Natural Polymer-Based Films and Coatings in Food Technology

Natural Polymer-Based Films and Coatings in Food Technology

Edited by

Ewelina Jamróz
Department of Chemistry, University of Agriculture,
Kraków, Poland
Department of Packaging and Logistics Processes, Krakow University of Economics, Kraków, Poland

Swarup Roy
Department of Food Technology and Nutrition, Lovely Professional University,
Phagwara, Punjab, India

ELSEVIER

Elsevier
Radarweg 29, PO Box 211, 1000 AE Amsterdam, Netherlands
125 London Wall, London EC2Y 5AS, United Kingdom
50 Hampshire Street, 5th Floor, Cambridge, MA 02139, United States

ISBN: 978-0-443-34244-8

For Information on all Elsevier publications
visit our website at https://www.elsevier.com/books-and-journals

Publisher: Matthew Deans
Acquisitions Editor: Jinia Dasgupta
Editorial Project Manager: Khyati Vat
Production Project Manager: Fizza Fathima
Cover Designer: Gopalakrishnan Venkatraman

Typeset by MPS Limited, Chennai, India

Contents

13. Gluten-based films and coatings

Jefferson Henrique Tiago Barros, Isamara Teodoro do Nascimento, Fernanda Ortolan, Luís Marangoni Júnior and Caroline Joy Steel

14. Zein films and coatings

Abhishek Parasar, Debabrata Nath and Pratap Kalita

15. Galactomannan films and coatings

Wilbert Rodriguez-Canto, Víctor Moo-Huchiin, Luis Chel-Guerrero, David Betancur-Ancona and Eduardo Castañeda-Pérez

30. Multilayer films and coatings based on biopolymers

Gourav Chakraborty, Shivani Desai, Srishti Upadhyay, Jeeva Kiran Banoth, Vijay Singh Sharanagat, Kshitiz Kumar and Yograj Bist

31. Migration from packaging materials to food

Violeta Ognenoska, Davor Daniloski and Anka Trajkovska Petkoska

32. Biodegradation aspect and life cycle assessment of biopolymer-based materials

Richert Agnieszka

33. Commercial applications of biopolymer film in industries

Avishikta Banerjee, Baru Venkata Naga Sahithi, Vasantha Veerappa Lakshmaiah, Jameel M. Al-Khayri and Praveen Nagella

List of contributors

Krishna Aayush Department of Food Science and Technology, Graphic Era (Deemed to be University), Dehradun, Uttarakhand, India

Reza Abedi-Firoozjah Student Research Committee, Kermanshah University of Medical Sciences, Kermanshah, Iran

Richert Agnieszka Department of Genetics, Faculty of Biology and Veterinary Science, Nicolaus Copernicus University in Toruń, Torun, Poland; Institute of Advanced Studies, Nicolaus Copernicus University in Toruń, Wileńska, Toruń, Poland

Iago Aguiar Dias Carmo School of Chemical Engineering, Universidade Estadual de Campinas (UNICAMP), Campinas, Brazil

Elaheh Ahmadi Institute for Sustainable Industries and Liveable Cities, Victoria University, Melbourne, VIC, Australia

Meraj Ahmed Department of Soil Science and Agricultural Chemistry, School of Agriculture, Lovely Professional University, Phagwara, Punjab, India

Lara Akay Department of Biology, Faculty of Science, Hacettepe University, Ankara, Turkey

K.P. Akhila Department of Microbiology, Central Coir Research Institute, University of Kerala, Alappuzha, Kerala, India

Jameel M. Al-Khayri Department of Agricultural Biotechnology, College of Agriculture and Food Sciences, King Faisal University, Al-Ahsa, Saudi Arabia

Avishikta Banerjee Department of Life Sciences, School of Sciences, CHRIST (Deemed to be University), Bengaluru, Karnataka, India

Jeeva Kiran Banoth Department of Food Engineering, National Institute of Food Technology Entrepreneurship and Management (NIFTEM-K), Sonipat, Haryana, India

Débora Parra Baptista Department of Food Engineering and Technology, School of Food Engineering (FEA), University of Campinas (Unicamp), Campinas, São Paulo, Brazil

Jefferson Henrique Tiago Barros Federal Institute of São Paulo (IFSP), Presidente Prudente, São Paulo, Brazil

David Betancur-Ancona Facultad de Ingeniería Química, Universidad Autónoma de Yucatán, Mérida, Yucatán, México

Yograj Bist Department of Biotechnology, Graphic Era (Deemed to be University), Dehradun, Uttarakhand, India

Paula Fernanda Janetti Bócoli Department of Food Engineering and Technology, School of Food Engineering (FEA), University of Campinas (Unicamp), Campinas, São Paulo, Brazil; Packaging Technology Center, Institute of Food Technology, Campinas, São Paulo, Brazil

Virginia Borroni Institute of Emerging Technologies and Applied Sciences (ITECA), National University of San Martín, San Martín, Argentina

Ana Caroline Ferreira Carvalho Department of Chemical Engineering, University of Blumenau, Blumenau, Santa Catarina, Brazil; Department of Food Engineering, Federal University of Santa Catarina, Florianópolis, Santa Catarina, Brazil

Eduardo Castañeda-Pérez Facultad de Ingeniería Química, Universidad Autónoma de Yucatán, Mérida, Yucatán, México

Nanci Castanha Packaging Technology Center (CETEA), Institute of Food Technology (ITAL), Campinas, São Paulo, Brazil

Gourav Chakraborty Department of Food Engineering, National Institute of Food Technology Entrepreneurship and Management (NIFTEM-K), Sonipat, Haryana, India

Neha Chandra School of Basic & Applied Sciences, K.R. Mangalam University, Gurugram, Haryana, India

Sahil Chaudhary Department of Food Science and Technology, I. K. Gujral Punjab Technical University, Kapurthala, Punjab, India

Luis Chel-Guerrero Facultad de Ingeniería Química, Universidad Autónoma de Yucatán, Mérida, Yucatán, México

Juan E. Cortés-Aguilera National Technological Institute of Mexico/Technological Institute of Tepic, Tepic, Nayarit, Mexico

Marlene J. Cran Institute for Sustainable Industries and Liveable Cities, Victoria University, Melbourne, VIC, Australia

Luan Ramos da Silva School of Food Engineering (FEA), University of Campinas (UNICAMP), Campinas, São Paulo, Brazil

Davor Daniloski School of Chemical Engineering, The University of Queensland, Brisbane, QLD, Australia

Eduarda Chiabai Rodrigues de Melo School of Chemical Engineering, Universidade Estadual de Campinas (UNICAMP), Campinas, Brazil

Carolina Krebs de Souza Department of Chemical Engineering, University of Blumenau, Blumenau, Santa Catarina, Brazil

S. Deotale Department of Chemical Engineering, Visvesvaraya National Institute of Technology (VNIT), Nagpur, Maharashtra, India

Shivani Desai Department of Food Engineering, National Institute of Food Technology Entrepreneurship and Management (NIFTEM-K), Sonipat, Haryana, India

V. Dhanya University of Kerala, Thiruvananthapuram, Kerala, India; Department of Microbiology, Sanatana Dharma College, University of Kerala, Alappuzha, Kerala, India

Krízia Emanuelle dos Santos Ferreira Department of Food Engineering and Technology, School of Food Engineering (FEA), University of Campinas (Unicamp), Campinas, São Paulo, Brazil

Larissa Graziele Rauber Duarte School of Food Engineering (FEA), University of Campinas (UNICAMP), Campinas, São Paulo, Brazil

Hadi Ebrahimi Student Research Committee, Yasuj University of Medical Sciences, Yasuj, Iran

Mariana Buranelo Egea Goiano Federal Institute of Education, Science, and Technology, Rio Verde, Goiás, Brazil

Farinaz Esmi Student Research Committee, Department of Food Science and Technology, Tabriz University of Medical Sciences, Tabriz, Iran

Han Lyn Foong Department of Food Technology, Faculty of Food Science and Technology, Universiti Putra Malaysia (UPM), Serdang, Malaysia; Department of Diagnostic & Allied Health Science, Faculty of Health and Life Sciences, Management and Science University, Shah Alam, Selangor, Malaysia

G. Ganga University of Kerala, Thiruvananthapuram, Kerala, India; Department of Biochemistry and Industrial Microbiology, Sree Ayyappa College, University of Kerala, Alappuzha, Kerala, India

Pragati Godara MMICT & BM (HM), Maharishi Markandeshwar Deemed to be University, Mullana, Ambala, Haryana, India

Vitor Emanuel de Souza Gomes Department of Food Engineering and Technology, School of Food Engineering (FEA), University of Campinas (Unicamp), Campinas, São Paulo, Brazil

Ramsés R. González-Estrada National Technological Institute of Mexico/Technological Institute of Tepic, Tepic, Nayarit, Mexico

Wiktoria Grzebieniarz Department of Chemistry, University of Agriculture, Kraków, Poland

Mandeep Gulati Food Packaging Technology Department, Polymer Materials Lab, CSIR-Central Food Technological Research Institute, Mysuru, Karnataka, India; Academy of Scientific and Innovative Research (AcSIR), Ghaziabad, Uttar Pradesh, India

Mohammad Hamidi Student Research Committee, Yasuj University of Medical Sciences, Yasuj, Iran

María Lidia Herrera Institute of Emerging Technologies and Applied Sciences (ITECA), National University of San Martín, San Martín, Argentina

Sıla Hoşoğlu Department of Biology, Faculty of Science, Hacettepe University, Ankara, Turkey

Divyashree J.S. Postharvest Research Lab, Department of Food Technology (Formerly Food Process Engineering), School of Bioengineering, SRM Institute of Science and Technology, Kattankulathur, Chengalpattu, Tamil Nadu, India

Ewelina Jamróz Department of Chemistry, University of Agriculture, Kraków, Poland; Department of Packaging and Logistics Processes, Krakow University of Economics, Kraków, Poland

S. Jaspin Department of Food Process Engineering, Entrepreneurship and Management, (NIFTEM-T), National Institute of Food Technology, Thanjavur, Tamil Nadu, India

Shivani Jaswal Department of Food Science and Technology, Dr. Y. S. Parmar University of Horticulture & Forestry, Solan, Himachal Pradesh, India

Pratap Kalita Pratiksha Institute of Pharmaceutical Sciences, Guwahati, Assam, India

Prabhjot Kaur Department of Food Technology and Nutrition, Lovely Professional University, Phagwara, Punjab, India

Arezou Khezerlou Department of Food Safety and Hygiene, School of Health, Gonabad University of Medical Sciences, Gonabad, Iran

Gözde Koşarsoy Ağçeli Department of Biology, Faculty of Science, Hacettepe University, Ankara, Turkey

Kshitiz Kumar Centre of Food Technology, University of Allahabad, Prayagraj, Uttar Pradesh, India

Veena P. Kumar Cashew Export Promotion Council of India (CEPCI) Laboratory & Research Institute, Kollam, Kerala, India; University of Kerala, Thiruvananthapuram, Kerala, India

Caroline Honaiser Lescano Department of Physiological Sciences (DCF), Institute of Biological Sciences (ICB), Federal University of Amazonas (UFAM), Manaus, Amazonas, Brazil

Caio Cesar de Lima Silva National Laboratory of Biosciences (LNBio), National Center for Research in Energy and Materials (CNPEM), Campinas, São Paulo, Brazil

Ulises M. López-García National Technological Institute of Mexico/Technological Institute of Tepic, Tepic, Nayarit, Mexico

Łukasz Łopusiewicz Department of Pharmaceutical Biology, Institute of Pharmacy, University of Greifswald, Greifswald, Germany; School of Medical & Health Sciences, VIZJA University, Warsaw, Poland

Fatemeh Madanifar Student Research Committee, Yasuj University of Medical Sciences, Yasuj, Iran

Amanda Alves Domingos Maia Packaging Technology Center, Institute of Food Technology, Campinas, São Paulo, Brazil

Gabriella Simon Maia School of Chemical Engineering, Universidade Estadual de Campinas (UNICAMP), Campinas, Brazil

Bianca Chieregato Maniglia São Carlos Institute of Chemistry, University of São Paulo (USP), São Carlos, São Paulo, Brazil

Karine Marafon Department of Food Engineering, Federal University of Santa Catarina, Florianópolis, Santa Catarina, Brazil

Luís Marangoni Júnior Department of Food Engineering and Technology, School of Food Engineering (FEA), University of Campinas (Unicamp), Campinas, São Paulo, Brazil

Luiz Henrique Capparelli Mattoso Brazilian Agricultural Research Corporation, Embrapa Instrumentation, São Carlos, São Paulo, Brazil

Giovana Maria Navarro Mendonça School of Pharmaceutical Sciences, São Paulo State University (UNESP), Araraquara, São Paulo, Brazil

Beatriz Montaño-Leyva Department of Research and Graduate Studies in Food, University of Sonora, Hermosillo, Sonora, Mexico

Víctor Moo-Huchiin Instituto Tecnológico de Mérida Campus Norte, Tecnológico Nacional de México, Mérida, Yucatán, México

Brenda Alessandra Munhoz Department of Food Engineering and Technology, School of Food Engineering (FEA), University of Campinas (Unicamp), Campinas, São Paulo, Brazil

Keshava Murthy P.S. Food Packaging Technology Department, Polymer Materials Lab, CSIR-Central Food Technological Research Institute, Mysuru, Karnataka, India; Academy of Scientific and Innovative Research (AcSIR), Ghaziabad, Uttar Pradesh, India

Shobha Muthukumaran Institute for Sustainable Industries and Liveable Cities, Victoria University, Melbourne, VIC, Australia; College of Sport, Health and Engineering, Victoria University, Melbourne, VIC, Australia

Amrutha N.R. Food Packaging Technology Department, Polymer Materials Lab, CSIR-Central Food Technological Research Institute, Mysuru, Karnataka, India; Academy of Scientific and Innovative Research (AcSIR), Ghaziabad, Uttar Pradesh, India

Praveen Nagella Department of Life Sciences, School of Sciences, CHRIST (Deemed to be University), Bengaluru, Karnataka, India

Isamara Teodoro do Nascimento Department of Food Engineering and Technology, School of Food Engineering (FEA), University of Campinas (Unicamp), Campinas, São Paulo, Brazil

Debabrata Nath Pratiksha Institute of Pharmaceutical Sciences, Guwahati, Assam, India

Pinku Chandra Nath Department of Bioengineering, School of Engineering and Architecture, Techno India University, Anandanagar, Tripura, India

S. Nimbkar Department of Food Process Engineering, National Institute of Technology (NIT), Rourkela, Odisha, India

Nikola Nowak-Nazarkiewicz Department of Chemistry, University of Agriculture, Kraków, Poland

Violeta Ognenoska Faculty of Technology and Technical Sciences, University St. Clement of Ohrid in Bitola, Veles, Macedonia

Josemar Gonçalves de Oliveira Filho Brazilian Agricultural Research Corporation, Embrapa Instrumentation, São Carlos, São Paulo, Brazil

Mircea Oroian Faculty of Food Engineering, Ştefan cel Mare University of Suceava, Suceava, Romania

Fernanda Ortolan Federal Institute of São Paulo (IFSP), Capivari, São Paulo, Brazil

Mohammadreza Ostovari Student Research Committee, Yasuj University of Medical Sciences, Yasuj, Iran

Vinay Kumar Pandey Department of FoodScience and Technology, Graphic Era (Deemed to be University), Dehradun, Uttarakhand, India

Abhishek Parasar Pratiksha Institute of Pharmaceutical Sciences, Guwahati, Assam, India

Anka Trajkovska Petkoska Faculty of Technology and Technical Sciences, University St. Clement of Ohrid in Bitola, Veles, Macedonia; Department of Materials Science and Engineering, Korea University, Seoul, South Korea

Rokalla Preethi Food Packaging Technology Department, Polymer Materials Lab, CSIR-Central Food Technological Research Institute, Mysuru, Karnataka, India; Academy of Scientific and Innovative Research (AcSIR), Ghaziabad, Uttar Pradesh, India

Amanda Alves Prestes Department of Food Engineering, Federal University of Santa Catarina, Florianópolis, Santa Catarina, Brazil

Elane Schwinden Prudêncio Department of Food Engineering, Federal University of Santa Catarina, Florianópolis, Santa Catarina, Brazil; Department of Food Science and Technology, Federal University of Santa Catarina, Federal University of Santa Catarina, Florianópolis, Santa Catarina, Brazil

Abhishek Biswal R. Postharvest Research Lab, Department of Food Technology (Formerly Food Process Engineering), School of Bioengineering, SRM Institute of Science and Technology, Kattankulathur, Chengalpattu, Tamil Nadu, India

Zahra Rahimi Motlagh Student Research Committee, Yasuj University of Medical Sciences, Yasuj, Iran

Rejish Ramakrishnan E Yuva Center, Anna University, Chennai, Tamil Nadu, India

María Regina Ramos Ramos Institute of Emerging Technologies and Applied Sciences (ITECA), National University of San Martín, San Martín, Argentina

Jeevan Prasad Reddy Food Packaging Technology Department, Polymer Materials Lab, CSIR-Central Food Technological Research Institute, Mysuru, Karnataka, India; Academy of Scientific and Innovative Research (AcSIR), Ghaziabad, Uttar Pradesh, India

Lina María Rodríguez Pineda Institute of Emerging Technologies and Applied Sciences (ITECA), National University of San Martín, San Martín, Argentina

Wilbert Rodriguez-Canto Facultad de Ingeniería Química, Universidad Autónoma de Yucatán, Mérida, Yucatán, México; Instituto Tecnológico de Mérida Campus Norte, Tecnológico Nacional de México, Mérida, Yucatán, México

Swarup Roy Department of Food Technology and Nutrition, Lovely Professional University, Phagwara, Punjab, India

Kevin Noe Ruiz-Hernández National Technological Institute of Mexico/Technological Institute of Tepic, Tepic, Nayarit, Mexico

Emmanuel Rotimi Sadiku Department of Chemical, Metallurgical and Materials Engineering, Institute of NanoEngineering Research (INER), Tshwane University of Technology, Pretoria, Republic of South Africa

Nitika Saini Department of Biotechnology, CT Institute of Pharmaceutical Sciences, Jalandhar, Punjab, India

Jonathan M. Sánchez-Silva Center for Research and Graduate Studies, Faculty of Chemical Sciences, Autonomous University of San Luis Potosí, San Luis Potosí, Mexico

Mahmood Alizadeh Sani Department of Food Science and Technology, School of Nutritional Sciences and Dietetics, Tehran University of Medical Sciences, Tehran, Iran

Luan Gustavo dos Santos Laboratory of Food Technology, School of Chemistry and Food Engineering, Federal University of Rio Grande, Rio Grande, Rio Grande do Sul, Brazil

Guilherme Frey Schütz School of Chemical Engineering, Universidade Estadual de Campinas (UNICAMP), Campinas, Brazil

Periyar Selvam Sellamuthu Postharvest Research Lab, Department of Food Technology (Formerly Food Process Engineering), School of Bioengineering, SRM Institute of Science and Technology, Kattankulathur, Chengalpattu, Tamil Nadu, India

Vijay Singh Sharanagat Department of Food Engineering, National Institute of Food Technology Entrepreneurship and Management (NIFTEM-K), Sonipat, Haryana, India

Gurvendra Pal Singh Department of Food Technology and Nutrition, Lovely Professional University, Phagwara, Punjab, India

Karthik Singirala Department of Food Technology and Nutrition, Lovely Professional University, Phagwara, Punjab, India

Mariana Spinei Faculty of Food Engineering, Ştefan cel Mare University of Suceava, Suceava, Romania

Caroline Joy Steel Department of Food Engineering and Technology, School of Food Engineering (FEA), University of Campinas (Unicamp), Campinas, São Paulo, Brazil

Divya Surisetti Department of Pharmacology, Kanpur Institute of Technology and Pharmacy, Kanpur, Uttar Pradesh, India

Zeba Tabassum School of Bioengineering and Biosciences, Lovely Professional University, Jalandhar, Punjab, India

Milad Tavassoli Cellular and Molecular Research Center, Yasuj University of Medical Sciences, Yasuj, Iran; Department of Nutrition, Faculty of Health and Nutrition Sciences, Yasuj University of Medical Sciences, Yasuj, Iran

Larissa Tessaro São Carlos Institute of Chemistry, University of São Paulo (USP), São Carlos, São Paulo, Brazil

P. Thivya Department of Food Process Engineering, National Institute of Technology (NIT), Rourkela, Odisha, India

Roshmi Thomas University of Kerala, Thiruvananthapuram, Kerala, India; Department of Microbiology, Sanatana Dharma College, University of Kerala, Alappuzha, Kerala, India

Gabriel Ferreira Toledo School of Chemical Engineering, Universidade Estadual de Campinas (UNICAMP), Campinas, Brazil

Srishti Upadhyay Department of Food Engineering, National Institute of Food Technology Entrepreneurship and Management (NIFTEM-K), Sonipat, Haryana, India

Vasantha Veerappa Lakshmaiah Department of Life Sciences, School of Sciences, CHRIST (Deemed to be University), Bengaluru, Karnataka, India

Baru Venkata Naga Sahithi Department of Life Sciences, School of Sciences, CHRIST (Deemed to be University), Bengaluru, Karnataka, India

Roniérik Pioli Vieira School of Chemical Engineering, Universidade Estadual de Campinas (UNICAMP), Campinas, Brazil

S.L. Vishnu Cashew Export Promotion Council of India (CEPCI) Laboratory & Research Institute, Kollam, Kerala, India; University of Kerala, Thiruvananthapuram, Kerala, India; Sree Narayana College for Women, University of Kerala, Kollam, Kerala, India

Nur Hanani Z.A. Department of Food Technology, Faculty of Food Science and Technology, Universiti Putra Malaysia (UPM), Serdang, Malaysia; Department of Diagnostic & Allied Health Science, Faculty of Health and Life Sciences, Management and Science University, Shah Alam, Selangor, Malaysia

Aida Zarshenas Student Research Committee, Yasuj University of Medical Sciences, Yasuj, Iran

Jianhua Zhang Institute for Sustainable Industries and Liveable Cities, Victoria University, Melbourne, VIC, Australia

Chapter 1

Introduction to biopolymer films and coatings

Ewelina Jamróz[1,2] and Swarup Roy[3]
[1]Departament of Packaging and Logistics Processes, Krakow University of Economics, Kraków, Poland, [2]Department of Chemistry, University of Agriculture, Kraków, Poland, [3]Department of Food Technology and Nutrition, Lovely Professional University, Phagwara, Punjab, India

1.1 Introduction

In the face of growing ecological challenges and the urgent need to minimize the negative impact of packaging on the environment, biopolymer films and coatings are becoming increasingly important in the food industry. Traditional plastics, while highly advantageous and functional, generate enormous amounts of problematic waste that contribute significantly to environmental pollution. Moreover, the micro and nano plastics generated from nonbiodegradable plastics also pose a serious threat to livestock and humans. As a result, the search for alternative, biodegradable materials capable of replacing conventional plastics has become a key priority for modern science and the packaging industry.

This book provides a comprehensive exploration of biopolymer films and coatings used in the food sector. It examines different types of films based on their source, including plant-based, animal-derived, microbial polymers, and synthetic materials. Their functional properties, such as mechanical strength, gas and moisture permeability, and their ability to create protective barriers for foods, as well as thermal and optical properties, are thoroughly discussed.

Particular attention is given to the role of biopolymer films in extending the freshness of food products. Their applications in protecting dairy products, fruits, vegetables, and other perishable foods are analyzed in detail. The book also explores innovative approaches, including active and intelligent films, which not only serve as protective packaging but also monitor food conditions in real time and interact with the storage environment.

A crucial aspect of this publication is the migration of components from films into food and its potential impact on food quality and safety. Additionally, issues related to the biodegradation of films and their environmental toxicity are thoroughly examined. The book presents research findings on biopolymer decomposition, as well as possibilities for recycling and reuse within a circular economy framework.

Legal regulations concerning biodegradable materials vary across countries and regions. Therefore, this book provides an overview of key standards and regulations currently in force in the European Union, the United States, Asia, and other parts of the world. Understanding these regulations is essential for the design and implementation of modern, compliant biopolymer packaging.

The book concludes with real-world examples of commercial applications of biopolymer films, highlighting their growing significance in the packaging market. It also identifies emerging technologies and trends that may shape the future of sustainable packaging. We believe this book serves as a valuable, up-to-date resource for scientists, students, packaging manufacturers, and anyone interested in sustainable development within the food industry.

Natural Polymer-Based Films and Coatings in Food Technology. DOI: https://doi.org/10.1016/B978-0-443-34244-8.00019-7

Section I

Biopolymers for production of films and coatings

Chapter 2

Starch films and coatings

Nanci Castanha[1], Caio Cesar de Lima Silva[2], Larissa Tessaro[3], Bianca Chieregato Maniglia[3], and Luís Marangoni Júnior[4]

[1]*Packaging Technology Center (CETEA), Institute of Food Technology (ITAL), Campinas, São Paulo, Brazil,* [2]*National Laboratory of Biosciences (LNBio), National Center for Research in Energy and Materials (CNPEM), Campinas, São Paulo, Brazil,* [3]*São Carlos Institute of Chemistry, University of São Paulo (USP), São Carlos, São Paulo, Brazil,* [4]*Department of Food Engineering and Technology, School of Food Engineering (FEA), University of Campinas (Unicamp), Campinas, São Paulo, Brazil*

2.1 Introduction and contextualization

The quality of food products is directly related to their organoleptic, nutritional, and microbiological properties, all of which can change dynamically during storage and distribution due to interactions with the environment or among the components of the food itself. To address these challenges and enhance food preservation, food packaging has evolved significantly, providing effective barriers against moisture, gases, and contaminants, thereby ensuring that products remain safe and fresh for longer periods (Debeaufort & Voilley, 2009).

Bio-based polymers, often referred to as biopolymers, are typically sourced from renewable materials. Common examples include polysaccharides like starch, cellulose, and chitosan; proteins such as casein and gluten; and lipids like waxes. Each chapter of this book focuses on one or more of these materials.

Films (used for wrapping food) and coatings (applied to or formed directly on food) made from biopolymers play a crucial role in addressing environmental concerns and food sustainability. They not only serve as effective barriers for food but also offer a sustainable alternative to traditional plastic packaging. It is important to emphasize that, in order to be classified as biodegradable and/or compostable, biopolymer-based packaging must undergo specific and rigorous testing to confirm its biodegradability. Moreover, to be classified as suitable for edible packaging—that is, a material to be consumed along with the food—the films and coatings made of biopolymers must meet several criteria, including desirable sensory attributes, biochemical and physicochemical stability, microbial resistance, and nontoxicity. Additionally, for biopolymers to be practical for industry use, they must be compatible with processing conditions similar to those used for petroleum-based plastics (Debeaufort & Voilley, 2009; Kramer, 2009).

Among various options, starch-based films and their blends with other biopolymers present a good alternative to petroleum-based plastics. The abundance, high biocompatibility, affordability, and biodegradability of starch make it a valuable material for sustainable food packaging. Ongoing research focuses on optimizing the design of starch-based films derived from diverse sources to meet sustainable packaging needs. Despite the technical challenges associated with their functionality, improvements can be achieved through starch modification, blending with other biopolymers or additives, and employing novel preparation techniques (Bangar et al., 2021). Starch-based films have demonstrated effectiveness in packaging a wide range of foods, including fruits, vegetables, baked goods, and many others, highlighting their commercial potential.

This chapter will provide a comprehensive overview of the properties and applications of starch-based films and coatings in food packaging, exploring both the challenges and opportunities associated with these materials while emphasizing their importance, functionality, and potential for enhancing food preservation using sustainable packaging solutions.

2.2 Starch: structure and properties

Starch is a natural biopolymer synthesized and stored in plants as an energy source, primarily found in various vegetables and grains. Starches are a versatile and widely used material, with undeniable economic importance. It has attracted significant academic and industrial interest due to its abundance, renewability, nontoxicity, and biodegradability, making it

Natural Polymer-Based Films and Coatings in Food Technology. DOI: https://doi.org/10.1016/B978-0-443-34244-8.00013-6

safe for consumption and environmentally convenient (Zhang et al., 2009). Starch's unique functionalities allow it to enhance food preservation while contributing to more sustainable packaging solutions in the food industry. The first step in developing new starch-based films and coatings is to understand the relationships among starch structure (including granules and molecules), processing methods, and the resulting properties.

2.2.1 Starch molecular structure

Starch is a biopolymer primarily composed of α-D-glucopyranosyl units, existing as water-insoluble granules that exhibit a unique and complex double-helical, semicrystalline, and birefringent structure. These granules are mainly composed of two polysaccharides: amylose and amylopectin. Amylopectin is highly branched, featuring chains of (1–4)-linked α-D-glucopyranosyl units along with (1–6)-linked branches, which create a treelike arrangement that constitutes the crystalline domains within the starch granule. In contrast, amylose is predominantly linear, composed mainly of (1–4)-linked α-D-glucopyranosyl units with fewer branches, and is typically found within the amorphous regions of the granule (Hizukuri et al., 2006; Liu, 2005; Tester et al., 2004). The granular structure of starch is highly organized, consisting of both amorphous and crystalline domains. The arrangement and distribution of these domains are influenced by genetic and environmental factors. In addition to amylose and amylopectin, starch granules may also contain minor components, such as lipids and phosphorus, which contribute to their overall properties. Different sources of starch exhibit varying molecular and granular characteristics, leading to a range of properties influenced by both the source and processing conditions. The key features of starch structure are well-documented in the literature and can be referenced for further information (Eliasson, 2006; Liu, 2005; Whistler et al., 1984). A detailed schematic representation of a starch granule structure showing its amorphous and crystalline regions, following the *blocklets* theory from the literature, is illustrated in Fig. 2.1.

2.2.2 Starch properties

The functional properties of various starchy products, including their texture and digestibility, are closely related to the processes of gelatinization and retrogradation. As a result, many industrial applications utilize gelatinized starches or depend on the gelatinization process itself. Gelatinizing starches is typically essential for the formation of films and coatings, with only a few exceptions (Averous & Halley, 2014). The ability of gelatinized starch to form films is a noteworthy characteristic that is exploited in development of biodegradable films (Bangar et al., 2021). Therefore, understanding the gelatinization process is crucial not only from a scientific perspective but also for its implications on the functional properties of the products. Gelatinization requires specific conditions, including adequate water (or other plasticizers) and temperatures above the starch's gelatinization threshold. Under these conditions, intermolecular bonds in starch granules weaken, allowing water to penetrate their rigid structure. As a result, the starch granule swells and loses its birefringence. If gelatinization continues, the starch transitions into the "pasting state," characterized by a two-phase system consisting of swollen granules and leached molecules, primarily amylose. Under specific conditions, such as cooling and storage, both amylose and amylopectin can undergo intermolecular reassociation, leading to the formation of a strong gel. This process is known as retrogradation (Liu et al., 2006; Liu, 2005). All these steps are illustrated in Fig. 2.2. The required water content and initial gelatinization temperature can vary depending on the starch source, meaning there is no universal set of gelatinization conditions applicable to all starches. As a result, each type of starch requires individual studies to determine its specific gelatinization conditions for optimal performance.

When analyzing the properties of starches, it is essential to recognize that naturally occurring starches might not possess all the desired characteristics. As a result, modifications are often necessary to meet the diverse and intricate demands of various industrial applications. Modified starches possess qualities that native starches lack, enabling new uses and providing competitive advantages through enhanced functional and value-added properties. Starch modification can be achieved using various methods, including chemical, physical, or enzymatic processes, and genetic improvement (Kaur et al., 2012). There is significant interest in optimizing these modification processes to develop new properties and applications. Current trends also emphasize the need for environmentally friendly practices and clean labels, leading to a reevaluation of traditional modification methods, particularly chemical ones, to better align with emerging demands (Maniglia et al., 2020, 2021).

2.3 Production of starch films and coatings

2.3.1 Definitions

In simple terms, coatings are either applied to or formed directly on foods, while films are self-supporting structures used to wrap food products. They can be applied to the surface of food or placed as thin layers between different components of

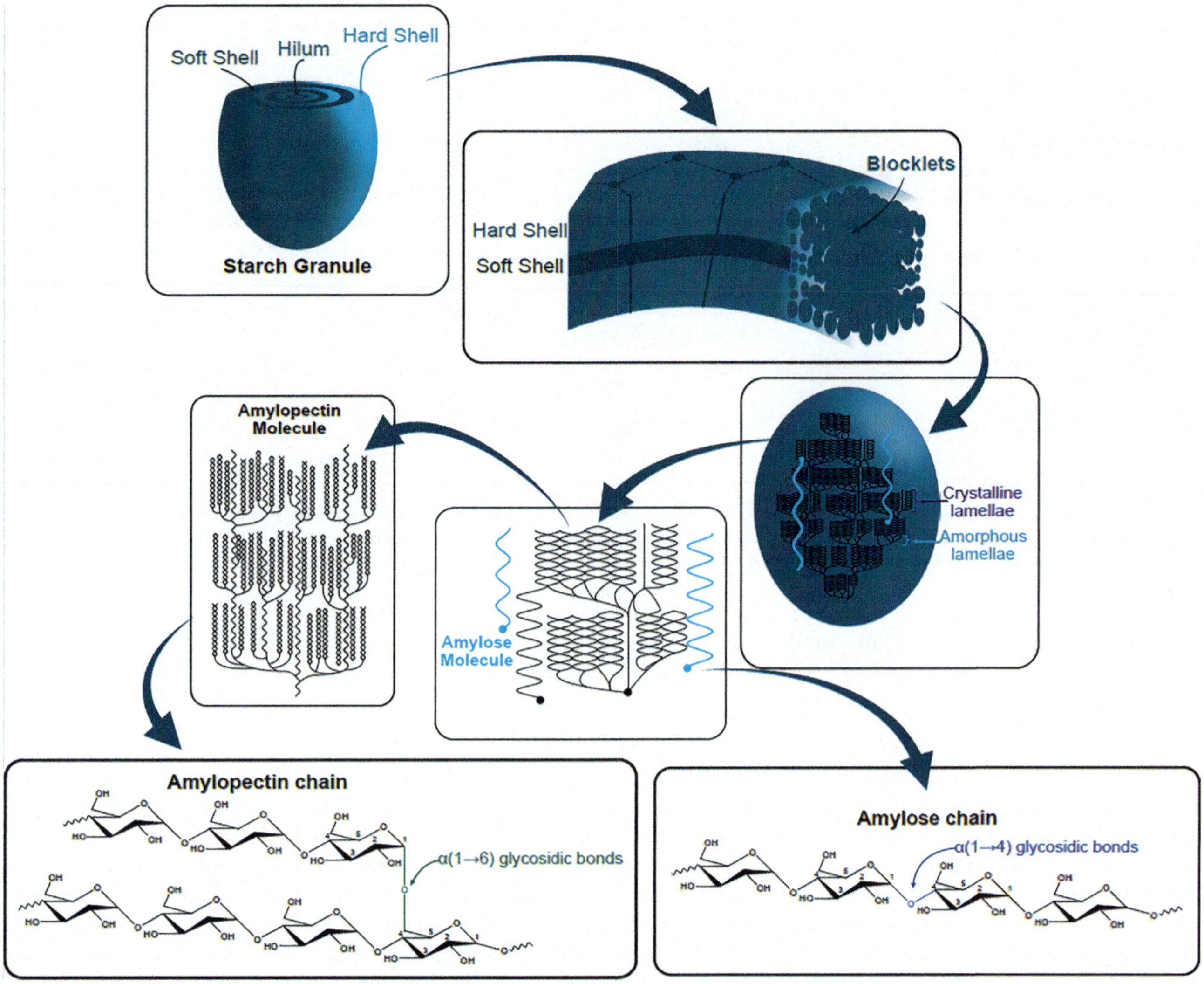

FIGURE 2.1 Schematic representation of a starch structure, from a starch granule to its molecular organization, emphasizing the alternation between the hard and the soft shells, the conformation of the blocklets in the shells and details of amylose and amylopectin molecules in the blocklets. Scheme inspired by the images and theories present in the literature.
Adapted from Gallant et al. (1992), Hizukuri et al. (2006), Liu (2005), Pérez et al. (2009), Tang et al. (2006), Tester et al. (2004), Eliasson & Gudmundsson (2006), and Castanha et al. (2017). Copyright 2024, with permission from Elsevier.

a food product. For instance, a film may be placed between fruit and the dough in a pie to provide protection against mass transfer and mechanical stress. Edible packaging refers to materials that are intended to be consumed alongside the food. This includes coatings applied to food and films that wrap around food products. These materials must fulfill specific criteria, including desirable sensory attributes, high barrier and mechanical efficiency, stability against biochemical, physicochemical, and microbial factors, nontoxicity, ease of use, low environmental impact, and cost-effectiveness (Debeaufort & Voilley, 2009) (Fig. 2.2).

2.3.2 Preparation techniques

Starch films and coatings are produced utilizing the gelatinization capacity of starch. In this irreversible process, starch granules swell and lose much of their internal molecular order, crystalline structure, and birefringence characteristics, ultimately leading to molecular solubilization (Bangar et al., 2021; Xie et al., 2006). Consequently, the preparation of starch films occurs in an aqueous solution, which is then transformed into a matrix under specific conditions involving applied force, heat, and the presence of plasticizers (Xie et al., 2006). The transformation of starch into films is most commonly achieved through solution casting or extrusion techniques (Bangar et al., 2021; Liyanapathiranage et al., 2023).

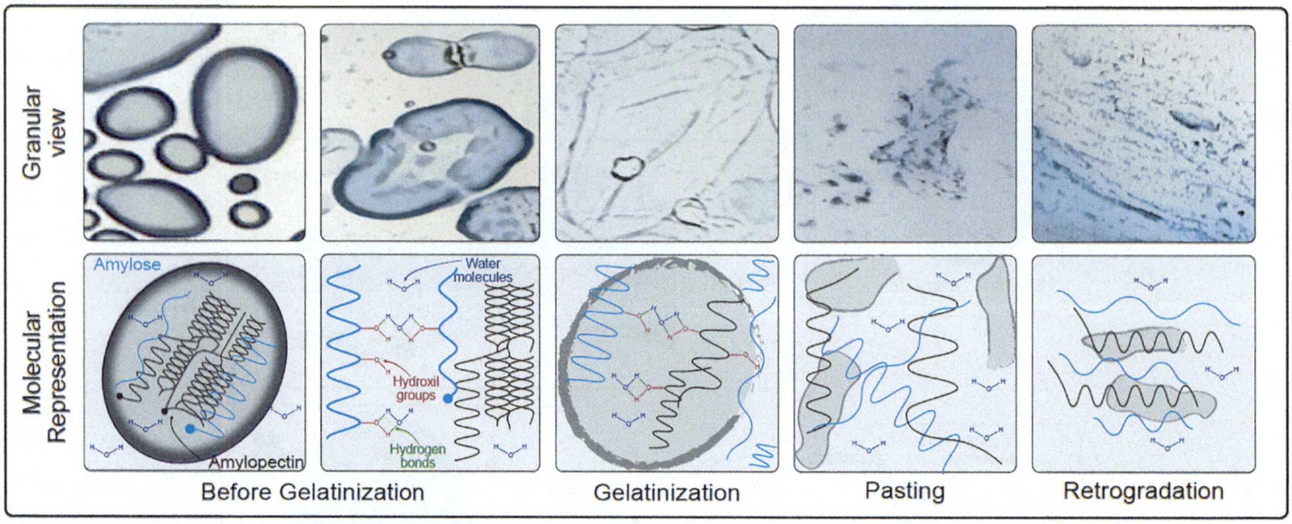

FIGURE 2.2 Schematic representation of starch pregelatinization, gelatinization, pasting, and retrogradation processes. Top row: granular view (microscopic images of potato starch). Bottom row: molecular view (illustration). Based on the theories presented in the literature. Adapted from Atwell et al. (1988), Liu (2005), Xie et al. (2007); and Castanha et al. (2019). Copyright 2024, with permission from Elsevier.

Conversely, coatings are typically applied using dipping or spray techniques (Gupta et al., 2022). A schematic representation of the main techniques for producing starch-based films and coatings is illustrated in Fig. 2.3.

2.3.2.1 *Films by casting*

The solvent-casting method is one of the most commonly used techniques for producing starch-based films, especially in laboratory and small-scale pilot projects (Suhag et al., 2020). The process involves three main steps (Fig. 2.3A): (1) dissolving or dispersing the biopolymer in a solvent and/or plasticizer usually involves applying heat to improve solubility and promote gelatinization; (2) pouring the mixture into a mold; and (3) drying the cast solution under controlled conditions (Bangar et al., 2021; Suhag et al., 2020). The film formation process begins with starch gelatinization. In this irreversible process, starch is mixed with water and/or a plasticizer and heated above its gelatinization temperature. The resulting gelatinized solution is then cast into a mold, where it undergoes drying to remove the solvent and form a polymer film. During the drying stage, retrogradation begins. This process involves the reassociation of amylose and amylopectin chains, which were disrupted during gelatinization, into a more ordered structure (Thakur et al., 2019; Wang et al., 2015). Drying ovens are commonly used during the drying phase of film formation. However, other equipment, such as microwaves, hot-air ovens, trays, and vacuum dryers, can also facilitate solvent evaporation and film peeling (Bangar et al., 2021; Velaga et al., 2018).

The removal of solvent is a critical factor influencing the properties of the films. For instance, rapid drying techniques can adversely affect the physical and structural characteristics of the films (Suhag et al., 2020; Velaga et al., 2018).

While water is an ideal plasticizer for starch, its evaporation can lead to brittleness in films. To mitigate this, it is recommended to combine water with other plasticizers (summarized in item 2.4 of this chapter) to create optimal conditions for solution casting. Plasticizers play a vital role in enhancing the flexibility of films by reducing intermolecular hydrogen bonding and increasing intermolecular spacing. This, in turn, improves the mechanical properties of starch-based films for packaging applications (Basiak et al., 2018; Thakur et al., 2019).

2.3.2.2 *Films by extrusion*

The extrusion process is a widely used and commercially preferred method for producing polymer films, valued for its ability to modify structural properties and enhance the physicochemical characteristics of the extruded materials. The process is conducted in an extruder that typically consists of one or two screws rotating along the axis of the barrel. The screw transports the material down the barrel to the die, which shapes the final product (Fig. 2.3B) (Giannakas, 2024). The extrusion process can be divided into three distinct zones: (1) the feeding zone, (2) the kneading zone, and (3) the heating zone (Bangar et al., 2021; Suhag et al., 2020). In a typical polymer extruder, the material melts due to frictional heating within the barrel, where temperatures are optimized to maintain proper viscosity, ensuring smooth flow without

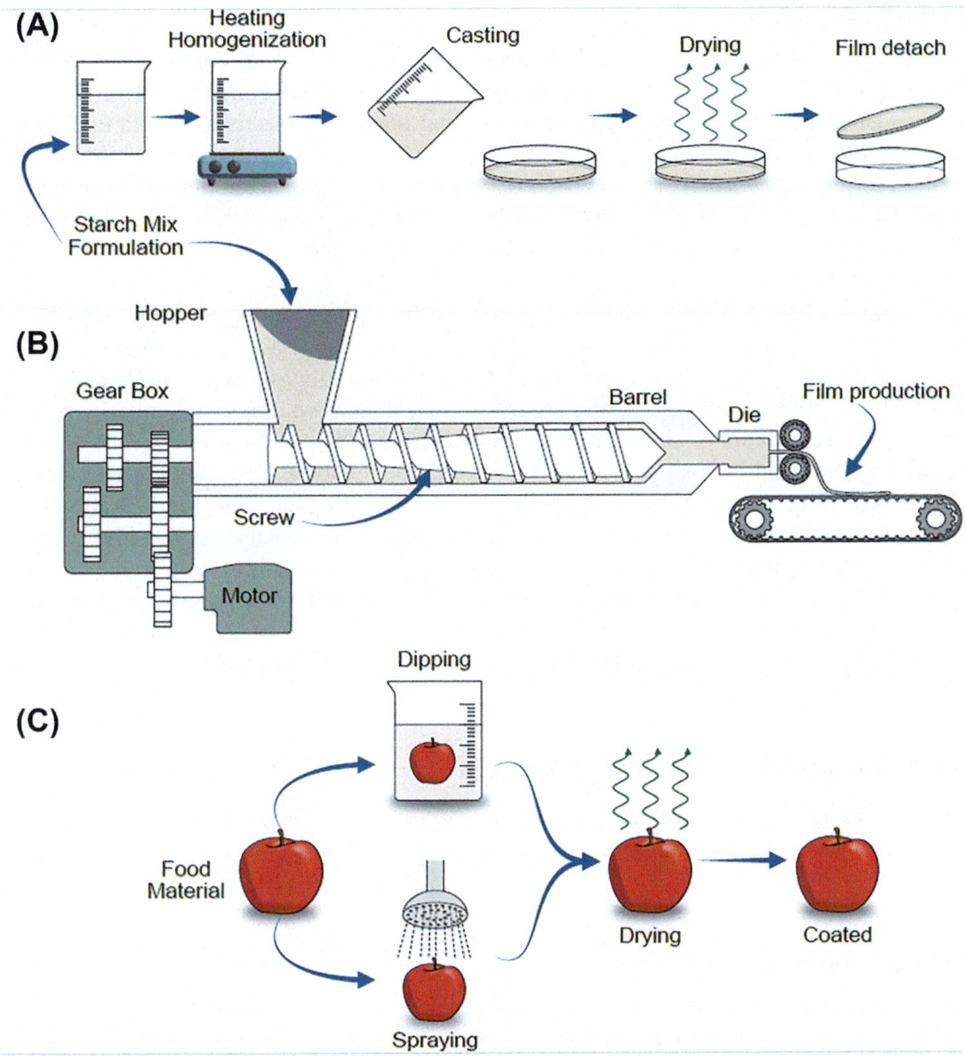

FIGURE 2.3 Techniques used to obtain films by (A) casting and (B) extrusion, and to obtain coatings by (C) dipping and spraying. Adapted from the literature Devi et al. (2024); V et al. (2022). Reproduction with permission from Creative Commons CC-BY license.

degradation. The molten polymer exits through the die and is subsequently cooled to solidify (Giannakas, 2024). Plasticizers can also be incorporated into the mixture during the extrusion process. Mechanical and thermal energies are the key to the extrusion technique, which work together to formulate films. The speed of the screw in the extruder significantly affects the specific mechanical energy applied. Variations in screw speed influence film properties such as homogeneity, shear rate, and stress, while also controlling the time of the film preparation (Bangar et al., 2021). As screw speed increases, the torque value required for the extrusion process decreases, which can enhance the specific properties of the resulting bio-based films (Suhag et al., 2020). Compared to casting, the extrusion process requires minimal space, operates faster, and is more energy-efficient, resulting in films with superior mechanical properties (such as elongation at break) and optical properties (transparency) (Bangar et al., 2021; Skurtys, no date). Additionally, extrusion is more attractive than casting because it eliminates the need for evaporation time, especially in industrial settings (Skurtys et al., 2010). However, alternative casting techniques, such as continuous casting, address several industrial challenges associated with using traditional casting methods for larger-scale projects (Otoni et al., 2017).

Overall, extrusion is recognized as a high-performance, cost-effective, and efficient method widely employed in the food industry, although it does have limitations regarding temperature tolerance and the moisture content of raw material blends (Bangar et al., 2021).

2.3.2.3 Application of coatings

While biopolymer films are formed separately and subsequently applied to food, coatings are prepared directly on the surface of the food material (Fig. 2.3C). Since film coatings come into direct contact with food, their composition and production must consider surface adhesion and the intended final quality of the food (Gupta et al., 2022).

Although several coating application techniques exist (Gupta et al., 2022), the two most commonly used methods are dipping and spraying coatings (Fig. 2.3C). The decision of application technique should consider factors such as the type of food being coated, its surface characteristics, and the coating's intended purpose. Elements like surface tension, density, shape, size, coating thickness, and the food's physical attributes all influence the selection of the application method (Debeaufort & Voilley, 2009).

In the dipping coating method, the food product is immersed in a film-forming solution, which involves three main steps: (1) immersion, where the food material is completely dipped into the film-forming solution; (2) evaporation, where the solvent is removed to form a film; (3) deposition, where a precursor solution forms a thin layer on the surface of the food material (Gupta et al., 2022). Dipping techniques often yield thicker coatings and are widely used for fresh fruits, vegetables, and meat products (Gupta et al., 2022; Tavassoli-Kafrani et al., 2016).

In contrast, the spraying method is more prevalent in industrial applications than dipping technique. In this approach, the coating is applied as a spray, dispersing the liquid solution into fine droplets. This method allows the smaller droplets to cover a larger surface area with the same amount of liquid, enhancing efficiency. The high pressure used enhances efficiency, making this method ideal for large-scale operations (Gupta et al., 2022).

The effectiveness of the coating is influenced by several factors, including pressure, viscosity, temperature, surface tension, and the spray nozzle type (Debeaufort & Voilley, 2009; Gupta et al., 2022). This form of application typically results in a more uniform coating but tends to produce a thinner layer compared to the dipping technique (Gupta et al., 2022; Tavassoli-Kafrani et al., 2016).

2.4 Properties of starch films and coatings

Before discussing the potential formulations and key properties of starch films, Table 2.1 provides a summary of the literature regarding starch-based film formulations and their characteristics.

2.4.1 Formulations

2.4.1.1 Plasticizers—types and quantity

Plasticizers are low molecular weight and nonvolatile compounds that are incorporated into films, coatings, and other polymeric materials to improve flexibility, malleability, and processing properties (Vieira et al., 2011). Plasticizers function by modifying the internal structure of the polymer, reducing intermolecular interactions between polymer chains, and thus facilitating their movement. This results in a more flexible and less rigid material (Thakur et al., 2019). The types of plasticizers used in starch-based films and coatings can include natural compounds such as polyols (e.g., glycerol and sorbitol), mono- and disaccharides (e.g., glucose and sucrose, respectively), and lipid derivatives (e.g., fatty acids and phospholipids) (Majeed et al., 2023; Maniglia et al., 2019), among other low molecular weight alcohols, in addition to water. In most preparations of starch-based films and coatings, starch is heated in water to its gelatinization temperature, which breaks the crystalline structure of the granules held together by hydrogen bonds (Lauer & Smith, 2020). Plasticizers are then added to prevent the intermolecular hydrogen bonds between starch chains from causing recrystallization during drying, which could lead to cracks and irregularities in the films. While water is always used as the primary plasticizer, glycerol is the most commonly used secondary plasticizer, as it is nontoxic and recognized as safe additive for use in the food industry (Thakur et al., 2019). The amount of plasticizer required varies depending on the characteristics of the starch and the desired application, but in general, the concentration of plasticizers ranges from 15% to 50% of the dry matter (Table 2.1).

The effect of different concentrations of glycerol on cassava starch-based films was evaluated (Bergo et al., 2010). Increasing the glycerol content improved the flexibility of the films but decreased their strength and stiffness, as measured by elongation at break, tensile strength, and elastic modulus. This observation was supported by a decrease in the glass transition temperature with higher glycerol concentrations. Glycerol interacts with starch chains, reducing intermolecular interactions, which makes the films more flexible and less resistant. Additionally, babassu starch-based films plasticized with glycerol or urea demonstrated greater flexibility and sensitivity to water, exhibiting lower crystallization and resistance compared to those plasticized with sorbitol or glucose (Maniglia et al., 2019). Glycerol and urea were identified as the most effective plasticizers due to their lower molecular weight.

TABLE 2.1 Summary of the literature information about starch-based films formulations and their properties.

Starch-based film formulations	Thickness (μm)	Tensile strength (MPa)	Elongation at break (%)	Elastic modulus (MPa)	Water vapor permeability	O_2 permeability	Optical properties	References
Cassava starch (5%) with glycerol (25%) at 75% relative humidity (RH)	80	2.4	43.8	51.8	23.8 g mm (kPa m^2 day)$^{-1}$	1.3 cm^3 (Pa m day)$^{-1}$	Opacity 9.3%	La Fuente et al. (2019)
Ozonized cassava starch (5%) with glycerol (25%) at 75% RH	70	5.5	39.4	71.2	30.0 g mm (kPa m^2 day)$^{-1}$	2.2 cm^3 (Pa m day)$^{-1}$	Opacity 3.7%	
Babassu starch (4%) with glycerol (19%) at 58% RH	77	24.5	1.9	18.7	5.5×10^{10} g (Pa m s)$^{-1}$	–	Opacity 4.1%	Maniglia et al. (2019)
Babassu starch (4%) with sorbitol (19%) at 58% RH	80	34.0	1.4	23.5	2.7×10^{10} g (Pa m s)$^{-1}$	–	Opacity 5.4%	
Rice starch (3%) with glycerol (25%) at 50% RH	128	15.0	28.0	7.7	4.0×10^{-14} kg m (Pa m^2 s)$^{-1}$	5.5×10^{-19} kg m (Pa m^2 s)$^{-1}$	Opacity 2.5%	Suriyatem et al. (2018)
Potato starch (5%) with glycerol (25%) at 75% RH	77	3.9	80.9	45.1	20.8 g mm (kPa m^2 day)$^{-1}$	–	Opacity 30.3%	La Fuente et al. (2020)
Ozonized potato starch (5%) with glycerol (25%) at 75% RH	61–65	4.2	28.4	61.1	22.5 g mm (kPa m^2 day)$^{-1}$	–	Opacity 9.2%	
Potato starch (2%) with glycerol (25%) at 53% RH	80	10.0	40.9	4.0	6.7 g mm (kPa m^2 h)$^{-1}$	15.7×10^{-14} cm^3 (Pa m s)$^{-1}$	Transmittance 86%	Moreno et al. (2015)
Wheat starch (1%) with glycerol (15%) at 53% RH	53	33.0	3.5	13.6	13×10^{10} g (Pa m s)$^{-1}$	0.061 cm^3 mm (atm m^2 day)$^{-1}$	–	Bonilla et al. (2013)
Turmeric starch (5%) with glycerol (25%) at 58% RH	89–101	9.2–3.6	1.9–3.3	7.5–11.4	4.6–8.0 $\times 10^{10}$ g (Pa m s)$^{-1}$	–	Opacity 29.5%–34.2%	Maniglia et al. (2022)
Potato starch (2.4%) with sorbitol (30%)	–	10	41	4.9	7.4 g mm (kPa m^2 day)$^{-1}$	1.3 cm^3 mm (atm m^2 day)$^{-1}$	–	Dong et al. (2024)
Potato starch (2.4%) with sorbitol (30%) and $Ti_3C_2T_x$ MXene nanoplatelets at 50% RH	–	15–18	3–33	8–20	0.6–3.0 g mm (kPa m^2 day)$^{-1}$	0.3–0.7 cm^3 mm (atm m^2 day)$^{-1}$	–	

(*Continued*)

TABLE 2.1 (Continued)

Starch-based film formulations	Thickness (μm)	Tensile strength (MPa)	Elongation at break (%)	Elastic modulus (MPa)	Water vapor permeability	O_2 permeability	Optical properties	References
Oxidized cassava starch (5%) with glycerol (1%), Tween 80 (1%) at 50% RH	73	2.1	3.5	–	1.04×10^{-12} g cm (Pa cm^2 s)$^{-1}$	0.30 g 100 g^{-1}	Transmittance 39%	Zhou et al. (2021)
Oxidized cassava starch (5%) with glycerol (1%), Tween 80 (1%), and oregano essential oil (0.5%–2.5%) at 50% RH	84–137	0.8–1.8	4.5–42	–	1.3–1.9 $\times 10^{-12}$ g cm (Pa cm^2 s)$^{-1}$	0.26–029 g (100 g)$^{-1}$	Transmittance 24%–36%	
Corn starch (5%) with glycerol (20%) and chitosan (1%–4%) at 75% RH	–	2.5–6.3	70–129	4–38	2–3 $\times 10^{-10}$ g (Pa m s)$^{-1}$	–	Opacity 1%–2%	Ren et al. (2017)
Corn starch (10%) with glycerol (30%), oxidized sucrose (1%–3%) and nanocellulose fibrillate (1%–5%) at 50% RH	22–32	1.2–5.6	22–40	-	0.038–0.074 g (m^2 min)$^{-1}$	10.3–29.4 cm^3 (m day)$^{-1}$	-	Othman et al. (2021)

2.4.1.2 Additives to improve functionality

The incorporation of additives into starch-based films and coatings is a widely adopted strategy aimed at enhancing the functionality, particularly in optimizing barrier, mechanical, and functional properties. Some additives, including inorganic compounds, nanocellulose, crosslinking agents, and chitosan (Table 2.1), significantly improve the performance of these films in terms of barriers to oxygen, water vapor, and mechanical properties (Lauer & Smith, 2020). The addition of fibrillated nanocellulose to corn starch-based films has been shown to enhance tensile strength and elastic modulus while reducing elongation at break and permeability to oxygen and water vapor (Othman et al., 2021). This occurs because cellulose nanofibrils act as reinforcing, rigid, and impermeable fillers. Conversely, the addition of thymol has been found to decrease the mechanical and water vapor barrier properties due to its plasticizing effect, which reduces cohesive forces within the matrix film. Furthermore, the addition of oxidized sucrose, a cross-linking agent, improves tensile strength and decreases water sensitivity by forming cross-links between the starch and fibrillated nanocellulose, resulting in a three-dimensional network that reinforces the material (Balakrishnan et al., 2019).

The addition of chitosan, a polysaccharide derived from the deacetylation of chitin, can also improve the properties of starch-based films, as it forms a blend with starch and has naturally antimicrobial properties. Interactions between the hydroxyl groups of corn starch and the amino groups of chitosan increase tensile strength and elongation at break while simultaneously reducing water vapor permeability and opacity (Ren et al., 2017).

2.4.1.3 Bioactive compounds

Bioactive compounds are natural substances present in foods, plants, or organisms that provide beneficial effects on human health. Generally, these compounds are secondary metabolites, including phenolic compounds, vitamins, terpenoids, and carotenoids, that can exert various effects in the body, for example, acting as antioxidants, antiinflammatories, or antimicrobials. When incorporated into starch-based films or coatings, bioactive compounds activate these materials, which can increase the shelf life of the food, inhibit the growth of microorganisms, and offer additional health benefits. Studies on the incorporation of bioactive compounds into starch-based films and coatings have shown that plant extracts and essential oils (EOs) rich in phenolic compounds and terpenoids, such as "siriguela" (*Spondias purpurea* L.) (Rodrigues et al., 2021) and oregano EO (Martins et al., 2021), can act as antioxidants, helping to extend the shelf life of foods such as fruits and meat. These compounds, mainly phenolics and terpenoids, reduce lipid oxidation, browning, and block UV/visible light, improving the oxygen barrier and absorbing light that degrades nutrients. They can capture oxygen to enhance the oxygen barrier or diffuse from the film/coating into the food, providing localized effects (Bonilla et al., 2012).

Additionally, bioactive compounds present in plant extracts and EOs can also act as natural antimicrobials. This biological activity is particularly valuable for food films and coatings, as the growth of spoilage microorganisms and pathogens is a primary cause of food spoilage (Han, 2003). For example, tapioca starch-based film containing pullulan and *Litsea cubeba* EO has been used as packaging for fresh beef meat, demonstrating significant antibacterial activity against foodborne pathogens like *Staphylococcus aureus* and *Escherichia coli*(Adame et al., 2024). Bioactive compounds can disrupt microbial cells through several mechanisms, including the destruction of the lipid bilayer of the cell membrane, the interruption of enzyme systems, the damage to bacterial genetic material, and the formation of hydroperoxides through the oxidation of unsaturated fatty acids (Atarés & Chiralt, 2016; Han, 2003). Antimicrobial food packaging films and coatings can either gradually release the antimicrobial compound into the food or inhibit microbial growth on the packaging surface without releasing substances. The first type gradually releases the agent or through coatings, while the second type acts upon direct contact with microorganisms (Bastarrachea et al., 2011). Bioactive compounds can be incorporated into films or coatings using various methods. The two main approaches involve either adding the compounds to the main formulation during production or including them in the food-contact layer of the packaging (Fig. 2.4) (Quintavalla & Vicini, 2002).

Some natural pigments, such as anthocyanins (phenolic compounds from the flavonoid group), can not only be incorporated into starch-based films and coatings to enhance their functionality but also be used to produce smart packaging (Luchese, Abdalla, et al., 2018). These smart packages function as pH indicators, as anthocyanins change their color depending on the pH of the environment. This feature allows for monitoring the acidity or alkalinity of foods and beverages during storage, helping to detect signs of spoilage or bacterial growth (Cheng et al., 2022). Packaging made from yam (*Dioscorea zingiberensis*) starch films, combined with anthocyanins extracted from butterfly pea (*Clitoria ternatea* L.) flowers, has been shown to effectively indicate the freshness and degree of deterioration of packaged chicken meat through color change (Wang et al., 2023).

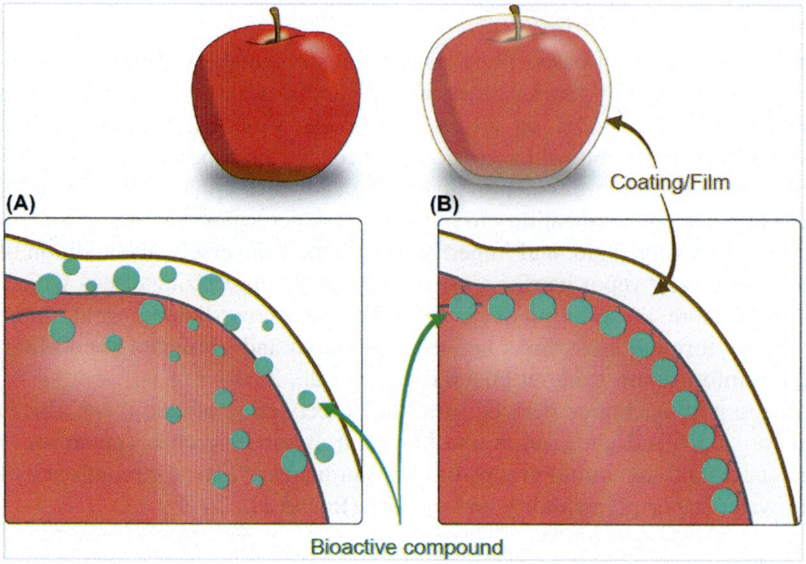

FIGURE 2.4 Bioactive compounds' availability in food packaging (A) by incorporating the compounds into the film or coating main formulation, or (B) by adding them to the food-contact layer of the packaging.
Adapted from Quintavalla & Vicini (2002). Copyright 2024, with permission from Elsevier.

2.4.2 Properties of starch films

The mechanical, barrier, and optical properties of starch-based films and coatings are fundamental for their application in food packaging, as they directly influence the protection and shelf life of products. Key attributes such as tensile strength, oxygen and water permeability, and transparency determine the functionality and efficiency of these materials. By modifying the composition and structure of these films, their properties can be optimized to meet various food preservation requirements and appearance preferences.

2.4.2.1 Mechanical properties

The mechanical properties of films are extremely important, especially in packaging and food protection applications, as they determine the effectiveness and functionality of the material under various conditions (Bangar et al., 2021). Key mechanical properties include tensile strength, elongation at break, and elastic modulus, which respectively indicate the maximum strength before rupture, flexibility, and stiffness of the material (Ren et al., 2017). These properties directly influence the performance, durability, and protection of the packaged product, and are typically evaluated using tensile test, where equipment applies a tensile load until the material breaks (Bonilla et al., 2013; Luchese, Benelli, et al., 2018; Suriyatem et al., 2018). In starch-based films, mechanical properties are directly related to the source and composition of starch, methods of preparation and modification, and the presence of other components, such as plasticizers (Table 2.1). The composition of starch, particularly the amylose and amylopectin content, differentiates one starch type from another, along with the size and shape of the starch granules (La Fuente et al., 2019). For instance, starch-based films made from high-amylose starch typically exhibit higher tensile strength compared to those made from waxy starches (Luchese, Benelli, et al., 2018). The linear amylose chains bind efficiently through strong bonds, predominantly hydrogen bonds and localized crystallization, resulting in more resistant and cohesive materials (Chaléat et al., 2014).

Starch-based films derived from different starch sources (cassava, corn, and wheat), despite having similar amylose content (~27.5%) and the same glycerol concentrations, exhibit varying mechanical properties (Luchese, Benelli, et al., 2018). While amylose content is important, other factors also influence the mechanical properties of the starch-based films, such as the molecular weight of amylose and amylopectin, the degree of amylopectin branching, and the shape, size, and crystallinity of starch granules (Chaléat et al., 2014; Thakur et al., 2019). The modification of starches can further affect the mechanical properties of the films. Different modification techniques, including chemical, physical, and enzymatic methods, aim to improve the processability, functionality, and applicability of starch-based films by improving their mechanical and barrier properties (Kumari & Sit, 2023). For instance, ozonation is a chemical modification that can break bonds within starch molecules, forming functional groups such as carbonyl and carboxyl, which in turn alter properties like tensile strength, elastic modulus, and elongation at break (La Fuente et al., 2019). The incorporation of nanoparticles as reinforcing fillers, using cross-linking as modifiers or blending with other biopolymers, can optimize the interactions between starch chains, forming a more cohesive and reinforced matrix with improved mechanical properties (Table 2.1).

2.4.2.2 *Barrier properties*

One of the main functions of a food packaging and coating is to create a barrier between the food and the external environment, helping to preserve the properties of the packaged or coated food and thus extend its shelf life (Trinh et al., 2023). Starch-based food packaging and coating are permeable, to varying degrees, to small molecules such as gases and vapors, as well as low molecular weight compounds like aromas and flavors (Xue et al., 2023). The oxygen and water vapor permeability are the widely measured barrier parameters of starch-based films. The permeation of gases and vapors in biopolymeric films follows the dissolution-diffusion model, which incorporates Henry's and Fick's laws and involves processes of absorption, diffusion, and desorption (Trinh et al., 2023). The permeability of these gases is influenced by factors such as temperature, moisture, relative humidity, pressure, crystallinity, film structure, and gas characteristics such as molecular weight and polarity (Kurek et al., 2014). Starch-based films typically exhibit excellent oxygen barrier properties, as the cohesive structure of the biopolymeric matrix creates a challenge for gas permeation (Bonilla et al., 2012). However, the incorporation of additional components, such as crosslinkers (glutaraldehyde), can diminish this property in starch films (Table 2.1). This may happen because oxygen has greater difficulty permeating cohesive matrices with strong bonds, low chain mobility, and/or the presence of lipids, as it is not soluble (Bangar et al., 2021; Trinh et al., 2023).

In contrast, the water vapor permeability of starch films is considerably higher than that of synthetic plastics (Table 2.1). For instance, a potato starch-based film (Moreno et al., 2015) exhibits water vapor permeability that is 670 times greater than that of high-density polyethylene with values of 6.7 e 0.01 g mm (kPa m^2 h)$^{-1}$, respectively (Bangar et al., 2021). The main reason for this disparity is that biopolymers like starch are water-soluble, hygroscopic, and possess an amorphous or partially crystalline structure, whereas synthetic plastics have a dense and impermeable structure. Many studies have explored the addition of various components to starch films to mitigate this limitation (Table 2.1). However, for practical applications, starch-based films remain highly permeable to water vapor, presenting a significant techno-scientific challenge that needs to be addressed.

2.4.2.3 *Optical properties*

Optical properties are essential for the application of starch-based films and coatings in food packaging. While the greater transparency of these materials is related to greater consumer acceptability (Maran et al., 2013), increased opacity can help protect the packaged food from exposure to light, especially spoilage caused by UV light. Light transmittance in the range of 200–800 nm is used to measure film transparency. The opacity of the film is defined by the absorbance at 600 nm divided by the thickness of the film, being an important parameter in evaluating its ability to block light (Singh et al., 2022). Several factors can influence the optical properties of films. For instance, glycerol can reduce the opacity of babassu starch-based films compared to other plasticizers (Maniglia et al., 2019). Increasing glycerol concentration has been shown to enhance the transparency of tapioca starch-based films, likely due to the dilution effect of glycerol within the biopolymeric matrix (Maran et al., 2013). Less opaque starch-based films are also an indication of greater compatibility between the components present, which can occur after a modification of the starch, for example (La Fuente et al., 2019, 2020).

Conversely, increasing opacity (and thus decreasing transparency) of starch-based films may be desirable for protecting light-sensitive foods. Starches from alternative sources rich in bioactive compounds, such as turmeric residue starch containing curcuminoids, can produce opaque (>30%) and colorful films, with the ability to absorb incident light and protect the packaged food (Maniglia et al., 2022). Additionally, increasing the opacity of starch-based films can also be achieved through the incorporation of various components, such as cellulose nanoparticles, metals, or EOs (Table 2.1), as these materials can disperse incident light and/or alter the refractive index.

2.4.3 Biodegradability

Most flexible food packaging is made from petroleum-derived polymers, offering effective barriers against gases, vapors, humidity, and odors, as well as good mechanical properties and low cost (Torres et al., 2011). However, these materials can cause significant environmental damage when disposed of incorrectly, accumulating in ecosystems and decomposing into microplastics, which disrupt natural habitats and affect food safety. In response to these issues, researchers are

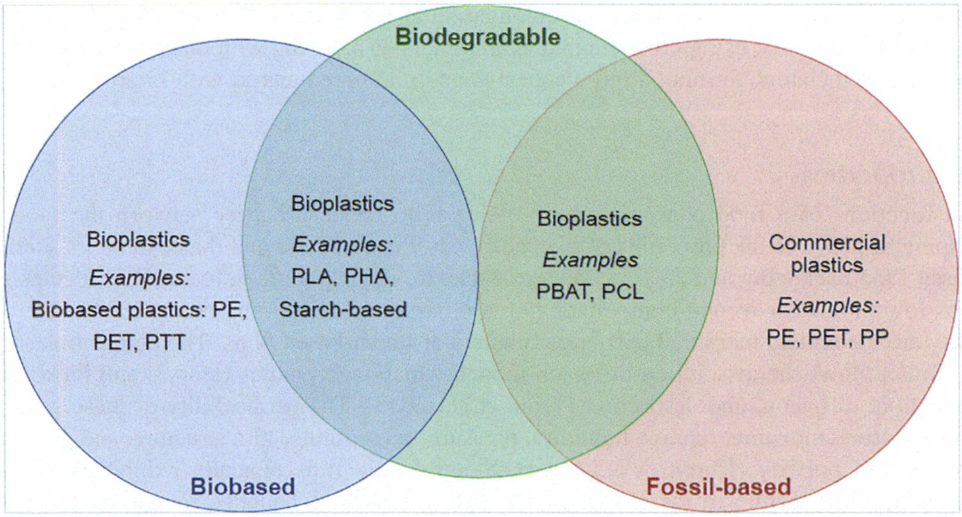

FIGURE 2.5 Indication of biodegradable packaging among bio-based and fossil-based materials. *PA*, Polyamides; *PBAT*, polybutyrate; *PBS*, polybutylene succinate; *PCL*, polycaprolactone; *PE*, polyethylene; *PET*, polyethylene terephthalate; *PHA*, polyhydroxyalkanoates; *PLA*, polylactic acid; *PP*, polypropylene; *PTT*, polytrimethylene terephthalate.
Adapted from Bioplastics (2022).

developing biodegradable biopolymer-based packaging to mitigate the environmental impacts of synthetic alternatives (Dong et al., 2024). It is important to note that bio-based materials are not always biodegradable or compostable, just as some synthetic materials can be (Fig. 2.5). A material's biodegradability depends on the presence and activity of microorganisms that break it down into smaller components, such as CO_2 and CH_4, using the material as a nutrient source (Koch, 2017). Compostability, which is distinct from biodegradability, requires a material to be biodegradable while also meeting additional criteria, including safety standards, since compost is intended to re-enter the food system. International standards for compostability have been established to assess these materials (Koch, 2017).

Pure starch-based material, without added plasticizers or processing aids, is typically both biodegradable and compostable. Consequently, films and coatings made from biopolymers derived from biomass, such as starch, represent a promising alternative to synthetic food packaging. These materials are abundant, renewable, cost-effective, and biodegradable, with excellent film-forming properties (Atarés & Chiralt, 2016). Under composting conditions, starch films based on Andean crops (including cassava, banana, sweet potato, gold potato, white carrot, and chickpea) exhibited over 80% weight loss after 15 days, with all tested films showing more than 90% weight loss after 30 days (Torres et al., 2011). Thermoplastic cassava starch-based films modified by esterification with citric acid and ascorbic acid displayed weight loss percentages of 61% and 65% after 30 days under soil burial conditions, respectively (Zain et al., 2018). In comparison, the weight loss for synthetic polymers was significantly lower; for example, polypropylene (a common synthetic polymer used in plastics production) lost only about 0.2% of its weight after 10 years (Albertsson, 1980).

2.5 Applications in food technology

Starch films and coatings have emerged as effective edible packaging solutions for food products, offering a protective barrier that helps retain quality and prevent spoilage. Research has demonstrated that these materials can significantly extend the shelf life of a variety of products, highlighting their promise as a sustainable alternative to synthetic packaging. Table 2.2 illustrates specific applications of starch films and coatings in the food industry, emphasizing their role in prolonging shelf life and preserving food quality (Fig. 2.6).

As evidenced by the studies presented in Table 2.2, starch-based biodegradable materials must possess specific characteristics to be effective in food packaging, thereby enhancing food quality, shelf life, and safety. Key properties such as mechanical strength, barrier protection (against water and oxygen), antibacterial, and antioxidant capabilities can be achieved by blending with other polymers or incorporating antioxidants and antimicrobial agents. Overall, starch-based films and coatings show significant promise for commercial use in packaging a variety of foods, including fruits, vegetables, bakery goods, and meats. As active packaging films, for example, starch films have been shown to improve the microbiological stability of refrigerated Argentinian Port Salut cheese when combined with natural antimicrobial agents

TABLE 2.2 Summary of the literature information about the food application of films and coatings.

Main formulation	Additives	Preparation technique	Main applications/ products	Major findings	Selected references
Film: cassava starch (5%) + glycerol (1.5%) + water (93.5%)	Natural extracts of green tea and basil (5 wt.% of the distilled water)	Casting: solutions were spread on plates and dried at 50°C for 48 h	Development of active and intelligent primary packaging	• The phenolic compounds in green tea and basil extracts significantly enhanced the antioxidant activity of the films • Films showed noticeable color changes when exposed to different pH environments • Maintained thermal stability up to 240°C	Medina-Jaramillo et al. (2017)
Film: cassava starch (3.4%) + polyvinyl alcohol (1.7%) + glycerol (25 wt.% on starch basis)	Betalains-rich red pitaya peel extract (0, 0.25, 0.50, and 1.00 wt.%)	Casting: solutions were spread on plates (24 cm diameter) and dried at 30°C for 48 h	Development of active and intelligent packaging films applied to monitor the freshness of shrimp	• Color alterations when exposed to alkaline environments (sensitivity to pH variations) • Extract improved the films' water vapor barrier, ultraviolet–visible light barrier, mechanical strength, antioxidant, and antimicrobial properties • The film containing 1.00 wt.% of the extract displayed visible color changes in response to the accumulated volatile nitrogen compounds during the shrimp spoilage process	Qin et al. (2020)
Film: tapioca starch (2.5 g) + glycerol (1 g) + water (46.5 g)	Natamycin (0.027%) + nisin (0.0068%)	Casting: solutions were spread on plates (7 cm diameter) and dried at 37°C for 48 h	Improve the microbiological stability of refrigerated Argentinian Port Salut cheese	• Inhibition of growth of yeasts and molds • Control of psychrotrophic bacteria originally present in the cheese stored at refrigeration temperature • Inhibition of a mixed culture (*Saccharomyces* • *cerevisiae and Listeria innocua*) present on the surface of cheese due to contamination, throughout the entire storage	Ollé Resa et al. (2016)
Film: cross-linked maize starch (3.61%) + glycerol (0.5%)	Grape juice (16.15:83.85 of grape juice: water ratio)	Casting: solutions were spread on plates (12 cm diameter) and dried at 50°C for 24 h	Protect the quality and microbial properties of chicken breast filets during cold storage	• Reduction of oxygen permeability and availability • Lower lipid oxidation and microbial (Total aerobic mesophilic, psychrophilic, and *Enterobacteriaceae*) growth during cold storage	Yıldırım-Yalçın et al. (2021)
Film: corn starch (5 g) + glycerol (1.5 g) + water	Red cabbage extract (64.18%) + sweet whey (4.36%)	Casting: solutions were spread on plates and dried at 35°C for 24 h	Ground beef packaging	• Production of colored films with good mechanical properties, high antioxidant capacity, low solubility and low water vapor permeability • The presence of anthocyanins contributed to the conservation of the ground meat	Ribeiro Sanches et al. (2021)

(Continued)

TABLE 2.2 (Continued)

Main formulation	Additives	Preparation technique	Main applications/ products	Major findings	Selected references
Film: potato starch (5%) + glycerin (5%) + water	Nano-SiO_2 (0.3%, 15–100 nm)	Casting: solutions were spread on plates (25 cm diameter) and dried at 50°C	White mushroom packaging	• Water resistance, mechanical properties, resistance to ultraviolet and thermal aging were improved with the addition of nano-SiO_2 • The films used for packaging white mushrooms exhibited exceptional preservation properties	Zhang et al. (2018)
Edible coating: cassava starch (2%) + glycerol (0.64%) + chitosan (2%) + acetic acid (1.5%) + glycerol (1.28%)	Essential oil (EO) of *Lippia gracilis* Schauer genotypes (1%–3%)	Sanitized guavas were immersed in the coating solution for 3 min and dried at room temperature	Impact of coating on the shelf life of guavas	• The coatings with EO effectively inhibited bacterial growth and reduced microbial counts, enhancing guava shelf life • Coatings maintained total soluble solids, and the presence of EO reduced titratable acidity and presented better color values • Details in Fig. 2.6	De Aquino et al. (2015)
Edible coating: gelatin (10%), 1:1 corn starch (3%–5%, native, waxy or modified waxy) + sorbitol (10%)	No additives	Sanitized grapes were immersed for 1 min in the coating solutions and dried at room temperature	Effect of coatings on quality of refrigerated Red Crimson grapes	• Coated grapes showed improved appearance and lower weight loss after 21 days of refrigeration, with no negative impact on sensory acceptability • Adding gelatin to the films increased mechanical strength, solubility, permeability to water vapor, and thickness, while decreasing opacity • Samples with sorbitol showed lower water vapor permeability and higher tensile strength than those with glycerol • Details in Fig. 2.6	Fakhouri et al. (2015)
Edible coating: corn starch (1.5%) + glycerol (20.2%) + tomato powder (4.0%–11.0% of starch weight)	Corn starch was modified using ascorbic acid (0.8%–2.0% of starch weight)	Molded raw doughs were immersed in the coating solutions for 3 min, being then proofed, baked (220°C, 20 min) and frozen (−18°C, 15–60 days)	Edible coatings for frozen bread dough	• Coated frozen dough resulted in breads with greater specific volume and less hardness compared to uncoated dough • The presence of K, S, Ca, and P minerals was confirmed, indicating effective coating mobility into the dough	Theóphilo Galvão et al. (2018)
Edible coating: rice starch (2.5%) + ι-carrageenan (1.5%) + sucrose fatty acid ester (2%) + glycerol (1.5%)	No additives	The coating (0.5 mL) was uniformly applied over the fruit surface and dried (35°C, 15 s)	Application of edible coating to improve the postharvest quality of "Cripps pink" apple	At low temperature, it was observed: • Reduced weight loss • Constant tissue firmness • No changes in the bioactive profile • Delay in skin color change • Reduction in fruit greasiness	Thakur et al. (2019)

Edible coating: rice starch (4% w/v) + glycerol (2% w/v)	Chitosan (0.5% and 1%), red palm oil (1%), sodium caseinate as emulsifier (0.2%)	The walnut kernels were immersed in the coating solution for 60 s, drained for 30 s, and then dried (25°C for 24 h)	Application of edible coating on dried walnut kernels	• The rice starch edible film alone and in combination with chitosan significantly enhanced shelf life and maintained quality of walnuts under high humidity and temperature	Aghazadeh et al. (2018)
Edible coating: cassava starch (3%)	Potassium sorbate (0.05%)	Fruits were immersed in the coating solutions for 3 min and drained for 3 h	Applied on minimally processed strawberries	• Cassava starch coatings extended strawberry shelf life up to 12 days compared to nine days for controls • Coatings reduced the respiration rate and delayed weight and firmness loss without affecting physicochemical properties or color	Garcia et al. (2012)

FIGURE 2.6 Examples of edible coatings for food applications: (A) Guavas coated with chitosan-cassava starch blend (left) and uncoated guavas (right), after eight days at room temperature; (B) red Crimson grapes uncoated (left) and coated with corn starch-gelatin blend (right), after 21 days of storage under refrigerated conditions.
Adapted from De Aquino et al. (2015); Fakhouri et al. (2015). Copyright 2024, with permission from Elsevier.

(Ollé Resa et al., 2016). They have also utilized anthocyanins to enhance the preservation of ground beef (Ribeiro Sanches et al., 2021) and protect the quality and microbial properties of chicken breast filets during cold storage when reinforced with grape juice (Yıldırım-Yalçın et al., 2021). Additionally, starch-based films have been effective in extending the shelf life of white mushrooms when combined with nanosilica (Zhang et al., 2018).

As active coatings, starch has been successfully used to extend the shelf life of various fruits, including guavas (De Aquino et al., 2015), grapes (Fakhouri et al., 2015), apples (Thakur et al., 2019), and strawberries (Garcia et al., 2012), with minimal or no additives. Starch coatings have also proven effective in preserving and prolonging the shelf life of other products, such as frozen bread dough when combined with ascorbic acid and tomato powder (Theóphilo Galvão et al., 2018), and dried walnut kernels when blended with chitosan, red palm oil, and sodium caseinate as an emulsifier (Aghazadeh et al., 2018).

Finally, as components of intelligent packaging systems, starch-based materials have been developed to respond to varying pH environments, utilizing natural extracts from green tea and basil (Medina-Jaramillo et al., 2017). Additionally, a packaging solution has been proposed that exhibits visible color changes in response to the accumulation of volatile nitrogen compounds during shrimp spoilage. This innovative approach combines cassava starch with betalain-rich extract from red pitaya peel to effectively monitor shrimp freshness (Qin et al., 2020).

Overall, the starch-based films and coatings exhibit significant potential in various applications due to their barrier properties, flexibility, and cost-effectiveness. These innovative materials present a promising alternative to conventional synthetic packaging, contributing to enhanced food safety and environmental sustainability.

2.5.1 Regulatory and safety considerations

When considering starch-based packaging for direct contact with food, it is essential to ensure that these materials comply with relevant regulatory standards to guarantee safety. In the European Union (EU), food contact materials (FCMs) are governed by Regulation (EC) No 1935/2004, which ensures that packaging materials do not endanger human health, do not release harmful substances into food, and do not alter the food's characteristics. In the United States, starch-based packaging is regulated as food-contact substances, with specific guidelines ensuring safety and compliance. In Brazil, the regulation of materials intended for food contact, including starch-based packaging, is overseen by Agência Nacional de Vigilância Sanitária (ANVISA). The legislation that governs the use of FCMs in Brazil includes Resolution RDC No. 18/

2008, which outlines the technical requirements for food packaging materials. According to this resolution, any material or substance that may come into contact with food is classified as an FCM and must comply with health and safety standards established by ANVISA and adhere to Good Manufacturing Practices (BPF). This includes ensuring that all substances used in packaging are safe, nontoxic, and do not pose a risk to food safety or human health. Starch-based films, particularly those made from native starch, are generally regarded as safe for food contact, as they do not pose health risks. However, if the starch undergoes chemical modifications or if plasticizers are used, additional safety assessments are necessary. Chemical modifications or the addition of plasticizers could potentially alter the properties of the material and lead to migration into food. While the risk is typically low, it is essential to use food-grade compounds in these cases to mitigate any potential health risks or impact on food quality. In both the EU and the United States, compliance with these safety standards is critical for the commercialization of starch-based packaging as a sustainable and safe alternative for food packaging (Koch, 2017).

2.6 Challenges, limitations, and future perspectives

Starch-based films and coatings offer an eco-friendly alternative to synthetic packaging, but their adoption faces significant challenges and limitations that must be addressed to improve functionality and scalability. As discussed, native starch films are prone to cracking, insufficient flexibility, and high hydrophilicity, which limit their application. Some alternatives have been explored such as the use of different additives, as well as the incorporation of nanofillers, such as nanocellulose, which has been shown to enhance mechanical strength and flexibility (Bangar et al., 2021; Cheng et al., 2022), cross-linking, blending with hydrophobic biopolymers (e.g., polycaprolactone), and the addition of water-resistant agents like waxes or EOs (Atarés & Chiralt, 2016; Dong et al., 2024). In addition, while starch films provide good oxygen barriers, their water vapor permeability and sealing capabilities remain suboptimal. Nanotechnology solutions, such as adding nanoparticles (e.g., silica or clay) or creating multilayer films, can improve these properties (Bangar et al., 2021). Nanomaterials like nanocellulose and nanosilica are being incorporated into starch films to enhance mechanical, thermal, and barrier properties while maintaining biodegradability (Cheng et al., 2022; Dong et al., 2024). Active coatings with antimicrobial, antioxidant, or pH-sensitive properties are being developed using bioactive compounds and natural extracts, broadening the scope of applications (Atarés & Chiralt, 2016; Cheng et al., 2022). Blending starch with other biopolymers or biodegradable synthetic polymers like polycaprolactone or cellulose acetate butyrate can overcome limitations related to water sensitivity and mechanical weakness (Debeaufort & Voilley, 2009; Dong et al., 2024). Research has shown that starch films blended with polycaprolactone and cellulose acetate butyrate achieve excellent biodegradability and mechanical properties, with weight loss ranging from 8% to 80% under composting conditions. These materials represent a viable option for reducing plastic waste while maintaining functional performance (Debeaufort & Voilley, 2009; Dong et al., 2024).

In future perspectives, the improvement of scalability and industrialization has been explored by developing efficient manufacturing processes, such as extrusion and continuous casting, which could facilitate large-scale production while maintaining high-quality films and coatings (Bangar et al., 2021). The economic feasibility of starch-based films and coatings for large-scale applications is a critical factor influencing their adoption in various industries, particularly food packaging. While these materials offer significant environmental benefits, transitioning from laboratory research to industrial-scale production presents several economic challenges and opportunities. In this sense, investments in green and energy-efficient technologies will support the industrial adoption of starch-based materials. For widespread industrial adoption, starch-based films must be compatible with existing packaging machinery. Challenges include the need for adjustments in sealing, printing, and filling systems to handle starch films with varying mechanical and thermal properties, as well as additional costs for retrofitting or replacing existing equipment. Overcoming these barriers requires collaboration between material scientists and industrial manufacturers to develop standardized, machine-compatible films.

In large-scale applications, the starch-based materials can be associated with for specific innovations such as smart films with controlled release of active agents or spoilage indicators that can revolutionize food preservation and reduce waste (Cheng et al., 2022). For agriculture, biodegradable coatings for seeds or agricultural films that degrade after use can address environmental concerns in farming. For biomedical applications, innovations in starch-based films for drug delivery systems, wound dressings, and biodegradable medical devices are promising areas of research.

Other points include improving regulatory and consumer acceptance by ensuring compliance with food safety regulations and demonstrating the nontoxic nature of starch-based coatings, which will be essential for broader acceptance (Koch, 2017). In addition, educating consumers on the environmental and health benefits of biopolymer-based packaging will drive demand.

Starch is widely available and inexpensive, as it can be sourced from abundant crops such as corn, potato, cassava, and wheat. However, the choice of starch source can impact the overall cost. Corn and potato starch are commonly used due to their global availability and relatively low cost. Specialty starches, such as those extracted from underutilized crops, may be more expensive but offer unique functional properties (e.g., higher amylose content or specific gelatinization behaviors). In regions where agricultural waste (e.g., cassava peels) can be utilized, the raw material cost can be further reduced, enhancing economic viability.

The future of starch-based films and coatings lies in innovative material designs, advanced processing technologies, and tailored applications, enabling these materials to overcome current limitations and unlock their full potential. By addressing challenges such as brittleness, water sensitivity, and scalability while embracing emerging technologies like nanotechnology, biocomposites, and intelligent packaging systems, starch-based materials can evolve into robust, sustainable solutions. These advancements will not only provide high-performance alternatives to synthetic packaging but also combine environmental benefits with enhanced functionality and adaptability across diverse industries, paving the way for a more sustainable future.

2.7 Conclusion

This chapter delved into the fundamental technological aspects associated with starch-based films and coatings. It covered a detailed examination of the structure and inherent properties of starch, alongside the various production and modification techniques utilized for creating starch-based films and coatings. Key material properties were discussed, as well as their diverse applications. Notably, the two predominant production methodologies—casting and extrusion—were described in detail, highlighting their distinctive processes, potential formulations, and the variety of applications within different food matrices, such as fruits, meats, and cheeses.

Furthermore, the chapter addressed several challenges and opportunities inherent in the field, including aspects of biodegradability and regulatory considerations. The investigation into starch films and coatings underscores their expansive potential as multifunctional materials within food technology, offering promising perspectives for future applications. With ongoing advancements in modification techniques, formulation development, and processing methods, starch-based films are well positioned to tackle numerous challenges currently faced by the food packaging industry. These innovations are paving the way for more sustainable and environmentally friendly solutions, contributing significantly to the evolution of packaging technologies. By enhancing the efficiency and functionality of these materials, starch films and coatings can lead to improvements in food preservation and safety, which are paramount concerns in today's eco-conscious marketplace.

Acknowledgments

The authors acknowledge the São Paulo Research Foundation (FAPESP) within the scope of the Center for Science and Development of Solutions for Post-Consumer Waste: Packaging and Products—CCD Circula (Grant #2021/11967-6). This study was financed, in part, by the São Paulo Research Foundation (FAPESP), Brazil. Process Numbers #2020/08727-8, #2024/02072-3, #2024/10994-8 and #2024/07555-2.

AI disclosure

During the preparation of this work, the author(s) used GPT-4o as a generative AI tool to improve the English language and overall readability of the text. After using this tool, the author(s) reviewed and edited the content as needed and take(s) full responsibility for the content of the publication.

References

Adame, M. Y., Shi, C., Li, C., Aziz, T., Alharbi, M., Cui, H., & Lin, L. (2024). Fabrication and characterization of pullulan/tapioca starch-based antibacterial films incorporated with *Litsea cubeba* essential oil for meat preservation. *International Journal of Biological Macromolecules, 268*. https://doi.org/10.1016/j.ijbiomac.2024.131775, https://www.sciencedirect.com/science/journal/01418130.

Aghazadeh, M., Karim, R., Sultan, M. T., Paykary, M., Johnson, S. K., & Shekarforoush, E. (2018). Comparison of starch films and effect of different rice starch-based coating formulations on physical properties of walnut during storage time at accelerated temperature. *Journal of Food Process Engineering, 41*(1). https://doi.org/10.1111/jfpe.12607, http://www.blackwellpublishers.co.uk/journal_new.asp?site=1.

Albertsson, A.-C. (1980). The shape of the biodegradation curve for low and high density polyethenes in prolonged series of experiments. *European Polymer Journal, 16*(7), 623–630. https://doi.org/10.1016/0014-3057(80)90100-7.

Atarés, L., & Chiralt, A. (2016). Essential oils as additives in biodegradable films and coatings for active food packaging. *Trends in Food Science & Technology, 48*, 51–62. https://doi.org/10.1016/j.tifs.2015.12.001.

Atwell, W. A., Hood, L. F., Lineback, D. R., Varriano-Marston, E., & Zobel, H. F. (1988). The terminology and methodology associated with basic starch phenomena. *Cereal Foods World, 33*, 306–311.

Averous, L., & Halley, P. J. (2014). Starch polymers: From the field to industrial products. In *Starch polymers: From genetic engineering to green applications*. France: Elsevier Inc, 3–10. http://www.sciencedirect.com/science/book/9780444537300, 10.1016/B978-0-444-53730-0.00018-X.

Balakrishnan, P., Sreekala, M. S., Geethamma, V. G., Kalarikkal, N., Kokol, V., Volova, T., & Thomas, S. (2019). Physicochemical, mechanical, barrier and antibacterial properties of starch nanocomposites crosslinked with pre-oxidised sucrose. *Industrial Crops and Products, 130*, 398–408. https://doi.org/10.1016/j.indcrop.2019.01.007.

Bangar, S. P., Purewal, S. S., Trif, M., Maqsood, S., Kumar, M., Manjunatha, V., & Rusu, A. V. (2021). Functionality and applicability of starch-based films: An eco-friendly approach. *Foods, 10*(9). https://doi.org/10.3390/foods10092181, https://www.mdpi.com/2304-8158/10/9/2181/pdf.

Basiak, E., Lenart, A., & Debeaufort, F. (2018). How glycerol and water contents affect the structural and functional properties of starch-based edible films. *Polymers, 10*(4), 412. https://doi.org/10.3390/polym10040412.

Bastarrachea, L., Dhawan, S., & Sablani, S. S. (2011). Engineering properties of polymeric-based antimicrobial films for food packaging: A review. *Food Engineering Reviews, 3*(2), 79–93. https://doi.org/10.1007/s12393-011-9034-8.

Bergo, P., Sobral, P. J. A., & Prison, J. M. (2010). Effect Of glycerol on physical properties of cassava starch films. *Journal of Food Processing and Preservation, 34*(2), 401–410. https://doi.org/10.1111/j.1745-4549.2008.00282.x.

Bioplastics, E. (2022). Fact Sheet - What are bioplastics? http://www.european-bioplastics.org/news/publications/.

Bonilla, J., Atarés, L., Vargas, M., & Chiralt, A. (2012). Edible films and coatings to prevent the detrimental effect of oxygen on food quality: Possibilities and limitations. *Journal of Food Engineering, 110*(2), 208–213. https://doi.org/10.1016/j.jfoodeng.2011.05.034.

Bonilla, J., Atarés, L., Vargas, M., & Chiralt, A. (2013). Properties of wheat starch film-forming dispersions and films as affected by chitosan addition. *Journal of Food Engineering, 114*(3), 303–312. https://doi.org/10.1016/j.jfoodeng.2012.08.005.

Castanha, Nanci, Matta Junior, Manoel Divino da, & Augusto, Pedro Esteves Duarte (2017). Potato starch modification using the ozone technology. *Food Hydrocolloids, 66*, 343–356. 10.1016/j.foodhyd.2016.12.001.

Castanha, Nanci, Santos, Débora Nascimento e, Cunha, Rosiane Lopes, & Augusto, Pedro Esteves Duarte (2019). Properties and possible applications of ozone-modified potato starch. *Food Research International, 116*, 1192–1201. 10.1016/j.foodres.2018.09.064.

Chaléat, C., Halley, P. J., & Truss, R. W. (2014). Mechanical properties of starch-based plastics. In *Starch polymers: From genetic engineering to green applications*. Australia: Elsevier B.V, 187–209. http://www.sciencedirect.com/science/book/9780444537300, 10.1016/B978-0-444-53730-0.00023-3.

Cheng, M., Yan, X., Cui, Y., Han, M., Wang, X., Wang, J., & Zhang, R. (2022). An eco-friendly film of pH-responsive indicators for smart packaging. *Journal of Food Engineering, 321*, 110943. https://doi.org/10.1016/j.jfoodeng.2022.110943.

De Aquino, A. B., Blank, A. F., & De Aquino Santana, L. C. L. (2015). Impact of edible chitosan-cassava starch coatings enriched with *Lippia gracilis* Schauer genotype mixtures on the shelf life of guavas (*Psidium guajava* L.) during storage at room temperature. *Food Chemistry, 171*, 108–116. https://doi.org/10.1016/j.foodchem.2014.08.077, www.elsevier.com/locate/foodchem.

Debeaufort, F., & Voilley, A. (2009). *Lipid-based edible films and coatings*. Springer Science and Business Media LLC, 135–168. 10.1007/978-0-387-92824-1_5.

Devi, L. S., Jaiswal, A. K., & Jaiswal, S. (2024). Lipid incorporated biopolymer based edible films and coatings in food packaging: A review. *Current Research in Food Science, 8*. https://doi.org/10.1016/j.crfs.2024.100720, https://www.sciencedirect.com/science/journal/26659271.

Dong, M., Mastroianni, G., Bilotti, E., Zhang, H., & Papageorgiou, D. G. (2024). Biodegradable starch-based nanocomposite films with exceptional water and oxygen barrier properties. *ACS Sustainable Chemistry and Engineering, 12*(29), 11056–11066. https://doi.org/10.1021/acssuschemeng.4c04198, http://pubs.acs.org/journal/ascecg.

Eliasson, A. (2006). In A. Eliasson (Ed.). *Carbohydrates in food*CRC Press, 2nd ed.

Eliasson, A., & Gudmundsson, M. (2006). Starch: Physicochemical and Functional Aspects. In A. Eliasson (Ed.), Carbohydrates in Food (2nd ed., pp. 391–469). CRC Press.

Fakhouri, F. M., Martelli, S. M., Caon, T., Velasco, J. I., & Mei, L. H. I. (2015). Edible films and coatings based on starch/gelatin: Film properties and effect of coatings on quality of refrigerated Red Crimson grapes. *Postharvest Biology and Technology, 109*, 57–64. https://doi.org/10.1016/j.postharvbio.2015.05.015, www.elsevier.com/inca/publications/store/5/0/3/3/1/3/index.htt.

Gallant, D. J., Bouchet, B., Buleon, A., & Perez, S. (1992). Nutrition 2 S3-s16S3 France Physical characteristics of starch granules and susceptibility to enzymatic degradation. *European Journal of Clinical, 46*.

Garcia, L. C., Pereira, L. M., De Luca Sarantópoulos, C. I. G., & Hubinger, M. D. (2012). Effect of antimicrobial starch edible coating on shelf-life of fresh strawberries. *Packaging Technology and Science, 25*(7), 413–425. https://doi.org/10.1002/pts.987.

Giannakas, A. E. (2024). Extrusion of biopolymers for food applications. In *Advances in biopolymers for food science and technology*. Greece: Elsevier, 137–169. https://www.sciencedirect.com/book/9780443190056, 10.1016/B978-0-443-19005-6.00007-4.

Gupta, V., Biswas, D., & Roy, S. (2022). A comprehensive review of biodegradable polymer-based films and coatings and their food packaging applications. *Materials, 15*(17). https://doi.org/10.3390/ma15175899, http://www.mdpi.com/journal/materials.

Han, J. H. (2003). Antimicrobial food packaging. In *Novel food packaging techniques*. Canada: Elsevier, 50–70. https://www.sciencedirect.com/book/9781855736757, 10.1533/9781855737020.1.50.

Hizukuri, S., Abe, J., & Hanashiro, I. (2006). Starch: Analytical aspects. In *Carbohydrates in food*. CRC Press, 305–390.

Kaur, B., Ariffin, F., Bhat, R., & Karim, A. A. (2012). Progress in starch modification in the last decade. *Food Hydrocolloids, 26*(2), 398–404. https://doi.org/10.1016/j.foodhyd.2011.02.016, https://www.journals.elsevier.com/food-hydrocolloids.

Koch, K. (2017). Starch-Based Films. In *Starch in food: Structure, function and applications*. Sweden: Elsevier, 747–767. https://www.sciencedirect.com/book/9780081008683, 10.1016/B978-0-08-100868-3.00019-6.

Kramer, M. E. (2009). *Structure and function of starch-based edible films and coatings*. Springer Science and Business Media LLC, 113–134. 10.1007/978-0-387-92824-1_4.

Kumari, B., & Sit, N. (2023). Comprehensive review on single and dual modification of starch: Methods, properties and applications. *International Journal of Biological Macromolecules, 253*(Pt 3), 126952. https://doi.org/10.1016/j.ijbiomac.2023.126952.

Kurek, M., Guinault, A., Voilley, A., Galić, K., & Debeaufort, F. (2014). Effect of relative humidity on carvacrol release and permeation properties of chitosan based films and coatings. *Food Chemistry, 144*, 9–17. https://doi.org/10.1016/j.foodchem.2012.11.132.

La Fuente, C. I. A., de Souza, A. T., Tadini, C. C., & Augusto, P. E. D. (2019). Ozonation of cassava starch to produce biodegradable films. *International Journal of Biological Macromolecules, 141*, 713–720. https://doi.org/10.1016/j.ijbiomac.2019.09.028, www.elsevier.com/locate/ijbiomac.

La Fuente, C. I. A., Castanha, N., Maniglia, B. C., Tadini, C. C., & Augusto, P. E. D. (2020). Biodegradable films produced from ozone-modified potato starch. *Journal of Packaging Technology and Research, 4*(1), 3–11. https://doi.org/10.1007/S41783-020-00082-0.

Lauer, M. K., & Smith, R. C. (2020). Recent advances in starch-based films toward food packaging applications: Physicochemical, mechanical, and functional properties. *Comprehensive Reviews in Food Science and Food Safety, 19*(6), 3031–3083. https://doi.org/10.1111/1541-4337.12627, http://onlinelibrary.wiley.com/journal/10.1111/(ISSN)1541-4337.

Liu, H., Yu, L., Xie, F., & Chen, L. (2006). Gelatinization of cornstarch with different amylose/amylopectin content. *Carbohydrate Polymers, 65*(3), 357–363. https://doi.org/10.1016/j.carbpol.2006.01.026.

Liu, Q. (2005). *Understanding starches and their role in foods food carbohydrates: Chemistry, physical properties, and applications*. Canada: CRC Press, 309–355. https://www.routledge.com/Food-Carbohydrates-Chemistry-Physical-Properties-and-Applications/Cui/p/book/9780849315749.

Liyanapathiranage, A., Dassanayake, R. S., Gamage, A., Karri, R. R., Manamperi, A., Evon, P., Jayakodi, Y., Madhujith, T., & Merah, O. (2023). Recent developments in edible films and coatings for fruits and vegetables. *Coatings, 13*(7). https://doi.org/10.3390/coatings13071177, www.mdpi.com/journal/coatings.

Luchese, C. L., Abdalla, V. F., Spada, J. C., & Tessaro, I. C. (2018). Evaluation of blueberry residue incorporated cassava starch film as pH indicator in different simulants and foodstuffs. *Food Hydrocolloids, 82*, 209–218. https://doi.org/10.1016/j.foodhyd.2018.04.010, https://www.journals.elsevier.com/food-hydrocolloids.

Luchese, C. L., Benelli, P., Spada, J. C., & Tessaro, I. C. (2018). Impact of the starch source on the physicochemical properties and biodegradability of different starch-based films. *Journal of Applied Polymer Science, 135*(33). https://doi.org/10.1002/app.46564, http://onlinelibrary.wiley.com/journal/10.1002/(ISSN)1097-4628.

Majeed, T., Dar, A. H., Pandey, V. K., Dash, K. K., Srivastava, S., Shams, R., Jeevarathinam, G., Singh, P., Echegaray, N., & Pandiselvam, R. (2023). Role of additives in starch-based edible films and coating: A review with current knowledge. *Progress in Organic Coatings, 181*. https://doi.org/10.1016/j.porgcoat.2023.107597, www.elsevier.com/locate/porgcoat.

Maniglia, B. C., Tessaro, L., Ramos, A. P., & Tapia-Blácido, D. R. (2019). Which plasticizer is suitable for films based on babassu starch isolated by different methods? *Food Hydrocolloids, 89*, 143–152. https://doi.org/10.1016/j.foodhyd.2018.10.038, https://www.journals.elsevier.com/food-hydrocolloids.

Maniglia, B. C., Castanha, N., Le-Bail, P., Le-Bail, A., & Augusto, P. E. D. (2020). Starch modification through environmentally friendly alternatives: A review. *Critical Reviews in Food Science and Nutrition, 0*(0), 1–24. https://doi.org/10.1080/10408398.2020.1778633.

Maniglia, B. C., Castanha, N., Rojas, M. L., & Augusto, P. E. (2021). Emerging technologies to enhance starch performance. *Current Opinion in Food Science, 37*, 26–36. https://doi.org/10.1016/j.cofs.2020.09.003.

Maniglia, B. C., Silveira, T. M. G., & Tapia-Blácido, D. R. (2022). Starch isolation from turmeric dye extraction residue and its application in active film production. *International Journal of Biological Macromolecules, 202*, 508–519. https://doi.org/10.1016/j.ijbiomac.2021.12.145, www.elsevier.com/locate/ijbiomac.

Maran, J. P., Sivakumar, V., Sridhar, R., & Thirugnanasambandham, K. (2013). Development of model for barrier and optical properties of tapioca starch based edible films. *Carbohydrate Polymers, 92*(2), 1335–1347. https://doi.org/10.1016/j.carbpol.2012.09.069.

Martins, P. C., Bagatini, D. C., & Martins, V. G. (2021). Oregano essential oil addition in rice starch films and its effects on the chilled fish storage. *Journal of Food Science and Technology, 58*(4), 1562–1573. https://doi.org/10.1007/s13197-020-04668-z, http://www.springerlink.com/content/121580/.

Medina-Jaramillo, C., Ochoa-Yepes, O., Bernal, C., & Famá, L. (2017). Active and smart biodegradable packaging based on starch and natural extracts. *Carbohydrate Polymers, 176*, 187–194. https://doi.org/10.1016/j.carbpol.2017.08.079.

Moreno, O., Atarés, L., & Chiralt, A. (2015). Effect of the incorporation of antimicrobial/antioxidant proteins on the properties of potato starch films. *Carbohydrate Polymers, 133*, 353–364. https://doi.org/10.1016/j.carbpol.2015.07.047.

Ollé Resa, C. P., Gerschenson, L. N., & Jagus, R. J. (2016). Starch edible film supporting natamycin and nisin for improving microbiological stability of refrigerated argentinian Port Salut cheese. *Food Control, 59*, 737–742. https://doi.org/10.1016/j.foodcont.2015.06.056.

Othman, S. H., Nordin, N., Azman, N. A. A., Tawakkal, I. S. M. A., & Basha, R. K. (2021). Effects of nanocellulose fiber and thymol on mechanical, thermal, and barrier properties of corn starch films. *International Journal of Biological Macromolecules, 183*, 1352–1361. https://doi.org/10.1016/j.ijbiomac.2021.05.082, www.elsevier.com/locate/ijbiomac.

Otoni, C. G., Avena-Bustillos, R. J., Azeredo, H. M. C., Lorevice, M. V., Moura, M. R., Mattoso, L. H. C., & McHugh, T. H. (2017). Recent advances on edible films based on fruits and vegetables—A review. *Comprehensive Reviews in Food Science and Food Safety, 16*(5), 1151–1169. https://doi.org/10.1111/1541-4337.12281, http://www.interscience.wiley.com/jpages/1541-4337.

Pérez, S., Baldwin, P. M., & Gallant, D. J. (2009). *Structural features of starch granules i starch*. France: Elsevier Inc, 149–192. http://www.sciencedirect.com/science/book/9780127462752, 10.1016/B978-0-12-746275-2.00005-7.

Qin, Y., Liu, Y., Zhang, X., & Liu, J. (2020). Development of active and intelligent packaging by incorporating betalains from red pitaya (*Hylocereus polyrhizus*) peel into starch/polyvinyl alcohol films. *Food Hydrocolloids, 100*, 105410. https://doi.org/10.1016/j.foodhyd.2019.105410.

Quintavalla, S., & Vicini, L. (2002). Antimicrobial food packaging in meat industry. *Meat Science, 62*(3), 373–380. https://doi.org/10.1016/s0309-1740(02)00121-3.

Ren, L., Yan, X., Zhou, J., Tong, J., & Su, X. (2017). Influence of chitosan concentration on mechanical and barrier properties of corn starch/chitosan films. *International Journal of Biological Macromolecules, 105*(Pt 3), 1636–1643. https://doi.org/10.1016/j.ijbiomac.2017.02.008.

Ribeiro Sanches, M. A., Camelo-Silva, C., da Silva Carvalho, C., Rafael de Mello, J., Barroso, N. G., Lopes da Silva Barros, E., Silva, P. P., & Pertuzatti, P. B. (2021). Active packaging with starch, red cabbage extract and sweet whey: Characterization and application in meat. *LWT, 135*. https://doi.org/10.1016/j.lwt.2020.110275, https://www.journals.elsevier.com/lwt.

Rodrigues, F. A. M., dos Santos, S. B. F., Lopes, M. Md. A., Guimarães, D. J. S., de Oliveira Silva, E., de Souza Filho, Md. S. M., Mattos, A. L. A., da Silva, L. M. R., de Azeredo, H. M. C., & Ricardo, N. M. P. S. (2021). Antioxidant films and coatings based on starch and phenolics from *Spondias purpurea* L. *International Journal of Biological Macromolecules, 182*, 354–365. https://doi.org/10.1016/j.ijbiomac.2021.04.012, www.elsevier.com/locate/ijbiomac.

Singh, G. P., Bangar, S. P., Yang, T., Trif, M., Kumar, V., & Kumar, D. (2022). Effect on the properties of edible starch-based films by the incorporation of additives: A review. *Polymers, 14*(10), 1987. https://doi.org/10.3390/POLYM14101987.

Skurtys, O. (n.d.). Food hydrocolloid edible films and coatings.

Skurtys, O., Acevedo, C., Pedreschi, F., Enrione, J., Osorio, F., & Aguilera, J. M. (2010). *Food hydrocolloid edible films and coatings food hydrocolloids: Characteristics, properties and structures*. Chile: Nova Science Publishers, Inc, 41–80. https://www.novapublishers.com/catalog/product_info.php?products_id=10994.

Suhag, R., Kumar, N., Petkoska, A. T., & Upadhyay, A. (2020). Film formation and deposition methods of edible coating on food products: A review. *Food Research International, 136*. https://doi.org/10.1016/j.foodres.2020.109582, www.elsevier.com/inca/publications/store/4/2/2/9/7/0.

Suriyatem, R., Auras, R. A., & Rachtanapun, P. (2018). Improvement of mechanical properties and thermal stability of biodegradable rice starch–based films blended with carboxymethyl chitosan. *Industrial Crops and Products, 122*, 37–48. https://doi.org/10.1016/j.indcrop.2018.05.047, www.elsevier.com/inca/publications/store/5/2/2/8/2/5.

Tang, H., Mitsunaga, T., & Kawamura, Y. (2006). Molecular arrangement in blocklets and starch granule architecture. *Carbohydrate Polymers, 63*(4), 555–560. https://doi.org/10.1016/j.carbpol.2005.10.016.

Tavassoli-Kafrani, E., Shekarchizadeh, H., & Masoudpour-Behabadi, M. (2016). Development of edible films and coatings from alginates and carrageenans. *Carbohydrate Polymers, 137*, 360–374. https://doi.org/10.1016/j.carbpol.2015.10.074, http://www.elsevier.com/wps/find/journaldescription.cws_home/405871/description#description.

Tester, R. F., Karkalas, J., & Qi, X. (2004). Starch - Composition, fine structure and architecture. *Journal of Cereal Science, 39*(2), 151–165. https://doi.org/10.1016/j.jcs.2003.12.001, http://www.elsevier.com/inca/publications/store/6/2/2/8/5/9/index.htt.

Thakur, R., Pristijono, P., Scarlett, C. J., Bowyer, M., Singh, S. P., & Vuong, Q. V. (2019). Starch-based edible coating formulation: Optimization and its application to improve the postharvest quality of "Cripps pink" apple under different temperature regimes. *Food Packaging and Shelf Life, 22*. https://doi.org/10.1016/j.fpsl.2019.100409, http://www.journals.elsevier.com/food-packaging-and-shelf-life/.

Theóphilo Galvão, A. M. M., de Oliveira Araújo, A. W., Carneiro, S. V., Zambelli, R. A., & do Socorro Rocha Bastos, M. (2018). Coating development with modified starch and tomato powder for application in frozen dough. *Food Packaging and Shelf Life, 16*, 194–203. https://doi.org/10.1016/j.fpsl.2018.04.003, http://www.journals.elsevier.com/food-packaging-and-shelf-life/.

Torres, F. G., Troncoso, O. P., Torres, C., Díaz, D. A., & Amaya, E. (2011). Biodegradability and mechanical properties of starch films from Andean crops. *International Journal of Biological Macromolecules, 48*(4), 603–606. https://doi.org/10.1016/j.ijbiomac.2011.01.026.

Trinh, B. M., Chang, B. P., & Mekonnen, T. H. (2023). The barrier properties of sustainable multiphase and multicomponent packaging materials: A review. *Progress in Materials Science, 133*. https://doi.org/10.1016/j.pmatsci.2023.101071, https://www.journals.elsevier.com/progress-in-materials-science.

V, A. K., Hasan, M., Mangaraj, S., M, P., Verma, D. K., & Srivastav, P. P. (2022). Trends in edible packaging films and its prospective future in food: A review. *Applied Food Research, 2*(1). https://doi.org/10.1016/j.afres.2022.100118, https://www.journals.elsevier.com/applied-food-research.

Velaga, S. P., Nikjoo, D., & Vuddanda, P. R. (2018). Experimental studies and modeling of the drying kinetics of multicomponent polymer films. *AAPS PharmSciTech, 19*(1), 425–435. https://doi.org/10.1208/s12249-017-0836-8, http://www.springerlink.com/content/1530-9932/.

Vieira, M. G. A., Da Silva, M. A., Dos Santos, L. O., & Beppu, M. M. (2011). Natural-based plasticizers and biopolymer films: A review. *European Polymer Journal, 47*(3), 254–263. https://doi.org/10.1016/j.eurpolymj.2010.12.011.

Wang, L., Yang, C., Deng, X., Peng, J., Zhou, J., Xia, G., Zhou, C., Shen, Y., & Yang, H. (2023). A pH-sensitive intelligent packaging film harnessing *Dioscorea zingiberensis* starch and anthocyanin for meat freshness monitoring. *International Journal of Biological Macromolecules, 245*, 125485. https://doi.org/10.1016/J.IJBIOMAC.2023.125485.

Wang, S., Li, C., Copeland, L., Niu, Q., & Wang, S. (2015). Starch retrogradation: A comprehensive review. *Comprehensive Reviews in Food Science and Food Safety, 14*(5), 568–585. https://doi.org/10.1111/1541-4337.12143.

Whistler, R. L., Bemiller, J. N., & Paschall, E. F. (1984). *Starch: Chemistry and technology*. Academic Press.

Xie, F., Liu, H., Chen, P., Xue, T., Chen, L., Yu, L., & Corrigan, P. (2006). Starch gelatinization under shearless and shear conditions. *In International Journal of Food Engineering, Vol. 2*(Issue 5). https://doi.org/10.2202/1556-3758.1162.

Xie, F., Liu, H., Chen, P., Xue, T., Chen, L., Yu, L., & Corrigan, P. (2007). Starch gelatinization under shearless and shear conditions. *International Journal of Food Engineering, 2*(5). https://doi.org/10.2202/1556-3758.1162.

Xue, W., Zhu, J., Sun, P., Yang, F., Wu, H., & Li, W. (2023). Permeability of biodegradable film comprising biopolymers derived from marine origin for food packaging application: A review. *Trends in Food Science & Technology, 136*, 295–307. https://doi.org/10.1016/j.tifs.2023.05.001.

Yıldırım-Yalçın, M., Sadıkoğlu, H., & Şeker, M. (2021). Characterization of edible film based on grape juice and cross-linked maize starch and its effects on the storage quality of chicken breast fillets. *LWT, 142*, 111012. https://doi.org/10.1016/j.lwt.2021.111012.

Zain, A. H. M., Ab Wahab, M. K., & Ismail, H. (2018). Biodegradation behaviour of thermoplastic starch: the roles of carboxylic acids on cassava starch. *Journal of Polymers and the Environment, 26*(2), 691–700. https://doi.org/10.1007/s10924-017-0978-5, http://www.kluweronline.com/issn/1566-2543/.

Zhang, R., Wang, X., & Cheng, M. (2018). Preparation and characterization of potato starch film with various size of Nano-SiO_2. *Polymers, 10*(10). https://doi.org/10.3390/POLYM10101172.

Zhang, Y. R., Zhang, S. D., Wang, X. L., Chen, R. Y., & Wang, Y. Z. (2009). Effect of carbonyl content on the properties of thermoplastic oxidized starch. *Carbohydrate Polymers, 78*(1), 157–161. https://doi.org/10.1016/j.carbpol.2009.04.023.

Zhou, Y., Wu, X., Chen, J., & He, J. (2021). Effects of cinnamon essential oil on the physical, mechanical, structural and thermal properties of cassava starch-based edible films. *International Journal of Biological Macromolecules, 184*, 574–583. https://doi.org/10.1016/j.ijbiomac.2021.06.067.

Chapter 3

Furcellaran films and coatings

Nikola Nowak-Nazarkiewicz[1], Wiktoria Grzebieniarz[1], and Ewelina Jamróz[1,2]

[1]*Department of Chemistry, University of Agriculture, Kraków, Poland,* [2]*Department of Packaging and Logistics Processes, Krakow University of Economics, Kraków, Poland*

3.1 Introduction

Plastics derived from fossil fuels have long been the industry standard for food packaging due to their durability and cost. However, their widespread use comes with a significant downside: the accumulation of nonbiodegradable waste, which many nations struggle to manage effectively. Despite their dominance in the market, concerns over environmental impact and the demand for safer, nontoxic materials have spurred interest in alternative solutions (Luu et al., 2025). Ensuring food quality and minimizing waste heavily depend on effective packaging systems. Traditionally, the food industry has favored plastic film packaging due to its lightweight nature, versatility, and affordability. Despite these advantages, the persistence of plastic waste in the environment has raised significant ecological concerns (Naderi Kalali et al., 2023). As a result, there is a growing movement toward biodegradable polymer-based alternatives that offer sustainable solutions without compromising packaging functionality. Biopolymers have emerged as a promising class of renewable materials capable of meeting sustainability goals while enhancing packaging functionality. Common bio-based film components include polysaccharides, proteins, and lipids, showing potential as eco-friendly materials. Additionally, alternative biopolymer sources and novel film formulations continue to be explored for their potential to enhance film properties. Materials derived from starch, chitosan, alginate, cellulose, and gelatin have demonstrated remarkable functional properties. Furthermore, advancements in composites, polymer blends, and active or intelligent films based on biopolymers have created new possibilities for improving mechanical and barrier properties (Kalita et al., 2025).

One particularly abundant resource with significant promise in sustainable packaging is seaweed. Found in vast quantities in marine ecosystems, seaweed contains a variety of functional metabolites, including polysaccharides, proteins, peptides, lipids, amino acids, polyphenols, and minerals. Based on their pigmentation and chemical composition, seaweed species fall into three main categories: brown (*Phaeophyceae*), green (*Chlorophyceae*), and red (*Rhodophyceae*) (Kajla et al., 2024). Various species of red seaweeds, including *Gracilaria* sp., *Porphyra capensis, Aedoesorbitosa, Notogenia stiriata,* and *Gelidium* sp., are widely recognized for their ability to generate sulfated galactans such as agar, carrageenan, and furcellaran. These compounds share a galactose backbone but differ in sulfate ester distribution and 3,6-anhydrogalactose content (Truus et al., 1997), resulting in distinct rheological properties that are beneficial to various industries, especially food technology (Marangoni et al., 2020).

Furcellaran, a polysaccharide derived from red algae, has recently gained attention for its superior film-forming capability (Jamróz et al., 2018b). Recognized under European Union food additive regulations alongside carrageenan, furcellaran has been increasingly utilized in the production of sustainable, active, and intelligent edible films. Its potential in creating antioxidant, antimicrobial, and food quality indicators has been highlighted in recent studies (Marangoni et al., 2020).

The aim of this study is to provide an in-depth overview of furcellaran-based films, exploring their production processes and applications in food packaging. A discussion of structural analysis will help elucidate the relationship between molecular composition and material properties. Additionally, key findings regarding film characteristics and their suitability for active and intelligent packaging will be presented. Finally, a critical evaluation of current challenges and future research directions will offer insight into the continued development of furcellaran films, potentially guiding future advancements in this field.

Natural Polymer-Based Films and Coatings in Food Technology. DOI: https://doi.org/10.1016/B978-0-443-34244-8.00005-7

3.2 Chemical structure of furcellaran

Furcellaran is a naturally occurring sulfated polysaccharide primarily extracted from red algae (*Furcellaria lumbricalis*). Its structural characteristics resemble those of both agar and carrageenan, making it a valuable component in various applications. This polysaccharide naturally exists as a mixture of sodium, potassium, magnesium, and calcium salts in a linear polymer, featuring repeating units of [→4)-3,6-anhydro-d-galactopyranose-(1→3)-galactopyranose-4′-sulfate-(1→]. The molecular weight of furcellaran varies based on extraction methods, with reported values ranging from 290 to 500 kDa (Hugerth & Sundelöf, 2001). Furcellaran is characterized by the presence of a single ester sulfate group per tetramer, specifically located at the fourth position of the galactose unit. Structurally akin to the algal polysaccharide κ-carrageenan, furcellaran primarily differs in its lower sulfate content. It can be regarded as a copolymer comprising both β- and κ-carrageenan components (Polášek et al., 2021). A comparative analysis showed that native Estonian furcellaran exhibited more intense peaks corresponding to (1→3) bound β-d-galactopyranose residues, whereas alkali-modified and commercial variants displayed stronger signals associated with (1→4) linked 3,6-anhydro-α-d-galactopyranose (Polášek et al., 2021).

Monosaccharide composition analysis, carried out through gas chromatography-mass spectrometry (GC-MS), revealed that furcellaran primarily consists of galactose and 3,6-anhydrogalactose, with smaller amounts of 6-O-methylgalactose, mannose, xylose, and glucose (Štěpánková et al., 2022). In compliance with European food additive regulations, carrageenan (classified as E407) must be derived from seaweed, contain less than 2% of an acid-insoluble material, and produce transparent gels. Although furcellaran was initially classified separately, regulatory reevaluation allowed for the recognition of its structural and functional similarities with carrageenan, ultimately categorizing both under the E407 designation (Marangoni et al., 2020).

3.3 Single-layer films based on furcellaran

Furcellaran-based films exhibit a range of functional properties that make them a promising alternative to conventional synthetic films. Due to the intrinsic characteristics of furcellaran, these films demonstrate unique structural and functional attributes that contribute to enhanced food preservation. While some bio-derived polymers require additional functional additives to achieve desirable properties, furcellaran-based films inherently possess beneficial traits such as antioxidant activity, antimicrobial effects, and UV-light resistance. These properties play a crucial role in extending the shelf life of packaged food products by reducing microbial growth, preventing oxidative deterioration, and shielding the contents from photooxidation caused by UV exposure. Additionally, furcellaran films can serve as an effective medium for incorporating bioactive compounds or nanomaterials, further enhancing their protective capabilities. By integrating functional reinforcements, furcellaran-based films can be tailored to diverse applications in active and intelligent food packaging. In numerous studies, the potential of such bio-functional packaging materials has been explored, highlighting their effectiveness in maintaining food quality and safety. The physical and functional properties of various types of single-layer films based on furcellaran and active additives are shown. For the first time, Jamróz et al. (2021) explored the physicochemical, structural, and mechanical properties of furcellaran-chitosan films at different ratios. The findings confirmed that CHIT/FUR films can be successfully developed by optimizing their proportions. However, thermal treatment weakened the mechanical strength and barrier properties compared to single-component films. FUR/CHIT films were produced using the casting method for potential packaging applications. Zeta potential analysis revealed that complex formation depends on the properties and concentration of both polysaccharides. Rheological tests showed that these solutions rapidly formed hydrogels at moderate temperatures. Fourier transform infrared (FTIR) confirmed intermolecular interactions between polymer functional groups, while differential scanning calorimetry (DSC) results indicated that the CHIT-FUR 9:1 film had the best thermal stability. The contact angle test showed that increasing furcellaran content made the material more hydrophilic, which may be unsuitable for high-humidity food packaging. Atomic force microscope (AFM) analysis indicated that chitosan-based films had high surface porosity, while furcellaran-based films were smoother. Adding furcellaran reduced chitosan's porosity, creating a more uniform surface. Thermal treatment influenced the films' functional properties, with characteristics falling between those of the individual components. Optimizing polysaccharide thermal treatment is crucial in the prevention of polymer degradation during the thermal-casting process. The developed films demonstrated strong thermal, mechanical, and water barrier properties, making them promising candidates for packaging materials. Films made from furcellaran (FUR) and chitosan (CHIT), enriched with gelatin hydrolysate (HGEL), were obtained without structural alterations in the CHIT–FUR matrix, as confirmed by FT-IR analysis. However, compared to the control film, AFM imaging revealed the formation of grooves and cracks on the surface after hydrolysate incorporation. The addition of HGEL reduced the thermal stability of the film but enhanced its antioxidant properties, as demonstrated by FRAP and DPPH tests (Janik et al., 2021). Pluta-Kubica et al. (2024) obtained films based on furcellaran and chitosan, which were enriched with bioactive peptides (LL-37 and RW4) having potential antimicrobial activity. The addition of peptides influenced the thickness of the furcellaran-based films. The control film without peptides measured 0.071±0.02 mm, while films incorporating LL-37 and RW4 peptides showed increased thickness values of 0.081±0.01 and 0.089±0.02 mm, respectively. This trend suggests that peptide incorporation contributes to structural modifications, likely due to molecular interactions affecting the polymer matrix. All tested films exhibited high water vapor transmission rate (WVTR) values, indicating significant permeability. The presence of peptides slightly reduced WVTR, with values of 519.59±19.72 g $(m^2{\cdot}d)^{-1}$ for the LL-37 film and 534.34±26.10 g $(m^2{\cdot}d)^{-1}$ for RW4, compared to 542.11±13.13 g $(m^2{\cdot}d)^{-1}$ for the control film. This minor reduction may be due to electrostatic interactions between furcellaran's negative charge and peptides' positive charges. However, the high WVTR level remains a concern for applications such as cheese packaging, where excessive moisture loss can negatively impact texture and microbial stability (Table 3.1).

Furcellaran-based films were reinforced with multi-walled carbon nanotubes (MWCNTs), graphene oxide (GO), and silver nanoparticles (AgNPs). AgNPs were synthesized in an eco-friendly manner using furcellaran as a stabilizing agent, confirmed by UV-vis spectroscopy with a surface plasmon resonance (SPR) peak at 400–500 nm. While nanofillers improved water content, solubility, and swelling, high WVTR values could limit their effectiveness for packaging applications. Among all nanofillers, AgNPs provided the most significant improvements in barrier, mechanical, and thermal properties, along with antimicrobial activity against *P*seudomonas *aeruginosa,* Enterococcus *faecalis,* and *S*taphylococcus *aureus*. Additional studies should be focused on the effects of these nanocomposite films with regard to food models, and the potential toxicity as well as migration of nanoparticles from the furcellaran matrix should be assessed (Jamróz et al., 2020).

Furcellaran films with carbon quantum dots (CQDs), MAN, and GO nanofillers were successfully developed. MAN and GO affected the color and mechanical properties, while CQDs did not. CQD films exhibited antioxidant properties and, along with MAN, provided UV barrier effects. The films showed antimicrobial activity in direct contact but did not release nanofillers into the environment (Jamróz, Kopel, et al., 2019).

Gelatin and furcellaran were used to create complexes, optimized based on ζ-potential and size measurements performed at varying pH and ionic strength. This was the first investigation of the physical, antimicrobial, and structural properties of furcellaran-gelatin films with selenium (SeNPs) and AgNPs. AgNPs were obtained using the ex situ method, that is, outside the film-forming matrix. SeNPs improved the films' mechanical properties, while both SeNPs and AgNPs influenced their morphology, as observed via scanning electron microscopy (SEM). The incorporation of AgNPs enhanced the films' UV-blocking ability. SeNP-loaded films exhibited antibacterial activity against *E*scherichia *coli, S. aureus,* and methicillin-resistant *S. aureus* (MRSA), demonstrated by clear inhibition zones (Jamróz, Kopel, et al., 2018).

Furcellaran-gelatin (FUR/GEL) biodegradable films with pu-erh (PE) and green tea (GTE) extracts were produced via casting. These films showed improved tensile strength (TS increased up to 24.14 MPa with GTE), while elongation at break remained unchanged. PE films had higher color parameters. GTE and PE enhanced antioxidant activity (DPPH up to 48.94%, ABTS up to 15.76%) and provided antimicrobial effects, with GTE films inhibiting *E. coli* and *S. aureus* (zones of 16 and 22.5 mm). Their pH sensitivity makes them suitable for intelligent food packaging (Jamróz, Kulawik, Krzyściak, et al., 2019).

Intelligent pH-indicator furcellaran (FUR) films based on extracts from beet root (BTR), elderberry (EB), blueberry (BB), green tea (GTE), and yerba mate (YM) were developed and tested. UV–Vis spectra and color responses varied across pH 1–14. The following color shifts were observed: red (BTR, BB, EB); orange (GT, YM); green in acidic pH (BTR, BB, EB); and brown in alkaline pH (GT, YM). These types of materials were designed to detect pH changes in fresh Atlantic mackerel packaging at 2°C (Jamróz, Kulawik, et al., 2019).

Active FUR/whey protein isolate (WPI) films with YM or WT extracts were successfully developed. The addition of YM improved the water vapor permeability, water content, solubility, elastic modulus, puncture resistance, and thermal stability of the films (Pluta-Kubica et al., 2019). Active packaging materials composed of furcellaran, whey protein isolate, and *B. officinalis* extract (BOE) were created using a simple solvent-casting method. The addition of BOE to the FUR/WPI matrix led to increased water content, solubility, elasticity, and antioxidant activity, while at the same time reducing tensile strength (Zając et al., 2021).

In the publication by Jancikova et al. (2019), active and intelligent properties of films based on furcellaran (FUR), gelatin hydrolysate (GELH), and rosemary extract (from fresh leaves [FRE] and dried leaves [DRE]) were determined at three different concentrations. The addition of rosemary extracts to FUR/GELH films resulted in increased values of parameters such as thickness, water content, and tensile strength. The UV barrier properties of the tested films were improved in FUR/GELH films due to the addition of rosemary extracts. The antioxidant activity (measured by DPPH and

TABLE 3.1 Physical and functional properties of single-layer films based on furcellaran.

Polymer	Active compounds	Type and name of single-layer film	Physical and functional properties of films	References
Furcellaran-chitosan	-	CHIT/FUR (ratio 9:1; 9:2; 9:3)	• Formation conditions were determined using zeta potential, component size, and mixture analysis. • Lowest water vapor transmission rate (WVTR) and best thermal stability were observed at a 9:1 FUR/CHIT ratio.	Jamróz et al. (2021)
Furcellaran-chitosan	Gelatin hydrolysate from carp skins (HGEL)	CHIT/FUR+HGEL	• Atomic force microscope results revealed HGEL addition created grooves and cracks, reducing surface roughness by ~21%. • Antioxidant activity improved with HGEL addition (DPPH by 2.318% and FRAP power by 444%). • Fourier transform infrared (FT-IR) analysis showed no matrix changes, indicating component compatibility.	Janik et al. (2021)
Peptides				
Furcellaran-chitosan	Peptides (LL-37 and RW4)	F-LL-37 and F-RW4	• The incorporation of LL-37 and RW4 peptides increased film thickness compared to the control. • Peptide addition slightly reduced WVTR.	Pluta-Kubica et al. (2024)
Plant extracts				
Furcellaran-gelatin	Yerba mate extract (YM) and green tea extract (GTE)	YM (5%, 10%, 20%) or GTE (5%, 10%, 20%)	• Tensile strength (up to 24.14 MPa with GTE) and antioxidant properties (DPPH up to 48.94%) improved. • GTE films showed antimicrobial activity against *E. coli* and *S. aureus* (16 and 22.5 mm inhibition zones).	Jamróz, Kulawik, Krzyściak, et al. (2019)
Furcellaran-whey protein isolate (WPI)	YM extract and white tea extract (WT)	FUR/WPI; FUR/WPI+YM; FUR/WPI+WT	• FTIR confirmed compatibility between components. • Films with YM improved thermal stability and reduced water solubility, water content, and vapor transition rate. • Films with WT had little effect on physical and mechanical properties.	Pluta-Kubica et al. (2019)
Furcellaran-WPI	*Borago officinalis* extract (BOE)	FUR/WPI+BOE	– FUR/WPI films without BOE: EAB 6.30%, TS 20.59 MPa.	Zając et al. (2021)

			– Films with BOE: EAB increased to 24.30%, TS decreased to 15.33 MPa. – BOE addition made the films darker and improved their antioxidant capacity.	
Furcellaran	Beetroot (BTR), elderberry (EB), blueberry (BB), green tea (GT), and YM extracts	FUR+BTR; FUR+EB;FUR+BB;FUR+GT;FUR+YM	– UV–Vis spectra and extract colors changed within pH 1–14. – Color shifts: Red (BTR, BB, EB); orange (GT, YM); green in acidic pH (BTR, BB, EB); brown in alkaline pH (GT, YM).	Jamróz, Kulawik, et al. (2019)
Essential oils				
Starch-furcellaran-gelatin	Lavender essential oil (OEL)	OEL (2%, 4%, 6%)	– Film thickness increased with OEL addition. – Solubility, water absorption, and swelling decreased as OEL concentration increased. – Tensile strength decreased, reducing mechanical strength, while elongation at break remained unchanged. – OEL affected the film's thermal behavior. – FTIR confirmed interactions between the film matrix and OEL. – Antioxidant properties improved with higher OEL concentration. – Films exhibited antimicrobial activity.	Jamróz et al. (2018b)
Starch-furcellaran-gelatin	Tea tree essential oil (TEO)	TEO (2%, 4%, 6%)	– Water content decreased as TEO concentration increased. – Tensile strength and Young's modulus significantly decreased at 4% and 6% TEO. – No significant effect on elongation at break. – TEO reduced the film's thermal stability. – Antimicrobial activity was observed against *S. aureus* (27.8 ± 0.29 mm) and *E. coli* (18.4 ± 0.20 mm).	Jamróz et al. (2018a)
Nanofillers				
Furcellaran	– Silver nanoparticles (AgNPs) – Graphene oxide (GO)	FUR+0.5GO (or MWCNTs);	– Environmentally friendly synthesis of AgNPs using furcellaran as a stabilizer, confirmed by UV-vis	Jamróz et al. (2020)

(*Continued*)

TABLE 3.1 (Continued)

Polymer	Active compounds	Type and name of single-layer film	Physical and functional properties of films	References
	– Multi-walled carbon nanotubes (MWCNTs)	FUR+1GO (or MWCNTs) and FUR+1.5GO (or MWCNTs); FUR+0.1AgNPs; FUR +0.15AgNPs; FUR +0.2AgNPs	spectroscopy (400–500 nm surface plasmon resonance peak). – Increased molecular weight of furcellaran, improved mechanical strength (TS and EAB) but reduced flexibility in water. – Scanning electron microscopy analysis showed even dispersion of GO and MWCNTs in FUR matrix, forming homogeneous films. – Films with AgNPs enhanced barrier, mechanical, and thermal performance, plus antimicrobial activity (against *P. aeruginosa*, *E. faecalis*, and *S. aureus*).	
Furcellaran-gelatin	Selenium nanoparticles (SeNPs) or AgNPs	FUR+GEL+SeNPs (5%, 10%, 15%); FUR+GEL+AgNPs (5%, 10%, 15%)	– AgNPs enhanced UV-blocking, and SeNP-loaded films demonstrated antibacterial activity against *E. coli*, *S. aureus*, and methicillin-resistant *S. aureus*.	Jamróz et al. (2018)
Furcellaran	– GO – Carbon quantum dots (CQDs) – Maghemite nanoparticles (MAN)	FUR+GO; FUR+CQDs; FUR+MAN	– MAN and GO increased tensile strength but reduced water content, solubility, and elongation at break. – FUR films with CQDs inhibited *Staphylococcus aureus* and *Escherichia coli* growth. – Films with nanofillers exhibited strong free radical scavenging ability.	Jamróz, Kopel, et al. (2019)

FRAP methods) significantly increased after the addition of DRE, reaching 88% DPPH inhibition and 207 μmol Trolox g^{-1} dried film in the FRAP value, while the addition of DRE did not significantly affect this biological property.

Starch, fish skin gelatin, and furcellaran (S/F/G) films with lavender essential oil (OEL) showed antioxidant and antimicrobial properties, potentially extending food shelf life. However, high solubility and low tensile strength may limit their use in high-moisture foods. FTIR confirmed OEL distribution, and DSC revealed altered thermal stability. Further research is needed on polysaccharide-protein films with essential oils, including toxicity, migration, barrier properties, and antifungal activity (Jamróz et al., 2018b). Tea tree essential oil (TEO) was incorporated into starch/furcellaran/gelatin (S/F/G) films at concentrations of 2%, 4%, and 6%. The solubility of the films significantly decreased at 6% TEO. Compared to the control sample (without TEO), water content declined as TEO concentration increased. Films containing 4% and 6% TEO showed a significant reduction in tensile strength and Young's modulus, while elongation at break remained unaffected. Fourier transform infrared spectroscopy analysis provided insight into potential interactions between polysaccharides, protein, and TEO. The addition of TEO reduced the thermal stability of the films but significantly enhanced their antioxidant properties. S/F/G films with TEO demonstrated antimicrobial effects against *S. aureus* and *E. coli*. These findings suggest that S/F/G films infused with TEO could serve as active packaging materials, improving microbial safety and extending food shelf life due to their strong antioxidant and antimicrobial properties (Jamróz, Juszczak, et al., 2018a).

3.4 Multilayer films based on furcellaran

Multilayer films utilizing furcellaran (FUR) possess sophisticated characteristics, positioning them as a compelling substitute for conventional synthetic packaging. By integrating furcellaran with biopolymers such as gelatin hydrolysates, chitosan, or carboxymethylcellulose in a layered configuration, they achieve superior mechanical, thermal, and barrier attributes, making them well-suited for preserving food. Naturally endowed with properties such as antioxidant and antibacterial effects, as well as resistance to UV radiation, these films contribute to prolonging food shelf life by curbing microbial proliferation, inhibiting oxidative processes, and providing protection against light-induced damage. Incorporating bioactive elements, including botanical extracts, peptides, or nanomaterials, enhances their performance, facilitating use in smart and active packaging solutions, such as indicators of food freshness. Research validates their ability to uphold food safety and quality, with improved water vapor barrier capabilities and thermal resilience, though elevated additive levels may compromise mechanical robustness. The characteristics of diverse multilayer furcellaran-based films incorporating active components are detailed in Table 3.2.

Double-layer biopolymer films based on furcellaran (FUR) and chitosan with gelatin hydrolysate (CHIT+HGEL) were enriched with AgNPs synthesized in extracts from water dock root (AgNPs+KB) and pu-erh tea (AgNPs+PE). These films are characterized by AgNPs with an average size below 100 nm, confirmed by SEM, X-ray diffraction (XRD), UV-VIS analyses (plasma resonance peak at 451.8 nm for AgNPs+KB and 440.8 nm for AgNPs+PE), and zeta potential. The addition of AgNPs enhances antibacterial activity against Gram-positive and Gram-negative pathogens, antioxidant activity, and thermal stability of the films, while reducing WVTR. The films have an orange-brown color, which may protect food products from UV radiation (Jamróz, Cabaj, et al., 2021).

Double-layer films based on furcellaran (FUR) and carboxymethylcellulose (CMC) with gelatin hydrolysate (HGEL) were successfully developed, exhibiting a distinct layered structure confirmed by SEM images. The first layer (FUR) was enriched with lingonberry extract (LBE), which imparts antioxidant activity (3.01–41.89 mM Trolox mg^{-1} in the FRAP assay, 0%–20.56% in the DPPH assay) and antimicrobial activity against Gram-positive and Gram-negative bacteria, particularly at 40% LBE addition. Films with 40% LBE demonstrate the highest thermal stability (improved by ~92%) compared to other tested films. The addition of LBE reduces mechanical properties but increases film flexibility (Jamróz, Tkaczewska, Zając, et al., 2022).

Active double-layer films based on furcellaran (FUR) and gelatin hydrolysate (HGEL), with the addition of the Ala-Tyr peptide, exhibit higher mechanical strength (TS = 13.4 MPa) and lower water solubility (WS = 62.8%) compared to films without the peptide. They show high antioxidant activity (FRAP = 6.6 mMol Trolox mg^{-1}) but lack radical scavenging ability (DPPH). The addition of the Ala-Tyr peptide improves the physicochemical, mechanical, and rheological properties of the films, making them more suitable for food packaging applications. These films exhibit active properties as packaging for Atlantic mackerel stored at 4°C for 15 days (Tkaczewska et al., 2021).

Jamróz et al. developed double-layer films based on furcellaran (FUR), protein hydrolysates from soybean bran (HSOY), and a layer of ethanolic soybean bran extract. The addition of the second layer increases the elastic modulus, antioxidant activity, thermal properties, hydrophobicity, and UV barrier properties, but it reduces tensile strength and elongation at break. These films were used as packaging for butter storage (Jamróz, Tkaczewska, Zając, et al., 2022).

Active double-layer films based on furcellaran (FUR) in the first layer and chitosan (CHIT) and gelatin hydrolysates (HGEL) in the second layer, enriched in the first layer with AgNPs (94.96 ± 3.33 nm, synthesized from YM extract), montmorillonite clay (MMT), ethanolic curcumin extract (CUR), and rosemary essential oil (REO), exhibit a yellow color and UV barrier properties. SEM and AFM images confirm the presence of AgNPs and a rougher, more porous microstructure at higher concentrations of active components. The films demonstrate rapid color changes (from orange to red at pH 2–10), indicating potential as freshness indicators for food products. Antioxidant activity increases with higher AgNPs and CUR concentrations, and composite No. 2 shows the highest antimicrobial activity against bacteria, although no antifungal activity is observed. The lowest concentration of active components reduces WVTR, but higher concentrations lead to a decline in mechanical (tensile strength) and thermal properties, lowering the elastic modulus and film stiffness (Jamróz, Janik, et al., 2022).

Double-layer films based on furcellaran (FUR) and gelatin hydrolysates (HGEL) from salmon skin, enriched with aqueous *Clitoria ternatea* flower extract at three concentrations in the first layer, have been developed. These films exhibit improved physicochemical properties, including better mechanical and thermal properties, lower WS, and reduced WVTR. The films show strong antioxidant activity (2.27–3.92 mM Trolox mg^{-1} in the FRAP assay, 36.67%–61.24% in the DPPH assay) and antimicrobial activity in vitro and in vivo against bacteria, though they lack fungicidal effects. The films were applied for the storage of salmon filets for 12 days at 4°C (Grzebieniarz, Tkaczewska, Juszczak, Kawecka, et al., 2023).

Based on furcellaran (FUR) and gelatin (GEL), bilayer films with the addition of *Phytolacca americana* (PA) berry extract in concentrations of 0%–6% were developed, characterized by good mechanical properties (TS 0.0011–0.0013 MPa, EAB 30.38%–33.51%) and water resistance (WVTR 574.74–588.49 g m^{-2}). SEM and AFM analyses confirm a uniform structure without cracks or air bubbles. The films exhibit high antioxidant activity: 4.77–20.70 mg GAE g^{-1} (Folin-Ciocalteu), 0.18–3.40 mM TE g^{-1} (FRAP), 48.63–53.99 mM TE g^{-1} (CUPRAC), and at 6% PA, they neutralize DPPH radicals (1.97%) and ABTS (17.34%). They are biodegradable, nontoxic to *Lepidium sativum* seeds, and show no ecotoxicity. These films, due to their physicochemical properties and antioxidant activity, are a promising alternative to synthetic packaging films in the food industry (Jasińska et al., 2024).

Furcellaran (FUR) and gelatin (GEL) were also used to develop double-layer films enriched with plant extracts, characterized by high biodegradability in vermicompost within several days and by environmental nontoxicity. They exhibit strong antioxidant properties, with the highest values recorded for the GE extract (76.64% in the DPPH test, 9.85% in metal chelating activity) and the dried tomatoes (DTE) extract (52.62% in the FRAP assay). The mechanical properties of the films improve during nine months of storage, indicating their durability (Nowak, Cholewa-Wójcik, et al., 2024). The addition of plant extracts significantly affects the functional parameters: films with garlic and black pepper (BPE) extracts exhibit the lowest solubility (58.40% and 59.26%, respectively) and water content (8.88%), while the film with garlic shows the lowest oxygen transmission rate (OTR) after six months of storage. All films provide good UV barrier properties, and the values of WVTR and OTR decrease over storage time, improving food storage conditions. The type of extract differentiates oxygen permeability but does not significantly affect WVTR, while water-related, color, and optical properties depend on the active substances as well as phenolic content of the extract. Positron annihilation lifetime spectroscopy (PALS) analysis confirms the correlation with gas barrier properties, especially for the garlic extract, indicating a new research direction in the design of biopolymer packaging materials (Nowak, Grzebieniarz, et al., 2024).

Triple-layer films based on furcellaran (FUR) and gelatin (GEL) enriched with *Curcuma longa* extract in citral are characterized by improved functional properties. SEM, XRD, and FTIR analyses confirm interactions between the film components and layers. The addition of the extract significantly decreases WS (from 74.79% to 57.25%) and increases thermal stability (melting temperature rises from 147.10°C to 158.80°C at the highest active compound concentration). At the same time, the films become more hydrophilic, as shown by a reduction in the water contact angle (WCA). Utilizing the film-forming ability of furcellaran and the good mechanical and barrier properties of the FUR/GEL complex, these films are a promising eco-friendly packaging material, particularly for the food industry. Due to the potential antimicrobial and intelligent properties of curcumin and citral, these films are developmental in nature but require further in vitro and in vivo studies to confirm their active and intelligent functionalities (Nowak, Grzebieniarz, Juszczak, et al., 2024).

Furcellaran-based (FUR) triple-layer films, used as eco-friendly packaging for vegan, sugar-free furcellaran gummies, with the addition of soy protein isolate and/or inulin, do not negatively affect the antioxidant activity, texture, or physicochemical properties of the product. Furcellaran demonstrates high potential both as a gelling agent, providing flexible hydrogels compatible with erythritol, soy protein, and inulin, and as a component of biodegradable packaging films (Stępień et al., 2023).

Triple-layer films based on furcellaran (FUR) and chitosan (CHIT), enriched with black seed by-product protein isolates (BBPI), comprise a first layer made of a BBPI-stabilized emulsion or BBPI dispersion, a second layer composed of a FUR

TABLE 3.2 Physical and functional properties of multilayer films based on furcellaran.

Polymer	Active compounds	Type and name of multilayer film	Physical and functional properties of films	References
Double-layer films				
Furcellaran (FUR); chitosan (CHIT)	– Silver nanoparticles (AgNPs) in extract from Lapathi radix 5%, 10%, and 15% (AgNP-PE 5%–15%) – Nanoparticles in extract from *Camellia sinensis* 5%, 10%, and 15% (AgNPs-KB 5%–15%) – Gelatin hydrolysate (HGEL)	FUR/CHIT+HGEL; FUR+AgNP-PE (5%–15%)/CHIT+HGEL; FUR+AgNP-KB (5%–15%)/CHIT+HGEL	– Improvement of thermal stability (only in certain types of films, depending on the type of active ingredient and its concentration). – Increment noted in water vapor transmission rate (WVTR) of up to ~5.51%–15.14%. – Reduction in water contact angle (WCA) by up to ~2.52%–11.15% (for second layer). – Increment in WCA by up to ~5.51%–15.14% (for first layer). – Improvement in antioxidant activity by up to ~52.90%–436.77% (FRAP method). – Improvement in antibacterial activity against *S. aureus, E. coli, Salmonella* spp., and *P. aeruginosa* bacteria.	Jamróz et al. (2021)
Carboxymethylcellulose (CMC); furcellaran (FUR)	• Gelatin hydrolysate (HGEL) • Lingonberry extract (LBE)	FUR/CMC+HGEL; FUR+10% LBE/CMC+HGEL; FUR+20%LBE/CMC+HGEL; FUR+40%LBE/CMC+HGEL	• Thermal stability enhancement. • Reduction in WVTR by up to ~0.19%–1.82% and WCA by up to ~4.13%–34.54%. • Improvement in EAB (by up to ~3.19%–32.36%) and PS (by up to ~2.05%–15.38%), and reduction in MBL (max. breaking load) (by up to ~6.28%–29.57%), in TS (by up to ~6.20%–29.46%) and in ME (modulus of elasticity) (by up to ~14.13%–76.28%). • Improvement in UV-blocking effects. • Improvement in antioxidant activity (by up to ~378.74%–1291.69% – FRAP method and by up to ~18.22%–24.61% – DPPH method). • Improvement in antibacterial activity against bacteria *S. aureus, E.coli, E. faecalis* and *P. aeruginosa*.	Jamróz, Tkaczewska, Juszczak, et al. (2022)

(Continued)

TABLE 3.2 (Continued)

Polymer	Active compounds	Type and name of multilayer film	Physical and functional properties of films	References
Furcellaran (FUR)	• Gelatin hydrolysate (HGEL) • Biopeptide (Ala-Tyr)	FUR+HGEL/Ala-Tyr	• Reduction in WCA by up to ~18.47% (for first layer). • Increment in WCA by up to ~47.74% (for second layer). • Reduction in water solubility by up to ~26.7% and increment in wood-based cellulose (WC) by up to ~2.7%. • Improvement in TS (by up to ~54.02%) and EAB (by up to ~30.17%). • Improvement in antioxidant activity (by up to 0.2%—DPPH method and by up to ~46.97%—FRAP method).	Tkaczewska et al. (2021)
Furcellaran (FUR)	– Protein hydrolysates from soybean bran (HSOY) – Ethanol extract from soybean bran (SOY EXTRACT)	FUR+HSOY/SOY EXTRACT	– Increment in WVTR by up to ~1.41%. – Improvement in thermal stability. – Reduction in MBL (max. breaking load) by up to ~25.89%, TS by up to ~27.02%, EAB by up to ~26.20% – Improvement in ME (modulus of elasticity) by up to ~256.30% – Improvement in UV-blocking effect. – Improvement in antioxidant activity by up to ~27.27% (FRAP method) and ~80.80% (DPPH method).	Jamróz, Tkaczewska, Zając, et al. (2022)
Furcellaran (FUR); chitosan (CHIT)	– Gelatin hydrolysate (HGEL) – Silver nanoparticles in yerba mate extract (AgNPs-YM) – Curcumin extract with rosemary essential oil (CUR-REO) – Montmorillonite solution (MMT)	FUR/CHIT+HGEL; FUR+6 ml CUR-REO+11 ml MMT+6 ml+AgNPs-YM/ CHIT+HGEL	– The average particle size of the nanoparticles was 94.96±3.33 nm (PDI - polydispersity index 0.240±0.009 nm). – Increment in thickness by up to 25%–50%. – Reduction in water content by up to 17.31%–25.78%. – Reduction in WC by up to ~72%. – Deterioration of the film's thermal properties. – Improvement in antioxidant activity. – Visible antimicrobial effect against *S. aureus*, *E. faecalis*, and *E. coli* and a slight effect against *S. enterica* and *P. aeruginosa*.	Jamróz et al. (2022)

Furcellaran (FUR)	– Gelatin hydrolysates (HGEL) – Aqueous extract of Clitoria flower (BP)	FUR+HGEL/FUR; FUR+HGEL/FUR+BP (5%, 7%, 5%, and 10%)	– Film thickness was 0.20 mm. – Reduction in water content, water contact angle, and WVTR. – Solubility was decreased. – In films with the addition of BP at 5% and 7.5%, the FRAP values did not differ significantly from the control. A significant increase was observed only with a 10% addition of Clitoria extract to the film (3.92 ± 0.03 mM Trolox mg^{-1}). – Antibacterial properties were observed of *Staphylococcus aureus, Enterococcus faecalis, Escherichia coli, Salmonella enterica,* and *Pseudomonas aeruginosa*. – Films have intelligent properties.	Grzebieniarz, Tkaczewska, Juszczak, Kawecka, et al. (2023)
Furcellaran (FUR); gelatin (GEL)	*Phytolacca americana* extract (PA)	FUR/FUR+GEL; FUR+PA/FUR+GEL	– Thickness values were around 0.15 m. – Reduction in thermal properties of up to 3.24%–12.33%. – Decrease in total reducing power by up to 93.29%–333.96%. – The values of mechanical properties, solubility, and water content did not differ statistically. – The extract reduced WVTR value.	Jasińska et al. (2024)
Furcellaran (FUR); gelatin (GEL)	– Lingonberry (LBE) – Granulated garlic (GE) – Dried tomatoes (DTE) – Tansy herb (TVE) Black pepper (BPE)	FUR/FUR+GEL; FUR+LBE/FUR+GEL; FUR+GE/FUR+GEL; FUR+DTE/FUR+GEL; FUR+TVE/FUR+GEL; FUR+BPE/FUR+GEL	– Composites are characterized by good mechanical properties. – The highest scavenging value of DPPH free radicals is characterized by the GE film (76%), BPE film (63%), and LBE film (57%). – The GE film also showed the highest metal chelation (9.85%). – None of the extracts showed statistically significant acceleration in the growth of cress seedlings. – The films are almost completely biodegradable after 14 days of incubation with vermicompost.	Nowak, Cholewa-Wójcik, et al. (2024)

(Continued)

TABLE 3.2 (Continued)

Polymer	Active compounds	Type and name of multilayer film	Physical and functional properties of films	References
Furcellaran (FUR); gelatin (GEL)	– Lingonberry (LBE) – Granulated garlic (GE) – DTE – Tansy herb (TVE) – BPE	FUR/FUR+GEL; FUR+LBE/FUR+GEL; FUR+GE/FUR+GEL; FUR+DTE/FUR+GEL; FUR+TVE/FUR+GEL; FUR+BPE/FUR+GEL	– Decreases in WVTR and oxygen transmission rates. – Good barrier properties against UV light. – The type of water extract does not significantly affect water vapor permeability of the film, but it differentiates the degree of oxygen permeability. – Research using the positron annihilation lifetime spectroscopy (PALS) method confirmed that the addition of plant extracts, in particular, those from garlic, affects the parameters of gas permeability after six months of storage.	Nowak, Grzebieniarz, et al. (2024)
Triple-layer films				
Furcellaran (FUR); gelatin (GEL)	*Curcuma longa* extract in citral (CUR)	FUR/FUR/FUR+GEL; FUR/FUR+CUR/FUR+GEL		Nowak, Grzebieniarz, Juszczak, et al. (2024)
Furcellaran (FUR); chitosan (CHIT); gelatin (GEL)	– Blackseed protein stabilized emulsion (EBSP) – Blackseed protein isolate solution in buffer (BSP)	FUR/FUR/CHIT+GEL; FUR+EBSP/FUR/CHIT+GEL; FUR+BSP/FUR/CHIT+GEL	– Addition of the blackseed emulsion and blackseed protein dispersion was determined. – Increased barrier properties were noted against UV radiation in BSP and EBSP. – Decreased antioxidant properties were observed for BSP and EBSP.	Kasprzak et al. (2024)
Furcellaran (FUR); chitosan (CHIT)	Gelatin hydrolysates (HGEL)	FUR+HGEL/FUR/CHIT+HGEL	– The thickness of the obtained films was 0.145 mm. – Film stored at $a_w = 0$ was characterized by the highest tensile strength. – The lowest value of water absorption recorded for this film was equal to 8.95% ± 1.82%. Moreover, the solubility value was at the level of 62.43% ± 3.70%. – The film stored in $a_w = 0$ conditions has the most desirable mechanical and aqueous properties.	Kruk et al. (2023)

Furcellaran (FUR); chitosan (CHIT)	– Gelatin hydrolysates (HGEL) – Curcumin ethanol extract enriched with lemongrass essential oil (CUR-LEO)	CHIT+HGEL/FUR/ FUR+HGEL; CHIT+HGEL/FUR+CUR;LEO (1 or 2 or 3)/FUR+HGEL	– Decreased water content by up to 13.63%–36.68%. – Decreased solubility by up to 47%–56%. – Decrease in the WVTR value was observed after the addition of CUR-LEO 2 by up to 50%. – Reduction in MBL (max. breaking load) by up to 4.16%–22.40%, TS 3.68%–21.32%, EAB 3.16%–18.93%. – Reduction of thermal properties by up to 2.49%–4.0%. – Increment in antioxidant activity (FRAP) by up to 233%–766% and DPPH by up to 88%–223%.	Jamróz et al. (2023)
Four-layer films				
Furcellaran (FUR); chitosan (CHIT); gelatin (GEL)	– Gelatin hydrolysate (HGEL) – AgNPs in tansy herb extract – Curcumin and capsaicin solution enriched with lemongrass oil (CUR/CAPS) – Montmorillonite solution (MMT)	FUR/FUR/CHIT+GEL/ CHIT+GEL; FUR+AgNPs /FUR+MMT/ CHIT+GEL+CAPSCUR/ CHIT+GEL+HGEL	– Nanosilver with a size of 110.5 nm and a low polydispersity coefficient (0.36). – All the produced composites were characterized by a barrier to UV light. – FILM II was characterized by the highest barrier properties. – Water content of 12%–15%. – Solubility of the composites oscillates around 50%. – The films did not show strong antioxidant properties measured by the FRAP method; however, they exhibited a high ability to scavenge free radicals, DPPH.	Nowak et al. (2024)
Furcellaran (FUR); chitosan (CHIT)	– Gelatin hydrolysate (HGEL) – YM solution with AgNPs Chitosan emulsion with capsaicin (CAPS-CHIT); curcumin (CUR); montmorillonite (MMT)	FUR/FUR/FUR/FUR; FUR+HGEL 0.5%/FUR +AgNPs (0.25%–1%)+CAPS-CHIT (0.25%–1%)/FUR +CUR (0.25%–1%)+MMT (0.25%–1%)/FUR	• Thickness values ranged between 0.19 and 0.20 mm. • Decreased water content by up to 7.73%–18.53%. • Reduction in solubility by up to 1.96%–3.48%, • Reduction in MBL (max. breaking load) by up to 2.52%–6.57%, TS – 11.04%–27.92%, ME (modulus of elasticity) 19.27%–30.28%. • Decrease in EAB by up to 5.94%–13.50%. • Reduction thermal properties by up to 5.55%–9.92%. • Decreased antioxidant activity FRAP by up to 105.77%–254.81% and DPPH 125.03%–427.90%.	Grzebieniarz, Tkaczewska, Juszczak, Krzyściak, et al. (2023)

and citric acid matrix, and a third layer made of a chitosan-gelatin matrix. BBPI, obtained with a 73% protein yield, is characterized by a high polyphenol content (1543 mg ml^{-1}), antioxidant activity (67% in the DPPH assay), and emulsifying capacity (encapsulation up to 96%), enabling the formation of stable emulsions at pH 10 and 12. The films show improved UV-Vis barrier properties, increased antioxidant activity (>70% in the DPPH assay), and reduced WS after the addition of the emulsion compared to the control. They are a promising packaging material for highly perishable food products (Kasprzak et al., 2024).

Triple-layer films based on furcellaran (FUR), with a middle layer enriched with a microemulsion of hazelnut oil, were also developed. They are characterized by good integration of the emulsion within the structure, as confirmed by SEM images. The addition of the microemulsion decreases the thermal and mechanical properties of the films, reducing tensile strength but increasing flexibility and surface hydrophilicity. The films, in the form of sachets, were used for storing cod liver oil for three months in refrigerated conditions (without light exposure) and at room temperature (with light exposure) (Pycia et al., 2024).

Kruk et al. developed triple-layer films based on furcellaran (FUR), chitosan (CHIT), and gelatin hydrolysates (HGEL) from carp skin. SEM and AFM analyses confirm the coherence of the layers and the uniform structure of the films, similar to single-layer films but with improved mechanical properties. The control sample shows a tensile strength of 7.76±0.35 MPa, elongation at break totaling 3.69%±0.33%, maximum breaking force of 15.86±0.58 N, water absorption equaling 25.80%±1.14%, and solubility at the level of 52.91%±1.31%. Reducing the water content in the film (storage at water activity $a_w = 0$) significantly improves mechanical properties (tensile strength up to 14.36±0.07 MPa, elongation at break up to 8.74%±0.69%) and lowers water absorption (to 8.95%±1.82%) as well as solubility (to 62.43%±3.70%). High water activity in the storage environment causes morphological and structural changes, deteriorating mechanical properties. Water sorption in the films follows the Brunauer–Emmett–Teller (BET) model, indicating physical multilayer sorption. Films stored in $a_w = 0$ conditions are characterized by optimal mechanical and water-related properties, suggesting that appropriate drying during production increases their potential applications as biodegradable packaging materials (Kruk et al., 2023).

Triple-layer films based on furcellaran (FUR), chitosan (CHIT), and gelatin hydrolysates (HGEL), with a middle layer enriched with an ethanolic extract of curcumin (CUR) and lemongrass essential oil (LEO), were developed. They are characterized by a uniform triple-layer structure, confirmed by SEM and AFM analyses, with even dispersion of CUR-LEO in the middle layer. The addition of CUR-LEO reduces film solubility due to the hydrophobicity of curcumin, improves UV barrier properties, and imparts antibacterial, antifungal properties, and antioxidant activity, which increases along with the rise in CUR-LEO concentration (Jamróz et al., 2023).

Four-layer biopolymer films based on furcellaran (FUR), chitosan (CHIT), and gelatin (GEL), enriched with active compounds (curcumin, capsaicin, AgNPs of ~110 nm synthesized with tansy extract, montmorillonite, and gelatin hydrolysate), exhibit strong antioxidant activity (enhanced against the DPPH radical) and improved physicochemical properties. The addition of active ingredients increases thermal stability (higher change in enthalpy ΔH), hydrophobicity, and UV barrier properties, while the presence of chitosan reduces WS, which is beneficial for packaging applications (Nowak, Tkaczewska, et al., 2024).

Other four-layer biopolymer films based on furcellaran, enriched with gelatin hydrolysate, curcumin, capsaicin, montmorillonite, and AgNPs in a YM extract, are characterized by a diversified structure, which is less homogeneous at higher concentrations of active ingredients, as confirmed by SEM and AFM analyses. The addition of active ingredients improves water properties (e.g., hydrophobicity) but does not significantly affect mechanical properties. The films exhibit antioxidant activity (1.04–2.74 mM Trolox mg^{-1} in FRAP assay, 7.67%–40.49% in DPPH assay) (Grzebieniarz, Tkaczewska, Juszczak, Krzyściak, et al., 2023).

3.5 Potential application of biopolymer films and coatings based on furcellaran

Due to its high antioxidant activity, ability to form stable multilayered systems, and excellent matrix properties enabling the incorporation of active compounds, furcellaran has emerged as one of the most promising biopolymers in the field of active and intelligent food packaging. Its previously described physicochemical properties, such as its anionic nature, ability to form gels, good mechanical strength, and high compatibility with a wide range of plant-based additives, proteins, and nanomaterials, make furcellaran-based materials extensively studied as modern packaging systems aimed at extending the shelf life of food products.

Furcellaran enables the formation of both homogeneous and complex multilayered structures, in which individual layers may serve barrier, antioxidant, antimicrobial, and indicator functions. Consequently, these materials not only protect products from external factors but can also actively influence food quality by, for example, delaying lipid oxidation processes, inhibiting microbial growth, or monitoring changes occurring during storage (e.g., changes in pH or the presence of volatile degradation compounds).

The application of furcellaran in the form of films, coatings, or emulsions, applied directly to food products or used as part of complex packaging systems, has been reported in numerous studies. Among the most commonly studied food models are products characterized by high water activity and short shelf life, such as fish, meat, cheese, or fresh fruits. The developed materials have demonstrated potential not only in extending the microbiological and oxidative shelf life of products but also in improving their sensory quality and reducing weight loss during storage (Table 3.3).

Atlantic salmon (*Salmo salar*) is widely recognized as a model food product in studies on active and intelligent packaging systems due to its high susceptibility to spoilage processes. The high water activity, considerable fat content, and presence of polyunsaturated fatty acids make salmon filets particularly sensitive to microbial growth, lipid oxidation, and deterioration of sensory properties during storage. For this reason, salmon is extensively used as a model for evaluating the effectiveness of biodegradable and active films, including those based on furcellaran.

The first description of salmon samples stored in furcellaran-based films was provided by Jamróz, Tkaczewska, Kopeć, et al. (2022). They developed a completely biodegradable packaging system for salmon storage, consisting of active trays and double-layer biopolymer-based films. The trays were produced from a mixture of potato starch, carboxymethyl cellulose, gelatin, cellulose, and dried green tea residues. The film consisted of a furcellaran layer with LBE (first layer) and a carboxymethyl cellulose layer with added gelatin hydrolysate (second layer). The aim of the study was to evaluate the effect of the applied packaging on the quality of salmon filets stored for 10 days at 4°C. Microbiological parameters, biogenic amine content, and lipid oxidation were analyzed. The use of the fully biodegradable packaging reduced microbial growth—after six days, the amount of aerobic mesophilic bacteria was 19% lower compared to the control group packed in low-density polyethylene (LDPE). A similar effect was observed for yeasts, molds, and psychrotrophic bacteria. Moreover, the packaging effectively limited the accumulation of biogenic amines—the total content of these compounds in the control sample at the end of storage was 311.89 mg kg^{-1}, while in the sample packed in the biodegradable packaging, it was only 248.10 mg kg^{-1}. Significantly lower concentrations of histamine, cadaverine, and tyramine were also recorded. The authors suggest that this effect may result from the antimicrobial properties of green tea and LBEs. The oxidative stability of lipids was also assessed in the study. Peroxide values did not differ significantly between the groups; however, the acid value after 10 days of storage was more than three times lower for the group packed in biodegradable trays (2.95 vs 8.43 mg KOH g^{-1}), indicating effective inhibition of lipid oxidation processes.

In earlier research, the authors also analyzed the use of the same film alone as packaging for cherry tomatoes (Jamróz, Tkaczewska, Juszczak, et al., 2022). In testing demonstrated the films' significant antioxidant and antibacterial activities; however, no antifungal activity was observed. In the application study, it was shown that the use of active films resulted in faster weight loss compared to synthetic LDPE film, which was attributed to the higher water vapor permeability and hydrophilic properties of the coatings. Additionally, a lower polyphenol content was observed in fruits covered with active films during the later days of storage (days 11 and 17), suggesting a deceleration of the ripening process in the presence of LBE. The color parameters of the fruits varied depending on the type of coating applied; however, no clear trend was observed regarding lightness (L*) and redness (a*).

Another important approach in the analysis of Atlantic salmon storage in a furcellaran-based matrix was the development by Grzebieniarz, Tkaczewska, Juszczak, Krzyściak, et al. (2023) of innovative, four-layer furcellaran films enriched with gelatin hydrolysate, curcumin, capsaicin, montmorillonite, and AgNPs in a YM extract. The use of these films for packaging salmon filets proved effective in limiting the growth of spoilage microorganisms—after 12 days of storage, the microbial count was 0.13 log CFU g^{-1} lower compared to the control. However, despite the high antioxidant activity of the films, no inhibition of lipid oxidation processes was noted in the stored samples.

Four-layer films containing these active ingredients were also developed by Nowak et al. (2024). The film was formulated based on furcellaran, gelatin, and chitosan, and was subsequently used to package Atlantic salmon. The films significantly inhibited microbial growth—after 12 days of storage, the bacterial count was 0.81 log CFU g^{-1} lower compared to samples packed in LDPE film, with the strongest antimicrobial effect observed during the first three days of storage. Despite the high antioxidant activity of the films confirmed by in vitro studies, it was not possible to inhibit lipid oxidation in the salmon.

In subsequent studies, in order to address the issue of increased lipid oxidation, the authors applied a fully active packaging system, combining the previously developed trays and four-layer films (Nowak et al., 2025). Atlantic salmon was again used as the model food product. The use of this packaging system extended the shelf life of salmon filets by six days compared to the packaging with LDPE. The tray, rich in polyphenols, further limited lipid oxidation in the salmon muscle. Thus, the combination of the previously developed materials enabled the creation of a fully biodegradable, active packaging system for Atlantic salmon.

Grzebieniarz, Tkaczewska, Juszczak, Kawecka, et al. (2023) developed double-layer active films based on furcellaran, enriched with an aqueous extract from butterfly pea flowers (*C. ternatea*), and a salmon skin gelatin hydrolysate. The active films were used to package salmon filets, which were stored for 12 days at 4°C. The application of the films extended the shelf life of the filets by six days and inhibited the growth of total viable counts (TVC) by approximately 0.5 log CFU g^{-1} compared to the control samples. Additionally, the films delayed lipid oxidation by three days, as confirmed by thiobarbituric acid reactive substances (TBARS) measurements.

In further studies, the authors created double-layer biopolymer coatings based on furcellaran (FUR) and salmon skin-derived gelatin (GEL), enriched with an aqueous extract from butterfly pea flowers (*C. ternatea*). The application of these active hydrogels as coatings for salmon filets (*S. salar*), stored under refrigerated conditions, resulted in a significant reduction in microbial growth throughout the storage period. Additionally, the coatings positively influenced the rate of lipid oxidation, slowing down oxidative processes in the fish. By the third day of storage, the total microbial count in samples coated with 1BP was 0.19 log CFU g^{-1} lower compared to samples wrapped in cling film, with the differences increasing over time, reaching 1.31 log CFU g^{-1} on day nine and 0.88 log CFU g^{-1} on day 12. A similar inhibitory effect was observed for yeast and mold counts—from day three to day 12 of storage, the colony counts in the 1BP samples were, on average, 0.63–1.10 log CFU g^{-1} lower than in the control samples. Importantly, only the 1BP coating effectively limited the growth of psychrotrophic bacteria during storage—on day 12, the psychrotrophic bacterial count in the 1BP samples was 9.18 log CFU g^{-1}, while in the other samples, the values were higher, reaching up to 9.88 log CFU g^{-1} (Grzebieniarz et al., 2024).

Atlantic salmon filets were also used in the study by Tkaczewska et al. (2024). Two types of innovative, active triple-layer films based on furcellaran, gelatin, and chitosan were developed. The first type (Film A) contained pure curcumin dissolved in LEO, while the second (Film B) contained curcumin derived from *C. longa* extract dissolved in citral. The aim of the trial was to use the developed films for packaging highly perishable products and to assess their impact on the quality and safety of stored products. Atlantic salmon filets, stored at 4°C for 14 days, were used for the study. Changes in color parameters, microbiological properties, lipid oxidation, biogenic amine accumulation, and sensory evaluation were analyzed. The application of the films showed that Film A partially inhibited the growth of microorganisms (total aerobic bacteria [TAB] as well as yeasts and molds) up to day eight of storage, whereas Film B did not exhibit antimicrobial activity against the tested pathogens. Both films effectively delayed the accumulation of biogenic amines during the initial storage period, particularly histamine, with better results observed for Film A. Studies on oxidation processes indicated that neither film inhibited lipid oxidation in the salmon filets. Moreover, Film A slightly accelerated oxidation, likely due to the pro-oxidative properties of curcumin when in direct contact with the food. Sensory evaluation revealed the negative impact of both films on the appearance and aroma of the salmon, mainly due to the migration of curcumin, which caused yellow discoloration of the product surface and unfavorable changes in aroma.

Furcellaran-based films and coatings have also been used with other fish species as model products—Atlantic mackerel and carp filets. In a study conducted by Jamróz, Kulawik, Guzik, et al. (2019), intelligent indicator films based on furcellaran, supplemented with natural plant extracts (beetroot, elderberry, blackberry, green tea, YM), were developed and applied as potential spoilage sensors for fresh Atlantic mackerel stored at 2°C. The aim was to assess whether changes in film color could reflect microbiological, chemical, and sensory changes occurring during fish storage. Although some films (particularly those containing green tea) showed a correlation between color change and increases in microbial contamination and lipid oxidation, the overall color change was too subtle to be clearly identified by a sensory panel. The films did not have a direct effect on the quality of the stored fish—they did not inhibit spoilage nor alter the physicochemical properties of the product, serving solely a diagnostic function. Ultimately, they were not sufficiently sensitive to effectively serve as freshness indicators under refrigerated conditions.

Tkaczewska et al. (2021) developed single- and double-layer films based on furcellaran and carp skin gelatin hydrolysate (CSGH), with the addition of the synthetic dipeptide Ala-Tyr in the outer layer (in the double-layer film). The aim of the study was to evaluate the impact of these films on the quality of Atlantic mackerel during 15 days of refrigerated storage (4°C). Microbiological parameters, biogenic amine content, lipid oxidation, fatty acid composition, and consumer perception were analyzed. The application of both single- and double-layer films limited the development of microorganisms in the mackerel. After 15 days of storage, a reduction in total microbial count by 2.12 log for the H film and 2.43 log for the P film was observed compared to the control sample. The growth of yeasts and molds was also inhibited, with the greatest effect noted for the P film (a reduction of 2.20 log). The amount of *Pseudomonas* spp. bacteria was also lower in the samples coated with the films, particularly in the H variant (a reduction of 1.67 log). The films also contributed to slowing down spoilage processes. The TVB-N value, an indicator of freshness, was significantly reduced in the samples coated with the films compared to the control—for both the single- and double-layer films. A delay in lipid oxidation processes was observed, especially during the early storage period—no increase in TBARS was recorded in the

TABLE 3.3 Applications of furcellaran-based films and coatings for food preservation and shelf-life extension.

Type of film/coating	Food product	Influence	References
Furcellaran-whey protein isolate films (FUR/WPI)/with yerba mate extract (FUR/WPI + YM)/ with white tea extract (FUR/WPI + WT)	Fresh soft rennet-curd cheese	• Reduction in water content and water activity in cheese compared to plastic film (LLDPE - linear low-density polyethylene). • Decrease in pH value of cheese for most of active films. • Total bacterial count decreased in cheese samples. • Inhibition of yeast and mold growth. • Improvement in overall sensory quality.	Pluta-Kubica et al. (2020)
Furcellaran and gelatin complex films with selenium nanoparticles (Se-AgNPs)	Mini kiwi	• Reduced weight loss due to lower WVTR in Se-AgNPs film. • Inhibited mold growth compared to control film (low-density polyethylene). • Extended shelf life of kiwi. • Decreased symptoms of fungal infection.	Jamróz et al. (2019)
Multilayer nano/mini, furcellaran/chitosan coatings with oregano essential oil and LL37 and RW4 bioactive peptides	Salmon nigiri	• Reduction of rice starch retrogradation, particularly at 4°C. • Improved texture through reduced hardness. • The pH value of the nigiri sushi sample with peptides slightly decreased during storage. • The pH of the hosomaki sushi sample with peptides remained stable. • Color intensity increased, and the natural appearance was better preserved.	Tadele et al. (2024)
Furcellaran films + beetroot (BTR)/elderberry (EB)/blueberry (BB)/green tea (GTE)/YM extracts	Atlantic mackerel (*Scomber scombrus*)	• The films functioned as pH-sensitive indicators without negatively affecting fish quality. • The colorimetric response was insufficient to accurately detect spoilage under refrigerated storage conditions.	Jamróz, Kulawik, et al. (2019)
Triple-layer furcellaran films: – first layer: furcellaran/gelatin hydrolysate complex (FUR+HGEL); second layer: furcellaran with curcumin extract and lemongrass essential oil (CUR-LEO); third layer: chitosan/gelatin hydrolysate complex (CHIT+HGEL).	Tomatoes, carp filets	• Weight loss of tomatoes stored in biopolymer films, compared to synthetic films, depended on the concentration of active components. • Reduced skin wrinkling compared to tomatoes stored in synthetic film.	Jamróz et al. (2023)
Single-layered film (H): – furcellaran/carp skin gelatin hydrolysate (FUR/CSGH). Double-layer film (P): first layer: furcellaran/carp skin gelatin hydrolysate (FUR/CSGH); second layer: antioxidant dipeptide Ala-Tyr.	Atlantic mackerel (*Scomber scombrus*)	• Reduction in total microbial count, yeasts and molds, and *Pseudomonas* spp. populations. • Extension of product shelf life. • Increase in the concentration of certain biogenic amines. • Inhibition of lipid oxidation during the initial storage period. • Improvement in consumer perception of the product.	Tkaczewska et al. (2021)

(Continued)

TABLE 3.3 (Continued)

Type of film/coating	Food product	Influence	References
Triple-layered coatings: first layer: chitosan mini-emulsion with oregano essential oil; second layer: furcellaran nanoemulsion; third layer: chitosan mini-emulsion with or without bioactive peptides LL37 or RW4.	Pork loins	• Inhibition of lipid and protein oxidization during refrigerated and frozen storage. • No negative impact on meat color attributes or texture profile. • No effect on fatty acid composition. • No effect on proximate composition (moisture, fat, protein, and ash). • Effective delay in the quality deterioration of pork loin.	Kulawik et al. (2025)
Furcellaran-gelatin coatings with green (GT) and pu-erh (PE) tea extracts.	Salmon nigiri	• No noticeable effect on the microbiological quality of the stored samples. • No influence on the product's shelf life. • No significant effect on sensory evaluation outcomes. • Substantial inhibition of lipid oxidization. • Significant influence on the color of the fish samples.	Kulawik et al. (2019)
Furcellaran and whey protein isolate (WPI) films with pu-erh (PE) or green tea (GTE) extracts.	Acid-curd cheese (quark)	• The number of *Lactococcus* and the total bacterial count decreased in almost all cheese samples during storage. • The number of yeast increased during storage in all samples, regardless of the type of packaging used. • Most of the tested films had negative impact on the sensory quality of the cheese.	Pluta-Kubica et al. (2021)
Active packaging (trays + films) film: first layer: furcellaran with lingonberry fruit extract; second layer: carboxymethyl cellulose and gelatin hydrolysate. Tray: a mixture of potato starch, carboxymethyl cellulose, gelatin, cellulose, and dried green tea grounds.	Salmon filets (*Salmo salar*)	• Potential inhibitory effect on biogenic amine accumulation and microbial growth was observed. • Filets stored in the tested packaging had a 19% lower total bacterial count on day six of storage. • The active packaging extended the shelf life of the product.	Jamróz, Tkaczewska, Kopeć, et al. (2022)
Double-layered film: first layer: furcellaran with lingonberry extract; second layer: carboxymethyl cellulose and gelatin hydrolysate.	Cherry tomatoes	• Faster weight loss observed in samples packed in biopolymer films. • Lower phenolic content in samples packed in biopolymer films.	Jamróz, Tkaczewska, Juszczak, et al. (2022)
Coatings based on chitosan, furcellaran, and gelatin hydrolysate with bioactive peptides RW4 and LL37.	Pork	• Inhibition of aerobic bacteria, mold, and yeast. • Extension of product shelf life. • No effect on water activity. • Sensory attributes related to extended shelf life had direct impact on sensory evaluation.	Kulawik et al. (2024)

(Continued)

TABLE 3.3 (Continued)

Type of film/coating	Food product	Influence	References
Furcellaran + AgNPs film	Rennet-curd (gouda) and an acid-curd (quark) cheese	• A decrease in yeast growth was observed in gouda and yeast growth was inhibited in quark. • A reduction in mold count was observed only in gouda. • A lower level of silver migration was detected in quark compared to gouda. • Improvement in the microbiological quality of cheeses during storage.	Pluta-Kubica et al. (2021)
Furcellaran, WPI, and *Borago officinalis* extract film.	Smoked pork ham	• Microbiological quality, accumulation of oxidation products, and sensory attributes were similar across all products.	Zając et al. (2021)
Single-layered film: furcellaran and protein hydrolysates from soybean bran (HSOY). Double-layered films: first layer furcellaran and protein hydrolysates from soybean bran; second layer: furcellaran.	Butter	• Increase in thiobarbituric acid reactive substances and peroxide value in samples packed in active double-layer films after 12 days of storage. • Decrease in polyunsaturated fatty acid content after 12 days of storage. • No inhibition of lipid oxidation despite high antioxidant activity of the films. • Quality of butter packed in biodegradable films comparable to that of synthetic film.	Jamróz, Tkaczewska, Zając, et al. (2022)
Films containing chitosan, furcellaran, gelatin hydrolysate, and bioactive peptides (Lfcin-β/ MBP-1, PLGG (Proline–Leucine–Glycine–Glycine)/ LL37/Cys-LL37/melittin/RW4/nisin).	Smoked pork ham	• Reduction in microbial counts in meat was observed. • The coatings with RW4 and LL37 peptides showed the highest effectiveness. • Effective inhibition of TVC growth, even at low peptide concentrations. • High antioxidant activity of biopolymer matrix.	Kulawik et al. (2023)
Four-layered films: first layer: furcellaran + gelatin hydrolysate (HGEL); second layer: furcellaran + capsaicin (CPAS) and silver nanoparticles (AgNPs); third layer: furcellaran + curcumin (CUR) and montmorillonite (MMT); fourth layer: furcellaran.	Salmon filets (*Salmo salar*)	• Inhibition of microbial growth responsible for filet spoilage was observed. • Reduction in microorganism count by 0.13 log CFU g^{-1} after 12 days of storage. • No inhibition of lipid oxidation in stored samples. • High antioxidant activity of the films did not translate into protection against lipid oxidation.	Grzebieniarz, Tkaczewska, Juszczak, Krzyściak, et al. (2023)
Four-layered films: – first layer: furcellaran + AgNPs; second layer: furcellaran + montmorillonite; third layer: chitosan/gelatin complex + capsaicin + curcumin; fourth layer: chitosan/gelatin complex + gelatin hydrolysate	Salmon filets (*Salmo salar*)	• The composite almost completely inhibited the growth of the total number of microorganisms and psychrotrophic bacteria during the first three days of storage. • Throughout the entire storage period, the film exhibited antimicrobial activity. • The level of substances reacting with TBA increased during the storage of samples.	Nowak et al. (2024)

(Continued)

TABLE 3.3 (Continued)

Type of film/coating	Food product	Influence	References
Active packaging (trays + film). Four-layered films: first layer: furcellaran + AgNPs; second layer: furcellaran + montmorillonite; third layer: chitosan/gelatin complex + capsaicin + curcumin; fourth layer: chitosan/gelatin complex + gelatin hydrolysate. Tray: a mixture of potato starch, carboxymethyl cellulose, gelatin, cellulose, and dried green tea grounds.	Salmon filets (*Salmo salar*)	Active packaging inhibited growth of microorganisms. Active packaging did not affect lipid oxidization.	Nowak et al. (2025)
Triple-layered film: first layer: furcellaran; second layer: furcellaran + *Oenothera* emulsion; third layer: furcellaran + spirulina.	Gummy jellies based on furcellaran	• The furcellaran film preserved the quality of the jellies during one week of storage. • Minimal water loss was attributed to drying, not packaging. • The jellies maintained good textural and physicochemical stability.	Stępień et al. (2023)
Furcellaran coatings with a gelatin extract obtained from carp skins and dried herbs (thyme or rosemary).	Carp filets	• Inhibition of biogenic amine formation in carp filets. • Slowdown of microbial spoilage. • Modification of sensory attributes—enhancement of the natural saltiness of the meat. • Improvement of sensory acceptance by consumers.	Tkaczewska et al. (2023)
Double-layered film: first layer: furcellaran + gelatin hydrolysate; second layer: furcellaran + butterfly pea flower extract.	Salmon filets (*Salmo salar*)	• Extension of salmon filet shelf life by six days. • Inhibition of total viable count (TVC) growth by approximately 0.5 log CFU g^{-1}. • Delay of lipid oxidation by three days. • Confirmation of antioxidant activity during storage.	Grzebieniarz, Tkaczewska, Juszczak, Kawecka, et al. (2023)
Double-layered coatings: first layer: furcellaran + gelatin; second layer: gelatin + butterfly pea flower extract.	Salmon filets (*Salmo salar*)	• Coatings effectively inhibited microbial growth throughout refrigerated storage. • Coated filets maintained better quality and showed a slower rate of oxidative changes compared to uncoated controls.	Grzebieniarz et al. (2024)
Chitosan/furcellaran (CHIT/FUR) coatings with bioactive peptides: Lfcinβ, PLGG (Proline–Leucine–Glycine–Glycine)/ MBP-1/ LL-37/Cys-LL-37/RW4/melittin.	Pork loins	• Almost complete inhibition of total viable count (TVC) growth in meat. • Extension of microbiological shelf life of meat products. • Effectiveness maintained even at low concentrations of RW4 and LL-37. • No antioxidant activity was observed in the coating.	Kulawik et al. (2022)
Film based on chitosan, furcellaran, and a gelatin hydrolysate.	American blueberries	• Inhibition of mold growth in blueberries. • Faster weight loss of fruits packed in biopolymer films. • Higher polyphenol content in fruits packed in HGEL-enriched films.	Janik et al. (2021)

(*Continued*)

TABLE 3.3 (Continued)

Type of film/coating	Food product	Influence	References
Double-layered films: first layer: furcellaran-gelatin complex; second layer: furcellaran + plant extract (garlic, lingonberry, tansy herb, black pepper, dried tomato extract).	Oil-based salad dressing	• No effect of the film on properties of the salad dressing.	Nowak, Cholewa-Wójcik et al. (2024)
Triple-layer film: first layer: furcellaran; second layer: furcellaran + hazelnut oil–water emulsion; third layer: furcellaran.	Liver oil	• Acid value remained stable during storage; however, iodine and peroxide values significantly increased, indicating fat rancidity. • Changes occurred in the fatty acid profile, depending on film type and storage conditions. The antioxidant protection provided by hazelnut oil was limited.	Pycia et al. (2024)
Triple-layered films: Film A: first layer: chit + HGEL; second layer: furcellaran + curcumin + lemongrass oil; third layer: furcellaran + HGel. Film B: first layer: furcellaran; second layer: furcellaran + *Curcuma longa* extract in citral; third layer: furcellaran-gelatin complex.	Salmon filets (*Salmo salar*)	• Both films negatively affected the sensory attributes of the product. • The films demonstrated potential in inhibiting the formation of biogenic amines.	Tkaczewska et al. (2024)

film-coated samples on day four, unlike in the unpackaged control samples. Moreover, a higher content of docosahexaenoic acid (DHA) and polyunsaturated fatty acids (PUFA) was found in the samples coated with the H film, suggesting greater effectiveness in protecting against lipid oxidation compared to the P film. In consumer evaluation, the use of the films did not diminish the product's attractiveness; on the contrary, it improved the perception of mackerel as an environmentally friendly and easily storable product. Additionally, an increase in the evaluation of taste and aroma was recorded for the single-layer film.

Fish products were also analyzed for the application of multilayer coatings based on furcellaran, chitosan, and gelatin. In a study involving carp filets, also conducted by Tkaczewska et al. (2023), edible coatings based on furcellaran enriched with carp gelatin extracts and dried herbs (thyme or rosemary) were developed. The aim of the research was to create simple, active coatings that could be used for packaging and stabilizing fresh fish products. Coatings containing rosemary extract exhibited the best protective properties—they effectively inhibited the formation of biogenic amines and slowed the growth of spoilage microflora (reductions of 0.53 and 0.29 log CFU g^{-1}, respectively). Coatings based on thyme also showed preservative effects, although they had a less favorable impact on the mechanical properties of the coatings (reduced strength and adhesion). The application of the coatings also altered the sensory characteristics of the filets—enhancing the natural saltiness of the meat and softening the typical fishy flavor, which was positively evaluated by the sensory panel.

Kulawik et al. (2019) developed edible protective coatings based on furcellaran and pork skin gelatin, enriched with aqueous extracts of green tea (GTE) and pu-erh tea (PE). The coatings were applied directly onto the surface of salmon nigiri using a gentle spraying technique, and their effect on product quality during refrigerated storage was analyzed. Four variants were evaluated: a control sample (without coating), a sample with only the furcellaran-gelatin matrix (FC), and samples enriched with green tea (GTE) or pu-erh tea (PE) extracts. The application of coatings containing tea extracts did not significantly affect the microbiological quality of the stored samples nor their shelf life. Similarly, sensory evaluation (including taste, aroma, and texture) revealed no statistically significant differences between the coated and control samples, suggesting the coatings' neutrality in terms of consumer acceptance. On the other hand, the coatings containing GT and PE extracts demonstrated a significant ability to inhibit lipid oxidation processes, which could substantially contribute to the protection of the fish's nutritional value during storage. Additionally, these extracts influenced the color of the fish, causing changes in color parameters, especially at higher additive concentrations.

In another study, Tadele et al. (2024) evaluated the impact of multilayer nano- and microemulsions based on furcellaran and chitosan, containing oregano essential oil and the bioactive peptides LL37 and RW4, on the quality of sushi during storage. The aim was to investigate whether the application of such emulsions could limit starch retrogradation in sushi rice, thereby improving the texture and other product properties. The results showed that the emulsions effectively reduced starch retrogradation, leading to decreased rice hardness, particularly during storage at 4°C. Furthermore, the coatings influenced changes in the color and pH of the sushi, especially in samples containing salmon. Sensory analysis confirmed that the application of the emulsions improved the overall sensory quality of sushi, including appearance, color, and aroma.

Coatings have also been used to evaluate their applicability to meat products, including pork loin and ham. Kulawik et al. (2025) developed triple-layer edible coatings based on furcellaran-chitosan mini- and nanoemulsions, enriched with oregano essential oil and bioactive peptides (LL37 or RW4). The coatings were applied to the surface of pork loin using an electrostatic spraying technique. The aim of the study was to assess the impact of the developed coatings on the quality, oxidative stability, and consumer perception of meat stored in refrigerated (4°C for 18 days) and frozen (−20°C for six months) conditions. The application of the coatings effectively inhibited lipid and protein oxidation processes in both storage conditions. TBARS values and thiol group content were significantly more favorable in the coated samples compared to the controls. At the same time, the coatings did not negatively affect meat color or texture profile, and slight changes in lightness, hue, as well as color saturation parameters were not significant in terms of sensory quality. No significant differences were found in proximate composition (moisture, fat, protein, ash) or fatty acid profiles between the coated and control samples. Consumer perception analysis showed that products with the applied coating were rated more favorably in terms of gloss, freshness, and purchase intent, regardless of storage time. Consumers also viewed the coating methods more positively, particularly when informed about benefits such as extended shelf life and the absence of the need for preservatives.

In a related investigation, antimicrobial coatings were developed based on chitosan and furcellaran (CHIT/FUR), and enriched with selected bioactive peptides—Lfcinβ, PLGG (Proline–Leucine–Glycine–Glycine), MBP-1, LL-37, Cys-LL-37, RW4, and melittin. An analysis of the antimicrobial activity of the peptide-enhanced coatings (5 μg mL^{-1}) demonstrated a synergistic antimicrobial effect in combination with the CHIT/FUR matrix. The coatings were applied to slices of smoked pork ham and pork tenderloins stored under refrigerated conditions. Analysis of total viable counts (TVC) showed that the CHIT/FUR complex alone almost completely inhibited bacterial growth, and the addition of RW4 and LL-37 peptides maintained or enhanced these properties. Other peptides reduced the antibacterial effectiveness of the coatings. In final tests, coatings containing RW4 and LL-37 were shown to effectively inhibit TVC growth even at low peptide concentrations (starting from 1.25 μg mL^{-1}). Although the effectiveness of LL-37 was slightly lower at the lowest doses, it remained significant. CHIT/FUR coatings enriched with RW4 and LL-37 demonstrated very strong antimicrobial properties, effectively extending the microbiological shelf life of the meat (Kulawik et al., 2022).

Building on these findings, the researchers also developed edible biopolymer coatings. The biopolymer coatings, based on chitosan, furcellaran, and carp skin gelatin hydrolysate, were additionally enriched with eight different bioactive peptides: Lfcin-β, MBP-1, PLGG (Proline–Leucine–Glycine–Glycine), LL-37, Cys-LL-37, melittin, RW4, and nisin. Under meat product storage conditions (smoked pork ham and pork tenderloin), the coatings demonstrated the ability to limit the growth of total viable aerobic bacteria (TVC), achieving reductions exceeding 3.5 log CFU g. Coatings containing the RW4 and LL-37 peptides were considered the most promising, as they effectively inhibited microbial growth even at low peptide concentrations (1.25–2.5 μg mL^{-1}). The addition of $CaCl_2$ in some coating variants did not produce clearly positive effects (Kulawik et al., 2023).

Continuing their research, Kulawik et al. (2024) conducted a study regarding the application of edible coatings based on chitosan, furcellaran, and fish gelatin hydrolysate (HGEL) enriched with the bioactive peptides LL-37 and RW4 to extend the shelf life of pork tenderloin stored in refrigerated (4°C) and frozen (−20°C) conditions. The coatings were prepared in three variants: without peptides (CC - control coating), with LL-37, and with RW4. The coatings were applied by immersing the meat samples in the coating solutions. The applied coatings exhibited antimicrobial activity, particularly during the first week of storage. A significant reduction in the counts of TAB, psychrotrophic bacteria, as well as yeasts and molds was observed compared to the control samples—by up to 1–2 log CFU g^{-1}. The strongest effect under refrigerated conditions was observed for the coating with LL-37, while in frozen storage, a delayed microbial growth during the thawing phase was noted. However, after prolonged storage, the antimicrobial activity was limited, and samples with RW4 even showed higher spoilage microflora counts than the control. pH analysis indicated that the coatings slowed the pH increase in meat samples, suggesting reduced intensity of protein degradation and proteolytic bacterial activity. Water activity remained unchanged across all variants, indicating coating stability and no impact on water migration within the product. Sensory analysis demonstrated that the coatings did not deteriorate the organoleptic characteristics of the meat; on the contrary, attributes such as color and aroma were rated higher in the coated samples compared to the control during the initial storage period. As a result, the sensory acceptability of coated samples was maintained for approximately one week longer, which was consistent with the microbiological findings.

Zając et al. (2021) also used smoked pork ham to evaluate the impact of furcellaran-based films on this food product. In the study, biodegradable films were prepared based on furcellaran (FUR) and WPI, enriched with *Borago officinalis* extract (BOE) serving as an antioxidant compound. The smoked hams were packaged using two film variants: meat coated before the smoking and cooking processes, and meat coated after these processes. Vacuum-packed ham in plastic film was implemented as the control sample. The application of furcellaran-whey protein films with BOE extract affected the quality of the stored ham. Products coated with the film exhibited a reduction in moisture content and water activity during storage, while these values remained stable in the control samples. Due to the darker color of the ham obtained with the extract, there was potential to shorten the smoking process time. The results of microbiological analyses, oxidation product accumulation levels, and sensory evaluations were comparable across all variants—both in samples packed with the film and in the control samples. The applied films did not negatively affect the microbiological or sensory quality of the ham.

Vegetables and fruits were also used as model products to evaluate the effects of furcellaran-based matrices. Due to their high water content, enzymatic activity, and susceptibility to oxidation and microbial spoilage, they represent particularly demanding models for testing the properties of active and biodegradable packaging materials.

Jamróz, Kopel, Juszczak, et al. (2019) developed double-layer biodegradable films based on furcellaran and gelatin (FUR/GEL), enriched with a mixture of selenium and silver nanoparticles (Se-AgNPs) at different concentrations (5%, 10%, and 15%). The aim of the study was not only to characterize the physicochemical, thermal, mechanical, and antibacterial properties of these films but, above all, to assess their suitability as active packaging materials for the storage of mini kiwi (*Actinidia arguta*). Storage testing demonstrated that kiwi packaged in FUR/GEL films containing nanoparticles exhibited lower weight loss, an absence of mold development, and a longer shelf life compared to fruits packed in standard LDPE film. As early as after four days of storage, brown spots appeared on the fruits packed in LDPE, followed by mold growth after six days, whereas fruits in the nanocomposite films remained free of visible disease symptoms. The strongest protective effect was observed with 10% and 15% Se-AgNPs content.

In further research, Jamróz et al. (2023) developed triple-layer biodegradable films based on furcellaran (FUR), chitosan (CHIT), and gelatin hydrolysates (HGEL) enriched with an ethanolic extract of curcumin and lemongrass essential oil (CUR-LEO). The active substance was incorporated into the middle (furcellaran) layer of the films at three different concentrations (1%, 2%, and 4%). The resulting materials exhibited both active properties (antioxidant and antimicrobial activity) and intelligent properties (the ability to signal pH changes through color variation). The functional properties of the obtained films were evaluated on two model food products: tomatoes and carp filets. Tomatoes were stored at room temperature for 33 days, during which weight loss and surface visual changes were periodically analyzed. Carp filets were packaged in CUR-LEO 3 films and stored for 36 hours at 22°C, with observations focused on color changes in the films as freshness indicators. The application of CUR-LEO-containing films positively affected the quality of the stored tomatoes. It was found that the weight loss of tomatoes packed in CUR-LEO 1 and CUR-LEO 2 films was significantly lower compared to unpackaged fruits and those packaged in synthetic (LDPE) film, especially during the mid-storage period (days 9–25). Moreover, tomatoes from these variants exhibited less skin wrinkling, indicating better moisture retention. In contrast, the CUR-LEO 3 film proved less effective—more pronounced wrinkling was observed on its surface, likely due to higher water vapor permeability. In the case of the carp filets, a distinct color change was observed in the CUR-LEO 3 film during storage—from yellow to orange. This shift was a response to the increase in pH caused by protein degradation and the release of volatiles.

In another study conducted by Janik et al. (2021), three-component biopolymer films based on furcellaran (FUR), chitosan (CHIT), and carp skin gelatin hydrolysate (HGEL) were developed. The films were used for packaging blueberries, which were stored for 10 days while monitoring quality changes. It was found that the use of CHIT–FUR/HGEL films inhibited mold growth on the blueberries compared to samples stored in synthetic film. However, fruits packaged in the biopolymer coatings exhibited faster weight loss. Additionally, fruits packaged in films containing gelatin hydrolysate had a higher polyphenol content, which may indicate an accelerated ripening process.

Cheese products of various types, including rennet cheeses such as gouda and acid-coagulated varieties, such as quark-type curd, were also analyzed in biodegradable furcellaran-based packaging. Due to their high water content, neutral to slightly acidic pH, and abundance of readily available nutrients, dairy products provide an ideal environment for microbial growth. Consequently, they are particularly challenging models for evaluating the effectiveness of active packaging materials.

Pluta-Kubica et al. (2020) developed and evaluated active edible films based on furcellaran and whey protein isolate (FUR/WPI), enriched with extracts from YM and white tea. The aim of the study was to assess their potential use as packaging materials for fresh rennet cheese. The films were characterized in terms of physicochemical, mechanical, and thermal properties, and their antioxidant activity was determined. They were subsequently applied to packaged cheese stored under refrigerated conditions for three weeks. The results demonstrated that the use of active FUR/WPI films contributed to reductions in moisture content and water activity in the cheese compared to plastic packaging (LLDPE - linear low-density polyethylene). Moreover, the pH value decreased in most cheese samples packaged in active films. The total bacterial count was significantly reduced ($P \leq 0.05$) in the cheese samples packaged in edible films, although no significant differences were observed between groups across specific weeks of storage. Importantly, the film containing white tea extract effectively inhibited the growth of yeasts and molds, which are the main spoilage organisms in fresh cheeses. From a sensory perspective, cheeses packaged in active films—particularly those with WT addition—were more often rated as desirable or highly desirable, indicating an improvement in overall sensory quality. Although the film with the YM extract had a slightly negative impact on the appearance and aroma of the cheese (due to intense coloration), it demonstrated beneficial antimicrobial properties and contributed to extending the product's shelf life.

In further studies, Pluta-Kubica, Jamróz, Juszczak, et al. (2021) developed edible biopolymer films based on furcellaran and WPI, enriched with aqueous extracts of green tea (GTE) or pu-erh tea (PE). The aim of the study was to characterize the mechanical, physicochemical, antioxidant, and antimicrobial properties of the obtained films and to evaluate their potential use as active packaging for acid-coagulated quark cheese. At the subsequent stage of the research, the films were used to package quark cheese, which was stored under refrigerated conditions for two weeks. It was found that the use of edible films with tea extracts did not improve the microbiological quality of the cheese—yeast counts increased in all variants and were significantly higher in samples packaged with GT and PE films compared to the control. A decrease in moisture content and water activity was also observed, attributed to the high water vapor permeability of the biopolymer films. The films did not have a clear effect on the pH of the cheese. In sensory evaluation, samples were assessed both immediately after packaging and after two weeks of storage. Edible films, particularly those with tea extracts, lowered the overall sensory quality of the cheese—negatively affecting appearance, taste, and texture. The engaged panelists also reported unacceptable color changes and surface discoloration. Although the study made it possible to demonstrate the potential of the films in terms of physicochemical and antioxidant properties, they were not suitable as packaging for acid-coagulated quark cheese, mainly due to their negative impact on sensory characteristics and limited antimicrobial effectiveness.

In parallel, Pluta-Kubica, Jamróz, Khachatryan, et al. (2021) developed a nanocomposite film based on furcellaran with the addition of silver nanoparticles (FUR + AgNPs), obtained via an in situ method, intended for use as biodegradable packaging material for cheeses. The aim of the study was to assess the impact of this film on the quality of two types of cheese: rennet cheese (gouda) and acid-coagulated cheese (quark-type curd). The moisture content, physicochemical properties, microbiological, and organoleptic quality of the cheeses, as well as the migration of AgNPs into the product, were analyzed in the study. It was found that the counts of *Lactococcus* bacteria and total bacterial counts in gouda remained unchanged during storage, regardless of packaging type. In the case of quark cheese, a decrease in *Lactococcus* counts was observed during storage. The application of FUR + AgNPs films slowed yeast growth in gouda and inhibited yeast growth in quark. Additionally, mold growth was hindered, although only in gouda samples. The level of silver migration was lower in quark than in gouda samples, and the amount of silver detected after storage did not pose a health risk to consumers. While the film improved the microbiological quality of the stored cheeses, the organoleptic quality of cheeses packaged in the active film was rated as less desirable compared to those packaged in traditional materials.

Oil-based products, due to their high susceptibility to oxidation processes, serve as an important model system for evaluating the effectiveness of biodegradable films enriched with active components. In several independent studies, attempts have been made to use furcellaran as a matrix for the development of active packaging materials intended to protect products with high fat content.

Nowak, Cholewa-Wójcik, et al. (2024) conducted research on double-layer films based on furcellaran and gelatin, enriched with aqueous plant extracts such as garlic, dried tomato, tansy, lingonberry, and pepper extracts. The aim was to assess their potential application for packaging food products, including an oil-based salad dressing. The analysis included the evaluation of acid value, peroxide value, induction time, and oxidative stability (Rancimat test). The results showed that, despite the high antioxidant activity of the films themselves, no significant differences were observed in the oxidative stability of the oil packaged in the active films compared to samples stored in synthetic LDPE film.

In another study, single-layer films based on furcellaran and HSOY were developed, along with double-layer variants, in which the second layer consisted of an ethanolic extract of soybean bran. The research objective was to evaluate the impact of these materials on the quality of stored butter. The addition of the second layer increased the antioxidant properties while improving the UV barrier and hydrophobicity of the films. However, despite the high antioxidant activity demonstrated under in vitro conditions, the films did not inhibit oxidation processes in the butter. After 12 days of storage, an increase in TBARS and peroxide value (PV), and a decrease in polyunsaturated fatty acid content were observed, suggesting potential pro-oxidative effects of the film matrix. The quality of butter packaged in active films was comparable to that of samples packed in conventional plastics (Jamróz, Tkaczewska, Zając, et al., 2022).

Other research focused on triple-layer furcellaran-based films, the middle layer of which was enriched with a hazelnut oil microemulsion. These materials showed good integration of the oil phase with the furcellaran matrix (confirmed by SEM imaging), although the addition of the microemulsion deteriorated the mechanical and thermal properties of the films. The films were used to package cod liver oil, which was stored for three months in different conditions (refrigerated and at room temperature, with light exposure). Analysis of acid, iodine, and peroxide values, as well as fatty acid profiles, indicated that rancidity processes progressed regardless of the type of film used. Despite the presence of α-tocopherol in the hazelnut oil, the protective effect was minimal, likely due to the sensitivity of vitamin E to light and oxygen (Pycia et al., 2024).

Furcellaran-based films were also applied for the storage of gummy jellies. In a study conducted by Stępień et al. (2023), vegan, sugar-free jelly candies were developed using furcellaran as a novel gelling agent, combined with soy protein isolate and/or inulin. The aim of the study was to determine the rheological, mechanical, antioxidant, and physicochemical properties of the jelly candies, as well as to evaluate the effect of a newly designed triple-layer furcellaran film as a packaging material on product quality during storage. Over one week of refrigerated storage, an increase in jelly candy hardness and a slight decrease in moisture content, as well as water activity, were observed. However, these changes were mainly attributed to insufficient drying of the product rather than the effect of the packaging film. The applied triple-layer furcellaran film did not negatively affect the physicochemical properties, texture, or antioxidant activity of the jelly candies during the one-week storage period.

Furcellaran, an anionic polysaccharide with exceptional gelling, mechanical, and compatibility properties with a wide range of bioactive compounds, has found broad application in the development of biodegradable films and coatings for active and intelligent food packaging. The numerous studies discussed in this subsection indicate that furcellaran-based matrices can be effectively used to extend the shelf life of food products with diverse characteristics, including fish, meat, vegetables, fruits, dairy products, oils, and confectionery products.

The developed furcellaran films and coatings exhibited multifunctional activity, ranging from antimicrobial and antioxidant properties to the ability to monitor quality changes through colorimetric responses. Particularly noteworthy was their positive impact on inhibiting microbial growth, reducing oxidative processes, and maintaining sensory parameters across various product categories. Research focused on enriching furcellaran with plant extracts, essential oils, AgNPs, and bioactive peptides further enhanced the functional properties of the developed packaging materials.

At the same time, some studies revealed limitations regarding the use of furcellaran as a standalone packaging matrix, particularly concerning its barrier properties against water vapor and its oxidative stability when applied to lipid-rich products. In some cases, negative effects of active additives on the sensory attributes of packaged products were observed, potentially limiting their commercial applicability.

The available data indicate that furcellaran represents a highly promising base material for the development of modern, biodegradable packaging systems. However, to fully realize its potential, further research is needed to enhance its barrier properties and mechanical durability.

Acknowledgements

Nikola Nowak-Nazarkiewicz work was financially supported by the Foundation for Polish Science START Award.

References

Grzebieniarz, W., Tkaczewska, J., Juszczak, L., Kawecka, A., Krzyściak, P., Nowak, N., ... Jamróz, E. (2023). The influence of aqueous butterfly pea (*Clitoria ternatea*) flower extract on active and intelligent properties of furcellaran Double-Layered films - In vitro and in vivo research. *Food Chemistry, 413*, 135612. https://doi.org/10.1016/j.foodchem.2023.135612.

Grzebieniarz, W., Tkaczewska, J., Juszczak, L., Krzyściak, P., Cholewa-Wójcik, A., Nowak, N., ... Jamróz, E. (2023). Improving the quality of multi-layer films based on furcellaran by immobilising active ingredients and impact assessment of the use of a new packaging material. *Food Chemistry, 428*, 136759. https://doi.org/10.1016/j.foodchem.2023.136759.

Grzebieniarz, W., Tkaczewska, J., Juszczak, L., Nowak, N., Krzyściak, P., Guzik, P., ... Jamróz, E. (2024). Design of active double-layer gel coatings based on furcellaran-gelatin and aqueous butterfly pea (*Clitoria ternatea*) flower extract for prolonging the shelf-life of salmon (*Salmo salar*). *Progress in Organic Coatings, 186*, 107945. https://doi.org/10.1016/j.porgcoat.2023.107945.

Hugerth, A., & Sundelöf, L. O. (2001). The effect of polyelectrolyte counterion specificity, charge density, and conformation on polyelectrolyte–amphiphile interaction: The carrageenan/furcellaran–amitriptyline system. *Biopolymers: Original Research on Biomolecules, 58*(2), 186–194.

Jamróz, E., Cabaj, A., Juszczak, L., Tkaczewska, J., Zimowska, M., Cholewa-Wójcik, A., ... Kopel, P. (2021). Active double-layered films enriched with AgNPs in great water dock Root and Pu-Erh extracts. *Materials, 14*(22), 6925. https://www.mdpi.com/1996-1944/14/22/6925.

Jamróz, E., Cabaj, A., Tkaczewska, J., Kawecka, A., Krzyściak, P., Szuwarzyński, M., ... Juszczak, L. (2023). Incorporation of curcumin extract with lemongrass essential oil into the middle layer of triple-layered films based on furcellaran/chitosan/gelatin hydrolysates – In vitro and in vivo studies on active and intelligent properties. *Food Chemistry, 402*, 134476. https://doi.org/10.1016/j.foodchem.2022.134476.

Jamróz, E., Janik, M., Juszczak, L., Kruk, T., Kulawik, P., Szuwarzyński, M., ... Khachatryan, K. (2021). Composite biopolymer films based on a polyelectrolyte complex of furcellaran and chitosan. *Carbohydrate Polymers*118627. https://doi.org/10.1016/j.carbpol.2021.118627.

Jamróz, E., Janik, M., Marangoni, L., Vieira, R. P., Tkaczewska, J., Kawecka, A., ... Juszczak, L. (2022). Double-layered films based on furcellaran, chitosan, and gelatin hydrolysates enriched with AgNPs in yerba mate extract, montmorillonite, and curcumin with rosemary essential oil. *Polymers, 14*(20), 4283. https://www.mdpi.com/2073-4360/14/20/4283.

Jamróz, E., Juszczak, L., & Kucharek, M. (2018a). Development of starch-furcellaran-gelatin films containing tea tree essential oil. *Journal of Applied Polymer Science, 135*(42), 46754. https://doi.org/10.1002/app.46754.

Jamróz, E., Juszczak, L., & Kucharek, M. (2018b). Investigation of the physical properties, antioxidant and antimicrobial activity of ternary potato starch-furcellaran-gelatin films incorporated with lavender essential oil. *International Journal of Biological Macromolecules, 114*, 1094–1101. https://doi.org/10.1016/j.ijbiomac.2018.04.014.

Jamróz, E., Khachatryan, G., Kopel, P., Juszczak, L., Kawecka, A., Krzyściak, P., ... Zimowska, M. (2020). Furcellaran nanocomposite films: The effect of nanofillers on the structural, thermal, mechanical and antimicrobial properties of biopolymer films. *Carbohydrate Polymers*116244. https://doi.org/10.1016/j.carbpol.2020.116244.

Jamróz, E., Kopel, P., Juszczak, L., Kawecka, A., Bytesnikova, Z., Milosavljevic, V., & Makarewicz, M. (2019). Development of furcellaran-gelatin films with Se-AgNPs as an active packaging system for extension of mini kiwi shelf life. *Food Packaging and Shelf Life, 21*, 100339. https://doi.org/10.1016/j.fpsl.2019.100339.

Jamróz, E., Kopel, P., Juszczak, L., Kawecka, A., Bytesnikova, Z., Milosavljević, V., ... Adam, V. (2018). Development and characterisation of furcellaran-gelatin films containing SeNPs and AgNPs that have antimicrobial activity. *Food Hydrocolloids, 83*, 9–16. https://doi.org/10.1016/j.foodhyd.2018.04.028.

Jamróz, E., Kopel, P., Tkaczewska, J., Dordevic, D., Jancikova, S., Kulawik, P., ... Adam, V. (2019). Nanocomposite furcellaran films—The influence of nanofillers on functional properties of furcellaran films and effect on linseed oil preservation. *Polymers, 11*(12), 2046. https://www.mdpi.com/2073-4360/11/12/2046.

Jamróz, E., Kulawik, P., Guzik, P., & Duda, I. (2019). The verification of intelligent properties of furcellaran films with plant extracts on the stored fresh Atlantic mackerel during storage at 2 °C. *Food Hydrocolloids, 97*, 105211. https://doi.org/10.1016/j.foodhyd.2019.105211.

Jamróz, E., Kulawik, P., Krzyściak, P., Talaga-Ćwiertnia, K., & Juszczak, L. (2019). Intelligent and active furcellaran-gelatin films containing green or pu-erh tea extracts: Characterization, antioxidant and antimicrobial potential. *International Journal of Biological Macromolecules, 122*, 745–757. https://doi.org/10.1016/j.ijbiomac.2018.11.008.

Jamróz, E., Tkaczewska, J., Juszczak, L., Zimowska, M., Kawecka, A., Krzyściak, P., & Skóra, M. (2022). The influence of lingonberry extract on the properties of novel, double-layered biopolymer films based on furcellaran, CMC and a gelatin hydrolysate. *Food Hydrocolloids, 124*, 107334. https://doi.org/10.1016/j.foodhyd.2021.107334.

Jamróz, E., Tkaczewska, J., Kopeć, M., & Cholewa-Wójcik, A. (2022). Shelf-life extension of salmon using active total biodegradable packaging with tea ground waste and furcellaran-CMC double-layered films. *Food Chemistry, 383*, 132425. https://doi.org/10.1016/j.foodchem.2022.132425.

Jamróz, E., Tkaczewska, J., Zając, M., Guzik, P., Juszczak, L., Kawecka, A., ... Wojdyło, A. (2022). Utilisation of soybean post-production waste in single- and double-layered films based on furcellaran to obtain packaging materials for food products prone to oxidation. *Food Chemistry, 387*, 132883. https://doi.org/10.1016/j.foodchem.2022.132883.

Jancikova, S., Jamróz, E., Kulawik, P., Tkaczewska, J., & Dordevic, D. (2019). Furcellaran/gelatin hydrolysate/rosemary extract composite films as active and intelligent packaging materials. *International Journal of Biological Macromolecules, 131*, 19–28. https://doi.org/10.1016/j.ijbiomac.2019.03.050.

Janik, M., Jamróz, E., Tkaczewska, J., Juszczak, L., Kulawik, P., Szuwarzyński, M., ... Kopel, P. (2021). Utilisation of carp skin post-production waste in binary films based on furcellaran and chitosan to obtain packaging materials for storing blueberries. *Materials, 14*(24), 7848. https://www.mdpi.com/1996-1944/14/24/7848.

Jasińska, J. M., Michalska, K., Szuwarzyński, M., Mazur, T., Cholewa-Wójcik, A., Kopeć, M., ... Jamróz, E. (2024). *Phytolacca americana* extract as a quality-enhancing factor for biodegradable double-layered films based on furcellaran and gelatin – Property assessment. *International Journal of Biological Macromolecules, 279*, 135155. https://doi.org/10.1016/j.ijbiomac.2024.135155.

Kajla, P., Chaudhary, V., Dewan, A., Bangar, S. P., Ramniwas, S., Rustagi, S., & Pandiselvam, R. (2024). Seaweed-based biopolymers for food packaging: A sustainable approach for a cleaner tomorrow. *International Journal of Biological Macromolecules, 274*, 133166. https://doi.org/10.1016/j.ijbiomac.2024.133166.

Kalita, P., Bora, N. S., Gogoi, B., Goswami, A., Pachuau, L., Das, P. J., ... Roy, S. (2025). Improving the hydrophobic nature of biopolymer based edible packaging film: A review. *Food Chemistry, 479*, 143793. https://doi.org/10.1016/j.foodchem.2025.143793.

Kasprzak, M., Jamróz, E., Nowak, N., Grzebieniarz, W., & Tkaczewska, J. (2024). Design of triple-layer films with blackseed protein as dispersion or emulsion. *Food Chemistry, 435*, 137533. https://doi.org/10.1016/j.foodchem.2023.137533.

Kruk, J., Tkaczewska, J., Szuwarzyński, M., Mazur, T., & Jamróz, E. (2023). Influence of storage conditions on functional properties of multilayer biopolymer films based on chitosan and furcellaran enriched with carp protein hydrolysate. *Food Hydrocolloids, 135*, 108214. https://doi.org/10.1016/j.foodhyd.2022.108214.

Kulawik, P., Jamróz, E., Janik, M., Tkaczewska, J., Krzyściak, P., Skóra, M., ... Tadele, W. (2022). Biological activity of biopolymer edible furcellaran-chitosan coatings enhanced with bioactive peptides. *Food Control, 137*, 108933. https://doi.org/10.1016/j.foodcont.2022.108933.

Kulawik, P., Jamróz, E., Janik, M., Tkaczewska, J., Krzyściak, P., Skóra, M., ... Tadele, W. (2023). Antimicrobial and antioxidant properties of chitosan-furcellaran-gelatin hydrolysate coatings enhanced with bioactive peptides. *Food Control, 153*, 109931. https://doi.org/10.1016/j.foodcont.2023.109931.

Kulawik, P., Jamróz, E., Tkaczewska, J., Vlčko, T., Zając, M., Guzik, P., ... Milosavljević, V. (2024). Application of antimicrobial chitosan-Furcellaran-hydrolysate gelatin edible coatings enriched with bioactive peptides in shelf-life extension of pork loin stored at 4 and −20 °C. *International Journal of Biological Macromolecules, 254*, 127865. https://doi.org/10.1016/j.ijbiomac.2023.127865.

Kulawik, P., Jamróz, E., Zając, M., Guzik, P., & Tkaczewska, J. (2019). The effect of furcellaran-gelatin edible coatings with green and pu-erh tea extracts on the microbiological, physicochemical and sensory changes of salmon sushi stored at 4 °C. *Food Control, 100*, 83–91. https://doi.org/10.1016/j.foodcont.2019.01.004.

Kulawik, P., Tadesse, E. E., Szymkowiak, A., Tkaczewska, J., Zając, M., Guzik, P., ... Jamróz, E. (2025). Effect of multilayer nano–/mini-furcellaran/chitosan emulsions with oregano essential oil on quality, oxidation and consumer perception of pork loins stored at 4 °C and −20 °C. *International Journal of Biological Macromolecules, 303*, 140491. https://doi.org/10.1016/j.ijbiomac.2025.140491.

Luu, T.-A., Le, V.-G., Thai, V.-A., Nguyen, M.-K., Lin, C., Pham, M.-T., ... Nguyen, D. D. (2025). Plastic wastes for carbon-based materials: Investigations on recent applications towards environmentally sustainable, carbon dioxide capture and green energy. *Environmental Pollution, 370*, 125899. https://doi.org/10.1016/j.envpol.2025.125899.

Marangoni, L., Vieira, R. P., Jamróz, E., & Anjos, C. A. R. (2020). Furcellaran: An innovative biopolymer in the production of films and coatings. *Carbohydrate Polymers*117221. https://doi.org/10.1016/j.carbpol.2020.117221.

Naderi Kalali, E., Lotfian, S., Entezar Shabestari, M., Khayatzadeh, S., Zhao, C., & Yazdani Nezhad, H. (2023). A critical review of the current progress of plastic waste recycling technology in structural materials. *Current Opinion in Green and Sustainable Chemistry, 40*, 100763. https://doi.org/10.1016/j.cogsc.2023.100763.

Nowak, N., Cholewa-Wójcik, A., Tkaczewska, J., Grzebieniarz, W., Tkacz, K., Modzelewska-Kapituła, M., ... Jamróz, E. (2024). The use of active compounds to shape the quality of active double-layer films based on furcellaran intended for packaging salad-dressing – Assessment of utilitarian and storage properties. *Food Chemistry, 438*, 137957. https://doi.org/10.1016/j.foodchem.2023.137957.

Nowak, N., Grzebieniarz, W., Cholewa-Wójcik, A., Juszczak, L., Konieczna-Molenda, A., Dryzek, E., ... Jamróz, E. (2024). Effects of selected plant extracts on the quality and functional properties of gelatin and furcellaran-based double-layer films. *Food and Bioprocess Technology, 17*(5), 1201–1214. https://doi.org/10.1007/s11947-023-03190-2.

Nowak, N., Grzebieniarz, W., Juszczak, L., Cholewa-Wójcik, A., Synkiewicz-Musialska, B., Huber, V., ... Jamróz, E. (2024). Influence of *Curcuma Longa* extract in citral addition on functional properties of thin films with triple-layer structure based on furcellaran and gelatin. *International Journal of Biological Macromolecules, 266*, 131344. https://doi.org/10.1016/j.ijbiomac.2024.131344.

Nowak, N., Tkaczewska, J., Cholewa-Wójcik, A., Grzebieniarz, W., Piasna-Słupecka, E., Drozdowska, M., & Jamróz, E. (2025). Multi-layer active biopolymer system as a perspective for improving the quality and scope of applying semi-rigid biodegradable packaging. *Food and Bioprocess Technology, 18*(3), 2682–2696. https://doi.org/10.1007/s11947-024-03617-4.

Nowak, N., Tkaczewska, J., Grzebieniarz, W., Juszczak, L., Mazur, T., Szuwarzyński, M., ... Jamróz, E. (2024). Active and intelligent four-layer films based on chitosan, gelatin, furcellaran and active ingredients—preparation, characterisation and application on salmon. *Food and Bioprocess Technology, 17*(7), 1862–1875. https://doi.org/10.1007/s11947-023-03238-3.

Pluta-Kubica, A., Jamróz, E., Guzik, P., Janik, M., Milosavljevic, V., Cholewa-Wójcik, A., & Kulawik, P. (2024). The effect of chitosan-furcellaran biopolymeric films and coatings incorporated with LL-37 and RW4 bioactive peptides on the microbiological and sensory quality of cheese. *International Dairy Journal, 157*, 106000. https://doi.org/10.1016/j.idairyj.2024.106000.

Pluta-Kubica, A., Jamróz, E., Juszczak, L., Krzyściak, P., & Zimowska, M. (2021). Characterization of furcellaran-whey protein isolate films with green tea or Pu-erh extracts and their application as packaging of an acid-curd cheese. *Food and Bioprocess Technology, 14*(1), 78–92. https://doi.org/10.1007/s11947-020-02570-2.

Pluta-Kubica, A., Jamróz, E., Kawecka, A., Juszczak, L., & Krzyściak, P. (2019). Active edible furcellaran/whey protein films with yerba mate and white tea extracts: Preparation, characterization and its application to fresh soft rennet-curd cheese. *International Journal of Biological Macromolecules*. https://doi.org/10.1016/j.ijbiomac.2019.11.102.

Pluta-Kubica, A., Jamróz, E., Kawecka, A., Juszczak, L., & Krzyściak, P. (2020). Active edible furcellaran/whey protein films with yerba mate and white tea extracts: Preparation, characterization and its application to fresh soft rennet-curd cheese. *International Journal of Biological Macromolecules, 155*, 1307–1316. https://doi.org/10.1016/j.ijbiomac.2019.11.102.

Pluta-Kubica, A., Jamróz, E., Khachatryan, G., Florkiewicz, A., & Kopel, P. (2021). Application of furcellaran nanocomposite film as packaging of cheese. *Polymers, 13*(9), 1428. https://www.mdpi.com/2073-4360/13/9/1428.

Polášek, Z., Salek, R. N., Vašina, M., Lyčková, A., Gál, R., Pachlová, V., & Buňka, F. (2021). The effect of furcellaran or κ-carrageenan addition on the textural, rheological and mechanical vibration damping properties of restructured chicken breast ham. *LWT, 138*, 110623. https://doi.org/10.1016/j.lwt.2020.110623.

Pycia, K., Grzebieniarz, W., Nowak, N., Juszczak, L., Kawecka, A., Szuwarzyński, M., ... Jamróz, E. (2024). Application possibilities of triple-layer furcellaran film with hazelnut oil microemulsion for packing cod liver oil. *Food Hydrocolloids, 147*, 109428. https://doi.org/10.1016/j.foodhyd.2023.109428.

Štěpánková, K., Ozaltin, K., Pelková, J., Pištěková, H., Karakurt, I., Káčerová, S., ... Mozetic, M. (2022). Furcellaran surface deposition and its potential in biomedical applications. *International journal of molecular sciences, 23*(13), 7439. https://www.mdpi.com/1422-0067/23/13/7439.

Stępień, A., Tkaczewska, J., Nowak, N., Grzebieniarz, W., Goik, U., Żmudziński, D., & Jamróz, E. (2023). Sugar-free, vegan, furcellaran gummy jellies with plant-based triple-layer films. *Materials, 16*(19), 6443. https://www.mdpi.com/1996-1944/16/19/6443.

Tadele, W., Kulawik, P., Stępień, A., Zając, M., Nowak, N., Grzebieniarz, W., ... Jamróz, E. (2024). Effects of multilayer Nano/mini Furcellaran/chitosan Emulsions with oregano essential oil and bioactive peptides on sensory, physicochemical properties and retrogradation in Sushi in cold storage conditions. *Innovative Food Science & Emerging Technologies, 96*, 103767. https://doi.org/10.1016/j.ifset.2024.103767.

Tkaczewska, J., Jamróz, E., Kasprzak, M., Zając, M., Pająk, P., Grzebieniarz, W., ... Juszczak, L. (2023). Edible coatings based on a furcellaran and gelatin extract with herb addition as an active packaging for carp fillets. *Food and Bioprocess Technology, 16*(5), 1009–1021. https://doi.org/10.1007/s11947-022-02952-8.

Tkaczewska, J., Kulawik, P., Jamróz, E., Guzik, P., Zając, M., Szymkowiak, A., & Turek, K. (2021). One- and double-layered furcellaran/carp skin gelatin hydrolysate film system with antioxidant peptide as an innovative packaging for perishable foods products. *Food Chemistry, 351*, 129347. https://doi.org/10.1016/j.foodchem.2021.129347.

Tkaczewska, J., Kulawik, P., Nowak, N., Grzebieniarz, W., Krzyściak, P., Tadele, W., ... Jamróz, E. (2024). Comparing the effects of duo-functional triple-layer films enriched with different sources of curcumin on the shelf-life of fish. *Foods, 13*(21), 3499. https://www.mdpi.com/2304-8158/13/21/3499.

Truus, K., Vaher, M., Usov, A. I., Pehk, T., & Kollist, A. (1997). Gelling galactans from the algal community of *Furcellaria lumbricalis* and *Coccotylus truncatus* (the Baltic Sea, Estonia): a structure-property study. *International Journal of Biological Macromolecules, 21*(1), 89–96. https://doi.org/10.1016/S0141-8130(97)00046-9.

Zając, M., Jamróz, E., Guzik, P., Kulawik, P., & Tkaczewska, J. (2021). Active biopolymer films based on furcellaran, whey protein isolate and Borago officinalis extract: Characterization and application in smoked pork ham production. *Journal of the Science of Food and Agriculture, 101*(7), 2884–2891. https://doi.org/10.1002/jsfa.10920.

Chapter 4

Carrageenan films and coatings

S. Jaspin[1], S. Nimbkar[2], S. Deotale[3], and P. Thivya[2]
[1]*Department of Food Process Engineering, Entrepreneurship and Management, (NIFTEM-T), National Institute of Food Technology, Thanjavur, Tamil Nadu, India,* [2]*Department of Food Process Engineering, National Institute of Technology (NIT), Rourkela, Odisha, India,* [3]*Department of Chemical Engineering, Visvesvaraya National Institute of Technology (VNIT), Nagpur, Maharashtra, India*

4.1 Introduction

Approximately 36% of food packaging materials worldwide are produced from petroleum-based plastics, such as polyethylene, polypropylene, and polystyrene. These materials have advantages due to their excellent barrier properties, ease of manufacture, and low cost, which contribute to an annual production increase of 8%, with a recycling rate of less than 5% (Cheng et al., 2022). These plastics are commonly disposed of through traditional methods like landfilling and incineration, leading to microplastics (particles less than 5 mm in diameter). These microplastics are ingested by living organisms, particularly aquatic animals, and pose significant risks to ecosystems and the food chain. This condition has prompted the exploration of alternative strategies for producing biopolymer-based packaging from available resources. Such materials have gained increasing research attention in recent years, driven by growing environmental concerns. Biopolymers are typically polysaccharides, proteins, and lipids derived from natural resources; they are nontoxic, biodegradable, biocompatible, and possess excellent film-forming properties. Polysaccharides are extensively employed to develop bio-based packaging films and coatings. These polysaccharides can be sourced from plants (e.g., starch, cellulose, pectin), animals (e.g., chitosan, keratin), microalgae/seaweeds (e.g., carrageenan, agar, sodium alginate [SA]), and microorganisms (e.g., pullulan, xanthan gum, cellulose, alginate). Seaweed is a readily available and cost-effective source of natural biopolymers, including carrageenan (CA), alginate, and agar. Among these, CA has gained significant research attention as a promising red seaweed-based polysaccharide due to its excellent film-forming ability (Sedayu et al., 2019).

CA is a sulfated polygalactan extracted from red seaweeds (*Chondrus crispus*) belonging to the class *Rhodophyceae*, with an ester sulfate content of 15%–40%. It consists of alternating units of D-galactose and 3,6-anhydrogalactose linked by α-1,3- and β-1,4-glycosidic bonds, forming a polymeric chain (Abud & Simon, 2022). CA has been used in Asia and Europe for centuries, with its commercial production beginning in the early 19th century as a powdered product. Initially, CA was utilized as a stabilizer in ice cream and chocolate milk, and later found applications in products like pudding, condensed milk, and toothpaste by the 1950s (Udo et al., 2023). In 2023, the global CA market was valued at $945.46 million and is forecasted to reach $1576.52 million by 2031, reflecting a compound annual growth rate (CAGR) of 6.6% from 2024 to 2031 ("Global Carrageenan Market - Industry Trends & Forecast to 2031," 2024). The Asia-Pacific region, mainly Indonesia, Philippines, China, and India, contributed to the major production of CA in the world (Carrageenan Market Size & Share Analysis - Growth Trends & Forecasts, 2023). Its distinctive properties and functional versatility have led to extensive exploration in various industries. Currently, CA is approved by the Food and Drug Administration (FDA) and classified as generally recognized as safe (GRAS), with an average molecular weight ranging from 200 to 800 kDa (Udo et al., 2023). It is a gelling, thickening, water retention, stabilizing, and emulsifying agent in numerous food products. The three primary types of CA used in food applications are kappa (κ-), iota (ι-), and lambda (λ-) CA, which are characterized by repeating D-galactose units and contain one, two, or three negatively charged organosulfate groups, respectively (Mendes et al., 2024). The specific degree of organosulfate substitution influences the classification and functional properties of CA. CA's ability to interact with food matrices influences its functional roles, including water retention, gelling, thickening, and stabilization (Udo et al., 2023). Over the years, it has been extensively incorporated into a variety of food products, such as meat, dairy, and flour-based items, while ongoing research continues to examine its mechanisms and applications across diverse food systems.

Natural Polymer-Based Films and Coatings in Food Technology. DOI: https://doi.org/10.1016/B978-0-443-34244-8.00014-8

κ-CA is widely recognized for its strong film-forming capabilities, enabling the production of transparent films with excellent mechanical and structural characteristics. These films exhibit comparatively higher tensile strength (TS) than those derived from other CA subtypes. However, κ-CA has limitations, including high hydrophilicity, poor water vapor barrier properties, and a lack of functional properties, which make it less effective at preventing food spoilage caused by microbial activity and oxidation, as well as limit its use in food packaging applications (Bhatia et al., 2024). To overcome these limitations, CA-based films and coatings are often enhanced by blending with other biopolymers, bioactive compounds (such as polyphenols, essential oils, and enzymes), and crosslinking agents, resulting in biocomposite films and coatings. These biocomposite films exhibit increased thickness, improved mechanical strength, and enhanced barrier properties against moisture (Wang, Liu et al., 2024). Another innovative approach involves the use of nanotechnology, where CA-based films and coatings are further enhanced by incorporating nanomaterials or nanocomposites (Aga et al., 2021). The addition of functional ingredients is crucial for developing active and smart/intelligent CA-based packaging materials capable of monitoring food quality and safety. This chapter highlights CA as a promising biopolymer for developing bio-based packaging films and coatings.

4.2 Structure, types, and properties of carrageenan

The term "carrageenin" (also known as "carrageenan") was initially used in 1840 to refer to a class of linear sulfated polysaccharides that were isolated from red algae, or red seaweeds. Different CA vary in their monosaccharide content, sulfation pattern (positions and quantities of sulfate groups), and molecular weights (MWs), even though their fundamental structure is made up of D-galactose units connected sequentially with α-(1–3) and β-(1–4) links (Michel et al., 2006). Their categorization is mainly determined by the precise positioning of the sulfate groups and the presence or absence of anhydrous moiety. Likely, CA can be classified into six fundamental types based on the number and location of sulfate groups: kappa (κ), lambda (λ-), iota (ι-), mu (μ-), theta (θ), nu (ν-), and xi (ξ) (Eha et al., 2021). Out of these, kappa (κ-), iota (ι-), and lambda (λ-) CA are the three main CA that are often utilized in the food industry, as shown in Fig. 4.1. They each include one, two, and three sulfate groups in their basic disaccharide units (Liu et al., 2021). Monosaccharide hydroxyl groups can be glycosylated, methylated, or sulfated. Furthermore, the structural diversity of

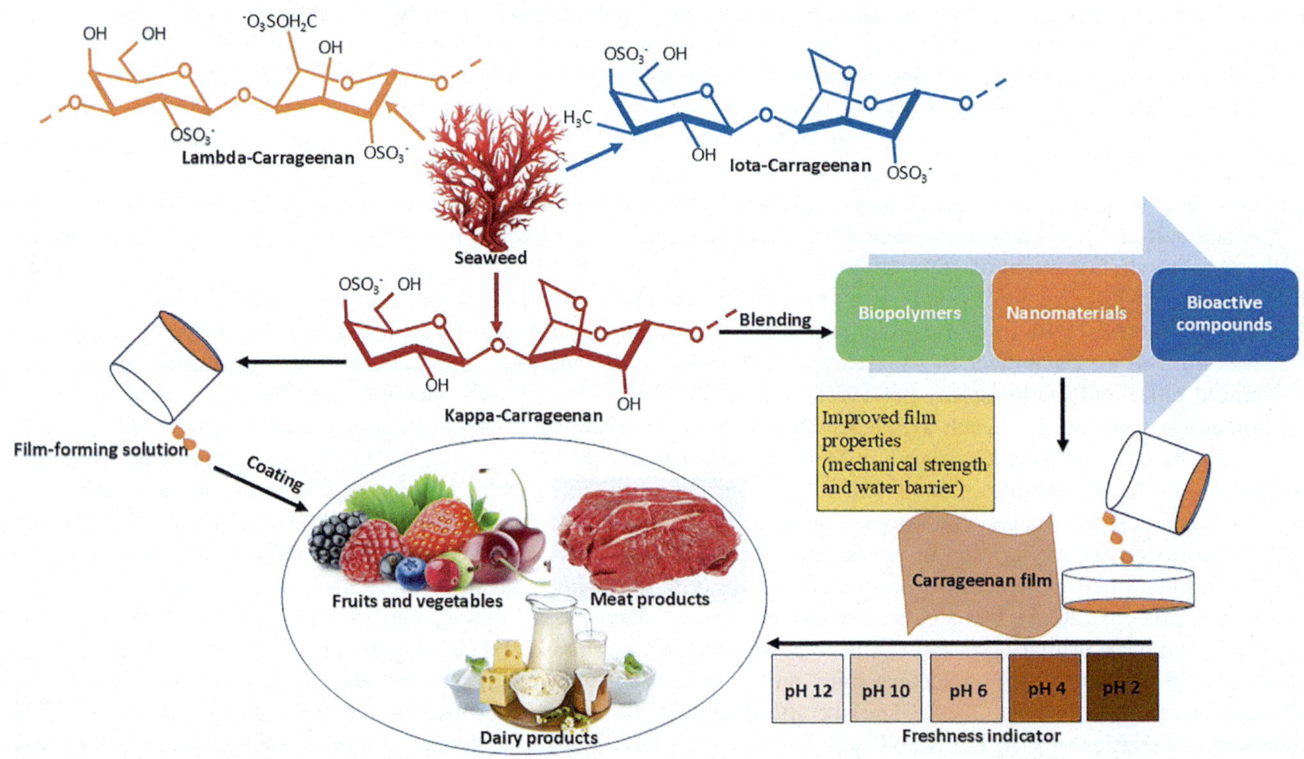

FIGURE 4.1 Structure and application of carrageenan-based coatings and films in food processing.

galactans can be further increased by the presence of numerous modified disaccharide units in a single molecule to create molecular hybrids (Ciancia et al., 2020). CA fundamental disaccharide units have different compositions, which correspond to different functions.

There are more potential fundamental repeating structural units for CA than there are letters in the Greek alphabet—42 to be exact. In place of the Greek letters previously used to identify at least 15 disaccharide units, a modified nomenclature scheme was suggested (Calvo et al., 2019). Carrageenose 4′-sulfate and carrageenose 2,4′-disulfate, for instance, are the suggested names for the κ- and ι-CA categories, respectively. More and more new CAs with distinct structural and bioactive characteristics have recently been discovered (Cotas et al., 2020). One of the primary factors influencing the viscosity, solubility, and gelation of CA is the MW distribution, which determines the prospective applications for these substances. MWs for food-grade CA are usually between 200 and 800 kDa. Nevertheless, upon acid hydrolysis at high temperatures (>80°C), degraded CA contains low MW fractions (20–40 kDa) (McKim et al., 2019). Poligeenan is one of the process's low MW fractions (MW = 10–20 kDa). Poligeenan and degraded CA have been shown to have detrimental gastrointestinal consequences, including neoplasms and intestinal ulcers. After receiving a modest dose of poligeenan (0.1%) orally for three months, rabbits developed colon lesions. Carcinogenicity and stomach ulcers are typically linked to the fraction with MW < 200 kDa (Younes et al., 2018). Notably, milder extraction techniques for CA also result in less MW fractions (20–50 kDa). The κ-type can have low MW fractions (<200 kDa) of up to 6.36% (David et al., 2018). The negative effects of these low MW fractions are widely thought to be controllable, even though they make up less than 5% of the overall amount of CA consumed. Furthermore, different preparations and varieties of CA contain varying amounts of low MW fractions, even for food-grade CA (David et al., 2020).

CA structure plays a major role in their biological activities and physicochemical characteristics, which in turn dictate how they may be used. The degree of sulfation affects the gel's elasticity, stiffness, and ability to form a gel. Generally, the types with a higher number of sulfate groups tend to have better solubility and weaker gelling/stabilizing properties. The percentages of sulfation in the κ-, ι-, and λ-types are 22%, 32%, and 38% (w/w), respectively (Campo et al., 2009). Because of this, the ι- and κ-types create gels, but the λ-type remains extremely viscous without forming gel-like structures. Remarkably, the CA isolated from algae in the *Gigartinaceae* family is a κ-/ι- hybrid, or a copolymer of ι- and κ-types, with special characteristics that set it apart from straightforward combinations of both types (Falshaw et al., 2001). Moreover, negatively charged CA interact with different molecules to acquire additional functionality in a synergistic or supportive way. For instance, κ-CA can combine with proteins or other hydrocolloids to create more appealing physical characteristics.

Random coils and other disordered conformations are present in the majority of nonstarch polysaccharides with heterogeneous structures in aqueous solutions (Goff & Guo, 2020). At every evaluated ionic strength, λ-CA maintains a random coil shape devoid of any secondary structure. Nonetheless, κ- and ι-CA experience a coil-helix transition brought on by ions, which increases the chain stiffness by two to three times (Schefer et al., 2015). Furthermore, ι-types have a larger ionic strength threshold for the transition than κ-types. At high temperatures and low salt concentrations, the secondary structure of the κ-type is identified as having a random coil shape, whereas at low temperatures and/or high ionic strengths, an ordered helical conformation forms. Depending on the quantity of salt, the κ-type exists in solution as a random coil at high temperatures (>50°C). However, it changes from a coil to a helix at low enough temperatures, and helical aggregation leads to gelation. In addition, when Na^+ or Li^+ is the only counterion present, ι-CA exhibits significant helix formation. Besides, a 10-fold increase in MW occurs together with the coil-helix transition of native ι-CA under nonaggregation circumstances. This conformational shift is also seen in a κ-/ι-type hybrid CA (de Velde et al., 2005). This hybrid type's gel strength naturally declines as the amount of the κ-type in the hybrid diminishes because the gelling temperature is independent of its composition; the gel strength can be controlled without bothering about the melting or gelation temperatures. Also, the chain composition of this hybrid determines its conformational order-disorder transition (Goff & Guo, 2020).

4.3 Extraction/manufacturing process of carrageenan

The first and most important stage in yielding seaweed is the extraction process. The seaweed species, the type of solvent, time, temperature, pH, ecological consequences, cost, size, and preferred qualities all affect the method selection and extraction efficiency. Proteins, unsaturated fatty acids, pigments, minerals, and polysaccharides are all included in seaweed extract. Flavonoid and polyphenolic substances, including phenolic terpenoids, phlorotannin, and bromophenol, are found in seaweed. Since phenol molecules frequently react with other chemicals, they are exceedingly challenging to quantitatively isolate. Nevertheless, it is possible to extract it on a laboratory basis (Dulanlebit & Hernani, 2023).

Seaweed is the primary raw material used to make CA extraction powder. Red seaweed species like *Eucheuma* and *Kappaphycus* are among the various types of seaweed that can be utilized. CA, the first ingredient in the extraction process, is abundant in these seaweeds. There are multiple stages involved in removing CA from seaweed, as shown in Fig. 4.2. CA is produced using two distinct processes, refined carrageenan (RC) and semirefined carrageenan (SRC), depending on its purity. Food-grade, CA or RC, is identified by the EU additive number E407, while SRC is designated E407a. According to the European Food Safety Authority (EFSA), when taken at the levels required to produce the appropriate food texture, both RC and SRC are safe and have not been shown to have any negative effects on humans (Mendes et al., 2024). The following discusses the procedures used in the manufacturing of both conventional and novel CA extraction methods.

4.3.1 Refined carrageenan

CA is found in the hydrosoluble component of the seaweed matrix's cell wall. CA is obtained by extracting the seaweed with hot water or an alkali solution of calcium hydroxide or potassium hydroxide (KOH). Additionally, polysaccharide enzymes like cellulose can be used to produce CA (Tarman et al., 2020). The process of RC consists of three main steps namely pretreatment, extraction, precipitation, drying, and grinding. In pretreatment, harvested seaweed is first properly cleaned to remove sand, mud, and other contaminants. This stage is crucial to guarantee the end product's purity. The seaweed may be partially or fully dried after washing, depending on the extraction technique. During the extraction process, drying can assist in reducing the water content and inactivating some enzymes that could break down CA (Rupert et al., 2022). Next, the prepared seaweed is immersed in an alkaline solution, typically KOH or sodium hydroxide (NaOH) (Al-Alawi et al., 2011). CA is released into the solution when the seaweed's cell walls are broken down by the alkaline environment. Temperature, extraction duration, and alkaline solution concentration are all crucial variables that must be

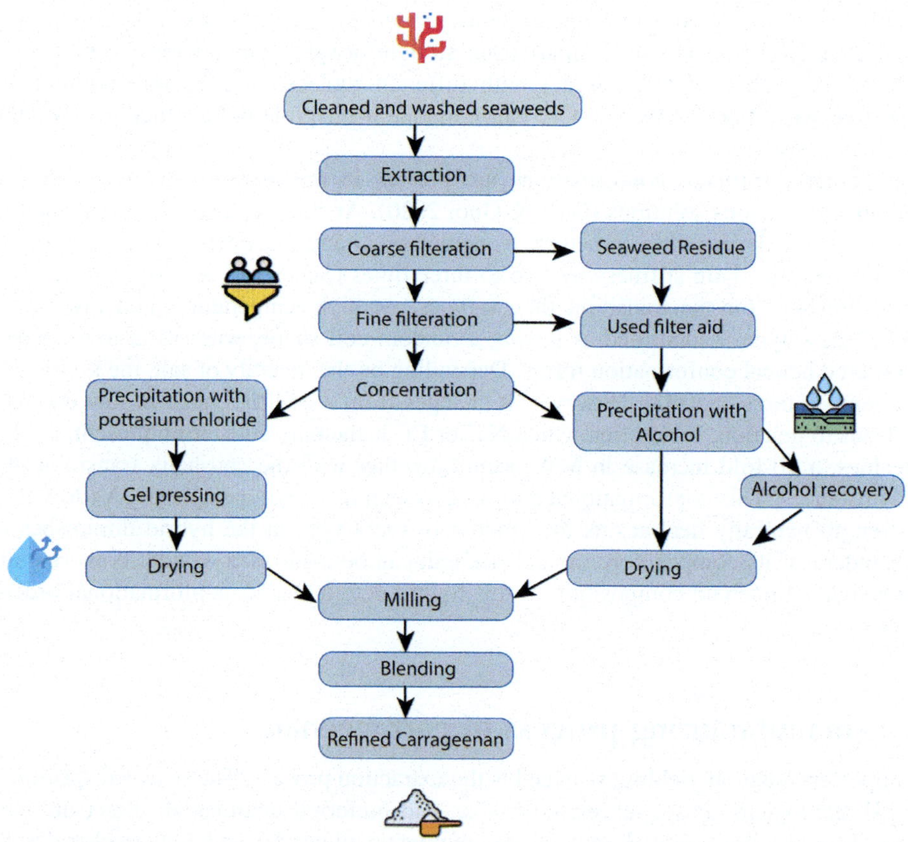

FIGURE 4.2 Flowchart of the manufacturing process of carrageenan.

effectively controlled (Vázquez-Delfín et al., 2014). To remove any leftover seaweed debris, the solution is filtered after the alkaline extraction. This filtration step is critical to achieve a clear solution that contains CA.

A precipitating agent, such as potassium chloride (KCl) for κ-CA or alcohol (ethanol or isopropanol) for ι-CA, is then added to the CA-containing solution. CA forms a solid precipitate when the precipitating agent is added. Usually, centrifugation or filtration is used to remove the precipitate from the solution. This separates the precipitating agent and other impurities from the CA in the residual liquid (Tarman et al., 2020). Drying eliminates any last traces of moisture from the separated CA precipitate. There are several techniques for drying, including oven, freeze, and air drying. For high-quality products, freeze-drying is frequently chosen because it better preserves the structure and characteristics of CA. The CA is dried and then milled into a fine powder. For a variety of uses, as well as for shipping and storage, the powder form is more practical (Li et al., 2014). Muthulakshmi et al. (2021) extracted CA from red seaweed (*Acanthophora spicifera*) using this method and obtained a 5 g yield of κ-CA powder (Muthulakshmi et al., 2021).

4.3.2 Semirefined carrageenan

SRC is a simpler technique than RC, and in this method, the CA from the seaweed is not removed. Seaweed, which is the raw material for SRC, is cleaned and cooked for a few hours in a hot alkali solution. The procedure results in the extraction of pigment, protein, fat, and salt from the alkali solution. Clean, dried seaweed is the end product of this technique. Furthermore, the reagent's hydroxide component enters the seaweed and lowers the carrageenan's sulfate content. Consequently, the CA's 3,6-anhydrogalactose and gel strength rise (Sedayu et al., 2018). The seaweed is cleaned and cut after being cooked in the alkali solution. The chopped seaweed is bleached with diluted sodium or calcium hypochlorite to produce white-colored SRC. Additionally, bleaching can aid in lowering the SRC's bacterial count (Tarman et al., 2020). The film-made SRC typically shows relatively poor mechanical properties and a water vapor barrier. This can be improved by blending SRC with hydrophobic materials, nanosized fillers, and a photosensitizer (solution of sodium benzoate) (Sedayu et al., 2021).

4.3.3 Novel extraction techniques for obtaining carrageenan

Red algae like *Eucheuma cottonii* (Balqis et al., 2017), *Mastocarpus stellatus* (Torres et al., 2017), and *Hypnea musciformis* (Rafiquzzaman et al., 2016) are used to extract CA using a variety of techniques, such as hydrolysis, microwave-based extraction, ultrasonic-assisted extraction (UAE), and supercritical fluid extraction methods. The most popular technique is hydrolysis extraction, which can yield up to 27% CA (Youssouf et al., 2017). Nevertheless, the extraction processes are difficult, solvent-consuming, and tedious. On the other hand, extraction techniques using microwaves, supercritical fluids, and ultrasonics have garnered more interest due to their better extraction yields, less energy usage, and quicker extraction times.

Sound waves that fall beyond the audible frequency range (usually >20 kHz) are referred to as ultrasound. UAE is a novel method where the plant cell wall can be physically destroyed by producing high pressure and high temperature quickly. This creates high-intensity shear stress, shock waves, and free radicals (Nie et al., 2021; Sharayei et al., 2021). A 150 W ultrasound power can produce a 50%–55% CA yield in 15 minutes (Youssouf et al., 2017). Aqueous UAE techniques have previously demonstrated that in the red algae *H. musciformis*, a higher extraction yield resulted in cell wall breakdown, reduced particle size, and greater mass transfer of cell contents (Mendes et al., 2024).

Microwave-assisted extraction (MAE) is also a new approach that can quickly heat the sample and solvent, since algae with high moisture concentrations are extremely vulnerable to microwave radiation. During extraction, variables like temperature, wave frequency, extraction time, solvent concentration, matrix properties, solid-to-liquid ratio, extraction pressure, and solvent properties must be optimized (Zhang et al., 2022). However, the CA yield is significantly impacted by the crucial parameter values. For instance, at 105°C, the yield of CA is 30.7% greater than the yield at a microwave temperature of 85°C (Cheng et al., 2022). Moreover, the yield of CA under neutral extraction circumstances is almost 37% greater than that under alkaline conditions (Boulho et al., 2017).

In the supercritical fluid extraction (SCFE) method, pure extracts were produced by rapidly permeating into seaweed while applying both pressure and temperature simultaneously. The extraction of bioactive substances using carbon dioxide and water as solvents works wonders to prevent CA from degrading at high temperatures. However, SCFE's effectiveness depends on the initial seaweed material's water content, solvent ratio, flow velocity, pressure, temperature, and biomass particle size (Dulanlebit & Hernani, 2023). SCFE offers a number of advantages, including quick extraction, high-quality results, and the use of ecologically friendly solvents (water) (Khalil et al., 2018). Additionally, the composition, bioactivity, and structure of CA may be altered by any of these techniques. Hence, future developments in CA extraction techniques will involve hybrid extraction techniques.

4.4 Methods used for the production of carrageenan-based coatings and films

CA-based coatings and films are commonly developed by combining CA with other biopolymers. Various methods, such as dipping/immersion, spray, emulsion, and sol-gel methods, are used to prepare CA coatings. The dipping method, followed by spray coating, is extensively used to prepare CA coatings. Techniques such as solvent casting, extrusion, electrospinning, and 3D printing are employed for CA film production. Among these, the solvent casting method is reported to be the most widely used. However, the hydrophilic nature and poor mechanical strength of CA limit its ability to withstand extensive force, pressure, and temperature. The methods of CA coating and film preparation are discussed in the following section, along with recent studies:

4.4.1 Methods of carrageenan-based coating

Coating is the process of applying a solution or film of material onto the surface of food to achieve specific properties or functions. This coated layer can aid various purposes, such as protecting the food surface from external damage, enhancing its appearance, improving shelf life, or imparting functional attributes like water and oil/grease resistance, and inhibiting the growth of microorganisms (Priya et al., 2023). Several coating techniques are used to produce CA coatings/films. These coating methods influence the effectiveness of CA-based coating materials applied to various food products, such as meat, fruits, and vegetables. The selection of coating methods depends on the nature, properties, and physical attributes of the foods being coated, such as size, shape, surface tension, density, viscosity, and thickness of coating materials (Bizymis & Tzia, 2022). The primary coating mechanism involves creating adhesion and diffusion between the film-forming solution and the surface of the food materials. In the case of fruits, two ripening patterns, climacteric and nonclimacteric, must be considered before the coating process. The adhesion between the coating materials and the surface of foods is formed through drying, warming, cooling, and coagulation processes. The different types of methods used for making CA-based coatings are explained in the following section:

4.4.1.1 Dipping method

Dipping is the most common method of coating food products and involves the immersion of a food sample in the CA-coating or film-forming solution (Priya et al., 2023). In the dipping method, firstly, CA is dissolved in water under moderate heat with constant stirring until the desired consistency is reached. The resultant mixture is called film-forming solution, which is further used for coating the food materials. In general, the dipping method involves three main steps: immersion and dwelling, deposition, and evaporation of solvents (Jafri & Jaafar, 2024). In the first step, the food material is immersed in the film-forming solution at a constant speed; the dwelling ensures enough quantity of solution for wetting substrate and complete interaction between food and coating matrix. The deposition process is used to develop thin layers of the precursor emulsion on the surface of food products. The excess surface liquid drains and is removed by deposition. During evaporation, the solvent and excess liquid are evaporated from the food products' surface using a heating and drying procedure. The product is dried either at room temperature or with the help of a dryer when the surplus coating is drained away (Jafri & Jaafar, 2024). In general, dip coating produces a thicker film than spray coating, affecting the other properties of the films. Although dip coating is a simple and cost-effective operation widely used, it requires a larger quantity of film-forming solution, leading to higher biopolymer requirements and, consequently, increased material costs for preparing the coating solution (Shigrekar & Amdoskar, 2024). Additionally, since the coating is in direct contact with food, it may influence the sensory characteristics of the films, which must align with consumer expectations (Cheng et al., 2022).

The multilayer or layer-by-layer coating or deposition can be formed by dipping the food in the coating solution several times. However, care must be taken when opting for a multilayer. That is, each layer should dry properly with minimal moisture before applying the next layer. This leads to time consumption and increases the thickness of the coated layer (Arnon-Rips & Poverenov, 2018). The dipping process is widely used to apply edible coatings to fresh produce. In general, fruits and vegetables are submerged for 5–30 seconds in the formulation of edible coating, and most fruits are easy to use (de Oliveira Filho et al., 2021). The most effective way to increase the microbial stability of freshly cut fruit is by dipping them in aqueous solutions that contain antimicrobials. The most common method to control the browning of fresh fruits and vegetables is by adding antioxidants to the dipping solution by immersing the fresh produce immediately after peeling and cutting (Pleşoianu & Nour, 2022). Dipping has proven less effective in direct applications on food surfaces of antimicrobial agents, as the loss of activity is caused by leaching on the food, enzymatic activity, and the reaction with additional food constituents (Patil et al., 2023). The dipping method is widely used at the laboratory level for coating applications due to its ease of operation and low cost. The dipping method is beneficial for the food product to coat its surface completely. It ensures good uniformity across a rough and complex shape on the surface of food products (Suhag et al., 2020). Several issues arise from the dipping method, including dilution of coating, waste or dirt accumulation, and microbial growth during drying. Furthermore, the drawback of the dipping mechanism is the dilution of the external layer and degraded functionality; that is, after dipping, the natural wax coating of fruit and vegetables might be detached (Mahmud et al., 2021).

4.4.1.2 Spray coating

Spray coating is a broadly used technique for applying coatings to food products. This method involves atomizing a film-forming solution under high pressure and dispersing it as droplets over the food surface using nozzles (Leya et al., 2024). It provides uniform coatings and efficient application, and various spraying techniques are employed in the industry. Air spray atomization is a cost-effective method that uses a high-velocity air stream to break the coating liquid into fine droplets. An air jet nozzle disperses the droplets evenly across the food surface, making it ideal for applying edible coatings (Gupta et al., 2022). Air-assisted airless atomization is another technique that combines partial fluid atomization with compressed air to achieve a finely atomized spray pattern. It is effective for high-viscosity and high-solids coatings, offering high production efficiency and a quality finish (Kumar et al., 2023). In pressure atomization or airless method, high pressure forces the coating solution through small nozzles, forming droplets via surface tension and viscosity control. Pressure is maintained below 3.5 bars to prevent damage to the film-forming system (Patil et al., 2023). Key parameters affecting spraying efficiency include pressure, viscosity, temperature, and the design of the spray nozzle. Droplet size, spray angle, flux rate, and drying conditions influence the final coating quality. Advanced spray systems, including electro-sprays, can produce uniform coatings with precise droplet sizes, enhancing the application of polymer and biopolymer-based solutions (Suhag et al., 2020). Spraying ensures uniform coatings with good mechanical and barrier properties, but controlling parameters like drying temperature and time is critical for achieving optimal results (Lakkannavar et al., 2024).

4.4.1.3 Emulsion and sol-gel coating

The emulsion coating is prepared by dissolving CA and other biopolymers in deionized water with the addition of a plasticizer (mostly glycerol) and is well homogenized. Ultrasound processors form the coarse emulsion coatings to obtain uniform, stable, and available emulsion coatings. The prepared emulsion coating is mostly applied directly on the food surface using the direct immersion/dip coating method. Examples of κ-CA-based composite emulsion coatings are CA/konjac glucomannan/camellia oil coated on chicken (Zhou et al., 2021) and κ-CA/cinnamon essential oils coated on pork meat (He & Wang, 2022). In the sol-gel method, CA is dissolved in water under constant stirring with moderate heat. For instance, the CA-based hybrid composite coating is prepared by the sol-gel method by dissolving CA powder and gelatin in water separately under constant stirring at 40°C and 50°C, followed by the addition of glutaraldehyde as a crosslinker in ethanol. The resultant film acts as a good corrosion barrier on medical-grade stainless steel (SS304) (Muthulakshmi et al., 2021).

4.4.2 Methods of carrageenan film formation

Edible films are solid matrices typically applied to the surface of food products, serving as primary packaging without altering the food's sensory or nutritional properties. These films are designed to be flavorless, colorless, and neutral to ensure they do not interfere with the sensory characteristics of the food. In addition to wrapping food products, edible films can be used as pouches or sachets for applications (Nair et al., 2023). These dissolve in hot or cold liquids, releasing their contents while providing convenience, portion control, and reducing solid waste. The development of edible films requires a thorough understanding of the chemical and structural properties of film-forming biopolymers and additives, tailored to specific applications. Films are typically produced using either wet or dry methods, known as solvent casting and extrusion processes, respectively. Solubility is crucial for the casting method, while the extrusion process relies on the thermoplasticity of biopolymers and characteristics such as phase transitions, glass transitions, and gelatinization (Butler et al., 2023). The solvent casting method is widely used for making CA films rather than extrusion due to their hydrophilicity. Edible films are increasingly used as delivery systems for active ingredients, offering a convenient and discreet method to incorporate colors, flavors, sweeteners, nutraceuticals, vitamins, drugs, or other functional components. These applications extend to various forms, including films, pouches, tablets, chews, gels, and other polymer-based structures.

4.4.2.1 Solvent casting method

Solvent casting is a commonly used method for producing films. This technique involves preparing a film-forming solution, spreading it onto a suitable mold, and then allowing it to dry. Several factors influence the film formation process. For example, the choice of mold significantly affects the ease of peeling off the film, with materials like Teflon and glass plates being ideal. In addition, the drying temperature and environmental humidity impact the film's thickness. Higher drying temperatures generally reduce the film's thickness and elongation at break (EAB), while increased ambient humidity enhances TS and water vapor permeability (WVP) (Cheng et al., 2022). However, films produced from high-concentration CA solutions often exhibit a spongy structure with large pores, which can hinder their practical use (Alam et al., 2019). This issue may arise from delayed delamination during the casting process. Although solvent casting is cost-effective, requires simple equipment, and is easy to operate, its long drying time limits its suitability for large-scale commercial production. The casting method, being a wet process, promotes better particle-particle interactions, resulting in more homogeneous particle distribution and fewer defects.

On the laboratory scale, the CA-based film-forming solutions are cast using different materials such as glass plate at size of (24 × 30 cm) (Roy & Rhim, 2019a) and (10 × 20 cm) (Mahajan et al., 2022), hydrophobic soda-lime glass petri dish (200 × 30 mm) (Bharti et al., 2021), petri dishes (polystyrene plates with diameters of 90, 150, and 100 mm) (Ali et al., 2023; Avila et al., 2020; da Rosa et al., 2020; de Lima Barizão et al., 2020; Jancikova et al., 2020; Kim et al., 2023; Santos et al., 2023), Teflon-coated glass plates (Park et al., 2024; Roy & Rhim, 2020) with diameters of 150 and 160 mm (Genecya, Adhika, & Wungu, 2023; Genecya, Adhika, Sutrisno et al., 2023; Sogut et al., 2022), size of 24 × 30 cm (Khan et al., 2024; Kim et al., 2022), squared molds (24 × 24 cm) (Liu et al., 2020), a polyethylene plastic mold (Gumus et al., 2024; Zhang et al., 2021), polypropylene plates (150 × 15 mm) (da Costa et al., 2024), standard polyacrylic plates (16 × 16 × 0.3 cm) (Farhan & Hani, 2020), a stainless-steel plate (35 × 21 cm) (Bian et al., 2024), a polytetrafluoroethylene plate (18 × 18 cm) (Zhou et al., 2024), acrylic plates (150 mm diameter) (Martiny, Pacheco, et al., 2020), and a plexiglass plate (Zhang et al., 2020).

After casting, the CA films should be dried under suitable conditions; for example, CA/melanin nanoparticles film dried at room temperature for 24 hours (Khan et al., 2024; Roy & Rhim, 2019a), 36 hours for CA/microcrystalline cellulose (MCC) from pistachio shells film (Park et al., 2024) and SRC/germinated fenugreek seeds extract film (Farhan & Hani, 2020), 48 hours for κ-CA/curcumin-β-cyclodextrin complex loaded film (Lai et al., 2022), and 70 hours for κ-CA/lignin film (Rukmanikrishnan et al., 2020). Similarly, κ-CA/*Ipomoea batatas* film dried at 40°C for 8 hours (Bharti et al., 2021), κ-CA/cassava starch film at 40°C for 16 hours (de Lima Barizão et al., 2020), and κ-CA/pomegranate flesh and peel extracts films dried at 30°C with 50% relative humidity (RH) for 48 hours (Liu et al., 2020), κ-CA/*Cymbopogon winterianus* essential oil film at 40°C for 6 hours (Santos et al., 2023), κ-CA/cassava starch/copper oxide particles film dried at 40°C ± 2°C for 16 hours (da Costa et al., 2024), κ-CA/cellulose nanocrystals (CNC) film dried in an oven at 45°C for 24 hours (Popescu et al., 2019), κ-CA/sodium carboxymethyl starch/carboxylated cellulose nanofibrils film dried at 50°C for 24 hours (Zhang et al., 2021b), arrowroot starch/iota-CA film and κ-CA/CNC film drying in an oven at 50°C for 6 hours (Abdillah & Charles, 2021; Gupta et al., 2021), CA/*Aloe vera* film at 55°C for 5 hours (Mahajan et al., 2022), sodium carboxymethyl starch/κ-CA/mulberry anthocyanins film and κ-CA/photoactivated proanthocyanidin film dried at 60°C for 24 hours (Santinon et al., 2024; Zhang et al., 2020), κ-and ι-CA/Lapacho tea extract film dried in a fume hood for 24 hours (Jancikova et al., 2020), κ-CA/fish gelatin/zein nanoparticle (ZNP)/turmeric essential oil (TEO) active films and CA/konjac glucomannan/gellan gum/mica nanosheets/anthocyanin intelligent films dried at 35°C overnight in a hot air oven (Ali et al., 2023; Bian et al., 2024), and CA/montmorillonite/silver and cerium oxide nanoparticles film dried at 60°C for 20 hours (Genecya, Adhika, & Wungu, 2023; Genecya, Adhika, Sutrisno et al., 2023).

After drying, the prepared composite films are peeled off and conditioned for 48 hours at 25°C ± 5°C with 50% ± 5% RH in a desiccator containing saturated salt solution (Bharti et al., 2021; Gumus et al., 2024; Khan et al., 2024; Kim et al., 2022; Lai et al., 2022; Li et al., 2020; Park et al., 2024; Roy & Rhim, 2020, 2019a). Notably, while preparing carrageenan film solution, there is a chance of producing air bubbles due to agitation, which should be removed by keeping the film-forming solution in a vacuum oven for 2 hours, or using ultrasonic treatment at 25°C for 3–10 minutes (Bharti et al., 2021; Lai et al., 2022; Sogut et al., 2022), or a thermostatic water bath at 55°C for 30 minutes (Abdillah & Charles, 2021), or by heating the solution at 75°C with constant stirring for 20 minutes (Park et al., 2024). If the air bubbles are not removed before casting, they can produce disordered films, leading to ease of breaking.

This method is also used to prepare the bilayer and multilayer carrageenan-based films by casting layer-by-layer and ensures the uniform volume of film solution, and each layer should dry enough prior to casting the next layer. For instance, chitosan/κ-CA film is made using a layer-by-layer solvent casting method (Zhuikova et al., 2020). This leads to an increase in the thickness of multilayer films compared to monolayer films due to high solid content. Koca and Bayramoğlu (2022) modified biaxially oriented polypropylene (BOPP) using layer-by-layer assembly with lysozyme, iota CA, and gum Arabic for food packaging. Five bilayer coatings of lysozyme/iota CA and lysozyme/gum Arabic reduced oxygen transmission rates by 66.15% and 56.89%, and water vapor transmission rates (WVTRs) by 28% and 33.52%, respectively. These edible coatings showed strong potential for preserving fresh produce with modified atmosphere packaging (Koca & Bayramoğlu, 2022). The lower processing temperature is an advantage of this method, which is particularly beneficial for blending CA with heat-sensitive materials and may undergo irreversible structural changes (Suhag et al., 2020). However, scaling up film production from the laboratory to the industrial level poses significant challenges.

4.4.2.2 Extrusion method

The extrusion method is widely utilized for producing bio-based packaging films due to its speed, efficiency, high energy utilization, and suitability for large-scale production. In this process, the material is extruded using a twin-screw system, passing through a die of a specific shape to form films under controlled parameter settings (Mount III, 2024). Unlike petrochemical materials, biopolymers are more sensitive to processing parameters such as screw speed, temperature, feed rate, and screw configuration. Research has shown that high extrusion temperatures and inappropriate pH levels can disrupt the polypeptide chains in proteins and alter their charge distribution (Cheng et al., 2022). Furthermore, even small adjustments in extrusion parameters can significantly influence the properties of the resulting films. Despite these challenges, the extrusion method has been successfully used to create biopolymer-based films, including those made from pectin and gelatin (Joye & Julian McClements, 2016). Studies indicate that consistent biopolymer films can be achieved by carefully managing factors such as extrusion speed, barrel temperature, and pH. However, the application of this method for producing CA films remains largely unexplored. Existing research has primarily focused on using extrusion to produce CA hydrogels and particles (Dong et al., 2021; Liu et al., 2023). Therefore, further research and development are needed to adapt the extrusion method for CA film production, thereby enhancing its commercial potential for creating CA-based packaging films.

4.4.2.3 Electrospinning

Electrospinning is an advanced technique used to produce CA films in the form of nanoscale fibers. This method employs a high-voltage electrostatic field to alter the surface of a polymer solution droplet, creating an electric potential between the droplet and a collector positioned at the end of a capillary (Sharma & James, 2022). When the electric field force exceeds the droplet's surface tension, a jet of charged polymer solution is ejected from the spinneret. As the jet is stretched during the process, the solvent gradually evaporates, resulting in partially or fully solidified electrospun fibers (Wang et al., 2019). The fibers produced through this method are characterized by an ultra-fine structure, high porosity, a large surface-to-volume ratio, and customizable morphology. These fibers can encapsulate bioactive compounds, enhancing the antimicrobial, oxidation resistance, and thermal stability properties of films. Due to its simplicity and cost-effectiveness, electrospinning is considered one of the most efficient techniques for fabricating one-dimensional materials (Banitaba & Ehrmann, 2021). However, the weak interactions between the electrospun fibers often result in films with poor mechanical strength (Rashid et al., 2021). For example, combining zein, CA, zinc oxide nanoparticles (ZnONPs), and rosemary essential oil has been shown to create maize protein/κ-CA electrospun fibers with improved surface hydrophobicity, antibacterial effectiveness, and antioxidant capabilities (Amjadi et al., 2020). Another example is the electrospun nanofibers produced from polyvinyl alcohol/Κ-CA/*Centaurea arvensis* anthocyanin, which showed high responsiveness of color changes (heather violet to green at pH 2–12) in 10 seconds compared to colorimetric films (15–40 minutes) indicators produced by the casting method (Forghani et al., 2022).

4.4.2.4 3D printing

3D printing is used in food packaging film production by enabling the creation of customized, efficient, and sustainable solutions. This technology allows for precise control over film design, incorporating features like multilayered structures, antimicrobial properties, and smart functionalities such as freshness indicators (Witek-Krowiak et al., 2024). It supports the use of biodegradable and edible materials, reducing environmental impact while maintaining food safety. Moreover, 3D printing minimizes material waste and offers potential for advanced innovations like pH-sensitive films and embedded sensors (Versino et al., 2023). Despite challenges like scalability and cost, advancements in high-speed printing and material science are enhancing their feasibility for large-scale applications, making it promising for sustainable and functional food packaging (Hussain et al., 2024). For instance, CA/collagen/allopurinol gel is prepared by ionic gelation, the obtained gel is used as ink, and the design and thickness of films are controlled by the 3D printing method (Vu et al., 2021). The gel/ink is pushed out and printed on a glass sheet during the printing process. After filling, the film is dried naturally at room temperature or at a controlled temperature in the oven. Furthermore, some factors, namely the thickness of the printing layer, the width of the print line, the filling area, the print speed, and the temperature of the print table, should be controlled during printing to produce an efficient film. There are not many studies on the application of 3D printing in the production of CA films, delivering an opportunity for future research to explore this technology for developing CA-based films.

4.5 Properties of carrageenan-based edible coatings and films

CA-based edible coatings and films exhibit a range of properties critical for their performance in food technology. These properties are categorized into physical, mechanical, barrier, and functional aspects, highlighting their suitability for specific applications and areas of improvement. CA-based edible coatings and films exhibit enhanced mechanical, thermal, and barrier properties through physical and chemical modifications. These modifications improve their strength and flexibility, making them suitable for food packaging applications as explained in Table 4.1. Additionally, the films demonstrate antibacterial properties, which help extend the shelf life of food products. The physicochemical properties, including gelation mechanisms and drying rates, are crucial for optimizing their performance in preserving food freshness and quality. Overall, these characteristics make CA a promising alternative to petroleum-based plastics.

4.5.1 Physical properties

The physical dimension of the film's cross-section is measured in micrometers (μm) or millimeters (mm). Thickness of the film affects barrier properties (e.g., water vapor and gas permeability). It also influences mechanical strength and transparency (Zhang et al., 2021). Transparency and opacity measurements evaluate the degree to which light passes through or is blocked by the film, respectively. Transparent films are desirable for food packaging to allow product visibility. Opaque films are useful for UV protection. Additives like fillers, pigments, or extracts (e.g., curcumin or tea extracts) can enhance opacity or provide UV-blocking capabilities (Sangeetha et al., 2024). The ability of the film to dissolve in water determines its suitability for various applications, such as instant food packaging or dissolvable wraps. The amount of water retained within the film matrix is crucial; high moisture content can lead to reduced mechanical strength and increased microbial activity. Controlled moisture is essential for stability. The swelling index measures the extent to which the film absorbs water and swells. The visual appearance of the film is assessed using colorimetric scales (L^*, a^*, b^* values), which influence consumer perception and product appeal. Films can be made clear, tinted, or opaque depending on the application. The UV-blocking property of the film refers to its ability to absorb or block ultraviolet radiation. Such films protect light-sensitive foods from oxidative degradation. Additives like tea extracts, nanoclays, or polyphenols can enhance this property (Cheng et al., 2022; Liang et al., 2023; Wang, Guo et al., 2024).

CA-based films exhibit several physical properties that enhance their suitability for food packaging. These films can form homogeneous film-forming solutions and are primarily fabricated through methods like direct coating, solvent casting, and electrospinning. However, pure CA films have inherent limitations, which can be addressed by incorporating other compounds to improve their physical and chemical properties. This enhancement allows CA-based films to effectively extend food shelf life by inhibiting microbial growth and reducing moisture loss (Cheng et al., 2022). The physical properties of these films, such as thickness, solubility, and the incorporation of additives, significantly influence their barrier and mechanical properties. These properties are crucial for determining the effectiveness of CA films in protecting and preserving food products. CA films are typically smooth, uniform, and transparent, which enhances their market appeal. The research paper highlights that the incorporation of copper-ferulic acid networks (CuFA NWs) into CA-based films significantly enhances their physical properties. The resulting Carr/CuFA films exhibit increased mechanical strength, improved surface hydrophobicity, and enhanced water vapor barrier properties. Additionally, these films possess UV-shielding capacity, making them more effective for food packaging applications. The interactions between CuFA NWs and the CA matrix are facilitated through hydrogen bonding and electrostatic interactions, contributing to these improved characteristics (Shi et al., 2024). As consumer preferences increasingly lean toward products that offer both sustainability and enhanced usability, developing such multifunctional packaging could reshape market offerings and foster broader acceptance of biodegradable alternatives.

4.5.1.1 *Mechanical properties*

TS is the maximum stress (force per unit area) a film can endure while being stretched before failure. It is usually expressed in MPa or pounds per square inch (psi). High TS is desirable for durability, structural integrity, and resistance to tearing during handling or application (Sun et al., 2021). EAB is the percentage increase in length a material undergoes

TABLE 4.1 Properties of carrageenan-based edible films and coatings.

Films/coatings	Biopolymers/ bioactives/ plasticizers used	Physical properties	Mechanical properties	Barrier properties	Functional properties	References
κ-CA and *Dillenia serrata* pectin	Curcumin	Reduced thickness and swelling	Tensile strength (TS): 0.03 N mm^{-2}	Lowest water vapor transmission rate (WVTR): 5.61 $g \cdot h^{-1} \cdot m^{-2}$	More effective against the Gram-negative (*E. coli*) than the Gram-positive (*L. monocytogenes*) bacteria	Sultan et al. (2023)
κ-CA and gluten	Garlic essential oil	Water solubility (40.57%)	Thickness: 0.73 mm and TS: 0.0009 N mm^{-2}	WVTR: 21.856 g h^{-1} m^{-2}	Antibacterial properties against *E. coli* bacteria of 26.93 mm and against *S. aureus* bacteria of 33.76 mm	Muflih and Syarifuddin (2023)
Konjac glucomannan (KGM)/κ-CA	Peppermint essential oil and thymol	Excellent UV-blocking activity	TS of 34.2 MPa and elongation at break (EB) of 21.9%	Water vapor permeability (WVP): 5.85 × 10^{-11} $g \cdot mm^{-1} \cdot h^{-1} \cdot Pa^{-1}$	Excellent antioxidant performance and antibacterial behavior against *Escherichia coli, Listeria monocytogenes,* and *Staphylococcus aureus*	Wu et al. (2023)
κ-CA	*Cymbopogon winterianus* essential oil	Transparency: 93.83%	TS and elongation at break (EAB): 2821.52 N m^{-1} and 3.07%	WVTR: 503.27 (g $m^{-2} \cdot$day), WVP: 1.68 (g $[Pa \cdot day \cdot m]^{-1}$) (×$10^{-5}$)	Antioxidant and antimicrobial against *Listeria monocytogenes*	Santos et al. (2023)
κ-CA and gelatin	Procyanidins	Thickness = 0.0362 mm	TS and EAB: 55.981 MPa and 8.527%	WVP: 2.094 ($g \cdot m \cdot m^{-2} \cdot s^{-1} \cdot Pa^{-1}$)	Controlled release of antioxidants	Wei et al. (2024)
κ-CA	Copper-ferulic acid networks (CuFA NWs)	Improved surface hydrophobicity, UV shielding	Enhanced mechanical properties	Improved water vapor barrier properties	Antibacterial capabilities	Shi et al. (2024)
κ-CA	Polyvinyl alcohol, spermidine carbon dots (CDs)	High transparency	Good mechanical performance	Good water vapor barrier ability and low migration	Antibacterial against *E. coli and S. aureus*	Li et al. (2024)
κ-CA and starch	Carrageenan from *Kappaphycus alvarezii*	Improved thermal stability	Enhanced mechanical strength	Increasing water solubility and water content and reducing the degradation rate and swelling degree	Not specified	Yap et al. (2024)
K-carrageenan	Rose petal-based carbon dots RP-CDs	Improved UV-blocking properties	Enhanced mechanical strength	Enhanced water contact angle and WVP	Higher total antioxidant activity, 2,2-Diphenyl-1-picrylhydrazyl (DPPH) and 2,2′-Azino-bis(3-ethylbenzothiazoline-6-sulfonic acid) (ABTS); higher antimicrobial properties	Wagh et al. (2024)

(Continued)

TABLE 4.1 (Continued)

Films/coatings	Biopolymers/ bioactives/ plasticizers used	Physical properties	Mechanical properties	Barrier properties	Functional properties	References
K- and ι-CA	Lapacho tea extract	Improved UV-blocking properties	TS: 0.42 ± 0.02 MPa and breaking strain: 77.6% ± 1.6%	-	TPC: 233.75 ± 0.104 mg gallic acid g^{-1}.DPPH: 87.63% ± 0.03%	Jancikova et al. (2020)
K-CA/ hydroxypropyl methylcellulose	*Prunus maackii* extract		43.20% ± 2.07% EAB: 1096.39 ± 51.29 N m^{-1}	2.23 ± 0.02 $cm^3 \cdot mm \cdot m^{-2}$ $day^{-1} \cdot atm^{-1}$ oxygen permeability	DPPH free radical scavenging rate of the films was 23.75% ± 0.84%	Sun et al. (2020)
κ-CA	Benzyl isothiocyanate	48.8% enhanced thermal degradation temperature	33.7% enhanced TS, decreased the EAB by 33.8%	Decreased the WVP by 60.1%, decreased the oxygen permeability by 42%	Enhanced antimicrobial activity against *Staphylococcus aureus, Salmonella typhimurium*, and *Pseudomonas fluorescens*	Du et al. (2024)
κ-CA	Tragacanth gum and clove essential oil (CEO)		Enhanced film EAB, whereas it reduced TS and Young's modulus	Reduced the films' moisture content and WVP	30% CEO exhibited a strong inhibitory effect against both *Escherichia coli* and *Bacillus* with higher antioxidant capacity	Tabassum et al. (2024)
κ-CA	Guar gum or locust bean gum	Good light transmittance, thermal stability improved	21.380 ± 1.536 MPa TS and 32.3% EAB	OP: 2.78 cm^3 mm m^{-2} day^{-1} atm^{-1}	Not specified	Wang et al. (2023)
κ-CA/poly(vinyl alcohol)	Gallic acid and stabilized with metallic ions	Suitable physical properties as a coating	TS: 10.28 ± 0.65 MPa	WVP: 5.36 ± 0.51 × 10^{-12} g $(Pa \times m \times s)^{-1}$	Good antimicrobial action against *Staphylococcus aureus* and *Escherichia coli*	de Jesus et al. (2023)
κ-CA	Grapefruit essential oil	67.10% transparency	TS: 98.21 ± 6.35 MPa EAB: 5.73% ± 1.78%	WVP: 0.532 (g·mm) $(m^2 \cdot h \cdot kPa)^{-1}$	11.36% DPPH inhibition, 30.73% ABTS inhibition	Bhatia et al. (2024)

before breaking under tensile stress. It indicates how much the material can elongate relative to its original length. It is expressed in a percentage. High EAB indicates that the material is flexible and can undergo significant deformation without breaking, whereas low EAB suggests the material is brittle and prone to breaking with little deformation (Sun et al., 2019). The mechanical properties of CA-based films can be enhanced through various modifications and the incorporation of additional biopolymers or active compounds. To achieve the highest TS and EAB in CA-based films and coatings, a combination of specific additives and conditions is essential. The incorporation of glycerol and montmorillonite has been shown to optimize these mechanical properties effectively. Specifically, a CA-based bioplastic with 3% glycerol and 20% montmorillonite achieved a TS of 3.43 MPa and an EAB of 47.01% (Genecya, Adhika, & Wungu, 2023; Genecya, Adhika, Sutrisno et al., 2023).

The mechanical properties of κ-CA-based edible films were significantly enhanced with the incorporation of grapefruit essential oil (GFO). The TS increased from 65.20 ± 4.71 to 98.21 ± 6.35 MPa in a concentration-dependent manner. This improvement is attributed to the intermolecular bonding between CA and GFO, resulting in the formation of more compact films. These enhanced mechanical properties indicate the potential of GFO-infused CA films for effective food packaging applications (Bhatia et al., 2024). Films combining κ-CA with poly(vinyl alcohol) and gallic acid showed varied TSs, with the κc3/GA10 film achieving a TS of 10.28 ± 0.65 MPa. These films also exhibited high EAB, indicating good flexibility. Lower κ-CA content promoted stiffer films with improved WVP, indicating a balance between flexibility and strength in these functional films (de Jesus et al., 2023). CA-based films exhibit favorable mechanical properties, including TS and flexibility, which are essential for food packaging applications. The films' structural integrity is influenced by their composition and preparation methods, allowing for customization to meet specific requirements. These films demonstrate good film-forming ability, contributing to their effectiveness as edible coatings that protect food while maintaining quality. Recent research highlights the potential of CA derivatives to enhance these mechanical properties further, making them suitable for smart active packaging solutions (Abdallah et al., 2024).

4.5.1.2 Barrier properties

Oxygen permeability (OP) quantifies the rate at which oxygen gas passes through a material over a given time under specific conditions (temperature, pressure, and RH). It is typically measured in $cm^3\ m^{-2}\ day^{-1}\ kPa^{-1}$. WVP measures the rate at which water vapor passes through a material under specific conditions, and it is typically measured in $g\ m^{-2}\ day^{-1}$. WVTR measures the amount of water vapor that passes through a material over a specific time, typically under standardized conditions of temperature and RH. It is a practical expression of a material's WVP and is commonly expressed as $g\ m^{-2}\ day^{-1}$ (Zena et al., 2024). CA-based edible films exhibit various barrier properties that are crucial for preserving food quality and extending shelf life. However, the inherent limitations of CA, such as its water solubility, necessitate modifications to enhance its barrier properties. CA films generally have high WVTR, which can be a limitation for certain applications. However, incorporating additives like pectin and curcumin can significantly reduce WVTR, enhancing the film's barrier properties against moisture (Sultan et al., 2023).

The study found that incorporating galactomannan (guar gum or locust bean gum) into κ-CA films significantly improved their barrier properties. The films exhibited lower OP and WVP as the galactomannan content increased, enhancing their effectiveness as edible coatings. The film with 15% galactomannan exhibited good light transmittance and enhanced thermal stability. The compatibility between κ-CA and galactomannan contributed to these improvements, making the κ-CA–galactomannan films suitable for food packaging applications due to their enhanced strength and barrier characteristics (Wang et al., 2023).

4.5.2 Functional properties

CA films, derived from marine red algae, exhibit a range of functional properties that make them valuable in various applications, particularly in wound care, drug delivery, and food packaging. These films are known for their biodegradability, biocompatibility, and film-forming abilities, which are enhanced by their unique structural properties. They exhibit significant antioxidant and antimicrobial properties, making them suitable for food packaging applications. The incorporation of essential oil improved the physical and mechanical properties of the films, such as increased film elongation and enhanced antimicrobial and antioxidant effects, while also reducing moisture content and WVP. This indicates that these biodegradable films are not only environmentally friendly but also suitable for packaging edible food items, addressing the issue of food spoilage and health risks associated with microbial contamination (Tabassum et al., 2024). Below, the antioxidant and antimicrobial properties of CA films are discussed in detail.

4.5.2.1 Antimicrobial properties

The use of antibacterial compounds in food packaging has evolved significantly, with a growing preference for natural over synthetic agents due to consumer awareness of potential health risks associated with synthetic compounds (Huat et al., 2023). CA, a natural polysaccharide derived from red algae, has been extensively studied for its potential as an antibacterial agent in food applications. Modified CA films, enhanced with essential oils such as orange and clove, have demonstrated significant antibacterial properties, particularly against Gram-positive bacteria like *Staphylococcus aureus* (Du et al., 2024). These films are developed using solvent casting methods, which allow for the incorporation of essential oils that enhance their functional properties, including increased thickness and improved barrier properties. The addition of essential oils not only improves the mechanical properties of the films but also enhances their antimicrobial efficacy, as seen in studies where clove essential oil significantly inhibited the growth of *Escherichia coli* (Tabassum et al., 2024). The development of low molecular weight κ-CA has also been explored to improve its antibacterial activity, with studies indicating that degradation of CA can enhance its efficacy against various bacterial strains by altering membrane permeability (Huang et al., 2024). Additionally, the incorporation of olive oil extracts into CA films has been shown to significantly reduce bacterial colony-forming units, further supporting the potential of CA-based films as effective antibacterial packaging solutions. These advancements highlight the potential of CA and its modified forms as sustainable and effective alternatives to synthetic antibacterial agents in the food industry, aligning with the increasing demand for natural and environmentally friendly packaging solutions (Huang et al., 2024).

A study reported that the CA film incorporated with jaboticaba peels extract (JPE) exhibits significant antioxidant and antimicrobial properties due to the high concentrations of phenolic compounds and anthocyanins, particularly cyanidin-3-glucoside (Cn-3-Glu). These bioactive compounds contribute to the film's ability to inhibit microbial growth, specifically against *E. coli*, enhancing food safety. A developed CA film with 50% JPE exhibited total phenolics of 121.16 mg g^{-1} GAE with 41.84% antioxidant activity. Similarly, a concentration of JPE (50 mg L^{-1}) shows 22.21% microbial inhibition. Furthermore, lowering the extract concentration did not significantly alter the inhibition levels. This evidence suggests that JPE may prevent the growth of the bacteria *E. coli* even at low levels. The extract's composition, which is high in phenolic compounds, is linked to its microbial suppression (Avila et al., 2022). CA films loaded with citric acid demonstrate enhanced antimicrobial effects. Citric acid, a known antimicrobial agent, increases the inhibition zones against various bacterial strains, including *S. aureus* and *E. coli*, by potentially lowering the pH and disrupting bacterial cell membranes (Krishnan et al., 2024). The incorporation of dual-ion crosslinking with Cu2+ and K+ in κ-CA/carboxymethyl chitosan membranes enhances antimicrobial properties. The ions may disrupt bacterial cell walls or interfere with essential ion channels, leading to bacterial cell death (Han et al., 2023). The combination of antimicrobial agents such as lysozyme, nisin, and grapefruit seed extract (GFSE) in CA films further enhances their antimicrobial efficacy. These agents work synergistically to inhibit bacterial growth, although the specific mechanisms may vary depending on the agent used (Manzoor et al., 2023).

By incorporating natural antimicrobial agents directly into the film matrix, these materials can actively extend the shelf life of food products while simultaneously providing health benefits (Wang et al., 2024). For instance, studies have shown that integrating essential oils such as thyme or oregano into CA films significantly enhances their antibacterial properties, effectively inhibiting microbial growth on packaged foods (Ramesh & Chellam, 2022). This dual functionality not only improves product safety but also aligns with consumer preferences for minimally processed and naturally preserved foods. As advancements in formulation techniques continue, the ability to tailor these films for specific applications could revolutionize how we approach food preservation, ultimately contributing to reduced food waste and improved sustainability in the industry.

4.5.2.2 Antioxidant properties

Modified CA films, when integrated with antioxidants, exhibit significant potential in enhancing the oxidative stability of food products by delaying the oxidation process. Antioxidants are crucial for scavenging free radicals, thereby mitigating oxidative stress caused by reactive oxygen species (ROS) in food systems (Bayram & Decker, 2023; Goyal et al., 2025). The incorporation of natural antioxidants into packaging films, such as those derived from food industry waste and natural extracts, has been shown to effectively prolong the shelf life of food by preventing lipid peroxidation and maintaining food quality (Barzan et al., 2024). This is particularly relevant in the context of CA films, which can be modified to include antioxidants that are released slowly into the food system, thereby providing a controlled and sustained protective effect against oxidation (Malekjani et al., 2024).

The use of natural antioxidants, such as tocopherols, polyphenols, and catechins, has been demonstrated to significantly improve the oxidative stability of food products, as seen in studies involving blended oils and chitosan-based films

(Wang et al., 2024, 2022). These antioxidants not only enhance the shelf life of food products by reducing rancidity and preserving nutritional quality but also offer a sustainable approach to food preservation by utilizing natural and renewable resources (Pruteanu et al., 2023). The development of such antioxidant packaging systems aligns with the principles of green chemistry and sustainability, offering a promising solution to the challenges of food spoilage and waste (Bhattacharya, 2025). Overall, the integration of antioxidants into modified CA films represents a viable strategy for extending the shelf life of food products while maintaining their safety and quality, thereby meeting the increasing industrial demand for effective and sustainable food packaging solutions.

The study demonstrated that incorporating GFO into κ-CA films significantly enhanced their antioxidant properties, as evidenced by increased radical scavenging activity measured through DPPH and ABTS assays. This study demonstrated that films loaded with GFO (0.3% v/v) exhibited the highest antioxidant activity, showing 11.36% inhibition against the DPPH radical and 30.73% inhibition against the ABTS cation radical (Bhatia et al., 2024). Another study concluded that the incorporation of *A. vera* extract into CA-based edible films significantly enhances the antioxidant capacity and antimicrobial properties of the films, particularly against *E. coli*, thereby improving the lipid oxidative and microbial stability of frozen dairy products like kulfi. The optimal formulation for the edible film was determined to be 1.5% CA, 14% glycerol, and 15% *A. Vera* extract, which resulted in the highest levels of antioxidant and antimicrobial potential, indicating its effectiveness for packaging frozen dairy products (Mahajan et al., 2021). Furthermore, it has been found that the eugenol-loaded CA-cellulose nanofiber film, particularly with 0.4% eugenol, exhibited significant antioxidant properties. This film demonstrated the highest release rate of polyphenols at 614.92 mg gallic acid L^{-1} sample over 28 days of storage and achieved a radical scavenging activity efficiency of 58%. These findings indicate that the 0.4% eugenol film is rich in antioxidants, making it a promising option for sustainable packaging applications that can enhance food shelf life (Wan Yahaya et al., 2024). The antioxidant-rich CA films help in extending the shelf life of food products by preventing oxidative degradation and microbial growth. These films are biodegradable, offering an eco-friendly alternative to conventional plastic packaging. Overall, the properties of CA-based films are explained in Table 4.1.

4.6 Nanotechnology in modification of carrageenan-based coatings and film properties

In food packaging, "nanotechnology" encompasses a diverse range of nanoparticles (NPs), and these materials are incorporated into CA solution due to their distinct physical, chemical, and biological characteristics. These nanoparticles are primarily integrated to enhance biopolymers' durability, abrasion resistance, and resistivity, including CA against moisture, water, light, and environmental gases (Gupta et al., 2022). The main drawbacks of CA are poor mechanical, water vapor barrier, and thermal properties, resulting in limited application in the food industry (Roy & Rhim, 2019a). To address these problems, different nanosized fillers have been reinforced with CA as explained in Table 4.2, resulting in improved resistance to gases, temperature, and humidity, as well as enhanced mechanical strength and flexibility. Both organic (lipid, polysaccharide, and protein-based NPs) and inorganic NPs (clays, salts, layered silicates, metal oxides, layered double hydroxides (LDH), metallic nanoparticles, gold nanorods, and cylindrical metal oxide nanotubes) and functional materials have been used to improve the properties of CA-based films. In general, organic nanofillers have significant advantages over inorganic nanofillers due to biodegradability, biocompatibility, nondosage limiting, and low cytotoxicity.

An example, the organic melanin nanoparticles (40–160 nm) are combined with CA to improve the film properties of CA film. The addition of melanin NPs up to 1.0% showed improvement in UV shielding, mechanical strength, heat resistance, hydrophobicity, antioxidant activity, and antimicrobial activity against Listeria *monocytogenes* and *E. coli* (Roy & Rhim, 2019a). Likewise, the blending of silver nanoparticle (AgNP) synthesized by melanin (reducing and capping agent) and montmorillonite along with silver and cerium oxide nanoparticles into CA film, provided strong UV-shielding properties and antibacterial activity against *E. coli* and *Listeria monocytogenes*, increased the thermal stability, mechanical strength, and resilience; however, it reduced the water contact angle and water vapor barrier properties (Genecya, Adhika, & Wungu, 2023; Genecya, Adhika, Sutrisno et al., 2023; Roy et al., 2019). Furthermore, the incorporation of CNC is reinforced to enhance the mechanical capability, reduce moisture content and water uptake, and increase the contact angle and water barrier of κ-CA film (Kassab et al., 2019; Popescu et al., 2019).

Another example is the inorganic nanoclay blended with κ-CA with the incorporation of *Zataria multiflora* plant extract (ZME) and glycerol (plasticizer). The study found that the addition of plasticizer with 3% ZME before the nanoclay addition showed higher TS (56%) and lower EAB (61%) compared to after nanoclay addition. Furthermore, the films exhibited excellent inhibition against *E. coli* and *P. aeruginosa* and improved UV barrier, microstructural, and thermal properties (Nouri et al., 2020). Zinc oxide (ZnO)-doped carbon dot combined with CA, along with the addition of

TABLE 4.2 Application of nanomaterials in the improvement of carrageenan-based films and coatings properties.

Types of nanomaterials	Carrageenan and other materials used	Key film properties	Functional properties	References
Zinc oxide nanoparticles (ZnONPs) synthesized by reacting zinc acetate with potassium hydroxide using melanin (capping agent)	κ-CA	Higher UV-blocking, thermal stability, mechanical, and water vapor barrier properties than the Carr/znonp *Monocytogenes*	Strong inhibition against *E. coli* and less inhibition against *L. monocytogenes*	Roy and Rhim (2019b)
ZnONPs	κ-CA, sodium alginate (SA), gelatin/oregano essential oil (OEO)	Improved the mechanical, thermal stability, UV resistance, and retention of OEO	Stronger antioxidant and antibacterial activity	Zhang, Zhang, et al. (2024)
Bacterial cellulose nanofibrils (BCN)	κ-CA and soy protein isolate	Significant decrease in water solubility (22.98% ± 0.78%), WVP (1.22 ± 0.14 g mm^{-1} S^{-1} Pa^{-1} 10^{-10}), and elongation at break (EAB) (57.77% ± 5.25%) properties	Showed antioxidant and antibacterial activity	Jancikova et al. (2020)
Zein nanoparticles (ZNP)	κ-CA/ gelatin/turmeric essential oil (TEO)	Improved physicochemical and mechanical properties	Showed excellent inhibition against *Salmonella enterica*	Ali et al. (2023)
Cellulose nanocrystals (CNCs) derived from Indian gooseberry pomace	κ-CA	5% CNC loading decreased WVP from 3.21 to 2.36 g mm (m^2 day $kPa)^{-1}$ and increased tensile strength (TS) from 23.28 to 39.75 Mpa	-	Gupta et al. (2021)
Silver nanoparticles (AgNPs) derived from pine needle extract	κ-CA	Significantly affected the surface color, water vapor barrier, surface wettability properties, and mechanical strength, and enhanced the UV-blocking properties	Showed strong antioxidant and potent antibacterial activity against *S. aureus* and *E. coli O157: H7*	Kim et al. (2022)
AgNPs	κ-CA and soluble soybean polysaccharide	Enhanced mechanical strength, thermal stability, and water resistance	Antibacterial activity against *Escherichia coli* and *Staphylococcus aureus*	Liu et al. (2022)
SiO_2–ZnO NPs	κ-semirefined iota CA	The addition of SiO_2–ZnO NPs increased TS, water vapor barrier performance, and UV-screening properties	Enhanced antimicrobial activity	Praseptiangga et al. (2021)
Cationic nanocellulose	κ-CA/OEO	The addition of OEO slightly decreases the TS and increases the EAB	Provides inhibitory effects against *S. aureus and E. coli*	Amanda et al. (2024)

Copper sulfide nanoparticles (CuS NPs)	κ-CA	The film showed high transparency with enhanced mechanical properties and thermal stability	Antibacterial activity against *Staphylococcus aureus* (99.9%) and *Escherichia coli* (99.2%)	Li et al. (2020)
Mica nanosheets	κ-CA/konjac glucomannan/ *Broussonetia papyrifera* fruit anthocyanin/gellan gum	The double-layer film WVP is reduced to 4.7 times, and TS is increased to 4.1 times with UV-blocking and higher water contact angle (102.4 degrees ± 1.18 degrees)	Film showed color changes (red to blue-gray) at different pH buffer solutions and volatile ammonia	Bian et al. (2024)
Cellulose nanofibers (CNFs)	κ-CA	Superior mechanical property (TS −99 ± 7 MPa)	Provides 85% DPPH inhibition and 0.06 mg GAE mg^{-1} total phenolic content	Lakshmi Balasubramaniam et al. (2021)
Electrospun nanofibers	κ-CA/polyvinyl alcohol/ *Centaurea arvensis*anthocyanin	Had adverse effect on mechanical properties	Displayed color changes from heather violet to green (2–12 pH)	Forghani et al. (2022)
AgNP adsorbed to halloysite nanotube (HNT) and treated with sodium dodecyl sulfate (SDS) or HNT-AgNP nanoparticles	κ-CA	Increased the mechanical strength and UV-barrier properties	Showed potent antibacterial activity against *L. monocytogenes*, and *E. coli*	Saedi, Shokri, and Rhim (2020)
ZNPs loaded with anthocyanin and cinnamaldehyde	κ-CA	Increasing the nanoparticle concentration resulted changes in the opacity (17.42 A mm^{-1}), microstructure, and thermal stability (65.3%)	Holds DPPH (48.39%) and ABTS (92.77%) radical scavenging activity and showed color changes (S_{Lab}: 16.98%–85.98%) as pH indicator	Zhou et al. (2024)
Comparison of three fillers: nanoclay, tetraethyl orthosilicate (TEOS), and 3-aminopropyltriethoxysilane (APTES)	κ-CA	Unfilled carrageenan film, nanoclay, TEOS, and APTES filler film properties exhibited: UV light transmission value: 30.9%, 23%, 18%, and 1%, respectively TS: 38.4, 38.6, 57, and 60 MPa TEOS films displayed higher thermal stability and nanoclay exhibited the lowest stability	-	Asadzadeh et al. (2023)
Carbon dots (CDs) derived from garlic (*Allium sativum*) cloves	κ-CA/SA	CDs-loaded film exhibited improved mechanical stability, good compatibility, excellent UV-blocking properties (~85.1% in the UV-A, and 99.0% in the UV-B), and surface hydrophobicity	Significant bacteriostatic effect against *E. coli* and *L. monocytogenes* Showed high antioxidant activity: ABTS (98.6% ± 0.3%) and DPPH (34.2% ± 0.3%)	Khan et al. (2023a)

(Continued)

TABLE 4.2 (Continued)

Types of nanomaterials	Carrageenan and other materials used	Key film properties	Functional properties	References
Hybrid nanoparticles (HNPs) prepared using modified Fe_3O_4nanoparticles by polyamidoamine (PAMAM) dendrimer and the decorated magnetic core-shell nanomaterials withAgNPs (Fe_3O_4@SiO_2@PAMAM@AgNP)	κ-CA	Exhibit high UV-barrier property and did not influence the mechanical and water vapor barrier properties	Clear antibacterial activity against *L. monocytogenes*, and *E. coli*	Saedi and Rhim (2020)
Titanium dioxide nanotube (TNT) and CuO-doped TNT (TNT–CuO)	κ-CA	Addition of NPs imparted UV-blocking properties and increased the mechanical strength, water vapor barrier, and surface hydrophobicity properties	The modified TiO_2 (TNT and TNT–CuO) films showed significantly higher antibacterial activity	Ezati et al. (2021)
Zinc sulfide nanoparticles (ZnSNP)/ nanocellulose fibers	κ-CA/agar/tea tree oil	The addition of ZnSNPs improved the mechanical strength and slightly improved the water vapor barrier, water resistance, and thermal stability of the film	Showed antioxidant and antibacterial activity	Roy and Rhim (2021a)
AgNPs	κ-CA	-	Higher bactericidal activity against *Staphylococcus aureus*	Dumas et al. (2024)
Bentonitenanoclay (BT)	κ-CA	With the increase the BT content, decreased the water uptake, water sorption ability, and water contact angle	-	Dogaru et al. (2020)
Cu-metal organic framework (Cu-MOF)	Gelatin/CA/anthocyanins from red cabbage	3 wt.% Cu-MOF showed excellent UV-barrier ability (98.5% in the UV-A and 99.9% in the UV-B),	Strong antioxidant efficacy against ABTS (100%) and DPPH (100%) and exhibited strong antibacterial activity after 3 h of incubation. Film exhibited pH-dependent color change under pH 2–12	Khan et al. (2023b)

antioxidants (derived from Kohlrabi peel), displayed excellent UV-blocking ability (85.2% of UV-A and 99.4% of UV-B), high antioxidant effect (~99% for ABTS; ~58.6% for DPPH radical scavenging activity), strong antibacterial activity (100%) against *L. monocytogenes,* and an 8.1 log CFU mL^{-1} reduction in the growth of *E. coli* after 12 hours of incubation (Khan et al., 2024). This film can be used as a smart/intelligent film to identify the freshness of perishable foods.

Roy and Rhim (2020) developed the bionanocomposite film from pullulan/CA/copper sulfide nanoparticles (CuSNP)/D-limonene (DL). They noted that the addition of DL and CuSNP significantly increased the UV-blocking properties without reducing transparency and thermal stability and improved the film's TS by 15% with a slight improvement in water barrier and antimicrobial activity. In addition, the use of copper nanoparticles (derived from *Argemone maxicana*) can improve the κ-CA films' thermal stability, elastic properties, water vapor resistance, UV resistance qualities, and antibacterial activity (against *S. aureus* and *E. coli*) (Kumari et al., 2024). Though nanotechnology has advanced significantly in food packaging at the research level, its commercial application remains limited due to health and environmental concerns, particularly during storage. Moreover, identifying and monitoring nanomaterials in food packaging film and the packed products, whether through release or interaction with the food constituents, remains a complex challenge. Researchers continue to explore various strategies to integrate nanoparticles with biopolymers like CA to improve film properties more effectively (Gupta et al., 2022).

4.7 Applications of carrageenan-based films for food preservation

4.7.1 Applications in fruit and vegetable processing

Fruits and vegetables form an important part of human diet as they serve as important sources of vitamins, minerals, and dietary fibers. Additionally, the presence of various bioactive compounds in them provides additional functional benefits. However, they have a short shelf life due to their perishable nature. Fruits and vegetables are living commodities and tend to respire even after harvest. Many fruits are climacteric in nature, which release ethylene and undergo ripening process after harvesting. Various external factors, such as temperature and humidity, govern the rate of respiration and ethylene release, affecting the shelf life of fruits and vegetables (Nair et al., 2020). Moreover, recently, the demand for high quality and minimally processed fruits and vegetables is also increasing. In case of minimally processed fruits and vegetables, as the natural protective barrier is breached, rapid deterioration is observed. Under these circumstances, there is a need for technological intervention to improve their shelf life with minimal loss of quality and nutritional properties. CA-based biodegradable films and coatings offer barrier for moisture and gaseous exchange, enhancing the shelf life of fruits and vegetables.

4.7.1.1 Carrageenan-based coating for active/intelligent packaging in fruit and vegetable processing

Coatings form a part of fruit or vegetable, forming an immediate barrier on the surface that can be consumed or removed prior to consumption. In a study on longan fruit, Lin et al. (2018) reported application of CA-based edible coating to control quality losses. The concentration of both polymers and glycerol was optimized using response surface methodology. The increasing concentration of polymer and plasticizer was found to significantly reduce water loss, weight loss, and respiratory rate. Based on the response optimization, 1.49% CA and 0.03% glycerol showed minimal quality loss. Combining CA with other biopolymers can offer beneficial effects in terms of permeability, mechanical properties, and overall performance of coatings. One such application of combining CA with chitosan and subsequent coating of dragon fruit is reported by Nguyen et al. (2020). Dragon fruit is popular due to its nutritional composition and sensory qualities; however, it is prone to several postharvest diseases, resulting in poor shelf life. Hot water treatment followed by composite coating of 1% chitosan and 0.2% CA emerged as an effective approach for maintaining the postquality of dragon fruit for 30 days at 10°C. Hot water prevented the accumulation of hydrogen peroxide, whereas the composite edible coating helped in maintaining the green color of bracts. Another work by the same group suggested that pretreatment with gibberellic acid or methyl jasmonate with same chitosan-CA composite film resulted in retained chlorophyll content and bract color of dragon fruit (Nguyen et al., 2021).

It has been observed that the incorporation of various active agents in CA-based coatings can improve functionality in terms of barrier properties, antimicrobial properties, or intelligent packaging applications. For instance, Wani et al. (2021) incorporated lemongrass oil as an antimicrobial agent in edible coatings of CA, xanthan gum, and gum Arabic individually and evaluated their performance in preserving strawberries. The edible coatings with lemongrass oil significantly restricted the growth of psychrophilic bacteria, yeast, and mold count of strawberries. Comparative study revealed that CA-based coating showed delayed weight loss and decay percentage, firmness, and higher retention of ascorbic acid, anthocyanins, and phenolic compounds during storage for up to 12 days under refrigeration. Another interesting study demonstrated ultraviolet blocking properties of CA-cassava starch-based film coatings incorporated with copper oxide particles. The films comprising 75% CA and 25% cassava starch were incorporated with varying levels of copper oxide particles (0%, 1%, 1.5%, and 2%) and were applied to pears. The coating with 2% copper oxide showed best results by maintaining the quality of pears for 30 days, showing better performance than conventional, nonbiodegradable polyvinyl chloride (PVC) film (da Costa et al., 2024). Several such applications of CA-based active coatings are reported in Table 4.2.

4.7.1.2 Carrageenan-based films for active/intelligent packaging

Films are typically molded into a form of a solid sheet that can be used to cover or wrap fruits and vegetables. Biodegradable edible films based on CA and other biopolymers have been developed recently. A study on development of edible films of iota-CA with arrowroot starch in different proportions was conducted by Abdillah and Charles (2021). Films prepared by combining CA with arrowroot starch showed relatively higher TS, barrier properties, water solubility, and swelling properties compared to the arrowroot starch films. When applied to cherry tomatoes, these composite films successfully delayed the weight loss, extending the shelf life to 10 days. The mechanical and barrier properties of developed biodegradable films play a strong role in enhancing the shelf life of the packed fruits and vegetables. Therefore, researchers are attempting to incorporate various active constituents into such films to improve these properties. One such study reported incorporation of ferric ion and polyphenols from the leaves of plant *Capsicum annum* into polyvinyl alcohol-CA films. This approach improved the mechanical and barrier properties, as well as imparted fire retardant, UV resistance, and antioxidant properties. The improved barrier properties prevented browning, overripening, and tissue softening ultimately delaying the spoilage process (Li et al., 2024). Using similar approach of metal-polyphenol complex, (Shi et al., 2024) incorporated CuFA NWs synthesized by solvothermal method into a CA matrix to improve the mechanical properties, hydrophobicity, and barrier properties. Based on the antimicrobial and antioxidant properties, films with 3% CuFA NWs were selected for application to blueberries and cherries. Fig. 4.3A indicates the change in appearance over 16-day storage. The unpacked blueberries showed shrinkage on eighth day, followed by complete

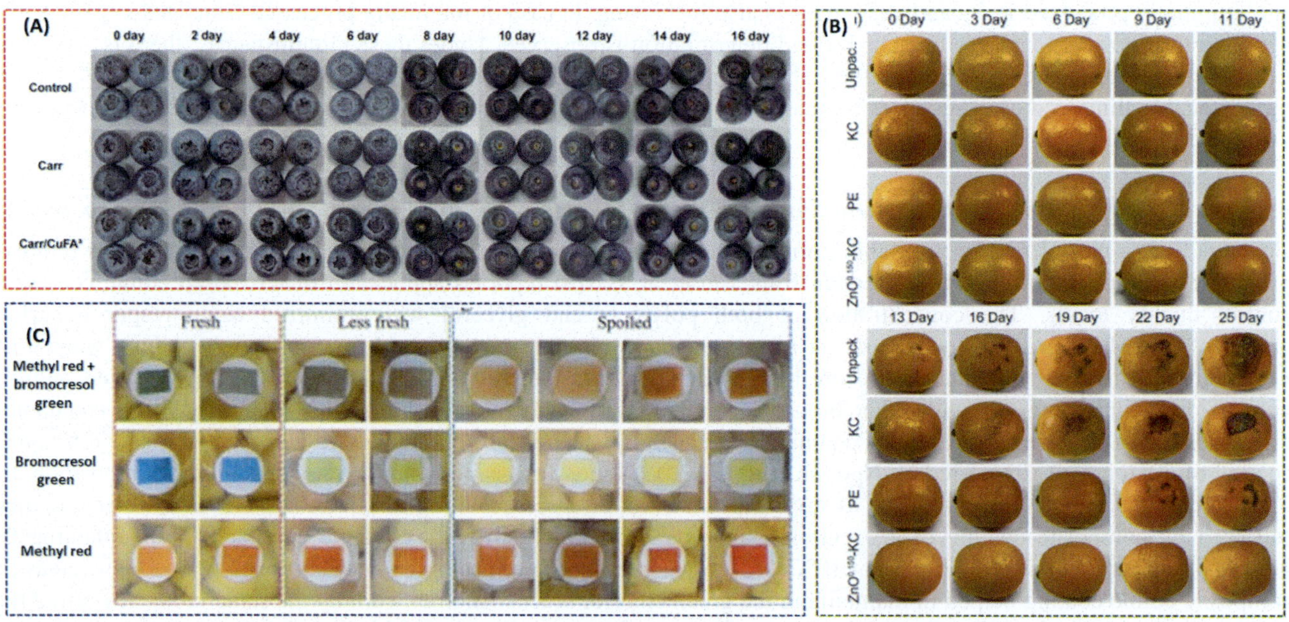

FIGURE 4.3 CA-based active/intelligent packaging of fruits: (A) effect of CA incorporated with copper-ferulic acid networks (CuFA NWs) on blueberries, (B) pseudo-pasteurization films comprising zinc oxide-doped carbon nanoparticles for preservation of kumquats, (C) mixed indicator-based intelligent films for monitoring freshness of mango.
From Shi, Shuo, Lan, Xi, Ding, Xiyan, Han, Ximei, Sun, Jing, Wang, Jianlong & Duan, Jinyou. (2024). Metal-phenolic networks spontaneously reinforced carrageenan-based packaging films with antibacterial and antioxidant properties. International Journal of Biological Macromolecules, 279(Pt 2). https://doi.org/10.1016/j.ijbiomac.2024.135143.

collapse on 14th day. Contrarily, the CA film with 3% copper-ferulic acid network was able to maintain freshness for up to 16 days without shrinkage. This is also confirmed by minimal changes in weight loss, total soluble solids, and firmness. Similar results were obtained in case of cherries, where the film was able to maintain good appearance and bright color (Shi et al., 2024). An interesting study on preservation of kumquat fruits reported pseudo-pasteurization films based on CA incorporated with ZnO-doped carbon nanoparticles as a filler material. Unique photothermal effect and antibacterial properties of the filler material enabled almost complete inactivation of *E. coli* and *L. monocytogenes* under near-infrared radiation (NIR) of 2 W cm^{-2} for 5 minutes at a low photothermal temperature (60°C–62°C). The effect of the pseudo-pasteurization film on storage of kumquat fruits can be observed in Fig. 4.3B. The fruits packed in CA films with filler material with intermittent NIR treatment (13th, 18th, and 23rd day) did not show spoilage for up to 25 days, suggesting an emerging method for extension of shelf life of kumquat fruits (Zhang et al., 2022).

As described before, ethylene is a plant hormone responsible for the ripening of fruits, which ultimately leads to aging and spoilage (Alonso-Salinas et al., 2024). Therefore, efforts are underway to reduce or absorb ethylene through the use of ethylene scavengers to delay senescence and extend shelf life. In their work with bananas, Ezati et al. (2021) integrated a well-known ethylene scavenger, TiO_2, in the form of nanotubes with CA-based films. Moreover, the TiO_2 nanotubes were modified with copper oxide to obtain antimicrobial activity through photocatalysis under visible light. The CA films with modified TiO_2 nanotubes maintained the freshness of bananas for up to 12 days compared to CA alone and CA with TiO_2, which showed complete spoilage.

As fruits and vegetables tend to respire during storage, they evolve carbon dioxide. Ethylene is also released during ripening process. CA-based biofilms can offer an intelligent feature that can help in monitoring the quality of packed fruits and vegetables by tracking these evolved gases. An intelligent label comprising CA, sodium carboxymethyl cellulose, and bromothymol blue as a pH-sensitive reagent was developed to monitor the freshness of freshly cut papaya. The presence of CA improved the water resistance and mechanical properties of the label. The label, being sensitive to pH, changed its color from blue to yellow-brown with increasing concentrations of carbon dioxide inside the package as a result of respiration effectively reflecting the quality of papaya (Rong et al., 2023). Of late, mixed indicators are being used to extend the color range and obtain response to wider pH. Considering the tedious experiments involved in optimization of concentrations of different indicators, a computational simulation algorithm was developed to arrive at the optimized composition of a mixed indicator. A pH-indicating intelligent label was developed by combining methylcellulose with kappa-CA and sodium tripolyphosphate for enhanced mechanical properties. Fig. 4.3C indicates the color response of labels comprising methyl red-bromocresol green, bromocresol green alone, and methyl red alone during storage of mango. From this, it can be concluded that the mixed indicator provided consistent color change compared to the individual counterparts. Furthermore, for accurate determination of freshness and classification, computer vision and convoluted neural networks were used. Experiments with mango, kiwi, and grape showed maximum accuracies of 96.4%, 96.3%, and 96.3%, respectively, for freshness determination (Jia et al., 2025).

4.7.2 Applications in fish processing

Fish are considered one of the best sources of protein and essential fatty acids. Being highly nutritious, they are also highly susceptible to autolytic and microbial spoilage, which significantly hinders the quality and safety (Nimbkar et al., 2023). Some of the spoilage products and pathogens can pose serious risk to human health (Sheng & Wang, 2021). This substantiates the importance of maintaining quality and safety of fish and fish products throughout the supply chain. Preservation of fish is achieved through conventional approaches such as drying, salting, fermentation, and novel techniques such as freezing, irradiation, and ozonization. Recently, CA-based films and coatings are emerging as novel, environmentally friendly approaches for packaging and preservation of fish and fish products. They act as barriers preventing moisture loss, lipid oxidation, and microbial contamination, further assisting in preservation.

4.7.2.1 Carrageenan-based coating for active/intelligent packaging in fish processing

A functional edible coating comprising fish gelatin and CA, loaded with pomegranate peel extract, was developed for preservation of Nile tilapia fish. The coating showed antimicrobial properties due to presence of phenolic compounds, alkaloids, and tannins in pomegranate peel extract, which inhibited the growth of spoilage microorganisms for 30 days under refrigerated conditions. The antimicrobial activity showed an increasing trend with increasing concentrations of the extract. Additionally, the coating also delayed the surge in chemical parameters such as total volatile basic nitrogen (TVB-N), thiobarbituric acid, reactive substances (TBARS), and peroxide value, which underline the antimicrobial and antioxidant properties (Khojah, 2020). Another work used dill essential oil as an antimicrobial agent incorporated in κ-CA coatings for maintaining the quality of rainbow trout filets. Additionally, the effect of a separate oxygen scavenging packaging system is also evaluated. Results revealed that the κ-CA-based edible coating with incorporation of 2% dill essential oil inhibited the growth of spoilage microbes and also delayed undesirable chemical changes. The coating with oxygen-absorbing packaging system showed best performance in maintaining the chemical composition and sensory quality of filets (Karimzadeh et al., 2024). It is noteworthy that the applications of CA-based coatings are not just restricted to preservation of fish freshness and extending shelf life. An interesting application of such coatings in reduction of oil uptake and protein oxidation in fried fish strips is reported by Gaurav et al. (2023). They studied the comparative effectiveness of different concentrations of biopolymers, carboxymethyl cellulose, CA, and xanthan gum on fat uptake, moisture, oiliness, texture, and protein oxidation of fish strips using multiobjective optimization. Though 1% carboxymethyl cellulose was found to be the most effective coating formulation, 0.75% CA was still an optimum formulation for reduced fat uptake, protein oxidation, and improved functionality of deep-fried fish strips.

4.7.2.2 Carrageenan-based films for active/intelligent packaging in fish processing

Recently, CA-based biodegradable films are emerging as a popular alternative for packaging of fish in order to extend shelf life. Additionally, considering the health concerns associated with improper packaging and storage of fish, monitoring quality throughout the supply chain has become indispensable. Consequently, there is an increasing interest toward development of multifunctional films that can perform the functions of active and intelligent packaging. Integration of various active and smart materials along with CA-based films achieves preservation as well as quality monitoring simultaneously. These smart materials, being sensitive to the metabolites produced during spoilage of fish, can effectively indicate the spoilage via a visual response. One such multifunctional film was developed by integrating TiO_2 nanoparticles and anthocyanins from saffron and red barberry into a crosslinked gelatin and CA network (Alizadeh Sani et al., 2022). The presence of TiO_2 nanoparticles imparted bacteriostatic properties against *S. aureus* and *E. coli*. The film with saffron-anthocyanins showed color change from violet-blue to green, whereas the one with barberry-anthocyanins changed from peach to yellow during 72-hour room temperature storage of trout fish. This is due to the change in pH as a result of ammonia release during spoilage of fish. In another similar approach, a bilayer CA-based film comprising an outer antimicrobial layer and an inner indicator layer was developed. Magnonol and purple sweet potato anthocyanin were used as natural antimicrobial active ingredient and freshness indicator, respectively. A bilayer film with 2% magnonol in outer layer and 0.2% purple sweet potato anthocyanin in inner layer was found to be the most optimum in terms of TS and water vapor transmission and showed distinguishable color change in 3–10 pH range. This film effectively reflected the quality and freshness of salmon and also prolonged its shelf life by two days at 4°C (Zhang et al., 2024).

Several studies have reported intelligent films by incorporating natural pigments; nevertheless, poor stability and dispersibility are concerns in their application. By using an encapsulation approach, the natural pigments can be effectively entrapped in suitable nanoparticles, followed by incorporation into a film. Han et al. (2025) developed ZNPs loaded with curcumin and anthocyanins using antisolvent precipitation, which were subsequently coated with chondroitin sulfate through electrostatic interactions and further incorporated into a CA-based film. The composite film with 40% nanoparticles was able to effectively monitor the quality of mandarin fish by changing its color from reddish-brown to dark brown over 24-hour storage. Additionally, this film demonstrated antimicrobial properties by controlling the TVB-N content of fish, extending its shelf life. The study clearly indicates the trends in TVB-N levels and visual differences in fish samples over storage time. In another work by the same research group, curcumin and epigallocatechin gallate were encapsulated in ZNPs and further incorporated into a CA-based film. Incorporation of epigallocatechin gallate along with curcumin synergistically improved the antioxidant activity and antimicrobial activity (Han et al., 2023). A multifunctional active/intelligent packaging film was developed by incorporating anthocyanins and carbon dots (CDs) from purple Kohlrabi doped with ZnO into a CA matrix. Apart from excellent antioxidant and antimicrobial properties, the films demonstrated visible color response to pH and ammonia concentration. During freshness monitoring of shrimp, the films with 3% of ZnO-doped CDs showed maximum change in color from red to green during storage at room temperature for 24 hours as observed in Fig. 4.4A. Moreover, the film was able to preserve the shrimp for up to 10 days at 4°C. After 10 days, the quality started to decline, which can be observed as the color of the film slightly changed (Fig. 4.4B) (Khan et al., 2024). The accuracy of quality determination strongly depends on the sensitivity of the intelligent packaging systems. However, difficulties in perceiving lower concentrations of ammonia create a hindrance in accurate determination of the quality of fish products. In order to address this, Bian et al. (2023) reported the use of diatomite, a highly porous material with strong adsorption properties. The film was developed by incorporating diatomite into a konjac glucomannan-CA matrix loaded with blueberry anthocyanin as an indicator. Upon application to shrimp quality monitoring, the film with 6% diatomite showed more pronounced color variation from bright pink to dark red to bluish violet over five days,

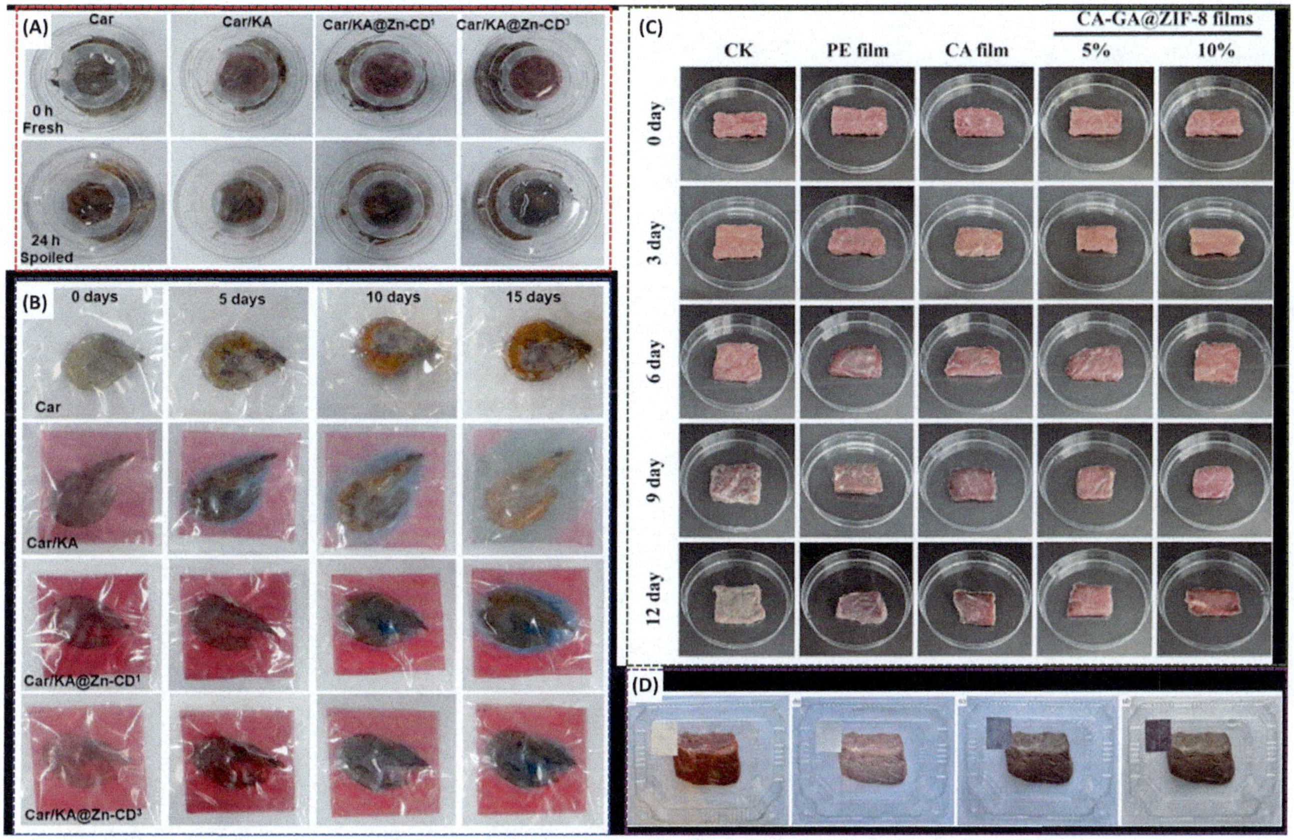

FIGURE 4.4 Carrageenan-based active/intelligent packaging of fish and meat (A) and (B) multifunctional active and intelligent CA-based films incorporated with zinc and carbon dots for shrimp preservation and freshness monitoring, (C) effect of gallic acid-metal organic framework incorporated CA-based film on beef shelf life, (D) nanofibrous intelligent mats for monitoring the freshness of beef.
From Khan, A., Riahi, Z., Tae Kim, J. & Rhim, J.W. (2024). Carrageenan-based multifunctional packaging films containing Zn-carbon dots/anthocyanin derived from Kohlrabi peel for monitoring quality and extending the shelf life of shrimps. Food Chemistry, 432. https://doi.org/10.1016/j.foodchem.2023.137215.

corresponding to increasing TVB-N concentration. This enabled relatively easier visible detection of spoilage of shrimp (Bian et al., 2023).

Table 4.3 lists diverse applications of CA-based films for active and intelligent packaging of fish.

4.7.3 Application in meat processing

Meat and meat products are largely consumed for their nutritional composition and are highly popular due to desirable sensory characteristics. However, being rich in moisture and essential nutrients, they are highly susceptible to spoilage and exhibit poor shelf life. Without appropriate measures of preservation, deterioration occurs due to autolysis of proteins, oxidation of lipids leading to rancidity, and the growth of various spoilage and pathogenic microbes (Ren et al., 2021). Unhygienically processed and improperly stored meat not only undergoes rapid spoilage but also may harbor various foodborne pathogens and lead to various health issues. With higher demand and processing of meat and meat products, the need for sustainable and green packaging material is highly relevant (Song et al., 2021). Biodegradable edible films and coatings based on CA can provide effective barrier and extend preservative effect for meat and meat products.

4.7.3.1 Carrageenan-based coatings for active/intelligent packaging in meat processing

Edible coatings are formed in direct contact with meat and can create barrier for oxygen and prevent moisture loss. In order to increase the hydrophobicity novel emulsion-based coatings for meat are becoming popular. In one such attempt, Zhou et al. (2021) developed an emulsion-based coating by incorporating camellia oil into a crosslinked CA and konjac

TABLE 4.3 Applications of carrageenan-based edible films and coatings.

Type of food examined	Film/coating	Active compounds used	Findings on food quality	Shelf life	References
Blueberry, raspberry	Coating	Green tea extract	Incorporation of green tea extract enhanced the antiviral activity of CA. Lota- and Lambda- CA showed higher antiviral activity due to lower viscosity compared to κ-CA		Falcó et al. (2019)
Strawberry	Coating	Carboxylated cellulose nanocrystals (C-CNC)	Coating of strawberries without formation of cracks. Coating with C-CNC showed antishrinkage properties	Five days	Zhang et al. (2021a)
Cherry tomato	Film	Kaolin clay, arbutin	Rapid in situ formation of film conforming the shape of cherry tomato within 30 s. Preservation of cherry tomato quality up to nine days at 28°C	Nine days	Yin et al. (2024)
Strawberry	Film	Melanin extract from Niger seed hulls	The film with 30 mg melanin showed better antibacterial activity than chloramphenicol against common foodborne pathogens The film extended the shelf life of strawberries up to five days at room temperature with effective control over heterotrophic bacteria	Five days at 25°C	Sunil Kumar et al. (2025)
Strawberry	Film	Aqueous Chinese hawthorn extract	The film with 15% hawthorn extract found to be most efficient in preservation of strawberries. The film demonstrated least increase in pH and acidity with least degree of deterioration and maintained shelf life of 10 days	10 days at 4°C	Yin et al. (2024); Cvetković et al. (2025)
Banana	Film	Quercetin	Integration of hydrophobic polyphenols into a film using borate ester bond and Schiff base bond. The hydrogel film effectively delayed spoilage of bananas for up to 10 days, demonstrating least weight loss (8.92%) and highest firmness	10 days at 20°C	Yin et al. (2024)

Cherry tomato and chicken	Film	Nanocellulose from *Enteromorpha prolifera*	A sustainable solution to "green tide" by value addition of *Enteromorpha prolifera* by extraction of nanocellulose. The composite film showed better performance in maintaining chicken freshness compared to polyethylene. In case of cherry tomatoes, the composite film showed highest retention of vitamin C, color, odor, and pH		Zhang et al. (2022)
Trout (*Oncorhynchus mykiss*) filet	Coating	Lemon essential oil	The filet coated with lemon essential oil incorporated CA coating showed better olfactory characteristics compared to uncoated one as confirmed by e-nose. The CA coating with lemon essential oil better preserved the muscle structure with compact muscle fibers	12 days at 4°C	Volpe et al. (2019)
Cyprinus carpio var. specularis	Film	Mulberry anthocyanin extract	Incorporation of mulberry anthocyanin extract had effect on moisture migration but limited impact on heat stability of films. Red color of films changed to dark blue in accordance with rising TVB-N levels		Zhang et al. (2020)
Shrimp	Film	Purple sweet potato and purple cabbage	Suitability of different polysaccharide matrices and anthocyanin combinations for determination of shrimp freshness is studied. Purple cabbage anthocyanin films showed better potential for monitoring shrimp freshness		Yong et al. (2022)
Chinese white shrimp	Film	Butterfly pea flower anthocyanin and TiO_2	A bilayer approach of film forming increased the improved the color stability and photostability without affecting the color response. The film showed visible color change sky blue to light yellowish green when the TVB-N concentration crossed regulatory limit of 30 mg kg^{-1}in 48–60 h		Zhang et al. (2023)
Silver carp filets	Film	*Malva sylvestris* anthocyanins	Successful encapsulation of *Malva sylvestris* anthocyanins in chitosan-CA electrospun nanofibers The intelligent label showed excellent color reversibility for up to three uses and stability for over a month.		Shavisi 2024)

(*Continued*)

TABLE 4.3 (Continued)

Type of food examined	Film/ coating	Active compounds used	Findings on food quality	Shelf life	References
			Nanofibrous mats with 4% extract showed remarkably visible change from white to green according to the chemical and microbial changes		
Shrimp	Hydrogel	Blueberry anthocyanins	Incorporation of polysaccharides resulted in increased hydrogen bonding, which showed enhanced gelling properties and color stability. Distinctive change in color of hydrogels from orange to dark brown corresponding to increasing TVB-N and total viable count		Yin et al. (2025)
Mandarin fish	Film	Anthocyanin and cinnamaldehyde	Optimized film could extend the shelf life of mandarin fish by reducing TVB-N levels by 13.3%. The nanocomposite films showed change in color from dark purple to black purple, amounting 14.8%	Seven days at 4°C	Zhou et al. (2024)
Shrimp	Film	Natural anthocyanin and ZnO	Films combining ZnO nanoparticles and anthocyanins showed antimicrobial activity against *S. aureus* and *E. coli.* The films containing black wolfberry anthocyanins and ZnO nanoparticles were found most optimum for monitoring shrimp freshness		Jiang, Yang, et al. (2024)
Hake (*Merluccius merluccius*)	Film	Blueberry and beetroot extract	Films with blueberry extract were found to have better physicochemical properties compared to the ones with beetroot extract. The film with blueberry extract showed visible color change from pink to light purple within four days of storage, which is verified by rising TVB-N and pH levels		Faria et al. (2024)
Fish (*Cyprinus carpio* var. specularis) and pork	Film	Bromothymol blue and rice straw fibers	Attachment of bromothymol blue to rice straw fiber particles to prevent the release of the former in food system. The film demonstrated evident color change from yellow to dark blue with reduction in meat freshness		Abdallah et al. (2024); Liu et al. (2022)

Sea bass	Film	Red grape skin anthocyanins and silver nanoparticles	During storage of sea bass, the composite film showed no color change in first three days and turned blackish purple and dark green on fourth and fifth day, respectively, corresponding to increasing TVB-N concentration. The film showed excellent resistance to UV exposure and stability up to 250 days at 25°C		Abdillah and Charles (2021); Chi et al. (2020)
Pork	Film	Grape skin powder	An environment friendly extraction-free process of direct incorporation of grape skin powder in the film. The color of the film changed from purple to green when the TVB-N content approached the regulatory limit		Chi et al. (2020)
Minced chicken	Film	SiO_2-ZnO nanoparticles	Incorporation of nanocomposites improved barrier properties, UV screening, and antimicrobial activity of films. The optimum nanocomposite film maintained the quality of minced chicken up to six days under refrigeration	Six days under refrigeration	Praseptiangga et al. (2021)
Pork	Coating	Cinnamon essential oil	The cinnamon oil functionalized coating showed reduction in total viable count, lactic acid bacterial and H_2S-producing bacteria count. The coating also significantly reduced the lipid oxidation in pork meat	Seven days at 4°C	He and Wang (2022)
Chicken breast	Coating	Allyl isothiocyanate	Coating with 200 ppm allyl isothiocyanate maintained aerobic bacterial population under threshold for 21 days. The coating with and without allyl isothiocyanate did not affect the rancidity of chicken breast	21 days at 4°C	Moller et al. (2023)
Chicken	Film	Salmonella phage PBSE191	A film with CA:glucomannan ratio of 7:3 with 40% sorbitol as plasticizer showed excellent adsorption of the phage. The films showed rapid release of phage with significant bacterial reduction, and the effect lasted for 48 and 96 h at 25°C and 4°C, respectively		Kim et al. (2023)

(*Continued*)

TABLE 4.3 (Continued)

Type of food examined	Film/ coating	Active compounds used	Findings on food quality	Shelf life	References
Chicken	Film	Benzyl isothiocyanate	Successful encapsulation of highly volatile benzyl isothiocyanate in β-cyclodextrin inclusion complex for incorporation in packaging film. Chicken packed in antimicrobial films showed significantly lower total viable count than the unpackaged sample after eight days of refrigerated storage	Eight days at 4°C	Huang et al. (2023)
Goat meat		Mekwiya date seed extract	A film with 0.88% incorporation of date seed extract showed highest phenolic content and antioxidant activity by DPPH and Ferric reducing antioxidant power (FRAP) assay. A film with 0.88% date seed extract extended the shelf life of meat beyond 14 days under refrigerated conditions	14 days at 4°C	Amokrane-Aidat et al. (2024)
Pork	Film	Anthocyanins and betacyanins from purple sweet potato and dragon fruit peels	Film with combination of anthocyanin and betacyanin in 3:1 ratio showed most significant color change during pork spoilage. Based on principal component analysis and correlation analysis, total color difference can indicate pork freshness		Gao et al. (2022)
Chicken	Film	Turmeric essential oil (TEO)	Encapsulation of TEO in zein nanoparticles achieves sustained release. The nanocomposite film reduced S. enterica population from 3.08 log CFU g^{-1} to 2.81 log CFU during 14-day storage	14 days at 4°C	Ali et al. (2023)
Beef	Film	*Oxalis triangularis* extract	Composite films with *Oxalis triangularis* extract possessed strong antioxidant activity and showed pink color at pH 2, light green color at pH 7–11 and yellow color at pH 12–13. A visible color transition from purple to pale purple in 24 h, and light green at 48 and 60 h suggested deteriorating beef quality		Jiang et al. (2023)

Pork	Film	Black corn seed powder	The concentration of black corn seed powder is directly proportional to the film haze and inversely proportional to transmittance. The color of the composite film transitioned from pink to grayish-green with increasing TVB-N content of pork		Chi et al. (2023)
Ready-to-eat ham	Film	Carvacrol	Incorporation of hydrophobic and volatile carvacrol successfully in a crosslinked film of gelatin and carrageenan. The crosslinked film effectively reduced the moisture loss, maintained the original texture and eliminated *L. monocytogenes* in ham up to seven days	Seven days at 4°C	Cui et al. (2023)
Meat patties	Film	Alpha-tocopherol	Delay in lipid oxidation was observed in meat patties packed in films containing alpha-tocopherol. Antioxidant-rich films can act as an alternative to synthetic antioxidants in processed meat products	12 days	Abd Hamid et al. (2019)
Pork	Film	*Rosa chinensis* flower extract	Composite film with anthocyanins showed strong antioxidant activity (71.72% of DPPH scavenging rate). The films showed significant color change from tan to olive over deteriorating quality of pork		Jiao et al. (2024)
Pork and shrimp	Film	Rose petal anthocyanins and carbon dots (CDs)	Incorporation of CDs (2%) and anthocyanins (5%) significantly improved the film properties and thermal stability. The composite film delivered both active and intelligent functions in preservation and freshness monitoring with visible color change from red to yellowish		Wagh et al. (2024)
Fish and pork	Film	Mulberry pomace particles	Increasing concentration of mulberry pomace particles increased the haze, which reduced the interference and improved the accuracy. The films showed obvious color change when the TVB-N content reached 14.60 and 19.04 mg 100 g^{-1}, for pork and fish, respectively		Zhang et al. (2023)

(*Continued*)

TABLE 4.3 (Continued)

Type of food examined	Film/ coating	Active compounds used	Findings on food quality	Shelf life	References
Pork and shrimp	Film	Curcumin	A simpler and easier direct dispersion of curcumin at lower concentrations (<3%) and controlled release from the film. A noticeable red shift in film color on third day of storage of pork and shrimp corresponding to increasing TVB-N levels		Liu et al. (2018)
Beef	Film	Benzyl isothiocyanate	Incorporation of benzyl isothiocyanate nanoemulsion in CA-based films significantly enhanced the antimicrobial activity against *S. aureus, Salmonella typhimurium,* and *Pseudomonas fluorescence.* The composite films effectively delayed the rise in total viable count, TVM-N, and TBARS, extending shelf life of raw beef up to 10 days	10 days at 4°C	Du et al. (2024)
Meat patties	Film	Eugenol and alpha-tocopherol	Films with eugenol and alpha-tocopherol had different modes of action and release behavior. Films with alpha-tocopherol are found to be better in prevention of lipid oxidation in meat patties		Wan Yahaya et al. (2024)
Beef	Film	Anthocyanin	The change in beef quality is related to the color change value of the indicator film. CA-carboxymethyl cellulose films with anthocyanin microcapsules changed color from pink to purple to green, corresponding to fresh, subfresh, and spoiled beef on zeroth, fourth, and seventh day, respectively		Cao et al. (2024)
Milk	Film	Mulberry polyphenolic extract	Change in color of films from purple to pink with spoilage of milk at 40°C. CA film with 4% mulberry polyphenolic extract showed most significant color change in response to the pH of milk		Liu et al. (2019)

Milk	Film	*Lycium ruthenicum* Murr. extract	CA film with 2.5% *Lycium ruthenicum* Murr. extract was found optimum for milk freshness monitoring. The color of the film changed from gray to grayish purple to pink when pH changed from 6.67 to 4.75 to 3.24 during spoilage		Liu et al. (2019)
Kalari (cheese)	Film	*Aloe vera* gel	Antimicrobial properties against *E. coli* and significant reduction in lipid oxidation due to polyphenols in the gel. The film with *Aloe vera* gel retained moisture, maintained juiciness, and flavor of cheese over 28 days storage	28 days	Kouser et al. (2023)
Milk	Film	Coconut husk-lignin-derived CDs	A film with 2% CDs showed excellent pH-dependent fluorescent color change in prescribed pH range. With spoilage of milk and subsequent decline in pH, the emission showed a redshift with noticeable changes in chromaticity values		Sangeetha et al. (2024)

glucomannan film-forming solution and observed its effect on quality and shelf life of chicken. The incorporation of camellia oil significantly reduced weight loss, pH, TBARS, and TVB-N, and also retarded lipid oxidation and microbial growth. A coating with 3.5% camellia oil was able to extend the shelf life of chicken for 10 days under refrigerated conditions without affecting odor and acceptability.

Improvement in the adhesive properties of edible coatings can ensure effective coating of meat products, leading to enhanced shelf life. One such work, inspired by the ability of mussels to attach to surfaces, resulted in the development of a highly adhesive CA-based edible coating. Herein, curcumin-loaded ZNPs were incorporated into a riboflavin-encapsulated CA-based coating. The most notable finding of the study is increased adhesive nature of the coatings due to interaction of CA with biogenic amines produced during storage of meat due to hydrogen bonding and electrostatic forces. The effectiveness of coating was tested for lamb and pork preservation, which revealed 1.8 and 2.3-fold increase in shelf life compared to the uncoated ones at 4°C. Apart from barrier properties, combination of curcumin and riboflavin had a strong role in preservation due to their antimicrobial activity under light irradiation (Li et al., 2025).

4.7.3.2 Carrageenan-based films for active/intelligent packaging in meat processing

Off late, the trend is toward development of packaging films where the concepts of food preservation and quality monitoring are combined, giving rise to active and intelligent packaging. Similar to fish, several studies have demonstrated CA-based films for meat and meat products. Considering that microbial growth is one of the major factors responsible for meat spoilage, antimicrobial films are under development. A CA-based active film was developed by incorporating olive leaf extract and evaluated for suitability for lamb meat packaging. It was observed that the film with olive leaf extracts was able to significantly reduce the coliform bacteria and aerobic mesophiles compared to the PVC and CA only films (Martiny et al., 2020). In another work by the same researchers, the active packaging film demonstrated decrease in psychrophiles in application for lamb meat compared to the conventional counterparts (Martiny, Raghavan, et al., 2020). Considering the higher cost associated with RC, Farhan and Hani (2020) developed an inexpensive SRC film functionalized with a water extract of germinated fenugreek seeds as an antioxidant and antimicrobial agent. During application studies on chicken breasts, the film with 20% germinated fenugreek extract showed 2.39 log reduction on seventh day of storage compared to control. Functionalization of biodegradable films with bioactive compounds and antioxidants significantly improves their performance; however, the susceptibility of these compounds to environmental stresses is a major concern. Therefore, encapsulation of such compounds in suitable matrices or nanostructures could be highly beneficial. Following this logic, gallic acid was encapsulated in a metal organic framework of zeolite imidazolium ester (ZIF-8) by solution impregnation method and was further incorporated into a CA-based film. The functionalized film showed improved mechanical strength, barrier properties, sustained release of gallic acid, antioxidant activity, and antimicrobial activity against *E. coli* O157:H7. A 12-day study on storage of beef revealed that the beef packed in CA films with encapsulated gallic acid showed better appearance compared to unpacked and polyethylene films (Fig. 4.4C). This is also confirmed by lower total viable count in CA-encapsulated gallic acid film-packed beef, which is attributed to the antimicrobial properties of the film (Zhang, Yi, et al., 2024).

Change in pH is a typical phenomenon associated with autolytic and microbial spoilage of meat. Therefore, in context of intelligent packaging, various pH-sensitive dyes and natural pigments are incorporated with CA films. Liu et al. (2024) integrated anthocyanins from *Padus virginiana* peel into a composite film of κ-CA and straw lignin for real-time freshness determination of chicken breasts. The anthocyanin-containing films changed their color from pink to grayish yellow in response to the release of nitrogenous compounds as a result of decreasing freshness. In another work on freshness monitoring of pork, black raspberry extract was incorporated into composite labels of pea starch and CA. During 60-hour storage study at 20°C, the TVB-N content was continuously rising and crossed the rejection limit of 15 mg 100 g^{-1} in first 36 hours. The composite labels did not show color change in first 24 hours; however, after 36 hours, the color changed from pink to brown, which clearly indicates the suitability of labels for clear identification of pork spoilage (Jiang, Liu, et al., 2024). In another work, *Rosa canina* petal extract and chitosan nanoparticles were incorporated into a CA-locust bean gum-based nanofibrous membranes prepared by electrospinning for monitoring beef freshness. Here, electrospinning is used as a technique for successful encapsulation of petal extracts, and chitosan nanoparticles were used to improve the mechanical strength, barrier properties, and thermal stability. The color of the intelligent nanofibrous membrane changed from white to blue on fourth day, which indicates the spoilage of beef and is confirmed by increased total viable count, pH, and TVB-N. Thus, the nanofibrous membrane with encapsulated *R. canina* petal extract proves to be a convenient indicator for monitoring beef freshness (Mohammadi et al., 2024).

Further, researchers are keen over development of multifunctional films where active and intelligent capabilities are combined in a single film. A composite film based on κ-CA and hydroxypropyl methylcellulose was integrated with an extract of *Prunus maackii* pomace and cork bark extract as a pH indicator and antimicrobial agent, respectively. The films possessed antibacterial activity against *E. coli* and *S. aureus* and also showed antioxidant activity, which is beneficial to prevent lard oxidation. The color of the films also significantly changed from reddish to yellow-green in accordance with the TVB-N content effectively representing the freshness of pork (Zhang et al., 2021b). Selected research works related to meat packaging are reported.

4.7.4 Applications in dairy processing

The milk and milk products are nutritionally rich and therefore susceptible to rapid microbial spoilage, creating challenges in handling and preservation. It is crucial to maintain the quality and safety of the dairy products at each stage of the supply chain, right from the production of feed for milking animals until it is consumed by the consumer. The packaging plays major role in protecting product during transport, distribution, sale, and consumption before the expiry. As the consumer awareness is increasing, the need for development of smart and biodegradable packaging that detects and indicates deterioration of quality and safety of product packed inside is essential (Mirza Alizadeh et al., 2021). The rising concerns about the degradability of huge plastic packaging material produced from routine domestic requirement of dairy products have generated the research necessity to develop and study biodegradable packaging solutions incorporating active and intelligent components. As CA derived from seaweed is known as food additive with applications in dairy products as well, it can also be used in developing biodegradable packaging material. Among the various biodegradable polysaccharides, CA has the ability to bind therapeutic agents and provide better packaging alternatives according to several studies (Mathew et al., 2024).

4.7.4.1 Carrageenan-based coatings for active/intelligent packaging in dairy processing

Various dairy products, such as butter, cheese, and paneer, are popular for their sensory properties; however, they tend to undergo microbial and chemical spoilage. Edible coatings functionalized with various active ingredients can significantly assist in shelf-life improvement. Studies have reported application of whey protein (Mileriene et al., 2021), alginate (Silva et al., 2022), chitosan (El-Sayed et al., 2021), agar (Guitián et al., 2019), and tapioca starch (Berti et al., 2019) with incorporation of various active ingredients aimed at preservation and shelf-life extension of cheese. Similar to above-mentioned biopolymers, CA also holds strong potential as a polymer for active edible coating for preservation of dairy products; therefore, it demands further research.

4.7.4.2 Carrageenan-based films for active/intelligent packaging in dairy processing

The studies have explored film-forming ability of CA by different methods, and it was observed that the CA can produce translucent films enabling visual inspection of the product packed in the film (Mathew et al., 2024). The CA-based biodegradable film was incorporated with *A. vera* extract to improve the antioxidant and antimicrobial properties, and it was further used for packaging of a frozen dairy product, kulfi. A six-month storage study at −18°C revealed that the kulfi packed in edible film with 15% *A. vera* extract reported lower microbial count, free fatty acids, and TBARS content compared to the control sample. With improved antimicrobial and antioxidant properties, the film was found to be suitable for packaging of frozen dairy products on commercial level (Mahajan et al., 2021). Another study on cottage cheese involved use of copper oxide nanoparticles incorporated into CA-based film for ultraviolet protection and antimicrobial activity. With application of this film, the cottage cheese was successfully stored for seven days under refrigeration without change in flavor, texture, and fragrance (Kumari et al., 2024).

Milk is primarily pasteurized to eliminate pathogenic microorganisms; however, spoilage microorganisms are not entirely eliminated. Presently, pasteurized milk is transported and marketed under low temperatures and is advised to be consumed before the expiry date. Under these conditions, monitoring freshness becomes critical. Roy and Rhim (2021b) developed a CA and gelatin-based indicator film functionalized with shikonin as an indicator and propolis as an antimicrobial agent. Upon application for milk freshness monitoring, the color of the film changes from purple to reddish pink while reducing pH from 6.6 (fresh milk) to 4.5 (spoiled milk). The consumers and retailers can clearly distinguish the fresh, spoiling, and spoiled milk by simply observing the color without opening the package. Another study on milk freshness monitoring used alizarin red S as a pH indicator dye and incorporated it into a guar gum-CA matrix to develop halochromic labels. While monitoring freshness of milk at 28°C, the color of the label changed from dark purple to lavender at 24 hours and to orange red at 48 hours, clearly identifying the levels of spoilage from fresh, subfresh, and spoiled, respectively (Chen et al., 2024). Overall, the application of carrageenan-based films and coatings is depicted, and the recent studies are explained in Table 4.3.

4.8 Challenges and opportunities of carrageenan-based coatings and films

CA-based coatings and films are extensively studied for their potential applications in food technology due to their unique properties. Despite the fact that the potential of CA-based coatings and films is often highlighted, it is crucial to critically examine the significant challenges and limitations that may impede their practical application in food packaging and other sectors. One of the foremost concerns is their mechanical weakness due to their hydrophilicity. Although ongoing research aimed at improving mechanical properties by blending with other biopolymers, CA films often exhibit brittleness. This fragility can lead to cracking and breakage during handling, altering the integrity of the packaging, the safety of the food products, and scalability for commercial use. To address this, various additives like plasticizers, nanofillers, essential oils, and other biopolymers can be explored in CA films and coatings.

CA films exhibit notable WVP and are less effective over gases like oxygen and carbon dioxide, limiting their use for products requiring extended preservation, such as meats and dairy. Their highly hydrophilic nature also results in poor moisture barrier properties, making them unsuitable for packaging moisture-sensitive products (Sedayu et al., 2019). The commercial viability of CA films and coatings is challenged by issues related to mechanical properties, moisture sensitivity, and the cost of functionalization. To enhance the mechanical strength and barrier properties of these films by blending with other biopolymers or incorporating nanomaterials or bioactive compounds has shown promise in expanding their applicability beyond just food packaging (Sedayu et al., 2019). Moreover, understanding the gelation mechanisms and drying rates can lead to optimized production processes that further improve performance characteristics (Wang et al., 2024).

The functionalization of CA films with bioactive compounds, such as anthocyanins and essential oils, enhances their antioxidant and antimicrobial properties, making them suitable for intelligent packaging applications (Wagh et al., 2024). However, the extraction and incorporation of these compounds can be costly and may not be economically viable for large-scale production. The use of metal-phenolic networks and other advanced materials to impart additional functionalities, like UV protection and antibacterial properties, further increases the cost and complexity of production. Moreover, while the exploration of smart packaging features is promising, the integration of such technologies into CA films is still in its early stages. The complexities involved in embedding sensors and ensuring their functionality can complicate the manufacturing process, potentially leading to higher costs and reduced scalability.

Although CA-based films offer environmental benefits due to their biodegradability, the economic feasibility of their production remains a challenge. The cost of raw materials, processing, and functionalization must be balanced against the environmental benefits to make these films a viable alternative to conventional packaging (Amin et al., 2021). The scalability of production techniques, such as solvent casting and electrospinning, also impacts the cost-effectiveness of CA films. These methods need to be optimized for large-scale production to reduce costs and improve efficiency (Mathew et al., 2024). Advancing research in this area aims to address these challenges and requires a balance between enhancing film properties and maintaining cost-effectiveness. Future research and development efforts should optimize production techniques and explore cost-effective additives to make CA-based films a feasible option for commercial sustainable food packaging applications.

4.9 Conclusion

Carrageenan is low-cost biopolymer obtained from red seaweed and has excellent film-forming properties. CA-based coatings and films present a promising solution for sustainable food packaging, offering key advantages such as biodegradability, renewability, and functional versatility. However, significant challenges remain in scaling their production, increasing cost, and optimizing their performance, which hinders commercial viability. This chapter emphasizes how integrating CA with other biopolymers, nanomaterials, and bioactive compounds is helping to enhance the mechanical, water barrier, antimicrobial, antioxidant, and UV-blocking capabilities, making them innovative alternatives to single-use plastics. These films are also integrated with various indicators and sensors, which further enhance functionality for real-time monitoring of freshness of fresh produce during supply chain, which helps to reduce postharvest losses and food waste. This offers opportunities for continued research and development of CA-based films that have the potential to transform the packaging industry by providing eco-friendly, efficient, and intelligent solutions for a wide range of applications.

References

Abdallah, R. B., Ghazouani, T., & Fattouch, S. (2024). Carrageenan based films. In *Polysaccharide based films for food packaging: Fundamentals, properties and applications*. Springer, 175–195.

Abdillah, A. A., & Charles, A. L. (2021). Characterization of a natural biodegradable edible film obtained from arrowroot starch and iota-carrageenan and application in food packaging. *International Journal of Biological Macromolecules, 191*, 618–626. https://doi.org/10.1016/j.ijbiomac.2021.09.141, www.elsevier.com/locate/ijbiomac.

Abud, E. M., Simon, R. (2022). Provisional chapter title: Food additives and reactions: part 1.

Aga, M. B., Dar, A. H., Nayik, G. A., Panesar, P. S., Allai, F., Khan, S. A., Shams, R., Kennedy, J. F., & Altaf, A. (2021). Recent insights into carrageenan-based bio-nanocomposite polymers in food applications: A review. *International Journal of Biological Macromolecules, 192*, 197–209. https://doi.org/10.1016/j.ijbiomac.2021.09.212, www.elsevier.com/locate/ijbiomac.

Al-Alawi, A. A., Al-Marhubi, I. M., Al-Belushi, M. S. M., & Soussi, B. (2011). Characterization of carrageenan extracted from *Hypnea bryoides* in Oman. *Marine Biotechnology, 13*(5), 893–899. https://doi.org/10.1007/s10126-010-9350-7.

Alam, J., Alhoshan, M., Shukla, A. K., Aldalbahi, A., & Ali, F. A. A. (2019). k-Carrageenan–A versatile biopolymer for the preparation of a hydrophilic PVDF composite membrane. *European Polymer Journal, 120*, 109219.

Ali, M. S., Haq, M., Roy, V. C., Ho, T. C., Park, J. S., Han, J. M., & Chun, B. S. (2023). Development of fish gelatin/carrageenan/zein bio-nanocomposite active-films incorporated with turmeric essential oil and their application in chicken meat preservation. *Colloids and Surfaces B: Biointerfaces, 226*. https://doi.org/10.1016/j.colsurfb.2023.113320, www.elsevier.com/locate/colsurfb.

Alizadeh Sani, M., Tavassoli, M., Salim, S. A., Azizi-lalabadi, M., & McClements, D. J. (2022). Development of green halochromic smart and active packaging materials: TiO_2 nanoparticle- and anthocyanin-loaded gelatin/κ-carrageenan films. *Food Hydrocolloids, 124*. https://doi.org/10.1016/j.foodhyd.2021.107324, https://www.journals.elsevier.com/food-hydrocolloids.

Alonso-Salinas, R., López-Miranda, S., Pérez-López, A. J., & Acosta-Motos, J. R. (2024). Strategies to delay ethylene-mediated ripening in climacteric fruits: Implications for shelf life extension and postharvest quality. *Horticulturae, 10*(8). https://doi.org/10.3390/horticulturae10080840, www.mdpi.com/journal/horticulturae.

Amanda, P., Ismadi, I., Ningrum, R. S., Nabila, S., & Prasetyo, K. W. (2024). Carrageenan functional film integrated with pickering emulsion of oregano oil stabilized by cationic nanocellulose for active packaging. *Food Science and Technology International, 30*(1), 61–72. https://doi.org/10.1177/10820132221132912, https://journals.sagepub.com/home/FST.

Amin, U., Khan, M. U., Majeed, Y., Rebezov, M., Khayrullin, M., Bobkova, E., Shariati, M. A., Chung, I. M., & Thiruvengadam, M. (2021). Potentials of polysaccharides, lipids and proteins in biodegradable food packaging applications. *International Journal of Biological Macromolecules, 183*, 2184–2198. https://doi.org/10.1016/j.ijbiomac.2021.05.182, www.elsevier.com/locate/ijbiomac.

Amjadi, S., Almasi, H., Ghorbani, M., & Ramazani, S. (2020). Reinforced ZnONPs/rosemary essential oil-incorporated zein electrospun nanofibers by κ-carrageenan. *Carbohydrate Polymers, 232*.

Amokrane-Aidat, R., Brahmi, F., Chennit, B., Smaoui, S., Elhadef, K., Chaari, M., Madani, K., & Boulekbache-Makhlouf, L. (2024). Sustainable gelatin-kappa carrageenan active packaging with Mekwiya date seeds to enhance goat meat quality and shelf life. *International Journal of Biological Macromolecules, 279*, 135285. https://doi.org/10.1016/j.ijbiomac.2024.135285.

Arnon-Rips, H., & Poverenov, E. (2018). Improving food products' quality and storability by using layer by layer edible coatings. *Trends in Food Science & Technology, 75*, 81–92. https://doi.org/10.1016/j.tifs.2018.03.003.

Asadzadeh, N., Ghorbanpour, M., & Sayyah, A. (2023). Effects of filler type and content on mechanical, thermal, and physical properties of carrageenan biocomposite films. *International Journal of Biological Macromolecules, 253*, 127551.

Avila, L. B., Barreto, E. R. C., de Souza, P. K., Silva, B. D. Z., Martiny, T. R., Moraes, C. C., Morais, M. M., Raghavan, V., & Rosa, G. S.da (2020). Carrageenan-based films incorporated with jaboticaba peel extract: An innovative material for active food packaging. *Molecules (Basel, Switzerland), 25*(23). https://doi.org/10.3390/molecules25235563, https://www.mdpi.com/1420-3049/25/23/5563/pdf.

Avila, L. B., Barreto, E. R. C., Moraes, C. C., Morais, M. M., & Rosa, G. S.da (2022). Promising new material for food packaging: An active and intelligent carrageenan film with natural jaboticaba additive. *Foods, 11*(6). https://doi.org/10.3390/foods11060792, https://www.mdpi.com/2304-8158/11/6/792/pdf.

Balqis, A. M. I., Khaizura, M. A. R. N., Russly, A. R., & Hanani, Z. A. N. (2017). Effects of plasticizers on the physicochemical properties of kappa-carrageenan films extracted from *Eucheuma cottonii*. *International Journal of Biological Macromolecules, 103*, 721–732.

Banitaba, S. N., & Ehrmann, A. (2021). Application of electrospun nanofibers for fabrication of versatile and highly efficient electrochemical devices: A review. *Polymers, 13*(11), 1741. https://doi.org/10.3390/polym13111741.

Barzan, G., Sacco, A., Giovannozzi, A. M., Portesi, C., Schiavone, C., Salafranca, J., Wrona, M., Nerín, C., & Rossi, A. M. (2024). Development of innovative antioxidant food packaging systems based on natural extracts from food industry waste and *Moringa oleifera* leaves. *Food Chemistry, 432*. https://doi.org/10.1016/j.foodchem.2023.137088, www.elsevier.com/locate/foodchem.

Bayram, I., & Decker, E. A. (2023). Underlying mechanisms of synergistic antioxidant interactions during lipid oxidation. *Trends in Food Science and Technology, 133*, 219–230. https://doi.org/10.1016/j.tifs.2023.02.003, http://www.elsevier.com/wps/find/journaldescription.cws_home/601278/description#description.

Berti, S., Ollé Resa, C. P., Basanta, F., Gerschenson, L. N., & Jagus, R. J. (2019). Edible coatings on Gouda cheese as a barrier against external contamination during ripening. *Food Bioscience, 31*. https://doi.org/10.1016/j.fbio.2019.100447, http://www.journals.elsevier.com/food-bioscience/.

Bharti, S. K., Pathak, V., Arya, A., Alam, T., Rajkumar, V., & Verma, A. K. (2021). Packaging potential of Ipomoea batatas and κ-carrageenan biobased composite edible film: Its rheological, physicomechanical, barrier and optical characterization. *Journal of Food Processing and Preservation, 45*(2). https://doi.org/10.1111/jfpp.15153, http://onlinelibrary.wiley.com/journal/10.1111/(ISSN)1745-4549.

Bhatia, S., Abbas Shah, Y., Al-Harrasi, A., Jawad, M., Koca, E., & Aydemir, L. Y. (2024). Enhancing tensile strength, thermal stability, and antioxidant characteristics of transparent kappa carrageenan films using grapefruit essential oil for food packaging applications. *ACS Omega, 9*(8), 9003–9012. https://doi.org/10.1021/acsomega.3c07366, pubs.acs.org/journal/acsodf.

Bhattacharya, S. (2025). Exploring the therapeutic and pharmaceutical potential of antioxidants. In *Antioxidants: Nature's defense against disease*. India: wiley, 587–612. https://onlinelibrary.wiley.com/doi/book/10.1002/9781394270576.

Bian, Z., Li, X., Zhang, J., Shi, M., Xin, Z., Wang, H., Komarneni, S., Zhang, K., Ni, Z., & Hu, G. (2024). Intelligent double-layer films based on gellan gum/mica nanosheets/anthocyanin/konjac glucomannan/carrageenan for food real-time freshness monitoring. *Food Hydrocolloids, 151*, 109767. https://doi.org/10.1016/j.foodhyd.2024.109767.

Bian, Z., Xu, W., Zhang, H., Shi, M., Ji, X., Dong, S., Chen, C., Zhao, G., Zhuo, X., Komarneni, S., Zhang, K., Ni, Z., & Hu, G. (2023). Simultaneously realizing enhancement of sensitivity for freshness monitoring and multinomial properties of carrageenan/konjac glucomannan/blueberry anthocyanin-based intelligent film by diatomite. *International Journal of Biological Macromolecules, 251*. https://doi.org/10.1016/j.ijbiomac.2023.126192, https://www.sciencedirect.com/journal/international-journal-of-biological-macromolecules.

Bizymis, A. P., & Tzia, C. (2022). Edible films and coatings: Properties for the selection of the components, evolution through composites and nanomaterials, and safety issues. *Critical Reviews in Food Science and Nutrition, 62*(31), 8777–8792. https://doi.org/10.1080/10408398.2021.1934652, https://www.tandfonline.com/loi/bfsn20.

Boulho, R., Marty, C., Freile-Pelegrín, Y., Robledo, D., Bourgougnon, N., & Bedoux, G. (2017). Antiherpetic (HSV-1) activity of carrageenans from the red seaweed *Solieria chordalis* (Rhodophyta, Gigartinales) extracted by microwave-assisted extraction (MAE). *Journal of Applied Phycology, 29*(5), 2219–2228. https://doi.org/10.1007/s10811-017-1192-5, www.wkap.nl/journalhome.htm/0921-8971.

Butler, I. P., Banta, R. A., Tyuftin, A. A., Holmes, J., Pathania, S., & Kerry, J. (2023). Pectin as a biopolymer source for packaging films using a circular economy approach: Origins, extraction, structure and films properties. *Food Packaging and Shelf Life, 40*. https://doi.org/10.1016/j.fpsl.2023.101224, https://www.sciencedirect.com/science/journal/22142894.

Calvo, G. H., Cosenza, V. A., Sáenz, D. A., Navarro, D. A., Stortz, C. A., Céspedes, M. A., Mamone, L. A., Casas, A. G., & Di Venosa, G. M. (2019). Disaccharides obtained from carrageenans as potential antitumor agents. *Scientific Reports, 9*(1). https://doi.org/10.1038/s41598-019-43238-y, www.nature.com/srep/index.html.

Campo, V. L., Kawano, D. F., Silva, D. Bd, & Carvalho, I. (2009). Carrageenans: Biological properties, chemical modifications and structural analysis - A review. *Carbohydrate Polymers, 77*(2), 167–180. https://doi.org/10.1016/j.carbpol.2009.01.020.

Cao, Y., Gou, Q., Song, Z., Zhang, L., Yu, Q., Zhu, X., & Li, S. (2024). Smart carrageenan/carboxymethyl cellulose films combined with zein/gellan gum microcapsules encapsulated by composite anthocyanins for chilled beef freshness monitoring. *Food Hydrocolloids, 153*, 110059. https://doi.org/10.1016/j.foodhyd.2024.110059.

Carrageenan Market Size & Share Analysis - Growth Trends & Forecasts. (2023). WWW Document, WWW Document.

Chen, J., Zhang, J., Liu, C., Sun, Y., Han, X., Sun, X., Pei, X., Huang, F., Li, X., & Chen, A. (2024). Development of halochromic labels based on binary systems of cationic guar gum and κ-carrageenan loaded with alizarin red S for monitoring milk and seafood freshness. *Food Hydrocolloids, 156*, 110272. https://doi.org/10.1016/j.foodhyd.2024.110272.

Cheng, C., Chen, S., Su, J., Zhu, M., Zhou, M., Chen, T., & Han, Y. (2022). Recent advances in carrageenan-based films for food packaging applications. *Frontiers in Nutrition, 9*. https://doi.org/10.3389/fnut.2022.1004588, journal.frontiersin.org/journal/nutrition.

Chi, W., Cao, L., Sun, G., Meng, F., Zhang, C., Li, J., & Wang, L. (2020). Developing a highly pH-sensitive κ-carrageenan-based intelligent film incorporating grape skin powder via a cleaner process. *Journal of Cleaner Production, 244*, 118862. https://doi.org/10.1016/j.jclepro.2019.118862.

Ciancia, M., Matulewicz, M. C., & Tuvikene, R. (2020). Structural diversity in galactans from red seaweeds and its influence on rheological properties. *Frontiers in Plant Science, 11*, 559986.

Cotas, J., Marques, V., Afonso, M. B., Rodrigues, C. M. P., & Pereira, L. (2020). Antitumour potential of gigartina pistillata carrageenans against colorectal cancer stem cell-enriched tumourspheres. *Marine Drugs, 18*(1). https://doi.org/10.3390/md18010050, https://www.mdpi.com/1660-3397/18/1/50/pdf.

Cui, H., Cheng, Q., Li, C., Khin, M. N., & Lin, L. (2023). Schiff base cross-linked dialdehyde β-cyclodextrin/gelatin-carrageenan active packaging film for the application of carvacrol on ready-to-eat foods. *Food Hydrocolloids, 141*, 108744. https://doi.org/10.1016/j.foodhyd.2023.108744.

Cvetković, K., Đorđević, N., Karabegović, I., Danilović, B., Dordevic, D., Dordevic, S., & Kushkevych, I. (2025). Production and Testing of Carrageenan-Based Films Enriched with Chinese Hawthorn Extract in Strawberry Packaging. *Processes, 13*(2), 379. https://doi.org/10.3390/pr13020379.

da Costa, J. C. M., Bruni, A. R. S., Jesus, G. A. M., Alves, E. S., de Oliveira S. Júnior, O., Martins, A. F., & Bonafe, E. G. (2024). Enhancing fresh pear preservation with UV-blocking film coatings based on κ-carrageenan, cassava starch, and copper oxide particles. *Journal of Food Engineering, 368*, 111853. https://doi.org/10.1016/j.jfoodeng.2023.111853.

da Rosa, G. S., Vanga, S. K., Gariepy, Y., & Raghavan, V. (2020). Development of biodegradable films with improved antioxidant properties based on the addition of carrageenan containing olive leaf extract for food packaging applications. *Journal of Polymers and the Environment, 28*(1), 123–130. https://doi.org/10.1007/s10924-019-01589-7, http://www.kluweronline.com/issn/1566-2543/.

David, S., Shani Levi, C., Fahoum, L., Ungar, Y., Meyron-Holtz, E. G., Shpigelman, A., & Lesmes, U. (2018). Revisiting the carrageenan controversy: Do we really understand the digestive fate and safety of carrageenan in our foods? *Food and Function, 9*(3), 1344–1352. https://doi.org/10.1039/c7fo01721a, http://pubs.rsc.org/en/journals/journal/fo.

David, S., Wojciechowska, A., Portmann, R., Shpigelman, A., & Lesmes, U. (2020). The impact of food-grade carrageenans and consumer age on the in vitro proteolysis of whey proteins. *Food Research International, 130*, 108964. https://doi.org/10.1016/j.foodres.2019.108964.

de Jesus, G. A. M., Berton, S. B. R., Simões, B. M., Zola, R. S., Monteiro, J. P., Martins, A. F., & Bonafé, E. G. (2023). κ-Carrageenan/poly (vinyl alcohol) functionalized films with gallic acid and stabilized with metallic ions. *International Journal of Biological Macromolecules, 253*, 127087.

de Lima Barizão, C., Crepaldi, M. I., Junior, Od. O. S., de Oliveira, A. C., Martins, A. F., Garcia, P. S., & Bonafé, E. G. (2020). Biodegradable films based on commercial κ-carrageenan and cassava starch to achieve low production costs. *International Journal of Biological Macromolecules, 165*, 582–590. https://doi.org/10.1016/j.ijbiomac.2020.09.150, www.elsevier.com/locate/ijbiomac.

de Oliveira Filho, J. G., Miranda, M., Ferreira, M. D., & Plotto, A. (2021). Nanoemulsions as edible coatings: A potential strategy for fresh fruits and vegetables preservation. *Foods, 10*(10), 2438. https://doi.org/10.3390/foods10102438.

de Velde, F., Antipova, A. S., Rollema, H. S., Burova, T. V., Grinberg, N. V., Pereira, L., Gilsenan, P. M., Tromp, R. H., Rudolph, B., & Grinberg, V. Y. (2005). The structure of κ/ι-hybrid carrageenans II. Coil–helix transition as a function of chain composition. *Carbohydrate Research, 340*, 1113–1129.

Dogaru, B.-I., Simionescu, B., & Popescu, M.-C. (2020). Synthesis and characterization of κ-carrageenan bio-nanocomposite films reinforced with bentonite nanoclay. *International Journal of Biological Macromolecules, 154*, 9–17.

Dong, Y., Wei, Z., & Xue, C. (2021). Recent advances in carrageenan-based delivery systems for bioactive ingredients: A review. *Trends in Food Science & Technology, 112*, 348–361.

Du, X. X., Ge, Z. T., Hao, H. S., Bi, J. R., Hou, H. M., & Zhang, G. L. (2024). An antibacterial film using κ-carrageenan loaded with benzyl isothiocyanate nanoemulsion: Characterization and application in beef preservation. *International Journal of Biological Macromolecules, 276*. https://doi.org/10.1016/j.ijbiomac.2024.133689, https://www.sciencedirect.com/science/journal/01418130.

Dulanlebit, Y. H., & Hernani, H. (2023). Overview of extraction methods for extracting seaweed and its applications. *Jurnal Penelitian Pendidikan IPA, 9*(2), 817–824. https://doi.org/10.29303/jppipa.v9i2.3053.

Dumas, L., de Souza, M. C., Bonafe, E. G., Martins, A. F., & Monteiro, J. P. (2024). Optimized incorporation of silver nanoparticles onto cotton fabric using k-carrageenan coatings for enhanced antimicrobial properties. *ACS Applied Bio Materials, 7*(10), 6908–6918. https://doi.org/10.1021/acsabm.4c01002, pubs.acs.org/journal/aabmcb.

Eha, K., Pehk, T., Heinmaa, I., Kaleda, A., & Laos, K. (2021). Impact of short-term heat treatment on the structure and functional properties of commercial furcellaran compared to commercial carrageenans. *Heliyon, 7*(4), e06640. https://doi.org/10.1016/j.heliyon.2021.e06640.

El-Sayed, H. S., El-Sayed, S. M., Mabrouk, A. M. M., Nawwar, G. A., & Youssef, A. M. (2021). Development of eco-friendly probiotic edible coatings based on chitosan, alginate and carboxymethyl cellulose for improving the shelf Life of UF soft cheese. *Journal of Polymers and the Environment, 29*(6), 1941–1953. https://doi.org/10.1007/s10924-020-02003-3, http://www.kluweronline.com/issn/1566-2543/.

Ezati, P., Riahi, Z., & Rhim, J.-W. (2021). Carrageenan-based functional films integrated with CuO-doped titanium nanotubes for active food-packaging applications. *ACS Sustainable Chemistry & Engineering, 9*, 9300–9307.

Falcó, I., Randazzo, W., Sánchez, G., López-Rubio, A., & Fabra, M. J. (2019). On the use of carrageenan matrices for the development of antiviral edible coatings of interest in berries. *Food Hydrocolloids, 92*, 74–85. https://doi.org/10.1016/j.foodhyd.2019.01.039.

Falshaw, R., Bixler, H. J., & Johndro, K. (2001). Structure and performance of commercial kappa-2 carrageenan extracts: I. Structure analysis. Food Hydrocolloids, *15*, 441–452.

Faria, C. S. V., Vieira, J. M., Vicente, A. A., & Martins, J. T. (2024). Locust bean gum/κ-carrageenan film containing blueberry or beetroot extracts as intelligent films to monitoring hake (merluccius merluccius) freshness. *Foods, 13*(19), 3088. https://doi.org/10.3390/foods13193088.

Farhan, A., & Hani, N. M. (2020). Active edible films based on semi-refined κ-carrageenan: Antioxidant and color properties and application in chicken breast packaging. *Food Packaging and Shelf Life, 24*, 100476. https://doi.org/10.1016/j.fpsl.2020.100476.

Forghani, S., Zeynali, F., Almasi, H., & Hamishehkar, H. (2022). Characterization of electrospun nanofibers and solvent-casted films based on *Centaurea arvensis* anthocyanin-loaded PVA/κ-carrageenan and comparing their performance as colorimetric pH indicator. *Food Chemistry, 388*, 133057. https://doi.org/10.1016/j.foodchem.2022.133057.

Gao, L., Liu, P., Liu, L., Li, S., Zhao, Y., Xie, J., & Xu, H. (2022). κ-carrageenan-based pH-sensing films incorporated with anthocyanins or/and betacyanins extracted from purple sweet potatoes and peels of dragon fruits. *Process Biochemistry, 121*, 463–480.

Gaurav, K., Mehta, N. K., Majumdar, R. K., Priyadarshini, M. B., Pal, P., Xavier, K. A. M., & Sharma, S. (2023). Carboxy methyl cellulose, xanthan gum, and carrageenan coatings reduced fat uptake, protein oxidation, and improved functionality in deep-fried fish strips: An application of the multiobjective optimization (MOO) approach. *ACS Omega, 8*(36), 32855–32866. https://doi.org/10.1021/acsomega.3c04072, pubs.acs.org/journal/acsodf.

Genecya, G., Adhika, D. R., Sutrisno, W., & Wungu, T. D. K. (2023). Characteristic improvement of a carrageenan-based bionanocomposite polymer film containing montmorillonite as food packaging through the addition of silver and cerium oxide nanoparticles. *ACS Omega, 8*(42), 39194–39202. https://doi.org/10.1021/acsomega.3c04575, pubs.acs.org/journal/acsodf.

Genecya, G., Adhika, D. R., & Wungu, T. D. K. (2023). Optimization of mechanical properties of carrageenan-based bioplastic as food packaging. *IOP Conference Series: Earth and Environmental Science,* 12079 IOP Publishing.

Global Carrageenan Market - Industry Trends and Forecast to 2031 [WWW Document], 2024. Data Bridg. Mark. Res. URL https://www.databridgemarketresearch.com/reports/global-carrageenan-market?srsltid=AfmBOoo1TagWCCg9pI-Yj7M-qWrz1V_vAqiH4Uush8EcWuetLwXmtOw0.

Goff, H. D., & Guo, Q. (2020). The role of hydrocolloids in the development of food structure. In *Handbook of food structure development*. Royal Society of Chemistry.

Goyal, D., Thirumal, S., Singh, D., Kumar, I., Singh, G., Kumar, R. K., & Sindhu (2025). Basics of antioxidants and their importance. *Antioxidants Nature's Def.*

Guitián, M. V., Ibarguren, C., Soria, M. C., Hovanyecz, P., Banchio, C., & Audisio, M. C. (2019). Anti-*Listeria monocytogenes* effect of bacteriocin-incorporated agar edible coatings applied on cheese. *International Dairy Journal, 97*, 92–98. https://doi.org/10.1016/j.idairyj.2019.05.016, www.elsevier.com/inca/publications/store/4/0/5/8/6/0/index.htt.

Gumus, T., Kaynarca, G. B., & Kamer, D. D. A. (2024). Optimization of an edible film formulation by incorporating carrageenan and red wine lees into fish gelatin film matrix. *International Journal of Biological Macromolecules, 258*, 128854.

Gupta, V., Biswas, D., & Roy, S. (2022). A comprehensive review of biodegradable polymer-based films and coatings and their food packaging applications. *Materials (Basel), 15*, 5899.

Gupta, V., Yadav, R., Tanwar, R., & Gaikwad, K. K. (2021). κ-Carrageenan-based bio-nanocomposite film reinforced with cellulose nanocrystals derived from amla pomace for food packaging. *Biomass Conversion Biorefinery*, 1–10.

Han, Y., Zhou, M., Deng, J., Cheng, C., Xu, Z., Hou, W., Yi, Y., McClements, D. J., & Chen, S. (2025). Intelligent carrageenan-based composite films containing color indicator-loaded nanoparticles for monitoring fish freshness. *Food Packaging and Shelf Life, 47*. https://doi.org/10.1016/j.fpsl.2024.101420, https://www.sciencedirect.com/science/journal/22142894.

Han, Y., Zhou, M., McClements, D. J., Liu, F., Cheng, C., Xiong, J., Zhu, M., & Chen, S. (2023). Investigation of a novel smart and active packaging materials: Nanoparticle-filled carrageenan-based composite films. *Carbohydrate Polymers, 301*. https://doi.org/10.1016/j.carbpol.2022.120331, http://www.elsevier.com/wps/find/journaldescription.cws_home/405871/description#description.

He, S., & Wang, Y. (2022). Antimicrobial and antioxidant effects of kappa-carrageenan coatings enriched with cinnamon essential oil in pork meat. *Foods, 11*(18), 2885. https://doi.org/10.3390/foods11182885.

Huang, Y., Liu, J., Li, Z., Cao, Z., Hao, H., Bi, J., Hou, H., Wu, H., & Zhang, G. (2023). Antibacterial film based on κ-carrageenan with benzyl isothiocyanate-β-cyclodextrin inclusion complex: Characterization and application in chicken preservation. *Food Hydrocolloids, 145*, 109063. https://doi.org/10.1016/j.foodhyd.2023.109063.

Huang, H., Wang, Q., Ning, Z., Ma, Y., Huang, Y., Wu, Y., Yang, Y., Xiao, M., & Ye, J. (2024). Preparation, antibacterial activity, and structure-activity relationship of low molecular weight κ-carrageenan. *International Journal of Biological Macromolecules, 266*. https://doi.org/10.1016/j.ijbiomac.2024.131021, https://www.sciencedirect.com/science/journal/01418130.

Huat, L. L., Alali, N. B., Ling, R. L. Z., Feng, S. S., & Teo, S.-S. (2023). Analysis of antimicrobial activity of carrageenan extracted from Kappaphycus Alvarezii. *Applied Microbiology: Theory & Technology*, 61–72.

Hussain, S., Akhter, R., & Maktedar, S. S. (2024). Advancements in sustainable food packaging: From eco-friendly materials to innovative technologies. *Sustainable Food Technology, 2*, 1297–1364.

Jafri, N. N. M., & Jaafar, J. (2024). Dip coating technique. In *Advanced ceramics for photocatalytic membranes: Synthesis methods, characterization and performance analysis, and applications in water and wastewater treatment*. Malaysia: Elsevier, 101–127. https://www.sciencedirect.com/book/9780323954181.

Jancikova, S., Dordevic, D., Jamroz, E., Behalova, H., & Tremlova, B. (2020). Chemical and physical characteristics of edible films, based on κ- and ι-carrageenans with the addition of lapacho tea extract. *Foods, 9*(3), 357. https://doi.org/10.3390/foods9030357.

Jia, H., Wu, C., Huang, M., & Zhu, Q. (2025). An intelligent fruit freshness monitoring system using hydrophobic indicator labels based on methylcellulose, k-carrageenan, and sodium tripolyphosphate, combined with deep learning. *International Journal of Biological Macromolecules, 291*, 140001. https://doi.org/10.1016/j.ijbiomac.2025.140001.

Jiang, C., Liu, T., Wang, S., Zou, Y., Cao, J., Wang, C., Hang, C., & Jin, L. (2023). Antioxidant and ammonia-sensitive films based on starch, κ-carrageenan and Oxalis triangularis extract as visual indicator of beef meat spoilage. *International Journal of Biological Macromolecules, 235*, 123698. https://doi.org/10.1016/j.ijbiomac.2023.123698.

Jiang, C., Liu, G., Zhang, Q., Wang, S., & Zou, Y. (2024). Preparation and characterization of intelligent packaging labels based on pea starch, κ-carrageenan and black raspberry extract for monitoring freshness of pork. *Food Chem. X, 24*, 101850.

Jiang, G., Yang, Y., Sheng, W. Y., Yang, L., Yang, H., Tang, T., Wang, C., & Tian, Y. (2024). Preparation and characterization of κ-carrageenan/dextran films blended with nano-ZnO and anthocyanin for intelligent food packaging. *International Journal of Biological Macromolecules, 282*, 137203.

Jiao, Y., Liu, T., Zhou, S., & Xu, Y. (2024). Evaluation of antioxidant and intelligent films for pork packaging and spoilage monitoring based on maize starch, carrageenan gum, and Rosa chinensis flower extracts. *Journal of Food Measurement and Characterization, 18*(9), 7455–7467. https://doi.org/10.1007/s11694-024-02740-2.

Joye, I. J., & Julian McClements, D. (2016). Biopolymer-based delivery systems: Challenges and opportunities. *Current Topics in Medicinal Chemistry, 16*(9), 1026–1039. https://doi.org/10.2174/1568026615666150825143130, http://www.eurekaselect.com/615/journal/current-topics-medicinal-chemistry.

Karimzadeh, Z., Fallah, A. A., Habibian Dehkordi, S., Shafiei, S., & Ghasemi, M. (2024). The active kappa-carrageenan coating and oxygen absorber packaging for the preservation of chilled rainbow trout fillets. *International Journal of Biological Macromolecules, 283*(Pt 1), 137572. https://doi.org/10.1016/j.ijbiomac.2024.137572.

Kassab, Z., Aziz, F., Hannache, H., Ben Youcef, H., & El Achaby, M. (2019). Improved mechanical properties of k-carrageenan-based nanocomposite films reinforced with cellulose nanocrystals. *International Journal of Biological Macromolecules, 123*, 1248–1256. https://doi.org/10.1016/j.ijbiomac.2018.12.030, www.elsevier.com/locate/ijbiomac.

Khalil, H. P. S., Lai, T. K., Tye, Y. Y., Rizal, S., Chong, E. W. N., Yap, S. W., Hamzah, A. A., Fazita, M. R., & Paridah, M. T. (2018). A review of extractions of seaweed hydrocolloids: Properties and applications. *Express Polymer Letters, 12*.

Khan, A., Priyadarshi, R., Bhattacharya, T., & Rhim, J.-W. (2023a). Carrageenan/alginate-based functional films incorporated with Allium sativum carbon dots for UV-barrier food packaging. *Food and Bioprocess Technology, 16*, 2001–2015.

Khan, A., Riahi, Z., Kim, J. T., & Rhim, J.-W. (2023b). Gelatin/carrageenan-based smart packaging film integrated with Cu-metal organic framework for freshness monitoring and shelf-life extension of shrimp. Food Hydrocolloids, *145*, 109180.

Khan, A., Riahi, Z., Tae Kim, J., & Rhim, J. W. (2024). Carrageenan-based multifunctional packaging films containing Zn-carbon dots/anthocyanin derived from Kohlrabi peel for monitoring quality and extending the shelf life of shrimps. *Food Chemistry, 432*. https://doi.org/10.1016/j.foodchem.2023.137215, www.elsevier.com/locate/foodchem.

Khojah, S. M. (2020). Bio-based coating from fish gelatin, K-carrageenan and extract of pomegranate peels for maintaining the overall qualities of fish fillet. *Journal of Aquatic Food Product Technology, 29*(8), 810–822. https://doi.org/10.1080/10498850.2020.1718261, http://www.tandfonline.com/toc/wafp20/current.

Kim, S., Kim, B. S., Bai, J., & Chang, Y. (2023). Antibacterial κ-carrageenan/konjac glucomannan-based edible hydrogel film containing *Salmonella* phage PBSE191 and its application in chicken meat. *LWT, 180*, 114707. https://doi.org/10.1016/j.lwt.2023.114707.

Kim, Y. H., Bang, Y. J., Yoon, K. S., Priyadarshi, R., & Rhim, J. W. (2022). Pine needle (*Pinus densiflora*) extract-mediated synthesis of silver nanoparticles and the preparation of carrageenan-based antimicrobial packaging films. *Journal of Nanomaterials, 2022*. https://doi.org/10.1155/2022/8395302, http://www.hindawi.com/journals/jnm/.

Koca, N., & Bayramoğlu, B. (2022). Layer-by-layer assembly of lysozyme with iota-carrageenan and gum Arabic for surface modification of food packaging materials with improved barrier properties. *Colloids and Surfaces A: Physicochemical and Engineering Aspects, 639*, 128391. https://doi.org/10.1016/j.colsurfa.2022.128391.

Kouser, F., Kumar, S., Bhat, H. F., Hassoun, A., Bekhit, A. E.-D. A., & Bhat, Z. F. (2023). Aloe barbadensis based bioactive edible film improved lipid stability and microbial quality of the cheese. *Foods, 12*(2), 229. https://doi.org/10.3390/foods12020229.

Krishnan, L., Ravi, N., Mondal, A. K., Aktar, F., Kumar, P., Ralph, P. J., & Kuzhiumparambil, U. (2024). Seaweed-based polysaccharides: Review of extraction, characterization, and bioplastic application.

Kumar, S., Reddy, A. R. L., Basumatary, I. B., Nayak, A., Dutta, D., Konwar, J., Purkayastha, M. D., & Mukherjee, A. (2023). Recent progress in pectin extraction and their applications in developing films and coatings for sustainable food packaging: A review. *International Journal of Biological Macromolecules, 239*. https://doi.org/10.1016/j.ijbiomac.2023.124281, www.elsevier.com/locate/ijbiomac.

Kumari, S., Kumari, A., Ahmed, J., Jasrotia, R., Sillanpää, M., Lakshmaiya, N., Kondal, N., Kandwal, A., & Sharma, R. (2024). Enhancing UV protection and antimicrobial properties in food packaging through the use of copper nanoparticles and κ-carrageenan based nanocomposite film. *Journal of Inorganic and Organometallic Polymers and Materials, 34*(11), 5538–5550. https://doi.org/10.1007/s10904-024-03231-z, https://www.springer.com/journal/10904.

Lai, D., Zhou, F., Zhou, A., Hamzah, S. S., Zhang, Y., Hu, J., & Lin, S. (2022). Comprehensive properties of photodynamic antibacterial film based on κ-Carrageenan and curcumin-β-cyclodextrin complex. *Carbohydrate Polymers, 282*, 119112. https://doi.org/10.1016/j.carbpol.2022.119112.

Lakkannavar, V., Yogesha, K. B., Prasad, C. D., Phanden, R. K., Srinivasa, G., & Prasad, S. C. (2024). Thermal spray coatings on high-temperature oxidation and corrosion applications – A comprehensive review. *Results in Surfaces and Interfaces, 16*, 100250. https://doi.org/10.1016/j.rsurfi.2024.100250.

Lakshmi Balasubramaniam, S., Singh Patel, A., & Nayak, B. (2021). Fabrication of antioxidative food packaging films using cellulose nanofibers, kappa-carrageenan, and gallic acid. *Journal of Food Processing and Preservation, 45*, e15480.

Leya, B., Nivetha, T. U., Freeda Blessie, R., & Pragalyaashree, M. M. (2024). Edible coating deposition methods: Dipping, spraying, fluidized bed, and panning. In *Food coatings and preservation technologies*. India: wiley, 485–514. https://onlinelibrary.wiley.com/doi/book/10.1002/9781394237623.

Li, F., Liu, Y., Cao, Y., Zhang, Y., Zhe, T., Guo, Z., Sun, X., Wang, Q., & Wang, Li (2020). Copper sulfide nanoparticle-carrageenan films for packaging application. *Food Hydrocolloids, 109*, 106094. https://doi.org/10.1016/j.foodhyd.2020.106094.

Li, F., Zhu, S., Du, Y., Zhe, T., Ma, K., Liu, M., & Wang, L. (2024). Carrageenan/polyvinyl alcohol composite film reinforced with spermidine carbon dots: An active packaging material with dual-mode antibacterial activity. *International Journal of Biological Macromolecules, 266*, 131343.

Li, L., Ni, R., Shao, Y., & Mao, S. (2014). Carrageenan and its applications in drug delivery. *Carbohydrate Polymers, 103*(1), 1–11. https://doi.org/10.1016/j.carbpol.2013.12.008.

Li, X., Fan, Y., Guo, J., Li, R., Liu, Z., Hou, Y., Qu, Z., & Liu, Q. (2024). Polyvinyl alcohol/kappa-carrageenan-based package film with simultaneous incorporation of ferric ion and polyphenols from *Capsicum annuum* leaves for fruit shelf-life extension. *International Journal of Biological Macromolecules, 266*, 131002.

Li, Y., Ni, Y., He, W., Li, H., Zhang, W., Tan, L., Zhao, J., & Xu, B. (2025). Mussel-inspired highly adhesive carrageenan-based coatings with self-activating enhanced activity for meat preservation. *Carbohydrate Polymers, 348*(Pt A), 122840. https://doi.org/10.1016/j.carbpol.2024.122840.

Liang, S., Yuan, X., Han, X., Han, C., Liu, Z., & Liang, M. (2023). Is seeing always good? The influence of organic food packaging transparency on consumers' purchase intentions. *International Journal of Food Science & Technology, 58*(12), 6517–6529. https://doi.org/10.1111/ijfs.16765.

Lin, M. G., Lasekan, O., Saari, N., & Khairunniza-Bejo, S. (2018). Effect of chitosan and carrageenan-based edible coatings on post-harvested longan (*Dimocarpus longan*) fruits. *CyTA - Journal of Food, 16*(1), 490–497. https://doi.org/10.1080/19476337.2017.1414078.

Liu, J., Wang, H., Wang, P., Guo, M., Jiang, S., Li, X., & Jiang, S. (2018). Films based on κ-carrageenan incorporated with curcumin for freshness monitoring. *Food Hydrocolloids, 83*, 134–142. https://doi.org/10.1016/j.foodhyd.2018.05.012.

Liu, F., Duan, G., & Yang, H. (2023). Recent advances in exploiting carrageenans as a versatile functional material for promising biomedical applications. *International Journal of Biological Macromolecules, 235*, 123787.

Liu, F., Hou, P., Zhang, H., Tang, Q., Xue, C., & Li, R. W. (2021). Food-grade carrageenans and their implications in health and disease. *Comprehensive Reviews in Food Science and Food Safety, 20*(4), 3918–3936. https://doi.org/10.1111/1541-4337.12790, http://onlinelibrary.wiley.com/journal/10.1111/(ISSN)1541-4337.

Liu, J., Dong, Y., Ma, Z., Rao, Z., Zheng, X., & Tang, K. (2022). Soluble soybean polysaccharide/carrageenan antibacterial nanocomposite films containing green synthesized silver nanoparticles. *ACS Applied Polymer Materials, 4*(8), 5608–5618. https://doi.org/10.1021/acsapm.2c00635.

Liu, J., Wang, H., Guo, M., Li, L., Chen, M., Jiang, S., Li, X., & Jiang, S. (2019). Extract from *Lycium ruthenicum* Murr. Incorporating κ-carrageenan colorimetric film with a wide pH–sensing range for food freshness monitoring. *Food Hydrocolloids, 94*, 1–10.

Liu, W., Ning, Y., Yun, Y., Wei, N., Pan, Z., & Wang, L. (2024). Development of pH-responsive intelligent films based on κ-carrageenan/straw lignin and anthocyanin from Padus virginiana peel for real-time monitoring of chicken. *International Journal of Biological Macromolecules, 270*(Pt 2), 132464. https://doi.org/10.1016/j.ijbiomac.2024.132464.

Liu, Y., Qin, Y., Bai, R., Zhang, X., Yuan, L., & Liu, J. (2019). Preparation of pH-sensitive and antioxidant packaging films based on $κ$-carrageenan and mulberry polyphenolic extract. *International Journal of Biological Macromolecules, 134*, 993–1001.

Liu, Y., Zhang, X., Li, C., Qin, Y., Xiao, L., & Liu, J. (2020). Comparison of the structural, physical and functional properties of κ-carrageenan films incorporated with pomegranate flesh and peel extracts. *International Journal of Biological Macromolecules, 147*, 1076–1088. https://doi.org/10.1016/j.ijbiomac.2019.10.075.

Mahajan, K., Kumar, S., Bhat, Z. F., Naqvi, Z., Mungure, T. E., & Bekhit, A. E. D. A. (2021). Functionalization of carrageenan based edible film using Aloe vera for improved lipid oxidative and microbial stability of frozen dairy products. *Food Bioscience, 43*. https://doi.org/10.1016/j.fbio.2021.101336, http://www.journals.elsevier.com/food-bioscience/.

Mahajan, K., Kumar, S., Bhat, Z. F., Singh, M., Bhat, H. F., Bhatti, M. A., & Bekhit, A. E. D. A. (2022). Aloe vera and carrageenan based edible film improves storage stability of ice-cream. *Applied Food Research, 2*(1). https://doi.org/10.1016/j.afres.2022.100128, https://www.journals.elsevier.com/applied-food-research.

Mahmud, N., Islam, J., & Tahergorabi, R. (2021). Marine biopolymers: Applications in food packaging. *Processes, 9*, 2245.

Malekjani, N., Karimi, R., Assadpour, E., & Jafari, S. M. (2024). Control of release in active packaging/coating for food products; approaches, mechanisms, profiles, and modeling. *Critical Reviews in Food Science and Nutrition, 64*(29), 10789–10811. https://doi.org/10.1080/10408398.2023.2228413, https://www.tandfonline.com/loi/bfsn20.

Manzoor, A., Khan, S., Dar, A. H., Pandey, V. K., Shams, R., Ahmad, S., Jeevarathinam, G., Kumar, M., Singh, P., & Pandiselvam, R. (2023). Recent insights into green antimicrobial packaging towards food safety reinforcement: A review. *Journal of Food Safety, 43*(4). https://doi.org/10.1111/jfs.13046, http://onlinelibrary.wiley.com/journal/10.1111/(ISSN)1745-4565.

Martiny, T. R., Pacheco, B. S., Pereira, C. M. P., Mansilla, A., Astorga–España, M. S., Dotto, G. L., Moraes, C. C., & Rosa, G. S. (2020). A novel biodegradable film based on κ-carrageenan activated with olive leaves extract. *Food Science and Nutrition, 8*(7), 3147–3156. https://doi.org/10.1002/fsn3.1554, onlinelibrary.wiley.com/journal/10.1002/%28ISSN%292048-7177.

Martiny, T. R., Raghavan, V., Moraes, C. C.de, Rosa, G. S.da, & Dotto, G. L. (2020). Bio-based active packaging: Carrageenan film with olive leaf extract for lamb meat preservation. *Foods, 9*, 1759.

Mathew, S. S., Jaiswal, A. K., & Jaiswal, S. (2024). Carrageenan-based sustainable biomaterials for intelligent food packaging: A review. *Carbohydrate Polymers*122267.

McKim, J. M., Willoughby, J. A., Blakemore, W. R., & Weiner, M. L. (2019). Clarifying the confusion between poligeenan, degraded carrageenan, and carrageenan: A review of the chemistry, nomenclature, and in vivo toxicology by the oral route. *Critical Reviews in Food Science and Nutrition, 59*(19), 3054–3073. https://doi.org/10.1080/10408398.2018.1481822, https://www.tandfonline.com/loi/bfsn20.

Mendes, M., Cotas, J., Gutiérrez, I. B., Gonçalves, A. M. M., Critchley, A. T., Hinaloc, L. A. R., Roleda, M. Y., & Pereira, L. (2024). Advanced extraction techniques and physicochemical properties of carrageenan from a novel *Kappaphycus alvarezii* cultivar. *Marine Drugs, 22*(11). https://doi.org/10.3390/md22110491, http://www.mdpi.com/journal/marinedrugs.

Michel, G., Nyval-Collen, P., Barbeyron, T., Czjzek, M., & Helbert, W. (2006). Bioconversion of red seaweed galactans: A focus on bacterial agarases and carrageenases. *Applied Microbiology and Biotechnology, 71*(1), 23–33. https://doi.org/10.1007/s00253-006-0377-7.

Mileriene, J., Serniene, L., Henriques, M., Gomes, D., Pereira, C., Kondrotiene, K., Kasetiene, N., Lauciene, L., Sekmokiene, D., & Malakauskas, M. (2021). Effect of liquid whey protein concentrate–based edible coating enriched with cinnamon carbon dioxide extract on the quality and shelf life of Eastern European curd cheese. *Journal of Dairy Science, 104*(2), 1504–1517. https://doi.org/10.3168/jds.2020-18732.

Mirza Alizadeh, A., Masoomian, M., Shakooie, M., Zabihzadeh Khajavi, M., & Farhoodi, M. (2021). Trends and applications of intelligent packaging in dairy products: A review. *Critical Reviews in Food Science and Nutrition, 62*(2), 383–397. https://doi.org/10.1080/10408398.2020.1817847, https://www.tandfonline.com/loi/bfsn20.

Mohammadi, Y., Shahbazi, Y., & Shavisi, N. (2024). Intelligent locust bean gum-k-carrageenan nanofibrous mats with *Rosa canina* petal anthocyanins and chitosan nanoparticles: Preparation, characterization, and application for monitoring of beef meat freshness. *LWT, 208*, 116707.

Moller, A., Leone, C., Kataria, J., Sidhu, G., Rama, E. N., Kroft, B., Thippareddi, H., & Singh, M. (2023). Effect of a carrageenan/chitosan coating with allyl isothiocyanate on microbial spoilage and quality of chicken breast. *Poultry Science, 102*(3), 102442. https://doi.org/10.1016/j.psj.2022.102442.

Mount III, E. M. (2024). Extrusion processes. In *Applied plastics engineering handbook*. Elsevier, 313–362.

Muflih, M. H., & Syarifuddin, A. (2023). The effect of carrageenan-gluten ratio and garlic essential oil concentration on physical, mechanical, and antimicrobial properties of edible film. *IOP Conference Series: Earth and Environmental Science, 1230*(1), 012165. https://doi.org/10.1088/1755-1315/1230/1/012165.

Muthulakshmi, L., Anand Kumar, B., Rajasekar, A., Annaraj, J., & Pruncu, C. I. (2021). The benefits of k-carrageenan-gelatin hybrid composite coating on the medical grade stainless steel (SS304) used as anticorrosive barrier. *Materials Chemistry and Physics, 258*, 123909. https://doi.org/10.1016/j.matchemphys.2020.123909.

Nair, M. S., Tomar, M., Punia, S., Kukula-Koch, W., & Kumar, M. (2020). Enhancing the functionality of chitosan- and alginate-based active edible coatings/films for the preservation of fruits and vegetables: A review. *International Journal of Biological Macromolecules, 164*, 304–320. https://doi.org/10.1016/j.ijbiomac.2020.07.083, www.elsevier.com/locate/ijbiomac.

Nair, S. S., Trafiałek, J., & Kolanowski, W. (2023). Edible packaging: A technological update for the sustainable future of the food industry. *Applied Sciences, 13*, 8234.

Nguyen, H. T., Boonyaritthongchai, P., Buanong, M., Supapvanich, S., & Wongs-Aree, C. (2021). Chitosan- and κ-carrageenan-based composite coating on dragon fruit (*Hylocereus undatus*) pretreated with plant growth regulators maintains bract chlorophyll and fruit edibility. *Scientia Horticulturae, 281*, 109916. https://doi.org/10.1016/j.scienta.2021.109916.

Nguyen, H. T., Boonyaritthongchai, P., Buanong, M., Supapvanich, S., & Wongs-Aree, C. (2020). Postharvest hot water treatment followed by chitosan- and κ-carrageenan-based composite coating induces the disease resistance and preserves the quality in dragon fruit (*Hylocereus undatus*). *International Journal of Fruit Science, 20*(3), S2030. https://doi.org/10.1080/15538362.2020.1851342, http://www.tandfonline.com/loi/wsfr20.

Nie, J., Chen, D., Ye, J., Lu, Y., & Dai, Z. (2021). Optimization and kinetic modeling of ultrasonic-assisted extraction of fucoxanthin from edible brown algae *Sargassum fusiforme* using green solvents. *Ultrasonics Sonochemistry, 77*, 105671. https://doi.org/10.1016/j.ultsonch.2021.105671.

Nimbkar, S., Auddy, M., Manoj, I., & Shanmugasundaram, S. (2023). Novel techniques for quality evaluation of fish: A review. *Food Reviews International, 39*(1), 639–662. https://doi.org/10.1080/87559129.2021.1925291.

Nouri, A., Tavakkoli Yaraki, M., Lajevardi, A., Rahimi, T., Tanzifi, M., & Ghorbanpour, M. (2020). An investigation of the role of fabrication process in the physicochemical properties of κ-carrageenan-based films incorporated with *Zataria multiflora* extract and nanoclay. *Food Packaging and Shelf Life, 23*. https://doi.org/10.1016/j.fpsl.2019.100435, http://www.journals.elsevier.com/food-packaging-and-shelf-life/.

Park, S.-Y., Kim, H.-L., & Her, J.-Y. (2024). Isolation of microcrystalline cellulose (MCC) from pistachio shells and preparation of carrageenan-based composite films. *Carbohydrate Polymer Technologies and Applications, 7*, 100423.

Patil, V., Shams, R., & Dash, K. K. (2023). Techno-functional characteristics, and potential applications of edible coatings: A comprehensive review. *Journal of Agriculture and Food Research*, 100886.

Pleşoianu, A. M., & Nour, V. (2022). Pectin-based edible coating combined with chemical dips containing antimicrobials and antibrowning agents to maintain quality of fresh-cut pears. *Horticulturae, 8*(5). https://doi.org/10.3390/horticulturae8050449, https://www.mdpi.com/2311-7524/8/5/449/pdf?version=1652781036.

Popescu, M. C., Dogaru, B. I., Sun, D., Stoleru, E., & Simionescu, B. C. (2019). Structural and sorption properties of bio-nanocomposite films based on κ-carrageenan and cellulose nanocrystals. *International Journal of Biological Macromolecules, 135*, 462–471. https://doi.org/10.1016/j.ijbiomac.2019.05.194, www.elsevier.com/locate/ijbiomac.

Praseptiangga, D., Mufida, N., Panatarani, C., & Joni, I. M. (2021). Enhanced multi functionality of semi-refined iota carrageenan as food packaging material by incorporating SiO_2 and ZnO nanoparticles. *Heliyon, 7*(5). https://doi.org/10.1016/j.heliyon.2021.e06963, http://www.journals.elsevier.com/heliyon/.

Priya, K., Thirunavookarasu, N., & Chidanand, D. V. (2023). Recent advances in edible coating of food products and its legislations: A review. *Journal of Agriculture and Food Research*, 100623.

Pruteanu, L. L., Bailey, D. S., Grădinaru, A. C., & Jäntschi, L. (2023). The biochemistry and effectiveness of antioxidants in food, fruits, and marine algae. *Antioxidants, 12*(4). https://doi.org/10.3390/antiox12040860, www.mdpi.com/journal/antioxidants.

Rafiquzzaman, S. M., Ahmed, R., Lee, J. M., Noh, G., Jo, Ga, & Kong, I. S. (2016). Improved methods for isolation of carrageenan from *Hypnea musciformis* and its antioxidant activity. *Journal of Applied Phycology, 28*(2), 1265–1274. https://doi.org/10.1007/s10811-015-0605-6, www.wkap.nl/journalhome.htm/0921-8971.

Ramesh, K., & Chellam, P. V. (2022). Carrageenan-based bionanocomposites for food packaging applications. In *Bionanocomposites for food packaging applications*. Elsevier, 295–322.

Rashid, T. U., Gorga, R. E., & Krause, W. E. (2021). Mechanical properties of electrospun fibers—a critical review. *Advanced Engineering Materials, 23*, 2100153.

Ren, B., Wu, W., Soladoye, O. P., Bak, K. H., Fu, Y., & Zhang, Y. (2021). Application of biopreservatives in meat preservation: A review. *International Journal of Food Science and Technology, 56*(12), 6124–6141. https://doi.org/10.1111/ijfs.15307, http://onlinelibrary.wiley.com/journal/10.1111/(ISSN)1365-2621.

Rong, L., Zhang, T., Ma, Y., Wang, T., Liu, Y., & Wu, Z. (2023). An intelligent label using sodium carboxymethyl cellulose and carrageenan for monitoring the freshness of fresh-cut papaya. *Food Control, 145*, 109420. https://doi.org/10.1016/j.foodcont.2022.109420.

Roy, S., & Rhim, J.-W. (2019a). Preparation of carrageenan-based functional nanocomposite films incorporated with melanin nanoparticles. *Colloids Surfaces B Biointerfaces, 176*, 317–324.

Roy, S., & Rhim, J.-W. (2019b). Carrageenan-based antimicrobial bionanocomposite films incorporated with ZnO nanoparticles stabilized by melanin. Food Hydrocolloids, *90*, 500–507.

Roy, S., & Rhim, J. W. (2020). Fabrication of copper sulfide nanoparticles and limonene incorporated pullulan/carrageenan-based film with improved mechanical and antibacterial properties. *Polymers, 12*(11), 1–14. https://doi.org/10.3390/polym12112665, https://www.mdpi.com/2073-4360/12/11/2665/pdf.

Roy, S., & Rhim, J.-W. (2021a). Carrageenan/agar-based functional film integrated with zinc sulfide nanoparticles and Pickering emulsion of tea tree essential oil for active packaging applications. *International Journal of Biological Macromolecules, 193*, 2038–2046.

Roy, S., & Rhim, J. W. (2021b). Preparation of gelatin/carrageenan-based color-indicator film integrated with shikonin and propolis for smart food packaging applications. *ACS Applied Bio Materials, 4*(1), 770–779. https://doi.org/10.1021/acsabm.0c01353, pubs.acs.org/journal/aabmcb.

Roy, S., Shankar, S., & Rhim, J.-W. (2019). Melanin-mediated synthesis of silver nanoparticle and its use for the preparation of carrageenan-based antibacterial films. Food Hydrocolloids, *88*, 237–246.

Rukmanikrishnan, B., Rajasekharan, S. K., Lee, J., Ramalingam, S., & Lee, J. (2020). K-Carrageenan/lignin composite films: Biofilm inhibition, antioxidant activity, cytocompatibility, UV and water barrier properties. *Materials Today Communications, 24*, 101346. https://doi.org/10.1016/j.mtcomm.2020.101346.

Rupert, R., Rodrigues, K. F., Thien, V. Y., & Yong, W. T. L. (2022). Carrageenan from *Kappaphycus alvarezii* (Rhodophyta, Solieriaceae): Metabolism, structure, production, and application. *Frontiers in Plant Science, 13*. https://doi.org/10.3389/fpls.2022.859635, https://www.frontiersin.org/journals/plant-science.

Saedi, S., & Rhim, J.-W. (2020). Synthesis of Fe_3O_4@ SiO_2@ PAMAM dendrimer@ AgNP hybrid nanoparticles for the preparation of carrageenan-based functional nanocomposite film. *Food Packaging Shelf Life. 24*.

Saedi, S., Shokri, M., & Rhim, J.-W. (2020). Preparation of carrageenan-based nanocomposite films incorporated with functionalized halloysite using AgNP and sodium dodecyl sulfate. Food Hydrocolloids, *106*, 105934.

Sangeetha, U. K., Sudhakaran, N., Parvathy, P. A., Abraham, M., Das, S., De, S., & Sahoo, S. K. (2024). Coconut husk-lignin derived carbon dots incorporated carrageenan based functional film for intelligent food packaging. *International Journal of Biological Macromolecules, 266*(Pt 1), 131005. https://doi.org/10.1016/j.ijbiomac.2024.131005.

Santinon, C., Borges, A., Simões, M., Gonçalves, A. S. C., Beppu, M. M., & Vieira, M. G. A. (2024). Visible-light photoactivated proanthocyanidin and kappa-carrageenan coating with anti-adhesive properties against clinically relevant bacteria. *International Journal of Biological Macromolecules, 263*. https://doi.org/10.1016/j.ijbiomac.2024.130611, https://www.sciencedirect.com/science/journal/01418130.

Santos, C., Ramos, A., Luís, Â., & Amaral, M. E. (2023). Production and characterization of k-carrageenan films incorporating cymbopogon winterianus essential oil as new food packaging materials. *Foods, 12*(11), 2169. https://doi.org/10.3390/foods12112169.

Schefer, L., Usov, I., & Mezzenga, R. (2015). Anomalous stiffening and ion-induced coil–helix transition of carrageenans under monovalent salt conditions. *Biomacromolecules, 16*, 985–991.

Sedayu, B. B., Cran, M. J., & Bigger, S. W. (2018). Characterization of semi-refined carrageenan-based film for primary food packaging purposes. *Journal of Polymers and the Environment, 26*, 3754–3761.

Sedayu, B. B., Cran, M. J., & Bigger, S. W. (2019). A review of property enhancement techniques for carrageenan-based films and coatings. *Carbohydrate Polymers, 216*, 287–302.

Sedayu, B. B., Cran, M. J., & Bigger, S. W. (2021). Effects of surface photocrosslinking on the properties of semi-refined carrageenan film. *Food Hydrocolloids, 111*, 106196.

Sharayei, P., Azarpazhooh, E., Zomorodi, S., Einafshar, S., & Ramaswamy, H. S. (2021). Optimization of ultrasonic-assisted extraction of astaxanthin from green tiger (*Penaeus semisulcatus*) shrimp shell. *Ultrasonics Sonochemistry, 76*, 105666. https://doi.org/10.1016/j.ultsonch.2021.105666.

Sharma, G. K., & James, N. R. (2022). Electrospinning: The technique and applications. In Recent developments in nanofibers research.

Shavisi, N. (2024). Electrospun fiber mats based on chitosan-carrageenan containing Malva sylvestris anthocyanins: Physic-mechanical, thermal, and barrier properties along with application as intelligent food packaging materials. *International Journal of Biological Macromolecules, 266*, 131077. https://doi.org/10.1016/j.ijbiomac.2024.131077.

Sheng, L., & Wang, L. (2021). The microbial safety of fish and fish products: Recent advances in understanding its significance, contamination sources, and control strategies. *Comprehensive Reviews in Food Science and Food Safety, 20*(1), 738–786. https://doi.org/10.1111/1541-4337.12671.

Shi, S., Lan, X., Ding, X., Han, X., Sun, J., Wang, J., & Duan, J. (2024). Metal-phenolic networks spontaneously reinforced carrageenan-based packaging films with antibacterial and antioxidant properties. *International Journal of Biological Macromolecules, 279*(Pt 2), 135143. https://doi.org/10.1016/j.ijbiomac.2024.135143.

Shigrekar, M., & Amdoskar, V. (2024). A review on recent progress and techniques used for fabricating superhydrophobic coatings derived from biobased materials. *RSC Advances, 14*(44), 32668–32699. https://doi.org/10.1039/d4ra04767b.

Silva, S. P. M., Ribeiro, S. C., Teixeira, J. A., & Silva, C. C. G. (2022). Application of an alginate-based edible coating with bacteriocin-producing *Lactococcus* strains in fresh cheese preservation. *LWT, 153*. https://doi.org/10.1016/j.lwt.2021.112486, https://www.journals.elsevier.com/lwt.

Sogut, E., Filiz, B. E., & Seydim, A. C. (2022). Whey protein isolate-and carrageenan-based edible films as carriers of different probiotic bacteria. *Journal of Dairy Science, 105*, 4829–4842.

Song, D. H., Hoa, V. B., Kim, H. W., Khang, S. M., Cho, S. H., Ham, J. S., & Seol, K. H. (2021). Edible films on meat and meat products. *Coatings, 11*(11). https://doi.org/10.3390/coatings11111344, https://www.mdpi.com/2079-6412/11/11/1344/pdf.

Suhag, R., Kumar, N., Petkoska, A. T., & Upadhyay, A. (2020). Film formation and deposition methods of edible coating on food products: A review. *Food Research International, 136*, 109582. https://doi.org/10.1016/j.foodres.2020.109582.

Sultan, R. A., Rahman, A. N. F., Dirpan, A., & Syarifuddin, A. (2023). Physical, mechanical, barrier, antibacterial properties, and functional group of carrageenan-based edible film as influenced by pectin from *Dillenia serrata* fruit peel and curcumin. *Current Research in Nutrition and Food Science, 11*(3), 1308–1321. https://doi.org/10.12944/CRNFSJ.11.3.32, www.foodandnutritionjournal.org/.

Sun, G., Chi, W., Xu, S., & Wang, L. (2020). Developing a simultaneously antioxidant and pH-responsive κ-carrageenan/hydroxypropyl methylcellulose film blended with *Prunus maackii* extract. *International Journal of Biological Macromolecules, 155*, 1393–1400. https://doi.org/10.1016/j.ijbiomac.2019.11.114.

Sun, H., Ji, T., Zhou, X., Bi, H., Xu, M., Huang, Z., & Cai, L. (2021). Mechanically strong, transparent, and biodegradable wood-derived film. *Materials Chemistry Frontiers, 5*(21), 7903–7909. https://doi.org/10.1039/d1qm00973g.

Sunil Kumar, B. T., Reddy, J. P., Vanajakshi, V., Dasalkar, A. H., Yannam, S. K., Hebbar, U. H., & Singh, S. A. (2025). Development and characterization of carrageenan-based antibacterial films incorporated with natural melanin pigment from niger seed hulls (Guizotia abyssinica) and their efficacy to enhance the shelf-life of strawberries. *Food Control, 174*, 111235. https://doi.org/10.1016/j.foodcont.2025.111235.

Sun, K.-qiang, Li, F.-yi, Li, J.-yong, Li, J.-feng, Zhang, C.-wei, Chen, S., Sun, X., & Cui, J.-feng (2019). Optimisation of compatibility for improving elongation at break of chitosan/starch films. *RSC Advances, 9*(42), 24451–24459. https://doi.org/10.1039/c9ra04053f.

Tabassum, A., Babu, H. S., Ahmed, T., Naik, K. P., Bhumika, D., Jeevan Prasad Reddy, P. R., Prabhakar, C. S., Vivek Babu, K., & Varaprasad, P. S. (2024). Keshava Murthy, development of antibacterial edible films for food packaging using tragacanth gum, carrageenan, and clove essential oil. *Journal of Applied Polymer Science, 141*(24). https://doi.org/10.1002/app.55495, http://onlinelibrary.wiley.com/journal/10.1002/(ISSN)1097-4628.

Tarman, K., Sadi, U., Santoso, J., & Hardjito, L. (2020). Carrageenan and its enzymatic extraction. *Encyclopedia of Marine Biotechnology, 1*, 147–159.

Torres, M. D., Chenlo, F., & Moreira, R. (2017). Thermal reversibility of kappa/iota-hybrid carrageenan gels extracted from *Mastocarpus stellatus* at different ionic strengths. *Journal of the Taiwan Institute of Chemical Engineers, 71*, 414–420.

Udo, T., Mummaleti, G., Mohan, A., Singh, R. K., & Kong, F. (2023). Current and emerging applications of carrageenan in the food industry. *Food Research International, 173*(Pt 2), 113369. https://doi.org/10.1016/j.foodres.2023.113369.

Vázquez-Delfín, E., Robledo, D., & Freile-Pelegrín, Y. (2014). Microwave-assisted extraction of the carrageenan from *Hypnea musciformis* (Cystocloniaceae, Rhodophyta). *Journal of Applied Phycology, 26*, 901–907.

Versino, F., Ortega, F., Monroy, Y., Rivero, S., López, O. V., & García, M. A. (2023). Sustainable and bio-based food packaging: A review on past and current design innovations. *Foods, 12*(5). https://doi.org/10.3390/foods12051057, www.mdpi.com/journal/foods.

Volpe, M. G., Coccia, E., Siano, F., Di Stasio, M., & Paolucci, M. (2019). Rapid evaluation methods for quality of trout (Oncorhynchus mykiss) fresh fillet preserved in an active edible coating. *Foods, 8*(4), 113. https://doi.org/10.3390/foods8040113.

Vu, Q. M., Nguyen, T. C., Dam, D. M. N., Vu, Q. T., Le, T. L., Hoang, T. D., Tran, T. K. N., Nguyen, T. A., Nguyen, P. H., & Thai, H. (2021). A novel method for preparation of carrageenan/fish scale collagen/allopurinol biocomposite film. *International Journal of Polymer Science, 2021*. https://doi.org/10.1155/2021/9960233, http://www.hindawi.com/journals/ijps/.

Wagh, R. V., Riahi, Z., Kim, J. T., & Rhim, J. W. (2024). Carrageenan-based functional films hybridized with carbon dots and anthocyanins from rose petals for smart food packaging applications. *International Journal of Biological Macromolecules, 272*. https://doi.org/10.1016/j.ijbiomac.2024.132817, https://www.sciencedirect.com/science/journal/01418130.

Wang, C., Wang, J., Zeng, L., Qiao, Z., Liu, X., Liu, H., Zhang, J., & Ding, J. (2019). Fabrication of electrospun polymer nanofibers with diverse morphologies. *Molecules (Basel, Switzerland), 24*(5), 834. https://doi.org/10.3390/molecules24050834.

Wang, G., Liu, L., Peng, F., Ma, Y., Deng, Z., & Li, H. (2024). Natural antioxidants enhance the oxidation stability of blended oils enriched in unsaturated fatty acids. *Journal of the Science of Food and Agriculture, 104*(5), 2907–2916. https://doi.org/10.1002/jsfa.13183.

Wang, R., Zhang, S., Liu, S., Sun, Y., & Xu, H. (2023). A contribution to improve barrier properties and reduce swelling ratio of κ-carrageenan film from the incorporation of guar gum or locust bean gum. *Polymers, 15*(7), 1751. https://doi.org/10.3390/polym15071751.

Wang, X., Guo, C., & Guo, H. (2024). Progress of carrageenan-based films and coatings for food packaging applications. *Packaging Technology and Science*.

Wang, Z., Huang, J., Yun, D., Yong, H., & Liu, J. (2022). Antioxidant packaging films developed based on chitosan grafted with different catechins: Characterization and application in retarding corn oil oxidation. *Food Hydrocolloids, 133*, 107970. https://doi.org/10.1016/j.foodhyd.2022.107970.

Wani, S. M., Gull, A., Ahad, T., Malik, A. R., Ganaie, T. A., Masoodi, F. A., & Gani, A. (2021). Effect of gum Arabic, xanthan and carrageenan coatings containing antimicrobial agent on postharvest quality of strawberry: Assessing the physicochemical, enzyme activity and bioactive properties. *International Journal of Biological Macromolecules, 183*, 2100–2108. https://doi.org/10.1016/j.ijbiomac.2021.06.008, www.elsevier.com/locate/ijbiomac.

Wan Yahaya, W. A., Azman, N. A., M. Krishnnan, P., Adam, F., & Almajano, M. P. (2024). Thermo-mechanical and antioxidant properties of eugenol-loaded carrageenan-cellulose nanofiber films for sustainable packaging applications. *Journal of Applied Polymer Science, 141*.

Wei, M., Shan, M., Zhang, L., Chen, N., Tie, H., Xiao, Y., & Li, Z. (2024). Preparation of gelatin/κ-carrageenan active films through procyanidins crosslinking: Physicochemical, structural, antioxidant and controlled release properties. *Food Hydrocolloids, 153*, 110023. https://doi.org/10.1016/j.foodhyd.2024.110023.

Witek-Krowiak, A., Szopa, D., & Anwajler, B. (2024). Advanced packaging techniques—A mini-review of 3D printing potential. *Materials (Basel), 17*, 2997.

Wu, X., Zhou, J., Liu, Z., Liu, J., He, S., & Shao, W. (2023). Constructing a biodegradable carrageenan based food packaging film according to the synergistic strategies between peppermint essential oil and thymol. *International Journal of Biological Macromolecules, 253*(Pt 8), 127537. https://doi.org/10.1016/j.ijbiomac.2023.127537.

Yap, X. Y., Khalid, M., Raju, G., Gew, L. T., & Yow, Y. Y. (2024). Synergistic effects of starch and carrageenan from *Kappaphycus alvarezii* in composite film formation: Physicochemical and degradable properties. *International Journal of Biological Macromolecules, 278*. https://doi.org/10.1016/j.ijbiomac.2024.135205, https://www.sciencedirect.com/science/journal/01418130.

Yin, C., Sun, Z., Yang, Y., Cui, M., Zheng, J., & Zhang, Y. (2024). Rapid in situ formation of κ-carrageenan-carboxymethyl chitosan-kaolin clay hydrogel films enriched with arbutin for enhanced preservation of cherry tomatoes. *International Journal of Biological Macromolecules, 273*, 132957. https://doi.org/10.1016/j.ijbiomac.2024.132957.

Younes, M., Aggett, P., Aguilar, F., Crebelli, R., Filipič, M., Frutos, M. J., Galtier, P., Gott, D., Gundert-Remy, U., Kuhnle, G. G., Lambré, C., Leblanc, J. C., Lillegaard, I. T., Moldeus, P., Mortensen, A., Oskarsson, A., Stankovic, I., Waalkens-Berendsen, I., Woutersen, R. A., ... Dusemund, B. (2018). Re-evaluation of carrageenan (E 407) and processed Eucheuma seaweed (E 407a) as food additives. *EFSA Journal, 16*(4). https://doi.org/10.2903/j.efsa.2018.5238, efsa.onlinelibrary.wiley.com/journal/18314732/.

Youssouf, L., Lallemand, L., Giraud, P., Soulé, F., Bhaw-Luximon, A., Meilhac, O., D'Hellencourt, C. L., Jhurry, D., & Couprie, J. (2017). Ultrasound-assisted extraction and structural characterization by NMR of alginates and carrageenans from seaweeds. *Carbohydrate Polymers, 166*, 55–63. https://doi.org/10.1016/j.carbpol.2017.01.041, http://www.elsevier.com/wps/find/journaldescription.cws_home/405871/description#description.

Zena, Y., Periyasamy, S., Tesfaye, M., Tumssa, Z., Mohamed, B. A., Karthik, V., Asaithambi, P., Getachew, D., & Aminabhavi, T. M. (2024). Trends on barrier characteristics improvement of emerging biopolymeric composite films using nanoparticles-A review. *Journal of the Taiwan Institute of Chemical Engineers*, 105488.

Zhang, C., Chi, W., Meng, F., & Wang, L. (2021a). Fabricating an anti-shrinking κ-carrageenan/sodium carboxymethyl starch film by incorporating carboxylated cellulose nanofibrils for fruit preservation. *International Journal of Biological Macromolecules, 191*, 706–713.

Zhang, C., Sun, G., Cao, L., & Wang, L. (2020). Accurately intelligent film made from sodium carboxymethyl starch/κ-carrageenan reinforced by mulberry anthocyanins as an indicator. *Food Hydrocolloids, 108*, 106012. https://doi.org/10.1016/j.foodhyd.2020.106012.

Zhang, C., Sun, G., Li, J., & Wang, L. (2021b). A green strategy for maintaining intelligent response and improving antioxidant properties of κ-carrageenan-based film via cork bark extractive addition. Food Hydrocolloids, *113*, 106470.

Zhang, H.-qiang, Zhou, J., Wang, H., Lan, D., Peng, Y., Li, Z., & Dai, F. (2023). Antimicrobial dual ionic-covalent k-carrageenan/carboxymethyl chitosan membranes with robust moisture retention property. *Polymers for Advanced Technologies, 34*(2), 691–701. https://doi.org/10.1002/pat.5920.

Zhang, L., Wang, W., Ni, Y., Yang, C., Jin, X., Wang, Y., yang, Y., Jin, Y., Sun, J., & Wang, J. (2022). ZnO/C-mediated k-carrageenan based pseudo-pasteurization films for kumquat preservation. *Food Hydrocolloids, 128*, 107582. https://doi.org/10.1016/j.foodhyd.2022.107582.

Zhang, L., Yu, D., Regenstein, J. M., Xia, W., & Dong, J. (2021). A comprehensive review on natural bioactive films with controlled release characteristics and their applications in foods and pharmaceuticals. *Trends in Food Science and Technology, 112*, 690–707. https://doi.org/10.1016/j.tifs.2021.03.053, http://www.elsevier.com/wps/find/journaldescription.cws_home/601278/description#description.

Zhang, M., Zhang, M., Zhao, Z., Zhu, J., Wan, X., Lv, Y., Tang, C., & Xu, B. (2024). Preparation and characterization of intelligent and active bi-layer film based on carrageenan/pectin for monitoring freshness of salmon. *International Journal of Biological Macromolecules, 276*, 133769.

Zhang, R., Han, Y., McClements, D. J., Xu, D., & Chen, S. (2022). Production, characterization, delivery, and cholesterol-lowering mechanism of phytosterols: A review. *Journal of Agricultural and Food Chemistry, 70*(8), 2483–2494. https://doi.org/10.1021/acs.jafc.1c07390, http://pubs.acs.org/journal/jafcau.

Zhang, S., Yi, J., Yuan, X., Zhang, Z., Shan, Z., & Wang, H. (2024). Fabrication and characterization of carrageenan-based multifunctional films integrated with gallic acid@ZIF-8 for beef preservation. *International Journal of Biological Macromolecules, 274*, 133319.

Zhang, X., Zhang, B., Mao, R., Huang, Z., Jing, K., Jin, C., Yang, B., Qi, J., Yu, M., & Xiong, G. (2024). A novel multilayer film based on sodium alginate/k-carrageenan-gelatin incorporated with ZnO nanoparticles and oregano essential oil for active food packing. *Progress in Organic Coatings, 187*, 108170.

Zhou, M., Han, Y., McClements, D. J., Cheng, C., & Chen, S. (2024). Co-encapsulation of anthocyanin and cinnamaldehyde in nanoparticle-filled carrageenan films: Fabrication, characterization, and active packaging applications. *Food Hydrocolloids, 149*. https://doi.org/10.1016/j.foodhyd.2023.109609, https://www.sciencedirect.com/science/journal/0268005X.

Zhou, X., Zong, X., Zhang, M., Ge, Q., Qi, J., Liang, J., Xu, X., & Xiong, G. (2021). Effect of konjac glucomannan/carrageenan-based edible emulsion coatings with camellia oil on quality and shelf-life of chicken meat. *International Journal of Biological Macromolecules, 183*, 331–339. https://doi.org/10.1016/j.ijbiomac.2021.04.165.

Zhuikova, Y. V., Zhuikov, V. A., Zubareva, A. A., Akhmedova, S. A., Sviridova, I. K., Sergeeva, N. S., & Varlamov, V. P. (2020). Physicochemical and biological characteristics of chitosan/κ-carrageenan thin layer-by-layer films for surface modification of nitinol. *Micron (Oxford, England: 1993), 138*, 102922. https://doi.org/10.1016/j.micron.2020.102922.

Chapter 5

Agar films and coatings

Karthik Singirala, Prabhjot Kaur, and Swarup Roy
Department of Food Technology and Nutrition, Lovely Professional University, Phagwara, Punjab, India

5.1 Introduction

In recent years, the health and environmental effects of plastic packaging have become more apparent, demonstrating the need for sustainable alternatives. Plastic-based food packaging materials are widely used because of their superior barrier properties, low cost, and durability. However, their applications in packaging have caused significant concerns, specifically about their nonbiodegradability and subsequent pollution. Their extensive use has raised significant concerns, particularly regarding their nonbiodegradability and associated pollution. Every year, millions of tons of plastic waste end up in landfills and oceans, contributing to environmental degradation and harming ecosystems (Cazón & Vázquez, 2020; Zhang et al., 2023). Furthermore, packaging materials made up of plastic often lead to the release of harmful chemicals into food products, such as phthalates and bisphenol A (BPA), which can cause serious health hazards, such as cancer, endocrine disruption, and developmental abnormalities (Priyadarshi et al., 2022; Yuan et al., 2020). These problems have intensified the demand for sustainable packaging solutions. Among various solutions being investigated, natural polymers made from renewable resources have gained significant attention (Roy & Rhim, 2022; Zhang et al., 2023). Agar, a polysaccharide extracted from red seaweed, has emerged as a promising solution for developing biodegradable and edible food packaging films. Agar is known for its excellent film-forming ability, nontoxic nature, transparency, and biodegradability (Mostafavi & Zaeim, 2020). It has been widely utilized in various industries, including food, pharmaceuticals, and biotechnology, for its unique properties (Madera-Santana et al., 2014; Mostafavi & Zaeim, 2020).

The use of agar in packaging is consistent with the global focus on decreasing environmental damage caused by plastic waste. Agar is a renewable and biodegradable material, which makes it environmentally friendly, and can help to reduce the effects of polymeric plastics on the environment (Roy and Rhim 2022; Bang et al., 2022). The FDA has classified agar as generally recognized as safe (GRAS), affirming its food safety, thereby making it ideal for use in food contact surfaces. The agar-based films under this current study are biodegradable and can be flexible and transparent; hence, they can be used in packaging various food products like vegetables, meats, and bakery products, among others (Dora, Ghosh et al., 2020; Dora, Singh, et al., 2020; Zhang et al., 2023).

Agar-based films are typically brittle, have low mechanical strength, and exhibit poor moisture barrier properties (Arham et al., 2016; Roy & Rhim, 2022). Such constraints have, however, limited their use in the general food packaging industry. Subsequently, in the last few years, significant efforts have been made to overcome these problems by altering the nature and morphology of agar-based films. Some of them are the use of plasticizers, which consist of glycerol and sorbitol, blending of agar with other biopolymers to improve the mechanical properties of the films, and the addition of bioactive functional components (Madera-Santana et al., 2014; Mostafavi & Zaeim, 2020).

Plasticizers are necessary for minimizing the brittleness of agar films by breaking intermolecular hydrogen bonding, which enhances flexibility and elongation at break (EAB) (Mostafavi & Zaeim, 2020; Uranga et al., 2020). Studies have shown that glycerol and sorbitol are excellent plasticizers, considerably increasing the elasticity of agar-based films (Arham et al., 2016; Madera-Santana et al., 2014). In contrast, biopolymer blends overcome mechanical limits by forming composite films with complementary characteristics. For example, combining agar with chitosan or lignin increases tensile strength (TS) while also adding functions like antibacterial activity and UV resistance (Giménez et al., 2013; Roy & Rhim, 2022). These improvements have increased the potential uses of agar-based films in active and smart packaging, where features such as pH sensitivity and antioxidant characteristics are used (Roy and Rhim, 2019; Zhang & Rhim, 2022).

Natural Polymer-Based Films and Coatings in Food Technology. DOI: https://doi.org/10.1016/B978-0-443-34244-8.00010-0

Another crucial area for improvement is to enhance the barrier qualities of agar films. Pure agar films often exhibit high water vapor permeability (WVP) and low moisture resistance, restricting their application in high-humidity situations (Roy & Rhim, 2022; Wang, Liu, et al., 2018). Researchers investigated the use of cross-linkers to improve the molecular structure of agar films while decreasing their permeability. Incorporating fillers such as nano-bacterial cellulose or essential oils has also shown potential in enhancing water resistance and generating films with antibacterial and antioxidant properties (Belay et al., 2020; Zhang et al., 2023). These improvements not only improve the actual performance of agar-based films but also provide new possibilities for their application in preserving and quality assurance of perishable foods.

Beyond improving the mechanical and barrier properties of agar-based films, recent research has also focused on enhancing their functional attributes to meet the evolving demands of modern food packaging. One significant direction is the development of active and smart packaging systems, where agar's versatility allows it to serve as a functional carrier of bioactive agents.

Active packaging incorporates components that interact with the food or its environment to extend shelf life (Giménez et al., 2013; Zhang & Rhim, 2022). For instance, agar-based films embedded with essential oils or antioxidants have demonstrated strong antibacterial and antioxidant activity, making them suitable for packaging fresh produce and other perishable items (Giménez et al., 2013; Roy and Rhim, 2019). Smart packaging, on the other hand, aims to provide real-time information regarding the condition of the packed food. pH-sensitive agar films, which change color in response to variations in food freshness, are an example of this application that is gaining popularity in the food sector (Roy & Rhim, 2022; Zhang & Rhim, 2022).

Despite these developments, large-scale commercialization of agar-based films presents economic and technological obstacles. The extraction and refining of agar from red seaweed can be costly, especially when manufacturing high-quality films with consistent attributes (Roy and Rhim, 2021a,b; Zhang et al., 2023). Furthermore, the mechanical characteristics of agar films must be improved to compete with traditional plastics in terms of durability and barrier performance (Mostafavi & Zaeim, 2020; Roy & Rhim, 2022). Addressing these problems needs ongoing innovation in film production processes, such as enhanced extrusion or thermo-compression procedures, as well as the search for cost-effective raw materials and additives (Swarup and Rim 2019; Wang, Liu, et al., 2018).

This chapter aims to provide a detailed examination of agar-based edible films and coatings, focusing on their fabrication, characteristics, and applications. It highlights current advances in increasing agar films' mechanical, barrier, and functional qualities, as well as their potential for active and smart packaging for food preservation. By addressing present problems and research gaps, this review emphasizes the potential for further innovation in the promising field of sustainable food packaging (Mostafavi & Zaeim, 2020; Zhang & Rhim, 2022). Additionally, agar-based films represent an important step in reducing the environmental effects of food packaging while maintaining food safety and quality (Cazón & Vázquez, 2020; Roy & Rhim, 2022).

5.2 Extraction process of agar

The extraction of agar from red seaweed has advanced significantly because of novel and sustainable approaches (Martínez-Sanz et al., 2021). There are several extraction methods used for the extraction of agar, which include traditional methods, modern methods, and combined techniques. Traditional methods often result in low yields and high energy usage. To overcome this, modern methods are used to improve efficiency, yield, and product quality (Chemat et al.,2020). The extraction process of agar is schematically represented in Fig. 5.1.

5.3 Traditional methods

5.3.1 Hot water extraction

Hot water extraction is an effective method for extracting agar from seaweed (Dulanlebit et al., 2023). In one approach, 50 g of dried seaweed powder was soaked in 500 mL of distilled water and heated to 90°C for 2 hours. The agar-rich fluid was then separated from the solid remains by muslin cloth filtration. To improve elastic characteristics, the filtrate was gelled, frozen overnight at −21°C, and then freeze-dried twice (−21°C/25°C). This technique resulted in a high-quality agar-based extract known as HW, which is notable for its increased gel strength (245 ± 10 g cm^{-2}) and moderate extraction yield ($11.9\% \pm 1.4\%$). Such procedures help refine the physical characteristics of agar for a variety of applications (Martínez-Sanz et al., 2021).

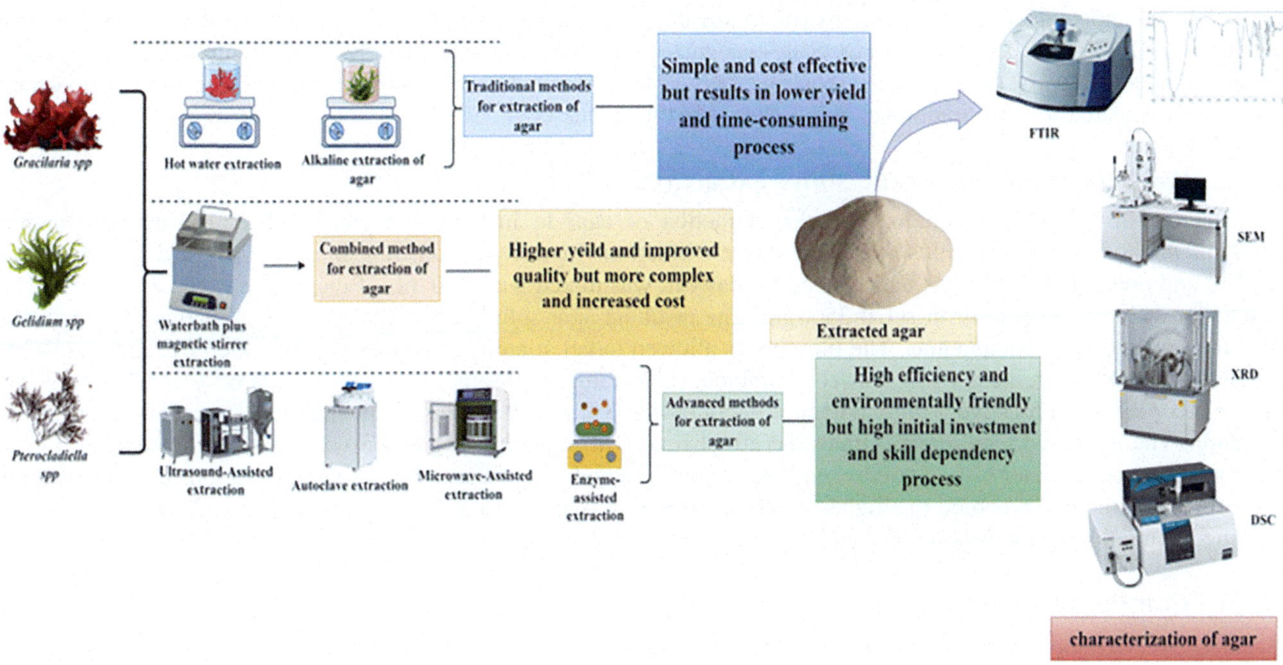

FIGURE 5.1 Extraction and characterization of agar.

5.3.2 Water bath extraction

Water bath extraction of agar is a process of extracting agar from seaweed with hot water, usually under controlled conditions, to maximize production and quality. In a study, 4 g of dried seaweed were measured and pretreated. Agar extraction was performed by boiling the dried seaweed sample in 300 mL of distilled water at 95°C for 6 hours in a water bath. The resultant agar solution was filtered using Whatman filter paper No. 541. The filtered extracts were allowed to gel at room temperature before being frozen at −25°C for 20 hours. After freezing, the gel was thawed at ambient temperature before drying in an oven at 60°C for 20 hours. This method resulted in a notably high extraction yield of 43.84% ± 0.21%. Despite its effectiveness in maximizing agar recovery, the method is considered to have low sustainability due to prolonged processing and energy use (Öğretmen & Duyar, 2018).

5.3.3 Alkaline extraction of agar

The traditional method for alkaline extraction of seaweeds, such as *Gracilaria lemaneiformis*, is an extensive process that includes alkaline, acidic, and bleaching treatments. Initially, seaweed is treated with a 7% NaOH solution at 90°C to eliminate contaminants, then washed repeatedly to achieve a neutral pH. The material is then acidified by using sulfuric acid and a mixed acid solution including EDTA and oxalate, with washing in between. Bleaching is done by using a 0.1% NaClO solution, followed by a thorough rinse. Finally, the treated seaweed is autoclaved at 95°C–102°C with steady stirring before being filtered under pressure. This method eliminates remaining contaminants, resulting in high-purity extracts suitable for subsequent uses. Despite its rigorous and time-intensive nature, taking approximately 12.33 hours, alkaline extraction yields a moderate agar output (12.32% ± 0.05%) with relatively lower gel strength (158.0 ± 1.5 g cm^{-2}) compared to other methods. The overall sustainability of this approach is considered low due to extensive chemical usage and long processing time (Xiao et al.,2019).

5.3.4 Combined methods

5.3.4.1 Combined heating-sonication treatment

A combined heating-sonication technique was used to improve the extraction of agar from seaweed. In this method, dried seaweed powder is soaked in distilled water, heated to 90°C, and then treated with ultrasound (e.g., UP-400S probe, 400 W, 24 kHz) for a set period, usually 30 minutes. The resultant agar solution is then filtered, freeze-thawed, and freeze-dried. This approach, which yields an extract classified as HW-US, is notable for its capacity to enhance extraction efficiency and improve the functional qualities of the agar by combining the effects of heat and ultrasound. The method achieves a moderate extraction yield (10.1% ± 0.1%) and is especially recognized for producing agar with high gel strength (275 ± 10 g cm^{-2}). Moreover, it is relatively time-efficient, requiring only about 30 minutes of sonication, and is considered moderately sustainable compared to more traditional approaches (Martínez-Sanz et al., 2021).

5.3.4.2 Water bath plus magnetic stirrer extraction

Advancement in combined techniques for the extraction of agar to increase the yield and quality includes another important method, water bath plus magnetic stirrer extraction. In this method, 4 g of dried seaweed were carefully weighed and pretreated. To extract agar, the dried seaweed was mixed with 300 mL of distilled water and boiled at 95°C for 6 hours with a magnetic stirrer set at 350 rpm. The resultant agar solution was filtered using Whatman No. 541 filter paper to eliminate any contaminants. The filtrate was allowed to gel at ambient temperature and then frozen at −25°C for 20 hours. After freezing, the gel was thawed at ambient temperature before drying in an oven at 60°C for 20 hours. This method recorded an impressively high extraction yield of 43.53% ± 0.61%, nearly equivalent to conventional water bath extraction. The increased yield demonstrates the effectiveness of integrating agitation with heating. As compared to the traditional method of water bath extraction and the modern method of autoclave extraction, the combined water bath plus magnetic stirrer method recorded the highest yield of extraction (Öğretmen & Duyar, 2018). This indicates the ability of combined extraction techniques to get high yields and improve the quality.

5.3.4.3 Combined sonication–autoclave

The effect of sonication pretreatment on agar extraction from *Eucheuma cottonii* and *Gelidium amansii* seaweeds was investigated using four variables: duration, concentration ratio, frequency, and intensity. The optimal agar production from *E. cottonii* was obtained using a 30-minute sonication period, a 1:20 seaweed-to-solvent ratio, a frequency of 35 Hz, and a power intensity of 30%. Similar conditions obtained the highest agar from *G. amansii.* Sonication had significant effects on agar quality and efficacy, with autoclaved seaweed providing agar that had a reduced sulfate concentration, making it excellent for gel electrophoresis applications. Additionally, autoclaving and sonication improved the gel strength of agar from both seaweeds, exceeding direct heating techniques. This combined technique achieved a remarkably high extraction yield of 49.10% and a gel strength of 840 g cm^{-2}, indicating its superior efficiency and functional performance (Din et al., 2019).

5.3.4.4 Ultrasound-microwave-assisted extraction

Ultrasound-microwave-assisted extraction (UMAE) is a novel technology that improves the extraction of agar from seaweed, notably *Gracilaria species.* This technology combines the advantages of ultrasonic and microwave technologies, resulting in larger yields and greater efficiency than standard extraction methods (Surendhiran et al., 2021). UMAE produced an agar yield of 26.68% and a gel strength of 345.30 ± 4.04 g cm^{-2}, placing it slightly behind microwave-assisted extraction (MAE) in performance. This hybrid approach combines the mechanical effects of ultrasound with the quick heating of microwaves to increase extraction efficiency. However, UMAE did not outperform ultrasound-assisted extraction (UAE) in terms of yield or quality measures, including gel strength, 3,6 AG content, and sulfate content. While UMAE provides a balanced approach, its lesser effectiveness compared to UAE indicates that more modification is needed for this method to compete with solo ultrasound-assisted approaches (Deepika et al., 2024). UMAE significantly increases the yield of agar, with studies showing yields of 12.45% at optimized conditions of 60°C for 60 minutes (Uju et al., 2018). Compared to conventional methods, UMAE reduces extraction time and energy consumption, making it a more sustainable option (Garcia-Vaquero et al., 2020; Uju et al., 2018). The combined effects of ultrasound and microwave energy lead to better extraction of bioactive compounds, enhancing the overall quality of the agar (Kwansang et al., 2022).

5.3.5 Advanced extraction methods

5.3.5.1 Microwave-assisted extraction

MAE is an efficient technique that improves agar yield and quality while reducing processing time. The extraction of carrageenan from red algae (*Gracilaria acerosa*) using microwave-assisted methods was enhanced through a response surface approach. Under ideal conditions, microwave power of 425 W, water-to-raw material ratio of 30 mL g^{-1}, and extraction period of 12 minutes, a maximum yield of 7.99% carrageenan was obtained. Gel permeation chromatography found that the isolated carrageenan had a molecular weight of 230 kDa. The principal monosaccharides discovered by composition analysis were galactose, xylose, mannose, glucose, fructose, and rhamnose. Furthermore, carrageenan exhibited strong antioxidant activity, indicating its potential as a natural component for medicinal and functional food applications (Maleki et al., 2023). MAE produced an agar yield of 27.04%. MAE was less efficient than UAE, but it outperformed traditional approaches. This technique leverages microwaves' rapid heating, enhancing extraction efficiency. However, the agar quality, including gel strength, 3,6 AG concentration, and sulfate content, was lower than that attained in the UAE. Despite these limitations, MAE is a feasible approach for industrial applications needing modest yields with lower energy usage and processing time when compared to previous methods (Deepika et al., 2024).

5.3.5.2 Ultrasound-assisted extraction

UAE is an advanced technique to improve the efficiency and sustainability of agar extraction from seaweed. This technique uses ultrasonic waves to break down cell walls, allowing agar to be released at lower temperatures and faster than other procedures (Ummat et al., 2021). UAE not only enhances agar output and quality, but it also saves energy, making it a viable option for industrial applications. In a study, a comparison between different extraction techniques was analyzed in which UAE produced the greatest agar extraction yield (32.99%), greatly outperforming other techniques. The UAE-extracted agar was of high quality, with a gel strength of (485.50 ± 4.27 g cm^{-2}), 3,6 anhydro-L-galactose (3,6 AG) content of (62.89% ± 2.89%), viscosity of (48.32 ± 0.03) centipoises, and sulfate content of (1.32% ± 0.02%). Under ideal circumstances (30 minutes, 60% amplitude), UAE was shown to be the most efficient green extraction method for obtaining high-quality agar while reducing environmental impact. Because of its efficiency and environmental friendliness, this approach has great potential for long-term manufacturing, notably in the pharmaceutical and beauty sectors (Deepika et al., 2024).

5.3.5.3 Autoclave extraction method

Another modern method of agar extraction is the autoclave extraction method. In this method, 4 g of dried seaweed were weighed and pretreated. Agar was extracted by boiling the seaweed sample (4 g) in 300 mL of distilled water in an autoclave at 110°C for 30 minutes. The resultant agar solution was filtered with Whatman filter paper 541. The filtrates were allowed to gel at room temperature before freezing at −25°C for 20 hours. The frozen gel was then thawed at ambient temperature and dried in a 60°C oven for 20 hours. This method yielded an agar extraction rate of 40.55% ± 1.52%, reflecting its effectiveness in maximizing yield within a relatively short processing time (Öğretmen & Duyar, 2018).

5.3.5.4 Enzyme-assisted extraction of seaweeds

Enzyme-assisted extraction (EAE) of agar from seaweed presents a promising alternative to traditional methods, enhancing yield and quality while minimizing environmental impact. EAE reduces the need for harsh chemicals and long processing times, aligning with sustainable practices. This method can be integrated into biorefinery concepts, promoting total biomass utilization (Terme et al., 2020). In a study on enzymatic extraction, 10 g of seaweed powder were mixed with 1 L of buffer solution in a 2 L reaction vessel. The vessel was supplied with an overhead stirrer and kept in a water bath heated to the ideal temperature for enzyme activity. When the temperature reached the correct level, 1 mL of a 0.1% enzyme solution was added, and the mixture was agitated at 350 rpm for 20 hours. After incubation, the enzyme was deactivated by heating the solution to 100°C for 10 minutes, then immediately chilling it on ice. The mixture was centrifuged at 1500×g for 10 minutes. The supernatant was then collected. The extract was freeze-dried, weighed to estimate yield, and then reconstituted in a determined volume of water for additional analysis. This study demonstrates the efficacy of enzymatic pretreatment in extracting bioactive chemicals from seaweed (Habeebullah et al., 2020). EAE can significantly improve agar yield compared to conventional methods. For instance, using enzyme-assisted techniques, agar yields of 24.7% with improved gel strength (706 g cm^{-2}) were achieved from *G. lemaneiformis* (Xiao et al., 2021).

The various extraction procedures investigated in the above studies represent substantial advances in the long-term production of high-quality agar from seaweed. Traditional techniques, such as hot water extraction and water bath procedures, provide fundamental concepts but are constrained by lower yields and longer processing durations. Innovative techniques, such as combined heating-sonication treatment and water bath with magnetic stirring, have increased efficiency and yield by capitalizing on heat and mechanical force synergy. Advanced methods, including MAE, UAE, and UMAE, improve agar quality and yield while lowering processing time and energy usage. Among them, the UAE stands out as the most efficient green extraction process, producing the largest yield and excellent agar quality while minimizing environmental effects.

Emerging technologies, such as EAE and autoclave-based procedures, offer new opportunities for refining extraction processes, focusing on specific bioactive chemicals, and satisfying industrial needs. These innovative techniques show promise for scalable and sustainable applications, notably in nutraceuticals, cosmetics, and medicines. These techniques emphasize the need to combine new technology with traditional methods to develop eco-friendly solutions that align with global sustainability goals while improving the functional qualities of agar for a variety of applications. Comparison of agar extraction techniques is tabulated in Table 5.1.

TABLE 5.1 Comparison of agar extraction techniques.

Methods	Extraction yield (%)	Gel strength (g cm^{-2})	Processing time	Sustainability	References
Hot water extraction	11.9 ± 1.4	245 ± 10	~2 h + freezing/drying	Low	Martínez-Sanz et al. (2021)
Water bath extraction	43.84 ± 0.21	Not specified	~6 h + freezing/drying	Low	Öğretmen and Duyar (2018)
Alkaline extraction	12.32 ± 0.05	158.0 ± 1.5	~12.33 h	Low	Xiao et al. (2019)
Heating-sonication treatment	10.1 ± 0.1	275 ± 10	~30 min sonication	Moderate	Martínez-Sanz et al. (2021)
Water bath + magnetic stirrer	43.53 ± 0.61	Not specified	~6 h + drying	Moderate	Öğretmen and Duyar (2018)
Sonication–autoclave	49.10	840	~30 min + autoclave	Moderate	Din et al. (2019)
Ultrasound-microwave-assisted extraction	26.68	345.30 ± 4.04	~60 min	High	Uju et al. (2018), Deepika et al. (2024)
Microwave-assisted extraction	27.04	425.30 ± 5.03	~12 min	Moderate–High	Deepika et al. (2024)
Ultrasound-assisted extraction	32.99	485.50 ± 4.27	~30 min	High	Deepika et al. (2024)
Autoclave extraction	40.55 ± 1.52	Not specified	~30 min + drying	Moderate	Öğretmen and Duyar (2018)
Enzyme-assisted extraction	24.7	706	~20 h	High (eco-friendly)	Xiao et al. (2021), Habeebullah et al. (2020)

5.4 Agar-based film/coating properties

Agar-based films have been significantly used for their potential in sustainable and functional food packaging. By incorporation of various compounds, mixes, and nanoparticles, these films' characteristics can be modified, exhibiting their excellent adaptability and nontoxic nature, thereby significantly influencing their use in packaging applications, particularly in the food industry. These films/coatings exhibit excellent mechanical, barrier, thermal, optical, antibacterial, and antioxidant properties, which are shown in Table 5.2.

5.4.1 Mechanical properties

The mechanical properties of agar-based films/coatings are significantly characterized by the incorporation of various biomacromolecules and the extraction process of agar. Films derived from Gelidium species, which contain higher agarose content, typically exhibit superior TS and elasticity compared to those extracted from Gracilaria species. For example, agar-chitosan (AG/CS) films show extensive improvements in TS and EAB, with TS increasing from 35.03 to 48.29 MPa and EAB rising from 14.79% to 22.62% as the content of chitosan increases. The strong hydrogen bonding between agar and chitosan molecules contributes to these enhancements (Huang et al., 2020). The incorporation of halloysite nanotubes (HNTs) into agar-chitosan films (AG/CS/HNT) further increases both TS and EAB, exhibiting the synergistic effects of these nanomaterials (Huang et al., 2020). Similarly, agar-lignin nanoparticle films exhibit enhanced TS and stiffness due to the stabilizing nature of lignin, although EAB is reduced slightly by maintaining a balance between strength and flexibility (Perera et al., 2021). Composite films with other additives, such as agar-melanin nanoparticles, agar-starch, and agar-cellulose nanocrystals, also exhibit enhanced mechanical properties, increased TS and stiffness while maintaining desirable flexibility, depending on the ratio of components (Guo et al., 2021; Martínez-Sanz et al., 2019; Roy & Rhim, 2019).

5.4.2 Barrier properties

The potential barrier properties of agar-based films are crucial for their use in food packaging. Agar films plasticized with sodium citrate and citric acid exhibit lower moisture absorption and WVP than films plasticized with glycerol, which makes them appropriate for managing moisture content (Shikov et al., 2017). Studies have shown that the incorporation of hydrophobic additives, such as essential oils or fatty acids, can significantly reduce WVP by introducing water-resistant domains within the film matrix. Sodium alginate-agar (SA-AG) coatings also provide strong oxygen and moisture barrier properties, significantly delaying lipid oxidation in refrigerated beef storage (Zhang et al., 2021). Agar-starch films reduce WVP and oxygen diffusion, particularly at higher agar concentrations, offering enhanced water resistance (Guo et al., 2021). The addition of HNTs into agar-chitosan films improves water resistance and reduces WVP, while lignin nanoparticle films exhibit enhanced UV and oxygen barrier properties (Huang et al., 2020; Perera et al., 2021). Further research on agar-silver nanoparticle composites shows that these films are suitable for packing perishable items since they reduce WVP and have excellent UV and oxygen barrier properties (Rhim et al., 2013). Agar/konjac glucomannan films also exhibit enhanced hydrophilicity and moisture barrier properties as glucomannan content increases (Qiao et al., 2019). Agar/carboxymethyl cellulose (CMC) films containing essential oils demonstrate improved water-barrier properties, although WVP increases with higher oil concentrations (Abdollahi et al., 2018).

5.4.3 Thermal stability

The thermal stability of agar-based films usually remains intact across most formulations, even with the inclusion of various strengthening agents. Under typical processing applications, agar-chitosan and agar-chitosan-halloysite films exhibit stable thermal degradation profiles, suggesting high performance (Huang et al., 2020). According to Zhang et al. (2021), beef quality can be maintained for up to 16 days in refrigerator storage due to the excellent thermal stability of SA-AG coatings containing ginger essential oil (GEO). The thermal degradation temperature rises when lignin is added to agar films, indicating improved thermal resistance (Perera et al., 2021). However, films produced using EAE methods, which preserve higher sulfate content, tend to have slightly reduced thermal stability compared to alkali-extracted agars. This is because sulfate groups hinder the formation of strong intermolecular interactions, thereby weakening the gel structure and reducing thermal resistance (Martínez-Sanz et al., 2019). Agar-lignin and agar-silver nanoparticle films are two other composites that exhibit enhanced heat stability, improving the packing material's longevity and durability (Shankar & Rhim et al., 2017a). Agar-starch films exhibit better thermal stability with higher agar content, contributing to greater structural integrity (Guo et al., 2021). Incorporating reinforcing agents, such as nanoparticles, has been shown to enhance the thermal properties of agar-based films, making them suitable for applications requiring heat resistance.

TABLE 5.2 Key properties of agar-based films/coatings.

S. no	Biopolymers used	Mechanical	Barrier	Thermal	Optical	Antioxidant	Antimicrobial	References
1.	Agar films with glycerol, sorbitol, and citric acid	Increased flexibility with glycerol and sorbitol; citric acid lowered strength.	Reduced water absorption and WVP with citric acid and sodium citrate; glycerol increased moisture uptake.	No significant impact on thermal stability.	Transparent films: no significant changes reported.			Shikov et al. (2017)
2.	Edible agar/maltodextrin films (A/M)	Tensile strength (TS) and Young's modulus (YM) enhanced with hydrophobic agents; highest TS is A/M-20% beeswax films (BW).	Hydrophobicity increased with all hydrophobic agents; A/M-liquid paraffin films have the highest water contact angle.	Enhanced thermal stability was observed A/M-beewax films; liquid paraffin reduced thermal stability.				Zhang et al., 2019
			Water vapor permeability (WVP) reduced in A/M-10% BW films ($6.86 \times 10^{-13}\ g \cdot m^{-1} \cdot s^{-1} \cdot Pa^{-1}$); beeswax showed the most significant reduction in WVP.					
		Elongation at break increased most significantly in A/M-20% films (240% higher than control).	Oxygen permeability (OP) reduced in A/M-10% films ($1.17 \times 10^{-17}\ cm^2 \cdot s^{-1} \cdot Pa^{-1}$); higher concentrations generally increased OP.					
3.	Agar-chitosan blends	Enhanced TS (35–42 MPa) and EAB (14.79%–22.62%) with higher chitosan concentration.	Decreased water permeability and oxygen diffusion due to compact matrix.		Increased UV blocking and light transmittance due to miscibility between agar and chitosan.			Huang et al. (2020)
4.	Agar-lignin composites	Increased TS and stiffness; EAB decreased.	Low WVP and enhanced oxygen barrier due to tortuous diffusion path; strong UV-barrier properties.	Reduced degradation temperature but increased stability at lower humidity.	Improved UV barrier due to chromophoric lignin groups; decreased light transmittance.	Enhanced due to lignin's antioxidant activity.	Effective against *E. coli* and *Listeria monocytogenes*.	Shankar et al. (2016)

5.	Agar-bacterial cellulose composites	TS enhanced linearly with bacterial cellulose incorporation (up to 10%); EAB decreased at higher concentrations.	Decreased WVP and OP; increased matrix density, reduced water and oxygen diffusion.	Improved thermal stability; degradation temperatures increased with cellulose concentration.	Enhanced opacity and reduced light transmittance due to cellulose dispersion in the matrix.			Wang et al. (2018)
6.	Agar-anthocyanin composites			Increased glass transition temperature (Tg) and stability due to polymer interactions.	Strong UV blocking; reduced UV and visible light transmittance by 85%–90%.	Increased antioxidant activity (ABTS scavenging increased up to 99%).		Roy et al. (2021)
7.	Agar-silver nanoparticles	Stiffness and TS increased due to nanoparticle reinforcement; EAB reduced.	Reduced WVP and OP due to decreased free volume for molecular transport; strong UV barrier.	Thermal degradation temperatures increased; char content increased at 600°C.	Enhanced UV blocking; darkened films reduce visible light transmittance.		Strong antimicrobial activity against *E. coli, L. monocytogenes*, and *Aeromonas hydrophila*.	Shankar and Rhim, 2017b
8.	Agar-melanin nanoparticles	TS increased by 34%, stiffness increased by 78%; higher concentration disrupted intermolecular interactions.	Slightly increased WVP at higher concentrations; improved UV- barrier properties; reduced OP.	Thermal stability remained unchanged; degradation patterns unaffected by nanoparticles.	UV- barrier properties enhanced; light transmittance reduced by 90%.	Enhanced antioxidant activity (DPPH and ABTS radical scavenging increased linearly with concentration).		Roy and Rhim (2019)
9.	Agar-based edible films	High elasticity and TS; enhanced by nanomaterial.	Reduced WVP, moderate oxygen barrier.	Moderate stability, enhanced by additives.	Transparent and glossy appearance.	Moderate, enhanced by essential oils and extract.	Limited.	Mostafavi and Zaeim (2020)
10.	Agar-chitosan blends	Increased stiffness and TS due to hydrogen bonding.	Reduced WVP; strong UV barrier.		Transparent and glossy films.	Limited.	Enhanced antimicrobial activity.	Cao et al. (2018)
11.	Agar-chitosan-halloysite nanotube films	TS and EAB increased with HNT addition; 48.29 MPa TS and 22.62% EAB with 40% chitosan and 8% HNT.	Improved moisture resistance and decreased OP with HNT incorporation.	Unchanged with HNT addition; consistent degradation patterns.	>70% transparency maintained in visible light.			Huang et al. (2020)

(*Continued*)

TABLE 5.2 (Continued)

S. no	Biopolymers used	Mechanical	Barrier	Thermal	Optical	Antioxidant	Antimicrobial	References
12.	Sodium alginate-agar coating	Enhanced mechanical adhesion to beef surface.	Improved oxygen and moisture barrier; delayed lipid oxidation.	Stability during refrigerated storage at 4°C for 10 days.	High transparency with no visible changes during storage.	Reduced thiobarbituric acid reactive substances (TBARS) and peroxide values during storage, indicating lipid oxidation resistance.	Moderate antimicrobial effects against *E. coli*, *S. aureus*, yeasts, and molds.	Zhang et al. (2021)
13.	Sodium alginate agar + ginger essential oil (SA-aga+GEO) coating	Improved flexibility and adhesion due to essential oil inclusion.	Enhanced oxygen and UV-barrier properties; effective lipid oxidation delay for 16 days.	Demonstrated stability during refrigerated storage (4°C) for 16 days, effectively preserving quality and extending the shelf life of beef.	Strong UV-blocking properties with GEO.	Significant antioxidant activity with reduced TBARS and peroxide values.	Significant reduction in microbial counts (total viable counts [TVC], *E. coli*, *S. aureus*, yeasts, molds).	Zhang et al. (2021)
14.	Agar-sodium alginate-silica nanoparticle films	Increased TS and reduced EAB due to nanoparticle stiffening effect.	Improved UV, oxygen, and moisture barriers due to dense network.	Increased thermal degradation temperature with silica nanoparticles.	Improved UV blocking and reduced visible light transmittance.		Enhanced with active agents.	Perera et al. (2021)
15.	Mulberry-infused seaweed-based films		pH-sensitive barrier for intelligent packaging; effective oxygen barrier at neutral pH.		Color changes at different pH levels for freshness indication.	.		Perera et al. (2021)
16.	Unpurified agar films	Improved TS due to interpenetrating polysaccharide/protein networks.	Enhanced water barrier; resistance to high humidity.	Stable under high-humidity conditions.	Greater UV-blocking capacity.			Martínez-Sanz et al. (2019)
17.	Agar/CMC biocomposite films with EO	Improved flexibility and surface hydrophobicity; reduced TS.	Increased WVP due to essential oil addition.		Reduced transparency with high essential oil loading.		Effective against *S. aureus* and *E. coli*.	Abdollahi et al. (2018)
18.	Agar-silver nanocomposites	Increased TS and YM; enhanced compatibility.	Reduced WVP and OP; excellent UV-barrier properties.	Improved thermal degradation resistance.	Strong UV-blocking properties.		Effective against *S. aureus* and *P. aeruginosa*.	Davidović et al. (2019)

19.	Starch/agar/maltodextrin blends	Improved TS due to agar; flexibility reduced with increased maltodextrin.	Moderate WVP; weaker barrier at higher maltodextrin concentrations.	Decreased stability due to maltodextrin.	High transparency in blends.			Wongphan et al. (2020)
20.	Pectin/agar-zinc sulfide (ZnS) nanocomposite films	Enhanced TS and stiffness.	Excellent UV protection; moderate water barrier.		Transparency unaffected by ZnS nanoparticles.		Effective against *E. coli* and *L. monocytogenes*.	Roy and Rhim (2021)
21.	CMC-agar active films	High TS; improved surface properties due to active agents.	Moderate water-barrier properties.	Enhanced thermal resistance with active components.	Moderate transparency.	Moderate antioxidant activity.	Effective against foodborne pathogens.	Gupta et al. (2022)
22.	Agar-based coatings for fruits	Moderate mechanical properties, sufficient to protect fruit integrity.	Reduced water loss; enhanced gas exchange barrier.	Stable under refrigerated conditions.	Transparent; retains fruit appearance.	Reduces oxidation of fruit components.	Delays microbial growth on coated fruits.	
23.	Agar and carrageenan coatings		Improves moisture retention and reduces OP.		Transparent and glossy.		Effective against spoilage microbes in fresh produce.	Sedayu et al. (2019)
24.	Agar films with sodium citrate plasticize	Increased TS compared to control; enhanced TS when combined with citric acid.	Lower water absorption compared to glycerol or sorbitol-plasticized films.		Transparent films; consistent clarity.			Shikov et al. (2017)
25.	Agar/konjac glucomannan films	Increased TS and elongation at 40°C drying temperature; reduced at 60°C due to phase separation.	Improved hydrophilicity with increasing konjac glucomannan (KGM) content.	Crystallinity reduced with higher KGM content, leading to amorphous structure.				Qiao et al. (2019)
26.	Starch/agar composite films	Improved TS and elongation; amorphous structure maintained in starch-dominant ratios.	Reduced WVP and OP at higher agar concentrations.	Increased thermal stability with enhanced structural integrity.	Moderate transparency.			Guo et al. (2021)
27.	Agar biopolymer coating on paper	Adequate TS for packaging; enhanced elasticity with glycerol addition.	Improved water and gas barrier potential; hydrophobic surface.	Stable under moisture exposure with gradual reduction in mechanical properties over time.	Transparent and smooth coating.			Dora, Ghosh et al. (2020); Dora, Singh, et al. (2020)

(Continued)

TABLE 5.2 (Continued)

S. no	Biopolymers used	Mechanical	Barrier	Thermal	Optical	Antioxidant	Antimicrobial	References
28.	Agar films with glycerol	Increased TS and flexibility, glycerol enhanced film ductility.	Reduced WVP due to glycerol; suitable for food packaging.	Improved stability under moisture exposure.	Transparent and glossy films.			Dora, Ghosh et al. (2020); Dora, Singh, et al. (2020)
29.	Agar/cellulose nanocrystal composite films	Enhanced mechanical properties; higher TS with cellulose nanocrystal addition.	Enhanced oxygen and water vapor barrier properties.	Thermal stability improved due to nanocrystal reinforcement.	Reduced transparency at higher nanocrystal concentrations.			Guo et al. (2021)
30.	Agar-based coatings for packaging	Sufficient mechanical strength for coating on paper substrates; ductility enhanced with glycerol addition.	Effective water and gas barrier properties; hydrophobic surface.	Maintained stability under varying moisture conditions.	Transparent and smooth appearance.			Dora, Ghosh et al. (2020); Dora, Singh, et al. (2020)

5.4.4 Optical properties

Agar-based films exhibit high transparency, making them ideal for applications requiring clear or see-through packaging. For instance, agar-chitosan and agar-halloysite films maintain excellent light transmittance (>70%) in the visible spectrum, supporting their use in transparent packaging systems (Huang et al., 2020). The transparency of these films can be affected by the presence of impurities, additives, or the method of extraction. For instance, films derived from sonication-assisted extractions tend to be more transparent due to the uniform distribution of polymer chains (Martínez-Sanz et al., 2019). When stored in a refrigerator, SA-AG coatings maintain their clarity and provide a visually appealing option for packaging perishable foods (Zhang et al., 2021). Agar-lignin nanoparticle films, on the other hand, have significant UV-blocking capabilities and lower light transmittance, which makes them appropriate for packaging that must be protected from photodegradation (Shankar et al., 2016; Perera et al., 2021). In addition, agar-silver nanoparticle composites show good UV-blocking properties; however, as nanoparticle concentrations increase, their transparency declines, balancing the demand for clarity and UV protection (Shankar and Rhim, 2017b). Furthermore, the addition of bioactive compounds, such as natural pigments or antioxidants, can impart specific colorations to the films, broadening their application in functional packaging.

5.4.5 Antimicrobial properties

To increase the shelf life of food products, antimicrobial agents are incorporated into agar-based films. The shelf life of refrigerated beef is greatly increased by SA-AG coatings containing GEO, which exhibit good antibacterial action *against Escherichia coli*, *Staphylococcus aureus*, yeasts, and molds (Zhang et al., 2021). Using the antimicrobial properties of silver, agar-based films that include silver nanoparticles (AgNPs) also exhibit strong antibacterial activity against *E. coli* and Listeria *monocytogenes* (Rhim et al., 2013). Essential oils are ideal for active food packaging because they improve antibacterial properties, as seen in agar/CMC films (Abdollahi et al., 2018). Additionally, it has been exhibited that agar-based fruit coatings suppress microbial development, thereby improving the shelf life of fresh food by preservation. Additionally, blending agar with polymers like chitosan, which possesses inherent antimicrobial properties, enhances the antibacterial activity of the films, particularly against Gram-positive and Gram-negative bacteria.

5.4.6 Antioxidant properties

The antioxidant properties of agar-based films are improved by adding natural components. Because lignin contains phenolic chemicals that are known to scavenge free radicals, agar-lignin composites show significant antioxidant activity (Shankar et al., 2016). Similarly, because melanin has natural antioxidant qualities, agar-melanin nanoparticle composites show enhanced radical scavenging ability, especially for DPPH and ABTS radicals (Roy & Rhim, 2019). Furthermore, thiobarbituric acid reactive substances (TBARS) and peroxide levels decrease when beef is chilled, indicating that SA-AG coatings with GEO have good lipid oxidation resistance (Zhang et al., 2021). Additionally, agar-cellulose nanocrystal composites exhibit potential antioxidant functionality when combined with bioactive agents (Guo et al., 2021). Agar's adaptability enables the addition of a wide range of additives, including chitosan, lignin, starch, essential oils, nanoparticles, and bioactive substances. Agar-based films are therefore a sustainable substitute for synthetic plastics due to their enhanced mechanical, barrier, thermal, optical, antibacterial, and antioxidant qualities. Moreover, the incorporation of antioxidants such as tocopherol or carotenoids into agar leads to films with improved functional properties, making them ideal for active packaging applications (Martínez-Sanz et al., 2019).

5.5 Agar-based active packaging films for food applications

Agar-based films are biodegradable, edible materials derived from seaweed. The gelling ability of agar is crucial for consistently producing these films. This property makes agar a sustainable alternative to artificial packaging materials and an effective way to preserve food. Agar films act as barriers to oxygen and moisture, which are the main causes of food spoilage. By creating these barriers, agar-based films help extend the shelf life of vegetables, meat, and baked goods, preserving food quality and maintaining freshness for longer periods.

The addition of antimicrobial compounds, essential oils, or natural antioxidants further improves the properties of agar-based films (Roy et al., 2023). Several studies have explored the use of agar-based films to enhance the quality and shelf life of food products. For example, an agar-based bioactive film containing probiotic strains (*Bifidobacterium lactis* B94 and *Lactobacillus paracasei* L26) and green tea extract was applied to hake filets. The film not only extended the shelf life of the filets but also increased the beneficial lactic acid bacteria in the fish (Pillai et al., 2024).

The efficiency and characteristics of agar films can be further improved by incorporating nanomaterials, blending with other biopolymers, adding plasticizers, hydrophobic elements, or incorporating antibacterial agents. Additionally, agar-based films with enhanced functionality have been extensively discussed (Table 5.3).

Fruits and vegetables: In apples, agar in combination with glycerol and water has been applied to extend their shelf life and reduce weight loss. The immersion method involves submerging apples in an agar-based solution containing bacterial cellulose, glycerol, and olive oil for 10 minutes. This process forms a thin protective layer without altering the fruit's sensory properties (Anh et al., 2024). In the case of bananas, agar blended with carrageenan and alginate gives antibacterial properties with improved shelf life and delayed ripening for up to 15 days (Jayakody et al., 2022). Numerous studies have shown that chitosan agar/alginate-based films have the potential to preserve the quality and shelf life of fruits such as guava, pear, and blueberries, along with vegetables such as cucumber, capsicum, and mushrooms (Nair et al., 2020). Similarly, agar films with zinc oxide nanoparticles (ZnONPs) improve the shelf life of green grapes. When green grapes are wrapped in plastic film after seven days, they develop several fungal spots on their surface and a noticeable mildew appearance along with a decay smell. Nanocomposite film containing 2% (w/w) and 4% (w/w) ZnO NPs persisted for 14 days. Therefore, the preservation efficiency of grapes was directly proportional to the concentration of nanoparticles. Therefore, So, it has been observed that the efficiency of preservation of grapes is directly proportional to the concentration of NPs (Kumar et al., 2019).

Agar-based nanocomposite films were developed using AgNPs as a filler, which improved the UV light protection, antibacterial capabilities, and thermal stability. With enhanced antibacterial property by the addition of AgNPs to the agar matrix, especially, against the Gram-negative foodborne pathogen *Aeromonas hydrophila* in *Lagestroemia speciosa* fruit. These advancements highlight the potential for commercial application in food preservation (Basumatary et al., 2018).

Ti-metal organic framework (Ti-MOF) doped with carbon dots (CDs) was synthesized and used for the fabrication of CMC/agar-based active packaging film for cherry tomatoes. It also exhibits antioxidant properties and shows strong antibacterial activity, further inhibiting the *E. coli* progress and completely stopping the growth of *L. monocytogenes*. Additionally, CMC/Agar/Ti-MOF film extends the shelf life of cherry tomatoes preserved at 4°C and delays the quality degradation (Khan et al., 2024). Another study demonstrates that natamycin-incorporated agar films act as antimicrobials and help preserve the quality of fresh strawberries. Natamycin-incorporated agar films prepared by the solvent casting method improve the antimicrobial activity of the films and lower the rate of water loss during storage (Wang et al., 2022).

Meat products: The oxidation of meat, containing polyunsaturated fatty acids, can lead to nutrient loss and spoilage due to the formation of aldehydes. Agar-based films were used for their application in both active and smart packaging, which plays a significant role in enhancing the shelf life of meat products by providing a barrier against oxidation and microbial contamination. Additionally, agar-based films can be customized and infused with different essential oils and nanoparticles to enhance the effectiveness of the packaging, preserving the quality and safety of meat during storage (Roy et al., 2023). Fresh beef loin was packaged with 1%–2% AgNPs incorporated in agar films to enhance antimicrobial properties and quality. Beef samples with AgNP film show a log reduction of the inoculated pathogenic bacteria during storage. Meat is highly susceptible to microbial contamination due to high water activity and nutritional content, which can cause off-flavor and taste, texture changes, and the development of slime after a few days. Further, metabolic processes, including lipid and myoglobin oxidation, are the primary causes of rancidity and discoloration. Silver-containing biopolymer films applied to fresh beef packaging, which are eco-friendly, maintain storage quality. Agar/AgNP composite film was incorporated for the packaging of refrigerated beef loin to increase the shelf life of beef by up to 15 days, eliminating the risk of contamination and rancidity as mentioned (Hong et al., 2021).

In a similar study, antimicrobial films were incorporated for the preservation of chicken based on agar/konjac glucomannan with carvacrol (1%, 1.5%, and 2%). The addition of 2% carvacrol showed strong antimicrobial activity against *S. aureus* and *E. coli*, prolonging the shelf life of refrigerated chicken breast from five to nine days (Peng et al., 2022). Another study of wrapping chicken meat with gelatin/chitosan-lactate/agar incorporated with Curcuma hydroethanolic extract films enhances the shelf life during storage and reduces bacterial growth as compared with unwrapped chicken meat (Thakur et al., 2024).

Seafood products: Seafood and poultry benefited from agar with silver or ZnONPs, which reduce lipid oxidation and enhance antimicrobial activity. Alginates/agar are naturally occurring indigestible polysaccharides that are useful for film-making. The cell wall of brown algae is a good source of them (Mahmud et al., 2021). Alginates are effectively used to preserve fresh-cut apples, potato strips, and fish filets. They play an important role in the shelf-life extension of seafood products while maintaining their sensorial and microbial quality. Because of perishability, fresh seafood has a short shelf life; thus, refrigeration or freezing alone is not enough for preservation. Alginate/agar films and coatings in hot-smoked rainbow trout increase the shelf life, with the coating of film containing 3% sodium alginate. Studies have shown that

TABLE 5.3 Agar-based films for food preservation.

S. no.	Composition	Food products	Outcome	Reference
1.	Agar, bacterial cellulose, glycerol, olive oil	Apples	Extend shelf life up to 20 days, reduce weight loss, antibacterial activity against *E. coli, M. catarrhalis* and *S. aureus*.	Anh et al. (2024)
2.	Agar, vegetable oil, egg protein	Bakery products (cake)	Sample did not show mold growth after 85 days, reduced microbial growth as water activity of edible film was 0.31.	De Pilli (2020)
3.	Agar/silver nanoparticles (AgNP) composite	Meat products (beef)	Stored up to 15 days at 5°C, one log reduction of pathogenic bacteria.	Hong et al. (2021)
4.	Agar/ZnO with 4% and 2% (W/W)	Green grape	Shelf-life extension and fresh appearance up to 14 and 21 days.	Kumar et al. (2019a)
5.	Agar, titanium oxide nanoparticles	Sour cream	Protect against fat oxidation with peroxide value 3.21 meq O_2 kg^{-1}, extend shelf life up to 90 days.	El-Sayed et al. (2024)
6.	Anthocyanin, shikonin with cellulose/agar	Packaged food product	UV-light barrier property, color-changing properties in pH range 2–12.	Roy et al. (2021)
7.	Seaweed-derived agar + alginate + carrageenan	Banana	Antibacterial, delay ripening.	Jayakody et al. (2022)
8.	Agar with silver or zinc oxide nanoparticles (ZnO NPs)	Sea food, poultry	Antimicrobial activity and reduce lipid oxidation up to 10 days at temp (4°C).	Mahmud et al. (2021)
9.	Sodium alginate agar with ginger essential oil	Beef, meat	Enhance preservation for up to three to six and nine days, retard oxidation with thiobarbituric acid reactive substances (TBARS) value 3.132, 1.868, 0.986 mg MDA kg^{-1}.	Zhang et al. (2021)
10.	Alginate/agar-based edible coating	Fresh cheese	Enhance shelf life throughout 10 days at 4°C, reduced migration of *L. monocytogenes*.	Silva et al. (2022)
11.	Silver agar-based films		Water vapor barrier, strong antimicrobial property.	Roy and Rhim, 2022
12.	Natamycin incorporated into agar films	Strawberry	Inhibition of *Aspergillus niger* and *Saccharomyces cerevisiae*, fungal colony level reduced by 0.51 and 1.05 log CFU g^{-1}.	Wang et al. (2022)
13.	Carbon dot dopped Ti-metal organic framework on agar film	Cherry tomatoes	Completely stopped *L. monocytogenes*, and inhibited *E. coli* progress, strong antioxidant activity.	Khan et al. (2024)
14.	*Aloe vera* oil with agar/gelatin edible film	Kashar cheese packaging	Enhance shelf life up to 20 days of storage temp at 4°C.	Islk et al. (2023)
15.	Carvacrol incorporation on agar/konjoc glucomannan	Chicken	Shelf-life extension up to five to nine days, antimicrobial activity against *Staphylococcus aureus* and *E. coli*.	Peng et al. (2022)
16.	Agar/alginate/collagen blend + AgNPs + grapefruit extract	Fresh potatoes	Prevent greening of potatoes, antibacterial activity against *Staphylococcus*sp., *Bacillus* sp., *Klebsiella* sp., and *Pseudomonas* sp.	Lieu et al. (2024)

(Continued)

TABLE 5.3 (Continued)

S. no.	Composition	Food products	Outcome	Reference
17.	Silver nanoparticles as fillers in agar-based nan composite film	Fruit (*Lagerstroemia speciosa*)	Antibacterial against *Aeromonas hydrophila* and Gram-negative bacteria.	Basumatary et al. (2018)
18.	Gelatin/agar blend + TiO_2 nanoparticles	Fish oil	Reduces the speed of photo-oxidation.	Vejdan et al. (2017)
19.	Agar/sugarcane wax with butterfly pea flower	Shrimp	Intelligent colorimetric pH sensoring film for shrimp freshness up to (2–6).	Hashim et al. (2022)
20.	Agar colloid-based edible film	Banana	Improved shelf life of fruit up to 15 days. Reduces weight loss through barrier against gasses.	Hussein Ziedan et al. (2018)

agar-based films on fish, shrimp, and pork with sodium alginate prolong the shelf life by at least three weeks, reduce thawing loss, weight loss, and cooking loss, and maintain functional properties (Erkan, 2015).

Bakery products: Agar based edible film made with vegetable oil and egg protein can be a good replacement for preservatives and primary plastic packaging of sweet baked goods. After 85 days of storage, this film does not exhibit mold growth, also sensory characteristics of the sweet goods, high water resistance property, and preserves the crispiness in bakery products such as cake. Certain essential oils, including those derived from plants, are known to possess antibacterial, antifungal, antioxidant, and anticarcinogenic properties. Dextrose agar was used as a growth medium to detect the presence of molds and yeasts. Chloramphenicol, a broad-spectrum antibiotic, was also added to prevent the growth of bacteria. Samples with edible film showed higher cohesion and hardness than samples without edible film. Additionally, during the compression test, the inclusion of edible film enhanced cohesiveness and reduced breaking (De Pilli, 2020; San et al., 2022).

Dairy products: The incorporation of agar and pectin in a protein mixture enhances the moisture barrier; further blending it with milk protein made of calcium caseinate (CC) and whey protein preserves the quality of milk products. Dairy items like mozzarella cheese packaged in biodegradable films made of protein preserve nutritional quality (Sandoval et al., 2016). *Streptococcus thermophilus* is one of the thermophilic bacteria used to produce cheese. Keshar cheese is one of the most consumed types of cheese in Turkey and can be subgrouped into semihard and pasta filet type cheeses. These are more susceptible to microbial and chemical spoilage. One of the main causes of cheese losses is microbial growth due to bacterial and fungal strains. Furthermore, agar–gelatin films enriched with *Aloe vera* oil were developed as edible, biodegradable packaging for Kashar cheese. The films showed antioxidant and antimicrobial activity, reducing bacterial and fungal growth by 2.30 and 3.37 log CFU g^{-1}, respectively, over 20 days at 4°C. They were effective against *E. coli*, *S. aureus*, *A. niger*, and *C. albicans*, suggesting their potential to extend shelf life and maintain dairy product quality (Islk et al., 2023; Sandoval et al., 2016). Overall, agar-based films have demonstrated significant potential for enhancing the shelf life of perishable foods. Their applications range from fresh produce to dairy and meat products, offering antimicrobial, antioxidant, and barrier properties. Further research should focus on optimizing formulations for commercial scalability and consumer acceptance.

5.5.1 Agar-based coatings

Edible coatings or films are a thin layer of biodegradable polymer used in primary packaging, with or without food-grade additives. Edible coatings are incorporated directly into food products and subsequently dried, whereas edible films are made as a thin layer and then applied to food products. Moreover, edible coatings with food-grade suspensions are applied in various ways to the food, such as spraying, dipping, spreading, etc. Agar-based coatings are used in various food products due to their gel-forming ability. Agar-based films and coatings are clear, transparent, homogeneous, flexible, and are used for the preservation of food (Shahidi & Hossain, 2022) (Table 5.4).

5.5.1.1 Agar-based coatings on fruits and vegetables

Agar-alginate coatings combined with citric and ascorbic acid effectively prevent browning and reduce microbial spoilage in fresh-cut apples with antioxidant benefits, which preserve the nutritional properties and extend the shelf life. Incorporation of antibrowning agents into the edible coatings is a conventional process. The main quality attributes, that is, appearance, color, taste, and texture, may be affected by coating formulation at different rates. Thus, coating agar-alginate with ascorbic and citric acid lowers weight loss by providing a barrier against moisture and a decline in browning index. Moreover, the addition of ascorbic acid acts as an antioxidant compound. After seven days of storage at 4°C, the lowest value of weight loss (0.33%), microbial load (2.95 log CFU g^{-1}), and antioxidant capacity (11.03 mmol Fe^{2+} kg^{-1}) were detected when ascorbic and citric acid were incorporated in the coating formulation (Najafi Marghmaleki et al., 2021).

For tomatoes (*Lycopersicon esculentum Mill.*), two seaweeds, namely red and brown algae, were selected due to their high gelling properties. Tomatoes are rich in phosphorus, potassium, folate, antioxidants, beta-carotene, and lycopene. They are available in all seasons and are easily perishable. Their spoilage can cause several changes in taste, smell, and texture. So, natural preservatives with a coating of seaweed gel are found to be a good choice, as they are cost-effective, with fewer or no side effects. The seaweeds were tested for phytonutrients, antibacterial, and antifungal activity through standard procedures. The potential of agar-based coatings at different concentrations to extend tomato shelf life was evaluated. Results showed that using seaweed-based gel enhanced tomato quality and significantly delayed spoilage, emphasizing its usefulness as a natural preservative for fruits and vegetables, and it can be used as a bio-preservative in fruit and vegetable preservation (Banu A et al., 2020).

TABLE 5.4 Agar-based coating for food preservation.

S. no.	Composition	Food product	Outcome	References
1.	Agar-alginate-based coating with citric and ascorbic acids	Fresh-cut apples	Lowest value of weight loss (0.33%), lowered browning index, storage life of seven days, antioxidant capacity (11.03 mmol Fe^{2+} kg^{-1}).	Najafi Marghmaleki et al. (2021)
2.	Agar/polyvinyl alcohol/maltodextrin + silver nanoparticles	Banana (*Musa acuminata*)	Maintain freshness up to five days at 25°C, slow decrease in respiration rate, antibacterial against *E. coli* and *Staphylococcus aureus*.	Nguyen et al. (2021)
3.	Agar-based seaweed coating	Tomato (*Lycopersicon esculentum Mill.*)	Antifungal and antibacterial activity, *K. alvarezii* maintains quality.	Banu A et al. (2020)
4.	Agar + sodium alginate + ginger essential oil	Beef	Retard oxidation, 18 days of shelf life, antibacterial property.	Zhang et al. (2021)
5.	Agar + anti-*Listeria monocytogenes*	Cheese	Antimicrobial at least 1 log unit to control *Listeria monocytogenes*, preserve quality.	Guitián et al. (2019)
6.	Agar + calcium alginate	Meat	Lower water activity, lower microbial count, reduce shrinkage.	Parreidt et al. (2018)
7.	Agar media + whey protein coating incorporating the lactoperoxidase system	Smoked salmon	Antimicrobial effect against *Listeria monocytogenes* up to 35 days of storage at 4°C and 10°C, inhibited 4.2 log colony-forming unit.	Min et al. (2005)
8.	Agar/alginates + sodium alginate solution	Fresh meat cuts	Lower water activity, reduce lipid oxidation.	Yb and Kola (2019)
9.	Agar + chitosan + lemongrass essential oil	Raspberries (*Rubus idaeus* L.)	Antimicrobial food packaging, extend the shelf life up to four to eight days, antibacterial against *E. coli*, *B. subtilis*, and *S. aureus*.	Jovanović et al. (2021)
10.	Agar/agarose + *Piper betle* L. leaf + gelatin	Duong (mandarin fruit)	Antitumor activity, antimicrobial, antiinflammatory.	Pham et al. (2024)
11.	Tryptic soy agar + green tea +hsian-tsao leaf gum+ tapioca starch	Fruit-based salads, romaine hearts, pork slices	Antimicrobial activity against Gram +ve bacteria.	Chiu and Lai (2010)
12.	Agar + locust bean gum + starch (RS4)	Rice	Lower starch digestibility, decreased glucose response, resistance to washing and cooking.	Choi et al. (2010)
13.	Agarose/gelatin with *Ocimum gratissimum* L. essential oil	Booth 7 avocado	Antioxidant, antimicrobial against *Listeria, Pseudomonas, Salmonella*, and *E. coli*, enhanced storability up to six days at 25°C.	Pham et al. (2023)
14.	Agar + carboxy methyl cellulose + starch	Eggs	Extend shelf life up to two to four weeks.	Qahtan Salman et al. (n.d.)
15.	Gallic acid-modified agarose coating	Grass carp (*Ctenopharyngodon idellus*)	100% inhibition of *E. coli* and *Staphylococcus aureus*, prevents fat oxidation, increases antioxidant property.	Fu et al. (2024)

A popular agricultural product in Vietnam for domestic consumption and export to other countries. However, bananas are prone to physical and biological deterioration, resulting in considerable postharvest losses during storage and transportation. Significant improvements were observed when the uncoated banana was compared to the coated one with PAM, which is a polymer blend of poly (vinyl alcohol) (PVOH), agar, and maltodextrin. It is used as a base film-forming matrix for the coating applied to banana preservation, incorporated with 0.2 AgNPs, exhibiting excellent antimicrobial properties, which show inhibition zones of 16 mm against *E. coli* and 13 mm against *S. aureus*. In terms of physical appearance, coated bananas have a better storage period of up to five days at 25°C. While b* value (yellowish) stayed higher at 54.11, which gives superior color retention, the L* value (Lightness) of PAM-0.2AgNPs coated bananas was 69.24 as compared to the uncoated one, which is 67.24. The b* value (yellowness) is higher at 54.11 in the coated one, indicating better color retention. Coated bananas lost just 6% of their weight by day 5, as compared to uncoated samples, indicating considerable weight loss reduction in coated fruits. Furthermore, firmness was retained better up to 1.75 kg cm^{-2}, but untreated bananas lost this property to 1 kg cm^{-2}. Respiration rate, as determined by CO_2, was significantly lower (105.30 mg CO_2 kg^{-1} $hour^{-1}$), indicating a delayed ripening process. In coated bananas, pH rises more gradually and titratable acidity falls more slowly, showing a slower metabolic process, which additionally lowers the TSS, which is correlated with starch breakdown during ripening, compared to bananas that were not coated, where TSS increased from 20% to 28%. Nanocomposite coating preserved the physical integrity, microbiological safety, and nutritional value over the storage period, in addition to extending the shelf life under ambient conditions (Nguyen et al., 2021). Doung (mandarin fruit) treated with agar, agarose, piper betel leaf, and gelatin coatings exhibited antimicrobial, antibacterial, and antioxidant properties (Pham, Tran, et al., 2023).

5.5.1.2 Meat products

Meat is composed of amino acids, fats, protein, water, etc., and its derived products make it susceptible to deterioration in physical, chemical, and organoleptic properties. Lipid and protein oxidation, together with microbial contamination, lead to the spoilage of meat products. To eliminate this, appropriate action should be taken to avoid degradation and increase shelf life. GEO contains bioactive compounds such as gingerol and shogaol, which exhibit strong antimicrobial and antioxidant properties. These compounds inhibit lipid oxidation and microbial proliferation, thereby extending the shelf life of fresh beef. Treating with coatings significantly reduces the oxidation of beef with TBARS values of 3.132, 1.868, 0.986 mg MDA kg^{-1} and peroxide value (PV) of 8.86, 7.46, and 4.26 meq kg^{-1} for uncoated; a similar reduction in microbial growth was also observed. The results have shown that AS-coated and AS+GEO-coated samples extend the shelf life of fresh beef by three to nine days. Furthermore, sensory characteristics were also improved during the storage period. The results have shown that coating with AS + GEO acts as an antibacterial coating for chilled beef storage (Zhang et al., 2021). Calcium alginate-based coatings in meat and poultry products were immersed in an aqueous dispersion of water-soluble algin and carbohydrate and then gelatinized with calcium chloride and CMC solution. Calcium alginate acts as a barrier and maintains lower water activity; coating helps to stabilize the meat color, reduces shrinkage, and increases meat tenderness (Parreidt et al., 2018). Similarly, fresh meat cuts coated with agar/alginate and sodium alginate solutions reduced the lipid oxidation and lowered the water activity (Yb & Kola, 2019).

5.5.1.3 Seafood products

Whey protein isolate coating acts as an antimicrobial and integrates with the lactoperoxidase system (LPOS) against *L. monocytogenes*. It was studied using agar as a growth medium for plate counting, disc covering, and disc surface-spreading tests. *L. monocytogenes* applied to smoked salmon before or after coating was examined after application and during storage at 40°C and 10°C for up to 35 days. Whey protein isolate (WPI) coating prepared with LPOS at 0.7% (w/w) in a coating solution initially reduced >3 and 1 log CFU g^{-1} of *L. monocytogenes* and total aerobic microorganisms in smoked salmon, respectively. The coatings prevented the growth of microorganisms (Min et al., 2005).

5.5.1.4 Dairy products

L. monocytogenes is a widely distributed foodborne pathogen that can cause many serious health issues and food safety problems. This can lead to listeriosis, a foodborne illness with a high mortality rate. So, controlling this foodborne pathogen is necessary. In cheeses, bacteriocins synthesized by lactic acid bacteria are natural alternatives to control this pathogen. Additionally, antimicrobial biopolymer coating has significant benefits in food packaging to preserve food products. The effect of anti-*L. monocytogenes enterocins* synthesized by *Enterococcus avium* DSMZ17511 was investigated when it was developed on agar edible films and applied as antimicrobial coatings to various cheese compositions; initially, the pathogens' viability was shown to have decreased by at least one log unit. Therefore, the application of agar coating in cheese significantly reduces the microbial growth, providing an effective, low-cost, and safe alternative for the preservation of cheese (Guitián et al., 2019).

5.5.1.5 Advanced applications

Raspberries packaged with agar, chitosan, and lemongrass essential oil act as an active component for the application in antimicrobial food packaging. Active coatings prolonged the shelf life of raspberries from four to eight days (Jovanović et al., 2021). For rice, coatings of agar, locust bean gum, and starch (RS4) decrease the starch digestibility and glucose response. The raw milled rice was soaked in prepared coating solutions, rolled in RS4 powder, and dried. Microscopic observation revealed that the RS4 coating on the rice surface with 0.1% LBG-agar mixture provides resistance to washing and cooking and covers the whole surface of rice to preserve the quality (Choi et al., 2010; Hu et al., 2024). Grass carp coated with gallic acid-modified agarose prevented fat oxidation, reduced moisture loss, and inhibited bacterial growth, resulting in enhanced shelf life (Fu et al., 2024).

5.6 Agar-based intelligent packaging

Agar-based food packaging materials represent a promising advancement in sustainable packaging technologies, offering a natural and eco-friendly alternative to conventional synthetic materials (Cebrián-Lloret et al., 2024). Their application is particularly valuable in addressing common issues in food systems, such as lipid oxidation. This process not only results in the formation of unpleasant odors and off-flavors but also promotes the growth of foodborne pathogens. Additionally, oxidation contributes to nutritional degradation by generating harmful aldehydes and depleting essential polyunsaturated fatty acids, compromising food quality and safety. To address these issues, agar-based edible films and functional coatings have been developed for both active and intelligent packaging applications. These films not only extend shelf life but also help monitor the quality and freshness of food products in real time.

Intelligent or smart packaging systems are designed to provide information about the product's condition, enabling consumers to assess its freshness and safety. These systems respond to changes in internal environmental factors such as temperature, pH, moisture, gas composition, and the presence of spoilage metabolites, which can impact the product's physical, chemical, and microbiological quality (Firouz et al., 2021). Agar-based films can function as biosensors by incorporating pH-sensitive color indicators that change in response to spoilage. For example, films with incorporated pH indicators can be used to monitor shrimp freshness, while the addition of natural dyes like shikonin enhances their ability to visually signal quality changes. These developments underscore the growing potential of agar-based materials in intelligent or smart packaging systems (Pillai et al., 2024; Roy et al., 2023). A schematic representation of agar-based intelligent packaging is shown in Fig. 5.2 and the application of agar based film in intelligent packaging represented in Table 5.5.

Fish filets: The potential use of anthocyanin-containing agar-agar-based films as intelligent packaging was studied for fish filets. It has been shown that the level of anthocyanins present in a film's composition has a major impact on structural properties of agar-based films. 30%–50% anthocyanin levels are the most appropriate. A more noticeable change in film color was observed when applied to minced fish, indicating a higher sensitivity of the packaging film to spoilage-related changes in this food matrix. Fat content in a raw fish material is an important factor influencing changes in film color, acidity, and oxidative processes during storage. This pH-sensitive color change property helps to detect shelf-life changes. These films are effective in monitoring freshness and ensuring quality (Chesnokova et al., 2024). Snapper fish: To detect food spoilage, intelligent packaging has integrated pH-responsive natural pigments such as anthocyanin. To create a pH biosensor for red snapper fish spoilage, anthocyanin extract is used in a multilayer chitosan gelatin-agar intelligent film for monitoring the freshness of fish during storage (How et al., 2024).

Lamb: Agar-chitosan composite films provide antibacterial and antioxidant properties, making them suitable for preserving the freshness of lamb. Chitosan-agar-based films used in smart packaging increase the shelf life of meat products (Contessa et al., 2023). Mutton: Creating an intelligent film with anthocyanins for real-time monitoring of freshness and quality is the most important area of packaging research. By using soybean isolate protein/xanthan gum/agar-*Lycium rithenicum* anthocyanin in an intelligent packaging for the preservation of mutton, SPI/XG/agar-5% provides great effectiveness toward the thickness of the film, permeability, extending the shelf life, and monitoring the freshness of the meat (Liu et al., 2024). From purple sweet potato, an intelligent pH indicator film was made, composed of agar/potato starch and anthocyanin extract. Agar and potato starch are both solid matrices that are employed to render anthocyanins, which are natural pigments. The pH indicator films showed pH changes and spoilage points of pork samples (Choi et al., 2017).

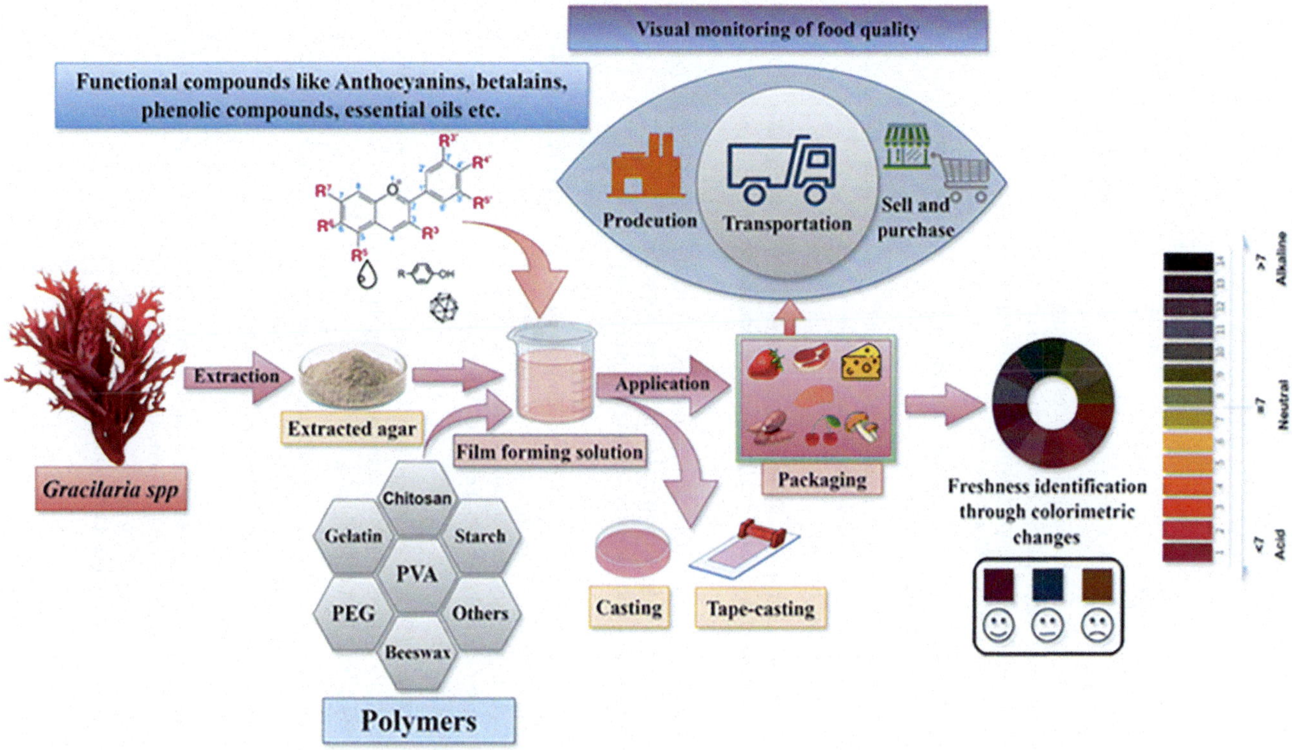

FIGURE 5.2 Application of agar-based packaging as a freshness indicator for food.

Shrimp: K-carrageenan-based intelligent packaging integrated with anthocyanin, as a colorant, with ZnO acting as an antibacterial agent in an agar plate, acts as a multifunctional intelligent packaging film. Furthermore, it functions as a freshness indicator in shrimp. The result of this study promotes the development of eco-friendly and sustainable packaging materials to replace traditional petroleum-based materials (Jiang et al., 2024). Sodium alginate and agar-based bilayer-structured indicators serve as a colorimetric film to modify quality parameters. It enhances the water-sustaining capacity and reduces dye migration. Freshness of shrimp can be revealed by significant color change, which signifies the freshness and spoilage of shrimp (An & Zhou, 2024). Real-time information of food quality provided by intelligent packaging imparts a protective effect to the food product by using active agents. A bilayer film made of alginate/agar with cellulose nanospheres from corncob was prepared using 1.5% (w/w) sodium alginate with 0.25% (w/v) butterfly pea extract incorporated, 2% agar, and 0.5% catechin to improve the film properties. The bilayer shows excellent pH sensitivity, excellent antibacterial properties, and effectively monitors shrimp freshness, shelf life, and quality (Romruen et al., 2022). From *Lithospermum erythrorhizom*, shikonin was extracted and employed in the shrimp to develop pH-indicating biodegradable starch-agar-based films. The composite film displays excellent color change properties from red to blue. In shrimp, it changes to blue-violet, indicating the onset of shrimp spoilage (Ezati & Rhim, 2021).

Kimchi: Smart colorimetric indicators for tracking kimchi freshness with aronia extract in agar, k-carrageenan, and cellulose nanofiber films was reported previously. AEP-films give high sensitivity to pH changes of ammonia and acetic acid. Alkalized AEP-indicator films are more effective in detecting color change during kimchi packaging tests with the most obvious color change from green, gray (fresh kimchi, pH 5.5, acidity 0.48%) to pale brown (optimal fermentation, pH 4.6, acidity 0.70%), and pale violet-brown (overfermented, pH 3.80, acidity 1.35%). Alkalized AEP-indicator films detect and give real-time information of fermented foods like kimchi (Oun et al., 2024).

Prawns: pH-indicator film made with semolina starch/agar-based intelligent films by incorporating butterfly pea flower anthocyanins to monitor freshness during storage has gained attention. From butterfly pea extract as a freshness indicator for tiger prawns, in which semolina/agar-based butterfly pea extract (S/A/BPE) films were developed with different concentrations. Film develops a darker color; results indicate that S/A/BPE film has the potential to be used as a pH indicator to detect the freshness of prawn samples (Saravanan et al., 2024).

Pork: Cochineal-loaded starch particles (2, 4, 6, and 8 wt.% on an agar basis) were added to an 8 wt.% polyvinyl alcohol solution and 0.2% agar solution to produce pH-responsive films using a casting technique. The addition of CSN reduced the pH-responsive films' water solubility but improved color and functional qualities such as antimicrobial and

TABLE 5.5 Application of agar-based films/coatings in intelligent packaging.

S. no.	Type of film/coating	Application	Function	References
1.	Agar-anthocyanin films	Fish filets	Agar film containing 30%–50% anthocyanin influences structural and mechanical properties. The greatest change in color observed in minced fish. Detect spoilage of fish by color change, film color shows important factor to the shift from active acidity to alkaline region.	Chesnokova et al. (2024)
2.	Bioindicator film from tapioca/agar/mulberry	Sausages	During storage time of sausage, color changes contributed to changes in pH ranged from light purple to purple green.	Wardana et al. (2018)
3.	K-carrageenan/dextran films + agar + nano-ZnO	Shrimp	ZnO enhances antibacterial and antioxidant properties, also pH sensitivity and color stability to detect freshness of shrimp. Film completely decomposed within 30 days.	Jiang et al. (2024)
4.	Agar+ starch + cellulose + chitin + guar gum	Meat products, sea food	Use of chemical dye as a spoilage and quality indicator. Change in pH is major indicator for spoilage of perishable food, real-time monitoring of freshness.	Bhargava et al. (2020)
5.	Sodium alginate/agar colorimetric film with methyl blue and bromothymol blue as indicators	Shrimp	Polypropylene fluoride/SiO_2, used as colorimetric film, enhances water retention property, and reduces dye migration. Monitoring freshness by color changes, serving as an indicator corresponding to freshness and spoilage of shrimp.	An and Zhou (2024)
6.	Aronia extract powder + agar + cellulose nanofiber	Kimchi (fermented food)	AEP containing film shows UV barrier, and reduced transmittance of light by 99.12% for agar, 98.86% for carrageenan, and 98.67% for CNF-based films. Exhibited high pH sensitivity, displays strong antioxidant property with 99% free radical scavengers. Enhanced color change detection from green, gray to pale violet-brown.	Oun et al. (2024)
7.	Samolina starch + agar-based intelligent film + butterfly pea flower anthocyanin	Prawns	Depending on the different BPE concentrations, the pH of the film changes (2.0–12.0). Color changes of prawn from dark blue at day 0 to intense green on day 2 beginning of spoilage.	Saravanan et al. (2024)
8.	Agar/potato starch and anthocyanin extract from purple sweet potato	Meat	Ultraviolet-visible spectrum of anthocyanin extract and agar/potato starch film shows color variations at different pH values are measured by colorimeter, used as a potential meat spoilage sensor. And spoilage of pork sample changing from red to green.	Choi et al. (2017)
9.	Linalool nanoparticles + cranberry extract (CEF) + agar/gelatin matrix	Pork	Encapsulation efficiency was 68.54%. pH indicator film based on CEF to develop smart film, enhanced antioxidant property (DPPH free radical scavenging rate increased from 26.8% to 36.35%) and antibacterial activity against *S. aureus* and *E. coli*. Retard spoilage of pork.	Song et al. (2024)

10.	Agar-starch composite films with shikonin	Shrimp	Addition of shikonin increases the film properties such as UV light barrier, mechanical strength, and water permeability (25% and 18%), shows antioxidant and antibacterial properties (*Listeria monocytogenes*), rapid color change when exposed to ammonia vapor, after some time shrimp packaging color shows reddish-pink to blue-violet, indicatiing spoilage.	Ezati and Rhim (2021)
11.	Alginate/agar + anthocyanin extracted from dried flowers of butterfly pea plant (*Clitoria ternatea)* + catechin-lysozyme	Shrimp	Shrimp freshness indicator with cellulose nitrate (CN). Film possesses excellent antimicrobial against Gram +ve and Gram-ve bacteria and antioxidant property with pH sensitive indicators, antibacterial against *Staphylococcus* and *E. coli*. After 24 h film color changes from blue to green and pH 6.7 to 7.6.	Romruen et al. (2022)
12.	Agarose solution + jamun fruit skin extract-based indicator	Chicken	Anthocyanin jamun skin extract was utilized as a quality indicator, smart indicator was used to identify the spoiling. Packaging material originally blue or dark violet turned yellow during course of storage, indicating variations in pH and meat's quality.	Talukder et al. (2020)
13.	Polyvinyl alcohol/agar containing cochineal film used as indicator extracted from female insect *Dactylopius coccus*	Pork	pH responsive from 2 to 12, improvement in functional properties such as antioxidant and antimicrobial properties. Agar/polyvinyl alcohol film 6% CSN exhibits sensitivity to ammonia, detection of freshness of pork by changing color.	Liu et al. (2023)
14.	Soybean isolate protein/Xanthan gum/agar-*Lycium ruthenicum* anthocyanins films	Mutton preservation	Real-time monitoring of freshness by pH-responsive intelligent indicator. 5% SPI/XG/agar exhibits great performance in terms of oxygen transmission rate, water vapor, and light. With 5% SPI/agar/XG extend shelf life.	Liu et al. (2024)
15.	Chickpea protein isolate electro-spun nanofibers with containing black chickpea peel anthocyanins	Seafood products	The active layer of the film controlled the pH, total volatile basic nitrogen, oxidation, and microbial load. Color change indicator during storage time of fish and shrimp was detected from red to pink then to olive.	Amjadi et al. (2024)
16.	Chitosan-gelatin-agar intelligent film with roselle anthocyanin extract	Snapper fish	Monitoring freshness by anthocyanin extract acts as a biosensor with concentrations (2%, 4%, and 6%), triple-layer film shows great barrier properties with lowest thickness, visible color variation from dark red to pale red. Storage up to seven days, used to monitor the freshness of fish during storage.	How et al. (2024)

antioxidant activities. It shows different color changes for pork freshness. Hence, pH-responsive film can be used as a potential packaging material for tracking the freshness of protein-rich food (Liu et al., 2023). To create multifunctional smart films, linalool nanoparticles (L-NPs) with an encapsulation effectiveness of 68.54% were created, and a pH indicator film based on cranberry extract (CEF) was added. CEF and L-NPs were evenly distributed in the gelatin/agar matrix. It enhances the antioxidant and antibacterial activity of the smart film (Song et al., 2024).

5.7 Conclusion

Agar-based films represent a significant step toward sustainable food packaging, providing a biodegradable and environmentally benign alternative to traditional plastics. Their flexible features, including superior film-forming ability, transparency, and compatibility with bioactive substances, make them ideal for a wide range of food packaging applications. Innovative improvements, such as plasticization, biopolymer mixing, and nanomaterial integration, have addressed critical constraints like brittleness and moisture sensitivity. These developments not only improve the mechanical and barrier qualities of agar-based films but also allow for active and intelligent capabilities like real-time freshness monitoring and antimicrobial protection. However, issues with cost-effective production and scalability persist, necessitating more research and innovation. Agar-based films can play a crucial role in decreasing plastic waste and boosting sustainable food systems.

AI disclosure

During the preparation of this work, the author(s) used ChatGPT and quillbot for paraphrasing. After using this tool/service, the author(s) reviewed and edited the content as needed and take(s) full responsibility for the content of the publication.

References

Abdollahi, M., Alboofetileh, M., Rezaei, M., & Behrooz, R. (2018). Comparing physicochemical and thermal properties of alginate nanocomposite films reinforced with organic/inorganic nanofillers. *Food Hydrocolloids, 74*, 101–111. https://doi.org/10.1016/j.foodhyd.2017.09.004.

Amjadi, S., Almasi, H., Gholizadeh, S., & Hamishehkar, H. (2024). Double-layer packaging based on active black chickpea protein isolate electrospun nanofibers and intelligent salep film containing black chickpea peel anthocyanins for seafood products. *International Journal of Biological Macromolecules, 278*, 134897. https://doi.org/10.1016/J.IJBIOMAC.2024.134897.

An, N., & Zhou, W. (2024). Sodium alginate/ager colourimetric film on porous substrate layer: Potential in intelligent food packaging. *Food Chemistry, 445*, 138790. https://doi.org/10.1016/J.FOODCHEM.2024.138790.

Anh, N. T., Nguyen, X. H., Ngo, T., & Van (2024). Research on development and characterization of composite membranes based on hybrid bacterial cellulose combined with glycerol and vegetable residues for the preservation of fresh fruit and food. *Journal of Engineering, 2024*(1), 4159277. https://doi.org/10.1155/2024/4159277.

Arham, A. F., Mulyati, S., Metusalach, & Salengke, S. (2016). Properties of agar-based edible films with added glycerol. *Food Hydrocolloids, 56*, 105–113.

Bang, Y. J., Roy, S., & Rhim, J. W. (2022). A facile in situ synthesis of resorcinol-mediated silver nanoparticles and the fabrication of agar-based functional nanocomposite films. *Journal of Composites Science, 6*(5), 124.

Banu A, T., Ramani P, S., & Murugan, A. (2020). Effect of seaweed coating on quality characteristics and shelf life of tomato (*Lycopersicon esculentum mill*). *Food Science and Human Wellness, 9*(2), 176–183. https://doi.org/10.1016/J.FSHW.2020.03.002.

Basumatary, K., Daimary, P., Das, S. K., Thapa, M., Singh, M., Mukherjee, A., & Kumar, S. (2018). *Lagerstroemia speciosa* fruit-mediated synthesis of silver nanoparticles and its application as filler in agar based nanocomposite films for antimicrobial food packaging. *Food Packaging and Shelf Life, 17*, 99–106. https://doi.org/10.1016/J.FPSL.2018.06.003.

Belay, Z. A., Tyeb, S., Rathore, R., Kumar, M., & Verma, M. (2020). Cross-linkers in biodegradable films: A review. *Journal of Polymer Research, 27*(3), 1–14.

Bhargava, N., Sharanagat, V. S., Mor, R. S., & Kumar, K. (2020). Active and intelligent biodegradable packaging films using food and food waste-derived bioactive compounds: A review. *Trends in Food Science & Technology, 105*, 385–401. https://doi.org/10.1016/J.TIFS.2020.09.015.

Cao, N., Fu, Y., & He, J. (2018). Preparation and physical properties of soy protein isolate and gelatin composite films. *Food Hydrocolloids, 33*, 181–187. https://doi.org/10.1016/j.foodhyd.2013.02.014.

Cazón, P., & Vázquez, M. (2020). Application of biopolymers in food packaging: A review. *Trends in Food Science & Technology, 104*, 198–206. https://doi.org/10.1016/j.tifs.2023.104198.

Cebrián-Lloret, V., Blanco-Díaz, M. T., García-Santiago, Y., & Aguilera-Alonso, P. (2024). Advances in sustainable packaging: The role of seaweed-based materials in food packaging. *Journal of Sustainable Food Systems, 45*(2), 345–360. https://doi.org/10.1016/j.jsfs.2024.01.001.

Chemat, F., Vian, M. A., Fabiano-Tixier, A. S., Nutrizio, M., Jambrak, A. R., Munekata, P. E., & Cravotto, G. (2020). A review of sustainable and intensified techniques for extraction of food and natural products. *Green Chemistry, 22*(8), 2325–2353.

Chesnokova, N., Kuznetsova, A., Levchuk, T., Kushnarenko, L., & Levochkina, L. (2024). Using antocyanins and agar-agar as a basis for smart packaging. *BIO Web of Conferences, 130*, 05003. https://doi.org/10.1051/BIOCONF/202413005003.

Chiu, P. E., & Lai, L. S. (2010). Antimicrobial activities of tapioca starch/decolorized hsian-tsao leaf gum coatings containing green tea extracts in fruit-based salads, romaine hearts and pork slices. *International Journal of Food Microbiology, 139*(1–2), 23–30. https://doi.org/10.1016/J.IJFOODMICRO.2010.01.010.

Choi, H. J., Lee, C. J., Cho, E. J., Choi, S. J., & Moon, T. W. (2010). Preparation, digestibility, and glucose response in mice of rice coated with resistant starch type 4 using locust bean gum and agar. *International Journal of Food Science & Technology, 45*(12), 2612–2621. https://doi.org/10.1111/J.1365-2621.2010.02434.X.

Choi, I., Lee, J. Y., Lacroix, M., & Han, J. (2017). Intelligent pH indicator film composed of agar/potato starch and anthocyanin extracts from purple sweet potato. *Food Chemistry, 218*, 122–128. https://doi.org/10.1016/J.FOODCHEM.2016.09.050.

Contessa, C. R., Rosa, G. S.da, Moraes, C. C., & Burkert, J. F.de M. (2023). Agar-agar and chitosan as precursors in the synthesis of functional film for foods: A review. *Macromol, 3*(2), 275–289. https://doi.org/10.3390/MACROMOL3020017.

Davidović, S., Novaković, A., Mitrić, M., Pavlović, V., & Stojanović, Z. (2019). Optimization of agar composite films for active food packaging. *Materials Chemistry and Physics, 228*, 217–223. https://doi.org/10.1016/j.matchemphys.2018.10.063.

Deepika, B., Ganesan, P., Sivaraman, B., Neethiselvan, N., & Padmavathy, P. (2024). Green extraction of agar from Hypnea pannosa seaweed: A comparative study of different techniques and optimization using response surface methodology. *Algal Research, 82*, 103618.

De Pilli, T. (2020). Development of a vegetable oil and egg proteins edible film to replace preservatives and primary packaging of sweet baked goods. *Food Control, 114*, 107273. https://doi.org/10.1016/J.FOODCONT.2020.107273.

Din, S. S., Chew, K. W., Chang, Y. K., Show, P. L., Phang, S. M., & Juan, J. C. (2019). Extraction of agar from *Eucheuma cottonii* and *Gelidium amansii* seaweeds with sonication pretreatment using autoclaving method. *Journal of Oceanology and Limnology, 37*(3), 871–880.

Dora, M., Ghosh, T., & Damodar, A. (2020). Food packaging and biodegradable polymers. *Food Science Advances, 24*(1), 12–19.

Dora, S., Singh, H., & Maurya, S. (2020). Edible packaging from plant-based raw materials for sustainable food packaging applications. *Journal of Food Processing and Preservation, 44*(12), e14999. https://doi.org/10.1111/jfpp.14999.

El-Sayed, S. M., El-Sayed, H. S., Hashim, A. F., & Youssef, A. M. (2024). Valorization of edible films based on chitosan/hydroxyethyl cellulose/olive leaf extract and TiO_2-NPs for preserving sour cream. *International Journal of Biological Macromolecules, 268*, 131727. https://doi.org/10.1016/J.IJBIOMAC.2024.131727.

Erkan, N. (2015). Use Of natural preservatives in seafood: Plant extracts, edible film and coating. *Journal of Food and Health Science*. https://doi.org/10.3153/jfhs15004.

Ezati, P., & Rhim, J. W. (2021). Starch and agar-based color-indicator films integrated with shikonin for smart packaging application of shrimp. *ACS Food Science and Technology, 1*(10), 1963–1969. https://doi.org/10.1021/ACSFOODSCITECH.1C00292/SUPPL_FILE/FS1C00292_SI_001.PDF.

Firouz, M. S., Mohi-Alden, K., & Omid, M. (2021). Intelligent packaging for food preservation: A review of innovations and emerging technologies. *Trends in Food Science & Technology, 112*, 203–217. https://doi.org/10.1016/j.tifs.2021.02.015.

Fu, L., Xiao, Q., Ru, Y., Hong, Q., Weng, H., Zhang, Y., Chen, J., & Xiao, A. (2024). Bio-based active packaging: Gallic acid modified agarose coatings in grass carp *(Ctenopharyngodon idellus)* preservation. *International Journal of Biological Macromolecules, 255*, 128196. https://doi.org/10.1016/J.IJBIOMAC.2023.128196.

Garcia-Vaquero, M., Ummat, V., Tiwari, B., & Rajauria, G. (2020). Exploring ultrasound, microwave and ultrasound–microwave assisted extraction technologies to increase the extraction of bioactive compounds and antioxidants from brown macroalgae. *Marine Drugs, 18*(3), 172.

Giménez, B., De Lacey, A. M. L., Pérez-Santín, E., López-Caballero, M. E., & Montero, P. (2013). Improving functional properties of agar-based films using antioxidants. *Food Hydrocolloids, 33*, 42–48.

Guitián, M. V., Ibarguren, C., Soria, M. C., Hovanyecz, P., Banchio, C., & Audisio, M. C. (2019). Anti-*Listeria monocytogenes* effect of bacteriocin-incorporated agar edible coatings applied on cheese. *International Dairy Journal, 97*, 92–98. https://doi.org/10.1016/J.IDAIRYJ.2019.05.016.

Guo, Z., Zhang, Y., & Yang, Y. (2021). Novel cellulose-based composite films with improved properties for intelligent food packaging. *Carbohydrate Polymers, 261*, 117847. https://doi.org/10.1016/j.carbpol.2021.117847.

Gupta, A., Ahmed, J., & Sivakumar, D. (2022). Recent advances in bio-based intelligent packaging: A review. *Trends in Food Science & Technology, 119*, 44–63. https://doi.org/10.1016/j.tifs.2021.10.008.

Habeebullah, S. F. K., Alagarsamy, S., Sattari, Z., Al-Haddad, S., Fakhraldeen, S., Al-Ghunaim, A., & Al-Yamani, F. (2020). Enzyme-assisted extraction of bioactive compounds from brown seaweeds and characterization. *Journal of Applied Phycology, 32*, 615–629.

Hashim, S. B. H., Elrasheid Tahir, H., Liu, L., Zhang, J., Zhai, X., Ali Mahdi, A., Nureldin Awad, F., Hassan, M. M., Xiaobo, Z., & Jiyong, S. (2022). Intelligent colorimetric pH sensoring packaging films based on sugarcane wax/agar integrated with butterfly pea flower extract for optical tracking of shrimp freshness. *Food Chemistry, 373*, 131514. https://doi.org/10.1016/J.FOODCHEM.2021.131514.

Hong, S. I., Cho, Y., & Rhim, J. W. (2021). Effect of Agar/AgNP composite film packaging on refrigerated beef loin quality. *Membranes, 11*(10), 750. https://doi.org/10.3390/MEMBRANES11100750.

How, Y. H., Wong, L. X., Kong, I., Nyam, K. L., & Pui, L. P. (2024). Development of multilayered pH-sensitive chitosan–gelatin-agar intelligent film incorporated with roselle anthocyanin extract for monitoring of the freshness of snapper fish. *Food and Bioprocess Technology, 17*(11), 4177–4194. https://doi.org/10.1007/S11947-024-03377-1/METRICS.

Hu, H., Jiang, H., Sang, S., McClements, D. J., Jiang, L., Wen, J., Jin, Z., & Qiu, C. (2024). Research advances in origin, applications, and interactions of resistant starch: Utilization for creation of healthier functional food products. *Trends in Food Science & Technology, 148*, 104519. https://doi.org/10.1016/J.TIFS.2024.104519.

Huang, T., Qian, Y., Wei, J., & Ismail, M. (2020). Enhancing mechanical, thermal, and barrier properties of agar films by adding chitosan and halloysite nanotubes. *Carbohydrate Polymers, 249*, 116813. https://doi.org/10.1016/j.carbpol.2020.116813.

Hussein Ziedan, E. S., El Zahaby, H. M., Maswada, H. F., & El Rafh Zoeir, E. H. A. (2018). Agar-agar a promising edible coating agent for management of postharvest diseases and improving banana fruit quality. *Journal of Plant Protection Research, 58*(3). https://doi.org/10.24425/122938.

Islk, I., Yenipazar, H., Saygun, A., Sahin Yesilcubuk, N., Ozkan Zayim, E., & Catalgil Giz, H. (2023). Aloe vera oil-added agar gelatin edible films for kashar cheese packaging. *ACS Omega, 8*(21), 18516–18522. https://doi.org/10.1021/ACSOMEGA.3C00147/ASSET/IMAGES/LARGE/AO3C00147_0002.JPEG.

Jayakody, M. M., Vanniarachchy, M. P. G., & Wijesekara, I. (2022). Seaweed derived alginate, agar, and carrageenan based edible coatings and films for the food industry: A review. *Journal of Food Measurement and Characterization, 16*(2), 1195–1227. https://doi.org/10.1007/S11694-021-01277-Y/METRICS.

Jiang, G., Yang, Y., Sheng, W. Y., Yang, L., Yang, H., Tang, T., Wang, C., & Tian, Y. (2024). Preparation and characterization of κ-carrageenan/dextran films blended with nano-ZnO and anthocyanin for intelligent food packaging. *International Journal of Biological Macromolecules, 282*, 137203. https://doi.org/10.1016/J.IJBIOMAC.2024.137203.

Jovanović, J., Ćirković, J., Radojković, A., Mutavdžić, D., Tanasijević, G., Joksimović, K., Bakić, G., Branković, G., & Branković, Z. (2021). Chitosan and pectin-based films and coatings with active components for application in antimicrobial food packaging. *Progress in Organic Coatings, 158*, 106349. https://doi.org/10.1016/J.PORGCOAT.2021.106349.

Khan, A., Riahi, Z., Kim, J. T., & Rhim, J. W. (2024). Effect of carbon dot-doped Ti-MOF on CMC/Agar film and active packaging application on storage quality of fruits. *Food Chemistry, 455*, 139911. https://doi.org/10.1016/J.FOODCHEM.2024.139911.

Kumar, S., Boro, J. C., Ray, D., Mukherjee, A., & Dutta, J. (2019). Bionanocomposite films of agar incorporated with ZnO nanoparticles as an active packaging material for shelf life extension of green grape. *Heliyon, 5*(6). https://doi.org/10.1016/j.heliyon.2019.e01867.

Kwansang, J., Chen, C. J., & Chaiprateep, E. O. (2022). Optimization of water-based ultrasonic-microwave assisted extraction (UMAE) of bioactive compounds from Garcinia mangostana pericarp. *Journal of Complementary and Integrative Medicine, 19*(2), 225–231.

Lieu, M. D., Dang, T. K. T., & Nguyen, T. H. (2024). Green synthesized silver nanoparticles, a sustainable approach for fruit and vegetable preservation: An overview. *Food Chemistry: X, 23*, 101664. https://doi.org/10.1016/J.FOCHX.2024.101664.

Liu, B., Gao, J., Liu, X., Zhang, X., Zeng, X., Zhang, X., & Zhao, P. (2024). Preparation of soybean isolate protein/xanthan gum/agar-Lycium ruthenicum anthocyanins intelligent indicator films and its application in mutton preservation. *International Journal of Biological Macromolecules, 283*, 137751. https://doi.org/10.1016/J.IJBIOMAC.2024.137751.

Liu, D., Zhong, Y., Pu, Y., Li, X., Chen, S., & Zhang, C. (2023). Preparation of pH-responsive films from polyvinyl alcohol/agar containing cochineal for monitoring the freshness of pork. *Foods, 12*(12), 2316. https://doi.org/10.3390/FOODS12122316/S1.

Madera-Santana, T., Freile-Pelegrín, Y., & Azamar-Barrios, J. A. (2014). Additives to agar-based films: A review. *Carbohydrate Polymers, 116*, 1–14.

Mahmud, N., Islam, J., & Tahergorabi, R. (2021). Marine biopolymers: Applications in food packaging. *Processes, 9*(12), 2245. https://doi.org/10.3390/PR9122245.

Maleki, N., Roomiani, L., & Tadayoni, M. (2023). Microwave-assisted extraction optimization, antimicrobial and antioxidant properties of carrageenan from red algae (*Gracilaria acerosa*). *Journal of Food Measurement and Characterization, 17*(2), 1156–1166.

Martínez-Sanz, M., Gomez-Barrio, L. P., Zhao, M., Tiwari, B., Knutsen, S. H., Ballance, S., & López-Rubio, A. (2021). Alternative protocols for the production of more sustainable agar-based extracts from *Gelidium sesquipedale*. *Algal Research, 55*, 102254.

Martínez-Sanz, M., López-Rubio, A., & Lagaron, J. M. (2019). Development of electrospun antimicrobial nanostructured biocomposites for active food packaging applications. *Food Hydrocolloids, 87*, 830–842. https://doi.org/10.1016/j.foodhyd.2018.07.044.

Min, S., Harris, L. J., & Krochta, J. M. (2005). *Listeria monocytogenes* inhibition by whey protein films and coatings incorporating the lactoperoxidase system. *Journal of Food Science, 70*(7), m317–m324. https://doi.org/10.1111/J.1365-2621.2005.TB11474.X.

Mostafavi, F. S., & Zaeim, D. (2020). Agar-based edible films for food packaging applications—A review. *International Journal of Biological Macromolecules, 156*, 605–616. https://doi.org/10.1016/j.ijbiomac.2020.05.123.

Nair, M. S., Tomar, M., Punia, S., Kukula-Koch, W., & Kumar, M. (2020). Enhancing the functionality of chitosan- and alginate-based active edible coatings/films for the preservation of fruits and vegetables: A review. *International Journal of Biological Macromolecules, 164*, 304–320. https://doi.org/10.1016/J.IJBIOMAC.2020.07.083.

Najafi Marghmaleki, S., Mortazavi, S. M. H., Saei, H., & Mostaan, A. (2021). The effect of alginate-based edible coating enriched with citric acid and ascorbic acid on texture, appearance and eating quality of apple fresh-cut. *International Journal of Fruit Science, 21*(1), 40–51. https://doi.org/10.1080/15538362.2020.1856018.

Nguyen, T. T., Huynh Nguyen, T. T., Tran Pham, B. T., Van Tran, T., Bach, L. G., Bui Thi, P. Q., & Ha Thuc, C. N. (2021). Development of poly (vinyl alcohol)/agar/maltodextrin coating containing silver nanoparticles for banana (*Musa acuminate*) preservation. *Food Packaging and Shelf Life, 29*, 100740. https://doi.org/10.1016/J.FPSL.2021.100740.

Öğretmen, Ö. Y., & Duyar, H. A. (2018). The effect of different extraction methods and pre-treatments on agar yield and physico-chemical properties of *Gelidium latifolium* (Gelidiaceae, Rhodophyta) from Sinop Peninsula Coast of Black Sea, Turkey. *Journal of Applied Phycology, 30*, 1355–1360.

Oun, A. A., Roy, S., Hong, S. J., Shin, G. H., Yoo, S. R., & Kim, J. T. (2024). Development of smart colorimetric indicators for tracking kimchi freshness by loading aronia extract in agar, κ-carrageenan, and cellulose nanofiber films. *International Journal of Biological Macromolecules, 270*, 132343. https://doi.org/10.1016/J.IJBIOMAC.2024.132343.

Parreidt, T. S., Müller, K., & Schmid, M. (2018). Alginate-based edible films and coatings for food packaging applications. *Foods, 7*(10), 170. https://doi.org/10.3390/FOODS7100170.

Peng, S., Zhang, J., Zhang, T., Hati, S., Mo, H., Xu, D., Li, H., Hu, L., & Liu, Z. (2022). Characterization of carvacrol incorporated antimicrobial film based on agar/konjac glucomannan and its application in chicken preservation. *Journal of Food Engineering, 330*, 111091. https://doi.org/10.1016/J.JFOODENG.2022.111091.

Perera, J. R., Arachchige, S. P., & Silva, S. R. (2021). Fabrication and characterization of agar-lignin composite films for food packaging. *International Journal of Polymer Science, 2021*, 6658917. https://doi.org/10.1155/2021/6658917.

Pham, B. T. T., Hoang, H. N. T., Trinh, C. D., Bui, Q. T. P., Phung, T. K., & Nguyen, T. T. (2023). Development of gelatin/agarose active coatings functionalized with *Ocimum gratissimum* L. essential oil for enhancing storability of 'Booth 7' avocado. *International Journal of Biological Macromolecules, 253*, 127516. https://doi.org/10.1016/J.IJBIOMAC.2023.127516.

Pham, B. T. T., Tran, N. Q., Huynh, N. Y. T., Bach, L. G., Ha Thuc, C. N., Nguyen, D. V., Ton That, Q., & Nguyen, T. T. (2024). Multifunctional coating films of gelatin/d-glucose/agarose reinforced by *Piper* L. leaf extract: Physicochemical properties and mandarin preservation. *Progress in Organic Coatings, 192*, 108509. https://doi.org/10.1016/J.PORGCOAT.2024.108509.

Pham, B. T. T., Tran, T. D., Phung, T. K., Bach, L. G., Nguyen, D. V., & Nguyen, T. T. (2023). Pectin/agarose-based coating functionalized with *Piper betel* L. extract for enhancing storability of postharvest bananas. *Progress in Organic Coatings, 185*, 107916. https://doi.org/10.1016/J.PORGCOAT.2023.107916.

Pillai, A. R. S., Eapen, A. S., Zhang, W., & Roy, S. (2024). Polysaccharide-based edible biopolymer-based coatings for fruit preservation: A review. *Foods, 13*(10). https://doi.org/10.3390/FOODS13101529.

Priyadarshi, R., Roy, S., Ghosh, T., Biswas, D., & Rhim, J. W. (2022). Antimicrobial nanofillers reinforced biopolymer composite films for active food packaging applications-A review. *Sustainable Materials and Technologies, 32*, e00353.

Qahtan Salman, K., Rabih Mohammed, M., & Al-Shadeedi, S.M. (n.d.). Effect of some natural coating of table egg on shelf life during refrigerator storage. Journal of Genetic and Environmental Resources Conservation, 2023(2), 103–113. www.gercj.com.

Qiao, C., Ma, X., & Zhang, J. (2019). Characterization and mechanical properties of agar/konjac glucomannan films. *International Journal of Biological Macromolecules, 124*, 898–904. https://doi.org/10.1016/j.ijbiomac.2018.11.222.

Rhim, J. W., Wang,, L. F., & Hong, S. I. (2013). Preparation and characterization of agar/silver nanoparticles composite films with antimicrobial activity. *Food Hydrocolloids, 33*, 327–335.

Romruen, O., Kaewprachu, P., Karbowiak, T., & Rawdkuen, S. (2022). Development of smart bilayer alginate/agar film containing anthocyanin and catechin-lysozyme. *Polymers, 14*(22), 5042. https://doi.org/10.3390/POLYM14225042.

Roy, S., Kim, H. J., & Rhim, J. W. (2021). Effect of blended colorants of anthocyanin and shikonin on carboxymethyl cellulose/agar-based smart packaging film. *International Journal of Biological Macromolecules, 183*, 305–315.

Roy, S., Chawla, R., Santhosh, R., Thakur, R., Sarkar, P., & Zhang, W. (2023). Agar-based edible films and food packaging application: A comprehensive review. In *Trends in food science and technology*, *Vol. 141,* Elsevier Ltd. https://doi.org/10.1016/j.tifs.2023.104198.

Roy, S., & Rhim, J. W. (2019). Agar-based composite films reinforced with melanin nanoparticles for active packaging applications. *International Journal of Biological Macromolecules, 123*, 647–655. https://doi.org/10.1016/j.ijbiomac.2018.11.147.

Roy, S., & Rhim, J. W. (2021a). Antioxidant and UV blocking properties of agar films incorporated with lignin. *Food Packaging and Shelf Life, 28*, 100664. https://doi.org/10.1016/j.fpsl.2021.100664.

Roy, S., & Rhim, J. W. (2021b). Preparation of pectin/agar-based functional films integrated with zinc sulfide nano petals for active packaging applications. *Colloids and Surfaces B: Biointerfaces, 207*, 111999.

Roy, S., & Rhim, J. W. (2022). Starch/agar-based functional films integrated with enoki mushroom-mediated silver nanoparticles for active packaging application. *Bioscience, 49*, 101867.

Remisha Jahan, K. T., Meenakshi, P., Kaur, Gurmeet, Mailk, Divya, Roy, Swarup, & Biswas, Deblina (2026). *Biomaterials for Sustainable Food Packaging*. Springer Nature Switzerland, 269–294. https://doi.org/10.1007/978-3-032-08027-1_9.

Roy, S., Rhim, J. W., & Jaiswal, L. (2019). Bioactive agar-based functional composite film incorporated with copper sulfide nanoparticles. *Food Hydrocolloids, 93*, 156–166. https://doi.org/10.1016/j.foodhyd.2019.02.034.

San, H., Laorenza, Y., Behzadfar, E., Sonchaeng, U., Wadaugsorn, K., Sodsai, J., Kaewpetch, T., Promhuad, K., Srisa, A., Wongphan, P., & Harnkarnsujarit, N. (2022). Functional polymer and packaging technology for bakery products. *Polymers, 14*(18), 3793. https://doi.org/10.3390/POLYM14183793.

Sandoval, L. N., López, M., Montes-Díaz, E., Espadín, A., Tecante, A., Gimeno, M., & Shirai, K. (2016). Inhibition of *Listeria monocytogenes* in fresh cheese using chitosan-grafted lactic acid packaging. *Molecules, 21*(4). https://doi.org/10.3390/MOLECULES21040469.

Saravanan, S., Abdullah, N. A., Hasanah, N. N., Azman, E. M., Abedin, N. H. Z., & Ismail-Fitry, M. R. (2024). development of semolina starch/agar-based intelligent films by incorporating butterfly pea flower anthocyanins to monitor the freshness of prawns. *Sains Malaysiana, 53*(1), 87–98. https://doi.org/10.17576/jsm-2024-5301-07.

Sedayu, B. B., Cran, M. J., & Bigger, S. W. (2019). Review of the mechanical and barrier properties of biodegradable films and composites in food packaging applications. *Polymers, 11*(9), 1437. https://doi.org/10.3390/polym11091437.

Shahidi, F., & Hossain, A. (2022). Preservation of aquatic food using edible films and coatings containing essential oils: A review. *Critical Reviews in Food Science and Nutrition, 62*(1), 66–105. https://doi.org/10.1080/10408398.2020.1812048 Taylor and Francis Ltd.

Shankar, S, & Rhim, J. W (2017a). Preparation and characterization of agar/lignin/silver nanoparticles composite films with ultraviolet light barrier and antibacterial properties. *Food Hydrocolloids, 71*, 76–84.

Shankar, S., & Rhim, J. W. (2017b). Properties of agar-based composite films reinforced with silver nanoparticles. *Carbohydrate Polymers, 146*, 261–269. https://doi.org/10.1016/j.carbpol.2016.12.040.

Shankar, S., Wang, L. F., & Rhim, J. W. (2016). Properties of agar/lignin/silver nanoparticle composite films. *International Journal of Biological Macromolecules, 86*, 570–577. https://doi.org/10.1016/j.ijbiomac.2016.01.059.

Shikov, V. S., Avdeenko, E. P., & Belogurova, V. A. (2017). Development of agar-based biodegradable films with enhanced mechanical properties. *Food Hydrocolloids, 70*, 53–62. https://doi.org/10.1016/j.foodhyd.2017.03.009.

Silva, S. P. M., Ribeiro, S. C., Teixeira, J. A., & Silva, C. C. G. (2022). Application of an alginate-based edible coating with bacteriocin-producing *Lactococcus strains* in fresh cheese preservation. *LWT, 153*, 112486. https://doi.org/10.1016/J.LWT.2021.112486.

Song, X., Wei, X., Liu, L., & Liu, Y. (2024). Gelatin/agar pH-indicator film based on cranberry extract loaded with linalool nanoparticle: Survey on physical, antimicrobial, and antioxidant properties. *International Journal of Biological Macromolecules, 268*, 131767. https://doi.org/10.1016/J.IJBIOMAC.2024.131767.

Surendhiran, D., Li, C., Cui, H., & Lin, L. (2021). Marine algae as efficacious bioresources housing antimicrobial compounds for preserving foods-A review. *International Journal of Food Microbiology, 358*, 109416.

Swarup, R., & Rhim, J. W. (2019). Melanin-mediated synthesis of copper oxide nanoparticles and preparation of functional agar/CuO NP nanocomposite films. *Journal of Nanomaterials, 2019*.

Talukder, S., Mendiratta, S. K., Kumar, R. R., Agrawal, R. K., Soni, A., Luke, A., & Chand, S. (2020). Jamun fruit (*Syzgium cumini)* skin extract based indicator for monitoring chicken patties quality during storage. *Journal of Food Science and Technology, 57*(2), 537–548. https://doi.org/10.1007/S13197-019-04084-Y/METRICS.

Terme, N., Hardouin, K., Cortès, H. P., Peñuela, A., Freile-Pelegrín, Y., Robledo, D., & Bourgougnon, N. (2020). Emerging seaweed extraction techniques: Enzyme-assisted extraction a key step of seaweed biorefinery. In *Sustainable seaweed technologies*. Elsevier, 225–256.

Thakur, R., Wickramarachchi, S., Pal, K., & Sarkar, P. (2024). Gelatin/chitosan-lactate/curcuma hydroethanolic extract-based antimicrobial films: Preparation, characterization, and application on chicken meat. *Food Hydrocolloids, 154*, 110075. https://doi.org/10.1016/J.FOODHYD.2024.110075.

Uju, U., Santoso, J., Ramadhan, W., & Abrory, M. F. (2018). Extraction of native agar from Gracillaria sp with ultrasonic acceleration at low temperature. *Jurnal Pengolahan Hasil Perikanan Indonesia, 21*(3), 414–422.

Ummat, V., Sivagnanam, S. P., Rajauria, G., O'Donnell, C., & Tiwari, B. K. (2021). Advances in pre-treatment techniques and green extraction technologies for bioactives from seaweeds. *Trends in Food Science & Technology, 110*, 90–106.

Uranga, J., Nguyen, B. T., Si, T., Guerrero, P., & De la Caba, K. (2020). Agar-based films and coatings: Enhancing barrier properties. *Food Packaging Advances, 20*(3), 342–356.

Vejdan, A., Ojagh, S. M., & Abdollahi, M. (2017). Effect of gelatin/agar bilayer film incorporated with TiO_2 nanoparticles as a UV absorbent on fish oil photooxidation. *International Journal of Food Science & Technology, 52*(8), 1862–1868. https://doi.org/10.1111/IJFS.13461.

Wang, L., et al. (2018). Nano-bacterial cellulose and agar nanocomposite films for improved barrier properties. *International Journal of Food Science, 53*(1), 65–74.

Wang, L., Liu, Y., & Xie, W. (2018). Properties and characterization of agar-based films with bacterial cellulose. *Carbohydrate Polymers, 195*, 252–260. https://doi.org/10.1016/j.carbpol.2018.04.033.

Wang, X. H., Song, X. J., Zhang, D. J., Li, Z. J., & Wang, H. J. (2022). Preparation and characterization of natamycin-incorporated agar film and its application on preservation of strawberries. *Food Packaging and Shelf Life, 32*, 100863. https://doi.org/10.1016/J.FPSL.2022.100863.

Wardana, A. A., Wiramukti, A., & Widyaningsih, T. D. (2018). Application of bioindicator film from tapioca/agar/mulberry (*Morus alba* L.) to detect sausage spoilage. *IOP Conference Series: Earth and Environmental Science, 195*(1), 012061. https://doi.org/10.1088/1755-1315/195/1/012061.

Wongphan, S., Yoksan, R., & Phromyothin, P. (2020). Biodegradable agar-starch composite films for active packaging. *Journal of Food Science, 85*(3), 621–631. https://doi.org/10.1111/1750-3841.15089.

Xiao, Q., Wang, X., Zhang, J., Zhang, Y., Chen, J., Chen, F., & Xiao, A. (2021). Pretreatment techniques and green extraction technologies for agar from *Gracilaria lemaneiformis*. *Marine Drugs, 19*(11), 617.

Xiao, Q., Weng, H., Ni, H., Hong, Q., Lin, K., & Xiao, A. (2019). Physicochemical and gel properties of agar extracted by enzyme and enzyme-assisted methods. *Food Hydrocolloids, 87*, 530–540.

Yb, B., & Kola, M. (2019). Influence of composite edible coating systems on preservation of fresh meat cuts and products: A brief review on their trends and applications. *Article in International Food Research Journal*. https://www.researchgate.net/publication/332832700.

Yuan, Y., Hao, J., Su, C., Qi, C., & He, G. (2020). Challenges in plastic waste management. *Journal of Environmental Research, 98*, 87–101.

Zhang, B., Liu, Y., Wang, H., Liu, W., Cheong, K., & Teng, B. (2021). Effect of sodium alginate-agar coating containing ginger essential oil on the shelf life and quality of beef. *Food Control, 130*, 108216. https://doi.org/10.1016/J.FOODCONT.2021.108216.

Zhang, S., Sani, M., Zhang, L., McClements, D. J., & Jafari, S. M. (2023). Sustainable food packaging innovations: A review. *Food Packaging and Shelf Life, 38*, 100926.

Zhang, X., & Rhim, J. W. (2022). Agar-based active and smart packaging. *Food Hydrocolloids, 47*, 89–101.

Zhang, Y., Wu, W., & Wei, J. (2019). Development of agar-based edible films incorporated with natural extracts. *Food Chemistry, 289*, 211–220. https://doi.org/10.1016/j.foodchem.2019.03.041.

Chapter 6

Chitosan films and coatings

Vitor Emanuel de Souza Gomes[1], Paula Fernanda Janetti Bócoli[1,2], Amanda Alves Domingos Maia[2], and Luís Marangoni Júnior[1]

[1]*Department of Food Engineering and Technology, School of Food Engineering (FEA), University of Campinas (Unicamp), Campinas, São Paulo, Brazil,* [2]*Packaging Technology Center, Institute of Food Technology, Campinas, São Paulo, Brazil*

6.1 Introduction

Plastics are widely used in the development of food packaging due to their properties, mainly because they act as a barrier against external contaminants and for their mechanical characteristics, which favor the processing, transport, and distribution stages along the supply chain (Arfelli et al., 2024). The most common plastics in the food packaging sector come from fossil sources, such as polyethylene (PE), polypropylene (PP), and polyethylene terephthalate (PET), among others. In addition to being more accessible and lightweight, these polymers have low cost and high performance; these factors influence the choice of these materials by the consumer goods industries (Cheng et al., 2024). However, the increased production and accumulation of plastic waste that is disposed of incorrectly cause significant environmental impacts, motivating the search for alternatives that reduce the carbon footprint and greenhouse gas emissions associated with the large-scale use of these materials (Gupta et al., 2024).

An alternative to minimize this impact is the development of packaging using polymers from renewable sources, which can promote sustainability by replacing the raw material of packaging. These compounds, originated from natural sources, can be derived from animals, plants, fungi, algae, and bacteria, and consist mainly of proteins, lipids, and carbohydrates (Kumar et al., 2024; Ribeiro et al., 2024). Among polysaccharide-based biopolymers, chitosan stands out as the second most abundant natural polysaccharide (Crini & Lichtfouse, 2019). It is obtained through alkaline hydrolysis (deacetylation) of chitin, composed of β-glucosamine molecules joined by β-1,4-glycosidic bonds, found in the exoskeletons of crustaceans and insects, in addition to the cell walls of bacteria and fungi (Wei et al., 2019). Residues of crustaceans, such as crabs, lobsters, and shrimp, are the most used by industry as chitin sources for chitosan production (Haghighi et al., 2020).

In terms of application, chitosan has become, over the last few decades, one of the most studied biopolymers. In the food industry, chitosan is promising as a replacement for synthetic polymers for the development of films and coatings due to its characteristics, such as nontoxicity, biodegradability, edibility, and film-forming ability (Abd El-Hack et al., 2023). In addition, it stands out for its antioxidant, antibacterial, and antifungal properties, acting against pathogenic and deteriorating microorganisms (Flórez et al., 2022). Compared to cellulose (more abundant biopolymer), chitosan has higher mechanical strength, higher crystallinity, stronger hydrogen bonds, and lower solubility (Stefanowska et al., 2023).

Pure chitosan films exhibit flexibility, transparency, and good gas barrier properties. However, their mechanical properties, thermal stability, capacity to interact with water, and production costs may limit their production on an industrial scale and their use in some applications (Haghighi et al., 2020). Thus, research is needed to improve the methods for producing and applying chitosan films, including combinations with other biopolymers (Crini & Lichtfouse, 2019), such as polysaccharides (pectin [Ngo et al., 2020], cellulose nanofiber [Xu et al., 2019], among others) and proteins (gelatin [Rezaee et al., 2018], collagen [Bhuimbar et al., 2019], whey protein [Yangilar, 2015], soy protein isolate [Xing et al., 2018], among others).

Thus, the objective of this chapter is to explore the development of chitosan-based films and coatings, emphasizing their properties, characteristics, and applicability, in addition to the challenges and limitations. Moreover, this chapter will address trends for the packaging sector, aiming at a more sustainable and innovative future aligned with food safety goals, which can provide a renewable and biodegradable alternative to traditional synthetic polymers.

Natural Polymer-Based Films and Coatings in Food Technology. DOI: https://doi.org/10.1016/B978-0-443-34244-8.00020-3

6.2 Chemical composition

Chitin, derived from the Greek word *chiton*, meaning "tunic" or "covering," was first isolated in 1811 from fungi as an alkali-resistant fraction (Ding et al., 2012; Muzzarelli et al., 2012). This natural biopolymer, produced through biosynthesis, is the second most abundant natural polymer in nature, only after cellulose (Piekarska et al., 2023; Silva et al., 2025). Chitin is usually considered a common natural biopolymer extensively used for decay protection and as a nontoxic, biodegradable, and effective oxygen barrier, in addition to having antimicrobial properties (Benavides & Franco, 2023; Silva et al., 2025; Zhang, Khan, et al., 2024; Zhang, Wang, et al., 2024). In addition, chitin presents a linear aminopolysaccharide structure that consists of N-acetyl-2-amino-2-deoxy-D-glucopyranose and 2-amino-2-deoxy-D-glucose in pyranose form and linked by 1–4 glycosidic bonds, with functional groups (-OH and -NH_2) (Khayrova et al., 2021).

The exoskeletons of marine organisms, by-products of fishing industry, invertebrates, insect cuticles, cell walls of fungi, green algae, and yeast are the most common sources of chitin (Alimi et al., 2023; Khayrova et al., 2021). Chitin content and properties such as yield, structure, crystallinity index, purity, morphology, and thermal stability differ, varying according to source; for example, crustacean shells, major source of this raw material, can contain around 20%–46% of chitin (Aoun et al., 2024; Khayrova et al., 2021). Chitin can be classified as crystalline microfibrils, presenting three isomeric structures: α-chitin, the most common and resistant chitin; β-chitin; and γ-chitin (Alimi et al., 2023; Isobe et al., 2022).

Chitosan is a cationic polysaccharide obtained by a chitin deacetylation process at higher temperatures (Barik et al., 2024) and is the most important chitin by-product, with functional groups, such as amino groups with chelating ligands capable of binding various metal ions properties and global annual production around 10 million tons (Lárez-Velásquez, 2023; Li et al., 2024). Chitosan can be represented by a straight-chain alkaline polysaccharide with 2-amine-2-deoxy-β-(1,4)-glucosamine and 2-acetamide-2-deoxy-β-(1,4)D-glucosamine, obtained by partial chitin deacetylation (55% or more) (Jantzen da Silva Lucas et al., 2021; Li et al., 2024; Xu et al., 2022).

Chitosan (pKa ~6.3) presents a positive charge at low pH, indicating the protonation of amine groups and acting as polyelectrolyte. Thus, chitosan presents higher water solubility than chitin and can be dissolved in weak acidic solutions (acetic, formic, hydrochloric, nitric, and lactic acids), adopting several conformations and structures, functionalized with a wide range of functional groups, allowing the modification of its surface composition for specific applications (Barik et al., 2024; Jiménez-Gómez & Cecilia, 2020; Khayrova et al., 2021; Picos-Corrales et al., 2023). These properties suggest chitosan as an important biomaterial for applications in industries (Alimi et al., 2023).

Chitin and chitosan present a similar chemical structure (Fig. 6.1), but they are insoluble in both aqueous and organic solvents (Barik et al., 2024; Jiménez-Gómez & Cecilia, 2020). However, chitosan exhibits higher solubility, high bacteriostatic, antimicrobial, biocompatibility, and biodegradability properties (Alimi et al., 2023).

Several chemical and biological chitin extraction methods are under study and development (Table 6.1). Typically, the main extraction process steps involve demineralization, deproteinization, and deacetylation of chitin, depending on source material and desired properties (Ben Aoun et al., 2024; Vo et al., 2024). Currently, chitosan can also be obtained by enzymatic reaction, microwave- and ultrasound-assisted reactions, and electrochemical processes, with ionic liquids and deep eutectic solvents (Ben Aoun et al., 2024; Felicia et al., 2024).

In order to have an efficient extraction process and to avoid the use of organic solvents and strong acid/alkaline solutions, enzymatic hydrolysis has been considered to remove proteins and calcium carbonates from chitosan and chitin, respectively (Lucas et al., 2021). Biological extraction methods, using proteolytic enzymes and microbiological fermentation, have some advantages, such as eco-friendly characteristics, lower energy consumption, and reduced solvent consumption (Aoun et al., 2024).

6.3 Preparation of chitosan-based films

To develop chitosan films for the food sector, various preparation methods are used to meet the requirements of this application. The solution casting method is the most viable on a laboratory and pilot scale, being widely applied in the manufacture of chitosan films. In turn, extrusion is a technique mainly explored to develop composite films combining chitosan with other biopolymers or even with synthetic polymers. In addition to these methods, electrospinning has been explored in the development of films based on chitosan, requiring its combination with other natural or synthetic polymers to improve its processability (Fig. 6.2).

6.3.1 Casting

In the development of chitosan-based films by the casting method, chitosan characteristics, such as source, molecular weight, and degree of deacetylation (DD), must be considered to ensure the good performance of the films obtained

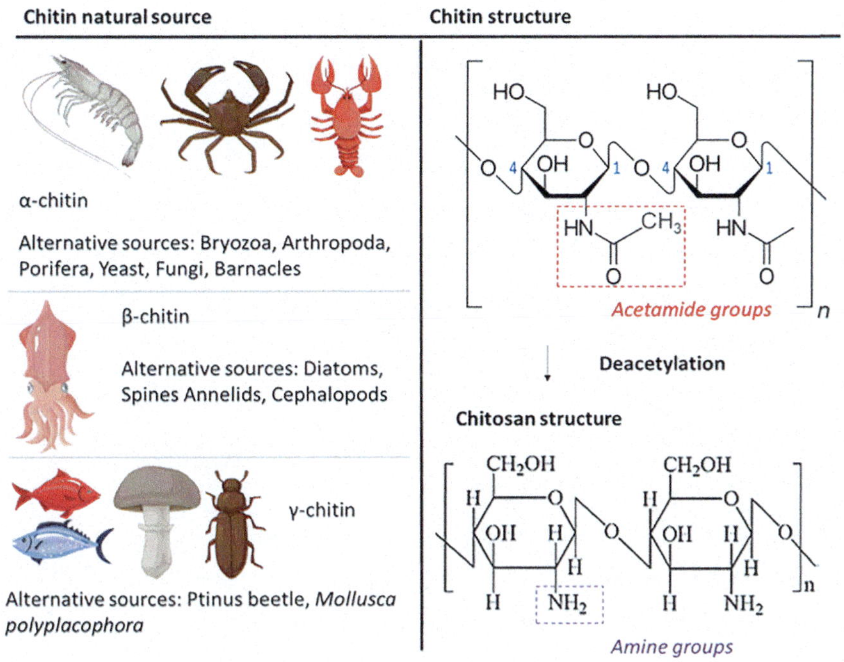

FIGURE 6.1 Natural sources of three allomorph structures (α, β, γ) and the chemical structure of chitin with the N-acetylglucosamine units that repeat to form long chains in the β-(1→4)-linkage and chemical structure of chitosan after the deacetylation process.
Reprinted with permission from Huang et al. (2025); Khayrova et al. (2021); Mohan et al. (2022). Elsevier, Copyright (2021).

(Barik et al., 2024). The first step involves dissolving chitosan (powder) in a weak acid solution, as chitosan is not soluble at pH 7. Acids such as acetic, lactic, citric, and formic acids are often used to obtain the filmogenic solution. Next, the biopolymer-solvent mixture is stirred and can be combined with heating to facilitate complete dissolution of chitosan. In general, temperatures between 24°C and 60°C are used. This step also has the addition of additives that contribute to the better processing and performance of the films, such as plasticizers, including polyols (glycerol, sorbitol, and polyethylene glycol [PEG]), sugars (glucose and sucrose), or lipids. After reaching a homogeneous filmogenic solution, it is filtered or centrifuged to remove insoluble particles and air bubbles formed during mixing. The solution is then deposited on a clean, flat surface, such as a Petri dish, and subjected to a drying step under controlled conditions of time, temperature, and humidity, which can occur in a vacuum oven, hot air oven, or climatic chamber. Finally, the film formed is removed from the surface, stored, and prepared for analyses, characterization, and application in foods (Barik et al., 2024; Zhao et al., 2023). In addition to the mentioned method, some laboratories have continuous casting pilot equipment, aiming to increase the scalability of film production. In addition, other techniques may be incorporated to enhance film properties, such as ultrasound and high-pressure homogenization.

6.3.2 Extrusion

Unlike the development of films by casting, the extrusion method for producing chitosan-based films does not require excessive use of solvents and provides a viable alternative for the development of biopolymer-based films on an industrial scale, making their commercialization viable. The first step consists of mixing several materials with a previously defined formulation, which is carried out with the aid of a mixer. Then, the mixed material passes through a twin-screw extruder, where it is subsequently cut into pellets and proceeds to the oven drying process. After drying, the dried pellets are extruded to produce the films, generally using a twin-screw extruder fitted with a flat die (Sarfraz et al., 2024).

The extrusion process is generally used to produce composites. Thus, cassava starch and chitosan composite films were produced. To this end, premixes containing native cassava starch, glycerol, water, and stearic acid were prepared at the ratios of 58:25:15:2, in addition to chitosan, glycerol, water, and acetic acid at the ratios of 34:39:22:5. The combination of the cassava starch and chitosan premixes at the ratios of 100/0, 75/25, and 50/50 was extruded in a corotating twin-screw extruder, with L/D ratio of 40, screw diameter of 16 mm, and die of 3 mm. Temperature conditions ranged between 120°C and 135°C, with screw speeds of 250 and 275 rpm, and feed rate between 0.2 and 0.9 kg $hour^{-1}$. The extrusion process resulted in films with promising properties, such as improved mechanical characteristics and reduced permeability to water vapor (Rodriguez Llanos et al., 2021).

TABLE 6.1 Major chitin extraction methods.

Sources	Extraction method	Route	Process	Impact	Advantages	Disadvantages	References
Crustacean waste, fish scales, insect exoskeletons, and microorganisms	Deacetylation	Hydrolysis of acetamide groups and cleavage of the acetyl group with the formation of an amino group	Chitin to chitosan through alkaline treatment, removing the acetyl group, while varying temperatures, time, and high concentrations of Na, KOH, and sodium hydroxide (NaOH)	Degree of deacetylation and the molecular weight of chitosan are related to deacetylation process, interfering with solubility, viscosity, and bioactivity	Short processing time; high deacetylation degree (%) of final product; industrial scale; efficient removal of organic salts	Environmentally unfriendly; solubilized minerals and proteins cannot be recovered and used as nutrients	Aoun et al. (2024); El Knidri et al. (2018); Isobe et al. (2022); Novikov et al. (2023); Xu et al. (2022)
	Demineralization	Chemical treatment with acid solvent	Removal of mineral content (calcium carbonate) using acid treatments (dilute hydrochloric acid)	Purity and quality			Aoun et al. (2024); El Knidri et al. (2018); Isobe et al. (2022); Novikov et al. (2023); Xu et al. (2022)
	Deproteinization	Chemical treatment with alkaline solvent and high temperatures (>100°C)	Removal of proteins using alkaline treatments with NaOH	Degree of deproteinization is related to the purity and structural integrity. Molecular weight and the degree of acetylation are related to the percent of solid/liquid ratio			Aoun et al. (2024); El Knidri et al. (2018); Isobe et al. (2022); Mohan et al. (2022); Novikov et al. (2023); Xu et al. (2022)
Crabs, insects, shrimp, crustaceans, and mollusks	Decolorization	Oxidants (hydrogen peroxide, sodium hypochlorite, phosphorus pentoxide, sulfur dioxide, potassium permanganate, and	Removal of pigment (astaxanthin and β-carotene) following several methods (alcohol washing, chloroform and alcohol solutions,	Purity and quality	High-quality chitin	Pigments can be economically important	Abidin et al. (2020); El Knidri et al. (2018); Lv et al. (2023); Marzieh et al. (2019)

		sodium carbonate) and organic/inorganic solvents (acetone, chloroform, ethyl acetate, and ethanol)	phosphorus pentoxide solutions, potassium permanganate, oxalic acid and sulfuric acid solutions, and sodium hypochlorite solutions)				
Crustacean shells	One-pot conversion	Converting chitin into chitosan oligosaccharides	One-pot synthesis of N,N-dimethyl chitosan oligosaccharide from chitin, considering acid-catalyzed tandem depolymerization-deacetylation-N-methylation with formaldehyde as the methylation reagent	Depolymerization, deacetylation, and N-methylation in a single step	Cost-effective	Depolymerization and deacetylation processes under strong acidic environment are environmentally unfriendly	Cheng et al. (2024)
Crab, Acetes, shrimp shells, squid pen, and mealworm's cuticles	Enzymatic	Filtration, drying, and precipitation processes with enzymes (chitinase, cellulase, protease, alcalase, esperase, neutrase, trypsin, ficin, and papain)	Deproteinization, deacetylation, and decrystallization of chitin by different enzymes (bacteria and fungi). Complex fungal (*Mucor circinelloides*) and bacterial (*Acetobacter pasteurianus and Bacillus subtilis*) fermentation for the deproteinization process, and calcium carbonate with protease and organic acids	Highest rate of deproteinization and demineralization, better yield, cost-effective, optimal activity depends on specific temperatures and pH levels	High quality and degree of deacetylation of the chitin; environmentally friendly; safe for food technology	High-time reaction and residual protein; expensive technology	Deng et al. (2020); Doan et al. (2019); El Knidri et al. (2018); Marzieh et al. (2019); Minguet-Lobato et al. (2024); Zhao et al. (2024)

(Continued)

TABLE 6.1 (Continued)

Sources	Extraction method	Route	Process	Impact	Advantages	Disadvantages	References
Shrimp waste (heads and shells)	Biological extraction	Three-step and single-step microbial fermentation techniques (microbial fermentation/ bacterial biotransformation)	*Bacillus cereus* converts chitin to chitosan through NaOH and acetic acid medium. *Paenibacillus mucilaginosus* and *Lactobacillus plantarum* (lactic acid bacteria)	The difference in morphology, structural modification, and surface charge	High quality of final product; environmentally safe; mineral and protein subproducts; maintenance of chitin structure; improved resource utilization	Long processing time; laboratory scale	Doan et al. (2021); El Knidri et al. (2018), Innocent and Padikasan (2023); Mohan et al. (2022)
	Chemical extraction		Major steps		Facultative step		
	Biological extraction						

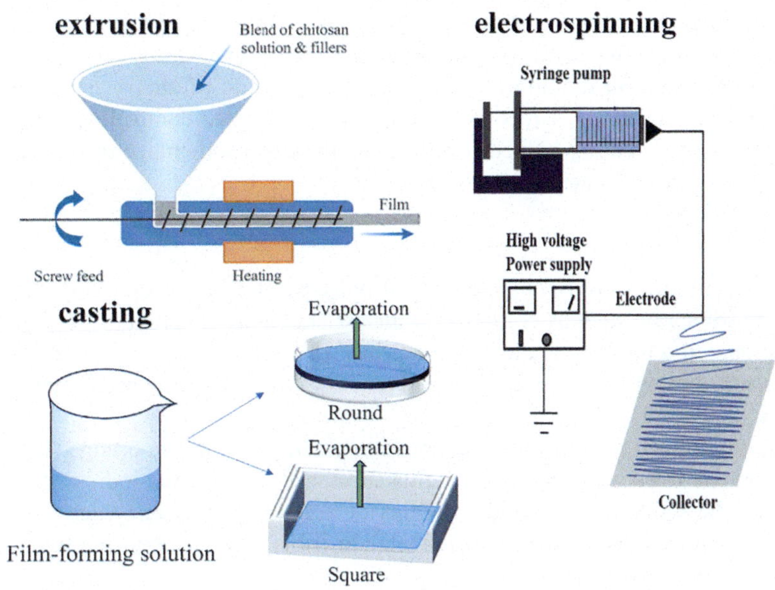

FIGURE 6.2 Methods for preparing chitosan films.
Reproduced with permission from Jin, J., Luo, B., Xuan, S., Shen, P., Jin, P., Wu, Z., & Zheng, Y. (2024). Degradable chitosan-based bioplastic packaging: Design, preparation and applications. International Journal of Biological Macromolecules, 266 (Pt) 1131253. https://doi.org/10.1016/j.ijbiomac.2024.131253. Elsevier, Copyright (2024).

6.3.3 Electrospinning

Electrospinning is a widely explored technique for the development of films using chitosan, standing out for its capacity to quickly process a variety of polymeric materials in the form of nano and microfibers (Zhao et al., 2023). The process occurs with the application of an electric field between the equipment's electrodes, which drives the movement of the polymer solution, forming a charged jet that is stretched, allowing solvent evaporation and fiber solidification. Several parameters influence the characteristics of the fibers produced, including applied voltage, distance between needle and collector, solution concentration, and polymer surface tension and viscosity (Vergara-Figueroa et al., 2020). In chitosan films, electrospinning is widely used to produce composite films. The development of nanofibers of poly-3-hydroxybutyrate (P(3HB)) and chitosan used a plastic syringe (external diameter = 1.3 mm and internal diameter = 0.79 mm), spinneret with a radius of 0.3 mm, a feed rate of 5 mL hour^{-1}, applied voltage of 25 kV, and a distance of 25 cm between the needle tip and collector (Sukhanova et al., 2020). This technique, when applied to chitosan films, has shown potential for microbial control in food. A study with electrospun nanofibers prepared with chitosan, gelatin, and moringa oil observed a significant antibacterial effect against *Listeria monocytogenes* and *Staphylococcus aureus* in cheese stored at 4°C and 25°C, with minimal impact on product color and sensory quality during four-day storage (Lin et al., 2019).

6.3.4 Coating

Different coating techniques, such as spread coating, spray coating, and layer-by-layer assembly (LBL), can be used to apply chitosan-based coatings to food. These coatings are applied directly to the food surface, acting as a primary packaging and forming a barrier that contributes, above all, to increasing food's shelf life.

6.3.4.1 Spread coating

An effective technique for coating food with chitosan films is spread coating. This method can be applied directly to the food surface, providing high protection against microorganisms due to direct contact (Jin et al., 2024). The first step consists of selecting food-grade materials to prepare the chitosan coating solution, which is then spread over the product with a brush or spatula, forming a uniform layer. After application, the film is dried, thus creating a coating on the product, which must then be stored under refrigeration or modified atmosphere, such as vacuum (Priyadarshi & Rhim, 2020).

In addition to the direct method, spread coating can also be carried out indirectly by applying chitosan to the surfaces of packaging materials (such as plastic film, paper, and cardboard), with the objective of increasing food protection during storage. Chitosan coatings not only significantly inhibit microorganism proliferation and oxidation, but they may also improve material performance (Jin et al., 2024). An active chitosan coating on kraft paper incorporating silver-impregnated zeolite was developed with the aim of extending the shelf life of cherry tomatoes. The results showed that the active packages developed tended to reduce the fruits' ripening rate, controlling ethylene concentration within the package. In addition, using this coating helped to minimize the environmental impact caused by the degradation of synthetic polymer coatings on paper (Soares et al., 2023).

6.3.4.2 Spray coating

Another chitosan coating technique is spray coating, which consists of directly applying the film solution to the food surface by means of an aerosol spray, such as compressed air-assisted coastal sprayers. As in the spread coating method, the solution needs to be prepared with food-grade raw materials. In addition, the spray coating method can be combined with different heat treatments, aiming to increase its effectiveness in inhibiting the microbial load in the product (Kumar et al., 2020; Muxika et al., 2017).

Oligochitosan spraying was applied to control *Colletotrichum gloeosporioides* in orange (laranja-baía) during storage at room temperature, where orange trees were treated with this substance. The results indicated that protopectin degradation was delayed during storage. In addition, increased activities of defense-related enzymes were observed in the peels of fruits treated with oligochitosan, with smaller lesion diameter in fruits that received the treatment (Deng et al., 2015).

6.3.4.3 Dip coating

The dip coating technique is an effective alternative to develop chitosan-based coatings, contributing to extend the shelf life of fresh food. This method assists in controlling microbial growth and preserving the sensory characteristics of the product, making it more resistant to cold damage. Compared to other methods for applying biopolymeric chitosan coatings, dip coating is simpler: properly sanitized and dried food is immersed directly in chitosan solutions with previously adjusted pH, remaining submerged for a certain time, which can range from 30 seconds to 30 minutes, depending on the type of application. This process may involve one, two, or even three immersions (Sarfraz et al., 2024). After treatment, excess coating is removed to ensure a uniform layer; then food is dried and stored under controlled conditions of temperature, pH, and relative humidity (Priyadarshi & Rhim, 2020). In studies on tomato coating, the use of different concentrations of chitosan delayed ripening and prolonged the shelf life of the product by up to 30 days, demonstrating the effectiveness of chitosan in reducing ethylene production (Sree et al., 2020).

6.3.4.4 Layer-by-layer assembly

Another technique that can be used to develop chitosan-based films is the LBL assembly method. This method requires no sophisticated equipment and uses polyelectrolytes obtained from polysaccharides, which capture liquid charges. Thus, there is attraction of opposite charges on the surface, as illustrated in Fig. 6.3. The coating is formed by means of spraying solutions or by immersion in a polyelectrolytic solution. pH variations significantly influence the development of films, as they change the electrical charge, which modifies the amount of polymer deposited (Barik et al., 2024).

A film composed of modified cellulose nanofiber, polylactic acid, and chitosan was developed using the LBL assembly method. In this process, chitosan served as booster and antimicrobial agent. For its preparation, a chitosan solution was dissolved in glacial acetic acid for 12 hours, forming a viscous solution. This solution was then applied over a flat glass and dried at 80°C. Polylactic acid was dissolved in dichloromethane at 6 mg mL^{-1} and mixed with the cellulose nanofibers at different concentrations (0%, 2%, 5%, 8%, and 10% by weight), and sonicated to improve dispersion. The resulting mixture was applied over the chitosan layer, air dried, and then subjected to vacuum drying at 40°C to remove residual solvents (Niu et al., 2018).

6.4 Chitosan film properties

Currently, the development and improvement of environmentally friendly and safe materials with functional properties are the most significant challenges. The application of films and coating materials can improve product shelf life and other properties. Several studies demonstrate the advancement of chitosan films and coatings. However, a thorough review of bio-based films and coatings that incorporate bioactive ingredients is essential (Gupta et al., 2024).

Chemical and molecular interactions developed by blending chitosan with active ingredients can enhance various properties. Important characteristics to consider include biocompatibility, film-forming ability, mechanical flexibility, and potential for functionalization (Guo et al., 2024; Gupta et al., 2024).

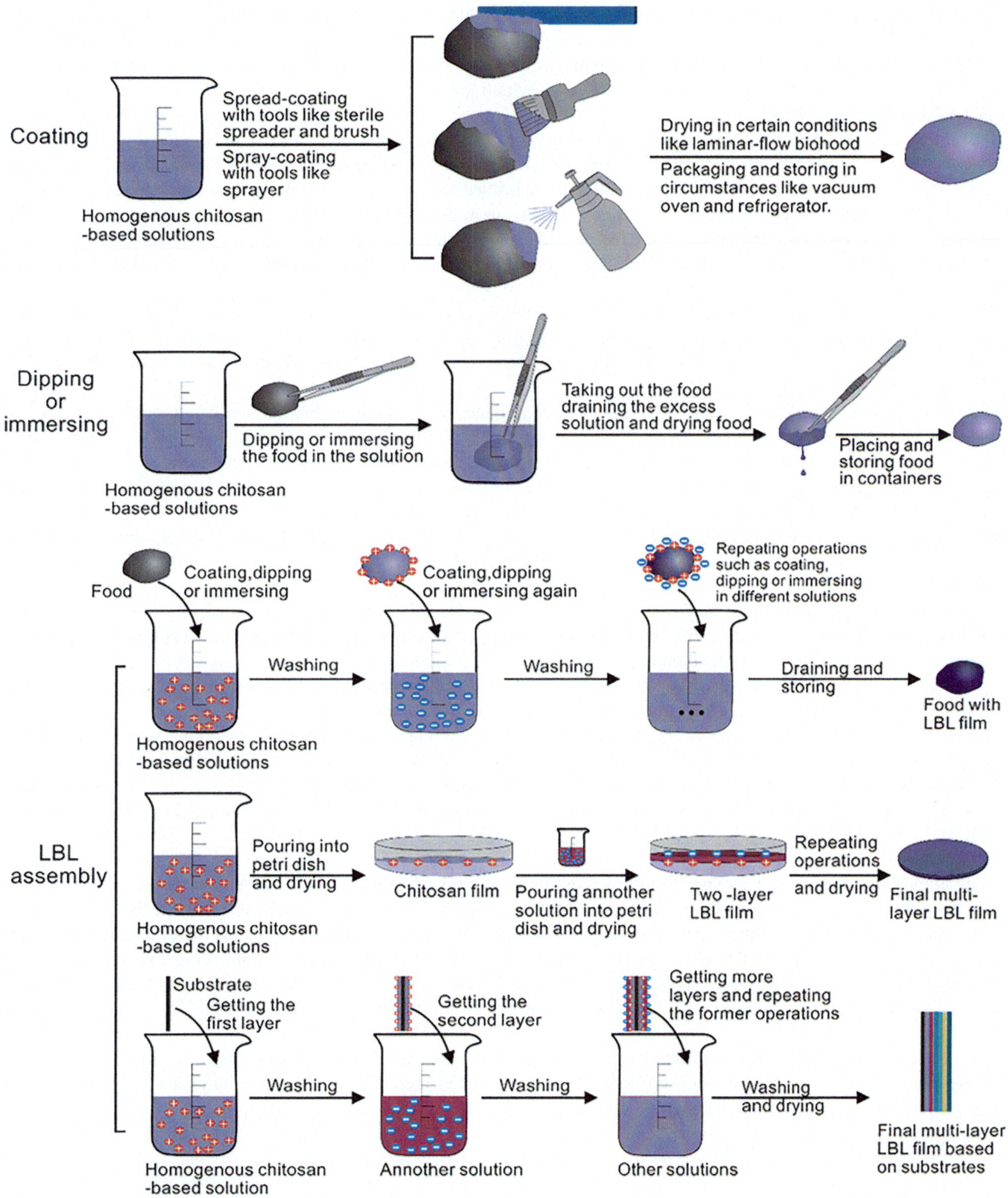

FIGURE 6.3 Chitosan coating methods (Wang et al., 2018)
Reproduced with permission from Wang, H., Qian, J., & Ding, F. (2018). Emerging Chitosan-Based Films for Food Packaging Applications. Journal of Agricultural and Food Chemistry, 66, (2), 395–413 https://doi.org/10.1021/acs.jafc.7b04528. American Chemical Society, Copyright (2018).

Chitosan is one of the most important, affordable, biocompatible, biodegradable, and eco-friendly resources. However, chitosan-based materials exhibit poor mechanical strength and water barrier properties (Flórez et al., 2022; Garreau et al., 2023; Mu et al., 2024; Vo et al., 2024). Thus, blending chitosan with active additives enhances its properties, enabling the use of the developed material in numerous applications, which are key characteristics explored in several studies (Aoun et al., 2024; Guo et al., 2024).

The improvement of mechanical and biodegradability properties of chitosan-based films can be related to the addition of plasticizers, most commonly blended with glycerin, sorbitol, and PEG, interfering in moisture sorption and antimicrobial activity. Plasticizers provide improved flexibility and reduced brittleness for chitosan-based films, resulting in higher elongation, lower tensile strength properties, films with lower solubility in water, and higher optical properties (Perera et al., 2023). Excessive use of plasticizers can compromise the effectiveness of chitosan-based films. Higher concentrations of glycerol, propylene glycol, and PEG can reduce mechanical and tensile strengths as well as barrier properties (Jin et al., 2024; Rojas-Lema et al., 2024).

However, this additive also decreases the antimicrobial activity of chitosan-based films, especially under high temperatures. Regarding biodegradability, chitosan films degrade faster in humid conditions, and while plasticizers enhance certain properties of chitosan films, it is crucial to balance these improvements with potential reductions in biodegradability (Jin et al., 2024; Rojas-Lema et al., 2024).

The noncovalent interactions between chitosan and alginate make them highly compatible biopolymers for food packaging and pharmaceutical applications (Perera et al., 2023). Sodium alginate, derived from brown algae, exhibits excellent water solubility, gel-forming ability, and film-forming properties. However, alginate–chitosan films typically show lower mechanical strength and water resistance. In contrast, the starch–chitosan blend is considered a promising material for film production due to its enhanced properties, including odorlessness, film-forming capability, low cost, renewability, and nontoxicity (Milbreta et al., 2024; Zhou et al., 2024). Additionally, blends that include pectin and nanoparticles improve transparency, mechanical and antimicrobial aspects, also extending the shelf life of food products (Perera et al., 2023).

Other important chitosan-based blending materials include lipids, oils, and fatty acids, attracting interest due to their potential applications in food packaging, drug delivery systems, and others, by improving the mechanical, thermal, and biological properties. Chitosan films blended with olive, clove, oregano, ginger, and corn oils exhibit enhanced surface morphology, thermal stability, tensile strength, elongation at break, water sorption, antimicrobial properties, and permeability, increasing their effectiveness as packaging materials (Jiang et al., 2023; Stefanowska et al., 2023).

Incorporation of cellulose nanocrystals, polyphenols, and transglutaminase enzymatic reaction increases mechanical strength. Quercetin reinforcement in chitosan-based materials decreases water vapor permeability, and the blending of chitosan with polyvinyl alcohol (PVA), clays, graphene oxide, and metal nanoparticles improves the structural integrity and water/gas permeability of the chitosan-based materials (Heras et al., 2024; Khanzada et al., 2024; Rostamabadi et al., 2024; Shi et al., 2024).

Chitosan's mechanical strength, flexibility, and barrier properties may be negatively impacted by the addition of incompatible substances, such as specific plasticizers or hydrophilic agents. These interactions can alter the structural integrity, water resistance, and biodegradability of the films, depending on the combinations of materials.

Chitosan crosslinking is an important process for improving the properties of chitosan-based materials for a wide range of applications, especially food packaging. This process alters the polymer structure, enhancing mechanical strength, biocompatibility, stability, hydrophobicity, and functionality by creating bonds between the polymer chains. Chitosan can be crosslinked through ionic bonding, covalent crosslinking, and interfacial polymerization methods (Bisla & Yoshitake, 2024; Spitzer et al., 2024; Yang et al., 2024).

However, challenges persist in optimizing crosslinking agents to achieve an ideal balance between mechanical strength and biodegradability. It is crucial to ensure that the materials meet the specific requirements of each application while maintaining environmental sustainability (Yang et al., 2024; Zhang, Khan, et al., 2024; Zhang, Wang, et al., 2024). Therefore, careful selection and optimization of additives and processes are essential to preserve the desired properties of chitosan-based materials.

6.5 Physical, mechanical, and barrier properties

6.5.1 Mechanical properties

The chemical, physical, mechanical, and biological properties of chitosan are determined by parameters such as molecular weight, DD—percentage of amino groups, polymer chain size, and the composition of chitosan itself, which is related to the distribution of the two components that constitute the polymer chain of the material—acetylated glucosamine and deacetylated glucosamine (Kołodziejska et al., 2021; Sarfraz et al., 2024; Shi et al., 2024).

Among the mechanical characteristics of chitosan-based films, studies show that better properties are related to increased or decreased DD. This type of performance may be due to the greater availability of reactive groups for interaction (Liu et al., 2012; Liu et al., 2020) and by increasing chain entanglement within and between chitosan crystals (Lazaridou et al., 2003).

As in the case of traditional synthetic polymers, information on tensile strength and percentage elongation, among others, is an important parameter for knowing the performance of chitosan films during their application. In addition to the intrinsic factors of the chitosan type used to manufacture the films, the amount of chitosan, the manufacturing process, the solvent type, and the storage time and temperature of the films are also critical factors for the mechanical performance of the materials (Barik et al., 2024; Stefanowska et al., 2023).

Compared to petroleum-based films, chitosan films exhibit a significant limitation regarding mechanical properties, particularly concerning elongation at break. The incorporation of plasticizers such as glycerol, sorbitol, and PEG can improve this characteristic, also facilitating the application of the film on an industrial scale (Shi et al., 2024).

The combination of chitosan with other materials or nanoparticles can make it an active substrate with often improved properties (Barik et al., 2024; Haghighi et al., 2020; Stefanowska et al., 2023). Films produced from chitosan and grape seed extract (GSE) showed higher moduli of elasticity and breaking strength when compared to films produced only with chitosan (Sogut & Seydim, 2018). Increased tensile strength in three chitosan films incorporating starch and glycerol; starch, gallic acid, and glycerol; and cane starch and olive oil was observed by Jiang et al. (2023). In this same study, when examining the incorporation of gelatin and other compounds into chitosan substrates, in addition to tensile strength, Young's modulus also increased in the presence of the combination of gelatin and nanocellulose. Table 6.2 shows some examples of chitosan formulations added with different compounds and their influence on the mechanical properties of the films produced.

The capacity to produce a package by means of heat sealing, which involves joining two layers of film through the pressure exerted between two plates heated for a certain time, is a fundamental requirement for the film to be viable on an industrial scale as a packaging material. Despite its technological potential, the inability to heat seal a chitosan film limits its application as a packaging material, and in this case, mixing it with a thermoplastic copolymer, such as starch or gelatin, is an alternative to circumvent this limitation.

Considering this context, Alves et al. (2022) studied the performance of films produced with different mixtures of chitosan and starch, along with the addition of reduced graphene oxide, in relation to sealing and to mechanical and water vapor barrier properties. As a result, the authors concluded that combining 75% starch with 25% chitosan led to higher seal strength (0.17 N mm^{-1}), indicating that this chitosan mass fraction acts as a plasticizer, helping to improve molecular entanglement during sealing.

Consistently, Prateepchanachai et al. (2019) sought to evaluate the effect of the combination of gelatin, glycerol, and CO_2 application on a chitosan substrate that promoted a better sealing capacity for the material, in addition to improving its mechanical properties. The combination of glycerol and gelatin promoted the highest seal strength among all samples, and the actual strength of the film prepared through the addition of glycerol and use of CO_2 to treat the film surface was lower than that of other films, except for the chitosan-only film. According to the authors, this type of behavior is due to the reaction between the molecules of glycerol and CO_2, since the presence of the three CHO bonds of glycerol helps in the formation of hydrogen bonds, making the bond stronger. The highest seal strength was obtained for the film prepared through the solution with 25% (w/w) glycerol and 10% (w/v) gelatin, with a value of 110.7 N m^{-1}. Even though this value is much lower than that normally obtained for a synthetic polymer, the authors considered this to be a good performance, even when compared to matrices of other biopolymers, such as starch and sago films presented in the literature.

Therefore, as demonstrated in this section, chitosan films still have limitations regarding their mechanical properties, but these can be improved through the incorporation of additives, reinforcing agents, crosslinking agents, and mixtures with other matrices from renewable and/or fossil sources.

6.5.2 Barrier properties: gases and water vapor

The intrinsic characteristics of the chitosan molecule, such as molecular weight and DD, as well as the use of additives and plasticizers, are important for the films' barrier properties, considering the hydrophilic nature of chitosan films, the barrier to water vapor is quite limited (Flórez et al., 2022; Shi et al., 2024). According to Kerch et al. (2010), with an increase in the molecular weight of chitosan, the water vapor transmission rate (WVTR) also increases.

TABLE 6.2 Different formulations of chitosan-based films and their influence on mechanical properties.

Formulation	Mechanical property	Barrier property	References
Starch (82–77/100 g) Chitosan (0–5/100 g) Glycerol (18/100 g) Oregano extract oil (0–1.0/100 g)	Influence on percentage elongation at break (%): 21.95 ± 1.98 to 48.40 ± 5.32	Water vapor permeability (WVP) × 10^{-10} (g [Pa m s]$^{-1}$): decreased from 1.39 ± 1.56 to 0.62 ± 0.15	Pelissari et al. (2009)
Chitosan and 1% salvia oil	Increase in tensile strength values (MPa): 20–26 on average	No results presented	Souza et al. (2017)
Chitosan and turmeric extract (1:2)	Increase in tensile strength values (MPa): 32.2 ± 0.9 to 47.9 ± 2.6	WVP × 10^{-4} (g mm [h m^2 Pa]$^{-1}$) increased slightly: 4.03 ± 0.15–5.51 ± 0.32	Kalaycıoğlu et al. (2017)
Chitosan (2% w/v), citric acid (1% v/v), glycerol (0.5:1)	Increase in elongation (%): 2.22 ± 0.15 to 26.43 ± 2.62 Decrease in tensile strength (MPa): 52.22 ± 0.15 to 9.48 ± 1.35 and tensile modulus (MPa): 12.40 ± 0.58 to 29.80 ± 0.57	Similar results to WVP (x 10^{12} g cm^{-1} s^{-1} Pa^{-1}): 3.27 ± 0.36 and 2.32 ± 0.21	Priyadarshi et al. (2018)
Starch (4% w/v) Chitosan (1.5% w/v) Citric acid (5%, 10%, 15%, and 20% w/w)	Increase in tensile strength (MPa) from 9.72 ± 0.87 to 10.49 ± 0.38 and elastic modulus (%) from 5.70 ± 0.1 to 8.96 ± 0.62	WVP (x 10^{12} g cm^{-1} s^{-1} Pa^{-1}) decreased from 3.03 to 2.05	Wu et al. (2019)
Chitosan and polyvinyl alcohol (PVA) with incorporation of ethyl vanillin (EV) at 1:3, 1:1, and 3:1 ratios	Increase in tensile strength (MPa) with increasing ratio of EV from 25 to 70	Water vapor transmission rate (g m^{-2}.d) decreased from 35 to 17 and oxygen transmission rate (cc [m^2.d]$^{-1}$) decreased from 1200 to 500	Narasagoudr et al. (2020)
Blend of chitosan with PVA using different agents: citric acid, succinic acid, and tetraethoxysilane (TEOS)	PVA favored tensile strength (from 17.9 ± 0.3 to 45.0 ± 0.7 MPa) and elongation at break (from 6.7% ± 0.7% to 86.4% ± 0.9%). Presence of the agents also increased the strength of the films for all formulations	No results presented	Wardhono et al. (2022)
Blend of chitosan with essential oils (EO)—10 wt.% and incorporation of silver nanoparticles (AgNP)—5 wt.%	Increase in tensile strength (MPa) with incorporation of AgNP from 60.4 ± 4.9 to 69.7 ± 3.9 and reduction with addition of EOs from 60.4 ± 4.9 to 51.3 ± 8.1	WVP (x 10^{-9} g m [m^2. Pa. s]$^{-1}$) increased with the presence of AgNP (from 1.26 ± 0.08 to 1.43 ± 0.05) and with EOs + AgNP (from 1.26 ± 0.08 to 1.42 ± 0.07)	Shankar et al. (2021)
Chitosan (2% w/v) EO (lemon, cinnamon, thyme)—1% Blend of two EOs—0.5% each	Increase in tensile strength in films with incorporation of cinnamon EO (12.31%) and the blend of cinnamon and lemon EO (19.25%) compared to the control film	WVP decreased with the addition of EOs (from 9.36 to 7.52 × 10^{-11} g m^{-1}s Pa^{-1})	Peng and Li (2014)

To improve this characteristic, lipid compounds, such as essential oils, are often added to chitosan. Studies show good results for water vapor barrier with the incorporation of essential oils of bergamot, citronella, cedarwood, and lemon, among others (Peng & Li, 2014; Sánchez-González et al., 2010; Shen & Kamdem, 2015). According to the authors, this effect may be due to the covalent interactions between the matrix and the constituent components of essential oils, mainly oxygenated terpenes, leading to a crosslinking reaction between them (Shen & Kamdem, 2015).

In the case of the addition of phenolic compounds, a reduction in water vapor permeability values is generally observed, due to the interactions between hydrogen and covalent bonds within the chitosan network and phenolic compounds, which reduces the availability of hydrophilic groups and, consequently, the affinity of the chitosan matrix for water molecules

(Balti et al., 2017; Kurek et al., 2018; Peng & Li, 2014). Nanoclays can also provide a gain in water barrier properties of chitosan films, due to the increase in the tortuous path formed in the polymer chain, making it difficult for water vapor molecules to pass through (Pires et al., 2021).

As for gas barrier, chitosan appears as a material with potential for a good barrier to CO_2 and O_2; however, the molecular weight and solvent type used in the formulation of chitosan films may influence their permeability (Shi et al., 2024). The application of reinforcing materials, such as nanofibers, cellulose nanocrystals, carbon nanotubes, and nanoclays, can be an important option for incorporation into chitosan, providing gains in the barrier properties of the material. This type of property is achieved due to the increased diffusion path that slows the permeation of gases through the biopolymer. The incorporation of zinc oxide nanoparticles (ZnO NPs) is efficient both for the antimicrobial action of chitosan films and for the good performance of barrier properties, in addition to being nontoxic when added to the films, being recognized as safe by the US Food and Drug Administration (FDA) as generally recognized as safe (GRAS) (Pires et al., 2021). According to Yadav et al. (2021), with the addition of ZnO NPs, oxygen permeability values decreased by up to 41% when compared to pure chitosan film.

6.5.3 Light and color barrier properties

Foods and beverages may be exposed to ultraviolet (UV) and visible radiation while being sold at points of sale. In both situations, in the case of light-sensitive products, this type of interaction can be detrimental to food shelf life, requiring a barrier usually provided by the packaging material.

Literature data indicate that chitosan films have almost no protection from the incidence of light in the visible range (400–700 nm), allowing a passage of up to 93%, and in the ultraviolet range (200–400 nm), transmittance values can reach 88% (Stefanowska et al., 2023).

Considering the importance of a light barrier for some applications of chitosan-based films, studies suggest that the application of natural additives rich in phenolic compounds, such as lychee pericarp powder or α-tocopherol, may increase their capacity to absorb UV radiation, improving the barrier property (Pal et al., 2021; Vilela et al., 2017). The same performance was observed by Mujtaba et al. (2019) by incorporating additives such as gallic acid, gentisic acid, and protocatechuic acid (PCA) into chitosan. The use of poly(limonene) (PLM) provided great barrier results for the chitosan film, especially in the UV region (Gonçalves et al., 2023). According to the authors, the incorporation of PLM, regardless of the concentration studied, introduced a very efficient light-blocking effect to the film, reaching 60% higher values than the control film in the region of 315–400 nm.

If, on the one hand, light protection can be provided to the chitosan film through structural modifications with other compounds, film color control also requires attention, considering the impact on the appearance of the product packaged or coated with the chitosan film. Studies have shown that the chitosan film's transparency is an intrinsic property of the material that can be influenced by its formulation and preparation method (Stefanowska et al., 2023).

The addition of GSE, at concentrations of 5%, 10%, and 15%, was studied by Sogut and Seydim (2018) in relation to chitosan film properties, and the authors found increase in film opacity with increasing GSE concentrations. According to them, this behavior can be attributed to the formation of an impenetrable matrix created by phenolic interactions, promoting light scattering through the film. In this case, light-sensitive foods become more protected against oxidation processes and/or color loss. Similar effects were also observed in the incorporation of turmeric and tea polyphenols, according to Wang et al. (2013) and Liu et al. (2017), respectively.

6.6 Active packages

6.6.1 Antimicrobial activity

Antimicrobial activity is an important biological property of chitosan, being active against several microorganisms, such as bacteria and fungi. Its action is correlated with the degree of polymerization of the molecule, the host, the composition of chitosan, and the substrate (Jiang et al., 2023). Studies evaluating the antimicrobial activity of chitosan in Gram-positive and Gram-negative bacteria have shown great results, especially for Gram-positive bacteria, such as *S. aureus* (Alimi et al., 2023).

Chitosan, in its pure form, has higher antimicrobial effectiveness at pH below 6.5, due to its characteristic structure with positively charged amino groups, which have affinity for the negatively charged membranes of microorganisms, promoting the binding of chitosan to pathogenic bacteria, thus increasing antimicrobial activity (Sarfraz et al., 2024).

There are several hypotheses on the form of action of chitosan in relation to antimicrobial activity. One hypothesis assumes that, in chitosan-based films, a protective layer is formed on food surface, blocking external microorganisms.

Another assumption is that the chitosan film restricts oxygen permeability, inhibiting microbiological growth, especially of aerobic microorganisms. A third antimicrobial mechanism includes adsorption of chitosan polymers on the surface of microorganisms, blocking their development (Jiang et al., 2023).

According to the authors (Fernández-Pan et al., 2015; Latou et al., 2014; Sudarshan et al., 1992), another hypothesis is that the NH_3^+ groups present in chitosan may interact with the negatively charged phosphoryl groups, leading to the leakage of intercellular electrolytes, resulting in bacterial death. Chitosan can also inhibit microbial replication by penetrating the nucleus, shielding DNA, and preventing RNA synthesis, or it can chelate essential nutrients and metals, making them unavailable to microorganisms (Sudarshan et al., 1992; Wang et al., 2019).

Chitosan, in addition to the traditional sources of crustacean exoskeletons, can also be extracted from fungi and yeasts, such as mushrooms. According to studies comparing the antimicrobial potential of chitosan from mushrooms and marine sources, chitosan from mushrooms showed greater action, and this characteristic is attributed to its lower molecular weight in relation to crustacean sources (Alimi et al., 2023; Savin et al., 2020).

The application of natural essential oils in edible films has major potential for food preservation. In chitosan-based films, these oils can regulate release rates of antimicrobial agents to maintain the concentration of active compounds over a long period. Studies with ginger oil, cinnamon oil, and thyme oil showed important results regarding microbiological action for different food matrices. In many cases, other properties of the films, such as light barrier and mechanical strength, were also improved with these combinations (Jiang et al., 2023; Stefanowska et al., 2023).

The incorporation of poly(β-pinene) oligomers, a monomer of natural origin with essential oil characteristics, widely found in nature, also provided great results regarding antimicrobial action in films (Rodrigues et al., 2021). According to the authors, for all microorganisms tested in the study, the blend of chitosan with the monomer was efficient against *S. aureus, Aspergillus niger*, *C. gloeosporioides*, and *Fusarium oxysporum*.

Similarly, the incorporation of phenolic compounds into chitosan films can positively influence microbiological inactivation, in addition to antioxidant action and modifications in mechanical properties, such as the application of GSE, carvacrol, and polyphenolic groups from kombucha-type tea (Jiang et al., 2023).

Another method for applying chitosan with antimicrobial potential is through its nanoparticle form. Several studies have documented its efficacy against different microorganisms through the synthesis of chitosan-based nanoparticles (Sarfraz et al., 2024). In addition, chitosan nanoparticle encapsulation with natural compounds has the potential to increase its activity spectrum (Hu et al., 2016).

Antifungal activity has also been found as a property of chitosan for different species, such as *Candida albicans*, after the incorporation of cellulose nanofibers into chitosan film (Stefanowska et al., 2023). Noshirvani et al. (2017) presented excellent results for chitosan films incorporating ginger and cinnamon oils regarding the antifungal action against the species *A. niger*.

6.6.2 Antioxidant activity

Antioxidant activity is another important aspect of chitosan that depends mostly on the presence of reactive groups in the chitosan biopolymer, which include the amino and hydroxyl groups, and on the molecular weight. The antioxidant effect of chitosan films occurs due to the capacity of free amino groups ($-NH_2$) to react with free radicals to form a stable macromolecule with ammonium groups ($-NH_3^+$), through reaction with hydrogen ions from the solution, resulting in effective antioxidant activity (Alves et al., 2022; Sarfraz et al., 2024).

In its natural form, chitosan already has antioxidant activity; however, this property can be improved by the incorporation of derivatives of the molecule itself into the structure or other compounds. There are several studies related to the mechanisms of action involved in the antioxidant activity of essential oil compounds, which include free radical scavenging, chain initiation prevention, reducing agents, peroxide termination, singlet oxygen quenching, and metal ion catalyst binding (Muñoz-Tebar et al., 2023).

Films enriched with olive pomace (de Moraes Crizel et al., 2018) or with mango leaf extract (Rambabu et al., 2019) showed effective protection against nut oxidation for 31 and 28 days, respectively. The positive effect of the antioxidant action of chitosan films combined with other compounds applied for the protection of fresh fruits and vegetables, such as apple, mango, and lychee, may be indicated by parameters such as color, firmness, weight loss, or respiration rate during food shelf life. The combination of antimicrobial and antioxidant action of chitosan films applied to fishery products—such as ground trout filet and carp filets—was assessed through the occurrence of late oxidation of lipids and proteins, demonstrating the efficiency of the films used in the studies (Flórez et al., 2022).

The incorporation of essential oils into chitosan films has shown promising results in delaying lipid oxidation in meat and fish products (Yuan et al., 2016). According to the authors, this type of coating preserved the characteristics of the products for an extended period under refrigerated conditions, and the sensory attributes were not affected. Similarly, chitosan films with essential oils were also effective when applied to fruits and vegetables, reducing respiration rates, gas exchange, and weight loss, as well as retaining some phenolic compounds, thus ensuring their nutritional quality.

Another study combined chitosan film with banana peel extract (BPE), which has antioxidant properties, aiming to expand its application as a coating to maintain the quality of apples during storage. The incorporation of BPE resulted in more yellowish films due to the lutein present in banana peels and contributed to a significant enhancement in the opacity of the chitosan film, which is considered an advantage for inhibiting lipid oxidation and preserving the organoleptic properties of the packaged food (Zhang et al., 2020).

A combined chitosan film with bio-vanillin (BV) and kaolin clay (KC) to produce an antifungal and biodegradable food packaging was studied, and the results showed that the film presented, besides strong antimicrobial activity with a 90% reduction in *Escherichia coli* and *S. aureus* growth and a 75% reduction in fungal growth, an antioxidant capacity of 80%, compared to chitosan film (Subramani & Manian, 2024).

The application of chitosan-based films incorporated with antioxidant agents represents a promising and sustainable strategy for active food packaging. These systems have demonstrated efficacy in mitigating oxidative processes, thereby preserving the physicochemical, nutritional, and sensory attributes of food during storage. Given their biodegradability and functional versatility, such films offer a valuable alternative to conventional synthetic materials, contributing to both food quality preservation and environmental sustainability.

6.7 Applications of chitosan films and coatings

The application of chitosan has major potential in the food industry, mainly due to its characteristics of being a nontoxic, biodegradable, biofunctional, biocompatible material that can be widely modified, with significant improvements in its active, mechanical, and barrier properties (Alves et al., 2022; Stefanowska et al., 2023). In addition, the formulation of chitosan films with other constituents is an economically favorable alternative, as it is a simpler process than synthesizing new polymers (Haghighi et al., 2020).

The varied chitosan options available on the market, with different molecular weights and DDs, significantly influence the properties obtained in the films. In addition, films can be produced from a matrix of pure or modified chitosan, or added with other substances, according to the objective to be achieved (Stefanowska et al., 2023). The functionalization of chitosan films with bioactive compounds, such as essential oils, plant extracts, nanoemulsions, or natural antioxidants, has further enhanced their technological applications. These additives provide antioxidant and/or antimicrobial action to the materials, which is particularly relevant in perishable foods, in which lipid oxidation and microbial growth are the main deterioration factors.

Chitosan-based films have potential applications for fresh products (fruits, vegetables, cheeses, meats, and fish) and foods with a short to medium shelf life (Haghighi et al., 2020). Foods such as tomatoes, carrots, bananas, papayas, pomegranates, fish, and others can be preserved with the use of pure chitosan films. Storing these foods with this type of coating results in greater stability, preserving antioxidant activity, and nutraceutical properties, which, in turn, prolongs the shelf life of the product (Cazón & Vázquez, 2020). Chitosan treatment in the fresh produce industry is safe for the consumer and the environment, and chitosan has been approved by the US Food and Drug Administration (USFDA) as a GRAS food additive (Romanazzi et al., 2017).

6.7.1 Applications in fruits and vegetables

The application of chitosan-based coatings to fruits and vegetables has shown positive results in reducing the respiration rate, weight loss through transpiration, and maintaining the firmness and color of the fruits. Such coatings act as semipermeable barriers, reducing gas exchange and oxidative stress, in addition to creating an environment less favorable for the growth of fungi and bacteria on the surface of the food. Fruits and vegetables treated with chitosan have a higher nutritional value, because chitosan can retain the contents of the ascorbic and phenolic compounds (Romanazzi et al., 2017).

The effectiveness of chitosan coatings on fruits and vegetables is influenced by concentration, pH, molecular weight, and deacetylation degree. Temperature and coating time also play a role, affecting the viscosity, adhesion, thickness, and quality of the coating. The type of fruit matters as well; different fruits require different chitosan formulations for optimal preservation (Saberi Riseh et al., 2024).

The deposition of a chitosan film layer on red fruits and grapes using the spray coating process was studied by Tessarin et al. (2016). The authors found a significant effect on flavonol and phenolic acid contents, without affecting fruit ripening and alcoholic fermentation.

Strawberries are fruits that possess a very short shelf life because of their high metabolic activity, water content, and susceptibility to contamination with fungal strains. Because of this, research was conducted to evaluate the impact on the shelf life and nutritional values of three newly developed chitosan-based edible coatings on strawberries and apple slices. The study showed that the treatments preserved the ascorbic acid, total polyphenol content, and antioxidant activity for all the tested samples compared to the control sample (Popescu et al., 2022).

Nowadays, chitosan/flavonoid films have been widely used in the preservation of fruits, vegetables, aquatic products, meat products, and edible oils, and are useful for indicating their freshness. The application of chitosan/anthocyanin films to fruits results in a gradual depletion of soluble solids and titratable acidity during storage, which promotes a progressive color shift of the films toward red hues (Liu et al., 2025). Still in this context, plasticizers and emulsifiers, when incorporated into chitosan films, result in improved flexibility and mobility of the biopolymer chain, which consequently also improves mechanical strength.

6.7.2 Applications in products of animal origin

Chitosan films and coatings have been widely used in the preservation of meat, fish, and processed products. Their antioxidant action directly inhibits lipid oxidation, one of the main causes of rancidity and changes in flavor in meat products. In addition, chitosan's antimicrobial activity helps control spoilage flora, slowing the growth of bacteria such as *L. monocytogenes*, Salmonella spp., and *E. coli*. When functionalized with bioactive compounds, these films promote a synergistic action that extends the shelf life of the product, maintaining its color, texture, and sensory acceptability during refrigerated storage.

Fishery products such as rainbow trout, hake, bass, and swordfish were stored by combining vacuum packaging with the application of pure chitosan films, yielding satisfactory results (Sarfraz et al., 2024).

In recent years, smart chitosan films with pH indicators have been widely studied. These films are used to monitor food freshness, as the indicator changes color when the product undergoes a pH variation. Thus, consumers can easily check if the food they are about to consume is in good condition (Flórez et al., 2022). For this type of application, there are several studies related to fishery products, considering the perishability of such foods and the importance of ensuring safety for consumers.

The effect of adding alizarin to chitosan film applied to fish was studied by Ezati and Rhim (2020). The results showed a film color change from light yellow to purple, which clearly indicated a pH change in the range of 4–10, demonstrating changes in the product. The application of butterfly pea (*Clitoria ternatea L.)* extract led chitosan films to change color from purple blue to dark green during fish preservation, also demonstrating potential modification in the product, according to Yan et al. (2021).

In this line of application, Yang et al. (2019) assessed the performance of chitosan films with black and purple rice extract, applied to pork, while also seeking to trace product pH changes based on film color change. Vo et al. (2019) conducted a study through the application of a film composed of chitosan, poly(vinyl alcohol), and anthocyanin extracted from red cabbage to pork belly slices, showing a light green color change when a slightly alkaline pH range is detected in the meat.

Similar studies have also been conducted on dairy products. Films formulated with cellulose and chitosan with addition of carrot anthocyanins have shown potential as food-grade biomaterials to control milk freshness and spoilage. In this study, after 48-hour storage at 20°C, a change from blue to pink-purple was observed, indicating a change in the milk (Ebrahimi Tirtashi et al., 2019).

Still in this context, a film with chitosan, melissa essence, and pomegranate peel extract was applied to creamy cheese. As the cheese's time and temperature increased, the samples became acidic, and this pH change resulted in a color change from blue to red (Pirsa et al., 2020).

Therefore, all examples presented highlight a range of application possibilities for chitosan films and coatings aimed at extending the shelf life of various foods, especially in relation to formulations with active and smart agents.

6.8 Trends and challenges in functional chitosan films

Currently, chitosan is one of the most explored biopolymers in the manufacture of films and coatings (Fig. 6.4. The application of chitosan during the development of new nontoxic, biocompatible, and renewable compounds with antioxidant, anticoagulant, antitumor, and antimicrobial properties can be related to the mitigation of environmental issues (Du et al., 2021; Saravanan et al., 2023).

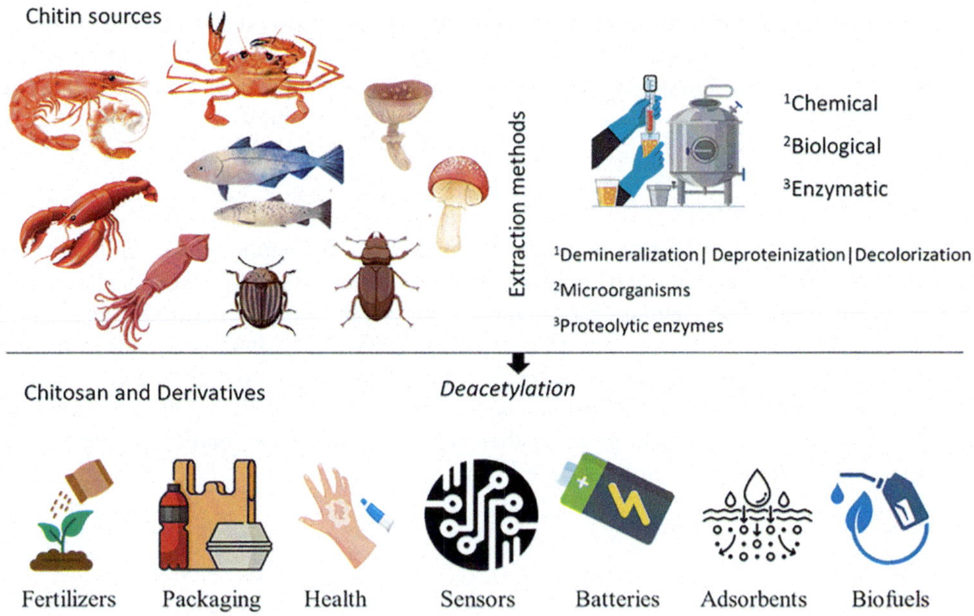

FIGURE 6.4 Chitin sources, extraction routes, and common chitin-based products developed.
From Aoun, R., Ben, Trabelsi, N., Abda lah, M., Mourtzinos, I., & Mhamdi, R. (2024). Towards a greener future: Exploring the cha lenges of extraction of chitin and chitosan as bioactive polysaccharides. Materials Today Communications, 39, https://doi.org/10.1016/j.mtcomm.2024.108761; Lárez-Velásquez, C. (2023). Chitosan: an overview of its multiple advantages for creating sustainable development poles. Polimeros, 33, (1). https://doi.org/10.1590/0104-1428.20220103 Available from http://w ww.scielo.br/scielo.php?script=sci_arttext&pid=S0104-14282023000100502&tlng=en.

These properties and characteristics of this raw material promoted an extensive application in multiple and different areas, such as biomedicine and pharmaceuticals, agriculture, food and packaging processing, waste and water treatment, textiles, and others (Vo et al., 2024). Chitosan properties enable the development of new materials with different concepts, technologies, and methodologies, such as monolayer, multilayer, and composite materials, polysaccharide concentration, additives, temperatures, and drying methodologies, and others (Díaz-montes & Castro-muñoz, 2021).

Innovative methods such as LBL assembly, nanoparticle incorporation, and the use of probiotics have also proven effective in extending the shelf life of various food products. These techniques not only enhance the viability of new materials but also improve food preservation through various mechanisms (Pandey et al., 2020).

However, the successful manufacturing process of chitosan is related to an alternative extraction method ensuring an efficient use, cost-effectiveness, and biocompatibility of the material (Aoun et al., 2024). Recently, to minimize environmental issues and to develop eco-friendly materials and green processes, several research efforts are under consideration (Jin et al., 2024; Mu et al., 2024; Vo et al., 2024).

Thus, chitosan has been studied in several areas, including food, environmental, cosmetics, agriculture, and biomedical fields, and can be processed as gels, nanoparticles, films, composites, and others (Garreau et al., 2023). However, chitosan raw material often exhibits low mechanical strength, high sensitivity to moisture and oxygen, and production and processing require optimization, such as incorporation of additives, plasticizers, and nanoparticles improving the chitosan materials' properties and ensuring the efficiency of the material (Barleany et al., 2020; Díaz-montes & Castro-muñoz, 2021; Garreau et al., 2023).

Chitosan presents both hydrophilic (amine and hydroxyl groups) and hydrophobic (acetyl groups) sites, containing both polar and nonpolar groups (Han et al., 2023; Wang et al., 2023). Chitosan's solubility in acidic media precludes its use in numerous applications (Garreau et al., 2023).

Nevertheless, the methods for processing and producing chitosan films require optimization to ensure consistency and scalability. While incorporating additives can enhance the films' properties, it also complicates the production process (Barleany et al., 2020; Díaz-montes & Castro-muñoz, 2021).

Novel and effective food preservation methods include irradiation, ultrasound, high-pressure treatment, modified atmosphere packaging, chemical materials, low temperatures, and others. However, these methods have technological and/or economic limitations, which limit their application (Heras et al., 2024).

Advanced methods require specialized equipment, qualified personnel, high initial costs, economic viability, and they can promote nutrient loss or changes in food texture and flavor, decreasing consumer acceptance (Felicia et al., 2024; Mu et al., 2024; Vo et al., 2024; Zhang, Khan, et al., 2024; Zhang, Wang, et al., 2024).

Therefore, in recent years, research on chitosan-based films and coatings has focused on the effective development and characterization of chitosan polymeric composites and their properties. Studies have also explored alternative blends and their integration into chitosan-based materials, in addition to the effects and impacts of nanotechnology on chitosan-based films and coatings (Aoun et al., 2024; Díaz-montes & Castro-muñoz, 2021; Jin et al., 2024; Vo et al., 2024).

6.9 Conclusion

Considering the content presented in this chapter, it can be concluded that the application of chitosan as a food packaging material demonstrates major potential in the development of more sustainable and efficient solutions for food product conservation, whether in the form of packaging or as a coating. Its origin from a renewable source and its antimicrobial properties are determining factors for reducing environmental impact while providing a barrier against oxidation, thus extending food shelf life. Furthermore, chitosan emerges as a promising alternative compared to conventional plastics, consistent with the growing demands for more sustainable practices. However, it is still necessary to conduct further studies to optimize its properties, in combination with other types of compounds, and to ensure the economic viability required to make its large-scale implementation feasible.

References

Abd El-Hack, M. E., Kamal, M., Alazragi, R. S., Alreemi, R. M., Qadhi, A., Ghafouri, K., Azhar, W., Shakoori, A. M., Alsaffar, N., Naffadi, H. M., Taha, A. E., & Abdelnour, S. A. (2023). Impacts of chitosan and its nanoformulations on the metabolic syndromes: A review. *Brazilian Journal of Biology, 83*. https://doi.org/10.1590/1519-6984.276530.

Abidin, N. A. Z., Kormin, F., Abidin, N. A. Z., Anuar, N. A. F. M., & Bakar, M. F. A. (2020). The potential of insects as alternative sources of chitin: An overview on the chemical method of extraction from various sources. *International Journal of Molecular Sciences, 21*(Issue 14), 1–25. https://doi.org/10.3390/ijms21144978.

Alimi, B. A., Pathania, S., Wilson, J., Duffy, B., & Frias, J. M. C. (2023). Extraction, quantification, characterization, and application in food packaging of chitin and chitosan from mushrooms: A review. *International Journal of Biological Macromolecules, 237*. https://doi.org/10.1016/j.ijbiomac.2023.124195, www.elsevier.com/locate/ijbiomac.

Alves, Z., Ferreira, N. M., Ferreira, P., & Nunes, C. (2022). Design of heat sealable starch-chitosan bioplastics reinforced with reduced graphene oxide for active food packaging. *Carbohydrate Polymers, 291*. https://doi.org/10.1016/j.carbpol.2022.119517, http://www.elsevier.com/wps/find/journaldescription.cws_home/405871/description#description.

Aoun, R., Ben, Trabelsi, N., Abdallah, M., Mourtzinos, I., & Mhamdi, R. (2024). Towards a greener future: Exploring the challenges of extraction of chitin and chitosan as bioactive polysaccharides. *Materials Today Communications, 39*. https://doi.org/10.1016/j.mtcomm.2024.108761.

Arfelli, F., Roguszewska, M., Torta, G., Iurlo, M., Cespi, D., Ciacci, L., & Passarini, F. (2024). Environmental impacts of food packaging: Is it all a matter of raw materials? *Sustainable Production and Consumption, 49*, 318–328. https://doi.org/10.1016/j.spc.2024.06.032.

Balti, R., Mansour, M. B., Sayari, N., Yacoubi, L., Rabaoui, L., Brodu, N., & Massé, A. (2017). Development and characterization of bioactive edible films from spider crab (*Maja crispata*) chitosan incorporated with Spirulina extract. *International Journal of Biological Macromolecules, 105*, 1464–1472. https://doi.org/10.1016/j.ijbiomac.2017.07.046, www.elsevier.com/locate/ijbiomac.

Barik, M., BhagyaRaj, G. V. S., Dash, K. K., & Shams, R. (2024). A thorough evaluation of chitosan-based packaging film and coating for food product shelf-life extension. *Journal of Agriculture and Food Research, 16*. https://doi.org/10.1016/j.jafr.2024.101164, https://www.sciencedirect.com/science/journal/26661543.

Barleany, D. R., Sari, D. K., & Lestari, R. S. D. (2020). Emerging trends in the development of chitosan films for active food packaging. *World Chemical Engineering Journal, 4*(Issue 2). http://jurnal.untirta.ac.id/index.php/WCEJ.

Ben Aoun, R., Trabelsi, N., Abdallah, M., Mourtzinos, I., & Mhamdi, R. (2024). Towards a greener future: Exploring the challenges of extraction of chitin and chitosan as bioactive polysaccharides. *Materials Today Communications, 39*. https://doi.org/10.1016/j.mtcomm.2024.108761, https://www.sciencedirect.com/science/journal/23524928.

Benavides, S., & Franco, W. (2023). Innovative integration of Arrayan (*Luma apiculata*) extracts in chitosan coating for fresh strawberry preservation. *International Journal of Molecular Sciences, 24*(19), 14681. https://doi.org/10.3390/ijms241914681.

Bhuimbar, M. V., Bhagwat, P. K., & Dandge, P. B. (2019). Extraction and characterization of acid soluble collagen from fish waste: Development of collagen-chitosan blend as food packaging film. *Journal of Environmental Chemical Engineering, 7*(2). https://doi.org/10.1016/j.jece.2019.102983, http://www.journals.elsevier.com/journal-of-environmental-chemical-engineering/.

Bisla, V., & Yoshitake, H. (2024). Control of mechanical and hydrophobic properties of silylated chitosan-starch films by cross-linking using carboxylic acids. *Carbohydrate Polymer Technologies and Applications, 7*, 100462. https://doi.org/10.1016/j.carpta.2024.100462.

Cazón, P., & Vázquez, M. (2020). Mechanical and barrier properties of chitosan combined with other components as food packaging film. *Environmental Chemistry Letters, 18*(2), 257–267. https://doi.org/10.1007/s10311-019-00936-3, https://www.springer.com/journal/10311.

Cheng, J., Gao, R., Zhu, Y., & Lin, Q. (2024). Applications of biodegradable materials in food packaging: A review. *Alexandria Engineering Journal, 91*, 70–83. https://doi.org/10.1016/j.aej.2024.01.080.

Crini, G., & Lichtfouse, E. (Vol. Eds.), (2019). *Sustainable agriculture reviews 36: Vol. 36*Springer International Publishinghttps://doi.org/10.1007/978-3-030-16581-9.

de Moraes Crizel, T., de Oliveira Rios, A., D. Alves, V., Bandarra, N., Moldão-Martins, M., & Hickmann Flôres, S. (2018). Active food packaging prepared with chitosan and olive pomace. *Food Hydrocolloids, 74*, 139–150. https://doi.org/10.1016/j.foodhyd.2017.08.007.

Deng, L., Zhou, Y., & Zeng, K. (2015). Pre-harvest spray of oligochitosan induced the resistance of harvested navel oranges to anthracnose during ambient temperature storage. *Crop Protection, 70*, 70–76. https://doi.org/10.1016/j.cropro.2015.01.016.

Deng, J. J., Mao, H. H., Fang, W., Li, Z. Q., Shi, D., Li, Z. W., Zhou, T., & Luo, X. C. (2020). Enzymatic conversion and recovery of protein, chitin, and astaxanthin from shrimp shell waste. *Journal of Cleaner Production, 271*. https://doi.org/10.1016/j.jclepro.2020.122655.

Díaz-montes, E., & Castro-muñoz, R. (2021). Trends in chitosan as a primary biopolymer for functional films and coatings manufacture for food and natural products. *Polymers, 13*(5), 1–28. https://doi.org/10.3390/polym13050767, https://www.mdpi.com/2073-4360/13/5/767/pdf.

Ding, F., Shi, X., Li, X., Cai, J., Duan, B., & Du, Y. (2012). Homogeneous synthesis and characterization of quaternized chitin in NaOH/urea aqueous solution. *Carbohydrate Polymers, 87*(1), 422–426. https://doi.org/10.1016/j.carbpol.2011.07.069.

Doan, C. T., Tran, T. N., Nguyen, V. B., Vo, T. P. K., Nguyen, A. D., & Wang, S. L. (2019). Chitin extraction from shrimp waste by liquid fermentation using an alkaline protease-producing strain, *Brevibacillus parabrevis*. *International Journal of Biological Macromolecules, 131*, 706–715. https://doi.org/10.1016/j.ijbiomac.2019.03.117.

Doan, V. K., Ly, K. L., Tran, N. M. P., Ho, T. P. T., Ho, M. H., Dang, N. T. N., Chang, C. C., Nguyen, H. T. T., Ha, P. T., Tran, Q. N., Tran, L. D., Vo, T. V., & Nguyen, T. H. (2021). Characterizations and antibacterial efficacy of chitosan oligomers synthesized by microwave-assisted hydrogen peroxide oxidative depolymerization method for infectious wound applications. *Materials, 14*(16). https://doi.org/10.3390/ma14164475.

Du, C., Zhao, X., Song, W., He, N., Jiang, S., Zhou, Y., & Zhang, G. (2021). Combined strategies to improve the expression of acidic mammalian chitinase in *Pichia pastoris* for the production of N, N'-diacetylchitobiose. *Biochemical Engineering Journal, 167*, 107907. https://doi.org/10.1016/j.bej.2020.107907.

Ebrahimi Tirtashi, F., Moradi, M., Tajik, H., Forough, M., Ezati, P., & Kuswandi, B. (2019). Cellulose/chitosan pH-responsive indicator incorporated with carrot anthocyanins for intelligent food packaging. *International Journal of Biological Macromolecules, 136*, 920–926. https://doi.org/10.1016/j.ijbiomac.2019.06.148.

El Knidri, H., Belaabed, R., Addaou, A., Laajeb, A., & Lahsini, A. (2018). Extraction, chemical modification and characterization of chitin and chitosan. *International Journal of Biological Macromolecules, 120*, 1181–1189. https://doi.org/10.1016/j.ijbiomac.2018.08.139.

Ezati, P., & Rhim, J. W. (2020). pH-responsive chitosan-based film incorporated with alizarin for intelligent packaging applications. *Food Hydrocolloids, 102*. https://doi.org/10.1016/j.foodhyd.2019.105629, https://www.journals.elsevier.com/food-hydrocolloids.

Felicia, W. X. L., Rovina, K., Mamat, H., Aziz, A. H. A., Lim, L. S., Jaziri, A. A., & Nurdiani, R. (2024). Advancements in fruit preservation technologies: Harnessing chitosan, aloe vera gel, and plant-based essential oils for coating applications. *Applied Food Research, 4*(2). https://doi.org/10.1016/j.afres.2024.100439, https://www.sciencedirect.com/science/journal/27725022.

Fernández-Pan, I., Maté, J. I., Gardrat, C., & Coma, V. (2015). Effect of chitosan molecular weight on the antimicrobial activity and release rate of carvacrol-enriched films. *Food Hydrocolloids, 51*, 60–68. https://doi.org/10.1016/j.foodhyd.2015.04.033.

Flórez, M., Guerra-Rodríguez, E., Cazón, P., & Vázquez, M. (2022). Chitosan for food packaging: Recent advances in active and intelligent films. *Food Hydrocolloids, 124*, 107328. https://doi.org/10.1016/j.foodhyd.2021.107328.

Garreau, C., Chiappisi, L., Micciulla, S., Morfin, I., Trombotto, S., Delair, T., & Sudre, G. (2023). Preparation of highly stable and ultrasmooth chemically grafted thin films of chitosan. *Soft Matter, 19*(8), 1606–1616. https://doi.org/10.1039/d3sm00003f.

Gonçalves, S. D.Á., Barbosa, M. H. R., Marangoni Júnior, L., Alves, R. M. V., & Vieira, R. P. (2023). Poly(limonene): A novel renewable oligomeric antioxidant and UV-light blocking additive for chitosan-based films. *Food Packaging and Shelf Life, 37*. https://doi.org/10.1016/j.fpsl.2023.101085, http://www.journals.elsevier.com/food-packaging-and-shelf-life/.

Guo, Y., Qiao, D., Zhao, S., Liu, P., Xie, F., & Zhang, B. (2024). Biofunctional chitosan–biopolymer composites for biomedical applications. *Materials Science and Engineering R: Reports, 159*. https://doi.org/10.1016/j.mser.2024.100775, https://www.sciencedirect.com/science/journal/0927796X.

Gupta, D., Lall, A., Kumar, S., Patil, T. D., & Gaikwad, K. K. (2024). Plant-based edible films and coatings for food-packaging applications: Recent advances, applications, and trends. *Sustainable Food Technology, 2*(5), 1428–1455. https://doi.org/10.1039/d4fb00110a, https://www.rsc.org/journals-books-databases/about-journals/sustainable-food-technology/.

Haghighi, H., Licciardello, F., Fava, P., Siesler, H. W., & Pulvirenti, A. (2020). Recent advances on chitosan-based films for sustainable food packaging applications. *Food Packaging and Shelf Life, 26*, 100551. https://doi.org/10.1016/j.fpsl.2020.100551.

Han, X., Xue, Y., Lou, R., Ding, S., & Wang, S. (2023). Facile and efficient chitosan-based hygroscopic aerogel for air dehumidification. *International Journal of Biological Macromolecules, 251*, 126191. https://doi.org/10.1016/j.ijbiomac.2023.126191.

Heras, M., Huang, C. C., Chang, C. W., & Lu, K. H. (2024). Trends in chitosan-based films and coatings: A systematic review of the incorporated biopreservatives, biological properties, and nanotechnology applications in meat preservation. *Food Packaging and Shelf Life, 42*. https://doi.org/10.1016/j.fpsl.2024.101259, https://www.sciencedirect.com/science/journal/22142894.

Hu, D., Wang, H., & Wang, L. (2016). Physical properties and antibacterial activity of quaternized chitosan/carboxymethyl cellulose blend films. *LWT, 65*, 398–405. https://doi.org/10.1016/j.lwt.2015.08.033.

Innocent, S., & Padikasan, I. A. (2023). Biotransformation of chitin to chitosan using *Bacillus cereus* JCM44 isolated from Kanyakumari, India and its characterization. *Journal of Pure and Applied Microbiology, 17*(1), 132–142. https://doi.org/10.22207/JPAM.17.1.02.

Isobe, N., Kaku, Y., Okada, S., Kawada, S., Tanaka, K., Fujiwara, Y., Nakajima, R., Bissessur, D., & Chen, C. (2022). Identification of chitin allomorphs in poorly crystalline samples based on the complexation with ethylenediamine. *Biomacromolecules, 23*(10), 4220–4229. https://doi.org/10.1021/acs.biomac.2c00714.

Jantzen da Silva Lucas, A., Quadro Oreste, E., Leão Gouveia Costa, H., Martín López, H., Dias Medeiros Saad, C., & Prentice, C. (2021). Extraction, physicochemical characterization, and morphological properties of chitin and chitosan from cuticles of edible insects. *Food Chemistry, 343*, 128550. https://doi.org/10.1016/j.foodchem.2020.128550, www.elsevier.com/locate/foodchem.

Jiang, A., Patel, R., Padhan, B., Palimkar, S., Galgali, P., Adhikari, A., Varga, I., & Patel, M. (2023). Chitosan based biodegradable composite for antibacterial food packaging application. *Polymers, 15*(10). https://doi.org/10.3390/polym15102235, http://www.mdpi.com/journal/polymers.

Jiménez-Gómez, C. P., & Cecilia, J. A. (2020). Chitosan: A natural biopolymer with a wide and varied range of applications. *Molecules (Basel, Switzerland), 25*(17). https://doi.org/10.3390/molecules25173981, https://www.mdpi.com/1420-3049/25/17/3981/pdf.

Jin, J., Luo, B., Xuan, S., Shen, P., Jin, P., Wu, Z., & Zheng, Y. (2024). Degradable chitosan-based bioplastic packaging: Design, preparation and applications. *International Journal of Biological Macromolecules, 266*(Pt 1), 131253. https://doi.org/10.1016/j.ijbiomac.2024.131253.

Kalaycıoğlu, Z., Torlak, E., Akın-Evingür, G., Özen, İ., & Erim, F. B. (2017). Antimicrobial and physical properties of chitosan films incorporated with turmeric extract. *International Journal of Biological Macromolecules, 101*, 882–888. https://doi.org/10.1016/j.ijbiomac.2017.03.174, www.elsevier.com/locate/ijbiomac.

Kerch, G., Zicans, J., & Meri, R. M. (2010). The effect of chitosan oligosaccharides on bread staling. *Journal of Cereal Science, 52*(3), 491–495. https://doi.org/10.1016/j.jcs.2010.08.007.

Khanzada, B., Akhtar, N., ul haq, I., Mirza, B., & Ullah, A. (2024). Polyphenol assisted nano-reinforced chitosan films with antioxidant and antimicrobial properties. *Food Hydrocolloids, 153*, 110010. https://doi.org/10.1016/j.foodhyd.2024.110010.

Khayrova, A., Lopatin, S., & Varlamov, V. (2021). Obtaining chitin, chitosan and their melanin complexes from insects. *International Journal of Biological Macromolecules, 167*, 1319–1328. https://doi.org/10.1016/j.ijbiomac.2020.11.086.

Kołodziejska, M., Jankowska, K., Klak, M., & Wszoła, M. (2021). Chitosan as an underrated polymer in modern tissue engineering. *Nanomaterials, 11*(11). https://doi.org/10.3390/nano11113019, https://www.mdpi.com/2079-4991/11/11/3019/pdf.

Kumar, A., Baranwal, J., Barse, B., Piga, I., Cincotti, A., & Gatto, G. (2024). Sustainability of food packaging. In *Advances in biopolymers for food science and technology*. Italy: Elsevier, 349–369. http://doi.org/10.1016/B978-0-443-19005-6.00014-1, https://www.sciencedirect.com/book/9780443190056.

Kumar, S., Mukherjee, A., & Dutta, J. (2020). Chitosan based nanocomposite films and coatings: Emerging antimicrobial food packaging alternatives. *Trends in Food Science & Technology, 97*, 196–209. https://doi.org/10.1016/j.tifs.2020.01.002.

Kurek, M., Garofulić, I. E., Bakić, M. T., Ščetar, M., Uzelac, V. D., & Galić, K. (2018). Development and evaluation of a novel antioxidant and pH indicator film based on chitosan and food waste sources of antioxidants. *Food Hydrocolloids, 84*, 238–246. https://doi.org/10.1016/j.foodhyd.2018.05.050, https://www.journals.elsevier.com/food-hydrocolloids.

Lárez-Velásquez, C. (2023). Chitosan: An overview of its multiple advantages for creating sustainable development poles. *Polimeros, 33*(1). https://doi.org/10.1590/0104-1428.20220103, http://www.scielo.br/scielo.php?script=sci_arttext&pid=S0104-14282023000100502&tlng=en.

Latou, E., Mexis, S. F., Badeka, A. V., Kontakos, S., & Kontominas, M. G. (2014). Combined effect of chitosan and modified atmosphere packaging for shelf life extension of chicken breast fillets. *LWT, 55*(1), 263–268. https://doi.org/10.1016/j.lwt.2013.09.010, https://www.journals.elsevier.com/lwt.

Lazaridou, A., Biliaderis, C. G., & Kontogiorgos, V. (2003). Molecular weight effects on solution rheology of pullulan and mechanical properties of its films. *Carbohydrate Polymers, 52*(2), 151–166. https://doi.org/10.1016/S0144-8617(02)00302-8.

Li, Z. C., Su, M. Y., Yuan, X. Y., Lv, H. Q., Feng, R., Wu, L. J., Gao, X. P., An, Y. X., Li, Z. W., Li, M. Y., Zhao, G. M., & Wang, X. P. (2024). Green fabrication of modified lignin/zeolite/chitosan-based composite membranes for preservation of perishable foods. *Food Chemistry, 460*. https://doi.org/10.1016/j.foodchem.2024.140713, https://www.sciencedirect.com/science/journal/03088146.

Lin, L., Gu, Y., & Cui, H. (2019). Moringa oil/chitosan nanoparticles embedded gelatin nanofibers for food packaging against *Listeria monocytogenes* and *Staphylococcus aureus* on cheese. *Food Packaging and Shelf Life, 19*, 86–93. https://doi.org/10.1016/j.fpsl.2018.12.005, http://www.journals.elsevier.com/food-packaging-and-shelf-life/.

Liu, J., Liu, S., Wu, Q., Gu, Y., Kan, J., & Jin, C. (2017). Effect of protocatechuic acid incorporation on the physical, mechanical, structural and antioxidant properties of chitosan film. *Food Hydrocolloids, 73*, 90–100. https://doi.org/10.1016/j.foodhyd.2017.06.035.

Liu, T., Wang, J., Chi, F., Tan, Z., & Liu, L. (2020). Development and characterization of novel active chitosan films containing fennel and peppermint essential oils. *Coatings, 10*(10), 936. https://doi.org/10.3390/coatings10100936.

Liu, X., Xu, F., Yong, H., Chen, D., Tang, C., Kan, J., & Liu, J. (2025). Recent advances in chitosan-based active and intelligent packaging films incorporated with flavonoids. *Food Chemistry: X, 25*. https://doi.org/10.1016/j.fochx.2025.102200.

Liu, Z., Ge, X., Lu, Y., Dong, S., Zhao, Y., & Zeng, M. (2012). Effects of chitosan molecular weight and degree of deacetylation on the properties of gelatine-based films. *Food Hydrocolloids, 26*(1), 311–317. https://doi.org/10.1016/j.foodhyd.2011.06.008.

Lv, J., Lv, X., Ma, M., Oh, D. H., Jiang, Z., & Fu, X. (2023). Chitin and chitin-based biomaterials: A review of advances in processing and food applications. *Carbohydrate Polymers, 299*. https://doi.org/10.1016/j.carbpol.2022.120142.

Marzieh, M. N., Zahra, F., Tahereh, E., & Sara, K. N. (2019). Comparison of the physicochemical and structural characteristics of enzymatic produced chitin and commercial chitin. *International Journal of Biological Macromolecules, 139*, 270–276. https://doi.org/10.1016/j.ijbiomac.2019.07.217.

Milbreta, U., Andze, L., Filipova, I., & Dortins, E. (2024). Effect of nanofibrillated cellulose on alginate and chitosan film properties as potential barrier coatings for paper food packaging. *BioResources, 19*(2), 3375–3389. https://doi.org/10.15376/biores.19.2.3375-3389.

Minguet-Lobato, M., Cervantes, F. V., Míguez, N., Plou, F. J., & Fernández-Lobato, M. (2024). Chitinous material bioconversion by three new chitinases from the yeast *Mestchnikowia pulcherrima*. *Microbial Cell Factories, 23*(1). https://doi.org/10.1186/s12934-024-02300-9.

Mohan, K., Ganesan, A. R., Ezhilarasi, P. N., Kondamareddy, K. K., Rajan, D. K., Sathishkumar, P., Rajarajeswaran, J., & Conterno, L. (2022). Green and eco-friendly approaches for the extraction of chitin and chitosan: A review. *Carbohydrate Polymers, 287*. https://doi.org/10.1016/j.carbpol.2022.119349.

Mu, L., Wu, L., Wu, S., Ye, Q., & Zhong, Z. (2024). Progress in chitin/chitosan and their derivatives for biomedical applications: Where we stand. *Carbohydrate Polymers, 343*. https://doi.org/10.1016/j.carbpol.2024.122233, https://www.sciencedirect.com/science/journal/01448617.

Mujtaba, M., Morsi, R. E., Kerch, G., Elsabee, M. Z., Kaya, M., Labidi, J., & Khawar, K. M. (2019). Current advancements in chitosan-based film production for food technology; A review. *International Journal of Biological Macromolecules, 121*, 889–904. https://doi.org/10.1016/j.ijbiomac.2018.10.109, www.elsevier.com/locate/ijbiomac.

Muñoz-Tebar, N., Pérez-Álvarez, J. A., Fernández-López, J., & Viuda-Martos, M. (2023). Chitosan edible films and coatings with added bioactive compounds: Antibacterial and antioxidant properties and their application to food products: A review. *Polymers, 15*(Issue 2). https://doi.org/10.3390/polym15020396.

Muxika, A., Etxabide, A., Uranga, J., Guerrero, P., & de la Caba, K. (2017). Chitosan as a bioactive polymer: Processing, properties and applications. *International Journal of Biological Macromolecules, 105*(Pt 2), 1358–1368. https://doi.org/10.1016/j.ijbiomac.2017.07.087.

Muzzarelli, R. A. A., Boudrant, J., Meyer, D., Manno, N., Demarchis, M., & Paoletti, M. G. (2012). Current views on fungal chitin/chitosan, human chitinases, food preservation, glucans, pectins and inulin: A tribute to Henri Braconnot, precursor of the carbohydrate polymers science, on the chitin bicentennial. *Carbohydrate Polymers, 87*(2), 995–1012. https://doi.org/10.1016/j.carbpol.2011.09.063.

Narasagoudr, S. S., Hegde, V. G., Vanjeri, V. N., Chougale, R. B., & Masti, S. P. (2020). Ethyl vanillin incorporated chitosan/poly(vinyl alcohol) active films for food packaging applications. *Carbohydrate Polymers, 236*. https://doi.org/10.1016/j.carbpol.2020.116049, http://www.elsevier.com/wps/find/journaldescription.cws_home/405871/description#description.

Ngo, T. M. P., Nguyen, T. H., Dang, T. M. Q., Tran, T. X., & Rachtanapun, P. (2020). Characteristics and antimicrobial properties of active edible films based on pectin and nanochitosan. *International Journal of Molecular Sciences, 21*(6). https://doi.org/10.3390/ijms21062224, https://www.mdpi.com/1422-0067/21/6/2224/pdf.

Niu, X., Liu, Y., Song, Y., Han, J., & Pan, H. (2018). Rosin modified cellulose nanofiber as a reinforcing and co-antimicrobial agents in polylactic acid /chitosan composite film for food packaging. *Carbohydrate Polymers, 183*, 102–109. https://doi.org/10.1016/j.carbpol.2017.11.079.

Noshirvani, N., Ghanbarzadeh, B., Gardrat, C., Rezaei, M. R., Hashemi, M., Le Coz, C., & Coma, V. (2017). Cinnamon and ginger essential oils to improve antifungal, physical and mechanical properties of chitosan-carboxymethyl cellulose films. *Food Hydrocolloids, 70*, 36–45. https://doi.org/10.1016/j.foodhyd.2017.03.015, https://www.journals.elsevier.com/food-hydrocolloids.

Novikov, V. Y., Derkach, S. R., Konovalova, I. N., Dolgopyatova, N. V., & Kuchina, Y. A. (2023). Mechanism of heterogeneous alkaline deacetylation of chitin: A review. *Polymers, 15*(Issue 7). https://doi.org/10.3390/polym15071729.

Pal, P., Pal, A., Nakashima, K., & Yadav, B. K. (2021). Applications of chitosan in environmental remediation: A review. *Chemosphere, 266*, 128934. https://doi.org/10.1016/j.chemosphere.2020.128934.

Pandey, V. K., Upadhyay, S. N., Niranjan, K., & Mishra, P. K. (2020). Antimicrobial biodegradable chitosan-based composite Nano-layers for food packaging. *International Journal of Biological Macromolecules, 157*, 212–219. https://doi.org/10.1016/j.ijbiomac.2020.04.149, www.elsevier.com/locate/ijbiomac.

Pelissari, F. M., Grossmann, M. V. E., Yamashita, F., & Pined, E. A. G. (2009). Antimicrobial, mechanical, and barrier properties of cassava starch-chitosan films incorporated with oregano essential oil. *Journal of Agricultural and Food Chemistry, 57*(16), 7499–7504. https://doi.org/10.1021/jf9002363Brazil, http://pubs.acs.org/doi/pdfplus/10.1021/jf9002363.

Peng, Y., & Li, Y. (2014). Combined effects of two kinds of essential oils on physical, mechanical and structural properties of chitosan films. *Food Hydrocolloids, 36*, 287–293. https://doi.org/10.1016/j.foodhyd.2013.10.013.

Perera, K. Y., Jaiswal, A. K., & Jaiswal, S. (2023). Biopolymer-based sustainable food packaging materials: Challenges, solutions, and applications. *Foods, 12*(12). https://doi.org/10.3390/foods12122422, www.mdpi.com/journal/foods.

Picos-Corrales, L. A., Morales-Burgos, A. M., Ruelas-Leyva, J. P., Crini, G., García-Armenta, E., Jimenez-Lam, S. A., Ayón-Reyna, L. E., Rocha-Alonzo, F., Calderón-Zamora, L., Osuna-Martínez, U., Calderón-Castro, A., De-Paz-Arroyo, G., & Inzunza-Camacho, L. N. (2023). Chitosan as an outstanding polysaccharide improving health-commodities of humans and environmental protection. *Polymers, 15*(3). https://doi.org/10.3390/polym15030526, http://www.mdpi.com/journal/polymers.

Piekarska, K., Sikora, M., Owczarek, M., Jóźwik-Pruska, J., & Wiśniewska-Wrona, M. (2023). Chitin and chitosan as polymers of the future—obtaining, modification, life cycle assessment and main directions of application. *Polymers, 15*(4). https://doi.org/10.3390/polym15040793, http://www.mdpi.com/journal/polymers.

Pires, J., Paula, C. D. D., Souza, V. G. L., Fernando, A. L., & Coelhoso, I. (2021). Understanding the barrier and mechanical behavior of different nanofillers in chitosan films for food packaging. *Polymers, 13*(5), 1–29. https://doi.org/10.3390/polym13050721, https://www.mdpi.com/2073-4360/13/5/721/pdf.

Pirsa, S., Karimi Sani, I., Pirouzifard, M. K., & Erfani, A. (2020). Smart film based on chitosan/*Melissa officinalis* essences/ pomegranate peel extract to detect cream cheeses spoilage. *Food Additives & Contaminants: Part A, 37*(4), 634–648. https://doi.org/10.1080/19440049.2020.1716079.

Popescu, P. A., Palade, L. M., Nicolae, I. C., Popa, E. E., Miteluț, A. C., Drăghici, M. C., Matei, F., & Popa, M. E. (2022). Chitosan-based edible coatings containing essential oils to preserve the shelf life and postharvest quality parameters of organic strawberries and apples during cold storage. *Foods, 11*(21). https://doi.org/10.3390/foods11213317.

Prateepchanachai, S., Thakhiew, W., Devahastin, S., & Soponronnarit, S. (2019). Improvement of mechanical and heat-sealing properties of edible chitosan films via addition of gelatin and CO_2 treatment of film-forming solutions. *International Journal of Biological Macromolecules, 131*, 589–600. https://doi.org/10.1016/j.ijbiomac.2019.03.067.

Priyadarshi, R., & Rhim, J.-W. (2020). Chitosan-based biodegradable functional films for food packaging applications. *Innovative Food Science & Emerging Technologies, 62*, 102346. https://doi.org/10.1016/j.ifset.2020.102346.

Priyadarshi, R., Sauraj, Kumar, B., & Negi, Y. S. (2018). Chitosan film incorporated with citric acid and glycerol as an active packaging material for extension of green chilli shelf life. *Carbohydrate Polymers, 195*, 329–338. https://doi.org/10.1016/j.carbpol.2018.04.089, http://www.elsevier.com/wps/find/journaldescription.cws_home/405871/description#description.

Rambabu, K., Bharath, G., Banat, F., Show, P. L., & Cocoletzi, H. H. (2019). Mango leaf extract incorporated chitosan antioxidant film for active food packaging. *International Journal of Biological Macromolecules, 126*, 1234–1243. https://doi.org/10.1016/j.ijbiomac.2018.12.196, www.elsevier.com/locate/ijbiomac.

Rezaee, M., Askari, G., EmamDjomeh, Z., & Salami, M. (2018). Effect of organic additives on physiochemical properties and anti-oxidant release from chitosan-gelatin composite films to fatty food simulant. *International Journal of Biological Macromolecules, 114*, 844–850. https://doi.org/10.1016/j.ijbiomac.2018.03.122.

Ribeiro, I. S., Maciel, G. M., Bortolini, D. G., Fernandes, I. D. A. A., Maroldi, W. V., Pedro, A. C., Rubio, F. T. V., & Haminiuk, C. W. I. (2024). Sustainable innovations in edible films and coatings: An overview. *Trends in Food Science and Technology, 143*. https://doi.org/10.1016/j.tifs.2023.104272, https://www.sciencedirect.com/science/journal/09242244.

Rodrigues, P. R., Junior, L. M., de Souza, W. F. C., Sato, H. H., Alves, R. M. V., & Vieira, R. P. (2021). O-ATRP synthesized poly(β-pinene) blended with chitosan for antimicrobial and antioxidant bio-based films production. *International Journal of Biological Macromolecules, 193*, 425–432. https://doi.org/10.1016/j.ijbiomac.2021.10.156, www.elsevier.com/locate/ijbiomac.

Rodriguez Llanos, J. H., Tadini, C. C., & Gastaldi, E. (2021). New strategies to fabricate starch/chitosan-based composites by extrusion. *Journal of Food Engineering, 290*. https://doi.org/10.1016/j.jfoodeng.2020.110224, http://www.sciencedirect.com/science/journal/02608774.

Rojas-Lema, S., Gomez-Caturla, J., Balart, R., Arrieta, M. P., & Garcia-Sanoguera, D. (2024). Development and characterization of thermoplastic zein biopolymers plasticized with glycerol suitable for injection molding. *Industrial Crops and Products, 218*, 119035. https://doi.org/10.1016/j.indcrop.2024.119035.

Romanazzi, G., Feliziani, E., Baños, S. B., & Sivakumar, D. (2017). Shelf life extension of fresh fruit and vegetables by chitosan treatment. *Critical Reviews in Food Science and Nutrition, 57*(3), 579–601. https://doi.org/10.1080/10408398.2014.900474.

Rostamabadi, H., Demirkesen, I., Colussi, R., Roy, S., Tabassum, N., de Oliveira Filho, J. G., Bist, Y., Kumar, Y., Nowacka, M., Galus, S., & Falsafi, S. R. (2024). Recent trends in the application of films and coatings based on starch, cellulose, chitin, chitosan, xanthan, gellan, pullulan, Arabic gum, alginate, pectin, and carrageenan in food packaging. *Food Frontiers, 5*(2), 350–391. https://doi.org/10.1002/fft2.342, onlinelibrary.wiley.com/journal/26438429.

Saberi Riseh, R., Vatankhah, M., Hassanisaadi, M., Shafiei-Hematabad, Z., & Kennedy, J. F. (2024). Advancements in coating technologies: Unveiling the potential of chitosan for the preservation of fruits and vegetables. *International Journal of Biological Macromolecules, 254*. https://doi.org/10.1016/j.ijbiomac.2023.127677.

Sánchez-González, L., González-Martínez, C., Chiralt, A., & Cháfer, M. (2010). Physical and antimicrobial properties of chitosan–tea tree essential oil composite films. *Journal of Food Engineering, 98*(4), 443–452. https://doi.org/10.1016/j.jfoodeng.2010.01.026.

Saravanan, A., Kumar, P. S., Yuvaraj, D., Jeevanantham, S., Aishwaria, P., Gnanasri, P. B., Gopinath, M., & Rangasamy, G. (2023). A review on extraction of polysaccharides from crustacean wastes and their environmental applications. *Environmental Research, 221*. https://doi.org/10.1016/j.envres.2023.115306, https://www.sciencedirect.com/journal/environmental-research.

Sarfraz, M. H., Hayat, S., Siddique, M. H., Aslam, B., Ashraf, A., Saqalein, M., Khurshid, M., Sarfraz, M. F., Afzal, M., & Muzammil, S. (2024). Chitosan based coatings and films: A perspective on antimicrobial, antioxidant, and intelligent food packaging. *Progress in Organic Coatings, 188*. https://doi.org/10.1016/j.porgcoat.2024.108235, https://www.sciencedirect.com/science/journal/03009440.

Savin, S., Craciunescu, O., Oancea, A., Ilie, D., Ciucan, T., Antohi, L. S., Toma, A., Nicolescu, A., Deleanu, C., & Oancea, F. (2020). Antioxidant, cytotoxic and antimicrobial activity of chitosan preparations extracted from *Ganoderma Lucidum* Mushroom. *Chemistry & Biodiversity, 17*(7). https://doi.org/10.1002/cbdv.202000175.

Shankar, S., Khodaei, D., & Lacroix, M. (2021). Effect of chitosan/essential oils/silver nanoparticles composite films packaging and gamma irradiation on shelf life of strawberries. *Food Hydrocolloids, 117*, 106750. https://doi.org/10.1016/j.foodhyd.2021.106750.

Shen, Z., & Kamdem, D. P. (2015). Development and characterization of biodegradable chitosan films containing two essential oils. *International Journal of Biological Macromolecules, 74*, 289–296. https://doi.org/10.1016/j.ijbiomac.2014.11.046, www.elsevier.com/locate/ijbiomac.

Shi, B., Hao, Z., Du, Y., Jia, M., & Xie, S. (2024). Mechanical and barrier properties of chitosan-based composite film as food packaging: A review. *BioResources, 19*(2), 4001–4014. https://doi.org/10.15376/biores.19.2.shi.

Silva, N. C., Chevigny, C., Domenek, S., Almeida, G., Assis, O. B. G., & Martelli-Tosi, M. (2025). Nanoencapsulation of active compounds in chitosan by ionic gelation: Physicochemical, active properties and application in packaging. *Food Chemistry, 463*. https://doi.org/10.1016/j.foodchem.2024.141129, https://www.sciencedirect.com/science/journal/03088146.

Soares, T. R. P., Reis, A. F., dos Santos, J. W. S., Chagas, E. G. L., Venturini, A. C., Santiago, R. G., Bastos-Neto, M., Vieira, R. S., Carvalho, R. A., da Silva, C. F., & Yoshida, C. M. P. (2023). NaY-Ag zeolite chitosan coating kraft paper applied as ethylene scavenger packaging. *Food and Bioprocess Technology, 16*(5), 1101–1115. https://doi.org/10.1007/s11947-022-02989-9, https://www.springer.com/journal/11947.

Sogut, E., & Seydim, A. C. (2018). The effects of Chitosan and grape seed extract-based edible films on the quality of vacuum packaged chicken breast fillets. *Food Packaging and Shelf Life, 18*, 13–20. https://doi.org/10.1016/j.fpsl.2018.07.006, http://www.journals.elsevier.com/food-packaging-and-shelf-life/.

Souza, V. G. L., Fernando, A. L., Pires, J. R. A., Rodrigues, P. F., Lopes, A. A. S., & Fernandes, F. M. B. (2017). Physical properties of chitosan films incorporated with natural antioxidants. *Industrial Crops and Products, 107*, 565–572. https://doi.org/10.1016/j.indcrop.2017.04.056, www.elsevier.com/inca/publications/store/5/2/2/8/2/5.

Spitzer, L., Adrien, A., Veettil, U. T., Olza, S., Alonso-Varona, A., & Fernandes, S. C. M. (2024). Exploring marine-based compounds as cross-linkers to improve the biocompatibility and sustainability of chitosan-based hydrogels. *Carbohydrate Polymer Technologies and Applications, 8*. https://doi.org/10.1016/j.carpta.2024.100541, https://www.sciencedirect.com/science/journal/26668939.

Sree, K. P., Sree, M. S., & Supriya, P. (2020). Samreen., Application of chitosan edible coating for preservation of tomato. *International Journal of Chemical Studies, 8*(4), 3281–3285. https://doi.org/10.22271/chemi.2020.v8.i4ao.10157.

Stefanowska, K., Woźniak, M., Dobrucka, R., & Ratajczak, I. (2023). Chitosan with natural additives as a potential food packaging. *Materials, 16*(4), 1579. https://doi.org/10.3390/ma16041579.

Subramani, G., & Manian, R. (2024). Bioactive chitosan films: Integrating antibacterial, antioxidant, and antifungal properties in food packaging. *International Journal of Biological Macromolecules, 278*. https://doi.org/10.1016/j.ijbiomac.2024.134596.

Sudarshan, N. R., Hoover, D. G., & Knorr, D. (1992). Antibacterial action of chitosan. *Food Biotechnology, 6*(3), 257–272. https://doi.org/10.1080/08905439209549838.

Sukhanova, A. A., Murzova, A. E., Boyandin, A. N., Kiselev, E. G., Sukovatyi, A. G., Kuzmin, A. P., & Shabanov, A. V. (2020). Poly-3-hydroxybutyrate/chitosan composite films and nonwoven mats. *International Journal of Biological Macromolecules, 165*(Pt B), 2947–2956. https://doi.org/10.1016/j.ijbiomac.2020.10.177.

Tessarin, P., Chinnici, F., Donnini, S., Liquori, E., Riponi, C., & Rombolà, A. D. (2016). Influence of canopy-applied chitosan on the composition of organic cv. Sangiovese and Cabernet Sauvignon berries and wines. *Food Chemistry, 210*, 512–519. https://doi.org/10.1016/j.foodchem.2016.04.137, www.elsevier.com/locate/foodchem.

Vergara-Figueroa, J., Alejandro-Martin, S., Cerda-Leal, F., & Gacitúa, W. (2020). Dual electrospinning of a nanocomposites biofilm: Potential use as an antimicrobial barrier. *Materials Today Communications, 25*, 101671. https://doi.org/10.1016/j.mtcomm.2020.101671.

Vilela, C., Pinto, R. J. B., Coelho, J., Domingues, M. R. M., Daina, S., Sadocco, P., Santos, S. A. O., & Freire, C. S. R. (2017). Bioactive chitosan/ellagic acid films with UV-light protection for active food packaging. *Food Hydrocolloids, 73*, 120–128. https://doi.org/10.1016/j.foodhyd.2017.06.037, https://www.sciencedirect.com/science/journal/0268005X.

Vo, T. S., Chit, P. P., Nguyen, V. H., Hoang, T., Lwin, K. M., Vo, T. T. B. C., Jeon, B., Han, S., Lee, J., Park, Y., & Kim, K. (2024). A comprehensive review of chitosan-based functional materials: From history to specific applications. *International Journal of Biological Macromolecules, 281*. https://doi.org/10.1016/j.ijbiomac.2024.136243, https://www.sciencedirect.com/science/journal/01418130.

Vo, T. V., Dang, T. H., & Chen, B. H. (2019). Synthesis of intelligent pH indicative films from chitosan/poly(vinyl alcohol)/anthocyanin extracted from red cabbage. *Polymers, 11*(7). https://doi.org/10.3390/polym11071088, https://res.mdpi.com/polymers/polymers-11-01088/article_deploy/polymers-11-01088-v2.pdf?filename=&attachment=1.

Wang, X. Y., Wang, J., Zhao, C., Ma, L., Rousseau, D., & Tang, C. H. (2023). Facile fabrication of chitosan colloidal films with pH-tunable surface hydrophobicity and mechanical properties. *Elsevier B.V., China Food Hydrocolloids, 137*. https://doi.org/10.1016/j.foodhyd.2022.108429, https://www.journals.elsevier.com/food-hydrocolloids.

Wang, H., Qian, J., & Ding, F. (2018). Emerging chitosan-based films for food packaging applications. *Journal of Agricultural and Food Chemistry, 66*(2), 395–413. https://doi.org/10.1021/acs.jafc.7b04528.

Wang, L., Dong, Y., Men, H., Tong, J., & Zhou, J. (2013). Preparation and characterization of active films based on chitosan incorporated tea polyphenols. *Food Hydrocolloids, 32*(1), 35–41. https://doi.org/10.1016/j.foodhyd.2012.11.034.

Wang, Y., Du, H., Xie, M., Ma, G., Yang, W., Hu, Q., & Pei, F. (2019). Characterization of the physical properties and biological activity of chitosan films grafted with gallic acid and caffeic acid: A comparison study. *Food Packaging and Shelf Life, 22*. https://doi.org/10.1016/j.fpsl.2019.100401.

Wardhono, E. Y., Pinem, M. P., Susilo, S., Siom, B. J., Sudrajad, A., Pramono, A., Meliana, Y., & Guénin, E. (2022). Modification of physio-mechanical properties of chitosan-based films via physical treatment approach. *Polymers, 14*(23). https://doi.org/10.3390/polym14235216, http://www.mdpi.com/journal/polymers.

Wei, L., Tan, W., Wang, G., Li, Q., Dong, F., & Guo, Z. (2019). The antioxidant and antifungal activity of chitosan derivatives bearing Schiff bases and quaternary ammonium salts. *Carbohydrate Polymers, 226*, 115256. https://doi.org/10.1016/j.carbpol.2019.115256.

Wu, H., Lei, Y., Lu, J., Zhu, R., Xiao, D., Jiao, C., Xia, R., Zhang, Z., Shen, G., Liu, Y., Li, S., & Li, M. (2019). Effect of citric acid induced crosslinking on the structure and properties of potato starch/chitosan composite films. *Food Hydrocolloids, 97*, 105208. https://doi.org/10.1016/j.foodhyd.2019.105208.

Xing, F., Zhang, S., Li, J., Li, L., & Shi, J. (2018). Crosslinked chitosan-based biocomposite films modified with soy protein isolate. *Polymer Composites, 39*(3), 942–949. https://doi.org/10.1002/pc.24024, http://onlinelibrary.wiley.com/journal/10.1002/(ISSN)1548-0569.

Xu, J., Xia, R., Zheng, L., Yuan, T., & Sun, R. (2019). Plasticized hemicelluloses/chitosan-based edible films reinforced by cellulose nanofiber with enhanced mechanical properties. *Carbohydrate Polymers, 224*, 115164. https://doi.org/10.1016/j.carbpol.2019.115164.

Xu, Y., Li, L., Cao, S., Zhu, B., & Yao, Z. (2022). An updated comprehensive review of advances on structural features, catalytic mechanisms, modification methods and applications of chitosanases. *Process Biochemistry, 118*, 263–273. https://doi.org/10.1016/j.procbio.2022.05.001, www.elsevier.com/inca/publications/store/4/2/2/8/5/7.

Yadav, S., Mehrotra, G. K., & Dutta, P. K. (2021). Chitosan based ZnO nanoparticles loaded gallic-acid films for active food packaging. *Food Chemistry, 334*, 127605. https://doi.org/10.1016/j.foodchem.2020.127605.

Yan, J., Cui, R., Qin, Y., Li, L., & Yuan, M. (2021). A pH indicator film based on chitosan and butterfly pudding extract for monitoring fish freshness. *International Journal of Biological Macromolecules, 177*, 328–336. https://doi.org/10.1016/j.ijbiomac.2021.02.137.

Yang, B., Liu, B., Gao, Y., Wei, J., Li, G., Zhang, H., Wang, L., & Hou, Z. (2024). PEG-crosslinked O-carboxymethyl chitosan films with degradability and antibacterial activity for food packaging. *Scientific Reports, 14*(1). https://doi.org/10.1038/s41598-024-61642-x.

Yang, J., Dahlström, C., Edlund, H., Lindman, B., & Norgren, M. (2019). pH-responsive cellulose–chitosan nanocomposite films with slow release of chitosan. *Cellulose, 26*(6), 3763–3776. https://doi.org/10.1007/s10570-019-02357-5.

Yangilar, F. (2015). Chitosan/whey protein (CWP) edible films efficiency for controlling mould growth and on microbiological, chemical and sensory properties during storage of Göbek Kashar cheese. *Korean Journal for Food Science of Animal Resources, 35*(2), 216–224. https://doi.org/10.5851/kosfa.2015.35.2.216.

Yuan, G., Chen, X., & Li, D. (2016). Chitosan films and coatings containing essential oils: The antioxidant and antimicrobial activity, and application in food systems. *Food Research International, 89*, 117–128. https://doi.org/10.1016/j.foodres.2016.10.004.

Zhang, W., Khan, A., Ezati, P., Priyadarshi, R., Sani, M. A., Rathod, N. B., Goksen, G., & Rhim, J. W. (2024). Advances in sustainable food packaging applications of chitosan/polyvinyl alcohol blend films. *Food Chemistry, 443*. https://doi.org/10.1016/j.foodchem.2024.138506, https://www.sciencedirect.com/science/journal/03088146.

Zhang, J., Wang, L., Mi, Y., Dong, F., & Guo, Z. (2024). Preparation and properties of crosslinked quaternized chitosan-based hydrogel films ionically bonded with acetylsalicylic acid for biomedical materials. *Marine Drugs, 22*(10). https://doi.org/10.3390/md22100450.

Zhang, W., Li, X., & Jiang, W. (2020). Development of antioxidant chitosan film with banana peels extract and its application as coating in maintaining the storage quality of apple. *International Journal of Biological Macromolecules, 154*, 1205–1214. https://doi.org/10.1016/j.ijbiomac.2019.10.275.

Zhao, H., Su, H., Sun, J., Dong, H., & Mao, X. (2024). Bioconversion of α-chitin by a lytic polysaccharide monooxygenase OsLPMO10A coupled with chitinases and the synergistic mechanism analysis. *Journal of Agricultural and Food Chemistry, 72*(13), 7256–7265. https://doi.org/10.1021/acs.jafc.3c08688.

Zhao, Y., Zhang, Y., Dong, H., Wu, W., Yang, X., & He, Q. (2023). Functional biopolymers for food packaging: Formation mechanism and performance improvement of chitosan-based composites. *Food Bioscience, 54*, 102927. https://doi.org/10.1016/j.fbio.2023.102927.

Zhou, T., Wang, H., Han, Q., Song, Z., Yu, D., Li, G., Liu, W., Dong, C., Ge, S., & Chen, X. (2024). Fabrication and characterization of an alginate-based film incorporated with cinnamaldehyde for fruit preservation. *International Journal of Biological Macromolecules, 274*(Pt 1), 133398. https://doi.org/10.1016/j.ijbiomac.2024.133398.

Chapter 7

Cellulose and its derivatives in films and coatings

Mariana Spinei and Mircea Oroian
Faculty of Food Engineering, Ştefan cel Mare University of Suceava, Suceava, Romania

7.1 Introduction

Cellulose is an unbranched bio-based polymer consisting of repeating glucose units $(C_6H_{10}O_5)_n$ and is considered the world's most abundant natural polymer, with a global production of ~1.5×10^{12} tons per year, in comparison with the main mineral and fossil sources of the worldwide reserves (Seddiqi et al., 2021). This polymer is generally found in nature in two forms: (1) as microfibrils in the cell walls (CWs) of wood and plants, the epidermal cell membranes of tunicates, and algae tissues; and (2) in the form of nanofiber networks synthesized by different bacterial species (e.g., *Acetobacter, Alcaligenes, Pseudomonas, Aerobacter,* etc.) (Seddiqi et al., 2021). Cellulose-based materials utilize a hierarchical framework that extends from nanoscale to macroscopic sizes in different shapes (e.g., nanocrystallite, fibrils, nanoscale disordered domains, and fibril aggregates) (Fig. 7.1) (Grzybek et al., 2024). Cellulose presents a complex structure, consisting of aggregates of very fine fibrils that are composed of multiple cellulose chains (Seddiqi et al., 2021). Each fibril contains alternating crystalline (large ordered) and amorphous (small disordered) regions with different cross-sectional dimensions (Wang, Qi, et al., 2024). A single cellulose chain traverses numerous amorphous and crystalline domains, while β-1,4 glycosidic bonds connect individual cellulose chain units. Therefore, cellulose chains are exceptionally aligned with the crystalline region of a cellulose fibril (Seddiqi et al., 2021).

The chemical modifications of cellulose, such as partial or total etherification and esterification of the three hydroxyls present on the anhydroglucose unit of the backbone, are carried out to improve the performance and processability, and more especially to obtain derivatives that are soluble in water and organic solvents (despite being highly hydrophilic, cellulose does not dissolve in water due to hydrogen bonding between its hydroxyl (–OH) groups) (Heinze et al., 2018). Historically, cellulose nitrate (CN) (inorganic cellulose derivative) is the oldest ester synthesized, but it has found only a few uses as a biopolymer, while the first organic ester that has been used since 1865 in various applications (e.g., membranes, films, or fibers) is cellulose acetate (CA) (Bakri & Rahman, 2021). Ether derivatives are more recent, since methylation was first documented in 1905 (Heinze et al., 2018). Therefore, cellulose derivatives represent a category of natural polymers in which cellulose is dissolved utilizing various solvents, including organic solvents, ionic liquids (ILs), and aqueous solvents prior to being used (Ursachi et al., 2024). Cellulose esters (e.g., CA, cellulose acetate phthalate [CAP], cellulose acetate butyrate [CAB], etc.) and cellulose ethers (e.g., hydroxyethyl cellulose [HEC], methyl cellulose, hydroxyethyl methylcellulose, ethyl cellulose [EC], etc.) (Seddiqi et al., 2021).

Biodegradability, worldwide abundance, biocompatibility, and renewability are the main features of cellulose and its derivatives as materials used in the food industry (Ursachi et al., 2024). Moreover, the various significant characteristics for using cellulose derivatives include hydrophobic and hydrophilic properties, organic solubility, high swelling, make mucoadhesive system, gel and film formation ability, act as a viscosity modifying agent, and thermoplastic effect in food systems (Sabbaghi, 2023). Consequently, cellulose derivatives are used in edible coatings, biodegradable films, encapsulation, emulsification, extrusion, rheology control, molding, diffusion barriers, as well as foam stabilizers, suspending agents, and stabilizing agents in the food industry (Sabbaghi, 2023). Recent innovations include the production of bacterial cellulose, the formulation of nanocellulose using eco-friendly processes, the development of smart packaging, and the use of different mixtures of cellulose derivatives to obtain a film or coating with desirable physicochemical characteristics (Sabbaghi, 2023; Spinei et al., 2024; Ursachi et al., 2024). Additionally, bio-composite cellulose, such as

Natural Polymer-Based Films and Coatings in Food Technology. DOI: https://doi.org/10.1016/B978-0-443-34244-8.00021-5

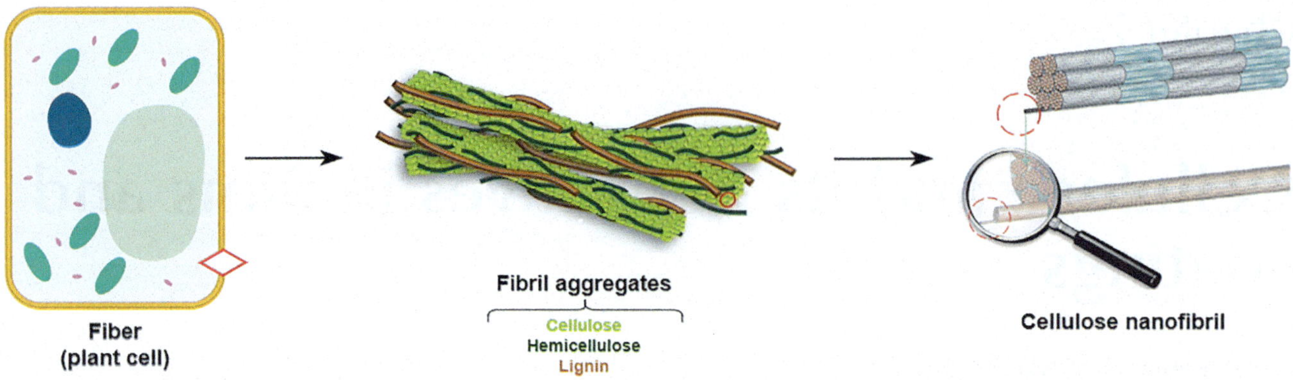

FIGURE 7.1 Schematic illustration of the cellulosic fiber. This figure presents a schematic representation of cellulosic fibers in various morphological forms.

nanocellulose, has a wide range of industrial applications, particularly in paper packaging, while cellulose-based molecules, commonly named microfibrillated cellulose, which are known as cellulose fibers, can accomplish the need for sustainable packaging due to their biodegradability (Liu et al., 2021). Therefore, cellulose and its derivatives can contribute to prolonging the shelf life of various food products due to their barrier and preservative characteristics (Liu et al., 2021). In this chapter, the application of cellulose and its various derivatives is investigated as materials used in films and coatings regarding innovations and principles of the food industry.

7.2 Cellulose

7.2.1 Cellulose structure

Cellulose is a high-molecular-weight, linear (unbranched) homopolymer composed of repeating *D*-glucose units interconnected by β-1,4 glycosidic bonds (Seddiqi et al., 2021). A hemiacetal is present at one end of the cellulose chain (Wang, Qi, et al., 2024). The cellulose configuration is complex as a result of variations in the aggregation and packing of the cellulose chains, which differ among organisms that produce cellulose (Seddiqi et al., 2021; Wang, Qi, et al., 2024). Each glucose residue in cellulose contains three hydroxyl groups. The degree of polymerization, which ranges between 1000 and 30,000, depends on the glucose units and is influenced by the source of cellulose and different treatments (Seddiqi et al., 2021). The intra-chain hydrogen bonding between oxygen of the contiguous ring of cellulose molecules and hydroxyl groups fortifies the connection and leads to an unbranched conformation of cellulose chain (Seddiqi et al., 2021).

Cellulose molecules organize into ordered parallel layers to form elementary fibrils, which can be found in plant CWs, tunicate epidermal cell membranes, algae CWs, or produced by bacteria. Linked by strong intra- and/or intermolecular hydrogen bonds and van der Waals forces, the cellulose chains are tightly packed together with a lateral size of 3–5 nm (Sabbaghi, 2023). Each elementary fibril represents a batch of cellulosic crystals aligned along the fibril axis, interspersed with disordered regions (Heinze, 2015). Bundles of elementary fibrils further form cellulose microfibrils, which have a cross-sectional width of 5–20 nm and can be several micrometers long, depending on their source (Heinze, 2015). Cellulose is a semicrystalline material, and its degree of crystallinity relies on a few factors, such as its source, pretreatment, and extraction technique (Seddiqi et al., 2021). The degree of crystallinity of plant- and wood-based cellulose (WC) generally ranges from 40% to 60%, whereas cellulose obtained from other sources, such as tunicin and bacteria, displays a higher degree of crystallinity with a range between 80% and 100% (Seddiqi et al., 2021).

Crystalline cellulose has different allomorphs, such as cellulose I, II, III, and IV (Seddiqi et al., 2021). Cellulose I consists of parallel chains in the crystalline configuration and is naturally obtained from various sources (plants, algae, trees, tunicates, and bacteria). The sheets formed by cellulose I, which are bonded together through van der Waals interactions and hydrogen bonds, play a major role in the rigidity and specific structure of cellulose (Seddiqi et al., 2021). Cellulose I is divided into two metastable structures: (1) triclinic (Iα) that is the allomorph known for most bacterial and algal cellulose, and (2) monoclinic (Iβ), which is generally found in tunicate- and plant-based cellulose (PC) (Seddiqi et al., 2021). Cellulose II is commonly obtained by regeneration (recrystallization and dissolution) or alkali treatment (mercerization) of cellulose I (Heinze, 2015). Throughout this transformation, the parallel chain configuration of cellulose I transitions into a more stable antiparallel chain configuration of cellulose II. Cellulose III can be produced from both cellulose I (referred to as cellulose III_I) and cellulose II (known as cellulose III_{II}) by immersing them in organic amine or liquid ammonia at low temperatures, while cellulose IV can be obtained from the thermal processing of cellulose III in glycerol (Seddiqi et al., 2021). Each of these allomorphs is identified by variations in their conformations, distinct packing arrangements, and, crucially, their specific patterns of intra- and intermolecular hydrogen bonding (Seddiqi et al., 2021).

7.2.2 Cellulose sources

Cellulose is distributed throughout nature and is classified depending on its source of origin, such as WC, PC, algae-based cellulose (AC), bacteria-based cellulose, and tunicate-based cellulose (TC) (Fig. 7.2) (Heinze, 2015; Seddiqi et al., 2021).

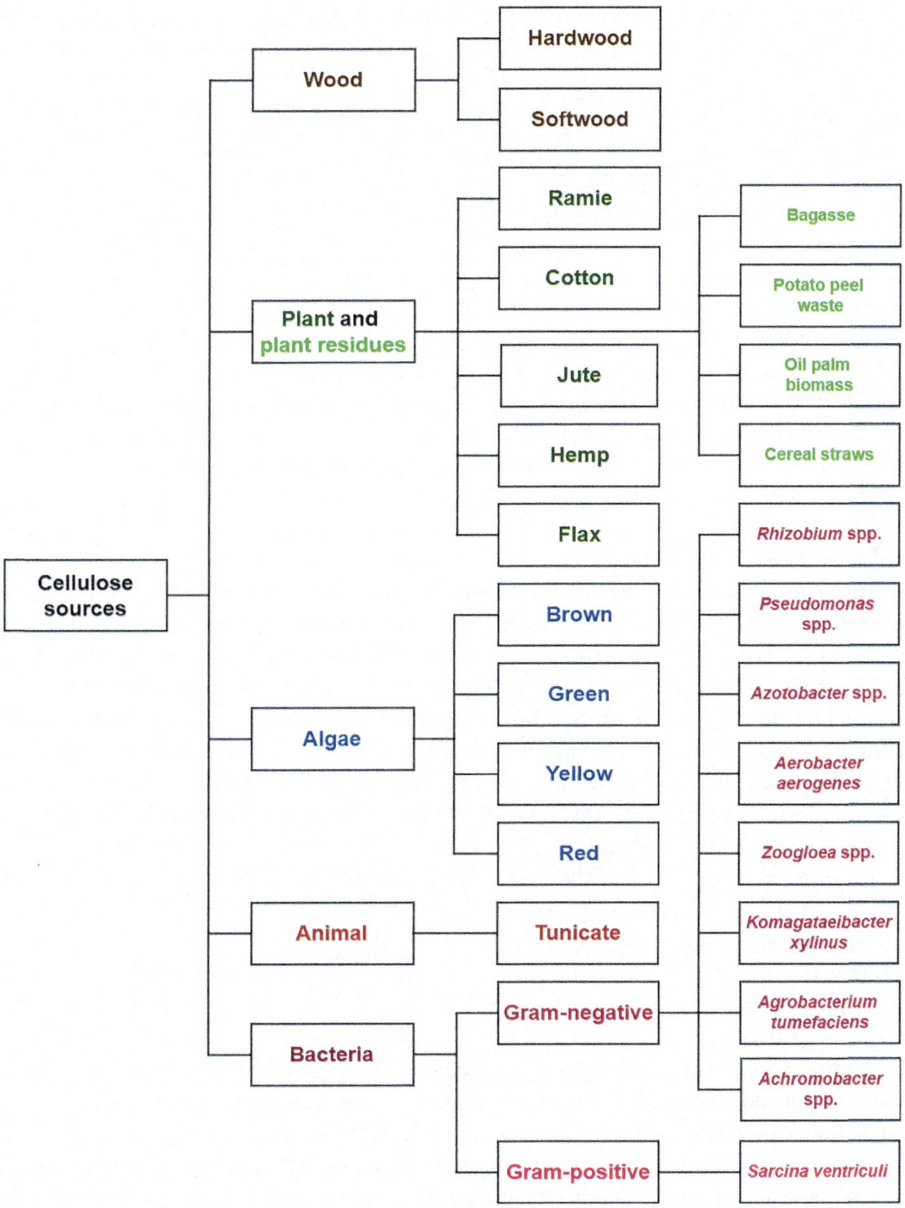

FIGURE 7.2 Classification of cellulose based on its source: wood, plant and plant residues, algae, animal (tunicate), and bacteria. This figure schematically presents the primary natural sources of cellulose.

In the long history of cellulose utilization, due to its availability and cost-effectiveness, PC and WC emerged as the most widely recognized types of cellulose, in contrast to other sources: BC, TC, and AC (Seddiqi et al., 2021).

7.2.2.1 *Plant and wood cellulose*

Cellulose has been extracted from softwood (e.g., pine, Monterey pine, Douglas fir, spruce, banyan fig, cotton tree, flame tree, kauri, burflower tree, and white leadtree) and hardwood (e.g., birch, beech, maple, sweet chestnut, tawa, alpine ash, English oak, karoi tree, gum Arabic tree, American mahogany, rain tree, southern blue gum, earleaf acacia, and oak) (Silvy et al., 2018; Tang et al., 2021). Additionally, a broad range of plant materials has been explored for cellulose extraction, including plants (e.g., cotton, ramie, hemp, kenaf, jute, and flax) and plant residues (e.g., bagasse, oil palm biomass, cereal straws, and other residues) (Heinze, 2015; Seddiqi et al., 2021). Fibers extracted from different sources, such as plants and wood, exhibit varying morphologies and dimensions, but share similarities in structure and chemical composition. They primarily consist of biodegradable polymers, including cellulose, hemicellulose, and lignin (Table 7.1). For example, spruce wood fibers are generally untwisted, while those from cotton are twisted (Heinze, 2015). In contrast to them, fibers from bast plants are straight and round. Notably, they all have an internal structure made up of multiple cell wall layers (Heinze, 2015).

WC and PC have a complex, multitiered structure consisting of parallel assemblies of elementary crystallites that arrange into fibers through hydrogen bonding (Heinze, 2015). Whereas regions of lower order or noncrystalline areas include bends between adjacent chains, organized crystallites arrange cellulose chains folded in a longitudinal manner (Heinze, 2015). The less organized areas show a relatively lower density and more irregular alignment (Seddiqi et al., 2021). The cellulose chains organize into a fundamental fibrillar unit, known as elementary fibrils, which are 100 nm in length and have a characteristic lateral dimension of 1.5–3.5 nm (Heinze, 2015). These elementary fibrils are subsequently combined into fibrillar bundles, referred to as microfibrils, which have widths ranging from 10 to 30 nm (Heinze, 2015). In plant CWs, a layer of amorphous cellulose, which is encased by hemicellulose, additionally envelops the microfibrils (Heinze, 2015). Although it is a product of PC, cotton linter, flax, and hemp fibers lack a significant presence of hemicellulose and lignin. The cellulose proportion in wood generally ranges from 40% to 50%, while in plants, its range is between 30% and 90%. Hardwood and softwood differ in chemical composition, especially in the levels of cellulose, hemicellulose, and lignin, along with their structural conformation (Seddiqi et al., 2021).

7.2.2.2 *Bacterial cellulose*

Cellulose produced by bacteria is known as "bio-cellulose," "bacterial nanocellulose," "microbial cellulose," or most commonly "bacterial cellulose" (BC) (Seddiqi et al., 2021). BC was first discovered by Adrian J. Brown in 1886 as a jelly membrane on the external part of a vinegar fermentation broth (Brown, 1886). Compared with PC and WC, BC is very pure (>90%), without binding to other polymers such as lignin and hemicellulose, and only a very low amount of carboxyl and carbonyl moieties is present (Heinze, 2015; Seddiqi et al., 2021). Consequently, the isolation and purification steps of BC are relatively straightforward and do not require extensive chemical or any other treatment, unlike PC and WC (Seddiqi et al., 2021). Since its discovery, BC has gained interest due to various advantages, such as extraordinary mechanical strength, high purity, biocompatibility, ultrafine fiber shapes, biodegradability, notably crystalline structure, chemical stability, high degree of polymerization, in situ moldability (namely shaping during biosynthesis), and great absorption capacity, resulting principally from the presence of nanofibrils of BC rather than microfibers of PC (Heinze, 2015; Seddiqi et al., 2021). Fundamentally, BC is recognized as a highly biocompatible material, nongenotoxic, and noncytotoxic, attracting interest in different domains (Seddiqi et al., 2021).

BC is generally produced extracellularly by Gram-negative bacteria (e.g., *Rhizobium* spp., *Pseudomonas* spp., *Zoogloea* spp., *Achromobacter* spp., *Azotobacter* spp., *Komagataeibacter xylinus*, *Aerobacter aerogenes*, and *Agrobacterium tumefaciens*) and only one genus of Gram-positive bacteria (i.e., *Sarcina ventriculi*) (Fig. 7.2) with an oxygen supply (air), nitrogen source, and carbon source (mainly *D*-glucose) (Seddiqi et al., 2021). *K. xylinus* is the most commonly utilized species of bacteria for the production of BC due to its ability to generate considerable quantities of BC from various nitrogen and carbon sources in liquid culture (Seddiqi et al., 2021). The most common carbon sources used for this purpose are glucose, fructose, mannitol, sucrose, maltose, and galactose, while nitrogen sources for the production of BC are yeast, peptone, peptone/yeast, ammonium sulfate, and urea (García-Sánchez et al., 2019; Volova et al., 2018). On the other hand, unconventional carbon and nitrogen sources are by-products and wastes from different industrial processing (e.g., pineapple canning industrial waste, fruit (pineapple, sweet lime, orange, and banana) peel waste extract, grape pomace hydrolysate, mango waste, etc.) (García-Sánchez et al., 2019; Gorgieva et al., 2023; Mardawati et al., 2023).

TABLE 7.1 Content of cellulose, hemicellulose, and lignin as primary constituents in the chemical composition of wood and plant fibers.

Type	Cellulose (%)	Hemicellulose (%)	Lignin (%)	References
Hardwood				
Aspen wood	52.7	21.7	19.5	Kostryukov et al. (2023)
Poplar wood	39.8	14.8	29.1	
Beech wood	40.0	23.0	21.0	
Oak wood	43.2	21.9	35.4	
Willow wood	42.4–45.3	20.6–22.9	16.9–18.9	
Cherry wood	46.0	29.0	18.0	
Gum arabic tree	43.9	17.7	18.4	Silvy et al. (2018)
Karoi tree	40.2	22.1	29.1	
Rain tree	49.0	27.1	21.1	
American mahogany	57.9	11.2	28.5	
Earleaf acacia	67.6	20.6	7.1	
Capirona tree	29.8–36.0	9.6–10.2	8.1–9.7	Javier-Astete et al. (2021)
Bolaina tree	20.4–42.4	14.2–19.3	11.2–12.2	
Softwood				
Spruce wood	45.5	22.9	27.9	Kostryukov et al. (2023)
Pine wood	42.0–50.0	24.0–27.0	20.0	
Larch wood	47.5	24.5	28.5	
White leadtree	51.70	14.00	18.40	Silvy et al. (2018)
Banyan fig	33.19	40.67	25.09	
Cotton tree	45.00	19.20	24.50	
Flame tree	67.60	10.87	8.13	
Burflower-tree	40.00	34.90	6.03	
Plants				
Cotton linter	89.7	1.0	2.7	Dorez et al. (2014)
Flax	80.0	13.0	2.0	
Hemp	74.1	7.6	2.2	
Sugar cane	51.8	27.6	10.7	
Bamboo	54.6	11.4	21.7	
Coconut	51.3	11.7	30.7	
Plant residues				
Barley straw	36.0–43.0	24.0–33.0	6.3–13.1	Kostryukov et al. (2023)
Wheat straw	35.0–39.0	23.0–30.0	12.0–16.0	
Oat straw	31.0–35.0	20.0–26.0	10.0–15.0	
Corn stalks	35.0–39.6	16.8–35.0	7.0–18.4	
Corn cobs	33.7–41.2	31.9–36.0	6.1–15.9	
Reed straw	49.4	31.5	8.7	
Sunflower stems	34.1–42.1	5.18–9.71	7.7–13.4	

The main features of BC include a high level of crystallinity (80%–90%), high mechanical strength linked to its inherent single cellulose nanofibers, and a high degree of polymerization (glucose units that range between 7000 and 16,000) (Seddiqi et al., 2021). BC can be transformed into different morphological forms, such as spheres, films, fleeces, and hollow particles, which enhances its adaptability for developing various films and coatings in food industry (Cazón & Vázquez, 2021). While PC, WC, and BC share an identical chemical structure, they exhibit varying mechanical characteristics due to differences in their structural organization and rate of degradation (Cazón & Vázquez, 2021; Seddiqi et al., 2021). BC is produced in several nonsynthetic and synthetic media by a group of acetic acid-producing bacteria (Cazón & Vázquez, 2021). Throughout biosynthesis, BC develops a pellicle consisting of an irregular nanofibrillar arrangement of cellulose chains featuring both disordered and crystalline areas (Fig. 7.3) (Cazón & Vázquez, 2021; Seddiqi et al., 2021). BC consists of <100 nm wide ribbon-shaped fibrils that are randomly assembled and consist of 7–8 nm wide nanofibrils accumulated in bundles (Heinze, 2015). Therefore, this particular nano-morphology results in a surface region that can hold water and presents high wet strength, excellent elasticity, and conformability, which is a great combination of advantageous characteristics for formulation of films and coatings (Cazón & Vázquez, 2021).

7.2.2.3 *Algal cellulose*

Various algal taxa were reported to contain cellulose, such as red algae Rhodophyta, green algae Chlorophyta (e.g., Ulvophyceae, Trebouxiophyceae, and Chlorophyceae), Charophyta (e.g., Charophyceae), brown algae (e.g., Phaeophyceae class), yellow-green algae (e.g., species from Xanthophyceae), and golden algae (e.g., Chrysophyceae class) (Zanchetta et al., 2021). Algal cellulose is not pure and is commonly associated with hemicellulose, pectin, protein, agars, alginates, and lignin (Seddiqi et al., 2021). It is generally known that brown algae consist of a dense CW composed mainly of fucoidans and alginates; the cellulose microfibrils are usually smaller but play a significant role by associating with alginates for structural functions (Zanchetta et al., 2021). The CW of red algae usually contains sulfated galactans, such as carrageenans and agars, which do not always correlate with cellulose (Zanchetta et al., 2021). It has been demonstrated that the CW of *Chondrus crispus* includes a layer of microfibrils set within a matrix of carrageenans, whereas *Gracilaria verrucosa* features two layers, consist of an inner agar layer and an outer layer predominantly composed of cellulose, along with a minor component of mannose and xylose (Zanchetta et al., 2021). Among green algae, Chlorella (e.g., Trebouxiophyceae) is known to have a primary fibrillar layer composed of polysaccharides (a mixture of cellulose, pectin, and hemicellulose) that may be protected by a robust outer layer of algaenan (Zanchetta et al., 2021).

According to current data in this field, *Nannochloropsis* sp. is the only microalgae studied for its potential to produce cellulose from which cellulose has been obtained (Lee et al., 2018). Nevertheless, there is limited information regarding other species; thus, some microalgae species might have great potential for cellulose production (Zanchetta et al., 2021). Therefore, Trebouxiophyceae class (e.g., *Chlorella* sp., *Oocystis* sp.), Chlorophyceae class (such as *Chlorococcum* sp., *Coelastrella* sp., *Selenastrum* sp., and *Scenedesmus* sp.), and Chrysophyceae class (e.g., *Dinobryon* sp. or *Dinophyta-phylum*) are other microalgae besides *Nannochloropsis* sp. that contain cellulose (Zanchetta et al., 2021). Among macroalgae, most studies on cellulose production concentrated on the Ulvophyceae class with species such as *Ulva* sp.,

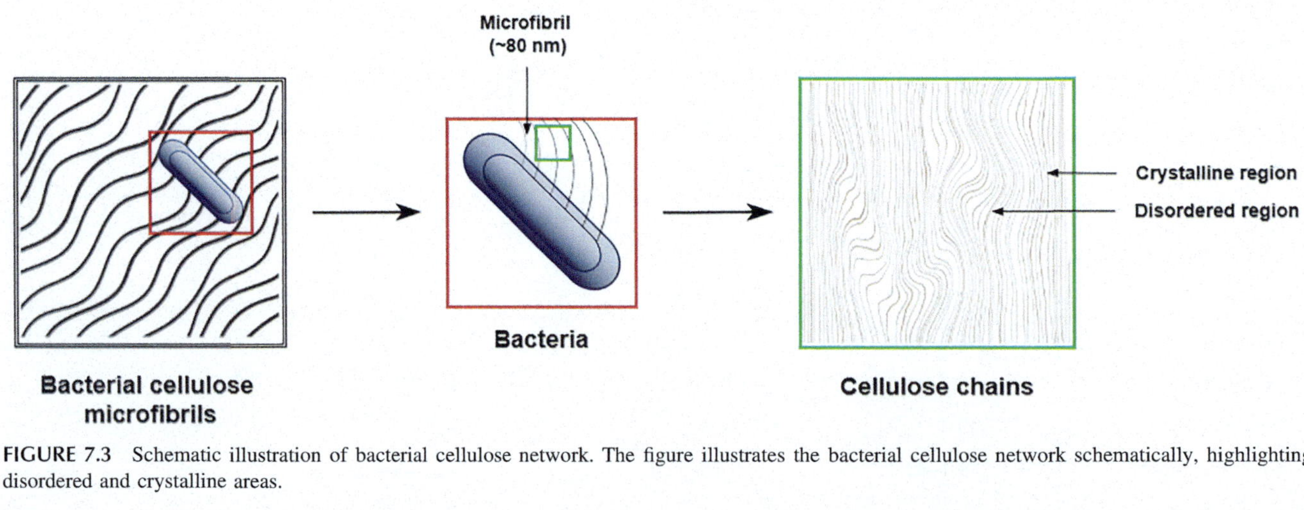

FIGURE 7.3 Schematic illustration of bacterial cellulose network. The figure illustrates the bacterial cellulose network schematically, highlighting disordered and crystalline areas.

Valonia sp., *Cladophora* sp., and *Boergesenia* sp. (Zanchetta et al., 2021). Many other species from the Cladophorales order, such as *Dictyosphaeria* sp., *Rhizoclonium* sp., *Apjohnia* sp., *Chaetomorpha* sp., *Siphonocladus* sp., and *Posidonia* sp., consist of highly crystalline and well-organized cellulose (Zanchetta et al., 2021). Cellulose extracted from green algae exhibits unique advantages compared to WC, BC, and PC because of its high crystallinity (>70%), high porosity within mesoporous range, low moisture adsorption capacity, and high specific surface area (Zanchetta et al., 2021).

7.2.2.4 Tunicate cellulose

Tunicates are invertebrate animals that live in the oceans and are the sole identified animal source of cellulose (Seddiqi et al., 2021). The term "Tunicata" is derived from its distinctive outer tissue known as the "tunic," which envelops the entire epidermis of the organism (Seddiqi et al., 2021). The cellulose serves as a framework in the tunic tissues. Complexes of enzymes that synthesize cellulose have been discovered in the plasma membrane of their epidermal cells and are responsible for cellulose production (Seddiqi et al., 2021). Additionally, the enzymes perform two functions: facilitating metamorphosis and ensuring proper tunic formation. Tunicates represent a highly abundant resource; the subphylum Tunicata consists of three classes, and every species in two of those classes, Ascidiacea and Thaliacea, has tunics (Chanthathamrongsiri et al., 2021). The third class, Appendicularia, does not have tunics but instead secretes cellulosic substances (Chanthathamrongsiri et al., 2021). There are more than 2300 species within Ascidiacea alone (Seddiqi et al., 2021). Tunicates have various shapes, such as flower-like, eggplant-like, and pineapple-like; thus, they are known by other names (e.g., "sea peach," "sea pineapple," etc.) (Lv et al., 2023). Based on functional organization and anatomical structure, tunicates are classified into solitary and colonial categories (Chanthathamrongsiri et al., 2021). Solitary tunicates live individually within their tunic, while colonial tunicates are interconnected by a common tunic from which they have budded (Chanthathamrongsiri et al., 2021). Most of the research on tunicate cellulose has concentrated on solitary tunicates, such as *Ascidia* sp., *Styela clava*, *Styela plicata*, *Halocynthia papillosa*, *Ciona intestinalis*, *Metandrocarpa uedai*, and *Halocynthia roretzi*, while studies on cellulose produced by colonial tunicates are limited; most of them are based on *Eudistoma* sp. (Chanthathamrongsiri et al., 2021).

Hundreds of cellulose nanofibrils (CNFs) are arranged together in the tunic. The form and size of a nanofibril bundle differ according to the species (Lv et al., 2023). Nanofibril bundles are grouped in a multilayered structure aligned with the epidermis surface (Seddiqi et al., 2021). The width of tunicate CNFs ranges from 10 to 30 nm, while the length ranges between 100 nm and 2 μm, and the aspect ratio values vary from 60 to 70 (Lv et al., 2023). In most cases, the dry tunic is composed of around 60% cellulose and 27% nitrogen-containing substances (Seddiqi et al., 2021). Nevertheless, after different treatments and extraction techniques, TC becomes highly crystalline (approximately 95%) and consists of nearly pure cellulose in the high aspect ratio fibril morphological form (Hong et al., 2024). The specific surface area of TC is high, ranging from 150 to 170 $m^2\,g^{-1}$ (Hong et al., 2024; Moon et al., 2021). Additionally, it exhibits a reactive surface as a result of the presence of hydroxyl groups and has values of polymerization degree ranging between 700 and 3500 (Moon et al., 2021).

7.2.3 Morphological forms of cellulosic particles

Cellulose is naturally found or extracted from different sources in specific forms and dimensions, which can be easily categorized into cellulose filaments, cellulose micro/nanofibrils, cellulose fibers, and cellulose crystals (Seddiqi et al., 2021). Each type of cellulosic particle possesses unique characteristics in terms of crystallinity, morphology, size, aspect ratio, and physicochemical properties (Seddiqi et al., 2021).

7.2.3.1 Cellulose fibers

Cellulose extracted from plants and wood is synthesized in form of microfibrils that are further organized to assemble cellulose fibers (Seddiqi et al., 2021). Commonly, cellulosic fibers are found in three different configurations, such as long fibers that have 20–100 cm length (strand fibers), short fibers of <60 mm length (staple fibers), and very short fibers that have 1–10 mm length (pulp fibers) (Seddiqi et al., 2021). A staple fiber is composed of a whole, but singular cell, whereas a strand fiber consists of numerous cells. Cotton fibers are classified as staple fibers, measuring 25–45 mm in length. The length of cotton linters, which are the fibers that cling to the cotton seed, is merely a few millimeters. Staple and strand fibers are obtained from wild or cultivated plants directly after a water retting procedure (Heinze, 2015). Pulp fibers are typically derived as separate cells from wood and occasionally from plant sources (e.g., fique, sisal, cotton linters, and banana) through a pulping procedure (Seddiqi et al., 2021). Pulping involves a thorough treatment with solutions that eliminate the lignin between the cells, leading to individual plant cells that measure only a few millimeters in length (Seddiqi et al., 2021).

7.2.3.2 Cellulose filaments

Cellulose nanofibers can be formed into filament shape through various techniques such as flow-focusing, wet-extrusion, or spinning (Seddiqi et al., 2021). These cellulose filaments may display various physicochemical characteristics (e.g., crystallinity, allomorph), but all filaments exhibit ductility, mechanical strength, and high aspect ratios (Tibolla et al., 2017). Additionally, there is another category of cellulose filaments, which is produced not by the aforementioned methods, but by separating cellulose fibers from plants and wood (Tibolla et al., 2017). This peeling technique is purely mechanical, requiring no enzymes or chemical reagents, and has no released effluent, making the resulting filament attractive for food industry uses. Filaments are longer, more continuous fibers with lower aspect ratios and fewer branches, resulting in mechanical and structural properties that differ from those of micro/nanofibrils, which are shorter, highly branched structures with a large surface area.

7.2.3.3 Cellulose micro/nanofibrils

Cellulose microfibrils, originally called microfibrillated cellulose (MFC), were first obtained by Herrick et al. (1983) and Turbak et al. (1983). Cellulose fibers (primarily from wood) are mechanically disintegrated into micro- and nanofibrils by being processed through a homogenizer. CNFs are generally obtained at a low concentration from water suspensions of plant and wood pulps and have a <100 nm width (Seddiqi et al., 2021). CNFs are not always produced as pure cellulose nanofibers, but they may be disintegrated from various sources that contain a significant quantity of lignin and hemicellulose (Heinze, 2015). Many techniques have been applied in order to isolate individual fibrils, including mechanical disintegration (e.g., homogenization, microfluidization), along with the use of enzymatic treatments (Seddiqi et al., 2021; Tibolla et al., 2017). Some chemical treatments were also applied in order to facilitate the disintegration of CNFs, such as carboxymethylation and 2,2,6,6-tetramethylpiperidine-1-oxyl radical (TEMPO)-catalyzed oxidation, which additionally impart various functionalities to the cellulose fibrils (Isogai, 2022). Commonly, the width of CNFs ranges between 2 and 100 nm, depending on various factors, such as cellulose source, pretreatment applied to cellulose material, and fibrillation method (Seddiqi et al., 2021). The CNFs have a length of 1 μm and are primarily composed of alternating disordered and crystalline regions (Seddiqi et al., 2021; Tibolla et al., 2017). Generally, CNFs exhibit an elastic modulus ranging from 14 to 36 GPa and a tensile strength of approximately 1 GPa under ambient conditions (Seddiqi et al., 2021), which is a great combination of advantageous features for formulation of films and coatings.

7.2.3.4 Crystalline cellulose

An accurate combination of chemical, mechanical, and/or enzymatic treatments applied to cellulose microfibrils results in the extraction of the highly crystalline regions of the cellulose microfibrils and further formation of microcrystalline cellulose (MCC) with a diameter on the micron scale or cellulose nanocrystals (CNCs) with a diameter on the nanoscale (Seddiqi et al., 2021). MCC and CNCs are rigid, rod-like particles consisting of cellulose chain segments in an excellent crystalline arrangement (Heinze, 2015). Industrial-scale MCC is produced by hydrolyzing wood and cotton cellulose with dilute mineral acids. MCC is identified by a significant level of crystallinity, and the values ranging between 55% and 80% (Bangar et al., 2023). MCC can be dried in order to obtain a fine-particle form or can be processed with a water-soluble polymer to create a colloidal form (Bangar et al., 2023; Seddiqi et al., 2021). Therefore, it has considerable potential to be utilized in various fields, such as food industry, pharmaceuticals, and cosmetics. Moreover, features like fibrous nature, stiffness, water insolubility, nontoxicity, strength, biodegradability, lightness, crystallinity, high internal porosity, and renewability make MCC increasingly attractive for use in diverse industrial sectors (Seddiqi et al., 2021). CNC particles are crystalline regions extracted from the cellulose microfibrils (Babaei-Ghazvini et al., 2024). Purified CNCs are often obtained by sulfuric acid hydrolysis of native cellulose, but ultrasonic treatment and aggressive sulfuric acid hydrolysis of bulk cellulose destroy highly crystalline CNC particles (Babaei-Ghazvini et al., 2024). CNCs have remarkable features, including a large specific area, high aspect ratio, high specific modulus (60–220 GPa), and strength (7500–7700 MPa), reactive surfaces, biodegradability, availability, and the capacity to produce stable suspensions in various low-polarity solvents when a surfactant is applied (Babaei-Ghazvini et al., 2024). The features of nanocrystals, such as length, shape, and width, are influenced by the cellulose source, extraction method, and extraction parameters (e.g., controlled temperature and time for acid hydrolysis, high-pressure homogenization conditions, and additional modifications like dialysis and neutralization) (Babaei-Ghazvini et al., 2024). CNCs can be extracted from different sources, such as PC, AC, WC, TC, and BC (Seddiqi et al., 2021). Moreover, CNC particles can be obtained by breaking down MCC (Seddiqi et al., 2021). After acid treatment of MCC, a stable CNC dispersion in organic solvents or water is created (Babaei-Ghazvini et al., 2024). Generally, CNCs exhibit a comparatively wide length distribution due to the diffusion-controlled mechanism of acid hydrolysis (Seddiqi et al., 2021). CNCs derived from PC and WC have a length between 100 and 500 nm and width values ranging between 3 and 30 nm, while CNCs extracted from TC and BC have a length of 100 nm to several microns and a width of 5–50 nm (Seddiqi et al., 2021). The adaptability and versatility of bio-nanocomposites permit these CNCs to be utilized for active food packaging (Gomri et al., 2022).

7.3 Cellulose derivatives

Cellulose derivatives are a category of natural polymers in which cellulose is dissolved or swollen using various solvents (e.g., ILs, aqueous, and organic solvents) before being added to films, filaments, or coatings (Ursachi et al., 2024). The swollen cellulose structure can be chemically modified in order to enhance the performance and processability for specific uses, while in the dissolved state, all the hydroxyl groups are available to the reactant molecules (Paunonen, 2013). Cellulose derivatives are advantageous due to their biodegradability, abundance, and renewability, making them useful across various industries. Cellulose derivatives are usually classified based on the type of chemical reaction occurring on the hydroxyl group (Liu et al., 2021):

1. substitution reactions on the hydroxyl groups:
   - esterification;
   - etherification;
   - other substitutions: (a) intra- and (b) intermolecular reactions.
2. oxidation reactions:
   - selective and controlled oxidation of the hydroxyl.
3. addition and substitution reactions:
   - electrophilic substitutions;
   - radical additions.
4. polymer derivatization:
   - graft copolymer synthesis;
   - block copolymer synthesis.
5. reactions with metal compounds:
   - interactions with organometallics.

The characteristics of cellulose derivatives are influenced not only by the type and level of substitution, but also by the pattern of functionalization throughout the polymer chain (Seddiqi et al., 2021). The regioselective production of cellulose derivatives is restricted by the low solubility of cellulose in organic solvents (e.g., ethanol, methanol, benzene, acetone, and hexane), as well as significant steric hindrance caused by the rigid and large cellulose backbone (Seddiqi et al., 2021). The hydroxyl group serves as the primary reactive site on the cellulose chain and can be modified with various reagents to obtain derivatives, such as cellulose ethers and cellulose esters (Liu et al., 2021). The cellulose derivatives may differ regarding key attributes (e.g., chemical substituent and structure, water interaction, solubility, and ionic character), which are summarized in Table 7.2.

7.3.1 Cellulose esters

Cellulose esters are a commercially accessible group of thermoplastic biopolymers that are derived from cellulose (Seddiqi et al., 2021). In comparison with cellulose, cellulose esters exhibit good solubility in typical solvents and melt before decomposition (Liu et al., 2021). Different morphological forms of cellulosic particles (e.g., fibrils, fibers, or crystals) can be esterified in order to obtain cellulose ester (Seddiqi et al., 2021). Based on chemical substituents, cellulose esters are classified into esters with single substituents (e.g., CA, cellulose sulfate [CS], CTA, and CN) and mixed substituents (e.g., CAP, CAB, cellulose acetate trimellitate [CAT], and cellulose acetate propionate [CAPr]). Therefore, some cellulose esters are described below.

7.3.1.1 Cellulose acetate

In 1865, CA was first synthesized by Paul Schutzenberger from the wood pulp using acetic anhydride (Liu et al., 2021). Initially, during the acetylation process, cellulose is treated with acetic acid for clarification. Afterwards, it is dehydrated and reacted with acetic anhydride utilizing an appropriate reagent (i.e., sulfuric acid-anhydrous acetic acid). The resulting material is cellulose triacetate (CTA) that can be hydrolyzed in an acid medium to a lower grade of substitution to obtain CA and cellulose diacetate (Liu et al., 2021). CA can also be extracted by the esterification of cellulosic sources (e.g., sugarcane, wood, and cotton) (Seddiqi et al., 2021). CA is broadly used in food packaging as a rigid wrapping film and

TABLE 7.2 Specific functional group, chemical structure, solvent, and ionic character of cellulose derivatives[a].

Cellulose derivatives	Substituent	Type	Specific functional group	Ionic character	Solvent	References
Cellulose esters	Single	CN	–OH, $–ONO_2$	Nonionic	Nitrobenzene, methanol, mixture of ethanol-ether	Abdellah Ali et al. (2020); Doelker (2005); Liu et al. (2021); Paunonen (2013); Sabbaghi (2023); Seddiqi et al. (2021); and Yekta et al. (2023)
		CA	–OH, $–O(C{=}O)CH_3$		Acetic acid, acetone, dimethyl acetamide	
		CS	–OH, $–OSO_3H$		Water	
		CTA	–OH, $–O(C{=}O)CH_3$		Acetic acid, chloroform, tetrachloroethane	
	Mixed	CAB	–OH, $–O(C{=}O)CH_3$, $–O(C{=}O)C_4H_9$		Ethanol, methanol, isopropanol, 1-propanol	
		CAPr	–OH, $–O(C{=}O)CH_3$, $–O(C{=}O)CH_2CH_3$		Acetic acid, acetone, dioxane	
		CAP	–OH, $–O(C{=}O)CH_3$, $–O(C{=}O)C_6H_4(CO)–$	Ionized above pK_a	Nonacidic medium, organic solvents	
		CAT	–OH, $–O(C{=}O)CH_3$, $–O(C{=}O)C_6H_4(CO)_2–$		Nonacidic medium, organic solvents	
Cellulose ethers	Single	MC	–OH, $–OCH_3$	Nonionic	Water (<50°C), acetic acid	
		EC	–OH, $–OCH_2CH_3$		Glycerol, propane-1,2-diol	
		HEC	–OH, $–OCH_2CH_2OH$		Hot water	
		HPC	–OH, $–OCH_2CH(OH)CH_3$		Water, anhydrous alcohol	
		NaCMC/CMC	–OH, $–OCH_2COONa$	Anionic	Water	
		CaCMC	–OH, $–OCH_2COOCa$		Insoluble in water	
	Mixed	MEC	–OH, $–OCH_3$, $–OCH_2CH_3$	Nonionic	Dimethyl sulfoxide, *N,N*-dimethylacetamide, *N*-methyl pyrrolidone, water	
		MHEC	–OH, $–OCH_3$, $–OCH_2CH_2OH$		Water, insoluble in most organic solvents	
		HPMC	–OH, $–OCH_3$, $–OCH_2CH(OH)CH_3$		Cold water, ethanol, methanol, acetone, isopropyl alcohol	
		HBMC	–OH, $–OCH_3$, $–OCH_2CH_2CH_2OH$		Organic solvents	
		EHEC			Water	

			–OH, $-OCH_2CH_3$, $-OCH_2CH_2OH$			
		NaCMHEC/CMHEC	–OH, $-OCH_2CH_2OH$, $-OCH_2COONa$	Anionic	Water	
		CMEC	–OH, $-OCH_2CH_3$, $-OCH_2COOH$		Insoluble in water, soluble in water-ethanol (70%–90%) solution	
Cellulose ether esters	Mixed	HPMCP	–OH, $-O(C{=}O)CH_3$, $-OCH_2CH(OH)CH_3$, –O $(C{=}O)C_6H_4(CO)-$	Ionized above pK_a	Insoluble in water, soluble in weak acid to neutral medium	
		HPMCAS	–OH, $-O(C{=}O)CH_3$, $-OCH_2CH(OH)CH_3$, –O $(C{=}O)C_6H_4(CO)_2-$		Nonacidic medium, organic solvents	
		HPCA	–OH, $-O(C{=}O)CH_3$, $-OCH_2CH(OH)CH_3$	Nonionic	Organic solvents	
		Polyquaternium-4	$-O(C{=}O)-$, $-N^+(CH_3)_2$, $-CH_2{=}CH-CH_2-$	Cationic	Water	
		Polyquaternium-10	$-O(C{=}O)-$, $-N^+(CH_3)_3$, $-CH_2-CH-OH-CH_3-$		Water	
		Polyquaternium-24	$-O(C{=}O)-$,–O–C–O–, $-N^+(CH_3)_2$, $-C_{12}H_{25}$		Water	
Cross-linked derivatives	Mixed	Croscarmellose sodium	–OH, $-OCH_2COOH$	Anionic	Insoluble in water, insoluble in organic solvents	
Grafted derivatives	Mixed	g-HEC	–OH, $-OCH_2CH_3$, $-C_{12}H_{25}$, $-C_{14}H_{29}$	Nonionic	Water	

[a] CN, *Cellulose nitrate;* CA, *cellulose acetate;* CS, *cellulose sulfate;* CTA, *cellulose triacetate;* CAB, *cellulose acetate butyrate;* CAPr, *cellulose acetate propionate;* CAP, *cellulose acetate phthalate;* CAT, *cellulose acetate trimellilate;* MC, *methyl cellulose;* EC, *ethyl cellulose;* HEC, *hydroxyethyl cellulose;* HPC, *hydroxypropyl cellulose;* NaCMC/CMC, *sodium carboxymethyl cellulose/carboxymethyl cellulose;* CaCMC, *calcium carboxymethyl cellulose;* MEC, *methyl ethyl cellulose;* MHEC, *methyl hydroxyethyl cellulose;* HPMC, *hydroxypropyl methyl cellulose;* HBMC, *hydroxybutyl methyl cellulose;* EHEC, *ethyl hydroxyethyl cellulose;* NaCMHEC/CMHEC, *sodium carboxymethyl hydroxyethyl cellulose/carboxymethyl hydroxyethyl cellulose;* CMEC, *carboxymethyl ethyl cellulose;* HPMCP, *hydroxypropyl methyl cellulose phthalate;* HPMCAS, *hydroxypropyl methyl cellulose acetate succinate;* HPCA, *hydroxypropyl cellulose acetate;* Polyquaternium-4, *esters of hydroxyethyl cellulose and diallyl dimethyl ammonium chloride;* Polyquaternium-10, *esters of hydroxyethyl cellulose and 2-hydroxypropyl trimethyl ammonium chloride;* Polyquaternium-24, *esters of hydroxyethyl cellulose and a lauryl dimethyl ammonium substituted epoxide;* Croscarmellose sodium, *crosslinked sodium carboxymethyl cellulose;* g-HEC, *hydroxyethyl cellulose grafted with alkyl C12-C14 chains.*

membrane (Paunonen, 2013). In cellulosic esters, the degree of substitution has a great impact on biodegradability and solubility (Liu et al., 2021). Therefore, CA has stable hydrolytic features, is insoluble in water, has lower water swelling, lower water content, and suitable mechanical characteristics, indicating that the hydrophilic nature is primarily caused by hydroxyl groups (Liu et al., 2021). Other characteristics of CA, such as nontoxicity, biocompatibility, and decomposability, make CA a significant and economical natural source (Liu et al., 2021). Gradual electrostatic assembly, electrospinning, and phase inversion by immersion precipitation and solvent evaporation are some of the most prevalent membrane manufacturing techniques based on CA (Liu et al., 2021). Some solvents, such as 1,3-dimethyl-2-imidazolidinone/lithium chloride, *N*-ethyl-pyridinium chloride, and *N,N*-dimethylacetamide/lithium chloride, dissolve CA but require prolonged pretreatment (Seddiqi et al., 2021). ILs have also been utilized as efficient solvents for CA, although their use in industry is limited because they are expensive (Seddiqi et al., 2021).

7.3.1.2 Cellulose nitrate

CN, or nitrocellulose, usually referred to as gun cotton, serves as a significant component in smokeless gunpowder due to its tendency to decompose explosively (Liu et al., 2021). CN is generated through the nitration process of cellulose obtained from the pulp of cotton linters and wood, using strong nitrating agents (i.e., nitric acid) (Seddiqi et al., 2021). During the nitration process, the hydroxyl groups present on the cellulose surface are replaced with nitrate esters as a result of the electrophilic attack by NO_2^+ ions on the –OH groups (Oprea & Voicu, 2020). This process was initially studied in relation to amine nitration and alcohol, and further research has indicated its relevance to cellulose as well (Liu et al., 2021). A complete nitration of cellulose is characterized by a final nitrogen content of 13.5% or higher than this value (Liu et al., 2021). The CN application is significantly influenced by the nitrogen content. Therefore, CN that has a nitrogen percentage ranging from 12.6% to 13.3% is classified as explosive, while those with less than 12.6% nitrogen exhibit remarkable physicochemical stability and biocompatibility, making them appropriate for applications in different fields (Liu et al., 2021; Seddiqi et al., 2021). CN membranes were identified as suitable for protein preservation due to the noncovalent interactions between the amine functional groups present in the protein structure and nitro groups of polymer (Liu et al., 2021). Additional uses of CN include osmosis and ultrafiltration membranes, as well as electrophoresis films (Oprea & Voicu, 2020).

7.3.1.3 Cellulose sulfate

CS is an ester of cellulose obtained by heterogeneous, quasihomogeneous, or homogeneous sulfation methods with different reagents, in particular sulfuric acid, dichloroethane, liquid sulfurous anhydride, and pyridine (Liu et al., 2021). CS possesses antibacterial, antiviral, and anticoagulant features and is soluble in water at high concentrations (Seddiqi et al., 2021). CS exhibits a high substitution degree (0.14–1.46), and the transitional reaction produces CN. The first report of cellulose being suspended in the SO_3-pyridine compound along with *N,N*-dimethylformamide as the cosolvent was published in 1979. The synthesis of CS with an even distribution of sulfate groups along the polymer chains requires a dipolar aprotic cosolvent that drastically reduces the solution viscosity and does not significantly influence the reactivity of the sulfating agent. The quasihomogeneous method relies on acetosulfation, described as a gradual dissolution of cellulose in *N,N*-dimethylformamide coupled with a blend of acetylating and sulfating agents (e.g., acetic anhydride/chlorosulfonic acid), succeeded by the removal of acetyl groups during precipitation, resulting in the transformation of CA-sulfate into CS (Liu et al., 2021). In addition to its low price, simplicity in preparation, and capacity for large-scale manufacturing, other features such as biodegradability, film-forming properties, and biocompatibility make it a great frontrunner for applications in the food industry (Liu et al., 2021; Seddiqi et al., 2021).

7.3.2 Cellulose ethers

Cellulose hydroxyl groups can be either partially or completely etherified with various reagents, such as α-halogenated carboxylic acids, epoxides, and halogenoalkanes (Seddiqi et al., 2021). The rate at which cellulose ethers dissolve is influenced by the alkalinity and acidity of the solution. Under acidic conditions, cellulose ethers dissolve quite slowly, whereas in alkaline solutions, they dissolve quickly (Seddiqi et al., 2021). Cellulose ethers can exhibit water solubility based on the substituent present in their chemical structure, pattern, and degree of substitution (Arai & Shikata, 2020). The majority of water-soluble ethers possess a degree of substitution ranging from 0.4 to 2 (Arai & Shikata, 2020). Among all cellulose ethers, hydroxypropyl methyl cellulose (HPMC), HEC, and carboxymethyl cellulose (CMC) are widely used in the formulation of different films and coatings due to their suitable mechanical and rheological properties, nontoxic nature, high thermo-gelling capability, and water retention ability (Heinze et al., 2018; Liu et al., 2021; Ursachi et al., 2024). Some relevant cellulose ethers are further described below.

7.3.2.1 *Ethyl cellulose and methyl cellulose*

Ethyl cellulose (EC) and methyl cellulose (MC) are both derivatives of cellulose and have similar comparative compositions (Liu et al., 2021). EC is a moderately ethylated cellulosic ether formed by the reaction of alkali cellulose with ethyl chloride (Ahmadi et al., 2022). EC is classified based on the degree of substitution or ethoxyl content, which is reported to range between 40% and 54.88%. Therefore, the solubility of EC in various solvents is influenced by its ethoxyl content (Ahmadi et al., 2022). EC dissolves in water with an ethoxyl content ranging between 19% and 29% or 0.8–1.3 ethyl substitutions in an anhydrous glucose unit; however, its solubility in water decreases with the enhancement of ethoxyl content (Ahmadi et al., 2022). EC with less than 46%–48% ethoxyl groups is completely soluble in polar and nonpolar organic solvents, such as methyl acetate, chloroform, and tetrahydrofuran, as well as in an aromatic hydrocarbon-ethanol mixture (Ahmadi et al., 2022). EC that contains 46%–48% or more ethoxyl groups is completely soluble in toluene, ethyl acetate, ethanol, chloroform, and methanol (Ahmadi et al., 2022; Seddiqi et al., 2021). MC is a methyl ether obtained via the reaction between alkali cellulose and methyl chloride (Liu et al., 2021). MC is soluble in cold water when the substitution values range between 1.4 and 2.0, and if this degree is 2.4–2.8, MC is typically soluble in water and some organic solvents (Paunonen, 2013; Seddiqi et al., 2021). It is nontoxic and forms transparent, continuous, flexible, and tasteless films that have great oxygen barrier properties, but it has an inferior characteristic in terms of water vapor (Paunonen, 2013). Therefore, MC can be utilized as an edible coating and film, and as a primary material in order to modify the barrier and mechanical features of blended or layered composites for food packaging (Paunonen, 2013). Crosslinking is one of the most applied methods used to alter MC hydrophilicity, which can be accomplished through radiation or various chemical crosslinking agents (e.g., epoxy compounds, aldehydes) (Liu et al., 2021; Oprea & Voicu, 2020).

7.3.2.2 *Hydroxyethyl cellulose and hydroxypropyl cellulose*

HEC and hydroxypropyl cellulose (HPC) are nonionic derivatives with superior water retention, biocompatibility, and nontoxicity that contain –OH groups in repeating glucose units in their structure (Trivedi, 2015). The chemical structure of HPC and HEC can be easily further modified due to the presence of –OH groups, which could offer novel properties that are relevant for film and coating development (Seddiqi et al., 2021). The high glass transition temperature of HPC provides substantial stability and limits recrystallization during storage, while free –OH groups of HEC can easily form hydrogen bonds with other groups of biopolymers, contributing to great stability in the solid state (Trivedi, 2015). Bochek et al. (2018) obtained composite films from mixtures of HEC and HPC aqueous solutions with poly-*N*-methyl-*N*-vinyl acetamide (PMVAA) and reported that HEC and HPC are compatible with PMVAA. Moreover, the introduction of one cellulose ether into the matrix of PMVAA inhibited the crystallization of the bulk biopolymer and resulted in the amorphization of all biopolymers within the blend (Bochek et al., 2018). On the other hand, Gökmen and Pekel Bayramgil (2022) reported that HEC gave the best cross-linked nanocomposite film formation among other biopolymers, such as HPC and HPMC. Therefore, HPC was not able to form cross-linked films, while HPMC did not provide sufficient compatibility regarding viscosity (Gökmen & Pekel Bayramgil, 2022).

7.3.2.3 *Carboxymethyl cellulose*

CMC is a water-soluble, anionic cellulose derivative, and a linear polysaccharide composed of anhydroglucose (Rahman et al., 2021). From a molecular perspective, the primary distinction between cellulose and CMC is the presence of certain anionic carboxymethyl groups (–CH_2COOH) in the CMC structure that substitute the hydrogen atoms from hydroxyl groups located in the cellulose framework (Rahman et al., 2021). CMC was initially synthesized in 1918. Cotton linters, corn husks, wood residue, rice, paper sludge, and mesquite trees were studied as sources of CMC (Liu et al., 2021). Based on its purity level, CMC may be categorized into semipurified, technical, and purified (Liu et al., 2021). CMC has particular characteristics, such as nontoxicity, hydrophilicity, high chemical stability, biodegradability, water stability, and biocompatibility that allow for the utilization of CMC in different fields as a binder, viscosity-increasing agent, stabilizer, rheological control agent, and film former (Seddiqi et al., 2021). Therefore, functional features, such as viscosity and water solubility of CMC, depend mainly on three factors: the position of carboxyl substituents in the polymer chain, degree of substitution, and molecular weight (Yildirim-Yalcin et al., 2022). CMC shows remarkable moisture absorption ability due to the presence of both carboxyl and hydroxyl groups (Yildirim-Yalcin et al., 2022). Furthermore, anionic character of CMC permits it to engage in ionic interactions with other ionic biopolymers, like polysaccharides and proteins (Yildirim-Yalcin et al., 2022).

7.4 Techniques for the fabrication of cellulose- and its derivatives-based films and coatings in food technology

Cellulose and its derivatives are film-forming materials that are typically used as structural substances in various polymeric mixtures (Paunonen, 2013). These materials are biodegradable and edible, recognized for their sensory and organoleptic characteristics, and can easily encapsulate or incorporate natural antioxidant and antimicrobial compounds (Liu et al., 2021). Moreover, cellulose and its derivatives provide good barrier properties, improve quality, and extend shelf life of food products (Liu et al., 2021). Although the terms coatings and films are frequently used with identical meanings, it is crucial to differentiate between them. Films are manufactured separately and are utilized to wrap the food product, whereas coatings are created directly on the surface of the food product (Martins et al., 2022). When selecting the type of packaging and food product that will be packaged, it is essential to take into account the method for the fabrication of films/coatings (Martins et al., 2022). There are various techniques for the fabrication of cellulose- and its derivatives-based films and coatings in food technology, each with its own advantages and disadvantages (Table 7.3). Some significant techniques are further described below.

Solution casting is the most commonly utilized technique for producing cellulose and its derivatives-based films due to its affordability and ease of handling (Martins et al., 2022). This method can be implemented as bench casting or continuous casting. The first method (bench casting) involves pouring the film-forming solution onto a Teflon plate or a solid surface, followed by the solvent evaporation (Liu et al., 2021; Martins et al., 2022). Consequently, this process leads to the rearrangement of molecules and the film formation. The drying procedure can take place at room temperature or in air circulation ovens at temperatures ranging from 35°C to 60°C (Martins et al., 2022; Suhag et al., 2020). The duration of the drying process is variable and is influenced by the type of dissolved substance, the volume of the filmogenic solution, and the drying temperature (Martins et al., 2022). During this process, the film thickness is controlled based on the quantity of solution poured onto the surface (Martins et al., 2022). The second method (continuous casting) is commonly applied in the conventional plastics sector, in addition to its growing use in the production of biodegradable films. This method involves the continuous application of the solution onto a moving level surface that has a blade; therefore the blade distributes the solution and regulates the film thickness (Martins et al., 2022; Suhag et al., 2020). In comparison with first casting method, continuous casting is more suitable for industrial uses due to its higher production capability and reduced drying time (Martins et al., 2022).

Extrusion represents a distinct method employed in the large-scale manufacture of polysaccharide-based films, especially cellulose-based films. In contrast to continuous casting, the films produced through extrusion exhibit different physical features (Martins et al., 2022). The film-forming solution is fed into a funnel and directed into a heated cylinder containing one or two screws (Martins et al., 2022). These screws are essential for mixing and compacting the solution, serving as a key component of the process. Factors such as screw type, temperature, and processing speed are adjusted based on the properties of the dissolved substance (Estevez-Areco et al., 2019; Suhag et al., 2020). Upon completion of the procedure, the extruded material is compressed, cooled, and ultimately shaped into film (Liu et al., 2021; Suhag et al., 2020). Considering these steps, it can be noted that extrusion is a continuous method that uses minimal solvent at high shear force and temperature while maintaining low cost and enhanced productivity (Estevez-Areco et al., 2019; Martins et al., 2022).

Regarding coatings, spraying (spray drying), immersion, and layer-by-layer (LBL) assembly are the most commonly used techniques (Martins et al., 2022; Suhag et al., 2020). In all of these methods, the mechanical and barrier characteristics primarily rely on the specific type of polysaccharide, its viscosity, concentration, and surface tension (Lara et al., 2020; Suhag et al., 2020). In the immersion technique, the food product is first submerged in a film-forming solution, forming a coating on its surface (Suhag et al., 2020). Afterwards, the excess solution is removed, and the solvent is either evaporated at ambient temperature or dried using air or nitrogen flow. The dissolved substance then polymerizes on the surface of food product, forming a thin film (Liu et al., 2021). One major advantage of the immersion method is its ability to coat irregular surfaces. However, its main drawbacks include the accumulation of contaminants and the growth of microorganisms in the immersion solution (Martins et al., 2022; Suhag et al., 2020).

Spraying (spray drying) is the most commonly utilized technique for coating food products on an industrial scale (Liu et al., 2021). A set of nozzles enhances the spraying technique by increasing the surface area of the film-forming solution through droplet formation, allowing it to evenly cover the surface of the food (Suhag et al., 2020). Therefore, homogeneous coatings are formed with regulated thickness. Additionally, the spraying apparatus prevents contamination of the filmogenic solution, and the capability to control temperature can enhance the automation of large-scale production (Lara et al., 2020; Martins et al., 2022; Suhag et al., 2020).

TABLE 7.3 Techniques for the fabrication of films and coatings in food technology, their advantages, and disadvantages.

Techniques	Advantages	Disadvantages	References
Solution casting	• The ease of production without the need for specialized equipment at low cost; • Results in improved particle-to-particle interactions making the particle packaging more uniform; • Possesses low optical retardation and low haze; • Exhibits remarkable transparency and flatness.	• Restricting the forms; • Possible trapping of toxic reagent within the polymer; • Restriction on the amount of film created; • Different evaporative levels and temperatures may result in the production of film with different properties; • Involves long drying time, which is not feasible for commercial production.	Suhag et al. (2020)
Layer-by-layer (LBL) assembly	• It is well-established and allows for conventional and unconventional assemblies; • Provides various characteristics to the films; • Produces more benefits for food preservation; • Various functions are encapsulated in one film.	• Nozzle shape effects and gravity draining can lead to inhomogeneity in the film; • It can be thermodynamically unstable; • It can be challenging to achieve for films assembled on the nanoscale.	Richardson et al. (2016)
Extrusion	• Improves physicochemical characteristics of extruded material; • It can be used to form multilayer films; • Short time of processing and no use of solvents; • Low energy consumption; • Enhanced mechanical and optical features (e.g., transparency and elongation).	• Limitation to process only low moisture and temperature tolerant raw material mixtures; • Restricts the use of specific polymers; • High cost of specialized equipment; • High maintenance cost.	Suhag et al. (2020)
Coating	• It is applied, sprayed, or placed at the surface of food products; • It is food grade, edible, and consists of bioactive compounds; • Offers thin films with low-surface roughness and good homogeneity.	• It should be selected based on the size and type of the substrate; • Requires sophisticated equipment and cleanroom facilities; • It is necessary to keep the thickness homogeneity of the substrate.	Butt (2022)
Polymeric hydrogel	• Increases the resistance of hydrogel to dissolution in water-based solutions.	• Low mechanical resistance; • Quick dissolution.	Batista et al. (2019)
Spraying (spray drying)	• Exhibits consistent coating with uniform thickness; • Allows automatic continuous production; • Permits to control the temperature of the coating solution.	• Needs to carefully adjust the viscosity of the coating solutions; • Highly viscous solutions are difficult to spray onto food products.	Suhag et al. (2020)
Electrospinning	• Cost-effective, affordable, and versatile technique; • Easy manipulation; • Reduces the amount of compounds required for the formulation of films and coatings.	• Low yield production; • High energy consumption; • Difficult to stabilize the process since it is a nonthermal technique.	Lopez-Polo et al. (2024)

(Continued)

TABLE 7.3 (Continued)

Techniques	Advantages	Disadvantages	References
Micro and nano encapsulation	• Exhibits superior functional and physicochemical features over traditional food films and coatings; • Improves food quality; • Prolongs shelf life of food products.	• Higher energy requirements (lyophilization as an encapsulation approach); • Long processing time (lyophilization as an encapsulation approach).	Saifullah et al. (2019); Xie et al. (2024)
Liposomes and nanoliposomes	• Experimental parameters can be easily controlled; • Control of physicochemical properties; • High miscibility in organic solvents; • Enhanced solvent power.	• High temperature deployment in liposomal preparation; • Phospholipid instability; • Heterogeneous size distribution; • Time consumption.	Alavi et al. (2022)
Nano micelles	• Biocompatibility; • Protect against oxidation; • Stable storage.	• Poor solubility of small-sized micelles; • Poor physical stability in vivo; • Poor loading capacity.	Perumal et al. (2022)
Nano emulsions	• Robust stability; • Apparatus simplicity.	• High energy consumption; • Increase of emulsion temperature during preparation.	Espitia et al. (2019)

The LBL technique is employed to achieve ultra-thin multilayer coatings, also referred to as nanolaminates (Richardson et al., 2016). This method depends on the sequential alternating application of various biomolecules on a surface, with a rinsing step occurring after each deposition (Liu et al., 2021). The selected biomolecules serve as building blocks, whose interactions (e.g., hydrogen bonds, electrostatic bonds, covalent bonds, acid-base interactions, or coordination interactions) facilitate the formation of a self-organizing matrix that can vary in composition, size, and shape (Richardson et al., 2016; Suhag et al., 2020).

7.5 Functional properties of films and coatings based on cellulose and its derivatives

Research and development in the food packaging sector aim to identify new alternatives that are not only compatible but also practical and suitable for commercial application. Consequently, polysaccharide polymers are currently being investigated for research, formulation, and development in the food packaging sector. Among these polymers, cellulose and its derivatives are gaining attention because they are among the most commonly used biopolymers. Cellulose derivatives, such as HEC, methyl cellulose, CMC, HPMC, and HPC, are renewable and natural polymers that are receiving growing attention (Liu et al., 2021). They exhibit a range of important properties, including mechanical strength, biocompatibility, porosity, solubility, thermal properties, biodegradability, crystallinity, water absorbency, and antimicrobial activity. In addition, they offer exceptional sensory qualities and various other biological functions (Liu et al., 2021). While antimicrobial and antioxidant characteristics are relatively low, the impregnation, incorporation, blending, and encapsulation of cellulose and its derivatives with numerous natural plant extracts and other biopolymers accomplish this requirement (Liu et al., 2021; Seddiqi et al., 2021).

7.5.1 Mechanical properties

The mechanical characteristics of a film are essential for wrapping and packaging food products of various shapes and sizes (Liu et al., 2021). Cellulose is primarily utilized alongside synthetic polymers since it facilitates blending and enhances their characteristics (Liu et al., 2021; Seddiqi et al., 2021). For food packaging materials, the key mechanical parameters include stress (σ), elongation at break, tensile strength, strain (ε), and Young's modulus (Ursachi et al., 2024). Cellulose blended with other polymers typically comprises two components: soft polymer and hard polymer (Liu et al., 2021). Many studies describe soft components as materials that deform irreversibly and eventually fracture, while hard polymers are more elastic and retain their shape (Seddiqi et al., 2021). HPMC films cross-linked with citric acid and β-cyclodextrin improved the mechanical, hygroscopic, and hydrophilic characteristics, and additionally, when they contained nisin, the films exhibited great antimicrobial properties (Wang, Wang, et al., 2024). Films formulated with methyl cellulose (less hydrophilic cellulose derivative) and impregnated with silver nanoparticles and *Lippiaalba* extract presented better mechanical and barrier characteristics (Nunes et al., 2018). Additionally, Teixeira et al. (2021) reported that the addition of different concentrations of triethyl citrate and glycerol in the matrix of CA films improved mechanical properties, light protection, and water vapor permeability. Therefore, the mechanical characteristics of developed films depend on their microstructure (specifically, their chemical composition and intermolecular interactions between their components), as well as the presence of various additives (such as lipid compounds, plasticizers, proteins, etc.) dispersed in the cellulose matrix (Martins et al., 2022).

7.5.2 Solubility and wetting properties

The solubility of cellulose is a significant factor for various applications, such as cellulose nano fiber spinning, bioethanol production, and cellulose gel preparation (Liu et al., 2021). Nonetheless, cellulose is not easy to dissolve in water and typical organic solvents due to its highly structured composition and strong hydrogen bonding, as well as hydrophobic and electrostatic interactions within the integrated fibrils, creating resilient bundles with many –OH groups located on the backbone (Seddiqi et al., 2021). The intricate inter and intramolecular hydrogen bonding arrangement of cellulose makes it insoluble in almost all organic and aqueous reagents (Liu et al., 2021). Therefore, in the early 20th century, it was initially reported that cellulose can be dissolved in other solvents, such as ILs (Liu et al., 2021). The dissolution of cellulose is fundamentally driven by the capacity of IL anions to form hydrogen bonds with cellulose, enabling its solubilization via a nonderivatizing process. However, only certain ILs possess the necessary interactions to effectively dissolve cellulose (Seddiqi et al., 2021). Further studies have indicated that the cellulose suspension in ILs is primarily influenced by the disruption of hydrogen bonding within cellulose chains, caused by the interaction with IL anions (Szabó et al., 2023). Moreover, cellulose solubility is greatly influenced by the anionic forms of the ILs (Szabó et al., 2023). Regarding cationic species, the impact on cellulose dissolution remains debated; however, several researchers agree that the cation has a crucial role, with the imidazolium cation considered more effective in dissolving cellulose in comparison with other cationic solvents (Szabó et al., 2023; Tomimatsu et al., 2019). One essential challenge in the use of ILs is their high viscosity, which affects the dispersion of cellulose in ILs (Tomimatsu et al., 2019). However, incorporating cosolvents into ILs is an efficient method for lowering their viscosity (Szabó et al., 2023; Tomimatsu et al., 2019).

7.5.3 Crystallinity

Crystallinity is an essential parameter that affects chemical reactivity, mechanical and physical features, and moisture-related properties of the polymeric fibers (Seki et al., 2022). Crystallinity-related parameters, such as crystal size, crystalline structure, characteristic crystalline peaks, and other complementary features in terms of crystallinity, should be determined in order to characterize a biopolymer (Seki et al., 2022). Cellulose crystallinity is influenced by its origin and cellulose source (Liu et al., 2021). The structure and shape of plant cells depend on the cellulose content and its tightly packed crystals, which are strong enough to hold the plant and are highly resistant to disintegration (Liu et al., 2021; Seddiqi et al., 2021). Unlike cellulose, which has a highly ordered crystalline structure, lignin and hemicellulose have an amorphous nature (Seki et al., 2022). The crystalline section of cellulose consists of numerous –OH groups with free ends that form hydrogen bonds (Heinze et al., 2018; Liu et al., 2021). Therefore, these hydrogen bonds contribute to the structural stability of crystalline cellulose by maintaining its tightly packed arrangement (Debnath et al., 2021). The degree of crystallinity in native cellulose varies depending on the measurement technique applied, including X-ray diffraction analysis, thermal analysis using differential scanning calorimetry (DSC), and infrared analysis (Debnath et al., 2021; Liu et al., 2021). Higher crystallinity can lead to enhanced tensile strength of cellulose (Debnath et al., 2021).

7.5.4 Thermal properties

Thermal decomposition of cellulosic fibers must be considered in composite production due to the high extrusion temperatures of the polymer matrix (Seki et al., 2022). The thermal degradation of plant fibers begins with the dehydration and breakdown of hemicellulose, followed by pyrolysis of lignin and cellulose decomposition (Seki et al., 2022). Consequently, the thermal decomposition of these fibers consists of four weight loss stages: (1) dehydration of cellulosic fibers occurring from room temperature to 110°C, (2) decomposition of hemicellulose (220°C–315°C), (3) breakdown of cellulose (300°C–450°C), and (4) degradation of lignin, which occurs between ambient temperature and 1000°C (Seki et al., 2022). The thermal behavior of cellulose is also influenced by various physicochemical pretreatments and cellulose sources (Liu et al., 2021). The thermal characteristics of natural polymers are examined through numerous analyses, such as DSC, dynamic mechanical analysis (DMA), thermogravimetry (TG), thermomechanometry (TMA), Fourier-transform infrared spectroscopy (FT-IR), and differential thermal analysis (DTA) (Liu et al., 2021). Therefore, Ursachi et al. (2024) reported that HPMC film incorporated with pectin at a concentration of 20% had the highest value of the maximum temperature for thermally induced weight loss in comparison with other cellulose derivatives (HEC and CMC) used in the formulation of the films. Moreover, they established that all developed films were thermally stable and might potentially serve as packaging for foods during transport and storage (Ursachi et al., 2024).

7.5.5 Barrier properties

Among the various factors that contribute to the formulation of effective packaging films, barrier properties are considered the key factor in preventing spoilage (Liu et al., 2021). To ensure proper food packaging, it is crucial to prevent oxygen transfer and moisture between food and the environment (Jali et al., 2024). The transfer of oxygen and moisture influences the appearance, shelf life, quality, odor, marketability, and taste of packaged food products (Jali et al., 2024). Consequently, the barrier features of food packaging materials are essential for extending and maintaining the shelf life of different packaged foods (Jali et al., 2024; Liu et al., 2021). The barrier requirements differ based on the product being packaged (Vijayakumar & Badwaik, 2022).

7.5.5.1 Oxygen barrier permeability

In the packaging sector, oxygen has the potential to degrade the quality of food, and food packaging materials that exhibit lower oxygen permeability values are more desirable (Jali et al., 2024). The oxygen permeability rate of packaging material is influenced by many factors, such as polymer characteristics (e.g., polarity, formulation, processing properties, molecular organization of the biopolymer, degree of crystallinity, and physical interaction between biopolymer and penetrant), environment (humidity and temperature), and penetrant properties (e.g., nature of the penetrant, biopolymer solubility, affinity to the biopolymer, and size of the penetrant) (Zabihzadeh Khajavi et al., 2020). Paunonen (2013) described cellophane and cellulose derivatives, such as MC, CA, and CMC, as the main materials used in food packaging. Therefore, Paunonen (2013) reported that cellophane films incorporated with cellulose extracted by an alkali/urea/water mixture were found to have superior oxygen barrier characteristics in comparison with traditional cellophane, while MC films have good barrier properties but poor water vapor barrier properties. Moreover, cellulose nanomaterials (CNFs, MCC, and CNCs) have been used to develop composite films and coatings by combining them with polymers that have poor oxygen barrier properties (Saedi et al., 2021).

7.5.5.2 Water vapor permeability

Cellulose and cellulose nanofibers display significant moisture resistance (Liu et al., 2021). This can be a result of porous composition of paper cellulose; it captures moisture and releases it with ease (Jali et al., 2024; Liu et al., 2021; Paunonen, 2013). These characteristics differ based on cellulose type, testing conditions, and density (Jali et al., 2024). Cellulose films exhibit better water vapor barrier characteristics at low relative humidity (RH) levels and temperatures (Liu et al., 2021). RH is not directly correlated with moisture absorption, which is described using distinct terminology (Jali et al., 2024). Ursachi et al. (2024) formulated films with commercial citrus pectin and different cellulose derivatives and reported that water vapor permeability (WVP) values for films developed with cellulose derivatives (control samples) were slightly higher than for films formulated with both pectin and cellulose derivatives. Similar tendency was observed when the water permeability of HPC film was improved with the addition of citric acid and β-cyclodextrin.

7.5.6 Sensory properties

The sensory characteristics of food products are correlated with flavor, color, aroma, texture, freshness, and appearance (Liu et al., 2021). Numerous studies have indicated and described that interconnection of long cellulose fibers permits the preservation of rheological features in food (Liu et al., 2021). This characteristic is associated with the viscosity. The linked microstructural fibers of cellulose contribute to better rheological characteristics, high water retention, and favorable structural and nutritional elements, though these fibers are less desirable for taste due to their insolubility or mouthfeel (Agarwal et al., 2018). However, if taste release is a priority for enhancing sensory perception, fibrillated fibers would be less desirable compared to more particulate materials. Furthermore, an aqueous solution of cellulose fibers exhibits a consistent viscoelastic gel-like structure that is essential for subsequent sensory evaluation (Agarwal et al., 2018).

Hence, cellulosic fibers are chosen as vectors to preserve the sensory attributes of food products (Agarwal et al., 2018; Liu et al., 2021).

7.5.7 Antimicrobial and antioxidant properties

Antimicrobial activity is one of the most significant characteristics of films and coatings. Oxidants, moisture, and availability of essential nutrients create an ideal environment for microorganisms to grow and survive (Liu et al., 2021). Several studies indicate that the cellulose matrix contributes to the antibacterial characteristic of developed films through its barrier property, interaction with moisture, and its role as a carrier for antimicrobial additives. Nevertheless, its antimicrobial activity is limited. Cellulose can be extracted from various sources, including bran, different fruit peels, wood, rice husk, wheat leftovers, and soft plant stems. These sources are abundant in phytochemicals and antimicrobial substances; therefore, cellulose is enriched with various properties (Liu et al., 2021). Cellulose and its derivatives serve as the most effective solutions when good thermal, barrier, and mechanical properties of the films are required, along with antimicrobial properties (Estevez-Areco et al., 2019). Films formulated with cellulose and its derivatives do not possess antioxidant properties; nonetheless, they are utilized for encapsulating various natural antioxidants, including essential oils (EOs) (Spinei et al., 2024). Moreover, Biswas et al. (2020) demonstrated that three cellulose esters (CA, CAB, and CAPr) can easily encapsulate different EOs (lime, rosemary, coffee, pimento berry, petitgrain, nutmeg, and anise essential oils) in film matrix, improving physicochemical properties and indicating that cellulose ester/essential oil system can be considered a bio-based food packaging material.

7.6 Application of cellulose and its derivatives in food preservation

Films and coatings are applied to numerous food products to maintain their quality by regulating biochemical, physiological, or oxidation processes. They are an accessible, inexpensive, and easy-to-apply method for preserving most vegetables and fruits compared to conventional packaging methods (Panahirad et al., 2021). The use of films and coatings would facilitate the implementation of less expensive packaging materials, consequently lowering the price of packaged products with significant economic profit (Panahirad et al., 2021). They protect various food products, such as vegetables and fruits, against physiological changes, chemical degradation, physical damage, and biological contamination (Liu et al., 2021). Films and coatings typically serve as a barrier to different gases and moisture transfers around foods, which prolongs the shelf life of perishable food products (Md Nor & Ding, 2020). Moreover, films and coatings also enhance the appearance of products by suppressing decay, hiding scars, providing shine, and addressing physiological issues (Panahirad et al., 2021). They may incorporate numerous active ingredients (e.g., colorings, flavorings, and sweeteners) that enhance the sensory and nutritional features of food products. Similarly, films and coatings can also encapsulate antioxidants, ions, and aroma compounds that, in turn, prevent browning reactions of fresh and fresh-cut vegetables and fruits (Liu et al., 2021; Md Nor & Ding, 2020; Panahirad et al., 2021). Furthermore, films and coatings decrease microbial spoilage, protect products from mechanical crushing, and prevent the release of volatile compounds. Various additives, mainly active ingredients, could be added to the formulation of films and coatings, which can enhance their functionality and efficiency (Panahirad et al., 2021). Liberating the active substances from film or coating materials to the surface of the product prevents undesirable reactions, such as oxidation, discoloration or browning, ethylene production, and microbial growth (Eyiz et al., 2020). Considering the final use of the film or coating, diverse compounds can be incorporated into film-forming solutions. In general, the film or coating material needs to be soluble or dispersible in specific solvents. At the same time, additional components like emulsifiers, plasticizers, and various additives may be incorporated into the film or coating solution in order to enhance functionality, stability, and integrity (Eyiz et al., 2020; Panahirad et al., 2021).

The fiber structure and peel chemistry play a crucial role in fruit packaging. They determine whether the packaging should be soft or hard and fibrous. Fruits contain a high amount of fructose and water, thereby creating an ideal environment for microorganisms. Ethylene is a natural substance emitted by fruits for ripening and is also a principal factor in their spoilage (Liu et al., 2021). Cellulose is frequently utilized to embed various antimicrobial agents, ensuring their controlled release within the matrix after activation. Besides microbial spoilage and ethylene, there are numerous other factors contributing to spoilage of fruits and vegetables, such as inadequate management of atmospheric temperature, bruising from poor transportation, moisture in storage areas, rodents and insects in storage facilities, improper grading during harvest and storage, ineffective cutting or harvesting methods, and contamination of packaging materials (Liu et al., 2021). Moreover, translucency, surface dehydration, texture loss, browning, off-flavor, water loss, bad aroma, softening, and microbes are the main reasons for deterioration of fresh and fresh-cut fruits and vegetables since they serve as a breeding ground for different types of bacteria (Liu et al., 2021). Therefore, cellulose and its derivatives are

TABLE 7.4 Blend of cellulose and its derivatives with other polymers for the preservation of different food products[a].

Food product	Polymers/additives	Main function of cellulose	Activity of film/coating	References
Fruits				
Apples (*"Red Delicious"* variety)	Bacterial cellulose nanocrystals-gelatin/CEOE	Decrease in the respiration rate	Increased firmness Retarding ripening and senescence during cold storage (60 days at 4°C)	Sadat Razavi et al. (2022)
Castile blackberry	Bacterial cellulose produced by *Komagataeibacter xylinus*	Water-retaining capacity	Inhibition of microbial growth during storage	Toscano Ávila et al. (2020)
Goldenberries	Carboxymethyl cellulose (CMC)/ polysaccharide-rich extract from SCG	Increased the resistance to gas diffusion	Shelf life increased to almost twofold when stored at 4°C and 95% RH	Ballesteros et al. (2022)
Mangoes	Cellulose nanocrystals (CNC)/ chitosan nanoparticles/PVA	Improved mechanical and barrier properties	Shelf life increased up to 20 days	Dey et al. (2021)
Raspberry	Cellulose nanofibers from RW-UP and RW-BP/chitosan	Improved mechanical and optical properties	Increased firmness during seven days of storage Protection against oxidation	Morcillo-Martín et al. (2023)
Strawberries	Cellulose nanofibers/chia seed mucilage	Reduced oxygen access	Increased shelf life Inhibited the activity of polyphenol oxidase and peroxidase enzymes	Mousavi et al. (2021)
Vegetables				
Black garlic	CNCs/chitosan/GL	Increased tensile strength, barrier properties, thermal stability	Improved antioxidant and antibacterial activities	Ding et al. (2023)
Cherry tomato	Cellulose nanofibers/gelatin	Lower barrier properties Improved mechanical properties Higher transparency	Maintained good luster and freshness Weight loss was only 16% after 14 days of storage at 25°C and 30% RH	Wang, Miao et al. (2024)
Fresh-cut potatoes, fried potatoes	CMC/flaxseed mucilage/NBE	Inhibited oil absorption	Reduced microbial load Lower browning index	Esmaeili et al. (2024)
Green chili	CMC/alginate/potato starch/GSE	Improved UV barrier properties and thermal stability	Enhanced shelf life up to 25 days Outstanding antimicrobial properties	Ramakrishnan et al. (2023)
Meat and meat products				
Chilled chicken meat	Bacterial cellulose produced by *Acetobacter xylinum*/TEO	Optimal fiber structure, thermal stability, crystallinity, and solution stability	Inhibition of microbial growth during storage	Xu et al. (2024)

Sausage (70% lean beef + 28% pork fat)	CNCs/chitosan/CIN and CL essential oils Pickering emulsions	Reduced water vapor permeability Reduced elongation at break Enhanced thermal stability	Prolonged shelf life Inhibited of the oxidation of beef fat Inhibited microbial growth and reproduction	Chen et al. (2024)
Vacuum-packaged beef	Bacterial cellulose nanocrystals/ NBCNs	Control of purge loss	Inhibition of lipid oxidation Inhibition of microbial growth	Gedarawatte et al. (2022)
Dairy products				
Fresh string cheese sticks	Hydroxyethyl cellulose (HEC) grafted by poly (AA-MBA)	Smooth surface Increasing hydrogen bonding	Extended shelf life of cheese from four to eight weeks Inhibition of microbial growth	Al-Moghazy et al. (2020)
High-moisture mozzarella cheese	CMC/natamycin	Good barrier against moisture, carbon dioxide, and oxygen	Extended shelf life for eight days Inhibition of microbial growth	Azhdari and Moradi (2022)
Koopeh cheese	Cellulose-starch aerogel/TDEO	Denser structure Reduction in aerogel porosity	Inhibition of microbial growth	Mirmoeini et al. (2023)
Nuts				
Brazil nuts	CMC/cassava starch/tocopherol mix	Excellent resistance to flow and stability	Minimization of oxidation up to 90 days of storage at 25°C	da Costa et al. (2024)
Pistachio kernel	CMC/*Astragalus* honey	Protection from oxygen and lipid oxidation	Increased shelf life Improved sensory properties	Khajeh-Ali et al. (2022)
Walnuts	Hydroxypropyl methyl cellulose (HPMC)/PEO/GA	Improved barrier properties	Decreased oxidation of walnuts during storage	Aydogdu et al. (2019)

[a] CEOE, *Cinnamon essential oil emulsion;* SCG, *spent coffee grounds;* RH, *relative humidity;* PVA, *poly-vinyl alcohol;* RW-UP, *unbleached pulp from raspberry waste;* RW-BP, *bleached pulp from raspberry waste;* GL, *garlic leaf extract;* NBE, *nano encapsulated burdock extract;* GSE, *grapefruit seed extract;* TEO, *thyme essential oil;* CIN *and* CL (EOs), *cinnamon essential oil combined with clove essential oil (composite essential oils);* NBCNs, *nisin-loaded bacterial cellulose nanocrystals;* AA-MBA, *acrylic acid and methyl butyl acrylate;* TDEO, *thymus daenensis essential oil;* PEO, *polyethylene oxide;* GA, *gallic acid.*

utilized as a stabilizer and also to encapsulate various antimicrobial components that help to resolve such challenges. Cellulose and its derivatives are not only used to preserve fruits and vegetables, but also for many other categories of food products, such as meat and meat products, dairy products, and nuts. Some examples of vegetables, fruits, and other types of food products that use cellulose and its derivatives as film or coating materials are shown in Table 7.4.

7.7 Conclusion and future considerations

Cellulose is probably the most abundant organic compound in the world, which is produced by various plants, woody plants, tunicates, bacteria, and algae. Cellulose presents a complex structure, consisting of aggregates of very fine fibrils that are composed of multiple cellulose chains, which can be easily categorized into cellulose filaments, cellulose micro/nanofibrils, cellulose fibers, and cellulose crystals. Cellulose derivatives are a category of natural polymers in which the cellulose is dissolved or swollen using various solvents (e.g., ILs, aqueous solvents, and organic solvents). Biodegradability, worldwide abundance, and renewability are the fundamental advantages of cellulose derivatives as materials used in different fields. The most common cellulose derivatives used in food packaging, especially in film and coating formulations, are cellulose ethers and cellulose esters. They display noticeable mechanical characteristics, biocompatibility, porosity, solubility, thermal properties, biodegradability, crystallinity, water absorbency, antimicrobial effects, and exceptional sensory features, as well as various other biological functionalities. Therefore, cellulose and its derivatives are great bio-composites and can easily encapsulate additional components like emulsifiers, plasticizers, and various additives in order to enhance functionality, stability, and integrity of film and coating formulations. However, the development of bio-composite films and coatings still requires more regulations and improvements for them to be applied at a commercial level. In order to gain further benefits from cellulose and its derivatives with relation to the food packaging industry, various aspects of their quality, function, and cost need to be addressed, along with the production of various bio-composites for future applications. Moreover, more research is needed for the blending cellulose and its derivatives with other biopolymers, as well as for evaluating the effectiveness of encapsulated polymers in order to formulate active and intelligent food packaging. Thus, this chapter provides useful insights regarding the development of films and coatings based on cellulose and its derivatives.

References

Abdellah Ali, S. F., William, L. A., & Fadl, E. A. (2020). Cellulose acetate, cellulose acetate propionate and cellulose acetate butyrate membranes for water desalination applications. *Cellulose, 27*(16), 9525–9543. https://doi.org/10.1007/s10570-020-03434-w, www.springer.com/journal/10570.

Agarwal, D., Hewson, L., & Foster, T. J. (2018). A comparison of the sensory and rheological properties of different cellulosic fibres for food. *Food and Function, 9*(2), 1144–1151. https://doi.org/10.1039/c7fo01495c, http://pubs.rsc.org/en/journals/journal/fo.

Ahmadi, P., Jahanban-Esfahlan, A., Ahmadi, A., Tabibiazar, M., & Mohammadifar, M. (2022). Development of ethyl cellulose-based formulations: A perspective on the novel technical methods. *Food Reviews International, 38*(4), 685–732. https://doi.org/10.1080/87559129.2020.1741007.

Alavi, M., Rai, M., Varma, R., Hamidi, M., & Mozafari, M. (2022). Conventional and novel methods for the preparation of micro and nanoliposomes. *Micro Nano Bio Aspects*. https://doi.org/10.22034/mnba.2022.150564.

Al-Moghazy, M., Mahmoud, M., & Nada, A. A. (2020). Fabrication of cellulose-based adhesive composite as an active packaging material to extend the shelf life of cheese. *International Journal of Biological Macromolecules, 160*, 264–275. https://doi.org/10.1016/j.ijbiomac.2020.05.217, www.elsevier.com/locate/ijbiomac.

Arai, K., & Shikata, T. (2020). Hydration/dehydration behavior of hydroxyethyl cellulose ether in aqueous solution. *Molecules (Basel, Switzerland), 25*(20), 4726. https://doi.org/10.3390/molecules25204726.

Aydogdu, A., Sumnu, G., & Sahin, S. (2019). Fabrication of gallic acid loaded Hydroxypropyl methylcellulose nanofibers by electrospinning technique as active packaging material. *Carbohydrate Polymers, 208*, 241–250. https://doi.org/10.1016/j.carbpol.2018.12.065.

Azhdari, S., & Moradi, M. (2022). Application of antimicrobial coating based on carboxymethyl cellulose and natamycin in active packaging of cheese. *International Journal of Biological Macromolecules, 209*(Pt B), 2042–2049. https://doi.org/10.1016/j.ijbiomac.2022.04.185.

Babaei-Ghazvini, A., Vafakish, B., Patel, R., Falua, K. J., Dunlop, M. J., & Acharya, B. (2024). Cellulose nanocrystals in the development of biodegradable materials: A review on CNC resources, modification, and their hybridization. *International Journal of Biological Macromolecules, 258*. https://doi.org/10.1016/j.ijbiomac.2023.128834, https://www.sciencedirect.com/science/journal/01418130.

Bakri, M. K. B., & Rahman, M. R. (2021). Extraction, types, and classification of cellulose. In *Fundamentals and recent advances in nanocomposites based on polymers and nanocellulose*. Malaysia: Elsevier, 19–40. https://doi.org/10.1016/B978-0-323-85771-0.00003-8, https://www.sciencedirect.com/book/9780323857710.

Ballesteros, L. F., Teixeira, J. A., & Cerqueira, M. A. (2022). Active carboxymethyl cellulose-based edible coatings for the extension of fresh goldenberries shelf-life. *Horticulturae, 8*(10). https://doi.org/10.3390/horticulturae8100936, www.mdpi.com/journal/horticulturae.

Bangar, S. P., Esua, O. J., Nickhil, C., & Whiteside, W. S. (2023). Microcrystalline cellulose for active food packaging applications: A review. *Food Packaging and Shelf Life, 36*. https://doi.org/10.1016/j.fpsl.2023.101048, http://www.journals.elsevier.com/food-packaging-and-shelf-life/.

Batista, R. A., Espitia, P. J. P., Quintans, Jd. S. S., Freitas, M. M., Cerqueira, M.Â., Teixeira, J. A., & Cardoso, J. C. (2019). Hydrogel as an alternative structure for food packaging systems. *Carbohydrate Polymers, 205*, 106–116. https://doi.org/10.1016/j.carbpol.2018.10.006, http://www.elsevier.com/wps/find/journaldescription.cws_home/405871/description#description.

Biswas, A., Do Socorro Rocha Bastos, M., Furtado, R. F., Kuzniar, G., Boddu, V., & Cheng, H. N. (2020). Evaluation of the properties of cellulose ester films that incorporate essential oils. *International Journal of Polymer Science, 2020*. https://doi.org/10.1155/2020/4620868, http://www.hindawi.com/journals/ijps/.

Bochek, A. M., Shevchuk, I. L., Gavrilova, I. I., Lebedeva, M. F., Lavrent'ev, V. K., & Panarin, E. F. (2018). Properties of composite films of hydroxyethyl cellulose and hydroxypropyl cellulose with poly-N-methyl-N-vinylacetamide. *Polymer Science, Series A, 60*(6), 788–795. https://doi.org/10.1134/s0965545x18060019.

Brown, A. J. (1886). XLIII. - On an acetic ferment which forms cellulose. *Journal of the Chemical Society, Transactions, 49*, 432–439. https://doi.org/10.1039/CT8864900432.

Butt, M. A. (2022). Thin-film coating methods: A successful marriage of high-quality and cost-effectiveness—A brief exploration. *Coatings, 12*(8), 1115. https://doi.org/10.3390/coatings12081115.

Cazón, P., & Vázquez, M. (2021). Bacterial cellulose as a biodegradable food packaging material: A review. *Food Hydrocolloids, 113*, 106530. https://doi.org/10.1016/j.foodhyd.2020.106530.

Chanthathamrongsiri, N., Petchsomrit, A., Leelakanok, N., Siranonthana, N., & Sirirak, T. (2021). The comparison of the properties of nanocellulose isolated from colonial and solitary marine tunicates. *Heliyon, 7*(8), e07819. https://doi.org/10.1016/j.heliyon.2021.e07819.

Chen, H., Xin, K., & Yu, Q. (2024). Sausage preservation using films composed of chitosan and a pickering emulsion of essential oils stabilized with waste-jujube-kernel-derived cellulose nanocrystals. *Foods, 13*(21), 3487. https://doi.org/10.3390/foods13213487.

da Costa, D. S., Furtado, K. F., Kluczkovski, A. M., Takeuchi, K. P., & Lopes, A. S. (2024). Cassava starch/carboxymethyl cellulose edible coating added of tocopherol: A strategy to preserve the oxidative stability of Brazil nuts. *Foods, 13*(17). https://doi.org/10.3390/foods13172732, www.mdpi.com/journal/foods.

Debnath, B., Haldar, D., & Purkait, M. K. (2021). A critical review on the techniques used for the synthesis and applications of crystalline cellulose derived from agricultural wastes and forest residues. *Carbohydrate Polymers, 273*, 118537. https://doi.org/10.1016/j.carbpol.2021.118537.

Dey, D., Dharini, V., Periyar Selvam, S., Rotimi Sadiku, E., Mahesh Kumar, M., Jayaramudu, J., & Nath Gupta, U. (2021). Physical, antifungal, and biodegradable properties of cellulose nanocrystals and chitosan nanoparticles for food packaging application. *Materials Today: Proceedings, 38*, 860–869. https://doi.org/10.1016/j.matpr.2020.04.885.

Ding, Y., Jiang, Y., Zhong, Y., Wang, D., Deng, Y., Meng, F., Li, Y., Zhang, M., & Zhang, C. (2023). Preparation of garlic stem cellulose nanocrystal/leaf extract/chitosan film for black garlic preservation by electrostatic spraying. *International Journal of Biological Macromolecules, 225*, 518–525. https://doi.org/10.1016/j.ijbiomac.2022.11.108.

Doelker, E. (2005). *Cellulose derivatives, 107*. Springer Science and Business Media LLC, 199–265. https://doi.org/10.1007/bfb0027554.

Dorez, G., Ferry, L., Sonnier, R., Taguet, A., & Lopez-Cuesta, J.-M. (2014). Effect of cellulose, hemicellulose and lignin contents on pyrolysis and combustion of natural fibers. *Journal of Analytical and Applied Pyrolysis, 107*, 323–331. https://doi.org/10.1016/j.jaap.2014.03.017.

Esmaeili, F., Mehrabi, M., Babapour, H., Hassani, B., & Abedinia, A. (2024). Active coating based on carboxymethyl cellulose and flaxseed mucilage, containing burdock extract, for fresh-cut and fried potatoes. *LWT, 192*, 115726. https://doi.org/10.1016/j.lwt.2024.115726.

Espitia, P. J. P., Fuenmayor, C. A., & Otoni, C. G. (2019). Nanoemulsions: Synthesis, characterization, and application in bio-based active food packaging. *Comprehensive Reviews in Food Science and Food Safety, 18*(1), 264–285. https://doi.org/10.1111/1541-4337.12405, http://onlinelibrary.wiley.com/journal/10.1111/(ISSN)1541-4337.

Estevez-Areco, S., Guz, L., Famá, L., Candal, R., & Goyanes, S. (2019). Bioactive starch nanocomposite films with antioxidant activity and enhanced mechanical properties obtained by extrusion followed by thermo-compression. *Food Hydrocolloids, 96*, 518–528. https://doi.org/10.1016/j.foodhyd.2019.05.054.

Eyiz, V., Tontul, İ., & Türker, S. (2020). The effect of edible coatings on physical and chemical characteristics of fruit bars. *Journal of Food Measurement and Characterization, 14*(3), 1775–1783. https://doi.org/10.1007/s11694-020-00425-0, http://rd.springer.com/journal/11694.

García-Sánchez, M. E., Robledo-Ortiz, J. R., Jiménez-Palomar, I., González-Reynoso, O., & González-García, Y. (2019). Production of bacterial cellulose by Komagataeibacter xylinus using mango waste as alternative culture medium. *Revista Mexicana de Ingeniería Química, 19*(2), 851–865. https://doi.org/10.24275/rmiq/bio743.

Gedarawatte, S. T. G., Ravensdale, J. T., Johns, M. L., Li, M., Al-Salami, H., Dykes, G. A., & Coorey, R. (2022). Evaluation of the water-holding and anti-spoilage effect of a bacterial cellulose nanocrystal coating for the storage of vacuum-packaged beef. *Food Packaging and Shelf Life, 31*. https://doi.org/10.1016/j.fpsl.2022.100818, http://www.journals.elsevier.com/food-packaging-and-shelf-life/.

Gökmen, F.Ö., & Pekel Bayramgil, N. (2022). Preparation and characterization of some cellulose derivatives nanocomposite films. *Carbohydrate Polymers, 297*, 120030. https://doi.org/10.1016/j.carbpol.2022.120030.

Gomri, C., Cretin, M., & Semsarilar, M. (2022). Recent progress on chemical modification of cellulose nanocrystal (CNC) and its application in nanocomposite films and membranes-A comprehensive review. *Carbohydrate Polymers, 294*, 119790. https://doi.org/10.1016/j.carbpol.2022.119790.

Gorgieva, S., Jančič, U., Cepec, E., & Trček, J. (2023). Production efficiency and properties of bacterial cellulose membranes in a novel grape pomace hydrolysate by Komagataeibacter melomenusus AV436T and Komagataeibacter xylinus LMG 1518. *International Journal of Biological Macromolecules, 244*, 125368. https://doi.org/10.1016/j.ijbiomac.2023.125368.

Grzybek, P., Dudek, G., & van der Bruggen, B. (2024). Cellulose-based films and membranes: A comprehensive review on preparation and applications. *Chemical Engineering Journal, 495*, 153500. https://doi.org/10.1016/j.cej.2024.153500.

Heinze, T. (2015). Cellulose: Structure and properties. *Advances in Polymer Science, 271*, 1–52. https://doi.org/10.1007/12_2015_319, http://www.springer.com/series/12.

Heinze, T., Seoud, & Koschella, A. (2018). *Cellulose Derivatives*. Springer International Publishing. https://doi.org/10.1007/978-3-319-73168-1.

Herrick, F. W., Casebier, R. L., Hamilton, J. K., & Sandberg, K. R. (1983). Microfibrillated cellulose: Morphology and accessibility. *Journal of Applied Polymer Science: Applied Polymer Symposia, 37*, 797–813.

Hong, S., Schnetzer, A., Lavoine, N., Lucia, L., & Eggleston, D. B. (2024). Innovating environmentally sustainable materials platforms by harnessing coastal marine tunicates. *ChemSusChem*. https://doi.org/10.1002/cssc.202401024, http://onlinelibrary.wiley.com/journal/10.1002/(ISSN)1864-564X.

Isogai, A. (2022). TEMPO-catalyzed oxidation of polysaccharides. *Polymer Journal, 54*(4), 387–402. https://doi.org/10.1038/s41428-021-00580-1.

Jali, S., Mohan, T. P., Mwangi, F. M., & Kanny, K. (2024). A review on barrier properties of cellulose/clay nanocomposite polymers for packaging applications. *Polymers, 16*(1). https://doi.org/10.3390/polym16010051, http://www.mdpi.com/journal/polymers.

Javier-Astete, R., Jimenez-Davalos, J., Zolla, G., & Cullen, D. (2021). Determination of hemicellulose, cellulose, holocellulose and lignin content using FTIR in *Calycophyllum spruceanum* (Benth.) K. Schum. and Guazuma crinita Lam. *PLoS One, 16*(10), e0256559. https://doi.org/10.1371/journal.pone.0256559.

Khajeh-Ali, S., Shahidi, F., & Sedaghat, N. (2022). Evaluation of the effect of carboxy methyl cellulose edible coating containing Astragalus honey (*Astragalus gossypinus*) on the shelf-life of pistachio kernel. *Food Control, 139*, 109094. https://doi.org/10.1016/j.foodcont.2022.109094.

Kostryukov, S. G., Matyakubov, H. B., Masterova, Y. Y., Kozlov, A. S., Pryanichnikova, M. K., Pynenkov, A. A., & Khluchina, N. A. (2023). Determination of lignin, cellulose, and hemicellulose in plant materials by FTIR spectroscopy. *Journal of Analytical Chemistry, 78*(6), 718–727. https://doi.org/10.1134/s1061934823040093.

Lara, G., Yakoubi, S., Villacorta, C. M., Uemura, K., Kobayashi, I., Takahashi, C., Nakajima, M., & Neves, M. A. (2020). Spray technology applications of xanthan gum-based edible coatings for fresh-cut lotus root (*Nelumbo nucifera*). *Food Research International, 137*. https://doi.org/10.1016/j.foodres.2020.109723, www.elsevier.com/inca/publications/store/4/2/2/9/7/0.

Lee, H. R., Kim, K. H., Mun, S. C., Chang, Y. K., & Choi, S. Q. (2018). A new method to produce cellulose nanofibrils from microalgae and the measurement of their mechanical strength. *Carbohydrate Polymers, 180*, 276–285. https://doi.org/10.1016/j.carbpol.2017.09.104, http://www.elsevier.com/wps/find/journaldescription.cws_home/405871/description#description.

Liu, Y., Ahmed, S., Sameen, D. E., Wang, Y., Lu, R., Dai, J., Li, S., & Qin, W. (2021). A review of cellulose and its derivatives in biopolymer-based for food packaging application. *Trends in Food Science and Technology, 112*, 532–546. https://doi.org/10.1016/j.tifs.2021.04.016, http://www.elsevier.com/wps/find/journaldescription.cws_home/601278/description#description.

Lopez-Polo, J., Muñoz-Shugulí, C., Patiño Vidal, M., & Patiño Vidal, C. (2024). Electrospun edible films and coatings: Development, functionality and food applications. *Trends in Food Science & Technology, 143*, 104253. https://doi.org/10.1016/j.tifs.2023.104253.

Lv, X., Han, J., Liu, M., Yu, H., Liu, K., Yang, Y., Sun, Y., Pan, P., Liang, Z., Chang, L., & Chen, J. (2023). Overview of preparation, modification, and application of tunicate-derived nanocellulose. *Chemical Engineering Journal, 452*, 139439. https://doi.org/10.1016/j.cej.2022.139439.

Mardawati, E., Rahmah, D. M., Rachmadona, N., Saharina, E., Pertiwi, T. Y. R., Zahrad, S. A., Ramdhani, W., Srikandace, Y., Ratnaningrum, D., Endah, E. S., Andriani, D., Khoo, K. S., Pasaribu, K. M., Satoto, R., & Karina, M. (2023). Pineapple core from the canning industrial waste for bacterial cellulose production by *Komagataeibacter xylinus*. *Heliyon, 9*(11). https://doi.org/10.1016/j.heliyon.2023.e22010, http://www.journals.elsevier.com/heliyon/.

Martins, B. A., de Albuquerque, P. B. S., & de Souza, M. P. (2022). Bio-based films and coatings: Sustainable polysaccharide packaging alternatives for the food industry. *Journal of Polymers and the Environment, 30*(10), 4023–4039. https://doi.org/10.1007/s10924-022-02442-0, https://www.springer.com/journal/10924.

Md Nor, S., & Ding, P. (2020). Trends and advances in edible biopolymer coating for tropical fruit: A review. *Food Research International, 134*, 109208. https://doi.org/10.1016/j.foodres.2020.109208.

Mirmoeini, S. S., Hosseini, S. H., Lotfi Javid, A., Esmaeili Koutamehr, M., Sharafi, H., Molaei, R., & Moradi, M. (2023). Essential oil-loaded starch/cellulose aerogel: Preparation, characterization and application in cheese packaging. *International Journal of Biological Macromolecules, 244*. https://doi.org/10.1016/j.ijbiomac.2023.125356, https://www.sciencedirect.com/journal/international-journal-of-biological-macromolecules.

Moon, S. M., Heo, J. E., Jeon, J., Eom, T., Jang, D., Her, K., Cho, W., Woo, K., Wie, J. J., & Shim, B. S. (2021). High crystallinity of tunicate cellulose nanofibers for high-performance engineering films. *Carbohydrate Polymers, 254*. https://doi.org/10.1016/j.carbpol.2020.117470, http://www.elsevier.com/wps/find/journaldescription.cws_home/405871/description#description.

Morcillo-Martín, R., Rabasco-Vílchez, L., Espinosa, E., Pérez-Rodríguez, F., & Rodríguez, A. (2023). Raspberry (*Rubus idaeus* L.) waste-derived nanocellulose for circular application in edible films and coatings. *LWT, 188*, 115438. https://doi.org/10.1016/j.lwt.2023.115438.

Mousavi, S. R., Rahmati-Joneidabad, M., & Noshad, M. (2021). Effect of chia seed mucilage/bacterial cellulose edible coating on bioactive compounds and antioxidant activity of strawberries during cold storage. *International Journal of Biological Macromolecules, 190*, 618–623. https://doi.org/10.1016/j.ijbiomac.2021.08.213, www.elsevier.com/locate/ijbiomac.

Nunes, M. R., de Souza Maguerroski Castilho, M., de Lima Veeck, A. P., da Rosa, C. G., Noronha, C. M., Maciel, M. V. O. B., & Barreto, P. M. (2018). Antioxidant and antimicrobial methylcellulose films containing Lippia alba extract and silver nanoparticles. *Carbohydrate Polymers, 192*, 37–43. https://doi.org/10.1016/j.carbpol.2018.03.014, http://www.elsevier.com/wps/find/journaldescription.cws_home/405871/description#description.

Oprea, M., & Voicu, S. I. (2020). Recent advances in composites based on cellulose derivatives for biomedical applications. *Carbohydrate Polymers, 247*, 116683. https://doi.org/10.1016/j.carbpol.2020.116683.

Panahirad, S., Dadpour, M., Peighambardoust, S. H., Soltanzadeh, M., Gullón, B., Alirezalu, K., & Lorenzo, J. M. (2021). Applications of carboxymethyl cellulose- and pectin-based active edible coatings in preservation of fruits and vegetables: A review. *Trends in Food Science and Technology, 110*, 663–673. https://doi.org/10.1016/j.tifs.2021.02.025, http://www.elsevier.com/wps/find/journaldescription.cws_home/601278/description#description.

Paunonen, S. (2013). Strength and barrier enhancements of cellophane and cellulose derivative films: A review. *BioResources, 8*(2), 3098–3121. https://doi.org/10.15376/biores.8.2.3098-3121, http://www.ncsu.edu/bioresources/BioRes_08/BioRes_08_2_Review_b_3098_Paunonen_Cellulose_Derivatives_Food_Packaging_Review_3308.pdf.

Perumal, S., Atchudan, R., & Lee, W. (2022). A review of polymeric micelles and their applications. *Polymers, 14*(12), 2510. https://doi.org/10.3390/polym14122510.

Rahman, M. S., Hasan, M. S., Nitai, A. S., Nam, S., Karmakar, A. K., Ahsan, M. S., Shiddiky, M. J. A., & Ahmed, M. B. (2021). Recent developments of carboxymethyl cellulose. *Polymers, 13*(8). https://doi.org/10.3390/polym13081345, https://www.mdpi.com/2073-4360/13/8/1345/pdf.

Ramakrishnan, R., Kulandhaivelu, S. V., & Roy, S. (2023). Alginate/carboxymethyl cellulose/starch-based active coating with grapefruit seed extract to extend the shelf life of green chilli. *Industrial Crops and Products, 199*, 116752. https://doi.org/10.1016/j.indcrop.2023.116752.

Richardson, J. J., Cui, J., Björnmalm, M., Braunger, J. A., Ejima, H., & Caruso, F. (2016). Innovation in layer-by-layer assembly. *Chemical Reviews, 116*(23), 14828–14867. https://doi.org/10.1021/acs.chemrev.6b00627, http://pubs.acs.org/journal/chreay.

Sabbaghi, H. (2023). *Perspective chapter: Cellulose in food production – principles and innovations*. IntechOpen. https://doi.org/10.5772/intechopen.109204.

Sadat Razavi, M., Golmohammadi, A., Nematollahzadeh, A., Ghanbari, A., Davari, M., Rovera, C., Carullo, D., & Farris, S. (2022). Impact of bacterial cellulose nanocrystals-gelatin/cinnamon essential oil emulsion coatings on the quality attributes of 'red delicious' apples. *Coatings, 12*(6), 741. https://doi.org/10.3390/coatings12060741.

Saedi, S., Garcia, C. V., Kim, J. T., & Shin, G. H. (2021). Physical and chemical modifications of cellulose fibers for food packaging applications. *Cellulose, 28*(14), 8877–8897. https://doi.org/10.1007/s10570-021-04086-0, www.springer.com/journal/10570.

Saifullah, M., Shishir, M. R. I., Ferdowsi, R., Tanver Rahman, M. R., & Van Vuong, Q. (2019). Micro and nano encapsulation, retention and controlled release of flavor and aroma compounds: A critical review. *Trends in Food Science and Technology, 86*, 230–251. https://doi.org/10.1016/j.tifs.2019.02.030, http://www.elsevier.com/wps/find/journaldescription.cws_home/601278/description#description.

Seddiqi, H., Oliaei, E., Honarkar, H., Jin, J., Geonzon, L. C., Bacabac, R. G., & Klein-Nulend, J. (2021). Cellulose and its derivatives: Towards biomedical applications. *Cellulose, 28*(4), 1893–1931. https://doi.org/10.1007/s10570-020-03674-w, www.springer.com/journal/10570.

Seki, Y., Selli, F., Erdoğan, Ü. H., Atagür, M., & Seydibeyoğlu, M.Ö. (2022). A review on alternative raw materials for sustainable production: Novel plant fibers. *Cellulose, 29*(9), 4877–4918. https://doi.org/10.1007/s10570-022-04597-4, www.springer.com/journal/10570.

Silvy, N., Reza, M. S., Uddin, M. N., & Akther, M. (2018). Comparison between different components of some available hardwood and softwood in Bangladesh. *IOSR Journal of Biotechnology and Biochemistry, 4*(1), 1–5.

Spinei, M., Oroian, M., & Ursachi, V.-F. (2024). Characterization of biodegradable films based on carboxymethyl cellulose and citrus pectin films enriched with bee bread oil and thyme oil. *LWT, 214*, 117088. https://doi.org/10.1016/j.lwt.2024.117088.

Suhag, R., Kumar, N., Petkoska, A. T., & Upadhyay, A. (2020). Film formation and deposition methods of edible coating on food products: A review. *Food Research International, 136*, 109582. https://doi.org/10.1016/j.foodres.2020.109582.

Szabó, L., Milotskyi, R., Sharma, G., & Takahashi, K. (2023). Cellulose processing in ionic liquids from a materials science perspective: Turning a versatile biopolymer into the cornerstone of our sustainable future. *Green Chemistry, 25*(14), 5338–5389. https://doi.org/10.1039/d2gc04730f.

Tang, X., Liu, G., Zhang, H., Gao, X., Li, M., & Zhang, S. (2021). Facile preparation of all-cellulose composites from softwood, hardwood, and agricultural straw cellulose by a simple route of partial dissolution. *Carbohydrate Polymers, 256*, 117591. https://doi.org/10.1016/j.carbpol.2020.117591.

Teixeira, S. C., Silva, R. R. A., de Oliveira, T. V., Stringheta, P. C., Pinto, M. R. M. R., & Soares, Nd. F. F. (2021). Glycerol and triethyl citrate plasticizer effects on molecular, thermal, mechanical, and barrier properties of cellulose acetate films. *Food Bioscience. 42*. https://doi.org/10.1016/j.fbio.2021.101202, http://www.journals.elsevier.com/food-bioscience/.

Tibolla, H., Pelissari, F. M., Rodrigues, M. I., & Menegalli, F. C. (2017). Cellulose nanofibers produced from banana peel by enzymatic treatment: Study of process conditions. *Industrial Crops and Products, 95*, 664–674. https://doi.org/10.1016/j.indcrop.2016.11.035, www.elsevier.com/inca/publications/store/5/2/2/8/2/5.

Tomimatsu, Y., Suetsugu, H., Yoshimura, Y., & Shimizu, A. (2019). The solubility of cellulose in binary mixtures of ionic liquids and dimethyl sulfoxide: Influence of the anion. *Journal of Molecular Liquids, 279*, 120–126. https://doi.org/10.1016/j.molliq.2019.01.093.

Toscano Ávila, J. A., Terán, D. A., Debut, A., Vizuete, K., Martínez, J., & Cerda-Mejía, L. A. (2020). Shelf life estimation of Blackberry (*Rubus glaucus* Benth) with bacterial cellulose film coating from *Komagataeibacter xylinus*. *Food Science & Nutrition, 8*(4), 2173–2179. https://doi.org/10.1002/fsn3.1525.

Trivedi, M. K. (2015). Influence of biofield treatment on physicochemical properties of hydroxyethyl cellulose and hydroxypropyl cellulose. *Journal of Molecular Pharmaceutics & Organic Process Research, 03*(02). https://doi.org/10.4172/2329-9053.1000126.

Turbak, A. F., Snyder, F. W., & Sandberg, K. R. (1983). Microfibrillated cellulose, a new cellulose product: Properties, uses and commercial potential. *Journal of Applied Polymer Science: Applied Polymer Symposia, 37*, 815–827.

Ursachi, V.-F., Oroian, M., & Spinei, M. (2024). Development and characterization of biodegradable films based on cellulose derivatives and citrus pectin: A comparative study. *Industrial Crops and Products, 219*, 119052. https://doi.org/10.1016/j.indcrop.2024.119052.

Vijayakumar, A., & Badwaik, L. S. (2022). Recent advancement in improvement of properties of polysaccharides and proteins based packaging film with added nanoparticles: A review. *International Journal of Biological Macromolecules, 203*, 515–525. https://doi.org/10.1016/j.ijbiomac.2022.01.181.

Volova, T. G., Prudnikova, S. V., Sukovatyi, A. G., & Shishatskaya, E. I. (2018). Production and properties of bacterial cellulose by the strain *Komagataeibacter xylinus* B-12068. *Applied Microbiology and Biotechnology, 102*(17), 7417–7428. https://doi.org/10.1007/s00253-018-9198-8, https://www.springer.com/journal/253.

Wang, M., Miao, X., Guo, F., Deng, Z., Bian, F., Xiao, T., & Chen, C. (2024). Optimized hybrid edible surface coating prepared with gelatin and cellulose nanofiber for cherry tomato preservation. *International Journal of Biological Macromolecules, 279*(Pt 1), 134822. https://doi.org/10.1016/j.ijbiomac.2024.134822.

Wang, Y., Qi, J., Zhang, M., Xu, T., Zheng, C., Yuan, Z., & Si, C. (2024). Cellulose-based aerogels, films, and fibers for advanced biomedical applications. *Chemical Engineering Journal, 497*, 154434. https://doi.org/10.1016/j.cej.2024.154434.

Wang, Y., Wang, T., Zhang, X., Wang, Q., Liu, S., Guo, Z., & Liu, H. (2024). The study of citric acid crosslinked β-cyclodextrin/hydroxypropyl cellulose food preservation film. *Nordic Pulp & Paper Research Journal, 39*(3), 435–446. https://doi.org/10.1515/npprj-2023-0088.

Xie, Q., Liu, G., & Zhang, Y. (2024). Edible films/coatings containing bioactive ingredients with micro/nano encapsulation: A comprehensive review of their fabrications, formulas, multifunctionality and applications in food packaging. *Critical Reviews in Food Science and Nutrition, 64*(16), 5341–5378. https://doi.org/10.1080/10408398.2022.2153794.

Xu, Y., Xin, J., Lyu, Y., & Zhang, C. (2024). Effects of bacterial cellulose/thyme essential oil emulsion coating on the shelf life of chilled chicken meat. *Journal of the Science of Food and Agriculture, 104*(9), 5577–5587. https://doi.org/10.1002/jsfa.13392.

Yekta, R., Abedi-Firoozjah, R., Azimi Salim, S., Khezerlou, A., & Abdolmaleki, K. (2023). Application of cellulose and cellulose derivatives in smart/intelligent bio-based food packaging. *Cellulose, 30*(16), 9925–9953. https://doi.org/10.1007/s10570-023-05520-1.

Yildirim-Yalcin, M., Tornuk, F., & Toker, O. S. (2022). Recent advances in the improvement of carboxymethyl cellulose-based edible films. *Trends in Food Science and Technology, 129*, 179–193. https://doi.org/10.1016/j.tifs.2022.09.022, http://www.elsevier.com/wps/find/journaldescription.cws_home/601278/description#description.

Zabihzadeh Khajavi, M., Ebrahimi, A., Yousefi, M., Ahmadi, S., Farhoodi, M., Mirza Alizadeh, A., & Taslikh, M. (2020). Strategies for producing improved oxygen barrier materials appropriate for the food packaging sector. *Food Engineering Reviews, 12*(3), 346–363. https://doi.org/10.1007/s12393-020-09235-y.

Zanchetta, E., Damergi, E., Patel, B., Borgmeyer, T., Pick, H., Pulgarin, A., & Ludwig, C. (2021). Algal cellulose, production and potential use in plastics: Challenges and opportunities. *Algal Research, 56*, 102288. https://doi.org/10.1016/j.algal.2021.102288.

Chapter 8

Alginate films and coatings

Elaheh Ahmadi[1], Jianhua Zhang[1], Shobha Muthukumaran[1,2], and Marlene J. Cran[1]
[1]*Institute for Sustainable Industries and Liveable Cities, Victoria University, Melbourne, VIC, Australia,* [2]*College of Sport, Health and Engineering, Victoria University, Melbourne, VIC, Australia*

8.1 Introduction

Alginates are linear anionic polysaccharides that are extracted from the cell walls of brown algae. The polysaccharide chains are composed of β-D-mannuronic acid (M) and α-L-guluronic acid (G) (Fig. 8.1), both 1–4 linked, in varying blocks of MM, GG, and MG subunits (Lee & Mooney, 2012; Rinaudo, 2008). These units contain carboxyl and hydroxyl groups aligned in specific conformations that are responsible for a range of important properties and functions. Alginate is commercially available as a sodium salt in a range of grades based on molecular weight, M:G ratios, and subunit block conformations. The GG blocks in the alginate polymers are associated with the formation of crosslinks, whereas the MG blocks facilitate lower gel solution viscosities due to higher flexibility of these blocks (Pournaki et al., 2024). These differences are due to the chemical configuration of the M and G units, and particularly the location of the carboxylic acid groups in the GG blocks shown in Fig. 8.1. Crosslinks are primarily formed when the carboxylic acid groups of the G block units in the polymer chains form a chelated structure in the presence of cations such as Ca^{2+} in a formation described as an "egg-box" configuration (Cao et al., 2020; Costa et al., 2018; Thiviya et al., 2023).

As a result of the unique structural variability, alginates can be engineered to produce a wide range of consumer and industrial materials. Examples include applications as food stabilizers and thickening agents, medicinal and pharmaceutical products, cosmetics, and textiles, among others (Gheorghita Puscaselu et al., 2020). A major application for alginate polymers is in the field of food packaging films (PFs) and coatings (Abdullah, Mohamad, et al., 2021). In comparison with conventional petroleum-based PFs, alginate films are inherently biodegradable. In addition, they are derived from raw materials that are obtained from abundant, renewable sources. The versatility of alginate films can be further enhanced by the addition of other polymers and additives that result in the formulation of active materials (Li et al., 2024). Many of these additives can impart antimicrobial and antioxidant activity to extend the shelf life of food products. They can also provide information to consumers in the case of colorimetric indicator films (IFs) (Li, Hu, et al., 2023). Alginate coatings can also be applied to fresh produce and fresh-cut foods to extend their shelf life and microbial safety, most commonly in conjunction with active agents (Tavassoli-Kafrani et al., 2016).

This chapter provides an overview of the use of alginates for food PFs and coatings. Techniques used to develop and optimize the films and coatings are discussed, and the use of additives to enhance both the alginate properties and the functionality of the materials is summarized. Examples of both academic studies and commercial alginate films and coatings are presented, and opportunities for further development in this field are discussed.

8.2 Alginate gels and films

8.2.1 Gels and crosslinking

The production of alginate films requires the preparation of a gel solution by dissolving relatively low amounts of sodium alginate in water. Alginate gels are formed when hydrogen bonds form between water molecules and the carboxylic acid groups in the polymer chains. Gel properties are highly dependent on factors such as the alginate's molecular weight, solution pH, and hardness (Lee & Mooney, 2012). The gels can be further stabilized by crosslinking using chemical agent (Klinkajon & Supaphol, 2014).

Natural Polymer-Based Films and Coatings in Food Technology. DOI: https://doi.org/10.1016/B978-0-443-34244-8.00016-1

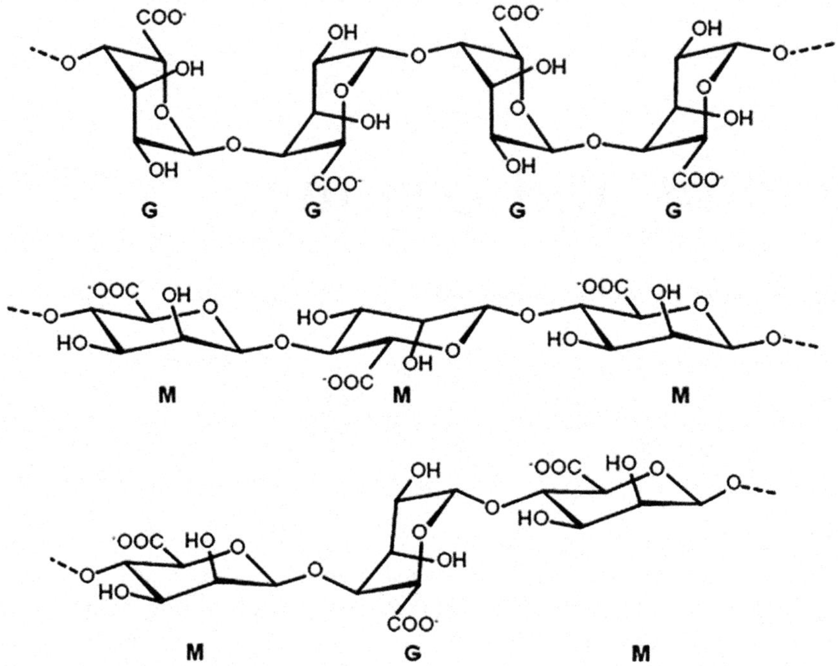

FIGURE 8.1 **Chemical structures of guar gum, MM, and MG block units.** Chemical structures of alginate block units, guluronic acid (G), and mannuronic acid (M).
From Lee, Kuen Yong & Mooney, David J (2012). Alginate: Properties and biomedical applications. Progress in Polymer Science, 37(1), 106–126. https://doi.org/10.1016/j.progpolymsci.2011.06.003.

8.2.2 Film production

Alginate films are formed when aqueous gels are dehydrated. In typical cases, alginate solutions are poured or cast into a vessel and allowed to dry under ambient conditions, or more quickly under mild heating. Depending on the grade and source of the alginate, clear or slightly colored films are formed when films are dry. These films are generally flexible but have poor mechanical properties and are highly soluble, which limits their use as food packaging materials unless modified. The previously described crosslinking reaction is used to improve the mechanical strength and inherent hydrophilicity of alginate films. Two main modes of crosslinking alginate are employed, namely internal and external gelation (Tavassoli-Kafrani et al., 2016). Internal gelation involves the use of a poorly soluble source of Ca^{2+} such as $CaCO_3$, which is activated by altering the solution pH (Tavassoli-Kafrani et al., 2016). Crosslinking occurs in situ, and films are formed by casting these solutions. External gelation is the more common approach, whereby dried or semidried cast films are exposed to Ca^{2+} solutions by immersion in a salt solution. The degree of crosslinking will depend on the concentration of the immersion solution and the exposure time (Li et al., 2016).

Although water remains in alginate gels and films following crosslinking, the drying process can result in shrinkage or contraction as water is removed in a process known as syneresis (Pournaki et al., 2024). This can result in poor mechanical strength and other undesirable properties. The main processing factors that influence whether a cross-linked alginate gel will undergo syneresis include the type of gelation and the Ca^{2+} concentration (Pournaki et al., 2024). Excess Ca^{2+}, which is often the case in external gelation, can exacerbate syneresis since only the GG-block units participate in crosslinking. The M:G ratio of the alginate is an important property, as high levels of MG blocks, and therefore more chain elasticity, induce greater syneresis (Pournaki et al., 2024). Alginate molecular weight also influences the propensity for syneresis (Draget et al., 2001). High molecular weight gels have greater chain flexibility, which can facilitate chain reorganization and contraction (Pournaki et al., 2024). It is therefore important to select the appropriate grade of alginate for a given film or coating application and then control the crosslinking reaction to minimize the possibility of syneresis.

Crosslinking can improve the tensile strength of alginate films (Girón-Hernández et al., 2021), but this can be to the detriment of film flexibility. Plasticizers are among the most common additives used to improve the flexibility and overall mechanical properties of alginate films (Bagheri et al., 2019; Paixão et al., 2019) and other biopolymer films in general (Imran et al., 2010; Jakubowska et al., 2023). Plasticizers are typically nonvolatile, low-molecular-weight compounds that are dispersed throughout the polymer network. Hydrogen bonding is disrupted, which reduces intermolecular forces and forces the polymer chains apart thereby improving their flexibility (Galdeano et al., 2009). When used in combination with crosslinking, plasticized films can be optimized to enhance the crucial properties required for PFs, including mechanical properties and water vapor permeability (Olivas & Barbosa-Cánovas, 2008).

8.2.3 Crosslinking agents

A wide range of chemical agents is used to crosslink biopolymers for PFs. They can be classified based on the type of cross-linked bonds formed (Garavand et al., 2017). These include ionic, covalent, and physical bonds that vary in the strength and stability of the crosslinks formed. They can also be more generally classified based on their action, which is either physical, chemical, or enzymatic, or combinations thereof. Ionic crosslinking using Ca^{2+} is the most common method and is achieved by immersing alginate films in $CaCl_2$ solutions. Other cations are also used, including Al^{3+}, Cu^{2+}, Fe^{3+}, Mn^{2+}, and Zn^{2+} with many of these alternative cations offering enhanced functionalities of the resulting films, including antimicrobial activity against foodborne microorganisms (Klinkajon & Supaphol, 2014). Organic acids and mixtures, including citric acid (Rangel-Marrón et al., 2019), tartaric acid (Singh et al., 2020), tannic acid (Yang et al., 2023), and ferulic acid (Yerramathi et al., 2021), are among other agents used to crosslink alginates. The crosslinking process is confirmed by the formation of carboxyl and ester bonds between the organic acid and alginate, typically using Fourier-transform infrared spectroscopy (Singh et al., 2020). Other organic chemicals, such as glutaraldehyde, are used for crosslinking, whereby acetal groups are formed between the aldehyde and alginate hydroxyl groups, often in the presence of an acid catalyst (Khapre et al., 2021; Sritweesinsub et al., 2015). However, aldehyde-based crosslinkers are no longer commonly used in food contact applications in favor of more food-safe agents. Combinations of crosslinking agents, such as ethylenediaminetetraacetic acid (EDTA) with Fe^{3+}(Jost & Reinelt, 2018) and tannic acid with Fe^{3+}(He et al., 2024), offer alternative strategies for enhanced crosslinking. In general, most films are cross-linked by immersion or dipping in aqueous solutions of one or more salts (Saswat & Maheshwari, 2023). Other approaches include immersion in complex multicomponent solutions (Bai et al., 2020) or via multiple immersion steps. Although not as common, crosslinking agents can be added to gel solutions prior to casting (internal gelation) (Jost & Reinelt, 2018) or by hybrid internal/external processes (Zhong et al., 2023). In situ crosslinking between layers of alginate and chitosan using sodium tripolyphosphate is another approach used to produce films with both improved mechanical and barrier properties and antioxidant activity (Bhatia et al., 2024).

Table 8.1 summarizes examples of crosslinking of alginate films and coatings used for food packaging and other general applications. The examples focus on ionic crosslinking with metallic anions, where the crosslinking process was the subject of optimization. Other examples show the use of alternative crosslinkers and processes such as internal gelation. The table demonstrates the broad approaches to crosslinking for enhancing film performance and functions, and, in general, immersion dipping in calcium chloride is the most common process used to crosslink alginate films for food packaging applications. Improved mechanical and barrier properties, such as water vapor permeability (WVP), are commonly reported, and the inclusion of other materials can enhance the functionality and ultimate properties of the films. For externally cross-linked films, $CaCl_2$ concentrations are typically in the range of 1%–10% w/w, and immersion times range from a few seconds to several minutes. The optimum $CaCl_2$ concentrations and immersion times depend on the type of alginate used, as discussed previously, and particularly on the nature of the M and G structural units. Some of the other metallic cations that successfully crosslink alginates can provide additional functionality, including Zn^{2+}, which can impart antibacterial properties to films (Liu et al., 2024a; Liu, Li, Niu, et al., 2025b).

8.2.4 Plasticizing agents

Plasticizers are almost always used during the preparation of alginate films, and a wide range of compounds can be used for this purpose. The most common additives used as plasticizers are polyols, including glycerol and sorbitol (Bagheri et al., 2019; Giz et al., 2020; Jost et al., 2014; Rangel-Marrón et al., 2019). Other plasticizers include natural and synthetic glycols, mono- and oligo-saccharides, seed oils, and other biopolymers (Eslami et al., 2023). The chemical nature and concentration of the plasticizer, as well as the other components of the alginate film and subsequent processing, all influence the final film properties (Silva et al., 2009). Plasticizing is commonly used to balance the physical properties of alginate films, particularly when crosslinking is also used, and the flexibility of the film is decreased, as depicted in Fig. 8.2 (Giz et al., 2020). Without glycerol, crosslinking results in a tighter network of polymer chains in the egg-box formation, whereas glycerol increases the spacing between polymer chains through hydrogen bonding. Although the main aim of incorporating plasticizers is to impart flexibility to the film, other desirable properties can be modulated, including barrier properties and grease resistance (Jost et al., 2014). Properties that influence the selection of plasticizers for alginate films include molecular weight, volatility, solubility, compatibility with other components, and food safety considerations

TABLE 8.1 Examples of various crosslinking optimization strategies for alginate films and coatings.

Composition and/or application	Crosslinker(s)	Process summary	Results	References
Alginate films for general applications	Al^{3+}, Ca^{2+}, Cu^{2+}, Fe^{2+}, Mg^{2+}, Zn^{2+}	Films dipped in 2%–10% cation solutions for 5–60 min	Films cross-linked with Ca^{2+} and Zn^{2+} were insoluble with higher tensile strength; films unaffected by Fe^{2+} and Mg^{2+} ions; intermediate effect with Cu^{2+} and Al^{3+}; fastest surface crosslinking with Cu^{2+}	Pavlath et al. (1999)
Alginate films with oregano essential oil	Ca^{2+}	Film solutions mixed with different amounts of $CaCO_3$ (internal crosslinking)	Degree of crosslinking enabled controlled release of essential oil	Benavides et al. (2012)
Alginate, cashew tree gum films	Ca^{2+}	Films with varying amounts of gum dipped in 2% $CaCl_2$ for different times up to 60 s	Best mechanical and barrier properties with low gum levels and longer immersion times	Azeredo et al. (2012)
Alginate, carboxymethyl cellulose (CMC) hydrogel films	Glutaraldehyde, Cu^{2+}	Films crosslinked in 2%–5% glutaraldehyde and/or 0.02 mol $CuSO_4$ for 30 min	Combination of crosslinkers decreased swelling and dissolution	Sritweesinsub et al. (2015)
Alginate/graphene oxide composite films for packaging, structural materials	Ca^{2+}, Ba^{2+}, Fe^{3+}	Films crosslinked for 6 h in 0.3 M solutions of cations	Films crosslinked with Fe^{3+} showed the best overall mechanical and thermal properties	Zheng et al. (2016)
Alginate films for general applications	Ca^{2+}	Films dipped in $CaCl_2$ solutions (0.375%–6% w/v) for 1 h	Optimum $CaCl_2$ concentration 1.5% w/v, improved mechanical properties and reduced swelling	Li et al. (2016)
Alginate/carrageenan packaging films (PFs)	Ca^{2+}	Films dipped in 0.5% $CaCl_2$ (in 20% v/v ethanol) for 30 min	Films containing 40% w/w alginate had best overall properties with good visual appearance; low solubility at all alginate levels	Ye et al. (2017)
Alginate (edible) films	Ca^{2+}	Films dipped in 1%–1.5% $CaCl_2$ for 5 min	Influence of ratio of M:G with higher G resulting in stronger films; properties also dependent on $CaCl_2$ concentration	Costa et al. (2018)
Alginate PFs	Ca^{2+}, ethylenediaminetetraacetic acid (EDTA), glucono-δ-lactone (GDL)	Films crosslinked by adding $CaCl_2$ with EDTA, $CaHPO_4$ and $CaCO_3$ and GDL for controlled hydrolysis, cast in plates, and continuous lamination	Films crosslinked using $CaHPO_4$ showed the best overall properties; demonstrated potential for continuous processing	Jost and Reinelt (2018)
Alginate PFs	Ca^{2+}	Films crosslinked in 0.2–0.8 M $CaCl_2$ for 2–8 min	Crosslinking in 0.8 M $CaCl_2$ for 8 min improved mechanical properties, reduced swelling	Bt Ibrahim et al. (2019)

Alginate PFs with different plasticizers	Ca^{2+}	Films dipped in 3% w/v Ca^{2+}, 3% v/v glycerol, and 3% v/v tributyl citrate (TC) for 15 min	Optimization of plasticizer/ mixtures; combination of glycerol and TC controlled solubility and hydrophilicity	Paixão et al. (2019)
Alginate/pectin PFs	Citric acid, tartaric acid	Alginate solutions heat treated with citric or tartaric acid with sodium hypophosphite catalyst, cast and dried	Films crosslinked with citric acid were stronger; both were readily compostable	Singh et al. (2020)
Alginate/ZnO nanoparticle antibacterial films	Ca^{2+}	Alginate internally gelled using 0.5%–2% w/w $CaSO_4$, mixing 24 h	Best overall properties with 1% $CaSO_4$, 2% ZnO	Satriaji et al. (2020)
Alginate films for packaging or wound dressing	Ca^{2+}	Films dipped in 0%–2% $CaCl_2$ for 30 min using different solvents	Synergistic effects of plasticizer and crosslinking on properties; nonlinearity with changes in concentrations	Giz et al. (2020)
Alginate, CMC PFs	Ba^{2+}, Ca^{2+}, Zn^{2+}	Films dipped in 2% w/v cation solutions for 1 min, additional films crosslinked in 1%–4% w/v Ca^{2+} for 1 min, pH 5.23–5.28	Optimum film properties using 2% w/v Ca^{2+} with highest tensile strength and good optical/thermal properties	Pinpru and Woramongkolchai (2020)
Alginate films for UV shielding	Fe^{3+}, EDTA	Films dipped in combinations of Fe^{3+} (1%–2.5% w/v) and/or EDTA (1%–1.5% w/v) for 6–12 min	Optimum mechanical and optical performance when cross-linked in Fe^{3+} 1.6% and EDTA 0.8% for 12 min	Bai et al. (2020)
Alginate PFs	Ca^{2+}	Films sprayed with 6% $CaCl_2$ solution, 0–32 mg cm^{-2}	Best overall film when sprayed with 3.2 mg Ca^{2+} cm^{-2}	Vieira et al. (2021)
Alginate films for water resistance	Ca^{2+}, Fe^{2+}, Fe^{3+}	Films dipped in 0.1 M $CaCl_2$, $FeCl_2$, $FeCl_3$, or $FeSO_4$ solutions for 5 min; further films dipped in 0.1 M $CaCl_2$ and $FeSO_4$ for 1–20 min	Films crosslinked with $FeSO_4$ had lowest water vapor permeability (WVP) and highest swelling index; WVP decreased with crosslinking time; films crosslinked with Fe were yellow in appearance	Choi et al. (2022)
Alginate, montmorillonite clay high performance films	Ca^{2+}, Ba^{2+}, Cd^{2+}, Co^{2+}, Cu^{2+}, Mn^{2+}, Ni^{2+}	Alginate/clay solutions crosslinked by addition of cation (X^{2+}) solutions (2:1 vol ratio Alg:X^{2+}) over 3 h	Weak crosslinking with Ca^{2+}, Co^{2+}, Mn^{2+}, and Ni^{2+} improved tensile properties, strong crosslinking with Ba^{2+}, Cd^{2+}, Cu^{2+} decreased properties due to competition with hydrogen bonding; best overall properties with Mn^{2+}; good light transmittance and UV barrier with different colored films	Wang, Song, et al. (2022)

(Continued)

TABLE 8.1 (Continued)

Composition and/or application	Crosslinker(s)	Process summary	Results	References
Alginate PFs	Ca^{2+}	Alginate gels cast onto glass, dried for 10 min, dipped into Ca^{2+} solutions 1%–13% until films released from slides; best film then optimized for dipping time 15–60 min	Best mechanical and water uptake properties when crosslinked 30 min in 3% Ca^{2+}	Saswat and Maheshwari (2023)
Alginate PFs	Ca^{2+}, putrescine	Alginate solutions precrosslinked with 0.01– 0.10 mM putrescine or 0.09 mM Ca^{2+} cast into films, then dipped for 15 min in 2% (w/v) $CaCl_2$, washed for 15 min in 5% (v/v) glycerol solution	Films precrosslinked with putrescine had better mechanical, improved water barrier properties, and good UV/vis light barrier properties	Zhong et al. (2023)
Alginate/HPMC PFs	Zn^{2+}	Films dipped in 5% $ZnCl_2$ for 60 s	Crosslinked 1:1 film had best mechanical and vapor properties; Zn^{2+} induced antibacterial properties	Liu et al. (2024b)
Alginate PFs	Cyclobutane-dicarboxylic acid (CBDA)	Alginate solutions mixed with varying levels of CBDA, films cast, and heat treated	Films with 4% CBDA solution had improved mechanical and UV shielding properties, with antibacterial and antimicrobial properties	Ruan et al. (2024)
Alginate coatings for polypropylene PFs	Ca^{2+}, GDL	Alginate solutions mixed with $CaHPO_4$ or $CaCO_3$, with GDL added to release Ca^{2+}	Coating provided good oxygen barrier properties; delayed oxidation of grape seed oil	Song et al. (2024)
Alginate PFs	Camellia seed cake protein (CSCP), citric acid	Alginate solutions mixed with CSCP, then citric acid, cast and dried	Films with 0.6% CSCP had best mechanical, antioxidant, UV, and barrier properties; good antibacterial activity	Chen et al. (2025)
Alginate coatings for fish preservation	Ca^{2+}	Alginate coatings dipped in 2% w/v $CaCl_2$ for 20 min	Crosslinked films more effectively controlled microbial growth and total volatile basic nitrogen (TVB-N) values, but increased lipid oxidation of fish	Vargas-Ramella et al. (2025)

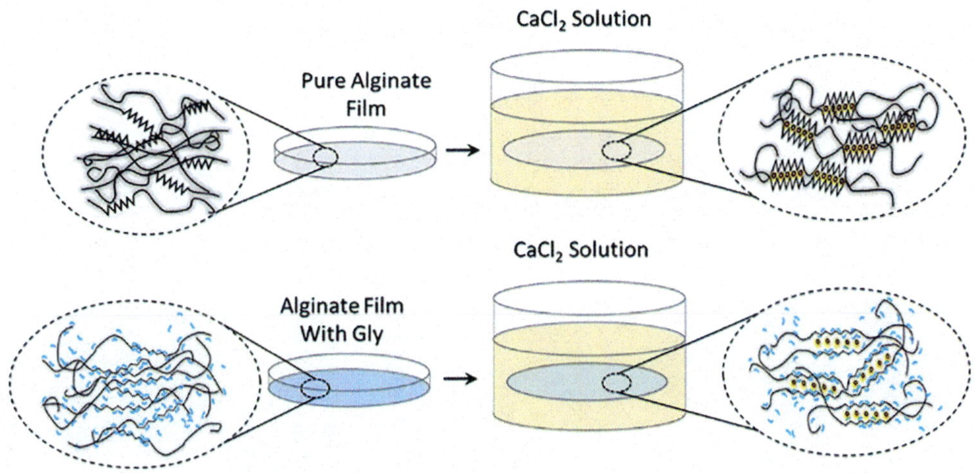

FIGURE 8.2 **Schematic representation of crosslinking without glycerol and with glycerol plasticizer.** Schematic representation of alginate crosslinking in calcium chloride without glycerol (top) and with glycerol (bottom).
From Giz, A.S., Berberoglu, M., Bener, S., Aydelik-Ayazoglu, S., Bayraktar, H., Alaca, B.E. & Catalgil-Giz, H. (2020). A detailed investigation of the effect of calcium crosslinking and glycerol plasticizing on the physical properties of alginate films. International Journal of Biological Macromolecules, 148, 49–55. https://doi.org/10.1016/j.ijbiomac.2020.01.103.

in the case of food packaging materials (Senturk Parreidt et al., 2018). When alginate films are produced in general, a nominal amount of plasticizer is often added to the formulation, and glycerol is among the most common at various levels, typically around 20%–50% relative to alginate (Hoque et al., 2024; Liu et al., 2023, 2024a; Santos et al., 2021; Silva de Farias et al., 2021).

In a study of calcium crosslinked alginate films, those plasticized with fructose, glycerol, sorbitol, and polyethylene glycol were more hydrophilic, and mechanical strength decreased with higher relative humidity (RH) compared to films without plasticizer (Olivas & Barbosa-Cánovas, 2008). There was a concurrent increase in elongation at break, and only fructose and sorbitol reduced the WVP at any RH level. For composite films of alginate and pectin, levels of glycerol were altered from 1% to 15% w/v in a simultaneous plasticizing and Ca^{2+} crosslinking process (Silva et al., 2009). In this case, optimized composite films with acceptable mechanical strength, low solubility, and swelling were obtained with 10% glycerol. A comparison of sorbitol (30%–50%) and glycerol (20%–40%) plasticizers for alginate demonstrated enhanced interactions between alginate and sorbitol, which improved barrier properties even with the higher levels of the plasticizer (Jost et al., 2014). In another study, hydrophilic (glycerol) and hydrophobic (tributyl citrate [TC]) plasticizers and their mixtures were investigated (Paixão et al., 2019). Films plasticized with TC alone were the least water-soluble with the highest tensile strength, whereas films with only glycerol were more homogenous and flexible. Mixed plasticizers imparted more compact film structures and higher elongation with increasing glycerol, but the composition of the mixtures did not significantly influence the overall film properties. A study of alginate films containing chestnut extracts compared commercial plasticizers (glycerol; epoxidized soybean oil; and epoxidized palm oil) with synthesized mixed plasticizers (mixed esters of propylene glycol and acetic acid; mixed esters of propylene glycol, oleic acid, and succinic acid; and epoxidized mixed esters of propylene glycol, oleic acid, and succinic acid) (Janik et al., 2023). Films containing each of the three synthesized plasticizers demonstrated superior mechanical and antibacterial properties compared to the commercial types with the latter observation attributed to the synergistic effects between chestnut extract and fatty acid-derived plasticizers.

8.3 Alginate films and coatings

Crosslinking and plasticizing are critical processes that are widely studied to optimize the properties of alginate films. A range of other additives and polymers can also be added to alginate films, including reinforcers, modifiers, natural extracts, various nanoparticles (NPs), and many others. Since alginate is comprised of hydroxyl and carboxylate groups, it provides an excellent matrix for these additives and can form bonds readily with their functional groups. Films can therefore be developed to impart more specific functional or active properties. Common approaches include the use of antioxidant and antimicrobial agents, colourimetric freshness indicators, and combinations of additives to produce PFs with wide-ranging applications. Table 8.2 presents examples of alginate films and coatings prepared with a range of additives from other

TABLE 8.2 Examples of food packaging alginate films and coatings.

Base film composition	Active agent(s)	Type[a]	Application and findings	References
Alginate	-	CF	Coating delaying ripening of sweet cherry fruits; maintained AO activity and total phenolics; coating increased storage from 8 to 16 days at 2°C	Díaz-Mula et al. (2012)
Alginate	-	CF	Coating reduced water loss of cantaloupe but increased for strawberries	Senturk Parreidt et al. (2019)
Alginate	*Arbutus unedo* L. leaf extracts	PF	Additive produced brittle films; water and acid simulants released more polyphenols with higher AO activity; films with 5% or 10% extract exhibited best oxidative stability	El Hammadi et al. (2024)
Alginate	Ascorbic and citric acids	PF AO	Antibrowning agents maintained color and increased antioxidant potential of coated fresh-cut mango cubes	Robles-Sánchez et al. (2013)
Alginate	Ascorbic and citric acids	PF AO	Combination of acids delayed film browning, proposed for AO packaging of confectionery/nuts	De'Nobili et al. (2016)
Alginate	Bilberry anthocyanin, bacterial cellulose nanocrystals, camellia oil Pickering emulsions	PF IF AO	Composite films had improved hydrophobicity, UV shielding, mechanical, thermal, WVP, and AO properties; nanocrystals stabilized camellia oil emulsions; color change from dark pink to pale yellow with increasing pH; color change from dark purple to pink on yogurt indicated pH changes and spoilage	Li, Wang, et al. (2023)
Alginate	Black soybean seed coat anthocyanins	PF IF	Additive reduced solubility and WVP, films sensitive to pH, NH_3, and H_2S; color change from orange to brown with increasing pH; change from light brown to light yellow/brown indicated spoilage of pork	Shi, Xu, Feng, et al. (2023)
Alginate	Carvacrol and methyl cinnamate	CF AM	Optimal concentrations of 0.98% (w/w) carvacrol and 1.45% (w/w) methyl cinnamate inhibited microbial growth	Peretto et al. (2014)
Alginate	Chestnut extracts, various plasticizers	PF AM	Films with synthesized plasticizers had superior mechanical and AM properties compared with commercial plasticizers	Janik et al. (2023)
Alginate	Chokeberry pomace, lemon balm extracts obtained with deep eutectic solvents	PF AO	Higher antioxidant capacity for films with lemon balm extract; other properties dependent on solvent	Kowalonek, Hamieau, et al. (2024)
Alginate	Cinnamaldehyde (xanthan gum stabilized), internal crosslinking with gluconolactone	PF AM AO	Coating on paper provided AO and AM activities; extended shelf life of strawberries to seven days (room temperature)	Zhou et al. (2024)

Alginate	Cinnamon essential oil (CEO), citric acid	PF AM AO	Combination of volatile and nonvolatile agents inhibited bacterial growth on sliced ham up to 12 days; essential oil also plasticized film	Lee et al. (2023)
Alginate	*Clitoria ternatea* (butterfly pea flower) extracts	PF IF AM	Films had improved tensile strength, light barrier properties, and AM activity; color change under heat sterilization and ammonia levels; color change in freshness of milk (blue to purple) and meat (blue to green); biodegradable in soil and beach sand within 15 days	Santos et al. (2022)
Alginate	Cobalt-based metal organic framework (MOF) nanoparticles (NPs)	PF AO	NPs improved mechanical strength, barrier properties, and UV shielding; AM films maintained freshness of shrimp under ambient temperature	Feng et al. (2023)
Alginate	Copper-molybdenum oxide NPs	PF AM	Films released Cu^{2+} ions, resulting in AM and antiviral properties with 5%–10% NPs	Assis et al. (2024)
Alginate	Crosslinking with crab shell $CaCO_3$ and gluconolactone	PF	Crosslinking provided superior thermo-mechanical properties, controlled release of drugs; proposed for active food packaging	Wang, Liu, et al. (2022)
Alginate	Hibiscetin, jujube pigment, laccaic acid	PF IF	Films with different pigments indicated color variations under different conditions; jujube pigment most sensitive to changes in quality of fresh salmon	Yan et al. (2024)
Alginate	Lactic, malic, citric acids	CF AM AO	Photodynamic therapy and citric acid provided best AM activity; maintained quality of cheese for seven days	Lisboa et al. (2022)
Alginate	Lemongrass essential oil	CF AM	Coating with 0.3% (w/v) oil maintained quality of fresh-cut pineapple; extended shelf life to 12 days	Azarakhsh et al. (2014)
Alginate	Lysine, curcumin, lignosulfonate acid NPs	PF IF AM AO	Alginate modification and additives improved mechanical properties, thermal stability, UV shielding, water resistance, AO, and AM properties; NPs stabilized curcumin; color change from yellow to light brown indicated shrimp spoilage	Yang et al. (2024)
Alginate	Metal oxide NPs	PF AM	Films inhibited growth of *Listeria* inoculated on salmon after four days refrigeration	Vizzini et al. (2020)
Alginate	MOF-based carrier, carvacrol	PF AM	Films were hydrophobic with good barrier and UV shielding properties; modified MOF exposed to alternating UV and visible light promoted release of carvacrol; films improved pork filet quality	Ning et al. (2025)
Alginate	Nanostructured lipid carriers	PF	Lipid carriers increased UV barrier and decreased WVP; proposed as carriers for hydrophobic bioactive agents	Karimi Khorrami et al. (2021)

(Continued)

TABLE 8.2 (Continued)

Base film composition	Active agent(s)	Type[a]	Application and findings	References
Alginate	Norbixin salts	PF AO	Additive improved barrier and UV shielding properties, lowered tensile strength; 0.03% norbixin salts reduced oxidation of sunflower oil during storage	de Farias et al. (2020)
Alginate	Pinolene (di-1-p-menthene)	CF AM AO	Layered coating of pinolene and alginate on fresh pistachios preserved appearance; increased shelf life up to 29 days	Sepasi et al. (2025)
Alginate	Polyphenolic truxillic acid (CBDA)	PF AM	Films with 10% CBDA had improved tensile strength with UV shielding and AM properties; delayed spoilage of pork tenderloin	Ruan et al. (2024)
Alginate	Propolis extracts, graphene oxide	PF AO	Films with additives had improved tensile strength and reduced WVP; graphene oxide controlled release of extracts; films prolonged shelf life of packaged blueberries, but films were dark in color	Guo et al. (2025)
Alginate	Protein hydrolysates	CF PF AM AO	Coated samples retained more moisture than films on chicken filets; coatings and films were AM and AO, films maintained color of filets	Merenkova and Zinina (2024)
Alginate	Purple onion skin and butterfly pea extracts	PF IF AM AO	Films with mixed extracts had AO and AM activity, improved film properties; color sensitivity to gases, pH, and temperature; color varied from red to blue, green to yellow, and shades of brown depending on composition and treatment	Santos and Martins (2024)
Alginate	Raspberry seed oil, black currant seed oil	PF AO	Oils and mixtures plasticized films and improved barrier properties and thermal stability; raspberry seed oil imparted high AO activity	Kowalonek et al. 2024; Kowalonek et al. (2025)
Alginate	Red cabbage anthocyanins, zinc oxide	PF IF AM	Additives improved mechanical, barrier, UV shielding, and AM properties; pH-sensitive color shift from violet to blue/green indicated shrimp spoilage	Anugrah et al. (2023)
Alginate	Rosemary and myrtle extracts, radiation treatment	CF AM AO	Irradiation and coatings with extracts inhibited lipid oxidation and microbial growth on beef meat; shelf life extension of up to six days	Smeti et al. (2025)
Alginate	Silver NPs, lemongrass essential oil	PF IF AM	Additives provided synergistic AM activity; films preserved color, softness, and surface texture of cheese for 14 days; color change due to oxidation of silver NPs indicated decline in cheese quality	Motelica et al. (2021)
Alginate	Sodium lactate (SL), sodium diacetate (SD), OptiForm (commercial formulation of SL and SD)	CF AM	Alginate coatings effective carriers of AM agents; coatings (with 2.4% SL plus 0.125% or 0.25% SD, and 2.5% OptiForm) on cold smoked salmon inoculated with *Listeria monocytogenes* delayed microbial growth	Neetoo et al. (2010)

Alginate	Sulfur NPs	PF AM	Best tensile, barrier, and UV shielding properties with 2 wt.% NPs but lower transparency; strong AM activity against *Listeria monocytogenes* after 12 h; film recommended for frozen meat products	Priyadarshi et al. (2021)
Alginate	Tannic acid	PF AM AO	Films with tannic acid had improved mechanical, UV shielding, barrier, AM, and AO properties; improvements due to strong hydrogen bonding with alginate	Li et al. (2022)
Alginate	Thyme oil (trehalose encapsulated), beta-cyclodextrin (βCD), Tween 20	PF AM	Most stable emulsions formed with Tween 20; better films color and opacity with trehalose and βCD; best encapsulation efficiency and AM activity with βCD	Navarro et al. (2016)
Alginate	Truxillic acid	PF AM AO	Films with 10% truxillic acid improved mechanical and barrier properties; imparted AM and AO properties on surface of chicken and sliced apples	Zhang, Zhang, Ruan, et al. (2024)
Alginate	Vanillin	CF AM	Coatings applied on grapes before and after harvest with vanillin reduced mold growth, maintained total phenolics, and AO capacity of grapes	Konuk Takma and Korel (2017)
Alginate	Yerba mate leaf extracts	PF AO	Additive increased mechanical strength and WVP; controlled release of total phenolics and chlorogenic acid; highest release in 10:90 ethanol:water	Silva de Farias et al. (2021)
Alginate	Zein	PF	Optimum alginate G:M ratio of 0.85 had best overall mechanical properties and hydrophobicity; zein enhanced film structure and properties	Luan et al. (2024)
Alginate	ZnO NPs	PF AM	ZnO NPs inhibited microbial growth in vitro and on ready-to-eat poultry meat	Akbar and Anal (2014)
Alginate	ZnO and Fe_2TiO_5 NPs	PF AM AO	Combined NPs reduced transparency, improved UV shielding, and thermal stability of film; lowered solubility in water; controlled spoilage of fresh-cut strawberries	Vasiljevic et al. (2024)
Alginate, agar, polyvinylidene fluoride (PVDF)/SiO_2	Methyl red and bromothymol blue dyes	IF	Layered film with synthetic dyes in alginate/agar layer; visible color change from red/orange to yellow/green indicating spoilage of shrimp	An and Zhou (2024)
Alginate, CMC	Phycocyanin	PF AO	Phycocyanin improved barrier and mechanical properties; AO capacity increased with increasing levels	Akhtar et al. (2024)
Alginate, CMC	Red-fleshed pitaya peel betalains obtained with deep eutectic solvents	PF IF AO	Films had improved tensile strength, WVP, solubility, UV shielding, and AO activity; color	Liu et al. (2025)

(*Continued*)

TABLE 8.2 (Continued)

Base film composition	Active agent(s)	Type[a]	Application and findings	References
			change from red/purple to yellow with increasing pH	
Alginate, CMC	Self-assembled hesperidin nanorods	PF AM AO	Nanorods improved barrier and mechanical properties, AM and AO activity; extended shelf life of grapes by 10 days compared to control films	Gao et al. (2025)
Alginate, carboxymethyl chitosan	Purple cauliflower extracts, Zn^{2+}, or Ca^{2+} crosslinking	PF IF	Crosslinking improved mechanical properties and WVP, reduced swelling, controlled extract release; Zn^{2+} crosslinked films were AM; color change from bright royal blue to yellow/green to brown indicated shrimp spoilage; Ca^{2+} crosslinked films more sensitive	Huang et al. (2023)
Alginate, carrageenan	Bilberry anthocyanins (BA), tannic acid (TA)	PF IF AM AO	TA imparted AM/AO activity and stabilized BA; films were UV shielding, color change from red to dark yellow indicated beef spoilage	Li, Feng, et al. (2024)
Alginate, carrageenan	Peanut shell extracts (flavonoids)	PF AO	Peanut shell extracts improved mechanical and barrier properties, thermal stability; extended shelf life of chilled pork cubes from 6 to 12 days	Wang, and Song, Cao, et al. (2024)
Alginate, cellulose	Betalains (red prickly pear extracts)	IF	Modified cellulose retained betalains; film color change from purple ($pH < 10$) to yellow ($pH > 10$) indicated fresh salmon spoilage	Halloub et al. (2023)
Alginate, cellulose nanocrystals	Black soybean seed coat anthocyanins	PF IF	Increasing nanocrystals improved mechanical properties, solubility, and barrier properties; films with 7% nanocrystals were color responsive with TVB-N and pH; color change from brown to brown/yellow indicated pork spoilage	Shi, Ren, et al. (2024)
Alginate, cellulose nanocrystals	Black soybean seed coat anthocyanins, nanosilica coating	PF IF	Films impregnated with 5% nanosilica were most water repellent; color change from red to yellow/green with increasing pH; indicated fresh carp spoilage storage over 14 days	Shi, Wu, et al. (2024)
Alginate, cellulose nanocrystals	Nisin	CF AM	Nanocrystals formed percolation network in Ca^{2+}-cross-linked film; reduced moisture loss and browning of coated litchi fruit; extended shelf life	Yin et al. (2024)
Alginate, cellulose nanofibers	-	PF	Layered films with nanofibers increased mechanical and thermal properties, reduced solubility, and WVP; improvements due to Ca^{2+}-carboxylate interactions	Xiao et al. (2021)
Alginate, cellulose nanofibers	Peanut red skin extracts	CF PF AM AO	Films had improved mechanical, barrier, UV shielding, AM, and AO properties; coating delayed decay of fresh fruit	Dai et al. (2022)

Alginate, cellulose nanofibrils	-	CF PF AM	Nanofibrils from cocoa by-products decreased transparency and WVP of film; coatings applied to Andean blueberries enhanced appearance and fruit quality for 21 days	Medina-Jaramillo et al. (2020)
Alginate, chitosan	Pterostilbene (blueberry extract)	PF AM AO	Improved mechanical, optical, water resistance, and barrier properties with additive; AO and AM activity enhanced; preserved shiitake mushrooms up to 10 days	Ruchika et al. (2025)
Alginate, chitosan, cellulose nanocrystals	Black soybean seed coat anthocyanins, beeswax coating	PF IF	Films with 5% beeswax coating were most water repellent; no testing of IF	Shi, Xu, Ren, et al. (2023)
Alginate, chitosan, ethyl cellulose	Ascorbic acid (mixed with alginate)	PF AO	Multilayered films with alginate/ascorbic acid in middle layer were water resistant with low WVP, high tensile strength, and high AO scavenging activity	Shah et al. (2025)
Alginate, gelatin	Onion peel extracts	IF	Color change from light pink to clear indicated milk spoilage by acidification from pH 6.5 to 5	Devi et al. (2024)
Alginate, gelatin	Plum peel extracts	PF IF	Extract improved mechanical, barrier, and water resistance properties; color change from red to yellow/green with increasing pH; indicated spoilage of fresh chicken breast	Chen et al. (2023)
Alginate, gellan gum	Modified black rice anthocyanins, curcumin, ZnO	PF IF AM AO	Bilayer film with AM/AO activities, extended shrimp shelf life by one day; color change from red/orange to dark green indicated shrimp spoilage	Li et al. (2024)
Alginate, gellan gum, CMC	Methyl red, bromothymol blue, silicon dioxide NPs, tannic acid	PF IF AM AO	Bilayer films with 1% SiO_2 NPs improved tensile, barrier, AM, and AO properties; color change with different pH and CO_2 levels indicated decline in strawberry quality	Zhang et al. (2025)
Alginate, gellan gum, CMC	Methyl red, bromothymol blue, tannic acid	PF IF AM AO	Bilayer films had improved tensile, barrier, AM and AO properties with increasing tannic acid; color change from red to yellow/green with variations in pH and CO_2 indicated change in strawberry freshness	Li et al. (2024)
Alginate, guar gum	Lemon peel-derived carbon quantum dots	PF AM AO	Quantum dots enhanced UV shielding, mechanical, and AO properties; delayed browning of blanched asparagus	Zhang et al. (2025)
Alginate, gum Arabic	*Syzygium cumini* seed extract	PF AO	Addition of extracts reduced thermal stability and mechanical properties; improved opacity, solubility, and WVP; films with high levels of extracts inhibited sunflower oil oxidation	Abdin et al. (2022)

(Continued)

TABLE 8.2 (Continued)

Base film composition	Active agent(s)	Type[a]	Application and findings	References
Alginate, hydroxypropyl methylcellulose	Curcumin, Zn^{2+} crosslinking	PF IF AM	Additives improved mechanical, barrier, thermal, and AM properties; color change from yellow to red with increasing pH; color change from yellow to orange indicated spoilage of pork and shrimp	Liu et al. (2025a)
Alginate, konjac glucomannan	*Lycium ruthenicum* anthocyanins, tea polyphenols	PF IF AM AO	Tea polyphenols enhanced stability of anthocyanins, improved mechanical, barrier, and AM/AO properties; color change from red to purple to dark green with increasing pH; film preserved milk and indicated spoilage	Wang et al. (2023)
Alginate, konjac glucomannan	Lycopene microcapsules	PF AM AO	Films with 2% microcapsules showed best AM and AO activities, high transparency, and good WVP; composite film prolonged shelf life of sweet cherries	Zeng et al. (2022)
Alginate, konjac glucomannan, cellulose nanocrystals	CEO, alizarin pigment	PF IF AM AO	Bilayer film, improved mechanical and light-shielding properties, AM and AO activity; color change from yellow/brown to purple/red indicates shrimp spoilage	Li, Hu, et al. (2024)
Alginate, konjac glucomannan, functionalized carbon nanotubes	Fe^{3+} crosslinking	PF AM	Dual-crosslinked network with high WVP, water resistance, thermal stability, and AM properties; extended strawberry shelf life up to six days	Chen, Wang, et al. (2024)
Alginate, methyl cellulose	Anthocyanins, Zn^{2+} crosslinking	PF IF AM	Zn^{2+} crosslinking imparted AM activity; pH-sensitive color shift from red/purple to dark blue/green indicated pork and shrimp spoilage	Wang and Liu (2024)
Alginate, methylcellulose	Dragon fruit peel extract, Zn^{2+} crosslinking	PF IF AM	Films had improved mechanical, thermal, UV shielding, AM properties; color change from purple to green with increased storage time indicating elevated TVB-N levels and spoilage of pork and shrimp	Liu et al. (2025a)
Alginate, pectin	*Aronia melanocarpa* anthocyanins, tea polyphenols	PF IF AM AO	Films with both additives had improved mechanical and barrier properties, extended shelf life of shrimp by 12 h; color change from red-pink-purple with increasing pH indicated freshness	Chen, Zhang, et al. (2024)
Alginate, pectin	Ascorbic, lactic acids	PF AM	Composite films were more hydrophobic with higher thermal stability and AM properties	Makaremi et al. (2019)
Alginate, pectin	Dragon fruit peel extract, Zn^{2+} crosslinking	PF IF AM	Films had low variability in WVP and tensile properties, better AM properties with Zn^{2+} crosslinking; color change from red to green with increasing pH used to indicate decline in shrimp freshness	Liu et al. (2025b)

Alginate, pectin	β-cyclodextrin (βCD) encapsulated trans-cinnamaldehyde	CF AM	βCD controlled release of cinnamaldehyde, layered film coating extended shelf life of fresh-cut watermelon from 7 to 12–15 days; maintained food quality and controlled microbial spoilage	Sipahi et al. (2013)
Alginate, pectin (apricot skin)	Rose anthocyanins	PF IF AO	Additive uniformly dispersed in film, altered mechanical and barrier properties, improved thermal, UV shielding, and AO properties; color sensitivity with pH changes, visible change from pink to dark yellow indicated spoilage of grass carp	Qiu et al. (2023)
Alginate, pectin, cellulose nanocrystals	Red cabbage anthocyanins	PF IF	Crosslinking and reinforcement with nanocrystals improved mechanical, thermal, and water resistance properties; pH-dependent color change from lilac to dark green to green/yellow indicated shrimp spoilage	Lei et al. (2023)
Alginate, pectin, cellulose nanofibrils	Black soybean seed coat anthocyanins	PF IF	Additive improved barrier, reduced moisture content, solubility, swelling; color change from red to brown indicated decline in freshness of goat meat with increasing pH and TVB-N	Wang et al. (2025)
Alginate, pectin, TiO_2 multilayer film	Dragon fruit peel betacyanins	PF IF	TiO_2 enhanced mechanical properties, reduced WVP, and stabilized betacyanins; pH-sensitive color change from red/violet to brown/yellow indicated increasing TVB-N levels and spoilage of mackerel filets	Du et al. (2024)
Alginate, pectin, xanthan gum	Blueberry anthocyanins	IF	Film monitored freshness of blueberries during storage and transportation; color change from purple to light pink indicated spoilage due to change in pH	Li, Hu, et al. (2023)
Alginate, polyvinyl alcohol (PVA)	Alizarin-MOF NPs	PF IF	Films had improved hydrophobicity; color of NPs sensitive to changes in pH; color change from yellow to purple corresponded with increasing TVB-N and indicated beef spoilage	Zhang, Zhang, Zhang, et al. (2024)
Alginate, PVA	Purple garlic peel extracts, TiO_2 NPs	PF IF AO	Addition of extracts and NPs improved mechanical and light barrier properties; color change from blue to red to yellow-green indicated spoilage of fresh beef	Jiang et al. (2023)
Alginate, PVA, PVDF support	Roselle anthocyanins	PF IF	Bilayer films were hydrophobic with good barrier properties (PVDF surface); color change from pink/red to green with increasing pH indicated change in pork freshness	Zhang et al. (2021)
Alginate, rice protein	Dialdehyde starch, anthocyanins	PF AO		Zhu et al. (2024)

(*Continued*)

TABLE 8.2 (Continued)

Base film composition	Active agent(s)	Type[a]	Application and findings	References
			Films with 10% starch improved mechanical, oxygen barrier, and UV shielding properties; 5% anthocyanins from *Lycium ruthenicum* fruit imparted antioxidant activity on fresh-cut cantaloupe	
Alginate, starch	Avocado seed extract	PF IF AM	Crosslinking improved water resistance, tensile, thermal, and AM properties (Gram+ only); color change from yellow to orange/red to dark brown with increasing pH; indicator of shrimp spoilage	Ngoc Le and Chen (2024)
Alginate, starch and fibers	-	PF	Fibers from babassu coconut improved solubility resistance, improved mechanical properties; proposed for PF	Raposo et al. (2021)
Alginate, tar NPs	Eugenol	PF AM	Tar NPs derived from rice husk pyrolysis encapsulated eugenol for controlled release; Ca^{2+} cross-linked films had improved mechanical properties, humidity resistance; lower release of eugenol and high AM activity	Taverna et al. (2022)

[a] *Types:* AO, *Antioxidant;* AM, *antimicrobial;* PF, *packaging film;* CF, *coating film;* IF, *indicator film;* WVP, *water vapor permeability;* TVB-N, *total volatile basic nitrogen.*

polymers and biopolymers, active compounds, and more complex combinations of agents. Examples include freshness IFs and coatings/films applied to real food samples, where alginate is the major or equally major component of the film or coating formulation. Unless otherwise specified, films are generally crosslinked with $CaCl_2$ and plasticized with glycerol.

8.3.1 General packaging films and coatings

Unlike some other marine-derived biopolymers, such as chitosan (Nasaj et al., 2024), alginate does not impart natural bioactivity. The majority of studies related to alginate films are therefore focused on the use of additives to enhance film properties. There are examples, however, of the development of alginate films and coatings for more primary purposes, such as general protective packaging and barrier films/coatings. Crosslinking and plasticizing are key processes frequently investigated to optimize alginate film properties (Bt Ibrahim et al., 2019; Choi et al., 2022; Jost & Reinelt, 2018; Li et al., 2016; Paixão et al., 2019; Pavlath et al., 1999; Saswat & Maheshwari, 2023; Vieira et al., 2021; Zhong et al., 2023). The main goal of these studies is to optimize the composition and processing of the film to obtain the best overall mechanical and barrier properties, hydrophobicity, and thermal stability. By optimizing these properties, alginate films can be more competitive with common primary PFs, such as polyethylene, which are inherently strong and hydrophobic.

Edible coatings are more unique to biopolymers such as alginate, and their optimization follows a similar approach to that of alginate films. It is also more common for studies related to alginate coatings to report real examples of the coating on food samples. One example reported the coating of mature sweet cherry fruits with sodium alginate at different concentrations (1%, 3%, or 5% w/v) with only glycerol added and no crosslinking applied (Díaz-Mula et al., 2012). The coatings effectively delayed postharvest ripening parameters and maintained total phenolics and AO activity of the fruit, with 3% and 5% slightly more effective. The storage stability was reported to double from 8 to 16 days at 2°C, both followed by two days at 20°C. In another study, fresh-cut cantaloupe and strawberries were coated with alginate containing plasticizers (glycerol, sunflower oil) and surfactants (span 80, tween 80), followed by crosslinking in Ca^{2+} (Senturk Parreidt et al., 2019). Variations in RH storage conditions were also studied with the main aim of investigating the effects of the coating on the water barrier properties. The coatings effectively reduced moisture loss from the cantaloupe with the opposite effect on strawberries. Differences in the hydrophilicity of the food surfaces were discussed as a contributing factor to the findings. In particular, the coating on the strawberries may have interacted with the natural protective surface waxes and decreased the resistance to water transfer. Lebranche mullet filets, uncoated or coated with Ca^{2+}-crosslinked alginate, were stored for 12 days at 4°C with and without vacuum packing (Vargas-Ramella et al., 2025). The coating reduced TVB-N and controlled microbial growth but increased lipid oxidation of the filets. The mechanism of oxidation was not investigated further, but it was implied that the use of $CaCl_2$ in the film may have a pro-oxidant effect. It was suggested that future studies could incorporate natural antioxidants in the coatings to reduce the lipid oxidation and generation of thiobarbituric acid reactive substances (TBARS) during storage.

8.3.2 Antimicrobial and antioxidant films

Alginate films often incorporate antimicrobial (AM) and antioxidant (AO) additives, with natural agents preferred over synthetic ones. Many natural compounds exhibit both AM and AO activities, which further expands the application of the films. The use of other naturally occurring and synthesized compounds is also widely reported, including basic metal salts that also facilitate crosslinking, and more advanced materials such as metal organic framework nanoparticles (MOF-NPs) (Feng et al., 2023). It is also common to combine additives for enhanced functionality with synergistic effects often reported. The wide range of natural agents that can be incorporated into alginate films includes essential oils and other plant extracts, organic acids and their derivatives, and other inorganic and organic compounds (Yildirim et al., 2018). The examples shown in Table 8.2 demonstrate the ability of alginates to effectively host these active agents and, in some cases, control the release of agents to further enhance the functionality of the film (Ning et al., 2025). The table categorizes alginate films by base polymer composition, which may include pure alginate or blends with polysaccharides such as pectin, gellan gum, guar gum, chitosan, starch, and cellulose derivatives. Although alginates are combined with a wide range of other polymers, they are commonly a secondary or minor component in many formulations. The alginate films can further be categorized based on the type of additive incorporated into the formulation or by the activity that they impart. However, there are many cases where the films are developed with multiple functions, which demonstrates the versatility of the additives. Anthocyanins are particularly useful AO additives that function to neutralize or inhibit reactive oxygen species (ROS) through scavenging. They can also damage the cell walls and membranes of certain bacteria, offering additional AM activity. In addition, many anthocyanins change color in response to changes in pH and are therefore studied as freshness IFs (Anugrah et al., 2023; Li, Hu, et al., 2023; Liu, Li, An, et al., 2025b; Shi, Xu, Ren, et al., 2023).

Many studies report the application of active alginate films on real foods, which can offer important insights into the functionality of the film and additives beyond in vitro studies. Common applications include films developed for whole and fresh-cut fruits and vegetables, dairy products, fresh and processed meat and fish, and other readily perishable food items. The main objectives of the film are to maintain freshness and/or increase the shelf life of the produce, with overall appropriate film properties. For example, alginate films containing Ag NPs and lemongrass essential oil were developed for packaging soft cheese (Motelica et al., 2021). The obtained films provided synergistic AM activity against Gram-positive and Gram-negative bacteria and were most effective against *Bacillus cereus*. It was suggested that the essential oil modified the stability of the bacterial cell wall and allowed the Ag^{+} released from the NPs to further damage the cell membranes, resulting in cell death. Further tests on cheese samples demonstrated preservation of the color, surface texture, and softness of samples for 14 days. It was also reported that the films darkened in color with temperature and light due to Ag NP oxidation, which was proposed for monitoring storage conditions. Another example reports the control of *Listeria* on salmon using alginate films containing Zn-MgO NPs (Vizzini et al., 2020). The release of Zn^{2+} from the NPs generates ROS, disrupts bacterial membranes, and interferes with metabolic processes, which results in cell death. Delayed browning of asparagus was achieved using alginate/guar gum films containing carbon quantum dots (QDs) (Zhang et al., 2025). The QDs prevented degradation of chlorophyll and delayed enzymatic browning of blanched asparagus by inhibiting peroxidase and polyphenol oxidase activities. The controlled release of carvacrol from light-activated alginate films improved pork filet quality by imparting AO and AM activity (Ning et al., 2025). Carvacrol inhibited bacterial growth, damaging cell membranes and providing AO activity by slowing the oxidation of myoglobin in the pork samples. Alginate films incorporating Fe_2TiO_5 and ZnO NPs were used to preserve the freshness of strawberries (Vasiljevic et al., 2024). The mechanism of Fe_2TiO_5 NP AO activity was not investigated, but the AM activity was due to the release of Zn^{2+} from the ZnO NPs, which penetrated the bacterial cell walls and produced ROS.

8.3.3 Indicator films

Other important types of active alginate films are those that serve to monitor the freshness of food products, primarily by the addition of color-sensitive agents. Natural bioactive compounds with color-changing pH sensitivity are useful additives for alginate-based IFs. These include flavonoids such as anthocyanins, which are the most common pigments used in IFs. Anthocyanins impart a broad range of bioactivities, so they are commonly used as indicators with AO and/or AM properties in alginate films. Table 8.2 demonstrates many examples of these hybrid films where anthocyanins and other pigments are used for multiple functions. In one example, alginate films containing purple sweet potato peel extracts were used to monitor the freshness of chicken filets at different temperatures (Zhao et al., 2022). The color of the anthocyanins in the extracts is sensitive to changes in pH and transitions from red (pH 2–3) to pink (pH 4–6), to purple (pH 7), to blue (pH 8), to green (pH 9–11), and yellow (pH 12–13). Due to this pH sensitivity, a color change of the film from pink to blue was an indicator of deterioration of the chicken proteins and release of volatile nitrogen compounds. In particular, the color change was a result of increasing pH, which corresponded with increasing TVB-N values. Inorganic additives can also offer color sensitivity and reactivity to environmental conditions to indicate specific types of food spoilage. These include Co-MOF NPs incorporated into alginate films that were used to detect ammonia and subsequent shrimp spoilage as evidenced by a distinct color change from pink to brown/black (Feng et al., 2023). In this case, Co-N bonds were formed, which resulted in a color transformation. The Co-MOF NPs also improved the mechanical strength, barrier properties, and UV shielding properties of the alginate film.

Many studies provide visual evidence of IFs reacting to pH changes or real food spoilage. As shown in Fig. 8.3, a distinct color change in IFs is evident in samples of beef mince and shrimp stored at 4°C and 25°C (Santos & Martins, 2024). In this study, alginate IFs containing purple onion skin and butterfly pea flower extracts were prepared with additives that improve key UV shielding, barrier, and thermal properties, and provide AO and AM activity. Color changes varied depending on the composition of the mixture of extracts and were more pronounced and perceptible than neat extracts. The films were claimed to provide both preservative properties and the ability to monitor the deterioration of protein-rich foods. Effective IFs require pigments with clear, high-contrast color changes at critical freshness thresholds. This ensures that food spoilage indication is clearly visible and easy to interpret for the majority of consumers. The broad range of plants and extracts that exhibit color sensitivity is also important to note. In many examples of IFs, small sections of films are used with clear film or other packaging to wrap or contain test food items. This ensures the food item is clearly visible and the natural color of the food is not masked. Other challenges exist for IFs such as pigment stability, which can be influenced by a range of factors including the extraction method and film processing. Environmental factors such as temperature, UV-light exposure, and storage can deteriorate pigments and adversely affect the response of IFs.

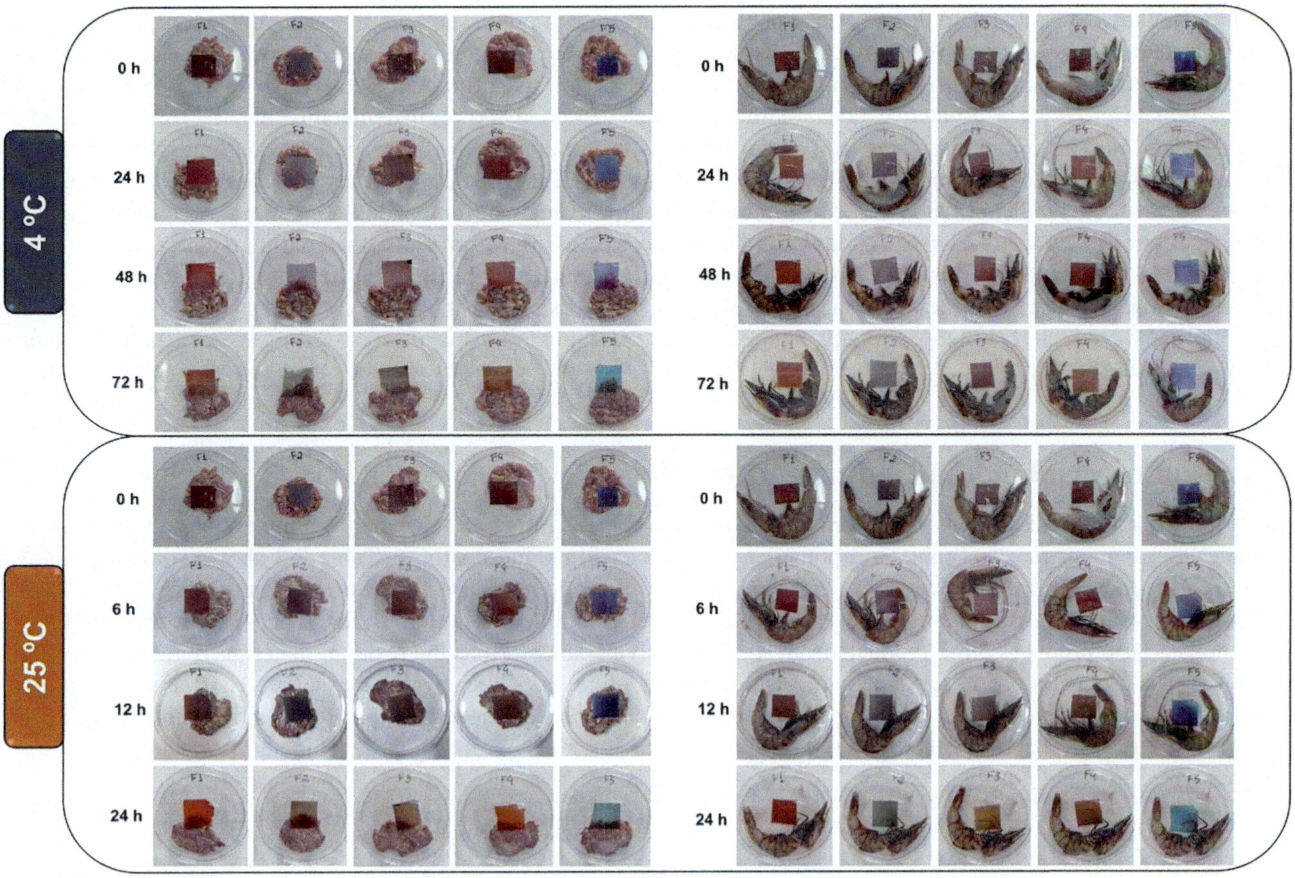

FIGURE 8.3 **Example of colorimetric alginate film with plant extracts to monitor beef mince and shrimp freshness.** Photographs of color-changing alginate film used to indicate beef mince and shrimp spoilage during storage at different temperatures.
From Santos, L.G. & Martins, V.G. (2024). Multifunctional alginate films blended with polyphenol-rich extract from unconventional edible sources: Bioactive properties, UV-light protection, and food freshness monitoring. International Journal of Biological Macromolecules, 262. https://doi.org/10.1016/j.ijbiomac.2024.130001.

8.3.4 Alginate coatings

Alginate coatings are applied directly to food products to extend their shelf life. In a typical process, food items are either dipped or sprayed with an alginate gel solution, which is then allowed to dry and/or further processed by immersion crosslinking. Alginate coatings are often labeled "edible," meaning they are safe for consumption. However, the presence of additives such as antimicrobial agents or plasticizers may affect their suitability. The term "edible" is frequently used interchangeably with "biodegradable" or "biobased," but these terms are not synonymous. There are many important considerations such as the nature and concentration of other additives that may impact the suitability of coatings for human consumption. The term is commonly used more generically to describe biobased and/or biodegradable coatings. There are countless examples of alginate coatings, which primarily include antioxidants and antimicrobial agents or their combinations. In addition to the more common additives, antibrowning agents can be used in coatings to minimize or delay browning due to enzymatic processes (Abdullah, Tong, et al., 2021; Robles-Sánchez et al., 2013; Rojas-Graü et al., 2007). Ascorbic acid is one such agent that acts both as a reducing agent and an enzymatic inhibitor.

Examples of preservative coatings include alginate containing citric or ascorbic acid applied to fresh-cut mangos that successfully preserved the mango color (Robles-Sánchez et al., 2013). In addition to the reducing capacity and enzyme-inhibiting action, ascorbic acid enhanced the antioxidant capacity by promoting gallic acid synthesis in the fruit. In another example, alginate was used as a coating in combination with chitosan and cinnamon essential oil (CEO) microcapsules for controlled release of the oil for mango preservation. After 14 days, mangos coated with five layers retained their appearance and commercial value, whereas the uncoated mangos deteriorated (Yin et al., 2019). The release of CEO inhibited chlorophyll hydrolysis and oxidation, reduced enzyme activity, and maintained the mango color. Alginate containing natamycin coated onto the calyx and pedicel of tomato fruit preserved both moisture and chlorophyll content of the coated parts. The optimum levels of alginate and natamycin were 0.5% and 0.3%, respectively, and the coating treatment maintained the overall quality and acceptance of the fruit and promoted a longer shelf life (Zhangyao et al., 2025). Alginate provided a protective layer inhibiting water loss, and the alginate/natamycin effectively reduced the peroxidase and polyphenol oxidase activities. Alginate films containing whey isolate protein fiber and glycyrrhizic acid temperature-sensitive emulsions improved UV-blocking, antioxidant, mechanical, and barrier properties with an optimum level of 30% of the emulsion when coated on grapes. Optimized alginate emulsions containing whey maintained the overall quality of the grapes for up to six days at 45°C (Shi et al., 2025). Specifically, the coating reduced respiration and water loss, and decreased the overall weight loss of the grapes. In addition, the coating slowed the degradation of organic acids in the grapes and reduced their respiratory metabolism. These examples show that coatings can successfully preserve whole and fresh-cut fruit, but there are many other examples that include fresh and processed meats, vegetables, and other readily perishable foods.

8.4 Commercial alginate films and coatings

Although academic studies show the potential for innovation and future development, commercial or industrial-scale examples can demonstrate the value and potential for alginate films and coatings. Table 8.3 shows some examples of commercial alginate films and coatings, as well as other developments or innovations in progress. These include NotPLA's paper coating for take-away foods and the innovative edible bubble for water, sauces, and other condiments. The company also produces other products from seaweed polymers, as well as the residual materials remaining after their extraction, such as cellulose to produce paper, among other products. Other examples include Evoware alginate films, which can be formed into sachets and wraps, Loliware who produce a seaweed-based drinking straw and polymer pellets that can be extruded. There are other key innovators in this field who are working to develop alginate, and other seaweed polymer-based PFs and coatings.

TABLE 8.3 Examples of commercial alginate films and coatings, and future innovations of seaweed polymers.

Company	Product(s)	Website
Notpla Limited	Food containers, gel pods, Ooho edible bubble	https://www.notpla.com/
Loliware Inc.	Blue Carbon Straw, seaweed resins	https://www.loliware.com/
Evoware	Seaweed-based film	https://rethink-plastic.com/
Kelpy[a]	Injection-moldable resins	https://www.kelpy.co/
Mujō	Algae-based packaging materials	https://mujolab.com/wordpress/
FlexSea	Seaweed packaging	https://flex-sea.com/
Sway Innovation Co.	Seaweed-based packaging	https://swaythefuture.com/
PakPlast	Seaweed wrap	https://pakplast.com.au/
EcoEnclose	Retail boxes with seaweed window	https://www.ecoenclose.com/
Searo Labs Ltd	Seaweed-based flexible film	https://searolabs.com/
B'Zeos	Drinking straws and food films	https://www.bzeos.com/
Carbonwave	Resins, packaging applications	https://carbonwave.com/
Oceanium[a]	Various products	https://oceanium.world/
Zerocircle Alternatives Pvt. Ltd.	Coatings, paper, resins	https://www.zerocircle.in/
PlantSea Ltd	Paper, punnets, pods, mulch film	https://plantsea.co.uk/

[a] *Indicates a start-up or company with products under development.*

8.5 Future perspectives

The continuing global movement toward more environmentally sound packaging opens many opportunities for innovative research. Alginates and other naturally sourced biopolymers and bio-derived materials represent a growing source of potential alternatives to conventional petroleum-based PFs and coatings. There are some key considerations and strategies needed to realize the full potential of alginates in this space. These include reducing the cost to make products competitive, improving film properties, appropriate use of additives, as well as commercial uptake and consumer acceptance. One approach to reducing raw material costs is the use of semirefined polymers, potentially combined with inexpensive fillers. Reducing costs can also be addressed by modifying raw materials so they can be processed using existing extrusion or production equipment.

It is also important to consider the regulatory environment related to food packaging, which can vary significantly globally and often within countries or regions. This is particularly important since the distribution of packaged foods occurs on a global scale, and packaging materials must meet specific import guidelines. The types and concentrations of additives, and their propensity to migrate from the packaging into foods must also meet regulatory guidelines. A common global priority in recent years is phasing out single-use plastic packaging and/or the responsible and management of waste plastics. This indirectly supports the further development of alginate and other biopolymer packaging materials that are needed to replace conventional packaging.

In the research environment, efforts focus on improving the performance properties of alginate films using reinforcing agents, fillers, crosslinkers, and plasticizers. Opportunities are unlimited in the scope of additives and processes that can be investigated, particularly in view of the versatility of alginate polymers. Since many of the highly active additives are plant-based, it is important to consider the sources of these additives to ensure they do not compete with food resources or encourage indiscriminate sourcing and/or farming practices. For both organic and inorganic additives, it is critical to ensure they are food safe (nontoxic and nonallergenic) and do not pose environmental concerns or excessive use of resources in extraction or synthesis in the case of more complex additives.

Alginate films, among most polysaccharide-based films, are invariably cast from hydrogel solutions prior to crosslinking or further treatment. Conventional commercial PFs are primarily produced using extrusion processing of polymer pellets, which offers rapid, efficient, and cost-effective film fabrication. To increase the viability of replacing or substituting petroleum-based films with natural alternatives, new processes are needed to enable producers to utilize existing equipment with or without modifications. Extrusion processing of alginates is not impossible but would require the development of alternative techniques to produce resins or powders suitable for extrusion. End-of-life options must also be taken into consideration such as the disposal and/or recovery of single-use PFs. These options must also meet regulatory requirements, but since waste management is generally a local or municipal responsibility, these regulations can vary more significantly than state or federally based guidelines. Although many alginate films are claimed as biodegradable or compostable, very few studies demonstrate this feature under composting (home and industrial) or landfill (anaerobic) conditions, or other environmental conditions (in soil, sand, seawater, etc.). Recovery and recyclability are potential disposal alternatives, but these are more complicated options since contamination of domestic recyclable materials can be a problem when organic-based materials are introduced.

New and frequent developments in the field of alginate packaging materials demonstrate the versatility and broad functionality of this biopolymer. With the ability to form films and direct coatings, with and without highly functional additives, alginates also offer the aspects of biodegradability from renewable and abundant resources. However, significant challenges remain, including limitations in mechanical and barrier properties and cost-effective large-scale production. Overcoming these challenges is crucial and can be achieved by conducting research into novel alginate blends, advanced processing techniques, and sustainable production methods. The widespread adoption of alginate packaging has the potential to significantly reduce plastic waste, create a more sustainable packaging industry, and offer environmentally friendly alternatives for various applications.

References

Abdin, M., El-Beltagy, A. E., El-sayed, M. E., & Naeem, M. A. (2022). Production and characterization of sodium alginate/gum arabic based films enriched with syzygium cumini seeds extracts for food application. *Journal of Polymers and the Environment, 30*(4), 1615–1626. https://doi.org/10.1007/s10924-021-02306-z, http://www.kluweronline.com/issn/1566-2543/.

Abdullah, N. A. S., Mohamad, Z., Khan, Z. I., Jusoh, M., Zakaria, Z. Y., & Ngadi, N. (2021). Alginate based sustainable films and composites for packaging: A review. *Chemical Engineering Transactions, 83*, 271–276. https://doi.org/10.3303/CET2183046, https://www.aidic.it/cet/21/83/046.pdf.

Abdullah, S. Z., Tong, W. Y., Leong, C. R., Rashid, S. A., Rozman, N. A. S., Hamid, N. H. M., Karim, S., Tumin, N. D., Muda, S. A., & Yacob, L. S. (2021). Development of cinnamaldehyde loaded-alginate based film for food packaging. *Asia-Pacific Journal of Molecular Biology and Biotechnology, 29*(3), 18–25. https://doi.org/10.35118/apjmbb.2021.029.3.03, http://www.msmbb.my/images/publication/volume_29/issue_3/03-Abdullah-et-al.pdf.

Akbar, A., & Anal, A. K. (2014). Zinc oxide nanoparticles loaded active packaging, a challenge study against *Salmonella typhimurium* and *Staphylococcus aureus* in ready-to-eat poultry meat. *Food Control, 38*(1), 88–95. https://doi.org/10.1016/j.foodcont.2013.09.065.

Akhtar, H. M. S., Zhao, Y., Li, L., & Shi, Q. (2024). Novel active composite films based on carboxymethyl cellulose and sodium alginate incorporated with phycocyanin: Physico-chemical, microstructural and antioxidant properties. *Food Hydrocolloids, 147*. https://doi.org/10.1016/j.foodhyd.2023.109440, https://www.journals.elsevier.com/food-hydrocolloids.

An, N., & Zhou, W. (2024). Sodium alginate/ager colourimetric film on porous substrate layer: Potential in intelligent food packaging. *Food Chemistry, 445*, 138790. https://doi.org/10.1016/j.foodchem.2024.138790.

Anugrah, D. S. B., Darmalim, L. V., Sinanu, J. D., Pramitasari, R., Subali, D., Prasetyanto, E. A., & Cao, X. T. (2023). Development of alginate-based film incorporated with anthocyanins of red cabbage and zinc oxide nanoparticles as freshness indicator for prawns. *International Journal of Biological Macromolecules, 251*. https://doi.org/10.1016/j.ijbiomac.2023.126203, https://www.sciencedirect.com/journal/international-journal-of-biological-macromolecules.

Assis, M., Cano-Vicent, A., Tuñon-Molina, A., Benzi-Chumachenco, R. R., Andrés, J., & Serrano-Aroca, A. (2024). Calcium alginate films loaded with copper-molybdenum oxide nanoparticles for antimicrobial applications. *Journal of Environmental Chemical Engineering, 12*(5), 113935. https://doi.org/10.1016/j.jece.2024.113935, https://www.sciencedirect.com/science/journal/22133437.

Azarakhsh, N., Osman, A., Ghazali, H. M., Tan, C. P., & Mohd Adzahan, N. (2014). Lemongrass essential oil incorporated into alginate-based edible coating for shelf-life extension and quality retention of fresh-cut pineapple. *Postharvest Biology and Technology, 88*, 1–7. https://doi.org/10.1016/j.postharvbio.2013.09.004.

Azeredo, H. M. C., Magalhães, U. S., Oliveira, S. A., Ribeiro, H. L., Brito, E. S., & De Moura, M. R. (2012). Tensile and water vapour properties of calcium-crosslinked alginate-cashew tree gum films. *International Journal of Food Science and Technology, 47*(4), 710–715. https://doi.org/10.1111/j.1365-2621.2011.02897.x.

Bagheri, F., Radi, M., & Amiri, S. (2019). Drying conditions highly influence the characteristics of glycerol-plasticized alginate films. *Food Hydrocolloids, 90*, 162–171. https://doi.org/10.1016/j.foodhyd.2018.12.001, https://www.journals.elsevier.com/food-hydrocolloids.

Bai, Y., Zhao, Y., Li, Y., Xu, J., Fu, X., Gao, X., Mao, X., & Li, Z. (2020). UV-shielding alginate films crosslinked with Fe^{3+} containing EDTA. *Carbohydrate Polymers, 239*. https://doi.org/10.1016/j.carbpol.2019.115480.

Benavides, S., Villalobos-Carvajal, R., & Reyes, J. E. (2012). Physical, mechanical and antibacterial properties of alginate film: Effect of the crosslinking degree and oregano essential oil concentration. *Journal of Food Engineering, 110*(2), 232–239. https://doi.org/10.1016/j.jfoodeng.2011.05.023.

Bhatia, S., Shah, Y. A., Al-Harrasi, A., Jawad, M., Khan, T. S., Koca, E., & Aydemir, L. Y. (2024). Tuning the structure and physiochemical properties of sodium alginate and chitosan composite films through sodium tripolyphosphate (STPP) crosslinking. *International Journal of Biological Macromolecules, 264*. https://doi.org/10.1016/j.ijbiomac.2024.130463, https://www.sciencedirect.com/science/journal/01418130.

Bt Ibrahim, S. F., Mohd Azam, N. A. N., & Amin, K. A. M. (2019). Sodium alginate film: The effect of crosslinker on physical and mechanical properties. *IOP Conference Series: Materials Science and Engineering, 509*. https://doi.org/10.1088/1757-899X/509/1/012063.

Cao, L., Lu, W., Mata, A., Nishinari, K., & Fang, Y. (2020). Egg-box model-based gelation of alginate and pectin: A review. *Carbohydrate Polymers, 242*. https://doi.org/10.1016/j.carbpol.2020.116389, http://www.elsevier.com/wps/find/journaldescription.cws_home/405871/description#description.

Chen, J., Zhang, Y., Liu, H., Lu, H., Xu, X., & Shen, M. (2025). Sodium alginate-camellia seed cake protein active packaging film cross-linked by electrostatic interactions for fruit preservation. *International Journal of Biological Macromolecules, 288*. https://doi.org/10.1016/j.ijbiomac.2024.138627, https://www.sciencedirect.com/science/journal/01418130.

Chen, K., Li, J., Li, L., Wang, Y., Qin, Y., & Chen, H. (2023). A pH indicator film based on sodium alginate/gelatin and plum peel extract for monitoring the freshness of chicken. *Food Bioscience, 53*. https://doi.org/10.1016/j.fbio.2023.102584, http://www.journals.elsevier.com/food-bioscience/.

Chen, S., Zhang, H., Jiang, Z., Ding, X., Chen, W., Ma, N., Xu, S., & Yang, L. (2024). Intelligent active packaging of sodium alginate and pectin mixed with *Aronia melanocarpa* anthocyanins and tea polyphenols for shrimp freshness monitoring and preservation. *International Journal of Biological Macromolecules, 283*. https://doi.org/10.1016/j.ijbiomac.2024.137754, https://www.sciencedirect.com/science/journal/01418130.

Chen, X., Wang, L., Zhang, D., Bu, N., Liu, W., Wu, Z., Mu, R., Tan, P., Zhong, Y., & Pang, J. (2024). Enhancing strawberry freshness: multifunction sustainable films utilizing two types of modified carbon nanotubes for photothermal food packaging. *ACS Applied Materials and Interfaces, 16*(46), 63964–63977. https://doi.org/10.1021/acsami.4c09955, http://pubs.acs.org/journal/aamick.

Choi, I., Lee, Y., Lyu, J. S., Lee, J. S., & Han, J. (2022). Characterization of ionically crosslinked alginate films: Effect of different anion-based metal cations on the improvement of water-resistant properties. *Food Hydrocolloids, 131*. https://doi.org/10.1016/j.foodhyd.2022.107785, https://www.journals.elsevier.com/food-hydrocolloids.

Costa, M. J., Marques, A. M., Pastrana, L. M., Teixeira, J. A., Sillankorva, S. M., & Cerqueira, M. A. (2018). Physicochemical properties of alginate-based films: Effect of ionic crosslinking and mannuronic and guluronic acid ratio. *Food Hydrocolloids, 81*, 442–448. https://doi.org/10.1016/j.foodhyd.2018.03.014, https://www.sciencedirect.com/science/journal/0268005X.

Dai, Q., Huang, X., Jia, R., Fang, Y., & Qin, Z. (2022). Development of antibacterial film based on alginate fiber, and peanut red skin extract for food packaging. *Journal of Food Engineering, 330*. https://doi.org/10.1016/j.jfoodeng.2022.111106, http://www.sciencedirect.com/science/journal/02608774.

de Farias, Y. B., Coutinho, A. K., Assis, R. Q., & Rios, Ad. O. (2020). Biodegradable sodium alginate films incorporated with norbixin salts. *Journal of Food Process Engineering, 43*(2), 13345. https://doi.org/10.1111/jfpe.13345, http://onlinelibrary.wiley.com/journal/10.1111/(ISSN)1745-4530.

De'Nobili, M. D., Soria, M., Martinefski, M. R., Tripodi, V. P., Fissore, E. N., & Rojas, A. M. (2016). Stability of L-(+)-ascorbic acid in alginate edible films loaded with citric acid for antioxidant food preservation. *Journal of Food Engineering, 175*, 1–7. https://doi.org/10.1016/j.jfoodeng.2015.11.015, http://www.sciencedirect.com/science/journal/02608774.

Devi, D., Kumar, S., & Mukherjee, A. (2024). Onion peel extract as milk freshness indicator in biopolymer-based intelligent packaging films. *Food and Humanity, 2*. https://doi.org/10.1016/j.foohum.2023.100223, https://www.sciencedirect.com/journal/food-and-humanity.

Díaz-Mula, H. M., Serrano, M., & Valero, D. (2012). Alginate coatings preserve fruit quality and bioactive compounds during storage of sweet cherry fruit. *Food and Bioprocess Technology, 5*(8), 2990–2997. https://doi.org/10.1007/s11947-011-0599-2.

Draget, K. I., Gåserød, O., Aune, I., Andersen, P. O., Storbakken, B., Stokke, B. T., & Smidsrød, O. (2001). Effects of molecular weight and elastic segment flexibility on syneresis in Ca-alginate gels. *Food Hydrocolloids, 15*(4-6), 485–490. https://doi.org/10.1016/s0268-005x(01)00046-7.

Du, H., Said, N. S., & Lee, W. Y. (2024). Development of pH-Sensitive intelligent films incorporating betacyanin from dragon fruit peel and TiO_2 nanoparticles for monitoring fish fillet freshness. *Sustainable Chemistry and Pharmacy, 42*. https://doi.org/10.1016/j.scp.2024.101839, https://www.sciencedirect.com/science/journal/23525541.

El Hammadi, N., Almajano, M. P., Pastor, M. V., & Codina-Torrella, I. (2024). *Artubuts unedo* L. leaves extract as a potential ingredient for the production of active-films and the preservation of oil-in-water (O/W) emulsions from oxidation. *LWT, 214*. https://doi.org/10.1016/j.lwt.2024.117110, https://www.sciencedirect.com/science/journal/00236438.

Eslami, Z., Elkoun, S., Robert, M., & Adjallé, K. (2023). A review of the effect of plasticizers on the physical and mechanical properties of alginate-based films. *Molecules, 28*(18). https://doi.org/10.3390/molecules28186637, http://www.mdpi.com/journal/molecules.

Feng, S., Tang, Q., Xu, Z., Huang, K., Li, H., & Zou, Z. (2023). Development of novel Co-MOF loaded sodium alginate based packaging films with antimicrobial and ammonia-sensitive functions for shrimp freshness monitoring. *Food Hydrocolloids, 135*. https://doi.org/10.1016/j.foodhyd.2022.108193, https://www.journals.elsevier.com/food-hydrocolloids.

Galdeano, M. C., Mali, S., Grossmann, M. V. E., Yamashita, F., & García, M. A. (2009). Effects of plasticizers on the properties of oat starch films. *Materials Science and Engineering C, 29*(2), 532–538. https://doi.org/10.1016/j.msec.2008.09.034.

Gao, Z., Xie, S., Chang, L., Tang, H., Sun, Z., Deng, Y., Hu, Y., Xu, Y., & Luan, M. (2025). Antioxidant and antimicrobial sodium alginate/sodium carboxymethyl cellulose films loaded with self-assembled hesperidin nanorods for fruits preservation. *Food Chemistry, 474*. https://doi.org/10.1016/j.foodchem.2025.143183, https://www.sciencedirect.com/science/journal/03088146.

Garavand, F., Rouhi, M., Razavi, S. H., Cacciotti, I., & Mohammadi, R. (2017). Improving the integrity of natural biopolymer films used in food packaging by crosslinking approach: A review. *International Journal of Biological Macromolecules, 104*, 687–707. https://doi.org/10.1016/j.ijbiomac.2017.06.093, www.elsevier.com/locate/ijbiomac.

Gheorghita Puscaselu, R., Lobiuc, A., Dimian, M., & Covasa, M. (2020). Alginate: From food industry to biomedical applications and management of metabolic disorders. *Polymers, 12*(10), 2417. https://doi.org/10.3390/polym12102417.

Girón-Hernández, J., Gentile, P., & Benlloch-Tinoco, M. (2021). Impact of heterogeneously crosslinked calcium alginate networks on the encapsulation of β-carotene-loaded beads. *Carbohydrate Polymers, 271*. https://doi.org/10.1016/j.carbpol.2021.118429, http://www.elsevier.com/wps/find/journaldescription.cws_home/405871/description#description.

Giz, A. S., Berberoglu, M., Bener, S., Aydelik-Ayazoglu, S., Bayraktar, H., Alaca, B. E., & Catalgil-Giz, H. (2020). A detailed investigation of the effect of calcium crosslinking and glycerol plasticizing on the physical properties of alginate films. *International Journal of Biological Macromolecules, 148*, 49–55. https://doi.org/10.1016/j.ijbiomac.2020.01.103, https://www.elsevier.com/locate/ijbiomac.

Guo, J., Khan, M. R., Ahmad, N., & Zhang, W. (2025). Enhancing fruit preservation with sodium alginate films incorporating propolis extract and graphene oxide. *International Journal of Biological Macromolecules, 288*. https://doi.org/10.1016/j.ijbiomac.2024.138778, https://www.sciencedirect.com/science/journal/01418130.

Halloub, A., Raji, M., Essabir, H., Nekhlaoui, S., Bensalah, M. O., Bouhfid, R., & Qaiss, Aek (2023). Stable smart packaging betalain-based from red prickly pear covalently linked into cellulose/alginate blend films. *International Journal of Biological Macromolecules, 234*. https://doi.org/10.1016/j.ijbiomac.2023.123764, https://www.sciencedirect.com/journal/international-journal-of-biological-macromolecules.

He, J., Yang, S., Goksen, G., Cong, X., Rizwan Khan, M., & Zhang, W. (2024). Functionalized pectin/alginate food packaging films based on metal-phenol networks. *Food Bioscience, 58*. https://doi.org/10.1016/j.fbio.2024.103635, https://www.sciencedirect.com/science/journal/22124292.

Hoque, M., Babu, R. P., McDonagh, C., Jaiswal, S., Tiwari, B. K., Kerry, J. P., & Pathania, S. (2024). Pectin/sodium alginate-based active film integrated with microcrystalline cellulose and geraniol for food packaging applications. *International Journal of Biological Macromolecules, 271*. https://doi.org/10.1016/j.ijbiomac.2024.132414, https://www.sciencedirect.com/science/journal/01418130.

Huang, H. L., Tsai, I. L., Lin, C., Hang, Y. H., Ho, Y. C., Tsai, M. L., & Mi, F. L. (2023). Intelligent films of marine polysaccharides and purple cauliflower extract for food packaging and spoilage monitoring. *Carbohydrate Polymers, 299*. https://doi.org/10.1016/j.carbpol.2022.120133.

Imran, M., El-Fahmy, S., Revol-Junelles, A. M., & Desobry, S. (2010). Cellulose derivative based active coatings: Effects of nisin and plasticizer on physico-chemical and antimicrobial properties of hydroxypropyl methylcellulose films. *Carbohydrate Polymers, 81*(2), 219–225. https://doi.org/10.1016/j.carbpol.2010.02.021.

Jakubowska, E., Gierszewska, M., Szydłowska-Czerniak, A., Nowaczyk, J., & Olewnik-Kruszkowska, E. (2023). Development and characterization of active packaging films based on chitosan, plasticizer, and quercetin for repassed oil storage. *Food Chemistry, 399*. https://doi.org/10.1016/j.foodchem.2022.133934, www.elsevier.com/locate/foodchem.

Janik, W., Nowotarski, M., Ledniowska, K., Shyntum, D. Y., Krukiewicz, K., Turczyn, R., Sabura, E., Furgoł, S., Kudła, S., & Dudek, G. (2023). Modulation of physicochemical properties and antimicrobial activity of sodium alginate films through the use of chestnut extract and plasticizers. *Scientific Reports, 13*(1). https://doi.org/10.1038/s41598-023-38794-3, https://www.nature.com/srep/.

Jiang, K., Li, J., Brennan, M., Brennan, C., Chen, H., Qin, Y., & Yuan, M. (2023). Smart indicator film based on sodium alginate/polyvinyl alcohol/TiO_2 containing purple garlic peel extract for visual monitoring of beef freshness. *Polymers, 15*(21). https://doi.org/10.3390/polym15214308, http://www.mdpi.com/journal/polymers.

Jost, V., Kobsik, K., Schmid, M., & Noller, K. (2014). Influence of plasticiser on the barrier, mechanical and grease resistance properties of alginate cast films. *Carbohydrate Polymers, 110*, 309–319. https://doi.org/10.1016/j.carbpol.2014.03.096.

Jost, V., & Reinelt, M. (2018). Effect of Ca^{2+} induced crosslinking on the mechanical and barrier properties of cast alginate films. *Journal of Applied Polymer Science, 135*(5). https://doi.org/10.1002/app.45754, http://onlinelibrary.wiley.com/journal/10.1002/(ISSN)1097-4628.

Karimi Khorrami, N., Radi, M., Amiri, S., & McClements, D. J. (2021). Fabrication and characterization of alginate-based films functionalized with nanostructured lipid carriers. *International Journal of Biological Macromolecules, 182*, 373–384. https://doi.org/10.1016/j.ijbiomac.2021.03.159, www.elsevier.com/locate/ijbiomac.

Khapre, M. A., Pandey, S., & Jugade, R. M. (2021). Glutaraldehyde-cross-linked chitosan–alginate composite for organic dyes removal from aqueous solutions. *International Journal of Biological Macromolecules, 190*, 862–875. https://doi.org/10.1016/j.ijbiomac.2021.09.026.

Klinkajon, W., & Supaphol, P. (2014). Novel copper (II) alginate hydrogels and their potential for use as anti-bacterial wound dressings. *Biomedical Materials, 9*(4), 045008. https://doi.org/10.1088/1748-6041/9/4/045008.

Konuk Takma, D., & Korel, F. (2017). Impact of preharvest and postharvest alginate treatments enriched with vanillin on postharvest decay, biochemical properties, quality and sensory attributes of table grapes. *Food Chemistry, 221*, 187–195. https://doi.org/10.1016/j.foodchem.2016.09.195.

Kowalonek, J., Hamieau, M., & Szydłowska-Czerniak, A. (2024). Influence of different deep eutectic solvents and plant extracts on antioxidant, mechanical, and color properties of alginate film. *Polymers, 16*(14). https://doi.org/10.3390/polym16142084, http://www.mdpi.com/journal/polymers.

Kowalonek, J., Łukomska, B., Łukomska, O., & Stachowiak-Trojanowska, N. (2024). Alginate films enriched in raspberry and/or black currant seed oils as active food packaging. *Molecules, 29*(9). https://doi.org/10.3390/molecules29092012, http://www.mdpi.com/journal/molecules.

Kowalonek, J., Łukomska, B., & Szydłowska-Czerniak, A. (2025). Color, structure, and thermal stability of alginate films with raspberry and/or black currant seed oils. *Molecules, 30*(2), 245. https://doi.org/10.3390/molecules30020245.

Lee, K. Y., & Mooney, D. J. (2012). Alginate: Properties and biomedical applications. *Progress in Polymer Science, 37*(1), 106–126. https://doi.org/10.1016/j.progpolymsci.2011.06.003.

Lee, M., Rüegg, N., & Yildirim, S. (2023). Evaluation of the antimicrobial activity of sodium alginate films integrated with cinnamon essential oil and citric acid on sliced cooked ham. *Packaging Technology and Science, 36*(8), 647–656. https://doi.org/10.1002/pts.2733, http://onlinelibrary.wiley.com/journal/10.1002/(ISSN)1099-1522.

Lei, Y., Yao, Q., Jin, Z., & Wang, Y. C. (2023). Intelligent films based on pectin, sodium alginate, cellulose nanocrystals, and anthocyanins for monitoring food freshness. *Food Chemistry, 404*. https://doi.org/10.1016/j.foodchem.2022.134528, www.elsevier.com/locate/foodchem.

Li, C., Song, A., Wu, Y., Gao, Y., & Li, C. (2024). Intelligent double-layer film based on gellan gum/modified anthocyanin/curcumin/sodium alginate/zinc oxide for monitoring shrimp freshness. *International Journal of Biological Macromolecules, 274*. https://doi.org/10.1016/j.ijbiomac.2024.132724, https://www.sciencedirect.com/science/journal/01418130.

Li, H., Liu, C., Sun, J., & Lv, S. (2022). Bioactive edible sodium alginate films incorporated with tannic acid as antimicrobial and antioxidative food packaging. *Foods, 11*(19). https://doi.org/10.3390/foods11193044, www.mdpi.com/journal/foods.

Li, H., Wang, Z., Zhu, F., & Li, G. (2024). Alginate-based active and intelligent packaging: Preparation, properties, and applications. *International Journal of Biological Macromolecules, 279*. https://doi.org/10.1016/j.ijbiomac.2024.135441, https://www.sciencedirect.com/science/journal/01418130.

Li, J., Wu, Y., He, J., & Huang, Y. (2016). A new insight to the effect of calcium concentration on gelation process and physical properties of alginate films. *Journal of Materials Science, 51*(12), 5791–5801. https://doi.org/10.1007/s10853-016-9880-0.

Li, R., Feng, H., Wang, S., Zhuang, D., & Zhu, J. (2024). A colorimetry-enhanced tri-functional film with high stability by polyphenol-anthocyanin co-pigmentation/conjugate: New prospect for active intelligent food packaging. *Food Chemistry, 447*. https://doi.org/10.1016/j.foodchem.2024.138927, https://www.sciencedirect.com/science/journal/03088146.

Li, S., Hu, X., Zhang, S., Zhao, J., Wang, R., Wang, L., Wang, X., Yuan, Y., Yue, T., Cai, R., & Wang, Z. (2024). A versatile bilayer smart packaging based on konjac glucomannan/alginate for maintaining and monitoring seafood freshness. *Carbohydrate Polymers, 340*. https://doi.org/10.1016/j.carbpol.2024.122244, https://www.sciencedirect.com/science/journal/01448617.

Li, S., Wang, X., Luo, Y., Chen, Z., Yue, T., Cai, R., Muratkhan, M., Zhao, Z., & Wang, Z. (2023). A green versatile packaging based on alginate and anthocyanin via incorporating bacterial cellulose nanocrystal-stabilized camellia oil Pickering emulsions. *International Journal of Biological Macromolecules, 249*. https://doi.org/10.1016/j.ijbiomac.2023.126134, https://www.sciencedirect.com/journal/international-journal-of-biological-macromolecules.

Li, Y., Hu, Z., Huo, R., & Cui, Z. (2023). Preparation of an indicator film based on pectin, sodium alginate, and xanthan gum containing blueberry anthocyanin extract and its application in blueberry freshness monitoring. *Heliyon, 9*(3). https://doi.org/10.1016/j.heliyon.2023.e14421, http://www.journals.elsevier.com/heliyon/.

Li, Y., Wu, Y., & Li, C. (2024). Development of CO_2-sensitive antimicrobial bilayer films based on gellan gum and sodium alginate/sodium carboxymethyl cellulose and its application in strawberries. *International Journal of Biological Macromolecules, 264*. https://doi.org/10.1016/j.ijbiomac.2024.130572, https://www.sciencedirect.com/science/journal/01418130.

Lisboa, M., Chagas, C., da Costa, J. C. M., Rossoni, D., Mikcha, J. M. G., Silva, J. Vd. O., Batista, A., Caetano, W., Madrona, G. S., Tonon, L. A. C., Pozza, M. Sd. S., Stafussa, A. P., & Scapim, M. Rd. S. (2022). Characterization and effect of edible alginate acidified coatings associated with photodynamics on cheese preservation. *Applied Food Research, 2*(1). https://doi.org/10.1016/j.afres.2022.100113, https://www.journals.elsevier.com/applied-food-research.

Liu, C., Li, N., An, X., Li, X., & Liu, Z. (2025a). A smart food freshness indicator film for shrimp and pork with the higher visuality, barrier, flame retardancy, and antibacterial effects. *Food and Bioprocess Technology*. https://doi.org/10.1007/s11947-024-03742-0, https://www.springer.com/journal/11947.

Liu, C., Li, N., An, X., Li, X., & Liu, Z. (2025b). Pectin/zinc alginate films containing anthocyanins from dragon fruit peel as intelligent pH indicators for shrimp freshness monitor. *International Journal of Biological Macromolecules, 285*. https://doi.org/10.1016/j.ijbiomac.2024.138317, https://www.sciencedirect.com/science/journal/01418130.

Liu, C., Li, N., Niu, L., Li, X., Feng, J., & Liu, Z. (2024a). Eco-friendly methylcellulose/zinc alginate film with multi-function properties: Thermal stability, flame retardancy and antibacterial activities. *International Journal of Biological Macromolecules, 281*. https://doi.org/10.1016/j.ijbiomac.2024.136237, https://www.sciencedirect.com/science/journal/01418130.

Liu, C., Li, N., Niu, L., Li, X., Feng, J., & Liu, Z. (2024b). Effect of Zn^{2+} crosslinked semi-interpenetrating network on thermal properties and antibacterial property of hydroxypropyl methylcellulose/sodium alginate blend film. *Journal of Materials Science, 59*(10), 4354–4365. https://doi.org/10.1007/s10853-023-08959-6, https://www.springer.com/journal/10853.

Liu, C., Li, N., Niu, L., Li, X., Feng, J., & Liu, Z. (2025a). Eco-friendly smart packaging film with high thermal stabilities, antibacterial activities, and food freshness monitoring. *Food and Bioprocess Technology, 18*(2), 1549–1562. https://doi.org/10.1007/s11947-024-03537-3, https://www.springer.com/journal/11947.

Liu, C., Li, N., Niu, L., Li, X., Feng, J., & Liu, Z. (2025b). The preparation, physicochemical characterization and antibacterial activity of Pectin/Zinc Alginate composite film. *Carbohydrate Research, 550*. https://doi.org/10.1016/j.carres.2025.109412, https://www.sciencedirect.com/science/journal/00086215.

Liu, Y., Mu, X., Wang, H., Wang, M., Zhang, L., Chen, G., & Wang, S. (2025). Extraction of betalaine from red pitaya peel using deep eutectic solvent and its research in films. *Food and Bioproducts Processing, 149*, 16–24. https://doi.org/10.1016/j.fbp.2024.11.005, https://www.sciencedirect.com/science/journal/09603085.

Liu, Y., Zhang, Y., Zhen, M., Wu, Y., Ma, M., Cheng, Y., & Jin, Y. (2023). Effect of catechin and tannins on the structural and functional properties of sodium alginate/gelatin/ poly(vinylalcohol) blend films. *Food Hydrocolloids, 135*. https://doi.org/10.1016/j.foodhyd.2022.108141, https://www.sciencedirect.com/science/journal/0268005X.

Luan, Q. Y., Wang, Y. S., Zhang, Y. X., Hu, X. T., & Chen, H. H. (2024). Mechanism of the G/M ratio and zein in enhancing the mechanical and hydrophobic properties of sodium alginate films. *International Journal of Biological Macromolecules, 280*. https://doi.org/10.1016/j.ijbiomac.2024.136079, https://www.sciencedirect.com/science/journal/01418130.

Makaremi, M., Yousefi, H., Cavallaro, G., Lazzara, G., Goh, C. B. S., Lee, S. M., Solouk, A., & Pasbakhsh, P. (2019). Safely dissolvable and healable active packaging films based on alginate and pectin. *Polymers, 11*(10). https://doi.org/10.3390/polym11101594, https://res.mdpi.com/d_attachment/polymers/polymers-11-01594/article_deploy/polymers-11-01594.pdf.

Medina-Jaramillo, C., Quintero-Pimiento, C., Gómez-Hoyos, C., Zuluaga-Gallego, R., & López-Córdoba, A. (2020). Alginate-edible coatings for application on wild andean blueberries (*Vaccinium meridionale* Swartz): Effect of the addition of nanofibrils isolated from cocoa by-products. *Polymers, 12*(4), 824. https://doi.org/10.3390/polym12040824.

Merenkova, S., & Zinina, O. (2024). Effect of bioactive packaging materials based on sodium alginate and protein hydrolysates on the quality and safety of refrigerated chicken meat. *Polymers, 16*(23), 3430. https://doi.org/10.3390/polym16233430.

Motelica, L., Ficai, D., Oprea, O. C., Ficai, A., Ene, V. L., Vasile, B. S., Andronescu, E., & Holban, A. M. (2021). Antibacterial biodegradable films based on alginate with silver nanoparticles and lemongrass essential oil–innovative packaging for cheese. *Nanomaterials, 11*(9). https://doi.org/10.3390/nano11092377, https://www.mdpi.com/2079-4991/11/9/2377/pdf.

Nasaj, M., Chehelgerdi, M., Asghari, B., Ahmadieh-Yazdi, A., Asgari, M., Kabiri-Samani, S., Sharifi, E., & Arabestani, M. (2024). Factors influencing the antimicrobial mechanism of chitosan action and its derivatives: A review. *International Journal of Biological Macromolecules, 277*. https://doi.org/10.1016/j.ijbiomac.2024.134321, https://www.sciencedirect.com/science/journal/01418130.

Navarro, R., Arancibia, C., Herrera, M. L., & Matiacevich, S. (2016). Effect of type of encapsulating agent on physical properties of edible films based on alginate and thyme oil. *Food and Bioproducts Processing, 97*, 63–75. https://doi.org/10.1016/j.fbp.2015.11.001, http://www.elsevier.com/wps/find/journaldescription.cws_home/713883/description#description.

Neetoo, H., Ye, M., & Chen, H. (2010). Bioactive alginate coatings to control *Listeria monocytogenes* on cold-smoked salmon slices and fillets. *International Journal of Food Microbiology, 136*(3), 326–331. https://doi.org/10.1016/j.ijfoodmicro.2009.10.003.

Ngoc Le, T. A., & Chen, W. N. (2024). Composite films produced from upcycling of tropical fruit seeds are capable of monitoring shrimp freshness. *Food and Humanity, 2*. https://doi.org/10.1016/j.foohum.2024.100234, https://www.sciencedirect.com/journal/food-and-humanity.

Ning, H., Lu, L., Zhang, Y., Pan, L., & Lu, L. (2025). Development of novel sodium alginate-based light-responsive controlled-release active packaging film. *International Journal of Biological Macromolecules, 304*. https://doi.org/10.1016/j.ijbiomac.2025.140780, https://www.sciencedirect.com/science/journal/01418130.

Olivas, G. I., & Barbosa-Cánovas, G. V. (2008). Alginate–calcium films: Water vapor permeability and mechanical properties as affected by plasticizer and relative humidity. *LWT - Food Science and Technology, 41*(2), 359–366. https://doi.org/10.1016/j.lwt.2007.02.015.

Paixão, L. C., Lopes, I. A., Barros Filho, A. K. D., & Santana, A. A. (2019). Alginate biofilms plasticized with hydrophilic and hydrophobic plasticizers for application in food packaging. *Journal of Applied Polymer Science, 136*(48). https://doi.org/10.1002/app.48263, http://onlinelibrary.wiley.com/journal/10.1002/(ISSN)1097-4628.

Pavlath, A. E., Gossett, C., Camirand, W., & Robertson, G. H. (1999). Ionomeric films of alginic acid. *Journal of Food Science, 64*(1), 61–63. https://doi.org/10.1111/j.1365-2621.1999.tb09861.x.

Peretto, G., Du, W. X., Avena-Bustillos, R. J., Berrios, J. D. J., Sambo, P., & McHugh, T. H. (2014). Optimization of antimicrobial and physical properties of alginate coatings containing carvacrol and methyl cinnamate for strawberry application. *Journal of Agricultural and Food Chemistry, 62*(4), 984–990. https://doi.org/10.1021/jf4042886.

Pinpru, N., & Woramongkolchai, S. (2020). Crosslinking effects on alginate/carboxymethyl cellulose packaging film properties. *Chiang Mai Journal of Science, 47*(4), 712–722. https://epg.science.cmu.ac.th/ejournal/journalDetail.php?journal_id=11100.

Pournaki, S. K., Aleman, R. S., Hasani-Azhdari, M., Marcia, J., Yadav, A., & Moncada, M. (2024). Current review: Alginate in the food applications. *J, 7*(3), 281–301. https://doi.org/10.3390/j7030016.

Priyadarshi, R., Kim, H. J., & Rhim, J. W. (2021). Effect of sulfur nanoparticles on properties of alginate-based films for active food packaging applications. *Food Hydrocolloids, 110*. https://doi.org/10.1016/j.foodhyd.2020.106155, https://www.sciencedirect.com/science/journal/0268005X.

Qiu, L., Zhang, M., Huang, M., Chitrakar, B., & Chang, L. (2023). Fabrication of sodium alginate/apricot peel pectin films incorporated with rose anthocyanin-rich extract for monitoring grass carp (*Ctenopharyngodon idellus*) freshness. *Food Packaging and Shelf Life, 39*. https://doi.org/10.1016/j.fpsl.2023.101150, http://www.journals.elsevier.com/food-packaging-and-shelf-life/.

Rangel-Marrón, M., Mani-López, E., Palou, E., & López-Malo, A. (2019). Effects of alginate-glycerol-citric acid concentrations on selected physical, mechanical, and barrier properties of papaya puree-based edible films and coatings, as evaluated by response surface methodology. *LWT, 101*, 83–91. https://doi.org/10.1016/j.lwt.2018.11.005.

Raposo, A. Kd. S., Paixão, L. C., Rocha, A. A., Lopes, I. A., Santos, G. A. S., Ribeiro, G. A. C., de Menezes, A. S., Barros Filho, A. K. D., & Santana, A. A. (2021). Characterization of biodegradable films produced from mixtures of alginate, starch and babassu fibers. *Journal of Polymers and the Environment, 29*(4), 1212–1226. https://doi.org/10.1007/s10924-020-01952-z, http://www.kluweronline.com/issn/1566-2543/.

Rinaudo, M. (2008). Main properties and current applications of some polysaccharides as biomaterials. *Polymer International, 57*(3), 397–430. https://doi.org/10.1002/pi.2378.

Robles-Sánchez, R. M., Rojas-Graü, M. A., Odriozola-Serrano, I., González-Aguilar, G., & Martin-Belloso, O. (2013). Influence of alginate-based edible coating as carrier of antibrowning agents on bioactive compounds and antioxidant activity in fresh-cut Kent mangoes. *LWT - Food Science and Technology, 50*(1), 240–246. https://doi.org/10.1016/j.lwt.2012.05.021.

Rojas-Graü, M. A., Tapia, M. S., Rodríguez, F. J., Carmona, A. J., & Martin-Belloso, O. (2007). Alginate and gellan-based edible coatings as carriers of antibrowning agents applied on fresh-cut Fuji apples. *Food Hydrocolloids, 21*(1), 118–127. https://doi.org/10.1016/j.foodhyd.2006.03.001.

Ruan, P., Zhang, K., Zhang, W., Kong, Y., Zhou, Y., Yao, B., Wang, Y., & Wang, Z. (2024). Polyphenolic truxillic acid crosslinked sodium alginate film with notable antimicrobial and biodegradable properties for food packaging. *International Journal of Biological Macromolecules, 279*. https://doi.org/10.1016/j.ijbiomac.2024.135184, https://www.sciencedirect.com/science/journal/01418130.

Ruchika, S. K., Sharma, R., Kumar, S. K., Yadav, A., & Saneja (2025). Development of a multifunctional and sustainable pterostilbene nanoemulsion incorporated chitosan-alginate food packaging film for shiitake mushroom preservation. *International Journal of Biological Macromolecules, 293*. https://doi.org/10.1016/j.ijbiomac.2024.139241, https://www.sciencedirect.com/science/journal/01418130.

Santos, L. G., Alves-Silva, G. F., & Martins, V. G. (2022). Active-intelligent and biodegradable sodium alginate films loaded with *Clitoria ternatea* anthocyanin-rich extract to preserve and monitor food freshness. *International Journal of Biological Macromolecules, 220*, 866–877. https://doi.org/10.1016/j.ijbiomac.2022.08.120, www.elsevier.com/locate/ijbiomac.

Santos, L. G., & Martins, V. G. (2024). Multifunctional alginate films blended with polyphenol-rich extract from unconventional edible sources: Bioactive properties, UV-light protection, and food freshness monitoring. *International Journal of Biological Macromolecules, 262*. https://doi.org/10.1016/j.ijbiomac.2024.130001, https://www.sciencedirect.com/science/journal/01418130.

Santos, L. G., Silva, G. F. A., Gomes, B. M., & Martins, V. G. (2021). A novel sodium alginate active films functionalized with purple onion peel extract (*Allium cepa*). *Biocatalysis and Agricultural Biotechnology, 35*. https://doi.org/10.1016/j.bcab.2021.102096, http://www.journals.elsevier.com/biocatalysis-and-agricultural-biotechnology/.

Saswat, O., & Maheshwari, P. H. (2023). Development of biodegradable calcium alginate films for packaging applications. *Emergent Materials, 6*(1), 197–209. https://doi.org/10.1007/s42247-022-00411-w, https://www.springer.com/journal/42247.

Satriaji, K. P., Garcia, C. V., Kim, G. H., Shin, G. H., & Kim, J. T. (2020). Antibacterial bionanocomposite films based on $CaSO_4$-crosslinked alginate and zinc oxide nanoparticles. *Food Packaging and Shelf Life, 24*. https://doi.org/10.1016/j.fpsl.2020.100510, http://www.journals.elsevier.com/food-packaging-and-shelf-life/.

Senturk Parreidt, T., Lindner, M., Rothkopf, I., Schmid, M., & Müller, K. (2019). The development of a uniform alginate-based coating for cantaloupe and strawberries and the characterization of water barrier properties. *Foods, 8*(6), 203. https://doi.org/10.3390/foods8060203.

Senturk Parreidt, T., Müller, K., & Schmid, M. (2018). Alginate-based edible films and coatings for food packaging applications. *Foods, 7*(10), 170. https://doi.org/10.3390/foods7100170.

Sepasi, S., Nouri, L., Shakerardekani, A., & Mohammadi-Moghadam, M. (2025). Effect of di-1-p-menthene (pinolene) and sodium alginate biodegradable natural polymers on the quality characteristics and shelf life of fresh pistachios. *Journal of Stored Products Research, 111*. https://doi.org/10.1016/j.jspr.2025.102558, https://www.sciencedirect.com/science/journal/0022474X.

Shah, Y. A., Bhatia, S., Al-Harrasi, A., & Khan, T. S. (2025). Chitosan/sodium alginate/ethyl cellulose-based multilayer film incorporated with L-ascorbic acid for improved barrier and antioxidant properties. *International Journal of Biological Macromolecules, 284*. https://doi.org/10.1016/j.ijbiomac.2024.138169, https://www.sciencedirect.com/science/journal/01418130.

Shi, S., Ren, Y., Zhang, H., Pan, N., Xu, X., & Xia, X. (2024). Sodium alginate-based indicator film with enhanced physicochemical properties induced by cellulose nanocrystals and monitor the freshness of chilled meat. *International Journal of Biological Macromolecules, 278*. https://doi.org/10.1016/j.ijbiomac.2024.134631, https://www.sciencedirect.com/science/journal/01418130.

Shi, S., Wu, X., Wang, Y., Li, W., Zhang, H., Lou, X., Xia, X., & Liang, W. (2024). Sodium-alginate-based indicator film containing a hydrophobic nanosilica layer for monitoring fish freshness. *International Journal of Biological Macromolecules, 265*. https://doi.org/10.1016/j.ijbiomac.2024.130714, https://www.sciencedirect.com/science/journal/01418130.

Shi, S., Xu, X., Feng, J., Ren, Y., Bai, X., & Xia, X. (2023). Preparation of NH3- and H2S-sensitive intelligent pH indicator film from sodium alginate/black soybean seed coat anthocyanins and its use in monitoring meat freshness. *Food Packaging and Shelf Life, 35*. https://doi.org/10.1016/j.fpsl.2022.100994, http://www.journals.elsevier.com/food-packaging-and-shelf-life/.

Shi, S., Xu, X., Ren, Y., Zhang, H., Du, X., Li, H., & Xia, X. (2023). Beeswax coating improves the hydrophobicity of sodium alginate/ anthocyanin/ cellulose nanocrystal indicator film. *Food Hydrocolloids, 144*. https://doi.org/10.1016/j.foodhyd.2023.108930, https://www.journals.elsevier.com/food-hydrocolloids.

Shi, Y., Li, B., Yuan, J., Qin, Y., Zhou, L., Zhong, J., Liu, W., & Liu, C. (2025). Controlled release active films prepared from temperature responded emulsions and their application in grape preservation. *Colloids and Surfaces A: Physicochemical and Engineering Aspects, 709*. https://doi.org/10.1016/j.colsurfa.2025.136190, https://www.sciencedirect.com/science/journal/09277757.

Silva, M. Ad, Bierhalz, A. C. K., & Kieckbusch, T. G. (2009). Alginate and pectin composite films crosslinked with Ca^{2+} ions: Effect of the plasticizer concentration. *Carbohydrate Polymers, 77*(4), 736–742. https://doi.org/10.1016/j.carbpol.2009.02.014.

Silva de Farias, N., Silva, B., de Oliveira Costa, A. C., & Müller, C. M. O. (2021). Alginate based antioxidant films with yerba mate (*Ilex paraguariensis* St. Hil.): Characterization and kinetics of phenolic compounds release. *Food Packaging and Shelf Life, 28*, 100548. https://doi.org/10.1016/j.fpsl.2020.100548.

Singh, P., Baisthakur, P., & Yemul, O. S. (2020). Synthesis, characterization and application of crosslinked alginate as green packaging material. *Heliyon, 6*(1). https://doi.org/10.1016/j.heliyon.2019.e03026, http://www.journals.elsevier.com/heliyon/.

Sipahi, R. E., Castell-Perez, M. E., Moreira, R. G., Gomes, C., & Castillo, A. (2013). Improved multilayered antimicrobial alginate-based edible coating extends the shelf life of fresh-cut watermelon (*Citrullus lanatus). LWT - Food Science and Technology, 51*(1), 9–15. https://doi.org/10.1016/j.lwt.2012.11.013.

Smeti, S., Tibaoui, S., Koubaier, H. B. H., Lakoud, A., & Atti, N. (2025). Combined effects of alginate based active edible coatings and irradiation treatment on the quality characteristics of Beef Meat at 2°C. *Applied Food Research, 5*(1). https://doi.org/10.1016/j.afres.2025.100743, https://www.sciencedirect.com/science/journal/27725022.

Song, H. G., Lee, S., & Han, J. (2024). Enhanced oxygen barrier properties of sodium alginate coatings in humid environments: Ionic crosslinking of sodium alginate by calcium ions released from calcium hydrogen phosphate and calcium carbonate. *Food Science and Biotechnology*. https://doi.org/10.1007/s10068-024-01711-z, https://www.springer.com/journal/10068.

Sritweesinsub, W., Charuchinda, S., Zhou, J., & Adiguzel, O. (2015). Alginate/carboxymethyl cellulose hydrogel films in relation to crosslinking with glutaraldehyde and copper sulfate. *MATEC Web of Conferences, 30*, 02005. https://doi.org/10.1051/matecconf/20153002005.

Tavassoli-Kafrani, E., Shekarchizadeh, H., & Masoudpour-Behabadi, M. (2016). Development of edible films and coatings from alginates and carrageenans. *Carbohydrate Polymers, 137*, 360–374. https://doi.org/10.1016/j.carbpol.2015.10.074, http://www.elsevier.com/wps/find/journaldescription.cws_home/405871/description#description.

Taverna, M. E., Busatto, C. A., Saires, P. J., Bertero, M. P., Sedran, U. A., & Estenoz, D. A. (2022). Bio-composite films based on alginate and rice husk tar microparticles loaded with eugenol for active packaging. *Waste and Biomass Valorization, 13*(6), 3061–3070. https://doi.org/10.1007/s12649-022-01679-z, http://www.springer.com/engineering/journal/12649.

Thiviya, P., Gamage, A., Liyanapathiranage, A., Makehelwala, M., Dassanayake, R. S., Manamperi, A., Merah, O., Mani, S., Koduru, J. R., & Madhujith, T. (2023). Algal polysaccharides: Structure, preparation and applications in food packaging. *Food Chemistry, 405*. https://doi.org/10.1016/j.foodchem.2022.134903, www.elsevier.com/locate/foodchem.

Vargas-Ramella, M., da Silva, D., Dilarri, G., Zortea, A. V. L., Mendes, C. R., de Souza Laurentino, G., Campagnol, P. C. B., de Oliveira, A. F., & da Silveira, C. B. (2025). Quality effects of sodium alginate coating cross-linked with $CaCl_2$ on Mugil liza fillets during storage. *Food Control, 170*. https://doi.org/10.1016/j.foodcont.2024.111048, https://www.sciencedirect.com/science/journal/09567135.

Vasiljevic, Z. Z., Vunduk, J., Dojcinovic, M. P., Miskovic, G., Tadic, N. B., Vidic, J., & Nikolic, M. V. (2024). ZnO and Fe_2TiO_5 nanoparticles obtained by green synthesis as active components of alginate food packaging films. *Food Packaging and Shelf Life, 43*. https://doi.org/10.1016/j.fpsl.2024.101280, https://www.sciencedirect.com/science/journal/22142894.

Vieira, T. M., Moldão-Martins, M., & Alves, V. D. (2021). Design of chitosan and alginate emulsion-based formulations for the production of monolayer crosslinked edible films and coatings. *Foods, 10*(7). https://doi.org/10.3390/foods10071654, https://www.mdpi.com/2304-8158/10/7/1654/pdf.

Vizzini, P., Beltrame, E., Zanet, V., Vidic, J., & Manzano, M. (2020). Development and evaluation of qpcr detection method and zn-mgo/alginate active packaging for controlling *Listeria monocytogenes* contamination in cold-smoked salmon. *Foods, 9*(10). https://doi.org/10.3390/foods9101353, https://www.mdpi.com/2304-8158/9/10/1353.

Wang, C., & Liu, C. (2024). A pH-sensitive intelligent packaging film harnessing anthocyanin for food freshness monitoring. *Food and Bioprocess Technology, 17*(12), 5312–5323. https://doi.org/10.1007/s11947-024-03431-y, https://www.springer.com/journal/11947.

Wang, C., Song, Z., Cao, Y., Han, L., Yu, Q., Han, G., & Zhu, X. (2024). Characterization of sodium alginate-carrageenan films prepared by adding peanut shell flavonoids as an antioxidant: Application in chilled pork preservation. *International Journal of Biological Macromolecules, 266*. https://doi.org/10.1016/j.ijbiomac.2024.131081, https://www.sciencedirect.com/science/journal/01418130.

Wang, H., Liu, J., Fan, X., Ren, J., Liu, Q., & Kong, B. (2022). Fabrication, characterisation, and application of green crosslinked sodium alginate hydrogel films by natural crab-shell powders to achieve drug sustained release. *LWT, 171*. https://doi.org/10.1016/j.lwt.2022.114147, https://www.journals.elsevier.com/lwt.

Wang, J., Song, T., Chen, H., Ming, W., Cheng, Z., Liu, J., Liang, B., Wang, Y., & Wang, G. (2022). Bioinspired high-strength montmorillonite-alginate hybrid film: The effect of different divalent metal cation crosslinking. *Polymers, 14*(12). https://doi.org/10.3390/polym14122433, https://www.mdpi.com/2073-4360/14/12/2433/pdf?version=1655353042.

Wang, S., Han, X., Chen, Y., Shao, Y., Song, J., Yang, C., Jiang, S., Bai, R., Ye, X., & Ding, W. (2025). Preparation of sodium alginate/pectin/cellulose nanofibrils films containing black soybean seed coat anthocyanins for monitoring goat meat freshness. *Food Control, 171*. https://doi.org/10.1016/j.foodcont.2024.111096, https://www.sciencedirect.com/science/journal/09567135.

Wang, S., Li, R., Han, M., Zhuang, D., & Zhu, J. (2023). Intelligent active films of sodium alginate and konjac glucomannan mixed by *Lycium ruthenicum* anthocyanins and tea polyphenols for milk preservation and freshness monitoring. *International Journal of Biological Macromolecules, 253*. https://doi.org/10.1016/j.ijbiomac.2023.126674, https://www.sciencedirect.com/journal/international-journal-of-biological-macromolecules.

Xiao, J., Zhao, K., Tian, X., Song, Q., Zhang, K., Ma, Y., & Wang, W. (2021). Impact of calcium-carboxylate interactions in cellulose nanofiber reinforced alginate based film with triple-decker-like structure. *LWT, 151*. https://doi.org/10.1016/j.lwt.2021.112197, https://www.journals.elsevier.com/lwt.

Yan, Z., Zhang, S., Lin, S., Ding, N., Qiang, J., Xie, T., & Zhang, S. (2024). An intelligent indicator film based on food derived materials for real-time visual monitoring of salmon freshness. *Journal of Aquatic Food Product Technology, 33*(9), 734–746. https://doi.org/10.1080/10498850.2024.2424843, http://www.tandfonline.com/toc/wafp20/current.

Yang, J., Fei, T., Zhang, W., & Cong, X. (2023). Tannic acid and Ca^{2+} double-crosslinked alginate films for passion fruit preservation. *Foods, 12*(21). https://doi.org/10.3390/foods12213936, www.mdpi.com/journal/foods.

Yang, M., Kou, X., Gao, Y., Zhang, B., Liu, Q., Ma, Y., Xia, Y., Qu, Y., & Tan, Y. (2024). Development of a novel curcumin loaded lysine modified alginate intelligent film reinforced by lignosulfonic acid with high color stability for monitoring shrimp freshness. *Food Packaging and Shelf Life, 43*. https://doi.org/10.1016/j.fpsl.2024.101278, https://www.sciencedirect.com/science/journal/22142894.

Ye, Z., Ma, P., Tang, M., Li, X., Zhang, W., Hong, X., Chen, X., & Chen, D. (2017). Interactions between calcium alginate and carrageenan enhanced mechanical property of a natural composite film for general packaging application. *Polymer Bulletin, 74*(8), 3421–3429. https://doi.org/10.1007/s00289-016-1894-x.

Yerramathi, B. B., Kola, M., Annem Muniraj, B., Aluru, R., Thirumanyam, M., & Zyryanov, G. V. (2021). Structural studies and bioactivity of sodium alginate edible films fabricated through ferulic acid crosslinking mechanism. *Journal of Food Engineering, 301*. https://doi.org/10.1016/j.jfoodeng.2021.110566, http://www.sciencedirect.com/science/journal/02608774.

Yildirim, S., Röcker, B., Pettersen, M. K., Nilsen-Nygaard, J., Ayhan, Z., Rutkaite, R., Radusin, T., Suminska, P., Marcos, B., & Coma, V. (2018). Active packaging applications for food. *Comprehensive Reviews in Food Science and Food Safety, 17*(1), 165–199. https://doi.org/10.1111/1541-4337.12322, http://www.interscience.wiley.com/jpages/1541-4337.

Yin, C., Huang, C., Wang, J., Liu, Y., Lu, P., & Huang, L. (2019). Effect of chitosan- and alginate-based coatings enriched with cinnamon essential oil microcapsules to improve the postharvest quality of mangoes. *Materials, 12*(13). https://doi.org/10.3390/ma12132039, https://res.mdpi.com/materials/materials-12-02039/article_deploy/materials-12-02039.pdf?filename=&attachment=1.

Yin, X., Zhou, Y., Tang, Y., Kong, D., Xiao, W., Gan, L., Huang, J., & Zhang, Y. (2024). A crosslinked and percolation network alginate coating for litchi prevention. *International Journal of Biological Macromolecules, 275*. https://doi.org/10.1016/j.ijbiomac.2024.133252, https://www.sciencedirect.com/science/journal/01418130.

Zeng, Y., Wang, Y., Tang, J., Zhang, H., Dai, J., Li, S., Yan, J., Qin, W., & Liu, Y. (2022). Preparation of sodium alginate/konjac glucomannan active films containing lycopene microcapsules and the effects of these films on sweet cherry preservation. *International Journal of Biological Macromolecules, 215*, 67–78. https://doi.org/10.1016/j.ijbiomac.2022.06.085, www.elsevier.com/locate/ijbiomac.

Zhang, J., Huang, X., Shi, J., Liu, L., Zhang, X., Zou, X., Xiao, J., Zhai, X., Zhang, D., Li, Y., & Shen, T. (2021). A visual bi-layer indicator based on roselle anthocyanins with high hydrophobic property for monitoring griskin freshness. *Food Chemistry, 355*. https://doi.org/10.1016/j.foodchem.2021.129573, www.elsevier.com/locate/foodchem.

Zhang, J., Zhang, J., Huang, X., Zhai, X., Li, Z., Shi, J., Sobhy, R., Khalifa, I., & Zou, X. (2025). Lemon-derived carbon quantum dots incorporated guar gum/sodium alginate films with enhanced the preservability for blanched asparagus active packaging. *Food Research International, 202*. https://doi.org/10.1016/j.foodres.2025.115736, https://www.sciencedirect.com/science/journal/09639969.

Zhang, J., Zhang, J., Zhang, X., Huang, X., Shi, J., Sobhy, R., Khalifa, I., & Zou, X. (2024). Ammonia-responsive colorimetric film of phytochemical formulation (Alizarin) grafted onto ZIF-8 carrier with poly(vinyl alcohol) and sodium alginate for beef freshness monitoring. *Journal of Agricultural and Food Chemistry, 72*(20), 11706–11715. https://doi.org/10.1021/acs.jafc.4c02227, http://pubs.acs.org/journal/jafcau.

Zhang, K., Zhang, W., Ruan, P., Zhou, Y., Yao, B., Wang, Y., & Wang, Z. (2024). Fabrication and characterization of truxillic alginate bioplastic with decent antibacterial, antioxidant and degradability properties in food packaging. *Materials Today Communications, 41*. https://doi.org/10.1016/j.mtcomm.2024.111022, https://www.sciencedirect.com/science/journal/23524928.

Zhang, X., Li, Y., Li, Y., & Li, C. (2025). Development of a high-barrier, antimicrobial, and pH-sensitive nanocomposite film based on gellan gum/sodium carboxymethylcellulose/sodium alginate with nano-SiO_2 for strawberry preservation and monitoring. *International Journal of Biological Macromolecules, 298*. https://doi.org/10.1016/j.ijbiomac.2025.139983, https://www.sciencedirect.com/science/journal/01418130.

Zhangyao, L., Hu, L., Ran, Y., Qingmin, C., Maorun, F., & Xiaofei, X. (2025). The composite edible film of sodium alginate and natamycin maintained color of pedicel and calyx of tomato fruit. *Postharvest Biology and Technology, 222*, 113413. https://doi.org/10.1016/j.postharvbio.2025.113413.

Zhao, M., Nuerjiang, M., Bai, X., Feng, J., Kong, B., Sun, F., Li, Y., & Xia, X. (2022). Monitoring dynamic changes in chicken freshness at 4 °C and 25 °C using pH-sensitive intelligent films based on sodium alginate and purple sweet potato peel extracts. *International Journal of Biological Macromolecules, 216*, 361–373. https://doi.org/10.1016/j.ijbiomac.2022.06.198, www.elsevier.com/locate/ijbiomac.

Zheng, H., Yang, J., & Han, S. (2016). The synthesis and characteristics of sodium alginate/graphene oxide composite films crosslinked with multivalent cations. *Journal of Applied Polymer Science, 133*(27). https://doi.org/10.1002/app.43616, http://onlinelibrary.wiley.com/journal/10.1002/(ISSN)1097-4628.

Zhong, Y., Yang, M., Chen, J., Mi, H., Ge, Y., Lv, J., & Li, J. (2023). Pre-crosslinking with putrescine improves mechanical and thermal properties of alginate film. *Journal of Food Engineering, 340*. https://doi.org/10.1016/j.jfoodeng.2022.111314, http://www.sciencedirect.com/science/journal/02608774.

Zhou, T., Wang, H., Han, Q., Song, Z., Yu, D., Li, G., Liu, W., Dong, C., Ge, S., & Chen, X. (2024). Fabrication and characterization of an alginate-based film incorporated with cinnamaldehyde for fruit preservation. *International Journal of Biological Macromolecules, 274*. https://doi.org/10.1016/j.ijbiomac.2024.133398, https://www.sciencedirect.com/science/journal/01418130.

Zhu, H., Qin, X., Wang, Q., & Zhong, J. (2024). Improved barrier, mechanical and antioxidant properties of pH-responsive film by incorporating dialdehyde starch and anthocyanins into rice protein/sodium alginate. *International Journal of Biological Macromolecules, 281*. https://doi.org/10.1016/j.ijbiomac.2024.136131, https://www.sciencedirect.com/science/journal/01418130.

Chapter 9

Pullulan based film coatings

Ana Caroline Ferreira Carvalho[1,2], Elane Schwinden Prudêncio[2,3], Karine Marafon[2], Amanda Alves Prestes[2], and Carolina Krebs de Souza[1]

[1]*Department of Chemical Engineering, University of Blumenau, Blumenau, Santa Catarina, Brazil,* [2]*Department of Food Engineering, Federal University of Santa Catarina, Florianópolis, Santa Catarina, Brazil,* [3]*Department of Food Science and Technology, Federal University of Santa Catarina, Federal University of Santa Catarina, Florianópolis, Santa Catarina, Brazil*

9.1 Introduction

Pullulan is a biodegradable polysaccharide produced by the fungus *Aureobasidium pullulans*, which was discovered in 1938 and characterized in the 1950s. With a chemical formula of $(C_6H_{10}O_5)_n$, its structure is composed of maltotriose units connected by α-(1,6) and α-(1,4) bonds, conferring properties such as water solubility, formation of transparent films, and an excellent barrier against oxygen (Fig. 9.1). These characteristics make pullulan attractive for several industrial applications, including the pharmaceutical and food sectors, with an emphasis on food packaging and preservation (de Souza et al., 2023; Thomas et al., 2024). Pullulan is flexible due to its glycosidic bonds, allowing for chemical modifications for specific applications. It can be acetylated or carboxymethylated to improve properties such as heat resistance or solubility. Its biocompatibility is also widely explored in areas such as biomedicine and cosmetics, in addition to uses in pharmaceutical products, including controlled-release capsules (Carrigy et al., 2019; Sabater et al., 2019; Sugumaran & Ponnusami, 2017).

This polymer has a powdery appearance ranging from white to light yellow, with a smooth texture. It dissolves easily in water and has neutral sensory characteristics, being tasteless and odorless. In addition, it is edible, biodegradable, and recognized for its safety, as it is considered nontoxic, noncarcinogenic, nonimmunogenic, and nonmutagenic. These properties give it the status of "generally recognized as safe" (GRAS), making it a versatile and reliable option for a wide range of commercial applications (de Souza et al., 2023; Ku et al., 2022; Tiwari et al., 2019; Wani et al., 2021).

Several strains of *Aureobasidium* species have been utilized to produce pullulan from diverse carbon sources, including glucose, inulin, sucrose, xylose, and agro-industrial residues. Among these, *A. pullulans* MTCC1991 and *A. melanogenum* TN1–2 have demonstrated the highest pullulan yields, showcasing their efficiency in leveraging these substrates for optimal polymer production (Wei et al., 2021). During the process, *A. pullulans* converts the substrates into polymers through specific metabolic pathways, such as glycolysis and the pentose phosphate pathway, which generate the metabolic precursors necessary for polysaccharide synthesis. Optimized fermentation methods have been developed to increase efficiency and economic viability, making the process more sustainable and accessible. Fermentation is adjusted to control parameters such as pH, temperature, and incubation time, promoting greater productivity (de Souza et al., 2023; Ghosh et al., 2022).

Pullulan is valued as a sustainable alternative to synthetic polymers in the packaging sector due to its biodegradability and functional characteristics. Recent studies highlight its use as an edible coating and biodegradable film in food, extending the shelf life of products by creating effective barriers against moisture and oxygen (de Souza et al., 2023; Ghosh et al., 2022). These materials not only reduce food waste but also meet the demand for environmentally friendly packaging, aligned with the Sustainable Development Goals. Furthermore, recent research shows that integrating pullulan with other materials, such as nanocomposites or natural additives, can further improve its mechanical and barrier properties (Moris et al., 2022; Mulla et al., 2023). These innovations broaden the scope of application, making it competitive in markets such as bioplastics and advanced packaging.

Finally, pullulan represents a promising and sustainable approach to addressing environmental and industrial challenges, offering innovative solutions in diverse sectors. Its structural versatility and unique physicochemical properties highlight its increasing relevance in modern technologies and in replacing conventional polymers.

Natural Polymer-Based Films and Coatings in Food Technology. DOI: https://doi.org/10.1016/B978-0-443-34244-8.00011-2

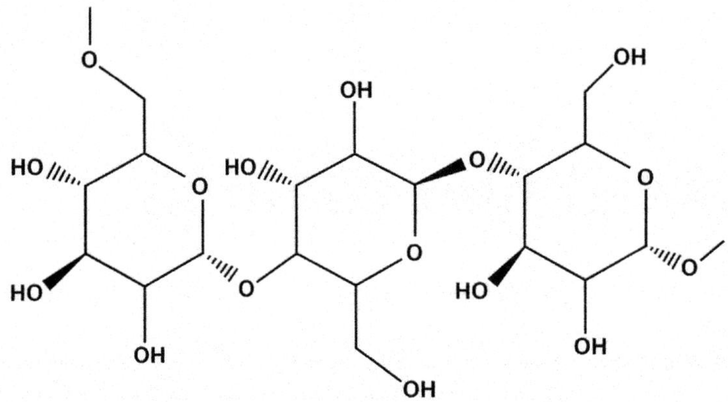

FIGURE 9.1 Chemical structure of pullulan.

9.2 Advantages and disadvantages of using pullulan in films and coatings

Pullulan is a biopolymer widely used in various applications, especially in producing films and coatings for food. This polymer stands out for its unique properties that make it ideal for such purposes. It exhibits high solubility, structural flexibility, transparency, fiber-forming ability, thermal resistance, and a broad viscosity range. Pullulan has excellent barrier properties against moisture, gases, and vapors, with a particular emphasis on its high oxygen resistance. By providing an oxygen barrier, pullulan protects food from oxidative processes that can affect color, flavor, odor, and nutritional properties. The use of pullulan not only enhances the sensory aspects of food but also extends its shelf life, preserving freshness, firmness, and flavor for longer periods. Its barrier against nonpolar substances further contributes to maintaining product quality. Although pullulan is excellent for film formation on its own, combining it with other materials can enhance its properties, resulting in films with improved mechanical strength and water vapor resistance. This makes it even more effective for preserving food products. Additionally, incorporating essential oils into pullulan films provides antimicrobial properties, further extending the shelf life of food and offering an innovative solution for the sustainable packaging market (de Souza et al., 2023; Li et al., 2020; Rashid et al., 2024; Sugumaran & Ponnusami, 2017; Tabasum et al., 2018).

Pullulan, in addition to being widely used as a coating and in producing films, stands out in an industrial market that is constantly growing. This biopolymer is especially valued as an additive in food products, significantly contributing to the improvement of quality aspects. It serves as a thickener, gelling agent, flocculant, stabilizer, texture enhancer, binding agent, and as a source of dietary fiber in low-calorie diets, thanks to its prebiotic properties.

Due to its excellent mechanical properties, pullulan enhances adhesion and gloss in food products without interfering with their visual, aromatic, or caloric characteristics. Furthermore, its water solubility facilitates both application and removal when necessary, making it a versatile and functional option for the food industry (Kraśniewska et al., 2019; Shanmugam et al., 2022; Tabasum et al., 2018; Yeasmin et al., 2020).

As a natural, biodegradable, and sustainable material, pullulan is widely recognized as safe. It possesses remarkable characteristics, such as being nontoxic, biocompatible, edible, tasteless, odorless, and free from carcinogenic, immunogenic, or mutagenic properties. These attributes have earned pullulan the status of GRAS by the Food and Drug Administration (FDA), making it approved for use in food products. In the European Union, it has been registered as a food additive under the number E1204. Pullulan stands out for its numerous advantages as a biopolymer in the production of films and coatings, serving as a sustainable and biodegradable alternative to conventional plastics. This substitution represents a significant step forward in the pursuit of more eco-friendly materials aligned with current environmental preservation demands (de Souza et al., 2023; Ghosh et al., 2022; Kanmani & Lim, 2013; Kraśniewska et al., 2019; Sugumaran & Ponnusami, 2017).

Although pullulan presents numerous advantages, its use still faces certain limitations that hinder its large-scale application. One of the main drawbacks is its high production cost, especially when compared to other polymers, which compromises its economic feasibility in industrial processes. When used alone, pullulan forms brittle films with low mechanical strength and rigidity, making them susceptible to breakage. Another relevant point is its high hydrophilicity, which results in films with high water vapor permeability. This can compromise the stability of food products, making them more vulnerable to deterioration. Additionally, pullulan-based films do not exhibit antimicrobial activity, which limits their ability to protect food against microbiological contamination. For these reasons, it is common to combine pullulan with other biopolymers or plasticizers to improve its flexibility, durability, and to optimize the film's mechanical and barrier properties (Adame et al., 2024; Farris et al., 2014; Goswami et al., 2021; Singh et al., 2019; Zheng et al., 2023).

9.3 Methods for preparing pullulan films

The primary purpose of using films and coatings in food preservation is to protect the product from environmental factors such as humidity, oxygen, and contaminants that can lead to deterioration. These protective layers create a barrier that slows down degradation processes, thus increasing the shelf life of the food. In addition, films and coatings can be formulated to gradually release antimicrobial agents or antioxidants, further enhancing the stability and quality of the food over time (Finardi et al., 2022).

Films are generally preformed, thin, and flexible layers that are applied to food products or used to wrap them. They are usually made from biodegradable polysaccharides (e.g., pullulan, pectin, chitosan, carrageenan, alginate, starch, cellulose, and xanthan gum), proteins (e.g., whey or soy protein, collagen, gelatin, gluten, and casein), and lipids (candelilla, paraffin, carnauba, beeswax, polyethylene, rice bran, ouricouri wax, and jojoba oil). In food packaging, films are physical barriers to external elements (moisture, gases, and contaminants) and can act as primary packaging. They can be enriched with active compounds, such as antimicrobials or antioxidants, to increase the shelf life of the food. The films are generally freestanding and can be removed from the surface of the food (Suhag et al., 2020).

Coatings are typically applied directly to the surface of food items by dipping, spraying, or brushing, forming an edible layer that adheres closely to the food. Unlike films, coatings are not freestanding and are often intended to be consumed with food. They protect the food from moisture loss, oxidative damage, and microbial contamination. Coatings can be formulated with bioactive ingredients to provide functional properties, such as delaying ripening or reducing spoilage, which are especially useful for fruits, vegetables, and meat products (Mueller et al., 2024; Santos et al., 2022).

9.3.1 Film techniques

The cohesion among film-forming materials is essential for achieving mechanical strength and a continuous, uniform structure; materials with weaker cohesion are more likely to produce films with reduced strength. The structural properties of film play a crucial role in guiding the film-forming process. Meanwhile, critical attributes of the final film (such as permeability to moisture and gases, elasticity, mechanical durability, rheological behavior, and color and light transmission) are shaped by the physical chemistry of the biopolymers. Although plasticizers are needed to enhance flexibility, they can also decrease cohesion among film components, leading to a structure with diminished durability. The method used to prepare pullulan films is casting (de Souza et al., 2023; Finardi et al., 2022; Hoffmann et al., 2022). The advantages and disadvantages of these techniques can be seen in Table 9.1.

9.3.1.1 Casting method

The casting method involves three steps to prepare a film from biopolymers. Pullulan is dissolved in water or another suitable solution, which may contain plasticizers and other additives. The first step is the solubilization of the film's raw materials in a suitable solvent, obtaining the film solution, and then drying the molten solution. In addition, this method has some applications and is used in industries, pilot-scale tests, and laboratory applications. The formulation of films starts with selecting a polymer or blend of raw materials to serve as the film's foundation. This polymer is then

TABLE 9.1 Advantages and disadvantages of film techniques.

Techniques	Advantages	Disadvantages	Reference
Casting method	Easy manufacturing. Low cost. No complex equipment is required. Formulates a cohesive matrix.	Temperature differences in solvent evaporation can interfere with material properties. The challenge in developing stable matrices for large-scale production.	Suhag et al. (2020)

solubilized, dissolved, or dispersed in a suitable solvent. The solution is poured into a specified mold, like a glass surface, through the casting method. During the drying phase, time is allowed for the solvent to evaporate entirely, creating a polymeric film that adheres to the mold. To aid in solvent evaporation during film casting, equipment such as hot air convection ovens, microwaves, air dryers, tray dryers, and vacuum dryers may be used, facilitating the removal of the film once the process is complete. It is essential to develop and study the formulation so that the matrix is cohesive and that the plasticizers used help to form an easily removable film with satisfactory and resistant mechanical characteristics, thermal stability, efficient barrier properties, and a uniform microstructure. In addition, the addition of plasticizers to the film matrix positively improves these properties (Suhag et al., 2020; Usman et al., 2016).

9.3.2 Coating techniques

The methods of coating applications on food depend on the purpose of the coating, the type of surface, and the nature of the food. The coating can be applied to meat, vegetables, fruit, etc. For this, it is necessary to know the nature of the food. Regarding vegetables and fruit, it is vital to understand and consider the degree of ripeness. The size of the food and the thickness, viscosity, and density of the coating must also be considered (Cazón et al., 2017).

Applying the coating solution involves adhesion and diffusion between the coating and the food surface. The wetting stage allows one to validate the coating's affinity for the product and determine the operating time (Finardi et al., 2022). The techniques applied are immersion, spraying, fluidized bed, and panning. Table 9.2 shows the advantages and disadvantages of these techniques.

9.3.2.1 *Immersion method*

This is a standard method for coating food. The food is first immersed in the pullulan coating solution, and at this stage, the speed is controlled (it must be constant), and the amount of solution coated must be controlled to ensure that the entire surface is covered. Then, the deposition step is carried out, which consists of removing the excess coating liquid and, finally, the evaporation of the coating solvents through heating and drying. The application of coatings imparts various characteristics to food. It can also help preserve and increase shelf life, improve flavor, and prevent moisture loss and undesirable external compounds from entering the food when desired. It also makes the handling process easier and quicker. One of the applications of coating is on fruits and vegetables. The aim is to stabilize the microbial load of fruit and vegetables (Santos et al., 2022; Suhag et al., 2020).

9.3.2.2 *Spraying method*

Spraying is a commonly applied technique that spreads droplets across the food's surface. The pullulan solution is sprayed onto the substrate using a spray system. This method is divided into three types: air spray atomization, air-assisted airless atomization, and pressure atomization. In air spray atomization, a high-speed air stream envelops a fluid that flows at a lower velocity from a tube (Huang et al., 2023; Suhag et al., 2020).

The interaction between the air and fluid increases the fluid's speed, disrupting its flow and resulting in atomization. This approach also employs air to disperse fine droplets onto the food's surface. Air jets break down the cylindrical water

TABLE 9.2 Advantages and disadvantages of coating techniques.

Techniques	Advantages	Disadvantage	Reference
Immersion method	Help with food preservation. Improves food characteristics. Common method. It makes it possible to create different functional formulations with specific characteristics for the foods to be coated.	Difficulties with food product respiration. Difficulty in dilution of coating.	Suhag et al. (2020)
Spraying method	High productivity. Quality coating. Temperature control, consistent coating, and automatic, continuous production.	Highly viscous solutions are complex to spray.	

stream into tiny droplets, enabling the coating to be applied to food items. Air-assisted airless atomization involves partially atomizing the fluid with a nozzle tip. Complete atomization is achieved with small injections of compressed air. Consequently, the result is a uniform spray pattern, making this technique suitable for coatings with high viscosity and high solids content. This technique allows heat and pressure in fluids with these characteristics. In pressure atomization, pressure is applied to the coating through small-diameter nozzles to provide surface tension and high coating viscosity onto the food. Parameters such as pressure, viscosity, and nozzle design affect spraying efficiency (Ahuja & Kumar Rastogi, 2023; Suhag et al., 2020).

The ideal pressure, up to 3.5 bars, prevents damage to the film-forming system and ensures the coating's thickness and mechanical properties. Drying times and temperatures influence film formation, as droplet size control and coating quality in spraying systems depend on the piston, nozzle, temperature, and air and fluid flow (Suhag et al., 2020).

9.4 Properties of pullulan films

Recognized as one of the most researched biopolymers in the field of sustainable packaging, pullulan's performance has been improved through various strategies (Das et al., 2023). This includes the integration of natural fillers for reinforcement, blending with other compounds, and the application of physical and chemical modifications.

These advances enable customized packaging solutions for specific needs (Thomas et al., 2024) and ensure that pullulan films maintain their barrier properties during handling, storage, and transportation, increasing their effectiveness in protecting products against environmental factors (Ezati et al., 2023).

Pullulan is particularly valued for its ability to form flexible, thin, and strong films with interesting mechanical integrity (Rashid et al., 2024), and it is also considered a polysaccharide of potential relevance for application in the food sector, as it is an edible, nontoxic, nonhygroscopic, and nonmutagenic polymer (Yang et al., 2020).

However, using pullulan can present some disadvantages in terms of its barrier properties, mechanical resistance, and biological functionality (Agrawal et al., 2022; Roy, & Min, Biswas, et al., 2023).

To reduce these limitations, research has been dedicated to studying chemical modifications of pullulan to improve or confer more advantageous properties in the postharvest preservation of fruits and vegetables (Ghosh et al., 2022). Although progress in research and development allows for an increase in the production scale of biodegradable packaging, the production and application of pullulan-based packaging materials remain challenges (Aquinas et al., 2024).

Pullulan has been widely exploited as an oxygen barrier coating in commercial packaging, in addition to its role as a primary packaging material. Its edible nature makes it particularly appealing for food applications, where it can help extend shelf life, further highlighting its effectiveness in sustainable packaging solutions (Ghosh et al., 2022).

The water solubility of pullulan positions it as a versatile polysaccharide with wide applications in the industrial, food, pharmaceutical, and packaging sectors (Agrawal et al., 2022; Oliveira et al., 2023). Its linear, unbranched structure allows water molecules to interact easily with its glucose chains, leading to rapid dissolution. Even at lower temperatures, pullulan dissolves efficiently, forming clear, colorless solutions. The viscosity of these solutions varies with concentration, offering flexibility for use in different formulations (Reddy et al., 2022).

Films and coatings developed with pullulan exhibit promising oxygen barrier properties due to their stability in aqueous solutions across a wide pH range, low viscosity, and nongelling properties (Hernandez-Tenorio et al., 2024). These films can block oxygen primarily due to the dense packing of polymer chains in the film matrix, which creates a structure that inhibits the passage of oxygen molecules (Xiao, Lim, et al., 2012). Compared to other biopolymers and synthetic materials, pullulan stands out for its superior oxygen barrier capability, even with relatively low film thickness. This property is particularly beneficial for packaging oxygen-sensitive foods, such as fruits, vegetables, and processed products, where oxidation can lead to spoilage and quality loss (Kraśniewska et al., 2019).

While pullulan-based materials excel at blocking oxygen and other gases, their water vapor barrier properties are less robust. Due to their hydrophilic nature, pullulan-based formulations tend to absorb moisture, which can lead to increased water vapor permeability over time (Dong & Tian, 2024).

To mitigate this, researchers have explored blending pullulan with other materials or applying surface treatments to improve its moisture resistance without compromising its biodegradability or transparency (Tabasum et al., 2018). However, pullulan's water solubility can still be advantageous in specific applications, such as dissolvable packaging for single-use products (Thomas et al., 2024), including coatings for highly edible foods that can be readily cooked and consumed.

In addition to oxygen, pullulan films have demonstrated the ability to effectively block the passage of carbon dioxide and nitrogen. This versatility makes them suitable for modified atmosphere packaging (MAP) techniques, where controlling the internal gas composition is crucial for maintaining product freshness (Tajeddin et al., 2018).

Pullulan's barrier against volatile compounds and aromas also contributes to its use in packaging applications where odor retention or prevention is important, such as packaging for strongly smelling foods (Vasile & Baican, 2021).

Pullulan films have garnered interest from the industry and researchers due to their mechanical properties, including strength, elasticity, and durability, which make them ideal for a variety of packaging applications (Rostamabadi et al., 2024). These properties contribute to the functionality of the films as reliable barriers, protecting products from physical damage and environmental factors, such as air and moisture.

Pullulan films exhibit excellent tensile strength, withstanding significant forces without tearing or damage. This strength is essential in packaging, as it ensures the film effectively protects products during handling, storage, and transport without compromising its integrity (Shah et al., 2023). In addition to strength, pullulan films are also highly elastic. This flexibility allows the films to stretch and adapt to various shapes and packaging formats without losing functionality. Their elasticity allows them to conform to irregularly shaped products and to withstand minor deformations that may occur during transportation (Roy, & Min, Biswas, et al., 2023).

The elongation and tensile strength at break are essential when describing the mechanical properties of thin films used in food applications (Tan et al., 2022). The tensile strength of pure pullulan films is approximately 50 MPa. Some studies show that plasticizers used in film formulations increase flexibility and extensibility by improving the mobility of polymer chains and reducing intermolecular forces (Dewan et al., 2024).

Commonly used plasticizers, such as potassium sorbate, glycerol, triacetin, and vitamin E, reduce the tensile strength of pullulan-based films, and increasing the concentration of these plasticizers further decreases the tensile strength of these materials (Vieira et al., 2011).

The durability of pullulan films means they maintain their structural integrity over time, even when exposed to external factors such as moisture or mechanical pressure (Rostamabadi et al., 2024).

Although pullulan is sensitive to water, researchers are applying modifications to increase its water resistance, thus improving its long-term durability. These combined mechanical properties make pullulan films a strong, flexible, and resilient choice for protective packaging (Wypij et al., 2023).

Pullulan films and coatings can provide a high level of transparency, allowing consumers to see the product without compromising protection. In the food industry, the application of this material enhances visual appeal and consumer trust by demonstrating the freshness and quality of the packaged or coated product (Rashid et al., 2024).

The sensory properties of pullulan films (odorless and tasteless) ensure that there is no interference with the flavor, aroma, or overall sensory experience of the food in contact with the material, including edible films (Gniewosz et al., 2022). Moreover, pullulan's ability to form thin, flexible films and coatings minimizes sensory impact, allowing consumers to enjoy the product without noticing the presence of the packaging material (Nunes et al., 2023).

The barrier properties of pullulan films, along with their strength, elasticity, and durability, ensure that they provide excellent protection for packaged products, while their optical and sensory properties, such as transparency and lack of flavor impact, enhance their appeal, especially in the food industry (Fig. 9.2). Together, these characteristics position pullulan as a versatile material for sustainable and functional packaging solutions (de Souza et al., 2023).

9.4.1 Research data: pullulan application in food packaging

Pullulan is a natural polysaccharide that is widely studied for applications in biodegradable packaging due to its remarkable properties.

It has been noted that pullulan enhances the mechanical properties of films and coatings used in packaging. For instance, pullulan films combined with geranium essential oil exhibited changes in mechanical properties, including increased weight and thickness, along with a decrease in tensile strength, elongation, and elastic modulus. This suggests that incorporating essential oils can modulate the mechanical characteristics of films (Simões et al., 2024). Similarly, integrating apple fibers into pullulan films resulted in a significant increase in both the tensile strength index and elastic modulus of the films, indicating an improvement in mechanical properties (Luís et al., 2021).

Regarding oxygen barrier properties, it was observed that pullulan nanocomposites with graphene oxide performed well as oxygen barriers, even with low graphene oxide loadings, due to the homogeneous distribution and unidirectional alignment of the nanoparticles. Consequently, the exceptional oxygen barrier properties of pullulan nanocomposites provided greater mechanical flexibility and good optical transparency, enhancing the value of food packaging materials (Unalan et al., 2015). Additionally, pullulan was applied to pullulan-silica coatings, demonstrating excellent oxygen barrier properties and suitability for food packaging applications (Farris et al., 2012).

Antimicrobial and antioxidant properties have also been observed in the application of pullulan materials in combination with other substances. Pullulan films infused with biosynthesized silver nanoparticles exhibited significant

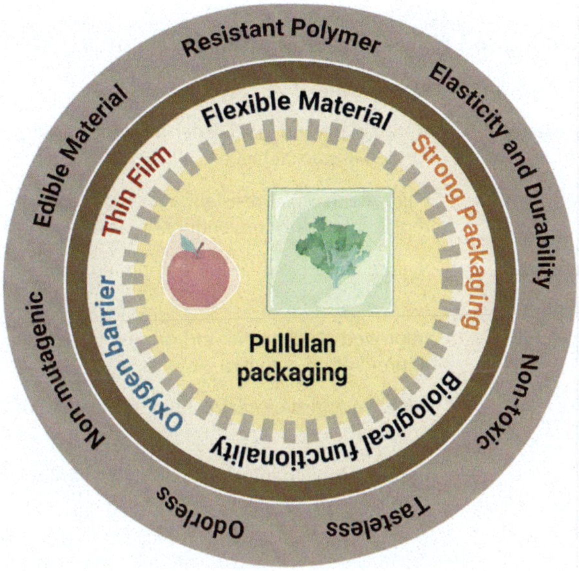

FIGURE 9.2 Properties of pullulan packaging.

antibacterial activity against food-borne pathogens, highlighting their potential to enhance the safety of packaged foods (Wypij et al., 2023). Incorporating *Cistus ladanifer* essential oil into pullulan films provided effective antimicrobial properties against both Gram-positive and Gram-negative bacteria, indicating a promising application in active food packaging (Luís et al., 2020). Additionally, pullulan and gelatin films infused with grape seed extract and sulfur nanoparticles exhibited significant antioxidant activity and antimicrobial properties, thereby broadening their applicability in food packaging (Roy, & Min, Biswas, et al., 2023).

In terms of sustainability, pullulan materials have stood out in food applications. Pullulan-based edible coatings applied to sliced cheese improve the product's shelf life, providing a sustainable alternative to traditional plastic packaging (Erceg et al., 2024).

These studies emphasize the versatility of pullulan as a material for biodegradable packaging, showcasing its strong mechanical properties, excellent oxygen barrier, and potential for functionalization with antimicrobial and antioxidant agents, making it a promising candidate for the development of sustainable packaging solutions.

9.5 Applications in food packaging

Recent innovations in food packaging have increasingly shifted toward the development of environmentally friendly and biodegradable alternatives to petroleum-based materials, which are major contributors to pollution. Pullulan, a natural polysaccharide produced through microbial fermentation, has shown considerable promise in applications (Table 9.3) such as edible films and coatings (Mahmud et al., 2024).

Originally used primarily as a stabilizer and thickener in food products, pullulan's role has expanded in recent years as its utility in the design of packaging materials has become more evident (de Souza et al., 2023). With attributes like biodegradability, odorlessness, tastelessness, transparency, homogeneity, and strong barrier properties—particularly in relation to oxygen and other gases—pullulan is being explored as a viable solution for modern and sustainable packaging technologies for food preservation (Wu et al., 2021).

In the development of pullulan-based packaging, this polysaccharide is typically combined with other active materials, such as plant extracts (polyphenols, terpenoids, and terpenes), essential oils (thymol, carvacrol), microorganisms (bacteriocins and functional enzymes), and animal extracts, to enhance the functionality and properties of the packaging (Rashid et al., 2024). In addition to its use as a packaging film, pullulan has also been applied in various other forms of food packaging materials for different applications and purposes (Ghosh et al., 2022).

Pullulan's preservation capacity has been demonstrated and improved when combined with other technologies. Strategies such as MAP, high hydrostatic pressure, edible coatings, active materials, the application of low temperatures, and storage with irradiation are alternatives to enhance the preservation of food products (Rashid et al., 2024).

TABLE 9.3 Application of pullulan in food films and coatings (Hirayama et al., 2022).

Types of materials	Main results	References
Pullulan/agar films with eucalyptus essential oil and rutin	Antioxidant and UV barrier properties and potential for active packaging.	Roy and Rhim (2023)
Pullulan films with starch nanoplatelets	High mechanical resistance and toughness, replacement for conventional plastics.	Zhang et al. (2023)
Pullulan/gelatin films with grape seed extract and sulfur nanoparticles	Significant antioxidant and antimicrobial activity.	Roy et al. (2022)
Pullulan films with titanium dioxide nanoparticles	Improved mechanical and oxygen barrier properties, antimicrobial activity.	Pereda et al. (2020)
Pullulan films with silver nanoparticles	Effective antimicrobial activity against *E. coli* and *S. aureus*.	Wypij et al. (2023)
Pullulan/CMC films with zinc oxide nanoparticles	Improved mechanical properties and antimicrobial activity.	Kanmani and Lim (2013)

Researchers have utilized pullulan as a reducing agent, stabilizer, and eco-friendly coating to synthesize silver nanoparticles (Kang et al., 2023). Ultrasound has been used to develop pullulan films for strawberry preservation. In addition, silver nanoparticles synthesized through ultrasonic irradiation and incorporated into the pullulan matrix have shown high effectiveness in preserving meat and other food products (Khan et al., 2019).

9.5.1 Pullulan-based edible coatings

An edible coating is a thin (<0.3 mm), consumable layer applied to the surface of food to protect it from spoilage, extend its shelf life, delay ripening, and preserve its freshness and nutritional content. These coatings can also enhance sensory qualities, like appearance and flavor (Nunes et al., 2023).

Made from natural compounds, which are often obtained from industrial waste, such as polysaccharides (e.g., pullulan, starch, or cellulose), lipids, proteins, or combinations thereof, and associated with different additives, edible coatings act as barriers to oxygen, moisture, and microbial contamination (Díaz-Montes & Castro-Muñoz, 2021). Edible coatings reduce moisture loss in fruits and vegetables, slow down oxidation in confectioneries, and improve the texture or appearance of processed foods. They provide a biodegradable, sustainable alternative to synthetic packaging, and are safe for consumption (Rios et al., 2022).

Edible coating is a material formed directly on food of plant or animal origin. Due to its unique properties, such as being tasteless, nontoxic, and biodegradable, it results in high-quality food without synthetic chemicals (Nunes et al., 2023). Because of the distinct characteristics of each type of food (fruit, vegetables, and meat), the quality and efficiency of each coating depend on the application methods used, such as spraying, brushing, foaming, wrapping, panning, dipping, and fluidized bed coating (Priya et al., 2023).

The incorporation of natural active compounds from agro-industrial waste into edible coatings is a significant trend in research on food packaging materials. Like other materials, edible coatings can contain active ingredients that not only help preserve food but also offer extended benefits to consumers (Yaashikaa et al., 2023).

For fresh or processed meats, pullulan-based coatings, mainly those containing antimicrobial or antioxidant agents, form a protective layer that reduces oxidation and microbial growth by limiting exposure to air and contaminants, thereby delaying spoilage and discoloration (Khan et al., 2019).

Modified pullulan is one of the polysaccharides with proven potential for developing edible coatings for food preservation (de Souza et al., 2023).

Several studies on the chemical modification of pullulan indicate significant improvements in the food packaging properties of this biopolymer (Chaari & Smaoui, 2024; Erceg et al., 2024; Hernandez-Tenorio et al., 2024). Results demonstrate the effective use of esterified pullulan mainly as an edible coating for fruit, while other types of chemical modifications can be used in nonedible packaging films (Rashid et al., 2024).

Zhang et al. (2019) applied a multilayer edible coating containing pullulan and chitosan, which preserved the quality attributes of fresh papayas longer than the control samples. Similarly, Treviño-Garza et al. (2017) used a multilayer edible coating combining mucilage, chitosan, and pullulan to enhance the quality of fresh-cut pineapples stored at room temperature, extending their shelf life by up to 14 days. Eroglu et al. (2014) developed a pullulan-based edible emulsion that increased the shelf life of chilled strawberries by five days.

Priyadarshi et al. (2022) applied an edible pectin-pullulan coating with grape seed extract to raw and roasted peanut seeds. This antioxidant coating reduced lipid oxidation by 75% over 30 days, an improvement attributed to the antioxidant capacity of grape seed extract and the oxygen barrier properties of pullulan.

In addition to edible coatings, pullulan-based films have emerged as a viable option for food packaging, offering greater sustainability and protection for various food groups (Rashid et al., 2024).

9.5.2 Pullulan-based films

Films based on pullulan are also considered an alternative for food packaging materials due to their biodegradability, natural origin, efficient film-forming capacity, and reduction of the environmental impact caused by synthetic films (Islam et al., 2024).

Pullulan-based films are transparent, odorless, and colorless, and they have a low permeability against oil (Farris et al., 2014) and oxygen (Kraśniewska et al., 2019).

These films can be applied in various food categories, including fresh produce, meats, bakery items, and dairy products, offering multiple benefits such as reducing oxidation, retaining moisture, and preventing microbial growth (Dewan et al., 2024).

In fresh produce, pullulan films slow down ripening and reduce moisture loss, while in meats and seafood, they protect against spoilage and oxidation (Goswami et al., 2021; Gounga et al., 2008). In bakery and confectionery items, pullulan films maintain freshness by locking in moisture and preventing staling (Farris et al., 2014). Additionally, pullulan films applied to nuts, dried fruits, and dairy products help retain flavor and prevent spoilage (Tang et al., 2025). Pullulan is also used for beverages and capsules, preserving sensitive ingredients until consumption (Goswami et al., 2021).

In addition to these applications, research has improved the durability and water resistance of pullulan-based films by blending them with other biopolymers (Agrawal et al., 2022). These advancements, along with the potential for active packaging that indicates the freshness of the packaged product and exhibits antimicrobial and antioxidant properties (essential oils or plant extracts), expand the possibilities for using pullulan films in the food sector (Soltani Firouz et al., 2021).

Pullulan-based films can be engineered to release active substances when needed, providing targeted protection based on the specific needs of the food product (Fig. 9.3).

These active films are developed not only to protect food but also to interact with the product or its environment, improving storage conditions and shelf life by incorporating functional compounds (Benbettaïeb et al., 2019).

For example, pullulan films can be combined with natural antimicrobial agents, such as essential oils (thymol, menthol, eucalyptol, and methyl salicylate), silver nanoparticles, or plant extracts, to inhibit microbial growth, thereby helping to prevent spoilage and contamination (Rashid et al., 2024). Additionally, antioxidant compounds like grape seed extract or vitamin E can be integrated into the films to reduce oxidative damage, particularly in high-fat foods such as nuts and meats (Gupta et al., 2020).

When engineered to respond to changes in environmental conditions, like temperature or humidity, pullulan films can function as "smart" packaging (Yildirim & Candan, 2023).

Zhao et al. (2019) demonstrated that the addition of lactoferrin to the formulation of pullulan-based edible films helped reduce the hydrophobicity of the material's surface and decreased water vapor permeability.

Xiao, Tong, et al. (2012) incorporated sodium alginate into pullulan-based edible films, increasing both the storage and loss moduli, with the effect becoming more pronounced as the sodium alginate concentration increased (40%–60%).

Gounga et al. (2008) developed edible films based on pullulan, whey protein isolate (WPI), and glycerol, which exhibited low solubility and high transparency when low concentrations of pullulan were utilized.

Morsy et al. (2015) presented results on incorporating oregano and rosemary essential oils, as well as zinc oxide nanoparticles, into pullulan-based edible films, which demonstrated an increase in the safety of fresh or processed meat and poultry products stored in a refrigerated environment.

Pullulan films present a flexible and eco-friendly solution for food packaging, with uses spanning from edible coatings on fresh produce and processed foods to biodegradable packaging for snacks and baked goods. Adding antimicrobial

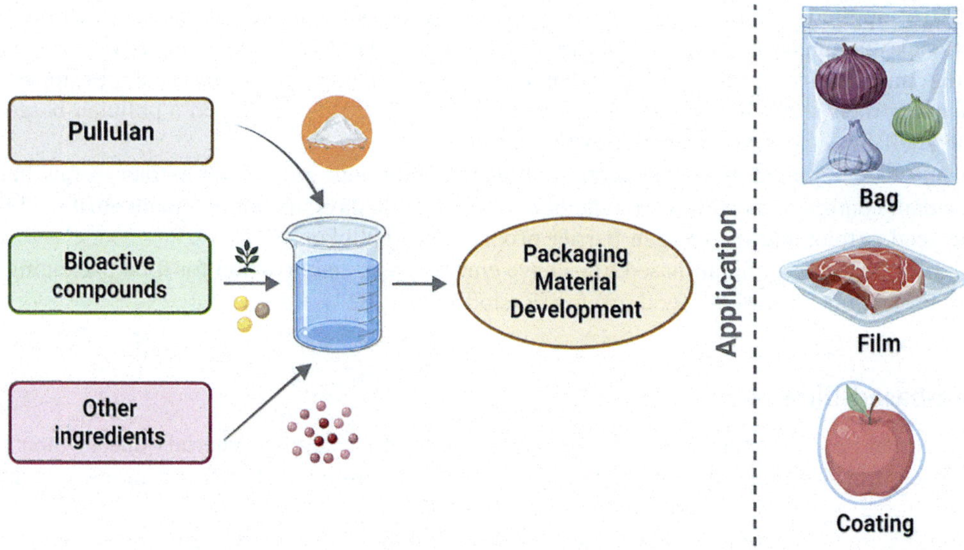

FIGURE 9.3 Schematic drawing of the application of pullulan and bioactive compounds in the development of materials used for food packaging.

agents, antioxidants, and other functional components within pullulan films boosts their preservative properties, helping to extend shelf life and enhance food safety (Dewan et al., 2024).

9.6 Development of active and functional films

Ensuring food quality and safety requires using suitable packaging materials and technologies. During storage, factors such as pathogenic and spoilage microorganisms, chemical and enzymatic reactions, physical damage, and environmental conditions can lead to undesirable changes in food. Beyond the conventional roles of packaging, including preservation, transportation, and providing information about food, recent decades have seen a growing focus on incorporating features such as biodegradability, sustainability, renewability, edibility, and active or intelligent functionalities (Naiaretti & Tita, 2021).

Active packaging (Fig. 9.4) is designed to modify the environmental conditions within the packaging to preserve or extend the product's shelf life while maintaining its quality and safety. Additionally, some active packaging incorporates bioactive compounds that provide properties such as antimicrobial, antioxidant, CO_2, and O_2 release, and ethylene regulation (Westlake et al., 2023).

Active packaging systems are generally categorized into absorbers and emitters. Absorber systems eliminate unwanted compounds from the packaging's headspace (such as oxygen, ethylene, carbon dioxide, and moisture) or the food's surroundings, which could otherwise compromise its quality. On the other hand, emitter systems incorporate substances into the packaging material, such as carbon dioxide, ethanol, antioxidants, antimicrobials, and preservatives, which are gradually released into the product over time (Gómez-Estaca et al., 2014).

Ethylene is a compound generated and released through plant metabolic processes. It plays a crucial role in plant tissue growth, ripening, aging, and eventual decay. In the case of climacteric produce, packaging can lead to an accumulation of ethylene, which increases the respiration rate of the vegetable and hastens the senescence process. Conversely, using an ethylene absorber to decrease its concentration within the packaging slows the senescence process, effectively extending the vegetable's shelf life (Dobrucka & Cierpiszewski, 2014).

Water constitutes a substantial portion of the mass in fruits and vegetables, and its movement drives the transfer from within the food to the external environment. A significant drawback of this mass transfer during preserving fruits and vegetables is dehydration, which alters their nutrient content, taste, and texture. Moreover, when the mass loss reaches 3%–10%, the produce starts to fade and shrink irreversibly, diminishing its visual appeal to consumers and leading to a decline in its commercial value (Carvalho et al., 2024).

Frequent condensation inside packaging for fruits and vegetables can hinder the product's visibility, a crucial factor influencing purchasing decisions. Active packaging equipped with moisture absorbers helps address this issue by reducing condensation and mitigating quality deterioration in these foods. Techniques to lower humidity levels within the

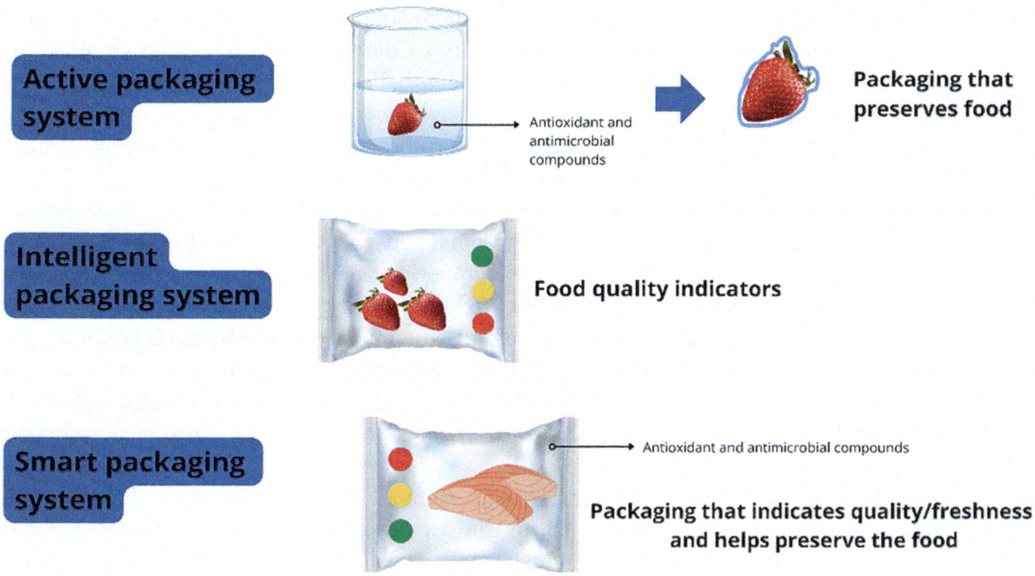

FIGURE 9.4 Illustration of the functionality of active, intelligent, and smart packaging systems.

packaging create a drier environment, ideal for preserving vegetables. Various antimicrobial compounds are utilized as alternatives for developing materials in or as part of food packaging. Antimicrobial agents incorporated into active packaging may exist in solid or gaseous forms. Such packaging can include films impregnated with antimicrobial substances, which, during storage, act to inhibit or eliminate microbial growth, thereby extending the food's shelf life (Han et al., 2018).

In addition to microbiological deterioration, lipid oxidation can also impact the sensory and nutritional qualities of foods, particularly those rich in unsaturated lipids, which are highly prone to oxidation. These reactions alter the food's taste, leading to rancidity and significantly reducing the product's appeal and consumer acceptance. Active packaging with antioxidant properties can be categorized into two types: one that releases antioxidant compounds into the food and another that eliminates unwanted substances, like oxygen, which triggers oxidation (Westlake et al., 2023).

Recent advancements have been explored in developing pullulan-based biopolymers for food packaging and preservation. Pullulan has gained significant attention as a biopolymer for packaging due to its potential for customization through natural fillers, blends, and physical or chemical modifications. These approaches enhance its properties for sustainable packaging. While creating biodegradable pullulan-based packaging on an industrial scale presents challenges, its use as a functional oxygen barrier in commercial active packaging is well studied. Given its edible nature, pullulan shows excellent promise as a coating material to extend food shelf life (Ghosh et al., 2022).

Erceg et al. (2024) added pullulan to sliced cheese by modifying it through an eco-friendly method for coating formulations combined with chitosan. The resulting coating exhibited significantly improved barrier properties, reducing moisture vapor permeability by 72.2 and 77.7 times, and enhanced mechanical strength, with tensile strength increasing by up to 3.9 and 9.4 times. A new strategy was introduced for controlling the microbiological growth of sliced cheese, utilizing a synergistic effect of *Helichrysum italicum* essential oil and its hydrolate in the coating. This method successfully preserved the cheese's sensory qualities, extending its shelf life by at least 50% through better barriers, mechanical and antimicrobial properties.

A study explored the impact of metal-organic frameworks loaded with cinnamon essential oil on the characteristics of gelatin/pullulan-based composite films. The results revealed that pullulan-based films exhibited notable antimicrobial properties against Staphylococcus *aureus*, *S. enterica*, Escherichia *coli*, and Listeria *monocytogenes*. Additionally, the films demonstrated a 98.16% reduction in the ABTS (ácido 2,2'-azino-bis(3-etilbenzotiazolina-6-sulfônico)) radical and effectively extended the shelf life of meat stored at 4°C, minimizing moisture loss by 3.35%, maintaining pH levels (6.2), and inhibiting bacterial growth by 99.9%. These findings highlight the promising potential of pullulan-based active packaging to enhance food safety and longevity (Hong et al., 2024).

Bian et al. (2024) manufactured active, pH-sensitive films based on gellan gum and pullulan with extracts from the fruits and leaves of *Broussonetia papyrifera* using a molding method. The films showed improved antioxidant properties, with maximum radical scavenging rates of 77.45% using 2,2-diphenyl-1-picrylhydrazyl and 66.21% using 2,2-azino-bis (3-ethylbenzothiazoline-6-sulfonic acid). The antibacterial capacity of the films also increased significantly.

Pullulan, chitosan, and carvacrol were used for the edible antibacterial film. The film showed satisfactory antibacterial activity against common bacteria in chilled meat, such as *Pseudomonas fluorescens*, *L. monocytogenes*, *E. coli*, *Pseudomonas putida*, *Enterobacter cloacae*, and *S. aureus*. Furthermore, when applied to preserve chilled goat meat, it increased the shelf life of goat meat to more than 15 days (Xiao et al., 2022).

The development of active biofilms based on chitosan/pullulan was incorporated with gallic acid using solvent casting. The films showed excellent antioxidant (~42.54%) and antimicrobial activities against food-borne pathogens. It was tested in practice for packaging green peppers at room temperature; comparing packaged green peppers with unpackaged green peppers, the shelf life was extended by up to 18 days without affecting quality. These findings indicate that the developed films meet promising conditions for active packaging applications for fresh food products (Gasti et al., 2023).

Intelligent packaging (Fig. 9.4) primarily monitors food changes during storage rather than directly extending its shelf life. Its principal function is to provide information about food quality, enhancing safety during transportation and storage. As sensors and indicators, intelligent packaging offers real-time updates on the product's freshness and preservation status (Hoffmann et al., 2022).

Certain fruits do not exhibit noticeable color changes or other signs of ripening, making it difficult to determine their maturity. However, they release volatile compounds during the ripening process, which can be detected by ripeness indicators that signal readiness through a color change. This enables consumers to select the fruit based on their preferred ripeness level (Fang et al., 2017).

High-protein foods such as fish and meat degrade over time during storage due to the action of enzymes and bacteria. This breakdown results in the buildup of volatile compounds like ammonia and di/trimethylamine (organic amines) inside the packaging. These compounds can be detected by pH and colorimetric indicators, which change color in response to the presence of these volatile substances, allowing consumers to assess the quality of the food (Fang et al., 2017).

Intelligent packaging equipped with time and temperature indicators provides real-time information about the storage conditions inside the packaging. These indicators are user-friendly, as they can be incorporated into the packaging, allowing consumers to assess food quality through a color change that reflects its condition at a specific temperature. Furthermore, this system helps monitor the food's transport conditions, ensuring its safety throughout its shelf life until consumption (Chen et al., 2017).

The quality of many foods is influenced by abrupt shifts in storage temperature, making this factor crucial for food distributors and manufacturers at every stage of the supply chain. Sensors are essential for tracking refrigerated foods' time and temperature history. Besides monitoring freshness, microbial growth, and chemical changes, these devices are essential for consumer safety. Smart packaging (Fig. 9.4) tracks alterations in the product or its surroundings by incorporating intelligent materials that respond to these changes. It employs biosensors or chemical sensors to assess the quality and safety of food throughout the entire supply chain, from production to consumption (Abedi-Firoozjah et al., 2023; Carvalho et al., 2024).

Food monitoring can be achieved by detecting and analyzing factors such as freshness, oxygen, carbon dioxide, pH, time, temperature, leaks, and pathogens. Smart packaging enables tracking a product throughout its entire lifecycle, assessing both internal and external conditions of the packaging, and providing real-time updates to the manufacturer and consumer about the product's status (Carvalho et al., 2024).

Integrating functional substances into packaging formulations can enhance the product's value and prolong food shelf life due to their active properties. These substances not only aid in preservation but also enable consumers to assess the freshness of food through intelligent packaging features. Additionally, using these functional substances boosts the packaging's effectiveness in maintaining food quality (Hoffmann et al., 2022).

9.7 Evaluation and testing of pullulan films

Mechanical properties are essential for forming films and coatings (Fig. 9.5), as they ensure the materials maintain their integrity even under stretching, preventing breakage. Pullulan exhibits good mechanical properties for film formation. However, certain limitations restrict its use in various food applications. It is considered brittle, as its high hydrophilicity reduces its mechanical strength. Additionally, it is costly and lacks antimicrobial properties, further limiting its potential applications (Agrawal et al., 2022; Chen et al., 2017; de Souza et al., 2023; Ghosh et al., 2022; Li et al., 2020).

Tensile strength and elongation at break are fundamental aspects for evaluating the quality of a polymer. As a hydrophilic material, water has a significant influence on its mechanical properties (Shanmugam et al., 2022; Tabasum et al., 2018). Thus, variations in the relative humidity of the environment where the film is applied can significantly alter

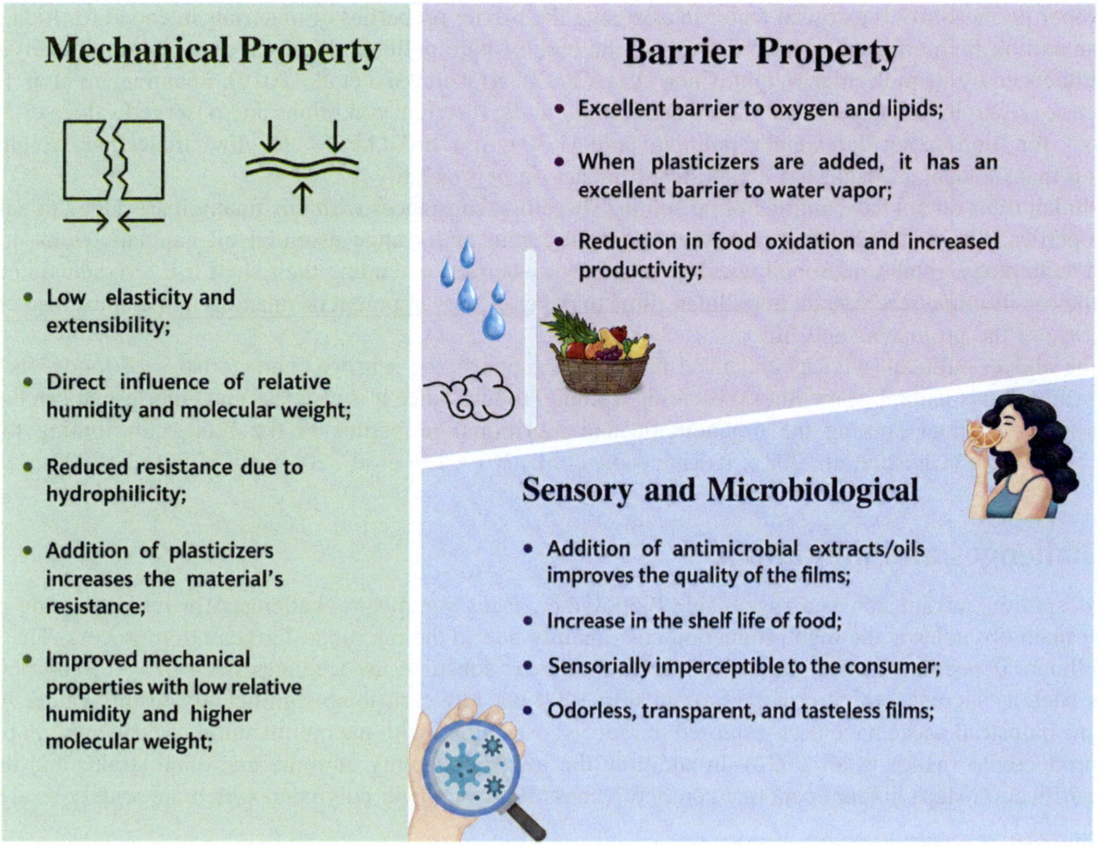

FIGURE 9.5 Mechanical, sensory, and barrier properties that are important in pullulan films.

its characteristics. Materials derived from pullulan exhibit varied mechanical properties, primarily influenced by molecular weight and relative humidity. In pullulan-based films, tensile strength can range from 5 to 85 MPa, depending on these conditions. Kraśniewska et al. (2019) and Xiao, Lim, et al. (2012) observed that pullulan films exhibit tensile strength between 55 and 70 MPa in environments with relative humidity ranging from 11% to 65%. In contrast, in environments with 85% relative humidity, this value decreases to approximately 30 MPa. Chen et al. (2017) reported tensile strength values ranging from 49 to 56 MPa, considering different molecular weights. These findings highlight the importance of controlling relative humidity and molecular weight in the quality of pullulan films. In general, the lower the relative humidity and the higher the molecular weight, the better the mechanical properties of these materials (de Souza et al., 2023; Kanmani & Lim, 2013; Kraśniewska et al., 2019).

Another very important parameter in the selection of a polymer is elongation at break. Pullulan films, for example, are characterized by low elasticity and extensibility, with values generally below 10%. Some studies report rates ranging from 3% to 5%, depending on molecular weight. These values indicate that the films are fragile, brittle, and not very extensible, making it necessary to add plasticizers to improve these properties (Kraśniewska et al., 2019; Li et al., 2020). Plasticizers enhance the elastic properties, providing the films with the necessary flexibility during and after production. This ensures that the products do not break during handling and storage. However, studies also show that while the addition of plasticizers improves the extensibility and elasticity of the material, it may cause a slight reduction in tensile properties (Ghosh et al., 2022; Kanmani & Lim, 2013; Sugumaran & Ponnusami, 2017).

Other important points include the barrier against gases and the permeability to water vapor. Pure pullulan films stand out for their excellent barrier properties to polar substances, such as oxygen and lipids. By limiting the entry of oxygen, the coating helps protect the product from processes that significantly compromise its flavor, odor, and color. Agrawal et al. (2022), de Souza et al. (2023), and Unalan et al. (2015) measured the oxygen permeability in pullulan films and identified a value of 6.337 $mL{\cdot}\mu m{\cdot}m^{-2}{\cdot}(24\ h)^{-1}$, higher than that of other polymers, such as polypropylene.

Water vapor permeability is a crucial factor in assessing the barrier properties of materials intended for food packaging, which must exhibit high efficiency in this regard. In the case of pure pullulan, these properties are relatively low, being strongly influenced by its molecular weight. Chen et al. (2017), Kraśniewska et al. (2019), Shanmugam et al. (2022), and Yeasmin et al. (2020) investigated water vapor permeability under different conditions and observed values of 21.40 g.mm $(m^2.d.kPa)^{-1}$ for high molecular weight pullulan and 41.30 g.mm $(m^2.d.kPa)^{-1}$ for low molecular weight pullulan, highlighting that molecular weight has a significant impact on permeability.

Pure pullulan films do not have antimicrobial action. Therefore, substances with this function are added to enhance their barrier properties. Liu et al. (2019) demonstrated that curcumin and orange essential oil nanoemulsions, treated with ultrasound technology, inhibit microbiological growth in strawberries, extending their shelf life. Priyadarshi et al. (2022) explored the use of grape seed extract in pullulan films to delay the development of rancidity in peanuts, also contributing to an increase in the product's shelf life.

Given the above, pullulan, with its enhanced properties, supports the sensory characteristics of foods. Its perception during consumption is natural, sometimes presenting a slight crunch. Since it is tasteless and colorless, it can be applied to various products without altering the organoleptic and nutritional properties of the food, contributing to increased consumer acceptance (Ghosh et al., 2022; Kraśniewska et al., 2019; Liu et al., 2019; Rashid et al., 2024).

9.8 Challenges and limitations

Pullulan, despite its advantages as a sustainable biopolymer, faces significant challenges for its large-scale production. One of the main obstacles is the high production cost, mainly due to the microbial fermentation process. The production cost of pullulan, depending on the scale of production and the substrate used, ranges from \$25 to \$100 per kilogram. Substrates such as sucrose, which costs approximately \$900 per ton, contribute significantly to this value. Alternatives such as agro-industrial waste have been explored to reduce costs, along with the optimization of fermentation processes to improve productivity (Moris et al., 2022). In addition, the low productivity of some microbial strains and the need for complex purification steps hinder economic competitiveness with synthetic polymers, which are widely used in industry (Wei et al., 2021).

Producing pullulan on an industrial scale requires innovative approaches. The use of agro-industrial waste as a substrate is a sustainable alternative. Still, the variable composition of these materials can lead to inconsistencies in the quality of the final product (Wani et al., 2021). Furthermore, the industrial infrastructure required for large-scale fermentation requires significant investment, which can limit economic viability, especially in highly competitive markets. The compatibility of pullulan with food is another challenge. Although it is safe for consumption, its interaction with different food matrices can alter sensory properties such as taste and texture, requiring fine-tuning for specific applications (Mulla et al., 2023). This requires formulation studies that increase development costs.

The stability of pullulan films is also a critical point. These materials, although efficient as barriers to gases and moisture, are sensitive to high relative humidity, which can compromise their functionality over time (Tiwari et al., 2019). Strategies such as the incorporation of additives or combination with other polymers can improve stability, but they increase production and development costs.

Finally, competitiveness with petroleum-based polymers, which are cheaper and scalable for various applications, represents a significant barrier. Despite the environmental benefits, the transition to materials such as pullulan depends on economic and regulatory incentives prioritizing sustainability (Carrigy et al., 2019). Overcoming these limitations requires continued research in biotechnology and materials engineering. Strategies such as the genetic modification of microorganisms to increase productivity and developing hybrid technologies to improve pullulan properties are promising (Sugumaran & Ponnusami, 2017). Furthermore, the growing market interest in sustainable solutions may drive their adoption, provided that costs are reduced and technical challenges are overcome.

9.9 Case studies and practical applications

This section deals with some application examples, consumer acceptability of pullulan, and the steel industry's use. Due to its unique properties, pullulan is a versatile polysaccharide with numerous applications across various industries. One of the applications of pullulan would be in films.

The use of pullulan films in food has expanded, contributing to the development of innovative solutions that meet a growing demand for sustainable alternatives to conventional plastics. Pullulan is a polysaccharide used in various applications by the food industry (Chauhan & Rao, 2024).

Fresh fruits such as apples, guava, peaches, strawberries, tomatoes, litchi, and pears are highly susceptible to water loss and spoilage. One of the main challenges food companies face is extending the shelf life of these products without resorting to artificial preservatives. In this scenario, using pullulan films has proven to be an effective solution (Niu et al., 2019). A pullulan coating has been developed for fresh fruit that acts as a barrier against water loss and oxygen penetration, slowing down the ripening process and spoilage caused by microorganisms. This coating is 100% edible and does not affect the flavor or texture of the fruit. By forming a protective layer, pullulan prevents internal moisture from escaping while reducing the amount of oxygen that comes into contact with the food, slowing down the aging process and preserving quality. In addition, pullulan films can be used as a substitute for conventional plastics, making the product more environmentally friendly (Niu et al., 2019; Rashid et al., 2024).

A study conducted by Zheng et al. (2023) showed that apples coated with pullulan films retained their crispness and freshness for up to two weeks longer than untreated apples. Furthermore, this study proved that pullulan effectively reduced losses due to bacteria and fungi, helping to extend shelf life without resorting to chemical additives. Another sector that has benefited from using pullulan films is the meat and sausage industry, where control of humidity, oxygen, and microorganisms is essential to ensure product safety and quality (de Souza et al., 2023).

A pullulan application has been developed for sausage coating (Cegiełka et al., 2016) and hams (Singh et al., 2019) to prevent moisture loss and increase the shelf life of products. Pullulan film is applied to the surface of products before they are packaged, forming a protective film. In meat products, this coating serves to preserve texture and freshness, and to prevent fat oxidation, which can lead to rancidity. Pullulan, a nonporous substance, prevents oxygen from entering and water from escaping, which is crucial for preventing dehydration and the growth of microorganisms. Meat products coated with pullulan film have shown a shelf life up to 30% longer than uncoated products (Khan et al., 2019).

The baking industry has also found benefits in using pullulan films. Breads, cakes, and other baked goods are the foods that suffer most from moisture loss as well as changes in texture and flavor. NithyaBalaSundari et al. (2020) conducted tests on applying pullulan in bread to improve its conservation. In the study by these authors, bakery products with pullulan presented better softness and appearance throughout their shelf life, in addition to less waste due to deterioration.

Pullulan also produces cereals and healthy snacks, adopting an innovative solution (Zang et al., 2024). The concept of edible packaging, which has gained popularity, is one of the most interesting approaches to reducing the use of plastics and other nonbiodegradable materials. Several types of packaging materials are used, but pullulan films have stood out for their resistance and flexibility, and they are perfectly compatible with the proposal of healthy and sustainable products. This approach reduces environmental impact and offers a differentiated consumption experience in which the consumer can choose to consume the packaging with the food. Thus, pullulan films are used as packaging for cereals and protein bars, which can be consumed with the product or discarded without impacting the environment (Zang et al., 2024).

Pullulan films have also been expanding in the functional food industry, where bioactive ingredients such as vitamins, minerals, and probiotics must be delivered efficiently and stably. Energy bars and dietary supplements that use pullulan films to encapsulate and protect active ingredients sensitive to heat, light, and moisture have been developed. In this case, the pullulan film serves as a protective barrier that helps maintain the efficacy of the bioactive ingredients until the moment of consumption. In this way, pullulan films have been shown to maintain the stability of the active ingredients during storage without losing their nutritional properties (Cheng et al., 2011).

Pullulan has also been used as a filler in low-calorie foods and beverages. Pullulan can be a prebiotic to protect bifidobacteria and stimulate their growth. Pullulan is also used in the food industry as a stabilizer, thickener, texturizer, and gelling agent to improve the sensory quality of foods (Rashid et al., 2024). Pullulan has also been classified as a safe food additive for various foods, including fermented vegetables, flours, fresh pasta and noodles, and frozen egg products. This recognition highlights the potential of pullulan as a safe food preservative (Chauhan & Rao, 2024).

As scientific research continues to advance, the use of pullulan is expanding, making its presence increasingly prominent in the food industry. Recently, pullulan has been used on food surfaces and in packaging preparations, providing a barrier against water and oxygen loss, as well as protection against pathogenic microorganisms and food oxidation. Thus, it can be summarized that pullulan can be an antioxidant and antimicrobial agent (Rashid et al., 2024).

Due to its enhancing properties, pullulan is widely used in soups, sauces, and beverages. It can also be used to stabilize the quality and texture of mayonnaise. Pullulan can be used to bind nuts to cookies. Pullulan can be used cautiously in producing low-calorie (diet) foods due to its presence on the GRAS list and its slow digestion (Oğuzhan & Yangılar, 2013).

Based on consumers' acceptance of pullulan, it would have no restrictions as it increases sensory quality. Although pullulan-based preservation has been studied extensively, practical application still faces some challenges. Most research work has been limited to observing apparent indicators, such as changes in physicochemical properties, bacterial enumeration, and sensory evaluation (Rashid et al., 2024). However, some disadvantages of pullulan include hydrophilicity, brittleness, and high cost. Combining it with other biopolymers, such as glucan, galactomannan, pectin, casein, chitosan, pea protein, gelatin, nanofibrillated cellulose, crosslinking, and plasticizers such as glycerol, can enhance this property to create edible packaging for the preservation of perishable products at a lower cost (Adame et al., 2024).

Using pullulan as a packaging or protective film on food products can help promote lifestyle changes. Many consumers are concerned about changing their eating patterns and lifestyle habits versus chronic diseases. For example, the increase in the intake of processed and high-calorie foods has led to an increase in overweight and obesity rates in countries around the world. Different food companies compete fiercely with each other to gain a large share of the market. One way to improve competition in the food market would be through packaging. Films can improve sensory acceptability and attributes, reduce lipid oxidation, and inhibit spoilage and pathogenic bacteria. The application of nanopackaging on food products is expected to prolong the products' shelf life. Pullulan packaging can improve sensory acceptability and attributes, reduce lipid oxidation, and inhibit spoilage and pathogenic bacteria counts (Anusha Siddiqui et al., 2024). It is worth mentioning that pullulan dry powder is white, tasteless, odorless, nontoxic, nonmutagenic, noncarcinogenic, biodegradable, edible (Trinetta, 2017), and only decomposes at high temperatures (250°C–280°C) (Trinetta, 2017), which has contributed to its acceptability and use.

Pullulan has been commercially available for over 25 years. Its commercial use has been encouraged by its unique mechanical and physical characteristics. The first proposed use of pullulan films was as edible pouches for premeasured food portions dissolved in water or hot liquid foods. Edible coatings applied directly to the surface of the food that can be eaten together with the product are the first commercial application of pullulan. Since then, considerable research has been conducted into the functional properties of pullulan-based edible films and coatings for food packaging applications (Trinetta, 2017).

According to Singh et al. (2019), the potential applications of pullulan in food items have been identified and recognized. Thus, the future challenge is to delineate the applications of pullulan at the industrial level and dictate the further successful penetration of pullulan in the food market (Singh et al., 2019).

Bio-based, disposable, biodegradable, and recyclable materials are finding their niche in the food market to replace petroleum-based packaging through multifunctionality and improved production. Pullulan is a unique polysaccharide that offers significant advantages over other bio-based materials due to its distinct physicochemical characteristics and antimicrobial properties. Despite these advantages, the main constraint in using pullulan is its cost. Engineering innovations, improved strains, and new strategies can reduce the production costs of pullulan, thus encouraging the use of this high-performance, environmentally friendly, and efficient material in food packaging applications (Trinetta, 2017).

9.10 Future trends and research

Pullulan is a natural, edible, nontoxic microbial polysaccharide with excellent film-forming capacity for edible coatings and food packaging. It has gained prominence as a sustainable alternative to conventional plastic films, especially in the food industry. Being biodegradable and produced from renewable sources, it represents an eco-friendly solution with unique properties such as moisture control, gas barrier capabilities, and highly transparent, colorless, and odorless films (de Souza et al., 2023; Kanmani & Lim, 2013; Rashid et al., 2024).

Despite offering numerous advantages for use, films made solely from pullulan have drawbacks related to mechanical properties and antimicrobial barriers, which can limit their application potential. Recent innovations involving chemical modifications and combinations with other biopolymers have enhanced their potential, making pullulan films thinner, more durable, and more versatile. These improvements have increased their mechanical strength and antimicrobial barriers, broadening their applications across various products. Their production requires less energy and reduces greenhouse gas emissions. Being recyclable and compostable, pullulan supports the circular economy, proving to be a conscious choice for the future and aligning with companies' sustainability strategies (de Souza et al., 2023; Ghosh et al., 2022; Rashid et al., 2024; Sugumaran & Ponnusami, 2017).

Pullulan has proven to be a versatile and promising ingredient in areas that extend beyond the food sector. Although widely used in edible films, its properties make it a valuable option in the fields of pharmaceuticals, cosmetics, oils, paints, textiles, and even medicine. Pullulan has been explored for applications such as encapsulation of active ingredients, controlled drug release, paint enhancement, stabilization and preservation of food products, skincare products (such as face masks and creams), wound dressings, bone repair, and more recently, in 3D printing. The material has demonstrated potential for creating customized structures, such as medical implants, thanks to its ability to be easily combined with other materials. These applications are expected to be further explored in the coming years, with advancements in the selection of polymers to combine with pullulan (Kraśniewska et al., 2019; Tabasum et al., 2018; Wang et al., 2025).

The growing demand for sustainable solutions and technological advancements positions pullulan as a strategic ingredient for the future. With new research and innovations, it is likely to become more accessible, affordable, and competitive, expanding its use across various sectors. However, there are still few commercial products based on pullulan available in the market. This underscores the need for more significant investment in research and development to increase its application. Such expansion will not only represent innovation but also a commitment to more sustainable practices, aligned with the demands of an increasingly conscious and future-oriented market (Agrawal et al., 2022; de Souza et al., 2023; Li et al., 2020; Shanmugam et al., 2022; Sugumaran & Ponnusami, 2017; Yeasmin et al., 2020).

9.11 Conclusion

Pullulan is a natural polymer that serves as a versatile material for developing films and coatings in the food packaging industry. Its unique properties, such as high transparency, biodegradability, and excellent gas barrier characteristics, make it a promising choice to meet the growing demands for sustainable and functional packaging solutions. Its ability to form thin, strong, and flexible films allows for various applications, from coatings for fruit to active packaging that extends the shelf life of food.

Among the greatest benefits of pullulan are its ability to create effective barriers against oxygen and its compatibility with bioactive compounds, such as antioxidants and antimicrobials. These properties not only improve food preservation but also allow the production of smart and functional packaging, capable of meeting the needs of the modern consumer. In addition, pullulan can be combined with other biopolymers or additives, further expanding its application potential.

Despite its advantages, the use of pullulan still faces challenges, such as sensitivity to moisture, which can compromise its mechanical properties, and the high production cost, which limits its commercialization on a large-scale. However, these limitations are being addressed through ongoing research, including the development of composites and chemical modifications to improve its stability under adverse conditions.

The future of pullulan in the packaging industry is promising, especially in the context of the growing demand for environmentally friendly and sustainable materials. Innovations in pullulan production, such as using low-cost substrates and more efficient fermentation processes, could reduce costs and make the material more accessible. In addition, advances in incorporating functional agents could expand its applications beyond food packaging to include sectors such as pharmaceuticals and cosmetics.

The growth potential of pullulan in the global packaging market aligns with trends toward sustainable consumption and the search for alternatives to conventional plastics. Governments, industries, and consumers are increasingly focused on reducing their environmental impact, which creates a favorable environment for the adoption of materials such as pullulan.

Therefore, pullulan is a remarkable example of how science and innovation can converge to meet the needs of the food packaging market. Its exceptional properties and customization possibilities provide a solid foundation for new applications and technological advancements. As research evolves, pullulan is expected to play an even more significant role in transitioning to more sustainable packaging solutions, contributing to a greener and more efficient future. The chapter reinforces the importance of pullulan as a standout material, presenting both its current achievements and future opportunities in the packaging sector.

Acknowledgments

The authors are grateful to the Foundation for Research Support and Innovation of Santa Catarina State (FAPESC: TO - 2025TR001604), the National Council for Scientific and Technological Development (CNPq n° 302867/2023-6; CNPq n° CNPq, 303069/2022-8) and the Ministry of Science, Technology, Innovation and Communications (MCTIC). This study was financed in part by the Coordenação de Aperfeiçoamento de Pessoal de Nível Superior e Brasil (CAPES) Finance Code 001. We would also like to thank the Krebs Research Group.

AI disclosure

During the preparation of this work, the author(s) used Grammarly software to help with grammar corrections. The AI of this software improved the concordance and readability of the text. After using this tool/service, the author(s) reviewed and edited the content as needed and take(s) full responsibility for the content of the publication.

References

Abedi-Firoozjah, R., Salim, S. A., Hasanvand, S., Assadpour, E., Azizi-Lalabadi, M., Prieto, M. A., & Jafari, S. M. (2023). Application of smart packaging for seafood: A comprehensive review. *Comprehensive Reviews in Food Science and Food Safety, 22*(2), 1438–1461. https://doi.org/10.1111/1541-4337.13117, http://onlinelibrary.wiley.com/journal/10.1111/(ISSN)1541-4337.

Adame, M. Y., Shi, C., Li, C., Aziz, T., Alharbi, M., Cui, H., & Lin, L. (2024). Fabrication and characterization of pullulan/tapioca starch-based antibacterial films incorporated with *Litsea cubeba* essential oil for meat preservation. *International Journal of Biological Macromolecules, 268*. https://doi.org/10.1016/j.ijbiomac.2024.131775, https://www.sciencedirect.com/science/journal/01418130.

Agrawal, S., Budhwani, D., Gurjar, P., Telange, D., & Lambole, V. (2022). Pullulan based derivatives: Synthesis, enhanced physicochemical properties, and applications. *Drug Delivery, 29*(1), 3328–3339. https://doi.org/10.1080/10717544.2022.2144544.

Ahuja, A., & Kumar Rastogi, V. (2023). Spray coating of edible insect waxes for liquid food packaging. *Applied Surface Science, 624*, 157150. https://doi.org/10.1016/j.apsusc.2023.157150.

Anusha Siddiqui, S., Mahmood Salman, S. H., Ali Redha, A., Zannou, O., Chabi, I. B., Oussou, K. F., Bhowmik, S., Nirmal, N. P., & Maqsood, S. (2024). Physicochemical and nutritional properties of different non-bovine milk and dairy products: A review. *International Dairy Journal, 148*. https://doi.org/10.1016/j.idairyj.2023.105790, www.elsevier.com/inca/publications/store/4/0/5/8/6/0/index.htt.

Aquinas, N., Chithra, C. H., & Bhat, M. R. (2024). Progress in bioproduction, characterization and applications of pullulan: A review. *Polymer Bulletin, 81*(14), 12347–12382. https://doi.org/10.1007/s00289-024-05300-2, https://www.springer.com/journal/289.

Benbettaïeb, N., Karbowiak, T., & Debeaufort, F. (2019). Bioactive edible films for food applications:Influence of the bioactive compounds on film structure and properties. *Critical Reviews in Food Science and Nutrition, 59*(7), 1137–1153. https://doi.org/10.1080/10408398.2017.1393384.

Bian, Z., Wu, X., Sun, X., Huang, X., Zhuo, X., Wang, H., Komarneni, S., Zhang, K., Ni, Z., & Hu, G. (2024). Gellan gum and pullulan-based films with triple functionalities of antioxidant, antibacterial and freshness indication properties for food packaging. *International Journal of Biological Macromolecules, 278*(Pt 2), 134825. https://doi.org/10.1016/j.ijbiomac.2024.134825.

Carrigy, N. B., Ordoubadi, M., Liu, Y., Melhem, O., Barona, D., Wang, H., Milburn, L., Ruzycki, C. A., Finlay, W. H., & Vehring, R. (2019). Amorphous pullulan trehalose microparticle platform for respiratory delivery. *International Journal of Pharmaceutics, 563*, 156–168. https://doi.org/10.1016/j.ijpharm.2019.04.004, www.elsevier.com/locate/ijpharm.

Carvalho, A. C. F., Bose, I., Ghosh, S., Hoffmann, T. G., Bertoli, S. L., Schwinden Prudêncio, E., Roy, S., & De Souza, C. K. (2024). Valorization of agrowaste derived materials for food packaging application. In *Sustainable materials for food packaging and preservation: Food security and sustainability*. Brazil: Elsevier, 199–221. https://doi.org/10.1016/B978-0-443-13567-5.00009-5, https://www.sciencedirect.com/book/9780443135675.

Cazón, P., Velazquez, G., Ramírez, J. A., & Vázquez, M. (2017). Polysaccharide-based films and coatings for food packaging: A review. *Food Hydrocolloids, 68*, 136–148. https://doi.org/10.1016/j.foodhyd.2016.09.009, https://www.journals.elsevier.com/food-hydrocolloids.

Cegiełka, A., Gniewosz, M., Hać-Szymańczuk, E., & Chlebowska-Śmigiel, A. (2016). Effect of the addition of pullulan on the quality of low-fat homogenized scalded sausages. *CyTA - Journal of Food, 15*(1), 1–8. https://doi.org/10.1080/19476337.2016.1225125.

Chaari, M., & Smaoui, S. (2024). Pullulan as a biopolymer from microorganisms: Role in food packaging. *Current Food Science and Technology Reports, 2*(2), 149–156. https://doi.org/10.1007/s43555-024-00023-x.

Chauhan, K., & Rao, A. (2024). Clean-label alternatives for food preservation: An emerging trend. *Heliyon, 10*(16), e35815. https://doi.org/10.1016/j.heliyon.2024.e35815.

Chen, C.-T., Chen, K.-I., Chiang, H.-H., Chen, Y.-K., & Cheng, K.-C. (2017). Improvement on physical properties of pullulan films by novel cross-linking strategy. *Journal of Food Science, 82*(1), 108–117. https://doi.org/10.1111/1750-3841.13577.

Cheng, K. C., Demirci, A., & Catchmark, J. M. (2011). Pullulan: Biosynthesis, production, and applications. *Applied Microbiology and Biotechnology, 92*(1), 29–44. https://doi.org/10.1007/s00253-011-3477-y.

Das, A., Ringu, T., Ghosh, S., & Pramanik, N. (2023). A comprehensive review on recent advances in preparation, physicochemical characterization, and bioengineering applications of biopolymers. *Polymer Bulletin, 80*(7), 7247–7312. https://doi.org/10.1007/s00289-022-04443-4.

de Souza, C. K., Ghosh, T., Lukhmana, N., Tahiliani, S., Priyadarshi, R., Hoffmann, T. G., Purohit, S. D., & Han, S. S. (2023). Pullulan as a sustainable biopolymer for versatile applications: A review. *Materials Today Communications, 36*. https://doi.org/10.1016/j.mtcomm.2023.106477, http://www.journals.elsevier.com/materials-today-communications/.

Dewan, M. F., Islam, M. N., & Nagase, K. (2024). Pullulan-based films: Unveiling its multifaceted versatility for sustainability. *Advances in Polymer Technology, 2024*(1). https://doi.org/10.1155/2024/2633384.

Dobrucka, R., & Cierpiszewski, R. (2014). Active and intelligent packaging food - research and development – A review. *Polish Journal of Food and Nutrition Sciences, 64*(1), 7–15. https://doi.org/10.2478/v10222-012-0091-3.

Dong, H., & Tian, Z. (2024). Preparation and characterization of pullulan-based packaging paper for fruit preservation. *Molecules (Basel, Switzerland), 29*(6), 1394. https://doi.org/10.3390/molecules29061394.

Díaz-Montes, E., & Castro-Muñoz, R. (2021). Edible films and coatings as food-quality preservers: An overview. *Foods, 10*(2), 249. https://doi.org/10.3390/foods10020249.

Erceg, T., Šovljanski, O., Tomić, A., Aćimović, M., Stupar, A., & Baloš, S. (2024). Comparison of the properties of pullulan-based active edible coatings implemented for improving sliced cheese shelf life. *Polymers, 16*(2), 178. https://doi.org/10.3390/polym16020178.

Eroglu, E., Torun, M., Dincer, C., & Topuz, A. (2014). Influence of pullulan-based edible coating on some quality properties of strawberry during cold storage. *Packaging Technology and Science, 27*(10), 831–838. https://doi.org/10.1002/pts.2077.

Ezati, P., Khan, A., Priyadarshi, R., Bhattacharya, T., Tammina, S. K., & Rhim, J. W. (2023). Biopolymer-based UV protection functional films for food packaging. *Food Hydrocolloids, 142*. https://doi.org/10.1016/j.foodhyd.2023.108771, https://www.journals.elsevier.com/food-hydrocolloids.

Fang, Z., Zhao, Y., Warner, R. D., & Johnson, S. K. (2017). Active and intelligent packaging in meat industry. *Trends in Food Science and Technology, 61*, 60–71. https://doi.org/10.1016/j.tifs.2017.01.002, http://www.elsevier.com/wps/find/journaldescription.cws_home/601278/description#description.

Farris, S., Introzzi, L., Fuentes-Alventosa, J. M., Santo, N., Rocca, R., & Piergiovanni, L. (2012). Self-assembled pullulan-silica oxygen barrier hybrid coatings for food packaging applications. *Journal of Agricultural and Food Chemistry, 60*(3), 782–790. https://doi.org/10.1021/jf204033d.

Farris, S., Unalan, I. U., Introzzi, L., Fuentes-Alventosa, J. M., & Cozzolino, C. A. (2014). Pullulan-based films and coatings for food packaging: Present applications, emerging opportunities, and future challenges. *Journal of Applied Polymer Science, 131*(13). https://doi.org/10.1002/app.40539, http://www.interscience.wiley.com/jpages/0021-8995.

Finardi, S., Hoffmann, T. G., Angioletti, B. L., Mueller, E., Lazzaris, R. S., Bertoli, S. L., Hlebová, M., Khayrullin, M., Nikolaeva, N., Shariati, M. A., & de Souza, C. K. (2022). Development and application of antioxidant coating on *Fragaria* Spp. stored under isothermal conditions. *Journal of Microbiology, Biotechnology and Food Sciences, 11*(4). https://doi.org/10.55251/jmbfs.5432, https://office2.jmbfs.org/index.php/JMBFS/index.

Gasti, T., Dixit, S., Chougale, R. B., & Masti, S. P. (2023). Fabrication of novel gallic acid functionalized chitosan/pullulan active bio-films for the preservation and shelf-life extension of green chillies. *Sustainable Food Technology, 1*(3), 390–403. https://doi.org/10.1039/d3fb00007a, www.rsc.org/journals-books-databases/about-journals/sustainable-food-technology/.

Ghosh, T., Priyadarshi, R., Krebs de Souza, C., Angioletti, B. L., & Rhim, J. W. (2022). Advances in pullulan utilization for sustainable applications in food packaging and preservation: A mini-review. *Trends in Food Science and Technology, 125*, 43–53. https://doi.org/10.1016/j.tifs.2022.05.001, http://www.elsevier.com/wps/find/journaldescription.cws_home/601278/description#description.

Gniewosz, M., Pobiega, K., Kraśniewska, K., Synowiec, A., Chaberek, M., & Galus, S. (2022). Characterization and antifungal activity of pullulan edible films enriched with propolis extract for active packaging. *Foods, 11*(15), 2319. https://doi.org/10.3390/foods11152319.

Goswami, S., Sharma, A., Anjana, & Bhattacharya, M. (2021). Pullulan films and natural compounds: Applications and perspectives. In *Biopolymer-based nano films: Applications in food packaging and wound healing*. India: Elsevier, 19–41. https://www.sciencedirect.com/book/9780128233818, https://doi.org/10.1016/B978-0-12-823381-8.00006-5.

Gounga, M. E., Xu, S.-Y., Wang, Z., & Yang, W. G. (2008). Effect of whey protein isolate–pullulan edible coatings on the quality and shelf life of freshly roasted and freeze-dried Chinese chestnut. *Journal of Food Science, 73*(4). https://doi.org/10.1111/j.1750-3841.2008.00694.x.

Gupta, M., Dey, S., Marbaniang, D., Pal, P., Ray, S., & Mazumder, B. (2020). Grape seed extract: Having a potential health benefits. *Journal of Food Science and Technology, 57*(4), 1205–1215. https://doi.org/10.1007/s13197-019-04113-w.

Gómez-Estaca, J., López-de-Dicastillo, C., Hernández-Muñoz, P., Catalá, R., & Gavara, R. (2014). Advances in antioxidant active food packaging. *Trends in Food Science & Technology, 35*(1), 42–51. https://doi.org/10.1016/j.tifs.2013.10.008.

Han, J.-W., Ruiz-Garcia, L., Qian, J.-P., & Yang, X.-T. (2018). Food packaging: A comprehensive review and future trends. *Comprehensive Reviews in Food Science and Food Safety, 17*(4), 860–877. https://doi.org/10.1111/1541-4337.12343.

Hernandez-Tenorio, F., Saez, A. A., Palacio, D. A., Galeano, E., Marin-Palacio, L. D., & Giraldo-Estrada, C. (2024). Formulations based on pullulan and a derivative as coating material for the food sector. *Carbohydrate Polymers, 342*. https://doi.org/10.1016/j.carbpol.2024.122393, https://www.sciencedirect.com/science/journal/01448617.

Hirayama, D., Nunnenkamp, L. A., Braga, F. H. G., & Saron, C. (2022). Enhanced mechanical properties of recycled blends acrylonitrile–butadiene–styrene/high–impact polystyrene from waste electrical and electronic equipment using compatibilizers and virgin polymers. *Journal of Applied Polymer Science, 139*(13). https://doi.org/10.1002/app.51873.

Hoffmann, T. G., Angioletti, B. L., Bertoli, S. L., & de Souza, C. K. (2022). Intelligent pH-sensing film based on jaboticaba peels extract incorporated on a biopolymeric matrix. *Journal of Food Science and Technology, 59*(3), 1001–1010. https://doi.org/10.1007/s13197-021-05104-6, http://www.springerlink.com/content/121580/.

Hong, S. J., Riahi, Z., Shin, G. H., & Kim, J. T. (2024). Development of innovative active packaging films using gelatin/pullulan-based composites incorporated with cinnamon essential oil-loaded metal-organic frameworks for meat preservation. *International Journal of Biological Macromolecules, 267*. https://doi.org/10.1016/j.ijbiomac.2024.131606, https://www.sciencedirect.com/science/journal/01418130.

Huang, S., Zou, S., & Wang, Y. (2023). Construction of compostable packaging with antibacterial property and improved performance using sprayed coatings of modified cellulose nanocrystals. *Carbohydrate Polymers, 305*, 120539. https://doi.org/10.1016/j.carbpol.2023.120539.

Islam, M. N., Ali, M. A., Ahiduzzaman, M., Khan, M. A. R., & Azam, M. S. (2024). Pullulan-based films. In *Polysaccharide based films for food packaging: Fundamentals, properties and applications*. Bangladesh: Springer Nature, 95–120. https://link.springer.com/book/10.1007/978-981-99-4898-7, https://doi.org/10.1007/978-981-99-4898-7_4.

Kang, L., Liang, Q., Chen, H., Zhou, Q., Chi, Z., Rashid, A., Ma, H., & Ren, X. (2023). Insights into ultrasonic treatment on the properties of pullulan/oat protein/nisin composite film:mechanical, structural and physicochemical properties. *Food Chemistry, 402*, 134237. https://doi.org/10.1016/j.foodchem.2022.134237.

Kanmani, P., & Lim, S. T. (2013). Development and characterization of novel probiotic-residing pullulan/starch edible films. *Food Chemistry, 141*(2), 1041–1049. https://doi.org/10.1016/j.foodchem.2013.03.103, www.elsevier.com/locate/foodchem.

Khan, M. J., Shameli, K., Sazili, A. Q., Selamat, J., & Kumari, S. (2019). Rapid green synthesis and characterization of silver nanoparticles arbitrated by curcumin in an alkaline medium. *Molecules (Basel, Switzerland), 24*(4). https://doi.org/10.3390/molecules24040719, https://www.mdpi.com/1420-3049/24/4/719/pdf.

Kraśniewska, K., Pobiega, K., & Gniewosz, M. (2019). Pullulan – Biopolymer with potential for use as food packaging. *International Journal of Food Engineering, 15*(9). https://doi.org/10.1515/ijfe-2019-0030.

Ku, T.-H., Nakatsubo, F., Kuboki, T., Yano, H., & Abe, K. (2022). Hindrance to nanofibrillation of undried pulp produced by the kraft cooking process. *Carbohydrate Polymers, 291*, 119481. https://doi.org/10.1016/j.carbpol.2022.119481.

Li, S., Yi, J., Yu, X., Wang, Z., & Wang, L. (2020). Preparation and characterization of pullulan derivative/chitosan composite film for potential antimicrobial applications. *International Journal of Biological Macromolecules, 148*, 258–264. https://doi.org/10.1016/j.ijbiomac.2020.01.080.

Liu, Y., Liu, Y., Han, K., Cai, Y., Ma, M., Tong, Q., & Sheng, L. (2019). Effect of nano-TiO_2 on the physical, mechanical and optical properties of pullulan film. *Carbohydrate Polymers, 218*, 95–102. https://doi.org/10.1016/j.carbpol.2019.04.073.

Luís, Â., Ramos, A., & Domingues, F. (2020). Pullulan films containing rockrose essential oil for potential food packaging applications. *Antibiotics, 9*(10), 681. https://doi.org/10.3390/antibiotics9100681.

Luís, Â., Ramos, A., & Domingues, F. (2021). Pullulan–apple fiber biocomposite films: Optical, mechanical, barrier, antioxidant and antibacterial properties. *Polymers, 13*(6), 870. https://doi.org/10.3390/polym13060870.

Mahmud, M. Z. A., Mobarak, M. H., & Hossain, N. (2024). Emerging trends in biomaterials for sustainable food packaging: A comprehensive review. *Heliyon, 10*(1), e24122. https://doi.org/10.1016/j.heliyon.2024.e24122.

Moris, H., Ghaee, A., Karimi, M., Nouri-Felekori, M., & Mashak, A. (2022). Preparation and characterization of Pullulan-based nanocomposite scaffold incorporating Ag-Silica Janus particles for bone tissue engineering. *Biomaterials Advances, 135*, 212733. https://doi.org/10.1016/j.bioadv.2022.212733.

Morsy, M. K., Sharoba, A. M., Khalaf, H. H., El-Tanahy, H. H., & Cutter, C. N. (2015). Efficacy of antimicrobial pullulan-based coating to improve internal quality and shelf-life of chicken eggs during storage. *Journal of Food Science, 80*(5), M1066. https://doi.org/10.1111/1750-3841.12855Egypt, http://onlinelibrary.wiley.com/journal/10.1111/(ISSN)1750-3841.

Mueller, E., Hoffmann, T. G., Schmitz, F. R. W., Helm, C. V., Roy, S., Bertoli, S. L., & de Souza, C. K. (2024). Development of ternary polymeric films based on cassava starch, pea flour and green banana flour for food packaging. *International Journal of Biological Macromolecules, 256*. https://doi.org/10.1016/j.ijbiomac.2023.128436, https://www.sciencedirect.com/science/journal/01418130.

Mulla, M. Z., Rostamabadi, H., Habibi, N., & Falsafi, S. R. (2023). Pullulan nanocomposites: Effect of nanoparticles and essential oil reinforcement on its performance and food packaging applications. *Food and Humanity, 1*, 887–894. https://doi.org/10.1016/j.foohum.2023.08.006, https://www.sciencedirect.com/journal/food-and-humanity.

Naiaretti, D. G., & Tita, O. (2021). Biopolymers used in food packaging: A review. *Food Technology, 16*(2), 3–19.

NithyaBalaSundari, S., Nivedita, V., Chakravarthy, M., Srisowmeya, G., Antony, U., & Nandhini Dev, G. (2020). Characterization of microbial polysaccharides and prebiotic enrichment of wheat bread with pullulan. *LWT, 122*, 109002. https://doi.org/10.1016/j.lwt.2019.109002.

Niu, B., Shao, P., Chen, H., & Sun, P. (2019). Structural and physiochemical characterization of novel hydrophobic packaging films based on pullulan derivatives for fruits preservation. *Carbohydrate Polymers, 208*, 276–284. https://doi.org/10.1016/j.carbpol.2018.12.070.

Nunes, C., Silva, M., Farinha, D., Sales, H., Pontes, R., & Nunes, J. (2023). Edible coatings and future trends in active food packaging–fruits' and traditional sausages' shelf life increasing. *Foods, 12*(17), 3308. https://doi.org/10.3390/foods12173308.

Oliveira, V. R. L., Coelho, V. V., Sérvulo, E. F. C., & Calado, V. Md. A. (2023). Study of the molecular weight of pullulan produced by *Aureobasidium pullulans* from industrial waste. *Materials Research, 26*. https://doi.org/10.1590/1980-5373-MR-2023-0060, https://www.scielo.br/j/mr/i/2023.v26suppl1/.

Oğuzhan, P., & Yangılar, F. (2013). Pullulan: Production and usage in food ındustry. *African Journal of Food Science and Technology, 4*(3), 57–63.

Pereda, M., Marcovich, N. E., & Villanueva, M. E. (2020). Antimicrobial and antioxidant properties of chitosan and pullulan edible films enriched with natural extracts: A comparative study. *Food Control, 112*. https://doi.org/10.1016/j.foodcont.2020.107089.

Priya, K., Thirunavookarasu, N., & Chidanand, D. V. (2023). Recent advances in edible coating of food products and its legislations: A review. *Journal of Agriculture and Food Research, 12*, 100623. https://doi.org/10.1016/j.jafr.2023.100623.

Priyadarshi, R., Riahi, Z., & Rhim, J.-W. (2022). Antioxidant pectin/pullulan edible coating incorporated with *Vitis vinifera* grape seed extract for extending the shelf life of peanuts. *Postharvest Biology and Technology, 183*, 111740. https://doi.org/10.1016/j.postharvbio.2021.111740.

Rashid, A., Qayum, A., Liang, Q., Kang, L., Ekumah, J.-N., Han, X., Ren, X., & Ma, H. (2024). Exploring the potential of pullulan-based films and coatings for effective food preservation: A comprehensive analysis of properties, activation strategies and applications. *International Journal of Biological Macromolecules, 260*(Pt 2), 129479. https://doi.org/10.1016/j.ijbiomac.2024.129479.

Reddy, C. N., Mishra, B., Mandal, S. K., Agrawal, D. C., & Kruthiventi, C. (2022). An insight into pullulan and its potential applications. In *Polysaccharides of microbial origin: Biomedical applications: With 257 figures and 85 tables*. India: Springer International Publishing, 247–277. https://link.springer.com/book/10.1007/978-3-030-42215-8, https://doi.org/10.1007/978-3-030-42215-8_15.

Rios, D. Ad. S., Nakamoto, M. M., Braga, A. R. C., & da Silva, E. M. C. (2022). Food coating using vegetable sources: Importance and industrial potential, gaps of knowledge, current application, and future trends. *Applied Food Research, 2*(1). https://doi.org/10.1016/j.afres.2022.100073, https://www.journals.elsevier.com/applied-food-research.

Rostamabadi, H., Demirkesen, I., Colussi, R., Roy, S., Tabassum, N., de Oliveira Filho, J. G., Bist, Y., Kumar, Y., Nowacka, M., Galus, S., & Falsafi, S. R. (2024). Recent trends in the application of films and coatings based on starch, cellulose, chitin, chitosan, xanthan, gellan, pullulan, Arabic gum, alginate, pectin, and carrageenan in food packaging. *Food Frontiers, 5*(2), 350–391. https://doi.org/10.1002/fft2.342, onlinelibrary.wiley.com/journal/26438429.

Roy, S., Ezati, P., & Rhim, J. W. (2022). Fabrication of antioxidant and antimicrobial pullulan/gelatin films integrated with grape seed extract and sulfur nanoparticles. *ACS Applied Bio Materials, 5*(5), 2316–2323. https://doi.org/10.1021/acsabm.2c00148, pubs.acs.org/journal/aabmcb.

Roy, S., Min, S. J., Biswas, D., & Rhim, J. W. (2023). Pullulan/chitosan-based functional film incorporated with curcumin-integrated chitosan nanoparticles. *Colloids and Surfaces A: Physicochemical and Engineering Aspects, 660*. https://doi.org/10.1016/j.colsurfa.2022.130898, www.elsevier.com/locate/colsurfa.

Roy, S., & Rhim, J.-W. (2023). Pullulan/agar-based functional film containing eucalyptus essential oil and rutin. *Coatings, 13*(2), 460. https://doi.org/10.3390/coatings13020460.

Sabater, C., Olano, A., Corzo, N., & Montilla, A. (2019). GC–MS characterisation of novel artichoke (*Cynara scolymus*) pectic-oligosaccharides mixtures by the application of machine learning algorithms and competitive fragmentation modelling. *Carbohydrate Polymers, 205*, 513–523. https://doi.org/10.1016/j.carbpol.2018.10.054.

Santos, S. P. D., Angioletti, B. L., Hoffmann, T. G., Gonçalves, M. J., Bertoli, S. L., Hlebová, M., Khayrullin, M., Gribkova, V., Shariati, M. A., & De Souza, C. K. (2022). Whey based biopolymeric coating as an alternative to improve quality of fresh fruits (Malpighia Emarginata D.C.) from Southern Brazil. *Journal of Microbiology, Biotechnology and Food Sciences, 11*(5). https://doi.org/10.55251/jmbfs.5433, https://office2.jmbfs.org/index.php/JMBFS/index.

Shah, Y. A., Bhatia, S., Al-Harrasi, A., Afzaal, M., Saeed, F., Anwer, M. K., Khan, M. R., Jawad, M., Akram, N., & Faisal, Z. (2023). Mechanical properties of protein-based food packaging materials. *Polymers, 15*(7). https://doi.org/10.3390/polym15071724, http://www.mdpi.com/journal/polymers.

Shanmugam, R., Mayakrishnan, V., Kesavan, R., Shanmugam, K., Veeramani, S., & Ilangovan, R. (2022). Mechanical, barrier, adhesion and antibacterial properties of pullulan/graphene bio nanocomposite coating on spray coated nanocellulose film for food packaging applications. *Journal of Polymers and the Environment, 30*(5), 1749–1757. https://doi.org/10.1007/s10924-021-02311-2.

Simões, A., Ramos, A., Domingues, F., & Luís, Â. (2024). Pullulan-tween 40 emulsified films containing geraniol: Production and characterization as potential food packaging materials. *European Food Research and Technology, 250*(6), 1721–1732. https://doi.org/10.1007/s00217-024-04514-y.

Singh, R. S., Kaur, N., & Kennedy, J. F. (2019). Pullulan production from agro-industrial waste and its applications in food industry: A review. *Carbohydrate Polymers, 217*, 46–57. https://doi.org/10.1016/j.carbpol.2019.04.050, http://www.elsevier.com/wps/find/journaldescription.cws_home/405871/description#description.

Soltani Firouz, M., Mohi-Alden, K., & Omid, M. (2021). A critical review on intelligent and active packaging in the food industry: Research and development. *Food Research International, 141*, 110113. https://doi.org/10.1016/j.foodres.2021.110113.

Sugumaran, K. R., & Ponnusami, V. (2017). Review on production, downstream processing and characterization of microbial pullulan. *Carbohydrate Polymers, 173*, 573–591. https://doi.org/10.1016/j.carbpol.2017.06.022.

Suhag, R., Kumar, N., Petkoska, A. T., & Upadhyay, A. (2020). Film formation and deposition methods of edible coating on food products: A review. *Food Research International, 136*, 109582. https://doi.org/10.1016/j.foodres.2020.109582.

Tabasum, S., Noreen, A., Maqsood, M. F., Umar, H., Akram, N., Nazli, Zi. H., Chatha, S. A. S., & Zia, K. M. (2018). A review on versatile applications of blends and composites of pullulan with natural and synthetic polymers. *International Journal of Biological Macromolecules, 120*, 603–632. https://doi.org/10.1016/j.ijbiomac.2018.07.154, www.elsevier.com/locate/ijbiomac.

Tajeddin, B., Ahmadi, B., Sohrab, F., & Ahmadi Chenarbon, H. (2018). Polymers for. In *Modified atmosphere packaging applications*. Elsevier BV, 457–499. https://doi.org/10.1016/b978-0-12-811516-9.00014-2.

Tan, S. X., Ong, H. C., Andriyana, A., Lim, S., Pang, Y. L., Kusumo, F., & Ngoh, G. C. (2022). Characterization and parametric study on mechanical properties enhancement in biodegradable chitosan-reinforced starch-based bioplastic film. *Polymers, 14*(2). https://doi.org/10.3390/polym14020278, https://www.mdpi.com/2073-4360/14/2/278/pdf.

Tang, B., Ma, J., Liu, L., Xu, J., Zhang, H., Li, K., Tu, X., & Guan, Q. (2025). Preparation of pullulan-shellac edible films with improved water-resistance and UV barrier properties for Chinese cherries preservation. *Journal of Future Foods, 5*(1), 107–118. https://doi.org/10.1016/j.jfutfo.2024.01.010.

Thomas, N., Puluhulawa, L. E., Cindana Mo'o, F. R., Rusdin, A., Mohd Gazzali, A., & Budiman, A. (2024). Potential of pullulan-based polymeric nanoparticles for improving drug physicochemical properties and effectiveness. *Polymers, 16*(15), 2151. https://doi.org/10.3390/polym16152151.

Tiwari, S., Patil, R., Dubey, S. K., & Bahadur, P. (2019). Derivatization approaches and applications of pullulan. *Advances in Colloid and Interface Science, 269*, 296–308. https://doi.org/10.1016/j.cis.2019.04.014.

Treviño-Garza, M. Z., García, S., Heredia, N., Alanís-Guzmán, M. G., & Arévalo-Niño, K. (2017). Layer-by-layer edible coatings based on mucilages, pullulan and chitosan and its effect on quality and preservation of fresh-cut pineapple (*Ananas comosus*). *Postharvest Biology and Technology, 128*, 63–75. https://doi.org/10.1016/j.postharvbio.2017.01.007, www.elsevier.com/inca/publications/store/5/0/3/3/1/3/index.htt.

Trinetta, V. (2017). *The application of edible and active pullulan coatings on foods*. Elsevier BV. https://doi.org/10.1016/b978-0-08-100596-5.21129-3.

Unalan, I. U., Wan, C., Figiel, Ł. F., Olsson, R. T., Trabattoni, S., & Farris, S. (2015). Exceptional oxygen barrier performance of pullulan nanocomposites with ultra-low loading of graphene oxide. *Nanotechnology, 26*(27). https://doi.org/10.1088/0957-4484/26/27/275703, http://iopscience.iop.org/0957-4484/26/27/275703/pdf/0957-4484_26_27_275703.pdf.

Usman, A., Hussain, Z., Riaz, A., & Khan, A. N. (2016). Enhanced mechanical, thermal and antimicrobial properties of poly(vinyl alcohol)/graphene oxide/starch/silver nanocomposites films. *Carbohydrate Polymers, 153*, 592–599. https://doi.org/10.1016/j.carbpol.2016.08.026, http://www.elsevier.com/wps/find/journaldescription.cws_home/405871/description#description.

Vasile, C., & Baican, M. (2021). Progresses in food packaging, food quality, and safety—Controlled-release antioxidant and/or antimicrobial packaging. *Molecules (Basel, Switzerland), 26*(5), 1263. https://doi.org/10.3390/molecules26051263.

Vieira, M. G. A., Da Silva, M. A., Dos Santos, L. O., & Beppu, M. M. (2011). Natural-based plasticizers and biopolymer films: A review. *European Polymer Journal, 47*(3), 254–263. https://doi.org/10.1016/j.eurpolymj.2010.12.011.

Wang, H., Chen, X., Yang, H., Wu, K., Guo, M., Wang, X., Fang, Y., & Li, L. (2025). A novel gelatin composite film with melt extrusion for walnut oil packaging. *Food Chemistry, 462*. https://doi.org/10.1016/j.foodchem.2024.141021, https://www.sciencedirect.com/science/journal/03088146.

Wani, S. M., Mir, S. A., Khanday, F. A., & Masoodi, F. A. (2021). Advances in pullulan production from agro-based wastes by *Aureobasidium pullulans* and its applications. *Innovative Food Science and Emerging Technologies, 74*. https://doi.org/10.1016/j.ifset.2021.102846, http://www.elsevier.com/locate/ifset.

Wei, X., Liu, G. L., Jia, S. L., Chi, Z., Hu, Z., & Chi, Z. M. (2021). Pullulan biosynthesis and its regulation in *Aureobasidium* spp. *Carbohydrate Polymers, 251*. https://doi.org/10.1016/j.carbpol.2020.117076, http://www.elsevier.com/wps/find/journaldescription.cws_home/405871/description#description.

Westlake, J. R., Tran, M. W., Jiang, Y., Zhang, X., Burrows, A. D., & Xie, M. (2023). Biodegradable biopolymers for active packaging: Demand, development and directions. *Sustainable Food Technology, 1*(1), 50–72. https://doi.org/10.1039/d2fb00004k, www.rsc.org/journals-books-databases/about-journals/sustainable-food-technology/.

Wu, F., Misra, M., & Mohanty, A. K. (2021). Challenges and new opportunities on barrier performance of biodegradable polymers for sustainable packaging. *Progress in Polymer Science, 117*, 101395. https://doi.org/10.1016/j.progpolymsci.2021.101395.

Wypij, M., Rai, M., Zemljič, L. F., Bračič, M., Hribernik, S., & Golińska, P. (2023). Pullulan-based films impregnated with silver nanoparticles from the *Fusarium culmorum* strain JTW1 for potential applications in the food industry and medicine. *Frontiers in Bioengineering and Biotechnology, 11*. https://doi.org/10.3389/fbioe.2023.1241739, http://journal.frontiersin.org/journal/bioengineering-and-biotechnology#archive.

Xiao, L., Kang, S., Lapu, M., Jiang, P., Wang, X., Liu, D., Li, J., & Liu, M. (2022). Preparation and characterization of chitosan/pullulan film loading carvacrol for targeted antibacterial packaging of chilled meat. *International Journal of Biological Macromolecules, 211*, 140–149. https://doi.org/10.1016/j.ijbiomac.2022.05.044.

Xiao, Q., Lim, L. T., & Tong, Q. (2012). Properties of pullulan-based blend films as affected by alginate content and relative humidity. *Carbohydrate Polymers, 87*(1), 227–234. https://doi.org/10.1016/j.carbpol.2011.07.040.

Xiao, Q., Tong, Q., & Lim, L. T. (2012). Pullulan-sodium alginate based edible films: Rheological properties of film forming solutions. *Carbohydrate Polymers, 87*(2), 1689–1695. https://doi.org/10.1016/j.carbpol.2011.09.077.

Yaashikaa, P. R., Kamalesh, R., Senthil Kumar, P., Saravanan, A., Vijayasri, K., & Rangasamy, G. (2023). Recent advances in edible coatings and their application in food packaging. *Food Research International, 173*(Pt 2), 113366. https://doi.org/10.1016/j.foodres.2023.113366.

Yang, X., Li, A., Li, X., Sun, L., & Guo, Y. (2020). An overview of classifications, properties of food polysaccharides and their links to applications in improving food textures. *Trends in Food Science & Technology, 102*, 1–15. https://doi.org/10.1016/j.tifs.2020.05.020.

Yeasmin, S., Yeum, J. H., & Yang, S. B. (2020). Fabrication and characterization of pullulan-based nanocomposites reinforced with montmorillonite and tempo cellulose nanofibril. *Carbohydrate Polymers, 240*. https://doi.org/10.1016/j.carbpol.2020.116307, http://www.elsevier.com/wps/find/journaldescription.cws_home/405871/description#description.

Yildirim, M., & Candan, Z. (2023). Smart materials: The next generation in science and engineering. *Materials Today: Proceedings*. https://doi.org/10.1016/j.matpr.2023.10.116.

Zang, J., Xiao, P., Chen, Y., Liu, Z., Tang, D., Liu, Y., Chen, J., Tu, Y., & Yin, Z. (2024). Hydrocolloid application in yogurt: Progress, challenges and future trends. *Food Hydrocolloids, 153*, 110069. https://doi.org/10.1016/j.foodhyd.2024.110069.

Zhang, L., Yu, X., Liu, L., Chen, Y., & Xu, Y. (2023). Bioinspired pullulan-starch nanoplatelets nanocomposite films with enhanced strength and toughness. *Carbohydrate Polymers, 299*. https://doi.org/10.1016/j.carbpol.2023.120190.

Zhang, L., Huang, C., & Zhao, H. (2019). Application of pullulan and chitosan multilayer coatings in fresh papayas. *Coatings, 9*(11), 745. https://doi.org/10.3390/coatings9110745.

Zhao, Z., Xiong, X., Zhou, H., & Xiao, Q. (2019). Effect of lactoferrin on physicochemical properties and microstructure of pullulan-based edible films. *Journal of the Science of Food and Agriculture, 99*(8), 4150–4157. https://doi.org/10.1002/jsfa.9645, http://onlinelibrary.wiley.com/journal/10.1002/(ISSN)1097-0010.

Zheng, M., Zhu, Y., Zhuang, Y., Tan, K. B., & Chen, J. (2023). Effects of grape seed extract on the properties of pullulan polysaccharide/xanthan gum active films for apple preservation. *International Journal of Biological Macromolecules, 241*, 124617. https://doi.org/10.1016/j.ijbiomac.2023.124617.

Chapter 10

Whey protein films and coatings

Brenda Alessandra Munhoz*, Krízia Emanuelle dos Santos Ferreira*, Luís Marangoni Júnior, and Débora Parra Baptista

Department of Food Engineering and Technology, School of Food Engineering (FEA), University of Campinas (Unicamp), Campinas, São Paulo, Brazil

10.1 Introduction

Recently, there has been a growing awareness among the population regarding sustainability, leading to increased concern about reducing waste, excess production, and the use of nonbiodegradable materials, particularly in the food sector. In this context, new film and coating technologies derived from biopolymers offer a promising alternative to packaging made from nonrenewable sources, such as plastic polymers, which may be associated with higher greenhouse gas emissions (Kandasamy et al., 2021).

Films and coatings are materials used in the food packaging sector to ensure the integrity and quality of the final product, protecting it from extrinsic factors (Kumar et al., 2022). Using biopolymers in packaging enables the use of renewable resources and the potential repurposing of industrial waste, contributing to more sustainable and eco-friendly solutions (Avramescu et al., 2020). Whey is a coproduct generated in large volumes during cheese production, and it has been processed using various technologies that enable the recovery of its constituents, such as lactose and whey proteins. Although the technologies for processing whey to obtain high-value-added coproducts are well established, the whey generated is not fully processed, which can result in wasted nutrients, environmental impacts, and financial losses.

Therefore, the use of whey in the production of edible films and coatings presents a promising application for this coproduct, as it has the potential to provide a functional barrier between food and the external environment, thereby ensuring its quality (Avramescu et al., 2020). Additionally, biopolymer production from whey can contribute to the United Nations' 2030 Sustainable Development Goals (SDGs), particularly goals 12 and 13, which focus on Responsible Consumption and Production and Climate Action, respectively (Fematt-flores et al., 2022; United Nations, 2024).

Whey contains significant amounts of proteins, with β-lactoglobulin and α-lactalbumin being the most abundant. These proteins can contribute to the development of films and coatings due to their favorable physicochemical, mechanical, optical, and sensory properties. As a result, whey protein shows great potential for use in this technology, either in the form of whey protein concentrate (WPC) or whey protein isolate (WPI). Furthermore, whey proteins can be combined with other biopolymers and various additives to enhance the desired properties (Kandasamy et al., 2021). In this context, this chapter aims to explore the characterization and production of whey, as well as the processing, composition/formulation, properties, applications, challenges, and emerging trends in the development of whey protein-derived films and coatings.

10.2 Whey

Whey is a coproduct of the dairy industry, obtained through acid or enzymatic milk coagulation, followed by curd drainage. Although it was once regarded as a waste product of cheese production, whey is now recognized as a valuable coproduct from which several high-value ingredients are obtained, including whey powder, lactose, WPC, and WPI (O'Mahony and Fox, 2014; Tsermoula et al., 2021).

During the cheese-making process, approximately 80%–90% of the milk volume is expelled as whey, a cloudy, yellowish liquid that contains about 50% of the milk solids (Tsermoula et al., 2021). The composition of whey is variable,

* These authors contributed equally to this work and share first authorship.

Natural Polymer-Based Films and Coatings in Food Technology. DOI: https://doi.org/10.1016/B978-0-443-34244-8.00007-0

depending on factors such as the processing method, type of coagulation (acid or enzymatic), cheese variety, cow's nutritional management practices, and stage of lactation. Despite this variability, lactose is the most abundant component (45–50 $g \cdot L^{-1}$), followed by proteins (6–11 $g \cdot L^{-1}$), minerals (8–10 $g \cdot L^{-1}$), lactic acid (0.5–9 $g \cdot L^{-1}$), and lipids (0.6–5 $g \cdot L^{-1}$). Among the components of whey, proteins are considered the most economically and technologically significant. This is due to their recognized functional properties, including acid stability, film formation, gelling, and emulsification. Additionally, whey proteins offer various nutritional benefits (Kandasamy et al., 2021; Ramos et al., 2015).

10.2.1 Whey production and processing

Obtaining whey involves destabilizing the milk's physicochemical system through acid or enzymatic coagulation, followed by the separation of a curd or precipitate primarily composed of caseins and an aqueous phase, which is the whey. Milk coagulation requires the destabilization of casein micelles, the structural form of caseins, the major proteins in milk. Casein micelles consist of clusters of thousands of protein molecules, including casein fractions (α_{s1}-, α_{s2}-, -, β- and κ-casein) and calcium phosphate. These micelles are stabilized by hydrophobic interactions, internal calcium-phosphate bonds, and surface properties that promote electrostatic repulsion due to the micelle's negative charge, as well as steric hindrance from the C-terminal region of κ-casein, known as "hairs," located on the micelle's surface (Dalgleish & Corredig, 2012).

Acidification of milk, either through the action of lactic acid bacteria that ferment lactose to produce lactic acid or by direct acid addition, leads to the neutralization of casein micelle charges, thereby reducing electrostatic repulsion. This process also causes the collapse of κ-casein hairs, reducing steric hindrance and facilitating micelle aggregation (Fox et al., 2015). Acid coagulation can lead to the formation of either a gel or a precipitate, depending on the rate of acidification. In the slow acidification process, as occurs during milk fermentation, aggregation forces gradually overcome the repulsive forces between micelles, forming a network characteristic of an acid gel. In contrast, rapid acidification, particularly when acid is directly added to milk with stirring and heating, causes abrupt destabilization, leading to extensive fusion between micelle surfaces. This results in dense aggregates that settle as precipitates (Fox et al., 2017).

Enzymatic coagulation, commonly used in cheese production, results from the action of the enzyme chymosin, which cleaves the bond between phenylalanine at position 105 and methionine at position 106 of κ-casein. This cleavage removes the macropeptide, known as the "hair" of κ-casein, which is solubilized in the whey, thereby exposing the para-κ-casein portion that is sensitive to free ionic calcium. As a result, steric hindrance and electrostatic repulsion are reduced, leading to the formation of an aggregate between micelles, a three-dimensional protein network that expels the aqueous phase from the curd (Fox et al., 2017; Hyslop, 2003).

As described above, milk's acid and enzymatic coagulation destabilizes casein micelles, which facilitates the expulsion of the milk's aqueous phase, or whey, enabling its separation and use as a coproduct in the development of new products. Thus, whey can be classified as acid or sweet, depending on the type of coagulation used to obtain it. Enzymatic coagulation produces sweet whey, which has a pH range of 5.3–6.6. In contrast, acid whey results from acid coagulation and has a pH ranging from 4.4 to 5.3 (Poppi et al., 2010). Besides the difference in acidity, acid whey has a higher content of calcium phosphate compared to sweet whey, due to solubilization induced by acidification (Rocha-Mendoza et al., 2021).

Whey proteins are not sensitive to acidification and remain soluble even as the pH decreases. They are also resistant to the action of proteolytic enzymes used in cheese production. As a result, regardless of the coagulation method applied to milk, destabilization leads to the separation of two protein fractions: the caseins, which aggregate, and the whey proteins, which remain soluble in the aqueous phase, are consequently considered coproducts of cheese production (Madureira et al., 2007; Subbiah Prabhakaran et al., 2024).

To separate and concentrate the protein fraction in the whey, membrane separation techniques are employed based on the molecular size of the components. This physical separation occurs under hydraulic pressure, with the feed solution flowing tangentially across the membrane surface (cross-flow filtration). Components larger than the membrane's pore size are retained as the retentate, while the permeate, consisting of smaller molecular weight components, passes through the membrane material (Damodaran & Parkin, 2017).

Commercially, various protein products derived from whey are available, including WPCs (containing 35%–80% protein), WPIs (containing over 90% protein), whey hydrolysates, lactose, lactose-reduced whey, and demineralized whey. The composition and purity of these products vary depending on the processing and filtration techniques employed (Kandasamy et al., 2021).

Ultrafiltration and nanofiltration are commonly used technologies for whey protein concentration. Nanofiltration is typically employed when partial desalination of whey or permeate, previously treated with ultrafiltration, is required.

Ultrafiltration is primarily used to concentrate whey proteins and is frequently applied in the production of WPCs (Britz & Robinson, 2008). This method utilizes membranes with pore sizes ranging from 0.01 to 0.1 μm, relying on the selective removal of high molecular weight components from the permeate. A concentration principle is also applied to the production of WPIs. In this case, proteins are concentrated using ion exchange chromatography or membrane filtration techniques (McSweeney and O'Mahony, 2016). To obtain the isolate, membrane filtration is followed by intensive diafiltration to enhance protein purification. In this process, water is added to the retentate, allowing solutes to be removed as they permeate the membrane (Britz & Robinson, 2008).

In summary, whey undergoes processing steps to obtain protein ingredients, including pasteurization, clarification, separation, concentration, and fractionation. After these steps, the protein concentrate can be dried for subsequent commercialization (McSweeney and O'Mahony, 2016).

10.2.2 Whey proteins

Whey proteins account for approximately 20% of milk proteins. They are globular, with compact and folded peptide chains, where hydrophilic amino acids are positioned on the surface, forming the interface between water and protein. In contrast, hydrophobic amino acids are located within the core of these structures, which also contain significant amounts of α-helix and β-sheet structures (Walstra et al., 2006).

Whey proteins include β-lactoglobulin, α-lactalbumin, lactoferrin, bovine serum albumin, immunoglobulins, and glycomacropeptides (Belitz et al., 2009; Haraguchi et al., 2006). The specificities of the main whey proteins are summarized in Table 10.1.

β-Lactoglobulin and α-lactalbumin are the major protein fractions of whey, representing approximately 80% of these proteins (Fig. 10.1) (Damodaran & Parkin, 2017).

10.2.2.1 β-Lactoglobulin

β-Lactoglobulin is the major whey protein, accounting for 55% of these proteins. Its relatively small structure consists of 162 amino acid residues and has a molecular weight of 18.3 kDa. This globular thermolabile protein exhibits secondary, tertiary, and quaternary structures and contains two intramolecular disulfide bonds (Cys 66–Cys 160 and Cys 106–Cys 119), as well as a free thiol group (Cys 121). β-Lactoglobulin is classified as a lipocalin protein due to its ability to bind small molecules within hydrophobic cavities (Damodaran & Parkin, 2017; Kandasamy et al., 2021).

β-Lactoglobulin exists in two main genetic variants, A and B, and can be found as monomers, dimers, or octamers, depending on factors such as pH, temperature, and ionic strength. The quaternary dimer structure, where molecules are strongly linked primarily by hydrophobic interactions, is the predominant form under physiological conditions. Its solubility is directly influenced by pH and ionic strength. Generally, it does not precipitate in acidified media (Damodaran & Parkin, 2017; Fox et al., 2015; Walstra et al., 2006).

TABLE 10.1 Major whey protein properties, concentrations, and physiological functions.

Protein	Molecular weight (kDa)	Concentration (g L^{-1})	Proportion (%)	Biological activity
β-Lactoglobulin	18.3	3.5	0.35	Retinol transport
α-Lactalbumin	14.2	1.2	0.12	Antimicrobial; apoptotic effect
Immunoglobulins	150–900	0.7	0.07	Transmission of passive immunity to neonates
Glycomacropeptide	7.0	0.01	0.001	–
Bovine serum albumin	65.0	0.4	0.04	Anticarcinogenic
Lactoferrin	86.1	0.02–0.35	0.003	Antimicrobial; antiinflammatory

Source: From Poppi, F.A., Costa, M., de, R., Rensis, C.M.V.B., & Sivieri, K. (2010). Whey and its Proteins: Composition and Functional Activity. Ciênc. Biol. Saúde., 12, (2), 31–37; Ramos, O.L., Pereira, R.N., Rodrigues, R.M., Teixeira, J.A., Vicente, A.A. & Malcata, F.X. (2015). Whey and Whey Powders: Production and Uses. In Encyclopedia of Food and Health (pp. 498–505). Elsevier Inc. https://doi.org/10.1016/B978-0-12-384947-2.00747-9.

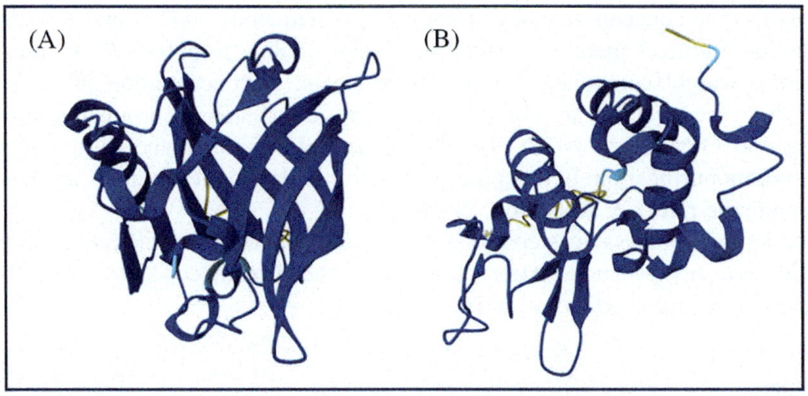

FIGURE 10.1 Structure of the main bovine whey proteins: β-lactoglobulin (A) and α-lactalbumin (B). Protein structures from the AlphaFold Protein Structure Database using UniProt codes P02754 and P00711 as entries for β-lactoglobulin and α-lactalbumin, respectively.
This figure uses images from the AlphaFold Protein Structure Database (AFDB; Varadi et al., 2022, 2024) under a Creative Commons 4.0 license (https://creativecommons.org/licenses/by/4.0/).

The structure of β-lactoglobulin includes eight antiparallel β-sheets, which form a calyx structure specific to this protein fraction. This structure houses the main hydrophobic binding site, which accommodates linear molecules. The conformation of the protein can change in response to pH variations. Under acidic conditions, the protein adopts a closed state, preventing molecules from binding to the specific site. In contrast, the protein adopts an open state under alkaline conditions, exposing the binding site (Sakurai & Goto, 2006; Chatterton et al., 2006).

β-Lactoglobulin exhibits good foaming and heat-induced gelation capacities, and also solubility across a wide pH range, making it a versatile ingredient for texturizing, water binding, and egg white replacement in various food applications (Chatterton et al., 2006).

10.2.2.2 α-Lactalbumin

α-Lactalbumin constitutes 20% of whey proteins and 3.5% of total milk proteins, with a molecular weight of 14.2 kDa. It exists in two main genetic variants, A and B (Belitz et al., 2009). Its structure consists of 123 amino acids and is rich in tryptophan, with approximately four residues per mole, and in sulfur, present in cystine, with four disulfide bonds per mole, and methionine. However, it lacks free sulfhydryl groups. The protein has a compact fold and globular structure, with 26% α-helix and 14% β-structure (Fox et al., 2015; Permyakov, 2020).

This protein fraction is classified as a metalloprotein due to its ability to bind one Ca^{2+} ion per mole through interactions with asparagine (Asp) residues. This characteristic enhances the heat stability of α-lactalbumin (Fox et al., 2015). It is worth noting that α-lactalbumin is recognized for its biological functions, including its antimicrobial activity. Additionally, its binding to minerals, particularly calcium and zinc, can enhance their absorption, providing consumers with increased nutritional benefits (Deeth & Bansal, 2019).

10.2.2.3 Minor whey proteins

Cow's milk contains small amounts of serum albumin (Poppi et al., 2010). Its molecular mass is approximately 65 kDa, consisting of 582 amino acid residues, with 17 disulfide bonds and one free sulfhydryl group (Fox et al., 2015). This protein has an elliptical shape and plays biological roles, such as regulating osmotic pressure (Deeth & Bansal, 2019; Walstra et al., 2006).

Immunoglobulins are glycoproteins derived from the blood, synthesized in response to antigen stimulation, and function as antibodies (Deeth & Bansal, 2019). They are characterized as monomers or polymers consisting of four polypeptide chains: two shorter chains, approximately 20 kDa each, and two longer chains with molecular weights ranging from 50 to 70 kDa, all linked by disulfide bonds (Fox et al., 2015; Walstra et al., 2006).

Lactoferrin is also a glycoprotein, with a molecular weight of approximately 86.1 kDa, and it strongly binds two moles of iron per mole of protein (Deeth & Bansal, 2019). It performs several bioactive functions, including high antibacterial activity (Walstra et al., 2006).

10.3 Whey protein films and coatings

Films and coatings are employed in the food industry to maintain the integrity and quality of the final product (Avramescu et al., 2020). Over the years, plastics derived from fossil sources have been widely used in polymeric packaging due to their low cost and strong performance. However, because of their complex decomposition and disposal processes, they have been identified as contributors to contamination and environmental risks, highlighting the need to explore alternative natural sources (Xu et al., 2024). Currently, only 20% of plastic materials are recycled globally each year. As a result, both industrial and academic sectors are focused on conducting research to address the issues associated with the improper disposal of packaging materials. The application of a circular economy to existing materials, along with the development of films and coatings derived from renewable sources, represents an important step toward enhancing sustainability in the food packaging production chain (Brown & Börkey, 2024; Kandasamy et al., 2021; Marangoni Júnior et al., 2022; Xu et al., 2024; Yang et al., 2024).

Although biopolymer-based films and coatings face several challenges, such as cost, processability, and performance properties, their applications in food packaging may improve the preservation of food's physical, chemical, and sensory attributes, ultimately extending product shelf life. Additionally, these materials provide other benefits, such as improved ecological outcomes, since they can either be consumed as edible films or are biodegradable and/or compostable (Lopez-Polo et al., 2024). In this context, the development of biopolymer-based films and coatings presents an attractive alternative for applications like food packaging, positively impacting manufacturing processes (Avramescu et al., 2020).

Biopolymer-based films must possess physical properties such as flexibility, low thickness, and satisfactory mechanical and barrier performance to effectively wrap food products. However, these characteristics still represent a challenge for the scientific and industrial community. In contrast, biopolymer coatings are applied directly to the food surface by immersing the product in a film-forming solution, followed by solvent evaporation (Kandasamy et al., 2021).

Films and coatings produced from whey proteins hold promising prospects, supported by research and industrial applications, due to their abundant source and biodegradability (Kandasamy et al., 2021). Whey protein biopolymers exhibit desirable properties, including mechanical and optical strength, and good oxygen permeability. These polymers possess a shiny, colorless, and translucent appearance, excellent flexibility, and are odorless (络Ематt-flores et al., 2022). Furthermore, their application can guarantee food quality, preserving the physical-chemical and sensory characteristics of the products (Schmid et al., 2012). Thus, whey protein-based films and coatings have been tested on various foods, including nuts, meats, fish, cheeses, and fruits, without affecting their sensory attributes (Badr et al., 2014; Reinoso et al., 2008; Seyfzadeh et al., 2013).

Compared to other natural sources of biopolymers, whey protein stands out due to its amphiphilic nature, potential for denaturation, electrostatic charges, and structural conformation. The practical development of whey protein-based materials relies on low-energy bonds, such as electrostatic interactions, hydrogen bonds, or van der Waals forces. In the production of adhesive polymers, the globular structure of whey proteins is reorganized to achieve linearity, forming irreversible aggregates through thiol-disulfide interactions. These physicochemical characteristics significantly influence the final properties of films and coatings (Henriques et al., 2016; Kandasamy et al., 2021).

Heat treatment is widely used to form gels, with denaturation being the most common technique for obtaining a uniform film-forming solution (Kim & Ustunol, 2001). The hydrophobic and -SH groups are located inside the whey protein molecules; however, denaturation exposes these functional and hydrophobic groups, allowing the formation of a three-dimensional network with intermolecular S-S bonds and hydrophobic interactions during solvent evaporation (Kandasamy et al., 2021; Krochta & Pérez-Gago, 2002; Ramos et al., 2012). Therefore, whey protein-based films and coatings exhibit distinct characteristics resulting from the interactions within this three-dimensional network (Kontogianni et al., 2022).

Thermal denaturation of whey proteins occurs in two stages. The first stage involves heating above 40°C and pH 6.0, which causes the dissociation of β-lactoglobulin dimers into monomers. As the temperature rises above 65°C, β-lactoglobulin unfolds, exposing hydrophobic and thiol groups, leading to aggregation with other smaller protein compounds. In the second stage, these smaller aggregates interact to form new, irreversible aggregates with high molecular weight. It is important to note that, for irreversible denaturation of the main whey proteins, temperatures of at least 69°C for β-lactoglobulin and 80°C for α-lactalbumin are required (Kandasamy et al., 2021).

Disulfide bonds, formed in the polymer matrix as a result of heat treatment, contribute to partial insolubility in water. Meanwhile, hydrogen bonds enhance water vapor permeability, as the polar nature of these films and coatings facilitates the mass transfer of water vapor molecules through the material (Schmid et al., 2012). The films and coatings formed from these interactions are characterized by transparency, flexibility, lack of color and taste, and resistance and insolubility in water (Mihalca et al., 2021).

The protein network that characterizes films and coatings is formed through the rearrangement of the conformation of these proteins, with the exposure of the hydrophobic and free SH groups, to create the three-dimensional network. However, this matrix is highly permeable to water vapor, which represents the main limitation in films and coatings. This challenge is often addressed by formulating film-forming solutions with various matrices and additives, thereby improving undesirable characteristics.

10.4 Composition of films and coatings

Film-forming solutions are developed from WPCs or isolates, consisting of three main components: the matrix, made up of biopolymers; additives (such as plasticizers, surfactants, and crosslinking agents) to enhance functional properties; and the solvent, typically water or ethanol (Ribeiro et al., 2021). Careful solvent selection is important to ensure compatibility among all the ingredients of the film-forming solution (Han & Aristippos, 2005).

The steps for preparing whey protein films or coatings include dissolving WPC or WPI, approximately 5%–12%, in water. The pH of the solution can then be adjusted to between 6 and 8, or it can be left unadjusted and heated to temperatures between 80°C and 90°C for 10–30 minutes to induce protein denaturation (Krochta & Pérez-Gago, 2002). Depending on their compatibility with the heating process, the remaining ingredients can be added before or after heating (Fernandes et al., 2020).

It is important to note that increasing the protein concentration in the solution can enhance resistance, as a higher number of sulfhydryl groups on the surface of the structures promotes the formation of more S-S bonds in the film-forming matrix (Oliveira et al., 2017).

Despite its advantages, this matrix has limitations, with water vapor permeability being the primary concern (Schmid & Müller, 2018). To address this issue, biopolymer blends are commonly used. These blends rely on physical interactions, such as intermolecular hydrogen bonds, and require polymer compatibility, which depends on adhesion and cohesion at the interface (Achari et al., 2024).

The interaction between the biopolymer and the substrate is termed adhesion, while cohesion refers to the interaction between the polymers in the solution. These two interactions are fundamental in determining the formation of films and coatings and their properties (Khwaldia et al., 2004). Cohesive forces in films or coatings made from a single biopolymer can limit strength, leading to undesirable brittleness. For this reason, additives—mainly plasticizers and other biopolymers (blends)—are added to reduce intermolecular forces, consequently increasing the mobility of the polymer chains (Kandasamy et al., 2021; Sothornvit & Krochta, 2005).

The use of biopolymer blends in whey protein-based films and coatings has already been reported, such as the mixture of WPI with biofiber gum (Jin et al., 2023) or pectin (Ke et al., 2025). In these studies, the authors discussed the limitations of using whey proteins individually and sought to overcome their disadvantages in blends, obtaining improvements in strength, flexibility, and water vapor permeability when using blends to formulate films and coatings (Jin et al., 2023; Ke et al., 2025).

In addition to blends with biopolymers, blends with synthetic polymers are also utilized for the same purpose. Biodegradable polymers are the most commonly explored, with a focus on sustainability. The most frequently used combinations include whey proteins with polyvinyl alcohol (PVA), polylactic acid (PLA), polybutylene succinate (PBS), polybutylene adipate-co-terephthalate (PBAT), or poly ε-caprolactone (PCL). These blends exhibit improved processing characteristics and, consequently, enhanced functional properties of films and coatings, resulting in more flexible, durable, resistant packaging, and capable of degrading more rapidly. However, the main challenge lies in the compatibility between whey proteins and synthetic polymers since biopolymers are hydrophilic and synthetic polymers are hydrophobic. This challenge can be addressed by incorporating plasticizers, which help improve compatibility and facilitate the formation of a more consistent matrix (Achari et al., 2024).

Additives can be incorporated into the preparation of whey protein films and coatings to enhance their functional properties, thereby improving processability, morphology, and mechanical characteristics. These additives include plasticizers, crosslinkers, thickeners, stabilizers, surfactants, and other compounds that contribute to the overall performance of the films and coatings (Ribeiro et al., 2021).

Plasticizing agents are added to the formulation to prevent the formation of a rigid and brittle structure, which can result from extensive bonding between the biopolymer chains. By reducing these bonds, plasticizers help to decrease tensile strength and increase the elongation of the film (Schmid et al., 2012). The addition of plasticizers can enhance the flexibility, extensibility, and dispensability of films. It also influences the permeability to water vapor, oxygen, and solutes in coatings, while facilitating the release of active compounds by increasing the mobility of the polymer chains (Ribeiro et al., 2021; Shimazu et al., 2007). Water is highly effective as a plasticizer; however, its effectiveness may be reduced in amphiphilic biopolymers due to the environmental conditions to which the material is exposed. Glycerol and sorbitol are also commonly used as plasticizers (Kumar et al., 2022).

Crosslinker agents are additives that reduce the mobility of biopolymer chains through the formation of intermolecular covalent bonds, thereby enhancing the functional properties and extending the shelf life of films and coatings. Various crosslinking methods can be employed, including the addition of aldehydes, heat treatments, irradiation, and acid-base treatments. Among these, glutaraldehyde is the most commonly used crosslinker in whey protein films and coatings due to its unique advantages, such as improving water resistance and preventing the premature degradation of bioactive compounds typically incorporated into active packaging (Ribeiro et al., 2021).

Other commonly used additives in whey protein-based films and coatings are essential oils derived from various natural sources. These oils offer valuable properties that enhance the shelf life of foods coated with protein films and coatings. Their primary benefits include antimicrobial and antioxidant activities, which help delay food deterioration and reduce the waste rate of perishable foods—approximately 25% of which are discarded due to microbiological contamination. Furthermore, incorporating antimicrobial agents into biopolymer formulations contributes to the creation of "clean label" foods by replacing synthetic additives with renewable, natural alternatives (Kumar et al., 2022). Çakmak et al. (2020) investigated edible films based on whey, incorporating lemon and bergamot essential oils, and found them to have a satisfactory antimicrobial effect against *Escherichia coli* and *Staphylococcus aureus*. Other essential oils, such as those derived from thyme, oregano, and cumin, have also been successfully applied in food films, offering additional antimicrobial properties to enhance food preservation (Catarino et al., 2017; Lanciu Dorofte et al., 2023; Nemati et al., 2023; Oliveira et al., 2017). Since protein films are generally neutral, adding essential oils can influence the final product's smell and taste. Moreover, these compounds are released gradually, helping maintain low microbial counts during the storage of the films. This characteristic can also be explored in pharmaceutical applications. It is important to note that the diffusion of essential oils within the biopolymer matrix is influenced by various parameters, including the source of the oil, processing conditions, formulation, temperature, pH, and water activity (Mohamed et al., 2022; Nemati et al., 2023).

Another class of additives that can be incorporated into whey protein films is surfactants. These compounds enhance the spreading ability of films and coatings by increasing surface energy and promoting molecular interactions within the film network. Surfactants improve wettability and adhesion between the food and the coating and help ensure a uniform surface (Ribeiro et al., 2021). This effect is achieved by reducing the surface tension of the film-forming solution, as the surfactant concentration can influence the colligative properties of the molecules present. Commonly used surfactants in formulations include Tween 80 and Tween 20 (Rodríguez et al., 2006).

Thickeners and stabilizers are also incorporated into formulations to enhance texture, act as barriers, and maintain the functional activity of films and coatings. Polysaccharides, such as gums, pectin, and starch, as well as proteins like gelatin and pectin, are used as thickeners. These ingredients help form more homogeneous biopolymers, improving their structural, mechanical, and organoleptic characteristics (Salehi, 2019). Additionally, stabilizers help maintain the structural integrity of the films and prevent phase separation during storage, thereby enhancing the strength, durability, and resistance of the formed network. Common stabilizers include cellulose derivatives, seaweed extracts, and vegetable and microbial gums. The stabilizer concentration may vary depending on the intended purpose (Lacroix & Le Tien, 2005; Salehi, 2019).

10.5 Methods for preparing films and coatings

The preparation of films and coatings must enable the formation of a dry polymer network that minimizes the interaction of the environment with the packaged food (Henriques et al., 2016). In film development, various techniques are utilized to produce an edible biopolymer for food packaging applications, including casting (wet), extrusion (dry), and electrospinning (Kumar et al., 2022).

The casting method is commonly employed at the laboratory scale. It is a low cost, wet-processing method in which the polymeric material and required additives are dissolved in a suitable solvent to prepare the film-forming solution. The resulting solution is then evenly distributed in a mold. After spreading, the solution is dried to evaporate the solvent, which can be carried out using a forced circulation drying oven, vacuum dryers, or microwave radiation (Kumar et al., 2022). After the evaporation of the solvent, the films are carefully removed from the mold (Fig. 10.2). The challenges of this method include the potential formation of surface defects, such as wrinkles and cracks, in the resulting film. Therefore, parameters such as film thickness, solution composition, and drying conditions must be carefully controlled to regulate molecular interactions and the mobility of the polymer chains (Ribeiro et al., 2021).

The extrusion technique is commonly used in commercial processes, leveraging the thermoplastic properties of polymers with minimal solvent usage (dry method) and few unit operations. It involves applying heat above the glass

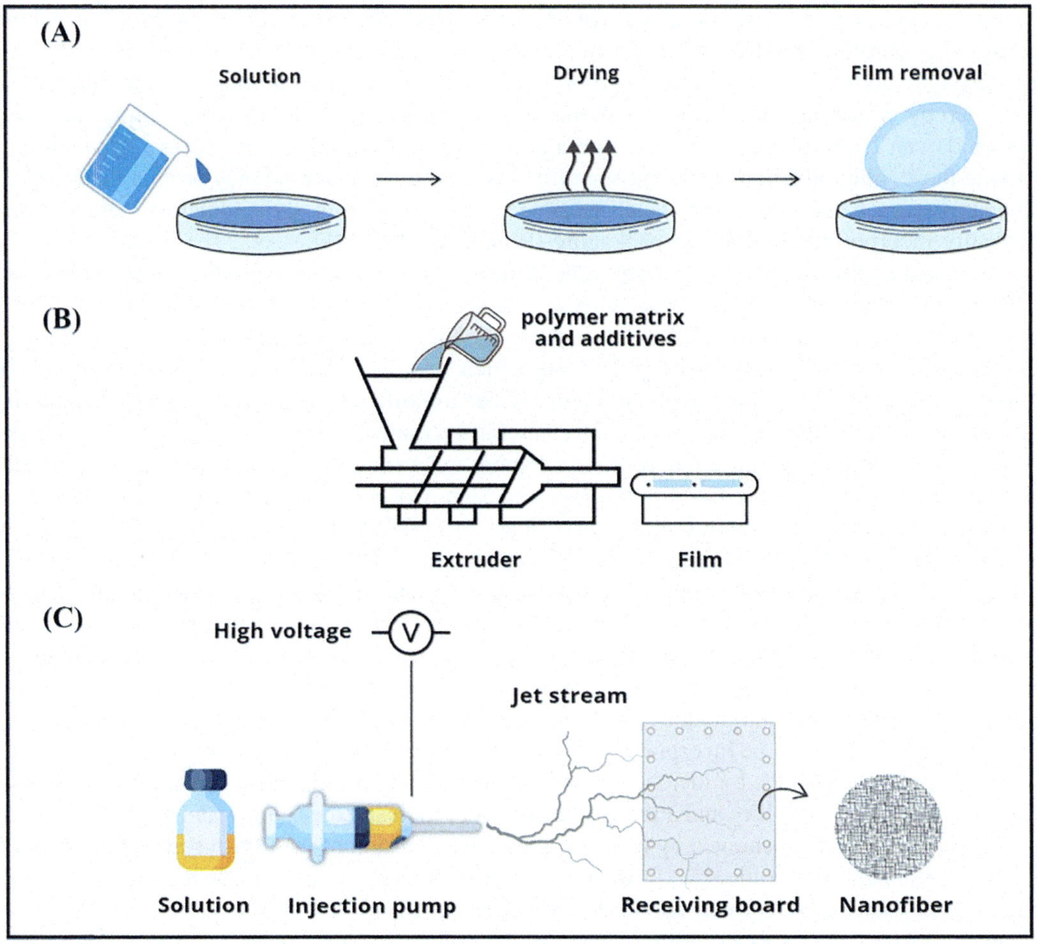

FIGURE 10.2 Illustrative scheme of the methods used for film preparation. Casting (A), extrusion (B), and electrospinning (C) methods.

transition temperature of the added plasticizer under conditions of low water content, which facilitates the development of viscoelastic properties (Ribeiro et al., 2021). In the extruder, the ingredients are introduced through the feed section and moved to the mixing and compression stage by the rotation of a screw. Heating is then applied to homogenize the mixture. At the end of the extruder, a die is used to define the shape and thickness of the film (Fig. 10.2B). Key parameters, such as temperature, screw speed, solvent content, and zone size, must be carefully controlled to ensure process efficiency and the proper development of the mechanical and optical properties. Multilayer films with enhanced properties can be produced using a twin-screw extruder, a process known as coextrusion. It is important to note that the biopolymer must be able to withstand higher temperatures due to the increased shear forces in this process (Kumar et al., 2022).

A technique recently employed in film production is electrospinning. This method produces ultrafine fibers using high voltages, eliminating the need for purification steps and the complex conditions (such as high temperatures or acidity) required by other methods. The applied voltage causes charge accumulation on the surface of the protein solution, resulting in Coulombic repulsion between the polymer particles. The equipment includes a voltage power supply (continuous or alternating) connected to a needle, where the particles are charged, initiating the Coulombic repulsive force. Once surface tension is overcome, ultrafine fibers are formed and collected on a plate (Wang et al., 2023). Fig. 10.2C illustrates the electrospinning method.

Some factors must be considered when producing films by electrospinning, such as protein solution concentration, molecular weight, viscosity, flow rate, and voltage. Forming these fibrous materials is considered simple and low cost, presenting advantageous characteristics, as they can be developed on a large scale with various thicknesses and biocompatible materials (Lopez-Polo et al., 2024).

Four distinct techniques are reported for applying edible coatings: immersion, spraying, fluidized bed, and panning. Among these, immersion and spraying are the most widely studied. The choice of application method depends on several factors, as it is crucial to analyze and understand the properties of the food to be coated, including its surface characteristics and the intended function of the coating. The application process consists of two main stages: the first involves spreading the solution over the product's surface, while the second focuses on adhering the coating material to the food's surface (Kumar et al., 2022).

The most commonly used method, particularly for fruits and vegetables, is immersion, which is characterized by the formation of uniform layers. The application of this technique can be divided into three main stages. First, the food product is fully immersed in the solution at a constant speed. The second stage involves keeping the product submerged for a specific duration (ranging from 5 seconds to 3 minutes), depending on the characteristics of the food. Finally, the food is removed from the film-forming solution and subjected to solvent evaporation, resulting in a thin layer on the product's surface (Priya et al., 2023). Fig. 10.3A illustrates the immersion process.

The immersion time depends on factors such as the density, surface tension, and viscosity of the film-forming solution. Prolonged exposure can lead to excessive moisture adsorption, which is undesirable. However, very short exposure times

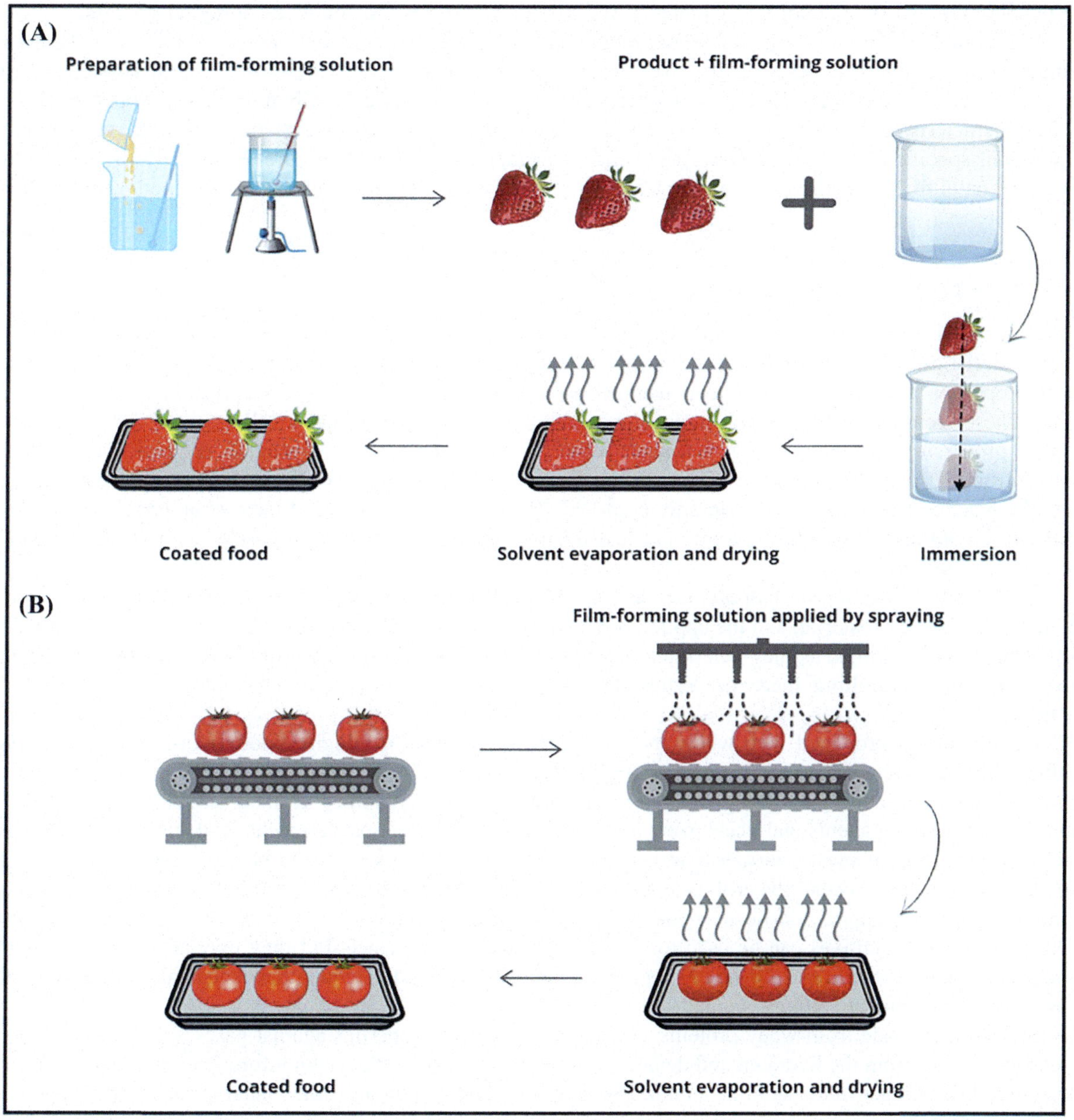

FIGURE 10.3 Illustrative scheme of methods used for edible coating application. Immersion (A) and spraying (B) methods.

are also problematic, as they may result in an uneven coating layer. The subsequent step, solvent evaporation, can occur at room temperature or with heat application. During this stage, the product must rest, allowing the drying process to form a uniform thin film (Jafri & Jaafar, 2024).

In the spraying method, the liquid film-forming solution is applied to the food as fine droplets, allowing for larger surface area coverage and a more uniform coating. This technique typically uses air spray atomization, pressure atomization, or air-assisted airless atomization. In air spray atomization, high air speed is used to transform the liquid into droplets, with the solution being dispersed at a slower rate alongside the fast-moving air, which induces atomization. High pressure serves the same function in pressure atomization, converting the liquid into droplets. The air-assisted airless atomization technique involves the use of compressed air after the coating solution is ejected, also generating small droplets (Andrade et al., 2012; Kumar et al., 2022).

Viscous solutions cannot be applied using this method, as they cannot be converted into small droplets. In such cases, immersion is the more appropriate technique. The tiny droplets and favorable rheological properties of the fluid enhance the process's efficiency, ensuring the products' complete coating. After spraying, the food must be left to dry under specific temperature and time conditions (Suhag et al., 2020). Fig. 10.3B illustrates the spraying process.

Posttreatment of films and coatings, regardless of the technique used, is essential. These coatings modify the internal atmosphere of the products, controlling the exchange of gases and moisture. Therefore, careful storage is crucial to enhance both the food's durability and the packaging's effectiveness, as they generally have active properties (Priya et al., 2023). The characteristics of each film or coating determine the required postprocessing care, making it important to understand their specific properties. Factors such as heat treatment, storage temperature, exposure to light, moisture, and gases must be carefully considered to improve the longevity of the packaging and extend the food's shelf life (Jafri & Jaafar, 2024).

10.6 Properties of films and coatings

Whey protein-based films and coatings exhibit specific functional properties, including optical, mechanical, active, and barrier characteristics, that must be analyzed and optimized for their intended application.

Optical properties are closely linked to consumer perception of the biopolymer's appearance when used in food packaging. These properties include color, transparency, and light barrier performance, all essential for achieving optimal results after processing and maintaining consistent physical-chemical and sensory quality (Fematt-flores et al., 2022). Films made from WPI exhibit better transparency than synthetic films, as they contain colorless compounds and glycerol plasticizer, enhancing transparency (Kandasamy et al., 2021). In a study by Fematt-Flores et al. (2022), films developed with WPI were colorless, shiny, transparent, and less opaque than films made from casein; overall, they demonstrated favorable optical properties.

The strength and flexibility of biopolymers are mechanical properties derived from the compounds used in the formulation, the base raw material, and the applied processing techniques. These properties are crucial for maintaining the integrity of films and coatings, as they determine the material's extensibility and its ability to withstand external forces. Therefore, carefully controlling these properties during development is essential (Avramescu et al., 2020; Schmid et al., 2012).

In films and coatings produced with whey protein, a firm network with a higher modulus of resistance is provided, originating from the three-dimensional protein network developed during the process. It is also worth noting that the higher the protein concentration, the greater the strength of the biopolymer formed (Fematt-flores et al., 2022). Additionally, crosslinking agents and plasticizers are commonly added to improve mechanical strength and flexibility. In the case of the addition of plasticizers, less molecular interaction is observed in the protein network, exhibiting better mechanical performance. When glycerol is applied, the crosslinking with β-lactoglobulin promotes an increase in elongation capacity and reduces tensile strength, and when sorbitol is applied, this resistance is even lower (Kandasamy et al., 2021). Chemical additives can be used to enhance physical performance, but they may also alter the organoleptic properties of foods. Perforation and tensile testing are commonly employed to evaluate mechanical properties, particularly in research and development settings (Fematt-flores et al., 2022).

Films and coatings made from whey proteins exhibit strong barrier properties against gases, particularly oxygen. This barrier capacity arises from the hydrated and denatured whey protein molecules, which form a compact three-dimensional structure through hydrogen and disulfide bonds (Song et al., 2022). On the other hand, protection against moisture permeability is limited due to the hydrophilic nature of whey proteins, and other biopolymers. As a result, film-forming solutions are typically created from blends to combine the strengths of different compounds and overcome their individual limitations (Zhang et al., 2024).

Whey protein films and coatings can exhibit active properties when antioxidant and antimicrobial compounds, such as essential oils and plant extracts, are incorporated during formulation (Lanciu Dorofte et al., 2023). Essential oils contain components such as terpenes, terpenoids, and aromatic molecules, which are effective in combating microbial activity. These components are released at the interface between the biopolymer and the food, thereby enhancing the product's shelf life. Bleoanca et al. (2020) evaluated the antimicrobial effects of films made with whey protein and thyme essential oil, demonstrating their activity against *Bacillus subtilis*, *Geotrichum candidum*, and *Torulopsis stellata*. Çakmak et al. (2020) observed antimicrobial activity against pathogens in whey protein films incorporating lemon and bergamot essential oils at concentrations between 0.05% and 0.1%. Additionally, whey protein formulations can contribute to antioxidant activity due to their metal ion-sequestering properties.608Ematt-Flores et al. (2022) reported antioxidant activity in edible whey protein films without the addition of other components, attributing this effect to the presence of bioactive peptides derived from whey protein. Furthermore, these packages can be classified as smart, containing indicators capable of altering their visual characteristics. These sensors detect changes in the quality, freshness, and safety of packaged foods. Typically, these indicators are colorimetric and respond to pH, gas concentration, temperature, or humidity variations, providing consumers with visual information about the product's condition (Yin et al., 2023). Biopolymer-based packaging has great potential for preserving food safety and can also signal changes in the product, reflecting shifts in the storage environment. As a result, it enables the monitoring of food deterioration during transportation, storage, and throughout its shelf life (Ke et al., 2025). For example, pH and ammonia-sensing intelligent films made from cellulose nanocrystals and WPI incorporated with curcumin and phycocyanin were developed by Sun et al. (2024). The results indicated that these films possess multifunctional potential as intelligent packaging for preserving the freshness of pork, as they change color in response to the meat's deterioration.

10.7 Applicability

Population growth is directly linked to increased food production and a significant rise in waste generation (Uribarrena et al., 2024). These factors emphasize the urgent need to develop sustainable alternatives to reduce food waste and minimize environmental impact.

To achieve this, maintaining food quality and safety is essential, requiring the exploration of new technologies. In this regard, reusing whey, a coproduct of the cheese industry, presents a promising strategy. Whey proteins are biopolymers that can serve important functions in food packaging, contributing to sustainability due to their excellent biodegradability and status as a coproduct. Films and coatings based on this biopolymer offer a sustainable solution, provide nutritional benefits, and extend the shelf life of products. Their applicability is particularly effective and has been most widely explored in dairy products, especially cheeses, as well as in fruits, vegetables, fish, and meat products (Catarino et al., 2017; Kumar & Saini, 2021; Nemati et al., 2023; Seyfzadeh et al., 2013; Zhang et al., 2024).

Whey protein-based coatings are used on fruits and vegetables to provide barriers against moisture and gases, factors that can affect the ripening and early senescence of the fruit or vegetable. They can also prevent enzymatic browning reactions through whey protein's antioxidant capacity, preventing the oxidation of phenolic compounds mediated by polyphenol oxidase in foods such as apples and potatoes (Kandasamy et al., 2021).

This application is particularly beneficial for highly perishable foods, such as meat and fish, aiming to extend product shelf life and reduce the need for chemical preservatives. The use of whey protein biopolymer acts as a barrier to the oxidation of myoglobin and lipids, while also inhibiting the growth of spoilage microorganisms (Fematt-flores et al., 2022). Catarino et al. (2017) observed an increase in the shelf life of processed meat products when coatings made from whey protein and *Origanum virens* essential oil were applied.

Fish generally have high moisture content and water activity, which promote a higher rate of chemical, enzymatic, and microbial reactions. Freezing is a commonly used preservation method, but it can lead to defects in the integrity of the fish. To address this challenge, whey protein biopolymers can be applied as coatings for fish and seafood (Rodriguez-Turienzo et al., 2011).

Dairy products are consumed worldwide and contain macronutrients such as proteins, carbohydrates, and lipids, which can undergo physical-chemical reactions during and after production. Cheese, in particular, is rich in fatty acids and may be susceptible to lipolytic enzymes, forming off-flavors. Additionally, some cheese varieties have high moisture content, making them vulnerable to microbial growth during storage and ripening (Vukić et al., 2017). Therefore, whey protein films and coatings can contribute to the preservation of these products mainly due to the gas barrier and protection against microbial contamination. Wagh et al. (2014) found no significant differences in the organoleptic characteristics of cheddar cheese wrapped in whey protein-based films compared to cheeses without or wrapped in casein-based films. However, they observed that using protein films reduced the rate of cheese deterioration. Additionally, incorporating antioxidant and

TABLE 10.2 Examples of the use of whey protein films and coatings in various food products.

Food product	Types of biopolymers	Composition	Objective	References
Beef	Film	Whey protein isolate (WPI) Essential oils (thyme, cinnamon, and cumin)	Reduction of microbial load	Badr et al. (2014)
Plum (*Prunus domestica* L.)	Coating	WPI Linseed oil Beeswax	Improve quality and nutritional value	Reinoso et al. (2008)
Kilka fish	Coating	WPI Sodium alginate	Reduction of *Staphylococcus* and increased shelf life	Seyfzadeh et al. (2013)
Freeze-dried strawberry	Coating	WPI	Improve rehydration speed	Huang et al. (2009)
Banana	Coating	WPI	Slow down the ripening speed	Cisneros-Zevallos and Krochta (1998)
Tomato	Coating	WPI Xanthan gum Glyceryl monostearate Clove oil	Extend shelf life	Kumar and Saini (2021)
Asian sea bass steak (*Lates calcarifer*)	Coating	WPI Ginger extract	Extend shelf life and maintain organoleptic characteristics	Chaijan et al. (2022)
Apple	Coating	WPI	Gas barrier (oxygen)	Cisneros-Zevallos and Krochta (2003)

antimicrobial substances into the biopolymer formulation for cheese coatings contributed to a longer shelf life by preserving desirable compounds and maintaining microbiological integrity. Similarly, Nemati et al. (2023) evaluated the preservation of Iranian white cheese coated with a whey protein biopolymer containing cumin essential oil and found improvements in durability, mechanical properties, microbial load, and moisture retention.

Table 10.2 presents studies that evaluated the application of whey-based films and coatings in various food matrices.

10.8 Opportunities and limitations

Several challenges may arise in the production, marketing, and development of whey protein-based films and coatings. Whey protein is sensitive to changes in pH, temperature, and moisture, which must be carefully considered during the formulation of the film-forming solution to ensure a uniform biopolymer. Additionally, the incorporation of additives and agents should be tested in advance, as they may compromise the integrity and functionality of the final product (Mohamed et al., 2022).

In terms of marketing, films and coatings face challenges in scaling up production and integrating them into the logistics chain without compromising quality. Additionally, ensuring cost-effectiveness and competitive pricing remains a key concern. Consumer acceptance may also pose a barrier, as there is limited awareness of this type of sustainable packaging (Kandasamy et al., 2021).

The application and storage conditions are key factors determining the quality of whey protein films and coatings, as they generally exhibit good gas barrier properties. However, a study by Çakmak et al. (2020) showed a lower oxygen permeability rate for whey protein films than for polysaccharide films. This could interfere with the cellular respiration process of fruits and vegetables, potentially affecting their ripening stages. Due to the hydrophilic nature of whey proteins, these biopolymers also have low moisture permeability, a limitation that has led to the development of formulations combined with other additives (Jin et al., 2023; Zhang et al., 2024).

Given the challenges associated with applying whey protein-based films and coatings, the most widely explored strategy to overcome these limitations is the use of blends with other biopolymers and additives to impart additional properties to the packaging (Schmid et al., 2012; Zhang et al., 2024). Therefore, preparing film-forming solutions involves a mixture of polymers, plasticizers, and essential oils, among other components. This approach enhances properties such as resistance, flexibility, and bioactivity, resulting in improved packaging that benefits consumers (Kandasamy et al., 2021; Kumar et al., 2022).

The food industry has been exploring new technologies to reduce environmental impacts as consumers increasingly take active roles in supporting sustainability (Zhang et al., 2024). Packaging production plays a key role in this context, as it is widely used and has the potential to benefit both the environment and nutrition, with an increasing amount of research and innovation focused on developing these materials.

Whey protein-based films and coatings offer promising food packaging qualities, with potential health and product quality benefits. For these reasons, further research is needed to explore their use on a larger production scale, as there is growing demand in the market. However, scientific and market research are still required to develop effective strategies and technologies for processing and distribution. It is also essential to establish partnerships and invest in marketing and advertising to reach a greater number of consumers (Xu et al., 2024).

Finally, it is essential for regulatory agencies to progress in the discussions and publication of specific legislation regarding biopolymers intended for food contact, with the goal of ensuring consumer safety.

10.9 Conclusion

Whey protein-based films and coatings represent excellent alternatives to meet consumers' demand for sustainable solutions in the food industry, as they offer strong properties for food packaging while contributing to sustainability. Research has shown promising formulations of these films and coatings, combining whey proteins with other biopolymers and additives, such as plasticizers, to enhance the packaging properties. While whey proteins alone help increase shelf life and provide active properties, such as excellent gas barrier capabilities, biodegradability, edibility, optical properties, and bioactivity—especially antimicrobial effect—the addition of other ingredients further boosts the value of these packages, benefiting both the packaged product and the consumer.

Another key advantage is their versatility across various products, carrying active and functional compounds that extend shelf life and enhance safety and organoleptic characteristics, particularly in cheeses, fruits, vegetables, meat products, and fish. From a cost-benefit perspective, this development is crucial for the industry, as it will shape the future large-scale production of these packages. However, industrial production still relies on further scientific research to optimize product performance, processing technologies, and manufacturing methods. Additionally, market research is necessary to understand consumer preferences and assess the sales potential of these packaging solutions.

AI disclosure

During the preparation of this work, the author(s) used Grammarly and ChatGPT in order to improve readability and language of the work. After using this tool/service, the author(s) reviewed and edited the content as needed and take(s) full responsibility for the content of the publication.

References

Achari, D. D., Kouser, S., Gasti, T., & Dixit, S. (2024). Biodegradable polymer blends in food packaging and preservations. In *Sustainable materials for food packaging and preservation: Food security and sustainability*. India: Elsevier, 171–197. 10.1016/B978-0-443-13567-5.00008-3, https://www.sciencedirect.com/book/9780443135675.

Andrade, R. D., Skurtys, O., & Osorio, F. A. (2012). Atomizing spray systems for application of edible coatings. *Comprehensive Reviews in Food Science and Food Safety, 11*(3), 323–337. https://doi.org/10.1111/j.1541-4337.2012.00186.x.

Avramescu, S. M., Butean, C., Popa, C. V., Ortan, A., Moraru, I., & Temocico, G. (2020). Edible and functionalized films/coatings-performances and perspectives. *Coatings, 10*(7). https://doi.org/10.3390/coatings10070687, https://res.mdpi.com/d_attachment/coatings/coatings-10-00687/article_deploy/coatings-10-00687-v2.pdf.

Badr, K. R., Ahmed, Z. S., & El Gamal, M. S. (2014). Evaluation of the antimicrobial action of whey protein edible films incorporated with cinnamon, cumin and thyme against spoilage flora of fresh beef. *International Journal of Agricultural Research, 9*(5), 242–250. https://doi.org/10.3923/ijar.2014.242.250, http://www.scialert.net/qredirect.php?doi=ijar.2014.242.250&linkid=pdf.

Belitz, H. D., Grosch, W., & Schieberle, P. (2009). *Food chemistry*. Berlin Heidelberg, Germany: Springer, 1–1070. http://springerlink.com/openurl.asp?genre=book&isbn=978-3-540-69933-0, http://10.1007/978-3-540-69934-7.

Bleoanca, I., Enachi, E., & Borda, D. (2020). Thyme antimicrobial effect in edible films with high pressure thermally treated whey protein concentrate. *Foods*, 9(7). https://doi.org/10.3390/foods9070855.

Britz, T. J., & Robinson, R. K. (2008). *Advanced dairy science and technology*. South Africa: South Africa Blackwell Publishing Ltd, 1–300. http://onlinelibrary.wiley.com/book/10.1002/9780470697634, http://10.1002/9780470697634.

Brown, A., & Börkey, P. (2024). *Plastics recycled content requirements, 236*. OECD Publishing.

Çakmak, H., Özselek, Y., Turan, O. Y., Fıratlıgil, E., & Karbancioğlu-Güler, F. (2020). Whey protein isolate edible films incorporated with essential oils: Antimicrobial activity and barrier properties. *Polymer Degradation and Stability*, 179, 109285. https://doi.org/10.1016/j.polymdegradstab.2020.109285.

Catarino, M. D., Alves-Silva, J. M., Fernandes, R. P., Gonçalves, M. J., Salgueiro, L. R., Henriques, M. F., & Cardoso, S. M. (2017). Development and performance of whey protein active coatings with *Origanum virens* essential oils in the quality and shelf life improvement of processed meat products. *Food Control, 80*, 273–280. https://doi.org/10.1016/j.foodcont.2017.03.054.

Chaijan, M., Chaijan, S., Panya, A., Nisoa, M., Cheong, L.-Z., & Panpipat, W. (2022). Combined effects of prior plasma-activated water soaking and whey protein isolate-ginger extract coating on the cold storage stability of Asian sea bass (*Lates calcarifer*) steak. *Food Control, 135*, 108787. https://doi.org/10.1016/j.foodcont.2021.108787.

Chatterton, D. E. W., Smithers, G., Roupas, P., & Brodkorb, A. (2006). Bioactivity of β-lactoglobulin and α-lactalbumin-Technological implications for processing. *International Dairy Journal, 16*(11), 1229–1240. https://doi.org/10.1016/j.idairyj.2006.06.001.

Cisneros-Zevallos, L.A., & Krochta, J. M. (1998). Book of abstracts, proceedings of the IFT annual meeting whey protein coatings delay ripening of banana fruits.

Cisneros-Zevallos, L., & Krochta, J. M. (2003). Whey protein coatings for fresh fruits and relative humidity effects. *Journal of Food Science, 68*(1), 176–181. https://doi.org/10.1111/j.1365-2621.2003.tb14136.x.

Dalgleish, D. G., & Corredig, M. (2012). The structure of the casein micelle of milk and its changes during processing. *Annual Review of Food Science and Technology, 3*(1), 449–467. https://doi.org/10.1146/annurev-food-022811-101214.

Damodaran, S., & Parkin, K. L. (2017). *Fennema's food chemistry* (5th e d). CRC Press.

Deeth, H., & Bansal, N. (2019). *Whey proteins: From milk to medicine*. Academic Press.

Fematt-flores, G. E., Aguiló-aguayo, I., Marcos, B., Camargo-olivas, B. A., Sánchez-vega, R., Soto-caballero, M. C., Salas-salazar, N. A., Flores-córdova, M. A., & Rodríguez-roque, M. J. (2022). Milk protein-based edible films: Influence on mechanical, hydrodynamic, optical and antioxidant properties. *Coatings, 12*(2). https://doi.org/10.3390/coatings12020196, https://www.mdpi.com/2079-6412/12/2/196/pdf.

Fernandes, L. M., Guimarães, J. T., Pimentel, T. C., Esmerino, E. A., Freitas, M. Q., Carvalho, C. W. P., Cruz, A. G., & Silva, M. C. (2020). *Edible whey protein films and coatings added with prebiotic ingredients agri-food industry strategies for healthy diets and sustainability: New challenges in nutrition and public health*. Brazil: Elsevier, 177–193. https://www.sciencedirect.com/book/9780128172261, 10.1016/B978-0-12-817226-1.00007-2.

Fox, P. F., Guinee, T. P., Cogan, T. M., & McSweeney, P. L. H. (2017). *Fundamentals of cheese science*. Springer.

Fox, P. F., Uniacke-Lowe, T., McSweeney, P. L. H., & O'Mahony, J. A. (2015). *Dairy chemistry and biochemistry*. Springer International Publishing. 10.1007/978-3-319-14892-2.

Han, J. H., & Aristippos, G. (2005). Edible films and coatings: A review. In *Innovations in food packaging*. Canada: Elsevier, 239–262. https://www.sciencedirect.com/book/9780123116321, 10.1016/B978-012311632-1/50047-4.

Haraguchi, F. K., Abreu, W. C.de, & Paula, H.de (2006). Proteínas do soro do leite: composição, propriedades nutricionais, aplicações no esporte e benefícios para a saúde humana. *Revista de Nutrição, 19*(4), 479–488. https://doi.org/10.1590/s1415-52732006000400007.

Henriques, M., Gomes, D., & Pereira, C. (2016). *Whey protein edible coatings: Recent developments and applications food engineering series*. Portugal: Springer, 177–196. springer.com/series/5996, 10.1007/978-3-319-24040-4_10.

Huang, L. L., Zhang, M., Yan, W. Q., Mujumdar, A. S., & Sun, D. F. (2009). Effect of coating on post-drying of freeze-dried strawberry pieces. *Journal of Food Engineering, 92*(1), 107–111. https://doi.org/10.1016/j.jfoodeng.2008.10.031.

Hyslop, D. B. (2003). *Advanced dairy chemistry, v. 1- proteins* (3rd.). Kluwer Academic/Plenum Publishers.

Jafri, N. N. M., & Jaafar, J. (2024). *Dip coating technique advanced ceramics for photocatalytic membranes: Synthesis methods, characterization and performance analysis, and applications in water and wastewater treatment*. Malaysia: Elsevier, 101–127. https://www.sciencedirect.com/book/9780323954181, 10.1016/B978-0-323-95418-1.00016-1.

Jin, T. Z., Yadav, M. P., & Qi, P. X. (2023). Antimicrobial and physiochemical properties of films and coatings prepared from bio-fiber gum and whey protein isolate conjugates. *Food Control, 148*. https://doi.org/10.1016/j.foodcont.2023.109666, https://www.journals.elsevier.com/food-control.

Kandasamy, S., Yoo, J., Yun, J., Kang, H. B., Seol, K. H., Kim, H. W., & Ham, J. S. (2021). Application of whey protein-based edible films and coatings in food industries: An updated overview. *Coatings, 11*(9). https://doi.org/10.3390/coatings11091056, https://www.mdpi.com/2079-6412/11/9/1056/pdf.

Ke, F., Yang, M., Ji, W., & Liu, D. (2025). Functional pH-sensitive film based on pectin and whey protein for grape preservation and shrimp freshness monitoring. *Food Chemistry, 463*(Pt 1), 141092. https://doi.org/10.1016/j.foodchem.2024.141092.

Khwaldia, K., Ferez, C., Banon, S., Desobry, S., & Hardy, J. (2004). Milk proteins for edible films and coatings. *Critical Reviews in Food Science and Nutrition, 44*(4), 239–251. https://doi.org/10.1080/10408690490464906.

Kim, S.-J., & Ustunol, Z. (2001). Solubility and moisture sorption isotherms of whey-protein-based edible films as influenced by lipid and plasticizer incorporation. *Journal of Agricultural and Food Chemistry, 49*(9), 4388–4391. https://doi.org/10.1021/jf010122q.

Kontogianni, V. G., Kasapidou, E., Mitlianga, P., Mataragas, M., Pappa, E., Kondyli, E., & Bosnea, L. (2022). Production, characteristics and application of whey protein films activated with rosemary and sage extract in preserving soft cheese. *LWT, 155*, 112996. https://doi.org/10.1016/j.lwt.2021.112996.

Krochta, J. M., & Pérez-Gago, M. B. (2002). *Formation and properties of whey protein films and coatings protein-based films and coatings*. Spain: CRC Press, 159–180. https://www.taylorfrancis.com/books/e/9781420031980.

Kumar, L., Ramakanth, D., Akhila, K., & Gaikwad, K. K. (2022). Edible films and coatings for food packaging applications: A review. *Environmental Chemistry Letters, 20*(1), 875–900. https://doi.org/10.1007/s10311-021-01339-z, https://www.springer.com/journal/10311.

Kumar, A., & Saini, C. S. (2021). Edible composite bi-layer coating based on whey protein isolate, xanthan gum and clove oil for prolonging shelf life of tomatoes. *Measurement: Food, 2*, 100005. https://doi.org/10.1016/j.meafoo.2021.100005.

Lacroix, M., & Le Tien, C. (2005). *Edible films and coatings from nonstarch polysaccharides innovations in food packaging*. Canada: Elsevier, 338–361. https://www.sciencedirect.com/book/9780123116321, 10.1016/B978-012311632-1/50052-8.

Lanciu Dorofte, A., Dima, C., Ceoromila, A., Botezatu, A., Dinica, R., Bleoanca, I., & Borda, D. (2023). Controlled release of β-CD-encapsulated thyme essential oil from whey protein edible packaging. *Coatings, 13*(3). https://doi.org/10.3390/coatings13030508, www.mdpi.com/journal/coatings.

Lopez-Polo, J., Muñoz-Shugulí, C., Patiño Vidal, M., & Patiño Vidal, C. (2024). Electrospun edible films and coatings: Development, functionality and food applications. *Trends in Food Science & Technology, 143*, 104253. https://doi.org/10.1016/j.tifs.2023.104253.

Madureira, A. R., Pereira, C. I., Gomes, A. M. P., Pintado, M. E., & Xavier Malcata, F. (2007). Bovine whey proteins - Overview on their main biological properties. *Food Research International, 40*(10), 1197–1211. https://doi.org/10.1016/j.foodres.2007.07.005.

Marangoni Júnior, L., Coltro, L., Dantas, F. B. H., & Vieira, R. P. (2022). Research on food packaging and storage. *Coatings, 12*(11). https://doi.org/10.3390/coatings12111714, www.mdpi.com/journal/coatings.

McSweeney, P. L. H., & O'Mahony, J. A. (2016). *Advanced dairy chemistry*. Springer.

Mihalca, V., Kerezsi, A. D., Weber, A., Gruber-traub, C., Schmucker, J., Vodnar, D. C., Dulf, F. V., Socaci, S. A., Fărcaş, A., Mureşan, C. I., Suharoschi, R., & Pop, O. L. (2021). Protein-based films and coatings for food industry applications. *Polymers, 13*(5), 1–24. https://doi.org/10.3390/polym13050769, https://www.mdpi.com/2073-4360/13/5/769/pdf.

Mohamed, A., El Galiou, O., Zantar, S., Arakrak, A., Laglaoui, A., & Zerrouk, M. H. (2022). *Edible films and coatings: Major challenges and potential applications in food packaging: A review Food Packaging: The Smarter Way*. Morocco: Springer Nature, 187–224. https://link.springer.com/book/10.1007/978-981-16-7196-8, 10.1007/978-981-16-7196-8_8.

Nemati, V., Hashempour-Baltork, F., Sadat Gharavi-Nakhjavani, M., Feizollahi, E., Marangoni Júnior, L., & Mirza Alizadeh, A. (2023). Application of a whey protein edible film incorporated with cumin essential oil in cheese preservation. *Coatings, 13*(8), 1470. https://doi.org/10.3390/coatings13081470.

Oliveira, S. P. L. F., Bertan, L. C., De Rensis, C. M. V. B., Bilck, A. P., & Vianna, P. C. B. (2017). Whey protein-based films incorporated with oregano essential oil. *Polimeros, 27*(2), 158–164. https://doi.org/10.1590/0104-1428.02016, http://www.scielo.br/pdf/po/v27n2/0104-1428-po-0104-142802016.pdf.

O'Mahony, J. A., & Fox, P. F. (2014). *Milk: An overview*. Elsevier BV, 19–73. 10.1016/b978-0-12-405171-3.00002-7.

Permyakov, E. A. (2020). α-Lactalbumin, amazing calcium-binding protein. *Biomolecules, 10*(9), 1–50. https://doi.org/10.3390/biom10091210, https://www.mdpi.com/2218-273X/10/9/1210/pdf.

Poppi, F. A., Costa, M., de, R., Rensis, C. M. V. B., & Sivieri, K. (2010). Whey and its proteins: Composition and functional activity. *Ciência Biológica e da Saúde. 12*(2), 31–37.

Priya, K., Thirunavookarasu, N., & Chidanand, D. V. (2023). Recent advances in edible coating of food products and its legislations: A review. *Journal of Agriculture and Food Research, 12*, 100623. https://doi.org/10.1016/j.jafr.2023.100623.

Ramos, O. L., Fernandes, J. C., Silva, S. I., Pintado, M. E., & Malcata, F. X. (2012). Edible films and coatings from whey proteins: A review on formulation, and on mechanical and bioactive properties. *Critical Reviews in Food Science and Nutrition, 52*(6), 533–552. https://doi.org/10.1080/10408398.2010.500528.

Ramos, O. L., Pereira, R. N., Rodrigues, R. M., Teixeira, J. A., Vicente, A. A., & Malcata, F. X. (2015). *Whey and whey powders: Production and uses encyclopedia of food and health*. Portugal: Elsevier Inc, 498–505. http://www.sciencedirect.com/science/book/9780123849533, 10.1016/B978-0-12-384947-2.00747-9.

Reinoso, E., Mittal, G. S., & Lim, L. T. (2008). Influence of whey protein composite coatings on plum (*Prunus domestica* L.) fruit quality. *Food and Bioprocess Technology, 1*(4), 314–325. https://doi.org/10.1007/s11947-007-0014-1.

Ribeiro, A. M., Estevinho, B. N., & Rocha, F. (2021). Preparation and incorporation of functional ingredients in edible films and coatings. *Food and Bioprocess Technology, 14*(2), 209–231. https://doi.org/10.1007/s11947-020-02528-4, http://www.springer.com/life+sci/food+science/journal/11947.

Rocha-Mendoza, D., Kosmerl, E., Krentz, A., Zhang, L., Badiger, S., Miyagusuku-Cruzado, G., Mayta-Apaza, A., Giusti, M., Jiménez-Flores, R., & García-Cano, I. (2021). Invited review: Acid whey trends and health benefits. *Journal of Dairy Science, 104*(2), 1262–1275. https://doi.org/10.3168/jds.2020-19038.

Rodriguez-Turienzo, L., Cobos, A., Moreno, V., Caride, A., Vieites, J. M., & Diaz, O. (2011). Whey protein-based coatings on frozen *Atlantic salmon* (*Salmo salar*): Influence of the plasticiser and the moment of coating on quality preservation. *Food Chemistry, 128*(1), 187–194. https://doi.org/10.1016/j.foodchem.2011.03.026.

Rodríguez, M., Osés, J., Ziani, K., & Maté, J. I. (2006). Combined effect of plasticizers and surfactants on the physical properties of starch based edible films. *Food Research International, 39*(8), 840–846. https://doi.org/10.1016/j.foodres.2006.04.002.

Sakurai, K., & Goto, Y. (2006). Dynamics and mechanism of the tanford transition of bovine β-lactoglobulin studied using heteronuclear NMR spectroscopy. *Journal of Molecular Biology, 356*(2), 483–496. https://doi.org/10.1016/j.jmb.2005.11.038.

Salehi, F. (2019). Characterization of new biodegradable edible films and coatings based on seeds gum: A review. *Journal of Packaging Technology and Research, 3*(2), 193–201. https://doi.org/10.1007/s41783-019-00061-0.

Schmid, M., Dallmann, K., Bugnicourt, E., Cordoni, D., Wild, F., Lazzeri, A., & Noller, K. (2012). Properties of whey-protein-coated films and laminates as novel recyclable food packaging materials with excellent barrier properties. *International Journal of Polymer Science, 2012*, 1–7. https://doi.org/10.1155/2012/562381.

Schmid, M., & Müller, K. (2018). *Whey protein-based packaging films and coatings whey proteins: From milk to medicine*. Germany: Elsevier, 407–437. http://www.sciencedirect.com/science/book/9780128121245, 10.1016/B978-0-12-812124-5.00012-6.

Seyfzadeh, M., Motalebi, A. A., Kakoolaki, S., & Gholipour, H. (2013). Chemical, microbiological and sensory evaluation of gutted kilka coated with whey protein based edible film incorporated with sodium alginate during frozen storage. *Iranian Journal of Fisheries Sciences, 12*(1), 140–153. http://www.jifro.ir/browse.php?a_id=878&slc_lang=en&sid=1&ftxt=1.

Shimazu, A. A., Mali, S., Victória, M., & Grossmann, E. (2007). Efeitos plastificante e antiplastificante do glicerol e do sorbitol em filmes biodegradáveis de amido de mandioca. *Semina: Ciências Agrárias, 28*, 79–88.

Song, H. G., Choi, I., Lee, J. S., Chang, Y., Yoon, C. S., & Han, J. (2022). Whey protein isolate coating material for high oxygen barrier properties: A scale-up study from laboratory to industrial scale and its application to food packaging. *Food Packaging and Shelf Life, 31*. https://doi.org/10.1016/j.fpsl.2021.100765, http://www.journals.elsevier.com/food-packaging-and-shelf-life/.

Sothornvit, R., & Krochta, J. M. (2005). *Plasticizers in edible films and coatings innovations in food packaging*. Thailand: Elsevier, 403–433. https://www.sciencedirect.com/book/9780123116321, 10.1016/B978-012311632-1/50055-3.

Subbiah Prabhakaran, G. Y., Molitor, M., Govindasamy-Lucey, S., & Lucey, J. A. (2024). Heat-stable whey protein isolate made using isoelectric precipitation and clarification. *Journal of Dairy Science, 107*(9), 6629–6642. https://doi.org/10.3168/jds.2024-24671.

Suhag, R., Kumar, N., Petkoska, A. T., & Upadhyay, A. (2020). Film formation and deposition methods of edible coating on food products: A review. *Food Research International, 136*, 109582. https://doi.org/10.1016/j.foodres.2020.109582.

Sun, H., Liu, X., Huang, Y., & Leng, X. (2024). Incorporating functional colorants in whey protein isolate–cellulose nanocrystal-blended edible films for pork freshness prediction. *International Journal of Biological Macromolecules, 283*(Pt 1), 137276. https://doi.org/10.1016/j.ijbiomac.2024.137276.

Tsermoula, P., Khakimov, B., Nielsen, J. H., & Engelsen, S. B. (2021). WHEY - The waste-stream that became more valuable than the food product. *Trends in Food Science and Technology, 118*, 230–241. https://doi.org/10.1016/j.tifs.2021.08.025, http://www.elsevier.com/wps/find/journaldescription.cws_home/601278/description#description.

United Nations. (2024). The sustainable development goals UN – United Nations UN. https://sdgs.un.org/goals

Uribarrena, M., Rovira-Cal, E., Urbina, L., Suárez, M. J., Aymerich, E., Guerrero, P., de la Caba, K., & Etxabide, A. (2024). Valorization of cheese whey: Closing the loop from protein extraction to whey protein film composting. *Green Chemistry, 26*(7), 4103–4111. https://doi.org/10.1039/d3gc04304e, http://pubs.rsc.org/en/journals/journal/gc.

Varadi, M., Anyango, S., Deshpande, M., Nair, S., Natassia, C., Yordanova, G., Yuan, D., Stroe, O., Wood, G., Laydon, A., Zídek, A., Green, T., Tunyasuvunakool, K., Petersen, S., Jumper, J., Clancy, E., Green, R., Vora, A., Lutfi, M., … Velankar, S. (2022). AlphaFold Protein Structure Database: Massively expanding the structural coverage of protein-sequence space with high-accuracy models. *Nucleic Acids Research*, 50(D1), D439 -D444. https://doi.org/10.1093/nar/gkab1061.

Varadi, M., Bertoni, D., Magana, P., Paramval, U., Pidruchna, I., Radhakrishnan, M., Tsenkov, M., Nair, S., Mirdita, M., Yeo, J., Kovalevskiy, O., Tunyasuvunakool, K., Laydon, A., Žídek, A., Tomlinson, H., Hariharan, D., Abrahamson, J., Green, T., Jumper, J., … Velankar, S. (2024). AlphaFold Protein Structure Database in 2024: providing structure coverage for over 214 million protein sequences. *Nucleic Acids Research*, 52(D1), D368 -D375. https://doi.org/10.1093/nar/gkad1011.

Vukić, M., Grujić, S., & Odzaković, B. (2017). *Application of edible films and coatings in food production advances in applications of industrial biomaterials*. Bosnia and Herzegovina: Springer International Publishing, 121–138. http://www.springer.com/in/book/9783319627663, 10.1007/978-3-319-62767-0_7.

Walstra, P., Wouters, J. T., & Geurts, T. J. (2006). *Dairy science and technology*. CRC.

Wang, Y., Khan, M. A., Chen, K., Zhang, L., & Chen, X. (2023). Electrospinning of natural biopolymers for innovative food applications: A review. *Food and Bioprocess Technology, 16*(4), 704–725. https://doi.org/10.1007/s11947-022-02896-z, https://www.springer.com/journal/11947.

Xu, X., Deng, S., Essawy, H., Lee, S. H., Lum, W. C., Zhou, X., Du, G., & Zhang, J. (2024). Chitosan-casein blended with condensed tannin and carnauba wax for the fabrication of antibacterial and antioxidant food packing films. *International Journal of Biological Macromolecules, 277*. https://doi.org/10.1016/j.ijbiomac.2024.133784, https://www.sciencedirect.com/science/journal/01418130.

Yang, J., Punia Bangar, S., Rizwan Khan, M., Hammouda, G. A., Alam, P., & Zhang, W. (2024). Biopolymer-based packaging films/edible coatings functionalized with ε-polylysine: New options for food preservation. *Food Research International, 187*, 114390. https://doi.org/10.1016/j.foodres.2024.114390.

Yin, W., Qiu, C., Ji, H., Li, X., Sang, S., McClements, D. J., Jiao, A., Wang, J., & Jin, Z. (2023). Recent advances in biomolecule-based films and coatings for active and smart food packaging applications. *Food Bioscience, 52*, 102378. https://doi.org/10.1016/j.fbio.2023.102378.

Zhang, R., Wang, B., Zhang, F., Zheng, K., & Liu, Y. (2024). Milk-derived antimicrobial peptides incorporated whey protein film as active coating to improve microbial stability of refrigerated soft cheese. *International Journal of Food Microbiology, 419*, 110751. https://doi.org/10.1016/j.ijfoodmicro.2024.110751.

Chapter 11

Pectin films and coatings

Gabriella Simon Maia, Guilherme Frey Schütz, Gabriel Ferreira Toledo, Eduarda Chiabai Rodrigues de Melo, Iago Aguiar Dias Carmo, and Roniérik Pioli Vieira
School of Chemical Engineering, Universidade Estadual de Campinas (UNICAMP), Campinas, Brazil

11.1 Introduction

The rising demand for sustainable alternatives to conventional plastics has led to increasing interest in biopolymers derived from renewable resources. Among them, pectin, a polysaccharide abundantly available as a by-product of the fruit processing industry, has attracted notable attention. Its biodegradability, film-forming ability, and natural compatibility with food systems make it an appealing candidate for packaging materials that address both environmental and functional needs (Roy & Rhim, 2021b; Roy et al., 2023).

Chemically, pectin is a heteropolysaccharide composed mainly of galacturonic acid units, with properties governed by its degree of esterification (DE). This structural versatility enables different gelling behaviors, solubility, and barrier functions that are critical for packaging applications (Nastasi et al., 2022; Priyadarshi et al., 2021; Roy & Rhim, 2021a; Santana et al., 2024). Industrially, pectin is already well established as a gelling agent, stabilizer, and thickener, but its transformation into edible coatings and films represents an expanding field that links traditional food technology with sustainable packaging innovation (Roy et al., 2023; Santana et al., 2024).

Beyond its intrinsic barrier and mechanical performance, pectin also offers functional advantages. Its antioxidant activity and compatibility with active additives, such as essential oils (Santana et al., 2024), polyphenols (Sharifi & Pirsa, 2021; Yang et al., 2021), and nanoparticles (Mallakpour & Mohammadi, 2022), allow the design of bioactive films with antimicrobial and antioxidant properties. These attributes extend food shelf life, improve safety, and create opportunities for smart or functional packaging solutions (Marangoni Júnior et al., 2022; Santana et al., 2024). Such multifunctionality is central to advancing packaging materials from passive barriers to active systems that interact beneficially with food products.

Finally, the sustainable value of pectin lies not only in its biodegradability but also in its alignment with circular economy principles. By valorizing agro-industrial residues, such as citrus peels and apple pomace, pectin-based materials reduce waste streams and support renewable resource utilization (Roy et al., 2023). This chapter will explore pectin's structure, sources, processing, film formation, and its role in advanced packaging applications. Emphasis will be placed on its dual function as both a structural and functional polymer, highlighting how pectin can help transition toward more environmentally responsible and innovative packaging solutions.

11.2 Structure, sources, and extraction methods

Pectin is a high molecular weight biopolymer, biocompatible and nontoxic, found in the middle lamella and cell walls of plants (Yadav et al., 2023). It consists mostly of galacturonic acid (GalA), the primary monomer, and an oxidized form of D-galactose, arabinose (Ara), galactose (Gal), rhamnose (Rha), xylose (Xyl), and fructose (Fru) (Mao et al., 2019). Their arrangements into four structures result in homogalacturonan (HG), which represents 60% of the mass, comprising α-1,4-linked galacturonic acid that may undergo esterification at C-6 and O-acetylation in positions C-2 and C-3; rhamnogalacturonan I (RG-I), containing a backbone of galacturonic acid and rhamnose with branches of other sugars; xylogalacturonan (XG), where galacturonic acid in the backbone is branched with xylose; and rhamnogalacturonan II (RGII), featuring a backbone of galacturonic acid with branches of rhamnose and other sugars (Stoffels et al., 2020).

Natural Polymer-Based Films and Coatings in Food Technology. DOI: https://doi.org/10.1016/B978-0-443-34244-8.00023-9

Considering that some carboxyl groups are methyl-esterified, the DE will be characterized as high methoxyl (HM) or low methoxyl (LM) based on whether the DE is above or below 50%, respectively (Cazón et al., 2017).

The obtention of pectin starts with its isolation from plant material, primarily produced from apples and citrus juice by-products (peels), as well as from sunflowers, pumpkin, carrot, and sugar beet roots, which are less commonly used sources (Munarin et al., 2012). Commercially, pectin is primarily produced from citrus juice by-products (85%) and apples (14%), with sunflowers, pumpkin, carrot, and sugar beet roots being less commonly used sources due to factors such as yield, extraction time, desired properties, cost, and raw material availability (Adetunji et al., 2017). The extraction process will also influence the molecular weight of pectin (60,000–130,000 g mol^{-1}); room temperature extractions result in high molecular weight but low yield, whereas higher temperature extractions (above 80°C) produce pectin with lower molecular weight but increased yield (Adetunji et al., 2017; Gentilini et al., 2014).

Pectin production process generally includes three steps: a pretreatment, the extraction, and a postextraction. The first may involve drying, washing, or blanching, and is intended to stabilize the raw material by deactivating bacteria and enzymes that could degrade pectin. The traditional extraction method employs an acid-mediated process using strong inorganic acids such as hydrochloric or nitric acid to maintain a pH between 1 and 3 at temperatures up to 90°C (Riyamol et al., 2023). A physical-chemical process involving hydrolysis of the plant tissue, extraction of pectin macromolecules, and solvent solubilization is highly influenced by temperature, pH, and time (Dranca & Oroian, 2018). Therefore, there are concerns about cost and environmental impact, along with the effects of extraction methods on their solubility, gelling properties, and molecular weight (Roy et al., 2023). The utilization of milder food-grade organic acids, such as acetic and citric acids, has been introduced with relative success (Cho et al., 2019). Additionally, in line with the concept of green chemistry, other emerging methods such as subcritical fluid extraction, ultrasound and enzyme-assisted techniques, and microwave-based processes are now being used to extract pectin from agro-industrial by-products (Adetunji et al., 2017).

Subcritical water, hot water, or superheated water extraction method refers to a procedure in which liquid water is kept at an elevated pressure, allowing it to reach temperatures above boiling point while avoiding phase change. Alternatively, if the solvent is other than water, the procedure is called accelerated solvent extraction (Adetunji et al., 2017). The elevated temperatures used provide benefits, including lower surface tension, reduced viscosity, higher vapor pressure, and improved mass transfer rates. In addition, the method could be employed for both ionic and nonionic compounds. Studies evaluating pectin extraction by subcritical fluid may be found in the literature (Benito-Román et al., 2022; Cui et al., 2022; Wang et al., 2022). The compound extracted by the method typically exhibits low molecular weight along with more pronounced hairy regions (RG-I) and side chains (Basak & Annapure, 2022).

Ultrasound-assisted extraction is an efficient "green" technique that can be completed in minutes, giving a final product with higher purity (Karbuz & Tugrul, 2021). Its mechanism involves sonication, which creates cavitation in the liquid solution. When these bubbles collapse near plant material, pressure and temperature rise, breaking down cell walls and releasing extractive compounds (Wang et al., 2017). The elevated mass transfer rate assures an elevated recovery of the compound with minimal degradation (Karbuz & Tugrul, 2021). This method has been employed in the extraction of pectin from pistachio, navel orange, tomato, sugar beet by-products, among others (Jafarzadeh-Moghaddam et al., 2021; Kazemi et al., 2020; Lasunon & Sengkhamparn, 2022; Patience et al., 2021; Wang et al., 2017).

A plant cell wall contains a complex network where pectin, proteins, and hemicellulose interact. Common enzymes used in pectin extraction include cellulase, α-amylase, pectin methyl esterase, and others (Yang et al., 2018). When the enzymatic method is used to extract pectin, it facilitates its release by degrading pectin or breaking down the plant cell wall, releasing the trapped pectin within the cellulose matrix through the hydrolysis of cellulose or hemicellulose (Wu, Chen, et al., 2022). This method offers several benefits for pectin recovery, including the elimination of acidic pH and high temperatures applied in the traditional extraction process, minimizes pretreatment steps, and shortens extraction time. It aligns with green chemistry by improving yield, reducing solvent use, and preserving the pectin's structural and functional properties for higher-quality extracts (Adetunji et al., 2017).

The microwave-assisted extraction involves an effective nonconventional method for extracting plant constituents from solid matrices (Karbuz & Tugrul, 2021). The electromagnetic radiation is converted into molecular motion, producing heat. This results in water vapor pressure within the cell walls of the heated sample, facilitating the dissolution of target components into the solvent (Liew et al., 2016). A few advantages of this method are the short procedure time that may range from seconds to an hour, also, considering that microwaves are nonionizing, they do not alter the molecular structure of the compound (Mao et al., 2019). When choosing the ideal method, key variables to consider include the solid-liquid ratio, type of eutectic solvent, temperature, pH, extraction time, pressure, intensity of ultrasound and microwaves, enzyme concentration, and scalability of the process (Colodel & Petkowicz, 2019).

11.3 Functional properties and film formation

Commonly used as a thickening and gelling agent, pectin's functional properties may change according to source, molecular weight, structure of the compound including different methylation degree, extraction method, and conditions (Reichembach & Lúcia de Oliveira Petkowicz, 2021). Its gelling properties will vary under specific conditions and differ between high-methoxyl pectin (HMP) and low-methoxyl pectin (LMP) (Cao et al., 2020). The gel formation in the former occurs through crosslinking interactions in HMP involving hydrophobic interactions among the methyl ester groups and hydrogen bonding between carboxyl and hydroxyl groups of the galacturonic acid chains. In contrast, in LMP, the formation results from interactions between ionized carboxyl groups and divalent cations (Wang et al., 2021). Similarly, the viscosity property of the polysaccharide increases with higher molar mass, the presence of side chains in RG-I, a higher degree of methyl esterification, and galacturonic acid content. Additionally, pectin solution viscosity will also be influenced by pH and the presence of salts (Cao et al., 2020).

A promising material that may be developed into films for food packaging with "green" characteristics such as sustainable, renewable, and biodegradable (Santana et al., 2024). Usually regarded as safe by the Food and Drug Administration, pectin cannot be made into films by itself due to its high hydrophilicity and frailty; therefore, it requires the introduction of additives to improve properties such as mechanical, barrier, antioxidant, and antimicrobial (Jovanović et al., 2021). The film's properties will be directly influenced by structure and, therefore, by source and extraction method. The presence of numerous carboxyl and hydroxyl groups will facilitate the interaction with additives and result in a film with desirable characteristics (Pacheco et al., 2019).

Solvent casting and extrusion are the most widely employed methods for producing pectin-based films. Solvent casting consists of dispersing pectin, plasticizers, and other additives in a chosen solvent, followed by casting and drying to obtain a continuous film. This approach is straightforward, highly adaptable at the laboratory scale, and enables homogeneous incorporation of bioactive agents, such as essential oils, polyphenols, or antimicrobial nanoparticles into the pectin matrix. However, it is less suitable for industrial production due to long drying times, dependence on solvent removal, and difficulties in achieving consistent thickness and mechanical strength (Singaram et al., 2024).

On the other hand, extrusion has greater relevance for industrial applications. In this process, pectin and additives are mixed under controlled temperature and shear inside an extruder, producing films via blow film extrusion or sheet extrusion (Fishman et al., 2004). Extrusion offers key advantages for pectin processing: it is solvent-free, continuous, and compatible with standard polymer-processing equipment, which favors scalability and cost-effectiveness. Despite these benefits, pectin's intrinsic hydrophilicity and thermal sensitivity can hinder its extrusion, often requiring blending with other biopolymers (e.g., starch, polylactic acid [PLA]), or the addition of plasticizers and compatibilizers to improve melt processability, flexibility, and barrier performance (Huang et al., 2021).

Recent studies demonstrate that pectin-based films prepared by either solvent casting or extrusion exhibit promising potential in food packaging, particularly as active films incorporating antimicrobial and antioxidant compounds (Huang et al., 2021). While solvent casting remains essential for fundamental investigations and proof-of-concept studies with pectin, extrusion is emerging as the preferred pathway for industrial translation and large-scale production. The next section will further explore specific applications and case studies where pectin films are tailored for sustainable food packaging solutions.

11.4 Characterization of pectin-based films

11.4.1 Mechanical properties

Pectin-based films and coatings have attracted significant attention in materials science due to their biodegradability and biocompatibility. Their mechanical properties are crucial for their performance in various applications, particularly in food packaging and biomedical sectors. Mechanical properties such as tensile strength (TS), elongation at break (EB), and elasticity modulus (EM) are commonly used to evaluate structural integrity and flexibility (Huang et al., 2021). While TS provides information about the film's resistance to tension forces, EB reflects the film's ductility. On the other hand, the EM describes the film's stiffness (Schutz et al., 2024). These properties can be influenced by factors such as the DE, molecular weight, crosslinking, incorporating additives such as plasticizers or fillers, and blending with other biopolymers. Understanding the mechanical behavior of pectin films is essential for optimizing their application in industries requiring specific durability and flexibility standards (Espitia et al., 2014; Roy et al., 2023).

Blending pectin with whey protein and anthocyanin-rich purple sweet potato extract led to films with increased tensile strength and elongation in the study by Ke et al. (2025). Interactions occurred between hydroxyl groups of anthocyanin and film matrices at concentrations of 300–500 mg 100 mL^{-1} of film-forming solution (Ke et al., 2025). However, the tensile strength and strain decreased at a concentration of 700–1000 mg 100 mL^{-1} of film-forming solution. A similar effect was observed by Dobrucka et al. (2025), in which ethanolic extracts of *Cannabis sativa* L. seeds at 0.5 wt.% increased tensile strength and EB. At 1.0 wt.%, however, a decrease in these mechanical properties was noted. This reduction was attributed to the fact that, at higher concentrations, the extract disrupts the polymeric chain continuity and accumulates within the matrix. This accumulation leads to points of concentrated stress, which are more susceptible to breakage (Dobrucka et al., 2025). Table 11.1 summarizes some recent findings regarding the mechanical properties of pectin films.

The improvement of the mechanical properties of pectin-based films is mostly due to the additives' compatibility and uniform dispersion within the film matrix, which facilitates the formation of strong hydrogen bonds between the additives and the film constituents. Developing films and coatings with desired mechanical performance requires achieving an ideal balance between TS, EB, and EM. The films are suitable for a variety of applications because of this balance, which ensures that they are not only durable and effective but also flexible enough to withstand mechanical stress without breaking.

11.4.2 Water vapor permeability

Water vapor permeability (WVP) and water vapor transmission rate (WVTR) are key indicators of how quickly water vapor can permeate through a film. Understanding these parameters is crucial for food packaging materials, as different types of food require specific levels of permeability to ensure optimal preservation and shelf life. As with mechanical properties, various factors such as blending, crosslinking, plasticizers, and other additives can significantly influence the water vapor barrier properties, either enhancing or diminishing the film's ability to resist water vapor transmission. Most studies on WVP or WVTR in bio-based films are conducted gravimetrically, following the American Society for Testing and Materials (ASTM) E96M/E96 method or its variations. This method measures the amount of water vapor that permeates through a specific area over a given time. Film thickness and the partial pressure gradient are critical factors that influence the results (Cazón et al., 2017).

Tea polyphenol nanoparticles were used as additives in pectin films in the work of Yang et al. (2024). Higher WVP values were observed in pure pectin films due to the abundance of hydroxyl groups. The incorporation of nanoparticles significantly reduced the WVP, suggesting that the free hydroxyl groups interacted with the nanoparticles, forming a more compact structure. This compacted matrix extended the diffusion path for vapor permeation through the film. Since WVP is a time-dependent parameter, this increased diffusion path resulted in lower WVP values (Yang et al., 2024).

Film processing methods can also contribute to different values of WVP. For instance, high-pressure processing (HPP) and $CaCl_2$ as crosslinking agents were explored as film modifiers. The film containing 70% pectin solution and 30% carboxymethyl cellulose (CMC) solution, submitted to treatment with 500 MPa, presented a reduction in WVP (1.55×10^{-10} $g{\cdot}m^{-1}{\cdot}s^{-1}{\cdot}Pa^{-1}$) compared to the control film without HPP (1.71×10^{-10} $g{\cdot}m^{-1}{\cdot}s^{-1}{\cdot}Pa^{-1}$). HPP could enhance the interaction between pectin and CMC, thus reducing WVP (Wang, Li, et al., 2024).

Different pectin extraction methods were used by Tristanto et al. (2024). Dragon fruit peel pectin was extracted using subcritical water or acid. The WVTR and WVP ranged from 926.90 to 1160.50 $g{\cdot}m^{-2}{\cdot}day^{-1}$ and 5.59×10^{-11} to 6.62×10^{-11} $g{\cdot}cm^{-1}{\cdot}s^{-1}{\cdot}Pa^{-1}$, respectively. Different extraction methods led to different compositions, for instance, pectin extracted with subcritical water presented a higher DE and galacturonic acid content, which increased its hydrophobicity when compared with acid-extracted pectin (Tristanto et al., 2024).

The incorporation of active agents such as essential oils or extracts can affect the barrier properties of the films. A pectin/sodium alginate-based film containing microcrystalline cellulose (MCC) and geraniol was tested for its WVP and WVTR at different relative humidity levels. The incorporation of MCC reduced WVP and WVTR by approximately 61% and 67%, respectively. This reduction is attributed to the formation of tortuous paths within the film matrix, as the well-dispersed MCC created a more compact structure, effectively limiting vapor transmission. However, the addition of geraniol had the opposite effect at lower concentrations. As the concentration of hydrophobic geraniol increased, both WVP and WVTR decreased, further enhancing the barrier properties (Hoque et al., 2024).

11.4.3 Oxygen permeability

Oxidation is one of the primary causes of food spoilage, making it essential for ideal packaging films/coatings to provide a balanced oxygen barrier. This barrier should prevent oxidation while allowing sufficient oxygen exchange to support respiration. Completely blocking oxygen can lead to anaerobic respiration, which can also cause issues with food preservation (Sistla & Mehraj, 2022). Films made from pectin extracted from different sources (citrus, apple, fig, and fig stalk) with or without crosslinking with $CaCl_2$ had their oxygen transmission rate (OTR) evaluated by Çavdaroğlu et al. (2023). The study

TABLE 11.1 Mechanical properties of pectin-based films.

Matrix	Modification/additives	Tensile strength (TS) (MPa)	Elongation at break (EB) (%)	Elasticity modulus (EM) (MPa)	Conclusion	References
Apple pomace pectin (3% w/v) + grass pea protein (3% w/v)	Propolis extract (PE) (0%–12% v/w)	1.97– 2.98	48.72– 81.28	-	Control film presented higher TS and lower EB due to strong interactions between pectin and protein. PE can act as a plasticizer, reducing TS and increasing EB.	Salimi et al. (2025)
Coffee pulp pectin (1–5 wt.%)	Honey, propolis, or glycerol (0–60 wt.%)	1.83–4.01	17.36– 10.26	12.36–37.00	An optimal film formulation was achieved with 2% pectin and 20% glycerol, honey, or propolis enhancing TS and EB.	Biratu et al. (2024)
Pectin + xanthan gum (1.5 wt.%)	Grapefruit essential oil (GFO) (0–40 μL)	2.91– 8.55	12.68– 52.84	-	TS increased and EB decreased with GFO incorporation, attributed to the enhanced crystallinity induced by GFO, which promoted the formation of a more compact structure within the films.	Bhatia et al. (2024)
Coffee pulp pectin (3–7 wt.%)	Microcrystalline cellulose (MCC) (3–7 wt.%)	2.42 –3.08	4.04–4.20	60.62–76.98	MCC significantly reinforced pectin films, likely due to the hydroxyl groups on the cellulose surface and its large surface area, which enhance interactions within the polymer matrix.	Dao et al. (2023)
Low-methoxy pectin (5 wt.%)	Calcium propionate (CP) (0.5 wt.%) + polyvinyl alcohol (PVA) (0–25 wt.%)	3.42– 12.71	29.0– 132.0	-	The addition of CP initially reduces TS by disrupting strong pectin chain interactions, but the presence of PVA increases TS by forming crystalline domains and strong hydrogen bonds.	Ren et al. (2022)
High-methoxy apple pectin (0.8% w/v)	Pectin/chitosan large or short fibers (0.8/0.3 w/v)	6.99– 10.76	45.42– 65.53	12.53– 17.26	The film containing short fibers presented higher TS and EB due to better compatibility and distribution of small fibers within the film matrix.	Younis et al. (2020)
High or low-methoxy pectin (3% wt.%)	Chitosan nanoparticles	30.81– 58.51	0.94–2.91	-	Interactions between chitosan amino groups and pectin carboxyl groups led to a significant improvement in TS and EB.	Lorevice et al. (2016)

demonstrated that the galacturonic acid content in pectin significantly influences oxygen permeability. Additionally, crosslinking had a notable effect, reducing oxygen permeability across all films by forming a more compact and denser network (Çavdaroğlu et al., 2023).

Guo, Ge, Yang, et al. (2021) investigated the effects of ultrasound treatment on films produced from watermelon peel pectin, demonstrating the significant impact of such treatments on film properties. Regarding oxygen permeability, a 10-minute ultrasound treatment (1.21×10^{-5} $cm^3 \cdot mm^{-2} \cdot day^{-1} \cdot atm^{-1}$) significantly reduced oxygen permeability compared to the control film (1.97×10^{-5} $cm^3 \cdot mm^{-2} \cdot day^{-1} \cdot atm^{-1}$). This reduction was attributed to ultrasound promoting increased hydrogen bonding between pectin chains, thereby inhibiting or extending oxygen permeation paths. However, longer ultrasound treatment times decreased the films' oxygen barrier properties due to structural defects. This highlights the importance of optimizing treatment conditions to ensure both processability and functionality in film production (Guo, Ge, Yang, et al., 2021).

As in the barrier to water vapor, the oxygen barrier strongly depends on compatibility between the matrix and additives and the good dispersion of those. As seen in another study conducted by Guo, Ge, Li, et al. (2021), beetroot extract powder was used as an additive in watermelon pectin films. The control film presented an oxygen permeability of 1.76×10^{-5} $cm^3 \cdot mm^{-2} \cdot day^{-1} \cdot atm^{-1}$, while the film containing 2% of beetroot extract had its permeability reduced to 0.38×10^{-5} $cm^3 \cdot mm^{-2} \cdot day^{-1} \cdot atm^{-1}$ (Guo, Ge, Li, et al., 2021). The effects of the incorporation of additives and the source of pectin in oxygen barriers are very similar to those in water vapor barriers, with key factors such as compatibility, dispersion, and molecular characteristics determining beneficial or diminishing effects. Understanding how these parameters affect a film's properties is crucial to determining its applicability.

11.4.4 Optical properties

Consumer approval of packaging materials is greatly influenced by the optical characteristics of pectin films and coatings. Transparency and color are examples of esthetic attributes that might influence how good and appealing a product is seen to be. Additionally, because these films can prevent oxidation, protect sensitive components from photodegradation, and preserve product color and nutritional content, their light barrier qualities are crucial for maintaining the quality and shelf life of the packaged food (Li et al., 2023; Reis et al., 2023; Roy et al., 2024). Therefore, improving these optical and barrier qualities is essential to enhancing the functionality and attractiveness of pectin-based packaging solutions for consumers.

The L*, a*, and b* values are commonly used to describe the color characteristics of films. In this case, L* stands for the film's brightness, and its values range from 0 (black) to 100 (white). The film's location on the green-red axis is indicated by the a* value, where positive values denote red color, and negative values denote green color. Similarly, the color variation along the blue-yellow axis is represented by the b* value, where positive values point toward yellow, and negative ones point toward blue (Anis et al., 2021).

Pectin films containing thymol were crosslinked with different cations (Fe^{3+}, Cu^{2+}, and Ca^{2+}), in which the color parameters vary from −24.59 to 2.58 for L*, −8.07 to 12.73 for a*, and 6.88 to 93.05 for b*. These results show that pectin films crosslinked with Fe^{3+} presented a yellowish color, indicated by the highest b* value, and Cu^{2+} showed a green-like color due to the highest a* value. The pectin/Fe^{3+} film also showed an impressive UV-blocking capacity with almost 0% light transmittance below 400 nm (Zhang et al., 2022). In the study by Han and Song (2021), watermelon rind pectin was incorporated with kiwi peel extract. The total color difference (ΔE) of the films with the extract varied from 4.56 to 13.76. Additionally, the a* value decreased, whereas the b* value increased, resulting in increased yellowness. These color changes are attributed to the natural pigments and phenolic compounds present in kiwi peel extract, which can increase the light barrier of the films (Han & Song, 2021).

Qiang et al. (2024) explored vanillin grafted onto pectin, followed by crosslinking with Fe^{3+} ions. This grafting process facilitated the formation of robust bonds between the hydroxyl groups in vanillin and Fe^{3+} ions, enhancing the film's structural integrity. Additionally, the presence of these phenolic groups imparted significant UV-blocking properties to the film, providing effective protection against UV radiation. Furthermore, the results of color analysis indicate that the presence of vanillin/Fe^{3+} in the films increased the yellowness with no significant reduction in the transparency of the films (Qiang et al., 2024).

Therefore, the aforementioned results indicate that additives can substantially alter both the color parameters and the UV-visible light barrier properties of the films. Functional groups within these additives that absorb UV light are critical for enhancing the light-blocking capacity of pectin films. However, it is equally important that these additives do not compromise the films' transparency, as visual clarity is essential for consumers when selecting packaged foods.

Nonetheless, these properties can be carefully adjusted to balance the protective benefits of the additives with the need for transparency, optimizing the films for both functionality and consumer appeal.

11.4.5 Thermal properties

The thermal properties of packaging films and coatings are crucial for maintaining product quality. Variations in temperature during storage and transportation can compromise a film's structural integrity, reducing its mechanical strength and barrier effectiveness. Ensuring that films are thermally stable helps preserve their functionality and the quality of the packaged products during storage and transportation (Shah et al., 2024). The thermal properties of pectin films are generally analyzed by thermogravimetric analysis (TGA), derivative thermogravimetry (DTG), and differential scanning calorimetry (DSC).

de Oliveira et al. (2021) showed that the method of processing pectin films did not affect their thermal stability. Films produced either by the casting technique or by extrusion followed by thermo-pressing presented the same TGA and DTG patterns, with three thermal events and the same residual mass after 600°C (de Oliveira et al., 2021). Blends of pectin/agar containing zinc sulfide nanoparticles were studied by Roy & Rhim (2021a, 2021b); the TGA results showed that the good compatibility between agar, pectin, and the nanoparticles did not cause changes in the thermal stability of the films. This good compatibility between film components is also confirmed by the DSC results (Roy & Rhim, 2021b).

The effects of honey, propolis, and glycerol on the thermal stability of coffee pulp pectin films were evaluated by TGA and DSC in the study by Biratul, Biratu et al. (2024). The DSC results showed a reduction in the glass transition temperature (T_g) for the films containing glycerol but did not significantly change in T_g for the films containing the other additives. Besides, honey and propolis can act as plasticizers; interactions with the pectin matrix enhance their thermal stability. These results are also confirmed by the TGA and DTG results (Biratu et al., 2024).

Chaichi et al. (2023) developed pectin films containing nanocellulose and three essential oils (clove, thyme, and cinnamon). The thermal changes were evaluated by DSC, and the melting temperature (T_m) and melting enthalpy were recorded. The results showed that the incorporation of the essential oils reduced the T_m of the composite films from 165.37°C in the control film to 156.09°C and 155.4°C in the thyme/cinnamon and thyme/cinnamon/clove films, respectively. The melting enthalpy followed the same trend. This effect is attributed to the plasticizing effect of essential oils (Chaichi et al., 2023).

Blends of pectin and kidney bean protein containing grape pomace polyphenols were studied by Amiri Samani et al. (2025). The TGA indicates that incorporating both biopolymers and grape pomace polyphenols enhances the thermal stability of the films compared to films made solely of pectin or protein. This improvement is evident from the higher maximum degradation temperature (T_{max}) observed in the blended films, suggesting that the addition of polyphenols reinforces the material structure, providing greater resistance to thermal degradation due to the formation of hydrogen bonds between film components (Amiri Samani et al., 2025).

11.5 Physical modifications of pectin films and coatings

11.5.1 Ultrasonication

Ultrasound treatment has been widely used in the food industry due to its beneficial effects on separation, extraction, and homogenization processes. As an eco-friendly approach, ultrasound avoids secondary pollution. It utilizes cavitation, where the formation, growth, and implosion of tiny gas bubbles in biopolymer suspensions generate micro-streaming and localized high-shear forces, promoting biopolymer modification. This technique also enhances blending and homogenization by reducing particle size, thus improving the uniformity of films. Additionally, ultrasound reduces aggregation in pectin solutions and can modify the molecular weight and galacturonic acid content of pectin (Guo, Ge, Yang, et al., 2021; Wang et al., 2018). As an example, Guo, Ge, Yang, et al. (2021) applied ultrasound treatment to watermelon peel pectin film-forming solutions with varying exposure times, finding that the treatment reduced solution viscosity and influenced the resulting film properties. Films produced with shorter exposure times (<10 minutes) displayed a more compact structure, likely due to better-dispersed pectin molecules and enhanced hydrogen bonding interactions (Guo, Ge, Yang, et al., 2021).

11.5.2 High-pressure processing

HPP is a nonthermal technique widely employed in the food industry to enhance product quality and extend shelf life (Devi et al., 2014). Pectin and carboxymethyl cellulose were combined and treated with high pressures varying from 400 to 600 MPa for 20 minutes (Wang, Li, et al., 2024). The results demonstrated that a 500 MPa 20 minutes^{-1} HPP treatment significantly enhanced the tensile strength and elongation at break of the films. Additionally, it improved the films' thermal stability without substantially affecting their thermal degradability.

11.6 Chemical modifications of pectin films and coatings

The presence of hydroxyl and carboxyl groups distributed along the backbone, in conjunction with a certain number of neutral sugars presented as side chains, endows pectin with the capacity to prepare a diverse array of derivatives. The formation of pectin derivatives allows for the modification of existing properties and the creation of new functional film properties (Chen et al., 2015). The application of chemical modification relies on the material's tendency to form lumps and agglomerates during dissolution, as well as its rapid hydration, swelling, and erosion. These characteristics are complemented by desirable properties such as gel formation, excellent biocompatibility, nontoxicity, and biodegradability (Bhatia et al., 2008; Chen et al., 2015; Kurita et al., 2012). Pectin modification has been accomplished using various approaches, including substitution (e.g., amidation, thiolation, sulfation), chain elongation (such as crosslinking and grafting), depolymerization (via acid or enzymatic hydrolysis, β-elimination, or mechanical degradation), and physical and biological methods (Roy & Rhim, 2021b).

11.6.1 Alkylation

Alkylation represents a specific chemical modification that can be employed to enhance water solubility (Sashiwa & Aiba, 2004), gel strength (Ghahramanpoor et al., 2011), emulsifying activity (Nilsson & Bergenståhl, 2006), or antibacterial activity (Xie et al., 2002), and increase hydrophobicity (Roy & Rhim, 2021b) of food coatings (Singaram et al., 2024). The introduction of long-chain alkylation can attach additional carbons to the carboxyl groups (– COOH) of pectin (Fig. 11.1). The substitution by alkylation can be divided into carboxylate and hydroxyl substitution, and different studies examine the molecular characterization, conformational changes, and gel properties of pectin alkylated with different alkyl chains (C_6, C_{12}, or C_{18}) at varying degrees of substitution (DS) (Liu et al., 2017).

The C-6 position of galacturonic acid in the pectin backbone is frequently naturally methoxylated, representing one of pectin's most distinctive structural features. The carboxylate alkylation converts the – COOH groups to – COO – alkyl, for example, by using tetrabutylammonium chloride (TBA) to initiate the reaction and obtain TBA-pectin with alkyl bromide in dimethyl sulfoxide. Subsequently, the effects of alkyl chain length and substitution ratios on the viscosities (apparent and intrinsic viscosity) and emulsifying properties were improved (Liang et al., 2015).

Alkylation of hydroxyl groups (C – OH) to form alkyl ether groups (C – O – alkyl) is another widely adopted approach for the hydrophobization of hydrocolloids (Fig. 11.1). Acetylation is one of the most important methods in the modification of pectin (Liang et al., 2015). For example, the presence of acetyl groups could reduce the cation's binding strength to individual galacturonic acid residues or cause the polymer to adopt a less favorable conformation for binding (Roy & Rhim, 2021b). Alkylated-pectin films exhibit significant hydrophobic properties, making them widely used as

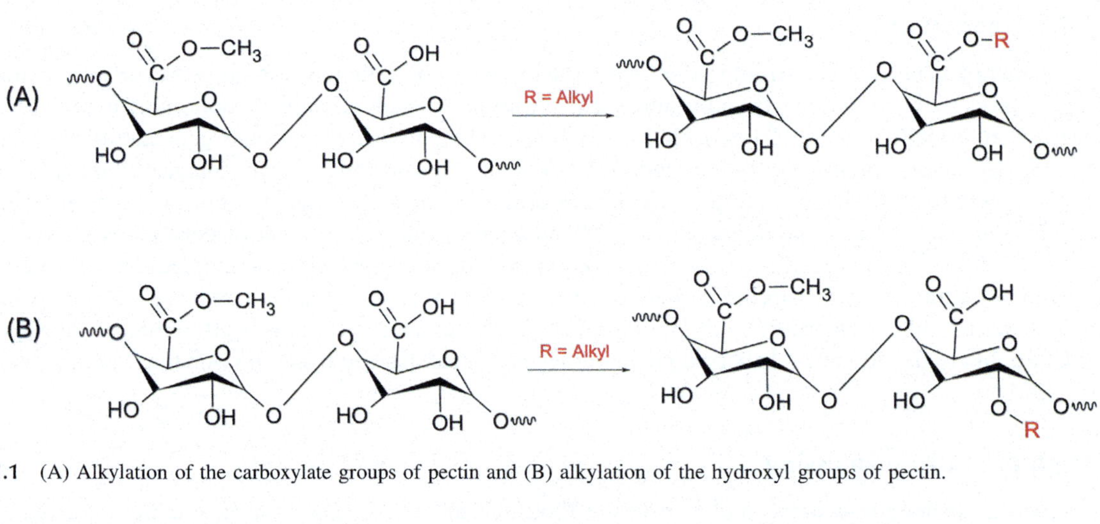

FIGURE 11.1 (A) Alkylation of the carboxylate groups of pectin and (B) alkylation of the hydroxyl groups of pectin.

matrices for controlled drug release. The release patterns of ibuprofen in the gastrointestinal tract are influenced by the reduced polarity and solubility of pectin, with studies reporting that acetylation enhances these properties (Miralles-Houzelle et al., 2001). Additionally, acylated pectin/gelatin-based films, combined with alkylated starch crystals, have been utilized for their antibacterial activity and coating applications. These composite films demonstrate antioxidant and antibacterial activities against *Escherichia coli* and *Staphylococcus aureus*. Furthermore, they inhibit lipid oxidation and improve the textural quality of *Trachinotus blochii* (golden pomfret) filets stored at 4°C, ensuring enhanced acceptability during storage (Song et al., 2023).

This strategy has already been adopted in the food industry, where acetylated pectin serves as a stabilizer and emulsifier. Chemical modification has proven to be an efficient, rapid, cost-effective, and environmentally friendly method for reducing the water absorption capacity of pectin, thereby enabling the exploration of new applications for this versatile polysaccharide (Roy & Rhim, 2021b). While the potential applications of modified pectin in food packaging are not yet fully understood, the observed improvements in this area could be adapted for biomedical applications with minimal modifications.

11.6.2 Amidation

The most commonly used amidated pectin derivatives contain primary amino groups ($-CO-NH_2$), with substitution primarily occurring in the methyl acetate groups ($CO-O-CH_3$) of the pectin (Fig. 11.2). These derivatives are commercially utilized in food technology due to their advantageous gelling properties. They demonstrate reduced sensitivity to calcium ions (Ca^{2+}), pH variations, and exhibit thermo-reversible behavior. A typical method for preparing amidated pectin involves ammonolysis of the methyl ester groups in highly methoxylated pectin using ammonia in anhydrous methanol, resulting in substitution of the hydrogen in the radical (Einhorn-Stoll et al., 2001; Wei et al., 2024).

Pectin amidation can also be achieved by reacting the polysaccharide with amino acids, followed by acetone precipitation and subsequent drying to produce the desired product. The combination of amidated pectin with various amino acids has shown potential to improve emulsifying properties and extend the shelf life of food products (Roy & Rhim, 2021b; Wang, Zhao, et al., 2024). Furthermore, amidated pectin has demonstrated excellent solubility in water across a pH range from acidic to neutral, even under conditions of high ionic strength. These properties make it an ideal candidate for producing pectin-based films for food packaging applications.

11.6.3 Quaternization

Quaternization is an effective method for enhancing the functional properties of polysaccharides, whereby ionic hydrocolloids are converted into their cationic derivatives. Quaternization is a chemical modification technique that is commonly employed to equip a compound with a cationic charge, primarily through the substitution of the quaternary ammonium group (Chan et al., 2024; de Almeida & da Silva, 2021). In general, under alkaline conditions, the demethoxylation of sodium hydroxide solutions is initiated, which subsequently leads to the quaternization process. It depends on the (DS%) of the quantity of R in the pectin chain.

For example, a well-established clay mineral, montmorillonite (MMT), is employed in quaternized pectin-montmorillonite to obtain hybrid films for sustained drug release, enhancing sustained release performance. Regardless of the simulated medium utilized, the cytotoxicity assay demonstrated that the material exhibited excellent biocompatibility (Meng et al., 2020). The quaternization of pectin improves biological activity, and the impact on polysaccharide toxicity needs more in vitro and in vivo tests (de Almeida & da Silva, 2021). Accordingly, this modification can be applied

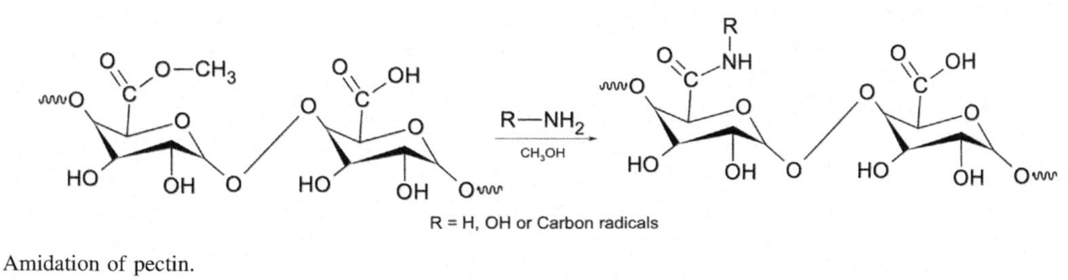

FIGURE 11.2 Amidation of pectin.

with comprehension and parsimony to food technology, given that the impact on packaging, for instance, using this technique remains incompletely understood.

11.6.4 Thiolation

Thiol-containing ligands are attached to the pectin chain via either ester or amide bonds. The synthesis of thiolated pectin involved the esterification of pectin with thioglycolic acid in the presence of hydrochloric acid (Fig. 11.3). This process results in the improvement of mucoadhesive properties without affecting the release profile (Roy & Rhim, 2022; Sharma & Ahuja, 2011). The preparation of thiolated pectin involves the formation of ester or amide bonds in some examples in Fig. 11.3A, thiolated pectin was synthesized by esterification of pectin with thioglycolic acid in the presence of hydrochloric acid (Sharma & Ahuja, 2011), and in Fig. 11.3B, thiolated pectin was synthesized through a covalent attachment with cysteine (Majzoob et al., 2006). Mucoadhesion of pectin on buccal tissue was found to be significantly enhanced and is observed to increase with the use of thiolated pectin (Hauptstein et al., 2014). In food applications, adding pectin reduces the network structure, weakening the gluten interactions, and thiolated pectin can maintain hydrophobicity or improve rheology, such as viscoelastic properties of gluten (Chen et al., 2023).

11.6.5 Crosslinking

Crosslinking involves forming chemical bonds between different molecular chains to create a stronger three-dimensional network (Garavand et al., 2017). The mechanism is defined as the ability to connect laterally between pectin molecules and uses metallic ions to form pectin-crosslinked films, and it can be divided into covalent bonds or electrostatic interactions. Metal ions such as calcium (Zhang et al., 2024), iron (Ghibaudo et al., 2018; Li, He, et al., 2024), zinc (Gawkowska et al., 2018), and copper (Zirak Hassan Kiadeh et al., 2021) can also crosslink with pectin to form an "egg-box" model similar to that of pectin and calcium ions (Zhang et al., 2022). These interactions lead to gel formation and improvement of water resistance (Zhang et al., 2022). For example, the alginate and pectin composite films crosslinked with Ca^{2+} ions require a two-step contact with Ca^{2+} using a plasticizer, first forming a low-structure prefilm that is further crosslinked, and second contact with a more concentrated Ca^{2+} solution containing plasticizer (Silva et al., 2009).

The use of epichlorohydrin as a crosslinker enables the preparation of novel hydrophilic pectin derivatives with significantly reduced water solubility (Semdé et al., 2003). Similarly, trisodium trimetaphosphate, a nontoxic crosslinker reacts with the hydroxyl groups of polysaccharides to strengthen the network, forming covalent gels (Prezotti et al., 2012). Crosslinked samples exhibit significantly enhanced thermal stability and network strength compared to noncrosslinked samples, highlighting the critical role of this method in food technology (Carbinatto et al., 2012).

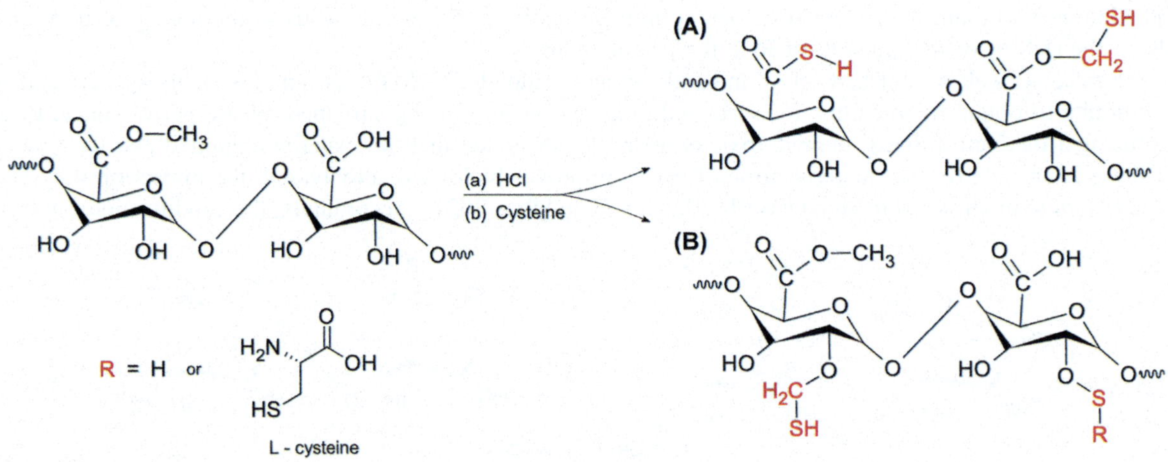

FIGURE 11.3 Thiolation of pectin using (A) esterification and (B) covalent attachment.

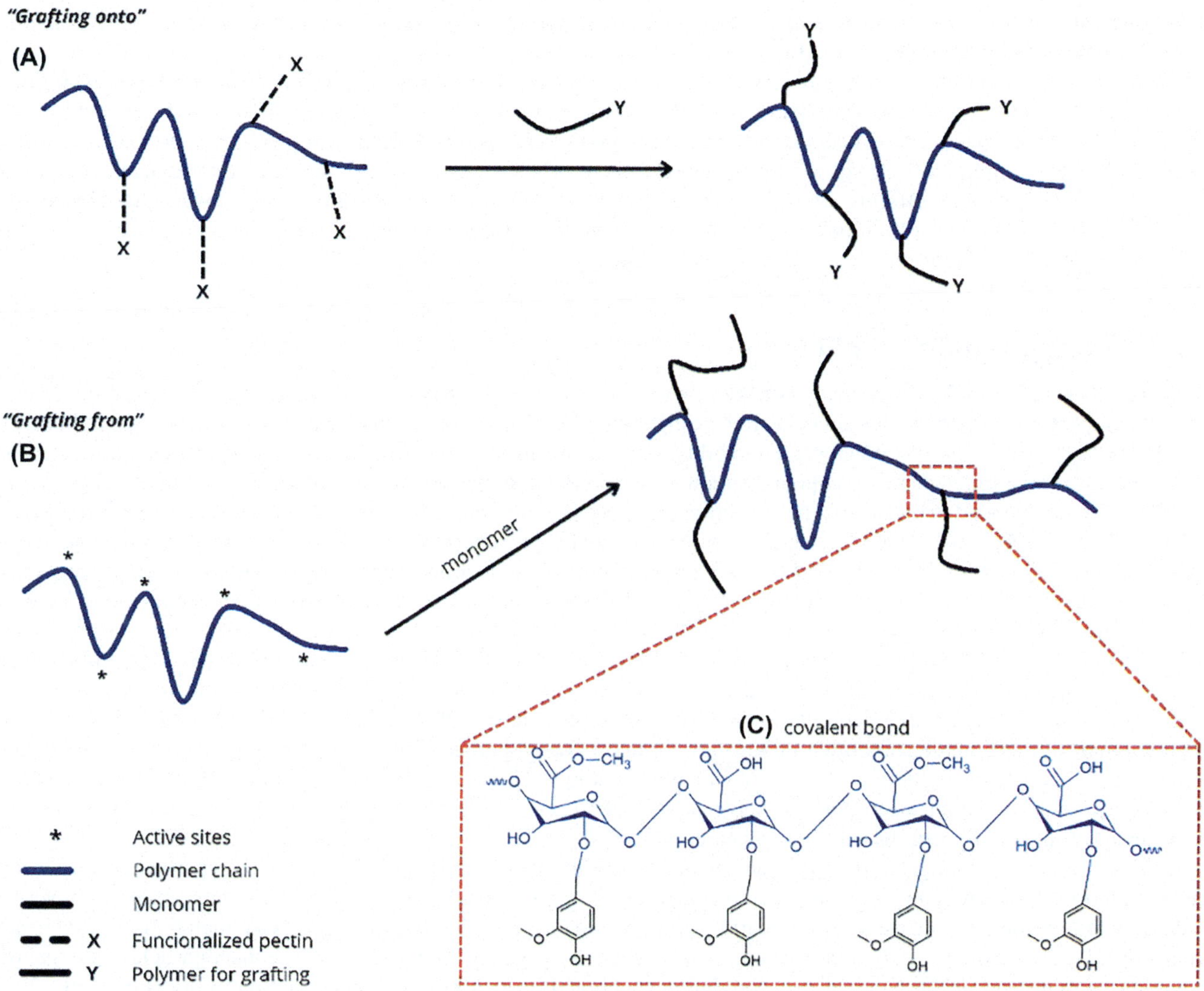

FIGURE 11.4 Schematic grafting examples (A) onto the pectin chain, (B) from the polymer pectin chain, and (C) covalent bond using vanillin as monomer grafted.
From Qiang, T., Ren, W., & Chen, L. (2024). Biodegradable, high mechanical strength, and eco-friendly pectin-based plastic film. Food Hydrocolloids, 149. https://doi.org/10.1016/j.foodhyd.2023.109539.

11.6.6 Grafting

Graft polymers consist of polymer segment branches covalently bonded to primary polymer chains. The process involves monomers and polymer chains, with radical polymerization offering a significant advantage due to its occurrence in aqueous media and high product yield. Grafting modifications are considered an attractive and promising method for polymer enhancement (Berlin & Kislenko, 1992; Vega-Hernández et al., 2021). Graft copolymerization can be categorized based on the initiation mechanism, including chemical (e.g., free-radical polymerization, ionic polymerization), radiation (e.g., free radical, ionic), photochemical, plasma radiation-induced, and enzymatic methods. Key factors influencing the grafting process include the nature of the backbone polymer, the type of monomer, solvent, initiator, and temperature (Bhattacharya & Misra, 2004).

The primary grafting mechanisms for pectin involve two approaches: the "onto" and "from" methods. The "onto" method involves functionalized pectin being added to the main chain (Fig. 11.4A), while the "from" method uses preexisting functional groups in the main chain, such as hydroxyl groups, as grafting sites (Fig. 11.4B). An example of covalent bond formation during grafting is the incorporation of vanillin (3-methoxy 4-hydroxybenzaldehyde), a natural flavor and antioxidant. This modification enhances the thermal stability, moisture resistance, and UV-blocking properties (200–400 nm) of the pectin film (Fig. 11.4C).

The grafting of vanillin to pectin can be induced with Fe^{3+} ions. Hydrogen bonds can form between the hydroxyl groups of pectin and the phenolic hydroxyl groups of vanillin, resulting in the combination of strong hydrogen bonds between the molecular chains of pectin. This process leads to the formation of outstanding mechanical properties and water resistance in the pectin. The grafting process allows for the enhancement of the level of control over the functionalized pectin with low polydispersity, thereby offering a range of advantages. Research in this field is progressing toward the objective of achieving biodegradability. It may prove to be an effective solution to the environmental pollution problems caused by components that resist biodegradation (Bhattacharya & Misra, 2004).

11.7 Incorporation of additives

Although polymeric biofilms have many benefits, they are often lower than traditional packaging in terms of mechanical strength and barrier properties. Researchers are investigating additives, including plasticizers, nanoparticles, and bioactive chemicals, to improve their functioning and durability to overcome these drawbacks and expand their uses. Plasticizers are low molecular weight, nonvolatile compounds frequently incorporated into polymeric materials to modify their properties according to the intended application. Their addition enhances the mobility of polymer chains by reducing intermolecular forces, leading to decreased rigidity, which is often reflected as a reduction in tensile strength. Simultaneously, this increased mobility imparts greater flexibility, resulting in improved elongation. Additionally, plasticizers effectively lower the second-order transition (T_g), thereby improving the processing capabilities of polymers and expanding their range of applications (Kokoszka et al., 2010; Vieira et al., 2011).

The most commonly used plasticizers in pectin films include glycerol ($C_3H_8O_3$) and sorbitol ($C_6H_{14}O_6$) (Aitboulahsen et al., 2020; Shahrampour et al., 2020). Glycerol, widely recognized as the most prevalent, is considered low in toxicity and poses minimal health risks, according to the European Food Safety Authority (Mortensen et al., 2017). However, the incorporation of these plasticizers often diminishes the bioactive properties of the films and provides limited nutritional value, which can negatively affect applications in edible coatings. This highlights the need to develop innovative natural plasticizers, such as natural deep eutectic solvents (NADES) and bio-based materials, to enhance the functional and nutritional attributes of pectin-based films (Chen et al., 2024; Gouveia et al., 2019; Nastasi et al., 2022).

Adding plasticizers like glycerol, while enhancing flexibility, can compromise UV barrier properties and accelerate food oxidation. To address this, inorganic and metallic nanoparticles have been investigated for their ability to block UV radiation and enhance the mechanical and barrier properties of food packaging films (do Nascimento et al., 2024; Sani et al., 2021). Furthermore, polymeric nanomaterials, such as chitosan and nanocellulose nanoparticles, are gaining attention as more sustainable alternatives, combining functionality with environmental compatibility (Lei et al., 2023; Lorevice et al., 2016).

11.8 Application in food preservation

The use of pectin in food packaging is generally associated with the development of active packaging solutions. Combining pectin with additives is widely promoted to extend the shelf life of stored food products. While alternative techniques such as extrusion, knife coating, and spraying have been explored in some studies, these films are most commonly produced using the solution-casting method. The solution-casting method is popular because it is the simplest technique for film production at the laboratory scale and does not require specialized equipment. Additionally, the lower processing temperatures of this method help preserve the structural integrity of many food-processing materials that might degrade under higher temperatures. However, this method is limited to sheet production, requires lengthy drying times, and is challenging to scale up for large-scale production (Roy et al., 2023; Wu, Wu, et al., 2022).

Pectin/gelatin films incorporated with curcumin and silver nanoparticles (AgNPs) demonstrated remarkable potential as multifunctional food packaging materials. The films exhibited excellent antioxidant activity, strong antibacterial performance with nearly 100% inhibition against *E. coli* and *S. aureus*, and good hydrophobicity, contributing to their effectiveness in food preservation. Importantly, they showed superior water vapor barrier properties, helping to extend product shelf life. Beyond protection, the films acted as intelligent packaging systems due to the pH-responsive behavior of curcumin, which enabled real-time monitoring of shrimp freshness. During storage, the films changed color from yellow to red as spoilage progressed, correlating with pH variations caused by the release of volatile amines. This visual indicator provided a simple and reliable tool for assessing food quality without opening the package. Together, these features highlight the films as sustainable, renewable biomass-based alternatives to petroleum-derived plastics, offering both active preservation and intelligent monitoring functions (Li, Wei, et al., 2024).

Edible films of pectin and carboxymethyl chitosan with roselle leaf extract (RLE) were developed as functional food packaging. The PCMCR-3 film (3% RLE) showed the best properties, including high tensile strength, elongation, hydrophobicity, and low WVP. RLE incorporation enhanced antioxidant and antibacterial activity, significantly extending fish filet shelf life at 4°C. While propolis extract (PE)-wrapped samples exceeded the spoilage limit (TVB-N > 20 mg 100 g^{-1}) after 8 days, PCMCR-3 films maintained freshness even after 12 days, effectively reducing protein degradation and microbial growth. This synergy between pectin/CMC and RLE demonstrates the potential of PCMCR-3 as a sustainable, active packaging solution for preserving perishable foods (Hlaing et al., 2025).

Pectin/gelatin nanocomposite films incorporating tannic acid (TA) and AgNPs were developed as antimicrobial, antioxidant, and UV-blocking packaging for strawberries. The films exhibited excellent UV protection, strong antibacterial activity against *E. coli* and *S. aureus*, and high antioxidant capacity (2,2-diphenyl-1-picrylhydrazyl (DPPH) 71.3%, 2,2′-azino-bis(3-ethylbenzothiazoline-6-sulfonic acid) (ABTS) 96.2%). In preservation tests, control and PE-packaged strawberries showed mold growth by days 4–6, while those wrapped in TA/AgNP-loaded films (P-G-T-A0.5) remained intact and free of contamination after eight days. These films also reduced weight loss, maintained higher hardness, and slowed the decline of total soluble solids, demonstrating superior freshness retention compared to conventional packaging. Overall, TA/AgNP nanocomposites present a sustainable strategy for extending the shelf life of highly perishable fruits (Li et al., 2025).

11.9 Biodegradability and environmental impact

With global plastic production exceeding 400 million tons annually, plastics are deeply embedded in modern society, playing a critical role in industries such as packaging, healthcare, construction, and electronics (Guicherd et al., 2024; Nayanathara Thathsarani Pilapitiya & Ratnayake, 2024). However, this widespread use poses significant environmental challenges, as only a small percentage is effectively recycled. Packaging, which typically has the shortest working life, is a major contributor to plastic waste, with up to 96.6% unrecycled (Rosenboom et al., 2022). In this scenario, pectin-based films emerge as an eco-friendly option to standard plastics. In addition to the possibility of different ways to obtain them, which transform agricultural waste into a valuable product, their biodegradability mitigates environmental impacts.

Pectin degradation primarily occurs through depolymerization, which proceeds via two main mechanisms. The first is β-elimination, where glycosidic bonds are cleaved, assisted by pectin lyase, resulting in the breakdown of the pectin structure. The second mechanism involves acid hydrolysis (pH < 3.0), where polygalacturonases specifically target and cleave the $\alpha(1 \rightarrow 4)$ bonds of the galacturonic acid backbone (Chen et al., 2015; Li, Peng et al., 2024; Norcino et al., 2020; Rehman et al., 2015). Both processes play a crucial role in the enzymatic and chemical degradation of pectin, influencing its biodegradability in various environments.

Among the methods studied in the literature, the biodegradation test in soil is the most commonly employed. This test is widely used due to the presence of microorganisms that facilitate the natural decomposition of pectin by producing pectinases, which efficiently degrade the material into simpler compounds (Kaul et al., 2024). Numerous studies have demonstrated that pectin exhibits excellent biodegradability, with more than 90% degradation within 30 days under soil conditions. Furthermore, research shows that the incorporation of additives into pectin-based films generally does not impede its degradability, maintaining its environmentally friendly profile even in the presence of stabilizers or other functional groups (do Nascimento et al., 2024; Jiang et al., 2022; Ren et al., 2022; Shivangi et al., 2021).

Ke et al. (2025) demonstrated that pectin and whey protein films, incorporating purple sweet potato extract as pH indicator, degraded rapidly, disintegrating within two days and reaching 50% degradation by the sixth day. This accelerated breakdown was attributed to the films' rapid soil moisture absorption, which enhanced microbial degradation (Ke et al., 2025). Similarly, Biratu, Woldemariam, and Gonfa reported that pectin-based edible films were completely degraded within nine days in soil due to their polygalacturonic acid content and high water solubility, which facilitated microbial access and accelerated degradation (Biratu et al., 2024). When applied to coated fertilizers, pectin-based materials can enhance nutrient release into the soil, benefiting plant growth and eliminating the possibility of contamination by microplastics (Xu et al., 2024).

These tests are generally performed at temperatures between 25°C and 30°C with humidity levels above 50% to simulate conditions conducive to biodegradation. Researchers typically use visual analysis, including morphological and optical microscopy, to monitor degradation progress. The degradation rate is often calculated based on the material's initial and final weight, while in some cases, CO_2 production is measured as an indicator of microbial activity and material breakdown, providing a more comprehensive evaluation of biodegradability (de Castro e Silva et al., 2020; Norcino et al., 2020; Shivangi et al., 2021).

Life-cycle assessment (LCA) is a technique for assessing the environmental impacts across each material production stage, from raw material extraction to disposal. Research on pectin production and its films highlights extraction and precipitation as major contributors to reduced sustainability, due to high energy demands and the use of ethanol and acetic acid (Duggal et al., 2024; Günkaya & Banar, 2016). Garcia-Garcia et al. demonstrated that microwave-assisted extraction significantly mitigates these impacts, achieving a 75% decrease in environmental impact compared to traditional thermal methods (Garcia-Garcia et al., 2019). Moreover, additional adjustments in the production process, such as incorporating bioethanol and renewable energy sources, can further reduce environmental impacts (Xu et al., 2024).

The rapid degradation of pectin films presents a promising alternative to conventional plastics, many of which are resistant to degradation and contribute significantly to environmental pollution. Single-use plastics, particularly in food packaging, are a major source of waste, and their disposal involves high costs and complex management. In that context, biodegradable materials, like pectin-based films, offer a viable solution, providing faster degradation and reduced environmental impact. Additionally, pectin's versatility, derived from various sources—some considered waste, enhances its attractiveness as an eco-friendly material for sustainable packaging applications.

11.10 Conclusions and future prospects

The increasing demand for innovative food packaging has made the development of new materials an important field of research. Pectin-based films and coatings offer an environmentally friendly alternative to synthetic polymers. Their mechanical and barrier capacities are influenced by pectin's inherent properties and modifications. Chemical alterations, including the addition of additives, are the most common strategies for improving performance and extending food shelf life. Despite progress, issues remain in optimizing large-scale production and ensuring sustainable raw material sources. Future research should focus on improving extraction procedures, finding innovative biocompatible additives, and enhancing the functional qualities of pectin films according to industry demands while maintaining environmental sustainability.

References

Adetunji, L. R., Adekunle, A., Orsat, V., & Raghavan, V. (2017). Advances in the pectin production process using novel extraction techniques: A review. *Food Hydrocolloids, 62*, 239–250. https://doi.org/10.1016/j.foodhyd.2016.08.015, https://www.journals.elsevier.com/food-hydrocolloids.

Aitboulahsen, M., El Galiou, O., Laglaoui, A., Bakkali, M., & Hassani Zerrouk, M. (2020). Effect of plasticizer type and essential oils on mechanical, physicochemical, and antimicrobial characteristics of gelatin, starch, and pectin-based films. *Journal of Food Processing and Preservation, 44*(6). https://doi.org/10.1111/jfpp.14480.

Amiri Samani, S., PourvatanDoust, S., Savarolyia, M., Aboutalebzadeh, S., Khezri, M., Kazemi, M., Khodaiyan, F., & Hosseini, S. S. (2025). Valorization of red grape pomace for sustainable food packaging: Development of pectin/kidney bean protein based biocomposite films enriched with grape pomace polyphenols. *Food Hydrocolloids, 160*, 110806. https://doi.org/10.1016/j.foodhyd.2024.110806.

Anis, A., Pal, K., & Al-Zahrani, S. M. (2021). Essential oil-containing polysaccharide-based edible films and coatings for food security applications. *Polymers, 13*(4), 1–32. https://doi.org/10.3390/polym13040575, https://www.mdpi.com/2073-4360/13/4/575/pdf.

Basak, S., & Annapure, U. S. (2022). The potential of subcritical water as a "green" method for the extraction and modification of pectin: A critical review. *Food Research International, 161*, 111849. https://doi.org/10.1016/j.foodres.2022.111849.

Benito-Román, Ó., Alonso-Riaño, P., Díaz de Cerio, E., Sanz, M. T., & Beltrán, S. (2022). Semi-continuous hydrolysis of onion skin wastes with subcritical water: Pectin recovery and oligomers identification. *Journal of Environmental Chemical Engineering, 10*(3), 107439. https://doi.org/10.1016/j.jece.2022.107439.

Berlin, A. A., & Kislenko, V. N. (1992). Kinetics and mechanism of radical graft polymerization of monomers onto polysaccharides. *Progress in Polymer Science, 17*(5), 765–825. https://doi.org/10.1016/0079-6700(92)90010-V.

Bhatia, M. S., Choudhari, P., Bhatia, N. M., & Deshmukh, R. (2008). Chemical modification of pectins, characterization and evaluation for drug delivery. *Scientia Pharmaceutica, 76*(4), 775–784. https://doi.org/10.3797/scipharm.0805-23India, http://www.scipharm.at/download.asp?id=306.

Bhatia, S., Al-Harrasi, A., Shah, Y. A., Saif Alrasbi, A. N., Jawad, M., Koca, E., Aydemir, L. Y., Alamoudi, J. A., Almoshari, Y., & Mohan, S. (2024). Structural, mechanical, barrier and antioxidant properties of pectin and xanthan gum edible films loaded with grapefruit essential oil. *Heliyon, 10*(3), e25501. https://doi.org/10.1016/j.heliyon.2024.e25501.

Bhattacharya, A., & Misra, B. N. (2004). Grafting: A versatile means to modify polymers: Techniques, factors and applications. *Progress in Polymer Science (Oxford), 29*(8), 767–814. https://doi.org/10.1016/j.progpolymsci.2004.05.002, http://www.sciencedirect.com/science/journal/00796700.

Biratu, G., Woldemariam, H. W., & Gonfa, G. (2024). Development of active edible films from coffee pulp pectin, propolis, and honey with improved mechanical, functional, antioxidant, and antimicrobial properties. *Carbohydrate Polymer Technologies and Applications, 8*, 100557. https://doi.org/10.1016/j.carpta.2024.100557.

Cao, L., Lu, W., Mata, A., Nishinari, K., & Fang, Y. (2020). Egg-box model-based gelation of alginate and pectin: A review. *Carbohydrate Polymers, 242*, 116389. https://doi.org/10.1016/j.carbpol.2020.116389.

Carbinatto, F. M., De Castro, A. D., Cury, B. S. F., Magalhães, A., & Evangelista, R. C. (2012). Physical properties of pectin-high amylose starch mixtures cross-linked with sodium trimetaphosphate. *International Journal of Pharmaceutics, 423*(2), 281–288. https://doi.org/10.1016/j.ijpharm.2011.11.042.

Çavdaroğlu, E., Büyüktaş, D., Farris, S., & Yemenicioğlu, A. (2023). Novel edible films of pectins extracted from low-grade fruits and stalk wastes of sun-dried figs: Effects of pectin composition and molecular properties on film characteristics. *Food Hydrocolloids, 135*, 108136. https://doi.org/10.1016/j.foodhyd.2022.108136.

Cazón, P., Velazquez, G., Ramírez, J. A., & Vázquez, M. (2017). Polysaccharide-based films and coatings for food packaging: A review. *Food Hydrocolloids, 68*, 136–148. https://doi.org/10.1016/j.foodhyd.2016.09.009, https://www.journals.elsevier.com/food-hydrocolloids.

Chaichi, M., Badii, F., Mohammadi, A., & Hashemi, M. (2023). Novel bioactive composite films based on pectin-nanocellulose-synergistic triple essential oils: Development and characterization. *Food and Bioprocess Technology, 16*(8), 1794–1805. https://doi.org/10.1007/s11947-023-03036-x.

Chan, S. Y., Owh, C., Yan, Y., Lim, Q. R. T., Chee, P. L., Zhang, Y., & Choo, W. S. (2024). Functionalization and structural evaluation of pectin. In *Handbook of natural polymers, volume 2: Functionalization, surface modification, and properties*. Singapore: Elsevier, 123–154. https://doi.org/10.1016/B978-0-323-99856-7.00005-7, https://www.sciencedirect.com/book/9780323998567.

Chen, H., Shang, K., Bian, X., Zhao, Z., Liu, Y., Lin, X., Wang, L., Zhang, W., Hu, X., & Guo, X. (2024). Enhanced functional pectin films incorporated with olive fruit extracts prepared by deep eutectic solvents. *Food Packaging and Shelf Life, 46*, 101361. https://doi.org/10.1016/j.fpsl.2024.101361.

Chen, J., Liu, W., Liu, C. M., Li, T., Liang, R. H., & Luo, S. J. (2015). Pectin modifications: A review. *Critical Reviews in Food Science and Nutrition, 55*(12), 1684–1698. https://doi.org/10.1080/10408398.2012.718722, www.tandf.co.uk/journals/titles/10408398.asp.

Chen, J., Cui, Y., Shi, W., Yang, F., Ma, Y., Zhang, S., & Wang, Y. (2023). Thiolation modification of pectin effectively alleviates the weakening of wheat gluten network caused by natural pectin. *Food Hydrocolloids, 144*, 109033. https://doi.org/10.1016/j.foodhyd.2023.109033.

Cho, E. H., Jung, H. T., Lee, B. H., Kim, H. S., Rhee, J. K., & Yoo, S. H. (2019). Green process development for apple-peel pectin production by organic acid extraction. *Carbohydrate Polymers, 204*, 97–103. https://doi.org/10.1016/j.carbpol.2018.09.086, http://www.elsevier.com/wps/find/journaldescription.cws_home/405871/description#description.

Colodel, C., & Petkowicz, C. L.de O. (2019). Acid extraction and physicochemical characterization of pectin from cubiu (*Solanum sessiliflorum* D.) fruit peel. *Food Hydrocolloids, 86*, 193–200. https://doi.org/10.1016/j.foodhyd.2018.06.013.

Cui, Y., Wang, S., Wang, S., Cao, S., Wang, X., & Lü, X. (2022). Extraction optimization and characterization of persimmon peel pectin extracted by subcritical water. *Food Chemistry: X, 16*, 100486. https://doi.org/10.1016/j.fochx.2022.100486.

Dao, D. N., Le, P. H., Do, D. X., Dang, T. M. Q., Nguyen, S. K., & Nguyen, V. (2023). Pectin and cellulose extracted from coffee pulps and their potential in formulating biopolymer films. *Biomass Conversion and Biorefinery, 13*(14), 13117–13125. https://doi.org/10.1007/s13399-022-02339-x, https://www.springer.com/journal/13399.

de Almeida, W. S., & da Silva, D. A. (2021). Does polysaccharide quaternization improve biological activity? *International Journal of Biological Macromolecules, 182*, 1419–1436. https://doi.org/10.1016/j.ijbiomac.2021.05.012, www.elsevier.com/locate/ijbiomac.

de Castro e Silva, P., de Oliveira, A. C. S., Pereira, L. A. S., Valquíria, M., Carvalho, G. R., Miranda, K. W. E., Marconcini, J. M., & Oliveira, J. E. (2020). Development of bionanocomposites of pectin and nanoemulsions of carnauba wax and neem oil pectin/carnauba wax/neem oil composites. *Polymer Composites, 41*(3), 858–870. https://doi.org/10.1002/pc.25416, http://onlinelibrary.wiley.com/journal/10.1002/(ISSN)1548-0569.

de Oliveira, A. C. S., Ferreira, L. F., de Oliveira Begali, D., Ugucioni, J. C., de Sena Neto, A. R., Yoshida, M. I., & Borges, S. V. (2021). Thermoplasticized pectin by extrusion/thermo-compression for film industrial application. *Journal of Polymers and the Environment, 29*(8), 2546–2556. https://doi.org/10.1007/s10924-021-02054-0, http://www.kluweronline.com/issn/1566-2543/.

Devi, A. F., Buckow, R., Hemar, Y., & Kasapis, S. (2014). Modification of the structural and rheological properties of whey protein/gelatin mixtures through high pressure processing. *Food Chemistry, 156*, 243–249. https://doi.org/10.1016/j.foodchem.2014.01.097.

Dobrucka, R., Dlugaszewska, J., Pawlik, M., & Szymański, M. (2025). Innovative active bio-based food packaging material with *Cannabis sativa* L. seeds extract as an agent to reduce food waste. *Colloids and Surfaces B: Biointerfaces, 245*, 114313. https://doi.org/10.1016/j.colsurfb.2024.114313.

do Nascimento, W. J., da Costa, J. C. M., Alves, E. S., de Oliveira, M. C., Monteiro, J. P., Souza, P. R., Martins, A. F., & Bonafe, E. G. (2024). Zinc oxide nanoparticle-reinforced pectin/starch functionalized films: A sustainable solution for biodegradable packaging. *International Journal of Biological Macromolecules, 257*. https://doi.org/10.1016/j.ijbiomac.2023.128461, https://www.sciencedirect.com/science/journal/01418130.

Dranca, F., & Oroian, M. (2018). Extraction, purification and characterization of pectin from alternative sources with potential technological applications. *Food Research International, 113*, 327–350. https://doi.org/10.1016/j.foodres.2018.06.065.

Duggal, M., Singh, D. P., Singh, S., Khubber, S., Garg, M., & Krishania, M. (2024). Microwave-assisted acid extraction of high-methoxyl kinnow (*Citrus reticulata*) peels pectin: Process, techno-functionality, characterization and life cycle assessment. *Food Chemistry: Molecular Sciences, 9*, 100213. https://doi.org/10.1016/j.fochms.2024.100213.

Einhorn-Stoll, U., Salazar, T., Jaafar, B., & Kunzek, H. (2001). Thermodynamic compatibility of sodium caseinate with different pectins. Influence of the milieu conditions and pectin modifications. *Nahrung/Food, 45*(5), 332. https://doi.org/10.1002/1521-3803(20011001)45:5<332::aid-food332>3.0.co;2-g.

Espitia, P. J. P., Du, W. X., Avena-Bustillos, Rd. J., Soares, Nd. F. F., & McHugh, T. H. (2014). Edible films from pectin: Physical-mechanical and antimicrobial properties - A review. *Food Hydrocolloids, 35*, 287–296. https://doi.org/10.1016/j.foodhyd.2013.06.005, https://www.journals.elsevier.com/food-hydrocolloids.

Fishman, M. L., Coffin, D. R., Onwulata, C. I., & Konstance, R. P. (2004). Extrusion of pectin and glycerol with various combinations of orange albedo and starch. *Carbohydrate Polymers, 57*(4), 401–413. https://doi.org/10.1016/j.carbpol.2004.05.014.

Garavand, F., Rouhi, M., Razavi, S. H., Cacciotti, I., & Mohammadi, R. (2017). Improving the integrity of natural biopolymer films used in food packaging by crosslinking approach: A review. *Journal of Biological Macromolecules, 104*, 687–707. https://doi.org/10.1016/j.ijbiomac.2017.06.093, www.elsevier.com/locate/ijbiomac.

Garcia-Garcia, G., Rahimifard, S., Matharu, A. S., & Dugmore, T. I. J. (2019). Life-cycle assessment of microwave-assisted pectin extraction at pilot scale. *ACS Sustainable Chemistry and Engineering, 7*(5), 5167–5175. https://doi.org/10.1021/acssuschemeng.8b06052, http://pubs.acs.org/journal/ascecg.

Gawkowska, D., Cybulska, J., & Zdunek, A. (2018). Cross-linking of sodium carbonate-soluble pectins from apple by zinc ions. *Carbohydrate Polymers, 196*, 1–7. https://doi.org/10.1016/j.carbpol.2018.05.024.

Gentilini, R., Bozzini, S., Munarin, F., Petrini, P., Visai, L., & Tanzi, M. C. (2014). Pectins from Aloe Vera: Extraction and production of gels for regenerative medicine. *Journal of Applied Polymer Science, 131*(2). https://doi.org/10.1002/app.39760.

Ghahramanpoor, M. K., Hassani Najafabadi, S. A., Abdouss, M., Bagheri, F., & Baghaban Eslaminejad, M. (2011). A hydrophobically-modified alginate gel system: Utility in the repair of articular cartilage defects. *Journal of Materials Science: Materials in Medicine, 22*(10), 2365–2375. https://doi.org/10.1007/s10856-011-4396-2.

Ghibaudo, F., Gerbino, E., Hugo, A. A., Simões, M. G., Alves, P., Costa, B. F. O., Dall′, V. C., Orto, Gómez-Zavaglia, A., & Simões, P. N. (2018). Development and characterization of iron-pectin beads as a novel system for iron delivery to intestinal cells. *Colloids and Surfaces B: Biointerfaces, 170*, 538–543. https://doi.org/10.1016/j.colsurfb.2018.06.052.

Gouveia, T. I. A., Biernacki, K., Castro, M. C. R., Gonçalves, M. P., & Souza, H. K. S. (2019). A new approach to develop biodegradable films based on thermoplastic pectin. *Food Hydrocolloids, 97*. https://doi.org/10.1016/j.foodhyd.2019.105175, https://www.sciencedirect.com/science/journal/0268005X.

Guicherd, M., Ben Khaled, M., Guéroult, M., Nomme, J., Dalibey, M., Grimaud, F., Alvarez, P., Kamionka, E., Gavalda, S., Noël, M., Vuillemin, M., Amillastre, E., Labourdette, D., Cioci, G., Tournier, V., Kitpreechavanich, V., Dubois, P., André, I., Duquesne, S., & Marty, A. (2024). An engineered enzyme embedded into PLA to make self-biodegradable plastic. *Nature, 631*(8022), 884–890. https://doi.org/10.1038/s41586-024-07709-1.

Günkaya, Z., & Banar, M. (2016). An environmental comparison of biocomposite film based on orange peel-derived pectin jelly-corn starch and LDPE film: LCA and biodegradability. *International Journal of Life Cycle Assessment, 21*(4), 465–475. https://doi.org/10.1007/s11367-016-1042-8, http://www.springerlink.com/content/0948-3349.

Guo, Z., Ge, X., Li, W., Yang, L., Han, L., & Yu, Q.-li (2021). Active-intelligent film based on pectin from watermelon peel containing beetroot extract to monitor the freshness of packaged chilled beef. *Food Hydrocolloids, 119*, 106751. https://doi.org/10.1016/j.foodhyd.2021.106751.

Guo, Z., Ge, X., Yang, L., Gou, Q., Han, L., & Yu, Q.-li (2021). Utilization of watermelon peel as a pectin source and the effect of ultrasound treatment on pectin film properties. *LWT, 147*, 111569. https://doi.org/10.1016/j.lwt.2021.111569.

Han, H. S., & Song, K. B. (2021). Antioxidant properties of watermelon (*Citrullus lanatus*) rind pectin films containing kiwifruit (*Actinidia chinensis*) peel extract and their application as chicken thigh packaging. *Food Packaging and Shelf Life, 28*. https://doi.org/10.1016/j.fpsl.2021.100636, http://www.journals.elsevier.com/food-packaging-and-shelf-life/.

Hauptstein, S., Hintzen, F., Müller, C., Ohm, M., & Bernkop-Schnürch, A. (2014). Development and in vitro evaluation of a buccal drug delivery system based on preactivated thiolated pectin. *Drug Development and Industrial Pharmacy, 40*(11), 1530–1537. https://doi.org/10.3109/03639045.2013.836213.

Hlaing, K. S. S., Fall, M., Tristanto, N. A., Carole, N. V. D., Kaharso, V. C., Golshany, H., Siddiquy, M., Yu, D., Yanshun, X., Qixing, J., & Xia, W. (2025). Preparation of edible films from pectin/carboxymethyl chitosan incorporating polyphenol-rich roselle leaf extracts for food packaging applications. *International Journal of Biological Macromolecules, 310*. https://doi.org/10.1016/j.ijbiomac.2025.143351, https://www.sciencedirect.com/science/journal/01418130.

Hoque, M., Babu, R. P., McDonagh, C., Jaiswal, S., Tiwari, B. K., Kerry, J. P., & Pathania, S. (2024). Pectin/sodium alginate-based active film integrated with microcrystalline cellulose and geraniol for food packaging applications. *International Journal of Biological Macromolecules, 271*. https://doi.org/10.1016/j.ijbiomac.2024.132414, https://www.sciencedirect.com/science/journal/01418130.

Huang, J., Hu, Z., Hu, L., Li, G., Yao, Q., & Hu, Y. (2021). Pectin-based active packaging: A critical review on preparation, physical properties and novel application in food preservation. *Trends in Food Science & Technology, 118*, 167–178. https://doi.org/10.1016/j.tifs.2021.09.026.

Jafarzadeh-Moghaddam, M., Shaddel, R., & Peighambardoust, S. H. (2021). Sugar beet pectin extracted by ultrasound or conventional heating: A comparison. *Journal of Food Science and Technology, 58*(7), 2567–2578. https://doi.org/10.1007/s13197-020-04763-1, http://www.springerlink.com/content/121580/.

Jiang, H., Zhang, W., Xu, Y., Cao, J., & Jiang, W. (2022). Properties of pectin-based films from white-fleshed pitaya (*Hylocereus undatus*) peel waste as affected by montmorillonite. *Food Packaging and Shelf Life, 34*, 100952. https://doi.org/10.1016/j.fpsl.2022.100952.

Jovanović, J., Ćirković, J., Radojković, A., Mutavdžić, D., Tanasijević, G., Joksimović, K., Bakić, G., Branković, G., & Branković, Z. (2021). Chitosan and pectin-based films and coatings with active components for application in antimicrobial food packaging. *Progress in Organic Coatings, 158*, 106349. https://doi.org/10.1016/j.porgcoat.2021.106349.

Karbuz, P., & Tugrul, N. (2021). Microwave and ultrasound assisted extraction of pectin from various fruits peel. *Journal of Food Science and Technology, 58*(2), 641–650. https://doi.org/10.1007/s13197-020-04578-0.

Kaul, K., Rajauria, G., & Singh, R. (2024). Valorization of agro-industrial waste for pectinase production and its influence on circular economy. *Food and Bioproducts Processing, 148*, 141–153. https://doi.org/10.1016/j.fbp.2024.09.008.

Kazemi, M., Khodaiyan, F., Labbafi, M., & Hosseini, S. S. (2020). Ultrasonic and heating extraction of pistachio by-product pectin: Physicochemical, structural characterization and functional measurement. *Journal of Food Measurement and Characterization, 14*(2), 679–693. https://doi.org/10.1007/s11694-019-00315-0, http://rd.springer.com/journal/11694.

Ke, F., Yang, M., Ji, W., & Liu, D. (2025). Functional pH-sensitive film based on pectin and whey protein for grape preservation and shrimp freshness monitoring. *Food Chemistry, 463*(Pt 1), 141092. https://doi.org/10.1016/j.foodchem.2024.141092.

Kokoszka, S., Debeaufort, F., Hambleton, A., Lenart, A., & Voilley, A. (2010). Protein and glycerol contents affect physico-chemical properties of soy protein isolate-based edible films. *Innovative Food Science & Emerging Technologies, 11*(3), 503–510. https://doi.org/10.1016/j.ifset.2010.01.006.

Kurita, O., Miyake, Y., & Yamazaki, E. (2012). Chemical modification of citrus pectin to improve its dissolution into water. *Carbohydrate Polymers, 87*(2), 1720–1727. https://doi.org/10.1016/j.carbpol.2011.09.081.

Lasunon, P., & Sengkhamparn, N. (2022). Effect of ultrasound-assisted, microwave-assisted and ultrasound-microwave-assisted extraction on pectin extraction from industrial tomato waste. *Molecules, 27*(4), 1157. https://doi.org/10.3390/molecules27041157.

Lei, Y., Yao, Q., Jin, Z., & Wang, Y.-C. (2023). Intelligent films based on pectin, sodium alginate, cellulose nanocrystals, and anthocyanins for monitoring food freshness. *Food Chemistry, 404*(Pt A), 134528. https://doi.org/10.1016/j.foodchem.2022.134528.

Li, J., Peng, C., Mao, A., Zhong, M., & Hu, Z. (2024). An overview of microbial enzymatic approaches for pectin degradation. *International Journal of Biological Macromolecules, 254*(Pt 1), 127804. https://doi.org/10.1016/j.ijbiomac.2023.127804.

Li, S., Fang, C., Wei, N., Wei, J., Feng, T., Liu, F., Liu, X., & Wu, B. (2025). Antimicrobial, antioxidative, and UV-blocking pectin/gelatin food packaging films incorporated with tannic acid and silver nanoparticles for strawberry preservation. *International Journal of Biological Macromolecules, 308*(Pt 3), 142445. https://doi.org/10.1016/j.ijbiomac.2025.142445.

Li, S., Wei, N., Wei, J., Fang, C., Feng, T., Liu, F., Liu, X., & Wu, B. (2024). Curcumin and silver nanoparticles loaded antibacterial multifunctional pectin/gelatin films for food packaging applications. *International Journal of Biological Macromolecules, 266*(Pt 1), 131248. https://doi.org/10.1016/j.ijbiomac.2024.131248.

Li, X., He, J., Zhang, W., Khan, M. R., Ahmad, N., & Tian, W. (2024). Pectin film fortified with zein nanoparticles and Fe3+-Encapsulated propolis extract for enhanced fruit preservation. *Food Hydrocolloids, 157*, 110405. https://doi.org/10.1016/j.foodhyd.2024.110405.

Li, X. L., Shen, Y., Hu, F., Zhang, X. X., Thakur, K., Rengasamy, K. R. R., Khan, M. R., Busquets, R., & Wei, Z. J. (2023). Fortification of polysaccharide-based packaging films and coatings with essential oils: A review of their preparation and use in meat preservation. *International Journal of Biological Macromolecules, 242*. https://doi.org/10.1016/j.ijbiomac.2023.124767, https://www.sciencedirect.com/journal/international-journal-of-biological-macromolecules.

Liang, R.-hong, Wang, L.-hua, Chen, J., Liu, W., & Liu, C.-mei (2015). Alkylated pectin: Synthesis, characterization, viscosity and emulsifying properties. *Food Hydrocolloids, 50*, 65–73. https://doi.org/10.1016/j.foodhyd.2015.04.007.

Liew, S. Q., Ngoh, G. C., Yusoff, R., & Teoh, W. H. (2016). Sequential ultrasound-microwave assisted acid extraction (UMAE) of pectin from pomelo peels. *International Journal of Biological Macromolecules, 93*, 426–435. https://doi.org/10.1016/j.ijbiomac.2016.08.065, www.elsevier.com/locate/ijbiomac.

Liu, C.-mei, Guo, X.-juan, Liang, R.-hong, Liu, W., & Chen, J. (2017). Alkylated pectin: Molecular characterization, conformational change and gel property. *Food Hydrocolloids, 69*, 341–349. https://doi.org/10.1016/j.foodhyd.2017.03.008.

Lorevice, M. V., Otoni, C. G., Moura, M. R.de, & Mattoso, L. H. C. (2016). Chitosan nanoparticles on the improvement of thermal, barrier, and mechanical properties of high- and low-methyl pectin films. *Food Hydrocolloids, 52*, 732–740. https://doi.org/10.1016/j.foodhyd.2015.08.003, https://www.journals.elsevier.com/food-hydrocolloids.

Majzoob, S., Atyabi, F., Dorkoosh, F., Kafedjiiski, K., Loretz, B., & Bernkop-Schnürch, A. (2006). Pectin-cysteine conjugate: Synthesis and in-vitro evaluation of its potential for drug delivery. *Journal of Pharmacy and Pharmacology, 58*(12), 1601–1610. https://doi.org/10.1211/jpp.58.12.0006.

Mallakpour, S., & Mohammadi, N. (2022). Development of sodium alginate-pectin/TiO_2 nanocomposites: Antibacterial and bioactivity investigations. *Carbohydrate Polymers, 285*, 119226. https://doi.org/10.1016/j.carbpol.2022.119226.

Mao, G., Wu, D., Wei, C., Tao, W., Ye, X., Linhardt, R. J., Orfila, C., & Chen, S. (2019). Reconsidering conventional and innovative methods for pectin extraction from fruit and vegetable waste: Targeting rhamnogalacturonan I. *Trends in Food Science and Technology, 94*, 65–78. https://doi.org/10.1016/j.tifs.2019.11.001, http://www.elsevier.com/wps/find/journaldescription.cws_home/601278/description#description.

Marangoni Júnior, L., Fozzatti, C. R., Jamróz, E., Vieira, R. P., & Alves, R. M. V. (2022). Biopolymer-based films from sodium alginate and citrus pectin reinforced with SiO_2. *Materials, 15*(11), 3881. https://doi.org/10.3390/ma15113881.

Meng, Y.-jie, Wang, S.-ya, Guo, Z.-wei, Cheng, M.-mei, Li, J., & Li, D.-qiang (2020). Design and preparation of quaternized pectin-Montmorillonite hybrid film for sustained drug release. *International Journal of Biological Macromolecules, 154*, 413–420. https://doi.org/10.1016/j.ijbiomac.2020.03.140.

Miralles-Houzelle, M. C., Hubert, P., & Dellacherie, E. (2001). Hydrophobic alkyl chains-pectin conjugates. Comparative study of some physicochemical properties in relation to covalent coupling vs ionic association. *Langmuir, 17*(5), 1384–1391. https://doi.org/10.1021/la0012081.

Mortensen, A., Aguilar, F., Crebelli, R., Di Domenico, A., Dusemund, B., Frutos, M. J., Galtier, P., Gott, D., Gundert-Remy, U., Leblanc, J.-C., Lindtner, O., Moldeus, P., Mosesso, P., Parent-Massin, D., Oskarsson, A., Stankovic, I., Waalkens-Berendsen, I., Woutersen, R. A., Wright, M., ... Lambré, C. (2017). Re-evaluation of glycerol (E 422) as a food additive. *EFSA Journal, 15*(3). https://doi.org/10.2903/j.efsa.2017.4720.

Munarin, F., Tanzi, M. C., & Petrini, P. (2012). Advances in biomedical applications of pectin gels. *International Journal of Biological Macromolecules, 51*(4), 681–689. https://doi.org/10.1016/j.ijbiomac.2012.07.002.

Nastasi, J. R., Kontogiorgos, V., Daygon, V. D., & Fitzgerald, M. A. (2022). Pectin-based films and coatings with plant extracts as natural preservatives: A systematic review. *Trends in Food Science and Technology, 120*, 193–211. https://doi.org/10.1016/j.tifs.2022.01.014, http://www.elsevier.com/wps/find/journaldescription.cws_home/601278/description#description.

Nayanathara Thathsarani Pilapitiya, P. G. C., & Ratnayake, A. S. (2024). The world of plastic waste: A review. *Cleaner Materials, 11*, 100220. https://doi.org/10.1016/j.clema.2024.100220.

Nilsson, L., & Bergenståhl, B. (2006). Adsorption of hydrophobically modified starch at oil/water interfaces during emulsification. *Langmuir, 22*(21), 8770–8776. https://doi.org/10.1021/la060870f.

Norcino, L. B., Mendes, J. F., Natarelli, C. V. L., Manrich, A., Oliveira, J. E., & Mattoso, L. H. C. (2020). Pectin films loaded with copaiba oil nanoemulsions for potential use as bio-based active packaging. *Food Hydrocolloids, 106*, 105862. https://doi.org/10.1016/j.foodhyd.2020.105862.

Pacheco, M. T., Villamiel, M., Moreno, R., & Moreno, F. J. (2019). Structural and rheological properties of pectins extracted from industrial sugar beet by-products. *Molecules, 24*(3), 392. https://doi.org/10.3390/molecules24030392.

Patience, N. A., Schieppati, D., & Boffito, D. C. (2021). Continuous and pulsed ultrasound pectin extraction from navel orange peels. *Ultrasonics Sonochemistry, 73*, 105480. https://doi.org/10.1016/j.ultsonch.2021.105480.

Prezotti, F. G., Meneguin, A. B., Evangelista, R. C., & Ferreira Cury, B. S. (2012). Preparation and characterization of free films of high amylose/pectin mixtures cross-linked with sodium trimetaphosphate. *Drug Development and Industrial Pharmacy, 38*(11), 1354–1359. https://doi.org/10.3109/03639045.2011.650863.

Priyadarshi, R., Kim, S.-M., & Rhim, J.-W. (2021). Pectin/pullulan blend films for food packaging: Effect of blending ratio. *Food Chemistry, 347*, 129022. https://doi.org/10.1016/j.foodchem.2021.129022.

Qiang, T., Ren, W., & Chen, L. (2024). Biodegradable, high mechanical strength, and eco-friendly pectin-based plastic film. *Food Hydrocolloids, 149*, 109539. https://doi.org/10.1016/j.foodhyd.2023.109539.

Rehman, H. U., Aman, A., Nawaz, M. A., & Qader, S. A. U. (2015). Characterization of pectin degrading polygalacturonase produced by *Bacillus licheniformis* KIBGE-IB21. *Food Hydrocolloids, 43*, 819–824. https://doi.org/10.1016/j.foodhyd.2014.08.018, https://www.journals.elsevier.com/food-hydrocolloids.

Reichembach, L. H., & Lúcia de Oliveira Petkowicz, C. (2021). Pectins from alternative sources and uses beyond sweets and jellies: An overview. *Food Hydrocolloids, 118*, 106824. https://doi.org/10.1016/j.foodhyd.2021.106824.

Reis, C. A., Gomes, A., & do Amaral Sobral, P. J. (2023). Films based on biopolymers incorporated with active compounds encapsulated in emulsions: Properties and potential applications—A review. *Foods, 12*(19). https://doi.org/10.3390/foods12193602, www.mdpi.com/journal/foods.

Ren, W., Qiang, T., & Chen, L. (2022). Recyclable and biodegradable pectin-based film with high mechanical strength. *Food Hydrocolloids, 129*, 107643. https://doi.org/10.1016/j.foodhyd.2022.107643.

Riyamol, N., Gada Chengaiyan, J., Rana, S. S., Ahmad, F., Haque, S., & Capanoglu, E. (2023). Recent advances in the extraction of pectin from various sources and industrial applications. *ACS Omega, 8*(49), 46309–46324. https://doi.org/10.1021/acsomega.3c04010, pubs.acs.org/journal/acsodf.

Rosenboom, J. G., Langer, R., & Traverso, G. (2022). Bioplastics for a circular economy. *Nature Reviews Materials, 7*(2), 117–137. https://doi.org/10.1038/s41578-021-00407-8, www.nature.com/natrevmats/.

Roy, S., & Rhim, J. W. (2022). Gelatin/cellulose nanofiber-based functional films added with mushroom-mediated sulfur nanoparticles for active packaging applications. *Journal of Nanostructure in Chemistry, 12*(5), 979–990. https://doi.org/10.1007/s40097-022-00484-3, https://www.springer.com/journal/40097.

Roy, S., Priyadarshi, R., Łopusiewicz, Ł., Biswas, D., Chandel, V., & Rhim, J.-W. (2023). Recent progress in pectin extraction, characterization, and pectin-based films for active food packaging applications: A review. *International Journal of Biological Macromolecules, 239*, 124248. https://doi.org/10.1016/j.ijbiomac.2023.124248.

Roy, S., Ramakrishnan, R., Goksen, G., Singh, S., & Łopusiewicz, Ł. (2024). Recent progress on UV-light barrier food packaging films – A systematic review. *Innovative Food Science & Emerging Technologies, 91*, 103550. https://doi.org/10.1016/j.ifset.2023.103550.

Roy, S., & Rhim, J.-W. (2021a). Fabrication of pectin/agar blended functional film: Effect of reinforcement of melanin nanoparticles and grapefruit seed extract. *Food Hydrocolloids, 118*, 106823. https://doi.org/10.1016/j.foodhyd.2021.106823.

Roy, S., & Rhim, J.-W. (2021b). Preparation of pectin/agar-based functional films integrated with zinc sulfide nano petals for active packaging applications. *Colloids and Surfaces B: Biointerfaces, 207*, 111999. https://doi.org/10.1016/j.colsurfb.2021.111999.

Salimi, A., Khodaiyan, F., Askari, G., & Amraei, A. (2025). Effect of propolis extract-loaded films made of apple pomace pectin and grass pea protein on the shelf life extension of black mulberry. *Food Hydrocolloids, 158*, 110549. https://doi.org/10.1016/j.foodhyd.2024.110549.

Sani, I. K., Geshlaghi, S. P., Pirsa, S., & Asdagh, A. (2021). Composite film based on potato starch/apple peel pectin/ZrO_2 nanoparticles/microencapsulated Zataria multiflora essential oil; investigation of physicochemical properties and use in quail meat packaging. *Food Hydrocolloids, 117*, 106719. https://doi.org/10.1016/j.foodhyd.2021.106719.

Santana, J. V., Marangoni Júnior, L., Cassol, G. Z., Sato, H. H., & Vieira, R. P. (2024). Advanced pectin-based films: Enhancing antioxidant, antibacterial, UV barrier, and physicochemical properties upon oligomeric limonene derivative incorporation. *Food Hydrocolloids, 149*. https://doi.org/10.1016/j.foodhyd.2023.109558, https://www.journals.elsevier.com/food-hydrocolloids.

Sashiwa, H., & Aiba, S.-ichi (2004). Chemically modified chitin and chitosan as biomaterials. *Progress in Polymer Science, 29*(9), 887–908. https://doi.org/10.1016/j.progpolymsci.2004.04.001.

Schutz, G. F., de Ávila Gonçalves, S., Alves, R. M. V., & Vieira, R. P. (2024). A review of starch-based biocomposites reinforced with plant fibers. *International Journal of Biological Macromolecules, 261*. https://doi.org/10.1016/j.ijbiomac.2024.129916, https://www.sciencedirect.com/science/journal/01418130.

Semdé, R., Moës, A. J., Devleeschouwer, M. J., & Amighi, K. (2003). Synthesis and enzymatic degradation of epichlorohydrin cross-linked pectins. *Drug Development and Industrial Pharmacy, 29*(2), 203–213. https://doi.org/10.1081/DDC-120016728.

Shah, Y. A., Bhatia, S., Al-Harrasi, A., Oz, F., Khan, M. H., Roy, S., Esatbeyoglu, T., & Pratap-Singh, A. (2024). Thermal properties of biopolymer films: Insights for sustainable food packaging applications. *Food Engineering Reviews, 16*(4), 497–512. https://doi.org/10.1007/s12393-024-09380-8, https://www.springer.com/journal/12393.

Shahrampour, D., Khomeiri, M., Razavi, S. M. A., & Kashiri, M. (2020). Development and characterization of alginate/pectin edible films containing Lactobacillus plantarum KMC 45. *LWT, 118*, 108758. https://doi.org/10.1016/j.lwt.2019.108758.

Sharifi, K. A., & Pirsa, S. (2021). Biodegradable film of black mulberry pulp pectin/chlorophyll of black mulberry leaf encapsulated with carboxymethylcellulose/silica nanoparticles: Investigation of physicochemical and antimicrobial properties. *Materials Chemistry and Physics, 267*, 124580. https://doi.org/10.1016/j.matchemphys.2021.124580.

Sharma, R., & Ahuja, M. (2011). Thiolated pectin: Synthesis, characterization and evaluation as a mucoadhesive polymer. *Carbohydrate Polymers, 85*(3), 658–663. https://doi.org/10.1016/j.carbpol.2011.03.034.

Shivangi, S., Dorairaj, D., Negi, P. S., & Shetty, N. P. (2021). Development and characterisation of a pectin-based edible film that contains mulberry leaf extract and its bio-active components. *Food Hydrocolloids, 121*, 107046. https://doi.org/10.1016/j.foodhyd.2021.107046.

Silva, M. Ad, Bierhalz, A. C. K., & Kieckbusch, T. G. (2009). Alginate and pectin composite films crosslinked with Ca2+ ions: Effect of the plasticizer concentration. *Carbohydrate Polymers, 77*(4), 736–742. https://doi.org/10.1016/j.carbpol.2009.02.014.

Singaram, A. J. V., Guruchandran, S., & Ganesan, N. D. (2024). Review on functionalized pectin films for active food packaging. *Packaging Technology and Science, 37*(4), 237–262. https://doi.org/10.1002/pts.2793, http://onlinelibrary.wiley.com/journal/10.1002/(ISSN)1099-1522.

Sistla, Y. S., & Mehraj, S. (2022). Molecular simulations to understand the moisture, carbon dioxide, and oxygen barrier properties of pectin films. *Journal of Molecular Modeling, 28*(4). https://doi.org/10.1007/s00894-022-05069-z, https://rd.springer.com/journal/894.

Song, Y., Zhang, F., Yu, R., Zheng, H., & Wang, P. (2023). Acylated pectin/gelatin-based films incorporated with alkylated starch crystals: Characterization, antioxidant and antibacterial activities, and coating preservation effects on golden pomfret. *International Journal of Biological Macromolecules, 241*, 124532. https://doi.org/10.1016/j.ijbiomac.2023.124532.

Stoffels, P., Müller, M. J., Stachurski, S., Terfrüchte, M., Schröder, S., Ihling, N., Wierckx, N., Feldbrügge, M., Schipper, K., & Büchs, J. (2020). Complementing the intrinsic repertoire of Ustilago maydis for degradation of the pectin backbone polygalacturonic acid. *Journal of Biotechnology, 307*, 148–163. https://doi.org/10.1016/j.jbiotec.2019.10.022, www.elsevier.com/locate/jbiotec.

Tristanto, N. A., Cao, W., Chen, N., Suryoprabowo, S., Soetaredjo, F. E., Ismadji, S., & Hua, X. (2024). Pectin extracted from red dragon fruit (*Hylocereus polyrhizus*) peel and its usage in edible film. *International Journal of Biological Macromolecules, 276*(Pt 1), 133804. https://doi.org/10.1016/j.ijbiomac.2024.133804.

Vega-Hernández, M.Á., Cano-Díaz, G. S., Vivaldo-Lima, E., Rosas-Aburto, A., Hernández-Luna, M. G., Martinez, A., Palacios-Alquisira, J., Mohammadi, Y., & Penlidis, A. (2021). A review on the synthesis, characterization, and modeling of polymer grafting. *Processes, 9*(2), 375. https://doi.org/10.3390/pr9020375.

Vieira, M. G. A., Da Silva, M. A., Dos Santos, L. O., & Beppu, M. M. (2011). Natural-based plasticizers and biopolymer films: A review. *European Polymer Journal, 47*(3), 254–263. https://doi.org/10.1016/j.eurpolymj.2010.12.011.

Wang, C., Qiu, W.-Y., Chen, T.-T., & Yan, J.-K. (2021). Effects of structural and conformational characteristics of citrus pectin on its functional properties. *Food Chemistry, 339*, 128064. https://doi.org/10.1016/j.foodchem.2020.128064.

Wang, J., Zhao, C., Chen, Y., Wang, F., Feng, L., Wang, Y., Li, C., Bao, Y., & Zheng, J. (2024). Pectins amidated with different amino acids via MMTM activation: Structural characteristics and emulsifying properties. *Food Hydrocolloids, 152*, 109894. https://doi.org/10.1016/j.foodhyd.2024.109894.

Wang, L., Li, Y., Osei, P. O., Gao, F., Wu, X., & Liao, X. (2024). Preparation of pectin/carboxymethyl cellulose composite films via high-pressure processing and enhancement of antimicrobial packaging. *International Journal of Biological Macromolecules, 281*(Pt 4), 136462. https://doi.org/10.1016/j.ijbiomac.2024.136462.

Wang, L., Zhao, Z., Zhao, H., Liu, M., Lin, C., Li, L., & Ma, B. (2022). Pectin polysaccharide from Flos Magnoliae (Xin Yi, Magnolia biondii Pamp. flower buds): Hot-compressed water extraction, purification and partial structural characterization. *Food Hydrocolloids, 122*, 107061. https://doi.org/10.1016/j.foodhyd.2021.107061.

Wang, W., Wu, X., Chantapakul, T., Wang, D., Zhang, S., Ma, X., Ding, T., Ye, X., & Liu, D. (2017). Acoustic cavitation assisted extraction of pectin from waste grapefruit peels: A green two-stage approach and its general mechanism. *Food Research International, 102*, 101–110. https://doi.org/10.1016/j.foodres.2017.09.087.

Wang, W., Chen, W., Zou, M., Lv, R., Wang, D., Hou, F., Feng, H., Ma, X., Zhong, J., Ding, T., Ye, X., & Liu, D. (2018). Applications of power ultrasound in oriented modification and degradation of pectin: A review. *Journal of Food Engineering, 234*, 98–107. https://doi.org/10.1016/j.jfoodeng.2018.04.016.

Wei, H., Qi, J., Liao, J., Zhuo, T., & Xiao, R. (2024). Co-influence of the degree of esterification and degree of amidation on the properties of acid-sugar-calcium amidated pectin gels. *Food Hydrocoll*, 110707.

Wu, D., Chen, S., Ye, X., Zheng, X., Ahmadi, S., Hu, W., Yu, C., Cheng, H., Linhardt, R. J., & Chen, J. (2022). Enzyme-extracted raspberry pectin exhibits a high-branched structure and enhanced anti-inflammatory properties than hot acid-extracted pectin. *Food Chemistry, 383*, 132387. https://doi.org/10.1016/j.foodchem.2022.132387.

Wu, W., Wu, Y., Lin, Y., & Shao, P. (2022). Facile fabrication of multifunctional citrus pectin aerogel fortified with cellulose nanofiber as controlled packaging of edible fungi. *Food Chemistry, 374*, 131763. https://doi.org/10.1016/j.foodchem.2021.131763.

Xie, W., Xu, P., Wang, W., & Liu, Q. (2002). Preparation and antibacterial activity of a water-soluble chitosan derivative. *Carbohydrate Polymers, 50*(1), 35–40. https://doi.org/10.1016/S0144-8617(01)00370-8.

Xu, F., Zhang, S., Wu, Q., An, C., Li, X., Chen, X., Chen, Y., Zhang, X., & Dong, Z. (2024). A biodegradable multifunctional pectin–montmorillonite fertilizer coating: Controlled-release, water-retention and soil-cementation. *International Journal of Biological Macromolecules, 277*(Pt 2), 134118. https://doi.org/10.1016/j.ijbiomac.2024.134118.

Yadav, K., Yadav, S., Anand, G., Yadav, P. K., & Yadav, D. (2023). Hydrolysis of complex pectin structures: Biocatalysis and bioproducts. In *Polysaccharide Degrading Biocatalysts*. India: Elsevier, 205–225. https://doi.org/10.1016/B978-0-323-99986-1.00011-9, https://www.sciencedirect.com/book/9780323999861.

Yang, J., Fan, Y., Cui, J., Yang, L., Su, H., Yang, P., & Pan, J. (2021). Colorimetric films based on pectin/sodium alginate/xanthan gum incorporated with raspberry pomace extract for monitoring protein-rich food freshness. *International Journal of Biological Macromolecules, 185*, 959–965. https://doi.org/10.1016/j.ijbiomac.2021.06.198.

Yang, W., Zhang, S., Feng, A., Li, Y., Wu, P., Li, H., & Ai, S. (2024). Water-insoluble tea polyphenol nanoparticles as fillers and bioactive agents for pectin films to prepare active packaging for fruit preservation. *Food Hydrocolloids, 156*, 110364. https://doi.org/10.1016/j.foodhyd.2024.110364.

Yang, Y., Wang, Z., Hu, D., Xiao, K., & Wu, J. Y. (2018). Efficient extraction of pectin from sisal waste by combined enzymatic and ultrasonic process. *Food Hydrocolloids, 79*, 189–196. https://doi.org/10.1016/j.foodhyd.2017.11.051, https://www.journals.elsevier.com/food-hydrocolloids.

Younis, H. G. R., Abdellatif, H. R. S., Ye, F., & Zhao, G. (2020). Tuning the physicochemical properties of apple pectin films by incorporating chitosan/pectin fiber. *International Journal of Biological Macromolecules, 159*, 213–221. https://doi.org/10.1016/j.ijbiomac.2020.05.060, www.elsevier.com/locate/ijbiomac.

Zhang, W., Cao, J., & Jiang, W. (2022). Effect of different cation in situ cross-linking on the properties of pectin-thymol active film. *Food Hydrocolloids, 128*, 107594. https://doi.org/10.1016/j.foodhyd.2022.107594.

Zhang, Y., Kong, Q., Niu, B., Liu, R., Chen, H., Xiao, S., Wu, W., Zhang, W., & Gao, H. (2024). The dual function of calcium ion in fruit edible coating: Regulating polymer internal crosslinking state and improving fruit postharvest quality. *Food Chemistry, 447*, 138952. https://doi.org/10.1016/j.foodchem.2024.138952.

Zirak Hassan Kiadeh, S., Ghaee, A., Farokhi, M., Nourmohammadi, J., Bahi, A., & Ko, F. K. (2021). Electrospun pectin/modified copper-based metal–organic framework (MOF) nanofibers as a drug delivery system. *International Journal of Biological Macromolecules, 173*, 351–365. https://doi.org/10.1016/j.ijbiomac.2021.01.058, www.elsevier.com/locate/ijbiomac.

Chapter 12

Gelatin films and coatings

Rejish Ramakrishnan
E Yuva Center, Anna University, Chennai, Tamil Nadu, India

12.1 Gelatin films and coatings

The global food packaging industry is transforming as consumer demand for environmentally friendly, safe, and innovative solutions continues to rise. Traditional petroleum-based materials dominate the market, yet their detrimental environmental impact, including substantial contributions to nonbiodegradable waste, has prompted a search for sustainable alternatives (Ramakrishnan et al., 2024). Among the various biodegradable materials explored, gelatin, a protein derived from animal collagen, has emerged as a promising candidate for food packaging applications. Known for its excellent film-forming properties, transparency, and biodegradability, gelatin provides effective oxygen barrier properties, making it suitable for extending the shelf life of perishable food products (Tonicíolli Rigueto et al., 2022).

However, gelatin-based films face limitations such as brittleness, poor mechanical strength, and sensitivity to moisture, which hinder their widespread application (Ying et al., 2024). To overcome these challenges, researchers have investigated strategies to enhance the properties of gelatin films, such as incorporating plasticizers like glycerol to improve flexibility and blending gelatin with other biopolymers for better mechanical and functional performance. The advancements in active and intelligent packaging systems utilizing gelatin films, such as embedding antimicrobial and antioxidant agents, have shown significant potential to reduce food waste and ensure quality preservation (Pulla-Huillca et al., 2021). This growing body of research highlights the role of gelatin-based films and coatings as an essential component in the transition toward sustainable packaging solutions.

Gelatin is a widely used natural polymer derived from the partial hydrolysis of collagen, a structural protein abundantly present in the connective tissues of animals such as skin, bones, and tendons (Alipal et al., 2021). This biopolymer is renowned for its versatility, biodegradability, and unique functional properties, making it indispensable in the food, pharmaceutical, and cosmetics industries. Collagen, the primary raw material for gelatin production, consists of amino acids arranged in a triple helix, stabilized by hydrogen bonds (Dille et al., 2021). The triple-helix structure comprises repeating units of glycine, proline, and hydroxyproline, which contribute to its rigidity and stability. The conversion of collagen to gelatin involves breaking down the triple-helix structure into smaller, water-soluble polypeptides. This process occurs through hydrolysis, which can be achieved using acidic, alkaline, or enzymatic methods. The type of hydrolysis significantly impacts the physicochemical properties of the resulting gelatin, such as gel strength, viscosity, and isoelectric point (Liao et al., 2021).

12.2 Gelatin synthesis, properties, and applications

Gelatin, as a biopolymer, is renowned for its versatility, biodegradability, and unique functional properties, making it indispensable in the food, pharmaceutical, and cosmetics industries. Collagen, the primary raw material for gelatin production, consists of amino acids arranged in a triple helix, stabilized by hydrogen bonds (Tang et al., 2022). The chemical structure of collagen consists of a repeating sequence of glycine-X-Y triplets, where X and Y are frequently proline and hydroxyproline (Masrouri & Qin, 2023). During hydrolysis, the triple helix is disrupted, and the hydrogen bonds are broken, yielding denatured collagen or gelatin with a linear polypeptide structure. The resulting gelatin exhibits functional groups, including amine (-NH_2) and carboxyl (-COOH) groups, which contribute to its solubility and reactivity. This chemical transformation enhances the polymer's functionality, making it suitable for various industrial applications.

Natural Polymer-Based Films and Coatings in Food Technology. DOI: https://doi.org/10.1016/B978-0-443-34244-8.00008-2

Gelatin can be extracted through three primary methods: acid treatment, alkaline treatment, and enzymatic treatment. Acid treatment, commonly used for porcine skin, involves subjecting collagen to acidic conditions for a short period, typically one to three days (Rather, Akhter, et al., 2022). This process yields Type A gelatin, characterized by an isoelectric point between 7 and 9, making it suitable for applications requiring higher pH compatibility. Alkaline treatment, often applied to bovine hides and bones, involves collagen undergoing prolonged exposure to alkaline conditions, which can extend up to several weeks. This method produces Type B gelatin with an isoelectric point between 4.7 and 5.3, making it ideal for acidic applications (Mushtaq et al., 2022). Enzymatic treatment employs proteolytic enzymes, such as papain or trypsin, to hydrolyze collagen under controlled conditions. Although enzymatically derived gelatin often exhibits superior functional properties, the process can be cost prohibitive.

The extraction process begins with raw material preparation, where animal tissues such as skin, bones, or tendons are thoroughly cleaned to remove impurities like fat, blood, and debris. This is typically achieved through washing, degreasing, and defatting steps. Once the raw material is cleaned, the pretreatment stage commences, involving acid or alkali solutions to weaken the collagen structure (Jayaprakash et al., 2024). The specific pretreatment depends on the source material; for example, an acid bath is preferred for porcine skin, while alkaline treatment is used for bovine hides. Following pretreatment, the materials are subjected to the hydrolysis phase, where the collagen is converted into gelatin through acid, alkaline, or enzymatic hydrolysis. The reaction conditions, including temperature, pH, and time, are meticulously controlled to achieve the desired physicochemical properties.

After hydrolysis, the gelatin solution is filtered to remove insoluble residues and other impurities (Cansu, 2023). Filtration is followed by concentration, which involves evaporating excess water to obtain a gelatin solution with a higher solid content. The concentrated gelatin solution is then sterilized to eliminate microbial contaminants. Once sterilized, the gelatin is dried using techniques such as drum drying or freeze-drying. The dried gelatin is then ground into powder or formed into sheets for industrial use.

The chemical structure of gelatin shows that it retains functional groups such as amine ($-NH_2$), hydroxyl (-OH), and carboxyl (-COOH), enabling it to form hydrogen bonds and ionic interactions. These properties contribute to its solubility, gel-forming ability, and reactivity, which are essential for its applications. Gelatin's functional properties include gel strength, viscosity, thermo-reversibility, and film-forming ability. Gel strength, measured in Bloom units, reflects the firmness of gelatin gels and is influenced by factors such as molecular weight and extraction conditions (Mushtaq et al., 2022). The viscosity of gelatin solutions is crucial for applications requiring specific flow properties, such as in food and pharmaceutical formulations. Thermo-reversibility, a hallmark of gelatin, enables it to form gels upon cooling and melt upon heating, a property extensively utilized in confectionery and drug encapsulation. Gelatin's film-forming ability allows it to produce transparent, flexible, and biodegradable films, making it suitable for packaging and coating applications (Wang, Ding, et al., 2021).

Gelatin-based films and coatings are widely used in various industries. In the food sector, gelatin coatings extend the shelf life of perishable items like fruits, vegetables, and meat by acting as barriers against oxygen and moisture (Chettri et al., 2023). The incorporation of antimicrobial and antioxidant agents into gelatin films has further enhanced their functionality in active packaging systems. In the pharmaceutical industry, gelatin is used to manufacture capsules and drug delivery systems due to its biocompatibility and film-forming properties (Ahmady & Abu Samah, 2021). Gelatin hydrogels serve as carriers for controlled drug release, demonstrating their potential in advanced pharmaceutical applications. In cosmetics, gelatin is utilized in face masks and hair care products for its protein content and film-forming capabilities (Gupta et al., 2022). Despite its numerous advantages, gelatin-based materials face challenges such as moisture sensitivity and brittleness. Researchers have developed strategies to overcome these limitations, including the use of plasticizers like glycerol and sorbitol to improve flexibility. Blending gelatin with other biopolymers, such as chitosan or starch, enhances mechanical strength and water resistance (Tiozon et al., 2021). The incorporation of nanoparticles, such as nanocellulose or silver nanoparticles, has shown promise in improving barrier properties and imparting antimicrobial activity (Ullah et al., 2024).

The future of gelatin lies in addressing these challenges and exploring new applications. Advances in cross-linking, blending, and nanotechnology are expected to enhance gelatin's properties, expanding its potential uses. The development of nonanimal gelatin alternatives derived from plant-based or microbial sources holds promise for addressing ethical and dietary concerns. This comprehensive understanding of gelatin synthesis and its diverse applications highlights its pivotal role in shaping sustainable materials across industries. Gelatin exhibits several critical physicochemical and functional properties that contribute to its wide range of applications. Its gel strength, measured in Bloom units, varies depending on the extraction process and source material. The viscosity of gelatin solutions, influenced by molecular weight and concentration, is another critical parameter for industrial use (Rather, Akhter, et al., 2022). A unique characteristic of gelatin is its thermo-reversibility, where it forms gels upon cooling and melts upon heating, a property extensively utilized in food and pharmaceutical industries. Gelatin demonstrates exceptional film-forming ability, enabling the production of transparent, flexible, and biodegradable films. It also acts as an emulsifier by stabilizing oil-in-water emulsions and enhances the texture of aerated food products like marshmallows. Gelatin's natural origin ensures its biodegradability and edibility, aligning with the increasing demand for sustainable materials.

Gelatin-based films and coatings have found extensive applications across various industries. In the food industry, gelatin coatings extend the shelf life of perishable items like fruits, vegetables, and meat by acting as barriers against oxygen and moisture (Rather, Makroo, et al., 2022). The incorporation of antimicrobial and antioxidant agents into gelatin films has further enhanced their functionality in active packaging systems, ensuring food safety and quality. In the pharmaceutical sector, gelatin's film-forming ability and biodegradability make it an ideal material for manufacturing capsules and drug delivery systems. Gelatin hydrogels serve as carriers for controlled drug release, demonstrating their potential in advanced pharmaceutical applications. Similarly, in the cosmetics industry, gelatin is utilized in face masks and hair care products due to its protein content and film-forming properties.

Despite its many advantages, gelatin-based films face challenges such as moisture sensitivity and brittleness, limiting their application in high-humidity environments. Researchers have explored various strategies to enhance gelatin's properties to address these limitations. Adding plasticizers like glycerol, sorbitol, or oleic acid improves the flexibility and mechanical properties of gelatin films (Ahmad et al., 2023). Blending gelatin with other polymers, such as chitosan or starch, enhances its mechanical strength and water resistance (Ramakrishnan et al., 2023). Recent advancements in nanotechnology have also shown promise, with the incorporation of nanoparticles like nanocellulose and silver nanoparticles improving barrier properties and imparting antimicrobial activity to gelatin-based materials (Yang et al., 2021). When compared to other biopolymers, such as starch, chitosan, and alginate, gelatin stands out for its excellent film-forming ability and biodegradability. However, its high moisture sensitivity and moderate cost are areas where other biopolymers may offer advantages. Gelatin's unique combination of properties and its potential for functional enhancements make it a valuable material in the pursuit of sustainable and eco-friendly alternatives to petroleum-based polymers.

Another challenge facing the gelatin industry is the growing ethical concern related to the use of animal-derived products (Duarte et al., 2023). As a biopolymer sourced primarily from animal collagen, gelatin is not suitable for vegetarians, vegans, and individuals adhering to specific religious dietary laws. This has spurred significant research into the development of alternative sources of gelatin, including plant-based and microbial alternatives. Plant-based gelling agents, such as agar-agar and pectin, have long been used as substitutes for gelatin, but they do not possess the same functional properties, particularly in terms of gel strength and thermo-reversibility (Ramli et al., 2023). Microbial fermentation offers another potential route for producing gelatin alternatives, where genetically engineered microorganisms could be used to synthesize gelatin-like proteins. These microbial sources have the advantage of being more sustainable and ethically acceptable, as they do not rely on animal products.

The future of gelatin-based materials looks promising, with growing interest in sustainable and renewable materials driving research and innovation. Efforts to enhance gelatin's properties through advanced techniques, such as cross-linking, blending, and nanotechnology, continue to expand its application potential. The exploration of nonanimal gelatin alternatives, derived from plant-based or microbial sources, holds promise for addressing ethical and dietary concerns, further broadening its scope of use. This comprehensive understanding of gelatin synthesis and its diverse applications highlights its pivotal role in shaping the future of sustainable materials across industries. Gelatin is a biopolymer that plays a crucial role in numerous industries, particularly in food, pharmaceuticals, and cosmetics, due to its unique combination of biodegradability, versatility, and functional properties (Aksun Tümerkan, 2021). As a derivative of collagen, gelatin's structural and chemical characteristics make it an invaluable material, yet it also poses various challenges that have sparked significant research into improving its properties. Collagen, the primary raw material for gelatin production, is a fibrous protein composed of long chains of amino acids arranged in a repeating sequence of glycine-X-Y triplets, where X and Y are often proline and hydroxyproline (Masrouri & Qin, 2023). This sequence is integral to the formation of the triple-helix structure that gives collagen its strength and stability. The process of converting collagen into gelatin involves hydrolysis, which breaks the triple helix down into smaller, water-soluble polypeptides, significantly altering its structure and functional properties. The hydrolysis of collagen can be accomplished using three primary methods: acidic, alkaline, and enzymatic treatments (Talha et al., 2024). Each of these methods has a profound impact on the final properties of the gelatin, including its gel strength, viscosity, and solubility.

The hydrolysis process is crucial in determining the physicochemical properties of gelatin, and it involves the disruption of the stable triple-helix structure. In acidic conditions, collagen undergoes partial hydrolysis, resulting in the production of Type A gelatin, which typically has an isoelectric point between 7 and 9, making it suitable for applications requiring higher pH stability (Tang et al., 2022). In contrast, alkaline hydrolysis, which is commonly applied to collagen sourced from bovine hides and bones, produces Type B gelatin, characterized by a lower isoelectric point, ranging from 4.7 to 5.3. This makes Type B gelatin more appropriate for acidic conditions (Samatra et al., 2022). Enzymatic hydrolysis, which employs specific proteolytic enzymes like papain or trypsin, enables a more controlled breakdown of collagen. Enzymatic hydrolysis often results in gelatin with superior functional properties, such as enhanced solubility and gel strength, although the method is generally more costly than the acidic and alkaline methods (Mirzapour-Kouhdasht et al., 2021). During the hydrolysis process, long collagen chains are cleaved into smaller polypeptides, and the hydrogen bonds that stabilize the triple helix are broken, leading to the formation of linear polypeptides. These polypeptides, which retain functional groups like amine (-NH_2), hydroxyl (-OH), and carboxyl (-COOH) groups, are responsible for the solubility, reactivity, and gel-forming ability of the resulting gelatin. These structural changes are essential for gelatin's role as a gelling agent, stabilizer, and emulsifier in various applications.

Gelatin's most notable physicochemical property is its gel strength, which is commonly measured in Bloom units. The gel strength is influenced by several factors, including the molecular weight of the gelatin, which is in turn affected by the extraction process and source material. Higher molecular weight gelatin typically results in firmer gels, while lower molecular weight gelatin forms softer gels (Cao et al., 2021). Viscosity is another critical parameter that affects the performance of gelatin in different applications. The viscosity of gelatin solutions, which is directly related to the concentration and molecular weight of the gelatin, is particularly important in food products where the flow properties of the gelatin solution influence the texture and consistency of the final product (Wang, Qiao, et al., 2021). Thermo-reversibility is another defining characteristic of gelatin, and it refers to the ability of gelatin to form a gel upon cooling and melt upon heating (Blanco-López et al., 2024). This property makes gelatin highly versatile in applications such as confectionery and pharmaceuticals, where the ability to form gels and melt at specific temperatures is essential. Furthermore, gelatin's film-forming ability is a key feature that has led to its use in biodegradable films and coatings. These films are particularly important in the food packaging industry, as they provide a sustainable alternative to petroleum-based plastics. Gelatin-based films can also act as barriers to moisture and oxygen, which helps extend the shelf life of perishable items (Luo et al., 2022).

Gelatin is widely used in the food industry, where it serves as a gelling agent, stabilizer, emulsifier, and thickener (M. I. Ahmad et al., 2024). In confectionery, gelatin is essential for the production of gummy candies, jellies, and marshmallows, where its ability to form stable gels is crucial. Gelatin is also employed in the production of clarified fruit juices, dairy products, and various desserts. Beyond its gelling properties, gelatin also plays a significant role in improving the texture and mouthfeel of food products, enhancing their sensory qualities. In the pharmaceutical industry, gelatin is used primarily for the production of capsules. Both soft and hard gelatin capsules are manufactured using gelatin due to its film-forming properties, which allow the encapsulation of active pharmaceutical ingredients (APIs) (Dille et al., 2021). The biodegradability and biocompatibility of gelatin make it an ideal material for drug delivery systems, where gelatin-based hydrogels are used for controlled drug release. These hydrogels can be designed to release drugs at specific rates, making gelatin an invaluable material in therapeutic applications. In the cosmetics industry, gelatin's protein content and film-forming ability are exploited in the production of face masks, hair care products, and moisturizing treatments. The ability of gelatin to form flexible films that adhere to the skin or hair, combined with its natural hydrophilic properties, makes it an ideal component in cosmetic formulations aimed at enhancing skin and hair health.

Gelatin is a versatile and valuable biopolymer with a broad range of applications across several industries. Its unique chemical structure, which involves the breakdown of collagen into smaller polypeptides, gives it a set of functional properties that make it suitable for various uses, including gelling, emulsifying, and film-forming. Although challenges related to moisture sensitivity and ethical concerns regarding its animal-derived origin remain, ongoing research in the areas of material enhancement and alternative sources of gelatin is expected to address these issues and expand its potential applications. The continued development of gelatin-based materials, especially in the context of sustainability and ethics, will likely shape the future of this biopolymer and ensure its continued relevance in a wide array of industrial sectors.

12.3 Sources and properties of gelatin

Gelatin has become a pivotal material in the development of sustainable and functional food packaging solutions, offering biodegradability and compatibility with edible coatings. Bovine (from cows, oxen, bison, buffalo, and other animals in the Bovidae family) and porcine (from pigs, *Sus scrofa domesticus*) gelatin are among the most commonly used in this domain due to their high gel strength (200–300 Bloom) and superior flexibility, making them ideal for applications such as edible films, protective coatings, and biodegradable wraps (Hassan et al., 2021). Bovine gelatin, derived from hides and bones, provides moderate oxygen and moisture barrier properties, enabling it to extend the shelf life of food products (Table 12.1). However, its cultural acceptability is limited in Hindu communities, while concerns over bovine spongiform

TABLE 12.1 Source of gelatin, applications, and properties.

Source	Food packaging applications	Gel strength (Bloom value)	Barrier properties	Mechanical properties	Bioactive properties	Sustainability
Bovine	Edible films, biodegradable packaging, and protective coatings for food	200–300	Moderate oxygen and water vapor barrier properties	High tensile strength and flexibility	Limited; can be functionalized with antimicrobial agents	Moderate sustainability; high water/energy use in processing
Porcine	Edible coatings for confectionery, meat, and dairy products; films for moisture retention	250–300	High moisture resistance; moderate oxygen barrier	Superior flexibility and elasticity	Limited; often combined with active components	Moderate sustainability depends on pig-farming practices
Fish	Biodegradable films for seafood, fruits, and vegetables	100–200	Excellent oxygen barrier; poor moisture barrier	Moderate strength; prone to brittleness	Intrinsic antioxidant and antimicrobial properties	High sustainability; derived from processing by-products
Chicken	Limited use in food coatings or packaging films	80–120	Low-barrier properties against moisture and gases	Weak tensile properties	Limited; niche functionalization possibilities	Moderate sustainability; derived from poultry waste
Marine	Edible packaging films with antioxidant and antimicrobial properties	50–150	Good oxygen barrier; bioactive functionality	Moderate tensile strength	Intrinsically high; antimicrobial and antioxidant effects	Excellent sustainability; renewable marine resources
Synthetic	Bioengineered edible packaging films; smart packaging (responsive to food spoilage)	Customizable	Superior tailored gas and moisture barrier properties	High strength; tunable elasticity	High; customizable for bioactive and intelligent responses	Sustainable if produced using renewable energy sources

encephalopathy (BSE) remain a challenge (Uddin et al., 2021). Porcine gelatin, known for its high tensile strength and moisture resistance, is a cost-effective alternative but is restricted in Halal, Kosher, and vegetarian markets (Osman & Abo-Dief, 2023).

Gelatin is widely used in food packaging due to its biocompatibility, biodegradability, and ability to form functional films. Bovine and porcine gelatin, the most extensively studied sources, exhibit excellent mechanical properties, with gel strengths ranging from 200 to 300 Bloom, making them ideal for robust packaging applications such as edible films and coatings (Zhu et al., 2023). Studies have shown that bovine gelatin films provide moderate oxygen and water vapor barrier properties, effectively slowing oxidative degradation in packaged foods. Porcine gelatin, known for its superior elasticity, offers enhanced moisture resistance and flexibility, which are critical for packaging products like confectionery and dairy. Both sources can be further functionalized with bioactive agents, such as essential oils or antimicrobial peptides, to improve food preservation, although intrinsic bioactivity is limited.

Fish and marine gelatin have emerged as eco-friendly alternatives, offering unique bioactive and barrier properties (Kim et al., 2025). Fish gelatin, derived from cold-water species, has been shown in studies to exhibit excellent oxygen barrier functionality, comparable to synthetic materials like polyvinyl alcohol (PVA). However, its films tend to be brittle, requiring plasticizers like glycerol to improve mechanical flexibility. Marine gelatin, particularly from sources like jellyfish or other invertebrates, provides an added advantage due to its intrinsic antioxidant and antimicrobial properties, as demonstrated in recent studies (Chiarelli et al., 2023). These bioactivities enable marine gelatin films to actively extend the shelf life of perishable foods by inhibiting microbial growth and oxidative damage. However, both fish and marine gelatins are less effective as moisture barriers compared to mammalian sources, which may limit their applications in high-humidity environments.

Synthetic gelatin, engineered through chemical synthesis or microbial processes, has opened new frontiers in food packaging. Scientific advancements have allowed for the development of gelatin-based smart packaging that can respond to environmental stimuli or indicate food spoilage (Gilchrist et al., 2019). Studies have highlighted synthetic gelatin's tunable barrier properties, enabling precise control over gas and moisture permeability to meet specific packaging needs. Furthermore, its ability to be customized with bioactive or intelligent components, such as pH-sensitive dyes or antimicrobial molecules, makes it a promising candidate for advanced applications. Despite its potential, challenges remain in scaling production cost-effectively while maintaining environmental sustainability.

Marine and fish gelatins hold significant promise as sustainable materials for food packaging, especially in applications where bioactivity is crucial. While mammalian gelatins dominate in mechanical performance and cost efficiency, their intrinsic bioactivity is limited, necessitating external modifications to meet modern food preservation standards. Synthetic gelatin offers unparalleled flexibility in tailoring properties but remains constrained by production costs. As scientific research continues to improve the functionality and scalability of these materials, gelatin-based packaging is poised to play a vital role in the shift toward sustainable and functional food packaging systems.

12.4 Gelatin-based active food packaging

The growing demand for sustainable, eco-friendly packaging solutions in the food industry has accelerated the development of biodegradable and bioactive packaging materials. Among these, gelatin-based films have emerged as a promising alternative to traditional plastic packaging, offering numerous benefits in terms of both functionality and environmental sustainability. Gelatin, a biopolymer derived from collagen, is widely regarded for its biodegradability, safety, and ability to form flexible, transparent films. Gelatin-based films not only provide effective physical protection for food but also possess the potential to be "active," meaning they can incorporate bioactive compounds that extend the shelf life of food, reduce spoilage, and preserve nutritional quality. These films are versatile, customizable, and can be enhanced with a range of bioactive agents, making them suitable for various food types. In this section, we will explore the properties, advantages, challenges, and applications of gelatin-based active food packaging, focusing on its mechanical properties, biodegradability, and incorporation of antimicrobial and antioxidant compounds to create functional food packaging solutions (Table 12.2).

One of the most significant advantages of gelatin-based films is their ability to be modified with bioactive compounds to enhance their functionality beyond basic protection. These bioactive films can provide antimicrobial, antifungal, and antioxidant properties, which are essential for preserving food quality, extending shelf life, and preventing contamination. The incorporation of bioactive compounds into gelatin films is a key feature of "active packaging," which not only protects the food but also interacts with it to improve its safety and quality during storage. These films are commonly enhanced with various active agents such as metal-organic frameworks (MOFs), antimicrobial peptides, essential oils, and nanoparticles to improve their antioxidant, antibacterial, and mechanical properties, making them viable alternatives to

TABLE 12.2 Base polymer, active ingredient, product test, and significant results.

S. no.	Base polymer	Active ingredient	Type of agent	Product tested	Significant results	References (DOI)
1.	Agar/gelatin	CD@metal-organic frameworks (MOF) (iron-based MOF + carbon quantum dots)	Antioxidant, antifungal, antibacterial	Tomato fruits	71.9% DPPH, 100% ABTS antioxidant, 100% UV-blocking, increased tensile strength	Riahi et al. (2024)
2.	Gelatin	Nisin A (chitin nanowhiskers)	Antimicrobial	Sea bream, cold-smoked salmon, chicken breast	Inhibition of spoilage pathogens, effective antimicrobial properties	Etxabide et al. (2024)
3.	Chitosan/gelatin	Curcumin (γ-CD-MOF)	Antimicrobial	Strawberry	Tensile strength increase (4.22–9.23 MPa), extended shelf life by eight days	Cai et al. (2025)
4.	Gelatin	Monascus red, phycocyanin, safflower yellow	Antioxidant	Blueberry	625% increase in tensile strength, 96% reduction in oxygen transmission, 92% antioxidant capacity	Liu et al. (2024)
5.	Fish skin gelatin	Tea polyphenols, kojic acid	Antimicrobial, antioxidant	Air-dried chicken	High tensile strength, UV-blocking, reduced bacterial populations	Zhang et al. (2024)
6.	Gelatin	Trehalose, *Pichia membranifaciens*	Antifungal	Cherry tomato	62% survival of yeast, 72.41% inhibition of *Botrytis cinerea*	Guo et al. (2024)
7.	Chitosan/gelatin	Carbon dots from banana peels	Antimicrobial, antioxidant	Packaged meat	74% DPPH, 99% ABTS antioxidant, 6 h inhibition of *Listeria monocytogenes*	Sul et al. (2023)
8.	Gelatin	Zn-MOF (zinc-based metal-organic frameworks)	Antioxidant, antibacterial	Cherry tomato	100% ABTS scavenging, improved water resistance, delayed texture deterioration	Riahi, Hong, et al. (2023)
9.	Chitosan/gelatin	Prussian blue nanoparticles	Antibacterial, photothermal, antioxidant	Strawberry, pork	99.99% reduction of *E. coli*, 15 days shelf life extension of strawberries	Hao et al. (2024)
10.	Gelatin/sodium alginate	Date pit extract	Antioxidant, antibacterial	Raw minced beef	94% DPPH, 88% ABTS antioxidant, 25 mm inhibition zone against *Listeria monocytogenes*	Elhadef et al. (2024)
11.	Soluble soybean polysaccharide/gelatin	Curcumin	Antioxidant, antibacterial	Edible oils	99.6% ABTS scavenging, 2.3–5.5 MPa tensile strength increase	Dong et al. (2023)

(Continued)

TABLE 12.2 (Continued)

S. no.	Base polymer	Active ingredient	Type of agent	Product tested	Significant results	References (DOI)
12.	Gelatin/zein	Cinnamaldehyde, thymol	Antioxidant, antibacterial	Strawberries	99.9% DPPH scavenging, 67.5% *E. coli* inhibition, seven days preservation	Wu et al. (2023)
13.	Chitosan/gelatin	Sulfur-functionalized chitin	Antioxidant, antimicrobial	Packaged chicken	100% ABTS, 95.9% DPPH scavenging, preserved chicken quality for 20 days	Khan et al. (2024)
14.	Gelatin/κ-carrageenan	Lactoferrin-loaded Cr-MOFs	Antimicrobial	Strawberries	Reduced swelling, antibacterial against *E. coli* and *S. aureus*	Khezerlou et al. (2023)
15.	Gelatin/serum plasma	Curcumin	Antioxidant, antibacterial	Pork loins	55 volatile aroma components identified, increased shelf life	Ibeogu et al. (2024)
16.	Gelatin/carrageenan	Cu-MOF, anthocyanins	Antioxidant, antibacterial	Shrimp	pH-responsive, UV-blocking, extended shelf life	Khan et al. (2023)
17.	Gelatin-based polyurethane	Quaternary ammonium groups	Antimicrobial	Strawberries	Extended shelf life of strawberries to six days	Zhong et al. (2023)
18.	Gelatin-based	Curcumin-loaded nanoencapsulation	Antimicrobial	Beef	95% inhibition, 90% antioxidant capacity, extended shelf life	da Silva et al. (2024), Zhuang et al. (2024)
19.	Fish gelatin	Açaí oil	Preservation agent	Tomatoes	Reduced mass loss by 8%, firmness improved to 25 N	da Silva et al. (2024)
20.	Chitosan/gelatin	Ginger essential oil, β-CD	Antioxidant, antifungal	Blueberry	Inhibited fungal growth, 16 days at 25°C, 40 days at 4°C	Hossen et al. (2024)
21.	Chitosan/furcellaran/gelatin	RW4 and LL37 peptides	Antimicrobial	Pork loins	Reduced microbial load, maintained sensory attributes	Kulawik et al. (2024)
22.	Gelatin-based	Zinc oxide nanoparticles (ZnO-NPs)	Antimicrobial	Talaga cheese, camel meat	Reduced *L. monocytogenes* by 2.76 log CFU g^{-1}, extended shelf life	Korany et al. (2024)

traditional plastic packaging. For instance, a study incorporating iron-based MOFs hybridized with carbon quantum dots (CDs) into agar/gelatin films demonstrated superior antioxidant and antibacterial activities, enhancing the shelf life of tomato fruits stored at low temperatures for up to 24 days (Riahi et al., 2024). Similarly, the incorporation of chitin nanowhiskers (CNWs) and riboflavin for photo-crosslinking in gelatin films showed improvements in mechanical properties and solubility, offering effective preservation for seafood products such as sea bream filets and smoked salmon (Etxabide et al., 2024). Moreover, curcumin-loaded γ-cyclodextrin-metal-organic frameworks (γ-CD-MOFs) in chitosan/gelatin composite films were developed for strawberry preservation, enhancing their tensile strength, UV protection, and antibacterial properties, which allowed strawberries to be preserved for up to eight days (Cai et al., 2025). The development of gelatin-based films with additional bioactive materials has also shown promising results in fruit preservation, such as bioactive gelatin films containing monascus red, phycocyanin, and safflower yellow, which were found to preserve blueberries by reducing weight loss and delaying hardness reduction (Liu et al., 2024).

In the realm of meat preservation, fish skin gelatin (FG) films combined with tea polyphenols and kojic acid-loaded nanoparticles have exhibited antimicrobial properties and improved shelf life for air-dried chicken (Zhang et al., 2024). Active films composed of yeast *Pichia membranifaciens* (PM) and trehalose also demonstrated significant antifungal activity against *Botrytis cinerea*, preserving the quality of cherry tomatoes (Guo et al., 2024). Another promising approach involved gelatin films infused with carbon dots derived from banana peels, which showed effective antibacterial activity against *Listeria monocytogenes* and extended the shelf life of minced meat (Sul et al., 2023). These findings highlight the versatility of gelatin films in providing antimicrobial solutions for a wide range of perishable foods, ultimately contributing to food safety and waste reduction. Moreover, eco-friendly solutions such as films incorporating zinc metal-organic frameworks (Zn-MOFs) demonstrated excellent barrier properties and antimicrobial effects, successfully preserving the freshness of cherry tomatoes and minimizing microbial growth (Riahi, Hong, et al., 2023). Prussian blue nanoparticles (PBNPs) integrated into chitosan/gelatin films also exhibited multifaceted sterilization mechanisms, including bacterial trapping, contact killing, and photothermal/photodynamic activities, which proved effective in preserving strawberries and pork (Hao et al., 2024). Similarly, gelatin-sodium alginate-based films with date pit extracts demonstrated significant antioxidant and antibacterial activities, preserving raw minced beef while delaying oxidation and spoilage (Elhadef et al., 2024).

In addition to the development of antimicrobial properties, gelatin-based films have also been integrated with photodynamic agents such as curcumin-loaded nanoencapsulation, which showed remarkable antioxidant and antibacterial activities when applied to beef preservation (Zhuang et al., 2024). These films, when combined with açaí oil, were effective in reducing mass loss and maintaining firmness during postharvest storage of tomatoes (da Silva et al., 2024). Films containing ginger essential oil and β-cyclodextrin were developed to preserve blueberries, demonstrating strong antimicrobial properties against fungal growth while extending the fruit's shelf life (Hossen et al., 2024). Additionally, gelatin films enhanced with sulfur-functionalized chitin exhibited superior mechanical properties and UV-blocking capabilities, preserving packaged chicken for up to 20 days under refrigerated conditions (Khan et al., 2024). The incorporation of lactoferrin-loaded chromium-based metal-organic frameworks (L@Cr-MOFs) into gelatin–κ-carrageenan films was another strategy explored for strawberry preservation, effectively delaying spoilage molds and bacteria (Khezerlou et al., 2023). Furthermore, serum plasma-infused gelatin films incorporating curcumin were shown to maintain the sensory quality and microbial safety of fresh pork, highlighting their potential as active packaging materials for a wide range of meat products (Ibeogu et al., 2024).

These studies underscore the growing importance of gelatin-based films in the food industry, demonstrating their potential as multifunctional, sustainable alternatives to conventional packaging materials. Whether for fruits, vegetables, meats, or oils, gelatin-based films offer not only enhanced preservation but also contribute to the increasing demand for eco-friendly, biodegradable solutions in food packaging.

12.5 Intelligent packaging: gelatin-based solutions for food preservation

Intelligent packaging has gained considerable attention as a promising approach to improving food safety, quality, and sustainability. Unlike traditional packaging, intelligent packaging systems incorporate functionalities that allow for real-time monitoring and active interaction with the product inside. Among the various materials explored for intelligent packaging, gelatin has emerged as a highly promising biopolymer due to its unique properties, such as biodegradability, film-forming ability, and the potential to incorporate bioactive agents. These films can be designed to provide additional functionalities, such as freshness indicators, antimicrobial effects, and shelf life prolongation, making them an ideal candidate for the development of intelligent packaging solutions for the food industry (Table 12.3).

TABLE 12.3 Base polymer, intelligent ingredient, product tested, and significant results.

S. no.	Base polymer	Intelligent ingredient	Product tested	Significant results	References (DOI)
1.	Chitosan-gelatin	Butterfly pea flower anthocyanin extract	Beef	The film demonstrated excellent color stability, antioxidant activity, and precise differentiation of beef freshness stages.	Li, Wang, et al. (2024)
2.	Pectin-gelatin	Curcumin and silver nanoparticles	Shrimp	Inhibition of *E. coli* (99.57%) and *S. aureus* (100%). pH-sensitive color changes from yellow to red across pH range 3–12.	Li, Wei, et al. (2024)
3.	Gelatin films cross-linked with oxidized alginate	Anthocyanin	Flesh foods	Tensile strength of 45 MPa, color change from pink to green, 40% DPPH radical scavenging activity.	Ding et al. (2024)
4.	Gelatin	Resazurin	Shrimp	Color change from yellow to pink with a $\Delta E = 15.4$, indicating spoilage through pH shifts.	Ezati et al. (2023)
5.	Gelatin/polyvinyl alcohol	Cu-MOFs and anthocyanins	Shrimp	pH-sensitive color changes from pink to green, 100% antioxidant activity, and bacterial inhibition.	Riahi, Khan, et al. (2023)
6.	Gelatin	Purple cabbage anthocyanins	Chilled fish	Color changes from pink to green, correlating with TVB-N levels.	Pang et al. (2023)
7.	Gelatin-chitosan-furcellaran	Curcumin, capsaicin, AgNPs, montmorillonite, gelatin hydrolysate	Salmon	Reduced microbial growth by 0.81 log CFU^{-1}, but no delay in lipid oxidation.	Nowak et al. (2024)
8.	Gelatin nanofibers	Black elderberry, Au nanoparticles, SnO_2	Hake fish	Significant color change in L value (52.29%), indicating freshness.	Cetinkaya et al. (2024)
9.	Gelatin-chitosan	β-cyclodextrin and corn poppy complex	Shrimp	Color changes from coral to goldenrod after seven days of storage, indicating spoilage.	Bakhshizadeh et al. (2023)
10.	Starch-gelatin	Purple sweet potato anthocyanins	*Flammulina velutipes* (mushrooms)	Color transitions from green to yellow, reflecting spoilage.	Zong et al. (2023)
11.	Starch-gelatin	Ligand for milk adulteration detection	Milk	Films monitor freshness with color changes and detect preservatives.	Fotouhi et al. (2023)
12.	CMC-gelatin	Red cabbage anthocyanins and Pistacia leaf extract	Shrimp	UV light barrier, 93% DPPH inhibition, and color changes indicate spoilage.	Sadi and Ferfera-Harrar (2023)
13.	Gelatin	Purple cabbage anthocyanin, chondroitin sulfate, tannic acid	Fish	Color change correlates with fish spoilage, with enhanced mechanical properties.	Hao et al. (2024)
14.	Gelatin-chitosan	β-cyclodextrin/*Viola odorata* anthocyanins	Kilka fish	Color change from pink to green as pH increased, monitored spoilage.	Amjadi et al. (2024)
15.	Fish gelatin	*Coleus scutellarioides* anthocyanin	Rainbow trout fish filet	Color change from purple to green correlating with TVB-N levels in spoilage.	Hematian et al. (2023)
16.	Furcellaran-chitosan-gelatin	Curcumin and lemongrass essential oil	Carp, tomatoes	Color changes from yellow to orange, extended shelf life.	Jamróz et al. (2023)

Gelatin-based intelligent packaging offers the ability to provide real-time feedback about the condition of food products. One of the most significant advantages of gelatin-based films is their capacity to incorporate natural compounds that respond to changes in environmental factors such as pH, moisture, or microbial activity. These bioactive compounds enable the films to act as indicators of the freshness or spoilage of the food, providing both manufacturers and consumers with valuable information about the state of the product. This functionality is particularly beneficial for perishable products, such as meat, seafood, and dairy, which are prone to microbial contamination and quality degradation.

For instance, gelatin-based films have been enhanced with chitosan and anthocyanin extracts to create intelligent packaging systems for beef (Li, Wang, et al., 2024). These films were designed to monitor freshness by changing color in response to pH variations associated with spoilage. The incorporation of chitosan, a biopolymer with antimicrobial properties, and anthocyanin, a natural pH-sensitive dye, enables the films to act as visual indicators of freshness, enhancing the convenience of monitoring food quality throughout the distribution chain. Curcumin-loaded gelatin films have also been investigated for their potential in intelligent packaging, particularly for seafood (Li, Wei, et al., 2024). The addition of curcumin, an antioxidant compound, and silver nanoparticles, which have antimicrobial properties, resulted in films that not only exhibit pH-sensitive color changes but also offer protection against microbial contamination. These films are capable of reducing bacterial growth on shrimp, thus improving product safety and shelf life. In another study, gelatin films were developed with anthocyanins to monitor meat spoilage (Ding et al., 2024). These films were designed to change color from light pink to green as the pH of the meat altered during spoilage, offering an easy-to-use freshness indicator. This color change is a direct result of the anthocyanins' sensitivity to pH fluctuations, which are typically associated with food degradation. By integrating such bioactive compounds into gelatin films, intelligent packaging can provide a direct and visual cue about the quality of the product.

Gelatin films can also incorporate resazurin, a dye used for microbial activity detection, which changes color in response to spoilage in shrimp packaging (Ezati et al., 2023). These films change color from yellow to pink as ammonia vapors are released during spoilage, providing an immediate visual indication of the product's condition. This technology is an example of how gelatin-based films can be designed to respond to specific spoilage-related biomarkers, enhancing their role in intelligent food packaging. Another innovative approach involves gelatin/PVA composite films for monitoring shrimp freshness (Riahi, Khan, et al., 2023). These films exhibit strong antioxidant properties and excellent mechanical performance, while also incorporating pH-responsive color changes. The combination of gelatin with PVA enhances the film's durability and versatility, making it suitable for use in intelligent packaging systems for a variety of seafood products. In the context of chilled fish storage, gelatin films have been subjected to ultrasound treatment to improve their properties (Pang et al., 2023). This treatment increased the films' hydrophobicity and mechanical strength, while incorporating purple cabbage anthocyanins provided a colorimetric response to pH changes. The color shift from red to yellow is indicative of spoilage, making these films an effective tool for real-time monitoring of food quality. Multilayered gelatin films have been developed for seafood packaging, combining curcumin, capsaicin, silver nanoparticles, and gelatin hydrolysate to create a multifunctional system (Nowak et al., 2024). These films provide significant antimicrobial protection, especially against bacterial contamination in salmon, while also displaying pH-responsive properties. While the films did not significantly affect lipid oxidation, their antimicrobial properties offer a promising solution for ensuring the safety and extending the shelf life of seafood products. Another breakthrough in intelligent packaging involves gelatin-based nanofibers loaded with bioactive compounds such as elderberry extract and gold nanoparticles (Cetinkaya et al., 2024). These nanofibers exhibited a color change from yellow to red as spoilage occurred in seafood products, such as hake fish filets. The inclusion of these bioactive agents not only enhanced the mechanical properties of the gelatin nanofibers but also provided a visual tool for monitoring food freshness.

In addition to seafood, gelatin-based films have been utilized for shrimp spoilage detection, incorporating chitosan nanofibers and β-cyclodextrin/corn poppy complexes (Bakhshizadeh et al., 2023). These films exhibited a distinct color transition from coral to goldenrod as spoilage progressed, allowing consumers and industry professionals to easily assess the freshness of shrimp. The films also demonstrated antimicrobial properties, further enhancing their functionality as intelligent packaging systems. For mushroom freshness monitoring, gelatin-based films incorporating purple sweet potato anthocyanins were developed (Zong et al., 2023). These films showed strong mechanical properties and exhibited a noticeable color change as spoilage progressed, offering a convenient method for monitoring the condition of mushroom products during storage. The combination of gelatin with anthocyanins provides a versatile solution for intelligent packaging in a variety of food categories. Furthermore, gelatin-based bio-films have been integrated with a progressive web application (PWA) for milk quality testing (Fotouhi et al., 2023). This integration allows for real-time monitoring and colorimetric analysis, with the films selectively detecting preservatives and pH changes. This system offers an innovative approach to milk quality control, enabling on-site testing without the need for specialized equipment. Finally, a cross-linked CMC/gelatin bio-nanocomposite film was developed for shrimp spoilage monitoring (Sadi & Ferfera-Harrar, 2023). These films exhibited significant colorimetric pH indication and antimicrobial properties, making them an effective tool for intelligent packaging systems. The films provided a simple and reliable method for monitoring shrimp freshness, combining the benefits of gelatin-based materials with bioactive agents to enhance food safety.

12.6 Conclusion

Gelatin-based films, with their ability to integrate bioactive compounds, offer numerous advantages for intelligent packaging applications. These films can be tailored to specific food products, providing real-time feedback on freshness, spoilage, and microbial contamination. The combination of gelatin's natural properties with bioactive agents makes these films not only effective in preserving food quality but also environmentally friendly due to their biodegradability. As the demand for sustainable and functional packaging solutions increases, gelatin-based intelligent packaging holds great promise in addressing the challenges faced by the food industry.

The continued development of gelatin-based films with enhanced functionalities for intelligent packaging systems will contribute to food safety, reduce waste, and improve sustainability in the food packaging sector. These innovations provide an opportunity for the food industry to meet consumer demand for transparent and environmentally conscious packaging solutions. As more bioactive agents are incorporated into gelatin films, their versatility and performance in intelligent packaging will only continue to expand, offering significant benefits for both manufacturers and consumers.

References

Ahmad, M. I., Li, Y., Pan, J., Liu, F., Dai, H., Fu, Y., Huang, T., Farooq, S., & Zhang, H. (2024). Collagen and gelatin: Structure, properties, and applications in food industry. *International Journal of Biological Macromolecules, 254*, 128037. https://doi.org/10.1016/j.ijbiomac.2023.128037.

Ahmad, M. S., Ali, R. R., Majid, R. A., & Mohamad, Z. (2023). Properties enhancement of packaging materials based on gelatin. *Environmental Quality Management, 33*(2), 277–284. https://doi.org/10.1002/tqem.21930, http://onlinelibrary.wiley.com/journal/10.1002/(ISSN)1520-6483.

Ahmady, A., & Abu Samah, N. H. (2021). A review: Gelatine as a bioadhesive material for medical and pharmaceutical applications. *International Journal of Pharmaceutics, 608*, 121037. https://doi.org/10.1016/j.ijpharm.2021.121037.

Aksun Tümerkan, E. T. (2021). Sustainable utilization of gelatin from animal-based agri-food waste for the food industry and pharmacology. In *Valorization of agri-food wastes and by-products: Recent trends, innovations and sustainability challenges*. Turkey: Elsevier, 425–442. https://doi.org/10.1016/B978-0-12-824044-1.00041-6, https://www.sciencedirect.com/book/9780128240441.

Alipal, J., Mohd Pu'ad, N. A. S., Lee, T. C., Nayan, N. H. M., Sahari, N., Basri, H., Idris, M. I., & Abdullah, H. Z. (2021). A review of gelatin: Properties, sources, process, applications, and commercialisation. *Materials Today: Proceedings, 42*, 240–250. https://doi.org/10.1016/j.matpr.2020.12.922.

Amjadi, S., Gholizadeh, S., Hamishehkar, H., Moghaddam, T. N., Haghi, P. B., & Bakhshizadeh, M. (2024). Development of an intelligent packaging system based on gelatin/chitosan nanofibers and containing β-cyclodextrin/*Viola odorata* L. Anthocyanins complex. *Journal of Polymers and the Environment, 32*(2), 884–897. https://doi.org/10.1007/s10924-023-03041-3, https://www.springer.com/journal/10924.

Bakhshizadeh, M., Moghaddam, T. N., Tavassoli, M., Ayaseh, A., & Bangar, S. P. (2023). Gelatin/chitosan nanofibres containing β-cyclodextrin complex and corn poppy *(Papaver rhoeas* L.) for intelligent packaging. *International Journal of Food Science and Technology, 58*(5), 2360–2368. https://doi.org/10.1111/ijfs.16369, http://onlinelibrary.wiley.com/journal/10.1111/(ISSN)1365-2621.

Blanco-López, M., Marcos-García, A., González-Garcinuño, Á., Tabernero, A., & Martín del Valle, E. M. (2024). Exploring the effect of experimental conditions on the synthesis and stability of alginate–gelatin coacervates. *Polymers for Advanced Technologies, 35*(8). https://doi.org/10.1002/pat.6554.

Cai, X., Chen, L., Yang, X., Wang, Y., Xu, J., Zhang, R., Ling, S., & Liu, Y. (2025). Active curcumin-loaded γ-cyclodextrin-metal organic frameworks as nano respiratory channels for reinforcing chitosan/gelatin films in strawberry preservation. *Food Hydrocolloids, 159*, 110656. https://doi.org/10.1016/j.foodhyd.2024.110656.

Cansu, Ü. (2023). Comparative evaluation of different separation and concentration procedures on some quality and functional properties of fish gelatin. *Innovative Food Science & Emerging Technologies, 83*, 103237. https://doi.org/10.1016/j.ifset.2022.103237.

Cao, W., Shi, L., Hao, G., Chen, J., & Weng, W. (2021). Effect of molecular weight on the emulsion properties of microfluidized gelatin hydrolysates. *Food Hydrocolloids, 111*, 106267. https://doi.org/10.1016/j.foodhyd.2020.106267.

Cetinkaya, T., Bildik, F., Altay, F., & Ceylan, Z. (2024). Gelatin nanofibers with black elderberry, Au nanoparticles and SnO_2 as intelligent packaging layer used for monitoring freshness of Hake fish. *Food Chemistry, 437*(Pt 1), 137843. https://doi.org/10.1016/j.foodchem.2023.137843.

Chettri, S., Sharma, N., & Mohite, A. M. (2023). Edible coatings and films for shelf-life extension of fruit and vegetables. *Biomaterials Advances, 154*, 213632. https://doi.org/10.1016/j.bioadv.2023.213632.

Chiarelli, P. G., Suh, J. H., Pegg, R. B., Chen, J., & Mis Solval, K. (2023). The emergence of jellyfish collagen: A comprehensive review on research progress, industrial applications, and future opportunities. *Trends in Food Science and Technology, 141*. https://doi.org/10.1016/j.tifs.2023.104206, http://www.elsevier.com/wps/find/journaldescription.cws_home/601278/description#description.

Dille, M. J., Haug, I. J., & Draget, K. I. (2021). Gelatin and collagen. In *Handbook of hydrocolloids*. Elsevier, 1073–1097. https://doi.org/10.1016/B978-0-12-820104-6.00028-0.

Ding, F., Wu, R., Huang, X., Shi, J., & Zou, X. (2024). Anthocyanin loaded composite gelatin films crosslinked with oxidized alginate for monitoring spoilage of flesh foods. *Food Packaging and Shelf Life, 42*, 101255. https://doi.org/10.1016/j.fpsl.2024.101255.

Dong, Y., Rao, Z., Liu, Y., Zheng, X., Tang, K., & Liu, J. (2023). Soluble soybean polysaccharide/gelatin active edible films incorporated with curcumin for oil packaging. *Food Packaging and Shelf Life, 35*, 101039. https://doi.org/10.1016/j.fpsl.2023.101039.

Duarte, A. C., Costa, E. C., Filipe, H. A. L., Saraiva, S. M., Jacinto, T., Miguel, S. P., Ribeiro, M. P., & Coutinho, P. (2023). Animal-derived products in science and current alternatives. *Biomaterials Advances, 151*. https://doi.org/10.1016/j.bioadv.2023.213428, https://www.sciencedirect.com/journal/biomaterials-advances/about/aims-and-scope.

Elhadef, K., Chaari, M., Akermi, S., Ennouri, K., Ben Hlima, H., Fourati, M., Chakchouk Mtibaa, A., Ennouri, M., Sarkar, T., Shariati, M. A., Gökşen, G., Pateiro, M., Mellouli, L., Lorenzo, J. M., & Smaoui, S. (2024). Gelatin-sodium alginate packaging film with date pits extract: An eco-friendly packaging for extending raw minced beef shelf life. *Meat Science, 207*. https://doi.org/10.1016/j.meatsci.2023.109371, http://www.elsevier.com/locate/meatsci.

Etxabide, A., Mojío, D., Guerrero, P., de la Caba, K., & Gómez-Estaca, J. (2024). Chitin nanowhisker-containing photo-crosslinked antimicrobial gelatin films. *Food Hydrocolloids, 147*, 109371. https://doi.org/10.1016/j.foodhyd.2023.109371.

Ezati, P., Khan, A., & Rhim, J.-W. (2023). Resazurin-impregnated gelatin-based indicator for intelligent packaging applications. *Colloids and Surfaces A: Physicochemical and Engineering Aspects, 675*, 131950. https://doi.org/10.1016/j.colsurfa.2023.131950.

Fotouhi, M., Nasihatkon, B., Solouki, S., & Seidi, S. (2023). A card instead of a lab: A ligand embedded in a bio-composite of starch/gelatin intelligent film for milk quality test followed by colorimetric analysis. *International Journal of Biological Macromolecules, 228*, 426–434. https://doi.org/10.1016/j.ijbiomac.2022.12.205.

Gilchrist, A. E., Lee, S., Hu, Y., & Harley, B. A. C. (2019). Soluble signals and remodeling in a synthetic gelatin-based hematopoietic stem cell niche. *Advanced Healthcare Materials, 8*(20). https://doi.org/10.1002/adhm.201900751, http://onlinelibrary.wiley.com/journal/10.1002/(ISSN)2192-2659.

Guo, H., Xu, Y., Chen, H., Si, X., Zhou, M., & Zhu, E. (2024). Antagonistic yeast and trehalose-enriched gelatin film: A bioactive antifungal packaging film for cherry tomato preservation. *Food Packaging and Shelf Life, 42*, 101258. https://doi.org/10.1016/j.fpsl.2024.101258.

Gupta, S., Sharma, S., Kumar Nadda, A., Saad Bala Husain, M., & Gupta, A. (2022). Biopolymers from waste biomass and its applications in the cosmetic industry: A review. *Materials Today: Proceedings, 68*, 873–879. https://doi.org/10.1016/j.matpr.2022.06.422.

Hao, R., Pang, S., Mraz, J., Geng, Y., Liu, Y., & Pan, J. (2024). Anthocyanin modified by chondroitin sulphate and tannic acid improved the quality-indicating properties of gelatin-based intelligent film. *Food Chemistry: X, 24*, 101779. https://doi.org/10.1016/j.fochx.2024.101779.

Hassan, N., Ahmad, T., Zain, N. M., & Awang, S. R. (2021). Identification of bovine, porcine and fish gelatin signatures using chemometrics fuzzy graph method. *Scientific Reports, 11*(1). https://doi.org/10.1038/s41598-021-89358-2, www.nature.com/srep/index.html.

Hematian, F., Baghaei, H., Mohammadi Nafchi, A., & Bolandi, M. (2023). Preparation and characterization of an intelligent film based on fish gelatin and *Coleus scutellarioides* anthocyanin to monitor the freshness of rainbow trout fish fillet. *Food Science & Nutrition, 11*(1), 379–389. https://doi.org/10.1002/fsn3.3068.

Hossen, M. A., Shimul, I. M., Sameen, D. E., Rasheed, Z., Tanga, W., Chen, M., & Liu, Y. (2024). Chitosan/gelatin coating loaded with ginger essential oil/β-cyclodextrin inclusion complex on quality and shelf life of blueberries. *International Journal of Biological Macromolecules, 279*. https://doi.org/10.1016/j.ijbiomac.2024.135026, https://www.sciencedirect.com/science/journal/01418130.

Ibeogu, I. H., Bako, H. K., Yar, M. S., Zhao, Q., Zhu, J., Zhao, D., Zhang, M., Ke, W., Shan, K., Zhou, G., & Li, C. (2024). Gelatin-serum plasma film incorporated with curcumin for improvement of antioxidant and antibacterial properties for fresh pork packaging application. *Food Hydrocolloids, 149*. https://doi.org/10.1016/j.foodhyd.2023.109617, https://www.sciencedirect.com/science/journal/0268005X.

Jamróz, E., Cabaj, A., Tkaczewska, J., Kawecka, A., Krzyściak, P., Szuwarzyński, M., Mazur, T., & Juszczak, L. (2023). Incorporation of curcumin extract with lemongrass essential oil into the middle layer of triple-layered films based on furcellaran/chitosan/gelatin hydrolysates – In vitro and in vivo studies on active and intelligent properties. *Food Chemistry, 402*, 134476. https://doi.org/10.1016/j.foodchem.2022.134476.

Jayaprakash, S., Mohamad Abdul Razeen, Z., Naveen Kumar, R., He, J., Milky, M. G., Renuka, R., & Sanskrithi, M. V. (2024). Enriched characteristics of poultry collagen over other sources of collagen and its extraction methods: A review. *International Journal of Biological Macromolecules, 273*(Pt 1), 133004. https://doi.org/10.1016/j.ijbiomac.2024.133004.

Khan, A., Riahi, Z., Kim, J. T., & Rhim, J. W. (2023). Gelatin/carrageenan-based smart packaging film integrated with Cu-metal organic framework for freshness monitoring and shelf-life extension of shrimp. *Food Hydrocolloids, 145*. https://doi.org/10.1016/j.foodhyd.2023.109180, https://www.journals.elsevier.com/food-hydrocolloids.

Khan, A., Riahi, Z., Kim, J. T., & Rhim, J. W. (2024). Chitosan/gelatin-based multifunctional films integrated with sulfur-functionalized chitin for active packaging applications. *Food Hydrocolloids, 149*. https://doi.org/10.1016/j.foodhyd.2023.109537, https://www.sciencedirect.com/science/journal/0268005X.

Khezerlou, A., Tavassoli, M., Alizadeh-Sani, M., Hashemi, M., Ehsani, A., & Bangar, S. P. (2023). Multifunctional food packaging materials: Lactoferrin loaded Cr-MOF in films-based gelatin/κ-carrageenan for food packaging applications. *International Journal of Biological Macromolecules, 251*, 126334. https://doi.org/10.1016/j.ijbiomac.2023.126334.

Kim, S., Lim, H., Park, S., Kim, D. S., & Kwak, H. W. (2025). Durable functional fish gelatin film using biomass-derived liquid hydrolysate as an eco-friendly crosslinker. *Industrial Crops and Products, 225*. https://doi.org/10.1016/j.indcrop.2025.120564, https://www.sciencedirect.com/science/journal/09266690.

Korany, A. M., Abdel-Atty, N. S., Zeinhom, M. M. A., & Hassan, A. H. A. (2024). Application of gelatin-based zinc oxide nanoparticles bionanocomposite coatings to control *Listeria monocytogenes* in Talaga cheese and camel meat during refrigerated storage. *Food Microbiology, 122*. https://doi.org/10.1016/j.fm.2024.104559, https://www.sciencedirect.com/science/journal/07400020.

Kulawik, P., Jamróz, E., Tkaczewska, J., Vlčko, T., Zając, M., Guzik, P., Janik, M., Tadele, W., Golian, J., & Milosavljević, V. (2024). Application of antimicrobial chitosan-furcellaran-hydrolysate gelatin edible coatings enriched with bioactive peptides in shelf-life extension of pork loin stored at 4 and −20 °C. *International Journal of Biological Macromolecules, 254*(Pt 2), 127865. https://doi.org/10.1016/j.ijbiomac.2023.127865.

Li, R., Wang, S., Feng, H., Zhuang, D., & Zhu, J. (2024). An intelligent chitosan/gelatin film via improving the anthocyanin-induced color recognition accuracy for beef sub-freshness differentiation monitoring. *Food Hydrocolloids, 146*, 109219. https://doi.org/10.1016/j.foodhyd.2023.109219.

Li, S., Wei, N., Wei, J., Fang, C., Feng, T., Liu, F., Liu, X., & Wu, B. (2024). Curcumin and silver nanoparticles loaded antibacterial multifunctional pectin/gelatin films for food packaging applications. *International Journal of Biological Macromolecules, 266*, 131248. https://doi.org/10.1016/j.ijbiomac.2024.131248.

Liao, W., Zhu, Y., Lu, Y., Wang, Y., Dong, X., Xia, G., & Shen, X. (2021). Effect of extraction variables on the physical and functional properties of tilapia gelatin. *LWT, 146*, 111514. https://doi.org/10.1016/j.lwt.2021.111514.

Liu, S., Qiao, S., Zhu, J., Yang, Y., Chen, H., Dai, H., Zhu, H., Yu, Y., Ma, L., Zhang, Y., & Wang, H. (2024). Enhanced barrier and antioxidant properties of gelatin films by structural-colored bioactive materials for food packaging. *Food Hydrocolloids, 150*, 109744. https://doi.org/10.1016/j.foodhyd.2024.109744.

Luo, Q., Hossen, M. A., Zeng, Y., Dai, J., Li, S., Qin, W., & Liu, Y. (2022). Gelatin-based composite films and their application in food packaging: A review. *Journal of Food Engineering, 313*, 110762. https://doi.org/10.1016/j.jfoodeng.2021.110762.

Masrouri, M., & Qin, Z. (2023). Effects of terminal tripeptide units on mechanical properties of collagen triple helices. *Extreme Mechanics Letters, 64*, 102075. https://doi.org/10.1016/j.eml.2023.102075.

Mirzapour-Kouhdasht, A., Moosavi-Nasab, M., Yousefi, R., & Eun, J.-B. (2021). Bio/multi-functional peptides derived from fish gelatin hydrolysates: Technological and functional properties. *Biocatalysis and Agricultural Biotechnology, 36*, 102152. https://doi.org/10.1016/j.bcab.2021.102152.

Mushtaq, F., Raza, Z. A., Batool, S. R., Zahid, M., Onder, O. C., Rafique, A., & Nazeer, M. A. (2022). Preparation, properties, and applications of gelatin-based hydrogels (GHs) in the environmental, technological, and biomedical sectors. *International Journal of Biological Macromolecules, 218*, 601–633. https://doi.org/10.1016/j.ijbiomac.2022.07.168.

Nowak, N., Tkaczewska, J., Grzebieniarz, W., Juszczak, L., Mazur, T., Szuwarzyński, M., Guzik, P., & Jamróz, E. (2024). Active and intelligent four-layer films based on chitosan, gelatin, furcellaran and active ingredients—preparation, characterisation and application on salmon. *Food and Bioprocess Technology, 17*(7), 1862–1875. https://doi.org/10.1007/s11947-023-03238-3.

Osman, N. A., & Abo-Dief, H. M. (2023). The origins, usage, and production methods of halal and kosher gelatin. In *Halal and kosher food: Integration of quality and safety for global market trends*. Saudi Arabia: Springer International Publishing, 119–130. https://doi.org/10.1007/978-3-031-41459-6_10, https://link.springer.com/book/10.1007/978-3-031-41459-6.

Pang, S., Wang, Y., Jia, H., Hao, R., Jan, M., Li, S., Pu, Y., Dong, X., & Pan, J. (2023). The properties of pH-responsive gelatin-based intelligent film as affected by ultrasound power and purple cabbage anthocyanin dose. *International Journal of Biological Macromolecules, 230*, 123156. https://doi.org/10.1016/j.ijbiomac.2023.123156.

Pulla-Huillca, P. V., Gomes, A., Quinta Barbosa Bittante, A. M., Lourenço, R. V., & Sobral, P. Jd. A. (2021). Wettability of gelatin-based films: The effects of hydrophilic or hydrophobic plasticizers and nanoparticle loads. *Journal of Food Engineering, 297*. https://doi.org/10.1016/j.jfoodeng.2021.110480, http://www.sciencedirect.com/science/journal/02608774.

Ramakrishnan, R., Kim, J. T., Roy, S., & Jayakumar, A. (2024). Recent advances in carboxymethyl cellulose-based active and intelligent packaging materials: A comprehensive review. *International Journal of Biological Macromolecules, 259*, 129194. https://doi.org/10.1016/j.ijbiomac.2023.129194.

Ramakrishnan, R., Kulandhaivelu, S. V., Roy, S., & Viswanathan, V. P. (2023). Characterisation of ternary blend film of alginate/carboxymethyl cellulose/starch for packaging applications. *Industrial Crops and Products, 193*. https://doi.org/10.1016/j.indcrop.2022.116114, www.elsevier.com/inca/publications/store/5/2/2/8/2/5.

Ramli, S. S., Nizar, N. N. A., Heng, J. Y. Y., Karde, V., Abidin, S. A. S. Z., & Taib, M. N. (2023). Gelatin substitute. In *Innovation of food products in halal supply chain worldwide*. Malaysia: Elsevier, 87–98. https://doi.org/10.1016/B978-0-323-91662-2.00010-7, https://www.sciencedirect.com/book/9780323916622.

Rather, J. A., Akhter, N., Ashraf, Q. S., Mir, S. A., Makroo, H. A., Majid, D., Barba, F. J., Khaneghah, A. M., & Dar, B. N. (2022). A comprehensive review on gelatin: Understanding impact of the sources, extraction methods, and modifications on potential packaging applications. *Food Packaging and Shelf Life, 34*, 100945. https://doi.org/10.1016/j.fpsl.2022.100945.

Rather, J. A., Makroo, H. A., Showkat, Q. A., Majid, D., & Dar, B. N. (2022). Recovery of gelatin from poultry waste: Characteristics of the gelatin and lotus starch-based coating material and its application in shelf-life enhancement of fresh cherry tomato. *Food Packaging and Shelf Life, 31*. https://doi.org/10.1016/j.fpsl.2021.100775, http://www.journals.elsevier.com/food-packaging-and-shelf-life/.

Riahi, Z., Hong, S. J., Rhim, J. W., Shin, G. H., & Kim, J. T. (2023). High-performance multifunctional gelatin-based films engineered with metal-organic frameworks for active food packaging applications. *Food Hydrocolloids, 144*. https://doi.org/10.1016/j.foodhyd.2023.108984, https://www.journals.elsevier.com/food-hydrocolloids.

Riahi, Z., Khan, A., Rhim, J.-W., Shin, G. H., & Kim, J. T. (2023). Gelatin/poly(vinyl alcohol)-based dual functional composite films integrated with metal-organic frameworks and anthocyanin for active and intelligent food packaging. *International Journal of Biological Macromolecules, 249*, 126040. https://doi.org/10.1016/j.ijbiomac.2023.126040.

Riahi, Z., Khan, A., Rhim, J. W., Shin, G. H., & Kim, J. T. (2024). Synergistic effect of iron-based metal-organic framework hybridized with carbon quantum dots in agar/gelatin films for fruit preservation. *Food Packaging and Shelf Life, 45*. https://doi.org/10.1016/j.fpsl.2024.101330, https://www.sciencedirect.com/science/journal/22142894.

Sadi, A., & Ferfera-Harrar, H. (2023). Cross-linked CMC/Gelatin bio-nanocomposite films with organoclay, red cabbage anthocyanins and pistacia leaves extract as active intelligent food packaging: Colorimetric pH indication, antimicrobial/antioxidant properties, and shrimp spoilage tests. *International Journal of Biological Macromolecules, 242*(Pt 3), 124964. https://doi.org/10.1016/j.ijbiomac.2023.124964.

Samatra, M. Y., Noor, N. Q. I. M., Razali, U. H. M., Bakar, J., & Shaarani, S. M. (2022). Bovidae-based gelatin: Extractions method, physicochemical and functional properties, applications, and future trends. *Comprehensive Reviews in Food Science and Food Safety, 21*(4), 3153–3176. https://doi.org/10.1111/1541-4337.12967, http://onlinelibrary.wiley.com/journal/10.1111/(ISSN)1541-4337.

da Silva, A. C. P., Barbosa, J. R., da Silva Araújo, C., Sousa Batista, J. T., Xavier Neves, E. M. P., Pereira Cardoso, D. N., Peixoto Joele, M. R. S., & de Fátima Henriques Lourenço, L. (2024). A new edible coating of fish gelatin incorporated into açaí oil to increase the post-harvest shelf life of tomatoes. *Food Chemistry, 438*. https://doi.org/10.1016/j.foodchem.2023.138047, https://www.sciencedirect.com/science/journal/03088146.

Sul, Y., Ezati, P., & Rhim, J.-W. (2023). Preparation of chitosan/gelatin-based functional films integrated with carbon dots from banana peel for active packaging application. *International Journal of Biological Macromolecules, 246*, 125600. https://doi.org/10.1016/j.ijbiomac.2023.125600.

Talha, M., Tanveer, M., Abid, A., Maan, A. A., Khan, M. K. I., Shair, H., Tanveer, N., & Mustafa, A. (2024). Valorization of poultry slaughter wastes via extraction of three structural proteins (gelatin, collagen and keratin): A sustainable approach for circular economy. *Trends in Food Science and Technology, 152*. https://doi.org/10.1016/j.tifs.2024.104667, https://www.sciencedirect.com/science/journal/09242244.

Tang, C., Zhou, K., Zhu, Y., Zhang, W., Xie, Y., Wang, Z., Zhou, H., Yang, T., Zhang, Q., & Xu, B. (2022). Collagen and its derivatives: From structure and properties to their applications in food industry. *Food Hydrocolloids, 131*, 107748. https://doi.org/10.1016/j.foodhyd.2022.107748.

Tiozon, R. J. N., Bonto, A. P., & Sreenivasulu, N. (2021). Enhancing the functional properties of rice starch through biopolymer blending for industrial applications: A review. *International Journal of Biological Macromolecules, 192*, 100–117. https://doi.org/10.1016/j.ijbiomac.2021.09.194, www.elsevier.com/locate/ijbiomac.

Toniciolli Rigueto, C. V., Rosseto, M., Alessandretti, I., de Oliveira, R., Wohlmuth, D. A. R., Ferreira Menezes, J., Loss, R. A., Dettmer, A., & Pizzutti, I. R. (2022). Gelatin films from wastes: A review of production, characterization, and application trends in food preservation and agriculture. *Food Research International, 162*. https://doi.org/10.1016/j.foodres.2022.112114, www.elsevier.com/inca/publications/store/4/2/2/9/7/0.

Uddin, S. M. K., Hossain, M. A. M., Sagadevan, S., Al Amin, M., & Johan, M. R. (2021). Halal and kosher gelatin: Applications as well as detection approaches with challenges and prospects. *Food Bioscience, 44*. https://doi.org/10.1016/j.fbio.2021.101422, http://www.journals.elsevier.com/food-bioscience/.

Ullah, M. W., Alabbosh, K. F., Fatima, A., Islam, S. U., Manan, S., Ul-Islam, M., & Yang, G. (2024). Advanced biotechnological applications of bacterial nanocellulose-based biopolymer nanohybrids: A review. *Advanced Industrial and Engineering Polymer Research, 7*(1), 100–121. https://doi.org/10.1016/j.aiepr.2023.07.004, https://www.sciencedirect.com/science/journal/25425048.

Wang, H., Ding, F., Ma, L., & Zhang, Y. (2021). Edible films from chitosan-gelatin: Physical properties and food packaging application. *Food Bioscience, 40*, 100871. https://doi.org/10.1016/j.fbio.2020.100871.

Wang, X., Qiao, C., Song, K., Jiang, S., & Yao, J. (2021). Hofmeister effect on the viscosity properties of gelatin in dilute solutions. *Colloids and Surfaces B: Biointerfaces, 206*, 111944. https://doi.org/10.1016/j.colsurfb.2021.111944.

Wu, X., Liu, Z., He, S., Liu, J., & Shao, W. (2023). Development of an edible food packaging gelatin/zein based nanofiber film for the shelf-life extension of strawberries. *Food Chemistry, 426*, 136652. https://doi.org/10.1016/j.foodchem.2023.136652.

Yang, Z., Chaieb, S., & Hemar, Y. (2021). Gelatin-based nanocomposites: A review. *Polymer Reviews, 61*(4), 765–813. https://doi.org/10.1080/15583724.2021.1897995.

Ying, Q., Zhan, S., Yu, H., Li, J., Jia, R., Wei, H., Roura, E., Tan, X., Qiao, Z., & Huang, T. (2024). Gelatin based preservation technologies on the quality of food: A comprehensive review. *Critical Reviews in Food Science and Nutrition*, 1–18. https://doi.org/10.1080/10408398.2024.2361298.

Zhang, A., Fan, X., Zeng, X., Xu, J., Zhou, C., Xia, Q., Sun, Y., Wu, Z., & Pan, D. (2024). Enhancing physicochemical, antimicrobial, and release properties of fish skin gelatin films using dual-layer nanoparticles loaded with tea polyphenols/kojic acid for air-dried chicken preservation. *Food Hydrocolloids, 149*, 109580. https://doi.org/10.1016/j.foodhyd.2023.109580.

Zhong, Y., Zhang, T., Zhang, W., Wang, G., Zhang, Z., Zhao, P., Liu, X., & Li, H. (2023). Antibacterial castor oil-based waterborne polyurethane/gelatin films for packaging of strawberries. *Food Packaging and Shelf Life, 36*, 101055. https://doi.org/10.1016/j.fpsl.2023.101055.

Zhu, X., Gu, S., Guo, D., Huang, X., Chen, N., Niu, B., & Deng, X. (2023). Determination of porcine derived components in gelatin and gelatin-containing foods by high performance liquid chromatography-tandem mass spectrometry. *Food Hydrocolloids, 134*, 107978. https://doi.org/10.1016/j.foodhyd.2022.107978.

Zhuang, D., Li, R., Zhang, M., Wang, S., Ahmad, H. N., & Zhu, J. (2024). Photodynamic-responsive gelatin-based coating with high utilization curcumin loaded bilayer nanoencapsulation: A promising environmental food packaging. *International Journal of Biological Macromolecules, 256*(Pt 2), 128476. https://doi.org/10.1016/j.ijbiomac.2023.128476.

Zong, Z., Liu, M., Chen, H., Farag, M. A., Wu, W., Fang, X., Niu, B., & Gao, H. (2023). Preparation and characterization of a novel intelligent starch/gelatin binary film containing purple sweet potato anthocyanins for *Flammulina velutipes* mushroom freshness monitoring. *Food Chemistry, 405*(Pt B), 134839. https://doi.org/10.1016/j.foodchem.2022.134839.

Chapter 13

Gluten-based films and coatings

Jefferson Henrique Tiago Barros[1], Isamara Teodoro do Nascimento[2], Fernanda Ortolan[3], Luís Marangoni Júnior[2], and Caroline Joy Steel[2]

[1]*Federal Institute of São Paulo (IFSP), Presidente Prudente, São Paulo, Brazil,* [2]*Department of Food Engineering and Technology, School of Food Engineering (FEA), University of Campinas (Unicamp), Campinas, São Paulo, Brazil,* [3]*Federal Institute of São Paulo (IFSP), Capivari, São Paulo, Brazil*

13.1 Introduction

The growing demand for sustainable solutions in the packaging industry has prompted the exploration of alternatives to conventional plastic materials, particularly due to environmental impact concerns (Geyer et al., 2017). In this context, biomaterial-based films and coatings represent a promising innovation that offers effective protection for food while reducing postconsumer waste (Assad et al., 2020). Notably, some of these materials are edible, allowing consumption along with food, thus supporting a circular economy and sustainability (Rocca-Smith, Marcuzzo, et al., 2016). Among the biopolymers utilized in the production of biomaterial-based films and coatings, wheat gluten stands out as a significant protein-based raw material (Assad et al., 2020). Its protein structure, which consists of gliadins and glutenins, facilitates the formation of a cohesive and flexible three-dimensional network.

This network-forming property endows gluten films with a balanced combination of mechanical strength and flexibility (Gennadios et al., 1993). The ability to structure such three-dimensional networks is essential for developing films that not only protect food but also endure mechanical stresses during transportation and storage, thereby maintaining their integrity over time (Mangavel et al., 2004). Unlike petroleum-based plastics, gluten films naturally decompose in the environment, helping to mitigate the environmental impact associated with packaging waste (Rocca-Smith, Marcuzzo, et al., 2016).

This chapter aims to provide a comprehensive analysis of gluten-based films and coatings, focusing on their chemical composition and the physicochemical properties that directly influence their performance. It will begin by exploring various preparation methods, followed by a detailed discussion of the processes used to modify their characteristics for different applications. Next, the practical applications of these materials will be examined, with particular emphasis on food preservation.

The chapter will also address the current limitations of gluten films and present innovative strategies that are being developed to overcome these challenges. Finally, emerging trends in gluten film research will be discussed, including the combination with other biopolymers and the application of nanotechnology, highlighting their potential for innovation and contributions to the field of sustainable packaging.

13.2 The chemistry of gluten

Gluten is a complex protein made up of two primary protein fractions: gliadins and glutenins. These proteins are vital in determining the viscoelastic properties of the gluten network, which are essential for food processing (e.g., bread production), as well as for the production of films and coatings. Gliadins are monomeric proteins that belong to the prolamin class and are characterized by their solubility in alcohol and low molecular weight (LMW), typically ranging from 28 to 55 kDa (Wieser, 2022). They contain a high proportion of hydrophobic amino acids, such as proline and glutamine, which enable the formation of helical structures (Zhang et al., 2020). This composition gives gliadins the extensibility of the dough.

This property is essential for producing films that require flexibility and malleability. In contrast, glutenins are polymeric proteins belonging to the glutelin class, and they are soluble in acidic or alkaline solutions (Wieser, 2022). These proteins exhibit a more rigid and ordered structure, with high molecular weights (HMWs) ranging from 500 kDa to

Natural Polymer-Based Films and Coatings in Food Technology. DOI: https://doi.org/10.1016/B978-0-443-34244-8.00017-3

1.5 million Daltons (Zhang et al., 2020). Glutenins consist of HMW and LMW subunits that polymerize through covalent bonds, primarily disulfide bonds, which are crucial for maintaining structural integrity and enhancing the elasticity and strength of the gluten network (Mangavel et al., 2004). These bonds create long interwoven chains that impart cohesion and elasticity to the gluten network.

In films and coatings, glutenins are primarily responsible for mechanical strength and structural stability, ensuring that the film maintains its integrity over time (Rovera et al., 2020). The ratio of gliadins to glutenins in gluten is a crucial factor in determining the properties of the resulting films and coatings. A proper balance between these two protein fractions enables the formation of films that combine flexibility and strength, essential characteristics for food packaging applications (Sharma et al., 2017). For instance, a higher proportion of gliadins tends to produce more extensible films, while a greater concentration of glutenins results in stiffer and more resilient films (Xu & Li, 2023).

The ability to adjust this ratio offers a significant advantage by allowing the modulation of the mechanical and barrier properties of gluten films, making them suitable for various applications, from food preservation to protective coatings for perishable products (Zhang et al., 2020).

13.3 Gluten network formation

The formation of the gluten network is a complex process that involves a series of chemical reactions and physical interactions, which are essential for achieving the viscoelastic properties characteristic of this biopolymer. These reactions primarily take place during mechanical processing steps such as kneading, extrusion, and mixing, creating optimal conditions for the rearrangement of gluten proteins, mainly glutenins and gliadins.

The formation of the gluten network requires water and mechanical action. Water plays a vital role by promoting the movement of protein chains, which supports efficient interactions. During kneading, the application of mechanical energy organizes these protein chains and promotes the development of a cohesive and elastic network. As gluten dries, water evaporation strengthens intermolecular interactions, particularly disulfide bonds, leading to the formation of a solid, stable material with improved mechanical strength and barrier properties, making it ideal for packaging applications (Ortolan & Steel, 2017).

The main interactions that occur during gluten network formation are disulfide bonds, hydrophobic interactions, hydrogen bonds, and dityrosine bonds.

Disulfide bonds: Disulfide bonds (–S–S–) play a crucial role in the formation of the gluten network. These covalent bonds are formed between the thiol (–SH) groups of cysteine residues found in glutenin subunits (Mangavel et al., 2004). During mechanical processing, such as dough kneading, the application of mechanical force in the presence of oxygen (and sometimes oxidants) promotes the oxidation of thiol groups, facilitating the formation of intra- and intermolecular disulfide bonds (Zhang et al., 2020). This interlinking of glutenin chains results in a complex three-dimensional structure that imparts elasticity and tensile strength to gluten. Consequently, disulfide bonds are essential for the mechanical strength and structural integrity of gluten, contributing to its elasticity and resistance to deformation (Xu & Li, 2023).

Hydrophobic interactions and hydrogen bonds: In addition to disulfide bonds, hydrophobic interactions and hydrogen bonds play significant roles in stabilizing the gluten protein network. Hydrophobic regions, which are rich in amino acids such as proline and glutamine, tend to aggregate in aqueous environments, thereby compacting the protein network and enhancing resistance to water solubility (Wieser, 2022). Meanwhile, hydrogen bonds form between the amino and carboxyl groups of amino acid side chains, providing additional cohesion and contributing to the structural integrity of the gluten network (Zhang et al., 2020).

Dityrosine bonds: Dityrosine bonds also act as an additional stabilization factor in the gluten network. These bonds form through the reaction of two adjacent tyrosine residues, creating a covalent linkage that enhances the strength and durability of the protein network. Dityrosine bonds are particularly significant under processing conditions involving high temperatures and/or pressures, where these reactions may be amplified, contributing to the rigidity and stability of the resulting material (Xu & Li, 2023).

The various bonds and interactions that occur contribute to the structural integrity of the gluten film, providing mechanical strength and elasticity. Additionally, they directly influence key properties such as water vapor permeability and solubility. The ability to modulate these interactions enables the development of films with specific characteristics tailored to various food applications. Enhancing these properties involves combining gluten with other biopolymers and employing nanotechnology to develop more efficient and sustainable films and coatings (Zhang et al., 2020).

13.4 Vital gluten

In the various studies on the formation of gluten-based films and coatings, an ingredient known as vital gluten is frequently used. This ingredient is a dehydrated form of gluten, widely utilized in the food industry due to its ability to impart elasticity and structure to products such as bread and pasta. Vital gluten differs from wet gluten that is extracted directly from wheat flour, as it undergoes a drying process that stabilizes its protein bonds and directly influences the formation and properties of films and coatings (Guilbert et al., 1998; Ortolan & Steel, 2017). Fig. 13.1 illustrates hydrated gluten (A) and vital gluten (B), highlighting the respective bonds and interactions within their structures.

When vital gluten is used to produce films or coatings, it is crucial to understand the implications of its preformed structure and the stabilized interactions between gliadin and glutenin proteins. Therefore, hydration facilitates the creation of additional molecular interactions, which contribute to films and coatings with enhanced mechanical strength and reduced initial flexibility. This occurs because vital gluten, when rehydrated, already possesses a partially established structure, requiring less mechanical manipulation to create the three-dimensional protein network (Sartori et al., 2018).

Furthermore, the hydrophobic interactions and hydrogen bonds that are essential for the cohesion of the gluten network may be less dynamic in vital gluten due to the presence of stabilized disulfide bonds. This can impact critical factors such as water vapor permeability and the solubility of the films, which are crucial for their application in food packaging. This stability can be beneficial in developing films with more consistent barrier properties, particularly in high-humidity environments (Zink et al., 2016). These characteristics confer both advantages and challenges to vital gluten during film formation, impacting its mechanical strength, elasticity, and barrier performance. Therefore, when selecting vital gluten as a base material, it is essential to understand its intrinsic characteristics to optimize the film formation process and enhance the final properties of the films.

13.5 Physicochemical properties of gluten films and coatings

To better understand the physicochemical properties of gluten film, it is essential to define and differentiate between gluten films and coatings. Both play significant roles in the food industry, but their applications and characteristics may vary.

Gluten films: Gluten films are thin, continuous layers of material that can be utilized independently or as part of a packaging solution. These films are created from gluten solutions applied to a surface, followed by the evaporation of the solvent to form a solid sheet. Gluten films are widely used as physical barriers. By fully enclosing foods, they avoid oxidation and moisture loss, extending their life and maintaining freshness and product quality (Ansorena et al., 2016; Sharma et al., 2017).

Gluten coatings: Unlike films, gluten coatings are applied directly to the surface of foods, typically in liquid form, and solidify into a protective layer after solvent evaporation. Coatings can be applied through methods such as dipping, spraying, or brushing and are often used to enhance appearance, control moisture loss, and prevent microbial spoilage.

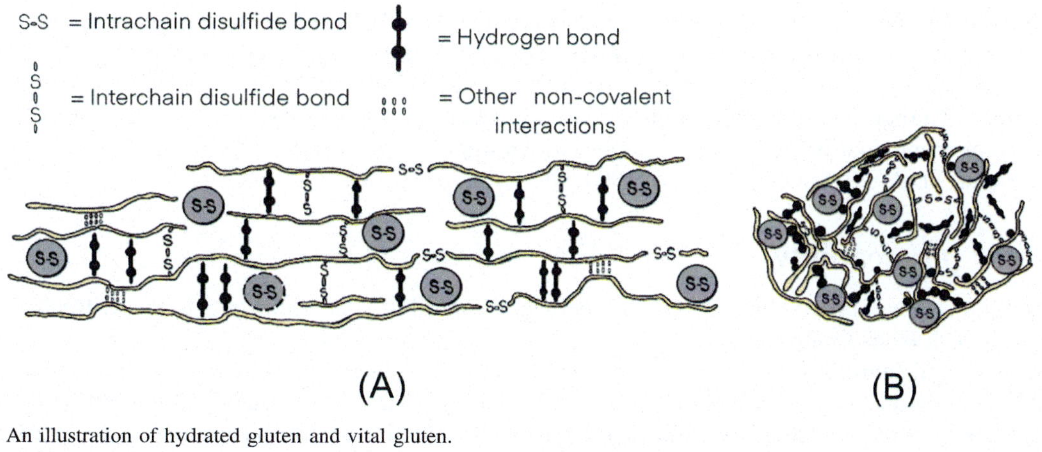

FIGURE 13.1 An illustration of hydrated gluten and vital gluten.
Modified from Ortolan, F., & Steel, C.J. (2017). Protein Characteristics that Affect the Quality of Vital Wheat Gluten to be Used in Baking: A Review. Comprehensive Reviews in Food Science and Food Safety, 16(3), 369–381. https://doi.org/10.1111/1541-4337.12259, http://www.interscience.wiley.com/jpages/1541-4337.

They are particularly beneficial for fruits, vegetables, and baked goods, where the coating acts as an additional barrier against oxidation and moisture loss (Guilbert et al., 1998; Han et al., 2005).

With this distinction in mind, we will now discuss the main physicochemical properties of gluten films and coatings and their significance in food packaging applications. For films, properties such as tensile strength, elongation at break, and oxygen barrier performance are important, whereas water vapor permeability and hydrophobicity are key for coatings.

13.6 Mechanical properties

The mechanical properties of gluten films are vital to maintain integrity during storage and distribution, effectively protecting food in the packaging industry. The mechanical properties evaluated are traction and stretching resistance in the interval.

Tensile strength: Tensile strength refers to the film's ability to withstand stress before breaking. This property is particularly critical for films used to package products that undergo handling, transportation, and storage, where the film must protect the food without tearing. The tensile strength of gluten films typically ranges from 0.5 to 4 MPa, depending on formulation and processing conditions. This strength can be modified by adding plasticizers, which increase flexibility (Gennadios et al., 1993; Han et al., 2005).

Elongation at break: Elongation at break measures the extent to which the film can stretch before breaking, making it an important property for applications where the film needs to conform to the shape of the packaged product. Gliadins contribute to the film's flexibility because of their monomeric nature and ability to form weak, noncovalent interactions, which promote molecular sliding and stretching. This property allows the film to adapt easily to different food shapes without rupturing. This characteristic is especially useful for flexible packaging that needs to enclose food effectively without compromising film integrity (Gontard et al., 1994). Gluten films generally exhibit elongation at break values ranging from 190% to 500%, depending on formulation and use of additives. Recent studies have investigated the effects of incorporating nanoparticles and other modifying agents to improve this property (Kayserilioğlu et al., 2001; Larré et al., 2000; Mojumdar et al., 2011; Rovera et al., 2020; Zuo & Dai, 2009).

13.7 Gas barrier properties

Gas permeability defines the film's ability to restrict the passage of gas molecules, such as oxygen and carbon dioxide, which are critical for food preservation.

Oxygen barrier: Oxygen is a primary contributor to lipid oxidation and the degradation of sensitive compounds, such as vitamins and pigments. Therefore, an effective oxygen barrier is essential for extending the shelf life of packaged foods. Gluten-based films have demonstrated low oxygen permeability, with values ranging from 1.5 to 6.0 $cm^3 \cdot \mu m \cdot m^{-2} \cdot day^{-1} \cdot kPa^{-1}$, making them highly effective for preserving oxidation-prone foods. Studies indicate that incorporating lipids or hydrophobic compounds into the gluten matrix can significantly reduce oxygen permeability, enhancing the film's barrier efficiency (Gontard et al., 1994; Chen et al., 2019; Rocca-Smith, Marcuzzo, et al., 2016).

Carbon dioxide barrier: The film's ability to regulate carbon dioxide permeability is also important, particularly in packaging that employs modified atmospheres to prolong the shelf life of fresh foods. Typically, gluten films exhibit carbon dioxide permeability values ranging from 2.0×10^{-8} to 1.0×10^{-7} $cm^3 \cdot m^{-2} \cdot s^{-1} \cdot Pa^{-1}$, depending on their composition and modifications. Carbon dioxide permeability can be enhanced by modifying the film composition, such as incorporating components like beeswax or fatty acids, which improve the barrier without compromising other essential properties (Fakhouri et al., 2018; Gontard et al., 1994; Sartori et al., 2018).

13.8 Water permeability

Water permeability is one of the most challenging properties to control in gluten films due to the hydrophilic nature of the gluten network, which tends to absorb water.

Hydrophobicity and solubility: Gluten films are naturally water-soluble, which can limit their applications, especially in high-humidity environments or with high-moisture foods (Gontard et al., 1994; Sharma et al., 2017). To overcome this limitation, incorporating lipids such as beeswax or fatty acids into the film matrix has proven to be an effective strategy (Rocca-Smith, Lullien-Pellerin, et al., 2016). These lipids create a hydrophobic barrier that prevents water penetration, thereby improving moisture resistance and enhancing the film's integrity under challenging conditions (Sartori et al., 2018). Recent research has also explored nanotechnology to develop nano-structured gluten films that enhance water resistance without compromising biodegradability (Zhang et al., 2020).

Water vapor permeability: Water vapor permeability is another critical measure, as it determines the film's ability to minimize moisture exchange between food and the external environment. Due to their hydrophilic nature, gluten films exhibit relatively high water vapor permeability, which may be a disadvantage for certain applications (Gontard et al., 1994). Reported values include 0.0048 $g{\cdot}mm{\cdot}m^{-2}{\cdot}day^{-1}{\cdot}mm\ Hg^{-1}$ for bilayer gluten films that incorporate lipid layers as barriers and 4.1×10^{-7} $g{\cdot}m\,(m^2{\cdot}s{\cdot}Pa)^{-1}$ for hydrophilic gluten films under high-humidity conditions (Roy et al., 2000).

Adding lipids like waxes and oils effectively reduces permeability, enhancing the film's suitability for moisture-sensitive food packaging (Fakhouri et al., 2018). Mechanical properties, gas permeability, and water vapor permeability are essential factors that determine the applicability of these films in various contexts. Technological innovations, such as the incorporation of nanoparticles and the development of multilayer films, are currently being investigated, providing new possibilities to enhance these properties and expand the use of gluten films and coatings in the food industry.

13.9 Preparation methods for gluten films and coatings

The production of gluten-based films and coatings involves various techniques that facilitate the formation of continuous and uniform structures. Each method relies on distinct chemical and physical principles, and each has its own advantages and disadvantages, which directly influence the choice of technique for particular applications. The following sections outline key gluten film preparation methods, detailing their processes and industrial applications.

13.9.1 Casting

The casting method, illustrated in Fig. 13.2, is a traditional technique widely employed to produce gluten films. The fundamental principle of casting involves creating a film from a liquid solution, which solidifies into a continuous layer upon solvent evaporation. This process begins with dissolving gluten in water to form a colloidal solution (Fig. 13.2A). Plasticizers, such as glycerol, along with other additives like antimicrobial agents or lipids, are incorporated to adjust the film's mechanical and barrier properties (GONTARD et al., 1994). The mixture is then poured onto a flat surface to ensure even spreading (Fig. 13.2B). Solvent evaporation can take place at room temperature or in an oven (Fig. 13.2C), resulting in the formation of the film (Fig. 13.2D) (Bourtoom, 2008; Chen et al., 2019).

13.9.1.1 Advantages

- Simplicity and low cost: The casting method is relatively straightforward and does not require complex equipment, making it accessible for laboratories and small-scale industries.
- Versatility: This method allows for the easy incorporation of various additives to modify the film's properties as needed.

13.9.1.2 Disadvantages

- Solvent evaporation time: The solvent evaporation process can be slow, particularly at room temperature, which may limit industrial efficiency.
- Limited uniformity: Achieving completely uniform films can be challenging, especially over large surfaces.
- Fragility: Films produced by casting tend to be more fragile and less resistant to mechanical stress.

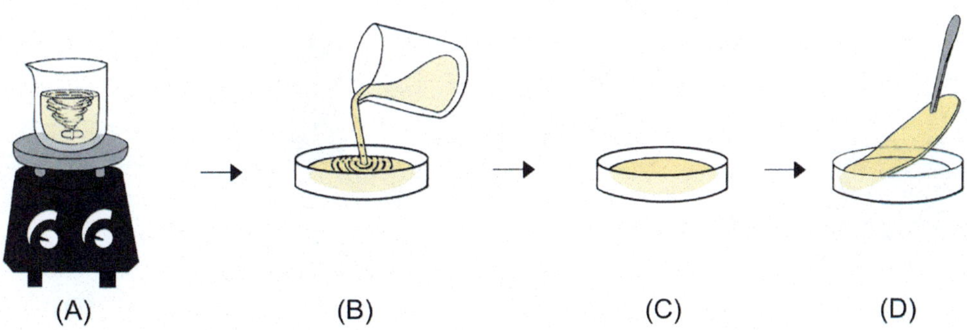

FIGURE 13.2 Casting method for gluten film production.

The casting method is especially suitable for producing thin films in applications where high precision in thickness control is not critical. It is widely employed for packaging dry products or as food coatings, where flexibility and mechanical strength are secondary considerations (Nur Hanani et al., 2014).

13.9.2 Compression molding

Compression molding is a technique that utilizes heat and pressure to transform a mixture of gluten and additives into films with enhanced mechanical properties, as illustrated in Fig. 13.3. This process capitalizes on gluten's ability to form a cohesive network when subjected to high temperatures and pressure. The mixture is placed in a mold, where the simultaneous application of heat (100°C–150°C) and pressure (3–10 MPa) promotes fusion, resulting in a dense and continuous film (Zubeldía et al., 2015). This method is commonly employed to produce thicker films that require greater strength and durability (Wang, 2022; Zhang et al., 2020).

13.9.2.1 Advantages

- High mechanical strength: The application of heat and pressure promotes the formation of a dense, cohesive protein network, resulting in films with superior tensile strength and durability.
- Suitable for thicker films: The method is ideal for producing films with greater thickness, which is necessary for applications that require enhanced robustness.
- Composition versatility: Compression molding allows for the incorporation of a wide range of additives to improve the film's mechanical and barrier properties.

13.9.2.2 Disadvantages

- Process complexity: This method requires specialized equipment for temperature and pressure control, making the process more complex and costly compared to casting.
- Format limitations: Compression molding may provide less versatility in terms of the shapes and dimensions of the films produced.
- Energy cost: The use of high heat and pressure increases the energy cost associated with the process, potentially limiting its feasibility for large-scale industrial applications.

This method is particularly well suited for producing films that require high mechanical strength and greater thickness, especially in situations where durability is critical, or the film must endure significant stress.

13.9.3 Extrusion

Extrusion is an industrial process utilized for large-scale film production, as illustrated in Fig. 13.4. This process involves pushing a heated gluten mixture through a die under high pressure, resulting in a continuous film that is then cooled and collected. This method allows for the production of films that exhibit good mechanical strength and uniformity while being both cost-effective and time-efficient for high-volume manufacturing (Tanada-Palmu et al., 2000). The application of heat and pressure during extrusion fosters the development of a cohesive gluten network, which enhances the film's barrier properties and durability (Guilbert et al., 1998).

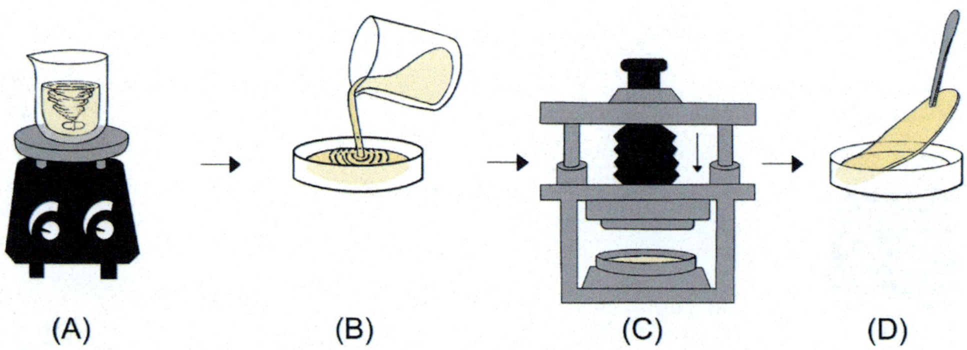

FIGURE 13.3 Compression molding for gluten film production.

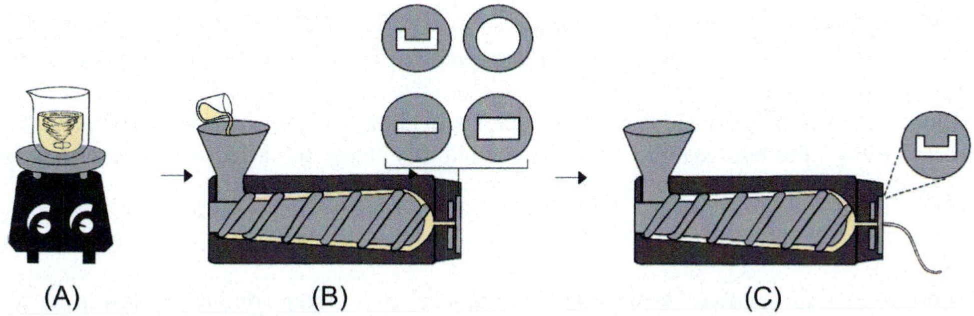

FIGURE 13.4 Gluten film production via extrusion.

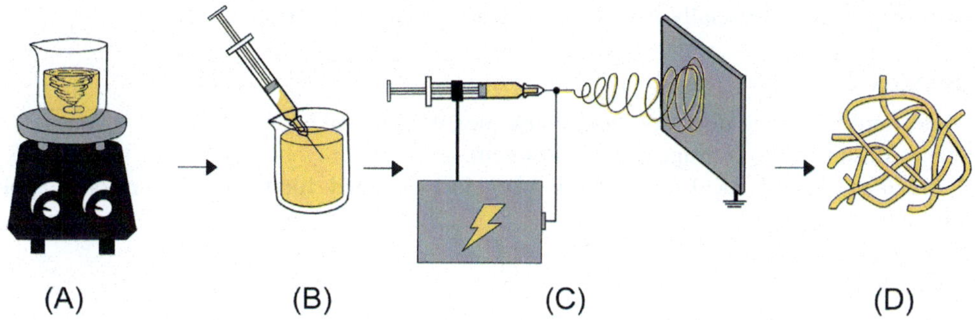

FIGURE 13.5 Gluten film production via electrospinning.

13.9.3.1 Advantages

- High productivity: Efficient for mass production, facilitating continuous, large-scale manufacturing.
- Property control: Offers precise control over film attributes such as thickness, composition, and texture.
- Versatility: Capable of producing films in variety of shapes and thicknesses.
- Application: This method is ideal for standardized food packaging, including bags, and sheets.

13.9.3.2 Disadvantages

- Complexity and cost: Requires expensive, intricate equipment that consumes a significant amount of energy.
- Parameter adjustments: Success hinges on precise adjustments of temperature, pressure, and extrusion speed.

13.9.4 Electrospinning

The electrospinning process, illustrated in Fig. 13.5, is an innovative technique for producing gluten films and coatings at the nanometric scale (Zhang et al., 2023). In this method, a gluten solution is electrically charged and ejected through a needle under high voltage toward a grounded collector. As the solution is expelled, it forms ultra-thin fibers that deposit on the collector, resulting in a fibrous matrix. Electrospinning enables the production of films with a high surface area, enhancing barrier properties and allowing for the incorporation of bioactive compounds, such as antioxidants and antimicrobials. This technique is particularly advantageous for creating functional coatings due to its ability to form porous, flexible structures that can be directly applied to fresh foods, thereby extending their shelf life (Zink et al., 2016).

13.9.4.1 Advantages

- Nanometric films: Enable the production of extremely thin films with enhanced barrier properties.
- Bioactive incorporation: The porous structure facilitates the incorporation of antioxidants and antimicrobials, enhancing functionality.
- Application flexibility: Capable of creating coatings that can be applied directly to food.

13.9.4.2 Disadvantages

- Technical complexity: Requires specialized equipment and precise control over operational parameters, such as voltage and solution viscosity.
- Limited scalability: Currently more suited for small-scale applications or research and development.
- Environmental sensitivity: Process stability and reproducibility may be affected by changes in humidity and temperature.

13.9.5 Dip coating

Dip coating, shown in Fig. 13.6, involves immersing the food item in a gluten solution, followed by the removal of the solvent by evaporation to create a continuous coating around the product. This straightforward and effective technique is particularly suitable for coating irregularly shaped foods, such as whole fruits or meat cuts. While it is not a method for producing films, it serves as an application technique utilizing casting-based coatings (Gennadios et al., 1993). Dip coating ensures uniform coverage and is especially beneficial for fresh produce (Bourtoom, 2008).

13.9.5.1 Advantages

- Simplicity: Easy to implement and does not require complex equipment.
- Complete coverage: Ensures full coverage of the food item.
- Suitability for irregular shapes: Ideal for coating irregularly shaped foods, where uniform application may be challenging with other methods.

13.9.5.2 Disadvantages

- Limited thickness control: Achieving precise control over coating thickness can be difficult.
- Drying time: Requires time for solvent evaporation in order to solidify the coating.
- Material waste: Some gluten solution may be wasted during the dipping process, particularly in large-scale operations.
- Dip coating is particularly well suited for whole or segmented food items, such as fruits and vegetables.

13.10 Modifying the properties of gluten films and coatings

A variety of strategies are employed to modify the mechanical, barrier, and functional properties of gluten-based films and coatings using additives such as plasticizers, lipids, crosslinking agents, and nanotechnology. Each modification technique is rooted in specific chemical principles that alter the structure and behavior of gluten proteins, leading to enhanced film properties. Below, we explore these modifications, focusing on their mechanisms of action, chemical effects, and the latest research in the field.

13.10.1 Use of plasticizers

Plasticizers are LMW molecules that embed themselves between gluten protein chains, reducing intermolecular interactions such as van der Waals forces and hydrogen bonds. This process increases the mobility of protein chains,

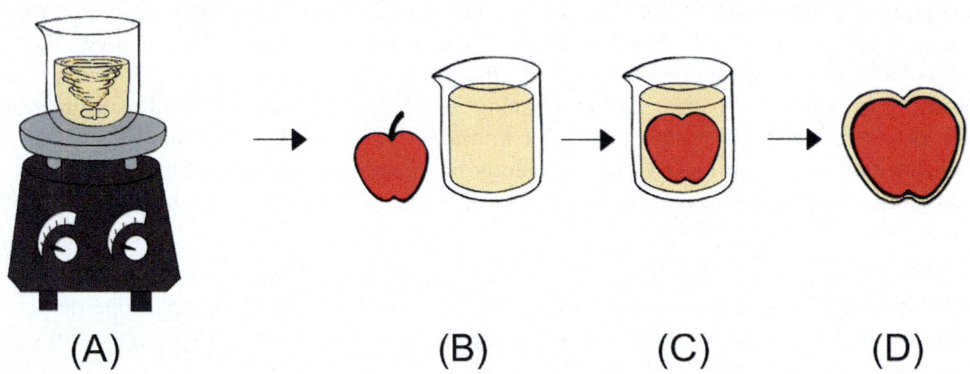

FIGURE 13.6 Dip coating application.

resulting in decreased film rigidity and enhanced flexibility. For instance, glycerol, a triol, demonstrates a plasticizing effect by disrupting intermolecular interactions through the formation of hydrogen bonds with the side chains of gluten amino acids. This modification enhances flexibility without significantly compromising the film's barrier properties (Zhang et al., 2020). Additionally, this interaction reduces the internal cohesion of the protein network, producing a more pliable and flexible material (Sartori et al., 2018).

Glycerol: Glycerol ($C_3H_8O_3$) is the most widely used plasticizer due to its effectiveness, low cost, and nontoxic nature. Incorporating glycerol into gluten films enhances flexibility, making it ideal for applications that require adaptable packaging, such as fresh food wraps. However, there is a maximum limit to the amount of glycerol that can be added; excessive amounts can lead to sticky films with reduced mechanical strength. Research indicates that finding the optimal glycerol dosage is crucial for balancing flexibility and strength while avoiding overly sticky material (Sharma et al., 2017).

Triethylene glycol: With a larger molecular structure, triethylene glycol ($C_6H_{14}O_4$) provides enhanced film stability over time, reducing rigidity and demonstrating a lower tendency for plasticizer migration. This characteristic is particularly valuable in applications where film durability is essential, such as long-term packaging. Triethylene glycol has also been shown to preserve film flexibility without compromising mechanical strength, making it a viable alternative to glycerol in certain packaging contexts (Gennadios et al., 1993).

Current research focuses on developing new plasticizers derived from natural sources, such as epoxidized vegetable oils, which can improve the biodegradability and functionality of gluten films. These innovative plasticizers not only enhance film flexibility but also improve barrier properties, making them more suitable for sustainable packaging applications. Additionally, there is increasing interest in exploring the synergistic effects of combining different plasticizers, such as glycerol and triethylene glycol, to enhance mechanical and barrier properties without sacrificing strength or durability (Assad et al., 2020; Bourtoom, 2008).

13.10.2 Addition of lipids

Incorporating lipids into gluten films aims to enhance their barrier properties, particularly by reducing water vapor permeability. Due to their hydrophobic nature, lipids disperse within the gluten protein matrix, forming physical barriers that impede the passage of water molecules (Chen et al., 2019; Rocca-Smith, Marcuzzo, et al., 2016). The hydrophobic nature of lipids creates microdomains within the film, repelling water and thereby reducing water vapor permeability. This effectively enhances the film's resistance in high-humidity environments (Rocca-Smith, Marcuzzo, et al., 2016).

Beeswax: Beeswax is a naturally occurring wax produced in the bee's hives by honeybees *A. mellifera* (Tinto et al., 2017). Beeswax is highly hydrophobic and, when incorporated into gluten films, forms a continuous barrier against water vapor. This property is particularly advantageous for packaging products with high-moisture content, such as cheese and fresh meats. Additionally, due to its solid nature at room temperature, beeswax contributes to the rigidity and structural stability of the film (Bourtoom, 2008).

Fatty acids (stearic acid): Fatty acids, such as stearic acid ($C_{18}H_{36}O_2$), are incorporated into gluten films to enhance flexibility and water resistance. The structure of fatty acids facilitates effective integration into the protein matrix, creating a barrier against water vapor permeability while preserving the film's elasticity. This characteristic is crucial for applications that require flexible packaging, such as wraps for fresh fruits and vegetables (Gontard et al., 1994).

Current research focuses on utilizing functional lipids, such as those derived from essential oils, to provide additional benefits, like antimicrobial and antioxidant properties (Cha & Chinnan, 2004). Furthermore, methods for dispersing lipids at the nanoscale are being explored to improve the uniformity of the hydrophobic barrier and enhance the efficiency of the film (Rovera et al., 2020).

13.10.3 Chemical treatments

Chemical treatments involve inducing crosslinking between protein molecules to enhance film cohesion and mechanical strength. This process can be represented by the interaction between functional groups in gluten proteins, such as hydroxyl (-OH) and carboxyl (-COOH) groups, mediated by crosslinking agents like tannic acid (Gennadios et al., 1993; Wang, 2022). This chemical reaction results in the formation of a denser, more durable three-dimensional network that is less prone to solubility in water and more resistant to adverse environmental conditions (Zhang et al., 2020).

Tannic acid: Tannic acid ($C_{76}H_{52}O_{46}$) is a polyphenol that can crosslink gluten proteins, significantly enhancing mechanical strength and film durability. These properties are particularly beneficial for packaging that requires additional protection against mechanical stress and oxidation, such as fresh foods and meat products (Zhang et al., 2020).

Gallic acid: Similar to tannic acid, gallic acid ($C_7H_6O_5$) contributes to the formation of stronger films with enhanced antioxidant properties, making it advantageous for packaging oxidation-prone foods, such as oils and baked goods (Zhang et al., 2020).

Research is ongoing to identify new crosslinking agents that can impart additional benefits, such as antimicrobial activity and pH sensitivity (Xu & Li, 2023; Zhang et al., 2020). Furthermore, there is increasing interest in physical treatments, such as UV irradiation or controlled thermal processes, which can induce crosslinking without the need for chemical additives (Zuo & Dai, 2009).

13.11 Nanotechnology in film applications

Nanotechnology involves incorporation of nanoparticles or creation of nanocomposites that can significantly enhance the functional properties of gluten films and coatings. These modifications introduce nanoscale particles into the protein matrix, allowing for molecular interactions that improve properties such as mechanical strength, permeability, and active functionalities, including antimicrobial and antioxidant activity (Rezaei et al., 2020; Zhang et al., 2020).

Silver nanoparticles (AgNPs): AgNPs are well-known for their antimicrobial properties. When integrated into gluten films, AgNPs interact with microbial cell membranes, leading to dysfunction and cell death. This characteristic is particularly valuable for active packaging solutions for perishable foods, where ensuring food safety is a priority (Xu & Li, 2023; Zhang et al., 2020).

Nanocomposites: The combination of gluten with other biopolymers at the nanoscale, such as nanoclays or carbon nanotubes, can yield materials with significantly enhanced mechanical and barrier properties. These nanocomposites form a denser, more cohesive structure that reduces water vapor and gas permeability while increasing both tensile and flexural strength (Rovera et al., 2020).

Current research is investigating the incorporation of additional nanoparticles, such as zinc oxide (ZnO) and titanium dioxide (TiO_2), which provide benefits like UV protection and antioxidant effects. Techniques are being developed to effectively disperse and distribute these nanoparticles within the protein matrix to maximize efficacy and minimize the risks of agglomeration (Zhang et al., 2020).

Moreover, applications of nanotechnology in gluten films are evolving to include advanced functionalities, such as self-healing properties and spoilage detection. These innovations have the potential to transform traditional food packaging into smart, active solutions.

13.12 Applications of gluten films and coatings

The effectiveness of gluten films and coatings is influenced by their physicochemical properties and the ability to tailor these properties for various applications. Below, we detail their primary applications along with a critical analysis of their advantages, limitations, and current research gaps.

13.12.1 Biodegradable packaging

One of the most promising applications for gluten films is their use in biodegradable packaging. Due to their natural origin and biodegradability, gluten films offer an eco-friendly alternative to conventional plastics, which are widely recognized for their environmental impact due to persistence and accumulation in ecosystems (Assad et al., 2020; Guilbert et al., 1998). However, using gluten films for food packaging presents significant challenges. Water vapor permeability and mechanical fragility are the primary obstacles. While the mechanical strength of these films is suitable for some applications, further enhancements are needed to meet the demands of more rigorous uses (Li et al., 2023).

Strategies such as lipid incorporation and nanotechnology applications are being explored to address these challenges; however, long-term studies validating these approaches under real-world conditions are still lacking (Xu & Li, 2023). Current research aims to enhance the barrier and strength properties of gluten films, but there is a notable shortage of studies assessing the economic feasibility of these modifications on an industrial scale (Sharma et al., 2017). Furthermore, the compatibility of these films with various food types, especially those with high-moisture content, requires additional investigation (Guilbert et al., 1998). Controlled degradation after use is another underexplored area, which is essential for ensuring that these films fulfill their environmental purpose without generating new waste problems (Zink et al., 2016).

13.13 Shelf life extension for fruits and vegetables

Gluten films and coatings have proven highly effective in extending the shelf life of fresh fruits and vegetables. These coatings serve as physical barriers that limit moisture loss and oxygen uptake, both of which contribute to postharvest deterioration (Assad et al., 2020; Han et al., 2005). When applied directly to the surfaces of fruits and vegetables, gluten coatings slow down dehydration and oxidation processes, thus preserving the freshness of the products for a longer period (Guilbert et al., 1998). The primary advantage of gluten coatings for postharvest preservation is their ability to form a barrier without significantly altering the appearance or taste of the product (Mangavel et al., 2004). However, the effectiveness of these coatings can vary depending on the type of product and storage conditions. For instance, in fruits with thin skins, such as strawberries, gluten coatings may be less effective due to their water vapor permeability (Rocca-Smith, Marcuzzo, et al., 2016). Additionally, enhancing the moisture resistance of these films is essential for their application in high-humidity or refrigerated storage conditions (Sartori et al., 2018). Despite advancements, there remains a lack of standardization in the application and evaluation methods for gluten coatings applied to fruits and vegetables. Long-term field studies are necessary to assess the real-world effectiveness of these coatings in various storage and transport environments (Wang, 2022). Furthermore, the interaction between gluten coatings and other preservation technologies, such as modified atmosphere packaging, warrants further investigation to improve the preservation of fresh products (Vargas et al., 2012).

Figs. 13.7 and 13.8 illustrate the application of gluten- and lignocellulose-based coatings on fruits, highlighting their impact on conservation and mechanical resistance.

Figs. 13.9 and 13.10 show the hybrid gluten and lignocellulose film, and lychees packaged in this film, respectively.

The images demonstrate how gluten-based coatings and films offer sustainable and effective solutions for preserving fresh fruits. These coatings and films not only enhance resistance to physical impacts but also provide moisture barriers and antimicrobial properties (Chen et al., 2022).

13.14 Applications in baking

In the baking industry, gluten films are primarily used as coatings for bread and cakes, helping to retain moisture and maintain the freshness of baked goods. These coatings form a barrier that prevents moisture loss, ensuring the texture of baked products remains soft for extended periods (Gennadios et al., 1993; Guilbert et al., 1998). Additionally, these

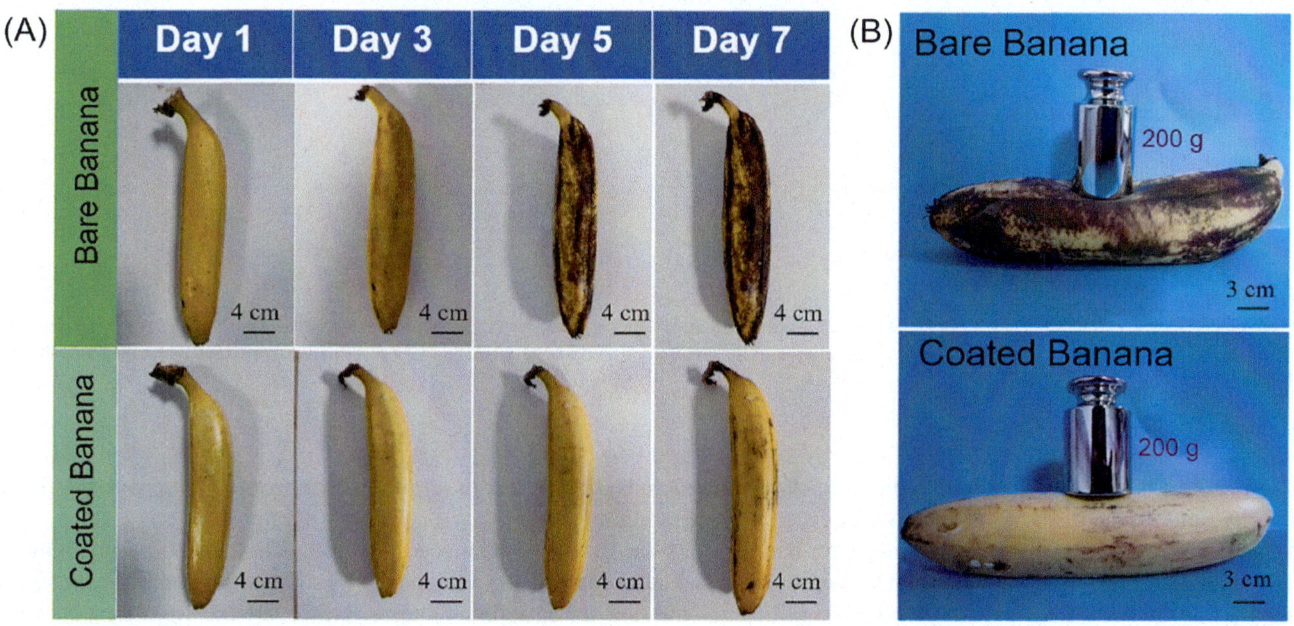

FIGURE 13.7 Gluten and lignocellulose coating on bananas.
From Chen, Y., Li, Y., Qin, S., Han, S., & Qi, H. (2022). Antimicrobial, UV blocking, water-resistant and degradable coatings and packaging films based on wheat gluten and lignocellulose for food preservation. Composites Part B: Engineering, 238, 109868. https://doi.org/10.1016/j.compositesb.2022.109868.

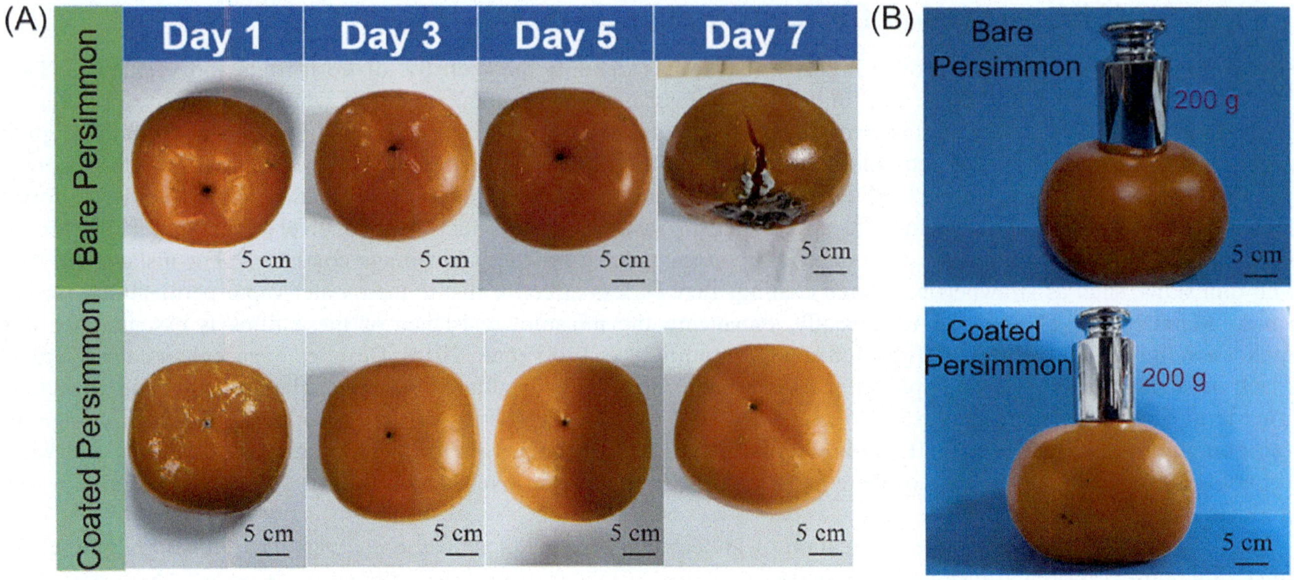

FIGURE 13.8 Gluten and lignocellulose coating on persimmons.
From Chen, Y., Li, Y., Qin, S., Han, S., & Qi, H. (2022). Antimicrobial, UV blocking, water-resistant and degradable coatings and packaging films based on wheat gluten and lignocellulose for food preservation. Composites Part B: Engineering, 238, 109868. https://doi.org/10.1016/j.compositesb.2022.109868.

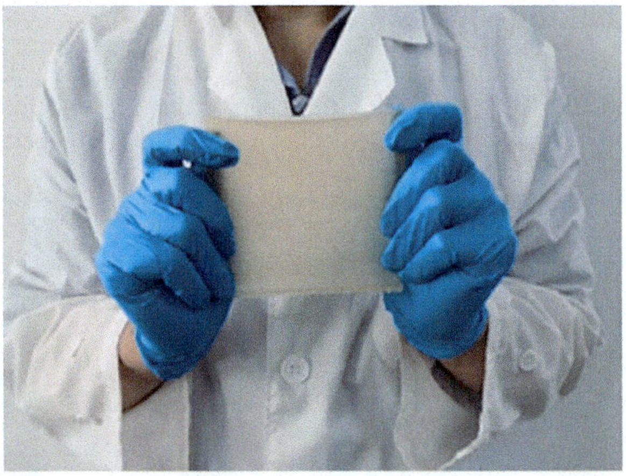

FIGURE 13.9 Photograph of the hybrid film (gluten and lignocellulose).
From Chen, Y., Li, Y., Qin, S., Han, S., & Qi, H. (2022). Antimicrobial, UV blocking, water-resistant and degradable coatings and packaging films based on wheat gluten and lignocellulose for food preservation. Composites Part B: Engineering, 238, 109868. https://doi.org/10.1016/j.compositesb.2022.109868.

coatings can be enhanced with ingredients such as antioxidants, flavorings, or antimicrobial agents to improve the flavor, quality, and safety of the products (Assad et al., 2020; Han et al., 2005).

Gluten films are particularly well suited for baked goods due to their proteinaceous origin, which is compatible with the wheat proteins (wheat flour) commonly used in most baked products. This compatibility allows the coating to integrate seamlessly into the final product without significantly altering its structure or taste. However, the fragility and water solubility of gluten films present challenges, particularly for products that need to be stored for extended periods or under high-humidity conditions (Guilbert et al., 1998).

Furthermore, the oxygen barrier provided by these films is limited, which may not adequately protect products containing fats sensitive to oxidation, such as cream-topped cakes (Kozłowska et al., 2019). There is a need for more detailed studies on the interactions between gluten films and various baking ingredients, as well as knowledge of how

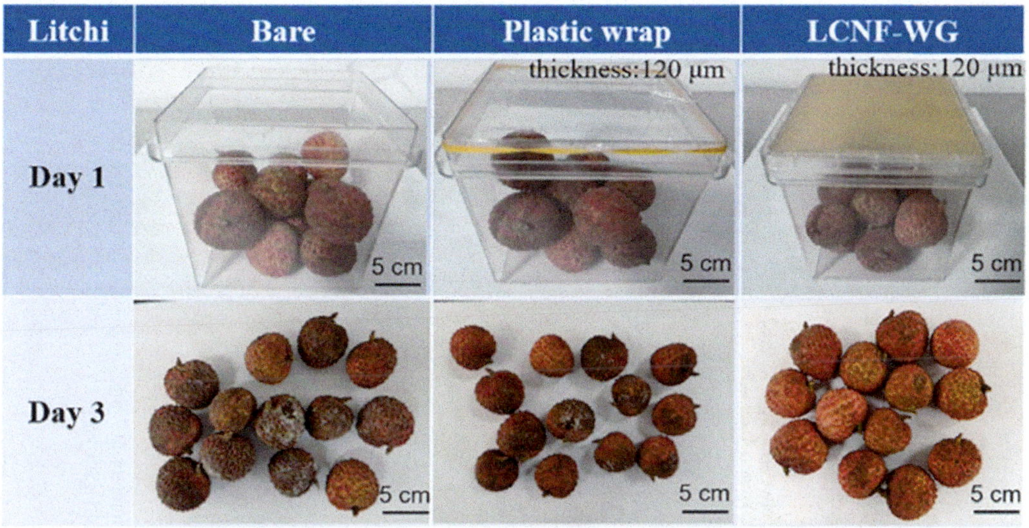

FIGURE 13.10 Images of lychees stored for three days.
From Chen, Y., Li, Y., Qin, S., Han, S., & Qi, H. (2022). Antimicrobial, UV blocking, water-resistant and degradable coatings and packaging films based on wheat gluten and lignocellulose for food preservation. Composites Part B: Engineering, 238, 109868. https://doi.org/10.1016/j.compositesb.2022.109868.

these interactions affect the overall quality of the final product (Ortolan & Steel, 2017). Additionally, the economic feasibility of implementing these approaches on a large scale requires further investigation (Wang, 2022).

Table 13.1 summarizes the key technical parameters and applications of gluten-based films and coatings, demonstrating that composition and fabrication methods significantly influence mechanical, thermal, and barrier properties. The use of phenolic compounds, such as tannic and gallic acids, has a direct impact on both mechanical and permeability properties. These compounds can function as plasticizers or crosslinkers, depending on their concentration, leading to variations in tensile modulus (ranging from 0.53 to 2.66 MPa) and elongation at break (from 3.59% to 296.8%).

Combining gluten with lipids and emulsifiers such as stearic acid, lecithin, and glycerol monostearate enhances tensile strength and elasticity, resulting in elongations of up to 150%. This characteristic makes the films particularly suitable for wrapping high-moisture foods or products with irregular surfaces (Rocca-Smith, Marcuzzo, et al., 2016). Additionally, the incorporation of lignocellulosic nanostructures has improved oxygen barrier properties of gluten films, making them effective in applications such as coatings and packaging for fruits, including bananas and lychees (Chen et al., 2022).

13.14.1 Active packaging

Gluten films can be modified to function as active packaging, serving not only as physical barriers but also interacting with the food to extend its shelf life (Ansorena et al., 2016). Incorporating antimicrobial and antioxidant agents into gluten films represents a promising strategy for inhibiting microbial growth and delaying the oxidation of oxygen-sensitive foods (Gómez-Heincke et al., 2016).

However, significant technical challenges remain, including achieving the uniform distribution of active agents within the film and ensuring their controlled release. Premature or excessive release of antimicrobial agents can reduce their efficacy and negatively impact the flavor and safety of packaged foods (Bourtoom, 2008).

More research is needed to improve encapsulation technologies for active agents within gluten films, ensuring their controlled and effective release over time (Chen et al., 2019). Additionally, further exploration of how these active films interact with various food types and storage conditions is necessary to verify their efficacy and safety (Wang, 2022).

13.14.2 Smart packaging

Smart packaging is an emerging application that combines gluten films with chemical sensors capable of indicating food freshness, pH changes, or the presence of gases such as ethylene (Rovera et al., 2020). This technology is particularly beneficial for fresh fruits and vegetables, where spoilage can often go undetected until visible signs appear (Zhang et al., 2020).

TABLE 13.1 Characteristics and applications of gluten-based films and coatings.

Film/coating composition	Application	Method	Properties	Water vapor permeability *(WVP)*	References
Gluten, glycerol, cinnamon essential oil, or turmeric essential oil	Baby carrots	Casting	Thermal degradation of the matrix beginsat 290°C	Between 1.44 and 1.75 (g·mm $[kPa \cdot h \cdot m^2]^{-1}$)	Rivera Leiva et al. (2022)
Gluten, glycerol, thyme essential oil	-	Extrusion	Tensile modulus: 7.8–30 MPa Tensile strength: 1.1–2.4 MPa Elongation at break: 22.7%–54.1%	Between 1.50 and 2.62 ($\times 10^{10}$) g $(Pa \cdot s \cdot m)^{-1}$	Ansorena et al. (2016)
Gluten, glycerol, lignocellulosic nanofibers (LCNF)	Banana, persimmon, and grape	Dip coating	Barrier (O_2): 15.9 $cm^{-3} \cdot \mu m \cdot m^{-2} \cdot day^{-1} \cdot kPa^{-1}$	-	Chen et al. (2022)
Gluten, glycerol, LCNF	Lychee	Casting	Tensile modulus: 10.2 MPa Elongation at break: 124%	10.3 $g \cdot mm^{-1} \cdot m^{-2} \cdot day^{-1}$	
Gluten, glycerol, phenolic compounds	Salmon	Casting	Tensile modulus: 6.70–8.30 MPa	Between 5.06 and 7.46 g·mm $(m^2 \cdot day \cdot kPa)^{-1}$	Yu and Xu (2024)
Gluten, glycerol, ethanol, and hydrochloric acid gluten film + lipid, mono- and diglyceride emulsifier, beeswax	-	Casting	Tensile modulus: 0.1–20 MPa Tensile strength: 0.8–30 MPa Elongation at break: 0.8%–150%	-	Rocca-Smith, Marcuzzo, et al. (2016)
Gluten, sodium sulfite, ethanol, glycerol, lipids with beeswax and/or DATEM	Hamburger bun	Casting	Tensile modulus: 3.51–8.96 MPa Elongation at break: 3.9%–60.10%	Between 1.24 and 3.25 (g·mm $[m^2 \cdot day \cdot mmHg]^{-1}$)	Mohammadi, Zoghi and Azizi (2023)
Gluten, banana fiber	Production of boxes, cups, and plates	Casting + hydraulic compression molding	Tensile modulus: 11–13 MPa	-	Nataraj et al. (2018)
Gluten, TEOS alkoxide, acetic acid	Paper (cellulosic substrates)	Casting	-	Between 9.7 and 881 g $(m^2 \cdot day)^{-1}$	Rivera Leiva et al. (2022)

Recent advances in this area include the development of a biodegradable gluten film infused with chlorophyll and polypyrrole, which exhibited changes in color and conductivity during storage, enabling real-time monitoring of food quality (Chavoshizadeh et al., 2020a). Similarly, films containing thyme essential oil and chlorophyll have proven effective as active packaging, inhibiting lipid oxidation in oils while indicating the validity of the food (Chavoshizadeh et al., 2020b). Another example involved the incorporation of *Heracleum persicum* essence, magnesium oxide, and polypyrrole into a gluten film, resulting in a material with antioxidant properties and electrical conductivity, making it suitable for use as active and intelligent packaging for perishable foods (Fazeli, 2021). Challenges in this field include effectively integrating chemical sensors into gluten films without compromising their physical or barrier properties (Wang, 2022).

Furthermore, ensuring the reliability and accuracy of sensor readings across varying storage conditions and for different types of food remains a significant technical challenge (Xu & Li, 2023).

13.15 Biomedical applications

Beyond packaging, gluten films are being investigated for biomedical applications, such as controlled drug delivery systems and wound dressings. The biodegradability and biocompatibility of gluten make it an ideal candidate for these applications (Assad et al., 2020). For instance, gluten films loaded with therapeutic agents can be engineered to release medications in a controlled manner over time, providing an innovative solution for medical treatments (Mangavel et al., 2004). However, research in this area is still in its early stages, and further studies are needed to ensure the efficacy and safety of these materials in clinical settings (Wang, 2022).

13.16 Challenges and limitations

Gluten films and coatings demonstrate significant potential for the sustainable packaging industry, but they also face challenges and limitations that must be addressed to facilitate broader adoption. Among the most critical challenges are the mechanical fragility and high water solubility of gluten films, which stem from the intrinsic characteristics of gluten's protein structure. The following sections will examine these challenges in detail and discuss strategies developed to mitigate them.

13.16.1 Mechanical fragility of gluten films and coatings

The fragility of gluten films is directly linked to the protein structure of the gluten network, which is primarily composed of gliadins and glutenins. The glutenin network provides strength and elasticity and is stabilized by covalent bonds, particularly disulfide bonds (–S–S–). Under low-humidity conditions, these bonds can lead to a rigid and brittle structure, as the lack of water reduces protein chain mobility and increases internal network tension (Rovera et al., 2020).

Moreover, insufficient levels of plasticizers can result in excessively rigid films. In the absence of agents that disrupt intermolecular interactions, the attractive forces between glutenin chains intensify, creating materials that are prone to breaking under mechanical stress, especially in dry conditions (Sartori et al., 2018). This fragility is a consequence of the strong internal cohesion of the protein network when adequate plasticizers are lacking (Patni et al., 2014). Adding plasticizers, such as glycerol, is a common strategy to reduce fragility. Glycerol inserts itself between protein chains, diminishing intermolecular interactions and enhancing film flexibility. Chemically, glycerol forms hydrogen bonds with the hydroxyl (-OH) groups of the proteins, disrupting direct interactions between glutenin chains (Chen et al., 2019).

However, the addition of plasticizers can also have adverse effects, such as reducing gas barrier properties. By increasing the mobility of protein chain, plasticizers can enhance the film's permeability to oxygen and other gases, which can be detrimental to food preservation applications (Sartori et al., 2018). This occurs because plasticizers like glycerol decrease the density of the protein network, facilitating greater diffusion of gas molecules (Chen et al., 2019). Therefore, achieving a balance between flexibility and barrier functionality is critical, as increased flexibility may compromise the film's effectiveness in protecting food from external factors.

13.16.2 Water solubility of gluten films and coatings

The high water solubility of gluten films poses a significant challenge, particularly in applications involving high-moisture foods. This solubility arises from the hydrophilic nature of gluten's protein components, especially gliadins, which contain numerous polar groups, such as -OH and $-NH_2$, that readily form hydrogen bonds with water molecules.

These functional groups promote interactions with water, resulting in increased moisture absorption by the film (Zuo & Dai, 2009). In high-humidity environments, these interactions lead to swelling and eventual dissolution of the protein matrix, compromising the film's structural integrity (Sartori et al., 2018). Thus, modifying the protein matrix is essential to enhance water resistance (Rovera et al., 2020). One common strategy to reduce water solubility is the incorporation of hydrophobic compounds, such as lipids and waxes, into the gluten matrix.

Lipids like beeswax and fatty acids are water-insoluble and can create a physical barrier within the film structure, preventing water penetration (Bourtoom, 2008). The hydrophobic properties of lipids generate microdomains within the film, significantly reducing water vapor permeability and enhancing water resistance (Guilbert et al., 1998). The addition of waxes, in particular, improves the film's integrity in humid environments without compromising its mechanical properties (Rocca-Smith, Marcuzzo, et al., 2016).

However, achieving uniform lipid distribution and preventing lipid migration to the film surface during storage remain significant challenges that require further investigation (Fakhouri et al., 2018). Another effective strategy involves inducing chemical crosslinking using agents such as tannic acid. This phenolic compound reacts with the amino and carboxyl groups of gluten proteins, forming crosslinks that enhance the internal cohesion of the protein network and reduce water solubility (Nur Hanani et al., 2014; Zink et al., 2016). Crosslinking also improves the mechanical strength of the film, resulting in a more robust material that is less prone to water dissolution (Zink et al., 2016). However, it is essential to carefully control chemical treatments to avoid compromising flexibility and other desirable properties (Gennadios et al., 1993; Mangavel et al., 2004).

13.16.3 Crosslinking

Using crosslinking agents such as tannic acid and gallic acid is another effective strategy for enhancing the mechanical strength and durability of gluten films (Esmaeili et al., 2021; Gennadios et al., 1993). These agents react with the functional groups in the proteins, forming a denser and more cohesive three-dimensional network (Nur Hanani et al., 2014). However, the introduction of crosslinks must be carefully controlled, as excessive crosslinking can result in a material that is overly rigid and brittle, thereby compromising the film's flexibility (Guilbert et al., 1998).

While gluten films and coatings have significant potential as sustainable packaging materials, challenges related to mechanical fragility and water solubility continue to limit their large-scale application. Solutions such as adding plasticizers, incorporating hydrophobic compounds, and inducing chemical crosslinking provide promising pathways; however, each approach carries its own limitations and challenges (Mangavel et al., 2004; Wang, 2022).

13.17 Current research and future trends

Recent research on gluten-based films and coatings focuses on addressing the material's intrinsic challenges, such as mechanical fragility and high water solubility, while also exploring new applications and combinations with other materials to enhance functionality. Below are the key innovations and emerging trends from the past five years.

13.18 Combining gluten with other biopolymers

A significant area of research involves combining gluten with other biopolymers, such as chitosan, alginate, and pectin. These biopolymers offer complementary properties that can help mitigate the limitations of gluten and expand its range of applications (Assad et al., 2020).

Chitosan: Chitosan is obtained by N-deacetylation of chitin (poly-β-(1→4)-N-acetyl-d-glucosamine), which is a major component of the shells of crustaceans and insects. It is a natural polysaccharide that has become of great interest to both the scientific community and the food industry due to its multiple possible applications, which include the formation of biodegradable films and coatings (Vargas et al., 2012). Known for its natural antimicrobial properties, chitosan is often combined with gluten to create films that not only protect food but also inhibit microbial growth (Han et al., 2005). Furthermore, chitosan enhances the mechanical strength and gas barrier properties of the films, making them more robust and functional (Patni et al., 2014).

Alginate: Alginates are naturally occurring polysaccharides commonly produced by, and refined from, various genera of brown algae. The molecular structure of alginates is composed of unbranched, linear binary copolymers of β-d-mannuronic acid (M) and α-l-guluronic acid (G) residues linked by 1–4 glycosidic bonds (Parreidt et al., 2018).

Another widely studied biopolymer, alginate, improves the flexibility and strength of gluten films (Rocca-Smith, Marcuzzo, et al., 2016). The combination of gluten and alginate produces films with enhanced gas and water vapor barrier properties, which are essential for preserving food in high-humidity environments (Zuo & Dai, 2009).

Emerging applications for gluten films and coatings are extending beyond the food industry, with research investigating their use in innovative areas such as smart packaging, functional edible films, and even biomedical applications (Assad et al., 2020; Wang, 2022).

13.19 Legal and regulatory issues

The regulation of gluten-based films and coatings is still under discussion in many regions worldwide.

United States: The Food and Drug Administration (FDA) evaluates materials used in food packaging for safety and efficacy. While gluten is generally recognized as safe (GRAS), gluten-based films incorporating new additives, such as nanoparticles or antimicrobial agents, may require additional assessment to ensure consumer safety.

European Union: The European Food Safety Authority (EFSA) regulates materials that come into contact with food, including biopolymer-based films and coatings. The introduction of new technologies, such as nanotechnology in gluten-based packaging, may necessitate updated regulatory guidelines to ensure that these materials do not cause risks to human health.

Future research should focus on strategies for scalability, environmental impact assessments, and functional improvements in the properties of gluten films. Additionally, exploring new applications in areas such as biomedicine and smart packaging can further expand the potential of these materials (Wang, 2022).

13.20 Conclusion

This chapter has examined the potential of gluten-based films and coatings as promising alternatives to conventional plastics, with a specific emphasis on their applications in the food industry. Gluten films provide significant advantages, including biodegradability, abundant availability of raw materials, and the capacity to modify their physicochemical properties. The mechanical strength and gas barrier properties of these films can be significantly enhanced through the addition of plasticizers, lipids, and crosslinking agents, while challenges associated with water solubility can be mitigated by incorporating hydrophobic compounds and utilizing nanotechnology. However, despite these benefits, considerable challenges remain, such as mechanical fragility in dry conditions and high water solubility, which need to be addressed to facilitate widespread commercial adoption. The modification strategies discussed in this chapter, including the combination of gluten with other biopolymers and the application of nanotechnology, offer promising solutions, but their practical implementation on an industrial scale still requires further research.

For gluten films to transition from a research topic to a commercial reality, continuous effort is needed on multiple fronts. Enhancing functional properties, such as water resistance and mechanical durability, is a priority. Additionally, addressing scalability and cost-effectiveness is essential to ensure these materials can be produced economically and sustainably on a large scale. While significant progress has been made, the field of gluten-based films and coatings still offers vast potential for exploration. Ongoing technological advancements and in-depth research hold the promise of transforming these materials from emerging solutions into foundational pillars for future sustainability. Readers are encouraged to engage more deeply in this field, contributing to the innovations that will make gluten films a tangible and impactful reality in the modern world.

Acknowledgments

This study was financed, in part, by the São Paulo Research Foundation (FAPESP), Brazil, Process Number *2024/07555-2*.

References

Ansorena, M. R., Giménez, B., Ibarguren, C., & Cian, R. (2016). Edible films based on whey proteins and fatty acids for packaging applications. *Journal of Food Science and Technology, 53*(5), 2033–2041.

Assad, M., Tien, M., Lacroix, & Vu, K. D. (2020). Protein-based packaging of plant origin: Sustainable solutions for food industry and biomedical applications. *Journal of Food Packaging and Preservation, 45*(3), 65–80.

Bourtoom, T. (2008). Edible films and coatings: Characteristics and properties. *International Food Research Journal, 15*(3), 237–248. http://www.ifrj.upm.edu.my.

Cha, D. S., & Chinnan, M. S. (2004). Biopolymer-based antimicrobial packaging: A review. *Critical Reviews in Food Science and Nutrition, 44*(4), 223–237. https://doi.org/10.1080/10408690490464276.

Chavoshizadeh, S., Pirsa, S., & Mohtarami, F. (2020a). Conducting/smart color film based on wheat gluten/chlorophyll/polypyrrole nanocomposite. *Food Packaging and Shelf Life, 24*. https://doi.org/10.1016/j.fpsl.2020.100501.

Chavoshizadeh, S., Pirsa, S., & Mohtarami, F. (2020). Sesame oil oxidation control by active and smart packaging system using wheat gluten/chlorophyll film to increase shelf life and detecting expiration date. *European Journal of Lipid Science and Technology, 122*(3). https://doi.org/10.1002/ejlt.201900385.

Chen, H., Wang, J., Cheng, Y., Wang, C., Liu, H., Bian, H., Pan, Y., Sun, J., & Han, W. (2019). Application of protein-based films and coatings for food packaging: A review. *Polymers, 11*(12). https://doi.org/10.3390/polym11122039.

Chen, Y., Li, Y., Qin, S., Han, S., & Qi, H. (2022). Antimicrobial, UV blocking, water-resistant and degradable coatings and packaging films based on wheat gluten and lignocellulose for food preservation. *Composites Part B: Engineering, 238*, 109868. https://doi.org/10.1016/j.compositesb.2022.109868.

Esmaeili, A., Hashemi, M., & Badii, F. (2021). Effect of tannic acid on functional properties of edible films based on whey protein and basil seed gum. *Journal of Food Science and Technology, 58*(9), 3631–3639.

Fakhouri, F. M., Martelli, S. M., Caon, T., Velasco, J. I., Buontempo, R. C., Bilck, A. P., & Innocentini Mei, L. H. (2018). The effect of fatty acids on the physicochemical properties of edible films composed of gelatin and gluten proteins. *LWT, 87*, 293–300. https://doi.org/10.1016/j.lwt.2017.08.056.

Fazeli, S. (2021). Nanocomposite film based on gluten modified with *Heracleum persicum* essence, magnesium oxide, and polypyrrole for food packaging. *Food Hydrocolloids, 119*.

Gontard, Nathalie, Duchez, Christophe, Cuq, Jean-Louis, & Guilbert, Stéphane (1994). Edible composite films of wheat gluten and lipids: water vapour permeability and other physical properties. *International Journal of Food Science & Technology. 29*(1), 39–50. https://doi.org/10.1111/j.1365-2621.1994.tb02045.x.

Gennadios, A., Gontard, N., & Guilbert, S. (1993). Physical properties of edible films made from caseinates, whey proteins, and soy protein isolates. *Journal of Food Science, 58*(2), 378–381.

Geyer, R., Jambeck, J. R., & Law, K. L. (2017). Production, use, and fate of all plastics ever made. *Science Advances, 3*(7). https://doi.org/10.1126/sciadv.1700782, http://advances.sciencemag.org/.

Gontard, N., Guilbert, S., & Cuq, J. L. (1994). Edible wheat gluten films: Influence of the main process variables on film properties using response surface methodology. *Journal of Food Science, 59*(5), 1135–1137.

Guilbert, S., Gontard, N., & Morel, M. H. (1998). Edible and/or biodegradable packaging: Challenges and opportunities. *Journal of Food Science and Technology, 33*(4), 111–118.

Gómez-Heincke, D., Martínez, I., Partal, P., Guerrero, A., & Gallegos, C. (2016). Development of antimicrobial active packaging materials based on gluten proteins. *Journal of the Science of Food and Agriculture, 96*(10), 3432–3438. https://doi.org/10.1002/jsfa.7525.

Han, J. H., Krochta, J. M., & Mulder-Johnston, D. (2005). Edible coatings and films based on milk proteins. *Food Technology, 49*(10), 62–70.

Kayserilioğlu, B. S., Stevels, W., Mulder, W., & Akkas, N. (2001). Mechanical and biochemical characterization of wheat gluten films as a function of pH and co-solvent. *European Journal of Plant Pathology, 107*(6), 637–647. https://doi.org/10.1023/A:1011224009913.

Kozłowska, M., Żbikowska, A., Szpicer, A., & Półtorak, A. (2019). Oxidative stability of lipid fractions of sponge-fat cakes after green tea extracts application. *Journal of Food Science and Technology, 56*(5), 2628–2638. https://doi.org/10.1007/s13197-019-03750-5.

Larré, C., Desserme, C., Barbot, J., & Gueguen, J. (2000). Properties of deamidated gluten films enzymatically cross-linked. *Journal of Agricultural and Food Chemistry, 48*(11), 5444–5449. https://doi.org/10.1021/jf0003505.

Li, L., Wang, Q., Liu, C., Hong, J., & Zheng, X. (2023). Effect of oven roasting on major chemical components in cereals and its modulation on flour-based products quality. *Journal of Food Science, 88*(7), 2740–2757. https://doi.org/10.1111/1750-3841.16625, http://onlinelibrary.wiley.com/journal/10.1111/(ISSN)1750-3841.

Mangavel, C., Barbot, J., Guilbert, S., & Popineau, Y. (2004). Properties and microstructure of thermoplastic films made from wheat gluten. *Journal of Applied Polymer Science, 91*(1), 620–628.

Mohammadi, M., Zoghi, A., & Azizi, M. H. (2023). Assessment of properties of gluten-based edible film formulated with beeswax and DATEM for hamburger bread coating. *Food Science & Nutrition, 11*(4), 2061–2068. https://doi.org/10.1002/fsn3.3242.

Mojumdar, S. C., Moresoli, C., Simon, L. C., & Legge, R. L. (2011). Edible wheat gluten (WG) protein films. *Journal of Thermal Analysis and Calorimetry, 104*(3), 929–936. https://doi.org/10.1007/s10973-011-1491-z.

Nataraj, D., Sakkara, S., Meenakshi, H. N., & Reddy, N. (2018). Properties and applications of citric acid crosslinked banana fibre-wheat gluten films. *Industrial Crops and Products, 124*, 265–272. https://doi.org/10.1016/j.indcrop.2018.07.076.

Nur Hanani, Z. A., Roos, Y. H., & Kerry, J. P. (2014). Use and application of gelatin as potential biodegradable packaging materials for food products. *International Journal of Biological Macromolecules, 71*, 94–102. https://doi.org/10.1016/j.ijbiomac.2014.04.027.

Ortolan, F., & Steel, C. J. (2017). Protein characteristics that affect the quality of vital wheat gluten to be used in baking: A review. *Comprehensive Reviews in Food Science and Food Safety, 16*(3), 369–381. https://doi.org/10.1111/1541-4337.12259, http://www.interscience.wiley.com/jpages/1541-4337.

Parreidt, T. S., Müller, K., & Schmid, M. (2018). Alginate-based edible films and coatings for food packaging applications. *Foods, 7*(10). https://doi.org/10.3390/foods7100170, https://www.mdpi.com/2304-8158/7/10.

Patni, N., Patel, A., & Shrimali, M. (2014). Biodegradable wheat gluten-based films: Effect of citric acid and glycerol as plasticizer on properties. *International Journal of Research in Environmental Science and Technology, 4*(1), 10–16.

Rezaei, M., Pirsa, S., & Chavoshizadeh, S. (2020). Photocatalytic/antimicrobial active film based on wheat gluten/ZnO nanoparticles. *Journal of Inorganic and Organometallic Polymers and Materials, 30*(7), 2654–2665. https://doi.org/10.1007/s10904-019-01407-6.

Rivera Leiva, A. F., Hernández-Fernández, J., & Ortega Toro, R. (2022). Active Films Based on Starch and Wheat Gluten (Triticum vulgare) for Shelf-Life Extension of Carrots. *Polymers, 14*(23), 5077. https://doi.org/10.3390/polym14235077.

Rocca-Smith, J. R., Lullien-Pellerin, V., Guilbert, S., Gontard, N., & Angellier-Coussy, H. (2016). Effect of lipid incorporation on performance of wheat gluten-based films. *Journal of Agricultural and Food Chemistry, 64*(40), 7654–7662.

Rocca-Smith, J. R., Marcuzzo, E., Karbowiak, T., Centa, J., Giacometti, M., Scapin, F., Venir, E., Sensidoni, A., & Debeaufort, F. (2016). Effect of lipid incorporation on functional properties of wheat gluten based edible films. *Journal of Cereal Science, 69*, 275–282. https://doi.org/10.1016/j.jcs.2016.04.001, http://www.elsevier.com/inca/publications/store/6/2/2/8/5/9/index.htt.

Rovera, D., Alves, A., Fakhouri, F. M., & Gonçalves, C. (2020). Water vapor barrier properties of edible films based on gluten and whey protein. *Food Research International, 128*(4).

Roy, S., Harris, P., & Stroud, J. (2000). Effect of humidity on the physical properties of hydrophilic wheat gluten films. *Journal of Membrane Science, 180*, 47–55.

Sartori, T., Caon, T., Martelli, S. M., Innocentini-Mei, L. H., & Fakhouri, F. M. (2018). Properties of films produced from blends of gelatin, starch, and glycerol. *Journal of Polymers and the Environment, 26*(6), 2451–2459.

Sharma, P., Singh, A. K., & Mishra, H. N. (2017). Isolation and development of edible films using whey proteins and antimicrobial properties of films enriched with nisin. *Journal of Food Science and Technology, 54*(1), 343–352.

Tanada-Palmu, P., Helén, H., & Hyvönen, L. (2000). Preparation, properties and applications of wheat gluten edible films. *Agricultural and Food Science in Finland, 9*(1), 23–35. https://doi.org/10.23986/afsci.5650.

Tinto, W. F., Elufioye, T. O., & Roach, J. (2017). *Waxes*. Elsevier BV, 443–455. https://doi.org/10.1016/b978-0-12-802104-0.00022-6.

Vargas, M., Sánchez-González, L., Cháfer, M., Chiralt, A., & González-Martínez, C. (2012). *Edible chitosan coatings for fresh and minimally processed foods*. Elsevier BV, 66–95. https://doi.org/10.1533/9780857095664.1.66.

Wang, e (2022). *RSC Medicinal Chemistry,* (1), 85–102.

Wieser, H. (2022). *Chemistry of gluten proteins wheat gluten: Structure and functionality*. Springer, 34–67.

Xu, J., & Li, Y. (2023). Wheat gluten–based coatings and films: Preparation, properties, and applications. *Journal of Food Science, 88*(2), 582–594. https://doi.org/10.1111/1750-3841.16454.

Yu, T., & Xu, J. (2024). Characteristics of films prepared from wheat gluten and phenolic extracts from Porphyra haitanensis and its application for salmon preservation. *Foods, 13*(15), 2442. https://doi.org/10.3390/foods13152442.

Zhang, H., Jin, C., Lv, S., Ren, F., & Wang, J. (2023). Study on electrospinning of wheat gluten: A review. *Food Research International, 169*. https://doi.org/10.1016/j.foodres.2023.112851.

Zhang, Y., Jin, X., & Zhang, W. (2020). Superior water stability and mechanical properties of wheat gluten-based films improved by tannic acid. *Journal of Materials Chemistry, 32*(2), 1463–1473.

Zink, I. A., Choudhury, N., & Hill, G. (2016). Physical, chemical, and biochemical properties of gluten-based films for food packaging. *Journal of Food Science, 45*(3), 133–142.

Zubeldía, F., Ansorena, M. R., & Marcovich, N. E. (2015). Wheat gluten films obtained by compression molding. *Polymer Testing, 43*, 68–77. https://doi.org/10.1016/j.polymertesting.2015.02.001.

Zuo, G., & Dai, Z. (2009). Preparation and properties of wheat gluten films. *Journal of Polymer Materials Science, 33*(2), 113–120.

Chapter 14

Zein films and coatings

Abhishek Parasar, Debabrata Nath, and Pratap Kalita
Pratiksha Institute of Pharmaceutical Sciences, Guwahati, Assam, India

14.1 Introduction

The environmental footprint is a global issue created by the food industry. Consequently, meeting the food demand for the growing global population also poses a challenge, as the global population is expected to reach 9.5 billion in the next 30 years, according to the worldwide population report. Over the last few decades, the increasing consumption of packaged foods has directly accelerated the production rate, leading to a corresponding increase in demand for food packaging. The purpose is to prevent food waste by maintaining food safety, extending shelf life, and preventing spoilage. According to an earlier report, Europe generates around 40% of packaged foods in various plastic products (i.e., containers, bags, foils), resulting in huge amounts of postconsumer plastic waste that directly contribute to environmental pollution and impact the planet (Sani et al., 2023). Packaging is a crucial component that prevents the deterioration of food materials, such as fruits, vegetables, and meat, while simultaneously improving the shelf life of these foods and maintaining their quality for a prolonged period (Lan et al., 2023). Petroleum-based plastics have been used in food packaging due to their unique physicochemical attributes, such as high output, low price, high degree of industrialization, and relatively complete production industry chain. A recent series of investigations has revealed that petroleum-based plastic materials harm the environment due to their slow degradation process. Therefore, several steps and attempts have been explored to replace traditional packaging materials with new ones due to the gradual depletion of nonrenewable resources, such as oil and coal. Recent scientific investigations have revealed that renewable, readily available, biocompatible, biodegradable, versatile, and environmentally friendly materials are the preferred choice to replace traditional petroleum-based plastic packaging materials, also helping to mitigate environmental pollution (Alirezalu et al., 2021; Amin et al., 2022; Lan et al., 2023). In addition, biopolymers are abundantly obtained from natural sources and used to fabricate novel, environmentally friendly, bio-based packaging materials for food applications. They can help solve the issues and problems generated by plastic-derived food packaging materials (Bhargava et al., 2020).

Zein is a natural water-insoluble macromolecule among the biopolymers obtained from plants in the form of protein. The unique physicochemical properties, including nontoxicity, biocompatibility, and biodegradability, of zein make it suitable for film preparation and coating applications. It was also thoroughly investigated that zein is an excellent moisture barrier, and its mechanical properties make it an appropriate and functional polymer for developing primary food packaging materials (Babayev et al., 2023; Yuan et al., 2022). The hydrophobic nature of the zein polymer is attributed to the presence of unique hydrophobic amino acids, which are insoluble in water but soluble in an aqueous ethanolic solution of 55%–90%, and are usually unstable in the pH range of 5–7. Considering the chemical structure of zein, it consists of at least four types of proteins: α-, β-, γ-, and δ-zein, which produce an unequal distribution of amino acids, molecular weights, solubility indices, and isoelectric points at pH 6.2 (Cheng & Jones, 2017; Yuan et al., 2022).

The term "edible films" refers to preformed films adhered to food substances, and "edible coatings" refers directly to the application of these coatings on the surface of food by dipping, spraying, or brushing. In recent decades, the applicability of zein-based edible films and coatings has become increasingly interesting due to their unique properties in preserving fruits, vegetables, and other food substances, such as bakery products. These edible zein films and coating materials (with a thickness generally less than 0.3–0.5 mm) possess mechanical and hydrophobic properties that prevent the entry of water molecules, oxygen, and air into the packed items (Lan et al., 2023; Mouzakitis et al., 2022; Yuan et al., 2022). Edible films and coatings were classified based on the materials applied during preparation and coating time. Zein-based edible films and coatings were found to significantly maintain the packed foods in their fresh form, preserving the aromatic flavor,

Natural Polymer-Based Films and Coatings in Food Technology. DOI: https://doi.org/10.1016/B978-0-443-34244-8.00003-3

color, and preventing contamination. Additionally, the zein-based films and coatings exhibited no moisture absorption, prevented the transfer of water vapor, adjusted gas permeation, and blocked the entry or contamination of pathogens, including bacteria and fungi (Jafarzadeh et al., 2020; Lan et al., 2023). Recently, several research investigations have also been conducted to engineer the zein-based functional films and coatings by incorporating bioactive compounds, plasticizers, nanomaterials, and other kinds of additives during the manufacturing time to make them more functional and versatile food packaging materials that keep the packed foods fresh, maintain quality, shelf life for a longer duration of time, and save from spoilage (Ahammed et al., 2021; Lan et al., 2023). Other studies have also demonstrated zein's versatile nature in advanced food preservation applications, such as zein-based nanoparticles and nanofibers loaded with antimicrobial or antioxidant agents, including thymol and curcumin. They have demonstrated remarkable potential in extending the shelf life of fruits by inhibiting microbial growth, which ultimately delays spoilage (Ren et al., 2025; Kunadu et al., 2025). Similarly, zein films incorporated with *Fructus Ligustri Lucidi* essential oil exhibit potent antioxidant and antibacterial activity, which helps maintain the quality of freshly cut fruits (Chen et al., 2025). These studies have demonstrated that zein is a natural candidate for edible films and coatings in sustainable food preservation. Considering all these scientific investigations, there is a need for a comprehensive literature review to compile the various sources of zein and their applications in films and coatings. Therefore, this chapter of the book is designed to profile several sources of zein and the multiple aspects of zein-based films and coatings.

14.2 Zein and its chemistry

Zein, a prolamin protein found predominantly in the endosperm of corn (*Zea mays*), constitutes a significant portion of the grain's protein content and is among the most studied cereal prolamins (Wu & Messing, 2014). Its name derives from the Latin word *Zea*, meaning "grain," reflecting its close association with maize. Zein is rich in hydrophobic amino acids, such as leucine, proline, and alanine, making it insoluble in water but soluble in alcohols like ethanol and isopropanol. This unique solubility profile is key to its functional properties (Tortorella et al., 2021). Zein is classified into four types: α-zein, β-zein, γ-zein, and δ-zein, based on their solubility and molecular weight, with α-zein being the most abundant and functionally significant (Tortorella et al., 2021; Wu et al., 2010). Known for its excellent film-forming ability, zein produces transparent, glossy, and water-resistant films, making it a versatile material for various applications. In the food industry, it is used for coatings and encapsulations, while in the pharmaceutical field, it serves as a component of drug delivery systems. Beyond food, zein holds potential in nonfood applications, such as biodegradable packaging, adhesives, and industrial coatings, aligning with the growing demand for sustainable and eco-friendly materials. Additionally, zein's valorization of agricultural by-products from corn processing highlights its role in waste reduction and the creation of value-added products, further underscoring its industrial and environmental significance (Lan et al., 2023; Tortorella et al., 2021).

14.2.1 Sources of zein

Zein is a storage protein derived exclusively from the seeds of corn (*Z. mays*), a globally cultivated cereal crop (Siddique et al., 2018). Predominantly localized in the endosperm, it accounts for 44%–79% of the total protein content in this starchy tissue, which provides nutrients during seed germination. Zein content varies among corn varieties, with dent corn and flint corn being particularly rich sources (Liu et al., 2023; Siddique et al., 2018). Dent corn, widely used in industrial processing, offers high yields and protein content, making it a primary source for zein extraction. Flint corn, although less common, also contains significant amounts of zein, particularly in traditional landraces (Liu et al., 2023). Industrially, zein is primarily obtained as a by-product of corn wet and dry milling processes, which produce corn gluten meal, a coproduct with a protein content of 60%–70%, serving as an economical and efficient raw material for zein production (Anderson & Lamsal, 2011; Uan-on et al., 2018). Speciality corn varieties, such as high-lysine and waxy corn, are occasionally explored in research due to their unique protein compositions, although they are not widely used for industrial purposes (Liu et al., 2023; Siddique et al., 2018). Regional variations in corn, influenced by factors such as soil, climate, and agricultural practices, also impact the zein content and quality, potentially affecting its functional properties and applications (Liu et al., 2023; Siddique et al., 2018; Uan-on et al., 2018).

14.2.2 Structure of zein

Zein belongs to the prolamin class of storage proteins, which are rich in proline and glutamine (Tortorella et al., 2021). These amino acids play a crucial role in determining the protein's structure and function (Xing et al., 2023). Zein is characterized by its hydrophobic nature, attributed to its high content of nonpolar amino acids, such as leucine, alanine, proline, and phenylalanine (Karthikeyan et al., 2014; Tortorella et al., 2021). This hydrophobicity, combined with a low proportion of charged amino acids, such as lysine, histidine, and arginine, makes zein insoluble in water but soluble in alcohols and aqueous ethanol solutions (Li et al., 2012). The lack of essential amino acids, particularly lysine and tryptophan, affects its nutritional profile but enhances its structural and functional versatility (Karthikeyan et al., 2014; Xing et al., 2023). Zein is categorized into four main types: α-zein, β-zein, γ-zein, and δ-zein, based on their molecular weight, amino acid composition, and solubility characteristics (Darie-Ion et al., 2021; Feng et al., 2009).

i). ***α-zein***: This is the most abundant form, constituting 70%–85% of total zein. Structurally, α-zein is predominantly composed of α-helices arranged into nine repeating helical segments interspersed with short loops (Feng et al., 2009). These helices are stabilized by intramolecular hydrophobic interactions, which form a compact, rod-like structure (Corradini et al., 2014). α-zein is soluble in 70%–95% ethanol and plays a pivotal role in imparting functional properties, such as film-forming ability and thermoplasticity, to the protein (Darie-Ion et al., 2021). Its abundance and solubility make it the primary focus of industrial and research applications (https://tinyurl.com/4pfcbm4b).

ii). ***β-zein***: β-zein is a smaller fraction of zein but is notable for its cysteine-rich composition, which facilitates additional disulfide bonding (Darie-Ion et al., 2021; Feng et al., 2009). Its structural properties and solubility closely resemble those of γ-zein, as it dissolves primarily in reducing environments. While less studied, β-zein contributes to the overall functionality and stability of zein in composite forms (Darie-Ion et al., 2021).

iii). ***γ-zein***: γ-zein constitutes 10%–20% of the total zein fraction. It is distinguished by its high cysteine content, which forms intramolecular and intermolecular disulfide bonds (Darie-Ion et al., 2021; Feng et al., 2009). These disulfide bonds contribute to a more compact, globular structure, increasing its resistance to thermal and chemical denaturation. γ-zein is soluble in reducing agents and provides rigidity and stability to zein films, enhancing their mechanical and barrier properties (Feng et al., 2009).

iv). ***δ-zein***: Representing the least abundant fraction at less than 5% of total zein, δ-zein has a unique amino acid composition (Darie-Ion et al., 2021; Feng et al., 2009). It contains fewer cysteine residues compared to β-zein and γ-zein, resulting in a more linear and less compact structure (Feng et al., 2009). δ-zein is soluble in higher concentrations of ethanol and is thought to contribute to the flexibility and ductility of zein-based films and materials (Darie-Ion et al., 2021; Feng et al., 2009).

The secondary structure of zein is dominated by α-helices, particularly in α-zein, which form stable helical bundles due to hydrophobic interactions. This structural arrangement enables zein to exhibit a thermoplastic nature, which is crucial to its ability to form films, fibers, and coatings (Corradini et al., 2014; Tihminlioglu et al., 2010). The high degree of hydrophobicity also contributes to its low water vapor permeability, making it suitable for barrier applications in food packaging and drug delivery systems (Tihminlioglu et al., 2010). Furthermore, the presence of disulfide bonds in β-zein and γ-zein adds mechanical strength and chemical resistance, while δ-zein contributes complementary flexibility, creating a balance of properties in zein-based materials (Corradini et al., 2014; Wang & Fan, 2019). The structural diversity among the four types of zein highlights its multifunctionality, enabling its application across industries (Tihminlioglu et al., 2010). The intricate interplay between the hydrophobic and disulfide-linked domains of zein provides insights into its remarkable film-forming capabilities, stability, and adaptability to various processing conditions (Corradini et al., 2014; Wang & Fan, 2019).

14.2.3 Properties of zein

Zein exhibits a remarkable array of physicochemical and functional properties, primarily stemming from its hydrophobic nature, specific amino acid composition, and secondary structural characteristics. These properties have positioned zein as a versatile biomaterial with diverse industrial applications (Singh et al., 2012; Wang & Fan, 2019).

14.2.3.1 ***Solubility***

Zein's solubility profile is one of its defining characteristics, dictated by its high content of hydrophobic amino acids such as leucine, proline, and alanine. These residues form a structure that is insoluble in water but dissolves readily in alcohols, such as ethanol, isopropanol, and aqueous ethanol solutions (50%–90%) (Singh et al., 2012). The solubility varies with factors such as alcohol concentration, pH, and temperature. For example, higher ethanol concentration typically increases zein's solubility, and the presence of salts or mild acidic conditions can further influence its dissolution (Reshma & Rekha, 2024). This property is crucial for its extraction and formulation in various applications, including food coatings and pharmaceutical excipients.

14.2.3.2 *Film-forming ability*

Zein's exceptional ability to form films is a hallmark of its utility. These films are highly transparent, glossy, and exhibit a natural water-resistant quality, primarily due to zein's hydrophobic amino acid composition (Tortorella et al., 2021). These α-helical structures within zein molecules arrange into compact formulations that stabilize the films and create low oxygen and water vapor permeability (Singh et al., 2012). These films also have inherent flexibility and durability, making them suitable for applications in food packaging, pharmaceutical encapsulation, and biodegradable coatings. Furthermore, plasticizers such as glycerol, sorbitol, or polyethylene glycol can be added to adjust the film's flexibility and mechanical strength, catering to specific functional requirements (Shukla & Cheryan, 2001).

14.2.3.3 *Thermal plasticity*

Zein demonstrates thermoplastic behavior, allowing it to soften and become moldable under heat. This arises from its ability to form noncovalent interactions, such as hydrophobic associations and hydrogen bonds, when heated and cooled (Reshma & Rekha, 2024). This thermoplastic nature enables zein to be processed into films, fibers, or molded shapes, often used in the production of biodegradable plastics, agricultural mulch films, and eco-friendly packaging materials (Tables 14.1 and 14.2).

14.2.3.4 *Mechanical strength and elasticity*

Zein-based materials exhibit impressive mechanical properties, striking a balance between tensile strength and elasticity. The mechanical strength arises from the compact packing of its hydrophobic domains. At the same time, its elasticity can be attributed to the flexible nature of its α-helical secondary structure (Tortorella et al., 2021). By incorporating plasticizers or blending with other polymers, these properties can be tailored, making zein suitable for use in durable and flexible materials for industrial applications (Shukla & Cheryan, 2001).

14.2.3.5 *Chemical resistance*

Zein exhibits resistance to many chemicals, particularly water and mild acidic or basic solutions. This chemical stability is a result of its low polar amino acid content, which minimizes its interaction with polar solvents (Reshma & Rekha, 2024). Its resistance to enzymatic degradation under certain conditions also enhances its durability, making it suitable for applications requiring prolonged stability, such as drug delivery systems and protective coatings (Tivano & Chiono, 2023).

14.2.3.6 *Biodegradability and environmental friendliness*

As a naturally occurring protein, zein is biodegradable and breaks down under natural environmental conditions, making it an environmentally friendly alternative to synthetic polymers. This aligns with global efforts to develop sustainable materials and reduce plastic waste, positioning zein as a key player in the development of eco-friendly packaging and agricultural films (Tivano & Chiono, 2023).

14.2.3.7 *Adhesiveness*

Zein exhibits strong adhesive properties, particularly under conditions of heat and moisture. This makes it highly effective as a binder in pharmaceutical formulations, where it can enhance tablet integrity and control drug release. Additionally, zein's adhesiveness is leveraged in industrial applications, such as coatings, glues, and composite materials (Aytac et al., 2020; Lin et al., 2011).

14.2.3.8 *Barrier properties*

Zein provides excellent barrier properties, especially against oxygen, grease, and oil. This is primarily due to its hydrophobic nature and tightly packed molecular structure, which limits the diffusion of gases and liquids. These barrier properties are highly advantageous for preserving the shelf life of food products and protecting sensitive pharmaceutical compounds (Helgeson et al., 2009; Tivano & Chiono, 2023).

14.2.3.9 *Thermal stability*

Zein demonstrates high thermal stability, capable of withstanding temperatures up to 200°C without significant degradation. This thermal resilience is crucial in processes that involve high heat, such as extrusion, film casting, and the production of thermoplastic materials. Zein's ability to retain its structural integrity under heat further enhances its functionality in a wide range of applications (Aytac et al., 2020; Lin et al., 2011).

TABLE 14.1 Overview of zein types, properties, and applications.

Parameter	Description		References
Class of protein	Zein belongs to the prolamin class of storage proteins, rich in proline and glutamine		Tortorella et al. (2021); Xing et al. (2023)
Hydrophobicity	High content of nonpolar amino acids (leucine, alanine, proline, phenylalanine) makes zein hydrophobic		Tortorella et al. (2021); Karthikeyan et al. (2014)
Water solubility	Zein is insoluble in water due to low levels of charged amino acids (lysine, histidine, arginine) but is soluble in alcohols and aqueous ethanol solutions		Li et al. (2012)
Nutritional profile	Deficient in essential amino acids such as lysine and tryptophan		Xing et al. (2023); Karthikeyan et al. (2014)
Types of zein	α-Zein, β-zein, γ-zein, and δ-zein—classified based on molecular weight, amino acid composition, and solubility		Darie-Ion et al. (2021); Feng et al. (2009)
	α-Zein	• Most abundant (70%–85%) • Predominantly α-helices in nine repeating segments with intramolecular hydrophobic interactions • Soluble in 70%–95% ethanol • Key role in film forming and thermal plasticity	Darie-Ion et al. (2021); Feng et al. (2009); Corradini et al. (2014)
	β-Zein	• Cysteine-rich, forms disulfide bonds • Soluble in reducing environments • Contributes to functionality and stability in composites	Darie-Ion et al. (2021); Feng et al. (2009)
	γ-Zein	• Accounts for 10%–20% • High cysteine content leads todisulfide bonds (inter/intramolecular) • Compact, globular structure, resistant to thermal and chemical denaturation • Enhances rigidity and stability in films	Darie-Ion et al. (2021); Feng et al. (2009)
	δ-Zein	• Least abundant (less than 5%) • Fewer cysteine residues; linear and less compact structure • Soluble in higher ethanol concentrations • Adds flexibility and ductility to zein-based films	Darie-Ion et al. (2021); Feng et al. (2009)
Secondary structure	Dominated by α-helices (stable helical bundles formed via hydrophobic interactions)		Corradini et al. (2014); Tihminlioglu et al. (2010)
Thermoplastic nature	Allows formation of films, fibers, and coatings		Corradini et al. (2014); Tihminlioglu et al. (2010)
Barrier properties	High hydrophobicity results in low water vapor permeability, useful for food packaging and drug delivery		Tihminlioglu et al. (2010)
Disulfide bond contribution	• β-zein and γ-zein: add mechanical strength and chemical resistance • δ-zein: provides flexibility and ductility • Creates a balance of properties in materials		Corradini et al. (2014), Wang and Fan (2019)
Applications	Versatile in forming films, coatings, and fibers due to hydrophobicity, stability, and adaptability under various conditions		Corradini et al. (2014); Tihminlioglu et al. (2010); Wang and Fan (2019)

TABLE 14.2 Summary of zein coating applications in food preservation.

Parameters	Application	Food items	Method	Results	References
Coating technique	Develop protective film	General food items	Dipping, brushing, spraying	Creates a barrier to enhance food safety	Otoni et al. (2017)
Shelf-life extension	Prevents hardness loss, weight loss	Tomatoes	Zein coating	Six-day shelf-life extension through sensory assessment	Li et al. (2020)
Ripening control	Slows ripening, reduces ethylene generation	Mangoes and tomatoes	Zein coating	Minimized color variation and weight reduction	Zapata et al. (2008)
Oxidative stability	Decrease lipid oxidation	Precooked pork chops	Zein film	Extended shelf life, preserved color, and flavor	Park et al. (2012)
Physical quality preservation	Maintains appearance and firmness	Peeled garlic cloves	Zein coating	Prevented oxygen entry, maintained color	Torun and Ozdemir (2022)
Chemical stability	Slows ripening, reduces oxidative damage	Guava fruits	Zein coating with tannic acid (ZTA)	Fewer changes in color and flavor, reduced weight fluctuation	Santos et al. (2018)
Active packaging	Enhances sensory characteristics	General food products	Bioactive packaging	Optimizes food storage, integrates bioactive features	Basavegowda and Baek (2021)
Antioxidant properties	Inhibit lipid oxidation	Fresh Kashar cheese	Zein films with polyphenolic molecules	Controlled delivery of antioxidants improves stability	Unalan et al. (2013)
Microbial inhibition	Reduces microbial counts	Beef specimens	Coating with essential oils and extract	Significant reduction in microbial counts	Sayadi et al. (2021)
Shelf-life extension with antibacterial agents	Restricts anthracnose development	Mango fruit	Zein nano-coating with curcumin	Extended shelf life, minimized anthracnose generation	Chuacharoen and Sabliov (2022)
Preservation of freshness	Prevents spoilage and softening	Strawberries	Electrospun zein film with thyme oil	Preserved damage and microbiological decomposition	Ansarifar and Moradinezhad (2021)
Stabilization of bioactive compounds	Improves stability of nutrients	Various food items	Zein-based nanocomposites	Enhanced bioavailability and bioaccessibility	Ren et al. (2019)

14.2.3.10 *Compatibility with additives and modifications*

Zein can be blended with various additives to enhance or customize its properties. Plasticizers, such as glycerol or polyethylene glycol, improve flexibility and reduce brittleness, while crosslinking agents and nanoparticles can enhance mechanical strength, thermal stability, and barrier properties. Additionally, zein can be combined with other biopolymers, such as starch or chitosan, to create composite materials with optimized functionality. This adaptability makes zein a versatile candidate for research and industrial innovation (Tivano & Chiono, 2023). Fig. 14.1 displays the properties of zein.

14.3 Zein-based films and coatings for food preservation

Zein is a hydrophobic and biodegradable protein extracted from various crops, including *Z. mays* L. (corn), broomcorn (sorghum), rye, wheat, and barley, all of which are highly productive and sustainable agricultural species (Kumar et al., 2024). The structure of the corn kernel includes the pericarp, germ, hull, endosperm, and fiber. The endosperm accounts for approximately 83% of the kernel's dry weight and is primarily composed of starch (88%) and residual proteins, including glutamine and proline, which together form prolamin (8%), with zein being the dominant component (Liu et al., 2023). Zein is characterized by a high content of cysteine and methionine but lacks essential amino acids such as lysine and tryptophan (Sharif et al., 2019). Chemical analysis of zein's amino acid composition reveals a high concentration of hydrophobic amino acid residues, which contribute to its solubility in various organic solvents and blends, making it inherently hydrophobic. Specifically, significant amounts of hydrophobic residues such as leucine, proline, alanine, and phenylalanine are present. Zein is classified into four main types: α, β, γ, and δ, based on their solubility properties. The α-zein types, typically found as disulfide-bridged dimers, have molecular weights of 19 and 22 kDa, with the 22 kDa form differing due to an amino acid insertion in its C-terminal region (Lai & Padua, 1997; Momany et al., 2006).

Zein is known for its excellent film-forming capabilities, which are attributed to its unique amino acid composition. Additionally, zein is an eco-friendly, renewable polymer that has attracted significant attention in the scientific community due to its tissue compatibility, biodegradability, and moderate water vapor permeability. It can polymerize, forming films and sheets that can replace synthetic polymers in various applications. Zein has also shown great potential for electrospinning, with the resulting nanofibers offering improved structural properties that are both more uniform and flexible (Aghababaei et al., 2024). Zein can be polymerized, creating films and sheets that can serve various applications as substitutes for synthetic polymers. The molecular structure, weight, and polymerization of zein enable it to form biodegradable films with excellent mechanical properties and water resistance in acetone or alcohol-based solutions. Zein films are shiny, oil-resistant biopolymers with natural free radical scavenging abilities, protecting oils and fats from oxidation and degradation (Ibrahim et al., 2019). As an active coating, zein helps

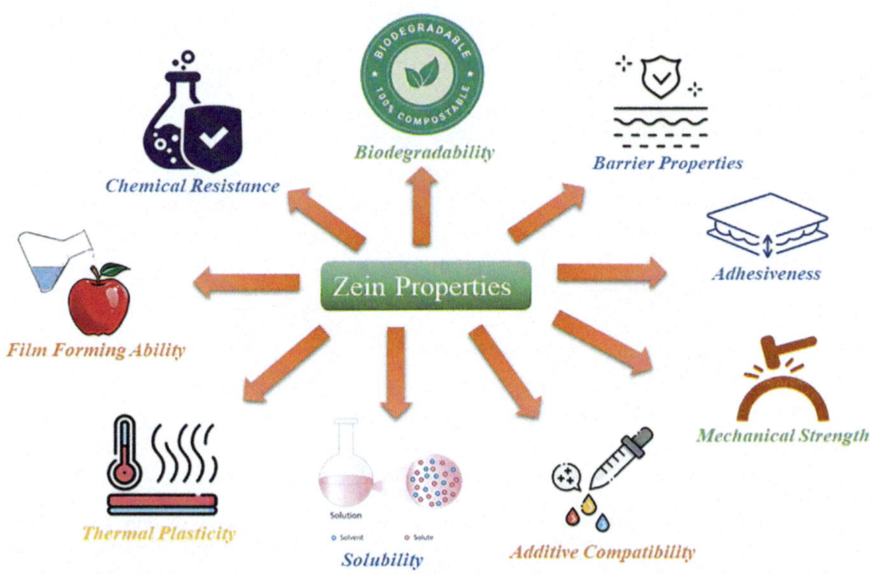

FIGURE 14.1 **Properties of zein.** The figure illustrates the major physicochemical and functional properties of zein protein in a radial format, highlighting its versatility as a biomaterial.

preserve food from oxidation, thereby maintaining its nutritional value. Its film-forming properties have led to its use in edible coatings and food preservation (Foong et al., 2024). Research has explored zein-based coatings for food packaging with hydrophobic or hydrophilic properties (Rahman et al., 2023). Zein films infused with flavonoids like ferulic acid and quercetin exhibit antibacterial and antioxidant effects (Bejenaru et al., 2024). Incorporating polyphenolic compounds, nanomaterials, and plasticizers helps address the issues of fragility, poor mechanical strength, and low elasticity in pure zein (Pei et al., 2024).

Given that most food packaging polymers and resins are produced through extrusion, it is essential to provide a brief overview of the extrusion process for zein and its composites. In short, extrusion processing involves melting the polymer through high temperatures and shear forces, then forcing it through a die. The resulting extrudate is drawn and cooled to form a fiber. However, the presence of degradation products, such as residual oxidants or monomers, can negatively affect the properties of various polymers. Extrusion is a commonly used method for processing zein (Selling & Utt, 2013; Selling, 2010; Bouman et al., 2015). Zein is typically extruded at around 90°C. The impact of temperature on zein has been studied using dynamic mechanical analysis (DMA) techniques, with tests conducted at temperatures up to 166°C (Zheng et al., 2014). Similar to starch, zein is usually plasticized with agents like polyethylene glycols before extrusion (Verbeek & van den Berg, 2010). The extrusion of zein sheets and films, often combined with oleic acid, has been demonstrated using slit die and blowing head techniques. However, extruded zein sheets and films tend to develop micro-voids due to moisture evaporation. The quantity and size of these micro-voids are influenced by both the extrusion process and any subsequent heat treatments (Wang & Padua, 2003).

The barrier properties of zein, along with those of other protein-, polysaccharide-, and lipid-based films, have been extensively studied. The performance of these materials can be improved by incorporating various active compounds, such as antimicrobial agents, antioxidants, nutrients, oxygen and carbon dioxide absorbers, and enzymes. For instance, edible coatings with antimicrobial agents can help prevent food spoilage caused by microbial growth on the surface. This chapter will, therefore, review the various methods and processes for fabricating zein films and examine their properties, particularly in relation to potential applications in food packaging (Fig. 14.2).

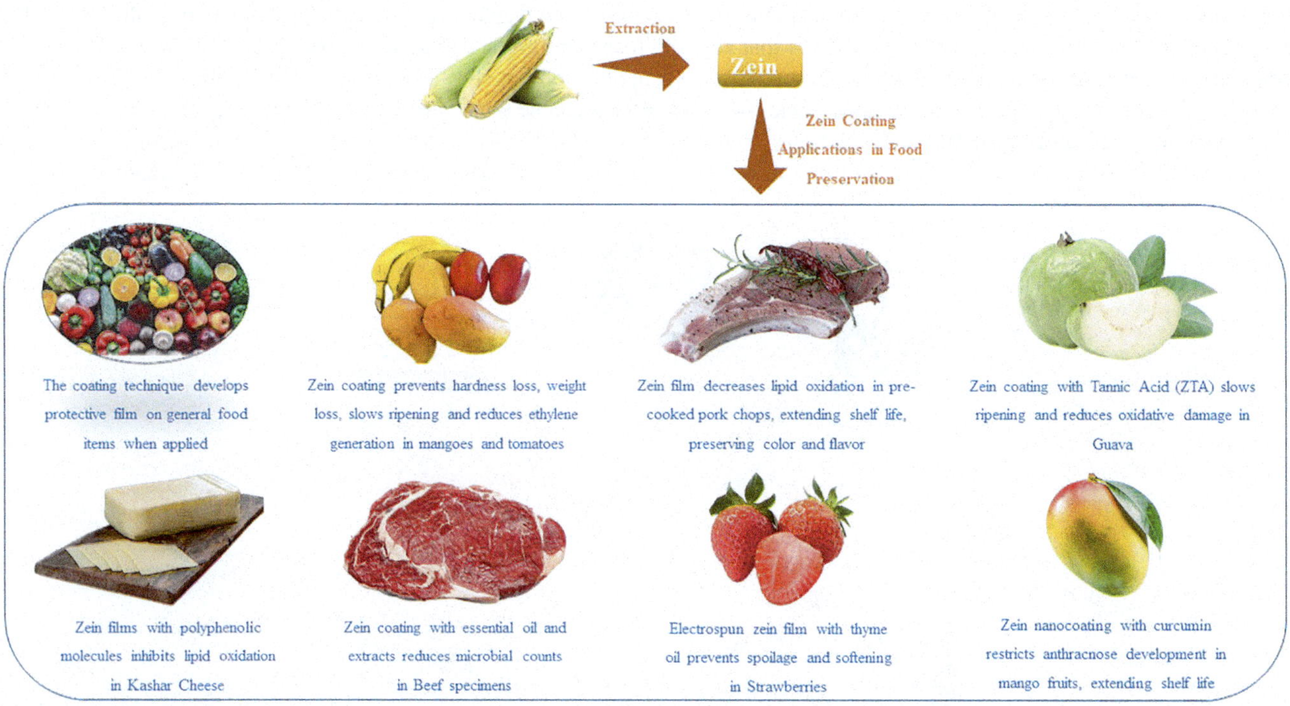

FIGURE 14.2 Zein coating applications in food preservation. The image illustrates the extraction of zein, a corn-derived protein, and its diverse applications in food preservation. Zein-based coatings and films are shown to enhance shelf life, inhibit lipid oxidation, reduce microbial growth, and delay ripening in various food products, such as fruits, vegetables, meat, cheese, and dairy. Specialized formulations, including zein with tannic acid, curcumin, or essential oils, demonstrate targeted preservation benefits like oxidative stability, microbial control, and firmness retention across a range of perishable items.

14.4 Zein-based film fabrication methods

Zein-based packaging materials can be produced using two primary methods: the wet process, which includes solvent casting, electrospaying, and electrospinning, and the dry process, which involves extrusion and blowing.

14.4.1 Solvent casting method

The solvent-casting process is the simplest method used for fabricating zein films. In this method, zein and plasticizer (e.g., glycerol, triethyl glycerol, and oleic acid) are solubilized in a suitable solvent (such as an aqueous ethanol solution), and the solution is dropped onto a substrate. Early studies by Evans and colleagues demonstrated that 20 wt.% solutions of zein in diethylene glycol, triethylene glycol, propylene glycol, and di-propylene glycol remained clear and fluid for up to two years when stored at 27°C. In contrast, a similar concentration of zein in 85% aqueous ethyl alcohol gelled within approximately 30 days at the same temperature. The researchers introduced the term "peptization" to describe the critical cloud points, which refer to the phenomenon where protein solutions become cloudy upon cooling. In essence, the cloud point is the temperature at which a solution turns turbid as it cools. For a substance to effectively dissolve zein, it must have a proper balance of polar and nonpolar groups; for example, propylene glycol is a good solvent for zein, whereas absolute propanol is not (Zhang et al., 2015). The interaction between protein and solvent is crucial in selecting an appropriate solvent for zein. For example, zein solutions prepared with aqueous ethanol undergo a transition from a Newtonian to a non-Newtonian shear-thinning behavior after being stored at room temperature for 24 hours (Souza et al., 2022). In addition to conventional primary and binary solvents, ionic liquids have also been found to solubilize zein, especially at higher temperatures. The superior solubilizing power of ionic liquids compared to traditional solvents is attributed to their low molecular volume, polarity, and ability to function as weak hydrogen bond acceptors. This suggests that advanced solvents, such as ionic liquids, could be beneficial for creating zein-based films (Tomlinson et al., 2014). Films made via the solvent-casting method exhibit consistent thickness, homogeneity, and isotropic optical clarity. Furthermore, this method typically does not require specialized equipment, although selecting an appropriate organic solvent for casting can be time-consuming. Additionally, the organic solvents used to dissolve zein may present significant health, safety, and environmental concerns (Kandasamy et al., 2021). Films produced through the solvent-casting technique for packaging are typically thin, transparent, glossy, and flexible, with a consistent thickness. Additionally, this method enables the incorporation of various heat-sensitive compounds into zein-based packaging films. However, there are certain drawbacks to solvent casting, including the use of toxic and unsustainable organic solvents, the need for the polymer and plasticizer to dissolve in the solvent, and the requirement for the solvent to have the appropriate viscosity for effective casting (Gaona-Sánchez et al., 2015).

14.4.2 Electrospraying method

Electrospraying, also known as electrohydrodynamic spraying, is a top-down fabrication technique that utilizes fine droplets as building blocks, which are deposited layer by layer to form a continuous coating on a substrate (Chinnappan et al., 2022). Electrospraying involves applying a high voltage to a polymer solution, which creates a charged jet that breaks up into fine droplets as it travels toward a collector. This process allows for the production of zein films with controlled thickness and morphology. In electrospraying, an electric field is applied to a zein-plasticizer solution, forcing it through a nozzle and onto a substrate, where it is evenly distributed in a controlled atmosphere at low temperatures. This process creates micron or nanosized polymer droplets that are regulated by the electric field (Patel & Haemmerich, 2024). The size distribution or dispersity of the droplets can be adjusted by varying factors such as the polymer solution concentration, the flow rate of the solution through the spraying nozzle, and the voltage applied to the nozzle (Zdraveva et al., 2023). Due to its ability to produce high-quality films, the electrospraying technique has been employed by many researchers to create zein-based films. To prepare zein for electrospraying, it is usually dissolved in an appropriate solvent, such as ethanol or acetic acid, to achieve a solution with the desired viscosity. The concentration of zein in the solution plays a crucial role in the process, as it needs to be carefully optimized. An ideal concentration ensures smooth film formation while preventing nozzle clogging during the spraying procedure (Gaona-Sánchez et al., 2015). The success of electrospraying largely depends on optimizing several key parameters, including the applied voltage (typically in the range of 20–30 kV), the flow rate, and the distance between the nozzle and the collector. These factors directly impact the size and morphology of the resulting zein particles or fibers. For example, higher voltages tend to produce finer fibers, whereas lower flow rates may result in thicker films. Careful adjustment of these conditions is essential for achieving the desired properties in zein-based materials (Gaona-Sánchez et al., 2015; Liu et al., 2018). As the charged droplets travel toward the collector, their high surface area-to-volume ratio causes them to evaporate quickly, resulting in the formation of solid films. The zein films produced through this process can possess valuable properties, such as enhanced barrier performance against moisture and oxygen, making them particularly suitable for food packaging applications. These characteristics are crucial for preserving the quality and shelf life of food products (Shi et al., 2009; Wang et al., 2021).

14.4.3 Electrospinning method

The electrospinning technique is a highly effective method for producing zein-based films, which are gaining popularity in food packaging due to their beneficial properties. This process involves the formation of ultrathin fibers from a zein solution, resulting in films with enhanced functional features. Electrospinning utilizes electrostatic forces to transform a polymer solution into fine fibers. The process begins with preparing a zein solution, typically dissolved in solvents like ethanol or acetic acid. The solution is then exposed to a high-voltage electric field, causing it to form a Taylor cone at the tip of the nozzle. As the voltage is increased, the cone elongates, and a jet of polymer solution is ejected, which solidifies into fibers as the solvent evaporates. This process enables the creation of zein fibers with improved properties suitable for various applications, particularly in the food packaging industry. The zein nanofibers produced by the electrospinning process exhibit high tensile strength, hydrophobicity, safety, and biocompatibility; however, they are poorly stretchable (Fabra et al., 2013; Wu et al., 2023).

14.4.4 Extrusion

Extrusion is a dry processing method that utilizes heat and shear forces to form fibers from a polymer; however, heat-induced degradation can impact film quality (Bayer, 2021). Zein extrusion, typically performed at 90°C, is commonly used in food packaging applications. The extrusion temperature is critical to the final product's quality. For example, heating zein to 120°C can cause protein crosslinking, while temperatures around 180°C lead to chain breakdown and color changes. Between 100°C and 220°C, the protein structure undergoes gradual changes, with rapid alterations occurring at 220°C and significant secondary structure changes at 240°C. Differential scanning calorimetry (DSC) and TGA show a shift in the amide I and II peaks, with abrupt changes above 220°C (Anderson & Lamsal, 2011). Higher extrusion temperatures can also disrupt protein bodies, converting them into fibrils (Beniwal et al., 2021). The extrusion process begins with dissolving zein in a solvent, typically a 70% ethanol aqueous solution, to form a zein resin that is easily processable. To improve flexibility and decrease brittleness, plasticizers such as oleic acid are frequently incorporated, making the resulting films more suitable for practical applications. The prepared zein resin is introduced into an extruder, where it undergoes controlled temperature and pressure conditions, typically ranging from 60°C to 110°C. A twin-screw extruder is commonly employed for efficient mixing and consistent processing of the material. Depending on the die configuration, the extrusion process can produce various forms, including sheets, films, or pellets. During extrusion, the zein resin is forced through a die to form continuous sheets or films. The thickness of the extruded product can be adjusted by modifying the die size and material flow rate. For example, slit dies can produce flat ribbons or sheets, whereas blown extrusion methods are used to create tubular films (Hong et al., 2024; Masanabo et al., 2022).

14.4.5 Blowing method

The blowing method for creating zein-based films is an innovative technique that leverages the distinct properties of zein, a protein sourced from maize. This approach enables the production of thermoplastic films with advantageous mechanical and physical characteristics, making them ideal for a range of applications, including biodegradable packaging (Oliviero et al., 2010). The fabrication of zein-based films begins with preparing a zein solution, typically by dissolving zein in an aqueous alcohol mixture, such as 70% ethanol. To improve flexibility and prevent brittleness, plasticizers like glycerol or oleic acid are often added, as pure zein films tend to be fragile. The zein solution is then fed into an extruder, where it undergoes heating and pressurization. This phase thermoplasticizes the zein, enabling it to flow through the extruder. The temperature is carefully regulated (usually between 60°C and 120°C) to avoid degradation of the zein. Following extrusion, the molten zein is passed through a die to form a tube-like shape. Air is introduced into the tube, causing it to expand and form a thin film. This blowing step ensures uniform film thickness and enhances the film's mechanical properties by aligning the molecular structure of the zein. After the film reaches the desired thickness, it is rapidly cooled to solidify. The cooling process helps preserve the film's structure and improves its barrier properties, particularly against moisture and oxygen (Azevedo et al., 2022; Liu et al., 2016; Luecha et al., 2010).

14.5 Application of zein-based films and coatings

Polyolefin plastics are commonly found in food packaging, but their degradation is incredibly slow, often taking decades or even centuries, which can lead to significant environmental pollution (Chang-Bravo et al., 2014). To address this issue, researchers are exploring the use of natural edible materials as alternatives to conventional plastic packaging (Shankar & Rhim, 2016). These edible films are typically made from natural biopolymers, such as proteins, polysaccharides, and lipids. As a result, edible composite films not only preserve freshness and safety but also help eliminate pollution, presenting promising opportunities for practical use (Xiao et al., 2021).

14.5.1 Zein-based coating application

In 1990, procedures for preparing zein coatings and films were developed to maintain the nutritional value of unprocessed foods within an eco-friendly packaging framework, aiming to reduce reliance on nonbiodegradable synthetic materials derived from petroleum that pose environmental risks (Asgher et al., 2020). Zein coatings can be produced through various methods, including extrusion, spin casting, and solvent casting, with solvent casting being the most commonly employed method. For film production, techniques such as 3D printing, spray drying, and electrospinning are employed. Chemical modifications to zein improve the thermal, mechanical, and surface-wetting properties of the films (Deng et al., 2018). Typically, these films and coatings have a thickness of less than 0.3 mm, and their effectiveness in food preservation depends on their mechanical, physical, and chemical properties (Tomadoni et al., 2020). Initial research on zein-based films and coatings focused on enhancing surface durability. For example, applying a zein film to tomatoes helped maintain firmness, minimize weight loss, and prevent color changes during storage, resulting in a six-day extension of shelf life, as indicated by sensory evaluations (Li et al., 2020).

Because 2-hydroxypropyl-β-cyclodextrin (HP-β-CD) interacts with hydrophobic amino acids, Vitiello et al. (2025) used it as a functional ingredient in zein bio-coatings. The study examined the influence of HP-β-CD on zein aggregation under varying conditions of temperature, ethanol concentration, and zein content. According to the results, the addition of HP-β-CD decreased zein aggregation and increased its solubility. Additionally, strawberries were coated in zein/HP-β-CD dispersion, which formed a protective layer that successfully extended their shelf life (Vitiello et al., 2025). Ren et al. (2025) developed a food-grade, biocompatible Pickering emulsion (H-ZSH/T) using sodium bicarbonate as a template to stabilize thymol-loaded hollow zein nanoparticles. These nanoparticles exhibited superior dispersibility and uniformity (92.24–95.18 nm, PDI 0.151–0.179), as well as effective thymol encapsulation through hydrogen bonding, electrostatic, and hydrophobic interactions. H-ZSH/T/T exhibited potent antibacterial activity, especially against *E*scherichia *coli*, by disrupting cell membranes. When applied as a coating, it reduced microbial growth and preserved firmness. It extended the shelf life of blueberries and cantaloupe, proving to be a stable, eco-friendly alternative to conventional food coatings (Ren et al., 2025).

Zein-coated mangoes and tomatoes showed a slower ripening process during storage, likely due to reduced ethylene production and lower respiration rates. This helped limit color changes and weight loss, allowing the fruits to remain effective during preservation (Zapata et al., 2008). The shelf life of peeled garlic cloves stored at 15°C was also evaluated, considering factors such as appearance, firmness, allicin, total sugar content, weight loss, and aroma. It was found that the zein-based coating provided excellent water vapor permeability, which prevented oxygen from penetrating and helped maintain the color of the garlic cloves (Torun & Ozdemir, 2022). Additionally, a zein-based coated film made from oleic acid (2.5%) and zein (4%) applied to the outer layer of macadamia nuts acted as an oxygen barrier, reducing oxidative degradation during storage without affecting the nuts' texture or color. This delay in fatty acid oxidation in macadamia nut oil helps maintain its economic value and prevents rancidity, which can spoil the flavor (Colzato et al., 2011).

14.5.2 Zein-based film application

Zein shows significant promise for creating edible packaging films due to its plentiful availability, biodegradable nature, and effective barrier against water vapor (Shi et al., 2012; Shukla & Cheryan, 2001). While zein is insoluble in water because of its high hydrophobic amino acid content, it can dissolve in alcohol concentrations ranging from 50% to 95% and in alkaline solutions with a pH above 11 (Anvar et al., 2025; Wu et al., 2023). However, zein films tend to have limited mechanical strength, and the addition of plasticizers can enhance their performance (Solano & de Gante, 2014; Xu et al., 2012). Research has demonstrated that incorporating glycerol and polyethylene glycol can improve the flexibility of zein films (Zhang et al., 2015). Furthermore, introducing the right amounts of oleic acid as a plasticizer can enhance the film's mechanical properties, decrease water vapor permeability, and boost its water resistance (Masamba et al., 2016; Turasan et al., 2018). For film production, techniques such as 3D printing, spray drying, and electrospinning are employed.

Zein-based films have shown the ability to reduce oxidative degradation in various food products by lowering lipid oxidation, which helps to prevent off-flavors in precooked pork chops and extends their shelf life during storage. Incorporating a plastic-like layered zein film containing natural polyphenols into the packaging of freshly chopped meat also effectively inhibits lipid oxidation while preserving color (Park et al., 2012).

In a study, Fan et al. (2025) developed zein-based two-dimensional network nanofiber films (T/N/Z films) designed for efficient ethylene removal through electrospinning. The incorporation of sodium chloride (NaCl) and Tween 80 facilitated fiber splitting, lowered surface tension, and reinforced hydrogen bonding, resulting in a stable network morphology. The optimized 0.7% T/N/Z films achieved an ethylene adsorption efficiency of 27.91 ± 1.55 mg $m^{-3} \cdot h^{-1}$, markedly higher than that of the control group. As a result, mango preservation trials revealed that fruits coated with the 0.7% T/N/Z films exhibited reduced browning (2.70% ± 2.42%) and improved firmness (804.23 ± 65.54 g), effectively extending storage life by more than 20 days (Fan et al., 2025).

Liang et al. (2025) developed a biodegradable zein-curcumin nanoparticles/sodium alginate (U-ZcNPs/SA) film through ultrasonic-assisted self-assembly, enhancing nanoparticle dispersion, mechanical strength, and oxygen barrier properties by improving electrostatic and hydrogen bonding interactions. These films exhibited potent antibacterial activity against *E. coli* and Staphylococcus *aureus*, effectively preserving grapes and strawberries by reducing spoilage and maintaining quality. Overall, U-ZcNPs/SA films present a sustainable and efficient approach to active food packaging (Liang et al., 2025). Chen et al. (2025) developed biodegradable zein films incorporated with Fructus *Ligustri Lucidi* essential oil (FEO) to enhance the preservation of fresh-cut fruits and vegetables. Using polyethylene glycol 400 and glucose as plasticizers, the optimized fructus zein film (FZF) film exhibited uniform FEO distribution, strong antioxidant activity (DPPH: 81.63%, ABTS: 91.44%), and effective antibacterial action against *S. aureus* and *E. coli*. It also demonstrated low cytotoxicity and significantly reduced browning in apple slices, extending their freshness. Overall, the FZF film offers a natural, safe, and effective approach for active food packaging (Chen et al., 2025).

Cui et al. (2020) investigated a zein-based active nanocomposite film loaded with impregnated chitosan nanoparticles, along with pomegranate extract, which inhibits the growth of *Listeria monocytogenes* and extends the shelf life of pork by up to 14 days during refrigerated storage (Cui et al., 2020). Another researcher, Li et al. found that electrospun zein/gelatin-based nanofiber mats/films with the bioactive molecule resveratrol show excellent antioxidant and antibacterial potential against *S. aureus* and *E. coli*, as well as better thermal stability and a longer shelf life of three days at 4°C (Li et al., 2020). Zein films containing polyphenolic compounds such as gallic acid and catechin function as antioxidants when combined with lysozyme, effectively inhibiting lipid oxidation and controlling the growth of *L. monocytogenes* in fresh Kashar cheese during 56 days of cold storage. The arrangement of these polyphenolic molecules within the zein influenced their release over the course of the study. The findings suggest that active packaging, which enables the controlled delivery of natural antioxidants, can improve the stability and functionality of fresh Kashar cheese (Ünalan et al., 2013). The use of electrospun zein film infused with 4% v/w thyme essential oil as bioactive packaging for polyethylene terephthalate effectively protected strawberries from damage, microbial decay, and texture softening (Ansarifar et al., 2021; Ansarifar & Moradinezhad, 2022). Additionally, various food materials were tested to evaluate the effectiveness of zein packaging created using different methods. In the past five years, numerous zein-based products have been developed in laboratories; however, there are only a limited number of commercial packages and nanocomposites available for active packaging, despite the increasing interest in nanotechnology-based food packaging in the agri-food production sector (Bumbudsanpharoke & Ko, 2015).

14.6 Conclusion and future perspective

Zein is a functional macromolecule that exhibits excellent film-forming, nontoxic, biodegradable, biocompatible, and soluble properties, making it applicable to food preservation. Applying and incorporating zein into the films and coatings enhances the commercial value, production, and cultivation rates. Therefore, the extraction technology must be developed to isolate zein from the bulk by-products generated in several food industries. The functional zein films and coatings produced by incorporating bioactive compounds during fabrication protect the products from spoilage. Therefore, the mechanisms by which the functional zein coatings and films preserve food must be understood.

References

Aghababaei, F., McClements, D. J., Martinez, M. M., & Hadidi, M. (2024). Electrospun plant protein-based nanofibers in food packaging. *Food Chemistry, 432*, 137236.

Ahammed, S., Liu, F., Wu, J., Khin, M. N., Yokoyama, W. H., & Zhong, F. (2021). Effect of transglutaminase crosslinking on solubility property and mechanical strength of gelatin-zein composite films. *Food Hydrocolloids, 116* https://doi.org/10.1016/j.foodhyd.2021.106649, https://www.sciencedirect.com/science/journal/0268005X.

Alirezalu, K., Pirouzi, S., Yaghoubi, M., Karimi-Dehkordi, M., Jafarzadeh, S., & Mousavi Khaneghah, A. (2021). Packaging of beef fillet with active chitosan film incorporated with ε-polylysine: An assessment of quality indices and shelf life. *Meat Science, 176*. https://doi.org/10.1016/j.meatsci.2021.108475, http://www.elsevier.com/locate/meatsci.

Anvar, A., Azizi, M. H., & Gavlighi, H. A. (2025). Exploring the effect of natural deep eutectic solvents on zein: Structural and functional properties. *Current Research in Food Science, 10*, 100965.

Amin, U., Khan, M. K. I., Maan, A. A., Nazir, A., Riaz, S., Khan, M. U., Sultan, M., Munekata, P. E. S., & Lorenzo, J. M. (2022). Biodegradable active, intelligent, and smart packaging materials for food applications. *Food Packaging and Shelf Life, 33*. https://doi.org/10.1016/j.fpsl.2022.100903, http://www.journals.elsevier.com/food-packaging-and-shelf-life/.

Anderson, T. J., & Lamsal, B. P. (2011). Zein extraction from corn, corn products, and coproducts and modifications for various applications: A review. *Cereal Chemistry, 88*(2), 159–173.

Anderson, T. J., & Lamsal, B. P. (2011). Development of new method for extraction of α-Zein from corn gluten meal using different solvents. *Cereal Chemistry, 88*(4), 356–362.

Ansarifar, E., & Moradinezhad, F. (2022). Encapsulation of thyme essential oil using electrospun zein fiber for strawberry preservation. *Chemical and Biological Technologies in Agriculture, 9*, 1–11.

Ansarifar, E., & Moradinezhad, F. (2021). Preservation of strawberry fruit quality via the use of active packaging with encapsulated thyme essential oil in zein nanofiber film. *International Journal of Food Science & Technology, 56*(9), 4239–4247. https://doi.org/10.1111/ijfs.15130.

Asgher, M., Qamar, S. A., Bilal, M., & Iqbal, H. M. (2020). Bio-based active food packaging materials: Sustainable alternative to conventional petrochemical-based packaging materials. *Food Research International, 137*, 109625.

Aytac, Z., Huang, R., Vaze, N., Xu, T., Eitzer, B. D., Krol, W., & Demokritou, P. (2020). Development of biodegradable and antimicrobial electrospun zein fibers for food packaging. *ACS Sustainable Chemistry & Engineering, 8*(40), 15354–15365.

Azevedo, A. G., Barros, C., Miranda, S., Machado, A. V., Castro, O., Silva, B., Saraiva, M., Silva, A. S., Pastrana, L., Carneiro, O. S., & Cerqueira, M. A. (2022). Active flexible films for food packaging: A review. *Polymers, 14*(12), 2442.

Babayev, A., Spasojević, L., Škrbić, J., Bučko, S., Kocić-Tanackov, S., Bulut, S., Fraj, J., Petrović, L., Milinković Budinčić, J., Sharipova, A., Aidarova, S., & Katona, J. (2023). Antimicrobial pseudolatex zein films with encapsulated carvacrol for sustainable food packaging. *Food Packaging and Shelf Life, 37*, 101076. https://doi.org/10.1016/j.fpsl.2023.101076.

Basavegowda, N., & Baek, K.-H. (2021). Synergistic antioxidant and antibacterial advantages of essential oils for food packaging applications. *Biomolecules, 11*(9), 1267. https://doi.org/10.3390/biom11091267.

Bayer, I. S. (2021). *Zein in food packaging sustainable food packaging technology*. Italy: Wiley, 199–224. https://onlinelibrary.wiley.com/doi/book/10.1002/9783527820078, http://10.1002/9783527820078.ch7.

Bejenaru, C., Radu, A., Segneanu, A. E., Biţă, A., Ciocîlteu, M. V., Mogoşanu, G. D., Bradu, I. A., Vlase, T., Vlase, G., & Bejenaru, L. E. (2024). Pharmaceutical applications of biomass polymers: Review of current research and perspectives. *Polymers, 16*(9), 1182.

Beniwal, A. S., Singh, J., Kaur, L., Hardacre, A., & Singh, H. (2021). Meat analogs: Protein restructuring during thermomechanical processing. *Comprehensive Reviews in Food Science and Food Safety, 20*(2), 1221–1249.

Bhargava, N., Sharanagat, V. S., Mor, R. S., & Kumar, K. (2020). Active and intelligent biodegradable packaging films using food and food waste-derived bioactive compounds: A review. *Trends in Food Science and Technology, 105*, 385–401. https://doi.org/10.1016/j.tifs.2020.09.015, http://www.elsevier.com/wps/find/journaldescription.cws_home/601278/description#description.

Bouman, J., Belton, P., Venema, P., et al. (2015). The development of direct extrusion-injection moulded zein matrices as novel oral controlled drug delivery systems. *Pharmaceutical Research, 32*(8), 2775–2786.

Bumbudsanpharoke, N., & Ko, S. (2015). Nano-food packaging: An overview of market, migration research, and safety regulations. *Journal of Food Science, 80*(5), R910–R923.

Chang-Bravo, L., Lopez-Cordoba, A., & Martino, M. (2014). Biopolymeric matrices made of carrageenan and corn starch for the antioxidant extracts delivery of Cuban red propolis and yerba mate. *Reactive & Functional Polymers, 85*, 11–19.

Chen, H., Yang, C., Zhuansun, X., Li, Y., Han, R., Wang, L., Ding, S., & Liu, Q. (2025). Zein film loaded with Fructus Ligustri Lucidi essential oil: Preparation, characterization and application in fresh-cut apple preservation. *Food Control, 167*, 110811.

Cheng, C. J., & Jones, O. G. (2017). Stabilizing zein nanoparticle dispersions with ι-carrageenan. *Food Hydrocolloids, 69*, 28–35. https://doi.org/10.1016/j.foodhyd.2017.01.022, https://www.journals.elsevier.com/food-hydrocolloids.

Chinnappan, B. A., Krishnaswamy, M., Xu, H., & Hoque, M. E. (2022). Electrospinning of biomedical nanofibers/nanomembranes: Effects of process parameters. *Polymers, 14*(18), 3719.

Chuacharoen, T., & Sabliov, C. M. (2022). Development of coating material by incorporating curcumin-loaded zein nanoparticles to maintain the quality of mango (*Mangifera indica* L. cv. Nam Dokmai). *Journal of Agriculture and Food Research, 10*, 100444. https://doi.org/10.1016/j.jafr.2022.100444.

Colzato, M., Scramin, J. A., Forato, L. A., Colnago, L. A., & Assis, O. B. G. (2011). 1h Nmr investigation of oil oxidation in macadamia nuts coated with zein-based films. *Journal of Food Processing and Preservation, 35*(6), 790–796.

Corradini, E., Curti, P. S., Meniqueti, A. B., Martins, A. F., Rubira, A. F., & Muniz, E. C. (2014). Recent advances in food-packing, pharmaceutical and biomedical applications of zein and zein-based materials. *International Journal of Molecular Sciences, 15*(12), 22438–22470.

Cui, H., Surendhiran, D., Li, C., & Lin, L. (2020). Biodegradable zein active film containing chitosan nanoparticle encapsulated with pomegranate peel extract for food packaging. *Food Packaging and Shelf Life, 24*, 100511.

Darie-Ion, L., Jayathirtha, M., Hitruc, G. E., Zaharia, M. M., Gradinaru, R. V., Darie, C. C., & Petre, B. A. (2021). A proteomic approach to identify zein proteins upon eco-friendly ultrasound-based extraction. *Biomolecules, 11*(12), 1838.

Deng, L., Kang, X., Liu, Y., Feng, F., & Zhang, H. (2018). Characterization of gelatin/zein films fabricated by electrospinning vs solvent casting. *Food Hydrocolloids, 74*, 324–332.

Fan, X., Sun, F., Chang, L., & Huang, J. (2025). Development of electrospun zein-based two-dimensional network nanofiber films and their application in mango preservation. *Food Bioscience, 73*, 107584.

Fabra, M. J., Lopez-Rubio, A., & Lagaron, J. M. (2013). High barrier polyhydroxyalcanoate food packaging film by means of nanostructured electrospun interlayers of zein. *Food Hydrocolloids, 32*(1), 106–114.

Feng, L., Zhu, J., Wang, G., Tang, Y., Chen, H., Jin, W., & Song, R. (2009). Expressional profiling study revealed unique expressional patterns and dramatic expressional divergence of maize α-zein super gene family. *Plant Molecular Biology, 69*, 649–659.

Foong, H. L., Ismail-Fitry, M. R., Noranizan, M. A., Tan, T. B., & Hanani, Z. N. (2024). Recent advances in extruded polylactic acid-based composites for food packaging: A review. *International Journal of Biological Macromolecules*, 131340.

Gaona-Sánchez, V. A., Calderon-Dominguez, G., Morales-Sanchez, E., Chanona-Perez, J. J., Velazquez-De La Cruz, G., Mendez-Mendez, J. V., Terrés-Rojas, E., & Farrera-Rebollo, R. R. (2015). Preparation and characterisation of zein films obtained by electrospraying. *Food Hydrocolloids, 49*, 1–10.

Helgeson, M. S., Graves, W. R., Grewell, D., & Srinivasan, G. (2009). Degradation and nitrogen release of zein-based bioplastic containers. *Journal of Environmental Horticulture, 27*(2), 123–127.

Hong, T., Tan, Z., Yang, T., Xu, D., Jin, Y., Wu, F., & Xu, X. (2024). Dynamic behavior of zein-gluten interaction during extruded noodle processing. *Food Hydrocolloids, 147*, 109320.

Ibrahim, S., Riahi, O., Said, S. M., Sabri, M. F., & Rozali, S. (2019). *Biopolymers from crop plants*. https://doi.org/10.1016/B978-0-12-803581-8.11573-5.

Jafarzadeh, S., Jafari, S. M., Salehabadi, A., Nafchi, A. M., Uthaya Kumar, U. S., & Khalil, H. P. S. A. (2020). Biodegradable green packaging with antimicrobial functions based on the bioactive compounds from tropical plants and their by-products. *Trends in Food Science and Technology, 100*, 262–277. https://doi.org/10.1016/j.tifs.2020.04.017, http://www.elsevier.com/wps/find/journaldescription.cws_home/601278/description#description.

Kandasamy, S., Yoo, J., Yun, J., Kang, H. B., Seol, K. H., Kim, H. W., & Ham, J. S. (2021). Application of whey protein-based edible films and coatings in food industries: An updated overview. *Coatings, 11*(9), 1056.

Karthikeyan, K., Vijayalakshmi, E., & Korrapati, P. S. (2014). Selective interactions of zein microspheres with different class of drugs: An in vitro and in silico analysis. *AAPS PharmSciTech, 15*, 1172–1180.

Kumar, P., Longmei, N., Choudhary, M., Gupta, M., Kumar, B., Jat, B. S., Bhushan, B., Dagla, M. C., & Aggarwal, S. K. (2024). Enhancement of nutritional quality in maize grain through QTL-based approach. *Cereal Research Communications, 52*(1), 39–55.

Kunadu, A. P.-H., Arcot, Y., Cisneros-Zevallos, L., Barouei, J., Akbulut, M. E. S., & Matthew Taylor, T. (2025). Nanoencapsulation of Curcumin and Quercetin in Zein-chitosan Shells for Enhanced Broad-spectrum Antimicrobial Efficacy and Shelf-life Extension of Strawberries. *Journal of Food Protection, 88*(6), 100517.

Lai, H. M., & Padua, G. W. (1997). Properties and microstructure of plasticized zein films. *Cereal Chemistry, 74*(6), 771–775.

Lan, X., Zhang, X., Wang, L., Wang, H., Hu, Z., Ju, X., & Yuan, Y. (2023). A review of food preservation based on zein: The perspective from application types of coating and film. *Food Chemistry, 424*, 136403. https://doi.org/10.1016/j.foodchem.2023.136403.

Li, Y., Li, J., Xia, Q., Zhang, B., Wang, Q., & Huang, Q. (2012). Understanding the dissolution of α-zein in aqueous ethanol and acetic acid solutions. *The Journal of Physical Chemistry B, 116*(39), 12057–12064.

Li, Y., Bai, Y., Huang, J., Yuan, C., Ding, T., Liu, D., & Hu, Y. (2020). Airglow discharge plasma treatment affects the surface structure and physical properties of zein films. *Journal of Food Engineering, 273*, 109813.

Lin, T., Lu, C., Zhu, L., & Lu, T. (2011). The biodegradation of zein in vitro and in vivo and its application in implants. *AAPS PharmSciTech, 12*, 172–176.

Liu, F., Avena-Bustillos, R. J., Woods, R., Chiou, B. S., Williams, T. G., Wood, D. F., & Zhong, F. (2016). Preparation of zein fibers using solution blow spinning method. *Journal of Food Science, 81*(12), N3015–N3025.

Liu, G., An, D., Li, J., & Deng, S. (2023). Zein-based nanoparticles: Preparation, characterization, and pharmaceutical application. *Frontiers in Pharmacology, 14*, 1120251.

Liu, Z. P., Zhang, Y. Y., Yu, D. G., Wu, D., & Li, H. L. (2018). Fabrication of sustained-release zein nanoparticles via modified coaxial electrospraying. *Chemical Engineering Journal, 334*, 807–816.

Liang, Q., Cao, P., Lu, H., Du, Y., Kang, L., Ma, H., & Ren, X. (2025). Ultrasonic engineering of zein-curcumin nanoparticles/sodium alginate composite films for active food packaging applications. *International Journal of Biological Macromolecules, 315*, 144268.

Luecha, J., Sozer, N., & Kokini, J. L. (2010). Synthesis and properties of corn zein/montmorillonite nanocomposite films. *Journal of Materials Science, 45*, 3529–3537.

Masamba, K., Li, Y., Hategekimana, J., Zehadi, M., Ma, J. G., & Zhong, F. (2016). Evaluation of mechanical and water barrier properties of transglutaminase cross-linked zein films incorporated with oleic acid. *IInternational Journal of Food Science & Technology, 51*, 1159–1167.

Masanabo, M. A., Ray, S. S., & Emmambux, M. N. (2022). Properties of thermoplastic maize starch-zein composite films prepared by extrusion process under alkaline conditions. *International Journal of Biological Macromolecules, 208*, 443–452.

Momany, F. A., Sessa, D. J., Lawton, J. W., et al. (2006). Structural characterization of α-zein. *Journal of Agricultural and Food Chemistry, 54*(2), 543–547.

Mouzakitis, C. K., Sereti, V., Matsakidou, A., Kotsiou, K., Biliaderis, C. G., & Lazaridou, A. (2022). Physicochemical properties of zein-based edible films and coatings for extending wheat bread shelf life. *Food Hydrocolloids, 132*. https://doi.org/10.1016/j.foodhyd.2022.107856, https://www.journals.elsevier.com/food-hydrocolloids.

Oliviero, M., Di Maio, E., & Iannace, S. (2010). Effect of molecular structure on film blowing ability of thermoplastic zein. *Journal of Applied Polymer Science, 115*(1), 277–287.

Otoni, C. G., Avena-Bustillos, R. J., Azeredo, H. M. C., Lorevice, M. V., Moura, M. R., Mattoso, L. H. C., & McHugh, T. H. (2017). Recent advances on edible films based on fruits and vegetables—a review. *Comprehensive Reviews in Food Science and Food Safety, 16*(5), 1151–1169. https://doi.org/10.1111/1541-4337.12281, http://www.interscience.wiley.com/jpages/1541-4337.

Park, H. Y., Kim, S. J., Kim, K. M., You, Y. S., Kim, S. Y., & Han, J. (2012). Development of antioxidant packaging material by applying corn-zein to LLDPE film in combination with phenolic compounds. *Journal of Food Science, 77*(10). https://doi.org/10.1111/j.1750-3841.2012.02906.x E273.

Patel, P. R., & Haemmerich, D. (2024). Review on electrospray nanoparticles for drug delivery: Exploring applications. *Polymers for Advanced Technologies, 35*(7), e6507.

Pei, J., Palanisamy, C. P., Srinivasan, G. P., Panagal, M., Kumar, S. S. D., & Mironescu, M. (2024). A comprehensive review on starch-based sustainable edible films loaded with bioactive components for food packaging. *International Journal of Biological Macromolecules,* 133332.

Rahman, M., Dip, T. M., Haase, T., Truong, Y. B., Le, T. C., & Houshyar, S. (2023). Fabrication of zein-based fibrous scaffolds for biomedical applications—A review. *Macromolecular Materials and Engineering, 308*(12), 2300175.

Ren, X., Hou, T., Liang, Q., Zhang, X., Hu, D., Xu, B., Chen, X., Chalamaiah, M., & Ma, H. (2019). Effects of frequency ultrasound on the properties of zein-chitosan complex coacervation for resveratrol encapsulation. *Food chemistry, 279*, 223–230. https://doi.org/10.1016/j.foodchem.2018.11.025.

Ren, M., Xie, T., Chen, L., Zhao, T., & Zhou, C. (2025). Pickering emulsion stabilized by hollow Zein/SSPS nanoparticles loaded with Thymol: Formation, characterization, and application in fruit preservation. *Food Research International, 201*, 115561.

Reshma, A. R., & Rekha, M. R. (2024). Zein—A plant protein as a promising biopolymer for biomedical applications: A perspective. *Journal of Polymer Science and Engineering, 7*(1), 3618.

Sani, I. K., Masoudpour-Behabadi, M., Sani, M. A., Motalebinejad, H., Juma, A. S., Asdagh, A., & Mohammadi, F. (2023). Value-added utilization of fruit and vegetable processing by-products for the manufacture of biodegradable food packaging films. *Food Chemistry, 405*, 134964.

Sayadi, M., Mojaddar Langroodi, A., & Jafarpour, D. (2021). Impact of zein coating impregnated with ginger extract and Pimpinella anisum essential oil on the shelf life of bovine meat packaged in modified atmosphere. *Journal of Food Measurement and Characterization, 15*(6), 5231–5244. https://doi.org/10.1007/s11694-021-01096-1.

Selling, G. W. (2010). The effect of extrusion processing on Zein. *Polymer Degradation and Stability, 95*(12), 2241–2249. https://doi.org/10.1016/j.polymdegradstab.2010.09.013.

Selling, G. W., & Utt, K. D. (2013). Effect of multiple extrusion passes on zein. *Polymer Degradation and Stability, 98*(1), 184–189.

Shankar, S., & Rhim, J. W. (2016). Preparation of nanocellulose from micro-crystalline cellulose: The effect on the performance and properties of agar-based composite films. *Carbohydrate Polymers, 135*, 18–26.

Sharif, N., Fabra, M. J., & López-Rubio, A. (2019). Nanostructures of zein for encapsulation of food ingredients. *Biopolymer Nanostructures for Food Encapsulation Purposes,* 217–245.

Shi, K., Kokini, J. L., & Huang, Q. (2009). Engineering zein films with controlled surface morphology and hydrophilicity. *Journal of Agricultural and Food Chemistry, 57*(6), 2186–2192.

Shi, K., Yu, H., Rao, S. L., & Lee, T.-C. (2012). Improved mechanical property and water resistance of zein films by plasticization with tributyl citrate. *Journal of Agricultural and Food Chemistry, 60*, 5988–5993.

Shukla, R., & Cheryan, M. (2001). Zein: The industrial protein from corn. *Industrial Crops and Products, 13*(3), 171–192.

Siddique, T., Asghar, A., Pasha, I., & Shahid, M. (2018). Evaluation of functional properties of Zein extracted from various maize varieties. *Journal of Biological Regulators and Homeostatic Agents, 32*(4), 899–903.

Singh, N., Singh, S., Kaur, A., & Singh Bakshi, M. (2012). *Zein: Structure, Production, film Properties and Applications.*

Solano, A. C. V., & de Gante, C. R. (2014). Development of biodegradable films based on blue corn flour with potential applications in food packaging. Effects of plasticizers on mechanical, thermal, and microstructural properties of flour films. *Journal of Cereal Science, 60*, 60–66.

Souza, E. M., Ferreira, M. R., & Soares, L. A. (2022). Pickering emulsions stabilized by zein particles and their complexes and possibilities of use in the food industry: A review. *Food Hydrocolloids, 131*, 107781.

Santos, T. M., Souza Filho, M.d. S. M., Silva, E.d. O., Silveira, M. R. S.d., Miranda, M. R. A.d., Lopes, M. M. A., & Azeredo, H. M. C. (2018). Enhancing storage stability of guava with tannic acid-crosslinked zein coatings. *Food Chemistry, 257*, 252–258.

Tihminlioglu, F., Atik, İ. D., & Özen, B. (2010). Water vapor and oxygen-barrier performance of corn–zein coated polypropylene films. *Journal of Food Engineering, 96*(3), 342–347.

Tivano, F., & Chiono, V. (2023). Zein as a renewable material for the preparation of green nanoparticles for drug delivery. *Frontiers in Biomaterials Science, 2*, 1156403.

Tomadoni, B., Capello, C., Valencia, G. A., & Gutiérrez, T. J. (2020). Self-assembled proteins for food applications: A review. *Trends in Food Science & Technology, 101*, 1–16.

Tomlinson, S. R., Kehr, C. W., Lopez, M. S., Schlup, J. R., & Anthony, J. L. (2014). Solubility of the corn protein zein in imidazolium-based ionic liquids. *Industrial & Engineering Chemistry Research, 53*(6), 2293–2298.

Tortorella, S., Maturi, M., Buratti, V. V., Vozzolo, G., Locatelli, E., Sambri, L., & Franchini, M. C. (2021). Zein as a versatile biopolymer: Different shapes for different biomedical applications. *RSC Advances, 11*(62), 39004–39026.

Torun, M., & Ozdemir, F. (2022). Milk protein and zein coatings over peeled garlic cloves to extend their shelf life. *Scientia Horticulturae, 291*, 110571.

Turasan, H., Barber, E. A., Malm, M., & Kokini, J. L. (2018). Mechanical and spectroscopic characterization of crosslinked zein films cast from solutions of acetic acid leading to a new mechanism for the crosslinking of oleic acid plasticized zein films. *Food Research International, 108*, 357–367.

Uan-on, T., Baibang, C., & Shuwisitkul, D. (2018). Extraction and characterization of zein protein from corn for controlled drug release. *Current Applied Science and Technology, 18*(3), 167–178.

Ünalan, İ. U., Arcan, I., Korel, F., & Yemenicioğlu, A. (2013). Application of active zein-based films with controlled release properties to control *Listeria monocytogenes* growth and lipid oxidation in fresh Kashar cheese. *Innovative Food Science & Emerging Technologies, 20*, 208–214.

Verbeek, C. J., & van den Berg, L. E. (2010). Extrusion processing and properties of protein-based thermoplastics. *Macromolecular Materials and Engineering, 295*(1), 10–21.

Vitiello, A., Catanzano, O., Guida, M., Scognamiglio, A., Gallo, A., Costabile, G., Ungaro, F., Miro, A., & Quaglia, F. (2025). Harnessing HP-β-CD in zein edible coatings for enhanced sensitive fruits preservation. *Carbohydrate Polymers, 367*, 123969.

Wang, D., Liu, Y., Sun, J., Sun, Z., Liu, F., Du, L., & Wang, D. (2021). Fabrication and characterization of gelatin/zein nanofiber films loading perillaldehyde for the preservation of chilled chicken. *Foods, 10*(6), 1277.

Wang, X., & Fan, M. (2019). Interaction behaviors and structural characteristics of zein/NaTC nanoparticles. *RSC Advances, 9*(10), 5748–5755.

Wang, Y., & Padua, G. W. (2003). Tensile properties of extruded zein sheets and extrusion blown films. *Macromolecular Materials and Engineering, 288*(11), 886–893.

Wu, X., Liu, Z., He, S., Liu, J., & Shao, W. (2023). Development of an edible food packaging gelatin/zein based nanofiber film for the shelf-life extension of strawberries. *Food Chemistry, 426*, 136652.

Wu, Y., Du, J., Zhang, J., Li, Y., & Gao, Z. (2023). pH effect on the structure, rheology, and electrospinning of maize zein. *Foods, 12*(7), 1395.

Wu, Y., & Messing, J. (2014). Proteome balancing of the maize seed for higher nutritional value. *Frontiers in Plant Science, 5*, 240. https://doi.org/10.3389/fpls.2014.00240.

Wu, Y., Holding, D. R., & Messing, J. (2010). -Zeins are essential for endosperm modification in quality protein maize. *Proceedings of the National Academy of Sciences, 107*(29), 12810–12815.

Xiao, Y., Liu, Y., Kang, S., Cui, M., & Xu, H. (2021). Development of pH-responsive antioxidant soy protein isolate films incorporated with cellulose nanocrystals and curcumin nanocapsules to monitor shrimp freshness. *Food Hydrocoll, 120*, 106893.

Xing, J., Li, Z., Zhang, W., & Wang, P. (2023). The composition, structure, and functionalities of prolamins from highland barley. *Molecules, 28*(14), 5334.

Xu, H., Chai, Y. W., & Zhang, G. Y. (2012). Synergistic effect of oleic acid and glycerol on zein film plasticization. *Journal of Agricultural and Food Chemistry, 60*, 10075–10081.

Yuan, Y., Ma, M., Xu, Y., & Wang, D. (2022). Surface coating of zein nanoparticles to improve the application of bioactive compounds: A review. *Trends in Food Science & Technology, 120*, 1–15. https://doi.org/10.1016/j.tifs.2021.12.025.

Zapata, P. J., Guillén, F., Martínez-Romero, D., Castillo, S., Valero, D., & Serrano, M. (2008). Use of alginate or zein as edible coatings to delay postharvest ripening process and to maintain tomato (*Solanum lycopersicon* Mill) quality. *Journal of the Science of Food and Agriculture, 88*(7), 1287–1293. https://doi.org/10.1002/jsfa.3220.

Zdraveva, E., G Srček, V., Kraljić, K., Škevin, D., Slivac, I., & Obranović, M. (2023). Agro-industrial plant proteins in electrospun materials for biomedical application. *Polymers, 15*(12), 2684.

Zhang, Y., Cui, L., Che, X., et al. (2015). Zein-based films and their usage for controlled delivery: Origin, classes and current landscape. *Journal of Controlled Release, 206*, 206–219.

Zheng, X. Q., Liu, X. L., Yu, S. F., et al. (2014). Effects of extrusion and starch removal pretreatment on zein proteins extracted from corn gluten meal. *Cereal Chemistry, 91*(5), 496–501.

Chapter 15

Galactomannan films and coatings

Wilbert Rodriguez-Canto[1,2], Víctor Moo-Huchiin[2], Luis Chel-Guerrero[1], David Betancur-Ancona[1], and Eduardo Castañeda-Pérez[1]

[1]*Facultad de Ingeniería Química, Universidad Autónoma de Yucatán, Mérida, Yucatán, México,* [2]*Instituto Tecnológico de Mérida Campus Norte, Tecnológico Nacional de México, Mérida, Yucatán, México*

15.1 Introduction

Pollution arising from the use of fossil-based plastics in packaging that is not properly disposed of after use is one of the main problems facing the environment on a global scale. Most plastic packaging ends up in the trash or in landfills as nondegradable waste. Despite efforts to reduce plastic waste, the problem persists. Therefore, it is up to modern societies to develop alternatives in order to preserve the environment and reduce pollution.

In food packaging, the use of plastic is popular and widely employed by humans due to its lightweight and effective barrier properties. However, its unsustainable origins and production processes do not favor ecological development or the practice of green chemistry. Due to recent concerns about the environmental threats associated with the massive use of petrochemical-derived plastic materials, increasing attention has been paid to scientific and technological research on the production of biodegradable and compostable packaging materials from biopolymers.

Edible coatings and films are known as environmentally friendly packaging materials, as they replace synthetic or plastic materials and are becoming a more sustainable option for food preservation. They are used to extend shelf life, prevent microbiological contamination, minimize lipid oxidation processes, and, in general, reduce the effects of food degradation. They are good carriers of functional ingredients and additives such as color, flavor, minerals, vitamins, and antioxidant agents to improve the functionality, nutritional value, and durability of food.

Polysaccharides such as starch, chitosan, alginates, carrageenans, cellulose, and their derivatives are among the most widely used materials to produce edible films and coatings. Galactomannan is a plant polysaccharide commonly found in the endosperm of legume seeds. The four main plant sources of galactomannans are carob (*Ceratonia siliqua*), guar (*Cyamopsis tetragonolobus*), tara (*Caesalpinia spinosa*), and fenugreek (*Trigonella foenum-graecum*).

Galactomannan is composed of a main chain of mannose units linked by β-(1→4) bonds and galactose units linked by α-(1→6) bonds as branches. It is a long-chain polymer with a high average molecular weight (>2,000,000 g mol^{-1}), and it has the potential to produce matrices with improved cohesive strength, making it a promising raw material for film formation. The galactose content and its distribution along the chain differ depending on the species, location, and environment. Furthermore, galactomannan is a nontoxic substance permitted for human consumption, which makes it particularly interesting for biomedical and food applications.

Differences in fine structure are also considered a determining factor in relation to the physicochemical and technological properties of galactomannans. The distribution of D-galactose groups along the mannan chain is a consequence of the in vivo synthetic mechanism, which varies from one species to another depending on structural differences around the active sites of the enzymes, and has a relevant effect on the functionality of the polysaccharide. In general, it is possible to generate edible films and coatings from galactomannans with good mechanical and rheological properties. However, due to their structure, they are hydrophilic materials, which have a low water vapor barrier and are sensitive to ambient humidity. This chapter explores the influence of the fine structure of galactomannans on the properties of edible films and coatings prepared from them.

Natural Polymer-Based Films and Coatings in Food Technology. DOI: https://doi.org/10.1016/B978-0-443-34244-8.00037-9

15.2 Galactomannans

Galactomannans are nonstarchy polysaccharides consisting of a linear chain of β-(1→4)-linked mannose units with α-(1→6)-linked D-galactose branches. These polysaccharides are primarily derived from the seeds of various leguminous plants, with the Leguminosae family being the main source of the most studied galactomannans (Geronço et al., 2021; Sharma et al., 2021). Within these plants and in fungal structures, galactomannans play crucial roles in maintaining structural integrity and functionality by enhancing mechanical strength and flexibility. In the endosperm of seeds, they also function as storage polysaccharides, providing energy during germination (Sharma et al., 2021).

The extraction of galactomannans generally involves obtaining the endosperm from ground, dehulled, and sifted seeds. The resulting flour often contains impurities, such as proteins and insoluble cellulosic polymers, which are also extracted during milling. Purification typically involves dissolution in water at various temperatures, followed by precipitation with ethanol, dialysis, and freeze-drying (Kontogiorgos, 2018).

Beyond their natural functions, galactomannans find extensive use in the food, pharmaceutical, and cosmetic industries. Their ability to form controlled-release drug delivery systems is particularly valuable for oral and topical formulations, enabling targeted delivery. In cosmetics, the moisturizing and emulsifying properties of galactomannans enhance the performance and sensory appeal of skincare products (Santos et al., 2015).

The unique structural composition of galactomannans—a linear mannan backbone with lateral galactose groups—dictates their diverse physicochemical properties and functionality. One key aspect is their interaction with water. Galactomannans are soluble in water, with their viscosity being influenced by temperature and the mannose-to-galactose ratio; higher galactose content generally increases solubility. Furthermore, their three-dimensional, entangled structure makes them highly hydrophilic, allowing them to absorb and retain significant amounts of water, leading to gel formation. The thermal and mechanical properties of galactomannans are also important for various applications. Their thermal stability is influenced by the degree of substitution (DS) and molecular weight, with higher DS often altering thermal transition temperatures. In terms of mechanical properties, galactomannans can form films and coatings, contributing to the structural integrity and barrier properties of these materials. Finally, galactomannans exhibit relatively good stability against enzymatic and chemical degradation, although this can be influenced by the specific source and processing conditions. The proportions of mannose and galactose in different sources of galactomannan vary, affecting their physical and functional properties. The seeds and powders of the most widely used galactomannans in industry are depicted in Fig. 15.1.

Fenugreek gum (FG) (*T. foenum-graecum* L.), with a mannose:galactose ratio of approximately 1:1 (Fig. 15.2), exhibits properties that make it useful in various pharmaceutical applications. FG has been extensively studied for its use

FIGURE 15.1 Seeds and powders of the gums: (A) guar, (B) tara, (C) fenugreek, and (D) locust bean. The various forms of the seeds used to extract guar, tara, locust bean, and fenugreek gums are presented, along with the galactomannan powder obtained from the seeds.
Modified from Prajapati, V. D. et al. (2013). Galactomannan: A versatile biodegradable seed polysaccharide. International Journal of Biological Macromolecules.

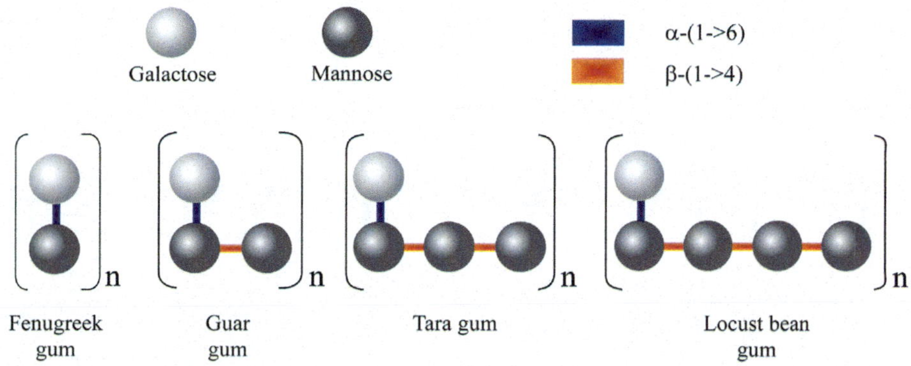

FIGURE 15.2 Structural representation of different galactomannans based on their mannose-to-galactose ratio.

as a rapidly dissolving matrix, disintegrant, binder, sustained-release agent, and suspending agent. Additionally, modified forms of FG, such as the octenyl succinate anhydride derivative and the amino derivative, have demonstrated potential in specialized applications, including bioadhesive intravaginal films (Kontogiorgos, 2018; Liu, Lei, et al., 2020).

Guar gum (GG) (*C. tetragonolobus*) has a mannose-to-galactose ratio of approximately 2:1 (Fig. 15.2). GG is widely utilized in the food industry. Its ability to thicken, stabilize, and emulsify makes it a valuable ingredient in a variety of products, including sauces, dressings, ice cream, and baked goods. It also improves texture and moisture retention. In pharmaceutical formulations, GG serves as a binder, disintegrant, suspending agent, and thickening agent. It has also been used to create matrix tablets, hydrogels, microspheres, and ocular inserts for controlled drug delivery. GG's susceptibility to microbial degradation in the large intestine makes it an ideal candidate for colon-targeted drug delivery. Research on the use of GG-based nanomaterials as drug carriers remains limited, highlighting an area for further study (Ghosh et al., 2018; Yadav & Maiti, 2020).

Tara gum (TG) (*C. spinosa*) has a mannose-to-galactose ratio of approximately 3:1 (Fig. 15.2). Its galactose substitution occurs approximately every third mannose unit, which is lower than that of fenugreek and GGs. This lower substitution allows TG to form stronger films. However, the industrial use of these films is limited by their poor mechanical and water vapor barrier properties. TG exhibits similar functional properties to guar and locust bean gums (LBGs), reaching high viscosity in water quickly and interacting synergistically with xanthan gum to increase gel strength and prevent syneresis. Due to its lower cost compared to guar and LBGs, TG serves as an alternative food additive in many countries, primarily as a thickener and stabilizer (Mukherjee et al., 2023).

LBG (*C. siliqua*), also known as carob gum, has a mannose and galactose ratio close to 4:1 (Fig. 15.2). LBG is a galactomannan derived from the seeds of the *C. siliqua* tree. It possesses biocompatible, bioabsorbable, biodegradable, nonteratogenic, and nonmutagenic properties. LBG has received Generally Recognized As Safe (GRAS) status from the US Food and Drug Administration (FDA), allowing its use in the food industry as a thickener, stabilizer, emulsifier, and gelling agent. It also finds application in the pharmaceutical industry as an excipient and in various biomedical applications, including superdisintegrants, matrix materials, buccal mucoadhesives, hydrogel microspheres, and floating tablets (Chen and Chen, 2018; Yadav & Maiti, 2020).

Other less-studied galactomannans are derived from *Delonix regia* (Rodriguez-Canto et al., 2019), *Punica granatum*, *Leucaena leucocephala*, *Gleditsia sinensis* Lam. (Liu, Lei, et al., 2020), *Morchella esculenta* (Zhang et al., 2022), *Astragalus gombiformis* Pomel (Fabaceae) (Bouziane et al., 2024), *Prosopis velutina* (Gómez-Rodríguez et al., 2024), and others. These galactomannans, despite being less researched, are also finding new and recent applications.

The mannose-to-galactose ratio, along with the distribution of galactose units along the mannose backbone (Fig. 15.3) (which can be regular, random, or in blocks), directly influences the ability of galactomannans to form intermolecular hydrogen bond networks. These networks, in turn, determine the material's viscosity, elasticity, rigidity, and gelling capacity (Dea & Morrison, 1975; Landin & Echezarreta, 2010; Liu, Lei, et al., 2020).

The mannose-to-galactose ratio in galactomannans, as well as their molecular weight, are characteristics that vary depending on the plant source. The Man/Gal ratio in legume seeds is fundamentally determined by the biosynthetic process, which involves the cooperative action of enzymes such as mannosyltransferases (responsible for constructing the mannan backbone) and galactosyltransferases (which add the galactose side branches). Differences in this mannose-to-galactose ratio among distinct plant sources can be attributed to the relative activity of mannosyltransferases and galactosyltransferases during galactomannan synthesis, the concentration of nucleotide-sugar substrates at the site of

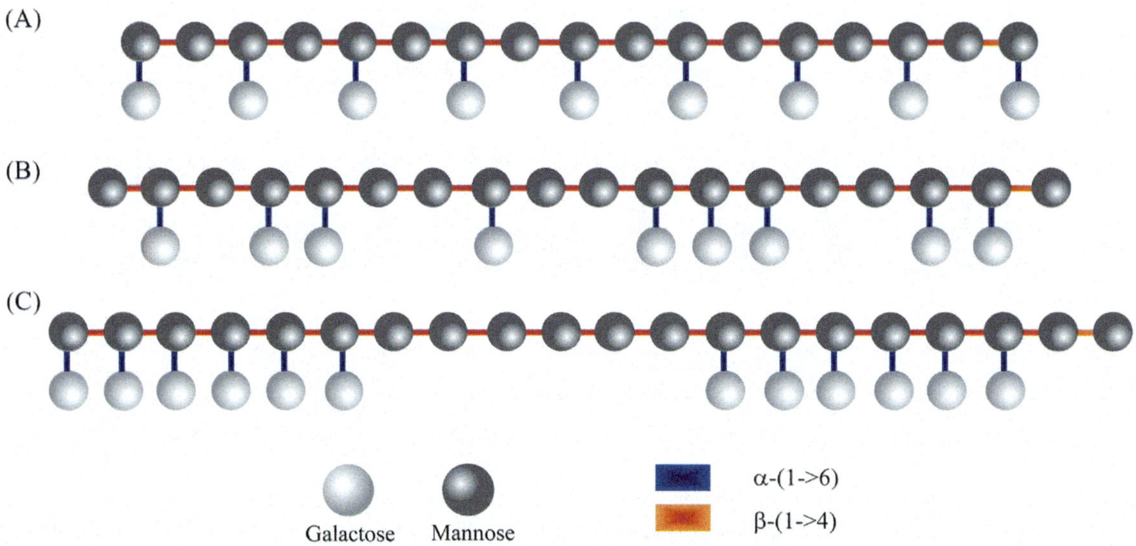

FIGURE 15.3 Types of galactose distribution along the galactomannan backbone: (A) regular, (B) random, and (C) block.

synthesis, and postdepositional modification by enzymes like α-galactosidase (Edwards et al., 1992). Furthermore, β-mannanases have also been found at the sites where galactomannans are synthesized; these enzymes could play a role in controlling the molecular weight of the chain.

Recent innovations have seen the emergence of galactomannans in the development of edible films and coatings. These films can improve food storage capacity while also providing microbial safety by acting as barriers against moisture and gases such as oxygen and carbon dioxide. Incorporating active compounds, including natural antimicrobials and antioxidants, into these films further expands their functionality, making them suitable for various food applications (Cerqueira et al., 2011).

The trend toward using renewable and biodegradable materials for packaging has positioned galactomannans as a viable alternative to traditional nonbiodegradable plastics. Several challenges and limitations affect galactomannans' efficacy and commercial viability in various applications. Therefore, formulation and packaging must be designed to ensure that galactomannans retain their properties throughout their shelf life, which poses additional challenges for manufacturers.

15.3 Preparation methods for edible films and coatings

Edible films and coatings are one of the primary forms of food packaging. The differences between them lie in their physical form and application process. Films are generally obtained as solid sheets and then used as a food wrap that can be removed or consumed with the food, while coatings are applied in liquid form by coating, spraying, or dipping. After drying on the food, an edible layer is formed that can be removed or consumed (Trajkovska Petkoska et al., 2021).

Basically, there are two methods for producing edible films and coatings: wet and dry processes (Barbosa et al., 2021). Wet processes use techniques such as casting, dipping, spraying, and spreading, while dry processes use thermopressing, thermoforming, and extrusion. The main difference between dry and wet processes is that the latter require a drying step. Films are typically prepared by casting, extrusion, and thermopressing/thermoforming, while coatings are prepared by dipping, spraying, and spreading (Umaraw et al., 2020).

15.3.1 Wet processes

Wet processes mainly comprise the application of the casting method or solvent casting, which is the most commonly used technique for film production at the laboratory or pilot scale. For this method, the film-forming components are homogenized with a solvent and dried to obtain a film. Conventionally, it involves three steps: (1) dissolving or dispersing the biopolymers in solvents is typically the first step that must be carried out to generate films; (2) additives such as plasticizers, colorants, flavorings, and antimicrobial agents (Moura-Alves et al., 2023) are added to the matrix material to increase the durability and flexibility of these materials (Umaraw et al., 2020). A possible requirement will be to adjust the pH and heat the solutions above their melting point to increase their solubility; (3) this film-forming solution is then poured onto a flat surface and dried at the appropriate temperature and relative humidity.

The casting method is the most commonly chosen for the preparation of edible active packaging and is a well-established technique at the laboratory scale. For this process, the dissolution of the selected biopolymers in a specific film solution is required, along with the chosen plasticizer and active compounds. Water or ethanol is normally the selected solvent, and after the pH is adjusted, the solution is heated above the melting temperature until complete homogenization. Finally, the solution is spread on a flat surface to dry (Suhag et al., 2020). Different drying processes can be applied, such as air drying, hot surface drying, infrared, and microwave techniques, and the film obtained after the drying step is used to wrap the food sample (Fig. 15.4). Despite its effectiveness in lab settings, the casting method faces challenges in industrial-scale production due to its long drying times and size limitations (Jeya Jeevahan et al., 2020).

Spraying, dipping, and spreading techniques follow the same steps as the casting technique up to the point of cooling the solution. The spraying technique is also a very common method that allows for a uniform coating around the food sample, as the process is easy to control. The solution is sprayed onto the food sample using nozzles, while the sample rotates at a controlled speed (Kurek et al., 2017). For the dipping technique, the food sample is immersed in the solution, and after it cools, the excess liquid is drained, and the sample is allowed to air dry. The dipping technique is low cost and easy to apply, but there may be microorganism growth in the dipping tank, in addition to generating a large amount of waste (Andrade et al., 2012). The spreading, or brushing, technique consists of spreading the solution over the surface of the food sample with the help of a brush or spatula. It is normally applied to small foods, such as fruits and vegetables, and results in homogeneous coatings, the quality of which depends on the efficiency of the operator (Yan et al., 2019).

These wet processes can be applied in a layer-by-layer mode, where they are used with different polymeric matrices sequentially. The layer-by-layer technique enhances polymeric properties in food packaging but requires additional production time, as each layer must dry before applying the next (Blanco-Pascual & Gómez-Estaca, 2016).

The most important advantage of wet methods is that they can be easily employed without special equipment. Furthermore, as many food processing materials cannot withstand higher temperatures, the lower temperatures during processing steps can be beneficial. They have disadvantages, such as a longer drying time, making them unsuitable for commercial manufacturing. In addition, scaling up film production from the laboratory to large-scale production presents a challenge due to the possibility of variations in quality and a risk of limiting the consistent development of commercial operations caused by factors such as heating, speed, and temperature (Suhag et al., 2020).

15.3.2 Dry processes

In dry processes, the added components are converted into films by taking advantage of the thermoplastic behavior of the mixture, using low moisture levels. The techniques are varied and consist of extrusion, compression molding, injection molding, etc. (Kumar et al., 2022).

The extrusion technique is preferred in the industry as it allows for obtaining food packaging with improved mechanical and optical properties in a very short period and with low energy consumption. Although extrusion is generally low cost, it

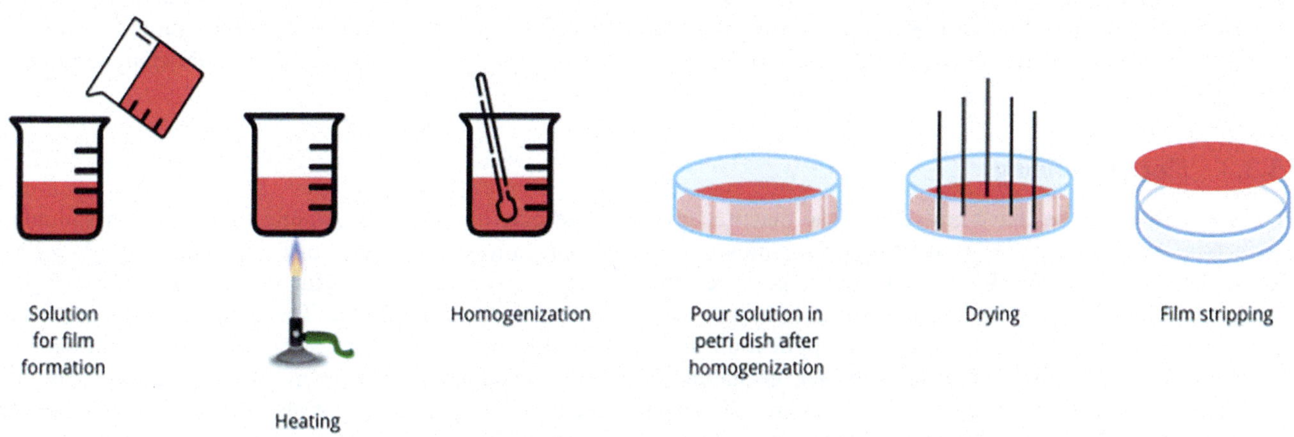

FIGURE 15.4 Wet casting method for edible film formation.

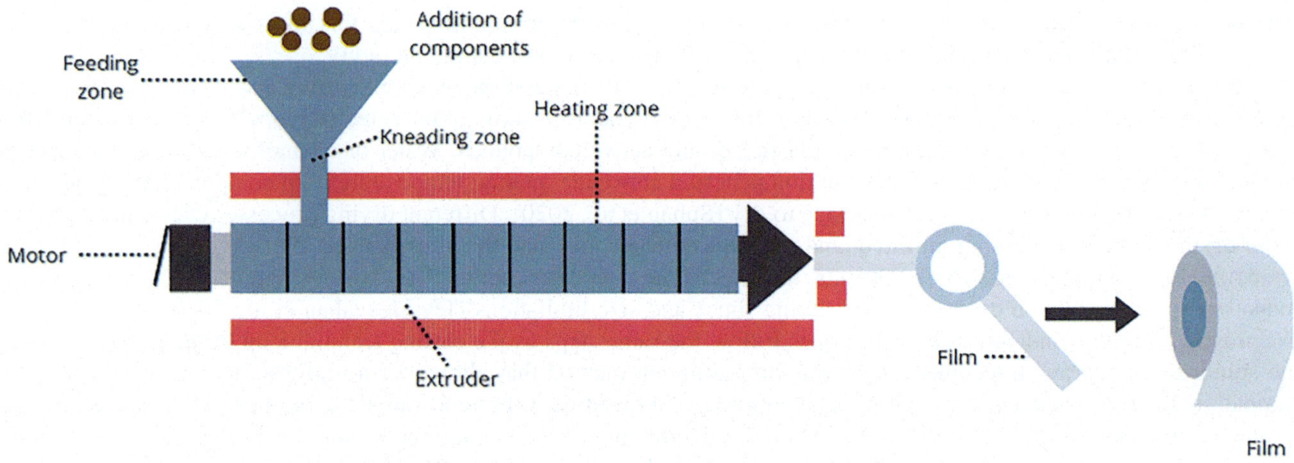

FIGURE 15.5 Dry process by extrusion technique for edible film formation.

requires a high initial investment for equipment and maintenance. This technique involves a minimal content of solvents and plasticizers, the possibility of using viscous polymers, and easy control of operating temperature and pressure (Fig. 15.5). However, some polymers cannot be used with this technique due to the limitation of using low moisture levels or because the materials have low-temperature tolerance. For example, chitosan is not a thermoplastic material and cannot be extruded. Extrusion can also be applied with multiple layers in a process known as coextrusion, which allows for improving and diversifying the properties of the films (Li & Ye, 2017; Suhag et al., 2020).

Other dry processes include compression molding, injection molding, and thermopressing (Nair et al., 2023). Compression molding is one of the oldest methods and is considered a sustainable process due to its low energy requirement and rapid film formation. The polymeric materials are heated under high pressure in the mold until they solidify, with processing time being a critical parameter in determining the mechanical and optical properties of the films. It is generally used in combination with the extrusion method to prepare the film-forming material before its main thermoforming process, and the films produced can improve the shelf life of food, providing a promising active food packaging material (Ceballos et al., 2020).

The injection molding technique is primarily used to manufacture plastic items, but it is also suitable for the mass production of edible films. It consists of three stages: filling, packing, and cooling, and its parameters are injection pressure, molding temperature, and preinjection pressure temperature (Lisitsyn et al., 2021). This method can be used in combination with the extrusion, and studies on the injection molding method are scarce (Kumar et al., 2022). Another dry process is thermopressing, also known as thermoforming, which is commonly used for the production of utensils, particularly plastic ones. Similar to extrusion, it involves high temperatures and pressures for film formation. It is used on an industrial scale and has recently been applied to the production of bioactive films (Blanco-Pascual & Gómez-Estaca, 2016).

15.3.3 Electrospinning

Recently, new processes have emerged in the manufacture of edible active packaging. Among them is electrospinning, which is a simple technique used to produce nanoscale fibers from various polymeric materials using a high-voltage electric field. The technique has several advantages, such as low cost for producing films with ultrafine structures, tailored morphology, high porosity, and a high surface area-to-volume ratio (Zhang et al., 2020). The temperature is controlled during the process, which allows for the application and preservation of the characteristics of sensitive compounds, making it advantageous for effectively encapsulating bioactive compounds. However, electrospinning has low productivity, which limits large-scale production. The characteristics of the polymers can also limit this technique due to their barrier properties and poor mechanical properties (Wen et al., 2021).

Some review articles have provided an overview of the electrospinning technique for a range of applications. For example, Torres-Giner et al. (2016) reviewed electrospun nanostructures as advanced bioactive platforms for tissue engineering, drug delivery, food packaging, and functional coatings. The functionalization of electrospun surfaces through the incorporation of bioactive substances has also been proposed, highlighting the encapsulation of compounds in electrospun nanofibers for potential food applications. New strategies are proposed to develop novel functional foods through the use of emulsion electrospinning and coaxial electrospinning techniques, using natural biopolymers based on polysaccharides and proteins (Wen et al., 2021).

Although a variety of electrospinning designs have been developed in the past few years, the current studies mainly focus on the formation, morphology, and structure of nanofibers by blending, emulsion, and coaxial electrospinning, whereas fewer applications in food packaging systems have been explored. Additionally, these works are still at the laboratory scale, and the commercial production in active food packaging is hindered by relatively low productivity (Zhang et al., 2020).

Another challenge is the strong plasticization caused by the hydrophilicity of natural biopolymers, which deteriorates the barrier property at higher humidity and therefore considerably limits their applications. Hence, the improvement of electrospinning-based fibrous mats in the stability, functionality, barrier, and mechanical properties is greatly significant for packaging applications (Tian et al., 2013).

15.3.4 Electrohydrodynamic atomization (electrospraying)

Films and coatings have traditionally been prepared by casting and extrusion. Electrohydrodynamic atomization (EHDA), also known as electrostatic atomization or electrospraying, is a novel processing route for forming thin films, as it presents significant advantages such as the relatively easy generation of droplets, the avoidance of coalescence due to the electric charge of the same polarity on the droplets, and a narrow droplet size distribution by obtaining the cone-jet mode (Watanabe et al., 2003).

Electrohydrodynamic spray coating technology is suitable for the additive manufacturing of thin films on the micro- and nanoscale. It is a process in which a liquid is forced through a capillary, and a potential difference on the order of kilovolts is applied between the capillary and the collecting electrode (Fig. 15.6). Electrosprays have a high intrinsic deposition efficiency due to their guided deposition on the collecting electrode, which is favored by the electric field. However, there are still some challenges in quality control and parameter adjustment during the coating process, especially when coating on nonconductive and nonhydrophilic substrates (Hu et al., 2024).

EHDA can occur in many modes, but the stable cone-jet mode is the most studied and desirable, as it generates nearly monodisperse droplets of a few micrometers in size. In this way, the droplet size can be controlled by varying the flow rate, the applied voltage, or the physical properties of the electrosprayed liquid. This phenomenon was first systematically studied by Zeleny between 1914 and 1917. Its technological importance increased in the last two decades after discoveries about its application in determining the mass of large biomolecules by mass spectrometry (Pareta & Edirisinghe, 2006).

15.3.5 Method of application of edible films and coatings on food products

An important factor to consider is the method of application of the films or coatings, since the form or technique used affects the cost and effectiveness and must be adapted to the food production process. A well-chosen application method ensures uniform coverage, maximizing the product's shelf life (Costa et al., 2018). The main techniques used for food coating are spraying, electrostatic spraying, and dipping. Other methods include brushing, individual wrapping, solvent casting, extrusion, spreading, thin-film hydration, encapsulation, and fluidized bed processing (Bizymis & Tzia, 2022). Edible films and coatings must be uniform and free of defects to optimize the functions for which they were designed (Ju et al., 2019).

15.3.6 Properties of films and coatings

Among the functions of films and coatings in the food industry, the protection of food from the external environment during distribution and storage can be highlighted, with the goal of extending its shelf life. For films and coatings to be able to fulfill this objective, they are required to have the appropriate properties, which will depend on the product they will contain. Among the main properties to consider are gas barrier, mechanical, and rheological properties, in addition to solubility, among others.

The rheological properties of film- or coating-forming solutions help to understand the possible interactions of the polymer chains, which will determine the properties of the films generated from them (Zhang et al., 2019). Furthermore, properties such as viscosity are important for the correct application of the forming solution in coatings, as well as for the handling the solutions in the production of films.

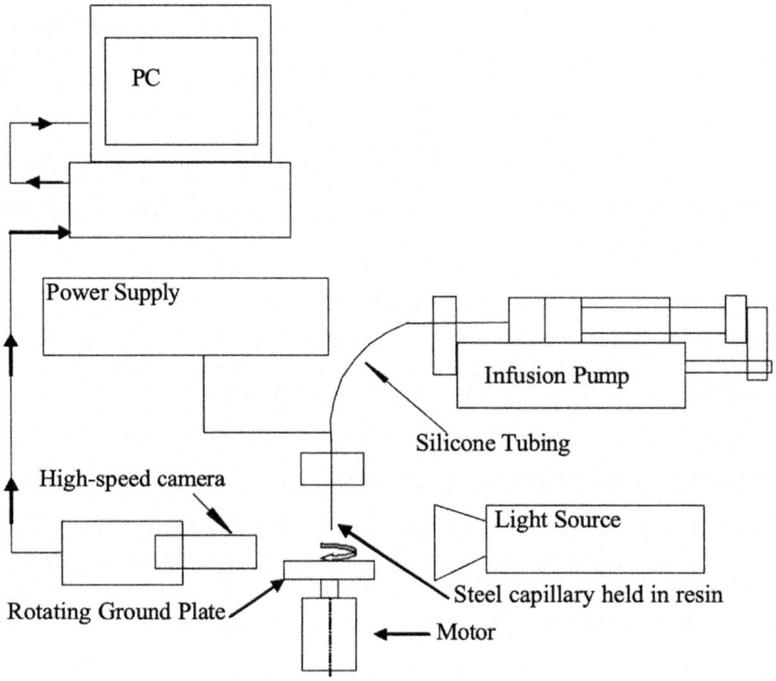

FIGURE 15.6 Schematic representation of equipment set-up used for the electrohydrodynamic experiments.
From Pareta, R., & Edirisinghe, M.J. (2006). A novel method for the preparation of starch films and coatings. Carbohydrate Polymers, 63(3), 425–431.

On the other hand, gas barrier properties are determining factors in the preservation of many edible products, since some require the absence of oxygen to prevent oxidation and rancidity (high fat content), while in vegetable products, for example, it will depend on the respiration rate (Yun et al., 2024). Water vapor permeability (WVP) is one of the most studied properties, since high permeability could affect the texture of a dry food due to moisture gain, in contrast to water loss in a product with high water content, which is undesirable.

It is expected that the mechanical properties of the film will have a high maximum tensile strength (TS), allowing it to withstand the stresses to which the product will be subjected during handling and transport, as well as a low elongation at break, but with a sufficient value so that it is not brittle (Niranjana Prabhu & Prashantha, 2018).

Therefore, it is essential to know these properties to determine the feasibility of using the polymer or to chart routes to improve properties that are deficient.

15.4 Solubility and rheological properties of galactomannans

15.4.1 Solubility of galactomannans

A fundamental property of galactomannan gums is solubility, since, according to their behavior in an aqueous environment, they can exhibit different functional properties. As a general rule, more soluble gums tend to form more viscous solutions, and factors such as galactose content, temperature, and particle size influence solubility (Table 15.1) and, therefore, the viscosity of the gum.

The ability of a galactomannan to disperse and completely dissolve in an aqueous medium allows the polymer molecules to hydrate and unfold, which increases the internal resistance to flow and, therefore, the viscosity. As can be seen in Table 15.1, in general, the solubility of galactomannan gums is strongly influenced by their composition, specifically by the mannose-to-galactose (M/G) ratio, since a greater amount of galactose generally leads to greater water solubility. Likewise, the arrangement of galactose side chains along the mannan backbone is influential, given that the presence of more galactosyl groups in the galactomannan structure causes chain bending when maintained in solution and reduces the intrinsic chain stiffness. This is because the galactose side chains create steric hindrance, which reduces the formation of hydrogen bonds between galactomannan molecules and, therefore, increases interaction with water, improving solubility (Sharma et al., 2021). For example, GG, with an M/G ratio of around 2, is more water-soluble than LBG, which has an M/G ratio of around 4. FG is the most soluble, while GG dissolves in water at room temperature and

TABLE 15.1 Solubility behavior of tara, locust bean, guar, and fenugreek gums at different temperatures.

Galactomannan	Cold water solubility	Hot water solubility	Notes	References
Tara	Fairly	High	Slowly disperses in cold water	Mukherjee et al. (2023)
Locust	Slightly	<90%	70%–85% when heated to 80°C 30 min^{-1}	Barak and Mudgil (2014)
Guar	High	High	A 1% aqueous dispersion exhibits a viscosity up to 10,000 cP	Sharma and Sharma, 2018)
Fenugreek	High	High	Higher solubility than guar gum	Gadkari et al. (2019); Liu, Lei, et al. (2020)

LBG requires temperatures close to 80°C for dissolution, while FG has a high dissolution and solubility capacity at room temperature. The distribution of galactose groups along the mannan backbone also affects the solution properties. Although the exact distribution can vary, a more uniform galactose distribution tends to improve solubility. Regions of the mannan backbone that are not substituted with galactose can associate with each other through hydrogen bonds, which reduces solubility. LBG, with more unsubstituted mannan regions, has a greater propensity to form gels as a result of the synergism when forming junction zones. Finally, the rate of galactomannan dissolution and viscosity development also depends on the salt concentration of the medium, but since they are uncharged molecules, they are independent of pH.

15.4.2 Rheological properties of galactomannans

Galactomannans are versatile polysaccharides that have gained relevance in the food and pharmaceutical industries, among others, due to their unique functional properties. The rheological properties of polymers are fundamental to understanding their behavior in various applications. These properties are influenced by factors such as molecular weight, concentration, and temperature. Polymers typically exhibit high viscosity due to the entanglement of their molecular chains, which increases with molecular weight and concentration. Likewise, they often show non-Newtonian behavior, such as shear thinning, where viscosity decreases with increasing shear rate. Furthermore, they have viscoelastic characteristics, combining properties of viscous liquids and elastic solids. This is reflected in their storage modulus (G'), which measures elastic energy, and loss modulus (G''), which measures energy dissipation; the ratio of which ($\tan \delta$) helps to determine whether elastic or viscous behavior predominates (Brady et al., 2017; Sharma et al., 2021).

The presence of galactose side groups separates the mannan chains; therefore, the most interactive galactomannans are those whose backbones have fewer galactose substitutions. This results in the viscosity of a galactomannan solution increasing as the number of galactose side groups is reduced, but only up to a certain point, which can be considered optimal for each galactomannan. After that point, the viscosity decreases due to the lower solubility of the resulting polymers (Liu, Chang, et al., 2020; Senarathna et al., 2023). This general behavior is observed in the viscosity of the different galactomannans in Fig. 15.7.

Fig. 15.7A shows how the apparent viscosity varies as a function of frequency at different concentrations (% w/w) and is measured at 25°C. The presence of galactose groups acts as a steric hindrance, which hinders interactions between the mannose chains. The reduction of these groups allows the mannose chains to come closer together and form stronger interactions, which increases viscosity, as observed in the cases of GG with a lower galactose content compared to that of fenugreek at the same concentration (Fig. 15.7A). However, an excessive reduction of galactose decreases solubility (Senarathna et al., 2023), which is the case with LBG, and therefore, viscosity, as observed in Fig. 15.7B. In all solutions, a clear pseudoplastic tendency is observed: the apparent viscosity decreases with increasing shear rate, a characteristic behavior of non-Newtonian fluids. This behavior is explained by the disentanglement and alignment of the polymer chains in the direction of flow, which reduces interaction between adjacent chains (Liu, Chang, et al., 2020). Concentration significantly affects viscosity; higher viscosities are observed at higher concentrations. At higher concentrations, the increased presence of polymer chains enhances molecular interactions and entanglement, leading to higher viscosity (Mukherjee et al., 2023). It can be observed that at high concentrations, LBG exhibits the highest viscosity, suggesting a lower interaction with water due to its structure with a higher proportion of mannose; conversely, fenugreek shows the lowest viscosity at similar concentrations due to its greater interaction with water, a product of its higher proportion of galactose. GG exhibits intermediate behavior in terms of viscosity. At high shear rates ($1000\ s^{-1}$), the viscosities of different galactomannans converge. For instance, fenugreek at 0.25% and GG at 1% exhibit viscosities ranging from 2 to 20 mPa·s. This suggests that, under strong flow forces, molecular interactions become negligible (Fig. 15.7B).

The intricate relationship between the structure of galactomannans and their viscosity and flow behavior significantly impacts their applications (Senarathna et al., 2023). In addition to the ratio of galactose to mannose units in galactomannans, their molecular configuration determines how these polysaccharides interact with water, influencing their ability to thicken and stabilize solutions in such a way that, at certain concentrations, they form viscous solutions or gels, and their interaction with other polysaccharides (Sharma et al., 2021). The ratio of galactose to mannose units and the molecular configuration of the galactomannan influence its interaction with water, determining its ability to thicken, stabilize solutions, and form films (Liu and Chang, et al., 2020).

15.5 Gas barrier properties

15.5.1 Permeability of gases in films and coatings

The gas barrier properties of a film or coating are determined by its permeability to O_2, CO_2, and H_2O. A simple model used to explain the phenomenon of gas passage through a dense polymeric film is the solution-diffusion model, in which

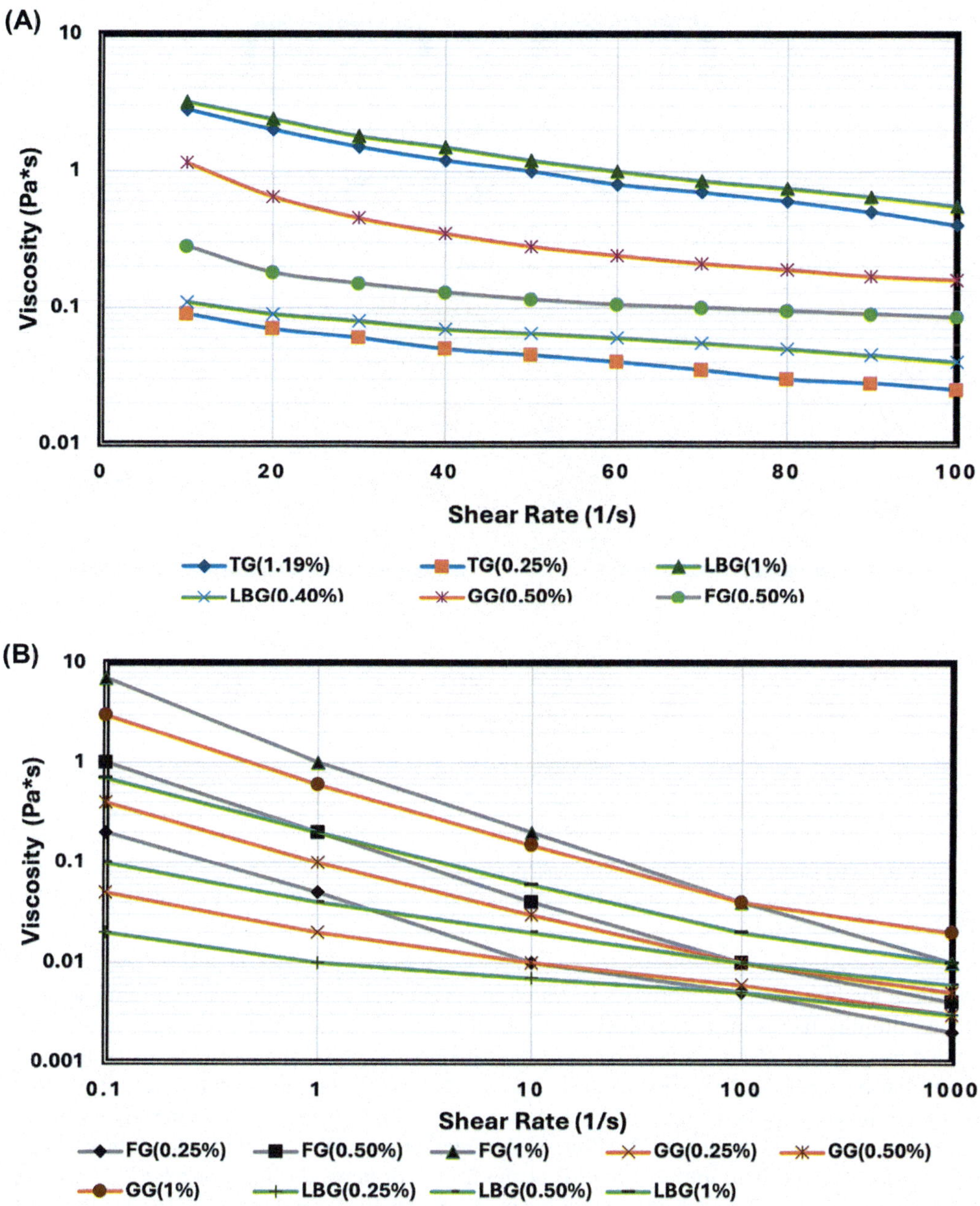

FIGURE 15.7 Flow curves of various galactomannans at different shear rates: (A) short range and (B) wide range. *TG*, Tara gum; *LBG*, locust bean gum; *GG*, guar gum; *FG*, fenugreek gum.
From Liu, F., Chang, W., Chen, M., Xu, F., Ma, J. & Zhong, F. (2020). Film-forming properties of guar gum, tara gum and locust bean gum. Food Hydrocolloids, 98. Sittikijyothin, Torres & Gonçalves. (2005). Modelling the rheological behaviour of galactomannan aqueous solutions. Carbohydrate Polymers, 59(3), 339–350.

permeability (P) is given by the product of the diffusivity (D) and the solubility (S) of the gas in the film. It is important to note that in this model, it is assumed that the process occurs in a steady state (Gökkaya Erdem et al., 2021). The flow of gas through the polymer matrix is generally driven by a concentration gradient of the gas on both surfaces of the film, for example, by the difference in partial pressures of the permeating gas between the outer and inner surfaces, as well as by the temperature difference. In Fig. 15.8, the solution-diffusion process is represented, which begins with the adsorption of the permeating gas on the surface of the film, which can be idealized as the solubilization process described by Henry's law. Subsequently, the gas diffuses through the film, obeying Fick's law, and finally exits by desorption on the opposite surface of the film, also represented by Henry's law. Solubility can be understood as the partitioning behavior of the gas in contact with the film surface, which will depend on the properties of the gas and the film material. On the other hand, the

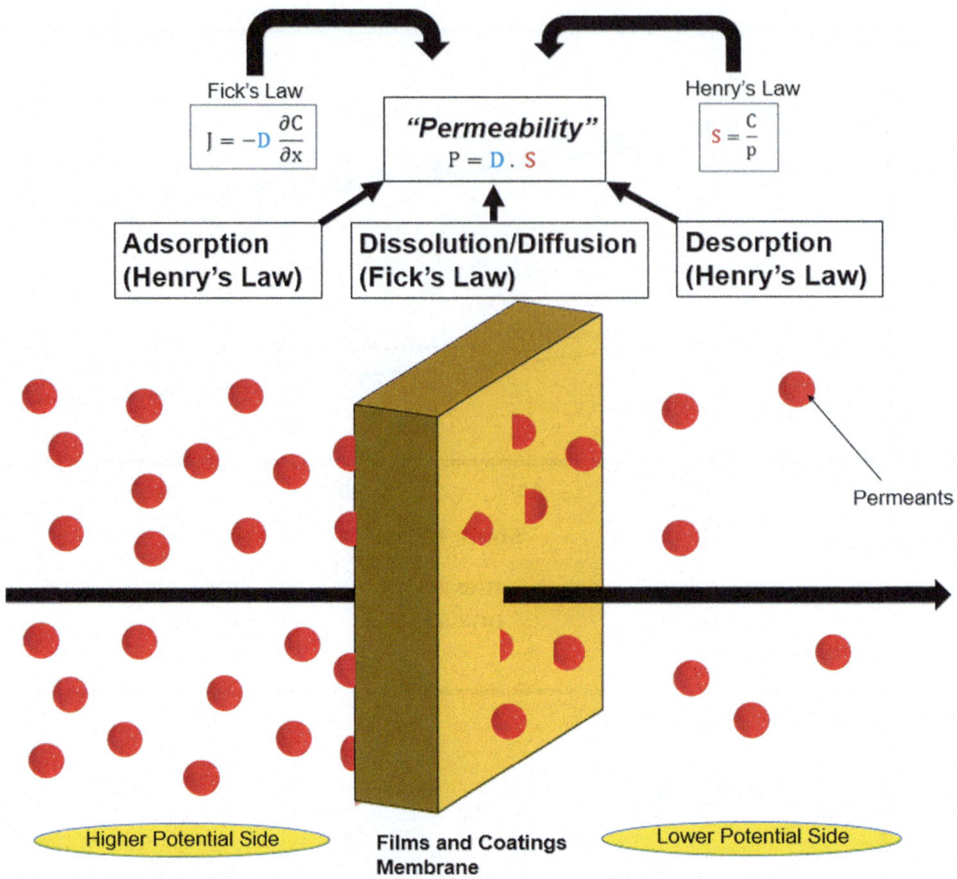

FIGURE 15.8 Solution-diffusion model of gas permeability through dense films.
From Trinh, B. M., Chang, B. P. & Mekonnen, T. H. (2023). The barrier properties of sustainable multiphase and multicomponent packaging materials: A review. Progress in Materials Science, 133.

passage of gas within the film is a kinetic process in which the properties of the film material influence its behavior and is generally the main limiting factor in permeability.

Among the main properties of the material, the chemical structure, the polymer shape (microstructure), crystallinity, and molecular weight can be mentioned (Agarwal, 2021; Xue et al., 2023). The chemical structure of the film or coating material results in the polarity of the chains, thus determining the affinity between the material and the gas. Polarity is a crucial factor for the solubility of the gas in the film, as the interaction of the gas with the surface chains, as well as its ability to make its way through them into the interior of the film, will depend on it. For example, polymers with a large number of -OH groups (such as polysaccharides) will have a high affinity for polar gas molecules like water vapor, while they will have a low affinity for nonpolar gas molecules like O_2 and CO_2. The morphology and architecture of the polymer are two factors that affect the diffusivity of the permeating gas within the film. Gas penetration and movement within this material are enabled by the free volume, which also permits the mobility of polymeric chains. This volume will depend on the cohesive forces that polymer chains have in inter and intramolecular interactions, as well as on the steric repulsions arising from some bulky groups. A similar effect is observed in films with semicrystalline structures, where the crystalline regions, in addition to restricting the movement of the chains in the amorphous phase, also act as impermeable barriers to gases, making their path more tortuous and thus decreasing gas permeability (Trinh et al., 2023).

15.5.2 Relationship of galactomannan characteristics to the gas barrier properties of their films or coatings

Differences in the characteristics of galactomannans extracted from various plant sources lead to variations in the gas barrier properties of films made from them, particularly due to the M/G ratio and the distribution of galactose units along the mannan backbone (Chel-Guerrero et al., 2024).

Water vapor barrier properties are typically quantified by parameters such as the water vapor transmission rate (WVTR), permeance, and WVP, which can be determined according to the American Society for Testing and Materials (ASTM) E96. WVTR represents the mass flow rate of water vapor per unit film area and is typically expressed in $g{\cdot}h^{-1}{\cdot}m^{-2}$. Permeance, on the other hand, accounts for the partial pressure difference across the film faces and is expressed in $g{\cdot}h^{-1}{\cdot}m^{-2}{\cdot}Pa^{-1}$. Both WVTR and permeance depend on film thickness and, therefore, cannot be considered intrinsic material properties. However, WVP normalizes the permeance by the film thickness (more precisely, WVP is typically calculated as permeance multiplied by thickness), and its units are typically $g{\cdot}m^{-1}{\cdot}Pa^{-1}{\cdot}s^{-1}$. Consequently, WVP is only a function of the temperature and humidity of the measurement. Galactomannans generally exhibit poor water vapor barrier properties due to the large number of -OH groups in their structure, which allows water molecules to interact via hydrogen bonds with the chains, enabling their diffusion through the film. Conversely, the presence of these polar groups causes the film to act as a good barrier against nonpolar gases such as CO_2 and O_2. In addition to polarity, the M/G ratio, as well as the distribution of galactose units along the chain, will influence the degree of chain packing due to potential inter and intramolecular interactions (Fig. 15.9). However, care must be taken when attributing the effect of each of these characteristics, as the gas barrier properties depend on all of these factors together. The use of plasticizers is common to produce flexible polysaccharide films, as without them, these films tend to be rigid and brittle. Generally, the addition of plasticizers increases the film's gas permeability since they increase free volume, thereby enhancing gas diffusion through the film. Furthermore, the plasticizer's polarity can alter gas solubility within the film.

The effect of the M/G ratio on the properties of films prepared from the galactomannans of *A. pavonina*, *C. tetragonolobus*, *C. pulcherrima*, *C. siliqua*, and *S. japonica*, with M/G ratios of 1.3, 1.7, 2.9, 3.4, and 5.6, respectively, has been studied (Santos et al., 2015). These films were generated from a solution containing 1.5% (w/v) galactomannan and 0.5% (w/v) glycerol. It was observed that the contact angle (CA) of the films increased from 65.22 degrees to 96.03 degrees as the M/G ratio increased from 1.3 to 2.9, while upon increasing the M/G ratio from 2.9 to 5.6, the CA decreased to 65.16 degrees. The CA is related to the hydrophilicity of the film, so it can be an indicator of the adsorption of water molecules on the film surface (solubilization). With respect to the WVP, the same behavior presented in CA was observed when M/G ratio increased, going from 6.78×10^{-11} to 10.69×10^{-11} $g\ m^{-1}\ Pa^{-1}\ s^{-1}$ and then decreasing to 8.6×10^{-11} $g\ m^{-1}\ Pa^{-1}\ s^{-1}$. This initial increase in WVP could be due to the decrease in galactose residues on the main chain, which allowed for better hydration between the chains, while upon increasing the M/G ratio from 2.9 to 5.6, the linear zones without galactose residues on the chain were able to form zones with a high packing density, hindering the diffusion of water molecules within the film. The CO_2 permeability showed the same behavior, increasing from 28.81×10^{-15} to 42.9×10^{-15} $g\ m\ (Pa\ s\ m^2)^{-1}$ and then decreasing to 28.74×10^{-15} $(Pa\ s\ m^2)^{-1}$. However, the O_2 permeability decreased as the M/G ratio increased from 1.3 to 3.4, from 2.79×10^{-15} to 1.75×10^{-15} $g\ m\ (Pa\ s\ m^2)^{-1}$. The same behavior was observed in films made with GG and LBG, using sorbitol as a plasticizing agent at 40% in relation to the galactomannan (Mikkonen et al., 2010), where the O_2 permeability decreased with increasing M/G ratio. Furthermore, in this research, the effect of relative humidity (RH) on WVP was also observed, finding that the guar and LBG films had a WVP of 3.94 ×

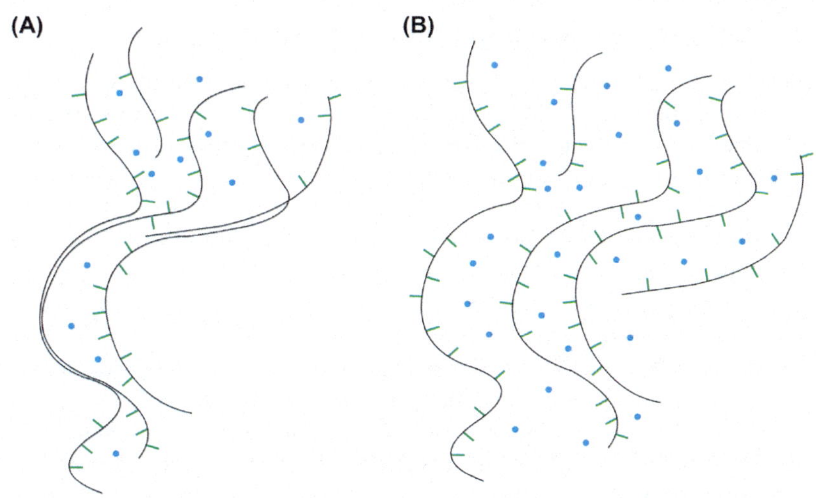

FIGURE 15.9 Interactions between galactose-unsubstituted regions (*green*) along the mannan backbone (*black*) and their effect on gas (*blue*) permeability. Galactomannan with (A) a high mannose-to-galactose ratio and (B) a low mannose-to-galactose ratio.

10^{-11} and 3.01×10^{-11} g m^{-1} Pa^{-1} s^{-1}, respectively, at relative humidity conditions of 0% inside and 54% outside the permeability cup. However, when changing the relative humidity conditions to 32% inside and 86% outside, the WVP showed no difference between the guar and LBG films, increasing at these %RH conditions to 4.34×10^{-11} g m^{-1} Pa^{-1} s^{-1}. Therefore, RH affects the diffusion and solubility mechanisms in the WVP of the films due to their swelling.

The reduction in gas permeability, due to the denser packing caused by the increase in the M/G ratio, is also supported by the increase in the glass transition temperature (Tg) of the galactomannan films as the M/G ratio rises. Ideally, an inverse relationship exists between the glass transition temperature (Tg) and free volume, as a larger free volume requires less thermal energy to initiate coordinated chain movement, leading to a lower Tg. However, factors such as interactions between pendant groups and the overall polymer structure can influence this behavior. This was observed in films made with GG and LBG, using glycerol as a plasticizer at 10% w/w in relation to the galactomannan, where the WVP decreased from 4.55×10^{-11} to 3.39×10^{-11} g m^{-1} Pa^{-1} s^{-1} and the Tg increased from −15.43°C to −12.14°C between GG and LBG, respectively. A lower Tg in the GG film indicates that there is greater chain mobility, due to a greater free space caused by the larger number of galactose residues, which prevent chain packing (Kurt & Kahyaoglu, 2014).

The interaction of the linear zones of the main chain without galactose residues in galactomannans can also be observed in their interaction with other polysaccharides. For example, the synergistic effect between the galactose-free sections of the LBG galactomannan chain (M/G = 4) and the double helices of κ-carrageenan in gel formation is well known, resulting in greater elasticity and decreased brittleness (Russo Spena et al., 2024). The production of films made with a mixture of κ-carrageenan, GG, and LBG has been reported, where the films obtained from the mixture of κ-carrageenan and LBG showed lower O_2 permeability, forming a denser film. This was attributed to the higher M/G ratio of LBG compared to that of GG, resulting in greater packing of the unbranched chains (Wang et al., 2023).

Furthermore, the properties of films and film-forming solutions have been related to the M/G ratio, using guar, tara, and LBGs. It was observed, via atomic force microscopy (AFM), that the microstructure of the gums in the film-forming solution varied depending on the M/G ratio, changing from a "bean-like" morphology to a "rod-like" morphology. This is attributed to the fact that, at low M/G ratios, the density of pendant galactose residues increases, which generates steric hindrance for the association of the galactomannan chains, while at higher M/G ratios, chain association increases. This result agrees with that observed in the transverse relaxation times of water protons, measured by low-field nuclear magnetic resonance (LF-NMR), which indicated that the reduction in hydrogen bonds between water/glycerol and the polymer is associated with the increase in the M/G ratio, possibly due to the lower availability of polar groups (e.g., -OH). This was also evident in the microstructure of the films, with a decrease in roughness on the surface of the tara and LBG films compared to those of GG. This relationship between the microstructure and the M/G ratio could be related to the decrease in the WVP of the films, going from 1.28×10^{-10} g m^{-1} Pa^{-1} s^{-1} for the GG film to 8.89×10^{-10} g m^{-1} Pa^{-1} s^{-1} for the LBG film (Lia and Chang, et al., 2020).

15.6 Mechanical properties

15.6.1 Factors affecting the mechanical properties of polymers

Polymers, with long molecular chains, are synthesized by methods such as addition and condensation polymerization, resulting in materials with diverse properties and applications. They are classified into thermoplastics, elastomers, and thermosets, each with mechanical characteristics that determine their use in sectors such as packaging and aerospace. For example, additive manufacturing has gained prominence for its ability to create complex designs with polymers, improving design freedom and reducing waste (Mehrpouya et al., 2019). Understanding the mechanical properties of these polymers is crucial, as it directly informs their suitability for specific applications, thus playing a fundamental role in the continuous innovation of polymer technology.

The mechanical properties of polymers are fundamental to their application in various fields, as they determine their response to external forces. These materials exhibit varied mechanical behaviors due to their viscoelastic nature, which combines characteristics of solids and viscous liquids. Understanding these properties is essential for selecting appropriate materials and designing components that can withstand different stress conditions. Polymers are viscoelastic, meaning that their mechanical properties depend on time and temperature. This behavior is characterized by elastic and viscous responses to stress, observable in stress-strain behavior (Liu et al., 2013; Seymour & Carraher, 1984).

The stress-strain behavior of polymers is influenced by their microstructure and morphology. Elastic moduli, such as Young's modulus (determined in tensile or compression tests) and shear modulus (determined in shear stress tests), are critical for understanding the stiffness and flexibility of polymers, which vary with temperature and the degree of crystallinity (Fakhrayi Mofrad et al., 2024; Landel & Nielsen, 1993).

The factors that affect the mechanical properties of polymers are diverse and significant. First, molecular weight plays a crucial role; as molecular weight increases, both the viscosity and the mechanical strength of the material increase. Crystallinity is also a determining factor, as its increase leads to greater stiffness, although it can reduce viscoelasticity. On the other hand, cross-links in the polymer structure tend to increase elasticity and decrease creep. Branching in the polymer chain alters the viscoelastic response of the material. Intermolecular forces not only enhance the rigidity and resistance to deformation of the material by strengthening the interactions between polymer chains (Fakhrayi Mofrad et al., 2024) but also affect the glass transition temperature (Tg). Studies indicate that a greater intensity of these forces is associated with a higher Tg. This implies that the material retains its rigid character up to a higher temperature; provided it operates below its Tg, it exhibits increased rigidity under these conditions (Santos et al., 2015).

In the context of polymeric films, three mechanical properties are fundamental to determining their performance (Funami, 2011): (1) TS, which describes the film's ability to resist stretching forces without fracturing. It is commonly measured in megapascals (MPa) and is a key indicator of the material's structural integrity during handling and storage. Higher TS implies that the film can withstand greater stresses before breaking. (2) Elongation at break, which represents the maximum deformation a film can tolerate before fracturing, expressed as a percentage of its original length. This parameter is essential for applications where flexibility and impact absorption capacity are required. Films with high elongation at break can stretch considerably without breaking. (3) Elastic modulus (EM), also known as Young's modulus, which quantifies the stiffness of the film. It is calculated as the slope of the stress-strain curve in the elastic region. A high EM indicates that the material is stiff and deforms little under stress, while a low value is associated with greater flexibility and deformation capacity. The correct evaluation of these parameters is essential for determining the performance and adaptability of the films in different applications.

The comparison between galactomannans is based on the correlation between their intrinsic chemical structure—-particularly the proportion and distribution of mannose and galactose units, molecular weight, and degree of branching—and the mechanical properties of the resulting films. Furthermore, it is evident that chemical or physical modifications applied to these polysaccharides allow for adjusting their characteristics for specific applications.

15.6.2 Mechanical properties of galactomannan films from various plant sources

To form a cohesive film, strong interactions between the polymer chains are required (Senarathna et al., 2023). These depend on the galactose-to-mannose ratio and the distribution of galactose groups along the mannan chain, which directly impacts the film's properties; a uniform galactose distribution favors film flexibility, while regions with blocks of unsubstituted mannose generate more rigidity (Santos et al., 2015). The presence of more galactose subunits can separate the mannan chains due to steric hindrance, thus weakening the interactions between the chains (Senarathna et al., 2023). The molecular composition and conformation of each galactomannan are key in the formation of polymeric networks. The ability to establish intermolecular interactions, such as hydrogen bonds and van der Waals forces, determines the network density and the mechanical properties of the films. A greater number of hydroxyl groups reinforces hydrogen bonds, resulting in a more cross-linked network (Mukherjee et al., 2023).

The different sources of galactomannans exhibit significant variations in their mechanical properties, making them suitable for various applications (Santos et al., 2015). These properties depend not only on the mannose/galactose ratio but also on the concentration, preparation method, and the process of obtaining the gum, among other factors (da Silva et al., 2021).

The results of a study (Liu and Lei, et al., 2020) in which the influence of the composition of three native gums on film preparation under the same conditions was evaluated are shown in Fig. 15.10.

The addition of glycerol to the formulation of the galactomannan films presented in Fig. 15.10 caused those of TG and LBG to be extensible, exhibiting ductile and flexible behaviors; however, the GG film was the most brittle and rigid. These differences in material fracture could be due to the more substituted and branched structure of GG, which has a lower mannose/galactose (M/G) ratio than TG and LBG, resulting in the greater galactose substitution in GG hindering hydrogen bonding, which is reflected in lower values of EM and TS. In contrast, TG and LBG exhibit higher values of both properties due to a greater density of mannose linkages.

Another study conducted with GG and LBG (Mikkonen et al., 2007) to evaluate the effect of the type of plasticizer, which, as has been mentioned, is necessary to improve functional properties, yielded the results shown in Fig. 15.11.

GG and LBG films exhibited significant differences in mechanical properties, with LBG films consistently showing higher TS regardless of plasticizer concentration. Additionally, LBG films demonstrate greater elongation at break compared to GG films. These findings suggest that galactomannan structure directly influences film formation, with the strongest and most flexible films developing from galactomannans with lower galactose substitution (LBG). The optimal

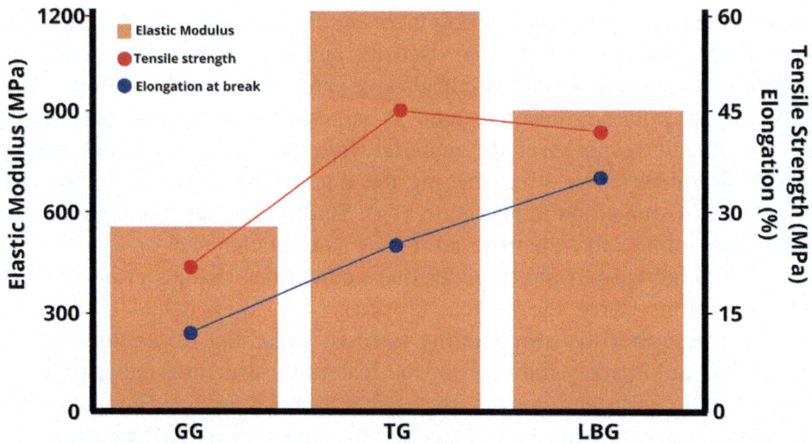

FIGURE 15.10 Mechanical properties of films made with different galactomannans.
From Liu, F., Chang, W., Chen, M., Xu, F., Ma, J. & Zhong, F. (2020). Film-forming properties of guar gum, tara gum and locust bean gum. Food Hydrocolloids, 98.

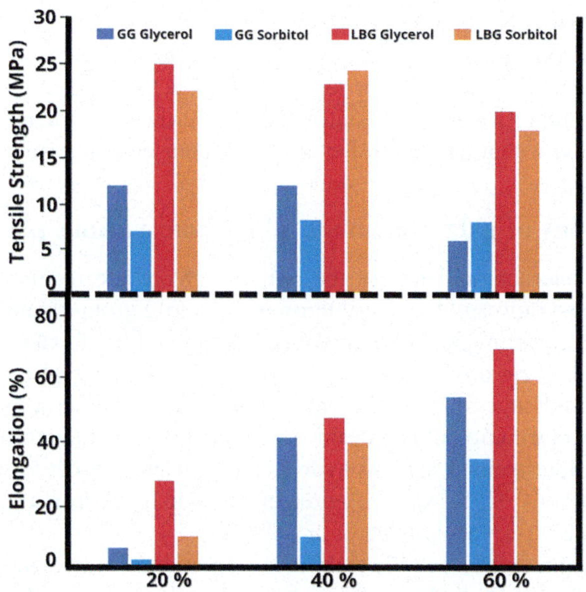

FIGURE 15.11 Breaking strength and elongation at break of guar gum (GG) and locust bean gum (LBG) films as a function of plasticizer type and incorporation level.
From Mikkonen, K. S., Rita, H., Helén, H., Talja, R. A., Hyvönen, L. & Tenkanen, M. (2007). Effect of polysaccharide structure on mechanical and thermal properties of galactomannan-based films. Biomacromolecules, 8(10), 3198–3205.

plasticizer concentration depends on the desired balance between flexibility and strength, as increasing plasticizer improves elongation at break while potentially compromising other structural properties.

The M:G ratio appears to be a key determinant of mechanical properties in galactomannan films, as evidenced by TG's intermediate structural parameters compared to other galactomannans. However, notable discrepancies exist in mechanical properties even when films are formulated under similar conditions. For instance, TS values varied considerably between experiments, with one study reporting approximately 45 MPa for both LBG and TG, while another reported maximum values of only 25 MPa. These variations stem largely from the lack of standardized testing methods for bioplastics, which, unlike conventional materials, require specialized techniques due to their sensitivity to environmental factors and delicate nature. The limitations of conventional testing equipment and the absence of uniform protocols further complicate the collection of reproducible and reliable data (Boudarel et al., 2018), as can be seen in Table 15.2, where the measurements and preparation methods of the films depend on various factors. LBG galactomannans, with their M/G ratio of 4, demonstrate mechanical properties particularly suitable for applications requiring high stiffness. They exhibit elevated EM

TABLE 15.2 Mechanical characteristics of films derived from different galactomannan sources.

Galactomannan	Tensile strength (MPa)	Elastic modulus (MPa)	Elongation at break (%)	Plasticizer	Information about the film and test	References
Fenugreek gum (FG)						
Native FG	3.03	N/R	~90	Glycerol (1.5% w/w)	FG 1.5% w/v; drying: oven 45°C, 12 h; standard: ASTM D882; specimen dimensions: 10 × 140 mm; crosshead speed: 12.5 mm min^{-1}	Memiş et al. (2017)
Guar gum (GG)						
Native GG	3.67	N/R	50.76	Glycerol (33.3% w/w)	GG 1.5% w/v; thickness ~70 μm; drying: oven 35°C, 16 h; standard: ASTM D882-91; specimen dimensions: 45 × 20 mm; crosshead speed: 5 mm min^{-1}	Dos Santos et al. (2015)
Cationic GG	1.16	2.96	40.38	Glycerol (5% w/v)	Cationic GG 6% w/v; thickness ~600 μm; drying: oven 40°C, 24 h; standard: ASTM D882-02; crosshead speed: 50 mm min^{-1}	Dallabrida and Braz, 2024)
Pea starch GG	6–34	26–41	68–25	Glycerol (15%–35% w/w)	GG: 0.1%–0.5% and starch: 2%–3% (w/v); thickness: 0.36–1.42 mm; drying: oven 40°C, 24 h; standard: ASTM D882; specimen dimensions: 45 × 15 mm; crosshead speed: 1 mm s^{-1}	Saberi et al. (2017)
Tara gum (TG)						
Native TG	45–90	N/R	16–44	Glycerol (7.5%–30% w/w)	TG 1% (w/v); drying: oven 35°C, 24 h; standard: ASTM D882; specimen dimensions: 20 × 80 mm; crosshead speed: 0.5 mm s^{-1}	Antoniou et al. (2014)
TG-chitosan nanoparticles	22.5–52.5	N/R	30–45	Glycerol (20% w/v)	TG 1% (w/v), variable concentrations of chitosan in bulk and particulate forms; drying: oven 35°C; standard: ASTM D882 with some modifications; specimen dimensions: 20 × 80 mm; crosshead speed: 0.5 mm s^{-1}	Antoniou et al. (2015)
Locust bean gum (LBG)						
Native LBG	~36	N/R	~32	Glycerol (0.2% w/v)	LBG 1% w/v; drying: oven 45°C, 12 h; standard: ASTM D882; specimen dimensions: 60 × 10 mm; test speed: 0.5 mm s^{-1}	Liu, Lei, et al. (2020)
Carbamoylethyl locust bean gum (CLBG)	18.55	N/R	6.11	Glycerol (1% w/w)	CLBG 5% w/v; drying: oven 45°C, 24 h; standard: ASTM D882-02	

and TS comparable to or exceeding that of guar galactomannans, though with more limited elongation at break. Like GG, LBG's network structure requires additional compounds to form robust films, with research confirming that both TS and elongation at break are significantly influenced by the concentration of plasticizer used during manufacturing.

15.7 Conclusion

Rising concern about pollution from fossil fuel-derived plastics has driven research into environmentally friendly alternatives for food packaging, highlighting the potential of edible coatings and films made from biopolymers. Among these, galactomannans, nonstarch polysaccharides present in legumes such as locust bean, guar, tara, and fenugreek, are emerging as promising materials due to their biodegradability, nontoxicity, and ability to form cohesive matrices.

Galactomannans can be processed by wet methods (casting, dipping, spraying). Wet methods are suitable for heat-sensitive materials and do not require specialized equipment, although they have long drying times. Galactomannan films can improve the shelf life of food by acting as barriers against oxygen and carbon dioxide, and by carrying functional ingredients such as antimicrobials and antioxidants.

The structure of galactomannans, characterized by a mannan backbone linked by β-(1→4) bonds and galactose branches linked by α-(1→6) bonds, directly influences their physicochemical and functional properties. This chapter has analyzed the relationship between the characteristics of galactomannans (mannose/galactose ratio (M/G), galactose distribution along the chain, the chemical structure of the galactomannan, as well as its molecular weight) and the properties of films and coatings made from them. In general, gas barrier properties, as well as mechanical properties, increase with increasing mannose:galactose ratio in the galactomannan. However, as the ratio increases, their solubility decreases, making it necessary to increase the temperature during processing to achieve dispersion of the galactomannan in water, as in the case of locust bean galactomannan. Due to the hydrophilic nature of galactomannans, they are sensitive to humidity conditions, so they have a low water vapor barrier, and the film structure can be lost due to the deterioration of mechanical properties. Because of this situation, the use of galactomannans in packaging applications presents challenges. To overcome these limitations, modification of the galactomannans, blending them with other biopolymers, and the incorporation of additives or fillers are necessary. For example, to overcome the brittleness of the films, plasticizers such as glycerol and sorbitol must be incorporated, although these additives can affect their solubility and gas diffusion. In addition, variability in the composition and structure of galactomannans from different plant sources and the lack of standardized methods for mechanical characterization make it difficult to obtain reproducible and reliable results.

Despite these challenges, galactomannans offer a promising and sustainable alternative as a basis for the development of materials that can replace conventional plastics in food packaging. A thorough understanding of the mechanical properties of these polymers and the appropriate use of additives is essential for their successful application in various industries. This understanding will enable researchers and engineers to design materials that not only meet strict performance criteria but are also environmentally sustainable.

AI disclosure

During the preparation of this work, the author(s) used Gemini 2.0 Pro for translation, grammar, and syntax review. After using this tool/service, the author(s) reviewed and edited the content as needed and take(s) full responsibility for the content of the publication.

References

Agarwal, S. (2021). Major factors affecting the characteristics of starch based biopolymer films. *European Polymer Journal, 160*. https://doi.org/10.1016/j.eurpolymj.2021.110788.

Andrade, R. D., Skurtys, O., & Osorio, F. A. (2012). Atomizing spray systems for application of edible coatings. *Comprehensive Reviews in Food Science and Food Safety, 11*(3), 323–337. https://doi.org/10.1111/j.1541-4337.2012.00186.x.

Antoniou, J., Liu, F., Majeed, H., Qazi, H. J., & Zhong, F. (2014). Physicochemical and thermomechanical characterization of tara gum edible films: Effect of polyols as plasticizers. *Carbohydrate Polymers, 111*, 359–365. https://doi.org/10.1016/j.carbpol.2014.04.005.

Antoniou, J., Liu, F., Majeed, H., & Zhong, F. (2015). Characterization of tara gum edible films incorporated with bulk chitosan and chitosan nanoparticles: A comparative study. *Food Hydrocolloids, 44*, 309–319. https://doi.org/10.1016/j.foodhyd.2014.09.023.

Barak, S., & Mudgil, D. (2014). Locust bean gum: Processing, properties and food applications-A review. *International Journal of Biological Macromolecules, 66*, 74–80. https://doi.org/10.1016/j.ijbiomac.2014.02.017.

Barbosa, C. H., Andrade, M. A., Vilarinho, F., Fernando, A. L., & Silva, A. S. (2021). Active edible packaging. *Encyclopedia, 1*(2), 360–373. Available from: https://doi.org/10.3390/encyclopedia1020030, https://www.mdpi.com/2673-8392/1/2/30.

Bizymis, A.-P., & Tzia, C. (2022). Edible films and coatings: Properties for the selection of the components, evolution through composites and nanomaterials, and safety issues. *Critical Reviews in Food Science and Nutrition, 62*(31), 8777–8792. https://doi.org/10.1080/10408398.2021.1934652.

Blanco-Pascual, N., & Gómez-Estaca, J. (2016). Edible food packaging: Materials and processing technologiesproduction and processing of edible packaging: Stability and applications. In M. A. Cerqueira, R. N. Pereira, O. L. Ramos, J. A. Teixeira, & A. A. Vicente (Eds.). *Taylor & Francis Group, Oxford, UK Taylor & Francis Group, Oxford, UK.*

Boudarel, H., Mathias, J.-D., Blaysat, B., & Grédiac, M. (2018). Towards standardized mechanical characterization of microbial biofilms: Analysis and critical review. *npj Biofilms and Microbiomes, 4*(1). https://doi.org/10.1038/s41522-018-0062-5.

Bouziane, G., Henni, A., Bouricha, M., Boual, Z., Belkhalfa, H., Bachari, K., & Ould El Hadj, M. D. (2024). A new galactomannan extracted from the seeds of *Astragalus gombiformis* Pomel (Fabaceae) and its utilization in the biosynthesis of silver nanoparticles. *Process Biochemistry, 136*, 48–59. https://doi.org/10.1016/J.PROCBIO.2023.11.009.

Brady, J., Drig, T., Lee, P. I., & Li, J.-X. (2017). Polymer properties and characterization. *Developing Solid Oral Dosage Forms: Pharmaceutical Theory and Practice: Second Edition,* 181–223. https://doi.org/10.1016/B978-0-12-802447-8.00007-8.

Ceballos, R. L., Ochoa-Yepes, O., Goyanes, S., Bernal, C., & Famá, L. (2020). Effect of yerba mate extract on the performance of starch films obtained by extrusion and compression molding as active and smart packaging. *Carbohydrate Polymers, 244*. https://doi.org/10.1016/j.carbpol.2020.116495.

Cerqueira, M. A., Bourbon, A. I., Pinheiro, A. C., Martins, J. T., Souza, B. W. S., Teixeira, J. A., & Vicente, A. A. (2011). Galactomannans use in the development of edible films/coatings for food applications. *Trends in Food Science and Technology, 22*(12), 662–671. https://doi.org/10.1016/J.TIFS.2011.07.002.

Chel-Guerrero, L., Betancur-Ancona, D., Aguilar-Vega, M., & Rodríguez-Canto, W. (2024). Films properties of QPM corn starch with *Delonix regia* seed galactomannan as an edible coating material. *International Journal of Biological Macromolecules, 255*. https://doi.org/10.1016/j.ijbiomac.2023.128408.

Chen, W. L., Chen, H. L., Guo, G. W., Huang, Y. C., Chen, C. Y., Tsai, Y., Huang, K. F., & Yang, C. H. (2018). Locust bean gum galactomannan hydrolyzed by thermostable β-D-mannanase may reduce the secretion of pro-inflammatory factors and the release of granule constituents. *International Journal of Biological Macromolecules, 114*, 181–186. https://doi.org/10.1016/J.IJBIOMAC.2018.03.097.

Costa, M. J., Maciel, L. C., Teixeira, J. A., Vicente, A. A., & Cerqueira, M. A. (2018). Use of edible films and coatings in cheese preservation: Opportunities and challenges. *Food Research International, 107*, 84–92. https://doi.org/10.1016/j.foodres.2018.02.013.

Dallabrida, K. G., Braz, W. C., Marchiori, C., Alves, T. M., Cruz, L. S., Trindade, G. A.de M., Machado, P., da Rosa, L. S., Khalil, N. M., Rego, F. G.de M., Fajardo, A. R., Ferreira, L. M., Sari, M. H. M., & Reolon, J. B. (2024). Exploring cationic guar gum: Innovative hydrogels and films for enhanced wound healing. *Pharmaceutics, 16*(9). https://doi.org/10.3390/pharmaceutics16091233.

Dea, I. C. M., & Morrison, A. (1975). Chemistry and interactions of seed galactomannans. *Advances in Carbohydrate Chemistry and Biochemistry, 31*(C), 241–312. https://doi.org/10.1016/S0065-2318(08)60298-X.

Edwards, M., Scott, C., Gidley, M. J., & Reid, J. S. G. (1992). Control of mannose/galactose ratio during galactomannan formation in developing legume seeds. *Planta, 187*(1), 67–74. https://doi.org/10.1007/BF00201625.

Fakhrayi Mofrad, N., Bahadori, P., & Raos, G. (2024). Ultimate molecular mechanical properties of polyolefin chains. *Macromolecules, 57*(9), 3901–3913. https://doi.org/10.1021/acs.macromol.3c02609.

Funami, T. (2011). Next target for food hydrocolloid studies: Texture design of foods using hydrocolloid technology. *Food Hydrocolloids, 25*(8), 1904–1914. https://doi.org/10.1016/j.foodhyd.2011.03.010.

Gadkari, P. V., Reaney, M. J. T., & Ghosh, S. (2019). Assessment of gelation behaviour of fenugreek gum and other galactomannans by dynamic viscoelasticity, fractal analysis and temperature cycle. *International Journal of Biological Macromolecules, 126*, 337–344. https://doi.org/10.1016/j.ijbiomac.2018.12.132.

Geronço, M. S., da Silveira Ramos, I. F., da Silva Filho, E. C., dos Santos Rizzo, M., Ribeiro, A. B., & da Costa, M. P. (2021). Are structurally modified galactomannan derivatives biologically active? *Polysaccharides, 2*(1), 1–15. https://doi.org/10.3390/polysaccharides2010001.

Ghosh, S. K., Das, A., Basu, A., Halder, A., Das, S., Basu, S., Abdullah, M. F., Mukherjee, A., & Kundu, S. (2018). Semi-interpenetrating hydrogels from carboxymethyl guar gum and gelatin for ciprofloxacin sustained release. *International Journal of Biological Macromolecules, 120*, 1823–1833. https://doi.org/10.1016/J.IJBIOMAC.2018.09.212.

Gökkaya Erdem, B., Dıblan, S., & Kaya, S. (2021). A comprehensive study on sorption, water barrier, and physicochemical properties of some protein- and carbohydrate-based edible films. *Food and Bioprocess Technology, 14*(11), 2161–2179. https://doi.org/10.1007/s11947-021-02712-0.

Gómez-Rodríguez, G. H., González-García, G., Álvarez-Bajo, O., García-Sifuentes, C. O., Argüelles-Monal, W. M., Lizardi-Mendoza, J., & López-Franco, Y. L. (2024). Enzyme-catalyzed transesterification of galactomannan extracted from mesquite seed (*Prosopis velutina*) with vinyl carboxylate esters. *Heliyon, 10*(10). https://doi.org/10.1016/J.HELIYON.2024.E31421.

Hu, C., Chen, J., Chen, W., Tang, W., Wang, G., Pan, F., & Yin, Z. (2024). Multiobjective optimization strategy for enhancing the efficiency and quality of organic thin-film manufacturing with electrohydrodynamic atomization coating. *ACS Omega, 9*(37), 38970–38988. https://doi.org/10.1021/acsomega.4c05402.

Jeya Jeevahan, J., Chandrasekaran, M., Venkatesan, S. P., Sriram, V., Britto Joseph, G., Mageshwaran, G., & Durairaj, R. B. (2020). Scaling up difficulties and commercial aspects of edible films for food packaging: A review. *Trends in Food Science and Technology, 100*, 210–222. https://doi.org/10.1016/j.tifs.2020.04.014.

Ju, J., Xie, Y., Guo, Y., Cheng, Y., Qian, H., & Yao, W. (2019). Application of edible coating with essential oil in food preservation. *Critical Reviews in Food Science and Nutrition, 59*(15), 2467–2480. https://doi.org/10.1080/10408398.2018.1456402.

Kontogiorgos, V. (2018). Galactomannans (guar, locust bean, fenugreek, tara). *Encyclopedia of Food Chemistry,* 109–113. https://doi.org/10.1016/B978-0-08-100596-5.21589-8.

Kumar, A., Hasan, M., Mangaraj, S., Pravitha, M., Verma, D. K., & Srivastav, P. P. (2022). Trends in edible packaging films and its prospective future in food: A review. *Applied Food Research, 2*. https://doi.org/10.1016/j.afres.2022.100118.

Kurek, M., Ščetar, M., & Galić, K. (2017). Edible coatings minimize fat uptake in deep fat fried products: A review. *Food Hydrocolloids, 71*, 225–235. https://doi.org/10.1016/j.foodhyd.2017.05.006.

Kurt, A., & Kahyaoglu, T. (2014). Characterization of a new biodegradable edible film made from salep glucomannan. *Carbohydrate Polymers, 104*(1), 50–58. https://doi.org/10.1016/j.carbpol.2014.01.003.

Landel, R. F., & Nielsen, L. E. (1993). *Mechanical properties of polymers and composites* (2). CRC Press. http://doi.org/10.1201/b16929.

Landin, M., & Echezarreta, M. (2010). *Polysaccharides: Development, properties and applicationsgalactomannans: Old and new pharmaceutical materials*. Nova Science Publishers Inc, 1–26.

Li, M., & Ye, R. (2017). *Biopackagingedible active packaging for food application: Materials and technology*. CRC, 1.

Lisitsyn, A., Semenova, A., Nasonova, V., Polishchuk, E., Revutskaya, N., Kozyrev, I., & Kotenkova, E. (2021). Approaches in animal proteins and natural polysaccharides application for food packaging: Edible film production and quality estimation. *Polymers, 13*(10). https://doi.org/10.3390/polym13101592.

Liu, D., Rehkopf, J., & Reitman, M. (2013). In Myer. Kutz. (Ed.). *Handbook of measurement in science and engineeringmechanical properties of polymers*John Wiley & Sonshttp://doi.org/10.1002/9781118436707.hmse038.

Liu, F., Chang, W., Chen, M., Xu, F., Ma, J., & Zhong, F. (2020). Film-forming properties of guar gum, tara gum and locust bean gum. *Food Hydrocolloids, 98*. https://doi.org/10.1016/j.foodhyd.2019.03.028.

Liu, Y., Lei, F., He, L., Xu, W., & Jiang, J. (2020). Physicochemical characterization of galactomannans extracted from seeds of *Gleditsia sinensis* Lam and fenugreek. Comparison with commercial guar gum. *International Journal of Biological Macromolecules, 158*, 1047–1054. https://doi.org/10.1016/J.IJBIOMAC.2020.04.208.

Mehrpouya, M., Dehghanghadikolaei, A., Fotovvati, B., Vosooghnia, A., Emamian, S. S., & Gisario, A. (2019). The potential of additive manufacturing in the smart factory industrial 4.0: A review. *Applied Sciences (Switzerland), 9*(18). https://doi.org/10.3390/app9183865.

Memiş, S., Tornuk, F., Bozkurt, F., & Durak, M. Z. (2017). Production and characterization of a new biodegradable fenugreek seed gum based active nanocomposite film reinforced with nanoclays. *International Journal of Biological Macromolecules, 103*, 669–675. https://doi.org/10.1016/j.ijbiomac.2017.05.090.

Mikkonen, K. S., Rita, H., Helén, H., Talja, R. A., Hyvönen, L., & Tenkanen, M. (2007). Effect of polysaccharide structure on mechanical and thermal properties of galactomannan-based films. *Biomacromolecules, 8*(10), 3198–3205.

Mikkonen, K. S., Heikkilä, M. I., Helén, H., Hyvönen, L., & Tenkanen, M. (2010). Spruce galactoglucomannan films show promising barrier properties. *Carbohydrate Polymers, 79*(4), 1107–1112. https://doi.org/10.1016/j.carbpol.2009.10.049.

Moura-Alves, M., Esteves, A., Ciríaco, M., Silva, J. A., & Saraiva, C. (2023). Antimicrobial and antioxidant edible films and coatings in the shelf-life improvement of chicken meat. *Foods, 12*(12). https://doi.org/10.3390/foods12122308.

Mukherjee, K., Dutta, P., Badwaik, H. R., Saha, A., Das, A., & Giri, T. K. (2023). Food industry applications of Tara gum and its modified forms. *Food Hydrocolloids for Health, 3*. https://doi.org/10.1016/j.fhfh.2022.100107.

Nair, S. S., Trafiałek, J., & Kolanowski, W. (2023). Edible packaging: A technological update for the sustainable future of the food industry. *Applied Sciences, 13*(14).

Niranjana Prabhu, T., & Prashantha, K. (2018). A review on present status and future challenges of starch based polymer films and their composites in food packaging applications. *Polymer Composites, 39*(7), 2499–2522. https://doi.org/10.1002/pc.24236.

Pareta, R., & Edirisinghe, M. J. (2006). A novel method for the preparation of starch films and coatings. *Carbohydrate Polymers, 63*(3), 425–431. https://doi.org/10.1016/j.carbpol.2005.09.018.

Rodriguez-Canto, W., Chel-Guerrero, L., Fernandez, V. V. A., & Aguilar-Vega, M. (2019). Delonix regia galactomannan hydrolysates: Rheological behavior and physicochemical characterization. *Carbohydrate Polymers, 206*, 573–582. https://doi.org/10.1016/j.carbpol.2018.11.028.

Russo Spena, S., Pasquino, R., & Grizzuti, N. (2024). K-carrageenan/locust bean gum gels for food applications—a critical study on potential alternatives to animal-based gelatin. *Foods, 13*(16). https://doi.org/10.3390/foods13162575.

Saberi, B., Thakur, R., Bhuyan, D. J., Vuong, Q. V., Chockchaisawasdee, Sl, Golding, J. B., Scarlett, C. J., & Stathopoulos, C. E. (2017). Development of edible blend films with good mechanical and barrier properties from pea starch and guar gum. *Starch/Staerke, 69*(1-2). https://doi.org/10.1002/star.201600227.

Santos, V. D., Souza, B., Teixeira, J. A., Vicente, A. A., & Cerqueira, M. (2015). Relationship between galactomannan structure and physicochemical properties of films produced thereof. *Journal of Food Science and Technology, 52*(12), 8292–8299. https://doi.org/10.1007/s13197-015-1961-6.

Senarathna, S., Navaratne, S., Wickramasinghe, I., & Coorey, R. (2023). Use of fenugreek seed gum in edible film formation: Major drawbacks and applicable methods to overcome. *Journal of Food Science and Technology, 60*(7), 1860–1869. https://doi.org/10.1007/s13197-022-05465-6.

Seymour, R. B., & Carraher, C. E. (1984). *Structure—Property relationships in polymersmechanical properties of polymers*. Springer. http://doi.org/10.1007/978-1-4684-4748-4_5.

Sharma, G., Sharma, S., Kumar, A., Al-Muhtaseb, A. H., Naushad, M., Ghfar, A. A., Mola, G. T., & Stadler, F. J. (2018). Guar gum and its composites as potential materials for diverse applications: A review. *Carbohydrate Polymers, 199*, 534–545. https://doi.org/10.1016/j.carbpol.2018.07.053.

Sharma, P., Sharma, S., Ramakrishna, G., Srivastava, H., & Gaikwad, K. (2021). A comprehensive review on leguminous galactomannans: Structural analysis, functional properties, biosynthesis process and industrial applications. *Critical Reviews in Food Science and Nutrition, 62*(2), 443–465. https://doi.org/10.1080/10408398.2020.1819196.

da Silva, D. A., Melo-Aires, G. C., & da Silva-Pena, R. (2021). In A. N. Barros, & I. Gouvinhas. (Eds.). *Innovation in the food sector through the valorization of food and agro-food by-productsgums-characteristics and applications in the food industry,* Intechopen, http://doi.org/10.5772/intechopen.95078.

Suhag, R., Kumar, N., Petkoska, A. T., & Upadhyay, A. (2020). Film formation and deposition methods of edible coating on food products: A review. *Food Research International, 136*. https://doi.org/10.1016/j.foodres.2020.109582.

Tian, L., Prabhakaran, M. P., Ding, X., & Ramakrishna, S. (2013). Biocompatibility evaluation of emulsion electrospun nanofibers using osteoblasts for bone tissue engineering. *Journal of Biomaterials Science, Polymer Edition, 24*(17), 1952–1968. https://doi.org/10.1080/09205063.2013.814096.

Torres-Giner, S., Pérez-Masiá, R., & Lagaron, J. M. (2016). A review on electrospun polymer nanostructures as advanced bioactive platforms. *Polymer Engineering and Science, 56*(5), 500–527. https://doi.org/10.1002/pen.24274.

Trajkovska Petkoska, A., Daniloski, D., D'Cunha, N. M., Naumovski, N., & Broach, A. T. (2021). Edible packaging: Sustainable solutions and novel trends in food packaging. *Food Research International, 140*. https://doi.org/10.1016/j.foodres.2020.109981.

Trinh, B. M., Chang, B. P., & Mekonnen, T. H. (2023). The barrier properties of sustainable multiphase and multicomponent packaging materials: A review. *Progress in Materials Science, 133*. https://doi.org/10.1016/j.pmatsci.2023.101071.

Umaraw, P., Munekata, P. E. S., Verma, A. K., Barba, F. J., Singh, V. P., Kumar, P., & Lorenzo, J. M. (2020). Edible films/coating with tailored properties for active packaging of meat, fish and derived products. *Trends in Food Science and Technology, 98*, 10–24. https://doi.org/10.1016/j.tifs.2020.01.032.

Wang, R., Zhang, S., Liu, S., Sun, Y., & Xu, H. (2023). A Contribution to improve barrier properties and reduce swelling ratio of κ-carrageenan film from the incorporation of guar gum or locust bean gum. *Polymers, 15*(7). https://doi.org/10.3390/polym15071751.

Watanabe, H., Matsuyama, T., & Yamamoto, H. (2003). Experimental study on electrostatic atomization of highly viscous liquids. *Journal of Electrostatics, 57*(2), 183–197. https://doi.org/10.1016/S0304-3886(02)00139-0.

Wen, Y., Kok, M. D. R., Tafoya, J. P. V., Sobrido, A. B. J., Bell, E., Gostick, J. T., Herou, S., Schlee, P., Titirici, M.-M., Brett, D. J. L., Shearing, P. R., & Jervis, R. (2021). Electrospinning as a route to advanced carbon fibre materials for selected low-temperature electrochemical devices: A review. *Journal of Energy Chemistry, 59*, 492–529. https://doi.org/10.1016/j.jechem.2020.11.014.

Xue, W., Zhu, J., Sun, P., Yang, F., Wu, H., & Li, W. (2023). Permeability of biodegradable film comprising biopolymers derived from marine origin for food packaging application: A review. *Trends in Food Science and Technology, 136*, 295–307. https://doi.org/10.1016/j.tifs.2023.05.001.

Yadav, H., & Maiti, S. (2020). Research progress in galactomannan-based nanomaterials: Synthesis and application. *International Journal of Biological Macromolecules, 163*, 2113–2126. https://doi.org/10.1016/j.ijbiomac.2020.09.062.

Yan, J., Luo, Z., Ban, Z., Lu, H., Li, D., Yang, D., Aghdam, M. S., & Li, L. (2019). The effect of the layer-by-layer (LBL) edible coating on strawberry quality and metabolites during storage. *Postharvest Biology and Technology, 147*, 29–38. https://doi.org/10.1016/j.postharvbio.2018.09.002.

Yun, X., Liu, L., Hu, J., Sun, T., Zhang, J., & Dong, T. (2024). Mechanical and gas permeability properties of poly(L-lactic acid)–based films and their application in fresh produce preservation—Review. *Packaging Technology and Science, 37*(4), 293–317. https://doi.org/10.1002/pts.2796.

Zhang, C., Li, Y., Wang, P., & Zhang, H. (2020). Electrospinning of nanofibers: Potentials and perspectives for active food packaging. *Comprehensive Reviews in Food Science and Food Safety, 19*(2), 479–502. https://doi.org/10.1111/1541-4337.12536.

Zhang, L., Liu, Z., Han, X., Sun, Y., & Wang, X. (2019). Effect of ethanol content on rheology of film-forming solutions and properties of zein/chitosan film. *International Journal of Biological Macromolecules, 134*, 807–814. https://doi.org/10.1016/j.ijbiomac.2019.05.085.

Zhang, N.-N., Ma, H., Zhang, Z.-F., Zhang, W.-N., Chen, L., Pan, W.-J., Wu, Q.-X., Lu, Y.-M., & Chen, Y. (2022). Characterization and immunomodulatory effect of an alkali-extracted galactomannan from *Morchella esculenta*. *Carbohydrate Polymers, 278*. https://doi.org/10.1016/J.CARBPOL.2021.118960.

Chapter 16

Films and coatings based on dairy product ingredients

Gözde Koşarsoy Ağçeli, Sıla Hoşoğlu, and Lara Akay
Department of Biology, Faculty of Science, Hacettepe University, Ankara, Turkey

16.1 Introduction

In the current era, there is a mounting interest in food packaging systems made from edible biopolymers. These include proteins, carbohydrates, and lipids. They must be biodegradable, safe for the environment, and provide an environmentally friendly alternative to petroleum-based materials commonly used in the food industry. Moreover, these packaging materials must be nontoxic, demonstrate stability in microbial state, possess good mechanical and barrier properties, exhibit excellent sensory characteristics, and be produced at minimal costs while being fully compatible with the packaged product (Rouf & Kokini, 2017).

Approximately 4–11 million tons of dairy waste are disposed of annually, given its elevated organic content, which represents a notable environmental risk. Furthermore, milk products frequently have a limited shelf life (one to two weeks). Additionally, the food industry generates a considerable quantity of by-products and waste, including unsold products that have exceeded their shelf life (Fig. 16.1). Nevertheless, milk proteins can be readily separated from dairy waste and transformed into bio-based polymeric materials, which could serve as a potential solution to both environmental challenges and food coating (Chen et al., n.d.; Ferreira et al., 2022).

Milk represents a primary source of protein, offering significant nutritional and functional benefits. The milk protein is composed of 80% casein and 20% whey protein, which can be further subdivided into distinct components. These proteins provide support for the digestive, cardiovascular, immune, endocrine, and central nervous systems. They also exhibit a range of properties, including antioxidant, antimicrobial, antihypertensive, and immunomodulatory activities (Gerna et al., 2023) (Fig. 16.2). The prevalence of whey and other by-products derived from dairy processing has considerable interest within the packaging industry, incorporating these substances as primary elements in edible films and coatings. It is well established that milk proteins, particularly casein and whey, comprise the essential components for the production of an optimal biodegradable and biocompatible film or coating formulation (Rouf & Kokini, 2017).

Natural biopolymer-based films for packaging have great potential for increasing the quality, stability, and safety of food products. Polymer nanomaterial composites, particularly those derived from natural biopolymers, have attracted considerable attention in recent years due to their significant advancements in packaging properties. These enhancements, including improved modulus and strength, reduced gas permeability, and better water resistance, are linked to their dispersion at the nanometer scale. Furthermore, the incorporation of biologically active components into these materials facilitates the development of packaging solutions with customized bio-functional properties, rendering them highly suitable for active food packaging applications (Xu et al., 2024). The eco-friendly nature of these materials positions them as a primary candidate for application in the food and packaging industries.

Biopolymers can be produced from a variety of sources, including animals and plants. Some of these can be synthesized chemically from bio-based monomers or by microorganisms (Rouf & Kokini, 2017).

For producing biodegradable films and coatings, proteins are highly suitable biopolymeric materials. When applied to food surfaces, edible films act as physical barriers to protect the food from environmental changes (temperature, light, and humidity). By preserving food qualities, including color, firmness, freshness, and nutritional value, these films help extend shelf life (Gerna et al., 2023).

Natural Polymer-Based Films and Coatings in Food Technology. DOI: https://doi.org/10.1016/B978-0-443-34244-8.00024-0

FIGURE 16.1 The acquisition of milk products from farm animals, their production processes, and short shelf life lead to significant amounts of waste.

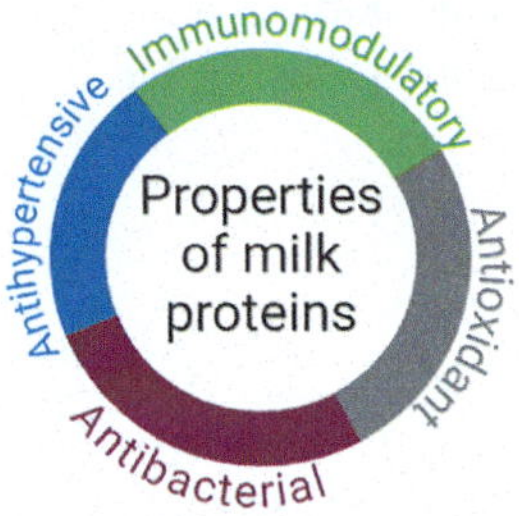

FIGURE 16.2 Various properties of milk proteins are antihypertensive, immunomodulatory, antioxidant, and antimicrobial.

Milk proteins, especially casein and whey, are among the natural biopolymers that possess unique properties for edible film production. Their three-dimensional network structures provide superior stability, endurance, and barrier properties in comparison to polysaccharide-based coatings. These proteins have additional functionalities, including antimicrobial and immunomodulatory effects, as discussed in the introduction, making them ideal for eco-friendly and active packaging materials (Elsabee & Abdou, 2013). Casein is easily processed into caseinates, which form films with minimal processing, utilizing its thermal stability, biodegradability, and strong binding capabilities. Correlatively, whey proteins, recovered from cheese whey, are excellent raw materials for films due to their ability to form polymeric structures through crosslinking. Collectively, these dairy proteins offer innovative, sustainable solutions for enhancing the quality and functionality of food packaging (Chen et al., n.d.).

16.2 Dairy products and film production

16.2.1 Lactose as a film and coating material

Lactose, a disaccharide from milk, has multifunctional roles in the field of food packaging, especially with a focus on edible films. Lactose does not serve as the main film-forming agent due to its poor film-forming properties, inclination to crystallize, low film quality, and mechanical properties, but it plays an essential indirect role in composite edible films. Lactose improves film performance when paired with other components and functional properties to maintain food quality. Lactose (in pure form) is not a prevailing film-forming agent. Due to its crystalline character, it is not appropriate for the formation of flexible and durable films, which are essential properties for food packaging. Nevertheless, lactose derivatives like lactose hydrolysates and oligosaccharides have shown potential when used as additives in protein or polysaccharide-based coating solutions to improve functionality (Ghanbarzadeh & Almasi, 2011). Their derivatives have decreased water solubility and increased film flexibility, making them appealing for food packaging purposes (Missaghi & Fassihi, 2004) (Fig. 16.3).

Protein-based films, such as whey protein isolate (WPI) or casein, often incorporate lactose as a plasticizer or filler. Lactose has been reported to increase the flexibility and processability of protein films without sacrificing their excellent oxygen barrier properties. Protein-lactose films are very useful in packaging applications where oxidative degradation must be minimized, such as fresh produce, dairy, and meat products.

In composite coatings, lactose can also be used with lipids, including stearic acid, beeswax, or glycerol monostearate (Khwaldia et al., 2004). Lipids are well known for their moisture barrier, and lactose helps improve the stability of the film and sensory properties. This combination is crucial in the packaging of perishable foods, where moisture management is

FIGURE 16.3 Advantages of lactose in food coatings and films.

key to extending the shelf life (Tharanathan, 2003). Coating fruits or baked goods, for instance, may use lactose incorporated into lipid layers for water transportation, mechanical stability, and permeability control. Moreover, lactose-derived oligosaccharides have been investigated for their antimicrobial polyfunctionality in food coatings. These derivatives can prevent the growth of microbes, changing the surface characteristics of the film to be unfavorable for spoilage organisms (Ghazali et al., 2012). Some important applications relate to more minimally processed foods, like fresh-cut fruits and vegetables, for which coatings with lactose derivatives can enhance protection levels and limit spoilage.

Indeed, in order to create multilayered coatings, lactose-based films are used with other biopolymers in practical applications. For example, lactose-coated casein coatings on fresh cheese or yogurt packaging improve barriers for oxygen and moisture while still being degradable. Lactose-modified alginate or starch films have also been evaluated as protective edible films to preserve fruits, vegetables, and bakery products (Metha et al., 2024).

Though not always direct, lactose's contribution to edible films and coatings is essential for the advancement of food packaging technologies. Utilized in coatings as a plasticizer, filler, or functional additive, lactose enables the formation of biodegradable packaging that also preserves food, prolongs shelf life, and ensures safety. Microbial fermentation has the potential to generate sustainable polymer-based food packaging materials that combine lactose with other macromolecules, such as proteins and lipids.

16.2.2 Whey: film and coating applications

Interest in edible films and coatings (especially whey based) is growing due to the increasing demand for sustainable and eco-friendly packaging solutions. As by-products of cheese, whey proteins are rich in functional properties that qualify them to replace plastic in food packaging. In a world where plastic waste and environmental degradation have taken center stage in the global food industry, whey protein films not only help mitigate these issues but also work to improve food preservation.

Whey protein-based edible films are among the most widely accepted due to their biodegradability (Elvira López-Caballero et al., 2024). Whey protein films decompose in nature, compared to traditional plastic packaging that takes hundreds of years to decompose. This may be a simple step to reduce plastic waste. This makes them an effective tool in the food industry's transition to more sustainable, eco-packing alternatives.

Whey is a by-product of cheese-making, which, in many situations, is underutilized. Whey protein films help to close the loop in dairy production and support a circular economy by converting waste into functional food packaging materials (Fig. 16.4). This enhances the value of whey while reducing the environmental footprint of dairy waste (Tiţa et al., 2024). Among other plastic materials, whey protein films and coatings are highly effective at preserving quality and extending the shelf life of food products. These coatings considerably outperform traditional polymers in barrier properties for oxygen, moisture, and volatile chemicals, that is, the three biggest agents of food spoilage. Whey protein coatings form a protective layer on food, reducing oxidation and microbial growth, and slowing enzymatic spoilage of food (Kandasamy et al., 2021).

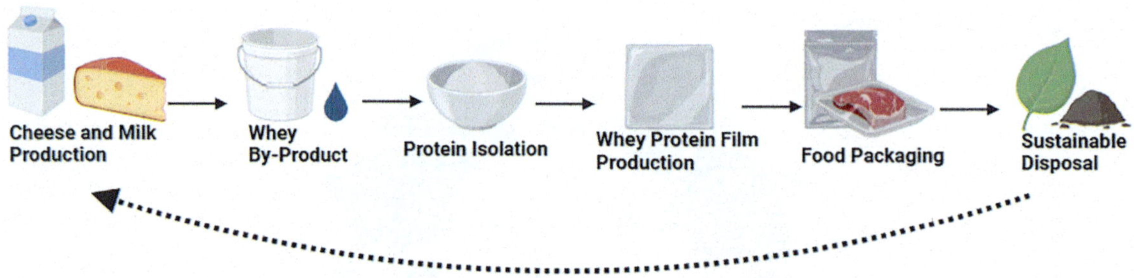

FIGURE 16.4 Circular economy of whey protein-based films in food packaging.

The ability of whey protein films to prevent oxidative browning and minimize moisture loss makes their preservation advantages especially noticeable in fresh produce. For instance, whey protein coatings help preserve the texture, color, and freshness of fresh-cut fruits and vegetables, which are particularly prone to spoiling. Likewise, in dairy products, these films also prevent surface moisture and protect against microbial contamination, thus contributing to improved product quality during storage. Besides protecting fresh produce, whey protein-based films can also prevent spoilage in bakery products, meat, and dairy items.

Additionally, different functional additives can be integrated into whey protein films to modify their properties. The most common type of modification involves incorporation of natural antimicrobial agents like essential oils or plant extracts. Before reviewing the antimicrobial effects of some of the additives; it is worth mentioning that these additives can provide antimicrobial activity to the films; thus preventing food products from microbial spoilage. This is especially helpful in the case of perishables such as meat and dairy, which can spoil due to bacterial growth (Li et al., 2024; Zhang et al., 2024a).

Besides incorporating antimicrobial agents, the addition of nanoparticles such as silver and zinc oxide has also enhanced barrier properties of the films against oxygen and moisture diffusion. These nanoparticles have other functions, like antimicrobial activity, which enhance food quality preservation by the film (Li et al., 2024; Zhang et al., 2024a).

Another important advantage of whey protein films includes clarity and edibility. The clear, shiny appearance of these films would allow consumers to view the quality of the product inside, while whey protein films are not toxic, in contrast to many synthetic polymers that could contain harmful chemicals or require unsafe additives. This safety aspect makes whey protein films even more intriguing in the food industry, where consumer health and food safety are of utmost importance. Continued research and innovation in this field will be required to overcome current obstacles, such as moisture sensitivity and production costs, fully realizing the potential of whey protein films in the food industry.

16.2.3 Casein-based films

Casein, the main protein present in milk, is the material used to construct edible films and coatings for food packaging. These unique features make it an excellent candidate for sustainability, food preservation, and quality-related applications. Casein films allow for the production of biodegradable and renewable materials that can replace synthetic plastics, as well as meet the growing market for sustainable packaging materials.

16.2.3.1 Structural properties of casein-based films

Casein exhibits unique structural properties that allow it to form cohesive, durable films. Its amphiphilic properties and ability to cross-link augment its effectiveness as a film-forming agent. It consists of three principal types: α-casein, β-casein, and κ-casein. These molecules, in their native state, assemble into micelles; hence, their disruption (e.g., using pH change or enzymatic treatment) allows the reorganization of these molecules into that of a film matrix. Usually, casein films are prepared by the solution casting method, which involves dissolving casein in water, adding a plasticizer (e.g., glycerol or sorbitol), and subsequently drying to obtain a thin film (Li et al., 2024; Zhang et al., 2024a).

Owing to their proteinaceous nature, casein-based films show even lesser tensile strength and elasticity than their gelatinous counterparts, yet harder species among biopolymers (such as gelatin) nowadays. These are effective oxygen barriers that are very important for preserving food. However, they are inherently hydrophilic, which limits their water vapor permeability (WVP). These challenges can be alleviated by using hydrophobic additives like beeswax and essential oils (Khan et al., 2021).

16.2.3.2 Functional properties of casein-based films

By preserving food and minimizing pollution simultaneously, casein-based films possess special functional properties. Additionally, casein films exhibit better oxygen barrier properties than many synthetic polymers, thanks to their dense protein matrix. This makes them ideal for packaging products sensitive to oxidation, such as meat, cheese, and fruits and vegetables.

The functionalization of casein films with additives can broaden their field of applications. Functional films are coated with natural antimicrobial agents such as essential oils (thyme, oregano), as well as antimicrobial nanoparticles (silver, zinc oxide), to endow these films with antimicrobial properties and promote food safety by inhibiting the growth of bacteria. These properties are especially beneficial for less-processed food, like fresh vegetables and dairy products, that can degrade quickly. Moreover, advanced modifications, such as the combination of casein with lipid-based ingredients or polysaccharides, improve moisture control and increase the durability of perishable products (Ponce et al., 2016).

However, casein films still face challenges regarding moisture sensitivity and relatively high production costs. They have lower mechanical stability compared with synthetic polymers, but this can be significantly improved with plasticizers and by producing nanocomposites. To date, new materials have been developed using biopolymers, which have gained popularity as edible and biodegradable food packaging due to their mechanical and thermal properties.

Food packaging with casein shows a promising step toward eco-friendly structures while keeping food safe and of high quality.

16.2.4 Biofilms from dairy by-products

Microorganisms that adhere to surfaces and are encased in a matrix of self-produced extracellular polymeric substances (EPS) are called biofilms, and biofilms have been studied for their implications in multiple fields, including the food industry. Biofilms in dairy can be a double-edged sword: some biofilms act as a substrate for probiotic cultures and bioactive elements; however, biofilms may also cause contamination and spoilage if not managed properly. The potential of using biofilms in combination with food fermentation by-products has been recently studied as a possible novel application for the preservation, packaging, and probiotic delivery of food products in the future, due to their flexible and protective nature.

By-products of dairy, such as cream and yogurt residues, are rich in nutrients (proteins, fats, lactose, minerals, etc.), making them an ideal substrate for the growth of microorganisms. Whey proteins, in particular, are excellent at forming films on lipid surfaces, and lactose and fats encourage the growth of many microorganisms, such as probiotics, bacteria, and yeasts. Probiotic strains have shown an ability to survive in various food matrices, and adhesion to those matrices is an essential factor for lactobacilli species, particularly *Lactiplantibacillus plantarum* and *Lacticaseibacillus rhamnosus,* to produce biofilms leading to stability and activity. Such biofilms can serve as protective layers, enhancing probiotic survivability in unfavorable gastrointestinal environments (Rezaei et al., 2023).

Biofilms from dairy by-products have significant potential as edible coatings and biodegradable packaging. There are many benefits of these biofilms, such as being biodegradable, nontoxic, and having an increased shelf life for food products.

Biofilms produced from dairy by-products (e.g., yogurt) can coat fresh produce, dairy products, and meats, acting as protective coatings. Conventional lipid-based edible coatings could further reduce water loss, lipid oxidation, and growth of spoilage microorganisms. Yogurt-based biofilms, for example, are effective in reducing microbial load on fruits and vegetables when combined with naturally occurring antimicrobial agents (such as essential oils), which would extend their shelf life. It has been reported that yogurt-derived biofilms, when combined with antimicrobial agents such as thyme essential oil, effectively inhibit the growth of foodborne pathogens, including *Listeria monocytogenes* and *Escherichia coli*.

In addition to edible coatings, biofilms from dairy by-products can be used to create sustainable packaging materials. Not only these biodegradable films minimize the amount of waste that could cause harm to the environment, but they can also help in the preservation of food by providing a barrier to O_2, moisture, and contaminants (Salama et al., 2022).

While the possible applications are promising, there are several challenges to be addressed before biofilms derived from dairy can be broadly used in food production. Biofilms can develop in dairy processing environments on surfaces like equipment, pipes, and tanks that shelter spoilage and pathogenic microorganisms. These biofilms can pose a serious threat to food safety and product quality. Pathogens such as Salmonella, *E. coli*, and Listeria can exist in biofilms, and eliminating them can be difficult, even with standard cleaning protocols.

Biofilms that may form on the surface of processing equipment also contribute to difficulties in sanitizing, since biofilms exhibit higher resistance against cleaning agents compared to planktonic microbes. This highlights the importance of regular monitoring and the use of advanced cleaning agents that are specifically designed for effective biofilm control to ensure safe and hygienic production of dairy products.

To industrialize biofilms from dairy residues, it is necessary to optimize certain parameters, such as raw material regularity and biofilm formation process. Inconsistencies in biofilm quality can occur due to variations in the composition of dairy by-products, including the protein and fat percentages. Standardization of the production process and the availability of quality by-products are required for the upscaling of this technology.

And although dairy-based biofilms have the ability to be utilized as an alternative to synthetic food coatings and packaging materials, the cost-effectiveness of large-scale production of such biofilms still seems to be an issue. This has to be competitive with the cost of extracting, processing, and applying conventional materials (Tennert et al., 2023).

16.3 Combinations with biopolymers

16.3.1 Reinforcement of milk-based films with tragacanth gum and chitosan

To enhance the mechanical, functional, and barrier qualities of milk-based films, natural biopolymers such as tragacanth gum and chitosan have been investigated for reinforcement in recent years. Tragacanth gum is a natural polysaccharide that can be extracted from the Astragalus plant. It consists of two main fractions: tragacanthin (water soluble) and bassorin (water swellable), representing 60%–70% of the total gum (Tonyali et al., 2018). This gum has long been recognized for its capacity to produce films and improve the characteristics of various biopolymer films. Tragacanth gum enhances the microstructural and physiochemical properties of whey protein-based films. Recent studies have shown that this gum can greatly increase the tensile strength of whey protein films, making them more suitable for use in food packaging applications (Ma et al., 2025). The interactions between the gum and the protein matrix cause this improvement, which makes the film more flexible and stable. Additionally, tragacanth gum lowers the film's WVP, which is an essential characteristic for stopping moisture transfer between the packaged food and its environment (Tonyali et al., 2018). Furthermore, tragacanth gum improves the mechanical qualities and uniformity of the film by contributing to its surface shape. This is particularly important for the development of materials for functional food packaging that preserve food quality and extend shelf life (Ma et al., 2025).

Chitosan is another commonly used reinforcement material in milk-based films. It is formed by the deacetylation of the chitin biopolymer. Chitosan copolymer is widely used in food packaging and edible film production due to its antimicrobial properties, biodegradability, and oxygen barrier capacity (Elsabee & Abdou, 2013).

Chitosan has the ability to enhance the functional features of milk-based films, particularly when combined with other biopolymers. Such as, the blending of chitosan with whey proteins and casein showed improvements in the mechanical and antimicrobial properties of edible films. Chitosan contributes to enhancing the tensile strength and flexibility of these films, thereby increasing their suitability for food packaging applications. Moreover, chitosan exhibits antimicrobial activity; when synergized with proteins like casein, it enables the films to effectively inhibit the growth of both bacterial and fungal contaminants, thereby extending the shelf life of packaged food products. Besides these advantages, challenges related to the moisture sensitivity and mechanical performance of chitosan films persist. To reduce these limitations, crosslinking with milk proteins such as casein and the incorporation of hydrophobic agents like carnauba wax have been studied to improve the hydrophobicity and water vapor resistance of the films (Xu et al., 2024). In addition, the inclusion of compounds such as condensed tannins has been shown to enhance the films' UV resistance, antioxidant, and antimicrobial activity. Collectively, these findings and developments emphasize chitosan's potential as a critical component in the formulation of high-performance, biodegradable food packaging materials (Xu et al., 2024).

16.3.2 Multilayer films: the compatibility of milk and natural polymers

The utilization of multilayer films combines the unique functionalities of various polymers with the intention of developing a package with superior performance in the food packaging industry (Kaiser et al., 2018). Commonly, multilayer packages consist of 3–12 layers, but food packaging typically has 3–7 layers (Butler & Morris, 2016) (Fig. 16.5). The most frequent multilayer film preparation procedures include coextrusion, coinjection with stretch blow molding, lamination, and coating (Ajitha et al., 2016). The choice of these is influenced by the type of polymer, interactions between polymers, layer compatibility, adhesiveness, and production scale. The compatibility of milk proteins in multilayer films with other natural polymers will be covered in this section.

The chitosan/whey protein edible film reduced the microbial contaminants and extended the product's shelf life. The coating improved texture preservation, postponed the development of undesirable acidity, and appeared to have no effect

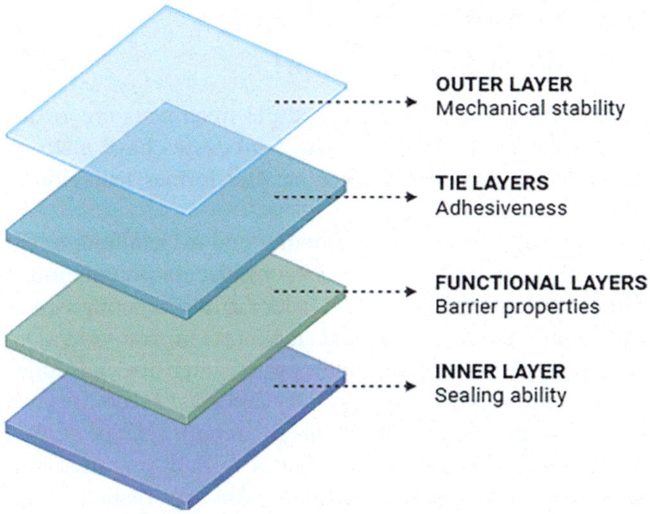

FIGURE 16.5 The general structural characteristics of multilayer films in food coatings.

on sensory characteristics of ricotta cheese. There is potential for the benefits derived from biopolymers to also be obtained with other dairy products (Di Pierro et al., 2011). It has been found that the multilayer edible food coating consisting of whey protein, pectin, and transglutaminase is effective in preventing the spoilage of fresh-cut apples, potatoes, and carrots over a 10-day storage period. A decrease in weight loss, inhibition of microbial growth, preservation of antioxidant activity, and no changes in the firmness and chewiness of the fruits and vegetables were all indicators of this effect. Compared to food coatings made from a protein-based film-forming solution alone, the performance of this enzymatically crosslinked protein/polysaccharide coating was superior (Di Pierro et al., 2011). According to other research, lamb meat's shelf life was extended by the WPI/cellulose nanofibre/TiO_2 nanoparticle/rosemary essential oil nanocomposite film when compared to the control samples. The best shelf life of the control samples was six days, according to the results of microbial and sensory evaluations; in contrast, the samples packed with nanocomposite films had a shelf life of 12–15 days (Alizadeh Sani et al., 2017). The compatibility of Furcellaran, a natural polysaccharide obtained from the red algae *Furcellaria lumbricalis*, with WPI and enhanced with extracts from yerba mate (*Ilex paraguariensis*) and white tea was shown in a study examining the effects of active edible films made of these components. When compared to the plastic packaging, the films significantly extended the shelf life of fresh soft rennet-curd cheese, decreased fungal contamination, and more effectively inhibited the growth of coliform bacteria (Pluta-Kubica et al., 2020). To find a viable alternative to complex emulsion-based coatings for fruit preservation, another study demonstrated that conjugating chitosan with WPI resulted in films with improved barrier, textural, and antioxidant properties. This edible coating delayed the senescence and ripening of strawberry fruits by up to five days (Muley & Singhal, 2020). Adding xylan and transglutaminase to WPI films to improve their characteristics. The films' barrier qualities were enhanced by xylan and transglutaminase, which decreased their oxygen permeability and WVP. Overall, WPI-xylan films, which are produced from waste by-products, provide a sustainable alternative for edible food packaging (Seiwert et al., 2021).

Research on casein shows that the mechanical and barrier properties of composite films significantly improve when they consist of pectin, casein, and gelatin. The best film's antimicrobial qualities are enhanced by adding clove essential oil (1.5% v/v), which reduces the microbial count and increases the zone of inhibition (Saikia & Badwaik, 2018). The incorporation of pectin into casein and egg albumin greatly enhances the composite film's physicochemical, structural, morphological, and thermal characteristics. When compared to films produced from pure casein or pure egg albumin, the composite films show better mechanical, structural, thermal, and physical properties (Sood & Saini, 2022). A comparable study observed that bitter orange oil-loaded casein-sodium alginate-based edible composite films showed superior morphological characteristics, thermal stability, and mechanical properties (Bhatia, Al-Harrasi, Al-Azri, Ullah, Bekhit, et al., 2022). When gallic acid is added as a natural crosslinker into gelatin-casein-based composite films, it enhances their physicochemical, mechanical, and barrier properties (Bhatia, Al-Harrasi, Al-Azri, Ullah, Makeen, et al., 2022).

It is clear from the examples given that a range of natural polymers and oils has been successfully tested in multilayer films combined with casein and whey protein, with compatibility observed in most cases.

16.4 Film formation mechanisms

16.4.1 Modification of milk proteins by heat and pH

The physicochemical transformations triggered by thermal and pH treatments are critical determinants of the structural and functional characteristics of milk protein-based films. These processes have a strong effect on protein denaturation, aggregation, and interactions, all of which are essential to the film-former behavior and the mechanical, optical, and barrier properties of films.

Heat treatment denatures whey proteins such as β-lactoglobulin and α-lactalbumin, exposing their hydrophobic amino acid residues and sulfhydryl groups. This exposure allows for intermolecular interactions such as disulfide bond formation and hydrophobic clustering that drive proteins to aggregate. Under elevated heating conditions, β-lactoglobulin intermixes with caseins, forming complexes that are vital for elastic and stable protein networks (Donato & Guyomarc'h, 2009). The extent of protein denaturation and aggregation is determined by the intensity and duration of heat treatment. Moderate heating can improve its film-forming properties (i.e., the viscosity and gelation ability), while excessive heating may cause overaggregation, leading to brittle films and reduced mechanical strength (Nayik et al., 2024).

pH modifications additionally have an effect on the behavior of milk during heat treatment. As pH decreases toward the isoelectric point of caseins (~4.6), the decrease in protein solubility results in aggregation and the formation of dense protein networks (Asaduzzaman et al., 2021). On the contrary, higher pH promotes electrostatic repulsion between proteins, resulting in more elastic but less cohesive networks (Svanborg et al., 2015). Under acidic pH conditions combined with heat, β-lactoglobulin–casein aggregates formed in a coacervate phase facilitate the formation of stronger and more cohesive films. In alkaline pH, soluble whey protein–casein complexes form and result in films with improved flexibility and transparency but may lead to a reduction in barrier properties (Marhons et al., 2023).

The synergistic action of heat and pH adjustment enables one to design milk protein-based films to meet defined functional criteria. For instance, the processing conditions can be adjusted to tune films intended for food packaging for tensile strength, elasticity, or WVP. This understanding can help unveil the potential for sustainable and performance-enhancing packaging materials in response to challenges faced in the packaging industry (Lewis, n.d.).

Film production largely relies on a precise modification of milk proteins through controlled heat treatment and pH adjustment. These modifications not only improve the functionality of milk protein-based films but also inspire for other food packaging industries to innovate.

16.4.2 Structural reinforcement with crosslinking agents

Biopolymer-based food coatings require the use of crosslinking agents to improve their structural and functional properties. These agents achieve covalent or noncovalent linkages between polymer chains to establish a 3D net structure, which enhances the mechanical strength, flexibility, and barrier properties of the coatings. Crosslinking is an important approach employed to improve the performance of food packaging materials concerning the safety and quality of food. The crosslinking in food coatings is typically performed using chemical agents (e.g., glutaraldehyde, citric acid, or genipin) or enzymatic processes (e.g., microbial transglutaminase). These agents amplify polymer engagements, for instance, crosslinking among proteins and polysaccharides results in enhanced integrity of the resulting film. Glutaraldehyde is one of the most researched crosslinking agents due to its potential to crosslink amino groups in proteins, which can tremendously increase both the tensile strength and water resistance of protein-based films (Azeredo & Waldron, 2016). Likewise, the growing use of natural cross-linkers such as citric acid is due to its eco-friendliness and the formation of ester linkages by its hydroxyl groups with polysaccharides. This makes starch-based films more effective as water vapor barriers.

Chemical crosslinking formations in the coatings significantly alter their mechanical and barrier characteristics. The film becomes tougher due to crosslinking, which creates a strong molecular grid. For example, an increase in elastic modulus and fracture toughness in crosslinked gelatin hydrogels has been reported in the literature (Serhan et al., 2019). Crosslinked coatings possess lower hydrophilicity and are suitable for moisture-sensitive food coatings. One such example is the enhancement of water resistance without loss of biodegradability in whey protein films by using microbial transglutaminase. Increased crosslinking also augments oxygen and lipid barrier properties necessary for food preservation. It was shown with starch-chitosan films that citric acid crosslinked the film, leading to lower oxygen permeability while still retaining flexibility (Pruksawan et al., 2023).

The current status of the use of crosslinking agents in food packaging remains extremely challenging in spite of the benefits. Some types of chemical crosslinkers can be a safety issue if residual agents are not effectively removed.

Consequently, there is an increasing attention on natural and enzymatic crosslinkers to fulfill the demands of both regulators and consumers for sustainability and food safety. More efficient crosslinking methods suitable for food-based coatings are now being developed as a key area of future research.

16.5 Functional features and applications

16.5.1 Moisture, gas, and oil barrier properties of milk-based films

Milk-based films, primarily derived from casein or whey proteins, are known for their moderate to low moisture barrier properties due to their hydrophilic nature (Dutta & Sit, 2024; Ke et al., 2025; Zhu et al., 2024). The protein structure plays a critical role in determining WVP. Crosslinking techniques, such as enzymatic treatments with transglutaminase or chemical agents like glutaraldehyde, can improve moisture resistance by creating a denser polymer network. However, the use of plasticizers like glycerol or sorbitol, while enhancing film flexibility, often compromises moisture barrier performance by introducing free volume in the matrix that facilitates water molecule transport (Devi et al., 2024; Ke et al., 2025; Xiao et al., 2019; Zhang et al., 2024b). To counteract this, hydrophobic additives such as beeswax, fatty acids, or essential oils can be incorporated, creating a water-resistant layer that significantly reduces WVP. Environmental factors, particularly relative humidity (RH), also affect moisture barrier efficiency, as higher RH levels increase water-protein interactions, leading to greater permeability (Dutta & Sit, 2024; Ke et al., 2025; Pino et al., 2024).

Milk-based films exhibit excellent gas barrier properties, particularly against oxygen and carbon dioxide, making them ideal for food packaging applications. The dense and compact protein structure limits gas diffusion, ensuring extended shelf life for packaged goods. Environmental conditions, such as temperature and humidity, can influence these properties, but overall, milk-based films outperform many synthetic polymers in gas impermeability. Enhancements such as crosslinking or the addition of nanoparticle fillers can further improve these properties by reinforcing the matrix structure (Innocente et al., 2023; Papadaki et al., 2024).

In terms of fat or oil barrier properties, milk-based films perform moderately due to the polar nature of proteins, which can allow limited interaction with nonpolar substances like oils. However, this can be improved by incorporating lipid-based components or waxes, which form a hydrophobic layer within or on the surface of the film. Such modifications reduce oil permeability and make these films suitable for applications involving greasy or oily foods. The combination of hydrophilic and hydrophobic characteristics in milk-based films can be tailored to achieve specific functional requirements, making them versatile materials for sustainable food packaging solutions (Devi et al., 2024; Innocente et al., 2023; Ke et al., 2025; Zhang et al., 2025; Zhang et al., 2024b).

16.5.2 Development of antimicrobial and antioxidant coatings using milk-based films

Milk-based films, particularly those derived from casein and whey proteins, have emerged as promising candidates for the development of antimicrobial and antioxidant coatings in food packaging (Fig. 16.6). These films are not only biodegradable and sustainable but also provide functional benefits such as active protection against microbial contamination and oxidative spoilage. By incorporating bioactive compounds, these films can serve as multifunctional coatings to prolong the shelf life of food products while reducing environmental impact.

Milk proteins (especially whey and casein) can be used as active and edible packaging materials. Milk proteins have the advantages of high biodegradability, nontoxicity, and superior gas barrier properties. In addition, thanks to their antimicrobial and antioxidant properties, they have significant potential to preserve food products and extend shelf life. These proteins, combined with various antimicrobial agents (e.g., essential oils, nisin, and natamycin), offer innovative packaging solutions that protect food products from microbial spoilage. In this way, they stand out as both an environmentally friendly and functional alternative (Chen et al., n.d.).

Milk proteins inherently possess some antimicrobial properties due to their bioactive components. For instance, peptides released during the enzymatic hydrolysis of casein and whey proteins, such as lactoferrin, exhibit bacteriostatic and bactericidal effects by binding to microbial cell membranes or interfering with enzyme activity (Singh et al., 2023). However, their natural antimicrobial activity can be significantly enhanced through the incorporation of additional antimicrobial agents.

Essential oils such as oregano, thyme, and clove oils have been widely used to enhance the antimicrobial activity of milk-based films. A study demonstrated that WPI films incorporated with oregano, rosemary, and garlic essential oils exhibited strong inhibitory effects on bacteria. The edible film enriched with oregano essential oil demonstrated significantly higher effectiveness against the tested bacteria at a 2% concentration compared to films containing garlic or rosemary extracts ($P < .05$). While WPI films with rosemary essential oil showed no antimicrobial activity,

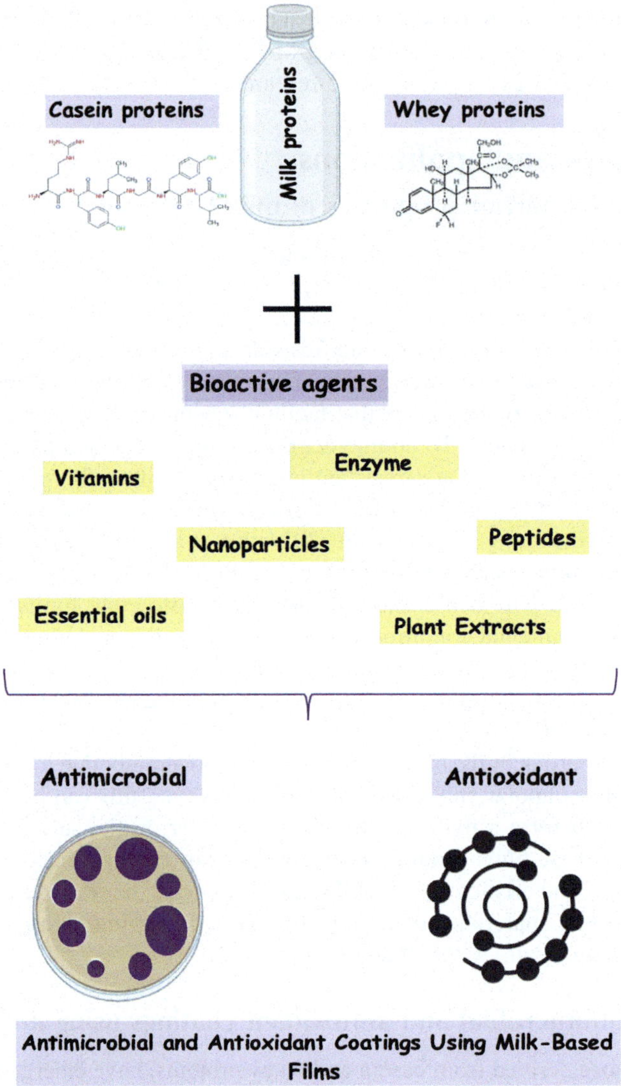

FIGURE 16.6 Antimicrobial and antioxidant coatings using milk-based films.

the incorporation of garlic essential oil into WPI films exhibited an inhibitory effect only at concentrations of 3% and 4% ($P < .05$). These findings indicate that the antimicrobial properties of certain spice extracts can be effectively expressed when incorporated into WPI-based edible films (Seydim & Sarikus, 2006).

The study conducted by Dong et al. investigated the applications of casein and silver nanoparticles (AgNPs) in biopolymer-based bilayer films. Casein provides both homogeneous distribution and content protection against UV rays and oxygen, thanks to its amphiphilic structure. Casein nanoparticles produced by the desolvation method played an important role in extending the shelf life of foods by preventing fat oxidation. In addition, AgNPs synthesized via casein showed strong antibacterial effects, increased the mechanical and barrier properties of bilayer films, and ensured that sensitive foods, such as almond oil, were preserved for longer periods (Dong et al., 2019).

In the study by Oussalah et al., milk protein-based edible films were prepared containing 1% (w/v) thyme, 1% (w/v) pimento, or 1% thyme-pimento (1:1) essential oil mixture and were applied to the surface of beef slices. These films were used to control the growth of pathogenic bacteria and to increase the shelf life of meat during storage at 4°C. Meat and films were subjected to regular microbial and biochemical analyses for seven days. The lipid oxidation potential of meat was assessed with thiobarbituric reactive substances (TBARS). The presence of phenolic compounds from essential oils was measured by determining total phenolic compounds. Antioxidant properties of the films were analyzed using a modified procedure with the N,N-diethyl-p-phenylenediamine color change method. Thyme-based films stabilized lipid oxidation in beef slices, while pimento-based films showed the highest antioxidant activity. Bioactive films were tested on meat surfaces containing *E. coli* O157:H7 and Pseudomonas spp. The films containing oregano were most effective against both bacteria, while the films containing pimento oils were least effective. At the end of storage, samples coated with oregano extract-containing films showed a 0.95 log reduction in Pseudomonas spp. and a 1.12 log reduction in *E. coli* O157:H7. These results indicate that oregano-based films are effective for both microbial control and oxidative stability (Oussalah et al., 2004).

Milk proteins, particularly casein and whey proteins, exhibit powerful antioxidant effects by scavenging reactive oxygen species (ROS), chelating metals, and modulating enzymes involved in ROS production and elimination. Casein fractions, which are precursors to antioxidant peptides, enhance antioxidant activity through their amino acid composition, including tryptophan, tyrosine, and histidine. Casein hydrolysis produces bioactive peptides like β-casomorphins and casein phosphopeptides (CPP), which bind metals, improve bioavailability, and protect against lipid peroxidation. Similarly, whey proteins, including lactoferrin and β-lactoglobulin, play key roles in antioxidant defense by scavenging free radicals, chelating metals, and enhancing the availability of essential ions. Lactoferrin also suppresses inflammation and reduces ROS production, although its antioxidant capacity diminishes with iron saturation. β-Lactoglobulin, rich in sulfur-containing amino acids (SAA) such as cysteine, is critical for protecting against oxidative stress, with peptides released during hydrolysis showing potent antioxidant properties. Additionally, SAA like methionine and cysteine, which are sensitive to ROS, contribute to antioxidant defenses by producing important metabolites like glutathione (GSH). The antioxidant capacity of milk is further supported by enzymes such as superoxide dismutase (SOD), catalase (CAT), glutathione peroxidase (GPx), and lactoperoxidase (LPx), which work together to neutralize ROS and protect cells from oxidative damage (Bielecka et al., 2022; Bottani et al., 2020; Li et al., 2019).

16.6 Sustainability and environmental impacts

Milk-based films offer significant ecological advantages, making them a promising alternative to traditional plastic packaging solutions in terms of sustainability and environmental protection. One of the most notable benefits of milk-based films is their biodegradability and compostability. Unlike conventional plastics, which can take hundreds of years to break down, milk-based films decompose naturally, significantly reducing long-term environmental pollution. This property helps prevent the accumulation of plastic waste in landfills and oceans, which can have devastating effects on ecosystems. Since these films are biodegradable, they break down into nontoxic substances, contributing to a cleaner environment (Singh et al., 2023; Zhu et al., 2024).

Milk-based films are derived from renewable resources, such as casein and whey proteins, which are natural components found in milk. This makes them an environmentally friendly alternative to petroleum-based plastics, which rely on nonrenewable fossil fuels. By utilizing milk proteins, these films support a more sustainable production cycle and reduce dependence on depleting natural resources. Additionally, the production process of milk-based films requires significantly less energy compared to the manufacturing of traditional plastics. This reduction in energy consumption lowers the overall carbon footprint of the packaging, contributing to a decrease in global greenhouse gas emissions.

Another important ecological advantage of milk-based films is their suitability for food packaging. They help extend the shelf life of food products while being eco-friendly. Traditional plastic packaging can take hundreds of years to decompose, causing long-lasting harm to the environment. Milk-based films, on the other hand, provide an effective barrier to protect food from spoilage, moisture loss, and contamination without causing long-term harm to the planet. They can also be safely disposed of, either by composting or breaking down naturally, thus avoiding pollution and waste (Bielecka et al., 2022; Nayik et al., 2024).

Furthermore, milk-based films help address the challenges associated with plastic recycling. Plastics are often difficult to recycle and require complex processes, which can further contribute to environmental strain. Milk-based films, however, are more easily compostable, reducing the strain on recycling systems and contributing to waste reduction. These films can decompose in organic waste facilities, where they are transformed into valuable compost that can be used to nourish soil, thus completing the natural cycle and minimizing waste accumulation.

In conclusion, milk-based films provide significant ecological advantages over traditional plastic packaging. Their biodegradability, compostability, and renewable nature make them a crucial component in the move toward sustainable and environmentally friendly food packaging. These films not only offer a solution to plastic pollution but also reduce energy consumption and support the use of renewable resources, making them a vital tool for addressing environmental challenges and fostering a circular economy in the packaging industry (Devi et al., 2024; Donato & Guyomarc'h, 2009; Singh et al., 2023).

16.7 The future of milk-based coatings and films

The future of milk-based coatings and films is marked by continued innovation, driven by the growing demand for sustainable and eco-friendly food packaging solutions. Derived from milk proteins like casein and whey, these films offer a biodegradable and environmentally friendly alternative to traditional plastic packaging (Alvarez et al., 2014; Innocente et al., 2023; Li et al., 2024). Researchers are focused on enhancing the mechanical properties, flexibility, and barrier performance of these films by combining milk proteins with polysaccharides such as chitosan, pectin, and cellulose. This improves their strength while maintaining their biodegradability. Additionally, the incorporation of natural antimicrobial agents, like essential oils or nanoparticles, is being explored to extend the shelf life of food products and improve food safety. These films are also being developed to include antioxidant properties, which can prevent oxidation in fatty foods, preserving both their nutritional and sensory qualities.

Milk-based films are also being designed with functional capabilities, such as controlled-release systems for bioactive compounds. This allows for a slow release of preservatives or nutrients, tailored to specific food types. These advancements promise to make milk-based films not only a protective layer for food but also an active participant in preserving their quality. Another key benefit of milk-based films is their environmental impact. As biodegradable and compostable materials, they offer a promising alternative to conventional plastics, which can take hundreds of years to degrade. In contrast, milk-based films naturally break down, reducing pollution and waste. Furthermore, the production of these films uses low-energy processes, minimizing the carbon footprint associated with food packaging. With sustainability and waste reduction becoming more pressing global issues, milk-based films provide a viable solution to help reduce plastic pollution and promote a circular economy in the food packaging industry (Innocente et al., 2023; Ke et al., 2025; Papadaki et al., 2024; Torun & Ozdemir, 2022; Zhang et al., 2025).

References

Ajitha, A. R., Aswathi, M. K., Maria, H. J., Izdebska, J., & Thomas, S. (2016). Multilayer polymer films. *Springer Series in Materials Science, 223*, 229–258. https://doi.org/10.1007/978-94-017-7324-9_8, http://www.springer.com/series/856?changeHeader.

Alizadeh Sani, M., Ehsani, A., & Hashemi, M. (2017). Whey protein isolate/cellulose nanofibre/TiO_2 nanoparticle/rosemary essential oil nanocomposite film: Its effect on microbial and sensory quality of lamb meat and growth of common foodborne pathogenic bacteria during refrigeration. *International Journal of Food Microbiology, 251*, 8–14. https://doi.org/10.1016/j.ijfoodmicro.2017.03.018.

Alvarez, M. V., Ortega-Ramirez, L. A., Gutierrez-Pacheco, M. M., Bernal-Mercado, A. T., Rodriguez-Garcia, I., Gonzalez-Aguilar, G. A., Ponce, A., Moreira, Md. R., Roura, S. I., & Ayala-Zavala, J. F. (2014). Oregano essential oil-pectin edible films as anti-quorum sensing and food antimicrobial agents. *Frontiers in Microbiology, 5*. https://doi.org/10.3389/fmicb.2014.00699, http://journal.frontiersin.org/Journal/10.3389/fmicb.2014.00699/full.

Asaduzzaman, M., Mahomud, M. S., & Haque, M. E. (2021). Heat-induced interaction of milk proteins: Impact on yoghurt structure. In *International Journal of Food Science, Vol. 2021*Hindawi Limited. https://doi.org/10.1155/2021/5569917.

Azeredo, H. M. C., & Waldron, K. W. (2016). Crosslinking in polysaccharide and protein films and coatings for food contact - A review. *Trends in Food Science and Technology, 52*, 109–122. https://doi.org/10.1016/j.tifs.2016.04.008, http://www.elsevier.com/wps/find/journaldescription.cws_home/601278/description#description.

Bhatia, S., Al-Harrasi, A., Al-Azri, M. S., Ullah, S., Bekhit, A. E. D. A., Pratap-Singh, A., Chatli, M. K., Anwer, M. K., & Aldawsari, M. F. (2022). Preparation and physiochemical characterization of bitter orange oil loaded sodium alginate and casein based edible films. *Polymers, 14*(18). https://doi.org/10.3390/polym14183855, http://www.mdpi.com/journal/polymers.

Bhatia, S., Al-Harrasi, A., Al-Azri, M. S., Ullah, S., Makeen, H. A., Meraya, A. M., Albratty, M., Najmi, A., & Anwer, M. K. (2022). Gallic acid crosslinked gelatin and casein based composite films for food packaging applications. *Polymers, 14*(19). https://doi.org/10.3390/polym14194065, http://www.mdpi.com/journal/polymers.

Bielecka, M., Cichosz, G., & Czeczot, H. (2022). Antioxidant, antimicrobial and anticarcinogenic activities of bovine milk proteins and their hydrolysates - A review. In *International Dairy Journal, Vol. 127*Elsevier Ltd. https://doi.org/10.1016/j.idairyj.2021.105208.

Bottani, M., Cattaneo, S., Pica, V., Stuknytė, M., De Noni, I., & Ferraretto, A. (2020). In vitro antioxidant properties of digests of hydrolyzed casein and caseinophosphopeptide preparations in cell models of human intestine and osteoblasts. *Journal of Functional Foods, 64*, 103673. https://doi.org/10.1016/j.jff.2019.103673.

Butler, T. I., & Morris, B. A. (2016). PE-based multilayer film structures. In *Multilayer flexible packaging: Second edition*. United States: Elsevier Inc, 281–310. https://doi.org/10.1016/B978-0-323-37100-1.00017-X, http://www.sciencedirect.com/science/book/9780323371001.

Chen, L., Goksen, G., Raju Panjagari, N., Salama, H.H., Punia Bangar, S., Rusu, A., Chaudhary, V., Kajla, P., Kumari, P., Trif, M., & Lorenzo, J.M. (n. d.). *OPEN ACCESS EDITED BY Milk protein-based active edible packaging for food applications: An eco-friendly approach.*

Devi, D., Kumar, S., & Mukherjee, A. (2024). Onion peel extract as milk freshness indicator in biopolymer-based intelligent packaging films. *Food and Humanity, 2*, 100223. https://doi.org/10.1016/j.foohum.2023.100223.

Di Pierro, P., Sorrentino, A., Mariniello, L., Giosafatto, C. V. L., & Porta, R. (2011). Chitosan/whey protein film as active coating to extend Ricotta cheese shelf-life. *LWT, 44*(10), 2324–2327. https://doi.org/10.1016/j.lwt.2010.11.031, https://www.journals.elsevier.com/lwt.

Donato, L., & Guyomarc'h, F. (2009). Formation and properties of the whey protein/κ-casein complexes in heated skim milk – A review. *Dairy Science and Technology, 89*(1), 3–29. https://doi.org/10.1051/dst:2008033.

Dong, X., Shannon, H. D., Amirsoleimani, A., Brion, G. M., & Escobar, I. C. (2019). Thiol-affinity immobilization of casein-coated silver nanoparticles on polymeric membranes for biofouling control. *Polymers, 11*(12). https://doi.org/10.3390/polym11122057, https://res.mdpi.com/d_attachment/polymers/polymers-11-02057/article_deploy/polymers-11-02057.pdf.

Dutta, D., & Sit, N. (2024). Casein-based films reinforced with bamboo shoot fibers modified by ultrasound and cellulase. *Industrial Crops and Products, 222*, 120112. https://doi.org/10.1016/j.indcrop.2024.120112.

Elsabee, M. Z., & Abdou, E. S. (2013). Chitosan based edible films and coatings: A review. *Materials Science and Engineering C, 33*(4), 1819–1841. https://doi.org/10.1016/j.msec.2013.01.010.

Elvira López-Caballero, M., Figueroa Pires, A., Díaz, O., Cobos, A., & Dias Pereira, C. (2024). *A review of recent developments in edible films and coatings-focus on whey-based materials*. https://doi.org/10.3390/foods1316.

Fematt-flores, G. E., Aguiló-aguayo, I., Marcos, B., Camargo-olivas, B. A., Sánchez-vega, R., Soto-caballero, M. C., Salas-salazar, N. A., Flores-córdova, M. A., & Rodríguez-roque, M. J. (2022). Milk protein-based edible films: Influence on mechanical, hydrodynamic, optical and antioxidant properties. *Coatings, 12*(2). https://doi.org/10.3390/coatings12020196, https://www.mdpi.com/2079-6412/12/2/196/pdf.

Gerna, S., D'Incecco, P., Limbo, S., Sindaco, M., & Pellegrino, L. (2023). Strategies for exploiting milk protein properties in making films and coatings for food packaging: A review. *Foods, 12*(6). https://doi.org/10.3390/foods12061271, www.mdpi.com/journal/foods.

Ghanbarzadeh, B., & Almasi, H. (2011). Physical properties of edible emulsified films based on carboxymethyl cellulose and oleic acid. *International Journal of Biological Macromolecules, 48*(1), 44–49. https://doi.org/10.1016/j.ijbiomac.2010.09.014.

Ghazali, H. M., Adzahan, N. M., Azarakhsh, N., Osman, A., Ghazali, H. M., Tan, C. P., & Adzahan, M. (2012). Optimization of alginate and gellan-based edible coating formulations for fresh-cut pineapples. *Article in International Food Research Journal, 19*.

Innocente, N., Calligaris, S., Di Filippo, G., Melchior, S., Marino, M., & Nicoli, M. C. (2023). Process design for the production of peptides from whey protein isolate with targeted antimicrobial functionality. *International Journal of Food Science and Technology, 58*(5), 2505–2517. https://doi.org/10.1111/ijfs.16393, http://onlinelibrary.wiley.com/journal/10.1111/(ISSN)1365-2621.

Kaiser, K., Schmid, M., & Schlummer, M. (2018). Recycling of polymer-based multilayer packaging: A review. In *Recycling*, *Vol. 3*MDPI AG. https://doi.org/10.3390/recycling3010001 Issue 1.

Kandasamy, S., Yoo, J., Yun, J., Kang, H. B., Seol, K. H., Kim, H. W., & Ham, J. S. (2021). Application of whey protein-based edible films and coatings in food industries: An updated overview. *Coatings, 11*(9). https://doi.org/10.3390/coatings11091056, https://www.mdpi.com/2079-6412/11/9/1056/pdf.

Ke, F., Yang, M., Ji, W., & Liu, D. (2025). Functional pH-sensitive film based on pectin and whey protein for grape preservation and shrimp freshness monitoring. *Food Chemistry, 463*(Pt 1), 141092. https://doi.org/10.1016/j.foodchem.2024.141092.

Khan, M. R., Volpe, S., Valentino, M., Miele, N. A., Cavella, S., & Torrieri, E. (2021). Active casein coatings and films for perishable foods: Structural properties and shelf-life extension. *Coatings, 11*(8). https://doi.org/10.3390/coatings11080899, https://www.mdpi.com/2079-6412/11/8/899/pdf.

Khwaldia, K., Ferez, C., Banon, S., Desobry, S., & Hardy, J. (2004). Milk proteins for edible films and coatings. *Critical Reviews in Food Science and Nutrition, 44*(4), 239–251. https://doi.org/10.1080/10408690490464906.

Lewis, G. (n.d.). *High pressure technologies in dairy processing*. https://app.biorender.com/biorender-.

Li, H., Pan, Y., Yang, Z., Rao, J., & Chen, B. (2019). Improving antioxidant activity of β-lactoglobulin by nature-inspired conjugation with gentisic acid. *Journal of Agricultural and Food Chemistry, 67*(42), 11741–11751. https://doi.org/10.1021/acs.jafc.9b05304.

Li, Y., Wu, H., Deng, S., Essawy, H., Brosse, N., Fan, M., Du, G., Chen, X., Zhou, X., & Liao, J. (2024). Novel hydroxyl-terminated hyperbranched polymer as a synergistic modifier with tannin for preparation of casein-based films with superior performance. *International Journal of Biological Macromolecules, 278*(Pt 1), 134672. https://doi.org/10.1016/j.ijbiomac.2024.134672.

Ma, K., Yang, J., Goksen, G., Alamri, A. S., Alhomrani, M., Xia, G., & Zhang, W. (2025). Application of tragacanth gum as a potential food packaging film and its performance enhancement strategy. *Food Hydrocolloids, 162*. https://doi.org/10.1016/j.foodhyd.2024.110894, https://www.sciencedirect.com/science/journal/0268005X.

Marhons, Š., Hyršlová, I., Stetsenko, V., Jablonská, E., Veselý, M., Míchová, H., Čurda, L., & Štětina, J. (2023). Properties of yoghurt treated with microbial transglutaminase and exopolysaccharides. *International Dairy Journal, 144*, 105701. https://doi.org/10.1016/j.idairyj.2023.105701.

Metha, C., Pawar, S., & Suvarna, V. (2024). Recent advancements in alginate-based films for active food packaging applications. Sustainable Food Technology, 2(5), 1246–1265. https://doi.org/10.1039/D3FB00216K.

Missaghi, S., & Fassihi, R. (2004). A novel approach in the assessment of polymeric film formation and film adhesion on different pharmaceutical solid substrates. *AAPS PharmSciTech, 5*(2), 32–39. https://doi.org/10.1208/pt050229.

Muley, A. B., & Singhal, R. S. (2020). Extension of postharvest shelf life of strawberries (*Fragaria ananassa*) using a coating of chitosan-whey protein isolate conjugate. *Food Chemistry, 329*. https://doi.org/10.1016/j.foodchem.2020.127213, www.elsevier.com/locate/foodchem.

Nayik, G. A., Gull, A., Masoodi, L., Navaf, M., Sunooj, K. V., Ucak, İ., Afreen, M., Kaur, P., Rehal, J., Jagdale, Y. D., Ramniwas, S., Singh, R., & Mugabi, R. (2024). Milk proteins: Chemistry, functionality and diverse industrial applications. *Cogent Food and Agriculture, 10*(1). https://doi.org/10.1080/23311932.2024.2377686, www.tandfonline.com/action/journalInformation?journalCode=oafa20.

Oussalah, M., Caillet, S., Salmiéri, S., Saucier, L., & Lacroix, M. (2004). Antimicrobial and antioxidant effects of milk protein-based film containing essential oils for the preservation of whole beef muscle. *Journal of Agricultural and Food Chemistry, 52*(18), 5598–5605. https://doi.org/10.1021/jf049389q.

Papadaki, A., Lappa, I. K., Manikas, A. C., Pastore Carbone, M. G., Natsia, A., Kachrimanidou, V., & Kopsahelis, N. (2024). Grafting bacterial cellulose nanowhiskers into whey protein/essential oil film composites: Effect on structure, essential oil release and antibacterial properties of films. *Food Hydrocolloids, 147*. https://doi.org/10.1016/j.foodhyd.2023.109374, https://www.journals.elsevier.com/food-hydrocolloids.

Pino, P., Vigani, B., Valentino, C., Ianev, D., Ruggeri, M., Boselli, C., Cornaglia, A. I., Grisoli, P., Onida, B., Bosco, F., Sandri, G., & Rossi, S. (2024). Sustainable whey proteins-nanostructured zinc oxide-based films for the treatment of chronic wounds: New insights from biopharmaceutical studies. *International Journal of Biological Macromolecules, 263*. https://doi.org/10.1016/j.ijbiomac.2024.130655, https://www.sciencedirect.com/science/journal/01418130.

Pluta-Kubica, A., Jamróz, E., Kawecka, A., Juszczak, L., & Krzyściak, P. (2020). Active edible furcellaran/whey protein films with yerba mate and white tea extracts: Preparation, characterization and its application to fresh soft rennet-curd cheese. *International Journal of Biological Macromolecules, 155*, 1307–1316. https://doi.org/10.1016/j.ijbiomac.2019.11.102.

Ponce, A., Roura, S. I., & Moreira, M. R. (2016). Casein and chitosan polymers: Use in antimicrobial packaging. In *Antimicrobial food packaging*. Elsevier Inc, 455–466. https://doi.org/10.1016/B978-0-12-800723-5.00037-1.

Pruksawan, S., Lim, J. W. R., Lee, Y. L., Lin, Z., Chee, H. L., Chong, Y. T., Chi, H., & Wang, F. K. (2023). Enhancing hydrogel toughness by uniform cross-linking using modified polyhedral oligomeric silsesquioxane. *Communications Materials, 4*(1). https://doi.org/10.1038/s43246-023-00402-2, https://www.nature.com/commsmat/.

Rezaei, Z., Salari, A., Khanzadi, S., Rhim, J. W., & Shamloo, E. (2023). Preparation of milk-based probiotic lactic acid bacteria biofilms: A new generation of probiotics. *Food Science and Nutrition, 11*(6), 2915–2924. https://doi.org/10.1002/fsn3.3273, onlinelibrary.wiley.com/journal/10.1002/%28ISSN%292048-7177.

Rouf, T. B., & Kokini, J. L. (2017). Natural biopolymer-based nanocomposite films for packaging applications. In *Bionanocomposites for packaging applications*. United States: Springer International Publishing, 149–177. https://doi.org/10.1007/978-3-319-67319-6_8, http://www.springer.com/in/book/9783319673196.

Saikia, M., & Badwaik, L. S. (2018). Characterization and antimicrobial property of casein, gelatin and pectin based active composite films. *Journal of Packaging Technology and Research, 2*(3), 233–242. https://doi.org/10.1007/s41783-018-0044-3.

Salama, H. H., Trif, M., Rusu, A. V., & Bhattacharya, S. (2022). Application of functional and edible coatings and films as promising strategies for developing dairy functional products—A review on yoghurt case. *Coatings, 12*(6). https://doi.org/10.3390/coatings12060838, https://www.mdpi.com/2079-6412/12/6/838/pdf?version=1655295261.

Seiwert, K., Kamdem, D. P., Kocabaş, D. S., & Ustunol, Z. (2021). Development and characterization of whey protein isolate and xylan composite films with and without enzymatic crosslinking. *Food Hydrocolloids, 120*. https://doi.org/10.1016/j.foodhyd.2021.106847, https://www.sciencedirect.com/science/journal/0268005X.

Serhan, M., Sprowls, M., Jackemeyer, D., Long, M., Perez, I. D., Maret, W., Tao, N., & Forzani, E. (2019). Total iron measurement in human serum with a smartphone. AIChE Annual Meeting, Conference Proceedings. 2019. 9780816911127. American Institute of Chemical Engineers. United States. https://www.aiche.org/conferences/, https://doi.org/10.1039/x0xx00000x.

Seydim, A. C., & Sarikus, G. (2006). Antimicrobial activity of whey protein based edible films incorporated with oregano, rosemary and garlic essential oils. *Food Research International, 39*(5), 639–644. https://doi.org/10.1016/j.foodres.2006.01.013.

Singh, A., Duche, R. T., Wandhare, A. G., Sian, J. K., Singh, B. P., Sihag, M. K., Singh, K. S., Sangwan, V., Talan, S., & Panwar, H. (2023). Milk-derived antimicrobial peptides: Overview, applications, and future perspectives. *Probiotics and Antimicrobial Proteins, 15*(1), 44–62. https://doi.org/10.1007/s12602-022-10004-y, https://www.springer.com/journal/12602.

Sood, A., & Saini, C. S. (2022). Utilization of peel of white pomelo for the development of pectin based biodegradable composite films blended with casein and egg albumen. *Food Chemistry Advances, 1*. https://doi.org/10.1016/j.focha.2022.100054, https://www.journals.elsevier.com/food-chemistry-advances.

Svanborg, S., Johansen, A. G., Abrahamsen, R. K., & Skeie, S. B. (2015). The composition and functional properties of whey protein concentrates produced from buttermilk are comparable with those of whey protein concentrates produced from skimmed milk. *Journal of Dairy Science, 98*(9), 5829–5840. https://doi.org/10.3168/jds.2014-9039, http://www.elsevier.com/wps/find/journaldescription.cws_home/721317/description#description.

Tennert, C., Sarra, G., Stähli, A., Sculean, A., & Eick, S. (2023). Bovine milk and yogurt affect oral microorganisms and biofilms in-vitro. *Oral Health and Preventive Dentistry, 21*, 69–76. https://doi.org/10.3290/j.ohpd.b3920023, https://www.quintessence-publishing.com/deu/en.

Tharanathan, R. N. (2003). Biodegradable films and composite coatings: Past, present and future. *Trends in Food Science and Technology, 14*(3), 71–78. https://doi.org/10.1016/S0924-2244(02)00280-7, http://www.elsevier.com/wps/find/journaldescription.cws_home/601278/description#description.

Tița, M. A., Moga, V. M., Constantinescu, M. A., Bătușaru, C. M., & Tița, O. (2024). Harnessing the potential of whey in the creation of innovative food products: Contributions to the circular economy. *Recycling, 9*(5). https://doi.org/10.3390/recycling9050079, www.mdpi.com/journal/recycling.

Tonyali, B., Cikrikci, S., & Oztop, M. H. (2018). Physicochemical and microstructural characterization of gum tragacanth added whey protein based films. *Food Research International, 105*, 1–9. https://doi.org/10.1016/j.foodres.2017.10.071.

Torun, M., & Ozdemir, F. (2022). Milk protein and zein coatings over peeled garlic cloves to extend their shelf life. *Scientia Horticulturae, 291*, 110571. https://doi.org/10.1016/j.scienta.2021.110571.

Xiao, G., Li, L., Yan, A., & He, X. (2019). Direct detection of melamine in infant formula milk powder solution based on SERS effect of silver film over nanospheres. *Spectrochimica Acta Part A: Molecular and Biomolecular Spectroscopy, 223*, 117269. https://doi.org/10.1016/j.saa.2019.117269.

Xu, X., Deng, S., Essawy, H., Lee, S. H., Lum, W. C., Zhou, X., Du, G., & Zhang, J. (2024). Chitosan-casein blended with condensed tannin and carnauba wax for the fabrication of antibacterial and antioxidant food packing films. *International Journal of Biological Macromolecules, 277*. https://doi.org/10.1016/j.ijbiomac.2024.133784, https://www.sciencedirect.com/science/journal/01418130.

Zhang, R., Liu, R., Han, J., Ren, L., & Jiang, L. (2024). Protein-based packaging films in food: Developments, applications, and challenges. In *Gels, Vol. 10*Multidisciplinary Digital Publishing Institute (MDPI). https://doi.org/10.3390/gels10070418 Issue 7.

Zhang, R., Wang, B., Zhang, F., Zheng, K., & Liu, Y. (2024). Milk-derived antimicrobial peptides incorporated whey protein film as active coating to improve microbial stability of refrigerated soft cheese. *International Journal of Food Microbiology, 419*. https://doi.org/10.1016/j.ijfoodmicro.2024.110751.

Zhang, Y., Tian, W., Deng, Y., Cao, Y., & Xiao, J. (2025). Regulation of whey protein microparticle adsorption and interfacial film formation by polysaccharides for enhanced Pickering emulsion stability. *Food Hydrocolloids, 162*, 110879. https://doi.org/10.1016/j.foodhyd.2024.110879.

Zhu, Y., Fu, B., Zhu, P., Shu, S., Wang, D., Xu, H., & Cai, D. (2024). Waste milk-derived biodegradable mulch film fabrication and nutrient slow-release performance evaluation. *Chemical Engineering Journal, 497*, 155008. https://doi.org/10.1016/j.cej.2024.155008.

Chapter 17

Polyhydroxyalkanoates films and coating

Zeba Tabassum
School of Bioengineering and Biosciences, Lovely Professional University, Jalandhar, Punjab, India

17.1 Introduction

For decades, the global food packaging industry has relied heavily on synthetic plastics, valued for their low cost, durability, versatility, and ease of processing. These materials have enabled advances in preservation, storage, and transportation, but their very advantages have also caused serious environmental challenges. Plastic packaging is a major contributor to solid waste and land and marine pollution, with its non-biodegradable nature leading to long-term accumulation and the formation of microplastics that threaten ecosystems and human health. Moreover, the linear production–consumption–disposal model of petroleum-based plastics drives unsustainable exploitation of resources. These issues underscore the urgent need for greener alternatives that combine functionality with sustainability. Biopolymers have emerged as promising substitutes or complements to conventional plastics in food packaging. Sourced from renewable feedstocks such as plants, microorganisms, and agricultural waste, they offer biodegradability, reduced carbon footprint, and compatibility with circular economy principles (Kabir et al., 2020; Shaikh et al., 2021; Tabassum et al., 2024, 2025; Tabassum, Girdhar, et al., 2023; Tabassum, Mohan, et al., 2023). Among biopolymers, starch, cellulose, polylactic acid (PLA), chitosan, and polyhydroxyalkanoates (PHAs), the latter stand out for their versatility and eco-friendliness, often matching conventional plastics in both barrier and mechanical performance.

17.2 Polyhydroxyalkanoate

PHAs are microbial polyesters synthesized as intracellular carbon and energy reserves. Their production can draw on diverse substrates, including sugars, vegetable oils, and even industrial waste streams, enhancing both economic feasibility and sustainability. Structural tunability further adds to their appeal: for example, polyhydroxybutyrate (PHB) is stiff and brittle, while copolymers such as PHBV and PHBH provide flexibility and toughness, enabling tailored solutions for specific packaging needs. One of the most critical roles of food packaging is to extend the shelf life of perishables by shielding them from oxygen, moisture, light, and microbial contamination. PHAs demonstrate excellent gas barrier properties, particularly against oxygen and carbon dioxide, making them suitable for fresh produce, dairy, meat, and other sensitive foods. Their tensile strength and modulus are comparable to polypropylene, and they display strong resistance to moisture and oil. Importantly, PHAs naturally degrade in soil, compost, and marine environments without leaving harmful residues, addressing a key drawback of conventional plastics. Biocompatibility and nontoxicity further strengthen the case for PHAs. Unlike some petroleum-based plastics that may leach additives, PHAs are safe for direct food contact and align with increasing regulatory and consumer demand for healthier packaging. Their ability to biodegrade under composting conditions also supports integration into circular waste management systems. Despite these advantages, PHA adoption is constrained by high production costs, scalability issues, and competition with cheaper synthetic plastics. Recent innovations, however, are steadily overcoming these barriers. Advances in microbial engineering, fermentation, and substrate optimization are improving yields and lowering costs. The use of inexpensive feedstocks such as agricultural residues, food processing waste, and even CO_2 as carbon sources holds promise for large-scale production. Additionally, blending PHAs with other biopolymers or reinforcing them with nanofillers like cellulose nanocrystals, chitosan nanoparticles, or montmorillonite nanoclay can enhance functional properties while reducing costs. Such composites and nanocomposites often display improved mechanical, thermal, and barrier performance, strengthening their competitiveness. PHAs also align with global sustainability agendas, particularly the UN Sustainable Development Goals (SDGs)

Natural Polymer-Based Films and Coatings in Food Technology. DOI: https://doi.org/10.1016/B978-0-443-34244-8.00006-9

related to responsible consumption, climate action, and ecosystem protection. By reducing fossil dependency and minimizing plastic pollution, they contribute to a circular bioeconomy. Policies promoting biodegradable alternatives and consumer awareness of environmental issues are further stimulating interest in PHA-based packaging. Applications of PHAs span films, trays, bottles, and coatings, with studies showing their capacity to extend food shelf life by reducing microbial growth and oxidation. Multilayer structures and active packaging systems enhanced with natural antimicrobials or antioxidants offer additional functionality, simultaneously reducing food spoilage and waste. Regionally, PHAs present opportunities for valorizing agricultural residues and agro-industrial by-products, fostering local value chains and supporting rural economies. In summary, PHAs represent a transformative step toward sustainable food packaging. They combine biodegradability, renewable sourcing, and functional performance with the potential to significantly extend the shelf life of perishables. Although challenges of cost and scalability remain, ongoing technological progress and supportive policies are driving their industrial relevance. With rising consumer preference for eco-friendly materials, PHAs are poised to play a central role in the next generation of food packaging. This chapter explores their production, properties, innovations, and applications, emphasizing their role in advancing a greener and more sustainable future. Polyhydroxyalkanoates (PHAs) are a family of naturally occurring polyesters synthesized by a wide variety of microorganisms as intracellular carbon and energy storage materials. These biopolymers can be produced using diverse substrates, ranging from sugars and vegetable oils to industrial waste streams, making their production pathway highly flexible and sustainable. The ability to valorize low-cost renewable and waste resources for PHA production further enhances their economic and ecological feasibility. Structurally, PHAs consist of hydroxyalkanoate monomers, and their properties can be tailored by altering monomer composition. For instance, polyhydroxybutyrate (PHB) is a stiff and brittle homopolymer, while copolymers such as polyhydroxybutyrate-co-hydroxyvalerate (PHBV) or polyhydroxybutyrate-co-hydroxyhexanoate (PHBH) offer greater flexibility and toughness. This tunability provides scope for customizing PHAs for specific food packaging needs. One of the most critical requirements for food packaging is its ability to extend the shelf life of perishable goods by acting as a protective barrier against external factors such as oxygen, moisture, light, and microbial contamination. PHAs exhibit favorable gas barrier properties, particularly against oxygen and carbon dioxide, which makes them suitable for packaging fresh produce, dairy, meat, and other sensitive products. Their mechanical performance, including tensile strength and modulus, is often comparable to widely used plastics like polypropylene (PP). Moreover, PHAs show high resistance to moisture and oil, which is particularly advantageous for packaging foods with high water activity or fat content. Importantly, PHAs degrade naturally in various environments, including soil, compost, and marine settings, without leaving behind harmful residues—addressing the end-of-life problem associated with traditional plastics. Another dimension of PHAs' potential lies in their biocompatibility and nontoxicity, which allows for safe contact with food. Unlike certain petroleum-based plastics that may leach harmful additives, PHAs are generally regarded as safe, aligning with increasing regulatory and consumer demand for healthier and safer food packaging materials. Their biodegradability under controlled composting conditions also supports integration into waste management systems, where packaging waste can be converted into biomass or CO_2 through microbial activity, thereby closing the material loop. Despite these advantages, the widespread adoption of PHAs in food packaging has been limited by factors such as high production costs, limited scalability, and competition with low-cost synthetic plastics. However, ongoing research and innovations are addressing these challenges. Advances in microbial engineering, fermentation technologies, and substrate optimization are driving down production costs and improving yields. The use of inexpensive feedstocks such as agricultural residues, food processing waste, and even CO_2 as carbon sources for microbial fermentation holds significant promise for large-scale sustainable production. Additionally, blending PHAs with other biopolymers or incorporating nanofillers such as cellulose nanocrystals, chitosan nanoparticles, or montmorillonite nanoclay has been shown to enhance their functional performance while reducing costs. Such PHA-based composites and nanocomposites can exhibit superior mechanical, thermal, and barrier properties, making them more competitive with conventional plastics. The integration of PHAs into food packaging also aligns with broader global sustainability goals, including the UN SDGs. Specifically, PHAs contribute to responsible consumption and production (SDG 12), climate action (SDG 13), life below water (SDG 14), and life on land (SDG 15). By reducing reliance on fossil-based plastics and minimizing environmental pollution, PHAs support the transition toward circular bioeconomy models. Consumer awareness and policy interventions, such as bans on single-use plastics and incentives for biodegradable alternatives, are further driving market interest in PHA-based solutions. From an application standpoint, PHAs have demonstrated strong potential in various food packaging formats, including films, trays, bottles, and coatings. Their inherent biodegradability ensures that after fulfilling their function, the packaging materials do not persist in the environment. Studies have shown that PHA-based films can significantly slow down microbial growth and oxidation in perishable foods, thus enhancing shelf life. For example, meat and dairy products stored in PHA packaging have exhibited lower microbial counts and slower spoilage compared to conventional packaging. The ability of PHAs to be processed into multilayer structures, combined with functional additives like natural antimicrobials or antioxidants, opens up avenues for developing active and intelligent packaging systems. These innovations not only preserve food quality but also contribute to reducing food waste, which is itself a critical sustainability challenge. Furthermore, PHAs present opportunities for regional resource valorization. Countries with abundant agricultural residues or agro-industrial by-products can harness these as feedstocks for PHA production, creating local value chains and reducing dependency on imported raw materials. This decentralization of production can stimulate rural economies while supporting sustainable waste management. Collaborative research and industry-academia partnerships are playing a pivotal role in scaling up PHA production and improving their cost-effectiveness for food packaging applications. In conclusion, PHAs represent a significant step forward in the development of sustainable food packaging solutions. They combine the benefits of biodegradability, renewable sourcing, and functional performance with the potential to extend the shelf life of perishable foods. While challenges related to cost and scalability remain, ongoing technological advances and supportive policy frameworks are steadily enhancing their viability. As consumer preferences shift toward environmentally responsible products, PHAs are well positioned to emerge as a cornerstone of the next generation of food packaging materials. This paper explores the diverse innovations surrounding PHAs, highlighting their eco-friendly attributes, potential for prolonging food shelf life, and role in shaping a greener tomorrow. Over the past decades, there has been a significant growth in the number of publications related to food packaging, including biodegradable, antimicrobial, active, and smart packaging, see Fig. 17.1. This chapter explores the potential of PHA-based packaging materials for prolonging the shelf life of perishable goods, addressing their production, properties, innovations, and industrial relevance.

Polyhydroxyalkanoates (PHAs) produced from waste-derived carbon sources present an intrinsically circular and commercially promising route to sustainable food-packaging materials when embedded within an industrial symbiosis framework. By valorizing heterogeneous organic residues—for example, food processing by-products (fruit/vegetable peels, dairy whey), lignocellulosic agro-residues (rice husk, sugarcane bagasse after hydrolysis), glycerol from biodiesel production, and carbon-rich streams from municipal wastewater treatment (volatile fatty acid-rich fermentates)—mixed microbial cultures or engineered PHA producers can accumulate high yields of short-chain and medium-chain PHAs under feast–famine and nutrient-limiting regimes without competing with food/feed markets. Integration into an industrial symbiosis network captures synergies at multiple scales: off-gases and waste heat from adjacent chemical or food processing plants can meet energy demands for pretreatment and downstream drying; concentrated organic wastes diverted from landfills reduce biogas plant load and provide consistent feedstock; glycerol or crude glycerin streams supply high-carbon substrates for fed-batch bioreactors; and solid residues from PHA extraction (e.g., biomass cake) can be anaerobically digested to produce biogas, closing the energy loop. Technically, upstream strategies focus on low-cost feedstock conditioning (pH control, hydrolysis to soluble sugars or acidogenic fermentation to volatile fatty acids

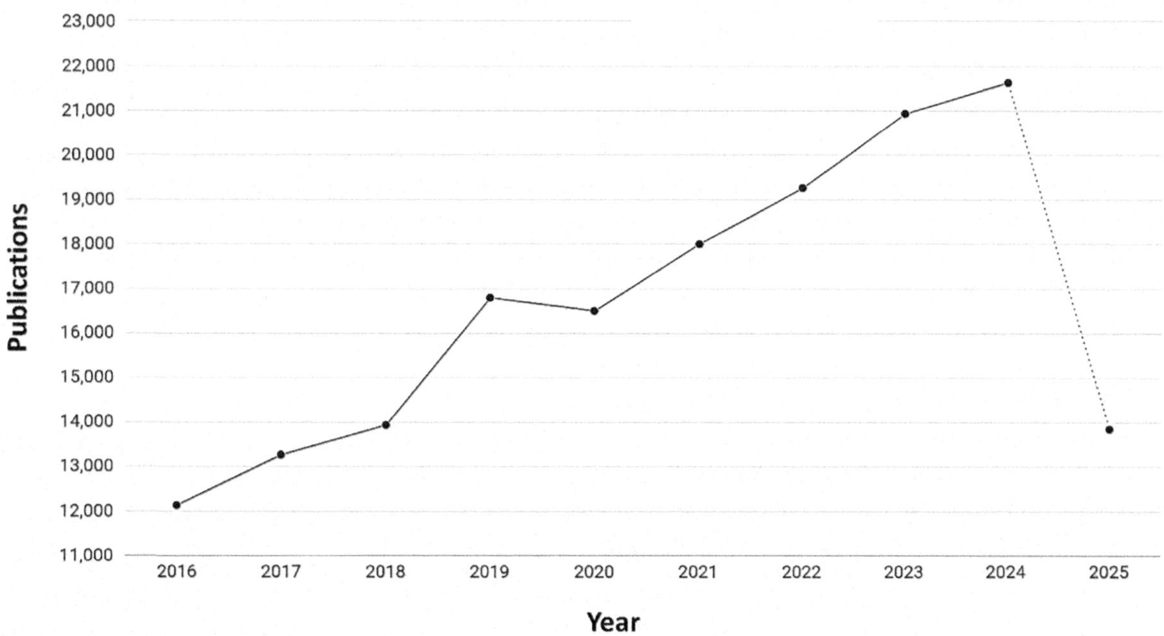

FIGURE 17.1 Recent publications on PHA.

[VFAs]), robust microbial consortia selection (native mixed microbial consortia [MMC] selected under open conditions or pure cultures such as Cupriavidus necator for homopolymeric P(3HB) or recombinant Pseudomonas for medium-chain length PHAs), and reactor configurations that balance productivity and robustness (sequencing batch reactors, continuous stirred tanks with cell retention). Downstream processing must mitigate solvent use and environmental impact—greener extraction methods (nonchlorinated solvents, supercritical CO_2, enzymatic digestion, or mechanical disruption combined with aqueous surfactant washes) are increasingly viable at pilot scale and preserve polymer molecular weight for film formation. Materially, waste-derived PHAs can match or exceed key barrier and mechanical properties required for certain food-packaging applications (stiffness and tensile strength comparable to low-density polyethylene for some PHA grades, and excellent oxygen barrier when oriented or blended), and they are amenable to modification (plasticization, copolymerization, nanofiller addition—for example, montmorillonite, ZnO nanoparticles, cellulose nanocrystals) to tune water vapor permeability, flexibility, and thermal transitions for targeted perishables. From a circularity and life-cycle perspective, coupling PHA production with existing industrial operations reduces feedstock transport and embodied emissions and creates opportunities for localized closed-loop packaging supply chains: manufacturers can convert their own process residues into film-grade PHA, thermoform or cast packaging on-site or at regional hubs, and return used packages for composting or controlled anaerobic digestion with energy recovery. However, scale-up barriers remain: feedstock heterogeneity affects fermentation stability and polymer consistency, capital expenditure for bioreactors and solvent-free recovery is substantial, and regulatory compliance for food contact requires rigorous migration testing and toxicological validation for residual extractables (especially when using waste streams). Economic viability is improved by coproduct valorization (energy generation from residuals, sale of residual biomass as fertilizer after appropriate treatment), policy incentives (extended producer responsibility, biodegradable packaging mandates), and industrial partnerships that internalize external waste management costs into material economics. In sum, waste-sourced PHAs embedded in an industrial symbiosis model offer a technically credible and environmentally beneficial pathway to produce food-contact materials that advance circular economy goals, but commercial deployment depends on integrated process design, greener downstream routes, consistent feedstock management, and regulatory pathways tailored to waste-derived biopolymers (De Donno Novelli et al., 2021; Elboughdiri et al., 2025; Gacem et al., 2025; Hong et al., 2025; Jannatiha & Gutiérrez, 2025; Liu et al., 2021; Rodriguez-Perez et al., 2018; Yukesh Kannah et al., 2022).

17.3 Polyhydroxyalkanoates and their derivatives

Polyhydroxyalkanoates (PHAs) are a diverse family of biodegradable polyesters produced by microorganisms as intracellular carbon and energy storage materials. Their structural diversity arises from variations in the monomer composition, primarily due to differences in the side chain (R group) attached to the hydroxyalkanoate unit. This structural variability governs their physicochemical properties, such as crystallinity, flexibility, thermal stability, and degradation rate, making them suitable for a wide range of applications (Fig. 17.2).

Masood (2017) classifies PHAs and their derivatives based on the carbon number and type of substituent group (R group) (Table 7.1).

Upstream processing of PHA is costly, contributing nearly 40% of the total production expenses. However, this cost can be significantly reduced by using low-cost raw materials such as agro-industrial residues and lignocellulosic waste. Recent studies have particularly emphasized the potential of hydrothermal processing of lignocellulosic waste, which offers an efficient and economical route for PHA synthesis. This approach not only enhances sustainability but also demonstrates promising applicability of PHAs in the food packaging industry (Tripathi et al., 2022).

17.4 Recent studies

Khosravi-Darani and Yazdian (2022) examine the deployment of polyhydroxyalkanoates (PHAs), microbially synthesized biodegradable polyesters, as promising candidates for sustainable food packaging. The authors emphasize PHAs' unique physicochemical properties that make them suitable substitutes for conventional plastics. They discuss microbial synthesis pathways, which can be enhanced under nutrient-limited conditions (e.g., with restricted nitrogen, oxygen, or phosphorus, and an excess of carbon), often leveraging renewable feedstocks like agro-wastes, vegetable oils, or starch. These strategies align both ecological goals and economic feasibility. Additionally, the chapter notes that mechanical disruption processing methods for PHA extraction are particularly promising from an environmental performance standpoint, reducing reliance on chemical-intensive recovery units (Khosravi-Darani & Yazdian, 2022).

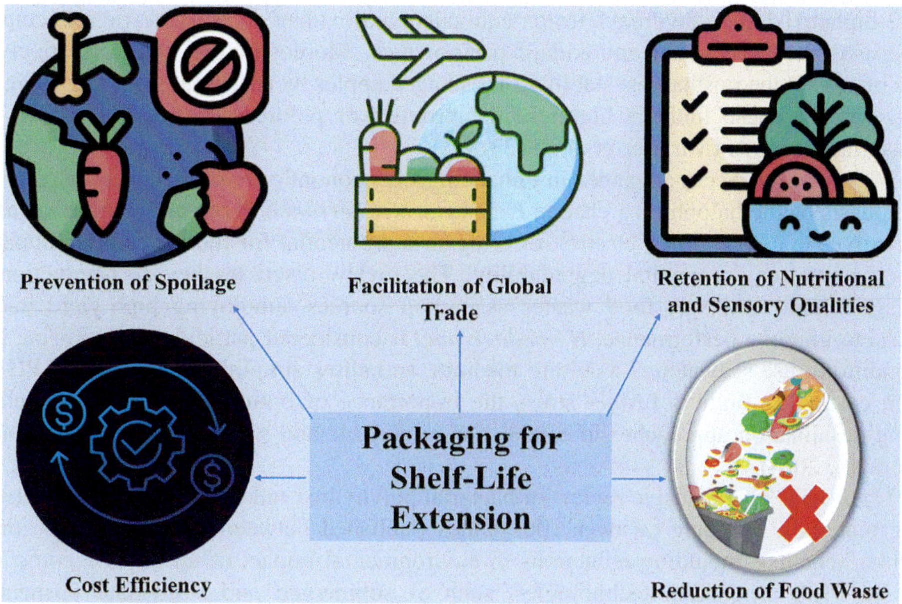

FIGURE 17.2 Need for proper packaging.

TABLE 17.1 PHAs and its derivatives.

n	R group	PHAs	Abbreviation
n = 1	R = hydrogen	Poly-(3-hydroxypropionate)	PHP
	R = methyl	Poly-(3-hydroxybutyrate)	PHB
	R = ethyl	Poly-(3-hydrovalerate)	PHV
	R = propyl	Poly-(3-hydroxyhexanoate)	PHHx
	R = pentyl	Poly-(3-hydroxyoctanoate)	PHO
	R = nonyl	Poly-(3-hydroxydodecanoate)	PHDD
n = 2	R = hydrogen	Poly-(4-hydroxybutyrate)	P4HB
n = 3	R = hydrogen	Poly-(5-hydrovalerate)	P5HV

Source: From Masood, F. (2017). Polyhydroxyalkanoates in the Food Packaging Industry. Elsevier BV, 153–17710.1016/b978-0-12-811942-6.00008-x.

Acharjee et al. (2023) review polyhydroxyalkanoates (PHAs) as a promising biodegradable alternative to petroleum-based plastics, particularly to mitigate microplastic pollution. They highlight PHAs' favorable attributes—including thermoplasticity, hydrophobicity, vapor-barrier capacity, and potential carbon neutrality due to biomass origins—making them suitable for applications such as food containers, trays, and foam utensils. The paper also addresses PHAs' expanding use beyond packaging in fields like biomedicine and agriculture, noting their biocompatibility and safe degradation in vivo. However, the authors acknowledge substantial obstacles to widespread implementation, particularly high production costs, poor mechanical properties (e.g., brittleness, hydrophilicity), and economic feasibility challenges. They note ongoing efforts to improve gas and water barrier characteristics as a key strategy to enhance PHA performance for broader adoption (Acharjee et al., 2023).

Recent work by Mirpoor et al. (2023) demonstrated the potential of polyhydroxyalkanoate (PHA)-based bioplastics functionalized with phloretin as active food packaging materials. The incorporation of phloretin (5–20 mg) imparted notable antioxidant and antimicrobial activity without significantly compromising the films' mechanical and hydrophilic properties. Spectroscopic analyses (Fourier Transform Infrared [FTIR], Raman) confirmed structural interactions, while bioassays (2,2-diphenyl-1-picrylhydrazyl, trolox equivalent antioxidant capacity, ferric reducing antioxidant power, metal chelation) verified dose-dependent antioxidant performance. Moreover, the films effectively inhibited *Listeria monocytogenes* and improved the postharvest stability of packaged apples by reducing browning, weight loss, and nutrient degradation during storage. These findings highlight the promise of phloretin-loaded PHA films as multifunctional, biodegradable packaging solutions (Mirpoor et al., 2023).

Atarés et al. (2024) reviewed recent advances in enhancing the economic feasibility of polyhydroxyalkanoate (PHA) production, with emphasis on the halophilic archaeon *Haloferax mediterranei*. PHAs are highlighted as biodegradable and biocompatible alternatives to conventional plastics, offering strong potential for food packaging applications due to their thermoplastic properties and environmental degradability. The review discusses key cost-reduction strategies such as utilizing low-cost agricultural and agro-food wastes as carbon sources, employing high-yield microbial strains, and functionalizing PHAs to improve performance. *H. mediterranei* is considered particularly promising because it can grow under nonsterile conditions, use seawater as a culture medium, and allow simplified, water-based PHA recovery, thereby reducing production costs. The authors further stress the importance of optimizing effluent recycling, valorizing side streams, and refining desalination approaches to establish a sustainable and scalable biotechnological platform for PHA production (Atarés et al., 2024).

Stublić et al. (2024) provide an expansive review of bacterial polyhydroxyalkanoates (PHAs) as sustainable alternatives for food packaging materials. Drawing on over 100 studies published between 2018 and 2023, the authors highlight notable environmental benefits—including reductions in environmental impact of up to 50% compared to conventional plastics. Advances in PHA production technologies, such as submerged and solid-state fermentation, have led to improvements in production efficiency of up to 60%, while refined extraction and purification methods have decreased energy use by approximately 30%. The review further emphasizes material innovations—blending PHAs with other biodegradable polymers like PLA, starch, and cellulose to enhance mechanical performance by up to 40%. Additionally, the incorporation of antimicrobial agents and essential oils extends shelf life by about 25% and improves barrier properties by around 35%, underscoring the significant potential of active PHA-based packaging solutions (Stublić et al., 2024).

Life cycle assessment (LCA) has been widely applied to evaluate the environmental performance of poly (hydroxyalkanoate) (PHA) plastics, particularly their carbon footprint. In this regard, Vea et al. conducted an LCA to examine how greenhouse gas (GHG) emissions from PHA materials may contribute to climate tipping points, considering 13 critical Earth system elements. To improve barrier properties against oxygen, water vapor, and aromas, the study investigated PHA films either laminated with PLA or metallized with aluminum (Al) and aluminum oxides (AlOx). The analysis also accounted for variations in the degradation kinetics of PHA at end-of-life, focusing on CO_2 and CH_4 emissions. Their findings indicated that PHA-based plastics laminated with Al or AlOx exhibited the lowest carbon footprint (0.05 kg $CO_2\,kg^{-1}$ PHA) when compared with PHA/PLA composites and conventional plastics. The authors concluded that such PHA-based materials hold strong promise as sustainable alternatives to traditional plastics, particularly within a circular economy framework (Yeo et al., 2024).

17.5 Future trends of polyhydroxyalkanoates in food technology

Recent advances in polyhydroxyalkanoate (PHA) research show a strong progression toward their integration into sustainable and functional food packaging systems. Current trends indicate a move beyond traditional biodegradable plastics toward bioactive and intelligent PHA-based films capable of enhancing food preservation and monitoring freshness in real time. Studies have demonstrated PHBV–tannin composites with improved antioxidant and mechanical performance, while the incorporation of natural extracts and polyphenols imparts antimicrobial and barrier functions. Parallelly, the valorization of food and agro-industrial waste as carbon sources for microbial PHA synthesis aligns with the circular bioeconomy and significantly reduces production costs—one of the main barriers to commercialization. Technological improvements such as ultrasonic-assisted blending, reactive extrusion, and copolymerization are enhancing ductility, transparency, and water resistance, making PHAs viable for edible coatings and flexible food wraps. Moreover, emerging "smart" PHA packaging incorporating biosensors or pH indicators reflects a paradigm shift toward interactive food safety systems. From my analysis of these studies, PHAs hold a strategically strong position for the future of food technology due to their biodegradability, renewability, and functional tunability, but their large-scale adoption will depend on achieving cost parity with conventional plastics, ensuring food-contact safety compliance, and developing robust postuse composting infrastructure. Given their biodegradation efficiency and compatibility with natural additives, PHAs are poised to become a cornerstone material for next-generation sustainable food packaging, bridging environmental responsibility with advanced food preservation functionality.

17.6 Conclusion

The eco-friendliness of PHA-based food packaging lies in its reduced carbon emissions, lower dependence on fossil fuels, and the utilization of agricultural by-products as renewable feedstocks. By extending the shelf life of perishable foods, PHAs not only help to minimize food waste but also generate economic benefits across the supply chain. Despite their great promise, challenges related to cost-effectiveness, consumer acceptance, and scalability continue to hinder widespread adoption. Among these, the high cost of carbon sources remains a major barrier to sustainable production. Strategies such as employing metabolically engineered strains and leveraging agro-industrial waste streams offer promising solutions to lower production costs. Overall, the integration of PHA biopolymers into food packaging represents a significant step forward in addressing plastic pollution, reducing food waste, and moving toward a more sustainable and greener future.

AI disclosure

During the preparation of this work, the author(s) used ChatGPT in order to improve research language. After using this tool/service, the author(s) reviewed and edited the content as needed and take(s) full responsibility for the content of the publication.

References

Acharjee, S. A., Bharali, P., Gogoi, B., Sorhie, V., Walling, B., & Alemtoshi (2023). PHA-based bioplastic: A potential alternative to address microplastic pollution. *Water, Air, and Soil Pollution, 234*(1). https://doi.org/10.1007/s11270-022-06029-2, https://www.springer.com/journal/11270.

Atarés, L., Chiralt, A., González-Martínez, C., & Vargas, M. (2024). Production of polyhydroxyalkanoates for biodegradable food packaging applications using haloferax mediterranei and agrifood wastes. *Foods, 13*(6), 950. https://doi.org/10.3390/foods13060950.

De Donno Novelli, L., Moreno Sayavedra, S., & Rene, E. R. (2021). Polyhydroxyalkanoate (PHA) production via resource recovery from industrial waste streams: A review of techniques and perspectives. *Bioresource Technology, 331*, 124985. https://doi.org/10.1016/j.biortech.2021.124985.

Elboughdiri, N., Kriaa, K., Maatki, C., Assadi, A. A., Tabassum, Z., Chopra, C., Sakran, K. A., & Wani, A. K. (2025). Eco-smart packaging: A green approach to prolonging freshness of chicken and chicken products. *Food and Bioprocess Technology*. https://doi.org/10.1007/s11947-025-04038-7, https://www.springer.com/journal/11947.

Gacem, A., Yadav, K. K., Tabassum, Z., Thajudeen, K. Y., Ahmed, M. M., Laeliocattleya, R. A., Khamidah, A., Antarlina, S. S., Alreshidi, M. A., Aldosari, F. M., Kaseem, M., Chopra, C., Wani, A. K., & Malik, T. (2025). Valorization of eggshell waste as sustainable mechanical reinforcement in biodegradable bio-packaging materials. *Journal of Agriculture and Food Research, 23*. https://doi.org/10.1016/j.jafr.2025.102273, https://www.sciencedirect.com/science/journal/26661543.

Hong, S. J., Ahn, K., Kim, J. T., & Kim, Y. T. (2025). Polyhydroxyalkanoates for sustainable food packaging: A comprehensive review on production, applications, and end-of-life scenarios for food packaging. *Trends in Food Science and Technology, 165*. https://doi.org/10.1016/j.tifs.2025.105294, https://www.sciencedirect.com/science/journal/09242244.

Jannatiha, N., & Gutiérrez, T. J. (2025). Recycling and revalorization of PLA and PHA-based food packaging waste: A review. *Sustainable Materials and Technologies, 44*, e01364. https://doi.org/10.1016/j.susmat.2025.e01364.

Kabir, E., Kaur, R., Lee, J., Kim, K. H., & Kwon, E. E. (2020). Prospects of biopolymer technology as an alternative option for non-degradable plastics and sustainable management of plastic wastes. *Journal of Cleaner Production, 258*. https://doi.org/10.1016/j.jclepro.2020.120536, https://www.journals.elsevier.com/journal-of-cleaner-production.

Khosravi-Darani, K., & Yazdian, F. (2022). *Polyhydroxyalkanoates (PHAs) in food packaging biodegradable polymer-based food packaging*. Iran: Springer Nature, 115–122. https://doi.org/10.1007/978-981-19-5743-7, 10.1007/978-981-19-5743-7_5.

Liu, H., Kumar, V., Jia, L., Sarsaiya, S., Kumar, D., Juneja, A., Zhang, Z., Sindhu, R., Binod, P., Bhatia, S. K., & Awasthi, M. K. (2021). Biopolymer poly-hydroxyalkanoates (PHA) production from apple industrial waste residues: A review. *Chemosphere, 284*. https://doi.org/10.1016/j.chemosphere.2021.131427, www.elsevier.com/locate/chemosphere.

Masood, F. (2017). *Polyhydroxyalkanoates in the food packaging industry*. Elsevier BV, 153–177. 10.1016/b978-0-12-811942-6.00008-x.

Mirpoor, S. F., Patanè, G. T., Corrado, I., Giosafatto, C. V. L., Ginestra, G., Nostro, A., Foti, A., Gucciardi, P. G., Mandalari, G., Barreca, D., Gervasi, T., & Pezzella, C. (2023). Functionalization of polyhydroxyalkanoates (PHA)-based bioplastic with phloretin for active food packaging: Characterization of its mechanical, antioxidant, and antimicrobial activities. *International Journal of Molecular Sciences, 24*(14). https://doi.org/10.3390/ijms241411628, http://www.mdpi.com/journal/ijms.

Rodriguez-Perez, S., Serrano, A., Pantión, A. A., & Alonso-Fariñas, B. (2018). Challenges of scaling-up PHA production from waste streams. A review. *Journal of Environmental Management, 205*, 215–230. https://doi.org/10.1016/j.jenvman.2017.09.083, https://www.sciencedirect.com/journal/journal-of-environmental-management.

Shaikh, S., Yaqoob, M., & Aggarwal, P. (2021). An overview of biodegradable packaging in food industry. *Current Research in Food Science, 4*, 503–520. https://doi.org/10.1016/j.crfs.2021.07.005.

Stublić, K., Ranilović, J., Ocelić Bulatović, V., & Kučić Grgić, D. (2024). Advancing sustainability: Utilizing bacterial polyhydroxyalkanoate for food packaging. *Processes, 12*(9), 1886. https://doi.org/10.3390/pr12091886.

Tabassum, Z., Mohan, A., Mamidi, N., Khosla, A., Kumar, A., Solanki, P. R., Malik, T., & Girdhar, M. (2023). Recent trends in nanocomposite packaging films utilising waste generated biopolymers: Industrial symbiosis and its implication in sustainability. *IET Nanobiotechnology, 17*(3), 127–153. https://doi.org/10.1049/nbt2.12122.

Tabassum, Z., Girdhar, M., Kumar, A., Malik, T., & Mohan, A. (2023). ZnO nanoparticles-reinforced chitosan–xanthan gum blend novel film with enhanced properties and degradability for application in food packaging. *ACS Omega, 8*(34), 31318–31332. https://doi.org/10.1021/acsomega.3c03763.

Tabassum, Z., Girdhar, M., Malik, T., Kumar, A., & Mohan, A. (2024). Advancing sustainability: A novel biopolymer-based degradable nanoclay composite film for next-generation packaging. *Materials Advances*. https://doi.org/10.1039/D4MA00476K.

Tabassum, Z., Girdhar, M., Anand, A., Kumari, N., Sood, B., Malik, T., Kumar, A., & Mohan, A. (2025). Trash to treasure: Advancing resource efficiency using waste-derived fillers as sustainable reinforcing agents in bioplastics. *Materials Advances*. https://doi.org/10.1039/D4MA01043D.

Tripathi, A. D., Mishra, P. K., Khosarvi-Darani, K., Agarwal, A., & Paul, V. (2022). Hydrothermal treatment of lignocellulose waste for the production of polyhydroxyalkanoates copolymer with potential application in food packaging. *Trends in Food Science and Technology, 123*, 233–250. https://doi.org/10.1016/j.tifs.2022.03.018, http://www.elsevier.com/wps/find/journaldescription.cws_home/601278/description#description.

Yeo, J. C. C., Muiruri, J. K., Fei, X., Wang, T., Zhang, X., Xiao, Y., Thitsartarn, W., Tanoto, H., He, C., & Li, Z. (2024). Innovative biomaterials for food packaging: Unlocking the potential of polyhydroxyalkanoate (PHA) biopolymers. *Biomaterials Advances, 163*. https://doi.org/10.1016/j.bioadv.2024.213929, https://www.sciencedirect.com/science/journal/27729508.

Yukesh Kannah, R., Dinesh Kumar, M., Kavitha, S., Rajesh Banu, J., Kumar Tyagi, V., Rajaguru, P., & Kumar, G. (2022). Production and recovery of polyhydroxyalkanoates (PHA) from waste streams – A review. *Bioresource Technology, 366*, 128203. https://doi.org/10.1016/j.biortech.2022.128203.

Chapter 18

Polylactic acid films and coatings

Nur Hanani Z.A.[1,2] and Han Lyn Foong[1,2]

[1]*Department of Food Technology, Faculty of Food Science and Technology, Universiti Putra Malaysia (UPM), Serdang, Malaysia,* [2]*Department of Diagnostic & Allied Health Science, Faculty of Health and Life Sciences, Management and Science University, Shah Alam, Selangor, Malaysia*

18.1 Introduction

Historically, petrochemical-based plastics have dominated the packaging industry since their commercialization in the late 19th century (Robertson, 2019). However, their environmental drawbacks, including nonbiodegradability and high carbon footprints, have driven the search for sustainable alternatives. In recent years, attention has shifted toward sustainable and smart packaging solutions.

Sustainable packaging refers to the utilization of renewable sources and often involves valorization of biomass or by-products of the food processing industry, converting them into polymers or additives for food packaging materials. This approach not only reduces reliance on fossil-based resources but also promotes the valorization of waste materials. Biopolymers have emerged as promising eco-friendly alternatives to conventional plastics, particularly for single-use applications. Derived from renewable resources or synthesized through microbial and chemical processes, biopolymers aim to mitigate environmental challenges posed by petrochemical-based plastics. Among these, polylactic acid (PLA) stands out due to its biodegradability, mechanical strength, and versatility in applications such as food packaging.

PLA is an aliphatic polyester produced either through the fermentation of carbohydrates or chemical synthesis of lactic acid. Microbial fermentation is the dominant commercial method for lactic acid production, favored for its cost-efficiency, reduced carbon dioxide emissions, minimal reliance on fossil-derived feedstocks, high product specificity, and relatively low environmental footprint (Han Lyn et al., 2024; Swetha et al., 2023). Renewable resources such as corn, sugar beet, tapioca, and wheat are used to derive lactic acid. The lactic acid monomer exists in two optically active configurations: L- and D-isomers. The ratio of these isomers affects key macromolecular properties, including the degree of crystallinity, melting temperature, and processability (Yu et al., 2023). PLA synthesis involves three main steps: lactic acid production, lactide formation, and polymerization, as illustrated in Fig. 18.1. Commercial PLA is primarily derived from L-lactide, yielding poly(L-lactic acid), a semicrystalline material with relatively high melting and glass transition temperatures (Auras et al., 2010).

PLA offers several advantages, including high mechanical strength, excellent thermoforming capability, biocompatibility, renewability, and compostability. Furthermore, its production results in 15%–60% lower carbon emissions and 25%–55% lower energy consumption compared to petrochemical-based polymers (Connolly et al., 2019; Zubairi et al., 2022). Its compatibility with conventional plastic processing methods has facilitated its commercial adoption in various applications, including biodegradable food packaging.

Despite its numerous advantages, PLA has notable limitations, including brittleness and high hardness. These characteristics stem primarily from its wide molecular weight distribution and relatively low molecular weight, which contribute to reduced thermal resistance and impact strength, as well as inadequate gas barrier properties, limiting its performance in certain food applications. The brittleness of PLA is further influenced by its molecular structure. Strong intermolecular forces, such as hydrogen bonding, restrict the mobility of polymer chains, reducing the material's ability to deform under stress without fracturing. These factors restrict the broader use of PLA in packaging, especially when compared to traditional plastics such as polyethylene terephthalate (PET). These challenges emphasize the need for continuous research to improve PLA's properties. Current strategies include the development of PLA-based nanocomposites, multilayer structures, and surface modifications, which aim to enhance its thermal stability, mechanical strength, and barrier properties.

Natural Polymer-Based Films and Coatings in Food Technology. DOI: https://doi.org/10.1016/B978-0-443-34244-8.00009-4

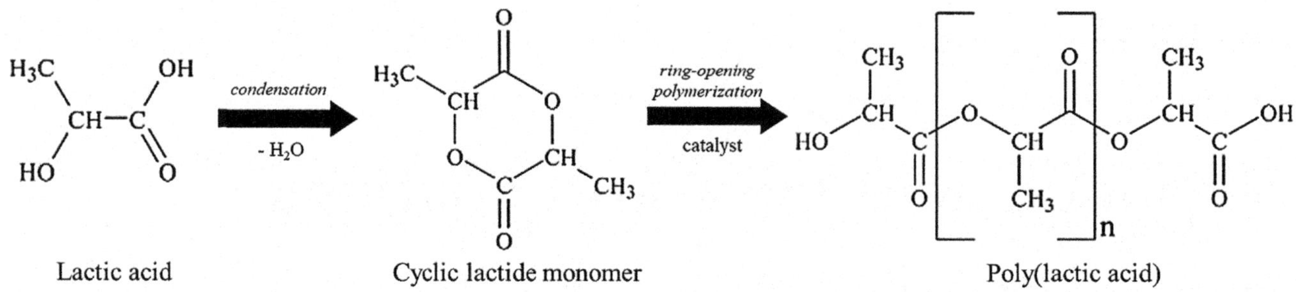

FIGURE 18.1 Polymerization process of lactic acid to form polylactic acid.
Modified from Yu, J., Xu, S., Liu, B., Wang, H., Qiao, F., Ren, X., & Wei, Q. (2023). PLA bioplastic production: From monomer to the polymer. European Polymer Journal, 193. https://doi.org/10.1016/j.eurpolymj.2023.112076.

From an environmental perspective, PLA offers several advantages over traditional petrochemical-based polymers, which will be discussed in subsequent sections. However, concerns related to the cultivation of starch-based crops, including deforestation, excessive water use, and pesticide application, highlight potential drawbacks in PLA production. Comprehensive life cycle assessment (LCA) studies are, therefore, crucial to evaluate the environmental trade-offs associated with PLA and guide its broader adoption.

18.2 Properties of polylactic acid films and coatings

18.2.1 Mechanical properties

Mechanical properties determine a material's ability to endure stresses during handling, transportation, and storage. Mechanical properties are generally represented by tensile and yield strength, elongation at break, and Young's modulus. They are assessed using equipment such as a universal testing machine or pull tester, which applies a pulling force to film samples. Key factors such as tensile strength, flexibility, and resistance to deformation play a vital role in determining a material's suitability for various packaging applications. Packaging with higher mechanical strength ensures product integrity, reduces waste caused by damage, and enhances consumer confidence in the packaged goods. As innovations in sustainable packaging materials advance, achieving a balance between mechanical strength and environmental sustainability has become a focal point in packaging design.

Despite its many advantages, PLA faces challenges related to its brittleness, limiting its applications in food packaging. Various strategies have been explored to mitigate this limitation and enhance the mechanical properties of PLA-based films and coatings.

In addition to modifying mechanical properties, essential oils (EOs) also impart antimicrobial and antioxidant properties, enhancing food preservation. The incorporation of additives, such as EOs and extracts, into PLA matrices has been shown to reduce tensile strength while increasing elasticity. This phenomenon is attributed to the plasticizing effect, in which the additive molecules disrupt the compact polymer matrix, creating discontinuities that reduce mechanical resistance. Simultaneously, these additives enhance the mobility of polymer chains, improving flexibility and ductility. The concentration of EOs or additives also plays a critical role in determining mechanical performance. For example, Zeid et al. (2019) demonstrated that incorporating thyme, rosemary, and oregano EOs into PLA films significantly reduced tensile strength ($p < 0.05$) and slightly lowered elongation at break ($p \geq 0.05$). The authors attributed these effects to the low EO concentration (10 wt.%) used in their study, which was insufficient to act as an effective plasticizer.

In extruded PLA composites, EOs can also influence the solidification behavior during calendaring. Their presence has been shown to reduce the solidification rates of PLA, thereby extending the time for polymer network orientation. This effect results in softer composites upon drying, as reported by a previous study (Gonon et al., 2023).

PLA is also used as a coating material to reinforce other biobased polymers. For instance, chitosan/nanocellulose composite films that were coated with a PLA (2 wt.%) solution exhibited significantly improved tensile strength and strain while demonstrating reduced water solubility (Sobhan et al., 2021).

The interaction of PLA with different plasticizers varies significantly, making it crucial to optimize plasticizer concentration for desired properties. For example, PLA films plasticized with epoxidized karanja oil showed a consistent increase in ductility at concentrations up to 5 wt.%. However, at higher concentrations (10 wt.%), ductility and elongation at break decreased due to plasticizer saturation within the polymer matrix, leading to phase separation (Garcia-Garcia et al., 2020).

By understanding and optimizing the interaction of PLA with plasticizers and additives, researchers can develop PLA-based materials that address its mechanical limitations, paving the way for broader applications in sustainable food packaging solutions.

18.2.2 Barrier properties

The shelf life and quality of food products are greatly influenced by external factors such as temperature, humidity, light, and oxygen. To enhance food shelf life and reduce food losses, interactions between food products and these environmental factors should be minimized (Roy et al., 2024).

Photooxidation is a key challenge, where light exposure—particularly ultraviolet (UV) light—catalyzes reactions in lipid-containing foods, leading to the formation of free radicals. These radicals can cause protein oxidation, aggregation, and the deterioration of nutritional and sensory qualities. Foods such as meat, dairy products, nuts, cooking oils, and wine are especially prone to photooxidation. Effective packaging materials must provide robust light and oxygen barrier properties to address these issues. Additionally, moisture plays a major role in food spoilage, contributing to increased water activity, microbial growth, and undesirable textural changes.

Pure PLA films and coatings are limited in their barrier properties, exhibiting high UV light transmittance (85% at 250 nm and 95% at 300 nm) and oxygen permeability (5.7×10^{-8} cm^2 s^{-1} at 30°C) (Bao et al., 2006; Ran et al., 2023). They also show a significantly higher water vapor transmission rate (WVTR) of 263.10 g $(m^2.24\ hour)^{-1}$ compared to petrochemical-based plastics such as low-density polyethylene (LDPE) (4.70–14.0 g $[m^2.24\ hour]^{-1}$) and polypropylene (0.05–1.55 g $[m^2.24\ hour]^{-1}$) (Matweb, 2024). Therefore, improving the light, oxygen, and moisture barrier properties of PLA is crucial for its application in food packaging.

18.2.2.1 Water vapor and oxygen barrier properties

Incorporating nanoparticles into PLA films has shown improvements in water vapor permeability (WVP). Shankar et al. (2018) demonstrated that adding 0.5 wt.% zinc oxide nanoparticles (ZnO NPs) significantly improved WVP. However, higher nanoparticle concentrations led to increased WVP. This shows that a low concentration of nanoparticles is effective as fillers to increase the tortuous path of the polymer matrix, whereas higher concentrations might modify the free volume at the interfaces of nanoparticles and polymer matrix, resulting in an increase in WVP.

Natural extracts have also been incorporated into PLA films, resulting in improvements in barrier properties. For example, incorporating a 2% ethanolic extract of green tea leaves resulted in a 35% reduction in the WVP of PLA films (Martins et al., 2018).

Hydrophobic additives, such as oils and lipids, improve the water vapor and oxygen barrier properties of PLA films. For instance, PLA/perilla EO nanofiber films exhibited the lowest WVP and oxygen permeability at a 2 wt.% EO concentration. Beyond this concentration, agglomeration of oil droplets increased permeability by creating micropores in the polymer matrix (Wang, Sun, et al., 2021).

It is important to note that the barrier performance of PLA films and coatings is sensitive to relative humidity. Karkhanis et al. (2021) found that WVP values for neat PLA and PLA/cellulose nanocrystals (CNC) composites increased with relative humidity (RH), being lowest at 50% relative humidity and highest at 85% RH. Incorporating 1% CNC decreased WVP by 41%–49%, and the effect was more prominent at lower humidity levels, as illustrated in Table 18.1.

The compatibility of hydrophobic compounds with PLA is a critical consideration. Sometimes, they weaken the barrier properties due to incompatibility of the film components. For instance, PLA films coated with chitosan containing clove or argan oil showed contrasting results. Films with clove oil had better barrier properties due to the strong interactions between clove oil and chitosan, forming a compact structure (Fig. 18.2). In contrast, argan oil droplets coalesced, creating micropores and increasing WVP and oxygen permeability (Criollo-Feijoo et al., 2024; Kang & Song, 2019; Stoleru et al., 2021).

18.2.2.2 Light barrier properties

Pure PLA films and coatings are highly transparent, with over 90% transmittance for visible light (wavelengths > 400 nm) and a significant reduction at wavelengths < 350 nm (Valerini et al., 2018). Enhancements in light barrier properties can be achieved through the incorporation of EOs or nanostructured metal oxides.

Many EOs contain polyphenols such as flavonoids, which exhibit strong UV absorbance due to their aromatic structures, allowing them to absorb UVA and UVB rays in the wavelength range of 200–400 nm.

Additionally, incorporation of EO might increase the opacity of PLA films, which can be attributed to the innate color of the EO or the nonhomogeneity that arises between the oils and the PLA matrix. Generally, EOs impart a slight brown

TABLE 18.1 Water vapor permeability values of polylactic acid (PLA) and PLA/cellulose nanocrystals (CNC) films at different relative humidities and a constant temperature of 23°C.

	% RH	Water vapor permeability (10^{-16} kg.m [m^2.s.Pa]$^{-1}$)	
	PLA	PLA/1% CNC	% Decrease
50	327.2 ± 16.1	166.9 ± 4.1	49
57	331.7 ± 20.6	182.7 ± 5.0	45
64	330.7 ± 27.3	187.7 ± 2.9	43
71	318.9 ± 21.3	181.6 ± 3.6	43
78	314.8 ± 43.9	196.5 ± 6.1	38
85	341.8 ± 50.0	200.3 ± 10.0	41

Source: From Karkhanis, S.S., Stark, N.M., Sabo, R.C., & Matuana, L.M. (2021). Potential of extrusion-blown poly(lactic acid)/cellulose nanocrystals nanocomposite films for improving the shelf-life of a dry food product. Food Packaging and Shelf Life, 29. https://doi.org/10.1016/j.fpsl.2021.100689.

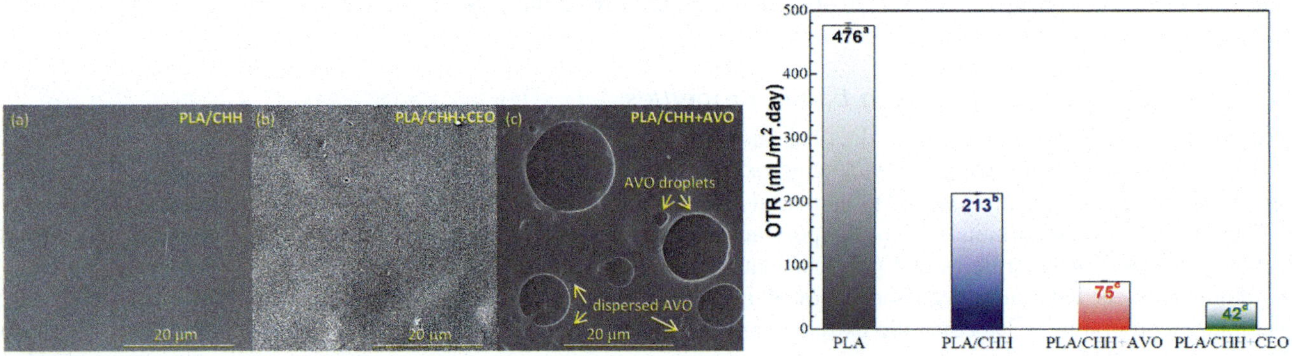

FIGURE 18.2 Scanning electron microscope micrographs and oxygen transmission rate of polylactic acid (PLA) samples surface modified with argan vegetal oil (AVO)- and clove essential oil (CEO)-loaded chitosan (CHH) coatings.
Modified from Stoleru, E., Vasile, C., Irimia, A., & Brebu, M. (2021). Towards a Bioactive Food Packaging: Poly(Lactic Acid) Surface Functionalized by Chitosan Coating Embedding Clove and Argan Oils. Molecules, 26(15). https://doi.org/10.3390/molecules26154500.

tint to PLA films, as shown in Fig. 18.3, and the intensity is often affected by the concentration of phenolic compounds in the EOs and the amount of oil incorporated.

To maintain high clarity while lowering UV light transmittance, colorless, nanostructured metal oxides have been incorporated into PLA films and coatings, including titanium dioxide (TiO_2) and zinc oxide (ZnO). These n-type semiconductors often exhibit good absorbance in the ultraviolet region due to their ability to absorb the energy of a UV photon during the excitation process of electrons from the valence band to the conduction band (Shehzad et al., 2024). Often, doping elements such as silver (Ag), aluminum (Al), and manganese (Mn) are incorporated to further improve their functional properties. For instance, in PLA films with an Al-doped ZnO coating, aluminum atoms promote the formation of alumina (Al_2O_3) nanograins, which improve the water vapor and oxygen barrier properties and antimicrobial properties while maintaining high optical transparency (Bala et al., 2011; Valerini et al., 2018).

18.2.3 Thermal properties

The incorporation of additives into PLA films plays a crucial role in tailoring their mechanical and thermal properties to meet the demands of various applications. These additives, including plasticizers and fillers, can modify the crystallization behavior, glass transition temperature (T_g), and overall thermal stability of PLA, offering both advantages and potential challenges.

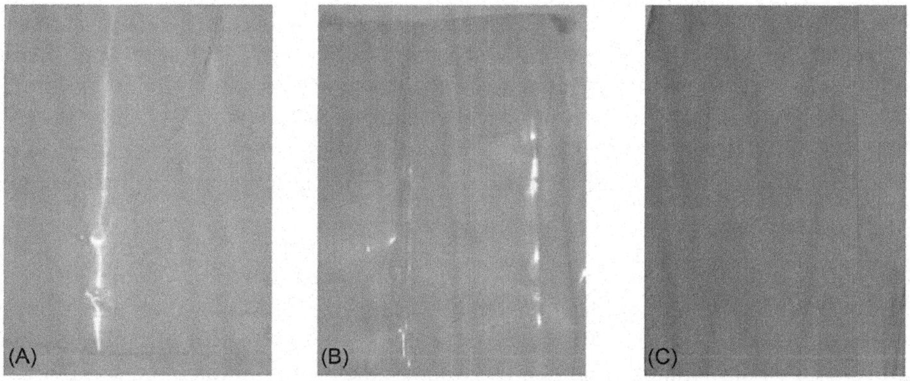

FIGURE 18.3 Neat polylactic acid films are clear and transparent and become increasingly brown with the increase in concentration of green tea extract.
From Martins, C., Vilarinho, F., Sanches Silva, A., Andrade, M., Machado, A.V., Castilho, M.C., Sá, A., Cunha, A., Vaz, M.F., & Ramos, F. (2018). Active polylactic acid film incorporated with green tea extract: Development, characterization and effectiveness. Industrial Crops and Products, 123, 100–110. https://doi.org/10.1016/j.indcrop.2018.06.056.

The incorporation of additives, such as plasticizers, typically increases the ductility of PLA films but is often accompanied by a decrease in the T_g. This phenomenon can be attributed to the enhanced segmental mobility of PLA chains. For example, the addition of capsicum oleoresin-impregnated nanoporous silica (SiCO) to PLA composite films decreased both the T_g and thermal stability, a change associated with the increased chain mobility induced by SiCO particles. Furthermore, the cold crystallization temperature (T_{cc}) was reduced, enabling crystallization to occur at lower temperatures during heating. Additionally, the SiCO particles acted as nucleating agents, lowering the energy required for cold crystallization. Meanwhile, the decreased thermal stability can be attributed to the lower thermal stability of capsicum oleoresin in SiCO (Techawinyutham et al., 2019).

Similarly, the incorporation of epoxidized soybean oil (ESO) as a plasticizer decreased both the T_g and T_{cc} of PLA films (Xing & Matuana, 2016). Interestingly, during the storage test, the T_g gradually increased over the first 40 days and then leveled off for the remainder of the storage period. This reversible change in thermal properties was attributed to the segregation of ESO from the film matrix during storage. Nevertheless, at the end of the 283-day storage period, the plasticized PLA films remained more flexible, with T_g values still lower than those of neat PLA films.

A decrease in the crystallization and melting temperatures of PLA has also been observed in nitrilotriacetic acid-grafted PLA films. This was attributed to intramolecular crosslinking of PLA chains (Herskovitz & Goddard, 2020). The authors noted that as the composites resulted in only a single phase on the thermogram, this indicates that there was no sign of immiscibility.

The thermal properties of PLA composite films and coatings can also be influenced by biodegradation. During soil degradation, PLA/starch films showed a more significant increase in T_{cc} compared to neat PLA films. This suggested that more polymeric chains participated in the crystallization process as degradation progressed. Moreover, starch/PLA composites degraded faster than neat PLA, further highlighting the impact of biodegradation on the thermal behavior of the films.

18.2.4 Biodegradability and compostability

The biodegradability of PLA composites has garnered significant attention due to their potential to reduce environmental pollution. PLA can undergo natural degradation under specific conditions, breaking down into eco-friendly by-products such as carbon dioxide and water, making it a promising material for sustainable applications. Several factors, including environmental conditions, additives, and structural modifications, influence the rate and efficiency of PLA degradation.

The biodegradability of PLA composites has been investigated using various methods, including natural weathering, accelerated weathering, and composting. PLA can undergo biodegradation under certain conditions without producing environmental pollution (Wu et al., 2023). Degradation of PLA can be divided into heterogeneous (surface degradation) and homogeneous (intramolecular polymer degradation), occurring through three primary chemical reactions: main-chain cleavage, side-chain cleavage, and cross-link cleavage. The degradation predominantly involves ester bond hydrolysis, splitting long polymer chains into shorter oligomers, dimers, or monomers. This process represents a critical, rate-limiting step in its overall breakdown (Siakeng et al., 2019). These smaller units, once reduced to a molecular weight below 10,000 g mol^{-1}, can penetrate microbial cell walls, serving as substrates for enzymatic degradation into carbon dioxide and water.

Temperature and humidity are critical factors influencing PLA degradation rates. Under soil or domestic composting conditions at 20°C, degradation may take up to a year, while temperatures above 25°C can reduce the timeframe to 12 weeks (Ho et al., 1999; Rudnik & Briassoulis, 2011). Industrial composting conditions, characterized by temperatures near PLA's glass transition temperature (~60°C) and high RH (60%–80%), can accelerate degradation to as little as 30 days (Emadian et al., 2017).

Incorporating natural fibers and polymers into PLA can significantly enhance its degradation rate, particularly with hydrophilic additives such as polysaccharides. For instance, PLA reinforced with chia seed flour degrades faster due to the hydrophilic lignocellulosic fillers facilitating the transfer of water, enzymes, and microorganisms into the composite (Dominguez-Candela et al., 2022). Similarly, the addition of corn (10 wt.%) to PLA improved biodegradability from 63.6% to 79.7% at 58°C over 90 days (Sarasa et al., 2009), whereas the incorporation of kenaf/HDPE composite enhanced the degradation of PLA by inducing micro-cracking on the surface, which reduced its structural integrity (Umar et al., 2012). Starch and cellulose also act as nutrient sources for microbes, increasing the degradation rate, as evidenced by the absence of a lag phase in carbon dioxide evolution during composting (Mayekar et al., 2023).

Natural fibers such as hemp, coir, and rice husk further enhance PLA biodegradability due to their high water affinities. PLA biocomposites reinforced with coir fibers showed a 34.0% weight loss in a soil burial test after 18 days, compared to 18% for neat PLA (Dong et al., 2014).

The structural configuration of PLA films also affects their degradation rate. Typically, monolayer PLA films degrade faster than bilayer or multilayer films. For example, PLA/Bio-Flex F 2110 bilayer films exhibited lower degradation rates than monolayer PLA films during hydrolytic degradation, which is attributed to the presence of stabilizers in the bilayer structure (Scaffaro et al., 2020).

18.2.5 Functional properties

The incorporation of active agents, even at low concentrations, is a promising strategy to enhance the antioxidant and antimicrobial properties of PLA films. Given the growing consumer preference for natural food products free from synthetic additives, there has been increasing interest in using EOs and extracts derived from natural sources such as flowers, leaves, stems, bark, fruit peels, and seeds as additives in active packaging. These bioactive compounds provide a sustainable alternative to traditional preservatives and may improve the functionality of PLA-based films.

One such natural extract that has been successfully incorporated into PLA films is green tea leaf extract (Andrade et al., 2023; Martins et al., 2018). In a previous study, PLA film containing 2% green tea extract exhibited 2,2-diphenyl-1-picrylhydrazy (DPPH) inhibition that was almost 70% higher than that of a PLA/rosemary extract film, highlighting the superior antioxidant activity of green tea extract (Andrade et al., 2023). However, when both green tea and rosemary extracts were combined, the resulting PLA films exhibited lower DPPH inhibition than the green tea extract-only formulation, suggesting that the synergy between different extracts can influence the functional properties of PLA films in unpredictable ways.

In another study, the incorporation of tea polyphenols and cinnamaldehyde into PLA/polybutylene adipate (PBAT) films was explored for their antimicrobial activities. Cinnamaldehyde demonstrated stronger inhibitory effects against *Escherichia coli* and *Staphylococcus aureus* than tea polyphenols, underlining the importance of selecting appropriate active agents for specific applications (Wang et al., 2022). Additionally, curcumin has been added to PLA films to enhance antioxidant properties, although a study (Roy & Rhim, 2020) found that the antioxidant activity of PLA/curcumin composite films was slightly lower than that of pure curcumin. This was attributed to the limited release of curcumin from the polymer matrix, which hinders its bioavailability due to interactions with the PLA structure.

The antioxidant and antimicrobial properties of PLA films can also be enhanced by incorporating a secondary layer containing active compounds. For example, PLA films coated with chitosan, alginate, or methylcellulose containing olive leaf extract have been reported to exhibit significant antioxidant and antimicrobial activities (Fiorentini et al., 2024). Additionally, a coating of the crude supernatant isolated from *Pediococcus pentosaceus* containing nisin has been shown to be effective in inhibiting growth of *Bacillus cereus*, *Salmonella enterica*, and *E. coli* (Todhanakasem et al., 2022).

The functional properties of PLA composites are influenced by several factors, including the pH of the food product. The efficiency of metal chelation and radical scavenging activities in PLA composites depends on the pKa of the active compounds. For example, nitrilotriacetic acid-grafted PLA composites effectively form metal chelation complexes at pH levels above 9, with radical scavenging activities peaking at pH 6.7 or higher (Kay et al., 2022). These findings suggest that PLA composites with the highest antioxidant activity are most suitable for slightly acidic to alkaline food systems, such as plant-based beverages and alkaline diets (Table 18.2).

TABLE 18.2 Additives in polylactic acid (PLA)-based films and coatings for food packaging.

Polymer blend	Additive	Results	Sources
PLA	Ag, graphene, titanium dioxide (TiO_2)	• Improved mechanical resistance by 30%. • With the incorporation of 3% nano-Ag-graphene-TiO_2 complex, water vapor and oxygen permeability decreased by 27% and 24.6% respectively. • Reduced solubility in food simulant.	Peter et al. (2021)
PLA	α-Tocopherol	• Absorption of UV light at 260–320 nm. • Incorporation of α-tocopherol has no effect on thermal properties.	Manzanarez-López et al. (2011)
PLA	Calcium carbonate, PLA-polyethylene glycol plasticizer	• Composite showed a lower cold crystallization temperature and glass transition temperature compared to pure PLA. • Composite exhibited lower tensile strength and higher elongation at break compared to pure PLA.	Cisar et al. (2023)
PLA	Capsicum oleoresin-impregnated nanoporous silica	• Reduced glass transition temperature and thermal stability. • Antimicrobial properties against Gram-positive and Gram-negative pathogenic bacteria.	Techawinyutham et al. (2019)
PLA	Carvacrol and cinnamaldehyde	• Reduced water vapor permeability but increased oxygen permeability. • PLA with carvacrol showed superior antimicrobial capacity against *Escherichia coli* and mold than cinnamaldehyde. • Increased flexibility and stretchability with the incorporation of essential oils (EOs).	Gonon et al. (2023)
PLA	Epoxidized karanja oil (EKO) as plasticizer	• Maximum decrease in tensile strength by 9%; maximum elongation of break by 77.5% compared to neat PLA. • Decreased glass transition temperature with increased EKO concentration.	Garcia-Garcia et al. (2020)
PLA	Epoxidized soybean oil (ESO) as plasticizer	• Zero shear viscosity decreased significantly from the range of 2395–2495 to 204–345 Pa-s with the incorporation of ESO. • Glass transition temperature decreased from 60.9°C to 45.8°C • Degree of crystallinity increased from 1.52% to 3.13%	Xing and Matuana (2016)
PLA	Graphene nanoplatelets, TiO_2	• Reduced hydrophilicity with the incorporation of TiO_2. • Increased light absorbance with increased TiO_2 concentration.	Shehzad et al. (2024)

(Continued)

TABLE 18.2 (Continued)

Polymer blend	Additive	Results	Sources
PLA	Green tea extract	• Increased DPPH radical scavenging activities and total phenolic content. • Tensile strength decreased by 12% with the incorporation of green tea extract. • Strain at break increased by 36% with 2% extract. • Decreased water vapor transmission rate.	Martins et al. (2018)
PLA	Green tea and rosemary extract	• Green tea extract presented the highest water vapor permeability and opacity. • Rosemary extract resulted in the highest oxygen permeability.	Andrade et al. (2023)
PLA	Nitrilotriacetic acid (NTA) as antioxidant; 1,3-bis(4,5-dihydro-2-oxazolyl)benzene (PBO) as crosslinker	• The bond between NTA active ligand and base PLA polymer is of covalent nature, making it unlikely to migrate to the food. • Radical scavenging activity is pH dependent and is effective at or above pH 6.7.	Kay et al. (2022)
PLA	NTA as chelating active precursor	• Increased film hydrophilicity from the polar carboxylic acids in NTA. • Decreased crystallization and melting temperature following grafting polymerization. • NTA-grafted PLA films exhibited 2,2'-azino-bis(3-ethylbenzothiazoline-6-sulfonic acid (ABTS) radical scavenging activity (trolox equivalence).	Herskovitz and Goddard (2020)
PLA	*Perilla* EO	• Improved mechanical properties as well as water vapor and oxygen barrier properties, thermal stability, and antibacterial and antioxidant properties. • Reduced water swelling rates.	Wang, Sun, et al. (2021), Wang, Shen, et al. (2021)
PLA	Sputter-coated zinc oxide (ZnO)/ Al_2O_3	• Reduced UV light transmittance while maintaining high visible transparency. • Strong antibacterial activity against *E. coli*.	Valerini et al. (2018)
PLA	Thyme, rosemary, and oregano EO	• Decreased tensile strength and percent elongation in all films. • All films with EOs exhibited radical scavenging activities. • All films were fully transparent and colorless, except oregano EO, which resulted in a slight yellow tint.	Zeid et al. (2019)
PLA	*Thymus vulgaris* EO and mediterranean propolis extract	• High inhibition activities against *Staphylococcus aureus* and *Penicillium* sp. • EO reduced brittleness and improved elasticity.	Ardjoum et al. (2021)

(*Continued*)

TABLE 18.2 (Continued)

Polymer blend	Additive	Results	Sources
PLA	ZnO nanoparticles	• Tensile strength increased by 37.5%. • 0.5% ZnO nanoparticles resulted in lowest water vapor permeability. • Light transmittance decreased with increasing nanoparticles concentration. • The composite films exhibited strong antimicrobial activities against *E. coli* and *Listeria monocytogenes*.	Shankar et al. (2018)
PLA/ cellulose nanocrystals (CNC)	Montmorillonite (MMT), nanosilica, alumina, carbon nanofibers, coconut oil, and olive oil as compatibilizers	• Improved oxygen and water vapor barrier properties. • Improved thermo-mechanical properties.	Dhar et al. (2017)
PLA/CNC	Reduced graphene oxide (rGO)	• Increased porosity, crystallinity, and thermal stability with the incorporation of rGO. • Hydrophilicity increased compared to neat PLA. • Antibacterial activities against both *S. aureus* and *E. coli* bacterial strains.	Verma et al. (2023)
PLA/CNC	Not determined (ND)	• Decreased water vapor permeability compared to neat PLA film, most effective at lower relative humidity (50% RH) than higher humidity (85% RH).	Karkhanis et al. (2021)
PLA/chia seed flour	(3-Glycidyloxypropyl) trimethoxysilane (GPS), Xibond, epoxidized chia oil, and maleinized chia oil as compatibilizers	• Elongation at break of film compatibilized using epoxidized chia oil was 10 times higher than uncompatibilized film. • Compatibilized films has lower glass transition temperature and higher cold crystallization temperature. • Decreased water uptake (wt.%) with all compatibilizers except maleinized chia oil. • All compatibilized films showed over 90% weight after 31 days of composting.	Dominguez-Candela et al. (2022)
PLA/diatomaceous earth	Maleinized linseed oil	• Maleinized linseed oil shows good miscibility at concentrations lower than 5 phr. • Impact strength and ductility increased significantly compared to uncompatibilized film.	Gonzalez et al. (2019)
PLA/grape pomace	ND	• Decreased glass transition, cold crystallization, and melting temperatures with increasing grape pomace concentration. • Tensile strength decreased; elastic modulus and elongation at break increased. • Antimicrobial activity against *E. coli* and *L. innocua*.	Bruna et al. (2023)

(*Continued*)

TABLE 18.2 (Continued)

Polymer blend	Additive	Results	Sources
PLA/ pineapple leaf fiber/corn powder	Supernatant from *Pediococcus pentosaceus*	• Increased melting temperatures and heat resistance. • Decreased tensile strength and strain at break compared to neat PLA. • Inhibited growth of *Bacillus cereus*, *Salmonella enterica*, and *E. coli*.	Todhanakasem et al. (2022)
PLA/ poly (butylene adipate-co-terephthalate) (PBAT-co-terephthalate)	Ginger and lime peel oil	• Increased elongation at break with incorporation of PBAT-co-terephthalate and EOs. • Incorporation of EO, particularly lemon peel oil, significantly lowered hydrophilicity. • High PLA film had lower oxygen permeability compared to high PBAT film.	Laorenza and Harnkarnsujarit (2023)
PLA/PBAT/corn starch	Joncryl additives that define rheology (ADR) 4370 F as chain extender; cinnamaldehyde and tea polyphenols as antioxidants	• Cinnamaldehyde effectively slowed the growth of *E. coli* and *S. aureus*. • Water vapor permeability increased with the incorporation of EOs. • Tea polyphenols effectively decreased oxygen permeability.	Wang et al. (2022)
PLA/poly(butylene succinate)	Lesser galangal EO	• Decreased tensile strength by up to 22%. • Increased water vapor and oxygen permeabilities. • Delayed mold and bacterial growth.	Wongphan et al. (2023)
PLA/thermoplastic starch (TPS)	Maleic anhydride as free radical reaction monomer and dicumyl peroxide as crosslinker	• TPS acted as initial nutrient source for microbes in the compost, accelerating degradation.	Mayekar et al. (2023)
PLA/TPS	Glycerol and sorbitol as plasticizers	• Increased hydrophilicity with increasing TPS concentrations. • Increased impact strength; decreased elongation at break with increasing PLA content.	Esmaeili et al. (2019)
PLA/TPS/PBSA	Tween 60 as compatibilizer	• Increased tensile strength, extensibility, and impact strength. • Crystallinity increased with increasing Tween 60 content. • Water vapor and oxygen permeability were reduced up to 30% and 62%, respectively, with the incorporation of Tween 60.	Yoksan and Dang (2023)
PLA/zeolite	Triethyl citrate (TEC) as plasticizer	• TEC increased hydrophilicity and accelerated PLA hydrolysis in high humidity. • Increased water vapor permeability with increasing zeolite and TEC concentrations.	Lee et al. (2020)
Bilayer/coating/laminate			
PLA	Chitosan coating	• Increased rigidity with increased chitosan content. • Reduced water vapor permeability with chitosan coating.	Kumbunleu et al. (2020)

(Continued)

TABLE 18.2 (Continued)

Polymer blend	Additive	Results	Sources
PLA	Chitosan, alginate, and methylcellulose coatings incorporated with olive leaf extract	• Effective antimicrobial activities against *S. aureus*, particularly with methylcellulose and chitosan. • Highest migration of phenolic compounds from chitosan coating, followed by methylcellulose and alginate.	Fiorentini et al. (2024)
PLA	Chitosan coating with argan vegetal oil and clove essential oil (CEO)	• Lowest oxygen transmission rate achieved with CEO. • CEO exhibited stronger antioxidant and antimicrobial activities.	Stoleru et al. (2021)
PLA	Chitosan/sodium caseinate blend enriched with rosemary EO	• Improved elasticity and water vapor barrier. • Incorporation of 2% rosemary EO resulted in significantly higher antioxidant activity.	Fiore et al. (2021)
PLA	Chitosan/nanocellulose/methyl red films	• PLA coating improved tensile strength and strain. • Water solubility decreased significantly.	Sobhan et al. (2021)

18.3 Applications of polylactic acid films and coatings for food packaging

Various food products possess unique characteristics such as pH, moisture content, and storage temperature, all of which can significantly influence the performance of PLA films and coatings. To ensure the packaging maintains its structural integrity and optimally preserves the food's quality, PLA films must be highly customizable in terms of mechanical properties. This flexibility allows for the development of packaging solutions that are tailored to the specific needs of different food products, addressing challenges such as varying moisture levels, sensitivity to oxygen, and the need for enhanced barrier properties.

In active packaging systems, PLA films and coatings act as carriers for active compounds, ensuring the sustained release of preservatives such as antimicrobial agents or antioxidants on the food surface. This not only extends the shelf life of food products but also ensures that the bioactive agents are effectively retained in the packaging. As the demand for sustainable and functional packaging solutions grows, PLA films are being increasingly researched and developed to meet both the practical and environmental needs of modern food packaging.

This section delves into the application of PLA films in packaging for a range of food products, including meat and seafood, dairy, and dried and baked goods. These sectors have seen promising results with the use of PLA packaging, demonstrating the material's versatility and feasibility in preserving the quality and safety of a diverse array of food products.

18.3.1 Meat and seafood

Meat and seafood are among the most perishable food products due to their high-moisture content, typically ranging from 60% to 80%, and their susceptibility to microbial contamination and oxidative spoilage. Common pathogens such as *Listeria monocytogenes*, *Salmonella* spp., and *Pseudomonas* spp. thrive in these conditions, leading to foodborne illnesses and quality degradation. Additionally, the oxidative deterioration of lipids in high-fat meats and seafood results in off-flavors, rancidity, and discoloration. Therefore, the primary objectives of packaging these products are to inhibit microbial growth, delay oxidative processes, and preserve the natural color, texture, and flavor.

PLA films are well suited for addressing these challenges due to their customizable barrier properties, antimicrobial potential, and environmental sustainability. To combat microbial contamination and spoilage, especially for high-risk items like sausages, PLA can be enhanced with antimicrobial and antioxidant agents, such as EOs, silver nanoparticles, or natural extracts. This is especially critical for meats with high-fat content, as they are more susceptible to oxidative rancidity, necessitating films with strong antioxidant properties. Cinnamaldehyde, an effective antimicrobial agent, has been incorporated into PLA films to extend the shelf life of meat analogs. In a study where meat analogs were stored for ten days at 4°C, PLA/PBAT films containing cinnamaldehyde effectively reduced the growth of *E. coli* and *S. aureus* by 3.6 and 4.1 log CFU g^{-1}, respectively. Additionally, the moisture evaporation was minimized, helping to maintain the texture of the meat analogs (Wang et al., 2022).

An alternative method to extend the shelf life of meat and seafood is the use of high-oxygen barrier or vacuum packaging. Modified or coated PLA films can significantly reduce oxygen permeability, which in turn inhibits aerobic bacterial growth and minimizes lipid oxidation. However, vacuum packaging also presents challenges, as it can promote the growth of anaerobic bacteria, such as *Clostridium* spp. (Dalla Rosa, 2019). Therefore, it is crucial to achieve an optimal balance between oxygen reduction and microbial control.

The visual appeal of packaged meat is another crucial factor influencing consumer perception. Bright red meat is associated with freshness due to the oxygenated form of myoglobin, oxymyoglobin, at the muscle surface. However, oxygen removal in vacuum-sealed or low-permeability films can lead to the formation of metmyoglobin, which imparts a brown hue and suggests diminished freshness. This color change, although reversible, poses a challenge for packaging design. Transparent PLA films with controlled oxygen permeability offer a solution, as they allow sufficient oxygen interaction to maintain the bright red appearance while minimizing spoilage risks (Dalla Rosa, 2019). This adaptability ensures that PLA films meet both consumer expectations and food safety requirements.

Packaging seafood presents additional challenges due to the rapid spoilage and textural changes associated with high water activity and protein degradation. Fresh fish, for instance, can rapidly develop off-odors and slimy textures due to bacterial activity. PLA films modified with antimicrobial agents effectively reduce bacterial growth on seafood surfaces, extending shelf life and preserving quality. Additionally, seafood often requires packaging with superior moisture barrier properties to prevent water loss and texture deterioration. PLA, when blended with hydrophobic additives or layered with coatings, can achieve this functionality. A previous study demonstrated that the incorporation of green tea extract into PLA films helped reduce lipid oxidation in smoked salmon, as evidenced by a significant decrease in the p-anisidine and thiobarbituric reactive substances (TBARS) values over 45 days of storage (Martins et al., 2018).

Freezing is an effective method for storing meat and seafood; however, storage below −18°C poses challenges for PLA films. Although their oxygen barrier properties improve at low temperatures due to the cold crystallization of PLA (Safandowska et al., 2022), which helps prevent oxidative spoilage, PLA's high glass transition temperature (~60°C) makes it susceptible to brittleness in freezing conditions. Below its glass transition temperature, PLA transitions into a glassy state where its polymer chains have limited mobility, causing the material to become stiff and brittle. This increases the likelihood of cracking or tearing, which can lead to freezer burn and compromised product texture. These issues can be mitigated by incorporating plasticizers such as polyethylene glycol (PEG), glycerol, or triethyl citrate (TEC) to lower the glass transition temperature and improve flexibility, as discussed earlier.

PLA films are also beneficial in the context of emerging trends in intelligent packaging, which work by monitoring freshness indicators such as pH, carbon dioxide levels, or microbial activity, providing real-time feedback on the condition of the product. This was demonstrated in a study in which methyl red was incorporated into PLA-coated chitosan/nanocellulose films to monitor the freshness of beef (Sobhan et al., 2021). As spoilage occurred, ammonia and other volatile basic amines were released, causing an increase in pH. This pH shift resulted in a distinct color change of the film from red to yellow, as illustrated in Fig. 18.4. Such smart packaging solutions can provide consumers and retailers with visual cues, helping to ensure the timely consumption of products and reduce food waste by accurately indicating product freshness.

18.3.2 Dairy

Dairy products, such as milk, cheese, yogurt, cream, ice cream, and butter, are highly perishable due to their moisture content, protein, fat, and carbohydrate composition. For instance, milk has an approximate moisture content of 87%, creating an environment conducive to microbial growth and nutrient degradation. Soft cheeses and yogurts, with their elevated water activity, are particularly vulnerable to spoilage by bacteria such as *L. monocytogenes*, *S. aureus*, and *Salmonella typhi* (Bastam et al., 2021). Proper packaging is essential for maintaining freshness and quality, minimizing moisture loss, inhibiting microbial growth, and delaying oxidation. Additionally, dairy products exhibit a range of pH values: milk typically has a pH of around 6.5–6.8, while yogurts and cheeses can range from 4.0 to 5.5 due to lactic acid production. Although PLA is fairly stable under weakly acidic and neutral conditions, its degradation might be accelerated

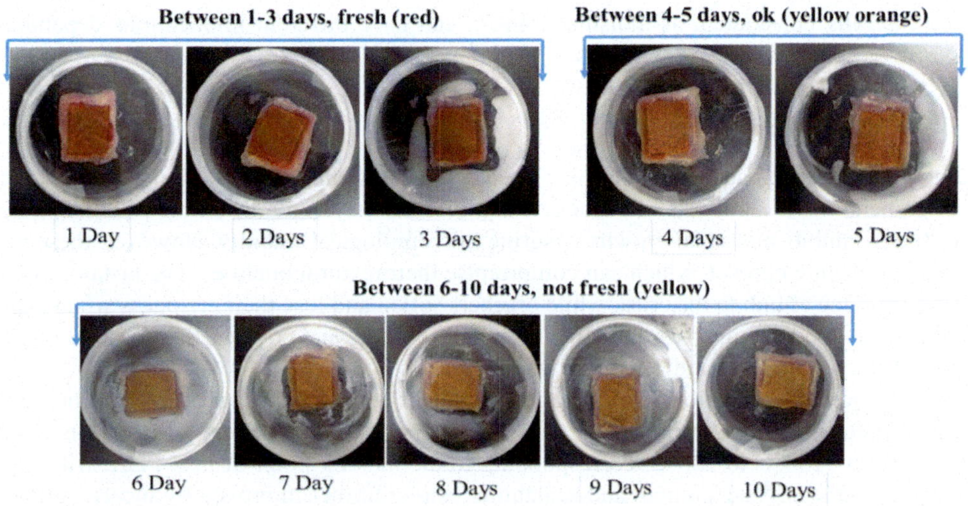

FIGURE 18.4 Color change of polylactic acid-coated chitosan/nanocellulose film with methyl red as a pH indicator, with respect to beef spoilage in chilled storage (4.0°C ± 0.2°C).
From Sobhan, A., Muthukumarappan, K., Wei, L., Zhou, R., & Tummala, H. (2021). Development of a polylactic acid-coated nanocellulose/chitosan-based film indicator for real-time monitoring of beef spoilage. Analytical Methods, 13(23), 2612–2623. https://doi.org/10.1039/d1ay00365h.

at lower pH due to the catalytic action of protons (H^+). Effective dairy packaging must address these challenges by minimizing moisture loss, inhibiting microbial growth, and preserving sensory qualities.

In chilled storage, typically maintained between 0°C and 4°C, the packaging material must provide a balance between moisture control and oxygen barrier properties to extend product shelf life. PLA's moderate oxygen barrier can help reduce oxidative rancidity in high-fat products like cream and cheese. However, its relatively high WVTR poses a challenge, leading to potential moisture imbalances that can affect texture and shelf life. For example, cheese may develop surface hardening due to moisture loss, while yogurt may experience syneresis, a condition where whey separates from the product. To mitigate these effects, PLA can be modified through coatings such as clay or wax. The inclusion of antimicrobials and antioxidants, such as natural plant polyphenols, can further enhance the shelf life of dairy products. For example, an anthocyanin-functionalized layered clay/PVA coating has been applied to PLA films to improve their gas barrier properties while delaying bacterial growth in pasteurized milk (Mao et al., 2023). However, for dairy products that require frozen storage, such as ice cream, the brittleness of PLA might result in compromised packaging. As mentioned in the previous section, plasticizers can be added to enhance its flexibility. Alternatively, PLA linings can be applied to paperboard ice cream tubs to improve the moisture resistance of paperboard while reinforcing the mechanical properties of PLA.

Transparency is another advantage of PLA films, allowing consumers to visually assess the quality of dairy products, such as milk and cheese. However, high UV penetration is not ideal because it promotes photooxidation, which accelerates the breakdown of lipids in dairy products and consequently leads to rancidity. This can be overcome by incorporating nanofillers or phenolic compounds, which enhance the light barrier properties of PLA films and coatings, particularly in the UV region.

The oxygen permeability of PLA is a key consideration, as exposure to oxygen can lead to lipid oxidation and rancidity in dairy products, such as butter, resulting in the formation of off-flavors. Enhancements like oxygen scavengers and multilayer constructions can bolster PLA's barrier capabilities, extending shelf life and preserving the nutritional content of products. Active packaging functionalities, which incorporate natural antimicrobial agents, can help further extend the freshness and safety of dairy items by controlling microbial growth.

Dairy products are prone to antimicrobial spoilage caused by bacteria, yeasts, and molds. As discussed in earlier sections, PLA-based active packaging can be embedded with natural antimicrobials (e.g., EOs, organic acids) or nanoparticles (e.g., silver ions, ZnO) that inhibit the growth of spoilage microbes, extending shelf life and maintaining quality. A slow release of embedded antimicrobial agents allows for sustained protection against microbial growth over time. This is particularly beneficial for dairy products with longer storage periods, such as cheese or butter.

Despite the potential benefits, PLA's high water vapor permeability presents particular challenges for dairy products with strict moisture requirements, such as cheese. Moreover, the performance of PLA under varying environmental conditions must be carefully considered, as moisture transfer can have different implications depending on the type of dairy product.

18.3.3 Dried and baked goods

Dried and baked goods, such as bread, cakes, biscuits, crackers, cereals, and various dried snacks, typically exhibit low-moisture levels, often below 10%. Products like crackers and biscuits may contain as little as 2%–3% moisture. This low water activity effectively inhibits microbial growth, ensuring microbiological stability. However, these products are highly sensitive to changes in moisture content, which can compromise their textural qualities. For instance, bread becomes stale as moisture migrates from the crumb to the crust, while crackers soften and lose their crispness when exposed to excessive moisture. Similarly, baked goods with a high-fat content, such as pastries and cookies, are prone to lipid oxidation, which leads to undesirable flavors, rancidity, and a reduced shelf life.

PLA films present a promising solution for the packaging of dried and baked goods. Their heatsealability provides secure packaging that protects against contamination during transportation and storage. PLA films can be heat sealed at temperatures ranging from ~150°C to 180°C, corresponding to the melting point of neat PLA, with higher temperatures requiring shorter sealing periods. Seal strength and sealability can be further improved by incorporating copolymers with faster crystallization rates, such as poly(butylene succinate) (PBS) (Chuakhao et al., 2024).

Despite these advantages, the relatively high WVTR of PLA poses challenges for moisture-sensitive products. While PLA's moderate oxygen barrier properties help delay oxidative rancidity in fat-rich items, they may not suffice for products requiring enhanced protection. To address these limitations, PLA films can be tailored to meet the specific needs of dried and baked goods. Strategies include blending PLA with hydrophobic additives, such as EOs, or integrating nanofillers like clay or silica nanoparticles to improve moisture and oxygen barrier properties. Additionally, coatings made from biopolymers like chitosan or natural waxes can significantly improve the water vapor barrier, preserving the crispness of crackers and the softness of bread. However, it is critical to ensure that additives, particularly EOs, do not impart undesirable odors to the packaged food. The porous nature of leavened baked goods makes them especially prone to odor adsorption.

Advanced packaging technologies can further enhance the performance of PLA films. Incorporating oxygen scavengers or vacuum-sealing methods can create a low-oxygen environment that protects products like cookies and pastries from oxidation and staling. Active packaging solutions, which include natural antioxidants such as tocopherols, EOs, and polyphenols, can further extend shelf life by mitigating lipid oxidation. Although dried and baked goods inherently have low water activity, incorporating antimicrobial agents into PLA films may provide additional protection against spoilage, especially in high-humidity conditions.

Consumer convenience is another important consideration. Resealable PLA-based packaging formats, such as zip-lock pouches, allow users to maintain freshness after opening, particularly for items like biscuits and chips that are consumed in multiple servings.

In summary, PLA films offer a sustainable and effective packaging solution for dried and baked goods by addressing critical factors such as moisture control, oxidative stability, and product freshness. Through advancements in material engineering and active packaging technologies, PLA can help manufacturers preserve the sensory and nutritional qualities of these products while contributing to the reduction of food waste and environmental impact (Table 18.3).

18.4 Life cycle assessment

As sustainability continues to be a priority for industries and consumers, evaluating the environmental impacts of PLA-based food packaging has become crucial. LCA offers a comprehensive method for assessing the environmental performance of packaging materials, covering all stages from raw material extraction to disposal or recycling. Building on the LCA framework, the product environmental footprint (PEF) provides a more integrated perspective on the overall sustainability of PLA packaging, offering further insight into its environmental implications (Table 18.4).

PLA, as a bioplastic, offers several advantages, such as lower greenhouse gas emissions during production and biodegradability under industrial composting conditions, helping to address plastic waste concerns. However, its reliance on agricultural feedstocks raises several environmental issues, including land use changes, fertilizer runoff, and resource depletion. Additionally, PLA's limited biodegradability in natural environments and the energy-intensive polymerization process contribute to sustainability challenges. To mitigate these issues and extend the material's life cycle, viable end-of-life (EoL) solutions, including mechanical recycling, solvent-based recycling, and chemical recycling, can be considered based on the quality of PLA waste and regional conditions.

TABLE 18.3 Applications of polylactic acid (PLA)-based films and coatings as active and smart food packaging.

Type of food	Additives/active compounds	Results/shelf life	Source
Active packaging			
Almonds and beef	Rosemary and green tea extract	Lower malonaldehyde (MDA) eq. value compared to control PLA films	Andrade et al. (2023)
Broccoli, banana, grapes, and tomato	Chitosan coating	Inhibited mold growth up to seven days of storage in room temperature	Kumbunleu et al. (2020)
Cake	Carvacrol and cinnamaldehyde	Delayed mold growth during storage, up to four days	Gonon et al. (2023)
Chicken	*Perilla* essential oil (EO)	Lower TVB-N content, thiobarbituric reactive substances, and total viable bacterial counts, extending shelf life for up to 12 days	Wang, Shen, et al. (2021)
Chicken	Chitosan/sodium caseinate/ rosemary EO coating	Lower average concentration of MDA over 21 days	Fiore et al. (2021)
Crackers	Cellulose nanocrystals	Delayed period to reach the critical moisture content, extending shelf life by 40%	Karkhanis et al. (2021)
Curd cheese	Ag-graphene-TiO_2	Delayed color change, whey leakage, weight loss, and rancidity	Peter et al. (2021)
Edible oil and potato chips	Reduced graphene oxide	Prolonged shelf life of edible oil by 13.5 times	Goh et al. (2016)
Fish (minced)	Thyme, rosemary, and oregano EOs	All PLA/EO films demonstrated antimicrobial and antioxidant activities; samples packed with PLA/ oregano films showed the lowest MDA concentration	Zeid et al. (2019)
Fish (paste)	Zinc oxide nanoparticles	Number of viable colonies of *E. coli* and *L. monocytogenes* was reduced to 0 in 10 days of storage	Shankar et al. (2018)
Margarine	TiO_2 and lycopene	Lower tetrabutylammonium chloride and peroxide value (PV) compared to control PLA film	Pirsa and Asadi (2021)
Milk	Anthocyanin-functionalized layered clay	Maintained pH and prolonged shelf life of milk from 60 to 85 h	Mao et al. (2023)
Mushroom	Polybutylene adipate	Effectively prevented condensation of water vapor inside the package; slow down the respiration rate and microbial growth	He et al. (2022)
Rice (cooked)	Lesser galangal EO	Films with 4% and 6% of lesser galangal EO showed significantly lower total viable counts (TVC) on days three and six of storage	Wongphan et al. (2023)
Sausage	Lemongrass	Reduced population of *L. monocytogenes* on sausage	Yang and Song (2016)
Soybean oil	α-Tocopherol	Lower PV compared to control PLA film throughout a storage period of 60 days	Manzanarez-López et al. (2011)
Soy protein-based meat	Cinnamaldehyde	Inhibited *S. aureus* growth up to 10 days	Wang et al. (2022)
Shrimp	Ginger and lemon peel EO	Lemon peel oil showed lower total viable count than ginger oil while ginger oil was more efficient in delaying melanosis	Laorenza and Harnkarnsujarit (2023)

(Continued)

TABLE 18.3 (Continued)

Type of food	Additives/active compounds	Results/shelf life	Source
Smoked salmon slices	Green tea extract	Low green tea extract concentration (1% w/w) was more effective in controlling lipid oxidation than a higher concentration (2% w/w)	Martins et al. (2018)
Tilapia slices	Gelatin/chitosan nanofibers with betel leaf extract	Delayed microbial growth and lipid oxidation	Tagrida et al. (2022)
Yogurt	-	Lower oxygen in headspace compared to polystyrene container due to lower oxygen transmission rate	Mikloskova et al. (2021)
Smart packaging			
Beef	Methyl red	Color change from red to yellow with pH increase which also corresponded to increase in total viable count	Sobhan et al. (2021)
Shrimp	Curcumin	Color change of curcumin from yellow to red-orange with the release of volatile amines	Cvek et al. (2022)

LCA comparisons between PLA and PET for applications like drinking water bottles have been extensively reported. Generally, PLA demonstrates a lower global warming potential (GWP) and carbon footprint compared to PET, along with a reduced environmental impact in terms of fossil fuel consumption. However, PLA's production is resource-intensive, involving significant land and water use, as well as the application of fertilizers, herbicides, and water for raw material cultivation. These inputs contribute to high levels of eutrophication and freshwater toxicity (Tamburini et al., 2021; Yanat et al., 2025).

In a study comparing PLA, PET, and aluminum drinking bottles, LCA was performed across the entire product lifecycle, including production, use, and disposal stages. Composting PLA, while aligned with the biogenic carbon cycle, offers limited net benefits in terms of reducing greenhouse gas emissions. In contrast, the closed-loop recycling of PET results in a 5% reduction in emissions, with open-loop recycling offering even greater savings of up to 24% by reducing the reliance on virgin material. While aluminum bottles are highly reusable and 100% recyclable, they come with substantial production energy costs due to the energy-intensive processes involved in metal extraction and bottle shaping. Additionally, their frequent cleaning with hot water and soap significantly increases environmental burdens, particularly in terms of water use and greenhouse gas emissions (Tamburini et al., 2021).

The incorporation of additives into PLA to improve its physical and functional properties requires careful consideration within the LCA. For example, an LCA of PLA biocomposites reinforced with agave fibers revealed notable environmental trade-offs. Replacing pure PLA with fiber-reinforced composites reduces the overall PLA content by approximately 20%, but the production process remains resource-intensive. PLA still accounts for over 68% of the environmental impacts across key indicators, including eutrophication and acidification. Among various molding methods, extrusion-injection molding exhibited the lowest environmental impact, with a GWP of 20,634.62 kg CO_2-eq per functional unit, compared to 51,296.30 kg for thermo-compression molding (Lara-Topete et al., 2024). However, the pretreatment steps for agave fibers contributed only minimally, accounting for an average of just 1.16% of impacts across impact categories. Composting agave bagasse emerged as the most environmentally friendly waste management option, with a GWP of only 51.95 kg CO_2-eq per ton managed, while incineration, though offsetting natural gas use, resulted in a higher GWP of 90.60 kg CO_2-eq. Despite the potential of agave fiber-reinforced PLA biocomposites, their environmental costs surpass those of conventional plastics such as LDPE in most impact categories, highlighting the need for improved production efficiency and region-specific evaluations to enhance the sustainability of PLA-based food packaging solutions.

Similarly, the incorporation of chitin nanocrystals (ChNC) into PLA films improves mechanical and barrier properties, making them a viable option for food packaging (Yanat et al., 2025). However, the production of ChNC is resource-intensive, involving high chemical usage and energy-demanding steps such as freeze-drying, which increases the environmental footprint. EoL strategies such as composting enhance the sustainability of PLA and ChNC/PLA composites, making them more environmentally friendly than PET, which typically ends up in landfills (Table 18.5).

TABLE 18.4 Product environmental footprint impact categories of life cycle assessment.

Impact category	Description
Acidification	Indicator of potential acidification of soils and water measures the environmental impact caused by the release of acidic gases, such as nitrogen oxides and sulfur oxides, which react with water and other substances in the atmosphere to form acidic compounds, leading to a decrease in pH and subsequent harm to ecosystems and biodiversity.
Climate change—fossil, biogenic, land use	Global warming potential measures a system's ability to trap greenhouse gases in the atmosphere over its entire life cycle, expressed in terms of the radiative forcing caused by the emission of a substance relative to an equivalent mass of carbon dioxide, ultimately indicating its contribution to climate change.
Depletion of abiotic resources—minerals and metals, fossil fuels	Indicator of the depletion of natural nonfossil resources measures the reduction in the availability of nonrenewable natural resources, such as metals, minerals, and other raw materials. Indicator of the depletion of natural fossil fuel resources tracks the consumption and exhaustion of nonrenewable fossil fuels, such as coal, oil, and natural gas.
Eco-toxicity (freshwater)	Indicator of the harmful effects of pollutants, such as heavy metals and chemicals, on aquatic species. These pollutants can cause physiological damage, disrupt reproduction, and reduce biodiversity, impacting ecosystem health and food chains. Long-term exposure can lead to bioaccumulation, posing risks to both aquatic life and human health.
Eutrophication—freshwater, marine, terrestrial	Indicator of the enrichment of the freshwater, marine, and terrestrial ecosystems with nutritional elements, due to the emission of nitrogen-containing compounds. The imbalance can disrupt ecosystems and reduce biodiversity.
Human toxicity—cancer, noncancer	Indicator of the impact of toxic substances emitted into the environment on humans. Noncancer toxic substances can cause a range of health issues, such as respiratory problems, organ damage, and neurological disorders. Cancer-related toxic substances have the potential to cause mutations that may lead to the development of cancer.
Ionizing radiation, human health	Indicator of the potential damage to human health and ecosystems linked to the emissions of radionuclides arise from their radioactive decay, which releases ionizing radiation that can cause cancer, genetic mutations, and other health issues in humans, while also harming plant and animal life by disrupting biological processes and damaging habitats.
Land use	The measure of changes in soil quality includes factors such as biotic production, erosion resistance, and mechanical filtration. These indicators are crucial for evaluating the long-term health and sustainability of soil ecosystems.
Ozone depletion	Indicator of the breakdown of the ozone layer in the stratosphere, leading to increased penetration of UV-B radiation to the Earth's surface.
Particulate matter emissions	Indicator of the potential incidence of disease due to particulate matter emissions reflects the health impact associated with exposure to airborne particles, including respiratory and cardiovascular conditions, as well as reduced life expectancy, caused by pollutants such as PM2.5 and PM10.
Photochemical ozone formation	Indicators of gas emissions that contribute to the formation of photochemical ozone in the lower atmosphere (smog) when catalyzed by sunlight.
Water use	Indicator of the amount of water used, evaluated in relation to regional water scarcity factors, reflecting the impact of water consumption on local water availability.

TABLE 18.5 End-of-life (EoL) scenarios and climate change impacts of 1 kg of polyethylene terephthalate (PET), polylactic acid (PLA), and chitin nanocrystal reinforced PLA (chitin nanocrystals [ChNC]/PLA) nanocomposites, including their EoL management.

	EoL scenarios	Material	Energy	Disposal	Total
PET	42% recycling, 26.1% energy recovery, 31.9% landfill	2.84	0.13	0.05	3.02 kg CO_2 eq.
PLA	42% recycling, 26.1% energy recovery, 31.9% landfill	2.70	0.17	0.01	2.88 kg CO_2 eq.
ChNC/PLA	100% composting	4.26	–	0.07	4.33 kg CO_2 eq.

Source: Reprinted with permission from Elsevier. Modified from Yanat, M., Voortman, E., Buijsse, C., & Schroën, K. (2025). Life cycle assessment (LCA) of chitin nanocrystal reinforced polylactic acid (ChNC/PLA) and polyethylene terephthalate (PET) packaging material. Carbohydrate Polymers, 348(Pt B). https://doi.org/10.1016/j.carbpol.2024.122927.

Optimizing ChNC production processes by reducing chemical usage and adopting more sustainable drying methods could further improve the environmental performance of ChNC/PLA composites.

In summary, while PLA-based food packaging offers notable environmental benefits, such as lower greenhouse gas emissions and biodegradability under industrial composting conditions, its sustainability is tempered by challenges associated with agricultural inputs, resource use, and its limited biodegradability in natural environments. With the current technology, the recycling of contaminated food-contact nonbiopolymeric packaging is rarely feasible, as most recycling facilities lack the capacity to handle such materials, resulting in the majority of plastics being sent to landfills. This challenge is further exacerbated by the incorporation of other polymers or composites during PLA manufacturing to enhance its properties, which complicates recycling processes. Consequently, composting of PLA-based packaging often becomes the most viable EoL option.

18.5 Challenges

18.5.1 Processing challenges

PLA presents several challenges during processing, which can affect its properties and suitability for food packaging applications. These challenges include thermal degradation, slow crystallization rates, depolymerization, and issues related to the incorporation of additives, such as nanofillers. Such limitations can lead to brittleness, low melt strength, and compromised barrier properties against water vapor and oxygen, thereby impacting the performance of PLA films and coatings.

One of the significant challenges in processing PLA is its susceptibility to thermal degradation at elevated temperatures, particularly above 200°C. This degradation manifests as depolymerization, discoloration, and racemization, which adversely affect the polymer's mechanical and optical properties (Wu et al., 2023). Additionally, side reactions, including ester exchange, can occur during processing, leading to the formation of cyclic oligomers. These degradation products can weaken the structural integrity of PLA, limiting its utility in demanding food packaging applications. The crystallization behavior of PLA further complicates its processing. PLA has a slow crystallization rate, which can hinder the development of a uniform crystalline structure during production. This slow rate affects the mechanical properties of the final material, often resulting in brittleness and reduced durability. Enhancing crystallization through the incorporation of nucleating agents or blending with other polymers is a common approach, but these modifications can introduce additional processing complexities.

The incorporation of nanofillers into PLA to enhance its barrier and mechanical properties also presents challenges. Achieving uniform dispersion of nanoparticles is critical, as poor dispersion can lead to the formation of aggregates that negatively impact the composite's performance. The immiscibility of nanofillers with PLA exacerbates this issue, making it difficult to achieve consistent interfacial compatibility. Certain nanofillers, such as layered silicates, carbon nanotubes, and hydrophilic nanoparticles like CNC, are often incompatible with PLA due to differences in their polarity and/or surface properties. This incompatibility leads to poor dispersion and weak interfacial bonding, adversely affecting the performance of PLA composites. Furthermore, the addition of nanofillers at higher concentrations can introduce processing inefficiencies and scalability issues, particularly when transitioning from laboratory-scale experiments to industrial production. To address these challenges, surface functionalization of fillers, such as grafting hydrophobic groups or using organophilic treatments, can improve compatibility with PLA. Additionally, the use of compatibilizers can effectively enhance interfacial bonding. Advanced dispersion techniques, including ultrasonication or high-shear mixing, are also effective in achieving a homogeneous filler dispersion.

Thermo-mechanical degradation during repeated processing cycles is another concern for PLA. Prolonged exposure to heat and shear stresses can result in molecular weight reduction, further compromising melt strength and processability. This degradation can lead to difficulties in achieving consistent film thickness and quality, which are critical for food packaging applications.

Addressing these challenges requires a multifaceted approach. Modifying PLA through copolymerization or blending with other biodegradable polymers can improve its melt strength and thermal stability. Similarly, the use of compatibilizers and surface treatments for nanofillers can enhance dispersion and interfacial compatibility, ensuring more uniform material properties. For industrial-scale applications, optimizing processing conditions such as extrusion temperature, cooling rates, and shear forces is essential to maintain material integrity and ensure reproducibility.

18.5.2 Cost considerations

A significant drawback of PLA compared to petrochemical-based plastics is its relatively high production cost, which poses a major challenge to its widespread commercialization. The primary contributor to this high cost is the purification of lactic acid, a critical step in PLA production. During fermentation, lactic acid must be separated from a complex broth containing pigments, cells, proteins, inorganic salts, by-product organic acids, and starch (Yu et al., 2023). This purification process accounts for approximately 50%–60% of the total production cost in industrial settings. The efficiency of purification technologies directly influences the overall yield, making it a pivotal yet challenging aspect of PLA manufacturing.

The higher price of PLA compared to conventional plastics limits its adoption, particularly for cost-sensitive applications such as packaging for low-value items like bread and crackers. However, advancements in production technologies are expected to gradually reduce PLA's cost, making it increasingly competitive with traditional plastics. For example, recent innovations have enabled the economical production of high molecular weight PLA, expanding its applications beyond niche medical uses, such as implants, tissue scaffolds, and sutures, to broader markets (Lim et al., 2008).

While the incorporation of nanofillers to enhance PLA's barrier and functional properties can improve its performance, it also contributes to increased production costs. These composites are often marketed as reusable to offset their higher cost; however, many current PLA composites remain suitable only for single-use applications due to limitations in processing technology. Similarly, blending PLA with other polymers offers a practical strategy for addressing cost concerns. Polymers such as acrylonitrile butadiene styrene (ABS), natural fibers, and biocarbons can be combined with PLA to improve flexibility and toughness, while also lowering material costs. These blends help mitigate the economic disadvantage of PLA by reducing the reliance on the relatively expensive biopolymer.

Production costs are further influenced by factors such as cooling times and sample removability during manufacturing. Extended cooling times or deformation of products during demolding can reduce efficiency, undermining the cost-effectiveness of PLA-based materials and processes. Optimizing these parameters is essential to streamline production and minimize waste, ultimately making PLA-based materials more economically viable for large-scale applications.

To achieve cost-effective solutions, researchers and manufacturers must focus on developing scalable, durable production techniques. Innovations that improve the feasibility of PLA production, such as advanced fermentation processes or improved nanofiller dispersion methods, hold promise for reducing costs while maintaining material quality. Additionally, balancing performance improvements with affordability will be critical to expanding PLA's adoption in the competitive packaging market.

18.5.3 Shelf life and storage conditions

PLA is hydrolytically degradable, making water-polymer interactions a critical factor in determining its general properties and stability. The presence of water in food and the surrounding atmosphere can lead to polymer degradation, which is a challenge for its application in food packaging. The rate of degradation increases under higher temperatures and humidity levels, as confirmed by studies showing a progressive decrease in PLA's molecular weight with an increase in RH from 50% to 100%. This degradation correlates with a reduction in the glass transition temperature, further affecting the material's structural integrity (Rocca-Smith et al., 2017).

Moisture uptake can plasticize the amorphous regions of PLA, weakening its structure and reducing mechanical strength. Additionally, an irreversible decrease in the degree of polymerization exacerbates this loss of strength. However, under drier conditions, some properties can be partially recovered. For instance, tensile strength measurements indicated that 94% of the initial tensile strength could be regained after exposure to low-humidity environments (Holm et al., 2006).

Surface modifications can also influence the stability and performance of PLA films. Plasma-activated PLA, for example, initially exhibits increased hydrophilicity, but this effect diminishes over time. Hydrophobic recovery occurs as polar groups on the surface either migrate into the polymer bulk or are removed entirely (Izdebska-Podsiadły & Dörsam, 2021). This phenomenon highlights the dynamic nature of PLA's interaction with environmental factors.

Photodegradation is a significant issue for PLA films, as exposure to ultraviolet (UV) light accelerates their degradation. Studies have shown that UV-treated PLA films lose their mechanical properties faster than untreated samples. For instance, research indicates that PLA films stored under high humidity and moderate temperatures experience a noticeable reduction in elongation at break within a similar timeframe, with UV exposure further accelerating this process. These observations highlight the need to address UV exposure to preserve the structural integrity of PLA-based materials.

Despite these challenges, PLA packaging is well suited for specific food applications. It performs effectively for perishable, high-moisture foods with short shelf lives, and semidry foods intended for longer storage. For long-term applications, incorporating opaque secondary packaging can shield PLA from UV exposure and delay photodegradation. Alternatively, enhancing the UV barrier properties of PLA films and coatings by adding EOs or nanostructured metal oxides, as previously discussed, offers a promising solution.

In summary, the shelf life and storage stability of PLA films and coatings depend heavily on environmental conditions such as humidity, temperature, and UV exposure. While these factors present challenges, practical strategies such as controlling environmental exposure and improving barrier properties can extend the functional lifespan of PLA packaging. These advancements reinforce PLA's potential as a sustainable material for food packaging applications.

18.5.4 Regulatory frameworks

Regulatory frameworks are essential in evaluating the suitability of PLA films and coatings for food packaging, ensuring the safety of both consumers and the environment. A critical concern is the potential release of particles, particularly when incorporating nanoparticles. Inorganic nanostructured metal oxides such as ZnO, titanium dioxide (TiO_2), copper oxide (CuO), and magnesium oxide (MgO) are frequently explored for their high surface area, functional properties, and selectivity (Perera et al., 2023). However, the potential toxicological effects of these nanomaterials remain under scrutiny, with ongoing debates in the scientific community (Valerini et al., 2018). To address these concerns, regulatory standards emphasize the assessment of migration and release of such particles into food to prevent human exposure to harmful levels.

The European Union (EU) has established comprehensive regulations for food contact materials (FCMs) to ensure consumer safety. Commission Regulation (EU) No. 10/2011 specifies stringent migration limits, setting an overall threshold of 10 mg dm^{-2} or 60 mg kg^{-1} of food for plastic materials in contact with food. These limits aim to minimize the transfer of substances from packaging to food, preserving its integrity and safety (European Commission, 2011). Additionally, Regulation (EC) No. 450/2009 provides guidelines for active materials, which release substances into the food matrix. Active components that are intrinsic to the material's passive structure are exempt from inclusion in the overall migration limit, provided their functionality complies with safety standards. These two regulations focus on passive and active contributions, highlighting the importance of rigorous compliance testing for PLA-based materials.

In the United States, the Food and Drug Administration (FDA) also governs the use of lactic acid, the primary building block of PLA, in food contact applications. Under Title 21 of the Code of Federal Regulations (CFR), Part 184.1061, lactic acid is classified as generally recognized as safe (GRAS) for direct human food use, with no limitations other than adherence to current good manufacturing practices. Furthermore, under Part 184.1(a), substances affirmed as GRAS for direct food use are also deemed GRAS as indirect FCMs. This regulatory endorsement ensures the safe integration of PLA into food packaging systems.

In the EU, lactic acid is similarly authorized as a food additive (E270) under Annex II and Annex III of Regulation (EC) No. 1333/2008. Its use, typically without specified maximum levels (quantum satis), is permitted across several food categories, with restrictions applied in specific categories such as infant formula and baby foods for infants and young children. For instance, lactic acid serves as both a polymer production aid and a monomer derived from microbial fermentation under Regulation (EU) No. 10/2011. Additional overarching requirements for FCMs are outlined in Regulation (EC) No. 1935/2004, which mandates the inertness of materials to prevent harmful substance transfer, and Regulation (EC) No. 2023/2006, which establishes good manufacturing practices for all FCMs.

In summary, the regulatory landscape for PLA as food packaging is robust, addressing both material composition and functional performance. These frameworks not only ensure consumer safety but also provide a pathway for PLA's sustainable adoption in the food packaging industry, balancing innovation with stringent compliance requirements.

18.6 Outlook

18.6.1 Emerging research areas and technological advancements

The landscape of food packaging is rapidly evolving, driven in part by the increasing commercialization of bioplastics. With the global plastics crisis and ambitious net-zero carbon emission goals, there is a growing focus on sustainable alternatives to conventional plastics. Single-use plastic items, including takeout containers, cups, cutlery, straws, wraps, and shopping bags, remain significant contributors to plastic pollution. Governments and industry stakeholders have identified these items as primary culprits, prompting a shift toward bioplastics, such as PLA.

Despite its advantages, PLA is not without limitations. Its poor thermal resistance, with a degradation threshold around 43°C (Collazo-Bigliardi et al., 2019), renders it unsuitable for packaging hot foods or beverages. Consequently, improving PLA's thermal stability is a priority in ongoing research. Recent advancements in material science, such as incorporating copolymers, blending PLA with other polymers, and introducing nanofillers, show promise in enhancing PLA's mechanical and thermal properties. These modifications aim to expand the applications of PLA while maintaining its biodegradability.

The high production cost of PLA remains a significant barrier to its widespread adoption. However, emerging technologies are addressing this issue. For instance, novel purification methods, such as those employing hydrophobic deep eutectic solvents (HDES), have demonstrated the potential to lower production costs while improving yields. HDES systems, composed of dialkyl amides and terpenes, have effectively separated organic acids, like lactic acid, from fermentation broths (Liu et al., 2020).

Consumer and industrial demand for home-compostable and marine-biodegradable bioplastics is also shaping the future of the bioplastics market. While PLA is industrially compostable, its limited biodegradability in home composting conditions presents a challenge. Future advancements in PLA could unlock its potential to degrade effectively in a wider range of environments, making it a strong contender for markets prioritizing home-compostable solutions.

Looking ahead, the development and commercialization of bioplastics will likely focus on materials with high renewable content. As advancements continue, future consumers will be better equipped to make environmentally responsible choices, selecting food packaging options that align with sustainability goals while reducing the global reliance on traditional plastics. By addressing cost, performance, and environmental challenges, PLA-based composites are set to play a crucial role in the transition to sustainable food packaging solutions.

18.7 Summary

PLA has emerged as a promising material for food packaging. Its environmental benefits, such as lower greenhouse gas emissions and industrial compostability, align with global sustainability goals. However, its inherent limitations, including brittleness, high water vapor permeability, and limited natural biodegradability, present challenges that must be addressed to fully realize its potential.

The performance of PLA as a food packaging material is influenced by its mechanical and barrier properties. These can be enhanced through the incorporation of additives like antimicrobial agents, plasticizers, and hydrophobic compounds. The careful optimization of additive concentrations is crucial to creating a homogeneous and continuous polymeric network. Imbalances in additive levels can lead to structural inconsistencies, compromising both the mechanical strength and functional properties of the material. For instance, excessive plasticizers may cause softening and reduce durability, while insufficient hydrophobic agents might fail to provide adequate moisture resistance. By tailoring PLA formulations to suit the pH, moisture content, and chemical composition of packaged foods, it is possible to extend freshness, improve product compatibility, and maximize packaging performance.

Hydrolytic degradation of PLA is a key factor in its shelf life and storage stability. Environmental conditions such as temperature, humidity, and UV exposure accelerate this process. Moisture uptake amplifies the plasticizing effect on PLA's amorphous regions, weakening its structure, while UV radiation accelerates molecular degradation, reducing mechanical strength. Studies indicate that protective strategies, including UV-filtering coatings or secondary opaque packaging, can mitigate these effects. Furthermore, PLA's functionality can be enhanced by formulations that match the specific requirements of the food product, such as reducing WVTRs for high-moisture items or improving oxygen barriers for fatty foods prone to oxidation.

The suitability of PLA for food packaging is regulated by stringent frameworks to ensure consumer safety. In the European Union, Commission Regulation No. 10/2011 sets migration limits for materials in contact with food, while the US FDA classifies lactic acid, PLA's primary building block, as GRAS. These regulations provide guidelines for both passive and active packaging systems, ensuring compliance while maintaining functional properties.

PLA's versatility enables its application across a wide range of food products, from high-moisture items like meat and dairy to low-moisture goods such as crackers and baked items. For instance, hydrophobic modifications protect dry foods from moisture absorption, while active packaging solutions incorporating antimicrobial or antioxidant agents extend the shelf life of perishable items. Despite these advantages, PLA's limited thermal resistance (~43°C) restricts its use for hot food packaging, highlighting an area for further improvement.

Emerging research is addressing these challenges, particularly regarding cost and functionality. The high production cost of PLA, driven by the expense of lactic acid purification, remains a barrier. However, advancements like HDES are improving purification efficiency and reducing costs. Future developments are also focusing on fully marine-biodegradable and home-compostable PLA variants, which align with consumer and industrial demand for more versatile bioplastics.

In summary, PLA offers significant promise as a sustainable food packaging material, but its adoption requires addressing key challenges, including cost, environmental stability, and functional limitations. By optimizing formulations, incorporating advanced additives at optimal concentrations, and leveraging emerging technologies, PLA can deliver enhanced performance while contributing to sustainable food packaging solutions. Continued innovation will be vital to expanding its role in reducing plastic waste and advancing global sustainability efforts.

References

Andrade, M. A., Barbosa, C. H., Cerqueira, M. A., Azevedo, A. G., Barros, C., Machado, A. V., Coelho, A., Furtado, R., Correia, C. B., Saraiva, M., Vilarinho, F., Silva, A. S., & Ramos, F. (2023). PLA films loaded with green tea and rosemary polyphenolic extracts as an active packaging for almond and beef. *Food Packaging and Shelf Life, 36*. https://doi.org/10.1016/j.fpsl.2023.101041, http://www.journals.elsevier.com/food-packaging-and-shelf-life/.

Ardjoum, N., Chibani, N., Shankar, S., Fadhel, Y. B., Djidjelli, H., & Lacroix, M. (2021). Development of antimicrobial films based on poly(lactic acid) incorporated with *Thymus vulgaris* essential oil and ethanolic extract of Mediterranean propolis. *International Journal of Biological Macromolecules, 185*, 535–542. https://doi.org/10.1016/j.ijbiomac.2021.06.194, www.elsevier.com/locate/ijbiomac.

Auras, R., Lim, L.-T., Selke, S. E. M., Tsuji, H., & Kumar, K. (2010). *Design and synthesis of different types of poly(lactic acid) poly(lactic acid: Synthesis, structures, properties, processing, and applications*. Wiley, 43–58 http://doi.org/10.1002/9780470649848.ch4.

Bala, T., Armstrong, G., Laffir, F., & Thornton, R. (2011). Titania–silver and alumina–silver composite nanoparticles: Novel, versatile synthesis, reaction mechanism and potential antimicrobial application. *Journal of Colloid and Interface Science, 356*(2), 395–403. https://doi.org/10.1016/j.jcis.2011.01.044.

Bao, L., Dorgan, J. R., Knauss, D., Hait, S., Oliveira, N. S., & Maruccho, I. M. (2006). Gas permeation properties of poly(lactic acid) revisited. *Journal of Membrane Science, 285*(1-2), 166–172. https://doi.org/10.1016/j.memsci.2006.08.021.

Bastam, M. M., Jalili, M., Pakzad, I., Maleki, A., & Ghafourian, S. (2021). Pathogenic bacteria in cheese, raw and pasteurised milk. *Veterinary Medicine and Science, 7*(6), 2445–2449. https://doi.org/10.1002/vms3.604, tinyurl.com/y7omcl8y.

Bruna, J. E., Castillo, M., López de Dicastillo, C., Muñoz-Shuguli, C., Lira, M., Guarda, A., Rodríguez-Mercado, F. J., & Galotto, M. J. (2023). Development of active biocomposite films based on poly(lactic acid) and wine by-product: Effect of grape pomace content and extrusion temperature. *Journal of Applied Polymer Science, 140*(38). https://doi.org/10.1002/app.54425, http://onlinelibrary.wiley.com/journal/10.1002/(ISSN)1097-4628.

Chuakhao, S., Rodríguez, J. T., Lapnonkawow, S., Kannan, G., Müller, A. J., & Suttiruengwong, S. (2024). Formulating PBS/PLA/PBAT blends for biodegradable, compostable packaging: The crucial roles of PBS content and reactive extrusion. *Polymer Testing, 132*. https://doi.org/10.1016/j.polymertesting.2024.108383, https://www.sciencedirect.com/science/journal/01429418.

Cisar, J., Drohsler, P., Pummerova, M., Sedlarik, V., & Skoda, D. (2023). Composite based on PLA with improved shape stability under high-temperature conditions. *Polymer, 276*, 125943. https://doi.org/10.1016/j.polymer.2023.125943.

Collazo-Bigliardi, S., Ortega-Toro, R., & Chiralt, A. (2019). Using lignocellulosic fractions of coffee husk to improve properties of compatibilised starch-PLA blend films. *Food Packaging and Shelf Life, 22*, 100423. https://doi.org/10.1016/j.fpsl.2019.100423.

Commission. (2011). On plastic materials and articles intended to come into contact with food Commission Regulation, 1–89.

Connolly, M., Zhang, Y., Brown, D. M., Ortuño, N., Jordá-Beneyto, M., Stone, V., Fernandes, T. F., & Johnston, H. J. (2019). Novel polylactic acid (PLA)-organoclay nanocomposite bio-packaging for the cosmetic industry; migration studies and in vitro assessment of the dermal toxicity of migration extracts. *Polymer Degradation and Stability, 168*. https://doi.org/10.1016/j.polymdegradstab.2019.108938, https://www.journals.elsevier.com/polymer-degradation-and-stability.

Criollo-Feijoo, J., Salas-Gomez, V., Cornejo, F., Auras, R., & Salazar, R. (2024). Cassava bagasse starch and oregano essential oil as a potential active food packaging material: A physicochemical, thermal, mechanical, antioxidant, and antimicrobial study. *Heliyon, 10*(16), e36150. https://doi.org/10.1016/j.heliyon.2024.e36150.

Cvek, M., Paul, U. C., Zia, J., Mancini, G., Sedlarik, V., & Athanassiou, A. (2022). Biodegradable films of PLA/PPC and curcumin as packaging materials and smart indicators of food spoilage. *ACS Applied Materials and Interfaces, 14*(12), 14654–14667. https://doi.org/10.1021/acsami.2c02181, http://pubs.acs.org/journal/aamick.

Dalla Rosa, M. (2019). Packaging sustainability in the meat industry. In *Sustainable meat production and processing*. Italy: Elsevier, 161–179. http://doi.org/10.1016/B978-0-12-814874-7.00009-2, https://www.sciencedirect.com/book/9780128148747.

Dhar, P., Gaur, S. S., Soundararajan, N., Gupta, A., Bhasney, S. M., Milli, M., Kumar, A., & Katiyar, V. (2017). Reactive extrusion of polylactic acid/cellulose nanocrystal films for food packaging applications: Influence of filler type on thermomechanical, rheological, and barrier properties. *Industrial and Engineering Chemistry Research, 56*(16), 4718–4735. https://doi.org/10.1021/acs.iecr.6b04699, http://pubs.acs.org/journal/iecred.

Dominguez-Candela, I., Gomez-Caturla, J., Cardona, S. C., Lora-García, J., & Fombuena, V. (2022). Novel compatibilizers and plasticizers developed from epoxidized and maleinized chia oil in composites based on PLA and chia seed flour. *European Polymer Journal, 173*, 111289. https://doi.org/10.1016/j.eurpolymj.2022.111289.

Dong, Y., Ghataura, A., Takagi, H., Haroosh, H. J., Nakagaito, A. N., & Lau, K. T. (2014). Polylactic acid (PLA) biocomposites reinforced with coir fibres: Evaluation of mechanical performance and multifunctional properties. *Composites Part A: Applied Science and Manufacturing, 63*, 76–84. https://doi.org/10.1016/j.compositesa.2014.04.003, https://www.journals.elsevier.com/composites-part-a-applied-science-and-manufacturing.

Emadian, S. M., Onay, T. T., & Demirel, B. (2017). Biodegradation of bioplastics in natural environments. *Waste Management, 59*, 526–536. https://doi.org/10.1016/j.wasman.2016.10.006, www.elsevier.com/locate/wasman.

Esmaeili, M., Pircheraghi, G., Bagheri, R., & Altstädt, V. (2019). Poly(lactic acid)/coplasticized thermoplastic starch blend: Effect of plasticizer migration on rheological and mechanical properties. *Polymers for Advanced Technologies, 30*(4), 839–851. https://doi.org/10.1002/pat.4517.

Fiore, A., Park, S., Volpe, S., Torrieri, E., & Masi, P. (2021). Active packaging based on PLA and chitosan-caseinate enriched rosemary essential oil coating for fresh minced chicken breast application. *Food Packaging and Shelf Life, 29*, 100708. https://doi.org/10.1016/j.fpsl.2021.100708.

Fiorentini, C., Leni, G., de Apodaca, E. D., Fernández-de-Castro, L., Rocchetti, G., Cortimiglia, C., Spigno, G., & Bassani, A. (2024). Development of coated PLA films containing a commercial olive leaf extract for the food packaging sector. *Antioxidants, 13*(5). https://doi.org/10.3390/antiox13050519, www.mdpi.com/journal/antioxidants.

Garcia-Garcia, D., Carbonell-Verdu, A., Arrieta, M. P., López-Martínez, J., & Samper, M. D. (2020). Improvement of PLA film ductility by plasticization with epoxidized karanja oil. *Polymer Degradation and Stability, 179*, 109259. https://doi.org/10.1016/j.polymdegradstab.2020.109259.

Goh, K., Heising, J. K., Yuan, Y., Karahan, H. E., Wei, L., Zhai, S., Koh, J. X., Htin, N. M., Zhang, F., Wang, R., Fane, A. G., Dekker, M., Dehghani, F., & Chen, Y. (2016). Sandwich-architectured poly(lactic acid)-graphene composite food packaging films. *ACS Applied Materials and Interfaces, 8*(15), 9994–10004. https://doi.org/10.1021/acsami.6b02498, http://pubs.acs.org/journal/aamick.

Gonon, H., Srisa, A., Promhuad, K., Chonhenchob, V., Bumbudsanpharoke, N., Jarupan, L., & Harnkarnsujarit, N. (2023). PLA thermoformed trays incorporated with cinnamaldehyde and carvacrol as active biodegradable bakery packaging. *Food Packaging and Shelf Life, 38*, 101123. https://doi.org/10.1016/j.fpsl.2023.101123.

Gonzalez, L., Agüero, A., Quiles-Carrillo, L., Lascano, D., & Montanes, N. (2019). Optimization of the loading of an environmentally friendly compatibilizer derived from linseed oil in poly(lactic acid)/diatomaceous earth composites. *Materials, 12*(10), 1627. https://doi.org/10.3390/ma12101627.

Han Lyn, F., Ismail-Fitry, M. R., Noranizan, M. A., Tan, T. B., & Nur Hanani, Z. A. (2024). Recent advances in extruded polylactic acid-based composites for food packaging: A review. *International Journal of Biological Macromolecules, 266*. https://doi.org/10.1016/j.ijbiomac.2024.131340, https://www.sciencedirect.com/science/journal/01418130.

He, X., Wang, X., Liu, Y., Fang, H., Zheng, S., Liu, H., Guan, W., & Yan, R. (2022). Biodegradable PBAT/PLA packaging maintained the quality of postharvest shiitake mushroom by modified humidity and atmosphere. *Food Packaging and Shelf Life, 34*, 100949. https://doi.org/10.1016/j.fpsl.2022.100949.

Herskovitz, J. E., & Goddard, J. M. (2020). Reactive extrusion of nonmigratory antioxidant poly(lactic acid) packaging. *Journal of Agricultural and Food Chemistry, 68*(7), 2164–2173. https://doi.org/10.1021/acs.jafc.9b06776, http://pubs.acs.org/journal/jafcau.

Ho, K. L. G., Pometto, A. L., Hinz, P. N., Gadea-Rivas, A., Briceño, J. A., & Rojas, A. (1999). Field exposure study of polylactic acid (PLA) plastic films in the banana fields of Costa Rica. *Journal of Environmental Polymer Degradation, 7*(4), 167–172.

Holm, V. K., Ndoni, S., & Risbo, J. (2006). The stability of poly(lactic acid) packaging films as influenced by humidity and temperature. *Journal of Food Science, 71*(2), E40, Available from https://doi.org/10.1111/j.1365-2621.2006.tb08895.x, http://www3.interscience.wiley.com/journal/118509799/issueyear?year=2008.

Izdebska-Podsiadły, J., & Dörsam, E. (2021). Storage stability of the oxygen plasma-modified PLA film. *Bulletin of Materials Science, 44*(2). https://doi.org/10.1007/s12034-021-02355-z.

Kang, J. H., & Song, K. B. (2019). Characterization of Job's tears (*Coix lachryma-jobi* L.) starch films incorporated with clove bud essential oil and their antioxidant effects on pork belly during storage. *LWT, 111*, 711–718. https://doi.org/10.1016/j.lwt.2019.05.102, https://www.journals.elsevier.com/lwt.

Karkhanis, S. S., Stark, N. M., Sabo, R. C., & Matuana, L. M. (2021). Potential of extrusion-blown poly(lactic acid)/cellulose nanocrystals nanocomposite films for improving the shelf-life of a dry food product. *Food Packaging and Shelf Life, 29*. https://doi.org/10.1016/j.fpsl.2021.100689, http://www.journals.elsevier.com/food-packaging-and-shelf-life/.

Kay, I. P., Herskovitz, J. E., & Goddard, J. M. (2022). Interfacial behavior of a polylactic acid active packaging film dictates its performance in complex food matrices. *Food Packaging and Shelf Life, 32*. https://doi.org/10.1016/j.fpsl.2022.100832, http://www.journals.elsevier.com/food-packaging-and-shelf-life/.

Kumbunleu, J., Rattanapun, A., Sapsrithong, P., Tuampoemsab, S., & Sritapunya, T. (2020). The study on chitosan coating on poly(lactic acid) film packaging to extend vegetable and fruit life. *IOP Conference Series: Materials Science and Engineering, 811*(1), 012021. https://doi.org/10.1088/1757-899x/811/1/012021.

Laorenza, Y., & Harnkarnsujarit, N. (2023). Ginger oil and lime peel oil loaded PBAT/PLA via cast-extrusion as shrimp active packaging: Microbial and melanosis inhibition. *Food Packaging and Shelf Life, 38*, 101116. https://doi.org/10.1016/j.fpsl.2023.101116.

Lara-Topete, G. O., Castanier-Rivas, J. D., Gradilla-Hernández, M. S., & González-López, M. E. (2024). Life cycle assessment of agave bagasse management strategies: PLA biocomposites versus conventional waste disposal practices. *Sustainable Chemistry and Pharmacy, 37*. https://doi.org/10.1016/j.scp.2024.101435, https://www.sciencedirect.com/science/journal/23525541.

Lee, M., Jung, B. N., Kim, G. H., Kang, D. H., Park, H. J., Shim, J. K., & Hwang, S. W. (2020). The effect of triethyl citrate on the dispersibility and water vapor sorption behavior of polylactic acid/zeolite composites. *Polymer Testing, 89*. https://doi.org/10.1016/j.polymertesting.2020.106571, https://www.journals.elsevier.com/polymer-testing.

Lim, L. T., Auras, R., & Rubino, M. (2008). Processing technologies for poly(lactic acid). *Progress in Polymer Science (Oxford), 33*(8), 820–852. https://doi.org/10.1016/j.progpolymsci.2008.05.004.

Liu, L., Wei, Q., Zhou, Y., & Ren, X. (2020). Using dialkyl amide via forming hydrophobic deep eutectic solvents to separate citric acid from fermentation broth. *Green Chemistry, 22*(8), 2526–2533. https://doi.org/10.1039/c9gc04401a.

Manzanarez-López, F., Soto-Valdez, H., Auras, R., & Peralta, E. (2011). Release of α-Tocopherol from Poly(lactic acid) films, and its effect on the oxidative stability of soybean oil. *Journal of Food Engineering, 104*(4), 508–517. https://doi.org/10.1016/j.jfoodeng.2010.12.029.

Mao, L., Wang, C., Yao, J., Lin, Y., Liao, X., & Lu, J. (2023). Design and fabrication of anthocyanin functionalized layered clay/poly(vinyl alcohol) coatings on poly(lactic acid) film for active food packaging. *Food Packaging and Shelf Life, 35*, 101007. https://doi.org/10.1016/j.fpsl.2022.101007.

Martins, C., Vilarinho, F., Sanches Silva, A., Andrade, M., Machado, A. V., Castilho, M. C., Sá, A., Cunha, A., Vaz, M. F., & Ramos, F. (2018). Active polylactic acid film incorporated with green tea extract: Development, characterization and effectiveness. *Industrial Crops and Products, 123*, 100–110. https://doi.org/10.1016/j.indcrop.2018.06.056, www.elsevier.com/inca/publications/store/5/2/2/8/2/5.

Matweb (2024). Overview of materials for polypropylene, Film Grade.

Mayekar, P. C., Limsukon, W., Bher, A., & Auras, R. (2023). Breaking it down: How thermoplastic starch enhances poly(lactic acid) biodegradation in compost—A comparative analysis of reactive blends. *ACS Sustainable Chemistry and Engineering, 11*(26), 9729–9737. https://doi.org/10.1021/acssuschemeng.3c01676, http://pubs.acs.org/journal/ascecg.

Mikloskova, H., Witte, F., Joeres, E., & Terjung, N. (2021). Storage stability of plain stirred whole milk yoghurt (3.7% fat) packed in polylactic acid and polystyrene. *International Dairy Journal, 120*, 105088. https://doi.org/10.1016/j.idairyj.2021.105088.

Perera, K. Y., Pradhan, D., Rafferty, A., Jaiswal, A. K., & Jaiswal, S. (2023). A comprehensive review on metal oxide-nanocellulose composites in sustainable active and intelligent food packaging. *Food Chemistry Advances, 3*. https://doi.org/10.1016/j.focha.2023.100436, https://www.journals.elsevier.com/food-chemistry-advances.

Peter, A., Mihaly Cozmuta, L., Nicula, C., Mihaly Cozmuta, A., Talasman, C. M., Drazic, G., Peñas, A., Calahorro, A. J., Sagratini, G., & Silvi, S. (2021). Chemical and organoleptic changes of curd cheese stored in new and reused active packaging systems made of Ag-graphene-TiO_2-PLA. *Food Chemistry, 363*. https://doi.org/10.1016/j.foodchem.2021.130341, www.elsevier.com/locate/foodchem.

Pirsa, S., & Asadi, S. (2021). Innovative smart and biodegradable packaging for margarine based on a nano composite polylactic acid/lycopene film. *Food Additives and Contaminants - Part A Chemistry, Analysis, Control, Exposure and Risk Assessment, 38*(5), 856–869. https://doi.org/10.1080/19440049.2021.1891299, http://www.tandfonline.com/toc/tfac20/current.

Ran, X., Qu, Y., Wang, Y., Cui, B., Shen, Y., & Li, Y. (2023). Enhanced UV-blocking capabilities of polylactic acid derived from renewable resources for food and drug packaging: A mini-review. *Journal of Composites Science, 7*(10). https://doi.org/10.3390/jcs7100410, www.mdpi.com/journal/jcs.

Robertson, G. L. (2019). *History of food packaging*. Elsevier BV. http://doi.org/10.1016/b978-0-08-100596-5.22535-3.

Rocca-Smith, J. R., Chau, N., Champion, D., Brachais, C.-H., Marcuzzo, E., Sensidoni, A., Piasente, F., Karbowiak, T., & Debeaufort, F. (2017). Effect of the state of water and relative humidity on ageing of PLA films. *Food Chemistry, 236*, 109–119. https://doi.org/10.1016/j.foodchem.2017.02.113.

Roy, S., Ramakrishnan, R., Goksen, G., Singh, S., & Łopusiewicz, Ł. (2024). Recent progress on UV-light barrier food packaging films – a systematic review. *Innovative Food Science & Emerging Technologies, 91*, 103550. https://doi.org/10.1016/j.ifset.2023.103550.

Roy, S., & Rhim, J. W. (2020). Preparation of bioactive functional poly(lactic acid)/curcumin composite film for food packaging application. *International Journal of Biological Macromolecules, 162*, 1780–1789. https://doi.org/10.1016/j.ijbiomac.2020.08.094, www.elsevier.com/locate/ijbiomac.

Rudnik, E., & Briassoulis, D. (2011). Degradation behaviour of poly(lactic acid) films and fibres in soil under Mediterranean field conditions and laboratory simulations testing. *Industrial Crops and Products, 33*(3), 648–658. https://doi.org/10.1016/j.indcrop.2010.12.031.

Safandowska, M., Makarewicz, C., Rozanski, A., & Idczak, R. (2022). Barrier properties of semicrystalline polylactide: The role of the density of the amorphous regions. *Macromolecules, 55*(22), 10077–10089. https://doi.org/10.1021/acs.macromol.2c01490.

Sarasa, J., Gracia, J. M., & Javierre, C. (2009). Study of the biodisintegration of a bioplastic material waste. *Bioresource Technology, 100*(15), 3764–3768. https://doi.org/10.1016/j.biortech.2008.11.049.

Scaffaro, R., Maio, A., Gulino, F. E., Di Salvo, C., & Arcarisi, A. (2020). Bilayer biodegradable films prepared by co-extrusion film blowing: Mechanical performance, release kinetics of an antimicrobial agent and hydrolytic degradation. *Composites Part A: Applied Science and*

Manufacturing, 132. https://doi.org/10.1016/j.compositesa.2020.105836, https://www.journals.elsevier.com/composites-part-a-applied-science-and-manufacturing.

Shankar, S., Wang, L. F., & Rhim, J. W. (2018). Incorporation of zinc oxide nanoparticles improved the mechanical, water vapor barrier, UV-light barrier, and antibacterial properties of PLA-based nanocomposite films. *Materials Science and Engineering C, 93*, 289–298. https://doi.org/10.1016/j.msec.2018.08.002.

Shehzad, A., Aslam, M., Basit, M., & Salman, M. (2024). Effect of titanium oxide on structural and optoelectrical properties of PLA/PEG/graphene nanocomposite films. *Synthetic Metals, 307*, 117696. https://doi.org/10.1016/j.synthmet.2024.117696.

Siakeng, R., Jawaid, M., Ariffin, H., Sapuan, S. M., Asim, M., & Saba, N. (2019). Natural fiber reinforced polylactic acid composites: A review. *Polymer Composites, 40*(2), 446–463. https://doi.org/10.1002/pc.24747, http://onlinelibrary.wiley.com/journal/10.1002/(ISSN)1548-0569.

Sobhan, A., Muthukumarappan, K., Wei, L., Zhou, R., & Tummala, H. (2021). Development of a polylactic acid-coated nanocellulose/chitosan-based film indicator for real-time monitoring of beef spoilage. *Analytical Methods, 13*(23), 2612–2623. https://doi.org/10.1039/d1ay00365h.

Stoleru, E., Vasile, C., Irimia, A., & Brebu, M. (2021). Towards a Bioactive Food Packaging: Poly(Lactic Acid) Surface Functionalized by Chitosan Coating Embedding Clove and Argan Oils. *Molecules (Basel, Switzerland), 26*(15), 4500. https://doi.org/10.3390/molecules26154500.

Swetha, T. A., Bora, A., Mohanrasu, K., Balaji, P., Raja, R., Ponnuchamy, K., Muthusamy, G., & Arun, A. (2023). A comprehensive review on polylactic acid (PLA) – Synthesis, processing and application in food packaging. *International Journal of Biological Macromolecules, 234*. https://doi.org/10.1016/j.ijbiomac.2023.123715, www.elsevier.com/locate/ijbiomac.

Tagrida, M., Gulzar, S., Nilsuwan, K., Prodpran, T., Zhang, B., & Benjakul, S. (2022). Polylactic acid film coated with electrospun gelatin/chitosan nanofibers containing betel leaf ethanolic extract: Properties, bioactivities, and use for shelf-life extension of tilapia slices. *Molecules (Basel, Switzerland), 27*(18), 5877. https://doi.org/10.3390/molecules27185877.

Tamburini, E., Costa, S., Summa, D., Battistella, L., Fano, E. A., & Castaldelli, G. (2021). Plastic (PET) vs bioplastic (PLA) or refillable aluminium bottles – What is the most sustainable choice for drinking water? A life-cycle (LCA) analysis. *Environmental Research, 196*. https://doi.org/10.1016/j.envres.2021.110974, http://www.elsevier.com/inca/publications/store/6/2/2/8/2/1/index.htt.

Techawinyutham, L., Siengchin, S., Parameswaranpillai, J., & Dangtungee, R. (2019). Antibacterial and thermomechanical properties of composites of polylactic acid modified with capsicum oleoresin-impregnated nanoporous silica. *Journal of Applied Polymer Science, 136*(31). https://doi.org/10.1002/app.47825.

Todhanakasem, T., Panjapiyakul, S., & Koombhongse, P. (2022). Novel pineapple leaf fibre composites coating with antimicrobial compound as a potential food packaging. *Packaging Technology and Science, 35*(1), 97–105. https://doi.org/10.1002/pts.2612.

Umar, A. H., Zainudin, E. S., & Sapuan, S. M. (2012). Effect of accelerated weathering on tensile properties of kenaf reinforced high-density polyethylene composites. *Journal of Mechanical Engineering and Sciences, 2*, 198–205. https://doi.org/10.15282/jmes.2.2012.7.0018.

Valerini, D., Tammaro, L., Di Benedetto, F., Vigliotta, G., Capodieci, L., Terzi, R., & Rizzo, A. (2018). Aluminum-doped zinc oxide coatings on polylactic acid films for antimicrobial food packaging. *Thin Solid Films, 645*, 187–192. https://doi.org/10.1016/j.tsf.2017.10.038.

Verma, K., Siddiki, S. H., Maity, C. K., Mishra, R. K., & Moniruzzaman, M. (2023). Development of reduced graphene oxide (rGO) reinforced poly(lactic) acid/cellulose nanocrystal composite through melt mixing: Effect of nanofiller on thermal, structural, biodegradation and antibacterial properties. *Industrial Crops and Products, 204*. https://doi.org/10.1016/j.indcrop.2023.117307, www.elsevier.com/inca/publications/store/5/2/2/8/2/5.

Wang, D., Sun, Z., Sun, J., Liu, F., Du, L., & Wang, D. (2021). Preparation and characterization of polylactic acid nanofiber films loading Perilla essential oil for antibacterial packaging of chilled chicken. *International Journal of Biological Macromolecules, 192*, 379–388. https://doi.org/10.1016/j.ijbiomac.2021.09.190.

Wang, L., Xu, J., Zhang, M., Zheng, H., & Li, L. (2022). Preservation of soy protein-based meat analogues by using PLA/PBAT antimicrobial packaging film. *Food Chemistry, 380*, 132022. https://doi.org/10.1016/j.foodchem.2021.132022.

Wang, S., Shen, Q., Guo, C., & Guo, H. (2021). Comparative study on water vapour resistance of poly(lactic acid) films prepared by blending, filling and surface deposit. *Membranes, 11*(12), 915. https://doi.org/10.3390/membranes11120915.

Wongphan, P., Nampanya, P., Chakpha, W., Promhuad, K., Laorenza, Y., Leelaphiwat, P., Bumbudsanpharoke, N., Sodsai, J., Lorenzo, J. M., & Harnkarnsujarit, N. (2023). Lesser galangal (*Alpinia officinarum* Hance) essential oil incorporated biodegradable PLA/PBS films as shelf-life extension packaging of cooked rice. *Food Packaging and Shelf Life, 37*. https://doi.org/10.1016/j.fpsl.2023.101077, http://www.journals.elsevier.com/food-packaging-and-shelf-life/.

Wu, Y., Gao, X., Wu, J., Zhou, T., Nguyen, T. T., & Wang, Y. (2023). Biodegradable polylactic acid and its composites: Characteristics, processing, and sustainable applications in sports. *Polymers, 15*(14). https://doi.org/10.3390/polym15143096, http://www.mdpi.com/journal/polymers.

Xing, C., & Matuana, L. M. (2016). Epoxidized soybean oil-plasticized poly(lactic acid) films performance as impacted by storage. *Journal of Applied Polymer Science, 133*(12). https://doi.org/10.1002/app.43201, http://onlinelibrary.wiley.com/journal/10.1002/(ISSN)1097-4628.

Yanat, M., Voortman, E., Buijsse, C., & Schroën, K. (2025). Life cycle assessment (LCA) of chitin nanocrystal reinforced polylactic acid (ChNC/PLA) and polyethylene terephthalate (PET) packaging material. *Carbohydrate Polymers, 348*(Pt B), 122927. https://doi.org/10.1016/j.carbpol.2024.122927.

Yang, H. J., & Song, K. B. (2016). Application of lemongrass oil-containing polylactic acid films to the packaging of pork sausages. *Korean Journal for Food Science of Animal Resources, 36*(3), 421–426. https://doi.org/10.5851/kosfa.2016.36.3.421, http://ocean.kisti.re.kr/downfile/volume/ksfsar/CSSPBQ/2016/v36n3/CSSPBQ_2016_v36n3_421.pdf.

Yoksan, R., & Dang, K. M. (2023). The effect of polyethylene glycol sorbitan monostearate on the morphological characteristics and performance of thermoplastic starch/biodegradable polyester blend films. *International Journal of Biological Macromolecules, 231*, 123332. https://doi.org/10.1016/j.ijbiomac.2023.123332.

Yu, J., Xu, S., Liu, B., Wang, H., Qiao, F., Ren, X., & Wei, Q. (2023). PLA bioplastic production: From monomer to the polymer. *European Polymer Journal, 193*, 112076. https://doi.org/10.1016/j.eurpolymj.2023.112076.

Zeid, A., Karabagias, I. K., Nassif, M., & Kontominas, M. G. (2019). Preparation and evaluation of antioxidant packaging films made of polylactic acid containing thyme, rosemary, and oregano essential oils. *Journal of Food Processing and Preservation, 43*(10). https://doi.org/10.1111/jfpp.14102, http://onlinelibrary.wiley.com/journal/10.1111/(ISSN)1745-4549.

Zubairi, S. I., Ab Kadir, I. A., Nurzahim, Z., & Lazim, A. (2022). Evaluation of poly(L-lactic acid) (PLLA) rapid indicator film on deterioration degree of Refined, Bleached and Deodorised Malaysian Tenera palm olein oil (RBDPO) during long-term repetitive deep-fat frying. *Arabian Journal of Chemistry, 15*(4), ScienceDirect https://doi.org/10.1016/j.arabjc.2022.103726, http://colleges.ksu.edu.sa/Arabic%20Colleges/CollegeOfScience/ChemicalDept/AJC/default.aspx, http://www.sciencedirect.com/science/journal/18785352.

Chapter 19

Poly(3-hydroxybutyrate) films and coatings

Łukasz Łopusiewicz[1,2]
[1]Department of Pharmaceutical Biology, Institute of Pharmacy, University of Greifswald, Greifswald, Germany, [2]School of Medical & Health Sciences, VIZJA University, Warsaw, Poland

19.1 Introduction

Contemporary packaging industry, dominated by petrochemical-based plastics, poses a serious challenge to the environment (Wani et al., 2025). Since their discovery in the 20th century, plastics have gained immense popularity due to their versatility, lightness, and low production costs. Between the 1950s and the present day, approximately one million tons of plastic have been produced, of which only 30% remains in active use, with the rest classified as waste (Jin et al., 2023). It is estimated that over 80 million tons of plastic are used annually in the packaging industry alone, much of which has a very short lifespan—from production to disposal (Iordanskii et al., 2018). As much as 66% of global plastic packaging production is accounted for by the food sector (Bairwan et al., 2024), which is dominated by the use of polyethylene (PE), polypropylene (PP), and polyethylene terephthalate (PET) (Atarés et al., 2024). The high durability of these materials, although desirable from a functional point of view, means that their degradation in the environment can take hundreds of years or may never occur (Jin et al., 2023). Plastic waste, especially from the food sector, contributes to soil and water contamination (Usurelu et al., 2022; Jin et al., 2023; Sun et al., 2024). This problem has been further exacerbated by the COVID-19 pandemic, which has drastically increased the consumption of single-use packaging materials and medical devices made of plastic (Usurelu et al., 2022; Choi et al., 2024). Only a small percentage of this waste is recycled—most of it is fragmented and ends up in the environment as microplastics. Microplastics enter the soil, groundwater, oceans, and then the food chain, where they can induce oxidative stress, cytokine release, immune responses, and cell damage (Usurelu et al., 2022; Jin et al., 2023). In the context of public health and food safety, this is an alarming phenomenon. Despite numerous campaigns promoting recycling and the implementation of circular economy systems, a significant proportion of waste still ends up in landfills (Arrieta et al., 2014; Iordanskii et al., 2018). Furthermore, plastic production currently accounts for 4%–8% of global oil consumption, and according to forecasts, this could rise to as much as 20% by 2050 (Vanheusden et al., 2024). Faced with rising raw material prices and limited fossil resources, the industry is facing the need to transform toward more sustainable solutions (Nanni & Messori, 2021; Zhang et al., 2024).

In response to these problems, the scientific community is focusing on developing alternative, environmentally friendly packaging materials based on renewable raw materials, such as polymers derived from microorganisms, plants, or waste (Bairwan et al., 2024; Bugnicourt et al., 2014; Zhang et al., 2024). The use of biopolymers in the packaging industry fits perfectly with the ideas of green chemistry, the circular economy, and sustainable eco-design of materials (Iordanskii et al., 2018; Nanni & Messori, 2021; Vanheusden et al., 2024). This makes it possible not only to reduce CO_2 emissions and fossil fuel consumption, but also to limit the generation of persistent waste and environmental pollution. Biopolymers such as poly(lactic acid) (PLA), poly(3-hydroxybutyrate) (PHB), poly(3-hydroxybutyrate-co-3-hydroxyvalerate) (PHBV), and poly(butylene adipate-co-terephthalate) (PBAT) are biodegradable, and some of them are already being produced on a commercial scale (Dhar et al., 2015; Ibrahim et al., 2021). Nevertheless, their share in total plastic production remains low (Vanheusden et al., 2024). Due to the diversity of their sources and chemical structures, biodegradable polymers are divided into three main categories: natural polymers, synthetic polymers, and microbiological polymers (Ibrahim et al., 2021). The first group includes compounds such as cellulose, starch, chitosan, collagen, and gelatin—all obtained from natural resources and easily biodegradable. The second group consists of synthetic biodegradable materials such as

Natural Polymer-Based Films and Coatings in Food Technology. DOI: https://doi.org/10.1016/B978-0-443-34244-8.00002-1

polyesters (e.g., poly(lactides)—PLA, polycaprolactone—PCL) and aliphatic-aromatic copolymers. The third group consists of polymers biosynthesized by microorganisms, such as polyhydroxyalkanoates (PHAs), including PHB and PHBV, which can be easily degraded by enzymes present in the environment (Bugnicourt et al., 2014; Zhang et al., 2024). The advantage of bioplastics is that they can be processed using methods used in the traditional plastics industry, such as extrusion, injection molding, and film extrusion (Vanheusden et al., 2024). However, the mass implementation of bioplastics poses technological challenges. Some natural materials (e.g., PLA, starch) are highly biodegradable but have low moisture resistance and poor mechanical properties, which limits their suitability for packaging products with high moisture content, such as meat or fish (Dhar et al., 2015; Choi et al., 2024). On the other hand, materials with higher strength often require special composting conditions, which may limit their degradation in the natural environment (Choi et al., 2024; Ibrahim et al., 2021). In addition to their susceptibility to moisture, many bioplastics also have limited mechanical and chemical durability compared to traditional plastics, so modifications are needed to increase the applicability of materials (Atarés et al., 2024).

In recent years, advanced biopolymer nanocomposites have attracted particular interest. Through the introduction of nanoparticles (e.g., nanosilica, nanocellulose, zinc oxide [ZnO], silver, or montmorillonite), they significantly improve the mechanical, barrier, and bioactive properties of packaging materials (Dhar et al., 2015). This approach makes it possible to increase moisture resistance, reduce gas migration, and extend the shelf life of food products. Furthermore, carefully selected additives can result in materials' antimicrobial properties, which are particularly important in the context of packaging fresh and minimally processed food. Although nanocomposite technology still needs to be optimized in terms of consumer safety and regulatory compliance, its potential in the context of the future of biodegradable packaging is widely recognized. In addition to the nanotechnological approach, chemical and enzymatic modification plays an important role in improving the functionality of bioplastics. Examples of such modifications include esterification, acetylation, oxyethylation, and grafting reactions, which allow the adjustment of the hydrophobic, barrier, and thermal and mechanical properties of materials (Zhou et al., 2021). Enzymatic functionalization of biopolymer surfaces, on the other hand, not only improves adhesion between composite layers but also allows for the incorporation of bioactive substances. This approach is gaining importance in the context of the development of active and intelligent packaging that responds to environmental factors. The introduction of functional chemical groups also increases the compatibility of biopolymers with natural additives (e.g., essential oils, plant extracts), allowing the development of materials with complex and controlled antimicrobial and antioxidant properties (Zhou et al., 2021). Another innovative direction in the field of biopolymer packaging is additive manufacturing techniques, including 3D printing. The use of additive methods enables the design of packaging structures with customized geometry, adapted to the shape and type of food product. Biopolymers such as PLA, PHA, or their blends with functional additives are already used as base materials in three-dimensional printing. This technology allows for precise placement of material layers, opening up new possibilities for the integration of active properties such as the gradual release of antimicrobial substances, permeability control, and visual detection of product quality changes. 3D printing also offers potential for smart packaging, enabling the production of packaging with built-in sensors or structures that respond to changes in temperature, pH, or the presence of gases (Kontárová et al., 2020; Melčová et al., 2020; Bittolo Bon et al., 2021). Although this technology is still in the early stages of implementation in the food industry, its potential is promising in the context of personalization and functionalization of sustainable packaging.

At the same time, there is growing interest in implementing the concept of biorefineries as a crucial element of the circular economy. Biorefineries enable the comprehensive processing of plant biomass and agri-food waste into valuable products, including biopolymers, bio-oils, biogases, and bioactive compounds (Bairwan et al., 2024). Such systems not only minimize waste but also increase the efficiency of natural resource use. In the context of packaging, biorefinery products can serve as substrates for the biosynthesis of polymers (e.g., PHA) or as functional additives (e.g., antioxidants, natural dyes, and antimicrobials). The implementation of biorefineries on a local scale—for example, at food processing plants—can significantly increase the sustainability of packaging production and reduce its carbon footprint. This approach fits into the broader circular economy paradigm, promoting synergistic links between agriculture, the food industry, and the packaging materials sector. In the face of these challenges, the further development of packaging materials should take into account both the improvement of the functional properties of biopolymers and integration with circular economy concepts, sustainable raw material sourcing, innovative polymer processing, and modification technologies (Nanni & Messori, 2021; Bairwan et al., 2024).

19.2 Polyhydroxyalkanoates

Polyhydroxyalkanoates (PHAs) are one of the most promising groups of biopolymers that could revolutionize the packaging industry and a number of other sectors, including biomedicine and tissue engineering. They are linear polyesters of aliphatic hydroxyacids synthesized by various species of bacteria and archaea, accumulated intracellularly as a carbon and energy reserve under conditions of carbon excess and deficiency of essential nutrients such as nitrogen or phosphorus (Bugnicourt et al., 2014; Dehouche et al., 2020; Ibrahim et al., 2021; Zhuikova et al., 2022). The first observations of PHA granules in microbial cells are attributed to Martinus W. Beijerinck in 1888, who, as a pioneer of environmental microbiology, observed cytoplasmic inclusions in bacterial cells (Bugnicourt et al., 2014). However, the real breakthrough came in 1926, when Maurice Lemoigne described polyhydroxybutyrate (PHB) as the first known member of this family identified in the bacterium *Bacillus megaterium* (Sosa-Hernández et al., 2020; Usurelu et al., 2022). In the following decades, research on PHA biosynthesis was mainly academic and exploratory in nature. It was not until the 1970s and 1980s that interest in these biopolymers resurfaced in the context of industrial applications, particularly thanks to the discovery that bacteria such as *Ralstonia eutropha* (now *Cupriavidus necator*) can produce not only PHB but also its copolymers, for instance with 3-hydroxyvalerate (Jin et al., 2023). The current approach focuses on the sustainable production of these materials and their use as substitutes for conventional plastics with a high environmental footprint.

PHAs are a family of biopolymers with significant structural diversity that can be classified according to the length of the monomer side chain. Three main classes are distinguished: (1) short chain (scl-PHA, 3–5 carbon atoms), (2) medium chain (mcl-PHA, 6–14 carbon atoms), and (3) long chain (lcl-PHA, >15 carbon atoms) (Bugnicourt et al., 2014; Xavier et al., 2015; Jin et al., 2023). At the structural level, PHAs can also be divided into homopolymers (made up of one type of monomer, e.g., PHB) and copolymers (containing two or more different monomers), such as PHBV (a copolymer of hydroxybutyrate and hydroxyvalerate), PHBHx (with hydroxyhexanoate), or P3HB3HV3HHx (a complex three-component copolymer) (Ibrahim et al., 2021; Jin et al., 2023; Atarés et al., 2024). The properties of these materials can be modified to a large extent by selecting the appropriate monomer composition, which affects the melting point, crystallinity, flexibility, and biodegradability. PHBV, for instance, has a lower melting point and greater flexibility than PHB, which increases its industrial usefulness and facilitates processing under industrial conditions. With an increase in the share of 3HV in the copolymer, an improvement in mechanical and thermoplastic properties is observed, making PHBV a particularly attractive material for packaging and medical applications (Jin et al., 2023; Atarés et al., 2024). The physicochemical properties of PHAs depend largely on their monomer composition, molecular weight (ranging from 50 to 10,000 kDa), polydispersity (1.1–6.0), degree of crystallinity (0%–70%), as well as glass transition temperature (from −52°C to 4°C), melting point (up to 177°C), and thermal degradation (227°C–256°C) (SOZA-HERNANDEZ, 2020). PHAs can have very different stiffness (elastic modulus from 0.008 to 3500 MPa) and elongation (from 2% to even 1000%), which allows them to be tailored to the requirements of specific applications. PHB, as the best-known member of the PHA family, is characterized by high crystallinity and brittleness, which makes it difficult to process. However, its copolymers, such as PHBV and PHBHx, have a lower melting point, greater flexibility, and a better processing profile. As a result, they are used, among other things, in medicine as materials for tissue scaffolding, drug carriers, surgical sutures, implants, and wound dressings (Kovalcik et al., 2020; Halevas et al., 2021; Ibrahim et al., 2021). In the industrial sector, PHAs are used as biodegradable packaging films, protective coatings, as well as internal components in electronic devices and car parts (Mostafa et al., 2020; Yin et al., 2020). Their advantage over PLA is their ability to biodegrade completely in aquatic environments and the lack of specific composting conditions (Atarés et al., 2024; Zhang et al., 2024). Furthermore, the biodegradation of PHAs does not cause the emission of toxic products or a significant increase in CO_2 concentration in the atmosphere, as it fits into the natural closed carbon cycle (Policastro et al., 2021). The end products of their decomposition are harmless metabolites, including 3-hydroxy acids, which are commonly found in living organisms (Kovalcik et al., 2020). As a result, PHAs not only do not generate waste but can even be considered an "organically recyclable" material. It is worth noting that other bioplastics, such as starch- or corn-based plastics, although formally recognized as biodegradable, have a significant impact on the environment due to the need for intensive agricultural cultivation, the use of artificial fertilizers and pesticides, and competition with the food sector (Bugnicourt et al., 2014; Policastro et al., 2021). PHAs, on the other hand, can be produced from organic waste and by-products, which significantly reduces their environmental footprint and supports the circular economy.

PHAs, due to their biodegradability, biocompatibility, and modifiability, are particularly attractive as an alternative to conventional plastics in the packaging sector. Their applications include single-use packaging, barrier films, biodegradable bags, trays, and disposable tableware. PHBV and its blends with PLA or PBAT have shown satisfactory mechanical and barrier properties, similar to traditional plastics such as PP and PE (Atarés et al., 2024; Vanheusden et al., 2024). From an application point of view, properties such as water vapor and gas (O_2, CO_2) barrier properties, grease resistance, transparency, printability, and heat sealability, as well as thermal stability, are of key importance. For fresh and processed food packaging (e.g., meat, dairy products, fruit), PHAs must also meet food contact standards and have adequate shelf life under refrigerated and high humidity conditions. Current research on PHA nanocomposites and their functional coatings (e.g., with essential oils or metal nanoparticles) focuses on improving their barrier and active properties, enabling the development of active or smart packaging materials (Zhou et al., 2021; Jin et al., 2023).

Despite its numerous advantages, the full implementation of PHA on a large scale in industry faces significant economic and technological barriers. The cost of PHA production is currently much higher than that of petrochemical-based plastics—it is estimated that the price per ton of PHA can be up to three to five times higher than that of conventional polymers, such as PE or PP, mainly due to raw material costs, low biosynthesis yields, and purification costs (Bairwan et al., 2024; Zhang et al., 2024). From a technological point of view, the limited thermal stability of some PHAs (especially PHB), their brittleness, and difficulties in industrial processing are also obstacles. In addition, the lack of widely accepted norms and standards for processing and material characteristics limits the possibilities of integrating PHAs into existing production lines. These requirements necessitate the continuous development of fermentation technologies (including metabolic engineering of production strains), optimization of bioreactors, and the use of alternative substrates and technologies based on the circular bioeconomy (e.g., biorefineries using lignocellulosic waste) (Dhar et al., 2015; Jin et al., 2023). Nevertheless, growing interest in sustainable development and legislative support (e.g., bans on single-use plastics and green public procurement) is creating favorable conditions for overcoming these barriers and enabling wider implementation of PHAs in a circular economy.

Although PHAs are most often mentioned in the context of biodegradable packaging, the application potential of this group of biopolymers goes far beyond the disposable materials sector. Due to their unique properties, such as biocompatibility, nontoxicity, controlled degradation, and chemical modification, PHAs are used in fields such as regenerative medicine, precision agriculture, and modern organic electronics (Kovalcik et al., 2020; Halevas et al., 2021; Zhang et al., 2024). In biomedicine, PHAs are used as tissue scaffolds, drug delivery systems (DDSs), surgical sutures, and materials for implants, stents, and biodegradable sutures (Bittolo Bon et al., 2021; Ibrahim et al., 2021). For instance, PHB and its copolymers (PHBV, P3HB-co-4HB) are used as matrices for bone and nerve engineering, as well as carriers for active compounds in controlled-release systems. Their biodegradability and lack of immune response mean that they do not require surgical removal, which reduces the risk of complications and treatment costs. In the agricultural sector, PHAs are being considered as a base for the production of coatings for slow-release fertilizers, biodegradable pots, and agrofilms that reduce the use of herbicides and pesticides. Furthermore, PHA biodegradation products—3-hydroxy acids—are metabolites naturally present in the soil, making them a material friendly to soil organisms (Zhou et al., 2021). In modern organic electronics and environmental sensor technology, PHAs are being tested as substrates for biodegradable sensors, radio frequency identification (RFID) systems, and electroconductive composites. Their optical transparency, good barrier properties, and ability to be combined with conductive materials (e.g., silver nanoparticles [AgNPs], polyaniline) open the way for the development of biodegradable electronic components (Jin et al., 2023; Bairwan et al., 2024). The growing interest in these nonstandard applications of PHAs is part of a broader trend toward the development of so-called "advanced biopolymers," in which the functionality of the material goes beyond the classic "packaging—biodegradation" framework, offering new functional properties tailored to the needs of advanced technological and medical sectors.

19.3 PHB—poly(3-hydroxybutyrate)

Poly(3-hydroxybutyrate), better known as PHB, is a natural polymer belonging to the PHAs family (Fig. 19.1). This compound was first discovered in 1926 by Lemoigne as a product accumulated inside bacterial cells under conditions of nutrient deficiency, with the exception of carbon (Bugnicourt et al., 2014). As a homopolymer of 3-hydroxybutyrate, PHB exhibits both thermoplastic and biodegradable properties, making it the subject of intensive research in the context of alternatives to petrochemical-based plastics (Díez-Pascual & Díez-Vicente, 2014; Cherpinski et al., 2018). PHB can be produced by various species of microorganisms, which store it as an energy and carbon reserve in the form of intracellular granules (Khattab et al., 2021; Low et al., 2021). Its industrial-scale production is based on fermentation using renewable carbon sources such as glucose, starch, and agricultural and industrial waste (Kontárová et al., 2020; Zhang et al., 2024). PHB synthesis occurs as a result of the condensation of two acetyl-CoA molecules, leading to the formation of acetoacetyl-CoA. This compound is then reduced to hydroxymethyl-CoA, which polymerizes to PHB (Bugnicourt et al.,

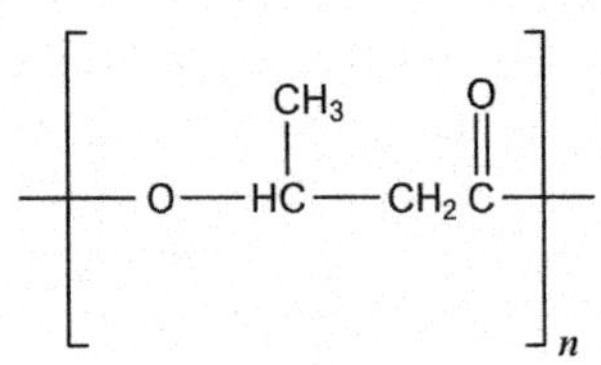

FIGURE 19.1 Chemical structure of poly(3-hydroxybutyrate).

2014; Suhazsini et al., 2020). PHB is an isotactic polyester with regular stereochemistry, consisting exclusively of R-3-hydroxybutyrate units, which gives it high crystallinity and significant thermal resistance (Dariš & Knez, 2020; Mierziak et al., 2021). The molecular weight of PHB can range from about 50,000 to over one million Da, depending on the microorganism used, the culture conditions, and the extraction method (Bugnicourt et al., 2014; Dariš & Knez, 2020). PHB exhibits a number of desirable properties from a materials engineering perspective: high melting point (approximately 175°C), significant elastic modulus (~3 GPa), and tensile strength (30–35 MPa), which brings it close to the properties of some polyolefins, such as polypropylene (Mlalila et al., 2018; Dariš & Knez, 2020). On the other hand, its high crystallinity (>50%–70%) causes brittleness and a narrow processing window, which limits its industrial applications (Díez-Pascual & Díez-Vicente, 2014; Yin et al., 2020). From a chemical point of view, the PHB monomer—3-hydroxybutyric acid—occurs naturally not only in microorganisms but also in higher organisms, where it functions as an energy metabolite and signal molecule. In animal organisms, it can influence gene expression, lipid metabolism, neuron function, and epigenetic activity by inhibiting histone deacetylases (Mierziak et al., 2021). Due to its structure and biological origin, PHB is a biodegradable and biocompatible material. Under natural conditions, it is broken down by enzymes such as lipases and esterases, and the degradation product is 3-hydroxybutyric acid, a nontoxic compound naturally found in human blood (Dariš & Knez, 2020; Sukhanova et al., 2020).

19.4 Physicochemical properties of poly(3-hydroxybutyrate)

As a thermoplastic material, PHB has a number of properties similar to those of classic plastics such as PP and PET, while remaining biodegradable and nontoxic (Mlalila et al., 2018; Melendez-Rodriguez et al., 2019). High crystallinity (>50%–70%) gives the material rigidity and mechanical resistance, but also causes brittleness and limited impact resistance (Atarés et al., 2024). With age, secondary crystallization occurs, leading to even greater stiffening of the material and deterioration of its elasticity (Dariš & Knez, 2020; Yin et al., 2020). Due to its high melting point (approximately 175°C) and susceptibility to thermal degradation already above 180°C, PHB has a very narrow processing window, which makes it difficult to use in conventional processing techniques such as extrusion or injection molding (Díez-Pascual & Díez-Vicente, 2014; Nanni & Messori, 2021). The processing of PHB is also hampered by its low solubility—it dissolves well only in a limited number of organic solvents, such as chloroform or dichloromethane (Dariš & Knez, 2020). To improve mechanical properties and expand the range of processing, structural modifications of PHB are used, such as copolymerization with hydroxybutyrate (3HV), formation of blends with other polymers (PLA, PCL, polyglycolic acid [PGA]), or additives in the form of plasticizers and nanofillers (Panaitescu et al., 2023; Bairwan et al., 2024). PHBV, a copolymer of PHB and 3-hydroxybutyrate, has a lower melting point, greater flexibility, and improved processing properties, making it a more promising material for packaging and biomedical applications (Suhazsini et al., 2020; Yin et al., 2020; Atarés et al., 2024). Importantly, the properties of PHB-based materials are highly dependent on the processing technique—films obtained by solution casting exhibit higher tensile strength and crack resistance than extrusion-molded materials, which is attributed to differences in spherulite morphology and nucleation density (Atarés et al., 2024).

As a biopolymer material, PHB offers a number of significant advantages that make it attractive to various industries. These include, above all, its complete biodegradability under natural conditions (soil, seawater, compost) and excellent biocompatibility, confirmed both in vitro and in vivo (Dariš & Knez, 2020; Sukhanova et al., 2020). The products of its biodegradation are compounds that are harmless to the environment and living organisms, including 3-hydroxybutyrate, which occurs naturally in the human body (Bugnicourt et al., 2014; Mierziak et al., 2021). From a processing point of view, PHB is a material with a high melting point, good mechanical properties, and barrier properties, especially against gases such as oxygen and carbon dioxide (Mlalila et al., 2018; Melendez-Rodriguez et al., 2019). These characteristics, combined with its renewable origin (fermentation with bacteria using, for example, agricultural waste), make it an attractive alternative to conventional plastics such as polyethylene and polypropylene (Kontárová et al., 2020; Zhang et al., 2024). On the other hand, a number of limitations significantly hinder the widespread use of PHB on an industrial scale. High crystallinity leads to excessive stiffness and brittleness, and also significantly narrows the processing window—the material undergoes thermal degradation just above its melting point (Díez-Pascual & Díez-Vicente, 2014). Furthermore, secondary crystallization occurring after processing results in further deterioration of mechanical properties during storage (Dariš & Knez, 2020; Atarés et al., 2024). Another significant problem is the high cost of PHB production. Compared to conventional plastics such as PP or PE, the cost of producing PHB can be up to 8–10 times higher, mainly due to the expensive carbon substrates used in the fermentation process (Khattab et al., 2021; Zhang et al., 2022, 2024). Although it is possible to use cheaper raw materials (e.g., used oils, waste from the agri-food industry), there is still a lack of stable and scalable technologies for the competitive production of PHB (Kontárová et al., 2020; Low et al., 2021).

The limited functionalization possibilities of the material are also significant—in its pure form, PHB is a hydrophobic material with no biological activity, which limits its application, for example, in tissue engineering (Dariš & Knez, 2020). Therefore, intensive work is underway to develop PHB-based blends, copolymers, and composites that eliminate its disadvantages and expand its areas of application (Bairwan et al., 2024; Panaitescu et al., 2023).

19.4.1 Mechanical properties and enhancement strategies

PHB belongs to the group of aliphatic polyesters, which, due to their biological origin and biodegradability, are considered an alternative to conventional plastics. Despite its high tensile strength (30–35 MPa), PHB is characterized by significant brittleness, low elongation at break (EB) (approximately 5%), and a narrow processing temperature range, which significantly limits its use in the packaging and biomedical industries. These limitations are caused by secondary crystallization of the amorphous phase, low nucleation density, and the formation of large spherulites that promote microcracks, as well as the proximity of the glass transition temperature to room temperature (Bugnicourt et al., 2014). In order to improve the mechanical properties of PHB, various modification strategies are used, including the addition of plasticizers, copolymerization, and the formation of composites with cellulose nanofibers and inorganic nanoparticles. The addition of biosynthetic poly(3-hydroxyoctanoate) (PHO) to PHB resulted in an increase in EB of up to 700%, compared to only 185% for the classic plasticizer tributyl 2-acetyl citrate (TAC), while maintaining better thermal stability (Frone et al., 2020). The use of plasticizers not only improves the elasticity of the material, but also affects its processing by lowering the glass transition temperature (T_g) and melting temperature. Lowering the T_g allows PHB to be processed at lower temperatures, which reduces its thermal degradation, particularly important due to the sensitivity of this polymer to temperatures above 180°C. However, plasticizers such as TAC can reduce thermal stability due to their low boiling point and migration within the material. Unlike TAC, the use of polymer modifiers such as PHO allows for the improvement of mechanical and processing properties without migration effects (Bugnicourt et al., 2014; Frone et al., 2020). During the processing of PHB, especially at high temperatures, a series of degradation reactions occur, leading to a significant loss of molecular weight. The most commonly observed degradation mechanism is the accidental breaking of chains through cis-elimination reactions, leading to the formation of low-molecular-weight products such as crotonic acid and its derivatives. This results not only in a reduction in the viscosity of the melt, but also in a shift in the crystallization temperature and an increase in the time required for crystallization to occur. This phenomenon, combined with the tendency of PHB to undergo secondary crystallization, significantly reduces its mechanical stability and makes it difficult to obtain reproducible material properties. The use of appropriate additives, such as lubricants, can partially limit this process, allowing PHB to be processed at 170°C–180°C without excessive degradation (Bugnicourt et al., 2014). The incorporation of nanocellulose (bacterial cellulose [BC], cellulose nanocrystals [CNC]), or metal nanoparticles (palladium nanoparticles [PdNPs], ZnO) results in a significant increase in elastic modulus and tensile strength, attributed to good interfacial adhesion resulting from hydrogen bonding and homogeneous dispersion in the PHB matrix (Cherpinski et al., 2018; Díez-Pascual & Díez-Vicente, 2014). In studies by Painaitescu et al., an increase in tensile strength of 28% was observed after the addition of 2% BC, and after additional plasma treatment—as much as 54% (Panaitescu et al., 2023). In turn, Albuquerque et al. showed that PHB/BC composites are characterized by significantly higher tensile strength compared to pure PHB, which was attributed to the effective network of cellulose microstructures (Albuquerque et al., 2021). It is worth noting that the processing technique used has a significant impact on the final mechanical properties of PHB and its composites. PHBV composites with cellulose nanofibers prepared by solution casting showed no significant changes in mechanical parameters, while the use of melt blending allowed for a clear improvement in tensile strength and elastic modulus. These differences result from the quality of nanofiber dispersion and the degree of their interaction with the matrix, which largely determine the efficiency of interphase stress transfer (Panaitescu et al., 2023). The addition of CNCs led to an increase in Young's modulus and tensile strength only when they were well dispersed in the polymer matrix. High concentrations of CNCs (above 5%) caused agglomeration, acting as stress concentrators, which reduced the mechanical parameters and EB (Usurelu et al., 2022). In the case of PHBV/CNCs composites, the reinforcing effect was visible up to a concentration of 2.3%, after which a decrease in strength was observed due to the formation of agglomerates and poor interphase adhesion. In the context of the deformation mechanics of PHB materials, it is also worth considering data from dynamic mechanical-thermal analysis (DMTA). These analyses revealed two characteristic relaxation processes: the first, occurring around the glass transition temperature (T_g), is associated with the movement of chain segments in the amorphous phase, while the second, at higher temperatures, reflects the mobility of segments between the crystalline and amorphous phases. The location and intensity of these peaks are strongly dependent on the thickness of the crystalline lamellae and the content of the rigid amorphous fraction (RAF), which makes DMTA a valuable diagnostic tool in the assessment of internal structural changes in PHB (Bugnicourt et al., 2014). In the analysis of the influence of different types of nanocellulose on the mechanical properties of PHB, different mechanisms of action of BC and CNCs are clearly outlined. BC, thanks to its three-dimensional microstructure network, acts as an effective reinforcing material, increasing strength without significantly reducing elasticity. In contrast, CNCs, as rigid and short crystalline structures, reinforce the material mainly by limiting the mobility of PHB segments and increasing the level of crystallinity. Their effectiveness largely depends on the quality of dispersion and chemical compatibility with the polymer matrix. Agglomeration of CNCs leads to stress concentration points and deterioration of mechanical properties, which is not observed in BC, whose network structure promotes even load distribution in the composite (Albuquerque et al., 2021; Usurelu et al., 2022). An important aspect in the context of the mechanical properties of PHB is the presence of the RAF, which limits the movement of chain segments and slows down the further development of crystallites. The presence of RAF has a significant impact on the formation of the spherulitic structure, and its amount depends on the molecular weight of PHB and the cooling conditions. Lowering the T_g by using plasticizers or controlling the crystallization process allows for increasing the elasticity of the material and extending the breaking length (Bugnicourt et al., 2014). In summary, despite its inherent brittleness and thermal limitations, the mechanical properties of PHB can be significantly improved by the use of natural modifiers, nanofillers, and techniques such as melt blending or surface activation. A key factor influencing the final result is the chemical compatibility of the components and their uniform dispersion in the polymer matrix, as confirmed by numerous experimental studies (Frone et al., 2020; Usurelu et al., 2022; Albuquerque et al., 2021; Panaitescu et al., 2023).

19.4.2 Crystallinity and modification approaches

The crystallinity of PHB is one of the pivotal parameters determining its processing properties, mechanical properties, and biodegradability (D'Amico et al., 2012; Laycock et al., 2013; Khasanah et al., 2017). It has been shown that the addition of nanofillers, such as ZnO nanoparticles, significantly affects the rate and mode of PHB crystallization. Díez-Pascual and Díez-Vicente observed an acceleration of the PHB crystallization process in the presence of ZnO, which results from the heterogeneous nucleation effect. In nanocomposites, the crystallization peak was sharpened, which clearly indicates the effective action of ZnO as a nucleating agent. The crystallinity level (X_c) increased from about 53% for the pure matrix to about 63% at a ZnO content of 5.0%, with no significant changes occurring with further increases in filler concentration. This mechanism is most likely related to the presence of numerous nucleation centers that form on the surface of ZnO nanoparticles, promoting the initiation of crystallization. Thanks to its high surface-to-volume ratio and potential interfacial interactions with polymer chains, ZnO can effectively lower the local energy barrier necessary for the formation of crystal nuclei (Díez-Pascual & Díez-Vicente, 2014). An interesting structural modification of PHB is its blending with other biopolymers, especially PLA. In studies on PLA-PHB-based electrospun nanocomposite fibers with the addition of CNCs, Arrieta et al. showed that the addition of CNC in amounts of 1%–5% improves both the thermal and mechanical resistance of the materials. Although the main objective of this study was to improve flexibility through the use of a plasticizer, acetyl tributyl citrate (ATBC), this modification also affected the preservation of the crystalline phase and degradation under composting conditions. By lowering the glass transition temperature (T_g), the ATBC plasticizer increased the mobility of polymer chain segments, which may promote the formation of ordered structures and support crystal growth under conditions conducive to crystallization. In turn, CNC, thanks to its rigidity and high specific surface area, can act as additional nucleation centers, enhancing the effect of heterogeneous nucleation. The synergistic action of ATBC and CNC therefore contributed to improving the performance parameters without simultaneously impairing the biodegradability of the material (Arrieta et al., 2014). Khasanah et al., on the other hand, attempted to gain an in-depth understanding of the crystalline behavior of PHB in confined spaces by analyzing ultra-thin PHB/poly(L-lactic acid) (PLLA) films with different thicknesses and molecular weights of the components (Khasanah et al., 2017). In films with a thickness of 30 nm, the presence of PLLA with an average molecular weight (Mw 13,100–23,000 g mol^{-1}) clearly inhibited PHB crystallization, which was not observed with PLLA with a higher molecular weight (Mw ≥ 50,000 g mol^{-1}). In films with a thickness of 13 nm, even stronger crystallization suppression was observed, regardless of the molecular weight of PLLA, confirming the important role of the confinement effect at the nanoscale. Reducing the film thickness led to a decrease in the mobility of polymer chain segments, which disrupted the reorganization necessary for crystal growth. In addition, the significant surface-to-volume ratio intensified the influence of surface physicochemical effects, such as adsorption or restriction of free volume for nucleation, which further hindered the development of the crystalline phase. Furthermore, differences in the crystal structure of PHB in the presence of PLLA were confirmed, showing that the crystal phase structure underwent significant modifications depending on the degree of order and the film thickness used. The authors suggest that phase separation at the free surface, chain entanglement, and PLLA particle size play a crucial role in this phenomenon. The authors also emphasized that mixtures of PHB with high-molecular-weight PLLA (Mw > 18,000 g mol^{-1}) exhibited phase separation, while low-molecular-weight PLLA was completely miscible with PHB in the molten state. Furthermore, the crystallization dynamics of both components and their mutual concentrations had a key influence on the final crystalline properties of the obtained materials (Khasanah et al., 2017). In summary, the crystallinity of PHB is influenced by a number of structural and environmental factors, including the presence of fillers (Díez-Pascual & Díez-Vicente, 2014), plasticizers, and reinforcing additives (Arrieta et al., 2014), as well as spatial constraints and compatibility of components in polymer blends (Khasanah et al., 2017). Understanding the mechanisms of crystallinity control is fundamental to the design of new PHB-based composite materials with optimized application properties (D'Amico et al., 2012).

19.4.3 Thermal properties and effects of additives

The thermal properties of PHB are crucial for both its processing and stability in end applications. As a highly crystalline polymer (crystallinity 60%–70%), PHB is characterized by a fast crystallization rate in the temperature range of 80°C–100°C, while below 60°C and above 130°C, this process slows down significantly, leading to an amorphous and viscous structure that persists for many hours. Due to the distinct liquid/solid phase transition, PHB can be processed very quickly, similar to polypropylene, with a typical melt flow index (MFI) of 30–40 g 10 minutes^{-1} (Bugnicourt et al., 2014). PHB contains not only surface water but also water bound structurally in the granules, which makes it susceptible to undesirable hydrolysis reactions during thermal processing. These reactions can lead to chain breakage and a significant reduction in molecular weight. Therefore, drying PHB prior to processing is essential—the best results are achieved using jet dryers at 80°C for at least 2 hours. Importantly, the granules regain moisture within a short time (approximately 30 minutes) after removal from the dryer, which emphasizes the need for immediate processing after drying (Bugnicourt et al., 2014). From the point of view of thermal degradation, PHB exhibits thermal instability in the range of 160°C–200°C, even before reaching its melting point. The main degradation mechanism is accidental chain breaking (cis-elimination), leading to the formation of terminal carboxyl and propenyl groups. These reactions occur through an ester decomposition mechanism with the temporary formation of a six-membered ring. In the initial phase, esterification of the end groups may also occur, which temporarily increases the molecular weight, but further degradation results in the formation of volatile oligomers and a decrease in Mw (Nanni & Messori, 2021). Thermogravimetric analysis (TGA) showed that the thermal decomposition of PHB occurs in one main stage, with a significant loss of mass in the range of 230°C–280°C. The presence of additives such as cellulose nanofibers or clove oil introduces further stages of decomposition—at approximately 330°C (cellulose degradation) and 121°C–188°C (eugenol oxidation), respectively. The presence of micro- and nanocomponents may affect thermal stability through mechanisms related to the limitation of chain mobility and barrier effects. In addition, ZnO as a filler not only exhibits a barrier effect but also high thermal conductivity (approximately 50 W mK^{-1}), which promotes even heat dissipation and prevents local areas of the material from overheating (Díez-Pascual & Díez-Vicente, 2014; Panaitescu et al., 2023). The addition of ZnO nanoparticles significantly improved the thermal resistance of the PHB matrix, which is attributed to the barrier effect that hinders the transport of decomposition products from the interior of the material to the gas phase. In addition, the high thermal conductivity of ZnO promotes uniform heat dissipation in the material, which also improves thermal stability (Díez-Pascual & Díez-Vicente, 2014). Plasma surface treatment also affects the thermal stability of the material. As a result of the action of reactive molecular forms, such as radicals and ions, chemical modification of the surface and an increase in the degradation temperature by up to 10°C occur. Panaitescu et al. noted that nanocomposites containing BC after plasma treatment show an increase in crystallization temperature (Tc) and a decrease in crystallinity, which may be the result of faster material aging and the formation of new functional chemical groups (Panaitescu et al., 2023). In turn, Perret et al. investigated the effect of thermal and mechanical stress on the properties of PHB fibers (Perret et al., 2020). Low-temperature stress relief (100°C–130°C) allowed the elasticity and viscoelastic properties of the material to be maintained, while stress relief at high stress led to an increase in mechanical strength and a decrease in EB. The presence of an oriented noncrystalline mesophase was observed, which developed between α crystallites during high-stress relaxation and was responsible for structural changes in the material. This mesophase could be restored by re-stretching, indicating the reversible nature of phase transformations occurring under the influence of temperature and stress. The ability of PHB to form and restore this phase demonstrates its susceptibility to dynamic regulation of mechanical and thermal properties already at the processing stage (Perret et al., 2020). In summary, the thermal properties of PHB depend on its chemical structure, the presence of additives, processing methods, and aging conditions. The key phenomena are: low thermal stability near the melting point (Nanni & Messori, 2021), the possibility of modifying Tc (Panaitescu et al., 2023), the barrier effect and thermal conductivity of fillers (Díez-Pascual & Díez-Vicente, 2014), the importance of drying before processing (Bugnicourt et al., 2014), and the influence of the mesocrystalline phase on mechanical and thermal behavior (Perret et al., 2020).

19.4.4 Gas and aroma barrier performance

Due to its high crystallinity, stereoregularity, and hydrophobic nature, PHB exhibits favorable barrier properties compared to other biopolymers used in the packaging industry (Koller, 2014; Usurelu et al., 2022). Its ability to limit the penetration of water vapor, oxygen, and volatile compounds is crucial in the context of packaging food that is sensitive to moisture and oxidation. The barrier properties of PHB are among the most important features determining its use in food packaging with extended shelf life. PHB exhibits excellent resistance to the permeation of certain gases and volatile substances, and its barrier properties can be modified over a wide range by using appropriate functional additives. Its barrier properties vary significantly depending on the compound. In the case of carbon dioxide, PHB has exceptionally low permeability, comparable to and, under certain conditions (4 atm), even better than PET (Koller, 2014). With regard to oxygen, pure PHB exhibits moderate barrier properties, which are significantly improved by additives such as ZnO or CNCs—for instance, 3% CNCs reduce the oxygen transmission rate threefold (Díez-Pascual & Díez-Vicente, 2014; Usurelu et al., 2022). On the other hand, PHB has a relatively high water vapor permeability (WVP) in relation to H_2O (water vapor), which can be problematic when packaging bakery products, but the addition of CNCs (e.g., 2%–3%) results in a threefold decrease in this value (Usurelu et al., 2022). When it comes to aromatic compounds (e.g., limonene), PHB exhibits higher barrier properties than PET, which is important for packaging products with an intense odor profile (Koller, 2014). In the case of organic solvents, however, the barrier properties depend on polarity: PHB insulates well against n-hexane and methanol, but is more permeable to compounds with similar polarity, such as acetone or toluene (Miguel & Iruin, 1999). Additives in the form of nanofillers, such as ZnO nanoparticles and CNCs, significantly improve barrier properties by creating a structure with increased tortuosity, limiting the diffusion of gases and water vapor (Díez-Pascual & Díez-Vicente, 2014; Usurelu et al., 2022). A 5 wt.% content of ZnO in the PHB composite led to a reduction in oxygen permeability of approximately 53%, which was attributed to both the good dispersion of the nanoparticles and their adhesion to the polyester matrix, as well as an increase in the crystallinity of the material. Similar phenomena were observed for nanoclay fillers, where the optimal barrier effect also occurred at a content of 5% (Díez-Pascual & Díez-Vicente, 2014). In the context of component migration into food, studies indicate that at low CNC concentrations (1%–2%), a reduction in total migration was observed, which resulted from good dispersion of the filler and strong interphase interactions limiting the mobility of polymer chains (Usurelu et al., 2022). However, higher CNC contents (≥3%) led to their agglomeration and deterioration of adhesion, resulting in an increase in migration levels—nevertheless, these values remained well below EU limits. The water vapor barrier properties are also improved by the use of CNCs. A threefold decrease in permeability was observed at a CNC content of 3%, which was due to the physical blocking of diffusion channels by well-distributed nanoparticles (Usurelu et al., 2022). Interestingly, further increases in the amount of CNCs did not bring additional benefits, which was explained by the formation of diffusion microspaces around agglomerates and a change in the porosity of the film (Cherpinski et al., 2018; Usurelu et al., 2022). Other studies have shown that the addition of 5% ZnO to the PHB matrix reduced WVP by 38% and water absorption by 66%, which was attributed to a more ordered crystal structure and a reduction in the amorphous phase. However, at a concentration of 10%, a deterioration of the effect was observed, associated with the formation of ZnO clusters creating preferential transport pathways for water and gases (Díez-Pascual & Díez-Vicente, 2014). PHB also exhibits excellent barrier properties against aromatic compounds such as limonene, achieving lower permeability values than PET (0.088×10^{-13} vs 1.17×10^{-13} $kg{\cdot}m{\cdot}s^{-1}{\cdot}m^{-2}{\cdot}Pa^{-1}$) (Sanchez-Garcia et al., 2007; Koller, 2014). Similarly, its effectiveness against water vapor and volatile solvents depends on their polar affinity to the matrix—compounds with extreme polarities were effectively blocked, while compounds with similar polarities (e.g., acetone, toluene) easily penetrated it (Miguel & Iruin, 1999). However, it should be emphasized that in some cases, the introduction of metallic fillers, such as PdNPs, may impair the barrier properties. In one study, PHB films containing PdNPs had more than twice the WVP value compared to pure PHB, which was attributed to the presence of unbound interfacial zones and agglomerates forming channels for gas transport (Cherpinski et al., 2018).

19.5 Poly(3-hydroxybutyrate)-based materials

Due to the limitations associated with the mechanical and thermal properties of PHB, numerous studies focus on modifying this polymer to improve its functionality. One of the most commonly used methods is the creation of copolymers, for example, with 3-hydroxybutyrate and 3-hydroxyvalerate (PHBV), which leads to a reduction in crystallinity, increased flexibility, and a lower melting point (Melendez-Rodriguez et al., 2019; Yin et al., 2020). Another strategy is to combine PHB with polymers such as PLA and PCL, which allows for the production of materials with diverse physicochemical properties (Sukhanova et al., 2020). The use of plasticizers such as propionic acid allows for the production of polymers with greater elongation and fracture resistance by changing the monomer ratio and creating PHBV with the desired property profile (Suhazsini et al., 2020). Another approach is to add nanofillers such as nanocellulose, layered silicates, or nano-carbon structures, which increase the mechanical strength, thermal stability, and barrier properties of PHB (Díez-Pascual, 2021; Panaitescu et al., 2023). Surface modifications of PHB, such as plasma treatment or combination with bioactive substances (e.g., essential oils, antimicrobial compounds), allow for its wider application in medicine and food packaging (Melendez-Rodriguez et al., 2019). Despite significant progress in modification, many challenges remain unresolved, including difficulties in thermal processing, the cost of raw materials, and the limited compatibility of PHB with other polymers. Nevertheless, the development of biorefinery technologies, advanced composites, and material engineering methods indicates the enormous potential for further adaptation of PHB in the industrial sector. Due to the significant limitations of pure PHB, such as high brittleness, low impact resistance, a narrow processing window, and low thermal stability, strategies for its modification through the creation of blends and composites are being intensively developed. The aim of these activities is to obtain materials with improved mechanical, thermal, barrier, and processing properties while maintaining their biodegradability and biocompatibility (Panaitescu et al., 2023; Popa et al., 2022).

19.6 Poly(3-hydroxybutyrate) blends and composite materials

Blending PHB with other biodegradable polymers, such as PLA, polybutylene succinate (PBS), PCL, or mcl-PHA, is one of the most commonly used methods to improve its performance properties (Laycock et al., 2013; Garcia-Garcia et al., 2017; Popa et al., 2022). Numerous studies have shown that the addition of PLA can increase oxygen barrier properties, stiffness, and thermal stability, although it does not completely eliminate the brittleness of PHB due to its own low EB (Arrieta et al., 2014; Iglesias-Montes et al., 2022; Patel et al., 2022; Pietrosanto et al., 2021). Better results are obtained with mcl-PHA, which acts as elasticity modifiers, lowers the glass transition temperature, and increases EB while maintaining the biodegradability of the material (Frone et al., 2020; Panaitescu et al., 2023). Another popular method is the use of low-molecular-weight or polymer plasticizers. Among the most commonly used are ATBC, glycerol and its derivatives, polyethylene glycol (PEG), triacetin, and tributyl citrate (TBC) (Frone et al., 2020; Kontárová et al., 2020; Quispe et al., 2021). They work by increasing the mobility of PHB chain segments, which leads to a reduction in the melting and glass transition temperatures and an improvement in EB. However, the effectiveness of many of them is limited by their incomplete compatibility with the PHB matrix and their tendency to migrate to the surface (Frone et al., 2020; Quispe et al., 2021). Among the numerous studies on improving the processing and mechanical properties of PHB, the evaluation of the effectiveness of low-molecular-weight plasticizers occupies a special place. Their main purpose is to reduce the brittleness of the material and extend the processing temperature range. The most commonly tested plasticizers include ATBC, triacetin (TA), TBC, tri(ethylene glycol)-bis(2-ethylhexyl) citrate, PEG400, PEG1500, and glycerol and its esters (Frone et al., 2020; Kontárová et al., 2020; Quispe et al., 2021). In the work of Quispe et al., triacetin and TBC showed a clear plasticizing effect, lowering the glass transition temperature (Tg) and increasing the deformation at break. The best mechanical properties were exhibited by samples with 30% glycerol tributyrate (GTB) added, but for pure glycerol and polyglycerols, a pro-degradation effect and migration of additives to the surface were observed (Quispe et al., 2021). Similar phenomena have also been described in the context of PEG—PEG1500 did not form a stable mixture with PHB, and its presence over a longer period of time led to precipitation from the matrix and loss of the plasticizing effect (Kontárová et al., 2020). Interesting conclusions are provided by studies comparing the effectiveness of classic low-molecular-weight plasticizers with their polymer counterparts. Frone et al. showed that biosynthesized poly(3-hydroxyoctanoate) (PHO), as a polymer modifier, is significantly more effective than traditional TAC. Although both additives lower the glass transition temperature of PHB, this effect was much more pronounced with PHO. Furthermore, EB increased by 700% for PHO, compared to only 185% for TAC. In addition, the use of X-ray photoelectron spectroscopy (XPS) and atomic force microscopy (AFM) techniques revealed a clear migration of TAC to the surface of the PHB film, which may negatively affect the stability of the material's properties over time. PHO did not exhibit this phenomenon, confirming its better compatibility with the matrix and more uniform distribution in the material structure (Frone et al., 2020). As a result, polymer modifiers, despite their higher synthesis costs, can offer significantly greater functional efficiency and durability in applications where mechanical and processing stability is crucial.

In the field of composites, natural fillers and nanofillers such as lignocellulosic fibers, nanocellulose, lignin, clays (MMT, talc), graphene and its derivatives, or carbon nanotubes are gaining increasing attention (D'Amico et al., 2012; Dan et al., 2018; Lugoloobi et al., 2020; Manikandan et al., 2020; Tanideh et al., 2020; Díez-Pascual, 2021). In the context of sustainable development and the circular economy, PHB composites containing fillers derived from biowaste and agricultural waste are attracting increasing interest. These fillers not only reduce the cost of the final material but also enable the reuse of waste raw materials, contributing to a reduction in the environmental footprint. An example is the use of agave fibers from the tequila industry as PHB reinforcement—their addition contributed to an increase in flexural strength, impact resistance, and heat deflection temperature (HDT) while reducing crystallinity (Smith et al., 2020). Similarly, lignin obtained in the form of nanoparticles and evenly dispersed in a PHB matrix showed reinforcing properties, increasing tensile strength by 13.2% and elastic modulus by 43.9%, while providing UV protection and improving thermal resistance. On the other hand, due to the presence of phenolic groups, in addition to its nucleating effect, it significantly reduced the transmission of UV and visible radiation, making PHB a potential material for light-protective applications (Lugoloobi et al., 2020). In turn, additives in the form of grape pomace or polyphenol extracts also had a positive effect on the thermal stability and microbial resistance of the material while maintaining its biodegradability (Nanni & Messori, 2021). The use of these cheap and widely available fillers can not only improve the properties of PHB-based materials but also enable the development of eco-friendly active packaging with antimicrobial and antioxidant properties. Nanocellulose, both in the form of nanofibers and nanocrystals, increases stiffness, improves barrier properties, and can act as a nucleating agent, accelerating the crystallization of PHB (Ruka et al., 2014; Dhar et al., 2015; Usurelu et al., 2022). However, its effectiveness is limited by its hydrophilic nature, which weakens compatibility with the hydrophobic PHB matrix, requiring surface treatment (TEMPO, silanes, compatible polymers) (Usurelu et al., 2022). In addition to improving mechanical and thermal parameters, the development of PHB composites also enables surface functionalization, giving materials new physicochemical properties. The use of additives such as graphene nanoparticles, carbon nanotubes, or lignin allows for the modification of electrical conductivity and UV resistance, as well as expanding the range of applications, including photoactive packaging and electronic materials (Dan et al., 2018; Lugoloobi et al., 2020; Manikandan et al., 2020).

The addition of reduced graphene oxide resulted in improved elastic modulus, tensile strength, and the emergence of electrical conductivity (Dan et al., 2018). Carbon nanotubes, depending on their diameter and degree of functionalization, improved not only the conductivity but also the stiffness and thermal resistance of composites. A common feature of these nanofillers is their ability to form ordered structures within the PHB matrix, which influences networked structure and extends the processing window of the material (Lugoloobi et al., 2020). Research on active composites indicates the possibility of functionalizing PHB mixtures with essential oils (e.g., clove oil), which improves their antimicrobial and thermal properties and increases resistance to microbial degradation (Albuquerque et al., 2021; Quispe et al., 2021). The use of additives such as grape pomace extracts also offers a thermally stabilizing effect and strengthens the PHB matrix while maintaining the ecological character of the material (Nanni & Messori, 2021). In recent years, there has been growing interest in PHB composites containing natural bioactive substances that act as antimicrobial and antioxidant additives. Their use not only extends the functionality of packaging materials but also fits in with the concept of "active packaging." Albuquerque et al. reported the development of a composite consisting of BC and PHB with the addition of clove oil, which improved mechanical properties (3.9-fold increase in tensile strength), resistance to microbial degradation (65% reduction in microbial growth), and thermal degradation (10°C increase in degradation temperature) (Albuquerque et al., 2021). In turn, Nanni et al. investigated the effect of additives such as lemon oil, cinnamon oil, and grape pomace extract. Although some of these compositions showed improved elasticity (e.g., EB increased from 2.5% to 6.5% after adding 30% clove oil), a deterioration in barrier properties (e.g., a threefold increase in WVP) was also observed due to the limited compatibility between the hydrophobic PHB matrix and the hydrophilic bioactive components. In addition, the use of essential oils such as cinnamon and citronella showed a positive effect on the elasticity of the material (increase in EB to 6.95%), but this was often at the expense of barrier properties, including increased WVP. The addition of 30% clove oil to the PHB/PEG mixture resulted in improved elasticity but also weakened the compatibility of the components of the system. Despite these limitations, such compositions can play an important role in the development of active, biodegradable food packaging with antibacterial and antioxidant properties, which improves their antimicrobial and thermal properties and increases resistance to microbial degradation. The use of additives such as grape pomace extracts also offers a thermally stabilizing effect and strengthens the PHB matrix while maintaining the ecological character of the material (Nanni & Messori, 2021).

Nevertheless, such solutions offer significant advantages in the context of the development of new-generation packaging materials—biodegradable, functional, and, at the same time, actively protecting the contents from undesirable environmental factors. In the context of tissue engineering applications, PHB composites with chitosan and hydroxyapatite show great promise. Increased hydrophilicity, the presence of free amino groups, and improved bioactivity enable the development of biostructures that support cell growth and proliferation (Sukhanova et al., 2020; Zhuikova et al., 2022). PHB composites with chitosan and hydroxyapatite are of particular importance in biomedical applications, as they allow the production of structures with increased bioactivity and tissue compatibility. Chitosan, as a natural polysaccharide with free amino groups, increases the hydrophilicity of the material and improves its ability to bind cells and drug carriers. The reduction of the contact angle from 84 to 42 degrees and the increase in compressive strength from 0.01 to 0.19 MPa after combining PHB with chitosan and nano-bioglass/titanium oxide (TiO_2) demonstrate the synergy of these components in the creation of scaffolds for tissue engineering (Zhuikova et al., 2022). In addition, the presence of hydroxyapatite in PHB/chitosan composites increases osteoinductivity and enables integration with natural bone tissue. The use of carbon nanotubes in such systems contributes to improved thermal and mechanical stability, paving the way for the production of scaffolds with optimized strength and susceptibility to surface modifications (Sukhanova et al., 2020; Zhuikova et al., 2022). Thus, these composites represent a promising platform for the development of smart biomaterials for applications in regenerative medicine.

The dynamic development of biodegradable material processing technologies creates new opportunities for specialized applications based on PHB blends and composites. One example is systems designed for 3D printing—Kontárová et al. developed PHB blends with PLA and selected plasticizers (including acetyl tris(2-ethylhexyl) citrate), which exhibited better mechanical and thermal properties compared to pure PHB–PLA blends. At the same time, the use of plasticizers slightly deteriorated the processing properties in layer printing technology, which requires further optimization (Kontárová et al., 2020). In another study, Arrieta et al. produced electrospun nonwovens from PLA/PHB blends with the addition of lactic acid oligomers and CNCs. The resulting mats were characterized by increased crystallinity, improved elongation, and enhanced barrier properties, making them attractive for applications in packaging or as biomedical scaffolds. In addition, some PHB composites exhibit UV-blocking properties, for example, through the use of lignin or cellulose nanofillers, which reduce light transmission and provide protection against photo-induced oxidation (Dhar et al., 2015; Lugoloobi et al., 2020). Such properties are desirable in the design of smart and active packaging materials, particularly in the context of packaging light-sensitive products. In addition to mechanical and functional properties, an important parameter determining the suitability of PHB composites is their biodegradability, especially in the context of diverse degradation environments. Studies have shown that the formation of immiscible blends, such as PHB with poly (ethylene adipate) or PHBV, leads to an accelerated degradation rate compared to pure components (Bugnicourt et al., 2014; Usurelu et al., 2022). This effect is explained by the presence of distinct phase boundaries in the material structure, which facilitate the penetration of water and microorganisms into the composite, initiating degradation at multiple points simultaneously. A similar phenomenon has been observed in PHB composites with natural fibers such as wood or lignin, where the porosity and structural diversity of the composite promote moisture absorption and facilitate enzymatic processes. Furthermore, the presence of natural components can attract soil microbiota, increasing the intensity of biological colonization of the material surface (Atarés et al., 2024). In this way, appropriate phase structure engineering can serve as a conscious strategy for designing PHB composites with increased biodegradation rate and efficiency. In summary, strategies for blending PHB with other polymers, the use of plasticizers, and the creation of composites with natural or functional fillers enable significant improvements in material properties and expand the range of applications—from packaging, through technical textiles, to regenerative medicine (Panaitescu et al., 2018; Briassoulis et al., 2021; Popa et al., 2022; Usurelu et al., 2022).

19.7 Poly(3-hydroxybutyrate)-based materials and their applications

The main area of application for PHB is the packaging industry, where PHB exhibits excellent barrier properties against oxygen and carbon dioxide, surpassing PET and polyvinyl chloride (PVC) in this respect (Sanchez-Garcia et al., 2007; Koller, 2014). Its use in food packaging extends product shelf life and reduces environmental impact, especially for fresh and perishable foods. In the medical industry, PHB has found application as a material for surgical sutures, orthopedic implants, tissue engineering scaffolds, and controlled DDSs (Zhuikova et al., 2022). Biocompatibility and complete biodegradability without leaving harmful metabolites make PHB attractive for biomedical applications (Dariš & Knez, 2020; Sukhanova et al., 2020). Another area of application is agriculture and horticulture—mulching films, seedling covers, biodegradable pots, and fertilizer capsules—where the degradation of PHB after use does not cause waste accumulation and does not interfere with the crop cycle (Low et al., 2021; Panaitescu et al., 2023). In addition, PHB is used as a matrix for encapsulating bioactive plant compounds, essential oils, and extracts with antimicrobial properties, which allows for the design of active packaging materials that support microbiological safety and extend the shelf life of food products (Díez-Pascual & Díez-Vicente, 2014; Melendez-Rodriguez et al., 2019). Fig. 19.2 presents representative applications of PHB-based films and coatings in various fields. Despite its numerous advantages, the mass application of PHB remains limited due to high production costs and a narrow processing window. However, with advances in production technology and growing legislative and social pressure to replace petrochemicals, the role of PHB in the materials economy is gradually increasing, and its applications will expand steadily in the coming decades.

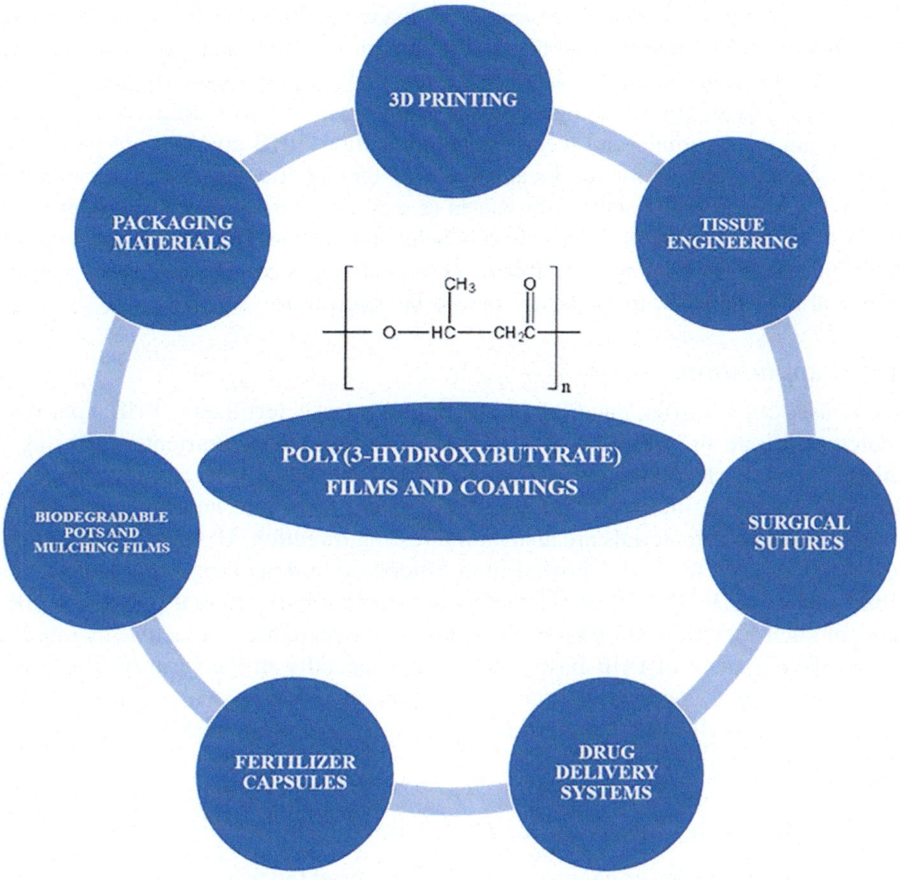

FIGURE 19.2 Selected applications of poly(3-hydroxybutyrate)-based films and coatings.

19.7.1 Poly(3-hydroxybutyrate) materials for biomedical engineering, drug delivery, and agricultural use

19.7.1.1 Biomedical engineering

PHB is increasingly used in advanced biomedical applications, DDSs, controlled fertilization, and photodegradable materials. Its biodegradability, biocompatibility, noncytotoxicity, and ability to be processed into films and coatings make it an attractive material for applications requiring precise adjustment of physicochemical properties to a specific biological or environmental context (Chang et al., 2018; Toalá et al., 2023). In the field of tissue engineering, PHB shows potential as a material for scaffolds for tissue regeneration, including skin, cartilage, bone, liver, blood vessels, and nerves (Kang et al., 2020; Mohammadalizadeh et al., 2020; Volkov et al., 2020). The challenges associated with its low hydrophilicity and bioactivity are being addressed through surface modifications, bioceramic additives (e.g., hydroxyapatite, bioglass), and by creating composites with chitosan, keratin, carbon fibers, or nanocellulose (Naderi et al., 2020; Parvizifard et al., 2020; Zhuikova et al., 2022). Plasma-treated PHB films and coatings show improved stem cell adhesion and cell growth due to surface modification or chemical functionalization, which increases hydrophilicity and the presence of functional groups, translating into better biointeractive properties (Chang et al., 2018), while fibers obtained by electrospinning allow the reproduction of the extracellular matrix structure, thus increasing their attractiveness for tissue regeneration (Korina et al., 2013; Toalá et al., 2023). It is worth noting that PHB films and coatings are also used as surfaces to support cell adhesion, differentiation, and migration. Furthermore, the use of multilayer coatings based on PHB, including the addition of ceramic materials or bioactive components, allows for the achievement of gradients of stiffness and bioactivity, which is particularly important in the regeneration of complex tissues such as tendon attachments or articular cartilage (Goonoo et al., 2017).

19.7.1.2 Drug delivery systems

In the area of DDSs, PHB and its copolymers, including PHBV, are used as matrices in the form of micro- and nanoparticles, capsules, and films for the controlled release of active substances, including anticancer drugs (doxorubicin, sorafenib), antiinflammatory compounds, and other bioactive substances (Babos et al., 2020; Sosa-Hernández et al., 2020; Halevas et al., 2021). Thanks to their bioresorbability and modifiability, PHB nanoparticles can be eliminated naturally through metabolism or filtration, and their degradation products, such as 3-hydroxybutyrate, are compatible with human metabolism (Murueva et al., 2013). PHB coatings also have potential as smart drug carriers, enabling selective drug release, response to environmental stimuli, and local therapeutic action. PHB coatings are particularly interesting when used as bioactive barriers or membranes for the local release of therapeutic substances, for example, at sites of tissue damage or implantation. By incorporating active substances (e.g., curcumin, antibiotics, anticancer compounds) into the film structure, a long-lasting, targeted therapeutic effect is achieved without the need for systemic drug administration (Tanideh et al., 2020). The ability to adjust the film thickness, porosity, and composition allows for the design of coatings tailored to specific clinical requirements, from release rate to bioresorption.

19.7.1.3 Agricultural applications

In agriculture, PHB is used as a carrier material in controlled-release fertilizers. PHB coatings that coat fertilizer granules, such as calcium nitrate or struvite, allow for the slow release of nutrients, reducing nitrogen losses and improving fertilization efficiency (Kontárová et al., 2022). PHB composites with cellulose nanofibers and starch have also been developed, which exhibit increased biodegradability and favorable nutrient release kinetics (França et al., 2021). PHB-based photodegradable materials are also an interesting direction. Hybrid films containing cobalt oxide and cobalt ferrite nanoparticles in a PHB matrix showed high efficiency in degrading organic dyes under sunlight, which opens up the possibility of using PHB in wastewater treatment and environmental protection (Bernal-Sánchez et al., 2024). The film form of such materials allows for their easy implementation as active membranes or biodegradable filters. In summary, the development of PHB-based materials, especially in the form of films and coatings, opens up wide possibilities in regenerative medicine, targeted therapy, agriculture, and environmental protection. Multi-directional modification strategies allow its natural limitations to be overcome, making PHB one of the most promising biopolymers of the new generation of materials.

19.7.2 Poly(3-hydroxybutyrate)-based materials for packaging applications

Food packaging plays a pivotal role in ensuring microbiological safety, extending product shelf life, and protecting against adverse environmental factors such as UV light, oxygen, moisture, and microbial and chemical contaminants (Díez-Pascual & Díez-Vicente, 2014; Bairwan et al., 2024). In an era of progressive environmental degradation caused by the excessive use of conventional plastics, the search for alternative, sustainable packaging materials has become one of the main research and technological challenges for the food industry (Jin et al., 2023; Kumari et al., 2023). Of particular interest are biopolymers of microbial origin, including PHB, which exhibit high biodegradability, biocompatibility, and compliance with EU and Food and Drug Administration (FDA) regulations on the migration of substances into food (Iordanskii et al., 2018). In order for a packaging material to be approved for use in the food industry, it must meet a number of technical and functional criteria. These include appropriate mechanical properties (tensile strength, impact resistance, flexibility), dimensional and thermal stability, barrier properties against gases and water vapor, low migration of components into food, and no impact on the organoleptic properties of packaged products (Bucci et al., 2005; Díez-Pascual & Díez-Vicente, 2014; Bairwan et al., 2024). PHB is characterized by high crystallinity and low gas permeability, which makes it ideal for passive applications. However, its brittleness and narrow processing window require structural modifications or appropriate additives to obtain a material with better performance properties (Bucci et al., 2005; Burgos et al., 2017; Iordanskii et al., 2018).

The primary role of food packaging is to protect its quality and safety and extend its shelf life by limiting access to factors that promote spoilage, such as microorganisms, oxygen, light, moisture, and volatile chemical compounds. Packaging material should therefore act as a barrier to the flow of gases and vapors, prevent the migration of contaminants, be mechanically, thermally, and chemically resistant, and, in light of current regulations, be recyclable or biodegradable (Díez-Pascual & Díez-Vicente, 2014; Bairwan et al., 2024). In the case of modern biopolymers such as PHB and its copolymers, barrier properties against oxygen and water vapor are of particular importance, as they directly affect the shelf life of food products. Mechanical parameters are also important: tensile strength, impact resistance, flexibility, and compression resistance. Pure PHB has been shown to have relatively high stiffness but limited resistance to dynamic loads (Bucci et al., 2005), which may hinder its use in packaging requiring flexibility, for example, in vacuum-packed products. In addition, packaging must meet strict migration requirements in accordance with EU regulations (including 10/2011/EC). PHB and PHBV materials enriched with active ingredients (e.g., essential oils, ZnO nanoparticles) show migration levels well below the permissible limits, confirming their safety in use (Díez-Pascual & Díez-Vicente, 2014; Kumari et al., 2023). Therefore, appropriate structural modification of PHB and proper processing engineering allows for the production of biopackaging that can meet modern technological and legislative requirements.

The use of PHB in food packaging requires detailed analysis of its behavior under conditions typical for the cold chain and storage at room temperature. Comparative studies of PHB and traditional polypropylene (PP) have shown that PHB has lower resistance to dynamic compression and impact—deformation was approximately 50% less than that of PP—which confirms its greater rigidity but lower flexibility (Bucci et al., 2005). Under refrigerated and frozen conditions, PHB exhibits poorer mechanical properties, while at higher temperatures, its performance is even better than that of PP. Despite this, PHB has excellent UV barrier properties, which may be important in the context of storing light-sensitive products. For example, PHB and PHB-copolymers packaging materials block, to some extent, light transmission in the 250–350 nm range, thus protecting products from photo-oxidative degradation (Bucci et al., 2005; Lugoloobi et al., 2020; Choi et al., 2024). Furthermore, modified forms of PHB can further limit the access of oxygen and moisture, which directly translates into a longer shelf life of packaged food.

The modern approach to food packaging is increasingly moving beyond the passive protection of the product—there is growing interest in active packaging systems, which release antimicrobial and antioxidant compounds in a controlled manner, extending the shelf life of products and improving microbiological safety. PHB, as a natural polymer, can be an effective matrix for the incorporation of many bioactive compounds. The literature describes a number of successful attempts to use PHB as a carrier for substances such as clove oil, eugenol, pediocin, vanillin, silver (AgNPs), ZnO, and curcumin—a compound with both antioxidant and indicator properties (Narayanan et al., 2013; Xavier et al., 2015; Díez-Pascual & Díez-Vicente, 2014; Choi et al., 2024). Active PHB films have shown high efficacy in combating pathogens such as *Escherichia coli*, *Staphylococcus aureus*, *Salmonella sp.*, *Bacillus cereus*, *Listeria monocytogenes,* as well as molds of the genus *Aspergillus* and *Penicillium* (Narayanan et al., 2013; Xavier et al., 2015; Iordanskii et al., 2018). Importantly, the release of active compounds often follows Fick's diffusion mechanism, which allows for gradual and controlled antimicrobial action. In addition, the use of ZnO and AgNPs nanopowders not only enhances the antimicrobial properties but also improves the barrier and mechanical properties of the material (Castro-Mayorga et al., 2018; Díez-Pascual & Díez-Vicente, 2014). The bioactive properties of PHB are therefore an important advantage in the context of smart packaging for food, combining protective functions with an active effect on microbiological and oxidative spoilage processes.

Due to its natural brittleness, low elasticity, and relatively high processing temperature, pure PHB requires modification to expand its range of packaging applications. One of the most common strategies is the addition of plasticizers such as PEG, ATBC, or lactic acid oligomers (OLA), which increase the material's stretchability and elasticity while lowering the processing temperature (Arrieta et al., 2014; Burgos et al., 2017; Kontárová et al., 2020). The literature has shown that the presence of plasticizers in the PHB–PLA system can lead to the production of continuous, elastic electrospun fibers, which significantly expand the application potential of this mixture in the production of biodegradable films and coatings (Arrieta et al., 2014). Moreover, modifications of PHB with nanoparticles (CNCs, ZnO, montmorillonite) not only improve the mechanical properties but also increase the gas barrier properties of the material, which is particularly important in the context of packaging products with high oxygen sensitivity (Díez-Pascual & Díez-Vicente, 2014; Kumari et al., 2023). The effectiveness of such modifications has also been confirmed in functional tests—PHB-based materials with additives showed increased tensile strength, better flexibility, and a significant reduction in oxygen and WVP, which translates into a real extension of the shelf life of packaged food (Albuquerque et al., 2021; Burgos et al., 2017).

19.7.2.1 Poly(3-hydroxybutyrate)-based smart indicators for food freshness monitoring

Alongside active packaging, the concept of smart packaging is rapidly developing, enabling real-time monitoring of the quality and freshness of food products. PHB, as a biodegradable and biocompatible material, is also used in this area, serving as a carrier for chemical indicators that react to changes occurring during food storage. An example of such a solution is PHB films containing curcumin, a natural dye that changes color under the influence of pH or the presence of ammonia, which is a product of protein metabolism during the spoilage of meat and fish (Choi et al., 2024). Studies have shown that such films can effectively signal deteriorating food quality at an early stage of degradation, which can significantly improve consumer safety and reduce food waste. In PHB-based systems with the addition of indicators, it is also possible to develop labels that monitor temperature, humidity, or volatile organic compounds, depending on the selection of the appropriate marker. Thanks to this, PHB can be used not only as a packaging material but also as a functional element of a new generation of quality control systems in the food supply chain (Choi et al., 2024).

19.7.2.2 Poly(3-hydroxybutyrate) integration into multilayer and composite structures for food packaging

In order to optimize its performance properties, PHB is increasingly used as a layer component in complex packaging structures, especially in the form of composites and laminates. Such systems allow the advantages of PHB (biodegradability, gas barrier properties, biocompatibility) to be combined with the functionality of other polymers or inorganic materials, such as thermoplastic starch (TPS), PLA, PCL, or nanofillers (ZnO, CNC, MMT) (Arrieta et al., 2014; Díez-Pascual & Díez-Vicente, 2014; Burgos et al., 2017; Kumari et al., 2023; Usurelu et al., 2022). PHB exhibits good processing compatibility with PLA, which allows the creation of blends or composite layers with increased flexibility, higher transparency, and reduced oxygen and WVP (Arrieta et al., 2014). Importantly, the addition of even small amounts of CNC significantly improves mechanical and thermal properties while maintaining the natural character of the material (Kumari et al., 2023). For commercial applications, processing techniques compatible with industrial production lines, such as coextrusion, electrospinning, and blow molding, are also of particular importance. PHB is highly compatible with these technologies, provided that the processing parameters and compatibilizers are selected appropriately.

19.7.2.3 Economic and industrial challenges of poly(3-hydroxybutyrate) in food packaging

Despite its many functional and environmental advantages, the widespread implementation of PHB as a packaging material in the food industry still faces significant economic and technological barriers. One of the main limitations is the high cost of PHB production, resulting from relatively low biosynthesis yields and the costs of substrates and polymer extraction processes from microbial cells (Khattab et al., 2021; Rami Reddy Tadi et al., 2021). It is estimated that the price of PHB is 3–5 times higher than that of traditional synthetic materials, which limits its market competitiveness. In response to this problem, intensive research is being conducted on the use of agri-food industry waste as alternative, inexpensive sources of carbon in the fermentation process, including whey, oil cakes, fruit wastes, and brans (Khattab et al., 2021; Rami Reddy Tadi et al., 2021; Sirohi et al., 2021a). These strategies not only significantly reduce the cost of PHB production, but also contribute to closing the organic material cycle in agriculture and the food industry. From a processing technology perspective, PHB has a narrow processing temperature range and high susceptibility to thermal degradation, which requires adjustments to the production conditions of films, injections, and protective coatings (Burgos et al., 2017). The integration of PHB into existing packaging industry production lines requires additional investment, but the development of coextrusion and hybrid polymer blend technologies increases the scalability and implementation potential of the material.

19.7.2.4 Regulatory compliance and food contact safety

The use of PHB in food packaging requires compliance with a number of international standards and regulations for materials that come into contact with food. PHB is a material recognized as safe (Generally Recognized As Safe [GRAS]) by the FDA, and its compliance with European requirements is confirmed, among others, by migration test results in accordance with Regulation (EU) No. 10/2011 (Díez-Pascual & Díez-Vicente, 2014; Bairwan et al., 2024). PHB is characterized by low global migration and no release of toxic degradation products, making it a safe material for packaging products with different chemical compositions—both dry and fatty, acidic, or containing alcohol. Cytotoxicity tests and toxicokinetic analyses show that PHB has no negative impact on model organisms and exhibits high microbiological stability (Zhuikova et al., 2022). It is worth noting that compliance with migration standards also applies to composites and mixtures of PHB with other biopolymers and active additives, which is important in the design of functional and smart materials. Due to these properties, PHB can be used not only in primary packaging but also in systems integrated with protective barriers or food quality monitoring labels.

In an era of growing consumer awareness and food market requirements, the concept of active and intelligent packaging has gained significant importance. In addition to basic functions such as protecting the product from mechanical damage, oxygen, and moisture, modern packaging must also prevent food spoilage and extend shelf life. One of the key developments in this field is the introduction of antimicrobial and antioxidant properties into packaging materials. The active compounds used in these types of materials can have various origins, ranging from natural essential oils (e.g., eugenol, carvacrol, vanillin) to metal nanoparticles (AgNPs, ZnO) and more complex systems such as mesoporous carriers or plasma modifications. These substances can be incorporated into a biopolymer matrix (e.g., PHB or PHBV) or located on the surface of the coating, and their release can be controlled.

19.8 Active poly(3-hydroxybutyrate) films and coatings

One of the most promising active packaging systems based on PHB is bionanocomposites containing ZnO nanoparticles. In studies conducted by Díez-Pascual and Díez-Vicente, a solution casting technique was used to obtain PHB films with the addition of ZnO without the need for surfactants or coupling agents. The use of ZnO led to a significant improvement in the mechanical and barrier properties of the material. An increase in Young's modulus, tensile strength, and impact resistance was observed even at low concentrations of the nanofiller. At the same time, a reduction in water absorption and an improvement in barrier properties against gases and water vapor were observed. Crucially, from a food safety perspective, these films exhibited strong antimicrobial activity against both Gram-positive (*S. aureus*) and Gram-negative (*E. coli*) bacteria, achieving 94% and 97% growth inhibition at 10% ZnO weight content, respectively. ZnO nanoparticles played a dual role—they acted as bactericidal agents and improved adhesion at the PHB-ZnO interface, which translated into better physical properties. At the same time, ZnO migration levels in various food simulants were well below the legal limits, indicating the potential for practical application of these materials as safe food packaging components (Díez-Pascual & Díez-Vicente, 2014). In another study, Panaitescu et al. proposed a three-component system consisting of PHBV, ZnO nanofibers, and surface activation by plasma treatment. This combination effectively increased surface wettability, improved the adhesion of bioactive compounds, and provided strong activity against *E. coli* and *L. monocytogenes* (Panaitescu et al., 2018).

Another effective solution in the field of active packaging is PHB composites containing AgNPs. In the study by Castro-Mayorga et al. (2018), AgNPs were biosynthesized during the fermentation of *C. necator* without the need to add a chemical reducing agent, which further increases the environmental attractiveness of this method. The resulting nanocomposites showed uniform dispersion of AgNPs in the PHB matrix and high antimicrobial activity against *Salmonella enterica* and *L. monocytogenes*. Importantly, the introduction of AgNPs did not adversely affect the thermal properties of the polymer, and the degree of crystallinity was only slightly reduced. PHB/AgNPs films also retained full biodegradability, which was confirmed under composting conditions—complete biodegradation of the material occurred within 40 days. The strong antibacterial activity, also confirmed in other reports (Salama et al., 2018; Łopusiewicz et al., 2022; Macieja et al., 2022), results from the ability of AgNPs to destabilize cell membranes and generate reactive oxygen species (ROS). Due to their broad spectrum of activity, high level of food safety, and environmental compatibility, PHB films modified with AgNPs are one of the most promising groups of active materials for packaging applications.

Natural plant-derived substances such as eugenol (found in clove oil), carvacrol (from oregano), and curcumin have found widespread use as active ingredients in PHB matrices (Choi et al., 2024; Melendez-Rodriguez et al., 2019; Narayanan et al., 2013; Requena et al., 2017). Their action is based on interactions with the cell membranes of microorganisms and inhibition of enzymatic activity. For instance, PHBV films with added eugenol reduced the population of *E. coli* and *S. aureus* by over 99% after 24 hours of contact with the material surface. In turn, the use of carvacrol in nanostructured PHB layers enabled the achievement of antifungal properties, which are particularly important in the packaging of some food products, such as bakery products and ripened cheeses. Curcumin, in addition to its antimicrobial properties, contributed to the system's ability to detect changes in product quality, showing potential for use as a smart freshness indicator. The advantage of such systems is their natural origin, low toxicity, and ability to synergize with antioxidant properties. These phenolic compounds slow down lipid oxidation, which is particularly important in the packaging of high-fat products, such as meat, dairy products, and nuts (Choi et al., 2024; Melendez-Rodriguez et al., 2019; Narayanan et al., 2013; Requena et al., 2017).

Chitosan, a natural polysaccharide with strong antimicrobial properties, is often used as a coating layer on PHB biopolymer films. The combination of PHB and chitosan results in a material with a much broader spectrum of activity—against both Gram-positive and Gram-negative bacteria, as well as yeast-like fungi. PHB coatings covered with a thin layer of chitosan significantly reduce the growth of L. *monocytogenes* and *Candida albicans* (Zhuikova et al., 2022). The mechanism of action of chitosan involves, among other things, interaction with negatively charged components of the cell wall of microorganisms, leading to their destabilization. Furthermore, chitosan can also act as a selective barrier, limiting the penetration of oxygen and moisture, thereby increasing the barrier properties of the entire system. The use of multilayer PHB/chitosan films indicates the high flexibility of this technology, including the possibility of controlled release of bioactive compounds and adaptation to different types of food.

In order to ensure controlled and prolonged release of antimicrobial and antioxidant compounds from packaging materials, mesoporous carriers such as silica (SiO_2), zeolites, and aluminum oxide are increasingly being used. These porous structures, characterized by a large specific surface area and appropriate pore diameter, enable the encapsulation and subsequent gradual release of active substances such as essential oils, plant phenols, or heavy metals in the form of nanoparticles. The use of such systems in a PHB matrix allows for the maintenance of antimicrobial activity for a longer period of time and reduces the burst release effect, which can lead to exceeding permissible migration limits. Kumari et al. in their study showed that PHB films containing mesoporous silica with adsorbed thymol inhibited the growth of *B. cereus* and *E. coli* for over 10 days of storage under refrigerated conditions while maintaining the sensory integrity of the food product. An additional advantage of using porous carriers is their effect on improving the mechanical and barrier properties of films, making them an attractive solution for packaging applications with extended shelf life (Kumari et al., 2023). PHB systems with mesoporous carriers are therefore the next step in the development of advanced new-generation active materials.

Other types of nanoparticles are also used to modify active PHB materials. For instance, Maureira et al. presented an innovative combination of PHB with MoS_2 layers, which effectively inhibited the growth of *L. monocytogenes*. The presence of MoS_2 not only increased the antimicrobial activity but also improved the barrier properties and biodegradability of the coatings (Maureira et al., 2023).

Another interesting area of research is antibacterial coatings based on the photocatalytic properties of TiO_2, particularly in the context of their activity under UV radiation (Yew et al., 2006; Venezia et al., 2023; Tasbihi et al., 2024; Korina et al., 2013). TiO_2, in the presence of UV light, generates ROS, which effectively destroy the cellular structures of bacteria, fungi, and viruses. Studies on TiO_2-modified PHB coatings have shown a significant reduction in the number of pathogenic microorganisms, such as *E. coli* and *S. aureus*, after UV exposure (Korina et al., 2013). TiO_2 nanocoatings applied by plasma or sol-gel methods not only preserve the transparency of the material but also enhance its mechanical properties and UV resistance, making them ideal for smart packaging for fresh and microbiologically sensitive foods. Another important aspect is the low level of metal ion migration and the photostability of the system, which allows for its safe use in contact with food. Further research focuses on combining TiO_2 with other functional components, such as silver or zinc, in order to increase the spectrum of activity and intensify synergistic effects.

Essential oils and their components, including those of citrus origin, such as limonene, linalool, citral, and geraniol, exhibit pronounced antimicrobial and antioxidant properties. They can be incorporated into PHB films by solution casting, coextrusion, or encapsulation in nanocarriers. Their volatility and sensitivity to light mean that optimization of process conditions is crucial for maintaining biological activity. In the study by Burgos et al., carvacrol was used as an active ingredient in PLA/PHB films with the addition of lactic acid oligomers (OLA). Increased antimicrobial efficacy, particularly against *S. aureus*, and strong antioxidant activity were demonstrated (Burgos et al., 2017). In another study, Moll and Chiralt investigated ferulic and *p*-coumaric acid additives, demonstrating prolonged film activity and controlled release of active compounds (Moll & Chiralt, 2023). The use of citrus oils and terpenoids in PHB films is particularly promising for packaging fresh products such as fruit, vegetables, and bread. They combine microbiological effectiveness with an acceptable sensory profile, making them an attractive solution for organic food producers and the convenience segment. In line with this trend, Narayanan et al. investigated the synergistic effect of pediocin, a bacteriocin produced by *Pediococcus acidilactici*, and eugenol. PHB/PHBV coatings with these ingredients effectively reduced the number of Gram-positive bacteria and yeasts while maintaining the flexibility and transparency of the material (Narayanan et al., 2013). Da Costa et al. investigated the use of oregano oil in combination with kaolinite and sepiolite additives. This modification improved mechanical properties, reduced component migration under acidic conditions, and showed high efficacy against Gram-positive and Gram-negative bacteria (da Costa et al., 2020).

On the other hand, Ahuja et al. incorporated a flavonol, chromone, into PHB films via solvent-casting method. The films showed high antioxidant activity and antimicrobial activity against mold (Aspergillus *niger*) as well as bacterial pathogens: *S. aureus* and *E. coli*. Packaging of bread samples with PHB/chromone films resulted in complete inhibition of microbial growth for up to 10 days of storage, thereby ensuring the microbiological safety of the food product (Ahuja et al., 2023). The study by Mittal et al. also has shown that PHB films containing clove essential oil (CEO) and nano-SiO_2 effectively inhibited the growth of *E. coli*, *S. aureus*, and *A. niger* and extended the shelf life of wheat bread to 10 days—a result comparable to polyethylene packaging. These effects were achieved through the synergistic action of oils and nanofillers, improving the mechanical and barrier properties of the matrix (Mittal et al., 2023). In the study by Iordanskii et al., PHB matrices containing chlorhexidine (CHD) were developed, which showed high antimicrobial activity while being able to release the active substance in a controlled manner. The use of CHD in PHB allowed for prolonging the activity period of the coating and reducing its toxicity, which makes this system a promising solution in the context of packaging for perishable products (Iordanskii et al., 2018). Solaiman et al. demonstrated that the biosurfactant sophorolipid, in combination with a PHB matrix, enables the formation of biodegradable films with activity against *Propionibacterium acnes* (Solaiman et al., 2015). This approach combines beneficial bioactive properties with compliance with food contact safety requirements.

19.9 Polyhydroxyalkanoates and poly(3-hydroxybutyrate) biosynthesis: technological pathways and sustainability aspects

19.9.1 Technological pathways and fermentation strategies

In response to the challenges associated with the high cost and production limitations of PHA, more research is focusing on innovative methods for producing these biopolymers. One of the key areas of development is the metabolic engineering of microbial strains—both bacteria and archaea—aimed at increasing biosynthetic yields and enabling the production of more complex copolymers (e.g., PHB-co-3HHx or PHBV-co-4HB). Genetic modifications include both the metabolic pathways of hydroxy acid synthesis and the transport and degradation systems of substrates, which allow for precise shaping of the monomer composition of the obtained polymers (Kovalcik et al., 2020; Ibrahim et al., 2021). At the same time, processes involving extremophile microorganisms, such as halophiles or thermophiles, are being developed that can produce PHA under extreme conditions (high salinity, elevated temperature), reducing the need for sterilization and simplifying biotechnological operations. This approach helps reduce costs and increase the process's resistance to environmental contamination. Continuous or semicontinuous fermentation systems are also increasingly being introduced to replace traditional batch processes, enabling more efficient substrate flow management and increasing the overall efficiency of PHA production (Dehouche et al., 2020; Mostafa et al., 2020).

The sustainable nature of PHB stems from its biological origin, biodegradability, and the possibility of using renewable raw materials and by-products from agriculture or the food industry as substrates for its production (Amadu et al., 2021; Briassoulis et al., 2021). PHB is produced by microbial biosynthesis and degrades under both aerobic and anaerobic conditions without generating toxic by-products (Sukhanova et al., 2020; Low et al., 2021). PHB production in a closed system (cradle-to-gate) shows significantly lower greenhouse gas emissions and nonrenewable energy use compared to petrochemical polymers such as PE and PP, especially when waste raw materials and appropriate energy allocation methods are used (Briassoulis et al., 2021). Comparative studies have also shown a lower impact of PHB on acidification potential and eutrophication potential, although PE performed slightly better for these two parameters (Briassoulis et al., 2021). Significant potential for the sustainable development of PHB is seen in the context of biorefineries, where organic waste (e.g., industrial wastewater rich in volatile fatty acids—[VFA]) can be converted into PHB while recovering valuable chemical compounds such as methyl crotonate or methacrylate (da Costa et al., 2020). The integration of PHB production with waste processing enables cost reduction, improved energy efficiency, and minimization of environmental impact. An important feature of PHB from a sustainable development perspective is its life cycle—the material can be recycled or completely degraded in the natural environment. The ability of PHB to degrade in marine conditions and its compostability confirm its potential as an alternative to durable plastics, contributing to the reduction of microplastic pollution (Bugnicourt et al., 2014; Kontárová et al., 2020).

The production of PHB using microorganisms has become an intensively developed alternative to petrochemical-based plastics in recent decades. PHB, as a biodegradable and biocompatible polyester, is synthesized under conditions of cellular stress, mainly with an excess of carbon and a deficiency of nitrogen, phosphorus, or sulfur, which promotes its accumulation as a reserve material in the form of cytoplasmic granules (Zhang et al., 2022; Atarés et al., 2024). At the biochemical level, PHB biosynthesis proceeds in three main enzymatic steps: condensation of two molecules of acetyl coenzyme A to acetoacetyl-CoA (catalyzed by β-ketothiolase—PhaA), reduction to (R)-3-hydroxybutyrate-CoA (with the participation of acetoacetyl-CoA reductase—PhaB), and polymerization of this monomer with the participation of PHB synthase (PhaC) (Sirohi et al., 2021a; Zhang et al., 2022; Sun et al., 2024). A wide range of microorganisms have been identified as capable of producing PHB. Strains such as *C. necator* (formerly known as *R. eutropha* or *Alcaligenes eutrophus*), *B. megaterium*, *Pseudomonas putida*, and *Halomonas* spp. are currently considered the most efficient producers (Kourilova et al., 2021; Zhang et al., 2022; Atarés et al., 2024). Features such as high dry cell weight, lack of an LPS layer (in *Bacillus* spp.), and tolerance to extreme environmental conditions (in halophiles) make these strains attractive in terms of both productivity and simplification of polymer extraction processes (Khattab et al., 2021; Low et al., 2021). Among those mentioned, *C. necator* deserves special attention. This strain exhibits exceptional metabolic versatility, the ability to grow in diverse environmental conditions, and the ability to utilize a wide range of carbon sources, including alcohols, polyols, and organic acids (Low et al., 2021; Zhang et al., 2022). It can achieve biomass densities of 140 $g\,L^{-1}$ and accumulate PHB at levels reaching 90% of dry cell weight, making it a leader in industrial applications (Zhang et al., 2022).

An increasingly popular strategy for increasing the efficiency and resilience of PHB production systems is the use of mixed microbial consortia (MMC). This approach allows for the effective use of diverse substrates, including agricultural and industrial waste, and increases the tolerance of fermentation processes to the presence of inhibitors. The use of MMC can also reduce the need for sterilization of culture media and lower the costs associated with controlling the microbiological purity of the process (Sirohi et al., 2021a; Atarés et al., 2024). In recent years, the range of waste substrates used for PHB biosynthesis has been significantly expanded. In addition to classic agricultural and food waste, such as molasses and whey, more unusual carbon sources are also being utilized. For example, extracts from carob pods were the only carbon source in a process involving *C. necator* and *B. megaterium*, achieving a yield of over 12 $g\,L^{-1}$ PHB (Manikandan et al., 2020). In turn, rice straw waste, subjected to enzymatic hydrolysis using an enzyme cocktail obtained from mold *Aspergillus nidulans*, enabled the production of high concentrations of reducing sugars, which were then effectively utilized by halophilic bacteria for PHB production (Naitam et al., 2022). Fermentation technologies used for PHB production include classic batch, fed-batch, and continuous fermentation processes. Fed-batch fermentation, characterized by controlled substrate feeding during the process, shows the greatest potential for scaling up from laboratory to pilot scale. Continuous fermentation, on the other hand, which enables stable production and high productivity with a constant supply of substrate and product removal, is most desirable in an industrial context (Jin et al., 2023; Atarés et al., 2024). In response to high production costs, numerous optimization strategies have been developed to improve the economic viability of the processes. These include the use of cheap carbon substrates (e.g., agri-food waste, waste glycerol, lignocellulose), the development of contaminant-resistant microorganisms (for nonsterile fermentation), and the use of two-step fermentation methods, in which the biomass growth phase is separated from the PHB accumulation phase (Manikandan et al., 2020; Naitam et al., 2022; Bairwan et al., 2024). The engineering of PHB producer metabolic pathways has also proved to be a significant breakthrough. The use of techniques such as CRISPR/Cas, phaCAB gene expression in *E. coli* strains, and increasing the nicotinamide adenine dinucleotide phosphate (NADPH) pool through modifications of the pentose phosphate pathway have enabled a significant increase in yield and control over the final properties of the polymer (Rami Reddy Tadi et al., 2021; Atarés et al., 2024; Sun et al., 2024). Furthermore, targeted modifications of extraction processes (e.g., ultrasound-assisted extraction) or copolymerization of PHB with other monomers (e.g., production of PHBV) enable the production of materials with reduced melting points and increased flexibility, which broadens their application possibilities, especially in the packaging sector (Briassoulis et al., 2021). It is also worth noting the prospects associated with the circular economy. In this context, the use of agri-food waste for PHB production not only reduces raw material costs but also contributes to greenhouse gas emission reductions, carbon recovery from biomass, and reduced dependence on fossil fuels. This approach is in line with the strategy for sustainable development of the chemical and plastics industry (Sirohi et al., 2021a; Atarés et al., 2024). One of the most promising directions for the development of the PHB fermentation process is its coupling with the production of high value-added coproducts. For example, Haloferax mediterranei strains not only accumulate PHB, but also simultaneously produce carotenoids, bacteriocins, archaeal lipids, and exopolysaccharides (EPS), which significantly improves the profitability of the entire biotechnological process (Atarés et al., 2024). Finally, a comparison of different types of fermentation (batch, fed-batch, continuous, two-stage) allows the selection of the optimal approach depending on the available substrates, process scale, and target product properties. Two-stage systems, which separate the growth phase from the PHB accumulation phase, allow for up to 90% polymer content in the dry cell mass, making them particularly attractive for industrial bioplastic production (Atarés et al., 2024). In summary, advanced fermentation technologies combined with modern metabolic engineering strategies and the use of waste substrates and low-cost processes make PHB production a promising area of industrial biotechnology. However, further research aimed at optimizing biosynthesis conditions, strain selection, and process integration will be crucial for the full implementation of PHB as an alternative to plastics on a global scale (Sirohi et al., 2021a; Zhang et al., 2022; Atarés et al., 2024).

19.9.2 Circular bioeconomy strategies: converting agro-industrial by-products into poly(3-hydroxybutyrate)

The use of waste as a substrate for the production of PHB is one of the most intensively developed approaches toward increasing the profitability and sustainability of biopolymers. The high cost of PHB production, mainly due to expensive, purified carbon substrates, is a major limitation to its wider commercialization. Therefore, an increasing number of studies are focusing on the use of cheap, readily available, and waste organic raw materials from various sectors of the agri-food, energy, and household industries (Sindhu et al., 2020; Kachrimanidou et al., 2021; Khattab et al., 2021; Sirohi et al., 2021a; Atarés et al., 2024). Among the most commonly used wastes are cheese whey (Carlozzi et al., 2021; Khattab et al., 2021), millet bran and rapeseed meal (Rami Reddy Tadi et al., 2021), grape pomace extracts (Kourilova et al., 2021), used cooking oil and fish oil (Sindhu et al., 2020; Loan et al., 2022), waste glycerin from biodiesel production (Kachrimanidou et al., 2021; Ta et al., 2023), kitchen waste and lignocellulose fractions (Zhang et al., 2022; Russo et al., 2024), as well as industrial fractions such as coffee grounds (Gigante et al., 2021), date syrup (Mostafa et al., 2020), banana leaf extracts, and palm oil waste (Mostafa et al., 2020). The processes of PHB production using the above-mentioned substrates include both direct bacterial fermentation (e.g., *B. flexus* Azu-A2, *B. cereus*, *C. necator*, *Neptunomonas concharum*, Ralstonia sp., *E. coli*) and more complex two-stage systems (e.g., production of lactic acid from whey and its further conversion to PHB by phototrophic bacteria) (Carlozzi et al., 2021). For many raw materials, preliminary enzymatic hydrolysis steps are used, as in the case of damaged wheat grains (Sirohi et al., 2021b) or fructose extracts from banana leaves (Low et al., 2021), in order to release simple sugars available to PHB-producing bacteria. It has also been shown that some fatty acid-rich wastes, such as used cooking oil or fish oil, allow some of the highest reported PHB production yields to be obtained (Loan et al., 2022), while fractions containing simple sugars, for example, date extracts and chicory root (Zhang et al., 2022), are particularly useful for halophilic and thermophilic bacteria. Microorganisms such as *Halomonas* sp. and *Tepidimonas taiwanensis* are capable of growing under high salinity or temperature conditions, which allows fermentation processes to be carried out under conditions that limit the risk of contamination (Kourilova et al., 2021; Park et al., 2021). In addition to classic substrates such as whey, lignocellulose, and oils, increasing attention is also being paid to the use of VFA as precursors in PHB production. *N. concharum* effectively converts acetate and butyrate to PHB under nonsterile conditions, suggesting potential for open bioprocesses with reduced operating costs (PU2020). From a process efficiency perspective, many studies have shown high levels of PHB accumulation in dry cell mass, often exceeding 60%–70% (Carlozzi et al., 2021; Khattab et al., 2021; Loan et al., 2022), proving that waste can be a valuable carbon source for PHB-synthesizing microorganisms. The implementation of such strategies can significantly reduce the unit cost of production, making it competitive with conventional plastics (Kachrimanidou et al., 2021). Ultimately, the use of waste for PHB production is not only in line with the principles of a circular economy, but also contributes to reducing the environmental impact of organic waste, making this direction particularly important both from the point of view of sustainable development and the commercial potential of biopolymers.

19.10 Biodegradability of poly(3-hydroxybutyrate)-based materials

The biodegradability of PHB is the subject of intensive research in the context of developing sustainable packaging materials. The key factor determining the rate and mechanism of PHB degradation remains its chemical structure and the environmental conditions in which the material functions. The literature repeatedly emphasizes the influence of parameters such as temperature, humidity, oxygen availability, microflora composition, pH, and the presence of nutrients (Briassoulis et al., 2021; Vaithanomsat et al., 2021; Iglesias-Montes et al., 2022; Popa et al., 2022). PHB is completely biodegradable in natural environments (soil, fresh water, and marine water) and controlled environments (aerobic composting, anaerobic composting, and activated sludge), as confirmed by numerous studies (Bugnicourt et al., 2014; Briassoulis et al., 2021; Popa et al., 2022; Usurelu et al., 2022). This process occurs without generating toxic products, mainly due to the enzymatic activity of microorganisms such as bacteria of the genus *Pseudomonas*, *Streptomyces*, *Bacillus*, and fungi of the genus *Aspergillus* (Bugnicourt et al., 2014; Thomas et al., 2020; Iglesias-Montes et al., 2022; Popa et al., 2022). These enzymes, depolymerases, catalyze the hydrolysis of ester bonds, leading to the formation of hydroxy acids, which are then metabolized to CO_2 and H_2O (aerobic conditions) or methane (anaerobic conditions) (Bugnicourt et al., 2014). Degradation occurs primarily in amorphous regions, which explains the slower decomposition rate observed in materials with higher crystallinity (Usurelu et al., 2022). PHB composites, such as PHBV or P(3HB-co-4HB), are characterized by increased susceptibility to degradation as the content of hydrophilic side units increases (Briassoulis et al., 2021). The influence of functional additives, such as plasticizers and nanofillers, is an important area of research. Tributyrin (TB), as a plasticizer, significantly accelerates the degradation of PHB, reducing the time of complete disintegration to as little as 14 days (Iglesias-Montes et al., 2022). On the other hand, additives with a higher molecular weight can limit water diffusion and enzyme availability, leading to slower decomposition (Usurelu et al., 2022). In the case of PHB/CNC nanocomposites, a low CNC content (2%) may increase the stiffness of the material, while a higher content (4%) contributes to increased hydrophilicity and accelerated degradation (Usurelu et al., 2022). The microbial degradation activity of PHB has also been confirmed under extreme conditions. The material decomposed in lake water at 6°C (Lugano) and under high-temperature composting conditions (above 50°C), where thermophilic microorganisms, such as *Aspergillus*, are capable of decomposing 90% of PHB within five days (Bugnicourt et al., 2014; Thomas et al., 2020). The addition of organic additives such as peat, clay, or wood flour allows the rate of biodegradation to be modified. For example, PHB composites with peat underwent a 43.6% reduction in mass, and those with wood flour a 33% reduction after 35 days of incubation in soil (Thomas et al., 2020). In addition, PHB degradation affects the physicochemical properties of the material, leading to a decrease in melting temperature and an increase in crystallinity (Iglesias-Montes et al., 2022). PHB meets the requirements of numerous biodegradability standards, such as ISO 14855, ISO 17556, ISO 15985, and ASTM D6691, making it an attractive material for the packaging industry (Usurelu et al., 2022). Even in

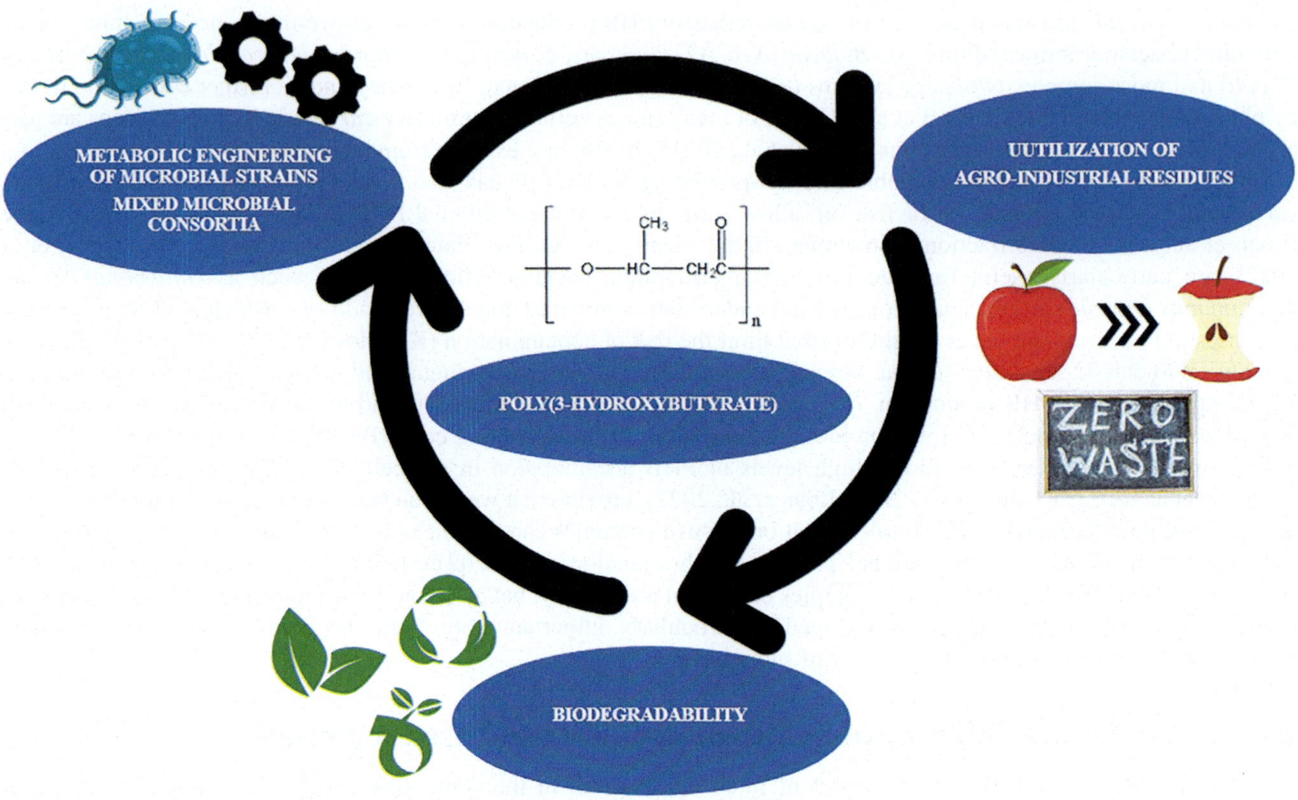

FIGURE 19.3 Key environmental benefits and green production strategies of poly(3-hydroxybutyrate)-based materials.

marine environments, which are considered less favorable for biodegradation, PHB shows the ability to degrade, albeit with less intensity (Briassoulis et al., 2021). In addition to biological degradation, chemical methods for converting PHB are also being considered. Thermochemical degradation in the presence of halogenated solvents allows the formation of derivatives such as crotonic acid residues or 3-bromobutyrate, which may be useful in chemical recycling (Boyandin et al., 2024). The biodegradability of PHB, its compatibility with various functional additives, and its compliance with international environmental standards confirm its high suitability in the context of reducing the negative impact of conventional plastics. As a result, PHB appears to be one of the key candidates for the further development of new-generation biopolymers (Bucci et al., 2005; Briassoulis et al., 2021; Iglesias-Montes et al., 2022). Certain aspects of sustainable production and eco-friendly features of PHB-based films and coatings are presented in Fig. 19.3.

19.11 Prospects and challenges ahead

Despite growing interest in PHB in the scientific community, its industrial-scale production remains limited. The high cost of production is a key obstacle, even when using relatively inexpensive substrates such as waste from the biodiesel industry or fruit processing. Sirohi et al. and Zhang et al. emphasized that it is necessary to develop more economical and energy-efficient methods for producing PHB using renewable energy sources and waste organic raw materials (Sirohi et al., 2021a; Zhang et al., 2022). In this context, it is also necessary to conduct full life cycle assessments (LCA) that can demonstrate a higher level of sustainability of PHB compared to traditional plastics, despite the current economic barriers (Zhang et al., 2022). In particular, the potential of lignocellulosic substrates is promising, but the presence of fermentation inhibitors such as furfural and acetic acid limits microbial performance. There is a need to develop inhibitor-resistant strains or to use adaptive evolution and selection strategies (Zhang et al., 2022). At the same time, future research directions should include metabolic modifications of microorganisms (e.g., *C. necator*) to increase PHB biosynthesis efficiency while reducing the production of side metabolites. Bairwan et al. indicate that another challenge is the processability of PHB and its copolymers, such as PHBV, whose low processing temperature range and tendency to thermal degradation hinder their industrial applications (Bairwan et al., 2024). Phase separation problems in polymer blends can result in reduced mechanical and barrier properties of the final products. Therefore, it is important to develop effective methods for compatibilization and improvement of phase dispersion in composite systems. The prospects for the development of PHB include advanced processing techniques such as electrospinning, high-barrier coatings on both sides, and multilayer structures. The combination of PHBV with nanomaterials—nanocellulose, nanoclays, or plant fibers—is becoming particularly interesting, as it may lead to materials with improved mechanical and functional properties (Bairwan et al., 2024). Despite the growing number of studies on PHBV composites, there are still no reports on the use of biodegradable and biocompatible nanocelluloses, which indicates a research gap worth further exploration. Another promising direction is smart packaging systems, which, in addition to their protective function, can act as freshness indicators, oxygen adsorbents, or carriers of antimicrobial substances (Bairwan et al., 2024). Their implementation in PHB-based structures could significantly increase the added value of this biopolymer in food industry applications. At the same time, research is needed to improve the compostability of these materials, especially in nonfood applications.

Conclusions

PHB is a promising alternative to conventional fossil fuel-based polymers, combining biodegradability, biocompatibility, and the ability to be biosynthesized from renewable sources under microbial fermentation conditions. Its physicochemical properties make it a suitable candidate for both short-term applications (e.g., disposable packaging) and specialized fields such as regenerative medicine and DDSs. Processing challenges (narrow thermal window, brittleness, mechanical instability) and production costs continue to limit the full exploitation of PHB's potential. However, with intensive work on modifications (copolymerization, nanofillers, plasticization), developing biorefinery technologies, and increasing access to waste substrates, PHB is expected to grow significantly in importance in the coming years. Further research should focus on: (1) developing cost-effective biosynthesis processes using organic waste, (2) developing PHB blends and composites with natural additives that have antibacterial or antioxidant properties, and (3) assessing the impact of long-term use of PHB in medical and environmental applications. Such targeted activities may contribute to establishing PHB as one of the key biopolymers of the 21st century.

AI statement: AI-based tools were used solely for language proofreading and editing purposes.

References

Ahuja, S., Bains, O., Mittal, M., Kamal, R., Aggarwal, N. K., & Arora, S. (2023). Multifunctional chromone-incorporated poly(hydroxybutyrate) luminescent film for active and intelligent food packaging. *International Journal of Biological Macromolecules, 246*, 125625. https://doi.org/10.1016/j.ijbiomac.2023.125625.

Albuquerque, R. M., Meira, H. M., Silva, I. D., Silva, C. J. G., Almeida, F. C. G., Amorim, J. D., Vinhas, G. M., Costa, A. F. S., & Sarubbo, L. A. (2021). Production of a bacterial cellulose/poly(3-hydroxybutyrate) blend activated with clove essential oil for food packaging. *Polymers and Polymer Composites, 29*, 259–270. https://doi.org/10.1177/0967391120912098.

Amadu, A. A., Qiu, S., Ge, S., Addico, G. N. D., Ameka, G. K., Yu, Z., Xia, W., Abbew, A.-W., Shao, D., Champagne, P., & Wang, S. (2021). A review of biopolymer (Poly-β-hydroxybutyrate) synthesis in microbes cultivated on wastewater. *Science of the Total Environment, 756*, 143729. https://doi.org/10.1016/j.scitotenv.2020.143729.

Arrieta, M. P., Samper, M. D., López, J., & Jiménez, A. (2014). Combined effect of poly(hydroxybutyrate) and plasticizers on polylactic acid properties for film intended for food packaging. *Journal of Polymers and the Environment, 22*, 460–470. https://doi.org/10.1007/s10924-014-0654-y.

Atarés, L., Chiralt, A., González-Martínez, C., & Vargas, M. (2024). Production of polyhydroxyalkanoates for biodegradable food packaging applications using haloferax mediterranei and agrifood wastes. *Foods, 13*. https://doi.org/10.3390/foods13060950.

Babos, G., Rydz, J., Kawalec, M., Klim, M., Fodor-Kardos, A., Trif, L., & Feczkó, T. (2020). Poly(3-Hydroxybutyrate)-based nanoparticles for sorafenib and doxorubicin anticancer drug delivery. *International Journal of Molecular Sciences, 21*. https://doi.org/10.3390/ijms21197312.

Bairwan, R. D., Yahya, E. B., Gopakumar, D., & H.P.S., A. K. (2024). Recent advances in poly(3-Hydroxybutyrate-co-3-Hydroxyvalerate) biocomposites in sustainable packaging applications. *Advanced Materials Letters, 15*. https://doi.org/10.5185/amlett.2024.011739, 2401-1739.

Bernal-Sánchez, L. J., Vázquez-Olmos, A. R., Sato-Berrú, R. Y., Mata-Zamora, E., Rivera, M., & Garibay-Febles, V. (2024). Multifunctional hybrid films made from CoT_3OTx_4 and $CoFeT_2OT_4$ nanoparticles inside a poly 3-hydroxybutyrate matrix and study of their impact in methyl orange photodegradation. *PLoS One, 19*, e0312611. https://doi.org/10.1371/journal.pone.0312611.

Bittolo Bon, S., Chiesa, I., Morselli, D., Degli Esposti, M., Fabbri, P., De Maria, C., Foggi Viligiardi, T., Morabito, A., Giorgi, G., & Valentini, L. (2021). Printable smart 3D architectures of regenerated silk on poly(3-hydroxybutyrate-co-3-hydroxyvalerate). *Materials & Design, 201*, 109492. https://doi.org/10.1016/j.matdes.2021.109492.

Boyandin, A. N., Bessonova, V. A., Sukhanova, A. A., & Ertiletskaya, N. L. (2024). Mechanism of thermochemical degradation of bacterial poly-3-hydroxybutyrate in solutions. *Russian Chemical Bulletin, 73*, 688–694. https://doi.org/10.1007/s11172-024-4179-9.

Briassoulis, D., Tserotas, P., & Athanasoulia, I.-G. (2021). Alternative optimization routes for improving the performance of poly(3-hydroxybutyrate) (PHB) based plastics. *Journal of Cleaner Production, 318*, 128555. https://doi.org/10.1016/j.jclepro.2021.128555.

Bucci, D. Z., Tavares, L. B. B., & Sell, I. (2005). PHB packaging for the storage of food products. *Polymer Testing, 24*, 564–571. https://doi.org/10.1016/j.polymertesting.2005.02.008.

Bugnicourt, E., Cinelli, P., Lazzeri, A., & Alvarez, V. (2014). Polyhydroxyalkanoate (PHA): Review of synthesis, characteristics, processing and potential applications in packaging. *Express Polymers Letter, 8*, 791–808. https://doi.org/10.3144/expresspolymlett.2014.82.

Burgos, N., Armentano, I., Fortunati, E., Dominici, F., Luzi, F., Fiori, S., Cristofaro, F., Visai, L., Jiménez, A., & Kenny, J. M. (2017). Functional properties of plasticized bio-based poly(Lactic Acid)_poly(Hydroxybutyrate) (PLA_PHB) films for active food packaging. *Food and Bioprocess Technology, 10*, 770–780. https://doi.org/10.1007/s11947-016-1846-3.

Carlozzi, P., Giovannelli, A., Traversi, M. L., & Touloupakis, E. (2021). Poly(3-hydroxybutyrate) bioproduction in a two-step sequential process using wastewater. *Journal of Water Process Engineering, 39*, 101700. https://doi.org/10.1016/j.jwpe.2020.101700.

Castro-Mayorga, J. L., Freitas, F., Reis, M. A. M., Prieto, M. A., & Lagaron, J. M. (2018). Biosynthesis of silver nanoparticles and polyhydroxybutyrate nanocomposites of interest in antimicrobial applications. *International Journal of Biological Macromolecules, 108*, 426–435. https://doi.org/10.1016/j.ijbiomac.2017.12.007.

Chang, C.-K., Wang, H.-M. D., & Lan, J. C. (2018). Investigation and characterization of plasma-treated poly(3-hydroxybutyrate) and Poly(3-hydroxybutyrate-co-3-hydroxyvalerate) biopolymers for an in vitro cellular study of mouse adipose-derived stem cells. *Polymers, 10*. https://doi.org/10.3390/polym10040355.

Cherpinski, A., Gozutok, M., Sasmazel, H. T., Torres-Giner, S., & Lagaron, J. M. (2018). Electrospun oxygen scavenging films of poly(3-hydroxybutyrate) containing palladium nanoparticles for active packaging applications. *Nanomaterials, 8*. https://doi.org/10.3390/nano8070469.

Choi, Y. H., Kim, J. T., Kim, M. H., & Park, W. H. (2024). Biodegradable poly(3-hydroxybutyrate-co-4-hydroxybutyrate)/curcumin composite film as a smart indicator of food spoilage. *Sensors and Actuators B: Chemical, 408*, 135511. https://doi.org/10.1016/j.snb.2024.135511.

da Costa, R. C., Daitx, T. S., Mauler, R. S., da Silva, N. M., Miotto, M., Crespo, J. S., & Carli, L. N. (2020). Poly(hydroxybutyrate-co-hydroxyvalerate)-based nanocomposites for antimicrobial active food packaging containing oregano essential oil. *Food Packaging and Shelf Life, 26*, 100602. https://doi.org/10.1016/j.fpsl.2020.100602.

D'Amico, D. A., Manfredi, L. B., & Cyras, V. P. (2012). Relationship between thermal properties, morphology, and crystallinity of nanocomposites based on polyhydroxybutyrate. *Journal of Applied Polymer Science, 123*, 200–208. https://doi.org/10.1002/app.34457.

Dan, L., Pope, M. A., & Elias, A. L. (2018). Solution-processed conductive biocomposites based on polyhydroxybutyrate and reduced graphene oxide. *The Journal of Physical Chemistry C, 122*, 17490–17500. https://doi.org/10.1021/acs.jpcc.8b02515.

Dariš, B., & Knez, Ž. (2020). Poly(3-hydroxybutyrate): Promising biomaterial for bone tissue engineering. *Acta Pharmaceutica, 70*, 1–15. https://doi.org/10.2478/acph-2020-0007.

Dehouche, N., Idres, C., Kaci, M., Zembouai, I., & Bruzaud, S. (2020). Effects of various surface treatments on Aloe Vera fibers used as reinforcement in poly(3-hydroxybutyrate-co-3-hydroxyhexanoate) (PHBHHx) biocomposites. *Polymer Degradation and Stability, 175*, 109131. https://doi.org/10.1016/j.polymdegradstab.2020.109131.

Dhar, P., Bhardwaj, U., Kumar, A., & Katiyar, V. (2015). Poly (3-hydroxybutyrate)/cellulose nanocrystal films for food packaging applications: Barrier and migration studies. *Polymer Engineering & Science, 55*, 2388–2395. https://doi.org/10.1002/pen.24127.

Díez-Pascual, A. M. (2021). Effect of graphene oxide on the properties of poly(3-Hydroxybutyrate-co-3-hydroxyhexanoate). *Polymers, 13*. https://doi.org/10.3390/polym13142233.

Díez-Pascual, A. M., & Díez-Vicente, A. L. (2014). Poly(3-hydroxybutyrate)/ZnO bionanocomposites with improved mechanical, barrier and antibacterial properties. *International Journal of Molecular Sciences, 15*, 10950–10973. https://doi.org/10.3390/ijms150610950.

França, D., Angelo, L. M., Souza, C. F., & Faez, R. (2021). Biobased poly(3-hydroxybutyrate)/starch/cellulose nanofibrils for nutrients coatings. *ACS Applied Polymer Materials, 3*, 3227–3237. https://doi.org/10.1021/acsapm.1c00418.

Frone, A. N., Nicolae, C. A., Eremia, M. C., Tofan, V., Ghiurea, M., Chiulan, I., Radu, E., Damian, C. M., & Panaitescu, D. M. (2020). Low molecular weight and polymeric modifiers as toughening agents in poly(3-Hydroxybutyrate) films. *Polymers, 12*. https://doi.org/10.3390/polym12112446.

Garcia-Garcia, D., Rayón, E., Carbonell-Verdu, A., Lopez-Martinez, J., & Balart, R. (2017). Improvement of the compatibility between poly(3-hydroxybutyrate) and poly(ε-caprolactone) by reactive extrusion with dicumyl peroxide. *European Polymer Journal, 86*, 41–57. https://doi.org/10.1016/j.eurpolymj.2016.11.018.

Gigante, V., Seggiani, M., Cinelli, P., Signori, F., Vania, A., Navarini, L., Amato, G., & Lazzeri, A. (2021). Utilization of coffee silverskin in the production of Poly(3-hydroxybutyrate-co-3-hydroxyvalerate) biopolymer-based thermoplastic biocomposites for food contact applications. *Composites Part A: Applied Science and Manufacturing, 140*, 106172. https://doi.org/10.1016/j.compositesa.2020.106172.

Goonoo, N., Bhaw-Luximon, A., Passanha, P., Esteves, S. R., & Jhurry, D. (2017). Third generation poly(hydroxyacid) composite scaffolds for tissue engineering. *Journal of Biomedical Materials Research, Part B: Applied Biomaterials, 105*, 1667–1684. https://doi.org/10.1002/jbm.b.33674.

Halevas, E., Kokotidou, C., Zaimai, E., Moschona, A., Lialiaris, E., Mitraki, A., Lialiaris, T., & Pantazaki, A. (2021). Evaluation of the hemocompatibility and anticancer potential of poly(ε-Caprolactone) and Poly(3-Hydroxybutyrate) microcarriers with encapsulated chrysin. *Pharmaceutics, 13*. https://doi.org/10.3390/pharmaceutics13010109.

Ibrahim, M. I., Alsafadi, D., Alamry, K. A., & Hussein, M. A. (2021). Properties and applications of poly(3-hydroxybutyrate-co-3-hydroxyvalerate) biocomposites. *Journal of Polymers and the Environment, 29*, 1010–1030. https://doi.org/10.1007/s10924-020-01946-x.

Iglesias-Montes, M. L., Soccio, M., Siracusa, V., Gazzano, M., Lotti, N., Cyras, V. P., & Manfredi, L. B. (2022). Chitin nanocomposite based on plasticized poly(lactic acid)/Poly(3-hydroxybutyrate) (PLA/PHB) blends as fully biodegradable packaging materials. *Polymers, 14*. https://doi.org/10.3390/polym14153177.

Iordanskii, A., Zhulkina, A., Olkhov, A., Fomin, S., Burkov, A., & Stilman, M. (2018). Characterization and evaluation of controlled antimicrobial release from petrochemical (PU) and biodegradable (PHB) packaging. *Polymers, 10*. https://doi.org/10.3390/polym10080817.

Jin, A., del Valle, L. J., & Puiggalí, J. (2023). Copolymers and blends based on 3-hydroxybutyrate and 3-hydroxyvalerate units. *International Journal of Molecular Sciences, 24*. https://doi.org/10.3390/ijms242417250.

Kachrimanidou, V., Ioannidou, S. M., Ladakis, D., Papapostolou, H., Kopsahelis, N., Koutinas, A. A., & Kookos, I. K. (2021). Techno-economic evaluation and life-cycle assessment of poly(3-hydroxybutyrate) production within a biorefinery concept using sunflower-based biodiesel industry by-products. *Bioresource Technology, 326*, 124711. https://doi.org/10.1016/j.biortech.2021.124711.

Kang, J., Hwang, J.-Y., Huh, M., & Yun, S. I. (2020). Porous Poly(3-hydroxybutyrate) scaffolds prepared by non-solvent-induced phase separation for tissue engineering. *Macromolecular Research, 28*, 835–843. https://doi.org/10.1007/s13233-020-8109-x.

Khasanah, Takahashi, I., Reddy, K. R., & Ozaki, Y. (2017). Crystallization of ultrathin poly(3-hydroxybutyrate) films in blends with small amounts of poly(l-lactic acid): Correlation between film thickness and molecular weight of poly(l-lactic acid). *RSC Advances, 7*, 52651–52660. https://doi.org/10.1039/C7RA10996B.

Khattab, A. M., Esmael, M. E., Farrag, A. A., & Ibrahim, M. I. A. (2021). Structural assessment of the bioplastic (poly-3-hydroxybutyrate) produced by *Bacillus flexus* Azu-A2 through cheese whey valorization. *International Journal of Biological Macromolecules, 190*, 319–332. https://doi.org/10.1016/j.ijbiomac.2021.08.090.

Koller (2014). n.d. Poly(hydroxyalkanoates) for food packaging: Application and attempts towards implementation. *Applied Food Biotechnology, 1*(1), 3–15. https://doi.org/10.22037/afb.v1i1.7127.

Kontárová, S., Přikryl, R., Melčová, V., Menčík, P., Horálek, M., Figalla, S., Plavec, R., Feranc, J., Sadílek, J., & Pospíšilová, A. (2020). Printability, mechanical and thermal properties of poly(3-Hydroxybutyrate)-Poly(Lactic Acid)-plasticizer blends for three-dimensional (3D) printing. *Materials, 13*. https://doi.org/10.3390/ma13214736.

Kontárová, S., Přikryl, R., Škarpa, P., Kriška, T., Antošovský, J., Gregušková, Z., Figalla, S., Jašek, V., Sedlmajer, M., Menčík, P., & Mikolajová, M. (2022). Slow-Release Nitrogen Fertilizers with Biodegradable Poly(3-hydroxybutyrate) Coating: Their Effect on the Growth of Maize and the Dynamics of N Release in Soil. *Polymers, 14*. https://doi.org/10.3390/polym14204323.

Korina, E., Stoilova, O., Manolova, N., & Rashkov, I. (2013). Multifunctional hybrid materials from poly(3-Hydroxybutyrate), TiO_2 nanoparticles, and chitosan oligomers by combining electrospinning/electrospraying and impregnation. *Macromolecular Bioscience, 13*, 707–716. https://doi.org/10.1002/mabi.201200410.

Kourilova, X., Pernicova, I., Vidlakova, M., Krejcirik, R., Mrazova, K., Hrubanova, K., Krzyzanek, V., Nebesarova, J., & Obruca, S. (2021). Biotechnological conversion of grape pomace to poly(3-hydroxybutyrate) by moderately thermophilic bacterium *Tepidimonas taiwanensis*. *Bioengineering, 8*. https://doi.org/10.3390/bioengineering8100141.

Kovalcik, A., Obruca, S., Kalina, M., Machovsky, M., Enev, V., Jakesova, M., Sobkova, M., & Marova, I. (2020). Enzymatic hydrolysis of poly(3-Hydroxybutyrate-co-3-Hydroxyvalerate) scaffolds. *Materials, 13*. https://doi.org/10.3390/ma13132992.

Kumari, S. V. G., Pakshirajan, K., & Pugazhenthi, G. (2023). Facile fabrication and characterization of novel antimicrobial and antioxidant poly (3-hydroxybutyrate)/essential oil composites for potential use in active food packaging applications. *International Journal of Biological Macromolecules, 252*, 126566. https://doi.org/10.1016/j.ijbiomac.2023.126566.

Laycock, B., Halley, P., Pratt, S., Werker, A., & Lant, P. (2013). The chemomechanical properties of microbial polyhydroxyalkanoates. *Progress in Polymer Science, 38*, 536–583. https://doi.org/10.1016/j.progpolymsci.2012.06.003.

Loan, T. T., Trang, D. T. Q., Huy, P. Q., Ninh, P. X., & Van Thuoc, D. (2022). A fermentation process for the production of poly(3-hydroxybutyrate) using waste cooking oil or waste fish oil as inexpensive carbon substrate. *Biotechnology Reports, 33*, e00700. https://doi.org/10.1016/j.btre.2022.e00700.

Łopusiewicz, Ł., Macieja, S., Śliwiński, M., Bartkowiak, A., Roy, S., & Sobolewski, P. (2022). Alginate biofunctional films modified with melanin from watermelon seeds and zinc oxide/silver nanoparticles. *Materials*, 15. https://doi.org/10.3390/ma15072381.

Low, T. J., Mohammad, S., Sudesh, K., & Baidurah, S. (2021). Utilization of banana (Musa sp.) fronds extract as an alternative carbon source for poly(3-hydroxybutyrate) production by Cupriavidus necator H16. *Biocatalysis and Agricultural Biotechnology, 34*, 102048. https://doi.org/10.1016/j.bcab.2021.102048.

Lugoloobi, I., Li, X., Zhang, Y., Mao, Z., Wang, B., Sui, X., & Feng, X. (2020). Fabrication of lignin/poly(3-hydroxybutyrate) nanocomposites with enhanced properties via a Pickering emulsion approach. *International Journal of Biological Macromolecules, 165*, 3078–3087. https://doi.org/10.1016/j.ijbiomac.2020.10.156.

Macieja, S., Środa, B., Zielińska, B., Roy, S., Bartkowiak, A., & Łopusiewicz, Ł. (2022). Bioactive carboxymethyl cellulose (CMC)-based films modified with melanin and silver nanoparticles (AgNPs)—The effect of the degree of CMC substitution on the in situ synthesis of agnps and films' functional properties. *International Journal of Molecular Sciences, 23*. https://doi.org/10.3390/ijms232415560.

Manikandan, N. A., Pakshirajan, K., & Pugazhenthi, G. (2020). A closed-loop biorefinery approach for polyhydroxybutyrate (PHB) production using sugars from carob pods as the sole raw material and downstream processing using the co-product lignin. *Bioresource Technology, 307*, 123247. https://doi.org/10.1016/j.biortech.2020.123247.

Maureira, J., Olate-Moya, F., Bastías, R., Farias, S., Alvarez, R., Rosales-Cuello, N., & Palza, H. (2023). Multifunctional poly(3-hydroxybutyrate) composites with MoS2 for food packaging applications. *European Polymer Journal, 188*, 111914. https://doi.org/10.1016/j.eurpolymj.2023.111914.

Melčová, V., Svoradová, K., Menčík, P., Kontárová, S., Rampichová, M., Hedvičáková, V., Sovková, V., Přikryl, R., & Vojtová, L. (2020). FDM 3D printed composites for bone tissue engineering based on plasticized poly(3-hydroxybutyrate)/poly(d,l-lactide) blends. *Polymers, 12*. https://doi.org/10.3390/polym12122806.

Melendez-Rodriguez, B., Figueroa-Lopez, K. J., Bernardos, A., Martínez-Máñez, R., Cabedo, L., Torres-Giner, S., & M. Lagaron, J. (2019). Electrospun antimicrobial films of poly(3-hydroxybutyrate-co-3-hydroxyvalerate) containing eugenol essential oil encapsulated in mesoporous silica nanoparticles. *Nanomaterials, 9*. https://doi.org/10.3390/nano9020227.

Mierziak, J., Burgberger, M., & Wojtasik, W. (2021). 3-Hydroxybutyrate as a metabolite and a signal molecule regulating processes of living organisms. *Biomolecules, 11*. https://doi.org/10.3390/biom11030402.

Miguel, O., & Iruin, J. J. (1999). Evaluation of the transport properties of Poly(3-hydroxybutyrate) and its 3-hydroxyvalerate copolymers for packaging applications. *Macromolecular Symposia, 144*, 427–438. https://doi.org/10.1002/masy.19991440140.

Mittal, M., Ahuja, S., Yadav, A., & Aggarwal, N. K. (2023). Development of poly(hydroxybutyrate) film incorporated with nano silica and clove essential oil intended for active packaging of brown bread. *International Journal of Biological Macromolecules, 233*, 123512. https://doi.org/10.1016/j.ijbiomac.2023.123512.

Mlalila, N., Hilonga, A., Swai, H., Devlieghere, F., & Ragaert, P. (2018). Antimicrobial packaging based on starch, poly(3-hydroxybutyrate) and poly (lactic-co-glycolide) materials and application challenges. *Trends in Food Science & Technology, 74*, 1–11. https://doi.org/10.1016/j.tifs.2018.01.015.

Mohammadalizadeh, Z., Karbasi, S., & Arasteh, S. (2020). Physical, mechanical and biological evaluation of poly (3-hydroxybutyrate)-chitosan/ MWNTs as a novel electrospun scaffold for cartilage tissue engineering applications. *Polymer-Plastics Technology and Materials, 59*, 417–429. https://doi.org/10.1080/25740881.2019.1647244.

Moll, E., & Chiralt, A. (2023). Polyhydroxybutyrate-co-hydroxyvalerate (PHBV) with phenolic acids for active food packaging. *Polymers, 15*. https://doi.org/10.3390/polym15214222.

Mostafa, Y. S., Alrumman, S. A., Alamri, S. A., Otaif, K. A., Mostafa, M. S., & Alfaify, A. M. (2020). Bioplastic (poly-3-hydroxybutyrate) production by the marine bacterium Pseudodonghicola xiamenensis through date syrup valorization and structural assessment of the biopolymer. *Scientific Reports, 10*, 8815. https://doi.org/10.1038/s41598-020-65858-5.

Murueva, A. V., Shishatskaya, E. I., Kuzmina, A. M., Volova, T. G., & Sinskey, A. J. (2013). Microparticles prepared from biodegradable polyhydroxyalkanoates as matrix for encapsulation of cytostatic drug. *Journal of Materials Science: Materials in Medicine, 24*, 1905–1915. https://doi.org/10.1007/s10856-013-4941-2.

Naderi, P., Zarei, M., Karbasi, S., & Salehi, H. (2020). Evaluation of the effects of keratin on physical, mechanical and biological properties of poly (3-hydroxybutyrate) electrospun scaffold: Potential application in bone tissue engineering. *European Polymer Journal, 124*, 109502. https://doi.org/10.1016/j.eurpolymj.2020.109502.

Naitam, M. G., Tomar, G. S., & Kaushik, R. (2022). Optimization and production of holocellulosic enzyme cocktail from fungi *Aspergillus nidulans* under solid-state fermentation for the production of poly(3-hydroxybutyrate). *Fungal Biology and Biotechnology, 9*, 17. https://doi.org/10.1186/s40694-022-00147-6.

Nanni, A., & Messori, M. (2021). Effect of the wine wastes on the thermal stability, mechanical properties, and biodegradation's rate of poly(3-hydroxybutyrate). *Journal of Applied Polymer Science, 138*, 49713. https://doi.org/10.1002/app.49713.

Narayanan, A., Neera, Mallesha, & Ramana, K. V. (2013). Synergized antimicrobial activity of eugenol incorporated polyhydroxybutyrate films against food spoilage microorganisms in conjunction with pediocin. *Applied Biochemistry and Biotechnology, 170*, 1379–1388. https://doi.org/10.1007/s12010-013-0267-2.

Panaitescu, D. M., Frone, A. N., Nicolae, C.-A., Gabor, A. R., Miu, D. M., Soare, M.-G., Vasile, B. S., & Lupescu, I. (2023). Poly(3-hydroxybutyrate) nanocomposites modified with even and odd chain length polyhydroxyalkanoates. *International Journal of Biological Macromolecules, 244*, 125324. https://doi.org/10.1016/j.ijbiomac.2023.125324.

Panaitescu, D. M., Ionita, E. R., Nicolae, C.-A., Gabor, A. R., Ionita, M. D., Trusca, R., Lixandru, B.-E., Codita, I., & Dinescu, G. (2018). Poly(3-hydroxybutyrate) modified by nanocellulose and plasma treatment for packaging applications. *Polymers, 10*. https://doi.org/10.3390/polym10111249.

Park, Y.-L., Song, H.-S., Choi, T.-R., Lee, S. M., Park, S. L., Lee, H. S., Kim, H.-J., Bhatia, S. K., Gurav, R., Park, K., & Yang, Y.-H. (2021). Revealing of sugar utilization systems in Halomonas sp. YLGW01 and application for poly(3-hydroxybutyrate) production with low-cost medium and easy recovery. *International Journal of Biological Macromolecules, 167*, 151–159. https://doi.org/10.1016/j.ijbiomac.2020.11.163.

Parvizifard, M., Karbasi, S., Salehi, H., & Soleymani Eil Bakhtiari, S. (2020). Evaluation of physical, mechanical and biological properties of bioglass/ titania scaffold coated with poly (3-hydroxybutyrate)-chitosan for bone tissue engineering applications. *Materials Technology, 35*, 75–91. https://doi.org/10.1080/10667857.2019.1658169.

Patel, M. K., Hansson, F., Pitkänen, O., Geng, S., & Oksman, K. (2022). Biopolymer blends of poly(lactic acid) and Poly(hydroxybutyrate) and their functionalization with glycerol triacetate and chitin nanocrystals for food packaging applications. *ACS Applied Polymer Materials, 4*, 6592–6601. https://doi.org/10.1021/acsapm.2c00967.

Perret, E., Reifler, F. A., Gooneie, A., Chen, K., Selli, F., & Hufenus, R. (2020). Structural response of melt-spun poly(3-hydroxybutyrate) fibers to stress and temperature. *Polymer, 197*, 122503. https://doi.org/10.1016/j.polymer.2020.122503.

Pietrosanto, A., Scarfato, P., Di Maio, L., & Incarnato, L. (2021). Development of Pla / Phb blown films with improved performance for food packaging applications. *Chemical Engineering Transactions, 87*, 91–96. https://doi.org/10.3303/CET2187016.

Policastro, G., Panico, A., & Fabbricino, M. (2021). Improving biological production of poly(3-hydroxybutyrate-co-3-hydroxyvalerate) (PHBV) co-polymer: A critical review. *Reviews in Environmental Science and Bio/Technology, 20*, 479–513. https://doi.org/10.1007/s11157-021-09575-z.

Popa, M. S., Frone, A. N., & Panaitescu, D. M. (2022). Polyhydroxybutyrate blends: A solution for biodegradable packaging? *International Journal of Biological Macromolecules, 207*, 263–277. https://doi.org/10.1016/j.ijbiomac.2022.02.185.

Quispe, M. M., Lopez, O. V., Boina, D. A., Stumbé, J.-F., & Villar, M. A. (2021). Glycerol-based additives of poly(3-hydroxybutyrate) films. *Polymer Testing, 93*, 107005. https://doi.org/10.1016/j.polymertesting.2020.107005.

Rami Reddy Tadi, S., Dutt Ravindran, S., Balakrishnan, R., & Sivaprakasam, S. (2021). Recombinant production of poly-(3-hydroxybutyrate) by *Bacillus megaterium* utilizing millet bran and rapeseed meal hydrolysates. *Bioresource Technology, 326*, 124800. https://doi.org/10.1016/j.biortech.2021.124800.

Requena, R., Vargas, M., & Chiralt, A. (2017). Release kinetics of carvacrol and eugenol from poly(hydroxybutyrate-co-hydroxyvalerate) (PHBV) films for food packaging applications. *European Polymer Journal, 92*, 185–193. https://doi.org/10.1016/j.eurpolymj.2017.05.008.

Ruka, D. R., Simon, G. P., & Dean, K. (2014). Harvesting fibrils from bacterial cellulose pellicles and subsequent formation of biodegradable poly-3-hydroxybutyrate nanocomposites. *Cellulose, 21*, 4299–4308. https://doi.org/10.1007/s10570-014-0415-z.

Russo, G., Scocca, P., Gelosia, M., Fabbrizi, G., Giannoni, T., Urbani, S., Esposto, S., & Nicolini, A. (2024). Poly(3-hydroxybutyrate) production for food packaging from biomass derived carbohydrates by cupriavidus necator DSM 545. *Enzyme and Microbial Technology, 181*, 110516. https://doi.org/10.1016/j.enzmictec.2024.110516.

Salama, H. E., Aziz, M. S. A., & Saad, G. R. (2018). Thermal properties, crystallization and antimicrobial activity of chitosan biguanidine grafted poly (3-hydroxybutyrate) containing silver nanoparticles. *International Journal of Biological Macromolecules, 111*, 19–27. https://doi.org/10.1016/j.ijbiomac.2017.12.153.

Sanchez-Garcia, M. D., Gimenez, E., & Lagaron, J. M. (2007). Novel PET nanocomposites of interest in food packaging applications and comparative barrier performance with biopolyester nanocomposites. *Journal of Plastic Film & Sheeting, 23*, 133–148. https://doi.org/10.1177/8756087907083590.

Sindhu, et al. (2020). Bioconversion of waste cooking oil for the production of poly-3-hydroxybutyrate using *Bacillus cereus* MPTDC. *IJEB*. https://doi.org/10.56042/ijeb.v58i08.39413.

Sirohi, R., Kumar Gaur, V., Kumar Pandey, A., Jun Sim, S., & Kumar, S. (2021a). Harnessing fruit waste for poly-3-hydroxybutyrate production: A review. *Bioresource Technology, 326*, 124734. https://doi.org/10.1016/j.biortech.2021.124734.

Sirohi, R., Pandey, J. P., Tarafdar, A., Agarwal, A., Chaudhuri, S. K., & Sindhu, R. (2021b). An environmentally sustainable green process for the utilization of damaged wheat grains for poly-3-hydroxybutyrate production. *Environmental Technology & Innovation, 21*, 101271. https://doi.org/10.1016/j.eti.2020.101271.

Smith, M. K. M., Paleri, D. M., Abdelwahab, M., Mielewski, D. F., Misra, M., & Mohanty, A. K. (2020). Sustainable composites from poly(3-hydroxybutyrate) (PHB) bioplastic and agave natural fibre. *Green Chemistry: an International Journal and Green Chemistry Resource: GC, 22*, 3906–3916. https://doi.org/10.1039/D0GC00365D.

Solaiman, D. K. Y., Ashby, R. D., Zerkowski, J. A., Krishnama, A., & Vasanthan, N. (2015). Control-release of antimicrobial sophorolipid employing different biopolymer matrices. *Biocatalysis and Agricultural Biotechnology, 4*, 342–348. https://doi.org/10.1016/j.bcab.2015.06.006.

Sosa-Hernández, J. E., Villalba-Rodríguez, A. M., Romero-Castillo, K. D., Zavala-Yoe, R., Bilal, M., Ramirez-Mendoza, R. A., Parra-Saldivar, R., & Iqbal, H. M. N. (2020). Poly-3-hydroxybutyrate-based constructs with novel characteristics for drug delivery and tissue engineering applications—A review. *Polymer Engineering & Science, 60*, 1760–1772. https://doi.org/10.1002/pen.25470.

Suhazsini, P., Keshav, R., Narayanan, S., Chaudhuri, A., & Radha, P. (2020). A study on the synthesis of poly (3-hydroxybutyrate-co-3-hydroxyvalerate) by *Bacillus megaterium* utilizing cheese whey permeate. *Journal of Polymers and the Environment, 28*, 1390–1405. https://doi.org/10.1007/s10924-020-01687-x.

Sukhanova, A. A., Murzova, A. E., Boyandin, A. N., Kiselev, E. G., Sukovatyi, A. G., Kuzmin, A. P., & Shabanov, A. V. (2020). Poly-3-hydroxybutyrate/chitosan composite films and nonwoven mats. *International Journal of Biological Macromolecules, 165*, 2947–2956. https://doi.org/10.1016/j.ijbiomac.2020.10.177.

Sun, X., Shang, Y., Zhang, B., Guo, P., Luo, Y., & Wu, H. (2024). Engineering of fast-growing Vibrio natriegens for biosynthesis of poly(3-hydroxybutyrate-co-lactate). *Bioresources and Bioprocessing, 11*, 86. https://doi.org/10.1186/s40643-024-00801-4.

Ta, D.-T., Chiang, C.-J., Huang, Z.-X., Luu, N.-L., & Chao, Y.-P. (2023). High production of poly(3-hydroxybutyrate) in *Escherichia coli* using crude glycerol. *Bioresource Technology, 384*, 129315. https://doi.org/10.1016/j.biortech.2023.129315.

Tanideh, N., Azarpira, N., Sarafraz, N., Zare, S., Rowshanghiyas, A., Farshidfar, N., Iraji, A., Zarei, M., & El Fray, M. (2020). Poly(3-Hydroxybutyrate)-multiwalled carbon nanotubes electrospun scaffolds modified with curcumin. *Polymers, 12*. https://doi.org/10.3390/polym12112588.

Tasbihi, M., Kwon, S., Kim, B., Brüggemann, D., Hou, H., Lu, J., Amitrano, R., Grimm, T., García-Antón, J., Strasser, P., Riedel, S. L., & Schwarze, M. (2024). Polyhydroxykanoate-assisted photocatalytic TiO_2 films for hydrogen production. *Langmuir: the ACS Journal of Surfaces and Colloids, 40*, 25800–25810. https://doi.org/10.1021/acs.langmuir.4c02727.

Thomas, S., Shumilova, A. A., Kiselev, E. G., Baranovsky, S. V., Vasiliev, A. D., Nemtsev, I. V., Kuzmin, A. P., Sukovatyi, A. G., Avinash, R. P., & Volova, T. G. (2020). Thermal, mechanical and biodegradation studies of biofiller based poly-3-hydroxybutyrate biocomposites. *International Journal of Biological Macromolecules, 155*, 1373–1384. https://doi.org/10.1016/j.ijbiomac.2019.11.112.

Toalá, C. U., Prokhorov, E., Barcenas, G. L., Landaverde, M. A. H., Limón, J. M. Y., Gervacio-Arciniega, J. J., de Fuentes, O. A., & Tapia, A. M. G. (2023). Electrostrictive and piezoelectrical properties of chitosan-poly(3-hydroxybutyrate) blend films. *International Journal of Biological Macromolecules, 250*, 126251. https://doi.org/10.1016/j.ijbiomac.2023.126251.

Usurelu, C. D., Badila, S., Frone, A. N., & Panaitescu, D. M. (2022). Poly(3-hydroxybutyrate) nanocomposites with cellulose nanocrystals. *Polymers, 14*. https://doi.org/10.3390/polym14101974.

Vaithanomsat, P., Kongsin, K., Trakunjae, C., Boonyarit, J., Jarerat, A., Sudesh, K., & Chollakup, R. (2021). Biosynthesized Poly(3-Hydroxybutyrate) on coated pineapple leaf fiber papers for biodegradable packaging application. *Polymers, 13*. https://doi.org/10.3390/polym13111733.

Vanheusden, C., Samyn, P., Vackier, T., Steenackers, H., D'Haen, J., Peeters, R., & Buntinx, M. (2024). Fabrication of poly(3-hydroxybutyrate-co-3-hydroxyhexanoate)/ZnO nanocomposite films for active packaging applications: Impact of ZnO type on structure–property dynamics. *Polymers, 16*. https://doi.org/10.3390/polym16131861.

Venezia, V., Prieto, C., Evtoski, Z., Marcoaldi, C., Silvestri, B., Vitiello, G., Luciani, G., & Lagaron, J. M. (2023). Electrospun hybrid TiO_2/humic substance PHBV films for active food packaging applications. *Journal of Industrial and Engineering Chemistry, 124*, 510–522. https://doi.org/10.1016/j.jiec.2023.05.006.

Volkov, A. V., Muraev, A. A., Zharkova, I. I., Voinova, V. V., Akoulina, E. A., Zhuikov, V. A., Khaydapova, D. D., Chesnokova, D. V., Menshikh, K. A., Dudun, A. A., Makhina, T. K., Bonartseva, G. A., Asfarov, T. F., Stamboliev, I. A., Gazhva, Y. V., Ryabova, V. M., Zlatev, L. H., Ivanov, S. Y., Shaitan, K. V., & Bonartsev, A. P. (2020). Poly(3-hydroxybutyrate)/hydroxyapatite/alginate scaffolds seeded with mesenchymal stem cells enhance the regeneration of critical-sized bone defect. *Materials Science and Engineering: C, 114*, 110991. https://doi.org/10.1016/j.msec.2020.110991.

Wani, K. M., Pandita, G., Pillai, A. R. S., Eapen, A. S., & Roy, S. (2025). Sustainable coatings and films for active food packaging. In S.ul Islam, & M. Shahid (Eds.). *Green materials for active food packaging* (pp. 253–291). Singapore: Springer Nature Singapore. https://doi.org/10.1007/978-981-96-0369-5_8.

Xavier, J. R., Babusha, S. T., George, J., & Ramana, K. V. (2015). Material properties and antimicrobial activity of polyhydroxybutyrate (PHB) films incorporated with vanillin. *Applied Biochemistry and Biotechnology, 176*, 1498–1510. https://doi.org/10.1007/s12010-015-1660-9.

Yew, S.-P., Tang, H.-Y., & Sudesh, K. (2006). Photocatalytic activity and biodegradation of polyhydroxybutyrate films containing titanium dioxide. *Polymer Degradation and Stability, 91*, 1800–1807. https://doi.org/10.1016/j.polymdegradstab.2005.11.011.

Yin, F., Li, D., Ma, X., Li, J., & Qiu, Y. (2020). Poly(3-hydroxybutyrate-3-hydroxyvalerate) production from pretreated waste lignocellulosic hydrolysates and acetate co-substrate. *Bioresource Technology, 316*, 123911. https://doi.org/10.1016/j.biortech.2020.123911.

Zhang, C., Liu, Y., Jin, F., & Xi, Z. (2024). Formulation and characterization of poly(3-hydroxybutyrate)-based microspheres containing acetamiprid. *Journal of Polymers and the Environment, 32*, 5036–5045. https://doi.org/10.1007/s10924-024-03292-8.

Zhang, L., Jiang, Z., Tsui, T.-H., Loh, K.-C., Dai, Y., & Tong, Y. W. (2022). A review on enhancing cupriavidus necator fermentation for poly(3-hydroxybutyrate) (PHB) production from low-cost carbon sources. *Frontiers Bioengineering Biotechnology, 10*, 946085. https://doi.org/10.3389/fbioe.2022.946085.

Zhou, Y., Katsou, E., & Fan, M. (2021). Interfacial structure and property of eco-friendly carboxymethyl cellulose/poly(3-hydroxybutyrate-co-3-hydroxyvalerate) biocomposites. *International Journal of Biological Macromolecules, 179*, 550–556. https://doi.org/10.1016/j.ijbiomac.2021.03.009.

Zhuikova, Y., Zhuikov, V., & Varlamov, V. (2022). Biocomposite materials based on poly(3-hydroxybutyrate) and chitosan: A review. *Polymers, 14*. https://doi.org/10.3390/polym14245549.

Chapter 20

Polycaprolactone films and coatings

Milad Tavassoli[1,2], Fatemeh Madanifar[3], Aida Zarshenas[3], Zahra Rahimi Motlagh[3], Mohammadreza Ostovari[3], Mohammad Hamidi[3], and Hadi Ebrahimi[3]

[1]*Cellular and Molecular Research Center, Yasuj University of Medical Sciences, Yasuj, Iran,* [2]*Department of Nutrition, Faculty of Health and Nutrition Sciences, Yasuj University of Medical Sciences, Yasuj, Iran,* [3]*Student Research Committee, Yasuj University of Medical Sciences, Yasuj, Iran*

20.1 Introduction

The field of materials science is undergoing a transformative evolution, particularly in plastics, which are among the most widely used materials for their adaptability and versatility. Plastics play a pivotal role in numerous applications, with packaging being one of the most prominent. Their extensive use in this sector not only meets the demands of modern consumerism but also drives continuous innovation and advancement. This highlights the indispensable role plastics play in shaping the materials science industry and the broader economy (Abd Al-Ghani et al., 2021; Abdellah Ali, 2016; Ahmadi, Ahari, et al., 2024). However, the convenience of plastics, particularly those derived from fossil fuels, has given rise to significant and enduring challenges, primarily related to plastic waste and its environmental implications. The proliferation of single-use plastics has led to alarming levels of waste, resulting in widespread environmental pollution that threatens ecosystems and human health alike. This pressing dilemma underscores the urgent need for comprehensive, holistic strategies to mitigate the global environmental consequences of plastic waste (Abdellah Ali, 2020; Abioye et al., 2019). In particular, the increasing reliance on petroleum-based packaging materials within the food industry has sparked considerable concern and scrutiny from both regulatory bodies and environmentally conscious consumers. This has prompted a concentrated investigation into the potential of polymers produced from renewable resources as sustainable alternatives (Ahmadi, Ahari, et al., 2024; Ahmadi, Khoshkalampour, et al., 2024; Bahramian et al., 2025). The shift toward these alternative materials highlights the growing recognition of the vital connection between industrial practices and environmental challenges, particularly within the food packaging sector. Accordingly, the development of sustainable and innovative food packaging solutions has become a central focus, driven by global concerns about environmental impact (Ahn et al., 2024; Ali et al., 2014; Khezerlou et al., 2023). Among the promising candidates in this arena is polycaprolactone (PCL), a biodegradable and biocompatible synthetic polymer belonging to the polyester family. PCL is distinguished by its appealing set of physical, chemical, thermal, and mechanical properties, which render it applicable across a diverse range of fields. Its characteristics can be tailored by manipulating both its chemical composition and crystallinity, further enhancing its versatility (Bahramian et al., 2025; Phadtare et al., 2025). Additionally, PCL shows remarkable compatibility with various other polymers, positioning it as a highly significant contender for food packaging applications. Ongoing research on PCL extends beyond traditional applications and emphasizes its development as an environmentally friendly material. Investigations into its potential uses highlight PCL's critical role in advancing sustainable and innovative approaches across multiple scientific domains. The quest for sustainable and efficient food packaging solutions has spurred extensive research to identify innovative materials that balance functionality, environmental impact, and consumer safety (Yang et al., 2024; Zhang, Zhu, et al., 2024). In this context, PCL has emerged as a particularly promising option, showcasing a unique combination of properties that makes it exceptionally well-suited for creating composite films tailored for food packaging applications. This chapter provides a detailed analysis of the application of PCL-based composite films in the food packaging arena. It will delve into recent advancements and discoveries in this field, addressing both scientific innovations and environmental considerations that underscore the importance of developing sustainable packaging solutions in our quest for a more eco-friendly future (Yousaf et al., 2024; Zang et al., 2024).

Natural Polymer-Based Films and Coatings in Food Technology. DOI: https://doi.org/10.1016/B978-0-443-34244-8.00012-4

The primary objective of this comprehensive review is to provide an in-depth, nuanced understanding of the use of PCL-based films and coatings in two pivotal domains: biomedical applications and the preservation and enhancement of food quality. This exploration is crafted by systematically dissecting the extraction and synthesis methods associated with PCL, elucidating various fabrication techniques, and detailing the functional attributes of these films. By doing so, the review aspires to contribute significantly to the burgeoning field of sustainable and effective food packaging solutions (Xu et al., 2024; Yang et al., 2024; Zhang, Bu, et al., 2024). This chapter is distinctively structured to reflect the multifaceted nature of PCL-based composite films, thereby providing clarity and organization to the discussion. It commences with a thorough analysis of PCL itself, exploring its chemical properties, synthesis methods, including ring-opening polymerization (ROP) and other relevant techniques, and highlighting the unique functional properties that make PCL an appealing candidate for food packaging applications. Subsequently, the chapter delves into the various fabrication methods used to create composite films. This section aims to serve as a detailed guide for researchers and industry professionals, outlining techniques such as electrospinning, solvent casting, and extrusion that are pivotal in producing PCL-based materials. Each method is scrutinized to present a clear understanding of its advantages and limitations, offering insights on selection criteria based on desired film properties. The discourse then progresses to an in-depth examination of the characteristics of PCL-based composite films, emphasizing not only their mechanical and physical properties, such as tensile strength, flexibility, and barrier properties, but also their microstructural and optical features. This analysis reveals how these characteristics influence the films' performance in real-world applications. A significant portion of the review is dedicated to the practical utilization of these materials in food packaging. This section categorizes applications into three main classifications: active-based, intelligent-based, and nanofiber-based packaging solutions. Each category is explored in detail, highlighting how PCL films can enhance food preservation, extend shelf life, and serve as indicators for quality monitoring, underscoring their multifaceted roles in improving food quality and safety. As the chapter approaches its conclusion, it synthesizes key insights from the discussion, offering a panoramic overview of the current research landscape on PCL-based composite films. This final section not only recaps the range of applications, properties, and methodologies discussed but also identifies critical knowledge gaps in the field. Moreover, it suggests novel research pathways that could spur further innovation in drug delivery systems, tissue engineering, and food packaging. Ultimately, this chapter aspires to be an invaluable resource for a diverse audience, including researchers, industry experts, and policymakers engaged in the rapidly evolving sectors of sustainable biomedical applications and food industries. By promoting a deeper understanding of the potential benefits and challenges of PCL-based solutions, it aims to inspire future advancements and collaborative efforts in these critical areas.

20.2 Overview of polycaprolactone

PCL is a highly versatile and biodegradable polyester that has garnered substantial interest in both academic and industrial circles due to its unique properties and broad spectrum of applications. Created through the process of ROP of ε-caprolactone, PCL features a distinct set of attributes that position it as a favorable material for various sectors, particularly in the realm of composite films suitable for food packaging (Bahramian et al., 2025; Li, Li, et al., 2024; Ma et al., 2024; Sundar et al., 2024; Veras et al., 2024; Viel et al., 2024). One of the primary appeals of PCL is its exceptional thermal stability coupled with mechanical flexibility. These characteristics make it an ideal candidate for a variety of manufacturing methods, including extrusion and injection molding. PCL's low melting point, typically ranging from 59°C to 64°C, streamlines processing, allowing manufacturers to mold and shape it with remarkable ease. This property also facilitates the production of films tailored to specific requirements, enhancing their effectiveness across a range of packaging applications (Shi et al., 2024; Syafiq et al., 2024; Tousignant et al., 2024). A defining feature of PCL that sets it apart from traditional packaging materials is its biodegradability. As industries increasingly grapple with the environmental challenges posed by conventional packaging, PCL has emerged as a sustainable alternative that addresses growing ecological concerns. Understanding the complex interplay between PCL's durability, critical for ensuring the safety and longevity of food products, and its environmentally friendly qualities is vital for assessing its overall ecological footprint (Santana et al., 2024; Shorey & Mekonnen, 2024; Wang et al., 2024). From a chemical standpoint, PCL is characterized by its hydrophobic nature and crystalline structure. It possesses a glass transition temperature of −60°C, which contributes to its flexibility at low temperatures. The molecular weight of PCL can vary considerably, ranging from 3000 to 90,000 g mol^{-1}. This variance allows categorization by molecular weight, which influences mechanical properties and biodegradation behavior. Various factors, such as crystallinity, molecular weight, and external degradation conditions (temperature, pH, and environmental exposure), significantly influence the biodegradation rate of PCL (Ruan et al., 2024; Suryani, Rihayat, Fitria, Safitri, et al., 2024). Notably, studies indicate that increases in molecular weight or crystallinity can impede degradation. PCL undergoes multiple degradation mechanisms depending on the environment in which it is placed. These environments can include compost, soil, and active sludge, as well as interactions with fungal cultures and enzymes, each of which exhibits different rates of degradation. This complexity underscores the importance of understanding PCL's behavior across various settings to optimize its applications and effectively assess its environmental impact (Ozer & San Keskin, 2024; Qin, Fang, et al., 2024; Qin, Jiang, et al., 2024; Syafiq et al., 2024).

20.2.1 Chemistry and structure

PCL is a notable biodegradable polyester that has attracted considerable interest in sustainable materials, particularly in the food packaging sector. Its distinct chemical architecture features repeating units of -CH_2-CH(OH)-CH_2-CH_2-COOH, which underpin a range of exceptional functional properties such as biocompatibility, biodegradability, and the versatility for chemical modification (Li, Li, et al., 2024; Nair et al., 2024; Pei et al., 2024; Piscitelli et al., 2024). One of the defining characteristics of PCL is its relatively low melting point, typically around 60°C, which influences its processing and applications across various environments. Additionally, it possesses a glass transition temperature of approximately −60°C, indicating its ability to remain flexible and malleable at lower temperatures, an essential attribute for packaging materials that must retain performance under varying storage conditions (Moll & Chiralt, 2024; Moshkbid et al., 2024; Muñoz-Suarez et al., 2024). The production of PCL involves the ROP of ε-caprolactone, a process often facilitated by catalysts such as stannous octanoate, which enhance efficiency and control over the polymerization. The resulting semicrystalline structure of PCL is key to its physical properties, providing high toughness and elasticity essential for packaging materials that must endure the mechanical stresses of transport, handling, and storage (Martinaud et al., 2024; Mohamady Hussein et al., 2024). Adjusting the molecular weight of PCL is crucial in tailoring its mechanical properties to suit specific applications. By fine-tuning this parameter, manufacturers can create packaging films that are both durable and flexible, effectively meeting the diverse and evolving requirements of the food industry. This adaptability is vital for ensuring the materials can adequately protect various food products, particularly those sensitive to external conditions (Li, Li, et al., 2024; Liang et al., 2024; Ma et al., 2024). Another significant advantage of PCL is its biodegradation profile, which occurs over an average period of two to three years. This degradation is primarily driven by microbial activity or hydrolysis of the aliphatic ester bonds under physiological conditions. Such a moderate degradation rate is beneficial for applications that require controlled release of the packaging material, particularly for food packaging intended to safeguard edibles over extended periods without introducing long-term environmental risks (Leite et al., 2024; Li, Tseng, et al., 2024; Matumba et al., 2024). PCL's biocompatibility is perhaps one of its most critical attributes, confirming that it does not leach harmful substances into food products. This feature is especially important for packaging that comes into direct contact with consumables like fresh produce or dairy items, ensuring an environmentally friendly and safe option in terms of food safety (Khoshboui Lahijani et al., 2024; Lawless et al., 2024; Li, Li, et al., 2024). Moreover, PCL's chemical structure allows for significant versatility. Its properties can be enhanced by incorporating bioactive compounds or by blending with other biopolymers, such as polylactic acid (PLA). This modification capability not only enhances the safety and shelf life of food products but also opens the door to innovative packaging solutions tailored to specific market needs and consumer preferences. By leveraging these attributes, PCL stands out as a robust candidate for the future of sustainable food packaging (Huang, Huang, et al., 2024; Jiang et al., 2024; Kenawy et al., 2024).

In summary, the functional characteristics of PCL position it as an exceptionally valuable material for food packaging applications. PCL is notable for its biodegradability, flexibility, and ability to form a barrier against moisture and gases, which are critical for maintaining the freshness and safety of food products. Recent studies have expanded its potential applications, exploring innovative ways to enhance its performance and broaden its use in various food packaging contexts. The unique adaptability of PCL allows it to be customized for specific packaging requirements, such as varying thickness, strength, and barrier properties, making it suitable for a wide range of food items, from perishable goods to dry snacks. This versatility, in conjunction with ongoing research to improve its properties and reduce production costs, underscores PCL's promising role in the development of sustainable packaging solutions. As consumers and industries increasingly prioritize environmental responsibility, PCL stands out as a preferred choice for the future, aligning with the growing demand for effective, environmentally friendly food packaging alternatives.

20.2.2 Physicochemical properties

The degradation of PCL primarily occurs through two significant mechanisms: enzymatic degradation, commonly referred to as surface erosion, and hydrolytic degradation, or bulk erosion. Enzymatic degradation is characterized by a reduction in PCL weight while maintaining a relatively stable molecular weight. This process occurs on the material's surface, allowing it to break down without significantly altering its overall structural integrity at the molecular level. Fig. 20.1 shows the different methods of PCL synthesis. In contrast, hydrolytic degradation results in a concurrent decrease in both the molecular weight and the physical weight of the PCL, indicative of a more profound internal breakdown of the

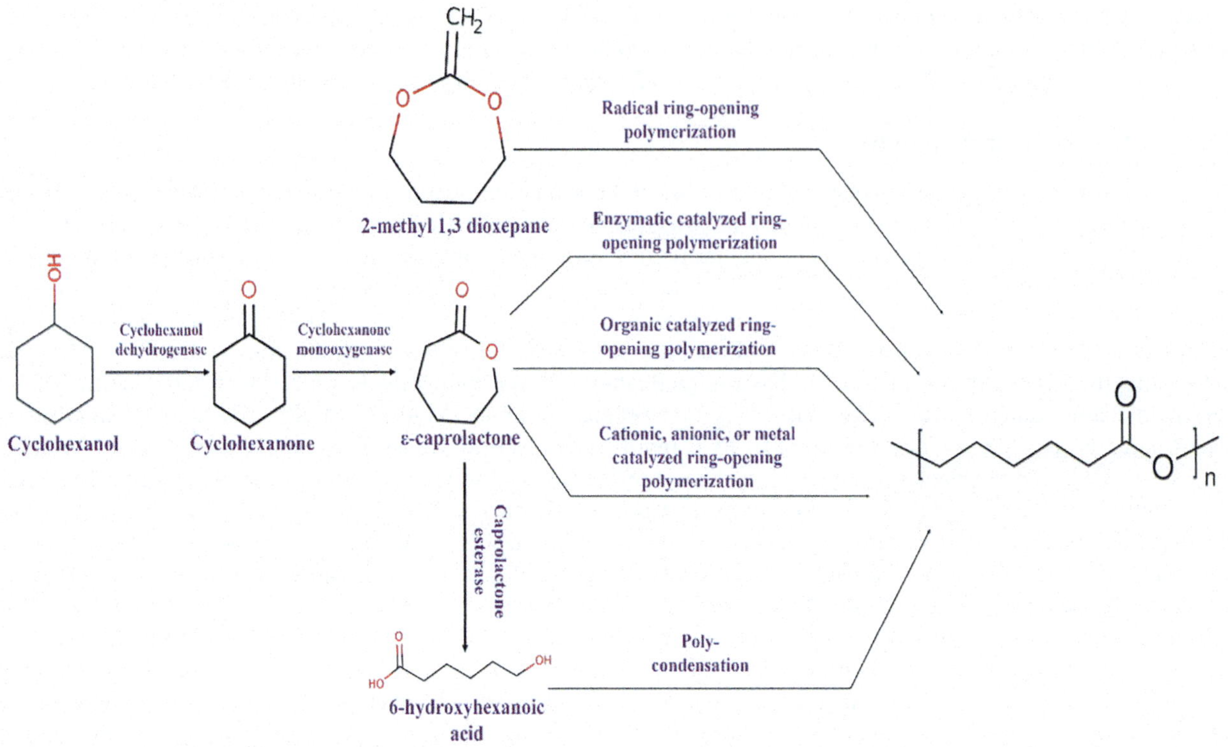

FIGURE 20.1 Schematic of different PCL synthesis methods.

polymer chains (Hu, Huang, et al., 2024; Hu, Li, et al., 2024; Huang, Fang, et al., 2024). PCL exhibits an optimal melting temperature of 58°C–60°C, which enhances its degradability, as composting environments typically exceed this range. The primary breakdown pathways involve enzymatic and chemical hydrolysis of PCL's ester bonds, with enzymatic hydrolysis emerging as the dominant mechanism driving degradation. Adding another layer of complexity, blending PCL with other biopolymers such as chitosan, starch, cellulose, and PLA can significantly enhance the tensile strength of the composites. This strengthening renders them less susceptible to microbial degradation, thereby extending the material's lifespan across various applications (Heiner et al., 2024; Mohamady Hussein et al., 2024; Ruan et al., 2024; Zhang, Zhu, et al., 2024). The degradation of PCL is influenced by temperature dynamics; elevated temperatures predominantly lead to end-chain scission, resulting in the breakdown of polymer chains starting from the terminal ends. Conversely, lower temperatures tend to favor random chain scission, where the breakdown occurs along the polymer backbone. Throughout the degradation process, an initial phase involves the disintegration of the polymer's amorphous regions, facilitated by nonenzymatic hydrolysis of ester groups. This phase results in increased PCL crystallinity while maintaining a stable molecular weight. During hydrolytic degradation, the presence of hydrolytically sensitive aliphatic ester linkages is crucial, as the polymer undergoes hydrolysis, releasing carboxylic acids, a byproduct of ester bond breakdown. Enzymatic catalysis can subsequently expedite the next phase of the decomposition process (Govindan et al., 2023; Ramos et al., 2023; Shahrampour et al., 2023). Despite its versatility, PCL's hydrophobic nature typically leads to slow degradation, often ranging from 2 to 4 years, depending on environmental conditions and the presence of other materials. However, the degradation rate of PCL can be modified by incorporating copolymers, such as lactide or glycolide, along with various natural fibrous materials. This ability to tailor degradation rates is particularly advantageous for applications that require specific timeframes before the material fully decomposes (Ghanem et al., 2024; Gu et al., 2024; Guicherd et al., 2024).

The synthesis of PCL leverages controlled ROP methods, which provide precise control over the molecular weight and properties of the resulting polymer. This level of precision is critical, as it directly influences the end-use characteristics of PCL, thereby broadening its practical applications across various fields. Beyond food packaging, PCL finds a range of applications in biomedical settings, particularly in drug delivery systems and tissue engineering scaffolds. Its compatibility with biological tissues and ability to degrade in a biological environment make it an invaluable material in these realms. Moreover, PCL's versatility extends into industrial applications, including the manufacturing of adhesives, coatings, and materials employed in 3D printing technologies (Ding et al., 2024; Dogra et al., 2024; Du et al., 2024; Elgharbawy et al., 2024).

In summary, this comprehensive exploration of PCL sets the groundwork for a deeper understanding of its application in composite films for food packaging. A thorough grasp of its intrinsic properties, synthesis techniques, and diverse applications enriches appreciation of PCL's multifaceted nature and its growing significance in the packaging sector. This insight not only underscores the material's flexibility but also highlights its potential impact on creating sustainable and biodegradable solutions in response to global environmental challenges (Clerici et al., 2024; Deng, Wu, et al., 2024; Deng, Yu, et al., 2024; Li, Li, et al., 2024). The production of PCL, a biodegradable polyester known for its versatility and useful properties, can be achieved through two primary synthetic methods: the polycondensation of 6-hydroxyhexanoic acid and the ROP of epsilon-caprolactone (ε-CL). The polycondensation method involves the direct reaction of 6-hydroxyhexanoic acid, which can occur under specific temperature and pressure conditions that favor the loss of water molecules and facilitate polymerization. This technique, while straightforward and cost-effective, presents challenges in finely tuning the molecular weight and achieving desired branching structures within the polymer. As a result, the control over the polymer's physical and mechanical properties may be limited, which can impact its functionality in practical applications. In contrast, the ROP method offers a more sophisticated approach for synthesizing PCL. This involves first synthesizing ε-caprolactone via a chemical ring-opening reaction, which then serves as the monomer for PCL production. The ROP method allows for enhanced precision in regulating both the molecular weight and the degree of branching of the resulting polymer chains. This is particularly beneficial in applications where specific material characteristics, such as mechanical strength, flexibility, and degradation rate, are critical (Ahn et al., 2024; Dogra et al., 2024; Elgharbawy et al., 2024; Elsayed et al., 2024). Research has demonstrated that the ROP technique is superior for producing high-quality PCL, as it allows meticulous control over reaction parameters and, consequently, the polymer's properties. This method employs various catalytic systems, including enzymatic, organic, and metal catalysts, each playing a crucial role in facilitating polymerization. The selection of a catalyst is pivotal, as it significantly affects not only the reaction conditions, such as temperature, pressure, and time, but also the yield and properties of the resultant PCL (Du et al., 2024; Li, Li, et al., 2024; Piscitelli et al., 2024; Zang et al., 2024). Enzymatic catalysts, for instance, are renowned for their ability to achieve high yields and specific selectivity in reactions, thereby enabling the formation of desired polymer architectures without the need for harsh reaction conditions. On the other hand, organic and metal catalysts may offer distinct advantages in reaction kinetics and can be tailored to optimize selectivity for specific polymerization outcomes. Ultimately, the choice of catalytic system is a critical factor in the development of PCL with tailored characteristics for diverse applications ranging from medical devices to biodegradable packaging materials (Chen, Qin, et al., 2022; Clerici et al., 2024; Ding et al., 2024; Jovanska et al., 2022). The synthesis of PCL involves several key factors that determine the quality and properties of the final product. Among these factors, temperature, reaction duration, solvent selection, and reactant concentration are of paramount importance. The optimal conditions for synthesizing PCL can vary significantly based on the specific synthesis method employed and the desired characteristics of the resultant polymer (Jiang et al., 2024; Nallathambi et al., 2024). One of the most commonly used methods for synthesizing PCL is ROP of ε-caprolactone (CL) monomers. This approach is favored for its ability to achieve higher degrees of polymerization and molecular weight, resulting in enhanced final product quality and lower polymer dispersity. The ROP method generally operates under higher temperatures and extended reaction times compared to traditional polycondensation techniques. This increase in temperature and reaction duration is necessary to facilitate opening of the ring structure of the CL monomers, thereby allowing the formation of long polymer chains (Hegde & Pathak, 2023; Pramanik, 2022; Sachan et al., 2023). In the ROP process, metal-based catalysts, specifically tin (II) octanoate ($Sn(Oct)_2$), are widely utilized. This catalyst has proven effective in producing high-molecular-weight PCL polymers with low polydispersity and minimal side reactions, such as transesterification. However, it is important to note that using $Sn(Oct)_2$ typically requires careful control of synthesis parameters, including higher temperatures and longer reaction times, to ensure a satisfactory polymerization rate (Bujok et al., 2021; Swain et al., 2017). Recent studies have underscored the significance of temperature and solvent selection in the synthesis of PCL. Different solvents can affect the solubility of the reactants, the polymerization rate, and the properties of the final polymer. As such, these factors must be thoughtfully optimized to achieve ideal synthesis conditions (El-Naggar et al., 2022; Hegde & Pathak, 2023; Pramanik, 2022). The precision offered by the ROP pathway allows chemists to finely tune various attributes of PCL, such as its molecular weight, branching properties, and crystallinity. This level of control makes PCL particularly suitable for a wide range of applications requiring specific mechanical properties or biodegradability. However, the complexity and higher costs of the ROP method compared with polycondensation can limit its use in certain settings, particularly where cost-effectiveness is a priority (Ali et al., 2014; Belkhir et al., 2024; Li et al., 2014). Furthermore, recent studies have reported innovative advancements in catalytic systems for the ROP of ε-caprolactone. Single-site metal alkoxide complexes have emerged as promising alternatives to traditional metal catalysts. These advanced catalysts exhibit the potential to achieve high conversion rates and selectivity, which are essential for producing PCL of superior quality. Using single-site metal alkoxide catalysts enables the synthesis of polymers or copolymers with precise regio- and stereoregularity, meeting the stringent performance standards of various applications. The benefits of these catalysts include the production of high-molecular-weight polymers, low polydispersity indices, and living characteristics, which foster the formation of structurally uniform polymer architectures (Belkhir et al., 2024; Lai et al., 2019; Swain et al., 2017). In conclusion, the synthesis of PCL is influenced by a multitude of critical parameters, including the synthesis method and the desired properties of the final product. By carefully optimizing these factors, researchers can create high-quality PCL that fits the needs of diverse applications.

The selection of catalysts is fundamental to influencing the properties and microstructure of polymers, significantly impacting their functionality and applicability. Among the advancements in this field, innovative single-site metal alkoxide catalysts have emerged as pivotal tools that empower chemists to synthesize regio- and stereoregular polymers or copolymers. These catalysts are essential for meeting the escalating performance requirements across various applications by enabling the production of polymers with elevated molecular weights, low polydispersity indices, and traits typical of living polymerization. This sophistication in polymer synthesis enhances not only the structural regularity of the resulting materials but also their overall performance (Belkhir et al., 2024; Jiang et al., 2024). Moreover, extensive research has been conducted on blending PCL with other polymers to enhance their mechanical, thermal, and biodegradation properties. Notably, the combination of PCL with polyglycolic acid (PGA) and poly(l-lactide) (PLA) has shown promising synergies. These blends have resulted in significant improvements in mechanical strength, flexibility, and biodegradability, making them suitable for diverse applications across multiple sectors, especially in food packaging. The resulting materials offer an advantageous balance of structural integrity, sustainability, and compatibility with food products, catering to consumer demands for eco-friendly packaging solutions (Ali et al., 2014; Li et al., 2014; Nallathambi et al., 2024). Additionally, the ROP technique for synthesizing PCL from ε-caprolactone (ε-CL) is recognized as the most thoroughly researched and preferred method in the field. This approach enables precise control over molecular weight and architecture, yielding high-quality PCL tailored for specific applications. The careful selection of catalysts and the optimization of reaction conditions are critical for customizing the physical and chemical properties of PCL, thereby enhancing its versatility and effectiveness (Hegde & Pathak, 2023; Sachan et al., 2023). Continued innovations in catalyst design and blending methodologies are anticipated to further improve PCL's functional capabilities, making it an even more effective candidate for environmentally sustainable packaging solutions. As the demand for biodegradable materials grows, the ongoing development of these technologies is poised to play a significant role in advancing the field of green chemistry and materials science (Jiang et al., 2024; Li et al., 2014; Nallathambi et al., 2024).

20.2.3 Biodegradation and biocompatibility

When discussing biopolymers, it is crucial to distinguish between the terms biodegradable and bioresorbable, as they represent distinct concepts. A material is biodegradable if it can be decomposed by environmental factors or biological organisms, breaking down into simpler compounds. However, this does not inherently imply that the material is bioresorbable. Bioresorbability refers to the ability of a substance to be fully absorbed and eliminated by a living organism, including its original structure and any by-products formed during degradation. In other words, bioresorbable materials can break down and be safely removed from the body without leaving harmful residues or side effects (Viel et al., 2024; Zhang, Zhu, et al., 2024). For example, PCLs are a class of poly(α-hydroxy) esters that can be degraded by naturally occurring microorganisms, such as bacteria and fungi, when exposed to outdoor environments. However, their degradation processes differ significantly between human and animal bodies. This difference is largely due to the absence of specific enzymes required to facilitate their breakdown in biological systems. It is important to note that this does not render PCLs nonbioresorbable; rather, the degradation process is slower, primarily relying on hydrolytic mechanisms (Li, Li, et al., 2024; Salević-Jelić et al., 2024; Shahsavar & Babaei, 2024; Suryani, Rihayat, Fitria, Sariadi, et al., 2024). Hydrolytic degradation of PCLs can occur through two primary pathways: surface degradation and bulk degradation. The predominant pathway is governed by the diffusion-reaction phenomenon, which dictates how water and degradation products interact with the polymer structure. In surface degradation or erosion, the hydrolytic cleavage of the polymer backbone occurs exclusively at the material's exterior. This situation arises when the rate at which the polymer undergoes chain scission to generate oligomers and monomers, which subsequently diffuse away from the surface, exceeds the rate of water penetration into the polymer mass (Muñoz-Suarez et al., 2024; Navarro Arrieta et al., 2024). As a result, the outer layer of the polymer gradually thins while the molecular weight of the internal bulk remains largely unchanged throughout the degradation process. This characteristic of surface erosion brings notable advantages, particularly in pharmaceutical applications where controlled drug release is essential. Because the degradation rate can be predicted from specific polymer characteristics and environmental conditions, researchers and developers can establish desired release profiles for drug delivery systems, thereby enhancing their efficacy and safety in therapeutic applications (Guicherd et al., 2024; Lawless et al., 2024; Moshkbid et al., 2024).

Bulk degradation is a complex process that occurs when water infiltrates the entire polymer volume, leading to hydrolysis as it penetrates throughout the polymer matrix. This infiltration initiates random hydrolytic chain scission, a reaction in which polymer chains are broken, resulting in a general decline in molecular weight throughout the material (Cintra et al., 2022; İlaslan et al., 2022). When water molecules effectively permeate the polymer bulk, they facilitate the hydrolysis of the polymer chains. This reaction allows smaller monomers or oligomers to escape from the bulk material, initiating gradual erosion and leading to an equilibrium state within this diffusion-reaction mechanism. However, if this equilibrium is disrupted, say, by rapid hydrolysis or uneven water distribution, the degradation process can trigger internal autocatalysis. This autocatalysis is often due to the presence of reactive end-group by-products such as carboxyl and hydroxyl groups that emerge during hydrolysis (Bhovi et al., 2022; Din et al., 2020; Duarah et al., 2023). In situations where water permeation is significant, oligomers and carboxyl groups produced can diffuse away from the polymer surface, leading to what is termed surface erosion. In contrast, bulk degradation results in the accumulated concentration of these autocatalysis by-products within the polymer, creating an acidic gradient. Specifically, carboxyl end groups generated during the cleavage of ester bonds tend to accumulate, which can greatly accelerate internal degradation processes compared to surface loss. This phenomenon results in a characteristic structure in which the outer layer of the polymer retains a higher molecular weight, while the interior becomes significantly degraded, yielding lower-molecular-weight regions. Thus, the overall degradation process often exhibits a bimodal molecular weight distribution, indicating the polymer's varied states throughout its volume (Awasthi et al., 2022; Chen, Oveissi, et al., 2022; Cintra et al., 2022). As the internal oligomers degrade to sufficiently small sizes, they can rapidly diffuse through the polymer's outer layer, leading to initial weight loss. This diffusion process can reduce the rate of continued chain scission, ultimately yielding a hollow structure that retains a higher molecular weight on the exterior (Chen, Oveissi, et al., 2022; Díaz-Pedraza et al., 2021; Kim et al., 2022). Moreover, the swift release of these oligomers and acidic by-products can elicit inflammatory responses in living organisms. This has been evidenced in the literature on bioresorbable medical devices, which often undergo varying physiological reactions due to degradation. Additionally, if the surrounding tissue lacks sufficient vascularization or exhibits low metabolic activity, it may struggle to buffer the pH changes caused by the accumulation of acidic by-products, leading to localized disturbances. A prime example of this can be observed with fiber-reinforced PGA pins used in orthopedic surgeries, which can cause increased osmotic pressure due to the build-up of localized fluid and subsequent rapid degradation, potentially resulting in adverse tissue responses. This interplay among degradation processes, molecular weight changes, and biological responses illustrates the intricate dynamics at work in polymer degradation in medical applications (El-Sayed & Youssef, 2023; Ravi et al., 2023). The homopolymer PCL demonstrates a complete degradation timeline of approximately two to four years. This period is influenced by several factors, including the initial molecular weight of the PCL-based device or implant. Importantly, the rate of hydrolysis, which is a key process in the degradation of PCL, can be adjusted by copolymerizing it with other lactones or various glycolides/lactides (Liu et al., 2023; Shahrampour et al., 2023; Vibha et al., 2023). In a series of thorough investigations, the research team examined the degradation behavior of PCL scaffolds under both in vitro and in vivo conditions. These studies revealed no evidence of internal catalytic processes driving degradation, as evidenced by the consistent molecular weight distribution over time and by cross-sectional analyses of the scaffold struts performed at 6- and 36-month intervals (Lam et al., 2009). Moreover, additional research on the hydrolytic degradation of PCL was conducted in two distinct settings: a controlled in vitro environment with saline and an in vivo model in rabbits. Findings from these studies indicated that degradation rates in both scenarios were similar, suggesting that enzymatic activity during the initial degradation phase (first 12 months) was not a significant factor in the overall degradation of the PCL material. The uniformity in degradation observed across different conditions suggests a predictable and robust degradation pathway for PCL scaffolds used in biomedical applications (Muñoz-Suarez et al., 2024; Richert et al., 2023; Sun et al., 2006).

PCL degradation has been extensively studied through numerous in vitro and in vivo experiments, employing various buffers and experimental conditions to better understand the underlying degradation processes. In vitro studies primarily use buffers such as phosphate-buffered saline (PBS), simulated body fluid (SBF), Ringer's solution, and Dulbecco's Modified Eagle's Medium (DMEM). These setups can be conducted either in the absence of living cells, allowing for a clearer focus on the material's intrinsic properties, or in the presence of cells, which provide insights into how biological factors interact with PCL.

On the other hand, in vivo studies predominantly investigate the subcutaneous (SC) implantation of PCL materials. These experiments often replicate the conditions under which capsular formations occur, a common phenomenon when biomaterials are introduced into living organisms. A variety of animal models, including rabbits, mice, and rats, are frequently employed for these studies. This approach enables researchers to evaluate PCL degradation within a physiological environment; however, it is crucial to acknowledge that degradation conditions within soft tissue may vary markedly from those experienced in typical orthopedic applications, affecting both the rate and mechanism of degradation

(Abdellah Ali, 2020; de Oliveira Gama et al., 2018). Pitt et al. investigated the degradation of PCL capsules implanted subcutaneously in rabbits over an impressive duration of 30 months. Their detailed findings revealed a gradual decrease in molecular weight (Mn), starting from a baseline of 50 kDa and diminishing to approximately 35 kDa after four months. This reduction continued as the study progressed, culminating in an Mn of about 17 kDa after one year. The observed decrease in molecular weight was primarily attributed to the preferential degradation of the polymer's more amorphous regions, which subsequently led to an increase in crystallinity. Interestingly, significant changes in mass loss were recorded around the 15–18-month mark, during which Mn further decreased from 13 to 9 kDa. This reduction in molecular weight corresponded with a 16% mass loss, indicating substantial material changes over time. Mechanical testing during this phase revealed the emergence of fractures within the PCL structure, coupled with ongoing increases in crystallinity, suggesting that the polymer was undergoing continuous structural transformations. The relationship between the decrease in Mn and the degradation time followed a logarithmic pattern, suggesting that the degradation process can be described by a first-order rate law for ester hydrolysis. Further investigations in the study included degradation analysis of tritium (3H)-labeled PCL capsules that were also implanted subcutaneously in rabbits. Remarkably, the complete clearance of these high-molecular-weight polymers, which initially had an Mn of around 130 kDa, was documented after a four-year period. At the 22-month evaluation, the researchers noted that only brittle fragments remained, illustrating the extent of degradation that had transpired over the study period. This comprehensive research underscores the complex interplay of factors influencing PCL degradation in biological systems and highlights the importance of understanding these mechanisms for future applications in biomedical engineering (Pitt et al., 1981). Research conducted by Sun and colleagues has provided significant insights into the degradation dynamics of low molecular weight 3H-labeled PCL with a molecular weight of 3 kDa following its implantation. In their study, the tracer was initially detected in biological samples as early as day 15 after implantation, with its concentration peaking by day 45. Fascinatingly, by day 135, an impressive 92% of the tracer was recovered from fecal and urinary excretions, highlighting the rapid and efficient breakdown of the PCL material in the body. The rapid degradation can be attributed to the low molecular weight of the PCL, which was introduced after reconstitution in a glycol ether, thereby facilitating its absorption and metabolism. Additionally, the research indicated that radioactivity levels in blood plasma and various organs returned to baseline within a certain time frame, further supporting the conclusion that there was minimal to no significant accumulation of PCL fragments in body tissues or organs. Moreover, in a manner congruent with the findings of Pitt et al., Sun's research on the subcutaneously implanted PCL capsules showed a marked reduction in molecular weight over time. Specifically, the molecular weight decreased from an initial value of 40 kDa to 26 kDa after three months, then continued to decline to 21.4 kDa after one year. This degradation pattern consistently exhibited first-order kinetics, indicating that the rate of degradation was proportional to the amount of PCL present at any given time. These observations underscore the effective biodegradability of PCL in biological systems and its potential implications for medical applications (Sun et al., 2006).

In a controlled in vitro study, Domingos et al. investigated the effects of SBF and PBS at 37°C on the degradation behavior of 3D-printed PCL scaffolds. Their research revealed that the molecular weight (Mn) of the PCL scaffolds began at 34 kDa. After six months of immersion in PBS, the Mn decreased to 32 kDa, indicating a modest degradation rate. In contrast, the scaffolds submerged in SBF exhibited a slight alteration in Mn, reducing to 33 kDa over the same six-month period. This outcome highlights the distinct effects of the two fluids on the degradation of PCL scaffolds (Domingos et al., 2010). These findings present a notable contrast to the results obtained by Yavuz et al., who specifically studied solvent-cast PCL films. Their research indicated a more pronounced reduction in Mn, with the molecular weight dropping from an initial 30 kDa to 21 kDa within just 30 days of exposure (Yavuz et al., 2002). Furthermore, degradation continued, resulting in a further decrease to 20 kDa after 60 days of immersion at 37°C in culture media, suggesting a faster degradation rate than that observed in the 3D-printed scaffolds. In a more extended analysis, Coombes and colleagues tracked the degradation of PCL over a significant period of 45 months while submerged in PBS at 37°C (Coombes et al., 2004). Their long-term study reported a consistent degradation trend that mirrored the earlier findings of Bolgen, where the initial Mn level of PCL started at a notably higher value of 78 kDa and declined to 40 kDa by the end of the 45-month period. This extended observation period underscores the gradual, sustained degradation of PCL under physiological conditions, providing valuable insights into the material's lifespan and suitability for biomedical applications.

Larranaga et al. conducted a comprehensive investigation into the degradation behavior of solid polycaprolactone (PCL) when immersed in PBS at a controlled temperature of 37°C over an extended period of 132 days (approximately 4.5 months). Throughout this prolonged exposure, the researchers observed only minimal mass loss, indicating a robust degradation resistance of the polymer under these specific conditions. They reported a decrease in the number-average molecular weight (Mn) of PCL, from 78 to 70 kDa by the end of the study. A key finding emphasized by the authors was that the highly crystalline regions of the PCL exhibit notable resilience to degradation. This characteristic is critical, as it may significantly influence the outcomes and interpretations of in vivo studies that evaluate the behavior of PCL in biological environments (Larrañaga et al., 2014). In a complementary study, Little and colleagues focused on the degradation of pressed PCL sheets, similarly immersed in PBS at 37°C, but over a slightly longer duration of 7 months. Their research demonstrated that during this extended period, the mechanical properties of PCL remained largely intact, showcasing the material's durability under environmental stressors. Like Larranaga et al., Little and associates noted virtually no mass loss from the polymer samples throughout the seven-month study, suggesting a highly stable degradation profile. They measured a modest decline in Mn, from 63 to 60 kDa by the end of the study, suggesting molecular weight changes without compromising overall structural integrity (Little et al., 2009). These collective findings from both studies offer valuable insights into the degradation dynamics of PCL under various environmental conditions. They underscore the complex interplay among molecular weight, polymer crystallinity, environmental factors, and underlying degradation mechanisms, which is crucial for predicting PCL behavior in practical applications, especially in biomedical contexts.

20.3 Application of polycaprolactone as food packaging coatings and films

PCL is a notable biodegradable polymer in the class of aliphatic polyesters. It is characterized by repeating units derived from hexanoate and is recognized for its diverse applications across fields such as food science, pharmaceuticals, and biomedicine. The significance of PCL stems from its unique properties, including tunable mechanical properties, compatibility with numerous other polymers, and high biocompatibility. One of the key features of PCL is its favorable solubility, which allows it to dissolve in various organic solvents. Additionally, its low glass transition temperature contributes to its semirigid behavior at ambient temperatures, making it suitable for applications that require flexibility and durability. The polymer's hydrophobic nature and semicrystalline structure further enhance its versatility, enabling researchers to explore its potential as nanocarriers for drug delivery, in food packaging materials, and in various biomedical and pharmaceutical applications. PCL's properties can be tailored through modifications in two crucial aspects: molecular weight and degree of crystallinity. By adjusting the molecular weight, one can alter the mechanical strength and degradation rate of the polymer, while variations in crystallinity can influence its overall thermal and physical characteristics. These modifications are important because they can directly affect the degradation of PCL, which occurs primarily through ester hydrolysis and enzymatic degradation. As researchers continue to investigate and refine PCL's capabilities, its applications are expected to expand, bolstered by its environmentally friendly and adaptable nature.

PCL has garnered significant attention in recent years for its potential applications in controlled drug delivery systems. Despite its popularity, PCL is characterized by a relatively slow degradation rate, typically ranging from two to three years. This prolonged degradation process, coupled with its hydrophobic nature and limited solubility, makes PCL more suitable for tissue engineering applications, particularly in bone tissue regeneration and repair. In these contexts, PCL is often utilized as an implant material or as a scaffold to support cell growth and tissue formation. To enhance the degradation characteristics of PCL and mitigate its inherent limitations, researchers have explored blending PCL with various other polymers. Notable examples include PLA, polylactic-co-glycolic acid (PLGA), and polyethylene glycol (PEG) (Uma Thanu Krishnan Neela et al., 2024; Zidar et al., 2024). These blends create block copolymers that improve PCL's degradation behavior by introducing amphiphilic structures, which lead to more favorable interactions with biological environments, thus promoting faster breakdown and integration into the body. From a regulatory perspective, PCL enjoys a favorable status, having been approved by the US Food and Drug Administration (FDA) and registered with the European Community for use in clinical applications. This regulatory support facilitates the incorporation of PCL into drug delivery systems and various medical devices, enhancing its reliability and safety profile for human use. The significance of PCL extends beyond its medical applications; it has become increasingly important in the development of biomaterials and eco-friendly materials. With a diverse range of compositions and properties, PCL can be tailored for various applications (Chakravarti & Annu, 2024; Ma et al., 2024; Raza et al., 2024). In the medical field, it is used in wound dressings, contraceptive devices, and dental products. Beyond healthcare, PCL's versatility allows it to be used in nonmedical sectors such as food packaging, where its biodegradability offers an environmentally friendly alternative, as well as in other environmental applications aimed at reducing plastic waste. The ongoing research and innovation surrounding PCL underscore its potential as a vital material in both medical and industrial domains (Obisesan et al., 2023).

The production of PCL composites can be accomplished through a range of processing methods, each offering distinct advantages and suitable for different applications. Some of the primary techniques include extrusion molding, compression molding, injection molding, electrospinning, 3D printing, melt spinning, film blowing, and ball mixing. Extrusion molding is the most prevalent method utilized in the polymer industry, primarily due to its remarkable adaptability and high production efficiency. This technique allows for continuous production, making it ideal for large-scale manufacturing. However, while methods like injection molding, compression molding, and blow molding are effective for creating various polymer products, they typically struggle to produce items that require high resolution or exhibit detailed features. These limitations can compromise the quality and intricate design of the finished product. Considering that PCL composites are increasingly favored for biomedical applications, the manufacturing processes must meet stringent quality standards and complex specifications. This requirement pushes researchers and manufacturers to select specific techniques to produce components with sophisticated geometric designs, ensuring they meet the necessary quality, accuracy, and precision. Among the array of available fabrication methods, electrospinning and 3D printing have emerged as particularly promising approaches for developing PCL-composite-based systems. Electrospinning is renowned for its ability to create nanofibers with high surface area-to-volume ratios, making it suitable for applications requiring enhanced drug delivery and tissue engineering scaffolds. Meanwhile, 3D printing offers exceptional versatility, enabling the fabrication of customized structures and shapes tailored to particular medical needs. Both of these methods are gaining traction in the biomedical and pharmaceutical fields, as they enable the production of complex, high-quality PCL composite constructs that meet the stringent standards required in these sensitive applications (Barnthip et al., 2022; Chukaew et al., 2023; Hsu et al., 2023).

PCL has emerged as a highly promising biodegradable polymer, particularly well-suited for the development of controlled drug delivery systems. This is largely attributed to its excellent biocompatibility and distinctive ability to modulate drug-release profiles. One of the standout features of PCL is its customizable properties, which can be finely tuned to meet specific pharmaceutical needs. These properties include biodegradation kinetics, which can be adjusted to control the rate at which the polymer breaks down in the body; crystallinity, which influences the material's mechanical strength and thermal behavior; and porosity, which affects drug diffusion rates. Furthermore, PCL exhibits favorable permeability and stability, ensuring that drugs remain encapsulated and can be released in a controlled manner over time. Its compatibility with various copolymers and drugs enhances its versatility, enabling uniform distribution of encapsulated drugs within the matrix and optimizing drug-loading efficiency (Chakravarti & Annu, 2024; Mary et al., 2023). A significant advantage of PCL is its gradual degradation, which enables the development of long-term drug-release carriers that sustain delivery over several months. This prolonged-release formulation is particularly beneficial for conditions requiring sustained therapeutic medication levels over an extended period. Another critical aspect of PCL is its minimal tendency to induce an acidic environment during degradation. This contrasts with other biodegradable polyesters, such as PLA, PGA, and PLGA, which can cause localized acidity that may compromise surrounding tissue and reduce overall efficacy. What makes PCL particularly intriguing is the ability to tailor its properties, encompassing physical, chemical, thermal, mechanical, and viscoelastic characteristics, through various modification techniques. These include altering synthesis processes or creating composites with both natural and synthetic polymers. Natural polymers such as starch, chitosan, gelatin, and collagen, as well as synthetic polymers like PEG, polyethylene oxide (PEO), polyurethane, and polyvinyl alcohol, can be incorporated to enhance the performance and functionality of PCL-based systems (Bahrani et al., 2024; Hsu et al., 2023; Ibrahim et al., 2023; Joy et al., 2023). This adaptability opens new avenues for innovation in drug delivery and biomedical applications. A study evaluated drug release profiles using polycaprolactone matrix and reservoir systems to derive empirical drug diffusion and release parameters. Strong correlations between these parameters and drug properties such as LogP, molecular weight, and solubility were observed. The effectiveness of the developed models was demonstrated using three prototypes: a matrix system for levonorgestrel, a reservoir system for timolol maleate, and a reservoir system for primaquine bisphosphate. The empirical equations accurately matched experimental data and can assist in designing tailored polycaprolactone thin-film devices for specific release profiles, providing valuable tools for controlled-release system design (Schlesinger et al., 2015). Fused deposition modeling (FDM) is an advanced 3D printing technique that enables the production of intricate drug delivery implants finely tuned to meet the unique requirements of individual patients. However, to achieve the intended drug activity, it is crucial to meticulously control specific processing temperatures and use appropriate filament preparation methods. Additionally, the materials used for these implants must meet rigorous biocompatibility standards to ensure safety and efficacy in clinical settings. In a study by Yang and colleagues, a novel drug-delivery implant using FDM printing technology was developed, demonstrating excellent biocompatibility and effective, controlled drug release for personalized medicine. The researchers created drug-loaded filaments via hot-melt extrusion (HME), selecting polycaprolactone for its desirable properties and using ibuprofen as the model drug. A significant aspect of the approach was the incorporation of chitosan, which enhanced filament structure and improved drug-release efficiency. The study investigated factors such as filament printability, variations in physical and chemical properties, and drug-release behavior, while ensuring that no significant degradation occurred during manufacturing. The results showed that while the tensile strength of the filament decreased with higher drug and chitosan concentrations, the drug-release efficiency was high, exceeding 99% over 120 hours, driven by diffusion and erosion mechanisms. Cell viability in the presence of a 72-hour implant extract was 75.3%, indicating adequate biocompatibility for clinical applications. This research highlights promising advancements in personalized drug delivery systems using FDM technology (Yang et al., 2022).

Nanocarrier-based drug delivery systems have become a pivotal area of research in the fight against cancer, characterized by their diverse structures designed to effectively impede cancer progression and improve the therapeutic index for cancer therapies. Despite the promise of these systems, many currently available nanostructures face significant limitations, including low stability and rapid drug release, which can undermine their therapeutic effectiveness. A recent study by Rashidy Ahmady and colleagues investigated a novel drug delivery system using capsaicin-loaded alginate nanoparticles combined with polycaprolactone-chitosan nanofiber mats (Fig. 20.2). Capsaicin, derived from chili peppers, is known for its potential anticancer properties. The study found that incorporating these nanoparticles into the nanofiber mats significantly extended capsaicin release from 120 hours to over 500 hours, enhancing its sustained delivery. In vitro tests showed that this system effectively inhibited the proliferation of MCF-7 human breast cancer cells while remaining nontoxic to human dermal fibroblasts (HDF). This indicates the potential for selective targeting of cancer cells without harming healthy tissue, underscoring the system's efficacy and safety for cancer therapy. Overall, the research suggests promising advancements in cancer treatment through improved nanocarrier systems (Ahmady et al., 2023).

Tissue engineering plays a critical role in regenerative medicine by focusing on developing biological substitutes to repair or replace damaged tissues and organs. A contemporary strategy in this field involves the use of injectable biodegradable polymers, which offer significant advantages in clinical settings. These materials are designed not only to fill anatomical defects but also to seamlessly integrate with therapeutic agents or living cells, facilitating their delivery to the target site. Furthermore, their injectable nature supports minimally invasive surgical procedures, which can reduce

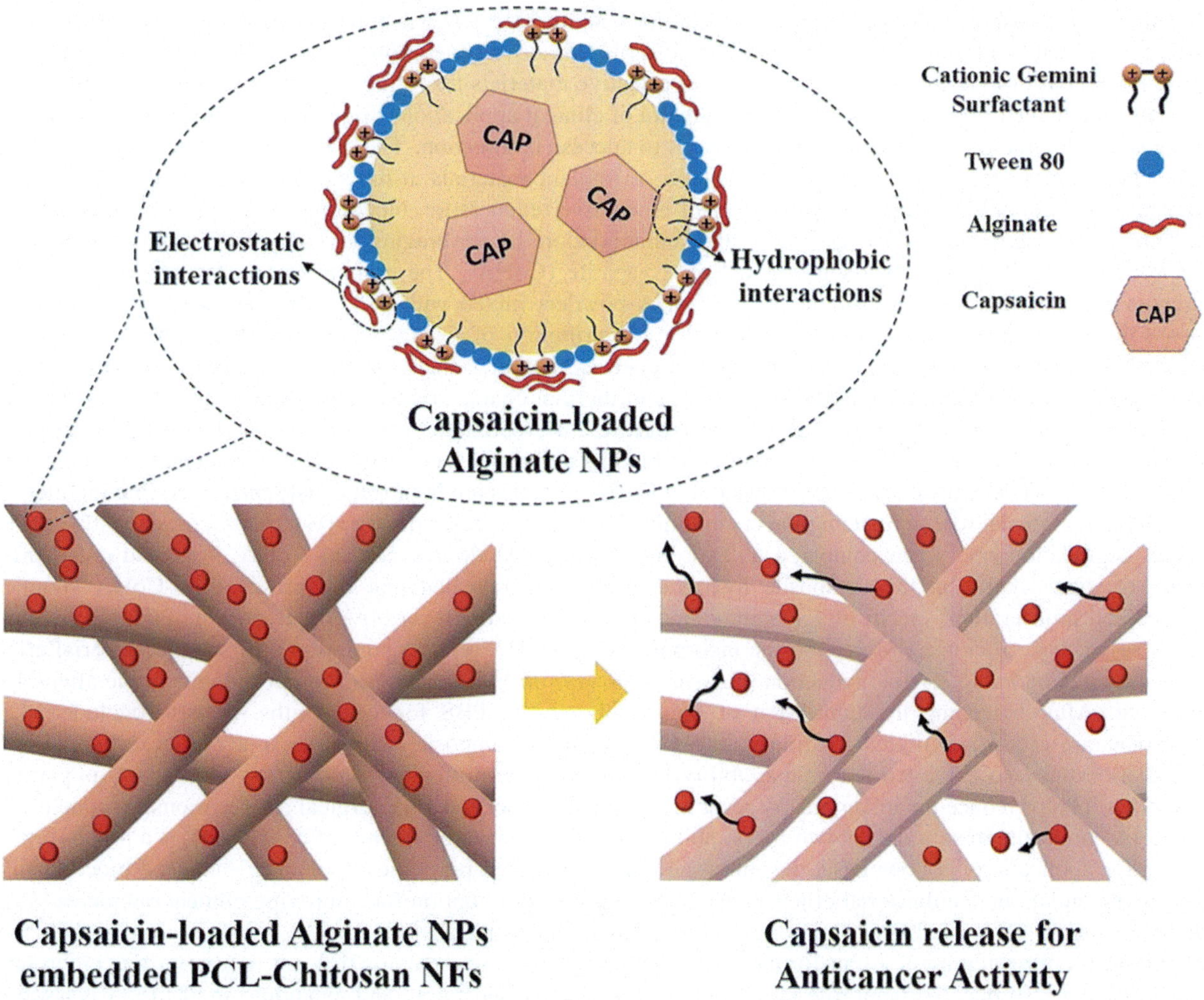

FIGURE 20.2 Schematic of a drug delivery system using alginate nanoparticles containing capsaicin combined with PCL-chitosan nanofiber mats. *From Ahmady, A.R., Solouk, A., Saber-Samandari, S., Akbari, S., Ghanbari, H., & Brycki, B.E. (2023). Capsaicin-loaded alginate nanoparticles embedded in polycaprolactone-chitosan nanofibers as a controlled drug delivery nanoplatform for anticancer activity. Journal of Colloid and Interface Science, 638, 616–628. https://doi.org/10.1016/j.jcis.2023.01.139, with reproducing permission from Elsevier under license number 6179270736435.*

recovery times and improve patient outcomes (Gupta et al., 2024; Kozielski & Sitti, 2017). Salgado and colleagues formulated a novel biodegradable injectable gel by combining PCL with sebacic acid (SA). The incorporation of SA was intended to enhance the degradation rate of PCL, a polymer known for its favorable mechanical properties and biocompatibility. While SA is recognized for its rapid degradation characteristics, caution is warranted regarding its concentration; elevated levels can lower pH, posing a risk to the biocompatibility of the resulting PCL blend. To evaluate the performance of the polycaprolactone-sebacic acid (PCL-SA) blend, degradation studies were conducted in PBS using a 5% concentration of SA. The stability and characteristics of the biomaterial following degradation were meticulously analyzed using a suite of diagnostic techniques, including differential scanning calorimetry (DSC), dynamic mechanical analysis (DMA), and scanning electron microscopy (SEM). These assessments provided insights into how the material's properties changed over time. Results from the degradation tests revealed that the elastic properties of the PCL-SA blend significantly decreased as it broke down. The material exhibited increased crystallinity and rigidity, although these changes were less pronounced compared to pure PCL, indicating that the presence of SA moderated some of the adverse effects associated with rapid degradation. Notably, the PCL-SA blend exhibited a more rapid degradation profile, losing its crystalline phase within just 30 days, yet it retained thermal and mechanical properties comparable to those of unmodified PCL. This finding suggests that the blend successfully met the predetermined objectives for both biodegradability and structural integrity. In addition to material characterization, subsequent cytotoxicity assays were conducted to assess the biocompatibility of the PCL-SA gel. The assays demonstrated that the gel supported high cell viability (> 90%), confirming its cytocompatibility. This aspect is crucial for potential applications in tissue engineering, where the ability to support cell growth and survival is essential for fostering tissue regeneration. Overall, the research highlights the promising potential of PCL-SA biodegradable gels as effective materials for future applications in tissue engineering and regenerative medicine (Salgado et al., 2012). In the realm of clinical applications of guided tissue regeneration and guided bone regeneration, one of the predominant challenges to success is infection. These infections often arise from foreign-body reactions influenced by the presence of bacteria or implant materials at the surgical wound site. To address this significant issue, Lin and colleagues have developed innovative regeneration membranes that specifically release cobalt ions. This design aims to reduce infection and inflammation by leveraging a composite material composed of polycaprolactone (PCL) and cobalt-substituted hydroxyapatite (CoHA). The composite membrane formulated by the researchers consists of 20 weight percent (wt.%) CoHA powders mixed with PCL. This membrane was manufactured using a solvent-casting method, resulting in a uniform distribution of CoHA within the PCL matrix. To thoroughly understand the membrane's physical and chemical properties, detailed analyses were conducted, focusing on surface morphology, crystal structure, chemical composition, and thermal characteristics. Furthermore, the researchers assessed the membrane's biocompatibility, potential for osteogenic differentiation, and inherent antibacterial properties. An additional critical evaluation involved determining the biodegradability of the membranes by immersing them in PBS for 6 months. The physicochemical analyses indicated that CoHA particles are homogeneously integrated into the membrane, thereby playing a crucial role in reducing the crystallinity of PCL. This reduction in crystallinity consequently leads to an enhanced degradation rate when compared to PCL membranes that do not contain CoHA. The study also involved culturing osteoblast cells on the composite membrane. Findings revealed that the presence of CoHA significantly stimulated cell proliferation; specifically, calcium deposition increased by over 90% relative to the PCL-only membrane after just seven days of culture. Moreover, the incorporation of CoHA powder conferred a robust antibacterial effect, as evidenced by a remarkable 2.4-fold reduction in *Escherichia coli* proliferation on the composite membrane after 24 hours of exposure. After extending the immersion of the membranes in PBS for six months, it was concluded that the polycaprolactone–cobalt-substituted hydroxyapatite (PCL–CoHA) composite membrane not only exhibited improved biodegradation but also exhibited the unique ability to continuously eliminate free radicals. This action is particularly beneficial, as it helps reduce the inflammatory response typically associated with surgical interventions. In conclusion, the PCL–CoHA composite membrane, known for its ability to enable controlled release of cobalt ions, is a highly promising candidate for applications in bone tissue regeneration. Its multifaceted properties, including enhanced biocompatibility, osteogenic potential, and antibacterial efficacy, establish it as a leading option for improving clinical outcomes in guided tissue regeneration and guided bone regeneration procedures (Lin et al., 2019).

Biodegradable films represent a transformative category of packaging materials that not only prioritize environmental sustainability but also play a crucial role in enhancing food quality and safety and extending shelf life (Tavassoli et al., 2023). Among these materials, PCL-based films stand out due to their unique properties and versatility. When combined with bioactive compounds or functional additives, PCL-based films offer a promising solution to tackle pressing food safety issues and align with increasingly health-conscious consumer preferences. The extensive evaluation and review of PCL-based packaging films indicate their effectiveness in real-world applications. The applications of PCL-based food packaging films for preserving food samples are outlined in Table 20.1. Numerous studies have demonstrated the

TABLE 20.1 Application of polycaprolactone-based food packaging films for food sample preservation.

Base material	Concentration of PCL	Other ingredients	Active compound	Preparation method	Food model	Key results	References
PCL	14%	Chitosan	Propolis extract	Electrospinning	Pork	PCL/chitosan-based nanofibrous films improved color stability, showed no difference in lipid oxidation, and a four-day delay in the onset of bacterial growth in pork during refrigerated storage.	Vargas Romero et al. (2021)
PCL	12%	Whey protein isolate	*Urtica dioica L.* extract	Electrospinning	Rainbow trout	PCL/whey protein isolate nanofibers containing *Urtica dioica* L. extract, controlled TVB-N, and tetrabutylammonium chloride, and the wrapped fish samples showed the lowest number of bacterial growth.	Alp Erbay et al. (2017)
PCL	18%	-	Linalool	Electrospinning	Salmon	PCL-based nanofibers containing linalool reduced the spoilage rate of salmon and effectively increased the shelf life of salmon.	Weligama Thuppahige and Karim, 2021
Chitosan/PCL	8%	Chitosan	Boron nitride nanosheets	Electrospinning	Tomato	Chitosan/PCL-based nanofibrous films containing boron nitride nanosheets showed better texture, lower temperature, and ultimately less weight loss in tomato samples.	Shen et al. (2022)
PCL	6.5%	Ethyl cellulose	Natamycin/trans-cinnamic acid	Electrospinning	Strawberry	PCL/ethyl cellulose-based nanofibers had good mechanical and physical properties and provided resistance to the fungus Botrytis cinerea in fruits.	Wu et al., 2024
PCL	12%	Polylactic acid (PLA)	Oregano essential oil @β-cyclodextrin	Electrospinning	Blackberry	PCL/PLA has been shown to delay postharvest decay, deterioration, and to reduce storage quality in blackberries.	Shi et al. (2022)
PCL	7%	Polyethylene glycol (PEG)	Commercially anthocyanin	Electrospinning	Chicken breast	PCL/PEG nanofibers caused a color change from bright red to bright blue-violet during storage of chicken breast samples.	Guclu et al. (2023)

(*Continued*)

TABLE 20.1 (Continued)

Base material	Concentration of PCL	Other ingredients	Active compound	Preparation method	Food model	Key results	References
PCL	7%	Polyethylene oxide	*Hibiscus rosa sinensis* anthocyanin	Electrospinning	Shrimp	The nanofiber films changed color from red to brown during shrimp storage.	Jovanska et al. (2022)
PCL	16, 18, 20 g mL^{-1}	Silk fibroin	Barberry anthocyanin	Electrospinning	Pork and shrimp	PCL/silk fibroin-based nanofibers caused a color change from pink to yellow during the storage of the samples. They also showed an increase in shelf life of shrimp by approximately 70% at 25°C and an increase in shelf life of pork by more than 100% at 4°C.	Qin et al. (2015)
PCL	4%	PLA	Cinnamaldehyde	Casting	Button mushroom	A reduction in the microbial counts of the sample, enhancing the general quality during storage.	Qin et al. (2015)
PCL	10%	-	Montmorillonite	Casting	Bread	No growth was observed in packaged bread	Seyrek et al. (2021)
PCL	-	PLA/PEG	Clove essential oil (CEO)/zinc oxide (ZO)	Casting	Egg	The addition of CEO/ZO to PCL-based films resulted in maximum reduction of pathogens during 21 days of storage at 4°C.	Ahmed et al. (2019)
PCL	5%	-	Ag-doped nanoclays	Casting	Ground beef	The addition of Ag-doped nanoclays to PCL-based films has been shown to have a bacteriostatic effect on the natural bacterial load of beef during storage and provides better microbial quality.	İlaslan et al. (2022)
PCL	6%	-	Almond skin extract	Casting	Fried almonds	PLC-based films with almond skin extract provided protection against lipid oxidation and improved their stability.	Valdés García et al. (2020)

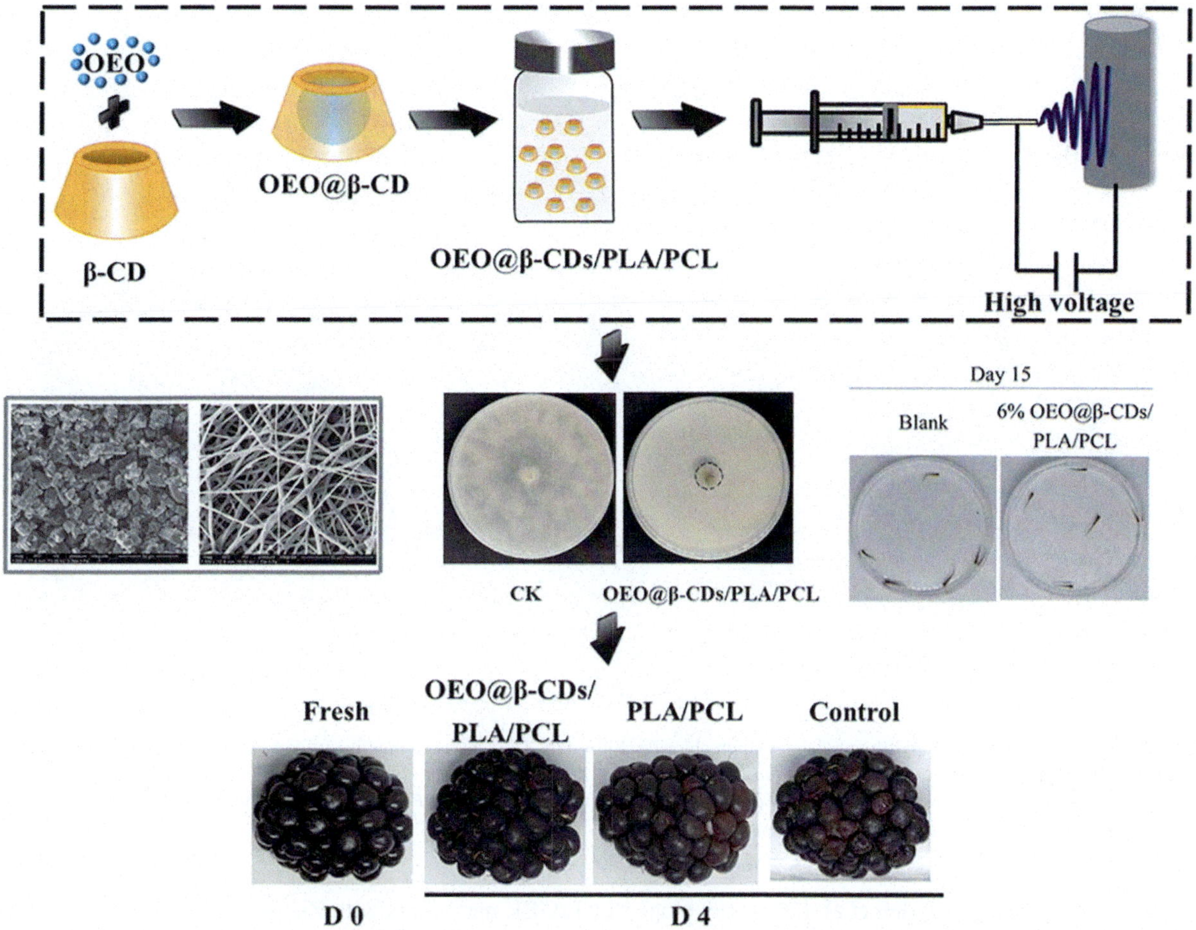

FIGURE 20.3 Effects of polylactic acid (PLA)/PCL/oregano essential oil (OEO)/β-CD encapsulated nanofiber films on blackberries' shelf life and quality over four days. *From Shi et al., 2022 Shi, C., Zhou, A., Fang, D., Lu, T., Wang, J., Song, Y., Lyu, L., Wu, W., Huang, C., & Li, W. (2022). Oregano essential oil/β-cyclodextrin inclusion compound polylactic acid/polycaprolactone electrospun nanofibers for active food packaging. Chemical Engineering Journal, 445, 136746. https://doi.org/10.1016/j.cej.2022.136746, with reproducing permission from Elsevier under license number 6179270934568.*

impressive capabilities of these films in active and intelligent packaging. Active packaging involves the integration of substances that interact with food products to improve preservation, while intelligent packaging incorporates sensors or indicators that provide real-time information about food freshness or spoilage. Research has highlighted the efficacy of PCL-based films enhanced with various active components, including antimicrobial agents and antioxidants, which significantly prolong the shelf life of food items. Additionally, incorporating nanofibers into these films further enhances their properties, providing a barrier against moisture and gases, which is essential for maintaining food quality. The application of these advanced PCL-based packaging films has yielded remarkable results across various food systems, demonstrating their potential to not only extend the shelf life of perishable goods but also to monitor their freshness effectively. This dual functionality not only addresses consumer concerns about food safety and waste reduction but also positions PCL-based biodegradable films as a key player in the future of sustainable packaging solutions.

Shi et al. (2022) examined the effects of PLA/PCL/oregano essential oil (OEO)/β-CD encapsulated nanofiber films (ENFs) on the shelf life and quality of blackberries (Fig. 20.3). Over four days, control, OEO, and PLA/PCL treatments showed mold infection and rotting, while blackberries treated with OEO@β-CD/PLA/PCL maintained a good appearance. Key quality indicators revealed that weight loss after four days was lower in OEO@β-CD/PLA/PCL (13.56 ± 0.46) compared to control (20.80 ± 0.55), OEO (17.07 ± 0.42), and PLA/PCL (16.63 ± 0.44). Firmness measurements also demonstrated greater preservation in OEO@β-CD/PLA/PCL (1.89 ± 0.96), and total solids were highest in this treatment (8.93 ± 0.33). The β-CD's controlled release of OEO contributed to prolonged antimicrobial activity and retention of quality (Shi et al., 2022).

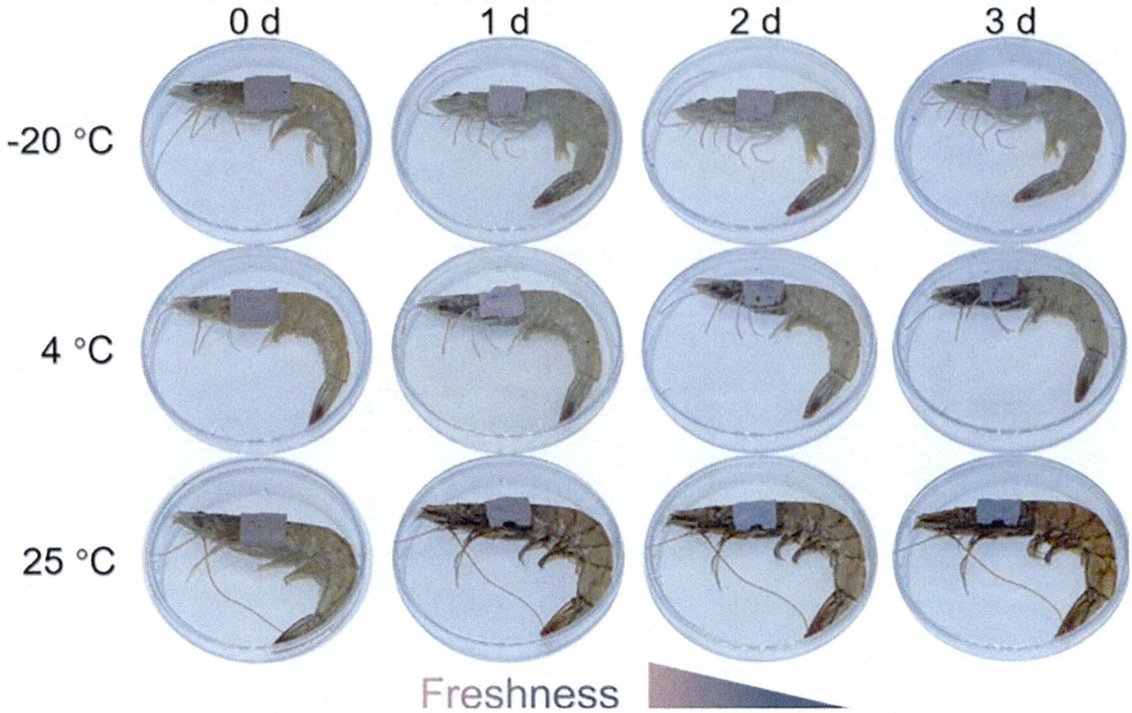

FIGURE 20.4 Applying hydroxypropyltrimethyl ammonium chloride chitosan/PCL/SKN-based intelligent food packaging to monitor shrimp freshness at −20°C, 4°C, and −25°C. *From Zou, Y., Sun, Y., Shi, W., Wan, B., & Zhang, H. (2023). Dual-functional shikonin-loaded quaternized chitosan/ polycaprolactone nanofibrous film with pH-sensing for active and intelligent food packaging. Food Chemistry, 399, 133962. https://doi.org/10.1016/j. foodchem.2022.133962, with reproducing permission from Elsevier under license number 6179271091797.*

In another study, Shanbehzadeh et al. (2022) developed PCL/GEL nanofibers with ZnNPs and cumin essential oil (EO) for cheese packaging, and PCL/Gel nanofibers containing ZO and cumin EO. They found that PCL/Gel/3% ZO/1% cumin EO, PCL/Gel/3% ZO, and PCL/Gel/3% ZO/3% cumin EO achieved higher scores for color, texture, and odor after 12 days compared to control samples. The inclusion of zinc oxide (ZO) and cumin EO minimized moisture loss and preserved cheese quality. The strongest antimicrobial effect against *Staphylococcus aureus* was seen in the PCL/Gel/3% ZO/3% cumin EO treatment, with a survival rate of 5.09 ± 0.29 on day 12 versus 8.82 ± 0.41 in controls, attributed to the synergistic effects of the compounds used (Shanbehzadeh et al., 2022).

Freshness indicator films represent an innovative approach to food packaging, specifically engineered to monitor and assess the quality and safety of food products throughout their storage life. These advanced films are composed of natural pigments extracted from various plant sources, such as curcumin (found in turmeric), alizarin (derived from madder root), and anthocyanins (which are responsible for the red, blue, and purple colors in many fruits and vegetables) (Bakhshizadeh et al., 2023; Tavassoli et al., 2022). Zou and coworkers developed significant strides in developing a new type of ENFs utilizing a combination of hydroxypropyltrimethyl ammonium chloride chitosan (HACC), PCL, and SKN (shikonin) materials (Fig. 20.4). These films not only exhibit enhanced antimicrobial and antioxidant properties but also serve as intelligent packaging that actively monitors the freshness of food products. The unique feature of these HACC/PCL/SKN films is their ability to undergo pH-responsive color changes, transitioning from red at pH 2 to a striking blue-purple at pH 12. This colorimetric response allows consumers and retailers alike to easily visually assess the freshness of the enclosed food items. In extensive testing, these films have demonstrated their effectiveness in indicating spoilage for shrimp stored under varying temperature conditions over a three-day period. As a result, they have promising potential for application across a diverse array of food products, enhancing safety and quality monitoring during storage and reducing food waste (Zou et al., 2023).

20.4 Conclusion and future perspectives

PCL has demonstrated remarkable success both independently and as a component of various composite materials across a broad spectrum of application fields, ranging from nonmedical to highly specialized biomedical contexts. The existing literature extensively documents PCL's compatibility with the development of innovative systems tailored to specific functions. One of the distinguishing characteristics of PCL is its relatively slow degradation rate, which can be precisely modulated, making it an attractive choice for researchers seeking to tailor the degradation kinetics of materials for different therapeutic applications. Furthermore, the ease with which PCL's crystallinity, porous architecture, and elastic modulus can be tuned enhances its suitability for the development of advanced drug delivery systems. In tissue engineering, PCL is particularly notable for its application in creating 3D porous scaffolds that provide an ideal environment for cellular attachment, growth, and differentiation. Extensive in vivo studies have shed light on PCL's degradation kinetics, its biocompatibility, and its biological behavior upon implantation, all contributing to its reputation as a reliable biomaterial. Importantly, PCL has received FDA approval for various clinical applications, facilitating its transition from laboratory research to real-world medical use. This regulatory endorsement not only validates the safety and efficacy of PCL-based systems but also motivates researchers and industry professionals to actively explore and commercialize PCL and its composites for diverse biomedical and pharmaceutical applications. Moreover, the versatility of PCL has led to its growing significance in the production of a wide array of medical products, including sutures, wound dressings, bone scaffolding, medical devices, and even food packaging solutions. With a wealth of ongoing research on PCL-based systems consistently yielding promising results, there is strong reason to anticipate that the market will increasingly adopt PCL-based products in the near future.

A key factor in the successful manipulation of PCL lies in its ability to achieve essential physical, chemical, viscoelastic, and mechanical properties tailored to specific applications. Conversely, it is critical to rigorously manage processing parameters during fabrication, as any deviations could adversely affect material properties and, consequently, render the polymer unsuitable for its intended application. It is therefore imperative to accurately identify optimal fabrication methods, meticulously control processing conditions, and conduct a thorough risk-benefit analysis of composites to avert potential failures in any PCL-based system. Given a comprehensive understanding of PCL and its inherent characteristics, there is substantial potential to harness its properties for the development of biomedical, pharmaceutical, and nonmedical devices or systems that meet stringent specifications. PCL's customizable attributes solidify its position as a pioneering polymer across an array of fields. The continued backing from regulatory authorities, as evidenced by approvals for clinical applications, suggests a promising trajectory toward higher success rates in upcoming innovations. Numerous devices and drug delivery systems currently in preclinical and clinical trials are expected to enter the market in the near future. Additionally, ongoing rigorous research is exploring the applicability of PCL across nearly all sectors, hinting at a wealth of opportunities for new product development in the years to come.

References

Abd Al-Ghani, M. M., Azzam, R. A., & Madkour, T. M. (2021). Design and development of enhanced antimicrobial breathable biodegradable polymeric films for food packaging applications. *Polymers, 13*(20). https://doi.org/10.3390/polym13203527, https://www.mdpi.com/2073-4360/13/20/3527/pdf.

Abdellah Ali, S.F. (2016). Mechanical and thermal properties of promising polymer composites for food packaging applications. IOP conference series: Materials science and engineering.

Abdellah Ali, S. F. (2020). Performance of cellulose acetate propionate in polycaprolactone and starch composites: Biodegradation and water resistance properties. *Biointerface Research in Applied Chemistry, 10*(3), 5382–5386. https://doi.org/10.33263/BRIAC103.382386, https://biointerfaceresearch.com/wp-content/uploads/2020/03/20695837103382386.pdf.

Abioye, A. A., Fasanmi, O. O., Rotimi, D. O., Abioye, O. P., Obuekwe, C. C., Afolalu, S. A., & Okokpujie, I. P. (2019). Review of the development of biodegradable plastic from synthetic polymers and selected synthesized nanoparticle starches. *Journal of Physics: Conference Series, 1378*(4), 042064. https://doi.org/10.1088/1742-6596/1378/4/042064.

Ahmadi, N., Ahari, H., Anvar, A., Khosravi-Darani, K., & Gharachorloo, M. (2024). Polycaprolactone (PCL)-based films integrated with hairy cellulose nanocrystals and silver nanoparticles for active Tilapia packaging applications. *Food Chemistry: X, 22*. https://doi.org/10.1016/j.fochx.2024.101490.

Ahmadi, S., Khoshkalampour, A., Ghorbani, M., Ramezani, S., Ghasempour, Z., & Ghareaghajlou, N. (2024). Development of active packaging material based on polycaprolactone/hydroxypropyl methylcellulose nanofibers containing carbon dot nanoparticles for meat preservation. *LWT, 197*. https://doi.org/10.1016/j.lwt.2024.115913.

Ahmady, A. R., Solouk, A., Saber-Samandari, S., Akbari, S., Ghanbari, H., & Brycki, B. E. (2023). Capsaicin-loaded alginate nanoparticles embedded polycaprolactone-chitosan nanofibers as a controlled drug delivery nanoplatform for anticancer activity. *Journal of Colloid and Interface Science, 638*, 616–628. https://doi.org/10.1016/j.jcis.2023.01.139, http://www.elsevier.com/inca/publications/store/6/2/2/8/6/1/index.htt.

Ahmed, J., Mulla, M., Jacob, H., Luciano, G., Bini, T. B., & Almusallam, A. (2019). Polylactide/poly(ε-caprolactone)/zinc oxide/clove essential oil composite antimicrobial films for scrambled egg packaging. *Food Packaging and Shelf Life, 21*, 100355. 10.1016/j.fpsl.2019.100355.

Ahn, K., Kambiz, S., Park, K., & Seo, J. (2024). Enhancing polyvinyl alcohol surface properties through pre-drying treatment and modified electrospun polycaprolactone nanofibers. *Progress in Organic Coatings, 194*, 108621. https://doi.org/10.1016/j.porgcoat.2024.108621.

Ali, F. B., Kang, D. J., Kim, M. P., Cho, C. H., & Kim, B. J. (2014). Synthesis of biodegradable and flexible, polylactic acid based, thermoplastic polyurethane with high gas barrier properties. *Polymer International, 63*(9), 1620–1626. https://doi.org/10.1002/pi.4662.

Awasthi, S. K., Kumar, M., Kumar, V., Sarsaiya, S., Anerao, P., Ghosh, P., Singh, L., Liu, H., Zhang, Z., & Awasthi, M. K. (2022). A comprehensive review on recent advancements in biodegradation and sustainable management of biopolymers. *Environmental Pollution, 307*. https://doi.org/10.1016/j.envpol.2022.119600, https://www.journals.elsevier.com/environmental-pollution.

Bahramian, B., Abedi-Firoozjah, R., Kiani-Salmi, N., Ebrahimi, A., Oladzadabbasabadi, N., Moradiyan, M. H., Khezerlou, A., Mazloomi, S. M., Ehsani, A., & Tavassoli, M. (2025). Development of polycaprolactone-based electrospun nanofiber incorporated lemon beebrush essential oil-loaded metal-organic frameworks as a novel active food packaging for meat preservation. *Food Control, 168*. https://doi.org/10.1016/j.foodcont.2024.110981, https://www.sciencedirect.com/science/journal/09567135.

Bahrani, S., Sadati Behbahani, E., Ghaedi, M., Amrollahi Miandeh, Y., & Asfaram, A. (2024). Molecular imprinted technology using in biosensing: A review. *Microchemical Journal, 203*, 110888. https://doi.org/10.1016/j.microc.2024.110888.

Bakhshizadeh, M., Moghaddam, T. N., Tavassoli, M., Ayaseh, A., & Bangar, S. P. (2023). Gelatin/chitosan nanofibres containing β-cyclodextrin complex and corn poppy (*Papaver rhoeas* L.) for intelligent packaging. *International Journal of Food Science and Technology, 58*(5), 2360–2368. https://doi.org/10.1111/ijfs.16369, http://onlinelibrary.wiley.com/journal/10.1111/(ISSN)1365-2621.

Barnthip, N., Teeka, J., Kantha, P., Teepoo, S., & Damjuti, W. (2022). Fabrication and characterization of polycaprolactone/cellulose acetate blended nanofiber mats containing sericin and fibroin for biomedical application. *Scientific Reports, 12*(1). https://doi.org/10.1038/s41598-022-26908-2.

Belkhir, N. L., Harrane, A., Bezzekhami, M. A., Belalia, M., Bellucci, S., & Hamed, D. (2024). Progress in nanocomposites based on starch nanocrystals graft polycaprolactone and modified clay: Synthesis, characterization, thermal and antibacterial properties. *Journal of Adhesion Science and Technology, 38*(22), 4099–4115. https://doi.org/10.1080/01694243.2024.2364818, http://www.tandfonline.com/toc/tast20/current.

Bhovi, V. K., Melinmath, S. P., & Gowda, R. (2022). Biodegradable Polymers and their applications: A review. *Mini-Reviews in Medicinal Chemistry, 22*(16), 2081–2101. https://doi.org/10.2174/1389557522666220128152847, http://www.eurekaselect.com/633/journal/mini-reviews-medicinal-chemistry.

Bujok, S., Peter, J., Halecký, M., Ecorchard, P., Machálková, A., Santos Medeiros, G., Hodan, J., Pavlova, E., & Beneš, H. (2021). Sustainable microwave synthesis of biodegradable active packaging films based on polycaprolactone and layered ZnO nanoparticles. *Polymer Degradation and Stability, 190*, 109625. https://doi.org/10.1016/j.polymdegradstab.2021.109625.

Chakravarti, A. R., & Annu (2024). Biomedical applications of polycaprolactone. In *Biopolymers for biomedical applications*. India: Wiley, 373–391. http://doi.org/10.1002/9781119865452.ch14, https://onlinelibrary.wiley.com/doi/book/10.1002/9781119865452.

Chen, H., Oveissi, F., Daly, S., Shahrbabaki, Z., Naficy, S., & Dehghani, F. (2022). A green and biodegradable plasticizer from copolymers of poly(β-hydroxybutyrate- co -ε-caprolactone). *Journal of Applied Polymer Science, 139*(22). https://doi.org/10.1002/app.52240.

Chen, S., Qin, J., & Du, J. (2022). Cross-linkable yet biodegradable polymer films. *Wuli Huaxue Xuebao/ Acta Physico - Chimica Sinica, 38*(8). https://doi.org/10.3866/PKU.WHXB202006029, http://www.whxb.pku.edu.cn/EN/Y2022/V38/I8/2006029.

Chukaew, S., Parivatphun, T., Thonglam, J., Khangkhamano, M., Meesane, J., & Kokoo, R. (2023). Biphasic scaffolds of polyvinyl alcohol/gelatin reinforced with polycaprolactone as biomedical materials supporting for bone augmentation based on anatomical mimicking; fabrication, characterization, physical and mechanical properties, and in vitro testing. *Journal of the Mechanical Behavior of Biomedical Materials, 143*, 105933. https://doi.org/10.1016/j.jmbbm.2023.105933.

Cintra, S. C., Braga, N. F., Morgado, G. F. D. M., Montanheiro, T. L. D. A., Marini, J., Passador, F. R., & Montagna, L. S. (2022). Development of new biodegradable composites materials from polycaprolactone and wood flour. *Wood Material Science and Engineering, 17*(6), 586–597. https://doi.org/10.1080/17480272.2021.1905712, http://www.tandf.co.uk/journals/titles/17480272.asp.

Clerici, N. J., Vencato, A. A., Helm Júnior, R., Daroit, D. J., & Brandelli, A. (2024). Electrospun Poly-ε-caprolactone nanofibers incorporating keratin hydrolysates as innovative antioxidant scaffolds. *Pharmaceuticals, 17*(8). https://doi.org/10.3390/ph17081016, https://www.mdpi.com/journal/pharmaceuticals.

Coombes, A. G. A., Rizzi, S. C., Williamson, M., Barralet, J. E., Downes, S., & Wallace, W. A. (2004). Precipitation casting of polycaprolactone for applications in tissue engineering and drug delivery. *Biomaterials, 25*(2), 315–325. https://doi.org/10.1016/s0142-9612(03)00535-0.

de Oliveira Gama, R., Bretas, R. E. S., & Oréfice, R. L. (2018). Control of the hydrophilic/hydrophobic behavior of biodegradable natural polymers by decorating surfaces with nano- and micro-components. *Advances in Polymer Technology, 37*(3), 654–661. https://doi.org/10.1002/adv.21706, http://onlinelibrary.wiley.com/journal/10.1002/(ISSN)1098-2329.

Deng, Z. A., Wu, M., Shen, C., Yang, X., Wang, D., Li, J., Wu, D., & Chen, K. (2024). Microfluidic-blow-spinning of carvacrol-loaded porphyrin metal – organic framework nanofiber films with synergistic antibacterial capabilities for food packaging. *Food Chemistry, 460*. https://doi.org/10.1016/j.foodchem.2024.140707, https://www.sciencedirect.com/science/journal/03088146.

Deng, J., Yu, W., Zhan, L., Zhu, X., Fang, Y., Zhang, P., ... Huang, C. (2024). Wettability regulation of membranes based on biodegradable aliphatic polyester. *Journal of Membrane Science, 711*. https://doi.org/10.1016/j.memsci.2024.123199.

Díaz-Pedraza, A., Piñeros-Castro, Y., & Ortega-Toro, R. (2021). Effect of potassium sorbate as an interface agent in biodegradable bi-layers polymers. *Revista Mexicana de Ingeniera Quimica, 20*(1), 345–354. https://doi.org/10.24275/rmiq/Alim1999, http://rmiq.org/ojs311/index.php/rmiq/article/download/1999/1114.

Din, M. I., Ghaffar, T., Najeeb, J., Hussain, Z., Khalid, R., & Zahid, H. (2020). Potential perspectives of biodegradable plastics for food packaging application-review of properties and recent developments. *Food Additives and Contaminants - Part A Chemistry, Analysis, Control, Exposure and Risk Assessment, 37*(4), 665–680. https://doi.org/10.1080/19440049.2020.1718219, http://www.tandfonline.com/toc/tfac20/current.

Ding, Y., Ikura, R., Yamaoka, K., Nishida, K., Sugawara, A., Uyama, H., Nara, S., & Takashima, Y. (2024). Reinforcement and controlling the stability of poly(ε-caprolactone)-based polymeric materials via reversible and movable cross-links employing cyclic polyphenylene sulfide. *ACS Macro Letters, 13*(10), 1265–1271. https://doi.org/10.1021/acsmacrolett.4c00495.

Dogra, V., Kishore, C., Mishra, A., Gaur, A., Gupta, J., Zafar, S., & Verma, A. (2024). Polymer based composite coatings in engineering applications. In *Hybrid composite materials: Experimental and theoretical analysis*. India: Springer Nature, 237–251. http://doi.org/10.1007/978-981-97-2104-7_9, https://link.springer.com/book/10.1007/978-981-97-2104-7.

Domingos, M., Chiellini, F., Cometa, S., De Giglio, E., Grillo-Fernandes, E., Bártolo, P., & Chiellini, E. (2010). Evaluation of in vitro degradation of PCL scaffolds fabricated via BioExtrusion. Part 1: Influence of the degradation environment. *Virtual and Physical Prototyping, 5*(2), 65–73. https://doi.org/10.1080/17452751003769440.

Du, Y., Zhang, Y., & Pang, J. (2024). Electrospun konjac glucomannan-gelatin/polycaprolactone bilayer nanofibrous films with improved hydrophobicity and mechanical properties for food packaging. *Food Bioscience, 61*, 104544. https://doi.org/10.1016/j.fbio.2024.104544.

Duarah, R., Aleksic, I., Milivojevic, D., Rameshkumar, S., Nikodinovic-Runic, J., & Padamati, R. B. (2023). Development of nystatin-based antifungal, biodegradable polymer composite materials for food packaging via melt processing approach. *Journal of Applied Polymer Science, 140*(45). https://doi.org/10.1002/app.54663, http://onlinelibrary.wiley.com/journal/10.1002/(ISSN)1097-4628.

Elgharbawy, A. S., El Demerdash, A. G. M., Sadik, W. A., Kasaby, M. A., Lotfy, A. H., & Osman, A. I. (2024). Synthetic degradable polyvinyl alcohol polymer and its blends with starch and cellulose—A comprehensive overview. *Polymers, 16*(10). https://doi.org/10.3390/polym16101356, http://www.mdpi.com/journal/polymers.

El-Naggar, M. E., Hasanin, M., & Hashem, A. H. (2022). Eco-friendly synthesis of superhydrophobic antimicrobial film based on cellulose acetate/polycaprolactone loaded with the green biosynthesized copper nanoparticles for food packaging application. *Journal of Polymers and the Environment, 30*(5), 1820–1832. https://doi.org/10.1007/s10924-021-02318-9, http://www.kluweronline.com/issn/1566-2543/.

Elsayed, H., Ibrahim, S., & Masoud, R. A. (2024). Novel nano biodegradable package films of polycaprolactone / extracted gelatin from white shavings. *Egyptian Journal of Chemistry, 67*(13), 537–548. https://doi.org/10.21608/ejchem.2024.257729.9075, https://ejchem.journals.ekb.eg/article_338972.html.

El-Sayed, S. M., & Youssef, A. M. (2023). Eco-friendly biodegradable nanocomposite materials and their recent use in food packaging applications: A review. *Sustainable Food Technology, 1*(2), 215–227. https://doi.org/10.1039/d2fb00021k, www.rsc.org/journals-books-databases/about-journals/sustainable-food-technology/.

Erbay, E. A., Dağtekin, B. B. G., Türe, M., Yeşilsu, A. F., & Torres-Giner, S. (2017). Quality improvement of rainbow trout fillets by whey protein isolate coatings containing electrospun poly (ε-caprolactone) nanofibers with Urtica dioica L. extract during storage. *LWT, 78*, 340–351.

Ghanem, A. F., Yassin, M. A., Cosquer, R., Gouanvé, F., Espuche, E., & Abdel Rehim, M. H. (2024). Polycaprolactone composite films infused with hyperbranched polyester/reduced graphene oxide: Influence on biodegradability, gas/water transport and antimicrobial properties for sustainable packaging. *RSC Advances, 14*(9), 5740–5753. https://doi.org/10.1039/d3ra08948g, http://pubs.rsc.org/en/journals/journal/ra.

Govindan, S., Ramos, M., & Al-Jumaily, A. M. (2023). On the effect of monomeric and polymeric plasticizer on polybutylene succinate (PBS), polyhydroxybutyrate (PHB), and polylactic acid (PLA) films with 20wt%PCL for flexible packaging application. *Materials Science Forum, 1087*, 3–12.

Gu, J., Xu, Z., Liu, Q., Tang, S., Zhang, W., Xie, S., Chen, X., Chen, J., Yong, K.-T., Yang, C., & Xu, G. (2024). Building a better silver bullet: Current status and perspectives of non-viral vectors for mRNA vaccines. *Advanced Healthcare Materials, 13*(3). https://doi.org/10.1002/adhm.202302409.

Guclu, N., Duzyer Gebizli, S., & Orhan, M. (2023). Development of polycaprolactone-based electrospun pH-sensitive sensors as instant colorimetric indicators for food packaging. *Coloration Technology, 139*(5), 527–543. 10.1111/cote.12701.

Guicherd, M., Ben Khaled, M., Guéroult, M., Nomme, J., Dalibey, M., Grimaud, F., Alvarez, P., Kamionka, E., Gavalda, S., Noël, M., Vuillemin, M., Amillastre, E., Labourdette, D., Cioci, G., Tournier, V., Kitpreechavanich, V., Dubois, P., André, I., Duquesne, S., & Marty, A. (2024). An engineered enzyme embedded into PLA to make self-biodegradable plastic. *Nature, 631*(8022), 884–890. https://doi.org/10.1038/s41586-024-07709-1.

Gupta, P., Sharma, S., Jabin, S., & Jadoun, S. (2024). Chitosan nanocomposite for tissue engineering and regenerative medicine: A review. *International Journal of Biological Macromolecules, 254*(Pt 1), 127660. https://doi.org/10.1016/j.ijbiomac.2023.127660.

Hegde, C., & Pathak, C. (2023). Synthesis, characterization and eco-friendly applications of biodegradable polycaprolactone-tapioca starch polymeric composite membrane. *Asian Journal of Chemistry, 36*(1), 18–24. https://doi.org/10.14233/ajchem.2024.28554.

Heiner, A. C., Bishop, K. M., Musgrave, G. M., Huber, A. K., Cushman, L. M., Bates, J. S., & Wang, C. (2024). Soil Compostability of ester-based thiol-ene photopolymer networks. *ACS Sustainable Chemistry and Engineering, 12*(24), 9067–9077. https://doi.org/10.1021/acssuschemeng.4c01018, http://pubs.acs.org/journal/ascecg.

Hsu, C. C., Hsu, C. C., Lin, C. M., Wang, D. M., & Hsieh, H. J. (2023). Fabrication of polycaprolactone scaffolds filled with cell-compatible porous matrix for enhanced cell growth for biomedical applications. *Journal of Applied Polymer Science, 140*(38). https://doi.org/10.1002/app.54432, http://onlinelibrary.wiley.com/journal/10.1002/(ISSN)1097-4628.

Hu, G., Huang, Q., Li, J., Wang, Z., Yu, Y., Yang, W., & Hu, Y. (2024). PCL/Fucoidan nanofiber membrane loaded HP-β-CD/EGC inclusion complexes for food packaging based on self-assembly strategy. *Food Hydrocolloids, 151*, 109836. https://doi.org/10.1016/j.foodhyd.2024.109836.

Hu, G., Li, J., Wang, Z., Yang, W., & Hu, Y. (2024). PCL/Yam polysaccharide nanofibrous membranes loaded with self-assembled HP-β-CD/ECG inclusion complexes for food packaging. *Food Chemistry, 438*, 138031. https://doi.org/10.1016/j.foodchem.2023.138031.

Huang, J., Fang, Y., Zhou, S., Tie, R., Zhou, S., Jin, Y., ... Kumar, R. (2024). Enhanced compatibility and toughness of poly(3-hydroxybutyrate-co-3-hydroxyvalerate)/polycaprolactone blends by ring-opening polymerization and hydrogen bonding of epoxy-terminated hyperbranched polyester. *Journal of Applied Polymer Science, 141*(41). https://doi.org/10.1002/app.56069.

Huang, Y., Huang, H., Wu, J., Feng, Q., Li, Y., Li, Q., & Sun, T. (2024). Preparation and properties of PCL coaxial electrospinning films with shell loaded with CEO and core coated LEO nanoemulsions. *Food Research International, 194*. https://doi.org/10.1016/j.foodres.2024.114817.

Ibrahim, S. W., Hamad, T. I., & Haider, J. (2023). Biological properties of polycaprolactone and barium titanate composite in biomedical applications. *Science Progress, 106*(4). https://doi.org/10.1177/00368504231215942, https://journals.sagepub.com/home/SCI.

İlaslan, K., Tornuk, F., & Durak, M. Z. (2022). Development of polycaprolactone biodegradable films reinforced with silver-doped organoclay and effect on the microbiological quality of ground beef meat. *Journal of Food Processing and Preservation, 46*(10). https://doi.org/10.1111/jfpp.16862, http://onlinelibrary.wiley.com/journal/10.1111/(ISSN)1745-4549.

Jiang, J., Qian, S., Song, T., Lu, X., Zhan, D., Zhang, H., & Liu, J. (2024). Food-packaging applications and mechanism of polysaccharides and polyphenols in multicomponent protein complex system: A review. *International Journal of Biological Macromolecules, 270*(Pt 2), 132513. https://doi.org/10.1016/j.ijbiomac.2024.132513.

Jovanska, L., Chiu, C. H., Yeh, Y. C., Chiang, W. D., Hsieh, C. C., & Wang, R. (2022). Development of a PCL-PEO double network colorimetric pH sensor using electrospun fibers containing Hibiscus rosa sinensis extract and silver nanoparticles for food monitoring. *Food Chemistry, 368*. https://doi.org/10.1016/j.foodchem.2021.130813, www.elsevier.com/locate/foodchem.

Joy, A., Unnikrishnan, G., Megha, M., Haris, M., Thomas, J., Kolanthai, E., & Muthuswamy, S. (2023). Design of biocompatible polycaprolactone-based nanocomposite loaded with graphene oxide/strontium nanohybrid for biomedical applications. *Applied Nanoscience (Switzerland), 13*(6), 4471–4484. https://doi.org/10.1007/s13204-022-02721-1, https://www.springer.com/journal/13204.

Kenawy, E. R., Hemdan, B. A., Abdel-Sattar, R., Khairallah, M. R., Shendy, S., Morsy, O. M., & El-Naggar, M. E. (2024). Assessing the bactericidal effectiveness of clove oil nanoemulsion-loaded polycaprolactone nanofibers for eco-friendly and endurable food packaging. *Polymers for Advanced Technologies, 35*(11). https://doi.org/10.1002/pat.6594, http://onlinelibrary.wiley.com/journal/10.1002/(ISSN)1099-1581.

Khezerlou, A., Tavassoli, M., Alizadeh Sani, M., Ehsani, A., & McClements, D. J. (2023). Smart packaging for food spoilage assessment based on *Hibiscus sabdariffa* L. anthocyanin-loaded chitosan films. *Journal of Composites Science, 7*(10). https://doi.org/10.3390/jcs7100404, www.mdpi.com/journal/jcs.

Khoshboui Lahijani, L., Ahari, H., Anvar, A., & Sharifan, A. (2024). pH-sensitive polycaprolactone/gelatin active packaging containing rosemary-loaded chitosan nanoparticles. *ACS Food Science and Technology, 4*(11), 2600–2613. https://doi.org/10.1021/acsfoodscitech.4c00428, https://pubs.acs.org/page/afsthl/about.html.

Kim, W., Han, T., Gwon, Y., Park, S., Kim, H., & Kim, J. (2022). Biodegradable and flexible nanoporous films for design and fabrication of active food packaging systems. *Nano Letters, 22*(8), 3480–3487. https://doi.org/10.1021/acs.nanolett.2c00246.

Kozielski, K. L., & Sitti, M. (2017). Gene delivery particle engineering strategies for shape-dependent targeting of cells and tissues. *Current Gene Therapy, 17*(2), 80–88. https://doi.org/10.2174/1566523217666170510161120, http://www.benthamdirect.org/pages/all_b_bypublication.php.

Lai, X., Liu, J., Liu, G., Wang, L., & Ren, Q. (2019). Study on synthesis and thermal stability of non-amine waterborne polyurethane-polyacrylate. *Journal of Applied Science and Engineering, 22*(1), 145–152. https://doi.org/10.6180/jase.201903_22(1).0014, http://www2.tku.edu.tw/~tkjse/22-1/14-CHE10707_0121.pdf.

Lam, C. X. F., Hutmacher, D. W., Schantz, J.-T., Woodruff, M. A., & Teoh, S. H. (2009). Evaluation of polycaprolactone scaffold degradation for 6 months in vitro and in vivo. *Journal of Biomedical Materials Research. Part A, 90A*(3), 906–919. https://doi.org/10.1002/jbm.a.32052.

Larrañaga, A., Aldazabal, P., Martin, F., & Sarasua, J. J. (2014). Hydrolytic degradation and bioactivity of lactide and caprolactone based sponge-like scaffolds loaded with bioactive glass particles. *Polymer Degradation and Stability, 110*, 121–128.

Lawless, S., Laycock, B., Lant, P., & Pratt, S. (2024). The extensional rheology of biodegradable polymers determines their propensity to be film blown. *Polymer Degradation and Stability, 227*, 110888. https://doi.org/10.1016/j.polymdegradstab.2024.110888.

Leite, L. S. F., Le Gars, M., Azeredo, H. M. C., Moreira, F. K. V., Mattoso, L. H. C., & Bras, J. (2024). Insights into the effect of carboxylated cellulose nanocrystals on mechanical and barrier properties of gelatin films for flexible packaging applications. *International Journal of Biological Macromolecules, 280*. https://doi.org/10.1016/j.ijbiomac.2024.135726, https://www.sciencedirect.com/science/journal/01418130.

Li, J., Tseng, J., Tang, D., Yong, Y., Sun, L., & Huo, Y. X. (2024). Upcycling C1 gas-derived resources in future food system. *Resources, Conservation and Recycling, 210*. https://doi.org/10.1016/j.resconrec.2024.107827, https://www.sciencedirect.com/science/journal/09213449.

Li, X., Li, S. C., & Huang, L. Q. (2014). Synthesis and property of water borne polyurethane/modified nano-ZnO composites packaging membranes. *Applied Mechanics and Materials, 469*, 167–170. https://doi.org/10.4028/www.scientific.net/AMM.469.167.

Li, J., Li, M., Wang, Z., Hu, G., Yang, W., & Hu, Y. (2024). Chitosan/PCL fibrous membranes loaded potassium cinnamate/β-CD clathrate compounds developed by ELS for fruit preservation. *Food Hydrocolloids, 156*. https://doi.org/10.1016/j.foodhyd.2024.110341.

Liang, X., Cha, D. K., & Xie, Q. (2024). Properties, production, and modification of polyhydroxyalkanoates. *Resources, Conservation and Recycling Advances, 21*. https://doi.org/10.1016/j.rcradv.2024.200206, https://www.sciencedirect.com/science/journal/26673789.

Lin, W.-C., Yao, C., Huang, T.-Y., Cheng, S.-J., & Tang, C.-M. (2019). Long-term in vitro degradation behavior and biocompatibility of polycaprolactone/cobalt-substituted hydroxyapatite composite for bone tissue engineering. *Dental Materials, 35*(5), 751–762. https://doi.org/10.1016/j.dental.2019.02.023.

Little, U., Buchanan, F., Harkin-Jones, E., McCaigue, M., Farrar, D., & Dickson, G. (2009). Accelerated degradation behaviour of poly(ε-caprolactone) via melt blending with poly(aspartic acid-co-lactide) (PAL). *Polymer Degradation and Stability, 94*(2), 213–220. https://doi.org/10.1016/j.polymdegradstab.2008.11.001.

Liu, Z., Qin, Z., Jia, H., Xu, J., Liu, M., & Hou, Z. (2023). Dual-crosslinked starch–poly(ester urethane)–oligochitosan films with high starch content: Application as biodegradable food packaging. *Food Packaging and Shelf Life, 37*, 101064. https://doi.org/10.1016/j.fpsl.2023.101064.

Ma, J., Liu, Y., Xu, J., Chen, Y., Liu, L., & Zhang, H. (2024). An insect lac blanket-mimetic and degradable shellac hydrogel/chitosan packaging film with controllable gas permeation for fresh-cut vegetables preservation. *International Journal of Biological Macromolecules, 275*(Pt 2), 133131. https://doi.org/10.1016/j.ijbiomac.2024.133131.

Martinaud, E., Hierro-Iglesias, C., Hammerton, J., Hadad, B., Evans, R., Sacharczuk, J., Lester, D., Derry, M. J., Topham, P. D., & Fernandez-Castane, A. (2024). Valorising cassava peel waste into plasticized polyhydroxyalkanoates blended with polycaprolactone with controllable thermal and mechanical properties. *Journal of Polymers and the Environment, 32*(8), 3503–3515. https://doi.org/10.1007/s10924-023-03167-4.

Mary, S. M. S., Malathi, S., Varadharaj, S., Arul, K. T., Verma, R. S., Ramya, J. R., Asokan, K., Krishna, J. B. M., & Kalkura, S. N. (2023). S. Moorthy Babu, Tuning the physiochemical properties of polycaprolactone-hydroxyapatite composite films by gamma irradiation for biomedical applications. *Biomaterials Advances, 155*. https://doi.org/10.1016/j.bioadv.2023.213679, https://www.sciencedirect.com/journal/biomaterials-advances/about/aims-and-scope.

Matumba, K. I., Mokhena, T. C., Ojijo, V., Sadiku, E. R., & Ray, S. S. (2024). Morphological characteristics, properties, and applications of polylactide/poly(ε-caprolactone) blends and their composites—A review. *Macromolecular Materials and Engineering, 309*(8). https://doi.org/10.1002/mame.202400056, http://onlinelibrary.wiley.com/journal/10.1002/(ISSN)1439-2054.

Mohamady Hussein, M. A., Alamri, E. S., Bayomy, H. M., Albalawi, A. N., Grinholc, M., & Muhammed, M. (2024). Developing novel hybrid bilayer nanofibers based on polylactic acid with impregnation of chamomile essential oil and gallic acid-stabilized silver nanoparticles. *International Journal of Biological Macromolecules, 282*. https://doi.org/10.1016/j.ijbiomac.2024.137262, https://www.sciencedirect.com/science/journal/01418130.

Moll, E., & Chiralt, A. (2024). Improving thermo-sealing of poly(3-hydroxybutyrate-co-3-hydroxyvalerate) by Blending with Polycaprolactone. *Polymers, 16*(23), 3255. https://doi.org/10.3390/polym16233255.

Moshkbid, E., Cree, D. E., Bradford, L., & Zhang, W. (2024). Biodegradable alternatives to plastic in medical equipment: Current state, challenges, and the future. *Journal of Composites Science, 8*(9). https://doi.org/10.3390/jcs8090342, www.mdpi.com/journal/jcs.

Muñoz-Suarez, A. M., Cortés-Rodríguez, M., & Ortega-Toro, R. (2024). Biodegradable films based on tilapia collagen (*Oreochromis* sp): Improvement of properties with PLA and PCL bilayers with potential use in sustainable food packaging. *Revista Mexicana de Ingeniera Quimica, 23*(3). https://doi.org/10.24275/rmiq/Alim24326, https://rmiq.org/iqfvp/Numbers/V23/No3/Alim24326.html.

Nair, A., Chandrashekhar H, R., Day, C. M., Garg, S., Nayak, Y., Shenoy, P. A., & Nayak, U. Y. (2024). Polymeric functionalization of mesoporous silica nanoparticles: Biomedical insights. *International Journal of Pharmaceutics, 660*. https://doi.org/10.1016/j.ijpharm.2024.124314, https://www.sciencedirect.com/science/journal/03785173.

Nallathambi, A., Thiyaku, T. T., & Aravinthan, R. S. M. V. (2024). Synthesis and characterization of biocomposite food packaging composite film using waste orange peel bio-plasticizer and PLA(Poly Lactic Acid) /PCL (Polycaprolactone) blend. *Waste and Biomass Valorization*. https://doi.org/10.1007/s12649-024-02724-9, https://www.springer.com/journal/12649.

Navarro Arrieta, C., Acevedo-Puello, V., Fuentes Ordoñez, E. G., Fonseca-Florido, H., & Ortega-Toro, R. (2024). Biodegradable monolayer film based on the collagen extracted from *Oreochromis* sp. processing byproducts blended with chitosan and assembled with PCL and PLA monolayers to form bilayers films. *Journal of Food Process Engineering, 47*(7). https://doi.org/10.1111/jfpe.14696, http://onlinelibrary.wiley.com/journal/10.1111/(ISSN)1745-4530.

Obisesan, O. S., Ajiboye, T. O., Mhlanga, S. D., & Mufhandu, H. T. (2023). Biomedical applications of biodegradable polycaprolactone-functionalized magnetic iron oxides nanoparticles and their polymer nanocomposites. *Colloids and Surfaces B: Biointerfaces, 227*. https://doi.org/10.1016/j.colsurfb.2023.113342, www.elsevier.com/locate/colsurfb.

Ozer, F. T., & San Keskin, N. O. (2024). Fabrication of stable dual-layer polymer-based nanofibers with applications for food preservation. *ChemistrySelect, 9*(10). https://doi.org/10.1002/slct.202304294, http://onlinelibrary.wiley.com/journal/10.1002/(ISSN)2365-6549.

Pei, J., Palanisamy, C. P., Srinivasan, G. P., Panagal, M., Kumar, S. S. D., & Mironescu, M. (2024). A comprehensive review on starch-based sustainable edible films loaded with bioactive components for food packaging. *International Journal of Biological Macromolecules, 274*. https://doi.org/10.1016/j.ijbiomac.2024.133332, https://www.sciencedirect.com/science/journal/01418130.

Phadtare, P., Viswapriya, V., Shinde, V., Mathure, D., & Salunke, M. R. (2025). Hybrid bionanocomposites as the advancements in biomedical utility. *Hybrid Advances, 8*. https://doi.org/10.1016/j.hybadv.2024.100365, https://www.sciencedirect.com/journal/hybrid-advances.

Piscitelli, E., Maya, I. C., Cocola, C., Martino, V., Abeni, E., Pelucchi, P., ... Reinbold, R. (2024). Long-term culture of patient-derived mammary organoids in non-biogenic electrospun scaffolds for identifying metalloprotein and motor protein activities in aging and senescence. *Advances in Protein Chemistry and Structural Biology, 141*, 331–360. https://doi.org/10.1016/bs.apcsb.2024.03.008.

Pitt, C., Chasalow, F., Hibionada, Y., Klimas, D., & Schindler, A. (1981). Aliphatic polyesters. I. The degradation of poly (ϵ-caprolactone) in vivo. *Journal of Applied Polymer Science, 26*(11), 3779–3787.

Pramanik, K. (2022). *Polycaprolactone: Applications, synthesis and characterization*. Nova Science Publishers. http://doi.org/10.52305/YNVU4463.

Qin, Y., Liu, D., Wu, Y., Yuan, M., Li, L., & Yang, J. (2015). Effect of PLA/PCL/cinnamaldehyde antimicrobial packaging on physicochemical and microbial quality of button mushroom (Agaricus bisporus). *Postharvest Biology and Technology, 99*, 73–79. 10.1016/j.postharvbio.2014.07.018.

Qin, Z., Fang, W., Jiang, Q., Li, J., & Zhang, H. (2024). The urchin-like gold nanoparticles/poly(ε-caprolactone)/chitosan electrospun nanofibers for antibacterial active packaging. *International Journal of Biological Macromolecules, 274*(Pt 1), 133287. https://doi.org/10.1016/j.ijbiomac.2024.133287.

Qin, Z., Jiang, Q., Zou, Y., Chen, M., Li, J., Li, Y., & Zhang, H. (2024). Synthesis of nanosized γ-cyclodextrin metal-organic frameworks as carriers of limonene for fresh-cut fruit preservation based on polycaprolactone nanofibers. *Small (Weinheim an der Bergstrasse, Germany), 20*(29). https://doi.org/10.1002/smll.202400399.

Ramos, M., Govindan, S., & Al-Jumaily, A. M. (2023). Property improvement of polybutylene succinate (PBS), polyhydroxybutyrate (PHB), and polylactic acid (PLA) films with PCL (Polycaprolactone) for flexible packaging application. *Materials Science Forum, 1087*, 41–50.

Ravi, G., Iriventi, P., & Osmani, R. A. M. (2023). Biodegradable polymers: A novel stride in drug delivery and therapeutics. *Research Journal of Pharmacy and Technology, 16*(6), 3047–3053. https://doi.org/10.52711/0974-360X.2023.00502, www.rjptonline.org.

Raza, M. A., Sharma, M. K., Nagori, K., Jain, P., Ghosh, V., Gupta, U., & Ajazuddin (2024). Recent trends on polycaprolactone as sustainable polymer-based drug delivery system in the treatment of cancer: Biomedical applications and nanomedicine. *International Journal of Pharmaceutics, 666*. https://doi.org/10.1016/j.ijpharm.2024.124734, https://www.sciencedirect.com/science/journal/03785173.

Richert, A., Olewnik-Kruszkowska, E., Malinowski, R., Kalwasińska, A., & Swiontek Brzezinska, M. (2023). Polycaprolactone-based films incorporated with birch tar—Thermal, physicochemical, antibacterial, and biodegradable properties. *Foods, 12*(23), 4244. https://doi.org/10.3390/foods12234244.

Ruan, J., Liao, C., Li, P., Li, X., & Zuo, Y. (2024). Synergistic preparation of a straw fiber/polylactic acid composite with high toughness and strength through interfacial compatibility enhancement and elastomer toughening. *International Journal of Biological Macromolecules, 275*(Pt 1), 133621. https://doi.org/10.1016/j.ijbiomac.2024.133621.

Sachan, R., Warkar, S. G., & Purwar, R. (2023). An overview on synthesis, properties and applications of polycaprolactone copolymers, blends & composites. *Polymer-Plastics Technology and Materials, 62*(3), 327–358. https://doi.org/10.1080/25740881.2022.2113890, https://www.tandfonline.com/loi/lpte21.

Salević-Jelić, A., Lević, S., Prieto, C., Jeremić, S., Stevanović, S., Rac, V., ... Lagaron, J. M. (2024). Polycaprolactone-based electrospun films incorporating sage extract: From active food packaging application to accelerated biodegradation by Pseudomonas. *Future Foods, 10*. https://doi.org/10.1016/j.fufo.2024.100465.

Salgado, C. L., Sanchez, E. M. S., Zavaglia, C. A. C., & Granja, P. L. (2012). Biocompatibility and biodegradation of polycaprolactone-sebacic acid blended gels. *Journal of Biomedical Materials Research - Part A, 100*(1), 243–251. https://doi.org/10.1002/jbm.a.33272.

Santana, I., Felix, M., & Bengoechea, C. (2024). Properties of biopolymer blends based on Rugulopteryx okamurae and hydrophobic polycaprolactone (PCL) and hydrophilic acylated soy protein isolated (SPIa). *Environmental Science and Pollution Research, 31*(25), 36615–36625. https://doi.org/10.1007/s11356-024-33659-2.

Schlesinger, E., Ciaccio, N., & Desai, T. A. (2015). Polycaprolactone thin-film drug delivery systems: Empirical and predictive models for device design. *Materials Science and Engineering C, 57*, 232–239. https://doi.org/10.1016/j.msec.2015.07.027.

Seyrek, M. E., Okur, M., & Saraçoğlu, N. (2021). Improvement of mechanical, thermal and antimicrobial properties of organically modified montmorillonite loaded polycaprolactone for food packaging. *Journal of Vinyl and Additive Technology, 27*(4), 894–908. 10.1002/vnl.21860.

Shahrampour, D., Razavi, S. M. A., & Sadeghi, A. (2023). Evaluation of green tea extract incorporated antimicrobial/antioxidant/biodegradable films based on polycaprolactone/polylactic acid and its application in cocktail sausage preservation. *Journal of Food Measurement and Characterization, 17*(1), 1058–1067. https://doi.org/10.1007/s11694-022-01670-1, https://www.springer.com/journal/11694.

Shahsavar, F., & Babaei, A. (2024). Investigating the effect of chitosan functionalized graphene oxide on the performance of biodegradable polycaprolactone. *Journal of Reinforced Plastics and Composites, 43*(7-8), 429–439. https://doi.org/10.1177/07316844231163909.

Shanbehzadeh, F., Saei-Dehkordi, S. S., & Semnani, D. (2022). Fabrication and characterization of electrospun nanofibrous mats of polycaprolactone/gelatin containing ZnO nanoparticles and cumin essential oil and their anti-staphylococcal potency in white cheese. *Food Bioscience, 49*. https://doi.org/10.1016/j.fbio.2022.101904, http://www.journals.elsevier.com/food-bioscience/.

Shen, C., Wu, M., Sun, C., Li, J., Wu, D., Sun, C., He, Y., & Chen, K. (2022). Chitosan/PCL nanofibrous films developed by SBS to encapsulate thymol/HPβCD inclusion complexes for fruit packaging. *Carbohydrate Polymers, 286*, 119267. 10.1016/j.carbpol.2022.119267.

Shi, C., Fang, D., Xia, S., Guo, Y., Wang, J., Lyu, L., Wu, W., Huang, C., & Li, W. (2024). Poly(lactic acid)/polycaprolactone nanofibrous packaging containing different functional agents for blackberry postharvest preservation. *International Journal of Biological Macromolecules, 279*(Pt 1), 134544. https://doi.org/10.1016/j.ijbiomac.2024.134544.

Shi, C., Zhou, A., Fang, D., Lu, T., Wang, J., Song, Y., Lyu, L., Wu, W., Huang, C., & Li, W. (2022). Oregano essential oil/β-cyclodextrin inclusion compound polylactic acid/polycaprolactone electrospun nanofibers for active food packaging. *Chemical Engineering Journal, 445*, 136746. https://doi.org/10.1016/j.cej.2022.136746.

Shorey, R., & Mekonnen, T. H. (2024). Oleic acid decorated kraft lignin as a hydrophobic and functional filler of cellulose acetate films. *International Journal of Biological Macromolecules, 268*. https://doi.org/10.1016/j.ijbiomac.2024.131672, https://www.sciencedirect.com/science/journal/01418130.

Sun, H., Mei, L., Song, C., Cui, X., & Wang, P. (2006). The in vivo degradation, absorption and excretion of PCL-based implant. *Biomaterials, 27*(9), 1735–1740. https://doi.org/10.1016/j.biomaterials.2005.09.019.

Sundar, N., Pavithra, A., & Ananda Kumar, S. (2024). Effect of crystallinity of bio-based polymers on compatibility and the tortuous path model of cellulose-based industrial products. *Biomass Conversion and Biorefinery, 14*(18), 22687–22702. https://doi.org/10.1007/s13399-023-04697-6.

Suryani, Rihayat, Tiku, Fitria, Sariadi, Yunus, M., Hasanah, U., Safitri, A., Anggoro, D. D., Kumoro, A. C., Dahnum, D., Restu, W. K., Sembiring, K. C., Indriyati, Ndruru, S. T. C. L., & Putri, A. M. H. (2024). The impact of chitosan incorporation on the mechanical characteristics of biodegradable packaging based on PLA/PCL blend. *E3S Web of Conferences, 503*, 08004. https://doi.org/10.1051/e3sconf/202450308004.

Suryani, Rihayat, T., Fitria, Safitri, A., & Izzati, I. A. (2024). Valorization of active agent nanochitosan-catechin in development mechanical properties based on polylactid acid (PLA)-polycaprolactone (PCL) mixture. *Journal of Advanced Research in Micro and Nano Engineering, 20*(1), 10–16. https://doi.org/10.37934/armne.20.1.1016, https://akademiabaru.com/submit/index.php/armne/article/download/5394/4157.

Swain, S. K., Sethy, P. K., & Pattanayak, A. J. (2017). Synthesis of soy protein based biocomposites for packaging applications. *Green Energy and Technology, 0*, 143–166. https://doi.org/10.1007/978-3-319-49382-4_7 9783319493817.

Syafiq, R. M. O., Ilyas, R. A., Rajeshkumar, L., AL-Oqla, F. M., Nukman, Y., Zuhri, M. Y. M., Atiqah, A., Thiagamani, S. M. K., Bangar, S. P., Barile, C., & Sapuan, S. M. (2024). Corn starch nanocomposite films reinforced with nanocellulose. *Physical Sciences Reviews, 9*(8), 2653–2681. https://doi.org/10.1515/psr-2022-0011, https://www.degruyter.com/journal/key/psr/html.

Tavassoli, M., Alizadeh Sani, M., Khezerlou, A., Ehsani, A., Jahed-Khaniki, G., & McClements, D. J. (2022). Smart biopolymer-based nanocomposite materials containing ph-sensing colorimetric indicators for food freshness monitoring. *Molecules (Basel, Switzerland), 27*(10), 3168. https://doi.org/10.3390/molecules27103168.

Tavassoli, M., Khezerlou, A., Firoozy, S., Ehsani, A., & Punia Bangar, S. (2023). Chitosan-based film incorporated with anthocyanins of red poppy (*Papaver rhoeas* L.) as a colorimetric sensor for the detection of shrimp freshness. *International Journal of Food Science & Technology, 58*(6), 3050–3057. https://doi.org/10.1111/ijfs.16432.

Tousignant, M. N., Tischler, V., Wagner, K., Lin, Z. S., Brusso, J., Izquierdo, R., & Lessard, B. H. (2024). Aerosol jet printed temperature sensors using an environmentally friendly bilayer dielectric. *Flexible and Printed Electronics, 9*(1). https://doi.org/10.1088/2058-8585/ad2ece, iopscience.iop.org/journal/2058-8585.

Uma Thanu Krishnan Neela, N., Szewczyk, P. K., Karbowniczek, J. E., Polak, M., Knapczyk-Korczak, J., & Stachewicz, U. (2024). Improving stability and mechanical strength of electrospun chitosan-polycaprolactone scaffolds using genipin cross-linking for biomedical applications. *Macromolecular Rapid Communications*. https://doi.org/10.1002/marc.202400869, http://onlinelibrary.wiley.com/journal/10.1002/(ISSN)1521-3927.

Valdés García, A., Juárez Serrano, N., Beltrán Sanahuja, A., & Garrigós, M. C. (2020). Novel Antioxidant Packaging Films Based on Poly(ε-Caprolactone) and Almond Skin Extract: Development and Effect on the Oxidative Stability of Fried Almonds. *Antioxidants, 9*(7), 629. 10.3390/antiox9070629.

Vargas Romero, E., Lim, L.-T., Suárez Mahecha, H., & Bohrer, B. M. (2021). The Effect of Electrospun Polycaprolactone Nonwovens Containing Chitosan and Propolis Extracts on Fresh Pork Packaged in Linear Low-Density Polyethylene Films. *Foods, 10*(5), 1110. 10.3390/foods10051110.

Veras, F. F., Clerici, N. J., Vencato, A. A., & Brandelli, A. (2024). Poly(ϵ-caprolactone) nanofibers functionalized with poultry feather hydrolysate as a novel antioxidant material. *Sustainable Food Technology, 2*(6), 1724–1734. https://doi.org/10.1039/d4fb00250d.

Vibha, C., Chandrajith, A. V., & Praveen, G. L. (2023). Biodegradable nanomaterials as antimicrobial agents. In *Antimicrobial nanosystems: Fabrication and development*. United Arab Emirates: Elsevier, 117–130. http://doi.org/10.1016/B978-0-323-91156-6.00022-1, https://www.sciencedirect.com/book/9780323911566.

Viel, T., Cocca, M., Esposito, R., Amato, A., Russo, T., Di Cosmo, A., Polese, G., Manfra, L., Libralato, G., Zupo, V., & Costantini, M. (2024). Effect of biodegradable polymers upon grazing activity of the sea urchin *Paracentrotus lividus* (Lmk) revealed by morphological, histological and molecular analyses. *Science of the Total Environment, 929*, 172586. https://doi.org/10.1016/j.scitotenv.2024.172586.

Wang, Z., Zang, C., Hu, G., Li, J., Yu, Y., Yang, W., & Hu, Y. (2024). PCL/Locust bean gum nanofibers loaded with HP-β-CD/Epicatechin clathrate compounds for fruit packaging. *International Journal of Biological Macromolecules, 276*(Pt 2), 133940. https://doi.org/10.1016/j.ijbiomac.2024.133940.

Weligama Thuppahige, V. T., & Karim, M. A. (2021). A comprehensive review on the properties and functionalities of biodegradable and semibiodegradable food packaging materials. *Comprehensive Reviews in Food Science and Food Safety, 21*(1), 689–718. 10.1111/1541-4337.12873.

Wu, M., Deng, Z. A., Shen, C., Yang, Z., Cai, Z., Wu, D., & Chen, K. (2024). Fabrication of antimicrobial PCL/EC nanofibrous films containing natamycin and trans-cinnamic acid by microfluidic blow spinning for fruit preservation. *Food Chemistry, 442*. https://doi.org/10.1016/j.foodchem.2024.138436.

Xu, J., Gao, Y., Chen, W., Sheng, M., Lin, Q., Shi, H., & Tan, Y. (2024). Properties of coaxial polycaprolactone cinnamaldehyde chitosan nanoparticles electrospun membrane and its effect on the shelf life of meat. *Food Control, 164*, 110555. https://doi.org/10.1016/j.foodcont.2024.110555.

Yang, Y., Wu, H., Fu, Q., Xie, X., Song, Y., Xu, M., & Li, J. (2022). 3D-printed polycaprolactone-chitosan based drug delivery implants for personalized administration. *Materials & Design, 214*, 110394. https://doi.org/10.1016/j.matdes.2022.110394.

Yang, Z., Zhang, D., & Cai, J. (2024). Robust design of starch composite nanofibrous films for active food packaging: Towards improved mechanical, antioxidant, and antibacterial properties. *International Journal of Biological Macromolecules, 260*(Pt 1), 129329. https://doi.org/10.1016/j.ijbiomac.2024.129329.

Yavuz, H., Babaç, C., Tuzlakoğlu, K., & Pişkin, E. J. (2002). Preparation and degradation of L-lactide and ε-caprolactone homo and copolymer films. *Polymer Degradation And Stability, 75*(3), 431–437.

Yousaf, A., Zhao, Z., Wu, L., Hanif, S., & Akhtar, N. (2024). Sensitive detection of bisphenol A and S in plastic-packaged frozen meat using trimetallic MOF-coated electrospun nanofibers. *Process Safety and Environmental Protection, 191*, 102–111. https://doi.org/10.1016/j.psep.2024.08.132.

Zang, C., Zhang, Y., Yang, W., & Hu, Y. (2024). Polycaprolactone/chitosan electrospun nanofibrous membranes loaded Chinese yam polysaccharide for active food packaging. *LWT, 198*, 115985. https://doi.org/10.1016/j.lwt.2024.115985.

Zhang, D., Bu, N., Zhou, L., Lin, L., Wen, Y., Chen, X., ... Pang, J. (2024). Quercetin-loaded melanin nanoparticle mediated konjac glucomannan/polycaprolactone bilayer film with dual-mode synergistic bactericidal activity for food packaging. *International Journal of Biological Macromolecules, 276*. https://doi.org/10.1016/j.ijbiomac.2024.133982.

Zhang, Y., Zhu, M., Huang, Z., Yang, F., Weng, Y., & Zhang, C. (2024). The effect of polylactic acid-based blend films modified with various biodegradable polymers on the preservation of strawberries. *Food Packaging and Shelf Life, 45*. https://doi.org/10.1016/j.fpsl.2024.101333.

Zidar, A., Zupančič, Š., Kristl, J., & Jeras, M. (2024). Impact of polycaprolactone, alginate, chitosan and zein nanofiber physical properties on immune cells for safe biomedical applications. *International Journal of Biological Macromolecules, 282*. https://doi.org/10.1016/j.ijbiomac.2024.137029.

Zou, Y., Sun, Y., Shi, W., Wan, B., & Zhang, H. (2023). Dual-functional shikonin-loaded quaternized chitosan/polycaprolactone nanofibrous film with pH-sensing for active and intelligent food packaging. *Food Chemistry, 399*, 133962. https://doi.org/10.1016/j.foodchem.2022.133962.

Chapter 21

Poly(butylene adipate-co-terephthalate) films and coatings

Amrutha N.R.[1,2], Rokalla Preethi[1,2], Mandeep Gulati[1,2], Keshava Murthy P.S.[1,2], and Jeevan Prasad Reddy[1,2]
[1]*Food Packaging Technology Department, Polymer Materials Lab, CSIR-Central Food Technological Research Institute, Mysuru, Karnataka, India,*
[2]*Academy of Scientific and Innovative Research (AcSIR), Ghaziabad, Uttar Pradesh, India*

Conventional plastics are among the most important materials of the 21st century. They have experienced rapid development and are extensively used by various industries owing to their excellent mechanical properties, versatility, durability, ease of processing, and cost-effectiveness. Regretfully, most of these traditional plastics, including polyethylene (PE), polyethylene terephthalate (PET), polystyrene (PS), and polypropylene (PP), are derived from petroleum resources, and their waste products are not biodegradable. Therefore, the rise in oil consumption and drastic environmental degradation are the effects of the increased production and consumption of conventional plastics, which should be resolved for future sustainable growth (Hamaide et al., 2014). Hence, recycling of used plastics and the use of biodegradable plastics should be notable approaches to address these challenges. However, the serious problems caused by conventional plastics cannot be resolved by recycling waste plastics, since it is not always feasible. Thus, using biodegradable polymers is an efficient solution to address these plastic issues in contrast to plastic recycling. Bioplastics are a promising alternative to nonbiodegradable plastics, which can break down easily by microorganisms present in nature, such as bacteria, fungi, and algae (Jian et al., 2020).

Currently, 2.5 million tons of bioplastics are produced annually. About 39% of the total market for biodegradable polymers accounts for packaging. Production of biodegradable polymers is predicted to triple over the next five years, and the demand for these materials is expanding annually (Bartolucci et al., 2023). Polyesters based polylactic acid (PLA), poly(butylene adipate-co-terephthalate) (PBAT), and poly(butylene succinate) (PBS) are widely used bioplastics for pacakging applications (Roy & Rhim, 2020a). PBAT is a well-known polyester-based polymer that produces high-quality packaging films because of its biocompatibility, high flexibility, and superior processability, making it a desirable option (Roy & Rhim, 2020b). PBAt-based materials are also used in other applications such as hygiene products (cotton swabs, diaper back sheets), biomedical fields, agriculture mulch films, and bin liners (de Castro et al., 2016).

21.1 Chemistry of poly(butylene adipate-co-terephthalate)

PBAT is produced by polymerizing adipic acid (AA), 1,4-butanediol, and terephthalic acid which combines aromatic and aliphatic polyester, using traditional polyester manufacturing equipment (Xing et al., 2019). It is prepared by a transesterification reaction under controlled conditions, which exhibit reproducible and predictable properties. The process includes several steps, such as melt polymerization, condensation, and esterification (Venkatesan et al., 2023). Schematic representation of PBAT synthesis is shown in Fig. 21.1. PBAT is prepared using a premixing, prepolymerization, and final-polymerization method. Long reaction periods, strong vacuum, and temperatures exceeding 190°C are necessary for the synthesis. These conditions are provided to promote condensation reaction and to eliminate byproducts (water). Organometallic compounds of tin, zinc, and titanium are employed as catalysts for polycondensation throughout the synthesis (Jian et al., 2020).

During the esterification process, at 200°C–220°C, 1,4-butanediols are reacted with AA and terephthalic acid (PTA) in a 1:1 ratio to produce PBAT oligomers with a molecular weight of 103 g mol^{-1}. The hydroxyl groups of alcohol react with the carboxyl groups of acids to form an ester bond during the esterification reaction. Under the influence of an acidic catalyst, such as p-toluenesulfonic acid or sulfuric acid, the reaction involves eliminating water from the reaction mixture

Natural Polymer-Based Films and Coatings in Food Technology. DOI: https://doi.org/10.1016/B978-0-443-34244-8.00004-5

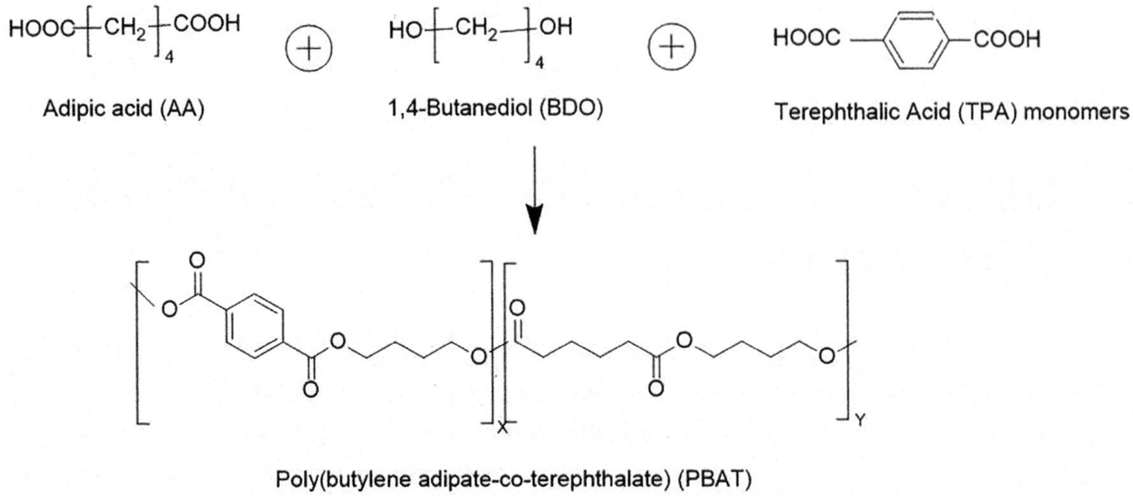

FIGURE 21.1 Schematic representation of poly(butylene adipate-co-terephthalate) (PBAT) synthesis.
From Venkatesan, R., Alagumalai, K., & Kim, S.C. (2023). Preparation and performance of biodegradable poly (butylene adipate-co-terephthalate) composites reinforced with novel $AgSnO_2$ microparticles for application in food packaging. Polymers, 15(3), 554. https://doi.org/10.3390/polym15030554.

(Nifant'ev et al., 2022). The residual water and unreacted reactants are eliminated using distillation, which purifies the resultant dibutyl adipate.

High molecular-weight polyester is the product of the process, which continues until the required molecular weight is achieved (Aversa et al., 2022). Melt polymerization, the last stage of PBAT production, involves high-temperature polyester heating under vacuum to eliminate any residual volatile contaminants. After cooling, the molten polymer is pelletized to use for various applications. Due to the presence of aromatic groups in the PBAT backbone, it has good mechanical and thermal properties.

Numerous factors affect the synthesis of PBAT, and altering the reaction conditions can change the PBAT characteristics. Additionally, PBAT molecular weight could be regulated by various parameters such as the ratio of reactants, reaction time, and temperature. The final properties of polymer could also be adjusted by adding branching agents (2,2-bis [hydroxymethyl]propionic acid) and copolymers (ethylene or propylene glycols) (Jian et al., 2020). Further, there are methods to improve the quantity and quality of PBAT in production at the industrial level. To regulate the crystallization behavior of the reaction system, for instance, various inorganic elements might be added throughout the synthesis process (Ferreira et al., 2019). To enhance the crystallization behavior of PBAT and to prevent tackiness, nucleating agents such as inorganic compounds, namely talc, silicon oxides, and chalk, are added at the final-polymerization step. As color stabilizers, phosphorus compounds like phosphorus acid and phosphoric acid could be added during the pre or postpolymerization process; however, this lowers the rate of condensation. Since industrial manufacturers protect the strategies of PBAT synthesis, these are rarely documented in the literature. About 460,000 tons of PBAT are produced annually worldwide (Bartolucci et al., 2023). Italy, Germany, and China are the three primary producers of PBAT worldwide (Aversa et al., 2022). The countries that produce PBAT and their brand names are tabulated in Table 21.1.

To reduce cost of PBAT, film thickness must be decreased, and strength should be increased. However, melt strength must be significantly increased to obtain stable ultra-thin films. An efficient way to achieve melt strength is the incorporation of long-chain branching into the polymer structure. Multifunctional monomers (≥3), such as multifunctional acids, alcohols, epoxides, isocyanates, and phenyl phosphites, could be introduced as branching agents to obtain long-chain branching copolyesters, which substantially expand the range of applications (Nifant'ev et al., 2022).

The biodegradability is one of PBAT's primary chemical characteristics. According to reports, PBAT, which is used commercially, has a 90% biodegradation rate in just three months. The organic waste system (OWS), a global organization, has tested PBAT's compostability by ASTM D6400 and EN 13432 standards. The findings demonstrate that PBAT satisfies the requirements outlined in this specification for assessing material characteristics, biodegradation, deterioration, and compost quality, making it a desirable choice for various applications, particularly in the food packaging sector (de Souza et al., 2022). The presence of the butylene adipate group affects the biodegradability of PBAT. There are two primary ways that PBAT might degrade: enzymatic (by bacteria, fungus, and algae found in the natural environment)

TABLE 21.1 Major commercially available copolyester poly(butylene adipate-co-terephthalate).

Company	Country	Brand name	Capacity (t/y)
BASF	Germany	ECOFLEX	60,000
KINGFA	China	ECOPOND	50,000
NOVAMONT	Italy	Origo-Bi	40,000
TUNHE	China	-	30,000
XINFU	China	-	20,000
JINHUI	China	ECOWORD	20,000

and nonenzymatic (such as chemical hydrolysis and heat degradation). The main cause of PBAT degradation in the absence of enzymes is that ester linkages exist in the polymer backbone, which can be easily hydrolyzed by microbes and break the polymer into smaller fragments that the microbes may utilize as sustenance. The biodegradation rate depends on several factors, such as microbes, temperature, humidity, and chemical properties such as degree of crystallinity and molecular weight, which are inversely related to degradation rate.

21.2 Poly(butylene adipate-co-terephthalate)-based film preparation methods

There are various ways to manufacture PBAT-based films: blown film extrusion, solution casting, compression molding, and hot pressing.

Firstly, solution casting, a laboratory-scale and straightforward method, wherein polymers are dissolved in a suitable solvent to make a film-forming solution, which is then cast over a mold to form a film. Solution casting is a technique that reliably disperses or dissolves additives by dissolving polymers in the proper solvent. After that, the material is poured onto a level surface and allowed to dry to create a film. Chloroform is the most widely used solvent in solution casting for creating PBAT films (Shankar & Rhim, 2018). The process involves dissolving PBAT granules in chloroform at 60°C under magnetic stirring for 24 hours (Venkatesan & Rajeswari, 2017). This method is an easy and oldest has advantages like being cost-effective and helps examine how experimental parameters affect the film's characteristics, but it is only appropriate for small laboratory scales. Other drawbacks of this approach include toxic solvent residues, shape imitation, and the long drying period.

Other film manufacturing processes are more dependable for large-scale film productions. Large batch samples and sophisticated equipment are needed for the extrusion and blown processes, which are frequently employed to produce film at an industrial scale that drives up the production cost of film (Suhag et al., 2020). The extrusion process can be used when large quantities need to be produced. This approach has several benefits, such as that solvents are not required, greater control over feed residence time, regulation of mixing degree, ease of handling high-viscosity polymers, and a broad range of process conditions, including temperatures between 70°C and 500°C and pressures between 0 and 500 bar (Gupta et al., 2022). This is the most effective method over the solution casting approach; the composition of polymer and filler is predetermined, and then large-scale manufacturing is done (Nambiar et al., 2019). Extrusion is among the most significant methods for industrially creating polymer blends and composites among the currently accessible polymer processing techniques (Ghosh et al., 2020). This procedure involves the extrusion of a molten polymer through an extrusion die. When the molten solution comes into contact with cooled air, it solidifies and forms a continuous polymer film tube (Kim et al., 2021). PBAT blown films prepare packaging material, covering films, and bags. A single screw extruder with film blowing was used to prepare blown films. The material was processed in single screw extruder using temperature profiles of 90°C, 120°C, 120°C, 120°C, and 120°C and then blowing (Gao et al., 2022).

Other PBAT-based materials of desired shapes can be obtained through compression molding, hot pressing, injection molding, and thermoforming. For the production of PBAT composite films, blends are prepared first by twin screw extruder and then by thermal compression (Xiong et al., 2020). Based on the preceding discussion, it can be concluded that the fabrication method can affect the attributes of the PBAT-based film.

21.3 Food packaging applications of poly(butylene adipate-co-terephthalate)-based films

Fat and lipid oxidation in food products produces disagreeable smells and odors, and the development of undesirable foodborne microorganisms renders food products intolerable. Food oxidation causes a shortage of polyunsaturated fatty acids and the production of harmful aldehydes, which leads to nutritional loss and food product deterioration. Active packaging is a cutting-edge technique whereby functional elements are purposefully added to the packing environment or the package material to improve the system's performance (Grzebieniarz et al., 2023).

Substances like antimicrobials (essential oils, chitosan, lysozyme, lactoferrin metals), antioxidants (plant extracts, phenolic compounds, lignin), ethylene scavengers (activated carbon, metal oxides, potassium permanganate), carbon dioxide emitters (sodium bicarbonate, ferrous carbonate, citric acid), and oxygen scavengers (gallic acid, iron, pyrogallol, glucose oxidase, ascorbic acid) are among active functional materials used in preparation of active food packaging. Releasing its bioactive substances from the packaging film is the mechanism behind active packaging. The antibacterial properties of bioactive fillers, free radical scavenging, and degassing (oxygen, carbon dioxide, and ethylene) are crucial for prolonging food shelf life in active packaging. To ensure prolonged effectiveness against foodborne pathogens and to delay food oxidation, bioactive compounds should be gradually and sustainably released (Perumal et al., 2022; Roy et al., 2023a).

According to Ahmed et al. (2022) and Kuai et al. (2021), bioactive antioxidants can scavenge free radicals directly or indirectly, reduce food oxidation, and prolong food shelf life (Ahmed et al., 2022; Kuai et al., 2021). However, in addition to direct interactions between microorganisms and bioactive chemicals, active compounds can also have an antibacterial effect by attaching to biomolecules like DNA, forming free radicals, or degrading membranes (Pandey et al., 2022). Several active agents, including nitrite (Katekhong et al., 2022), lignin-zinc oxide (LZn) (Xiao et al., 2022), curcumin (de Campos et al., 2019), and essential oils encapsulated in essential oils (da Silva et al., 2019), were used to create the PBAT-based active packaging films.

Blending, creating nanocomposites, using compatibilizers, activators, and chain extenders are some methods to enhance the qualities of PBAT-based films for food packaging. In contrast to PLA, which is hard and brittle, PBAT is flexible and elastic. With compatibilizers, blown films for food packaging applications can be made by combining these two bioplastics (Pietrosanto et al., 2020). Bio composites based on calcium phosphate glass microparticles and PBAT can be utilized as mulch films in agri-food packaging applications (Sciancalepore et al., 2022). According to da Silva et al. (2019), blend composites of PBAT, thermoplastic starch, cellulose nano whiskers, and compatibilizers have also been used to create flexible films (da Silva et al., 2019). Consequently, modified PBAT-based blends or composites are extensively utilized to prolong the shelf life of different food products.

In light of this, several active packaging materials enhance the quality of food products and their packaging qualities. Accordingly, meat can be kept fresher in the refrigerator by utilizing blown films made from nanocomposite materials consisting of zinc oxide nanoparticles, thermoplastic starch, and PBAT (Phothisarattana & Harnkarnsujarit, 2022). Meat is a perishable item that oxidizes easily and spoils quickly. The antibacterial action of the coating is enhanced by zinc oxide nanoparticles, which prolong the meat's shelf life. Pork meat can also be packaged using the active packaging method based on blown films of PBAT, thermoplastic starch, nisin, and ethylenediaminetetraacetic acid (EDTA) (Leelaphiwat et al., 2022).

Antimicrobial peptides are added to the PBAT-based packaging to prolong the shelf life of swine meat by preventing bacterial development. Chia oil can be packed in a blown film made of PBAT, thermoplastic starch, and curcumin to increase oxidative stability (Mücke et al., 2021). The banana's shelf life can be increased by the composite films made with zeolite and PBAT (Threepopnatkul et al., 2020).

Active antioxidant packaging based on poly(lactic acid), poly(butylene adipate-co- terephthalate), thermoplastic starch (TPS), and essential oils (peppermint and clove oil) has been shown to preserve straw mushrooms and extend their postharvest shelf life by up to four days (Gui et al., 2022). Films made from PBAT, poly (butylene succinate), and polyethylene oxide have been used to package white mushrooms (Chen et al., 2022). It is also said that PBAT and PLA-based packaging is utilized to preserve the quality of shiitake mushrooms (He et al., 2022). The browning and deterioration of mushrooms can be postponed by using active food packaging material based on PBAT. The overall quality was enhanced in terms of rate of respiration, appearance, texture, phenolic content, microbial count, and organoleptic qualities, even if packaging was unable to support the mushrooms' weight reduction.

Additionally, food packaging based on PBAT, thermoplastic starch, and nitrite is used for active packaging of meat products, where the active films can stabilize the lipids of meat and prevent microbial growth. PBAT and clay-based nanocomposite and PBAT, as well as titanium dioxide-based active packaging, can be used for banana packaging

(Katekhong et al., 2022). Because nitrite can have a plasticization effect and make starch compatible in the formulation of polymer blends, it is advantageous to use it in PBAT and thermoplastic starch-based films.

Extrusion may be used to create active antibacterial and antioxidant films based on PBAT, thermoplastic cassava starch, and curcumin. The antimicrobial qualities of these films were evaluated against *Staphylococcus aureus, Escherichia coli, and Pseudomonas aeruginosa* (de Campos et al., 2019). The blown extrusion-produced antimicrobial films based on starch, PBAT, and antimicrobial peptides (polyline and nisin) also have antibacterial properties against *S. aureus* and *E. coli* (Xiong et al., 2020). According to reports, the presence of PBAT-based functional packaging considerably extends the shelf life of fresh peaches. Commercial antimicrobial peptides added to the packaging matrix improve the antibacterial activity of the film, thereby extending the shelf life of fresh peaches.

In a recent study, Paulsen et al. (2022) utilized a PLA/PBAT blend film as wrapping material for broccoli head packaging and compared its performance to linear low-density polyethylene (LDPE) (Paulsen et al., 2022). They found that using the PLA/PBAT film extended the broccoli head's shelf life to three weeks when stored at 2°C. Additionally, the PLA/PBAT-based functional film was applied to packaging soy protein-based meat analogs (Wang et al., 2022). It was found that incorporating cinnamaldehyde and tea polyphenols into the PLA/PBAT film effectively inhibited bacterial growth in meat analogs. These potent antioxidant and antibacterial compounds in PBAT packaging were also reported to prolong the shelf life of vegan meat.

Yang et al. (2022) employed PBAT-based functional packaging film with added modified clay and quercetin for packaging bananas and blueberries (Yang et al., 2022). Their research demonstrated that this functional film packaging significantly delayed the decay of both fruits by at least one week, attributed to its excellent gas and water vapor barrier properties and robust antioxidant activity. Hence, it can be inferred that PBAT, either alone or combined with other polymers, effectively enhances the shelf life of packaged food. Additionally, PBAT can serve as a carrier for numerous bioactive ingredients. The functional components of PBAT-based packaging films further contribute to extending the shelf life of packaged foods.

21.4 Poly(butylene adipate-co-terephthalate)-based coating applications in food

Nowadays, the bioplastic PBAT is being evaluated as a viable option for food packaging to overcome the difficulties posed by fossil fuel–derived packaging materials. Greenhouse gas emissions, solid waste creation, and climate change are the environmental issues caused by packaging materials. Due to its superior packaging qualities and good biodegradability, PBAT is a good substitute for conventional packaging. Mulch films, carrier bags, cutlery, and packaging are the few items that are PBAT-based. Packaging made of PBAT effectively extends the shelf life of food.

To achieve food safety, foods must be packaged in high-barrier materials. The search for the type of packaging materials is being led by nanocomposite technology. The impact of zinc oxide loadings on the structure of PBAT was examined, along with the packaging's mechanical, thermal, and barrier properties. Zinc oxide (ZnO) nanoparticles and PBAT are used to create antimicrobial films. Superior antibacterial activity against *S. aureus and E. coli* is demonstrated by the resultant PBAT/ZnO nanofilms (Venkatesan et al., 2018).

Melt processing was used to create biodegradable PBAT films that contained varying concentrations of the antibacterial peptide nisin. According to these findings, PBAT/nisin films made through melt processing have a lot of promise for active food packaging materials applications to improve food safety (Zehetmeyer et al., 2016).

According to current study, biodegradable PBAT can be bio-reinforced using torrefied coffee grounds (CG) generated from agricultural commodities without the need for a compatibilizer. The outcomes of these solvent-free bio composites' preparation indicate that they may be suitable options for use in food packaging (Moustafa et al., 2017).

Lignin nanoparticles (LNP) and ZnO were combined to create LZn hybrid particles using a straightforward hydrothermal process at varying pH levels. It was thought that the PBAT/LZn composite films were used for active food packing (Xiao et al., 2022). Melt blending and blowing processes were used to create antimicrobial films based on starch, PBAT, and a commercial silver nanoparticles (AgNPs)@SiO_2 antibacterial composite product. The starch/PBAT/AgNPs@SiO_2 composite films exhibit robust antibacterial properties and a high-throughput production process, which makes them promising as antimicrobial packaging materials for commercial use (Zhou et al., 2021).

PBAT/AgNPs composite films were made using AgNPs. The nanocomposite films made from PBAT and AgNPs exhibited strong antibacterial effects against *E. coli* and *Listeria monocytogenes*, but the chemical structure and thermal stability of PBAT films remained largely unchanged. The produced films have a great chance of being employed as antimicrobial and UV-screening packaging films (Shankar & Rhim, 2016).

In addition to their exceptional physical qualities, packaging films' functional qualities are also sought after for active food packaging applications that aim to extend the shelf life of packed foods. By preventing the formation of undesirable foodborne pathogens and preventing oxidation and organoleptic changes in packed foods, the addition of antioxidants and antimicrobials to packaging systems enhances food quality (Roy et al., 2023b). To improve the functional qualities and suitability of PBAT-based packaging films for food packing and storage, a variety of antibacterial and antioxidants are added, such as nanofillers, essential oils, peptides, plant extracts, and phytochemicals.

A different study explored the antibacterial effectiveness of PLA/PBAT-based films against *L. monocytogenes* and *E. coli* in the presence of sulfur and sulfur nanoparticles (Kim et al., 2021). The research revealed that sulfur nanoparticles will enhance the antibacterial activity of PBAT-based packaging. Furthermore, active packaging composed of PLA, PBAT, grapefruit seed extract, and zinc oxide nanoparticles demonstrated a range of antioxidant properties (Kim et al., 2022).

According to Venkatesan et al. (2018), integrating silver oxide nanoparticles (0%–7%) into PBAT-based film exhibited strong antibacterial properties against foodborne pathogens *K. pneumoniae* and *S. aureus*. Furthermore, the addition of oregano essential oil microparticles significantly enhanced the antioxidant activity and total phenolic content of PBAT/flour-based packaging films (Jian et al., 2020).

Ferreira et al. examined the antimicrobial efficacy of PBAT-based packaging film loaded with cinnamon essential oil nanocapsules (Ferreira et al., 2021). They found that the functional film's antibacterial activity against *E. coli* increased over time. Similarly, another study by Sharma et al. (2020) revealed that incorporating eucalyptus and cinnamon essential oils into PLA/PBAT-based films exhibited strong antibacterial effects against foodborne pathogens, which could be beneficial for active food packaging (Sharma et al., 2020).

Numerous bioactive substances included in essential oils, including flavonoids, terpenes, esters, aldehydes, and flavonols, provide the packaging system with strong antioxidant potential (Nair et al., 2021). Overall, antioxidant and antibacterial properties of film are enhanced by the addition of functional substances to PBAT-based polymer matrix. This could potentially extend the shelf life of foods wrapped in PBAT film.

21.5 Poly(butylene adipate-co-terephthalate) coated paper and biopolymer films

Paper coatings made of petroleum-based substances, including waxes, polyethylene, and fluoro-compounds, are used to change the wettability and water and oil transport of paper. Despite their capacity to increase surface water resistance, these polymers have declined due to their drawbacks, which include poor recycling and a lack of biodegradability. However, employing such nonbiodegradable and nonrenewable materials generates greenhouse gases and endangers human health and the environment. Furthermore, recycling becomes challenging when plastics and materials from fossil fuels are used as barrier materials in paper-based packaging boxes. This raises difficulties for recycling, increases recycling costs, and raises ecological concerns about the waste produced because of the absence of biodegradation. Investigations into alternatives for petroleum-based packaging are becoming more and more popular as consumer demands for products with nutritional quality and extended shelf life rise (Basak et al., 2024).

One study examined the efficacy of esterified lignin as an additive for composite paper coatings made from PBAT. In order to determine whether the addition of esterified lignin in the PBAT framework gave the coated paper sheets the desired hydrophobicity and oleophobicity, the water and oil Cobb tests were conducted. About 40% of lesser water absorption was observed for 50% esterified lignin incorporation with a thickness of 60 μm, which could be assigned to substitution of hydroxyl groups with long hydrocarbon chains (Shorey & Mekonnen, 2022).

21.6 Other applications

PBAT and its derivatives are currently employed in a variety of fields, including industrial uses, environmental protection, and biomedical food packaging. Antimicrobial PBAT fabrics were created and utilized in hospitals to prevent infections (Wei et al., 2015) . Fukushima et al. investigated the biocompatibility of PBAT nanocomposites made of fluorohectorites, sepiolite, and MMTs. According to their report, nanocomposites are more in vitro hemocompatible than plain PBAT and show noncytotoxicity and positive hemostatic effects after coming into contact with blood (Fukushima et al., 2013). Mondal et al. used cetyltrimethylammonium bromide (CTAB) to modify montmorillonite (MMT) and created PBAT/CMMT nanocomposites by the solution intercalation method. Since the CMMT of nanocomposites contains a quaternary ammonium cation and has an intercalated shape, they found that PBAT/CMMT exhibits antibacterial properties (Mondal et al., 2014). Recently, PBAT-based nanocomposites incorporated with nanoclays, including bentonite and MMTs, were prepared by Bittman et al., wherein bentonite-modified PBAT demonstrated the optimal mechanical and thermal properties (Bittmann et al., 2012).

21.7 Biodegradation

The important, inevitable factor considered for sustainable packaging system is biodegradability. In contrast to biopolymers, synthetic bioplastics break down more quickly than plastic polymers made from petrochemicals. PBAT-based packaging films are typically tested in landfills to determine their biodegradability, and the findings indicate that they are biodegradable. According to earlier studies, PBAT/starch-based films showed outstanding biodegradation qualities in landfill tests after three months (Wang et al., 2015). Another study found that montmorillonite hinders the biodegradation of PLA/PBAT/MMT-based films in soil (Freitas et al., 2017). *Thermobacillus composti* KWC4 carboxylesterase enzyme was shown to degrade PBAT in another recently published study (Wu et al., 2023). The addition of catalytic enzymes has been shown to improve PBAT hydrolytic activity, depending on the enzyme's source. Enzymatic digestion's efficacy can be verified using scanning electron microscopy (SEM), thermogravimetric analysis (TGA), X-ray diffraction (XRD), and Fourier transform infrared (FTIR) characterization (Bartolucci et al., 2023).

Recent reports have indicated that the usage of photolysis and biodegradation in thermophilic environments can accelerate the deterioration of PLA/PBAT-based mulching films. The use of suitable strains of bacteria in the soil system was also found to be a significant contributing element to the quick breakdown of packing films made of polyester. The biodegradability of PBAT with organoclays (PBAT/natureplast [PBE 003]) composite films in freshwater as well as marine environments was documented by another investigation. According to the authors, over a 20-day period, the degradation of PBAT/PBE film was around 1% marginally higher in freshwater relative to seawater, presumably because there were more bacteria and algae that may break the polymer chain (Moustafa et al., 2021). A three-week soil burial test was also used to examine the biodegradability of the PBAT/soy protein isolate (SPI) blend. The pure PBAT film showed minimal deterioration; however, the SPI-added blend film had slightly deteriorated as verified by SEM images (Gaiping Guo et al., 2015). Recently, Diao et al. investigated the breakdown of PBAT/calcium carbonate layer in soil and simulated seawater. According to the authors, soil water systems were inferior to soil degradation. Additionally, the scientists coated the film with polyethylene glycol, which promoted the diffusion of the polymers and improved the film's biodegradation (Diao et al., 2022). A number of variables, including the compatibility between fillers and polymers, affect the biodegradability of PBAT film. Consequently, it can be concluded that the PBAT film has a slow rate of biodegradability, and while the addition of fillers or bioactive ingredients occasionally accelerates biodegradation, the opposite effect could also be possible.

21.8 Conclusions

As a synthetic polyester bioplastic, PBAT has significant potential to replace nonbiodegradable plastics made from petrochemicals, such as PE, PP, PS, and PET. PBAT-based packaging materials are biodegradable and incredibly flexible. PBAT-based flexible films have potential to use in food packaging, such as packaging films, trash bags, pouches, paper laminations, and mulch film. This book chapter examined the chemistry, PBAT manufacturers, film manufacturing techniques, and the application of PBAT as coatings, paper laminates, and biodegradation. PBAT films can be manufactured via solvent casting, cast film extrusion, and blown film extrusion methods. The mechanical strength and barrier qualities of PBAT-based films are inadequate, and they are nonfunctional. Consequently, adding bioactive fillers and other polymers improved the functional and physical efficiency of PBAT films. Furthermore, in soil and marine systems, the PBAT-based films demonstrated biodegradability. PBAT-based films can also be utilized for active food packaging applications to extend the shelf life of contained food. Overall, PBAT can be used in food packaging applications and can replace LDPE films, although it has several drawbacks, such as being almost three times more expensive than PE or PP and having lower barrier (vapor and gas) qualities than traditional polymers. Further study is required to enhance their qualities and increase their manufacturing volume for use in food packaging.

References

Ahmed, M. W., Haque, M. A., Mohibbullah, M., Khan, M. S. I., Islam, M. A., Mondal, M. H. T., & Ahmmed, R. (2022). A review on active packaging for quality and safety of foods: Current trends, applications, prospects and challenges. *Food Packaging and Shelf Life, 33*. https://doi.org/10.1016/j.fpsl.2022.100913 Article 100913.

Aversa, C., Barletta, M., Cappiello, G., & Gisario, A. (2022). Compatibilization strategies and analysis of morphological features of poly(butylene adipate-co-terephthalate) (PBAT)/poly(lactic acid) PLA blends: A state-of-art review. *European Polymer Journal, 173*. https://doi.org/10.1016/J.EURPOLYMJ.2022.111304 Article 111304.

Bartolucci, L., Cordiner, S., De Maina, E., Kumar, G., Mele, P., Mulone, V., ... Piechota, G. (2023). Sustainable valorization of bioplastic waste: A review on effective recycling routes for the most widely used biopolymers. *International Journal of Molecular Sciences, 24*(9), Article 7696. https://doi.org/10.3390/ijms24097696.

Basak, S., Dangate, M. S., & Samy, S. (2024). Oil-and water-resistant paper coatings: A review. *Progress in Organic Coatings, 186*, 107938.

Bittmann, B., Bouza, R., Barral, L., González-Rodríguez, M. V., & Abad, M.-J. (2012). Nanoclay-reinforced poly(butylene adipate-co-terephthalate) biocomposites for packaging applications. *Polymer Composites, 33*(11), 2022–2028. https://doi.org/10.1002/pc.22344.

de Campos, S. S., de Oliveira, A., Moreira, T. F. M., da Silva, T. B. V., da Silva, M. V., Pinto, J. A., Bilck, A. P., Gonçalves, O. H., Fernandes, I. P., Barreiro, M. F., Yamashita, F., Valderrama, P., Shirai, M. A., & Leimann, F. V. (2019). TPCS/PBAT blown extruded films added with curcumin as a technological approach for active packaging materials. *Food Packaging and Shelf Life, 22*. https://doi.org/10.1016/j.fpsl.2019.100424 Article 100424.

de Castro, J. G., Rodrigues, B. V. M., Ricci, R., Costa, M. M., Ribeiro, A. F. C., Marciano, F. R., & Lobo, A. O. (2016). *RSC Advances, 6*, 32615.

Chen, C., Chen, W., Dai, F., Yang, F., & Xie, J. (2022). Development of packaging films with gas selective permeability based on poly(butylene adipate-co-terephthalate)/ poly(butylene succinate) and its application in the storage of white mushroom (*Agaricus bisporus*). *Food and Bioprocess Technology, 15*, 1268–1283. https://doi.org/10.1007/S11947-022-02794-4.

Diao, X., Zhang, C., & Weng, Y. (2022). Properties and degradability of poly(butylene adipate-co-terephthalate)/calcium carbonate films modified by polyethylene glycol. *Polymers, 14*(3). https://doi.org/10.3390/polym14030484 Article 484.

Ferreira, F. V., Cividanes, L. S., Gouveia, R. F., & Lona, L. M. F. (2019). An overview on properties and applications of poly(butylene adipate-co-terephthalate)-PBAT based composites. *Polymer Engineering & Science, 59*(s2), E7–E15. https://doi.org/10.1002/PEN.24770.

Ferreira, R. R., de Souza, A. G., & Rosa, D. S. (2021). Essential oil-loaded nanocapsules and their application on PBAT biodegradable films. *Journal of Molecular Liquids, 337*(38), 116488. https://doi.org/10.1016/j.molliq.2021.116488.

Freitas, A. L. P., de, L., Carvalho, A. J. F., Gandini, A., Rosa, D. S., & Curvelo, A. A. S. (2017). Effect of montmorillonite and chain extender on rheological, morphological and biodegradation behavior of PLA/PBAT blends. *Polymer Testing, 62*, 189–195. https://doi.org/10.1016/j.polymertesting.2017.07.001.

Fukushima, K., Rasyida, A., & Yang, M.-C. (2013). Characterization, degradation and biocompatibility of PBAT-based nanocomposites. *Applied Clay Science, 80–81*, 291–298. https://doi.org/10.1016/j.clay.2013.04.015.

Gao, S., Zhai, X., Cheng, Y., Zhang, R., Wang, W., & Hou, H. (2022). Starch/PBAT blown antimicrobial films based on the synergistic effects of two commercial antimicrobial peptides. *International Journal of Biological Macromolecules, 204*, 457–465. https://doi.org/10.1016/J.IJBIOMAC.2022.01.183.

Gui, H., Zhao, M., Zhang, S., Yin, R., Hu, C., Fan, M., & Li, L. (2022). Active antioxidant packaging from essential oils incorporated polylactic acid/poly(butylene adipate-coterephthalate)/thermoplastic starch for preserving straw mushroom. *Foods, 11*(15), 2252. https://doi.org/10.3390/FOODS11152252.

Ghosh, T., Hazarika, D., Dhar, P., & Katiyar, V. (2020). Processing techniques for nanocellulose-based value-added products. *Cellulose Nanocrystals: An Emerging Nanocellulose for Numerous Chemical Processes,* 119–156. https://doi.org/10.1515/9783110648010-005/MACHINEREADABLECITATION/RIS.

Grzebieniarz, W., Biswas, D., Roy, S., & Jamroz, E. (2023). Advances in biopolymerbased multi-layer film preparations and food packaging applications. *Food Packaging and Shelf Life, 35*. https://doi.org/10.1016/j.fpsl.2023.101033 Article 101033.

Gupta, V., Biswas, D., & Roy, S. (2022). A comprehensive review of biodegradable polymer-based films and coatings and their food packaging applications. *Materials, 15*(17), 5899. https://doi.org/10.3390/MA15175899.

Guo, G., Zhang, C., Du, Z., Zou, W., Tian, H., Xiang, A., & Li, H. (2015). Structure and property of biodegradable soy protein isolate/PBAT blends. *Industrial Crops and Products, 78*, 141–148. https://doi.org/10.1016/j.indcrop.2015.06.009.

Hamaide, T., DeterreJ, R., & Feller, F. (2014). *Environmental impact of polymers*. John Wiley & Sons, Inc.

He, X., Wang, X., Liu, Y., Fang, H., Zheng, S., Liu, H., Guan, W., & Yan, R. (2022). Biodegradable PBAT/PLA packaging maintained the quality of postharvest shiitake mushroom by modified humidity and atmosphere. *Food Packaging and Shelf Life*. https://doi.org/10.1016/j.fpsl.2022.100949.

Jian, J., Xiangbin, Z., & Xianbo, H. (2020). An overview on synthesis, properties and applications of poly(butylene-adipate-co-terephthalate)-PBAT. *Advanced Industrial and Engineering Polymer Research, 3*, 19–26. https://doi.org/10.1016/j.aiepr.2020.01.001.

Katekhong, W., Wongphan, P., Klinmalai, P., & Harnkarnsujarit, N. (2022). Thermoplastic starch blown films functionalized by plasticized nitrite blended with PBAT for superior oxygen barrier and active biodegradable meat packaging. *Food Chemistry, 374*. https://doi.org/10.1016/J.FOODCHEM.2021.131709 Article 131709.

Kim, H. J., Lee, C. L., Yoon, K. S., & Rhim, J.-W. (2022). Synergistic effect of UV-C LED irradiation and PLA/PBAT-based antimicrobial packaging film on fresh-cut vegetables. *Food Control, 138*. https://doi.org/10.1016/J.FOODCONT.2022.109027 Article 109027.

Kim, S., Roy, S., Yoon, K. S., & Rhim, J.-W. (2021). Preparation of low-density polyethylene- and poly (lactide)/poly(butylene adipate-co-terephthalate)-based antibacterial films integrated with elemental sulfur and sulfur nanoparticles. *Packaging Technology and Science, 34*, 505–516. https://doi.org/10.1002/pts.2577.

Kuai, L., Liu, F., Chiou, B. S., Avena-Bustillos, R. J., McHugh, T. H., & Zhong, F. (2021). Controlled release of antioxidants from active food packaging: A review. *Food Hydrocolloids, 120*. https://doi.org/10.1016/j.foodhyd.2021.106992 Article 106992.

Leelaphiwat, P., Pechprankan, C., Siripho, P., Bumbudsanpharoke, N., & Harnkarnsujarit, N. (2022). Effects of nisin and EDTA on morphology and properties of thermoplastic starch and PBAT biodegradable films for meat packaging. *Food Chemistry, 369*. https://doi.org/10.1016/J.FOODCHEM.2021.130956 Article 130956.

Mondal, D., Bhowmick, B., Mollick, M. M. R., Maity, D., Saha, N. R., Rangarajan, V., Rana, D., Sen, R., & Chattopadhyay, D. (2014). Antimicrobial activity and biodegradation behavior of poly(butylene adipate-co-terephthalate)/clay nanocomposites. *Journal of Applied Polymer Science, 131*(7), Article 40079. https://doi.org/10.1002/app.40079.

Moustafa, H., El-Sayed, S. M., & Youssef, A. M. (2021). Synergistic impact of cumin essential oil on enhancing UV-blocking and antibacterial activity of biodegradable poly(butylene adipate-co-terephthalate)/clay platelets nanocomposites. *Journal of Thermoplastic Composite Materials*. https://doi.org/10.1177/0892705721989771.

Moustafa, H., Guizani, C., Dupont, C., Martin, V., Jeguirim, M., & Dufresne, A. (2017). Utilization of torrefied coffee grounds as reinforcing agent to produce high-quality biodegradable PBAT composites for food packaging applications. *ACS Sustainable Chemistry & Engineering, 5*(2), 1906–1916.

Mücke, N., da Silva, T. B. V., de Oliveira, A., Moreira, T. F. M., Venancio, C. D. S., Marques, L. L. M., Valderrama, P., Gonçalves, O. H., da Silva-Buzanello, R. A., Yamashita, F., Shirai, M. A., Genena, A. K., & Leimann, F. V. (2021). Use of watersoluble curcumin in TPS/PBAT packaging material: Interference on reactive extrusion and oxidative stability of chia oil. *Food and Bioprocess Technology, 14*, 471–482. https://doi.org/10.1007/s11947-021-02584-4.

Nair, U. A., Selvam, S. P., Dharini, V., Nambiar, R. B., Babu, P. A., Sadiku, E. R., & Jayaramudu, J. (2021). Development of antifungal biocomposite film against postharvest pathogens *Colletotrichum gloeosporioides* and *Lasiodiplodia theobromae*. *Materials Today: Proceedings, 38*, 1113–1120. https://doi.org/10.1016/j.matpr.2020.06.356.

Nambiar, R. B., Sellamuthu, P. S., Perumal, A. B., Sadiku, E. R., & Adeyeye, O. A. (2019). The use of chitosan in food packaging applications. In *Green biopolymers and their nanocomposites*. Singapore: Springer, 125–136. https://doi.org/10.1007/978-981-13-8063-1_5.

Nifant'ev, I. E., Bagrov, V. V., Komarov, P. D., Ilyin, S. O., & Ivchenko, P. V. (2022). The use of branching agents in the synthesis of PBAT. *Polymers, 14*(9), 1720. https://doi.org/10.3390/polym14091720.

Pandey, S., Sharma, K., & Gundabala, V. (2022). Antimicrobial bio-inspired active packaging materials for shelf life and safety development: A review. *Food Bioscience, 48*. https://doi.org/10.1016/j.fbio.2022.101730 Article 101730.

Paulsen, E., Lema, P., Martínez-Romero, D., & García-Viguera, C. (2022). Use of PLA/ PBAT stretch-cling film as an eco-friendly alternative for individual wrapping of broccoli heads. *Scientia Horticulturae, 304*. https://doi.org/10.1016/j.scienta.2022.1112601 Article 111260.

Perumal, A. B., Huang, L., Nambiar, R. B., He, Y., Li, X., & Sellamuthu, P. S. (2022). Application of essential oils in packaging films for the preservation of fruits and vegetables: A review. *Food Chemistry, 375*. https://doi.org/10.1016/j.foodchem.2021.131810 Article 131810.

Phothisarattana, D., & Harnkarnsujarit, N. (2022). Migration, aggregations and thermal degradation behaviors of TiO_2 and ZnO incorporated PBAT/TPS nanocomposite blown films. *Food Packaging and Shelf Life, 33*. https://doi.org/10.1016/J.FPSL.2022.100901 Article 100901.

Pietrosanto, A., Scarfato, P., Di Maio, L., & Incarnato, L. (2020). Development of ecosustainable PBAT-based blown films and performance analysis for food packaging applications. *Materials, 13*(23), 5395. https://doi.org/10.3390/MA13235395.

Roy, S., & Rhim, J.-W. (2020a). Preparation of bioactive functional poly(lactic acid)/ curcumin composite film for food packaging application. *International Journal of Biological Macromolecules, 162*, 1780–1789. https://doi.org/10.1016/j.ijbiomac.2020.08.094.

Roy, S., & Rhim, J.-W. (2020b). Curcumin incorporated poly(butylene adipate-coterephthalate) film with improved water vapor barrier and antioxidant properties. *Materials, 13*(19), 4369. https://doi.org/10.3390/ma13194369.

Roy, S., Ezati, P., Khan, A., & Rhim, J.-W. (2023a). New opportunities and advances in quercetin-added functional packaging films for sustainable packaging applications: A mini-review. *Critical Reviews in Food Science and Nutrition*. https://doi.org/10.1080/10408398.2023.2200553 Online.

Roy, S., Zhang, W., Biswas, D., Ramakrishnan, R., & Rhim, J.-W. (2023b). Grapefruit seed extract-added functional films and coating for active packaging applications: A review. *Molecules, 28*(2), 730. https://doi.org/10.3390/molecules28020730.

Sciancalepore, C., Togliatti, E., Giubilini, A., Pugliese, D., Moroni, F., Messori, M., & Milanese, D. (2022). Preparation and characterization of innovative poly(butylene adipate terephthalate)-based biocomposites for agri-food packaging application. *Journal of Applied Polymer Science, 139*(24), 52370. https://doi.org/10.1002/APP.52370.

Shankar, S., & Rhim, J. W. (2016). Tocopherol-mediated synthesis of silver nanoparticles and preparation of antimicrobial PBAT/silver nanoparticles composite films. *LWT-Food Science and Technology, 72*, 149–156.

Shankar, S., & Rhim, J. W. (2018). Preparation of antibacterial poly(lactide)/poly (butylene adipate-co-terephthalate) composite films incorporated with grapefruit seed extract. *International Journal of Biological Macromolecules, 120*, 846–852. https://doi.org/10.1016/j.ijbiomac.2018.09.004.

Sharma, S., Barkauskaite, S., Jaiswal, S., Duffy, B., & Jaiswal, A. K. (2020). Development of essential oil-incorporated active film based on biodegradable blends of poly (lactide)/poly (butylene adipate-co-terephthalate) for food packaging application. *Journal of Packaging Technology and Research, 4*, 235–245. https://doi.org/10.1007/s41783-020-00099-5.

Shorey, R., & Mekonnen, T. H. (2022). Sustainable paper coating with enhanced barrier properties based on esterified lignin and PBAT blend. *International Journal of Biological Macromolecules, 209*, 472–484.

da Silva, J. B. A., Santana, J. S., de Almeida Lucas, A., Passador, F. R., de Sousa Costa, L. A., Pereira, F. V., & Druzian, J. I. (2019). PBAT/TPS-nanowhiskers blends preparation and application as food packaging. *Journal of Applied Polymer Science, 136*(26), 47699. https://doi.org/10.1002/APP.47699.

de Souza, A. G., Komatsu, L. G. H., Barbosa, R. F. S., Parra, D. F., & Rosa, D. S. (2022). The effect of ZnO nanoparticles as Ag-carrier in PBAT for antimicrobial films. *Polymer Bulletin, 79*(6), 4031–4048. https://doi.org/10.1007/S00289-021-03681-2/FIGURES/10.

Suhag, R., Kumar, N., Petkoska, A. T., & Upadhyay, A. (2020). Film formation and deposition methods of edible coating on food products: A review. *Food Research International, 136*. https://doi.org/10.1016/j.foodres.2020.109582 Article 109582.

Threepopnatkul, P., Wongsuton, K., Jaiaue, C., Rakkietwinai, N., Stittatrakul, A., & Kulsetthanchalee, C. (2020). Effect of zeolite on mechanical and barrier properties of PBAT films for life extension of agricultural products. *Key Engineering Materials, 861*, 176–181. https://doi.org/10.4028/WWW.SCIENTIFIC.NET/KEM.861.176.

Venkatesan, R., & Rajeswari, N. (2017). ZnO/PBAT nanocomposite films: Investigation on the mechanical and biological activity for food packaging. *Polymers for Advanced Technologies, 28*(1), 20–27. https://doi.org/10.1002/PAT.3847.

Venkatesan, R., Rajeswari, N., & Tamilselvi, A. (2018). Antimicrobial, mechanical, barrier, and thermal properties of bio-based poly(butylene adipate-co-terephthalate) (PBAT)/Ag 2O nanocomposite films for packaging application. *Polymers for Advanced Technologies, 29*(1), 61–68. https://doi.org/10.1002/pat.4089.

Venkatesan, R., Alagumalai, K., & Kim, S. C. (2023). Preparation and performance of biodegradable poly (butylene adipate-co-terephthalate) composites reinforced with novel $AgSnO_2$ microparticles for application in food packaging. *Polymers, 15*(3), 554. https://doi.org/10.3390/polym15030554.

Wang, H., Wei, D., Zheng, A., & Xiao, H. (2015). Soil burial biodegradation of antimicrobial biodegradable PBAT films. *Polymer Degradation and Stability, 116*, 14–22. https://doi.org/10.1016/j.polymdegradstab.2015.03.007.

Wang, L., Xu, J., Zhang, M., Zheng, H., & Li, L. (2022). Preservation of soy protein-based meat analogues by using PLA/PBAT antimicrobial packaging film. *Food chemistry, 380*. https://doi.org/10.1016/J.FOODCHEM.2021.132022 Article 132022.

Wei, D., Wang, H., Ziaee, Z., Chibante, F., Zheng, A., & Xiao, H. (2015). Non-leaching antimicrobial biodegradable PBAT films through a facile and novel approach. *Materials Science and Engineering: C, 58*, 986–994. https://doi.org/10.1016/j.msec.2015.09.023.

Wu, P., Li, Z., Gao, J., Zhao, Y., Wang, H., Qin, H., & Han, X. (2023). Characterization of a PBAT degradation carboxylesterase from *Thermobacillus composti* KWC4. *Catalysts, 13*(2), Article 340. https://doi.org/10.3390/catal13020340.

Xiao, L., Yao, Z., He, Y., Han, Z., Zhang, X., Li, C., & Ma, P. (2022). Antioxidant and antibacterial PBAT/lignin-ZnO nanocomposite films for active food packaging. *Industrial Crops and Products, 187*, 115515.

Xing, Q., Buono, P., Ruch, D., Dubois, P., Wu, L., & Wang, W. J. (2019). Biodegradable UV-blocking films through core-shell lignin-melanin nanoparticles in poly(butylene adipate- co-terephthalate). *ACS Sustainable Chemistry and Engineering, 7*(4), 4147–4157. https://doi.org/10.1021/acssuschemeng.8b05755.

Xiong, S. J., Pang, B., Zhou, S. J., Li, M. K., Yang, S., Wang, Y. Y., Shi, Q., Wang, S. F., Yuan, T. Q., & Sun, R. C. (2020). Economically competitive biodegradable PBAT/ lignin composites: Effect of lignin methylation and compatibilizer. *ACS Sustainable Chemistry and Engineering, 8*(13), 5338–5346. https://doi.org/10.1021/ACSSUSCHEMENG.0C00789/SUPPL_FILE/SC0C00789_SI_001.PDF.

Yang, F., Chen, G., Li, J., Zhang, C., Ma, Z., Zhao, M., Yang, Y., Han, Y., Huang, Z., & Weng, Y. (2022). Effects of quercetin and organically modified montmorillonite on the properties of poly(butylene adipate-co-terephthalate)/thermoplastic starch active packaging films. *ACS Omega, 8*, 663–672.

Zehetmeyer, G., Meira, S. M. M., Scheibel, J. M., de Oliveira, R. V. B., Brandelli, A., & Soares, R. M. D. (2016). Biodegradable nisin-PBAT films for active food packaging applications: The impact of melt processing. *Journal of Applied Polymer Science, 133*(13) 42.

Zhou, S., Zhai, X., Zhang, R., Wang, W., Lim, L. T., & Hou, H. (2021). High-throughput fabrication of antibacterial starch/PBAT/AgNPs@ SiO_2 films for food packaging. *Nanomaterials, 11*(11), 3062.

Section II

Properties of biopolymer films and coatings

Chapter 22

Properties of biopolymer films and coatings

Juan E. Cortés-Aguilera[1], Jonathan M. Sánchez-Silva[2], Kevin Noe Ruiz-Hernández[1], Ulises M. López-García[1], Beatriz Montaño-Leyva[3], and Ramsés R. González-Estrada[1]

[1]*National Technological Institute of Mexico/Technological Institute of Tepic, Tepic, Nayarit, Mexico,* [2]*Center for Research and Graduate Studies, Faculty of Chemical Sciences, Autonomous University of San Luis Potosí, San Luis Potosí, Mexico,* [3]*Department of Research and Graduate Studies in Food, University of Sonora, Hermosillo, Sonora, Mexico*

22.1 Introduction

Biopolymer films and coatings have emerged as innovative and sustainable solutions in various industries, especially in food packaging. Understanding their properties is crucial because these materials offer a range of benefits that traditional synthetic films cannot. Biopolymers are derived from renewable resources, making them an eco-friendly alternative that reduces reliance on fossil fuels and decreases environmental pollution (Han, 2013; Pavlath & Orts, 2009). The properties of biopolymer films and coatings, such as their mechanical strength, barrier properties, thermal stability, and biodegradability, determine their effectiveness and applicability. For example, their ability to provide a strong barrier against moisture, gases, and contaminants is essential for extending the shelf life of perishable products and maintaining food safety (Ribeiro et al., 2021). Additionally, their mechanical properties, including flexibility and tensile strength (TS), ensure that they can withstand handling and transportation without compromising product integrity (Bourtoom, 2008). Furthermore, the biodegradability of biopolymer films and coatings addresses the growing concern over plastic waste accumulation (Tharanathan, 2003). As they break down naturally in the environment, they contribute to reducing the overall carbon footprint and support a circular economy. By exploring and optimizing the properties of biopolymer films and coatings, researchers and industries can develop sustainable, high-performance materials that meet the demands of modern applications while promoting environmental conservation.

22.2 Water characterization

The water-related properties of a material are fundamental to its suitability for use in food packaging and coatings. These properties determine the material's functionality, stability, and performance, ensuring optimal food protection. Key water-related characterizations include solubility percentage, water content, and swelling degree (Tan et al., 2024). As previously discussed, water solubility is one of the crucial factors in the utilization of films and coatings on food. This property indicates the material's suitability for packaging high-water-content foods and is also used to assess its biodegradability (Zhao et al., 2023). Another important property to consider in the development and application of food packaging is the material's moisture content, which relates to its stability when interacting with different food matrices (Yang, Punia Bangar, et al., 2024). Finally, the swelling degree of films measures the material's ability to absorb water or liquid solutions. This absorption capacity primarily depends on the material's components, as these define its behavior under varying atmospheric conditions and in contact with food (Ngasotter et al., 2025). The interaction of the material with water can occur in two ways: first, with water present in a solution, and second, with water resulting from contact with food or the environment. This interaction primarily occurs through molecular bonds formed by the functional groups present in the material, which determine its degree of hydrophobicity. These bonds—whether intermolecular or intramolecular—typically arise from the polarity of functional groups capable of forming hydrogen bonds, such as hydroxyl (-OH), carboxyl (-COOH), and amine ($-NH_2$) groups (Denise R et al., 2024).

Natural Polymer-Based Films and Coatings in Food Technology. DOI: https://doi.org/10.1016/B978-0-443-34244-8.00001-X

In the field of polymers, the interaction with water is significantly influenced by their physical structure, particularly the degree of crystallinity, which reflects the level of structural order of polymer chains. This structure is divided into two principal regions: the crystalline region, characterized by a high degree of order, and the amorphous region, where the chains exhibit a disordered arrangement (Uzun, 2023). The goal of food packaging materials is to achieve a partially amorphous structure. Reducing crystallinity through intermolecular interactions has been shown to improve the material's compatibility with foodstuffs and processability. However, this strategy also increases the material's hygroscopicity—its ability to absorb moisture—thereby reducing its moisture resistance (Pang et al., 2024). In light of the aforementioned evidence, the hydrophilic or hydrophobic nature of materials used in food packaging depends on the quantity and spatial arrangement of functional groups capable of interacting with water molecules. These interactions are influenced not only by the intrinsic properties of the material but also by external environmental conditions and the moisture content of the food.

The moisture content of food matrices varies considerably. For instance, fresh produce often contains a high-water content, which contributes to its freshness, facilitates microbial growth, and accelerates deterioration. Excess water increases the risk of contamination, posing a significant challenge to food preservation (Chen & Zhu, 2024; Zhou et al., 2023). Characterizing water is critically important for determining a film's suitability or coating for specific foodstuffs. Relative humidity directly influences food products, particularly regarding their hygroscopicity. Therefore, proper humidity control is essential to reduce or prevent product damage (Moustafa et al., 2023). Furthermore, water characterization allows for the prediction of the functional behavior of materials under various external conditions, including biodegradability. This process depends on environmental factors such as oxygen availability, temperature, and humidity. Moisture, in particular, significantly influences biodegradability, as the presence of hydrophilic groups in a material enhances this property (Nilsen-Nygaard et al., 2021; Plaeyao et al., 2025). In this type of material, components such as essential oils, plant extracts, phenolic compounds, and nanoparticles are integrated to enhance their functional performance (Thakur et al., 2023). Water characterization is paramount to adapting the material's application to different food matrices in this context. Moreover, precise moisture control is essential to ensure food quality and optimize the efficiency of these solutions.

The material's water-related properties can be influenced by several factors, which may either enhance or degrade its performance as a food protection solution. One significant factor is the material's chemical composition, determined by its constituent components. This composition can be modified in polymeric matrices by incorporating specific chemical compounds. This process introduces functional groups from active compounds that are covalently attached to the polymer chain, primarily via amide, ether, ester, and thioester bonds (Liu et al., 2024; Ojogbo et al., 2020; Zhang, Wei, et al., 2024).

Another fundamental factor is the material's morphology and structure, which directly influence its water-related properties. As previously mentioned, the degree of crystallinity or amorphism determines the availability of hydrophilic groups that can migrate and interact with water molecules (Maghsoud et al., 2018). In this context, a key structural aspect affecting water-related properties is the material's cross-linking, which stabilizes its three-dimensional structure. This assembly occurs through interactions between functional groups present in the polymer chains, either inter or intra-catenary. Cross-linking agents play a particularly significant role in this process, as they restrict the accessibility of functional groups capable of forming noncovalent interactions, such as hydrogen bonds (Zhang, Roy et al., 2023).

The performance of packaging materials is significantly influenced by external factors, particularly the environment in which the material and the food are situated. Temperature, light, and ambient humidity can accelerate degradation: (1) Elevated temperature: high temperatures can cause dehydroxylation of the material's structural bonds, compromising its integrity (Ullah et al., 2024); (2) UV light exposure: UV radiation can break chemical bonds, such as C-C, C-H, and C-O, whose functional groups may interact with water molecules, thereby accelerating material degradation (Roy, Ramakrishnan, et al., 2024); (3) High relative humidity: materials prone to hydrolytic degradation are particularly vulnerable, as their structural stability and functionality can be compromised (Holm et al., 2006). The combined effect of these environmental conditions reduces the shelf life of the packaging and impairs its ability to effectively protect the food. Given the diverse water-related characteristics of food packaging materials, their importance, and the factors that significantly influence them, it is essential to understand the techniques and equipment used to determine these properties. Table 22.1 presents the main analytical techniques for characterizing water in films and coatings.

The water-related properties of food packaging materials, whether films or coatings, determine their interaction with the environment and directly affect the behavior and performance of the food they protect. Generally, excess water in food is considered detrimental, as it promotes microbial growth and accelerates spoilage. Additionally, water can trigger chemical reactions and contribute to the degradation of water-soluble vitamins, such as vitamin C and B-complex (Perumal et al., 2024). In this context, the food's moisture content and capacity for moisture exchange with the environment are crucial for

TABLE 22.1 Water-related characterization test.

Test	Basis	Water-related characteristics	Properties that influence	Results	References
Water vapor permeability (WVP)	Diffusion and solubility of H_2O molecules in the material matrix.	Solubility, moisture content.	Crystallinity, cross-linking, hydrophobic-hydrophilic equilibrium.	By increasing the hydrophobicity of the material, the diffusion and dispersion of H_2O molecules are hindered, which decreases the water vapor permeability.	Rezaei et al. (2024)
Water adsorption and desorption isotherms	Relationship between the moisture content of a material (adsorbate) and the water activity (a_w) at constant temperature.	Moisture content, swelling degree.	Hydrophobic-hydrophilic equilibrium.	Characterization of the performance, efficiency, and adsorption and diffusion behavior of water in the material.	Samadi et al. (2015)
Thermogravimetric analysis	Changes in the mass of the material as a function of temperature and time.	Solubility, moisture content.	Chemical composition, crystallinity.	Determination of thermal stability until decomposition and water content.	Jiang et al. (2024)
Differential scanning calorimetry	Difference in heat flow between a sample and a reference during heating and cooling events.	Solubility, moisture content.	Chemical composition.	Thermal behavior of a substance, transition temperatures, crystallization temperature, melting, and evaporation of compounds.	Gharanjig et al. (2020)
Fourier transform infrared	Absorption of infrared energy in molecular bonds for the identification of chemical compounds.	Solubility, moisture content.	Chemical composition.	Absorption bands of -OH groups and hydrogen bonds are identified.	Dutta (2017)
Nuclear magnetic resonance	Electromagnetic radiation is absorbed by an atomic nucleus, such as hydrogen (H^+), keeping it in a static magnetic field that generates resonance.	Moisture content.	Chemical composition.	The technique can measure water diffusion or translational displacement of water molecules at intermolecular distances, as well as quantify bound water molecules.	Balthazar et al. (2021)
Energy dispersive X-ray spectroscopy (EDX)	Interaction of X-rays with matter and emission of characteristic X-rays.	Moisture content.	Chemical composition.	Identification and differentiation of elements present in the material.	Abd Mutalib et al. (2017)
Contact angle of water	Determines the geometry that a drop of H_2O acquires on the solid surface of the material.	Solubility, swelling degree.	Surface roughness and chemical composition.	The angle describes the level of wettability of a material, being (1) hydrophilic (<90°), (2) hydrophobic (>90°) and superhydrophobic (>150°).	Mohammad Karimi et al. (2024)

maintaining organoleptic properties by regulating moisture rejection or containment (Deng et al., 2017; Yang, Shang et al., 2024).

In conclusion, the water-related properties of films and coatings used in food packaging are crucial for ensuring materials' functionality, stability, and performance, and for optimizing food preservation. A material's water solubility, moisture content, and degree of swelling directly affect its interaction with food and the environment, thereby influencing the protection it offers against external factors such as humidity, temperature, and UV rays. Moreover, the material's chemical composition, physical structure, and cross-linking significantly impact its behavior toward water. Therefore, it is essential to control these properties to tailor the packaging to different food matrices and extend the product's shelf life. Accurate characterization of these properties will enhance the quality of the protected food and support the development of sustainable and biodegradable solutions that meet both environmental and market demands.

22.3 Thermal properties

The thermal properties of biopolymer films and coatings have a direct impact on their application in the area of food technology. It is essential to know the thermal stability of biopolymer films and coatings in order to safeguard the structural integrity and preservation of food at various stages of production (processing, transport, and storage). Therefore, it is indispensable that biopolymer films and coatings possess the ability to safeguard food safety and quality (Shah et al., 2024). Some of the specific properties obtained from the thermal analysis of biopolymer films and coatings are thermal stability, melting temperature, thermal degradation rate, glass transition temperature, and the relationship between mechanical and thermal properties (Hu et al., 2024). These properties and characteristics are obtained from characterization techniques, such as thermogravimetric analysis (TGA), differential scanning calorimetry (DSC), dynamic mechanical analysis (DMA), and thermomechanical analysis (TMA), and will be discussed below.

Thermogravimetric analysis (TGA): TGA is an indispensable characterization technique for the research and creation of biopolymer films and coatings because it is essential to know the thermal stability of these materials. In addition, it is necessary to determine exact decomposition temperatures and the effects of chemical or physical modifications made on biopolymer films and coatings (decomposition temperature shifts). The TGA technique uses a constant and controlled heating ramp, a heating range of 25°C–900°C, heating is performed in an inert atmosphere (nitrogen or argon), 5–10 mg of material is used, and the weight loss as a function of time (α, mass loss percentage) is recorded (Sanchez-Silva et al., 2023). The thermogravimetric curves of biopolymer films and coatings such as chitosan, gelatin, pectin, sodium alginate, and cellulose present specific weight loss zones that can be related to their compositional structure. For example, the first weight loss zone is related to the loss of moisture or solvents present, the second zone is usually the highest weight decay and is related to the structural decomposition of the biopolymer and is the most important zone for defining the thermal stability (Shah et al., 2024). In addition to the TGA curves, it is possible to determine the first derivative of TG curve with respect to either temperature or time, known as the derivative thermogravimetry (DTG); this DTG curve allows us to better visualize specific decomposition temperatures, for example, the initial thermal degradation temperature (T_{onset}) and the maximum temperature of degradation (T_{max}). Alasalvar et al. (2023) performed TGA and DTG curves of pectin films and found a first stage of decomposition between 41°C and 143°C due to water evaporation and loss of volatile compounds. The second stage of decomposition took place between 143°C and 390°C, which was attributed to the decomposition of the pectin structure. These values are important to define the treatment conditions to which these films and coatings will be subjected so that they will not be damaged and have direct repercussions on the integrity of the food. Table 22.2 shows the T_{onset} and T_{max} of films with a single biopolymer used in edible films.

DSC: DSC analysis is a characterization technique that quantifies heat transfer during heating or cooling of a sample in a wide range of temperatures (−90°C–550°C) and is an important tool for the characterization of biopolymer films and coatings, as it is essential to understand modifications made to the material according to the different environments to which it will be subjected, for example, in production or storage stages. The DSC technique is often used to estimate phase change enthalpy (ΔH), crystallization temperature (T_c), melting temperature (T_m), and glass transition temperature (T_g) (Leyva-Porras et al., 2020; Mishra et al., 2024). The T_g is associated with the molecular mobility of the biopolymer depending on the conditions to which it is exposed; the T_g is defined as the temperature where the material changes from a glassy hard material to a soft and rubbery material, or the opposite, from a rubbery state to a glassy state, and this is closely related to the thermal and mechanical properties of the material. Therefore, it is important to know this property of

TABLE 22.2 Characteristic temperatures in thermogravimetric analysis/derivative thermogravimetry analysis of films with single biopolymer.

Biopolymer	T_{onset} (°C)	T_{max} (°C)	References
Apple pectin	225	240	Šešlija et al. (2018)
Chitosan	241	294	Teixeira-Costa et al. (2023)
Cellulose	150	238	Gomaa et al. (2022)
Apple pectin	150	229	Alasalvar et al. (2023)
Citrus pectin	191	223	Nesic et al. (2022)
Sodium alginate	198	230	Li et al. (2022)
Bovine-hide gelatin	150	305	Wang et al. (2017)

the material when it is destined to be used in any food technology application, where the environments can damage some food. Generally, conventional DSC equipment is used to estimate the melting temperature (T_m) and to estimate the glass transition temperature (T_g); recently, more sensitive techniques such as modulated differential scanning calorimetry (MDSC) have been employed, which improve the accuracy to correctly estimate the T_g, as reported by Kadzińska et al. (2020) in the characterization of sodium alginate films, where they estimate the melting temperature (T_m = 134.39°C) by DSC and the glass transition temperature (T_g = −21.96°C) by MDSC. On the other hand, Meerasri and Sothornvit (2020) found the glass transition temperature of citrus pectin films to be 3.3°C, and when glycerol or gamma-aminobutyric acid was incorporated into these films, the T_g value decreased and influenced the flexibility of the film. Therefore, the DSC and MDSC are important tools for optimizing and improving the creation of biopolymer films and coatings, which increases the probability that these materials will be easily applicable in food technology, such as food packaging.

DMA: This characterization technique is widely used to evaluate the combined effect of temperature, time, mechanical stress, and even atmosphere on viscoelastic performance. Therefore, DMA equipment is used, which has the ability to investigate the direct impact of temperature on the resilience and rigidity of the biopolymer while subjected to low levels of mechanical force (oscillating force) (Shah et al., 2024). In other words, the DMA technique provides a comparison between the thermal-mechanical behavioral relationship, which is crucial for mechanical energy dispersion and breakage prevention during its application in food technology.

TMA: The TMA technique is the most important technique in the area of material science, specifically in thermal analysis. It determines dimensional changes in films as a function of temperature and/or time under a defined mechanical force in accordance with the following standards (DIN 51 005, ASTM D 3386, ASTM E 831, ASTM D 696, ISO 11359—Part 1–3) (Saba & Jawaid, 2018). From this characterization technique, it is possible to obtain information such as T_g, thermal expansion, and softening or hardening points. From TMA, the most important thermal property obtained is the coefficient of thermal expansion (CTE), which describes how much a material expands or contracts when its temperature changes; this has a direct impact on its application. Therefore, the data provided by the TMA is crucial for the creation of biopolymer films and coatings applied in food packaging, so that they have the ability to withstand thermal cycling, mainly during storage. In conclusion, for film design, the TMA technique is a useful method to estimate the combined thermal-mechanical behavior (Mishra et al., 2024).

Considering the main objective of the biopolymer films and coatings, which is to protect the integrity of food, it is especially important to know the thermal stability of the films and their structural performance in different environmental conditions. Since it is required to minimize decomposition or damage during the various stages of application, such as processing, transportation, and storage, thermal properties obtained from various characterization techniques, such as TGA, DSC, DMA, and TMA, are important to establish efficient applications in food technology. Similarly, during the search for modifications in biopolymer films and coatings, such modifications exhibit higher thermal stability and improved thermal properties, especially with the increasing necessity for sustainable and environmentally friendly packaging solutions.

22.4 Mechanical properties

Edible films and coatings are extensively applied to fruits and vegetables to enhance appearance, preserve quality, and extend shelf life. These materials are biodegradable and nontoxic and can be formulated from different components, including natural biopolymers (e.g., polysaccharides, proteins, and waxes), synthetic polymers, or a combination of them (composites) (Suhag et al., 2020). The chemical composition, physicochemical properties, and microstructure of films define their possible applications (Galus et al., 2020). To determine the suitability of films and coatings for fruit preservation, it is essential to assess properties such as mechanical performance, water solubility, thickness, crystallinity, thermal behavior, and vapor and gas permeability.

The mechanical properties of films and coatings play an essential role in determining their protective capacity, handling efficiency, and overall durability. Examining the mechanical properties is indispensable for evaluating their resilience, functionality, and efficiency in maintaining fruit quality. Various experimental approaches are employed to analyze the mechanical performance of these coatings under diverse environmental and operational conditions. Tensile tests are the most common. The parameters evaluated in tensile tests are TS, elongation at break (EAB), and Young's modulus (YM). These parameters are influenced by the nature and molecular structure of their components, as well as the compatibility and interaction between the polymers (Hammam, 2019; Silva-Weiss et al., 2013; Villagómez-Zavala et al., 2008). Table 22.3 shows different formulations of edible films and coatings, and their impact on mechanical properties.

TS represents the maximum stress the film can resist before breaking under tension. TS is important for maintaining the integrity of the coating. High TS allows the coating to withstand handling, transport, and storage without cracking or tearing. EAB refers to the film's ability to stretch before breaking. Films with high EAB values better withstand impact and deformation, reducing the risk of damage to both the fruit and the coating. YM measures the stiffness and flexibility of the material. Films with high YM are stiffer, which may be desirable for fruit protection; however, these films may be less flexible. On the other hand, lower YM materials are more flexible, making the coating more adaptable to the fruit's shape and movement during transport. In general, TS and EAB are the mechanical properties of principal interest because the aim is to obtain a coating or edible film capable of maintaining its structure despite external stress, without neglecting the elongation necessary to avoid being a brittle material.

Most edible films and coating formulations use plasticizers to improve the flexibility and extensibility of the materials (Hammam, 2019). Plasticizers such as glycerol and sorbitol are nonvolatile molecules with a low molecular weight that reduce interactions between polymer chains, thus facilitating the intermolecular spacing and sliding of the chains. Therefore, the film's flexibility and stretchability are improved (Jafarizadeh et al., 2011; Pavinatto et al., 2020). However, plasticizers may also decrease the mechanical resistance and barrier properties of the films (Jafarzadeh et al., 2018). In a study, the physico-mechanical and microstructural properties of edible films based on semolina flour with different plasticizers (sorbitol/glycerol) were evaluated. The results showed that the addition of plasticizer in semolina film has a great effect on the mechanical and barrier properties, such as an increase in elongation and a decrease in water vapor and oxygen permeability, TS, and YM (Jafarzadeh et al., 2018). Similarly, Anker et al. (2001) studied the effect of glycerol and sorbitol on the mechanical and barrier properties of whey films. They reported that high plasticizer concentrations reduce mechanical strength, YM, and glass transition temperature, but barrier properties were not affected.

Another way to improve the mechanical properties of edible films is by combining the functional properties of different types of polymers in a composite film. Likewise, nanoparticles (NPs) are widely used to enhance the properties of biopolymers (Pires et al., 2024). Due to their nanometric scale, their incorporation into biomaterials enhances mechanical and barrier properties. The mechanical characteristics of nanocomposites result in significant increases in the material's stiffness YM. Barikloo and Ahmadi (2018) examined the evolution of gas exchange and mechanical properties of strawberries coated with nanocomposite films composed of chitosan, clay, and silica during storage. The incorporation of clay and silica into the chitosan-based coating enhanced the physical, chemical, and mechanical properties of strawberries during the storage period. Cyras et al. (2008) developed glycerol-plasticized starch/clay nanocomposites and investigated the effect of incorporating clay on the barrier, thermal, and mechanical properties of plasticized starch. The nanocomposites resulted in a significant enhancement in mechanical performance, particularly in YM. Nevertheless, a decrease in elongation was observed, attributed to the reinforcing effect of MMT. They concluded that the increase in YM resulted from a strong interfacial interaction between the starch matrix and the clay, which altered the matrix morphology.

The mechanical properties of edible films and coatings are influenced by the structure of the polymers and the interactions between their components. These properties must be suitable for preserving the quality and integrity of food products during handling and distribution. Poor mechanical properties can directly affect the barrier properties and alter permeability, leading to microbial contamination and decreased shelf life.

TABLE 22.3 Composition of edible films and coatings and their effect on mechanical properties.

Coating/film composition	Effect on mechanical properties	Application	References
Gellan gum, glycerol, as a plasticizer	Increased elongation at break (EAB) with higher glycerol concentrations and decreased tensile strength (TS)	Applied to mango slices	Gómez et al. (2024)
Carboxymethyl chitosan and gelatin	Improved mechanical and barrier properties; optimization achieved through formulation adjustments	Can be used as an edible coating for fruits	Zhang, Cui et al. (2023)
Pectin with chitosan nanoparticles (CSNP), polysorbate 80, and garlic essential oil	The films showed low elongation values (reinforcing effect) linked to high interactions among CSNP surface groups (NH^{3+}) and pectin monomers (-COOH)	Active edible films	dos Santos et al. (2023)
Calcium caseinate, whey protein isolate, and vitamin E	Vitamin E incorporation improved the EAB and reduced the TS of the films because of its hydrophobic nature and the plasticizer effect	Used for wrapping or coating to enhance the nutritional value of food	Mei and Zhao (2003)
Starch, gelatin, and natural waxes (beeswax, candelilla and carnauba)	The addition of beeswax and candelilla reduced the TS of films, indicating the network destruction of the film matrix	Highly hydrophobic edible films	Cheng et al. (2023)
Soybean protein isolate (SPI) and hydroxypropyl methyl cellulose (HPMC)	The TS of the films increased slightly with the increase of HPMC, while the EAB gradually decreased. The interaction of HPMC hydrogen bonds interacted with amino groups on the SPI improves the internal network structure of the film		Zhang et al. (2020)
Chitosan, nano-titanium dioxide (nano-TiO_2), and litchi peel extract (LPE)	The addition of nano-TiO_2 increased the TS with and EAB decreased because of the reinforcing filler effect. Conversely, TS decreased and EAB increased with the addition of LPE, increasing film flexibility	As coating on watercored apples	Liu et al. (2021)
Chitosan films plasticized with 30% glycerol	Maintained bactericidal properties, enhanced elasticity, and hydrophobicity	Coating on strawberries	Pavinatto et al. (2020)
Chitosan and banana peel extract (BPE)	The mechanical properties of the films were improved when low amounts of BPE (4%) were added, increasing the TS and EAB. When higher amounts were added (8% and 12%), TS and EAB were reduced. The excess BPE disrupted the molecular structure of the chitosan, decreasing the mechanical properties	Coating on apples	Zhang et al. (2020)
Pectin and konjac glucomannan with tea polyphenols (TP)	The incorporation of TP remarkably increased the mechanical (especially EAB) and water-resistant properties of the films	Bioactive materials in food packaging	Lei et al. (2019)
Banana starch, chitosan, and Aloe vera gel (AV)	The inclusion of AV gel disrupts the starch-chitosan interaction, thus TS and EAB decrease	Strawberry coating	Pinzon et al. (2020)
Sago starch and carboxymethyl cellulose nanoparticles (CMCNPs)	TS and Young's modulus (YM) increased and elongation significant reduced when CMCNPs were incorporated. TS increase was due to surface interaction between starch matrix and CMCNPs due to chemical similarity between structures of polysaccharides	Edible films and coatings	Tabari (2018)

(Continued)

TABLE 22.3 (Continued)

Coating/film composition	Effect on mechanical properties	Application	References
Semolina flour with glycerol and sorbitol (3:1)	As plasticizer content increased, TS and the elastic modulus of the semolina films decreased, while EAB increased significantly	Edible films	Jafarzadeh et al. (2018)
High amylose corn starch modified by extrusion	Extrusion process improve mechanical properties of native starch films	As coating on mango (*Mangifera indica* L.)	Calderón-Castro et al. (2018)
Cross-linked arabinoxylans and *D. hansenii*/Gly	TS, EAB, and YM dropped when *D. hansenii* was added	As a functional entrapping film	González-Estrada et al. (2015)
Gelatin and Zataria multiflora essential oil (ZMO)	Incorporation of ZMO caused significant decrease in TS and YM and an increase in EAB of the films. Gelatin films were mainly stabilized by weak bonds, including hydrogen bonds and hydrophobic interactions	Antioxidant and antimicrobial films	Kavoosi et al. (2014)
Cassava starch, chitosan, and oregano essential oil (OEO)	The presence of OEO in starch-chitosan films reduced TS and YM, and increased EAB, achieving fewer rigid films and more flexible films	Antimicrobial films	Pelissari et al. (2009)
Potato starch and montmorillonite aqueous suspensions (MMT)	The addition of MMT into the film showed an enhancement in the mechanical properties, such as YM and TS. Though, a diminution in the deformation was observed due to the reinforcing action of MMT	Food packaging	Cyras et al. (2008)

22.5 Morphological behavior

The analysis of the morphological behavior of edible films and coatings plays a crucial role in understanding their potential for food applications (Paidari et al., 2021). The study of their morphological behavior—that is, their structure, texture, and properties at the microscopic and macroscopic levels—can provide insights into their performance and functionality in food systems (Galus et al., 2020). These materials are increasingly utilized in the food industry for extending shelf life, maintaining quality, and reducing environmental impact by offering biodegradable alternatives to synthetic packaging (Díaz-Montes & Castro-Muñoz, 2021; Kumar et al., 2022). The morphological properties can be assessed using scanning electron microscopy (SEM), transmission electron microscopy (TEM), atomic force microscopy (AFM), X-ray diffraction (XRD), profilometry, and Fourier transform infrared (FT-IR) techniques.

22.5.1 Scanning electron microscopy

SEM is the most powerful tool used for morphological analysis in materials science, including the characterization of edible films and coatings (Han, 2013). SEM can be utilized to assess film uniformity, layer composition, pore and crack presence (defects), surface texture (additive distribution), thickness (cross-sectional image), adhesion (interface between the food and coating), and coating integrity (degradation) (García et al., 2009; Liu et al., 2018). The thickness property is associated with functional properties of films, such as mechanical resistance, transparency, and water vapor permeability; micrographs of the cross-sectional film permit the evaluation of this parameter with high precision (Jeevahan et al., 2020; Kumari et al., 2017). The presence of defects such as pores and cracks can compromise the mechanical resistance of films as well as the barrier properties, such as water vapor permeability and CO_2/O_2 permeability (de Moura et al., 2009). The analysis of the surface texture gives information about the compatibility of the additives used in the formulation of film-forming solutions by observing smooth or rough surfaces; it provides crucial information about the additive distribution in the polymeric matrix (Zhang et al., 2021). Analyzing the coating adhesion of a biopolymer in food applications is critical for ensuring the coating's effectiveness and functionality. Proper adhesion is essential because it directly impacts the coating's ability to perform its intended roles, such as extending shelf life, preserving quality, and protecting against environmental factors (Martins et al., 2024; Sengupta & Han, 2014). Poor adhesion can lead to peeling or flaking, reducing the coatings' protective barrier property and compromising food quality and safety (Frabetti et al., 2021). The coating integrity is essential to maintain the quality, safety, and functional benefits of the food product. Proper integrity ensures the barrier properties, the appearance, and the presence of additives in the matrix (Basumatary et al., 2022; Tyagi et al., 2021).

22.5.2 Transmission electron microscopy

TEM is a critical tool for analyzing edible films and coatings due to its ability to provide high-resolution, detailed insights into the microstructure, morphology, and interactions within these materials. This information is vital for optimizing the functional and mechanical properties of edible films and coatings, especially those used in food packaging. This analysis can provide useful information about the following properties (Kumar et al., 2022):

Microstructural analysis, TEM allows for high-resolution imaging of the film or coating's microstructure, revealing details about the arrangement and size of nanoparticles, polymers, and other components.
Layer thickness, it helps in measuring the thickness of individual layers within multilayer films or coatings, which is crucial for understanding barrier properties.
Distribution of additives, TEM can visualize the distribution and dispersion of additives (like antimicrobials or antioxidants) within the film or coating matrix.
Crystallinity, it can provide insights into the crystalline structure of biopolymers used in the films, which affects mechanical and barrier properties.
Defects and imperfections, TEM helps identify any defects or imperfections in the film or coating that could impact its performance.

22.5.3 Atomic force microscopy

AFM is widely used for analyzing the microstructure and properties of edible films and coatings. This technique provides high-resolution images of the surface topography and roughness of edible films and coatings, which are crucial for understanding their barrier properties (de Paula Herrmann et al., 2004). By studying the surface characteristics, researchers can gain insights into the interactions between different components of the films and coatings, leading to improvements in their design and functionality. Additionally, AFM can measure mechanical properties such as stiffness, YM, and viscoelastic properties, providing valuable information about the strength and flexibility of the films (Ghanbarzadeh & Oromiehi, 2008).

AFM also plays a crucial role in determining the thickness of individual layers within multilayer films or coatings. This information is essential for optimizing the performance of edible films, especially regarding their barrier properties against gases and moisture. Furthermore, AFM can assess functional properties like electrical conductivity, surface potential, and piezoelectric responses, which are important for specific applications, such as incorporating active ingredients that require precise control over their release (Cheong et al., 2019; Denning et al., 2016; Durkan et al., 2000).

22.5.4 X-ray diffraction

XRD is a valuable technique useful for studying the structural properties of edible films and coatings. XRD helps identify the crystalline structure and phase composition of the materials used in these films. By analyzing the diffraction patterns, researchers can determine the degree of crystallinity, which affects the mechanical and barrier properties of the films (Abdillah & Charles, 2021; García et al., 2000; Zhang et al., 2016). This information is crucial for optimizing the formulation and improving the performance of edible films and coatings.

For instance, a study by Chen, Wang et al. (2024) utilized XRD to evaluate the effects of different plasticizers on the structure and physical properties of curdlan-based edible films. The results showed that the inclusion of plasticizers altered the crystalline structure of the curdlan matrix, leading to changes in the film's mechanical properties and water resistance. This highlights the importance of XRD in understanding how various additives influence the structural integrity of edible films.

22.5.5 Profilometry

This technique is useful for analyzing the surface topography and roughness of edible films and coatings (Porter & Felton, 2010). By providing high-resolution measurements of surface features, profilometry helps researchers understand the physical characteristics that influence the performance of these films (Porter & Felton, 2010). For example, the thickness and uniformity of the coatings can be accurately measured, which is crucial for ensuring consistent barrier properties and mechanical strength. One study utilized profilometry to evaluate the surface roughness of chitosan-based edible coatings applied to fruits. The results indicated that smoother surfaces were more effective in reducing water loss and extending the shelf life of the coated fruits (Silue & Fawole, 2024). This highlights the importance of surface characterization in optimizing the protective properties of edible films.

22.5.6 O_2/CO_2 barrier properties

The primary function of food packaging materials is to protect foodstuffs from physical harm and to preserve their organoleptic properties by delaying the natural deterioration that would otherwise occur. This is primarily achieved by preventing microbial contamination and serving as an effective barrier to gases (Moeini et al., 2022). In this context, the ability of food films and coatings to act as a barrier is crucial for food preservation.

The ability of a food packaging material to function as a barrier depends on its capacity to prevent the passage of molecules from the external environment into the food. By limiting the transmission of molecules such as O_2 and CO_2, oxidation is prevented, and the growth of aerobic microorganisms is restricted in the absence of oxygen (Wang et al., 2024). In this context, "barrier properties" are defined as protecting the food inside the packaging by preventing the entry or exit of low molecular weight permeant chemicals (Marano et al., 2022; Mujtaba et al., 2022).

Therefore, barrier properties are of paramount importance for protecting foodstuffs, particularly given that they are exposed to various atmospheric factors, including oxygen, carbon dioxide, and water vapor. These factors significantly influence the food's respiratory metabolism. By reducing direct contact with the external environment, nutrient consumption and water loss are minimized, thereby preventing changes in the food's visual and functional properties (Hu et al., 2024).

Oxygen is a key factor that significantly affects food shelf life, as its presence promotes the oxidation of lipids and vitamins, accelerates enzymatic reactions, and stimulates the growth of microorganisms such as aerobic bacteria, yeasts, and molds, contributing to food spoilage. Additionally, oxygen can degrade product quality and considerably reduce shelf life (Dey & Neogi, 2019; Rozo et al., 2024; Zhang, Wei, et al., 2024).

On the other hand, carbon dioxide is another important gas in food preservation, particularly in fresh foods such as fruits, where aerobic respiration produces CO_2 through the oxidation of sugars. However, in anaerobic conditions with excess CO_2, it triggers anaerobic metabolic reactions that decompose sugars, leading to softening and loss of texture (Sistla & Mehraj, 2022).

The barrier properties of materials used in food packaging are closely related to product spoilage. Maintaining an appropriate balance in the gas exchange between the food and the surrounding atmosphere is essential. This exchange, known as "gas permeation," involves several stages: first, gas molecules are adsorbed onto the surface of the barrier material (film or coating); then, they disperse on this surface and diffuse from areas of higher concentration to areas of lower concentration, following the concentration gradient. Finally, the molecules are desorbed from the opposite surface of the material (Xue et al., 2023; Trinh et al., 2023).

To estimate the permeability of gases in food packaging materials, the solution-diffusion model (described above) is used, following Fick's first law. Permeability is expressed as the molar flux of gas through the material, denoted as P_i, where the flux per unit area through the film is related to the partial pressure gradient, as shown in equation "*n*":

$$J_i = P_i \frac{\Delta P_i}{l}$$

f plasticizer content increased, Where the molar density flux of the component is given by J_i, the partial pressure difference of the component at the two sides of the material is ΔP_i, and the thickness is l. The dimensions of the permeability are given based on the amount of gas (moles or mass). Hence, the units in the international system would be reported as mol * m * m^{-2} * s^{-1} * Pa^{-1} (Giacinti Baschetti & Minelli, 2020; Souza et al., 2024).

The gas permeation rate, and therefore the barrier performance of materials, is determined by several factors, including film thickness, temperature, relative humidity, and the composition and structural properties of the material, such as crystallinity and morphology. Thickness is an essential property for packaging materials because uniformity is required for proper performance. It directly affects physicochemical performance and is critical in determining gas and water permeability (Kurabetta et al., 2024; Lade et al., 2023). Temperature also influences permeability and barrier capacity through thermal stability, which is observed as the degradation of functional and structural components that regulate the passage of gases (Alghamdi, 2024). As for relative humidity mainly affects hydrophilic materials, as an increase in internal humidity can modify the structure, reducing its stability. On the other hand, low relative humidity levels maintain the complex structure, allowing controlled gas transmission (Ma et al., 2025). Regarding the material's structure, cross-linking and the degree of crystallinity play a role, as this molecular arrangement limits or allows the transmission of molecules (Du et al., 2024; Rivera et al., 2024).

The gas barrier properties of food packaging materials depend mainly on the intrinsic properties of the raw materials used in their manufacture. Within this context, a wide variety of materials can act as a barrier to food. Historically, plastics derived from nonrenewable fossil sources, such as polyethylene, polypropylene, and polystyrene, have been used. Although these materials provide excellent barrier properties, their nonbiodegradable nature significantly impacts the environment negatively at the time of disposal (Bashir et al., 2018; Dang et al., 2025).

Bio-based polymers have been adopted to address the problem of nonrenewable packaging and reduce its environmental impact. These materials are characterized by their biodegradability, nontoxicity, economic viability, and ability to form compelling films (Huang et al., 2024). The most commonly used bio-based packaging polymers in the food industry include proteins, such as gelatin and casein, and polysaccharides, including chitosan, cellulose, starch, and polyvinyl alcohol. In addition, lipid matrices, such as waxes and oils, are also utilized (Garavand et al., 2017).

Despite each material's unique properties, environmental factors and the intrinsic characteristics of the food can trigger chemical reactions that depend on the heat sensitivity and stability of the vitamins. These reactions can reduce the barrier performance of the packaging material and accelerate food spoilage. To address this issue, the goal is to optimize material properties by developing more effective barriers against these factors (Parhi et al., 2020).

A relevant example is polysaccharides, which have good physical properties and water vapor transmission characteristics but are limited in their gas barrier capacity. To overcome this limitation, several strategies have been implemented, such as the addition of copolymers, plasticizers, cross-linking agents, nanomaterials, or the application of treatments that modify their structure, such as the use of cold plasma and high pressure (Roy, Malik, et al., 2024; Thomas et al., 2024).

Taking into account the different aspects that influence the gas barrier capacity, the design of the material must be adapted to the needs of the food matrix to which it will be applied; in the case of fresh and processed meats, reasonable control of oxidation and prevention of microbial growth must be achieved; the oxidation of meat involves the release of volatile amines and organic acids that act as indicators of freshness, so limiting their contact with oxygen delays decomposition (Nogueira et al., 2024). Fruits and vegetables must have an effective barrier to O_2 and CO_2 to reduce their respiration rate and metabolism and slow down their senescence, thus obtaining a higher quality fruit with a longer shelf life (Chen, Zhang et al., 2024; Xie et al., 2024).

To ensure the proper barrier performance and gas permeability of food films and coatings, it is essential to evaluate their O_2 permeability. This is typically measured according to standards such as ASTM D-3985 (based on carrier gas and Fick's law) and DIN 53380-5 (optical method), which relies on the dynamic quenching of luminescence caused by molecular oxygen. For CO_2, the manometric method DIN 53380-2 is commonly employed (Turan et al., 2017). Additionally, water vapor permeability is frequently determined, as it is closely related to gas transmission. This relationship provides insight into the chemical structure and morphology of the film, as well as the hydrophilic/hydrophobic ratio, steric hindrance, and the degree of ordering of the components within the film matrix (Li et al., 2025).

Barrier materials have gained a prominent role in the food industry due to technological advances in food preservation. These innovations extend the shelf life of products by delaying spoilage, thereby reducing losses and wastage of perishable foods such as fruit, vegetables, meat, seafood, dairy, and bakery products (Chew et al., 2024). In this context, as mentioned above, the current focus is on developing materials that provide comprehensive protection against microorganisms, act as a barrier against gas, moisture, and mechanical damage, and help preserve freshness and extend shelf life. In addition, using sustainable and biodegradable materials is being prioritized to reduce environmental impact (Xue et al., 2024).

Functionalization of materials is a key strategy for developing packaging that meets all required characteristics. This is achieved by incorporating compounds that compensate for deficiencies in the base material and reduce potential impacts on the food (Hu et al., 2024). In this context, three main innovative packaging types have been developed: active, multilayer, and biodegradable/compostable. (1) Active packaging: this type incorporates active compounds into the packaging material, allowing direct interaction with the food and its environment. This helps retard processes such as enzymatic activity, oxidation, and microbial growth, thereby preserving product quality (Salević-Jelić et al., 2024); (2) Multilayer packaging: this packaging consists of at least three layers (outer, middle, and inner) and is designed to provide enhanced barrier capacity against gases and moisture and increased mechanical strength. These structures can be further functionalized by adding active compounds between the layers (Vieira et al., 2022); (3) Biodegradable/compostable packaging: this provides a sustainable solution for food packaging, designed to decompose at the end of its shelf life. It can be broken down naturally by microorganisms or through composting, a process that converts organic waste into humic substances beneficial to the soil (Nizamuddin et al., 2024). These developments not only improve the functionality of packaging but also contribute to environmental sustainability and reduce the ecological impact associated with conventional packaging waste.

In summary, the barrier properties of food packaging materials are essential for extending product shelf life, maintaining food quality, and reducing food waste. These properties depend on factors such as material composition, structure, and design, and must be tailored to the specific needs of each food product. With this in mind, innovations in biodegradable and functional materials are transforming the packaging industry, providing sustainable solutions that protect food from external factors and help reduce environmental impact. The development of reactive, multilayer, and biodegradable packaging demonstrates how this technology can meet food preservation requirements while addressing sustainability issues.

22.6 Conclusions and future perspectives

The analysis of the properties of edible films and coatings is of paramount importance in the development and optimization of sustainable packaging solutions. By thoroughly understanding the mechanical strength, barrier properties, thermal stability, and biodegradability of these materials, researchers can design high-performance films that meet the demands of modern applications while minimizing environmental impact. The ability to control and enhance the functional properties of edible films and coatings contributes to the extension of shelf life, preservation of food quality, and reduction of plastic waste. Moreover, detailed analysis using advanced techniques like XRD, AFM, and TEM provides valuable insights into the microstructural and functional aspects of these materials. This knowledge enables the fine-tuning of formulations to achieve optimal performance and addresses the challenges associated with their application.

Future trends in the analysis and development of biopolymer films and coatings need to focus on enhancing their performance and sustainability. Researchers need to explore the incorporation of nanomaterials, crosslinkers, and plasticizers to improve mechanical, physical, and thermal properties. Advancements in technology are expected to lead to the development of intelligent packaging solutions that can monitor food quality and provide real-time information about spoilage. Additionally, there is a growing emphasis on scaling up production processes to make biopolymer films and coatings more economically viable and environmentally friendly. As consumer awareness and demand for sustainable packaging solutions increase, the industry is likely to see more innovations in biopolymer-based materials. These trends will contribute to reducing plastic waste, promoting a circular economy, and ensuring food safety and quality. In summary, the analysis of the properties of edible films and coatings is essential for advancing sustainable packaging technologies. It supports the development of environmentally friendly materials that offer effective protection and preservation, ultimately contributing to a more sustainable and resilient food system.

References

Abdillah, A. A., & Charles, A. L. (2021). Characterization of a natural biodegradable edible film obtained from arrowroot starch and iota-carrageenan and application in food packaging. *International Journal of Biological Macromolecules, 191*, 618–626. Available from https://doi.org/10.1016/j.ijbiomac.2021.09.141, www.elsevier.com/locate/ijbiomac.

Abd Mutalib, M., Rahman, M. A., Othman, M. H. D., Ismail, A. F., & Jaafar, J. (2017). Scanning electron microscopy (SEM) and energy-dispersive x-ray (EDX) spectroscopy. In *Membrane characterization*. Malaysia: Elsevier Inc, 161–179. Available from http://www.sciencedirect.com/science/book/9780444637765.

Alasalvar, H., Yildirim, Z., & Yildirim, M. (2023). Development and characterization of sustainable active pectin films: The role of choline chloride/glycerol-based natural deep eutectic solvent and lavender extracts. *Heliyon, 9*(11). https://doi.org/10.1016/j.heliyon.2023.e21756.

Alghamdi, A. M. (2024). Optical, thermal, mechanical, and antibacterial properties of polyvinyl alcohol/sodium alginate/$ZnMn_2O_4$ nanocomposites films for various optical devices and food packaging applications. *International Journal of Biological Macromolecules, 271*(Pt 2), 132689. https://doi.org/10.1016/j.ijbiomac.2024.132689.

Anker, M., Stading, M., & Hermansson, A.-M. (2001). Aging of whey protein films and the effect on mechanical and barrier properties. *Journal of Agricultural and Food Chemistry, 49*(2), 989–995.

Balthazar, C. F., Guimarães, J. T., Rocha, R. S., Pimentel, T. C., Neto, R. P. C., Tavares, M. I. B., Graça, J. S., Alves Filho, E. G., Freitas, M. Q., Esmerino, E. A., Granato, D., Rodrigues, S., Raices, R. S. L., Silva, M. C., Sant'Ana, A. S., & Cruz, A. G. (2021). Nuclear magnetic resonance as an analytical tool for monitoring the quality and authenticity of dairy foods. *Trends in Food Science and Technology, 108*, 84–91. Available from https://doi.org/10.1016/j.tifs.2020.12.011, http://www.elsevier.com/wps/find/journaldescription.cws_home/601278/description#description.

Barikloo, H., & Ahmadi, E. (2018). Effect of nanocomposite-based packaging and chitosan coating on the physical, chemical, and mechanical traits of strawberry during storage. *Journal of Food Measurement and Characterization, 12*(3), 1795–1817. https://doi.org/10.1007/s11694-018-9795-3.

Bashir, A., Jabeen, S., Gull, N., Islam, A., Sultan, M., Ghaffar, A., Khan, S. M., Iqbal, S. S., & Jamil, T. (2018). Co-concentration effect of silane with natural extract on biodegradable polymeric films for food packaging. *International Journal of Biological Macromolecules, 106*, 351–359. Available from https://doi.org/10.1016/j.ijbiomac.2017.08.025, www.elsevier.com/locate/ijbiomac.

Basumatary, I. B., Mukherjee, A., Katiyar, V., & Kumar, S. (2022). Biopolymer-based nanocomposite films and coatings: Recent advances in shelf-life improvement of fruits and vegetables. *Critical Reviews in Food Science and Nutrition, 62*(7), 1912–1935. Available from https://doi.org/10.1080/10408398.2020.1848789, https://www.tandfonline.com/loi/bfsn20.

Bourtoom, T. (2008). Edible films and coatings: Characteristics and properties. *International Food Research Journal, 15*(3), 237–248. Available from http://www.ifrj.upm.edu.my.

Calderón-Castro, A., Vega-García, M. O., de Jesús Zazueta-Morales, J., Fitch-Vargas, P. R., Carrillo-López, A., Gutiérrez-Dorado, R., Limón-Valenzuela, V., & Aguilar-Palazuelos, E. (2018). Effect of extrusion process on the functional properties of high amylose corn starch edible films and its application in mango *(Mangifera indica* L.) cv. Tommy Atkins. *Journal of Food Science and Technology, 55*(3), 905–914. Available from https://doi.org/10.1007/s13197-017-2997-6, http://www.springerlink.com/content/121580/.

Chen, J., Zhang, Y., Liu, H., Lu, H., Xu, X., & Shen, M. (2024). Sodium alginate-camellia seed cake protein active packaging film cross-linked by electrostatic interactions for fruit preservation. *International Journal of Biological Macromolecules*138627. https://doi.org/10.1016/j.ijbiomac.2024.138627.

Chen, W., & Zhu, Z. (2024). Ultralong luminescence lifetime imaging of edible plant tissue for humidity sensing in food packaging by a smartphone. *Food Chemistry, 454*, 139778. https://doi.org/10.1016/j.foodchem.2024.139778.

Chen, Y., Wang, J., Xu, L., Nie, Y., Ye, Y., Qian, J., Liu, F., & Zhang, L. (2024). Effects of different plasticizers on the structure, physical properties and film forming performance of curdlan edible films. *Foods, 13*(23), 3930. https://doi.org/10.3390/foods13233930.

Cheng, Y., Zhai, X., Wu, Y., Li, C., Zhang, R., Sun, C., Wang, W., & Hou, H. (2023). Effects of natural wax types on the physicochemical properties of starch/gelatin edible films fabricated by extrusion blowing. *Food Chemistry, 401*, 134081. https://doi.org/10.1016/j.foodchem.2022.134081.

Cheong, L. Z., Zhao, W., Song, S., & Shen, C. (2019). Lab on a tip: Applications of functional atomic force microscopy for the study of electrical properties in biology. *Acta Biomaterialia, 99*, 33–52. Available from https://doi.org/10.1016/j.actbio.2019.08.023, http://www.journals.elsevier.com/acta-biomaterialia.

Chew, Z. L., Koh, Q. Q., Chu, E. E., Kua, Y. L., Gan, S., Tan, K. W., & Lee, T. Z. E. (2024). Tunable durian seed gum-derived eutectogel as a novel coating material: Rheological, thermal, textural and barrier properties for enhanced food preservation. *International Journal of Biological Macromolecules, 267*. Available from https://doi.org/10.1016/j.ijbiomac.2024.131201, https://www.sciencedirect.com/science/journal/01418130.

Cyras, V. P., Manfredi, L. B., Ton-That, M. T., & Vázquez, A. (2008). Physical and mechanical properties of thermoplastic starch/montmorillonite nanocomposite films. *Carbohydrate Polymers, 73*(1), 55–63. https://doi.org/10.1016/j.carbpol.2007.11.014.

Dang, X., Han, S., Du, Y., Fei, Y., Guo, B., & Wang, X. (2025). Engineered environment-friendly multifunctional food packaging with superior nonleachability, polymer miscibility and antimicrobial activity. *Food Chemistry, 466*, 142192. https://doi.org/10.1016/j.foodchem.2024.142192.

de Moura, M. R., Aouada, F. A., Avena-Bustillos, R. J., McHugh, T. H., Krochta, J. M., & Mattoso, L. H. C. (2009). Improved barrier and mechanical properties of novel hydroxypropyl methylcellulose edible films with chitosan/tripolyphosphate nanoparticles. *Journal of Food Engineering, 92*(4), 448–453. https://doi.org/10.1016/j.jfoodeng.2008.12.015.

Deng, Z., Jung, J., & Zhao, Y. (2017). Development, characterization, and validation of chitosan adsorbed cellulose nanofiber (CNF) films as water resistant and antibacterial food contact packaging. *LWT - Food Science and Technology, 83*, 132–140. https://doi.org/10.1016/j.lwt.2017.05.013.

Denise R, M., Usharani, N., Saravanan, N., & Kanth, S. V. (2024). In vitro and in silico approach towards antimicrobial and antioxidant behaviour of water-soluble chitosan dialdehyde biopolymers. *Carbohydrate Research, 542*, 109192. https://doi.org/10.1016/j.carres.2024.109192.

Denning, D., Guyonnet, J., & Rodriguez, B. J. (2016). Applications of piezoresponse force microscopy in materials research: From inorganic ferroelectrics to biopiezoelectrics and beyond. *International Materials Reviews, 61*(1), 46–70.

de Paula Herrmann, P. S., Yoshida, C. M. P., Antunes, A. J., & Marcondes, J. A. (2004). Surface evaluation of whey protein films by atomic force microscopy and water vapour permeability analysis. *Packaging Technology and Science, 17*(5), 267–273. https://doi.org/10.1002/pts.662.

Dey, A., & Neogi, S. (2019). Oxygen scavengers for food packaging applications: A review. *Trends in Food Science & Technology, 90*, 26–34. https://doi.org/10.1016/j.tifs.2019.05.013.

Díaz-Montes, E., & Castro-Muñoz, R. (2021). Edible films and coatings as food-quality preservers: An overview. *Foods, 10*(2), 249. https://doi.org/10.3390/foods10020249.

dos Santos, V. S., Lorevice, M. V., Baccarin, G. S., da Costa, F. M., da Silva Fernandes, R., Aouada, F. A., & de Moura, M. R. (2023). Combining chitosan nanoparticles and garlic essential oil as additive fillers to produce pectin-based nanocomposite edible films. *Polymers, 15*(10), Available from https://doi.org/10.3390/polym15102244, http://www.mdpi.com/journal/polymers.

Du, Y., Zhang, Y., & Pang, J. (2024). Electrospun konjac glucomannan-gelatin/polycaprolactone bilayer nanofibrous films with improved hydrophobicity and mechanical properties for food packaging. *Food Bioscience, 61*, 104544. https://doi.org/10.1016/j.fbio.2024.104544.

Durkan, C., Chu, D. P., Migliorato, P., & Welland, M. E. (2000). Investigations into local piezoelectric properties by atomic force microscopy. *Applied Physics Letters, 76*(3), 366–368. https://doi.org/10.1063/1.125756.

Dutta, A. (2017). Fourier transform infrared spectroscopy. *Spectroscopic Methods for Nanomaterials Characterization, 2*, 73–93. Available from https://doi.org/10.1016/B978-0-323-46140-5.00004-2, http://www.sciencedirect.com/book/9780323461405.

Frabetti, A. C. C., de Moraes, J. O., Jury, V., Boillereaux, L., & Laurindo, J. B. (2021). Adhesion of food on surfaces: Theory, measurements, and main trends to reduce it prior to industrial drying. *Food Engineering Reviews, 13*(4), 884–901. Available from https://doi.org/10.1007/s12393-021-09286-9, http://www.springer.com/life+sci/food+science/journal/12393.

Galus, S., Arik Kibar, E. A., Gniewosz, M., & Kraśniewska, K. (2020). Novel materials in the preparation of edible films and coatings—A review. *Coatings, 10*(7), 674. https://doi.org/10.3390/coatings10070674.

Garavand, F., Rouhi, M., Razavi, S. H., Cacciotti, I., & Mohammadi, R. (2017). Improving the integrity of natural biopolymer films used in food packaging by crosslinking approach: A review. *International Journal of Biological Macromolecules, 104*, 687–707. Available from https://doi.org/10.1016/j.ijbiomac.2017.06.093, www.elsevier.com/locate/ijbiomac.

García, M. A., Martino, M. N., & Zaritzky, N. E. (2000). Microstructural characterization of plasticized starch-based films. *Starch-Stärke, 52*(4), 118–124.

García, M. A., Pinotti, A., Martino, M. N., & Zaritzky, N. E. (2009). *Characterization of starch and composite edible films and coatings*. Springer Science and Business Media LLC, 169–209. http://doi.org/10.1007/978-0-387-92824-1_6.

Ghanbarzadeh, B., & Oromiehi, A. R. (2008). Biodegradable biocomposite films based on whey protein and zein: Barrier, mechanical properties and AFM analysis. *International Journal of Biological Macromolecules, 43*(2), 209–215. https://doi.org/10.1016/j.ijbiomac.2008.05.006.

Gharanjig, H., Gharanjig, K., Hosseinnezhad, M., & Jafari, S. M. (2020). Differential scanning calorimetry (DSC) of nanoencapsulated food ingredients. In *Characterization of Nanoencapsulated Food Ingredients*. Iran: Elsevier, 295–346. Available from https://www.elsevier.com/books/characterization-of-nanoencapsulated-food-ingredients/jafari/978-0-12-815667-4.

Giacinti Baschetti, M., & Minelli, M. (2020). Test methods for the characterization of gas and vapor permeability in polymers for food packaging application: A review. *Polymer Testing, 89*, 106606. https://doi.org/10.1016/j.polymertesting.2020.106606.

Gomaa, M., Al-Badaani, A. A., Hifney, A. F., & Adam, M. S. (2022). Utilization of cellulose and ulvan from the green seaweed Ulva lactuca in the development of composite edible films with natural antioxidant properties. *Journal of Applied Phycology, 34*(5), 2615–2626. Available from https://doi.org/10.1007/s10811-022-02786-z, https://www.springer.com/journal/10811.

Gómez, H., Madera-Santana, T. J., Quintana-Owen, P., Vázquez, Sierra, & Vargas y Vargas, L. de. (2024). *Propiedades fisicoquímicas y bioactivas de películas comestibles basadas en alginato y miel melipona*.

González-Estrada, R., Calderón-Santoyo, M., Carvajal-Millan, E., De Jesús Ascencio Valle, F., Ragazzo-Sánchez, J.A., Brown-Bojorquez, F., & Rascón-Chu, A. (2015). Covalently cross-linked arabinoxylans films for Debaryomyces hansenii entrapment. Molecules, 20(6), 11373–11386. Available fromAvailable from:http://www.mdpi.com/1420-3049/20/6/11373/pdf; https://doi.org/10.3390/molecules200611373.

Hammam, A. R. A. (2019). Technological, applications, and characteristics of edible films and coatings: A review. *SN Applied Sciences, 1*(6), Available from https://doi.org/10.1007/s42452-019-0660-8, springer.com/snas.

Han, J. H. (2013). *Edible films and coatings: A review innovations in food packaging* (Second Edition). United States: Elsevier Ltd, 213–255. Available from http://www.sciencedirect.com/science/book/9780123946010.

Holm, V. K., Ndoni, S., & Risbo, J. (2006). The stability of poly(lactic acid) packaging films as influenced by humidity and temperature. *Journal of Food Science, 71*(2). https://doi.org/10.1111/j.1365-2621.2006.tb08895.x.

Hu, J., Yun, X., Zheng, Y., Sun, T., Song, L., Pan, P., & Dong, T. (2024). Development of ultra-thin poly(L-lactic acid)-based films integrating toughness, barrier properties, and gas selectivity: Towards gas-permeation controllable green food packaging. *Food Chemistry, 449*, 139218. https://doi.org/10.1016/j.foodchem.2024.139218.

Huang, X., Li, J., He, J., Luo, J., Cai, J., Wei, J., Li, P., & Zhong, H. (2024). Preparation of curcumin-loaded chitosan/polyvinyl alcohol intelligent active films for food packaging and freshness monitoring. *International Journal of Biological Macromolecules, 276*(Pt 1), 133807. https://doi.org/10.1016/j.ijbiomac.2024.133807.

Jafarizadeh, M. H., Osman, A., Tan, C. P., & Abdul, R. (2011). Development of an edible coating based on chitosan-glycerol to delay'Berangan'banana (Musa sapientum cv. Berangan) ripening process. *International Food Research Journal, 18*(3).

Jafarzadeh, S., Alias, A. K., Ariffin, F., & Mahmud, S. (2018). Physico-mechanical and microstructural properties of semolina flour films as influenced by different sorbitol/glycerol concentrations. *International Journal of Food Properties, 21*(1), 983–995. Available from https://doi.org/10.1080/10942912.2018.1474056, www.tandf.co.uk/journals/titles/10942912.asp.

Jeevahan, J. J., Chandrasekaran, M., Venkatesan, S. P., Sriram, V., Joseph, G. B., Mageshwaran, G., & Durairaj, R. B. (2020). Scaling up difficulties and commercial aspects of edible films for food packaging: A review. *Trends in Food Science & Technology, 100*, 210–222.

Jiang, G., Yang, Y., Sheng, W. Y., Yang, L., Yang, H., Tang, T., Wang, C., & Tian, Y. (2024). Preparation and characterization of κ-carrageenan/dextran films blended with nano-ZnO and anthocyanin for intelligent food packaging. *International Journal of Biological Macromolecules, 282*(Pt 4), 137203. https://doi.org/10.1016/j.ijbiomac.2024.137203.

Kadzińska, J., Bryś, J., Ostrowska-Ligęza, E., Estéve, M., & Janowicz, M. (2020). Influence of vegetable oils addition on the selected physical properties of apple–sodium alginate edible films. *Polymer Bulletin, 77*(2), 883–900. https://doi.org/10.1007/s00289-019-02777-0.

Kavoosi, G., Rahmatollahi, A., Mohammad Mahdi Dadfar, S., & Mohammadi Purfard, A. (2014). Effects of essential oil on the water binding capacity, physico-mechanical properties, antioxidant and antibacterial activity of gelatin films. *LWT - Food Science and Technology, 57*(2), 556–561. https://doi.org/10.1016/j.lwt.2014.02.008.

Kumar, L., Ramakanth, D., Akhila, K., & Gaikwad, K. K. (2022). Edible films and coatings for food packaging applications: A review. *Environmental Chemistry Letters, 20*(1), 875–900. Available from https://doi.org/10.1007/s10311-021-01339-z, https://www.springer.com/journal/10311.

Kumari, M., Mahajan, H., Joshi, R., & Gupta, M. (2017). Development and structural characterization of edible films for improving fruit quality. *Food Packaging and Shelf Life, 12*, 42–50. https://doi.org/10.1016/j.fpsl.2017.02.003.

Kurabetta, L. K., Masti, S. P., Gunaki, M. N., Hunashyal, A. A., Chougale, R. B., Dalbanjan, N. P., & Praveen Kumar, S. K. (2024). Exploration of physicochemical and biological properties of phenylalanine incorporated carboxymethyl cellulose/poly(vinyl alcohol) based bioactive films for food packaging applications. *Food Bioscience, 61*. Available from https://doi.org/10.1016/j.fbio.2024.104869, https://www.sciencedirect.com/science/journal/22124292.

Lade, B. D., Dhowlaghar, N., Pillai, S. S., & Patil, B. S. (2023). Physicochemical, mechanical, and antimicrobial properties of sodium alginate films as carriers of zein emulsion with pelargonic acid and eugenol for active food packing. *Food Packaging and Shelf Life, 40*. Available from https://doi.org/10.1016/j.fpsl.2023.101202, http://www.journals.elsevier.com/food-packaging-and-shelf-life/.

Lei, Y., Wu, H., Jiao, C., Jiang, Y., Liu, R., Xiao, D., Lu, J., Zhang, Z., Shen, G., & Li, S. (2019). Investigation of the structural and physical properties, antioxidant and antimicrobial activity of pectin-konjac glucomannan composite edible films incorporated with tea polyphenol. *Food Hydrocolloids, 94*, 128–135. https://doi.org/10.1016/j.foodhyd.2019.03.011.

Leyva-Porras, C., Cruz-Alcantar, P., Espinosa-Solís, V., Martínez-Guerra, E., Piñón-Balderrama, C. I., Martínez, I. C., & Saavedra-Leos, M. Z. (2020). Application of differential scanning calorimetry (DSC) and modulated differential scanning calorimetry (MDSC) in food and drug industries. *Polymers, 12*(1), Available from https://doi.org/10.3390/polym12010005, https://res.mdpi.com/d_attachment/polymers/polymers-12-00005/article_deploy/polymers-12-00005-v2.pdf.

Li, H., Liu, C., Sun, J., & Lv, S. (2022). Bioactive edible sodium alginate films incorporated with tannic acid as antimicrobial and antioxidative food packaging. *Foods, 11*(19), 3044. https://doi.org/10.3390/foods11193044.

Li, T., Guan, L., Zeng, M., Li, T., Lei, S., Tang, X., Pan, Y., Li, S., Zhou, M., Yuan, X., Zhang, Z., & Wu, H. (2025). Fabrication of corn starch-based composite films functionalized by Rosa roxburghii Tratt fruit pomace via extrusion compression molding for active food packaging. *International Journal of Biological Macromolecules, 284*(Pt 1), 138091. https://doi.org/10.1016/j.ijbiomac.2024.138091.

Liu, X., Wang, L., Qiao, Y., Sun, X., Ma, S., Cheng, X., Qi, W., Huang, W., & Li, Y. (2018). Adhesion of liquid food to packaging surfaces: Mechanisms, test methods, influencing factors and anti-adhesion methods. *Journal of Food Engineering, 228*, 102–117. https://doi.org/10.1016/j.jfoodeng.2018.02.002.

Liu, Z., Du, M., Liu, H., Zhang, K., Xu, X., Liu, K., Tu, J., & Liu, Q. (2021). Chitosan films incorporating litchi peel extract and titanium dioxide nanoparticles and their application as coatings on watercored apples. *Progress in Organic Coatings, 151*, 106103. https://doi.org/10.1016/j.porgcoat.2020.106103.

Liu, Z., Zhang, M., Hao, Y., Hu, W., Zhu, W., Wang, H., & Li, L. (2024). Application of surface-modified functional packaging in food storage: A comprehensive review. *Comprehensive Reviews in Food Science and Food Safety, 23*(3), Available from https://doi.org/10.1111/1541-4337.13343, http://onlinelibrary.wiley.com/journal/10.1111/(ISSN)1541-4337.

Ma, K., Yang, J., Goksen, G., Alamri, A. S., Alhomrani, M., Xia, G., & Zhang, W. (2025). Application of tragacanth gum as a potential food packaging film and its performance enhancement strategy. *Food Hydrocolloids, 162*, 110894. https://doi.org/10.1016/j.foodhyd.2024.110894.

Maghsoud, Z., Rafiei, M., & Famili, M. H. N. (2018). Effect of processing method on migration of antioxidant from HDPE packaging into a fatty food simulant in terms of crystallinity. *Packaging Technology and Science, 31*(3), 141–149. https://doi.org/10.1002/pts.2359.

Marano, S., Laudadio, E., Minnelli, C., & Stipa, P. (2022). Tailoring the barrier properties of PLA: A state-of-the-art review for food packaging applications. *Polymers, 14*(8), 1626. https://doi.org/10.3390/polym14081626.

Martins, V. F. R., Pintado, M. E., Morais, R. M. S. C., & Morais, A. M. M. B. (2024). Recent highlights in sustainable bio-based edible films and coatings for fruit and vegetable applications. *Foods, 13*(2), Available from https://doi.org/10.3390/foods13020318, www.mdpi.com/journal/foods.

Meerasri, J., & Sothornvit, R. (2020). Characterization of bioactive film from pectin incorporated with gamma-aminobutyric acid. *International Journal of Biological Macromolecules, 147*, 1285–1293. https://doi.org/10.1016/j.ijbiomac.2019.10.094.

Mei, Y., & Zhao, Y. (2003). Barrier and mechanical properties of milk protein-based edible films containing nutraceuticals. *Journal of Agricultural and Food Chemistry, 51*(7), 1914–1918. https://doi.org/10.1021/jf025944h.

Mishra, B., Panda, J., Mishra, A. K., Nath, P. C., Nayak, P. K., Mahapatra, U., Sharma, M., Chopra, H., Mohanta, Y. K., & Sridhar, K. (2024). Recent advances in sustainable biopolymer-based nanocomposites for smart food packaging: A review. *International Journal of Biological Macromolecules, 279*. Available from https://doi.org/10.1016/j.ijbiomac.2024.135583, https://www.sciencedirect.com/science/journal/01418130.

Moeini, A., Pedram, P., Fattahi, E., Cerruti, P., & Santagata, G. (2022). Edible polymers and secondary bioactive compounds for food packaging applications: Antimicrobial, mechanical, and gas barrier properties. *Polymers, 14*(12), 2395. https://doi.org/10.3390/polym14122395.

MohammadKarimi, S., Ershad-Langroudi, A., & Alizadegan, F. (2024). Surface analysis and water contact angle of modified natural biopolymers. In *Handbook of natural polymers*, *Volume 2*, Elsevier, 473–500. https://doi.org/10.1016/B978-0-323-99856-7.00010-0.

Moustafa, H., Hemida, M. H., Shemis, M. A., Dufresne, A., & Morsy, M. (2023). Functionalized GO nanoplatelets with folic acid as a novel material for boosting humidity sensing of chitosan/PVA nanocomposites for active food packaging. *Surfaces and Interfaces, 41*. Available from https://doi.org/10.1016/j.surfin.2023.103229, http://www.journals.elsevier.com/surfaces-and-interfaces.

Mujtaba, M., Lipponen, J., Ojanen, M., Puttonen, S., & Vaittinen, H. (2022). Trends and challenges in the development of bio-based barrier coating materials for paper/cardboard food packaging; A review. *Science of the Total Environment, 851*(Pt 2), 158328. https://doi.org/10.1016/j.scitotenv.2022.158328.

Nesic, A., Meseldzija, S., Cabrera-Barjas, G., & Onjia, A. (2022). Novel biocomposite films based on high methoxyl pectin reinforced with zeolite Y for food packaging applications. *Foods, 11*(3), 360. https://doi.org/10.3390/foods11030360.

Ngasotter, S., Xavier, K. A. M., Sagarnaik, C., Sasikala, R., Mohan, C. O., Jaganath, B., & Ninan, G. (2025). Evaluating the reinforcing potential of steam-exploded chitin nanocrystals in chitosan-based biodegradable nanocomposite films for food packaging applications. *Carbohydrate Polymers, 348*. Available from https://doi.org/10.1016/j.carbpol.2024.122841, https://www.sciencedirect.com/science/journal/01448617.

Nilsen-Nygaard, J., Fernández, E. N., Radusin, T., Rotabakk, B. T., Sarfraz, J., Sharmin, N., Sivertsvik, M., Sone, I., & Pettersen, M. K. (2021). Current status of biobased and biodegradable food packaging materials: Impact on food quality and effect of innovative processing technologies. *Comprehensive Reviews in Food Science and Food Safety, 20*(2), 1333–1380. https://doi.org/10.1111/1541-4337.12715.

Nizamuddin, S., Baloch, A. J., Chen, C., Arif, M., & Mubarak, N. M. (2024). Bio-based plastics, biodegradable plastics, and compostable plastics: Biodegradation mechanism, biodegradability standards and environmental stratagem. *International Biodeterioration and Biodegradation, 195*. Available from https://doi.org/10.1016/j.ibiod.2024.105887, https://www.sciencedirect.com/science/journal/09648305.

Nogueira, G. F., Meneghetti, B. B., Soares, I. H. B. T., Soares, C. T., Bevilaqua, G., Fakhouri, F. M., & de Oliveira, R. A. (2024). Multipurpose arrowroot starch films with anthocyanin-rich grape pomace extract: Color migration for food simulants and monitoring the freshness of fish meat. *International Journal of Biological Macromolecules, 265*. Available from https://doi.org/10.1016/j.ijbiomac.2024.130934, https://www.sciencedirect.com/science/journal/01418130.

Ojogbo, E., Ogunsona, E. O., & Mekonnen, T. H. (2020). Chemical and physical modifications of starch for renewable polymeric materials. *Materials Today Sustainability, 7–8*, 100028. https://doi.org/10.1016/j.mtsust.2019.100028.

Paidari, S., Zamindar, N., Tahergorabi, R., Kargar, M., Ezzati, S., shirani, N., & Musavi, S. H. (2021). Edible coating and films as promising packaging: A mini review. *Journal of Food Measurement and Characterization, 15*(5), 4205–4214. Available from https://doi.org/10.1007/s11694-021-00979-7, http://rd.springer.com/journal/11694.

Pang, Y., Zhang, N., Liu, Y., & Zhang, J. (2024). High-performance biodegradable poly(vinyl alcohol)/starch composites via alkaline-regulated crystalline engineering. *Industrial Crops and Products, 220*, 119143. https://doi.org/10.1016/j.indcrop.2024.119143.

Parhi, A., Tang, J., & Sablani, S. S. (2020). Functionality of ultra-high barrier metal oxide-coated polymer films for in-package, thermally sterilized food products. *Food Packaging and Shelf Life, 25*, 100514. https://doi.org/10.1016/j.fpsl.2020.100514.

Pavinatto, A., de Almeida Mattos, A. V., Malpass, A. C. G., Okura, M. H., Balogh, D. T., & Sanfelice, R. C. (2020). Coating with chitosan-based edible films for mechanical/biological protection of strawberries. *International Journal of Biological Macromolecules, 151*, 1004–1011. Available from https://doi.org/10.1016/j.ijbiomac.2019.11.076, www.elsevier.com/locate/ijbiomac.

Pavlath, A. E., & Orts, W. (2009). *Edible films and coatings: Why, what, and how?*. Springer Science and Business Media LLC, 1–23. http://doi.org/10.1007/978-0-387-92824-1_1.

Pelissari, F. M., Grossmann, M. V. E., Yamashita, F., & Pined, E. A. G. (2009). Antimicrobial, mechanical, and barrier properties of cassava starch-chitosan films incorporated with oregano essential oil. *Journal of Agricultural and Food Chemistry, 57*(16), 7499–7504. Available from https://doi.org/10.1021/jf9002363Brazil, http://pubs.acs.org/doi/pdfplus/10.1021/jf9002363.

Perumal, T., Krebs de Souza, C., Nihues, T. C., Jain, P., Gaikwad, K. K., & Roy, S. (2024). A review on biopolymer-based oil and water-resistant functional paper coating for food packaging. *Food Bioscience*105656. https://doi.org/10.1016/j.fbio.2024.105656.

Pinzon, M. I., Sanchez, L. T., Garcia, O. R., Gutierrez, R., Luna, J. C., & Villa, C. C. (2020). Increasing shelf life of strawberries *(Fragaria* ssp) by using a banana starch-chitosan-Aloe vera gel composite edible coating. *International Journal of Food Science and Technology, 55*(1), 92–98. Available from https://doi.org/10.1111/ijfs.14254, http://onlinelibrary.wiley.com/journal/10.1111/(ISSN)1365-2621.

Pires, A. F., Díaz, O., Cobos, A., & Pereira, C. D. (2024). A review of recent developments in edible films and coatings-focus on whey-based materials. *Foods, 13*(16), Available from https://doi.org/10.3390/foods13162638, www.mdpi.com/journal/foods.

Plaeyao, K., Talodthaisong, C., Yingyuen, W., Kaewbundit, R., Tun, W. S. T., Saenchoopa, A., Kayunkid, N., Wiwattananukul, R., Sakulsombat, M., & Kulchat, S. (2025). Biodegradable antibacterial food packaging based on carboxymethyl cellulose from sugarcane bagasse/cassava starch/chitosan/gingerol extract stabilized silver nanoparticles (Gin-AgNPs) and vanillin as cross-linking agent. *Food Chemistry, 466*. Available from https://doi.org/10.1016/j.foodchem.2024.142102, https://www.sciencedirect.com/science/journal/03088146.

Porter, S. C., & Felton, L. A. (2010). Techniques to assess film coatings and evaluate film-coated products. *Drug Development and Industrial Pharmacy, 36*(2), 128–142. https://doi.org/10.3109/03639040903433757.

Rezaei, A., Sadeghi, E., Assadpour, E., Moradiyan, M.hadi, Khaledian, S., Rezaei, N., Dehnad, D., Zhang, F., Azizi-lalabadi, M., & Jafari, S. M. (2024). Preparation and characterization of gelatin/Kappa-carrageenan halochromic films containing curcumin for active and smart food packaging. *Food Bioscience, 62*, 105304. https://doi.org/10.1016/j.fbio.2024.105304.

Ribeiro, A. M., Estevinho, B. N., & Rocha, F. (2021). Preparation and incorporation of functional ingredients in edible films and coatings. *Food and Bioprocess Technology, 14*, 209–231.

Rivera, P., Torres, A., Romero, J., Rodríguez, F., Arrieta, M. P., Olea, F., Silva, T., Maldonado, P., Quijada-Maldonado, E., & Tapia, A. (2024). Experimental and theoretical characterization of the release kinetic of carvacrol as inclusion complexes with β-cyclodextrin in poly(lactic acid) and Mater-Bi® processed by supercritical impregnation. *International Journal of Biological Macromolecules, 278*(Pt 1), 133946. https://doi.org/10.1016/j.ijbiomac.2024.133946.

Roy, S., Malik, B., Chawla, R., Bora, S., Ghosh, T., Santhosh, R., Thakur, R., & Sarkar, P. (2024). Biocompatible film based on protein/polysaccharides combination for food packaging applications: A comprehensive review. *International Journal of Biological Macromolecules, 278*(Pt 1), 134658. https://doi.org/10.1016/j.ijbiomac.2024.134658.

Roy, S., Ramakrishnan, R., Goksen, G., Singh, S., & Łopusiewicz, Ł. (2024). Recent progress on UV-light barrier food packaging films – a systematic review. *Innovative Food Science & Emerging Technologies, 91*, 103550. https://doi.org/10.1016/j.ifset.2023.103550.

Rozo, D. F., Alvarado, J. F., Chaparro, L. M., Medina, J. A., & Salcedo, F. (2024). Modeling oxidation kinetics of linseed oil in oxygen scavenger nanocapsules to be potentially used in active food packaging. *Food Packaging and Shelf Life, 42*. Available from https://doi.org/10.1016/j.fpsl.2024.101256, https://www.sciencedirect.com/science/journal/22142894.

Saba, N., & Jawaid, M. (2018). A review on thermomechanical properties of polymers and fibers reinforced polymer composites. *Journal of Industrial and Engineering Chemistry, 67*, 1–11. https://doi.org/10.1016/j.jiec.2018.06.018.

Salević-Jelić, A., Lević, S., Prieto, C., Jeremić, S., Stevanović, S., Rac, V., Vukašinović, I., Nedović, V., & Lagaron, J. M. (2024). Polycaprolactone-based electrospun films incorporating sage extract: From active food packaging application to accelerated biodegradation by *Pseudomonas*. *Future Foods, 10*, 100465. https://doi.org/10.1016/j.fufo.2024.100465.

Samadi, N., Hasanzadeh, R., & Rasad, M. (2015). Adsorption isotherms, kinetic, and desorption studies on removal of toxic metal ions from aqueous solutions by polymeric adsorbent. *Journal of Applied Polymer Science, 132*(11). https://doi.org/10.1002/app.41642.

Sanchez-Silva, J. M., Ocampo-Pérez, R., Padilla-Ortega, E., Sangaré, D., Escobedo-Bretado, M. A., Domínguez-Arvizu, J. L., Hernández-Majalca, B. C., Salinas-Gutiérrez, J. M., López-Ortiz, A., & Collins-Martínez, V. (2023). Pyrolysis kinetics of byrsonima crassifolia stone as agro-industrial waste through isoconversional models. *Molecules, 28*(2), Available from https://doi.org/10.3390/molecules28020544, http://www.mdpi.com/journal/molecules.

Sengupta, T., & Han, J. H. (2014). *Surface chemistry of food, packaging, and biopolymer materials*. Elsevier BV, 51–86. http://doi.org/10.1016/b978-0-12-394601-0.00004-7.

Šešlija, S., Nešić, A., Ružić, J., Kalagasidis Krušić, M., Veličković, S., Avolio, R., Santagata, G., & Malinconico, M. (2018). Edible blend films of pectin and poly(ethylene glycol): Preparation and physico-chemical evaluation. *Food Hydrocolloids, 77*, 494–501. https://doi.org/10.1016/j.foodhyd.2017.10.027.

Shah, Y. A., Bhatia, S., Al-Harrasi, A., Oz, F., Khan, M. H., Roy, S., Esatbeyoglu, T., & Pratap-Singh, A. (2024). Thermal properties of biopolymer films: Insights for sustainable food packaging applications. *Food Engineering Reviews, 16*(4), 497–512. Available from https://doi.org/10.1007/s12393-024-09380-8, https://www.springer.com/journal/12393.

Silue, Y., & Fawole, O. A. (2024). Global research network analysis of edible coatings and films for preserving perishable fruit crops: Current status and future directions. *Foods, 13*(15), 2321. https://doi.org/10.3390/foods13152321.

Silva-Weiss, A., Ihl, M., Sobral, P. J. A., Gómez-Guillén, M. C., & Bifani, V. (2013). Natural additives in bioactive edible films and coatings: Functionality and applications in foods. *Food Engineering Reviews, 5*(4), 200–216. https://doi.org/10.1007/s12393-013-9072-5.

Sistla, Y. S., & Mehraj, S. (2022). Molecular simulations to understand the moisture, carbon dioxide, and oxygen barrier properties of pectin films. *Journal of Molecular Modeling, 28*(4), Available from https://doi.org/10.1007/s00894-022-05069-z, https://rd.springer.com/journal/894.

Souza, V. G. L., Rodrigues, C., Pires, J. R. A., Fernando, A. L., Alves, V., & Coelhoso, I. (2024). *Permeation of oxygen and carbon dioxide through food packaging materials*. Springer Science and Business Media LLC, 219–232. http://doi.org/10.1007/978-1-0716-3613-8_12.

Suhag, R., Kumar, N., Petkoska, A. T., & Upadhyay, A. (2020). Film formation and deposition methods of edible coating on food products: A review. *Food Research International, 136*. Available from https://doi.org/10.1016/j.foodres.2020.109582, www.elsevier.com/inca/publications/store/4/2/2/9/7/0.

Tabari, M. (2018). Characterization of a new biodegradable edible film based on Sago Starch loaded with carboxymethyl cellulose nanoparticles. *Nanomedicine Research Journal, 3*(1), 25–30. Available from https://doi.org/10.22034/NMRJ.2018.01.004, http://www.nanomedicine-rj.com/article_31897_6300444ffaa95eebcf7c52566a285baa.pdf.

Tan, W., McClements, D. J., Chen, J., & Ma, D. (2024). Novel biopolymer-based active packaging material for food applications: Cinnamaldehyde-loaded calcium nanoparticles incorporated into alginate-carboxymethyl cellulose films. *Food Packaging and Shelf Life, 45*, 101351. https://doi.org/10.1016/j.fpsl.2024.101351.

Teixeira-Costa, B. E., Ferreira, W. H., Goycoolea, F. M., Murray, B. S., & Andrade, C. T. (2023). Improved antioxidant and mechanical properties of food packaging films based on chitosan/deep eutectic solvent, containing açaí-filled microcapsules. *Molecules, 28*(3), Available from https://doi.org/10.3390/molecules28031507, http://www.mdpi.com/journal/molecules.

Thakur, R., Santhosh, R., Kumar, Y., Suryavanshi, V. R., Singhi, H., Madhubabu, D., Wickramarachchi, S., Pal, K., & Sarkar, P. (2023). Characteristics and application of animal byproduct-based films and coatings in the packaging of food products. *Trends in Food Science & Technology, 140*, 104143. https://doi.org/10.1016/j.tifs.2023.104143.

Tharanathan, R. N. (2003). Biodegradable films and composite coatings: Past, present and future. *Trends in Food Science & Technology, 14*(3), 71–78. https://doi.org/10.1016/s0924-2244(02)00280-7.

Thomas, J., Selvin Thomas, P., & Stephen, R. (2024). Improved mechanical, thermal, and barrier properties of halloysite nanotubes and nanocellulose incorporated PVA-PEO films: For food packaging applications. *Food Packaging and Shelf Life, 46*, 101373. https://doi.org/10.1016/j.fpsl.2024.101373.

Trinh, B. M., Chang, B. P., & Mekonnen, T. H. (2023). The barrier properties of sustainable multiphase and multicomponent packaging materials: A review. *Progress in Materials Science, 133*, 101071. https://doi.org/10.1016/j.pmatsci.2023.101071.

Turan, D., Sängerlaub, S., Stramm, C., & Gunes, G. (2017). Gas permeabilities of polyurethane films for fresh produce packaging: Response of O_2 permeability to temperature and relative humidity. *Polymer Testing, 59*, 237–244. https://doi.org/10.1016/j.polymertesting.2017.02.007.

Tyagi, P., Salem, K. S., Hubbe, M. A., & Pal, L. (2021). Advances in barrier coatings and film technologies for achieving sustainable packaging of food products – A review. *Trends in Food Science and Technology, 115*, 461–485. Available from https://doi.org/10.1016/j.tifs.2021.06.036, http://www.elsevier.com/wps/find/journaldescription.cws_home/601278/description#description.

Ullah, A., Yang, H., Takemae, K., Wang, F., Lee, S., & Kim, I. S. (2024). Sustainable bioactive food packaging based on electrospun zein-polycaprolactone nanofibers integrated with aster yomena extract loaded halloysite nanotubes. *International Journal of Biological Macromolecules, 267*(Pt 2), 131375. https://doi.org/10.1016/j.ijbiomac.2024.131375.

Uzun, İ. (2023). Methods of determining the degree of crystallinity of polymers with X-ray diffraction: A review. *Journal of Polymer Research, 30*(10), Available from https://doi.org/10.1007/s10965-023-03744-0, https://www.springer.com/journal/10965.

Vieira, D. M., Andrade, M. A., Vilarinho, F., Silva, A. S., Rodrigues, P. V., Castro, M. C. R., & Machado, A. V. (2022). Mono and multilayer active films containing green tea to extend food shelf life. *Food Packaging and Shelf Life, 33*. Available from https://doi.org/10.1016/j.fpsl.2022.100918, http://www.journals.elsevier.com/food-packaging-and-shelf-life/.

Villagómez-Zavala, D. L., Gómez-Corona, C., San Martín Martínez, E., Pérez-Orozco, J. P., Vernon-Carter, E. J., & Pedroza-Islas, R. (2008). Comparative study of the mechanical properties of edible filmsmade from single and blended hydrophilic biopolymer matrices. *Revista Mexicana de Ingeniera Quimica, 7*(3), 263–273. Available from http://www.rmiq.org/Pdfs/Vol%207%20No%203/10_RMIQ_Vol7No3_100708.pdf.

Wang, S., Tang, K., Zhang, Z., Liu, H., Yao, Y., & Liao, X. (2024). PBAT/lignin-ZnO composite film for food packaging: Photo-stability, better barrier and antibacterial properties. *International Journal of Biological Macromolecules, 275*(Pt 2), 133651. https://doi.org/10.1016/j.ijbiomac.2024.133651.

Wang, K., Wang, W., Ye, R., Xiao, J., Liu, Y., Ding, J., Zhang, S., & Liu, A. (2017). Mechanical and barrier properties of maize starch–gelatin composite films: Effects of amylose content. *Journal of the Science of Food and Agriculture, 97*(11), 3613–3622. https://doi.org/10.1002/jsfa.8220.

Xie, D., Huang, G., Yang, M., Yi, S., Zheng, X., Zhu, M., & Zhai, H. (2024). Bionic modified atmosphere film Inspired by plant stomata with curcumin as controllable gas "Switches" for fruit preservation. *Chemical Engineering Journal, 500*, 156741. https://doi.org/10.1016/j.cej.2024.156741.

Xue, M., Huang, R., Wen, Z., Li, W., Chai, X., & Huang, L. (2024). Adsorption and barrier properties of bio-based coating on paper against MOSH for improving food packaging safety. *Progress in Organic Coatings, 194*, 108590. https://doi.org/10.1016/j.porgcoat.2024.108590.

Xue, W., Zhu, J., Sun, P., Yang, F., Wu, H., & Li, W. (2023). Permeability of biodegradable film comprising biopolymers derived from marine origin for food packaging application: A review. *Trends in Food Science & Technology, 136*, 295–307. https://doi.org/10.1016/j.tifs.2023.05.001.

Yang, J., Punia Bangar, S., Rizwan Khan, M., Hammouda, G. A., Alam, P., & Zhang, W. (2024). Biopolymer-based packaging films/edible coatings functionalized with ε-polylysine: New options for food preservation. *Food Research International, 187*, 114390. https://doi.org/10.1016/j.foodres.2024.114390.

Yang, S., Shang, P., Zhang, K., Wang, J., Zhang, B., Gao, X., Waterhouse, G. I. N., Xie, J., Zhang, L., & Xu, J. (2024). PBAT/PLA food packaging film containing sodium dehydroacetate-loaded diatomite as an antibacterial agent: Fabrication, water-gas regulation and long-acting antimicrobial mechanism. *Food Chemistry, 446*. Available from https://doi.org/10.1016/j.foodchem.2024.138880, https://www.sciencedirect.com/science/journal/03088146.

Zhang, P., Zhao, Y., & Shi, Q. (2016). Characterization of a novel edible film based on gum ghatti: Effect of plasticizer type and concentration. *Carbohydrate Polymers, 153*, 345–355.

Zhang, T., Wei, W., Wang, B., Sun, D., Sheng, S., Xu, H., & Wu, M. (2024). A new two-step combined modification method for preparing amylopectin biofilms with high transparency and ideal properties as foods packaging materials. *Food Packaging and Shelf Life, 42*, 101248. https://doi.org/10.1016/j.fpsl.2024.101248.

Zhang, W., Li, X., & Jiang, W. (2020). Development of antioxidant chitosan film with banana peels extract and its application as coating in maintaining the storage quality of apple. *International Journal of Biological Macromolecules, 154*, 1205–1214.

Zhang, W., Roy, S., Ezati, P., Yang, D. P., & Rhim, J. W. (2023). Tannic acid: A green crosslinker for biopolymer-based food packaging films. *Trends in Food Science and Technology, 136*, 11–23. Available from https://doi.org/10.1016/j.tifs.2023.04.004, http://www.elsevier.com/wps/find/journaldescription.cws_home/601278/description#description.

Zhang, W., Zhang, Y., Cao, J., & Jiang, W. (2021). Improving the performance of edible food packaging films by using nanocellulose as an additive. *International Journal of Biological Macromolecules, 166*, 288–296. https://doi.org/10.1016/j.ijbiomac.2020.10.185.

Zhang, Y.-L., Cui, Q.-L., Wang, Y., Liu, J.-L., & Zhang, Y.-Q. (2023). Mechanical and barrier properties optimization of carboxymethyl chitosan-gelatin-based edible film using response surface methodology. *Coatings, 13*(9), 1529.

Zhao, Y., Li, H., Chen, J., & Wang, Y. (2023). A novel high water-soluble antibacterial films-based guar gum incorporated with Aloe vera gel and ε-polylysine. *Food Chemistry, 427*, 136686. https://doi.org/10.1016/j.foodchem.2023.136686.

Zhou, L., Hao, M., Min, T., Bian, X., Du, H., Sun, X., Zhu, Z., & Wen, Y. (2023). Kaolin incorporated with thyme essential oil for humidity-controlled antimicrobial food packaging. *Food Packaging and Shelf Life, 38*. Available from https://doi.org/10.1016/j.fpsl.2023.101106, http://www.journals.elsevier.com/food-packaging-and-shelf-life/.

Chapter 23

UV barrier properties

Reza Abedi-Firoozjah[1], Arezou Khezerlou[2], Farinaz Esmi[3], and Mahmood Alizadeh Sani[4]
[1]*Student Research Committee, Kermanshah University of Medical Sciences, Kermanshah, Iran,* [2]*Department of Food Safety and Hygiene, School of Health, Gonabad University of Medical Sciences, Gonabad, Iran,* [3]*Student Research Committee, Department of Food Science and Technology, Tabriz University of Medical Sciences, Tabriz, Iran,* [4]*Department of Food Science and Technology, School of Nutritional Sciences and Dietetics, Tehran University of Medical Sciences, Tehran, Iran*

23.1 Introduction

Global food production struggles to meet the nutritional demands of the world's population, with approximately 815 million people experiencing hunger. This crisis is exacerbated by the annual loss of about 1.3 billion tons of food (Roy et al., 2024). A major contributor to food spoilage is the interaction of food products with environmental factors, including temperature, humidity, light exposure, and oxygen reactivity, all of which significantly shorten shelf life (Siemińska-Kuczer et al., 2022). Throughout the supply chain—from harvesting, storage, and transportation to processing, retail display in grocery stores and restaurants, and final consumption—food is constantly exposed to natural and artificial light (Fig. 23.1A) (Vijayakumar et al., 2022).

Sunlight reaching Earth consists of infrared, visible, and ultraviolet (UV) rays. Over recent decades, anthropogenic activities have caused massive depletion of the ozone layer, resulting in a sharp increase in UV radiation penetrating the atmosphere (Siemińska-Kuczer et al., 2022; Zhao et al., 2018). While UV radiation has some positive effects on human health, it is predominantly harmful, contributing to skin aging, cancer, erythema, DNA damage, and cataracts (Hamouda et al., 2022; Robinson et al., 2023). In food systems, consistent exposure to UV triggers photolysis and photooxidation reactions that generate active oxygen and free radicals. These lead to unpleasant odors and flavors, reduced nutritional value, discoloration, and accelerated degradation (Pandiselvam et al., 2022; Tripathi et al., 2024). Foods high in lipids or proteins, such as bakery items, meat, fish, nuts, dairy, and wines, are particularly susceptible (He et al., 2021).

Traditional packaging materials like metal, brown glass, plastic, paper, cardboard, and aluminum foil offer some protection against UV light (Chandran et al., 2024; Nayak & Mishra, 2025). However, commercially available UV blockers, including coatings, filters, and metal foils, often suffer from poor recyclability and leaching concerns (Ezati et al., 2023). The remarkable growth in plastic use has revolutionized food packaging due to its flexibility, lightweight nature, and ease of production. Yet, nondegradable petroleum-based plastics (e.g., polyethylene terephthalate, polyethylene, polypropylene, and polyvinyl chloride) persist in the environment for centuries, contributing to pollution, wildlife endangerment, and ecosystem disruption (Aboelela & Bassyouni, 2024). Inadequate waste management and low recycling rates have intensified public and environmental concerns, underscoring the urgent need for sustainable alternatives (Gore & Prajapat, 2022).

Biopolymers, derived from natural sources such as plants, animals, and microbes, or chemically synthesized from biological monomers, present a promising solution (Shoba et al., 2025). Due to their nontoxicity, low cost, and environmental compatibility, biopolymers are gaining traction for industrial applications (Sharma et al., 2025). Recent advances have focused on incorporating organic, inorganic, and synthetic UV-absorbing fillers into polymer matrices to create effective UV-protective films (Danmatam et al., 2023; Li et al., 2025; Silva et al., 2023).

By effectively controlling UV exposure through advanced natural polymer-based films and coatings, food shelf life can be significantly extended, reducing waste and supporting global food security. This chapter provides a comprehensive overview of UV-barrier properties in natural polymer-based systems for food packaging. It covers ultraviolet radiation and its evaluation, classification of UV-blocking materials (inorganic nanoparticles, organic compounds, biopolymers,

Natural Polymer-Based Films and Coatings in Food Technology. DOI: https://doi.org/10.1016/B978-0-443-34244-8.00018-5

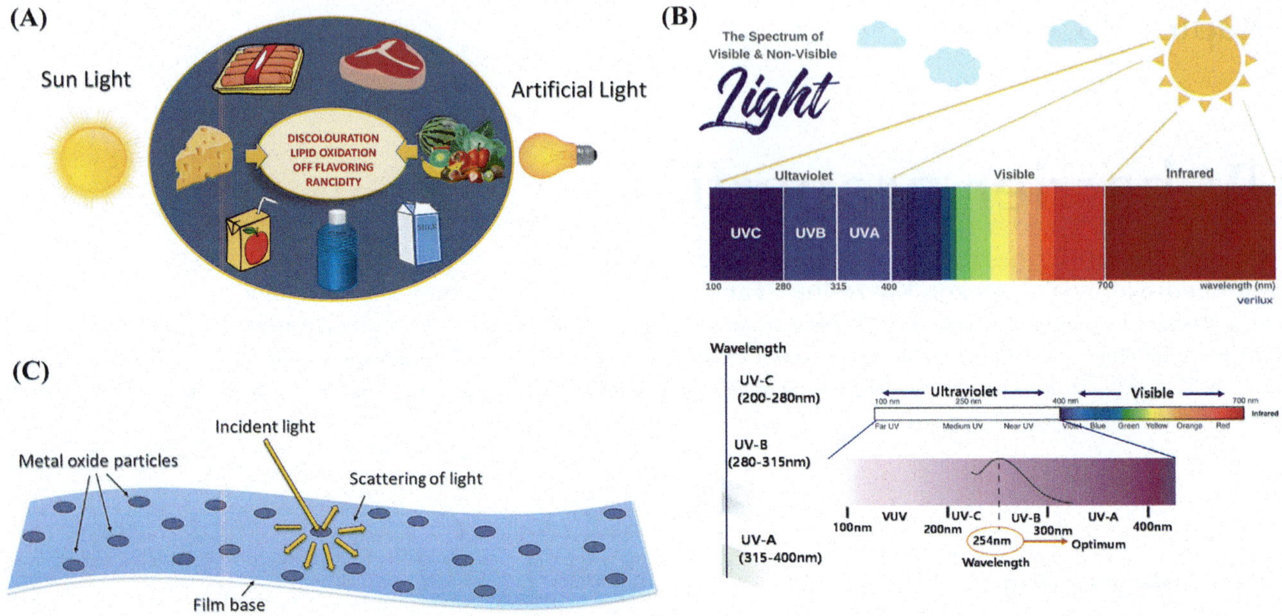

FIGURE 23.1 Effect of natural and artificial light on various food products. (A) Effect of natural and artificial light on various food products; (B) The position, wavelength, and frequency of UV radiation within the electromagnetic spectrum; (C) schematic image of reflection, refraction, and scattering of light on metal oxide particles.
Adapted from Roy, Swarup, Ramakrishnan, Rejish, Goksen, Gulden, Singh, Sunita & Łopusiewicz, Łukasz. (2024). Recent progress on UV-light barrier food packaging films – a systematic review. Innovative Food Science & Emerging Technologies, 91. https://doi.org/10.1016/j.ifset.2023.103550.

phytochemicals, and hybrid systems), mechanisms of UV blocking, practical applications in food preservation, limitations, and finally, challenges and future perspectives.

23.2 UV radiation and its evaluation

UV radiation spans the wavelength range of 100–400 nm and is classified into UVC (200–280 nm), UVB (280–320 nm), and UVA (320–400 nm) (Fig. 23.1B) (Manickam et al., 2025). The broader UV–visible spectrum (200–800 nm) includes UVC (mostly from artificial sources), the highly energetic UVB, lower-energy UVA, and visible light (Patil et al., 2021; Salian et al., 2021; Silva et al., 2023). Although UV light constitutes only about 5% of solar radiation reaching Earth, its energy is sufficient to break critical chemical bonds such as C–C, C–O, and C–H, causing widespread damage (Nikvarz et al., 2020).

As mentioned earlier, sunlight contains UV rays that can penetrate foods and promote undesirable chemical reactions (Chaudhary et al., 2022). The majority of the UV rays that cause these adverse effects are UV-A and UV-B, since most of the UV-C is absorbed in the upper atmosphere. Both UV-A and UV-B radiation can penetrate foods and promote degradation, thereby reducing their quality, decreasing their nutritional value, shortening their shelf life, and increasing food waste (Ezati et al., 2023). For this reason, UV-protective substances need to be incorporated into packaging materials to improve food quality (Gore & Prajapat, 2022; Zhao et al., 2023).

Evaluating the UV-barrier properties of food packaging films is essential to ensure effective protection against photodegradation while maintaining desirable optical clarity for consumer appeal. Concepts like the UV protection factor (UPF) for fabrics or sun protection factor (SPF) for sunscreens provide a foundation, though they are adapted differently in packaging contexts (Zhao et al., 2023). The SPF value indicates how much UV radiation reaches the skin with protection compared to without; for instance, SPF 30 means the skin absorbs about 1/30th of the UV rays, blocking approximately 96.7%. It is calculated using the formula:

$$\mathrm{SPF} = \frac{\sum_{290}^{400} E_\lambda S_\lambda}{\sum_{290}^{400} E_\lambda S_\lambda T_\lambda}$$

where E_λ is the erythema spectral effectiveness, S_λ is solar spectral irradiance, and T_λ is the spectral transmittance of the sample (Zhang & Naebe, 2021).

In food packaging, however, the primary tool is UV–visible (UV-Vis) spectrophotometry, which directly measures absorbance and transmittance in the 200–800 nm range to assess UV-blocking efficiency and visible light transparency (Khezerlou et al., 2025; Zhang & Naebe, 2021). Light transmittance at around 660 nm is often reported to evaluate visible transparency, with higher values indicating better clarity. UV-barrier performance is quantified by calculating the percentage blocking in specific ranges:

$$\text{UVA blocking}(\%) = 100 - \frac{\int_{315}^{400} T(\lambda)d\lambda}{\int_{315}^{400} d\lambda}$$

$$\text{UVB blocking}(\%) = 100 - \frac{\int_{280}^{315} T(\lambda)d\lambda}{\int_{280}^{315} d\lambda}$$

where $T(\lambda)$ is the average transmittance at wavelength λ.

Practical examples illustrate this approach.

These measurements highlight a common trade-off: stronger UV blocking often slightly reduces transparency, but optimized natural polymer composites can achieve excellent shielding without compromising packaging appeal (Robinson et al., 2023). Overall, UV-VIS transmittance spectra provide a reliable, standardized way to evaluate and compare UV-barrier properties in biopolymer-based films for food applications.

23.3 Classification of UV-blocking materials

Food spoilage is significantly influenced by oxidation triggered by UV radiation. Consequently, ensuring protection against UV exposure and preventing direct contact between UV sources and food ingredients are crucial considerations in packaging design. Although the food industry extensively employs chemical preservatives to inhibit spoilage, growing consumer awareness and concern about potential long-term health effects are prompting a shift in attitudes toward their use (Ganguly et al., 2018). Despite these additives being approved as safe and edible, many people worldwide are becoming increasingly cautious about consuming chemically preserved foods, driven by worries over possible adverse side effects in certain individuals (Alizadeh Sani et al., 2024; Ezati, Rhim, et al., 2022). The rise of this shift has led to the development and innovation of active packaging, where active substances are integrated into the packaging materials themselves rather than being directly added to the food ingredients (Moustafa et al., 2023; Quilez-Molina et al., 2020; Störmer et al., 2017). Increasingly, antioxidants like ascorbic acid are being incorporated into packaging materials instead of being directly added to food, as this method is gaining popularity for preventing oxidation. For many years, aluminum foil and metallized aluminum foil have been the preferred materials for blocking UV and visible light in packaging applications (Du & Zheng, 2024; Ezati et al., 2021; Ezati et al., 2023). Nonetheless, these materials present several drawbacks, such as limited transparency, elevated costs, and challenges related to recyclability (Kwon et al., 2018; Lazaridi et al., 2021). As research into UV-protective materials expands and their application in food packaging grows, opportunities have arisen to create alternative solutions that overcome the limitations associated with aluminum and metallized foils. Given that each UV absorber is specifically designed for a particular purpose, these limitations can be addressed effectively through the thoughtful selection and application of appropriate polymers and additives available today (Bintsis et al., 2000; Lazaridi et al., 2021; Pandiselvam et al., 2022; Rhim & Kim, 2014). This section highlights key categories of UV protection materials that are currently utilized or hold potential for use in food packaging, providing an overview of their significance and applications. Classified into nanomaterials, polymer-based substances, and polyphenol compounds, UV-blocking materials employed in food packaging play a crucial role in protecting products from ultraviolet radiation (Bintsis et al., 2000; Quilez-Molina et al., 2020; Tan et al., 2021).

23.3.1 Nanomaterials

In the realm of sustainable packaging, nanotechnology is essential, particularly in the development of active and intelligent packaging solutions that not only prevent UV-light penetration but also inhibit bacterial growth in food products. At the nanoscale, the physicochemical characteristics of bulk materials undergo significant transformation, leading to enhancements such as increased surface-to-volume ratio, reduced toxicity, and improved biocompatibility. Additionally, their optical and electrical properties can be optimized to boost UV-shielding capabilities (Lino et al., 2022; Prodpran et al., 2012). In sustainable packaging, nanomaterials contribute significantly by providing improved mechanical strength and thermal stability to composites, making them highly suitable for various applications (Kwon et al., 2018; Lino et al., 2022; Tupuna-Yerovi et al., 2025). Serving as a protective gas barrier between the food interior and the external environment, nanomaterials play a crucial role. Furthermore, certain nanomaterials possess the capacity to exhibit antimicrobial effects, effectively targeting both gram-positive and gram-negative bacterial strains. The application of nanotechnology in antimicrobial packaging has garnered significant interest, driven by its vital function in enhancing microbial safety and prolonging the shelf life of food items (Ezati, Priyadarshi, et al., 2022; Shi et al., 2020). As this field continues to develop, the integration of nanotechnological advancements plays an increasingly important role in food preservation. Concerns regarding health and food safety arise from the long-term use of chemically derived UV-shielding agents incorporated into bio-based polymer composites, as these substances may pose risks over extended periods (Ganguly et al., 2018; Paidari et al., 2021; Störmer et al., 2017). Sustainable packaging applications incorporate a variety of nanomaterials, including both organic and inorganic types, chosen for their distinctive physical and chemical characteristics (Paidari et al., 2021). A wide range of inorganic UV-absorbing materials, whether in the form of particles or thin film coatings, has been created. Research indicates that the UV protection capabilities of biopolymer films can be enhanced through the addition of different organic/inorganic nanomaterials, which function by absorbing and scattering light (Assis et al., 2020; Li, Guan, et al., 2023). Cellulose films containing melanin nanoparticles, sodium alginate embedded with nanoencapsulated lycopene and β-carotene, methylcellulose with nanoencapsulated β-carotene, and cellulose acetate incorporating nanoencapsulated norbixin, lycopene, or zeaxanthin are among the various examples. Furthermore, chitosan films containing melanin nanoparticles loaded with epigallocatechin-3-gallate, along with bacterial cellulose nanowhiskers, are also included (Zhao et al., 2022). It has been documented that the use of chitosan and cellulose nanofibers can improve UV protection. This enhanced shielding results from the potent light-scattering and absorption properties inherent to these organic nanomaterials (Haghighi et al., 2021; Li, Yang, et al., 2023). Commercial UV protection often relies on materials like metal-organic frameworks (MOFs), metals, and metal oxides, including zinc oxide (ZnO), ZrO_2, iron oxide, CeO_2, and titanium dioxide, that are widely utilized in various applications. Due to their broad absorption of UVA and UVB rays, affordability, and straightforward manufacturing process, ZnO nanoparticles are the most prevalent among these materials (Paidari et al., 2021). Using an in situ approach, Zhao et al. synthesized ZnO nanomaterials embedded within a polyvinyl alcohol (PVA) film specifically for UV-shielding purposes (Zhao et al., 2018).

23.3.2 Biopolymer and synthetic polymer nanocomposites

At present, the predominant materials used for food packaging are synthetic polymers derived from petroleum sources. Ultimately, these petroleum-based polymers remain the primary choice for such applications (Kwon et al., 2018; Zhang et al., 2023). Despite their advantages, including robust mechanical strength, thermal stability, ease of processing, and cost-effectiveness, these plastics raise serious environmental issues because they do not readily degrade, can be toxic, and negatively affect ecosystems (Roy & Rhim, 2021c; Shankar & Rhim, 2018). Ultimately, their environmental footprint outweighs the benefits they provide. Biopolymers, derived from natural sources such as plants and animals, are employed across diverse organic and mechanical fields. For widespread application, an optimal biopolymer must be nontoxic, nonallergenic, noncarcinogenic, biodegradable, and easily accessible (Kwon et al., 2018; Zou et al., 2022). Ultimately, these qualities make biopolymers a promising sustainable alternative to conventional plastics. Sources from plants and animals, such as starch, cellulose, chitosan, and proteins, serve as prominent examples (Rukmanikrishnan et al., 2020). Ultimately, these natural materials exemplify the diverse range of biopolymers utilized in various applications. Due to their abundance, affordability, environmental friendliness, biodegradability, and biocompatibility, biopolymers represent compelling substitutes for nonbiodegradable plastics derived from petroleum. In the end, their sustainable nature makes them highly appealing options in reducing ecological impact. Primarily because of their amine functional groups, chitosan stands out among polysaccharides for its effective UV-blocking capabilities across the full spectrum: UVA, UVB, and UVC (Ezati & Rhim, 2020; Ganguly et al., 2018; Kainulainen et al., 2018). Ultimately, this distinctive property highlights its potential as a protective agent against ultraviolet radiation. The level of UV protection offered by chitosan is influenced by both its degree of deacetylation and molecular weight, where higher deacetylation and lower molecular weights result in more effective shielding against ultraviolet radiation (Kainulainen et al., 2018; Xing et al., 2017; Zou et al., 2022). The ability of pectin, a plant-derived polysaccharide, to block approximately 70% of UV radiation is mainly due to the presence of carbonyl groups within its molecular structure. Its application alongside synthetic polymers, such as poly (methyl methacrylate), has led to the formation of UV-protective, transparent complexes (Shankar et al., 2015; Zou et al., 2022). In summary, pectin's UV-blocking properties, combined with its compatibility with synthetic materials, make it a valuable component in creating clear, protective coatings (Kainulainen et al., 2018; Shankar & Rhim, 2018). Demonstrating notable UV-shielding capabilities, protein-based biopolymers like gelatin can absorb approximately 70%–75% of ultraviolet radiation, a property primarily attributed to the amine groups present in their amino acid compositions. To improve the mechanical strength of eco-friendly packaging solutions, a range of biopolymers, mainly proteins and polysaccharides, are commonly employed. Ultimately, their incorporation not only provides UV protection but also enhances the durability of sustainable packaging materials. UV screening in biopolymers is largely due to the presence of functional groups such as carbonyl, hydroxyl, amino, and aromatic groups (Kainulainen et al., 2018; Min et al., 2023; Zou et al., 2022). These chemical features play a crucial role in providing protective properties. In conclusion, the inherent functional groups within biopolymers are key contributors to their ability to block ultraviolet radiation (Philp et al., 2013; Priyadarshi, Roy, et al., 2022). Enhancing UV protection can be achieved by raising the concentration of these functional groups within the biopolymers. Notably, gelatin and chitosan contain elevated amounts of such groups, which contribute to their strong ability to absorb ultraviolet radiation (Priyadarshi, Riahi, et al., 2022). Ultimately, the presence and abundance of these functional groups directly influence the UV-shielding effectiveness of these materials. The UV-blocking capability of gelatin is further enhanced by the presence of aromatic amino acids, such as tyrosine and tryptophan.

23.3.3 Polyphenolic compounds

Amid growing environmental and sustainability concerns, polyphenolic compounds have become prominent as eco-friendly and functional materials. Their primary sources are naturally derived from plants and animals. In conclusion, the interest in these compounds continues to rise, driven by their natural origin and potential for sustainable applications. Lignin stands out as a significant byproduct generated during the pulp, paper, and biorefining industries (Ezati et al., 2023). Its ability to absorb a wide range of UV radiation, specifically between 250 and 400 nm, is attributed to its chromophore functional groups. Ultimately, this property highlights lignin's potential as a valuable material for UV protection and related applications. Primarily composed of three phenyl propanoid units—sinapyl alcohol (S type), coniferyl alcohol (G type), and para-coumaryl alcohol (H type)—lignin's complex structure is stabilized by arylglycerol ether bonds and additional linkages. To conclude, these interconnected units form the intricate polymer that characterizes lignin's molecular architecture. Products like sunscreens, transparent films, paints, and varnishes incorporate this substance, which is currently the subject of ongoing research for its potential in microbial protection. While traditionally valued for its applications in these materials, new studies are expanding its use toward safeguarding against microbial contamination. Depending on the plant source and extraction technique, lignin's composition can differ significantly. In commercial applications, it is commonly available as either alkali lignin or sulfonated lignin. Unlike lignosulfonates, which contain water-attracting sulfonate groups and dissolve readily, alkali lignins, such as soda and kraft lignins, are insoluble in water (Ezati et al., 2023; Garrido-Romero et al., 2022; Roy & Rhim, 2021a).

The ability of lignin to block almost all UV radiation varies based on its concentration and type. To improve the UV-protective qualities of biopolymer-based packaging, different phytochemicals can be integrated, further enhancing their effectiveness (Alizadeh-Sani et al., 2021). Various additives have been incorporated into different films to enhance their properties: catechol has been added to chitosan/PVA films, quercetin to chitosan/gelatin and gelatin-based films, lignin to cellulose films, and anthocyanins to methylcellulose/chitosan nanofiber films. Additionally, essential oils (EOs) like Santalum album, thyme, cinnamon, and clove have been infused into chitosan, gelatin, or gellan gum matrices. When incorporated into gellan gum-chitosan films, emulsified thyme oil notably enhanced UV protection (Xiao et al., 2020). Likewise, the addition of cinnamon and clove oils increased the UV-shielding effectiveness of chitosan/acetylated starch films. The potential of phytochemicals as natural additives to improve UV protection in biopolymer-based packaging materials is underscored by these findings. Their use demonstrates a promising avenue for developing more effective, eco-friendly packaging solutions (Alizadeh-Sani et al., 2021).

Various plant EOs and extracts are rich in polyphenolic compounds, which have been studied for their potential as natural UV-barrier agents. Notably, carvacrol, present in thyme, sage, and oregano oils, along with thymol found in thyme and oregano, and eugenol found in clove, cinnamon leaf, and pimento oils, are among the key polyphenols identified for this purpose. Extensive research has been conducted on natural compounds such as quercetin and catechins found in kiwi peel, as well as catechins, epicatechins, procyanidins, and gallic acid present in grape seed extracts, for their ability to block UV radiation in food packaging films (Priyadarshi & Rhim, 2020; Priyadarshi, Purohit, et al., 2022; Roy & Rhim, 2021b, 2022; Xiao et al., 2020). Moreover, plant extracts like kiwi peel and grape seed extracts are notable sources of these bioactive substances, highlighting their potential in enhancing UV protection. While natural polyphenols offer promising benefits, their application is hindered primarily by inconsistent composition resulting from variations in plant sources, impacting reliability (Priyadarshi & Rhim, 2020; Prodpran et al., 2012). Additionally, their inherent coloration can alter the visual appeal and optical transparency of packaging materials, thereby restricting their suitability in contexts where clear and color-neutral packaging is essential (Alizadeh-Sani et al., 2021). Synthetic polyphenols are effective UV blockers due to their colorless appearance and strong absorption, whereas natural polyphenols are limited by inconsistent composition and coloration that can affect packaging clarity (Kwon et al., 2018; Priyadarshi & Rhim, 2020). Phenolic UV absorbers are popular because they efficiently block UV rays, convert harmful energy into safer forms, and are highly durable, making them a preferred choice across various industries.

23.3.4 Hybrid systems

To improve the properties of biopolymer-based packaging materials, various UV screening agents have been integrated into hybrid systems (Roy & Rhim, 2021a). These innovative combinations aim to boost UV protection and overall functionality (Braga et al., 2019; Li et al., 2021; Sani et al., 2022). Ultimately, such approaches represent a significant advancement in the development of more effective biopolymer packaging solutions. Incorporating inorganic nanomaterials like TiO_2, zinc sulfide, and silver nanoparticles (AgNPs) into biopolymer matrices exemplifies how both components work together to provide UV protection (Ananikov, 2019; Kwon et al., 2018). Such integration leverages the synergistic effects of nanomaterials and biopolymers to enhance the overall shielding capabilities. Ultimately, this approach demonstrates how the combination of these materials can effectively improve UV resistance in biopolymer-based systems (Roy & Rhim, 2021a; Yang et al., 2022). Integrating inorganic nanoparticles with phytochemicals, such as TiO_2 with tea tree oil or zinc sulfide with tea tree oil, enhances UV protection in biopolymer films (Sani et al., 2022). These synergistic combinations demonstrate how combining nanomaterials and natural compounds can significantly improve UV-shielding capabilities (Roy & Rhim, 2021a; Sheng et al., 2023). Combining inorganic nanoparticles with phytochemicals, such as TiO_2, zinc sulfide, and anthocyanins, enhances the UV protection of biopolymer films. These hybrid materials demonstrate that integrating nanomaterials with natural compounds can significantly improve the UV-resistant properties of biopolymer-based systems (Wang et al., 2021; Xie et al., 2019). The primary mechanism behind the improved UV-blocking capabilities of these hybrid systems is their enhanced light scattering and absorption, which results in markedly superior protective qualities in the final films (Bian et al., 2024; Shi et al., 2012). Ultimately, it is this combination of increased scattering and absorption that underpins their superior UV-shielding performance.

23.4 Mechanism of UV-blocking materials

The chemical composition of biopolymer films largely determines their UV protection capabilities, which are closely linked to their specific chemical makeup. The process of UV-light absorption within these films and coatings is primarily influenced by the types of biopolymers and nanomaterials incorporated (Ahmed et al., 2016; Yang et al., 2024). Ultimately, the effectiveness of UV shielding in biopolymer-based films is rooted in their unique chemical structures and the materials used. Primarily made up of carbohydrate units, most biopolymers feature various functional groups such as amino, hydroxy, alcohol, acid, ketone, and aldehyde (Riahi, Khan, Hong, et al., 2026; Xu et al., 2025). These groups are characteristic components that define their chemical nature. In summary, the presence of these functional groups is a common feature among biopolymers predominantly composed of carbohydrate structures (Barden & Decker, 2016; Liu, 2020). Because they lack aromatic structures, their capacity for π-π interactions is minimal, leading to weak or nonexistent absorption in the UV spectrum and permitting UV, visible, and near-infrared radiation to penetrate (Alizadeh Sani et al., 2024; Chung et al., 2020). Conversely, when nanomaterials are incorporated into biopolymeric films, they can serve as efficient barriers against UV light by actively absorbing it. When excited by UV radiation, these nanomaterials have the ability to emit visible light via fluorescence or photoluminescence, effectively transforming harmful UV rays into visible illumination (Patil et al., 2021; Singh et al., 2021). This capability introduces an eco-friendly dimension to UV protection, enhancing its sustainability. Influences such as size, shape, crystallinity, color, and surface characteristics of nanomaterials play a significant role in governing their primary mechanisms, namely absorption, backscattering, and reflection of UV rays (Fig. 23.1C). These processes are affected by these physical and chemical properties, which determine how effectively UV radiation is managed. The spectroscopic properties of the embedded nanomaterials, along with these influencing factors, ultimately determine the overall efficiency of UV protection in biopolymer nanocomposites (Delorme et al., 2020; Patil et al., 2020; Tripathi et al., 2024). The specific spectroscopic features of the nanomaterials embedded within biopolymer nanocomposites, combined with various influencing factors, play a crucial role in shaping the overall UV-shielding performance. Ultimately, these elements determine the level of protection offered by the nanocomposites against ultraviolet radiation (Ezati et al., 2023; Tripathi et al., 2024). The incorporation of UV-absorbing nanomaterials, both in terms of their type and concentration, significantly influences the UV protection offered by composite films. Ultimately, the effectiveness of these films in safeguarding against UV radiation hinges on these specific factors. Initially, pristine biopolymer films allow approximately 90% of UV radiation to pass through, providing only limited protection

(Hao et al., 2021; Rhim & Kim, 2014; Yan & Chai, 2021). However, the addition of even small quantities of nanomaterials can enable these films to start absorbing UV-C and some UV-B wavelengths. As the concentration of nanomaterials increases, the absorption capacity extends across the entire UV spectrum, UV-A, UV-B, and UV-C, potentially resulting in the near-complete blocking of UV radiation (Ezati et al., 2023; Feng et al., 2017; Wang et al., 2022).

The UV-shielding efficiency of these films is thus closely related to the amount and type of nanomaterials incorporated. Achieving broader UV absorption across the entire spectrum can be accomplished by blending multiple nanomaterials in precise ratios during the film-casting process (Feng et al., 2017; Rabani et al., 2021). An additional benefit is provided by incorporating photoresponsive, fluorescent, or photoluminescent nanomaterials, which not only absorb UV radiation but also emit visible light. The emission properties of these nanomaterials are influenced by the excitation wavelength, offering a versatile approach to enhancing UV protection with functional optical features (Badry et al., 2023). The ability to both protect against UV radiation and provide additional functional benefits opens up possibilities for smart or active packaging solutions. This combination of features not only strengthens UV shielding but also paves the way for innovative applications that leverage these active properties (Badry et al., 2023; Ran et al., 2023).

Chitosan and gelatin films have excellent UV protection because their amino, hydroxyl, and carbonyl groups provide free π-electrons that effectively absorb UV radiation, giving them strong shielding abilities among biopolymer materials (Patil et al., 2021; Qamar et al., 2020). Thanks to their high concentration of amino groups, chitosan possesses enhanced properties, including antioxidant activity, antimicrobial effects, and the ability to penetrate tissues. Owing to its chemical composition, which includes nitrogen-containing groups and UV-active compounds, chitosan stands out as an effective biopolymer for blocking UV radiation. Similarly, gelatin, a protein-derived biopolymer, demonstrates remarkable UV protection because of its amino acids, such as tryptophan and tyrosine, that contain carbonyl and amino functionalities. The UV-active elements within the material improve both its ability to absorb UV radiation and its barrier effectiveness (Ezati et al., 2023; Phan et al., 2021; Shankar et al., 2015; Shankar et al., 2016). Research indicates that gelatin-based films are capable of effectively blocking a substantial amount of UV light, allowing only about 29.4% to pass through while absorbing approximately 70.6% of UV radiation (Rhim & Kim, 2014). When doped with silver and copper nanoparticles, the UV-blocking ability of the material is greatly enhanced, decreasing transmittance to nearly 0.01% and markedly boosting its protective effectiveness. In contrast, films made from carboxymethyl cellulose (CMC) exhibit relatively poor UV shielding, permitting approximately 62% of UV light to pass through at 280 nm (Yan & Chai, 2021). The absence of significant UV absorption in CMC films and their high light transmission throughout the UV spectrum can be attributed to the lack of UV-active components within their chemical makeup. This limited capacity for UV protection might stem from the presence of solely carbonyl groups in their structure, which do not contribute effectively to UV shielding. While pure agar gel lacks a distinct UV absorption band, a subtle increase at lower wavelengths indicates that some light scattering occurs within the gel. To achieve substantial UV-blocking capabilities in films, it is crucial to dope or modify them with functional groups or active ingredients that feature aromatic structures, which can undergo π-π^* transitions when exposed to UV radiation (Assimopoulou & Papageorgiou, 2005; Sansonetti et al., 2021). Adding aromatic compounds to CMC and cellulose nanofibers films improves their UV protection by about 50% through absorption, while nanomaterials with fluorescence or photoluminescence further enhance UV shielding by converting UV energy into visible light. Demonstrating effective UV-blocking through these mechanisms, CQDs-loaded chitosan/gelatin biopolymer exhibited the highest UV absorbance, with a transmittance as low as 0.03% at 280 nm. This example highlights the significant potential of such biopolymer composites in UV protection applications (Ezati et al., 2021; Ezati, Rhim, et al., 2022).

23.5 Applications of UV-blocking agents in food packaging

UV-blocking packaging systems are engineered to block harmful UV rays, reflecting or absorbing them while transmitting visible light to maintain food quality. These materials, infused with UV protectants, are essential for safeguarding photosensitive foods from deterioration. This section highlights key product categories where such films can be effectively applied.

23.5.1 Meat and meat products

Meat, rich in essential nutrients, is a key dietary component but is vulnerable to oxidation and microbial contamination, which packaging can influence. Raw poultry's short shelf life necessitates advanced preservation techniques to meet safety standards. Transparent packaging offers an appealing presentation but lacks UV protection, while opaque options shield against light but hinder visibility (Fang et al., 2017; Song et al., 2021). Developing films that are clear in visible light but block UV rays is crucial.

Numerous studies have demonstrated that integrating UV-protective agents into packaging materials enhances their resistance to light-induced degradation. It is well established that such additives can significantly boost the photostability of packaging. Khan et al. (2023) developed garlic-derived carbon dots (CDs) to create carrageenan/sodium alginate films with strong UV-blocking capabilities (~85.1% in UV-A and 99.0% in UV-B). These films effectively protected raw meat from UV exposure, maintaining its color even after 30 hours of direct UV light. The UV-blocking properties helped extend the shelf life and preserve the visual quality of meat. Khan et al. (2023) and Jiang et al. (2024) developed a Swan-feather-inspired CMC/PVA@quercetin-based film that offers UV-triggered photochromic and anticounterfeiting features (Fig. 23.2A). Jiang et al. (2024) prepared films, utilizing quercetin's fluorescence and anticounterfeit properties, to extend food shelf life (Fig. 23.2B). The strong yellow-green fluorescence (peaking at 528 nm), 21.32 ns lifetime, and uniform distribution enhance food safety and protection against counterfeiting. The film extends minced pork freshness to five days at 4°C and 40% humidity, outperforming packaging by two days due to its antimicrobial, antioxidant, and UV-barrier properties (Fig. 23.2C).

Adding ZIF-67-loaded microcrystalline cellulose nanofillers to starch/pectin composite film significantly improves UV protection, reducing transmittance at 280 nm by up to 31.48% and at 660 nm by 59.10%, which helps prevent lipid oxidation in food (Zhao et al., 2025). This enhancement in UV shielding, despite decreased transparency, enhances the film's effectiveness in preserving food quality and safety. Accordingly, adding nanofiller (5%) to film significantly

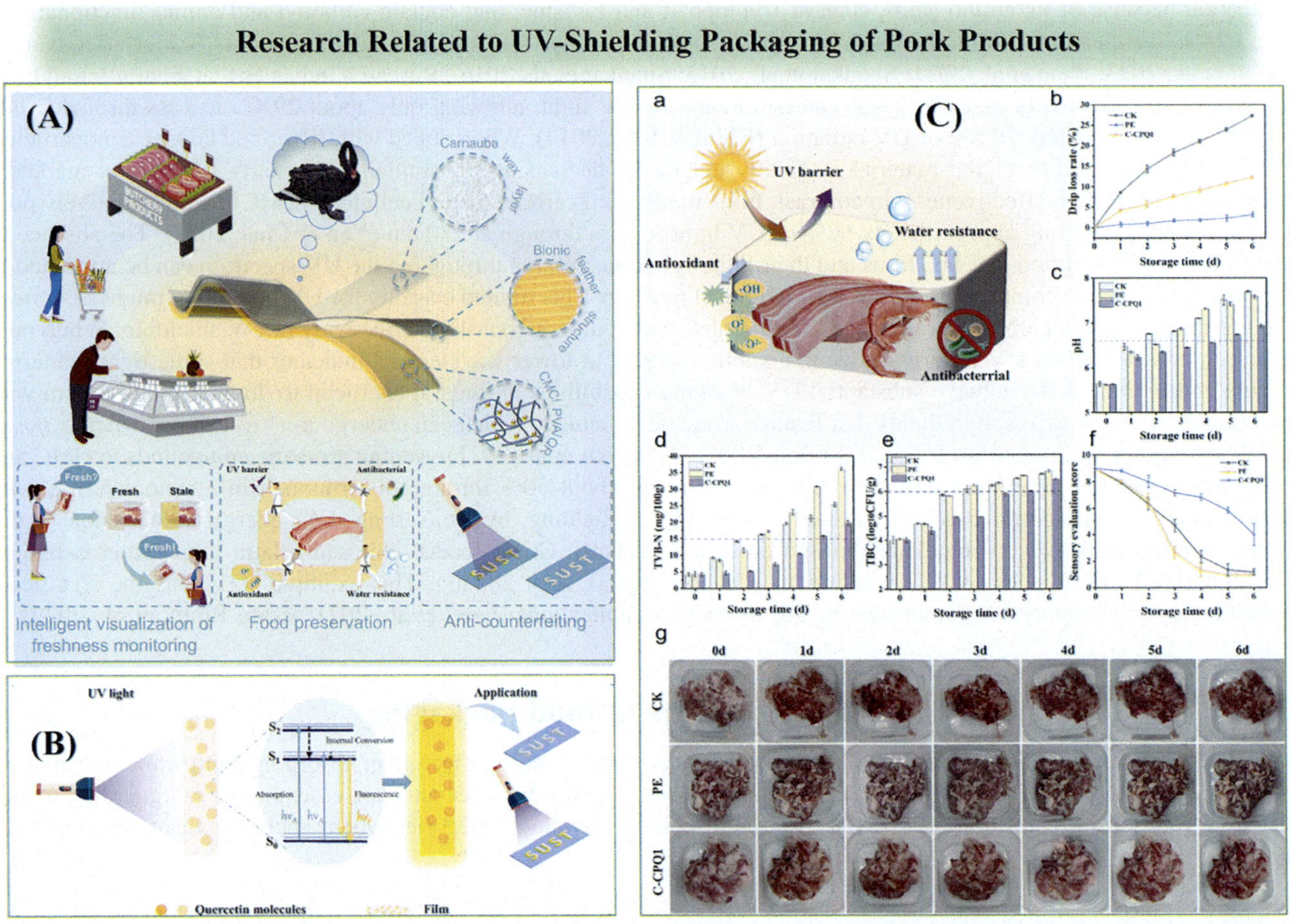

FIGURE 23.2 (A) Structure, functions, and applications diagram of bionic feather multifunctional films, (B) mechanism of fluorescence anticounterfeiting in the film, and (C) mechanism of freshness preservation and its visual images during storage.
Adapted from with permission. Jiang, H., Zhao, S., Li, Z., Chen, L., Mo, H., & Liu, X. (2024). Swan-feathers inspired smart-responsive sustainable carboxymethyl cellulose/polyvinyl alcohol based food packaging film for robustly integrated Intelligent and Active Packaging. Nano Today, 56. https://doi.org/10.1016/j.nantod.2024.102272.

improved pork preservation by reducing weight loss to 8.09% (vs 18.36% in control), delaying spoilage indicators like TVB-N (14.63 mg 100 g^{-1} vs over 16.52 mg 100 g^{-1} in controls), and maintaining acceptable pH and microbial levels for up to eight days. This packaging extended pork shelf life by two days compared to conventional methods, effectively preserving freshness and safety. Incorporation of anthocyanins can significantly reduce light transmission across UV and visible spectra, enhancing the film's protective barrier. In a recent study by Shahi et al. (2026), PVA/chitosan/starch-based smart films with ghangaru-anthocyanins blocked UV light up to 450 nm, while proposed films extended protection to 500 nm, effectively shielding food from harmful UV exposure and protecting the chicken meat from photodegradation (Shahi et al., 2026). Pure gelatin film shows the highest UV transmittance, offering minimal protection, but incorporating cellulose nanocrystals (CNCs) slightly improves UV shielding by forming a denser network (Sun et al., 2026). Adding nanolignin-stabilized Pickering emulsion (NPE) significantly enhances UV shielding, due to scattering and absorption by lipid droplets and bioactive compounds, making the film more effective in protecting food from UV-induced oxidation and quality degradation. The film loaded with CNC/NPE extended pork shelf life to eight days at 4°C, with significantly better preservation than control and PE groups, which spoiled by day 6. It maintained appearance, delayed pH increase (from 6.2 to 6.4), kept TVB-N below 15 mg 100 g^{-1}, slowed lipid oxidation, and kept microbial counts below 6 log CFU g^{-1}. Increasing (tea polyphenol [TP]; 1, 2, 4, and 8 mg mL^{-1}) content in chitosan films also reduces UV transmittance, with all films showing low transmittance between 200 and 400 nm (Wang et al., 2025). This enhances meat preservation by delaying microbial growth, pH increase (from 6.2 to 6.4 by day 5), and lipid oxidation, extending its shelf life from five to eight days. The addition of encapsulated beet leaf extract in mesoporous silica nanoparticles (BLE@MSNPs) into chitosan/cellulose nanocrystals (CS/CNCs) film improved UV-shielding properties and protected chicken filets from oxidation, thereby maintaining their freshness for up to eight days (Soufiani et al., 2025). Here, the untreated groups showed a 2.5-fold increase in protein oxidation, with carbonyl levels rising from 0.4151 to 1.0350 over eight days. However, BLE@MSNPs-loaded CS/CNCs films limited oxidation to 0.611, reducing damage by approximately 41%, thereby significantly enhancing meat preservation during storage.

23.5.2 Seafood and seafood products

Seafood's high perishability leads to rapid freshness loss, risking spoilage and health issues if not properly stored. To meet consumer demand for fresh, minimally processed products, innovative packaging methods have been developed (Kontominas et al., 2021; Siddiqui et al., 2024). UV-shielding and sustainable packaging show promise in reducing spoilage and maintaining quality.

Many seafood items are rich in polyunsaturated fatty acids, making them especially vulnerable to damage from light-induced oxidation. This issue can be mitigated through the use of packaging materials that effectively block UV light (Abedi-Firoozjah, Azimi-Salim, et al., 2025; Yue et al., 2024). For example, the neat κ-carrageenan films had high transmittance (~66.5%) and poor UV protection but remained transparent. Incorporating dextran reduced transmittance to 18.35%, and adding anthocyanins and ZnONPs further improved UV shielding, with the best films showing the highest UV barrier (decreased transmissivity to 2.73%), essential for protecting shrimp from UV-induced spoilage. Adding AgNPs/CDs enhances the chitosan/PVA film's light-shielding ability, and increasing its content (1, 2, 3, 4, and 5 wt.%) further reduces transmittance, with 3% NPs film achieving the highest UV shielding in the 200–400 nm range (Du & Zheng, 2024). This low transmittance effectively prevents photooxidation, helping to extend the shelf life of stored shrimp at 4°C for seven days. In a study by Murugan et al. (2025), hydrothermally processed Bambara groundnut pericarp powder produced BGP-CDs with sizes ranging from 1.34 to 15.02 nm, showing excellent optical properties (Murugan et al., 2025). Gelatin/PLA bilayer films containing BGP-CDs showed strong UV-shielding properties, with the BGP-CDs/3 sample (synthesized for 3 hours at 100 μg mL^{-1}) achieving up to 92.64% UV-B protection, enhancing preservation of packed Asian seabass slices by preventing photooxidation. Ponnusamy et al. (2025) found that the incorporating 1% tannin acid (TA) into chitosan/fish gelatin blend films doubled UVA protection from 33.02% to 68.27%, with higher concentrations (3% and 5%) achieving up to 90.08% UVA and 100% UVB shielding, significantly enhancing UV protection (Ponnusamy et al., 2025). This high UV barrier is due to TA's polyphenolic structure, which effectively absorbs UV radiation, making these films ideal for extending food shelf life. TA-rich packaging effectively suppressed most bacteria, resulting in a less diverse but more dominant microbial community, thereby enhancing preservation of fresh baby clams by inhibiting spoilage bacteria during refrigeration. The cellulose-based composite film's UV-shielding ability mainly comes from ZnO NPs, which absorb UV light effectively due to their 3.2 eV band gap, increasing absorbance at 280 nm nearly fivefold (Yue et al., 2024). This provides nearly 100% protection against UV-A and UV-B (Fig. 23.3). This robust UV protection reduces oxidation and discoloration in shrimp, significantly delaying the spoilage process (Fig. 23.3B).

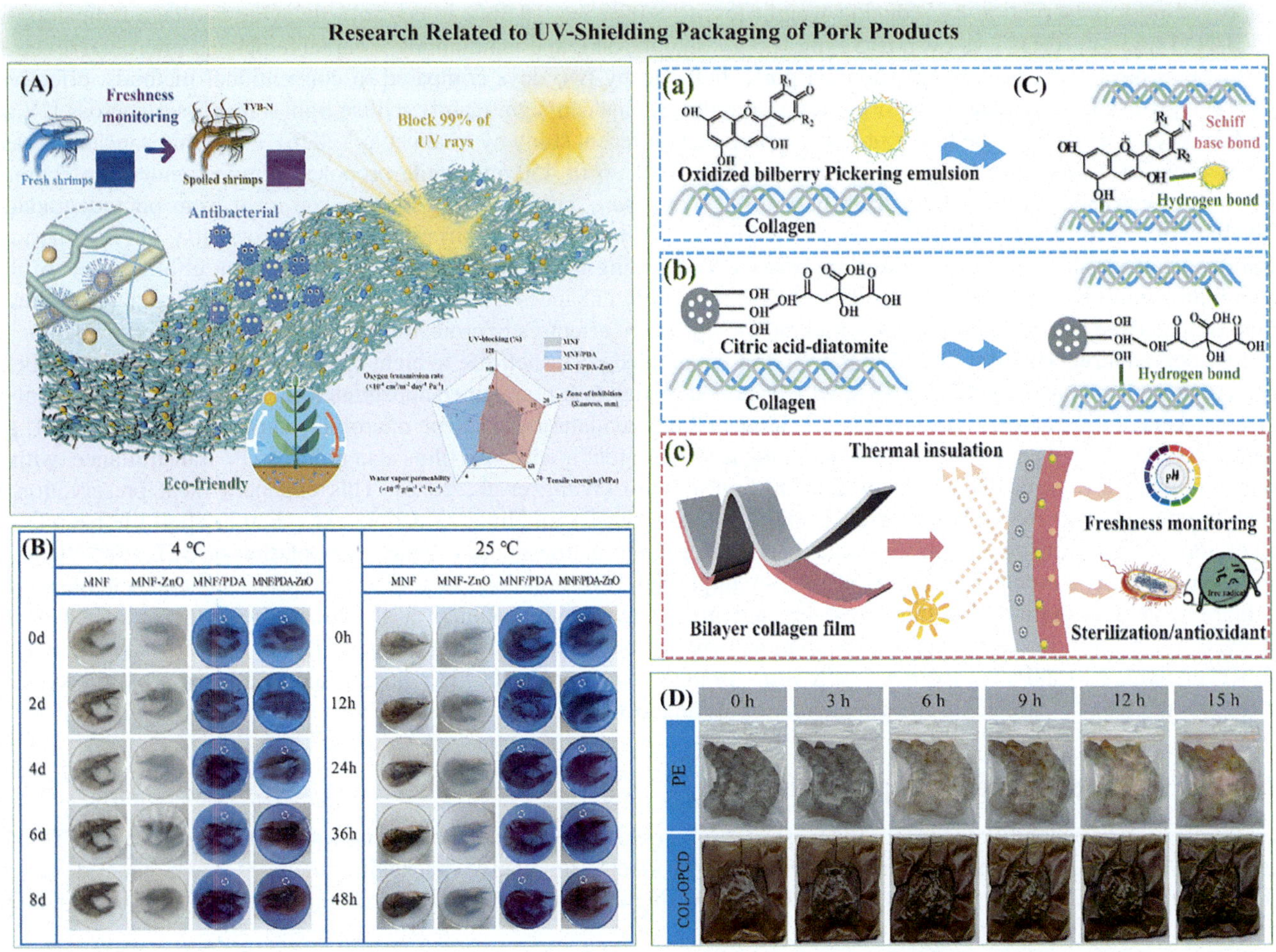

FIGURE 23.3 (A) Schematic diagram of the preparation of cellulose-based films loaded with zinc oxide nanoparticles (ZnO NPs), (B) application of the films for shrimp preservation, (C) preparation of the double-layer collagen-based films, and (D) shrimp preservation.
Reprinted with permission. Ran, R., Luo, X., Zhai, D., He, Y., Guo, X., He, L., & Li, G. (2025). Development of a dual-layer collagen film with insulating, antioxidant, antimicrobial, and pH-sensitive properties for the maintenance and visual monitoring of shrimp freshness. Food Hydrocolloids, 169, 111603. https://doi.org/10.1016/j.foodhyd.2025.111603; Yue, C., Wang, M., Zhou, Z., You, Y., Wang, G., & Wu, D. (2024). Cellulose-based intelligent packaging films with antibacterial, UV-blocking, and biodegradable properties for shrimp freshness monitoring. Chemical Engineering Journal, 488. https://doi.org/10.1016/j.cej.2024.150975.

Adding 1%, 2.5%, and 5% diosmin to chitosan films significantly increases UV absorbance, with the 5% diosmin film showing the highest absorbance, indicating improved UV-barrier properties (Xu et al., 2025). The enhancements are due to hydrogen bonding between diosmin's hydroxyl groups and chitosan's amino/hydroxyl groups, along with π–π stacking of diosmin's aromatic rings, which create a denser network that boosts UV absorption and decreases transparency; additionally, diosmin's conjugated structure naturally absorbs UV light, acting as an effective UV filter. During shrimp packaging, adding diosmin reduced pH increase and lipid oxidation, helping preserve shrimp quality by inhibiting enzyme activity and oxidative spoilage, with diosmin's antioxidant properties playing a key role. Collagen films loaded with citric acid-diatomite (CD)/oxidized bilberry/Pickering emulsion were also shown to have strong UV-shielding ability, which extended packaged shrimp shelf life (Fig. 23.3C and D) (Ran et al., 2025). The neat film transmitted 84.30% of UV light at 365 nm, but incorporating emulsion reduced UV transmittance to 1.30%, significantly improving UV protection while keeping the film transparent.

23.5.3 Milk and dairy products

The packaging of milk and dairy products, designed to protect against damage and contamination, varies based on their specific properties. UV-shielding and sustainable materials help extend shelf life by reducing nutrient loss and weight.

These solutions also facilitate handling, storage, and legal labeling to inform consumers and regulators (Biegańska, 2022; Rejeesh & Anto, 2023).

As shown in Fig. 23.4, the PVA/nanocellulose-AgNPs composite film loaded with Chun Jian peel extract demonstrated superior UV- shielding capabilities, shielding over 95% of UVB (200–275 nm) and most UVA (320–400 nm) wavelengths (Fig. 23.4A and B) (Yang et al., 2023). The dense polymer network formed by nanocellulose crosslinking and the UV-absorbing properties of phenolics and flavonoids contributed to this excellent UV shielding, making the film highly suitable for milk packaging applications that require effective UV protection. The corn straw core/cellulose nanofiber films achieved UV-protection factor values over 50, with the highest reaching 101.15, indicating excellent UV protection for food preservation. The improved UV-barrier mechanism involves corn straw core's light reflectivity and scattering, which enhance the film's ability to block UV rays (Li, Guan, et al., 2023). Thus, milk stored (25°C and 40%–50% relative humidity [RH]) with these films showed less microbial growth and spoilage after 14 days compared to unprotected samples (Fig. 23.4C and D). UV-protective packaging can enhance cheese quality and extend its shelf life by preventing oxidation of lipids, proteins, and pigments throughout storage (Jafarzadeh et al., 2021).

In another study, Qu et al. (2025) found that the ZnO-incorporated Na-caseinate/CMC films showed extremely low transmittance, nearly 0% at 280 nm and only 1.8% at 660 nm, indicating superior UV and light-blocking ability (Qu et al., 2025). This enhanced UV protection mechanism involves ZnO's strong light absorption and scattering properties, effectively preventing light-induced cheese spoilage. Moreover, the sustained antimicrobial and barrier properties of ZnO help inhibit pathogen growth and improve cheese safety over time, preventing external *L. monocytogenes* contamination for the entire 20-day storage period. Active packaging for butter utilizes bioactive compounds to inhibit oxidation and

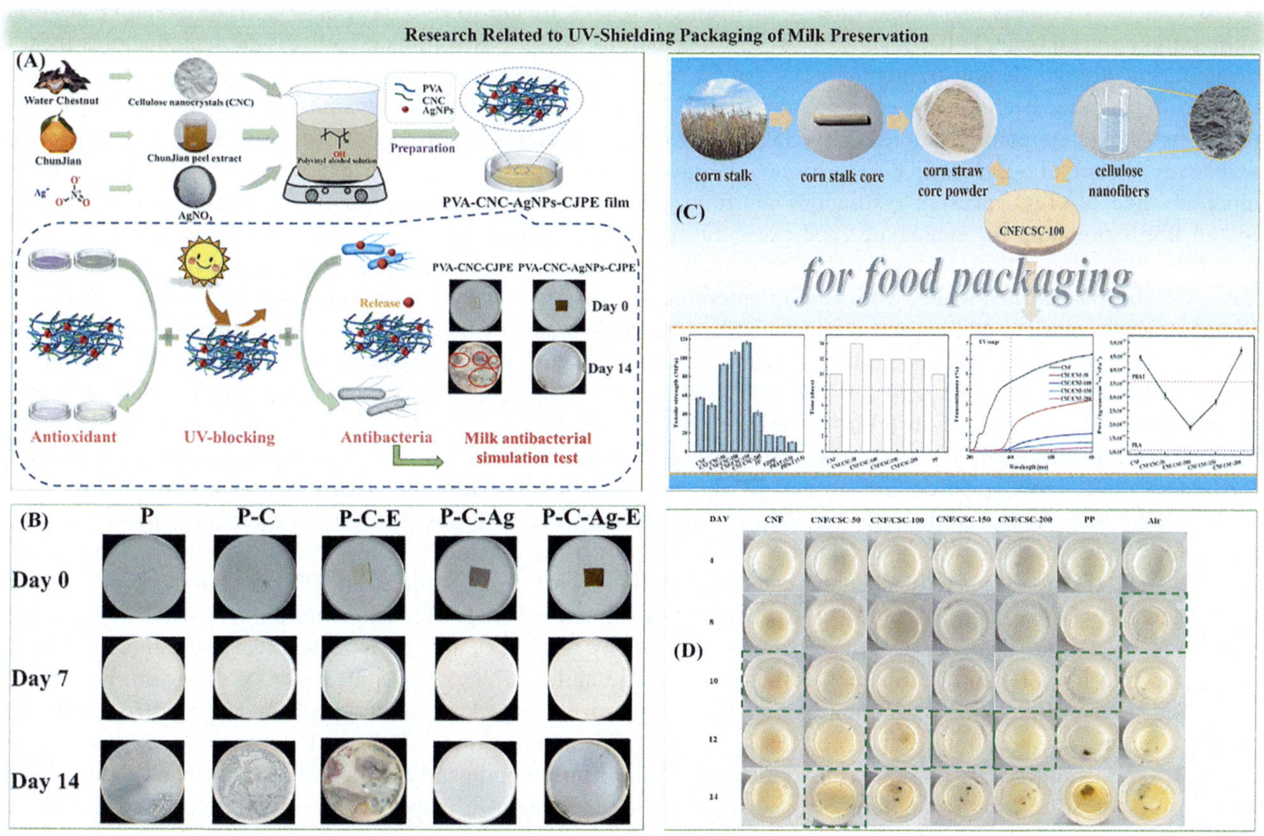

FIGURE 23.4 (A) Schematic drawing for fabrication of ChunJian peel extract@PVA/nanocellulose-silver nanoparticles composite film and (B) photos of the apparent changes and milk preservation during storage for a period of the 14th day. (C) Fabrication of the corn straw core/cellulose nanofibers composite films and (D) images of milk during the experimental period of the 14th day.

Reproduced with permission. Yang, D., Fan, B., Sun, G., He, Y.-C., & Ma, C. (2023). Ultraviolet blocking ability, antioxidant and antibacterial properties of newly prepared polyvinyl alcohol-nanocellulose-silver nanoparticles-ChunJian peel extract composite film. International Journal of Biological Macromolecules, 252, 126427. https://doi.org/10.1016/j.ijbiomac.2023.126427; Li, Z., Guan, J., Yan, C., Chen, N., Wang, C., Liu, T., & Shao, Z. (2023). Corn straw core/cellulose nanofibers composite for food packaging: Improved mechanical, bacteria blocking, ultraviolet and water vapor barrier properties. Food Hydrocolloids, 143. https://doi.org/10.1016/j.foodhyd.2023.108884.

prolong freshness, effectively extending shelf life. Additionally, UV-shielding materials help protect butter from light-induced spoilage, preserving its quality and flavor. Together, these properties ensure safer, longer-lasting butter products with maintained sensory qualities. For instance, Flórez et al. (2022) developed an active packaging film made from chitosan combined with *Santalum album* EOs (SAEOs), which was characterized and applied as a butter sachet to slow down lipid oxidation (Flórez et al., 2022). SAEOs-chitosan films exhibited very low transmittance, with values as low as 0.16% in UV-C, 1.14% in UV-B, and 3.50% in UV-A regions, indicating excellent UV-shielding properties of proposed films for butter preservation. The UV protection mechanism relies on SAEO's ability to absorb and scatter UV light, preventing light-induced spoilage and preserving food quality.

23.5.4 Fruits and vegetables

To ensure freshness and prevent spoilage, UV-shielding and sustainable materials improve fruit and vegetable packaging, supporting eco-friendly practices. Proper storage conditions, varying by type in temperature, light, humidity, and proximity to other foods, are crucial for maintaining quality (Sahu & Shrivas, 2024; Sequino et al., 2022). Fruits and vegetables, rich in nutrients, face challenges from microbial spoilage and short shelf life, impacting safety and sustainability (Abedi-Firoozjah, Bahramian, et al., 2025; Shi et al., 2024). UV light can degrade fruit and vegetable juices, causing color fading, flavor changes, and nutrient loss, but UV-shielding packaging can help mitigate these effects (Bahramian et al., 2025; Shahi et al., 2025). Postharvest losses from moisture loss, damage, microbes, and climate threaten fruit quality, which modern storage methods like low temperature, ethylene inhibitors, and UV-shielding packaging aim to prevent (Sganzerla et al., 2020). In a study by Bahramian et al. (2025), the addition of functionalized turnip peel bio-waste (TPCDs; 1, 3, and 5 wt.%) into methylcellulose/chitosan nanofiber films was shown to protect against UV radiation (96%) and maintain the freshness and visual quality of fresh-cut apples for up to five days at 4°C (Bahramian et al., 2025). Additionally, poly(limonene) (10% and 20%, w/w)-loaded citrus pectin-based films effectively protected apples from UV damage during storage, demonstrating their UV-barrier properties (Santana et al., 2024). The chitosan/pectin bilayer film loaded with epigallocatechin gallate/natamycin (EGCG/NATA) showed a UV transmittance of less than 1% in the 200–350 nm range at concentrations above 5%, providing effective UV protection for food packaging, especially strawberries (Fu et al., 2024). EGCG's strong UV-absorbing properties help prevent photodegradation of sensitive compounds like NATA, ensuring continuous antifungal activity during storage (Fig. 23.5A). This highlights the as-prepared film's potential for enhancing food preservation by blocking harmful UV rays, reducing weight loss by 30%, maintaining 50% higher firmness, and slowing the decline of vitamin C content by 20% compared to the control during storage. Gelatin-based films loaded with zein-quercetin nanoparticles also indicated proper UV protection effects, which help maintain the nutrient quality and retain an acceptable appearance of the strawberries for approximately eight days (Fig. 23.5B) (Fu et al., 2024). The addition of iron-based MOF with carbon quantum dots (CD@MOF) into the agar/gelatin matrix provided the film with significant UV protection, blocking 89.6% of UV-A and 99.5% of UV-B radiation, primarily due to the strong UV-absorbing properties of CDs and MOFs (Fig. 23.5C). Using composite film for tomato packaging significantly slowed degradation, with fruits showing negligible decay after 24 days at 4°C (Fig. 23.5D), compared to rapid spoilage in control groups (Riahi et al., 2024). The dispersion of clove essential oil into sodium alginate-gelatin films improved their UV screening properties and delayed the decay of bananas, making them a promising solution for food packaging with excellent UV-shielding properties (Upadhye et al., 2025). Food spoilage caused by photolysis and oxidation leads to blackening, bad taste, and nutritional loss. Adding nanofillers to polymer films enhances their oxygen and UV barriers, helping to slow degradation and extend shelf life. The addition of ZnONPs (0.2, 0.4, and 0.8 wt.%) to PVA films increases UV absorbance at 350 nm by 5%, 20%, and 40%, respectively (Channa et al., 2022). These composite films remain transparent in the visible range and can block up to 40% of UV radiation, making them suitable for UV-sensitive food packaging. As such, the optimized films kept apples fresh and unoxidized longer than unpackaged ones, demonstrating their effective preservative ability (Fig. 23.5E). Incorporating 2–4 wt.% TiO_2-functionalized bismuth-MOF particles into chitosan/gelatin films significantly enhanced UV-shielding performance, reducing UV-A and UV-B transmittance by up to 97.88% and 99.46%, respectively, while decreasing visible light transmittance to 18.5% (Riahi, Khan, Hong, et al., 2026). This improvement, driven by TiO_2's UV absorption and light-scattering effects, effectively delays browning index, declines weight loss, and maintains bananas' freshness for up to 10 days (Fig. 23.5F).

23.5.5 Edible oils

Edible oils rich in monounsaturated and polyunsaturated fats are highly prone to oxidation under UV exposure, affecting their color, flavor, and nutritional quality. Traditional packaging often offers insufficient protection, and synthetic

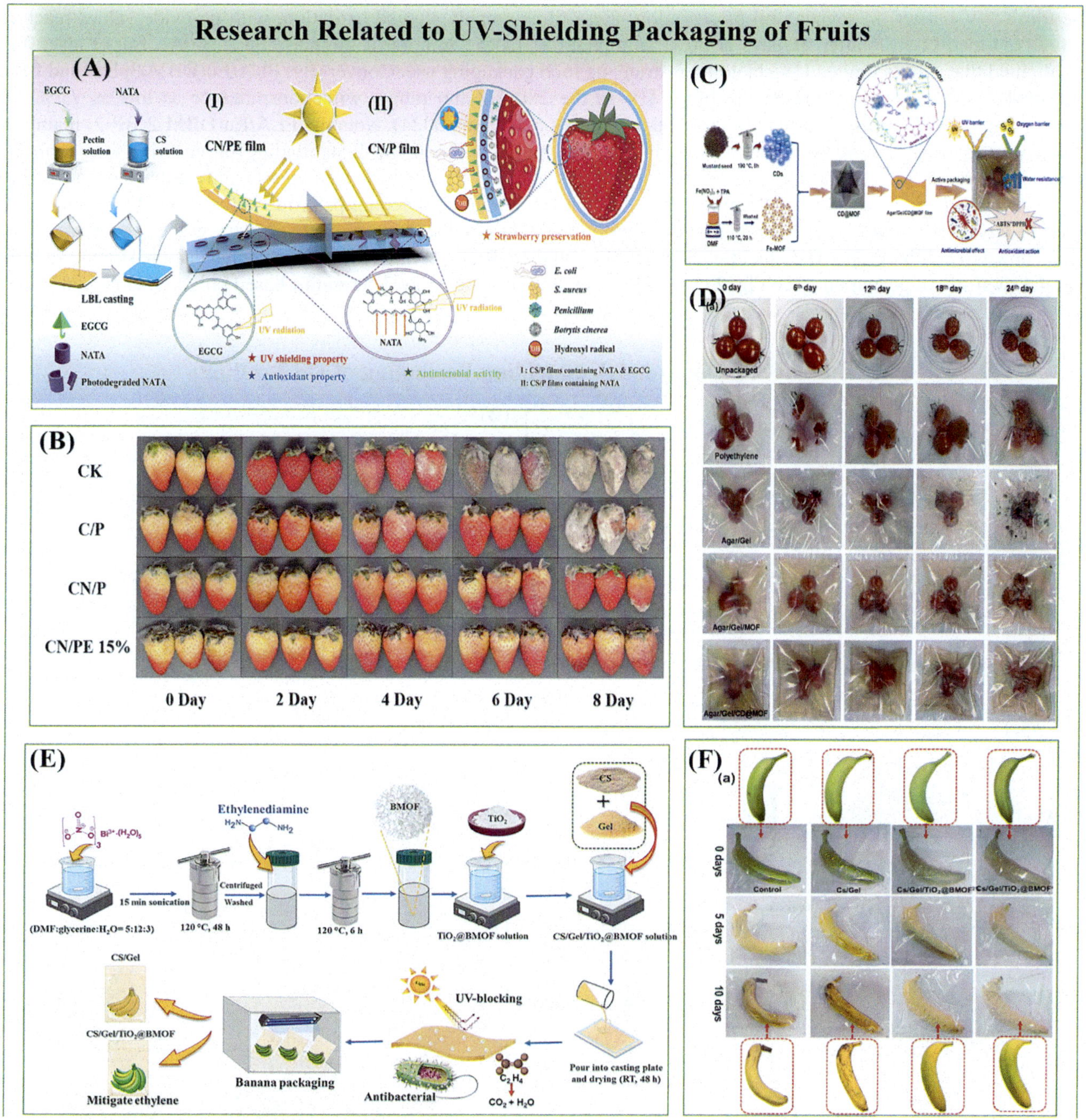

FIGURE 23.5 (A) Preparation diagram of chitosan/pectin bilayer film loaded with EGCG/NATA and (B) strawberry packaging at 25°C for eight days. (C) Schematic drawing of the development of CD@MOF@agar/gelatin matrix and (D) digital images of tomatoes stored at 4°C for 24 days. (E) Schematic illustration of polyvinyl alcohol films containing ZnONPs (0.2, 0.4, and 0.8 wt.%) and (F) digital images of packed bananas.
Reproduced with permission. (Fu et al., 2024; Riahi et al., 2024; Riahi, Khan, Hong, et al., 2026). Riahi, Z., et al., UV-barrier properties of carbon dots for food packaging applications, in Carbon Dots in Food Packaging and Preservation. 2026, Elsevier. p. 189–208. (n.d.).

antioxidants raise health concerns (Khemkaew et al., 2025; Liu et al., 2024). UV-shielding films can effectively prevent these detrimental changes and improve preservation. For example, Khemkaew et al. (2025) created chitosan films functionalized with 1.5% cashew leaf extract (CE), a natural preservative, and 0.5% TiO_2 nanoparticles, used as UV blockers (Khemkaew et al., 2025). The addition of CE and TiO_2 significantly decreased light transmission from 200 to 800 nm, mainly due to UV absorption by CE's polyphenols and light scattering by TiO_2, enhancing UV protection. The key mechanism involves phenolic compounds absorbing UV rays while TiO_2 NPs scatter and block UV light, improving the film's UV barrier. In soybean oil preservation, these films notably slowed oxidation, with peroxide values reaching only 10.54 meq O_2 kg^{-1} after 21 days at 50°C, compared to 145.51 in control samples (Fig. 23.6A). Liu et al. (2024) highlighted that PLA's limited UV shielding restricts its food packaging use. Conversely, PLA/Eu(DBM)3phen and PLA/Eu(DBM)2(BP-2)phen films effectively block UV in the 280–400 nm range, with transmittance as low as 7.7% and 10.6%, thanks to absorption and scattering by Eu complexes (Liu et al., 2024). Notably, PLA/Eu(DBM)2(BP-2) maintains minimal transmittance increase (from 7.7% to 11.6% over 12 days) due to BP-2 stabilizing Eu, ensuring long-term UV protection through phenolic absorption and nanoparticle scattering. These films also significantly extended flaxseed oil shelf life by reducing oxidation and pigment loss through UV shielding, keeping peroxide values low (13.73 mmol kg^{-1} after 21 days) comparable to dark storage. Pectin/pullulan edible films containing *Vitis vinifera* grape seed extract have been shown to exhibit proper UV-shielding abilities (Priyadarshi, Riahi, et al., 2022), which enhanced the storage stability and the shelf life of peanut oils (see Fig. 23.6B). Zhang et al. (2022) developed curcumin-based polyurethane-gelatin films

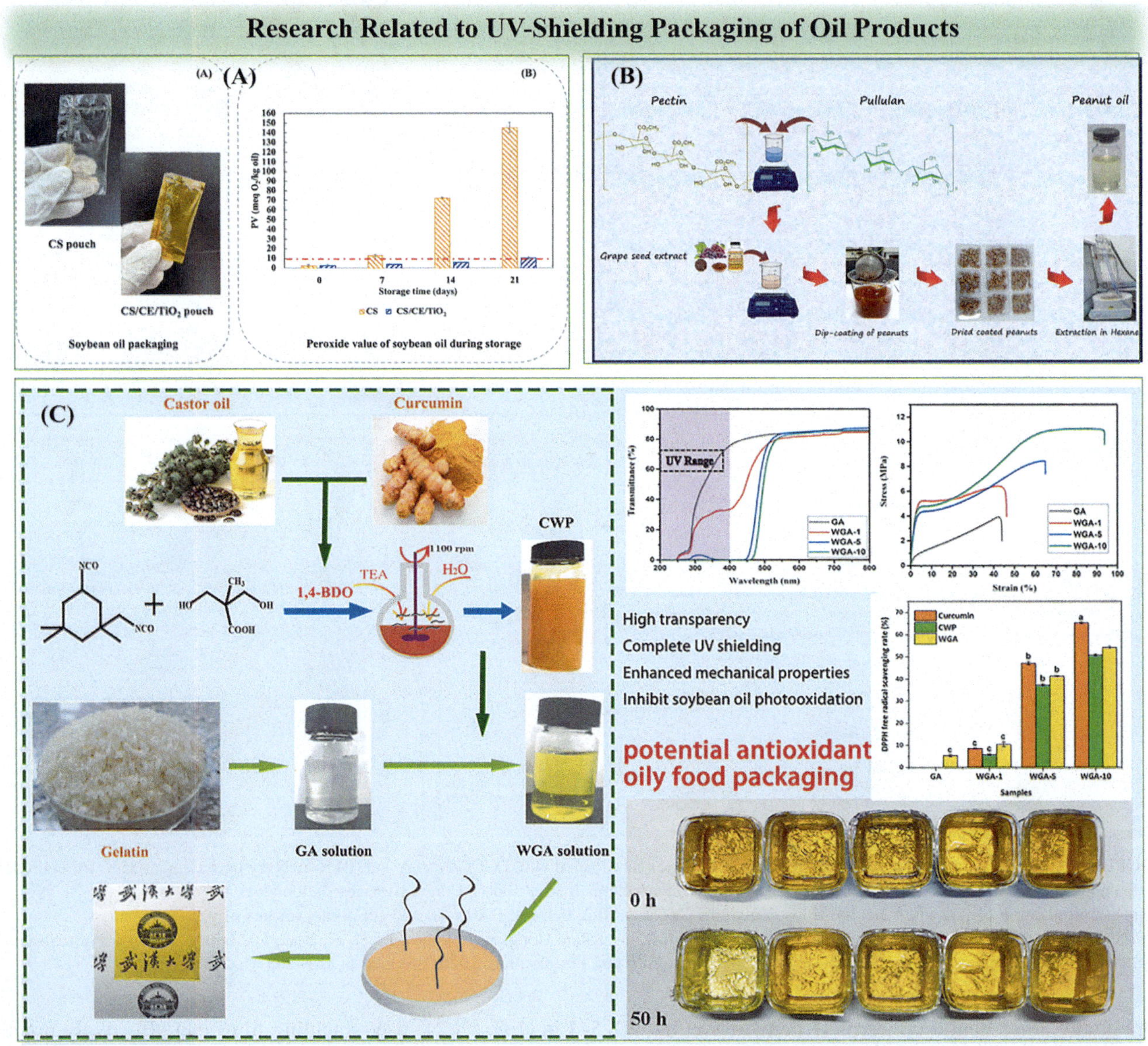

FIGURE 23.6 Application of UV-shielding packaging films based on (A) chitosan and cashew leaf extract + TiO_2, PLA/Eu(DBM)3phen, and curcumin-based polyurethane-gelatin for oil packaging.
(Khemkaew et al., 2025; Priyadarshi et al., 2022). Zhang, T., Zhang, W., Deng, Y., Chu, Y., Zhong, Y., Wang, G., & Li, H. (2022).Curcumin-based waterborne polyurethane-gelatin composite bioactive films for effective UV shielding and inhibition of oil oxidation. Food Control, 141. https://doi.org/10.1016/j.foodcont.2022.109199.

that effectively block UV rays and inhibit soybean oil oxidation. Incorporating 10 wt.% curcumin-based waterborne polyurethane maintained high transparency (83.83% at 600 nm) and increased antioxidant activity by 52.10%, reducing peroxide values by 87.84% and thiobarbituric acid values by 62.03%, thus significantly extending oil shelf life under UV exposure (see Fig. 23.6C) (Zhang et al., 2022). Citral-loaded pectin-beeswax edible films were also exhibited to provide strong UV-shielding properties to hazelnut oil (Kashiri, 2025).

23.5.6 Bakery products

Bakery products are susceptible to UV-induced damage, such as photodegradation and lipid oxidation, driving the need for UV-shielding packaging (Qian et al., 2021). These eco-friendly materials boost protection, extend shelf life, and support sustainable practices. Moreover, they facilitate easier handling, transportation, and compliance with labeling requirements (Schifferstein, 2024). In a study by Yang et al. (2024), the PVA/lignin-rich alkaline black liquor containing AgNPs (10 g L^{-1}) film achieved near-complete UV shielding, preventing 100% of UV radiation in the 200–400 nm range, significantly better than commercial PE films (Fig. 23.7A) (Yang et al., 2024). Its UV protection mechanism involves lignin's phenolic groups and AgNPs' absorption and scattering effects. The proposed film effectively preserved the freshness of bread and bananas by reducing weight loss, delaying browning, and preventing microbial growth over seven days (Fig. 23.7C). Its mechanism involves water retention, antimicrobial activity, and antioxidant properties, which inhibit spoilage and maintain food quality during storage (Fig. 23.7B). In a recent study, Sul et al. (2025) developed active bread packaging films based on chitosan/pullulan incorporating nitrogen-doped CDs from tangerine peel (Fig. 23.7D). Films loaded with N-CD nearly fully blocked UV-A and UV-B rays by reflecting and absorbing UV radiation, enhanced by polyphenols. The films delayed mold growth, with no mold after 12 days compared to 4 days in traditional packaging, while also maintaining moisture, reducing weight loss from 41% to 17.9%, and limiting microbial growth (Fig. 23.7E).

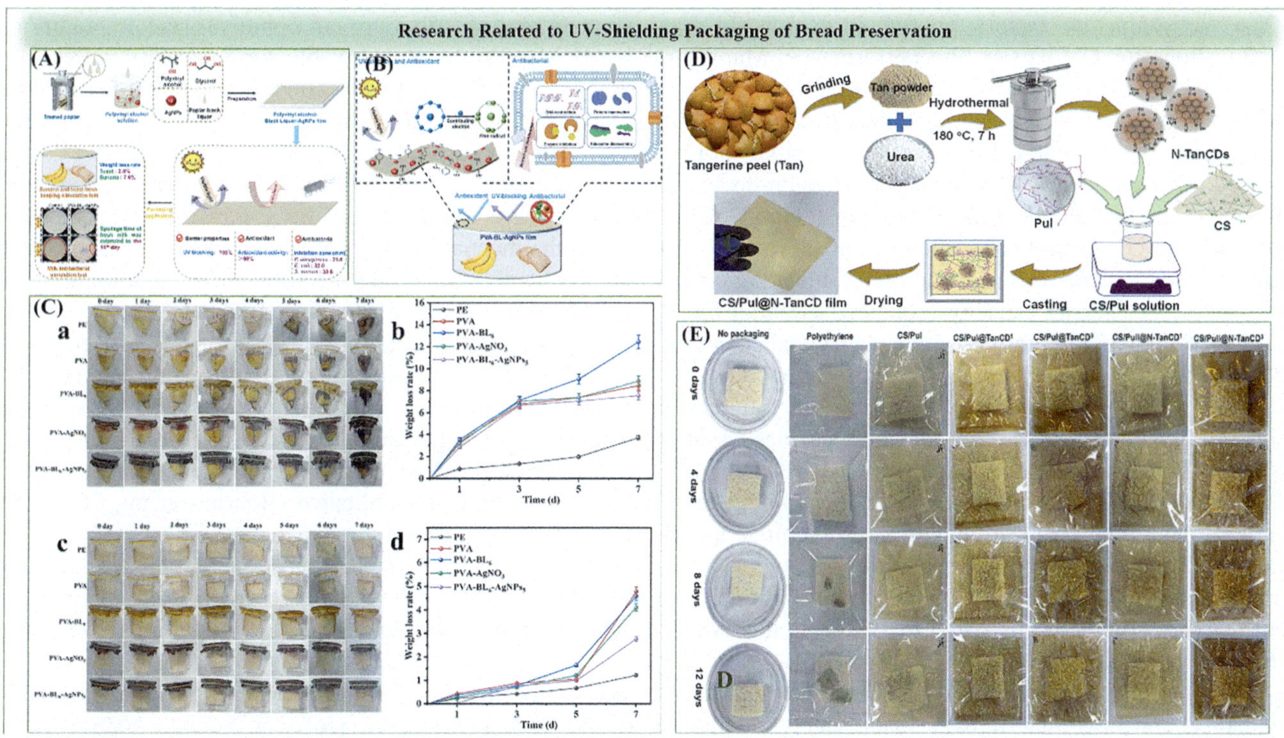

FIGURE 23.7 (A) Schematic drawing of the development of polyvinyl alcohol (PVA)/lignin-rich alkaline black liquor films containing AgNPs, (B) antibacterial/antiradical effects and UV-shielding mechanism of proposed film, and (C) assessing the freshness of bananas/bread. (D) nitrogen-doped CDs@chitosan/pullulan development process, and (E) visual images of sliced bread wrapped with the proposed films during storage for 12 days at room temperature.

Reprinted with permission. Yang, D., Fan, B., & He, Y.-C. (2024). UV-blocking, antibacterial, corrosion resistance, antioxidant, and fruit packaging ability of lignin-rich alkaline black liquor composite film. International Journal of Biological Macromolecules, 275, 133344. 2024; Sul, Y., Khan, A., Kim, J.T., & Rhim, J.-W. (2025). Tangerine peel-derived nitrogen-doped carbon dots incorporated chitosan/pullulan-based active packaging film for bread packaging. Colloids and Surfaces B: Biointerfaces, 245. https://doi.org/10.1016/j.colsurfb.2024.114339.

Riahi et al. (2022) and Sul et al. (2025) developed alginate films with sulfur quantum dots (SQD), significantly improving UV shielding by lowering T280 from 61.9% to 11.8%. This is closely due to the small size (5.3 nm) of SQD dispersing within the matrix to absorb UV radiation (Riahi et al., 2022). The prepared films also effectively inhibited mold growth on bread, with no mold after 14 days, extending shelf life compared to other packaging films. Chitosan films containing TiO_2 NPs have been shown to exhibit good UV-shielding abilities, which prolonged the shelf life of rye bread (Soltani et al., 2023).

23.6 Challenges and future perspective

The development of packaging films and coatings with strong UV-barrier properties represents one of the most significant challenges in modern food packaging industry (Tripathi et al., 2024). Although significant progress has been made in adding UV-shielding agents to biopolymer matrices, several technical, functional, and economic barriers continue to slow their large-scale commercial use (Patil et al., 2021). Unmodified natural polymers are biodegradable and biocompatible but offer poor inherent protection against UVA (315–400 nm) and UVB (280–315 nm) radiation. This creates a main problem: shifting to eco-friendly materials is essential for sustainability, yet these alternatives often fall short in barrier performance compared to traditional petroleum-based plastics, potentially reducing food protection effectiveness (Gore & Prajapat, 2022; Zhang et al., 2024).

A major technical difficulty lies in achieving good compatibility and stability when incorporating UV-blocking agents. Inorganic fillers like TiO_2 and ZnO are highly effective but tend to aggregate in biopolymer matrices, leading to opaque films that obscure product visibility, a key factor for consumer appeal (Alizadeh Sani et al., 2024; Channa et al., 2022). Poor interaction between many UV absorbers and the polar nature of natural polymers can cause phase separation and uneven distribution, weakening overall performance (Wang et al., 2025; Zhang et al., 2024). Ensuring uniform dispersion of nanoparticles at practical, cost-effective loadings remains a significant obstacle to scaling up from lab prototypes to industrial production.

The environmental factors present in real-world food storage and distribution environments create additional complications for UV-barrier film performance. Many biopolymer-based films with UV-blocking capacity suffer from instability when exposed to the combined effects of moisture, heat, oxygen, and light conditions ubiquitous in retail environments and consumer storage situations (Gore & Prajapat, 2022; Periyasamy et al., 2025). The mechanical and barrier properties of these films often deteriorate more rapidly than those of conventional plastic alternatives when exposed to accelerated UV aging, a phenomenon documented in comprehensive photodegradation studies of biodegradable polymers (Briassoulis et al., 2017; Guo & Guo, 2022). This photodegradation vulnerability is particularly problematic given that the UV protection is often forfeited precisely to maintain the mechanical integrity and barrier properties of the films themselves. The incorporation of UV-blocking compounds frequently reduces tensile strength, elongation at break, and water vapor barrier capacity properties essential for effective food preservation (Scaffaro et al., 2019).

Cost-effectiveness represents a major practical constraint limiting the implementation of advanced UV-barrier natural polymer films in commercial food packaging applications. The incorporation of nanofillers, lignin nanoparticles, CDs, MOFs, or other UV-blocking agents substantially increases the material production cost, often pricing these films beyond economic competitiveness with conventional plastics in price-sensitive food packaging markets (Alizadeh Sani et al., 2024). While research has demonstrated that incorporating PLA-grafted lignin nanoparticles can effectively block transmission of UV wavelengths while maintaining reasonable visible light transmittance (Boarino et al., 2022), the synthesis and functionalization processes required to achieve uniform dispersion remain expensive and not readily scalable. The development of extraction and processing methods for plant-derived UV absorbers such as quercetin, vanillic acid, or natural anthocyanins provides a more economical route, yet these compounds frequently exhibit lower UV-blocking efficiency compared to synthetic alternatives, requiring higher concentrations that compromise other film properties (Alizadeh Sani et al., 2024).

Migration and leaching of UV-blocking compounds from packaging into food products represent critical safety and regulatory concerns that remain inadequately addressed in the current literature (Jose & Salam, 2026). Lignin nanoparticles, zinc oxide nanoparticles, and other inorganic fillers can potentially leach into food products, particularly fatty foods or those with elevated acidity, creating safety risks that necessitate comprehensive migration studies under varied storage conditions and food simulants (Anushikha & Gaikwad, 2024; Wang et al., 2023). Regulatory frameworks governing the use of nanoparticles and other novel UV-absorbing agents in food contact materials remain underdeveloped, creating significant uncertainty regarding permissible materials and maximum allowable migration limits. This regulatory ambiguity substantially complicates commercial development strategies, as manufacturers must engage in extensive safety testing and regulatory approval processes with uncertain outcomes.

The challenge of maintaining optical properties while achieving UV protection remains technically challenging. Most effective UV-blocking agents, whether organic compounds or inorganic nanoparticles, inherently reduce film transparency; the latter substantially diminishes consumer appeal and restricts applications where product visibility is essential for marketing and consumer acceptance (Channa et al., 2022; Le et al., 2025). Films that successfully block UV-B and UV-A radiation while maintaining transparency represent exceptional achievements that have been demonstrated only under laboratory conditions with limited practical scalability (Patil et al., 2020). Furthermore, the photostability of many natural colorants and UV-absorbing compounds, while superior to synthetic alternatives in some respects, still proves insufficient under the prolonged storage and retail display conditions encountered in real food supply chains.

Future research and development in UV-barrier properties of natural polymer-based films must prioritize the development of multifunctional approach strategies that simultaneously address UV protection, mechanical performance, optical clarity, cost-effectiveness, and regulatory compliance. Biomass-derived CDs or MOFs emerge as particularly promising candidates, offering exceptional UV-absorbing capacity combined with antioxidant, antimicrobial, and pH-responsive properties, thereby enabling development of truly multifunctional packaging systems (Alizadeh Sani et al., 2024). These carbon-based nanomaterials can achieve significant UV blockage with blue-emitting CDs while maintaining high transparency and reduced opacity compared to traditional inorganic nanofillers. The sustainability profile of CDs synthesized from natural precursors such as plant biomass or waste materials further enhances their appeal for food packaging applications (Riahi et al., 2025; Riahi, Khan, Ebrahimi, et al., 2026).

Layer-by-layer assembly and multilayer film design strategies offer promising approaches for achieving UV protection without compromising other essential film properties (Borges et al., 2014). By strategically placing UV-blocking components in outer protective layers while maintaining mechanical and barrier properties in internal layers, researchers can optimize overall film performance. Composite films featuring with UV-blocking layers of titanium dioxide, ZnO, or other inorganic compounds have demonstrated superior performance compared to homogeneous films with similar overall composition. This stratified approach enables customization of functional properties while minimizing the overall quantity of UV-absorbing agents required, thereby reducing costs and migration concerns.

Regulatory progress is essential. Clear, harmonized guidelines and standardized migration tests covering diverse food types and storage conditions are necessary to build confidence and streamline approvals (Dai et al., 2025; Ilyas et al., 2021; Versino et al., 2023).

Consumer education highlighting sustainability benefits and improved food quality will also drive acceptance, especially if supported by successful pilot applications. Finally, integrating computational tools like machine learning for material screening, along with smart sensors for real-time monitoring, will accelerate innovation. These advances can lead to next-generation packaging that not only blocks UV effectively but also enhances overall food safety and reduces waste, paving the way for broader adoption of sustainable natural polymer films in the food industry.

23.7 Conclusion

In conclusion, the development of effective UV-barrier properties in natural polymer-based films and coatings represents a significant advancement in sustainable food packaging technology, directly addressing the vulnerabilities of light-sensitive foods to ultraviolet radiation. As highlighted throughout this chapter, UV exposure accelerates lipid oxidation, vitamin degradation, discoloration, and off-flavor development, compromising both the nutritional quality and shelf life of products such as dairy, meats, oils, fruits, and beverages.

Natural polymers offer inherent advantages of biodegradability, biocompatibility, and renewability, positioning them as eco-friendly alternatives to conventional petroleum-based materials. However, their limited intrinsic UV-blocking capability necessitates the strategic incorporation of specialized agents to achieve robust photoprotection.

The reviewed UV-blocking agents fall into distinct categories: inorganic nanoparticles, organic absorbers, inherently protective biopolymers, plant-based phytochemicals (including polyphenols and carotenoids), and combined hybrid formulations. These materials work through various principles, primarily absorbing, reflecting, or scattering harmful UV wavelengths. When integrated into biopolymer matrices, they improve light-blocking efficiency. Practical applications illustrate how these packaging materials extend shelf life and preserve sensory and nutritional attributes in various food categories. Still, hurdles remain, including possible loss of film transparency, concerns over additive migration into food, and the need to meet strict safety regulations.

Moving forward, opportunities lie in advanced solutions like hybrid nanocomposites with multiple functions, intelligent systems that respond to UV exposure, and the use of farm waste as an affordable source of natural blockers. By combining nanotechnology with sustainable practices, these natural polymer films can rival or surpass conventional packaging in UV protection while supporting environmental goals. Ultimately, such progress supports the broader implementation of biodegradable materials that prioritize food quality preservation and contribute to more sustainable food systems worldwide.

AI disclosure

During the preparation of this work, the author(s) used Grok and Deep in order to summarize, find synonyms, make grammatical and spelling corrections, and paraphrase. After using this tool/service, the author(s) reviewed and edited the content as needed and take(s) full responsibility for the content of the publication.

References

Abedi-Firoozjah, R., Azimi-Salim, S., Afrah, A., Moradiyan, M. H., Assadpour, E., Sarlak, Z., & Jafari, S. M. (2025). Comparison of the pH- and NH3-sensitivity of chitosan/polyvinyl alcohol smart films containing anthocyanins or betacyanins for monitoring fish freshness. *Carbohydrate Polymer Technologies and Applications, 9*, 100639. https://doi.org/10.1016/j.carpta.2024.100639 2025.

Abedi-Firoozjah, R., Bahramian, B., Tavassoli, M., Majlesi, M., Ghaderi, S., Assadpour, E., & Jafari, S. M. (2025). Alginate-based edible films/coatings/nanofibers in food packaging: A comprehensive review of recent advances. *Carbohydrate Polymer Technologies and Applications, 11*, 100955. https://doi.org/10.1016/j.carpta.2025.100955 2025.

Aboelela, D., & Bassyouni, M. (2024). Biodegradable polymers converting environmental challenges into eco-friendly solutions. *Egyptian Journal of Chemistry, 67*, 2015–2032 2024.

Ahmed, M., Pickova, J., Ahmad, T., Liaquat, M., Farid, A., & Jahangir, M. (2016). Oxidation of lipids in foods. *Sarhad Journal of Agriculture, 32*, 230–238 2016.

Alizadeh Sani, M., Khezerlou, A., Tavassoli, M., Abedini, A. H., & McClements, D. J. (2024). Development of sustainable UV-screening food packaging materials: A review of recent advances. *Trends in Food Science & Technology, 145*, 104366. https://doi.org/10.1016/j.tifs.2024.104366 2024.

Alizadeh-Sani, M., Tavassoli, M., Mohammadian, E., Ehsani, A., Khaniki, G. J., Priyadarshi, R., & Rhim, J.-W. (2021). pH-responsive color indicator films based on methylcellulose/chitosan nanofiber and barberry anthocyanins for real-time monitoring of meat freshness. *International Journal of Biological Macromolecules, 166*, 741–750 2021.

Ananikov, V.P. (2019). Organic–inorganic hybrid nanomaterials. In (Vol. 9, pp. 1197): MDPI.

Anushikha, & Gaikwad, K. K. (2024). Lignin as a UV blocking, antioxidant, and antimicrobial agent for food packaging applications. *Biomass Conversion and Biorefinery, 14*, 16755–16767 2024.

Assimopoulou, A., & Papageorgiou, V. (2005). Radical scavenging activity of *Alkanna tinctoria* root extracts and their main constituents, hydroxynaphthoquinones. *Phytotherapy Research: An International Journal Devoted to Pharmacological and Toxicological Evaluation of Natural Product Derivatives, 19*, 141–147 2005.

Assis, R. Q., Rios, P. D. A., de Oliveira Rios, A., & Olivera, F. C. (2020). Biodegradable packaging of cellulose acetate incorporated with norbixin, lycopene or zeaxanthin. *Industrial Crops and Products, 147*, 112212 2020.

Badry, R., El-Nahass, M. M., Nada, N., Elhaes, H., & Ibrahim, M. A. (2023). Structural and UV-blocking properties of carboxymethyl cellulose sodium/CuO nanocomposite films. *Scientific Reports, 13*, 1123 2023.

Bahramian, B., Abedi-Firoozjah, R., Khezerlou, A., Kiani-Salmi, N., Ahmadi, N., Alizadeh Sani, M., & Jafari, S. M. (2025). Biodegradable active packaging films based on methyl cellulose/chitosan nanofibers and turnip peel carbon dots for perishable food preservation. *Carbohydrate Polymer Technologies and Applications, 11*, 100950. https://doi.org/10.1016/j.carpta.2025.100950 2025.

Barden, L., & Decker, E. A. (2016). Lipid oxidation in low-moisture food: A review. *Critical Reviews in Food Science and Nutrition, 56*, 2467–2482 2016.

Bian, H., Luo, D., Shi, J., Zhou, X., Xu, T., Xiao, H., & Ragauskas, A. J. (2024). Cellulose nanofibrils-based packaging films synergistically reinforced by lignin and tea polyphenols for strawberry preservation. *Chemical Engineering Journal, 502*, 157843 2024.

Biegańska, M. (2022). Packaging of dairy products: Emerging strategies. In *Food packaging: The smarter way*. Springer, 127–164.

Bintsis, T., Litopoulou-Tzanetaki, E., & Robinson, R. K. (2000). Existing and potential applications of ultraviolet light in the food industry–A critical review. *Journal of the Science of Food and Agriculture, 80*, 637–645 2000.

Boarino, A., Schreier, A., Leterrier, Y., & Klok, H.-A. (2022). Uniformly dispersed poly (lactic acid)-grafted lignin nanoparticles enhance antioxidant activity and UV-barrier properties of poly (lactic acid) packaging films. *ACS Applied Polymer Materials, 4*, 4808–4817 2022.

Borges, J., Rodrigues, L. C., Reis, R. L., & Mano, J. F. (2014). Layer-by-layer assembly of light-responsive polymeric multilayer systems. *Advanced Functional Materials, 24*, 5624–5648 2014.

Braga, L. R., Pérez, L. M., Soazo, M.d V., & Machado, F. (2019). Evaluation of the antimicrobial, antioxidant and physicochemical properties of Poly (Vinyl chloride) films containing quercetin and silver nanoparticles. *LWT, 101*, 491–498 2019.

Briassoulis, D., Tserotas, P., & Hiskakis, M. (2017). Mechanical and degradation behaviour of multilayer barrier films. *Polymer Degradation and Stability, 143*, 214–230 2017.

Chandran, G. U., Kumar, A. A., Menon, S. K., Sambhudevan, S., & Shankar, B. (2024). The potential role of flavonoids in cellulose-based biopolymeric food packaging materials for UV radiation protection. *Cellulose, 31*, 4733–4773 2024.

Channa, I. A., Ashfaq, J., Gilani, S. J., Shah, A. A., Chandio, A. D., & Jumah, M. N. (2022). UV blocking and oxygen barrier coatings based on polyvinyl alcohol and zinc oxide nanoparticles for packaging applications. *Coatings, 12*, 897. 2022 https://www.mdpi.com/2079-6412/12/7/897.

Chaudhary, B. U., Lingayat, S., Banarjee, A. N., & Kale, R. D. (2022). Preparation and characterization of antioxidant, antimicrobial, and UV-light protection film based on poly (vinyl alcohol) and garlic peel extract. *Waste and Biomass Valorization, 13*, 4717–4734 2022.

Chung, Y. J., Kim, J., & Park, C. B. (2020). Photonic carbon dots as an emerging nanoagent for biomedical and healthcare applications. *ACS Nano, 14*, 6470–6497 2020.

Dai, K., Cao, S., Yuan, J., Wang, Z., Li, H., Yuan, C., & Xing, R. (2025). Recent advances of sustainable UV shielding materials: Mechanisms and applications. *ACS Applied Materials & Interfaces, 17*, 30402–30422 2025.

Danmatam, N., Pearce, J., & Pattavarakorn, D. (2023). UV-blocking properties of carboxymethyl cellulose film integrated with oolong tea extracts as eco-friendly packaging film. *Materials Today: Proceedings, 77*, 1052–1058 2023.

Delorme, M. M., Guimarães, J. T., Coutinho, N. M., Balthazar, C. F., Rocha, R. S., Silva, R., & Freitas, M. Q. (2020). Ultraviolet radiation: An interesting technology to preserve quality and safety of milk and dairy foods. *Trends in Food Science & Technology, 102*, 146–154 2020.

Du, Y., & Zheng, M. (2024). Silver nanoparticles/carbon dots nanocomposite with potent antibiofilm and long-term antimicrobial activity doped into chitosan/polyvinyl alcohol film for food preservation. *International Journal of Biological Macromolecules, 283*, 137744. https://doi.org/10.1016/j.ijbiomac.2024.137744 2024.

Ezati, P., & Rhim, J.-W. (2020). pH-responsive pectin-based multifunctional films incorporated with curcumin and sulfur nanoparticles. *Carbohydrate Polymers, 230*, 115638 2020.

Ezati, P., Priyadarshi, R., Bang, Y.-J., & Rhim, J.-W. (2021). CMC and CNF-based intelligent pH-responsive color indicator films integrated with shikonin to monitor fish freshness. *Food Control, 126*, 108046 2021.

Ezati, P., Priyadarshi, R., & Rhim, J.-W. (2022). Prospects of sustainable and renewable source-based carbon quantum dots for food packaging applications. *Sustainable Materials and Technologies, 33*, e00494 2022.

Ezati, P., Rhim, J.-W., Molaei, R., & Rezaei, Z. (2022). Carbon quantum dots-based antifungal coating film for active packaging application of avocado. *Food Packaging and Shelf Life, 33*, 100878 2022.

Ezati, P., Khan, A., Priyadarshi, R., Bhattacharya, T., Tammina, S. K., & Rhim, J.-W. (2023). Biopolymer-based UV protection functional films for food packaging. *Food Hydrocolloids, 142*, 108771 2023.

Fang, Z., Zhao, Y., Warner, R. D., & Johnson, S. K. (2017). Active and intelligent packaging in meat industry. *Trends in Food Science & Technology, 61*, 60–71 2017.

Feng, X., Zhao, Y., Jiang, Y., Miao, M., Cao, S., & Fang, J. (2017). Use of carbon dots to enhance UV-blocking of transparent nanocellulose films. *Carbohydrate Polymers, 161*, 253–260 2017.

Flórez, M., Cazón, P., & Vázquez, M. (2022). Active packaging film of chitosan and *Santalum album* essential oil: Characterization and application as butter sachet to retard lipid oxidation. *Food Packaging and Shelf Life, 34*, 100938. https://doi.org/10.1016/j.fpsl.2022.100938 2022.

Fu, X., Chang, X., Xu, S., Xu, H., Ge, S., Xie, Y., & Ding, S. (2024). Development of a chitosan/pectin-based multi-active food packaging with both UV and microbial defense functions for effectively preserving of strawberry. *International Journal of Biological Macromolecules, 254*, 127968. https://doi.org/10.1016/j.ijbiomac.2023.127968 2024.

Ganguly, P., Breen, A., & Pillai, S. C. (2018). Toxicity of nanomaterials: Exposure, pathways, assessment, and recent advances. *ACS Biomaterials Science & Engineering, 4*(2018), 2237–2275.

Garrido-Romero, M., Aguado, R., Moral, A., Brindley, C., & Ballesteros, M. (2022). From traditional paper to nanocomposite films: Analysis of global research into cellulose for food packaging. *Food Packaging and Shelf Life, 31*, 100788 2022.

Gore, A. H., & Prajapat, A. L. (2022). Biopolymer nanocomposites for sustainable UV protective packaging. *Frontiers in Materials, 9*, 855727 2022.

Guo, C., & Guo, H. (2022). Progress in the degradability of biodegradable film materials for packaging. *Membranes, 12*, 500 2022.

Haghighi, H., Gullo, M., La China, S., Pfeifer, F., Siesler, H. W., Licciardello, F., & Pulvirenti, A. (2021). Characterization of bio-nanocomposite films based on gelatin/polyvinyl alcohol blend reinforced with bacterial cellulose nanowhiskers for food packaging applications. *Food Hydrocolloids, 113*, 106454 2021.

Hamouda, S. A., Alshawish, N. K., Abdalla, Y. K., & Ibrahim, M. K. (2022). Ultraviolet radiation: Health risks and benefits. *Saudi Journal of Engineering and Technology, 7*, 533–541 2022.

Hao, G., Hu, Y., Shi, L., Chen, J., Cui, A., Weng, W., & Osako, K. (2021). Physicochemical characteristics of chitosan from swimming crab (*Portunus trituberculatus*) shells prepared by subcritical water pretreatment. *Scientific Reports, 11*, 1646 2021.

He, J., Evans, N. M., Liu, H., Zhu, Y., Zhou, T., & Shao, S. (2021). UV treatment for degradation of chemical contaminants in food: A review. *Comprehensive Reviews in Food Science and Food Safety, 20*, 1857–1886 2021.

Ilyas, R., Sapuan, S., Megashah, L., Ibrahim, R., Atikah, M., Ainun, Z., & Lee, C. (2021). Regulations for food packaging materials. *Bio-based Packaging: Material, Environmental and Economic Aspects*, 467–494 2021.

Jafarzadeh, S., Salehabadi, A., Mohammadi Nafchi, A., Oladzadabbasabadi, N., & Jafari, S. M. (2021). Cheese packaging by edible coatings and biodegradable nanocomposites; improvement in shelf life, physicochemical and sensory properties. *Trends in Food Science & Technology, 116*, 218–231. https://doi.org/10.1016/j.tifs.2021.07.021 2021.

Jiang, H., Zhao, S., Li, Z., Chen, L., Mo, H., & Liu, X. (2024). Swan-feathers inspired smart-responsive sustainable carboxymethyl cellulose/polyvinyl alcohol based food packaging film for robustly integrated Intelligent and Active Packaging. *Nano Today, 56*, 102272. https://doi.org/10.1016/j.nantod.2024.102272 2024.

Jose, J., & Salam, J. A. (2026). Food packaging with UV-shielding functional materials. *Novel Applications of Functionally Graded Materials,* 115–136 2026.

Kainulainen, T. P., Sirviö, J. A., Sethi, J., Hukka, T. I., & Heiskanen, J. P. (2018). UV-blocking synthetic biopolymer from biomass-based bifuran diester and ethylene glycol. *Macromolecules, 51,* 1822–1829 2018.

Kashiri, M. (2025). Effect of citral incorporation on the antimicrobial, barrier, mechanical, and UV-protective properties of pectin-beeswax films for food packaging applications. *Applied Food Research, 5,* 101051. https://doi.org/10.1016/j.afres.2025.101051 2025.

Khan, A., Priyadarshi, R., Bhattacharya, T., & Rhim, J.-W. (2023). Carrageenan/alginate-based functional films incorporated with *Allium sativum* carbon dots for UV-barrier food packaging. *Food and Bioprocess Technology, 16,* 2001–2015 2023.

Khemkaew, P., Jaisan, C., Kingwascharapong, P., Rawdkuen, S., Karbowiak, T., Degraeve, P., & Kaewprachu, P. (2025). Biobased antioxidant packaging from chitosan incorporating cashew leaf extract and TiO_2 nanoparticles for soybean oil preservation. *LWT, 228,* 118053. https://doi.org/10.1016/j.lwt.2025.118053 2025.

Khezerlou, A., Tavassoli, M., Abedi-Firoozjah, R., Mcclements, D. J., Ehsani, A., Bangar, S. P., & Sani, M. A. (2025). Metal–Organic Frameworks (MOFs): Innovative materials for food packaging applications. *Food and Bioprocess Technology,* 1–22 2025.

Kontominas, M. G., Badeka, A. V., Kosma, I. S., & Nathanailides, C. I. (2021). Recent developments in seafood packaging technologies. *Foods, 10,* 940 2021.

Kwon, S., Orsuwan, A., Bumbudsanpharoke, N., Yoon, C., Choi, J., & Ko, S. (2018). A short review of light barrier materials for food and beverage packaging. *Korean Journal of Packaging Science & Technology, 24,* 141–148 2018.

Lazaridi, E., Janssen, H.-G., Vincken, J.-P., Pirok, B., & Hennebelle, M. (2021). A comprehensive two-dimensional liquid chromatography method for the simultaneous separation of lipid species and their oxidation products. *Journal of Chromatography. A, 1644,* 462106. https://doi.org/10.1016/j.chroma.2021.462106 2021.

Le, G. H., Thanh, D. A., Pham, T. T., Tran, Q. V., Dao, N. N., Nguyen, K. T., & Quan, T. T. (2025). UV-blocking and mechanically reinforced starch films incorporating Ce-UiO-66 nanoparticles for food packaging applications. *RSC Advances, 15,* 30415–30426 2025.

Li, S., Fang, C., Wei, N., Wei, J., Feng, T., Liu, F., & Wu, B. (2025). Antimicrobial, antioxidative, and UV-blocking pectin/gelatin food packaging films incorporated with tannic acid and silver nanoparticles for strawberry preservation. *International Journal of Biological Macromolecules, 308,* 142445 2025.

Li, X., Yang, M., Zhang, W., Chen, K., & Wang, C. (2021). High concentration acid-induced discoloration polymeric dyes fabricated with UV-curable azobenzene-lignin-based waterborne polyurethane. *International Journal of Biological Macromolecules, 182,* 1953–1965. https://doi.org/10.1016/j.ijbiomac.2021.05.173 2021.

Li, Y., Yang, J., Sun, L., Liu, B., Li, H., & Peng, L. (2023). Crosslinked fish scale gelatin/alginate dialdehyde functional films incorporated with carbon dots derived from pomelo peel waste for active food packaging. *International Journal of Biological Macromolecules, 253,* 127290 2023.

Li, Z., Guan, J., Yan, C., Chen, N., Wang, C., Liu, T., & Shao, Z. (2023). Corn straw core/cellulose nanofibers composite for food packaging: Improved mechanical, bacteria blocking, ultraviolet and water vapor barrier properties. *Food Hydrocolloids, 143,* 108884. https://doi.org/10.1016/j.foodhyd.2023.108884 2023.

Lino, R. C., de Carvalho, S. M., Noronha, C. M., Sganzerla, W. G., da Rosa, C. G., Nunes, M. R., & Barreto, P. L. M. (2022). Production of methylcellulose films functionalized with poly-ε-caprolactone nanocapsules entrapped β-carotene for food packaging application. *Food Research International, 160,* 111750 2022.

Liu, L., Wu, Y., Yun, X., Wang, X., Li, J., Chen, L., & Song, L. (2024). UV-barrier poly(lactic acid) film with light-stabilized Eu complexes as filler. *International Journal of Biological Macromolecules, 271,* 132529. https://doi.org/10.1016/j.ijbiomac.2024.132529 2024.

Liu, M. (2020). Optical properties of carbon dots: A review. *Nanoarchitectonics, 2020,* 1–12.

Manickam, S., Haromuchadi, S. R., & Naik, R. (2025). Application of ultraviolet light in solid and semi-solid foods. In *Nonthermal light-based technologies in food processing.* Apple Academic Press, 107–161.

Min, S., Priyadarshi, R., Ezati, P., Rhim, J.-W., & Kim, J. T. (2023). Chitosan-based multifunctional coating combined with sulfur quantum dots to prevent Listeria contamination of enoki mushrooms. *Food Packaging and Shelf Life, 35,* 101014 2023.

Moustafa, H., El-Sayed, S. M., & Youssef, A. M. (2023). Synergistic impact of cumin essential oil on enhancing of UV-blocking and antibacterial activity of biodegradable poly (butylene adipate-co-terephthalate)/clay platelets nanocomposites. *Journal of Thermoplastic Composite Materials, 36,* 96–117 2023.

Murugan, G., Nilsuwan, K., Kim, J. T., Rhim, J. W., Yin, T., Zhang, B., & Benjakul, S. (2025). Bambara groundnut pericarp–derived carbon dots: Reinforcement of gelatin/PLA bilayer films and their function as preservative in packed asian seabass slices. *Journal of Food Science, 90*(2025), e70469.

Nayak, B., & Mishra, A. A. (2025). *Food packaging materials: Types and properties. concepts of food packaging* 2025.

Nikvarz, N., Khayati, G. R., & Sharafi, S. (2020). Preparation of UV absorbent films using polylactic acid and grape syrup for food packaging application. *Materials Letters, 276,* 128187 2020.

Paidari, S., Tahergorabi, R., Anari, E. S., Nafchi, A. M., Zamindar, N., & Goli, M. (2021). Migration of various nanoparticles into food samples: A review. *Foods, 10,* 2114 2021.

Pandiselvam, R., Barut, G. S., Yüksel, A. N., Tekgül, Y., Çalişkan, K. G., & Kothakota, A. (2022). Evaluation of the impact of UV radiation on rheological and textural properties of food. *Journal of Texture Studies, 53,* 800–808 2022.

Patil, A. S., Waghmare, R. D., Pawar, S. P., Salunkhe, S. T., Kolekar, G. B., Sohn, D., & Gore, A. H. (2020). Photophysical insights of highly transparent, flexible and re-emissive PVA@ WTR-CDs composite thin films: A next generation food packaging material for UV blocking applications. *Journal of Photochemistry and Photobiology A: Chemistry, 400,* 112647 2020.

Patil, A. S., Nille, O. S., Kolekar, G. B., Sohn, D., & Gore, A. H. (2021). UV-shielding biopolymer@ nanocomposites for sustainable packaging applications. In *Sustainable packaging*. Springer, 177–202.

Periyasamy, T., Asrafali, S. P., & Lee, J. (2025). Recent advances in functional biopolymer films with antimicrobial and antioxidant properties for enhanced food packaging. *Polymers, 17*, 1257 2025.

Phan, T. T. V., Phan, D. T., Cao, X. T., Huynh, T.-C., & Oh, J. (2021). Roles of chitosan in green synthesis of metal nanoparticles for biomedical applications. *Nanomaterials, 11*, 273 2021.

Philp, J. C., Bartsev, A., Ritchie, R. J., Baucher, M.-A., & Guy, K. (2013). Bioplastics science from a policy vantage point. *New Biotechnology, 30*, 635–646 2013.

Ponnusamy, A., Palamae, S., Prodpran, T., Kim, J. T., Zhang, B., Ma, L., & Benjakul, S. (2025). Fabrication of a chitosan–gelatin/polylactic acid bilayer active film loaded with tannic acid for enhancing shelf-life of refrigerated baby clams. *Foods, 14*, 3934. 2025 https://www.mdpi.com/2304-8158/14/22/3934.

Priyadarshi, R., & Rhim, J.-W. (2020). Chitosan-based biodegradable functional films for food packaging applications. *Innovative Food Science & Emerging Technologies, 62*, 102346 2020.

Priyadarshi, R., Purohit, S. D., Roy, S., Ghosh, T., Rhim, J.-W., & Han, S. S. (2022). Antiviral biodegradable food packaging and edible coating materials in the COVID-19 era: A mini-review. *Coatings, 12*, 577 2022.

Priyadarshi, R., Riahi, Z., & Rhim, J.-W. (2022). Antioxidant pectin/pullulan edible coating incorporated with Vitis vinifera grape seed extract for extending the shelf life of peanuts. *Postharvest Biology and Technology, 183*, 111740. https://doi.org/10.1016/j.postharvbio.2021.111740 2022.

Priyadarshi, R., Roy, S., Purohit, S.D., & Ghosh, T. (2022). Biopolymers for food packaging and biomedical applications: options or obligations? In (Vol. 12, pp. 1261): MDPI.

Prodpran, T., Benjakul, S., & Phatcharat, S. (2012). Effect of phenolic compounds on protein cross-linking and properties of film from fish myofibrillar protein. *International Journal of Biological Macromolecules, 51*, 774–782 2012.

Qamar, S. A., Ashiq, M., Jahangeer, M., Riasat, A., & Bilal, M. (2020). Chitosan-based hybrid materials as adsorbents for textile dyes–A review. *Case Studies in Chemical and Environmental Engineering, 2*, 100021 2020.

Qian, M., Liu, D., Zhang, X., Yin, Z., Ismail, B. B., Ye, X., & Guo, M. (2021). A review of active packaging in bakery products: Applications and future trends. *Trends in Food Science & Technology, 114*, 459–471 2021.

Qu, B., Xiao, Z., & Luo, Y. (2025). Facile in situ synthesis of ZnO-infused caseinate/carboxymethyl cellulose films with antibacterial properties for cheddar cheese packaging. *Carbohydrate Polymer Technologies and Applications, 10*, 100779. https://doi.org/10.1016/j.carpta.2025.100779 2025.

Quilez-Molina, A. I., Marini, L., Athanassiou, A., & Bayer, I. S. (2020). UV-blocking, transparent, and antioxidant polycyanoacrylate films. *Polymers, 12*, 2011. 2020 https://www.mdpi.com/2073-4360/12/9/2011.

Rabani, I., Lee, S.-H., Kim, H.-S., Yoo, J., Hussain, S., Maqbool, T., & Seo, Y.-S. (2021). Engineering-safer-by design ZnO nanoparticles incorporated cellulose nanofiber hybrid for high UV protection and low photocatalytic activity with mechanism. *Journal of Environmental Chemical Engineering, 9*, 105845. https://doi.org/10.1016/j.jece.2021.105845 2021.

Ran, R., Luo, X., Zhai, D., He, Y., Guo, X., He, L., & Li, G. (2025). Development of a dual-layer collagen film with insulating, antioxidant, antimicrobial, and pH-sensitive properties for the maintenance and visual monitoring of shrimp freshness. *Food Hydrocolloids, 169*, 111603. https://doi.org/10.1016/j.foodhyd.2025.111603 2025.

Ran, X., Qu, Y., Wang, Y., Cui, B., Shen, Y., & Li, Y. (2023). Enhanced UV-blocking capabilities of polylactic acid derived from renewable resources for food and drug packaging: A mini-review. *Journal of Composites Science, 7*, 410 2023.

Rejeesh, C., & Anto, T. (2023). Packaging of milk and dairy products: Approaches to sustainable packaging. *Materials Today: Proceedings, 72*, 2946–2951 2023.

Rhim, J.-W., & Kim, Y.-T. (2014). Biopolymer-based composite packaging materials with nanoparticles. In *Innovations in food packaging*. Elsevier, 413–442.

Riahi, Z., Priyadarshi, R., Rhim, J.-W., Lotfali, E., Bagheri, R., & Pircheraghi, G. (2022). Alginate-based multifunctional films incorporated with sulfur quantum dots for active packaging applications. *Colloids and Surfaces B: Biointerfaces, 215*, 112519. https://doi.org/10.1016/j.colsurfb.2022.112519 2022.

Riahi, Z., Khan, A., Rhim, J.-W., Shin, G. H., & Kim, J. T. (2024). Synergistic effect of iron-based metal-organic framework hybridized with carbon quantum dots in agar/gelatin films for fruit preservation. *Food Packaging and Shelf Life, 45*, 101330. https://doi.org/10.1016/j.fpsl.2024.101330 2024.

Riahi, Z., Khan, A., Ebrahimi, M., Rhim, J. W., Shin, G. H., & Kim, J. T. (2025). Exploring Sustainable Carbon Dots as UV-Blocking Agents for Food Preservation. *Comprehensive Reviews in Food Science and Food Safety, 24*, e70192 2025.

Riahi, Z., Khan, A., Ebrahimi, M., Rhim, J.-W., & Kim, J. T. (2026). UV-barrier properties of carbon dots for food packaging applications. In *Carbon dots in food packaging and preservation*. Elsevier, 189–208.

Riahi, Z., Khan, A., Hong, S. J., Rhim, J.-W., Shin, G. H., Min, S. C., & Kim, J. T. (2026). Photocatalytic ethylene scavenging by chitosan/gelatin packaging films loaded with TiO_2-functionalized bismuth metal organic frameworks for postharvest preservation of fruits. *Food chemistry, 499*, 147380. https://doi.org/10.1016/j.foodchem.2025.147380 2026.

Robinson, J., Begum, R., & Maqbool, M. (2023). Ultraviolet radiation: Benefits, harms, protection. *An Introduction to Non-Ionizing Radiation, 2*, 62 2023.

Roy, S., & Rhim, J.-W. (2021a). Carrageenan/agar-based functional film integrated with zinc sulfide nanoparticles and pickering emulsion of tea tree essential oil for active packaging applications. *International Journal of Biological Macromolecules, 193*, 2038–2046 2021a.

Roy, S., & Rhim, J.-W. (2021b). Fabrication of cellulose nanofiber-based functional color indicator film incorporated with shikonin extracted from *Lithospermum erythrorhizon* root. *Food Hydrocolloids, 114*, 106566. https://doi.org/10.1016/j.foodhyd.2020.106566 2021b.

Roy, S., & Rhim, J.-W. (2021c). Fabrication of chitosan-based functional nanocomposite films: Effect of quercetin-loaded chitosan nanoparticles. *Food Hydrocolloids, 121*, 107065 2021c.

Roy, S., & Rhim, J.-W. (2022). New insight into melanin for food packaging and biotechnology applications. *Critical Reviews in Food Science and Nutrition, 62*, 4629–4655 2022.

Roy, S., Ramakrishnan, R., Goksen, G., Singh, S., & Łopusiewicz, Ł. (2024). Recent progress on UV-light barrier food packaging films–A systematic review. *Innovative Food Science & Emerging Technologies, 91*, 103550 2024.

Rukmanikrishnan, B., Ramalingam, S., Rajasekharan, S. K., Lee, J., & Lee, J. (2020). Binary and ternary sustainable composites of gellan gum, hydroxyethyl cellulose and lignin for food packaging applications: Biocompatibility, antioxidant activity, UV and water barrier properties. *International Journal of Biological Macromolecules, 153*, 55–62 2020.

Sahu, M., & Shrivas, A. (2024). Exploring microbial spoilage in fruits: Causes, detection methods, and preventive measures. *International Journal of Innovative Research in Technology and Science, 12*, 470–477 2024.

Salian, A., Dutta, S., & Mandal, S. (2021). A roadmap to UV-protective natural resources: Classification, characteristics, and applications. *Materials Chemistry Frontiers, 5*, 7696–7723 2021.

Sani, M. A., Tavassoli, M., Salim, S. A., Azizi-lalabadi, M., & McClements, D. J. (2022). Development of green halochromic smart and active packaging materials: TiO_2 nanoparticle-and anthocyanin-loaded gelatin/κ-carrageenan films. *Food Hydrocolloids, 124*, 107324 2022.

Sansonetti, A., Bertasa, M., Corti, C., Rampazzi, L., Monticelli, D., Scalarone, D., & Canevali, C. (2021). Optimization of copper stain removal from marble through the formation of Cu (II) complexes in agar gels. *Gels, 7*, 111 2021.

Santana, J. V., Marangoni Júnior, L., Cassol, G. Z., Sato, H. H., & Vieira, R. P. (2024). Advanced pectin-based films: Enhancing antioxidant, antibacterial, UV barrier, and physicochemical properties upon oligomeric limonene derivative incorporation. *Food Hydrocolloids, 149*, 109558. https://doi.org/10.1016/j.foodhyd.2023.109558 2024.

Scaffaro, R., Maio, A., Sutera, F., Gulino, E. F., & Morreale, M. (2019). Degradation and recycling of films based on biodegradable polymers: A short review. *Polymers, 11*, 651 2019.

Schifferstein, H. N. (2024). Changes in appearance during the spoilage process of fruits and vegetables: Implications for consumer use and disposal. *Cleaner and Responsible Consumption, 12*, 100184 2024.

Sequino, G., Valentino, V., Torrieri, E., & De Filippis, F. (2022). Specific microbial communities are selected in minimally-processed fruit and vegetables according to the type of product. *Foods, 11*, 2164 2022.

Sganzerla, W. G., Rosa, G. B., Ferreira, A. L. A., da Rosa, C. G., Beling, P. C., Xavier, L. O., & Barreto, P. L. M. (2020). Bioactive food packaging based on starch, citric pectin and functionalized with *Acca sellowiana* waste by-product: Characterization and application in the postharvest conservation of apple. *International Journal of Biological Macromolecules, 147*, 295–303 2020.

Shahi, D. K., Awasthi, G. P., Rajendra Bahadur, G. C., Panthi, K. P., Chand, A. B., Shin, M., & Joshi, M. K. (2025). *Rhododendron arboreum* Sm. anthocyanin-infused starch, chitosan, and polyvinyl alcohol based composite films: Comparative analysis of physical, UV barrier, antioxidant and intelligent behavior. *International Journal of Biological Macromolecules, 302*, 140532. https://doi.org/10.1016/j.ijbiomac.2025.140532 2025.

Shahi, D. K., Awasthi, G. P., Shin, M., Achhami, H., Rai, E., Regmi, S. R., & Neupane, B. B. (2026). Sustainable PVA/chitosan/starch based smart packaging films with ghangaru-anthocyanins for real-time chicken meat freshness detection. *Food Analytical Methods, 19*, 13 2026.

Shankar, S., & Rhim, J.-W. (2018). Preparation of sulfur nanoparticle-incorporated antimicrobial chitosan films. *Food Hydrocolloids, 82*, 116–123 2018.

Shankar, S., Teng, X., Li, G., & Rhim, J.-W. (2015). Preparation, characterization, and antimicrobial activity of gelatin/ZnO nanocomposite films. *Food Hydrocolloids, 45*, 264–271 2015.

Shankar, S., Tanomrod, N., Rawdkuen, S., & Rhim, J.-W. (2016). Preparation of pectin/silver nanoparticles composite films with UV-light barrier and properties. *International Journal of Biological Macromolecules, 92*, 842–849 2016.

Sharma, C., Kundu, S., Singh, S., Saxena, J., Gautam, S., Kumar, A., & Pathak, P. (2025). From concept to shelf: engineering biopolymer-based food packaging for sustainability. *RSC Sustainability, 2025*.

Sheng, J., Ding, S., Liao, H., Yao, Y., Zhai, Y., Zhan, J., & Wang, X. (2023). Polyacrylonitrile/UV329/titanium oxide composite nanofibrous membranes with enhanced UV protection and filtration performance. *RSC Advances, 13*, 17622–17627 2023.

Shi, L., Shan, J., Ju, Y., Aikens, P., & Prud'homme, R. K. (2012). Nanoparticles as delivery vehicles for sunscreen agents. *Colloids and Surfaces A: Physicochemical and Engineering Aspects, 396*, 122–129 2012.

Shi, S., Lan, X., Ding, X., Han, X., Sun, J., Wang, J., & Duan, J. (2024). Metal-phenolic networks spontaneously reinforced carrageenan-based packaging films with antibacterial and antioxidant properties. *International Journal of Biological Macromolecules, 279*, 135143. https://doi.org/10.1016/j.ijbiomac.2024.135143 2024.

Shi, Y.-e, Zhang, P., Yang, D., & Wang, Z. (2020). Synthesis, photoluminescence properties and sensing applications of luminescent sulfur nanodots. *Chemical Communications, 56*, 10982–10988 2020.

Shoba, H., Jain, S., Panwar, N., Lakhawat, S., Rajpurohit, D., & Jain, N. (2025). A review on sources of biopolymers, antimicrobial agents, and biodegradable packaging. *Discover Food, 5*, 369 2025.

Siddiqui, S. A., Singh, S., Bahmid, N. A., & Sasidharan, A. (2024). Applying innovative technological interventions in the preservation and packaging of fresh seafood products to minimize spoilage-A systematic review and meta-analysis. *Heliyon, 10*, 2024.

Siemińska-Kuczer, A., Szymańska-Chargot, M., & Zdunek, A. (2022). Recent advances in interactions between polyphenols and plant cell wall polysaccharides as studied using an adsorption technique. *Food Chemistry, 373*, 131487 2022.

Silva, K. F., Marques, C. S., de Freitas Junior, A., Dias, M. V., & Mori, F. A. (2023). Whey protein isolate and kraft lignin multifunctional films for potential food packaging application: UV block and antioxidant potential. *Food Bioscience, 53*, 102581 2023.

Singh, H. K., Kumar, S., Bamne, J., Taiwade, K., Singh, N., Chandel, V., & Haque, F. Z. (2021). Analyzing the synthesis of various inorganic nanoparticles and their role in UV-shielding. *Materials Today: Proceedings, 46*, 5607–5618 2021.

Soltani, Z., Tavakolipour, H., & Tabari, M. (2023). The influence of chitosan and titanium dioxide nanoparticles incorporated with polylactic acid on prolonging rye bread shelf life. *Journal of Food Measurement and Characterization, 17*, 1806–1816 2023.

Song, D.-H., Hoa, V. B., Kim, H. W., Khang, S. M., Cho, S.-H., Ham, J.-S., & Seol, K.-H. (2021). Edible films on meat and meat products. *Coatings, 11*, 1344 2021.

Soufiani, S. P., Yousefi, S., Honarvar, M., Weisany, W., & Asadi, G. (2025). Bioactive chitosan packaging films with encapsulated beet leaf extract for preserving chicken fillets. *Food Chemistry: X*, 103335 2025.

Störmer, A., Bott, J., Kemmer, D., & Franz, R. (2017). Critical review of the migration potential of nanoparticles in food contact plastics. *Trends in Food Science & Technology, 63*, 39–50 2017.

Sul, Y., Khan, A., Kim, J. T., & Rhim, J.-W. (2025). Tangerine peel-derived nitrogen-doped carbon dots incorporated chitosan/pullulan-based active packaging film for bread packaging. *Colloids and Surfaces B: Biointerfaces, 245*, 114339. https://doi.org/10.1016/j.colsurfb.2024.114339 2025.

Sun, F., Zhao, W., Zhao, J., Wang, K., Peng, L., & Li, H. (2026). A sustainable gelatin-based film reinforced with nanolignin-stabilized pickering emulsion and cellulose nanocrystals from coffee leaves for food preservation. *Food Hydrocolloids, 172*, 112156. https://doi.org/10.1016/j.foodhyd.2025.112156 2026.

Tan, D., Ma, A., Wang, S., Zhang, Q., Jia, M., Kamal-Eldin, A., & Chen, G. (2021). Effects of the oxygen content and light intensity on milk photooxidation using untargeted metabolomic analysis. *Journal of Agricultural and Food Chemistry, 69*, 7488–7497 2021.

Tripathi, S., Kumar, L., Deshmukh, R. K., & Gaikwad, K. K. (2024). Ultraviolet blocking films for food packaging applications. *Food and Bioprocess Technology, 17*, 1563–1582 2024.

Tupuna-Yerovi, D. S., Schmidt, H., & Rios, A.d O. (2025). Biodegradable sodium alginate films incorporated with lycopene and β-carotene for food packaging purposes. *Food Science and Technology International, 31*, 23–35 2025.

Upadhye, S., Mujawar, S. S., & Kashte, S. B. (2025). Eco-friendly, antibacterial, antioxidant, ultraviolet blocking sodium alginate-gelatin films loaded with clove essential oil for food packaging. *Journal of Food Measurement and Characterization, 19*, 2803–2817 2025.

Versino, F., Ortega, F., Monroy, Y., Rivero, S., López, O. V., & García, M. A. (2023). Sustainable and bio-based food packaging: A review on past and current design innovations. *Foods, 12*, 1057 2023.

Vijayakumar, R., Sivaraman, Y., Pavagada Siddappa, K. M., & Dandu, J. P. R. (2022). Synthesis of lignin nanoparticles employing acid precipitation method and its application to enhance the mechanical, UV-barrier and antioxidant properties of chitosan films. *International Journal of Polymer Analysis and Characterization, 27*, 99–110 2022.

Wang, D., Gu, Y., Feng, S., Yang, W., Dai, H., Xiao, H., & Han, J. (2023). Lignin-containing biodegradable UV-blocking films: a review. *Green Chemistry, 25*, 9020–9044 2023.

Wang, H., Liao, Y., Zhu, G., Wang, L., Chen, Z., Li, X., & Song, P. (2025). Development of double-film composite food packaging with UV protection and microbial protection for cherry preservation. *Foods, 14*, 2283 2025.

Wang, L., Liu, X., Qi, P., Sun, J., Jiang, S., Li, H., & Zhang, S. (2022). Enhancing the thermostability, UV shielding and antimicrobial activity of transparent chitosan film by carbon quantum dots containing N/P. *Carbohydrate Polymers, 278*, 118957 2022.

Wang, X., Sun, X., Guan, X., Wang, Y., Chen, X., & Liu, X. (2021). Tannic interfacial linkage within ZnO-loaded fabrics for durable UV-blocking applications. *Applied Surface Science, 568*, 150960 2021.

Wang, Y., Guo, X., Qin, Z., Liu, R., Li, S., Du, M., & Bai, Y. (2025). Self-assembly of tea polyphenol onto the high strength chitosan film fabricated via a new way to enhance the antioxidant and antibacterial activity for meat preservation. *Food Packaging and Shelf Life, 49*, 101482. https://doi.org/10.1016/j.fpsl.2025.101482 2025.

Xiao, M., Shawkey, M. D., & Dhinojwala, A. (2020). Bioinspired melanin-based optically active materials. *Advanced Optical Materials, 8*, 2000932 2020.

Xie, W., Pakdel, E., Liu, D., Sun, L., & Wang, X. (2019). Waste-hair-derived natural melanin/TiO_2 hybrids as highly efficient and stable UV-shielding fillers for polyurethane films. *ACS Sustainable Chemistry & Engineering, 8*, 1343–1352 2019.

Xing, Q., Ruch, D., Dubois, P., Wu, L., & Wang, W.-J. (2017). Biodegradable and high-performance poly (butylene adipate-co-terephthalate)–lignin UV-blocking films. *ACS Sustainable Chemistry & Engineering, 5*, 10342–10351 2017.

Xu, H., Wu, P., & Cui, D. (2025). Diosmin-functionalized chitosan films as sustainable packaging materials for shrimp shelf-life extension. *International Journal of Biological Macromolecules, 329*, 147777. https://doi.org/10.1016/j.ijbiomac.2025.147777 2025.

Yan, N., & Chai, X.-S. (2021). Rapid determination of the content of carboxymethyl cellulose sodium in aqueous solution by a color indicator-assisted spectroscopy. *Polymer Testing, 93*, 106990 2021.

Yang, D., Fan, B., Sun, G., He, Y.-C., & Ma, C. (2023). Ultraviolet blocking ability, antioxidant and antibacterial properties of newly prepared polyvinyl alcohol-nanocellulose-silver nanoparticles-ChunJian peel extract composite film. *International Journal of Biological Macromolecules, 252*, 126427. https://doi.org/10.1016/j.ijbiomac.2023.126427 2023.

Yang, D., Fan, B., & He, Y.-C. (2024). UV-blocking, antibacterial, corrosion resistance, antioxidant, and fruit packaging ability of lignin-rich alkaline black liquor composite film. *International Journal of Biological Macromolecules, 275*, 133344 2024.

Yang, Z., Li, M., Zhai, X., Zhao, L., Tahir, H. E., Shi, J., & Xiao, J. (2022). Development and characterization of sodium alginate/tea tree essential oil nanoemulsion active film containing TiO_2 nanoparticles for banana packaging. *International Journal of Biological Macromolecules, 213*, 145–154 2022.

Yue, C., Wang, M., Zhou, Z., You, Y., Wang, G., & Wu, D. (2024). Cellulose-based intelligent packaging films with antibacterial, UV-blocking, and biodegradable properties for shrimp freshness monitoring. *Chemical Engineering Journal, 488*, 150975. https://doi.org/10.1016/j.cej.2024.150975 2024.

Zhang, H., Cheng, X., Liu, C., Liu, Z., Liu, L., Feng, C., & Yao, X. (2024). Ultraviolet-blocking polymers and composites: recent advances and future perspectives. *Journal of Materials Chemistry A* 2024.

Zhang, T., Zhang, W., Deng, Y., Chu, Y., Zhong, Y., Wang, G., & Li, H. (2022). Curcumin-based waterborne polyurethane-gelatin composite bioactive films for effective UV shielding and inhibition of oil oxidation. *Food Control, 141*, 109199. https://doi.org/10.1016/j.foodcont.2022.109199 2022.

Zhang, W., Sani, M. A., Zhang, Z., McClements, D. J., & Jafari, S. M. (2023). High performance biopolymeric packaging films containing zinc oxide nanoparticles for fresh food preservation: A review. *International Journal of Biological Macromolecules, 230*, 123188 2023.

Zhang, Y., & Naebe, M. (2021). Lignin: A review on structure, properties, and applications as a light-colored UV absorber. *ACS Sustainable Chemistry & Engineering, 9*, 1427–1442 2021.

Zhao, J., Zhao, W., Wang, K., Zhuang, Y., Li, H., & Peng, L. (2025). Development of multifunctional starch/pectin composite film engineered with ZIF-67-loaded microcrystalline cellulose for real-time monitoring and preserving of pork freshness. *Carbohydrate Polymers, 355*, 123358. https://doi.org/10.1016/j.carbpol.2025.123358 2025.

Zhao, L., Zhang, M., Mujumdar, A. S., Adhikari, B., & Wang, H. (2022). Preparation of a novel carbon dot/polyvinyl alcohol composite film and its application in food preservation. *ACS Applied Materials & Interfaces, 14*, 37528–37539 2022.

Zhao, L., Zhang, M., Mujumdar, A. S., & Wang, H. (2023). Application of carbon dots in food preservation: a critical review for packaging enhancers and food preservatives. *Critical Reviews in Food Science and Nutrition, 63*, 6738–6756 2023.

Zhao, Z., Mao, A., Gao, W., & Bai, H. (2018). A facile in situ method to fabricate transparent, flexible polyvinyl alcohol/ZnO film for UV-shielding. *Composites Communications, 10*, 157–162 2018.

Zou, F., Li, H., Dong, Y., Tewari, G. C., & Vapaavuori, J. (2022). Optically transparent pectin/poly (methyl methacrylate) composite with thermal insulation and UV blocking properties based on anisotropic pectin cryogel. *Chemical Engineering Journal, 439*, 135738 2022.

Chapter 24

Active and intelligent materials based on biopolymers

Josemar Gonçalves de Oliveira Filho[1], Luan Gustavo dos Santos[2], Luan Ramos da Silva[3], Caroline Honaiser Lescano[4], Larissa Graziele Rauber Duarte[3], Giovana Maria Navarro Mendonça[5], Luiz Henrique Capparelli Mattoso[1], and Mariana Buranelo Egea[6]

[1]*Brazilian Agricultural Research Corporation, Embrapa Instrumentation, São Carlos, São Paulo, Brazil,* [2]*Laboratory of Food Technology, School of Chemistry and Food Engineering, Federal University of Rio Grande, Rio Grande, Rio Grande do Sul, Brazil,* [3]*School of Food Engineering (FEA), University of Campinas (UNICAMP), Campinas, São Paulo, Brazil,* [4]*Department of Physiological Sciences (DCF), Institute of Biological Sciences (ICB), Federal University of Amazonas (UFAM), Manaus, Amazonas, Brazil,* [5]*School of Pharmaceutical Sciences, São Paulo State University (UNESP), Araraquara, São Paulo, Brazil,* [6]*Goiano Federal Institute of Education, Science, and Technology, Rio Verde, Goiás, Brazil*

24.1 Introduction

The growing demand for sustainable and efficient solutions for food preservation and reducing environmental impact has driven the development of innovative technologies in the packaging sector (Trajkovska Petkoska et al., 2021). In this scenario, active and intelligent materials stand out as a relevant study area, combining advanced functional properties with the potential to meet contemporary sustainability challenges (Drago et al., 2020). Unlike conventional packaging, these materials can interact dynamically with food or the environment, performing specific functions such as the controlled release of active compounds or the detection of changes in product quality (Du et al., 2023; Hou et al., 2023).

In this context, scientific and technological advances in developing biopolymer-based materials are explored, emphasizing their unique characteristics and practical applications. Biopolymers, obtained from renewable sources such as proteins, polysaccharides, and lipids, offer significant advantages in biodegradability and reduction of plastic waste (Cheng et al., 2024). When enriched with bioactive compounds or integrated with detection systems, these materials become multifunctional, playing crucial roles in food preservation, quality monitoring, and shelf-life extension (de Oliveira Filho et al., 2024; Sahraeian et al., 2024).

Initially, the approach focuses on incorporating active compounds into polymer matrices, highlighting categories such as plant extracts, lipids, vitamins, and polyphenols. Each of these classes is analyzed based on their antioxidant, antimicrobial, and stabilizing properties, and the challenges associated with their application. For example, plant extracts are rich in bioactive compounds, such as terpenes and flavonoids, which contribute to food preservation by inhibiting microbial growth and reducing oxidation (Filgueiras et al., 2024; Legesse et al., 2024). Lipids offer structural and functional advantages, while vitamins and polyphenols are crucial in maintaining nutritional value of packaged foods and can contribute to their sensory quality.

Then, we present advances in intelligent packaging, which integrates indicators to monitor food quality in real time. These systems, often based on colorimetric or chemical reactions, allow the detection of changes in pH, gas concentration, or other parameters that indicate product deterioration (Palanisamy et al., 2025). Recent studies on natural pigments, such as anthocyanins and curcumin, illustrate promising solutions for developing environmentally friendly and safe intelligent packaging (Rezaei et al., 2024; Song et al., 2024).

The content presented offers a comprehensive and up-to-date theoretical basis for researchers and students interested in biopolymers and their applications in multifunctional packaging. Throughout the information, it is hoped to promote an in-depth understanding of the contributions of these materials to technological innovation and sustainability in the food sector.

Natural Polymer-Based Films and Coatings in Food Technology. DOI: https://doi.org/10.1016/B978-0-443-34244-8.00026-4

24.2 Active materials based on biopolymers and bioactive compounds

Active biopolymeric materials derived from natural and renewable sources represent a sustainable and forward-looking alternative for the food packaging industry. These materials exhibit multifunctionality, extending the shelf life of food products through the incorporation of natural active compounds with antimicrobial and antioxidant properties. In addition, they can be engineered to promote the controlled release of bioactive agents, ensuring prolonged effectiveness and targeted action. Among the most relevant compounds incorporated are plant extracts, lipid-based compounds, vitamins, and polyphenols. Beyond functionality, these biopolymers strengthen environmental sustainability, enhance consumer safety, and foster the integration of advanced active features into modern packaging systems (Reis et al., 2023). In the following sections, we provide a detailed discussion on the incorporation of plant extracts, lipids, vitamins, and polyphenols into biopolymeric packaging materials.

24.2.1 Plant extracts

Plant extracts are concentrated preparations derived from various parts of plants, including roots, stems, leaves, flowers, and fruits. These extracts are obtained through specific extraction processes that separate the desired bioactive components from the raw plant material, typically using polar and/or nonpolar solvents. The solvents used depend on the nature of the compounds present in the plant material and result in differences in the properties of the extracts obtained (Bucić-Kojić et al., 2009; Thoo et al., 2010). These extracts can exist in different forms, such as liquid, solid, or viscous, and are categorized into (1) simple extracts, (2) standardized extracts with consistent levels of active ingredients, and (3) purified extracts containing isolated compounds. The extracts serve a wide range of applications in various sectors, including pharmaceuticals, nutraceuticals, cosmetics, agriculture, and packaging systems. The extraction process is crucial to preserve the integrity and efficacy of bioactive compounds, with factors such as pH, temperature, and solvent type significantly influencing yield and quality, and some plant extracts exhibit antioxidant and antimicrobial properties.

Antioxidant capacity refers to the ability of a substance to neutralize free radicals, which are unstable molecules that can cause cell damage and contribute to the development of various diseases, such as cancer and cardiovascular diseases. Antioxidant compounds donate electrons to free radicals, stabilizing them and thus preventing oxidative stress. This property may be associated with several compounds present in plant species, which are associated with the defense mechanism of these species. Some species stand out for their high antioxidant capacity (Table 24.1), presenting IC50 values of 35.83 mg L^{-1} (Rosmarinus *officinalis*), 28.68 mg L^{-1} (Azadirachta *indica*) (Alhaithloul et al., 2023), 52.15 mg L^{-1} (Verbascum *sinaiticum*) (Legesse et al., 2024), and 710.0 mg L^{-1} (Malus. *domestica*) (Pires et al., 2018).

The antimicrobial capacity of extracts refers to the ability of a substance to kill or inhibit the growth of microorganisms, including bacteria, viruses, fungi, and protozoa. This capacity is crucial in several applications, particularly in medicine, where antimicrobial agents treat infections and prevent diseases. Understanding the antimicrobial capacity is essential for applying extracts, considering the microorganisms that will be affected, thus making it possible to use this material in various processes, such as food packaging (Filgueiras et al., 2024). According to Table 24.1, the species have both antioxidant and antimicrobial capacities, and these act in synergy, preventing degradation by oxygen, as well as by the action of microorganisms.

The effect of an edible coating based on maltodextrin, enriched with *Pyracantha coccinea* extract and potassium nanocarbonate, was investigated for the postharvest preservation of refrigerated grapes. The results indicated that the coating preserved secondary metabolites and maintained the fruits' antioxidant capacity compared to uncoated samples. The *P. coccinea* extract exhibited both antimicrobial and antioxidant properties, effectively preventing the growth of molds and yeasts while reducing the degradation of phenolic compounds. The authors concluded that this coating represents a promising strategy to minimize losses and extend the shelf life of fruits (Ebrahimi et al., 2024).

Hamann et al. (2024) produced biodegradable films based on cellulose acetate, cornstarch, and chitosan, enriched with green tea extract for interleaving sliced meat products. The films demonstrated a controlled release of antioxidant activity for up to 60 days of storage, significantly delaying lipid oxidation in cooked ham. The authors recommend cornstarch-based films with 1.0% green tea extract as an effective solution to enhance the quality of meat products during storage. This approach supports environmental sustainability, promotes consumer health, and reduces reliance on chemical additives.

Malherbi et al. (2019) developed cassava starch-based sachets for olive oil packaging, investigating the incorporation of *Euterpe oleracea* Mart. extract into film properties and product shelf life. The sachets exhibited excellent extrusion performance and were well suited for olive oil storage, with açaí extract playing a vital role in maintaining product quality for up to 120 days.

TABLE 24.1 Plant extracts with antioxidant and antimicrobial properties.

Plants	Properties		References
	Antioxidant	Antimicrobial	
Rosmarinus officinalis	2,2-Diphenyl-1-picrylhydrazyl (DPPH) antioxidant activity	Antibacterial activity against *Bacillus cereus, Escherichia coli, Staphylococcus aureus, and Pseudomonas aeruginosa*.	Alhaithloul et al. (2023), Boukhari and Fatimi (2023), Wahba et al.
Azadirachta indica	DPPH antioxidant activity	Antibacterial activity against *Alcaligenes faecalis, Bacillus anthracis, Bacillus subtilis, Enterococcus faecalis, Escherichia coli, Escherichia coli O157:H7, Helicobacter pylori, Klebsiella pneumoniae, Lactobacillus acidophilus, Listeria monocytogenes, Methicillin-resistant S. aureus (MRSA), Porphyromonas gingivalis, Pseudomonas aeruginosa, Pseudomonas gingivalis, Salmonella* spp., *Salmonella typhi, Shigella* spp., *Staphylococcus epidermidis, Streptococcus mutans, Streptococcus viridans, Vibrio cholerae*, and *Vibrio parahaemolyticus*.	Alhaithloul et al. (2023), Biney et al. (2021), Wylie and Merrell (2022)
Origanum vulgare	Antioxidant activity through the downregulation of oxidant enzymes, such as hepatic NOX_4 and NOS_2	Antibacterial activity against *Streptococcus mutans, Streptococcus salivarius, Fusobacterium nucleatum, Porphyromonas gingivalis, Prevotella intermedia, Parvimonas micra, Staphylococcus aureus* (methicillin- and vancomycin-resistant), *Staphylococcus aureus* (methicillin-resistant) in raw milk, *S. aureus* (methicillin-resistant) in poultry meat, and *S. aureus* subsp. *aureus*.	Antoniadou et al. (2024), Madkour et al. (2024)
Allium sativum	Photochemiluminescence antioxidant activity	Antibacterial activity against Pseudomonas, Proteus, *S. aureus, Escherichia coli, Salmonella, Klebsiella, Micrococcus, Bacillus subtilis, Clostridium, Mycobacterium, Helicobacter*.	Bedouhene et al. (2024), Harris et al. (2001), Lu et al. (2024)
Campomanes-ia pubescens	DPPH antioxidant activity	Antibacterial activity against *S. aureus*.	Catelan et al. (2018), Pradella et al. (2021)
Campomanes-ia adamantium	DPPH antioxidant activity	Antimicrobial activity against *Candida glabrata, Cryptococcus neoformans, Saccharomyces cerevisiae, Candida krusei, Candida parapsilosis*, and *Candida tropicalis*.	Silva et al. (2022), Sá et al. (2018)
Malus domestica Borkh	DPPH antioxidant activity	Antibacterial activity against *S. aureus* (MRSA), *Listeria monocytogenes, Enterococcus faecalis, E. coli, Morganella morganii*.	Cendrowski et al. (2024), Pires et al. (2018), Vasile et al. (2021)
Psidium myrtoides	DPPH antioxidant activity	Antibacterial activity against *Streptococcus mitis, Streptococcus sanguinis, Streptococcus sobrinus, Streptococcus salivarius*, and *Streptococcus mutans*.	Dias et al. (2019), Ribeiro et al. (2024), Martins de Freitas et al. (2021)
Verbascum sinaiticum	DPPH antioxidant activity	Antibacterial activity against *S. aureus* ATCC 2592, *Streptococcus agalactiae* ATCC 12386, *Acinetobacter baumannii* ATCC 19606, and *Pseudomonas aeruginosa* ATCC 27853.	Geyesa et al. (2024), Legesse et al. (2024), Tegegn et al. (2024)
Thymus vulgaris	Ferric Reducing Antioxidant Power (FRAP) antioxidant activity	Antimicrobial activity against *Candida albicans, S. aureus, Enterococcus faecalis, Streptococcus mutans*, and *Pseudomonas aeruginosa*.	Oliveira et al. (2017), Gedikoğlu et al. (2019), Soares Mateus et al. (2024)

(*Continued*)

TABLE 24.1 (Continued)

Plants	Properties		References
	Antioxidant	Antimicrobial	
Cosmos sulphureus	2,2′-Azino-bis(3-ethylbenzothiazoline-6-sulfonic acid) (ABTS) antioxidant activity	-	Camiletti et al. (2024), Malaka et al. (2015), Ou et al. (2024)
Stevia rebaudiana	DPPH antioxidant activity	*Bacillus subtilis, Staphylococcus aureus, Micrococcus luteus, Serratia marcescens, Pseudomonas aeruginosa, Bacillus megaterium, Escherichia coli, Proteus vulgaris, Saccharomyces cerevisiae, Aspergillus niger, and Rigidoporus oligoporus.*	Shukla et al. (2009), Tadhani and Subhash (2006), Wen et al. (2024)

Incorporating grape pomace extract, an industrial by-product, into biodegradable arrowroot-based films can significantly enhance the antioxidant properties of packaging. Rich in anthocyanins, the extract influences film coloration and makes it suitable for photosensitive foods (Nogueira, Soares, et al., 2022). Additionally, the films function as smart materials, as the color shifts of anthocyanins respond to pH variations in food (Nogueira et al., 2024).

The method used to obtain extracts can impact their quality and effectiveness. Supercritical extraction, regarded as an environmentally friendly technique, yields extracts rich in bioactive compounds with antimicrobial and antioxidant properties (Kessler et al., 2024). Using this approach, Alloh et al. (2024) produced rosemary, thyme, and coriander extracts, obtaining significant amounts of terpenoids and potent antimicrobial agents. These extracts were incorporated into whey protein concentrate emulsions to create edible coatings for ultrafiltered cheeses, extending their shelf life to 40 days.

24.2.2 Lipids

Lipids are substances with structural and energetic functions, categorized into simple, compound, and derivative (Akoh, 2017). Furthermore, they are natural components of the diet and are well tolerated by the body. The functional characteristics associated with their structure indicate that these molecules have the potential to be incorporated into biomaterials. The benefits of the biological activities of specific lipids are also considerable attributes in terms of human health and improving the formulation of the biopolymer (Shishir et al., 2018).

The lipid molecules commonly used in polymeric systems are waxes, essential oils, triglycerides, and phospholipids. They are widely applied in food formulations as additives (Akhter et al., 2019), emulsifiers (Pereda et al., 2012), edible films, and coatings, and in lipid-based delivery systems with controlled active release (liposomes and solid lipid nanoparticles) (Perdones et al., 2016). Although lipids have multiple applications, they may present disadvantages, such as unpleasant taste, low gas permeability, and low mechanical properties. Incorporating or mixing lipids into polymeric materials can overcome these challenges (Qurat-ul-Ain et al., 2016).

Adding essential oils confers antioxidant and antimicrobial properties to the biopolymer and increases the surface's hydrophobicity. Antioxidants and antimicrobial agents help preserve food quality, freshness, and safety attributes by allowing food to be transported and stored safely. This is because they can inhibit the growth of microorganisms and prolong the shelf life of food by slowing down oxidative processes (Hosseini et al., 2019).

Essential oils are considered highly effective as antimicrobials, as described by researchers. For example, lemongrass (*Cymbopogon citratus*) essential oil incorporated into nanoemulsions showed antimicrobial activity against Escherichia *coli* (Salvia-Trujillo et al., 2015). Essential oils of oregano (*Origanum minutiflorum*) and sage (*Salvia officinalis*) added to pullulan films have demonstrated efficacy against pathogenic microorganisms, such as Staphylococcus *aureus*, Listeria *monocytogenes*, *E. coli* O157:H7, and *S. typhimurium* (Morsy et al., 2014).

The justification for the effects is mainly related to the chemical composition of essential oils. Bioactive compounds such as menthol, eugenol, carvacrol, geraniol, and cinnamaldehyde are reported (Solórzano-Santos & Miranda-Novales, 2012). Natural substances have been an alternative for cost savings and potentially eliminating food safety risks (Domínguez et al., 2018). However, the unstable composition of essential oils makes it challenging to preserve their properties over the long term.

Although they can be incorporated directly into the packaging to improve barrier performance, they are exposed to degradation due to light, oxidation, and heat (Sharma et al., 2021). Another critical point is that adding lipophilic compounds to packaging requires overcoming challenges, such as the possibility of interaction with substances present in the active compound itself. This can be seen, for example, in essential oils, where there is an increase in the release of phenolics due to the presence of lipids, which increases the migration of compounds from the packaging to the food (Sharma et al., 2021). Furthermore, the migration rate depends on the plant origin and the essential oil concentration (Ribeiro et al., 2024).

In this sense, new technologies are emerging, intending to improve the stability of essential oils by creating nanocapsules such as liposomes or lipid nanoparticles (Fernández-López & Viuda-Martos, 2018). Simple and compound lipids, such as triglycerides, phospholipids, and waxes, are widely used in lipid-based delivery systems. Their molecular structures aid in forming a stable matrix that makes it possible to encapsulate different bioactive ingredients, ensuring stability, bioavailability, and functionality (Fathi et al., 2012). Several lipid-based delivery systems can be designed in this sense, such as nanostructured lipid carriers, niosomes, phytosomes, liposomes, colloidal emulsions, and solid lipid nanoparticles (Gali et al., 2023).

The review by Christaki et al. (2021) discusses some aspects of essential oils' use and distribution system in cheese preservation. For example, incorporating microcapsules of essential oils in different types of cheese helps preserve the oxidation, volatilization, and interactions of the bioactive compounds present in the essential oil with other constituents in the food matrix. In addition, prolonging their antimicrobial and antioxidant effects through microencapsulation leads to their sustained release (Christaki et al., 2021). Another lipid molecule used to create a moisture barrier is beeswax. A film prepared from guar gum, and beeswax has been shown to be helpful in preserving cheese, its texture, color, and pH for up to two months from the point of packaging. This is because beeswax can restrict moisture migration and consequently extend the shelf life of products (Diyana et al., 2021).

24.2.3 Vitamins

Antioxidant compounds, even when administered in small amounts, prevent lipid oxidation and protect pigments, flavors, proteins, and vitamins from oxidative damage. They directly neutralize reactive oxygen species (ROS) or inhibit their formation, in addition to stabilizing free radicals and forming new, more stable radicals. For their action to be more efficient, they must be incorporated into foods at the beginning of processing, as they do not reverse oxidation that has already started (Bensid et al., 2022; Ribeiro et al., 2021).

Food industries face a major challenge with food spoilage, which can occur due to various factors. Among them are oxidation reactions, degradation of compounds, growth of microorganisms, oxidative rancidity, and browning caused by enzymatic processes (Valencia et al., 2019). These problems compromise the quality, flavor, and safety of products, requiring effective solutions to preserve them.

Vitamins with low molecular weight are of great importance in the cellular defense mechanism against oxidizing agents; these antioxidant compounds will end radical chain reactions, and some of these compounds may be soluble in water and will act in the cytosol or cytoplasmic matrix. Other groups of antioxidants are soluble in lipids and will act on membranes. The vitamins with low molecular weight that have the most representative capacity for antioxidant action are vitamins C, E, and A (Jomova et al., 2023).

L-ascorbic acid (Fig. 24.1), known as vitamin C, is commonly used in the food industry. It stands out for its antioxidant action, protecting sensitive components and preventing changes in color and flavor. It is commonly used as a color fixer, preservative, and acidity regulator in processed meats, baked goods, canned fruits, beers, jellies, and juices. Its preservative function also prevents the growth of microorganisms, prolonging shelf life and ensuring food quality and safety (Dodevska et al., 2023).

Vitamin C, present in large quantities in citrus fruits, is a powerful antioxidant. It helps neutralize free radicals in the body, protecting cells against oxidative stress and reducing inflammation. Thanks to these properties, vitamin C is important for preventing chronic diseases such as cancer and heart problems (Konstantinidi & Koutelidakis, 2019; Vignesh et al., 2024).

Vitamin E (Fig. 24.2), a fat-soluble vitamin discovered almost a century ago, has been widely studied and applied in the food, pharmaceutical, and cosmetic industries, mainly for its antioxidant properties. Its therapeutic characteristics have boosted research and the development of vitamin E products (Ribeiro et al., 2021). Vitamin E is a fat-soluble nutrient that prevents various diseases, including cardiovascular disease, diabetes, and cancer (Geng et al., 2023; Orucha et al., 2011).

Vitamin E is known to be a natural antioxidant. This compound is commonly used as a nutraceutical food, with the characteristic of increasing the shelf life and quality of food. An important characteristic of these molecules is that they

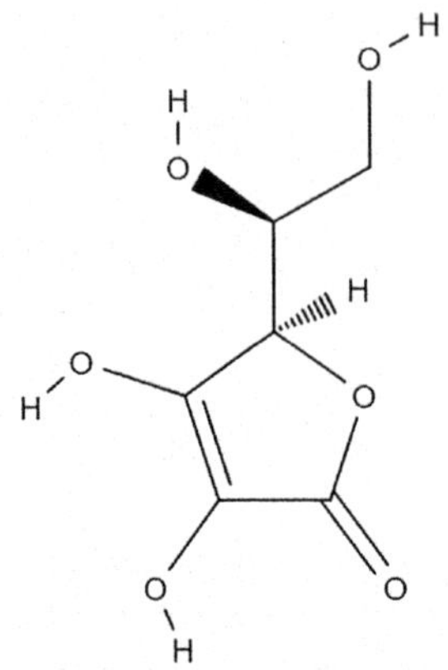

FIGURE 24.1 Molecular structure of vitamin C (L-ascorbic acid or ascorbate).

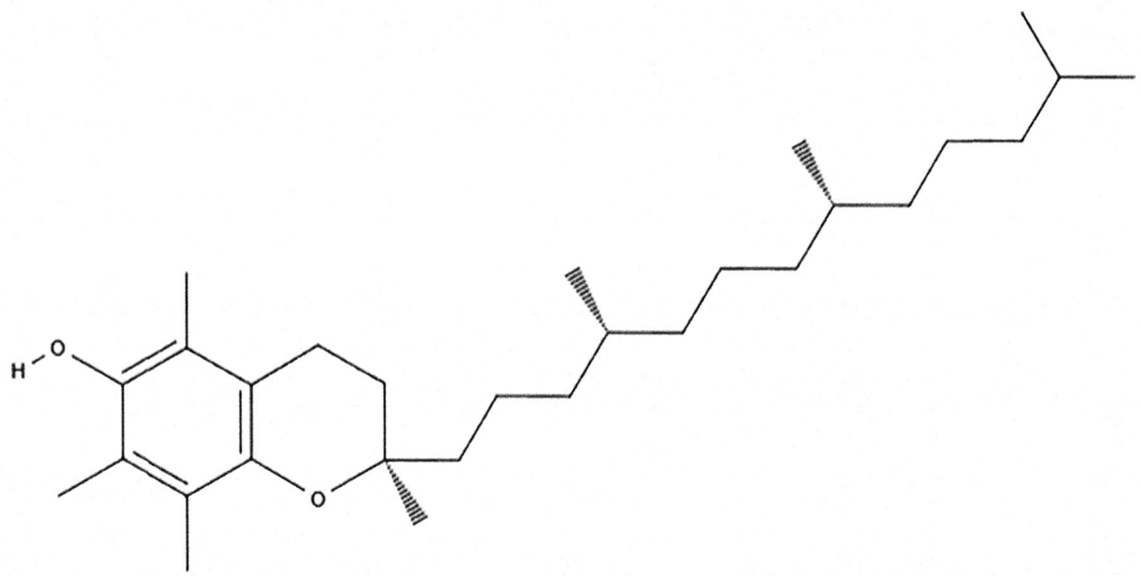

FIGURE 24.2 Molecular structure of vitamin E (tocopherol).

can be immobilized in matrix packaging or incorporated into foods as additive agents to improve their resistance to factors such as oxidation and antimicrobial activity (Lin & Pascall, 2014). Vitamins C and E, antioxidant compounds, are classified into two main classes, as shown in Fig. 24.3; they play an essential role as agents that neutralize free radicals, helping to preserve the nutritional quality of foods.

These groups of antioxidant compounds play important roles in preserving food quality and protecting health by neutralizing the harmful effects of free radicals. They act in different ways, performing specific functions that ensure the protection and stability of these foods. Table 24.2 shows some studies that used biopolymers with vitamin C applied to perishable and processed foods.

Vitamin C has been incorporated into biopolymer films using matrices such as carrageenan, alginate, carboxymethyl cellulose, gellan gum, pectin, apple fiber, and soy protein to preserve foods like papaya, fresh-cut apples, fresh-cut pineapple, tomatoes, and bram fish, as described in Table 24.2. For example, carboxymethyl cellulose films combined with calcium and citric acid were able to delay browning, maintain pulp firmness, reduce soluble solids content, titratable acidity, and changes in pH of fresh apple slices (Koushesh Saba & Sogvar, 2016). In the work of Song et al. (2011), films developed based on sodium alginate containing vitamin C and tea polyphenols reduced chemical deterioration, delayed mass loss, and increased the overall sensory quality of fish during storage.

Films based on vitamin E and biopolymers such as poly(lactic acid), basil seed gum, chitosan, hydroxypropyl methylcellulose (HPMC), or soy protein isolate (SPI) have been developed for use as active packaging for products such

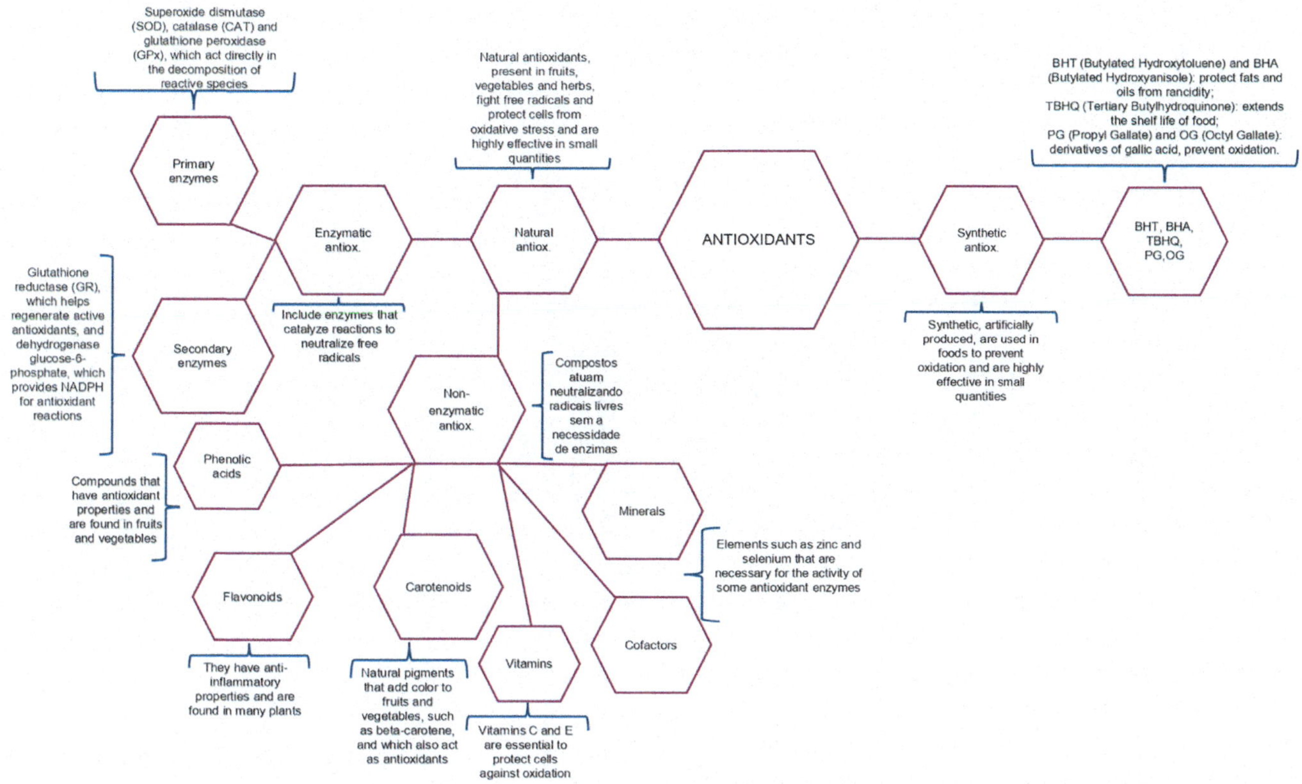

FIGURE 24.3 Different types of antioxidant compounds, including natural and synthetic types.

as corn oil, safflower oil, mushrooms, and salmon (Table 24.2). For example, Rosenbloom and Zhao (2021) applied films of SPI or HPMC that were packaged in soybean oil at 35°C for 60 days. They observed that SPI films containing tocopherol effectively minimized product oxidation. Similarly, Zhang et al. (2020) used chitosan and chitosan/zein films to package mushrooms at 4°C for 12 days. They observed less mass loss, less browning, and greater firmness in the mushrooms packed with tocopherol-enriched chitosan/zein films.

Therefore, incorporating vitamins, such as vitamins C and E, into biopolymeric films has shown great potential in developing active packaging for food preservation. These materials can contribute to food safety and develop more eco-friendly and functional packaging.

24.2.4 Polyphenols

Polyphenols are organic compounds widely distributed in the plant kingdom, chemically characterized by aromatic rings linked to hydroxyl groups (-OH) (Zagoskina et al., 2023). The secondary metabolic pathways of plants produce these compounds through the shikimic acid and malonic acid pathways or by converting amino acids into phenolic compounds through the enzymatic action of phenylalanine ammonia-lyase (PAL) (Sahraeian et al., 2024). The synthesis of polyphenols is directly related to essential functions of survival, development, and interaction of plants with the environment; that is, the phenolic compounds play essential roles in mitigating impacts caused by abiotic factors, such as extreme temperatures, high exposure to light, and nutrient deficiency, and biotic factors, such as pests, herbivores, and pathogens, in addition to acting in the attraction of pollinators, dispersers, and nitrogen-fixing microorganisms, contributing significantly to plant protection, reproduction, and development (El-Saadony et al., 2024). As polyphenols play a protective role in plants, they are found in greater concentrations in inedible parts, such as leaves, stems, bark, flowers, and pulp, often discarded as food processing waste (Santos & Martins, 2022).

For human health, foods rich in polyphenols are widely recommended due to their potent antioxidant and antiinflammatory effects, which regulate metabolism, prevent chronic and degenerative diseases, and delay premature aging (El-Saadony et al., 2024). On the other hand, polyphenols can be used as natural additives in packaging, coatings, and films, aligning with the trends of innovative materials in the food industry (Martins et al., 2022). Polyphenol-functionalized materials not only help extend the shelf life of foods through their antioxidant and antimicrobial properties

TABLE 24.2 Production of biopolymeric films with vitamins C and E for food preservation.

Application	Biopolymer	Agent additives	Results	References
Fresh-cut apples	Carboxymethyl cellulose	Glycerol, calcium, and ascorbic acid	The film suppressed browning, preserved pulp firmness, and reduced the soluble solids content, titratable acidity, and pH variation in apple slices.	Koushesh Saba and Sogvar (2016)
Fresh-cut apples	Alginate, gellan gum, pectin, and apple fiber	Glycerol, ascorbic acid, and inulin	Fresh-cut apples' quality parameters (browning, firmness, soluble solids content, titratable acidity, and pH) were enhanced.	Moreira et al. (2015)
Fresh-cut pineapple	Alginate	Glycerol, sunflower oil, calcium chloride, ascorbic acid, and citric acid	The browning process was delayed, resulting in an extended shelf life.	Robles-Sánchez et al. (2013)
Bream fish	Alginate	Glycerol, ascorbic acid, and tea polyphenols	The coatings proved effective in delaying fish decomposition, reducing chemical deterioration, minimizing water loss, and improving sensory quality compared to uncoated fish.	Song et al. (2011)
Corn oil	Poly(lactic acid)	Vitamin E	The distribution of vitamin E in the films had minimal impact on color, transparency, tensile strength, and barrier properties. Films with higher vitamin E in the outer layer released it more slowly, and the film with 4% vitamin E in the outer layer and 2% in the inner layer provided the best oxidative stability for corn oil.	Wang et al. (2024)
-	Basil seed gum	Vitamin E in free form, encapsulated in liposomes, and nanoliposomes	The incorporation of vitamin E in free form, encapsulated in liposomes, and in nanoliposomes into the films improved flexibility, extensibility, mechanical strength, UV barrier properties, and antioxidant activity, with emphasis on vitamin E in nanoliposomes, which exhibited higher antioxidant efficacy and vitamin E retention during storage.	Salehi et al. (2024)
Salmon	Chitosan	Montmorillonite and vitamin E	Chitosan films with 15% vitamin E and 1% montmorillonite offered a good vapor barrier, controlled release of tocopherol, and nutritionally enriched food. After eight days, magnesium was migrated from the film to the salmon.	da Costa et al. (2024)
Safflower oil	Hydroxypropyl methylcellulose (HPMC) or soy protein isolate (SPI)	Vitamin E and cellulose nanocrystals	Incorporating vitamin E into the films improved their properties, such as heat sealability, elongation, mechanical strength, and antioxidant capacity. SPI films with vitamin E showed better oxidation control and oxygen barrier results, while HPMC films with vitamin E exhibited good antioxidant capacity. Additionally, the films enhanced the preservation of safflower oil.	Rosenbloom and Zhao (2021)
Agaricus bisporus	Chitosan/zein	Vitamin E	Mushrooms packaged with the films displayed the highest firmness, increased catalase and superoxide dismutase activities, and higher total phenolic content and DPPH radical scavenging activity.	Zhang et al. (2020)

but may also have the potential to monitor and indicate the product's condition simultaneously, offering greater safety and convenience to consumers (Martins et al., 2022). Due to the broad advantages of using polyphenols, several studies have been improving extraction techniques, both conventional (e.g., maceration and reflux) and advanced (e.g., enzymatic, ultrasound, microwave, supercritical), aiming to obtain this compound in a sustainable way and with high added value from agro-industrial waste (Jiang et al., 2021; Santos et al., 2024).

24.2.4.1 Polyphenols classification

It is estimated that there are more than 8000 types of polyphenols that can perform different functions. Despite the immense quantity and subclasses found in nature, polyphenols can be categorized based on the organization of the phenolic rings and bonds with carbon, hydrogen, and oxygen atoms (El-Saadony et al., 2024). In this sense, polyphenols can be divided into two main categories: flavonoids and nonflavonoids. Composed of two simple arenes (A and B) linked by a heterocyclic ring containing oxygen (C), flavonoids correspond to the largest class of polyphenols known, with their chemical skeleton known as C6-C3-C6 (Martins et al., 2022). Flavonoids are synthesized from p-coumaric acid by the shikimic acid metabolic pathway, generating basic phenylpropanoid units that are catalyzed by a series of enzymes and converted into different subclasses of flavonoids, such as flavanones, flavonols, flavonols, and anthocyanidins (Zou et al., 2025), as shown in Fig. 24.4.

Flavanones (e.g., naringenin and hesperidin) have a saturated heterocyclic ring (C ring) with a C6-C3-C6 structure and are found in citrus fruits (Sahraeian et al., 2024). Flavonols (e.g., kaempferol and quercetin) are characterized by a double bond in the C ring between carbons 2 and 3 and are found mainly in onions and green tea (Zou et al., 2025). Flavanols (e.g., catechin and epicatechins) have a saturated C ring and a hydroxyl group in position 3 and are abundant in cocoa, grapes, and apples (Zagoskina et al., 2023). Anthocyanidins have an oxygenated flavylic nucleus (C ring). They are responsible for the vibrant colors of fruits, vegetables, and flowers, being able to vary their color from red, blue, green, and yellow depending on pH and temperature conditions (Martins et al., 2022).

Nonflavonoid phenolic compounds are formed by simple or complex phenolic units, which may include hydroxyl (-OH), carboxyl (-COOH), methoxyl (-OCH3) groups, or polymeric structures, not following the typical C6-C3-C6 skeleton of flavonoids (Santos et al., 2024). These compounds are synthesized mainly by the metabolic pathways of shikimic acid and malonic acid, with the participation of the enzyme PAL, converting phenylalanine into cinnamic acid, a precursor of many nonflavonoid phenolics (Mahmutović et al., 2024; Zou et al., 2025), such as phenolic acids, tannins, lignans, stilbenes, and curcuminoids (Fig. 24.5).

Phenolic acids (e.g., gallic acid, ferulic acid, and vanillic acid) are one of the most abundant groups of nonflavonoid polyphenols in nature and consist of a benzene ring linked to hydroxyl (-OH) and carboxyl (-COOH) groups (El-Saadony

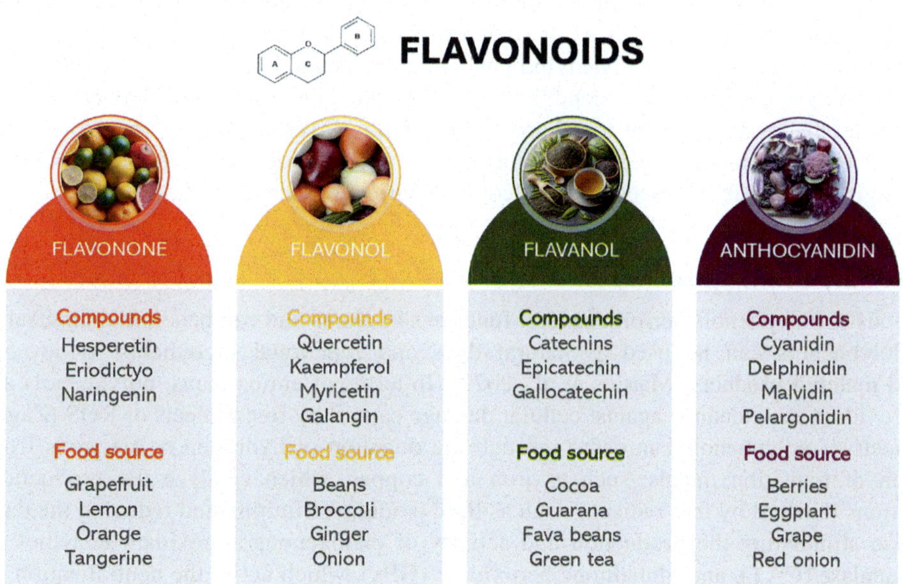

FIGURE 24.4 Types and sources of flavonoids.

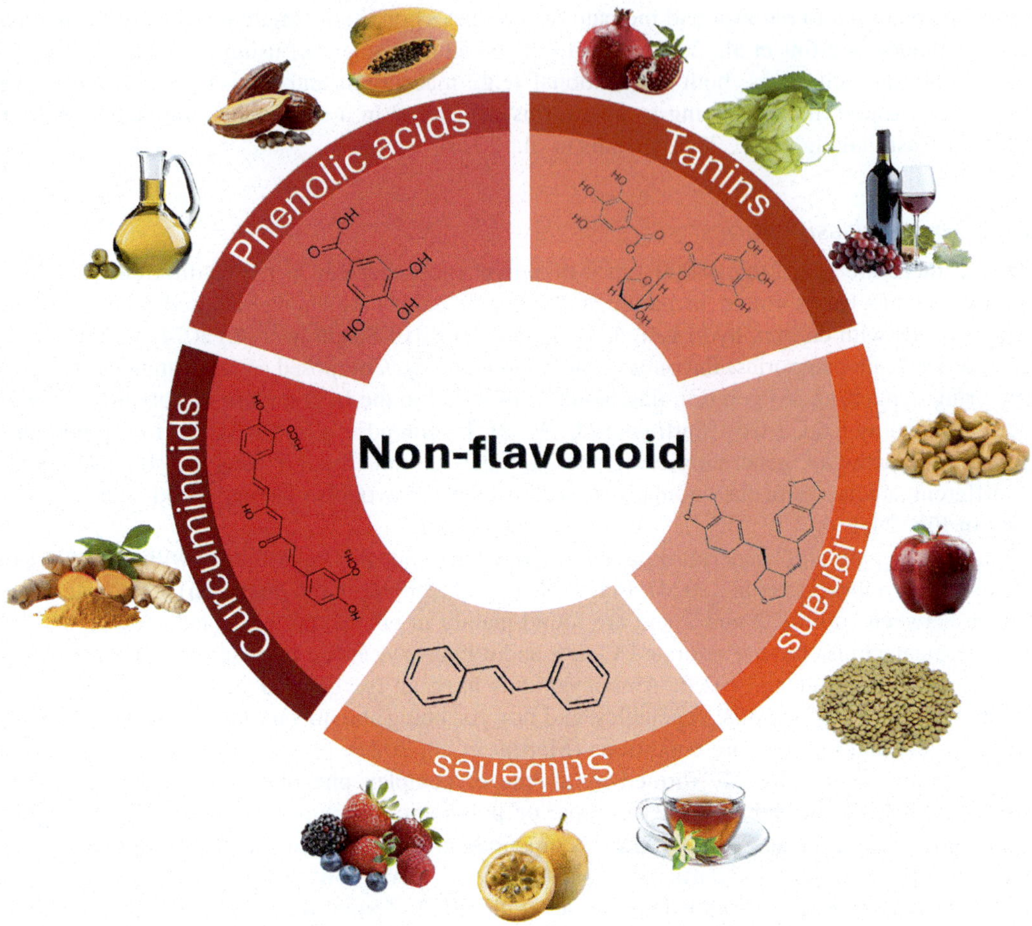

FIGURE 24.5 Chemical structure and sources of nonflavonoids.

et al., 2024). Tannins are high molecular weight polymerized polyphenols categorized as hydrolyzable (e.g., tannin gallate and ellagic acid derivatives) and condensed (e.g., proanthocyanidins) (Verma et al., 2024). Lignans (e.g., secoisolariciresinol and matairesinol) are derived from phenylpropane monomers, and their structure differs from that of other polyphenols by presenting a carbon-carbon bond between two phenolic rings (Rinaldo et al., 2010). Stilbenes are characterized by two aromatic rings linked to ethene (C=C), with resveratrol being the group's main representative (Bahramian et al., 2024). Curcuminoids (e.g., curcumin) have a structure of two phenolic rings linked to a conjugated carbon chain, which gives turmeric its intense yellow color (Rodriguez-Amaya et al., 2019).

24.2.4.2 Functionality of polyphenols

As mentioned previously, polyphenols perform several functions in plants and can benefit human health when included in the animal diet. Polyphenols can be used as natural dyes and functional ingredients to develop food, cosmetic, pharmaceutical, and material products (Martins et al., 2022). In terms of antioxidants, polyphenols act as scavengers of oxidizing agents, providing maintenance against cellular damage caused by free radicals or ROS (Zagoskina et al., 2023). The antioxidant effects of polyphenols can occur through the donation of hydrogen or electrons from hydroxyl groups (-OH), sequestration of transition metals, such as iron and copper, which catalyze the production of free radicals, interruption of reactions initiated by free radicals, such as lipid oxidation, limiting and reducing the damage caused to cell membranes, and also stimulating the production and activity of endogenous antioxidant enzymes such as superoxide dismutase (SOD), catalase (CAT), and glutathione peroxidase (GPx), which act in the neutralization of free radicals (El-Saadony et al., 2024). Thus, polyphenols play an essential role in cellular protection against damage caused by oxidative stress, preventing premature aging, chronic diseases, and deterioration of food products.

Polyphenols demonstrate remarkable antimicrobial capacity, acting as inhibitors of the growth of several pathogenic microorganisms, including bacteria, fungi, and viruses (Dai & Mumper, 2010). This action can occur through different mechanisms, such as altering the permeability of the cell membrane of microorganisms and allowing the entry of toxic compounds that result in cell death. Furthermore, polyphenols can inhibit essential enzymes and interfere with synthesizing proteins fundamental to microbial metabolism (Mahmutović et al., 2024). In the case of viruses, their antimicrobial activity includes the inhibition of viral replication by reducing the adhesion of the virus to host cells, making it challenging to initiate the infectious cycle (El-Saadony et al., 2024; Rinaldo et al., 2010). Their multiple pathways of antimicrobial action reinforce the potential of polyphenols as therapeutic agents and natural preservatives, expanding their relevance in several industrial areas.

Furthermore, polyphenols stand out as bioactive dyes, offering antioxidant and antimicrobial properties, as well as the ability to produce a wide range of colors in plants and products. Compounds such as anthocyanins and curcumins are the main polyphenols used as natural dyes, whose colors can be modulated by several factors, including pH, temperature, light exposure, and the presence of metals (Martins et al., 2022). These attributes make them safe and sustainable alternatives for cosmetics, food, textiles, and materials applications.

24.2.4.3 Polyphenols in food materials

The versatility of polyphenols has driven their application in the field of materials, especially in the development of packaging, coatings, and films for food. Research indicates that incorporating extracts or essential oils rich in polyphenols obtained from various sources offers a sustainable, ecological, and innovative solution for food preservation (Table 24.3). Materials functionalized with polyphenols can present better structural and functional properties; that is, the mechanical characteristics and barriers against gases and moisture are improved by adding bioactive compounds. Polyphenols act as active agents, providing antioxidant and antimicrobial effects, and light protection action to materials, helping preserve the packaged product's quality (Sahraeian et al., 2024).

Alginate-based sachets incorporated with polyphenol-rich macaúba extract demonstrated a 2.3-fold increase in tensile strength and an 82% reduction in water vapor permeability compared to sachets without the extract. When the functionalized sachet was used to store olive oil, it was observed that the oxidation rate of the product was significantly lower compared to storage in glass bottles or alginate sachets without the extract, evidencing its potential as functional packaging for food preservation (Alves-Silva et al., 2022). Films developed from polyvinyl alcohol (PVA), gelatin, and polyphenol-rich extract from Amaranth leaf showed high antioxidant and antimicrobial capacities, being able to extend the shelf life of chilled fish and chicken filets from 3 to 12 days compared to the film without the addition of the extract (Kanatt, 2020).

Edible coatings based on furcellara-gelatin loaded with green tea and puerh (*Camellia sinensis* L.) polyphenols were applied to salmon filets and, due to their antioxidant and antimicrobial effects, reduced the proliferation of microorganisms, in addition to inhibiting lipid oxidation compared to filets without the coating application (Kulawik et al., 2019). Minimally processed apples showed a lower browning rate when immersed in an edible coating based on starch/protein functionalized with polyphenol-rich extract from pink pepper, indicating greater conservation of the quality attributes of the processed fruit (Romani et al., 2018).

In this context, polyphenols are multifunctional compounds for the materials sector, offering a sustainable alternative to synthetic additives. Their extraction from food waste promotes the circular economy, adds value to processing by-products, and demonstrates their potential as a health-benefiting ingredient and a practical component for developing smart and environmentally responsible materials. Thus, incorporating polyphenols into materials improves food quality and safety and drives more innovative and sustainable industrial solutions aligned with the demands for greater efficiency and environmental respect.

24.3 Intelligent packaging based on biopolymers

Intelligent packaging systems monitor food and environmental conditions, providing consumers with information about product quality (Zhu et al., 2024). These systems typically include labels or tags with sensors or indicators that change appearance, such as color, in response to variations in temperature, pH, or gases like oxygen and carbon dioxide (Panwar et al., 2024).

pH-sensitive and chemically responsive colorimetric indicators are especially useful for food packaging, such as for meat and fish, allowing consumers to quickly identify fresh or spoiled food without opening the package. This enhances food safety and reduces waste (Alizadeh Sani et al., 2024).

TABLE 24.3 Functionalized materials with polyphenols applied in food products.

Source	Active compound	Material	Application	References
Açaí seeds and red beans	Anthocyanin and acid phenolic extracts	Bean flour-based films	Extend the shelf life of olive oil	Nogueira, Marasca, et al. (2022)
Olive leaf	Polyphenol-rich extract	Fish gelatin films	Reduction of *L. monocytogenes* growth in smoked salmon	Albertos et al. (2017)
Pink pepper residue	Polyphenol-rich extract	Chitosan films	Preserving the freshness of salmon filets	Merlo et al. (2019)
Clove	Polyphenol essential oil	Cassava starch films	Shelf-life extension of strawberries	Alves-Silva et al. (2022)
Cinnamon	Polyphenol essential oil	Edible chitosan coating	Shelf-life extension of minimally processed pineapple	Basaglia et al. (2021)
Helichrysum italicum flower	Polyphenol essential oil	Alginate-based coatings and films	Extend the shelf life of cherry tomatoes	Nkede et al. (2024)
Coffee by-product	Polyphenol-rich extract	Gelatine-based coatings	Reduction of lipid oxidation in hazelnuts	Turan et al.(2024)
Black chokeberry pomace	Polyphenol-rich extract	Gelatine-based coatings	Preservation effects on chilled pork	Yang et al. (2025)
Butterfly pea flower	Anthocyanin-rich extract	Sodium alginate films	Monitoring the freshness of pork, shrimp, and milk	Santos and Martins (2022)
Purple tomato pomace	Anthocyanin-rich extract	Chitosan films	Monitoring the quality of fish and milk	Li et al. (2021)
Black rice and turmeric	Anthocyanins and curcumin extract	Gellan gum/sodium alginate films	Monitoring shrimp freshness	Li et al. (2024)
Turmeric	Curcumin pickering emulsion	Starch/chitosan/ nanocellulose films	Meat freshness monitoring	Yang et al. (2025)

Using biodegradable polymers, such as proteins and polysaccharides, to produce intelligent packaging offers environmental advantages over petroleum-based polymers (Oliveira Filho, Bertolo, et al., 2021). Moreover, integrated indicators can monitor and record food conditions during storage, working with analytical sensors for greater precision (Luo, 2022).

24.3.1 Quality indicators

Polymeric films and nanofibers have been extensively explored to develop food quality indicators due to their unique properties and high efficiency. Biodegradable films are a sustainable alternative to synthetic materials, resistant to microbial degradation, and environmentally persistent. These films, made from biodegradable polymers, decompose into natural components like water and carbon dioxide. Classified into proteins (gelatin, collagen), polysaccharides (cellulose, chitosan), lipids (oils, waxes), and polyesters (PLA), they promote sustainability and meet the growing demand for eco-friendly food packaging (Bhowmik et al., 2024).

Nanofibers are promising materials for intelligent food packaging due to their high porosity and large specific surface area. These properties allow for the development of highly sensitive colorimetric indicators with uniform color responses and rapid detection capabilities using biopolymers (Almasi et al., 2022; Forghani et al., 2021).

Synthetic acid-base indicators have been widely used in intelligent packaging materials for their advantages, including low cost, stability, versatility, and intensity. These indicators belong to four main groups of organic dyes: triphenylmethanes, azo compounds, phthaleins, and sulfonphthaleins (Almasi et al., 2022). Recently, bromocresol purple, methyl red, bromocresol green, and bromothymol blue have been among the most-used synthetic dyes for freshness indicators in food (Oliveira Filho, Braga, et al., 2021).

Despite their advantageous properties, synthetic dyes raise increasing concerns regarding their biological incompatibility, toxicity, and environmental persistence, which may compromise consumer safety and sustainability. Consequently, attention has shifted toward natural, food-grade pigments that can act as intelligent ingredients in packaging systems (Azman et al., 2022).

Natural pH-sensitive compounds, such as anthocyanins, curcumin, alizarin, and phycocyanin, combine biodegradability, safety, and bioactivity with the ability to exhibit visible color changes in response to environmental variations. Anthocyanins, water-soluble pigments found in fruits, vegetables, and flowers, display various colors, ranging from red in acidic conditions to blue or green in alkaline environments. These changes occur due to alterations in their molecular structure, making them excellent for applications as colorimetric indicators in food and packaging (Oliveira Filho et al., 2021).

Curcumin, extracted from the root of turmeric (*Curcuma longa*), is widely used for its ability to detect acid-base reactions. This pigment changes color from yellow in neutral or acidic conditions to red in alkaline environments, making it particularly useful in systems requiring direct visual monitoring (Oliveira Filho et al., 2021). Alizarin, chemically known as 1,2-dihydroxyanthraquinone, is a natural dye derived from plant roots such as *Rubia tinctorum*. Its color varies from yellow at low pH to purple at high pH, making it a versatile alternative for detection systems (Ezati & Rhim, 2020). Phycocyanin, a blue pigment extracted from cyanobacteria such as spirulina (*Arthrospira platensis*), has antioxidant properties, is sensitive to pH changes, and is safe and nutritionally beneficial (Chentir et al., 2019).

Taken together, these compounds illustrate how natural pigments can function not only as safe and biodegradable alternatives to synthetic dyes, but also as multifunctional, intelligent ingredients, enabling food quality monitoring while simultaneously providing bioactivity and sustainability benefits.

According to Table 24.4, colorimetric indicators based on films and nanofibers functionalized with natural pigments, such as anthocyanins, curcumin, and betacyanins, have been successfully applied to monitor the quality of animal- and plant-based foods.

Gelatin and kappa-carrageenan films incorporating curcumin demonstrated pH sensitivity. When applied to monitor the freshness of chicken breast, they showed color changes correlated with deterioration (Rezaei et al., 2024). Esterified starch films with anthocyanins from black rice responded to different pH values, changing from purple to brownish yellow as the deterioration of pork belly progressed (Song et al., 2024). Cellulose films with alizarin demonstrated sensitivity to pH changes during fish deterioration (Xia et al., 2024).

Nanofibers were also explored (Table 24.4), such as gelatin and zein fibers with anthocyanins from purple sweet potato. These exhibited color changes at different pH levels and in the presence of volatile ammonia. These nanofibers also indicated the freshness of shrimp during deterioration (Wu et al., 2024). As pH increased, PLA/PEO nanofibers with phycocyanin changed from blue to bluish-green and white (Moreira et al., 2018).

Another application involved PVA nanofibers with purple cabbage extract anthocyanins, which displayed reversible color changes according to pH, allowing the monitoring of transient variations in fresh dates (Rutab). Tests showed a good

TABLE 24.4 Examples of films and nanofibers functionalized with natural indicators for monitoring food quality.

Material	pH-sensitive polymer matrix	Dye used	Food product	Main results	References
Film	Gelatin/kappa-carrageenan	Curcumin	Chicken breast meat	The incorporation of curcumin into the films resulted in improved functional and structural properties, with a significant reduction in solubility and permeability, as well as an increase in tensile strength, elongation at break, and Young's modulus. The films demonstrated pH sensitivity. When applied to monitor the freshness of chicken breast meat, the films exhibited color changes correlated with spoilage.	Rezaei et al. (2024)
Film	Esterified starch	Black rice anthocyanins	Pork belly	The incorporation of black rice anthocyanin into the films increased thickness and color difference, reduced transparency, enhanced elongation at break, and decreased tensile strength. The film responded to different pH values and exhibited color changes in various buffer solutions. The film changed from purple to yellowish-brown as the pork belly spoiled.	Song et al. (2024)
Film	Polyvinyl alcohol	Betacyanin from red beetroot extract	-	The incorporation of betacyanin increased film thickness, reduced water solubility, improved barrier properties (lower water vapor permeability), and significantly altered color parameters (lower lightness, higher a and b values), although all films exhibited low tensile strength. The films were able to detect pH changes.	Akter et al. (2024)
Film	Cellulose	Alizarin	Fish	The incorporation of alizarin into cellulose films led to homogeneous structures through hydrogen bonding, induced distinct pH-responsive color changes (more sensitive at low alizarin content), enhanced UV-shielding and antioxidant capacity, maintained good mechanical properties with tensile strength up to 31 MPa, and enabled their application as intelligent indicator films for fish freshness monitoring.	Xia et al. (2024)
Nanofiber	Gelatin/zein	Anthocyanin from purple sweet potatoes	Shrimp	The nanofibers exhibited a color change in the presence of different pH solutions and volatile ammonia. Additionally, they demonstrated a colorimetric response indicating the freshness of shrimp during spoilage.	Wu et al. (2024)
Nanofiber	PLA/PEO	Phycocyanin	-	The incorporation of phycocyanin into ultrafine fibers reduced fiber diameter, increased thickness, and enhanced color change perception (ΔE), demonstrating clear pH responsiveness and confirming their potential as innovative pH indicators for intelligent packaging.	Moreira et al. (2018)

Nanofiber	Polyvinyl alcohol	Anthocyanin from red cabbage extract	Date fruit (Rutab)	The nanofiber mat displayed reversible color changes in response to pH, enabling the monitoring of transient variations. Tests with fresh dates (Rutab) demonstrated that the mat's responses were well correlated with pH changes in the packaging, allowing for real-time monitoring.	Maftoonazad and Ramaswamy (2019)
Nanofiber	Zein/gelatin	Anthocyanin from red cabbage extract and purple sweet potatoes	Fish and milk	The nanofibers exhibited visible color changes, responding rapidly to pH variations (3–10) and ammonia vapor, proving effective in monitoring the spoilage of pasteurized milk and fish filets.	Oliveira Filho et al. (2024)

correlation between color changes and pH variations in the packaging, enabling real-time monitoring (Maftoonazad & Ramaswamy, 2019).

Finally, zein/gelatin nanofibers produced by solution-blown spinning technique with anthocyanins from purple cabbage and purple sweet potato exhibited visible color changes, responding rapidly to pH (3–10) and ammonia vapor. They effectively monitored the deterioration of pasteurized milk and fish filets (Oliveira Filho, Silva et al., 2024).

Therefore, films and nanofibers formulated with biopolymers and natural plant pigments have shown promise as smart materials for packaging, capable of real-time monitoring of the quality of various foods.

24.4 Conclusion

The growing demand for sustainable solutions in the food packaging sector has driven the development of innovative materials, emphasizing biopolymers derived from renewable sources such as proteins, polysaccharides, and lipids. These materials offer significant advantages, such as biodegradability, reducing plastic waste and promoting more sustainable development for the food industry.

In this context, active and intelligent materials combining advanced functional properties are gaining prominence. These materials interact dynamically with food and the environment, performing functions such as the controlled release of bioactive compounds and detecting changes in product quality. Compounds such as plant extracts, lipids, vitamins, and polyphenols, when integrated into biopolymers, offer essential benefits, such as antioxidant, antimicrobial, and stabilizing properties, contributing to preserving food and maintaining its nutritional and sensory quality.

In addition, intelligent packaging, which integrates real-time quality indicators based on chemical or colorimetric reactions, is an important innovation. These systems can detect changes in parameters such as pH and gas concentration, ensuring early detection of product deterioration. Using natural pigments, such as anthocyanins and curcumin, offers an environmentally friendly and safe alternative for developing these packaging products.

Therefore, combining biopolymers with bioactive compounds and intelligent systems represents a significant trend in technological innovation in the food packaging sector. These innovations contribute to sustainability, reduce environmental impact, and improve food safety and quality, meeting the demands of conscious consumers. Integrating these technologies can transform the food industry, promoting a more sustainable, efficient, and safe future.

Acknowledgments

The authors are grateful to FAPESP (processes 2023/02038-7, 2021/13260–7, and 2018/22214–6) and CNPq (141413/2019-0).

References

Akhter, R., Masoodi, F. A., Wani, T. A., & Rather, S. A. (2019). Functional characterization of biopolymer based composite film: Incorporation of natural essential oils and antimicrobial agents. *International Journal of Biological Macromolecules, 137*, 1245–1255. https://doi.org/10.1016/j.ijbiomac.2019.06.214, www.elsevier.com/locate/ijbiomac.

Akoh, C. C. (2017). *Food lipids: Chemistry, nutrition, and biotechnology, fourth edition food lipids: Chemistry, nutrition, and biotechnology, fourth edition*. United States: CRC Press, 1–1029. https://doi.org/10.1201/9781315151854, http://www.tandfebooks.com/doi/book/10.1201/9781315151854.

Akter, M., Islam, M. N., Yasmin, S., & Mahomud, M. S. (2024). Development of an intelligent packaging material incorporating betacyanin from red beetroot extract into polyvinyl alcohol films. *Future Foods, 10*. https://doi.org/10.1016/j.fufo.2024.100497, https://www.sciencedirect.com/science/journal/26668335.

Albertos, I., Avena-Bustillos, R. J., Martín-Diana, A. B., Du, W. X., Rico, D., & McHugh, T. H. (2017). Antimicrobial olive leaf gelatin films for enhancing the quality of cold-smoked Salmon. *Food Packaging and Shelf Life, 13*, 49–55. https://doi.org/10.1016/j.fpsl.2017.07.004, http://www.journals.elsevier.com/food-packaging-and-shelf-life/.

Alhaithloul, H. A. S., Alqahtani, M. M., Abdein, M. A., Ahmed, M. A. I., Hesham, A. E. L., Aljameeli, M. M. E., Al Mozini, R. N., Gharsan, F. N., Hussien, S. M., & El-Amier, Y. A. (2023). Rosemary and neem methanolic extract: Antioxidant, cytotoxic, and larvicidal activities supported by chemical composition and molecular docking simulations. *Frontiers in Plant Science, 14*. https://doi.org/10.3389/fpls.2023.1155698, https://www.frontiersin.org/journals/plant-science.

Alizadeh Sani, M., Zhang, W., Abedini, A., Khezerlou, A., Shariatifar, N., Assadpour, E., Zhang, F., & Jafari, S. M. (2024). Intelligent packaging systems for the quality and safety monitoring of meat products: From lab scale to industrialization. *Food Control, 160*, 110359. https://doi.org/10.1016/j.foodcont.2024.110359.

Alloh, P. B. R. A., El-Said, M. M., El-Sayed, H. S., Baranenko, D. A., & El-Messery, T. M. (2024). Extension of ultrafiltered cheese shelf life using edible coatings containing supercritical rosemary, thyme and coriander extracts as antimicrobial agents. *Federation Food Control, 163*. https://doi.org/10.1016/j.foodcont.2024.110479, https://www.sciencedirect.com/science/journal/09567135.

Almasi, H., Forghani, S., & Moradi, M. (2022). Recent advances on intelligent food freshness indicators; an update on natural colorants and methods of preparation. *Food Packaging and Shelf Life, 32*, 100839. https://doi.org/10.1016/j.fpsl.2022.100839.

Alves-Silva, G. F., Santos, L. G., Martins, V. G., & Cortez-Vega, W. R. (2022). Cassava starch films incorporated with clove essential oil and nanoclay as a strategy to increase the shelf life of strawberries. *International Journal of Food Science and Technology, 57*(10), 6690–6698. https://doi.org/10.1111/ijfs.16014, http://onlinelibrary.wiley.com/journal/10.1111/(ISSN)1365-2621.

Antoniadou, M., Rozos, G., Vaou, N., Zaralis, K., Ersanli, C., Alexopoulos, A., Dadamogia, A., Varzakas, T., Tzora, A., & Voidarou, C.(C. (2024). Comprehensive bio-screening of phytochemistry and biological capacity of oregano (*Origanum vulgare*) and *Salvia triloba* extracts against oral cariogenic and food-origin pathogenic bacteria. *Biomolecules, 14*(6), 619. https://doi.org/10.3390/biom14060619.

Azman, N. H., Khairul, W. M., & Sarbon, N. M. (2022). A comprehensive review on biocompatible film sensor containing natural extract: Active/intelligent food packaging. *Food Control, 141*. https://doi.org/10.1016/j.foodcont.2022.109189, https://www.journals.elsevier.com/food-control.

Bahramian, B., Abedi-Firoozjah, R., Ebrahimi, A., Tavassoli, M., Ehsani, A., & Naebe, M. (2024). Resveratrol-biopolymer materials: A sustainable approach to food packaging. *Trends in Food Science & Technology, 154*, 104761. https://doi.org/10.1016/j.tifs.2024.104761.

Basaglia, R. R., Pizato, S., Santiago, N. G., Maciel de Almeida, M. M., Pinedo, R. A., & Cortez-Vega, W. R. (2021). Effect of edible chitosan and cinnamon essential oil coatings on the shelf life of minimally processed pineapple (Smooth cayenne). *Food Bioscience, 41*. https://doi.org/10.1016/j.fbio.2021.100966, http://www.journals.elsevier.com/food-bioscience/.

Bedouhene, S., Senani, N., Rekeb, T., Chabane, M. D., Dermeche, S., & Messaoudi, D. (2024). Dual protection of aqueous garlic extract biomolecules against hemolysis and its oxidation products in preventing inflammation. *Cellular and Molecular Biology, 70*(7), 29–37. https://doi.org/10.14715/cmb/2024.70.7.5, https://www.cellmolbiol.org/index.php/CMB/article/view/5541.

Bensid, A., El Abed, N., Houicher, A., Regenstein, J. M., & Özogul, F. (2022). Antioxidant and antimicrobial preservatives: Properties, mechanism of action and applications in food–A review. *Critical Reviews in Food Science and Nutrition, 62*(11), 2985–3001. https://doi.org/10.1080/10408398.2020.1862046, https://www.tandfonline.com/loi/bfsn20.

Bhowmik, S., Agyei, D., & Ali, A. (2024). Smart chitosan films as intelligent food packaging: An approach to monitoring food freshness and biomarkers. *Food Packaging and Shelf Life, 46*, 101370. https://doi.org/10.1016/j.fpsl.2024.101370.

Biney, E. E., Nkoom, M., Darkwah, W. K., & Puplampu, J. B. (2021). High-performance liquid chromatography analysis and antioxidant activities of extract of *Azadirachta indica* (Neem) leaves. *Pharmacognosy Research*(1).

Boukhari, A., & Fatimi (2023). Extract of rosemary as food additive: The landmark patents. *In Biology and Life Sciences Forum, 26*.

Bucić-Kojić, A., Planinić, M., Tomas, S., Jakobek, L., & Šeruga, M. (2009). Influence of solvent and temperature on extraction of phenolic compounds from grape seed, antioxidant activity and colour of extract. *International Journal of Food Science & Technology, 44*(12), 2394–2401. https://doi.org/10.1111/j.1365-2621.2008.01876.x.

Camiletti, O. F., Prieto, M. C., Bergesse, A. E., Vera, L. G., & Grosso, N. R. (2024). A new active packaging system based on chickpea-based edible coatings added with microcapsules of *Cosmos sulphureus* Cav. flower extract. *Journal of the Science of Food and Agriculture, 104*(14), 8887–8896. https://doi.org/10.1002/jsfa.13715, http://onlinelibrary.wiley.com/journal/10.1002/(ISSN)1097-0010.

Catelan, T. B. S., Brum, C. C. S., Heredia-Vieira, S. C., Crispim, B. A., Grisolia, A. B., Santos, R. C. S., & Cardoso, C. A. L. (2018). Cytotoxicity, genotoxicity, antioxidant potential and chemical composition of leaves of Campomanesia pubescens (Mart. ex DC.) O.Berg. *Current Pharmaceutical Biotechnology, 19*(5), 416–421. https://doi.org/10.2174/1389201019666180626102443, http://www.eurekaselect.com/163213.

Cendrowski, A., Jakubowska, Z., & Przybył, J. L. (2024). Apple tree leaves (*Malus domestica* Borkh) as a valuable source of polyphenolic compounds with a high antioxidant capacity. *Applied Sciences, 14*(8), 3252. https://doi.org/10.3390/app14083252.

Cheng, J., Gao, R., Zhu, Y., & Lin, Q. (2024). Applications of biodegradable materials in food packaging: A review. *Alexandria Engineering Journal, 91*, 70–83. https://doi.org/10.1016/j.aej.2024.01.080.

Chentir, I., Kchaou, H., Hamdi, M., Jridi, M., Li, S., Doumandji, A., & Nasri, M. (2019). Biofunctional gelatin-based films incorporated with food grade phycocyanin extracted from the Saharian cyanobacterium *Arthrospira* sp. *Food Hydrocolloids, 89*, 715–725. https://doi.org/10.1016/j.foodhyd.2018.11.034.

Christaki, S., Moschakis, T., Kyriakoudi, A., Biliaderis, C. G., & Mourtzinos, I. (2021). Recent advances in plant essential oils and extracts: Delivery systems and potential uses as preservatives and antioxidants in cheese. *Trends in Food Science and Technology, 116*, 264–278. https://doi.org/10.1016/j.tifs.2021.07.029, http://www.elsevier.com/wps/find/journaldescription.cws_home/601278/description#description.

da Costa, D. S., dos Santos, L. N., Ferreira, N. R., Takeuchi, K. P., & Lopes, A. S. (2024). Properties of films and coatings added of tocopherol for food packaging: Tool-based review for systematic reviews and bibliometric analysis. *British Food Journal. 126*(13), 226–268. https://doi.org/10.1108/BFJ-04-2023-0292, https://www.emeraldinsight.com/loi/bfj.

Dai, J., & Mumper, R. J. (2010). Plant phenolics: Extraction, analysis and their antioxidant and anticancer properties. *Molecules (Basel, Switzerland), 15*(10), 7313–7352. https://doi.org/10.3390/molecules15107313, http://www.mdpi.com/1420-3049/15/10/7313/pdf.

Dias, A. L. B., Batista, H. R. F., Estevam, E. B. B., Alves, C. C. F., Forim, M. R., Nicolella, H. D., Furtado, R. A., Tavares, D. C., Silva, T. S., Martins, C. H. G., & Miranda, M. L. D. (2019). Chemical composition and in vitro antibacterial and antiproliferative activities of the essential oil from the leaves of *Psidium myrtoides* O. Berg (Myrtaceae). *Natural Product Research, 33*(17), 2566–2570. https://doi.org/10.1080/14786419.2018.1457664, www.tandf.co.uk/journals/titles/14786419.asp.

Diyana, Z. N., Jumaidin, R., Selamat, M. Z., & Suan, M. S. M. (2021). Thermoplastic starch/beeswax blend: Characterization on thermal mechanical and moisture absorption properties. *International Journal of Biological Macromolecules, 190*, 224–232. https://doi.org/10.1016/j.ijbiomac.2021.08.201.

Dodevska, T., Hadzhiev, D., & Shterev, I. (2023). A review on electrochemical microsensors for ascorbic acid detection: Clinical, pharmaceutical, and food safety applications. *Micromachines, 14*(1), 41. https://doi.org/10.3390/mi14010041.

Domínguez, R., Barba, F. J., Gómez, B., Putnik, P., Bursać Kovačević, D., Pateiro, M., Santos, E. M., & Lorenzo, J. M. (2018). Active packaging films with natural antioxidants to be used in meat industry: A review. *Food Research International, 113*, 93–101. https://doi.org/10.1016/j.foodres.2018.06.073, www.elsevier.com/inca/publications/store/4/2/2/9/7/0.

Drago, E., Campardelli, R., Pettinato, M., & Perego, P. (2020). Innovations in smart packaging concepts for food: An extensive review. *Foods, 9*(11), 1628. https://doi.org/10.3390/foods9111628.

Du, H., Sun, X., Chong, X., Yang, M., Zhu, Z., & Wen, Y. (2023). A review on smart active packaging systems for food preservation: Applications and future trends. *Trends in Food Science and Technology, 141*. https://doi.org/10.1016/j.tifs.2023.104200, http://www.elsevier.com/wps/find/journaldescription.cws_home/601278/description#description.

Ebrahimi, M., Karimi, R., & Garmakhany, A. D. (2024). The effect of maltodextrin edible coating containing pyracantha extract and potassium nanocarbonate on secondary metabolites, antioxidant capacity and microbiological properties of grape during cold storage. *Heliyon, 10*(13). https://doi.org/10.1016/j.heliyon.2024.e34123, https://www.sciencedirect.com/science/journal/24058440.

El-Saadony, M. T., Yang, T., Saad, A. M., Alkafaas, S. S., Elkafas, S. S., Eldeeb, G. S., Mohammed, D. M., Salem, H. M., Korma, S. A., Loutfy, S. A. (2024). Chemistry, bioavailability, bioactivity, nutritional aspects and human health benefits of polyphenols: A comprehensive review.

Ezati, P., & Rhim, J.-W. (2020). pH-responsive chitosan-based film incorporated with alizarin for intelligent packaging applications. *Food Hydrocolloids, 102*, 105629. https://doi.org/10.1016/j.foodhyd.2019.105629.

Fathi, M., Mozafari, M. R., & Mohebbi, M. (2012). Nanoencapsulation of food ingredients using lipid based delivery systems. *Trends in Food Science & Technology, 23*(1), 13–27. https://doi.org/10.1016/j.tifs.2011.08.003.

Fernández-López, J., & Viuda-Martos, M. (2018). Introduction to the special issue: Application of essential oils in food systems. *Foods, 7*(4), 56. https://doi.org/10.3390/foods7040056.

Filgueiras, C. T., Fakhouri, F. M., Garcia, V. Ad. S., Velasco, J. I., Nogueira, G. F., Ramos da Silva, L., & Oliveira, R. Ad (2024). Effect of adding red propolis to edible biodegradable protein films for coating grapes: Shelf life and sensory analysis. *Polymers, 16*(7). https://doi.org/10.3390/polym16070888, http://www.mdpi.com/journal/polymers.

Forghani, S., Almasi, H., & Moradi, M. (2021). Electrospun nanofibers as food freshness and time-temperature indicators: A new approach in food intelligent packaging. *Innovative Food Science & Emerging Technologies, 73*, 102804. https://doi.org/10.1016/j.ifset.2021.102804.

Martins de Freitas, S., Rodrigues, Fakhouri, Silva, Cardoso, A. L., Velasco, I., Filgueiras, & Garcia, A.dos S. (2021). Application of the Box–Behnken experimental design for the extraction of phenolic compounds from araçá-roxo (*Psidium myrtoides*). *Journal of Food Processing and Preservation, 45*(3).

Gali, L., Pirozzi, A., & Donsì, F. (2023). Biopolymer-and lipid-based carriers for the delivery of plant-based ingredients. *Pharmaceutics, 15*(3).

Gedikoğlu, A., Sökmen, M., & Çivit, A. (2019). Evaluation of *Thymus vulgaris* and *Thymbra spicata* essential oils and plant extracts for chemical composition, antioxidant, and antimicrobial properties. *Food Science & Nutrition, 7*(5), 1704–1714. https://doi.org/10.1002/fsn3.1007.

Geng, M., Feng, X., Wu, X., Tan, X., Liu, Z., Li, L., Huang, Y., Teng, F., & Li, Y. (2023). Encapsulating vitamins C and E using food-grade soy protein isolate and pectin particles as carrier: Insights on the vitamin additive antioxidant effects. *Food Chemistry, 418*, 135955. https://doi.org/10.1016/j.foodchem.2023.135955.

Geyesa, J. M., Esho, T. B., Legesse, B. A., & Wotango, A. S. (2024). Antibacterial and antioxidant potential analysis of *Verbascum sinaiticum* leaf extract and its synthesized silver nanoparticles. *Heliyon, 10*(2). https://doi.org/10.1016/j.heliyon.2024.e24215, https://www.sciencedirect.com/science/journal/24058440.

Hamann, D., Wlodarkievicz, M. E., Puton, B. M. S., Fischer, B., Colet, R., Paroul, N., Valduga, E., Zeni, J., Mignoni, M. L., Junges, A., Backes, G. T., & Cansian, R. L. (2024). Evaluation biodegradable films with green tea extract for interleafing sliced meat products. *Food Chemistry, 458*. https://doi.org/10.1016/j.foodchem.2024.140159, https://www.sciencedirect.com/science/journal/03088146.

Harris, J., Cottrell, S., Plummer, S., & Lloyd, D. (2001). Antimicrobial properties of Allium sativum (garlic). *Applied Microbiology and Biotechnology, 57*(3), 282–286. https://doi.org/10.1007/s002530100722.

Hosseini, H., Tajiani, Z., & Jafari, S. M. (2019). Improving the shelf-life of food products by nano/micro-encapsulated ingredients. In *Food quality and shelf life*. Iran: Elsevier, 159–200. https://doi.org/10.1016/B978-0-12-817190-5.00005-7, https://www.sciencedirect.com/book/9780128171905.

Hou, T., Ma, S., Wang, F., & Wang, L. (2023). A comprehensive review of intelligent controlled release antimicrobial packaging in food preservation. *Food Science and Biotechnology, 32*(11), 1459–1478. https://doi.org/10.1007/s10068-023-01344-8.

Jiang, T., Ghosh, R., & Charcosset, C. (2021). Extraction, purification and applications of curcumin from plant materials-A comprehensive review. *Trends in Food Science & Technology, 112*, 419–430. https://doi.org/10.1016/j.tifs.2021.04.015.

Jomova, K., Raptova, R., Alomar, S. Y., Alwasel, S. H., Nepovimova, E., Kuca, K., & Valko, M. (2023). Reactive oxygen species, toxicity, oxidative stress, and antioxidants: Chronic diseases and aging. *Archives of Toxicology, 97*(10), 2499–2574. https://doi.org/10.1007/s00204-023-03562-9, https://www.springer.com/journal/204.

Kanatt, S. R. (2020). Development of active/intelligent food packaging film containing *Amaranthus leaf* extract for shelf life extension of chicken/fish during chilled storage. *Food Packaging and Shelf Life, 24*, 100506. https://doi.org/10.1016/j.fpsl.2020.100506.

Kessler, J. C., Martins, I. M., Manrique, Y. A., Rodrigues, A. E., Barreiro, M. F., & Dias, M. M. (2024). Advancements in conventional and supercritical CO_2 extraction of Moringa oleifera bioactives for cosmetic applications: A review. *Journal of Supercritical Fluids, 214*. https://doi.org/10.1016/j.supflu.2024.106388, https://www.sciencedirect.com/science/journal/08968446.

Konstantinidi, M., & Koutelidakis, A. E. (2019). Functional foods and bioactive compounds: A review of its possible role on weight management and obesity's metabolic consequences. *Medicines, 6*(3), 94. https://doi.org/10.3390/medicines6030094.

Koushesh Saba, M., & Sogvar, O. B. (2016). Combination of carboxymethyl cellulose-based coatings with calcium and ascorbic acid impacts in browning and quality of fresh-cut apples. *LWT - Food Science and Technology, 66*, 165–171. https://doi.org/10.1016/j.lwt.2015.10.022.

Kulawik, P., Jamróz, E., Zając, M., Guzik, P., & Tkaczewska, J. (2019). The effect of furcellaran-gelatin edible coatings with green and pu-erh tea extracts on the microbiological, physicochemical and sensory changes of salmon sushi stored at 4 °C. *Food Control, 100*, 83–91. https://doi.org/10.1016/j.foodcont.2019.01.004.

Legesse, A. B., Emire, S. A., Tadesse, M. G., Dadi, D. W., Kassa, S. K., Oyinloye, T. M., & Yoon, W. B. (2024). Optimization of ultrasound-assisted extraction of *Verbascum sinaiticum* Leaves: Maximal phenolic yield and antioxidant capacity. *Foods, 13*(8). https://doi.org/10.3390/foods13081255, www.mdpi.com/journal/foods.

Li, C., Song, A., Wu, Y., Gao, Y., & Li, C. (2024). Intelligent double-layer film based on gellan gum/modified anthocyanin/curcumin/sodium alginate/zinc oxide for monitoring shrimp freshness. *International Journal of Biological Macromolecules, 274*(Pt 1), 132724. https://doi.org/10.1016/j.ijbiomac.2024.132724.

Li, Y., Wu, K., Wang, B., & Li, X. (2021). Colorimetric indicator based on purple tomato anthocyanins and chitosan for application in intelligent packaging. *International Journal of Biological Macromolecules, 174*, 370–376. https://doi.org/10.1016/j.ijbiomac.2021.01.182.

Lin, S. J., & Pascall, M. A. (2014). Incorporation of vitamin E into chitosan and its effect on the film forming solution (viscosity and drying rate) and the solubility andthermal properties of the dried film. *Food Hydrocolloids, 35*, 78–84. https://doi.org/10.1016/j.foodhyd.2013.04.015, https://www.journals.elsevier.com/food-hydrocolloids.

Lu, M., Pan, J., Hu, Y., Ding, L., Li, Y., Cui, X., Zhang, M., Zhang, Z., & Li, C. (2024). Advances in the study of vascular related protective effect of garlic (*Allium sativum*) extract and compounds. *The Journal of Nutritional Biochemistry, 124*, 109531. https://doi.org/10.1016/j.jnutbio.2023.109531.

Luo, X. (2022). Application of inkjet-printing technology in developing indicators/sensors for intelligent packaging systems. *Current Opinion in Food Science, 46*, 100868. https://doi.org/10.1016/j.cofs.2022.100868.

Madkour, M., Aboelenin, M. M., Habashy, W. S., Matter, I. A., Shourrap, M., Elolimy, A. A., & Aboelazab, O. (2024). Effects of oregano and/or rosemary extracts on growth performance, digestive enzyme activities, cecal bacteria, tight junction proteins, and antioxidants-related genes in heat-stressed broiler chickens. *Poultry Science, 103*(9). https://doi.org/10.1016/j.psj.2024.103996, https://www.sciencedirect.com/science/journal/00325791.

Maftoonazad, N., & Ramaswamy, H. (2019). Design and testing of an electrospun nanofiber mat as a pH biosensor and monitor the pH associated quality in fresh date fruit (Rutab). *Polymer Testing, 75*, 76–84. https://doi.org/10.1016/j.polymertesting.2019.01.011.

Mahmutović, L., Sezer, A., Bilajac, E., Hromić-Jahjefendić, A., Uversky, V. N., & Glamočlija, U. (2024). Polyphenol stability and bioavailability in cell culture medium: Challenges, limitations and future directions. *International Journal of Biological Macromolecules, 279*. https://doi.org/10.1016/j.ijbiomac.2024.135232, https://www.sciencedirect.com/science/journal/01418130.

Malaka, R., Hema, J. A., Muthukumarasamy, N. P., Sambandam, A., Subramanian, S., & Sevanan, M. (2015). Green synthesis of silver nanoparticles using cosmos sulphureus and evaluation of their antimicrobial and antioxidant properties. *Nano Biomedicine and Engineering, 7*(4), 160–168. https://doi.org/10.5101/NBE.V7I4.P160-168, http://nanobe.org/Assets/userfiles/sys_eb538c1c-65ff-4e82-8e6a-a1ef01127fed/files/7(4)_p160-168.pdf.

Malherbi, N. M., Schmitz, A. C., Grando, R. C., Bilck, A. P., Yamashita, F., Tormen, L., Fakhouri, F. M., Velasco, J. I., & Bertan, L. C. (2019). Corn starch and gelatin-based films added with guabiroba pulp for application in food packaging. *Food Packaging and Shelf Life, 19*, 140–146. https://doi.org/10.1016/j.fpsl.2018.12.008, http://www.journals.elsevier.com/food-packaging-and-shelf-life/.

Martins, V. G., Palezi, S. C., Alves-Silva, G. F., & Santos, L. G. (2022). Biodegradable packaging materials and techniques to improve their performance. In *Food packaging: The smarter way*. Brazil: Springer Nature, 61–105. https://doi.org/10.1007/978-981-16-7196-8_4, https://link.springer.com/book/10.1007/978-981-16-7196-8.

Soares Mateus, A. R., Serrano, C., Almeida, C., Soares, A., Rolim Lopes, V., & Sanches-Silva, A. (2024). Beyond thymol and carvacrol: Characterizing the phenolic profiles and antioxidant capacity of portuguese oregano and thyme for food applications. *Applied Sciences, 14*(19), 8924. https://doi.org/10.3390/app14198924.

Merlo, T. C., Contreras-Castillo, C. J., Saldaña, E., Barancelli, G. V., Dargelio, M. D. B., Yoshida, C. M. P., Ribeiro Junior, E. E., Massarioli, A., & Venturini, A. C. (2019). Incorporation of pink pepper residue extract into chitosan film combined with a modified atmosphere packaging: Effects on the shelf life of salmon fillets. *Food Research International, 125*. https://doi.org/10.1016/j.foodres.2019.108633, www.elsevier.com/inca/publications/store/4/2/2/9/7/0.

Moreira, J. B., Terra, A. L. M., Costa, J. A. V., & Morais, M. Gd (2018). Development of pH indicator from PLA/PEO ultrafine fibers containing pigment of microalgae origin. *International Journal of Biological Macromolecules, 118*, 1855–1862. https://doi.org/10.1016/j.ijbiomac.2018.07.028, www.elsevier.com/locate/ijbiomac.

Moreira, M. R., Cassani, L., Martín-Belloso, O., & Soliva-Fortuny, R. (2015). Effects of polysaccharide-based edible coatings enriched with dietary fiber on quality attributes of fresh-cut apples. *Journal of Food Science and Technology, 52*(12), 7795–7805. https://doi.org/10.1007/s13197-015-1907-z, http://www.springerlink.com/content/121580/.

Morsy, M. K., Khalaf, H. H., Sharoba, A. M., El-Tanahi, H. H., & Cutter, C. N. (2014). Incorporation of essential oils and nanoparticles in pullulan films to control foodborne pathogens on meat and poultry products. *Journal of Food Science, 79*(4), M675-M684 https://doi.org/10.1111/1750-3841.12400, http://www3.interscience.wiley.com/journal/118509799/issueyear?year=2008.

Nkede, F. N., Wardak, M. H., Fanze, M., Kondo, N., Wardana, A. A., Jothi, J. S., Tanaka, F., & Tanaka, F. (2024). The potential of helichrysum italicum essential oil-infused alginate coatings and film for prolonging the shelf-life of cherry tomatoes. *Food Packaging and Shelf Life, 46*. https://doi.org/10.1016/j.fpsl.2024.101381, https://www.sciencedirect.com/science/journal/22142894.

Nogueira, D., Marasca, N. S., Latorres, J. M., Costa, J. A. V., & Martins, V. G. (2022). Effect of an active biodegradable package made from bean flour and açaí seed extract on the quality of olive oil. *Polymer Engineering and Science, 62*(4), 1070–1080. https://doi.org/10.1002/pen.25907, http://onlinelibrary.wiley.com/journal/10.1002/(ISSN)1548-2634.

Nogueira, G. F., Soares, I. H. B. T., Soares, C. T., Fakhouri, F. M., & de Oliveira, R. A. (2022). Development and characterization of arrowroot starch films incorporated with grape pomace extract. *Polysaccharides, 3*(1), 250–263. https://doi.org/10.3390/polysaccharides3010014, https://www.mdpi.com/journal/polysaccharides.

Nogueira, G. F., Meneghetti, B. B., Soares, I. H. B. T., Soares, C. T., Bevilaqua, G., Fakhouri, F. M., & de Oliveira, R. A. (2024). Multipurpose arrowroot starch films with anthocyanin-rich grape pomace extract: Color migration for food simulants and monitoring the freshness of fish meat. *International Journal of Biological Macromolecules, 265*. https://doi.org/10.1016/j.ijbiomac.2024.130934, https://www.sciencedirect.com/science/journal/01418130.

Oliveira, J. Rd, de Jesus Viegas, D., Martins, A. P. R., Carvalho, C. A. T., Soares, C. P., Camargo, S. E. A., Jorge, A. O. C., & de Oliveira, L. D. (2017). *Thymus vulgaris* L. extract has antimicrobial and anti-inflammatory effects in the absence of cytotoxicity and genotoxicity. *Archives of Oral Biology, 82*, 271–279. https://doi.org/10.1016/j.archoralbio.2017.06.031, http://www.journals.elsevier.com/archives-of-oral-biology/.

de Oliveira Filho, J. G., Bertolo, M. R. V., Fernandes, S. S., Lemes, A. C., da Cruz Silva, G., Junior, S. B., de Azeredo, H. M. C., Mattoso, L. H. C., & Egea, M. B. (2024). Intelligent and active biodegradable biopolymeric films containing carotenoids. *Food Chemistry, 434*. https://doi.org/10.1016/j.foodchem.2023.137454, www.elsevier.com/locate/foodchem.

Oliveira Filho, J. Gd, Bertolo, M. R. V., Rodrigues, M.Á. V., Marangon, C. A., Silva, Gd. C., Odoni, F. C. A., & Egea, M. B. (2021). Curcumin: A multifunctional molecule for the development of smart and active biodegradable polymer-based films. *Trends in Food Science and Technology, 118*, 840–849. https://doi.org/10.1016/j.tifs.2021.11.005, http://www.elsevier.com/wps/find/journaldescription.cws_home/601278/description#description.

Oliveira Filho, J. Gd, Silva, Cd. O., Egea, M. B., Tonon, R. V., Paschoalin, R. T., Rios, Ad. O., Azeredo, H. M. Cd, & Mattoso, L. H. C. (2024). Efficient and sustainable production of intelligent nonwovens as indicators of food spoilage through solution blow spinning of proteins and natural pigments from agri-food waste. *Food Control, 166*. https://doi.org/10.1016/j.foodcont.2024.110707, https://www.sciencedirect.com/science/journal/09567135.

Oliveira Filho, J. Gd, Braga, A. R. C., Oliveira, B. Rd, Gomes, F. P., Moreira, V. L., Pereira, V. A. C., & Egea, M. B. (2021). The potential of anthocyanins in smart, active, and bioactive eco-friendly polymer-based films: A review. *Food Research International, 142*. https://doi.org/10.1016/j.foodres.2021.110202, www.elsevier.com/inca/publications/store/4/2/2/9/7/0.

Orucha, R., Prymeb, I. F., & Holmsenb, H. (2011). The fat soluble antioxidant vitamin E: Its metabolism, and biological and physiological significance. *Global Journal of Biochemistry| 2*(1).

Ou, H., Zuo, J., Gregersen, H., & Liu, X.-Y. (2024). Combination of supercritical CO_2 and ultrasound for flavonoids extraction from *Cosmos sulphureus:* Optimization, kinetics, characterization and antioxidant capacity. *Food Chemistry, 435*, 137598. https://doi.org/10.1016/j.foodchem.2023.137598.

Palanisamy, Y., Kadirvel, V., & Ganesan, N. D. (2025). Recent technological advances in food packaging: Sensors, automation, and application. *Sustainable Food Technology*. https://doi.org/10.1039/d4fb00296b.

Panwar, A., Kumar, S., Dhiman, A., Kumar, V., Gupta, D., & Sharma, A. (2024). pH sensitive indicator film based intelligent packaging systems of perishables: Developments and challenges of last decade. *Microchemical Journal, 207*, 111732. https://doi.org/10.1016/j.microc.2024.111732.

Perdones, Á., Chiralt, A., & Vargas, M. (2016). Properties of film-forming dispersions and films based on chitosan containing basil or thyme essential oil. *Food Hydrocolloids, 57*, 271–279. https://doi.org/10.1016/j.foodhyd.2016.02.006.

Pereda, M., Amica, G., & Marcovich, N. E. (2012). Development and characterization of edible chitosan/olive oil emulsion films. *Carbohydrate Polymers, 87*(2), 1318–1325. https://doi.org/10.1016/j.carbpol.2011.09.019.

Pires, T. C. S. P., Dias, M. I., Barros, L., Alves, M. J., Oliveira, M. B. P. P., Santos-Buelga, C., & Ferreira, I. C. F. R. (2018). Antioxidant and antimicrobial properties of dried Portuguese apple variety (*Malus domestica* Borkh. cv Bravo de Esmolfe). *Food Chemistry, 240*, 701–706. https://doi.org/10.1016/j.foodchem.2017.08.010, www.elsevier.com/locate/foodchem.

Pradella, L. C., Dourado, D. M., Santos, C. H. M. D., Fermiano, M. H., Matias, R., & Silva, B. A. K. D. (2021). Evaluation of the scientific activity of the leaf of the species campomanesia pubescens in wound model infected by *Staphylococcus aureus*. *Bioscience Journal, 37*. https://doi.org/10.14393/BJ-V37N0A2021-42182, http://www.seer.ufu.br/index.php/biosciencejournal/article/view/42182/32189.

Qurat-ul-Ain, Zia, K. M., Zia, F., Ali, M., Rehman, S., & Zuber, M. (2016). Lipid functionalized biopolymers: A review. *International Journal of Biological Macromolecules, 93*, 1057–1068. https://doi.org/10.1016/j.ijbiomac.2016.09.071, www.elsevier.com/locate/ijbiomac.

Reis, C. A., Gomes, A., & do Amaral Sobral, P. J. (2023). Films based on biopolymers incorporated with active compounds encapsulated in emulsions: Properties and potential applications—A review. *Foods, 12*(19). https://doi.org/10.3390/foods12193602, www.mdpi.com/journal/foods.

Rezaei, A., Sadeghi, E., Assadpour, E., Moradiyan, Mh, Khaledian, S., Rezaei, N., Dehnad, D., Zhang, F., Azizi-lalabadi, M., & Jafari, S. M. (2024). Preparation and characterization of gelatin/Kappa-carrageenan halochromic films containing curcumin for active and smart food packaging. *Food Bioscience, 62*. https://doi.org/10.1016/j.fbio.2024.105304, https://www.sciencedirect.com/science/journal/22124292.

Ribeiro, A. M. R., Fernandes, C. C., Menezes, Oliveira, D. S., Gonçalves, C. H. G., & Martins, M. L. D. (2024). Miranda, antibacterial screening of hexane extracts from *Psidium myrtoides*, a Brazilian native plant. *Ciência e Natura*.

Ribeiro, A. M., Estevinho, B. N., & Rocha, F. (2021). The progress and application of vitamin E encapsulation – A review. *Food Hydrocolloids, 121*. https://doi.org/10.1016/j.foodhyd.2021.106998, https://www.journals.elsevier.com/food-hydrocolloids.

Rinaldo, D., Mbéguié-A-Mbéguié, D., & Fils-Lycaon, B. (2010). Advances on polyphenols and their metabolism in sub-tropical and tropical fruits. *Trends in Food Science & Technology, 21*(12), 599–606. https://doi.org/10.1016/j.tifs.2010.09.002.

Robles-Sánchez, R. M., Rojas-Graü, M. A., Odriozola-Serrano, I., González-Aguilar, G., & Martin-Belloso, O. (2013). Influence of alginate-based edible coating as carrier of antibrowning agents on bioactive compounds and antioxidant activity in fresh-cut Kent mangoes. *LWT - Food Science and Technology, 50*(1), 240–246. https://doi.org/10.1016/j.lwt.2012.05.021.

Rodriguez-Amaya, D. B. (2019). Update on natural food pigments - A mini-review on carotenoids, anthocyanins, and betalains. *Food Research International, 124*, 200–205. https://doi.org/10.1016/j.foodres.2018.05.028, www.elsevier.com/inca/publications/store/4/2/2/9/7/0.

Romani, V. P., Hernández, C. P., & Martins, V. G. (2018). Pink pepper phenolic compounds incorporation in starch/protein blends and its potential to inhibit apple browning. *Food Packaging and Shelf Life, 15*, 151–158. https://doi.org/10.1016/j.fpsl.2018.01.003, http://www.journals.elsevier.com/food-packaging-and-shelf-life/.

Rosenbloom, R. A., & Zhao, Y. (2021). Hydroxypropyl methylcellulose or soy protein isolate-based edible, water-soluble, and antioxidant films for safflower oil packaging. *Journal of Food Science, 86*(1), 129–139. https://doi.org/10.1111/1750-3841.15543, http://onlinelibrary.wiley.com/journal/10.1111/(ISSN)1750-3841.

Sá, S., Chaul, L. T., Alves, V. F., Fiuza, T. S., Tresvenzol, L. M. F., Vaz, B. G., Ferri, P. H., Borges, L. L., & Paula, J. R. (2018). Phytochemistry and antimicrobial activity of *Campomanesia adamantium*. *Revista Brasileira de Farmacognosia, 28*(3), 303–311. https://doi.org/10.1016/j.bjp.2018.02.008, https://www.springer.com/journal/43450.

Sahraeian, S., Abdollahi, B., & Rashidinejad, A. (2024). Biopolymer-polyphenol conjugates: Novel multifunctional materials for active packaging. *International Journal of Biological Macromolecules, 280*(Pt 2), 135714. https://doi.org/10.1016/j.ijbiomac.2024.135714.

Salehi, S., Hashemi, H., Ghiasi, F., Mirzapour-Kouhdasht, A., Garcia-Vaquero, M., & Niakousari, M. (2024). Generation and evaluation of novel active biodegradable films based on modified basil seed gum and vitamin E nanoliposomes. *Food Hydrocolloids, 157*, 110422. https://doi.org/10.1016/j.foodhyd.2024.110422.

Salvia-Trujillo, L., Rojas-Graü, A., Soliva-Fortuny, R., & Martín-Belloso, O. (2015). Physicochemical characterization and antimicrobial activity of food-grade emulsions and nanoemulsions incorporating essential oils. *Food Hydrocolloids, 43*, 547–556. https://doi.org/10.1016/j.foodhyd.2014.07.012.

Santos, L. G., & Martins, V. G. (2022). Recovery of phenolic compounds from purple onion peel using bio-based solvents: Thermal degradation kinetics and color stability of anthocyanins. *Journal of Food Processing and Preservation, 46*(12). https://doi.org/10.1111/jfpp.17161, http://onlinelibrary.wiley.com/journal/10.1111/(ISSN)1745-4549.

Santos, L. G., de Paula, M., Gomes, B. M., Latorres, J. M., & Martins, V. G. (2024). High-power ultrasound as a trigger for extraction of proteins from brewery waste: Optimization and techno-functional properties. *Biomass Conversion and Biorefinery*. https://doi.org/10.1007/s13399-024-06109-9.

Sharma, S., Barkauskaite, S., Jaiswal, A. K., & Jaiswal, S. (2021). Essential oils as additives in active food packaging. *Food Chemistry, 343*. https://doi.org/10.1016/j.foodchem.2020.128403, www.elsevier.com/locate/foodchem.

Shishir, M. R. I., Xie, L., Sun, C., Zheng, X., & Chen, W. (2018). Advances in micro and nano-encapsulation of bioactive compounds using biopolymer and lipid-based transporters. *Trends in Food Science and Technology, 78*, 34–60. https://doi.org/10.1016/j.tifs.2018.05.018, http://www.elsevier.com/wps/find/journaldescription.cws_home/601278/description#description.

Shukla, S., Mehta, A., Bajpai, V. K., & Shukla, S. (2009). In vitro antioxidant activity and total phenolic content of ethanolic leaf extract of Stevia rebaudiana Bert. *Food and Chemical Toxicology, 47*(9), 2338–2343. https://doi.org/10.1016/j.fct.2009.06.024.

Silva, P. H.dos S., Abrão, F. Y., Paula, J. R.de, Ferreira, H. D., & Fiuza, T.de S. (2022). Evaluation of the antimicrobial activity of the crude ethanol extract, essential oil, and fractions from *Campomanesia pubescens* leaves. *Research, Society and Development, 11*(5). https://doi.org/10.33448/rsd-v11i5.28622 e56911528622.

Solórzano-Santos, F., & Miranda-Novales, M. G. (2012). Essential oils from aromatic herbs as antimicrobial agents. *Current Opinion in Biotechnology, 23*(2), 136–141. https://doi.org/10.1016/j.copbio.2011.08.005.

Song, W., Wu, N., He, Y., Zhao, H., Xu, J., & Ren, L. (2024). Intelligent color changing packaging film based on esterified starch and black rice anthocyanins. *Food Chemistry: X, 24*, 101930. https://doi.org/10.1016/j.fochx.2024.101930.

Song, Y., Liu, L., Shen, H., You, J., & Luo, Y. (2011). Effect of sodium alginate-based edible coating containing different anti-oxidants on quality and shelf life of refrigerated bream (*Megalobrama amblycephala*). *Food Control, 22*(3-4), 608–615. https://doi.org/10.1016/j.foodcont.2010.10.012.

Tadhani, M. B., & Subhash, R. (2006). In vitro antimicrobial activity of Stevia rebaudiana Bertoni leaves. *Tropical Journal of Pharmaceutical Research, 5*(1), 557–560.

Tegegn, G., Melaku, Y., Aliye, M., Abebe, A., Abdissa, N., Meresa, A., Degu, S., Hunsen, M., Hussein, A. A., & Endale, M. (2024). In vitro antimicrobial and antioxidant activities, essential oil composition, and in silico molecular modeling analysis of secondary metabolites from roots of *Verbascum sinaiticum*. *Zeitschrift für Naturforschung C, 79*(1-2), 25–39. https://doi.org/10.1515/znc-2023-0157.

Thoo, Y. Y., Ho, S. K., Liang, J. Y., Ho, C. W., & Tan, C. P. (2010). Effects of binary solvent extraction system, extraction time and extraction temperature on phenolic antioxidants and antioxidant capacity from mengkudu (*Morinda citrifolia*). *Food Chemistry, 120*(1), 290–295. https://doi.org/10.1016/j.foodchem.2009.09.064.

Trajkovska Petkoska, A., Daniloski, D., D'Cunha, N. M., Naumovski, N., & Broach, A. T. (2021). Edible packaging: Sustainable solutions and novel trends in food packaging. *Food Research International, 140*. https://doi.org/10.1016/j.foodres.2020.109981, www.elsevier.com/inca/publications/store/4/2/2/9/7/0.

Turan, D., Wang, Y., Grundmann, D., Paillart, M., Dieleman, R., & Rahn, A. (2024). Coffee by-product cascara as an edible active coating for enhancing hazelnut preservation and packaging. *Food Packaging and Shelf Life, 45*, 101350. https://doi.org/10.1016/j.fpsl.2024.101350.

Valencia, G. A., Luciano, C. G., & Monteiro Fritz, A. R. (2019). Smart and active edible coatings based on biopolymers. In *Polymers for agri-food applications*. Brazil: Springer International Publishing, 391–416. https://doi.org/10.1007/978-3-030-19416-1_20, https://link.springer.com/book/10.1007/978-3-030-19416-1.

Vasile, M., Bunea, A., Ioan, C. R., Ioan, B. C., Socaci, S., & Viorel, M. (2021). Phytochemical content and antioxidant activity of *Malus domestica* borkh peel extracts. *Molecules (Basel, Switzerland), 26*(24). https://doi.org/10.3390/molecules26247636, https://www.mdpi.com/1420-3049/26/24/7636/pdf.

Verma, P., Sen, R., Bamanna, A., Elhindawy, M., Nagpal, K., & Krishnan, V. (2024). Structural chemistry to therapeutic functionality: A comprehensive review on proanthocyanidins. *Biocatalysis and Agricultural Biotechnology, 55*, 102963. https://doi.org/10.1016/j.bcab.2023.102963.

Vignesh, A., Amal, T. C., Sarvalingam, A., & Vasanth, K. (2024). A review on the influence of nutraceuticals and functional foods on health. *Food Chemistry Advances, 5*, 100749. https://doi.org/10.1016/j.focha.2024.100749.

Wahba, R. H., El-Sonbati, A. Z., Diab, M. A., Gomaa, E. A., & AbouElleef, E. M. (2024). Electrochemical sensing of strontium ions cyclic voltammetry and investigation of rosemary extract's effects on their behavior: Antibacterial properties of rosemary extract and molecular docking analysis for potential COVID-19 therapeutic benefits. Microchemical Journal, 200, 110398.

Wang, H., Li, S., Wang, S., Zhou, J., Liu, C., Chen, C., & Xie, J. (2024). Development of controlled-release antioxidant poly (lactic acid) bilayer active film with different distributions of α-tocopherol and its application in corn oil preservation. *Food Chemistry, 439*, 138094. https://doi.org/10.1016/j.foodchem.2023.138094.

Wen, K., Zhang, K., Gao, W., Bai, S., Wang, J., Song, W., Zeng, Q., Peng, H., Lv, L., Xuan, Y., Li, S., Xu, M., & Ding, X. (2024). Effects of stevia extract on production performance, serum biochemistry, antioxidant capacity, and gut health of laying hens. *Poultry Science, 103*(1), 103188. https://doi.org/10.1016/j.psj.2023.103188.

Wu, X., Liu, Z., Liu, J., He, S., & Shao, W. (2024). Development of novel intelligent and edible dual-functional electrospun nanofiber films for shrimp preservation and freshness monitoring in real-time. *Food Bioscience, 61*, 104828. https://doi.org/10.1016/j.fbio.2024.104828.

Wylie, M. R., & Merrell, D. S. (2022). The antimicrobial potential of the neem tree *Azadirachta indica*. *Frontiers in Pharmacology, 13*. https://doi.org/10.3389/fphar.2022.891535, http://www.frontiersin.org/Pharmacology.

Xia, G., Ma, Y., Ma, Q., Yao, X., Xu, Z., Ji, X., & Zhang, F. (2024). Anti-ultraviolet, antioxidant and pH-responsive cellulose-based composite film incorporated with alizarin for intelligent packaging applications. *Food Packaging and Shelf Life, 46*, 101413. https://doi.org/10.1016/j.fpsl.2024.101413.

Yang, L., Chen, S., Ma, N., Chen, W., Zhang, Z., & Zhang, H. (2025). Effect of gelatin edible coating with *Aronia melanocarpa* pomace polyphenols on the cold storage of chilled pork. *Meat Science, 219*, 109677. https://doi.org/10.1016/j.meatsci.2024.109677.

Zagoskina, N. V., Zubova, M. Y., Nechaeva, T. L., Kazantseva, V. V., Goncharuk, E. A., Katanskaya, V. M., Baranova, E. N., & Aksenova, M. A. (2023). Polyphenols in plants: Structure, biosynthesis, abiotic stress regulation, and practical applications (Review). *Federation International Journal of Molecular Sciences, 24*(18). https://doi.org/10.3390/ijms241813874, http://www.mdpi.com/journal/ijms.

Zhang, L., Liu, Z., Sun, Y., Wang, X., & Li, L. (2020). Combined antioxidant and sensory effects of active chitosan/zein film containing α-tocopherol on *Agaricus bisporus*. *Food Packaging and Shelf Life, 24*, 100470. https://doi.org/10.1016/j.fpsl.2020.100470.

Zhu, Y., Gao, X., Gao, X., Jiang, Z., Alhomrani, M., Alamri, A. S., Alsanie, W. F., & Cui, H. (2024). Development of polysaccharide based intelligent packaging system for visually monitoring of food freshness. *International Journal of Biological Macromolecules, 277*. https://doi.org/10.1016/j.ijbiomac.2024.134588, https://www.sciencedirect.com/science/journal/01418130.

Zou, P., Otero, P., Mejuto, J. C., Simal-Gandara, J., Xiao, J., Cameselle, C., Chen, S., Lin, S., & Cao, H. (2025). Exploring the mechanism of flavonoids modification by dimerization strategies and their potential to enhance biological activity. *Food Chemistry, 467*. https://doi.org/10.1016/j.foodchem.2024.142266, https://www.sciencedirect.com/science/journal/03088146.

Chapter 25

Peptides

Pragati Godara[1], Divya Surisetti[2], Meraj Ahmed[3], Pinku Chandra Nath[4], Vinay Kumar Pandey[5], and Neha Chandra[6]

[1]*MMICT & BM (HM), Maharishi Markandeshwar Deemed to be University, Mullana, Ambala, Haryana, India,* [2]*Department of Pharmacology, Kanpur Institute of Technology and Pharmacy, Kanpur, Uttar Pradesh, India,* [3]*Department of Soil Science and Agricultural Chemistry, School of Agriculture, Lovely Professional University, Phagwara, Punjab, India,* [4]*Department of Bioengineering, School of Engineering and Architecture, Techno India University, Anandanagar, Tripura, India,* [5]*Department of FoodScience and Technology, Graphic Era (Deemed to be University), Dehradun, Uttarakhand, India,* [6]*School of Basic & Applied Sciences, K.R. Mangalam University, Gurugram, Haryana, India*

25.1 Introduction

There is a major shift occurring in the global food industry toward safer, more sustainable packaging solutions as environmental concerns, consumer awareness, and demand pressures increase. A key driver of this transformation is innovation in food packaging technologies, a crucial element for ensuring food safety, extending shelf life, maintaining nutritional quality, and making products more attractive (Olawade et al., 2024). Most of the currently used food packaging materials, including plastics derived from petrochemicals, have been favored in the market for several years because of their versatility, durability, and low cost, but are now subject to criticism for being nonbiodegradable and polluting the environment, especially causing the largest problems in the form of microplastic accumulation (Hosseini et al., 2023). This has led to the paradigm shift toward the development and adoption of sustainable and biodegradable substitutes. Within these, the edible films and coatings based on natural polymers are attracting growing interest. However, the increasing use of peptides as bioactive and functional agents to enhance the performance of these biodegradable systems offers even more exciting potential (Moghadas et al., 2024).

Many forms of food packaging do much more than just contain their contents. Flexible, semirigid, and rigid packaging systems are like the ones invented to provide mechanical protection, barrier properties to gases and moisture, microbial protection, and in some cases, active contact action in food to prolong its freshness (Nath, Ojha, et al., 2023). Traditionally, packaging materials can be categorized into plastics (e.g., polyethylene, polypropylene, polyethylene terephthalate [PET]), glass, metals (aluminum, tin), and paper-based composites. Synthetic plastics are the most common because they can be engineered to be flexible, cost-effective, and resistant to moisture. However, the benefits come at the expense of environmental harm when those items are discarded, since many of those materials are single-use and seldom end up being recycled properly (Nath, Debnath, et al., 2023).

Recent decades have seen technological advancements resulting in such innovations as the development of active packaging (which interacts with food to extend shelf life), intelligent packaging (which can monitor food condition or environment), and edible coatings (which can be applied in direct contact with food surfaces and consumed with the product) (Ali et al., 2025). These newer methods are not just aimed at improving preservation; some are also designed to meet health, safety, and environmental objectives (see Fig. 25.1). At the forefront of packaging technology, one of the most promising strategies for reducing dependence on conventional plastic packaging is the extensive use of natural and biodegradable materials to replace or minimize the use of synthetic packaging materials (Ibarra et al., 2025). Several convergent issues have led to the demand for sustainable food packaging. The first concern is the environmental impact caused by synthetic packaging waste, particularly plastic, that lasts in ecosystem for hundreds of years (Ibarra et al., 2025). Plastic waste mismanagement directly drives marine pollution, soil degradation, and threatens wildlife. Furthermore, increasing awareness among customers regarding the carbon footprint and toxicity risks caused by synthetic chemicals in packaging materials has led to obvious trends in favor of environmentally friendly and health-safe alternatives.

Natural Polymer-Based Films and Coatings in Food Technology. DOI: https://doi.org/10.1016/B978-0-443-34244-8.00027-6

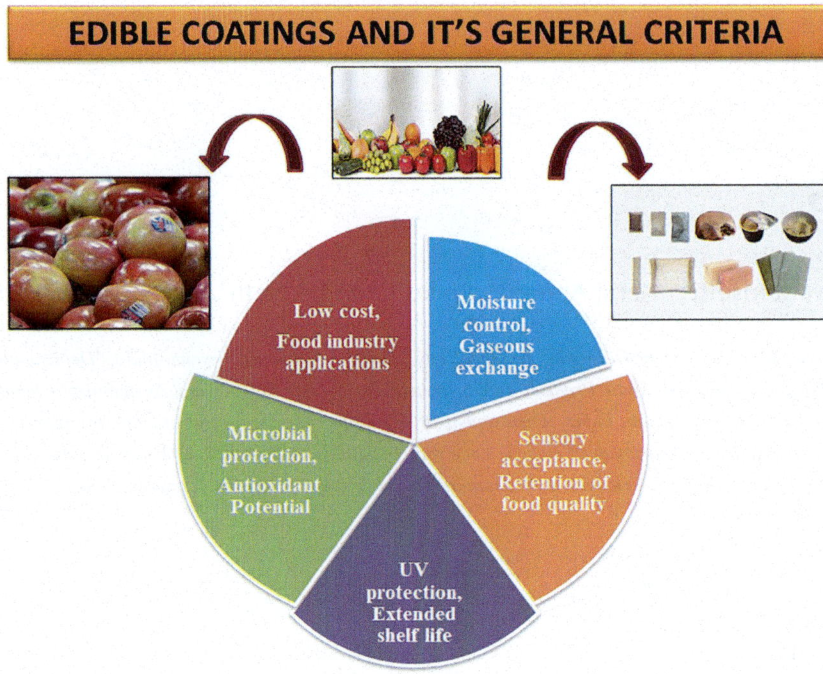

FIGURE 25.1 Edible coatings and their general criteria.

Natural polymers, which are derived from renewable biological sources such as plants, animals, and microorganisms, are a suitable alternative to these challenges. These factors include polysaccharides (starch, cellulose, chitosan), proteins (casein, soy protein, gelatin), and lipids. When formed into films and coatings, these materials also possess biodegradability, edibility, and in a number of cases, good compatibility with food systems (Cheng et al., 2025). However, they also exhibit limitations, including insufficient mechanical strength, poor moisture barrier properties, and vulnerability to degradation, which need to be solved for wider acceptance. Water-soluble polysaccharides such as starch, alginate, and pectin have good film-forming capacity, so they can be used for fresh fruits and vegetables. They have good oxygen barrier properties, and proteins like whey protein, gelatin, and soy protein isolate (SPI) are used for introducing bioactive compounds (Khanda et al., 2025). Lipids are also added to improve the moisture barrier but are often not used on their own due to their brittleness. Among promising agents, peptides, which are short chains of amino acids (AAs), emerge in the development of next-generation food packaging systems (Amin et al., 2021). These are excellent candidates for improving the structural and functional properties of biodegradable films and coatings due to their functionality, bioactivity, and compatibility with natural polymers. Peptides are obtained from natural proteins (e.g., via enzymatic hydrolysis) or synthesized (with defined sequences) to provide the desired properties.

Most peptides have intrinsic antimicrobial, antioxidant, and antiinflammatory activities. Their incorporation in edible coatings can serve as a bioactive barrier to actively inhibit spoilage microorganisms and oxidative degradation of food, helping to extend shelf life as well as retention of food quality (Kazemzadeh-Narbat et al., 2021). The intermolecular interactions between polysaccharides or proteins with antifungal peptides can enhance the mechanical strength, barrier properties, and stability of the natural polymer films (Holmberg et al., 2013). This solves the fundamental problem of natural films, which are brittle and have weak waterproofness. They can serve as carriers for other bioactive compounds and, as a result, provide a controlled-release mechanism in response to certain triggers (pH or temperature). This enables intelligent packaging solutions that engage with the food environment in a dynamic way. As peptides are derived from food-grade sources, they typically have a low risk of toxicity and good acceptance from regulatory and consumer aspects (Drexelius & Neundorf, 2021).

25.2 Natural polymers used in food films and coatings

25.2.1 Polysaccharides

Polysaccharides are among the most widely used biopolymers for development of edible films and coatings since they are abundant, renewable, and compatible with food products (Nath, Sharma, Debnath, Sharma et al., 2023). These are long-

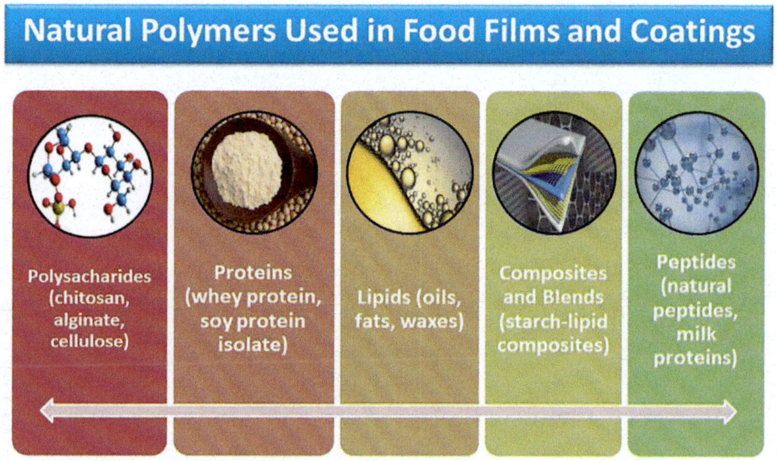

FIGURE 25.2 Natural polymers used in food films and coatings.

chain carbohydrates from diverse natural sources such as plants, animals, algae, and microbial fermentation. Some examples include chitosan, alginate, starch, and cellulose, which can exhibit different functional properties in food applications (Mishra et al., 2024). Natural polymers used in food films and coatings are shown in Fig. 25.2. Chitosan, a polysaccharide derived from chitin in crustacean shells, is noted for its antimicrobial action as well as its ability to form films and its biodegradability. It has superb oxygen barrier properties and can provide flexible and transparent films (Sharma et al., 2024). Its intrinsic antimicrobial activity has made it particularly suitable for use as packaging for perishable foods, such as meat, fish, and produce. Chitosan is only soluble under acidic conditions, which restricts its applications in neutral or basic food systems.

Natural polysaccharides serve as the basis for sodium alginate, a biopolymer derived from brown algae. It is an exceptional film-forming biopolymer that is hydrophilic. It has a very good oxygen barrier (Yadav, Mahapatra et al., 2025). Meat products are now wrapped in these alginate films, which were initially used to control the respiration rate of fruits and vegetables. These edible films help extend the shelf life of meat products when paired with antibacterial and active substances (Li et al., 2025). Black cumin oil is used as an antibacterial agent in an edible film made of alginate to package chicken breast meat. This film prevents *Escherichia coli* from growing and the color of meat from changing for approximately five days when kept at 40°C.

Sources such as corn, potato, and cassava are among the most common and inexpensive polysaccharides used in food films; starch is one of the polymers in this group. It can form transparent, tasteless, and biodegradable films (Nath, Nandi, et al., 2023; Pandey, Islam, et al., 2022). Although native starch films exhibit excellent gas barrier properties, they are brittle and have poor water resistance owing to the hydrophilic nature of starch components, amylose and amylopectin. It often requires modification methods to improve its mechanical and barrier properties, for example, plasticization or blending with other biopolymers. Apart from gelatin, cellulose and its derivatives, like carboxymethyl cellulose (CMC) or hydroxypropyl methylcellulose (HPMC), are the other important polysaccharides for film formation. Gas barrier properties of cellulose-based films are significant for fruits and vegetables, as cellulose-based films are the most commonly used. However, the moisture sensitivity of cellulose films, like that of other polysaccharides, constrains their widespread application (Pandey, 2022; Sharma et al., 2023).

25.2.2 Proteins

Proteins are valuable biopolymers with specific structural characteristics that allow them to form continuous networks with gaseous barrier properties, which makes them ideal for edible films and coatings applications. Protein-based films are biodegradable and nontoxic materials derived from either animal sources or plant sources, with reasonable mechanical integrity (Bashir et al., 2023; Roy et al., 2023). Common natural proteins used in food packaging films include gelatin, soy protein, and whey protein, all of which play unique roles in film formation and functional performance (Nath, Hazarika, Ojha et al., 2024). Peptide-based natural polymer films and coatings in food technology are shown in Table 25.1.

One of the most extensively studied protein polymers in the context of film-making applications is gelatin, a polymer obtained by the partial hydrolysis of collagen and composed predominantly of animal connective tissues (Ong et al., 2021). It is widely preferred because of its excellent film-forming properties, clarity, flexibility, and high compatibility

TABLE 25.1 Different types of peptide-based natural polymer films/coatings, preparation methods, action mechanisms and their applications.

Type of peptide–polymer film/coating	Preparation methodology	Mechanism of action	Key properties	Food applications	References
Antimicrobial peptide–chitosan film	Solvent casting with chitosan and antimicrobial peptides (AMPs) (e.g., nisin)	Disruption of microbial cell membranes	Antimicrobial, biodegradable, good film-forming ability	Surface coating for meats, dairy, fruits	Li et al. (2020)
Whey protein–bioactive peptide film	Film casting from whey protein isolate with enzymatic hydrolysates	Reactive oxygen species scavenging, microbial inhibition	Antioxidant, antimicrobial, flexible	Packaging for cheese, poultry products	Nath, Debnath, et al. (2023)
Gelatin–peptide composite film	Dissolution of gelatin with bioactive peptides (e.g., fish-skin peptides)	Cross-linking with gelatin matrix, microbial interaction	Good mechanical strength, antioxidant	Wrapping for seafood, baked goods	Nath et al. (2024)
Zein–peptide coating	Ethanol-based solution casting with hydrophobic peptides	Barrier formation, surface interaction with microbes	Water resistance, UV protection, bioactivity	Coating for fruits, nuts, confectioneries	
Alginate–peptide coating	Ionic gelation (e.g., with $CaCl_2$) incorporating AMPs	Ion-mediated gelation, release of bioactive agents	Edible, biocompatible, active release	Coating for fresh-cut fruits, vegetables	Hurteaux et al. (2005)
Starch–peptide film	Heat-induced gelatinization and peptide mixing, followed by casting	Peptide-starch interaction via hydrogen bonding	Transparency, flexibility, active bio-functionality	Edible films for snacks, bakery items	Gao et al. (2022)
Pullulan–peptide edible film	Blending pullulan with peptides, casting, and drying	Sustained release of peptides, surface barrier	Glossy, oxygen barrier, antimicrobial	Edible coatings for candies, fresh produce	Gan et al. (2022)
Pectin–functional peptide coating	Homogenization and casting with peptide inclusion	Electrostatic and hydrogen-bonding interactions	Biodegradable, hydrophilic, antioxidant	Fresh fruit preservation, yogurt packaging	Wang et al. (2022)
Soy protein–peptide blend film	Thermal denaturation and casting with soy peptides	Enhanced cross-linking, enzyme inhibition	Mechanical strength, water vapor barrier	Films for tofu, plant-based meats	Gao et al. (2025), Gimranov et al. (2025)

with various food matrices. Gelatin films possess good oxygen barrier properties and weak mechanical properties (Hamidi, 2022). However, their hydrophilic nature restricts their suitability as moisture barriers. This is commonly resolved through plasticization of gelatin using compounds such as glycerol or through the impregnation of hydrophobic materials (lipids).

Soy protein, notably SPI, is fast becoming an attractive plant-based option in the field of edible film technology. It is made from a blend of globular proteins, such as glycinin and β-conglycinin, that unfold and bond as a film is formed to produce a robust network (Qiu et al., 2025). The tensile strength and oxygen barrier properties of SPI films have been shown to be good, and they are regarded as suitable for packaging applications in fruits, vegetables, and bakery products. Soy protein films, however, are brittle and sensitive to moisture. Adding plasticizers or mixing them with polysaccharides can facilitate their flexibility and stability (Xu et al., 2025).

Another high-quality film-forming protein is whey protein, which is a by-product of cheese manufacturing. Whey proteins, including β-lactoglobulin (BLG) and α-lactalbumin (ALA), are abundant in whey protein isolate (WPI) and have been known to unfold and subsequently crosslink to yield films through heat treatment (Kheto et al., 2025). WPI films have the advantages of high clarity, good mechanical properties, and low oxygen permeability, which are particularly well suited for oxygen-sensitive food products such as nuts and oily snacks. Despite these benefits, whey protein films are hydrophilic and may need to have hydrophobic components or coating layers added to minimize moisture transmission (Zhu, 2024).

25.2.3 Lipids

With respect to edible films and coatings, lipids (such as fats, oils, waxes, etc.) are important because they are intended to improve moisture barrier properties. Due to their brittleness and low mechanical strength, lipids are mostly not used alone to create flexible, free-standing films; they are often combined with polysaccharides or proteins in composite systems to provide hydrophobicity and increase water resistance (Devi et al., 2024). Including them in edible coatings minimizes water loss and retention of texture, as well as prolonging the shelf life of moisture-sensitive food products (fruits, vegetables, meats, and baked products) (Wu et al., 2024).

Waxes are one of the best lipid-based materials to fortify moisture barrier properties; they not only have high hydrophobicity but also exhibit low water vapor permeability (Wang et al., 2024). Beeswax, for example, has long been widely used to coat fresh produce, inhibiting respiration and water loss. Plant-based Carnauba wax is also used for glossy finishes and to enhance appearance. These materials create relatively nonpermeable layers of insulation that are capable of stopping significant amount of water vapor flow. However, they tend to be brittle and inelastic, and often need additives such as plasticizers or may be used in multilayer systems in order to perform better (Islam et al., 2024).

Hydrophobic layers may also be provided by means of fatty acids and their salts. Edible coating formulations are usually applied by dipping, spraying, or brushing. Stearic acid, for example, has proven to decrease moisture migration in confectionery and bakery products. Due to their liquid form at room temperature, oils and glycerides are usually in emulsions with proteins or polysaccharides. They assist in altering the surface characteristics of coatings to obtain even more hydrophobic surfaces. Examples include sunflower oil entrapment in starch matrix, which has proven to lower moisture transference and increase surface gloss (Gupta, 2024).

25.2.4 Composites and blends

Composite and blend formulations are a very efficient strategy for obtaining advanced edible films and coatings, especially in terms of mechanical, barrier, and functional properties-focused improvement of the natural polymer systems (Manzoor, no date). Although each biopolymer class (polysaccharides, proteins, and lipids) can have specific beneficial properties, each type of biopolymer also presents certain limitations. Polysaccharides and proteins, for example, usually have good oxygen barrier properties but are hydrophilic, making them poor moisture resistors (Manzoor, no date). Lipids, on the other hand, provide a great barrier to moisture but offer little mechanical strength. This approach allows for the piecing together of various bio-based polymers to create films and coatings that not only utilize the benefits of both in one system but also have a compensating effect on their drawbacks.

Composites are defined as multiphase systems in which polymers with different physical or chemical properties are incorporated, as well as additional components such as plasticizers, emulsifiers, essential oils, nanoparticles, or bioactive peptides. Blends, in contrast, are generally mixtures of compatible polymers (usually class-specific, for example, two polysaccharides or proteins) processed together to create a homogeneous matrix (Udaipuria & Bhattacharya, 2025). A well-known and widely studied group of materials is starch-lipid composites, which are mixtures in which, after the starch matrix, dispersed lipid phases (waxes, oils, etc.) are embedded. These systems create emulsion-based films that exhibit superior water barrier properties with minimal strength loss. A favored blend of chitosan-protein composites combines the antimicrobial and film-forming properties of chitosan with the structural and gas barrier attributes of gelatin or whey protein. Such combinations produce coatings with enhanced flexibility, antifouling activity, and resistance to oxidation (Iñiguez-Moreno & González-González, 2025). Blends of polysaccharides and proteins (alginate or pectin with soy or whey protein) yield matrices with balanced mechanical and barrier properties. These films are perfect for semimoist and fresh food items, including fruits, vegetables, and dairy products. When properly designed, these blends can exhibit enhanced intermolecular interactions (e.g., hydrogen bonding, ionic interactions) that stabilize the matrix, thus conferring impermeability in the aqueous milieu (Stefanowska et al., 2025).

25.3 Peptides: structure, function, and sources

25.3.1 Definition and classification

Peptides are defined as short chains of AAs held together by peptide bonds and usually contain 2–50 AA residues. They represent certain free biomolecules that together form an intermediate between free AAs and full-length proteins, with special structural and functional characteristics. In food biotechnology and food science, high interest has been directed toward peptides as bioactive compounds with antimicrobial, antioxidant, antihypertensive, and immunomodulatory functions (Tandel & Florence, 2025). Peptides have great potential in natural polymer-based films and coatings, mainly because they enhance the mechanical and functional properties of the films, but they also promote the bio-preservative properties of food packaging systems. These possess broad-spectrum activity against bacteria, fungi, and viruses. Diverse mechanisms of antimicrobial action include alterations of bacterial membranes culminating in lysis. Antimicrobial peptides (AMPs) can be derived from food proteins through the processes of enzymatic hydrolysis, or they can even be naturally occurring, like defensins and cathelicidins in humans and animals. Due to their cationic and amphiphilic properties, they can interact with negatively charged microbial membranes. In food coatings, AMPs are incorporated in order to prevent the growth of spoilage and pathogenic organisms on the surface of food products (yadav, Makkar et al., 2025).

These peptides have the ability to scavenge free radicals, chelate pro-oxidant metal ions, and inhibit lipid peroxidation. Intrinsic antioxidant peptides have been linked to specific AA residues (e.g., histidine, proline, tyrosine, and methionine) with the potential to neutralize radicals (Thery et al., 2019). Antioxidant peptides are responsible for the maintenance of the oxidative stability of lipids in a food matrix, which, in turn, enhances shelf life and improves nutritional quality. A general term for any peptide that has a positive physiological effect aside from basic nutrition encompasses peptides exhibiting effects such as antihypertensive (e.g., angiotensin-converting enzyme [ACE]-inhibitory), antiinflammatory, opioid, and immunomodulatory. Incorporation of antimicrobials, whether in the form of AMPs or antioxidants, is most commonly associated with edible films, though there are some instances in the literature on the addition of phenolic compounds for nutraceutical or functional food packaging applications (Kulawik et al., 2022).

Natural peptides with biological functions and bioactive properties can be obtained from animal, plant, and microbial systems as well as by chemical synthesis. A wide variety of sources offers versatility in producing peptide functions for use in food preservation, safety, and quality, which is particularly applicable in the integration of peptide-enriched films and coatings based on natural polymers (Nithya et al., 2013). Source selection is often based on the intended bioactivity, the availability of the raw material, consumer preference (for instance, vegetarian or halal products), and the production cost.

25.3.1.1 Animal-derived peptides

Milk proteins (casein and whey) are abundant sources of bioactive peptides. Proteins from these sources undergo enzymatic hydrolysis to produce peptides with an array of functional properties, such as antioxidant, antimicrobial, antihypertensive, and immunomodulatory activities. Casein-derived peptides, such as casocidin and isracidin, exhibit antimicrobial properties, making them applicable for prolonging the shelf stability of perishable products. Marine proteins and by-products from fish (like skin, bones, and viscera) constitute excellent sources of marine peptides with great antioxidant and antimicrobial activity (Zatta Cassol et al., 2025). Fish skin and bone collagen hydrolysates are very interesting for their functional benefits as well as their biofilm formation (Daroit & Brandelli, 2021). Meat by-products, like blood, skin, and connective tissue, are an untapped but potential source. Hydrolyzed hemoglobin and myofibrillar proteins provide peptides that can reduce microbial load and delay oxidation (Vargas-Serna et al., 2025).

25.3.1.2 Plant-based peptides

Plant-based diets, clean-label trends, and the demand for natural and sustainable absorption in the manufacturing process are increasing the application of plant-based peptides. Storage proteins, mainly globulins, in legumes including soybeans, chickpeas, and lentils provide bioactive peptides through enzymatic hydrolysis with antioxidant and antimicrobial properties (Ashaolu & Suttikhana, 2023). Peptides derived from soy protein have been incorporated into edible coatings on fruits and vegetables, providing protection against oxidation and microorganisms. Profound diet studies have indicated that cereal grains such as wheat, rice, and oats contain proteins (e.g., gluten, zein) that undergo hydrolysis and yield peptides with antioxidative and health-promoting activities (Ying et al., 2021). Zein, a corn-derived film-forming protein, and bio-functional peptides are rich in the edible coatings. Bioactivities in oilseeds and nuts (e.g., sunflower and pumpkin seeds) also produce bioactive peptides with free radical scavenging prowess, which are frequently used in coatings designed for lipid-rich food preservation (Lee et al., 2024).

25.3.1.3 Microbial and fermentation-derived peptides

Microorganisms can produce peptides via fermentation or by genetically engineered expression systems. These peptides are frequently potent antimicrobials and have a variety of applications as food preservatives and pharmaceuticals. Bacteriocins, including nisin (*Lactococcus lactis*) and pediocin (*Pediococcus acidilactici*), are small AMPs that prevent the growth of foodborne pathogens such as *Listeria monocytogenes* and *Clostridium botulinum* (Farid et al., 2024). Due to their prior approvals as food preservatives, these peptides are promising candidates for application in active edible films. Lactic acid bacteria or yeast can be used in fermentation processes, which can also release bioactive peptides from food substrates, improving their bioactivity and stability. Such peptides are commonly added to fermented dairy, soy, and cereal products (Zhao et al., 2024).

25.3.1.4 Synthetic peptides

They can be synthesized either via solid-phase peptide synthesis (SPPS) or recombinant DNA technology. These methods provide exact control over AA sequence, making it possible to engineer peptides with tailored characteristics (e.g., target antimicrobial spectra, antioxidant capacities) (Gimranov et al., 2025). Though costly, synthetic peptides offer unmatched purity and reproducibility, which are critical for elucidating structure-function relationships and customizing coatings for niche applications. Synthetic peptides are likely to gain a growing role in functional packaging, as production costs decline and synthesis technologies improve.

25.4 Functional roles in food applications

Peptides have become increasingly important in their application to food, embodying a number of functions when applied to films and coatings based on natural polymers (Mamdouh et al., 2025). However, the main role of these peptides is their antimicrobial activity, with many of them, particularly nisin and other bacteriocins, having the ability to strongly inhibit a broad range of spoilage and pathogenic microorganisms by either lysing the microbial cell membranes and disrupting metabolism or by preventing the formation of biofilms (Roque-Borda et al., 2025). Embedded in edible films, these peptides act to optimize active packaging systems, extending food shelf life as well as food safety. Additionally, due to their low molecular weight, some peptides have good antioxidant activity, most notably those rich in AAs such as histidine, tyrosine, and cysteine, that scavenge reactive oxygen species (ROS) and prevent lipid oxidation (Peng et al., 2025).

This is especially advantageous in maintaining sensory and nutritional quality of lipid-rich or fresh-cut food products. Apart from bioactivity, peptides also contribute to providing structural properties to the film matrices that enhance mechanical strength, flexibility, and oxygen barrier properties (Asghar et al., 2025). Thanks to their amphiphilic characteristics, some peptides have been employed as natural emulsifiers; they allow the stabilization of lipid-in-water emulsions, promoting the incorporation of hydrophobic bioactives, such as essential oils, into biopolymer coatings (dos Santos Lima et al., 2025). From a nutritional point of view, the incorporation of peptides into edible films improves their quality by providing health-beneficial effects (e.g., antihypertensive or immunomodulatory effects) alongside essential AAs.

Their impact on sensory characteristics such as gloss, texture, and even taste may enhance consumer appeal further. Advanced applications additionally focus on utilizing peptides for their function within controlled-release systems or smart packaging technologies in which the bioactivity is activated by variations in environmental stimuli, including pH or temperature (Kasaai, 2025). Altogether, peptides are an important class of multifunctional agents that could greatly improve the performance, safety, and functionality of bio-based food packaging systems, particularly those derived from natural polymers.

25.5 Integration of peptides into natural polymer films and coatings

25.5.1 Methods of incorporation

Integrating peptides into polymeric-based films and coatings derived from natural polymers is crucial for creating functional and active packaging systems in food science (Khanda et al., 2025). Some peptides have bioactive properties, especially antimicrobial and antioxidant functions, which can greatly improve the preservative and protective potential of biopolymer matrices. However, their effectiveness relies on how the peptides are integrated into the system since functional activity needs to be retained while maintaining a homogeneous distribution and compatibility with the film-forming system (Mohammadi et al., 2025). There are numerous strategies for peptide integration, including blending, surface coating, and encapsulation, each with its own unique benefits and limitations relative to the end-use application and characteristics of the material. Integration of peptides into biodegradable films and coatings is illustrated in Fig. 25.3.

Mixing is the simplest and most widely used method for incorporating peptides into edible films. Peptides are incorporated into film-forming solutions, including a natural polymer (e.g., chitosan, gelatin, alginate), prior to casting or extrusion (Shirai et al., 2025). In the blending approach, peptides are homogeneously distributed in the polymer network and act uniformly in their bioactive function throughout the entire film. For example, AMPs in combination with chitosan or starch matrices have been reported to extend microbial inhibition in coated meats, fruits, and cheeses (Oliveira et al., 2025). Blending is especially effective for water-soluble peptides compatible with the polymer's solvent system. However, special precautions need to be taken to avoid heat or pH-induced degradation of the peptide during processing.

Surface coating is the process of applying a peptide solution to the surface of a preformed film or directly onto the food surface (Doan & et al., 2025). This approach is appropriate when high surface bioactivity is desired, such as when contact with pathogenic or oxidative agents is most probable. Surface coating is done by spraying, dipping, or brushing, which provides means for the introduction and immobilization of bioactive components after film formation during postprocessing, keeping their bioactivity (Tavassoli, Zhang, et al., 2025). It allows local changes in the peptide concentrations without affecting the bulk composition of the thin film. Challenges remain in achieving high adhesion to the underlying film and preventing fast desorption or inactivation during storage or food handling.

Encapsulation is an advanced method involving the encapsulation of peptides within a carrier system, which can be liposomes, nano-emulsions, polymeric nanoparticles, and protein-based microcapsules, followed by dispersion of these encapsulated systems into the film matrix (Barbosa-Nuñez et al., 2025). This protects the peptide from environmental factors like light, oxygen, and moisture, as well as from degradation by enzymes. It can also provide controlled or targeted

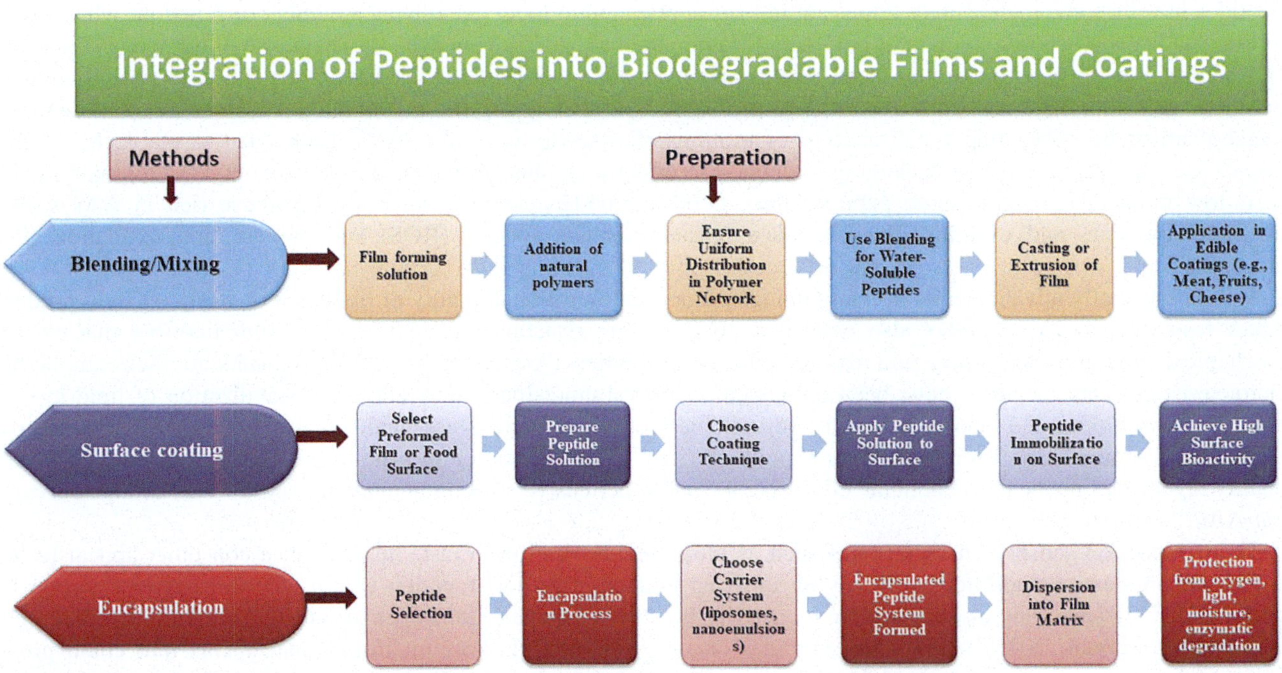

FIGURE 25.3 Integration of peptides into biodegradable films and coatings.

release of bioactive peptides in a storage or triggered/stimuli (e.g., pH) dependent manner. Composite encapsulated AMPs within pectin or alginate coatings can also be used to extend efficacy on fresh produce.

25.5.2 Compatibility and stability

Compatibility and stability of the peptides are crucial for their incorporation in various applications, including as biodegradable thin films and coatings based on natural polymers, as they affect the performance and efficiency of the packaging system in the end application (Al-Khalili et al., 2025). Compatibility relates to the physicochemical interactions of the peptides with the film-forming biopolymer matrix and encompasses stability, which includes the preservation of the peptide's structural integrity and functional activity throughout processing, storage, and application (Ahmed & Janaswamy, 2025). These two features must be considered in order to provide optimum functionalization of films with peptides for their use in food applications.

Depending on their specific AA content and the physicochemical conditions of the system (e.g., pH, ionic strength), peptides may establish hydrogen bonds, electrostatic, or hydrophobic interactions with these polymers (Pk, 2025). As an illustration, positively charged AMPs usually present better interaction with negatively charged polysaccharides, such as alginate, improving the cohesion and functional properties of the films. To provide maximal compatibility, researchers have been known to alter their film formulation by adjusting the pH, adding plasticizers (e.g., glycerol, sorbitol), or introducing surfactants that stabilize the mixture (Brasil et al., 2025). To improve the dispersion and incorporation of proteins into polymer matrices, cross-linking agents or compatible copolymers are often added. Moreover, compatibility could depend on the method of incorporation, whether it is blending, surface coating, or encapsulation, as mentioned in the previous section.

Stability is another important issue because peptides can degrade or be denatured under unfavorable conditions often experienced in food packaging film fabrication. Heat during drying, shear forces when mixing, and exposure to light, oxygen, or different humidity all tend to degrade or inactivate sensitive peptides. As an example, certain antioxidant peptides have their activity compromised after exposure to heat or oxidative stress involved in the formation of the film. AMPs, for example, may also exhibit undesirable folding or aggregation when subjected to suboptimal biophysical conditions, decreasing their capacity to engage microbial membranes (Reay et al., 2025).

Multiple strategies can be used to improve peptide stability. One of the most efficient ways is encapsulation, which shields peptides from various environmental stresses while allowing the slow release of peptides over time. Freeze-drying (lyophilization) of the peptides prior to incorporation can also mitigate structural degradation. Antioxidants or stabilizing agents can be coadded to mitigate oxidative degradation (Donno et al., 2025). Additionally, storage conditions of the final film or coating, including temperature, humidity, and light exposure, need to be controlled to ensure no degradation occurs over the target shelf life. In addition, cold (refrigerated) storage and vacuum sealing can often do wonders for peptide retention and activity.

25.6 Functional enhancements

25.6.1 Antimicrobial activity

Antimicrobial activity has been shown to be one of the most relevant functional improvements within active packaging, with the incorporation of peptides into both natural polymer coatings and films providing significant contributions to food safety (Castellanos et al., 2025; Manzoor et al., 2023). AMPs and protein hydrolysate-derived peptides can inhibit or kill a wide range of foodborne pathogens and spoilage microorganisms, increasing product shelf life and maintaining microbiological quality. The antimicrobial actions of peptides are diverse and distinct but often converge on the microbial cell membrane. The majority of AMPs have cationic and amphiphilic properties, which enable them to target the negatively charged elements of bacterial membranes, including lipoteichoic acid (LTA) in Gram-positive bacteria or lipopolysaccharides (LPS) in Gram-negative bacteria (Achi & Halami, 2025; Bose et al., 2023). They can insert into the membrane bilayer upon binding and create pores or channels, leading to membrane permeabilization and an imbalance of ions with the leakage of cytoplasm. As a result, this leads to cell lysis and cell death. Meanwhile, other AMPs can translocate across the membrane and affect inner activities like DNA replication, protein synthesis, or enzymatic activity. Certain peptides similarly interfere with bacterial signaling systems, such as quorum sensing, thereby blocking biofilm formation (Fig. 25.4).

Target pathogens can vary from Gram-positive to Gram-negative bacteria, although the latter group may require amphipathic properties due to significant differences in the structure of their outer membranes, and Gram-positive bacteria are typically more targeted with AMPs. Peptide-functionalized films might inhibit major foodborne pathogens, such as

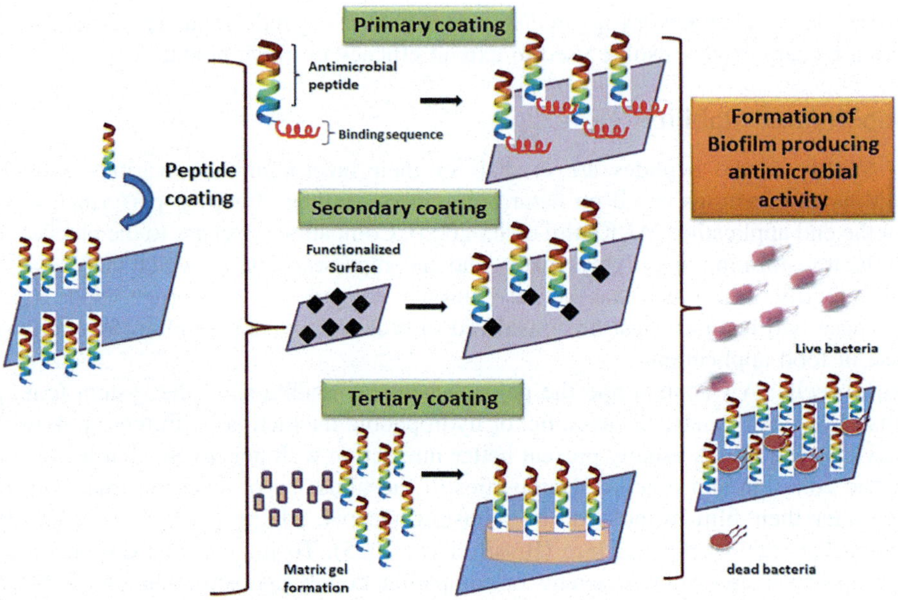

FIGURE 25.4 Formation of biofilm producing antimicrobial activity.

L. monocytogenes, Staphylococcus aureus, E. coli, Salmonella spp., and *Pseudomonas* spp., while some peptides might also exert antifungal effects against fungal contaminants, such as *Candida albicans* and *Aspergillus* species (Pack et al., 2025).

25.6.2 Antioxidant properties

The use of natural polymer-based films and coatings enriched with peptides not only provides important antioxidant activities essential for food quality preservation by delaying or preventing oxidative spoilage. The lipid oxidation and ROS production are among the most contributable factors for the spoilage of fatty foods and perishable products that result in rancidity, off-flavors, color changes, and loss of nutritional value (Oliveira et al., 2025). They have been defined as an effective solution to these challenges, and their incorporation into edible films allows for the efficient delivery of protective functionality in active food packaging systems. Antioxidant peptides are small to medium-sized fragments and are generally rich in their specific redox-active AAs, which include histidine, cysteine, tyrosine, tryptophan, methionine, and proline. These peptides endow antioxidant properties mainly through their free radical scavenging, metal ion chelation, and lipid peroxidation inhibition. These peptides exhibit antioxidant properties by neutralizing ROS like superoxide radicals (O_2^-), hydroxyl radicals (OH·), and peroxyl radicals (ROO·) through electron donation or hydrogen atom transfer mechanisms, helping to alleviate oxidative stress in the food matrix (Alam et al., 2025). A considerable number of antioxidant peptides can decrease the concentration of pro-oxidant metallic ions like iron (Fe^{2+}) and copper (Cu^{2+}), which are known to trigger lipid oxidation reactions. Peptides restrain these metals and hence prevent generation of highly reactive hydroxyl radicals through Fenton-type reactions. Peptides may also regenerate other antioxidants (tocopherols, ascorbic acid) or protect lipid-soluble compounds against oxidative degradation (Beristain-Bauza et al., 2025).

25.6.3 Nutritional and health benefits

Food packaging involves not only classic plastic and polymer materials, but also the incorporation of functional bioactive substances like peptides into natural polymer-derived films and coatings, thereby providing functional nutritional and health benefits that transform the packaging design into an active contributor to dietary health (Tavassoli, Zhang, et al., 2025). These peptides may serve not only as structural or preservative agents but also as bioavailable health-promoting additives when consumed. By incorporating them into edible films, a new strategy is formed for the delivery of functional nutrition, providing a structured and targeted output. Peptides applied in food packaging are related to the enzymatic hydrolysis of food-grade proteins with potential bioactive activities, such as casein, whey, soy, fish collagen, and plant proteins, generating specific peptides or mixtures of peptides (Tavassoli, Abedi-Firoozjah, et al., 2025). The biological activities of these bioactive peptides are well-studied, showing various physiological effects, for example, antihypertensive, immunomodulatory, antidiabetic, antithrombotic, and mineral-binding activities. Some peptides, for instance, have ACE inhibitory activity for controlling hypertension, while others work as immune enhancers or increase calcium absorption via soluble mineral-peptide complexes (Shirai et al., 2025). Peptide-functionalized films can also be utilized to combat nutritional deficiencies by fortifying foods with crucial AAs, particularly in vegetarian or low-protein diets. Peptic films derived from soy protein hydrolysates, for example, can add additional lysine and methionine, while fish-derived peptides supply high amounts of glycine, proline, and hydroxyproline. This functional enrichment is especially beneficial to vulnerable groups, such as children, the elderly, or people with certain metabolic disorders.

25.6.4 Mechanical and barrier improvements

In addition to their biological activities, peptides can also serve as cross-linking agents, which may improve the mechanical strength and barrier properties of natural polymer films and coatings, and increase the efficiency of packaging materials (Alam et al., 2025). Additives are essential features that assure the integrity of the films during handling, transporting, and storage, while preventing moisture and gas exchange to keep the food away from quality degradation. The effect of peptides on the mechanical properties (tensile strength), water vapor permeability (WVP), and gas permeability (mainly O_2 and CO_2) demonstrates their multidimensional effects on active and intelligent packaging systems (Soto et al., 2025). Depending on their sequence, structure, and polarity, peptides can interact with the polymer matrix through hydrogen bonding, electrostatic interactions, and hydrophobic associations. These interactions can also alter the film's microstructure to strengthen the polymer network and limit free volume between chains, thereby enhancing tensile strength and flexibility (Srivastava & Garg, 2025). The addition of short-chain peptides to chitosan or starch films, for example, leads to more compact and cohesive structures that demonstrate improved mechanical resistance to tearing and stretching. Films containing gelatin hydrolysates also show enhanced film-forming properties from homogeneous distribution of the peptides and chain alignment.

Peptides can modify the ambiphilic properties of the film matrix, which affect barrier properties. Numerous peptides are amphiphilic, so they can modulate the water transport pathways across the polymer framework (Zheng et al., 2025). This means that they have lower permeability to water vapor, which is an advantage in packaging high-moisture or lipophilic food products that are prone to evaporation and/or oxidation. Moreover, peptides contribute to decreasing gas permeability of the films, especially concerning oxygen, limiting oxidative spoilage of sensitive foods like meats, oils, and cut fruits (Wang et al., 2025).

25.7 Applications in food systems

25.7.1 Fresh produce

Natural polymer-based films and coatings incorporated with peptides are excellent candidates for improved control of qualities and performance characteristics of fresh fruits and vegetables, as the perishability of fresh produce is attributed to microbial attack, loss of moisture, and respiration. Incorporation of AMPs into polysaccharide or protein matrices, such as chitosan, alginate, or cellulose, can dramatically reduce the surface microbial load and limit the growth of common spoilage organisms, such as *Botrytis cinerea* and *Penicillium* spp. (Armghan Khalid et al., 2022). These coatings create a semipermeable barrier that not only regulates gas exchange (minimizing ethylene accumulation) but also provokes senescence and chlorophyll degradation, maintaining color and firmness during storage. The peptide-functionalized coatings also retain moisture, which prevents wilting and loss of weight, a problem with leafy greens and berries (Armghan Khalid et al., 2022). Moreover, a few antioxidant peptides can prevent the enzymatic browning found in sliced fruits like apples or bananas by displacing free radicals and inhibiting activity of the enzyme polyphenol oxidase. As these films are edible, they need not be removed and can add a further health benefit.

25.7.2 Meat and seafood

The high protein and lipid content in meat and seafood makes them susceptible to microbial spoilage and oxidative rancidity. Natural polymer-based peptide-infused coatings provide a good strategy for such perishable products. AMPs, including nisin, lactoferrin-derived peptides, and synthetic AMPs, effectively target *L. monocytogenes*, *Salmonella*, and *Pseudomonas* spp. when introduced into films made from gelatin, soy protein, or chitosan (Gutiérrez, 2017). The films can be wrapped around meat and fish or applied directly to create an active barrier against microbial infiltration. Antioxidant peptides derived from fish protein hydrolysates or milk casein are applied against lipid oxidation in meats, particularly when stored under cold or display conditions. This provides desirable color (e.g., bright red in beef), flavor, and nutritional quality. In seafood, especially oily species such as salmon and mackerel, coatings enriched with peptide molecules can extend the development of off-flavors and texture loss caused by lipid peroxidation.

25.7.3 Dairy products

Dairy products, mainly soft cheeses, yogurts, and high-moisture products, are susceptible to microbial contamination, mold growth, and oxidative alterations. Biopolymer films and coatings integrated with bioactive peptides provide multifunctional versatility to prolong the shelf life and ensure the safety of dairy products (Das et al., 2025). Preservatives, including AMPs nisin and lysozyme, are routinely used in cheese preservation and can be incorporated into casein or whey protein-based films to inhibit spoilage organisms (e.g., *Clostridium tyrobutyricum* and *Lactobacillus* spp.). Such coatings infused with these peptides may also be directly applied to cheese rinds or soft dairy products to inhibit fungal growth, limit surface dehydration, and maintain structural integrity during aging and storage (Nath, Nandi, et al., 2023). Coatings have been shown to increase the stability of probiotics in yogurts and fermented milk products and to provide protection against contamination from environmental sources.

25.7.4 Ready-to-eat and snack foods

Convenience drives increasing consumer demand for ready-to-eat (RTE) products and snack foods with diverse ingredients and expectations for extended shelf life, making preservation challenging (Ambrosio et al., 2022). Peptide-functionalized natural polymeric coatings form a tailored solution for combination foods by improving microbial safety, retarding oxidative deterioration, and maintaining structural and organoleptic features. Protein- or starch-based coatings with incorporated antimicrobials can prevent fungal and pathogenic growth on RTE products, including sandwiches, deli meats, and sushi (Zolqadri et al., 2025). These films can be used as wraps or internal barriers to reduce spoilage risk and cross-contamination at the points of consumption, particularly in minimally processed products. Crafted coverings (also known as coatings), particularly fatty and oily ones with some important fraction of antioxidant peptides, can help maintain baked snacks and bars from lack of flavor and rancidity while in storage. Restoration of mechanical properties of the packaging can also be achieved by use of peptide coating, enabling better moisture proofing for crispy snacks or oxygen impermeability in vacuum-packed RTE products (Nath et al., 2022). FSW-based peptides can even act as film-forming or emulsifying agents, which help coatings to be more homogeneous and adhere better to food surfaces of complex shapes.

25.8 Conclusion and future perspectives

The use of peptides in natural polymer-based films and coatings is a major innovation in the food logistics sector that responds to the increasing interest in sustainable, safe, and functional food systems. Active and intelligent packaging material is an emerging trend in food and pharmaceutical packaging, and peptides play an integral role in this area due to their antimicrobial, antioxidative, and nutritional qualities. Integrating peptides with polysaccharide, protein, and lipid-based biodegradable matrices enables the development of packaging materials that prolong shelf life and reduce spoilage while also promoting the value of the final food product. Peptides are also advantageous as they can interact with biological systems in low concentrations and are compatible with natural polymers, making them effective and desirable agents to improve food quality as well as safety. The prospect for peptide-based films and coatings is promising but heavily reliant on the resolution of multiple challenges. The stability, controlled release, and bioavailability of peptides during processing and storage are important; enabling the stability, controlled release, and functional integrity of peptides during processing and storage is critical; the development of encapsulation technologies, nano carrier systems, and stimuli-responsive materials may also provide better strategies for peptide delivery when incorporated into polymer matrices. Scaling production and ensuring that it is commercially viable, consistent, and fits within the regulatory framework will be key to their mass commercial use. Further investigation should focus on the fabrication of multifunctional and intelligent packaging systems by using designed peptide sequences that are responsive to environmental stimuli or interact with specific food components. Better knowledge of peptide–polymer interactions, consumer safety, and environmental effects will enable responsible and effective use of these technologies. In all instances, peptides represent a promising new frontier in the development of natural polymer-based food packaging, bringing together the principles of bioactivity and sustainability in response to the requirements of contemporary food systems and consumer demand.

References

Achi, S., & Halami (2025). Antimicrobial peptides from *Bacillus* spp.: Use in antimicrobial packaging. In *Antimicrobial food packaging*. Elsevier, 731–742.

Ahmed, S., & Janaswamy, S. (2025). Green fabrication of biodegradable films: Harnessing the cellulosic residue of oat straw. *International Journal of Biological Macromolecules, 303*, 140656. https://doi.org/10.1016/j.ijbiomac.2025.140656.

Alam, M. W., Saravanan, P., Al-Sowayan, N. S., Almutairi, H. H., Rosaiah, P., Prakash, N. G., & Ko, T. J. (2025). Polysaccharides and proteins based edible coatings for food protection: Classification, properties, & public demands (2020–2024). *Journal of Food Measurement and Characterization, 19*(3), 1533–1556. https://doi.org/10.1007/s11694-024-03090-9.

Ali, M., Ali, A., Ali, S., Chen, H., Wu, W., Liu, R., Chen, H., Ahmed, Z. F. R., & Gao, H. (2025). Global insights and advances in edible coatings or films toward quality maintenance and reduced postharvest losses of fruit and vegetables: An updated review. *Comprehensive Reviews in Food Science and Food Safety, 24*(1). https://doi.org/10.1111/1541-4337.70103.

Al-Khalili, M., Rahman, S., & Al-Habsi, N. (2025). Date seed–added biodegradable films and coatings for active food packaging applications: A review. *Packaging Technology and Science*.

Ambrosio, R. L., Gogliettino, M., Agrillo, B., Proroga, Y. T. R., Balestrieri, M., Gratino, L., Cristiano, D., Palmieri, G., & Anastasio, A. (2022). An active peptide-based packaging system to improve the freshness and safety of fish products: A case study. *Foods, 11*(3), 338. https://doi.org/10.3390/foods11030338.

Amin, U., Khan, M. U., Majeed, Y., Rebezov, M., Khayrullin, M., Bobkova, E., Shariati, M. A., Chung, I. M., & Thiruvengadam, M. (2021). Potentials of polysaccharides, lipids and proteins in biodegradable food packaging applications. *International Journal of Biological Macromolecules, 183*, 2184–2198. https://doi.org/10.1016/j.ijbiomac.2021.05.182.

Armghan Khalid, M., Niaz, B., Saeed, F., Afzaal, M., Islam, F., Hussain, M., Mahwish, Khalid, H. M. S., Siddeeg, A., & Al-Farga, A. (2022). Edible coatings for enhancing safety and quality attributes of fresh produce: A comprehensive review. *International Journal of Food Properties, 25*(1), 1817–1847. https://doi.org/10.1080/10942912.2022.2107005.

Asghar, W., Ubaid ur Rahman, H., & Khalid, N. (2025). Waste to wellness: Valorizing plant peel waste for functional and nutritional improvement of restructured meat products. *Nutrition and Health, 31*(1), 5–7. https://doi.org/10.1177/02601060251326803.

Ashaolu, T. J., & Suttikhana, I. (2023). Plant-based bioactive peptides: A review of their relevant production strategies, in vivo bioactivities, action mechanisms and bioaccessibility. *International Journal of Food Science and Technology, 58*(5), 2228–2235. https://doi.org/10.1111/ijfs.16384, http://onlinelibrary.wiley.com/journal/10.1111/(ISSN)1365-2621.

Barbosa-Nuñez, J. A., Espinosa-Andrews, H., Cardona, A. A. V., & Haro-González, J. N. (2025). Polymer-based encapsulation in food products: A comprehensive review of applications and advancements. *Journal of Future Foods, 5*(1), 36–49. https://doi.org/10.1016/j.jfutfo.2024.01.003, https://www.sciencedirect.com/science/journal/27725669.

Bashir, I., Dar, A. H., Dash, K. K., Pandey, V. K., Fayaz, U., Shams, R., Srivastava, S., & Singh, R. (2023). Deep eutectic solvents for extraction of functional components from plant-based products: A promising approach. *Sustainable Chemistry and Pharmacy, 33*, 101102. https://doi.org/10.1016/j.scp.2023.101102.

Beristain-Bauza, S.del C., Hernández-Carranza, P., Ruiz-López, I. I., & Ochoa-Velasco, C. E. (2025). *Synbiotic and combiotic edible films and coatings: A review of their properties, human health benefits, and practical application*. Elsevier BV, 1–16. https://doi.org/10.1016/b978-0-443-21874-3.00001-5.

Bose, I., Roy, S., Pandey, V. K., & Singh, R. (2023). A comprehensive review on significance and advancements of antimicrobial agents in biodegradable food packaging. *Antibiotics, 12*(6), 968. https://doi.org/10.3390/antibiotics12060968.

Brasil, Y. L., Cruz-Tirado, J. P., & Barbin, D. F. (2025). Analysis of biodegradable films made from cassava starch and oregano essential oil using hyperspectral imaging and portable NIR spectroscopy. *Spectrochimica Acta - Part A: Molecular and Biomolecular Spectroscopy, 332*. https://doi.org/10.1016/j.saa.2025.125857, https://www.sciencedirect.com/science/journal/13861425.

Castellanos, D. A., Rodríguez, B., Garavito, J., Garzon, L., Reinoso-Carvalho, F., & Porras, A. (2025). *Packaging, long-term stability, and usability of antimicrobial peptide products*. Elsevier BV, 393–404. https://doi.org/10.1016/b978-0-443-15393-8.00016-6.

Cheng, H., Huang, W., Zeng, S., Li, G., Gong, Y., & Ren, D. (2025). Optimization of the thermoforming process for straw-based biodegradable materials through biological mechanical processing and simulation. *Materials Today Communications, 43*, 111739. https://doi.org/10.1016/j.mtcomm.2025.111739.

Daroit, D. J., & Brandelli, A. (2021). In vivo bioactivities of food protein-derived peptides – A current review. *Current Opinion in Food Science, 39*, 120–129. https://doi.org/10.1016/j.cofs.2021.01.002, http://www.journals.elsevier.com/current-opinion-in-food-science/.

Das, P., Altemimi, A. B., Nath, P. C., Katyal, M., Kesavan, R.krishnan, Rustagi, S., Panda, J., Avula, S. K., Nayak, P. K., & Mohanta, Y. K. (2025). Recent advances on artificial intelligence-based approaches for food adulteration and fraud detection in the food industry: Challenges and opportunities. *Food Chemistry, 468*, 142439. https://doi.org/10.1016/j.foodchem.2024.142439.

Devi, L. S., Jaiswal, A. K., & Jaiswal, S. (2024). Lipid incorporated biopolymer based edible films and coatings in food packaging: A review. *Current Research in Food Science, 8*. https://doi.org/10.1016/j.crfs.2024.100720, https://www.sciencedirect.com/science/journal/26659271.

Doan, et al. (2025). Exploring the potential of chitosan–phytochemical composites in preventing the contamination of antibiotic-resistant bacteria on food surfaces: A review. *Molecules (Basel, Switzerland), 30*(3).

Donno, D., Neirotti, G., Fioccardi, A., Razafindrakoto, Z. R., Tombozara, N., Mellano, M. G., Beccaro, G. L., & Gamba, G. (2025). Freeze-drying for the reduction of fruit and vegetable chain losses: A sustainable solution to produce potential health-promoting food applications. *Plants, 14*(2), 168. https://doi.org/10.3390/plants14020168.

Drexelius, M. G., & Neundorf, I. (2021). Application of antimicrobial peptides on biomedical implants: Three ways to pursue peptide coatings. *International Journal of Molecular Sciences, 22*(24), 13212. https://doi.org/10.3390/ijms222413212.

Farid, M. S., Anjum, R., Yang, Y., Tu, M., Zhang, T., Pan, D., Sun, Y., & Wu, Z. (2024). Recent trends in fermented plant-based analogues and products, bioactive peptides, and novel technologies-assisted fermentation. *Trends in Food Science & Technology, 149*, 104529. https://doi.org/10.1016/j.tifs.2024.104529.

Gan, L., Jiang, G., Yang, Y., Zheng, B., Zhang, S., Li, X., Tian, Y., & Peng, B. (2022). Development and characterization of levan/pullulan/chitosan edible films enriched with ε-polylysine for active food packaging. *Food Chemistry, 388*, 132989. https://doi.org/10.1016/j.foodchem.2022.132989.

Gao, S., Zhai, X., Cheng, Y., Zhang, R., Wang, W., & Hou, H. (2022). Starch/PBAT blown antimicrobial films based on the synergistic effects of two commercial antimicrobial peptides. *International Journal of Biological Macromolecules, 204*, 457–465. https://doi.org/10.1016/j.ijbiomac.2022.01.183.

Gao, T., Yuan, C., Cui, B., Liu, P., Wu, Z., Fang, Y., & Lu, L. (2025). Effect of soybean peptides on structure and functional properties of soluble soybean polysaccharide edible films. *Food Hydrocolloids, 167*, 111405. https://doi.org/10.1016/j.foodhyd.2025.111405.

Gimranov, E., Santos, J., Regalado, L., Teixeira, C., Gomes, P., Santos, C., & Pereira-Dias, L. (2025). Synthetic peptides bioactive against phytopathogens have lower impact on some beneficial bacteria: An assessment of peptides biosafety in agriculture. *Journal of Environmental Management, 374*, 123942. https://doi.org/10.1016/j.jenvman.2024.123942.

Gupta, D. (2024). Plant based edible films and coatings for food packaging applications: Recent advances, applications, and trends.

Gutiérrez, T. J. (2017). Chitosan applications for the food industry. In *Chitosan: Derivatives, composites and applications*. Argentina: wiley, 183–232. https://doi.org/10.1002/9781119364849.ch8, https://onlinelibrary.wiley.com/doi/book/10.1002/9781119364849.

Hamidi, N. A. S. M. (2022). Potential application of plant-based derivatives as green components in functional coatings: A review. *Cleaner Materials, 4*.

Holmberg, K. V., Abdolhosseini, M., Li, Y., Chen, X., Gorr, S. U., & Aparicio, C. (2013). Bio-inspired stable antimicrobial peptide coatings for dental applications. *Acta Biomaterialia, 9*(9), 8224–8231. https://doi.org/10.1016/j.actbio.2013.06.017, http://www.journals.elsevier.com/acta-biomaterialia.

Hosseini, S. F., Mousavi, Z., & McClements, D. J. (2023). Beeswax: A review on the recent progress in the development of superhydrophobic films/coatings and their applications in fruits preservation. *Food Chemistry, 424*. https://doi.org/10.1016/j.foodchem.2023.136404, www.elsevier.com/locate/foodchem.

Hurteaux, R., Edwards-Lévy, F., Laurent-Maquin, D., & Lévy, M.-C. (2005). Coating alginate microspheres with a serum albumin-alginate membrane: Application to the encapsulation of a peptide. *European Journal of Pharmaceutical Sciences, 24*(2-3), 187–197. https://doi.org/10.1016/j.ejps.2004.10.010.

Ibarra, V. G., Sendón, R., R, A., & De Quirós, B. (2025). Antimicrobial food packaging based on biodegradable materials. In *Antimicrobial food packaging*. Elsevier, 493–529.

Iñiguez-Moreno, M., & González-González, R. B. (2025). Effect of gelatin and salicylic acid incorporated in chitosan coatings on strawberry preservation. *International Journal of Biological Macromolecules, 305*(Pt 1), 140918. https://doi.org/10.1016/j.ijbiomac.2025.140918.

Islam, R., Khan, T., Junaid, P., & Naseem, S. (2024). Recent developments in sustainable bio-based wax nanoemulsions as edible coatings for food commodities: A comprehensive review. *Packaging Technology and Science, 37*(12), 1175–1194. https://doi.org/10.1002/pts.2854, http://onlinelibrary.wiley.com/journal/10.1002/(ISSN)1099-1522.

Kasaai, M. R. (2025). *Biopolymers as fillers of biopolymer matrices to fabricate green nano-composites for food coating and packaging: A review*. Springer Science and Business Media LLC, 1–14. https://doi.org/10.1007/978-981-99-3516-1_152-1.

Kazemzadeh-Narbat, M., Cheng, H., Chabok, R., Alvarez, M. M., de la Fuente-Nunez, C., Phillips, K. S., & Khademhosseini, A. (2021). Strategies for antimicrobial peptide coatings on medical devices: A review and regulatory science perspective. *Critical Reviews in Biotechnology, 41*(1), 94–120. https://doi.org/10.1080/07388551.2020.1828810.

Khanda, M., Seal, P., Mohan, A. J., Arya, N., & Boda, S. K. (2025). Antimicrobial peptide coatings on implant surfaces – Current status and future considerations. *Nanoscale*. https://doi.org/10.1039/d5nr00953g.

Kheto, A., Adhikary, U., Dhua, S., Sarkar, A., Kumar, Y., Vashishth, R., Shrestha, B. B., & Saxena, D. C. (2025). A review on advancements in emerging processing of whey protein: Enhancing functional and nutritional properties for functional food applications. *Food Safety and Health, 3*(1), 23–45. https://doi.org/10.1002/fsh3.12067.

Kulawik, P., Jamróz, E., Janik, M., Tkaczewska, J., Krzyściak, P., Skóra, M., Guzik, P., Milosavljević, V., & Tadele, W. (2022). Biological activity of biopolymer edible furcellaran-chitosan coatings enhanced with bioactive peptides. *Food Control, 137*, 108933. https://doi.org/10.1016/j.foodcont.2022.108933.

Lee, C. C., Suttikhana, I., & Ashaolu, T. J. (2024). Techno-functions and safety concerns of plant-based peptides in food matrices. *Journal of Agricultural and Food Chemistry, 72*(22), 12398–12414. https://doi.org/10.1021/acs.jafc.4c02464, http://pubs.acs.org/journal/jafcau.

Li, C., Pei, J., Zhu, S., Song, Y., Xiong, X., & Xue, F. (2020). Development of chitosan/peptide films: Physical, antibacterial and antioxidant properties. *Coatings, 10*(12), 1193. https://doi.org/10.3390/coatings10121193.

Li, Y., Zheng, S., Song, Y., Jiang, Q., & Zhang, H. (2025). Preparation, characterization, and application of composite oleogels based on whey protein isolate and sodium alginate. *International Journal of Biological Macromolecules, 300*, 140317. https://doi.org/10.1016/j.ijbiomac.2025.140317.

Mamdouh, F., Hassabo, A. G., & Othman, H. (2025). Improving the performance properties of polyester fabrics through treatments with natural polymers. *Journal of Textiles, Coloration and Polymer Science, 22*(1), 219–231.

Manzoor, A. Plant-derived active substances incorporated as antioxidant, antibacterial or antifungal components in coatings/films for food packaging applications. Food Bioscience.

Manzoor, A., Khan, S., Dar, A. H., Kumar Pandey, V., Shams, R., Ahmad, S., Jeevarathinam, G., Kumar, M., Singh, P., & Pandiselvam, R. (2023). Recent insights into green antimicrobial packaging towards food safety reinforcement: A review. *Journal of Food Safety, 43*(4). https://doi.org/10.1111/jfs.13046.

Mishra, B., Panda, J., Mishra, A. K., Nath, P. C., Nayak, P. K., Mahapatra, U., Sharma, M., Chopra, H., Mohanta, Y. K., & Sridhar, K. (2024). Recent advances in sustainable biopolymer-based nanocomposites for smart food packaging: A review. *International Journal of Biological Macromolecules, 279*(Pt 4), 135583. https://doi.org/10.1016/j.ijbiomac.2024.135583.

Moghadas, H. C., Chauhan, R., & Smith, J. S. (2024). Application of plant oils as functional additives in edible films and coatings for food packaging: A review. *Foods, 13*(7). https://doi.org/10.3390/foods13070997, www.mdpi.com/journal/foods.

Mohammadi, B., Pirsa, S., & Mahmudi, M. (2025). Animal and vegetable proteins: Applications as film and coating in food industry. *Journal of Polymers and the Environment, 33*(3), 1233–1253. https://doi.org/10.1007/s10924-025-03487-7, https://www.springer.com/journal/10924.

Nath, P. C., Nandi, N. B., Tiwari, A., Das, J., & Roy, B. (2022). Applications of nanotechnology in food sensing and food packaging. In *Nanotechnology applications for food safety and quality monitoring*. India: Elsevier, 321–340. https://doi.org/10.1016/B978-0-323-85791-8.00006-9, https://www.sciencedirect.com/book/9780323857918.

Nath, P. C., Hazarika, T. K., Ojha, A., Sharma, R., Bhunia, B., & Nayak, P. K. (2024). Microbial electrochemical technologies for valorization of food wastes. In *Development in waste water treatment research and processes: Applied technologies for clean up of environmental contaminants*. India: Elsevier, 323–343. https://doi.org/10.1016/B978-0-443-13615-3.00030-3, https://www.sciencedirect.com/book/9780443136153.

Nath, P. C., Nandi, N. B., Khan, S. A., Bhunia, B., Bandyopadhyay, T. K., & Roy, B. (2023). Smart packaging for commercial food products. In *Green sustainable process for chemical and environmental engineering and science: Methods for producing smart packaging*. India: Elsevier, 197–209. https://doi.org/10.1016/B978-0-323-95644-4.00016-4, https://www.sciencedirect.com/book/9780323956444.

Nath, P. C., Debnath, S., Sridhar, K., Inbaraj, B. S., Nayak, P. K., & Sharma, M. (2023). A comprehensive review of food hydrogels: Principles, formation mechanisms, microstructure, and its applications. *Gels, 9*(1), 1. https://doi.org/10.3390/gels9010001.

Nath, P. C., Ojha, A., Debnath, S., Sharma, M., Sridhar, K., Nayak, P. K., & Inbaraj, B. S. (2023). Biogeneration of valuable nanomaterials from agro-wastes: A comprehensive review. *Agronomy, 13*(2), 561. https://doi.org/10.3390/agronomy13020561.

Nath, P. C., Debnath, S., Sharma, M., Sridhar, K., Nayak, P. K., & Inbaraj, B. S. (2023). Recent advances in cellulose-based hydrogels: Food applications. *Foods, 12*(2), 350. https://doi.org/10.3390/foods12020350.

Nath, P. C., Sharma, R., Debnath, S., Nayak, P. K., Roy, R., Sharma, M., Inbaraj, B. S., & Sridhar, K. (2024). Recent advances in production of sustainable and biodegradable polymers from agro-food waste: Applications in tissue engineering and regenerative medicines. *International Journal of Biological Macromolecules, 259*(Pt 1), 129129. https://doi.org/10.1016/j.ijbiomac.2023.129129.

Nath, P. C., Sharma, R., Debnath, S., Sharma, M., Inbaraj, B. S., Dikkala, P. K., Nayak, P. K., & Sridhar, K. (2023). Recent trends in polysaccharide-based biodegradable polymers for smart food packaging industry. *International Journal of Biological Macromolecules, 253*(Pt 8), 127524. https://doi.org/10.1016/j.ijbiomac.2023.127524.

Nithya, V., Murthy, P. S. K., & Halami, P. M. (2013). Development and application of active films for food packaging using antibacterial peptide of *Bacillus licheniformis* Me1. *Journal of Applied Microbiology, 115*(2), 475–483. https://doi.org/10.1111/jam.12258.

Olawade, D. B., Wada, O. Z., & Ige, A. O. (2024). Advances and recent trends in plant-based materials and edible films: A mini-review. *Frontiers in Chemistry, 12*. https://doi.org/10.3389/fchem.2024.1441650, http://journal.frontiersin.org/journal/chemistry.

Oliveira, I., Pinto, T., Afonso, S., Karaś, M., Szymanowska, U., Gonçalves, B., & Vilela, A. (2025). Sustainability in bio-based edible films, coatings, and packaging for small fruits. *Applied Sciences, 15*(3), 1462. https://doi.org/10.3390/app15031462.

Ong, G., Kasi, R., & Subramaniam, R. (2021). A review on plant extracts as natural additives in coating applications. *Progress in Organic Coatings, 151*, 106091. https://doi.org/10.1016/j.porgcoat.2020.106091.

Pack, B. W., Manheim, J., Chahrour, O., Wood, B., Zhang, L., Vogel, U., & Sheng, G. (2025). Therapeutic peptides control strategy: Perspective on current industry practices. *Organic Process Research and Development, 29*(2), 270–280. https://doi.org/10.1021/acs.oprd.4c00386, http://pubs.acs.org/journal/oprdfk.

Pandey (2022). A comprehensive review on clove (*Caryophyllus aromaticus* L.) essential oil and its significance in the formulation of edible coatings for potential food applications. *Frontiers in Nutrition, 9*.

Pandey, V. K., Islam, R. U., Shams, R., & Dar, A. H. (2022). A comprehensive review on the application of essential oils as bioactive compounds in Nano-emulsion based edible coatings of fruits and vegetables. *Applied Food Research, 2*(1), 100042. https://doi.org/10.1016/j.afres.2022.100042.

Peng, X., Oral, C. M., Urso, M., Ussia, M., & Pumera, M. (2025). Active microrobots for dual removal of biofilms via chemical and physical mechanisms. *ACS Applied Materials & Interfaces, 17*(2), 3608–3619. https://doi.org/10.1021/acsami.4c18360.

Pk, M. E. H. (2025). Carboxymethyl chitosan-based biodegradable films preparation and prospects for food preservation. *ChemistrySelect, 10*(1).

Qiu, C., Meng, Y., Zhang, Z., Li, X., McClements, D. J., Li, G., Jiang, L., Wen, J., Jin, Z., & Ji, H. (2025). Enhancement of soy protein functionality by conjugation or complexation with polysaccharides or polyphenols: A review. *Comprehensive Reviews in Food Science and Food Safety, 24*(1). https://doi.org/10.1111/1541-4337.70095.

Reay, M. K., Graf, M., Murphy, M., Li, G., Yan, C., Bhattacharya, M., Osbahr, H., Ma, J., Chengtao, W., Shi, X., Ren, S., Cui, J., Collins, C., Chadwick, D., Jones, D. L., Evershed, R. P., & Lloyd, C. E. M. (2025). Higher potential leaching of inorganic and organic additives from biodegradable compared to conventional agricultural plastic mulch film. *Journal of Hazardous Materials, 488*, 137147. https://doi.org/10.1016/j.jhazmat.2025.137147.

Roque-Borda, C. A., Primo, L. M. D. G., Medina-Alarcón, K. P., Campos, I. C., Nascimento, Cd. F., Saraiva, M. M. S., Berchieri Junior, A., Fusco-Almeida, A. M., Mendes-Giannini, M. J. S., Perdigão, J., Pavan, F. R., & Albericio, F. (2025). Antimicrobial peptides: A promising alternative to conventional antimicrobials for combating polymicrobial biofilms. *Advanced Science, 12*(1). https://doi.org/10.1002/advs.202410893, http://onlinelibrary.wiley.com/journal/10.1002/(ISSN)2198-3844.

Roy, B., Saikia, D., Nayak, P. K., Chandra Biswas, S., Bandyopadhyay, T. K., Bhunia, B., & Chandra Nath, P. (2023). Innovations in smart packaging technologies for monitoring of food quality and safety. In *Green sustainable process for chemical and environmental engineering and science: Methods for producing smart packaging*. India: Elsevier, 39–58. https://doi.org/10.1016/B978-0-323-95644-4.00013-9, https://www.sciencedirect.com/book/9780323956444.

dos Santos Lima, K., de Matos Fonseca, J., Monteiro, A., & Valencia, G. (2025). Progress in biopolymer coatings for achieving sustainable paper and paperboard packaging for food applications: A brief review. *Packaging Technology and Science, 38*(2), 109–116. https://doi.org/10.1002/pts.2861, http://onlinelibrary.wiley.com/journal/10.1002/(ISSN)1099-1522.

Sharma, R., Nath, P. C., & Seth, D. (2023). Optimization of spray-drying conditions using response surface methodology, physico-chemical characterization and shelf-life estimation of pineapple powder. *Sustainable Food Technology, 1*(5), 750–761. https://doi.org/10.1039/d3fb00008g, www.rsc.org/journals-books-databases/about-journals/sustainable-food-technology/.

Sharma, R., Nath, P. C., Das, P., Rustagi, S., Sharma, M., Sridhar, N., Hazarika, T. K., Rana, P., Nayak, P. K., & Sridhar, K. (2024). Essential oil-nanoemulsion based edible coating: Innovative sustainable preservation method for fresh/fresh-cut fruits and vegetables. *Food Chemistry, 460*(Pt 1), 140545. https://doi.org/10.1016/j.foodchem.2024.140545.

Shirai, M. A., Baú, T. R., Zanela, J., & Pimentel, T. C. (2025). Microalgae as an innovative active ingredient for edible films and coatings for food applications. *Algal Research, 86*, 103959. https://doi.org/10.1016/j.algal.2025.103959.

Soto, C., Rodriguez, J. C., Jimenez, W., Perez, I., Contreras, M., & Quiñones, C. E. (2025). Development and characterization of edible films from resistant polymers of native potato starch as a food coating. *Italian Association of Chemical Engineering - AIDIC, Peru Chemical Engineering Transactions, 115*, 7–12. https://doi.org/10.3303/CET25115002, https://www.cetjournal.it/index.php/cet/issue/view/vol115.

Srivastava, V., & Garg, H. (2025). Edible packaging: Novel approaches for food protection and sustainability. *Engineering Materials, 21*, 391–416. https://doi.org/10.1007/978-981-96-0369-5_12, https://www.springer.com/series/4288.

Stefanowska, K., Woźniak, M., Sip, A., Biegańska-Marecik, R., Dobrucka, R., & Ratajczak, I. (2025). Kombucha as a solvent for chitosan coatings: A new strategy to extend shelf life of red peppers. *Materials, 18*(7), 1605. https://doi.org/10.3390/ma18071605.

Tandel, H., & Florence, K. (2025). *Oral delivery of proteins and peptides: Concepts and applications*. Elsevier BV, 287–343. https://doi.org/10.1016/b978-0-443-27416-9.00011-3.

Tavassoli, M., Abedi-Firoozjah, R., Bahramian, B., Hashemi, M., Noori, S. M. A., Oladzadabbasabadi, N., Nagdalian, A., & Jafari, S. M. (2025). Glutaraldehyde cross-linking for improving the techno-functional properties of biopolymeric food packaging films; a comprehensive review. *Food Chemistry, 478*, 143740. https://doi.org/10.1016/j.foodchem.2025.143740.

Tavassoli, M., Zhang, W., Assadpour, E., Zhang, F., & Jafari, S. M. (2025). Self-healing packaging films/coatings for food applications; an emerging strategy. *Advances in Colloid and Interface Science, 339*, 103423. https://doi.org/10.1016/j.cis.2025.103423.

Thery, T., Lynch, K. M., & Arendt, E. K. (2019). Natural antifungal peptides/proteins as model for novel food preservatives. *Comprehensive Reviews in Food Science and Food Safety, 18*(5), 1327–1360. https://doi.org/10.1111/1541-4337.12480, http://onlinelibrary.wiley.com/journal/10.1111/(ISSN)1541-4337.

Udaipuria, N., & Bhattacharya, S. (2025). Novel carbohydrate polymer-based systems for precise drug delivery in colon cancer: Improving treatment effectiveness with intelligent biodegradable materials. *Biopolymers, 116*(1). https://doi.org/10.1002/bip.23632, http://onlinelibrary.wiley.com/journal/10.1002/(ISSN)1097-0282.

Vargas-Serna, C. L., Pineda-Osorio, A. N., Gallego-Ocampo, H. L., Plaza-Dorado, J. L., & Ochoa-Martínez, C. I. (2025). Transforming coffee and meat by-products into protein-rich meal via black soldier fly larvae *(Hermetia illucens)*. *Sustainability, 17*(2), 460. https://doi.org/10.3390/su17020460.

Wang, P., Gao, Y., Wang, D., Huang, Z., & Fei, P. (2022). Amidated pectin with amino acids: Preparation, characterization and potential application in Hydrocolloids. *Food Hydrocolloids, 129*, 107662. https://doi.org/10.1016/j.foodhyd.2022.107662.

Wang, X., Cao, Z., Su, J., Ge, X., & Zhou, Z. (2025). Oral barriers to food-derived active peptides and nano-delivery strategies. *Journal of Food Science, 90*(1). https://doi.org/10.1111/1750-3841.17672, http://onlinelibrary.wiley.com/journal/10.1111/(ISSN)1750-3841.

Wang, Z., Chandrapala, J., Truong, T., & Farahnaky, A. (2024). Binary wax oleogels: Improving physical properties and oxidation stability through substitution of carnauba wax with beeswax. *Journal of Food Science, 89*(7), 4372–4388. https://doi.org/10.1111/1750-3841.17140.

Wu, Y., Wu, H., & Hu, L. (2024). Recent advances of proteins, polysaccharides and lipids-based edible films/coatings for food packaging applications: A review. *Food Biophysics, 19*(1), 29–45. https://doi.org/10.1007/s11483-023-09794-7, https://www.springer.com/journal/11483.

Xu, J., Chen, Q., Qiu, Y., Wang, Z., Zeng, M., Qin, F., Chen, J., & He, Z. (2025). Competitive binding of flavors in the preparation of soy protein: Screening based on molecular docking. *Food Chemistry, 475*, 143279. https://doi.org/10.1016/j.foodchem.2025.143279.

yadav, P., Makkar, N., Jain, M., & Prakash, R. (2025). Peptide-based nanoparticles from agricultural and food waste: Applications in smart food packaging. *International Journal of Peptide Research and Therapeutics, 31*(3). https://doi.org/10.1007/s10989-025-10699-x.

Yadav, P., Mahapatra, U., Sharma, R., Rustagi, S., Katyal, M., Sharma, M., Nayak, P. K., Nath, P. C., & Sridhar, K. (2025). Recent advances in plant protein-based sustainable edible film and coatings for applications in the food-pharma industry – Opportunities and challenges: A review. *International Journal of Biological Macromolecules, 296*, 139698. https://doi.org/10.1016/j.ijbiomac.2025.139698.

Ying, X., Agyei, D., Udenigwe, C., Adhikari, B., & Wang, B. (2021). Manufacturing of plant-based bioactive peptides using enzymatic methods to meet health and sustainability targets of the sustainable development goals. *Frontiers in Sustainable Food Systems, 5*. https://doi.org/10.3389/fsufs.2021.769028.

Zatta Cassol, G., Rezende-de-Souza, J. H., & Savay-da-Silva, L. K. (2025). Fish byproducts as a protein source: A critical review of fish protein concentrate. *Food Reviews International, 41*(1), 201–217. https://doi.org/10.1080/87559129.2024.2392154, www.tandf.co.uk/journals/titles/87559129.asp.

Zhao, L., Liu, X., Wang, S., Yin, Z., An, T., Zhang, J., & Liu, Y. (2024). Research progress on fermentation-produced plant-derived bioactive peptides. *Frontiers in Pharmacology, 15*. https://doi.org/10.3389/fphar.2024.1438947.

Zheng, S., Tu, Y., Li, B., Qu, G., Li, A., Peng, X., Li, S., & Shao, C. (2025). Antimicrobial peptide biological activity, delivery systems and clinical translation status and challenges. *Journal of Translational Medicine, 23*(1). https://doi.org/10.1186/s12967-025-06321-9.

Zhu, F. (2024). Modifications of whey proteins for emulsion based applications: Current status, issues and prospectives. *Food Research International, 178*, 113935. https://doi.org/10.1016/j.foodres.2024.113935.

Zolqadri, R., Akbarbaglu, Z., Ramezani, A., Mazloomi, N., & Sarabandi, K. (2025). Microencapsulated red-garlic peptides with biopolymers: Influence on the ACE-inhibitory, biological activity, textural structures, and sensory properties of pan-breads. *Future Foods, 11*, 100550. https://doi.org/10.1016/j.fufo.2025.100550.

Chapter 26

Nanostructures

Lina María Rodríguez Pineda, María Regina Ramos Ramos, Virginia Borroni, and María Lidia Herrera
Institute of Emerging Technologies and Applied Sciences (ITECA), National University of San Martín, San Martín, Argentina

26.1 Introduction

Food packaging plays a crucial role in maintaining product quality and extending shelf life, particularly for items prone to microbiological spoilage, chemical oxidation, and physical deterioration. This is especially important for products like aquatic foods (Gürdal & Çetinkaya, 2024) and for fruits and vegetables (Kocira et al., 2021; Liyanapathiranage et al., 2023), which are highly vulnerable to these issues. Proper packaging helps preserve the freshness and safety of these products, preventing contamination and degradation while ensuring that they remain in optimal condition during storage and transportation. By protecting against environmental factors such as moisture, air, and light, effective food packaging plays a key role in reducing waste and improving consumer acceptance (Gürdal & Çetinkaya, 2024).

Petroleum-based plastics have been used for this purpose for many years. However, they are a significant concern in terms of environmental contamination due to their nonbiodegradable nature and their negative impact on marine life and various ecosystems. For these reasons, biopolymers have attracted considerable attention over the past two decades, as they are renewable, derived from natural sources, and cause less environmental harm. Lipids, carbohydrates, and proteins, either alone or in combination, have been successfully utilized to produce edible films with the potential to protect food products from microbiological deterioration, physical damage, and chemical damage.

To be viable alternatives to plastic films, biopolymer-based films must have acceptable mechanical and tensile properties, along with effective barriers to water vapor and oxygen. Depending on the specific use, they should also be capable of blocking UV light and displaying thermal stability. It's equally important that they offer strong resistance to foodborne microorganisms. In general, biopolymer-based films exhibit lower performance than conventional plastic materials. To address these limitations, a variety of strategies have been developed to enhance their mechanical, barrier, and functional properties. One such strategy is the incorporation of nanoparticles—either inorganic or natural—and using of nanoemulsions as starting systems, which have been widely studied. Fig. 26.1 summarizes the main strategies to improve the film's physical properties. The addition of nanoreinforcement has been used in all types of systems. For two-phase systems, reducing droplet size and using Pickering materials have proven successful in enhancing the film's performance.

Films prepared from two-phase systems have been reported to exhibit inferior mechanical and barrier properties compared to those made from solution systems. The presence of a dispersed phase can disrupt the connections between the building blocks in the structure, thereby compromising the physical properties of the films. However, strategies have been proposed to address this issue (Hoyos-Merlano et al., 2022) and will be discussed later in this chapter. In addition to providing structural benefits, the presence of a dispersed lipid phase with nanodroplets enables the inclusion of antioxidant and antimicrobial agents, as well as nutraceutical bioactive compounds, within the films, leading to beneficial effects on the packaged food products. The droplet size of the dispersed phase may be related to the accessibility of bioactive agents, such as antimicrobials and/or antioxidants, and it is a topic worth investigating.

In this chapter, the inclusion of nanocomponents in edible films to improve their potential to replace plastic materials and their impact on film structure and physical properties is reviewed. Relevant examples of film performance for food packaging are also discussed.

Natural Polymer-Based Films and Coatings in Food Technology. DOI: https://doi.org/10.1016/B978-0-443-34244-8.00025-2

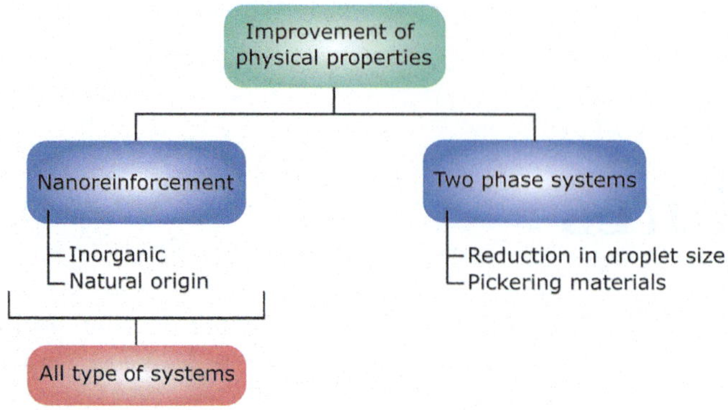

FIGURE 26.1 Strategies to improve the physical properties of films that include the use of nanomaterials.

26.2 Nanoreinforcements

Nanoreinforcement, using both inorganic and natural fillers, has become one of the main strategies to enhance the physical properties of edible films. To effectively improve these properties, in addition to the reinforcement's chemical nature, two key factors must be considered: the concentration and the geometry (shape and dimensions) of the selected nanoreinforcement. Table 26.1 provides examples of various reinforcements, both natural and mineral, that have been used to enhance the properties of films. All of these serve as successful examples of using nanostructures to create nanocomposite films. Fig. 26.2 shows two examples of TiO_2 nanoreinforcement, in the form of spheres and nanotubes. It is expected that the geometry of the nanoreinforcement will have a significant impact on mechanical and tensile properties. The shape will influence the interaction between the polymer and reinforcement, as well as their distribution in the matrix, which is highly relevant to the film's performance.

The shape of the reinforcement is crucial for enhancing the physical properties of films. However, in most reports published in the literature, nanoreinforcements are spherical (Hoyos-Merlano et al., 2022). One of the main reasons for using spherical reinforcements, such as TiO_2 P25, to improve the physical properties of biopolymer-based films is their commercial availability. Nevertheless, the geometry of the reinforcement is a key issue, and only a few reports address this subject.

Studies using whey protein nanofibrils have shown that TiO_2 in the form of nanotubes not only exhibits higher antibacterial activity than spherical TiO_2 nanoparticles but also increases interactions with whey protein nanofibrils, highlighting the importance of reinforcement geometry (Feng et al., 2019). Our previous studies, which involved formulating films with sodium caseinate, support these conclusions. The performance of TiO_2 nanotubes added to sodium caseinate solution-based films was significantly superior to that of TiO_2 nanospheres (Hoyos Merlano et al., 2023). As the nanotubes were longer, there were more points of interaction between the matrix and the reinforcement, making the reinforcement more effective in modifying the structure and improving the physical properties. The spatial distribution of nanoreinforcements is also highly relevant. However, the structural description of films is scarce. Other geometries, such as silk nanodiscs, successfully improved the performance of chitosan-based films. These were prepared using acid hydrolysis synthesis from silk fibroin. The discs exhibited enhanced crystallinity (95%) and superior thermal stability. The edible coatings improved the bananas' shelf life (weight, appearance, and firmness) over seven days at 25°C (Ghosh et al., 2021).

The importance of fiber length and its impact on physical properties has been emphasized in several studies. To enhance the mechanical strength of a collagen-based film, fibers of varying sizes (macro, micro, and nano) were prepared using high-pressure homogenization, and the resulting fiber-based edible films were analyzed. The films made with smaller-sized fibers exhibited a more uniform and denser structure, along with improved mechanical and water barrier properties, compared to those made with longer fibers. The Fourier-transform infrared (FTIR) spectrum showed no changes in the positions or intensities of the signals, suggesting that the improvement in physical properties was due to the noncovalent bonds and physical entanglement of the fibers (Ma et al., 2020). The smaller size allowed for a more uniform distribution within the film matrix, boosting the film's performance. This example highlights the significance of fiber length in film structure, which plays a crucial role in the film's mechanical behavior. It also suggests that there must be an optimal

TABLE 26.1 Different nanoreinforcements used to enhance physical properties.

Film	Reinforcement	Geometry	Preparation method	References
Whey protein isolate/cinnamon essential oil	Chitosan	Commercial nanofibers	Casting	Mohammadi et al. (2020)
Salvia macrosiphon seed mucilage	Nanoclay	Commercial powder	Casting	Davachi and Shekarabi (2018)
Salep-mucilage	Zinc oxide (ZnO)	Flower-like ZnO	Casting	Ekrami et al. (2021)
Pullulan	Liposomes	Spheres	Casting	Najafi et al. (2021)
Thermoplastic starch	Crystalline cellulose	Nanofibers	Casting	Midhun Dominic et al. (2021)
Chitosan	From silk fibroin	Silk nanodiscs	Casting	Ghosh et al. (2021)
Sodium caseinate	TiO_2	Spheres and nanotubes	Casting	Hoyos Merlano et al. (2023)
Pectin	Melanin nanoparticle	Spherical	Casting	Liang et al. (2023)
Chitosan/poly-vinyl-alcohol	Copper and titanium nanostructures	Spherical	Casting	Salmas et al. (2021)
Sodium alginate	Thyme-zeolite nanostructures	Spherical	Casting	Giannakas et al. (2022)
Chitosan/poly-vinyl-alcohol	Halloysite nanostructures	Tubular	Casting	Salmas et al. (2022)
Gelatin	Casein phosphopeptides	Nanofiber	Casting	Khedri et al. (2021)

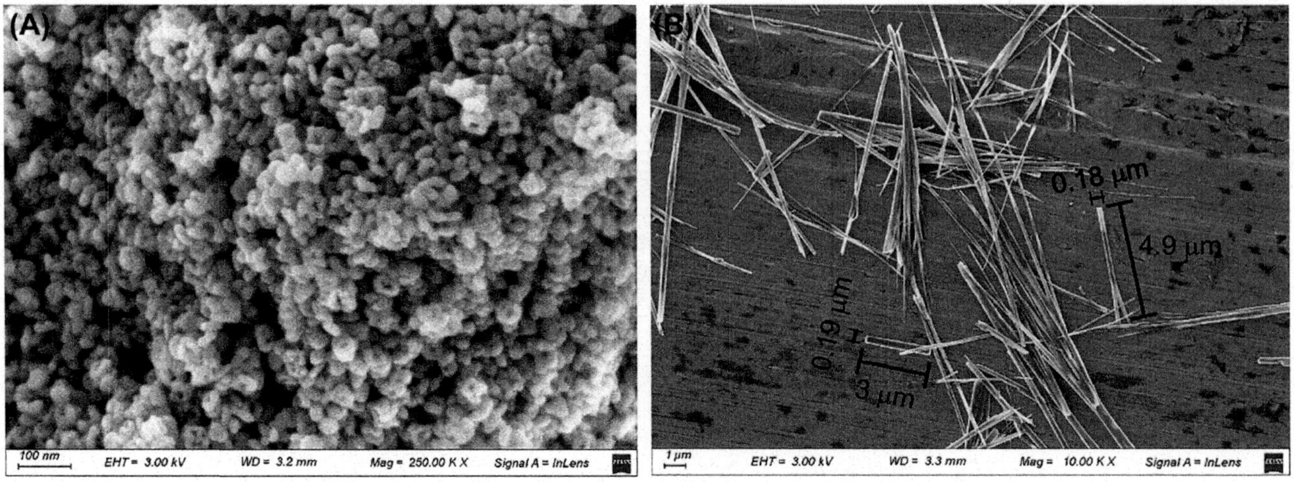

FIGURE 26.2 Different geometries of nanoreinforcement: (A) spheres, (B) nanotubes.

length, as tubes performed better than spheres due to having more points of interaction; however, if reinforcements are too long, they can lead to inhomogeneities in the matrix structure.

Another important aspect of the nanoreinforcement strategy is determining the optimal concentration of the reinforcement, which can vary depending on the nature and geometry of the nanoreinforcement. The reinforcement may increase hydrophobicity, thereby improving water barrier properties, and the strength of the polymer/reinforcement interactions may also depend on the physical interactions, which vary with the geometry and concentration of the reinforcement. If the reinforcement chemically reacts with the matrix, the effects on physical properties may be more pronounced, depending on the degree of crosslinking (Liu et al., 2021). The processing method is also crucial. The application of ultrasound treatment may expose functional groups in proteins and promote crosslinking or facilitate strong interactions (Cheng & Cui, 2021). For example, the random coil structure of gelatin was modified and transformed into a covalent network by adding pullulan dialdehyde, an oxidized polysaccharide, which improved physical properties. The molecular weight of pullulan dialdehyde may play a significant role by affecting the network structure and, therefore, the properties of the gelatin films (Liu et al., 2021).

In addition to the structural advantages, the incorporation of nanoreinforcement into films may also provide antimicrobial or antioxidant properties. These effects can help improve the overall quality of the product by extending its shelf life, while preserving the texture and appearance of food. By preventing the growth of harmful microorganisms or delaying oxidation, nanoreinforced films can significantly enhance food preservation, offering both functional and protective benefits.

26.2.1 Most common fillers used for reinforcement

Among the natural nanofibers used as reinforcement, chitosan is one of the most extensively studied nanofillers for reinforcing biopolymer films, owing to its ease of production, cost-effectiveness, and biodegradable nature (Table 26.1). It exhibits excellent properties, including its geometry (nanoscale in one dimension, such as diameter, and microscale in the other, such as length), which makes it effective in enhancing the mechanical strength of reinforced films. Furthermore, its antimicrobial activity helps extend the shelf life of products. Numerous studies have demonstrated the beneficial effects of incorporating chitosan nanofibers into various types of biopolymeric films, such as those based on whey protein isolate (Mohammadi et al., 2020) and bacterial cellulose (Almasi et al., 2018), to name just a few. For chitosan nanofibers, the optimum concentration reported was 6% based on the dry weight of whey protein isolate (Mohammadi et al., 2020).

The Cloisite 15A nanoclay, derived from natural montmorillonite clay and modified with a quaternary ammonium salt, has been highlighted in the literature as an effective natural reinforcement for films based on *Salvia macrosiphon* seed mucilage (Davachi & Shekarabi, 2018). According to scanning electron microscope (SEM) images, the nanoclay consists of spherical particles of varying sizes at the nanoscale. Its high hydrophobicity contributes significantly to improving the water vapor barrier properties of the films. When 2% Cloisite 15A nanoclay was incorporated into *S. macrosiphon* mucilage-based films, significant improvements were observed in various properties. These included enhanced mechanical and thermal properties, a higher water vapor barrier, and improved tensile properties such as elongation at break and tensile strength. The incorporation of this nanoclay, which is highly hydrophobic, played a key role in these enhancements, making the films more suitable for applications requiring resistance to moisture and improved strength (Davachi & Shekarabi, 2018).

Other types of nanostructures based on montmorillonite were also prepared through an ion-exchange process with Cu^+ and Ti^{4+} ions (Salmas et al., 2021). These nanostructures were added at concentrations of 3% and 6% to a chitosan/polyvinyl alcohol matrix. The results showed that the nanocomposite film with copper nanostructures had a more intercalated structure than the one with titanium. Due to its more connected structure, it exhibited better mechanical and barrier properties than the titanium-based film, highlighting the importance of nanoreinforcement distribution and its effect on film structure. Both copper and titanium montmorillonite nanostructures displayed strong antibacterial activity against *Staphylococcus aureus*, *Escherichia coli*, *Listeria monocytogenes*, and *Salmonella enterica*, making them promising materials for food packaging applications.

The beneficial effects of zinc oxide (ZnO) nanoparticles, an inorganic reinforcement, on physical properties were dependent on their surface area and the interactions between the nanoreinforcement and polymer branches. 3D ZnO structures demonstrate high stability against aggregation and a favorable surface-to-volume ratio. A significantly enhanced performance, driven by morphology, was observed in flower-like ZnO. These nanoparticles were able to notably improve UV protection efficiency within the biopolymers, compared to the 1D/2D ZnO nanosheet and nanorod structures. The report highlights the importance of the reinforcement's geometry, which affects the physical properties in different ways. The transparency of the films significantly decreased (~80%) with increasing concentrations of ZnO flower-like (from 1% to 4%). The nanocomposite films based on Salep-mucilage/ZnO flower-like exhibited excellent UV-protective properties and efficacy in reducing photo-oxidation in kilka fish oil under fluorescent light over 12 days of storage (Ekrami et al., 2021).

Eco-friendly materials for food packaging have been developed using cellulose nanofibers to strengthen a potato starch-based matrix. Starches from various sources have been used to produce films and coatings with a variety of properties because amylose has excellent film-forming abilities (Majeed et al., 2023). Cellulose has also garnered significant attention due to being the most abundant polysaccharide on Earth, found in the plant cell wall. It exists in microfibrils that contain tightly bound noncellulosic components such as hemicellulose, lignin, and pectin. These noncellulosic components are removed to obtain nanocellulose. The resulting nanostructures demonstrated a very high crystallinity index (~65%), which is essential, as the crystallinity of nanoreinforcements improves the thermal stability of films. The potato starch-based films were reinforced with varying concentrations (0%, 1%, 2%, 3%, 4%) of nanofibers. The film with 3% nanoreinforcement exhibited exceptional tensile properties (tensile strength ~161% higher than the control and tensile modulus ~167% higher than the control) and excellent thermal stability, making it a suitable alternative for sustainable packaging (Midhun Dominic et al., 2021).

In addition to improving physical properties, nanostructures made from generally recognized as safe (GRAS) nanomaterials have been developed to enhance the antibacterial properties of the films. An edible nanocomposite-based antibacterial packaging film was prepared by adding natural melanin nanoparticles extracted from cuttlefish ink in concentrations of 0.25, 0.75, 1.25, and 2.0 wt.% to a pectin matrix. It was found to be highly effective in controlling *L. monocytogenes*, with a reduction in bacterial content of over 90% after just 5 minutes. This nanocomposite film also exhibited improved mechanical properties and thermal stability compared to the control pectin film (Liang et al., 2023).

Bioactive casein phosphopeptides, protein components containing 2–20 amino acids, were used at concentrations of 0.1%, 0.2%, 0.3%, and 0.5% w/v to improve the physical and structural properties, as well as the antimicrobial and antioxidant capacities, of gelatin-based films (Khedri et al., 2021). The effects of nanoreinforcement concentrations on the film characteristics were complex. Film permeability and solubility decreased as nanoreinforcement concentrations increased from 0.1% to 0.3% w/v, along with the mechanical properties, compared to the control treatment. This effect was due to the addition of bioactive casein phosphopeptides, which increased the elongation at break, elastic modulus, and tensile strength of the films. However, further increases in concentration did not improve the films' properties. There was an optimal concentration that enhanced the films' physical behavior, made them effective against the tested Gram-positive bacteria *S. aureus* and *Bacillus cereus*, and provided antioxidant properties.

Complex reinforcements, created by modifying natural nanofillers, have also been used to improve the physical properties of films. One example is the modification of halloysite nanotube (HNT) nanoclay by adding thymol essential oil, with thymol loading reaching approximately 30%. The halloysite nanostructures, which have a hollow tubular shape, were incorporated into a hydrogel matrix made of chitosan and polyvinyl alcohol at concentrations of 5%, 10%, and 15% to create nanocomposite films (Salmas et al., 2022). These modified halloysite nanostructures were more effectively dispersed than pure HNT due to the presence of thymol essential oil, which enhanced the nanoreinforcement's dispersion within the matrix and improved the film's transparency. The addition of thymol showed promising effects on the performance of the films.

To test the films' ability to preserve food, the developed films were used to coat kiwi fruits. 15 uncoated and coated kiwi fruits were stored for 21 days at room humidity and 25°C. The results showed that thymol was primarily adsorbed onto the external surface of the HNT nanoclay, allowing it to be easily released and protect the kiwi fruits against foodborne pathogens. The uncoated kiwi fruits began to deteriorate by the third day, while the kiwi fruits coated with 15% HNT showed signs of decline by the 12th day. In contrast, the kiwi fruits coated with 15% HNT began to show signs of spoilage on the 15th day, indicating the effectiveness of the modified nanoclay films in preserving food quality and extending shelf life.

Another example of a complex form of nanoreinforcement involves the performance of a hybrid nanostructure, created by modifying a natural zeolite with thyme essential oil (Giannakas et al., 2022). This hybrid nanoreinforcement was then incorporated directly into a sodium alginate hydrogel at varying concentrations of 5%, 10%, and 15% to produce films intended for use in food packaging applications. To verify the distribution of the nanostructure powder within the polymer matrix, advanced techniques such as X-ray diffraction (XRD) and SEM were employed. The XRD analysis was used to determine the degree of crystallinity of the nanomaterial, a property that plays a crucial role in its reinforcement ability. Meanwhile, the SEM images provided valuable insights into the positioning and distribution of the nanostructures within the film.

The addition of zeolites (as a control) or the hybrid nanoreinforcement (thymol/zeolite) resulted in increased aggregation of spherical particles of alginate hydrogel on the film surface, as observed by SEM, compared to the film without reinforcement. SEM images also showed that the nanohybrids were evenly distributed across the film surface, indicating excellent compatibility with the polymer matrix. Notably, the distribution was more uniform for the hybrid nanoparticles than for the pure zeolites, suggesting significantly better incorporation of the hybrid nanoparticles into the polymer matrix.

This effective incorporation of the hybrid nanostructure significantly enhanced the properties of the film for its intended application in food packaging. The hybrid-reinforced films proved effective in extending the shelf life of soft cheese, particularly when the nanoreinforcement concentration ranged from 5% to 10%. This was mainly due to the antibacterial properties imparted by the thyme essential oil in the hybrid nanostructures, which helped protect the cheese from spoilage.

All of these studies highlighted the importance of nanoreinforcement distribution in edible films, which depends on the chemical nature of the reinforcement, its concentration, and its geometry (size and morphology). The presence of reinforcements modified the film's structure, which significantly impacted the film's mechanical properties and barrier capabilities. Reinforcement also plays a crucial role in protecting against foodborne microorganisms and chemical deterioration. Adding reinforcements to various matrixes resulted in formulations suitable for food packaging.

In most reports in the literature, the structures of nanoreinforcements and films were analyzed using SEM, which provided detailed surface images or cryofracture analyses. Additionally, the energy dispersive X-ray (EDS) detector attached to the SEM was utilized to obtain information on the distribution of elements on the surface and to demonstrate the homogeneity of the reinforcement within the film. However, despite its potential, using small-angle X-ray scattering (SAXS) techniques is relatively rare, even though it offers a more in-depth description of the structure at the nanoscale. By measuring the SAXS signal at various points within the film, it is possible to generate X-ray transmission images. These images are created by plotting the transmitted intensity at each sample location, divided by the transmitted intensity measured in a background region where no sample is present. This background region is often easily identifiable as a blue triangle (indicating low or no scattering) in the scattering maps. From these transmission microscopy images, a more quantitative distribution of the reinforcement within the film and an evaluation of the homogeneity of the matrix can be obtained. Furthermore, modeling the SAXS curves allows the calculation of structural parameters that describe the material's building block size or correlation distances. Although this approach offers powerful insights into the film structure, there are only a limited number of studies that provide this level of structural analysis (Montes-de-Oca-Ávalos et al., 2018, Montes-de-Oca-Ávalos et al., 2020).

Most of the research conducted on the properties of edible films has been carried out on a laboratory scale. However, the casting method, commonly used for film preparation, presents challenges when it is intended to be scaled up for industrial production. As a result, there is a need for more studies that utilize industrial-scalable methods, such as a twin-screw extruder, to better assess the performance of the films in real-world applications. By conducting these studies, researchers could gain a deeper understanding of how the material behaves during large-scale production and its effectiveness under more practical conditions.

26.3 Dispersed phase

A dispersed phase facilitates the encapsulation and controlled delivery of lipophilic compounds. Various systems have been employed for these purposes, including conventional emulsions (Arnon-Rips et al., 2019), nanoemulsions (Gul et al., 2018; Hanan et al., 2024), and nanoliposomes (Najafi et al., 2021), to name a few that are among the most commonly studied. This approach has proven highly successful in food products, as it allows for the addition of valuable compounds, such as vitamins and carotenoids, to beverages and juices, for instance. The same strategy has been chosen for edible coatings in food packaging to enhance the quality and extend the shelf life of fresh and processed foods. Reports on chicken breast filets and low-fat cut or homemade cheese have shown the potential of the approach to extend durability (Arnon-Rips et al., 2019; Mahcene et al., 2021).

The size of the droplets in the dispersed phase plays a significant role in the capability of the film to encapsulate and deliver different compounds. Research comparing the effects of droplet size on packaging performance has emphasized the advantages of reducing droplet size through high-energy methods. Specifically, coarse (conventional) and nanoemulsions of citral, a natural monoterpene derived from essential oils, were prepared and incorporated into matrixes made from either chitosan or carboxymethyl cellulose polysaccharides. These matrixes were used to create active edible coatings designed for solid-state delivery (Arnon-Rips et al., 2019).

The films made from nanoemulsions exhibited a more organized and dense structure compared to those made from coarse emulsions, which led to significant improvements in water vapor barrier properties. Additionally, the nanoemulsion-based films demonstrated superior mechanical and tensile strength. When applied to fresh-cut melons as a food model, the nanoemulsified edible coatings provided enhanced antimicrobial protection and extended the product's shelf life by up to 13 days. This highlights the superior performance of nanoemulsions in food preservation.

Similarly, corn starch-based films incorporating *Zataria multiflora* essential oil as the dispersed phase in conventional or nanoemulsion forms showed comparable behavior to films based on chitosan or carboxymethyl cellulose. Nanoemulsions, especially those containing cinnamaldehyde dissolved in the lipid phase, were also effective in preventing oxidation in ground beef patties during a 20-day storage period at $4°C \pm 1°C$ (Amiri et al., 2019).

The homogeneity of the structure and the small size of the dispersed phase, which was so well dispersed that it was not detectable by SEM, contributed to the excellent physical properties and performance of these films as edible coatings for fruits and vegetables. This was especially evident in composite films made from natural jicama starch, agarwood Aetoxylon bouya essential oil, and calcium propionate (Wigati et al., 2022). These films demonstrated enhanced durability and functionality, making them highly suitable for food preservation applications.

Other reports have indicated that another strategy to enhance the performance of edible films based on emulsions is to increase the stability of the initial coarse emulsion systems. This can be achieved by preparing Pickering emulsions. Beeswax-in-water emulsions with droplet sizes around 10 μm were stabilized with cellulose nanofibrils and carboxymethyl chitosan (Xie et al., 2020). The resulting films demonstrated improved properties despite the large droplet sizes of the dispersed phase, as the coverage of interphase droplets/water with cellulose nanofibrils provided the system with the desired stability.

Among other uses, nanoliposomes have been investigated for their potential applications as delivery systems in food products. They consist of a bilayer membrane made from food-grade phospholipids that encapsulates an aqueous-phase core. As a result, nanoliposomes are capable of encapsulating both hydrophobic molecules within the bilayer walls and hydrophilic compounds in their central aqueous phase.

Rapeseed lecithin has been selected to prepare food-grade liposomes due to its high polyunsaturated fatty acid composition, which is desirable from a nutritional perspective. Additionally, the content of residual tocopherol, with its antioxidant capability, provides added value to food products (Najafi et al., 2021). Pullulan-based films loaded with nanoliposomes as carriers of saffron extract, known as the world's most expensive spice, were prepared and characterized. Analysis by FTIR spectroscopy indicated that saffron extract was successfully encapsulated in the nanoliposomes and film matrix. The incorporation of nanoliposomes enhanced the oxygen barrier properties of the films, although it did not significantly affect the water vapor permeability (WVP). The resulting edible films or coatings can offer additional benefits due to the unique flavor and color of saffron. Furthermore, the incorporation of saffron extract can provide health benefits because of its antioxidant capacity.

Other systems have been stabilized by incorporating nanoscale colloidal materials based on lipids, proteins, or polysaccharides. Lipid-based nanoparticles are among the most commonly used because they can encapsulate a wide range of hydrophobic bioactives, such as nutraceuticals, pigments, antimicrobials, antioxidants, and drugs. Lipids can also minimize moisture loss and reduce the complexity of packaging due to their highly hydrophobic nature, which results from their high content of long-chain fatty acids or alcohols (Chen et al., 2022). Various lipid materials have been successfully prepared, containing either solid (Karimi Khorrami et al., 2021) or liquid (Mohammadi et al., 2020) cores.

The food-grade lipid-based nanoparticles have the potential to be uniformly incorporated into the film matrix, thereby enhancing the barrier and mechanical properties of the films. These colloidal nanoparticles also modify the structure of the films, as analyzed by SEM and atomic force microscopy (AFM). In alginate-based films, it was reported that the films containing nanolipid carriers exhibited a significantly different structure than the control film without nanoparticle addition (Karimi Khorrami et al., 2021). SEM analyses revealed a more porous structure, which increased with the concentration of lipid nanoparticles. Specifically, for the film containing 0.35% nanoparticles, the structure became highly porous and spongy, with the nanolipid carriers homogeneously distributed throughout the interior and on the surfaces of the films. AFM data also showed that the control sample had the smoothest surface, which can result from the formation of a more uniform biopolymer network within the films. Parameters such as the roughness average and root-mean-square roughness of the films increased significantly with the concentrations of nanolipid carriers, indicating that the presence of the nanoparticles strongly impacted the structure.

The formation of pores in the films has also been reported in chitosan-based systems containing Tween 20 (Cárdenas et al., 2010). According to the report, the structural changes were attributed to the ability of this small surfactant to inhibit the crosslinking of the calcium alginate chains. However, the mechanical properties of the films were not reported in the manuscript; therefore, the hypothesis was not verified. A different interpretation was provided for sodium caseinate films reinforced with TiO_2 nanotubes (Hoyos Merlano et al., 2023). In this case, the presence of Tween 20 enhanced the mechanical and tensile properties because it was able to interact through covalent bonds, forming an ester with the carboxylic groups of sodium caseinate. As a result, structural changes in the films were observed, which impacted other physical properties. Both interpretations shared that the interactions between Tween 20 and the matrix were more than just electrostatic interactions, as they interfered with the connections between the building blocks of the matrix.

26.3.1 Addition of essential oils

Nowadays, many consumers seek natural microbial and antioxidant agents, primarily due to safety concerns regarding the presence of residual chemicals in food. Although some of the most potent poisons known are of natural origin, there is a belief that natural compounds are safer than chemical additives. Therefore, incorporating natural preservatives, such as plant extracts and essential oils, pure or combined with antimicrobial and antioxidant properties, into bio-based packaging materials could make products more appealing to consumers (Rech et al., 2021).

Essential oils are clear, colorless liquids that are highly volatile and aromatic. They consist of 85%–99% volatile compounds, including terpenes, alcohols, phenols, aldehydes, ketones, amides, and amines. The composition of these compounds is specific to each essential oil, depending on the particular plant source, and can also vary based on the geographical origin of the oil. The bioactive components within essential oils are typically divided into two main categories: terpenes/terpenoids and aromatic/aliphatic compounds. Among these, terpenes and terpenoids usually make up the majority of the content. Common terpenes and terpenoids found in essential oils include citronellol, camphor, geraniol, terpinolene, limonene, carvacrol, carvone, cineole, thymol, and linalool (Anis et al., 2021).

The low molecular weight terpenes and terpenoids are the key constituents responsible for the antimicrobial properties of essential oils. While the precise mechanisms of their antimicrobial action are not yet fully understood, researchers have proposed several theories. One widely accepted explanation is that essential oils can increase the permeability of bacterial cell walls. This is believed to be related to their ability to dissolve the lipid components in cell membranes, which, in turn, compromises the structural integrity of the microbial cell wall (Bhavaniramya et al., 2019).

Essential oils are never added directly to biopolymer-based packaging materials, as most matrices have a hydrophilic nature, which leads to issues with solubility and a loss of activity. Furthermore, the amount of essential oils required to impart effective antimicrobial properties exceeds the organoleptic acceptance levels. For this reason, they are incorporated into the biopolymer matrix as conventional emulsions, nanoemulsions, liposomes, or lipid-based nanoparticles (Kong et al., 2022). Consequently, the particle size of the dispersed phase—namely, the emulsified essential oil—is expected to have significant effects on the functional properties of the edible film.

Different types of essential oils have been incorporated into various films to effectively prevent microbial spoilage and enhance the shelf life of food products (Shahidi & Hossain, 2022). Table 26.2 provides a summary of common essential oils frequently used in the formulation of films for food packaging. These examples represent successful applications that can play a significant role in helping prevent food spoilage, ensuring the preservation of freshness and safety.

One of the most commonly utilized essential oils for this purpose is cinnamon essential oil. It has been recognized by the US Food and Drug Administration (FDA) as GRAS, which makes it a trusted ingredient for many applications.

TABLE 26.2 Different essential oils used to improve the formulation of the films.

Film	Essential oil	Preparation method	References
Chitosan or carboxymethyl cellulose	Citral	Casting	Arnon-Rips et al. (2019)
Corn starch	*Zataria multiflora*	Casting	Amiri et al. (2019)
Whey protein isolate	Cinnamon	Casting	Mohammadi et al. (2020)
Hazelnut meal protein	Clove	Casting	Gul et al. (2018)
Gelatin	Thymol	Casting	Li et al. (2020)
Quinoa/chitosan	Thymol	Casting	Robledo et al. (2018)
Sodium alginate	Sweet orange	Casting	Das et al. (2020)
Basil gum	*Zataria multiflora*	Casting	Hashemi Gahruie et al. (2017)
Sodium alginate	Oregano	Coating	Lee et al. (2022)
Chitosan or whey protein	Garlic	Casting	Esmaeili et al. (2020)
Banana starch	Lemongrass/rosemary	Casting	Restrepo et al. (2018)
Carrageenan	Orange	Casting	Simona et al. (2021)
Chitosan/nisin	Perilla	Casting	Fang et al. (2025)
Carboxymethyl chitosan/pullulan	Galangal	Casting	Zhou, He, et al. (2021)

Cinnamon essential oil has gained considerable attention in recent years due to its powerful, broad-spectrum antimicrobial properties, making it effective against a wide range of harmful pathogens (Mohammadi et al., 2020).

Despite its notable antimicrobial benefits, adding cinnamon essential oil to whey protein isolate-based films has been shown to cause morphological irregularities. SEM analysis revealed changes in the structure of the films. Additionally, incorporating cinnamon essential oil led to a decrease in the tensile strength of the films, making them weaker and less durable. A similar effect was observed in films made from oxidized hydroxypropyl cassava starch (Zhou, Wu, et al., 2021). However, the reduction in strength could be overcome by adding chitosan nanofibers to the film composition. The chitosan nanofibers serve as nanoreinforcements, enhancing the overall mechanical properties of the films and counteracting the negative impact of cinnamon oil on their tensile properties (Mohammadi et al., 2020).

Another commonly used essential oil, clove, has long been utilized in various applications, including antimicrobial effects, pain relief, digestive issues, and respiratory conditions such as cough and asthma. As a food ingredient, it is used in amounts that adhere to good manufacturing practices, typically not exceeding 0.06%. Clove essential oil has a strong, spicy aroma, which may not always be suitable for food formulations. However, incorporating clove essential oil into edible films could help reduce its intense flavor compared to direct application in products. Films made from hazelnut meal protein and clove essential oil have been successfully prepared starting from nanoemulsions. The addition of clove essential oil to the films imparted antibacterial activity against *E. coli*, *L. monocytogenes*, *S. aureus*, *B. subtilis*, and *P. aeruginosa*, as well as antioxidant activity (Gul et al., 2018). The small size of the nanodroplets improved the homogeneity of the films, resulting in better performance. The encapsulation efficiency for the nanodroplets was higher, which enhanced the activity of the essential oil.

Thymol, the primary component of essential oils from thyme (*Thymus vulgaris*), is another plant oil classified as GRAS and known for its excellent antimicrobial and antioxidant properties against a wide range of pathogenic and spoilage bacteria. Due to its hydrophobic nature, thymol is effective in reducing the WVP of films and has been used to prepare nanoemulsion-based films with various formulations (Li et al., 2020). Although the addition of 1% thymol increased the hydrophobicity of the films, it did not affect the smooth and uniform surface structure of gelatin films (Li et al., 2020). However, when thymol nanoemulsions were incorporated into quinoa-chitosan films, the film microstructure became porous and heterogeneous (Robledo et al., 2018). These changes in microstructure were dependent on the interactions between the matrix and thymol, which varied according to the chemical nature of the matrix. Nevertheless, the microstructural changes of the film did not affect its antibacterial performance. When the quinoa-chitosan films were applied to extend the shelf life of cherry tomatoes, the coated tomatoes inoculated with *B. cinerea* showed a significant reduction in fungal growth after seven days at 5°C, compared to control samples (tomatoes without coating or those coated with quinoa protein/chitosan). This demonstrated the coating's effectiveness in prolonging the shelf life of cherry tomatoes (Robledo et al., 2018).

A citrus sinensis essential oil containing sweet orange essential oil was incorporated into sodium alginate-based films to create an edible nanoemulsion-based coating to protect tomatoes from contamination by both *Salmonella* and *Listeria*. The small size (43.23 nm) of the nanoemulsion droplets imparted transparency to the products, which positively influenced their marketability and appeal. Due to their small size, these nanodroplets, evenly distributed throughout the film, were also highly effective at controlling *Salmonella* and *Listeria* in both single and multispecies culture conditions, improving the firmness of tomatoes by up to 33% and reducing weight loss by up to three times compared to uncoated tomatoes. Sensory analysis revealed that using the edible coating increased the overall acceptance scores of the tomatoes (Das et al., 2020). Thus, coating efficiency was enhanced by reducing the size of the dispersed phase particles.

Orange essential oil, combined with trehalose, was also used to formulate carrageenan-based films. FTIR spectroscopy was used to study the interactions among the film components, while SEM was used to describe the structure. The results from both techniques confirmed that the amount of carrageenan used (0.3 and 0.5 g in the final volume, approximately 50 g) and the presence of orange essential oil influenced the compatibility of trehalose with the film matrix. The film with 0.3 g of carrageenan showed a more homogeneous surface than the one formulated with 0.5 g of carrageenan, which had a more heterogeneous surface. The data showed a synergistic effect between trehalose and orange essential oil that enhanced the physical properties and protective effects of the films against microorganisms, thereby extending shelf life and improving food quality (Simona et al., 2021).

The genus *Z. multiflora* Boiss., part of the Lamiaceae family, is a valuable medicinal and culinary plant commonly found in Iran, Pakistan, and Afghanistan. The essential oil's main components, carvacrol and thymol, contribute to its strong antimicrobial properties against a wide range of microorganisms, which have been well documented. Basil (*Ocimum basilicum* L.) is grown in various regions across Asia, Africa, and Central and South America. The basil seeds are often used in traditional beverages and desserts, cherished for their therapeutic properties and high fiber content. When soaked in water, the outer pericarp of basil seeds swells, forming a gelatinous mass. The hydrocolloids in basil seeds are similar to those in other commercial hydrocolloid sources. However, basil seed gum offers several notable advantages, such as low production costs, hydrophilicity, biocompatibility, biodegradability, and desirable rheological properties. These characteristics make basil seed gum an ideal candidate for the development of edible films and coatings. Due to its natural origin and beneficial properties, basil seed gum is increasingly being explored as an alternative to synthetic materials in food packaging. It offers a sustainable and effective solution for extending the shelf life of food products while maintaining their quality and safety. *Z. multiflora* essential oil was incorporated into a basil seed gum-based film in the form of a nanoemulsion. The antibacterial effect of the films increased as the droplet size decreased, and an increase in nanoemulsion concentration up to 3% enhanced the mechanical properties. The addition of nanoemulsion to the basil-based film improved its degree of compactness, resulting in a finer film network with higher density. SEM micrographs revealed good compatibility between the incorporated nanoemulsion and the basil-based film network, suggesting that nanoemulsions could function as compatible nanofillers within the matrix (Hashemi Gahruie et al., 2017). *Z. multiflora* essential oil was also incorporated into corn starch-based films (Amiri et al., 2019). Due to the small size of the droplets in the dispersed phase, the antioxidant capabilities, measured by peroxide value, thiobarbituric acid reactive substances, and carbonyl content, were excellent. The sensory analysis revealed an overall acceptability score of 5.85, indicating that the essential oil was not detectable when encapsulated.

Oregano essential oil contains the monoterpene carvacrol and is renowned for its exceptional antimicrobial properties. It was utilized to create single- and multipolysaccharide edible coating suspensions, incorporating bagasse cellulose nanocrystals, chitosan nanofibers, and sodium alginate as polysaccharide components. These suspensions were applied to strawberries via spray coating to protect the fruit. The incorporation of essential oils into polysaccharide coatings helps retain moisture and preserves other desirable sensory qualities of fresh fruit. Coatings made with multiple polysaccharides were more effective in inhibiting fungal growth than single-polysaccharide coatings. In contrast, uncoated strawberries and those with single-polysaccharide coatings exhibited higher weight loss than those with multipolysaccharide coatings. The multipolysaccharide coating effectively retained the desired moisture, firmness, and appearance of the strawberries, owing to its gas barrier properties resulting from the entangled matrix structure. The coatings successfully extended the shelf life of the fruit (Lee et al., 2022).

In recent years, perilla essential oil has attracted significant attention due to its beneficial chemical composition. This essential oil is particularly appreciated for its high content of unsaturated fatty acids, along with a variety of other valuable active compounds, including sterols, phenols, and ketones (Wang et al., 2021). Recent studies have explored the potential of composite edible films that utilize perilla essential oil as both a water-blocking agent and an antibacterial component. These films are made by combining chitosan, a naturally occurring cationic polysaccharide, with nisin, a peptide derived from *Lactococcus lactis* subspecies, to form the matrix. These films were prepared through a casting process, with the formulation optimized using response surface methodology.

The effectiveness of these films was tested in ready-to-eat fish balls. The fish balls were stored in freshness bags at room temperature for 12 days, and the films' ability to maintain quality was evaluated. The emulsion-based film, which contained droplets on a conventional scale, demonstrated significantly reduced total volatile basic nitrogen levels and bacterial counts compared to the control groups (Fang et al., 2025). These results highlight the potential of perilla essential oil-based films as an effective way to enhance the preservation and safety of fish products.

Galangal essential oil, the primary active ingredient of the galangal plant, has aroused the interest of researchers due to its beneficial properties and chemical composition. This essential oil contains oxygen-rich monoterpenes, such as 1,8-cineole, which are responsible for many of its health-promoting effects. Extensive research has demonstrated that galangal essential oil possesses a variety of therapeutic properties, including antioxidant, antibacterial, anticoagulant, antiinflammatory, and antitumor effects (Dong & Wang, 2018). Several studies have specifically highlighted the antimicrobial activity of galangal essential oil, showing its effectiveness against bacteria such as *S. aureus*, *E. coli*, and *L. monocytogenes* (Zhou, He, et al., 2021).

The use of essential oils, particularly for active packaging, has become increasingly popular due to their natural origin and well-documented antibacterial activity, which makes them an ideal choice for enhancing food safety and preservation (Biddeci et al., 2016). In line with this approach, galangal essential oil was incorporated as a hydrophobic phase in the formulation of composite films. These films were prepared by casting and combining two natural polysaccharides—carboxymethyl chitosan and pullulan—to form a stable matrix. The performance was evaluated by testing their effectiveness in preserving mangoes stored at 25°C ± 1°C for 15 days (Zhou, He, et al., 2021).

Several key parameters were assessed during the evaluation, including fruit weight loss, firmness, soluble solids content, and titratable acidity. The results indicated that the composite film was a promising eco-friendly packaging material, particularly for use in the fruit preservation industry. In addition to its preservation capabilities, the films also exhibited strong mechanical and barrier properties. These properties were likely attributed to the compatibility of the polysaccharides forming the matrix, as demonstrated by FTIR and XRD analysis. The results showed significant interactions between the functional groups of the polysaccharides—specifically, the hydroxyl groups of pullulan and the carboxyl groups of carboxymethyl chitosan. This interaction led to a homogeneous matrix structure, contributing to the excellent performance of the films in terms of mechanical strength and barrier functionality (Zhou, He, et al., 2021).

Garlic (*Allium sativum*) is an aromatic plant native to Central Asia, widely used as a flavoring agent. It belongs to the Alliaceae family, and both its fresh bulbs and essential oil are commonly used in savory foods. Due to its flavoring, antimicrobial, and antioxidant properties, garlic essential oil is particularly well-suited for use in processed meat products, such as hamburgers and sausages. Garlic essential oil was added to chitosan or whey protein-based films at a concentration of 2%, either by preparing a coarse emulsion with an IKA T 25 ultra-turrax homogenizer or by nanoencapsulating it in cholesterol/soybean phosphatidylcholine liposomes, to extend the shelf life of refrigerated vacuum-packed sausages (Esmaeili et al., 2020). The sausages were stored under refrigeration for 50 days. Based on microbiological and lipid stability analysis, all active films slowed lipid oxidation and the growth of main spoilage bacterial groups compared to the control. The best results were obtained with the chitosan film containing nanoencapsulated garlic essential oil, as measured by peroxide value, thiobarbituric acid reactive substances, and aerobic plate count. However, there were no significant differences in sensory properties among the samples, indicating that although garlic essential oil demonstrated microbiological activity against foodborne pathogens and antioxidant capabilities, its strong flavor could not be fully masked by encapsulation.

Lemongrass and rosemary, both of which possess remarkable antimicrobial and antifungal properties, are highly effective against common pathogens typically found during the postharvest stages of fruits and vegetables. These essential oils were combined and used as a nanoemulsion to improve the mechanical and barrier properties of films made from banana starch (Restrepo et al., 2018). The nanoemulsions were prepared using the emulsion-phase inversion method, a low-energy technique. The selected surfactant-to-oil ratios varied between 0.5 and 2. The concentration of essential oil nanoemulsions in the films ranged from 0.3% to 1% (w/v). The results of the study demonstrated that incorporating these essential oil nanoemulsions into the banana starch-based films had a plasticizing effect, which in turn increased the film's WVP, improved its transparency, and enhanced its elongation at break. However, the hydrophobic nature of the essential oils contributed to a reduction in the water solubility of the films. The use of nanoemulsions proved to be an effective method for encapsulating essential oils, ultimately leading to the development of a film suitable for food packaging purposes.

All these reports highlighted the importance of droplet size in the discontinuous phase. Reducing droplet size proved to be an effective strategy for improving structure and physical properties, with the nanoemulsion content playing a crucial role in ensuring structural homogeneity and mechanical behavior of the systems. Nanodroplets cause less disruption to the film structure and allow for better connections between the building blocks of the film. Smaller droplets exhibited high encapsulation efficiency, and the delivery of bioactive compounds was highly effective, enhancing the films' antibacterial and antioxidant properties. In this way, complex films behave similarly to simpler systems, like films made from solutions, in terms of mechanical and tensile properties but with improved functionality due to the addition of antibacterial and/or antioxidant components.

26.4 Conclusion

Although bio-based polymers generally have poorer physical properties than petroleum-based plastics, their performance was successfully enhanced by incorporating nanomaterials into formulations. Depending on the chemical nature, concentration, and geometry, nanoreinforcements were able to improve both mechanical and barrier properties. The effects were related to the interactions between the reinforcement and the matrix, and how they were distributed within the matrix. Nanoemulsions made it possible to create two-phase systems with the mechanical properties of one-phase systems, as the small size of the droplets in the dispersed phase did not disrupt the matrix structure or the connections among its components. When the lipid phase was an essential oil, the films also exhibited antimicrobial and antioxidant properties. The films discussed in this chapter improved the appearance and shelf life of various food products when used as packaging materials.

References

Almasi, H., Jafarzadeh, P., & Mehryar, L. (2018). Fabrication of novel nanohybrids by impregnation of CuO nanoparticles into bacterial cellulose and chitosan nanofibers: Characterization, antimicrobial and release properties. *Carbohydrate Polymers, 186*, 273–281. https://doi.org/10.1016/j.carbpol.2018.01.067.

Amiri, E., Aminzare, M., Azar, H. H., & Mehrasbi, M. R. (2019). Combined antioxidant and sensory effects of corn starch films with nanoemulsion of *Zataria multiflora* essential oil fortified with cinnamaldehyde on fresh ground beef patties. *Meat Science, 153*, 66–74. https://doi.org/10.1016/j.meatsci.2019.03.004, http://www.elsevier.com/locate/meatsci.

Anis, A., Pal, K., & Al-Zahrani, S. M. (2021). Essential oil-containing polysaccharide-based edible films and coatings for food security applications. *Polymers, 13*(4), 575. https://doi.org/10.3390/polym13040575.

Arnon-Rips, H., Porat, R., & Poverenov, E. (2019). Enhancement of agricultural produce quality and storability using citral-based edible coatings; the valuable effect of nano-emulsification in a solid-state delivery on fresh-cut melons model. *Food Chemistry, 277*, 205–212. https://doi.org/10.1016/j.foodchem.2018.10.117.

Bhavaniramya, S., Vishnupriya, S., Al-Aboody, M. S., Vijayakumar, R., & Baskaran, D. (2019). Role of essential oils in food safety: Antimicrobial and antioxidant applications. *Grain & Oil Science and Technology, 2*(2), 49–55. https://doi.org/10.1016/j.gaost.2019.03.001.

Biddeci, G., Cavallaro, G., Di Blasi, F., Lazzara, G., Massaro, M., Milioto, S., Parisi, F., Riela, S., & Spinelli, G. (2016). Halloysite nanotubes loaded with peppermint essential oil as filler for functional biopolymer film. *Carbohydrate Polymers, 152*, 548–557. https://doi.org/10.1016/j.carbpol.2016.07.041.

Cárdenas, G., Anaya, P., del Rio, R., Schrebler, R., von Plessing, C., & Schneider, M. (2010). Scanning electron microscopy and atomic force microscopy of chitosan composite films. *Journal of the Chilean Chemical Society, 55*(3), 352–354. https://doi.org/10.4067/S0717-97072010000300017.

Chen, H., Wu, C., Feng, X., He, M., Zhu, X., Li, Y., & Teng, F. (2022). Effects of two fatty acids on soy protein isolate/sodium alginate edible films: Structures and properties. *LWT, 159*, 113221. https://doi.org/10.1016/j.lwt.2022.113221.

Cheng, J., & Cui, L. (2021). Effects of high-intensity ultrasound on the structural, optical, mechanical and physicochemical properties of pea protein isolate-based edible film. *Ultrasonics Sonochemistry, 80*, 105809. https://doi.org/10.1016/j.ultsonch.2021.105809.

Das, S., Vishakha, K., Banerjee, S., Mondal, S., & Ganguli, A. (2020). Sodium alginate-based edible coating containing nanoemulsion of *Citrus sinensis* essential oil eradicates planktonic and sessile cells of food-borne pathogens and increased quality attributes of tomatoes. *International Journal of Biological Macromolecules, 162*, 1770–1779. https://doi.org/10.1016/j.ijbiomac.2020.08.086.

Davachi, S. M., & Shekarabi, A. S. (2018). Preparation and characterization of antibacterial, eco-friendly edible nanocomposite films containing *Salvia macrosiphon* and nanoclay. *International Journal of Biological Macromolecules, 113*, 66–72. https://doi.org/10.1016/j.ijbiomac.2018.02.106, www.elsevier.com/locate/ijbiomac.

Dong, F., & Wang, X. (2018). Guar gum and ginseng extract coatings maintain the quality of sweet cherry. *LWT, 89*, 117–122. https://doi.org/10.1016/j.lwt.2017.10.035.

Ekrami, M., Emam-Djomeh, Z., Joolaei-Ahranjani, P., Mahmoodi, S., & Khaleghi, S. (2021). Eco-friendly UV protective bionanocomposite based on Salep-mucilage/flower-like ZnO nanostructures to control photo-oxidation of kilka fish oil. *International Journal of Biological Macromolecules, 168*, 591–600. https://doi.org/10.1016/j.ijbiomac.2020.12.013.

Esmaeili, H., Cheraghi, N., Khanjari, A., Rezaeigolestani, M., Basti, A. A., Kamkar, A., & Aghaee, E. M. (2020). Incorporation of nanoencapsulated garlic essential oil into edible films: A novel approach for extending shelf life of vacuum-packed sausages. *Meat Science, 166*. https://doi.org/10.1016/j.meatsci.2020.108135, http://www.elsevier.com/locate/meatsci.

Fang, Z., Yang, Y., Lin, S., Xu, L., Chen, S., Lv, W., Wang, N., Dong, S., Lin, C., Xie, Y., Liu, J., Meng, M., Wen, W., & Yang, Y. (2025). Development and antimicrobial activity of composite edible films of chitosan and nisin incorporated with perilla essential oil-glycerol monolaurate emulsions. *Food Chemistry, 462*, 141006. https://doi.org/10.1016/j.foodchem.2024.141006.

Feng, Z., Li, L., Wang, Q., Wu, G., Liu, C., Jiang, B., & Xu, J. (2019). Effect of antioxidant and antimicrobial coating based on whey protein nanofibrils with TiO_2 nanotubes on the quality and shelf life of chilled meat. *International Journal of Molecular Sciences, 20*(5), 1184. https://doi.org/10.3390/ijms20051184.

Ghosh, T., Mondal, K., Giri, B. S., & Katiyar, V. (2021). Silk nanodisc based edible chitosan nanocomposite coating for fresh produces: A candidate with superior thermal, hydrophobic, optical, mechanical and food properties. *Food Chemistry, 360*, 130048. https://doi.org/10.1016/j.foodchem.2021.130048.

Giannakas, A. E., Salmas, C. E., Moschovas, D., Zaharioudakis, K., Georgopoulos, S., Asimakopoulos, G., Aktypis, A., Proestos, C., Karakassides, A., Avgeropoulos, A., Zafeiropoulos, N. E., & Nychas, G. J. (2022). The increase of soft cheese shelf-life packaged with edible films based on novel hybrid nanostructures. *Gels, 8*(9). https://doi.org/10.3390/gels8090539, www.mdpi.com/journal/gels.

Gul, O., Saricaoglu, F. T., Besir, A., Atalar, I., & Yazici, F. (2018). Effect of ultrasound treatment on the properties of nano-emulsion films obtained from hazelnut meal protein and clove essential oil. *Ultrasonics Sonochemistry, 41*, 466–474. https://doi.org/10.1016/j.ultsonch.2017.10.011, www.elsevier.com/inca/publications/store/5/2/5/4/5/1.

Gürdal, A. A., & Çetinkaya, T. (2024). Advancements in edible films for aquatic product preservation and packaging. *Reviews in Aquaculture, 16*(3), 997–1020. https://doi.org/10.1111/raq.12880, http://onlinelibrary.wiley.com/journal/10.1111/(ISSN)1753-5131.

Hanan, E., Dar, A. H., Shams, R., & Goksen, G. (2024). New insights into essential oil nano emulsions loaded natural biopolymers recent development, formulation, characterization and packaging applications: A comprehensive review. *International Journal of Biological Macromolecules, 280*(Pt 2), 135751. https://doi.org/10.1016/j.ijbiomac.2024.135751.

Hashemi Gahruie, H., Ziaee, E., Eskandari, M. H., & Hosseini, S. M. H. (2017). Characterization of basil seed gum-based edible films incorporated with *Zataria multiflora* essential oil nanoemulsion. *Carbohydrate Polymers, 166*, 93–103. https://doi.org/10.1016/j.carbpol.2017.02.103, http://www.elsevier.com/wps/find/journaldescription.cws_home/405871/description#description.

Hoyos-Merlano, N. T., Borroni, V., Rodriguez-Batiller, M. J., Candal, R. J., & Herrera, M. L. (2022). Nanoreinforcement as a strategy to improve physical properties of biodegradable composite films based on biopolymers. *Food Research International, 162*. https://doi.org/10.1016/j.foodres.2022.112178, www.elsevier.com/inca/publications/store/4/2/2/9/7/0.

Hoyos Merlano, N. T., Guz, L., Borroni, V., Candal, R. J., & Herrera, M. L. (2023). Effects of the geometry of reinforcement on physical properties of sodium caseinate/TiO_2 nanocomposite films for applications in food packaging. *Biopolymers, 114*(2). https://doi.org/10.1002/bip.23531, http://onlinelibrary.wiley.com/journal/10.1002/(ISSN)1097-0282.

Karimi Khorrami, N., Radi, M., Amiri, S., & McClements, D. J. (2021). Fabrication and characterization of alginate-based films functionalized with nanostructured lipid carriers. *International Journal of Biological Macromolecules, 182*, 373–384. https://doi.org/10.1016/j.ijbiomac.2021.03.159, www.elsevier.com/locate/ijbiomac.

Khedri, S., Sadeghi, E., Rouhi, M., Delshadian, Z., Mortazavian, A. M., de Toledo Guimarães, J., Fallah, M., & Mohammadi, R. (2021). Bioactive edible films: Development and characterization of gelatin edible films incorporated with casein phosphopeptides. *LWT, 138*, 110649. https://doi.org/10.1016/j.lwt.2020.110649.

Kocira, A., Kozłowicz, K., Panasiewicz, K., Staniak, M., Szpunar-Krok, E., & Hortyńska, P. (2021). Polysaccharides as edible films and coatings: Characteristics and influence on fruit and vegetable quality—A review. *Agronomy, 11*(5), 813. https://doi.org/10.3390/agronomy11050813.

Kong, I., Degraeve, P., & Pui, L. P. (2022). Polysaccharide-based edible films incorporated with essential oil nanoemulsions: Physico-chemical, mechanical properties and its application in food preservation—A review. *Foods, 11*(4), 555. https://doi.org/10.3390/foods11040555.

Lee, D., Shayan, M., Gwon, J., Picha, D. H., & Wu, Q. (2022). Effectiveness of cellulose and chitosan nanomaterial coatings with essential oil on postharvest strawberry quality. *Carbohydrate Polymers, 298*, 120101. https://doi.org/10.1016/j.carbpol.2022.120101.

Li, X., Yang, X., Deng, H., Guo, Y., & Xue, J. (2020). Gelatin films incorporated with thymol nanoemulsions: Physical properties and antimicrobial activities. *International Journal of Biological Macromolecules, 150*, 161–168. https://doi.org/10.1016/j.ijbiomac.2020.02.066.

Liang, Y., Zhao, Y., Sun, H., Dan, J., Kang, Y., Zhang, Q., Su, Z., Ni, Y., Shi, S., Wang, J., & Zhang, W. (2023). Natural melanin nanoparticle-based photothermal film for edible antibacterial food packaging. *Food Chemistry, 401*, 134117. https://doi.org/10.1016/j.foodchem.2022.134117.

Liu, J., Zhang, L., Liu, C., Zheng, X., & Tang, K. (2021). Tuning structure and properties of gelatin edible films through pullulan dialdehyde crosslinking. *LWT, 138*, 110607. https://doi.org/10.1016/j.lwt.2020.110607.

Liyanapathiranage, A., Dassanayake, R. S., Gamage, A., Karri, R. R., Manamperi, A., Evon, P., Jayakodi, Y., Madhujith, T., & Merah, O. (2023). Recent developments in edible films and coatings for fruits and vegetables. *Coatings, 13*(7). https://doi.org/10.3390/coatings13071177, www.mdpi.com/journal/coatings.

Ma, Y., Teng, A., Zhao, K., Zhang, K., Zhao, H., Duan, S., Li, S., Guo, Y., & Wang, W. (2020). A top-down approach to improve collagen film's performance: The comparisons of macro, micro and nano sized fibers. *Food Chemistry, 309*, 125624. https://doi.org/10.1016/j.foodchem.2019.125624.

Mahcene, Z., Khelil, A., Hasni, S., Bozkurt, F., Goudjil, M. B., & Tornuk, F. (2021). Home-made cheese preservation using sodium alginate based on edible film incorporating essential oils. *Journal of Food Science and Technology, 58*(6), 2406–2419. https://doi.org/10.1007/s13197-020-04753-3, http://www.springerlink.com/content/121580/.

Majeed, T., Dar, A. H., Pandey, V. K., Dash, K. K., Srivastava, S., Shams, R., Jeevarathinam, G., Singh, P., Echegaray, N., & Pandiselvam, R. (2023). Role of additives in starch-based edible films and coating: A review with current knowledge. *Progress in Organic Coatings, 181*. https://doi.org/10.1016/j.porgcoat.2023.107597, www.elsevier.com/locate/porgcoat.

Midhun Dominic, C. D., dos Santos Rosa, D., Camani, P. H., Kumar, A. S., Neenu, K. V., Begum, P. M. S., Dinakaran, D., John, E., Baby, D., Thomas, M. M., Joy, J. M., Parameswaranpillai, J., & Saeb, M. R. (2021). Thermopslastic starch nanocomposites using cellulose-rich *Chrysopogon zizanioides* nanofibers. *International Journal of Biological Macromolecules, 191*, 572–583. https://doi.org/10.1016/j.ijbiomac.2021.09.103, http://www.elsevier.com/locate/ijbiomac.

Mohammadi, M., Mirabzadeh, S., Shahvalizadeh, R., & Hamishehkar, H. (2020). Development of novel active packaging films based on whey protein isolate incorporated with chitosan nanofiber and nano-formulated cinnamon oil. *International Journal of Biological Macromolecules, 149*, 11–20. https://doi.org/10.1016/j.ijbiomac.2020.01.083.

Montes-de-Oca-Ávalos, J. M., Altamura, D., Candal, R. J., Scattarella, F., Siliqi, D., Giannini, C., & Herrera, M. L. (2018). Relationship between nano/micro structure and physical properties of TiO_2-sodium caseinate composite films. *Food Research International, 105*, 129–139. https://doi.org/10.1016/j.foodres.2017.11.011, www.elsevier.com/inca/publications/store/4/2/2/9/7/0.

Montes-de-Oca-Ávalos, J. M., Altamura, D., Herrera, M. L., Huck-Iriart, C., Scattarella, F., Siliqi, D., Giannini, C., & Candal, R. J. (2020). Physical and structural properties of whey protein concentrate - Corn oil - TiO_2 nanocomposite films for edible food-packaging. *Food Packaging and Shelf Life, 26*. https://doi.org/10.1016/j.fpsl.2020.100590, http://www.journals.elsevier.com/food-packaging-and-shelf-life/.

Najafi, Z., Kahn, C. J. F., Bildik, F., Arab-Tehrany, E., & Şahin-Yeşilçubuk, N. (2021). Pullulan films loading saffron extract encapsulated in nanoliposomes; preparation and characterization. *International Journal of Biological Macromolecules, 188*, 62–71. https://doi.org/10.1016/j.ijbiomac.2021.07.175, www.elsevier.com/locate/ijbiomac.

Rech, C. R., Brabes, K. C. S., Silva, B. E. B., Martines, M. A. U., Silveira, T. F. S., Alberton, J., Amadeu, C. A. A., Caon, T., Arruda, E. J., & Martelli, S. M. (2021). Antimicrobial and physical–mechanical properties of polyhydroxybutyrate edible films containing essential oil mixtures. *Journal of Polymers and the Environment, 29*(4), 1202–1211. https://doi.org/10.1007/s10924-020-01943-0, http://www.kluweronline.com/issn/1566-2543/.

Restrepo, A. E., Rojas, J. D., García, O. R., Sánchez, L. T., Pinzón, M. I., & Villa, C. C. (2018). Mechanical, barrier, and color properties of banana starch edible films incorporated with nanoemulsions of lemongrass (*Cymbopogon citratus*) and rosemary (*Rosmarinus officinalis)* essential oils. *Food Science and Technology International, 24*(8), 705–712. https://doi.org/10.1177/1082013218792133.

Robledo, N., Vera, P., López, L., Yazdani-Pedram, M., Tapia, C., & Abugoch, L. (2018). Thymol nanoemulsions incorporated in quinoa protein/chitosan edible films; antifungal effect in cherry tomatoes. *Food Chemistry, 246*, 211–219. https://doi.org/10.1016/j.foodchem.2017.11.032.

Salmas, C. E., Giannakas, A. E., Baikousi, M., Kollia, E., Tsigkou, V., & Proestos, C. (2021). Effect of copper and titanium-exchanged montmorillonite nanostructures on the packaging performance of chitosan/poly-vinyl-alcohol-based active packaging nanocomposite films. *Foods, 10*(12). https://doi.org/10.3390/foods10123038, https://www.mdpi.com/2304-8158/10/12/3038/pdf.

Salmas, C. E., Giannakas, A. E., Moschovas, D., Kollia, E., Georgopoulos, S., Gioti, C., Leontiou, A., Avgeropoulos, A., Kopsacheili, A., Avdylaj, L., & Proestos, C. (2022). Kiwi fruits preservation using novel edible active coatings based on rich thymol halloysite nanostructures and chitosan/polyvinyl alcohol gels. *Gels, 8*(12). https://doi.org/10.3390/gels8120823, www.mdpi.com/journal/gels.

Shahidi, F., & Hossain, A. (2022). Preservation of aquatic food using edible films and coatings containing essential oils: a review. *Critical Reviews in Food Science and Nutrition, 62*(1), 66–105. https://doi.org/10.1080/10408398.2020.1812048.

Simona, J., Dani, D., Petr, S., Marcela, N., Jakub, T., & Bohuslava, T. (2021). Edible films from carrageenan/orange essential oil/trehalose—Structure, optical properties, and antimicrobial activity. *Polymers, 13*(3), 1–19. https://doi.org/10.3390/polym13030332.

Wang, D., Sun, Z., Sun, J., Liu, F., Du, L., & Wang, D. (2021). Preparation and characterization of polylactic acid nanofiber films loading Perilla essential oil for antibacterial packaging of chilled chicken. *International Journal of Biological Macromolecules, 192*, 379–388. https://doi.org/10.1016/j.ijbiomac.2021.09.190.

Wigati, L. P., Wardana, A. A., Tanaka, F., & Tanaka, F. (2022). Edible film of native jicama starch, agarwood *Aetoxylon bouya* essential oil and calcium propionate: Processing, mechanical, thermal properties and structure. *International Journal of Biological Macromolecules, 209*, 597–607. https://doi.org/10.1016/j.ijbiomac.2022.04.021, www.elsevier.com/locate/ijbiomac.

Xie, B., Zhang, X., Luo, X., Wang, Y., Li, Y., Li, B., & Liu, S. (2020). Edible coating based on beeswax-in-water Pickering emulsion stabilized by cellulose nanofibrils and carboxymethyl chitosan. *Food Chemistry, 331*, 127108. https://doi.org/10.1016/j.foodchem.2020.127108.

Zhou, W., He, Y., Liu, F., Liao, L., Huang, X., Li, R., Zou, Y., Zhou, L., Zou, L., Liu, Y., Ruan, R., & Li, J. (2021). Carboxymethyl chitosan-pullulan edible films enriched with galangal essential oil: Characterization and application in mango preservation. *Carbohydrate Polymers, 256*, 117579. https://doi.org/10.1016/j.carbpol.2020.117579.

Zhou, Y., Wu, X., Chen, J., & He, J. (2021). Effects of cinnamon essential oil on the physical, mechanical, structural and thermal properties of cassava starch-based edible films. *International Journal of Biological Macromolecules, 184*, 574–583. https://doi.org/10.1016/j.ijbiomac.2021.06.067.

Section III

Application of biopolymer films and coatings

Chapter 27

Meat preservation

Sahil Chaudhary[1], Shivani Jaswal[2], and Nitika Saini[3]
[1]*Department of Food Science and Technology, I. K. Gujral Punjab Technical University, Kapurthala, Punjab, India,* [2]*Department of Food Science and Technology, Dr. Y. S. Parmar University of Horticulture & Forestry, Solan, Himachal Pradesh, India,* [3]*Department of Biotechnology, CT Institute of Pharmaceutical Sciences, Jalandhar, Punjab, India*

27.1 Introduction

Humans have consumed meat for millennia, and it has long constituted a significant source of dietary protein. In addition to its protein content, meat provides essential vitamins and minerals, some of which are either exclusively available in animal-derived products or exhibit greater bioavailability in such sources. Over the past three decades, global meat consumption has approximately doubled, and current projections indicate that it may double again by the middle of this century (Li et al., 2024), attributable to the global population, which is envisioned to grow to 9.7 billion by 2050, representing an increment of about one-third compared to 2015 (FAO, 2018). Any meat (flesh, organ, poultry, or fish) is a highly perishable material due to its rich, yet sensitive biological profile. Multiple conjugated facets, namely, temperature, O_2 presence, endogenous enzymes, moisture, light, and microorganisms affect the overall quality, leading to food waste and economic losses. The envisioned economic impact of meat spoilage and extensive product recalls is estimated to reach 10 billion US dollars each year on a global scale, leading to the wastage of valuable meat resources (Chen et al., 2024). Meat is a nutrient-dense matrix containing a good amount of protein, micronutrients, available energy-producing nutrients (glucose, ribose, amino acids, and nucleosides), moisture content, and an appropriate pH (5.5–6.5) that facilitate microbial proliferation. Meat-based foods are also susceptible to biochemical and physiological changes due to endogenous enzymes. All these factors justify the relatively short shelf life of meat, fish, and derived products and their vulnerability to quality degradation. The deterioration of meat and its subsequent waste contributes to a burdensome stream characterized by severe eco-toxicological effects. Additionally, the environmental impact of worldwide meat production is considerable, manifesting in land use, greenhouse gas emissions, and water consumption. In both scenarios—meat production and the management of meat waste—there is a substantial carbon footprint, which poses a greater economic challenge than the loss of valuable food resources in any context (Antonino et al., 2024). Meat, fish, and derived products could lose their quality and start to spoil at any point during the production to consumption hierarchy. There can be many reasons for the quality decline. Meat goes through several stages in the production line, so quality can be compromised at any stage, such as slaughtering, processing instruments and environments, storage units, animal intestines and blood, unhygienic laborers, and storage boxes. These elements, in conjunction with internal factors such as elevated moisture, fat, and protein levels, create an optimal environment for spoilage to commence (Fig. 27.1).

There are many important reasons, one of which is microbes. Microbes can arise from anywhere, and there is a domain of microbiota that thrives on meat and fish-based foods (Comi & Iacumin, 2025). Meat from livestock is susceptible to *Pseudomonas*, *Clostridium*, *Serratia*, *Lactobacillus*, *Leuconostoc*, *LAB*, *Staphylococcus*, and *Enterococcus*, while fish-based foods are more affected by *Streptococcus*, *Staphylococcus*, *Vibrio*, *Serratia, and Salmonella* spp. (Zhu et al., 2022). Apart from microbial causes, autolytic spoilage is also an important cause of meat spoilage. Different enzymes (proteases, lipoxygenase, lipase, phospholipase, and others) that are inherently present in the muscle cells of the carcass are responsible for the breakdown of proteins and fats into simpler units, bringing about meat softening, color alterations, and slime development. Oxidation of lipids also yields rancid smells and off-tastes. The production of free radicals enhances the interaction between the unsaturated fatty acids and molecular oxygen, producing lipid hydroperoxides, which lack a smell but are quite unstable, and their breakdown elicits various secondary compounds such as hydrocarbons, ketones,

Natural Polymer-Based Films and Coatings in Food Technology. DOI: https://doi.org/10.1016/B978-0-443-34244-8.00028-8

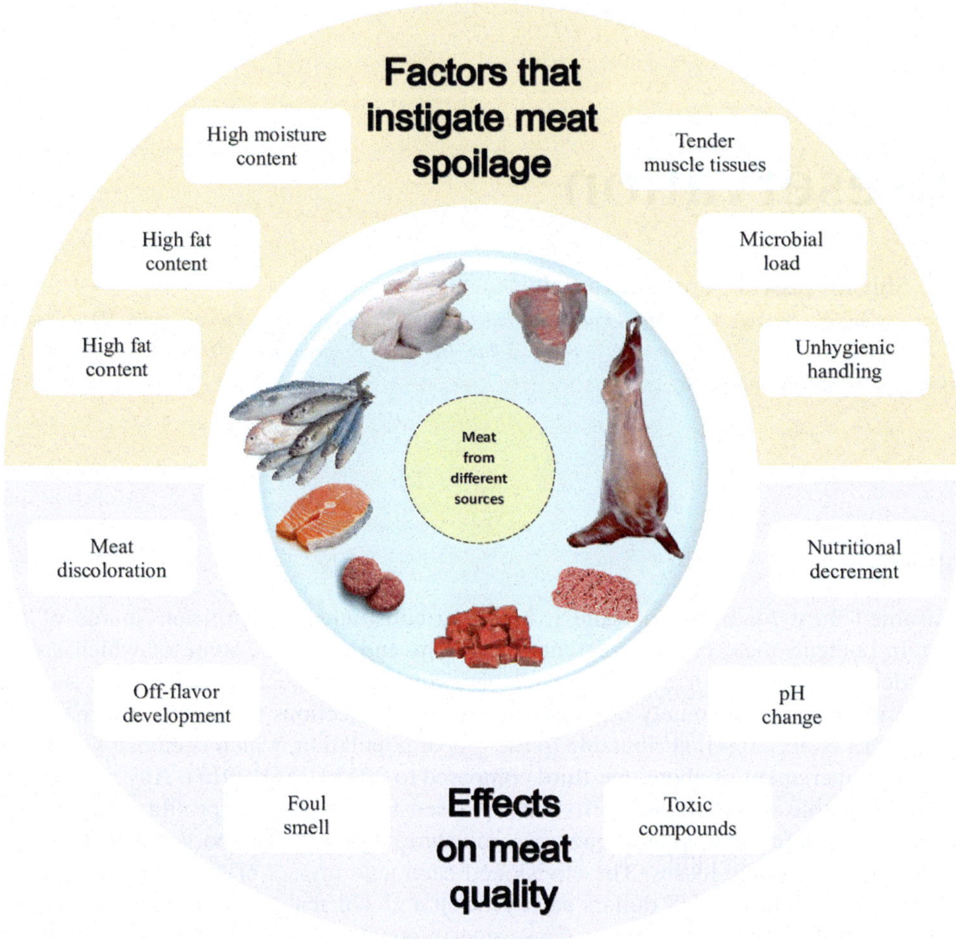

FIGURE 27.1 Different aspects associated with meat spoilage and quality deterioration.

aldehydes, alcohols, esters, and acids that give rise to unpleasant flavors and odors. Proteins present in meat are also vulnerable to oxidative chemical changes triggered by reactive species or as a result of interactions with by-products derived from oxidative stress (Ajaykumar & Mandal, 2020; Odeyemi et al., 2020).

Ensuring food safety is pivotal for businesses to flourish, uphold their reputation, and comply with food regulations. Freezing, chilling, and canning are available but have higher energy requirements. Moreover, these methods are not significant in meat transportation, as it is tortuous to maintain the appropriate environmental conditions during transportation. Chemical methods have been widely used in the past for meat quality maintenance, but due to increased awareness and the harmful effects of chemicals on health, this approach is no longer relevant in today's modern times, since people are getting more concerned about their intake. Also, resistance development by pathogens against chemicals is gaining attention among the scientific community. Certain bacteria, such as *E. coli*, *S. typhimurium*, and *L. monocytogenes*, have acquired the capability to thrive in cells by triggering acid tolerance response (ATR) mechanisms at reduced pH levels following initial adaptation to a mildly acidic environment. These challenges, exacerbated by consumer demand for safer and more sustainable options, necessitate a shift toward innovative preservation methods that not only confront these issues but also adhere to environmental management and resource conservation principles.

In recent decades, the concept of utilizing edible coatings and films has gained momentum for meat, fish, and poultry-based foods. Conceptually, an edible coating or film represents a primary protective packaging developed to maintain the quality and shelf life of perishable food products. The primary difference between the terms "coating" and "film" is that a coating is recognized as a direct application onto the food surface, while the film is fabricated separately and then wrapped around the comestible (Chavan et al., 2023). The materials most commonly used for making coatings and films are biopolymers. They are a sustainable, eco-friendly, and cost-effective alternative to petroleum-based counterparts. Natural biopolymers offer a range of options for developing coatings and films, including polysaccharides, proteins, and lipids.

These materials not only serve as excellent barriers to moisture, gas, and solute transfer but are also safe, biodegradable, and generally sourced from renewable resources. Furthermore, bioactive compounds can easily be incorporated into these edible coatings to improve their functionality (Jurić et al., 2024). Since these materials come from food waste and by-products, they are eco-friendly and do not induce harmful impacts on the environment; therefore, they connote a circular economy. This process enables materials that are not fully used to be reintroduced into the manufacturing process, adhering to the principles of a circular economy. Edible coatings and films for extending the shelf life of food contribute positively to the circular economy because, in contrast to traditional coatings, they are decomposable, nonharmful, and possess strong protective qualities (Gaspar & Braga, 2023). Also, the concept of quality preservation of meat and meat-based products through biopolymer coatings and films contributes to different Sustainable Development Goals (SDGs) launched in 2015 under "Agenda 2030." In particular, biopolymer coatings and films can extend meat shelf life, contributing to more food availability and thereby zero hunger (SDG 2), can benefit new industry, innovation, and infrastructure (SDG 9), and responsible consumption and production (SDG 12). These coatings and films can be incorporated with a variety of bioactive compounds, which could indirectly contribute to healthier consumption, contributing to improved health and well-being (SDG 3). Since these coatings and films are appropriate substitutes for conventional packaging, they therefore participate in global plastic reduction (SDG 15) (Gururani et al., 2023; Michel et al., 2024). In modern settings, biopolymer coatings have been employed beyond cosmetic goals to modify surfaces for probing properties, enhancing biocompatibility, and inducing functional characteristics. An exposition on routinely used biopolymers, their use for coatings and films, application for meat safety, and resultant effects have been stated below.

27.2 Biopolymers for better meat preservation

In 2021, the edible packaging market was \$0.84 billion, estimated to reach \$2.8 billion by 2030, compounding at a 14.31% annual growth rate from 2022 to 2030 (Sharma et al., 2024). The research toward biopolymers is flourishing, as these are suitable candidates to replace chemicals and synthetic counterparts. There are various renewable sources of biopolymers, such as plants, animals, and microbes, which render these materials sustainable and eco-friendly. The biopolymers comprise active functional groups on their side chains, which are responsible for their effectivity. The selection of specific biopolymers for different uses is determined by these essential functional groups, though primarily found ones include—the carboxyl group (C–O double bond), the hydroxyl group (-OH), and the amide group ($-NH_2$). Biopolymer-based coatings and films are increasingly adopted as solutions for the quality and shelf-life preservation of meat and meat-based products. It would not be surprising, which is why it is anticipated that the biopolymer market will surpass the USD 35.25 billion mark by 2030 at an 11.07% annual growth rate (Dey et al., 2024). Literature lays out substantial examples highlighting the utilization of biopolymers for effective meat preservation. Biopolymers of interest include polysaccharides (chitosan, alginate, cellulose, pectin, and starch), proteins (gelatin, zein, soy protein, and casein), lipids (beeswax, carnauba wax, and vegetable oils), and composites (Fig. 27.2).

27.2.1 Polysaccharides

Chitosan is an important biopolymer that is used in many fields. This biopolymer is obtained from the wastes and by-products of shrimps, crabs, and crustaceans, as these materials contain a valuable component called chitin, the deacetylation of which generates chitosan, which is of prime interest to the food sector. Chitosan is nontoxic, eco-friendly, and offers multiple biological properties namely, antioxidant, antimicrobial, antiinflammatory, anticoagulant, antitumoral, hypocholesterolemic, and antidiabetic activity (Tardy et al., 2023). United States Food and Drug Administration (USFDA), EU Regulation (No. 749/2012), and China National Standards (GB 29941–2013) have recognized chitosan as generally recognized as safe (GRAS) that endorses this biopolymer as a sound option to fabricate coatings and films (Chaudhary, 2023). Chitosan exhibits antimicrobial activity, attributable to its positively charged amino groups, which, when interacting with oppositely charged microbial membranes, affect the integrity, increasing permeation, outflow of intra-cellular liquids, and ultimate death (Cao et al., 2023; Fan et al., 2023).

Starch can be used to fabricate coatings and films. It can be obtained from numerous sources: rice, corn, potato, oats, rye, sorghum, soy, tapioca, arrowroot, and others. It is composed of amylose (linear-chain polysaccharide) (~25%) and amylopectin (branched-chain structure formed by glucose) (~75%). Starch-based coatings and films are colorless, bland in taste, and biodegradable. Since starch-based coatings are highly hydrophilic, they offer veneering with weaker water vapor barricade characteristics. It is important to know that native starch comprises a strong molecular structure of inter and intra-molecular hydrogen bonding between the hydroxyl groups; thus, disruption of starch granules occurs in a heated aqueous suspension. Firstly, the expansion of starch granules occurs as they absorb water in the amorphous region when

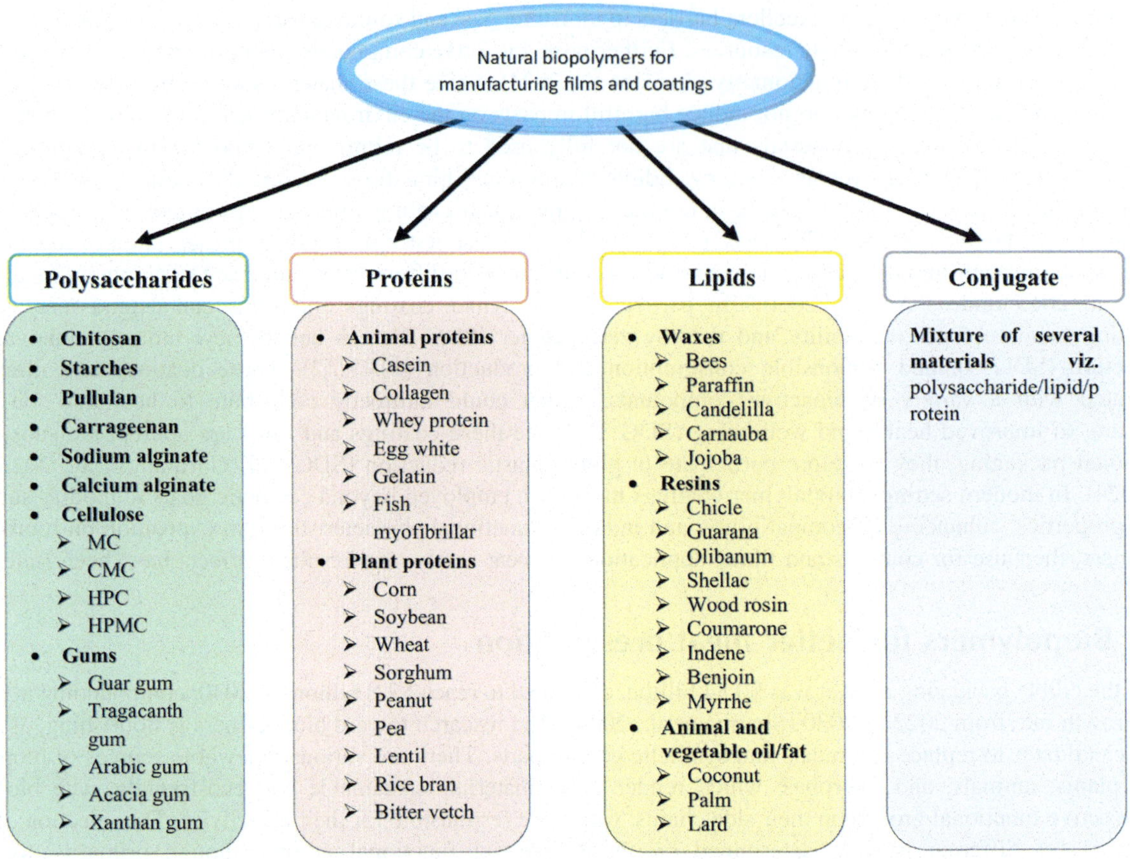

FIGURE 27.2 Different material sources for edible coating and film preparation.

the temperature remains below the gelatinization point (T_{gel}). Once the temperature escalates beyond the threshold level, the breakdown of the semicrystalline structure emanates, which is irreversible. Diffusing through the semicrystalline amylopectin clusters, water disrupts the hydrogen bonds and caters to the solubilization of amylose and amylopectin. Over time, amylose and amylopectin interweave, begetting starch film (Kupervaser et al., 2023).

Alginate is a multifunctional biopolymer that is procured from seaweeds, like *Laminaria* spp. and *Ascophyllum* spp., and finds applications in various realms such as pharma, textile, biomedical, and food (Khanzadi et al., 2020). Alginate is a glycuronan and an acidic linear polysaccharide consisting of residues of β-D-mannuronic acid (M block) and α-L-guluronic acid (G block), which are arranged in a block-wise fashion along the polymer chain (Kumar et al., 2023). There are different alginates, such as sodium alginate, potassium alginate, ammonium alginate, and calcium alginate, which are monovalent salts of alginic acid. Among these, sodium alginate is quite commonly reported to fabricate coatings of packaging and preservation interests (Khanzadi et al., 2020; Zhang et al., 2021). USFDA has also classified sodium alginate as a GRAS (generally recognized as safe) substance in the Code of Federal Regulations (CFR) (Title 21).

Pectin is a type of dietary fiber found in fruits and vegetables. This soluble fiber has many health-promoting benefits, including the potency to fight cancer. Pectin binds to carcinogens and toxins in the body, aiding in their removal from the human system. Pectin is used as a thickener, gelling, and stabilizing agent in a variety of foods. Apart, pectin has also been recognized to develop coatings and films of packaging interest, since pectin-based veneering are nontoxic and biodegradable. Structurally, pectin is a heteropolysaccharide with three domains—homogalacturonan (HG), rhamnogalacturonan-I (RG-I), and rhamnogalacturonan-II (RG-II). HG and RG-I account for the major concentrations (Nastasi et al., 2022).

Cellulose is ubiquitously found in plants and animals. Cellulose is a polymer of β-glucose. Cellulose forms a straight chain without any branches, which permits cellulose molecules to get close together. Hydrogen bonds synthesize in large numbers among the neighboring chains, which makes cellulose extremely strong. Cellulose chains grouped are often termed "microfibrils," while these microfibrils merge to yield "macrofibrils." These macrofibrils grouped grow into cellulose fiber, which forms the plant cell wall (Kumar et al., 2023; Maciel et al., 2020). Cellulose can be used to develop coatings and films, as they provide valuable functionalities such as high thermal resistance, nontoxicity, biodegradability, and conform sound physical and mechanical properties (Liu et al., 2021). Cellulose has many derivatives, among which methylcellulose is widely appreciated for its high water resistance. Methylcellulose has been explored as a barrier to oil absorption in deep-frying foods and for coatings and films for lipid-barrier properties

27.2.2 Proteins

Proteins from plant and animal sources are probed for coatings and film development. Proteins are easily processable and can be formulated into biodegradable films. Coatings and films developed with protein-based biopolymers trots out strong gas r̥ properties and adequate mechanical strength.

Collagen is obtained from the skin, tendons, connective tissues, ligaments, and bones of mammals, especially from cows and pigs. Partial hydrolysis of collagen produces a dispersible form of protein called gelatin (Mihalca et al., 2021). Gelatin is used to manufacture marshmallows, gummies, jellies, and many more. Apart from the food industry, this versatile component finds applications in pharma, photography, and even the paper industry. It is a colorless, odorless, and tasteless component (Liu et al., 2022; Rather et al., 2024). Gelatin can be used to develop coatings and films. It is a polypeptide with a higher molecular weight (MW) and is composed of oxygen, hydrogen, carbon, and nitrogen at 25.2%, 6.8%, 50.5%, and 17%, respectively (Toniciolli Rigueto et al., 2022). The viscosity of gelatin solutions relies upon certain factors such as concentration, MW of gelatin, pH, and temperature. The gelatin in aqueous solution above 40°C exhibits Newtonian viscosity behavior.

Milk proteins are another option to explore as coatings and films. Mainly classified as caseins and whey proteins, they form transparent and tasteless coatings. In particular, casein is deemed to fabricate films that are stable at different pH, temperature, and salt levels (Mihalca et al., 2021). Talking about whey, it is a yellowish liquid with a tint of green. Whey is a water-soluble protein by-product obtained postcasein extraction in cheese processing. Whey protein caseinate and whey protein isolate (WPI) are increasingly utilized to develop coatings. Whey protein-based films mete out highly interacting polymer networks with a 3D gel-type structure (Kandasamy et al., 2021).

Wheat gluten comprises >72% protein and is composed of two distinct protein fractions, called gliadin and glutenin. Wheat gluten is obtained through the process of washing wheat with water until all starch particles are expunged. The film formation property and integrity of gluten are traceable to its cohesiveness and elasticity. Wheat gluten offers good oxygen and CO_2 barricade properties (Kumar et al., 2023). Being a coproduct of the wheat starch industry, wheat gluten emerges as the preferable protein source to develop coatings, as it offers several benefits such as cross-linking ability upon heating, viscoelastic properties, and lower water solubility. However, associated limitations spring up upon transformation into coatings/films as brittleness and poor water barrier properties. However, amendments can be made using plasticizers possessing reactive groups (NH_2, –COOH, –OH, and –SH), and/or compounding with other biopolymers.

Zein is the premier component of corn protein, comprising 45%–50% (Maciel et al., 2020). Zein is abundant in hydrophobic amino acids that serve as a storage protein within the corn endosperm. Numerous studies have demonstrated that zein films exhibit significant water barricade properties, which can effectively restrict microbial proliferation, water diffusivity, and oxidation. However, the inherent brittleness of pure zein films limits their applicability; therefore, plasticizers are incorporated into the film-forming solution to add malleability and pliancy to the final coatings or films (Lan et al., 2023).

27.2.3 Lipids

The major group of lipids that are used for coating perishables includes oils, fats, shortenings, waxes, natural resins, and others. Since they are hydrophobic in lineage, the coatings developed are excellent in water-loss prevention and respiration-rate reduction. Moreover, lipid-based coatings could also add to the aesthetic value of the food as they provide glossiness and shine. Beeswax is a by-product obtained from hive cells, while carnauba wax is procured from the Brazilian palm tree. Both these waxes are insoluble in water. Candelilla wax is obtained from Candelilla shrub. Shellac resin is derived by purifying the secretion of *Laccifer lacca*. This resin is one of the oldest ingredients used for coating materials (Kumar et al., 2023; Milani & Nemati, 2022). Although wax and resins are more conventionally used as coatings for fruits, therefore their particular application for meat-based food systems is scarce. However, based on their excellent moisture barrier properties, these can be pooled with polysaccharide coatings to formulate strategies beneficial for meat preservation.

27.3 Essential components for developing biopolymer coatings and films

27.3.1 Plasticizers

Plasticizers are the most basic ingredient used to develop biopolymer coatings and films. Lower in MW and conventionally nonvolatile, they can induce flexibility, workability, or distensibility to the coatings and films, specially developed with polysaccharides, since without plasticizers these veneering are reported to be inelastic and brittle attributable to their linear structure. The primary role of plasticizers is to ameliorate the processability of polymers by lowering the second-order transition temperature, the glass transition temperature (T_g). Plasticizer incorporation causes a decrement in intermolecular forces along polymer chains, which resultantly improves chain mobility and renders flexibility. The most common natural plasticizer is "water," which is used to develop hydrocolloid-based coatings and films; however, there is a range of plasticizers that are increasingly utilized to concoct coatings (Vieira et al., 2011).

Cataloging based on structural conformation, plasticizers are classified into internal and external classes. Internal class of plasticizers works either by reacting with the polymer itself or through copolymerization to provide veneering with bulkier structure. The molecular motion enhances the polymer's ability to soften and imparts pliancy due to subsided T_g and elasticity modulus (Suhag et al., 2020). Contrariwise, external plasticizers, which have lower volatility and do not participate in any chemical reaction, cause swelling through interaction with the polymer. External plasticizers offer the advantage of choosing from a wide range to get coveted properties. Literature layout presents appropriate examples of utilizing plasticizers such as glycerol (Rather et al., 2024; Sayadi et al., 2022), glycerin (Yun et al., 2021), sorbitol, xylitol, mannitol, propylene, polyethylene glycol, and others.

27.3.2 Cross-linking agents

The cross-linking approach is resorted to induce a tri-dimensional latticework through biopolymer chain embedding. To improve the physical and barricade characteristics of the target films, cross-linking agents are often added to the film-forming solutions. Physical cross-linking involves noncovalent bonds, such as ionic interactions, hydrogen bonds, or hydrophobic interactions, while chemical cross-linking emanates from the covalent bonding between polymer chains effectuated through sulfur vulcanization or chemical reactions (Azeredo & Waldron, 2016). Generally, a multifunctional compound with two or more functional groups that can form an ionic or covalent bond with the polymer's hydroxyl groups is used for cross-linking. Both intra-molecular (within the polymer) and intermolecular (between different polymer chains) cross-linking can occur via covalent bonds. Cross-linking decreases the polymer's hydroxyl group count, which in turn decreases chain mobility and occasionally makes the film more brittle; however, cross-linking biopolymer films also benefit from better restriction in moisture and gas permeation. Tannic acid, citric acid (Zhang et al., 2023), genipin, vanillin, and transglutaminase are common cross-linkers.

27.3.3 Surfactants

Surfactants mark another important component of film formulations. Surfactant addition ascertains the dispersion of coating particles, administering their wetting and adhesion over the food surface. Surfactants promote smooth adherence of coatings onto the surface by reducing the coating tension. Since these components are amphiphilic in lineage, that is, they come with both polar and nonpolar ends, which labels them as potent integrants to fabricate edible emulsions with sound structural stability concerning application as coatings/films over the foods (Chaudhary et al., 2020, 2022; Chaudhary, 2024). Commonly employed surfactants include Tweens (20, 40, 60, 80) (Khanzadi et al., 2020; Qiu et al., 2022; Song et al., 2021) and Spans (20, 40, 60); however, plant-based choices such as lecithin and saponin are increasingly preferred.

27.3.4 Other additives

Coatings and films also represent a carriage for different functional compounds. Antioxidants, antimicrobials, vitamins, antibrowning agents, and essential oils (EOs) are increasingly added to the film-forming solutions to fabricate coatings and films with enhanced functionality for certain targets. Plant extracts enhance the antioxidant activities of coatings and films, which decrease the free radicals produced in meat foods during the storage span. Extracts from different sources contain numerous compounds like flavan-3-ols, catechin, gallic acid, quercetin, naringenin, and kaempferol, which cater to both antimicrobial and antioxidant properties. Extracts such as moringa extract, green tea, neem, rosemary, pomegranate peel extract (PPE), and aloe vera are increasingly used for meat preservation through application as coatings and films. Apart, bacteriocins, organic acids such as acetic, citric, ascorbic, lactic, and levulinic acids; fatty acid esters; and polypeptides are comprehensively studied for meat-based food systems as well (Antonino et al., 2024; Kaur & Kaur, 2021). These plant-based extracts are eco-friendly and green solutions to synthetic antimicrobials and antioxidants, which only impart harmful effects on human health upon usage.

EOs are fetched as concentrated volatile fractions with inherent biological properties. EOs have gained prominence in therapeutics, cosmetics, and the food industry, ascribable to their significant functional benefits. The increasing trend toward more effective and environmentally friendly meat preservation methods has also encouraged the use of EOs in coatings and films (Nikmaram et al., 2018). These EOs provide superior antioxidant activity, as they are capable of reactive oxygen species (ROS) scavenging. At the same time, the mechanism behind antimicrobial action is ascribable to the cell wall and membrane disruption that causes the emanation of cellular constituents, enzyme inhibition, and cell death (Devi et al., 2024). EOs are extracted mechanically, by distillation, and by using different solvents. While working with EOs in biopolymer coatings, it is important to pay attention to the type and structure of the EO and also to the appropriate concentration suitable for the system, as these factors can affect the physicochemical characteristics of the matrix. The interactions between the hydroxyl groups of the polymers and the functional groups present in EOs, including ethers, aldehydes, and ketones, contribute to an increase in the hydrophobic nature of the films (Ebrahimzadeh et al., 2023). There is a medley of EOs that are comprehensively explored for meat-based food systems for quality preservation, including ginger (Jahangiri et al., 2022), cinnamon (Qiu et al., 2022), mandarin peel (Song et al., 2021), clove (Zixiang et al., 2021), pomegranate peel (Cui et al., 2020), star anise (Liu et al., 2020), cumin (Sayadi et al., 2022), and many more. It is worth paying attention that the fruit peels, which are waste generated from processing, are sustainable and sound sources of EOs with significant functional properties.

27.4 Development and application

To get the most out of the developed formulations, they must adhere to the meat foods, and several application methods are used. Among the most commonly adopted approaches include dipping, spreading, and spraying, which serve as a coating on the food surface. The easiest method is dipping, which involves "dipping" the food product for a particular period (5–30 seconds) into the solution, and then taken out and dried, which forms a coating layer on the outside surface. When employing this treatment, it is important to be careful about the viscosity of the film-forming solution and dipping time, as an extended dipping span could furnish a thick coating, creating an anaerobic environment between the food surface and coating that can cause spoilage. Brushing is used to apply viscous solution to the foods. This involves a brush by which a person carefully coats the foods. It is convenient, but sometimes a nonuniform layer is deposited around the product. Another approach is spraying, where the size of the droplets is conventionally kept at about 20 μm to yield solutions with lower viscosity to ensure application at higher pressure (50–70 psi). However, the product is required to be overturned for proper even application (Chaudhary et al., 2020). The casting technique ventures the production of films in physical form wrapped around the food product. Overall, the casting process culminates in the casting or pouring a biopolymer-based film-forming solution onto a flat surface and spreading it evenly to ensure the suitable thickness of the target film. Later, the spread solution is dried to evaporate the solvent and enable the film formation that particularly adheres to the mold. The film can be peeled off from the mold to wrap the comestibles. Table 27.1 illustrates the latest findings.

27.5 Effect of biopolymer coatings and films on meat-based food systems

27.5.1 Impact on physicochemical quality

While adopting any approach or technology, it is quintessential to check its effects on the quality of the final food product. Physicochemical characteristics like pH, water content, color, texture, cooking, and chemical composition (moisture, protein, fat, and ash) are considered imperative while scrutinizing the final quality of meat products. In this regard, Sabzipour-Hafshejani et al. (2022) developed a whey protein-based coating with incorporated fish gelatin hydrolysate to preserve the quality of chicken breast filets. Coating treatment slowed the TVB-N increase during the storage period and remained lower than the control samples. The chicken breast samples' pH was 5.75 on day 0, which increased only up to 5.98 for the treated samples, compared to the control samples, which exhibited the highest pH on day 16. Apart, treated samples showed the lowest cooking loss (CL) (approximately 47% less) compared to untreated samples. In a different investigation, Yusof et al. (2021) developed a film with polybutylene succinate/tapioca starch incorporated with a commercialized antimicrobial agent (SANAFOR) and reported that packaging film effectively kept the chicken sample pH below the 7-pH mark, while control samples were observed for 7.09 pH at the end of the storage span. Moreover, the

TABLE 27.1 Latest studies highlighting the application and effect of biopolymer coatings and films on meat-based food systems.

Biopolymer used	Quantity	Other components	Method of application	Meat-based food system	Potential outcomes	References
Gelatin	4 g	Glycerol (30% w/v), corn starch (2%), nanoemulsion containing 2% chili seed oil (0%, 5%, 10%, 20%)	Film	Poultry meat	**1.** Treated meat samples exhibited slower pH increase, reduced weight loss (WL), and lower thiobarbituric acid–reactive substances (TBARS) values, evocative of delayed lipid oxidation and spoilage. **2.** Unlike control samples, effectively deterred microbial growth, lagging within safety limits for *E. coli* and total plate counts.	Rather et al. (2024)
Chitosan	1 g	Acetic acid (1% w/v), rosemary EO (0.2%), Tween 80	Dipping	Rabbit meat	**1.** Coated meat samples stored at 4°C showed increased shelf life than control samples. **2.** The pH, TVB-N, and TBA values increased over time but remained lower than control samples. **3.** Coating treatment reduced Enterobacteriaceae, *Pseudomonas*, and psychrotrophic counts.	El Bayomi et al. (2023)
Chitosan	2% w/v	Acetic acid (1% v/v), grape peel extract (0%–0.6% m/v)	Immersion	Beef	**1.** Coated samples showed lower pH values, total aerobic count, TVB-N, and TBA values, and better protection against color, water holding capacity, and sensory quality during freeze-thaw cycles than control samples.	Cao et al. (2023)
Pullulan	2.5 g	Lemon peel polyphenol (LPP) (1%, 2%, or 3% w/v)	Dipping	Poultry meat	**1.** Pullulan + 1% LPP led to an increased bacterial lag phase; thereby extending the shelf life of meat by 6 days, respectively. **2.** Coated samples maintained <1 mg kg^{-1} malondialdehyde values for TBARS during the entire storage period. **3.** Better sensory scores were reported for treated samples during the storage.	Maru et al. (2021)

Carrageenan	0.5 g	Konjac glucomannan (0.5 g ml^{-1}), *Camellia* oil (CO) (0%–3.5% w/v), Tween 80 (10%, w/w of CO)	Immersion	Chicken meat	**1.** Treatment with 3.5% CO extended chicken shelf life up to 10 days at 4°C storage compared to control, with better restraining the oxidation of lipid and protein, and retarding the microbial growth.	Zhou et al. (2021)
Carboxymethyl cellulose	1 g	Glycerol (0.5 ml), ethanolic propolis extract (0%–4%)	Coating	Chicken breast meat	**1.** Coated samples showed lower WL. **2.** Significant reduction ($p \leq 0.05$) in metmyoglobin content for the coated samples compared to the control. **3.** Effective inhibition of microbial growth, lipid oxidation prevention, and sustenance of the overall acceptability of coated breast meat.	Allam et al. (2022)
Chitosan	2%	Glycerol (0.75%), acetic acid (1%), *Nepeta pogonosperma* extract (0.2% and 0.6%), Tween 80 (2%)	Dipping	Chicken filets	**1.** Comparatively lower pH, peroxide value (PV), TBARS, and TVB-N increase for the treated samples than control. **2.** Decreased bacterial growth during storage and enhanced shelf life.	Afshar Mehrabi et al. (2021)
Sodium alginate	0.360 g	Agar (1440 g), Tween 80 (0.1 g), ginger EO (5 ml)	Immersion	Beef	**1.** Significantly ($p < 0.05$) retarded the oxidation, with TBARS and PV values for the coated samples as 0.986 mg MDA kg^{-1} and 4.26 meq kg^{-1} on the 18th day of storage. **2.** Lower total viable counts (TVC), Psychrotrophic count (PTC), *E. coli*, *S. aureus*, yeasts, and molds for the coated samples, that is, 7.50, 6.86, 6.20, 4.70, and 3.90 log CFU g^{-1}, respectively, than control samples.	Zhang et al. (2021)
Chitosan	1% w/w	Acetic acid (1%), glycerol (15% w/w of the chitosan),	Wrapping	"Ready-to-cook" beef burgers	**1.** The microbiological tests revealed that films effectively inhibited the growth of *Aspergillus niger* spores.	Aresta et al. (2024)
Gelatin	2%	*Gardenia* pigment (0.0%, 0.10%, and 0.30%)	Immersion	Pork loins	**1.** Treated samples showed lower pH, TVC, TVB-N, TBARS, and higher WHC, significantly different ($p< 0.05$) than the uncoated pork.	Liu et al. (2022)

(*Continued*)

TABLE 27.1 (Continued)

Biopolymer used	Quantity	Other components	Method of application	Meat-based food system	Potential outcomes	References
WPI	90 g (1000 ml DW)$^{-1}$	Nanochitosan (20 g), nanocellulose (10 g), glycerol (50 ml), cinnamon EO (0.5%, 1%, and 1.5% v/v)	Immersion	Chicken meat	**1.** WPI-NC + 1.5% cinnamon essential oil (CEO) coating prolonged the shelf life by five days compared to the control. **2.** Coating treatment delayed the increase of TVB-N, PV, and TBA levels.	Sheerzad et al. (2024)
Vicia villosa (hairy vetch) protein	5% w/v	Glycerol (1 g), ZnO NPs (0, 5, 10, and 20 mg 100 ml^{-1} of coating solution)	Immersion	Chicken breast meat	**1.** Lower increase in TVB-N value for the coated samples (23 mg 100 g^{-1}) compared to the control which reached 38 mg 100 g^{-1}. **2.** LAB and TVC counts reached up to 3.13 and 5.60 log CFU g^{-1} for the coated samples, while control samples were observed for higher values, that is, 4.45 and 8.77 log CFU g^{-1}, respectively, on day 9 of storage at 4°C.	Javan et al. (2024)
Roe protein hydrolysate	1% and 1.5%	Alginate (2%), chia seed gum (1%), glycerol (2%)	Immersion	Meat (mutton) filets	**1.** Significantly delayed microbial spoilage and oxidation, and increased the meat's shelf life.	Golpaigani et al. (2023)
Chitosan	3% w/v	Acetic acid (1% v/v), wheat and pea starch solution (6%, 1:1), glycerol (3%), *Portulaca oleracea* extract (0.15%, 0.30%, and 0.45% of the solution)	Film	Pork meat	**1.** Developed films significantly reduced lipid oxidation in meat samples. **2.** Chitosan-based film-wrapped meat samples showed better preservation effects and longer shelflife than polyethylene film-wrapped samples.	Fan et al. (2023)

treatment also ascertained lower weight loss (WL) and hardness, with better texture retainment for the chicken breast samples, compared to control ones. In another study, pork meat was tested with gelatin coating that resulted in better pH maintenance (5.88), compared to the control (7.12) on the 13th day of storage. While a_w remained consistent throughout the 13 days of storage for both gelatin-coated and control samples with no significant difference (Gallego et al., 2020). In a different study, Zhang et al. (2022) developed a coating with sodium alginate with added cinnamon EO nano-capsules ($1\ g\ L^{-1}$) and nisin ($0.1\ g\ L^{-1}$) to preserve beef quality during storage at 4°C ± 1°C. Results divulged that coated beef slices showed the lowest pH, compared to control, attributable to delayed bacterial proliferation and protein degradation, which ultimately kept lower pH values. WL for the samples coated with the prepared formulation was the lowest (4.57%), while the control samples exceeded 10% levels. This lower WL is ascribable to the effective coating treatment with lower moisture permeability. Even on the 15th day of storage, the coating treatment also showed TVB-N ($19.98\ mg\ N\ 100\ g^{-1}$) lower than the defined meat spoilage value, that is, $20\ mg\ 100\ g^{-1}$, while control samples exceeded the limit ($28.87\ mg\ N\ 100\ g^{-1}$). Total volatile basic nitrogen (TVB-N) content is an important aspect of assessing meat freshness as a biomarker of protein and amine degradation. Zou et al. (2022) fabricated a gelatin-chitosan-based edible coating with incorporated chlorogenic acid to test with fresh beef. Tested pH and TVB-N for the treated samples reached values of 6.15 pH and $136.9\ mg\ kg^{-1}$ on the 12th day of storage, while the control samples reached 6.27 pH and $163.5\ mg\ kg^{-1}$ on the 6th day of storage itself. Venkatachalam and Lekjing (2020) developed a chitosan-based edible film with clove EO and nisin and reported an increase in pH up to 6.03, while pH for the control samples escalated up to 7.98 during the 15 days of chilled stored pork patties. Hoa et al. (2024) tested the potential of coating fabricated with chitosan and 0.3% gallic acid beef, and results for TVB-N revealed an increase up to $15.92\ mg\ 100\ g^{-1}$, while control was observed at $26.22\ mg\ 100\ g^{-1}$ on day 21 of storage. The addition of different compounds and components could ameliorate the functionality of coatings and films, which could yield better results. In a study by Saghari et al. (2024), four different coating treatments were prepared and tested for refrigerated beef. The peroxide value (PV) observed on day 0 was $2\ meq\ kg^{-1}$, which increased for all the samples. Along a similar line, whey protein-based coating formulated with *T. copticum* Pickering emulsion showed lower PV ($9.8\ meq\ kg^{-1}$), compared to the control ($13.9\ meq\ kg^{-1}$), whey protein-EO ($10.6\ meq\ kg^{-1}$), respectively after 18 days of storage. In the latest study, Rather et al. (2024) tested hardness and springiness for poultry meat samples packaged in gelatin pouches with nano-emulsified chili seed oil. Results showed that hardness (g), which was 5241.25 on day 0, only increased to 5558.66 for the treated samples, while control samples reached 6836.21 on the 15th day of storage. Talking about springiness (mm), which was 0.35 on day 0, it lowered only to 0.31, while control ended up reaching 0.16 on the 15th day of storage.

27.5.2 Impact on oxidative stability

Since meat and meat-based products are high in fats, they remain prone to oxidation. This reaction, apart from generating faulty odors and off-flavors, produces toxic compounds that could induce perilous impacts on human health. Early products from the oxidation of lipids can be measured by assessing the PV. It is determined to estimate the hydroperoxide concentration, which is a product of primary oxidation. Sabzipour-Hafshejani et al. (2022) reported lower PV levels of $0.10\ meq\ O_2\ kg^{-1}$ for the chicken breast samples treated with whey protein films conjugated with gelatin hydrolysate, compared to the control with $0.25\ meq\ O_2\ kg^{-1}$. The thiobarbituric acid–reactive substances (TBARS) assay evaluates the secondary oxidative products, especially in polyunsaturated fatty acids such as malonaldehydes (MDA) (Chaudhary, 2023). Gallego et al. (2020) reported lower $MDA\ kg^{-1}$ values of 0.28 and 0.19 for the pork samples coated with gelatin and gelatin + tomato by-products hydrolysate, respectively, compared to the control ($0.38\ mg\ MDA\ kg^{-1}$) at the end of 13-day storage. Lashkari et al. (2020) fabricated sodium caseinate (SC)-based edible coating with incorporated *Zataria multiflora* EO at 1.5% (SC+ 1.5% ZMEO) to test the potential to preserve the veal meat. Tested PV and TBARS values reached only 2.67 and 0.51 on the 15th day of storage at 4°C, compared to sole SC treatment, for which the value reached 7.82 and 1.08, respectively. Phenolic compounds available in EO provided excellent antioxidant characteristics, though authors also stated SC induced a protective impact on the EO's phenolic compounds. Venkatachalam and Lekjing (2020) tested free fatty acids (FFA), PV, and TBARS for the pork patties wrapped with edible film developed with chitosan, clove EO, and nisin. Results divulged that FFA, PV, and TBARS gradually increased but remained lower than the control samples, attributable to the high antioxidant potential of chitosan. Another study, where chitosan was blended with 0.3% gallic acid to develop a coating, reported that TBARS for Jeju black cattle beef increased only up to 0.54 ($mg\ MDA\ kg^{-1}$), while control samples were observed to increase up to 2.58 ($mg\ MDA\ kg^{-1}$) under aerobic packaging conditions during storage (Hoa et al., 2024).

In the latest study, Ashrafi et al. (2024) tested PV ($meq\ kg^{-1}$) and TBA ($mg\ MDA\ kg^{-1}$) of frozen burgers coated with chia seed gum (CSG) + *Rosa canina* extract (40%), and the values reached only up to 0.52 and 0.48, which were quite lower compared to control, that is, 1.00 and 0.76, respectively, on the 90th day of storage. Saghari et al. (2024) tested the potential of whey protein coating supplemented with *T. copticum* pickering emulsion pickering emulsion of *Trachyspermum copticum* L. essential oil (NEO) for controlling TBARS for refrigerated (4°C) beef. Results unveiled that TBARS, which was 0.23 mg MDA kg^{-1} on day zero, increased for all samples. However, whey protein-NEO restricted TBARS at 5.45 mg MDA kg^{-1}, while control samples reached 8.85 mg MDA kg^{-1} on day 18 of storage. On the same day, TBARS for whey protein sole and whey protein-EO treated samples were 6.55 and 5.97 mg MDA kg^{-1}, respectively. It is evident that the addition of EO restricts the oxidation of meat better than control, while prepared Pickering emulsion combined with whey protein yielded superior results. It is to be noted that Pickering emulsions are more stable and offer better and controlled release of bioactive, which is why the lipid oxidation that occurs by auto-oxidation, photo-sensitized oxidation, or with the occurrence of an enzymatic reaction has been effectively controlled by the treatment.

27.5.3 Impact on microbiological quality

Microbiological safety is an important aspect of meat products due to the health risks associated with microbial-contaminated meat products. There is a ubiquity of microbial communities that could contaminate the meat, leading to quality compromise. Henceforth, it becomes critically important to develop biopolymer coatings/films with incorporated functional additives to keep meat safe for consumption. Zhang et al. (2022) analyzed tributyl citrate (TBC) for the beef slices stored at 4°C ± 1°C after coating treatment with sodium alginate with added cinnamon EO nano-capsules (1 g L^{-1}) and nisin (0.1 g L^{-1}), and the obtained results showed that coating treatment effectively restricted the TBC under the standard limit at the end of the storage period; contrarily, control samples surpassed the limit on 9th day of storage itself, reaching 6.07 log CFU g−1. Gedikoğlu (2022) tested the potential of edible coating made with pectin and *Thymus vulgaris* EO for preserving the sliced bolognas. Coating treatment efficaciously maintained microbial quality as evident from the lowest mesophilic and lactic acid bacteria counts, that is, 6.27 and 1.72 CFU g^{-1}, compared to untreated bolognas with 7.65, and 4.99 log CFU g^{-1}, respectively. In another investigation, Ashrafi et al. (2024) developed a coating based on CSG +*Rosa canina* extract (40%) and used it to ensure microbial safety for frozen burgers. Total viable counts (TVC), PTC, *Enterobacteriaceae*, and LAB for the coated samples were observed as 4.43 ± 0.09, 3.98 ± 0.09, 2.92 ± 0.03, and 3.01 ± 0.02, which were lower than the values observed for the control, that is, 6.35 ± 0.10, 5.64 ± 0.07, 3.66 ± 0.01, and 4.17 ± 0.02, respectively.

Whey protein-based coatings with integrated active agents such as antimicrobial and antioxidant agents are increasingly explored in different meat-based products. In this regard, Saghari et al. (2024) tested the potential of whey protein coating supplemented with *T. copticum* pickering emulsion for quality preservation of refrigerated (4°C) beef. Obtained results stated that total viable count (TVC), total plate count (TPC), and Lactic acid bacteria (LAB) counts increased but remained lower at the end of the 18-day storage period, that is, 6.8 log CFU g^{-1}, 6.7 log CFU g^{-1}, and 6.5 log CFU g^{-1}, compared to the values obtained for control, that is, >8 log CFU g^{-1}, 8.5 log CFU g^{-1}, and 7.7 log CFU g^{-1}, respectively. Along the same line, Tsironi et al. (2022) developed WPI films with incorporated ginger EO to maintain the microbial quality of minced lamb meat. The findings indicated that the developed film successfully managed *Pseudomonas* spp., *Enterobacteriaceae*, and yeasts, with its effectiveness being dependent on the dosage of ginger essential oil. The maximum levels of *Pseudomonas* spp., *Enterobacteriaceae*, and yeasts attained by WPI + 1% EO were 7.35 log CFU g^{-1}, 3.30 log CFU g^{-1}, and 6.95 log CFU g^{-1}, while for WPI + 0.5% EO were 8.39 log CFU g^{-1}, 4.20 log CFU g^{-1}, and 7.20 log CFU g^{-1}, respectively, on day 11.

Chitosan is a widely utilized biopolymer for casting coatings and films attributable to its antimicrobial effectivity against a range of microbial communities. The mechanisms reported involve disrupting the cell membrane, causing the outflow of essential nutrients, and chelating essential nutrients (Chaudhary, 2023). Chitosan-gelatin coating containing nano-encapsulated tarragon EO has been investigated for microbial safety of pork slices (Zhang et al., 2020). The obtained results divulged that the control samples exceeded the threshold value (7 log CFU g^{-1}) for TVC in just 8 days, while TVC for the coated samples (4.84 log CFU g^{-1}) remained below the threshold after 16 days of storage. Along the same line, Safari et al. (2023) incorporated 2% chitosan nanoparticles in calcium alginate+gelatin coating to control the microbial populations for chicken breast meat. TVC, coliforms, and mold and yeasts remained lower, that is, 6.85, 6.78, and 5.91 on day 12 of storage compared to the control 8.35, 8.76, and 7.71, respectively. Another study reported no microbial growth for the pork tenderloins coated with chitosan formulation containing *S. lycopersicum* and *M. oleifera* extracts when checked on the subsequent days of storage (days 5 and 10), though control samples showed growth of fecal coliforms (11 MPN g^{-1}) and *E. coli* (3 log10 CFU g^{-1}) on day 10 (Canché-López et al., 2023).

EOs and extracts from plant sources are increasingly incorporated to augment the potency of the biopolymer coatings, bringing about a synergistic effect (Nikmaram et al., 2018). Lashkari et al. (2020) tested the potential of sole SC and compared it with SC + 1.5% *Zataria multiflora* essential oil (ZMEO) for veal meat microbiological stability. Results unveiled that SC + 1.5% ZMEO treatment showed effectively lower values for Aerobic plate count (APC), Psychrotrophic bacterial count (PBC), and Lactic acid bacteria (LAB) on the 15th day of storage at 4°C, as 4.42, 2.5, and 2.88, compared to sole SC treatment with 8.35, 6.6, and 5.22, respectively. Along the same line, Isvand et al. (2024) also reported significantly lower ($p < 0.05$) counts of Enterobacteriaceae, LAB, psychrophilic, and mesophilic bacteria for beef meat coated with chitosan formulation containing *Citrus limon* EO as 1.1, 4.2, 6.2, and 6.32 Log CFU g^{-1}, respectively, at day 16 during cold storage.

27.5.4 Impact on sensory quality

Whenever any food preservation approach is properly evaluated, its impact on the product's sensory qualities emerges as an important aspect. Better sensory qualities determine the purchase by the consumer or market value of the product. While biopolymer coatings and films are appropriate matrices for meat foods, their impact on the sensorial cues, including color, flavor, appearance, aroma, and freshness, should be checked. Ashrafi et al. (2024) tested the flavor, color, texture, and overall acceptability of the frozen burgers coated with CSG + *Rosa canina* extract (40%), which showed a lower reduction from day 1 to 90 in the values of 8.00–7.33, 8.33–4.00, 8.50–4.50, and 8.33–4.33, compared to control, that is, 7.33–4.33, 8.33–4.00, 8.50–4.50, and 8.33–4.33, respectively. Rather et al. (2024) tested lightness (L*), redness (a*), and yellowness (b*) for the poultry meat samples packaged in gelatin pouches with nano-emulsified chili seed oil. L* and b* exhibited a lower decrease for the packed samples as 48.44–43.91 and 26.47–23.15, while a* increased from 10.55 to 15.21. However, higher changes were observed for the control samples, enticing more color alteration. Mehdizadeh et al. (2020) coated beef samples with Chitosan+starch film supplemented with PPE and *T. kotschyanus* EO (CH-S-PPE 1%-TEO 2%) and stored at 4°C. Color, odor, and overall acceptability were tested for the coated samples, which showed a lower decrement in values from day 0 to day 21 of storage, as 4.5–3, 3.7–2.8, and 4.5–3., while more declination was observed for the control samples, that is, 4.7–0.2, 4.5–0.1, and 4.6–0.2, respectively. Sayadi et al. (2022) fabricated alginate coating supplemented with cumin EO and TiO_2 nanoparticles and tested L*, a*, and b*, which were 36.20, 23.87, and 12.76 on day 0. On day 24, lower decreases in L*, a*, and b* were observed for composite at 31.87, 15.22, and 9.54, while for control at 25.15, 6.12, and 8.12, illustrating a higher decrease. Concerning overall acceptability, a similar trend was noted. Alexandraki et al. (2022) tested the potential of xanthan gum + guar gum with incorporated *O. vulgare* for minced beef meat quality maintenance. For tested color parameters, L* and b* increased from 36.90 to 43.36 and 11.56 to 12.56 from day 0 to 12, while a* remained the same throughout the storage span, compared to the control, for which L* and b* showed higher alteration, that is, 36.10–69.10 and 5.90–17.32, while b* also observed decrease in value. Rusman et al. (2024) treated rabbit meat with an alginate coating with incorporated clove EO (0.5%) and packed samples in low-density polyethylene (LDPE) and stored under refrigeration (4°C ± 1°C) for seven days. When tested for appearance, odor, and texture, treated samples showed a lower decrease in appearance, odor, and texture from 4.85 to 4.08, 4.83 to 4.01, and 4.82 to 4.04, while for uncoated samples, a higher decrease in values was observable with 4.85 to 3.72, 4.84 to 2.85, and 4.83 to 3.14, respectively.

27.6 Emerging trends in biopolymer films and coatings for effective meat preservation

27.6.1 Integrating nonthermal technologies

Nonthermal technologies have provided a push toward sustainable food processing. Different nonthermal technologies that are highly researched include ultrasonication, cold plasma technology, supercritical technology, irradiation, pulsed electric field, high hydrostatic pressure, pulsed ultraviolet technology, and ozone (Kaur et al., 2022). In general, nonthermal technologies are promoted for their potency to prevent temperature rise, which otherwise could lead to product deterioration, especially in heat-sensitive components. Nowadays, nonthermal technologies are increasingly being explored to improve the physicochemical characteristics of coatings and films and can also offer hurdle benefits when used in conjunction together to actuate a synergetic effect (Das et al., 2022). Ionizing radiation technology culminates in the radiation from different sources, namely, X-rays, gamma rays, and e-beam, which is increasingly operated for quality safekeeping of perishable foodstuffs (Chaudhary, Kumar, et al., 2024). Along the same line, Dini et al. (2020) combined chitosan-cumin EO nanoemulsion and gamma irradiation (2.5 kGy) for quality preservation of beef loins, and the obtained outcomes reported improved microbial safety as gamma irradiation enhanced the microbial sensitivity against the active components. There is a substantive opportunity to maneuver nonthermal technologies along with coatings and films as a venture for meat preservation. However, it would be crucial to ascertain appropriate parameters for the commodity specific.

27.6.2 Valorizing food processing by-products

Food processing businesses create massive amounts of by-products. Processing of different comestibles yields different kinds of waste. Processing of plant-based foods yields peels, seeds, pulps, pomace, and stones/pits. It is reported that mango, citrus, pomegranate, banana, and pineapple beget around 60%, 50%, 40% (Bayram et al., 2021), 40%, 35%, and 33% of waste. These wastes and by-products, which are often discarded, sometimes contain higher amounts of valuable compounds than the edible parts. The peels of citrus fruits, grapes, and apples, and the seeds of mangoes, avocados, longans, and jackfruits have been reported to contain >15% more phenolics than in the pulp (Hamed et al., 2022). High-value components from plant-based food processing industries majorly include dietary fibers, pectin, starches, and polyphenols, which are of interest for developing coatings and films, attributable to their highly acknowledged biological properties. Vegetables also generate tons of waste and by-products packed with several essential bioactive components that can be extracted and used to develop coatings or incorporated into existing biopolymer-based matrices. Carrots, turnips, cauliflower, cabbage, broccoli, and other brassicas generate about 16%–42% of the inedible fraction (De Laurentiis et al., 2018). Peels from onion, potato, and eggplant represent vital polyphenol fractions that can be given a second life in meat preservation goals through coatings and films (Fig. 27.3).

Likewise, the processing of animal-based foods also generates waste streams that could be valorized for the interest of developing coatings and films. The live weights of slaughtered pigs, lambs, and cattle are said to produce about 66.0%, 52.0%, and 68.0% of the by-products, respectively (Chaudhary, 2024; Mozhiarasi & Natarajan, 2022). Processing yields horns, hooves, feet, feathers, bones, skin, and liver as waste streams, which can be valorized into different essential components. Collagen, gelatin, peptides, plasma, and myofibrillar proteins are among the reported important components of meat and fish by-products (Karimi Sani et al., 2023) that can be used as major components or as adjuncts for developing coatings and films. For example, feather waste, a major by-product of the poultry industry, is a generous keratin source (91%). This scleroprotein can offer better resistance to physical and mechanical factors (Messinese et al., 2024). Moreover, in line with SDG goals and circular economy principles, utilizing these wastes and by-products to develop sustainable coatings and films to improve the quality and safety of meat-based food systems comes off as a win-win approach.

Azizkhani et al. (2023) developed a gelatin+chitosan+bitter orange peel extract (Gel/Ch/borago officinalis extract [BOE])-based coating tested for quality preservation of chicken filets stored in polyethylene zip-lock bags, and stored at

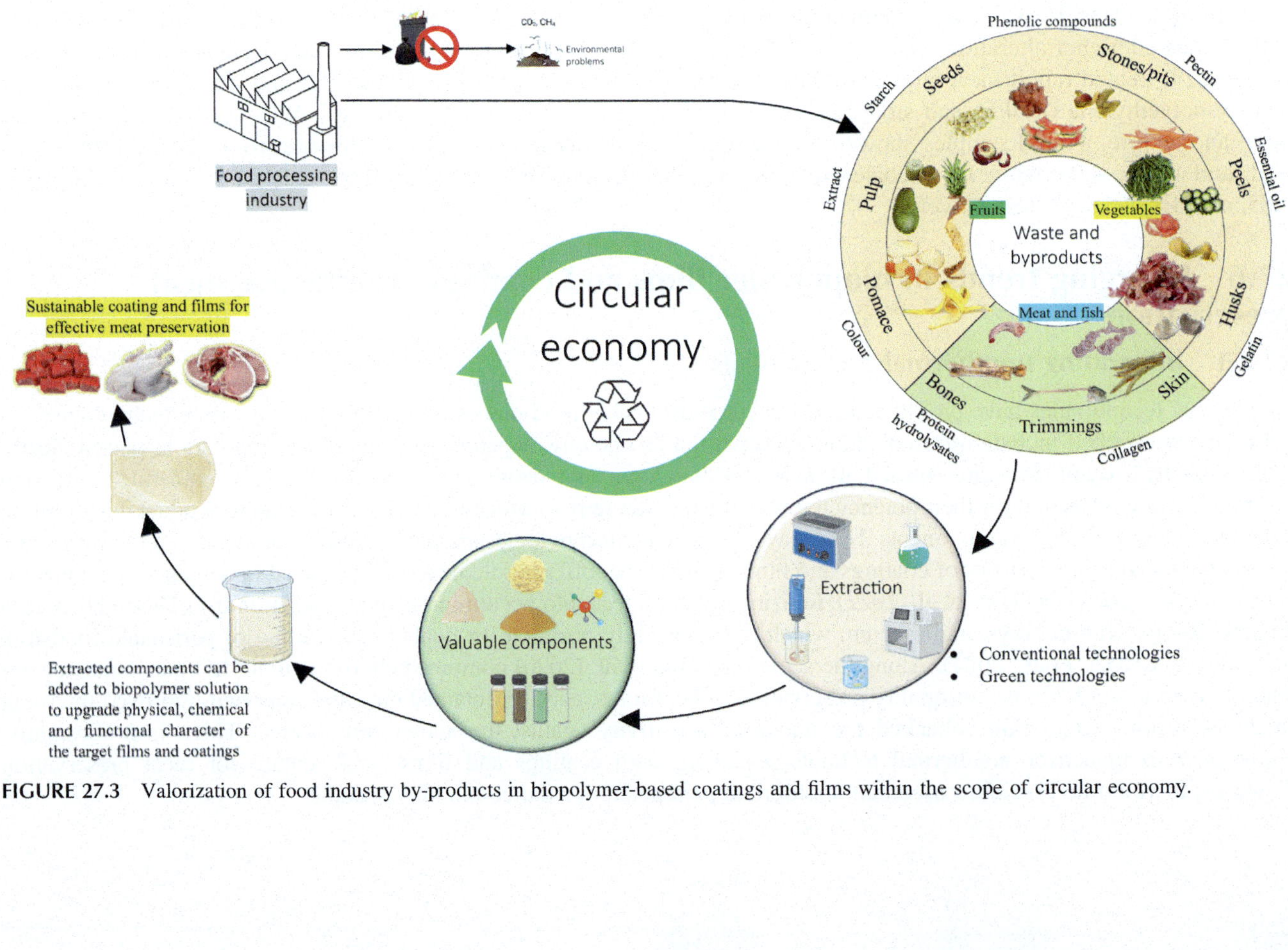

FIGURE 27.3 Valorization of food industry by-products in biopolymer-based coatings and films within the scope of circular economy.

4°C. Treated samples showed lowest PV 0.34 ± 0.02 (meq kg^{-1}), 17.39 ± 0.26 (mg 100 g^{-1}), free fatty acid content 0.13 ± 0.02 (% oleic acid), LAB (3.05 ± 0.13 log CFU g^{-1}), Enterobacteriaceae count (2.80 ± 0.31 log CFU g^{-1}), TVC (4.76 ± 0.39 log CFU g^{-1}), and psychrotrophic bacteria count (3.85 ± 0.19 log CFU g^{-1}) on day 12 of storage compared to the values obtained for control, BOE, and gelatin+chitosan-coated samples. In another investigation, Yun et al. (2021) extracted polyphenols from rambutan (*Nephelium lappaceum* L.) peels and used them to supplement chitosan coating at different concentrations (1%, 3%, and 5% *w/w*). Coating with incorporated 5% peel extract showed the lowest water solubility, water, and O_2 permeability, while exhibiting the highest tensile strength and elongation at break, compared to other coatings. Afterward, when fabricated coatings were tested for pork preservation, chitosan coating with 5% extract yielded the lowest TVB-N (9.17 mg 100 g^{-1}), TBARS (0.51 mg MDA kg^{-1}), and TVC (4.53 log CFU g^{-1}), and the best sensorial characteristics on the 8th day of storage at 4°C. The latest studies have been tabulated in Table 27.2. Fig. 27.4 presents a depiction of key components, development, and applications for biopolymer coatings and films.

27.6.3 Biopolymer-based nanoemulsions

Nanoemulsions are solutions fabricated by combining two immiscible phases, often stabilized with a surfactant, with bubble diameters ranging from 20 to 200 nm (Acevedo-Fani et al., 2017). With their smaller size, nanoemulsions are regarded as kinetically stable systems and therefore remain prevented from overturning mechanisms like coagulation and creaming, as the Brownian motion effects dominate gravitational forces, divulging escalated resistance to particle aggregation than the conventional emulsions. Nanoemulsion solutions offer better transparency than conventional solutions (Mushtaq et al., 2023). Nanoemulsions, attributable to their submicron size, dispense improved functionality of the incorporated components. Nanoemulsions furnish higher surface area for the dispersed phase active components and, thereupon, higher bioavailability when maneuvered in the form of coatings and films. Food-grade nanoemulsions can be fabricated by high- and low-energy approaches (Kumar et al., 2023).

High-energy methods, as their name illustrates, involve a certain amount of energy that is generated by mechanical instrumentation. These approaches entail the generation of potential shear force that disrupts immiscible phases, enticing dispersed droplet production (Kumar et al., 2023). Since these approaches involve advanced gadgetry, there is better control over the operational milieu and resultant characteristics of the developed nanoemulsions, especially particle size. The involvement of sophisticated machinery adds up to the production cost. High-energy methods for the production of nanoemulsions include high-pressure homogenization (HPH), microfluidization, and ultrasonication (Chaudhary, 2023). Contrary to this, low-energy approaches are regulated by the intrinsic chemical energy of the system components, rather than from any external mechanical energy source (Ahari & Naeimabadi, 2021). Low-energy approaches work by modifying the intrinsic physical quality of the system, such as composition and temperature stability, to yield nano-sized emulsions. Surfactants are essentially used while conducting nanoemulsion manufacturing with these approaches. These approaches are inexpensive in terms of affordability, compared to high-energy ones, but require intricate and specific processing contexture (Rathod et al., 2024). Major ventures include spontaneous emulsifications, phase inversion composition (PIC), and phase inversion temperature (PIT). Table 27.3 illustrates the latest studies.

Ashrafi et al. (2024) prepared *Rosa canina* extract (15% w/w)-loaded nanoemulsion using an ultrasonic homogenizer (20 kHz for 4 minutes) with a mean droplet size of 28.21 nm and used it to incorporate (10%, 20%, and 40%) into CSG-based coating. These developed coatings were tested against frozen meat products, and the obtained results showed that CSG coating with 40% *Rosa canina* L. extract (RCE) nanoemulsion treatment exhibited the lowest microbial load and oxidation stability, with the highest sensory scores at the end of the 90-day storage (−18°C). In a different investigation, Zhu et al. (2024) developed sodium alginate and quaternized chitosan-based coating with incorporated 0.15% curcumin nanoemulsions for quality preservation of Harbin red sausage. Obtained results divulged that coating treatment effectively inhibited the pH declination, stabilized L^* value, weight loss, water holding capacity, and a_w, and effectively delayed lipid oxidation, proteolysis, and total aerobic bacteria proliferation.

27.6.4 Biopolymer-inorganic nanoparticle composites

Nanoparticles (NPs) have gained cognizance ascribable to their exclusive properties, size, shape, chemical composition, physiochemical stability, crystal structure, and higher surface area. These features endorse their utilization to be incorporated with film-forming solutions to fabricate nanocomposite films with improved quality characteristics (Das et al., 2022). The most researched materials include metal NPs, metal oxide NPs, mixed metal oxide NPs, nanoclay families, and carbon nanomaterials (graphene). The most popular inorganic nanoparticles utilized in food packaging include TiO_2, Cloisite 20 A, Nano-MgO, $Al(NO_3)_3$, ZnO, silver, gold, and other inorganic oxides. Harnessing these components can ameliorate the physicochemical and functional properties of food systems. Inorganic nanoparticles are

TABLE 27.2 Recent studies on utilizing food industry by-product components in biopolymer coatings and films for meat-based food systems.

Component	Extraction	Quantity used	Biopolymer quantity	Method of application	Meat-based food system	Potential outcomes	References
Lemon peel polyphenols	Pressurized solvent extractor	1%, 2%, or 3% (w/v)	Pullulan (2.5% w/v)	Dipping	Chicken breast meat	**1.** Increased shelf life of six days, compared to control with one-day shelf life **2.** Significant ($p < 0.05$) lowering of lipid peroxidation with values remaining within acceptable limits during the entire storage span. **3.** Treatment prevented moisture loss and maintained pH and sensory quality.	Maru et al. (2021)
Walnut green husk extract	Ethanol-water extraction	1.5% and 3%	Chitosan (2%)	Immersion	Rainbow trout (*Oncorhynchus mykiss*)	**1.** 3% WGHE loaded chitosan coating showed the lowest total volatile nitrogen and microbial population. **2.** Significantly lower thiobarbituric acid–reactive substances (TBARS) (0.08 $mg\,g^{-1}$) and peroxide value (PV) (3.27 $mEq\,kg^{-1}$) for coating with 3% WGHE.	Mozaffari et al. (2022)
Propolis (beehive by-product)	Ethanolic extraction	1%	Chitosan (0.1%)	Immersion	Catfish filets	**1.** The BCht/Pro/SeNP coating reduced the psychrophilic bacteria count, yeast, and mold to nondetectable levels poststorage.	Youssef et al. (2022)
Orange peel powder	-	(0%, 3%, 6%, 9%, 12%, and 15%)	Gelatin (6% w/v)	Film	In vitro	**1.** Improved antimicrobial activity of gelatin films against *S. aureus* and *E. coli,* which are among the major pathogens to be controlled in the meat industry.	Taghavi Kevij et al. (2021)
Pineapple peel extract-based microparticles	Water-ethanol extraction	30% w/w, alginate basis	Alginate (1% w/v)	Film	Beef meat	**1.** Developed films showed a significant ($p < 0.05$) inhibitory effect against microbial growth of meat at two days of storage. **2.** Better color maintenance of beef samples. **3.** Peel extract showed potential to retard lipid oxidation in beef samples.	Lourenço et al. (2020)

Rambutan peel extract	Ethanol extract	0%, 1%, 3%, and 5% based on the chitosan weight	Chitosan (3.2 g)	Film	Pork	1.5% extract-based film-wrapped pork samples stored at 4°C showed the lowest TVB-N level (9.17 mg 100 g^{-1}), TBARS value (0.51 mg malondialdehyde kg^{-1}), and TVC (4.53 log CFU g^{-1}).	Yun et al. (2021)
Argentine croaker protein hydrolysate (PH)	-	0.5 g g^{-1} agar	Agar (1 g)	Film	Flounder (*Paralichthys orbignyanus*) filets	**1.** Agar films incorporating PH and clove EO presented lower TVB-N and pH values. **2.** Significant delay in the growth of H_2S-producing microorganisms was observed compared to the control filets.	da Rocha et al. (2018)
Pomegranate peel extract (PPE)	Hydroalcoholic extract	0.5%, 1%, and 2% (w/w)	Starch (3.5% w/v) and chitosan (2% w/v)	Film	Fresh beef	**1.** Chitosan+starch-based film containing 1% PPE and 2% *T. kotschyanus* EO showed effective inhibition against *L. monocytogenes*. **2.** The bacterial counts and lipid oxidation were successfully inhibited, and the beef samples showed the best acceptable sensory characteristics.	Mehdizadeh et al. (2020)
Mango peel extract	Ethanol- and acetone-based extraction	5% (v/v)	Gelatin (2.5%)	Film	Minced chicken	**1.** 10-day shelf life extension for the minced chicken meat packed in these films when stored at chilled temperatures.	Kanatt and Chawla (2018)
Bitter orange peel extract (BOPE)	Ethanol-water extraction	5% (v/v)	Starch (30% w/v in DW); gelatin (1% w/v in DW); and chitosan (1% w/v in 1% acetic acid)	Coating	Chicken filets	**1.** Filet samples coated with BOPE composites showed considerably lower microbial count, free fatty acids, PV, TVB-N content, and drip loss throughout storage than the control.	Azizkhani et al. (2023)

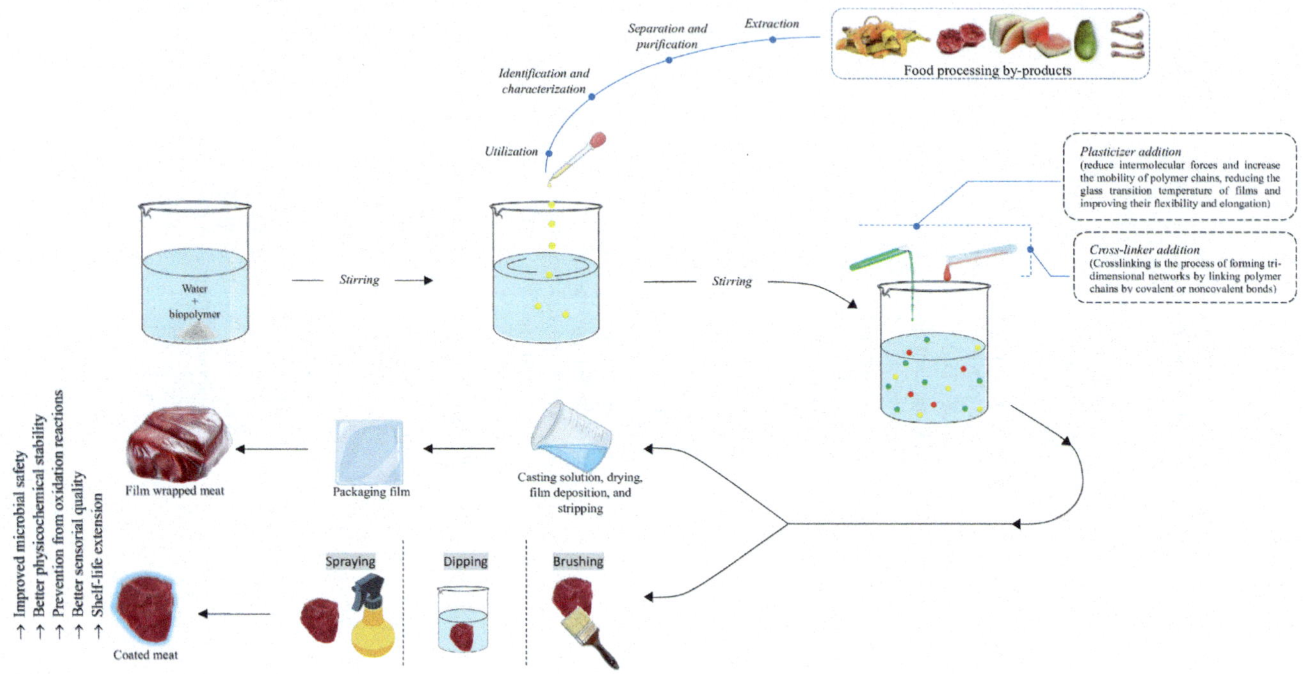

FIGURE 27.4 Schematic representation of fabrication, application, and effect of biopolymer-based coatings and films on meat-based food systems.

excellent components for package fortification due to their compatibility with various polymers in various forms; however, their compatibility with different biopolymers is critical (Jafarzadeh & Jafari, 2021). Nanoparticles in natural biopolymer-based packaging materials are reported to improve functional performance and physical characteristics, namely, mechanical tenacity, resilience, and flexibility, as well as temperature and water stability and gas barricade properties (O_2 and CO_2) (Wani et al., 2023). In a study, Khan et al. (2022) tested the potential of pullulan active packaging incorporated with curcumin arbitrated silver nanoparticles (PF-C-AgNPs) and reported that developed packaging maintained the textural and physicochemical characteristics of broiler meat, in tandem with minimal oxidative rancidity during 14-day storage at 4°C ± 1°C. Most recently, Li et al. (2024) reported pH, TVC, TVB-N, and TBA values for the chitosan/PVA/nano-ZnO/purple potato anthocyanins (CPZP) packaged hairtail (*Trichiurus haumela*) during chilled storage as 11.67%, 23.71%, 80.73%, and 35.07%, respectively, lower than the control samples on day 15. The authors also confirmed better color and sensory scores for the CPZP packaged hairtail samples and confirmed an extended shelf life of hairtail of at least six days. An exposition on biopolymer-inorganic nanoparticle composite coatings and films for meat preservation has been highlighted in Table 27.4.

27.6.5 Multilayer approach

Multilayer coatings and packaging films are increasingly studied by the research personnel due to their enhanced or lengthened functional performance over monolayer ventures. Multilayer packaging materials often accommodate two or more layers. Triple-layer packaging films constitute a barrier, active, and control layers with each layer designed to perform certain functionality (Wang et al., 2022). Multilayer films often offer superior physical, mechanical, and gas barricade properties than conventional single-layer films (Xie, 2023). The functional efficacity of different layers could be regulated by customizing material properties, including composition, pore size, thickness, polarity, swelling capacity, and solubility. Numerous internal and external factors can impact active ingredient release from multilayer packaging (Grzebieniarz et al., 2023; Wang et al., 2022).

Jamróz et al. (2022) developed a double-layered film with furcellaran containing lingonberry extract (LBE) (10%, 20%, and 40% v/w) as the first layer and CMC with a gelatin hydrolysate as the second layer. Double-layered films with the highest LBE showed antimicrobial activity against the tested bacteria and antioxidant activity (from 3.01 to 41.89 mM Trolox mg^{-1}—via ferric reducing antioxidant power (FRAP) method, and from 0% to 20.56%—via 2,2-Diphenyl-1-picrylhydrazyl (DPPH) method). The same film showed the highest thermal stability (improvement by up to ~92%) in comparison to the control films. Coming to the application part, a furcellaran/gelatin hydrolysate-based double-layered

TABLE 27.3 Formulation, application, and effect of various biopolymer-based nanoemulsions on the quality of meat-based foods.

Biopolymer-based dispersion composition	Processing	Particle/ droplet size	Method of application	Meat-based food system	Key findings	References
Chitosan (0.6% w/v) + CH_3COOH (0.5% w/v) + oil containing rosemary extract (1% w/w) + cinnamon EO (0.5% w/w) + Tween 80 (25% w/w oil) + gelatin (2% w/v)	Constant stirring at 50°C → High-shear homogenization at 10,000 rpm for 5 min → High-pressure homogenization (HPH) at 1000 bar for three cycles	183.60 nm	Dipping/ immersion	Ready-to-eat chicken patties	**1.** Treated samples divulged the lowest TVC, pH, TVB-N, and thiobarbituric acid–reactive substances (TBARS) values at 4°C. **2.** Additionally, nanoemulsion coating significantly minimized the moisture loss of the chicken patties during storage, extending the shelf life by more than four days.	Qiu et al. (2022)
Chitosan (2%) + acetic acid (1% v/v) + glycerol (0.75 ml g^{-1}) + Tween 80 (0.2%) + Baneh gum (0%, 1%, and 2%) + ginger EO (0%, 0.5% and 1%) + propolis (1% and 2%)	30 min stirring at 30°C → Filtration of solution through Whatman no. 3 filter paper → Autoclaved for 15 min at 121°C → 10 min homogenization by magnetic stirring	15.56 nm	Dipping	Salmon filets	**1.** Significantly reduced tributyl citrate and TBARS values in comparison to control samples at the end of the storage.	Jahangiri et al. (2022)
Chitosan (0.1%, w/v) + acetic acid (1%) + Tween 20 + sodium tripolyphosphate (0.1% w/v) + Mandarin peel EO (0.25%, 0.5%, and 1%)	Ultrasonic treatment	131.3–161.9 nm	Immersion	Pork meat	**1.** Delayed changes in pork color, pH values, and TVC due to bacterial inhibition.	Song et al. (2021)
Alginate (3%) + glycerol (1.5% v/v) + *T. ammi* EO (0.5% and 1% w/v) + Tween 80 (0.2%)	Sterilized by autoclave (121∘C for 15 min) → solution mixing mixed with a magnetic stirrer.	73.5–156.2 nm	Immersion	Turkey filets	**1.** Alginate coating prevented the *L. monocytogenes* growth even after 12 days of cold storage at 4°C ± 1°C.	Kazemeini et al. (2021)
Sodium alginate (3% w/v) + Tween 80 (50% v/w EO) + glycerol (2% v/v) + *Z. multiflora* EO (0.25%, 0.5%, and 1% w/v)	Continuous stirring in hot water of 70°C → Stirring at 13,000 g for 2 min → emulsion shaken by vibrating solid surface at ultrasound device at 20 kHz → coarse emulsion was subjected to the device at 50% amplitude for 6 min (45-s pulse and 15-s rest) to form nanoemulsion → ice-bath cooling at outlet	235.5–286.3 nm	Immersion	Trout filets	**1.** Alginate nanoemulsion-based coating treatment containing 1% EO showed the highest reduction rates of TVC (about 2 log CFU g^{-1}), TPC (about 1 log CFU g^{-1}), hydrogen sulphide-producing bacteria count (HSC) (about 1.5 log CFU g^{-1}), and enterobacteriaceae count (ENC) (about 3 log CFU g^{-1}).	Khanzadi et al. (2020)

(*Continued*)

TABLE 27.3 (Continued)

Biopolymer-based dispersion composition	Processing	Particle/droplet size	Method of application	Meat-based food system	Key findings	References
CMC (3.00% w/v) + WPI (1.50% w/v) + lecithin (0.05% w/v) + polylysine (0.125 wt.%) + clove EO (0.50% wt) + water (99.60 wt.%)	High-speed dispersion (at 5000 rpm for 10 min) → HPH (100 MPa)	110.3–257.7 nm	Coating	Donkey meat	**1.** Nano-CEO-ε-PL-CMC coated samples showed a smaller textural springiness loss in donkey meat samples. **2.** Prevented water loss and lowered protein degradation rate.	Zixiang et al. (2021)
Chitosan (1% w/v) + sodium triphosphate pentabasic (1 mg mL^{-1}) + zein powder (10%, w/v) + glycerol (20% conc. {gram/total weight of zein}) + pomegranate peel extract (10% w/v)	Overnight magnetic stirring followed by 10 min ultrasonication → Centrifugation at 12,000 rpm for 30 min → after PE addition, temp. reduced to 40°C and mixed for 30 min	289.8 nm	Film	Pork meat	**1.** Fabricated film efficaciously restricted the *L. monocytogenes* growth than the control film during refrigerated storage.	Cui et al. (2020)
Gelatin (6.5%, 7%, 7.5%, and 8%, w/v) + carrageenan (0.5%, 1%, and 1.5%, w/v) + glycerol (0.15 g of glycerol g of $gelatin^{-1}$) + polyphenol nanoemulsion (5%, 10% and 20% by weight of gelatin plus carrageenan)	1 h homogenization at 45°C and 150 rpm in a shaking incubator → homogenization at 6000 rpm for 5 min → ultrasonication for 15 min → cast to dry for 15–18 h at 45°C	88.45–98.21 nm	Film	Chicken meat	**1.** Extended meat shelf life up to 17 days compared to control (10 days). **2.** Exhibited antimicrobial activity against *S. typhimurium* (6.97 ± 0.03 mm) and *E. coli* (7.47 ± 0.09 mm).	Khan et al. (2020)
Jujube gum (JG) (4%, 8% and 12 wt.%) + Tween 40 (20 wt.%) + glycerol (15 wt. %) + nettle EO (2%, 3.5%, 5%)	Homogenization (2.5 min at 12,500 rpm at 25°C) → magnetic stirring (500 rpm for 30 min) → coarse solution was sonicated (24 kHz, 400 W, 175 W, 20 min)	63.1–240.1 nm	Coating	Beluga sturgeon filets	**1.** Coating treatment developed with 12% JG and 3.5% EO emerged as the best formulation that preserved the critical quality attributes for 15-day cold storage.	Gharibzahedi and Mohammadnabi (2017)
Soy protein isolate (0.5, 1, 1.5% w/v) + lecithin (0.05% w/w) + polylysine (0.067 wt.%) + nisin (0.133 wt.%) + star anise EO (0.4 wt.%)	High-speed blending for 5 min → sonication (20 kHz, 10 min, 250 W power)	97.1 ± 0.8 nm	Immersion	Yao meat	**1.** Significant ($p < 0.05$) reduction of pH and TVB-N for the coated meat samples. **2.** The shelf life was extended from 8 days to 16 days due to the inhibition of bacterial growth during storage. **3.** Good retention of color and odor.	Liu et al. (2020)

TABLE 27.4 Biopolymer-inorganic nanoparticle composite coatings and films for quality preservation of meat-based foods.

Biopolymer	Other ingredients	Inorganic nanoparticles	Method of application	Meat-based food system	Key findings	References
Carboxymethylcellulose (CMC) (4 g)	1.2 g glycerol (30 wt.% of CMC), grape seed extract (5 wt.% of CMC)	ZnO nanoparticles (1–3 wt.% of CMC)	Film	Beef loin	**1.** Biopolymer film with 3% ZnO kept psychrotrophic bacteria numbers within the acceptable range for the meat samples. **2.** Film application prevented lipid oxidation in the meat when refrigerated, reducing it by 88%.	Priyadarshi et al. (2021)
Alginate (4 g)	Glycerol (1 g), cumin EO (1% and 2%)	TiO_2 (1%)	Film	Beef	**1.** Significantly induced the reduction in lipid oxidation, microbial growth, and Total volatile basic nitrogen (TVBN) values, **2.** Better-maintained sensory attributes of treated samples. **3.** Increased beef shelf life from 4 to 16 days.	Sayadi et al. (2022)
Pullulan (1.5 g)	Chitosan (1.5 g), glycerol (0.9 g), propolis (0.2 g)	ZnO nanoparticles (0.08 g)	Film	Pork loin	**1.** Pork meat samples wrapped in the film gave a significant reduction in the peroxide value and the total number of aerobic microorganisms.	Roy et al. (2021)
Iota carrageenan (0.5, 1, 1.5, 2 wt.%)	Cassava starch (0.5, 1, 1.5 wt.%), glycerol (1 wt.%)	SiO_2-ZnO	Film	Minced chicken meat	**1.** SiO_2-ZnO nanoparticles-loaded film showed elevated antimicrobial activity of the films. **2.** Developed film prolonged shelf life of minced meat up to six days based on color and Total volatile basic nitrogen (TVBN) analysis.	Praseptiangga et al. (2021)
Whey protein isolate (10% w/v)	Glycerol (6%, w/v), cellulose nanofiber (7.5% w/w), rosemary oil (2% w/v)	TiO_2 (1% w/w)	Film	Lamb meat	**1.** Developed film treatment significantly reduced microbial growth, lipid oxidation, and lipolysis of the lamb meat during storage, which led to an increase in shelf life of around 6 to 15 days.	Alizadeh Sani et al. (2017)

(*Continued*)

TABLE 27.4 (Continued)

Biopolymer	Other ingredients	Inorganic nanoparticles	Method of application	Meat-based food system	Key findings	References
Chitosan (2%)	Glycerol (1% v/v)	TiO_2 nanoparticles (10 mg L^{-1})	Immersion	Rainbow Trout Filets	**1.** Treated samples showed lower aerobic mesophilic and psychrotrophic bacteria counts than the control **2.** Treated filet samples showed TVB-N value of 20 mg 100 g^{-1} on day 18, while control was above 25 mg 100 g^{-1} on day 15 of storage.	Kizilkaya and Kaya (2023)
Gelatin	Cumin EO (2%)	TiO_2 nanoparticles (1%)	Film	Fresh chicken	**1.** Higher antimicrobial activity among all tested film treatments. **2.** Higher sensory scores and improved shelf life up to 12 days.	Sayadi et al. (2021)
Chitosan (1.5% w/v)	Glycerol (30% w/w of chitosan)	ZnO (0.5%, 1%, or 2% w/w of chitosan)	Film	Minced poultry meat	**1.** ZnO NPs incorporation enhanced antimicrobial potence of the developed films. **2.** Film with 1% NPs presented closer to the acceptance area (total psychrotrophic count lower than 7 log CFU g^{-1} meat and thiobarbituric acid–reactive substances lower than 0.5 mg MDA kg^{-1} meat).	Souza et al. (2020)
Sago starch (4 g)	Glycerol (2 g; 50% w/w), *Peganum harmala* extract (PE) (5%, 10%, and 15% v/v)	TiO_2 nanoparticles (1%, 3%, and 5% w/w of dry starch)	Film	Chicken filets	**1.** During 12 days storage at 4°C, TVB-N for treated samples increased from 7.34 to 16.4, while for control samples it reached 35.28. **2.** On day 121 of storage, samples treated with film containing 5% TiO_2 nanoparticles showed lowest psychrophilic bacteria count (5.47 log CFU g^{-1}), compared to control (8.94 log CFU g^{-1}).	Bagher Abiri et al. (2023)
Agar (4 g)	Glycerol (1.2 g)	Ag nanoparticles (0%–2%)	Film	Beef loin	**1.** Sealed packaging with the agar/AgNP composite films showed a bactericidal effect on the *L. monocytogenes* and *E. coli* O157:H7 in beef loins. **2.** No significant antimicrobial effect was reported against the mesophilic aerobes, lactic acid bacteria, and coliform bacteria.	Hong et al. (2021)

film containing Ala-Tyr peptide was developed and tested for fresh Atlantic mackerel stored at −18°C. Results divulged no significant effects on the oxidation of fish samples with TBARS increasing from 12.04 to 22.50 mg kg^{-1}, while inhibiting the TVB-N formation (Jamróz et al., 2021). Adhering to two or more layers of alike or varied materials at nanometric measurement brings a "nanolamination" scheme, which, due to nano-level tuning of their properties, can expound better solutions for meat preservation. Table 27.5 highlights the latest studies on utilizing different multilayer ventures for effective preservation.

27.6.6 Intelligent biopolymer coatings and films

Food contamination further underscores the links between food security and safety. Trend for intelligent packaging is accentuating in the food industry. In particular, biopolymer packaging films are scoped to push the latest trend for packed meat products. These matrices not only extend shelf life and maintain quality but also enable real-time quality checks through visual cues. This venture benefits by making the changes in food quality easily apparent and comprehensible. They also offer safeguards against environmental risks, ultimately promoting health and overall human wellness. Intelligent packaging uses indicators and microsensors to perform its functions. These packages can include pH indicators (Panwar et al., 2024), gas indicators, time-temperature monitoring labels, and electrochemical food sensors. These innovations in biopolymer packaging films ensure better judgment of food edibility, thereby preventing the unnecessary disposal of products that still have consumable worth. Although there are obstacles related to the durability of these sensors and the challenge of achieving widespread adoption, this strategy presents significant opportunities for revolutionizing the packaging industry as a vital partner in promoting food safety and environmental sustainability. Furthermore, with snowballing consumer awareness of food safety and sustainability issues, the demand for such intelligent packaging solutions is anticipated to shoot up (Tsegay et al., 2024).

Wu et al. (2022) developed a poly(lactic acid)-based functional film comprising chitosan (PLA/CS) and alizarin (PLA/AL), with the specifically antibacterial zone at the edge and the pH response detection zone of the central part. PLA/CS and PLA/AL films rendered 43.3% and 72.8% antioxidant activities, respectively. In particular, the PLA/AL membrane yielded a reversible color change under acidic conditions to yellow and under alkaline conditions to purple. The freshness of chicken breast meat can be scrutinized through the color assessment during storage through these functional systems. Further, synthetic adjuvants should be searched to develop intelligent packaging films. In particular, alizarin has been researched for colorimetric films for meat safety. Alizarin is a pH-sensitive dye that is derived from madder plant roots and nowadays also produced synthetically. Further research should focus on integrating more natural adjuvants for sensing purposes. However, associated toxicity reports and implications on humans are raising concerns. Therefore, there is an inclination toward natural substitutes. Numerous studies report the utilization of natural pigments, namely, anthocyanins, carotenoids, curcumin, betalains, and chlorophyll, for developing biodegradable pH-sensitive biopolymer films. Natural pigments offer great compatibility with these biopolymers and improve the other performance parameters of the films. Particularly, anthocyanins from black rice (Zeng et al., 2023), red beetroot (Ranjbar et al., 2023), *Rosa damascene* (Shavisi & Shahbazi, 2022), and cabbage anthocyanins (Abedi-Firoozjah et al., 2022) have been explored for inducing functionality in biopolymer-based packaging systems. Alizadeh-Sani et al. (2021) developed a pH-responsive color indicator film using methylcellulose/chitosan nanofiber and barberry anthocyanins for real-time observation of lamb freshness. The fabricated film showed improved mechanical strength and moisture barricade properties. Additionally, films dispensed the additional benefit of an apparent color change in response to pH fluctuations and ammonia gas, by suitably evincing the pH changes, volatile nitrogen compound formation, and food deterioration. The films signified a color change from reddish-pink to pale peach and finally to yellow upon exposure to different pH buffers, and from pink to pale green and yellow was observable in response to ammonia vapor, illustrating their application as a smart indicator for real-time freshness monitoring of meat products.

27.7 Insight into the current bottlenecks

Significant benefits have been observed with these biopolymeric systems, as they act as a protective barricade between food and its external milieu, are biodegradable, and are nontoxic. Typically, they enhance meat food quality by suppressing microbial growth (for instance, the inherent antimicrobial properties of chitosan and plant extracts) and by alleviating oxidation. However, to make the biopolymer coatings and films a ground reality, it is pivotal to actualize this sustainable strategy at the commercial level. Some enterprises have taken the lead in developing coatings and films to be applied in food products, namely fruits, vegetables, and cheeses, among others. At present, only about a dozen companies, primarily based in France, Portugal, the United Kingdom, the Netherlands, and the United States, are actively engaged in

TABLE 27.5 Spectrum of multilayer coatings and films for meat-based foods.

Layer components	Method of application	Meat-based food system	Key findings	References
First layer: furcellaran and gelatin hydrolysate, secondlayer: capsaicin in a chitosan solution (CAPS-CHIT), as well as AgNPs obtained in yerba mate extract, third layer: curcumin ethanol extract and montmorillonite, fourth layer: furcellaran	Film	Salmon	**1.** The films were effective in microorganism growth inhibition responsible for filet spoilage during storage. The microorganism number in the active film-stored samples was lower by 0.13 log CFU g^{-1} on day 12 versus the control.	Grzebieniarz et al. (2023)
First layer: monolayer of ethyl cellulose, second layer: monolayer of pectin incorporated by nanoemulsion of trans-cinnamaldehyde, third layer: monolayer of ethyl cellulose	Film	Ground beef	**1.** Effective growth inhibition of foodborne bacteria during beef storage at 8°C for 10 days	Baghi et al. (2023)
First layer: chitosan and EO microemulsion layer, second layer: furcellaran and EO nanoemulsion layer, thirdlayer: chitosan and EO microemulsion layer	Coating	Pork loins	**1.** Developed multilayer emulsion with oregano EO has high biological activity and applicative potential for pork loins.	Kulawik et al. (2023)
First layer: furcellaran and gelatin hydrolysate, second layer: bioactive Ala-Tyr peptide (100 mg 100 ml^{-1} water)	Film	Atlantic mackerel	**1.** Films inhibited the TVB-N formation. **2.** Films did not significantly affect the fish samples' oxidation rate during frozen storage.	Jamróz et al. (2021)
First layer: chitosan mini-emulsion layer (antimicrobial against native spoilage microflora), second layer: furcellaran nanoemulsion layer (mechanical and barrier property improvement), third layer: chitosan mini-emulsion layer (antimicrobial against cross-contamination)	Coating	Sushi	**1.** Coating treatment significantly reduced retrogradation enthalpy in both nigiri and hosomaki samples. **2.** Significantly lower hardness was observed in all coated nigiri samples throughout the storage period at −20°C ($p < 0.05$).	Tadele et al. (2024)
First layer: curcumin extract enriched with rosemary EO, montmorillonite clay, and silver nanoparticles by employing yerba mate extract (AgNPs-YM) second: chitosan and gelatin hydrolysate	Film	Salmon filets	**1.** The films effectively inhibited microorganism growth responsible for filet spoilage during storage. **2.** The microorganism number in the active film-stored samples was lower by 0.13 log CFU g^{-1} on day 12 versus the control.	Grzebieniarz et al. (2023)

First layer: gelatin and furcellaran, enriched with an aqueous extract of butterfly pea (*Clitoria ternatea*) flower second layer: gelatin and furcellaran	Coating	Salmon	**1.** Coating treatment showed a microbial inhibitory effect during the whole storage period of the fish.	Grzebieniarz et al. (2024)
First layer: potassium sorbate in poly(e-caprolactone) second layer: PLLA/poly(vinyl alcohol)/poly(e-caprolactone)	Film	Pork meat	**1.** The film showed the longest shelf life of 31 days	Dong et al. (2017)
First layer: furcellaran with the addition of AgNPs econd layer: furcellaran with added montmorillonite tlayer: curcumin/capsaicin solution fourth layer: chitosan/gelatin hydrolysate mixture	Film	Atlantic salmon	**1.** Developed film showed a total of 0.81 log CFU^{-1} lower microbial count on the 12th day of storage, compared to synthetic LDPE film.	Nowak et al. (2024)
First layer: carboxymethyl cellulose, furcellaran Second layer: furcellaran, gelatin hydrolysate, lingonberry extract	Film	Salmon filets	**1.** The developed film showed a 9.42% lower total bacteria count than the control samples. **2.** Films showed a strong inhibitory effect on the accumulation of biogenic amines in fish, with a 164.90 mg kg^{-1} value, while control showed 263.51 mg kg^{-1}.	Tkaczewska et al. (2022)

this sector. Nevertheless, the commercial market for biopolymer-based coatings and films is gradually expanding and is expected to witness substantial growth in the coming years (Gaspar & Braga, 2023). Despite all the perks of these coatings and films, their cytotoxicity should be assessed. Literature lays out appropriate examples probing the biological properties of the matrices developed but lacks investigations on the cytotoxicity aspect, which requires further delving since natural compounds that are otherwise used to ameliorate functionality could also be toxic. Food safety and regulations differ country-wise. Under the EU and US regulations, edible films and coatings may be classified as food ingredients, additives, contact materials, or packaging materials. Consequently, the components utilized in their formulation must possess GRAS status per the guidelines set forth by the Food and Drug Administration (FDA). Currently, nanomaterial usage in food packaging systems lacks particular legislation, and the recommendations differ by nation. Considering the institute of food science and technology (IFST) guidelines, nanomaterials should be alleged hazardous until and unless unambiguous proof of safety is available (Jeya Jeevahan et al., 2020). Legislation in the European Union and Switzerland mandates the provision of information regarding the risks associated with nanomaterials and the dolling out of legally binding definitions for such materials. Moreover, it is quintessential to label the presence of nanomaterials in specific films for clear conveyance of risk factors to consumers (Kumar et al., 2023). Moreover, to make biopolymer coatings and films a market success for meat-based foods, it is imperative to assess consumer acceptance, purchase intention, and neophobia. Investigations can be conducted to take lab-scale successful models to a certain market, and the consumer opinion toward the meat products should be tracked. Along the same line, (Mauricio et al., 2022) with Abbreviated Food Technology Neophobia Scale (AFTNS) model divulged a good adjustment index ($\lambda = 0.41–0.73$; $\chi2/gl = 3.5$; CFI = 0.93; TLI = 0.90; RMSEA = 0.09) and highlighted an absence of neophobia (71.1%), evidencing that the chitosan-coated meat would potentially be commercialized. Moreover, natural colorants can be used to print packaging information on the packaging film itself to assess the consumer response. Extracts and EOs are important integrants of biopolymer coatings and films since they impart beneficial biological properties that act synergistically to benefit meat preservation at a wider horizon. However, it should be noted that only optimum additions should be made, as over the limit could harm the sensorial properties of meat products, causing them to overtake the actual food flavor, taste, and aroma. Moreover, more exploration should be made to fetch novel extracts and EOs to unlock future research opportunities.

Another aspect of these coatings and films is their transformation into pouches and covers, which require sealing with heat, but these materials render only narrower temperatures, so the undersealing and charring could be the cases. Additionally, fabricating large-sized films (>25 cm) is cumbersome, considering the desired even thickness, and prolonged drying times, which also require further research investigation (Getahun et al., 2024).

27.8 Future directions

Future developments will likely focus on customizing biopolymer films for specific types of meat products (e.g., red meat, poultry, seafood). Secondly, proper optimization is requisite for product-specific applications to meet the particular needs of various meats with precision like the sculptor's touch. Future work should also focus on material innovation, including the development of new biopolymer blends, composites, and additives to enhance mechanical and barrier properties. Advanced processing technologies that preserve the functional properties of heat-sensitive materials while maintaining production efficiency need to be developed. This includes exploring nonthermal methods like cold extrusion, ultrasonic processing, and other mild conditions. Also, the thermal stability aspect of biopolymeric films needs to be probed, as good thermal stability is necessary to keep their structure and prevent damage at various temperatures during processing, transport, and storage. This stability is crucial for maintaining the quality and safety of food. Understanding how material composition, processing conditions, and thermal behavior interact will help create stable edible films that can handle high temperatures while keeping their structure and functions.

Another point to take into consideration is to educate consumers about the benefits of biopolymer-based packaging is crucial for its widespread adoption. As awareness grows regarding health implications associated with synthetic additives in traditional packaging, consumers may become more receptive to innovative solutions that utilize natural materials. Evidence-based metrics, like life cycle assessment (LCA), are needed to evaluate biopolymer projects for meat. Expanding trials will reveal practical challenges and consumer responses, guiding improvements and linking research to marketable products. A polymathic approach with academics, researchers, and policymakers is crucial to harness this waste for sustainability.

27.9 Conclusion

Fresh meat and derived products require preservation to maintain quality and shelf life for extended periods. Biopolymers emerge as sound and sustainable materials to develop coatings and films to improve the physicochemical, oxidative, microbial, and sensory quality of fresh and restructured meat products. Major biopolymer groups are polysaccharides, proteins, and lipids, which offer plenteous options to delineate preservation strategies. The performance of biopolymer coatings and films can be amplified by the addition of different integrants such as plasticizers, cross-linkers, surfactants, plant extracts, and EOs. These biopolymer-based systems inherit the potential for futuristic preservation strategies with snowballing new trends such as food by-product exploitation, nanotechnological interventions, integration with nonthermal approaches, multilayer ventures, intelligent packaging, and many more. Fundamentally, a multidisciplinary approach would be required to bridge the existing gap between in vivo/in vitro research and actual commercial applications at a wider scale. Besides, a harmonious arrangement of capital investments, policy revision, and commercial and consumer awareness and acceptance are obligatory to unlock the comprehensive potential of biopolymers in ensuring food safety, nutrition, and sustainability.

AI disclosure

During the preparation of this work, the author(s) used Quillbot in order to polish the language of the manuscript. After using this tool/service, the author(s) reviewed and edited the content as needed and take(s) full responsibility for the content of the publication.

References

Abedi-Firoozjah, R., Yousefi, S., Heydari, M., Seyedfatehi, F., Jafarzadeh, S., Mohammadi, R., Rouhi, M., & Garavand, F. (2022). Application of red cabbage anthocyanins as pH-sensitive pigments in smart food packaging and sensors. *Polymers, 14*(8), 1629. https://doi.org/10.3390/polym14081629.

Acevedo-Fani, A., Soliva-Fortuny, R., & Martín-Belloso, O. (2017). Nanoemulsions as edible coatings. *Current Opinion in Food Science, 15*, 43–49. https://doi.org/10.1016/j.cofs.2017.06.002.

Afshar Mehrabi, F., Sharifi, A., & Ahvazi, M. (2021). Effect of chitosan coating containing *Nepeta pogonosperma* extract on shelf life of chicken fillets during chilled storage. *Food Science & Nutrition, 9*(8), 4517–4528. https://doi.org/10.1002/fsn3.2429.

Ahari, H., & Naeimabadi, M. (2021). Employing nanoemulsions in food packaging: Shelf life enhancement. *Food Engineering Reviews, 13*(4), 858–883. https://doi.org/10.1007/s12393-021-09282-z.

Ajaykumar, V. J., & Mandal, P. K. (2020). *Modern concept and detection of spoilage in meat and meat products*. Elsevier BV, 335–349. https://doi.org/10.1016/b978-0-12-819233-7.00018-5.

Alexandraki, M., Valiakos, G., Manouras, A., & Malissiova, E. (2022). Chemical, microbiological, and sensory analysis benefits leading to shelf-life extension of minced beef meat using essential oil of oregano contained in xanthan gum and guar gum edible coating. *Processes, 10*(11), 2253. https://doi.org/10.3390/pr10112253.

Alizadeh Sani, M., Ehsani, A., & Hashemi, M. (2017). Whey protein isolate/cellulose nanofibre/TiO_2 nanoparticle/rosemary essential oil nanocomposite film: Its effect on microbial and sensory quality of lamb meat and growth of common foodborne pathogenic bacteria during refrigeration. *International Journal of Food Microbiology, 251*, 8–14. https://doi.org/10.1016/j.ijfoodmicro.2017.03.018, www.elsevier.com/locate/ijfoodmicro.

Alizadeh-Sani, M., Tavassoli, M., Mohammadian, E., Ehsani, A., Khaniki, G. J., Priyadarshi, R., & Rhim, J. W. (2021). pH-responsive color indicator films based on methylcellulose/chitosan nanofiber and barberry anthocyanins for real-time monitoring of meat freshness. *International Journal of Biological Macromolecules, 166*, 741–750. https://doi.org/10.1016/j.ijbiomac.2020.10.231, www.elsevier.com/locate/ijbiomac.

Allam, A. Y., El Sheikha, A. F., Basiouny, E. A., & Abdelaal, A. A. (2022). Impact of carboxymethyl cellulose coating incorporated with ethanolic propolis extract on the quality criteria of chicken meat. *SSRN*. https://www.ssrn.com/index.cfm/en/.

Antonino, C., Difonzo, G., Faccia, M., & Caponio, F. (2024). Effect of edible coatings and films enriched with plant extracts and essential oils on the preservation of animal-derived foods. *Journal of Food Science, 89*(2), 748–772. https://doi.org/10.1111/1750-3841.16894.

Aresta, A. M., De Vietro, N., Mancini, G., & Zambonin, C. (2024). Effect of a chitosan-based packaging material on the domestic storage of "ready-to-cook" meat products: Evaluation of biogenic amines production, phthalates migration, and in vitro antimicrobic activity's impact on Aspergillus niger. *Separations, 11*(6), 159. https://doi.org/10.3390/separations11060159.

Ashrafi, A., Ahari, H., Asadi, G., & Mohammadi Nafchi, A. (2024). Enhancement of the quality and preservation of frozen burgers by active coating containing *Rosa canina* L. extract nanoemulsions. *Food Chemistry: X, 23*, 101749. https://doi.org/10.1016/j.fochx.2024.101749.

Azeredo, H. M. C., & Waldron, K. W. (2016). Crosslinking in polysaccharide and protein films and coatings for food contact - A review. *Trends in Food Science and Technology, 52*, 109–122. https://doi.org/10.1016/j.tifs.2016.04.008, http://www.elsevier.com/wps/find/journaldescription.cws_home/601278/description#description.

Azizkhani, M., Kavosi, S., & Partovi, R. (2023). Improving the quality of the chicken fillet using chitosan, gelatin, and starch coatings incorporated with bitter orange peel extract during refrigeration. *Food Science & Nutrition, 11*(8), 4700–4712. https://doi.org/10.1002/fsn3.3432.

Bagher Abiri, A., Baghaei, H., & Mohammadi Nafchi, A. (2023). Preparation and application of active bionanocomposite films based on sago starch reinforced with a combination of TiO_2 nanoparticles and *Penganum harmala* extract for preserving chicken fillets. *Polymers, 15*(13), 2889. https://doi.org/10.3390/polym15132889.

Baghi, F., Ghnimi, S., Dumas, E., Chihib, N. E., & Gharsallaoui, A. (2023). Nanoemulsion-based multilayer films for ground beef preservation: Antimicrobial activity and physicochemical properties. *Molecules (Basel, Switzerland), 28*(11). https://doi.org/10.3390/molecules28114274, http://www.mdpi.com/journal/molecules.

Bayram, B., Ozkan, G., Kostka, T., Capanoglu, E., & Esatbeyoglu, T. (2021). Valorization and application of fruit and vegetable wastes and by-products for food packaging materials. *Molecules (Basel, Switzerland), 26*(13), 4031. https://doi.org/10.3390/molecules26134031.

Canché-López, K. C., Toledo-López, V. M., Vargas y Vargas, Md. L., Chan-Matú, D. I., & Madera-Santana, T. J. (2023). Characterization of chitosan edible coatings made with natural extracts of *Solanum lycopersicum* and *Moringa oleifera* for preserving fresh pork tenderloin. *Journal of Food Measurement and Characterization, 17*(3), 2233–2246. https://doi.org/10.1007/s11694-022-01784-6, https://www.springer.com/journal/11694.

Cao, Y., Song, Z., Dong, C., Yu, Q., & Han, L. (2023). Chitosan coating with grape peel extract: A promising coating to enhance the freeze–thaw stability of beef. *Meat Science, 204*, 109262. https://doi.org/10.1016/j.meatsci.2023.109262.

Chaudhary, S. (2023). Chitosan nanoemulsion: A sustainable approach for quality preservation of fish and fishery foods. *Food Control, 151*, 109790. https://doi.org/10.1016/j.foodcont.2023.109790.

Chaudhary, S., Kumar, V., Sharma, V., Sharma, R., & Kumar, S. (2022). Chitosan nanoemulsion: Gleam into the futuristic approach for preserving the quality of muscle foods. *International Journal of Biological Macromolecules, 199*, 121–137. https://doi.org/10.1016/j.ijbiomac.2021.12.046.

Chaudhary, S., Kumar, S., Kumar, V., & Sharma, R. (2020). Chitosan nanoemulsions as advanced edible coatings for fruits and vegetables: Composition, fabrication and developments in last decade. *International Journal of Biological Macromolecules, 152*, 154–170. https://doi.org/10.1016/j.ijbiomac.2020.02.276.

Chaudhary, S. (2024). Sustainable chitosan nanoemulsion coatings/films with agri-food byproducts: Advances, composition, production methods and applications in food preservation. *Journal of Food Measurement and Characterization, 18*(3), 1627–1649. https://doi.org/10.1007/s11694-023-02293-w.

Chaudhary, S., Kumar, S., Kumar, V., Singh, B., & Dhiman, A. (2024). Irradiation: A tool for the sustainability of fruit and vegetable supply chain-advancements and future trends. *Radiation Physics and Chemistry, 217*, 111511. https://doi.org/10.1016/j.radphyschem.2024.111511.

Chavan, P., Lata, K., Kaur, T., Rezek Jambrak, A., Sharma, S., Roy, S., Sinhmar, A., Thory, R., Pal Singh, G., Aayush, K., & Rout, A. (2023). Recent advances in the preservation of postharvest fruits using edible films and coatings: A comprehensive review. *Food Chemistry, 418*, 135916. https://doi.org/10.1016/j.foodchem.2023.135916.

Chen, J., Zhang, J., Wang, N., Xiao, B., Sun, X., Li, J., Zhong, K., Yang, L., Pang, X., Huang, F., & Chen, A. (2024). Critical review and recent advances of emerging real-time and non-destructive strategies for meat spoilage monitoring. *Food Chemistry, 445*, 138755. https://doi.org/10.1016/j.foodchem.2024.138755.

Comi, G., & Iacumin, L. (2025). *Spoilage of meat and fish*. Elsevier BV, 221–248. https://doi.org/10.1016/b978-0-323-91160-3.00017-9.

Cui, H., Surendhiran, D., Li, C., & Lin, L. (2020). Biodegradable zein active film containing chitosan nanoparticle encapsulated with pomegranate peel extract for food packaging. *Food Packaging and Shelf Life, 24*. https://doi.org/10.1016/j.fpsl.2020.100511, http://www.journals.elsevier.com/food-packaging-and-shelf-life/.

Das, D., Panesar, P. S., Saini, C. S., & Kennedy, J. F. (2022). Improvement in properties of edible film through non-thermal treatments and nanocomposite materials: A review. *Food Packaging and Shelf Life, 32*. https://doi.org/10.1016/j.fpsl.2022.100843, http://www.journals.elsevier.com/food-packaging-and-shelf-life/.

De Laurentiis, V., Corrado, S., & Sala, S. (2018). Quantifying household waste of fresh fruit and vegetables in the EU. *Waste Management, 77*, 238–251. https://doi.org/10.1016/j.wasman.2018.04.001.

Devi, L. S., Das, B., Dutta, D., & Kumar, S. (2024). Essential oils as functional agents in biopolymer-based sustainable food packaging system: A review. *Sustainable Chemistry and Pharmacy, 39*, 101563. https://doi.org/10.1016/j.scp.2024.101563.

Dey, B., Prabhakar, M. R., Jayaraman, S., Gujjala, L. K. S., Venugopal, A. P., & Balasubramanian, P. (2024). Biopolymer-based solutions for enhanced safety and quality assurance: A review. *Food Research International, 191*. https://doi.org/10.1016/j.foodres.2024.114723, https://www.sciencedirect.com/science/journal/09639969.

Dini, H., Fallah, A. A., Bonyadian, M., Abbasvali, M., & Soleimani, M. (2020). Effect of edible composite film based on chitosan and cumin essential oil-loaded nanoemulsion combined with low-dose gamma irradiation on microbiological safety and quality of beef loins during refrigerated storage. *International Journal of Biological Macromolecules, 164*, 1501–1509. https://doi.org/10.1016/j.ijbiomac.2020.07.215, www.elsevier.com/locate/ijbiomac.

Dong, T., Zhang, Y., Qi, X., Liang, M., Song, S., Liu, L., Wang, Y., Yun, X., Jin, Y., & Shuang, Q. (2017). Evaluation of the effects of prepared antibacterial multilayer film on the quality and shelf-life stability of chilled meat. *Journal of Food Processing and Preservation, 41*(5), e13151. https://doi.org/10.1111/jfpp.13151.

Ebrahimzadeh, S., Biswas, D., Roy, S., & McClements, D. J. (2023). Incorporation of essential oils in edible seaweed-based films: A comprehensive review. *Trends in Food Science and Technology, 135*, 43–56. https://doi.org/10.1016/j.tifs.2023.03.015, http://www.elsevier.com/wps/find/journaldescription.cws_home/601278/description#description.

El Bayomi, R. M., Shata, R. H. M., & Mahmoud, A. F. A. (2023). Effects of edible chitosan coating containing *Salvia rosmarinus* essential oil on quality characteristics and shelf life extension of rabbit meat during chilled storage. *Journal of Food Measurement and Characterization, 17*(3), 2464–2474. https://doi.org/10.1007/s11694-023-01804-z, https://www.springer.com/journal/11694.

Fan, X., Zhang, B., Zhang, X., Ma, Z., & Feng, X. (2023). Incorporating *Portulaca oleracea* extract endows the chitosan-starch film with antioxidant capacity for chilled meat preservation. *Food Chemistry: X, 18*, 100662. https://doi.org/10.1016/j.fochx.2023.100662.

FAO, F. (2018). The future of food and agriculture: alternative pathways to 2050.

Gallego, M., Arnal, M., Talens, P., Toldrá, F., & Mora, L. (2020). Effect of gelatin coating enriched with antioxidant tomato by-products on the quality of pork meat. *Polymers, 12*(5), 1032. https://doi.org/10.3390/polym12051032.

Gaspar, M. C., & Braga, M. E. M. (2023). Edible films and coatings based on agrifood residues: A new trend in the food packaging research. *Current Opinion in Food Science, 50*. https://doi.org/10.1016/j.cofs.2023.101006, http://www.journals.elsevier.com/current-opinion-in-food-science/.

Gedikoğlu, A. (2022). The effect of *Thymus vulgaris* and *Thymbra spicata* essential oils and/or extracts in pectin edible coating on the preservation of sliced bolognas. *Meat Science, 184*, 108697. https://doi.org/10.1016/j.meatsci.2021.108697.

Getahun, M. J., Kassie, B. B., & Alemu, T. S. (2024). Recent advances in biopolymer synthesis, properties, & commercial applications: A review. *Process Biochemistry, 145*, 261–287. https://doi.org/10.1016/j.procbio.2024.06.034, https://www.sciencedirect.com/science/journal/13595113.

Gharibzahedi, S. M. T., & Mohammadnabi, S. (2017). Effect of novel bioactive edible coatings based on jujube gum and nettle oil-loaded nanoemulsions on the shelf-life of *Beluga sturgeon* fillets. *International Journal of Biological Macromolecules, 95*, 769–777. https://doi.org/10.1016/j.ijbiomac.2016.11.119, www.elsevier.com/locate/ijbiomac.

Golpaigani, M. H., Ariaii, P., Ahmadi, M., & Safari, R. (2023). Preservation effect of protein hydrolysate of rainbow trout roe with a composite coating on the quality of fresh meat during storage at 4 ± 1°C. *Journal of Food Measurement and Characterization, 17*(3), 2416–2428. https://doi.org/10.1007/s11694-022-01783-7, https://www.springer.com/journal/11694.

Grzebieniarz, W., Tkaczewska, J., Juszczak, L., Nowak, N., Krzyściak, P., Guzik, P., Kasprzak, M., Zimowska, M., & Jamróz, E. (2024). Design of active double-layer gel coatings based on furcellaran-gelatin and aqueous butterfly pea (*Clitoria ternatea*) flower extract for prolonging the shelf-life of salmon (*Salmo salar*). *Progress in Organic Coatings, 186*, 107945. https://doi.org/10.1016/j.porgcoat.2023.107945.

Grzebieniarz, W., Tkaczewska, J., Juszczak, L., Krzyściak, P., Cholewa-Wójcik, A., Nowak, N., Guzik, P., Szuwarzyński, M., Mazur, T., & Jamróz, E. (2023). Improving the quality of multi-layer films based on furcellaran by immobilising active ingredients and impact assessment of the use of a new packaging material. *Food Chemistry, 428*, 136759. https://doi.org/10.1016/j.foodchem.2023.136759.

Gururani, P., Bhatnagar, P., Dogra, P., Chandra Joshi, H., Chauhan, P. K., Vlaskin, M. S., Chandra Joshi, N., Kurbatova, A., Irina, A., & Kumar, V. (2023). Bio-based food packaging materials: A sustainable and Holistic approach for cleaner environment- a review. *Current Research in Green and Sustainable Chemistry, 7*. https://doi.org/10.1016/j.crgsc.2023.100384, https://www.sciencedirect.com/science/journal/26660865.

Hamed, I., Jakobsen, A. N., & Lerfall, J. (2022). Sustainable edible packaging systems based on active compounds from food processing byproducts: A review. *Comprehensive Reviews in Food Science and Food Safety, 21*(1), 198–226. https://doi.org/10.1111/1541-4337.12870, http://onlinelibrary.wiley.com/journal/10.1111/(ISSN)1541-4337.

Hoa, V.-B., Song, D.-H., Seol, K.-H., Kim, Y.-S., Kim, H.-W., Bae, I.-S., & Cho, S.-H. (2024). Effect of coating with combined chitosan and gallic acid on shelf-life stability of Jeju black cattle beef. *Animal Bioscience, 37*(1), 123–130. 10.5713/ab.23.0180.

Hong, S. I., Cho, Y., & Rhim, J. W. (2021). Effect of agar/agnp composite film packaging on refrigerated beef loin quality. *Membranes, 11*(10). https://doi.org/10.3390/membranes11100750, https://www.mdpi.com/2077-0375/11/10/750/pdf.

Isvand, A., Karimaei, S., & Amini, M. (2024). Assessment of chitosan coating enriched with *Citrus limon* essential oil on the quality characteristics and shelf life of beef meat during cold storage. *International Journal of Food Microbiology, 423*, 110825. https://doi.org/10.1016/j.ijfoodmicro.2024.110825.

Jafarzadeh, S., & Jafari, S. M. (2021). Impact of metal nanoparticles on the mechanical, barrier, optical and thermal properties of biodegradable food packaging materials. *Critical Reviews in Food Science and Nutrition, 61*(16), 2640–2658. https://doi.org/10.1080/10408398.2020.1783200.

Jahangiri, R., Hamedi, H., & Ahari, H. (2022). The effect of edible coating of chitosan and Baneh gum (*Pistacia atlantica*) containing propolis extract and ginger nanoemulsion.

Jamróz, E., Kulawik, P., Tkaczewska, J., Guzik, P., Zając, M., Juszczak, L., Krzyściak, P., & Turek, K. (2021). The effects of active double-layered furcellaran/gelatin hydrolysate film system with Ala-Tyr peptide on fresh Atlantic mackerel stored at −18°C. *Food Chemistry, 338*. 127867 https://doi.org/10.1016/j.foodchem.2020.127867.

Jamróz, E., Tkaczewska, J., Juszczak, L., Zimowska, M., Kawecka, A., Krzyściak, P., & Skóra, M. (2022). The influence of lingonberry extract on the properties of novel, double-layered biopolymer films based on furcellaran, CMC and a gelatin hydrolysate. *Food Hydrocolloids, 124*, 107334. https://doi.org/10.1016/j.foodhyd.2021.107334.

Javan, A. J., Baktash, S., Yancheshmeh, B. S., Parsaeimehr, M., Madanchi, H., Abdolshahi, A., Marvdashti, L. M., & Shriatifar, N. (2024). Effect of *Vicia villosa* protein isolate-based edible coating incorporated with ZnO nanoparticles on the shelf-life of chicken breast meat during cold storage. *Food Hydrocolloids for Health, 5*. https://doi.org/10.1016/j.fhfh.2024.100176, https://www.sciencedirect.com/science/journal/26670259.

Jeya Jeevahan, J., Chandrasekaran, M., Venkatesan, S. P., Sriram, V., Britto Joseph, G., Mageshwaran, G., & Durairaj, R. B. (2020). Scaling up difficulties and commercial aspects of edible films for food packaging: A review. *Trends in Food Science & Technology, 100*, 210–222. https://doi.org/10.1016/j.tifs.2020.04.014.

Jurić, M., Maslov Bandić, L., Carullo, D., & Jurić, S. (2024). Technological advancements in edible coatings: Emerging trends and applications in sustainable food preservation. *Food Bioscience, 58*, 103835. https://doi.org/10.1016/j.fbio.2024.103835.

Kanatt, S. R., & Chawla, S. P. (2018). Shelf life extension of chicken packed in active film developed with mango peel extract. *Journal of Food Safety, 38*(1). https://doi.org/10.1111/jfs.12385, http://www.wiley.com/bw/editors.asp?ref=0149-6085&site=1.

Kandasamy, S., Yoo, J., Yun, J., Kang, H. B., Seol, K. H., Kim, H. W., & Ham, J. S. (2021). Application of whey protein-based edible films and coatings in food industries: An updated overview. *Coatings, 11*(9). https://doi.org/10.3390/coatings11091056, https://www.mdpi.com/2079-6412/11/9/1056/pdf.

Karimi Sani, I., Masoudpour-Behabadi, M., Alizadeh Sani, M., Motalebinejad, H., Juma, A. S. M., Asdagh, A., Eghbaljoo, H., Khodaei, S. M., Rhim, J.-W., & Mohammadi, F. (2023). Value-added utilization of fruit and vegetable processing by-products for the manufacture of biodegradable food packaging films. *Food Chemistry, 405*(Pt B), 134964. https://doi.org/10.1016/j.foodchem.2022.134964.

Kaur, R., Shekhar, S., Chaudhary, S., Singh, B., & Prasad, K. (2022). Non-thermal food preservation technologies. In *Smart and sustainable food technologies*. India: Springer Nature, 157–195. https://doi.org/10.1007/978-981-19-1746-2_6, https://link.springer.com/book/10.1007/978-981-19-1746-2.

Kaur, R., & Kaur, L. (2021). Encapsulated natural antimicrobials: A promising way to reduce microbial growth in different food systems. *Food Control, 123*, 107678. https://doi.org/10.1016/j.foodcont.2020.107678.

Kazemeini, H., Azizian, A., & Adib, H. (2021). Inhibition of *Listeria monocytogenes* growth in turkey fillets by alginate edible coating with *Trachyspermum ammi* essential oil nano-emulsion. *International Journal of Food Microbiology, 344*, 109104. https://doi.org/10.1016/j.ijfoodmicro.2021.109104.

Khan, M. J., Ramiah, S. K., Selamat, J., Shameli, K., Sazili, A. Q., & Mookiah, S. (2022). Utilisation of pullulan active packaging incorporated with curcumin and pullulan mediated silver nanoparticles to maintain the quality and shelf life of broiler meat. *Italian Journal of Animal Science, 21*(1), 244–262. https://doi.org/10.1080/1828051X.2021.2012285, http://www.tandfonline.com/loi/tjas20#.VpSlwLfUhFo.

Khan, M. R., Sadiq, M. B., & Mehmood, Z. (2020). Development of edible gelatin composite films enriched with polyphenol loaded nanoemulsions as chicken meat packaging material. *CYTA - Journal of Food, 18*(1), 137–146. https://doi.org/10.1080/19476337.2020.1720826, http://www.tandfonline.com/toc/tcyt20/current.

Khanzadi, S., Keykhosravy, K., Hashemi, M., & Azizzadeh, M. (2020). Alginate coarse/nanoemulsions containing *Zataria multiflora* Boiss essential oil as edible coatings and the impact on microbial quality of trout fillet. *Aquaculture Research, 51*(3), 873–881. https://doi.org/10.1111/are.14418, http://onlinelibrary.wiley.com/journal/10.1111/(ISSN)1365-2109.

Kizilkaya, P., & Kaya, M. (2023). The effect of a chitosan/TiO_2-nanoparticle/rosmarinic acid-based nanocomposite coating on the preservation of refrigerated rainbow trout fillets (*Oncorhynchus mykiss*). *Food Science of Animal Resources, 43*(6).

Kulawik, P., Jamróz, E., Kruk, T., Szymkowiak, A., Tkaczewska, J., Krzyściak, P., Skóra, M., Guzik, P., Janik, M., Vlčko, T., & Milosavljević, V. (2023). Active edible multi-layer chitosan/furcellaran micro/nanoemulsions with plant essential oils and antimicrobial peptides: Biological properties and consumer acceptance. *Food Control, 150*, 109767. https://doi.org/10.1016/j.foodcont.2023.109767.

Kumar, S., Kumar, V., Sharma, R., Manhas, A., Chaudhary, S., & Sharma, S. (2023). *Edible nanoemulsion coating materials: Potential for fresh fruits and vegetables*. Informa UK Limited, 267–321. https://doi.org/10.1201/9781003305385-10.

Kupervaser, M. G., Traffano-Schiffo, M. V., Dellamea, M. L., Flores, S. K., & Sosa, C. A. (2023). Trends in starch-based edible films and coatings enriched with tropical fruits extracts: A review. *Food Hydrocolloids for Health, 4*. https://doi.org/10.1016/j.fhfh.2023.100138, https://www.journals.elsevier.com/food-hydrocolloids-for-health.

Lan, X., Zhang, X., Wang, L., Wang, H., Hu, Z., Ju, X., & Yuan, Y. (2023). A review of food preservation based on zein: The perspective from application types of coating and film. *Food Chemistry, 424*, 136403. https://doi.org/10.1016/j.foodchem.2023.136403.

Lashkari, H., Halabinejad, M., Rafati, A., & Namdar, A. (2020). Shelf life extension of veal meat by edible coating incorporated with zataria multiflora essential oil. *Journal of Food Quality, 2020*, 1–8. https://doi.org/10.1155/2020/8871857.

Li, C., Pradhan, P., Wu, X., Li, Z., Liu, J., Hubacek, K., & Chen, G. (2024). Livestock sector can threaten planetary boundaries without regionally differentiated strategies. *Journal of Environmental Management, 370*, 122444. https://doi.org/10.1016/j.jenvman.2024.122444.

Liu, Q., Zhang, M., Bhandari, B., Xu, J., & Yang, C. (2020). Effects of nanoemulsion-based active coatings with composite mixture of star anise essential oil, polylysine, and nisin on the quality and shelf life of ready-to-eat Yao meat products. *Food Control, 107*, 106771. https://doi.org/10.1016/j.foodcont.2019.106771.

Liu, Y., Ahmed, S., Sameen, D. E., Wang, Y., Lu, R., Dai, J., Li, S., & Qin, W. (2021). A review of cellulose and its derivatives in biopolymer-based for food packaging application. *Trends in Food Science and Technology, 112*, 532–546. https://doi.org/10.1016/j.tifs.2021.04.016, http://www.elsevier.com/wps/find/journaldescription.cws_home/601278/description#description.

Liu, Y., Liu, Z. H., Luo, C. Q., Xiao, C. T., Zhou, W. Y., & Xie, W. J. (2022). Effect of gelatin nano-coating containing *Gardenia pigment* on the preservation of pork slices. *Food Science and Biotechnology, 31*(4), 505–514. https://doi.org/10.1007/s10068-022-01048-5, http://www.springerlink.com/content/n3031612t262/.

Lourenço, S. C., Fraqueza, M. J., Fernandes, M. H., Moldão-Martins, M., & Alves, V. D. (2020). Application of edible alginate films with pineapple peel active compounds on beef meat preservation. *Antioxidants, 9*(8), 1–15. https://doi.org/10.3390/antiox9080667, https://www.mdpi.com/2076-3921/9/8/667/pdf.

Maciel, V. B. V., Contini, L. R. F., Yoshida, C. M. P., & Venturini, A. C. (2020). Application of edible biopolymer coatings on meats, poultry, and seafood. In *Biopolymer membranes and films: Health, food, environment, and energy applications*. Brazil: Elsevier, 515–533. https://doi.org/10.1016/B978-0-12-818134-8.00021-3, https://www.elsevier.com/books/biopolymer-membranes-and-films/de-moraes/978-0-12-818134-8.

Maru, V. R., Gupta, S., Ranade, V., & Variyar, P. S. (2021). Pullulan or chitosan based active coating by incorporating polyphenols from lemon peel in raw poultry meat. *Journal of Food Science and Technology, 58*(10), 3807–3816. https://doi.org/10.1007/s13197-020-04841-4, http://www.springerlink.com/content/121580/.

Mauricio, R. A., Campos, J. A. D. B., & Nassu, R. T. (2022). Meat with edible coating: Acceptance, purchase intention and neophobia. *Food Research International, 154*.

Mehdizadeh, T., Tajik, H., Langroodi, A. M., Molaei, R., & Mahmoudian, A. (2020). Chitosan-starch film containing pomegranate peel extract and *Thymus kotschyanus* essential oil can prolong the shelf life of beef. *Meat Science, 163*, 108073. https://doi.org/10.1016/j.meatsci.2020.108073.

Messinese, E., Pitirollo, O., Grimaldi, M., Milanese, D., Sciancalepore, C., & Cavazza, A. (2024). By-products as sustainable source of bioactive compounds for potential application in the field of food and new materials for packaging development. *Food and Bioprocess Technology, 17*(3), 606–627. https://doi.org/10.1007/s11947-023-03158-2.

Michel, M., Eldridge, A. L., Hartmann, C., Klassen, P., Ingram, J., & Meijer, G. W. (2024). Benefits and challenges of food processing in the context of food systems, value chains and sustainable development goals. *Trends in Food Science and Technology, 153*. https://doi.org/10.1016/j.tifs.2024.104703, https://www.sciencedirect.com/science/journal/09242244.

Mihalca, V., Kerezsi, A. D., Weber, A., Gruber-Traub, C., Schmucker, J., Vodnar, D. C., Dulf, F. V., Socaci, S. A.ța, Fărcaş, A., Mureşan, C. I., Suharoschi, R., & Pop, O. L. (2021). Protein-based films and coatings for food industry applications. *Polymers, 13*(5), 769. https://doi.org/10.3390/polym13050769.

Milani, J. M., & Nemati, A. (2022). Lipid-based edible films and coatings: A review of recent advances and applications. *Journal of Packaging Technology and Research, 6*(1), 11–22. https://doi.org/10.1007/s41783-021-00130-3.

Mozaffari, P., Pashangeh, S., Berizi, E., Majlesi, M., Hosseinzadeh, S., Salehi, S. O., Derakhshan, Z., & Giannakis, S. (2022). Potential of nanochitosan coating combined with walnut green husk to improve the preservation of rainbow trout (*Oncorhynchus mykiss*) during refrigerated storage. *Environmental Research, 214*. https://doi.org/10.1016/j.envres.2022.114019, http://www.elsevier.com/inca/publications/store/6/2/2/8/2/1/index.htt.

Mozhiarasi, V., & Natarajan, T. S. (2022). Slaughterhouse and poultry wastes: Management practices, feedstocks for renewable energy production, and recovery of value added products. *Biomass Conversion and Biorefinery*. https://doi.org/10.1007/s13399-022-02352-0, http://www.springer.com/engineering/energy+technology/journal/13399.

Mushtaq, A., Mohd Wani, S., Malik, A. R., Gull, A., Ramniwas, S., Ahmad Nayik, G., Ercisli, S., Alina Marc, R., Ullah, R., & Bari, A. (2023). Recent insights into nanoemulsions: Their preparation, properties and applications. *Food Chemistry: X, 18*. https://doi.org/10.1016/j.fochx.2023.100684, https://www.journals.elsevier.com/food-chemistry-x.

Nastasi, J. R., Kontogiorgos, V., Daygon, V. D., & Fitzgerald, M. A. (2022). Pectin-based films and coatings with plant extracts as natural preservatives: A systematic review. *Trends in Food Science and Technology, 120*, 193–211. https://doi.org/10.1016/j.tifs.2022.01.014, http://www.elsevier.com/wps/find/journaldescription.cws_home/601278/description#description.

Nikmaram, N., Budaraju, S., Barba, F. J., Lorenzo, J. M., Cox, R. B., Mallikarjunan, K., & Roohinejad, S. (2018). Application of plant extracts to improve the shelf-life, nutritional and health-related properties of ready-to-eat meat products. *Meat Science, 145*, 245–255. https://doi.org/10.1016/j.meatsci.2018.06.031, http://www.elsevier.com/locate/meatsci.

Nowak, N., Tkaczewska, J., Grzebieniarz, W., Juszczak, L., Mazur, T., Szuwarzyński, M., Guzik, P., & Jamróz, E. (2024). Active and intelligent four-layer films based on chitosan, gelatin, furcellaran and active ingredients—Preparation, characterisation and application on Salmon. *Food and Bioprocess Technology, 17*(7), 1862–1875. https://doi.org/10.1007/s11947-023-03238-3.

Odeyemi, O. A., Alegbeleye, O. O., Strateva, M., & Stratev, D. (2020). Understanding spoilage microbial community and spoilage mechanisms in foods of animal origin. *Comprehensive Reviews in Food Science and Food Safety, 19*(2), 311–331. https://doi.org/10.1111/1541-4337.12526, http://onlinelibrary.wiley.com/journal/10.1111/(ISSN)1541-4337.

Panwar, A., Kumar, S., Dhiman, A., Kumar, V., Gupta, D., & Sharma, A. (2024). pH sensitive indicator film based intelligent packaging systems of perishables: Developments and challenges of last decade. *Microchemical Journal, 207*, 111732. https://doi.org/10.1016/j.microc.2024.111732.

Praseptiangga, D., Widyaastuti, D., Panatarani, C., & Joni, I. M. (2021). Development and characterization of semi-refined iota carrageenan/SiO_2-ZnO bionanocomposite film with the addition of cassava starch for application on minced chicken meat packaging. *Foods, 10*(11), 2776. https://doi.org/10.3390/foods10112776.

Priyadarshi, R., Kim, S. M., & Rhim, J. W. (2021). Carboxymethyl cellulose-based multifunctional film combined with zinc oxide nanoparticles and grape seed extract for the preservation of high-fat meat products. *Sustainable Materials and Technologies, 29*. https://doi.org/10.1016/j.susmat.2021.e00325, https://www.journals.elsevier.com/sustainable-materials-and-technologies.

Qiu, L., Zhang, M., Chitrakar, B., Adhikari, B., & Yang, C. (2022). Effects of nanoemulsion-based chicken bone gelatin-chitosan coatings with cinnamon essential oil and rosemary extract on the storage quality of ready-to-eat chicken patties. *Food Packaging and Shelf Life, 34*, 100933. https://doi.org/10.1016/j.fpsl.2022.100933.

Ranjbar, M., Azizi Tabrizzad, M. H., Asadi, G., & Ahari, H. (2023). Investigating the microbial properties of sodium alginate/chitosan edible film containing red beetroot anthocyanin extract for smart packaging in chicken fillet as a pH indicator. *Heliyon, 9*(8). https://doi.org/10.1016/j.heliyon.2023.e18879, http://www.journals.elsevier.com/heliyon/.

Rather, J. A., Kaur, G., Shah, I. A., Majid, D., Makroo, H. A., & Dar, B. N. (2024). Sustainable gelatin-based packaging with nanoemulsified chilli seed oil for enhancing poultry meat preservation: An eco-friendly approach. *Food Chemistry Advances, 5*. https://doi.org/10.1016/j.focha.2024.100761, https://www.sciencedirect.com/science/journal/2772753X.

Rathod, N. B., Meral, R., Siddiqui, S. A., Nirmal, N., & Ozogul, F. (2024). Nanoemulsion-based approach to preserve muscle food: A review with current knowledge. *Critical Reviews in Food Science and Nutrition, 64*(19), 6812–6833. https://doi.org/10.1080/10408398.2023.2175347, https://www.tandfonline.com/loi/bfsn20.

da Rocha, M., Alemán, A., Romani, V. P., López-Caballero, M. E., Gómez-Guillén, M. C., Montero, P., & Prentice, C. (2018). Effects of agar films incorporated with fish protein hydrolysate or clove essential oil on flounder (*Paralichthys orbignyanus*) fillets shelf-life. *Food Hydrocolloids, 81*, 351–363. https://doi.org/10.1016/j.foodhyd.2018.03.017, https://www.journals.elsevier.com/food-hydrocolloids.

Roy, S., Priyadarshi, R., & Rhim, J.-W. (2021). Development of multifunctional pullulan/chitosan-based composite films reinforced with ZnO nanoparticles and propolis for meat packaging applications. *Foods, 10*(11), 2789. https://doi.org/10.3390/foods10112789.

Rusman, A. H., Kumar, P., Hayat, M. N., Abdul, M., Adamu, G., Ismail-Fitry, M. R., & Sazili, A. Q. (2024). Effect of sodium alginate coating containing clove essential oil on pH, cooking loss, colour and sensory attributes of refrigerated rabbit meat. *Journal of Meat Science, 19*(1). https://doi.org/10.48165/jms.2024.19.01.01, https://acspublisher.com/journals/index.php/jms/article/download/12118/12115.

Sabzipour-Hafshejani, F., Mirzapour-Kouhdasht, A., Khodaei, D., Taghizadeh, M. S., & Garcia-Vaquero, M. (2022). Impact of whey protein edible coating containing fish gelatin hydrolysates on physicochemical, microbial, and sensory properties of chicken breast fillets. *Polymers, 14*(16). https://doi.org/10.3390/polym14163371, http://www.mdpi.com/journal/polymers.

Safari, R., Yaghoubi, M., Marcinkowska-Lesiak, M., Paya, H., Sun, X., Rastgoo, A., Rafiee, M., & Alirezalu, K. (2023). The effects of double gelatin containing chitosan nanoparticles–calcium alginate coatings on the stability of chicken breast meat. *Food Science & Nutrition, 11*(12), 7673–7685. https://doi.org/10.1002/fsn3.3686.

Saghari, V., Jalali, H., Shariatifar, N., & Ziaolhagh, S. (2024). Evaluation of whey protein coating containing free and Pickering emulsion forms of *Trachyspermum copticum* L. essential oil on quality of refrigerated beef. *Food Science & Nutrition, 12*(11), 9187–9197. https://doi.org/10.1002/fsn3.4454.

Sayadi, M., Amiri, S., & Radi, M. (2021). Active packaging nanocomposite gelatin-based films as a carrier of nano TiO_2 and cumin essential oil: The effect on quality parameters of fresh chicken. *Journal of Food Measurement and Characterization,* 1–11.

Sayadi, M., Mojaddar Langroodi, A., Amiri, S., & Radi, M. (2022). Effect of nanocomposite alginate-based film incorporated with cumin essential oil and TiO_2 nanoparticles on chemical, microbial, and sensory properties of fresh meat/beef. *Food Science & Nutrition, 10*(5), 1401–1413. https://doi.org/10.1002/fsn3.2724.

Sharma, S., Nakano, K., Kumar, S., & Katiyar, V. (2024). Edible packaging to prolong postharvest shelf-life of fruits and vegetables: A review. *Food Chemistry Advances, 4*, 100711. https://doi.org/10.1016/j.focha.2024.100711.

Shavisi, N., & Shahbazi, Y. (2022). Chitosan-gum Arabic nanofiber mats encapsulated with pH-sensitive *Rosa damascena* anthocyanins for freshness monitoring of chicken fillets. *Food Packaging and Shelf Life, 32*, 100827. https://doi.org/10.1016/j.fpsl.2022.100827.

Sheerzad, S., Khorrami, R., Khanjari, A., Gandomi, H., Basti, A. A., & Khansavar, F. (2024). Improving chicken meat shelf-life: Coating with whey protein isolate, nanochitosan, bacterial nanocellulose, and cinnamon essential oil. *LWT, 197*. https://doi.org/10.1016/j.lwt.2024.115912, https://www.sciencedirect.com/science/journal/00236438.

Song, X., Wang, L., Liu, T., Liu, Y., Wu, X., & Liu, L. (2021). Mandarin *(Citrus reticulata* L.) essential oil incorporated into chitosan nanoparticles: Characterization, anti-biofilm properties and application in pork preservation. *International Journal of Biological Macromolecules, 185*, 620–628. https://doi.org/10.1016/j.ijbiomac.2021.06.195, www.elsevier.com/locate/ijbiomac.

Souza, V. G. L., Rodrigues, C., Valente, S., Pimenta, C., Pires, J. R. A., Alves, M. M., Santos, C. F., Coelhoso, I. M., & Fernando, A. L. (2020). Eco-friendly ZnO/chitosan bionanocomposites films for packaging of fresh poultry meat. *Coatings, 10*(2). https://doi.org/10.3390/coatings10020110, https://res.mdpi.com/d_attachment/coatings/coatings-10-00110/article_deploy/coatings-10-00110-v2.pdf.

Suhag, R., Kumar, N., Petkoska, A. T., & Upadhyay, A. (2020). Film formation and deposition methods of edible coating on food products: A review. *Food Research International, 136*. https://doi.org/10.1016/j.foodres.2020.109582, www.elsevier.com/inca/publications/store/4/2/2/9/7/0.

Tadele, W., Kulawik, P., Stępień, A., Zając, M., Nowak, N., Grzebieniarz, W., Jasińska, J. M., Vlcko, T., Szymkowiak, A., Milosavljević, V., Debnarova, D., & Jamróz, E. (2024). Effects of multilayer Nano/Mini Furcellaran/Chitosan Emulsions with oregano essential oil and bioactive peptides on sensory, physicochemical properties and retrogradation in Sushi in cold storage conditions. *Innovative Food Science & Emerging Technologies, 96*, 103767. https://doi.org/10.1016/j.ifset.2024.103767.

Taghavi Kevij, H., Salami, M., Mohammadian, M., Khodadadi, M., & Emam-Djomeh, Z. (2021). Mechanical, physical, and bio-functional properties of biopolymer films based on gelatin as affected by enriching with orange peel powder. *Polymer Bulletin, 78*(8), 4387–4402. https://doi.org/10.1007/s00289-020-03319-9.

Tardy, B. L., Richardson, J. J., Greca, L. G., Guo, J., Bras, J., & Rojas, O. J. (2023). Advancing bio-based materials for sustainable solutions to food packaging. *Nature Sustainability, 6*(4), 360–367. https://doi.org/10.1038/s41893-022-01012-5, https://www.nature.com/natsustain/.

Tkaczewska, J., Jamróz, E., Guzik, P., & Kopeć, M. (2022). Attempt to extend the shelf-life of fish products by means of innovative double-layer active biodegradable films. *Polymers, 14*(9), 1717. https://doi.org/10.3390/polym14091717.

Toniciolli Rigueto, C. V., Rosseto, M., Alessandretti, I., de Oliveira, R., Wohlmuth, D. A. R., Ferreira Menezes, J., Loss, R. A., Dettmer, A., & Pizzutti, I. R. (2022). Gelatin films from wastes: A review of production, characterization, and application trends in food preservation and agriculture. *Food Research International, 162*. https://doi.org/10.1016/j.foodres.2022.112114, www.elsevier.com/inca/publications/store/4/2/2/9/7/0.

Tsegay, Z. T., Hosseini, E., Varzakas, T., & Smaoui, S. (2024). The latest research progress on polysaccharides-based biosensors for food packaging: A review. *International Journal of Biological Macromolecules, 282*(Pt 4), 136959. https://doi.org/10.1016/j.ijbiomac.2024.136959.

Tsironi, M., Kosma, I. S., & Badeka, A. V. (2022). The effect of whey protein films with ginger and rosemary essential oils on microbiological quality and physicochemical properties of minced lamb meat. *Sustainability, 14*(6), 3434. https://doi.org/10.3390/su14063434.

Venkatachalam, K., & Lekjing, S. (2020). A chitosan-based edible film with clove essential oil and nisin for improving the quality and shelf life of pork patties in cold storage. *RSC Advances, 10*(30), 17777–17786. https://doi.org/10.1039/d0ra02986f.

Vieira, M. G. A., Da Silva, M. A., Dos Santos, L. O., & Beppu, M. M. (2011). Natural-based plasticizers and biopolymer films: A review. *European Polymer Journal, 47*(3), 254–263. https://doi.org/10.1016/j.eurpolymj.2010.12.011.

Wang, Q., Chen, W., Zhu, W., McClements, D. J., Liu, X., & Liu, F. (2022). A review of multilayer and composite films and coatings for active biodegradable packaging. *npj Science of Food, 6*(1). https://doi.org/10.1038/s41538-022-00132-8, https://www.nature.com/npjscifood/about.

Wani, N. R., Dar, A. H., Dash, K. K., Pandey, V. K., Srivastava, S., Jan, S. Y., Deka, P., & Sabahi, N. (2023). Recent advances in the production of bionanomaterials for development of sustainable food packaging: A comprehensive review. *Environmental Research, 237*. https://doi.org/10.1016/j.envres.2023.116948, https://www.sciencedirect.com/journal/environmental-research.

Wu, Y., Ma, Y., Gao, Y., Liu, Y., & Gao, C. (2022). Poly (lactic acid)-based pH responsive membrane combined with chitosan and alizarin for food packaging. *International Journal of Biological Macromolecules, 214*, 348–359. https://doi.org/10.1016/j.ijbiomac.2022.06.039.

Xie, F. (2023). Biopolymer-based multilayer films and coatings for food preservation: An update of the recent development. *Current Food Science and Technology Reports, 1*(1), 1–12. https://doi.org/10.1007/s43555-023-00002-8.

Youssef, D. M., Alshubaily, F. A., Tayel, A. A., Alghuthaymi, M. A., & Al-Saman, M. A. (2022). Application of nanocomposites from bees products and nano-selenium in edible coating for catfish fillets biopreservation. *Polymers, 14*(12). https://doi.org/10.3390/polym14122378, https://www.mdpi.com/2073-4360/14/12/2378/pdf?version=1655023752.

Yun, D., Qin, Y., Zhang, J., Zhang, M., Qian, C., & Liu, J. (2021). Development of chitosan films incorporated with rambutan (*Nephelium lappaceum* L.) peel extract and their application in pork preservation. *International Journal of Biological Macromolecules, 189*, 900–909. https://doi.org/10.1016/j.ijbiomac.2021.08.171.

Yusof, N. L., Mutalib, N. A. A., Nazatul, U. K., Nadrah, A. H., Aziman, N., Fouad, H., Jawaid, M., Ali, A., Kian, L. K., & Sain, M. (2021). Efficacy of biopolymer/starch based antimicrobial packaging for chicken breast fillets. *Foods, 10*(10). https://doi.org/10.3390/foods10102379, https://www.mdpi.com/2304-8158/10/10/2379/pdf.

Zeng, F., Ye, Y., Liu, J., & Fei, P. (2023). Intelligent pH indicator composite film based on pectin/chitosan incorporated with black rice anthocyanins for meat freshness monitoring. *Food Chemistry: X, 17*, 100531. https://doi.org/10.1016/j.fochx.2022.100531.

Zhang, B., Liu, Y., Wang, H., Liu, W., Cheong, K.-l, & Teng, B. (2021). Effect of sodium alginate-agar coating containing ginger essential oil on the shelf life and quality of beef. *Food Control, 130*, 108216. https://doi.org/10.1016/j.foodcont.2021.108216.

Zhang, M., Luo, W., Yang, K., & Li, C. (2022). Effects of sodium alginate edible coating with cinnamon essential oil nanocapsules and nisin on quality and shelf life of beef slices during refrigeration. *Journal of Food Protection, 85*(6), 896–905. 10.4315/jfp-21-380.

Zhang, H., Liang, Y., Li, X., & Kang, H. (2020). Effect of chitosan-gelatin coating containing nano-encapsulated tarragon essential oil on the preservation of pork slices. *Meat Science, 166*, 108137. https://doi.org/10.1016/j.meatsci.2020.108137.

Zhang, W., Roy, S., Assadpour, E., Cong, X., & Jafari, S. M. (2023). Cross-linked biopolymeric films by citric acid for food packaging and preservation. *Advances in Colloid and Interface Science, 314*, 102886. https://doi.org/10.1016/j.cis.2023.102886.

Zhou, X., Zong, X., Zhang, M., Ge, Q., Qi, J., Liang, J., Xu, X., & Xiong, G. (2021). Effect of konjac glucomannan/carrageenan-based edible emulsion coatings with camellia oil on quality and shelf-life of chicken meat. *International Journal of Biological Macromolecules, 183*, 331–339. https://doi.org/10.1016/j.ijbiomac.2021.04.165.

Zhu, Y., Wang, W., Li, M., Zhang, J., Ji, L., Zhao, Z., Zhang, R., Cai, D., & Chen, L. (2022). Microbial diversity of meat products under spoilage and its controlling approaches. *Frontiers in Nutrition, 9*. https://doi.org/10.3389/fnut.2022.1078201.

Zhu, Y.-ao, Li, F., Wang, M., Cao, Y., Kong, B., Liu, Q., Wang, H., & Chen, Q. (2024). Improving the storage quality of Harbin red sausages by quaternized chitosan/sodium alginate coating curcumin nano-emulsion. *Meat Science, 216*, 109585. https://doi.org/10.1016/j.meatsci.2024.109585.

Zixiang, W., Jingjing, Z., Huachen, Z., Ning, Z., Ruiyan, Z., Lanjie, L., & Guiqin, L. (2021). Effect of nanoemulsion loading a mixture of clove essential oil and carboxymethyl chitosan-coated ε-polylysine on the preservation of donkey meat during refrigerated storage. *Journal of Food Processing and Preservation, 45*(9). https://doi.org/10.1111/jfpp.15733.

Zou, J., Liu, X., Wang, X., Yang, H., Cheng, J., Lin, Y., & Tang, D. (2022). Influence of gelatin-chitosan-glycerol edible coating incorporated with chlorogenic acid, gallic acid, and resveratrol on the preservation of fresh beef. *Foods, 11*(23), 3813. https://doi.org/10.3390/foods11233813.

Chapter 28

Application and future trends in biopolymer films and coatings for effective fish preservation

Veena P. Kumar[1,2], S.L. Vishnu[1,2,3], V. Dhanya[2,4], Roshmi Thomas[2,4], K.P. Akhila[5], and G. Ganga[2,6]

[1]*Cashew Export Promotion Council of India (CEPCI) Laboratory & Research Institute, Kollam, Kerala, India,* [2]*University of Kerala, Thiruvananthapuram, Kerala, India,* [3]*Sree Narayana College for Women, University of Kerala, Kollam, Kerala, India,* [4]*Department of Microbiology, Sanatana Dharma College, University of Kerala, Alappuzha, Kerala, India,* [5]*Department of Microbiology, Central Coir Research Institute, University of Kerala, Alappuzha, Kerala, India,* [6]*Department of Biochemistry and Industrial Microbiology, Sree Ayyappa College, University of Kerala, Alappuzha, Kerala, India*

28.1 Introduction

It is widely known that fish and fish products are more perishable compared to meat or meat products (Rahman et al., 2024). Due to their perishable nature, the shelf life of fish products is minimal. To overcome this, proper packaging and aseptic conditions are required during the storage and transport of fish. Conventionally, petrochemical-based materials preserve fish from physical, chemical, and biological damage. Almost 70% of global waste is produced from packaging, especially petroleum-based materials (Vibha et al., 2022). The escalating global demand for seafood is juxtaposed against the inherent challenges of preserving its quality and safety, with fish being particularly susceptible to rapid spoilage. This vulnerability stems from its high moisture content, neutral pH, and the presence of enzymes and microorganisms that accelerate decomposition, resulting in significant economic losses and potential health risks (Díaz-Montes & Castro-Muñoz, 2021). Traditional packaging methods, often relying on synthetic plastics, have raised environmental concerns due to their nonbiodegradable nature and contribution to pollution (Kumar et al., 2021). These conventional materials, while providing barrier properties, fail to address the active preservation needs of fish products, such as inhibiting microbial growth and preventing oxidation, further shortening shelf life and increasing food waste (Alim et al., 2022). As consumers increasingly demand fresher, minimally processed foods with extended shelf lives, the need for innovative and sustainable packaging solutions becomes paramount. Biopolymer-based films and coatings have emerged as promising alternatives, offering a dual benefit of environmental compatibility and enhanced preservation capabilities, addressing both the ecological and food safety concerns associated with fish packaging (Asgher et al., 2020). The development of biodegradable or edible films from biopolymers presents an environmentally conscious alternative to synthetic plastic packaging (Rocha et al., 2018). Furthermore, consumer preferences for safe, healthy, and high-quality food products are driving the need for innovative food packaging systems designed to maintain the integrity of packaged food (Zhang et al., 2021).

Biopolymer films and coatings present a compelling alternative by leveraging natural polymers derived from renewable sources, offering biodegradability and biocompatibility (Galus et al., 2020). These materials can be engineered to provide tailored barrier properties against oxygen, moisture, and microbial intrusion, while also acting as carriers for antimicrobial and antioxidant agents that actively combat spoilage mechanisms. The utilization of biopolymers in food packaging not only addresses environmental concerns but also aligns with consumer demand for natural and sustainable products (Yuvaraj et al., 2021). The growing environmental consciousness and the need to reduce reliance on petroleum-based plastics have spurred research into biopolymers for food packaging applications (Mangaraj et al., 2018). Biopolymer-based packaging materials present a sustainable and efficient solution for preserving fish, reducing food waste, and

Natural Polymer-Based Films and Coatings in Food Technology. DOI: https://doi.org/10.1016/B978-0-443-34244-8.00030-6

meeting consumer demands for environmentally friendly products. Biopolymers, sourced from renewable resources, offer a sustainable alternative to traditional petroleum-based polymers, reducing environmental impact and promoting a circular economy (Donkor et al., 2023). Therefore, the application of biopolymer films and coatings in fish packaging represents a multifaceted approach to address the challenges of spoilage, shelf-life limitations, and environmental sustainability.

28.2 Biopolymer films and coatings: definitions and types

Biopolymer films and coatings are thin layers of material derived from natural, renewable sources, designed to encapsulate food products, providing a protective barrier against environmental factors and enhancing preservation. These films and coatings are composed of polymers extracted from biomass, including plants, animals, and microorganisms, offering a biodegradable and often edible alternative to synthetic packaging materials (Popescu et al., 2020). These materials offer a compelling alternative to conventional synthetic polymers, addressing environmental concerns and offering unique functionalities for food preservation. Biopolymers can be broadly classified into three major categories: proteins, polysaccharides, and lipids, each possessing distinct properties and functionalities that make them suitable for various packaging applications. The selection of the appropriate biopolymer depends on the specific requirements of the food product, including its sensitivity to moisture, oxygen, and microbial spoilage, as well as the desired mechanical and barrier properties of the packaging (Rydz et al., 2018).

Proteins, such as gelatin, collagen, and chitosan, are derived from animal sources and offer excellent film-forming properties, gas barrier characteristics, and potential for incorporating antimicrobial agents. Gelatin and collagen, obtained from animal by-products, provide good mechanical strength and flexibility, making them suitable for coatings that require structural integrity. Chitosan, derived from crustacean shells, exhibits inherent antimicrobial activity, inhibiting the growth of spoilage bacteria and extending the shelf life of fish products (Sathivel et al., 2007). Polysaccharides, including pectin, alginate, starch, and cellulose, are derived from plant and algal sources, offering biodegradability, biocompatibility, and the ability to form transparent and flexible films. Pectin and alginate, extracted from fruits and seaweed, respectively, form gels and films with good barrier properties, preventing moisture loss, and maintaining the freshness of fish. Starch and cellulose, abundant in plant biomass, can be modified to enhance their film-forming properties and produce sustainable packaging materials with reduced environmental impact. Lipid-based films, composed of waxes, fats, and oils, provide excellent moisture barrier properties, preventing dehydration and maintaining the quality of fish products (Paola et al., 2021).

28.2.1 Edible and nonedible coatings

Edible coatings can be consumed along with the food product, while nonedible coatings must be peeled off or removed before consumption. Edible coatings and films, composed of generally recognized as safe (GRAS) materials, offer a unique advantage by becoming an integral part of the food product, enhancing its texture, flavor, and nutritional value (Azizah et al., 2023; Kumar et al., 2021). Edible films, formulated from food-grade components, offer a unique approach to food packaging, providing a consumable layer that enhances the product's appeal and functionality (Freitas et al., 2022). Edible films and coatings present a pioneering technique in food preservation, differing from conventional packaging through their inherent edibility, allowing them to be ingested alongside the food product without posing any health risks (Mohamed et al., 2020). Edible coatings are applied directly onto the food surface as a liquid, forming a thin, protective layer upon drying, while films are preformed separately and then wrapped around the food product (Avramescu et al., 2020) (Fig. 28.1. These coatings not only extend shelf life but can also incorporate flavors, colors, and nutrients, enhancing the overall eating experience (Hassan et al., 2017). They are applied directly to the food surface, offering a seamless integration with the product and eliminating the need for removal before consumption (Bizymis & Tzia, 2021).

The selection between edible and nonedible coatings hinges on factors like the intended consumption method, desired sensory attributes, and regulatory considerations. Nonedible coatings, while providing similar protective functions, are designed to be removed before consumption, focusing primarily on extending shelf life and maintaining product integrity during storage and transportation. The choice between edible and nonedible coatings is determined by the specific application, considering factors such as consumer preferences, regulatory requirements, and the desired level of integration with the food product. The application of edible coatings represents a significant advancement in food technology, aligning with consumer demand for natural, sustainable, and convenient food solutions (Nagarajan et al., 2021). The development and application of edible films and coatings are dependent on creating materials that are durable, thin, and homogeneous (Kocira et al., 2021).

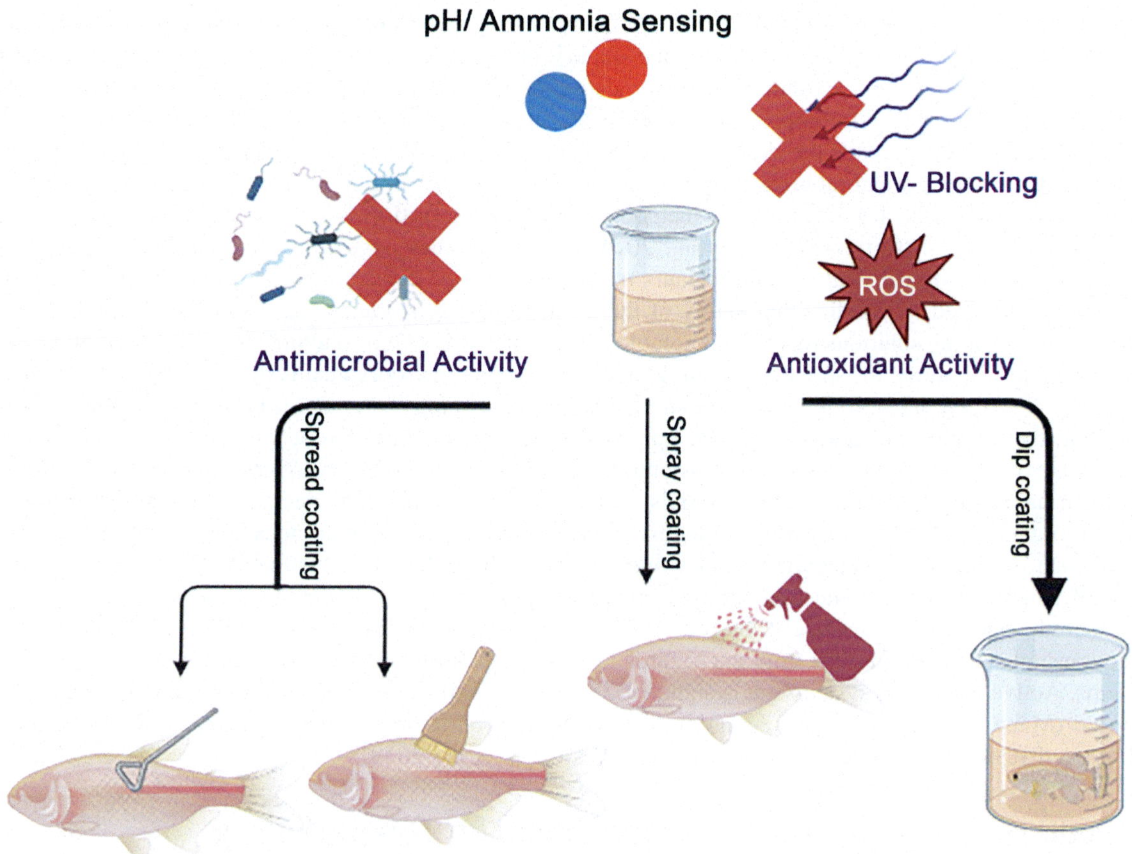

FIGURE 28.1 Types of edible coating methods and applications.

28.3 Mechanisms of fish preservation using biopolymer-based packaging

Biopolymer-based packaging employs a range of mechanisms to preserve fish, primarily focusing on physicochemical and microbiological protection.

28.3.1 Physicochemical protection

Physicochemical protection involves creating a barrier against moisture loss, purge accumulation, oxygen permeation, and water vapor transmission, all of which contribute to fish spoilage. Biopolymer films and coatings act as a barrier, reducing moisture loss from the fish, preventing dehydration, and maintaining its succulent texture (Gennadios et al., 1997). The films also minimize purge accumulation, which is the release of fluids from the fish tissue during storage, creating an unappealing appearance and promoting microbial growth. Additionally, they create an oxygen barrier, slowing down oxidation reactions that cause rancidity and discoloration, thus preserving the sensory attributes of the fish. Biopolymer coatings and films are effective in preventing moisture loss and purge accumulation, thereby maintaining the freshness and extending the shelf life of fish products (Hamzeh & Rezaei, 2012). By controlling these factors, biopolymer packaging can significantly extend the shelf life and maintain the quality of fish products. Lipids play an essential role in preventing moisture loss and preserving the freshness of the food product (Amin, Khan, Khan, et al., 2021).

28.3.2 Microbiological protection

Microbiological protection involves inhibiting the growth of spoilage and pathogenic microorganisms through the delivery of bioactive compounds like antimicrobials and antioxidants. Biopolymer films and coatings can be infused with natural antimicrobials, such as essential oils, plant extracts, or bacteriocins, which inhibit the growth of spoilage bacteria and extend the shelf life of fish (Nagarajan et al., 2021). By incorporating antimicrobial agents, biopolymer films can actively combat microbial growth, maintaining the safety and quality of the fish product. The release of antioxidants from the biopolymer matrix prevents lipid oxidation, a major cause of rancidity and off-flavors in fish (Dutta et al., 2008). Incorporating such compounds into biopolymer films creates an active packaging system that inhibits microbial growth, reduces lipid oxidation, and extends the shelf life of fish. The increasing health concerns related to synthetic compounds have triggered the demand for natural bioactive compounds for the preservation of meat and fish products (Hashemi et al., 2023). The use of lactic acid bacteria or their metabolites in biopreservation is considered an alternative to chemical substances for preservation (Cortés-Sánchez et al., 2024). Several studies have demonstrated that the incorporation of active compounds like antimicrobials or antioxidants in edible films and coatings can improve the microbial safety and shelf life, and can be used to stabilize the quality of the products during storage (Lacroix & Le Tien, 2005).

Antimicrobial packaging solutions represent a notable progression in active packaging technologies, aiming to eradicate or suppress the proliferation of spoilage-inducing and pathogenic microorganisms that contaminate food products (Tripathi et al., 2008). By integrating antimicrobial agents into packaging materials, it becomes feasible to bolster food safety, prolong the duration of preservation, and uphold the superior quality of food items (Umiyati et al., 2019). The sustained release of preservatives from these films onto the food surface ensures prolonged antimicrobial action, which is vital for ensuring lasting protection. The materials employed for antimicrobial packaging encompass a spectrum of organic and inorganic substances, such as bacteriocins, enzymes, organic acids, metal ions, and nanoparticles, each chosen for their distinct antimicrobial capabilities and compatibility with packaging constituents (Bangar et al., 2021). The integration of nanomaterials or metal ions into packaging designs signifies a deliberate strategy to impede or eliminate microbes, thereby guaranteeing the safety and preserving the integrity of packaged food products (Mai et al., 2023). This method demonstrates promise in mitigating the hazards linked to foodborne illnesses and enhancing the overall durability of food items. Active packaging systems are designed to extend the shelf life of foods and maintain their high quality for a longer period (Mane, 2016). Active packaging interacts with the food to change the conditions, increasing its shelf life, safety, and quality, or improving its sensory characteristics (Cruz et al., 2018).

28.4 Sources and preparation of biopolymers for fish packaging

28.4.1 Sources of biopolymers

Biopolymers for fish packaging are sourced from a variety of raw materials, including seafood by-products and plant-based sources, providing sustainable and renewable alternatives to traditional synthetic polymers (Amin, Khan, Majeed, et al., 2021). Seafood by-products, such as fish skin, scales, and crustacean shells, are rich in proteins like gelatin and collagen, as well as polysaccharides like chitosan (Butnaru et al., 2019). Gelatin and collagen, derived from fish processing waste, can be processed into films and coatings with excellent mechanical and barrier properties (Zou et al., 2022). Chitosan, extracted from crustacean shells, exhibits antimicrobial properties, making it ideal for preserving fish products. Plant-based biopolymers, including starch, cellulose, pectin, and alginate, are also widely used in fish packaging applications (Hari et al., 2021). Starch, derived from sources like corn, potatoes, and tapioca, can be processed into films and coatings with good mechanical strength and barrier properties (Brodnjak & Tihole, 2020). Cellulose, the main structural component of plant cell walls, can be used to form films with high tensile strength and gas barrier properties. Pectin and alginate, extracted from fruits and seaweed, respectively, can be used to form edible coatings that protect fish from dehydration and oxidation. The utilization of cassava starch in edible films is advantageous due to its cost-effectiveness, biocompatibility, and biodegradability (Pudjiastuti et al., 2021). These polysaccharides create a barrier against oxygen and moisture, which is vital for maintaining the quality of packed fresh produce by preventing dehydration, microbial development, and softening (Ismillayli et al., 2020).

28.4.2 Preparation methods of biopolymer matrix

The method of film formation is a critical step in determining the properties and performance of biopolymer packaging, with techniques like casting, extrusion, multilayering, and nanoengineering offering versatile options.

Casting involves dissolving the biopolymer in a solvent, pouring the solution onto a flat surface, and allowing the solvent to evaporate, resulting in a thin film (Khodaei et al., 2021). Extrusion involves melting the biopolymer and forcing it through a die to create a continuous film or sheet. The casting method can be used to create films with tailored thickness and composition, while extrusion is suitable for high-volume production of films with uniform properties (Silva et al., 2019). Multilayering involves combining different biopolymers or layering biopolymers with synthetic polymers to create films with enhanced properties. Nanoengineering involves incorporating nanoparticles into the biopolymer matrix to improve its mechanical, barrier, and antimicrobial properties. An electric field is a promising technology for preparation of edible films (Kocira et al., 2021). The strong point of the electrohydrodynamic technique is the fine control of the coating thickness, as well as the droplet size and distribution. The mechanical properties of films are enhanced by blending, layering, or filling cellulose derivatives with other bio or synthetic polymers, which strengthens their position in the packaging market (Paunonen, 2013). The incorporation of nanomaterials in biopolymer matrices can significantly enhance the performance of the resulting films, leading to improved mechanical strength, barrier properties, and antimicrobial activity (Bizymis & Tzia, 2021; Susmitha et al., 2021).

28.5 Active and intelligent packaging strategies

28.5.1 Active packaging

Active packaging strategies enhance the preservation of fish products by incorporating natural preservatives such as essential oils and plant extracts into biopolymer films. Essential oils derived from herbs and spices like thyme, oregano, rosemary, and clove contain antimicrobial and antioxidant compounds that inhibit the growth of spoilage bacteria and prevent lipid oxidation in fish (Rojas et al., 2021). Plant extracts, such as green tea extract, grape seed extract, and cranberry extract, are rich in polyphenols and other bioactive compounds with antimicrobial and antioxidant properties (Kumar & Neeraj, 2019). The incorporation of natural preservatives into biopolymer films allows for the controlled release of these compounds onto the fish surface, providing sustained antimicrobial and antioxidant activity throughout the storage period. The integration of active compounds, like antimicrobials and antioxidants, into edible films and coatings has been shown to enhance microbial safety, extend shelf life, and stabilize product quality during storage. The film-forming biopolymers can act as carriers for the active compounds, facilitating their controlled release and interaction with the packaged food. Antimicrobial films are designed to minimize or prevent microbial development within the food, making antimicrobial activity vital for food packaging (Gowsia et al., 2022). Antioxidant films have been developed to extend the shelf life of food products that are susceptible to oxidation. Active packaging systems absorb or release chemicals to extend the shelf life of food by, for example, providing antibacterial and antioxidant capabilities (Bhat et al., 2012; Silva et al., 2020).

One of the studies conducted by Bremenkamp and Sousa-Gallagher (2024) reveals the perfect use of edible coatings as a freshness enhancer and active material for preserving fish and fish products. Here, in this study, the researchers coated the baked fish products with a chitosan-alginate-$CaCl_2$ mixture at various concentrations. In their study, it is evident that the chitosan-coated fish product shows an extended shelf life compared to the uncoated one (Fig. 28.2).

28.5.2 Intelligent packaging

Intelligent packaging solutions provide real-time information on the quality and safety of fish products by incorporating sensors and indicators into biopolymer films. These sensors can detect changes in temperature, pH, gas composition, and microbial growth, providing valuable information about the freshness and safety of the fish (Siracusa & Lotti, 2019). Intelligent packaging helps in minimizing food waste, food poisoning, and contamination concerns impacting the environment and health (Jafarzadeh et al., 2024). The data collected by the sensors can be displayed on the packaging or transmitted wirelessly to a central monitoring system, allowing retailers and consumers to track the condition of the fish throughout the supply chain. These films can also be used to detect and report information on the state of packaged food to extend shelf life, maintain quality, and guarantee safety (Srisa et al., 2022). Intelligent packaging offers real-time monitoring of environmental changes, ensuring food safety and quality throughout the supply chain (Lee & Rahman, 2014). Intelligent packaging uses sensors and indicators to track parameters including humidity and spoilage chemicals, providing data on food quality without the need to open the package (Pramitasari et al., 2022). Intelligent packaging is used for monitoring the environmental conditions of the food product and communicating information about the product quality (Boonsiriwit et al., 2022). Unlike active packaging, intelligent packaging does not directly interact with the food product; instead, it focuses on monitoring and communicating information.

28.5.3 Biopolymers infused with nanoparticles for fish preservation

The integration of nanoparticles into biopolymer films enhances their functionalities, leading to improved mechanical strength, barrier properties, and antimicrobial activity. Nanoparticles, such as silver nanoparticles, titanium dioxide nanoparticles, and zinc oxide nanoparticles, have antimicrobial properties and can inhibit the growth of spoilage bacteria on fish surfaces. The incorporation of nanoparticles into biopolymer films can also improve their barrier properties,

FIGURE 28.2 Application of the chitosan-alginate combination for fish preservation. (A) Chitosan alone, (B) alginate-$CaCl_2$ combination, (C) chitosan-alginate-$CaCl_2$, and (D) uncoated stored at 4°C.
From Bremenkamp, Ina & Sousa-Gallagher, Maria J. (2024). Design and Development of an Edible Coating for a Ready-to-Eat Fish Product. Polymers, 16(3). https://doi.org/10.3390/polym16030346.

reducing the permeation of oxygen and water vapor, and preventing lipid oxidation and dehydration of fish. The combination of nanotechnology and biopolymers provides a promising avenue for developing high-performance packaging materials for fish products.

The use of biopolymers with nanoparticles represents an innovative approach to enhance the functionality of fish packaging. Nanoparticles, such as metal nanoparticles (e.g., silver, zinc oxide), metal oxide nanoparticles (e.g., titanium dioxide, silicon dioxide), and nanoclays, can be incorporated into biopolymer films to improve their mechanical, barrier, and antimicrobial properties. Metal nanoparticles, such as silver nanoparticles, exhibit broad-spectrum antimicrobial activity and can effectively inhibit the growth of spoilage bacteria in fish (Perera et al., 2023). Metal oxide nanoparticles, such as titanium dioxide and zinc oxide nanoparticles, can enhance the UV-blocking properties of biopolymer films, protecting fish from light-induced degradation. Nanoclays, such as montmorillonite, can improve the mechanical strength and gas barrier properties of biopolymer films, extending the shelf life of fish products (Odobašić et al., 2015). Nanocomposite films not only enhance food preservation but also meet industrial and consumer demands for effective packaging solutions (Janjarasskul & Krochta, 2010). The incorporation of nanoparticles into biopolymer films requires careful consideration of their potential migration into the food product and their impact on human health and the environment (Primožič et al., 2021).

28.5.4 Biopolymers integrated with plant extraction

The integration of barberry, abundant in anthocyanins and polyphenols, into color indicator films presents a novel approach, leveraging its antibacterial and antioxidant attributes to enhance food safety and shelf life (Sani et al., 2020). Such advancements underscore the potential of natural resources in developing intelligent packaging solutions that actively contribute to food preservation and safety. Silver nanoparticles can be synthesized using agricultural waste, and these nanoparticles can then be integrated into biopolymer coatings to impart antimicrobial properties (Flores-Contreras et al., 2024). One study explored doping poly(vinyl alcohol) with graphene oxide decorated with silver nanoparticles, showcasing an eco-friendly method for synthesizing nanocomposites with enhanced antibacterial capabilities (Gautam et al., 2021). Additionally, incorporating essential oils into chitosan coatings has shown promise in combating postprocessing contamination of ready-to-eat meat products, with studies demonstrating the effectiveness of clove bud oil and cinnamon bark oil in reducing microbial loads (Wrońska et al., 2021). In addition to their antimicrobial and antioxidant properties, plant extracts can also enhance the mechanical and barrier properties of biopolymer films.

28.6 Performance characteristics and properties

The performance of biopolymer films and coatings in fish packaging is determined by several key characteristics, including mechanical properties, barrier properties, antimicrobial and antioxidative activity, and sensory impact. Mechanical properties, such as tensile strength, elongation at break, and puncture resistance, determine the film's ability to withstand physical stress during handling and storage. Barrier properties, such as water vapor permeability, oxygen permeability, and UV-light transmission, determine the film's ability to protect fish from moisture loss, oxygen exposure, and light-induced degradation. The interplay between food and packaging is significantly influenced by environmental factors such as humidity, temperature variations, and pressure fluctuations inherent in the supply chain, thereby emphasizing the necessity for comprehensive testing and adaptation to diverse conditions to ensure optimal film performance (Shah et al., 2024).

The tensile strength and elongation at break of biopolymer films are influenced by factors such as the type of biopolymer, the presence of additives, and the film formation method. Chitosan, a polysaccharide derived from crustacean shells, exhibits inherent antimicrobial properties, inhibiting the growth of a broad spectrum of microorganisms, including bacteria, yeasts, and molds (Oleksy et al., 2023; Tripathi et al., 2008). The antimicrobial activity of chitosan is attributed to its polycationic nature, which disrupts the negatively charged microbial cell membranes, leading to cell lysis and death (Dutta et al., 2008). The antimicrobial and antioxidative activity of biopolymer films and coatings is crucial for inhibiting spoilage and pathogenic microorganisms and preventing lipid oxidation in fish products (Dutta et al., 2008).

The sensory impact of biopolymer films and coatings on fish products is an important consideration, as consumers may be sensitive to changes in appearance, odor, and texture. The incorporation of certain additives, such as essential oils or plant extracts, may impart desirable flavors or aromas to the fish product. However, some additives may also cause undesirable changes in sensory attributes, such as off-odors or discoloration. The inclusion of natural compounds in films may affect the mechanical characteristics of the films (Souza et al., 2020). It is worth mentioning that biopolymer films are biodegradable and translucent and possess high mechanical strength (Barik et al., 2024).

Antimicrobial and antioxidative activity determine the film's ability to inhibit the growth of spoilage bacteria and prevent lipid oxidation in fish. Sensory impact, such as appearance, odor, and texture, determines the film's effect on the consumer's perception of the fish product (Ezati et al., 2021). The effectiveness of edible films is significantly influenced by ambient relative humidity, with higher humidity levels diminishing their efficacy as barriers against water vapor and gases (Olivas & Barbosa-Cánovas, 2008). Beeswax, when integrated into edible films, not only improves the film's moisture resistance and mechanical integrity but also introduces a subtle flavor profile that can positively influence the sensory attributes of the packaged food (Hosseini et al., 2023).

For instance, the addition of glycerol and *Aloe vera* can improve the tensile strength, elongation, and transparency of edible films (Siskawardani et al., 2020). The development and optimization of biopolymer films and coatings with tailored performance characteristics are crucial for ensuring the quality, safety, and shelf life of fish products.

The careful tailoring of biopolymer composition and the incorporation of functional additives are essential for achieving the desired balance of performance characteristics in fish packaging applications. The optimization of biopolymer film properties, such as mechanical strength, barrier properties, and antimicrobial activity, is achieved through strategic selection of raw materials, modification of film-forming techniques, and incorporation of functional additives. The assessment of biopolymer film performance requires the use of standardized testing methods to ensure accurate and reliable results. The mechanical properties of biopolymer films are assessed using tensile testing, puncture testing, and tear resistance testing to determine their ability to withstand physical stress during handling and storage. Water vapor permeability is assessed using gravimetric methods, while oxygen permeability is assessed using gas permeability testers to determine the film's ability to protect fish from moisture loss and oxygen exposure. Moreover, advanced spectroscopic techniques are harnessed to investigate the interaction between film components and their influence on barrier characteristics, offering insights into film performance. Table 28.1 provides details of different biopolymers and the application of these materials by the infusion of other active components.

28.7 Application examples and case studies

Edible coatings comprising alginates, plasticizers like glycerol, and olive oil exemplify the application of biopolymer films (Racmayani et al., 2020). Alginate films exhibit transparency and good quality, providing barrier properties against external factors and hindering the permeation of oils, fats, moisture, and oxygen (Anis et al., 2021). Gelatin and chitosan-based edible coatings have been successfully applied to preprocessed and fresh fish filets to extend their shelf life and improve their quality. Gelatin coatings provide a barrier against oxygen and moisture, reducing lipid oxidation and dehydration, while chitosan coatings exhibit antimicrobial activity, inhibiting the growth of spoilage bacteria.

For instance, nano-engineered films composed of chitosan and proanthocyanidin have demonstrated enhanced antioxidant and antimicrobial properties, extending the shelf life of fish products. Additionally, nano-engineered films incorporating essential oils have shown promise in inhibiting the growth of spoilage bacteria and extending the shelf life of fish products. The incorporation of rosemary essential oil into chitosan coatings, facilitated by nanoemulsion technology, not only enhances antimicrobial efficacy but also mitigates potential organoleptic alterations, thereby preserving the sensory integrity of the fish product.

Multilayer systems combining bio and synthetic polymers offer a synergistic approach to fish packaging, leveraging the advantages of both types of materials. The design of multilayer films involves strategic selection of individual layers with complementary properties, such as mechanical strength, barrier properties, and antimicrobial activity, to achieve optimal packaging performance. In addition, incorporating nanomaterials and essential oils contributes to heightened antimicrobial properties, thus prolonging the duration of freshness. The integration of sodium alginate films with essential oils prevents microbial contamination, showcasing the potential of biopolymers to replace plastic packaging in food applications (Gheorghiţă et al., 2020). Chitosan, derived from chitin, is recognized for its biocompatibility, biodegradability, and nontoxicity, positioning it as a versatile material for biomedical and food packaging applications.

Chitosan-based coatings, potentially enhanced with chitooligosaccharides and glutathione, demonstrate promise as an effective method for inhibiting bacterial growth, reducing total volatile basic nitrogen and malondialdehyde, and maintaining the sensory properties of fish during storage (Wu et al., 2014). Research indicates that applying chitosan-based solutions as edible coatings can protect frozen fish from deterioration, effectively preventing weight loss and maintaining microbial growth within acceptable limits (Soares et al., 2013). The application of chitosan coatings extends the shelf life of fresh produce, meat products, bread, and dairy products.

Alginate films can be used as edible coatings that can be consumed along with the food product. These coatings can provide additional benefits, such as improving the texture, flavor, or appearance of the fish. They can also be used to deliver functional ingredients, such as antioxidants or omega-3 fatty acids, to the consumer (Fig. 28.3).

28.8 Challenges and limitations

Despite the promising potential of biopolymer films and coatings for fish packaging, there are several challenges and limitations that need to be addressed. One major challenge is the scalability and industrial adoption of biopolymer-based packaging technologies. The transition from laboratory-scale production to large-scale manufacturing requires optimization of film-forming processes, development of cost-effective production methods, and ensuring consistent product quality. Cost considerations, stemming from raw material procurement and processing methodologies, wield considerable influence over the economic viability of biopolymer films compared to their conventional counterparts. The cost of biopolymers can be higher than that of synthetic polymers, which may limit their adoption in price-sensitive markets. Life-cycle assessments are also required to comprehensively evaluate the environmental impacts of biopolymer production, use, and disposal, considering factors such as energy consumption, greenhouse gas emissions, and waste generation.

Moreover, the effectiveness of biopolymer films is intricately linked to environmental factors, encompassing temperature and humidity, which may impact their mechanical and barrier attributes. Additionally, the mechanical properties of biopolymer films, such as their strength and flexibility, may not be as good as those of synthetic polymers, which can limit their application in certain packaging uses (Cao et al., 2020). The long-term stability and durability of

TABLE 28.1 List of some relevant research articles on biopolymer-based food packaging and coating materials.

Martix/biopolymer	Fish/fish products	Active/component	Application	References
Cellulose	Tilapia filets	Cyanidin-3-glucoside	Intelligent packaging	Shi et al. (2022)
Pectin	Yellow croaker *(Larimichthys polyactis)*	Oregano essential oil (OEO)	Smart packaging	Lan et al. (2022)
Carboxymethyl cellulose (CMC)/furcellaran	Salmon filets	Tea ground waste	Active packaging	Jamróz et al. (2022)
Pullulan and sodium carboxymethylcellulose	Sea bass (*Lateolabrax maculatus*) filets	Gallic acid and ε-polylysine hydrochloride (PL)	Edible coating	Li et al. (2021)
Chitosan	Catfish filets	Alginate	Edible coating	Basarah et al. (2021)
Polylactic acid (PLA)	Indian anchovy (*Stolephorus indicus*)	Nanocellulose (NC)	Active packaging	Thomas et al. (2021)
Chitosan (CS)/methylcellulose (MC)	Fish filets	*Phyllanthus reticulatus* ripen fruit anthocyanin	Smart packaging	Gasti et al. (2021)
Chitosan/CMC	Fish patty	Glutaraldehyde and cinnamon essential oil	Active packaging	Valizadeh et al. (2020)
CMC/starch (S)	Grass crap	Purple sweet potato anthocyanins	Intelligent packaging	Jiang et al. (2020)
Gelatin/oxidized chitin nanocrystals	Hairtail fish (*Trichiurus haumela*)	Black rice bran anthocyanins	Intelligent packaging	Ge et al. (2020)
κ-carageenan	Rainbow trout filets (*Oncorhynchus mykiss*)	Red cabbage extract (*Brassica oleracea*)	Smart packaging	Rostamzad et al. (2019)
Gelatin and water-soluble chitosan	Fresh tuna	Squid Maillard peptides	Active/intelligent packaging	Jiang et al. (2019)
PLA and sawdust particles	Pangasius fish filets	Bacteriocin—Bac7293	Active packaging	Woraprayote et al. (2018)
Agar	Wuchang bream (*Megalobrama amblycephala*)	*Arnebia euchroma* root extracts	Intelligent packaging	Huang et al. (2019)
Starch	Fresh silver carp	Polyvinyl alcohol/roselle anthocyanins	Intelligent packaging	Zhai et al. (2017)
Chitosan/silver carp skin gelatin	Fish muscle—cultured grass carp (*Ctenopharyngodon idella*)	OEO	Active packaging	Wu et al. (2014)
Chitosan	Japanese sea bass (*Lateolabrax japonicus*)	Ergothioneine	Active packaging	Cai et al. (2014)
Chitosan	Fresh lingcod (*Ophiodon elongatues*) filets	Krill oil	Modified atmospheric packaging	Duan et al. (2010)
Chitosan	Rainbow trout (*Oncorhynchus mykiss*)	Cinnamon oil	Edible coating	Ojagh et al. (2010)

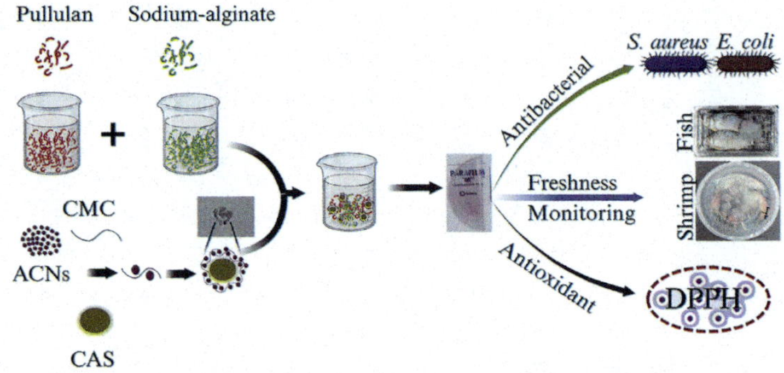

FIGURE 28.3 Pullulan/alginate film incorporated with anthocyanin-loaded casein/cellulose composite for real-time fish freshness monitoring (Rashid et al., 2024).
From Rashid, A., Qayum, A., Shah Bacha, S.A., Liang, Q., Liu, Y., Kang, L., Chi, Z., Chi, R., Han, X., Ekumah, J.N., Virk, M.S., Ren, X. & Ma, H. (2024). Novel pullulan-sodium alginate film incorporated with anthocyanin-loaded casein-carboxy methyl cellulose nanocomplex for real-time fish and shrimp freshness monitoring. Food Hydrocolloids, 156. https://doi.org/10.1016/j.foodhyd.2024.110356. (License no. 5937480623012).

biopolymer films under different storage conditions also need to be evaluated to ensure that they can provide adequate protection to the packaged fish throughout its shelf life.

Furthermore, consumer acceptance of biopolymer-based packaging is crucial for its successful adoption. Consumer perception of biopolymers is influenced by factors such as their appearance, texture, and odor, as well as their perceived safety and environmental friendliness. Addressing these challenges and limitations through research and development efforts is essential to unlock the full potential of biopolymer films and coatings for sustainable fish packaging (Bordón et al., 2021; Mahmud et al., 2024; Perera et al., 2023). Innovations in genetic engineering, catalyst development, and process optimization are expected to play critical roles in advancing these materials (Harun-Ur-Rashid & Imran, 2024).

Moreover, the lack of established regulatory frameworks and standards for biopolymer-based packaging materials can hinder their widespread adoption. Life-cycle assessment studies are needed to comprehensively evaluate the environmental impacts of biopolymer films, including energy consumption, greenhouse gas emissions, and waste generation, to ensure that they offer a net environmental benefit compared to conventional plastics. While biopolymers offer a promising alternative to traditional plastics in fish packaging, their widespread adoption necessitates overcoming challenges related to cost, scalability, and regulatory frameworks (Aider, 2010). Addressing these challenges will pave the way for the greater adoption of biopolymer films and coatings in the fish packaging industry, contributing to a more sustainable and environmentally friendly food system.

References

Aider, M. (2010). Chitosan application for active bio-based films production and potential in the food industry: Review. *LWT, 43*(6), 837–842. https://doi.org/10.1016/j.lwt.2010.01.021, https://www.journals.elsevier.com/lwt.

Alim, A. A. A., Shirajuddin, S. S. M., & Anuar, F. H. (2022). A review of nonbiodegradable and biodegradable composites for food packaging application [Review of a review of nonbiodegradable and biodegradable composites for food packaging application]. *Journal of Chemistry, 2022*, 1. https://doi.org/10.1155/2022/7670819 Hindawi Publishing Corporation.

Amin, U., Khan, M. K. I., Khan, M. U., Akram, M. E., Pateiro, M., Lorenzo, J. M., & Maan, A. A. (2021). Improvement of the performance of chitosan—Aloe vera coatings by adding beeswax on postharvest quality of mango fruit. *Foods, 10*(10), 2240. https://doi.org/10.3390/foods10102240.

Amin, U., Khan, M. U., Majeed, Y., Ребезов, M, Khayrullin, M., Bobkova, E. Y., Shariati, M. A., Chung, I. M., & Thiruvengadam, M. (2021). Potentials of polysaccharides, lipids and proteins in biodegradable food packaging applications [Review of Potentials of polysaccharides, lipids and proteins in biodegradable food packaging applications]. *International Journal of Biological Macromolecules, 183*, 2184. https://doi.org/10.1016/j.ijbiomac.2021.05.182 Elsevier BV.

Anis, A., Pal, K., & Al-Zahrani, S. M. (2021). Essential oil-containing polysaccharide-based edible films and coatings for food security applications [Review of Essential oil-containing polysaccharide-based edible films and coatings for food security applications]. *Polymers, 13*(4), 575. https://doi.org/10.3390/polym13040575 Multidisciplinary Digital Publishing Institute.

Asgher, M., Qamar, S. A., Bilal, M., & Iqbal, H. M. N. (2020). Bio-based active food packaging materials: Sustainable alternative to conventional petrochemical-based packaging materials. *Food Research International, 137*, 109625. https://doi.org/10.1016/j.foodres.2020.109625.

Avramescu, S. M., Butean, C., Popa, C. V., Ortan, A., Moraru, I., & Temocico, G. (2020). Edible and functionalized films/coatings-performances and perspectives. *Coatings, 10*(7), 20796412. https://doi.org/10.3390/coatings10070687, https://res.mdpi.com/d_attachment/coatings/coatings-10-00687/article_deploy/coatings-10-00687-v2.pdf.

Azizah, F., Nursakti, H., Ningrum, A., & Supriyadi (2023). Development of edible composite film from fish gelatin–pectin incorporated with lemongrass essential oil and its application in chicken meat. *Polymers, 15*(9), 2075. https://doi.org/10.3390/polym15092075.

Bangar, S. P., Chaudhary, V., Thakur, N., Kajla, P., Kumar, M., & Trif, M. (2021). Natural antimicrobials as additives for edible food packaging applications: A review. *Foods, 10*(10), 23048158. https://doi.org/10.3390/foods10102282, https://www.mdpi.com/2304-8158/10/10/2282/pdf.

Barik, M., BhagyaRaj, G. V. S., Dash, K. K., & Shams, R. (2024). A thorough evaluation of chitosan-based packaging film and coating for food product shelf-life extension. *Journal of Agriculture and Food Research, 16*. https://doi.org/10.1016/j.jafr.2024.101164, https://www.sciencedirect.com/science/journal/26661543.

Basarah, A. R., Pujiastuti, D. Y., & Waiprib, Y. (2021). Effect of deacetylation conditions on physicochemical properties of chitosan derived from shrimp shell and squid pen. *IOP Conference Series: Earth and Environmental Science, 679*(1). https://doi.org/10.1088/1755-1315/679/1/01203.

Bhat, R., Alias, A. K., & Paliyath, G. (2012). *Active and intelligent packaging of food progress in food preservation*. Wiley, 23–48. 10.1002/9781119962045.ch2.

Bizymis, A.-P., & Tzia, C. (2021). Edible films and coatings: Properties for the selection of the components, evolution through composites and nanomaterials, and safety issues [Review of Edible films and coatings: Properties for the selection of the components, evolution through composites and nanomaterials, and safety issues]. *Critical Reviews in Food Science and Nutrition, 62*(31), 8777. https://doi.org/10.1080/10408398.2021.1934652 Taylor & Francis.

Boonsiriwit, A., Itkor, P., Sirieawphikul, C., & Lee, Y. S. (2022). Characterization of natural anthocyanin indicator based on cellulose bio-composite film for monitoring the freshness of chicken tenderloin. *Molecules, 27*(9), 14203049. https://doi.org/10.3390/molecules27092752, https://www.mdpi.com/1420-3049/27/9/2752/pdf.

Bordón, P., Paz, R., Peñalva, C., Vega, G., Monzón, M., & García, L. (2021). Biodegradable polymer compounds reinforced with banana fiber for the production of protective bags for banana fruits in the context of circular economy. *Agronomy, 11*(2), 242. https://doi.org/10.3390/agronomy11020242.

Bremenkamp, I., & Sousa-Gallagher, M. J. (2024). Design and development of an edible coating for a ready-to-eat fish product. *Polymers, 16*(3), 346. https://doi.org/10.3390/polym16030346.

Brodnjak, U. V., & Tihole, K. (2020). Chitosan solution containing zein and essential oil as bio based coating on packaging paper. *Coatings, 10*(5). https://doi.org/10.3390/COATINGS10050497, https://www.mdpi.com/2079-6412/10/5/497.

Butnaru, E., Stoleru, E., Brebu, M. A., Darie-Nita, R. N., Bargan, A., & Vasile, C. (2019). Chitosan-based bionanocomposite films prepared by emulsion technique for food preservation. *Materials, 12*(3), 373. https://doi.org/10.3390/ma12030373.

Cai, L., Li, X., Wu, X., Lv, Y., Liu, X., & Li, J. (2014). Effect of chitosan coating enriched with ergothioneine on quality changes of Japanese sea bass (*Lateolabrax japonicas*). *Food and Bioprocess Technology, 7*(8), 2281–2290. https://doi.org/10.1007/s11947-013-1215-4.

Cao, W., Yan, J., Liu, C., Zhang, J., Wang, H., Gao, X., Yan, H., Niu, B., & Li, W. (2020). Preparation and characterization of catechol-grafted chitosan/gelatin/modified chitosan-AgNP blend films. *Carbohydrate Polymers, 247*, 116643. https://doi.org/10.1016/j.carbpol.2020.116643.

Cortés-Sánchez, A. D. J., Jaramillo-Flores, M. E., Díaz-Ramírez, M., Espinosa-Chaurand, L. D., & Torres-Ochoa, E. (2024). Biopreservation and the safety of fish and fish products, the case of lactic acid bacteria: A basic perspective. *Fishes, 9*(8). https://doi.org/10.3390/fishes9080303, www.mdpi.com/journal/fishes/.

Cruz, R. M. S., Alves, V., Khmelinskii, I., & Vieira, M. C. (2018). *New food packaging systems*. Elsevier BV, 63–85 10.1016/b978-0-12-811516-9.00002-6.

Díaz-Montes, E., & Castro-Muñoz, R. (2021). Edible films and coatings as food-quality preservers: An overview. *Foods, 10*(2), 249. https://doi.org/10.3390/foods10020249.

Donkor, L., Kontoh, G., Yaya, A., Bediako, J. K., & Apalangya, V. (2023). Bio-based and sustainable food packaging systems: Relevance, challenges, and prospects. *Applied Food Research, 3*(2), 100356. https://doi.org/10.1016/j.afres.2023.100356.

Duan, J., Jiang, Y., Cherian, G., & Zhao, Y. (2010). Effect of combined chitosan-krill oil coating and modified atmosphere packaging on the storability of cold-stored lingcod (*Ophiodon elongates*) fillets. *Food Chemistry, 122*(4), 1035–1042. https://doi.org/10.1016/j.foodchem.2010.03.065.

Dutta, P. K., Tripathi, S., Mehrotra, G. K., & Dutta, J. (2008). Perspectives for chitosan-based antimicrobial films in food applications. *Food Chemistry, 114*(4), 1173. https://doi.org/10.1016/j.foodchem.2008.11.047.

Ezati, P., Priyadarshi, R., Bang, Y.-J., & Rhim, J.-W. (2021). CMC and CNF-based intelligent pH-responsive color indicator films integrated with shikonin to monitor fish freshness. *Food Control, 126*, 108046. https://doi.org/10.1016/j.foodcont.2021.108046.

Flores-Contreras, E. A., González-González, R. B., Pablo Pizaña-Aranda, J. J., Parra-Arroyo, L., Rodríguez-Aguayo, A. A., Iñiguez-Moreno, M., González-Meza, G. M., Araújo, R. G., Ramírez-Gamboa, D., Parra-Saldívar, R., & Melchor-Martínez, E. M. (2024). Agricultural waste as a sustainable source for nanoparticle synthesis and their antimicrobial properties for food preservation. *Frontiers in Nanotechnology, 6*. https://doi.org/10.3389/fnano.2024.1346069, www.frontiersin.org/journals/nanotechnology#.

Freitas, J. A. M., Mendonça, G. M. N., Santos, L. B., Alonso, J. D., Mendes, J. F., Barud, H. S., & Azeredo, H. M. C. (2022). Bacterial cellulose/tomato puree edible films as moisture barrier structures in multicomponent foods. *Foods, 11*(15). https://doi.org/10.3390/foods11152336, www.mdpi.com/journal/foods.

Galus, S., Kibar, E. A. A., Gniewosz, M., & Kraśniewska, K. (2020). Novel materials in the preparation of edible films and coatings-A review. *Coatings, 10*(7). https://doi.org/10.3390/coatings10070674, https://res.mdpi.com/d_attachment/coatings/coatings-10-00674/article_deploy/coatings-10-00674.pdf.

Gasti, T., Dixit, S., D'souza, O. J., Hiremani, V. D., Vootla, S. K., Masti, S. P., Chougale, R. B., & Malabadi, R. B. (2021). Smart biodegradable films based on chitosan/methylcellulose containing *Phyllanthus reticulatus* anthocyanin for monitoring the freshness of fish fillet. *International Journal of Biological Macromolecules, 187*, 451–461. https://doi.org/10.1016/j.ijbiomac.2021.07.128, www.elsevier.com/locate/ijbiomac.

Gautam, S., Sharma, S., Sharma, B., & Jain, P. (2021). Antibacterial efficacy of poly (vinyl alcohol) nanocomposites reinforced with graphene oxide and silver nanoparticles for packaging applications. *Polymer Composites, 42*(6), 2829–2837. https://doi.org/10.1002/pc.26017.

Ge, Y., Li, Y., Bai, Y., Yuan, C., Wu, C., & Hu, Y. (2020). Intelligent gelatin/oxidized chitin nanocrystals nanocomposite films containing black rice bran anthocyanins for fish freshness monitorings. *International Journal of Biological Macromolecules, 155*, 1296–1306. https://doi.org/10.1016/j.ijbiomac.2019.11.101.

Gennadios, A., Hanna, M. A., & Kurth, L. B. (1997). Application of edible coatings on meats, poultry and seafoods: A review [Review of Application of edible coatings on meats, poultry and seafoods: A review]. *LWT, 30*(4), 337. https://doi.org/10.1006/fstl.1996.0202 Elsevier BV.

Gheorghiţă, R., Gutt, G., & Amariei, S. (2020). The use of edible films based on sodium alginate in meat product packaging: An eco-friendly alternative to conventional plastic materials. *Coatings, 10*(2), 166. https://doi.org/10.3390/coatings10020166.

Gowsia, I., Mir, F. A., & Banday, J. A. (2022). Preparation and characterization of polyvinyl alcohol–piperic acid composite film for potential food packaging applications. *Progress in Biomaterials, 11*(3), 281. https://doi.org/10.1007/s40204-022-00195-6.

Hamzeh, A., & Rezaei, M. (2012). The effects of sodium alginate on quality of rainbow trout (*Oncorhynchus mykiss*) fillets stored at 4 ± 2°C. *Journal of Aquatic Food Product Technology, 21*(1), 14–21. https://doi.org/10.1080/10498850.2011.579384.

Hari, K. D., Garcia, C. V., Shin, G. H., & Kim, J. T. (2021). Improvement of the UV barrier and antibacterial properties of crosslinked pectin/zinc oxide bionanocomposite films. *Polymers, 13*(15). https://doi.org/10.3390/polym13152403, https://www.mdpi.com/2073-4360/13/15/2403/pdf.

Harun-Ur-Rashid, M., & Imran, A. B. (2024). Emerging trends in engineering polymers: A paradigm shift in material engineering. *Recent Progress in Materials, 06*(03), 1–37. https://doi.org/10.21926/rpm.2403024.

Hashemi, S. M. B., Kaveh, S., Abedi, E., & Phimolsiripol, Y. (2023). Polysaccharide-based edible films/coatings for the preservation of meat and fish products: Emphasis on incorporation of lipid-based nanosystems loaded with bioactive compounds. *Foods, 12*(17). https://doi.org/10.3390/foods12173268, www.mdpi.com/journal/foods.

Hassan, B., Chatha, S. A. S., Hussain, A. I., Zia, K. M., & Akhtar, N. (2017). Recent advances on polysaccharides, lipids and protein based edible films and coatings: A review [Review of Recent advances on polysaccharides, lipids and protein based edible films and coatings: A review]. *International Journal of Biological Macromolecules, 109*, 1095. https://doi.org/10.1016/j.ijbiomac.2017.11.097 Elsevier BV.

Hosseini, S. F., Mousavi, Z., & McClements, D. J. (2023). Beeswax: A review on the recent progress in the development of superhydrophobic films/coatings and their applications in fruits preservation [Review of Beeswax: A review on the recent progress in the development of superhydrophobic films/coatings and their applications in fruits preservation]. *Food Chemistry, 424*, 136404. https://doi.org/10.1016/j.foodchem.2023.136404 Elsevier BV.

Huang, S., Xiong, Y., Zou, Y., Dong, Q., Ding, F., Liu, X., & Li, H. (2019). A novel colorimetric indicator based on agar incorporated with *Arnebia euchroma* root extracts for monitoring fish freshness. *Food Hydrocolloids, 90*, 198–205. https://doi.org/10.1016/j.foodhyd.2018.12.009.

Ismillayli, N., Hadi, S., Tri Dharmayani, N. K., Sanjaya, R. K., & Hermanto, D. (2020). Characterization of alginate-chitosan membrane as potential edible film. *IOP Conference Series: Materials Science and Engineering, 833*(1). https://doi.org/10.1088/1757-899X/833/1/012073, https://iopscience.iop.org/journal/1757-899X.

Jafarzadeh, S., Yildiz, Z., Yildiz, P., Strachowski, P., Forough, M., Esmaeili, Y., Naebe, M., & Abdollahi, M. (2024). Advanced technologies in biodegradable packaging using intelligent sensing to fight food waste. *International Journal of Biological Macromolecules, 261*(Pt 1), 129647. https://doi.org/10.1016/j.ijbiomac.2024.129647.

Jamróz, E., Tkaczewska, J., Kopeć, M., & Cholewa-Wójcik, A. (2022). Shelf-life extension of salmon using active total biodegradable packaging with tea ground waste and furcellaran-CMC double-layered films. *Food Chemistry, 383*, 132425. https://doi.org/10.1016/j.foodchem.2022.132425.

Janjarasskul, T., & Krochta, J. M. (2010). Edible packaging materials. *Annual Review of Food Science and Technology, 1*(1), 415–448. https://doi.org/10.1146/annurev.food.080708.100836.

Jiang, G., Hou, X., Zeng, X., Zhang, C., Wu, H., Shen, G., Li, S., Luo, Q., Li, M., Liu, X., Chen, A., Wang, Z., & Zhang, Z. (2020). Preparation and characterization of indicator films from carboxymethyl-cellulose/starch and purple sweet potato (*Ipomoea batatas* (L.) lam) anthocyanins for monitoring fish freshness. *International Journal of Biological Macromolecules, 143*, 359–372. https://doi.org/10.1016/j.ijbiomac.2019.12.024.

Jiang, W., Hu, S., Li, S., & Liu, Y. (2019). Evaluation of the preservation effect of gelatin-water soluble chitosan film incorporated with maillard peptides on bluefin tuna (*Thunnus thynnus*) slices packaging. *LWT, 113*, 108294. https://doi.org/10.1016/j.lwt.2019.108294.

Khodaei, D., Álvarez, C., & Mullen, A. M. (2021). Biodegradable packaging materials from animal processing co-products and wastes: An overview [Review of Biodegradable packaging materials from animal processing co-products and wastes: An overview]. *Polymers, 13*(15), 2561. https://doi.org/10.3390/polym13152561 Multidisciplinary Digital Publishing Institute.

Kocira, A., Kozłowicz, K., Panasiewicz, K., Staniak, M., Szpunar-Krok, E., & Hortyńska, P. (2021). Polysaccharides as edible films and coatings: Characteristics and influence on fruit and vegetable quality—A review. *Agronomy, 11*(5), 813. https://doi.org/10.3390/agronomy11050813.

Kumar, N., & Neeraj (2019). Polysaccharide-based component and their relevance in edible film/coating: A review. *Nutrition & Food Science, 49*(5), 793–823. https://doi.org/10.1108/nfs-10-2018-0294.

Kumar, N., Pratibha, Trajkovska Petkoska, A., Khojah, E., Sami, R., & Al-Mushhin, A. A. M. (2021). Chitosan edible films enhanced with pomegranate peel extract: Study on physical, biological, thermal, and barrier properties. *Materials, 14*(12). https://doi.org/10.3390/ma14123305, https://www.mdpi.com/1996-1944/14/12/3305/pdf.

Lacroix, M., & Le Tien, C. (2005). *Edible films and coatings from nonstarch polysaccharides,* 338–361. https://doi.org/10.1016/b978-012311632-1/50052-8.

Lan, W., Lang, A., Chen, M., & Xie, J. (2022). Combined effects of pectin–plant essential oil coating with vacuum packaging on the quality of large yellow croaker (*Pseudosciaena crocea*) during iced storage. *Journal of Food Safety, 42*(2). https://doi.org/10.1111/JFS.12960;PAGE:STRING:ARTICLE/CHAPTER.

Lee, S. J., & Rahman, A. T. M. M. (2014). *Intelligent packaging for food products*. Elsevier BV, 171–209. 10.1016/b978-0-12-394601-0.00008-4.

Li, Q., Zhang, J., Zhu, J., Lin, H., Sun, T., & Cheng, L. (2021). Effects of gallic acid combined with epsilon-polylysine hydrochloride incorporated in a pullulan–CMC edible coating on the storage quality of sea bass. *RSC Advances, 11*(47), 29675–29683. https://doi.org/10.1039/d1ra02320a.

Mahmud, M. Z. A., Mobarak, M. H., & Hossain, N. (2024). Emerging trends in biomaterials for sustainable food packaging: A comprehensive review [Review of Emerging trends in biomaterials for sustainable food packaging: A comprehensive review]. *Heliyon, 10*(1). https://doi.org/10.1016/j.heliyon.2024.e24122 Elsevier BV.

Mai, H., Dao, K. Q., Hoang, H. C., Nguyen, T. T., & Hoang Nguyen, D. (2023). Zinc ions implantation on the surface of commercial papers for antibacterial and antifungal applications. *VNU Journal of Science: Natural Sciences and Technology*. https://doi.org/10.25073/2588-1140/vnunst.5545.

Mane, K. A. (2016). A review on active packaging: An innovation in food packaging. *International Journal of Environment, Agriculture and Biotechnology, 1*(3), 544–549. https://doi.org/10.22161/ijeab/1.3.35.

Mangaraj, S., Yadav, A., Bal, L. M., Dash, S. K., & Mahanti, N. K. (2018). Application of biodegradable polymers in food packaging industry: A comprehensive review [Review of Application of biodegradable polymers in food packaging industry: A comprehensive review]. *Journal of Packaging Technology and Research, 3*(1), 77. https://doi.org/10.1007/s41783-018-0049-y Springer Science+Business Media.

Mohamed, S. A. A., El-Sakhawy, M., & El-Sakhawy, M. A. (2020). Polysaccharides, protein and lipid -based natural edible films in food packaging: A review [Review of Polysaccharides, protein and lipid -based natural edible films in food packaging: A review]. *Carbohydrate Polymers, 238*, 116178. https://doi.org/10.1016/j.carbpol.2020.116178 Elsevier BV.

Nagarajan, M., Rajasekaran, B., Benjakul, S., & Venkatachalam, K. (2021). Influence of chitosan-gelatin edible coating incorporated with longkong pericarp extract on refrigerated black tiger Shrimp (*Penaeus monodon*). *Current Research in Food Science, 4*, 345–353. https://doi.org/10.1016/j.crfs.2021.05.003.

Odobašić, A., Ćatić, S., Šestan, & Bratovčić, A. (2015). Application of polymer nanocomposite materials in food packaging. *Croatian Journal of Food Science and Technology, 7*(2). https://doi.org/10.17508/cjfst.2015.7.2.06.

Ojagh, S. M., Rezaei, M., Razavi, S. H., & Hosseini, S. M. H. (2010). Effect of chitosan coatings enriched with cinnamon oil on the quality of refrigerated rainbow trout. *Food Chemistry, 120*(1), 193–198. https://doi.org/10.1016/j.foodchem.2009.10.006.

Oleksy, M., Dynarowicz, K., & Aebisher, D. (2023). Advances in biodegradable polymers and biomaterials for medical applications—A review [Review of Advances in biodegradable polymers and biomaterials for medical applications—A review]. *Molecules, 28*(17), 6213. https://doi.org/10.3390/molecules28176213 Multidisciplinary Digital Publishing Institute.

Olivas, G. I., & Barbosa-Cánovas, G. V. (2008). Alginate-calcium films: Water vapor permeability and mechanical properties as affected by plasticizer and relative humidity. *Academic Press, Mexico LWT, 41*(2), 359–366. https://doi.org/10.1016/j.lwt.2007.02.015, https://www.journals.elsevier.com/lwt.

Paola, M. G. D., Paletta, R., Lopresto, C. G., Lio, G. E., Luca, A. D., Chakraborty, S., & Calabrò, V. (2021). Stability of film-forming dispersions: Affects the morphology and optical properties of polymeric films. *Polymers, 13*(9), 1464. https://doi.org/10.3390/polym13091464.

Paunonen, S. (2013). Strength and barrier enhancements of cellophane and cellulose derivative films: A review. *BioResources, 8*(2), 3098–3121. https://doi.org/10.15376/biores.8.2.3098-3121, http://www.ncsu.edu/bioresources/BioRes_08/BioRes_08_2_Review_b_3098_Paunonen_Cellulose_Derivatives_Food_Packaging_Review_3308.pdf.

Perera, K. Y., Jaiswal, A. K., & Jaiswal, S. (2023). Biopolymer-based sustainable food packaging materials: Challenges, solutions, and applications [Review of Biopolymer-based sustainable food packaging materials: Challenges, solutions, and applications]. *Foods, 12*(12), 2422. https://doi.org/10.3390/foods12122422 Multidisciplinary Digital Publishing Institute.

Popescu, P. A., Popa, E. E., Mitelut, A. C., & Popa, M. E. (2020). Development of recyclable and biodegradable food packaging materials – Opportunities and risks. *Current Trends in Natural Sciences, 9*(17), 142–146. https://doi.org/10.47068/ctns.2020.v9i17.016.

Pramitasari, R., Gunawicahya, L. N., & Anugrah, D. S. B. (2022). Development of an indicator film based on cassava starch–chitosan incorporated with red dragon fruit peel anthocyanin extract. *Polymers, 14*(19), 4142. https://doi.org/10.3390/polym14194142.

Primožič, M., Knez, Ž., & Leitgeb, M. (2021). Review of (Bio)Nanotechnology in food science—Food packaging. *Nanomaterials, 11*(2), 292. https://doi.org/10.3390/nano11020292 Multidisciplinary Digital Publishing Institute.

Pudjiastuti, L., Sugianto, N. N., Hamzah, A., Zuchrillah, D. R., Puspita, N. F., & Rosalya, A. (2021). Effect of antimicrobial addition from lime extract on edible film as food packaging. *IOP Conference Series: Earth and Environmental Science, 913*(1). https://doi.org/10.1088/1755-1315/913/1/012031, https://iopscience.iop.org/journal/1755-1315.

Racmayani, N., Husni, A., Isnansetyo, A., Jayanti, A. D., Sari, D. W. K., Puspita, I. D., Putra, M. M. P., Huda, N., Bhujel, R. C., & Adharini, R. I. (2020). Effect of different formulations on characteristic of biobased alginate edible films as biodegradable packaging. *E3S Web of Conferences, 147*, 03003. https://doi.org/10.1051/e3sconf/202014703003.

Rahman, S., Gogoi, J., Dubey, S., & Chowdhury, D. (2024). Animal derived biopolymers for food packaging applications: A review. *International Journal of Biological Macromolecules, 255*, 128197. https://doi.org/10.1016/j.ijbiomac.2023.128197.

Rashid, A., Qayum, A., Shah Bacha, S. A., Liang, Q., Liu, Y., Kang, L., Chi, Z., Chi, R., Han, X., Ekumah, J. N., Virk, M. S., Ren, X., & Ma, H. (2024). Novel pullulan-sodium alginate film incorporated with anthocyanin-loaded casein-carboxy methyl cellulose nanocomplex for real-time fish and shrimp freshness monitoring. *Food Hydrocolloids, 156*. https://doi.org/10.1016/j.foodhyd.2024.110356, https://www.sciencedirect.com/science/journal/0268005X.

Rocha, M.da, Souza, M. M.de, & Prentice, C. (2018). Biodegradable films: An alternative food packaging. *Elsevier eBooks*, 307. https://doi.org/10.1016/b978-0-12-811516-9.00009-9 Elsevier BV.

Rojas, A., Velásquez, E., Piña, C., Galotto, M. J., & López de Dicastillo, C. (2021). Designing active mats based on cellulose acetate/polycaprolactone core/shell structures with different release kinetics. *Carbohydrate Polymers, 261*. https://doi.org/10.1016/j.carbpol.2021.117849, http://www.elsevier.com/wps/find/journaldescription.cws_home/405871/description#description.

Rostamzad, H., Kamali, N., & Sabeti, A. (2019). Babakhani, Production and evaluation of smart biodegradable film based on carrageenan for fish fillet packaging. *Journal of Fisheries, 72*(1), 85–95. https://doi.org/10.22059/JFISHERIES.2019.74392.

Rydz, J., Musioł, M., Zawidlak-Węgrzyńska, B., & Sikorska, W. (2018). *Present and future of biodegradable polymers for food packaging applications*. Elsevier BV, 431–467. 10.1016/b978-0-12-811449-0.00014-1.

Sani, M. A., Tavassoli, M., Mohammadian, E., Ehsani, A., Khaniki, G. J., Priyadarshi, R., & Rhim, J.-W. (2020). pH-responsive color indicator films based on methylcellulose/chitosan nanofiber and barberry anthocyanins for real-time monitoring of meat freshness. *International Journal of Biological Macromolecules, 166*, 741. https://doi.org/10.1016/j.ijbiomac.2020.10.231.

Sathivel, S., Liu, Q., Huang, J., & Prinyawiwatkul, W. (2007). The influence of chitosan glazing on the quality of skinless pink salmon (*Oncorhynchus gorbuscha*) fillets during frozen storage. *Journal of Food Engineering, 83*(3), 366–373. https://doi.org/10.1016/j.jfoodeng.2007.03.009.

Shah, Y. A., Bhatia, S., Al-Harrasi, A., & Khan, T. S. (2024). Advancements in the biopolymer films for food packaging applications: A short review. *Biotechnology for Sustainable Materials, 1*(1). https://doi.org/10.1186/s44316-024-00002-1.

Shi, C., Ji, Z., Zhang, J., Jia, Z., & Yang, X. (2022). Preparation and characterization of intelligent packaging film for visual inspection of tilapia fillets freshness using cyanidin and bacterial cellulose. *International Journal of Biological Macromolecules, 205*, 357–365. https://doi.org/10.1016/j.ijbiomac.2022.02.072.

Silva, F. A. G. S., Dourado, F., Gama, M., & Poças, F. (2020). Nanocellulose bio-based composites for food packaging. *Portugal Nanomaterials, 10*(10), 1–29. https://doi.org/10.3390/nano10102041, https://www.mdpi.com/2079-4991/10/10/2041/pdf.

Silva, O. A., Pellá, M. G., Pellá, M. G., Caetano, J., Simões, M. R., Bittencourt, P. R. S., & Dragunski, D. C. (2019). Synthesis and characterization of a low solubility edible film based on native cassava starch. *International Journal of Biological Macromolecules, 128*, 290–296. https://doi.org/10.1016/j.ijbiomac.2019.01.132, www.elsevier.com/locate/ijbiomac.

Siracusa, V., & Lotti, N. (2019). *Intelligent packaging to improve shelf life*. Elsevier BV, 261–279. 10.1016/b978-0-12-817190-5.00008-2.

Siskawardani, D. D., Warkoyo, W., & Siwi, A. A. P. (2020). The effect of *Aloe vera* and glycerol addition on edible film of lesser yam starch (*Dioscorea esculenta* L. burkill). *Food Technology and Halal Science Journal, 3*(1), 26. https://doi.org/10.22219/fths.v3i1.13057.

Soares, N., Mendes, T. S., & Vicente, A. A. (2013). Effect of chitosan-based solutions applied as edible coatings and water glazing on frozen salmon preservation – A pilot-scale study. *Journal of Food Engineering, 119*(2), 316. https://doi.org/10.1016/j.jfoodeng.2013.05.018.

Souza, V. G. L., Pires, J. R. A., Rodrigues, C., Coelhoso, I. M., & Fernando, A. L. (2020). Chitosan composites in packaging industry-current trends and future challenges. *Polymers, 12*(2). https://doi.org/10.3390/polym12020417, https://res.mdpi.com/d_attachment/polymers/polymers-12-00417/article_deploy/polymers-12-00417.pdf.

Srisa, A., Promhuad, K., San, H., Laorenza, Y., Wongphan, P., Wadaugsorn, K., Sodsai, J., Kaewpetch, T., Tansin, K., & Harnkarnsujarit, N. (2022). Antibacterial, antifungal and antiviral polymeric food packaging in post-COVID-19 era. *Polymers, 14*(19), 4042. https://doi.org/10.3390/polym14194042.

Susmitha, A., Sasikumar, K., Rajan, D., Padmakumar M, A., & Nampoothiri, K. M. (2021). Development and characterization of corn starch-gelatin based edible films incorporated with mango and pineapple for active packaging. *Food Bioscience, 41*, 22124306. https://doi.org/10.1016/j.fbio.2021.100977, http://www.journals.elsevier.com/food-bioscience.

Thomas, G., Kumar, K. S., Panda, S. K., & Bindu, J. (2021). Nanocellulose incorporated polylactic acid films for chilled preservation of Indian Anchovy (*Stolephorus indicus*) (van Hasselt, 1823). *Fishery Technology, 58*(2). https://doi.org/10.56093/FT.V58I2.115424.

Tripathi, S., Mehrotra, G. K., & Dutta, P. K. (2008). Chitosan based antimicrobial films for food packaging applications. *E-Polymers, 8*(1). https://doi.org/10.1515/epoly.2008.8.1.1082.

Umiyati, R., Hidayat, C., Millati, R., Ariyanto, T., & Hadiyanto, B. (2019). Warsito, maryono, physical and antimicrobial properties of hydroxypropyl starch bio-plastics incorporated with nyamplung (*Calophyllum inophyllum*) cake extract as an eco-friendly food packaging. *E3S Web of Conferences, 125*, 07011. https://doi.org/10.1051/e3sconf/201912507011.

Valizadeh, S., Naseri, M., Babaei, S., & Hosseini, S. M. H. (2020). Shelf life extension of fish patty using biopolymer-coated active paper sheets. *Iran Food Packaging and Shelf Life, 26*. https://doi.org/10.1016/j.fpsl.2020.100603, http://www.journals.elsevier.com/food-packaging-and-shelf-life/.

Vibha, C., Parameswaranpillai, J., Krishnasamy, S., Siengchin, S., Jayakumar, A., Radoor, S., Mavinkere Rangappa, S., V. Salim, N., Hameed, N., Praveen, G. L., & Midhun Dominic, C. D. (2022). Biodegradable polymers and green-based antimicrobial packaging materials. *Biodegradable Polymers, Blends and Composites*, 717–733. https://doi.org/10.1016/B978-0-12-823791-5.00005-3 Elsevier.

Woraprayote, W., Pumpuang, L., Tosukhowong, A., Zendo, T., Sonomoto, K., Benjakul, S., & Visessanguan, W. (2018). Antimicrobial biodegradable food packaging impregnated with Bacteriocin 7293 for control of pathogenic bacteria in pangasius fish fillets. *LWT, 89*, 427–433. https://doi.org/10.1016/j.lwt.2017.10.026.

Wrońska, N., Katir, N., Miłowska, K., Hammi, N., Nowak, M., Kędzierska, M., Anouar, A., Zawadzka, K., Bryszewska, M., Kadib, A. E., & Lisowska, K. (2021). Antimicrobial effect of chitosan films on food spoilage bacteria. *International Journal of Molecular Sciences, 22*(11), 14220067. https://doi.org/10.3390/ijms22115839, https://www.mdpi.com/1422-0067/22/11/5839/pdf.

Wu, J., Ge, S., Liu, H., Wang, S., Chen, S., Wang, J., Li, J., & Zhang, Q. (2014). Properties and antimicrobial activity of silver carp (*Hypophthalmichthys molitrix*) skin gelatin-chitosan films incorporated with oregano essential oil for fish preservation. *Food Packaging and Shelf Life, 2*(1), 7–16. https://doi.org/10.1016/j.fpsl.2014.04.004.

Yuvaraj, D., Iyyappan, J., Gnanasekaran, R., Ishwarya, G., Harshini, R. P., Dhithya, V., Chandran, M., Kanishka, V., & Gomathi, K. (2021). Advances in bio food packaging – An overview. *Heliyon, 7*(9), e07998. https://doi.org/10.1016/j.heliyon.2021.e07998.

Zhai, X., Shi, J., Zou, X., Wang, S., Jiang, C., Zhang, J., Huang, X., Zhang, W., & Holmes, M. (2017). Novel colorimetric films based on starch/polyvinyl alcohol incorporated with roselle anthocyanins for fish freshness monitoring. *Food Hydrocolloids, 69*, 308–317. https://doi.org/10.1016/j.foodhyd.2017.02.014.

Zhang, L., Yu, D., Regenstein, J. M., Xia, W., & Dong, J. (2021). A comprehensive review on natural bioactive films with controlled release characteristics and their applications in foods and pharmaceuticals. *Trends in Food Science & Technology, 112*, 690–707. https://doi.org/10.1016/j.tifs.2021.03.053.

Zou, J., Liu, X., Wang, X., Yang, H., Cheng, J., Lin, Y., & Tang, D. (2022). Influence of gelatin-chitosan-glycerol edible coating incorporated with chlorogenic acid, gallic acid, and resveratrol on the preservation of fresh beef. *Foods, 11*(23), 3813. https://doi.org/10.3390/foods11233813.

Chapter 29

Emerging trends and innovations in biopolymer films and coatings for the preservation of fruits

Abhishek Biswal R.[1], Divyashree J.S.[1], Periyar Selvam Sellamuthu[1], and Emmanuel Rotimi Sadiku[2]

[1]*Postharvest Research Lab, Department of Food Technology (Formerly Food Process Engineering), School of Bioengineering, SRM Institute of Science and Technology, Kattankulathur, Chengalpattu, Tamil Nadu, India,* [2]*Department of Chemical, Metallurgical and Materials Engineering, Institute of NanoEngineering Research (INER), Tshwane University of Technology, Pretoria, Republic of South Africa*

29.1 Introduction

Fruit preservation faces various challenges, particularly related to microbial spoilage, moisture loss, and enzymatic degradation, which can quickly lead to reduced quality and shorter shelf life. Microbial contamination is a primary issue, as microorganisms such as bacteria, yeasts, and molds thrive on the sugars and nutrients present in fruit, accelerating spoilage. Strategies to combat microbial spoilage often involve adjusting storage conditions like lowering temperature and reducing oxygen levels, as each type of microorganism responds differently, complicating preservation efforts. Additionally, methods such as the use of modified atmospheres (e.g., low oxygen and high CO_2) can help control the microbial growth and delay the ripening; however, they must be finely tuned to avoid damage to the fruit itself, particularly for temperature-sensitive varieties such as tropical fruits. Another key challenge is the management of moisture loss and enzymatic browning. As fruits lose moisture, they become susceptible to wilting and shriveling, significantly impacting their appearance and texture (Peng, 2022). Technologies like edible coatings and controlled humidity storage can help retain moisture; however, these techniques are not always universally effective across different fruit types. Enzymatic browning, caused by polyphenol oxidase, is another issue that can lead to undesirable discoloration, especially in cut or bruised fruit. Common solutions include cold storage, which slows enzymatic activity, and chemical treatments, though there is growing consumer demand for natural and chemical-free preservation methods. Innovative preservation technologies are emerging to address these issues, such as high-pressure processing, UV-C light treatments, and the application of natural antimicrobials like essential oils (EOs). Each technique offers benefits but also poses specific challenges. For instance, high-pressure processing can effectively reduce microbial load without heat, maintaining freshness; yet, it is often costly. Similarly, UV-C treatments help sanitize the fruit surface, though their penetration is limited, meaning effectiveness depends on thorough coverage and application (Amit et al., 2017; Rosenthal et al., 2018).

Biopolymer films and coatings are essential for extending the shelf life of fresh produce by providing a protective barrier against environmental factors, moisture loss, and microbial growth. These materials, often derived from natural sources like polysaccharides, proteins, and lipids, are increasingly popular due to their biodegradable and nontoxic nature, aligning with sustainable food preservation practices. Edible coatings, for instance, form a thin layer on the fruit's surface, reducing respiration rates, delaying ripening, and enhancing texture retention. When infused with natural antimicrobials like EOs or organic acids, these films can also inhibit bacterial and fungal growth, effectively reducing spoilage. Additionally, biopolymer films can improve the fruit's antioxidant retention by limiting oxidation reactions, which helps maintain nutritional quality and sensory characteristics. Compared to conventional synthetic packaging, biopolymer-based materials allow for tailored modifications, such as controlling gas permeability and incorporating bioactive compounds, to meet specific preservation needs for different types of fruits and vegetables. As a result, biopolymer films and coatings not

Natural Polymer-Based Films and Coatings in Food Technology. DOI: https://doi.org/10.1016/B978-0-443-34244-8.00029-X

only extend shelf life but also support environmental sustainability by reducing dependency on plastics and improving overall food quality (Baldwin et al., 2011; Blancas-Benitez et al., 2022; Dhall, 2013).

This review explores the latest developments and innovations in biopolymer films and coatings specifically designed to enhance the preservation and shelf life of fruits. It examines various types of biopolymers, including polysaccharides, proteins, and lipids, which are increasingly used in the food industry due to their biodegradability, biocompatibility, and nontoxic properties. The chapter will also analyze the integration of active agents, such as antimicrobials, antioxidants, and other functional additives, into biopolymer matrices to improve their preservation capabilities. Additionally, this review aims to cover current applications, the effectiveness of different biopolymer-based treatments, and the advantages and challenges associated with using these sustainable materials over conventional synthetic packaging. Another objective is to identify current challenges, such as limitations in mechanical strength and biodegradability, and to discuss emerging solutions and future trends, such as the integration of nanotechnology and active packaging techniques. By examining these aspects, this review aims to contribute valuable insights into the role of biopolymer films and coatings as sustainable alternatives to conventional packaging methods, aligning with consumer demand for environmentally friendly and health-conscious food preservation practices.

29.2 Fundamentals of biopolymer films and coatings

29.2.1 Definition and characteristics of biopolymers

Biopolymers are naturally occurring polymers derived from renewable biological sources such as plants, animals, and microorganisms. They are composed of monomeric units linked by covalent bonds, forming long chains with unique chemical and physical properties (Ghuttora, 2016; Vinod et al., 2020). Unlike synthetic polymers, which are often petroleum-based, biopolymers are biodegradable and typically more environmentally friendly, breaking down into harmless by-products over time through natural processes (Pinaeva & Noskov, 2024). This characteristic makes them highly valuable in applications that prioritize sustainability, such as food packaging, agriculture, and biomedical devices. Biopolymers can be classified into three primary categories based on their sources and chemical compositions: polysaccharides, proteins, and polyesters. Polysaccharides like starch, cellulose, and chitosan are carbohydrate-based polymers that exhibit good film-forming abilities and moisture resistance, making them popular in edible coatings for fruits and vegetables (Cruz-Monterrosa et al., 2023; Di et al., 2022). Proteins, derived from sources like soy, wheat, and casein, offer excellent barrier properties and tensile strength. Polyesters, such as polylactic acid (PLA) and polyhydroxyalkanoates (PHA), are produced by bacterial fermentation and are known for their durability and biodegradability (Choi et al., 2020). The functional properties of biopolymers, including their flexibility, mechanical strength, barrier properties, and resistance to environmental conditions, can be tailored by modifying their chemical structures or blending them with other compounds. For instance, adding plasticizers can enhance the flexibility of a biopolymer, while incorporating antimicrobial agents can improve its preservation capabilities. Because of these adaptable characteristics, biopolymers are increasingly used as eco-friendly alternatives to conventional synthetic polymers, particularly in industries focused on reducing environmental impact and enhancing sustainability (Liu et al., 2011; Suderman et al., 2018).

29.2.2 Types of biopolymers used in fruit preservation

29.2.2.1 Polysaccharides

Polysaccharides are commonly used in edible coatings and films due to their natural abundance, biodegradability, and excellent film-forming properties. Derived from renewable sources such as plants, algae, and crustaceans, polysaccharides like starch, cellulose, alginate, and chitosan provide a protective barrier on food surfaces, reducing moisture loss, oxygen permeability, and microbial growth. Starch, sourced from corn or potatoes, is widely used for its transparency and low cost, whereas it often requires modification to improve its mechanical properties (Díaz-Montes, 2022; Nair et al., 2020). Alginate, extracted from brown seaweed, is valued for its gel-forming ability in the presence of calcium ions, making it effective for coating fruits and vegetables. Cellulose derivatives, such as carboxymethyl cellulose (CMC), offer strong barrier properties and are used to extend shelf life. Chitosan, derived from chitin, provides antimicrobial activity and is commonly applied to fresh produce to inhibit spoilage. These polysaccharides are instrumental in developing sustainable and functional edible coatings and films (Dutta et al., 2012).

29.2.2.2 Proteins

Proteins are widely used in edible coatings and films due to their excellent film-forming properties, barrier functionality, and biodegradability. Derived from both plant and animal sources, proteins such as soy, whey, casein, gelatin, and zein are commonly employed to create coatings that help extend the shelf life of various foods by reducing moisture loss, oxygen transfer, and microbial contamination (Dangaran et al., 2009; Floret et al., 2023). Soy protein, derived from soybeans, is highly abundant and offers good film-forming abilities and strong mechanical properties. Its films are relatively resistant to oils and are commonly used for fruits, vegetables, and nuts. Whey protein, a by-product of cheese production, forms transparent, flexible films with excellent oxygen barrier properties, which makes it ideal for use on fresh-cut fruits and vegetables to prevent oxidative browning (Dou et al., 2014). Casein, the main protein in milk, provides a water-soluble film with strong adhesive properties and high nutritional value. It is often combined with other ingredients to improve its moisture barrier (Bhat et al., 2016; Liu et al., 2021). Gelatin, obtained from animal collagen, is highly flexible and widely used for its gel-forming capabilities, which makes it suitable for protecting fragile foods like berries. Zein, a corn protein, is particularly effective for coating oily foods due to its hydrophobic nature and high resistance to grease. Each of these proteins brings unique properties that can be enhanced through modifications or by blending with other biopolymers, plasticizers, or antimicrobial agents. Due to their natural composition and functional versatility, protein-based films and coatings are increasingly favored as eco-friendly alternatives to synthetic materials in food preservation, contributing to sustainability and improved food quality (Coltelli et al., 2015; Yuan et al., 2022).

29.2.2.3 Lipid-based coatings

Lipid-based coatings and films are widely used in food preservation due to their excellent water barrier properties, which help reduce moisture loss and extend the shelf life of various food products (Milani & Nemati, 2022). Lipids such as waxes, fatty acids, and triglycerides are commonly utilized in these coatings, either alone or in combination with other biopolymers, to create thin layers that effectively protect the surfaces of fruits, vegetables, meats, and other perishable items (Devi et al., 2024). Waxes, such as beeswax, carnauba wax, and paraffin wax, are frequently applied to fruits like apples, citrus, and cucumbers to enhance their glossy appearance while forming a protective layer that limits moisture evaporation (Thavasilingam et al., 2023). Carnauba wax, for example, is valued for its durability and is commonly used on products requiring a harder coating. Fatty acids like stearic acid and palmitic acid are often incorporated into coatings for their hydrophobic nature, which significantly enhances water resistance (Chiumarelli & Hubinger, 2012; Ngamakeue & Chitprasert, 2016). When combined with proteins or polysaccharides, these fatty acids can improve the moisture barrier without compromising the coating's flexibility. Triglycerides, which include oils like palm oil and coconut oil, also contribute to water resistance and can provide a softer texture, making them suitable for delicate items such as berries or cut fruits. In addition to their water-repellent properties, lipid-based coatings can also serve as carriers for flavorings, colorants, and antimicrobial agents, further enhancing the protective and sensory qualities of the coating. Although lipid-based coatings are less effective as oxygen barriers compared to protein or polysaccharide films, they are particularly beneficial in applications where moisture control is a priority. As natural, biodegradable materials, lipid-based coatings offer a sustainable option for extending the shelf life of fresh and minimally processed foods while enhancing their appearance and quality (Nasirifar et al., 2018; Seyedi et al., 2015; Simovic et al., 2012).

29.2.3 Advantages of biopolymers over conventional packaging materials

Biopolymers offer several advantages over conventional packaging materials, making them an increasingly popular choice in the food packaging industries. One of the most significant benefits is their biodegradability (Asgher et al., 2020; Tang et al., 2012). Unlike traditional plastics, which can take hundreds of years to decompose, biopolymers break down naturally into nontoxic by-products within a much shorter time frame. This characteristic helps mitigate environmental pollution and the growing problem of plastic waste (Rendón-Villalobos et al., 2016). Another advantage is the use of renewable resources. Biopolymers are derived from natural materials such as plants, starches, and proteins, reducing reliance on fossil fuels (Klein & Poverenov, 2020). This shift supports sustainable practices and contributes to a circular economy by utilizing agricultural waste and by-products, which can lower overall production costs. Biopolymers also possess excellent barrier properties, providing effective protection against moisture, oxygen, and UV light, which can degrade food quality (Sanchez-Garcia et al., 2010). These materials can be engineered to meet specific packaging requirements, such as enhancing shelf life or preserving freshness. Furthermore, biopolymers can be tailored to include functional properties, such as antimicrobial activity or enhanced mechanical strength, through blending with other biopolymers or incorporating natural compounds like EOs (Bustamante-Torres et al., 2022). This versatility allows for the development of smart packaging solutions that can actively respond to environmental changes, thus improving food safety. Finally, consumer preference is shifting toward eco-friendly products. Biopolymer packaging aligns with this trend, appealing to environmentally conscious consumers and brands looking to reduce their carbon footprint. Overall, biopolymers represent a sustainable and innovative alternative to conventional packaging materials, offering benefits for both the environment and food preservation (Mishra et al., 2024).

29.3 Mechanisms of action in fruit preservation

29.3.1 Barrier properties against moisture and gases

Packaging films play a critical role in preserving the quality by extending the shelf life of fruits and vegetables by offering barrier properties against gases and moisture. These properties mainly prevent dehydration and inhibit the microbial growth, during storage and transportation of fresh produce. Moisture barrier films limits the transfer of water vapor between the produce and its surrounding environment that helps to maintain the freshness of fruits and common materials, such as polyethylene (PE) and polypropylene (PP), exhibit excellent water vapor resistance due to their nonpolar nature. Advanced multilayer films incorporating hydrophobic coatings like PLA or polyvinylidene chloride (PVDC) further enhance moisture resistance. Fruits continue to respire after harvesting by consuming oxygen and carbon dioxide. The gas-permeable films mainly regulate gas exchange to maintain an optimal atmosphere within the packaging, thereby slowing down respiration and delaying ripening. Incorporating active packaging solutions, such as oxygen scavengers or moisture absorbers, can further optimize barrier performance (Han, 2005; Robertson, 2005) (Fig. 29.1).

29.3.2 Antimicrobial and antioxidant properties

Biopolymers encapsulated with active ingredients like Essential oils, nanoparticles, and antimicrobial peptides exhibit antimicrobial action. Polysaccharides derived from chitin, chitosan exhibits intrinsic antimicrobial activity due to its possibility charged amino acids. Chitosan mainly interacts with negatively charged microbial cell membranes that disrupt cell integrity and lead to cell death. Biopolymers incorporated with metal nanoparticles (e.g., silver or zinc oxide) enhance antimicrobial efficiency by releasing ions that damage microbial DNA and proteins. The antimicrobial properties of biodegradable starch films infused with varying concentrations of EO and surfactants. Infusion of EO and surfactant influences the structure and antibacterial properties of the films. The active packaging films are made from starch and glycerol, reinforced with cellulose nanocrystals (CNC), and activated with extracted red propolis (ERP). These films possess better antimicrobial efficacy in packaging cheese, curds, and antioxidant efficacy in packaging butter. Incorporation of CNC improved the mechanical strength of films by reducing water permeability and water activity (Costa et al., 2014; Rabea et al., 2003; Song et al., 2018a) (Fig. 29.2).

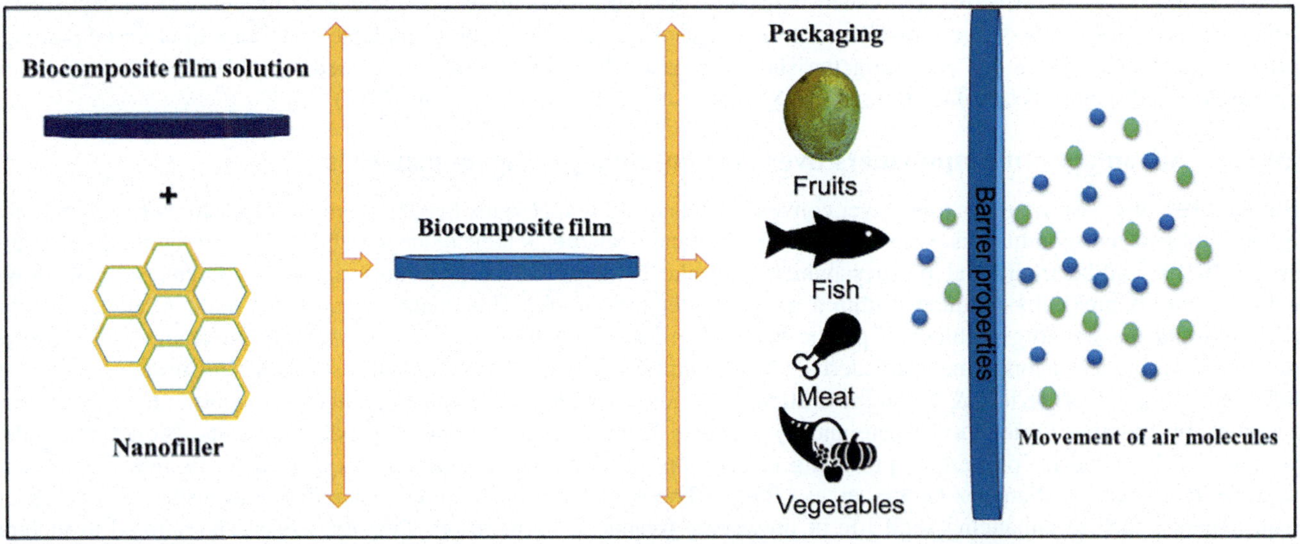

FIGURE 29.1 **Barrier properties of a biocomposite film.** Biocomposite films exhibiting enhanced barrier properties against moisture and gases, crucial for maintaining food quality and extending shelf life.

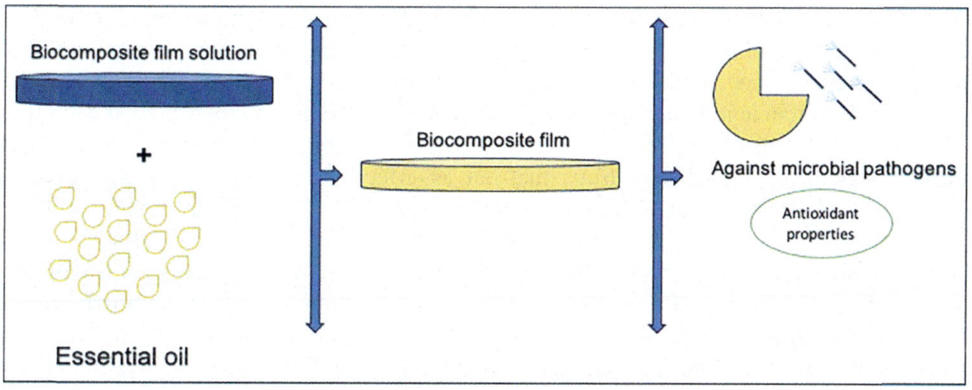

FIGURE 29.2 **Antimicrobial and antioxidant properties of biopolymer film.** Nature's shield: biopolymer films with antimicrobial and antioxidant action.

29.4 Recent trends in biopolymer film and coating solutions development

The food sector uses cellulose, the most prevalent natural polymer, because it is flexible and sustainable. Cellulose and its derivatives, derived from plant materials, have been widely studied as food biopolymer wrappers. Nontoxic, biocompatible, and biodegradable, cellulose is ideal for culinary applications while reducing environmental impact. Cellulose sheets are strong, flexible, and transparent, which resist moisture and gases, and are helpful for preserving the food quality. Regulatory bodies have classified cellulose and its derivatives as generally recognized as safe (GRAS), enabling them to be used as edible wrappers on food items. Cellulose films protect food with great mechanical properties. Water-soluble methylcellulose makes effective moisture-blocking films. It often covers fruits, vegetables, and baked items. Hydroxypropyl methylcellulose makes flexible, fat-, and oil-resistant coatings due to its film-forming capacity. Clear films are made from water-soluble CMC. In food coatings, it improves moisture retention and texture. Cellulose-based coatings can be customized with additives like antimicrobials, antioxidants, and colorants. Incorporating nanocellulose enhances their mechanical strength, transparency, and barrier properties. Nanocomposites improve protection against moisture and gases, while natural extracts like EO boost antimicrobial and antioxidant efficacy for better food preservation (Bangar et al., 2023; Nizardo et al., 2024; Singh et al., 2024).

Techniques like electrospinning and layer-by-layer assembly are being developed to create uniform functional coatings. Sustainable cellulose production methods, such as those using agricultural waste, are gaining focus. For fruits and vegetables, cellulose-based coatings like hydroxypropyl methyl cellulose (HPMC) reduce weight loss, maintain firmness, and extend the shelf life. Coating using CMC extracted from chicken filets and fish enhances moisture retention that prevents oxidation and reduces microbial contamination. In dairy sector, cellulose-based coatings on cheese prevent mold and moisture loss, while biodegradable films for yogurt packaging extend the shelf life. For baked goods, methylcellulose coatings prevents the staling and mold growth in bread (Makam et al., 2024; Yang et al., 2024a).

Starch, a natural polysaccharide extracted from the sources like corn, potatoes, and wheat, is used as a sustainable material for food coatings due to its biocompatibility, film-forming properties, and abundance. It forms strong, transparent, and flexible films that can act as an effective barrier to gases and moisture, preserving food quality and extending the shelf life. Combining starch with other biopolymers, such as proteins or lipids, enhances barrier and mechanical properties. Starch-based coatings reduce moisture loss, slow respiration, and prevent microbial growth, maintaining food's sensory and nutritional quality. These biodegradable coatings are environmentally friendly and avoid the health risks of synthetic preservatives. Additionally, they can be tailored with antioxidants, antimicrobials, and colorants to meet specific needs (de Oliveira Filho et al., 2022; Garcia et al., 2010; Versino et al., 2016).

Chitosan, derived from chitin in crab and insect exoskeletons, is a sustainable biopolymer gaining popularity in food applications. It is biocompatible, biodegradable, and safe for use as an edible coating, offering environmental benefits. Chitosan's intrinsic antibacterial and antifungal properties help extend shelf life of food by suppressing spoilage microbes. Its excellent film-forming ability creates flexible, translucent films that act as effective barriers to moisture and gases, preserving the food quality and delaying spoilage. Chitosan coatings can also be enriched with antioxidants, minerals, and flavor enhancers to boost food's nutritional value and appeal. Its compatibility with other biopolymers allows for the creation of composite films, making it a versatile material for sustainable food packaging solutions. Chitosan-based biopolymer coatings offer a promising and sustainable solution for food coating applications. Their natural antimicrobial properties, biodegradability, and ability to form effective barrier films make them ideal for extending the shelf life and preserving the quality of various food products (Hajji et al., 2018; Zhang et al., 2021; Chen et al., 2023a).

Gelatin derived from collagen in animal tissues is a versatile and sustainable material for food coatings. Its film-forming properties, biocompatibility, and biodegradability make it ideal for extending shelf life and preserving the food quality. Gelatin forms strong, pliable, and translucent films that act as effective barriers to moisture and gases, minimizing moisture loss, reducing respiration, and inhibiting microbial growth. These coatings are helpful to maintain the sensory and nutritional qualities of food without the health risk of synthetic preservatives. Gelatin coatings also reduce the environmental pollution due to their biodegradability. Applications include extending the shelf life of bread by preventing staling and mold growth, maintaining softness and freshness, and protecting fruit like strawberries and apples from microbial growth, oxidation, and moisture loss. Gelatin coatings on seafood and dairy products reduce spoilage by preventing oxidation and retain moisture by enhancing quality and shelf life. Edible gelatin films can be used for biodegradable, single-serving yogurt containers. Gelatin's adaptability allows for the incorporation of additives like antioxidants, antimicrobials, and colorants, further enhancing its protective and functional properties. Nanocomposite gelatin coatings offer superior moisture and gas protection, making gelatin a promising material for sustainable food packaging (Chandrababu et al., 2024; Ştefănescu et al., 2022).

Fatty acids and waxes are natural substances that are extensively used as food coatings that help extend the shelf life by forming protective barriers that reduce moisture loss, gas exchange, and microbial contamination. Fats are primarily composed of triglycerides, key components approved by regulatory bodies like the Food and Drug Administration (FDA) and European Food Safety Authority (EFSA). Oleic acid, mainly found in olive oil, creates moisture-resistant coatings. Stearic acid, extracted from animal fats and cocoa butter is valued for its film-forming properties. Beeswax also plays a major role in reducing moisture loss by delaying fruit ripening. Carnauba wax is known for its excellent barrier properties and a glossy finish. Similar to carnauba, candelilla wax provides protection for food products. Paraffin wax is used on cheese packaging to prevent moisture loss and microbial growth. Nanoparticles enhance the mechanical strength, barrier properties, and antibacterial activity of fatty acid and wax coatings, with nanocomposites providing excellent defence against gases and moisture. Adding natural extracts, EOs, and bioactive compounds further boosts their antibacterial and antioxidant properties, improving food preservation. Wax coatings protect fish from oxidation and spoilage while storing by ensuring freshness, while paraffin wax coatings on dairy products prevent moisture loss and mold growth. Fatty acid coatings on bread prevent staling and mold, preserving texture and flavor, while wax coatings on bakery products maintain freshness and reduce moisture loss (de Freitas et al., 2019; Díaz-Montes & Castro-Muñoz, 2021; Khwaldia, 2010; Prudnikov et al., 2023).

29.5 Innovations in application methods for biopolymer coatings

29.5.1 Dipping and spraying techniques

Dipping and spraying techniques are widely used for applying coatings on fruits and vegetables. These methods are cost-efficient and adaptable to various types of coatings, including waxes, biopolymers, and emulsions. Dipping involves immersing of fruits in coating solution for a specific duration, ensuring even coverage of the coating material. Some of the research demonstrated the effectiveness of dipping in alginate-based coatings for extending the freshness of strawberries by reducing moisture loss and microbial contamination. Applying chitosan coatings via dipping to citrus fruits significantly delayed microbial spoilage and extended shelf life (Gupta et al., 2024; Wu et al., 2024).

Spraying uses atomized nozzles to apply a fine mist of the coating solution over fruit surfaces. This technique is faster, uses less coating material, and is suitable for large-scale applications. Edible wax coating is applied to apples through spraying techniques that enhance visual appeal and reduce rehydration during storage. Nanoemulsion coating of mangoes demonstrates improved barrier properties during storage. Spraying allows a great control over coating thickness and reduces waste, but achieving uniformity may be challenging for irregularly shaped fruits. Both the techniques are essential in fruit coating industry and are selected based on the type of coating material, scale of operation, and desired results. Ongoing advancements in coating formulations and application methods promise further improvements in efficiency and effectiveness (Ali et al., 2020; Lee et al., 2024) (Fig. 29.3).

29.5.2 Electrospinning and electrostatic deposition

Electrospinning and electrostatic deposition are innovative techniques increasingly explored for applying coatings on fruits. These methods leverage electrical forces to achieve uniform, nano-scale coatings that offer enhanced barrier

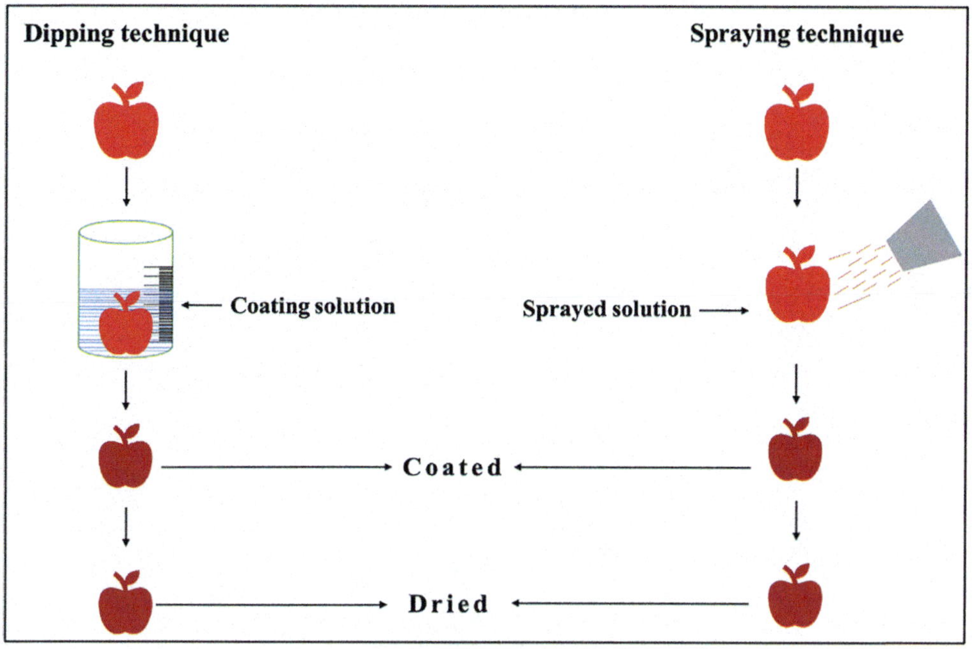

FIGURE 29.3 **Dipping and spraying technique of fruits.** Using dipping and spraying techniques to apply edible coatings, extending shelf life and maintaining fruit quality.

properties, antimicrobial activity, and controlled release functionality. Electrospinning uses a high-voltage electric field to create ultrafine polymer fibers that form a nanofibrous mat over the fruit surface. These mats provide excellent mechanical strength, gas permeability, and bioactive compound delivery. Electrospun using chitosan—polyethylene oxide nanofibers with EOs was coated on strawberries, in which the nanofibers significantly reduced microbial load and maintained fruit firmness during storage. Electrospun zein-based fibers were also applied to apples, demonstrating improved antioxidant activity and reduced weight loss. Electrospinning is particularly advantageous for creating bioactive coatings with controlled release properties, ensuring prolonged preservation effects (Yilmaz et al., 2016; Zhang et al., 2021).

Electrostatic deposition technique involves applying coating solutions to food material using electrically charged particles that are attracted to the fruit's surfaces. The main advantage of this technique ensures even coverage, reduces material waste, and enhances coating adherence. Polysaccharide-based coatings of mangoes using electrostatic deposition improve moisture retention and reduce spoilage rates. Nanoemulsion incorporated with EO-based coatings on tomatoes achieves uniform coatings with enhanced antimicrobial properties. This technique is advantageous for industrial-scale applications due to its precision, material efficiency, and compatibility with various coating formulations. However, optimizing the electrical parameters and coating material properties is critical to ensure consistent results. Both electrospinning and electrostatic deposition offer significant potential for advancing fruit coating technologies, with their ability to provide uniform, functional coatings. Ongoing research and technological advancements aim to address scalability and cost challenges, making these methods viable for commercial use (Barradas & de Holanda e Silva, 2021; Shiekh et al., 2021) (Fig. 29.4).

29.5.3 Layer-by-layer assembly

Layer-by-layer (LbL) assembly is a cutting-edge technique developed for multilayer coatings of fruits that offers precise control over the coating's structure and functionality. This method involves alternating deposition of oppositely charged materials such as polyelectrolytes, biopolymers, and nanoparticles that create highly organized and functional coatings. The LbL process mainly relies on electrostatic interactions, hydrogen bonding, or hydrophobic forces to subsequently deposit thin layers of materials. Each layer is applied by immersing the fruit in a coating solution or by spraying. The LbL coatings provide enhanced barrier properties that reduce moisture loss by controlling oxygen and carbon dioxide exchange. The LbL coatings using chitosan and alginate for strawberries effectively minimized weight loss and delayed

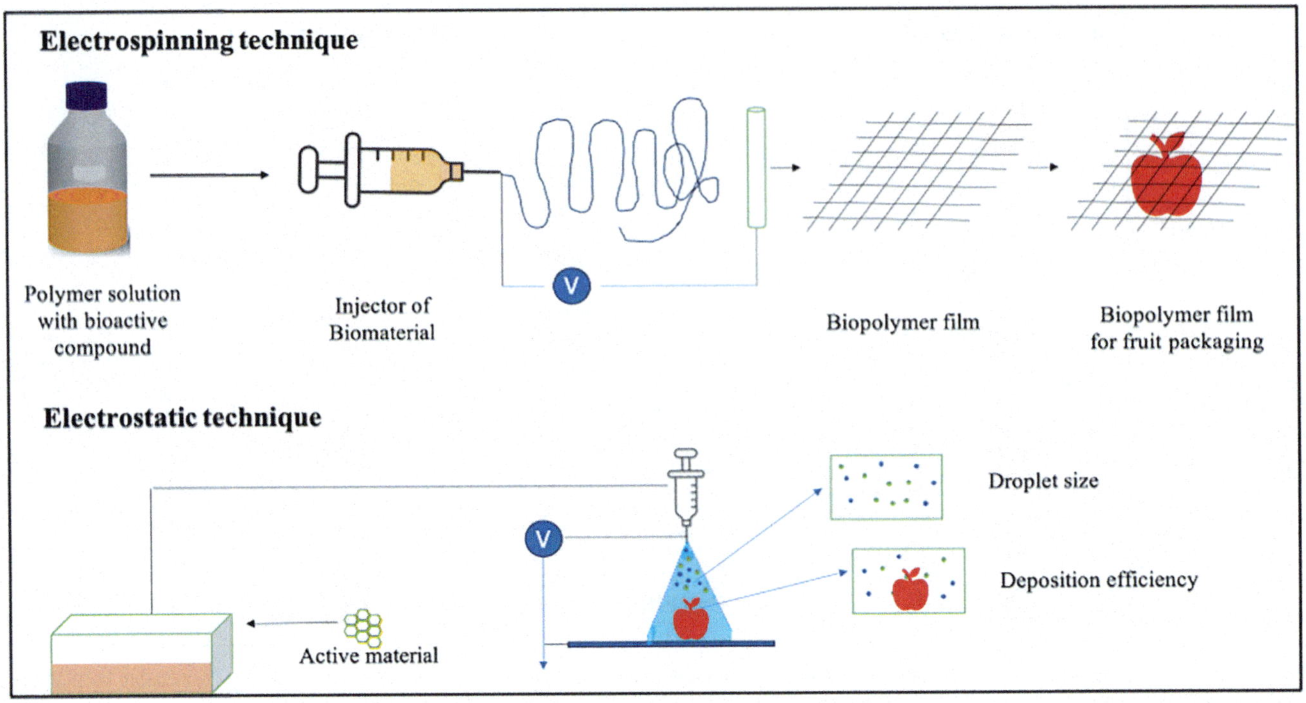

FIGURE 29.4 **Electrostatic and electrospinning techniques for packaging of fruits.** Extending freshness, one nanofiber at a time: electrospinning technology for advanced fruit packaging.

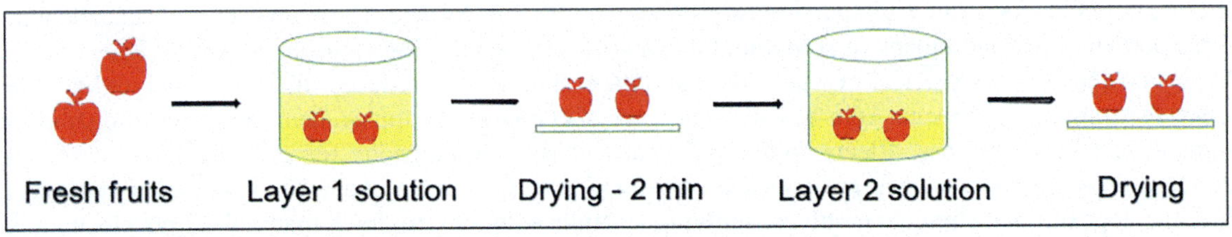

FIGURE 29.5 **Layer-by-layer coating solution of fruits.** A multilayered approach to fruit preservation, using edible coatings to extend freshness.

ripening. Coatings incorporated with silver nanoparticles and CMC for blueberries resulted in extended shelf life and reduced microbial contamination. Incorporated LbL films enable controlled release of antioxidants, vitamins, and antimicrobial agents that enhance fruit preservation. Coatings enriched with green tea extract on apples prolong the shelf life by enhancing the antioxidant activity. The LbL assembly is a versatile and promising technique for fruit coating, combining advanced functional properties with sustainable materials. As research progresses, this method holds significant potential for revolutionizing postharvest fruit preservation (Arnon-Rips & Poverenov, 2016; Gómez-Mascaraque et al., 2022; Yu et al., 2022) (Fig. 29.5).

29.6 Impact of biopolymer films and coatings on fruit quality

Biopolymer films and coatings have emerged as sustainable solutions to extend the shelf life and maintain the quality of fruits as shown in Tables 29.1 and 29.2. Derived natural products such as polysaccharides, proteins, and lipids create a protective barrier that helps in preserving the sensory, nutritional, and physiochemical properties of fruits during storage and transportation. Biopolymer films and coatings are transforming the way fruits are preserved by enhancing quality, reducing waste, and promoting sustainability. Continued research and development in this field hold immense potential for improving global food security and reducing environmental impact.

TABLE 29.1 Biopolymer coating on fruits and vegetables and its applications.

S. no.	Formulation	Fruits/veg	Interpretation	References
1.	Chitosan/fruit waste extract	Tomatoes/grapes	Coated fruits had a control over the decay index for more than 10 days, the physiochemical parameters were retained than uncoated fruits.	Thakur et al. (2024)
2.	Zucchini pectin	Tomatoes	Antioxidant value retained by 34.32%, minimal weight loss, and antimicrobial activity against *Staphylococcus aureus, E. coli, Aspergillus niger*. The firmness extended up to 11 days in coated, whereas 9 days in uncoated samples.	Jhanani et al. (2024)
3.	Pectin/beeswax/lime	Tomatoes/okra/eggplants	The shelf-life extended for 23, 21 and 11 days for tomatoes, okra and eggplants respectively.	Obiaocha-Nwaogwugwu et al. (2024)
4.	Guar gum/chitosan/oil (coconut/clove bud/cinnamic bark)	Khasi mandarins	Retention of firmness 5.6N, weight loss (38%), other quality parameters were retained for 20 days.	Goswami et al. (2024)
5.	Beeswax/moringa oil	Cucumber	Over a period of 27 days, shelf-life was retained. Vitamin C content of 37.7 mg/L, weight loss and color values for coated samples were at lower changes compared to uncoated samples.	Al-Rashdi et al. (2024)
6.	Chitosan/nano titanium oxide/sodium tripolyphosphate	Cucumber	Postharvest qualities were retained for 21 days, the antimicrobial activity against *Salmonella spp.*, retained highest crispness value.	Helal et al. (2022)
7.	Chitosan/beeswax	Mango	Minimum weight loss was observed in coated fruits (12.92%) on 30th day, TSS, firmness, pH, titratable acidity were maintained, disease incidence was recorded as 0% in treated fruits till 15 days of storage.	Eshetu et al. (2019)
8.	Beeswax/sodium nitroprusside	Pear	Fruit firmness and other physiochemical parameters such as TSS, titratable acidity was measured, freshness retained for 60 days.	Chouhan and Singh (2023)
9.	Aloe vera gel/beeswax	Salak Madu	Aloe vera coating reduced the decay incidence and colour changes of fruit up to 12 days of storage.	Sari et al. (2020)
10.	Cassava starch/*Moringa oleifera* seed oil	Mango	Physiochemical properties were retained for 12 days of storage, oil extract slowed down the degradation of pigments and retained the fruit shininess.	Silva et al. (2021)
11.	Beeswax/coconut oil	Lemon	Greenness of lemons were retained, reduced respiration, weight loss and ripening for 18 days of storage whereas uncoated retained the qualities for 6 days.	Nasrin et al. (2020)
12.	Alginate/chitosan nano multilayer/aloe vera	Tomatoes	The total colonies of yeast/molds were higher in uncoated than coated, retained the weight of the samples for 15 days through coating.	Flores-López et al. (2023)

(*Continued*)

TABLE 29.1 (Continued)

13.	Gum Arabic/ cactus pear stem/ moringa/ henna leaf extracts	Guava	Reduced weight loss, inhibition against *Rhizopus* rot infection, delayed the accumulation of fruit contents such as TSS and TA, extended the shelf-life for 21 days.	El-Gioushy et al. (2022)
14.	Gum Arabic/carboxy methyl cellulose/ moringa leaf extract	Avocado	Inhibition against *C. gleosporioides* for 33% and reduced weight loss and ripening, retaining the firmness were the major advantages on moringa extract.	Kubheka et al. (2020)
15.	Chitosan/pullulan/pomegranate peel extract	Green bell pepper	Physiochemical and organoleptic properties were retained for 18 days in c coated samples than uncoated, including phenolic content and total soluble solids.	Kumar et al. (2021)
16.	Guar gum/chitosan/tamarind seed starch/jackfruit seed starch	Kinnow fruits	Inhibited the decay incidence at large rate and reached 0.83% for coated fruits on day 25. Fruit firmness maintained for 6.77 N and TPC of 213.05 mg GAE/100 g due to tamarind seed starch incorporation.	Bhan et al. (2022)
17.	Chitosan/gelatin/ginger essential oil/ beta-cyclodextrin	Blueberries	Narrow particle distribution index 0.299 explains on the uniform mixture of the costing solution. Reduction in decaying, regulating anthocyanin, control in physiochemical activities lasted for 16 days at 25°C and 40 days at 4°C due to coating.	Hossen et al. (2024)
18.	Guar gum/almond gum/oregano essential oil	Okra	The freshness maintained for 4 days at 23°C for coated okra, antimicrobial inhibition against growth of *E. coli, S. aureus* and *A. niger* noted as 19 mm, 29 mm and 35 mm respectively by the coating solution	Shinde et al. (2024)

TABLE 29.2 Biopolymer packaging film of fruits and vegetables and its applications.

Film	Formulation	Target	Application	Reference
Edible	Starch/Iron yam/ Lemon essential oil	Instant food/ flavoring packaging	Better elongation, enhanced antibacterial effects.	Song et al. (2018b)
Biofilm	Sweet potato/Tapioca	Bioplastics	Higher amylose content, 96% soil biodegradability in a month, greater tensile strength up to load of 2.4 kg.	Patkar et al. (2020)
Bilayer	Pea starch/Polylactic acid	Cherry tomatoes	Higher thermal stability and barrier capacity. Weak interfacial adhesion between bilayers. Reduced weight loss and retention of organic acids in Cherry tomatoes.	Zhou et al. (2019)
Biodegradable	PLA-Polycaprolactone (PCL)	Strawberry	Higher toughness and small dispersion size of PCL-PLA led to better preservation of the colour and luster of the fruits.	Zhang et al. (2024)
Edible	Tilapia fish bone gelatin/passion fruit peel	Food wrap	20.38 MPa was the tensile strength for the film with 1.5 ml of the passion fruit peel extract, which gave better characteristics in comparison with other concentration of the extract.	Nasution et al. (2024)
Biodegradable	CMC/PVA/Pyrazoic acid hydrogels	Banana	Natural Biodegradability completed in 50 days. The browning, ripening, senescence had been delayed through these hydrogel wraps.	Gang et al. (2024)
Biodegradable	Gelatin/sodium alginate/NOPE (Neem oil Pickering emulsion)	Cherry tomatoes	Oxygen and water barrier capacity increased by 32.32% and 37.02%, reduction in *Botrytis cinerea* spores leading to better disease control of about 72.4%	Yang et al. (2024b)
Biodegradable	Chitosan/neem leaf extract/lemon peel-pectin	Apricots	SEM analyzed confirmed that uniform layer of film, potential control over the *Staphylococcus aureus* and *Aspergillus niger*. The shelf-life of film wrapped apricots increased by 50% than control fruits.	Firdaus et al. (2024)
Biodegradable	Modified atmosphere/ nonperforated	Raspberries	The film with 5.7 kPa O_2 and 31.6 kPa CO_2 had maintained the color of the fruit with a reduced change from harvest period.	Briano et al. (2015)
Bio-nano composite	Chitosan/CuO nanoparticles/ Hydroxypropyl cellulose	Jamun	SEM, FTIR revealed the uniform structure and the inter-bonds of the film, the tensile strength of 33 MPa and 4% transmittance at 360 nm enhanced UV-blockage on the fruit shelf-life, the antimicrobial activity against the Gram positive and Gram-negative bacteria led to the advantageous characteristics of the film, the film stored fruit had retained their antioxidant properties and the physical parameters compared to the control fruits.	Gunaki et al. (2024)

(Continued)

TABLE 29.2 (Continued)

Bio-nano composite	Jamun seed starch/tamarind kernel xyloglucan/chitosan nanoparticles	Sapota	FTIR/XRD analysis proved strong intermolecular interactions and crystallinity. Tensile strength of 20.42 MPa, contact angle of 89°, WVTR of 13.26g/h.m^2, antimicrobial activity against *Bacillus* and *E. coli* are the enhanced properties of film. Shelf-life of sapota retained for 12 days by retarded weight loss and color change.	Santhosh and sarkar (2024)
Biodegradable	Chitosan/curcumin/Tenebrio molitor larvae protein	Blueberries	Water vapor barrier increased by 37.04%, reduced decay index of the fruit, maintained quality parameters such as weight and aging.	Liu et al. (2024)
Bio-nano composite	Starch/selenium nanoparticles	Litchi	Biodegradability was confirmed by 11 days of testing, the film enhanced the biosafety and quality of fruits by retainment of antioxidants and other nutrients.	Yu et al. (2024)
Biodegradable	Chitosan/pullulan/*Artemisia annua* essential oil	Grape berries	Elongation at break increased to 71.25% and UV-transmittance was 0% and the storage time was increased through the incorporated with essential oil.	Chen et al. (2023b)
Active	Chitosan/cornstarch/grapefruit seed extract (GFSE)	Grapes	Antimicrobial activity of the GFSE incorporated film was notably higher than the control which was 36.34 mm and 40.33 mm against *E. coil* and *Staphylococcus aureus* respectively. Overall acceptability for the film wrapped grapes was higher than the unwrapped samples on the day 10.	Ghoshal and singh (2024)

29.7 Conclusion

The continuous exploration and application of biopolymer-based films and coatings have showcased their potential as sustainable alternatives to conventional plastic packaging for preserving fruits. Emerging trends in biopolymer technology, particularly the integration of active ingredients like EOs, nanoparticles, and other functional additives, have enhanced the antimicrobial, antioxidant, and mechanical properties of these films. These advancements not only extend the shelf life of fruits but also maintain their quality during storage and transportation, offering substantial benefits to the agro-food industry.

Biopolymers such as polysaccharides, proteins, and lipids have been extensively studied for their film-forming capabilities, biodegradability, and eco-friendliness. Innovations in combining biopolymers with advanced techniques, such as response surface methodology (RSM), have optimized the formulation and processing conditions to achieve desirable functional properties. The development of nanocomposites and multilayer films has further broadened the scope of biopolymer applications, improving their barrier and mechanical characteristics while maintaining environmental compatibility.

The incorporation of EOs and bioactive compounds into biopolymer matrices represents a significant leap in active packaging technology. These components provide antimicrobial and antioxidant functionalities, which are crucial for delaying microbial spoilage and oxidative degradation. However, challenges remain in ensuring the controlled release of these active agents to maintain their efficacy throughout the storage period. Techniques such as encapsulation within hydrogel systems and the use of nanostructured materials are being explored to address these issues, demonstrating promising results in extending fruit shelf life and reducing food waste.

Despite these advancements, the commercial adoption of biopolymer films faces several hurdles, including high production costs, scalability challenges, and limited consumer awareness. The development of cost-effective production methods and the use of renewable resources could significantly enhance their market feasibility. Additionally, regulatory frameworks and standardization of biopolymer-based packaging are essential to ensure their safety, functionality, and widespread acceptance.

Future research should focus on interdisciplinary approaches that integrate material science, food technology, and environmental studies to develop next-generation biopolymer films. Emphasis should be placed on scaling up production processes, improving the cost efficiency of raw materials, and exploring innovative applications in diverse fruit preservation scenarios. Furthermore, the integration of IoT and smart packaging solutions could revolutionize the monitoring and management of fruit quality during storage and distribution.

In conclusion, biopolymer-based films and coatings represent a sustainable and innovative solution for fruit preservation, aligning with global efforts to reduce plastic waste and promote environmental sustainability. The synergistic combination of advanced materials, cutting-edge technologies, and strategic research directions holds the promise of overcoming existing challenges and unlocking the full potential of biopolymers in food packaging. By addressing technical, economic, and regulatory barriers, biopolymer films can play a pivotal role in transforming the agro-food industry toward a more sustainable future.

AI disclosure

During the preparation of this work, the author(s) used ChatGPT in order to correct grammar errors. After using this tool/service, the author(s) reviewed and edited the content as needed and take(s) full responsibility for the content of the publication.

References

Ali, U., Basu, S., & Mazumder, K. (2020). Improved postharvest quality of apple (Rich Red) by composite coating based on arabinoxylan and β-glucan stearic acid ester. *International Journal of Biological Macromolecules, 151*, 618–627.

Al-Rashdi, S., Al-Subhi, N., Al-Dairi, M., & Pathare, P. B. (2024). Effect of a moringa oil–beeswax edible coating on the shelf-life and quality of fresh cucumber. *Processes, 12*(6), 1148.

Amit, S. K., Uddin, M. M., Rahman, R., Islam, S. R., & Khan, M. S. (2017). A review on mechanisms and commercial aspects of food preservation and processing. *Agriculture & Food Security, 6*, 1–22.

Arnon-Rips, H., & Poverenov, E. (2016). Biopolymers-embedded nanoemulsions and other nanotechnological approaches for safety, quality, and storability enhancement of food products: Active edible coatings and films. In *Emulsions*. Academic Press, 329–363.

Asgher, M., Qamar, S. A., Bilal, M., & Iqbal, H. M. (2020). Bio-based active food packaging materials: Sustainable alternative to conventional petrochemical-based packaging materials. *Food Research International, 137*, 109625.

Baldwin, E. A., Hagenmaier, R., & Bai, J. (Eds.). (2011). *Edible coatings and films to improve food quality*. CRC press.

Bangar, S. P., Scott Whiteside, W., Suri, S., Barua, S., & Phimolsiripol, Y. (2023). Native and modified biodegradable starch-based packaging for shelf-life extension and safety of fruits/vegetables. *International Journal of Food Science & Technology, 58*(2), 862–870.

Barradas, T. N., & de Holanda e Silva, K. G. (2021). Nanoemulsions of essential oils to improve solubility, stability and permeability: A review. *Environmental Chemistry Letters, 19*(2), 1153–1171.

Bhan, C., Asrey, R., Meena, N. K., Rudra, S. G., Chawla, G., Kumar, R., & Kumar, R. (2022). Guar gum and chitosan-based composite edible coating extends the shelf life and preserves the bioactive compounds in stored Kinnow fruits. *International Journal of Biological Macromolecules, 222*, 2922–2935.

Bhat, M. Y., Dar, T. A., & Singh, L. R. (2016). *Casein proteins: Structural and functional aspects. Milk proteins-From structure to biological properties and health aspects, 10*, 64187.

Blancas-Benitez, F. J., Montaño-Leyva, B., Aguirre-Güitrón, L., Moreno-Hernández, C. L., Fonseca-Cantabrana, Á., del, C., Romero-Islas, L., & González-Estrada, R. R. (2022). Impact of edible coatings on quality of fruits: A review. *Food Control, 139*, 109063.

Briano, R., Giuggioli, N. R., Girgenti, V., & Peano, C. (2015). Biodegradable and compostable film and modified atmosphere packaging in postharvest supply chain of raspberry fruits (cv. G randeur). *Journal of Food Processing and Preservation, 39*(6), 2061–2073.

Bustamante-Torres, M., Arcentales-Vera, B., Estrella-Nuñez, J., Yánez-Vega, H., & Bucio, E. (2022). Antimicrobial activity of composites-based on biopolymers. *Macromol, 2*(3), 258–283.

Chandrababu, V., Parameswaranpillai, J., Gopi, J. A., Pathak, C., Dominic, C. M., Feng, N. L., & Ganguly, S. (2024). Progress in food packaging applications of biopolymer-nanometal composites—A comprehensive review. *Biomaterials Advances*, 213921.

Chen, Y., Liu, Y., Dong, Q., Xu, C., Deng, S., Kang, Y., & Li, L. (2023a). Application of functionalized chitosan in food: A review. *International Journal of Biological Macromolecules, 235*, 123716.

Chen, W., Liu, H., Chai, Y., Guo, C., Luo, C., Chen, D., & Huang, C. (2023b). Chitosan–pullulan films enriched with Artemisia annua essential oil: Characterization and application in grape preservation. *International Journal of Biological Macromolecules, 243*, 125216.

Chiumarelli, M., & Hubinger, M. D. (2012). Stability, solubility, mechanical and barrier properties of cassava starch–Carnauba wax edible coatings to preserve fresh-cut apples. *Food Hydrocolloids, 28*(1), 59–67.

Choi, S. Y., Cho, I. J., Lee, Y., Kim, Y. J., Kim, K. J., & Lee, S. Y. (2020). Microbial polyhydroxyalkanoates and nonnatural polyesters. *Advanced Materials, 32*(35), 1907138.

Chouhan, N. K., & Singh, V. (2023). Effect of postharvest edible coatings on quality and shelf life of pear fruit cv. *Punjab Beauty*.

Coltelli, M. B., Wild, F., Bugnicourt, E., Cinelli, P., Lindner, M., Schmid, M., & Lazzeri, A. (2015). State of the art in the development and properties of protein-based films and coatings and their applicability to cellulose based products: An extensive review. *Coatings, 6*(1), 1.

Costa, S. S., Druzian, J. I., Machado, B. A. S., de Souza, C. O., & Guimarães, A. G. (2014). Bi-functional biobased packing of the cassava starch, glycerol, licuri nanocellulose and red propolis. *PLoS ONE, 9*(11), e112554.

Cruz-Monterrosa, R. G., Rayas-Amor, A. A., González-Reza, R. M., Zambrano-Zaragoza, M. L., Aguilar-Toalá, J. E., & Liceaga, A. M. (2023). Application of polysaccharide-based edible coatings on fruits and vegetables: Improvement of food quality and bioactivities. *Polysaccharides, 4*(2), 99–115.

Dangaran, K., Tomasula, P. M., & Qi, P. (2009). Structure and function of protein-based edible films and coatings. *Edible Films and Coatings for Food Applications*, 25–56.

de Freitas, C. A. S., de Sousa, P. H. M., Soares, D. J., da Silva, J. Y. G., Benjamin, S. R., & Guedes, M. I. F. (2019). Carnauba wax uses in food–A review. *Food Chemistry, 291*, 38–48.

Devi, L. S., Jaiswal, A. K., & Jaiswal, S. (2024). Lipid incorporated biopolymer based edible films and coatings in food packaging: A review. *Current Research in Food Science*, 100720.

Dhall, R. K. (2013). Advances in edible coatings for fresh fruits and vegetables: A review. *Critical Reviews in Food Science and Nutrition, 53*(5), 435–450.

de Oliveira Filho, J. G., Albiero, B. R., Calisto, Í. H., Bertolo, M. R. V., Oldoni, F. C. A., Egea, M. B., & Ferreira, M. D. (2022). Bio-nanocomposite edible coatings based on arrowroot starch/cellulose nanocrystals/carnauba wax nanoemulsion containing essential oils to preserve quality and improve shelf life of strawberry. *International Journal of Biological Macromolecules, 219*, 812–823.

Di, X., Liang, X., Shen, C., Pei, Y., Wu, B., & He, Z. (2022). Carbohydrates used in polymeric systems for drug delivery: From structures to applications. *Pharmaceutics, 14*(4), 739.

Dou, Y., Xu, S., Liu, X., Han, J., Yan, H., Wei, M., & Duan, X. (2014). Transparent, flexible films based on layered double hydroxide/cellulose acetate with excellent oxygen barrier property. *Advanced Functional Materials, 24*(4), 514–521.

Dutta, J., Tripathi, S., & Dutta, P. K. (2012). Progress in antimicrobial activities of chitin, chitosan and its oligosaccharides: A systematic study needs for food applications. *Food Science and Technology International, 18*(1), 3–34.

Díaz-Montes, E. (2022). Polysaccharides: Sources, characteristics, properties, and their application in biodegradable films. *Polysaccharides, 3*(3), 480–501.

Díaz-Montes, E., & Castro-Muñoz, R. (2021). Edible films and coatings as food-quality preservers: An overview. *Foods, 10*(2), 249.

El-Gioushy, S. F., Abdelkader, M. F., Mahmoud, M. H., Abou El Ghit, H. M., Fikry, M., Bahloul, A. M., & Gawish, M. S. (2022). The effects of a gum arabic-based edible coating on guava fruit characteristics during storage. *Coatings, 12*(1), 90.

Eshetu, A., Ibrahim, A. M., Forsido, S. F., & Kuyu, C. G. (2019). Effect of beeswax and chitosan treatments on quality and shelf life of selected mango (*Mangifera indica* L.) cultivars. *Heliyon, 5*(1).

Firdaus, S., Ahmad, F., & Zaidi, S. (2024). Preparation and characterization of biodegradable food packaging films using lemon peel pectin and chitosan incorporated with neem leaf extract and its application on apricot fruit. *International Journal of Biological Macromolecules, 263*, 130358.

Flores-López, M. L., Vieira, J. M., Rocha, C. M., Lagarón, J. M., Cerqueira, M. A., Jasso de Rodríguez, D., & Vicente, A. A. (2023). Postharvest quality improvement of tomato (*Solanum lycopersicum* L.) fruit using a nanomultilayer coating containing *Aloe vera. Foods, 13*(1), 83.

Floret, C., Monnet, A. F., Micard, V., Walrand, S., & Michon, C. (2023). Replacement of animal proteins in food: How to take advantage of nutritional and gelling properties of alternative protein sources. *Critical Reviews in Food Science and Nutrition, 63*(7), 920–946.

Gang, F., Xu, M., Zhang, S., Zhang, C., He, J., Xiao, Y., & Zhang, J. (2024). Biodegradable active composite hydrogel packaging for postharvest climacteric bananas preservation. *Food Chemistry, 442*, 138494.

Garcia, L. C., Pereira, L. M., de, L., Sarantópoulos, C. I., & Hubinger, M. D. (2010). Selection of an edible starch coating for minimally processed strawberry. *Food and Bioprocess Technology, 3*, 834–842.

Ghoshal, G., & Singh, J. (2024). Study of coating effectiveness of grape fruit seed extract incorporated Chitosan/cornstarch based active packaging film on grapes. *Food Chemistry Advances, 4*, 100651.

Ghuttora, N. (2016). Increase the usage of biopolymers and biodegradable polymers for sustainable environment.

Gunaki, M. N., Masti, S. P., D'souza, O. J., Eelager, M. P., Kurabetta, L. K., Chougale, R. B., & Kumar, S. P. (2024). Fabrication of CuO nanoparticles embedded novel chitosan/hydroxypropyl cellulose bio-nanocomposites for active packaging of jamun fruit. *Food Hydrocolloids, 152*, 109937.

Gupta, V., Meena, N. K., Kumar, A., Bhatnagar, P., Jain, S. K., Jat, S. K., & Choudhary, K. (2024). Hydrocolloid-based bilayer edible coatings maintain bioactive compounds and nutritional quality of 'Dashehari'mango (*Mangifera indica* L.) fruits. *Food and Bioprocess Technology, 17*(8), 2483–2498.

Gómez-Mascaraque, L. G., Balanč, B., Djordjević, V., & Bugarski, B. (2022). Encapsulation techniques for food purposes. *Encapsulation in Food Processing and Fermentation*, 37–80.

Goswami, M., Mondal, K., Prasannavenkadesan, V., Bodana, V., & Katiyar, V. (2024). Effect of guar gum-chitosan composites edible coating functionalized with essential oils on the postharvest shelf life of Khasi mandarin at ambient condition. *International Journal of Biological Macromolecules, 254*, 127489.

Hajji, S., Younes, I., Affes, S., Boufi, S., & Nasri, M. (2018). Optimization of the formulation of chitosan edible coatings supplemented with carotenoproteins and their use for extending strawberries postharvest life. *Food Hydrocolloids, 83*, 375–392.

Han, J. H. (2005). *Innovations in food packaging*. Elsevier.

Helal, M., Sami, R., Algarni, E., Alshehry, G., Aljumayi, H., Al-Mushhin, A. A., & Almasoudi, A. (2022). Active bionanocomposite coating quality assessments of some cucumber properties with some diverse applications during storage condition by chitosan, nano titanium oxide crystals, and sodium tripolyphosphate. *Crystals, 12*(2), 131.

Hossen, M. A., Shimul, I. M., Sameen, D. E., Rasheed, Z., Tanga, W., Chen, M., & Liu, Y. (2024). Chitosan/gelatin coating loaded with ginger essential oil/β-cyclodextrin inclusion complex on quality and shelf life of blueberries. *International Journal of Biological Macromolecules, 279*, 135026.

Jhanani, G. K., AlSalhi, M. S., Naveena, T., & Shanmuganathan, R. (2024). As assessment of shelf life increasing competence of pectin (Zucchini) based edible coating on tomatoes. *Environmental Research*, 119368.

Khwaldia, K. (2010). Water vapor barrier and mechanical properties of paper-sodium caseinate and paper-sodium caseinate-paraffin wax films. *Journal of Food Biochemistry, 34*(5), 998–1013.

Klein, M., & Poverenov, E. (2020). Natural biopolymer-based hydrogels for use in food and agriculture. *Journal of the Science of Food and Agriculture, 100*(6), 2337–2347.

Kubheka, S. F., Tesfay, S. Z., Mditshwa, A., & Magwaza, L. S. (2020). Evaluating the efficacy of edible coatings incorporated with moringa leaf extract on postharvest of 'Maluma'avocado fruit quality and its biofungicidal effect. *HortScience, 55*(4), 410–415.

Kumar, N., Ojha, A., Upadhyay, A., Singh, R., & Kumar, S. (2021). Effect of active chitosan-pullulan composite edible coating enrich with pomegranate peel extract on the storage quality of green bell pepper. *LWT, 138*, 110435.

Lee, Y. C., Yu, M. C., Tsay, J. S., Hou, C. Y., Huang, P. H., & Liang, Y. S. (2024). Investigation of different nanoemulsion coatings to prevent anthracnose and maintain quality parameters during storage of Irwin mango. *LWT, 205*, 116501.

Liu, M., Zou, X., Wu, X., Li, X., Chen, H., Pan, F., & Peng, W. (2024). Preparation of chitosan/Tenebrio molitor larvae protein/curcumin active packaging film and its application in blueberry preservation. *International Journal of Biological Macromolecules, 275*, 133675.

Liu, H., Chaudhary, D., Ingram, G., & John, J. (2011). Interactions of hydrophilic plasticizer molecules with amorphous starch biopolymer—An investigation into the glass transition and the water activity behavior. *Journal of Polymer Science Part B: Polymer Physics, 49*(14), 1041–1049.

Makam, R. M. M., Omar, W. N. N. W., Bin Jihat, D. A., Nor, N. U. M., Shamjuddin, A., & Amin, N. A. S. (2024). The potential of carboxylmethyl cellulose from empty fruit bunch as versatile material in food coating: A review. *Carbohydrate Polymers, 338*, 122194.

Liu, K., Xu, X., Liu, H., Liu, Z., Zhao, K., Ma, Y., & Zhang, K. (2021). Mechanical properties and water sensitivity of soybean protein isolate film improved by incorporation of sodium caseinate and transglutaminase. *Progress in Organic Coatings, 153*, 106154.

Milani, J. M., & Nemati, A. (2022). Lipid-based edible films and coatings: A review of recent advances and applications. *Journal of Packaging Technology and Research, 6*(1), 11–22.

Mishra, B., Panda, J., Mishra, A. K., Nath, P. C., Nayak, P. K., Mahapatra, U., & Sridhar, K. (2024). Recent advances in sustainable biopolymer-based nanocomposites for smart food packaging: A review. *International Journal of Biological Macromolecules*, 135583.

Nair, M. S., Tomar, M., Punia, S., Kukula-Koch, W., & Kumar, M. (2020). Enhancing the functionality of chitosan-and alginate-based active edible coatings/films for the preservation of fruits and vegetables: A review. *International Journal of Biological Macromolecules, 164*, 304–320.

Nasirifar, S. Z., Maghsoudlou, Y., & Oliyaei, N. (2018). Effect of active lipid-based coating incorporated with nanoclay and orange peel essential oil on physicochemical properties of *Citrus sinensis*. *Food Science & Nutrition, 6*(6), 1508–1518.

Nasrin, T. A. A., Rahman, M. A., Arfin, M. S., Islam, M. N., & Ullah, M. A. (2020). Effect of novel coconut oil and beeswax edible coating on postharvest quality of lemon at ambient storage. *Journal of Agriculture and Food Research, 2*, 100019.

Nasution, H., Harahap, H., Yustira, A., Julianti, E., & Jaafar, M. (2024). Innovative edible food wraps from tilapia fish bone gelatin and passion fruit peel extract. *Case Studies in Chemical and Environmental Engineering, 10*, 100990.

Ngamakeue, N., & Chitprasert, P. (2016). Encapsulation of holy basil essential oil in gelatin: Effects of palmitic acid in carboxymethyl cellulose emulsion coating on antioxidant and antimicrobial activities. *Food and Bioprocess Technology, 9*, 1735–1745.

Nizardo, N. M., Sugandi, N. A. D., & Handayani, A. S. (2024). Poly (vinyl alcohol)/carboxymethyl cellulose/cellulose nanofibrils nanocomposite as coating for food packaging paper. *Polymer-Plastics Technology and Materials, 63*(5), 447–458.

Obiaocha-Nwaogwugwu, C. N., Eruteya, O. C., & Ahaotu, I. (2024). Shelf-life extension of tomatoes (*Solanumlycopersicum*), okra (*Abelmoschus esculentus*) and eggplant (*Solanum melongena*) using edible coating. *Scientia Africana, 23*(2), 384–397.

Patkar, M. S., Prasad, R. R., Rahate, K. S., Shetye, P. A., Sinha, R. A., & Chitnis, K. S. (2020). Extraction and characterization of starch from tubers and its application as bioplastic. *World Journal of Environmental Biosciences, 9*(3-2020), 1–5.

Peng, J. (2022). Latest advances in preservation technology for fresh fruit and vegetables. *Foods, 11*(20), 3236.

Pinaeva, L. G., & Noskov, A. S. (2024). Biodegradable biopolymers: Real impact to environment pollution. *Science of The Total Environment*, 174445.

Prudnikov, E., Polishchuk, I., Sand, A., Hamad, H. A., Massad-Ivanir, N., Segal, E., & Pokroy, B. (2023). Self-assembled fatty acid crystalline coatings display superhydrophobic antimicrobial properties. *Materials Today Bio, 18*, 100516.

Rabea, E. I., Badawy, M. E. T., Stevens, C. V., Smagghe, G., & Steurbaut, W. (2003). Chitosan as antimicrobial agent: Applications and mode of action. *Biomacromolecules, 4*(6), 1457–1465.

Rendón-Villalobos, R., Ortíz-Sánchez, A., Tovar-Sánchez, E., & Flores-Huicochea, E. (2016). The role of biopolymers in obtaining environmentally friendly materials. *Composites from renewable and sustainable materials*, 151.

Robertson, G. L. (2005). *Food packaging: Principles and practice*. CRC press.

Rosenthal, A., Deliza, R., Welti-Chanes, J., & Barbosa-Cánovas, G. V. (2018). *Fruit preservation: Novel and conventional technologies*. Springer.

Sanchez-Garcia, M. D., Lopez-Rubio, A., & Lagaron, J. M. (2010). Natural micro and nanobiocomposites with enhanced barrier properties and novel functionalities for food biopackaging applications. *Trends in Food Science & Technology, 21*(11), 528–536.

Santhosh, R., & Sarkar, P. (2024). Fabrication of jamun seed starch/tamarind kernel xyloglucan bio-nanocomposite films incorporated with chitosan nanoparticles and their application on sapota (*Manilkara zapota*) fruits. *International Journal of Biological Macromolecules, 260*, 129625.

Sari, P. R. P., Darmawati, E., & Ahmad, U. (2020). Aloe vera and beeswax based coating to maintain shelf life of salak cv. Madu. In *IOP conference series: Earth and environmental science*. IOP Publishing 012014 Vol. 542, No. 1.

Seyedi, S., Koocheki, A., Mohebbi, M., & Zahedi, Y. (2015). Improving the physical and moisture barrier properties of *Lepidium perfoliatum* seed gum biodegradable film with stearic and palmitic acids. *International Journal of Biological Macromolecules, 77*, 151–158.

Shiekh, K. A., Ngiwngam, K., & Tongdeesoontorn, W. (2021). Polysaccharide-based active coatings incorporated with bioactive compounds for reducing postharvest losses of fresh fruits. *Coatings, 12*(1), 8.

Shinde, M. M., Malik, M., Kaur, K., Gahlawat, V. K., Kumar, N., Chiraang, P., & Upadhyay, A. (2024). Formulization and characterization of guar gum and almond gum based composite coating and their application for shelf-life extension of okra (Hibiscus esculentus). *International Journal of Biological Macromolecules, 262*, 129630.

Silva, S., Sousa, A. P., Gomes, J., Andrade, R., Lima, G., Siqueira, E., & Campos, A. R. (2021). Postharvest conservation of Tommy atkins mango fruits during storage using'*Moringa oleifera*'oil-based coating. *Australian Journal of Crop Science, 15*(2), 278–283.

Simovic, S., Barnes, T. J., Tan, A., & Prestidge, C. A. (2012). Assembling nanoparticle coatings to improve the drug delivery performance of lipid based colloids. *Nanoscale, 4*(4), 1220–1230.

Singh, A., Singh, J., Singh, N., & Kaur, N. (2024). Development and characterization of gelatin/carboxymethyl cellulose based polymeric film with inclusion of ionic liquid to enhance the shelf life of food. *Food Packaging and Shelf Life, 42*, 101251.

Song, X., Zuo, G., & Chen, F. (2018a). Effect of essential oil and surfactant on the physical and antimicrobial properties of corn and wheat starch films. *International Journal of Biological Macromolecules, 107*, 1302–1309.

Song, X., Cheng, L., & Tan, L. (2018b). Edible iron yam and maize starch convenient food flavoring packaging films with lemon essential oil as plasticization. *Food Science and Technology, 39*(4), 971–979.

Ştefănescu, B. E., Socaciu, C., & Vodnar, D. C. (2022). Recent progress in functional edible food packaging based on gelatin and chitosan. *Coatings, 12*(12), 1815.

Suderman, N., Isa, M. I. N., & Sarbon, N. M. (2018). The effect of plasticizers on the functional properties of biodegradable gelatin-based film: A review. *Food bioscience, 24*, 111–119.

Tang, X. Z., Kumar, P., Alavi, S., & Sandeep, K. P. (2012). Recent advances in biopolymers and biopolymer-based nanocomposites for food packaging materials. *Critical Reviews in Food Science and Nutrition, 52*(5), 426–442.

Thakur, R. K., Biswas, P. K., & Singh, M. (2024). Biovalorization of fruit wastes for development of biodegradable antimicrobial chitosan-based coatings for fruits (tomatoes and grapes). *Applied Biochemistry and Biotechnology, 196*(3), 1175–1193.

Thavasilingam, K., Senthil Kumar, A., Sakthimurugan, D., & Meenatchisundaram, P. (2023). Synthesis and characterization of fumed silica dispersed bees/carnauba wax-epoxy biocomposite coating: A comparative study. *Biomass Conversion and Biorefinery, 13*(8), 7375–7383.

Versino, F., Lopez, O. V., Garcia, M. A., & Zaritzky, N. E. (2016). Starch-based films and food coatings: An overview. *Starch-Stärke, 68*(11-12), 1026–1037.

Vinod, A., Sanjay, M. R., Suchart, S., & Jyotishkumar, P. (2020). Renewable and sustainable biobased materials: An assessment on biofibers, biofilms, biopolymers and biocomposites. *Journal of Cleaner Production, 258*, 120978.

Wu, J., Zhang, L., & Fan, K. (2024). Recent advances in polysaccharide-based edible coatings for preservation of fruits and vegetables: A review. *Critical Reviews in Food Science and Nutrition, 64*(12), 3823–3838.

Yang, Y., Zhang, J., & Li, C. (2024a). Delivery of probiotics with cellulose-based films and their food applications. *Polymers, 16*(6), 794.

Yang, Z., Chen, B., Tahir, H. E., Li, Z., Huang, X., Li, M., & Xiao, J. (2024b). Gelatin/sodium alginate-based biodegradable films functionalized by persimmon pectin/ovalbumin-stabilized neem essential oil Pickering emulsion: Application for cherry tomato preservation. *Progress in Organic Coatings, 192*, 108448.

Yilmaz, A., Bozkurt, F., Cicek, P. K., Dertli, E., Durak, M. Z., & Yilmaz, M. T. (2016). A novel antifungal surface-coating application to limit postharvest decay on coated apples: Molecular, thermal and morphological properties of electrospun zein–nanofiber mats loaded with curcumin. *Innovative Food Science & Emerging Technologies, 37*, 74–83.

Yu, D., Yu, Z., Zhao, W., Regenstein, J. M., & Xia, W. (2022). Advances in the application of chitosan as a sustainable bioactive material in food preservation. *Critical Reviews in Food Science and Nutrition, 62*(14), 3782–3797.

Yu, Y., Zhou, J., Chen, Q., Xie, F., Zhang, D., He, Z., & Cai, J. (2024). Self-reinforced multifunctional starch nanocomposite film for litchi fruit postharvest preservation. *Chemical Engineering Journal, 486*, 150262.

Yuan, Y., Ma, M., Xu, Y., & Wang, D. (2022). Surface coating of zein nanoparticles to improve the application of bioactive compounds: A review. *Trends in Food Science & Technology, 120*, 1–15.

Zhang, X., Ismail, B. B., Cheng, H., Jin, T. Z., Qian, M., Arabi, S. A., & Guo, M. (2021). Emerging chitosan-essential oil films and coatings for food preservation-A review of advances and applications. *Carbohydrate Polymers, 273*, 118616.

Zhang, Y., Zhu, M., Huang, Z., Yang, F., Weng, Y., & Zhang, C. (2024). The effect of polylactic acid-based blend films modified with various biodegradable polymers on the preservation of strawberries. *Food Packaging and Shelf Life, 45*, 101333.

Zhou, X., Yang, R., Wang, B., & Chen, K. (2019). Development and characterization of bilayer films based on pea starch/polylactic acid and use in the cherry tomatoes packaging. *Carbohydrate Polymers, 222*, 114912.

Chapter 30

Multilayer films and coatings based on biopolymers

Gourav Chakraborty[1], Shivani Desai[1], Srishti Upadhyay[1], Jeeva Kiran Banoth[1], Vijay Singh Sharanagat[1], Kshitiz Kumar[2], and Yograj Bist[3]

[1]*Department of Food Engineering, National Institute of Food Technology Entrepreneurship and Management (NIFTEM-K), Sonipat, Haryana, India,* [2]*Centre of Food Technology, University of Allahabad, Prayagraj, Uttar Pradesh, India,* [3]*Department of Biotechnology, Graphic Era (Deemed to be University), Dehradun, Uttarakhand, India*

30.1 Introduction

The negative environmental impacts associated with the production, indiscriminate use, and inadequate disposal management of synthetic plastics, scarcity of nonrenewable petroleum resources, and increasing door-to-door food delivery lead to a considerable increase in the consumption of packaged foods. This lifestyle change has boosted attention toward the greener, safer, and biodegradable solutions for synthetic plastics. The renewable source, wide availability, and exceptional degradability of biopolymers make them a suitable alternative in the creation of eco-friendly films (Grzebieniarz et al., 2023). The films produced from biopolymers not only solve environmental problems, such as the accumulation of waste caused by petroleum-based materials (plastics) and microplastics accumulation in aquatic organisms, but also improve the shelf life, quality, and safety of packaged food. Biopolymers with oxygen in their molecular chains undergo enzymatic hydrolysis by microorganisms in soil, leading to the rapid degradation of these materials.

In the last decades, various biopolymers have been explored for the development of biodegradable packaging films and coatings. Polysaccharides are mostly used as substrates for the preparation of packaging films and coatings, sourced from animal materials (chitosan [CS]), plant materials (starch, pectin, agar, sodium alginate [SA], carrageenan [CAR], furcellaran [FUR], cellulose, and its derivatives). The next group of biopolymers is proteins (gelatin, zein, soy protein isolate [SPI], whey protein, gluten, casein, keratin, silk fibroin, etc.), which are also ideal materials for biodegradable packaging and coatings. Among proteins, gelatin, a water-soluble protein with high-molecular weight, is the most widely used in film matrix due to its characteristics such as low oxygen and oil permeability, gelling, thickening, emulsifying capacity, and encapsulating ability (Grzebieniarz et al., 2023). Other alternative options are the lipids extracted from plants, animals, and insects, and the bio-based polymers produced from microbial synthesis (polyhydroxyalkanoates, bacterial cellulose, gellan gum, xanthan gum, guar gum, acacia gum, curdlan, etc.). These biopolymers are the backbone of biopolymer-based films and coatings due to their wide adaptability, sustainability, cost-effectiveness, biocompatibility, film-forming ability, and functional versatility. However, poor functional, antimicrobial, antioxidant, mechanical, and barrier properties to oxygen and water, compared to available packaging, limit their application in food packaging.

Monolayered film structures based on biopolymers exhibit great potential to replace synthetic polymers due to faster decomposition compared to multilayer films (MFs). However, achieving the desired properties for packaging applications is challenging with monolayer biopolymer-based films due to their poor barrier properties, lower mechanical strength, and high-water solubility (WS). The operational drawbacks of monolayered biopolymer-based films can be minimized by incorporating plasticizers, cross-linkers, and fillers in MFs through different methods, which provide complementary properties such as better barrier properties, higher mechanical properties, or broad functions by minimizing the disadvantages of a single layer. MFs are composed of thin, independent sheets of materials of lower thickness, each layer with a specific role in making a single sandwich-like structure, whereas biopolymer coatings are either applied directly to products for immediate consumption or added to films to improve their functional properties. The MFs, typically composed of three to seven layers with thicknesses ranging from 20 to 400 μm, integrate the unique characteristics of various materials, providing a range of functionalities for flexible packaging (Alias et al., 2022). Each layer in an MF is strategically designed in a

Natural Polymer-Based Films and Coatings in Food Technology. DOI: https://doi.org/10.1016/B978-0-443-34244-8.00031-8

sequence as top, bottom, outer, and inner, to combine the unique properties of various polymers, resulting in packaging with superior protection and durability (Fig. 30.1). The outermost layer (barrier layer) is in direct contact with the external environment; therefore, it must have superior barrier properties against moisture, gases, and microorganisms to prevent spoilage of packed food from external environment and microbial access, and to protect the active ingredients within the inside layers. The intermediate layer (active layer) contains the bioactive ingredients (antioxidants and antimicrobials); therefore, it must have properties that retain the ingredients in stable form and must have good diffusion properties to diffuse the ingredients through the innermost layer at a controlled rate. Unlike the outer barrier layer, this layer should retain moisture to maintain the desired humidity for packaged food. Furthermore, innermost layer (control layer) is in direct contact with food, so it must be designed appropriately for the release of active ingredients into the food product at a controlled rate (Wang et al., 2022). However, the layer properties may vary with the application of MFs with varying food products.

These approaches modulate the mechanical and barrier properties and control the release of active ingredients through the films, addressing sustainability and suitability challenges for food packaging applications. The introduction of active substances (bioactive compounds from plant materials, nanofillers) and antimicrobial ingredients (essential oils, organic acids, enzymes, and bacteriocins) in a specific layer of film composites helps to eliminate the limitations associated with the biopolymers or monolayered films. Incorporating active materials into MFs can also serve as intelligent indicators, enabling the monitoring and assessment of the food product's freshness.

MFs are typically produced through the solution casting method (wet method), the layer-by-layer (LbL) extrusion method (dry method), and electrohydrodynamic methods. Among these methods, extrusion and solution casting are the most widely used to produce films from solids and solutions, respectively. The structure of the films and their properties vary with the method of fabrication. By changing the ratio of active ingredients in the intermediate layer (active layer), the barrier properties and release time of the active ingredients can be controlled. One or two methods in combination are also in practice for creating MFs with better performance. Furthermore, dipping is the most commonly used technique for coating on various substrates; though simpler in practice, it has poor control over the thickness. Therefore, more state-of-the-art techniques have evolved, such as spray coating, electrospraying, LbL electrophoretic deposition, vacuum impregnation, etc. Furthermore, the selection of the best combinations of film-forming composition, optimization of the production process, nature of food material, type of contact of food with package, nature of active ingredients (concentration, molecular weight, size, polarity, etc.), and their controlled release through packaging are important variables to improve functional performance of films for their use in food applications. While bio-based films are sustainable, ensuring they also meet the stringent safety standards of packaged food is challenging. This is the major concern while producing the film, which is maintaining the integrity of packaged food products and extending their shelf life, along with safeguarding health.

This chapter comprehensively discusses the methods of preparation of MFs and coatings, the properties of MFs, and their potential applications in different sectors of the food industry. Additionally, the challenges associated with the biopolymer-based MFs and perspectives to overcome these challenges to make them the best, efficient, and inexpensive are also discussed.

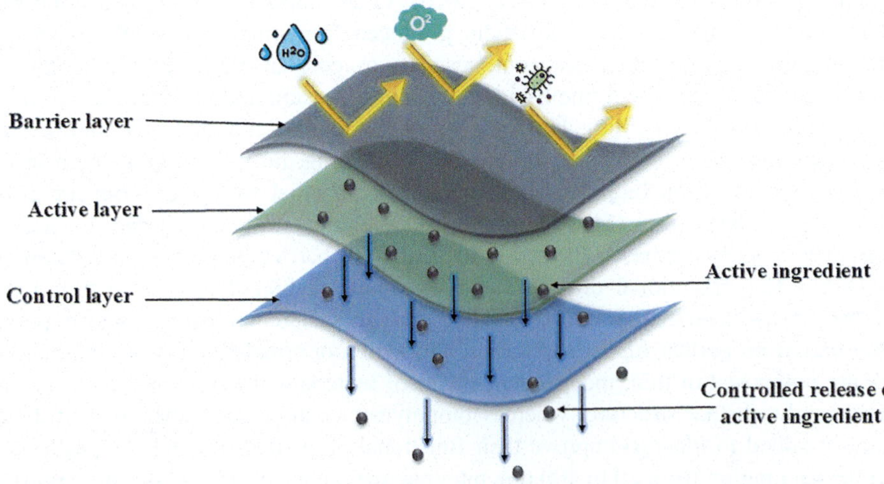

FIGURE 30.1 Multilayer packaging film.

30.2 Different biopolymers used for multilayer films

30.2.1 Polysaccharide

30.2.1.1 Starch

Starch is the predominant carbohydrate found in nature, comprising around 80% of total carbohydrate production. It is made up of two glucose-based polymers, that is, amylose and amylopectin, and is commonly synthesized by green plants like corn, wheat, potatoes, rice, and cassava. Starch, due to its abundant natural availability, has attracted significant attention for the development of biodegradable films (Singh et al., 2023). Although natural starch offers several benefits, its application in food packaging is limited by certain drawbacks, such as high-moisture sensitivity, poor mechanical strength, low barrier efficiency, and minimal resistance to elevated temperatures. To overcome these limitations, plasticizers like glycerol, sorbitol, and xylitol are commonly incorporated to improve the flexibility and reduce the brittleness of starch (Talja et al., 2007). Additionally, blending starch with synthetic biodegradable polymers such as polylactic acid (PLA), polyvinyl alcohol (PVA), and polycaprolactone (PCL) can significantly enhance its mechanical strength, water resistance, and processability (Cheng et al., 2021). Moreover, various modification techniques, including physical, chemical, and enzymatic, are also employed to further improve the functional properties of starch-based biodegradable materials.

Heidemann et al. (2019) explored the use of atmospheric cold plasma to enhance the bonding properties of PCL and PLA films for improved integration with starch layers. The plasma treatment modified the surface characteristics of PCL and PLA, increasing their roughness and reducing the water contact angle, thereby making the surfaces more receptive to adhesion. Following this surface modification, bilayer films combining PCL or PLA with starch were fabricated. The study found that plasma-treated films exhibited significantly better adhesion between the starch and polymer layers, with improved resistance to delamination that correlated with longer treatment times.

30.2.1.2 Pectin

Pectin is a naturally occurring polysaccharide derived from plants, classified as a heteropolysaccharide. Its molecular structure is primarily composed of galacturonic acid units, along with minor amounts of other sugars such as rhamnose, galactose, arabinose, and mannose. The physicochemical characteristics of pectin can vary significantly depending on its source, largely due to differences in molecular structure. These variations are influenced by factors such as the proportion of constituent sugars, the arrangement of side chains, and the degree of esterification (Roy et al., 2023). Pectin-based films have gained widespread use as sustainable alternatives for food packaging applications. These functional films have proven effective in extending the shelf life of products such as soybean oil, cheese, meat, and fish. Additionally, pectin films incorporated with color indicators have been utilized as smart packaging tools to monitor the freshness of perishable items like meat and fish (Ezati & Rhim, 2020; Guo et al., 2021; Rodsamran & Sothornvit, 2019). Jridi et al. (2020) developed bioactive composite films using pectin extracted from blood orange peels combined with fish gelatin (FG), specifically for cheese packaging applications. The authors revealed that blending pectin with gelatin resulted in films with enhanced physical properties compared to films made from pure gelatin. Various gelatin–pectin ratios were explored, and the 50:50 blend emerged as the most effective, showing notable improvements in both mechanical strength and barrier performance.

30.2.1.3 Agar

Agar consists mainly of two key components, that is, agarose and agaropectin. Agarose is the component that imparts strong gelling capabilities, making it ideal for use in the formulation of films and coatings. In contrast, agaropectin contains various anionic groups, such as sulfate, pyruvate, and glycuronate, which tend to interfere with and reduce the gel strength of agar (Chen et al., 2021). Agar readily dissolves in hot water and forms a gel upon cooling, a property that has made it widely used in the food industry as a thickening, gelling, and stabilizing agent. Beyond its culinary applications, agar has also gained interest in the field of packaging due to the strong mechanical properties exhibited by films made from it. In a study by Diop et al. (2023), an LbL casting method was employed to fabricate innovative bilayer plastic composites, combining distinct layers of agar/chitosan and PVA/agar. The physical and mechanical behavior of these composite films was found to be highly dependent on the relative thicknesses of the individual layers, indicating that variations in layer composition directly impacted the overall film performance.

30.2.1.4 Sodium alginate

Alginates are naturally occurring polysaccharides predominantly present in the cell walls of brown seaweeds like kelp and other marine algae. In addition to these plant-based sources, alginates can also be biosynthesized by certain bacteria, resulting in what are known as bacterial alginates. These microbially derived alginates often exhibit more uniform and well-defined chemical and physical characteristics (Yerramathi et al., 2025). Alginate-based biopolymer films have gained recognition as an eco-friendly and sustainable packaging option. These films stand out for their biodegradability and compostability, while also offering excellent mechanical strength, barrier effectiveness, and favorable sensory characteristics, making them highly suitable for use in food packaging (Dave et al., 2023). Zhang, Guan, et al. (2024) and Zhang, Zhang, et al. (2024) recently introduced an innovative MF composed of SA and κ-carrageenan, with gelatin incorporated into the inner layer. This advanced layered configuration demonstrated notable benefits for food packaging, including superior mechanical performance and increased effectiveness in blocking moisture and gases.

30.2.1.5 Chitosan

Chitosan, a naturally alkaline polysaccharide, is produced through the N-deacetylation of chitin, primarily sourced from marine crustaceans, and ranks as the second most abundant natural polymer. Unlike chitin, chitosan exhibits significantly improved solubility, which broadens its potential for use across various applications (Sarfraz et al., 2024). Advancements in science and technology have brought increased focus to chitosan's potential in food packaging, largely due to its excellent film-forming capabilities. In a study by Li et al. (2019), films were developed using an LbL assembly technique involving chitosan and alginate. This method resulted in films with enhanced tensile strength (TS), reduced water vapor permeability (WVP), and improved hydrophobic and light-blocking properties compared to conventional film structures.

30.2.1.6 Cellulose

Cellulose and its derivatives, derived from a variety of natural sources, are among the most extensively utilized biopolymers, particularly in packaging applications. Structurally, cellulose is a polysaccharide made up of repeating D-glucose units arranged in a linear chain, with the chemical formula $(C_6H_{10}O_5)_n$ (Hu et al., 2019). In a recent study, Cazón et al. (2024) designed and evaluated multilayer composite films made from bacterial cellulose and chitosan, incorporating grape bagasse extract as a natural antioxidant along with glycerol as a plasticizer. The results demonstrated that these films hold strong potential for food packaging, offering added functional benefits through the integration of naturally derived antioxidants.

30.2.2 Protein

30.2.2.1 Gelatin

Gelatin, a fibrous protein derived from the partial hydrolysis of collagen, is water-soluble and commonly used in the production of edible films. Its linear molecular structure and relatively short polymer chains play a key role in its excellent film-forming properties (Kanatt, 2020). The mechanical strength, functionality, and barrier performance of gelatin films are largely influenced by their formulation and processing techniques. To enhance their moisture resistance and vapor barrier properties, additives such as plasticizers, cross-linking agents, or complementary biopolymers, including SPI, various oils, fatty acids, and polysaccharides, can be incorporated into the film matrix (Suderman et al., 2018). In a study conducted by Jiang et al. (2022), multilayer active films were developed using a combination of gelatin and myofibrillar protein. Compared to their single-layer counterparts, these multilayer structures demonstrated significantly improved resistance to water penetration, highlighting their potential for advanced packaging applications.

30.2.2.2 Soy protein isolates

SPIs are produced from soy flour through a process called isoelectric precipitation, which involves removing nonprotein components such as carbohydrates, fats, and fibers to obtain a highly purified protein product (Meganaharshini et al., 2023). SPI is a highly purified form of soy protein, comprising more than 90% protein content. Film-forming solutions using SPI can be prepared under either acidic or alkaline conditions; however, films cast from acidic solutions tend to exhibit superior physical and mechanical characteristics (Qi et al., 2024). Yong et al. (2023) developed bilayer antimicrobial films by cross-linking SPI and chitosan (CS) with curcumin, incorporating both antibacterial and antioxidant functions. The resulting films showed marked improvements in TS, UV-blocking capability, and surface hydrophobicity. Additionally, WVP, thermal shrinkage, and opacity were significantly minimized, enhancing the overall performance of the films.

30.2.2.3 Zein

Zein is a water-insoluble, alcohol-soluble thermoplastic protein known for its strong film-forming capabilities and hydrophobic nature. It has been commercially utilized for years, particularly in the edible coating of products like nuts and confectionery items (García Ibarra et al., 2016). Zein, a type of prolamin protein, is primarily located in the endosperm of corn (maize) and is chiefly made up of three major components: α-zein, β-zein, and γ-zein (Ezati et al., 2023). Chen et al. (2025) investigated the physicochemical and structural characteristics of gelatin–zein films, comparing those created by blending with those formed using an LbL self-assembly technique. As the proportion of zein increased, both types of films exhibited reductions in elongation at break (EAB), WVP, and melting temperature. However, the WVP of the bilayer films remained relatively stable despite the varying zein content.

30.2.2.4 Casein

Casein, the primary phosphoprotein found in milk, is widely utilized as a food additive and has increasingly been explored for use in the development of biodegradable food packaging films. These films are noted for their effective water vapor barrier properties, likely attributed to the hydrophobic segments within the casein molecular structure. Szyk-Warszyńska et al. (2014) constructed multilayer films using α-casein or β-casein combined with poly-L-arginine through an LbL assembly method. The study focused on how the type of casein influenced the film properties, revealing that films incorporating α-casein were marginally thicker than those made with β-casein.

30.3 Methods of preparation for multilayer films and coating techniques

The preparation of MFs with desirable functional properties necessitates careful selection of packaging material and other ingredients, film-forming methods, and preparation conditions (Uz & Altınkaya, 2011; Wang et al., 2022). The selection of materials and design for these films is influenced by factors such as the type of polymer, interactions between polymers, layer adhesiveness, compatibility of individual layers, and the intended production scale (Anukiruthika et al., 2020). A variety of techniques have been employed to create MFs. The widely used methods for preparing MFs are casting, LbL assembly, electrospinning, and electrospraying (Grzebieniarz et al., 2023). These methods of preparation of MFs and coating techniques are discussed below.

30.3.1 Wet or casting method

Solvent or solution casting is the most commonly employed technique for preparing biodegradable films in laboratory settings. This method involves three main steps (Fig. 30.2): preparation of solution, casting it into a mold, and drying (Wang et al., 2022). To make MFs, this approach can be adapted by sequentially pouring and layering each film-forming solution one at a time. For the preparation of second and subsequent layers, the first layer must achieve a gel-like state. However, not all biopolymers are suitable for this purpose, as some do not form a gel during the drying process (Grzebieniarz et al., 2023). The primary advantage of the casting method lies in its simplicity. However, the high-liquid content in the film-forming solution makes the drying process both time-intensive and energy-demanding. In addition to its time-consuming drying process, the casting technique faces other challenges, such as poor control over thickness, limited format options typically restricted to simple plates and tubes, and difficulties in scaling up production. These scaling challenges arise primarily from the sensitivity of film characteristics due to variations in factors such as temperature and air speed during drying (Avila et al., 2023). Despite its limitations, the casting method has been widely employed by researchers to develop bio-based MFs. For example, Andrade et al. (2016) utilized this technique to produce MFs composed of polyhydroxyalkanoates and hydrocolloids. Similarly, Bora et al. (2016) made bilayer films by casting a pectin–SA composite layer onto a casein film. In another study, Jamróz et al. (2021) demonstrated a casting approach to produce MFs, completing the entire process within a few hours. Successive layers were applied at intervals of 1–2 hours. There are also some studies that involve a layering process that spans several days. In such cases, the first layer must be fully dried before the second layer can be applied (Fu et al., 2022; Sun et al., 2022).

30.3.2 Dry or extrusion method (layer by layer)

MFs can also be fabricated using the LbL method, also known as the Langmuir-Blodgett deposition technique. Renowned for its cost-effectiveness, simplicity, and versatility, this method has become an important option for making MFs (Avila et al., 2023). This technique relies on interactions such as electrostatic forces, hydrogen bonding, hydrophobic attractions, or molecular entanglements between adjacent layers. Typically, it involves creating MFs or coatings by sequentially

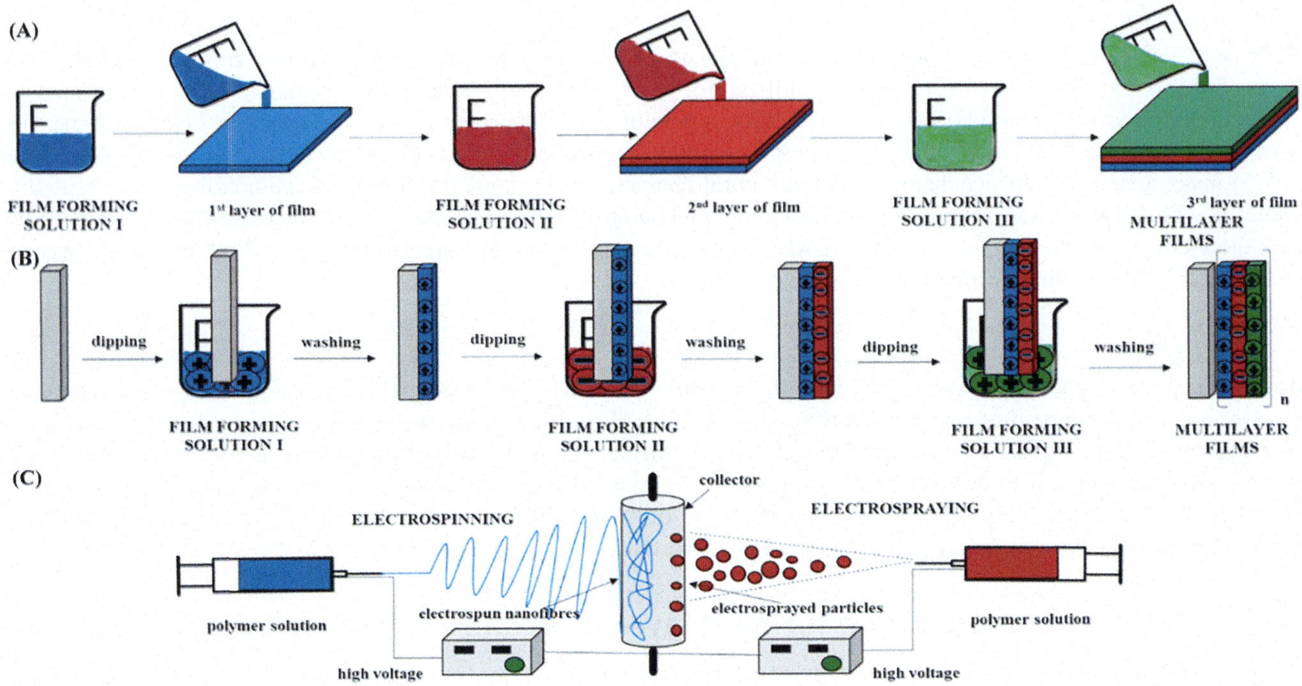

FIGURE 30.2 Methods for obtaining multilayer films: (A) casting solution; (B) layer-by-layer; and (C) electrospinning and electrospraying. From Grzebieniarz et al. (2023). Advances in biopolymer-based multi-layer film preparations and food packaging applications. Food Packaging and Shelf Life, 35. https://doi.org/10.1016/j.fpsl.2023.101033.

depositing multiple film-forming materials onto a substrate (Wang et al., 2022). The most common mechanism driving LbL assembly in solutions is the electrostatic attraction between ions. This process involves the alternate deposition of oppositely charged polyelectrolytes or nanoparticles onto a chosen template (Fig. 30.2). Once the desired multilayer structure of specific thickness is achieved, the template is removed, leaving behind the MF (Zhang et al., 2019). Electrostatic LbL assembly offers precise control over the structure and thickness of MFs while enabling the seamless incorporation of functional substances, such as conductive polymers, photopolymers, or biological macromolecules, into the film.

Hydrogen bonding is another key driving force in LbL assembly. For example, to enhance the durability of conventional superhydrophilic films, Hwangbo et al. (2016) incorporated branched polyethyleneimine into a macromolecule, creating superhydrophilic antibacterial nanocoatings through hydrogen bond assembly. These bonds formed between the hydroxyl groups of silsesquioxane and the amine groups of branched polyethyleneimine contribute to the coating's functionality and stability. Often referred to as a push-pull system, LbL charge transfer interactions involve an electron donor and an electron acceptor positioned at opposite ends of a conjugated structure.

This mechanism can serve as a driving force for assembly in polymer solutions, enabling the construction of multilayered polymer films in nonionic polymer systems (Bathula et al., 2017). MFs created using electrostatic interactions, hydrogen bonding, or charge transfer interactions often demonstrate limited stability in polar solvents and high-concentration salt solutions. In recent years, LbL assembly utilizing covalent bonding has advanced significantly, relying on chemical reactions between polymers to facilitate the assembly process. Due to the higher energy of covalent bonds compared to electrostatic forces and hydrogen bonds, MFs constructed through covalent bonding exhibit superior stability and durability (Zhang et al., 2019). These interactions largely depend on the characteristics of the materials used in creating biodegradable packaging. For example, proteins typically form cross-links with adjacent molecules through a combination of hydrophobic interactions, electrostatic forces, covalent bonds, and hydrogen bonding (Calva-Estrada et al., 2019; Hassan et al., 2018). LbL films offer the capability for controlled drug release while also exhibiting antimicrobial and antifungal properties, making them highly versatile and widely utilized in tissue engineering applications (Borges et al., 2015; Kumorek et al., 2020; Li et al., 2020). This technique allows for the creation of nano, micro, and macro composites while enabling the integration of diverse components into their structures, including active or intelligent elements (Arnon-Rips & Poverenov, 2018). Various techniques, including dipping, spin-coating, spray-coating, and the more recent brushing method, have been employed to develop LbL films (Kharlampieva et al., 2009; Park et al., 2018;

Schlenoff et al., 2000). Among these, the dip-coating method is particularly versatile, accommodating substrates of nearly any shape while minimizing material usage, as the same solutions can be reused for multiple deposition cycles. This approach allows for the deposition of a wide range of materials, from synthetic compounds to biological molecules, and provides precise control over the film's structure, thickness, and functionality. Despite these advantages, the process is time-intensive, with each deposition or rinsing step typically requiring 5 to 20 minutes (Iqbal et al., 2024). Achieving the desired physicochemical and mechanical properties of MFs created through LbL technology depends on various factors, such as the type of absorbent used, the total number and sequence of layers, and the processing conditions. For successful film production, two key requirements must be met: the polyelectrolytes should be water-soluble and possess opposite electrostatic charges. The overall film thickness is influenced by the total number of layers and the thickness of individual layers. The layer thickness is influenced by several factors, including the molecular weight of the compounds, pH, ionic strength, type of counterions, deposition method, and ambient relative humidity. With the LbL technique, individual layer thicknesses can be precisely controlled, ranging from 1 to 100 nm (Lindström & Österberg, 2020; Richardson et al., 2015). The development of bio-modified paper with enhanced properties using the LbL deposition of CS and a dopamine-grafted carboxymethyl cellulose (CMC) polymer (CMC-g-DA) revealed that a six-layer CS/CMC-g-DA composite significantly improved the TS of the paper by 71.6% (Yu et al., 2022). Additionally, it reduced air permeability by 69.2% and WVP by 56.3%, while achieving 90% antibacterial effectiveness against *Staphylococcus aureus* and *Escherichia coli*. Yan, Wang, et al. (2022) and Yan, Sun, et al. (2022) created bilayer composite films using the LbL assembly of SA and CS layers. The resulting films demonstrated exceptional TS, barrier properties, and oxidation resistance, highlighting their potential as effective food preservation materials with diverse applications. Zhang, Guan, et al. (2024) and Zhang, Zhang, et al. (2024) utilized the LbL technique to fabricate high-barrier flexible films composed of poly (L-lactic acid) (PLLA), CS, and graphene oxide (GO) for meat packaging applications. It was found that the films outperformed commercial polyamide (PA)/polyethylene (PE) films in maintaining meat freshness, owing to their enhanced barrier properties and notable antibacterial activity. These findings highlight the promising potential of these films for use in fresh meat preservation. Wang, Qu, et al. (2024) and Wang, Wang, et al. (2024) developed an LbL assembly system using carboxymethyl chitosan (CMCS) and SA with varying concentrations of GO. It was found that increasing both the number of layers and the GO concentration significantly enhanced the mechanical strength and barrier properties of the self-assembled films. These improvements contributed to the effective preservation of freshness in fresh-cut apples, showcasing the potential of this approach for extending the shelf life of perishable produce. The LbL method offers several advantages over other techniques for fabricating multilayer packaging. It is a straightforward and cost-effective process that does not require specialized equipment, making it accessible for a wide range of applications. By selecting different types, quantities, and sequences of film-forming materials, it is possible to tailor the film's physicochemical and structural properties, such as composition, thickness, and charge to achieve various functional attributes. Furthermore, this method is highly versatile, as it is not restricted to flat surfaces and can be adapted to create coatings for food items of diverse shapes (Wang et al., 2022).

30.3.3 Electrohydrodynamic method

The electrohydrodynamic process utilizes a high-voltage electric field to transform polymer solutions into microscale or nanoscale fibers or particles. This technique is primarily carried out using two widely adopted methods: electrospinning and electrospraying (Wang et al., 2022). The standard configuration for both electrospinning and electrospraying includes four key components, as shown in Fig. 30.2: (1) a high-voltage power supply (ranging from 1 to 30 kV), typically operating in direct current mode, although alternating current mode can also be used; (2) a stainless steel needle or capillary with a blunt tip; (3) a syringe pump; and (4) a grounded collector, which can be a flat plate or a rotating drum to gather the resulting fibers or particles (Anu Bhushani & Anandharamakrishnan, 2014).

30.3.3.1 Electrospinning

Electrospinning is a technique used to generate polymer fibers by applying a high voltage to a polymer solution, creating an electrically charged jet. In this process, the polymer solution is held within a capillary tube, with a voltage applied between the tube and a collector plate. Under the influence of the electric field, the solution is expelled as a fluid stream, forming a structure known as a Taylor Cone before stretching into fibers (Grzebieniarz et al., 2023). Needleless electrospinning technology offers an alternative approach, which is to generate polymer jets directly from the liquid surface using a spinning electrode. This approach simplifies maintenance, enhances operational stability, and significantly improves production efficiency (Omer et al., 2021). Packaging materials created through electrospinning are characterized by their high porosity, well-aligned polymer structure, and extensive specific surface area, making them particularly beneficial for certain food-related applications (Zhao, Duan, et al., 2020; Zhao, Wang, et al., 2020). It shows great potential for developing active packaging materials and creating nanostructured layers for food packaging. The unique properties of electrospun fibers, such as their nanoscale diameter, high surface-to-volume ratio, responsiveness to atmospheric changes, and ability to encapsulate heat-sensitive active compounds, make them ideal for use in advanced packaging solutions (Anu Bhushani & Anandharamakrishnan, 2014). It also enables the creation of materials from a wide variety of polymers chosen for their unique barrier properties, mechanical durability, and compatibility with various food products. These materials can be infused with active additives such as antioxidants, antimicrobials, and oxygen scavengers, enhancing their effectiveness in preserving food. By reducing oxidation and inhibiting microbial activity, they significantly extend the shelf life of packaged foods (Topuz & Uyar, 2024). In a study conducted by Akinalan Balik et al. (2019), the electrospinning technique was employed to fabricate multilayer structures aimed at enhancing the barrier properties against volatile aromatic compounds. By utilizing pectin and polyethylene oxide, they successfully produced uniform fibers. The resulting films demonstrated excellent homogeneity, transparency, reduced porosity, and notably improved barrier performance against water vapor and limonene. Yao et al. (2023) used the electrospinning technique to develop nanofiber membranes composed of tetracycline hydrochloride (TCH), peanut protein isolate (PPI), and PLLA. Antibacterial testing revealed that the membranes incorporating TCH exhibited clear inhibition zones against *E. coli* and *S. aureus*. Moreover, the size of the inhibition zones increased notably with higher concentrations of TCH, demonstrating enhanced antibacterial efficacy. Zhang et al. (2022) used coaxial electrospinning technology to produce nanofibers with thymol as the core and ethylene-vinyl alcohol copolymer (EVOH) as the outer shell. The findings revealed that EVOH nanofibers without thymol exhibited no antibacterial activity, whereas the inclusion of thymol significantly enhanced antibacterial properties. The antibacterial effectiveness increased proportionally with the thymol concentration within the nanofibers. Thymol/EVOH nanofibrous films were shown to effectively extend the shelf life of strawberries, highlighting their potential for application in fruit preservation. The process of preparing electrospun fibers also comes with several challenges. One significant issue is ensuring the compatibility between antimicrobial agents and polymer matrices, as composites of inorganic and organic antimicrobial agents with commonly used polymers often face difficulties with even dispersion due to interface-related problems. Additionally, during usage, antimicrobial agents can be lost or inactivated due to washing, friction, or other factors. Fixing these agents securely within the fiber structure, through methods like chemical bonding, remains a challenging task (Xu et al., 2025).

30.3.3.2 Electrospraying

Electrospraying operates similarly to electrospinning, but with specific adjustments to the polymer solution and processing conditions. These modifications ensure that the jet formed from the Taylor cone remains stable and does not elongate, resulting in the production of fine droplets rather than continuous fibers (Jaworek & Sobczyk, 2008). The key principle behind electrospraying involves applying an electric charge to the polymer solution, which is crucial for the process. When a high voltage is applied, it deforms the surface of the droplet, causing the formation of micro- or nanoscale droplets, depending on the specific process conditions. The underlying mechanism suggests that the electric field induces a Coulomb force within the droplet, which competes with the cohesive forces holding the droplet together, ultimately leading to the production of fine droplets (Bock et al., 2011). When the Coulomb force applied exceeds the cohesive forces within the droplet, it disrupts the surface tension, leading to the formation of micro- or nanodroplets. The critical point at which the droplet breaks up into these smaller droplets is known as the Rayleigh limit (LR), and this occurs when the electrostatic force surpasses the surface tension, causing the droplet to destabilize (Xie et al., 2006). In a study conducted by Khan et al. (2013), a water-in-oil emulsion was electrosprayed to create thin, edible barrier films. The inclusion of an emulsifier, polyglycerol polyricinoleate, helped stabilize the jet, while whey protein isolates significantly improved the water vapor barrier properties of the resulting film.

Electrohydrodynamic techniques for packaging offer several key advantages over traditional methods. These methods enable the direct, continuous, and large-scale production of polymer nanofibers, making them highly suitable for commercial applications. Additionally, the conditions required for film or coating formation are relatively mild, allowing for the use of a wide variety of raw materials. The equipment for production is both cost-effective and user-friendly, though care is needed due to the high voltages involved. Moreover, the properties and structure of the resulting films can be easily tailored by adjusting the materials and processing parameters. Finally, these methods facilitate the creation of fibers and particles at the nanometer scale, resulting in films or coatings with large surface areas and high porosity, which can enhance their functionality (Wang et al., 2022) (Fig. 30.3).

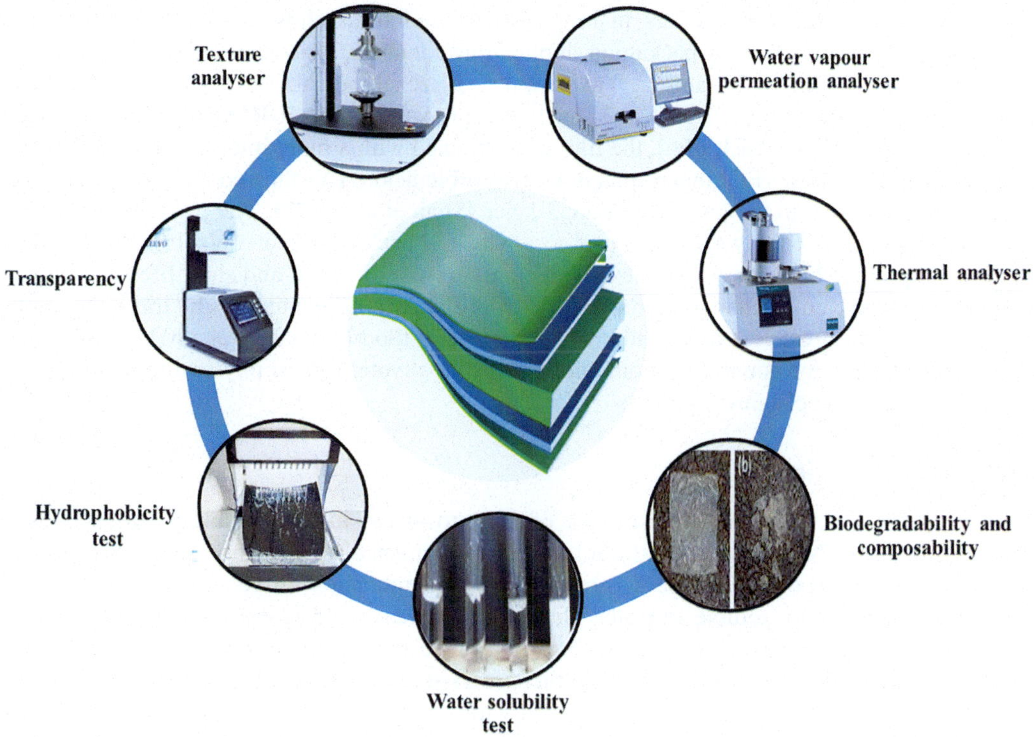

FIGURE 30.3 Properties of multilayer films based on biopolymers.

30.4 Properties of multilayer films

Biopolymer-based MFs are emerging as a sustainable alternative in food packaging due to their biodegradable nature. However, they have certain limitations, such as poor barrier properties, lower mechanical strength, and high WS. The properties of MF depend on the physical, chemical, and mechanical integrity of the composite materials present in it and the sequence of layers. Thus, the design of MF can be improved through material modifications, incorporating plasticizers, cross-linkers, and fillers, as well as innovative layering techniques, combining the advantages of each monolayer to achieve enhanced barrier properties, mechanical strength, and functional attributes (Grzebieniarz et al., 2023). This approach addresses existing limitations and significantly improves functional requirements, making them a promising solution for sustainable food packaging for commercial applications.

30.4.1 Physical properties

The type of compounds, fillers, and methods used for preparing biopolymer-based MFs significantly influence their physical properties. They play a crucial role in their performance and applicability in food packaging. The thickness of MF is influenced by the film-forming components and their interaction with one another. Zhao, Wang, et al. (2020) and Zhao, Duan, et al. (2020) reported that the polysaccharide films (CS or Alg) reinforced with cellulose nanofibers (CNF) (SA-CNF or CS-CNF) by blending or LbL method showed higher thickness compared to the monolayered film. The thickness was further increased with the concentration of CNF. Furthermore, the findings showed that the thickness of Alg-CNF films prepared using the blending (casting) method achieved a lower thickness than the LbL method, whereas CS-CNF films prepared by the LbL method achieved a sharp reduction in thickness compared to casting, which might be attributed to the agglomeration of CS and CNF due to electrostatic interaction in blending. Li et al. (2019) studied the synergistic effect of cross-linking and LbL method on film properties. The film prepared using cross-linking (Ferulic acid [FA]) in CS-SA biopolymer by LbL method (LbL-CS-FA-SA) had a significant reduction of 74.5%, 63.6%, and 54.8% in film thickness compared to the film prepared by direct mixing (CS-SA), cross-linking (CS-FA-SA), and LbL without cross-linking (LBL-CS-SA), respectively. This was due to the increased interactions between the cross-linkers.

The color attribute allows visual assessment of the appearance of MFs, which influences the functional performance of film, the appeal of packed food products, and consumer acceptance (Alias et al., 2022). It varies with the source of the biopolymer, a combination of raw materials and their compatibility, additives, processing conditions, environmental factors, etc. The induction of functional fillers in biopolymer-based films showed excellent physical and functional performance, which can be used for food packaging applications. Jamróz, Tkaczewska, Kopec, et al. (2022), Jamróz, Tkaczewska, Zając, et al. (2022), and Jamróz, Tkaczewska, Juszczak, et al. (2022) reported that with increasing concentration of the lingonberry extract (LBE) in bilayer film prepared from FUR/ CMC with the addition of gelatin hydrolysate (HGEL) (FUR/CMC + HGEL film), the film was obtained with a more intense and darker color. MF prepared with the CS, SA, and ethyl cellulose (EC) incorporated with ascorbic acid (AA) showed a more intricate surface structure with a light brown coloration due to the presence of a CS layer (Shah et al., 2025). The physical properties of MFs are a critical determinant of their functionality and consumer acceptance as packaging materials. It not only enhances the visual appeal of the packaged product but also contributes significantly to its protection and shelf life. The integration of active ingredients with pH-responsive color-changing properties further elevates these materials by transforming them into smart films capable of indicating the freshness and quality of the packaged food (Nowak et al., 2024). This dual functionality, combining protection and real-time communication, underscores the pivotal role of physical properties in advancing MFs as next-generation sustainable packaging solutions.

30.4.2 Functional properties

WS is a physical property that indicates the resistance of films to water; higher solubility corresponds to lower water resistance (Arham et al., 2016). MFs generally exhibit less WS performance compared to single-layer and natural polymer films. Nagarajan et al. (2017) reported that the WS of a PLA MF combined with gelatin was lower than that of a pure gelatin film, which was completely soluble in water. Similarly, González and Igarzabal (2013) observed that bilayered films made from SPI and PLA showed improved water resistance compared to single-layer films. In another study, Sanyang et al. (2015) found that MFs incorporating synthetic-derived biomass had lower WS, indicating enhanced water resistance. The reduction in WS for biomass films combined with synthetic components was attributed to the hydrophobic nature of PLA, which limits water absorption. This low WS helps reduce water activity in packaging materials, thereby contributing to the preservation of food products. Further, the WS of neat PLA films was found to be lower than that of bilayer films containing varying concentrations of zein particles with reduced thickness. Increasing the thickness of the zein coating and the PLA layer enhanced hydrophobic interactions between the PLA and zein layers, improving water resistance. In contrast, Martucci and Ruseckaite (2010) reported that bilayer films incorporating gelatin facilitated the migration of water molecules into the unprotected inner edges of the multilayered film due to the hydrophilic nature of gelatin. Upon contact with water, gelatin swelled and dissolved, causing delamination of the external PLA layers.

30.4.3 Barrier properties

Packaging materials must exhibit excellent barrier properties against water vapor, oxygen, and carbon dioxide to ensure optimal food preservation. The goal is to reduce film permeability to the lowest possible level, as this directly affects the condition and shelf life of packaged food. It is especially critical when dealing with freshly harvested fruits and vegetables, which continue their respiratory activity even after being packed and stored (Grzebieniarz et al., 2023). Generally, MFs offer superior barrier properties compared to monolayer films due to their extended diffusion path in multilayers, significantly improving the shelf life of packed food (Cerqueira et al., 2010).

30.4.3.1 Water vapor permeability

WVP is a critical property of food packaging films, as it determines the ease with which water vapor passes through the material, impacting moisture control and shelf life. Effective moisture transfer control is essential in packaging applications to minimize interaction between the food product and the surrounding atmosphere, requiring films with low WVP (Suderman et al., 2018). The rate of water vapor permeation can be theoretically determined by applying Fick's Law of diffusion in conjunction with Henry's Law of solubility. Factors influencing WVP in biopolymer films include film thickness, where thinner films are generally more permeable (Mora-Palma et al., 2024), and the molecular structure and crystallinity of the biopolymer itself (Trifol et al., 2020). The use of plasticizers, which enhance flexibility, can further affect water vapor barrier properties (Eslami et al., 2023). Protein-based films, while offering excellent gas barrier properties and mechanical strength, exhibit poor water vapor resistance due to their hydrophilic nature (Cheng et al., 2024). Laminating biopolymer films with water-resistant polymers can improve moisture-blocking performance, as demonstrated in MFs, which typically show significantly lower WVP compared to single-layer films (Pasquier et al., 2022). For instance, multilayer gelatin and PLA films achieved a WVP of $3.50 \pm 0.05 \times 10^{-10}$ $g{\cdot}m{\cdot}m^{-2}{\cdot}s^{-1}{\cdot}Pa^{-1}$, much lower than single-layer gelatin films at $38.00 \pm 0.05 \times 10^{-10}$ $g{\cdot}m{\cdot}m^{-2}{\cdot}s^{-1}{\cdot}Pa^{-1}$ (Hosseini et al., 2016). Biomass MFs demonstrated

WVP values ranging from 0.53 to 3.26 × 10^{-10} g·m·m^{-2}·s^{-1}·Pa^{-1}, with microbial-derived MFs performing better (0.40 to 0.09 × 10^{-10} g·m·m^{-2}·s^{-1}·Pa^{-1}) than biomass-synthesis or biomass-biomass films, due to a more cohesive matrix that reduces micropores and enhances barrier properties (Andrade et al., 2016). In contrast, biomass-biomass films exhibited minimal WVP improvements, due to thermodynamic incompatibility between biopolymers creating inconsistent matrices (Pereda et al., 2011). Various studies confirm the potential of bilayer and MFs in enhancing water vapor resistance, with examples such as agar-gelatin (3.26 ± 0.05 × 10^{-10} g·m·m^{-2}·s^{-1}·Pa^{-1}) and gelatin-PLA films (2.65 ± 0.04 × 10^{-10} g·m·m^{-2}·s^{-1}·Pa^{-1}) showcasing improved barrier properties (Nilsuwan et al., 2018; Vejdan et al., 2016). These findings highlight the promise of innovative multilayer designs in optimizing moisture barrier performance for food packaging applications (Table 30.1).

30.4.3.2 Gas barrier properties

Gas barrier properties refer to the ability of a film to resist the flow of gases. Gas permeation in biopolymeric films is governed by the processes of sorption and diffusion, both of which are significantly influenced by the size and structural properties of the migrating molecules. In rubbery polymers, those functioning above their glass transition temperature, gas sorption follows the principles of Fick's first law of diffusion combined with Henry's law of solubility (Shah et al., 2024). Oxygen permeability (OP) is a key parameter used to evaluate the capacity of a film to inhibit oxidation reactions, directly impacting the preservation of packaged food (Giteru et al., 2015). Minimizing OP is particularly crucial when packaging fresh fruits and vegetables, which continue respiratory activities postharvest (Arnon-Rips & Poverenov, 2018). As moisture significantly influences packaging performance, OP is often assessed under both dry and humid conditions (Carosio et al., 2014). The effectiveness of gas barrier properties plays a vital role in protecting the packaged contents from degradation, contributing to extended shelf life and quality maintenance (George & Venugopal, 2023). In MFs, the alternating arrangement of polymers in nanoscale layers modifies the crystallization patterns compared to bulk materials, enhancing barrier properties. The enhancement arises from the confined environment within each layer, which alters polymer chain behavior and improves gas resistance (Carr et al., 2012). Specifically, the formation of oriented lamellae in nanolayered films increases the tortuosity of the gas diffusion pathway, resulting in superior barrier performance (Greco et al., 2017; Ming et al., 2020). Protein-based films inherently exhibit excellent oxygen barrier properties, outperforming polysaccharide films in resistance to oxygen and carbon dioxide permeability due to their dense molecular structure (Suderman et al., 2018). Conversely, synthesis-derived films, characterized by their hydrophobic nature, tend to be less effective as oxygen barriers (Ghaani et al., 2016). The application of MFs has demonstrated significant improvements in barrier performance over single-layer counterparts. For example, a triple-layer film composed of PLA and FG (PLA/FG/PLA) reduced OP by 87% and WVP by 91% compared to monolayer films of the same materials (Hosseini et al., 2016). Similarly, electrospun MFs combining starch/poly(ε-caprolactone) (PCL) with carvacrol showed a ~200% decrease in WVP (Tampau et al., 2018). Bilayer films, such as those made from gelatin and epigallocatechin gallate (EGCG), also exhibited lower OP compared to their monolayer equivalents (Nilsuwan et al., 2019). However, other studies have observed variations, such as bilayer films of olive flounder gelatin (OSG) and PLA demonstrating higher OP than their monolayer counterparts, potentially due to differences in interfacial compatibility or structural integrity (Lee & Song, 2017).

30.4.4 Thermal properties

The thermal stability of films is a vital property for optimizing the potential applications of MFs, particularly in improving their performance. Thermal stability refers to the ability of a film to resist heat without degrading or altering its structure. The melting temperature (Tm) is the key aspect, which corresponds to the helix-coil transition involving the release of water molecules that stabilize the helix through hydrogen-bonded bridges (Staroszczyk et al., 2012). Protein-based films are generally unstable due to their sensitivity to moisture, low melting, and dissolving temperatures, which make them prone to structural degradation under heat. However, integrating microbial- or synthesis-derived films in multilayer structures can enhance the thermal properties of these films. The interaction between different polymer layers in MFs can influence the Tm, often leading to a phenomenon known as melting point depression. This occurs when the molecules formed at the polymer interfaces disrupt the crystal lattice, hindering the alignment and packing of polymer chains during melting (Vega et al., 2022). Biomass-synthesis MFs have shown Tm values ranging from 145.54°C to 148.12°C, while biomass-microbial films exhibit higher Tm values, ranging from 166.40°C to 174.00°C. Some examples include gelatin-polyhydroxyalkanoate (PHA) multilayer films with a Tm of 174.00°C ± 0.05°C (Andrade et al., 2016), gelatin-PLA films at 165.50°C ± 0.05°C (Hosseini et al., 2016), and gelatin–zein films at 111.02°C ± 0.28°C (Deng et al., 2018). Notably, gelatin-PLA MFs with a Tm of 148.12°C ± 0.05°C demonstrate significantly higher thermal stability compared to gelatin

TABLE 30.1 Properties of biopolymer-based multilayer films (MFs).

Names of compounds	Barrier properties	Mechanical properties	Optical properties	Physical properties	Thermal properties	References
Polylactic acid (PLA) and fish gelatin (FG)	• The water vapor permeability (WVP) of the multilayer film (0.125 g mm kPa h m^{-2}) decreased 11-fold compared with that of the FG film. • Oxygen permeability (OP) of MF (5.02 $cm^3\ m^{-2}$ day bar) decreased eight times compared to PLA film.	• The tensile strength (TS) of the triple-layer film (25 ± 2.13 MPa) was greater than that of the FG film (7.48 ± 1.70 MPa).	• The light transmittance of pure PLA film, FG, and triple-layer film was 78.43%, 91.83%, and 85.48%, respectively, at 600 nm. • Opacity decreased from 2.90 ± 0.36 to 1.54 ± 0.08.	-	• Multilayer film showed only one distinguishing T_g value, ranging between those of the PLA (44.3°C) and FG (29.8°C).	Hosseini et al. (2016)
Bioactive MF loaded with l-ascorbic acid (AA) containing CS, sodium alginate (SA), and ethyl cellulose (EC)	• The MF, which incorporates CS, SA, and EC layers, showed the lowest water vapor transmission rate (WVTR) (54.99 $g \cdot d^{-1} \cdot m^{-2}$).	• MF exhibited a higher TS of 0.56 MPa. • MF and EC films exhibited lower elongation at breaks (EABs) of 96.46% and 86.28%, respectively, compared to the CS film (107.18%).	• The MF had a transmittance of 81.85% (lower than the individual) and 27.23% (lower) glossiness.	• MF had a rough surface. • The MF exhibited the highest thickness value at 0.118 mm, followed by the EC film at 0.102 mm.	• The AA-loaded MF film exhibited a higher percentage (66.20%) of inhibition of 2,2-diphenyl-1-picrylhydrazyl (DPPH) radical.	Shah et al. (2025)
PLA with cellulose nanocrystals (CNCs)15 wt.% polyvinyl alcohol (PVA) or kappa-CAR (CG)	• The WVTR values of MF drop to around 23 g m^{-2} day^{-1}, 20 times lower than that of the pure CNC film.	• The oxygen transmission rate of Laminate PVA (2.2 $cm^3\ m^{-2}$ day^{-1}) was found to be comparable to the Laminate CNC film (1.7 $cm^3\ m^{-2}\ day^{-1}$).	• The TS and stiffness increased to 78 MPa and 3.5 GPa, respectively • The strain value decreased to 3.33%.	-	-	Chen et al. (2024)

Gelatin, CS, and carboxymethyl cellulose (CMC) with different layers, incorporating selenium nanoparticles (SeNPs) and beetroot extract (BTE)	• WVP and OP decreased with the addition of BTE.	-	• Multilayer films had enhanced antioxidant activity with BTE and BTE combined with SeNPs having 78.81 ± 0.42% and 83.44 ± 0.52%, respectively.	• The incorporation of SeNPs and BTE led to a significant increase in the thickness of the films, ranging from 0.32 ± 0.09 to 1.22 ± 0.10 mm.	• The Tg and Tm values were observed to increase to 73.89°C and 171.27°C, respectively, after the addition of SeNPs.	Ahmad et al. (2025)
Sage seed gum-gelatin, TiO_2, and electrospun zein fibers encapsulating pomegranate peel extract	• The addition of the sage seed gum-gelatin composite film layer containing TiO_2 imparts a brighter appearance to the film.	-	• The opacity of the multilayer film containing 80 $g.m^{-2}$ of electrospun zein-pomegranate flower extract (SG-Ge/Ze 8/TiO_2) is nearly double that of the control.	• The thickness of the film-containing extract and the film without extract was 0.23 –0.17 mm, respectively. • The DPPH free radical scavenging activity and total phenolic content of the film-containing extract increased by 547% and 899%, respectively. • Without extract, films exhibited the highest biodegradability.	-	Valizadeh et al. (2024)
CMC and PLA via incorporation of tannin and ferric chloride	-	• Multilayer films showed an excellent TS of about 70 MPa, which was higher than that of other bio-based PLA MFs. • The CMC/TA–PLA/Fe film showed an	-	• Multilayer films exhibited water contact angles (WCAs) around 110°, indicative of a highly hydrophobic surface.	• A T_g of 66.91°C for CMC–PLA and CMC/TA–PLA films, with T_m values of 161.72°C and 162.32°C, respectively, comparable to standalone PLA films.	Xiao et al. (2024)

(Continued)

TABLE 30.1 (Continued)

Names of compounds	Barrier properties	Mechanical properties	Optical properties	Physical properties	Thermal properties	References
		increased EAB of 8%, and its Young's modulus decreased from 47.38 ± 2.03 MPa to 44.10 ± 2.11 MPa compared with the CMC–PLA/Fe film.			• The incorporation of Fe^{3+} elevated the T_g to 88.67°C for CMC–PLA/Fe and 79.68°C for CMC/TA–PLA/Fe.	
CS and mussel adhesive protein-mimetic polymer (dopamine-grafted CMC, CMC-g-DA) on the paper surface and subsequent oxidative cross-linking by sodium periodate	• The multilayer-modified paper sample exhibited a 69.2% reduction in air permeability and a 56.3% decrease in WVP.	• The Tensile index (TI), EAB, and elasticity modulus values of the $(CS/CMC\text{-}g\text{-}DA)_6$ multilayer-modified paper improved by 88.7%, 45.4%, and 24.3%, respectively.	-	-	• The inhibition percentages of the cross-linked $(CS/CMC\text{-}g\text{-}DA)_6$ multilayer-modified paper were very comparable to those of the reference antibiotic, with inhibition percentages of 98.3% and 96.6% against *S.* aureus and *E. coli*.	Yu et al. (2022)
Nanofibers (CNF), oxidized nanofibers (CNF/2,2,6,6-Tetramethylpiperidinyloxyl (TEMPO)) and nanocrystals (CNC) in CS	• CNF and CNF/TEMPO coatings decreased the OP of the PLA-based samples by one order of magnitude, and CNC coatings even by two orders of magnitude.	• Coated PLA(CNC) films performed better (3.2×10^{-20} $m^3.m\ [m^2.s.Pa]^{-1}$) than PLA(CNF) and PLA (CNF/TEMPO) (1.1×10^{-19} at best).	-	-	-	Guivier et al. (2023)

single-layer films at 93.03°C ± 0.05°C (Nagarajan et al., 2017). Thus, MFs were found to improve the thermal performance of protein-based materials, broadening their application scope.

30.4.5 Optical properties

Transparency is a key property of films, as it determines their visual appeal and can significantly influence consumer decisions. It is defined as the ability of a film to allow light to pass through without altering its direction (Zhao et al., 2022). Transparency affects the appearance of food products and plays a crucial role in defining the performance of films and coatings for food wrapping and container applications (Nazmi & Sarbon, 2020). A lower transparency value indicates higher transparency, meaning the film allows more light to pass through, while a higher value denotes lower transparency or greater opacity (Guzman-Puyol et al., 2022). MFs can enhance the color and transparency of packaging by combining polymers with complementary properties. For example, a zein-gelatin MF has a transparency of 3.55% ± 0.11%, which is significantly lower than the 12.28% ± 0.02% of a zein single-layer film (Xia et al., 2019). Transparency values for biomass-biomass MFs range from 0.84% to 3.55%, as seen in a starch-zein bilayer film (0.84%) (Chen et al., 2019), while biomass-synthesis films, such as fish gelatin-PLA, exhibit even lower values, ranging from 0.46% to 1.54% (Nilsuwan et al., 2018). Notably, the fish gelatin-PLA film, with 0.46% transparency, demonstrates the highest clarity. Biomass films combined with synthesis-derived polymers generally have superior transparency compared to biomass-biomass films, likely due to the ordered arrangement, partial crystallization, and compact structure of PLA molecules within the film matrix (Nagarajan et al., 2017). The improved clarity of MFs provides consumers with a clearer visual signal of the packaged product, enhancing its appeal and marketability.

30.4.6 Mechanical properties

To ensure the practical application of MFs, it is essential that these films remain stable in various suspending media, and they can resist different stresses/external forces. Characterizing and modulating film mechanical properties has emerged as a significant challenge in recent years, particularly for films composed of polysaccharides and polyamino acids, which are rather soft and less robust compared to films made of synthetic polyelectrolytes. Recent advancements have shown that the mechanical properties of biopolymer-based MFs can be improved through various approaches, such as cross-linking, incorporating additives, and processing techniques.

Mechanical properties like TS and EAB are key factors for packaging films, as they define the film's strength and flexibility, ensuring durability and resistance during handling and use (Castilla et al., 2022). TS refers to the maximum force a film can withstand before breaking, making it a key factor in selecting edible films for packaging products that require high protection (Khairunnisa et al., 2018; Katili et al., 2013). EAB is expressed as the percentage of film length change relative to its original length and measures the film's ability to stretch before failure. A higher EAB indicates greater flexibility and energy absorption capacity (Nazmi & Sarbon, 2020). These mechanical properties are influenced by the composition of the biopolymer matrix, the method of film preparation, and the incorporation of active ingredients (Bhaskar et al., 2023).

The reported studies demonstrate that the MFs possess comparable or even higher mechanical properties than single-layer films. The films developed from the combination of biopolymers with the incorporation of organic substances addressed the mechanical issues associated with the single-layer films. Wang et al. (2023) reported a slight increase in TS and EAB of CS film with an additional alginate layer due to hydrogen bonding and oppositely charged polymers. The introduction of inorganic particles, zinc oxide (ZnO), into the above film matrix further improved the mechanical properties compared to a monolayer CS film. Gelatin films containing 30% sucrose demonstrate a TS of approximately 3.03 MPa and an EAB of 152.02%, while the addition of CS to FG enhances both TS and elastic modulus, resulting in stronger films (Hosseini et al., 2013). Protein-based films typically display superior mechanical properties compared to lipid and polysaccharide-based films, although their TS and EAB values remain limited (Zubair et al., 2025). Generally, protein and polysaccharide films have low TS values but high EAB values, indicating limited mechanical robustness. The application of MFs has shown potential in enhancing the mechanical properties of single-layer films, particularly EAB. Studies have demonstrated that MFs often outperform single-layer films in mechanical performance. For example, gluten-PHB MFs exhibit significantly lower EAB values (31.00% ± 5.00%) compared to single-layer gluten films (136.00% ± 12.00%) (Fabra et al., 2013). MFs combining gelatin and EGCG achieve a TS of 40.70 ± 1.66 MPa and an EAB of 7.50% ± 0.30%, outperforming single-layer gelatin films (Nilsuwan et al., 2019). However, in starch-based films, combining cassava starch and CS in a bilayer structure decreases both TS and EAB compared to single-layer films of the same materials (Valencia-Sullca et al., 2018). Triple-layer films composed of kefiran (KAF), gelatin (GEL), and their blend at intermediate layer (KAF/KAF+GEL/GEL) displayed higher TS values than their single-layer counterparts as well as blend film (Wang et al., 2018). Similarly, introducing graphene into CS/rGO@silver nanoparticles (AgNPs)/CS triple-layer films significantly enhances mechanical properties, yielding TS values 312.5% higher than commercial polyethylene (PE) films (Gu et al., 2021). The EAB of films also varies with polymer combinations. OSG and PLA films achieve a high EAB of 72.21% ± 5.80% (Lee and Song, 2017). In contrast, combinations like gelatin-PLA (11.60% ± 0.80%) (Nilsuwan et al., 2018), starch-CS (2.28% ± 0.14%) (Valencia-Sullca et al., 2018), and zein-Polyhydroxy butyrate-co-valerate (1.80% ± 0.10%) films exhibit lower EAB values. The compact structure and higher crystallinity of zein-Polyhydroxy butyrate-co-valerate films are likely responsible for their high strength but limited flexibility (Fabra et al., 2013). Despite advances in MF development, challenges remain. For instance, Guiga et al. (2010) observed partial layer mixing in triple-layer films made from EC and hydroxypropyl methylcellulose (HPMC) with nisin or Nisaplin. While this mixing posed challenges in composite layer thickness, the films successfully delayed nisin desorption from 0.5 to 20 hours and demonstrated excellent microbiological properties. Hence, as evidenced by the studies, MFs have high-mechanical properties, leading to more robustness in the area of food packaging.

30.4.7 Biodegradability and compostability

Biodegradability refers to the ability of materials to naturally break down by microorganisms, transforming into CO_2 and organic matter. González and Igarzabal (2013) conducted biodegradability tests at 21°C ± 2°C and a relative humidity of 39.5% ± 4% for 75 days in a closed environment. They reported that native SPI films degraded completely, while SPI films with PLA exhibited 50%–60% degradation after 55 days. After 75 days, degradation reached 65%–80%, with the PLA layer degrading only partially, demonstrating a slower degradation rate compared to SPI films. Similar findings were reported by Wang, Qu, et al. (2024) and Wang, Wang, et al. (2024) for corn starch and PLA bilayer films, wherein a 10-week degradation experiment demonstrated good degradability of the bilayer films. Scaffaro et al. (2020) tested bilayer films (comprising PLA, Bio-flex, carvacrol, and nanoclay) under different pH conditions (4, 7, and 10) at 37°C for up to 864 hours. Both the bilayer and monolayer films showed stability in acidic and neutral conditions but degraded in alkaline environments. In another study, Wu et al. (2021) prepared trilayer films of polyester, starch, and polyester with varying starch and polyester layer thickness ratios using coextrusion blowing. The trilayer films exhibited a higher degradation rate compared to polyester films, losing up to 80% of their mass after a 120-day degradation test.

Scaffaro et al. (2018) also fabricated monolayer and bilayer films using PLA and Mater-Bi through coextrusion blow molding. Degradation tests revealed that the bilayer films degraded faster than the respective monolayers under alkaline conditions. This behavior was attributed to the interfacial delamination of the two layers, which allowed the pH solution to accumulate at the interface, synergistically accelerating degradation. The delamination in blended or composite systems was due to absorbed water, which tended to concentrate at the interface between the two phases.

30.4.8 Antioxidant and antimicrobial properties

MFs can be enhanced with antibacterial and antioxidant properties by incorporating various functional agents into their formulation. These agents help combat bacterial growth and prevent oxidation reactions, extending the films' functional performance. A wide range of antimicrobial and antioxidant agents can be incorporated into films, including ZnO, AgNPs, copper oxides (Cu_2O and CuO), carvacrol, tea polyphenol (TP), LBE, green tea extract (GTE), citric acid (CA), essential oils from cinnamon and oregano, and lauroyl arginate ethyl (LAE) (Xie 2023). When these agents are added to one or more layers of MFs, they can be released gradually over time, thereby prolonging the films' antimicrobial and antioxidant functions. (Hasheminya et al., 2018) reported that CMC films containing *Zataria multiflora* essential oil effectively prevent oxidative damage. Additionally, biocomposite films made of kefiran and CMC, with copper oxide nanoparticles (CuONPs), demonstrate strong antibacterial properties, effectively killing *S. aureus* and *E. coli*. Incorporating ZnO particles into polymer matrices enhances antioxidant capacity by scavenging free radicals and reducing oxidative stress during degradation. ZnO particles also exhibit antimicrobial properties, such as breaking down methylene blue dye under UV light, highlighting their effectiveness in combating microbial growth (Castilla et al., 2022). Similarly, incorporating essential oil from *Cinnamodendron dinisii* into zein-CS films enhances both antioxidant and antimicrobial properties. Films infused with thymol and cinnamaldehyde create a microbe-resistant environment, thereby improving food safety (Bhaskar et al., 2023). Gelatin-based films are particularly effective in preventing food oxidation caused by UV light, helping to preserve nutrients and prevent off-flavors. Combining CS with gelatin imparts bacteriostatic and fungistatic properties, making the films highly effective against microbial growth (Hosseini et al., 2013). Furthermore, CMC films enriched with resveratrol and eugenol exhibit enhanced antimicrobial properties, demonstrating their suitability for active food packaging applications (Khin et al., 2024).

30.4.9 Surface properties

30.4.9.1 Hydrophobicity

MFs incorporating natural biopolymers can be designed with outer layers made of highly water-resistant materials to enhance their moisture barrier properties. For example, a paper treated with a wax-coated layer of CAR, CS, and montmorillonite achieved a high-water contact angle of 151.4°, demonstrating excellent water resistance (Xie, 2023). Protein-based films can be made more water-resistant through cross-linking, a process that reduces their solubility and increases their flexibility. For instance, Bhaskar et al. (2023) reported that adding CS derivatives to whey protein films enhances their hydrophobic properties through noncovalent interactions, where molecules adhere without forming strong chemical bonds. Similarly, incorporating Cu_2ONCs particles into the film matrix increases the water contact angle from 83.33° to 87°, further improving hydrophobicity (Yan et al., 2021). In kefiran films, the addition of oleic acid enhances both flexibility and water resistance (Exarhopoulos et al., 2022). Additionally, Hosseini et al. (2013) found that FG films exhibit greater hydrophobicity than mammalian gelatin films. This difference is attributed to the lower levels of proline and hydroxyproline in FG, which affect its structural properties and water resistance.

30.4.10 Controlled release of active compounds

Biopolymer films are specialized materials designed to release active compounds over time. These active compounds, distributed throughout the films, determine their performance. The release rate of these substances depends on several factors, including the shape of the film, the type of polymers used, and the specific properties of the active compounds. The film's shape can influence the release pattern, while different polymers interact with active compounds in various ways, impacting their release rates. Additionally, the inherent characteristics of the active compounds play a significant role in determining their release behavior (Khandeparkar et al., 2024). Studies have shown that using triple-layer films can significantly slow the release of nisin, a bactericidal compound. Typically, nisin is released within about half an hour, but with a triple-layer film, the release can extend up to 20 hours. This demonstrates the effectiveness of such films in controlling the release rate of active compounds (Grzebieniarz et al., 2023). Incorporating active compounds into food packaging films helps maintain food freshness and safety by ensuring a slow and controlled release, thereby preserving food quality for a longer duration (Xie, 2023). For example, adding Cu_2O and nano-CS to MFs enhances their functional properties, allowing for the controlled release of antibacterial compounds (Yan et al., 2021). Similarly, packaging films made from kefiran and CMC can regulate the release of active compounds. The inclusion of nanoparticles, such as CuONPs, in the polymer matrix further influences the release rates of these compounds (Hasheminya et al., 2018).

30.5 Application of multilayer films in food packaging

Food packaging requires ensuring safety and quality, and extending the shelf life of enclosed food materials. In order to be used in food packaging, multilayer biopolymer needs to demonstrate the qualities such as high-mechanical (strength, flexibility, stiffness) and barrier (for water vapor and gas) properties, ease of processing, and low cost, as possessed by popular plastic polymers such as polyethylene (PE), polypropylene (PP), polystyrene (PS), and polyethylene terephthalate (PET). Double and multilayer biopolymer films exhibit improved mechanical strength, lower water vapor permeability, and high-water resistance (or low WS). By a suitable combination of films, many MFs also exhibit antimicrobial and antioxidant properties. MFs and coatings have been used in the packaging of a wide range of food products, including fruits and vegetables, meat, bakery items, snacks, and frozen meals. The MFs used in food packaging consist of at least one layer of bio-based material, while the second and subsequent layers may either be bio-based or synthetic. Composite films made of polysaccharide-protein complexes are very common for food applications (Table 30.2).

30.5.1 Fresh produce (fruits and vegetables)

A double-layer film consisting of PLA and PVA- quaternary ammonium CS demonstrated antimicrobial properties due to the dispersion of antimicrobial agent CuO by the suitable carrier for stored cherry tomatoes. The packed tomatoes also had considerably fewer surface wrinkles and water loss (Fu et al., 2022). Zhou et al. (2019) used PLA as a second layer over pea starch as the first layer and used the composite film for the storage of cherry tomatoes. The bilayer film created a low-oxygen environment, reducing the respiration rate and acid metabolism of the tomatoes. The rate of decrease of vitamin C in the stored tomatoes was less in the bilayer film due to the inhibition of the oxidation reaction, owing to the oxygen barrier property of the film.

TABLE 30.2 Application of bio-based multilayer packaging in food.

Food product	First layer	Second layer	Properties	References
Cherry tomatoes	Polylactic acid (PLA)	Polyvinyl alcohol (PVA) (quaternary ammonium CS)	Inhibition rate (%) ↑ ([41.1–99.6] [*E. coli*], [51.8–99.7] [*S. aureus*]), low surface wrinkle, and water loss.	Fu et al. (2022)
	Pea starch	PLA	PLA addition increase water resistance properties of film. Reduced respiration rate, acid metabolism, and vitamin C oxidation during storage. Oxygen permeability ↓(0.234×10^{-12}–0.942×10^{-12} $cm^3 \cdot m^{-1} \cdot s^{-1} \cdot Pa^{-1}$)	Zhou et al. (2019)
	Furcellaran (FUR)/ lingonberry extract (LBE)	Carboxymethyl cellulose/gelatin hydrolysate (HGEL)	Antioxidant activity ↑ (FRAP-3.01–41.89 mM Trolox mg^{-1}) (DPPH-0%–20.56%) Elongation at break ↑ (40.67%–73.03%) Weight loss ↑	Jamróz, Tkaczewska, Kopec, et al. (2022), Jamróz, Tkaczewska, Zając, et al. (2022), Jamróz, Tkaczewska, Juszczak, et al. (2022)
	Chitosan	PVA	Increased shelf life With the addition of Cu2O@NCs, the film inhibited the microbial growth on cherry tomatoes	Yan et al. (2021)
	Chitosan	Sodium Alginate (SA)	Decrease in the concentration of H_2O_2 and an increase in the activity of catalase.	Zhu et al. (2019)
Red grape	Chitosan	SA	Films were still relatively full, and the surface maintained smooth.	Wang et al. (2018)
Strawberries	Zein (Z)/gelatin (G)	Zein (Z)/ gelatin (G) zein (Z)/ gelatin (G) oregano oil (OEO)	By comparison, no moisture condensation occurred in the 1 L film, due to the higher water vapor permeability (1.234×10^{-10} $gm^{-1}s^{-1}$ Pa^{-1}) and water absorption capacity of protein-based matrix. Trilayer films (3L7.5%, 3L15%U, 3L15%I) with OEO in upper/ intermediate layers showed better performance.	Cai et al. (2020)
Fish	FUR	HGEL	Antioxidant properties (DPPH- 2.9%–3.1 %) (FRAP-3.5–6.6 mM Trolox mg^{-1}) High-mechanical properties (tensile strength- 8.7–13.4 MPa), elongation at break – 28.5%–37.1% Cadaverine levels also remained slow throughout storage, with a maximum concentration of 0.22 mg kg^{-1}.	Jamróz et al. (2021)
	FUR	Carp skin gelatin hydrolysates (CSGH)	One-layer films with the hydrolysate showed no significant increase in TVB-N observed on day four of storage. Slowed lipid oxidation. Despite 15 days of storage, the levels of all biogenic amines samples were low, not exceeding 7 mg kg^{-1}.	Tkaczewska et al. (2021)
	Rice protein hydrolysate	Chitosan	Reduced TVB-N, oxidation, and bacterial colony count	Yan et al. (2022)

Meat	Cassava starch	Chitosan	Controlled bacterial growth Bilayer films were slightly less effective than CH monolayers at controlling the bacterial growth.	Valencia-Sullca et al. (2018)
	Carrageenan	Konjac glucomannan (KGM)	Indicate real-time meat freshness The TVB-N value of chicken meat increased significantly from 6.9 to 51.43 after four days. With the increase of storage time, Carrageenan-Anthocyanin-Konjac glucomannan/camellia oil and Carrageenan-Anthocyanin-Curcumin-Konjac glucomannan/camellia oil labels appeared a noticeable color change.	Zhou et al. (2021)
	Methylcellulose	Chitosan nanofiber	Indicates a change in pH and ammonia gas. The increase in pH caused the color indicator film to turn from reddish to pale peach.	Alizadeh-Sani et al. (2021)
Chili	Cassava starch	Chitosan	Excellent antifungal activity C/CS/LEO coated significantly reduced weight loss ($P < .05$) after four days of storage while C and C/CS coated significantly reduced weight loss ($P < .05$) after 12 days of storage.	Perdana et al. (2021)
	PLA	KGM/wheat gluten	Decrease in the number of *Aspergillus* and *Penicillium* species. The specific volume of bread samples increased with longer fermentation times. Among the tested samples, bread stored using PLA/KGM/WG film (ES group) and those exposed to open air (S group) showed a linear increase in specific volume during storage, ranging from approximately 151 g to 241 g.	Wang et al. (2023)

Double-layer film consisting of FUR (a negatively charged polysaccharide extracted from red algae Furcellaria lumbricalis) as the first layer and CMC along with HGEL as the second layer demonstrated antimicrobial and antioxidant activity due to the addition of water extract from lingonberries in the FUR layer. However, the developed biodegradable film had low mechanical strength and high-water permeability, which resulted in weight loss of tomatoes that were wrapped with the developed film during the storage period of 17 days (Jamróz et al., 2021). Jamróz et al. (2018) also developed a MF consisting of starch, FUR, and gelatin, which exhibited high solubility and low TS, which limits their application in high-moisture food. The mechanical properties were improved by the incorporation of Se-Ag nanoparticles in the film consisting of FUR and gelatin (Jamróz et al., 2018). The added nanoparticles also improved the thermal stability and UV-light barrier property of the developed film, which helped to reduce the weight loss and increase the shelf life of stored kiwi fruit (Jamróz, Kopel, et al., 2019; Jamróz, Kulawik, et al., 2019). In grapefruit, the AgNPs were successful in increasing the shelf life by 14 days (Kumar et al., 2018). Another approach to reduce the solubility in water and improve the mechanical properties of FUR/GEL-based films was by incorporation of green tea and Pu-erh tea extract (Jamróz, Kopel, et al., 2019; Jamróz, Kulawik, et al., 2019). As these films exhibit good antioxidant and antimicrobial properties, they can be used as an intelligent packaging system in fresh fruit and vegetables.

CS has been used as the first layer of many developed double-layer films for the storage of vegetables and has been found to extend the shelf life. The MF containing CS and PVA was developed and used in the storage of cherry tomatoes, where it was observed that weight loss and decrease in the hardness were minimal, while the smooth and glossy appearance of the tomatoes was retained over 15 days. CS is also used as an effective coating material to extend the shelf life of fruits such as mangoes, pears, sweet cherries, and kiwifruit. Owing to its hydrophilic nature, CS alone is not able to impart the desired packaging functionality; it is used with fillers such as nano-TiO_2 and nano-Ag crystals. A bilayer film of CS and alginate was used in preserving cherry tomatoes, found to regulate the ascorbate-glutathione cycle in cherry tomatoes and alleviate the accumulation of oxidative damage during postharvest storage (Hu et al., 2019). Multilayer edible coating consisting of hydrophobic zein as outer layer, hybrid zein/ gelatin as middle layer, and hydrophilic gelatin as inner layer, when applied to freshly cut fruit, demonstrated a preservative effect by preventing weight loss, rapid browning, and inhibiting bacterial growth (Xia et al., 2019).

30.5.2 Meat, poultry, and seafood

Multilayer packaging has been used in the literature numerous times for the storage of meat, poultry, and seafood. For example, pork meat packaged using bilayer films of cassava starch with or without essential oils in the CS layer was effective in controlling the bacterial growth (Valencia-Sullca et al., 2018). Meat spoilage or freshness detection can be achieved with bilayer films developed from biopolymers. For this, the pH-sensitive dyes obtained from plant extracts such as curcumin, pomegranate peel extract, anthocyanin, and chlorophyll are incorporated into the films to develop the intelligent packaging system. For example, a bilayer film consisting of an emulsified outer layer of konjac glucomannan (KGM) and camellia oil, and an inner indicative layer of CAR, curcumin (CUR), and/or anthocyanin (ATH) was fabricated to monitor the chicken meat freshness in real time at 25°C (Zhou et al., 2021). Similarly, pH-responsive color indicator films are also developed by blending ATH with methylcellulose (MC)/CS nanofiber, which showed a color change in response to pH, ammonia gas, and were suitable to indicate the formation of TVB-N (Alizadeh-Sani et al., 2021).

Multilayer trays based on starch, CMC, gelatin, cellulose, and waste from brewing green tea grounds were used in the storage and packaging of salmon filets. The trays exhibited an inhibitory effect on biogenic amine accumulation, simultaneously inhibiting the growth of spoilage bacteria (Jamróz, Tkaczewska, Juszczak, et al., 2022; Jamróz, Tkaczewska, Kopec, et al., 2022; Jamróz, Tkaczewska, Zając, et al., 2022). Atlantic mackerel (*Scomberscombrus*), when packed in double-layer film FUR, HGEL, and Ala-Tyr dipeptide as antimicrobial/antioxidant ingredients, exhibited a lack of significant increases in total volatile basic nitrogen (TVB-N) throughout the storage period of four months (Jamróz et al., 2021). Mackerel has also been stored in a coating consisting of FUR and Carp skin gelatin hydrolysates (CSGH), which proved to be effective in slowing the lipid oxidation, inhibiting the growth of microorganisms, and extending the shelf life. Here, though, some biogenic amine levels increased, but the levels did not exceed the safety limit of 7 mg kg^{-1} (Tkaczewska et al., 2021). Rice protein hydrolysate/CS composite film was used to store fresh salmon at 4°C for seven days, where the oxidation properties, such as TVB-N and peroxide value, were found to be lower than those of the sample kept in PE and PP films, showing that MFs are also effective in controlling oxidation. The antibacterial nature of the film also controlled the oxidation of salmon and reduced the bacterial colony count, thus improving the sensory properties of the stored salmon (Yan et al., 2022).

30.5.3 Dairy products

In some cases, multilayer packaging and coatings have been utilized for dairy-based products, particularly to monitor food freshness effectively, as documented in the literature. Li et al. (2023) developed a dual-layer intelligent film using CMCS for milk preservation and freshness monitoring. The outer layer, composed of CMCS, was cross-linked with oxidized SA to enhance water resistance. Anthocyanin extract from *Oxalis triangularis* ssp. was used to achieve colorimetric sensitivity. *Papilionacea* was incorporated into the structure. The film's inner layer, consisting of a gelatin–zein blend and linalool as an antibacterial agent, was electrospun onto the outer layer. During storage at 25°C, the film's color shifted from grass green to light brown due to the production of volatile acids, such as acetic and butyric acids, resulting from milk spoilage. The controlled release of linalool from the inner layer helped slow this color change by inhibiting the growth of *S. aureus* and *E. coli*, highlighting its dual functionality in preservation and spoilage indication. Coating solutions derived from hemicellulose have been employed to enhance the shelf life of cheese. This was achieved by chemically altering CS and galactomannan in combination with a plasticizer and corn oil (Cerqueira et al., 2010).

Single-layer films using FUR combined with protein hydrolysates from soybean bran were also developed. In the double-layer films, the first layer consisted of FUR and soybean protein hydrolysates, while the second layer included an ethanol extract of soybean bran. The impact of the second layer on the storage stability of butter was examined. Although the double-layer films demonstrated high-antioxidant activity, single-layer films proved unsuitable for butter storage. Butter stored in single-layer films showed reduced polyunsaturated fatty acid content, along with increased thiobarbituric acid reactive substances (TBARS) and peroxide values, indicating notable pro-oxidative effects. However, the quality of butter stored in both single- and double-layer films was comparable to that of butter stored in synthetic packaging (Jamróz, Tkaczewska, Juszczak, et al., 2022; Jamróz, Tkaczewska, Kopec, et al., 2022; Jamróz, Tkaczewska, Zając, et al., 2022).

30.5.4 Bakery and confectionery

Composite film consisting of cassava starch and CS incorporated with lemongrass essential oil exhibited excellent antifungal activity against *Aspergillus flavus* and antiaflatoxigenic properties, which have been exploited for the packaging of bakery products (Perdana et al., 2021). A bilayer film prepared with cinnamaldehyde-loaded PLA/KGM/wheat gluten was used for the packaging of bread samples fermented at different times. A marked decrease in the number of the main spoilage fungi (*Aspergillus* and *Penicillium*) in samples packed with the MF was noted. A reduction in fungal diversity was also observed during storage. Texture of the bread was well maintained, and film also prevented the moisture evaporation from the packaging environment, thus increasing the shelf life (Wang et al., 2023). Starch and other polysaccharide-based MFs have been widely explored for cereal product packaging. These are reinforced with suitable fillers to improve their barrier and mechanical properties.

30.6 Conclusion

MFs and coatings based on biopolymers have revolutionized the field of sustainable food packaging by offering an effective alternative to traditional synthetic materials. Their exceptional functional properties, including enhanced mechanical strength, superior barrier performance, and the ability to incorporate active agents, position them as a pivotal innovation for maintaining food quality and extending shelf life. The diverse preparation techniques, such as LbL assembly, electrospinning, and electrospraying, enable precise customization, allowing these films to cater to specific requirements across various food sectors, including meat, fish, poultry, fruits, vegetables, dairy products, and bakery items.

Despite their numerous advantages, challenges such as scalability, cost efficiency, and the compatibility of different biopolymers and active components remain critical areas for further research and development. Addressing these limitations will pave the way for broader industrial adoption and increased commercial viability of multilayer biopolymer films. As the demand for environmentally friendly and functional food packaging solutions continues to rise, multilayer biopolymer films and coatings hold immense potential to contribute to global sustainability goals while ensuring food safety and quality. By fostering interdisciplinary collaborations and advancing technological innovations, these materials can redefine the future of sustainable packaging.

References

Ahmad, H. N., Yong, Y., Tang, Z., Li, R., Munawar, N., & Zhu, J. (2025). Multifunctional layer-by-layer smart film with betalains and selenium nanoparticles for intelligent meat freshness monitoring and preservation. *Food Chemistry, 471*, 142737. https://doi.org/10.1016/j.foodchem.2024.142737.

Akinalan Balik, B., Argin, S., Lagaron, J. M., & Torres-Giner, S. (2019). Preparation and characterization of electrospun pectin-based films and their application in sustainable aroma barrier multilayer packaging. *Applied Sciences, 9*(23), 5136. https://doi.org/10.3390/app9235136.

Alias, A. R., Wan, M. K., & Sarbon, N. M. (2022). Emerging materials and technologies of multi-layer film for food packaging application: A review. *Food Control, 136*. https://doi.org/10.1016/j.foodcont.2022.108875, https://www.journals.elsevier.com/food-control.

Alizadeh-Sani, M., Tavassoli, M., Mohammadian, E., Ehsani, A., Khaniki, G. J., Priyadarshi, R., & Rhim, J. W. (2021). pH-responsive color indicator films based on methylcellulose/chitosan nanofiber and barberry anthocyanins for real-time monitoring of meat freshness. *International Journal of Biological Macromolecules, 166*, 741–750. https://doi.org/10.1016/j.ijbiomac.2020.10.231, www.elsevier.com/locate/ijbiomac.

Andrade, C. S., Fonseca, G. G., Innocentini Mei, L. H., & Fakhouri, F. M. (2016). Development and characterization of multilayer films based on polyhydroxyalkanoates and hydrocolloids. *Journal of Applied Polymer Science, 134*(6). https://doi.org/10.1002/app.44458, http://onlinelibrary.wiley.com/journal/10.1002/(ISSN)1097-4628.

Anu Bhushani, J., & Anandharamakrishnan, C. (2014). Electrospinning and electrospraying techniques: Potential food based applications. *Trends in Food Science & Technology, 38*(1), 21–33. https://doi.org/10.1016/j.tifs.2014.03.004.

Anukiruthika, T., Sethupathy, P., Wilson, A., Kashampur, K., Moses, J. A., & Anandharamakrishnan, C. (2020). Multilayer packaging: Advances in preparation techniques and emerging food applications. *Comprehensive Reviews in Food Science and Food Safety, 19*(3), 1156–1186. https://doi.org/10.1111/1541-4337.12556, http://onlinelibrary.wiley.com/journal/10.1111/(ISSN)1541-4337.

Arham, R., Mulyati, M. T., Metusalach, M., & Salengke, S. (2016). Physical and mechanical properties of agar based edible film with glycerol plasticizer. *International Food Research Journal, 23*(4), 1669–1675. http://www.ifrj.upm.edu.my.

Arnon-Rips, H., & Poverenov, E. (2018). Improving food products' quality and storability by using Layer by Layer edible coatings. *Trends in Food Science & Technology, 75*, 81–92. https://doi.org/10.1016/j.tifs.2018.03.003.

Avila, L. B., Schnorr, C., Silva, L. F. O., Morais, M. M., Moraes, C. C., da Rosa, G. S., Dotto, G. L., Lima, É. C., & Naushad, M. (2023). Trends in bioactive multilayer films: Perspectives in the use of polysaccharides, proteins, and carbohydrates with natural additives for application in food packaging. *Foods, 12*(8). https://doi.org/10.3390/foods12081692, www.mdpi.com/journal/foods.

Bathula, C., Buruga, K., Lee, S. K., Khazi, I. A. M., & Kang, Y. (2017). Microwave assisted synthesis of bithiophene based donor-acceptor-donor oligomers and their optoelectronic performances. *Journal of Molecular Structure, 1139*, 125–129. https://doi.org/10.1016/j.molstruc.2017.03.033, www.elsevier.com/inca/publications/store/5/0/0/8/5/0/index.htt.

Bhaskar, R., Zo, S. M., Narayanan, K. B., Purohit, S. D., Gupta, M. K., & Han, S. S. (2023). Recent development of protein-based biopolymers in food packaging applications: A review. *Polymer Testing, 124*. https://doi.org/10.1016/j.polymertesting.2023.108097, https://www.journals.elsevier.com/polymer-testing.

Bock, N., Woodruff, M. A., Hutmacher, D. W., & Dargaville, T. R. (2011). Electrospraying, a reproducible method for production of polymeric microspheres for biomedical applications. *Polymers, 3*(1), 131–149. https://doi.org/10.3390/polym3010131Australia, http://www.mdpi.com/2073-4360/3/1/131/pdf.

Borges, A. F., Silva, C., Coelho, J. F. J., & Simões, S. (2015). Oral films: Current status and future perspectives: I-Galenical development and quality attributes. *Journal of Controlled Release, 206*, 1–19. https://doi.org/10.1016/j.jconrel.2015.03.006, www.elsevier.com/locate/jconrel.

Cai, J., Xiao, J., Chen, X., & Liu, H. (2020). Essential oil loaded edible films prepared by continuous casting method: Effects of casting cycle and loading position on the release properties. *Food Packaging and Shelf Life, 26*, 100555. https://doi.org/10.1016/j.fpsl.2020.100555.

Calva-Estrada, S. J., Jiménez-Fernández, M., & Lugo-Cervantes, E. (2019). Protein-based films: Advances in the development of biomaterials applicable to food packaging. *Food Engineering Reviews, 11*(2), 78–92. https://doi.org/10.1007/s12393-019-09189-w, https://link.springer.com/content/pdf/10.1007/s12393-019-09189-w.pdf.

Carosio, F., Colonna, S., Fina, A., Rydzek, G., Hemmerlé, J., Jierry, L., Schaaf, P., & Boulmedais, F. (2014). Efficient Gas and Water Vapor Barrier Properties of Thin Poly(lactic acid) Packaging Films: Functionalization with Moisture Resistant Nafion and Clay Multilayers. *Chemistry of Materials, 26*(19), 5459–5466. https://doi.org/10.1021/cm501359e.

Carr, J. M., Langhe, D. S., Ponting, M. T., Hiltner, A., & Baer, E. (2012). Confined crystallization in polymer nanolayered films: A review. *Journal of Materials Research, 27*(10), 1326–1350. https://doi.org/10.1557/jmr.2012.17.

Castilla-Caballero, D., Sadak, O., Martínez-Díaz, J., Martínez-Castro, V., Colina-Márquez, J., Machuca-Martínez, F., Hernandez-Ramirez, A., Vazquez-Rodriguez, S., & Gunasekaran, S. (2022). Solid-state photocatalysis for plastics abatement: A review. *Materials Science in Semiconductor Processing, 149*, 106890. https://doi.org/10.1016/j.mssp.2022.106890.

Cazón, P., Puertas, G., & Vázquez, M. (2024). Characterization of multilayer bacterial cellulose-chitosan films loaded with grape bagasse antioxidant extract: Insights into spectral and water properties, microstructure, and antioxidant activity. *International Journal of Biological Macromolecules, 268*(Pt 2), 131774. https://doi.org/10.1016/j.ijbiomac.2024.131774.

Cerqueira, M. A., Sousa-Gallagher, M. J., Macedo, I., Rodriguez-Aguilera, R., Souza, B. W. S., Teixeira, J. A., & Vicente, A. A. (2010). Use of galactomannan edible coating application and storage temperature for prolonging shelf-life of "Regional" cheese. *Journal of Food Engineering, 97*(1), 87–94. https://doi.org/10.1016/j.jfoodeng.2009.09.019.

Chen, C., Wang, L., Shams Es-haghi, S., Tajvidi, M., Wang, J., & Gardner, D. J. (2024). Biodegradable and recyclable bio-based laminated films of poly (lactic acid) and cellulose nanocrystals for food barrier packaging. *Food Packaging and Shelf Life, 42*, 101244. https://doi.org/10.1016/j.fpsl.2024.101244.

Chen, X., Cui, F., Zi, H., Zhou, Y., Liu, H., & Xiao, J. (2019). Development and characterization of a hydroxypropyl starch/zein bilayer edible film. *International Journal of Biological Macromolecules, 141*, 1175–1182.

Chen, X., Fu, X., Huang, L., Xu, J., & Gao, X. (2021). Agar oligosaccharides: A review of preparation, structures, bioactivities and application. *Carbohydrate Polymers, 265*, 118076. https://doi.org/10.1016/j.carbpol.2021.118076.

Cheng, H., Chen, L., McClements, D. J., Yang, T., Zhang, Z., Ren, F., Miao, M., Tian, Y., & Jin, Z. (2021). Starch-based biodegradable packaging materials: A review of their preparation, characterization and diverse applications in the food industry. *Trends in Food Science and Technology, 114*, 70–82. https://doi.org/10.1016/j.tifs.2021.05.017, http://www.elsevier.com/wps/find/journaldescription.cws_home/601278/description#description.

Cheng, J., Gao, R., Zhu, Y., & Lin, Q. (2024). Applications of biodegradable materials in food packaging: A review. *Alexandria Engineering Journal, 91*, 70–83.

Dave, P. K., Ramana Rao, T. V., & Thakkar, V. R. (2023). Development of films and coatings from alginates. In *Biopolymer-based films and coatings: Trends and challenges*. India: CRC Press, 101–120. http://doi.org/10.1201/9781003303671-5, https://www.taylorfrancis.com/books/edit/10.1201/9781003303671.

Deng, L., Kang, X., Liu, Y., Feng, F., & Zhang, H. (2018). Characterization of gelatin/zein films fabricated by electrospinning vs solvent casting. *Food Hydrocolloids, 74*, 324–332. https://doi.org/10.1016/j.foodhyd.2017.08.023.

Diop, C. I. K., Beltran, S., Sanz, M. T., Garcia-Tojal, J., & Trigo-lopez, M. (2023). Designing bilayered composite films by direct agar/chitosan and citric acid-crosslinked PVA/agar layer-by-layer casting for packaging applications. *Food Hydrocolloids, 144*. https://doi.org/10.1016/j.foodhyd.2023.108987, https://www.journals.elsevier.com/food-hydrocolloids.

Eslami, Z., Elkoun, S., Robert, M., & Adjallé, K. (2023). A review of the effect of plasticizers on the physical and mechanical properties of alginate-based films. *Molecules (Basel, Switzerland), 28*(18), 6637. https://doi.org/10.3390/molecules28186637.

Exarhopoulos, S., Goulas, A., & Dimitreli, G. (2022). Biodegradable films from kefiran-based cryogel systems. *Macromology, 2*(3), 324–345. https://doi.org/10.3390/macromol2030021.

Ezati, P., Khan, A., Priyadarshi, R., Bhattacharya, T., Tammina, S. K., & Rhim, J. W. (2023). Biopolymer-based UV protection functional films for food packaging. *Food Hydrocolloids, 142*. https://doi.org/10.1016/j.foodhyd.2023.108771, https://www.journals.elsevier.com/food-hydrocolloids.

Ezati, P., & Rhim, J.-W. (2020). pH-responsive pectin-based multifunctional films incorporated with curcumin and sulfur nanoparticles. *Carbohydrate Polymers, 230*, 115638. https://doi.org/10.1016/j.carbpol.2019.115638.

Fabra, M. J., Lopez-Rubio, A., & Lagaron, J. M. (2013). High barrier polyhydroxyalcanoate food packaging film by means of nanostructured electrospun interlayers of zein. *Food Hydrocolloids, 32*(1), 106–114. https://doi.org/10.1016/j.foodhyd.2012.12.007, https://www.journals.elsevier.com/food-hydrocolloids.

Fu, D., Ding, Y., Guo, R., Zhang, J., Wang, H., Niu, B., & Yan, H. (2022). Polylactic acid/polyvinyl alcohol-quaternary ammonium chitosan double-layer films doped with novel antimicrobial agent CuO@ZIF-8 NPs for fruit preservation. *International Journal of Biological Macromolecules, 195*, 538–546. https://doi.org/10.1016/j.ijbiomac.2021.12.022.

García Ibarra, V., Sendón, R., & Rodríguez-Bernaldo De Quirós, A. (2016). Antimicrobial food packaging based on biodegradable materials. In *Antimicrobial food packaging*. Spain: Elsevier Inc, 363–384. http://doi.org/10.1016/B978-0-12-800723-5.00029-2, http://www.sciencedirect.com/science/book/9780128007235.

George, N., & Venugopal, B. (2023). Gas barrier properties of biopolymers. In *Handbook of biopolymers*. Singapore: Springer Nature Singapore, 297–321.

Giteru, S. G., Coorey, R., Bertolatti, D., Watkin, E., Johnson, S., & Fang, Z. (2015). Physicochemical and antimicrobial properties of citral and quercetin incorporated kafirin-based bioactive films. *Food Chemistry, 168*, 341–347. https://doi.org/10.1016/j.foodchem.2014.07.077.

González, A., & Igarzabal, C. I. A. (2013). Soy protein–Poly (lactic acid) bilayer films as biodegradable material for active food packaging. *Food Hydrocolloids, 33*(2), 289–296.

Greco, A., Esposito Corcione, C., & Maffezzoli, A. (2017). Diffusion in oriented lamellar nanocomposite: Numerical analysis of the effects of dispersion and intercalation. *Computational Materials Science, 133*, 45–51. https://doi.org/10.1016/j.commatsci.2017.03.011.

Grzebieniarz, W., Biswas, D., Roy, S., & Jamróz, E. (2023). Advances in biopolymer-based multi-layer film preparations and food packaging applications. *Food Packaging and Shelf Life, 35*, 101033. https://doi.org/10.1016/j.fpsl.2023.101033.

Gu, B., Jiang, Q., Luo, B., Liu, C., Ren, J., Wang, X., & Wang, X. (2021). A sandwich-like chitosan-based antibacterial nanocomposite film with reduced graphene oxide immobilized silver nanoparticles. *Carbohydrate Polymers, 260*, 117835. https://doi.org/10.1016/j.carbpol.2021.117835.

Guiga, W., Swesi, Y., Galland, S., Peyrol, E., Degraeve, P., & Sebti, I. (2010). Innovative multilayer antimicrobial films made with Nisaplin® or nisin and cellulosic ethers: Physico-chemical characterization, bioactivity and nisin desorption kinetics. *Innovative Food Science & Emerging Technologies, 11*(2), 352–360. https://doi.org/10.1016/j.ifset.2010.01.008.

Guivier, M., Almeida, G., Domenek, S., & Chevigny, C. (2023). Resilient high oxygen barrier multilayer films of nanocellulose and polylactide. *Carbohydrate Polymers, 312*, 120761. https://doi.org/10.1016/j.carbpol.2023.120761.

Guo, Z., Ge, X., Li, W., Yang, L., Han, L., & Yu, Q.-li (2021). Active-intelligent film based on pectin from watermelon peel containing beetroot extract to monitor the freshness of packaged chilled beef. *Food Hydrocolloids, 119*, 106751. https://doi.org/10.1016/j.foodhyd.2021.106751.

Guzman-Puyol, S., Benítez, J. J., & Heredia-Guerrero, J. A. (2022). Transparency of polymeric food packaging materials. *Food Research International, 161*. https://doi.org/10.1016/j.foodres.2022.111792, www.elsevier.com/inca/publications/store/4/2/2/9/7/0.

Hasheminya, S. M., Rezaei Mokarram, R., Ghanbarzadeh, B., Hamishekar, H., & Kafil, H. S. (2018). Physicochemical, mechanical, optical, microstructural and antimicrobial properties of novel kefiran-carboxymethyl cellulose biocomposite films as influenced by copper oxide nanoparticles (CuONPs). *Food Packaging and Shelf Life, 17*, 196–204. https://doi.org/10.1016/j.fpsl.2018.07.003, http://www.journals.elsevier.com/food-packaging-and-shelf-life/.

Hassan, B., Chatha, S. A. S., Hussain, A. I., Zia, K. M., & Akhtar, N. (2018). Recent advances on polysaccharides, lipids and protein based edible films and coatings: A review. *International Journal of Biological Macromolecules, 109*, 1095–1107. https://doi.org/10.1016/j.ijbiomac.2017.11.097, www.elsevier.com/locate/ijbiomac.

Heidemann, H. M., Dotto, M. E. R., Laurindo, J. B., Carciofi, B. A. M., & Costa, C. (2019). Cold plasma treatment to improve the adhesion of cassava starch films onto PCL and PLA surface. *Colloids and Surfaces A: Physicochemical and Engineering Aspects, 580*. https://doi.org/10.1016/j.colsurfa.2019.123739, www.elsevier.com/locate/colsurfa.

Hosseini, S. F., Javidi, Z., & Rezaei, M. (2016). Efficient gas barrier properties of multi-layer films based on poly(lactic acid) and fish gelatin. *International Journal of Biological Macromolecules, 92*, 1205–1214. https://doi.org/10.1016/j.ijbiomac.2016.08.034, www.elsevier.com/locate/ijbiomac.

Hosseini, S. F., Rezaei, M., Zandi, M., & Ghavi, F. F. (2013). Preparation and functional properties of fish gelatin–chitosan blend edible films. *Food Chemistry, 136*(3-4), 1490–1495.

Hu, L., Du, H., Liu, C., Zhang, Y., Yu, G., Zhang, X., Si, C., Li, B., & Peng, H. (2019). Comparative evaluation of the efficient conversion of corn husk filament and corn husk powder to valuable materials via a sustainable and clean biorefinery process. *ACS Sustainable Chemistry & Engineering, 7*(1), 1327–1336. https://doi.org/10.1021/acssuschemeng.8b05017.

Hwangbo, S., Jeong, H., Heo, J., Lin, X., Kim, Y., Chang, M., & Hong, J. (2016). Antibacterial nanofilm coatings based on organosilicate and nanoparticles. *Reactive and Functional Polymers, 102*, 27–32. https://doi.org/10.1016/j.reactfunctpolym.2016.03.004.

Iqbal, M. H., Kerdjoudj, H., & Boulmedais, F. (2024). Protein-based layer-by-layer films for biomedical applications. *Chemical Science, 15*(25), 9408–9437. https://doi.org/10.1039/d3sc06549a, http://pubs.rsc.org/en/journals/journal/sc.

Jamróz, E., Tkaczewska, J., Kopec, M., & Cholewa-W´ojcik, A. (2022). Shelf-life extension of salmon using active total biodegradable packaging with tea ground waste and furcellaran-CMC double-layered films. *Food Chemistry, 383*. Article 132425.

Jamróz, E., Kulawik, P., Tkaczewska, J., Guzik, P., Zając, M., Juszczak, L., Krzyściak, P., & Turek, K. (2021). The effects of active double-layered furcellaran/gelatin hydrolysate film system with Ala-Tyr peptide on fresh Atlantic mackerel stored at −18 °C. *Food Chemistry, 338*, 127867. https://doi.org/10.1016/j.foodchem.2020.127867.

Jamróz, E., Tkaczewska, J., Zając, M., Guzik, P., Juszczak, L., Kawecka, A., Turek, K., Zimowska, M., & Wojdyło, A. (2022). Utilisation of soybean post-production waste in single- and double-layered films based on furcellaran to obtain packaging materials for food products prone to oxidation. *Food Chemistry, 387*, 132883. https://doi.org/10.1016/j.foodchem.2022.132883.

Jamróz, E., Juszczak, L., & Kucharek, M. (2018). Investigation of the physical properties, antioxidant and antimicrobial activity of ternary potato starch-furcellaran-gelatin films incorporated with lavender essential oil. *International Journal of Biological Macromolecules, 114*, 1094–1101. https://doi.org/10.1016/j.ijbiomac.2018.04.014.

Jamróz, E., Kopel, P., Juszczak, L., Kawecka, A., Bytesnikova, Z., Milosavljevic, V., & Makarewicz, M. (2019). Development of furcellaran-gelatin films with Se-AgNPs as an active packaging system for extension of mini kiwi shelf life. *Food Packaging and Shelf Life, 21*, 100339.

Jamróz, E., Kulawik, P., Krzyściak, P., Talaga-Ćwiertnia, K., & Juszczak, L. (2019). Intelligent and active furcellaran-gelatin films containing green or pu-erh tea extracts: Characterization, antioxidant and antimicrobial potential. *International Journal of Biological Macromolecules, 122*, 745–757.

Jamróz, E., Tkaczewska, J., Juszczak, L., Zimowska, M., Kawecka, A., Krzyściak, P., & Skóra, M. (2022). The influence of lingonberry extract on the properties of novel, double-layered biopolymer films based on furcellaran, CMC and a gelatin hydrolysate. *Food Hydrocolloids, 124*, 107334.

Jaworek, A., & Sobczyk, A. T. (2008). Electrospraying route to nanotechnology: An overview. *Journal of Electrostatics, 66*(3-4), 197–219. https://doi.org/10.1016/j.elstat.2007.10.001.

Jiang, J., Watowita, P. S. M. S. L., Chen, R., Shi, Y., Geng, J. T., Takahashi, K., Li, L., & Osako, K. (2022). Multilayer gelatin/myofibrillar films containing clove essential oil: Properties, protein-phenolic interactions, and migration of active compounds. *Food Packaging and Shelf Life, 32*. https://doi.org/10.1016/j.fpsl.2022.100842, http://www.journals.elsevier.com/food-packaging-and-shelf-life/.

Jridi, M., Abdelhedi, O., Salem, A., Kechaou, H., Nasri, M., & Menchari, Y. (2020). Physicochemical, antioxidant and antibacterial properties of fish gelatin-based edible films enriched with orange peel pectin: Wrapping application. *Food Hydrocolloids, 103*, 105688. https://doi.org/10.1016/j.foodhyd.2020.105688.

Kanatt, S. R. (2020). Development of active/intelligent food packaging film containing Amaranthus leaf extract for shelf life extension of chicken/fish during chilled storage. *Food Packaging and Shelf Life, 24*, 100506. https://doi.org/10.1016/j.fpsl.2020.100506.

Katili, S., Harsunu, B. T., & Irawan, S. (2013). *Effect of plasticizer concentration of glycerol and chitosan compositions in the solvent on the physical properties of chitosan edible film. Jurnal Teknologi, 6*(1), 29–38.

Khairunnisa, S., Junianto, J., Zahidah, Z., & Rostini, I. (2018). The effect of glycerol concentration as a plasticizer on edible films made from alginate towards its physical characteristic. *World Scientific News,* 130–141.

Khan, M. K. I., Maan, A. A., Schutyser, M., Schroën, K., & Boom, R. (2013). Electrospraying of water in oil emulsions for thin film coating. *Journal of Food Engineering, 119*(4), 776–780. https://doi.org/10.1016/j.jfoodeng.2013.05.027.

Khandeparkar, A. S., Paul, R., Sridhar, A., Lakshmaiah, V. V., & Nagella, P. (2024). Eco-friendly innovations in food packaging: A sustainable revolution. *Sustainable Chemistry and Pharmacy, 39*. https://doi.org/10.1016/j.scp.2024.101579, https://www.sciencedirect.com/science/journal/23525541.

Kharlampieva, E., Kozlovskaya, V., Chan, J., Ankner, J. F., & Tsukruk, V. V. (2009). Spin-assisted layer-by-layer assembly: Variation of stratification as studied with neutron reflectivity. *Langmuir: the ACS Journal of Surfaces and Colloids, 25*(24), 14017–14024. https://doi.org/10.1021/la9014042, http://pubs.acs.org/doi/pdfplus/10.1021/la9014042.

Khin, M. N., Ahammed, S., Kamal, M. M., Saqib, M. N., Liu, F., & Zhong, F. (2024). Investigating next-generation edible packaging: Protein-based films and coatings for delivering active compounds. *Food Hydrocolloids for Health, 6*. https://doi.org/10.1016/j.fhfh.2024.100182, https://www.sciencedirect.com/science/journal/26670259.

Kumar, S., Shukla, A., Baul, P. P., Mitra, A., & Halder, D. (2018). Biodegradable hybrid nanocomposites of chitosan/gelatin and silver nanoparticles for active food packaging applications. *Food Packaging and Shelf Life, 16*, 178–184. https://doi.org/10.1016/j.fpsl.2018.03.008, http://www.journals.elsevier.com/food-packaging-and-shelf-life/.

Kumorek, M., Minisy, I. M., Krunclová, T., Voršiláková, M., Venclíková, K., Chánová, E. M., Janoušková, O., & Kubies, D. (2020). pH-responsive and antibacterial properties of self-assembled multilayer films based on chitosan and tannic acid. *Republic Materials Science and Engineering C, 109*. https://doi.org/10.1016/j.msec.2019.110493, https://www.journals.elsevier.com/materials-science-and-engineering-c.

Lee, K. Y., & Song, K. B. (2017). Preparation and characterization of an olive flounder (*Paralichthys olivaceus*) skin gelatin and polylactic acid bilayer film. *Journal of Food Science, 82*(3), 706–710.

Li, K., Zhu, J., Guan, G., & Wu, H. (2019). Preparation of chitosan-sodium alginate films through layer-by-layer assembly and ferulic acid crosslinking: Film properties, characterization, and formation mechanism. *International Journal of Biological Macromolecules, 122*, 485–492.

Li, L., Wang, X., Li, D., Qin, J., Zhang, M., Wang, K., Zhao, J., & Zhang, L. (2020). LBL deposition of chitosan/heparin bilayers for improving biological ability and reducing infection of nanofibers. *International Journal of Biological Macromolecules, 154*, 999–1006. https://doi.org/10.1016/j.ijbiomac.2020.03.152.

Li, L., Xia, L., Xiao, F., Xiao, Y., Liu, L., Jiang, S., & Wang, H. (2023). Colorimetric active carboxymethyl chitosan/oxidized sodium alginate-*Oxalis triangularis* ssp. papilionacea anthocyanins film@ gelatin/zein-linalool membrane for milk freshness monitoring and preservation. *Food Chemistry, 405*, 134994.

Lindström, T., & Österberg, F. (2020). Evolution of biobased and nanotechnology packaging – A review. *Nordic Pulp & Paper Research Journal, 35*(4), 491–515. https://doi.org/10.1515/npprj-2020-0042.

Martucci, J. F., & Ruseckaite, R. A. (2010). Three-layer sheets based on gelatin and poly(lactic acid), part 1: Preparation and properties. *Journal of Applied Polymer Science, 118*(5), 3102–3110. https://doi.org/10.1002/app.32751.

Meganaharshini, M., Sudhakar, V., Dhivya Bharathi, N., & Deepak, S. (2023). Review on recent trends in the application of protein concentrates and isolates – A food industry perspective. *Food and Humanity, 1*, 308–325. https://doi.org/10.1016/j.foohum.2023.05.022.

Ming, Y., Zhou, Z., Zhang, S., Wei, Y., Hao, T., & Nie, Y. (2020). Molecular simulation of crystallization of polymers confined in cylindrical nanodomain. *Polymer, 206*, 122818. https://doi.org/10.1016/j.polymer.2020.122818.

Mora-Palma, R. M., Martinez-Munoz, P. E., Contreras-Padilla, M., Feregrino-Perez, A., & Rodriguez-Garcia, M. E. (2024). Evaluation of water diffusion, water vapor permeability coefficients, physicochemical and antimicrobial properties of thin films of nopal mucilage, orange essential oil, and orange pectin. *Journal of Food Engineering, 366*. https://doi.org/10.1016/j.jfoodeng.2023.111865, https://www.sciencedirect.com/science/journal/02608774.

Nagarajan, M., Prodpran, T., Benjakul, S., & Songtipya, P. (2017). Properties and characteristics of multi-layered films from tilapia skin gelatin and poly (Lactic Acid). *Food Biophysics, 12*(2), 222–233. https://doi.org/10.1007/s11483-017-9478-3.

Nazmi, N. N. M., & Sarbon, N. M. (2020). Characterization on antioxidant and physical properties of gelatin based composite films with incorporation of *Centella asiatica* (Pegaga) extract. *Food Research, 4*(1), 224–233. https://doi.org/10.26656/fr.2017.4(1).243, https://www.myfoodresearch.com/uploads/8/4/8/5/84855864/_29__fr-2019-243_nazmi.pdf.

Nilsuwan, K., Benjakul, S., Prodpran, T., & de la Caba, K. (2019). Fish gelatin monolayer and bilayer films incorporated with epigallocatechin gallate: Properties and their use as pouches for storage of chicken skin oil. *Food Hydrocolloids, 89*, 783–791. https://doi.org/10.1016/j.foodhyd.2018.11.056.

Nilsuwan, K., Benjakul, S., & Prodpran, T. (2018). Physical/thermal properties and heat seal ability of bilayer films based on fish gelatin and poly(lactic acid). *Food Hydrocolloids, 77*, 248–256. https://doi.org/10.1016/j.foodhyd.2017.10.001.

Nowak, N., Tkaczewska, J., Cholewa-Wójcik, A., Grzebieniarz, W., Piasna-Słupecka, E., Drozdowska, M., & Jamróz, E. (2024). Multi-layer active biopolymer system as a perspective for improving the quality and scope of applying semi-rigid biodegradable packaging. *Food and Bioprocess Technology*, 1–15.

Omer, S., Forgách, L., Zelkó, R., & Sebe, I. (2021). Scale-up of electrospinning: Market overview of products and devices for pharmaceutical and biomedical purposes. *Pharmaceutics, 13*(2), 286. https://doi.org/10.3390/pharmaceutics13020286.

Park, K., Choi, D., & Hong, J. (2018). Nanostructured polymer thin films fabricated with brush-based layer-by-layer self-assembly for site-selective construction and drug release. *Scientific Reports, 8*(1). https://doi.org/10.1038/s41598-018-21493-9.

Pasquier, E., Mattos, B. D., Koivula, H., Khakalo, A., Belgacem, M. N., Rojas, O. J., & Bras, J. (2022). Multilayers of renewable nanostructured materials with high oxygen and water vapor barriers for food packaging. *ACS Applied Materials and Interfaces, 14*(26), 30236–30245. https://doi.org/10.1021/acsami.2c07579, http://pubs.acs.org/journal/aamick.

Perdana, M. I., Ruamcharoen, J., Panphon, S., & Leelakriangsak, M. (2021). Antimicrobial activity and physical properties of starch/chitosan film incorporated with lemongrass essential oil and its application. *LWT, 141*, 110934. https://doi.org/10.1016/j.lwt.2021.110934.

Pereda, M., Ponce, A. G., Marcovich, N. E., Ruseckaite, R. A., & Martucci, J. F. (2011). Chitosan-gelatin composites and bi-layer films with potential antimicrobial activity. *Food Hydrocolloids, 25*(5), 1372–1381. https://doi.org/10.1016/j.foodhyd.2011.01.001.

Qi, W., Tong, X., Wang, M., Liu, S., Cheng, J., & Wang, H. (2024). Impact of soybean protein isolate concentration on chitosan-cellulose nanofiber edible films: Focus on structure and properties. *International Journal of Biological Macromolecules, 255*, 128185. https://doi.org/10.1016/j.ijbiomac.2023.128185.

Richardson, J. J., Björnmalm, M., & Caruso, F. (2015). Technology-driven layer-by-layer assembly of nanofilms. *Science (New York, N.Y.), 348*(6233). https://doi.org/10.1126/science.aaa2491, http://www.sciencemag.org/content/348/6233/aaa2491.full.pdf.

Rodsamran, P., & Sothornvit, R. (2019). Lime peel pectin integrated with coconut water and lime peel extract as a new bioactive film sachet to retard soybean oil oxidation. *Food Hydrocolloids, 97*, 105173. https://doi.org/10.1016/j.foodhyd.2019.105173.

Roy, S., Priyadarshi, R., Łopusiewicz, Ł., Biswas, D., Chandel, V., & Rhim, J.-W. (2023). Recent progress in pectin extraction, characterization, and pectin-based films for active food packaging applications: A review. *International Journal of Biological Macromolecules, 239*, 124248. https://doi.org/10.1016/j.ijbiomac.2023.124248.

Sanyang, M. L., Sapuan, S. M., Jawaid, M., Ishak, M. R., & Sahari, J. (2015). Effect of plasticizer type and concentration on tensile, thermal and barrier properties of biodegradable films based on sugar palm (*Arenga pinnata*) starch. *Polymers, 7*(6), 1106–1124.

Sarfraz, M. H., Hayat, S., Siddique, M. H., Aslam, B., Ashraf, A., Saqalein, M., Khurshid, M., Sarfraz, M. F., Afzal, M., & Muzammil, S. (2024). Chitosan based coatings and films: A perspective on antimicrobial, antioxidant, and intelligent food packaging. *Progress in Organic Coatings, 188*. https://doi.org/10.1016/j.porgcoat.2024.108235, https://www.sciencedirect.com/science/journal/03009440.

Scaffaro, R., Maio, A., Gulino, E. F., Morreale, M., & La Mantia, F. P. (2020). The effects of nanoclay on the mechanical properties, carvacrol release and degradation of a PLA/PBAT blend. *Materials, 13*(4), 983. https://doi.org/10.3390/ma13040983.

Scaffaro, R., Sutera, F., & Botta, L. (2018). Biopolymeric bilayer films produced by co-extrusion film blowing. *Polymer Testing, 65*, 35–43. https://doi.org/10.1016/j.polymertesting.2017.11.010.

Schlenoff, J. B., Dubas, S. T., & Farhat, T. (2000). Sprayed polyelectrolyte multilayers. *Langmuir: The ACS Journal of Surfaces and Colloids, 16*(26), 9968–9969. https://doi.org/10.1021/la001312i.

Shah, Y. A., Bhatia, S., Al-Harrasi, A., & Khan, T. S. (2025). Chitosan/sodium alginate/ethyl cellulose-based multilayer film incorporated with L-ascorbic acid for improved barrier and antioxidant properties. *International Journal of Biological Macromolecules, 284*. https://doi.org/10.1016/j.ijbiomac.2024.138169, https://www.sciencedirect.com/science/journal/01418130.

Shah, Y. A., Bhatia, S., Al-Harrasi, A., & Khan, T. S. (2024). Advancements in the biopolymer films for food packaging applications: A short review. *Biotechnology for Sustainable Materials, 1*(1), 2.

Singh, S., Sharanagat, V. S., Desai, S., Kumar, K., Upadhyay, S., & Chakraborty, G. (2023). Interaction of pulsed electric modified elephant foot yam (*Amorphophallus paeoniifolius)* starch and polyvinyl alcohol to enhance the mechanical and barrier properties of film. *Starch - Stärke, 75*(5-6). https://doi.org/10.1002/star.202200229.

Staroszczyk, H., Pielichowska, J., Sztuka, K., Stangret, J., & Kołodziejska, I. (2012). Molecular and structural characteristics of cod gelatin films modified with EDC and TGase. *Food Chemistry, 130*(2), 335–343. https://doi.org/10.1016/j.foodchem.2011.07.047.

Suderman, N., Isa, M. I. N., & Sarbon, N. M. (2018). The effect of plasticizers on the functional properties of biodegradable gelatin-based film: A review. *Food Bioscience, 24*, 111–119. https://doi.org/10.1016/j.fbio.2018.06.006.

Sun, Y., Zhang, M., Adhikari, B., Devahastin, S., & Wang, H. (2022). Double-layer indicator films aided by BP-ANN-enabled freshness detection on packaged meat products. *Food Packaging and Shelf Life, 31*, 100808. https://doi.org/10.1016/j.fpsl.2021.100808.

Szyk-Warszyńska, L., Kilan, K., & Socha, R. P. (2014). Characterization of casein and poly-l-arginine multilayer films. *Journal of Colloid and Interface Science, 423*, 76–84. https://doi.org/10.1016/j.jcis.2014.02.031, http://www.elsevier.com/inca/publications/store/6/2/2/8/6/1/index.htt.

Talja, R. A., Helén, H., Roos, Y. H., & Jouppila, K. (2007). Effect of various polyols and polyol contents on physical and mechanical properties of potato starch-based films. *Carbohydrate Polymers, 67*(3), 288–295. https://doi.org/10.1016/j.carbpol.2006.05.019.

Tampau, A., González-Martínez, C., & Chiralt, A. (2018). Release kinetics and antimicrobial properties of carvacrol encapsulated in electrospun poly-(ε-caprolactone) nanofibres. Application in starch multilayer films. *Food Hydrocolloids, 79*, 158–169. https://doi.org/10.1016/j.foodhyd.2017.12.021.

Tkaczewska, J., Kulawik, P., Jamróz, E., Guzik, P., Zając, M., Szymkowiak, A., & Turek, K. (2021). One- and double-layered furcellaran/carp skin gelatin hydrolysate film system with antioxidant peptide as an innovative packaging for perishable foods products. *Food Chemistry, 351*, 129347. https://doi.org/10.1016/j.foodchem.2021.129347.

Topuz, F., & Uyar, T. (2024). Electrospinning of sustainable polymers from biomass for active food packaging. *Sustainable Food Technology, 2*(5), 1266–1296. https://doi.org/10.1039/d4fb00147h, https://www.rsc.org/journals-books-databases/about-journals/sustainable-food-technology/.

Trifol, J., Plackett, D., Szabo, P., Daugaard, A. E., & Giacinti Baschetti, M. (2020). Effect of crystallinity on water vapor sorption, diffusion, and permeation of PLA-based nanocomposites. *ACS Omega, 5*(25), 15362–15369. https://doi.org/10.1021/acsomega.0c01468, pubs.acs.org/journal/acsodf.

Uz, M., & Altınkaya, S. A. (2011). Development of mono and multilayer antimicrobial food packaging materials for controlled release of potassium sorbate. *LWT - Food Science and Technology, 44*(10), 2302–2309. https://doi.org/10.1016/j.lwt.2011.05.003.

Valencia-Sullca, C., Vargas, M., Atarés, L., & Chiralt, A. (2018). Thermoplastic cassava starch-chitosan bilayer films containing essential oils. *Food Hydrocolloids, 75*, 107–115. https://doi.org/10.1016/j.foodhyd.2017.09.008.

Valizadeh, R., Zandi, M., Ganjloo, A., & Dardmeh, N. (2024). Bioactive multilayer film based on the sage seed gum-gelatin, TiO_2 and electrospun zein fibers encapsulating pomegranate peel extract. *International Journal of Biological Macromolecules, 283*(Pt 2), 137826. https://doi.org/10.1016/j.ijbiomac.2024.137826.

Vega, J. F., Souza-Egipsy, V., Expósito, M. T., & Ramos, J. (2022). Melting temperature depression of polymer single crystals: Application to the eco-design of tie-layers in polyolefinic-based multilayered films. *Polymers, 14*(8). https://doi.org/10.3390/polym14081622, https://www.mdpi.com/2073-4360/14/8/1622/pdf.

Vejdan, A., Ojagh, S. M., Adeli, A., & Abdollahi, M. (2016). Effect of TiO_2 nanoparticles on the physico-mechanical and ultraviolet light barrier properties of fish gelatin/agar bilayer film. *LWT - Food Science and Technology, 71*, 88–95. https://doi.org/10.1016/j.lwt.2016.03.011.

Wang, Q., Chen, W., Zhu, W., McClements, D. J., Liu, X., & Liu, F. (2022). A review of multilayer and composite films and coatings for active biodegradable packaging. *npj Science of Food, 6*(1). https://doi.org/10.1038/s41538-022-00132-8, https://www.nature.com/npjscifood/about.

Wang, L., Zhang, Y., Xing, Q., Xu, J., & Li, L. (2023). Quality and microbial diversity of homemade bread packaged in cinnamaldehyde loaded poly (lactic acid)/konjac glucomannan/wheat gluten bilayer film during storage. *Food Chemistry, 402*, 134259.

Wang, W., Qu, W., Zhang, C., Fan, Q., Xue, W., Wang, J., Yu, M., Wu, H., & Zhu, J. (2024a). Graphene-oxide-infused marine biopolymers: Layer-by-layer assembled nanocomposite packaging for prolonged fresh-cut apple preservation. *ACS Sustainable Chemistry & Engineering, 12*(8), 3012–3024. https://doi.org/10.1021/acssuschemeng.3c06363.

Wang, W., Xiao, J., Chen, X., Luo, M., Liu, H., & Shao, P. (2018). Fabrication and characterization of multilayered kafirin/gelatin film with one-way water barrier property. *Food Hydrocolloids, 81*, 159–168.

Wang, Y., Wang, K., Chen, M., Zhao, P., Wang, Y., Wang, X., & Wang, J. (2024b). Development and characterization of biodegradable bilayer packaging films based on corn starch-polylactic acid as raw material. *Journal of Food Measurement and Characterization, 18*(1), 625–639.

Wu, S., Wang, W., Zhang, R., Zhai, X., & Hou, H. (2021). Preparation and characterization of biodegradable trilayer films based on starch and polyester. *International Journal of Biological Macromolecules, 183*, 1058–1066. https://doi.org/10.1016/j.ijbiomac.2021.05.051.

Xia, C., Wang, W., Wang, L., Liu, H., & Xiao, J. (2019). Multilayer zein/gelatin films with tunable water barrier property and prolonged antioxidant activity. *Food Packaging and Shelf Life, 19*, 76–85. https://doi.org/10.1016/j.fpsl.2018.12.004.

Xiao, J., Liu, T., Chu, Q., Yu, C., Yin, Y., Xuan, L., & Wu, S. (2024). Development of an UV–resistant multilayer film with enhanced compatibility between carboxymethyl cellulose and polylactic acid via incorporation of tannin and ferric chloride. *Molecules, 29*(12), 2822. https://doi.org/10.3390/molecules29122822.

Xie, J., Lim, L. K., Phua, Y., Hua, J., & Wang, C. H. (2006). Electrohydrodynamic atomization for biodegradable polymeric particle production. *Journal of Colloid and Interface Science, 302*(1), 103–112. https://doi.org/10.1016/j.jcis.2006.06.037.

Xie, F. (2023). Biopolymer-based multilayer films and coatings for food preservation: an update of the recent development. *Current Food Science and Technology Reports, 1*(1), 1–12. https://doi.org/10.1007/s43555-023-00002-8.

Xu, W., Sun, K., Hou, S., & Chen, A. (2025). Research progress of advanced polymer composite antibacterial materials based on electrospinning. *European Polymer Journal, 222*, 113623. https://doi.org/10.1016/j.eurpolymj.2024.113623.

Yan, J., Li, M., Wang, H., Lian, X., Fan, Y., Xie, Z., & Li, W. (2021). Preparation and property studies of chitosan-PVA biodegradable antibacterial multilayer films doped with Cu2O and nano-chitosan composites. *Food Control, 126*, 108049.

Yan, Q., Wang, L., Sun, X., Fan, F., Ding, J., Li, P., & Fang, Y. (2022). Improvement in the storage quality of fresh salmon (Salmo salar) using a powerful composite film of rice protein hydrolysates and chitosan. *Food Control, 142*, 109211.

Yan, W., Sun, H., Liu, W., & Chen, H. (2022). Preparation and properties of blended composite film manufactured using walnut-peptide–chitosan–sodium alginate. *Foods, 11*(12). https://doi.org/10.3390/foods11121758. Article 12.

Yan, Q., Wang, L., Sun, X., Fan, F., Ding, J., Li, P., Zhu, Y., Xu, T., & Fang, Y. (2022). Improvement in the storage quality of fresh salmon (Salmo salar) using a powerful composite film of rice protein hydrolysates and chitosan. *Food Control, 142*, 109211. https://doi.org/10.1016/j.foodcont.2022.109211.

Yao, F., Zheng, Y., Gao, Y., Du, Y., & Chen, F. (2023). Electrospinning of peanut protein isolate/poly-L-lactic acid nanofibers containing tetracycline hydrochloride for wound healing. *Industrial Crops and Products, 194*, 116262. https://doi.org/10.1016/j.indcrop.2023.116262.

Yerramathi, B. B., Muniraj, B. A., Donadi, J., Bandi, K., & Kola, M. (2025). Transformation of alginate based films into food packaging material - confines and possibilities as a renewable resource: A review. *Reactive and Functional Polymers, 215*. https://doi.org/10.1016/j.reactfunctpolym.2025.106358, https://www.sciencedirect.com/science/journal/13815148.

Yong, Y., Wang, S., Li, L., Li, R., Ahmad, H. N., Munawar, N., & Zhu, J. (2023). A curcumin-crosslinked bilayer film of soy protein isolate and chitosan with enhanced antibacterial property for beef preservation and freshness monitoring. *International Journal of Biological Macromolecules, 247*, 125778. https://doi.org/10.1016/j.ijbiomac.2023.125778.

Yu, F., Shi, H., Wang, K., Li, H., & Peng, L. (2022). Preparation of robust and fully bio-based modified paper via mussel-inspired layer-by-layer assembly of chitosan and carboxymethyl cellulose for food packaging. *International Journal of Biological Macromolecules, 222*(Pt A), 1238–1249. https://doi.org/10.1016/j.ijbiomac.2022.09.243.

Zhang, W., Liu, R., Sun, X., An, H., Min, T., Zhu, Z., & Wen, Y. (2022). Leaf-stomata-inspired packaging nanofibers with humidity-triggered thymol release based on thymol/EVOH coaxial electrospinning. *Food Research International, 162*. https://doi.org/10.1016/j.foodres.2022.112093, www.elsevier.com/inca/publications/store/4/2/2/9/7/0.

Zhang, J., Guan, B., Zhang, Y., Hu, J., Sun, T., Dong, T., & Yun, X. (2024). Development of high barrier poly(l-actic acid)/chitosan/graphene oxide flexible films for meat packaging by layer-by-layer. *Food Bioscience, 60*, 104304. https://doi.org/10.1016/j.fbio.2024.104304.

Zhang, X., Xu, Y., Zhang, X., Wu, H., Shen, J., Chen, R., Xiong, Y., Li, J., & Guo, S. (2019). Progress on the layer-by-layer assembly of multilayered polymer composites: Strategy, structural control and applications. *Progress in Polymer Science, 89*, 76–107. https://doi.org/10.1016/j.progpolymsci.2018.10.002.

Zhang, X., Zhang, B., Mao, R., Huang, Z., Jing, K., Jin, C., Yang, B., Qi, J., Yu, M., Xiong, G., & Hu, Y. (2024). A novel multilayer film based on sodium alginate/*k*-carrageenan-gelatin incorporated with ZnO nanoparticles and oregano essential oil for active food packing. *Progress in Organic Coatings, 187*, 108170. https://doi.org/10.1016/j.porgcoat.2023.108170.

Zhao, J., Wang, Y., & Liu, C. (2022). Film transparency and opacity measurements. *Food Analytical Methods, 15*(10), 2840–2846.

Zhao, K., Wang, W., Teng, A., Zhang, K., Ma, Y., Duan, S., & Guo, Y. (2020). Using cellulose nanofibers to reinforce polysaccharide films: Blending vs layer-by-layer casting. *Carbohydrate polymers, 227*, 115264.

Zhao, L., Duan, G., Zhang, G., Yang, H., He, S., & Jiang, S. (2020). Electrospun functional materials toward food packaging applications: A review. *Nanomaterials (Basel, Switzerland), 10*(1), 150. https://doi.org/10.3390/nano10010150.

Zhou, X., Yang, R., Wang, B., & Chen, K. (2019). Development and characterization of bilayer films based on pea starch/polylactic acid and use in the cherry tomatoes packaging. *Carbohydrate Polymers, 222*, 114912.

Zhou, X., Yu, X., Xie, F., Fan, Y., Xu, X., Qi, J., & Zhang, F. (2021). pH-responsive double-layer indicator films based on konjac glucomannan/camellia oil and carrageenan/anthocyanin/curcumin for monitoring meat freshness. *Food Hydrocolloids, 118*, 106695.

Zhu, J., Wu, H., & Sun, Q. (2019). Preparation of crosslinked active bilayer film based on chitosan and alginate for regulating ascorbate-glutathione cycle of postharvest cherry tomato (*Lycopersicon esculentum*). *International Journal of Biological Macromolecules, 130*, 584–594.

Zubair, M., Mujahid, M., Shahzad, S., Rauf, Z., Hussain, A., Ayyash, M., & Ullah, A. (2025). Exploring protein-based films and coatings for active food packaging applications: A comprehensive review. *International Journal of Biological Macromolecules, 320*, 146070. https://doi.org/10.1016/j.ijbiomac.2025.146070.

Chapter 31

Migration from packaging materials to food

Violeta Ognenoska[1], Davor Daniloski[2], and Anka Trajkovska Petkoska[1,3]
[1]*Faculty of Technology and Technical Sciences, University St. Clement of Ohrid in Bitola, Veles, Macedonia,* [2]*School of Chemical Engineering, The University of Queensland, Brisbane, QLD, Australia,* [3]*Department of Materials Science and Engineering, Korea University, Seoul, South Korea*

31.1 Introduction

The migration of substances from packaging materials into food is a critical issue that has garnered increasing attention in recent years due to its implications for food safety, quality, and public health (Gupta et al., 2024). Packaging materials are essential for protecting food, extending its shelf life, and maintaining its quality during storage and transportation. However, these materials can also serve as a source of chemical contaminants, which may migrate into food under certain conditions (Ščetar et al., 2022). This migration is influenced by various factors, including the type of packaging material, the nature of the food, and storage conditions such as temperature and time (Omer et al., 2025). Recent research showed that the substances migrating from packaging materials can include monomers, additives, residual solvents, and degradation products (Franz & Welle, 2022). These compounds may pose health risks if their migration levels exceed safety thresholds. For instance, the European Union has established specific migration limits (SML) for certain substances in food-contact materials (FCMs) to ensure consumer safety (Regulation [EU] No 10/2011). Similarly, regulatory bodies such as the US Food and Drug Administration (FDA) and the World Health Organisation (WHO) have provided guidelines for the use of food-packaging materials and the permissible levels of migrated substances (Gupta et al., 2024).

Plastics, which dominate the packaging industry, have been widely studied for their potential to release chemical substances into food (Yashwanth et al., 2025). Studies have identified a wide range of migrating substances, including phthalates, bisphenol A (BPA), and other plasticizers, which are known to disrupt endocrine functions and pose long-term health risks (Groh et al., 2019; Muncke et al., 2017). Similarly, migration from other packaging materials, such as metals, paper, and glass, has also been investigated. For example, heavy metal migration from can coatings and printed paperboard into food has raised significant concerns (Mertoğlu-Elmas & Çınar, 2018; Vázquez-Loureiro et al., 2023). Emerging research underscores the need to better understand the complex interactions between packaging materials and food components. The chemical composition of the food, particularly its fat, water, and acid content, plays a significant role in determining the extent of migration (Arvanitoyannis & Kotsanopoulos, 2014). Moreover, advancements in analytical techniques, including gas chromatography-mass spectrometry (GC-MS) and high-performance liquid chromatography (HPLC), have enabled more precise identification and quantification of migrants in food (Rodríguez-Ramos et al., 2024).

The environmental impact of packaging materials further complicates the issue, as biodegradable and recycled materials are increasingly being used to reduce waste and promote global sustainability trends (Cheng et al., 2024). However, these materials may introduce new challenges related to chemical migration, necessitating comprehensive risk assessments (Mahmud et al., 2024). Therefore, this chapter aims to review recent developments in the study of migration from packaging materials to food, focusing on key chemical contaminants, their health implications, and the regulatory framework governing FCMs. Additionally, it is possible that certain compounds from the food matrix may leach into food-packaging materials. This chapter synthesizes recent research to provide a comprehensive understanding of the challenges and opportunities in ensuring food-packaging safety.

Natural Polymer-Based Films and Coatings in Food Technology. DOI: https://doi.org/10.1016/B978-0-443-34244-8.00032-X

31.2 Interaction between food-packaging materials and food

When food is packaged, it comes into direct contact with the inner surface of packaging, allowing interactions between the food and the material (Guazzotti et al., 2022). Components may transfer from the packaging to the food or vice versa, influenced by mass transfer mechanisms like migration, permeation, and sorption (Fig. 31.1). These processes, driven by concentration/pressure differences, can alter the quality of both food and packaging over time (Garba, 2023). This mass transfer is governed by Fick's and Henry's laws of diffusion.

31.2.1 Mass transfer through the food-packaging material

Permeation of low molecular weight gases/vapors through polymeric walls (as packaging materials) is measured by the ability of the polymer matrix to absorb and diffuse the penetrant—generally described as a "solubilization-diffusion" mechanism. It usually happens as a three-stage process: (1) absorption of penetrant on the (polymer) surface/packaging film; (2) diffusion of penetrant through the material(s); (3) desorption of penetrant from the opposite surface of the packaging material/film (Fig. 31.2). Under steady-state conditions, gases or vapors will diffuse through polymer films at a constant rate if a constant pressure difference is maintained across the film (Alibekov et al., 2024; Trinh et al., 2023; Wu et al., 2021).

The diffusive flux, J, of a permeant through a polymer can be defined as the amount passing through a plane surface of unit area normal to the direction of flow during unit time (Eq. 31.1):

$$J = \frac{Q}{A \cdot t} \tag{31.1}$$

where, Q is the total amount of permeant that has passed through certain area (A) during certain time (t).

The First Fick's Law is defined as (Eq. 31.2):

$$J = -D\frac{dc}{dx} \tag{31.2}$$

where D is diffusivity coefficient and $\frac{dc}{dx}$ concentration gradient in the x-direction.

Combining Eqs. (31.1) and (31.2):

$$Q = \frac{D(c_1 - c_2)}{x} A \cdot t \tag{31.3}$$

For the case of very small concentration differences of the permeant, then Henry's Law could be applied:

$$c = S \cdot p \tag{31.4}$$

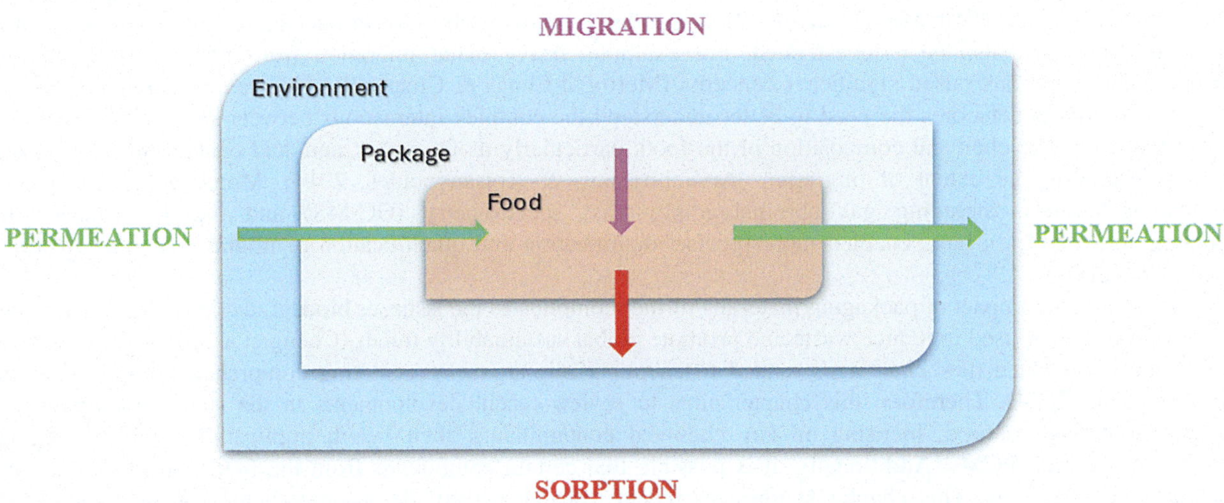

FIGURE 31.1 Migration, sorption, and permeation through the packaging material.

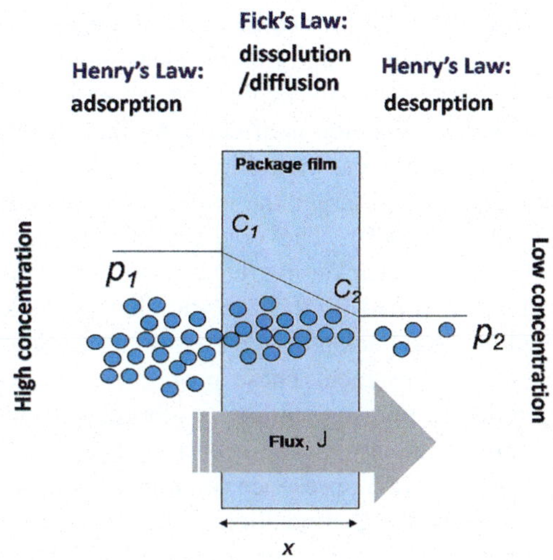

FIGURE 31.2 Mass transfer through the packaging film/wall.

where S is the gas solubility coefficient (Henry's constant), and p is the pressure of diffused gas. In general, solubility leads the rate of molecules entering the membrane, while diffusivity dictates the flow rate through the membrane. Combination of previous Eqs. (31.3) and (31.4) leads to:

$$Q = \frac{D \cdot S \cdot (p_1 - p_2)}{x} A \cdot t \tag{31.5}$$

At the thermodynamic equilibrium, the gas permeability coefficient (P) is given as (Eq. 31.6):

$$P = D \cdot S \tag{31.6}$$

That means that permeability depends on diffusivity and solubility of permeant in the polymer /packaging film, or combination of Eqs. (31.5) and (31.6) provides:

$$P = \frac{Q \cdot x}{A \cdot \Delta p \cdot t} \tag{31.7}$$

31.2.2 Migration

Chemicals present in FCMs—such as monomers, additives, processing agents, or reaction by-products—can migrate into food (Geueke et al., 2023). This process, known as migration, is influenced by several factors, including elevated temperatures, storage duration, the chemical properties of both the FCMs and the food, and the physical attributes of the FCMs, such as pore size, thickness, and surface area (Muncke et al., 2014). Diffusion, adsorption, desorption, and evaporation are the primary mechanisms by which substances from FCMs migrate into food (Balaji & Mayuri, 2024). Migration primarily occurs through diffusion, where molecules move from areas of higher concentration to lower concentration until equilibrium is reached. The mass diffusion process is usually governed by Fick's law (Alamri et al., 2021). Chemical migration into food typically occurs from primary packaging in direct contact with the food. However, contamination can also occur, though less frequently, from secondary, tertiary, or quaternary packaging, including cartons, pallets, or containers (Urbelis & Cooper, 2021). Chemicals like printing inks (e.g., benzophenone) may transfer to food-contact surfaces through processes like "*set-off*," thus external packaging surfaces transfer chemicals during stacking and storage. Migration can also happen via evaporation, with chemicals leaching into food through the gaseous phase (Ardic et al., 2015).

31.2.3 Permeation

Permeability is the ability of a substance, such as gas or water vapor, to pass through a material that offers resistance to its flow. When there are no cracks, pinholes, or defects in the material, the main mechanism for this flow is activated diffusion. In this process, the permeate first dissolves into the material at the side with a higher concentration, then moves through the material driven by a concentration gradient, and finally releases/evaporates from the opposite side (Robertson, 1993). The ease with which specific gases permeate a material depends on their solubility in the material, which influences how readily they dissolve and diffuse. The diffusion step is affected by factors including the size, shape, and polarity of the gas molecules, as well as the material's properties, including its crystallinity, degree of cross-linking, and polymer chain flexibility (Favvas et al., 2017). Generally, the larger the molecule of the permeate, the lower its permeability rate through the polymer material (Gajdoš et al., 2000). The concept of permeability is normally associated with the quantitative evaluation of the barrier properties of a packaging (e.g., plastic) material. For example, the choice of polymers in the food-packaging industry is largely determined by the specific characteristics of the food product (Bianchi et al., 2024). Additionally, manufacturing, handling, and packaging processes can significantly impact the final properties of the packaging material, particularly its barrier performance. These properties are closely linked to the polymer's intrinsic structure, including its degree of crystallinity, ratio of crystalline to amorphous phases, polymer type, and any thermal or mechanical treatments applied before or after contact with food (Farrell et al., 2024). Other influencing factors might also include the chemical composition of the polymer (e.g., presence of polar or nonpolar groups), the degree of cross-linking, and the glass transition temperature. Moreover, the composition of the food itself (e.g., protein and fat content, aroma compounds) can influence the material's sorption behavior, while external factors like temperature and pH, and for certain polymers, relative humidity can affect their barrier properties (Siracusa, 2012).

31.2.4 Sorption

Sorption, commonly referred to as scalping, involves the absorption or adsorption of elements such as flavors, aromas, lipids, and moisture from the food into the packaging material. In essence, sorption can encompass both absorption (elements being taken up into the material) and adsorption (elements adhering to the surface of the material) (Caner, 2018). This process often leads to the diffusion of these components through the packaging, causing a reduction in food flavor (Gnanasekharan & Floros, 1997). Aroma, a critical quality factor for many foods, can be lost through packaging material, leading to flavor degradation, scalping, or contamination, ultimately compromising product quality (Caner, 2018). The processes of sorption and migration often occur simultaneously, as substances can adhere to or penetrate packaging materials while concurrently moving into the food product (Garba, 2023).

31.3 Factors influencing migration

The migration of substances from the packaging material toward the packed food items is affected by several factors like the nature of the packaging material, the nature of the food (composition), external factors, and intentionally added and nonintentionally added substances. Fig. 31.3 summarizes the factors that influence migration processes in food-packaging systems.

31.3.1 Material composition

The composition of the packaging material plays a significant role in the migration of substances into food. Different types of packaging materials exhibit varying migration tendencies due to their chemical and physical properties (Alamri et al., 2021). The selection of packaging materials must consider their compatibility with the food product to minimize potential health risks and preserve food quality.

Plastic dominates the packaging industry because it has numerous food-packaging advantages compared to its disadvantages, which have enormously contributed to its preference as a food-contact material for packing various food items (Ščetar, 2021). Polypropylene (PP), polyethylene (PE), polyvinyl chloride (PVC), polystyrene (PS), polyethylene terephthalate (PET), and polyurethane (PU) are the packaging materials globally dominating the plastic food-packaging industry (Muzeza et al., 2023). In most products made from plastic materials, the core polymer is combined with a formulation known as a plastic compound. This formulation includes various chemical compounds, referred to as additives, which are added to enhance the polymer's performance, functionality, and durability over time (Trajkovska Petkoska et al., 2021a). The most frequently used additives in different types of polymer-based packaging materials include functional additives: stabilizers, antistatic agents, flame retardants, plasticizers, lubricants, slip agents, curing agents, foaming agents, biocides, to name a few; colorants: pigments and soluble azo colorants; fillers: mica, talc, kaolin, clay, calcium carbonate, and barium sulfate; and reinforcements: glass and carbon fibers (Hahladakis et al., 2018; Muzeza et al., 2023). However, the additives in plastics may migrate into food under certain conditions, such as chemical

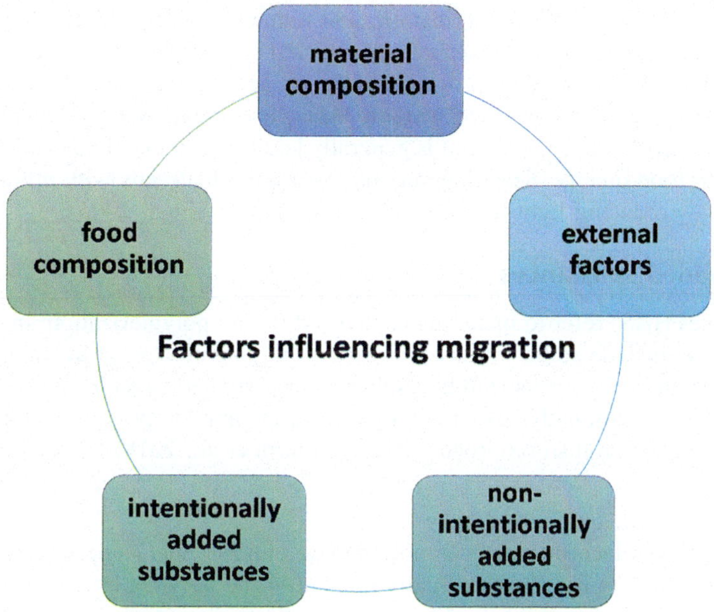

FIGURE 31.3 Factors that influence migration process in food-packaging system.

properties of the packaging materials, the composition of the food, the temperature, and the duration of contact between the packaging and food (Fang & Vitrac, 2017).

Paper-based materials are often coated with waxes or plastic films to enhance their barrier properties. However, untreated paper and cardboard can release contaminants, including mineral oils from printing inks and adhesives (Grob, 2018). Metals like aluminum, tin plate, tin-free steel, and stainless steel, as well as metal-based packaging materials in both rigid and semirigid forms (cans, foil wraps, and retort pouches), are commonly used in food packaging. However, their interaction with food depends on the presence of coatings or lacquers. If these coatings degrade, metal ions can migrate into the food, especially in an acidic environment (Deshwal & Panjagari, 2020). Glass is considered one of the most inert packaging materials. Nevertheless, the sealing materials, such as gaskets, can contribute to migration, particularly under high temperatures and storage times (Qureshi et al., 2024).

31.3.2 External factors

Temperature significantly influences the migration process by affecting both the diffusivity of substances within the packaging material and their solubility in food (Gupta et al., 2024). Elevated temperatures accelerate molecular movement, increasing the rate at which substances migrate. Studies indicate that substances like phthalates and BPA exhibit increased migration rates at higher temperatures and prolonged storage times (Agarwal et al., 2022; Khanniri et al., 2024). In a study conducted by Alin and Hakkarainen (2012), it is noted that in microwave heating of polycarbonate packaging with the use of food simulants (water, ethanol, and isooctane), the migration rates of m-tert-butyl-phenol, 4-ethoxy-benzoic acid ethyl ester, benzophenone, and 9,9-dimethylxanthene are higher than in conventional heating. In a review by Bhunia et al. (2013), it is concluded that microwave heating degrades the antioxidants in food-contact plastics. Further, during food processing, such as sterilization or pasteurization, packaging materials are subjected to high temperatures. These conditions can lead to polymer degradation, releasing monomers and other substances into the food. Another study by Paraskevopoulou et al. (2012) found that increasing the time of exposure along with increased temperature showed higher values of styrene migration into the food.

Multiple factors influence migration, including food composition, type of contact, duration, temperature, packaging material, and the chemical properties of migrating substances (Arvanitoyannis & Kotsanopoulos, 2014).

Exposure of packaged foods to light, including ultraviolet (UV) and visible wavelengths, can cause oxidative deterioration of lipids, vitamins, proteins, and colorants, leading to nutrient loss, off-flavors, and discoloration (Chemat et al., 2023). Factors such as light intensity, exposure duration, and packaging material properties influence these effects.

Oxygen availability is crucial for light-induced oxidation, and reducing oxygen levels (e.g., via vacuum packaging or modified atmospheres) can mitigate damage (Duncan & Chang, 2012). Noteworthy, the effectiveness of light protection depends on material factors like thickness, coloration, and additives. Metals provide the best protection, followed by glass, plastics, and paper. Additives like titanium oxide, carbon black, and cavitation techniques can improve light barrier properties in plastics (Andrady et al., 2023). UV light is generally less harmful than blue-violet visible light (400–500 nm), which is more detrimental to food quality. Selecting suitable packaging materials with optimal light-blocking properties helps preserve food quality by reducing light-induced damage (Singh & Singh, 2005).

31.3.3 Intentionally added substances

Intentionally added substances (IAS) refer to materials introduced during polymerization, including monomers, catalysts, and their impurities. They also include additives incorporated during manufacturing to enhance material properties, such as plasticizers, antioxidants, light and thermal stabilizers, lubricants, antistatic agents, and slip additives, along with their impurities (Nerin et al., 2013). Additionally, IAS may arise from incomplete polymerization during polymer formation and residual solvents left from the production process (García Ibarra et al., 2018; Muzeza et al., 2023).

31.3.3.1 Plasticizers

Plasticizers are chemical additives incorporated into polymers to enhance mechanical properties such as flexibility and ease of processing. When plasticizers are not chemically bonded to the polymer chains, they can migrate under certain conditions, resulting in unwanted changes to the polymer's properties and contamination of the surrounding environment (Muobom et al., 2020). In the context of food packaging, this migration allows plasticizers to diffuse into the food, potentially causing food safety issues, including food poisoning (Eslami et al., 2023). To mitigate plasticizer migration, more complex, high molecular weight plasticizers have been developed that better integrate into the polymer matrix, reducing their mobility. However, over time and under specific conditions, even these plasticizers can leach out, posing environmental and health risks (Li et al., 2022). High molecular weight phthalates, such as di(2-ethylhexyl) phthalate (DEHP) and diisononyl phthalate (DiNP), are widely used as plasticizers in PVC materials, including food-packaging and food-contact products (Wang & Qian, 2021). Over the past two decades, regulations in the European Union and the United States have restricted the use of several ortho-phthalates, including DEHP and di-n-butyl phthalate (DnBP), in consumer products. As a result, alternative plasticizers like di(2-ethylhexyl) adipate (DEHA), 1,2-cyclohexane dicarboxylic acid diisononyl ester (DINCH), and dioctyl terephthalate (also called di(2-ethylhexyl) terephthalate, DEHT) have been adopted for use in PVC materials. Notably, DEHT differs structurally from ortho-phthalates despite its similar naming convention (Edwards et al., 2022).

31.3.3.2 Solvents

Residual solvents such as toluene, methanol, ethanol, ethyl acetate, methyl ethyl ketone, isopropyl alcohol, acetone, and cyclohexane can be retained in multilayer packaging materials or introduced during printing and lamination processes. These solvents can alter the flavor of food and potentially pose health risks to humans (Seo & Shin, 2010). Directive 2009/32/EC regulates permissible extraction solvents in food processing to ensure safety and compliance with good manufacturing practices (GMPs) (https://eur-lex.europa.eu/EN/legal-content/summary/safe-food-processing-common-standards-for-extraction-solvents.html).

31.3.4 Nonintentionally added substances

Nonintentionally added substances (NIAS) present a unique challenge in FCMs made from synthetic, polymer-based materials like plastics or metal coatings. During polymerization, free radicals are generated to facilitate the process, but these radicals can react with other compounds in the mixture, resulting in randomly formed by-products (Kato & Conte-Junior, 2021). Additionally, oligomers (short-chain polymers with 2–10 monomer units) are created, which may migrate into food. Identifying these random compounds in synthetic polymers is difficult, complicating chemical risk assessments, as only known substances can undergo toxicological evaluation (Muncke, 2016). Oligomers, polymers, by-products, and degradation products are referred to as NIAS (Coniglio et al., 2020).

31.3.5 Food composition

Food composition significantly affects migration behavior by altering the solubility and reactivity of migrating substances. Lipophilic substances migrate preferentially into high-fat foods due to their greater solubility in fat (Singh et al., 2023). In a review by Etxabide et al. (2022), it was found that the presence of higher fat content increased the migration rate of migratable substances such as oligomers and oligomeric saturated hydrocarbons, and consequently, their total migrated concentration. Another study by Bomfim et al. (2011) confirms that residues of 3-caprolactam (CAP), a precursor monomer of nylon 6 used in food-packaging materials, migrated into bologna sausage, turkey blanquete, poultry breast, and pate, respectively, presented specific migration levels exceed the limit established in Europe (15 mg kg^{-1} of food). The results were further confirmed by Đurić et al. (2024), where diisobutyl phthalate (DIBP) and dibutyl phthalate (DBP) largely migrated into salmon meat from the packaging (polyamide polyethylene) after six months of storage, reaching 73.77 and 78.45 μg kg^{-1}, respectively. Additionally, the highest concentrations of di(2-ethylhexyl) phthalate (DEHP) after six months of storage were present in the salmon meat packed in the above-mentioned packaging type (253.56 μg kg^{-1}). In contrast, the lowest amount of DEHP was observed in the carp meat that was packed in PP (157.72 μg kg^{-1}).

Acidic foods, such as citrus juices and tomato products, can corrode metal packaging, enhancing the release of metal ions. Metal cans are widely used for packaging processed foods due to their excellent barrier properties, mechanical strength, and recyclability (Deshwal & Panjagari, 2020). They are impermeable to moisture, gases, and flavors, and degrade faster than glass or plastic when exposed to the environment. However, metal cans can corrode when exposed to certain food components like oxygen, pigments, nitrates, sulfur compounds, and salts. Foods such as tomatoes, peppers, and fish are particularly aggressive in causing corrosion (Dhuey et al., 2021). To prevent this, polymeric coatings are applied as protective linings inside cans. The thermal breakdown of amino acids in tomatoes can produce volatile methyl sulfides, which are absorbed by the can's protective lining (Kontominas et al., 2006). This absorption leads to damage in the lining, including moisture penetration, delamination, and exposure of the can's base metal. As a result, the electrolytes from the tomatoes interact with the exposed metal, causing corrosion on the can walls (Dhuey et al., 2022). The water content of food affects the migration of hydrophilic substances. The study by Schmid and Welle (2020) explained and reviewed the migration of different substances such as antimony trioxide, acetaldehyde, 2-aminobenzamide, and PET oligomers from the bottles into the water.

31.4 Migration tests and regulations

All FCMs must undergo testing before being marketed to ensure their safety during intended use and throughout the product's shelf life. Different countries have specific regulations outlining the required tests, including food simulants and test conditions (Ščetar et al., 2022).

31.4.1 Simulant testing

Food simulants mimic the chemical and physical behavior of different food types during migration testing. In Europe, six simulants are used to represent foods with different properties, such as low-fat, acidic, high-fat, or dry foods. However, in some cases, standard simulants may not be suitable for specific tests (Nerín et al., 2023). For long-shelf-life foods, migration studies are conducted at higher temperatures over shorter durations using simulants that mimic food properties. The choice of simulants depends on the packaging's expiry time, shelf life, and food type (Nerin, 2016). However, migration traces often go undetected in taste and odor tests and require advanced techniques like liquid chromatography or mass spectrometry for precise analysis (Yenidoğan et al., 2023). The guidelines from US FDA recommend the use of three types of food simulants for migration testing: 10% (v/v) ethanol for aqueous and acidic food, 10% or 50% ethanol for low and high-alcoholic beverages, and 50% ethanol for fatty foods (FDA, 2020-D-1925) (US FDA, 2007). In some cases, such as when testing for rigid poly(vinyl chloride), polystyrene, rubber-modified polystyrene, and poly(ethylene terephthalate), 50% ethanol is recommended as an effective fatty-food simulant. The guidelines also provide specific recommendations for temperature and duration (US FDA, 2007).

31.4.2 Analytical methods for monitoring migration processes through the packaging materials

Various techniques are used for identifying and analyzing the chemical composition of microplastics, including scanning electron microscopy-energy dispersive (SEM-EDS), Fourier Transform Infrared (FTIR), near infrared (NIR), Raman, and nuclear magnetic resonance (NMR) spectroscopies. Namely, SEM-EDS provides detailed information on microplastic surfaces and additives. FTIR and Raman spectroscopy are highly effective for chemical characterization, offering nondestructive analysis by measuring interactions between molecules and light. FTIR is suitable for particles up to 20 μm, while Raman can analyze particles as small as 1 μm (Guo et al., 2024; Tirkey & Upadhyay, 2021). Additionally, techniques like GC-MS and HPLC enable precise identification and quantification of migrants (Nerín et al., 2022). The contents of residual solvent levels in food-packaging materials can be determined using headspace extraction-gas chromatograph/mass spectrometry (HS-GC/MS) (Sapozhnikova & Nuñez, 2022; Seo & Shin, 2010). Additionally, high-resolution mass spectrometry (HRMS) allows the detection of previously unidentified compounds, providing a more comprehensive understanding of migration phenomena (Hollender et al., 2017).

31.4.3 Food-packaging regulations and standards

Food-packaging regulations and standards play a critical role in ensuring the safety, quality, and sustainability of materials used in direct or indirect contact with food. These legislations are designed to prevent contamination, protect consumer health, and promote environmentally friendly practices (Gupta et al., 2024). A central focus of modern packaging laws is migration, which refers to the transfer of substances from packaging materials into food products. Regulations have established SML to ensure that materials used in packaging do not release harmful chemicals above permissible levels. Table 31.1 provides an updated list of the most important global and EU legislations and standards for food packaging.

TABLE 31.1 Global and EU regulations and guidelines for food packaging.

Regulation/guideline	Region	Purpose	Key focus areas
EU regulations			
Regulation (EC) No 1935/2004	EU	General requirements for materials intended to come into contact with food	Safety, inertness, labeling, traceability, and compliance declaration
Regulation (EC) No 2023/2006		Good manufacturing practices for food-contact materials (FCMs)	Quality assurance systems and processes to ensure compliance with food safety standards
Regulation (EU) No 10/2011		Specific rules for plastic materials and articles in contact with food	Migration limits, permitted substances, and testing requirements for plastics
Regulation (EU) No 450/2009		Active and intelligent packaging materials for food	Conditions for packaging that actively interacts with food or provides information about its condition
Regulation (EU) No 282/2008		Recycled plastics intended for FCMs	Criteria for the use of recycled plastics, including decontamination processes
Regulation (EC) No 1907/2006		Registration, evaluation, authorization, and restriction of chemicals	Chemical substances used in packaging materials and safety compliance
Regulation (EC) No 1333/2008		Additives used in FCMs	Defines permitted additives, limits, and compliance testing
Council Directive 85/374/EEC		Liability for defective products	Holds manufacturers accountable for noncompliant or unsafe packaging materials
International regulations and guidelines			
Codex Alimentarius (CAC/RCP 1–1969)	Global (FAO/World Health Organisation)	International food standards, including hygiene and safety requirements for packaging materials	Safety standards, material suitability, and contaminant prevention
ISO 22000:2018	Global (ISO)	Food safety management systems, including packaging safety requirements	Comprehensive food safety processes, traceability, and risk management
ISO 18602:2013		Packaging and environment—optimization of the packaging system	Sustainability, source reduction, and recyclability

Source: https://eur-lex.europa.eu, https://www.fao.org, https://www.iso.org

31.5 Health implications of chemical migration from food packaging

While packaging materials play a key role in prolonging the shelf life of food and maintaining its quality, certain elements within these materials have been linked to a range of health concerns through the transfer of chemical substances into the food (Robertson, 2016). According to a study by Stevens et al. (2024), many plastics used in FCMs release substances that affect nuclear receptors linked to endocrine and metabolic functions, posing potential health risks. PVC, PU, and low-density polyethylene (LDPE) exhibited the strongest endocrine-disrupting activities, while PET and high-density polyethylene (HDPE) released fewer harmful compounds (Stevens et al., 2024).

Plasticizers often used in food packaging, such as phthalates and bisphenols, are highly functional chemicals that have become a concern due to their toxicity (i.e., endocrine/hormone modulators) and ability to migrate from FCMs into food matrices and the environment, resulting in human and environmental health risks (Muzeza et al., 2023; Tumu et al., 2023). For example, BPA, also known by its chemical name, 2,2-(4,4′-dihydroxydiphenyl) propane, is frequently used in the production of tin cans. It functions as both a plasticizer and an agent that inhibits PVC from fully polymerizing (Blankenship & Coady, 2005; Muzeza et al., 2023). Research indicates that BPA can leach from packaging into food, disrupting the hormone systems involved in reproduction. Over time, these disruptions may contribute to miscarriages, infertility, and increase the risk of cancer (Mukhi et al., 2024). Furthermore, paper and cardboard packaging often contain phthalates, paint residues, stickers, labels, adhesives, and other chemicals with potential endocrine-disrupting activity, posing a risk of food contamination. Sealing lids or internal linings in glass containers often incorporate plastic or metal components that may introduce endocrine-disrupting substances (de Paula & Alves, 2024). Although metal-based packaging offers excellent barriers and is widely used for food, there is concern over the migration of BPA and heavy metals (such as lead, cadmium, and mercury), which can lead to carcinogenic and mutagenic toxicity (Deshwal & Panjagari, 2020).

31.6 Novel trends in food packaging, materials, and technologies

31.6.1 Biodegradable and biobased polymers

Biobased materials offer a promising alternative to conventional materials due to their nontoxicity, biocompatibility, and biodegradability. They can notably reduce energy use and carbon emissions associated with traditional packaging materials (Upadhyay et al., 2024). Research in this area is expanding, with examples including wood-based biopolymers (cellulose, hemicellulose, starch, lignin), protein-derived films (gelatin, keratin, wheat gluten, soy protein, whey protein isolates), and microbial polymers (polyhydroxyalkanoate [PHA], polylactic acid [PLA]) (Nizamuddin et al., 2024). These have all been investigated as sustainable options for food packaging. Biopolymers can be classified into two groups: those sourced naturally, particularly proteins, polysaccharides, lipids, and their composites, as well as materials made from renewable sources (plant or animal feedstock), and those synthesized from biomass (Upadhyay et al., 2024). "Bio-based" refers to renewable, agriculture-derived materials used in plastic production, while "biodegradable" indicates a material's capacity to be fully broken down by environmental microorganisms. However, not all biobased materials are biodegradable. Biodegradable and biocompostable bioplastics are biobased materials that meet established standards for biodegradability and compostability (Trajkovska Petkoska et al., 2021b).

Various standardized testing protocols exist at national, continental, and international levels, providing harmonized criteria for confirming these properties (Stoica et al., 2022). Biodegradable materials and edible films, developed from proteins, polysaccharides, and lipids, present eco-friendly alternatives that minimize migration risks and environmental impact. Additionally, bioactive components, such as essential oils and phenolic compounds, can enhance antimicrobial and antioxidant properties, offering safer solutions for food preservation and extending their shelf life (Trajkovska Petkoska et al., 2021a; Trajkovska Petkoska, Daniloski, D'Cunha, et al., 2021).

Rising crude oil prices, increasing consumer demand for eco-friendly products, the growth of convenience packaging, and the development of composting facilities are driving the adoption of biobased/biodegradable packaging (Bauer et al., 2024). As production scales up, costs decrease, further boosting its appeal. Biopolymers like starch, cellulose, chitosan, polylactic acid (PLA), polycaprolactone (PCL), and polyhydroxybutyrate (PHB) offer benefits such as breathability, low-temperature sealability, and affordability. Current trends include blending various biopolymers (e.g., starch-PLA, starch-PCL) to produce bottles, containers, films, bags, and other packaging items. Overall, these factors are making biodegradable packaging increasingly popular (Pawar & Purwar, 2013). Thus, selecting the right feedstock material is critical, as it influences the final polymer's properties. Lignocellulosic fibers from plants like pineapple, sisal, and jute strengthen biodegradable/biobased polymers, while agricultural waste (e.g., postharvest residues, vegetable-processing by-products) can serve as both polymer precursors and sources of plasticizers that enhance flexibility as well as an active component in novel types of packaging. To improve economic viability and move toward a circular economy, second- and third-generation biopolymers rely on alternative inputs—such as nonfood crops, residues of first-generation feedstock (e.g., corn stover, sugarcane bagasse, wood, palm fruit bunches, switchgrass), algae, and even industrial or municipal waste (Jha et al., 2024).

However, they often lack the strength, flexibility, and barrier qualities of traditional plastics (Porta et al., 2020). In addition, biodegradable polymers still face obstacles in practical packaging, particularly their insufficient gas and moisture barriers that limit their use in food packaging. Strategies like modifying polymer chains, adjusting crystallinity, using melt blending and multilayer coextrusion, employing nanotechnology, and applying surface coatings have all been explored to address these challenges (Wu et al., 2021). Nowadays, several biobased polymers are used either as standalone materials or in blends/compositions with other biobased or nonbiobased materials in food applications. Nevertheless, their native form often requires modification to improve its structure and physicochemical properties for broader usability. For instance, while cellulose presents strong mechanical and thermal stability, its water sensitivity often needs improvement. In fact, cellulose acetate binds less water and swells less than unmodified cellulose, whereas cellulose sulfate has antimicrobial effects (Bauer et al., 2024). The most used biobased materials along with their main sources and key properties, are given in Table 31.2.

31.6.2 Nanocomposites

Concerns about food spoilage due to inadequate barrier and active properties have driven the incorporation of nanotechnology into food packaging. Common nanomaterials such as zinc oxide (ZnO), TiO_2, SiO_2, Ag, and halloysite nanoclays are used; they improve the properties of composite materials due to their small size and/or proper distribution in the polymer matrix (Nath et al., 2022). Polymeric nanocomposites exhibit significantly enhanced packaging capabilities, such as reduced oxygen and/or moisture permeability, superior barrier performance, heightened resistance to solvents, improved thermal durability, advanced biological properties, antimicrobial effectiveness, fire resistance, and/or exceptional mechanical strength (Kausar, 2020). In general, nano-based packaging materials are derived from polysaccharides (starch, chitosan), proteins (whey), lipids (wax, resin), or their compositions, and then biopolymers (PLA, polyvinyl alcohol [PVA], etc.) (Trajkovska Petkoska et al., 2025; Trinh et al., 2023). Commonly used nanomaterials in food packaging, typical polymer matrices, and key benefits are given in Table 31.3.

31.6.3 Layered structures

In recent years, multilayer polymer films have become popular in flexible packaging due to their superior barrier properties, durability, transparency, sealability, and cost-effectiveness. While these films address the limitations of single-layer polymers, such as poor resistance to gases, moisture, and light, they also raise concerns about recyclability and migration of materials (Ščetar, 2021). Multilayer films combine the strengths of different polymers, enhancing shelf life and enabling features like antimicrobial and antioxidant properties. The multilayering approach integrates multiple materials or layers within a single package, each fulfilling a distinct role to improve overall functionality (Robertson, 2016; Wang et al., 2022). For instance, the outer layer may deliver structural durability and protection against environmental influences. Meanwhile, the core layer can incorporate active substances, such as antioxidants, antimicrobials, or moisture regulators, which are gradually released to preserve the food (Siracusa, 2012). The innermost layer, which comes into direct contact with the food, can be engineered to provide a regulated release of beneficial compounds (Akhrib et al., 2024), or to protect against migration issues.

31.6.4 Surface modification: coatings

Applying coatings to conventional packaging substrates has emerged as a practical strategy for reducing the migration of potentially harmful substances into food. By adding a functional layer, manufacturers create a barrier between the packaging material and the product, curtailing the movement of plasticizers, residual monomers, or other undesired compounds (Daniloski et al., 2019). Coating technology is key to boosting barrier properties, especially for biodegradable substrates. Hydrophobic coatings (e.g., lipids, waxes, PHAs) excel at moisture resistance, while proteins and polysaccharides offer better gas barriers. Protein-based coatings generally provide stronger mechanical properties, but good adhesion is essential to avoid delamination (Kusuma et al., 2024). Coatings can also deliver antimicrobial or antioxidant functions via components like essential oils, chitosan, pectin, or certain nanoparticles. However, issues related to adhesion, heat sealability, and potential coating migration into food still limit widespread application (Jahangiri et al., 2024).

TABLE 31.2 Commonly used biobased materials in food packaging.

Material	Main source	Key properties	References
Cellulose and derivatives	Wood pulp, cotton linters	Good mechanical strength; transparent and flexible; biodegradable; moderate oxygen barrier, poor moisture barrier	Romão et al. (2022)
Starch-based films	Corn, potato, tapioca	Good film-forming ability; biodegradable; poor barrier to moisture	Żołek-Tryznowska and Kałuża (2021)
Polylactic acid	Fermented plant sugars (corn, sugarcane)	Biodegradable; transparent; good mechanical strength, though can be brittle; moderate oxygen and moisture barrier	O'Loughlin et al. (2023), Gerometta et al. (2019)
Polyhydroxyalkanoates (PHAs) (e.g., PHB)	Microbial fermentation of sugars, oils, or waste biomass	Fully biodegradable; high crystallinity; good barrier properties	Ragaert et al. (2019)
Chitosan	Chitin from crustacean shells or fungi	Antimicrobial, good film-forming ability; excellent oxygen barrier, moderate moisture barrier	Li et al. (2022), Liu et al. (2022)
Protein-based films (e.g., gelatin, whey, soy protein)	Animal or plant proteins	Biodegradable; good oxygen barrier; sensitive to moisture	Zhang and Rhim (2022)
Alginate	Brown seaweed	Gel-forming ability, biodegradable, antimicrobial; moderate oxygen barrier, poor moisture barrier	Kontominas (2020), Theagarajan et al. (2019)
Pectin	Citrus peels, apple pomace	Perfect film-forming capability, biodegradable, edible; excellent oxygen barrier capacity, high water permeability	Younis et al. (2022)
Lignin-based materials	Lignocellulosic biomass	Tensile strength; antioxidant, antibacterial, and UV barrier; good thermal properties	Zubair et al. (2024)
Agar-based materials	Red seaweed	Biodegradable, edible, brittle; high moisture permeability and poor thermal stability	Mostafavi and Zaeim (2020)
Polybutylene succinate	Fermentation of sugar-based monomers	Thermal stability and a high crystallization; biodegradable; moderate oxygen and moisture barrier	Aliotta et al. (2022)
Carnauba, candelilla, rice bran wax	Plant waxes	Water repellent; biodegradable, edible; excellent moisture barrier	Pashova (2023)

31.6.5 Active and intelligent packaging options in context of beneficial migration

Packaging technologies have evolved in response to consumer demand for more advanced and creative solutions that not only protect food but also enhance its quality and safety (Daniloski et al., 2021). Unlike conventional packaging, which serves primarily as a barrier against external conditions, active and intelligent packaging systems engage in continuous interaction with the food product and its environment, leading to what can be termed as "beneficial migration." This beneficial or intentional migration, unlike the "negative" migration, when harmful chemical compounds transfer from the packaging to food, involves the purposeful movement of active components to or from the packaged product, helping

TABLE 31.3 Commonly used nanomaterials in food packaging.

Nanomaterial	Polymer matrix	Key benefits	Typical foodpackaging applications	References
Zinc oxide	Polylactic acid (PLA), PHB, PE, PP, etc.	Antimicrobial properties; improved mechanical strength; enhanced barrier against gases and moisture	Fresh produce packaging, dairy products, bakery items	Salsabila and Nandiyanto (2023), Smaoui et al. (2023)
Titanium dioxide (TiO_2)	PLA, PCL, PE, polyethylene terephthalate	UV protection; antibacterial effects; improved tensile strength and elasticity	Light-sensitive products (e.g., dairy), extended shelf-life packaging	Zhang and Rhim (2022)
Silicon dioxide (SiO_2)	PLA, PHB, polybutylene adipate-co-terephthalate, PE, etc.	Enhanced mechanical properties; reduced water vapor and gas permeability; improved clarity	Wide range of food packaging	Koreshkov et al. (2024)
Silver (Ag)	PLA, PE, PP	Strong antimicrobial activity; enhanced shelf life; increased product safety; good moisture barrier; moderate gas barrier	Meat, poultry, and fish packaging	Simbine et al. (2019), Trotta et al. (2023)
Halloysite nanoclay	PLA, PCL, starch-based polymers	Enhanced mechanical strength and stiffness; improved barrier properties to O_2 and CO_2; possible controlled release of active agents	High-barrier films, active packaging for fresh and processed foods	Deshmukh et al. (2023)

maintain and/or improve the overall food quality during extended shelf time. In most cases, terms like migration, release, or diffusion have been used by many authors to determine the quantity of the available active compounds (e.g., antioxidants, antimicrobials, or bioactives) that can be transferred to the food surface; their release from the packaging material to the packaged food is designed to be fast/slow, but with kinetics that compromise the food quality and safety (García-Anaya et al., 2025). In general, active packaging integrates (natural) active agents into eco-friendly materials (like a reservoir system) that either release or absorb specific compounds. These agents may migrate in a controlled way into the headspace or onto the surface of the food, effectively delaying oxidation, controlling respiration rates, limiting microbial growth, or managing moisture migration (Mane, 2016). The release rate from the reservoir system depends upon the nature, thickness, area, and the permeability of the packaging material as well as the nature of the active compound that migrates and its diffusion properties (Li et al., 2024; Westlake et al., 2022). Depending on the system's function, active packaging can be categorized into "releasing" or "absorbing" types. Releasing systems disperse active substances such as antioxidants, antimicrobials, or flavor enhancers, thereby safeguarding quality or imparting desirable characteristics to the packaged product. Absorbing (scavenging) systems capture unwanted substances like excess oxygen, moisture, or ethylene that accumulate over time, thereby preserving the product's quality (Akhrib et al., 2024). These active systems can be developed through various methods, including incorporation, coating, immobilization, or

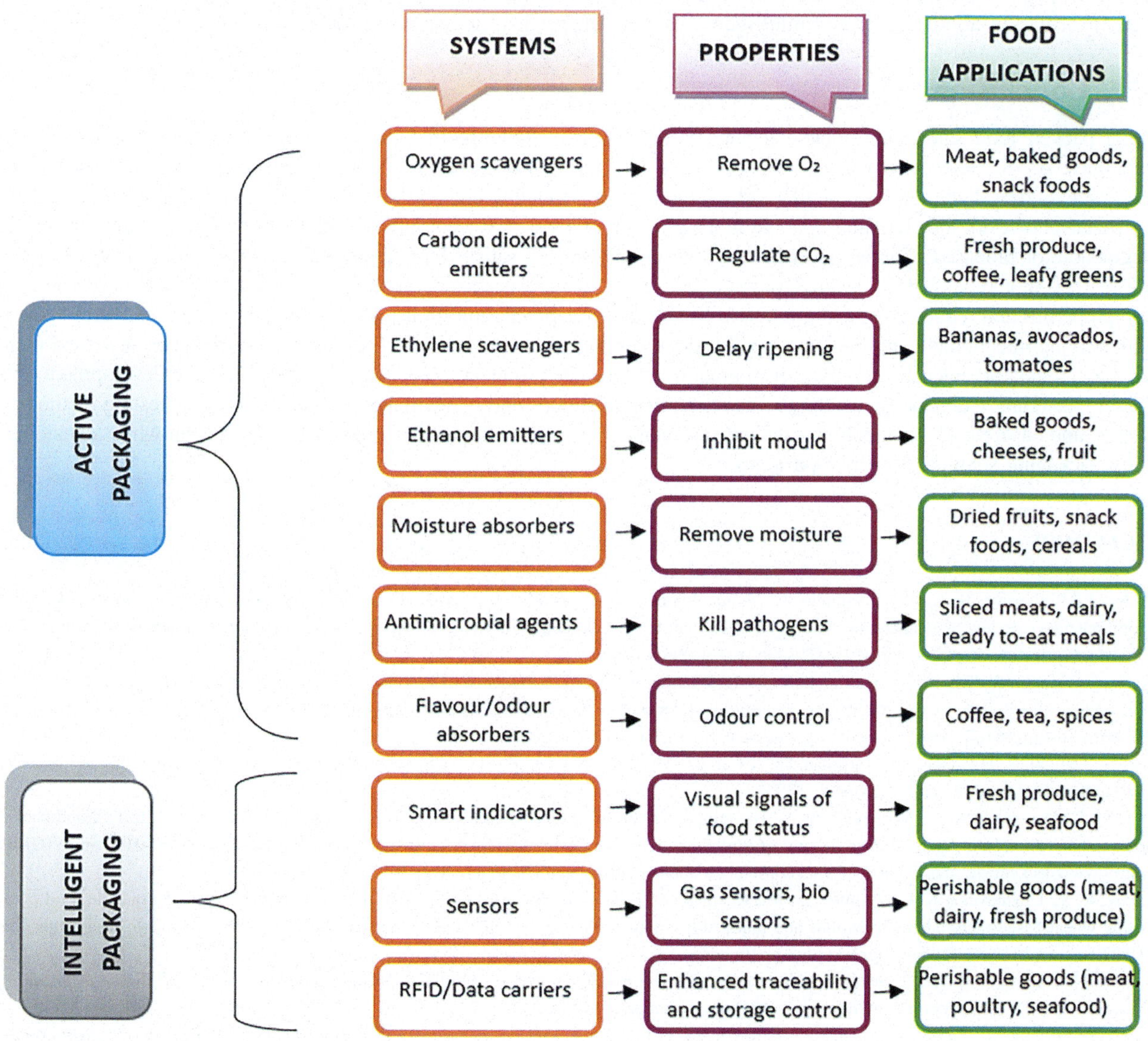

FIGURE 31.4 Active and intelligent packaging systems in context of beneficial migration and its monitoring.

surface modification of the packaging material (Vilela et al., 2018). The intentional release or scavenging of specific compounds thus constitutes a strategic form of beneficial migration, one that is carefully designed to prolong shelf life and maintain food safety.

On the other hand, intelligent packaging focuses on monitoring the condition of the product or its surrounding environment, relaying real-time information to both producers and consumers without directly interacting with the food (Müller & Schmid, 2019). By employing indicators, sensors, and data carriers (e.g., barcodes or radio frequency identification (RFID) tags), intelligent packaging tracks factors such as temperature, freshness, and integrity (Zhao et al., 2023). Although these systems generally do not engage in chemical or microbial processes, their capability to signal adverse changes or deterioration complements active packaging's protective role. Through user-friendly cues like color change in an indicator, intelligent packaging can facilitate timely interventions, thereby indirectly ensuring that beneficial migration processes within the packaging environment remain effective (Abekoon et al., 2024; Thirupathi Vasuki et al., 2023).

By merging active and intelligent strategies in food-packaging options, packaging can both act (through controlled release or absorption of substances) and inform (through real-time indicators and data), creating a powerful, dynamic defence against spoilage and ensuring the highest quality experience for consumers. This synergistic approach to beneficial migration is shaping the future of food packaging—sustainable, safe, and smart (Versino et al., 2023). Fig. 31.4 depicts information about active and intelligent systems, properties, and food applications.

31.7 Conclusions

Keeping packed food safe, with suitable quality, is an ongoing challenge that involves understanding materials and related technologies. As discussed, migration issues depend on the type of packaging, the kind of the packed food, and environmental factors, such as heat, light, and storage time. Plastic and metal packaging can release substances, including phthalates, BPA, and heavy metals; however, they remain important in the global food market. Recent innovations focus on safer, greener materials like biodegradable or biobased options; nevertheless, they usually do not satisfy barrier properties as packaging materials for food; therefore, they should be blended, layered, or made in forms of composites/nanocomposites, or suitably coated in order to improve barrier properties and reduce chemical migration. Yet, high costs, processing limits, and recyclability issues can slow their wider use. Identifying intentionally and nonintentionally added substances also remains a challenge, requiring better analytical methods. Worldwide, regulatory agencies set rules for migration limits, guide production practices, and encourage the development of better packaging materials. Collaboration among researchers, manufacturers, and policymakers is crucial to update testing methods, align standards, and ensure packaging is both safe and sustainable. This joint effort will protect consumers health while advancing eco-friendly packaging solutions for the food industry.

References

Abekoon, T., Buthpitiya, B. L. S. K., Sajindra, H., Samarakoon, E. R. J., Jayakody, J. A. D. C. A., Kantamaneni, K., & Rathnayake, U. (2024). A comprehensive review to evaluate the synergy of intelligent food packaging with modern food technology and artificial intelligence field. *Discover Sustainability, 5*(1). https://doi.org/10.1007/s43621-024-00371-7.

Agarwal, A., Gandhi, S., Tripathi, A. D., Iammarino, M., & Homroy, S. (2022). Analysis of Bisphenol A migration from microwaveable polycarbonate cups into coffee during microwave heating. *International Journal of Food Science and Technology, 57*(12), 7477–7485. Available from https://doi.org/10.1111/ijfs.16103, http://onlinelibrary.wiley.com/journal/10.1111/(ISSN)1365-2621.

Akhrib, S., Djellali, S., Haddaoui, N., Karimian, D., & Carraro, M. (2024). Biocomposites and poly (lactic acid) in active packaging: A review of current research and future directions. *Polymers, 17*(1).

Alamri, M. S., Qasem, A. A. A., Mohamed, A. A., Hussain, S., Ibraheem, M. A., Shamlan, G., Alqah, H. A., & Qasha, A. S. (2021). Food packaging's materials: A food safety perspective. *Journal of Biological Sciences, 28*(8), 4490–4499. Available from https://doi.org/10.1016/j.sjbs.2021.04.047, http://www.ksu.edu.sa/sites/Colleges/CollegeofScinces/BotanyDepartment/SJBS/Default.aspx.

Alibekov, R. S., Urazbayeva, K. U., Azimov, A. M., Rozman, A. S., Hashim, N., & Maringgal, B. (2024). Advances in biodegradable food packaging using wheat-based materials: Fabrications and innovations, applications, potentials, and challenges. *Foods, 13*(18), Available from https://doi.org/10.3390/foods13182964, www.mdpi.com/journal/foods.

Alin, J., & Hakkarainen, M. (2012). Migration from polycarbonate packaging to food simulants during microwave heating. *Polymer Degradation and Stability, 97*(8), 1387–1395. https://doi.org/10.1016/j.polymdegradstab.2012.05.017.

Aliotta, L., Seggiani, M., Lazzeri, A., Gigante, V., & Cinelli, P. (2022). A brief review of poly (butylene succinate) (PBS) and its main copolymers: Synthesis, blends, composites, biodegradability, and applications. *Polymers, 14*(4), 844. https://doi.org/10.3390/polym14040844.

Andrady, A. L., Heikkilä, A. M., Pandey, K. K., Bruckman, L. S., White, C. C., Zhu, M., & Zhu, L. (2023). Effects of UV radiation on natural and synthetic materials. *Photochemical & Photobiological Sciences, 22*(5), 1177–1202. https://doi.org/10.1007/s43630-023-00377-6.

Ardic, M., Kahve, H. I., & Duran, A. (2015). Chemical migration in food technology. *Academic Journal of Science, 4*(02), 163–168.

Arvanitoyannis, I. S., & Kotsanopoulos, K. V. (2014). Migration phenomenon in food packaging. Food-package interactions, mechanisms, types of migrants, testing and relative legislation-A review. *Food and Bioprocess Technology, 7*(1), 21–36. https://doi.org/10.1007/s11947-013-1106-8.

Balaji, J., & Mayuri, T. (2024). Migration study of food packaging material. In *Food Packaging* (pp. 37–41). Selfypage Developers Pvt. Ltd. https://doi.org/10.58532/v3bcagp1ch2

Bauer, M. G., Henkel, F., Gürer, U., & Lieleg, O. (2024). Bio-based and degradable food packaging materials: Where are they? *Advanced Materials Interfaces*. Available from https://doi.org/10.1002/admi.202400645, http://onlinelibrary.wiley.com/journal/10.1002/(ISSN)2196-7350.

Bhunia, K., Sablani, S. S., Tang, J., & Rasco, B. (2013). Migration of chemical compounds from packaging polymers during microwave, conventional heat treatment, and storage. *Comprehensive Reviews in Food Science and Food Safety, 12*(5), 523–545. https://doi.org/10.1111/1541-4337.12028.

Bianchi, E., Papadopoulos, L., Soccio, M., Siracusa, V., Gazzano, M., Robert, T., Bikiaris, D. N., & Lotti, N. (2024). Mechanical properties, gas permeability and biodegradation mechanism of biobased poly(ester amide)s from 2,5-furandicarboxylic acid and amido diols for sustainable food packaging. *Polymer Degradation and Stability, 230*. Available from https://doi.org/10.1016/j.polymdegradstab.2024.111049, https://www.sciencedirect.com/science/journal/01413910.

Blankenship, A. L., & Coady, K. (2005). *Bisphenol a encyclopedia of toxicology*. United States: Elsevier, 314–317. https://doi.org/10.1016/B0-12-369400-0/00139-3, http://doi.org/10.1016/B0-12-369400-0/00139-3.

Bomfim, M. V. J., Zamith, H. P. S., & Abrantes, S. M. P. (2011). Migration of ε-caprolactam residues in packaging intended for contact with fatty foods. *Food Control, 22*(5), 681–684. https://doi.org/10.1016/j.foodcont.2010.09.017.

Caner, C. (2018). Fundamentals of the sorption (scalping) phenomena in packaged foods. *Reference Module in Food Science,* 1–12.

Chemat, A., Song, M., Li, Y., & Fabiano-Tixier, A.-S. (2023). Shade of innovative food processing techniques: Potential inducing factors of lipid oxidation. *Molecules (Basel, Switzerland), 28*(24), 8138. https://doi.org/10.3390/molecules28248138.

Cheng, J., Gao, R., Zhu, Y., & Lin, Q. (2024). Applications of biodegradable materials in food packaging: A review. *Alexandria Engineering Journal, 91*, 70–83. https://doi.org/10.1016/j.aej.2024.01.080.

Coniglio, M. A., Fioriglio, C., & Laganà, P. (2020). *Non-intentionally added substances*. Springer Science and Business Media LLC, 43–58. http://doi.org/10.1007/978-3-030-39134-8_4.

Daniloski, D., Petkoska, A. T., Galić, K., Ščetar, M., Kurek, M., Vaskoska, R., Kalevska, T., & Nedelkoska, D. N. (2019). The effect of barrier properties of polymeric films on the shelf-life of vacuum packaged fresh pork meat. *Meat Science, 158*. Available from https://doi.org/10.1016/j.meatsci.2019.107880, http://www.elsevier.com/locate/meatsci.

Daniloski, D., Petkoska, A. T., Lee, N. A., Bekhit, A. E. D., Carne, A., Vaskoska, R., & Vasiljevic, T. (2021). Active edible packaging based on milk proteins: A route to carry and deliver nutraceuticals. *Trends in Food Science and Technology, 111*, 688–705. Available from https://doi.org/10.1016/j.tifs.2021.03.024, http://www.elsevier.com/wps/find/journaldescription.cws_home/601278/description#description.

de Paula, L. C. P., & Alves, C. (2024). Food packaging and endocrine disruptors. *Jornal de Pediatria, 100*, S40–S47. https://doi.org/10.1016/j.jped.2023.09.010.

Deshmukh, R. K., Hakim, L., & Gaikwad, K. K. (2023). Active packaging materials. *Current Food Science and Technology Reports, 1*(2), 123–132. https://doi.org/10.1007/s43555-023-00004-6.

Deshwal, G. K., & Panjagari, N. R. (2020). Review on metal packaging: Materials, forms, food applications, safety and recyclability. *Journal of Food Science and Technology, 57*(7), 2377–2392. Available from https://doi.org/10.1007/s13197-019-04172-z, http://www.springerlink.com/content/121580/.

Dhuey, E., Castada, H. Z., Barringer, S., Joseph, J., Hadad, C. M., Ruffley, K., & Pascall, M. A. (2021). Heat-induced compounds development in processed tomato and their influence on corrosion initiation in metal food cans. *Food Science & Nutrition, 9*(8), 4134–4145. https://doi.org/10.1002/fsn3.2376.

Dhuey, E., Ruffley, K., & Pascall, M. A. (2022). Characterization of morphological changes to an epoxy-based polymer used as a corrosion-preventative lining in retorted canned tomatoes. *Journal of Food Engineering, 313*, 110753. https://doi.org/10.1016/j.jfoodeng.2021.110753.

Duncan, S. E., & Chang, H. H. (2012). Implications of light energy on food quality and packaging selection. *Advances in Food and Nutrition Research, 67*, 25–73. Available from https://doi.org/10.1016/B978-0-12-394598-3.00002-2, http://www.elsevier.com/wps/find/bookdescription.cws_home/704624/description#description.

Đurić, B., Kartalović, B., Habschied, K., Novakov, N., Vranešević, J., Brkić, B., & Mastanjević, K. (2024). Effects of packaging material type, storage time and lipid content on phthalate migration in smoked fish meat. *Applied Sciences, 14*(4), 1660. https://doi.org/10.3390/app14041660.

Edwards, L., McCray, N. L., VanNoy, B. N., Yau, A., Geller, R. J., Adamkiewicz, G., & Zota, A. R. (2022). Phthalate and novel plasticizer concentrations in food items from U.S. fast food chains: A preliminary analysis. *Journal of Exposure Science and Environmental Epidemiology, 32*(3), 366–373. Available from https://doi.org/10.1038/s41370-021-00392-8, http://www.nature.com/jes/index.html.

Eslami, Z., Elkoun, S., Robert, M., & Adjallé, K. (2023). A review of the effect of plasticizers on the physical and mechanical properties of alginate-based films. *Molecules (Basel, Switzerland), 28*(18), 6637. https://doi.org/10.3390/molecules28186637.

Etxabide, A., Young, B., Bremer, P. J., & Kilmartin, P. A. (2022). Non-permanent primary food packaging materials assessment: Identification, migration, toxicity, and consumption of substances. *Comprehensive Reviews in Food Science and Food Safety, 21*(5), 4130–4145. Available from https://doi.org/10.1111/1541-4337.13001, http://onlinelibrary.wiley.com/journal/10.1111/(ISSN)1541-4337.

Fang, X., & Vitrac, O. (2017). Predicting diffusion coefficients of chemicals in and through packaging materials. *Critical Reviews in Food Science and Nutrition, 57*(2), 275–312. Available from https://doi.org/10.1080/10408398.2013.849654, www.tandf.co.uk/journals/titles/10408398.asp.

Farrell, R., Cortese, Y. J., Devine, D. M., Gately, N., Rueda, M., Rodriguez, L., & Pezzoli, R. (2024). The function and properties of common food packaging materials and their suitability for reusable packaging: The transition from a linear to circular economy. *Current Research in Green and Sustainable Chemistry, 9*. Available from https://doi.org/10.1016/j.crgsc.2024.100429, https://www.sciencedirect.com/science/journal/26660865.

Favvas, E. P., Katsaros, F. K., Papageorgiou, S. K., Sapalidis, A. A., & Mitropoulos, A. C. (2017). A review of the latest development of polyimide based membranes for CO_2 separations. *Reactive and Functional Polymers, 120*, 104–130. https://doi.org/10.1016/j.reactfunctpolym.2017.09.002.

Franz, R., & Welle, F. (2022). Recycling of post-consumer packaging materials into new food packaging applications—Critical review of the European approach and future perspectives. *Sustainability, 14*(2), 824. https://doi.org/10.3390/su14020824.

Gajdoš, J., Galić, K., Kurtanjek, Ž., & Ciković, N. (2000). Gas permeability and DSC characteristics of polymers used in food packaging. *Polymer Testing, 20*(1), 49–57. https://doi.org/10.1016/s0142-9418(99)00078-1.

Garba, A. I. (2023). Food preservation packaging. In *Food processing and packaging technologies – Recent advances*.

García Ibarra, V., Rodríguez Bernaldo de Quirós, A., Paseiro Losada, P., & Sendón, R. (2018). Identification of intentionally and non-intentionally added substances in plastic packaging materials and their migration into food products. *Analytical and Bioanalytical Chemistry, 410*(16), 3789–3803. https://doi.org/10.1007/s00216-018-1058-y.

García-Anaya, M. C., Sepúlveda, D. R., & Acosta-Muñiz, C. H. (2025). Contributing factors to the migration of antimicrobials in active packaging films. *Food Research International, 200*. Available from https://doi.org/10.1016/j.foodres.2024.115514, https://www.sciencedirect.com/science/journal/09639969.

Gerometta, M., Rocca-Smith, J. R., Domenek, S., & Karbowiak, T. (2019). *Physical and chemical stability of PLA in food packaging*. Elsevier BV. http://doi.org/10.1016/b978-0-08-100596-5.22471-2.

Geueke, B., Groh, K. J., Maffini, M. V., Martin, O. V., Boucher, J. M., Chiang, Y. T., Gwosdz, F., Jieh, P., Kassotis, C. D., Łańska, P., Myers, J. P., Odermatt, A., Parkinson, L. V., Schreier, V. N., Srebny, V., Zimmermann, L., Scheringer, M., & Muncke, J. (2023). Systematic evidence on migrating and extractable food contact chemicals: Most chemicals detected in food contact materials are not listed for use. *Critical Reviews in Food Science and Nutrition, 63*(28), 9425–9435. Available from https://doi.org/10.1080/10408398.2022.2067828, https://www.tandfonline.com/loi/bfsn20.

Gnanasekharan, V., & Floros, J. D. (1997). Migration and sorption phenomena in packaged foods. *Critical Reviews in Food Science and Nutrition, 37*(6), 519–559. Available from https://doi.org/10.1080/10408399709527788, www.tandf.co.uk/journals/titles/10408398.asp.

Grob, K. (2018). Mineral oil hydrocarbons in food: A review. *Food Additives & Contaminants: Part A, 35*(9), 1845–1860. https://doi.org/10.1080/19440049.2018.1488185.

Groh, K. J., Backhaus, T., Carney-Almroth, B., Geueke, B., Inostroza, P. A., Lennquist, A., Leslie, H. A., Maffini, M., Slunge, D., Trasande, L., Warhurst, A. M., & Muncke, J. (2019). Overview of known plastic packaging-associated chemicals and their hazards. *Science of the Total Environment, 651*, 3253–3268. Available from https://doi.org/10.1016/j.scitotenv.2018.10.015, www.elsevier.com/locate/scitotenv.

Guazzotti, V., Hendrich, V., Gruner, A., Fiedler, D., Störmer, A., & Welle, F. (2022). Migration of styrene in yogurt and dairy products packaged in polystyrene: Results from market samples. *Foods, 11*(14), 2120. https://doi.org/10.3390/foods11142120.

Guo, X., Dai, H., & He, L. (2024). Migration testing of microplastics from selected water and food containers by Raman microscopy. *Journal of Hazardous Materials, 462*, 132798. https://doi.org/10.1016/j.jhazmat.2023.132798.

Gupta, R. K., Pipliya, S., Karunanithi, S., Eswaran U, G. M., Kumar, S., Mandliya, S., Srivastav, P. P., Suthar, T., Shaikh, A. M., Harsányi, E., & Kovács, B. (2024). Migration of chemical compounds from packaging materials into packaged foods: Interaction, mechanism, assessment, and regulations. *Foods, 13*(19), Available from https://doi.org/10.3390/foods13193125, www.mdpi.com/journal/foods.

Hahladakis, J. N., Velis, C. A., Weber, R., Iacovidou, E., & Purnell, P. (2018). An overview of chemical additives present in plastics: Migration, release, fate and environmental impact during their use, disposal and recycling. *Journal of Hazardous Materials, 344*, 179–199. Available from https://doi.org/10.1016/j.jhazmat.2017.10.014, www.elsevier.com/locate/jhazmat.

Hollender, J., Schymanski, E. L., Singer, H. P., & Ferguson, P. L. (2017). Nontarget screening with high resolution mass spectrometry in the environment: Ready to go? *Environmental Science and Technology, 51*(20), 11505–11512. Available from https://doi.org/10.1021/acs.est.7b02184, http://pubs.acs.org/journal/esthag.

Jahangiri, F., Mohanty, A. K., & Misra, M. (2024). Sustainable biodegradable coatings for food packaging: Challenges and opportunities. *Green Chemistry, 26*(9), 4934–4974. Available from https://doi.org/10.1039/d3gc02647g, http://pubs.rsc.org/en/journals/journal/gc.

Jha, S., Akula, B., Enyioma, H., Novak, M., Amin, V., & Liang, H. (2024). Biodegradable biobased polymers: A review of the state of the art, challenges, and future directions. *Polymers, 16*(16), 2262. https://doi.org/10.3390/polym16162262.

Kato, L. S., & Conte-Junior, C. A. (2021). Safety of plastic food packaging: The challenges about non-intentionally added substances (NIAS) discovery, identification and risk assessment. *Polymers, 13*(13), Available from https://doi.org/10.3390/polym13132077, https://www.mdpi.com/2073-4360/13/13/2077/pdf.

Kausar, A. (2020). A review of high performance polymer nanocomposites for packaging applications in electronics and food industries. *Journal of Plastic Film & Sheeting, 36*(1), 94–112. https://doi.org/10.1177/8756087919849459.

Khanniri, E., Bayanati, M., Koushki, M. R., Ferdosi, R., Sohrabvandi, S., Esmaeili, S., Akbari, M. E., & Forouhar, P. (2024). Migration of Bisphenol A and several phthalate acid contaminants into bottled drinking water: Influence of storage conditions and their health risks. *International Journal of Environmental Analytical Chemistry, 104*(20), 8672–8689. Available from https://doi.org/10.1080/03067319.2023.2209869, www.tandf.co.uk/journals/titles/03067319.asp.

Kontominas, M. G. (2020). Use of alginates as food packaging materials. *Foods, 9*(10), 1440. https://doi.org/10.3390/foods9101440.

Kontominas, M. G., Prodromidis, M. I., Paleologos, E. K., Badeka, A. V., & Georgantelis, D. (2006). Investigation of fish product-metal container interaction using scanning electron microscopy-X-ray microanalysis. *Food Chemistry, 98*(2), 225–230. https://doi.org/10.1016/j.foodchem.2005.06.004.

Koreshkov, M., Antreich, S. J., Bismarck, A., Fritz, I., Reimhult, E., Takatsuna, Y., & Zirbs, R. (2024). Sustainable food packaging using modified SiO_2 nanofillers in biodegradable polymers. *Materials Chemistry Frontiers, 8*(16), 2754–2763. Available from https://doi.org/10.1039/d4qm00206g, rsc.li/frontiers-materials.

Kusuma, H. S., Sabita, A., Putri, N. A., Azliza, N., Illiyanasafa, N., Darmokoesoemo, H., Amenaghawon, A. N., & Kurniawan, T. A. (2024). Waste to wealth: Polyhydroxyalkanoates (PHA) production from food waste for a sustainable packaging paradigm. *Food Chemistry: Molecular Sciences, 9*. Available from https://doi.org/10.1016/j.fochms.2024.100225, https://www.sciencedirect.com/science/journal/26665662.

Li, H., Aguirre-Villegas, H. A., Allen, R. D., Bai, X., Benson, C. H., Beckham, G. T., Bradshaw, S. L., Brown, J. L., Brown, R. C., Cecon, V. S., Curley, J. B., Curtzwiler, G. W., Dong, S., Gaddameedi, S., García, J. E., Hermans, I., Kim, M. S., Ma, J., Mark, L. O., ... Huber, G. W. (2022). Expanding plastics recycling technologies: Chemical aspects, technology status and challenges. *Green Chemistry, 24*(23), 8899–9002. Available from https://doi.org/10.1039/d2gc02588d, http://pubs.rsc.org/en/journals/journal/gc.

Li, J., Sun, H., & Weng, Y. (2024). Natural extracts and their applications in polymer-based active packaging: A review. *Polymers, 16*(5), 625. https://doi.org/10.3390/polym16050625.

Liu, T., Li, J., Tang, Q., Qiu, P., Gou, D., & Zhao, J. (2022). Chitosan-based materials: An overview of potential applications in food packaging. *Foods, 11*(10), 1490. https://doi.org/10.3390/foods11101490.

Mahmud, M. Z. A., Mobarak, M. H., & Hossain, N. (2024). Emerging trends in biomaterials for sustainable food packaging: A comprehensive review. *Heliyon, 10*(1), Available from https://doi.org/10.1016/j.heliyon.2024.e24122, https://www.sciencedirect.com/science/journal/24058440.

Mane, K. A. (2016). A Review on active packaging: An innovation in food packaging. *International Journal of Environment, Agriculture and Biotechnology, 1*(3), 544–549. https://doi.org/10.22161/ijeab/1.3.35.

Mertoğlu-Elmas, G., & Çınar, G. (2018). Toxic metals in paper and paperboard food packagings. *BioResources, 13*(4), 7560–7580. Available from https://doi.org/10.15376/BIORES.13.4.7560-7580, http://www.ncsu.edu/bioresources/Back_Issues.htm.

Mostafavi, F. S., & Zaeim, D. (2020). Agar-based edible films for food packaging applications - A review. *International Journal of Biological Macromolecules, 159*, 1165–1176. Available from https://doi.org/10.1016/j.ijbiomac.2020.05.123, www.elsevier.com/locate/ijbiomac.

Mukhi, S., M., S., Reghupathi, D. R., Manjrekar, & Hegde, S. (2024). Effect of food packaging on human health – A review. *Journal of Chemical Health Risks, 14*(1), 2089–2097.

Müller, P., & Schmid, M. (2019). Intelligent packaging in the food sector: A brief overview. *Foods, 8*(1), 16. https://doi.org/10.3390/foods8010016.

Muncke, J. (2016). *Chemical migration from food packaging to food*. Elsevier BV. https://doi.org/10.1016/b978-0-08-100596-5.03311-4.

Muncke, J., Backhaus, T., Geueke, B., Maffini, M. V., Martin, O. V., Myers, J. P., Soto, A. M., Trasande, L., Trier, X., & Scheringer, M. (2017). Scientific challenges in the risk assessment of food contact materials. *Public Health Services, US Dept of Health and Human Services, Switzerland Environmental Health Perspectives, 125*(9), Available from https://doi.org/10.1289/EHP644, https://ehp.niehs.nih.gov/wp-content/uploads/2017/09/EHP644.alt_.pdf.

Muncke, J., Myers, J. P., Scheringer, M., & Porta, M. (2014). Food packaging and migration of food contact materials: Will epidemiologists rise to the neotoxic challenge? *Journal of Epidemiology and Community Health, 68*(7), 592–594. Available from https://doi.org/10.1136/jech-2013-202593, http://jech.bmj.com/content/68/7/592.full.pdf.

Muobom, S. S., Umar, A.-M. S., Soongseok, Y., & Brolin, A.-P. (2020). A review on plasticizers and eco-friendly bioplasticizers: Biomass sources and market. *International Journal of Engineering Research & Technology, 9*(5), 1138–1144.

Muzeza, C., Ngole-Jeme, V., & Msagati, T. A. M. (2023). The mechanisms of plastic food-packaging monomers' migration into food matrix and the implications on human health. *Foods, 12*(18), Available from https://doi.org/10.3390/foods12183364, www.mdpi.com/journal/foods.

Nath, D., Chetri, R., Santhosh, R., & Sarkar, P. (2022). Nanocomposites in food packaging. In *Food, medical, and environmental applications of nanomaterials*. India: Elsevier, 167–203. http://doi.org/10.1016/B978-0-12-822858-6.00007-8, https://www.sciencedirect.com/book/9780128228586.

Nerin, C. (2016). *Plastics and polymers for food packaging manufacturing*. Elsevier BV. https://doi.org/10.1016/b978-0-08-100596-5.03195-4.

Nerin, C., Alfaro, P., Aznar, M., & Domeño, C. (2013). The challenge of identifying non-intentionally added substances from food packaging materials: A review. *Analytica Chimica Acta, 775*, 14–24. https://doi.org/10.1016/j.aca.2013.02.028.

Nerín, C., Bourdoux, S., Faust, B., Gude, T., Lesueur, C., Simat, T., Stoermer, A., Van Hoek, E., & Oldring, P. (2022). Guidance in selecting analytical techniques for identification and quantification of non-intentionally added substances (NIAS) in food contact materials (FCMS). *Food Additives & Contaminants: Part A, 39*(3), 620–643. https://doi.org/10.1080/19440049.2021.2012599.

Nerín, C., Canellas, E., & Vera, P. (2023). *Migration of packaging and labeling components and advances in analytical methodology supporting exposure assessment*. Elsevier BV, 218–239. https://doi.org/10.1016/b978-0-12-819470-6.00005-6.

Nizamuddin, S., Baloch, A. J., Chen, C., Arif, M., & Mubarak, N. M. (2024). Bio-based plastics, biodegradable plastics, and compostable plastics: Biodegradation mechanism, biodegradability standards and environmental stratagem. *International Biodeterioration and Biodegradation, 195*. Available from https://doi.org/10.1016/j.ibiod.2024.105887, https://www.sciencedirect.com/science/journal/09648305.

O'Loughlin, J., Doherty, D., Herward, B., McGleenan, C., Mahmud, M., Bhagabati, P., Boland, A. N., Freeland, B., Rochfort, K. D., Kelleher, S. M., Fahy, S., & Gaughran, J. (2023). The potential of bio-based polylactic acid (PLA) as an alternative in reusable food containers: A review. *Sustainability, 15*(21), 15312. https://doi.org/10.3390/su152115312.

Omer, E., Bakiri, A., Hammel, Y. A., Sanders, M. J., Koster, S., & Ciclet, O. (2025). Deciphering the complexity of the chemicals in food packaging materials using molecular networks. *Food Chemistry, 462*. Available from https://doi.org/10.1016/j.foodchem.2024.140853, https://www.sciencedirect.com/science/journal/03088146.

Paraskevopoulou, D., Achilias, D. S., & Paraskevopoulou, A. (2012). Migration of styrene from plastic packaging based on polystyrene into food simulants. *Polymer International, 61*(1), 141–148. https://doi.org/10.1002/pi.3161.

Pashova, S. (2023). Application of plant waxes in edible coatings. *Coatings, 13*(5), 911. https://doi.org/10.3390/coatings13050911.

Pawar, P., & Purwar, A. H. (2013). Biodegradable polymers in food packaging. *American Journal of Engineering Research, 2*(5), 151–164.

Porta, R., Sabbah, M., & Di Pierro, P. (2020). Biopolymers as food packaging materials. *International Journal of Molecular Sciences, 21*(14), 4942. https://doi.org/10.3390/ijms21144942.

Qureshi, I., Habib, M., Singh, S., Jan, K., McClements, D. J., & Bashir, K. (2024). *Introduction to food packaging*. Elsevier BV, 1–37. http://doi.org/10.1016/b978-0-323-95624-6.00001-1.

Ragaert, P., Buntinx, M., Maes, C., Vanheusden, C., Peeters, R., Wang, S., & Cardon, L. (2019).

Robertson, G. (1993). Permeability of thermoplastic polymers. *Food Packaging: Principles and Practice*, 73–110.

Robertson, G. L. (2016). Packaging and food and beverage shelf life. In *The stability and shelf life of food*. Australia: Elsevier, 77–106. http://doi.org/10.1016/B978-0-08-100435-7.00003-4, http://www.sciencedirect.com/science/book/9780081004357.

Rodríguez-Ramos, R., Santana-Mayor, A., Herrera-Herrera, A. V., Socas-Rodríguez, B., & Rodríguez-Delgado, M. A. (2024). Recent advances in the analysis of plastic migrants in food. *TrAC Trends in Analytical Chemistry, 178*, 117847. https://doi.org/10.1016/j.trac.2024.117847.

Romão, S., Bettencourt, A., & Ribeiro, I. A. C. (2022). Novel features of cellulose-based films as sustainable alternatives for food packaging. *Polymers, 14*(22), 4968. https://doi.org/10.3390/polym14224968.

Salsabila, A. R., & Nandiyanto, A. B. D. (2023). The biodegradable food packaging with the addition of ZnO nanoparticles. *Fullerene Journal of Chemistry, 7*(2), 111–119.

Sapozhnikova, Y., & Nuñez, A. (2022). Non-targeted analysis with liquid chromatography - high resolution mass spectrometry for the identification of food packaging migrants. *Journal of Chromatography. A, 1676*, 463215. https://doi.org/10.1016/j.chroma.2022.463215.

Ščetar, M., Daniloski, D., Tinjić, M., Kurek, M., & Galić, K. (2022). Effect of ultrasound treatment on barrier changes of polymers before and after exposure to food simulants. *Polymers, 14*(5), 990. https://doi.org/10.3390/polym14050990.

Ščetar, M. (2021). *Plastics packaging materials and processing for food, pharmaceuticals and cosmetics*. Wiley, 105–130. http://doi.org/10.1002/9781119825081.ch5.

Schmid, P., & Welle, F. (2020). Chemical migration from beverage packaging materials—A review. *Beverages, 6*(2), 37. https://doi.org/10.3390/beverages6020037.

Seo, I., & Shin, H. S. (2010). Determination of toluene and other residual solvents in various food packaging materials by gas chromatography/mass spectrometry (GC/MS). *Food Science and Biotechnology, 19*(6), 1429–1434. https://doi.org/10.1007/s10068-010-0204-x.

Simbine, E. O., Rodrigues, L. D. C., Lapa-Guimarães, J., Kamimura, E. S., Corassin, C. H., & Oliveira, C. A. F. D. (2019). Application of silver nanoparticles in food packages: A review. *Food Science and Technology, 39*(4), 793–802. https://doi.org/10.1590/fst.36318.

Singh, A. K., Pal, P., Pandey, B., Goksen, G., Sahoo, U. K., Lorenzo, J. M., & Sarangi, P. K. (2023). Development of "Smart Foods" for health by nanoencapsulation: Novel technologies and challenges. *Food Chemistry: X, 20*. Available from https://doi.org/10.1016/j.fochx.2023.100910, https://www.sciencedirect.com/science/journal/25901575.

Singh, R. K., & Singh, N. (2005). Quality of packaged foods. In *Innovations in food packaging*. United States: Elsevier, 24–44. http://doi.org/10.1016/B978-012311632-1/50035-8, https://www.sciencedirect.com/book/9780123116321.

Siracusa, V. (2012). Food packaging permeability behaviour: A report. *International Journal of Polymer Science, 2012*, 1–11. https://doi.org/10.1155/2012/302029.

Smaoui, S., Chérif, I., Ben Hlima, H., Khan, M. U., Rebezov, M., Thiruvengadam, M., Sarkar, T., Shariati, M. A., & Lorenzo, J. M. (2023). Zinc oxide nanoparticles in meat packaging: A systematic review of recent literature. *Food Packaging and Shelf Life, 36*. Available from https://doi.org/10.1016/j.fpsl.2023.101045, http://www.journals.elsevier.com/food-packaging-and-shelf-life/.

Stevens, S., Bartosova, Z., Völker, J., & Wagner, M. (2024). Migration of endocrine and metabolism disrupting chemicals from plastic food packaging. *Environment International, 189*, 108791. https://doi.org/10.1016/j.envint.2024.108791.

Stoica, D., Alexe, P., Ivan, A. S., Moraru, D. I., Ungureanu, C. V., Stanciu, S., & Stoica, M. (2022). *Biopolymers: Global carbon footprint and climate change*. Springer Science and Business Media LLC, 35–54. http://doi.org/10.1007/978-3-030-98392-5_3.

Theagarajan, R., Dutta, S., Moses, J. A., & Anandharamakrishnan, C. (2019). Alginates for food packaging applications. In *Alginates: Applications in the* biomedical and food industries (pp. 205–232). https://doi.org/10.1002/9781119487999.ch11

Thirupathi Vasuki, M., Kadirvel, V., & Pejavara Narayana, G. (2023). Smart packaging—An overview of concepts and applications in various food industries. *Food Bioengineering, 2*(1), 25–41. https://doi.org/10.1002/fbe2.12038.

Tirkey, A., & Upadhyay, L. S. B. (2021). Microplastics: An overview on separation, identification and characterization of microplastics. *Marine Pollution Bulletin, 170*. Available from https://doi.org/10.1016/j.marpolbul.2021.112604, www.elsevier.com/locate/marpolbul.

Trajkovska Petkoska, A., Kumar, N., & Pratibha (2025). Sustainable materials for food packaging and preservation. In *Bionanocomposites in food packaging and preservation, 149*. Elsevier doi.org/10.1016/B978-0-443-13567-5.00007-1.

Trajkovska Petkoska, A., Daniloski, D., D'Cunha, N. M., Naumovski, N., & Broach, A. T. (2021). Edible packaging: Sustainable solutions and novel trends in food packaging. *Food Research International, 140*. Available from https://doi.org/10.1016/j.foodres.2020.109981, www.elsevier.com/inca/publications/store/4/2/2/9/7/0.

Trajkovska Petkoska, A., Daniloski, D., Kumar, N., Pratibha, & Broach, A. T. (2021a). Active edible packaging: A sustainable way to deliver functional bioactive compounds and nutraceuticals. In *Environmental footprints and eco-design of products and processes*. North Macedonia: Springer, 225–264. http://doi.org/10.1007/978-981-16-4609-6_9, springer.com/series/13340.

Trajkovska Petkoska, A., Daniloski, D., Kumar, N., Pratibha, & Broach, A. T. (2021b). Biobased materials as a sustainable potential for edible packaging. In *Environmental footprints and eco-design of products and processes*. North Macedonia: Springer, 111–135. http://doi.org/10.1007/978-981-16-4609-6_5, springer.com/series/13340.

Trinh, B. M., Chang, B. P., & Mekonnen, T. H. (2023). The barrier properties of sustainable multiphase and multicomponent packaging materials: A review. *Progress in Materials Science, 133*. Available from https://doi.org/10.1016/j.pmatsci.2023.101071, https://www.journals.elsevier.com/progress-in-materials-science.

Trotta, F., Da Silva, S., Massironi, A., Mirpoor, S. F., Lignou, S., Ghawi, S. K., & Charalampopoulos, D. (2023). Silver bionanocomposites as active food packaging: Recent advances & future trends tackling the food waste crisis. *Polymers, 15*(21), Available from https://doi.org/10.3390/polym15214243, http://www.mdpi.com/journal/polymers.

Tumu, K., Vorst, K., & Curtzwiler, G. (2023). Endocrine modulating chemicals in food packaging: A review of phthalates and bisphenols. *Comprehensive Reviews in Food Science and Food Safety, 22*(2), 1337–1359. https://doi.org/10.1111/1541-4337.13113.

Upadhyay, A., Agbesi, P., Arafat, K. M. Y., Urdaneta, F., Dey, M., Basak, M., Hong, S., Umeileka, C., & Argyropoulos, D. (2024). Bio-based smart packaging: Fundamentals and functions in sustainable food systems. *Trends in Food Science and Technology, 145*. Available from https://doi.org/10.1016/j.tifs.2024.104369, https://www.sciencedirect.com/science/journal/09242244.

US Food and Drug Administration (US FDA). (2007). Guidance for Industry: Preparation of Premarket Submissions for Food Contact Substances: Chemistry Recommendations. U.S. Department of Health and Human Services.

Urbelis, J. H., & Cooper, J. R. (2021). Migration of food contact substances into dry foods: A review. *Food Additives & Contaminants: Part A, 38*(6), 1044–1073. https://doi.org/10.1080/19440049.2021.1905188.

Vázquez-Loureiro, P., Lestido-Cardama, A., Sendón, R., Bustos, J., Cariou, R., Paseiro-Losada, P., & Rodríguez-Bernaldo de Quirós, A. (2023). Investigation of migrants from can coatings: Occurrence in canned foodstuffs and exposure assessment. *Food Packaging and Shelf Life, 40*, 101183. https://doi.org/10.1016/j.fpsl.2023.101183.

Versino, F., Ortega, F., Monroy, Y., Rivero, S., López, O. V., & García, M. A. (2023). Sustainable and bio-based food packaging: A review on past and current design innovations. *Foods, 12*(5), Available from https://doi.org/10.3390/foods12051057, www.mdpi.com/journal/foods.

Vilela, C., Kurek, M., Hayouka, Z., Röcker, B., Yildirim, S., Antunes, M. D. C., Nilsen-Nygaard, J., Pettersen, M. K., & Freire, C. S. R. (2018). A concise guide to active agents for active food packaging. *Trends in Food Science and Technology, 80*, 212–222. Available from https://doi.org/10.1016/j.tifs.2018.08.006, http://www.elsevier.com/wps/find/journaldescription.cws_home/601278/description#description.

Wang, Q., Chen, W., Zhu, W., McClements, D. J., Liu, X., & Liu, F. (2022). A review of multilayer and composite films and coatings for active biodegradable packaging. *npj Science of Food, 6*(1), Available from https://doi.org/10.1038/s41538-022-00132-8, https://www.nature.com/npjscifood/about.

Wang, Y., & Qian, H. (2021). Phthalates and their impacts on human health. *Healthcare, 9*(5), 603. https://doi.org/10.3390/healthcare9050603.

Westlake, J. R., Tran, M. W., Jiang, Y., Zhang, X., Burrows, A. D., & Xie, M. (2022). Biodegradable active packaging with controlled release: Principles, progress, and prospects. *ACS Food Science and Technology, 2*(8), 1166–1183. Available from https://doi.org/10.1021/acsfoodscitech.2c00070, https://pubs.acs.org/page/afsthl/about.html.

Wu, F., Misra, M., & Mohanty, A. K. (2021). Challenges and new opportunities on barrier performance of biodegradable polymers for sustainable packaging. *Progress in Polymer Science, 117*, 101395. https://doi.org/10.1016/j.progpolymsci.2021.101395.

Yashwanth, A., Huang, R., Iepure, M., Mu, M., Zhou, W., Kunadu, A., Carignan, C., Yegin, Y., Cho, D., Oh, J. K., Taylor, M. T., Akbulut, M. E. S., & Min, Y. (2025). Food packaging solutions in the post-per- and polyfluoroalkyl substances (PFAS) and microplastics era: A review of functions, materials, and bio-based alternatives. *Comprehensive Reviews in Food Science and Food Safety, 24*(1), Available from https://doi.org/10.1111/1541-4337.70079, http://onlinelibrary.wiley.com/journal/10.1111/(ISSN)1541-4337.

Yenidoğan, S., Aydemir, C., & Doğan, C. E. (2023). Packaging-food interaction and chemical migration. *Cellulose Chemistry & Technology*.

Younis, H., Zhao, G., & Abdellatif, H. (2022). *Pectin and its applicability in food packaging*. IntechOpen. http://doi.org/10.5772/intechopen.101614.

Zhang, W., & Rhim, J.-W. (2022). Titanium dioxide (TiO_2) for the manufacture of multifunctional active food packaging films. *Food Packaging and Shelf Life, 31*, 100806. https://doi.org/10.1016/j.fpsl.2021.100806.

Zhao, Y., Li, C., Xia, X., Tan, M., Wang, H., Lv, Y., Cheng, Y., Tao, Y., Lu, J., Li, D., & Du, J. (2023). Eco-friendly and intelligent cellulosic fibers-based packaging system for real-time visual detection of food freshness. *Chemical Engineering Journal, 474*, 146013. https://doi.org/10.1016/j.cej.2023.146013.

Żołek-Tryznowska, Z., & Kałuża, A. (2021). The influence of starch origin on the properties of starch films: Packaging performance. *Materials, 14*(5), 1146. https://doi.org/10.3390/ma14051146.

Zubair, M., Rauf, Z., Fatima, S., & Ullah, A. (2024). Lignin-derived bionanocomposites as functional food packaging materials. *Sustainable Food Technology, 2*(4), 945–966. https://doi.org/10.1039/d4fb00105b.

Chapter 32

Biodegradation aspect and life cycle assessment of biopolymer-based materials

Richert Agnieszka[1,2]

[1]*Department of Genetics, Faculty of Biology and Veterinary Science, Nicolaus Copernicus University in Toruń, Torun, Poland,* [2]*Institute of Advanced Studies, Nicolaus Copernicus University in Toruń, Wileńska, Toruń, Poland*

32.1 Problem statement

Polymeric materials have found wide application in almost every field of life, including construction, food, packaging, and medical industries, and are widely used due to their special properties (Andrady & Neal, 2009; Hopewell et al., 2009; Sivan, 2011). In the case of traditionally used plastics, their remarkable durability and resistance to microbial degradation result in the accumulation of unnecessary waste in the environment (Geyer et al., 2017; Vroman & Tighzert, 2009). The increasing amount of plastic waste is one of today's major environmental challenges. Synthetic polymers are a potential threat not only to humans but also to all other forms of life (Ali et al., 2021; Jambeck et al., 2015; Rochman et al., 2016). The answer to the situation is to turn to natural materials. However, it is important to remember that not all of them are subject to biological degradation. Abiotic degradation occurs with any type of material. Environmental conditions cause polymeric materials to break down into smaller and smaller fragments. This causes waste to disappear from our sight, but the truth is that these fragments become smaller and smaller. We are talking about the issue of microplastics, a recently discovered new threat to humans and nature (Gewert et al., 2015; Koelmans et al., 2022). Therefore, there is an urgent need to find effective degradation methods that stand in opposition to traditional methods of waste disposal and landfilling (Ali et al., 2021). Biodegradation may be the answer, as a mechanism for managing unwanted waste so that it is no longer a threat to the environment (Hartmann et al., 2019; Wright & Kelly, 2017). To this end, the ability of microorganisms to degrade polymeric substances is being studied under various near-natural conditions (Mistry et al., 2023). The efficiency of a given microorganism is influenced by its enzymatic activity, so it is also important to know the mechanism of reaction catalysis (Lucas et al., 2008; Mohanan et al., 2020; Shah et al., 2008).

The term biodegradation consists of the word "degradation," which in biological terms means the complete decomposition of organic material to its basic components, resulting in the production of carbon dioxide and water, and the prefix "bio-," indicating a process that occurs with the participation of living organisms: bacteria, fungi, and other organisms capable of synthesizing the appropriate enzymes (Geyer, 2020; Ignatyev et al., 2014; Wei & Zimmermann, 2017). In the context of polymers, biodegradation refers to the deterioration of their physical and chemical properties caused by the action of microorganisms under aerobic and anaerobic conditions, leading to a decrease in the molecular weight of polymers down to the level of simple compounds such as CO_2, H_2O, CH_4, and other low-molecular products (Rogovina et al., 2011; Singh & Sharma, 2008).

Biodegradation research is of great interest nowadays because of concern for the environment and the desire to invent ways to remove any pollutants found in it, such as xenobiotics (drugs, pesticides, and environmental pollutants of organic and inorganic origin). Among the most toxic are aromatic hydrocarbons, among which we find many phenolic derivatives. These organic compounds are biodegraded and detoxified mainly by microorganisms, which, over the course of trillions of years, have developed a variety of mechanisms, metabolic pathways, and specialized enzymes through which they are able to carry out biodegradation. This process requires finding cultures of microorganisms that can use harmful compounds as

Natural Polymer-Based Films and Coatings in Food Technology. DOI: https://doi.org/10.1016/B978-0-443-34244-8.00035-5

a source of carbon and energy and convert them into products that are safe for the environment (as it assimilates them) (Kyrikou & Briassoulis, 2007; Łabużek et al., 2002).

Due to the crystalline and relatively solid structure of "plastics," their biological degradation is a selective process. Among the many types of polymers, the physicochemical properties of each affect their biodegradability. Lower molecular weight and density of the polymer correlate positively with the rate of biodegradation, as does the increased availability of functional groups that increase hydrophobicity. Structural complexity, such as the presence or absence of branching and easily broken bonds, the molecular composition of the polymer, and whether it is a composite. The degradation process is also influenced by the physical form of the polymer waste, that is, film, pellet, or fibers, in which it is deposited in the environment and degraded (Gu, 2003; Kjeldsen et al., 2018; Restrepo-Flórez et al., 2014; Richert et al., 2023; Tokiwa et al., 2009). Treatments used to improve the degradation process and properties of polymers sometimes involve mixing two polymers. This has a positive effect on certain properties such as flexibility or strength of the material, but also significantly accelerates the degradation process under ambient conditions (Mayekar et al., 2023).

Other methods that accelerate the degradation process of plastic materials involve subjecting the material to chemical pretreatment, thermodegradation, or photo-oxidation. These methods reduce the molecular weight of the polymer so that low-molecular-weight organic acids are produced, which are a food source for microorganisms. Shorter polymer chains biodegrade first. Modifying the structure of widely used films is also promising. Addition of pro-oxidant substances pro-oxidants that initiate and accelerate free radical reactions under aerobic conditions results in the polymer chain being split into products more susceptible to biodegradation. This solution is widely used for polyethylene films and is a promising solution to the problem of environmental pollution from polyethylene waste (Chiellini et al., 2003; Mierzwa-Hersztek et al., 2019).

Nowadays, an equally common method to facilitate the biodegradation of artificial materials is to combine them with natural ingredients and natural polymers. The assumption is that the resulting product, based on a synthetic polymer resistant to biological degradation, with a biodegradable additive, will be partially degraded under the influence of biological factors in the environment (Boryniec et al., 2004). The most common natural raw material used to modify plastic structures is starch. It is a cheap and readily available polymer that can be used to produce high-molecular-weight materials and as an additive. Nowadays, a lot of research is being conducted on one of the most widely used plastics in the industry polyethylene modified with the addition of starch, focusing on the properties and degradation of this plastic. Modification of polymers is accompanied at the same time by a change in the structure and properties of the resulting product, often negatively affecting its durability. Nevertheless, such modification significantly accelerates the degradation of the polymer, and decomposition by microorganisms is facilitated by polyethylene and the natural additive (Łabużek et al., 2005).

Biodegradable polymers used in the production of plastics are easily obtained and, thanks to their properties, are utilized in many industries just like plastics. To increase their use, products are easily modified to impart biocidal properties. Substances used for this include natural plant extracts and oils, as well as substances produced by microorganisms (Gibas & Richert, 2018). However, such modifications of the polymer material can also negatively affect the biodegradation of the polymer material.

Biofilm plays a very important role in biodegradation processes. A biofilm is defined as a multicellular structure surrounded by a layer of organic and inorganic substances produced by the microorganisms that form it, showing adhesion to biotic and abiotic surfaces (Costerton et al., 1995; Czaczyk & Wojciechowska, 2003). These cells adhere to each other and to solid surfaces. Bacterial cells living in the biofilm environment show resistance to a significant proportion of antimicrobial agents and are about 1000 times more resistant to these agents than microorganisms remaining in suspension (Trafny, 2000). The formation of a biofilm is a complex process that depends on the type and characteristics of the microorganisms that form it, as well as the structure, composition, and properties of the colonized material. In the process of biofilm formation, several stages are distinguished: (1) initial (reversible) adhesion, (2) irreversible adhesion of microorganisms, (3) formation of microcolonies (growth and maturation), (4) mature form, and (5) biofilm dispersion (Kołwzan, 2011). Stage one is adhesion. An important role in this stage is played by extracellular polymers produced by microorganisms, as well as lipopolysaccharides and proteins of their cell wall, and extracellular structures: cilia and fimbriae. The geometric structure of the surface (smoothness, roughness) and mechanical damage (Czaczyk & Wojciechowska, 2003) have a great influence on the colonization of the material.

Bacterial deposition on the substrate occurs in a basic and fundamental manner. In the basal phase, microorganisms are not permanently attached to the substrate, making it relatively easy to remove them. Some microorganisms are detached from the substrate either freely or by physical or chemical means (Celina, 2013; Czaczyk & Wojciechowska, 2003).

At a distance of about 3 nm from the substrate, the basal phase begins, in which chemical forces play a major role: hydrogen bonds, the formation of ionic pairs and complexes, covalent bonds, and exopolymers produced by bacteria. This is an irreversible phase, in which the removal of microorganisms from the substrate requires complex chemical treatments (bacterial cell death) or mechanical removal from the surface. Extrinsic structures of bacteria, such as cilia or fimbriae, are extremely important in the colonization of a given surface (Kunicki-Goldfinger, 2008).

In the irreversible phase, microorganisms multiply, form microcolonies, and differentiate. There is static cell division at a given site, as well as migration of individual cells across the surface, resulting in the formation of larger clusters. The first microcolonies consist of a layer of 3–5 cells, often dispersed as a layer of microorganisms adjacent to the substrate. The mature biofilm, on the other hand, is surrounded by a thick layer of glycocalyx, to which minerals, organic compounds, and other organisms are adsorbed (Trafny, 2000).

The biofilm can be both single- and multilayered, and can be formed by microorganisms of one or many species. There are three types of biofilm structure: (flat, otherwise known as two-dimensional characteristic in an environment with a strong flow of aqueous solution), columnar, and the so-called fungus model, noticeable with slow water flow (Flemming & Wingender, 2010). The spatial structure of the biofilm depends on many factors, such as hydrodynamic conditions, nutrient content, bacterial motility, intercellular communication, exopolysaccharide, and protein content.

Cells of microorganisms that form a biofilm are connected to each other by an extracellular polymeric substance composed mainly of exopolysaccharides (EPS). Its components play an important role initially in the formation and later in the function of the biofilm. Extracellular biopolymers also provide protection to the biological membrane by preventing direct access of the disinfectant to bacterial cells (Hall-Stoodley et al., 2004).

The excretion of metabolic products, as well as the supply of nutrients to microorganisms, is enabled by a network of channels found in the biofilm. Rhamnolipids are responsible for their proper functioning. This system allows only metabolically active cells to function properly, enabling the biofilm to grow in thickness, and these cells are found in the surface layer of the microcolony. In the inner layer of the microcolony, there are cells in a state of anabiosis, besides, in the deeper layers, there are extracellular substances that interfere with its proper functioning. Intercellular signaling, which involves the production of signaling molecules and their free diffusion between cells, is responsible for the proper functioning of the biofilm. The phenomenon of communicating and recognizing the number of cells (sensing cell density) in a biofilm is called quorum sensing (QS). The ability to transmit intercellular information means that biofilms can function like primitive multicellular organisms.

The layer most vulnerable to damage, both mechanical and by biocides, is the surface of the biofilm. Many bacteria have an envelope or mucus on the outer surface of the cell wall. It forms a barrier, protecting the cell from desiccation and phagocytosis, reduces the rate of biocide diffusion through the biofilm layers, and plays an important role in cell adhesion processes to the surface of solid materials. The large number of polysaccharides in the envelope increases the number of free functional groups, which also results in high resistance of microorganisms. Bacterial resistance to biocides may be due to the activity of cellular pump-like proteins that recognize various types of threats and remove them from the cytoplasm. Environmental factors are also not insignificant. Cells in the deeper layers of the biofilm have significantly fewer nutrients and oxygen, leading to a reduced growth rate and reduced sensitivity to toxic substances. The slowed growth protects the cell from the effects of antimicrobial substances. Metabolic changes in biofilm cells may be the reason for the high resistance of biological membranes to antimicrobial agents. In addition, some cellular enzymes are able to modify the structure of bactericidal substances and thus change their properties and potency. Some scientists also believe that biosurfactants (e.g., rhamnolipids) increase biofilm resistance to phagocytosis and participate in infectious processes (Flemming & Wingender, 2010). In addition, long-term exposure of bacterial biofilms to high-concentration bactericides can lead to point mutations in genes, which can result in increased resistance of microorganisms.

From an ecological point of view, packaging made from biodegradable polymeric materials is a good alternative to those made from conventional materials, for example, polyolefins (polyethylenes, polypropylenes), as they do not linger as postconsumer waste and allow the circulation of elements in nature. When assessing suitability for composting, uniform criteria for this evaluation are needed.

In Germany, the basis for developing certification criteria for biodegradable products has been established. This was done by the Association of Manufacturers of Biodegradable Materials (Interessengemeinschaft Biologisch Abbaubare Werkstoffe [IBAW]). The organization developed DIN 54900, the "compostability" label, mechanisms for selective collection and segregation of biodegradable waste and an information system on these issues (Czaja-Jagielska, 2008; Żakowska, 2003).

DIN 54900, EN 13432:2002, and ASTM 6400 have become the basis for the DIN CERTCO unit (a member of the German DIN Standards Organization), which certifies biodegradable products (Czaja-Jagielska, 2008). The certification is based on strictly defined rules and detailed procedures. It aims to confirm the compostability of polymeric materials, as well as their mixtures and composites containing additional ingredients such as paints, varnishes, or dyes (Czaja-Jagielska, 2008; Żakowska, 2003).

In order to inform recipients that packaging is subject to collection along with bio-waste, it is marked with a sign. This mark is recognized in many European countries, that is, Denmark, the Netherlands, the United Kingdom, Italy, and Poland. Belgium, Finland, and Norway have their own logos in this regard. AIB-Vincotte Group in Belgium stashes two marks: "OK biodegradable" and "OK compost" (Żakowska, 2003).

In the United States, certification is based on ASTM D 6400:2004 (Czaja-Jagielska, 2008).

In addition to biodegradation, aging is a crucial process that reduces the performance of polymeric materials.

Aging is a process of structural changes occurring in the polymer under the influence of long-term external factors, leading to a deterioration of the plastic's performance. The first noticeable sign is a change in surface coloration, such as yellowing and tarnishing. Aging (after a certain period of use) is undoubtedly one of the few disadvantages that plastic products have. On the one hand, we require that plastics retain certain properties for as long as possible within the accepted period of time, regardless of the conditions and environment in which they are used, on the other hand—in accordance with ecological trends—plastic products should be environmentally friendly and, depending on their purpose, show good recyclability.

32.2 Method description

Due to the lack of complete knowledge of all interdependent factors and their mutual correlation in the environment, the place where the biodegradation process occurs, the exact decomposition time of biodegradable plastics is difficult to estimate. Therefore, tests that are used to quickly analyze whether a plastic will biodegrade under specific conditions provide a lot of valuable information. Furthermore, research is being conducted on such microorganisms, whose presence in the environment can significantly accelerate the biological degradation of polymeric materials. In the long run, such research can greatly facilitate and even solve the problem of lingering plastic waste in the environment. However, quantitative methods that determine plastic degradation may differ from one another, making it more difficult for researchers to compare their findings.

To date, many tests have been developed to study the progress of biodegradability of materials. Some of the methods are based on evaluating only a specific stage of the process rather than the total degradation. Weight loss, changes in molar mass of the polymer, and elimination of small particles are studied. Some methods are based on monitoring physical and chemical changes in the structure of the polymer, as well as macroscopic and microscopic changes on surfaces and in the microstructure of the materials under study. A separate category is formed by analyses of microbial growth properties. Biodegradability tests are performed in accredited laboratories and require the use of various analytical techniques. Microscopic techniques, chromatography, and spectrophotometric methods, which complement each other, are used to observe changes in the structure of materials (Tiwari et al., 2018).

Evaluation of visible changes appearing on plastics is possible when performing most tests. Observations of surface changes in polymers are conducted using microscopic optical techniques. High-end microscopes equipped with a digital camera and lenses of various magnifications are utilized. The defragmentation of the surface, the formation of holes, cracks, and pits, the formation of a biofilm on the surface of the plastic, and color changes allow for the determination of the occurrence of the biodegradation process. However, the presence of these changes can indicate any interaction between microorganisms present in the environment and the plastic. More advanced observations by scanning electron microscope (SEM) and atomic force microscopy (AFM) are used to obtain information about the mechanism of degradation taking place (Ikada, 1999; Shah et al., 2008).

An AFM is used to analyze the surface of the tested materials. Observations are based on interatomic forces, which allow for obtaining three-dimensional images. The roughness parameters of the film surface are calculated using special software. This technique also allows for determining the cavities formed in the film after material modification or degradation processes. The AFM method is easy to use, does not require special preparation of samples for testing, and does not damage the surface of the material (Tiwari et al., 2018). This technique also allows for imaging samples with phase contrast; this imaging takes place in tapping mode and provides better visualization of small samples. Phase imaging is characterized by greater sensitivity, which allows for collecting data on micro changes on the surface of the tested structure (Tang et al., 2016). Changes in the morphology of the material, both before and after the degradation of the film by microorganisms, can be examined using a scanning electron microscope (SEM). In contrast to AFM, this technique requires prior sample preparation, usually by sputtering a gold layer. SEM images are taken at high magnification because a small-scale image does not provide sufficient information about the surface structure and precise morphology of the material (Shah et al., 2008; Teng et al., 2020).

Visual methods, despite being easy and quick to perform, are objective and dependent on the person conducting the analysis. For this reason, they should be performed automatically, and standardized computer software should be used for the analysis. Commonly used biodegradability tests are based on measurements of oxygen consumption and carbon dioxide release (Kjeldsen et al., 2018). The reaction efficiency can be calculated based on the change in the released carbon dioxide converted from the amount of plastic, that is, the carbon source. However, these methods do not take into account the enzymatic activity of the microorganisms used in the study, which is a certain limitation, because there are enzymes that decompose plastic, accumulating biomass in the environment rather than producing carbon dioxide. Therefore, the reaction conditions should be selected in such a way that they allow for the complete conversion of the tested material into carbon dioxide. The final fate of any additives that are not ultimately converted into gas is also not included in the results. In this case, soil analysis methods are useful, which allow for the consideration of the fate of additives in the research material. This analytical method also allows for the determination of the potential formation of microplastics. However, these studies are more time-consuming and generate higher costs.

There are some limitations to the studies, for example, standard methods are not able to realistically represent the natural environment in which the reaction takes place, which is associated with predicting biodegradability in this environment. Laboratory tests create much more controlled conditions than the natural environment. This can cause discrepancies; plastic that decomposed in the laboratory may show lower degradation properties in nature due to changing conditions at the place of decomposition or the lack of a suitable microorganism correlating with the degradation process. Here, these reports mainly concern the aquatic environment (Harrison et al., 2018). Biomass growth studies are used when the only source of carbon is plastic and the reaction conditions are strictly controlled. The simplest method at the initial stages of degradation is to determine the mass loss of the tested material. After incubation, the plastic should be free of the developing bacterial biofilm. For this purpose, distilled water and an ethyl alcohol solution are used to ensure that the material is completely free of bacterial cells before weighing (Das & Kumar, 2015a).

Current methods lack general standardization, which not only makes it difficult for researchers to establish a uniform time scale for biodegradability, but also creates problems with comparing the results of tests on the biodegradability of materials. Problems with the research methodology include general reaction conditions such as temperature, which can range from a dozen to 55°C, oxygen conditions, and incubation time. Most studies require at least three months of incubation, but the maximum time in many studies is up to 24 months. The difference in results is influenced by the source of the microorganism inoculum. In the case of methods based on the same inoculum, for example, from soil or compost, studies conducted in different laboratories may differ significantly due to the diversity of microorganisms present in the sample (Kjeldsen et al., 2018; Richert et al., 2023). It is also worth mentioning the form of plastic whose biodegradability is being tested, its molecular weight, and source. Any pretreatment with physical factors such as light will have an impact, but also whether it was a product just after production, used appropriately at the end of its life cycle, or extracted from the environment.

Microorganisms currently play an important role in combating the problem of environmental pollution. Bacteria and fungi naturally produce enzymes that are commonly used in biotechnology. The presence or addition of a microorganism can have a positive effect on the processes related to the bioremediation of soils contaminated with heavy metals and plant growth, which also accelerates soil remediation. Bacterial strains are also used as a biopreparation to combat plastic waste pollution. Due to the large diversity of microorganisms in the natural environment, researchers often encounter difficulties in classifying isolated microorganisms, which affects the determination of the ecological role of a given organism. As in the case of techniques for testing the susceptibility of polymeric materials to degradation, there are differences in the methodology used for the extraction of genetic material and subsequent sequencing of DNA of environmental origin. To understand the role played by microorganisms in ecosystems, traditional studies used so far are supplemented with molecular biology techniques. Commonly used biochemical methods for determining the composition of environmental samples may be insufficient. Ribosomal RNA (rRNA) sequencing is a standard method for identifying and characterizing the diversity of microorganisms in the environment (Ibrahim et al., 2016).

Sequencing of the 16S rRNA gene can reduce the number of microorganisms incorrectly identified by traditional methods due to greater objectivity. Classical methods are exposed to subjective assessment or error by the examiner, due to phenotypic variability, recognition of morphologically appropriate colonies and correct classification is difficult. One of the methods used to identify bacterial strains is gene sequencing, however, this requires prior amplification of genetic material by means of polymerase chain reaction using universal primers. Sequencing means reading the order of base pairs in the DNA chain, which allows one to learn the structure of the gene. Currently, modern DNA sequencing, called next-generation sequencing, is used. It is based on the classic Sanger sequencing method, which is not limited to individual DNA fragments. A frequently used method is the Illumina system, which is based on the attachment of special adapters to DNA fragments. The process takes place in a flow chamber with a microplate covered with oligonucleotides, to which single-stranded fragments of genetic material are attached after denaturation. DNA fragments are amplified using a polymerase chain reaction of the "cat's back" type, during which the DNA fragment bends and binds complementarily to a region on the plate, creating a bridge nearby. The denaturation and amplification processes are repeated many times, as a result of which the plate is covered with so-called clusters. In the next stage, the nucleotides supplied to the system are labeled with fluorochromes, which, after excitation, give an image of which terminator is currently attached to the cluster. Based on the images generated at the moment of nucleotide attachment, their order in the tested DNA sequence can be determined (Orłowska & Sobczyk, 2017).

The 16S rRNA gene sequence is a widely used tool for the classification and identification of bacterial strains due to its ubiquity in all known bacterial taxa. Involved in the process of protein synthesis in living organisms, analyses based on the sequence of this gene cannot miss any bacteria known to researchers. The length of the 16S rRNA gene is about 1500 base pairs, including variable and conserved regions. This is an appropriate length for bioinformatic analysis. The genes are conserved but differ depending on the taxon (Woese, 1987). Analysis of the 16S rRNA sequence allows for the measurement and determination of relationships between bacterial isolates isolated from the environment. The 16S rRNA sequence is an ideal tool for classification of bacterial genera and species, especially those from the same family. It allows for comparison of results obtained from different environments by many different researchers. Currently, this sequence is used to identify new bacterial species (Ibrahim et al., 2016; Thompson et al., 2017; Willis et al., 2019).

Thanks to sequencing, we are able to determine the microorganisms useful in biodegradation processes. With this knowledge, we can use another research tool, crucial in biodegradation processes, namely bioaugmentation. Bioaugmentation involves the use of selected strains or mixed cultures of microorganisms to enrich the local microflora to effectively reduce pollutants by transforming them into simple compounds (Herrero & Stuckey, 2014; Tondera et al., 2021). It is a promising and practical technique that has been used to solve the problem of large amounts of polymer waste (Mistry et al., 2023; Mohanan et al., 2020). In order to properly carry out bioaugmentation, attention should be paid to various factors, including the physicochemical properties of the environment to be purified and the microbiological availability of pollutants. The selection of appropriate microorganisms that can effectively decompose pollutants without being harmful to the native microbiota is crucial (Pal et al., 2021). Bioaugmentation can be performed both in situ, in contaminated areas, and ex situ, in restricted areas or in bioreactors (Angelim et al., 2013). To create effective bioreactors for waste treatment, the presence and stability of suitable microbial consortia that have sufficient activity to degrade the contaminants present inside play a key role (Herrero & Stuckey, 2014). The most commonly used options in bioaugmentation are the introduction of preadapted, pure bacterial or fungal strains into the bioreactor or the addition of preadapted microbial consortia. Alternatively, genetically modified organisms whose adaptive capabilities have been improved can also be used (Herrero & Stuckey, 2014; Tondera et al., 2021).

Microbial biomass used as specialized inoculum in bioaugmentation can consist of homogeneous cell suspension produced in bioreactors under optimal conditions. Selected strains may not grow or show any activity in the bioreactor due to antagonistic interactions between them. After in situ inoculum application, microorganisms must adapt to the conditions prevailing in the contaminated environment. Therefore, microorganisms introduced into the environment may not adapt properly to difficult conditions such as periods of starvation, changes in low temperatures, and consequently be unable to colonize it and compete with native microflora. Thus, the preparation of inoculum from microorganisms that show high activity and are resistant to unfavorable environmental conditions is essential and constitutes a bottleneck in bioaugmentation strategies (Angelim et al., 2013; Herrero & Stuckey, 2014). To avoid problems with introducing microorganisms into the environment, an important step in research on bioaugmented microorganisms is to check under which conditions close to natural they perform best. For this purpose, their catabolic activity in soil, compost, and sometimes also in aquatic environments is studied (Mohanan et al., 2020).

To determine effective environmental remediation strategies, it is necessary to clearly define whether the goal is to reduce a specific contaminant or to reduce total organic carbon (TOC). A specific contaminant may only constitute a small part of the total TOC load, but it may also be the main cause of toxicity to microorganisms in the biosystem. Effective bioaugmentation should lead to a rapid reduction of toxicity to native microflora. Therefore, specific consortia of microorganisms, rather than single strains, may be more useful in achieving effective bioaugmentation. (Herrero & Stuckey, 2014; Yao et al., 2023) The use of consortia of microorganisms rather than pure cultures in bioaugmentation has gained popularity due to the synergy between them, which in turn improves the efficiency of hydrocarbon and complex chemical degradation (Angelim et al., 2013; Mao et al., 2012). One of the studies indicating biodegradability is the determination of bacterial biofilm on the surface of the film. It is performed using the spectrophotometric method of measuring the amount of crystal violet bound by bacterial cells forming the biofilm (Kroupitski et al., 2009; Sela et al., 2006; Stepanovic et al., 2000).

In this test, any strains of microorganisms can be used, which are grown in a liquid medium, which is a nutrient broth with the composition (g dm^{-3}): peptone—5.0, meat extract—3.0, distilled water—1 dm^3, pH 7.4. Incubation is carried out at 37°C for 24 hours. From the obtained cultures, 2 cm^3 of suspension is sterilely collected and transferred to Eppendorf tubes. The tubes, together with the culture, are centrifuged at 10,000 rpm for 6–7 minutes using a MiniSpin centrifuge (Eppendorf). The supernatant is removed, and the sediment is suspended in 1 cm^3 of diluent, which is peptone water with the composition (g dm^{-3}): peptone—1, distilled water—1 dm^3. The cell suspension obtained in this way is transferred to a densitometer (Densi-La-MeterII, Lachema) in order to bring its optical density to the value of 1, which, according to the McFarland scale, corresponds to 3 × 108 bacterial cells in 1 cm^3. Then, 20 cm^3 of nutrient broth and 0.2 cm^3 of previously boiled bacterial suspension are poured into sterile dishes. A sterilized (80% ethanol) and dried with sterile filter paper fragment of the tested foil measuring 0.5 × 1.5 cm is placed in the prepared dishes. The whole is incubated at 35°C for 48 hours. Then, the foil fragments are gently washed three times with distilled water to remove unattached cells and dried in air for 30–45 minutes. After this time, the foils are flooded with 1% crystal violet and left for 45 minutes—binding of the violet by the resulting biofilm. Then, to remove unbound violet, it is washed again three times with distilled water and dried for 30 minutes. Next, the foil fragments are placed in chemically clean test tubes, and 2.5 cm^3 of ethyl alcohol (96%) is poured. The whole is sealed with parafilm to prevent alcohol evaporation and vortexed—extraction of violet bound by the biofilm. After washing off the crystal violet, the resulting alcohol solution is placed in a spectrophotometer (Marcel Pro), and the absorbance value is measured at a wavelength of $\lambda = 595$. Measurements should be taken in triplicate for each sample.

Another indirect method related to biodegradation analyses is the determination of the activity of extracellular bacterial hydrolases. Test bacterial strains are multiplied in a nutrient broth medium with the following composition (g dm^{-3}): peptone—5.0, meat extract—3.0, distilled water—1 dm^3, pH 7.4. After inoculation of the medium, incubation is carried out at 37°C for 24 hours. In order to separate cells from the postculture fluid, the 24-hour culture of the test strains in the nutrient broth should be centrifuged at 10,000 rpm for 10 minutes at 4°C. The obtained postculture fluid containing hydrolases is used for further studies. The general activity of hydrolytic enzymes after contact with the foil should be determined using a nonspecific substrate—fluorescein diacetate (Adam & Duncan, 2001). For this purpose, pieces of foil (5 × 5 cm) are placed in sterile Petri dishes, onto which 3 cm^3 of the obtained postculture fluid is applied. The whole should be covered with another piece of foil (4.5 × 4.5 cm). The samples are preincubated for 1 hour at 4°C, and then the postculture fluid is collected from the tested foils into Eppendorf tubes. To the collected postculture fluid, for each 1, 0.01 cm^3 of fluorescein diacetate solution at a concentration of 1000 μg cm^{-3} is added. Then the samples are incubated for an hour at 30°C.

The concentration of fluorescein released by hydrolases should be measured using a HITACHI F-2500 spectrofluorimeter, relative to the standard curve, at an excitation wavelength of 480 nm and an emission wavelength of 505 nm. Inhibition or stimulation (in %) should be determined relative to the control sample, which consists of appropriate standard materials. The standard curve for fluorescein concentration is prepared as follows: 0.2265 g of fluorescein sodium salt is transferred to a 100 cm^3 volumetric flask and made up to 100 cm^3 with double-distilled water. The final concentration of fluorescein in this stock solution is 2000 μg cm^{-3}. Then, 1 cm^3 is taken from the obtained stock solution and transferred to another 100 cm^3 volumetric flask and made up to 100 cm^3 with double-distilled water. In this way, a working solution is obtained in which the fluorescein concentration is 20 μg cm^{-3}. From the working solution, 5, 10, 15, 20, and 25 cm^3 are taken successively and transferred to the next five measuring flasks with a capacity of 100 cm^3, which are filled to a volume of 100 cm^3 with bi-distilled water. In this way, standard solutions with fluorescein concentrations of 1, 2, 3, 4, and 5 μg cm^{-3} are obtained. The fluorescence of the prepared standard solutions is measured on a Hitachi F-2500 fluorimeter at an excitation wavelength of $\lambda = 480$ nm and an emission wavelength of $\lambda = 505$ nm. Based on the obtained results, a standard curve of fluorescein concentration is prepared.

Biodegradability can be tested by determining the oxygen consumption by microorganisms in specific environments in the presence of the tested polymeric materials. Analyses are performed using the Oxi Top Control WTW Wissenschaftlich-Technische Werkstätten GmbH (WTW, Germany) measuring system. The tests are conducted in the presence of microorganisms found, for example, in mature compost, activated sludge, river, and seawater. Polymer materials are used as respiratory substrates. A sample of a film with a given surface area should be cut into smaller pieces and placed in a glass measuring vessel. Then, 200 cm^3 of an appropriately prepared mixture with microorganisms and NT600 nitrification inhibitor is poured into the vessel, and a stirring rod is placed in it. A rubber quiver with CO_2 absorbent (0.4 g NaOH) is placed in each measuring vessel, and then the Oxi Top measuring heads are screwed on. The measuring vessels prepared in this way should be placed on a mixing platform and in a thermostatic cabinet. Incubation should be carried out for seven days at a temperature of 20°C. The measurement values are recorded every 24 hours using the OC 110 controller in the "Special Biochemical Oxygen Demand (BOD)" mode.

The biodegradability test of the films is performed in triplicate. The control (reference) value is determined based on the measurement of oxygen consumption by the mixture of microorganisms during incubation in the presence of polymeric materials. The biodegradation environments may have different natures. Biodegradation tests often use mature compost, from which an extract containing a mixture of microorganisms is prepared. For this purpose, 100 g of mature compost is added to 900 cm^3 of Ringer's solution (composition [g dm^{-3}]: NaCl—2.250, KCl—0.105, $CaCl_2$—0.120, $NaHCO_3$—0.050, distilled water—1 dm^3). Then, the whole is homogenized using a mechanical homogenizer. The obtained suspension should be filtered through a cloth with a mesh size of 50 μm. The filtrate is used for tests as a suspension of microorganisms present in the compost. Activated sludge can also be used in the tests. The sludge should be centrifuged at 10,000 rpm for 5 minutes at 20°C. The supernatant is removed, and the sludge is suspended in sterile physiological saline. The suspension should then be centrifuged again and suspended in physiological saline. The resulting suspension of washed activated sludge is added to Ringer's solution in a ratio of 10:190. The mixture is poured into Oxi Top measuring bottles, and the appropriate samples of polymeric materials are introduced into them.

River water or seawater can be used in the tests. Both types of water should be directly introduced into Oxi Top measuring bottles without any treatment. Then, film samples are introduced into the bottles, and the test procedure is carried out in accordance with the OxiTop system. After the microbiological degradation tests, tests of changes in selected properties should be performed. It is worth determining changes in the material structure using the SEM method and parameters such as water vapor permeability of oxygen and carbon dioxide for individual film samples before and after the biodegradation processes.

32.3 Discussion and evaluation

Polymers can become a potential carbon source for heterotrophic microorganisms (Danso et al., 2019; Gu, 2003). It is believed that chitosan produced by fungi is decomposed and used as a source of carbon, nitrogen, and energy by microorganisms capable of its degradation (Sawaguchi et al., 2015). The process of polymer degradation can be divided into two main parts. The first is the fragmentation of polymers leading to the formation of compounds with a lower molecular weight. This can take place through abiotic reactions, that is, oxidation, photodegradation, and in the further stages of biotic reactions, that is, degradation by microorganisms (Vroman & Tighzert, 2009). The large surface area of the material makes it difficult for microorganisms to adsorb, and the high molecular weight makes it unfavorable for microbial decomposition, as it becomes impossible to assimilate through the membrane and subject it to the action of cellular enzymes. Therefore, biotic reactions are the final steps of polymer degradation (Gu, 2003).

The first stages of plastic degradation are exposure to UV radiation with accompanying mechanical damage, which ultimately leads to the breakdown of larger fragments into smaller pieces. The resulting particles are susceptible to further degradation (Danso et al., 2019). Biodegradation is a slow process that includes environmental factors and the action of naturally occurring microbial species (Srikanth et al., 2022). Plastics were created due to human activity, and many of them did not exist in nature before. Many enzymes that are highly specific to individual polymer substances have not evolved in nature. The high resistance of polymers to biodegradation is a consequence of many of their characteristic features. It results mainly from the high molecular weight of the carbon chain, which often does not have side chains, making it less frequently assimilated. Strong C-C bonds and an extremely hydrophobic surface make polymers less susceptible to microbial colonization of their surface. Polymers have both amorphous and crystalline forms, and each has different levels of degradation. Those containing amorphous domains are more susceptible to microbial enzymes. The crystalline nature of polymer materials makes them less accessible (Danso et al., 2019; Jaiswal et al., 2020). In addition to all of the above-mentioned factors, additional stabilizers and other chemical agents that change mechanical and physical properties are also a problem, as they can interfere with enzyme activity (Danso et al., 2019). It is believed that the closer the structure of a polymer is to a natural molecule, the easier it is to degrade and mineralize. Polymers such as cellulose or chitin, compared to synthetic polymers, can be completely biodegraded by heterotrophic microorganisms in the natural environment (Gu, 2003). The rate at which the biodegradability of a material occurs depends not only on the origin of the polymer, its chemical composition and structure, but also on environmental conditions, such as the microbial community present, humidity, and temperature (Geyer, 2020; Vroman & Tighzert, 2009). In the case of chitosan, the critical factors controlling the biodegradation rate are the degree of deacetylation and molecular weight, followed by chemical modifications (Matica et al., 2017).

Among all microorganisms, fungi have attracted special attention from researchers. Among the many common and thoroughly studied representatives, there are those that have proven their effectiveness in degrading natural and artificial polymers. These include fungi from the genera Aspergillus, Penicillium, or Cladosporium, and some types of saprotrophic fungi, such as Pleurotus, Agaricus, or Trichoderma. They play a key role, acting on the polymer material by secreting many different types of enzymes with a lysing effect, for example, chitinases, lipases, and proteases, or lignocellulolytic enzymes (Jaiswal et al., 2020; Srikanth et al., 2022). In addition to the production of biopolymers that will be able to replace petroleum-derived materials on the market, scientists are focusing on identifying organisms that can produce enzymes capable of degrading plastics present in the natural environment. Despite the biodegradability of biopolymers, this process is still not 100% effective, which creates a field for identifying microorganisms that would serve as a natural preparation accelerating the biodegradation of plastics. Plastics are polymers made of molecules with a high molecular weight, so they cannot penetrate the cell membranes of microorganisms. Bacteria involved in the degradation process secrete enzymes called exoenzymes outside the cell (Shah et al., 2014). Bacteria produce enzymes such as lactase, protease, and catalase intracellularly and then secrete them into the external environment via exocytosis to break down large polymers into smaller molecules soluble in water. These molecules are able to penetrate the semipermeable bacterial membrane and can be a source of energy.

Metabolic activity varies depending on the organism from which it originates, so different enzymes break down different types of polymers into oligomers and monomers. Degradation can be carried out anaerobically, with the end products being carbon dioxide, water, and methane. In contrast, aerobic metabolism produces water and carbon dioxide (Starnecker & Menner, 1996; Gu et al., 2000; Singh and Sharma, 2008). The mechanism of polymer biodegradation consists of three basic steps. Microorganisms first attach to the polymer surface, then begin to degrade the material by secreting enzymes. This is done to use the polymer as a source of carbon and energy needed for bacterial growth. The next step is biofragmentation, that is, the breakdown of the polymer chain into shorter chains of oligomers and low-molecular-weight molecules. The final step is the assimilation of smaller forms of molecules by bacterial cells (Danso et al., 2019).

The use of microorganisms allows us to eliminate the problem of pollution in a safe and environmentally sustainable way. Enzymes produced by microorganisms ensure rapid and effective degradation of plastic waste. In the literature, more and more types of polymers, both biodegradable and nonbiodegradable, are assigned to specific types of microorganisms and enzymes capable of decomposing them (Mohanan et al., 2020; Ahmed et al., 2018). Species that degrade polymers have been identified as *Streptococcus*, *Klebsiella*, *Micrococcus*, *Staphylococcus*, and *Pseudomonas*, among others (Das & Kumar, 2015b).

Enzymes are biological catalysts, which means that their role is to increase the rate of biochemical reactions in living organisms. At the same time, they do not undergo any transformations themselves. It can therefore be stated that without enzymes, most reactions of cellular metabolism would not be possible (Azevedo et al., 2004). During degradation, an important role is played by the cooperation of exo- and endolytic hydrolases (Gu, 2003; Poshina et al., 2018). Chitinases and chitosanases can have both endo- and exolytic activity (Aam et al., 2010). Endoenzymes randomly hydrolyze bonds along the entire length of the polymer chain, creating smaller molecules, that is, oligosaccharides, in the form of dimers or trioses (Poshina et al., 2018). Exoenzymes catalyze the release of dimers from the reducing or nonreducing end of the polysaccharide chain, yielding short chains or smaller molecules, for example, oligomers, dimers, and monomers. These are small enough to pass through the outer bacterial membrane and then be used as a source of carbon and energy. This process is the actual depolymerization (Gu, 2003; Poshina et al., 2018). The degradation is called mineralization when the end products are inorganic compounds, for example, CO_2, H_2O, or CH_4 (Banerjee et al., 2014; Vroman & Tighzert, 2009).

In addition, chitosan-lytic enzymes exhibit so-called processivity. Processive enzymes do not release the substrate after only one cleavage, but remain bound to the substrate, allowing new cleavage to occur when the polymeric substrate slides through the substrate-binding cleft. Processivity has been shown to reduce enzyme efficiency relative to soluble and more accessible polymeric substrates such as chitosan (Aam et al., 2010). Enzymatic hydrolysis of the glycosidic bond is a nucleophilic substitution reaction at the anomeric carbon atom. This hydrolysis occurs via two main mechanisms, resulting in complete retention or inversion of the anomeric configuration. Both hydrolysis reactions occur via acid catalysis, that is, they require a pair of carboxylic acids in the active site of the enzyme. One carboxylic acid acts as a proton donor, facilitating the abstraction of the rest of the atoms from the central functional group atom, which forms the cation. The second carboxylic acid acts as a base (reversing mechanism) or as a nucleophile (retaining mechanism). In both mechanisms, the proton donor position is located within a hydrogen bond of the glycosidic oxygen (Aam et al., 2010; Davies & Henrissat, 1995). Chitosanases belonging to families GH46, GH75, and GH80, and family 19 chitinases use the reversing mechanism. Chitinases from family 18 use the retaining mechanism (Aam et al., 2010).

Fungi of the genus Trichoderma are classified as filamentous ascomycetes belonging to the order Hypocreales (Ascomycota) and include a large number of representatives commonly found throughout the world (Sandoval-Denis et al., 2014; Zeilinger et al., 2016). They include species such as *T. harzianum*, *T. hamatum*, *T. asperellum*, *T. atroviride*, *T. koningii*, *T. virens*, and *T. viride* (Zhang et al., 2021). These saprophytic fungi occur abundantly in soil and play an important role in the decomposition of organic matter, especially plant residues. Isolated Trichoderma strains can be easily recognized by their characteristic green, rapidly growing colonies, as well as by the typical branching patterns of conidiophores (Sandoval-Denis et al., 2014).

Representatives of this genus have found application in biotechnology due to their ability to produce a wide range of bioactive compounds in the form of secondary metabolites (Sandoval-Denis et al., 2014). In recent years, more than 1000 diverse metabolites have been isolated and identified, with a number of potential applications (Zeilinger et al., 2016; Zhang et al., 2021). This indicates their promising industrial, agricultural, and medicinal potential. Many of the synthesized secondary metabolites are used commercially due to their ability to accelerate plant growth and to inhibit the growth of pathogens (Wang et al., 2009; Zhang et al., 2021). Trichoderma uses β-1,3-glucanase, chitinase, peroxidase, and polyphenol oxidase to inhibit bacterial growth in plants (Tyśkiewicz et al., 2022). Among the representatives of Trichoderma spp., the ability for mycoparasitism, that is, parasitic antagonism toward other fungi, is also widespread (Zeilinger et al., 2016). One of the representatives exhibiting mycoparasitism is *T. atroviride* (Mukherjee et al., 2022). The cascade of reactions leading to parasitism on hyphae or killing of the host is triggered by the response to characteristic stimuli. These are physical contact with the host hyphae and response to proteins located on the surface of its cells (e.g., lectins). This is a signal to release enzymes that degrade the cell wall (Loc et al., 2020). The parasite also responds to various compounds released as a result of cell wall degradation, and so, for example, chitosan will enhance the growth and even stimulate the spore formation of *T. atroviride* (Mukherjee et al., 2022; Lopez-Moya et al., 2019).

Fungi of the genus Trichoderma are characterized by their ability to secrete a set of extracellular enzymes, including enzymes capable of degrading the cell wall: chitinases, β-glucanases, and proteases (Tyśkiewicz et al., 2022). The cell wall of fungi consists mostly of 90% polysaccharides: chitin, β-(1,3)-, β-(1,4)-, and β-(1,6)-glucans (D-glucose units linked by β-glycosidic bonds), α-glucans, chitosan, mannan, and galactomannan, as well as proteins (Sood et al., 2020; Tyśkiewicz et al., 2022). Chitinase is considered the most important enzyme, because chitin is the most abundant component of fungal cell walls (Loc et al., 2020). Some fungal plant pathogens produce chitosan during early infection to avoid host defense response (Lopez-Moya &Suárez-Fernández et al., 2019). The tolerance of fungi to chitosan within this genus is very diverse and depends on the structure of the cell membrane. The membranes of chitosan-sensitive fungi (*T. harzianum* or *T. neocrassum*) are very fluid (rich in unsaturated free fatty acids [FFA]). In contrast, chitosan-resistant fungi (*T. koningiopsis*) have low cell membrane fluidity (enriched in saturated fatty acids such as palmitic or stearic acid) (Lopez-Moya &Suárez-Fernández et al., 2019). Currently, there are studies that combine the need for chitosan degradation with the biological properties of fungi involving Trichoderma. Studies conducted by Wang et al. (2009) prove that Trichoderma is able to digest chitosan to oligosaccharides. In turn, the chitinolytic activity of Trichoderma results from the presence of many genes encoding the synthesis of chitinases, the high diversity of which is responsible for effective biocontrol against a wide range of fungal phytopathogens (Li et al., 2006). The experiment conducted by Gueye et al. (2020) showed that the addition of colloidal chitin and yeast extract as sources of carbon and nitrogen in the culture medium significantly increased the chitinolytic activity of *T. asperellum* strains. This indicates an increased production of hydrolytic enzymes in the presence of the fungal pathogen.

Contamination of polymeric materials is the beginning of biodegradation (Kania-Surowiec, 2013). Material destruction processes that occur under the influence of biological activity are called biodeterioration or microbiologically influenced corrosion (MIC). The colonization of material surfaces is easier when the structure is damaged and disrupted by wind, water, and air pollution (Kołwzan, 2011). Water easily enters the damage, which contributes to the settlement of microorganisms on the surface of the material. Initially, it is covered with single cells, but over time, they begin to multiply and form colonies, from which a mature form of biofilm is created (Czaczyk & Wojciechowska, 2003). Its presence is not without significance for biodegradation processes. Various groups of microorganisms can decompose or transform organic compounds into simpler substances. The processes and rate of decomposition of a given organic material are influenced by microorganisms and their enzymes as well as external factors such as oxygen concentration or biofilm density (Kania-Surowiec, 2013).

The ever-increasing amount of industrial and municipal waste, characterized by low susceptibility to biotic and abiotic factors, is a serious threat to the environment. One method to reduce waste is the production and use of degradable plastics. Degradation can be defined as an irreversible process of gradual decomposition of polymer material molecules, during which polymer chains are shortened and their fragments are eliminated, the degree of polymerization and molecular weight are reduced, and strength is diminished. In favorable conditions, this process is completed by depolymerization and the formation of simple chemical compounds (Shah et al., 2008). It is worth asking how we recognize that plastic has aged. Most often, these are changes noticeable in the appearance of the material, and then in mechanical properties, which can be determined using specific tests. A commonly known visual effect is a change in color, increased brittleness, loss of gloss, and surface cracking, which ultimately make the polymer product unfit for use. In chemical terms, these are changes in the molecular structure—the appearance of functional groups indicating material degradation, changes in the supramolecular structure—recrystallization and polymorphic changes, and consequently a decrease in functionality through reduced mechanical strength. Knowledge of the nature and mechanisms of the aging process and the factors influencing this process is necessary in order to obtain products that are required to be used correctly and safely in specific conditions throughout their assumed period of use. Phenomena related to the aging process and the conditions of use of the material affect its properties, determining primarily its appropriate selection, as well as designing the right shape or using the right method of its processing (Campbell & White, 2003; Malikb et al., 2006).

In relation to polymeric materials, the most important destructive processes include thermal and photodegradation. Most often, they occur simultaneously, and their effect depends on the type of polymer, the presence of impurities, auxiliary agents (such as dyes or fillers), and the type of products made from them. Many polymers are subject to degradation caused by external atmospheric conditions, in which photochemical reactions occurring under the influence of photons from solar ultraviolet radiation and atmospheric oxygen cause the disintegration of macromolecule chains as a result of breaking chemical bonds. In the natural environment, many physical, chemical, biological, and other factors affect the material, the most important of which are heat, oxygen, moisture, UV radiation, chemical impurities, and mechanical stresses. Polymer materials are exposed to atmospheric conditions during the use of a given product. Atmospheric factors and conditions can act simultaneously, and their share and intensity depend on, among others, the degree of environmental pollution, climate zone, or season (Chapiro et al., 1997; Wiles & Scott, 2006; Fieldman, 2002). The cognitive process of aging materials in natural conditions is impossible due to time, which in this case is a factor limiting the performance of such analyses. Therefore, the solution and necessity is to learn about the impact of the environment on materials in fully controlled, reproducible, and repeatable conditions, providing the possibility of assessing their impact on individual parameters of the material. It should be remembered that laboratory conditions are not able to fully simulate the real environment, due to differences in conditions, such as variable UV radiation intensity, spraying time, humidity and its fluctuations, or variable temperatures. Nevertheless, several main directions of research on phenomena related to the aging process of materials can be distinguished. The first of them includes research on resistance to microclimate in places located at different latitudes. The second group is focused on research on the impact of stabilizers and aging retarders on the rate of this process. Yet another range of issues concerns the study of the course of aging processes occurring in unconventional measurement conditions, for example, at elevated temperature or in an artificial environment. Another group includes the analysis of aging mechanisms under the influence of selected external factors and the prediction of changes in material properties based on short-term measurements (Wypych, 2008).

Aging tests are conducted in so-called aging chambers, where aging conditions (humidity, temperature, light, atmosphere) can be controlled. Aging conditions for individual materials are described in the form of ISO and American Society for Testing and Materials (ASTM) standards and norms. Changes in parameters are measured using mechanical methods such as stretching or tearing, visible light spectroscopy, and microscopic methods. Each of the structural symptoms of degradation can be assessed using analytical methods: recrystallization X-ray diffraction, degree of polymerization viscometric viscosity measurement, liquid chromatography; oxidation infrared spectroscopy. All effects indicating the ongoing degradation process may be the first sign that the material is not fully capable of continuing to perform its intended function, and its continued use may be dangerous (Jakubowicz et al., 1999). The resistance of materials to adverse phenomena causing aging depends on many factors, primarily temperature and time. In order to determine the criteria characterizing the change in the properties of a material, two cognitive algorithms are most often used (ISO 472). The first one experimentally determines the degree of intensity and kinetics of changes in the tested property, and then determines an equation describing, for example, the aging rate. Such an equation allows for calculating at any time -approximately determined by the accuracy of the approximation -the current value of the tested quantity, interesting in a given case. The second algorithm determines the critical aging time, that is, the time after which the tested property changes by a certain fixed value. The measure of aging resistance can be assumed as the loss of specific physicochemical properties (e.g., 50% loss of a given property) as a result of the action of a specific factor constituting the source of aging, under established conditions. Reaching the critical aging time most often means excluding the product from use.

Regardless of the type of method used to study the aging process, the time for conducting measurements is usually much shorter than the assumed time of use of the plastic product. The results obtained from aging tests constitute the basis for further analysis based on selected mathematical methods, allowing for an adequate description of the occurring phenomena. Among the forecasting methods, the most frequently mentioned in the literature are the analysis of the distribution of a random variable, regression analysis, and time series analysis. The correctness of selecting the appropriate mathematical model, based on which the calculation algorithm is adopted and the process simulation is carried out, is verified experimentally in relation to the actual results. The use of selected forecasting methods allows for the simulation determination of the degree of changes in the tested properties of the product occurring over a period longer than the assumed test time (ISO 175; ISO 4892; ISO 2440). Forecasting is a recognized scientific method of analyzing issues related to determining future changes in the properties of used materials. The process of predicting and assessing the modifications that may occur in the research object is based on theoretical analyses, analytical considerations, logical premises, and practical experiences. The degradation process can occur both in the use cycle of the finished product and during the production process. Most often, changes occur in the structure of polymers, during production, changes are determined by temperature and atmospheric oxygen. For some materials, the processing and degradation temperatures are very similar, so this is a significant technological problem, consisting of the difficulties of selecting the right parameters.

Thanks to industrial development, new technologies are being implemented, including specialist additives and thermal stabilizers, which ensure process thermal stability. They can be divided into primary and secondary. Temperature and atmospheric oxygen cause polymer chains to crack with the release of free radicals (R• and ROO•). Primary stabilizers are designed to capture free radicals and transform them into the semiactive form of ROOH. This group of stabilizers includes salts of higher fatty acids and amine or phenolic compounds. They are selected depending on the type of material used. Adding primary stabilizers to polymers slows down the degradation process. In the next stage, it is necessary to use secondary stabilizers. They are responsible for transforming ROOH into inactive forms of ROH. This is necessary because ROOH forms are not stable and can easily decompose back into free radicals, which would start the degradation process all over again. However, by obtaining ROH forms, we completely stop the degradation process (ISO 472; ISO 175; EN ISO 4892; ISO 2440).

32.4 Summary and conclusions

- Depending on the stages and degree of polymer biodegradation, various parameters change. The first to be observed are visual changes, for example, color change, cracking, pore formation, or biofilm formation on the surface. They do not always prove that the degradation process is underway, but they can be an indicator suggesting microorganism attack or the influence of the aquatic environment (Shah et al., 2008; Tokiwa et al., 2009).
- The parameter that clearly indicates the biodegradation of polymer materials is their molecular weight. Its reduction is observed from the beginning of the process and is associated with the shortening of the main polymer chains (Kabir et al., 2020). The reduction in molecular weight is accompanied by a decrease in polymer viscosity. The methods that allow for the assessment of structural changes include gel permeation chromatography, Fourier-transform infrared spectroscopy (FT-IR), and TG/DSC thermal analysis (Shah et al., 2008; Tsuji & Ikada, 2000a, 2000b).
- A decrease in the strength of materials is also observed, which is an obvious consequence of breaking polymer chains. In the case of plastics, the most frequently measured parameter is tensile strength, but hardness, deformability, and modulus of elasticity also change (Kale et al., 2007a, 2007b; Tsuji & Ikada, 2000a, 2000b).
- Further changes can be observed in the range of polymer crystallinity, which also affects the rate of additional degradation. The first areas to undergo degradation are the crystalline regions of polymer materials. As the process progresses, the polymer structure loosens, which facilitates the reorganization and crystallization of the remaining chains. Any changes in polymer crystallinity can be detected using the following methods: X-ray analysis, infrared spectroscopy (FT-IR), and thermal analysis (TG/DSC) (Tsuji & Ikada, 2000a, 2000b; Kale et al., 2007a, 2007b).
- The last measurable parameter is mass. It decreases as a result of material defragmentation and the removal of degradation products from the environment (Andrady, 2017; Niaounakis, 2015).
- During degradation, changes in the roughness and wetting angle of the material surface can also be observed. Reducing wettability (increasing hydrophilicity) promotes the progression of degradation (Tokiwa et al., 2009).
- The assessment of the degradation rate can be carried out not only on the basis of changes occurring in the material itself but also by analyzing the released degradation products and their impact on the change in environmental parameters.
- It is important to understand the mechanisms of the aging process and to assess its impact on the properties of materials. The gradual and inevitable increase in demand for polymer materials means that the research and development work conducted for years focuses on explaining the complex mechanisms of the aging process, determining its impact on the properties -mainly physical and chemicalas well as assessing the impact of processing conditions on the course of this phenomenon.
- Testing polymer materials under the influence of environmental parameters and determining the degree of their aging allows for the detection of early stages of degradation and provides valuable information about the aging process, which should be taken into account during the design and construction of various products made from these materials. Only a detailed description of the phenomenon, as well as an understanding of the mechanisms responsible for polymer degradation, allows for the design of materials with specific durability and strength. Research into the aging process of products is also very important for the recirculation of materials and, above all, for their safe use.
- It should be noted that all plastics are susceptible to a process called aging. Due to the factors that determine the aging of plastics, we can distinguish the following types of degradation: photochemical, thermal, biological, chemical, oxidative, mechanical, and radiation. The impact of external factors can cause dulling, discoloration, and cracking, as well as loss of brittleness, elasticity, or stiffness of the material.

- Fungi of the Trichoderma genus produce hydrolytic enzymes associated with the degradation of polymers, especially high activity of chitinolytic enzymes (Tyśkiewicz et al., 2022).
- Bioaugmentation with Trichoderma fungi accelerates the biodegradation of polymeric material films.
- The activity of microbiological enzymes in the soil after the introduction of biodegradable films varied depending on the type of film (Kale et al., 2007a, 2007b; Tsuji & Ikada, 2000a, 2000b).
- The inclusion of pure bacterial isolates in the biodegradation process can both accelerate and enhance the efficiency of the biodegradation of the analyzed materials.
- Methods for determining biodegradability may be based on the description of physicochemical, barrier, and structural changes in the film under the influence of a biological factor, for example, a pure bacterial isolate (Shah et al., 2008; Tokiwa et al., 2009).

References

Aam, B. B., Heggset, E. B., Norberg, A. L., Sørlie, M., Vårum, K. M., & Eijsink, V. G. H. (2010). Production of chitooligosaccharides and their potential applications in medicine. *Marine Drugs, 8*(5), 1482–1517. https://doi.org/10.3390/md8051482.

Adam, G., & Duncan, H. (2001). Development of sensitive and rapid method for the measurement of total microbial activity using fluorescein diacetate (FDA) in a range of soil. *Soil Biology and Biochemistry, 33*, 943–951.

Ahmed, I., et al. (2018). Biodegradation of plastics: Current scenario and future prospects for environmental safety. *Environmental Pollution, 25*, 7287–7298.

Ali, S., et al. (2021). Microbial degradation of synthetic plastics: A review. *Environmental Science and Pollution Research, 28*(42), 58926–58943. https://doi.org/10.1007/s11356-021-14774-0.

Andrady, A. L. (2017). The plastic in microplastics: A review. *Marine Pollution Bulletin, 119*(1), 12–22. https://doi.org/10.1016/j.marpolbul.2017.01.082.

Andrady, A. L., & Neal, M. A. (2009). Applications and societal benefits of plastics. *Philosophical Transactions of the Royal Society B: Biological Sciences, 364*(1526), 1977–1984. https://doi.org/10.1098/rstb.2008.0304.

Angelim, A. S., et al. (2013). An innovative bioremediation strategy using a bacterial consortium entrapped in chitosan beads. *Journal of Environmental Management, 127*, 10–17.

Azevedo, H. S., & Reis, R. L. (2004). Understanding the enzymatic degradation of biodegradable polymers and strategies to control their degradation rate. In R. L. Reis, & J. San Román (Eds.). *Biodegradable systems in tissue engineering and regenerative medicine* (pp. 177–201). CRC Press. https://doi.org/10.1201/9780203491232.ch12.

Banerjee, I., et al. (2014). Mineralization of synthetic polymers by microorganisms. *Bioresource Technology, 168*, 10–22. https://doi.org/10.1016/j.biortech.2014.04.049.

Boryniec, A., et al. (2004). Application of spectroscopic methods to study the biodegradation process of packaging materials. Zastosowanie metod spektroskopowych do badania procesu biodegradacji materiałów opakowaniowych *Polymers, 49*(9), 656–661.

Campbell, G. A., & White, J. R. (2003). Polymer aging and lifetime prediction: Physico-chemical concepts and industrial practice. *Polymer Degradation and Stability, 82*(1), 257–268.

Celina, M. C. (2013). Review of polymer oxidation and its relationship with materials performance and lifetime prediction. *Polymer Degradation and Stability, 98*(12), 2419–2429. https://doi.org/10.1016/j.polymdegradstab.2013.06.024.

Chapiro, A., et al. (1997). Environmental effects on polymer aging. *Polymer Degradation and Stability, 58*(2–3), 219–229. https://doi.org/10.1016/S0141-3910(97)00085-2.

Chiellini, E., Corti, A., & Swift, G. (2003). Biodegradation of thermally oxidized, fragmented low-density polyethylenes. *Polymer Degradation and Stability, 81*(2), 341–351. https://doi.org/10.1016/S0141-3910(03)00105-8.

Costerton, J. W., Lewandowski, Z., Caldwell, D. E., Korber, D. R., & Lappin-Scott, H. M. (1995). Microbial biofilms. *Annual Review of Microbiology, 49*, 711–745. https://doi.org/10.1146/annurev.mi.49.100195.003431.

Czaczyk, K., & Wojciechowska, K. (2003). Formation of bacterial biofilms: The nature of the phenomenon and mechanisms of interaction. Tworzenie biofilmów bakteryjnych - istota zjawiska i mechanizmy oddziaływania *Biotechnology (Reading, Mass.), 62*, 180–192.

Czaja-Jagielska, N. (2008). Biodegradacja modyfikowanych folii opakowaniowych z wytworzeniem glebowych bakterii denitryfikacyjnych. In *Wydawnictwo Akademii Ekonomicznej w Poznaniu. Biodegradation of modified packaging films with the use of soil denitrifying bacteria*. Poznań University of Economics Press in Polish.

Danso, D., Chow, J., & Streit, W. R. (2019). Plastics: Environmental and biotechnological perspectives on microbial degradation. *Applied and Environmental Microbiology, 85*(19), e01095-19. https://doi.org/10.1128/AEM.01095-19.

Das, M., & Kumar, A. (2015a). Bacterial strains degrading plastics: Diversity and mechanisms. *International Biodeterioration & Biodegradation, 105*, 8–16. https://doi.org/10.1016/j.ibiod.2015.08.015.

Das, M., & Kumar, A. (2015b). Mass loss and bacterial biofilm studies for biodegradation assessment. *International Biodeterioration & Biodegradation, 105*, 17–24. https://doi.org/10.1016/j.ibiod.2015.08.016.

Davies, G., & Henrissat, B. (1995). Structures and mechanisms of glycoside hydrolases. *Structure (London, England: 1993), 3*(9), 853–859. https://doi.org/10.1016/S0969-2126(01)00220-9.

Fieldman, D. (2002). Polymer aging under environmental stress. *Polymer Testing, 21*(3), 283–288. https://doi.org/10.1016/S0142-9418(01)00116-1.

Flemming, H. C., & Wingender, J. (2010). The biofilm matrix. *Nature Reviews. Microbiology, 8*, 623–633.

Gewert, B., Plassmann, M. M., & MacLeod, M. (2015). Pathways for degradation of plastic polymers floating in the marine environment. *Environmental Science: Processes & Impacts, 17*(9), 1513–1521. https://doi.org/10.1039/C5EM00207A.

Geyer, R. (2020). Production, use, and fate of synthetic polymers. *Nature Reviews Materials, 5*, 437–438. https://doi.org/10.1038/s41578-020-00247-0.

Geyer, R., Jambeck, J. R., & Law, K. L. (2017). Production, use, and fate of all plastics ever made. *Science Advances, 3*(7), e1700782. https://doi.org/10.1126/sciadv.1700782.

Gibas, E., & Richert, A. (2018). Impact of bacteria on oxy-degradable films containing essential oils and silver, and copper nanoparticles. *Przemysł Chemiczny, 97*(2), 291–293.

Gu, J. D., et al. (2000). Aerobic and anaerobic biodegradation of polyvinyl alcohol in marine sediment. *Journal of Environmental Sciences, 12*(2), 129–135.

Gu, J. D. (2003). Microbial degradation of polymers and chitosan. *International Biodeterioration & Biodegradation, 52*, 69–91.

Gueye, N., Ndiaye, M., Diop, T. A., & Kane, A. (2020). Factors affecting the chitinase activity of Trichoderma asperellum isolated from agriculture field soils. *Journal of Applied Biology & Biotechnology, 8*(2), 41–44. https://doi.org/10.7324/JABB.2020.80208.

Hall-Stoodley, L., Costerton, J. W., & Stoodley, P. (2004). Bacterial biofilms: From the natural environment to infectious diseases. *Nature Reviews. Microbiology, 2*(2), 95–108. https://doi.org/10.1038/nrmicro821.

Harrison, J. P., et al. (2018). Plastic pollution in the aquatic environment: Evidence and knowledge gaps. *Environmental Science & Technology, 52*(12), 6079–6093. https://doi.org/10.1021/acs.est.8b00571.

Hartmann, N. B., Hüffer, T., Thompson, R. C., Hassellöv, M., Verschoor, A., Daugaard, A. E., Rist, S., Karlsson, T., Brennholt, N., Cole, M., Herrling, M. P., Hess, M. C., Ivleva, N. P., Lusher, A. L., & Wagner, M. (2019). Are we speaking the same language? Recommendations for a definition and categorization framework for plastic debris. *Environmental Science & Technology, 53*(3), 1039–1047. https://doi.org/10.1021/acs.est.8b05297.

Herrero, M., & Stuckey, D. C. (2014). Bioaugmentation and its application in wastewater treatment: A review. *Biotechnology Advances, 33*(3–4), 495–511. https://doi.org/10.1016/j.biotechadv.2015.02.009.

Hopewell, J., Dvorak, R., & Kosior, E. (2009). Plastics recycling: Challenges and opportunities. *Philosophical Transactions of the Royal Society B: Biological Sciences, 364*(1526), 2115–2126. https://doi.org/10.1098/rstb.2008.0311.

Ibrahim, M., et al. (2016). Analysis of microbial communities using rRNA gene sequencing. *Journal of Microbiological Methods, 125*, 27–35. https://doi.org/10.1016/j.mimet.2016.03.007.

Ignatyev, I. A., Thielemans, W., & Vander Beke, B. (2014). Recycling of polymers: A review. *ChemSusChem, 7*(6), 1579–1593. https://doi.org/10.1002/cssc.201300898.

Ikada, Y. (1999). Degradability of polymers as influenced by the environmental conditions. *Polymer Degradation and Stability, 64*(2), 127–132. https://doi.org/10.1016/S0141-3910(98)00130-2.

Jaiswal, P., et al. (2020). Fungal enzymes for polymer biodegradation: Current status and future prospects. *Frontiers in Microbiology, 11*, 590183. https://doi.org/10.3389/fmicb.2020.590183.

Jakubowicz, I., Yarahmadi, N., & Gevert, T. (1999). Effects of accelerated and natural ageing of plasticized polyvinyl chloride (PVC). *Polymer Degradation and Stability, 66*(3), 415–421. https://doi.org/10.1016/S0141-3910(99)00094-2.

Jambeck, J. R., Geyer, R., Wilcox, C., Siegler, T. R., Perryman, M., Andrady, A., Narayan, R., & Law, K. L. (2015). Plastic waste inputs from land into the ocean. *Science (New York, N.Y.), 347*(6223), 768–771. https://doi.org/10.1126/science.1260352.

Kabir, E., Kaur, R., Lee, J., Kim, K.-H., & Kwon, E. E. (2020). Prospects of biopolymer technology as an alternative option for non-degradable plastics and sustainable management of plastic wastes. *Journal of Cleaner Production, 258*, 120536. https://doi.org/10.1016/j.jclepro.2020.120536.

Kale, G., Auras, R., Singh, S. P., & Narayan, R. (2007a). Biodegradability of polylactide bottles in real and simulated composting conditions. *Polymer Testing, 26*(8), 1049–1061. https://doi.org/10.1016/j.polymertesting.2007.07.006.

Kale, G., Auras, R., Singh, S. P., & Narayan, R. (2007b). Degradation of commercial biodegradable packages under real composting and ambient exposure conditions. *Journal of Polymers and the Environment, 15*, 125–138. https://doi.org/10.1007/s10924-007-0055-6.

Kania-Surowiec, I. (2013). Biological beds in wastewater treatment from plastics recycling. Złoża biologiczne w oczyszczaniu ścieków z recyklingu tworzyw sztucznych. Inżynieria Ekologiczna, *Ecological Engineering, 32*, 74–84.

Kjeldsen, P., et al. (2018). Biodegradability assessment of polymers in landfill simulation tests. *Waste Management, 75*, 197–204. https://doi.org/10.1016/j.wasman.2018.02.025.

Koelmans, A. A., Redondo-Hasselerharm, P. E., Mohamed Nor, N. H., de Ruijter, V. N., Mintenig, S. M., & Kooi, M. (2022). Risk assessment of microplastic particles. *Nature Reviews Materials, 7*, 138–152. https://doi.org/10.1038/s41578-021-00411-Y.

Kołwzan, B. (2011). Analysis of the biofilm phenomenon: Conditions of its formation and functioning. Analiza zjawiska biofilmu - warunki jego powstawania i funkcjonowania. Ochrona Środowiska, *Environmental Protection, 4*, 3–14.

Kroupitski, Y., Pinto, R., Brandl, M. T., Belausov, E., & Sela, S. (2009). Interactions of Salmonella enteric with lettuce leaves. *Journal of Applied Microbiology, 106*, 1876–1885.

Kunicki-Goldfinger, W. J. H. (2008). *The life of bacteria*. PWN Scientific Publishers Życie bakterii. Wydawnictwo Naukowe PWN.

Kyrikou, I., & Briassoulis, D. (2007). Biodegradation of agricultural plastic films: A critical review. *Journal of Polymers and the Environment, 15*, 125–150. https://doi.org/10.1007/s10924-007-0053-8.

Łabużek, S., Pająk, J., Nowak, B., Majduk, E., & Karcz, J. (2002). Microbiological degradation of LDPE/PBSU·AD film. *Polymers, 47*, 256–261.

Łabużek, S., Pająk, J., & Nowak, B. (2005). Biodegradation of modified polyethylene in soil under laboratory conditions. Biodegradacja modyfikowanego polietylenu w glebie w warunkach laboratoryjnych. *Polymers, 50*(9), 675–681.

Li, B., Kennedy, J. F., Peng, J. L., Yie, X., & Xie, B. J. (2006). Preparation and performance evaluation of glucomannan-chitosan-nisin ternary antimicrobial blend film. *Carbohydrate Polymers, 65*, 488–494.

Loc, N. H., Huy, N. D., Quang, H. T., Lan, T. T., & Ha, T. T. T. (2020). Characterisation and antifungal activity of extracellular chitinase from a biocontrol fungus, *Trichoderma asperellum* PQ34. *Mycology, 11*(1), 38–48. https://doi.org/10.1080/21501203.2019.1703839.

Lopez-Moya, F., et al. (2019). Chitosan inhibits growth and affects development in phytopathogenic fungi. *Fungal Genetics and Biology, 130*, 18–25. https://doi.org/10.1016/j.fgb.2019.04.001.

Lopez-Moya, F., Suárez-Fernández, M., & Lopez-Llorca, L. V. (2019). Molecular mechanisms of chitosan interactions with fungi and plants. *International Journal of Molecular Sciences, 20*(2), 332. https://doi.org/10.3390/ijms20020332.

Lucas, N., Bienaime, C., Belloy, C., Queneudec, M., Silvestre, F., & Nava-Saucedo, J. E. (2008). Polymer biodegradation: Mechanisms and estimation techniques - A review. *Chemosphere, 73*(4), 429–442. https://doi.org/10.1016/j.chemosphere.2008.06.064.

Malikb, A., et al. (2006). Factors influencing polymer degradation under environmental stress. *Journal of Polymer Science, 44*(10), 1223–1236.

Mao, J., Luo, Y., Teng, Y., Li, Z., & Christie, P. (2012). Bioremediation of polycyclic aromatic hydrocarbon-contaminated soil by a bacterial consortium and associated microbial community changes. *International Biodeterioration & Biodegradation, 70*, 141–147. https://doi.org/10.1016/j.ibiod.2012.03.002.

Matica, M. A., et al. (2017). Chitosan biodegradability and factors influencing it. *Carbohydrate Polymers, 176*, 1–9. https://doi.org/10.1016/j.carbpol.2017.07.082.

Mayekar, P. C., Limsukon, W., Bher, A., & Auras, R. A. (2023). Breaking it down: How thermoplastic starch enhances poly(lactic acid) biodegradation in compost - A comparative analysis of reactive blends. *ACS Sustainable Chemistry & Engineering, 11*(26), 9729–9737. https://doi.org/10.1021/acssuschemeng.3c01676.

Mierzwa-Hersztek, M., Gondek, K., & Kopeć, M. (2019). Degradation of polyethylene and biocomponent-derived polymer materials: An overview. *Journal of Polymers and the Environment, 27*(3), 600–611. https://doi.org/10.1007/s10924-019-01368-4.

Mistry, A. N., Kachenchart, B., Pinyakong, O., Assavalapsakul, W., Jitpraphai, S. M., Somwangthanaroj, A., & Luepromchai, E. (2023). Bioaugmentation with a defined bacterial consortium: A key to degrade high molecular weight polylactic acid during traditional composting. *Bioresource Technology, 367*, 128237. https://doi.org/10.1016/j.biortech.2022.128237 Article.

Mohanan, N., Montazer, Z., Sharma, P. K., & Levin, D. B. (2020). Microbial and enzymatic degradation of synthetic plastics. *Frontiers in Microbiology, 11*, 580709. https://doi.org/10.3389/fmicb.2020.580709.

Mukherjee, P. K., Mendoza-Mendoza, A., Zeilinger, S., & Horwitz, B. A. (2022). Mycoparasitism as a mechanism of Trichoderma-mediated suppression of plant diseases. *Fungal Biology Reviews, 39*, 15–33.

Niaounakis, M. (2015). *Biopolymers: Processing and products*. William Andrew Publishing. https://doi.org/10.1016/C2014-0-00663-3.

Orłowska, A., & Sobczyk, B. (2017). The role of next-generation sequencing in microbial ecology. *Molecular Biology Reports, 44*(4), 343–350. https://doi.org/10.1007/s11033-017-3984-3.

Pal, S., et al. (2021). Selection criteria of microorganisms for bioaugmentation. *Biodegradation, 32*, 233–246. https://doi.org/10.1007/s10532-021-09968-0.

Poshina, D. N., et al. (2018). Role of exo- and endoenzymes in microbial degradation of polysaccharides. *Biotechnology Advances, 36*(2), 306–312. https://doi.org/10.1016/j.biotechadv.2017.11.001.

Restrepo-Flórez, J. M., Bassi, A., & Thompson, M. R. (2014). Microbial degradation and deterioration of polyethylene - A review. *International Biodeterioration & Biodegradation, 88*, 83–90. https://doi.org/10.1016/j.ibiod.2013.12.014.

Richert, A., et al. (2023). Comparative analysis of biodegradability test methods. *Polymer Testing, 122*, 107463. https://doi.org/10.1016/j.polymertesting.2023.107463.

Rochman, C. M., Browne, M. A., Underwood, A. J., van Franeker, J. A., Thompson, R. C., & Amaral-Zettler, L. A. (2016). The ecological impacts of marine debris: Unraveling the demonstrated evidence from what is perceived. *Ecology, 97*(2), 302–312. https://doi.org/10.1890/14-2070.1.

Rogovina, S. Z., Grachev, A. V., Aleksanyan, K. V., & Prut, E. V. (2011). Study of the thermal stability of blends based on synthetic polymers and natural polysaccharides. *Russian Journal of Bioorganic Chemistry, 37*, 791–795. https://doi.org/10.1134/S1068162011070211.

Sandoval-Denis, M., et al. (2014). Identification and diversity of Trichoderma species using molecular tools. *Mycologia, 106*(3), 450–465. https://doi.org/10.3852/13-215.

Sawaguchi, A., Ono, S., Oomura, M., Inami, K., Kumeta, Y., Honda, K., Sameshima-Saito, R., Sawa, Y., & Sameshima-Yamashita, Y. (2015). Chitosan degradation and associated changes in bacterial community structures in two contrasting soils. *Soil Science and Plant Nutrition, 61*(3), 471–480. https://doi.org/10.1080/00380768.2014.1003965.

Sela, S., Frank, S., Belausov, E., & Pinto, R. (2006). A mutation in the luxS gene influences Listeria monocytogenes biofilm formation. *Applied and Environmental Microbiology, 72*(8), 5653–5658. https://doi.org/10.1128/AEM.00048-06.

Shah, A. A., Hasan, F., Hameed, A., & Ahmed, S. (2008). Biological degradation of plastics: A comprehensive review. *Biotechnology Advances, 26*(3), 246–265. https://doi.org/10.1016/j.biotechadv.2007.12.005.

Shah, A. A., et al. (2014). Microbial enzymes for biodegradation of synthetic plastics. *Applied Microbiology and Biotechnology, 98*, 8447–8461. https://doi.org/10.1007/s00253-014-5884-4.

Singh, B., & Sharma, N. (2008). Mechanistic implications of plastic degradation. *Polymer Degradation and Stability, 93*(3), 561–584. https://doi.org/10.1016/j.polymdegradstab.2007.11.008.

Sivan, A. (2011). New perspectives in plastic biodegradation. *Current Opinion in Biotechnology, 22*(3), 422–426. https://doi.org/10.1016/j.copbio.2011.01.013.

Sood, M., et al. (2020). Fungal cell wall structure and role in enzymatic degradation. *Frontiers in Microbiology, 11*, 1032. https://doi.org/10.3389/fmicb.2020.01032.

Srikanth, K., et al. (2022). Microbial degradation of polymers: Current perspectives. *Environmental Science & Technology, 56*(4), 2230–2242. https://doi.org/10.1021/acs.est.1c05640.

Starnecker, T., & Menner, M. (1996). Microbial degradation of biodegradable polymers in soil. *Applied Microbiology and Biotechnology, 46*(4), 319–324.

Stepanovic, S., Vukovic, D., Dakic, I., Savic, B., & Svabic-Vlahovic, M. (2000). A modified microtiter-plate test for quantification of staphylococcal biofilm formation. *Journal of Microbiological Methods, 40*, 175–179.

Tang, F. G., Bao, P., & Su, Z. H. (2016). Analysis of nanodomain composition in high-impact polypropylene by atomic force microscopy-infrared. *Analytical Chemistry, 88*(9), 4926–4930. https://doi.org/10.1021/acs.analchem.6b00798.

Teng, J., et al. (2020). SEM-based morphological analysis of degraded polymers. *Materials Characterization, 169*, 110567. https://doi.org/10.1016/j.matchar.2020.110567.

Thompson, L. R., Sanders, J. G., McDonald, D., Amir, A., Ladau, J., Locey, K. J., Prill, R. J., Tripathi, A., Gibbons, S. M., Ackermann, G., Navas-Molina, J. A., Janssen, S., Kopylova, E., Vázquez-Baeza, Y., González, A., Morton, J. T., Mirarab, S., Zech Xu, Z., Jiang, L., Haroon, M. F., et al. (2017). A communal catalogue reveals Earth's multiscale microbial diversity. *Nature, 551*, 457–463. https://doi.org/10.1038/nature24621.

Tiwari, A., et al. (2018). Imaging techniques for studying polymer biodegradation. *Journal of Environmental Chemical Engineering, 6*(2), 2858–2865. https://doi.org/10.1016/j.jece.2018.04.012.

Tokiwa, Y., Calabia, B. P., Ugwu, C. U., & Aiba, S. (2009). Biodegradability of plastics. *International Journal of Molecular Sciences, 10*(9), 3722–3742. https://doi.org/10.3390/ijms10093722.

Tondera, K., et al. (2021). Bioaugmentation using specialized microbial consortia. *Journal of Hazardous Materials, 403*, 123579. https://doi.org/10.1016/j.jhazmat.2020.123579.

Trafny, E. A. (2000). Formation of *Pseudomonas aeruginosa* biofilm and its significance in the pathogenesis of chronic infections. Powstawanie biofilmu Pseudomonas aeruginosa i jego znaczenie w patogenezie zakażeń przewlekłych. Postępy Mikrobiologii *Advances in Microbiology, 39*, 55–71.

Tsuji, H., & Ikada, Y. (2000a). Properties and morphology of poly(L-lactide). II. Hydrolysis in alkaline solution. *Journal of Polymer Science Part A: Polymer Chemistry, 38*(8), 1522–1530.

Tsuji, H., & Ikada, Y. (2000b). Properties and morphology of poly(L-lactide). IV. Effects of structural parameters on long-term hydrolysis of poly(L-lactide) in phosphate-buffered solution. *Polymer Degradation and Stability, 67*(1), 179–189. https://doi.org/10.1016/S0141-3910(99)00115-8.

Tyśkiewicz, R., Nowak, A., Ozimek, E., & Jaroszuk-Ściseł, J. (2022). Trichoderma: The current status of its application in agriculture for the biocontrol of fungal phytopathogens and stimulation of plant growth. *International Journal of Molecular Sciences, 23*(4), 2329. https://doi.org/10.3390/ijms23042329.

Vroman, I., & Tighzert, L. (2009). Biodegradation of polymers: Mechanisms and factors. *Materials, 2*(2), 307–344. https://doi.org/10.3390/ma2020307.

Wang, X., et al. (2009). Microbial consortia and degradation of pollutants. *Bioresource Technology, 100*(14), 3439–3446. https://doi.org/10.1016/j.biortech.2009.02.030.

Wei, R., & Zimmermann, W. (2017). Microbial enzymes for the recycling of recalcitrant petroleum-based plastics: How far are we? *Microbial Biotechnology, 10*(6), 1308–1322. https://doi.org/10.1111/1751-7915.12710.

Wiles, D. M., & Scott, G. (2006). Polyolefins with controlled environmental degradability. *Polymer Degradation and Stability, 91*(7), 1581–1592. https://doi.org/10.1016/j.polymdegradstab.2005.09.010.

Willis, C. D., et al. (2019). 16S rRNA gene sequencing in microbiology. *Frontiers in Microbiology, 10*, 3044. https://doi.org/10.3389/fmicb.2019.03044.

Woese, C. R. (1987). Bacterial evolution and the use of 16S rRNA sequences in systematics. *Microbiological Reviews, 51*(2), 221–271. https://doi.org/10.1128/mr.51.2.221-271.1987.

Wright, S. L., & Kelly, F. J. (2017). Plastic and human health: A micro issue? *Environmental Science & Technology, 51*(12), 6634–6647. https://doi.org/10.1021/acs.est.7b00423.

Wypych, G. (2008). *Handbook of material weathering* (4th ed.). ChemTec Publishing.

Yao, X., Guo, H., Zhang, K., Zhao, M., Ruan, J., & Chen, J. (2023). Trichoderma and its role in biological control of plant fungal and nematode disease. *Frontiers in Microbiology, 14*, 1160551. https://doi.org/10.3389/fmicb.2023.1160551.

Żakowska, H. (2003). *Biodegradable packaging*. Central Research and Development Centre for Packaging Opakowania biodegradowalne. Centralny Ośrodek Badawczo-Rozwojowy Opakowań.

Zeilinger, S., et al. (2016). Trichoderma spp. in industry and agriculture. *Fungal Biology Reviews, 30*(2), 74–90. https://doi.org/10.1016/j.fbr.2016.01.001.

Zhang, X., et al. (2021). Secondary metabolites of Trichoderma and their applications. *Biotechnology Advances, 47*, 107676. https://doi.org/10.1016/j.biotechadv.2020.107676.

Chapter 33

Commercial applications of biopolymer film in industries

Avishikta Banerjee[1], Baru Venkata Naga Sahithi[1], Vasantha Veerappa Lakshmaiah[1], Jameel M. Al-Khayri[2], and Praveen Nagella[1]

[1]*Department of Life Sciences, School of Sciences, CHRIST (Deemed to be University), Bengaluru, Karnataka, India,* [2]*Department of Agricultural Biotechnology, College of Agriculture and Food Sciences, King Faisal University, Al-Ahsa, Saudi Arabia*

33.1 Introduction

The application of films, especially those derived from natural biopolymers, has gained increasing attention from researchers and industrialists owing to their versatility and environmentally friendly nature. Biofilms formed from naturally occurring biopolymers—such as polysaccharides, proteins, and lipids—are considered viable alternatives to synthetic films for various industrial and commercial uses (Mamun & Chen, 2022; Niaounakis, 2015). Biopolymer applications range from protective films that cover different packaging to those used in the pharmaceuticals industry. These biopolymer films are thus a sustainable substitute that is crucial for the progression of present technologies without reducing or affecting their performance in any way (Olatunji, 2015). The increasing concerns arising from the degradation of synthetic packaging materials during disposal have posed a significant threat to the environment. Furthermore, the growing demand for renewable, sustainable and biodegradable packaging material had led to innovations in identifying biopolymers as best alternative materials (Mamun & Chen, 2022). Natural biopolymer films are thin individual layers of macromolecules derived from plant, animal, and microbial sources. They are created through practices like casting, extrusion, or coating and apply characteristic features like mechanical strength, elasticity, impermeability to gases, and water resistance depending on the kind of biopolymer used (Balart et al., 2021). Common examples of these biopolymers include cellulose, a plant-derived polysaccharide known for its high tensile strength and barrier properties (Balart et al., 2021); gelatin, a protein derived from collagen, which is valued for its film-forming capacity in food and pharmaceutical applications (Tyuftin & Kerry, 2021); and chitosan, a biopolymer derived from chitin, which has excellent antimicrobial properties (Jiménez-Gómez & Cecilia, 2020). Most of these biofilms are already being employed and are currently in the pipeline to be employed in areas such as food packaging and drug delivery systems. This demonstrates the versatility of the films and their technological affiliations for better industrial applications (Olatunji, 2015).

The application of natural biopolymer films has gained attention in various sectors over time. For instance, in the food industry, polysaccharide-based films such as starch and pectin are used to create edible coatings for perishable products to increase their shelf life and protect them from moisture and oxygen (Klein & Poverenov, 2020). In the medical field, biopolymer films like alginate dressings are employed in wound care since the material is biocompatible and has hydrogel-forming capability to support healing and new tissue formation (Baranwal et al., 2022; Udayakumar et al., 2021). Globally, proteins such as silk fibroin films are being applied in the cosmetic industry for the persistent release of active ingredients in skin care products (Aggarwal et al., 2020; Alves et al., 2020). These diverse applications demonstrate the importance of natural biopolymer films in catering to emerging and increasing global needs for high-performance and environmentally friendly films in multiple sectors. This chapter focuses on the industrial and commercial applications of biofilms, particularly those made from natural biopolymers. It also explores the various aspects related to the biopolymer films, including their raw material sources, production methods, environmental benefits, and practical uses. By examining the types, properties, and advantages of these materials, the chapter aims to highlight how natural biopolymers can address current challenges in material science and enhance sustainability. To illustrate their real-world applications, the chapter also includes case studies and examples from fields such as environmental engineering, pharmaceuticals, and food

Natural Polymer-Based Films and Coatings in Food Technology. DOI: https://doi.org/10.1016/B978-0-443-34244-8.00036-7

packaging. Natural biopolymers are macromolecules produced by living organisms and can be broadly classified into three main categories, as shown in Fig. 33.1

1. Polysaccharides, including cellulose, chitosan, and starch, are well-known for their film-forming ability, barrier functionality, and biodegradability. For example, cellulose acetate (CA), a cellulose derivative, is utilized in packaging films due to its transparency and mechanical durability (Liu et al., 2021).
2. Proteins, including whey protein, collagen, and gelatin, produce films with high oxygen barrier properties, making them ideal for use in the food and biomedical industries. For example, gelatin-based films are used as edible films in food coatings (Aaliya et al., 2021; Tyuftin & Kerry, 2021).
3. Lipids, such as wax and fatty acids, can generate hydrophobic films that act as moisture barriers and are frequently coupled with polysaccharides or proteins to enhance mechanical properties and efficiency. Lipid-based films are used extensively in the packaging sector to preserve moisture-sensitive products (Abdullah et al., 2022). Each of these biopolymers is characterized by distinct physicochemical properties that enable them to be suited for various applications in industries.

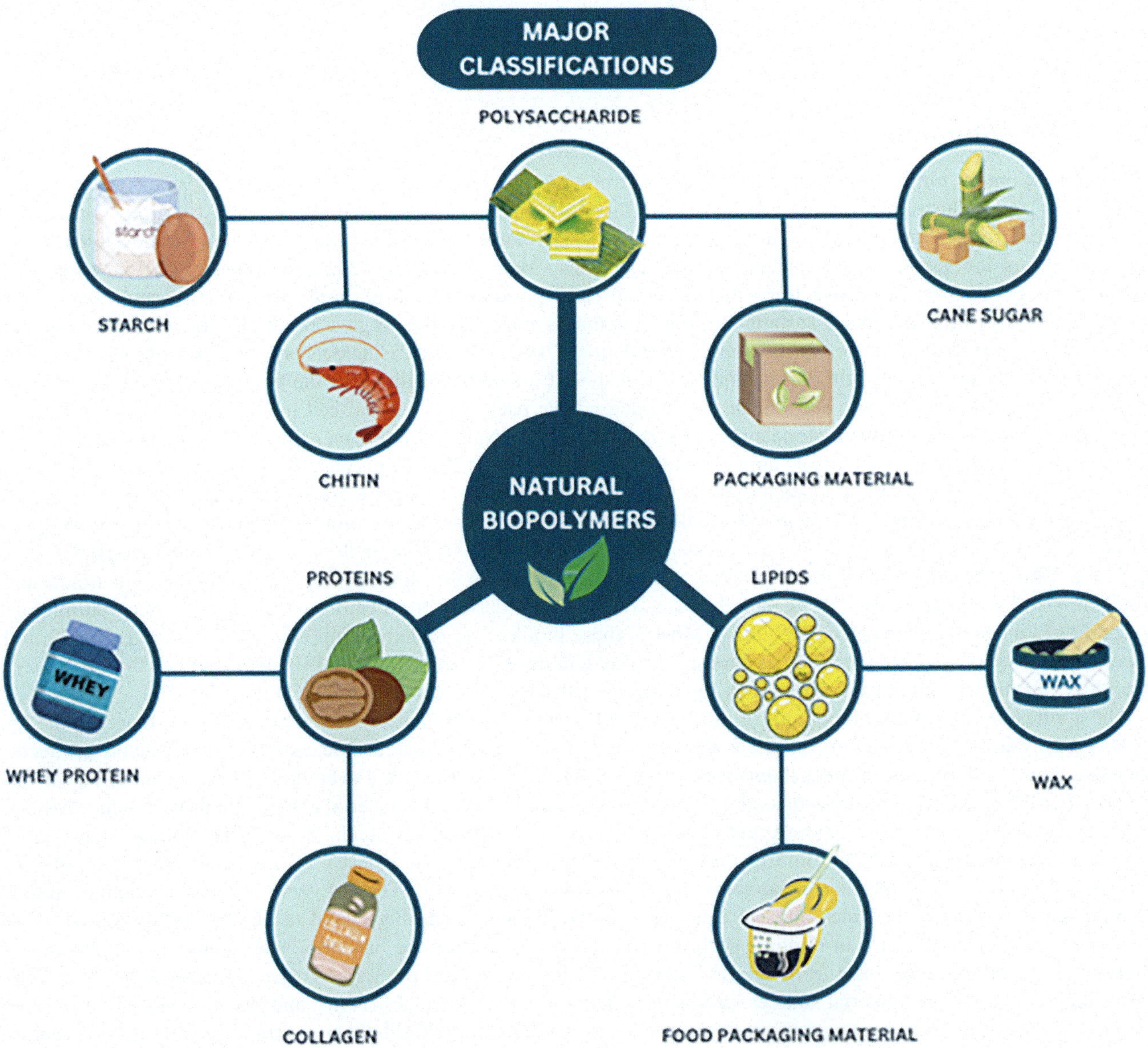

FIGURE 33.1 Major classifications of natural biopolymers.

Natural biopolymer films have various advantages over conventional synthetic polymers like polyethylene and polypropylene (Moeini et al., 2022). The most significant aspect is their biodegradability. Biopolymer films can be easily broken down in natural environments, minimizing the long-term adverse impacts and waste management concerns associated with synthetic plastics (Gowthaman et al., 2021; Teixeira-Costa & Andrade, 2021). Furthermore, because they are made from renewable resources such as plants (cellulose, starch), animals (gelatin, casein), and marine sources (chitosan), biopolymer films facilitate the use of sustainable, nonpetrochemical feedstocks (Mamun & Chen, 2022). Additionally, their nontoxic and biocompatible properties make them ideal for sensitive applications like food packaging and medical devices, where consumer safety is crucial (Gowthaman et al., 2021; Mamun & Chen, 2022; Vinod et al., 2020). For instance, films based on casein are utilized as edible coatings that do not jeopardize consumers' health in any way (Khan et al., 2021).

Natural biopolymer films have been adopted in industrial and commercial sectors due to their widespread environmental benefits, in addition to their general usefulness. Firstly, biopolymer films can be broken down by natural processes like microbial degradation, compared to synthetic polymers, which persist in the environment for decades, producing toxic by-products (Gowthaman et al., 2021). The usage of biopolymers significantly reduces plastic waste accumulation in landfills and oceans. Secondly, biopolymer production often involves the use of agricultural by-products or industrial waste materials, which conform to the principle of a circular economy by promoting waste valorization and reducing fossil fuel dependency (Mellinas et al., 2020). For example, chitosan is derived from shrimp and crab shell waste, wherein a waste product is transformed into a valuable resource (Seenuvasan et al., 2021). Thus, biodegradability and sustainability of these biofilms make them a promising solution for reducing the environmental carbon footprint of industries that rely heavily on single-use plastics.

33.2 Industrial applications of natural biopolymer films

33.2.1 Packaging industry

In the post-COVID-19 era, consumer preferences have shifted toward food safety, hygiene, and environmentally friendly solutions, driving interest in biodegradable packaging materials. Biopolymer films derived from renewable sources such as plants, animals, and microorganisms offer a sustainable alternative to conventional packaging. These films, composed of natural polymers like polysaccharides, proteins, and lipids, provide biodegradability and enhance food preservation. Active packaging, achieved by incorporating antimicrobial agents into biopolymer films, prevents microbial spoilage and extends the shelf life of perishable products while ensuring food quality and safety. Plant extracts, such as those from *Moringa oleifera* and durian leaves, are rich in phenolic compounds with strong antimicrobial and antioxidant properties. When incorporated into biopolymer films, these compounds disrupt microbial form, growth, and nutrient uptake, making them effective for packaging fresh meats and fruits. Similarly, essential oils (EOs) from garlic, thyme, cumin, and cinnamon, known for their amphiphilic properties and bioactive molecules, are incorporated into biopolymer films to inhibit bacterial and fungal growth, further enhancing food preservation. Some examples of biopolymers are listed in Table 33.1.

Case studies of companies using biopolymer films as sustainable packaging materials (e.g., biodegradable snack bags, compostable cutlery wrappers)

1. **BioBag International**
   Biopolymer used: Mater-Bi
   Processing method: Film extrusion
   Application: Biodegradable bags
   BioBag International produces compostable bags using Mater-Bi, a biopolymer derived from renewable resources like starches and vegetable oils. These are mixed in the appropriate proportions to manufacture biofilm through blowing, casting, and extrusion techniques. These bags are ideal for food packaging and carrier bag collection. They are bioeconomically certified compostable biofilms. These biopolymers decompose into organic matter within 90–180 days in industrial composting, offering an eco-friendly alternative to plastic bags (Khoo & Tan, 2010).
2. **Green Dot Bioplastics**
   Biopolymer used: EcoPure
   Processing method: Injection molding and film extrusion
   Application: Compostable cutlery and food packaging
   Green Dot Bioplastics produces compostable products using EcoPure, a biopolymer made from renewable plant materials. Through injection molding and film extrusion, they create sustainable cutlery, straws, and packaging films

TABLE 33.1 Examples of starch and cellulose-based biopolymer packaging materials.

S. no.	Material	Composition	Properties	Applications	References
Starch-based biopolymers					
1.	Cassava starch-based foams with grape stalks	Cassava starch, grape stalks (reinforcement)	High flexibility, mechanical strength, resilience	Bakery product packaging (e.g., English cakes), maintains integrity over nine days at 55% relative humidity, 23°C	Engel et al. (2019)
2.	Starch films with citric pectin and Feijoa peel flour	Starch, citric pectin, Feijoa peel flour	Antioxidant, antimicrobial, enhances barrier properties	Active packaging for fruits like apples, preserves freshness for 15 days, controls oxidative damage	Sganzerla et al. (2020)
3.	Cowpea starch films with maqui berry extract (MBE)	Cowpea starch, MBE	High antioxidant activity, delays lipid oxidation	Packaging for high-fat foods like fish (e.g., salmon), reduces lipid oxidation during cold storage (4°C)	López De Dicastillo et al. (2016)
4.	Yam starch films with eugenol	Yam starch, eugenol (3% concentration)	Antimicrobial, extends shelf life by over 50%, reduces microbial growth	Meat packaging (e.g., pork), continuous antibacterial action for perishable items	Cheng et al. (2019)
5.	Pea starch/polylactic acid (PLA) bilayer films	Pea starch, PLA	Improved moisture and oxygen barrier, reduced weight loss	Fresh produce packaging (e.g., cherry tomatoes), minimizes dehydration during storage	Zhou et al. (2019)
6.	Rye starch films with rosehip extract (RHE)	Rye starch, RHE	Antioxidant, inhibits lipid oxidation, improves light barrier, and flexibility	Packaging for meats (e.g., chicken breast), preserves quality during storage	Go and Song (2019)
7.	Acetylated cassava starch and green tea extract blended with linear low-density polyethylene (LLDPE)	Acetylated cassava starch, LLDPE, green tea extract	Enhanced mechanical strength, antioxidant release reduces oxidative rancidity	Packaging for cured meats (e.g., bacon), slows oxidative spoilage during storage	Panrong et al. (2019)
8.	Corn starch-based films with bacterial cellulose coating	Corn starch, bacterial cellulose nano whiskers	Improved tensile strength, moisture, and gas barrier properties	Perishable food packaging, better moisture resistance for extended shelf life	Fabra et al. (2016)
9.	Thermoplastic starch with carboxymethyl chitosan and propolis extract	Thermoplastic starch, carboxymethyl chitosan, propolis extract	Antimicrobial, antioxidant, improved mechanical strength	Suitable for oil-based foods and meats, inhibits microbial growth, extends shelf life	Valencia-Sullca et al. (2018)
10.	Potato starch films reinforced with TiO_2 nanoparticles	Potato starch, TiO_2 nanoparticles	Antimicrobial, improved water resistance, mechanical strength	Packaging for moisture-sensitive items (e.g., mushrooms), prevents microbial growth and moisture loss	Dash et al. (2019)

Cellulose-based biopolymers					
1.	Cellulose acetate	Acetylated cellulose	Hydrophobic, high mechanical strength, flexible, controls moisture transfer	Fresh produce and dairy product packaging, preserves quality and freshness	Teixeira et al. (2022)
2.	Carboxymethyl cellulose	Water-soluble cellulose derivative	Transparent, excellent barrier against fats and oils, antimicrobial when additives included	Packaging for meats and cheese, food coatings, reduces food spoilage and waste	Hu et al. (2016)
3.	Hydroxypropyl methylcellulose (HPMC)	HPMC cellulose derivative	Excellent gas barrier (oxygen, CO_2), poor water resistance (can be improved with blends)	Fresh fruit and vegetable packaging reduces spoilage by preserving humidity	Farina et al. (2020)
4.	Methylcellulose (MC)	Methylcellulose (heat stable)	Strong oil and fat barrier, heat stability	Baked goods and fried food coatings, reduces moisture loss, maintains texture and quality	García et al. (2002)
5.	Bacterial cellulose	Cellulose produced by bacteria	High tensile strength, excellent water-retention capacity	Fresh produce packaging, ideal for high-moisture foods (e.g., fruits and vegetables), durable and protective	Chen et al. (2024)
6.	Cellulose nitrate	Modified cellulose	High durability, resistant to moisture and light	Packaging for pharmaceuticals and cosmetics, protects sensitive formulations from environmental damage	Ahmadi et al. (2022)
7.	Ethyl cellulose	Water-insoluble cellulose derivative	Moisture barrier, flexible, lightweight	Pharmaceutical coatings (tablets, capsules), protect active ingredients and moisture-sensitive food items	Ahmadi et al. (2022)
8.	Nanocellulose (NCC/NFC)	Nanocrystalline or nanofibrillated cellulose	Enhanced mechanical strength, improved barrier properties	Packaging requiring mechanical strength, smart packaging (environment responsive)	Azeredo et al. (2017)
9.	Cellophane	Regenerated cellulose film	Transparent, biodegradable, excellent gas and moisture barrier	Food packaging (candies, baked goods), wraps for perishable items, eco-friendly alternative to plastic	Azeredo et al. (2017)
10.	Cellulose-PLA composites	Cellulose blended with PLA	Biodegradable, improved tensile strength, barrier against gases	Packaging for fruits, vegetables, and other food items requiring extended shelf life	Ferrer et al. (2017)

that degrade in industrial and home compost environments. These provide commitment to sustainable, eco-friendly, and uncompromising alternative packing material (2022).

3. **TIPA Corp (TIPA Corp Ltd)**
 Biopolymer used: Bio-based polyethylene (PE) and PLA
 Processing method: Coextrusion
 Application: Compostable packaging films
 TIPA Corp develops flexible, compostable packaging using bio-based PE and PLA through coextrusion. Their multilayer films mimic conventional plastics while being compostable within 180 days, leaving no harmful residues. Ideal for food packaging, TIPA's solutions enhance shelf life and support sustainability (Bianchi et al., 2023).
4. **Nature Works Limited Liability Company (LLC)**
 Biopolymer used: Ingeo PLA
 Processing method: Injection molding and film extrusion
 Application: Biodegradable snack bags
 NatureWorks LLC produces Ingeo, a biopolymer from renewable cornstarch, using injection molding and film extrusion. Ingeo creates biodegradable snack bags with excellent mechanical and barrier properties, breaking down in industrial composting within 60–90 days, offering a sustainable alternative to traditional packaging (Vink et al., 2010).
5. **Mondi Group**

 Biopolymer used: Biomass-based polyethylene
 Processing method: Extrusion coating
 Application: Compostable wrappers
 Mondi Group uses biomass-based polyethylene from renewable resources to create compostable food wrappers through extrusion coating. These wrappers decompose in industrial composting, supporting sustainable packaging and reducing reliance on fossil fuels, exemplifying the shift toward biodegradable alternatives in the packaging industry (Dirpan et al., 2023).

33.2.2 Agriculture

Biopolymer films, such as poly(butylene adipate-co-terephthalate) (PBAT) and PLA, are increasingly used in agriculture for applications like mulch films and seed coatings due to their eco-friendly and functional properties. Unlike conventional polyethylene films, biopolymer-based films decompose into nontoxic by-products like water, CO_2, and biomass, eliminating disposal challenges (Mistretta et al., 2021). These films can be customized for specific needs by adjusting thickness, biodegradability, and mechanical properties. Nanocomposites like organo-modified clay and calcium carbonate are added to PBAT/PLA blends to enhance mechanical properties such as stiffness, tear strength, and UV resistance while maintaining biodegradability. In seed coatings, biopolymer films regulate water absorption and nutrient release, improving germination and seedling vigor (Abdul Khalil et al., 2018). Beyond environmental sustainability, biopolymer films protect soil from erosion, improve water retention, create a favorable microclimate for crops, and enable controlled fertilizer and pesticide release, reducing chemical use and boosting the yields. Derived from renewable sources like starch and cellulose, these films enhance soil health and lower agriculture's carbon footprint (Gamage et al., 2023). A list of commercial biopolymers and their applications is presented in Table 33.2.

These innovations represent a broad range of biodegradable, compostable, and environmentally friendly solutions to address the challenges of synthetic agricultural films. Each product offers specific benefits, such as improved mechanical strength, moisture retention, and UV resistance, while contributing to sustainability through biodegradation and reduced soil pollution.

33.2.3 Application in biomedical industry

The role of biopolymers in wound care and cosmetics is profoundly increasing due to their bioactivity, antimicrobial, antioxidant, antiinflammatory properties, and structural similarity to extracellular matrix (ECM) components such as collagen, chitosan, alginate, cellulose, and hyaluronic acid (HA).

33.2.3.1 Wound healing and tissue engineering applications

Biopolymers have demonstrated remarkable potential in accelerating wound healing and supporting tissue regeneration. Their unique biochemical properties, such as the ability to mimic the ECM, enhance cell adhesion, proliferation, and differentiation, which are critical for tissue repair (Sahana & Rekha, 2018).

TABLE 33.2 Examples of commercial products and innovations in agricultural films.

S. no.	Product	Type	Applications	Details	References
1.	ECOVIO by Badische Anilin- und Soda-Fabrik (BASF)	PLA/poly(butylene adipate-co-terephthalate) (PBAT) blend	Mulch films, seed coatings, packaging	Compostable polymer blend of PBAT and PLA with excellent flexibility, tensile strength, and biodegradability. Used as an eco-friendly alternative to polyethylene films	Souza et al. (2020)
2.	Mater-Bi by Novamont	Starch-based biopolymer	Mulching films	Made from renewable feedstocks like starch, fully biodegradable, reducing soil pollution. Widely used in European sustainable agriculture	Merino et al. (2022)
3.	BioFlex by Forschungsinstitut Kunststoff und Recycling (FKuR)	PLA/PBAT blend	Mulch films, agricultural packaging	Biodegradable and compostable, balancing flexibility and durability. Decomposes under composting conditions, leaving no toxic residues	Merino et al. (2022)
4.	BioBag mulch film	Compostable biopolymer (PBAT, PLA)	Mulching	Designed to biodegrade within the growing season. Used in organic and conventional farming for weed suppression and moisture conservation. Certified for organic farming	Fourati et al. (2021)
5.	Cereplast Compostables	PLA-based biopolymer	Agricultural films, seed coatings	Engineered for quick biodegradability, offering sustainable alternatives to synthetic films. Suitable for applications requiring rapid decomposition	Fourati et al. (2021)
6.	AGROFILM by Polysack	PLA/cellulose blend	Greenhouse films, mulching films	Biodegradable with enhanced strength and UV resistance. Designed for horticulture, degrading after the growing season, eliminating disposal needs	Fourati et al. (2021)
7.	Solanyl by Rodenburg	Potato starch-based biopolymer	Mulch films, seedling pots	Derived from nonfood grade potato starch, biodegradable, disintegrates naturally, and enriches soil with organic content	Souza et al. (2020)
8.	Biolice by Limagrain	Corn starch-based biopolymer	Mulch films	Made from corn starch, biodegradable, reduces weed growth, conserves soil moisture, and leaves no harmful residues. Popular in sustainable European agriculture	Souza et al. (2020)
9.	Plantic by Kuraray	Starch-based biopolymer	Agricultural films, packaging	Known for excellent barrier properties, used for greenhouse films and soil mulching. Biodegradable and compostable under industrial conditions	Fourati et al. (2021)
10.	PLA mulch films by NatureWorks	Polylactic acid	Mulching	PLA-based films certified for composting. Eco-friendly alternative to polyethylene, enhancing soil quality after degradation while conserving moisture	Souza et al. (2020)

Collagen, a protein ubiquitous in mammalian ECM, is one of the most extensively studied biopolymers for wound healing. Its triple-helical structure, formed from amino acid chains, provides both tensile strength and a scaffold-like environment for cell migration and proliferation. Collagen-based materials, such as hydrogels and films, have been shown to reduce inflammation and support re-epithelialization in chronic wounds, as they promote granulation tissue formation through the interaction of amino acid sequence arginine-glycine-aspartic acid (RGD)sequences with integrins on cell surfaces. Studies incorporating collagen with bioactive compounds, such as silver nanoparticles, have further enhanced its antimicrobial properties, making it suitable for infected wounds (Gobi et al., 2021).

Similarly, chitosan, derived from the deacetylation of chitin, is valued for its antimicrobial activity, hemostatic properties, and ability to form hydrogels and nanofibers. These properties make it a versatile candidate for wound dressings and tissue engineering scaffolds. For instance, chitosan-alginate composite films have been developed to provide a moist healing environment while effectively absorbing exudates. Their use has been associated with improved fibroblast activity and angiogenesis, crucial steps in the repair of chronic and acute wounds (Sahana & Rekha, 2018).

Cellulose, particularly bacterial cellulose (BC), is another biopolymer that has gained prominence due to its high water-retention capacity and structural similarity to human ECM. Functionalized cellulose dressings, such as oxidized cellulose, have demonstrated excellent hemostatic and antimicrobial properties, significantly enhancing the healing of partial-thickness burns and chronic ulcers (Gobi et al., 2021).

33.2.3.2 *Drug delivery system applications*

The integration of biopolymers into drug delivery systems leverages their ability to form smart-release matrices, ensuring targeted and sustained delivery of therapeutic agents. This characteristic is particularly valuable for reducing systemic side effects and improving the bioavailability of drugs.

Silk fibroin, extracted from the cocoons of *Bombyx mori*, has emerged as an excellent matrix for drug delivery due to its mechanical robustness, biodegradability, and tunable hydrophilic-hydrophobic balance. Silk fibroin nanoparticles and hydrogels have been employed for the delivery of chemotherapeutic agents, like doxorubicin, achieving controlled-release profiles and enhanced therapeutic efficacy. Crosslinked silk fibroin scaffolds have also shown promise in tissue engineering applications, demonstrating their multifunctional utility (Sahana & Rekha, 2018).

Chitosan nanoparticles, prepared through techniques such as ionic gelation, are widely used for encapsulating hydrophilic and hydrophobic drugs. Their mucoadhesive properties ensure prolonged retention at the target site, improving drug absorption and therapeutic outcomes. Chitosan-based systems have been successfully applied for delivering anticancer drugs and antibiotics, with studies highlighting their potential to overcome multidrug-resistant bacterial infections (Opriş et al., 2024).

Alginate, a polysaccharide derived from brown algae, forms hydrogels through ionotropic gelation with divalent cations, making it suitable for oral and localized drug delivery. Alginate-based systems have been investigated for encapsulating proteins and small molecules, with promising results in achieving controlled and site-specific drug release (Opriş et al., 2024). Some examples are mentioned in Table 33.3.

33.2.3.3 *Exploring the properties that make biopolymer films suitable for biomedical applications*

Biopolymer films, derived from natural sources like polysaccharides, proteins, and glycosaminoglycans, possess unique properties that make them indispensable for biomedical applications. Their suitability is attributed to their inherent characteristics, such as biocompatibility, nontoxicity, biodegradability, and tunable physicochemical properties. These features enable their use in areas such as wound healing, tissue engineering, and drug delivery systems (Fazal et al., 2023).

33.2.3.3.1 Biocompatibility

Biocompatibility ensures that biopolymer films integrate seamlessly with human tissues without triggering adverse immune responses. This property is critical for medical devices, implants, and therapeutic applications. For instance, chitosan—a polysaccharide derived from chitin—is widely recognized for its capacity to promote cellular interactions and tissue regeneration. When used as a scaffold or dressing, it facilitates fibroblast adhesion and migration, which are crucial processes in wound healing (Opriş et al., 2024).

33.2.3.3.2 Nontoxicity and safety

A defining feature of biopolymer films is their nontoxic nature, both during application and after degradation. Unlike synthetic materials, which may release harmful residues, biopolymers degrade into nontoxic by-products. HA, commonly found in connective tissue, exemplifies this property. It is metabolized into harmless sugars, making it an excellent candidate for applications ranging from cosmetic fillers to joint repair (Wu et al., 2022).

TABLE 33.3 Examples of commercial biopolymer-based medical products.

S. no.	Product	Type	Applications	Details/innovations	References
1.	RESOMER by Evonik	PLA, poly ε-caprolactone, PLA-PEG blends	Sutures, bone fixation screws, surgical meshes, drug delivery systems, 3D printing implants	Biodegradable polymers with customizable degradation periods. Used for complex medical implants like cranial scaffolds. Innovations include osteoconductive composites with calcium phosphate for enhanced bone regeneration	Song et al. (2018)
2.	ACTIVHEAL by advanced medical solutions	Collagen and bioactive polymers	Wound care for diabetic ulcers, pressure sores, burns	Hydrophilic hydrogel dressings maintain a moist environment for faster healing. Variants impregnated with antimicrobial agents, like silver ions, enhance infection control	Gobi et al. (2021)
3.	CHITOGUARD by LifeBond	Chitosan-based materials	Surgical wound management, trauma care	Chitosan's hemostatic and antimicrobial properties are effective in vascular surgeries, emergency wound closure, and combat injuries. Forms a rapid hemostatic barrier	Opriş et al. (2024)
4.	HYALOFILL by Anika Therapeutics	Hyaluronic acid (HA)-based dressings	Chronic wounds, diabetic ulcers,cosmetic/orthopedic applications	Supports cell proliferation, modulates inflammation, and accelerates extracellular matrix remodeling. Crosslinked HA molecules extend therapeutic effect and provide structural integrity	Song et al. (2018)
5.	GELFOAM by Pfizer	Gelatin-based sponge	Hemostatic agent for neurosurgery, cardiovascular surgery, organ transplantation	Porcine-derived sponge assists in clot formation and controls bleeding. Absorbed naturally by the body within four tosix weeks	Sahana and Rekha (2018)
6.	ORTHOCORD by DePuy Synthes	Polydioxanone and PLA-based sutures	Ligament and tendon repair	Combines tensile strength and controlled degradation. Coated with bioactive compounds that promote tissue integration and faster healing	Leong et al. (2020)
7.	EPIFIX by MiMedx	Amniotic membrane tissue	Hard-to-heal wounds (e.g., venous leg ulcers, complex surgical sites)	Acts as a bioactive scaffold rich in collagen, growth factors, and cytokines. Promotes cell proliferation, reduces inflammation, and is FDA-approved for regenerative medicine	Gobi et al. (2021)
8.	DERMABOND by Ethicon	Cyanoacrylate-based adhesive	Wound closure in emergency rooms and minor surgeries	Forms a flexible, antimicrobial seal over wounds, reducing infection risk and improving esthetic healing outcomes	Leong et al. (2020)
9.	BIOVISC by Biomatrix	HA-based injectable	Osteoarthritis treatment	Improves joint lubrication, reduces pain, and delays joint degeneration. Crosslinked HA formulations provide prolonged therapeutic effects	Sahana and Rekha (2018)
10.	INNOVASIVE by Arthrex	PLA and polyglycolic acid copolymers	Bioresorbable interference screws, orthopedic implants	Naturally degrading devices eliminate the need for removal surgeries. Structural enhancements support tissue regeneration and withstand mechanical stress	da Silva et al. (2018)

33.2.3.3.3 Biodegradability

The ability to degrade naturally is a hallmark of biopolymers, aligning with the need for temporary biomedical applications. Alginate, for example, breaks down into its constituent sugars under physiological conditions, facilitating its use in transient wound dressings or drug carriers. This ensures that the material does not require surgical removal or pose long-term environmental risks (Fasiku et al., 2019).

33.2.3.3.4 Mechanical versatility and customizability

The mechanical properties of biopolymers, such as elasticity, tensile strength, and porosity, can be tailored to meet specific biomedical needs. Cellulose-based films, with their high tensile strength and barrier properties, are ideal for wound dressings that require durability. Similarly, protein-based films, like gelatin, can be modified to form hydrogels or scaffolds that mimic the ECM, promoting tissue repair (Wu et al., 2022).

33.2.3.3.5 Antimicrobial properties

Some biopolymers, such as chitosan, exhibit intrinsic antimicrobial activity, making them particularly effective in infection-prone applications. Chitosan disrupts bacterial membranes and inhibits biofilm formation, making it a preferred choice for wound dressings and drug delivery systems where sterility is paramount (Nathanael & Oh, 2020).

33.2.3.3.6 Hydrophilicity and moisture regulation

Biopolymers like alginate and BC are hydrophilic, allowing them to maintain a moist environment conducive to wound healing. This moisture retention accelerates re-epithelialization and minimizes scarring. In drug delivery systems, hydrophilic films ensure controlled drug release by regulating diffusion rates (Kanjanamosit et al., 2009).

33.2.3.3.7 Functionalization and hybridization

The chemical structures of biopolymers allow for extensive functionalization, enhancing their performance for specialized applications. For instance, functional groups in HA enable the attachment of therapeutic agents or crosslinking to create hydrogels with specific degradation rates. Collagen can be blended with synthetic polymers to improve its mechanical stability, making it suitable for load-bearing applications in tissue engineering (Sulaeva et al., 2020).

33.2.3.3.8 Environmental sustainability

Biopolymers are renewable and biodegradable, addressing the dual challenge of medical waste and environmental sustainability. Their production from natural sources like algae, plants, or animal by-products ensures a reduced carbon footprint compared to petroleum-based materials. The unique combination of biocompatibility, nontoxicity, biodegradability, and customizability makes biopolymer films invaluable in biomedical science. These properties not only enhance patient outcomes but also align with global sustainability goals, positioning biopolymers as critical components in the future of medical technology (Sulaeva et al., 2020).

33.2.4 Cosmetics and personal care

Biopolymer films are increasingly used in skincare and cosmetics due to their multifunctionality, eco-friendliness, and compatibility with human skin. Derived from natural polymers like polysaccharides, proteins, and glycoproteins, they provide hydration, deliver active ingredients, and enhance skin health and appearance (Mendoza-Muñoz et al., 2023). hyaluronic acid (HA) known for its hydrating properties, binds water molecules and reduces trans-epidermal water loss, improving skin elasticity and radiance (Sahana & Rekha, 2018). Chitosan and alginate form breathable, protective layers on the skin, shielding it from pollutants and UV rays. Chitosan's antimicrobial properties are ideal for acne-prone skin, while alginate's gel-forming ability soothes and enhances barrier function (Opriş et al., 2024). Collagen-based films address aging by promoting fibroblast activity and improving skin elasticity. They deliver peptides directly to the dermis, enhancing structural integrity and reducing wrinkles (Gobi et al., 2021). After cosmetic procedures, HA and alginate films mimic the ECM, aiding cell repair and tissue regeneration. Biopolymers like silk fibroin encapsulate and deliver active ingredients, ensuring sustained release and prolonged efficacy, particularly in targeted applications like undereye patches (Sahana & Rekha, 2018). Innovations in cellulose, xanthan gum, and pectin hydrogels are widely used in sheet masks for superior hydration. Additionally, biopolymers' biodegradable nature supports sustainability in skincare formulations and packaging. Few examples are mentioned in Table 33.4.

TABLE 33.4 Examples of biopolymer film products in the cosmetics market.

S. no.	Product	Type	Applications	Details/innovations	References
1.	Verdessence by BASF	Biopolymer-based ingredients	Cosmetic creams, serums, masks, and powders	Includes five biopolymers: RiceTouch (matte finish), glucomannan (texturizer for serums), alginate (sensory enhancer), xanthan (stabilizer), tara (sun care texture enhancer). Sustainable, non-GMO, and cold-processing compatible	Serra et al. (2024)
2.	HyaCare by Evonik	Hyaluronic acid-based biopolymer	Moisturizers, serums, and antiaging creams	Exceptional water retention provides long-lasting hydration and improves skin elasticity. Reduces fine lines and wrinkles. Used in high-end and medical-grade skincare	Mitura et al. (2020)
3.	Chitoclear by Primex	Marine chitosan-based biopolymer	Acne-control formulations, serums, and scalp treatments	Provides antimicrobial and hydrating effects. Forms a breathable protective film to soothe sensitive and inflamed skin. Reduces bacterial activity effectively	Mitura et al. (2020)
4.	Seafilm by Marinova	Fucoidan-based polysaccharide (brown seaweed)	Sheet masks, hydrating and antiaging products	Offers antioxidant and antiinflammatory properties, enhancing skin repair and regeneration. Effective for environmentally stressed skin	Olsen (2015)
5.	Collactive by Copalis	Marine collagen and elastin blend	Antiaging creams, firming serums	Supports skin's structural integrity, reduces wrinkle depth, and enhances skin elasticity. Ideal for luxury skincare targeting mature skin	Rahman et al. (2024)
6.	SilkGel by Givaudan	Silk protein-based biopolymer	Primers, setting sprays, and antiaging serums, hair care	Forms a silky protective film that locks in moisture and delivers actives. Provides a smooth, lustrous finish for both skin and hair care products	Liu et al. (2024)
7.	Biophilic by Lubrizol	Phospholipid-based biopolymer	Emulsions, creams, and serums for dry and sensitive skin	Improves emulsification and enhances sensory properties. Hydrates effectively and controls delivery of active ingredients, making it ideal for sensitive skin	Salvatore et al. (2020)
8.	AquaColl by De Nederlandse Staatsmijnen (DSM)	Soluble collagen derivative	Premium moisturizing, antiaging, and repair creams	Rebuilds the skin's natural barrier, enhances hydration, and reduces irritation. Focuses on rejuvenation and repair in high-end skincare products	Rahman et al. (2024)

33.2.5 Electronics and engineering

Biopolymer films have emerged as transformative materials in engineering applications, particularly in flexible electronics, advanced sensors, and functional coatings. These materials, derived from natural sources such as cellulose, silk fibroin, chitosan, and PLA, exhibit properties such as high mechanical strength, lightweight structures, biocompatibility, and biodegradability, making them highly suitable for next-generation technologies (Sun et al., 2018).

33.2.5.1 Biopolymer films in flexible electronics

Flexible electronics require materials that combine mechanical flexibility with environmental sustainability, and biopolymer films meet these demands effectively. Silk fibroin, derived from silkworm cocoons, is valued for its optical transparency, mechanical durability, and adaptability to nonplanar surfaces, making it ideal for wearable devices, implantable sensors, and biodegradable circuits. When combined with conductive materials like graphene or carbon nanotubes, silk fibroin forms bio-friendly piezoresistive sensors for applications such as fitness trackers and robotic skins (Sahana & Rekha, 2018). Nanocellulose-based films, known for their tensile strength, thermal stability, and low thermal expansion, are used as substrates in flexible displays, organic light emitting diodes (OLEDs), and photovoltaic cells. These films, when integrated with conductive nanomaterials, enable foldable supercapacitors and wearable biosensors. PLA, derived from renewable sources like corn starch, offers biodegradability and is used in flexible circuit boards and energy-harvesting devices through the addition of piezoelectric nanoparticles (da Silva et al., 2018). PHAs, microbial polyesters with excellent thermal properties, are increasingly used in wearable and implantable electronics for medical monitoring due to their biodegradability and versatility (Sun et al., 2018).

33.2.5.2 Biopolymer films in advanced sensors

Biopolymer films are extensively used in the development of sensors that require precise and consistent signal transduction capabilities. Their biocompatibility and functional groups allow them to immobilize bioactive molecules for sensitive detection. Derived from chitin, chitosan exhibits excellent film-forming capabilities and antimicrobial properties. These films serve as matrices in biosensors to detect glucose, cholesterol, and pathogens. Electrochemical sensors utilizing chitosan-polymer composites have demonstrated improved sensitivity due to their high ionic conductivity and ability to immobilize enzymes effectively (Jiménez-Gómez & Cecilia, 2020; Vasudevan et al., 2024). HA-based biopolymer films are employed in electrochemical sensors and optical biosensors. Their hydrophilicity and biocompatibility enhance the stability of immobilized enzymes and antibodies, ensuring accurate detection of biomolecules in medical diagnostics. Pectin, a polysaccharide extracted from plant cell walls, is used in environmental sensors to detect heavy metals and toxins. Pectin-based films, when integrated with conductive polymers, provide cost-effective solutions for environmental monitoring while ensuring minimal ecological impact (Vasudevan et al., 2024).

33.2.5.3 Biopolymer films in functional coatings

The coating industry benefits immensely from biopolymer films, which provide eco-friendly solutions for improving surface properties such as hydrophobicity, antimicrobial activity, and wear resistance. CA and nanocellulose are used in coatings for electronic devices to enhance durability and thermal stability. These coatings offer protection against mechanical wear and environmental factors while maintaining lightweight structures suitable for portable devices (Haniffa et al., 2016; Sun et al., 2018). Biodegradable coatings made from PLA are widely applied in electronics packaging and corrosion-resistant layers for metal components. Their hydrophobic nature protects sensitive components from moisture, ensuring the longevity of devices. Silk fibroin-based coatings are utilized in implantable devices to enhance biocompatibility and reduce inflammation. They form a protective barrier that enables seamless integration of devices with biological tissues, making them ideal for medical-grade sensors and electrodes (Sahana & Rekha, 2018; Sun et al., 2018).

33.2.5.4 Advantages of biopolymer films in engineering applications

- **Lightweight and flexible**: Biopolymer films, such as silk fibroin and nanocellulose, combine low density with mechanical strength, making them ideal for portable and wearable devices.
- **Biocompatibility:** Their ability to integrate with biological systems without causing adverse effects ensures their widespread use in healthcare technologies, like biosensors and implantable electronics.
- **Eco-friendly nature:** Derived from renewable resources, biopolymers such as PLA and PHAs align with sustainable engineering practices, reducing environmental impact.

- **Tailorability:** Biopolymers can be chemically or physically modified to achieve specific properties, like conductivity, piezoelectricity, and moisture resistance, expanding their application scope.
- **Biodegradability:** Unlike synthetic materials, biopolymers degrade naturally, addressing issues of electronic waste and pollution.

33.2.5.4.1 Examples of innovative uses in electronics and commercial products

1. **BioFlex films for flexible electronics:** BioFlex is a PLA/PBAT blend adapted for use in flexible and stretchable electronic components. These biodegradable films provide the necessary mechanical strength and flexibility for developing eco-friendly circuits and sensors used in wearable technologies, like health monitors and fitness trackers (Merino et al., 2022; Souza et al., 2020).
2. **Cellulose nanofiber displays:** Nanocellulose-based films are being used as substrates for flexible displays in devices like e-readers and foldable smartphones. These films offer transparency, thermal stability, and biodegradability, reducing electronic waste compared to traditional plastic alternatives. Companies like Fujitsu have explored nanocellulose films for lightweight yet durable display substrates (Liu et al., 2021; Souza et al., 2020).
3. **Chitosan-based biosensors:** Chitosan films are used in biosensors to detect biomolecules such as glucose or cholesterol. Their biocompatibility and antimicrobial properties make them ideal for wearable devices in medical diagnostics. These sensors often incorporate chitosan blended with conductive materials like graphene for enhanced performance (Jiménez-Gómez & Cecilia, 2020).
4. **Silk fibroin electronics:** Silk fibroin films are employed in biodegradable electronics, particularly in implantable devices such as temporary pacemakers and bioresorbable sensors. Their optical transparency and compatibility with organic semiconductors enable applications in bio-integrated circuits and photodetectors (Opriş et al., 2024).
5. **Smart packaging films with pH indicators:** Biopolymer films embedded with anthocyanin dyes have been developed as pH-responsive indicators for monitoring the freshness of perishable goods, like meat. These intelligent films, crafted from PLA and chitosan blends, also have potential in electronic skins to monitor environmental conditions, such as humidity and acidity (Haniffa et al., 2016; Merino et al., 2022).

33.2.5.4.2 Commercial products

1. **DuPont's Kapton alternatives:** Biopolymer-based alternatives to Kapton, typically used in flexible circuits, are being developed for sustainable electronics. These products utilize CA films, which offer similar dielectric properties while remaining biodegradable (Dias et al., 2019).
2. **Flexible conductive films by nanocellulose companies:** Several startups, such as GranBio, are commercializing nanocellulose conductive films for use in flexible solar panels and supercapacitors. These films support energy storage applications in portable and wearable devices (Dias et al., 2019).
3. **Paptic in flexible packaging:** While primarily developed for packaging, Paptic, a cellulose-based film, is being adapted for protective coatings in electronics due to its durability and ability to resist wear and tear, offering a sustainable alternative to synthetic films (Liu et al., 2021).

33.2.6 Food and beverage industry

Biopolymer films have become an essential innovation in food packaging, particularly for the preservation of fresh produce and other perishable goods. These materials, derived from renewable sources such as plant polysaccharides, animal proteins, and microbial biopolymers, are increasingly recognized for their role in reducing food spoilage, enhancing safety, and offering sustainable packaging alternatives. Through the integration of bioactive agents, these films not only protect food but also improve its quality, making them vital in modern food preservation technologies.

33.2.6.1 Polysaccharide-based biopolymer films

Polysaccharides such as cellulose, starch, pectin, and alginate are among the most studied materials for biopolymer films due to their abundant availability, film-forming ability, and excellent barrier properties against oxygen and oils. These materials are often used in coatings and standalone films for fresh produce, baked goods, and minimally processed foods.

1. **Cellulose and nanocellulose:** Cellulose-based films, particularly those reinforced with nanocellulose, demonstrate superior tensile strength, thermal stability, and gas barrier properties. Their high crystallinity and ability to form dense structures make them effective in reducing oxygen permeability, thereby slowing oxidation and spoilage. Nanocellulose films have been shown to extend the shelf life of fruits like strawberries and apples by minimizing respiration and moisture loss (Basumatary et al., 2022).

2. **Starch films:** Starch-based films are widely used due to their transparency, biodegradability, and ability to form a continuous matrix. These films are particularly effective in controlling moisture migration, a critical factor in preventing shriveling and wilting in vegetables. However, due to their hydrophilic nature, starch films are often blended with other biopolymers or reinforced with nanoparticles, such as montmorillonite clay, to enhance water resistance (Basumatary et al., 2022).
3. **Pectin and alginate films**: Pectin, a polysaccharide extracted from fruits, forms flexible and transparent films that are excellent carriers for antioxidants and antimicrobials. Similarly, alginate films, derived from algae, are utilized for their ability to form water-insoluble gels in the presence of calcium ions. These films are effective in packaging fresh-cut fruits and vegetables by creating a moisture barrier while maintaining firmness (food preservation).

33.2.6.2 Protein-based biopolymer films: applications and properties

Protein-based biopolymer films are extensively studied for their excellent gas barrier properties and their ability to serve as carriers for bioactive agents. These films are biodegradable, renewable, and often exhibit superior mechanical strength compared to polysaccharide or lipid-based films. Proteins such as gelatin, casein, soy protein, and zein are commonly employed in food packaging applications.

1. **Gelatin films:** Gelatin, derived from collagen, is a flexible, transparent material used in packaging oxygen-sensitive foods like meat, dairy, and baked goods due to its low oxygen permeability. While gelatin films have high water vapor transmission, they can be improved for moisture regulation by blending with other biopolymers or crosslinking with agents like glutaraldehyde. Incorporating EOs (e.g., oregano or thyme) or extracts (e.g., green tea polyphenols) boosts their antimicrobial properties. Gelatin films enhanced with silver nanoparticles are particularly effective in reducing bacterial growth, such as in raw chicken packaging (Basumatary et al., 2022).
2. **Casein films:** Casein, a milk protein, is widely utilized for its film-forming capacity and compatibility with active agents. It creates strong, flexible films with good adhesion properties. Casein-based films are commonly used as coatings for bakery products to prevent staling and as protective barriers for cheese packaging. Their ability to incorporate antimicrobials like nisin makes them effective against spoilage microorganisms (Picchio et al., 2018).
3. **Soy protein isolate (SPI):** SPI, a plant-derived protein, forms strong, gas-impermeable films. When blended with biopolymers like pectin or nanocellulose, its functionality is enhanced. Recent studies show its use in composite films for improved water resistance, particularly in high-moisture foods like fresh-cut fruits (Zhen et al., 2022). SPI films are also used as barriers for frozen foods and as protective coatings for fruits, reducing microbial contamination and oxidative degradation (Basumatary et al., 2022).
4. **Zein films:** Zein, a corn protein, is particularly useful in coating applications due to its hydrophobicity. These films are used to coat nuts, seeds, and confectionery to prevent rancidity and maintain freshness. Zein coatings are also employed for pharmaceutical-grade food products, offering controlled-release mechanisms for active ingredients (Huang et al., 2021).

33.2.6.3 Lipid-based biopolymer films: applications and properties

Lipid-based films are known for their excellent water vapor barrier properties due to the hydrophobic nature of lipids. However, their use as standalone films is limited due to their brittleness. Instead, lipids are often incorporated into composite films with polysaccharides or proteins to combine the strengths of various materials (Castro-Rosas et al., 2016).

1. **Beeswax and carnauba wax coatings:** Beeswax and carnauba wax are commonly used lipids in food coatings due to their natural origin and hydrophobic characteristics. Wax coatings provide an effective barrier against moisture loss while maintaining the product's sensory qualities. They also impart a glossy appearance, enhancing consumer appeal. Carnauba wax coatings are widely applied to fruits like apples, citrus, and avocados to prevent dehydration and delay ripening. Beeswax coatings are used for artisanal cheese preservation, maintaining freshness and preventing microbial contamination (Basumatary et al., 2022).
2. **Fatty acids and monoglycerides:** Fatty acids such as stearic acid and lauric acid are used to improve the water resistance of polysaccharide and protein-based films. These lipid components are often blended into hydrocolloid matrices (e.g., starch or gelatin) to reduce water vapor permeability while retaining flexibility. Fatty acid-enriched coatings are employed for minimally processed vegetables and bakery items to extend shelf life and preserve texture (Basumatary et al., 2022; Vijayendra & Shamala, 2014).
3. **Shellac:** Shellac is a natural resin secreted by the Laccifer lacca insect and is extensively used as an edible coating. Its glossy finish and excellent water barrier properties make it ideal for coating candies, nuts, and pharmaceutical-grade food items. Shellac coatings are commonly applied to confectioneries and processed fruits to protect against moisture absorption and maintain structural integrity during storage (Basumatary et al., 2022; Souto et al., 2021).

33.2.7 Impact on food safety and reduction of food waste

Biopolymer films enhance food safety by preventing microbial contamination and oxidative spoilage. Chitosan-based films, for instance, inhibit pathogens like Salmonella and Escherichia *coli*, reducing the need for chemical preservatives (Basumatary et al., 2022). These films also reduce food waste by extending shelf life, with coatings extending freshness of items like tomatoes and bananas by four to seven days (Basumatary et al., 2022). Advances in protein and lipid-based films, incorporating bioactive agents and nanoparticles, offer intelligent packaging solutions that align with sustainability goals by reducing plastic waste in the food industry.

33.2.7.1 Case studies of commercial adoption in the food industry

Biopolymer films are reshaping the food packaging landscape, offering sustainable solutions that extend shelf life and maintain food quality. Let us explore some notable case studies highlighting their development, properties, and commercial applications.

33.2.7.1.1 Hemicellulose-based biopolymer films for bakery products and produce

The USDA developed hemicellulose-based films in 2018, combining agricultural residues with methylcellulose and carboxymethyl cellulose (CMC). The films, enhanced with glycerol and polyethylene glycol, offered improved tensile strength, moisture resistance, and oxygen barrier properties. These films were introduced into the US market to package fresh produce and bakery items. Their application in extending shelf life by 20%–30% proved essential in minimizing spoilage during transport and storage (Xu et al., 2019). There is a need for stringent guidelines and regulations governing the application of hemicellulose at commercial level. The focus of current research should be towards optimizing the cost effective and sustainable method in extraction and purification of hemicellulose from agricultural waste to meet the global demand (Mathura et al., 2024).

33.2.7.1.2 Starch-based edible films for fresh-cut fruits and vegetables

Introduced commercially in 2017, starch-based edible films were developed to coat fresh-cut fruits and vegetables, leveraging advancements in polysaccharide processing and nanoparticle reinforcement. The films, enhanced with EOs, provided antimicrobial and antioxidant properties. Major retailers in Europe and Latin America adopted these films for products like fresh-cut apples, mangoes, and leafy greens. These coatings reduced microbial contamination and maintained product freshness (Kocira et al., 2021). By 2021, the use of starch-based films reduced fresh produce spoilage in Latin America by nearly 18%, benefiting an estimated 25 million consumers annually. Retailers reported cost savings of approximately $2 billion due to reduced wastage along the supply chain.

33.2.7.1.3 Chitosan films for seafood packaging

Commercially introduced in 2019, chitosan films utilized crustacean shell waste to produce antimicrobial packaging for seafood. Innovations included embedding nanoparticles like zinc oxide and EOs to enhance microbial inhibition. Commercial application in frozen shrimp and salmon packing in markets of Japan and the United States. These films extended the shelf life of seafood by an average of five to seven days. They also reduced risks of contamination from *Listeria monocytogenes* and *E. coli* (Jiménez-Gómez & Cecilia, 2020; Priyadarshi & Rhim, 2020). By 2023, these films helped decrease spoilage rates in the global seafood supply chain by approximately 12%, saving an estimated 300,000 metric tons of seafood annually. This reduction significantly impacted food security for over 30 million consumers worldwide.

33.2.7.1.4 Polylactic acid-based transparent films for snack packaging

PLA-based films gained commercial traction in 2018. These films were developed to improve heat resistance and flexibility by blending PLA with other biodegradable polymers like PBAT (Palai et al., 2020). Commercial application of this polymer finds its wide adoption in snack packaging for chips, granola bars, and chocolates. These transparent films maintain product visibility and offer compostability, making them attractive for eco-conscious brands. By 2022, PLA films replaced approximately 25% of synthetic plastics in the snack packaging sector in Europe, preventing around 500,000 tons of plastic waste annually. Consumer surveys showed that 60% of shoppers preferred products with PLA-based eco-packaging, reflecting a significant shift in market trends.

33.2.7.1.5 Gelatin-based films for meat preservation

Year and development: Gelatin-based films were first introduced in 2016 for vacuum-sealed packaging of processed meats. These films incorporated natural antioxidants, like green tea polyphenols, to prevent lipid oxidation and discoloration.

Commercial application: These films were adopted by meat processors in North America and Europe to package high-value products such as sausages, ham, and bacon. Their oxygen barrier properties helped maintain product quality for up to 20 days under refrigeration (Chaari et al., 2022).

Impact: By 2021, the use of gelatin-based films reduced meat wastage in retail by approximately 10%, equating to 400,000 tons of saved meat annually. This impact benefited around 40 million consumers and contributed to cost savings of nearly $1.5 billion throughout the supply chain.

33.2.8 Application in textiles and apparel

The textile industry is increasingly adopting biopolymers to address sustainability concerns and enhance the functional properties of textiles. Biopolymers such as chitosan, cellulose, silk fibroin, PLA, and alginate offer distinct advantages, including biodegradability, renewability, and specialized functionalities like antimicrobial activity, moisture management, and improved dyeability. Some examples are mentioned in Table 33.5

33.2.8.1 Chitosan in the textile industry

Chitosan, derived from chitin, is gaining popularity in the textile industry due to its biodegradability, antimicrobial activity, and environmental sustainability. It is used as a finish on textiles, providing antibacterial and antifungal properties ideal for medical, sports, and hygiene applications (Antunes et al., 2021; Jiménez-Gómez & Cecilia, 2020). Chitosan enhances dyeability by interacting with dye molecules, improving color fixation on fibers such as cotton, polyester, and nylon, which is beneficial for eco-friendly textiles. Additionally, its water-repellent properties make it suitable for outdoor and protective clothing, while studies suggest it also improves fabric durability and strength (Jiménez-Gómez & Cecilia, 2020).

33.2.8.2 Cellulose-based fibers in textiles

Cellulose, the most abundant natural polymer, is central to sustainable textiles. Lyocell, made through a closed-loop process, is more eco-friendly than traditional viscose, offering strength, moisture-wicking, and biodegradability (Antunes et al., 2021; Liu et al., 2021). Cellulose fibers are breathable, soft, and comfortable, making them ideal for reducing synthetic fiber use. Modified cellulose fibers, like rayon, provide excellent dye uptake and color retention, and can be tailored for specific applications, like enhanced water absorption or quick-drying properties for activewear (Antunes et al., 2021; Sayyed et al., 2019).

33.2.8.3 Silk fibroin in textiles

Silk fibroin, a natural protein from silkworms, is valued in the textile industry for its strength, flexibility, and biocompatibility. It is used in high-end fashion, medical textiles (e.g., wound dressings, surgical sutures), and orthopedic textiles (Koh et al., 2015; Opriş et al., 2024). Beyond traditional textiles, silk fibroin's electric conductivity and biodegradability make it ideal for bioelectronics, wearable sensors, and smart textiles. It is also used in tissue engineering and biodegradable implants for drug delivery (Koh et al., 2015).

33.2.8.4 Polylactic acid in textiles

PLA, a biodegradable thermoplastic derived from renewable resources like corn starch or sugarcane, is increasingly used in textiles for its sustainability. PLA fibers are soft, lustrous, and moisture-wicking, making them ideal for activewear, casual wear, and eco-friendly fashion (da Silva et al., 2018; Merino et al., 2022). PLA fibers are known for their thermal and mechanical stability. PLA fibers contribute to reducing the textile industry's carbon footprint, while biodegradability presents challenges for outdoor applications. Current research is enhancing PLA's resistance to UV degradation and environmental conditions (Merino et al., 2022).

33.2.8.5 Alginate in textiles

Alginate, derived from brown seaweed, is gaining attention in the textile industry for its biodegradability and ability to form gels and fibers. This biopolymer is used as a coating or finishing agent. It improves moisture management, water retention, and antimicrobial properties (Kanjanamosit et al., 2009). Alginate is also being explored for medical textiles, including wound dressings and tissue engineering scaffolds, offering controlled drug release and biodegradability. Its film-

TABLE 33.5 Examples of biopolymer-based textile products on the market.

S. no.	Biopolymer textile	Source	Properties	Applications	References
1.	Lyocell (Tencel)	Wood pulp (sustainably sourced)	Biodegradable, soft, breathable, and highly absorbent. Closed-loop production with 99% solvent recovery	Clothing, home textiles, and medical textiles	Jadhav et al. (2021)
2.	Piñatex (pineapple leaf fiber)	Pineapple leaf fibers	Lightweight, durable, versatile, and biodegradable. Requires less water and fewer chemicals compared to leather	Fashion, accessories, footwear, and upholstery	Sayyed et al. (2019)
3.	Castor oil-based biopolymers	Castor oil seeds	Hydrophobic, heat-resistant, strong, durable, and biodegradable	High-performance apparel, automotive upholstery, and technical textiles	Kunduru et al. (2015)
4.	Polylactic acid fabrics	Fermented plant starch (corn/sugarcane)	Biodegradable, strong, lightweight, compostable, and moisture-wicking	Sportswear, activewear, casual wear, and medical textiles (e.g., surgical drapes)	Merino et al. (2022)
5.	Hemp and flax fibers	Hemp and flax plants	Strong, durable, naturally antimicrobial, low water, and chemical requirement	Clothing, upholstery, and industrial textiles	Muzyczek (2012)

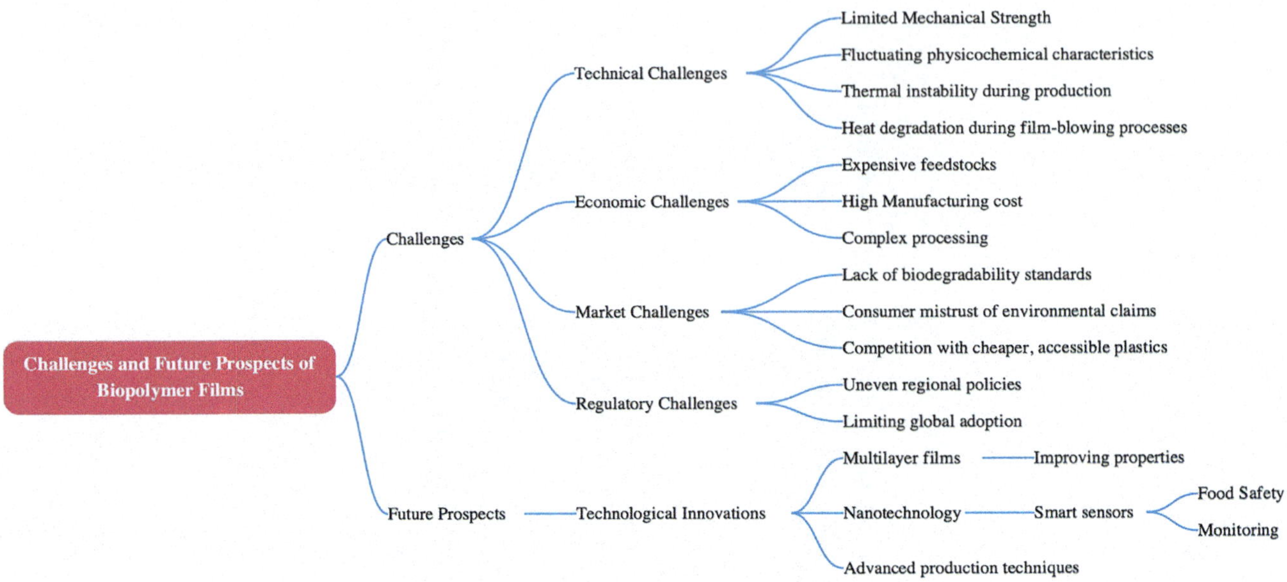

FIGURE 33.2 Overview of challenges and future prospects of biopolymer films.

forming properties make it suitable for eco-friendly food packaging. As demand for sustainable textiles grows, alginate's use is expected to expand even into consumer apparel and fashion (Teixeira-Costa & Andrade, 2021).

33.3 Challenges and future prospects

Biopolymer films face several challenges that hinder their wider use, such as limited mechanical strength, fluctuating properties, and thermal instability, which affect their performance compared to traditional plastics. High manufacturing costs, due to expensive feedstocks and complex processes, further limit their economic viability. While methods like film blowing and coextrusion techniques promise improvements in film properties, issues like poor interlayer adhesion and heat degradation remain. Market and regulatory barriers, including inconsistent biodegradability standards and consumer mistrust, also limit adoption. Despite these obstacles, advancements in biopolymer films are focused on improving their mechanical and thermal properties, integrating nanotechnology and smart sensors for enhanced functionality. The application of agricultural or industrial waste can reduce the costs and improve scalability. Research focusing on bioconversion of waste to biopolymers needs more attention and financial support. The future of biopolymer films lies in addressing these challenges through research, innovation, and investment in infrastructure, paving the way for their contribution to sustainable development and the green economy. An overview of challenges and future prospects have been mentioned in Fig. 33.2

33.4 Conclusion

Biopolymer films, derived from renewable sources like polysaccharides, proteins, and lipids, are gaining traction across industries such as biomedical, cosmetics, textiles, and electronics due to their biodegradability and biocompatibility. In biomedicine, they are used in wound dressings, drug delivery, and tissue engineering, while in cosmetics, they offer sustainable packaging and controlled ingredient release. In textiles, they provide eco-friendly waterproofing and antibacterial properties, and in electronics, they are used for flexible, bio-integrated devices. Despite challenges in scalability, performance, and cost, emerging technologies like coextrusion and hybrid composites are enhancing their mechanical strength and durability. Future advancements in nanotechnology and material science are expected to improve biopolymer film properties, enabling them to replace petroleum-based plastics and contribute to a circular economy. These innovations will support sustainable, resource-efficient solutions across various industries.

Acknowledgments

The authors express appreciation for the support provided by the Deanship of Scientific Research, Vice Presidency for Graduate Studies and Scientific Research, King Faisal University, Saudi Arabia [Grant No. KFU264092].

References

(2022). Green Dot Bioplastics launches expanded line of compostable resins and teams up with Mayco to reclaim fibers. Addit Polym: https://doi.org/10.12968/s0306-3747(22)70138-8

Mistretta, M. C., Botta, L., La Mantia, F. P., Di Fiore, A., & Cascone, M. (2021). Film Blowing of Biodegradable Polymer Nanocomposites for Agricultural Applications. *Macromolecular Materials and Engineering, 306*(9). https://doi.org/10.1002/mame.202100177.

Aaliya, B., Sunooj, K. V., & Lackner, M. (2021). Biopolymer composites: A review. *International Journal of Biobased Plastics, 3*, 40–84. https://doi.org/10.1080/24759651.2021.1881214.

Abdullah, Cai, J., Hafeez, M. A., Wang, Q., Farooq, S., Huang, Q., Tian, W., & Xiao, J. (2022). Biopolymer-based functional films for packaging applications: A review. *Frontiers in Nutrition, 9*. https://doi.org/10.3389/fnut.2022.1000116, journal.frontiersin.org/journal/nutrition.

Abdul Khalil, H. P. S., Banerjee, A., Saurabh, C. K., Tye, Y. Y., Suriani, A. B., Mohamed, A., Karim, A. A., Rizal, S., & Paridah, M. T. (2018). Biodegradable Films for Fruits and Vegetables Packaging Application: Preparation and Properties. *Food Engineering Reviews, 10*(3), 139–153. 10.1007/s12393-018-9180-3.

Aggarwal, J., Sharma, S., Kamyab, H., & Kumar, A. (2020). The realm of biopolymers and their usage: An overview. *Journal of Environmental Treatment Techniques, 8*(3), 1005–1016. http://www.jett.dormaj.com/docs/Volume8/Issue%203/The%20Realm%20of%20Biopolymers%20and%20Their%20Usage%20-%20An%20Overview.pdf.

Ahmadi, P., Jahanban-Esfahlan, A., Ahmadi, A., Tabibiazar, M., & Mohammadifar, M. (2022). Development of ethyl cellulose-based formulations: A perspective on the novel technical methods. *Food Reviews International, 38*(4), 685–732. https://doi.org/10.1080/87559129.2020.1741007.

Alves, T. F. R., Morsink, M., Batain, F., Chaud, M. V., Almeida, T., Fernandes, D. A., da Silva, C. F., Souto, E. B., & Severino, P. (2020). Applications of natural, semi-synthetic, and synthetic polymers in cosmetic formulations. *Cosmetics, 7*(4), 75. https://doi.org/10.3390/cosmetics7040075.

Antunes, J. C., Domingues, J. M., Miranda, C. S., Silva, A. F. G., Homem, N. C., Amorim, M. T. P., & Felgueiras, H. P. (2021). Bioactivity of Chitosan-Based Particles Loaded with Plant-Derived Extracts for Biomedical Applications: Emphasis on Antimicrobial Fiber-Based Systems. *Marine Drugs, 19*(7), 359. 10.3390/md19070359.

Azeredo, H. M. C., Rosa, M. F., & Mattoso, L. H. C. (2017). Nanocellulose in bio-based food packaging applications. *Industrial Crops and Products, 97*, 664–671. https://doi.org/10.1016/j.indcrop.2016.03.013.

Balart, R., Garcia-Garcia, D., Fombuena, V., Quiles-Carrillo, L., & Arrieta, M. P. (2021). Biopolymers from natural resources. *Polymers, 13*(15), 2532. https://doi.org/10.3390/polym13152532.

Baranwal, J., Barse, B., Fais, A., Delogu, G. L., & Kumar, A. (2022). Biopolymer: A sustainable material for food and medical applications. *Polymers, 14*(5), 983. https://doi.org/10.3390/polym14050983.

Basumatary, I. B., Mukherjee, A., Katiyar, V., & Kumar, S. (2022). Biopolymer-based nanocomposite films and coatings: Recent advances in shelf-life improvement of fruits and vegetables. *Critical Reviews in Food Science and Nutrition, 62*(7), 1912–1935. https://doi.org/10.1080/10408398.2020.1848789.

Bianchi, E., Guidotti, G., Soccio, M., Siracusa, V., Gazzano, M., Salatelli, E., & Lotti, N. (2023). Biobased and compostable multiblock copolymer of poly(l -lactic acid) containing 2,5-furandicarboxylic acid for sustainable food packaging: The role of parent homopolymers in the composting kinetics and mechanism. *Biomacromolecules, 24*(5), 2356–2368. https://doi.org/10.1021/acs.biomac.3c00216.

Castro-Rosas, J., Cruz-Galvez, A. M., Gomez-Aldapa, C. A., Falfan-Cortes, R. N., Guzman-Ortiz, F. A., & Rodríguez-Marín, M. L. (2016). Biopolymer films and the effects of added lipids, nanoparticles and antimicrobials on their mechanical and barrier properties: A review. *International Journal of Food Science and Technology, 51*(9), 1967–1978. https://doi.org/10.1111/ijfs.13183.

Chaari, M., Elhadef, K., Akermi, S., Ben Akacha, B., Fourati, M., Chakchouk Mtibaa, A., Ennouri, M., Sarkar, T., Shariati, M. A., Rebezov, M., Abdelkafi, S., Mellouli, L., & Smaoui, S. (2022). Novel active food packaging films based on gelatin-sodium alginate containing beetroot peel extract. *Antioxidants, 11*(11), 2095. https://doi.org/10.3390/antiox11112095.

Chen, G., Wang, K., Chen, P., Cai, D., Shao, Y., Xia, R., Li, C., Wang, H., Ren, F., Cheng, X., & Yu, Y. (2024). Fully biodegradable packaging films for fresh food storage based on oil-infused bacterial cellulose. *Advanced Science, 11*(23). https://doi.org/10.1002/advs.202400826.

Cheng, J., Wang, H., Kang, S., Xia, L., Jiang, S., Chen, M., & Jiang, S. (2019). An active packaging film based on yam starch with eugenol and its application for pork preservation. *Food Hydrocolloids, 96*, 546–554. https://doi.org/10.1016/j.foodhyd.2019.06.007.

Dash, K. K., Ali, N. A., Das, D., & Mohanta, D. (2019). Thorough evaluation of sweet potato starch and lemon-waste pectin based-edible films with nano-titania inclusions for food packaging applications. *International Journal of Biological Macromolecules, 139*, 449–458. https://doi.org/10.1016/j.ijbiomac.2019.07.193, www.elsevier.com/locate/ijbiomac.

da Silva, D., Kaduri, M., Poley, M., Adir, O., Krinsky, N., Shainsky-Roitman, J., & Schroeder, A. (2018). Biocompatibility, biodegradation and excretion of polylactic acid (PLA) in medical implants and theranostic systems. *Chemical Engineering Journal, 340*, 9–14. https://doi.org/10.1016/j.cej.2018.01.010.

Dias, O. A. T., Konar, S., Leão, A. L., & Sain, M. (2019). Flexible electrically conductive films based on nanofibrillated cellulose and polythiophene prepared via oxidative polymerization. *Carbohydrate Polymers, 220*, 79–85. https://doi.org/10.1016/j.carbpol.2019.05.057, http://www.elsevier.com/wps/find/journaldescription.cws_home/405871/description#description.

Dirpan, A., Ainani, A. F., & Djalal, M. (2023). A review on biopolymer-based biodegradable film for food packaging: Trends over the last decade and future research. *Polymers (Basel), 15*, 2781. https://doi.org/10.3390/polym15132781.

Engel, J. B., Ambrosi, A., & Tessaro, I. C. (2019). Development of biodegradable starch-based foams incorporated with grape stalks for food packaging. *Carbohydr Polym, 225*, 115234. https://doi.org/10.1016/j.carbpol.2019.115234.

Fabra, M. J., López-Rubio, A., Ambrosio-Martín, J., & Lagaron, J. M. (2016). Improving the barrier properties of thermoplastic corn starch-based films containing bacterial cellulose nanowhiskers by means of PHA electrospun coatings of interest in food packaging. *Food Hydrocolloids, 61*, 261–268. https://doi.org/10.1016/j.foodhyd.2016.05.025, https://www.journals.elsevier.com/food-hydrocolloids.

Farina, V., Passafiume, R., Tinebra, I., Palazzolo, E., & Sortino, G. (2020). Use of aloe vera gel-based edible coating with natural anti-browning and anti-oxidant additives to improve post-harvest quality of fresh-cut 'fuji' apple. *Agronomy, 10*(4), 515. https://doi.org/10.3390/agronomy10040515.

Fasiku, V. O., Owonubi, S. J., Mukwevho, E., Aderibigbe, B. A., Sadiku, E. R., Lemmer, Y., Babu Reddy, A., Manjula, B., Nkuna, C., Dludlu, M. K., Adeyeye, O. A., Varaprasad, K., & Tippabattini, J. (2019). *Bionanopolymers for drug delivery materials horizons: From nature to nanomaterials*. South Africa: Springer Nature, 191–220. www.springer.com/series/16122.

Fazal, T., Murtaza, B. N., Shah, M., Iqbal, S., Rehman, M. U., Jaber, F., Dera, A. A., Awwad, N. S., & Ibrahium, H. A. (2023). Recent developments in natural biopolymer based drug delivery systems. *RSC Advances, 13*(33), 23087–23121. https://doi.org/10.1039/d3ra03369d, http://pubs.rsc.org/en/journals/journal/ra.

Ferrer, A., Pal, L., & Hubbe, M. (2017). Nanocellulose in packaging: Advances in barrier layer technologies. *Industrial Crops and Products, 95*, 574–582. https://doi.org/10.1016/j.indcrop.2016.11.012.

Fourati, Y., Tarrés, Q., Delgado-Aguilar, M., Mutjé, P., & Boufi, S. (2021). Cellulose nanofibrils reinforced PBAT/TPS blends: Mechanical and rheological properties. *International Journal of Biological Macromolecules, 183*, 267–275. https://doi.org/10.1016/j.ijbiomac.2021.04.102.

Gamage, A., Gangahagedara, R., Gamage, J., Jayasinghe, N., Kodikara, N., Suraweera, P., & Merah, O. (2023). Role of organic farming for achieving sustainability in agriculture. *Farming System, 1*(1), 100005. https://doi.org/10.1016/j.farsys.2023.100005.

García, M. A., Ferrero, C., Bértola, N., Martino, M., & Zaritzky, N. (2002). Edible coatings from cellulose derivatives to reduce oil uptake in fried products. *Innovative Food Science & Emerging Technologies, 3*(4), 391–397. https://doi.org/10.1016/s1466-8564(02)00050-4.

Go, E. J., & Song, K. B. (2019). Antioxidant properties of rye starch films containing rosehip extract and their application in packaging of chicken breast. *Starch/Staerke, 71*(11-12). https://doi.org/10.1002/star.201900116, http://onlinelibrary.wiley.com/journal/10.1002/(ISSN)1521-379X.

Gobi, R., Ravichandiran, P., Babu, R. S., & Yoo, D. J. (2021). Biopolymer and synthetic polymer-based nanocomposites in wound dressing applications: A review. *Polymers, 13*(12), 1962. https://doi.org/10.3390/polym13121962.

Gowthaman, N. S. K., Lim, H. N., Sreeraj, T. R., Amalraj, A., & Gopi, S. (2021). *Advantages of biopolymers over synthetic polymers*. Elsevier BV, 351–372.

Haniffa, M. M., Ching, Y., Abdullah, L., Poh, S., & Chuah, C. (2016). Review of bionanocomposite coating films and their applications. *Polymers, 8*(7), 246. https://doi.org/10.3390/polym8070246.

Hu, D., Wang, H., & Wang, L. (2016). Physical properties and antibacterial activity of quaternized chitosan/carboxymethyl cellulose blend films. *LWT, 65*, 398–405. https://doi.org/10.1016/j.lwt.2015.08.033.

Huang, X., Zhou, X., Dai, Q., & Qin, Z. (2021). Antibacterial, antioxidation, UV-blocking, and biodegradable soy protein isolate food packaging film with mangosteen peel extract and ZnO nanoparticles. *Nanomaterials, 11*(12), 3337. https://doi.org/10.3390/nano11123337.

Jadhav, S., Lidhure, A., Thakre, S., & Ganvir, V. (2021). Modified Lyocell process to improve dissolution of cellulosic pulp and pulp blends in NMMO solvent. *Cellulose, 28*(2), 973–990. https://doi.org/10.1007/s10570-020-03580-1.

Jiménez-Gómez, C. P., & Cecilia, J. A. (2020). Chitosan: A natural biopolymer with a wide and varied range of applications. *Molecules, 25*. https://doi.org/10.3390/molecules25173981.

Kanjanamosit, N., Muangnapoh, C., & Phisalaphong, M. (2009). Biosynthesis and characterization of bacteria cellulose–alginate film. *Journal of Applied Polymer Science, 115*(3), 1581–1588. https://doi.org/10.1002/app.31138.

Khan, M. R., Volpe, S., Valentino, M., Miele, N. A., Cavella, S., & Torrieri, E. (2021). Active casein coatings and films for perishable foods: Structural properties and shelf-life extension. *Coatings, 11*(8). https://doi.org/10.3390/coatings11080899, https://www.mdpi.com/2079-6412/11/8/899/pdf.

Khoo, H. H., & Tan, R. B. H. (2010). Environmental impacts of conventional plastic and bio-based carrier bags. *The International Journal of Life Cycle Assessment, 15*(4), 338–345. https://doi.org/10.1007/s11367-010-0163-8.

Klein, M., & Poverenov, E. (2020). Natural biopolymer-based hydrogels for use in food and agriculture. *Journal of the Science of Food and Agriculture, 100*(6), 2337–2347. https://doi.org/10.1002/jsfa.10274.

Kocira, A., Kozłowicz, K., Panasiewicz, K., Staniak, M., Szpunar-Krok, E., & Hortyńska, P. (2021). Polysaccharides as edible films and coatings: Characteristics and influence on fruit and vegetable quality—A review. *Agronomy, 11*(5), 813. https://doi.org/10.3390/agronomy11050813.

Koh, L.-D., Cheng, Y., Teng, C.-P., Khin, Y.-W., Loh, X.-J., Tee, S.-Y., Low, M., Ye, E., Yu, H.-D., Zhang, Y.-W., & Han, M.-Y. (2015). Structures, mechanical properties and applications of silk fibroin materials. *Progress in Polymer Science, 46*, 86–110. https://doi.org/10.1016/j.progpolymsci.2015.02.001.

Kunduru, K. R., Basu, A., Haim Zada, M., & Domb, A. J. (2015). Castor oil-based biodegradable polyesters. *Biomacromolecules, 16*(9), 2572–2587. https://doi.org/10.1021/acs.biomac.5b00923, http://pubs.acs.org/journal/bomaf6.

Leong, N. L., Kator, J. L., Clemens, T. L., James, A., Enamoto-Iwamoto, M., & Jiang, J. (2020). Tendon and ligament healing and current approaches to tendon and ligament regeneration. *Journal of Orthopaedic Research, 38*(1), 7–12. https://doi.org/10.1002/jor.24475, http://onlinelibrary.wiley.com/journal/10.1002/(ISSN)1554-527X.

Liu, T., Wang, Y., Liu, J., Han, X., Zou, Y., Wang, P., Xu, R., Tong, L., Liu, J., Liang, J., Sun, Y., Fan, Y., & Zhang, X. (2024). An injectable photocuring silk fibroin-based hydrogel for constructing an antioxidant microenvironment for skin repair. *Journal of Materials Chemistry B, 12*(9), 2282–2293. https://doi.org/10.1039/d3tb02214e.

Liu, Y., Ahmed, S., Sameen, D. E., Wang, Y., Lu, R., Dai, J., Li, S., & Qin, W. (2021). A review of cellulose and its derivatives in biopolymer-based for food packaging application. *Trends in Food Science & Technology, 112*, 532–546. https://doi.org/10.1016/j.tifs.2021.04.016.

López De Dicastillo, C., Rodríguez, F., Guarda, A., & Galotto, M. J. (2016). Antioxidant films based on cross-linked methyl cellulose and native Chilean berry for food packaging applications. *Carbohydrate Polymers, 136*, 1052–1060. https://doi.org/10.1016/j.carbpol.2015.10.013, http://www.elsevier.com/wps/find/journaldescription.cws_home/405871/description#description.

Mamun, A. A., & Chen, J. Y. (2022). *Industrial applications of biopolymers and their environmental impact*. CRC Press.

Mellinas, C., Ramos, M., Jiménez, A., & Garrigós, M. C. (2020). Recent trends in the use of pectin from agro-waste residues as a natural-based biopolymer for food packaging applications. *Materials, 13*(3). https://doi.org/10.3390/ma13030673, https://res.mdpi.com/d_attachment/materials/materials-13-00673/article_deploy/materials-13-00673-v2.pdf.

Mendoza-Muñoz, N., Leyva-Gómez, G., Piñón-Segundo, E., Zambrano-Zaragoza, M. L., Quintanar-Guerrero, D., Del Prado Audelo, M. L., & Urbán-Morlán, Z. (2023). Trends in biopolymer science applied to cosmetics. *International Journal of Cosmetic Science, 45*(6), 699–724. https://doi.org/10.1111/ics.12880.

Merino, D., Zych, A., & Athanassiou, A. (2022). Biodegradable and Biobased Mulch Films: Highly Stretchable PLA Composites with Different Industrial Vegetable Waste. *ACS Applied Materials & Interfaces, 14*(41), 46920–46931. https://doi.org/10.1021/acsami.2c10965.

Mitura, S., Sionkowska, A., & Jaiswal, A. (2020). Biopolymers for hydrogels in cosmetics: review. *Journal of Materials Science: Materials in Medicine, 31*(6). 10.1007/s10856-020-06390-w.

Moeini, A., Pedram, P., Fattahi, E., Cerruti, P., & Santagata, G. (2022). Edible polymers and secondary bioactive compounds for food packaging applications: Antimicrobial, mechanical, and gas barrier properties. *Polymers, 14*(12), 2395. https://doi.org/10.3390/polym14122395.

Muzyczek, M. (2012). The use of flax and hemp for textile applications. *Handbook of Natural Fibres, 2*, 312–328. https://doi.org/10.1016/B978-1-84569-698-6.50010-3, http://www.sciencedirect.com/science/book/9781845696986.

Nathanael, A. J., & Oh, T. H. (2020). Biopolymer coatings for biomedical applications. *Polymers, 12*(12), 1–26. https://doi.org/10.3390/polym12123061, https://www.mdpi.com/2073-4360/12/12/3061/pdf.

Niaounakis, M. (2015). *Biopolymers: Applications and trends biopolymers: Applications and Trends*. Elsevier Inc, 1–604. undefined http://www.sciencedirect.com/science/book/9780323353991.

Olatunji, O. (2015). *Natural polymers: Industry techniques and applications*. Nigeria: Springer International Publishing, 1–370. http://doi.org/10.1007/978-3-319-26414-1.

Olsen, I. (2015). Biofilm-specific antibiotic tolerance and resistance. European Journal of Clinical Microbiology & Infectious Diseases, *34*, 877–886. https://doi.org/10.1007/s10096-015-2323-z.

Opriş, O., Mormile, C., Lung, I., Stegarescu, A., Soran, M.-L., & Soran, A. (2024). An Overview of biopolymers for drug delivery applications. *Applied Sciences, 14*(4), 1383. https://doi.org/10.3390/app14041383.

Palai, B., Mohanty, S., & Nayak, S. K. (2020). Synergistic effect of polylactic acid(PLA) and Poly(butylene succinate-co-adipate) (PBSA) based sustainable, reactive, super toughened eco-composite blown films for flexible packaging applications. *Polymer Testing, 83*, 106130. https://doi.org/10.1016/j.polymertesting.2019.106130.

Panrong, T., Karbowiak, T., & Harnkarnsujarit, N. (2019). Thermoplastic starch and green tea blends with LLDPE films for active packaging of meat and oil-based products. *Food Packaging and Shelf Life, 21*, 100331. 10.1016/j.fpsl.2019.100331.

Picchio, M. L., Linck, Y. G., Monti, G. A., Gugliotta, L. M., Minari, R. J., & Alvarez Igarzabal, C. I. (2018). Casein films crosslinked by tannic acid for food packaging applications. *Food Hydrocolloids, 84*, 424–434. https://doi.org/10.1016/j.foodhyd.2018.06.028, https://www.journals.elsevier.com/food-hydrocolloids.

Priyadarshi, R., & Rhim, J.-W. (2020). Chitosan-based biodegradable functional films for food packaging applications. *Innovative Food Science & Emerging Technologies, 62*, 102346. https://doi.org/10.1016/j.ifset.2020.102346.

Rahman, A., Rehmani, R., Pirvu, D. G., Huang, S. M., Puri, S., & Arcos, M. (2024). Unlocking the therapeutic potential of marine collagen: A scientific exploration for delaying skin aging. *Marine Drugs, 22*(4), 159. https://doi.org/10.3390/md22040159.

Sahana, T. G., & Rekha, P. D. (2018). Biopolymers: Applications in wound healing and skin tissue engineering. *Molecular Biology Reports, 45*(6), 2857–2867. https://doi.org/10.1007/s11033-018-4296-3.

Salvatore, L., Gallo, N., Natali, M. L., Campa, L., Lunetti, P., Madaghiele, M., Blasi, F. S., Corallo, A., Capobianco, L., & Sannino, A. (2020). Marine collagen and its derivatives: Versatile and sustainable bio-resources for healthcare. *Materials Science and Engineering: C, 113*, 110963. https://doi.org/10.1016/j.msec.2020.110963.

Mathura, S. R., Landázuri, A. C., Mathura, F., Andrade Sosa, A. G., & Orejuela-Escobar, L. M. (2024). Hemicelluloses from bioresidues and their applications in the food industry – towards an advanced bioeconomy and a sustainable global value chain of chemicals and materials. *Sustainable Food Technology, 2*(5), 1183–1205. https://doi.org/10.1039/d4fb00035h.

Sayyed, A. J., Deshmukh, N. A., & Pinjari, D. V. (2019). A critical review of manufacturing processes used in regenerated cellulosic fibres: Viscose, cellulose acetate, cuprammonium, LiCl/DMAc, ionic liquids, and NMMO based lyocell. *Cellulose, 26*, 2913–2940. https://doi.org/10.1007/s10570-019-02318-y.

Seenuvasan, M., Malar, C. G., & Growther, L. (2021). Production of a biopolymer film from biological wastes and its statistical analysis. *Bioresour Technol Rep, 13*, 100610. https://doi.org/10.1016/j.biteb.2020.100610.

Serra, M., Gudina, E., Botelho, C., Teixeira, J. A., & Barros, A. N. (2024). Polyglutamate: unleashing the versatility of a biopolymer for cosmetic industry applications. *Cosmetics, 11*(3), 76. https://doi.org/10.3390/cosmetics11030076.

Sganzerla, W. G., Rosa, G. B., Ferreira, A. L. A., da Rosa, C. G., Beling, P. C., Xavier, L. O., Hansen, C. M., Ferrareze, J. P., Nunes, M. R., Barreto, P. L. M., & de Lima Veeck, A. P. (2020). Bioactive food packaging based on starch, citric pectin and functionalized with *Acca sellowiana* waste by-product: Characterization and application in the postharvest conservation of apple. *International Journal of Biological Macromolecules, 147*, 295–303. https://doi.org/10.1016/j.ijbiomac.2020.01.074, www.elsevier.com/locate/ijbiomac.

Song, R., Murphy, M., Li, C., Ting, K., Soo, C., & Zheng, Z. (2018). Current development of biodegradable polymeric materials for biomedical applications. *Drug Design, Development and Therapy, 12*, 3117–3145. https://doi.org/10.2147/DDDT.S165440, https://doi.org/10.2147/DDDT.S165440.

Souto, E. B., Yoshida, C. M. P., Leonardi, G. R., Cano, A., Sanchez-Lopez, E., Zielinska, A., Viseras, C., Severino, P., da Silva, C. F., & Barbosa, Rd. M. (2021). Lipid-polymeric films: Composition, production and applications in wound healing and skin repair. *Pharmaceutics, 13*(8). https://doi.org/10.3390/pharmaceutics13081199, https://www.mdpi.com/1999-4923/13/8/1199/pdf.

Souza, P. M. S., Sommaggio, L. R. D., Marin-Morales, M. A., & Morales, A. R. (2020). PBAT biodegradable mulch films: Study of ecotoxicological impacts using *Allium cepa*, *Lactuca sativa* and HepG2/C3A cell culture. *Chemosphere, 256*. https://doi.org/10.1016/j.chemosphere.2020.126985, www.elsevier.com/locate/chemosphere.

Sulaeva, I., Hettegger, H., Bergen, A., Rohrer, C., Kostic, M., Konnerth, J., Rosenau, T., & Potthast, A. (2020). Fabrication of bacterial cellulose-based wound dressings with improved performance by impregnation with alginate. *Materials Science and Engineering: C, 110*, 110619. https://doi.org/10.1016/j.msec.2019.110619.

Sun, H., Lv, F., Liu, L., Gu, Q., & Wang, S. (2018). Conjugated polymer materials for photothermal therapy. *Advanced Therapeutics, 1*(6). https://doi.org/10.1002/adtp.201800057.

Teixeira, S. C., Oliveira, T. Vd, Batista, L. F., Silva, R. R. A., Lopes, Md. P., Ribeiro, A. R. C., Rigolon, T. C. B., Stringheta, P. C., & Soares, Nd. F. F. (2022). Anthocyanins of açaí applied as a colorimetric indicator of milk spoilage: A study using agar-agar and cellulose acetate as solid support to be applied in packaging. *Polysaccharides, 3*(4), 715–727. https://doi.org/10.3390/polysaccharides3040041, https://www.mdpi.com/journal/polysaccharides.

Teixeira-Costa, B. E., & Andrade, C. T. (2021). Natural polymers used in edible food packaging—history, function and application trends as a sustainable alternative to synthetic plastic. *Polysaccharides, 3*, 32–58. https://doi.org/10.3390/polysaccharides3010002.

Tyuftin, A. A., & Kerry, J. P. (2021). Gelatin films: Study review of barrier properties and implications for future studies employing biopolymer films. *Food Packaging and Shelf Life, 29*, 100688. https://doi.org/10.1016/j.fpsl.2021.100688.

Udayakumar, G. P., Muthusamy, S., Selvaganesh, B., Sivarajasekar, N., Rambabu, K., Banat, F., Sivamani, S., Sivakumar, N., Hosseini-Bandegharaei, A., & Show, P. L. (2021). Biopolymers and composites: Properties, characterization and their applications in food, medical and pharmaceutical industries. *Journal of Environmental Chemical Engineering, 9*(4), 105322. https://doi.org/10.1016/j.jece.2021.105322.

Valencia-Sullca, C., Vargas, M., Atarés, L., & Chiralt, A. (2018). Thermoplastic cassava starch-chitosan bilayer films containing essential oils. *Food Hydrocolloids, 75*, 107–115. https://doi.org/10.1016/j.foodhyd.2017.09.008.

Vasudevan, M., Perumal, V., Karuppanan, S., Ovinis, M., Bothi Raja, P., Gopinath, S. C. B., & Immanuel Edison, T. N. J. (2024). A comprehensive review on biopolymer mediated nanomaterial composites and their applications in electrochemical sensors. *Critical Reviews in Analytical Chemistry, 54*(7), 1871–1894. https://doi.org/10.1080/10408347.2022.2135090.

Vijayendra, S. V. N., & Shamala, T. R. (2014). Film forming microbial biopolymers for commercial applications-A review. *Critical Reviews in Biotechnology, 34*(4), 338–357. https://doi.org/10.3109/07388551.2013.798254, http://www.tandfonline.com/loi/ibty20.

Vink, E. T. H., Davies, S., & Kolstad, J. J. (2010). ORIGINAL RESEARCH: The eco-profile for current Ingeo® polylactide production. *Ind Biotechnol (New Rochelle N Y), 6*, 212–224. https://doi.org/10.1089/ind.2010.6.212.

Vinod, A., Sanjay, M. R., Suchart, S., & Jyotishkumar, P. (2020). Renewable and sustainable biobased materials: An assessment on biofibers, biofilms, biopolymers and biocomposites. *Journal of Cleaner Production, 258*, 120978. https://doi.org/10.1016/j.jclepro.2020.120978.

Wu, J., Shaidani, S., Theodossiou, S. K., Hartzell, E. J., & Kaplan, D. L. (2022). Localized, on-demand, sustained drug delivery from biopolymer-based materials. *Expert Opinion on Drug Delivery, 19*(10), 1317–1335. https://doi.org/10.1080/17425247.2022.2110582, http://www.tandfonline.com/loi/iedd20.

Xu, J., Xia, R., Zheng, L., Yuan, T., & Sun, R. (2019). Plasticized hemicelluloses/chitosan-based edible films reinforced by cellulose nanofiber with enhanced mechanical properties. *Carbohydrate Polymers, 224*, 115164. https://doi.org/10.1016/j.carbpol.2019.115164.

Zhen, Y., Wang, K., Wang, J., Qiao, D., Zhao, S., Lin, Q., & Zhang, B. (2022). Increasing the pH value during thermal processing suppresses the starch digestion of the resulting starch-protein-lipid complexes. *Carbohydrate Polymers, 278*, 118931. https://doi.org/10.1016/j.carbpol.2021.118931.

Zhou, X., Yang, R., Wang, B., & Chen, K. (2019). Development and characterization of bilayer films based on pea starch/polylactic acid and use in the cherry tomatoes packaging. *Carbohydrate Polymers, 222*, 114912. https://doi.org/10.1016/j.carbpol.2019.05.042.

Chapter 34

Legal regulations and consumer acceptances

Krishna Aayush[1,*] and Gurvendra Pal Singh[2,*]
[1]*Department of Food Science and Technology, Graphic Era (Deemed to be University), Dehradun, Uttarakhand, India,* [2]*Department of Food Technology and Nutrition, Lovely Professional University, Phagwara, Punjab, India*

34.1 Introduction

The global food industry continually develops innovations for better food safety, extended shelf life, and environmental sustainability. Some of the advancements discovered and developed in this field are edible nanobased packaging systems. These are active and smart packaging systems with all the advantages of nanotechnology and may ensure multi-dimensional preservation of food quality while meeting consumer demands for environmentally friendly materials (Sekhon, 2010; Singh et al., 2023). However, the successful commercialization of these technologies depends not only on their technical feasibility but also on regulatory compliance and consumer acceptance. Edible nanobased packaging systems are a new category of food packaging, integrating nanoscale materials into edible films or coatings. Such materials can be natural polymers, nanoparticles, or bioactive compounds that enhance the physical, chemical, and functional properties of the packaging. Essentially, they create packaging that is not only biodegradable but also consumable, adding value to the packaged food without generating waste (Aayush et al., 2022; Singh et al., 2023). These materials, at nano dimensions, exhibit unique features such as barrier properties, strength, and antimicrobial activity. For example, silver, zinc oxide, or titanium dioxide nanoparticles are known for their growth inhibitory activities against microorganisms, which can lengthen the shelf life of perishable food products such as fruits and vegetables, dairy products, and seafood. This includes adding nanostructured lipid carriers or encapsulated essential oils into the packaging as a delivery system for the bioactive compounds that could supply antioxidant and antimicrobial functions directly at the food surface (Nikolic et al., 2021; Singh et al., 2023). Edible nanobased systems also play a crucial role in intelligent packaging applications. The above-mentioned applications comprise indicators or sensors that detect environmental changes in food, such as temperature, humidity, or gas components, and, as a result, inform consumers about the spoilage-quality fact or the contamination-safety fact. In this manner, edible nanobased packaging presents solutions that not only enhance the safety and quality of foods but also address issues relating to environmental concerns posed by traditional synthetic plastic packaging (Chavan et al., 2023; Wongphan et al., 2022). In short, though much has been touted about edible nanobased packaging systems, these must comply strictly with legal regulations to gain widespread acceptance. Food packaging is subject to complex, region-specific regulations, and nanotechnology-based packaging is no exception. Thus, the packaging materials need to be nontoxic to the environment and meet established quality standards without posing any harm to humans (Dainelli et al., 2008). Regulatory bodies such as the US Food and Drug Administration (FDA), the European Food Safety Authority (EFSA), and Codex Alimentarius establish protocols for evaluating and approving nanomaterials in food packaging. These protocols often require migration testing to assess the degree to which nanoparticles are transferred into the food during processing and toxicological studies to determine whether a possible health impact may arise (Rojas-Graü et al., 2009). The edible packaging is more thorough and requires more studies to determine digestibility, allergenicity, and interactions with the food product among its components. The main issue here is that no globally established rule governs nanotechnology in food packaging. Even when some guiding rules are in place in

* Both authors contributed equally.

Natural Polymer-Based Films and Coatings in Food Technology. DOI: https://doi.org/10.1016/B978-0-443-34244-8.00033-1

certain countries, most others are still developing a comprehensive framework. The uncertainty in regulation is likely a major obstacle to the rapid pace of the innovation activities and the process of scaling technologies for international markets by manufacturers. On the other hand, nanotechnology is a rapidly evolving concept that mandates regular changes to existing regulatory standards, which, on one hand, will cope with scientific progress and, on the other hand, consider the public safety aspects. Besides the above, consumer acceptance will be the critical factor in the success of edible nanobased packaging systems. No matter how technological or efficient, it may be, it cannot continue without the confidence of the users and their willingness to adopt it.

The idea of edible packaging, especially when combined with nanoscale materials, is a relatively new concept to most consumers, and acceptance depends on many factors, such as awareness, perceived safety, and compatibility with consumer values. Most consumers are now beginning to develop awareness about the impacts their choices have on the environment and thus prefer to select products in accordance with sustainability goals (Li et al., 2020). Edible packaging systems that avoid plastic waste and comprise natural, biodegradable ingredients will surely appeal to this eco-sensitive consumer group. However, adding nanotechnology may sometimes raise skepticism since the long-term health and environmental impacts of nanotechnology are not known (Roy & Rhim, 2022). Transparency must be ensured in the creation of consumer trust between manufacturers and policymakers. Labels should be transparent, stating that nanotechnology is safe and beneficial for edible packaging and certified by regulatory bodies. This will further bridge the knowledge gap through educational campaigns on how nanotechnology enhances food safety and quality and poses minimal risks if applied responsibly. The market success of edible nanobased packaging will depend on consumer preferences for convenience, affordability, and sensory appeal. For example, packaging that is simple to use, not overly expensive, and does not change taste or texture will become popular very quickly. Industries have to execute market surveys to know their customers' choices and redesign their products correspondingly.

34.2 Global regulatory frameworks

The adoption and commercialization of edible nanobased packaging systems are crucially based on the regulations of the world. Such regulatory frameworks ensure that materials in edible nanobased packaging systems are nontoxic for human consumption, sustainable, and effective in food preservation. Nevertheless, the existence of great variability in standards at the country and regional levels leads to opportunities for innovation, but it is also challenging (Kumar et al., 2018; Sorrentino et al., 2007). The focal point of this chapter is the regulatory scenario worldwide with a focus on the most important international standards, the policy of specific regions, and the intricacies of harmonizing these frameworks.

34.2.1 Key international standards

34.2.1.1 International Organization for Standardization

The International Organization for Standardization stands for the initials ISO, which is a world body well known to everyone for producing and distributing safety, quality, efficiency, and/or- environmental management standards. For instance, in the food-packaging sector, which includes high-end technologies like nanotechnology, ISO is instrumental in the editing a network of standards as a mode of evaluation for the different packaging materials where safety, performance, and function are key tasks of packaging. Those ratings are obligatory for companies of any size to build trust and maintain the consistency of their products across the international trade arena (Robertson, 2016; Wirth, 2013). ISO has introduced several standards specific to nanotechnology, including ISO/TS 80004—defines standardized terminology for nanomaterials, ISO/TR 12885—provides general guidance on occupational safety when handling engineered nanomaterials. These standards are useful to manufacturers and regulators dealing with the classification and characterization of nanomaterials for innovative packaging systems involving edible nanobased packaging. The use of nanotechnology in food packaging is ensured to be safe, efficient, and to meet worldwide expectations by providing clear guidelines through ISO (Coles et al., 2003). ISO's influence extends to food safety management systems beyond packaging as well. Along the same lines, the 22000 ISO standard presents requirements that need to be met by all organizations engaged in food production from the farm to the customer. It is a documented system that manages food safety through the logical examination of food safety hazards and taking measures to mitigate risks, that is, ensuring food safety, thus ensuring that products are safe to eat. Its aim is to aid regulation and guarantee the quality of food by preventing products that are of poor safety standards from entering the market. The standard is particularly significant in the current world where products move from one border to another to be processed. This ISO document introduces a way for enterprises to conduct this type of work more efficiently, which leads to new bridging trust among the consumers and the responsible bodies (Coles et al., 2003; Trienekens & Zuurbier, 2008). The ISO standards are not mandatory, yet most industries comply with them as they are credible and cover the widest areas, opening the way for general acceptance. Therefore, ISO, being a certification agency, is in a unique position to help companies that want to verify their products and personnel certifications in one of the countries, as the required standard for each country is common. Besides, it has a dominant role in the development of global public health, the streamlining of international trade, and the new food packaging and safety systems. The ISO infrastructure will accurately adhere to the fact that the introduction of newer technologies, such as nanotechnology, into the global market in a controlled manner and the standards of quality and safety should be followed.

34.2.1.2 Codex Alimentarius

The Codex Alimentarius, also known as "Food Code," is a comprehensive collection of internationally accepted food standards, guidelines, and codes of practice that is drawn up by the Codex Alimentarius Commission, an international entity run jointly by the Food and Agriculture Organization (FAO) and the World Health Organization (WHO) since the early 1960s. The primary objectives of Codex are to safeguard consumer health and ensure fair practices in the food trade by providing a uniform set of guidelines that countries can adopt. Although Codex standards are only recommendations with voluntary application, they have been extensively used as a reference point for national legislation and, under the World Trade Organization's (WTO) Agreement on Sanitary and Phytosanitary Measures (SPS Agreement), play a crucial role in facilitating international trade (Organization, 2018). Codex standards are based on a wide variety of aspects related to food safety, such as food hygiene, food additives, residues of pesticides and veterinary drugs, contaminants, labeling and presentation, and methods of analysis and sampling. These standards cover the critical areas in which food safety should be put into practice, and food products must be safe to be consumed and conform to uniform standards for facilitating just trade between nations van der (Meulen, 2019).

The Codex framework also extends its guidance on food contact materials to include the principles for evaluating the safety of additives, contaminants, and packaging components. While Codex lacks specific standards for nanotechnology in food packaging, it strongly suggests the main uses of these technologies, such as risk assessment, toxicological evaluation, and adherence to good manufacturing practices (GMP). The invention-based principles are well suited to nanotechnology, a field that requires even more rigorous safety evaluations to secure regulatory approval and gain consumer acceptance. The Codex framework serves as the starting point for various other countries in creating a common environment for international trade and food safety. This will not only be a vital tool in ensuring proper food safety through these technologies but will also support their involvement in the global market (Meulen, 2019; Sothornvit & Krochta, 2005) (van der).

34.2.2 Region-specific regulations

34.2.2.1 United States: Food and Drug Administration guidelines for food contact materials

In the United States, the FDA plays a critical role in regulating the safety of food contact materials under the Federal Food, Drug, and Cosmetic Act (FFDCA). The FDA ensures that materials intended to come into contact with food, including packaging, utensils, and processing equipment, are safe for use and do not pose risks to public health. The FDA does not conduct direct testing; manufacturers are responsible for providing safety data, which the FDA evaluates to ensure that there is no contamination and to protect the consumer (Darrow et al., 2020). For edible nanobased packaging with advanced nanotechnology, the FDA evaluates nanotechnology applications based on existing safety frameworks but has not established specific regulations exclusively for nanotechnology in food packaging. Manufacturers must go through the premarket notification (PMN) process, where all new materials will be submitted for safety evaluation by the FDA prior to introduction. Testing migration is one of the essentials, as it will ensure that neither nanoparticles nor other substances in the package will migrate into food, which could be harmful to humans. Moreover, the generally recognized as safe (GRAS) designation allows substances with scientific consensus as safe entry into the marketplace without premarket approval (Heckman, 2005). The FDA has also addressed issues of guidance about nanotechnology, insisting that manufacturers are more transparent in terms of what materials they are using. In particular, such information may include particle size, surface properties, chemical composition, and interactions with food. This transparency is required by the FDA to ensure a holistic safety assessment of nanotechnology-enabled materials (Meral et al., 2022). The FDA's guidelines are based on thorough scientific assessments and risk evaluation. The FDA seeks to balance two important objectives: the need to protect public health and support innovation and trade in food products. Manufacturers can provide assurance that their food contact materials are safe while complying with US standards if they follow these regulations, thus helping create consumer confidence and promoting safe commercialization of new packaging technologies (Darrow et al., 2020; Heckman, 2005).

34.2.2.2 European Union: EFSA standards and registration, evaluation, authorization, and restriction of chemicals regulation

In the European Union (EU), the EFSA, as the leading scientific advisory body, is the principal authority responsible for food contact material legislation and safety assessments. Its work is conducted on a framework basis under the auspices of Regulation (EC) No 1935/2004, thus setting detailed guidelines to create and guarantee that materials for contact with food are safe for human consumption (Heckman, 2005). Complementing this framework, Commission Regulation (EU) No 10/2011 of January 14, 2011, on plastic materials and articles intended to come into contact with food provides detailed provisions regarding positive lists of authorized substances, migration limits (overall and specific), and compliance testing. This Regulation applies directly and uniformly across all Member States, thereby creating binding obligations upon manufacturers, importers, and users of plastic food contact materials within the internal market (Commission Regulation [EU], 2011). Food-packaging material, among others containing nanotechnology, shall not pose a hazard to health and shall not affect food composition, flavor, aroma, or any other sensory property. The EFSA demands very high levels of risk assessments for such materials, such as characterization of nanoparticles in terms of physical, chemical, and morphological properties; toxicological studies on the possible adverse effects such as cytotoxicity and genotoxicity; and migration and exposure assessment to assess nanoparticle transfer into food and the levels of consumer exposure (Alfei et al., 2020). To complement these efforts, the EU enforces the Registration, Evaluation, Authorisation, and Restriction of Chemicals (REACH) Regulation, governing all chemical use, including nanomaterials. According to REACH, manufacturers and importers are responsible for registering nanomaterials, submitting thorough safety data about their impacts on the environment and human health, and complying with demanding standards in the name of safety and sustainability—protecting not only human health but also the environment. Together, the standards set by EFSA, reinforced by Commission Regulation (EU) No 10/2011, along with the REACH regulation, provide a robust framework for the management of the safe use of food contact material in the EU. Focusing on scientific risk assessments, EFSA works to ensure that chemicals are put to safe use and safely documented, thus promoting strong consumer protection and environmental sustainability (Heckman, 2005; Miettinen, 2019). EFSA's approach to risk assessment for nanomaterials is built upon a detailed framework requiring full physicochemical characterization, hazard identification, and exposure assessment, coupled with rigorous migration and toxicity testing for food contact applications. The documentation emphasizes strict adherence to established guidance and a tiered, science-based review process. In contrast, the US FDA evaluates nanomaterials within existing regulatory frameworks specific to each product type, applying product-focused risk assessment methodologies and requiring manufacturers to fully characterize nanomaterials and ensure compliance with applicable safety, efficacy, and quality standards, as outlined in FDA guidance documents. Although both EFSA and FDA recognize the unique behavior of nanomaterials, EFSA imposes more explicit and structured requirements related to food contact materials, whereas FDA adopts a more flexible approach, governed by the statutory requirements relevant to the specific product category, often relying on industry guidance and general scientific principles for risk assessment (FDA, 2024; More et al., 2021).

34.2.2.3 Canada: Health Canada and Canadian Food Inspection Agency policies

Canada's food contact material health security regulation is under the Health Canada and another agency, the Canadian Food Inspection Agency (CFIA). Though there is no dedicated specific legislation in Canada regarding food packaging with nanotechnology, these agencies work under the Food and Drugs Act to provide a general framework for food safety security and public health. As complements to each other, they regulate, monitor, and enforce food safety standards (Bietlot & Kolakowski, 2012). Health Canada is responsible for developing policies on food safety and guidelines about the use of nanotechnology within food contact materials. The guidelines suggest a case-by-case risk assessment of possible risks presented by the application of nanomaterials. Notably, Health Canada's Guidelines for Novel Foods influence the assessment of nanotechnology applications, requiring case-by-case risk evaluations based on possible risks presented by the nanomaterials. Among the steps for evaluating the possible risks are the following: Evaluation of Stability and Safety of Nanoparticles: This includes determining the stability of nanomaterials in different conditions as well as determining whether they can present toxicity toward consumers. Migration Testing: Determine whether nanoparticles from packaging materials migrate into food and, if so, whether such migration is at levels that could pose a risk to consumer health (Bietlot & Kolakowski, 2012; Racicot et al., 2020). The CFIA focuses on food safety regulation and enforcement in compliance with the standards of food production, processing, packaging, and distribution. Food contact materials and their labeling meet consumer expectations for safety and environmental sustainability as part of the CFIA's key roles. This ensures that products meet both domestic and international food safety standards, thus enhancing public confidence in Canadian food products (Racicot et al., 2020). In integrating these efforts, Health Canada and CFIA collaborate to prevent foodborne illnesses by adhering to high standards of food safety. Assessments of nanotechnology are integrated with rigorous inspection protocols that ensure the innovative solutions involving nanomaterials applied in these packaging systems are safe for use while providing consumers with assurance and health for the public. The regulation supports leadership in Canada for food safety and sustainability and accommodates the developments in food-packaging technology. Additionally, the Canadian Environmental Protection Act, 1999 (CEPA), administered by Environment and Climate Change Canada in collaboration with Health Canada, establishes a comprehensive framework to manage and control new substances, including nanomaterials, before their entry into the Canadian market. CEPA aims to protect the environment and human health through pollution prevention and sustainable development. Under CEPA, new substances, including nanomaterials, must undergo notification, assessment, and authorization processes to evaluate potential environmental and health risks, applying the precautionary principle to minimize harm from uncertain risks (Canadian Environmental Protection Act, 1999, S.C. 1999, c. 33) (Canadian Environmental Protection Act, 1999).

34.2.2.4 Asia-Pacific: key countries such as India, Japan, and China

The Asia-Pacific region is seen to have dynamic and diverse regulatory landscapes, with countries like India, Japan, and China making significant strides in nanotechnology that is implemented in food packaging while taking precautions to preserve food safety and public health. All these nations have their regulatory frameworks, which serve to exhibit the approaches and priorities with which each country manages its own food contact material safety and innovation (Ravichandran, 2010).

34.2.2.4.1 India

The regulation covers various packaging materials, including plastics, metals, paper, and glass, and it also discusses the chemical interaction of these with different types of foodstuffs. The regulation still does not directly consider nanotechnology, but it is used as a basis to study new technologies, such as nanomaterials used in food contact applications. Some of the emerging technologies, like edible nanobased packaging, gain high limelight because they involve increased shelf life and the preservation of food quality, as well as waste avoidance in the environment (Han, 2003; Ravichandran, 2010). Hence, edible packaging represents a transformative innovation in the food-packaging line. These are derived from polysaccharides, proteins, and lipids, and they are also biodegradable and food-grade materials. Edible packaging reduces the amount of waste that occurs and, at the same time, serves as an active barrier against microorganisms, oxidation, and moisture loss within food products. Infusion of nanotechnology into edible packaging material will improve functionality in the form of enhanced mechanical strength and heat resistance with antimicrobial properties. For instance, nanoparticles might be used in edible films to form smart packaging systems that could track food freshness or release active agents such as antioxidants or preservatives (Joshi et al., 2024; Singh et al., 2023). The regulatory framework set by India, which still focuses on traditional materials, can be molded to the new paradigm. Emphasis on protocols for safety and migration testing will ensure that new creative solutions, like edible and nanobased packaging, find their way into the marketplace responsibly. India, in its course of cultivating a regulatory environment blending innovation with public safety, is steering the way toward sustainable advanced packaging solutions that will find harmony with emerging global trends and consumer preferences.

34.2.2.4.2 Japan

Japan has a science-based approach to food safety through the strict application of the Food Sanitation Act administered by the Ministry of Health, Labour, and Welfare of Japan (MHLW). The nation views the application of nanotechnology in food contact materials on a case-by-case basis with detailed assessments of safety to meet standards both within and outside the country's borders. Such an approach reflects Japan's commitment to consumer health while promoting innovation in food packaging. The rigorous assessment of domestic and imported food-packaging materials in Japan reveals that the country is serious about high-quality and safe food products (Walravens, 2017).

34.2.2.4.3 China

China has developed general guidelines for the use of nanotechnology in food packaging under the National Health Commission. It offers guidelines with conditions such as requirements for migration testing and toxicological evaluations that closely comply with international standards. Moreover, traceability and quality control measures ensure that food contact materials are at the highest levels of safety and sustainability. This proactive approach reflects the seriousness of China in harmonizing its domestic regulation with the rest of the world, building confidence in the safety and quality of its food-packaging materials (Idumah, 2023).

34.3 Challenges in harmonizing international regulations

34.3.1 Variability in standards

One of the significant challenges encountered in the harmonization of international regulation is regional standard differences. Some countries require exhaustive premarket testing for nanomaterials, whereas others rely on general safety laws that do not specifically address nanotechnology. This variation creates confusion among manufacturers and reduces the scalability of edible nanobased packaging systems (Charlebois et al., 2014).

34.3.2 Rapid technological advancements

The dynamic nature of nanotechnology is ahead of its regulatory frameworks. New materials and applications emerge every day, which challenges the regulators to update policies that may address the risks without stifling innovation (Charlebois et al., 2014).

34.3.3 Limited scientific consensus

The long-term effects of nanomaterials on human health and the environment have yet to be agreed upon by scientists. This has posed a challenge in establishing standard safety assessments and managing risk practices (Dainelli et al., 2008).

34.3.4 Trade barriers

It gets quite challenging for multinationals to market edible nanobased packaging systems worldwide since products, meeting the requirements of one country, would fail to meet those of another country. Different rules and regulations across countries can limit trade (Charlebois et al., 2014).

34.3.5 Public perception and misinformation

There are different perceptions of nanotechnology across cultures and regions. While some view it as a revolution, others have taken a rather skeptical stance due to fears of safety and ethical concerns. Harmonization of regulation must take into account the public sentiment and include provisions for educating and engaging consumers in the process (Wirth, 2013).

34.4 Safety assessments and compliance

The safety of edible packaging systems is ensured by a multilateral process involving scientific evaluation and regulation. Edible systems designed to be more functional by using nanoscale materials are subject to strict standards concerning human health and environmental compliance.

34.4.1 Risk assessment procedures for nanomaterials in food packaging

Safety and effectiveness are considered in edible packaging systems, which may include nanomaterials, essential oils, plant extracts, or other functional additives. It is only by this means that new materials used to improve food quality and shelf life, and reduce the environmental footprint, may be introduced into food use after systematic assessment of potential hazards, probability of exposure, and resultant risks. These assessments are crucial for the regulatory and manufacturing parties to ensure that these new materials are safe for human consumption but cause minimal damage to the environment (Honarvar et al., 2016). Material identification and properties are some of the steps in performing risk assessment. In the case of edible packaging, this encompasses understanding the chemical characterization of materials to be applied, such as nanoparticles and bio-based polymers, in addition to essential oils and their plant extracts. Impurities, including possible contaminants, are highly relevant to nanoparticles because trace contamination could have safety implications. Analogous factors for the physical characterization of nanoparticles are size, shape, surface area, and surface chemistry, because such properties have a dramatic influence on their behavior, stability, and interaction with food. In edible films and coatings, natural antimicrobials and antioxidants most widely used are essential oils and plant extracts, which have to be characterized by their chemical stability and their potential for interaction with other components in the packaging matrix (Onyeaka et al., 2022). Exposure assessment follows, taking into account how components of edible packaging might come into contact with food or be ingested by consumers. Studies on migration are particularly relevant in systems that contain nanoparticles, as the materials could transfer from the packaging to food under conditions such as high temperature or long-term storage. Essential oils and plant extracts, being volatile in nature, can diffuse into the food matrix, so understanding their migration dynamics is essential. Factors determining the degree of exposure are the type of food, whether it is solid, liquid, or fatty; acidity; and storage conditions. Consumer demographics by age, diet, and region are also considerations since these influence both the likelihood and the magnitude of exposure. Thus, children and the elderly might be more susceptible to some materials' effects owing to physiological differences (Hayden et al., 2012; Mura, 2020). Hazard identification is one of the main aspects of risk evaluation, and it concerns interactions between edible packaging materials used and biological systems. For instance, nanoparticles may exhibit effects that are not present with the same materials at the nanoscale; such materials can be cytotoxic, genotoxic, or even immunotoxic. In most studies, the generation of reactive oxygen species (ROS) by nanoparticles is compared with oxidative stress damage to cellular components, such as DNA, proteins, or lipids. Similarly, essential oils and plant extracts should be assessed for potential allergenic or toxic effects, especially at higher concentrations. Although these natural additives are GRAS at certain levels of exposure, their interactions with food constituents and long-term cumulative exposure should be properly evaluated (Organization, 2005; Wirth, 2013). The final step, risk characterization, brings together data from hazard identification and exposure assessment to conclude the safety of the edible packaging system. This step is to know how the extent of adverse effects changes with the amount of the material taken in by dose-response relationship analysis. In the case of nanoparticles, thresholds such as acceptable daily intake (ADI) or tolerable upper intake levels (UL) are established that can be used to define safe limits of exposure. Similarly, essential oils and plant extracts are assessed such that functional benefits, which include antimicrobial activity, are achieved without exceeding toxic thresholds. These assessments are dynamic and hence require periodic revision based on new scientific data and technological inputs (Onyeaka et al., 2022). The safety of individual components should be complemented by the assessment of synergistic effects resulting from the combination of nanoparticles, essential oils, plant extracts, and other additives within the edible packaging matrix. The interaction of materials may influence migration behavior, stability, and overall efficacy. For example, nanoparticles may act as carriers for essential oils, improving their stability and controlled release. Such combinations may, however, change the toxicity profiles, thus requiring comprehensive assessment in realistic conditions (Álvarez-Martínez et al., 2021).

Overall, risk assessment for edible packaging systems with nanomaterials and functional additives represents a rapidly emerging multidisciplinary field. This, therefore, requires consideration of the effects of individual components on other individual components and interactions between food, its environment, and its components. This is key to the consistency and reliability of these evaluations, which would then ensure that these edible packaging solutions build consumer trust and are widely accepted. With rigorous risk assessment and further improvement of safety protocols, edible packaging systems can unlock their full potential as sustainable, functional alternatives to conventional packaging materials.

34.4.2 Toxicological evaluations and migration limits

This development has brought a revolution in food packaging that can include edible packaging systems consisting of films and coatings with nanomaterials, essential oils, plant extracts, and other functional additives. Materials such as these not only enhance the protective and functional attributes of packaging but also reduce dependence on synthetic plastics to contribute to sustainability. Safe consumption, however, necessitates rigorous toxicological assessments and migration tests. These processes are crucial to establishing how parts of packaging interact with the body and food; they provide information that allows the establishment of regulatory limits that would safeguard the consumer (Singh et al., 2023; Szczepańska et al., 2018).

34.4.2.1 Toxicological testing

Toxicological evaluations form the backbone of assessing edible packaging materials for safety. Such tests involve studying the potential health effects of individual constituents, as well as their combined interactions within the packaging matrix.

34.4.2.1.1 In vitro studies

In vitro studies can be regarded as the first stepping stone in toxicological research since they aid in deducing the potential hazards introduced by packaging materials, be they nanoparticles, essential oils, or bio-based polymers. This study will involve cytotoxicity through culture systems that will determine whether the materials presented are hazardous to cells. Some of the notable assessments include genotoxic studies, which investigate whether exposure to these materials may cause genetic damage associated with mutations or carcinogenicity. The researchers also study cellular uptake and interaction to confirm whether nanoparticles or other additives can penetrate cell membranes and accumulate inside the cells, thus leading to harmful effects. Other in vitro experiments also identify potential inflammatory reactions, such as the induction of pro-inflammatory cytokines, which would imply activation of the immune system. Another factor that is crucial to evaluate is the oxidative stress that materials may induce and whether it causes the generation of ROS, leading to cellular damage. Generally, such assessments serve as a solid foundation for what could be in wait regarding novel food-packaging materials prior to advancing such into studies on animals or on human beings, or indeed for any approvals by regulating agencies (Reščeк et al., 2016; Szczepańska et al., 2018).

34.4.2.1.2 In vivo studies

Although in vitro studies are fundamental for generating initial safety information, understanding how packaging materials interact in a living organism requires in vivo studies. Researchers test the biodistribution of materials, particularly nanoparticles, for their absorption, distribution, metabolism, and excretion in the body using animal models. Long-term evaluations are performed in these studies to check for possible chronic effects, including potential risks such as carcinogenicity, reproductive toxicity, or any other adverse health outcomes. The digestive aspect is vital in edible packaging. The material's breakdown and the absorption processes in the gastrointestinal tract must, therefore, be evaluated. The utilization of such packaging, therefore, has to involve the use of natural polymers, essential oils, and plant extracts that have safe metabolic profiles, are nontoxic, and do not amass harmful effects in the body. These extensive in vivo evaluations complement in vitro studies and, therefore, present an integrated view of food-packaging material safety (Mohanraj and Chen, 2006; Onyeaka et al., 2022).

34.4.2.1.3 Migration testing

Migration testing is an important part of understanding how the active components of edible packaging actually interact with food, potentially transferring materials like nanoparticles or volatile compounds from essential oils under a variety of conditions. These studies mimic storage as well as usage under appropriate conditions to provide a full-range safety evaluation. Static testing places food simulants in contact with packaging materials under static conditions, whereas dynamic testing studies migration rates under varying conditions such as temperature, acidity of the food, and moisture. Standardized simulants of water, ethanol, and acetic acid are used to represent aqueous, alcoholic, and acidic foods, respectively, in their behavior. Then, through advanced analytical techniques, one can precisely detect and measure migration. For instance, gas chromatography-mass spectrometry (GC-MS) is among the techniques used to identify volatile compounds from essential oils. Additionally, Inductively coupled plasma mass spectrometry (ICP-MS) is used to determine the migration of nanoparticles and metal ions. Techniques have reached a level such that even trace amounts of migrating substances are detected; thus, the safety profile becomes more complete for edible packaging materials (Paidari et al., 2021; Szczepańska et al., 2018).

34.4.2.1.4 Setting migration limits

The toxicological data allow regulatory agencies to set limits for migration, ensuring safety for consumers. These limits establish the maximum amount of a substance that can migrate from packaging into food. Major regulatory frameworks include:

EFSA: This establishes SMLs for food contact materials so that exposure to potentially hazardous substances remains below the level of potentially harmful amounts.

US FDA: requires migration testing as part of the premarket notification process for new food contact materials. The FDA evaluates safety based on data from toxicological studies and migration studies.

For edible packaging, there is an added consideration: whether it is biodegradable, digestible, or interacts with food constituents as part of comprehensive safety.

34.4.3 Special considerations for edible packaging

Edible packaging systems are not like other materials in that they will directly contact food and potentially be ingested. Such contact and potential ingestion require a higher level of review for safety. The digestibility and metabolism of these materials are also very important. Films and coatings based on polysaccharides, proteins, or lipids must be safely digested and metabolized. For example, the assessment of additives such as essential oils with antimicrobial and antioxidant activity is carried out considering their bioavailability and health effects (Singh et al., 2023). In other words, some essential oils may cause gastrointestinal irritation and allergic reactions if taken in excessive dosages, so proper dosing is crucial. Interaction with food is also very important since packaging materials must not compromise organoleptic properties like flavor, texture, or odor. While materials such as nanoparticles, essential oils, or plant extracts will likely have a little negative impact regarding imparting off-flavors and alterations to the nutritional profile, functional components like antimicrobial properties or oxygen scavenging should not be detrimentally affected, as they should still be effective without detracting from the food (Aayush et al., 2022).

34.4.4 Comprehensive safety framework

Toxicological assessment and migration testing are part of ensuring the safety and consumer acceptability of edible packaging systems. Manufacturers, with help from strict scientific methodologies and observance of regulations, can guarantee consumers that such novel materials are safe for consumption by humans and the environment at large. Nanotechnology and other additives, such as essential oils, once integrated with edible packaging, must always be monitored, and advanced analytical techniques and harmonized regulations must be adopted for their use in sustainable and functional food-packaging solutions. In this way, edible packaging will meet its promise to revolutionize food preservation while safeguarding public health (Idumah, 2023; Onyeaka et al., 2022).

34.4.5 Environmental impact regulations

Sustainable consumption and use of nanomaterials applied in edible packaging can be achieved when such environmental impact is critically considered and mitigated, usually conducted by undertaking a life-cycle assessment for the various activities along the life-cycle pathway, including raw material extraction, processes in manufacture, and ultimate end-of-life disposal of a package. The evaluation also consists of sourcing materials, whether natural polymers, nanoparticles, etc., and energy consumption during manufacturing, greenhouse gases emitted, and waste produced as part of the nanobased packaging consideration (Saha et al., 2020). In addition, concern about the environmental fate includes nanoparticle persistence, ecosystem bioaccumulation, hazards to wildlife, and accidental release through degradation of packaging. The EU's REACH Regulation and the US Toxic Substances Control Act (TSCA) require manufacturers to assess and report on the environmental impact of nanomaterials. Many countries even have guidelines for handling these materials safely and for disposal at the end. In addition, sustainability standards promote the use of renewable resources, such as plant-based polymers, closed-loop recycling systems, and the development of biodegradable nanoparticles to minimize ecological harm and support sustainable practices in edible packaging (Miettinen, 2019).

34.4.6 Integrated approach to safety and compliance

An integrated approach to safety and compliance would combine risk assessments, toxicological evaluations, and considerations of environmental impacts to create a safe and sustainable outcome. This requires collaboration on the part of scientists with regulatory agencies and industry. There will be a need for constant monitoring and research to address this gap, including the development of predictive models for behavior in biological and environmental systems, as well as the refinement of analytical techniques that may be sensitive enough to allow the detection of nanoparticles (Honarvar et al., 2016). The global harmonization of standards through entities such as Codex Alimentarius and ISO is critical for the acceptance of edible nanobased packaging in the international market. Public involvement and openness can help build consumer trust. Proper labeling, certifications, and education campaigns can clear the misconceptions, while the safe aspects, along with the benefits of nanotechnology, are presented in the solutions provided in packaging (Wirth, 2013).

34.5 Consumer acceptance

34.5.1 Consumer acceptance of edible packaging systems

Consumer acceptance is the biggest dependency for the successful adoption of edible packaging systems, particularly those containing nanotechnology, active components, and natural additives. It is a function of public perception, which is based on awareness, safety issues, environmental concerns, and economic factors. A thorough understanding of these factors and barriers is essential to make such advanced packaging solutions common.

34.5.2 Awareness and perception

34.5.2.1 Public perception of nanotechnology and active packaging

Nanotechnology in food packaging is a new advancement, but its technical complexity limits the public's understanding. Although some consumers benefit from nanotechnology and active packaging innovations, which prolong shelf life, preserve food quality, and reduce waste, others are wary. The term "nanoparticles" can be intimidating because people are not familiar with it or associate it with adverse industrial applications. Similarly, active packaging, which interacts with food in order to increase safety or quality, may be perceived as altering the food itself (Sharma et al., 2021). Further demystification of the purpose and benefits of these products will come from educating consumers about the science behind these technologies. For example, the acceptance of nanoparticles that may act as antimicrobial agents or oxygen scavengers in edible coatings may increase. It also relates to the desire for more natural and organic products, where some additives have a natural origin, such as essential oils and plant extracts (Liu et al., 2016).

34.5.2.2 Misinformation and skepticism

Misinformation is a significant challenge to consumer acceptance. Quite often, it is due to the media or the lack of clarity from the regulator that negative perceptions are amplified concerning safety regarding edible packaging. Many people fear that these nanoparticles might migrate into food and become toxic or cause damage in the long term. Therefore, to overcome such fears, scientific research on these materials must be clearly communicated, along with safety assessments and regulatory approvals. It is helpful for consumer acceptance if such materials undergo extensive testing—toxicological evaluations and migration studies (Paidari et al., 2021).

34.5.2.3 Factors influencing acceptance of health and safety concerns

Consumer confidence depends on the safety of edible packaging. Though regulatory approvals and scientific evidence assure the safety of such materials, individual concerns about nanoparticles and chemical interactions persist. To alleviate these concerns, one needs to highlight the fact that the product development of edible packaging systems undergoes rigorous safety evaluations. This is to ensure that the materials used are safe for health and safety standards (Onyeaka et al., 2022). In addition, natural additives, including essential oils, are recommended since they are well established regarding their health benefits, with antimicrobial and antioxidant properties, among others. The biodegradability and digestibility of edible films and coatings are consistent with consumer preferences for eco-friendly and safe products, and this alternative option to traditional packaging is in line with sustainability without endangering safety and functionality. Highlighting success stories where edible packaging is being safely introduced into food systems can further reassure consumers (Joshi et al., 2024).

34.5.2.4 Sustainability and environmental friendliness

Consumer acceptance is largely driven by sustainability. With growing awareness of environmental issues, edible packaging systems are considered to be one of the most eco-friendly alternatives to single-use plastics, which are often made from renewable resources such as polysaccharides, proteins, and lipids. Consumers are increasingly attracted to products that support waste reduction and circular economies. Another positive aspect of nanotechnology is that it enhances sustainability, improving material efficiency and minimizing resource consumption. However, these benefits must be communicated effectively to dispel myths about the environmental impact of nanomaterials. Demonstrating how these systems reduce food spoilage and waste can increase consumer interest (Henrique et al., 2013).

34.5.2.5 Cost implications and willingness to pay

Cost is a critical factor in influencing consumer choices. Edible packaging systems, particularly those based on advanced technologies such as nanoparticles or bio-based materials, are likely to be more expensive than conventional ones. Although some consumers are willing to pay extra for sustainable and innovative products, others may not be. Affordability needs to be addressed by manufacturers and policymakers to improve market penetration. Economies of scale, technological advancements, and incentives for sustainable practices can reduce costs over time. In addition, the long-term benefits, such as extended shelf life, reduced food waste, and environmental sustainability, can be used to justify the initial investment for consumers (Madhusudan et al., 2018).

34.6 Bridging gaps between regulations and consumers

Advanced technologies, including nanotechnology in food packaging, open new frontiers for innovation. The advancement comes with challenges: aligning the regulatory framework with consumer expectations and trust. The gap between regulations and consumers must be closed to achieve acceptance and safety of novel packaging solutions. Transparency, labeling, certifications, and public engagement are important for bridging the gap.

34.6.1 Contribution of transparency to consumer trust

Transparency is the very foundation of trust among manufacturers, regulators, and consumers. Consumers tend to fear the safety, sustainability, and health impacts of food packages, especially those with cutting-edge technologies. Being transparent about the design of the product, its safety evaluations, and compliance with regulations will therefore greatly help improve consumer confidence (Charlebois et al., 2014). For instance, information related to the materials used for packaging can be shared in terms of natural polymers or essential oils with functional benefits such as antimicrobial activity or extended shelf life. In the same way, safety protocols that are followed diligently, migration testing, and toxicological evaluation can be cited as proof that these materials are safe to use. The more consumers are informed as to how, for instance, the FDA or EFSA evaluates and approves these products, the more they are likely to rely on such innovative packaging solutions (Sengul & Asmatulu, 2020).

34.6.2 Importance of labeling and certifications

Labeling and certification serve as a physical indicator that a product meets all safety and quality standards. Proper and transparent labeling helps the consumers understand what material is used in the packaging and what kind of impact that might have on the food's quality and safety. For example, if the label claims the material is biodegradable or edible, that's quite an advantage, as it would attract environmentally conscious customers. Certification from well-known regulatory bodies or third-party independent sources also enhances consumer confidence. Examples include ISO 22000 for food safety management systems or specific biodegradable material certifications that assure that a product has passed stringent tests and meets international standards. For example, in nanotechnology-based packaging, compliance with the REACH Regulation in the EU or the TSCA in the United States may provide an indication of conformance with guidelines and alleviate some of the concerns related to novel materials (Wirth, 2013). In addition, standardized icons or claims such as "GRAS-certified" or "nanotechnology-safe" could render complex regulatory details simpler and easier for consumers to understand. Therefore, labeling must be informative as well as readable to link technical regulations to consumer understanding.

34.6.3 Public engagement and educational campaigns

Public engagement and education campaigns must be undertaken to demystify emerging technologies and the misconceptions consumers have developed. Most consumers are less aware of the benefits associated with edible nanobased packaging and safety measures integrated into the packaging. Consumer skepticism or resistance to using these innovations arises from that lack of awareness. Through education, this gap is bridged: clear, science-based information is provided. Manufacturer-conducted, regulator-led, or consumer advocacy group-initiated workshops, seminars, and online campaigns could all potentially educate consumers about the added shelf life for food products through nanotechnology packaging, minimizing food loss, and improving food safety. It can also depict the compatibility of such technologies with other sustainability objectives, such as biodegradability and diminished impact on the environment, thereby appealing to eco-friendly consumers (Onyeaka et al., 2022). Moreover, direct interaction with consumers through focus groups or feedback surveys helps manufacturers and regulators respond to specific concerns and refine their communication strategies. Public forums where scientists, policymakers, and consumers can interact foster an environment of transparency and mutual understanding. Such initiatives not only educate but also empower consumers to make informed decisions (Miao et al., 2023).

34.6.4 Addressing consumer concerns through dialog

Consumer concerns about new technologies are usually related to health, safety, and the environment. An example would be fears about nanoparticles' migration into food or the long-term effects of consuming edible packaging materials. All such concerns must be proactively addressed through transparent communication supported by solid scientific data.

This would further intensify dialog through the involvement of consumer representatives in regulatory debates. Such representation ensures consideration of consumer perspectives during the development of guidelines and standards. An opportunity would also be present to demonstrate how stringent safety evaluation is for innovative solutions in packaging (Honarvar et al., 2016).

34.6.5 The role of digital tools in bridging the gap

Information can be passed through various digital media, such as mobile apps and QR codes on packaging. Scanning a QR code can reveal all the details of the composition of the packaging material, safety evaluation, and certifications. Educational content, including videos or infographics, can also be provided on interactive platforms to explain complex ideas, such as nanotechnology and food packaging. Social media campaigns, blogs, and webinars will help spread the information even more widely. These means provide the opportunity for the manufacturers and regulators to address various sections of society through two-way communication, so the queries and concerns of consumers can be addressed instantly.

34.6.6 The need for harmonized standards and global collaboration

One of the challenges of bridging the gap between regulations and consumers is inconsistency in safety and quality standards between regions. Global cooperation in trying to harmonize these standards would ensure that products conform to uniform safety criteria, irrespective of where they are manufactured or sold. Programs such as Codex Alimentarius and ISO are also meant to develop universally acceptable guidelines to facilitate the global trade of innovative packaging solutions. Harmonized standards also make consumer communications easier, providing a uniform framework for marking and certifications. For instance, the "safe for food contact" certification mark, unified across the world, could reduce confusion and further increase confidence (Meulen, 2019; Wirth, 2013) (van der).

34.6.7 Fostering long-term consumer acceptance

It is not a one-time process but an ongoing process as technology advances. Continuous engagement, transparent communication, and commitment to safety and sustainability are the requirements for the long-term acceptance of edible nanobased packaging among consumers. As consumers gain knowledge and confidence, they will be more willing to embrace and promote innovative packaging solutions, which will push the industry toward a more sustainable and technologically advanced future. Consequently, this gap between legislation and consumers should be closed only by a multidimensional transparency practice of clear labeling, public education, and international cooperation. In relation to these strategic practices, one would effectively facilitate the stakeholders in ensuring all forms of innovative technologies are included in food packaging in a manner that has effective consumptive benefits to the ultimate benefit of the consumer or environment (Kumari et al., 2022; Sharma et al., 2021).

34.7 Future trends and challenges

Edible packaging systems containing nanotechnology and sustainable additives would greatly improve food safety, quality, and environmental sustainability. However, there are trends and challenges that are likely to influence the future of these innovations. For instance, areas include trends of expected changes in regulation, consumer acceptance, and strategies to enhance compliance, among others. These trends should be understood in order for edible packaging to be fully integrated into global markets.

Regulatory frameworks for nanotechnology in food packaging are constantly updated to address the complexities surrounding advanced materials. While current regulations are mainly based on toxicological evaluations, migration limits, and material labeling, future updates will focus more on the harmonization of global standards and the inclusion of emerging safety concerns. Presently, there are still disparities in regulatory approaches; for example, the US FDA, European EFSA, and Asia-Pacific guidelines create problems for manufacturers operating at international levels. A consensus across the globe on definitions, testing protocols, and safety thresholds is expected to help streamline compliance processes. Scientific understanding of nanotechnology will also continue to deepen; consequently, regulatory bodies will begin to address emerging risks in the form of long-term impacts, bioaccumulation, and interactions with complex food matrices. Increased transparency in regulations, with clearer labeling of nanomaterials, will surely address consumer skepticism over the long term. There will also be further movements toward standardizing certifications and logos to indicate the safe application of nanotechnology, fostering informed decision-making among consumers.

Consumer acceptance is the greatest fundamental success factor in implementation, and innovation to trust has much to do with the gap bridged by strategies. Public awareness campaigns for edible packaging can develop safety issues for such types of packaging and benefits such as nanotechnology enhancing performance through nanoscale. The simplest explanations combined with tangible examples would prove an antidote against myth-mongering and counter all misinformation in due course. Transparency in product labeling is another critical factor, as consumers increasingly value certifications indicating adherence to stringent safety and sustainability standards. Labels such as "Eco-Friendly," "Safe Nano," or "Biodegradable" can resonate with environmentally conscious buyers, fostering trust and encouraging adoption. Moreover, making edible packaging systems cost-competitive with traditional materials is crucial for widespread acceptance. Investments in research, economies of scale, and government subsidies can help lower production costs. This may further expand the accessibility and acceptability of these systems, especially in developing regions, due to tailored solutions for various consumer markets.

Global acceptance of edible packaging systems will depend on the concerted efforts made by stakeholders all over the world. International research collaborations between academia, industries, and governments can fast-track the research on nanotechnology and bio-based packaging. Such associations are necessary for driving innovation in material development, safety assessments, and scalability. Sustainability initiatives on plastic waste and carbon footprints also play a central role. Developing and using biodegradable and renewable materials are encouraged by various programs such as the United Nations' Sustainable Development Goals and the circular economy initiatives. More importantly, governments and international organizations should focus more on the funding for research and development of sustainable packaging. The policy is what will influence the change to greener alternatives, for example, tax incentives, subsidies on green technologies, and bans on plastics.

Thus, it is the regulatory evolution, consumer acceptance, and cooperation from a global perspective that will push edible packaging systems toward revolutionizing the food-packaging industry. Harmonization of standards is an imperative factor, along with raising public awareness of the packaging systems, to facilitate greater acceptance and integration into the transformed industry.

34.8 Conclusion

The development and adoption of edible packaging systems, especially those based on nanotechnology and sustainable additives, depend largely on the regulatory interface and consumer acceptance. Regulation would help ensure that innovations in such areas are safe, of good quality, and environmentally compatible to address issues of toxicological impact, migration limits, and sustainability standards. Such consumer trust and adoption, however, would depend on understanding benefits, transparent information on products, and assurance of their safety. These, of course, have to go hand in hand to succeed in the implementation of edible packaging systems in the world market.

Moving forward, there is a need for regulatory frameworks to be geared toward consumer needs. The approach is multifaceted, and regulators must take it when adopting proactive policies that not only address the current safety concerns but also predict the risks emerging from advanced materials. For instance, this would encompass global harmonization of standards, allowing manufacturers to function across markets without constraints, and introducing transparent labeling systems empowering consumers. At the same time, education on the benefits and safety of edible packaging systems should be conducted among the public. Such a message could help dispel false information and increase public trust in these innovations and their contributions to sustainability, food quality, and safety.

Global collaboration among stakeholders, including governments, academia, and industry, is essential to foster sustainable innovation. Combining robust legal frameworks with consumer-centric strategies will ensure the edible packaging industry achieves widespread acceptance while addressing environmental challenges and public health issues. This integrated path ensures that both regulatory and consumer perspectives are at the forefront, paving the way for long-term success and adoption of edible packaging technologies.

AI disclosure

During the preparation of this work, the author(s) used Paperpalai in order to search references and improve the text language. After using this tool/service, the author(s) reviewed and edited the content as needed and take(s) full responsibility for the content of the publication.

References

Aayush, K., McClements, D. J., Sharma, S., Sharma, R., Singh, G. P., Sharma, K., & Oberoi, K. (2022). Innovations in the development and application of edible coatings for fresh and minimally processed Apple. *Food Control, 141*. https://doi.org/10.1016/j.foodcont.2022.109188, https://www.journals.elsevier.com/food-control.

Alfei, S., Marengo, B., & Zuccari, G. (2020). Nanotechnology application in food packaging: A plethora of opportunities versus pending risks assessment and public concerns. *Food Research International, 137*, 109664. https://doi.org/10.1016/j.foodres.2020.109664.

Álvarez-Martínez, F. J., Barrajón-Catalán, E., Herranz-López, M., & Micol, V. (2021). Antibacterial plant compounds, extracts and essential oils: An updated review on their effects and putative mechanisms of action. *Phytomedicine: International Journal of Phytotherapy and Phytopharmacology, 90*, 153626. https://doi.org/10.1016/j.phymed.2021.153626.

Bietlot, H. P., & Kolakowski, B. (2012). Risk assessment and risk management at the Canadian Food Inspection Agency (CFIA): A perspective on the monitoring of foods for chemical residues. *Drug Testing and Analysis, 4*(1), 50–58. https://doi.org/10.1002/dta.1352, http://onlinelibrary.wiley.com/journal/10.1002/(ISSN)1942-7611.

Canadian Environmental Protection Act. (1999). https://laws-lois.justice.gc.ca/eng/acts/c-15.31/FullText.html.

Charlebois, S., Sterling, B., Haratifar, S., & Naing, S. K. (2014). Comparison of global food traceability regulations and requirements. *Comprehensive Reviews in Food Science and Food Safety, 13*(5), 1104–1123. https://doi.org/10.1111/1541-4337.12101, http://onlinelibrary.wiley.com/journal/10.1111/(ISSN)1541-4337.

Chavan, P., Lata, K., Kaur, T., Rezek Jambrak, A., Sharma, S., Roy, S., Sinhmar, A., Thory, R., Pal Singh, G., Aayush, K., & Rout, A. (2023). Recent advances in the preservation of postharvest fruits using edible films and coatings: A comprehensive review. *Food Chemistry, 418*. https://doi.org/10.1016/j.foodchem.2023.135916, www.elsevier.com/locate/foodchem.

Coles, R., McDowell, D., & Kirwan, M. J. (2003). *Food packaging technology, 5*. CRC Press.

Commission Regulation (EU). (2011). http://data.europa.eu/eli/reg/2011/10/oj.

Considerations for Drug Products that Contain Nanomaterials FDA. (2024). https://www.fda.gov/drugs/cder-small-business-industry-assistance-sbia/considerations-drug-products-contain-nanomaterials.

Dainelli, D., Gontard, N., Spyropoulos, D., Zondervan-van den Beuken, E., & Tobback, P. (2008). Active and intelligent food packaging: Legal aspects and safety concerns. *Trends in Food Science & Technology, 19*(1), S103–S112. https://doi.org/10.1016/j.tifs.2008.09.011.

Darrow, J. J., Avorn, J., & Kesselheim, A. S. (2020). FDA approval and regulation of pharmaceuticals, 1983-2018. *JAMA - Journal of the American Medical Association, 323*(2), 164–176. https://doi.org/10.1001/jama.2019.20288, http://jama.jamanetwork.com/journal.aspx.

Han, J. H. (2003). *Antimicrobial food packaging novel food packaging techniques*. Canada: Elsevier Ltd, 50–70. http://doi.org/10.1016/B978-1-85573-675-7.50008-0, http://www.sciencedirect.com/science/book/9781855736757.

Hayden, S. C., Zhao, G., Saha, K., Phillips, R. L., Li, X., Miranda, O. R., Rotello, V. M., El-Sayed, M. A., Schmidt-Krey, I., & Bunz, U. H. F. (2012). Aggregation and interaction of cationic nanoparticles on bacterial surfaces. *Journal of the American Chemical Society, 134*(16), 6920–6923. https://doi.org/10.1021/ja301167y.

Heckman, J. H. (2005). Food packaging regulation in the United States and the European Union. *Regulatory Toxicology and Pharmacology, 42*(1), 96–122. https://doi.org/10.1016/j.yrtph.2005.02.002, http://www.elsevier.com/inca/publications/store/6/2/2/9/3/9/index.htt.

Henrique, M. A., Silvério, H. A., Flauzino Neto, W. P., & Pasquini, D. (2013). Valorization of an agro-industrial waste, mango seed, by the extraction and characterization of its cellulose nanocrystals. *Journal of Environmental Management, 121*, 202–209. https://doi.org/10.1016/j.jenvman.2013.02.054, https://www.sciencedirect.com/journal/journal-of-environmental-management.

Honarvar, Z., Hadian, Z., & Mashayekh, M. (2016). Nanocomposites in food packaging applications and their risk assessment for health. *Electronic Physician, 8*(6), 2531–2538. https://doi.org/10.19082/2531.

Idumah, C. I. (2023). Global status of nanotechnology policies in the packaging sector. *Informa UK Limited,* 368–395. https://doi.org/10.1201/9781003142287-18.

Joshi, M., Aayush, K., Sharma, K., Bose, I., Khan, A. A., Atanassova, M., Yang, T., Murariu, O. C., Sharma, S., & Caruso, G. (2024). Fiber and nanofiber based edible packaging for enhancing the shelf life of food: A review. *Food Bioscience, 59*. https://doi.org/10.1016/j.fbio.2024.103970, https://www.sciencedirect.com/science/journal/22124292.

Kumar, S., Shukla, A., Baul, P. P., Mitra, A., & Halder, D. (2018). Biodegradable hybrid nanocomposites of chitosan/gelatin and silver nanoparticles for active food packaging applications. *Food Packaging and Shelf Life, 16*, 178–184. https://doi.org/10.1016/j.fpsl.2018.03.008, http://www.journals.elsevier.com/food-packaging-and-shelf-life/.

Kumari, S., Yadav, B. S., & Yadav, R. B. (2022). Effect of nano-conversion on morphological, rheological and thermal properties of barley starch. *Journal of Food Science and Technology, 59*(2), 467–477. https://doi.org/10.1007/s13197-021-05029-0, http://www.springerlink.com/content/121580/.

Li, Q., Gao, R., Wang, L., Xu, M., Yuan, Y., Ma, L., Wan, Z., & Yang, X. (2020). Nanocomposites of bacterial cellulose nanofibrils and zein nanoparticles for food packaging. *ACS Applied Nano Materials, 3*(3), 2899–2910. https://doi.org/10.1021/acsanm.0c00159.

Liu, J., Wang, Z., Sheng, A., Liu, F., Qin, F., & Wang, Z. L. (2016). In situ observation of hematite nanoparticle aggregates using liquid cell transmission electron microscopy. *Environmental Science and Technology, 50*(11), 5606–5613. https://doi.org/10.1021/acs.est.5b06305, http://pubs.acs.org/journal/esthag.

Madhusudan, P., Chellukuri, N., & Shivakumar, N. (2018). Smart packaging of food for the 21st century – A review with futuristic trends, their feasibility and economics. *Materials Today: Proceedings, 5*(10), 21018–21022. https://doi.org/10.1016/j.matpr.2018.06.494.

Meral, R., Kose, Y. E., Ceylan, Z., & Cavidoglu, İ. (2022). The potential use of agro-industrial by-products as sources of bioactive compounds: A nanotechnological approach. *Studies in Natural Products Chemistry, 73*, 435–466. https://doi.org/10.1016/B978-0-323-91097-2.00014-5, http://www.elsevier.com/wps/find/bookdescription.cws_home/715507/description#description.

Meulen (2019). Impact of the codex alimentarius. *European Food and Feed Law Review, 14*(1), 29–50.

Miao, X., Magnier, L., & Mugge, R. (2023). Switching to reuse? An exploration of consumers' perceptions and behaviour towards reusable packaging systems. *Resources, Conservation and Recycling, 193*, 106972. https://doi.org/10.1016/j.resconrec.2023.106972.

Miettinen, M. (2019). Empirical analysis of regulative risk assessment processes of nanomaterials under the Toxic Substances Control Act (TSCA) and European Union regulation concerning the Registration, Evaluation, Authorization and restriction of Chemicals (REACH). *Journal of Physics: Conference Series, 1323*(1), 012023. https://doi.org/10.1088/1742-6596/1323/1/012023.

Mohanraj, V. J., & Chen, Y. (2006). Nanoparticles-a review. *Tropical Journal of Pharmaceutical Research, 5*(1), 561–573.

More, S., Bampidis, V., Benford, D., Bragard, C., Halldorsson, T., Hernández-Jerez, A., Hougaard Bennekou, S., Koutsoumanis, K., Lambré, C., Machera, K., Naegeli, H., Nielsen, S., Schlatter, J., Schrenk, D., Silano, V., Turck, D., Younes, M., Castenmiller, J., Chaudhry, Q., ... Schoonjans, R. (2021). Guidance on risk assessment of nanomaterials to be applied in the food and feed chain: Human and animal health. *EFSA Journal, 19*(8). https://doi.org/10.2903/j.efsa.2021.6768.

Mura, P. (2020). Advantages of the combined use of cyclodextrins and nanocarriers in drug delivery: A review. *International Journal of Pharmaceutics, 579*, 119181. https://doi.org/10.1016/j.ijpharm.2020.119181.

Nikolic, M. V., Vasiljevic, Z. Z., Auger, S., & Vidic, J. (2021). Metal oxide nanoparticles for safe active and intelligent food packaging. *Trends in Food Science and Technology, 116*, 655–668. https://doi.org/10.1016/j.tifs.2021.08.019, http://www.elsevier.com/wps/find/journaldescription.cws_home/601278/description#description.

Onyeaka, H., Passaretti, P., Miri, T., & Al-Sharify, Z. T. (2022). The safety of nanomaterials in food production and packaging. *Current Research in Food Science, 5*, 763–774. https://doi.org/10.1016/j.crfs.2022.04.005, https://www.journals.elsevier.com/current-research-in-food-science.

Organization, W. H. (2005). *Enhancing participation in Codex activities: An FAO/WHO training package*. Food & Agriculture Org.

Organization, W. H. (2018). *Understanding the codex alimentarius*. Food & Agriculture Org.

Paidari, S., Tahergorabi, R., Anari, E. S., Nafchi, A. M., Zamindar, N., & Goli, M. (2021). Migration of various nanoparticles into food samples; A review. *Foods, 10*(9). https://doi.org/10.3390/foods10092114, https://www.mdpi.com/2304-8158/10/9/2114/pdf.

Racicot, M., Comeau, G., Tremblay, A., Quessy, S., Cereno, T., Charron-Langlois, M., Venne, D., Hébert, G., Vaillancourt, J.-P., Fravalo, P., Ouckama, R., Mitevski, D., Guerin, M. T., Agunos, A., DeWinter, L., Catford, A., Mackay, A., & Gaucher, M.-L. (2020). Identification and selection of food safety-related risk factors to be included in the Canadian Food Inspection Agency's Establishment-based Risk Assessment model for Hatcheries. *Zoonoses and Public Health, 67*(1), 14–24. https://doi.org/10.1111/zph.12650.

Ravichandran, R. (2010). Nanotechnology applications in food and food processing: Innovative green approaches, opportunities and uncertainties for global market. *International Journal of Green Nanotechnology: Physics and Chemistry, 1*(2), P72–P96. https://doi.org/10.1080/19430871003684440.

Reščček, A., Ščetar, M., Hrnjak-Murgić, Z., Dimitrov, N., & Galić, K. (2016). Polyethylene/polycaprolactone nanocomposite films for food packaging modified with magnetite and casein: Oxygen barrier, mechanical, and thermal properties. *Polymer-Plastics Technology and Engineering, 55*(14), 1450–1459. https://doi.org/10.1080/03602559.2016.1163606.

Robertson, G. L. (2016). *Food packaging: Principles and practice* (Third Edition). Australia: CRC Press, 1–688. http://doi.org/10.1201/b21347, http://www.tandfebooks.com/doi/book/9781439862421.

Rojas-Graü, M. A., Soliva-Fortuny, R., & Martín-Belloso, O. (2009). Edible coatings to incorporate active ingredients to fresh-cut fruits: A review. *Trends in Food Science and Technology, 20*(10), 438–447. https://doi.org/10.1016/j.tifs.2009.05.002.

Roy, S., & Rhim, J.-W. (2022). Starch/agar-based functional films integrated with enoki mushroom-mediated silver nanoparticles for active packaging applications. *Food Bioscience, 49*, 101867. https://doi.org/10.1016/j.fbio.2022.101867.

Saha, T., Hoque, M. E., & Mahbub, T. (2020). Biopolymers for sustainable packaging in food, cosmetics, and pharmaceuticals. In *Advanced processing, properties, and applications of starch and other bio-based polymers*. Bangladesh: Elsevier, 197–214. http://doi.org/10.1016/B978-0-12-819661-8.00013-5, https://www.sciencedirect.com/book/9780128196618.

Sekhon, B. S. (2010). Food nanotechnology–An overview. *Nanotechnology, Science and Applications, 3*.

Sengul, A. B., & Asmatulu, E. (2020). Toxicity of metal and metal oxide nanoparticles: A review. *Environmental Chemistry Letters, 18*(5), 1659–1683. https://doi.org/10.1007/s10311-020-01033-6, https://www.springer.com/journal/10311.

Sharma, S., Barkauskaite, S., Jaiswal, A. K., & Jaiswal, S. (2021). Essential oils as additives in active food packaging. *Food Chemistry, 343*, 128403. https://doi.org/10.1016/j.foodchem.2020.128403.

Singh, G. P., Aayush, K., & Yang, T. (2023). Nanotechnology enabled packaging materials (Second Edition). In *Encyclopedia of Food Safety1-4*, Canada: Elsevier. V3-V3-719. https://doi.org/10.1016/B978-0-12-822521-9.00154-4, https://www.sciencedirect.com/book/9780128225219.

Sorrentino, A., Gorrasi, G., & Vittoria, V. (2007). Potential perspectives of bio-nanocomposites for food packaging applications. *Trends in Food Science and Technology, 18*(2), 84–95. https://doi.org/10.1016/j.tifs.2006.09.004.

Sothornvit, R., & Krochta, J. M. (2005). Plasticizers in edible films and coatings. In *Innovations in Food Packaging*. Thailand: Elsevier, 403–433. http://doi.org/10.1016/B978-012311632-1/50055-3, https://www.sciencedirect.com/book/9780123116321.

Szczepańska, N., Kudłak, B., & Namieśnik, J. (2018). Recent advances in assessing xenobiotics migrating from packaging material – A review. *Analytica Chimica Acta, 1023*, 1–21. https://doi.org/10.1016/j.aca.2018.03.045.

Trienekens, J., & Zuurbier, P. (2008). Quality and safety standards in the food industry, developments and challenges. *International Journal of Production Economics, 113*(1), 107–122. https://doi.org/10.1016/j.ijpe.2007.02.050.

Walravens, T. (2017). Food safety and regulatory change since the 'mad cow' in Japan: Science, self-responsibility, and trust. *Contemporary Japan, 29*(1), 67–88. https://doi.org/10.1080/18692729.2017.1256987.

Wirth, D. A. (2013). The international organization for standardization: Private voluntary standards as swords and shields. In *Research handbook on environment, health and the WTO*. United States: Edward Elgar Publishing Ltd, 139–163. http://doi.org/10.4337/9781781006146.00011, http://www.elgaronline.com/view/9781847208972.00011.xml.

Wongphan, P., Khowthong, M., Supatrawiporn, T., & Harnkarnsujarit, N. (2022). Novel edible starch films incorporating papain for meat tenderization. *Food Packaging and Shelf Life, 31*, 100787. https://doi.org/10.1016/j.fpsl.2021.100787.

Chapter 35

Trends and challenges in biopolymer-based films and coatings

Ewelina Jamróz[1,2]
[1]*Department of Chemistry, University of Agriculture, Kraków, Poland,* [2]*Department of Packaging and Logistics Processes, Krakow University of Economics, Kraków, Poland*

35.1 Introduction

The modern world is facing serious challenges related to environmental degradation, climate change, and the excess of plastic waste that is currently piling up in landfills, oceans, and land ecosystems. According to estimates, more than 400 million tons of plastics are produced worldwide each year, a significant portion of which comprises single-use materials such as packaging films, bags, and protective coatings. Traditional polymers, derived from petroleum, are highly durable and chemically resistant, making them very difficult to decompose—full degradation can take hundreds of years. In response to these issues, the industry and scientific communities are constantly and increasingly looking for alternatives that would be more environmentally friendly.

In this context, biopolymers are gaining particular importance; that is, polymers produced from renewable biological sources such as starch, cellulose, chitosan, alginates, plant proteins, or microorganisms (e.g., bacteria producing PHA). Biopolymers can be biodegradable, compostable, and, in some cases, also recyclable, which creates new possibilities in the design of packaging, films, and protective coatings. More and more companies are implementing them in their production processes, especially in the industries of, for example, food, pharmaceuticals, cosmetics, or agriculture. Biopolymer-based films and coatings are one of the most dynamically developing categories of ecological materials. They are a real alternative to traditional plastic films such as polyethylene (PE), polypropylene (PP), or polyethylene terephthalate (PET). Depending on the application, biopolymers can be processed into thin layers with different properties, from transparent, flexible films to rigid and resistant coatings with protective and/or bioactive effects. The development of this group of materials is driven not only by growing social and environmental pressure but also by changing legal regulations, especially in the European Union. For example, Single-Use Plastics Directive (SUP) prohibits the use of many types of disposable plastic products, which forces manufacturers to look for alternatives. At the same time, incentives are emerging in the form of tax breaks or subsidies for research on and development of sustainable materials. However, despite their intensive development and numerous advantages, biopolymer films and coatings also encounter a number of technological, economic, and logistical challenges. The most significant include high production costs, limited mechanical durability, low resistance to moisture and atmospheric factors, as well as problems with unifying biodegradability standards. Additionally, some biopolymers may be difficult to process in existing production lines, which requires investment in new infrastructure (Fig. 35.1).

The aim of this chapter is to present the most important trends and challenges related to the development and implementation of biopolymer-based films and coatings. In this study, current directions of research and innovations are discussed, and the technological and economic barriers that stand in the way of the widespread use of these materials are analyzed. Current scientific data and applicable legal regulations comprise the basis of the conducted analysis, which allows for a comprehensive look at the potential and limitations of biopolymer material solutions.

Natural Polymer-Based Films and Coatings in Food Technology. DOI: https://doi.org/10.1016/B978-0-443-34244-8.00034-3

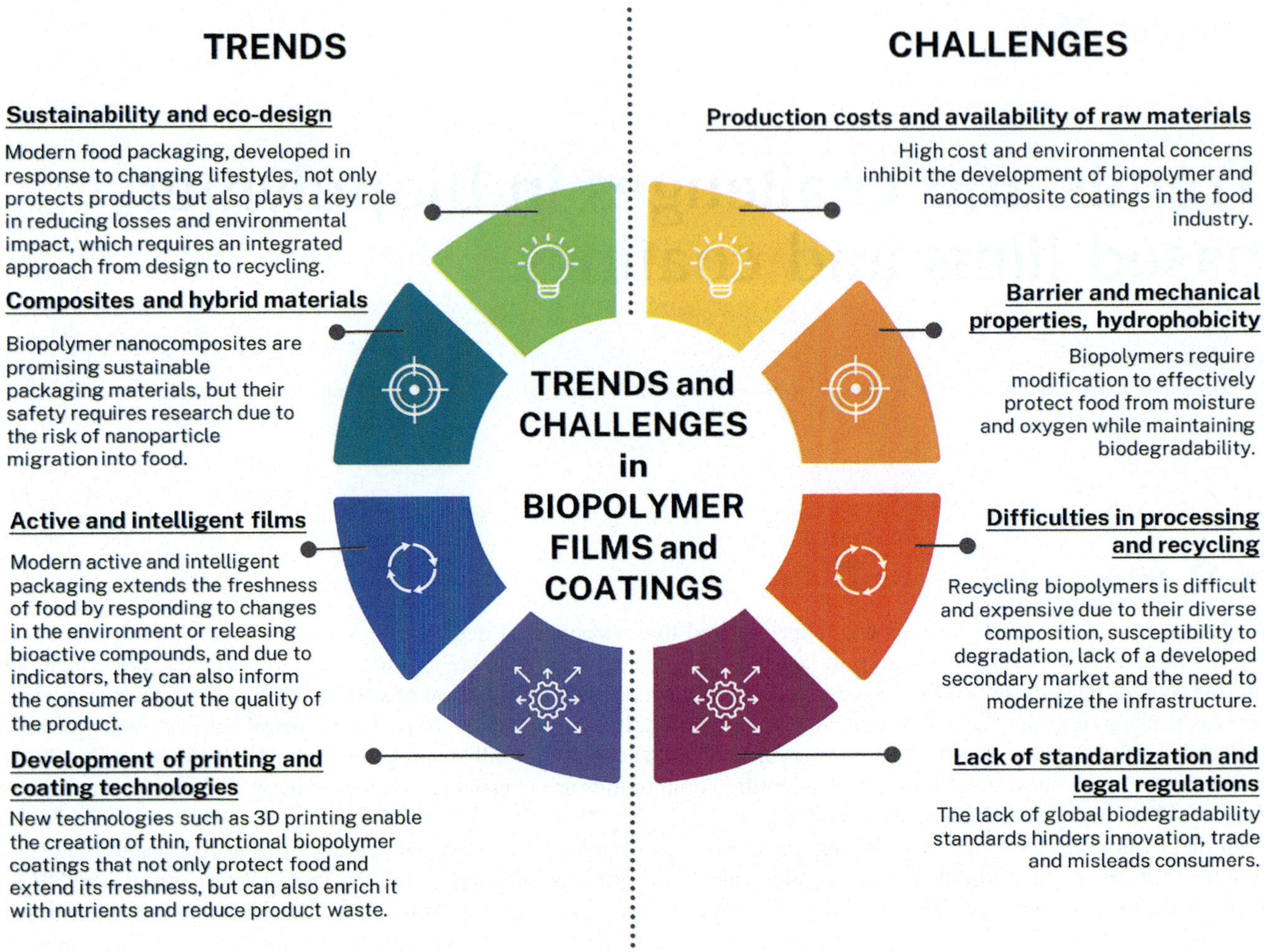

FIGURE 35.1 Schematic summary of trends and challenges in the area of biopolymer materials.

35.2 Current trends in the development of biopolymer films and coatings

35.2.1 Sustainable development and ecodesign

Lifestyle changes have significantly affected what and how people eat, which has forced the food industry to develop new packaging solutions. Today, almost all food is sold in packaging that not only allows it to be stored but also protects it throughout the journey from production to transport and finally to sale and consumption. How we produce and consume food shapes the environment, economy, and society, which is why it is necessary to implement sustainable solutions: from more efficient land use, through modern processing and packaging methods, to improved distribution and sales systems. The goal is to achieve a balance in which the benefits outweigh the costs.

Nowadays, packaging plays a key role in protecting food from contamination and harmful factors such as moisture, oxygen, UV radiation, and microorganisms. The development of modern packaging technologies allows for the extension of the shelf life of products through active systems that slow down oxidation and the growth of bacteria. Intelligent packaging goes a step further by monitoring the quality of products, helping to reduce losses resulting from improper storage, transport, or sale. Investing in such solutions is also an opportunity to reduce the total impact of the food sector on the environment, especially by minimizing food waste. However, in order to choose the best solutions, comprehensive life-cycle analyses (LCA) and up-to-date qualitative and quantitative data are needed. Educational activities and support from experts in the industry, science, and administration are also necessary to promote a culture of sustainable development among all participants in the supply chain, that is, from producers to consumers. Among the available possibilities are better protection of fresh produce, recovery of surpluses from farms, use of agro-food industry waste for packaging production, design of reusable and recyclable packaging, development of active and intelligent packaging materials, harmonization of regulations on expiration dates, and implementation of systems enabling better synchronization of supplies and reduction of losses.

Nonetheless, there is currently a serious gap between product and packaging design, their production, and reuse of materials from waste, which makes it difficult to close the raw material loop. The transition to a circular economy, therefore, requires a holistic approach that considers constraints and opportunities at every stage of the supply chain. The design phase is of particular significance here because it is at this stage that more than 80% of the product's environmental impact is determined. Well-designed food packaging can contribute to reducing both food losses and packaging waste while ensuring that food protection remains the most important factor in minimizing any environmental impact throughout the product's life cycle (Versino et al., 2023).

35.2.2 Composites and hybrid materials

In recent years, there has been a growing interest in biopolymer nanocomposites used for intelligent food packaging due to their ability to support sustainable development and improve food safety. Biopolymers obtained from natural sources, such as plants, microorganisms, or animal organisms, are an attractive alternative to traditional petroleum-based plastics. Their combination with nanostructured additives, such as nanoparticles or nanofibres, allows for significant improvements in mechanical, barrier, and functional properties, making them ideal materials for use in modern packaging systems (Mishra et al., 2024).

One of the major challenges associated with the use of nanomaterials in edible coatings is the risk of their migration to food. Due to their microscopic size and large specific surface area, nanoparticles have an increased ability to penetrate from the coating material to the food product, which may pose potential health risks for consumers (Karnwal et al., 2025). An example of this is titanium dioxide nanoparticles, used in the food industry, among others, as a component of coatings with a whitening effect. There is a risk that such particles will get into the food and be unknowingly consumed by consumers. Although research on the safety of these materials is still ongoing, food safety supervisors, such as the European Food Safety Authority (EFSA), are adopting a very cautious approach. In response to scientific uncertainty and possible health risks, strict regulations have been implemented on the use of nanomaterials in packaging and coatings that come into contact with food. This has been done in order to limit the possibility of their migration (Authority, 2011). The definition of potential risks resulting from the use of nanotechnology should be refined, taking the unique properties of nanomaterials and the dynamics of their migration into account. There is a risk that food in contact with packaging containing nanoparticles may become a source of exposure for the consumer, especially in the event of improper functioning of such packaging, which may lead to the migration of these particles to the product. Therefore, it is crucial to conduct research on the possibility of using nanomaterials in food packaging and to thoroughly assess their safety through migration tests carried out in strictly controlled conditions (Mishra et al., 2024).

35.2.3 Active and intelligent coatings

A current trend is the development of "smart" films and coatings that respond to changing environmental conditions, for example, temperature, pH (intelligent packaging), or release bioactive substances, for example, antioxidants and preservatives (active packaging). Such systems are particularly valued in the food industry, where they can prolong the freshness of products.

Food packaging can be made of materials with active properties that continuously absorb or emit specific chemical compounds, thus helping to extend the shelf life of stored products. Depending on the mode of action, they are divided into two main types: migratory and nonmigratory. In migratory packaging, active substances are released directly into the interior of the packaging, affecting the product or its environment. In the latter systems, on the other hand, these components are permanently bound to the packaging material, for example, by chemically attaching them to the surface (Amin et al., 2022). Modern solutions in the field of active packaging most often use elements that absorb specific substances, such as oxygen, ethylene, carbon dioxide, or undesirable aromas. In parallel, systems are used that release selected compounds; for instance, they emit carbon dioxide or gradually release flavor substances. The properties of the film play a key role here, that is, its ability to control the presence and movement of gases has a direct impact on the effectiveness of protecting the packaged food (Xue, Zhu, Sun, Yang, Wu, & Li, 2023).

During storage, packaged food undergoes various changes that lead to certain deterioration in its quality. These changes affect the composition of the atmosphere inside the packaging, as they can include pH fluctuations, modification of gas proportions, and the appearance of decomposition products resulting from chemical and microbiological processes, such as amines, ammonia, or hydrogen sulfide. Intelligent packaging acts as a kind of "sensor," reacting to changes in the product and providing information regarding its freshness. The appropriate selectivity of the packaging material, that is, the ability to let specific molecules, such as ammonia, oxygen, volatile acids, or carbon dioxide through, is of great significance. It is these properties that enable assessment of the food's condition as well as early detection of product spoilage processes (Xue et al., 2023). For early detection of these changes to be enabled, special indicators—compounds that react to the presence of specific substances or groups of compounds—are added to biopolymer-based packaging materials. An example is natural dyes that change color in response to specific environmental conditions. Such a visual color change acts as a simple and intuitive signal for the consumer, informing him/her as to whether the product is still suitable for consumption (Amin et al., 2022).

35.2.4 Development of printing and coating application technologies

Innovative methods used to apply biopolymer coatings, such as 3D printing, plasma spraying, or vacuum coating, allow for obtaining thin, uniform layers with a specific structure. This further permits the creation of very light and, at the same time, functional materials.

The 3D printing technology used in food production is becoming an increasingly promising tool in the hands of engineers and scientists involved in food industry innovation. Its application goes far beyond creating visually attractive forms—it allows for the precise shaping of the composition of products, their structure, sensory properties, and nutritional value. One of the more dynamically developing applications of this technology is the production of edible protective coatings, which respond to the growing needs in the area of shelf-life extension for products, improving their quality and reducing food losses (Andriani & Handayani, 2023; Karnwal et al., 2025). In the food industry, 3D printing involves the gradual deposition of material—in this case, food ingredients—layer-by-layer, which allows for the creation of complex structures with unprecedented accuracy. This process is based on three key components: a properly prepared printing material (e.g., vegetable or fruit puree, plant proteins), a printing device, and a control system that allows for modeling of the product structure (Hakim et al., 2024). For the process to be effective, the ingredients must have appropriate rheological properties, including viscosity and flowability (Karnwal et al., 2025). An example of applying this technology is chitosan coatings, which alone have properties that inhibit the growth of microorganisms. In combination with 3D printing, it becomes possible to precisely apply them to fruits and vegetables in a way that minimizes moisture loss and prolongs freshness of the product (Deng et al., 2020). Moreover, additional functional ingredients, such as vitamins, antioxidants, or microelements, can be introduced into edible protective layers, so that the coating not only protects the food but also enriches its nutritional profile. Xie et al. (2023) proved that printed coatings can effectively combine protective and health-promoting functions, becoming a carrier of valuable substances. This technology can also facilitate the reduction of food waste. The technology of 3D printing enables the use of surplus food products, for example, unsaleable fruit or vegetables, as raw materials for the production of edible coatings (Karnwal et al., 2025).

Despite its numerous advantages, implementing 3D printing technology on a larger scale also poses challenges. Equipment costs, the need for highly specialized knowledge in food technology, and difficulties in maintaining balance between the functionality of coatings and their sensory characteristics (such as taste or texture) are factors that limit the implementation rate of this technology. Nevertheless, the dynamic development of research and innovation in this area indicates the enormous potential of 3D printing in shaping the future of the food industry (Wang et al., 2024). In coming years, we can expect increasingly advanced solutions combining technological precision with consumer expectations and requirements of sustainable development. Edible coatings printed via 3D technology have the potential to become not only an effective way to protect food, but also a tool to improve its nutritional value and attractiveness—and thus, an important element of innovative packaging and preservation systems for food products (Karnwal et al., 2025).

35.3 Challenges associated with biopolymer implementation

35.3.1 Production costs and raw material availability

Compared to conventional plastics, biopolymers are still more expensive, which limits their widespread use. The production of renewable raw materials requires suitable agricultural conditions and may compete with food crops, raising ethical and economic controversies.

One of the main factors limiting the broad implementation of edible coatings in the food industry is the cost associated with their production, especially those resulting from the extraction of natural, high-quality active ingredients. The process of extracting plant-derived compounds is often expensive and difficult to standardize at an industrial level, which leads to problems with the repeatability of the properties in the final product (Costa et al., 2023). Despite edible coatings having great potential as natural preservatives, not all application methods, such as spraying, are cost-effective enough, especially in the case of products with a specific surface or structure (Karnwal et al., 2025).

What is more, the effective performance of such coatings often requires the use of complex mixtures of polymers and active substances, which together create a protective layer with appropriate mechanical and barrier properties. This, in turn, translates into higher manufacturing costs and increased complexity of the technological process, especially in large-scale production. For example, in the case of vegetables such as tomatoes, the commonly used dipping method can lead to uneven coverage, which causes variable coating thickness and, as a result, unstable performance (Poonia & Mishra, 2021).

This type of inhomogeneity is a serious obstacle in ensuring repeatable effects, and it further hinders the development of scalable solutions. For this reason, the need to develop more efficient, economical methods for coating application that will enable their even distribution and improve compatibility between different polymer components is increasingly emphasized. Only by improving application techniques and optimizing formulations based on natural components will it be possible to transform edible coatings into a realistically cost-effective and widely applied method for food protection (Karnwal et al., 2025).

The key challenge standing in the way of broadly implementing nanocomposite films is the standardization of their mass production processes. Transitioning methods developed in laboratory conditions to an industrial scale, while maintaining improved material properties, requires advanced technological solutions and large financial outlays. Although the addition of metal nanoparticles significantly enhances the properties of biopolymer films, it also raises concerns about their safety for human health and the environment. Therefore, in-depth research on the toxicity and mechanisms regarding the impact of these nanoparticles is necessary to dispel doubts and meet regulatory requirements. Another significant limitation is the economic aspect of the project—the production of nanomaterials is still associated with higher costs than the production of conventional plastic packaging. Such a cost difference is a significant barrier to their market development and mass availability. Overcoming this obstacle is crucial for the further development and popularization of ecological packaging solutions (Jogaiah et al., 2025). In addition to technological and economic challenges, limited public awareness of nanomaterials remains a significant barrier. Consumers are still cautious about modern solutions based on nanotechnology, which hinders their acceptance in the market. Therefore, it is crucial to conduct broad education that promotes the benefits of using these materials and dispels concerns related to their safety.

To fully exploit the potential of nanocomposites, it is necessary to focus future research on eliminating existing barriers. Optimizing production, both in terms of scalability and cost-effectiveness, is particularly imperative. Developments in biotechnology and nanotechnology will play a key role here, enabling more advanced and affordable solutions. Close cooperation between scientists, industries, and regulatory institutions is also essential. Only together can clear guidelines on the safety of using metal nanoparticles in packaging materials be developed. Biogenic and nano-biogenic nanoparticles (MNPs) have the potential to radically change the food sector, offering not only superior protective barriers but also antimicrobial properties, detection functions, and extended shelf life of food products (Jogaiah et al., 2025).

At the same time, doubts about their impact on the environment cannot be ignored, both during production and use, and in the final stages of the life cycle, especially at the stage of disposal. Therefore, thorough research is necessary concerning the biodegradability, ecotoxicity, and long-term impact of these materials on human health and ecosystems. Comparative LCA between nanomaterials and conventional plastic packaging will allow for a better evaluation of their actual ecological value and sustainable nature (Gupta et al., 2024). Along with advances in technology and increases in production efficiency, the cost differences between nanomaterials and traditional plastics can be expected to decrease, paving the way for their wider use in everyday life (Jogaiah et al., 2025).

35.3.2 Barrier and mechanical properties, hydrophobicity

Many biopolymers have poorer barrier properties (e.g., gas and water vapor permeability) and lower mechanical resistance than their synthetic counterparts. This necessitates the use of additives or structural modifications that can affect the biodegradability of the material. Properly matching the packaging to the requirements of good practice (GP) is crucial for maintaining the quality and durability of food. The properties of the packaging material, including its resistance to water vapor transmission (WVP), determine whether the product will be effectively protected in specific storage conditions. Different food groups have different needs—for example, nuts, bread, or cold cuts are best stored in a dry environment, while fresh vegetables, fruit, or meat require protection against moisture loss. The packaging should also allow for control of internal atmosphere, which is of particular significance in the case of plant products. After harvest, fruit, vegetables, and mushrooms continue to breathe, which reduces the oxygen levels and increases the carbon dioxide concentration inside the packaging. As a result, a gas gradient is created, which leads to the exchange of gases through the packaging film until equilibrium is reached, slowing down the respiratory processes and thus prolonging freshness (Badillo & Segura-Ponce, 2020).

It is therefore crucial that the barrier properties of the material are matched to the breathing intensity of the product in question; too high permeability can lead to excessively rapid gas exchange, which disrupts balance.

In the case of products such as meat, which do not breathe, the oxygen content in the packaging plays a significant role. A concentration that is too high accelerates the oxidation of fats and myoglobin, leading to a change in color and deterioration of quality. Therefore, meat requires packaging with a very high barrier, limiting the access to oxygen, minimizing visual and flavor degradation, and hindering the growth of microorganisms (McSharry et al., 2020).

Natural polymers used in the production of packaging materials are characterized by a hydrophilic nature, which is associated with numerous difficulties regarding their use in the food packaging sector. Substances such as polysaccharides or proteins easily absorb water, which negatively affects their ability to create an effective moisture barrier (Manzoor et al., 2022). Protective abilities against moisture are crucial for maintaining food freshness and durability (Kalita et al., 2025; Moschopoulou et al., 2019).

One of the key problems in designing biopolymer packaging is reconciling mechanical properties with the required resistance to moisture. Materials such as starch, cellulose, or proteins tend to swell in conditions of increased humidity, which leads to a decrease in their mechanical stability and usability. For this reason, it is difficult to obtain packaging that is both durable and moisture resistant. As an example, although films based on polysaccharides are environmentally friendly and biodegradable, their low resistance to water limits their use in packaging food that is particularly sensitive to moisture, unless they are appropriately modified (Kalita et al., 2025).

35.3.3 Difficulties in processing and recycling

Biopolymers often have different processing requirements than traditional plastics, requiring special temperatures, pressures, or additives. In addition, some of them are problematic to separate from other waste, which makes material recycling difficult. The issue of recycling is a significant problem in the context of food packaging materials. In order to adapt the barrier properties of packaging to the various needs of food, polymer blends are often used, or multilayer structures are created. Although products made from such processed materials are of high quality, their reprocessing is difficult. Problems mainly arise at the sorting stage due to the complex and often nonstandard composition of packaging and differences in production technologies. For this reason, there is a growing need for advanced material tracking and identification systems that would allow for improving the recycling process (Xue et al., 2023).

Recycling biopolymers is associated with a number of technological challenges resulting from their specific natural properties. One of the main difficulties is the diversity of the composition of these materials. Unlike traditional plastics, which are characterized by more uniform properties, biopolymers are derived from various biological raw materials. This diverse origin means that they exhibit different chemical and physical properties, which significantly complicates their processing. As a result, different recycling methods may be necessary for each type of biopolymer. An additional significant challenge is the susceptibility of biopolymers to degradation during recycling. These materials were developed to decompose more easily compared to traditional plastics, which works well in composting but poses a serious problem when it comes to their recovery. Especially in the case of mechanical recycling, further disintegration of polymer chains occurs, which leads to a deterioration of mechanical properties and limits the possibilities of reusing the recovered material (Shen et al., 2010).

Another significant issue is the profitability of recycling biopolymers and the processes related to their collection. Sorting and processing are associated with high costs. Unlike traditional plastics, biopolymers have not yet reached mass production scale, which translates into higher expenditures on infrastructure and recycling operations. In addition, the market for secondary raw materials from biopolymers is developing quite slowly. This is mainly due to the fact that recycled biopolymers are often inferior in quality to new materials, which discourages producers from using them. The situation is worsened by the fact that the prices of new biopolymers are falling, which further reduces the profitability of investments in their recovery technologies. Moreover, the existing network of recycling plants is mainly adapted to the processing of classic synthetic plastics, and adapting it to the specificity of biopolymers requires significant expenditures on modernization or the construction of new installations (Akinsemolu et al., 2024).

35.3.4 Lack of standardization and legal regulations

There is still a lack of clear standards for the certification of biodegradability or compostability of materials. For this reason, manufacturers and consumers are often unclear as to which products are truly ecological and which only appear to be so (so-called "greenwashing"). Despite the global interest in sustainable development and the growing social pressure to reduce plastic waste, regulations on the biodegradability of materials, including biopolymer films and coatings, remain largely inconsistent and unharmonized at the international level. Different countries use various legislative approaches, often based more on recommendations than precise legal requirements. This situation creates a number of barriers for manufacturers who want to introduce their solutions to foreign markets while, at the same time, making it difficult for consumers to make informed decisions. The United States is an example of a country where there are no clear, nationwide legal standards concerning the biodegradability of packaging. Although many states and cities introduce local regulations supporting ecological solutions, the lack of central regulations means that manufacturers must follow general guidelines for environmental protection. In practice, this leads to a wide range of interpretations regarding the concept of "biodegradability" and the risk of using unclear marketing labels that may mislead consumers. In the European Union, the situation is somewhat different, with environmental issues being strongly integrated into the overall packaging policy. Although the EU encourages the use of biodegradable materials, current regulations are mainly focused on requirements for compostability (compliance with the EN 13432 standard). However, even in this area, there is a lack of consistent and uniform definitions, especially with regard to materials not intended for industrial composting, such as thin-film coatings or materials used locally (e.g., for packaging fresh produce). This makes it problematic to implement innovative solutions and classify them under the current regulations. India is one of the countries that publicly promotes the use of biodegradable packaging, especially in the context of reducing plastic waste. Despite this, there are no clearly formulated regulations regarding the biodegradability of coatings or packaging films. Instead, general environmental guidelines are being introduced, which, although desirable from a long-term perspective, do not provide manufacturers with a clear legal framework for the composition, labeling, or certification of biodegradable materials. A similar situation occurs in China, where the central government promotes environmental solutions, including biodegradable coatings, but there is a lack of detailed legal regulations specifying which materials can be classified as biodegradable and under what conditions. In practice, this means that the implementation of such materials is often based on voluntary initiatives by producers and industry consortia, without the obligation of going through standardized assessment procedures. In Australia, too, there is a growing interest in biodegradable materials, both at the government and consumer levels. However, here as well, the recommendations and GPs dominate, rather than clear statutory regulations. Guidelines on the environmental impact of packaging, such as the Australian "Sustainable Packaging Guidelines," suggest the use of readily degradable materials, but do not include mandatory certification standards or inspection procedures for biopolymer-based coatings (Karnwal et al., 2025).

Legal regulations have huge impact on the direction of innovation in packaging materials, especially those intended for contact with food. These regulations, increasingly rigorous, require manufacturers to eliminate certain chemical substances and conduct comprehensive safety tests before allowing the material to be marketed. For example, the European Union, known for its very restrictive approach to chemical substances in packaging, requires compliance with a number of standards, which can significantly extend the process of implementing new biopolymer solutions (Azam et al., 2022). Depending on the region of the world, the approach to regulating packaging materials can significantly differ. In the US, the Food and Drug Administration (FDA) oversees this area, in Europe—EFSA, in India—Food Safety and Standards Authority of India (FSSAI), in China—China Food and Drug Administration (CFDA), and in Australia—Food Standards Australia New Zealand (FSANZ). Each of these bodies is guided by a slightly different legislative philosophy. Despite differences, the common denominator is concern for consumer health, the impact of materials on the environment, and the reliability of manufacturers' declarations concerning the properties and functions of the offered products (Karnwal et al., 2025). Such regulatory diversity is not only a technological barrier but also a strategic challenge for companies that want to commercialize their solutions on a global scale. The need for adaptation to many different dynamically changing legal standards requires large expenditures on research, certification, and documentation, which can significantly slow down the pace of innovation.

35.4 Summary and future perspectives for development

Biopolymer films and coatings are a promising direction for the development of packaging and functional materials that respond to the challenges of the modern world, including the needs related to environmental protection, waste reduction, and the circular economy. Although there are still many technological, economic, and legislative barriers, intensive research and investment in this area lead to systematic improvements in the properties of biopolymers and their increased availability. In the future, we can expect even greater specialization of films and coatings—they will be not only biodegradable but also intelligent, adaptive, and multifunctional. However, the key factor for success will be the development of solutions with competitive production costs, as well as the implementation of clear recycling and certification systems.

References

Akinsemolu, A. A., Idowu, A. M., & Onyeaka, H. N. (2024). Recycling technologies for biopolymers: Current challenges and future directions. *Polymers, 16*(19). https://doi.org/10.3390/polym16192770, http://www.mdpi.com/journal/polymers.

Amin, U., Khan, M. K. I., Maan, A. A., Nazir, A., Riaz, S., Khan, M. U., Sultan, M., Munekata, P. E. S., & Lorenzo, J. M. (2022). Biodegradable active, intelligent, and smart packaging materials for food applications. *Food Packaging and Shelf Life, 33*. https://doi.org/10.1016/j.fpsl.2022.100903, http://www.journals.elsevier.com/food-packaging-and-shelf-life/.

Andriani, V., & Handayani, N. A. (2023). Recent technology of edible coating production: A review. *Materials Today: Proceedings, 87*, 200–206. https://doi.org/10.1016/j.matpr.2023.02.397, https://www.sciencedirect.com/science/journal/22147853.

Authority, E. F. S. (2011). Use of the EFSA comprehensive European food consumption database in exposure assessment. *EFSA journal, 9*(3).

Azam, S. E., Yasmeen, F., Rashid, M. S., & Latif, M. F. (2022). Physical factors affecting the antibacterial activity of Silver (Ag) and Zinc Oxide (ZnO) Nanoparticles (NPs), their application in edible and inedible food packaging, and regulation in food products. *International Journal of Agriculture and Biosciences, 11*(3), 181–193. https://doi.org/10.47278/journal.ijab/2023.025, http://www.ijagbio.com/pdf-files/volume-11-no-3-2022/181-193.pdf.

Badillo, G. M., & Segura-Ponce, L. A. (2020). Classic and reaction-diffusion models used in modified atmosphere packaging (MAP) of fruit and vegetables. *Food Engineering Reviews, 12*(2), 209–228. https://doi.org/10.1007/s12393-020-09214-3, http://www.springer.com/life+sci/food+science/journal/12393.

Costa, B. P., Carpiné, D., Ikeda, M., Pazzini, I. A. E., da Silva Bambirra Alves, F. E., de Melo, A. M., & Ribani, R. H. (2023). Bioactive coatings from non-conventional loquat (*Eriobotrya japonica* Lindl.) seed starch to extend strawberries shelf-life: An antioxidant packaging. *Progress in Organic Coatings, 175*. https://doi.org/10.1016/j.porgcoat.2022.107320, www.elsevier.com/locate/porgcoat.

Deng, Z., Wang, T., Chen, X., & Liu, Y. (2020). Applications of chitosan-based biomaterials: A focus on dependent antimicrobial properties. *Marine Life Science & Technology, 2*(4), 398–413. https://doi.org/10.1007/s42995-020-00044-0.

Gupta, R. K., Guha, P., & Srivastav, P. P. (2024). Investigating the toxicological effects of nanomaterials in food packaging associated with human health and the environment. *Journal of Hazardous Materials Letters, 5*. https://doi.org/10.1016/j.hazl.2024.100125, https://www.sciencedirect.com/science/journal/26669110.

Hakim, L., Deshmukh, R. K., Lee, Y. S., & Gaikwad, K. K. (2024). *Edible ink for sustainable food printing and packaging applications: A review*.

Jogaiah, S., Mujtaba, A. G., Mujtaba, M., Archana, De Britto, S., Geetha, N., Belorkar, S. A., & Shetty, H. S. (2025). Chitosan-metal and metal oxide nanocomposites for active and intelligent food packaging; a comprehensive review of emerging trends and associated challenges. *Carbohydrate Polymers, 357*, 123459. https://doi.org/10.1016/j.carbpol.2025.123459.

Kalita, P., Bora, N. S., Gogoi, B., Goswami, A., Pachuau, L., Das, P. J., Baishya, D., & Roy, S. (2025). Improving the hydrophobic nature of biopolymer based edible packaging film: A review. *Food Chemistry, 479*. https://doi.org/10.1016/j.foodchem.2025.143793, https://www.sciencedirect.com/science/journal/03088146.

Karnwal, A., Kumar, G., Singh, R., Selvaraj, M., Malik, T., & Al Tawaha, A. R. M. (2025). Natural biopolymers in edible coatings: Applications in food preservation. *Food Chemistry: X, 25*. https://doi.org/10.1016/j.fochx.2025.102171, https://www.sciencedirect.com/science/journal/25901575.

Manzoor, A., Dar, A. H., Pandey, V. K., Shams, R., Khan, S., Panesar, P. S., Kennedy, J. F., Fayaz, U., & Khan, S. A. (2022). Recent insights into polysaccharide-based hydrogels and their potential applications in food sector: A review. *International Journal of Biological Macromolecules, 213*, 987–1006. https://doi.org/10.1016/j.ijbiomac.2022.06.044, www.elsevier.com/locate/ijbiomac.

McSharry, S., Koolman, L., Whyte, P., & Bolton, D. (2020). The microbiology of beef steaks stored aerobically or anaerobically in vacuum pack films with different oxygen barrier properties. *Food Packaging and Shelf Life, 26*, 100597. https://doi.org/10.1016/j.fpsl.2020.100597.

Mishra, B., Panda, J., Mishra, A. K., Nath, P. C., Nayak, P. K., Mahapatra, U., Sharma, M., Chopra, H., Mohanta, Y. K., & Sridhar, K. (2024). Recent advances in sustainable biopolymer-based nanocomposites for smart food packaging: A review. *International Journal of Biological Macromolecules, 279*. https://doi.org/10.1016/j.ijbiomac.2024.135583, https://www.sciencedirect.com/science/journal/01418130.

Moschopoulou, E., Moatsou, G., Syrokou, M. K., Paramithiotis, S., & Drosinos, E. H. (2019). Food quality changes during shelf life. In *Food quality and shelf life*. Greece: Elsevier, 1–31. https://doi.org/10.1016/B978-0-12-817190-5.00001-X, https://www.sciencedirect.com/book/9780128171905.

Poonia, A., & Mishra, A. (2021). Edible nanocoatings: Potential food applications, challenges and safety regulations. *Nutrition & Food Science, 52*(3), 497–514. https://doi.org/10.1108/nfs-07-2021-0222.

Shen, L., Worrell, E., & Patel, M. K. (2010). Open-loop recycling: A LCA case study of PET bottle-to-fibre recycling. *Resources, Conservation and Recycling, 55*(1), 34–52. https://doi.org/10.1016/j.resconrec.2010.06.014.

Versino, F., Ortega, F., Monroy, Y., Rivero, S., López, O. V., & García, M. A. (2023). Sustainable and bio-based food packaging: A review on past and current design innovations. *Foods, 12*(5). https://doi.org/10.3390/foods12051057, www.mdpi.com/journal/foods.

Wang, Y., McClements, D. J., Bai, C., Xu, X., Sun, Q., Jiao, B., Miao, S., Wang, Q., & Dai, L. (2024). Application of proteins in edible inks for 3D food printing: A review. *Trends in Food Science & Technology, 153*, 104691. https://doi.org/10.1016/j.tifs.2024.104691.

Xie, Y., Liu, Q., Zhang, W., Yang, F., Zhao, K., Dong, X., Prakash, S., & Yuan, Y. (2023). Advances in the potential application of 3D food printing to enhance elderly nutritional dietary intake. *Foods, 12*(9), 1842. https://doi.org/10.3390/foods12091842.

Xue, W., Zhu, J., Sun, P., Yang, F., Wu, H., & Li, W. (2023). Permeability of biodegradable film comprising biopolymers derived from marine origin for food packaging application: A review. *Trends in Food Science & Technology, 136*, 295–307. https://doi.org/10.1016/j.tifs.2023.05.001.

Author Index

Note: Page numbers followed by "*f*" and "*t*" refer to figures and tables, respectively.

A

B

D

E

H

I

J

K

L

M

N

O

P

Q

R

S

T

U

Subject Index

Note: Page numbers followed by "*f*" and "*t*" refer to figures and tables, respectively.

D

G

H

I

J

K

L

Q

R

Printed and bound by CPI Group (UK) Ltd, Croydon, CR0 4YY
29/09/2026
02224671-0001